BERGEY'S MANUAL OF

Systematic Bacteriology

Second Edition

Volume Five

The *Actinobacteria*, Part B

BERGEY'S MANUAL OF

Systematic Bacteriology

Second Edition

Volume Five

The *Actinobacteria*, Part B

Michael Goodfellow, Peter Kämpfer, Hans-Jürgen Busse, Martha E. Trujillo, Ken-ichiro Suzuki, Wolfgang Ludwig and William B. Whitman
EDITORS, VOLUME FIVE

William B. Whitman
DIRECTOR OF THE EDITORIAL OFFICE

Aidan C. Parte
MANAGING EDITOR

WITH CONTRIBUTIONS FROM 116 COLLEAGUES

Springer

William B. Whitman
Bergey's Manual Trust
Department of Microbiology
527 Biological Sciences Building
University of Georgia
Athens, GA 30602-2605
USA

ISBN 978-0-387-95043-3 e-ISBN 978-0-387-68233-4
DOI 10.1007/978-0-387-68233-4
Springer New York Dordrecht Heidelberg London

Library of Congress Control Number: 2012930836

Printed on acid-free paper.

Springer is part of Springer Science+Business Media (www.springer.com)

This volume is dedicated to our colleagues
Michael Goodfellow and Peter H.A. Sneath.

Michael retired from the Board of Trustees of Bergey's Manual during preparation of this volume. His tremendous efforts as an editor, author and officer of the Trust are truly appreciated.

Our late eminent colleague Peter H.A. Sneath (1923–2011) was a Trustee (1978–1994) and Chairman of the Trust (1990–1994). He also served as an editor of previous editions of The Manual and made many other remarkable contributions to the systematics of prokaryotes.

Preface to volume 5 of the second edition of *Bergey's Manual of Systematic Bacteriology*

Prokaryotic systematics remains a vibrant and exciting field of study, one of challenges and opportunities, great discoveries and gradual advances. To honor one of the leaders of our field, the Trust presented the Bergey Award in recognition of outstanding contributions to the taxonomy of prokaryotes to Antonio Ventosa in 2010. The Bergey Medal, in recognition of life-long contributions to the field of systematic bacteriology, was also awarded to Michael Goodfellow, Zhiheng Liu, Ji-Sheng Ruan, and James Tiedje in 2011.

Volume 5 will be the last volume to be edited by Michael Goodfellow, who served on the Trust for many years and has continued to be active during his retirement. Mike contributed to volumes 2 and 4 of the first edition of *Bergey's Manual of Systematic Bacteriology* and has made an enormous contribution to the present volume as an author, editor, and mentor. As a leader in actinobacterial research for many decades, he is also directly responsible for much of the wealth of information about this fascinating group of microorganisms described in the current volume. Mike served as the Vice-Chairman of the Trust for many years and Chairman for the last 3 years. During his tenure, the Trust underwent important transitions for the future beyond the second edition. Adept at saying the most difficult things in the nicest way and a master of the telling omission, he was the right person at the right time.

Acknowledgements

The Trust is indebted to all of the contributors and reviewers, without whom this work would not be possible. The Editors are grateful for the time and effort that each has expended on behalf of the entire scientific community. We also thank the authors for their good grace in accepting comments, criticisms, and editing of their manuscripts.

The Trust recognizes its enormous debt to Dr Aidan Parte, whose enthusiasm and professionalism have made this *Manual* possible. The completion of the second edition is due in great measure to his dedication, good judgment, and hard work. His vision for excellent science has made the *Manual* more than it would have been.

We also recognize the special efforts of Dr Jean Euzéby in checking and correcting where necessary the nomenclature and etymology of every described taxon in this volume.

The Trust also thanks its Springer colleagues, especially Editorial Director Andrea Macaluso and Production Manager Susan Westendorf, for all of their efforts. As this will be the last volume of the *Manual* published in collaboration with Springer, the Trust also wishes to acknowledge the tremendous support and understanding that Springer has demonstrated over the last 13 years in helping us to publish this comprehensive synthesis of the systematics of prokaryotes.

In addition, we thank Amina Ravi, our manager at our typesetters, SPi, for her work in the proofing and production of this and the previous two volumes.

We thank our current copyeditors, proofreaders and other staff, including Susan Andrews, Joanne Auger, Hannah Berle, Robert Gutman, Judy Leventhal, Linda Sanders, Tyler Sgro, Dana Schneider, and Mohammed Waqar, without whose hard work and attention to detail the production of this volume would have been impossible. Lastly, we thank Dale Boyer and the other members of the Department of Microbiology at the University of Georgia for their unfailing support of this endeavor.

William B. (Barny) Whitman

Contents

Contributors

Hiroshi Akasaka
Creative Research Initiative "Sousei" (CRIS), Hokkaido University, Kita 21, Nishi 10, Kita-ku, Sapporo 001-0021, Japan

Vladimir N. Akimov
VKM – All-Russian Collection of Microorganisms, G.K. Skryabin Institute of Biochemistry and Physiology of Microorganisms, Russian Academy of Sciences, Pushchino, Moscow Region 142290, Russia
akimov@ibpm.pushchino.ru

Robin C. Anderson
Research Microbiologist/Lead Scientist, United States Department of Agriculture, Southern Plains Agricultural Research Center, Food and Feed Safety Research Unit, 2881 F&B Road, College Station, TX 77845, USA
robin.anderson@ars.usda.gov

Elena V. Ariskina
VKM – All-Russian Collection of Microorganisms, G.K. Skryabin Institute of Biochemistry and Physiology of Microorganisms, Russian Academy of Sciences, Pushchino, Moscow Region 142290, Russia
lena@ibpm.pushchino.ru

Brian Austin
Institute of Aquaculture, School of Natural Sciences, University of Stirling, Stirling FK9 4LA, Scotland, UK
brian.austin@stir.ac.uk

Undine Behrendt
Leibniz Centre for Agricultural Landscape Research (ZALF), Institute of Landscape Matter Dynamics, Eberswalder Straße 84, D-15374 Müncheberg, Germany
ubehrendt@zalf.de

Yoshimi Benno
Benno Laboratory, Innovation Center, RIKEN, 2-1 Hirosawa, Wako, Saitama 351-0198, Japan
benno828@riken.jp

David R. Benson
Department of Molecular & Cell Biology, University of Connecticut, Storrs, CT 06269-3125, USA
david.benson@uconn.edu

Kathryn A. Bernard
National Microbiology Laboratory, Public Health Agency of Canada, 1015 Arlington Street, Winnipeg, Manitoba R3E 3R2, Canada
kathy.bernard@phac-aspc.gc.ca

Alison M. Berry
Department of Plant Sciences, One Shields Avenue, University of California, Davis, CA 95616, USA
amberry@ucdavis.edu

Bruno Biavati
Department of Agricultural Sciences, Bologna University, Viale Fanin 42, 40127 Bologna, Italy
bruno.biavati@unibo.it

Sandra Buczolits
Institut für Bakteriologie, Mykologie und Hygiene, Veterinärmedizinische Universität Wien, Veterinärplatz 1, A-1210 Wien, Austria
sandra.buczolits@vetmeduni.ac.at

Hans-Jürgen Busse
Institut für Bakteriologie, Mykologie und Hygiene, Veterinärmedizinische Universität Wien, Veterinärplatz 1, A-1210 Wien, Austria
hans-juergen.busse@vetmeduni.ac.at

W. Ray Butler
Division of Tuberculosis Elimination, National Center for HIV, STD and Tuberculosis Prevention, Centers for Disease Control and Prevention, Atlanta, GA, USA
wrb1@cdc.gov

Lorena Carro
Departamento de Microbiología y Genética, Edificio Departamental, Lab. 205, Campus Unamuno, Universidad de Salamanca, 37007 Salamanca, Spain
lcg@usal.es

Linda Cavaletti
Fondazione Istituto Insubrico di Ricerca per la Vita, via Robert Lepetit 34, 21040 Gerenzano, Italy
lindacavaletti@ricercaperlavita.it

Wen-Feng Chen
State Key Laboratories of Agrobiotechnology, College of Biological Sciences, China Agricultural University, Beijing 100193, P.R. China
chenwf@cau.edu.cn

Matthew D. Collins
The University of Reading, Whiteknights, Reading RG6 6AP, UK

Milton S. da Costa
Department of Life Sciences, University of Coimbra, 3001-401, Coimbra, Portugal
milton@ci.uc.pt

Xiao-Long Cui
Yunnan Institute of Microbiology, Yunnan University, Cuihu Beilu 2, Kunming, Yunnan 650091, P.R. China
xlcui@ynu.edu.cn

Ewald B. M. Denner
Institut für Bakteriologie, Mykologie und Hygiene, Veterinärmedizinische Universität Wien, Veterinärplatz 1, A-1210 Wien, Austria
ewald.denner@vetmeduni.ac.at

Floyd E. Dewhirst
Department of Molecular Genetics,
The Forsyth Institute, 245 First Street,
Cambridge, MA 02142, USA
fdewhirst@forsyth.org

Stefano Donadio
KtedoGen Srl and NAICONS Scrl, Via Fantoli 16/15,
20138 Milano, Italy
stefano.donadio@ktedogen.com, sdonadio@naicons.com

Lubov V. Dorofeeva
VKM – All-Russian Collection of Microorganisms,
G.K. Skryabin Institute of Biochemistry and Physiology
of Microorganisms, Russian Academy of Sciences,
Pushchino, Moscow Region 142290, Russia
dorofeeva@ibpm.pushchino.ru

Jean P. Euzéby
Ecole Nationale Veterinaire, 23 chemin des Capelles,
B.P. 87614, 31076 Toulouse cedex 3, France
jean.euzeby@gmail.com

Lyudmila I. Evtushenko
VKM – All-Russian Collection of Microorganisms,
G.K. Skryabin Institute of Biochemistry and Physiology
of Microorganisms, Russian Academy of Sciences, Pushchino,
Moscow Region 142290, Russia
evtushenko@ibpm.pushchino.ru

José F. Fernández-Garayzábal
Departamento de Sanidad Animal,
Facultad de Veterinaria, Universidad Complutense,
28040 Madrid, Spain
garayzab@vet.ucm.es

Christopher Franco
Department of Medical Biotechnology, School of Medicine,
Flinders University, Bedford Park, South Australia 5042,
Australia
chris.franco@flinders.edu.au

Guido Funke
Department of Medical Microbiology & Hygiene,
Gärtner & Colleagues Laboratories, Elisabethenstrasse 11,
D-88212 Ravensburg, Germany
ldg.funke@t-online.de

George M. Garrity
Department of Microbiology, Michigan State University,
6162 Biomedical and Physical Sciences Building,
East Lansing, MI 48824-4320, USA
garrity@msu.edu

Olga Genilloud
Fundación MEDINA, Avda del Conocimiento 3,
Parque Tecnológico Ciencias de la Salud, 18100 Armilla,
Granada, Spain
olga.genilloud@medinaandalucia.es, olga_genilloud@wanadoo.es

Michael Goodfellow
School of Biology, Ridley Building, University of Newcastle,
Newcastle upon Tyne NE1 7RU, UK
m.goodfellow@newcastle.ac.uk

Anthony C. Greene
Microbial Gene Research and Resources Facility,
School of Biomolecular and Physical Sciences,
Griffith University, Brisbane, Queensland 4111,
Australia
t.greene@griffith.edu.au

Ingrid Groth
Molekulare und Angewandte Mikrobiologie,
Leibniz-Institut für Naturstroff-Forschung
und Infektionsbiologie e. V., Hans-Knöll-Institut,
Beutenbergstrasse 11a, D-07745 Jena, Germany
ingrid.groth@gmx.net

Val Hall
Anaerobe Reference Unit, Public Health Wales Microbiology,
University Hospital of Wales, Cardiff CF14 4XW, UK
hallv@cardiff.ac.uk

Satoshi Hanada
Institute for Biological Resources and Functions,
National Institute of Advanced Industrial Science
and Technology (AIST), Tsukuba Central 6, 1-1-1 Higashi,
Tsukuba 305-8566, Japan
s-hanada@aist.go.jp

Lesley Hoyles
Microbial Ecology & Health Group,
Department of Food & Nutritional Sciences,
University of Reading, Whiteknights Campus, Reading, UK
lesley_hoyles@hotmail.com

Wael N. Hozzein
Botany Department, Faculty of Science, Beni-Suef University,
Beni-Suef, Egypt
hozzein29@yahoo.com

Ying Huang
State Key Laboratory of Microbial Resources,
Institute of Microbiology, Chinese Academy of Sciences,
No. 1 West Beichen Road, Chaoyang District, Beijing 100101,
P.R. China
huangy@im.ac.cn

Paul R. Jensen
Scripps Institution of Oceanography, University of California
San Diego, La Jolla, CA 92093-0204, USA
pjensen@ucsd.edu

Amanda L. Jones
Department of Biology, Food and Nutritional Sciences,
School of Life Sciences, A314 Ellison Building, Northumbria
University, Newcastle upon Tyne NE1 8ST, UK
amanda.l.jones@northumbria.ac.uk

Akiko Kageyama
Kitasato Institute for Life Sciences, Kitasato University,
5-9-1 Shirokane, Minato-ku, Tokyo 108-8641, Japan
kageyama@nihs.go.jp

Peter Kämpfer
Institut für Angewandte Mikrobiologie,
Justus-Liebig-Universität Giessen, Heinrich-Buff-Ring
26-32 (IFZ), D-35392 Giessen, Germany
peter.kaempfer@umwelt.uni-giessen.de

Hiroaki Kasai
Marine Biotechnology Institute, 3-75-1, Heita, Kamashi,
Iwate 026-0001, Japan
hkasai@kitasato-u.ac.jp

Ellen M. Kerr
Science and Advice for Scottish Agriculture,
Roddinglaw Road, Edinburgh, EH12 9FJ, Scotland, UK

Seung Bum Kim
Department of Microbiology and Molecular Biology,
Chungnam National University, 220 Gung-dong, Yuseong,
Daejeon 305-764, Republic of Korea
sbk01@cnu.ac.kr

Helmut König
Institute of Microbiology and Wine Research,
Johannes Gutenberg University, Becherweg 15,
55128 Mainz, Germany
hkoenig@uni-mainz.de

Valentina I. Krausova
VKM – All-Russian Collection of Microorganisms,
G.K. Skryabin Institute of Biochemistry and Physiology
of Microorganisms, Russian Academy of Sciences, Pushchino,
Moscow Region 142290, Russia
vikrau_z@mail.ru

Noel R. Krieg
617 Broce Drive, Blacksburg, VA 24060-2801, USA
nrk@vt.edu

Takuji Kudo
Japan Collection of Microorganisms, RIKEN, Hirosawa, Wako,
Saitama 351-0198, Japan
kudo@jcm.riken.jp

Yashawant Kumar
NCIMB Ltd, Ferguson Building,
Craibstone Estate, Bucksburn, Aberdeen AB21 9YA,
Scotland, UK

David P. Labeda
National Center for Agricultural Utilization Research,
U.S. Department of Agriculture, Peoria, IL 61604-3999, USA
david.labeda@ars.usda.gov

Paul A. Lawson
Department of Botany and Microbiology, George Lynn
Cross Hall, 770 Van Vleet Oval, The University of Oklahoma,
Norman, OK 73019-0245, USA
paul.lawson@ou.edu

Soon Dong Lee
School of Biological Sciences and Research Center
for Molecular Microbiology, Seoul National University,
Republic of Korea, Seoul, Korea
sdlee@cheju.ac.kr

Wen-Jun Li
Key Laboratory of Microbial Diversity in Southwest China,
Ministry of Education and Laboratory for Conservation
and Utilization of Bio-Resources, Yunnan Institute
of Microbiology, Yunnan University, Kunming 650091,
P.R. China
wjli@ynu.edu.cn, liact@hotmail.com

Zhi-Heng Liu
Institute of Microbiology, Chinese Academy of Sciences,
P.O. Box 2714, Beijing 100190, P.R. China
zhliu@sun.im.ac.cn

Nicole Lodders
Institut für Angewandte Mikrobiologie,
Justus-Liebig-Universität Giessen, Heinrich-Buff-Ring
26-32 (IFZ), D-35392 Giessen, Germany
nicole.lodders@umwelt.uni-giessen.de

Wolfgang Ludwig
Lehrstuhl für Mikrobiologie, Technische Universität
München, Emil-Ramann-Str. 4, D-85350 Freising, Germany
ludwig@mikro.biologie.tu-muenchen.de

John G. Magee
HPA Microbiology Services Newcastle Laboratory,
Level 2, Freeman Hospital High Heaton,
Newcastle upon Tyne NE7 7DN, UK
john.magee@hpa.org.uk

Matthias Maiwald
Department of Pathology and Laboratory Medicine,
KK Women's and Children's Hospital,
100 Bukit Timah Road, Singapore 229899
matthias.maiwald@flinders.edu.au, matthias_maiwald@yahoo.com

Luis A. Maldonado
Instituto de Ciencias del Mar y Limnología (ICMyL),
Universidad Nacional Autónoma de México (UNAM),
04510 Mexico, DF, Mexico
lamaldo@icmyl.unam.mx

Célia Manaia
Escola Superior de Biotecnologia, Universidade Católica
Portuguesa, R. Dr António Bernardino de Almeida,
4200-072 Porto, Portugal
cmmanaia@esb.ucp.pt

Eustoquio Martínez-Molina
Departamento de Microbiología y Genética,
Edificio Departamental, Lab. 209, Universidad de Salamanca,
Salamanca, Spain
emm@usal.es

Abdul M. Maszenan
Environmental Engineering Research Centre,
School of Civil and Environmental Engineering,
Nanyang Technological University, 639798, Singapore
cmaszenan@ntu.edu.sg

Pedro F. Mateos
Departamento de Microbiología y Genética,
Edificio Departamental, Campus Unamuno,
Universidad de Salamanca, 37007 Salamanca, Spain
pfmg@usal.es

Atsuko Matsumoto
Kitasato Institute for Life sciences, Kitasato University,
5-9-1 Shirokane, Minato-ku, Tokyo 108-8641, Japan
amatsu@lisci.kitasato-u.ac.jp

Paola Mattarelli
Department of Agricultural Sciences, Bologna University,
Viale Fanin 42, 40127 Bologna, Italy
paola.mattarelli@unibo.it

Anne L. McCartney
Microbial Ecology and Health Group, Department
of Food and Nutritional Sciences, University of Reading,
Whiteknights, P.O. Box 226, Reading, RG6 6AP, UK
a.l.mccartney@reading.ac.uk

Andrew McDowell
Centre for Infection and Immunity, School of Medicine,
Dentistry and Biomedical Sciences, Queen's University Belfast,
Medical Biology Centre, 97 Lisburn Road, Belfast BT9 7BL, UK
a.mcdowell@qub.ac.uk

Simon J. McIlroy
Biotechnology Research Centre, La Trobe University,
P.O. Box 199, Bendigo, Victoria 3550, Australia
s.mcilroy@latrobe.edu.au

Patricia Messenberg Guimarães
Embrapa Genetic Resources and Biotechnology,
PqEB Final W3 Norte, P.O. Box 02372, 70770-900 Brasília-DF,
Brazil
messenbe@cenargen.embrapa.br

William M. Moe
Department of Civil and Environmental Engineering,
Louisiana State University, Baton Rouge, LA 70803, USA
moemwil@lsu.edu

Paolo Monciardini
KtedoGen srl, via Fantoli 16/15, 20138 Milano, Italy
paolo.monciardini@ktedogen.com

Kazunori Nakamura
Institute for Biological Resources and Functions, National Institute of Advanced Industrial Science and Technology (AIST), Tsukuba, Ibaraki, Japan
k.nakamura@aist.go.jp

Futoshi Nakazawa
Department of Oral Microbiology, School of Dentistry, Health Sciences University of Hokkaido, 1757 Kanazawa, Tobetsu-Ishikari, Hokkaido, 061-0293, Japan
nakazawa@hoku-iryo-u.ac.jp

Olga I. Nedashkovskaya
Pacific Institute of Bioorganic Chemistry, of the Far-Eastern Branch of the Russian Academy of Sciences, Pr. 100 let Vladivostoku 159, 690022, Vladivostok, Russia
olganedashkovska@piboc.dvo.ru, olganedashkovska@yahoo.com

Balbina Nogales
Area de Microbiologia, Dept. Biologia, Universitat de les Illes Balears, Crtra. Valldemossa km 7.5, 07122 Palma de Mallorca, Spain
bnogales@uib.es

Philippe Normand
IFR41 CNRS Ecologie Genetique Evolution, UMR 5557 CNRS Ecologie Microbienne, Université Claude-Bernard Lyon1, Bat G. Mendel, 43 Blvd du 11 novembre 1918, 69622 Villeurbanne Cedex, France
normand@biomserv.univ-lyon1.fr

Paul R. Norris
School of Life Sciences, University of Warwick, Coventry CV4 7AL, UK
p.r.norris@warwick.ac.uk

Olga C. Nunes
LEPAE-Departamento de Engenharia Química, Faculdade de Engenharia, Universidade do Porto, R. Dr Roberto Frias, Porto, 4200-465, Portugal
opnunes@fe.up.pt

Cristina Pascual
National Documentation Centre, 48 Vas. Constantinou Avenue, GR11635 Athens, Greece
cpascual@ekt.gr, cristina_93@hotmail.com

Bharat K. C. Patel
Microbial Discovery Research Unit, School of Biomolecular and Physical Sciences, Griffith University, Nathan Campus, Kessels Road, Brisbane, Queensland 4111, Australia
b.patel@griffith.edu.au

Sheila Patrick
Centre for Infection and Immunity, School of Medicine, Dentistry and Biomedical Sciences, Queen's University Belfast, Medical Biology Centre, 97 Lisburn Road, Belfast BT9 7BL, UK
sheila.patrick@qub.ac.uk

Jerome J. Perry (Deceased)
3125 Eton Road, Raleigh, NC 27608, USA

Rüdiger Pukall
DSMZ – Deutsche Sammlung von Mikroorganismen und Zellkulturen, Inhoffenstraße 7 B, 38124 Braunschweig, Germany
rpu@dsmz.de

Erika T. Quintana
Instituto Politécnico Nacional (IPN), Escuela Nacional de Ciencias Biológicas (ENCB), Department of Microbiology, General Microbiology Laboratory, Prolongación de Carpio y Plan de Ayala s/n, Col. Santo Tomás, Deleg. Miguel Hidalgo, C.P. 11340, Mexico City, Mexico
erika_quintana@hotmail.com

Fred A. Rainey
Department of Biological Sciences, University of Alaska Anchorage, Providence Drive, Anchorage, AK 99508, USA
farainey@gmail.com

David A. Relman
Departments of Medicine, and of Microbiology & Immunology, Stanford University School of Medicine, Stanford, CA 94305-5124, USA
relman@stanford.edu

Raúl Rivas
Departamento de Microbiología y Genética, Edificio Departamental, Lab. 209, Campus Unamuno, Universidad de Salamanca, 37007 Salamanca, Spain
raulrg@usal.es

Gerard S. Saddler
Science and Advice for Scottish Agriculture, Roddinglaw Road, Edinburgh, EH12 9FJ, Scotland, UK
gerry.saddler@sasa.gsi.gov.uk

Klaus P. Schaal
Institut für Medizinische Mikrobiologie, Immunologie und Parasitologie, Universitätsklinikum Bonn, Sigmund-Freud-Strasse 25, D-53105 Bonn, Germany
schaal@microbiology-bonn.de, kpschaal@t-online.de

Jenny Schäfer
Bundesanstalt für Arbeitsschutz und Arbeitsmedizin, "Biologische Arbeitsstoffe", Nöldnerstrasse 40–42, 10317 Berlin, Germany
schaefer.jenny@baua.bund.de

Peter Schumann
DSMZ – Deutsche Sammlung von Mikroorganismen und Zellkulturen, Inhoffenstraße 7 B, 38124 Braunschweig, Germany
psc@dsmz.de

Susmitha Seshadri
Department of Microbiology, University of Georgia, 527 Biological Sciences Building, Cedar Street, Athens, GA 30602, USA
ssusmita@uga.edu

Robert J. Seviour
Biotechnology Research Centre, La Trobe University, P.O. Box 199, Bendigo, Victoria 3550, Australia
r.seviour@latrobe.edu.au

Peter P. Sheridan
Department of Biological Sciences, Idaho State University, P.O. Box 8007, Pocatello, ID 83209, USA
sherpete@isu.edu

Jacques A. Soddell
Cajid Media, 21 Wirth Street, Bendigo, Victoria 3550, Australia
jacques@cajid.com

Cathrin Spröer
DSMZ – Deutsche Sammlung von Mikroorganismen und Zellkulturen, Inhoffenstraße 7 B, 38124 Braunschweig, Germany
ckc@dsmz.de

Erko Stackebrandt
40 Rue des Ecoles, 75005 Paris, France
erko@dsmz.de

Thaddeus B. Stanton
Agricultural Research Service – Midwest Area,
National Animal Disease Center, United States Department of Agriculture, P.O. Box 70, 1920 Dayton Avenue,
Building 24, Ames, IA 50010-0070,
USA
thad.stanton@ars.usda.gov

Virginie Storms
Lab. voor Microbiologie en Microbiele Genetica,
Faculteit Wetenschappen, Univeristeit of Gent,
K.L. Ledeganckstraat 35, B-9000 Ghent, Belgium
virginie.storms@ugent.be

Ken-ichiro Suzuki
NITE Biological Resource Center (NBRC),
National Institute of Technology and Evaluation,
2-5-8, Kazusakamatari 2-chome, Kisarazu-shi, Chiba 292-0818,
Japan
suzuki-ken-ichiro@nite.go.jp

Yōko Takahashi
Kitasato Institute for Life Sciences, Kitasato University,
5-9-1 Shirokane, Minato-ku, Tokyo 108-8641, Japan
ytakaha@lisci.kitasato-u.ac.jp

Mariko Takeuchi
Takeda Chemical Industries, 17-85, Juso-honmachi, 2-chome,
Yodogawa-ku, Osaka 532-8686, Japan

Tomohiko Tamura
NITE Biological Resource Center (NBRC),
National Institute of Technology and Evaluation,
2-5-8, Kazusakamatari, Kisarazu-shi,
Chiba 292-0818, Japan
tamura-tomohiko@nite.go.jp

Geok Yuan Annie Tan
Microbiology Division, Institute of Biological Sciences,
University of Malaya, 50603 Kuala Lumpur, Malaysia
gyatan@um.edu.my

Shu-Kun Tang
The Key Laboratory for Microbial Resources of the Ministry of Education, P.R.China, and Laboratory for Conservation and Utilization Bio-Resources, Yunnan Institute of Microbiology, Yunnan University, Kunming 650091,
P.R. China

Tian-shen Tao
College of Life Sciences, Wuhan University, Wuhan 430072,
Hubei Province, P.R. China
taotianshen@126.com

Martha E. Trujillo
Departamento de Microbiología y Genética,
Edificio Departamental, Lab. 205, Campus Unamuno,
Universidad de Salamanca, 37007 Salamanca,
Spain
mett@usal.es

Takanori Tsukamoto
Plant Protection Division, Food Safety and
Consumer Affairs Bureau, Ministry of Agriculture,
Forestry and Fisheries (MAFF) 1-2-1 Kasumigaseki,
Chiyoda-ku, Tokyo 100-8950, Japan
tsukamotot@pps.maff.go.jp

Atsuko Ueki
Faculty of Agriculture, Yamagata University,
Wakaba-machi 1-23, Tsuruoka 997-8555, Japan
uatsuko@tds1.tr.yamagata-u.ac.jp

Katuji Ueki
Faculty of Agriculture, Yamagata University,
Wakaba-machi 1-23, Tsuruoka 997-8555, Japan
kueki@tds1.tr.yamagata-u.ac.jp

Andreas Ulrich
Institute of Landscape Matter Dynamics, Leibniz-Centre
for Agricultural Landscape Research (ZALF) e.V.,
Eberswalder Strasse 84, D-15374 Müncheberg, Germany
aulrich@zalf.de

Peter Vandamme
Laboratorium voor Microbiologie, Faculteit Wetenschappen,
Universiteit Gent, Ledeganckstraat 35, B-9000 Gent, Belgium
peter.vandamme@ugent.be

Encarna Velázquez
Departmento de Microbiología y Genética, Lab 209,
Edificio Departamental, Universidad de Salamanca,
Campus Miguel de Unamuno, 37007 Salamanca, Spain
evp@usal.es

António Veríssimo
Centro de Neurociências e Biologia Celular and Department of Life Sciences, University of Coimbra, Apartado 3046,
3001-401 Coimbra, Portugal
averissimo@uc.pt

Gernot Vobis
Departamento de Botánica, Centro Regional Universitario
Bariloche, Universidad Nacional del Comahue, Quintral,
1250 8400 San Carlos de Bariloche, Prov. de Río Negro,
Argentina
agavobis@bariloche.com.ar

William G. Wade
Microbiology, King's College London Dental Institute,
Floor 17, Tower Wing, Guy's Campus, London SE1 9RT, UK
william.wade@kcl.ac.uk

Alan C. Ward
School of Biology, University of Newcastle upon Tyne,
Newcastle upon Tyne NE1 7RU, UK
alan.ward@ncl.ac.uk

William B. Whitman
Department of Microbiology, University of Georgia,
527 Biological Sciences Building, Cedar Street, Athens,
GA 30602-2605, USA
whitman@uga.edu

Monika Wieser
Institut für Bakteriologie, Mykologie und Hygiene,
Veterinärmedizinische Universität Wien, Veterinärplatz 1,
A-1210 Wien, Austria
monika.wieser@vetmeduni.ac.at

Atteyet F. Yassin
Institut für Medizinische Mikrobiologie und Immunologie
der Universität Bonn, Sigmund-Freud-Straße 25, 53127 Bonn,
Germany
yassin@mibi03.meb.uni-bonn.de, atteyet-alla.yassin@ukb.uni-bonn.de

Akira Yokota
Institute of Molecular and Cellular Biosciences,
The University of Tokyo, 1-1-1, Yayoi, Bunkyo-Ku,
Tokyo 113-0032, Japan
uayoko@gmail.com

Jung-Hoon Yoon
Laboratory of Microbial Function, Korea Research Institute of Bioscience and Biotechnology (KRIBB), P.O. Box 115, Yusong, Taejon, South Korea
jhyoon@kribb.re.kr

Xiao-Yang Zhi
The Key Laboratory for Microbial Resources of the Ministry of Education, P.R.China, and Laboratory for Conservation and Utilization Bio-Resources, Yunnan Institute of Microbiology, Yunnan University, Kunming 650091, P.R. China

On using the *Manual*

Noel R. Krieg and George M. Garrity

Citation

The *Systematics* is a peer-reviewed collection of chapters, contributed by authors who were invited by the Trust to share their knowledge and expertise of specific taxa. Citations should refer to the author, the chapter title, and inclusive pages rather than to the editors.

Arrangement of the *Manual*

As in the previous volumes of this edition, the *Manual* is arranged in phylogenetic groups based upon the analyses of the 16S rRNA presented in the introductory chapter "Road map of the phylum *Actinobacteria*". This phylum has been substantially modified since the publication of volume 1 in 2001, reflecting both the availability of more experimental data and a different method of analysis. Since volume 5 includes only the phylum *Actinobacteria*, taxa are arranged by class, order, family, genus and species. Within each taxon, the nomenclatural type is presented first. Other taxa are presented in alphabetical order without consideration of degrees of relatedness.

Articles

Each article dealing with a bacterial genus is presented wherever possible in a definite sequence as follows:

a. Name of the genus. Accepted names are in boldface, followed by "defining publication(s)", i.e. the authority for the name, the year of the original description, and the page on which the taxon was named and described. The superscript AL indicates that the name was included on the Approved Lists of Bacterial Names, published in January 1980. The superscript VP indicates that the name, although not on the Approved Lists of Bacterial Names, was subsequently validly published in the *International Journal of Systematic and Evolutionary Microbiology* (or the *International Journal of Systematic Bacteriology*). Names given within quotation marks have no standing in nomenclature; as of the date of preparation of the *Manual* they had not been validly published in the *International Journal of Systematic and Evolutionary Microbiology*, although they may have been "effectively published" elsewhere. Names followed by the term "nov." are newly proposed but will not be validly published until they appear in a Validation List in the *International Journal of Systematic and Evolutionary Microbiology*. Their proposal in the *Manual* constitutes only "effective publication", not valid publication.

b. Name of author(s). The person or persons who prepared the Bergey's article are indicated. The address of each author can be found in the list of Contributors at the beginning of the *Manual*.

c. Synonyms. In some instances a list of some synonyms used in the past for the same genus is given. Other synonyms can be found in the *Index Bergeyana* or the *Supplement to the Index Bergeyana*.

d. Etymology of the name. Etymologies are provided as in previous editions, and many (but undoubtedly not all) errors have been corrected. It is often difficult, however, to determine why a particular name was chosen, or the nuance intended, if the details were not provided in the original publication. Those authors who propose new names are urged to consult a Greek and Latin authority before publishing in order to ensure grammatical correctness and also to ensure that the meaning of the name is as intended.

e. Salient features. This is a brief resume of the salient features of the taxon. The most important characteristics are given in boldface. The DNA G+C content is given.

f. Type species. The name of the type species of the genus is also indicated along with the defining publication(s).

g. Further descriptive information. This portion elaborates on the various features of the genus, particularly those features having significance for systematic bacteriology. The treatment serves to acquaint the reader with the overall biology of the organisms but is not meant to be a comprehensive review. The information is normally presented in the following sequence:

Colonial morphology and pigmentation
Growth conditions and nutrition
Physiology and metabolism
Genetics, plasmids, and bacteriophages
Phylogenetic treatment
Antigenic structure
Pathogenicity
Ecology

h. Enrichment and isolation. A few selected methods are presented, together with the pertinent media formulations.

i. Maintenance procedures. Methods used for maintenance of stock cultures and preservation of strains are given.

j. Procedures for testing special characters. This portion provides methodology for testing for unusual characteristics or performing tests of special importance.

k. Differentiation of the genus from other genera. Those characteristics that are especially useful for distinguishing the genus from similar or related organisms are indicated here, usually in a tabular form.

l. Taxonomic comments. This summarizes the available information related to taxonomic placement of the genus and indicates the justification for considering the genus a distinct taxon. Particular emphasis is given to the methods of molecular biology used to estimate the relatedness of the genus to other taxa, where such information is available. Taxonomic information regarding the arrangement and status of the various species within the genus follows. Where taxonomic controversy exists, the problems are delineated and the various alternative viewpoints are discussed.

m. Further reading. A list of selected references, usually of a general nature, is given to enable the reader to gain access to additional sources of information about the genus.

n. Differentiation of the species of the genus. Those characteristics that are important for distinguishing the various species within the genus are presented, usually with reference to a table summarizing the information.

o. List of species of the genus. The citation of each species is given, followed in some instances by a brief list of objective synonyms. The etymology of the specific epithet is indicated. Descriptive information for the species is usually presented in tabular form, but special information may be given in the text. Because of the emphasis on tabular data, the species descriptions are usually brief. The type strain of each species is indicated, together with the collection(s) in which it can be found. (Addresses of the various culture collections are given in the article in volume 1 entitled *Culture Collections: An Essential Resource for Microbiology*.) The 16S rRNA gene sequence used in phylogenetic analysis and placement of the species into the taxonomic framework is given, along with the GenBank (or other database) accession number. Additional comments may be provided to point the reader to other well-characterized strains of the species and any other known DNA sequences that may be relevant.

p. Species *incertae sedis*. The List of Species may be followed in some instances by a listing of additional species under the heading "Species *Incertae sedis*" or "Other organisms", etc. The taxonomic placement or status of such species is questionable, and the reasons for the uncertainty are presented.

q. References. All references given in the article are listed alphabetically at the end of the family chapter

Tables

In each article dealing with a genus, there are generally three kinds of table: (a) those that differentiate the genus from similar or related genera, (b) those that differentiate the species within the genus, and (c) those that provide additional information about the species (such information not being particularly useful for differentiation). The meanings of symbols are as follows:

+, 90% or more of the strains are positive
d, 11–89% of the strains are positive
−, 90% or more of the strains are negative
D, different reactions occur in different taxa (e.g., species of a genus or genera of a family)
v, strain instability (NOT equivalent to "d")
w, weak reaction.
nd, not determined or no data.
nr, not reported.

These symbols, and exceptions to their use, as well as the meaning of additional symbols, are given in footnotes to the tables.

Use of the *Manual* for determinative purposes

Many chapters have keys or tables for differentiation of the various taxa contained therein. For identification of species, it is important to read both the generic and species descriptions because characteristics listed in the generic descriptions are not usually repeated in the species descriptions.

The index is useful for locating the articles on unfamiliar taxa or in discovering the current classification of a particular taxon. Every bacterial name mentioned in the *Manual* is listed in the index. In addition, an up-to-date outline of the taxonomic framework is provided in the introductory chapter "Road map of the phylum *Actinobacteria*".

Errors, comments, and suggestions

As in previous volumes, the editors and authors earnestly solicit the assistance of all microbiologists in the correction of possible errors in *Bergey's Manual of Systematic Bacteriology*. Comments on the presentation will also be welcomed as well as suggestions for future editions. Correspondence should be addressed to:

Editorial Office
Bergey's Manual Trust
Department of Microbiology
University of Georgia
Athens, GA 30602-2605, USA
Tel: +1-706-542-4219; fax +1-706-542-6599
e-mail: bergeys@uga.edu

Order XI. **Micromonosporales** ord. nov.

Olga Genilloud

Mi.cro.mo.no.spo′ra.les. N.L. fem. n. *Micromonospora* type genus of the order; suff. *-ales* ending to denote a order; N.L. fem. pl. n. *Micromonosporales* the *Micromonospora* order.

The order was formed by elevation of the suborder *Micromonosporineae* Stackebrandt, Rainey and Ward-Rainey 1997, 486[VP] emend. Zhi, Li and Stackebrandt 2009, 599. It contains the family *Micromonosporaceae.* The recognition of this taxon is based on the distinct phylogenetic position of its constituent members as determined by 16S rRNA gene sequence analysis and by the presence of a family-specific pattern of 16S rRNA nucleotide positions. The pattern of 16S rRNA gene signature nucleotides for the order is as indicated for the family by Zhi et al. (2009).

Type genus: **Micromonospora** Ørskov 1923, 156[AL].

References

Ørskov, J. 1923. Investigations into the Morphology of the Ray Fungi. Levin and Munksgaard, Copenhagen, Denmark.

Stackebrandt, E., F.A. Rainey and N.L. Ward-Rainey. 1997. Proposal for a new hierarchic classification system, *Actinobacteria* classis nov. Int. J. Syst. Bacteriol. *47*: 479–491.

Zhi, X.-Y., W.-J. Li and E. Stackebrandt. 2009. An update of the structure and 16S rRNA gene sequence-based definition of higher ranks of the class *Actinobacteria,* with the proposal of two new suborders and four new families and emended descriptions of the existing higher taxa. Int. J. Syst. Evol. Microbiol. *59*: 589–608.

Family I. **Micromonosporaceae** Krasil'nikov 1938, 272[AL] emend. Zhi, Li and Stackebrandt 2009, 599

Olga Genilloud

Mi.cro.mo.no.spo.ra.ce′a.e. N.L. fem. n. *Micromonospora* type genus of the family; suff. *-aceae* ending to denote a family; N.L. fem. pl. n. *Micromonosporaceae* the *Micromonospora* family.

Aerobic, Gram-stain-positive, non-acid-fast organisms that form a non-fragmenting, branched, and septate mycelium that rarely carry scant aerial hyphae. The family is a member of the order *Micromonosporales* and cannot be distinguished from other suprageneric groups of *Actinomycetales* by using a set of exclusive phenotypic characters. The family encompasses a chemotaxonomically and morphologically diverse group of filamentous bacteria that, at the time of writing, included 17 phylogenetically closely related genera that present distinctive morphological and chemotaxonomic characteristics. **The organisms are mesophilic, showing optimum growth between 20 and 30°C, and grow well at pH 7;** growth can be observed in some genera from10 to 40°C and from pH 4.5 to 12. Colonies on agar media are flat to elevated with smooth or wrinkled surfaces and show a large variety of pigments. **Most strains contain carotenoid mycelial pigments**, but some produce blue-green and maroon to purple pigments. Diffusible pigments may be formed depending on the media. **Single spores, short spore chains, or sporangia borne directly from the substrate hyphae may be formed**. Spores may be motile with tufts of polar flagella. With few exceptions, cell walls contain ***meso*-2,6-diaminopimelic acid and glycolated muramic acid**. Predominant menaquinones may include all types of the MK-9 and MK-10 series. **Phosphatidylethanolamine is the diagnostic phospholipid**. Mycolic acids are absent. Members of the family *Micromonosporaceae* are widely distributed in the environment and have been found in sediments, soils, rhizospheres, colonizing living and decaying plant material, and in freshwater and marine habitats.

DNA G+C content (mol%): 69–73 (T_m, HPLC).

Type genus: **Micromonospora** Ørskov 1923, 156[AL].

Further descriptive information

The family *Micromonosporaceae* currently comprises the genera *Micromonospora, Actinocatenispora, Actinoplanes, Asanoa, Catellatospora, Catenuloplanes, Couchioplanes, Dactylosporangium, Krasilnikovia, Longispora, Luedemannella, Pilimelia, Polymorphospora, Salinispora, Spirilliplanes, Verrucosispora,* and *Virgisporangium.* These genera can be distinguished on the basis of their chemotaxonomic, phenotypic, and phylogenetic characteristics.

Phylogeny. Members of the family *Micromonosporaceae* are phylogenetically related as determined by 16S rRNA gene sequence analysis and form a coherent cluster within the actinomycete subphylum of Gram-stain-positive bacteria that is well separated from other families classified in the order *Actinomycetales* (Embley and Stackebrandt, 1994). The 16S rRNA gene sequence similarity values for members within the family *Micromonosporaceae* are above 91% (Matsumoto et al., 2003). The distribution of genus-specific properties and signature nucleotides that define the genera correlate with the phylogenetic distinctness of each genus. The relationships between the constituent genera of the family are shown in Figure 208.

Cell morphology. Members of the family *Micromonosporaceae* show three major types of sporulating structures, namely single spores, spore chains, and sporangia. The type genus *Micromonospora* Ørskov 1923 is characterized by the production of nonmotile spores that are borne singly, sessile, or on short or long sporophores, often in clusters on extensively branched substrate hyphae. Spores are spherical, ovoid, or ellipsoidal with a thick wall that may carry blunt spiny ornamentations (Kawamoto, 1989). The genus *Salinispora* Maldonado et al. 2005 cannot be distinguished morphologically from members of the genus *Micromonospora. Salinispora* strains do not produce aerial hyphae, but form spores singly or in clusters; the spores have smooth surfaces and are sessile or borne on short sporophores. Salinisporae are abundant and widespread in marine sediments and other marine substrates and will only grow on 20–25% seawater or on a sodium-enriched medium. The genus *Verrucosispora* Rheims et al. 1998 does not produce aerial hyphae, but does form granular structures on the mycelial surface. Nonmotile spores are borne singly and may be sessile or carried on short sporophores. Initially, the spores are warty, but turn hairy in older cultures.

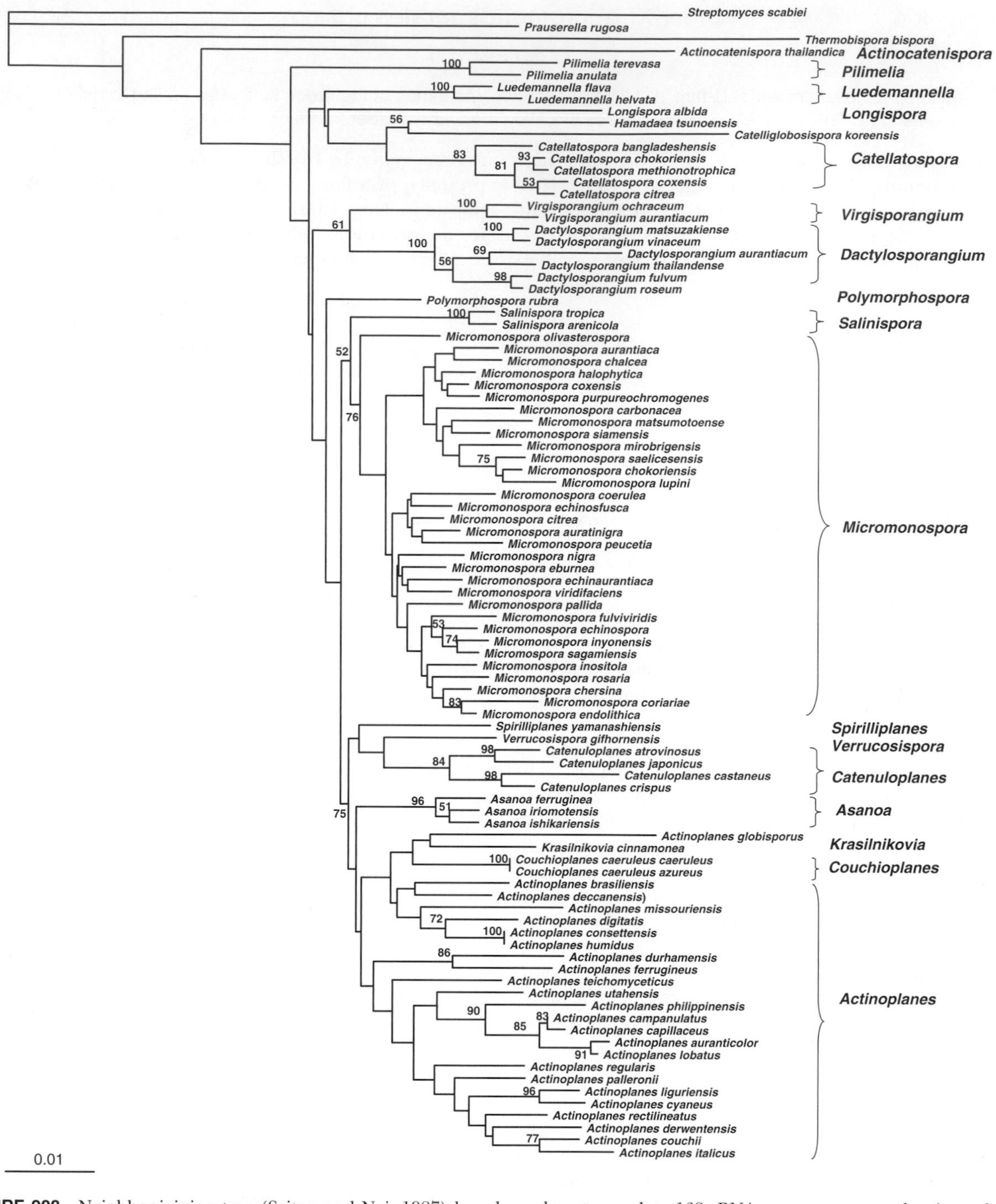

FIGURE 208. Neighbor-joining tree (Saitou and Nei, 1987) based on almost complete 16S rRNA gene sequences showing relationship between type strains classified in the family *Micromonosporaceae*. The tree is based on the Kimura two-parameter method with the confidence values of the branches determined by bootstrap analyses (Felsenstein, 1985) based on 1000 replicates. Only values >50% are shown at the nodes. Sequence accession numbers of type strains are given in the text. Bar = 0.01 substitutions per nucleotide position.

The genera *Actinoplanes* Couch 1950 emend. Stackebrandt et al. 1997, *Dactylosporangium* Thiemann et al. 1967, *Pilimelia* Kane 1966, and *Virgisporangium* Tamura et al. 2001 produce spore vesicles (sporangia) in which motile spores develop at the tips of short or long sporangiophores on the surface of substrate mycelia. The spores are formed within a sporangial envelope by fragmentation of branched or unbranched, straight or coiled sporogenous hyphae. The spore vesicles of the genera *Actinoplanes* and *Pilimelia* may be campanulate, cylindrical, digitate, lobate, ovoid, spherical, or irregular. Members of the genus *Pilimelia* can be differentiated from *Actinoplanes* strains by their keratinophilic character and their slow growth.

The genus *Dactylosporangium* produces two different types of spores: motile spores inside spore vesicles (sporangia) and non-motile spores borne singly on substrate hyphae. Motile spores are formed within finger-shaped to claviform, oligosporous spore vesicles that contain a single row of 2–5 spores. The sporangiospores are spherical, rod-shaped or oblong, ellipsoidal, or ovoid to pyriform. The spores are motile with polar or lateral tufts of flagella. The spore walls are single-layered and are not ornamented. The nonmotile spores are spherical and are formed terminally on short branches of the substrate hyphae (Vobis, 1989b, 1991). Members of the genus *Virgisporangium* Tamura et al. 2001 produce rod-shaped spore vesicles that develop singly or in clusters on the substrate mycelium. Each spore vesicle contains a single row of six or more oval to short-rod-shaped motile spores. In contrast, members of the genus *Luedemannella* Ara and Kudo 2007d form spherical spore vesicles on short sporangiophores that arise singly or in clusters from the substrate mycelium; each sporangiophore carries several nonmotile spherical to oval spores.

Members of the genera *Actinocatenispora* Thawai et al. 2006a, *Asanoa* Lee and Hah 2002, *Catellatospora* Lee and Hah 2002, *Catenuloplanes* Yokota et al. 1993, *Couchioplanes* Tamura et al. 1994, *Krasilnikovia* Ara and Kudo 2007a, *Longispora* Matsumoto et al. 2003, *Polymorphospora* Tamura et al. 2006, and *Spirilliplanes* Tamura et al. 1997 are non-sporangiate and produce chains of spores. Members of the genus *Catellatospora* do not form aerial hyphae, but produce short chains of nonmotile spores that emerge singly or in tufts from the vegetative mycelium (Asano and Kawamoto, 1986; Koch et al., 1996a). *Couchioplanes* strains form chains of motile spores that, together with aerial mycelia, are often aggregated into clusters that resemble sporangia, though true spore vesicles are not observed. Members of the genus *Krasilnikovia* also form globose pseudosporangial structures (Ara and Kudo, 2007a) that contain nonmotile spores.

Catenuloplanes strains form motile arthrospores in short chains that are arranged in one or two spirals that are often aggregated with normally sparse aerial mycelia. The spores are rod-shaped, straight, or curved with smooth surfaces (Kudo et al., 1999; Tamura et al., 1995; Yokota et al., 1993). Members of the genus *Spirilliplanes* produce long spore chains that are in spirals of five to ten turns. Spores are oval or rod-shaped with smooth surfaces and are motile in the presence of 10% soil extract or distilled water (Tamura et al., 1997).

The genus *Asanoa* was described to accommodate two species that were initially assigned to the genus *Catellatospora*, but it was found subsequently that these species could be clearly differentiated from members of the genus *Catellatospora* and other genera with validly published names classified in the family *Micromonosporaceae*. *Asanoa* strains sporulate poorly on tapwater and glycerol/calcium malate agars (Lee and Hah, 2002). Members of the taxon do not produce aerial mycelia or globose bodies, which are observed in some *Catellatospora* species. Members of the genus *Longispora* Matsumoto et al. 2003 produce a sparse aerial mycelium and only sporulate when grown on gellan gum; spores are borne on short sporophores that arise from the vegetative mycelium. The spores are cylindrical, have smooth surfaces, and are carried in short straight chains of 20 or more spores. Members of the genus *Polymorphospora* Tamura et al. 2006 develop short chains of nonmotile spores of varying shapes that differentiate into rods. Finally, members of the genus *Actinocatenispora* Thawai et al. 2006a produce aerial hyphae on oatmeal-nitrate agar; these develop into cylindrical spore chains containing more than 10 spores.

Chemotaxonomy. The phenotypic heterogeneity of the family *Micromonosporaceae* is further increased when chemotaxonomic markers are considered, namely the amino acids of the peptidoglycan, whole-organism sugar patterns, phospholipid types, and menaquinone and fatty acid composition (Table 190). Most members of the family are characterized by a cell-wall chemotype II *sensu* Lechevalier and Lechevalier (1970a); the wall peptidoglycan contains *meso*-diaminopimelic acid (*meso*-A_2pm) and/or 3-hydroxy-A_2pm. The first amino acid of the peptide side chain is glycine in all members of the family, apart from *Pilimelia* which contains acetate. The genus *Pilimelia* can be differentiated from other members of the family as it contains *N*-acetylated, as opposed to *N*-glycolated, muramic acid (Koch et al., 1996a). Deviations from this wall chemotype have been observed in *Micromonospora* species that contain LL-A_2pm (Kawamoto et al., 1981) and in *Couchioplanes* and *Catenuloplanes* species that have L-lysine and D- and L-serine as diagnostic amino acids in the peptidoglycan rather than *meso*-A_2pm acid (Horan and Brodsky, 1986a; Tamura et al., 1994; Yokota et al., 1993).

Arabinose, galactose, and xylose are the major sugars present in whole-organism hydrolysates [sugar pattern D *sensu* Lechevalier and Lechevalier (1970a)] with variable amounts of other sugars. The diagnostic phospholipid of the cell membrane is phosphatidylethanolamine [phospholipid type II *sensu* Lechevalier et al. (1977)], but diphosphatidylglycerol, phosphatidylglycerol, and phosphatidylinositol may be present. However, *Catenuloplanes* strains also contain phosphatidylcholine (Yokota et al., 1993) and, hence, have a phospholipid type III (Table 190).

Members of the family *Micromonosporaceae* contain complex and highly variable fatty acids, although saturated iso- and anteiso-branched fatty acids are predominant in nearly all genera, apart from the genus *Pilimelia*. Unsaturated fatty acids may or may not be present, although they are major components in *Longispora* strains (Matsumoto et al., 2003). 10-Methyl-branched fatty acids are found in certain strains of *Micromonospora*; cyclic fatty acids are absent from all strains, as are mycolic acids. The menaquinone composition is heterogeneous. Major amounts of tetra-, hexa- and/or octahydrogenated menaquinones with 9, 10, and 12 isoprene units are distributed among genera and species (Koch et al., 1996a; Vobis, 1989b) (Table 190).

Genetics. A family-specific pattern of 16S rRNA gene signatures was defined by Stackebrandt et al. (1997) to distinguish members of the family *Micromonosporaceae* from all other actinomycetes. The 11 16S rRNA gene signatures detected in representatives of *Actinoplanes, Catellatospora, Couchioplanes, Catenuloplanes, Dactylosporangium, Micromonospora*, and *Pilimelia* consist of nucleotides at positions 66:103 (G–C), 127:234 (A–U), 153:168 (C–G), 502:543 (G–C), 589:650 (C–G), 747 (A), 811 (U), 840:846 (C–G), 952:1229 (C–G), 1116:1184 (C–G), and 1133:1141 (G–C) (Stackebrandt et al., 1997). Since then, genus-specific signature nucleotides have been described for *Actinocatenispora* [positions 502:543 (U–G), 747 (G), and 811 (C)], *Krasilnikovia* [positions 445:489 (C–G), 446:488 (C–G), 1011:1018 (U–A), and 1263:1272 (G–U)], *Longispora* [position 502 (A); Matsumoto et al., 2003], *Luedemannella* [positions 139:224 (U–A), 381 (A), 656:750 (A/G–C), and 999:1041 (C–U)], *Polymorphospora* at position 1244 (U) (Tamura et al., 2006), *Verrucosispora* [positions 1133:1141 (A–U); Rheims et al., 1998], and *Virgisporangium* [positions 502:543 (G–C) and 1116:1184 (C–G); Tamura et al., 2001].

Acknowledgements

I am especially grateful to Dr Oscar Salazar for the updated phylogenetic tree of the family *Micromonosporaceae*.

TABLE 190. Morphological and chemotaxonomic characteristics of members of the family *Micromonosporaceae*[a]

Characteristic	*Micromonospora*	*Actinocatenispora*	*Actinoplanes*	*Asanoa*	*Catellatospora*	*Catenuloplanes*	*Couchioplanes*	*Dactylosporangium*	*Krasilnikovia*	*Longispora*	*Luedemannella*	*Polymorphospora*	*Pilimelia*	*Salinispora*	*Spirilliplanes*	*Verrucosispora*	*Virgisporangium*
Single spores	+	–	–	–	–	–	–	+	–	–	–	–	–	+	–	+	–
Sporangia	–	–	+	–	–	–	–	+	–	–	+	–	+	–	–	–	+
Spore chains	–	+	–	+	+	+	+	–	+	+	–	+	–	–	+	–	–
Motile spores	–	–	+	–	–	+	+	+	–	–	–	–	+	–	+	–	+
Salt required	–	–	–	–	–	–	–	–	–	–	–	–	–	+	–	–	–
Cell-wall chemotype[b]	II	II	II	II	II	VI	VI	II	II	II	II	II	II	II	II	II	II
Whole-cell sugars[c]	Ara, Xyl	Gal, Xyl, Ara, Glu, Man, Rib	Ara, Xyl	Ara, Xyl, Gal	Ara, Xyl, Gal	Xyl	Ara, Xyl, Gal	Ara, Xyl	Gal, Man, Xyl, Ara, Rib, Glu	Ara, Xyl, Gal	Xyl, Gal, Man, Rha, Rib, Ara	Xyl	Ara, Xyl	Ara, Xyl, Gal	Xyl, Gal	Man, Xyl	Ara, Gal, Man, Rha, Xyl
Fatty acid type[d]	3b	3b	2d	2d	3b	2c	2c	3b	2d	2d	2d	2a	2d	3a	2d	2b	2d
Major menaquinones (MK-)	10(H_4, H_6) 9(H_4,H_6)	9(H_4, H_6)	9(H_4) 10(H_4)	10(H_6, H_8)	10(H_8, H_6) 9(H_4, H_6)	9(H_8)[e], 10(H_8)	9(H_4)	9(H_4, H_6, H_8)	9(H_6)	10(H_4, H_6)	9(H_4, H_6)	10(H_4, H_6) 9(H_4, H_6)	9(H_2, H_4)	10(H_4)	10(H_4)	9(H_4)	10(H_4, H_6, H_8)
Phospholipid type[f]	II	II	II	II	II	III	II	II	II	II	II	II	II	II	II	II	II
DNA G+C content (mol%)	71–73	72	72–73	71–72	71–72	71–73	70–72	72–73	71	70	71	70–71	70–72	70–73	69	70	71–72

[a]Except where marked, data are from Stackebrandt and Kroppenstedt (1987), Vobis (1989b), Goodfellow et al. (1990b), Yokota et al. (1993) Tamura et al. (2006, 1997, 2001, 1994), Rheims et al. (1998), Kudo et al. (1999), Lee et al. (2000b), Lee and Hah (2002), Matsumoto et al. (2003), Maldonado et al. (2005), Thawai et al. (2006a), and Ara and Kudo (2007a, 2007d). +, Positive; –, negative.

[b]Data from Lechevalier and Lechevalier (1970a).

[c]Ara, arabinose; Gal, galactose; Glu, glucose; Man, mannose; Rha, rhamnose; Rib, ribose; Xyl, xylose.

[d]Data from Kroppenstedt (1985).

[e]Reported to be MK-10(H_4) and MK-11(H_4) by Kudo et al. (1999).

[f]Data from Lechevalier et al. (1977).

Genus I. **Micromonospora** Ørskov 1923, 156[AL]

Olga Genilloud

Mi.cro.mo.no.spo′ra. Gr. adj. *mikros* small; Gr. adj. *monos* single, solitary; Gr. fem. n. *spora* a seed and in biology a spore; N.L. fem. n. *Micromonospora* small, single-spored (organism).

Well-developed, branched, substrate mycelium (0.2–0.6 μm diameter). **Nonmotile spores are borne singly, sessile, or terminally on short sporophores**. Sporophore development is monopodial or in some cases sympodial. Spores are spherical to oval in shape (0.7–1.5 μm) and in most species have blunt spiny projections. The spores are often carried in branched clusters on short hyphae of the substrate mycelium. **Aerial mycelium is usually absent**, but some cultures develop sterile short aerial hyphae. Gram-stain-positive, mesophilic, non-acid-fast. Aerobic to microaerophilic. Chemo-organotrophic. Sensitive to pH below 5.0. Growth usually occurs between 20 and 40°C, but not above 50°C. **Cell wall contains *meso*-diaminopimelic acid and/or 3-OH-diaminopimelic acid. Arabinose and xylose are the characteristic sugars present in whole-organism hydrolysates of most species**, but variable amounts of galactose, glucose, mannose, and rhamnose can also be found depending on the species. **The major phospholipids are phosphatidylethanolamine, phosphatidylinositol, and phosphatidylinositol mannosides**. Saturated and unsaturated fatty acids are present, mostly $C_{15:0}$ iso, $C_{17:0}$ iso, $C_{15:0}$ anteiso, $C_{17:0}$ anteiso, $C_{17:0}$ 10-methyl, and $C_{17:0}$ 10-methyl. Mycolic acids are absent. Major menaquinones are MK-9(H_4), MK-9(H_6), MK-10(H_4), MK-10(H_6), or MK-12(H_6). Isolated from soil, plant materials, freshwater, and marine environments.

DNA G+C content (mol%): 68–75 (T_m, HPLC).

Type species: **Micromonospora chalcea** (Foulerton 1905) Ørskov 1923, 156[AL] ("*Streptothrix chalcea*" Foulerton 1905, 1199).

Further descriptive information

The genus *Micromonospora* is well defined in terms of morphological, chemotaxonomic, and phylogenetic criteria (Kawamoto, 1989; Koch et al., 1996a; Kroppenstedt, 1985) that distinguish it from other members of the family *Micromonosporaceae* (Table 191). *Micromonospora* species are defined based on chemotaxonomic markers, physiological characteristics, and on phylogenetic relationships.

Phylogeny. The intrageneric structure of the genus *Micromonospora* was described first by Koch et al. (1996b) in an analysis of 16S rRNA gene sequences of representatives of 15 species with validly published names, four subspecies, and 19 species with names that have not been validly published. This and later phylogenetic studies based on 16S rRNA gene sequences of the type strains of the genus have shown that *Micromonospora* species form a tight clade within the family *Micromonosporaceae*; they show 16S rRNA similarity values within the range 96.7–99% (Koch et al., 1996b). These high values cannot be used to confirm the species status of closely related strains, not least because nucleotide substitutions in variable regions of RNA genes may induce errors when establishing their relative phylogenetic position. However, *Micromonospora* species tend to group consistently into subclusters, the relative positions of which are generally conserved in the topology of phylogenetic trees, although *Micromonospora olivasterospora* represents an unrelated line of descent (Koch et al., 1996b; Kroppenstedt et al., 2005a; Thawai et al., 2005b). Nevertheless, few internal relationships can be confirmed by bootstrap values higher than 50%. The phylogenetic relationships between *Micromonospora* species are presented in Figure 209.

Cell morphology. *Micromonospora* strains frequently form raised and folded colonies on agar media. Colonies are initially pale yellow to light orange and, depending on the strain, they develop a blue-green, brown, dark orange, purple, or red pigmentation with age. Soluble pigments may be formed. Older colonies tend to be covered by a black, brown-black, or green-black mucous mass of spores (Kawamoto, 1989). Spores often occur in dense clusters and may be produced in distinct areas on the surface or be completely embedded in the substrate. Mycelial pigments sensitive to pH changes have been observed in *Micromonospora coerulea* (blue-green) and *Micromonospora echinospora* (maroon-purple) (Kawamoto, 1989). Other species characterized by soluble pigments are *Micromonospora chalcea* (yellow), *Micromonospora halophytica* (red-brown), *Micromonospora olivasterospora* (olive-green), *Micromonospora purpureochromogenes* (dark-brown), and *Micromonospora rosaria* (wine-red) (Horan and Brodsky, 1986b; Kawamoto, 1989). Nevertheless, mycelial pigmentation, mycelial development, and sporophore morphology have little diagnostic value for the identification of *Micromonospora* species.

Micromonosporae often produce short sporophores which occur in branched clusters on short hyphae of the substrate mycelium (Luedemann and Casmer, 1973). The sympodial sporophore of *Micromonospora carbonacea* (Luedemann and Brodsky, 1964) has also been observed in other species. The formation of single spores on the substrate mycelium is the main morphological characteristic of the genus *Micromonospora*. Occasionally, spores have been found longitudinally in pairs or in larger groups of spores (Luedemann and Casmer, 1973) and a sporulation process resembling sporangial formation has been described in *Micromonospora purpureochromogenes* NRRL B-2671 (Stevens, 1975).

The spore surface ornamentations formed on the outer layer of the spore wall differ from those observed in the sheath of some *Streptomyces* spores (Luedemann and Casmer, 1973). Spore ornamentations of *Micromonospora* strains described originally on the basis transmission electron microcopy as "smooth", "warty", or "blunt-spiny" have been shown by scanning electron microscopy to have blunt-spiny surfaces with variable spine sizes, but this characteristic is not a diagnostic characteristic for the differentiation of *Micromonospora* species (Kawamoto, 1989) (Figure 210).

Sporogenesis involves the formation of a sporulation septum that is followed by spore maturation with a thickening of wall layers (Hardisson and Suarez, 1979; Luedemann and Casmer, 1973). Spore dehiscence is favored by a stretching of the hyphal wall near the sporulation septum. The thickened walls may explain the high refractility and the relative differential resistance of mature spores to physical and chemical treatments, such as sonication or heat treatment up to 75°C

TABLE 191. Differential characteristics of the genus *Micromonospora* and related taxa classified in the family *Micromonosporaceae*

Characteristic	*Micromonospora*	*Actinocatenispora*	*Actinoplanes*	*Asanoa*	*Catellatospora*	*Catenuloplanes*	*Couchioplanes*	*Dactylosporangium*	*Krasilnikovia*	*Longispora*	*Luedemannella*	*Pilimelia*	*Polymorphospora*	*Salinispora*	*Spirilliplanes*	*Verrucosispora*	*Virgisporangium*
Single spores	+	−	−	−	−	−	−	+	−	−	−	−	−	+	−	+	−
Sporangia	−	−	+	−	−	−	−	+	−	−	+	+	−	−	−	−	+
Spore chains	−	+	−	+	+	+	+	−	+	+	−	−	+	−	+	−	−
Motile spores	−	−	+	−	−	+	+	+	−	−	−	+	−	−	+	−	+
Salt requirement	−	−	−	−	−	−	−	−	−	−	−	−	−	+	−	−	−
Cell-wall chemotype[a]	II	II	II	II	II	VI	VI	II	II	II	II	II	II	II	II	II	II
Whole-organism sugars	Ara, Xyl	Gal, Xyl, Ara, Glu, Man, Rib	Ara, Xyl	Ara, Xyl, Gal	Ara, Xyl, Gal	Xyl	Ara, Xyl, Gal	Ara, Xyl	Gal, Man, Xyl, Ara, Rib, Glu	Ara, Xyl, Gal	Xyl, Gal, Man, Rha, Rib, Ara	Ara, Xyl	Xyl	Ara, Xyl, Gal	Xyl, Gal	Man, Xyl	Ara, Gal, Man, Rha, Xyl
Fatty acid type[b]	3b	3b	2d	2d	3b	2c	2c	3b	2d	2d	2d	2d	2a	3a	2d	2b	2d
Major menaquinones (MK-)	$10(H_4, H_6)$ $9(H_4, H_6)$	$9(H_4, H_6)$	$9(H_4)$ $10(H_4)$	$10(H_6, H_8)$	$10(H_8, H_6)$ $9(H_4, H_6)$	$9(H_8)$, $10(H_8)$	$9(H_4)$	$9(H_4, H_6, H_8)$	$9(H_6)$	$10(H_4, H_6)$	$9(H_4, H_6)$	$9(H_2, H_4)$	$10(H_4, H_6)$ $9(H_4, H_6)$	$10(H_4)$	$10(H_4)$	$9(H_4)$	$10(H_4, H_6, H_8)$
Phospholipid type[c]	PII	PII	PII	PII	PII	PIII	PII	PII	PII	PII	PII	PII	PII	PII	PII	PII	PII

[a]Lechevalier and Lechevalier (1970a, 1970b).

[b]Kroppenstedt (1985).

[c]Lechevalier et al. (1977).

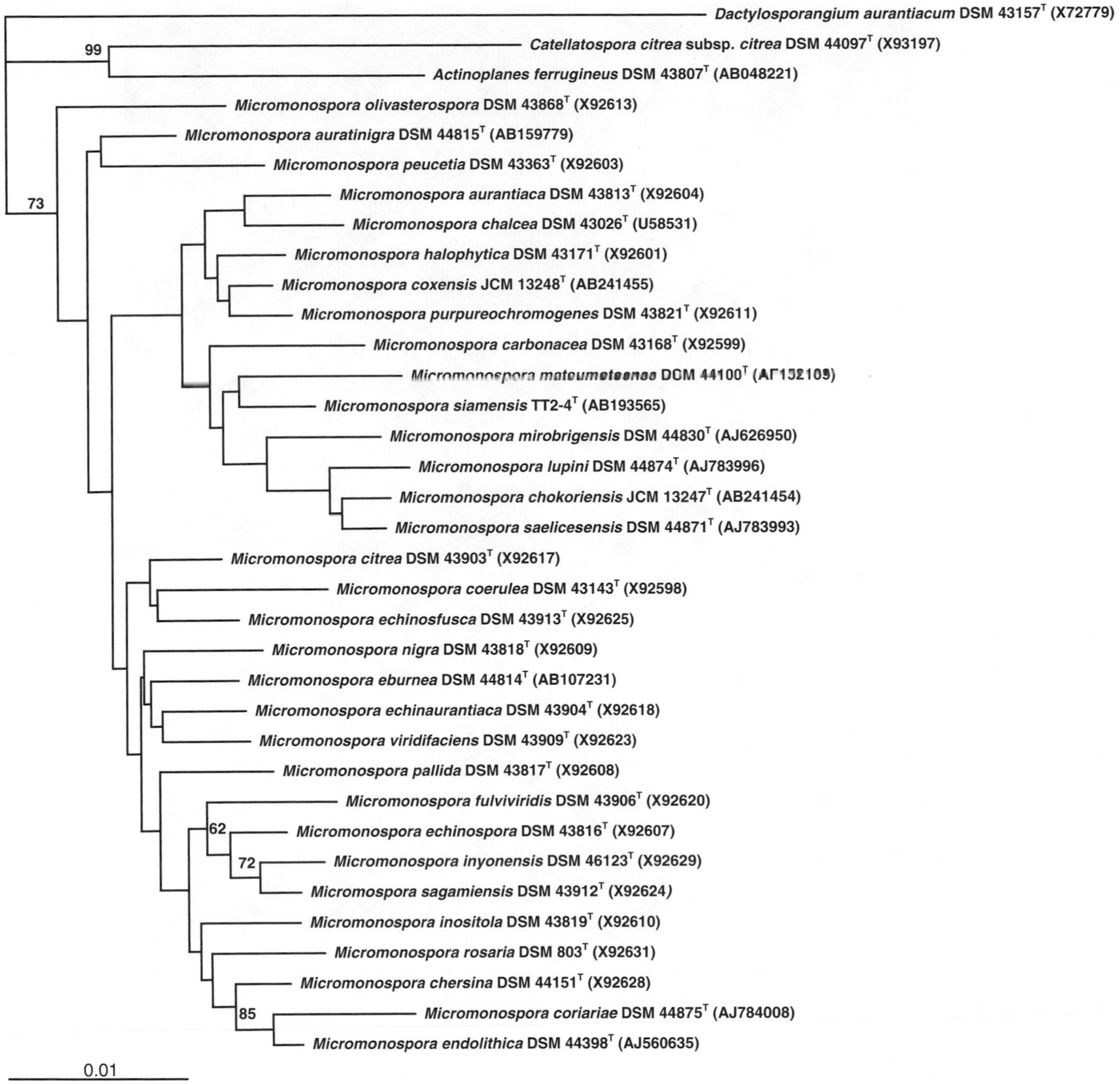

FIGURE 209. Neighbor-joining tree (Saitou and Nei, 1987) based on almost-complete 16S rRNA gene sequences showing relationships between type strains of the genus *Micromonospora*. The tree is based on the Kimura two-parameter method and confidence values of branches were determined by bootstrap analysis (Felsenstein, 1985) based on 1000 replicates. Only values >50% are shown at nodes. Bar = 0.01 substitutions per nucleotide position.

(Johnston and Cross, 1976; Kawamoto et al., 1982; Suárez et al., 1980). Spore viability is only markedly reduced at temperatures above 75°C, as shown in *Micromonospora chalcea* (Suárez et al., 1980). Spores are resistant to desiccation and remain viable in soil or in mud for extended periods (Cross and Attwell, 1974). Heat shock above 70°C induces spore germination (Ensign, 1982; Hoskisson et al., 2000) and dry heat treatments at 120°C clearly select for strains of this genus (Hayakawa et al., 1991b). *Micromonospora* spores show some resistance to treatment with different chemical agents, such as acetone (Kawamoto et al., 1982), dimethylformamide, formamide, *tert*-butyl alcohol, and phenol (Hayakawa et al., 1991b). Spore viability is not affected within the pH range 6.0–8.0, but decreases at acidic pH values (Kawamoto et al., 1982). In general, the spore germination process in *Micromonospora* is slower but similar to that observed in other actinomycetes (Kawamoto, 1989).

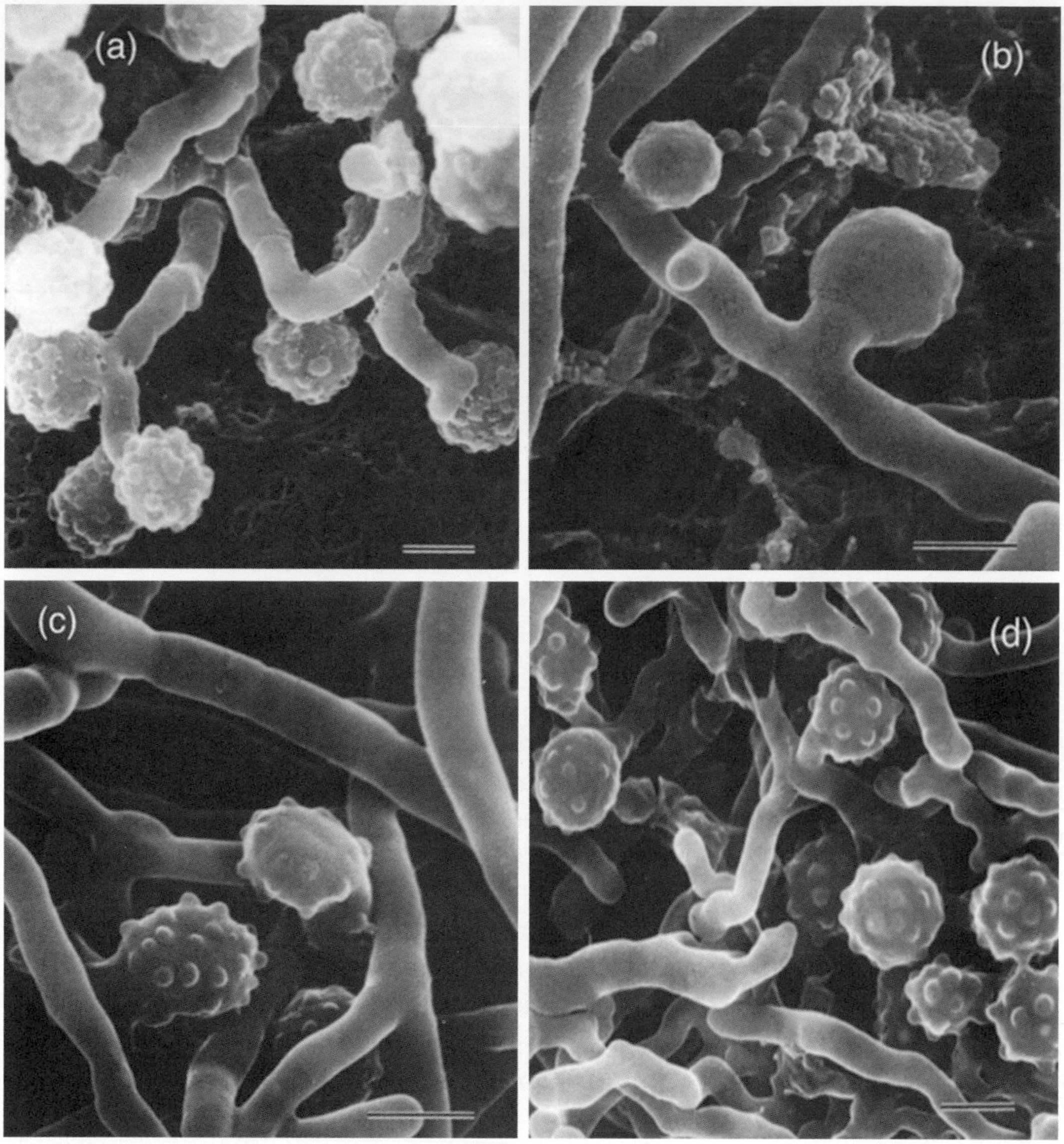

FIGURE 210. Scanning electron micrographs of: (a) *Micromonospora carbonacea* NRRL 2972^T; (b) *Micromonospora chalcea* ATCC 12452^T; (c) *Micromonospora purpureochromogenes* ATCC 27007^T; and (d) *Micromonospora echinospora* NRRL 2985^T. Bar = 0.5 µm.

Cell-wall composition. The genus *Micromonospora* is characterized by a cell-wall type II (Lechevalier and Lechevalier, 1970a, 1970b), a whole-organism sugar pattern D (Lechevalier and Lechevalier, 1970a, 1970b), a phospholipid type PII (Lechevalier et al., 1977), and a fatty acid pattern 3b (Kroppenstedt, 1985). Cell-wall hydrolysates may contain glycine, glutamic acid, diaminopimelic acid (A_2pm), including *meso*- and 3-OH-A_2pm, and D-alanine (Kawamoto et al., 1981). The presence of 3-OH-A_2pm is limited to certain species and its presence/absence divides *Micromonospora* species into two groups (Kawamoto, 1989). The acyl type of the cell-wall muramic acid is glycolyl and there is an almost equimolar ratio of glycolic acid and A_2pm (Kawamoto et al., 1981). Glycine is always present in the first position of the peptide subunit that is attached to *N*-glycolylmuramic acid. However, the peptidoglycan cannot always be hydrolyzed with lysozyme (β-*N*-acetylmuramidase) as observed in *Micromonospora olivasterospora* and *Micromonospora sagamiensis* (Kawamoto et al., 1981) or in strains where protoplasts cannot be effectively obtained (Szvoboda et al., 1980). The cell-wall sugars arabinose and xylose are constituents of the cell wall of most *Micromonospora* species, albeit in variable amounts. Glucose, galactose, mannose, and rhamnose can also be found in whole-organism hydrolysates, although amounts differ from species to species (Kawamoto et al., 1981); galactose, glucose, and rhamnose are found instead of arabinose in certain species (Kroppenstedt et al., 2005a; Trujillo et al., 2005).

Lipids and menaquinone composition. Characteristic phospholipids include diphosphatidylglycerol, phosphatidylethanolamine, phosphatidylinositol, and phosphatidylinositol mannosides (Lechevalier et al., 1977). Dassain et al. (1983) identified the glycolipids, monoglucosyldiglycerides, diglucosyldiglycerides, esters of trehalose, and fatty acids. Predominant fatty acids are iso- and anteiso-branched components. Fatty acids patterns are of type 3b with saturated and unsaturated fatty acids including $C_{15:0}$ anteiso, $C_{17:0}$ anteiso, $C_{17:0}$ 10-methyl, $C_{18:0}$ 10-methyl, $C_{16:0}$ iso, $C_{15:0}$ iso, and $C_{17:0}$ iso acids (Kroppenstedt, 1985). Mycolic acids and cyclic fatty acids are not present. The menaquinone composition of members of the genus *Micromonospora* is complex and heterogeneous; strains can be divided into three groups based on the predominant menaquinone. The two major groups encompass species that have menaquinones with either nine (MK-9) or ten isoprene units (MK-10); the exception, *Micromonospora pallida*, has menaquinones with 12 units (MK-12) (Collins et al., 1984; Hirsch et al., 2004a; Kawamoto, 1989; Kroppenstedt et al., 2005a; Thawai et al., 2004b, 2005a, 2005b; Tomita et al., 1992b; Trujillo et al., 2005).

Growth conditions. The identification of *Micromonospora* species is supported by physiological characteristics, such as growth on special media, carbon utilization profiles, glycosidase activity, nitrate reduction, and NaCl tolerance (Kawamoto, 1989). Fructose, glucose, sucrose, and starch are utilized as sole carbon sources by almost all strains and most strains utilize cellobiose, galactose, mannose, trehalose, and xylose, but little, if any, growth occurs with glycerol, inositol, mannitol, rhamnose, ribose, or salicin. Carbohydrate utilization patterns can be affected by the basal medium and this can account for discrepancies between results reported by different authors (Kawamoto, 1989). *Micromonospora* strains show α- and β-glucosidase, β-galactosidase, and β-*N*-acetylglucosaminidase activities, but variable results are obtained for α-galactosidase, α-mannosidase, and β-xylosidase (Kawamoto, 1989). Inorganic ammonium salts and acidic and basic amino acids are better nitrogen sources than nitrate salts (Kawamoto et al., 1983). NaCl tolerance ranges from 1.5 to 5% (w/v). Micromonosporae do not grow below pH 5.0 or above pH 9.5 and most strains do not grow at 45°C.

Genetics. *Micromonospora* species are defined genomically by having total genomic DNA–DNA reassociation values lower than 70%; this contrasts with their high 16S rRNA similarity values. Indeed, 16S rRNA gene sequences are almost identical with similarities of 99.7–99.9% for members of *Micromonospora* species and subspecies that were later reclassified as synonyms in DNA–DNA hybridization studies (Koch et al., 1996b). These workers also noted that the composition of the different groups of *Micromonospora* species are supported by signature nucleotides between 16S rRNA gene positions 603 and 627 (Koch et al., 1996b). DNA–DNA reassociation studies have confirmed that the genus is overspeciated (Kasai et al., 2000). The protein-encoding gene *gyrB* has been used as an alternative gene to determine the phylogenetic position and relationships of *Micromonospora* type strains given the high consistency found in their 16S rRNA gene sequences (Kasai et al., 2000). The branching patterns in the *gyrB*-based phylogenetic tree of *Micromonospora* type strains representing 14 species were similar to those of the corresponding 16S rRNA-based tree (Kasai et al., 2000).

Genetics, phages, and plasmids. Various lytic and temperate phages, and phages with no characterized infection cycles have been described in *Micromonospora* species (Alexander et al., 2003; Caso et al., 1990; Kikuchi and Perlman, 1977, 1978; Li et al., 2004; Tilley et al., 1990). Plasmids occur frequently, with many of them being cryptic. Numerous host vectors and shuttle vector systems have been developed from *Micromonospora* plasmids and have allowed genetic recombination studies of important biosynthetic gene clusters (Hosted et al., 2005; Inouye et al., 1994; Li et al., 2003; Parag and Goedeke, 1984; Takada et al., 1994; Vukov and Vasiljevic, 1998).

Ecology. *Micromonospora* strains can be isolated easily from soil and aquatic environments, but almost all described type strains have a terrestrial origin. Micromonosporae are very common in alkaline and neutral soils (Jensen, 1930; Vobis, 1992) and although many strains have been shown to be sensitive to acid pH (Kawamoto, 1989), they have also been isolated from acid soils and plant substrates on acidified media (Zenova et al., 2004). *Micromonospora* spores are hydrophilic, highly resistant to heat treatment and to different chemical agents, and can survive for years in the environment (Cross, 1981b).

It was widely accepted for a long time that the principal habitats of micromonosporae were aquatic ecosystems (Cross, 1981a; Goodfellow and Haynes, 1984). *Micromonospora* strains have been isolated extensively from freshwater and marine aquatic environments (Cross, 1981a, 1981b; Goodfellow and Williams, 1983), are frequently present in water samples of streams and rivers (Burman, 1973), and are amongst the predominant actinomycetes in lake sediments and mud samples (Johnston and Cross, 1976). An active role in the degradation of biopolymers in sediments has been proposed given their ability to decompose cellulose, chitin, and xylan (Erikson, 1941; Hunter et al., 1981). Their tolerance to low oxygen tensions suggests that they may grow under the microaerophilic conditions found in alluvial soils, floodplain meadows, and wet soils of river ecosystems (Goodfellow and Williams, 1983; Vobis, 1992; Zenova and Zviagintsev, 2002). *Micromonospora* strains predominate in actinomycete communities isolated from plant litter, lichens, roots, and organic soil horizons (González et al., 2005; Zenova et al., 1994). Their widespread occurrence even in vegetated arid zones has been explained by their ability to degrade biopolymers (Zenova et al., 1994).

Micromonospora strains have been found to be constituents of endophytic actinobacterial populations recovered from plant tissues; they have been described as colonizing roots of *Casuarina* and *Triticum* species (Coombs and Franco, 2003; Valdés et al., 2005). A selective advantage has been suggested for their association with plants given their ability to suppress the fungal pathogens *Pythium* and *Phytophthora* (Coombs and Franco, 2003). An isolate most closely related to "*Micromonospora yulongensis*" (name not validly published) has been described from surface-sterilized wheat root tissue and culture-independent methods have identified clones of endophytic actinobacteria closely related to *Micromonospora endolithica* and *Micromonospora peucetica* in wheat roots suggesting the presence of a large diversity of *Micromonospora* species in plants (Conn and Franco, 2004; Coombs and Franco, 2003). More recently, surface-sterilized legume nodules of *Lupinus angustifolius*, *Pisum sativum*, and *Vicia sativa* have been shown to be extremely rich reservoirs for the isolation of novel *Micromonospora* strains (Trujillo et al., 2006, 2007).

Micromonospora species have been isolated from diverse marine habitats, including coastal areas (Watson and Williams, 1974), salt marshes (Hunter et al., 1981), and deep-sea sediments (Weyland, 1969, 1981). It was widely accepted for many years that *Micromonospora* spores were washed from soils and accumulated as dormant spores in lake and sea sediments (Cross, 1981a, 1981b; Goodfellow and Haynes, 1984; Johnston and Cross, 1976; Weyland, 1969, 1981). Despite prior evidence suggesting the adaptation of actinomycetes to the marine environment and the existence of an indigenous marine population (Jensen et al., 1991; Magarvey et al., 2004; Mincer et al., 2002; Okami and Okazaki, 1978; Takizawa et al., 1993), it was only recently that active mycelial growth was demonstrated in sand particles (Jensen et al., 2005). In spite of this, many actinomycetes isolated using seawater-based media grow well in media prepared using distilled water.

Secondary metabolism. Micromonosporae, after streptomycetes, are one of the most prolific producers of bioactive secondary metabolites; they produce an extremely large diversity of chemical structures. Bérdy (2005) cited more than 740 *Micromonospora* strains as producers of different bioactive metabolites. They are particularly important producers of aminoglycosides (fortimicins, gentamicins, kanamycin, mannosidostreptomycin, neomycin B, paromamine, sagamicin, sisomicin, and verdamicin), ansamycins (halomicins, rifamycins), anthracyclines (daunorubicin, sibanomicin), anthraquinones (dynemicin), macrolides (antibiotic XK 41-B-2, erythromycins, juvenimicins, megalomycins, rosaramicin, and rustmicin), oligosaccharides (antlermicin and everninomicin), enediyne antitumor antibiotics (calicheamycin), and peptide antibiotics (actinomycin, bottromycin, microsporin, and thiocoraline) (Horan, 1999; Vobis, 1992; Wagman and Weinstein, 1980).

Enrichment and isolation procedures

Micromonosporae are easily isolated from soils and sediments by plating serial dilutions onto suitably nutritionally poor selective media that limit the growth of members of fast-growing bacterial species. The media most commonly used are arginine-glycerol-salts agar (Hunter et al., 1984), arginine-vitamin agar, starch-casein-nitrate agar supplemented with B vitamins (Shearer, 1987), cellulose-asparagine agar (Goodfellow and Haynes, 1984), colloidal chitin agar (Hsu and Lockwood, 1975), humic acid-vitamins agar (Hayakawa and Nonomura, 1987a), Kodoka's cellulose benzoate medium (Sandrak, 1977), and M3 medium (Rowbotham and Cross, 1977); inoculated plates are incubated at 28–30°C for at least 2–3 weeks. The addition of antibiotics such as gentamicin (1–10 µg/ml) (Ivanitskaia et al., 1978), nalidixic acid (20–30 µg/ml), novobiocin (25–50 µg/ml) (Sveshnikova et al., 1976), and tunicamycin (20–50 µg/ml) (Nonomura and Hayakawa, 1988; Wakisaka et al., 1982) to media promote the isolation of *Micromonospora* species. Different pretreatment regimes can also be used such as heating soil suspensions at 70°C for 10–30 min (Rowbotham and Cross, 1977; Sandrak, 1977), and dry heating of soil samples at 120°C for 60 min (Shearer, 1987). The pretreatment of soil suspensions with ammonia then chlorine for 10–30 min (Burman et al., 1969; Willoughby, 1969a) or with 1.5% (w/v) phenol followed by plating onto humic acid-vitamins agar supplemented with nalidixic acid (20 µg/ml) and tunicamycin (20 µg/ml) have been shown to be highly effective isolation methods (Hayakawa et al., 1991b).

The isolation of micromonosporae from plant materials, including roots, requires a preliminary sterilization of the material surface by successive treatments in 99% ethanol, 3.1% NaOCl, and 99% ethanol prior to plating (Coombs and Franco, 2003). Selective media prepared with artificial seawater are used in combination with specific methods to isolate micromonosporae from marine sediments and marine invertebrates. Dilution and heat shock treatments, stamping methods, and enrichments based on the incubation of sediments plated on membrane filters in humid chambers have been effective for the isolation of *Micromonospora* strains from sediments (Magarvey et al., 2004; Mincer et al., 2002). The slow growth of isolates frequently requires extended incubation for up to 10 weeks at lower temperatures (Goodfellow and Haynes, 1984).

Maintenance procedures

Sporulated cultures can be maintained for months on agar slopes or sealed plates at 4°C. Long-term storage can be achieved by lyophilization, by liquid drying, or by maintaining spores or liquid seed cultures in 10–15% glycerol at −80°C.

Differentiation of the genus *Micromonospora* from other genera

The morphological and chemotaxonomic characteristics that distinguish *Micromonospora* from other genera classified in the family *Micromonosporaceae* are given in Table 191. The genus can be differentiated from other members of the family on the basis of cell morphology, chemotaxonomy, and 16S rRNA gene sequence analysis.

Taxonomic comments

The description of *Micromonospora* species began with taxonomic studies on gentamicin-producing strains (Luedemann and Brodsky, 1964). The *Approved Lists of Bacterial Names* (Skerman et al., 1980) included 12 species (*Micromonospora aurantiaca*, *Micromonospora brunnea*, *Micromonospora carbonacea*, *Micromonospora chalcea*, *Micromonospora coerulea*, *Micromonospora echinospora*, *Micromonospora gallica*, *Micromonospora halophytica*, *Micromonospora inositola*, *Micromonospora purpurea*, *Micromonospora purpureochromogenes*, and *Micromonospora rhodorangea*) and seven subspecies as members of the genus *Micromonospora*, though the type strain of *Micromonospora gallica* is not extant (Kawamoto, 1989). *Micromonospora olivasterospora* (Kawamoto et al., 1983), *Micromonospora rosaria* (Horan and Brodsky, 1986b), and *Micromonospora chersina* (Tomita et al., 1992b) were described later and *Catellatospora matsumotoense* was transferred to the genus as *Micromonospora matsumotoense* (Lee et al., 1999).

DNA–DNA hybridization experiments and the study of the intrageneric relationships deduced from 16S rRNA and *gyrB* phylogenetic studies of *Micromonospora* type strains have confirmed the reclassification of many species and subspecies of the genus (Kasai et al., 2000; Koch et al., 1996b). *Micromonospora echinospora* subsp. *pallida* was reclassified as the novel species *Micromonospora pallida*, and the subspecies *Micromonospora halophytica* subsp. *nigra* as *Micromonospora nigra*. DNA–DNA reassociation studies on strains of *Micromonospora* species have confirmed that the taxa *Micromonospora echinospora* subsp. *ferruginea*, *Micromonospora purpurea*, and *Micromonospora rhodorangea* are synonyms of *Micromonospora echinospora* and that *Micromonospora brunnea* is a synonym of *Micromonospora purpureochromogenes* (Kasai et al., 2000; Szabó and Fernandez, 1984). In addition, the division of *Micromonospora carbonacea* into two subspecies, *Micromonospora carbonacea* subsp. *carbonacea* and *Micromonospora carbonacea* subsp. *aurantiaca*, originally established on the basis of the morphology of the spores, is not supported by DNA relatedness data; hence, the subspecies are considered as synonyms of *Micromonospora carbonacea* (Kasai et al., 2000). The phylogenetic position of the type strain of *Micromonospora aurantiaca* within the genus has been confirmed by 16S rRNA analysis (Kasai et al., 2000), in spite of an early association of this taxon with the genus *Actinoplanes* (Kawamoto, 1989).

Descriptions of 18 novel *Micromonospora* species have been published recently: *Micromonospora auratinigra, Micromonospora chokoriensis, Micromonospora citrea, Micromonospora coriariae, Micromonospora coxensis, Micromonospora eburnea, Micromonospora echinaurantiaca, Micromonospora echinofusca, Micromonospora endolithica, Micromonospora fulviviridis, Micromonospora inyonensis, Micromonospora lupini, Micromonospora mirobrigensis, Micromonospora peucetia, Micromonospora saelicesensis, Micromonospora sagamiensis, Micromonospora siamensis,* and *Micromonospora viridifaciens* (Ara and Kudo, 2007b; Hirsch et al., 2004a; Kroppenstedt et al., 2005a; Thawai et al., 2004b, 2005a, 2005b; Trujillo et al., 2005, 2006, 2007).

Since submission of this chapter, the following *Micromonospora* species have been described: *Micromonospora chaiyaphumensis* (Jongrungruangchok et al., 2008b); *Micromonospora krabiensis* (Jongrungruangchok et al., 2008a); *Micromonospora marina* (Tanasupawat et al., 2010); *Micromonospora narathiwatensis* (Thawai et al., 2007); *Micromonospora pattaloongensis* (Thawai et al., 2008); *Micromonospora pisi* (Garcia et al., 2010); *Micromonospora rifamycinica* (Huang et al., 2008); and *Micromonospora tulbaghiae* (Kirby and Meyers, 2010).

Differentiation of species of the genus *Micromonospora*

Characteristics which can be used to differentiate between *Micromonospora* species are shown in Table 192 and Table 193.

Acknowledgements

Many thanks to Dr Oscar Salazar for his contribution to updating the phylogenetic analysis of *Micromonospora* species.

List of species of the genus *Micromonospora*

1. **Micromonospora chalcea** (Foulerton 1905) Ørskov 1923, 156[AL] ("*Streptothrix chalcea*" Foulerton 1905, 1199)

chal'ce.a. L. fem. adj. *chalcea* brazen, of brass.

Raised and folded reddish orange colonies that turn brown, olive-brown to dark-brown with age and eventually black upon sporulation. The substrate mycelium is covered by a moist to dry spore layer; a pale yellow fluorescent diffusible pigment is produced in yeast starch media. Does not form melanin pigments. Spores are produced abundantly on short or long sporophores. Dark brown occasionally sessile, oval to spherical spores (0.7–1.0 µm in diameter) are formed. Phase-contrast microscopy has shown that the spores have smooth surfaces.

Diagnostic carbohydrate utilization pattern is good growth on melibiose, but growth does not occur on mannitol or rhamnose. Grows on arabinose, cellobiose, fructose, glucose, galactose, lactose, levulose, mannose, soluble starch, sucrose, trehalose, and xylose as sole carbon sources, but not on glycerol, inositol, salicin, or ribose. Decomposes cellulose; negative for nitrate reduction; liquefies gelatin and digests milk. Good growth occurs between 27 and 37°C, but does not grow at 45°C. Maximum NaCl tolerance is 5% (w/v). Peptidoglycan contains *meso*-A_2pm; major menaquinones are MK-10(H_4) and MK-10(H_6).

Source: isolated from air, soil, and aquatic environments.

DNA G+C content (mol%): 71.9 (HPLC).

Type strain: ATCC 12452, DSM 43026, NBRC 13503, JCM 3031, NRRL B-2344, VKM Ac-822.

Sequence accession nos: U58531, X92594 (16S rRNA gene); AB014148 (*gyrB*).

2. **Micromonospora aurantiaca** Sveshnikova, Maksimova and Kudrina 1969, 758[AL]

au.ran.ti.a'ca. N.L. fem. adj. *aurantiaca* orange-colored.

Colonies grown on synthetic media produce light yellow, orange, or dark orange mycelium without a spore layer. On organic media, colonies present a pale orange to bright orange pigmentation of the substrate mycelium that turns grayish on sporulation on some media. Does not produce melanin pigments. Spores are formed in clusters, albeit moderately so.

Arabinose, cellobiose, fructose, galactose, melibiose, and xylose are used as sole carbon sources, but not glycerol, lactose, mannitol, rhamnose, ribose, or salicin. Decomposes cellulose; positive for nitrate reduction, milk peptonization, and decomposition of tyrosine; negative for starch hydrolysis.

Source: soil.

DNA G+C content (mol%): 71.6 (HPLC).

Type strain: ATCC 27029, DSM 43813, NBRC 16125, NBRC 16155, JCM 10878, NRRL B-16091, VKM Ac-1936.

Sequence accession nos: X92604 (16S rRNA gene); AB015621 (*gyrB*).

Additional remarks: according to Rule 12b of the *Bacteriological Code* (1990 Revision), Euzéby and Tindall (2004) requested the replacement of the specific epithet "*aurantiaca*" in *Micromonospora aurantiaca* by "*sandarakina*".

3. **Micromonospora auratinigra** corrig. Thawai, Tanasupawat, Itoh, Suwanborirux and Kudo 2004a, 1425[VP] [*Micromonospora aurantionigra* (*sic*)] (Effective publication: Thawai, Tanasupawat, Itoh, Suwanborirux and Kudo 2004b, 13.)

au.ra.ti.ni'gra. L. adj. *auratus -a -um* gold-colored; L. adj. *niger -gra -grum* black; N.L. fem. adj. *auratinigra* gold- and black-colored, referring to the color of colonies.

Colonies are vivid orange on yeast extract-malt extract agar (ISP medium 2) turning to brownish black or black after sporulation; a brown soluble pigment is produced. Aerial mycelium is absent. Does not produce melanin pigments. Single spores are formed on sporophores produced on substrate hyphae. Growth is observed between 20 and 30°C, but not above 40°C. Maximum NaCl concentration for growth is 2% (w/v); pH range for growth is 5–10.

Arabinose, cellobiose, fructose, galactose, glucose, lactose, melibiose, raffinose, ribose, salicin, and xylose are used as sole carbon sources, but not glycerol, mannitol, or

TABLE 192. Morphological and chemotaxonomic characteristics of *Micromonospora* species[a]

Characteristic	*M. chalcea* DSM 43026^{T}	*M. aurantiaca* DSM 43813^{T}	*M. auratinigra* DSM 44815^{T}	*M. carbonacea* DSM 43168^{T}	*M. chersina* DSM 44151^{T}	*M. chokoriensis* JCM 13247^{T}	*M. citrea* DSM 43903^{T}	*M. coerulea* DSM 43143^{T}	*M. coriariae* DSM 44875^{T}	*M. coxensis* JCM 13248^{T}	*M. eburnea* DSM 44814^{T}	*M. echinaurantiaca* DSM 43904^{T}	*M. echinofusca* DSM 43913^{T}	*M. echinospora* DSM 43816^{T}
Substrate mycelium	Red-orange	Yellow orange to dark orange	Bright orange to brown black	Orange to black	Light orange-yellow	Golden brown to dark brown	Salmon to yellow orange	Blue-green	Orange	Light brown to cinnamon brown	Yellowish orange	Orange to light yellow	Brown orange	Dark brown to purple/red brown to orange
Diffusible pigment	Light yellow	None	Brown	None	Yellow	None	Yellow orange	None	None	None	Pale yellow	None	None	None
Diaminopimelic acid isomer	*meso*	*meso*	*meso* + 3-OH	*meso* + 3-OH	*meso*	*meso*	*meso*	*meso*	*meso*	*meso*	*meso*	*meso*	*meso*	*meso* + 3-OH
Whole-organism sugars[b]	nd	Xyl, Ara	Glu, Xyl, Ara, Gal, Man, Rib	nd	Glu, Man, Xyl, Ara	Rib, Man, Xyl, Gal, Glu, Ara	Xyl, Glu	nd	Glu, Man, Ara, Xyl, Rib	Glu, Rha, Man, Ara, Xyl, Gal, Rib	Xyl, Ara	Xyl, Glu, Rham	Xyl, Glu	nd
Major menaquinones	MK-10	nd	MK-9, MK-10	MK-9	MK-9, MK-10	MK-10	MK-9	MK-10	MK-10	MK-10	MK-9	MK-9	MK-9	MK-10
DNA G+C content (mol%)	71.9	71.6	72.8	73.3	72.9	71	nd	71.7	70.2	73	71.5	nd	nd	71.7
Maximum NaCl tolerance (%, w/v)	5	4	2	3	3	3	nd	1.5	1	3	4	nd	nd	3
Temperature growth range (°C)	27–45	nd	25–30	27–37	18–49	20–37	nd	24–41	12–37	15–37	25–45	nd	nd	27–37
Antibiotic produced				Everninomicin complex	Dynemicins									Gentamicin complex

[a]Data from: Ara and Kudo (2007b); Asano et al. (1989a); Hirsch et al. (2004a); Horan and Brodsky (1986b); Kasai et al. (2000); Kawamoto (1989); Kawamoto et al. (1974, 1983); Kroppenstedt et al. (2005a); Lee et al. (1999); Luedemann (1971); Luedemann and Brodsky (1964, 1965); Nara et al. (1977); Thawai et al. (2004b, 2005a, 2005b); Tomita et al. (1992b); Trujillo et al. (2005, 2006, 2007); Weinstein et al. (1968, 1970). nd, Not determined.

[b]Ara, Arabinose; Gal, galactose; Glu, glucose; Man, mannose; Rha, rhamnose; Rib, ribose; Xyl, xylose.

M. endolithica DSM 44398^{T}	*M. fulviviridis* DSM 43906^{T}	*M. halophytica* DSM 43171^{T}	*M. inositola* DSM 43819^{T}	*M. inyonensis* DSM 46123^{T}	*M. lupini* DSM 44874^{T}	*M. matsumotoense* DSM 44100^{T}	*M. mirobrigensis* DSM 44830^{T}	*M. nigra* DSM 43818^{T}	*M. olivasterospora* DSM 43868^{T}	*M. pallida* DSM 43817^{T}	*M. peucetia* DSM 43363^{T}	*M. purpureochromogenes* DSM 43821^{T}	*M. rosaria* DSM 803^{T}	*M. saelicesensis* DSM 44871^{T}	*M. sagamiensis* DSM 43912^{T}	*M. siamensis* TT2-4^{T}	*M. viridifaciens* DSM 43909^{T}
Orange, olive to black	Salmon orange to light yellow	Orange brown	Bright orange	Yellow olive to red brown to yellow	Light orange	Red to brown orange	Orange	Orange olive to brown black	light brown to olive	light ivory brown to black	Deep orange to green	Dark brown	Orange brown to purple black	Orange	Coral red	Vivid orange	Pastel yellow to nut brown
None	None	Red-brown	None	Brown yellow	None	Red-brown	None	None	Olive green	None	None	Dark brown	Wine red	Orange brown-brown	None	Pale yellow	None
meso	*meso*	*meso* + 3-OH	*meso* + 3-OH	*meso*	*meso*	*meso* + 3-OH	*meso*	*meso* + 3-OH	*meso* + 3-OH	*meso* + 3-OH	*meso*	*meso*	*meso* + 3-OH	*meso*	*meso*	*meso*	*meso*
Xyl, Ara, Gal, Rib, Rham	Xyl, Glu	nd	nd	Xyl, Glu, Ara	Glu, Man, Ara, Xyl, Rha	Ara, Xyl, Gal, 3-OH-Met-Rham	Glu, Gal, Man, Xyl	nd	Xyl, Ara, Glu	nd	Xyl, Ara, Glu	nd	Xyl, Ara, Glu	Glu, Man, Ara, Xyl, Rib, Rha	Xyl, Glu	Xyl, Ara	Xyl, Glu, Rham
MK-10	MK-9	MK-9	MK-10	MK-9	MK-10	MK-10	MK-10	MK-9	MK-10	MK-12	MK-9	MK-10	MK-10	MK-10	MK-9	MK-10	MK-9
70	nd	72.5	71.4	nd	70.9	71	68.6	71.7	71.9	71.1	nd	73	72.9	71.6	nd	73	nd
2.5	nd	4	1.5	nd	2	2	3	4	3	3	nd	1.50	2	2	nd	5	nd
8–39	nd	18–40	25–40	nd	20–37	nd	20–37	18–40	28–38	nd	nd	nd	35–40	20–37	nd	20–40	nd
		Halomycin complex	Antibiotic XK-41 complex	Sisomicin				Halomycin complex	Fortimicin complex	Gentamicin complex	Adriamycin		Rosaramicin		Sagamicin (XK62-2)		Antibiotic 37505

TABLE 193. Utilization of carbohydrates and other physiological characteristics of *Micromonospora* type strains[a]

Characteristic	*M. chalcea* DSM 43026^{T}	*M. aurantiaca* DSM 43813^{T}	*M. auratinigra* DSM 44815^{T}	*M. carbonacea* DSM 43168^{T}	*M. chersina* DSM 44151^{T}	*M. chokoriensis* JCM 13247^{T}	*M. citrea* DSM 43903^{T}	*M. coerulea* DSM 43143^{T}	*M. coriariae* DSM 44875^{T}	*M. coxensis* JCM 13248^{T}	*M. eburnea* DSM 44814^{T}	*M. echinaurantiaca* DSM 43904^{T}	*M. echinofusca* DSM 43913^{T}	*M. echinospora* DSM 43816^{T}	*M. endolithica* DSM 44398^{T}	*M. fulviviridis* DSM 43906^{T}	*M. halophytica* DSM 43171^{T}	*M. inositola* DSM 43819^{T}	*M. inyonensis* DSM 46123^{T}	*M. lupini* DSM 44874^{T}	*M. matsumotoense* DSM 44100^{T}	*M. mirobrigensis* DSM 44830^{T}	*M. nigra* DSM 43818^{T}	*M. olivasterospora* DSM 43868^{T}	*M. pallida* DSM 43817^{T}	*M. peucetia* DSM 43363^{T}	M *purpureochromogenes* DSM 43821^{T}	*M. rosaria* DSM 803^{T}	*M. saelicesensis* DSM 44871^{T}	*M. sagamiensis* DSM 43912^{T}	*M. siamensis* TT2-4^{T}	*M. viridifaciens* DSM 43909^{T}
Nitrate reduction	v	+	−	+	v	nd	nd	−	+	nd	+	nd	nd	v	−	nd	+	−	nd	−	+	−	+	+	+	nd	v	−	−	+	−	−
Peptonization of milk	+	+	w	+	-	nd	nd	−	nd	nd	+	nd	nd	+	nd	nd	+	−	nd	nd	v	nd	v	+	+	nd	w	+	nd	+	+	+
Starch hydrolysis	+	−	+	+	+	nd	nd	+	+	nd	+	nd	nd	+	+	nd	+	+	nd	+	+	+	+	+	+	nd	−	+	v	+	+	nd
Decomposition of tyrosine	−	+	−	nd	−	nd	nd	−	−	nd	−	nd	nd	−	+	nd	−	−	nd	−	−	nd	−	−	−	nd	−	+	−	nd	nd	−
Liquefaction gelatin	+	+	+	+	+	nd	nd	nd	nd	nd	+	nd	nd	+	+	nd	w	+	nd	+	+	+	v	+	+	nd	w	+	+	w	+	+
Carbohydrate utilization:																																
L-Arabinose	w	w	+	+	+	+	+	−	+	+	−	+	−	+	+	+	+	−	+	−	+	+	+	+	+	−	−	+	+	−	+	+
Cellobiose	+	+	+	+	+	nd	nd	+	+	nd	+	nd	nd	+	+	nd	+	+	nd	+	+	+	+	+	+	nd	+	+	+	nd	+	nd
D-Fructose	w	+	+	+	+	+	−	+	+	+	−	+	+	−	+	+	+	+	+	nd	+	nd	w	+	+	+	+	+	nd	−	w	+
D-Galactose	+	+	+	+	+	+	nd	+	−	+	w	nd	nd	−	+	nd	+	+	nd	+	+	+	+	+	−	nd	+	w	+	+	+	nd
Glycerol	−	−	−	−	-	−	nd	−	nd	+	+	nd	nd	−	nd	nd	−	−	nd	nd	−	nd	−	−	−	nd	+	−	nd	−	-	nd
L-Inositol	−	nd	nd	−	-	−	−	−	nd	−	nd	−	−	−	−	−	−	w	+	−	−	nd	−	−	−	−	−	nd	+	−	nd	+
Lactose	+	−	+	+	w	+	nd	+	nd	+	+	nd	nd	−	+	nd	+	+	nd	nd	+	nd	+	−	−	nd	w	+	nd	−	+	nd
D-Mannitol	−	−	−	−	−	−	−	+	nd	−	w	+	−	−	−	+	−	v	+	+	−	−	−	−	−	−	−	+	+	−	-	+
D-Mannose	+	nd	nd	+	+	+	nd	nd	+	+	nd	nd	nd	+	+	nd	+	+	nd	+	+	+	+	+	+	nd	nd	+	+	nd	nd	nd
Melezitose	−	nd	nd	−	-	nd	nd	nd	−	nd	nd	nd	nd	−	nd	nd	nd	nd	nd	nd	−	−	nd	−	−	nd	−	−	nd	nd	nd	nd
Melibiose	+	+	+	+	+	+	nd	+	+	nd	+	nd	nd	−	+	nd	+	+	nd	+	−	nd	+	−	−	nd	+	−	+	nd	+	nd
Raffinose	+	w	+	−	v	+	−	+	+	+	+	+	−	−	+	+	+	+	+	+	+	+	+	−	−	−	+	−	+	−	+	+
L-Rhamnose	−	−	−	−	−	−	−	−	+	w	+	−	−	+	−	−	−	−	+	−	−	−	−	−	+	−	−	+	−	−	+	+
D-Ribose	−	−	+	−	v	w	nd	−	+	+	−	nd	nd	v	−	nd	−	−	nd	nd	w	nd	−	+	w	nd	−	+	nd	nd	-	+
Salicin	−	−	+	v	nd	+	nd	w	+	+	+	nd	nd	−	+	nd	+	−	nd	nd	−	nd	+	−	−	nd	−	−	nd	nd	+	nd
Soluble starch	+	−	+	+	+	nd	nd	nd	+	nd	w	nd	nd	+	nd	nd	+	+	nd	nd	+	+	+	+	+	nd	nd	+	nd	nd	nd	+
Sucrose	+	nd	nd	+	+	+	+	nd	−	+	nd	+	+	+	+	+	+	+	+	+	+	+	+	+	+	+	+	+	−	+	nd	+
Trehalose	+	nd	nd	+	+	+	nd	nd	+	+	nd	nd	nd	+	+	nd	+	nd	nd	+	+	+	+	+	+	nd	nd	+	−	nd	nd	nd
D-Xylose	+	+	+	+	+	+	−	+	+	+	+	−	−	+	+	−	+	+	+	nd	+	nd	+	+	+	−	+	+	nd	+	+	+

[a]Data from: Ara and Kudo (2007b); Asano et al. (1989a); Hirsch et al. (2004a); Horan and Brodsky (1986b); Kasai et al. (2000); Kawamoto (1989); Kawamoto et al. (1974, 1983); Kroppenstedt et al. (2005a); Lee et al. (1999); Luedemann and Brodsky (1964, 1965); Luedemann (1971); Nara et al. (1977); Thawai et al. (2004b, 2005a, 2005b); Tomita et al. (1992b); Trujillo et al. (2005, 2007, 2006); Weinstein et al. (1968, 1970). nd, Not determined; v, variable response; w, weak response.

rhamnose. Positive for starch hydrolysis and gelatin liquefaction; weakly positive for milk peptonization; negative for nitrate reduction and for decomposition of adenine, hypoxanthine, tyrosine, and xanthine. Hydrogen sulfide is not produced.

Cell wall contains glutamic acid, glycine, alanine, and *meso*-A_2pm and 3-OH-A_2pm. Arabinose and xylose are the characteristic whole-organism sugars, but galactose, glucose, mannose, and ribose are also present. The phospholipid profile contains diphosphatidylglycerol, phosphatidylethanolamine, phosphatidylinositol, and phosphatidylinositol mannosides. Predominant menaquinones are MK-10(H_4), MK-9(H_4), and MK-10(H_6); the major fatty acids include $C_{15:0}$ iso and $C_{16:0}$ iso with small amounts of $C_{17:0}$ iso, $C_{17:0}$ anteiso, and $C_{15:0}$ anteiso. Does not contain mycolic acids. DNA–DNA relatedness with other *Micromonospora* type strains ranges between 32 and 53%.

Source: isolated from soil collected in a peat swamp forest, Thailand.

DNA G+C content (mol%): 72.8 (HPLC).

Type strain: TT1-11, JCM 12357, PCU 239, DSM 44815, TISTR 1532.

Sequence accession no. (16S rRNA gene): AB159779.

4. **Micromonospora carbonacea** Luedemann and Brodsky 1965, 51[AL]

car.bo.na′ce.a. L. n. *carbo -onis* coal, charcoal; L. suff. *-aceus* suffix of various meanings, but signifying in general made of or belonging to; N.L. fem. adj. *carbonacea* charcoal-like (referring to color of spores).

Colonies are raised, folded, and initially orange, but turn brown to black on sporulation. Spores are generally abundant. Characteristic blackish sporulating peripheral sectors are formed. Spore layer is moist or dry, but not viscid. Spores are oval to spherical (0.7–1.0 μm in diameter), and smooth-walled by light microscopy. Sporulation is observed in liquid media as sparsely dispersed clumps of mycelium, which consist mostly of unbranched, long, loosely woven, fine mycelial strands. Spores remain firmly attached to the sympodial type of sporophore, and are only free in older cultures. The mycelium does not degenerate into polymorphic bodies. Fair to good growth occurs on Czapek's sucrose agar supplemented with 0.1% $CaCO_3$. Melanin pigments are not produced. Good growth occurs between 27 and 37°C, but does not grow at 45°C.

Arabinose, cellobiose, fructose, glucose, galactose, lactose, levulose, mannose, melibiose, soluble starch, sucrose, trehalose, and xylose are used as sole carbon sources, but not mannitol or rhamnose. Poor growth is observed on arabinose, dulcitol, glycerol, inositol, raffinose, ribose, salicin, sorbose, and sorbitol. Gelatin is liquefied. Milk is digested. Positive for nitrate reduction, hydrolysis of tyrosine, and oxidase reaction. Tolerates salt up to 3% (w/v).

Major fatty acids are $C_{15:0}$ iso and $C_{17:1}$ ω8 and the predominant menaquinone is MK-9(H_4). Produces antibiotics of the everninomicin complex.

Source: isolated from a soil sample collected from Olean, New York.

DNA G+C content (mol%): 73.3 (HPLC).

Type strain: ATCC 27114, DSM 43168, NBRC 14108, JCM 3139, NRRL 2972.

Sequence accession nos: X92599 (16S rRNA gene); AB014147 (*gyrB*).

Additional remarks: in the original description, poor growth was reported on raffinose and salicin, but these sugars are utilized in a chemically defined agar medium (Kawamoto, 1989). *Micromonospora carbonacea* subsp. *carbonacea* Luedemann and Brodsky 1965[AL] and *Micromonospora carbonacea* subsp. *aurantiaca* Luedemann and Brodsky 1965[AL] are synonyms of *Micromonospora carbonacea* Luedemann and Brodsky 1965, 51[AL] (Kasai et al., 2000).

5. **Micromonospora chersina** Tomita, Hoshino, Ohkusa, Tsuno and Miyaki 1992a, 656[VP] (Effective publication: Tomita, Hoshino, Ohkusa, Tsuno and Miyaki 1992b, 25.)

cher.si'na. Gr. adj. *khersinos* living in dry land; N.L. fem. adj. *chersina* referring to the savanna vegetation from which this organism was isolated.

Light orange to yellow colonies turning light olive gray or black upon sporulation. Fluorescent yellow diffusible pigment produced on Czapek's sucrose-nitrate agar and inorganic salts-starch agar (ISP medium 4). Melanin pigments are not produced. Single spherical spores (1.2–1.8 μm) are sessile or borne on short or long monopodial sporophores developed from well-branched substrate hyphae (0.5 μm diameter). The spore surface carries short blunt spines. Short sterile aerial hyphae occasionally developed on some media. Moderate growth occurs on inorganic salts starch and yeast-malt extract agars, but poor growth occurs on many media, including Czapek's agar, and glycerol-asparagine and tyrosine agars. Grows between 18 and 49°C, and optimally between 37 and 44°C. pH range for growth is 5.5–10.5. Maximum NaCl tolerance is 3% (w/v).

Arabinose, cellobiose, fructose, galactose, glucose, mannose, melibiose, sucrose, trehalose, and xylose are used as sole carbon sources, but not arabinose, inositol, glycerol, mannitol, rhamnose, or ribose. Utilization of raffinose is variable depending on the medium. Positive for gelatin liquefaction and starch hydrolysis. Nitrate reduction is variable depending on the medium. Negative for tyrosinase activity. Positive for α- and β-glucosidases, but negative for α-mannosidase and β-xylosidase.

Cell wall contains *meso*-A_2pm and glycine. Whole-organism sugars include glucose and mannose, and traces of arabinose and xylose. Predominant menaquinones are MK-9(H_4), MK-9(H_6), MK-10(H_4), and MK-10(H_6). Major phospholipids are phosphatidylinositol and phosphatidylethanolamine. Produces antitumor antibiotic dynemicin.

Source: isolated from an Indian soil.

DNA G+C content (mol%): 72.9 (HPLC).

Type strain: M956-1, ATCC 53710, DSM 44151, NBRC 15963, JCM 9459.

Sequence accession nos: X92628 (16S rRNA gene); AB015622 (*gyrB*).

6. **Micromonospora chokoriensis** Ara and Kudo 2007, 1372[VP] (Effective publication: Ara and Kudo 2007b, 35.)

cho.ko.ri.en'sis. N.L. fem. adj. *chokoriensis* of or pertaining to Chokoria, Bangladesh, the origin of the soil from which the type strain was isolated

Golden brown to dark brown colonies are formed on yeast extract-starch agar without soluble pigments. Melanin pigments are not produced. Well-developed branched substrate mycelium bears single sessile spores. Spores are spherical with a rough to nodular spore surface. Grows at pH 5–9, at 20–37°C, and in 3% NaCl.

Arabinose, fructose, galactose, glucose, lactose, maltose, mannose, α-melibiose, raffinose, salicin, sucrose, trehalose, and xylose are used as sole carbon sources, but not adonitol, erythritol, fructose, glycerol, rhamnose, *myo*-inositol, or mannitol. Ribose is weakly utilized.

Cell wall contains *meso*-A_2pm. Whole-organism sugars include arabinose, galactose, glucose, mannose, ribose, and xylose. Contains diphosphatidylglycerol, phosphatidylethanolamine, phosphatidylinositol, and phosphatidylinositol mannosides as major polar lipids. Major fatty acids are $C_{15:0}$ iso, $C_{16:0}$ iso, $C_{17:0}$ iso, and $C_{15:0}$ anteiso. The predominant menaquinone is MK-10(H_4); minor amounts of MK-10(H_6), MK-9(H_4), and MK-9(H_6) are formed.

Source: isolated from sandy soil in forest-side waterfall.

DNA G+C content (mol%): 71 (HPLC).

Type strain: 2-19(6), JCM 13247.

Sequence accession no. (16S rRNA gene): AB241454.

7. **Micromonospora citrea** Kroppenstedt, Mayilraj, Wink, Kallow, Schumann, Secondini and Stackebrandt 2005b, 1743[VP] (Effective publication: Kroppenstedt, Mayilraj, Wink, Kallow, Schumann, Secondini and Stackebrandt 2005a, 333.)

ci'tre.a. L. fem. adj. *citrea* of or pertaining to the citrus-tree.

Colonies are pastel orange on yeast-malt extract agar (ISP medium 2), pastel yellow on inorganic salts-starch agar (ISP medium 4), and yellow orange on peptone-yeast extract-iron agar (ISP medium 6). A yellow orange soluble pigment is formed on peptone-yeast extract-iron agar. Aerial mycelium is absent. Single spores are formed on substrate mycelium.

Arabinose, glucose, and sucrose are used as sole carbon sources, but not fructose, inositol, mannitol, raffinose, rhamnose, or xylose. Positive for alkaline phosphatase, *N*-acetyl-β-glucosaminidase, esterase, α-and β-galactosidases, α-glucosidase, and acid phosphatase, but negative for α-fucosidase, β-glucosidase, β-glucuronidase, lipase, and α-mannosidase.

Peptidoglycan contains *meso*-A_2pm. Whole-organism diagnostic sugars are glucose and xylose. Major menaquinone is MK-9(H_4); phosphatidylethanolamine is the diagnostic phospholipid. Major fatty acids are $C_{15:0}$ iso, $C_{16:0}$ iso, and $C_{17:0}$ 10-methyl.

Source: isolated from lake mud, China.

DNA G+C content (mol%): not determined.

Type strain: DSM 43903, ATCC 35571, JCM 3256, NBRC 14025, NRRL B-16101.

Sequence accession no. (16S rRNA gene): X92617.

8. **Micromonospora coerulea** Jensen 1932, 177[AL]

coe.ru'le.a. L. fem. adj. *coerulea* dark blue, dark green.

Slow-growing colonies that require 3–5 weeks to develop. Pale-yellow-orange colonies turn yellow-green then dark blue-green to greenish black with age. pH-sensitive soluble pigment is produced. Spherical spores (0.8–1.5 µm in diameter) are borne on short or lateral sporophores. Mycelium fragmentation is observed rarely in old liquid cultures. Good growth occurs between 24 and 37°C.

Fructose, galactose, and lactose are used as sole carbon sources, but poor growth is seen on arabinose and arabinose. Starch is hydrolyzed.

Source: isolated from soil from Mt Haleakala, Maui Island, Hawaii.

DNA G+C content (mol%): 71.7 (HPLC).

Type strain: ATCC 27008, DSM 43143, NBRC 13504, JCM 3175, VKM Ac-661.

Sequence accession nos: X92598 (16S rRNA gene); AB014151 (*gyrB*).

9. **Micromonospora coriariae** Trujillo, Kroppenstedt, Schumann, Carro and Martínez-Molina 2006, 2384[VP]

co.ri.a.ri'a.e. N.L. gen. n. *coriariae* of *Coriaria*, pertaining to the isolation of the type strain from root nodules of *Coriaria myrtifolia.*

Raised, folded, intensively orange colonies turn darker with age, does not form diffusible pigments. Abundant growth on Bennett's SA1 and yeast extract-malt extract (ISP medium 2) agars. Single spores form at the tips of well-developed substrate hyphae. Optimum growth is at 28°C. Grows at pH 7–9, does not grow below pH 6.5.

Alanine, arabinose, arginine, cellobiose, fructose, gluconate, glucose, histidine, mannose, melibiose, pyruvate, raffinose, rhamnose, salicin, serine, starch, trehalose, and xylose are used as sole carbon sources, but not ascorbic acid, galactose, glutaric acid, lysine, melezitose, proline, propionic acid, sorbitol, sorbose, sucrose, tyrosine, valine, or xylitol. Positive for acid and alkaline phosphatases, esterases, α- and β-galactosidases, α- and β-glucosidases, and *N*-acetyl-β-glucosaminidase, but not for α-chymotrypsin, β-glucuronidase, or α-mannosidase. Oxidase- and catalase-positive. Degrades arbutin, casein, esculin, gelatin, starch, and Tween 80, but not tyrosine, urea, or Tween 20.

Whole-organism hydrolysates contain *meso*-A_2pm, and arabinose, glucose, mannose, ribose, and, xylose as characteristic sugars. Fatty acid profile is characterized by significant amounts of $C_{15:0}$ iso, $C_{16:0}$ iso, $C_{17:0}$, and $C_{17:0}$ 10-methyl. Major menaquinone is MK-10(H_4), with minor amounts of MK-10(H_6) and MK-9(H_4).

Source: the type strain was isolated from root nodules of *Coriaria myrtifolia.*

DNA G+C content (mol%): 70.2 (T_m).

Type strain: NAR01, DSM 44875, LMG 23557.

Sequence accession no. (16S rRNA gene): AJ784008.

10. **Micromonospora coxensis** Ara and Kudo 2007, 1372[VP] (Effective publication: Ara and Kudo 2007b, 36.)

cox.en'sis. N.L. fem. adj. *coxensis* pertaining to Cox's Bazaar, Bangladesh, the origin of the soil from which the type strain was isolated.

Colonies on yeast extract-starch agar are light brown to cinnamon brown without diffusible pigment. Melanin pigments are not produced. Well-developed substrate mycelium is formed with spherical single spores borne directly on substrate hyphae. Spore surface is nodular to warty. Grows at pH 5–9, at 15–37°C, and in 3% (w/v) NaCl.

Adonitol (weak), arabinose, fructose, galactose, glycerol, lactose, mannose, maltose, α-melibiose, raffinose, rhamnose, ribose, salicin, sucrose, trehalose, and xylose are used as sole carbon sources, but not erythritol, *myo*-inositol, or mannitol.

Whole-organism hydrolysates contain *meso*-A_2pm, and arabinose, galactose, glucose, mannose, rhamnose, ribose, and xylose as characteristic sugars. Fatty acid profile is characterized by significant amounts of $C_{17:0}$, $C_{17:1}$ iso, $C_{15:0}$ iso, $C_{17:0}$ anteiso, and $C_{16:1}$ iso. Major menaquinones are MK-10(H_6) and MK-10(H_8), with minor amounts of MK-10(H_4), MK-9(H_4), MK-9(H_6), and MK-9(H_8).

Source: isolated from sandy soil next to a forest-side waterfall.

DNA G+C content (mol%): 73 (HPLC).

Type strain: 2-30-b(28), JCM 13248, MTCC 8093.

Sequence accession no. (16S rRNA gene): AB241455.

11. **Micromonospora eburnea** Thawai, Tanasupawat, Itoh, Suwanborirux, Suzuki and Kudo 2005b, 55[VP]

e.bur'ne.a. L. fem. adj. *eburnea* of ivory, white as ivory, referring to the color of colonies.

Well-developed and branched substrate mycelium. Colonies have a characteristic yellowish white or dull orange color depending on the media. They turn grayish black upon sporulation on yeast extract-malt extract agar (ISP medium 2); a pale yellow soluble pigment is produced on this medium and on oatmeal (ISP medium 3) and nutrient agars. Nonmotile single spores (0.45 µm in diameter) are borne on substrate hyphae. Aerial mycelium is not produced. Optimal temperature for growth is 25–30°C, does not grow above 45°C. Maximum NaCl concentration tolerated for growth is 4% (w/v).

Cellobiose, glucose, glycerol, lactose, melibiose, raffinose, rhamnose, salicin, and xylose are used as sole carbon sources, but not arabinose, fructose, or ribose. Positive for milk peptonization, starch hydrolysis, and nitrate reduction.

Cell-wall hydrolysates contain alanine, glycine, glutamic acid, and *meso*-A_2pm. Muramic acid is glycolated. Arabinose and xylose are the characteristic whole-organism sugars. Whole-organism hydrolysates contain galactose, glucose, mannose, and ribose. The characteristic phospholipids are phosphatidylethanolamine, phosphatidylglycerol, phosphatidylinositol, and phosphatidylinositol mannosides. Predominant menaquinones are MK-9(H_4), MK-10(H_4), and MK-9(H_6). Major fatty acids include $C_{15:0}$ iso, $C_{16.0}$, $C_{17:0}$ iso, $C_{15:0}$ anteiso, $C_{17:0}$, and $C_{17:0}$ anteiso. Mycolic acids are absent.

Source: isolated from a peat swamp forest soil in southern Thailand.

DNA G+C content (mol%): 71.5 (HPLC).

Type strain: LK2-10, DSM 44814, JCM 12345, PCU 238, TISTR 1531.

Sequence accession no. (16S rRNA gene): AB107231.

12. **Micromonospora echinaurantiaca** Kroppenstedt, Mayilraj, Wink, Kallow, Schumann, Secondini and Stackebrandt 2005b, 1743[VP] (Effective publication: Kroppenstedt, Mayilraj, Wink, Kallow, Schumann, Secondini and Stackebrandt 2005a, 333.)

e.chin.au.ran.ti.a′ca. Gr. n. *echinos* hedgehog, sea-urchin, N.L. fem. adj. *aurantiaca* orange-colored; N.L. fem. adj. *echinaurantiaca* spiny and orange-colored.

Orange-colored colonies. Aerial mycelium is not formed. Pigment of substrate mycelium ranges from bright orange on inorganic salts-starch agar (ISP medium 4), peptone-yeast extract-iron agar (ISP medium 6), and yeast extract-malt extract agar (ISP medium 2) to deep orange on oatmeal agar (ISP medium 3). Soluble pigments are not produced.

Arabinose, fructose, glucose, mannitol, sucrose, and raffinose are used as sole carbon sources, but not inositol, rhamnose, or xylose. Positive for alkaline phosphatase, *N*-acetyl-β-glucosaminidase, esterase, α-glucosidase, α- and β-galactosidase, and phosphatase acid, but not for β-glucosidase, β-glucuronidase, lipase, α-mannosidase, or α-fucosidase.

Peptidoglycan contains *meso*-A_2pm. Whole-organism diagnostic sugars are glucose and rhamnose. Major menaquinone is MK-9(H_4). Phosphatidylethanolamine is the diagnostic phospholipid. Major fatty acids include $C_{15:0}$ iso, $C_{16.0}$, and $C_{17:0}$ iso.

Source: isolated from a Chinese soil.

DNA G+C content (mol%): not determined.

Type strain: DSM 43904, ATCC 35572, JCM 3257, NBRC 14022, NRRL B-16102.

Sequence accession no. (16S rRNA gene): X92618.

13. **Micromonospora echinofusca** Kroppenstedt, Mayilraj, Wink, Kallow, Schumann, Secondini and Stackebrandt 2005b, 1743[VP] (Effective publication: Kroppenstedt, Mayilraj, Wink, Kallow, Schumann, Secondini and Stackebrandt 2005a, 333.)

e.chi.no.fus′ca. Gr. n. *echinos* hedgehog, sea-urchin; L. fem. adj. *fusca* brown-colored, N.L. fem. adj. *echinofusca* spiny and brown-colored.

Colonies are orange-colored ranging from deep brownish-orange on inorganic-salts-starch agar (ISP medium 4), and glucose-asparagine agar (ISP medium 5), to pastel orange on oatmeal agar (ISP medium 3), tyrosine agar (ISP medium 7), and yeast extract-malt extract agar (ISP medium 2). Aerial mycelium is absent. Poor growth occurs on peptone-yeast extract-iron agar (ISP medium 6). Soluble pigments are not formed.

Fructose, glucose, and sucrose are used as sole carbon sources, but not arabinose, inositol, mannitol, raffinose, rhamnose, or xylose. Positive for *N*-acetyl-β-glucosaminidase, alkaline phosphatase, α- and β-galactosidases, α- and β glucosidases, and acid phosphatase, but not for α-fucosidase, glucuronidase, lipase, or α-mannosidase.

Peptidoglycan contains *meso*-A_2pm. Whole-organism diagnostic sugars are glucose and xylose. Major menaquinone is MK-9(H_4). Phosphatidylethanolamine is the diagnostic phospholipid. Major fatty acids are $C_{15:0}$ iso, $C_{16:0}$ iso, and $C_{17:1}$ ω8*c*.

Source: isolated from excrement of a chukar at Beijing Zoological Garden, China.

DNA G+C content (mol%): not determined.

Type strain: DSM 43913, JCM 3327, NBRC 14267.

Sequence accession no. (16S rRNA gene): X92625.

14. **Micromonospora echinospora** Luedemann and Brodsky 1964, 121[AL] emend. Kasai, Tamura and Harayama 2000, 131

e.chi.no.spo′ra. Gr. n. *echinos* hedgehog, sea-urchin; Gr. n. *spora* seed and in biology a spore; N.L. fem. n. *echinospora* spiny spore.

Colonies are orange-brown, maroon to dark purple and folded. Maroon to dark-purple pigments are pH-sensitive (red in the acid range and blue-green and precipitable in the basic range). Black spore layer is waxy to dry. Does not form melanin pigments. Sporulation is moderate, requiring 2–3 weeks for completion. Occasionally forms a sterile aerial mycelium in the form of a gray bloom. Spores are spherical (1.0–1.5 μm) and rough walled under phase-contrast microscopy. Sporophores are single or in small clusters on the same hyphae. Grows between 27 and 37°C, but not at 45°C.

Arabinose, cellobiose, glucose, mannose, sucrose, trehalose, soluble starch, and xylose are used as sole carbon sources, but not glycerol, inositol, raffinose, sorbitol, or salicin. Poor to fair growth occurs on galactose, lactose, and levulose. Diagnostic carbohydrate utilization pattern: good growth on rhamnose; poor growth on melibiose, mannitol, and ribose. Slow decomposition of cellulose. Gelatin is weakly liquefied. Milk is weakly digested.

Most strains produce antibiotics of the gentamicin complex.

Source: isolated from a soil sample collected from Jamesville, New York.

DNA G+C content (mol%): 71.7 (HPLC).

Type strain: ATCC 15837, DSM 43816, NBRC 13149, JCM 3073, NRRL 2985, VKM Ac-669.

Sequence accession nos: X92607 (16S rRNA gene); AB014154 (*gyrB*).

Additional remarks: earlier heterotypic synonyms are *Micromonospora rhodorangea* Wagman et al. 1974, *Micromonospora echinospora* subsp. *ferruginea* Luedemann and Brodsky 1964, and *Micromonospora purpurea* Luedemann and Brodsky 1964 (Skerman et al., 1980).

15. **Micromonospora endolithica** Hirsch, Mevs, Kroppenstedt, Schumann and Stackebrandt 2004b, 631[VP] (Effective publication: Hirsch, Mevs, Kroppenstedt, Schumann and Stackebrandt 2004a, 172.)

en.do.li'thi.ca. L. prep. *endo* in, within; Gr. n. *lithos* stone; L. suff. *-icus -a –um*, suffix used in adjectives with the sense of belonging to; N.L. fem. adj. *endolithica* (growing) within stone.

Orange, folded colonies that turn grayish-black, olive, or black upon sporulation. Soluble pigments are not produced. Vegetative hyphae (0.5–1.0 µm in diameter) often produce terminal "lemon-shaped" bodies. Spores (0.8–1.1 µm) are produced singly on short hyphae side branches and have short warty surfaces. Temperature range for growth is 8–39°C, with an optimal temperature of 27–29°C. Tolerates up to 7% salt, but optimum for growth is 2%.

Arabinose, cellobiose, fructose, galactose, glucose, lactose, maltose, mannitol (weak), mannose, melibiose, raffinose (weak), sucrose, trehalose, and xylose are used as sole carbon sources, but not *N*-acetylglucosamine, citrate, dextrin, inositol, oxalate, rhamnose, ribose, sorbitol, or sorbose. Catalase- and cytochrome oxidase-positive. Degrades adenine, casein, gelatin, starch, tyrosine, and xanthine, but not cellulose, chitin, or xylan. Hydrolyzes esculin.

Cell-wall amino acids include alanine, glycine, glutamic acid, and *meso*-A_2pm, and whole-organism sugars are arabinose, galactose, rhamnose, ribose, and xylose. Characteristic cell-wall sugars are arabinose and xylose. The characteristic phospholipids are phosphatidylethanolamine, phosphatidylglycerol, and phosphatidylinositol. Predominant menaquinones are MK-10(H_4) and MK-10(H_6). Major fatty acids include $C_{16:0}$ iso, $C_{17:1}$ ω8*c*, $C_{17:1}$ iso ω9*c*, and $C_{15:0}$ iso; $C_{18:0}$ 10-methyl is also present. Sensitive to 50 µg/ml lysozyme.

Source: isolated from sandstone in Antarctica.

DNA G+C content (mol%): 70.0 (T_m).

Type strain: AA-459, DSM 44398, JCM 12677, NRRL B-24248.

Sequence accession no. (16S rRNA gene): AJ560635.

16. **Micromonospora fulviviridis** Kroppenstedt, Mayilraj, Wink, Kallow, Schumann, Secondini and Stackebrandt 2005b, 1743[VP] (Effective publication: Kroppenstedt, Mayilraj, Wink, Kallow, Schumann, Secondini and Stackebrandt 2005a, 335.)

ful.vi.vi'ri.dis. L. adj. *fulvus* tawny, brown; L. adj. *viridis* green; N.L. fem. adj. *fulviviridis* brown-green.

Salmon orange to salmon pink colonies are formed on inorganic salts-starch (ISP medium 4), oatmeal (ISP medium 3), and yeast extract-malt extract (ISP medium 2) agars. Pastel yellow substrate mycelium is formed on glycerol-asparagine (ISP medium 5) and peptone-yeast extract (ISP medium 6) agars. Soluble pigments are not formed.

Arabinose, fructose, glucose, mannitol, raffinose, and sucrose are used as sole carbon sources, but not inositol, rhamnose, or xylose. Positive for alkaline phosphatase, α- and β-galactosidases, α- and β-glucosidases, and acid phosphatase, but not for *N*-acetyl-β-glucosaminidase, α-fucosidase, β-glucuronidase, or α-mannosidase.

Peptidoglycan contains *meso*-A_2pm. Whole-organism diagnostic sugars are glucose and xylose. Major menaquinone is MK-9(H_4). Phosphatidylethanolamine is the diagnostic phospholipid. Major fatty acids are $C_{15:0}$ iso, $C_{16.0}$, and $C_{17:0}$ iso.

Source: isolated from soil.

DNA G+C content (mol%): not determined.

Type strain: DSM 43906, ATCC 35574, JCM 3259, NRRL B-6104.

Sequence accession no. (16S rRNA gene): X92620.

17. **Micromonospora halophytica** Weinstein, Luedemann, Oden and Wagman 1968, 436[AL]

ha.lo.phy'ti.ca. N.L. fem. adj. *halophytica* growing under the influence of seawater.

Forms folded orange colonies that turn orange-brown with age. Sporulation is abundant, especially in older colonies. Light reddish brown diffusible pigment is produced in media containing galactose, lactose, levulose, mannose, raffinose, or trehalose. Melanin pigments are not produced. Grows between 18 and 40°C, but not at 50°C. Good growth within pH range 6.8–7.8. Spores are produced randomly along branching mycelium on short or long sporophores, but are occasionally sessile. Abundant dark colored spores occur in older cultures. Spores are spherical to ellipsoidal (up to 1.2 µm in diameter).

Arabinose, fructose, galactose, glucose, lactose, levulose, mannose, melibiose, raffinose, starch, sucrose, trehalose, and xylose are used as sole carbon sources. Poor growth occurs on adonitol, cellulose, rhamnose, inositol, mannitol, ribose, and sorbitol. Hydrolyzes starch, decomposes cellulose, reduces nitrate, liquefies gelatin, and digests milk. Produces antibiotics of the halomicin antibiotic complex.

Source: isolated from a salt pool in Syracuse, New York.

DNA G+C content (mol%): 72.5 (HPLC).

Type strain: ATCC 27596, DSM 43171, NBRC 14112, JCM 3125, NRRL 2998.

Sequence accession nos: X92601 (16S rRNA gene); AB014157 (*gyrB*).

18. **Micromonospora inositola** Kawamoto, Okachi, Kato, Yamamoto, Takahashi, Takasawa and Nara 1974, 495[AL]

i.no.si'to.la. N.L. fem. adj. *inositola* intended to mean that the bacterium is able to utilize inositol.

Folded, bright orange to orange colonies are formed on Bennett's, nutrient, and potato agars. Soluble pigments are not produced. Does not form a spore layer. Poor growth occurs on glucose-asparagine, inorganic starch, and oatmeal agars. Poor sporulation occurs on most media. Spores are oval to spherical (0.8–1.0 µm in diameter) and borne on short sporophores. Grows between 25 and 40°C and at pH 5.5–8.5.

Cellobiose, fructose, galactose, glucose, inositol (weak), lactose, mannose, raffinose, and xylose are used as sole carbon sources, but not arabinose, glycerol, mannitol, rhamnose, ribose, salicin, or sorbitol. Liquefies gelatin. Milk is slightly coagulated, but not peptonized. Negative for nitrate reduction. Cellulose is slightly decomposed. Starch is hydrolyzed. Produces antibiotics of the macrolide XK-41 complex.

Source: isolated from a soil sample from Hokkaido, Japan.

DNA G+C content (mol%): 71.4 (HPLC).

Type strain: ATCC 21773, DSM 43819, JCM 6239, NRRL B-16095.

Sequence accession nos: X92610 (16S rRNA gene); AB014158 (*gyrB*).

19. **Micromonospora inyonensis** Kroppenstedt, Mayilraj, Wink, Kallow, Schumann, Secondini and Stackebrandt 2005b, 1743[VP] (Effective publication: Kroppenstedt, Mayilraj, Wink, Kallow, Schumann, Secondini and Stackebrandt 2005a, 337.)

in.yo.nen'sis. N.L. fem. adj. *inyonensis* of or pertaining to Inyo County, California.

Viscid to poorly plicate colonies with fair to poor growth of substrate mycelium. Aerial mycelium is not formed. Substrate mycelium ranges from copper brown on inorganic salts-starch agar (ISP medium 4), to red-green on oatmeal agar (ISP medium 3), yellow on peptone-yeast extract-iron (medium ISP 6) and tyrosine (ISP medium 7), and yellow-olive on yeast extract-malt extract agar (ISP medium 2). Soluble brown yellow pigment is produced on ISP medium 6. Sparse growth occurs on glycerol-asparagine agar (ISP medium 5). Long branched mycelium is formed that is regular and nonseptate and with a mean diameter of 0.5 µm. Ovoid to spherical spores (1.0–1.5 µm in diameter) are rough walled and occasionally borne singly on sporophores.

Poor utilization of arabinose, fructose, glucose, inositol, mannitol, rhamnose, raffinose, sucrose, and xylose is observed. Positive for alkaline phosphatase, esterase, *N*-acetyl-β-glucosaminidase, α- and β-glucosidases, and acid phosphatase, but not for α-fucosidase, lipase, α- or β-galactosidase, β-glucuronidase, or α-mannosidase.

Peptidoglycan contains *meso*-A_2pm. Whole-organism diagnostic sugars are arabinose, glucose, and xylose. Major menaquinone is MK-9(H_4). Phosphatidylethanolamine is the diagnostic phospholipid. Major fatty acids are $C_{15:0}$ iso, $C_{16:0}$ iso, and $C_{17:1}$ iso. Produces the aminoglycoside sisomicin (also known as rickamicin or antibiotic 6640).

Source: isolated from soil.

DNA G+C content (mol%): not determined.

Type strain: DSM 46123, JCM 3188, ATCC 27600, NBRC 13156, NRRL 3292.

Sequence accession no. (16S rRNA gene): X92629.

20. **Micromonospora lupini** Trujillo, Kroppenstedt, Fernández-Molinero, Schumann and Martínez-Molina 2007, 2803[VP]

lu'pi.ni. L. gen. n. *lupini* of a lupin, referring to the isolation of the first strains from *Lupinus angustifolius.*

Light orange, folded, raised colonies without diffusible pigments are formed on yeast extract-malt extract agar. Produces well-developed branched hyphae (0.3–0.6 µm in diameter) with smooth-surfaced spores produced at hyphal tips. Grows at 20–37°C. Good growth occurs in 1% (w/v) NaCl, but growth is variable in 2% (w/v) NaCl.

Alanine, cellobiose, galactose, maltose, mannose, melibiose, raffinose, and trehalose are assimilated as carbon sources, but not arginine, histidine, lysine, proline, rhamnose, serine, sorbitol sorbose, tyrosine, valine, or xylitol. Catalase- and oxidase-positive. Nitrate is not reduced. Degrades arbutin, casein, esculin, gelatin, starch, and xylan, but not tyrosine. Positive for alkaline phosphatase, esterases and lipases, α- and β-galactosidases, α- and β-glucosidases, and *N*-acetyl-β-glucosaminidase, but not for urease, or β-glucuronidase.

Whole-organism hydrolysates contain *meso*-A_2pm and arabinose, glucose, mannose, rhamnose, and xylose as characteristic sugars. Diagnostic phospholipid is diphosphatidylethanolamine. Major fatty acids are $C_{16:0}$ iso and $C_{15:0}$ iso. Predominant menaquinone is MK-10(H_4). One of the strains of this species produces the antitumoral lupinamicins A and B.

Source: isolated from root nodules of *Lupinus angustifolius.*

DNA G+C content (mol%): 70.9 (T_m).

Type strain: Lupac 14N, DSM 44874, LMG 24055.

Sequence accession no. (16S rRNA gene): AJ783996.

21. **Micromonospora matsumotoense** (Asano, Masunaga and Kawamoto 1989a) Lee, Goodfellow and Hah 2000a, 3[VP] (Effective publication: Lee, Goodfellow and Hah 1999, 353.) (*Catellatospora matsumotoense* Asano, Masunaga and Kawamoto 1989a, 313)

mat.su.mo.to.en'se. L. deriv. *matsumotoense* (*sic*) of Matsumoto Nagago, Japan, the location of the soil sample from which the type strain was isolated.

Orange colonies are formed on oatmeal and yeast extract-malt extract agars and yellow colonies are formed on glucose-yeast extract and nutrient agars. A reddish-brown soluble pigment is produced on oatmeal agar. Does not require thiamine for growth. Grows in the presence of 2% (w/v) NaCl, but not in 3% (w/v) NaCl.

L-Arabinose, cellobiose, fructose, glucose, galactose, lactose, maltose, mannose, raffinose, ribose, starch, sucrose, trehalose, and xylose are used as sole carbon sources, but not adonitol, D-arabinose, dextran, gluconate, glycerol, mannitol, melezitose, melibiose, rhamnose, salicin, sorbitol, sorbose, or xylitol. Decomposes casein, elastin, gelatin, and starch. Resistant to neomycin (5 µg/ml), novobiocin (50 µg/ml), tetracycline (20 µg/ml), and vancomycin (20 µg/ml).

Cell walls contain *meso*- and 3-OH-A_2pm. Muramic acid is glycolated. Diagnostic cell-wall sugars are arabinose, galactose, 3-*O*-methylrhamnose, and xylose. Major menaquinones are MK-10(H_4), MK-10(H_6), and MK-10(H_8). Diphosphatidylglycerol, phosphatidylethanolamine, phosphatidylinositol, and phosphatidylglycerol are the predominant polar lipids. Major fatty acids include $C_{16:0}$ iso, $C_{18:1}$, $C_{17:1}$, $C_{16:1}$ iso, and $C_{15:0}$ iso.

Source: isolated from soil.

DNA G+C content (mol%): 71.0 (HPLC).

Type strain: 6393-C, ATCC 49364, CIP 106812, DSM 44100, NBRC 14550, JCM 9104, NRRL B-16490, VKM Ac-2009, MSNU 22003.

Sequence accession no. (16S rRNA gene): AF152109.

22. **Micromonospora mirobrigensis** Trujillo, Fernández-Molinero, Velázquez, Kroppenstedt, Schumann, Mateos and Martínez-Molina 2005, 879[VP]

mi.ro.bri.gen'sis. N.L. fem. adj. *mirobrigensis* of or belonging to Mirobriga, the region in Spain where the type strain was isolated.

Small, raised, and folded colonies. Colonies grow slowly on Bennett's, glucose-yeast extract, and nutrient agars, and are 2–3 mm in diameter after incubation for 2 weeks on SA1 medium. Initially, colonies are orange, but turn brownish black with age and upon sporulation. Produces a well developed substrate mycelium with single warty spores. Grows between 20–37°C, with an optimal temperature of 28°C, but does not grow at 45°C. Grows at pH 7.0 and tolerates NaCl up to 3% (w/v).

Arabinose, cellobiose, galactose, glucose, maltose, mannose, raffinose, sucrose, and trehalose are used as sole carbon sources, but not citrate, malate, mannitol, melezitose, rhamnose, sorbitol, sorbose, or xylitol. Hydrolyzes arbutin and esculin. Degrades casein, gelatin, starch, and xylan. Nitrate is not reduced. Shows resistance to ampicillin (2 μg). Positive for acid phosphatase, alkaline phosphatase, esterase lipase, α- and β-galactosidases, β-glucuronidase, α- and β-glucosidases, and *N*-acetyl-β-glucosaminidase.

Cell wall contains *meso*-A_2pm. Whole-organism sugars are galactose, glucose, mannose, and xylose. Major fatty acids are $C_{15:0}$ iso, $C_{16:0}$ iso, $C_{17:0}$ iso, and $C_{17:0}$ anteiso. Predominant menaquinones are MK-10(H_4) and MK-10(H_6). Major phospholipids are diphosphatidylglycerol, phosphatidylethanolamine, phosphatidylglycerol, and phosphatidylinositol.

Source: isolated from a water sample taken from a pond in the region of Mirobriga, Ciudad Rodrigo, Spain.

DNA G+C content (mol%): 68.6 (T_m).

Type strain: WA201, DSM 44830, LMG 22229.

Sequence accession no. (16S rRNA gene): AJ626950.

23. **Micromonospora nigra** (Weinstein, Luedemann, Oden and Wagman 1968) Kasai, Tamura and Harayama 2000, 131[VP] (*Micromonospora halophytica* subsp. *nigra* Weinstein, Luedemann, Oden and Wagman 1968, 437; Kawamoto 1989, 2446)

ni'gra. L. fem. adj. *nigra* black.

Orange colonies turn olive-brown to black upon sporulation. Diffusible pigments are not produced. A black spore layer is observed on media containing fructose, galactose, lactose, raffinose, or trehalose. Melanin pigments are not produced. Grows between 18 and 40°C, but not at 50°C.

Adonitol (weak), arabinose, cellulose (weak), fructose, galactose, lactose, salicin, and raffinose are used as sole carbon sources, but not glycerol, mannitol, or rhamnose. Starch is hydrolyzed, cellulose is decomposed, and nitrate is reduced. Gelatin liquefaction and milk digestion are variable. Produces antibiotics of the halomycin complex.

Source: isolated from a salt pool in Syracuse, New York.

DNA G+C content (mol%): 71.7 (HPLC).

Type strain: ATCC 33088, DSM 43818, NBRC 16103, JCM 8973, NCIMB 2225, NRRL 3097.

Sequence accession nos: X92609 (16S rRNA gene); AB014156 (*gyrB*).

24. **Micromonospora olivasterospora** Kawamoto, Yamamoto and Nara 1983, 110[VP]

o.li.va.ste.ro.spo'ra. L. n. *oliva* an olive; Gr. n. *aster* a star; Gr. n. *spora* a seed and in biology a spore; N.L. fem. n. *olivasterospora* olive-colored spore that looks like a star.

Light brown to dark yellow colonies that become covered with an olive to dark green waxy layer of spores with age. Olive-green soluble, non-pH-sensitive pigments are formed on oatmeal and yeast extract-malt extract agars. Good growth occurs on most organic media. Melanin pigments are not formed. Well-developed branched and septate substrate hyphae (0.5 μm diameter) are formed. Spores are oval to spherical (approx 1.0 μm diameter), and rough-surfaced as seen by phase-contrast microscopy. Spores are borne on short sporophores or sessile, occurring randomly or in clusters. Terminally or intercalary chlamydospore-like swellings are sometimes present. Grows at 28–38°C and pH 6.8–7.8.

Fructose, galactose, glycerol, mannose, maltose, ribose, starch, sucrose, trehalose, and xylose are used as sole carbon sources, but not arabinose, dulcitol, inositol, lactose, mannitol, raffinose, melezitose, α-melibiose, rhamnose, salicin, sorbitol, or sorbose. NH_4NO_3, $(NH_4)_2SO_4$, NH_4Cl, arginine, aspartic acid, glutamic acid, histidine, and serine are used as sole nitrogen sources. Starch is hydrolyzed. Skim milk is peptonized, but not coagulated. Weak decomposition of cellulose. Positive for α- and β-glucosidases, and *N*-acetyl-β-glucosaminidase, but not for α-fucosidase, α- or β-galactosidases, or α-mannosidase.

Cell walls contain D-alanine, 3-OH-A_2pm, glycine, and glutamic acid. Arabinose, glucose, and xylose are the predominant whole-organism sugars. Major polar lipids are diphosphatidylglycerol, phosphatidylethanolamine, phosphatidylinositol, and phosphatidylinositol mannosides. Major fatty acids are $C_{15:0}$ iso and $C_{16:0}$ iso. Produces antibiotics of the fortimicin complex.

Source: isolated from a soil sample from Hiroshima, Japan.

DNA G+C content (mol%): 71.9 (HPLC).

Type strain: MK-70, ATCC 21819, DSM 43868, NBRC 14304, JCM 7348, NRRL 8178, VKM Ac-1317.

Sequence accession nos: X92613 (16S rRNA gene); AB014159 (*gyrB*).

25. **Micromonospora pallida** (Luedemann and Brodsky 1964) Kasai, Tamura and Harayama 2000, 131[VP] (*Micromonospora echinospora* subsp. *pallida* Luedemann and Brodsky 1964, 116)

pal'li.da. L. fem. adj. *pallida* pale.

Light ivory to light melon-yellow colonies, which turn brown to black upon sporulation. Does not produce purple mycelial pigments, or melanin pigments. Spores are spherical, dark brown to black. Sporophores are solitary or in small clusters.

Arabinose, cellobiose, glucose, levulose, mannose, soluble starch, sucrose, trehalose, and xylose are used as sole carbon sources. Good growth occurs on rhamnose, whereas slight growth is found on ribose with abundant sporulation. Poor growth occurs on salicin. Growth on and decomposition of cellulose is slow. Positive for nitrate reduction, gelatin liquefaction, and milk digestion. Good growth is observed at 27–37°C.

Possesses a different menaquinone (MK-12) to other *Micromonospora* species. Produces antibiotics of the gentamicin complex.

Source: isolated from soil collected in Jamesville, New York.

DNA G+C content (mol%): 71.1 (HPLC).

Type strain: ATCC 15838, DSM 43817, NBRC 16070, JCM 3133, NRRL 2996.

Sequence accession nos: X92608 (16S rRNA gene); AB014153 (*gyrB*).

26. **Micromonospora peucetia** Kroppenstedt, Mayilraj, Wink, Kallow, Schumann, Secondini and Stackebrandt 2005b, 1743[VP] (Effective publication: Kroppenstedt, Mayilraj, Wink, Kallow, Schumann, Secondini and Stackebrandt 2005a, 337.)

peu.ce'ti.a. L. fem. adj. *peucetia* Peucetian, of Peucetia, Latin name of landscape in Apulia (Puglia), Southern Italy.

Bottle green colonies are formed on inorganic salts-starch agar (ISP medium 4), pine green colonies are formed on glycerol-asparagine (ISP medium 5) and tyrosine agars (ISP medium 7), pale green to beige red colonies are formed on oatmeal agar (ISP medium 3), deep orange colonies are formed on yeast extract-malt extract agar (ISP medium 2), and chrome green colonies are formed on peptone-yeast extract-iron agar (ISP medium 6). Aerial hyphae are absent. Single spores are produced on substrate mycelium. Soluble pigments are not formed.

Fructose, glucose, and sucrose are used as sole carbon sources. Positive for alkaline phosphatase, *N*-acetyl-β-glucosaminidase, esterase (C4), α-glucosidase, and acid phosphatase, but not for α-fucosidase, α- or β-galactosidases, lipase, β-glucosidase, β-glucuronidase, or α-mannosidase.

Peptidoglycan contains *meso*-A_2pm. Whole-organism diagnostic sugars are arabinose, glucose, and xylose. Major menaquinone is MK-9(H_4). Phosphatidylethanolamine is the diagnostic phospholipid. Major fatty acids are $C_{15:0}$ iso, $C_{16:0}$, and $C_{17:1}$ ω8.

Source: isolated from soil.

DNA G+C content (mol%): not determined.

Type strain: DSM 43363, JCM 12820.

Sequence accession no. (16S rRNA gene): X92603.

27. **Micromonospora purpureochromogenes** (Waksman and Curtis 1916) Luedemann 1971, 244[VP] ("*Actinomyces purpureochromogenes*" Waksman and Curtis 1916, 113; *Micromonospora fusca* Jensen 1932, 173; *Micromonospora brunnea* Sveshnikova, Maksimova and Kudrina 1969, 762)

pur.pur.e.o.chro.mo'ge.nes. L. adj. *purpureus* purple colored; Gr. n. *chroma* color; Gr. v. *gennaio* to produce; N.L. part. adj. *purpureochromogenes* producing purple color (relating to the color of the diffusible pigment).

Dark brown colonies are formed on yeast extract-glucose agar after 21 days; a dark brown, diffusible, and water-soluble pigment is produced in most media. Very slow growth is seen on Czapek's sucrose agar. Spores (0.8–1.0 μm in diameter) are formed singly or in clusters. A monopodial branching system of sporulating hyphae is found, best observed at the periphery of colonies. Apparent dichotomous branching of sporulating hyphae is observed. Fragmentation of mycelium occurs in liquid cultures. Grows at 26–37°C, but not at 45°C.

Fructose, galactose, glucose, glycerol, lactose, levulose, melibiose, raffinose, sucrose, and xylose are used as sole carbon sources. Poor growth occurs on D- and L-arabinose. Milk is poorly peptonized. Limited hydrolysis of casein occurs and nitrate reduction is variable.

Source: isolated from adobe soil in California.

DNA G+C content (mol%): 73 (HPLC).

Type strain: ATCC 27007, IMRU 3343, ATCC 27334, DSM 43821, NBRC 13324, JCM 3156, NRRL B-2101, NRRL B-16094, VKM Ac-937.

Sequence accession nos: X92611 (16S rRNA gene); AB014161 (*gyrB*).

28. **Micromonospora rosaria** (*ex* Wagman, Waltz, Marquez, Murawski, Oden, Testa and Weinstein 1972) Horan and Brodsky 1986b, 478[VP] ("*Micromonospora rosaria*" Wagman, Waltz, Marquez, Murawski, Oden, Testa and Weinstein 1972, 641)

ro.sa'ri.a. L. fem. adj. *rosaria* of roses, rose-. The epithet refers to the wine red diffusible pigment produced by the strain.

Raised, folded colonies are formed on most media with orange brown to purple black vegetative mycelial pigments. Produces characteristic wine red diffusible pigment on most media. Short, sterile aerial hyphae produce a gray white bloom in some media. Single, warty spores (1.5–1.8 μm) are formed on short sporophores or are sessile. Good growth occurs on most rich organic media. Good growth is found at 35–40°C; grows poorly at 42°C.

D- and L-Arabinose, fructose, galactose (weak), glucose, lactose, maltose, mannose, mannitol, rhamnose, ribose, sucrose, trehalose, and xylose are used as sole carbon sources, but not adonitol, dulcitol, glycerol, melibiose, melezitose, or raffinose. Degrades chitin, casein, starch, gelatin, tyrosine, and xylan. Hydrolyzes esculin. Nitrate reduction is negative.

Sensitive to acid. Resistant to clindamycin, erythromycin, everninomicin, gentamicin, lincomycin, novobiocin, penicillin G, and sisomicin, at concentrations of 50 μg/ml, but is sensitive to aminoglycosides, rifamycin, and tetracycline at this concentration.

Whole-organism hydrolysates contain *meso*- and 3-OH-A_2pm. Arabinose, galactose, and xylose are the characteristic sugars. Produces the macrolide antibiotic rosaramicin.

Source: isolated from soil.

DNA G+C content (mol%): 72.9 (HPLC).

Type strain: SCC 957, 67694, NRRL 3718, ATCC 29337, NBRC 13697, DSM 803, JCM 3159.

Sequence accession nos: X92631 (16S rRNA gene); AB014163 (*gyrB*).

29. **Micromonospora saelicesensis** Trujillo, Kroppenstedt, Fernández-Molinero, Schumann and Martínez-Molina 2007, 801[VP]

sa.e.li.ces.en'sis. N.L. fem. adj. *saelicesensis* of or pertaining to Saelices, the place where the plants from which the first strains were isolated were collected.

Orange colonies are formed on yeast extract-malt extract agar (ISP medium 2). Orange-brown to brown diffusible pigments are produced on oatmeal agar (ISP medium 3). Well-developed branched hyphae (0.3–0.6 µm diameter) are observed with smooth-surfaced spores formed at the tips of hyphae. Good growth is seen at 20–37°C. Grows in the presence of 2% (w/v) NaCl.

Arabinose, cellobiose, galactose, glutarate, histidine, inositol, maltose, mannose, melibiose, and raffinose are used as sole carbon sources, but not alanine, arginine, gluconate, lysine, proline, rhamnose, serine, sorbitol sorbose, sucrose, trehalose, valine, or xylitol. Catalase-positive and oxidase-variable. Nitrate is not reduced. Degrades arbutin, casein, gelatin, and xylan; starch degradation is variable. Esculin is hydrolyzed. Positive for acid and alkaline phosphatases, esterases and lipases, α- and β-galactosidase, α- and β-glucosidases, and *N*-acetyl-β-glucosaminidase, but not for β-glucuronidase or urease.

Whole-organism hydrolysates contain *meso*-A_2pm, and arabinose, glucose, mannose, ribose, rhamnose, and xylose as characteristic sugars. Diagnostic phospholipid is diphosphatidylethanolamine. Major fatty acids are $C_{15:0}$ iso, $C_{16:0}$, and $C_{17:1}$ ω8. Predominant menaquinone is MK-10(H_4).

Source: isolated from root nodules of *Lupinus angustifolius.*

DNA G+C content (mol%): 71.6 (T_m).

Type strain: Lupac 09, DSM 44871, LMG 24056.

Sequence accession no. (16S rRNA gene): AJ783993.

30. **Micromonospora sagamiensis** Kroppenstedt, Mayilraj, Wink, Kallow, Schumann, Secondini and Stackebrandt 2005b, 1743[VP] (Effective publication: Kroppenstedt, Mayilraj, Wink, Kallow, Schumann, Secondini and Stackebrandt 2005a, 338.)

sa.ga.mi.en'sis. N.L. fem. adj. *sagamiensis* of or belonging to Sagami Bay.

Salmon pink colonies are formed on inorganic salts-starch (ISP medium 4) oatmeal agar (ISP medium 3) and tyrosine agar (ISP medium 7), red colonies are formed on glycerol-asparagine agar (ISP medium 5), and pastel coral red colonies are formed on yeast extract-malt extract agar (ISP medium 2). Does not grow on peptone-yeast extract-iron agar (ISP medium 6). Aerial mycelium is not formed. Diffusible pigments are not produced. Oval or spherical spores with rough surfaces are produced singly on the substrate mycelium. Grows optimally at 30–40°C; optimum pH for growth is between 7.0 and 8.0.

Galactose, glucose, levulose, sucrose, raffinose (weak), and xylose are used as sole carbon sources, but not arabinose, glycerol, lactose, inositol, mannitol, or rhamnose. Positive for starch hydrolysis, nitrate reduction, and milk coagulation. Cellulose decomposition is weak. Positive for alkaline phosphatase, α- and β-galactosidases, α- and β-glucosidases, *N*-acetyl-β-glucosaminidase, and acid phosphatase, but not for α-fucosidase, β-glucuronidase, lipase, or α-mannosidase.

Peptidoglycan contains *meso*-A_2pm. Whole-organism diagnostic sugars are glucose and xylose. Major menaquinone is MK-9(H_4). Phosphatidylethanolamine is the diagnostic phospholipid. Major fatty acids are $C_{15:0}$ iso, $C_{16:0}$, $C_{17:1}$ iso, and $C_{17:0}$.

Produces sagamicin (XK62-2) and antibiotics of the gentamicin C complex.

Source: isolated from forest soil, Japan.

DNA G+C content (mol%): not determined.

Type strain: MK-65, DSM 43912, JCM 3310, ATCC 21826, NRRL 11334.

Sequence accession no. (16S rRNA gene): X92624.

31. **Micromonospora siamensis** Thawai, Tanasupawat, Itoh, Suwanborirux and Kudo 2006b, 2[VP] (Effective publication: Thawai, Tanasupawat, Itoh, Suwanborirux and Kudo 2005a, 233.)

si.am.en'sis. N.L. fem. adj. *siamensis* of or pertaining to Siam, the old name for Thailand, the origin of the soil from which the type strain was isolated.

Forms vivid orange colonies that turn brownish black upon sporulation, and a pale yellow soluble pigment on yeast extract-malt extract agar (ISP medium 2). Aerial mycelium is absent. Melanin pigments are not formed. Develops abundant branched substrate hyphae which carry single spores. Spores are nonmotile and smooth-surfaced. Grows optimally at 25–30°C, but not above 40°C. Tolerates up to 5% (w/v) NaCl.

L-Arabinose, cellobiose, fructose (weak), galactose, glucose, lactose, melibiose, raffinose, rhamnose, salicin, and xylose are used as sole carbon sources, but not glycerol, mannitol, or ribose. Positive for milk peptonization, starch hydrolysis, and gelatin liquefaction, but negative for nitrate reduction and H_2S production.

Peptidoglycan contains *meso*-A_2pm. Whole-organism diagnostic sugars are arabinose and xylose. Predominant menaquinones are MK-10(H_4) and MK-10(H_6). Diphosphatidylglycerol, phosphatidylethanolamine, phosphatidylinositol, and phosphatidylinositol mannosides are the major polar lipids. Rich in $C_{15:0}$ iso, $C_{16:0}$ iso, $C_{17:0}$ anteiso, and $C_{17:0}$ fatty acids.

Source: isolated from a peat swamp forest soil, Thailand.

DNA G+C content (mol%): 73.0 (HPLC).

Type strain: TT2-4, JCM 12769, PCU 266, TISTR 1554.

Sequence accession no. (16S rRNA gene): AB193565.

32. **Micromonospora viridifaciens** Kroppenstedt, Mayilraj, Wink, Kallow, Schumann, Secondini and Stackebrandt 2005b, 1744[VP] (Effective publication: Kroppenstedt, Mayilraj, Wink, Kallow, Schumann, Secondini and Stackebrandt 2005a, 338.)

vi.ri.di.fa'ci.ens. L. adj. *viridis* green; L. part. adj. *faciens* making, N.L. part. adj. *viridifaciens* green making.

Beige red colonies are formed on oatmeal agar (ISP medium 3), yellow colonies are formed on peptone-yeast extract-iron agar (ISP medium 6) and yellow to brown are formed colonies on yeast extract-malt extract-iron agar (ISP medium 2). Sparse growth is seen on inorganic salts-starch (ISP medium 4), glycerol-asparagine agar (ISP medium 5), and tyrosine agar (ISP medium 7).

Aerial mycelium is absent. Soluble pigments, including melanin, are not produced. Spores with smooth surfaces are formed singly on the substrate mycelium. Good growth occurs between 28 and 45°C.

Good utilization of ribose and starch is observed, but poorly utilizes arabinose, fructose, glucose, inositol, mannitol, rhamnose, raffinose, and sucrose. Positive for liquefaction of gelatin, and milk coagulation and peptonization, but negative for nitrate reduction and tyrosine degradation. Positive for alkaline phosphatase, α-glucosidase, and acid phosphatase; negative for α-fucosidase, lipase, *N*-acetyl-β-glucosaminidase, α- and β-galactosidase, β-glucosidase, β-glucuronidase, and α-mannosidase.

Peptidoglycan contains *meso*-A_2pm. Whole-organism diagnostic sugars are glucose, rhamnose, and xylose. Major menaquinone is MK-9(H_4). Phosphatidylethanolamine is the diagnostic phospholipid. Major fatty acids are $C_{15:0}$ iso and $C_{17:0}$ 10-methyl. Produces antibiotic 37505.

Source: isolated from soil sample, Japan.

DNA G+C content (mol%): not determined.

Type strain: DSM 43909, JCM 3267, ATCC 31146.

Sequence accession no. (16S rRNA gene): X92623.

Species *incertae sedis*

1. **Micromonospora gallica** (Erikson 1935; Waksman 1961) Kawamoto 1989

gal'li.ca. L. fem adj. *gallica* of or belong to the Gauls.

This species was included in the *Approved Lists of Bacterial Names* (Skerman et al., 1980), but the type strain NCTC 4582^{T} is not extant (Kawamoto, 1989).

Genus II. **Actinocatenispora** Thawai, Tanasupawat, Itoh and Kudo 2006a, 1792^{VP}

THE EDITORIAL BOARD

Ac.ti.no.ca.te.ni.spo'ra. Gr. n. *aktis -inos* ray; L. n. *catena* chain; Gr. n. *spora* seed and in biology a spore; N.L. fem. n. *Actinocatenispora* spore chain-producing ray (fungus).

Gram-stain-positive. Nonmotile. **Aerobic**. **Sporeforming** branching substrate hyphae; mycelia are yellow to vivid orange in color. Aerial hyphae or vegetative mycelium bear spore chains consisting of more than 10 spores. Cylindrical spores (0.3–0.4 × 0.5–1.0 μm) have a smooth surface. Cell wall contains glutamic acid, glycine, alanine, and *meso*-diaminopimelic acid. The *N*-acyl group of the cell-wall muramic acid is glycolyl. Characteristic whole-cell sugars are arabinose and xylose. Possess phospholipid pattern type II; cellular phospholipids are phosphatidylethanolamine, diphosphatidylglycerol, phosphatidylinositol, phosphatidylinositol mannosides, phosphatidylglycerol and unidentified ninhydrin-negative phospholipids. The predominant fatty acids are iso- and anteiso-methyl branched acids. Major menaquinone is MK-9(H_4). Mycolic acids are not detected.

DNA G+C content (mol%): 72.0–74.3.

Type species: **Actinocatenispora thailandica** Thawai, Tanasupawat, Itoh and Kudo 2006a, 1793^{VP}.

Further descriptive information

Phylogenetic analysis of the 16S rRNA gene positions the genus within the family *Micromonosporaceae*. The closest phylogenetic neighbor is *Phytohabitans suffuscus* (93.4%, accession no. AB490769) (Inahashi et al., 2010). So far, all the described species have all been isolated from soil. Environmental clones with high 16S rRNA gene sequence similarity have been detected in cellulosic waste from a simulated low-level-radioactive-waste site (99.2 %, accession no. GQ263629 and 99.0%, accession no. GQ263525). Environmental isolates identified as *Micromonospora* sp. and *Solwaraspora* sp. but with high 16S rRNA gene sequence similarity to *Actinocatenispora* have been cultivated from marine sediments (94% accession no. DQ448714 and ~93% accession nos AY552766–AY552773, respectively) (Gontang et al., 2007; Magarvey et al., 2004).

Enrichment and isolation procedures

Actinocatenispora thailandica strain TT2-10^{T} was isolated from peat swamp forest soil in Pattaloong Province, Thailand. Samples were taken from the soil surface were cultivated on starch-casein nitrate agar for 21 d at 30°C. Isolated colonies were purified on yeast extract-malt extract agar (ISP medium no. 2) and maintained on the same medium as a working culture (Thawai et al., 2004b), and the pure culture was kept at 4–10°C on yeast extract-malt extract agar (ISP 2 medium) slants.

Maintenance procedures

Strains are maintained yeast extract-malt extract agar (ISP medium no. 2) or in 20% (v/v) glycerol at −20 °C or −70°C.

Differentiation of the genus *Actinocatenispora* from closely related genera

Spore motility and true aerial mycelium production differentiates *Actinocatenispora* from other genera of the family *Micromonosporaceae*, except for *Longispora*. *Actinocatenispora* can be differentiated from *Longispora* by differences in the menaquinone and fatty acid composition.

List of species of the genus *Actinocatenispora*

1. **Actinocatenispora thailandica** Thawai, Tanasupawat, Itoh and Kudo 2006a, 1793[VP]

thai.lan'di.ca. N.L. fem. adj. *thailandica* of or belonging to Thailand, where the type strain was isolated.

White aerial mycelia are formed on oatmeal-nitrate agar. Soluble yellow pigment present on oatmeal agar. Temperature range for growth 25–30°C, does not grow above 40°C. Does not grow below pH 4.5, maximum NaCl tolerance is 7%. Major menaquinones are MK-9(H_4) and MK-9(H_6); MK-9(H_2) and MK-9(H_8) are also produced. Whole-cell sugars are galactose, xylose, arabinose, glucose, mannose, and ribose. Utilizes D-glucose, D-mannitol, D-melibiose, D-raffinose, glycerol, *myo*-inositol, salicin, and cellobiose. Does not utilize D-ribose, L-rhamnose, lactose, D-galactose, L-arabinose or D-fructose. Reduces nitrate, weakly positive for peptonization of milk and gelatin liquefaction. Does not hydrolyze starch. Melanin and H_2S are not produced. The major cellular fatty acid components are $C_{16:0}$ iso (26%), $C_{17:0}$ anteiso (19%), $C_{15:0}$ iso (20%) and $C_{17:0}$ iso (11%).

For all other characteristics, refer to the genus description.

DNA G+C content (mol%): 72.0 (HPLC).

Type strain: TT2-10, JCM 12343, PCU 235, DSM 44816.

Sequence accession no. (16S rRNA gene): AB107233.

2. **Actinocatenispora rupis** Seo and Lee 2009, 3082[VP]

ru'pis. L. gen. n. *rupis* of a cliff, referring to the site from which the type strain was isolated.

Vegetative mycelium is well developed and pale to strong yellow in color. Hyphal swellings are observed on the tips of the vegetative mycelium. Growth temperature optimum is 37°C, range is 25–42°C; and pH optimum is 5.1–9.1, range is 5.1–12.1; NaCl tolerance up to 4%. Degrades DNA, elastin, esculin, casein, Tween 80 and chitin, but not cellulose, starch, hypoxanthine, tyrosine, urea or xanthine. Catalase- and oxidase-negative. Utilizes D-arabinose, dextran, D-galactose, lactose, maltose, D-mannose, methyl α-D-glucoside, sucrose, adonitol, *meso*-erythritol, D-sorbitol and citrate as sole carbon sources. Does not utilize inulin, melezitose, methyl α-D-mannoside, salicin, L-sorbose, dulcitol, D-xylitol, acetate, benzoate, formate, malate, succinate or tartrate. Does not reduce nitrate to nitrite. Major menaquinones are MK-9(H_4), MK-9(H_6) and MK-9(H_8). The whole-cell sugars are glucose, rhamnose, ribose, arabinose and xylose. The diagnostic polar lipids are phosphatidylethanolamine, phosphatidylinositol and phosphatidylglycerol. The predominant fatty acids are $C_{16:0}$ iso (39%), $C_{17:0}$ anteiso (15%) and $C_{16:1}$ iso (13%).

For all other characteristics, refer to the genus description.

DNA G+C content (mol%): 74.3 (HPLC).

Type strain: CS5-AC17, DSM 45178, NRRL B-24660.

Sequence accession no. (16S rRNA gene): AM980986.

3. **Actinocatenispora sera** Matsumoto, Takahashi, Fukumoto and Ōmura 2007, 2653[VP]

se'ra. L. fem. adj. *sera* late.

Growth temperature optimum is 18–25°C, range is 13–37°C. Growth occurs at pH 6–9. No growth at 5% NaCl. Melanoid pigments are not produced. Negative for the liquefaction of gelatin. Hydrolyzes starch. Positive for coagulation and peptonization of milk. Reduces nitrate. Utilizes adonitol, D-glucose, L-rhamnose, xylitol and D-xylose. Does not utilize L-arabinose, D-cellobiose, D-fructose, glycerol, *myo*-inositol, maltose, D-mannose, D-mannitol, melibiose, raffinose, D-ribose, trehalose and sucrose. Cellulose is not decomposed. Predominant cellular fatty acids are $C_{17:0}$ anteiso (25%), $C_{16:0}$ iso (22%), $C_{17:0}$ iso (16%) and $C_{15:0}$ iso (14%).

Isolated from soil in Niigata Prefecture, Japan.

For all other characteristics, refer to the genus description.

DNA G+C content (mol%): 72.0 (HPLC).

Type strain: KV-744, NRRL B-24477, NBRC 101916.

Sequence accession no. (16S rRNA gene): AB263096.

Genus III. **Actinoplanes** Couch 1950, 89[AL] emend. Stackebrandt and Kroppenstedt 1987, 112

Gernot Vobis, Jenny Schäfer and Peter Kämpfer

Ac.ti.no.pla'nes. Gr. n. *aktis aktinos* ray, beam; Gr. masc. n. *planes* a wanderer, roamer; N.L. masc. n. *Actinoplanes* literally, a ray wanderer (intended to signify an actinomycete with swimming spores).

Substrate mycelium is developed on various agar media. Hyphae are 0.2–1.2 µm in diameter, branched, and septated; fragmentation is very rare. Gram-stain-positive, although parts may be Gram-stain-negative. Non-acid-fast. Aerial mycelium is absent or scanty.

Sporangia are produced on the surface of the substrate, sessile on the agar or detached on short sporangiophores. In many cases, they are formed at the tip of so-called palisade hyphae. **The shape of the sporangia varies from spherical, subspherical to irregular** (mean diameter 7.0–16.0 µm), **or cylindrical, bell-shaped, lobate, or digitate** (mean dimensions: 5.0–12.0 µm wide and 8.0–20.0 µm long). Numerous spores are produced within the sporangium, arranged in coiled, parallel, or irregular chains. **Sporangiospores are globose, subglobose, oval** (mean diameter 1.1–1.6 µm), **oblong** (mean 0.8 × 1.6 µm) or **rod-shaped** (mean 0.8 × 3.0 µm); **motile by a polarly inserted tuft of flagella**.

Grows under aerobic conditions. Colonies on various complex agar media are flat, elevated, or convoluted with a smooth, wrinkled, or ridged surface. The characteristic color of substrate mycelium is orange; by loss or additional formation of pigments, nearly all modifications occur from ivory, pale yellow to red, violet, green, brown and black. Most strains do not require organic growth factors. Chemo-organotrophic, mesophilic, in some cases moderately psychrophilic or thermotolerant. Grows well between 20 and 28°C.

The peptidoglycan of the cell wall contains glycine, *meso*-diaminopimelic acid and/or hydroxy-diaminopimelic acid. The diagnostic sugar of whole-cell hydrolysates is xylose; galactose and/or arabinose are also present. Phosphatidylethanolamine is the predominant phospholipid. Major menaquinone is MK-9(H_4). Iso/anteiso-branched and monounsaturated fatty acids and/or *cis*-9,10-octadecanoic acid (oleic acid) are the predominant fatty acids.

DNA G+C content (mol%): 69.0–73.0 (T_m).

Type species: **Actinoplanes philippinensis** Couch 1950, 89[AL].

Further descriptive information

Phylogeny. A phylogenetic study based on 16S rRNA gene sequence analysis assigned the genus *Actinoplanes* to the family *Micromonosporaceae.* This has been confirmed by other 16S rRNA gene sequence studies of this genus (Stackebrandt and Kroppenstedt, 1987; Stackebrandt et al., 1997; Tamura and Hatano, 2001). The taxon currently encompasses 28 species with validly published names. Representatives of species with validly published names show a high similarity in the 16S rRNA *Actinoplanes* gene tree (Figure 211), with the exception of *Actinoplanes globisporus.* Similarity values of the different species within the genus *Actinoplanes* range between 95.0 and 100%. The 16S rRNA gene sequence of *Actinoplanes globisporus* differs by 3.1–5.0% from those of other species of the genus *Actinoplanes* (similarity values between the sequences of *Actinoplanes globisporus* and those of other species of the genus *Actinoplanes* range from 95.0 to 96.9%).

Similarity values between *Actinoplanes* and the other genera within the family *Micromonosporaceae* (based on the type species) range from 97.7 to 94.4%. Based on distance calculations of the type species of the different genera, the nearest relatives are the genera *Spirilliplanes* (97.0%), *Polymorphospora* (96.9%), and *Micromonospora* (96.8%).

Cell morphology. Strains of *Actinoplanes* produce substrate mycelium on solid agar media. Substrate hyphae are fine, 0.2–1.0 µm in diameter, branched, and septated. Fragmentation is very rare, e.g, observable in *Actinoplanes couchii*, developing substrate mycelium that fragments in irregular rod-shaped cells when cultured on DSMZ medium 65 (Kämpfer et al., 2007). Strains may be morphogenetically characterized by specialized, vertically orientated hyphae directly under the surface of agar media (Couch, 1963). They are two to three times thicker than the usual hyphae and strongly arranged in parallel positions. In general, these so called palisade hyphae are identical to sporangiophores (Figure 213A), first described from the species *Actinoplanes philippinensis* (Couch, 1950).

The development of aerial mycelium in the genus *Actinoplanes* is not frequent and is observed only in some species, e.g. *Actinoplanes couchii, Actinoplanes ferrugineus, Actinoplanes liguriensis,*

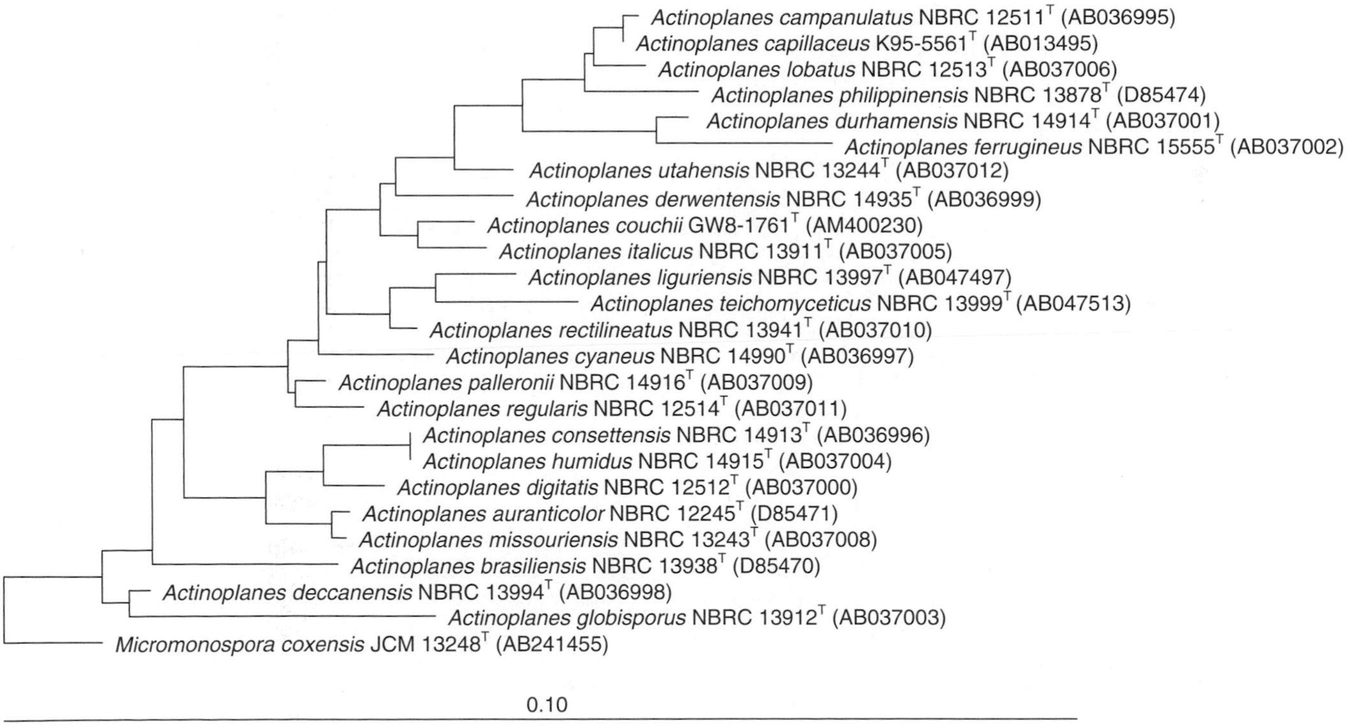

FIGURE 211. Phylogenetic analysis of *Actinoplanes* species based on 16S rRNA gene sequences available from EMBL (accession numbers are given in parentheses). Multiple alignment, distances (distance options according to the Kimura-2 model; Kimura, 1980) and clustering with the neighbor-joining method were performed by using the software packages MEGA (Molecular Evolutionary Genetics Analysis) version 4 (Tamura et al., 2007). Bootstrap values based on 1000 replications are given as percentages at the branching points. *Micromonospora coxensis* was used as an outgroup.

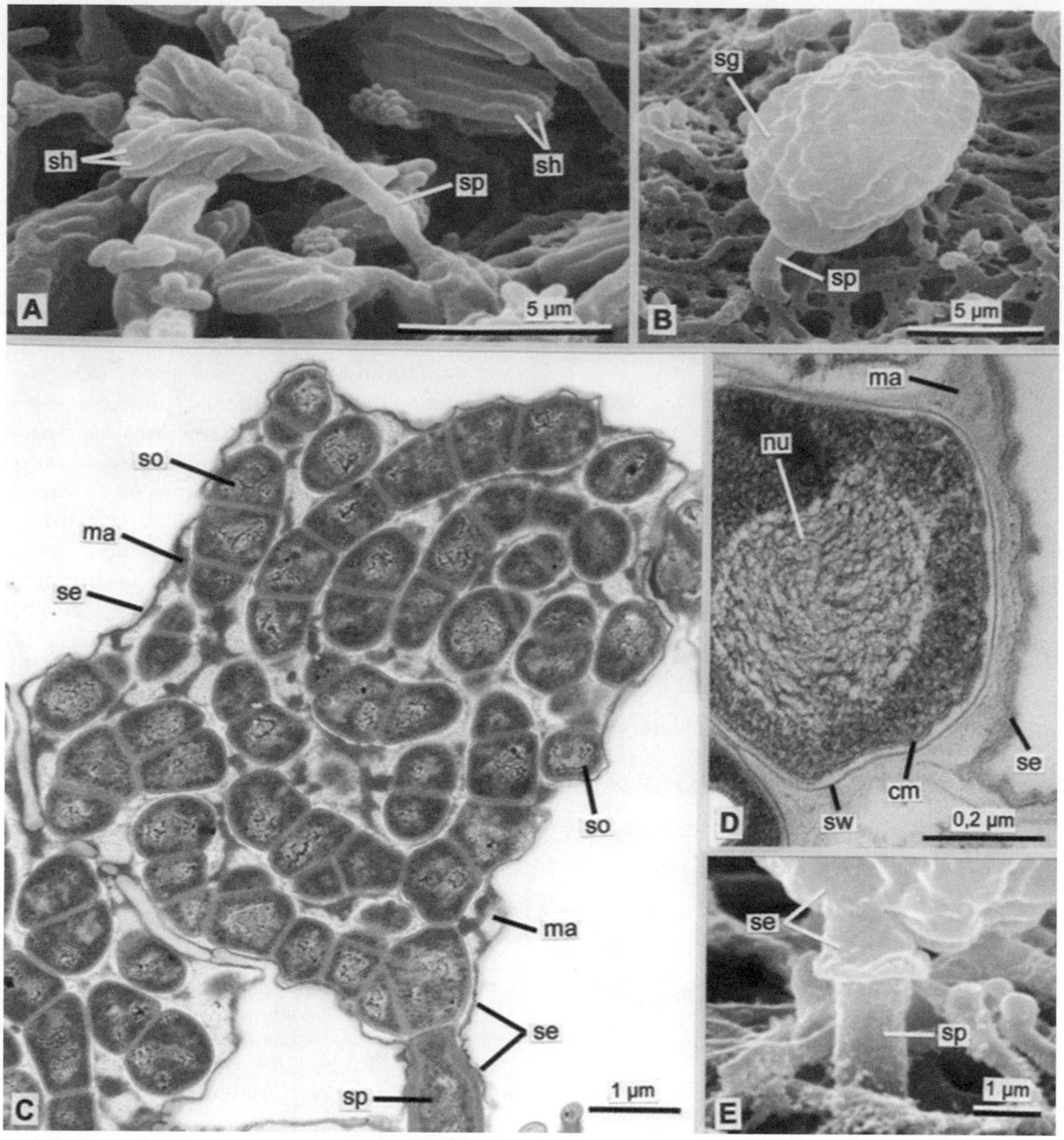

FIGURE 212. Multispored sporangia of *Actinoplanes*. A, "*Ampullariella*"-type with rows of sporogenous hyphae orientated in parallel (strain MB-VE 1144, SEM); B, subglobose sporangium developed on substrate mycelium (strain MB-SE 50, SEM); C, longitudinal section of a sporangium containing spores in chains (strain MB-SE 50, TEM); D, spore on the inner side of a thin, "membranous" sporangial envelope (strain ATCC 21983, TEM); E, sporangiophore in transition between humid substrate and air-exposed sporangium (strain MB-SE 50, SEM). Abbreviations: SEM, scanning electron microscope; TEM, transmission electron microscope; ma, intrasporangial matrix; se, sporangial envelope; sg, sporangium; sh, sporogenous hypha; so, spore; sp, sporangiophore; sw, spore wall.

Actinoplanes rectilineatus, and *Actinoplanes teichomyceticus*, if cultured on special media. In general, this kind of aerial mycelium is named rudimentary or sterile because it does not reach the dimensions of typical aerial mycelium and its hyphae are not involved in any production of sporangia or spores. Aerial mycelium is white and imparts to the colonies a powdery appearance (Lechevalier and Lechevalier, 1975; Parenti et al., 1978).

Sporangia are the characteristic reproductive structures of *Actinoplanes*, which are developed by substrate mycelium (Figure 212A, B, E). The sporangiophores are either more or less undifferentiated hyphae or palisade hyphae. They transgress the aqueous surroundings of the substrate. Once air-exposed, sporangia are additionally covered by a hydrophobic sheath, which passes over without interruption into the sporangial envelope (Figure 212C, E). The single steps of sporangial development are described elsewhere (Lechevalier and Holbert, 1965; Lechevalier et al., 1966; Vobis, 1997; Vobis and Kothe, 1985): inside the sporangial envelope, sporogenous hyphae are septated into spore-shaped segments, which are rounded off into individual spores (Figure 212C). In most species, the spore chains are arranged in coils or, in the case of the former "*Ampullariella*" strains, in parallel rows. Evidently, the external form of a sporangium depends on the internal arrangement of the sporogenous hyphae/spore chains (Figure 212A,

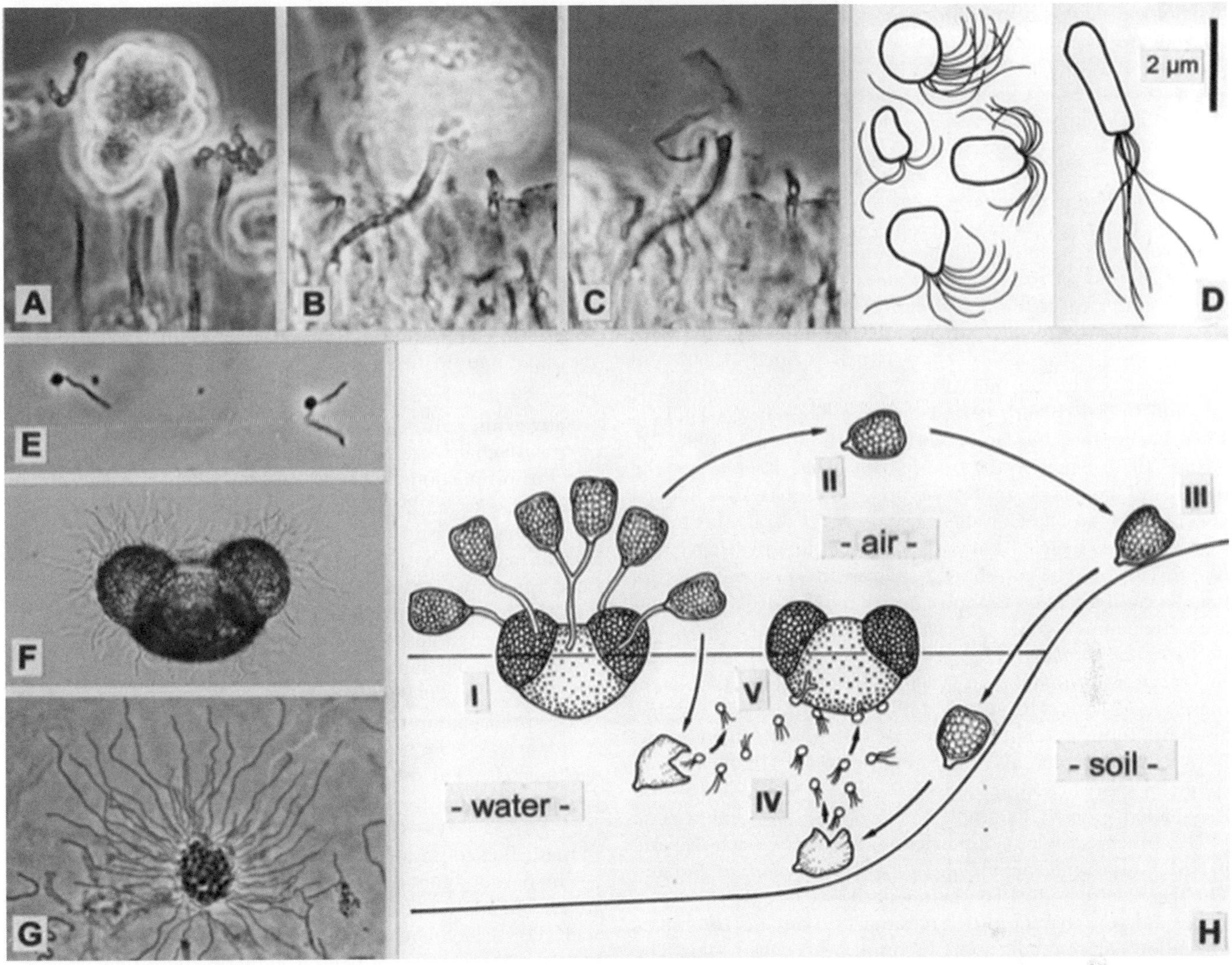

FIGURE 213. Release of sporangiospores (A, B, C), and aero-aquatic life cycle of *Actinoplanes* (H). A, Sporangium originated on palisade hypha, submersed in water; B, burst sporangium and swarming spores; C, left empty sporangial envelope; D, sporangiospores with a polarly inserted tuft of flagella, varying in size and shape; E, germinating spores; F, growth of mycelium on a pollen grain of *Pinus* sp.; G, germinated sporangium, developing new mycelium. H (I) Floating pollen grain with mycelium and sporangia; (II) dispersal of sporangia by air; (III) dry-resistant sporangia on soil; (IV) flagellated sporangiospores are released in water; (V) new substrate is colonized by germinating spores. Note: A, B, C, E, F, G (strain MB-SE 165; light microscope, phase-contrast; for dimensions compare Figure 212).

B). Very rarely, chains orientated in spirals can be observed (G. Vobis, unpublished). The surface of the sporangial envelope is generally smooth (Figure 212A–D). External ornamentation is described only for *Actinoplanes capillaceus*, including four invalidly proposed species of "*Ampullariella*" (Matsumoto et al., 2000; Miyadoh et al., 1997; Seino, 1983). In all these cases, the sporangia are covered with hair-like structures, resembling the hairy spores of *Streptomyces*.

The polysporous sporangia vary in shape and size (Vobis, 1997). In most species, the sporangia are spherical (globose), subspherical (subglobose), oval to irregular, occasionally wider than they are long, and rarely lobed. The smallest sporangia have a diameter of 4 µm, with the biggest being 25 µm (mean diameter of about 11 µm). As an extreme exception, a strain with sporangia of 47 µm in diameter has been observed (Vobis, 1992). The former "*Ampullariella*" species, including *Actinoplanes capillaceus* and *Actinoplanes rectilineatus*, have sporangial shapes from cylindrical or bottle-shaped to bell-shaped and digitate, frequently lobed and papillate (Vobis and Kothe, 1989). Extremely small sporangia may be found in *Actinoplanes digitatis* (3 µm wide and 6 µm long), and the biggest are found in *Actinoplanes regularis* (14 × 30 µm). Typical sporangia with parallel spore chains have a mean size of 5–12 × 8–20 µm.

The spores produced within the sporangia have a smooth surface and are actively motile by flagella (zoospores or planospores). With regard to their shape, they can be subdivided into two categories (Figure 213D). The typical globose spores vary from spherical (globose), subspherical (subglobose), to oval with a mean diameter of 1.35 µm (range: 0.8–2.0 µm). The sporangiospores of the former "*Ampullariella*" species are distinct rod-shaped or bacilliform, four times longer than they are wide, with a mean width of 0.75 μm and mean length of 3.0 μm (range: from 0.5 × 2.0 μm to 1.0 × 4.0 μm). Oblong spores are twice as long as they are wide and are generally described as short rod-shaped; with a mean width of 0.8 μm and a mean length of 1.63 μm (range: from 0.5 × 1.0 μm to 1.2 × 2.3 μm), they may belong to the last category. Variations in the shape of spores of one strain (e.g. globose, subglobose, oval) are explained by slight distance deviations during the segmentation process of the branched sporogenous hyphae.

The motility of the spores is caused by means of a tuft of flagella, inserted at the apical end (Figure 213D). This polytrichous type of flagellation was first documented by light microscopy in *Actinoplanes philippinensis* (Couch, 1950) and confirmed by electron microscopy preparations in other species and strains (Bland, 1970; Hanton, 1968; Karwowski et al., 1988; Lechevalier and Holbert, 1965; Miyadoh et al., 1997; Palleroni, 1979; Schäfer, 1973). In *Actinoplanes rectilineatus*, the number of flagella ranges from 17 to 40; their bases are hooked (Lechevalier and Lechevalier, 1975). In general, the tuft is formed by less flagella. They are 2–6 μm in length (Couch and Bland, 1974b). The rod-shaped spores with their polar tuft of flagella can be named lophotrichous (Schäfer, 1973) and they are well documented in *Actinoplanes regularis* (Kane, 1966), *Actinoplanes campanulatus* (Higgins et al., 1967), and *Actinoplanes capillaceus* (Matsumoto et al., 2000). The number of flagella on these species ranges from 1 to 12 (Higgins et al., 1967) and they are 3.5–6.0 μm in length (Couch and Bland, 1974b). Peritrichous flagellation has also been reported occasionally, both in spherical spores (Ruan et al., 1976; Willoughby, 1968) and in rod-shaped spores (Ruan and Zhang, 1974). Nonomura et al. (1979) observed polar (lophotrichous and amphitrichous) and peritrichous flagella arrangement in 12 "*Ampullariella*" isolates. The flagella are up to 19.0 μm in length (mean of 6–12 μm).

Besides the characteristic sporangia producing zoospores, formation of nonmotile spores (conidia) may also occur. The conidiophores develop in a similar manner to the sporangia, but in the absence of the sporangial envelope, the sporogenous (conidiogenous) hyphae appear as brushlike bunches (Willoughby, 1966). Conidiophores originate at the margin of the colonies of *Actinoplanes regularis* (Couch, 1963). In "*Actinoplanes arizonaensis*", conidiogenous hyphae occur as digitate hyphae on the surface of the colony (Karwowski et al., 1988). They are interpreted as abortive sporangia. *Actinoplanes campanulatus* develops microconidia and *Actinoplanes lobatus* develops oval conidia in moniliform arrangement, which are produced by substrate mycelium (Couch, 1963). Microconidia are also formed in the substrate mycelium of *Actinoplanes auranticolor* (Hanton, 1968) and chlamydospores are formed in *Actinoplanes globisporus* (Thiemann, 1967).

Cell-wall composition. The peptidoglycan of the cell wall contains glycine and *meso*- and/or hydroxy-diaminopimelic acid, according to wall chemotype II of Lechevalier and Lechevalier (1970a). It has been demonstrated that the acyl groups of the muramyl residue in the cell-wall peptidoglycan are of the glucolyl type (Uchida and Seino, 1997).

The principal wall sugars are arabinose, galactose, glucose, mannose, and xylose. The latter is the diagnostic sugar of whole-cell hydrolysates (Tamura and Hatano, 2001). The presence of xylose and arabinose corresponds to the whole-cell sugar pattern type D of Lechevalier and Lechevalier (1970a). Galactose was not detected in *Actinoplanes regularis*, *Actinoplanes campanulatus*, or *Actinoplanes digitatis* (Stackebrandt and Kroppenstedt, 1987).

The polar lipid profile is characterized by the presence of phosphatidylethanolamine as diagnostic phospholipid and the absence of phosphatidylcholine and amino-containing phosphoglycolipid, corresponding to phospholipid type PII of Lechevalier et al. (1981, 1977). Many species also contain phosphatidylinositol, phosphatidylinositol mannoside, and phosphatidylglycerol (Table 194). The type strain of *Actinoplanes regularis* is of particular note as it lacks phosphatidylethanolamine (Stackebrandt and Kroppenstedt, 1987). The whole polar lipid composition may also contain a highly hydrophilic glycolipid and further unknown or uncharacterized lipids and glycolipids (Goodfellow et al., 1990b; Kämpfer et al., 2007).

Fatty acids. *Actinoplanes* species contain complex mixtures of straight-chain, branched-chain, and unsaturated fatty acids and the proportions of these vary considerably between the strains (Ara et al., 2010; Goodfellow et al., 1990b; Sun et al., 2009). Characteristic fatty acid patterns for *Actinoplanes* species include major amounts of iso/anteiso branched- and monosaturated fatty acids, as well as *cis*-9,10-octadecanoic acid (oleic acid, n-$C_{18:1}$), 14-methylpentadecanoic ($C_{16:0}$ iso), and 14-methylhexadecanoic ($C_{17:0}$ anteiso) acids (Ara et al., 2010; Kämpfer et al., 2007; Matsumoto et al., 2000; Stackebrandt and Kroppenstedt, 1987; Sun et al., 2009; Wink et al., 2006).

The predominant saturated straight-chain fatty acids of the *Actinoplanes* species are penta- ($C_{15:0}$), hexa- ($C_{16:0}$), and heptadecanoic ($C_{17:0}$) acids. Major saturated branched fatty acids are 13-methyltetradecanoic ($C_{15:0}$ iso) and 14-methylpentadecanoic ($C_{16:0}$ iso), as well as 12-methyltetradecanoic ($C_{15:0}$ anteiso) and 14-methylhexadecanoic ($C_{17:0}$ anteiso) acids.

The predominant components of the unsaturated fatty acids are *cis*-9-heptadecenoic acid ($C_{17:1}$ ω9*c*) and *cis*-9-octadecenoic acid ($C_{18:1}$ ω9*c*, oleic acid). However, some differences have been detected by Kämpfer et al. (2007), where only *cis*-8-heptadecenoic acid ($C_{17:1}$ ω8*c*) instead of *cis*-9-heptadecenoic acid were detected in the analyzed strains, e.g. *Actinoplanes philippinensis*, *Actinoplanes couchii*, *Actinoplanes italicus*, and *Actinoplanes rectilineatus*. Minor amounts of tetra- and octadecanoic ($C_{14:0}$, $C_{18:0}$) acids, as well as 15-methylhexadecanoic- ($C_{17:0}$ iso) and *cis*-9-hexadecenoic acid ($C_{16:1}$ ω9*c*) have been found in most species. In a few cases, 12-methyltridecanoic acid ($C_{14:0}$ iso), 10-methylheptadecanoic acid ($C_{17:0}$ 10-methyl), and *cis*-9,14-methylhexadecenoic ($C_{17:1}$ iso ω9*c*) acids (Ara et al., 2010; Kämpfer et al., 2007; Sun et al., 2009) were detected in *Actinoplanes* species.

No fatty acid data are available for *Actinoplanes cyaneus* and *Actinoplanes digitatis*. Most analyses of the fatty acid components have been made according to Sasser (1990); fatty acid analyses described by Goodfellow et al. (1990b) were done according to Miwa et al. (1960). The distribution of cellular fatty acids amongst *Actinoplanes* species is summarized in Table 194.

Actinoplanes species contain tetrahydrogenated menaquinones with nine isoprene units [MK-9(H_4)] as predominant isoprenologue, with moderate amounts of MK-9(H_2) and MK-9(H_6) (Goodfellow et al., 1990b; Stackebrandt and Kroppenstedt, 1987). Minor proportions of MK-9(H_8) can also be found (Goodfellow et al., 1990b; Kämpfer et al., 2007). Tetrahydrogenated menaquinones with 10 side chains [MK-10(H_4)] may occur additionally as characteristic components of the quinone

TABLE 194. Differential characteristics of *Actinoplanes* species[a]

Characteristic	1. *A. philippinensis*	2. *A. auranticolor*	3. *A. brasiliensis*	4. *A. campanulatus*	5. *A. capillaceus*	6. *A. consettensis*	7. *A. couchii*	8. *A. cyaneus*	9. *A. deccanensis*	10. *A. derwentensis*	11. *A. digitatis*	12. *A. durhamensis*	13. *A. ferrugineus*	14. *A. globisporus*	15. *A. humidus*	16. *A. italicus*	17. *A. liguriensis*	18. *A. lobatus*	19. *A. missouriensis*	20. *A. palleronii*	21. *A. rectilineatus*	22. *A. regularis*	23. *A. sichuanensis*	24. *A. teichomyceticus*	25. *A. toevensis*	26. *A. tereljensis*	27. *A. utahensis*	28. *A. xinjiangensis*
Shape of sporangia:																												
Globose (subglobose, oval, pyriform)	+	–	(+)	–	–	+	+	+	+	+	–	+	+	–	+	(+)	+	–	+	+	–	–	+	+	+	–	(+)	+
Irregular	–	+	+	+	–	–	–	–	–	–	–	–	+	+	–	+	–	+	(+)	–	–	–	–	–	–	+	+	–
Lobed	–	+	–	+	–	–	–	–	–	–	(+)	–	–	–	–	–	–	+	–	–	–	–	–	–	–	–	+	–
Umbrella	–	–	+	–	–	–	–	–	–	–	–	–	–	–	–	–	–	–	–	–	–	–	–	–	–	–	–	–
Campanulate	–	–	–	+	+	–	–	–	–	–	–	–	–	–	–	–	–	–	–	–	–	–	–	–	–	–	–	–
Digitate	–	–	–	–	–	–	–	–	–	–	+	–	–	–	–	–	–	–	–	–	–	–	–	–	–	–	+	–
Cylindrical	–	–	–	(+)	–	–	–	–	–	–	–	–	–	–	–	–	–	(+)	–	–	+	+	–	–	–	–	–	–
Spore arrangement within sporangia:																												
Coils	+	–	–	–	–	–	–	nd	+	–	–	–	+	+	–	+	+	–	+	–	–	–	nd	+	nd	nd	+	nd
Irregular	–	+	+	–	–	+	+	nd	–	+[r]	–	+	–	–	+	–	–	–	–	+	–	–	nd	–	nd	nd	–	nd
Parallel rows	–	–	–	+	+	–	–	nd	–	–	+	–	–	–	–	–	–	+	–	–	+	+	nd	–	nd	nd	–	nd
Shape of spores:																												
Globose (subglobose, oval)	+	–	(+)	–	–	nd	+	+	+	+	–	nd	+	+	nd	+	+	–	+	nd	+	–	nd	+	+	+	+	nd
Rods (oblong, short rod-like)	–	+	+	+	+	nd	–	–	–	–	+	nd	–	–	nd	–	–	+	–	nd	–	+	nd	–	–	–	–	nd
Aerial mycelium:																												
Absent	+	+	+	+	+	+	–	+	+	+	+	+	–	+	+	+	–	+	+	+	–	+	+	–	+	+	+	+
Rudimentary	–	–	–	–	–	–	+	–	–	–	–	–	+	–	–	–	+	–	–	–	–	–	–	–	–	–	–	–
Well developed	–	–	–	–	–	–	–	–	–	–	–	–	–	–	–	–	–	–	–	–	+	–	–	+	–	–	–	–
Color of substrate mycelium:																												
White	–	–	–	–	–	–	–	nd	–	–	–	–	–	+	–	–	–	+	–	–	+	–	–	–	–	–	–	+
Yellow	+	–	–	–	+	+	(+)	nd	–	–	(+)	–	(+)	–	(+)	–	(+)	–	(+)	+	+	(+)	–	+	(+)	–	–	–
Orange	+	+	+	+	–	–	+	nd	+	+	–	+	–	–	+	+	+	–	+	–	+	+	+	+	(+)	(+)	+	+
Red	+	–	(+)	+	+	–	(+)	nd	(+)	–	+	–	–	–	–	+	–	+	(+)	–	–	+	+	–	(+)	(+)	(+)	–
Brown	(+)	–	–	+	–	+	–	nd	–	–	+	–	+	–	+	+	–	–	–	+	–	(+)	–	(+)	(+)	+	–	–
Black	–	–	–	+	–	–	–	nd	–	–	(+)	–	–	–	–	–	–	–	–	–	–	–	–	–	–	–	–	–
Production of soluble pigments:																												
Yellow	(+)	+	–	(+)	–	–	–	–	–	–	+	–	–	–	–	(+)	+	(+)	–	–	(+)	(+)	–	+	–	–	(+)	–
Red	–	–	–	–	–	–	+	–	–	–	–	–	(+)	–	–	+	+	–	–	–	–	–	–	+	(+)	(+)	–	–
Green	–	–	–	(+)	–	–	–	–	–	–	(+)	–	–	–	–	–	–	(+)	–	–	–	(+)	–	–	–	–	–	–
Blue	–	–	–	–	–	–	–	+	–	–	–	–	–	–	–	–	–	–	–	–	–	–	–	–	–	–	–	–
Brown	+	–	–	(+)	–	–	+	–	–	–	+	–	+	–	+	–	–	(+)	–	–	–	+	–	–	+	+	(+)	–
Physiological tolerance to (%, w/v):																												
Brilliant green (0.001)	–	–	nd	+	nd	–	nd	nd	nd	–	–	–	nd	nd	–	–	nd	+	+	–	nd	–	nd	nd	nd	nd	–	nd

(continued)

TABLE 194. (continued)

Characteristic	1. *A. philippinensis*	2. *A. aurantiicolor*	3. *A. brasiliensis*	4. *A. campanulatus*	5. *A. capillaceus*	6. *A. consettensis*	7. *A. couchii*	8. *A. cyaneus*	9. *A. deccanensis*	10. *A. derwentensis*	11. *A. digitatis*	12. *A. durhamensis*	13. *A. ferrugineus*	14. *A. globisporus*	15. *A. humidus*	16. *A. italicus*	17. *A. liguriensis*	18. *A. lobatus*	19. *A. missouriensis*	20. *A. palleronii*	21. *A. rectilineatus*	22. *A. regularis*	23. *A. sichuanensis*	24. *A. teichomyceticus*	25. *A. toevensis*	26. *A. tereljensis*	27. *A. utahensis*	28. *A. xinjiangensis*
Lysozyme (0.005)	+	+	nd	–[b], +[j]	nd	d	nd	nd	nd	d	+	+	+	nd	d	–[b], +[j]	nd	+	+	d	+	+	nd	nd	nd	nd	+	nd
Lysozyme (0.01)	+	+	nd	–	nd	d	nd	nd	nd	d	+	+	nd	nd	d	–	nd	–	+	–	nd	–	nd	nd	nd	nd	+	nd
Potassium tellurite (0.001)	+	+	nd	+	nd	+	nd	nd	nd	d	+	+	nd	nd	+	–	nd	–	+	+	nd	+	nd	nd	nd	nd	+	nd
Potassium tellurite (0.005)	+	+	nd	+	nd	–	nd	nd	nd	d	+	+	nd	nd	d	–	nd	–	+	–	nd	–	nd	nd	nd	nd	+	nd
Potassium tellurite (0.01)	+	+	nd	–	nd	–	nd	nd	nd	d	+	+	nd	nd	–	–	nd	–	+	–	nd	–	nd	nd	nd	nd	–	nd
Sodium azide (0.0001)	+	+	nd	+	nd	d	nd	nd	nd	d	+	d	nd	nd	d	+	nd	+	+	d	nd	+	nd	nd	nd	nd	+	nd
Sodium azide (0.001)	+	–	nd	+	nd	d	nd	nd	nd	+	+	d	nd	nd	–	+	nd	–	+	d	nd	–	nd	nd	nd	nd	–	nd
Resistance to antibiotics (µg/ml):																												
Ampicillin (0.5)	+	+	nd	+	nd	+	nd	nd	nd	+	+	d	nd	nd	+	–	nd	+	+	+	nd	+	nd	nd	nd	nd	+	nd
Ampicillin (2.0)	+	+	nd	+	nd	+	nd	nd	nd	–	–	–	nd	nd	d	–	nd	+	+	d	nd	+	nd	nd	nd	nd	+	nd
Ampicillin (8.0)	–	+	nd	+	nd	–	nd	nd	nd	–	–	–	nd	nd	–	–	nd	–	+	–	nd	–	nd	nd	nd	nd	–	nd
Ampicillin (100.0)	+[j]	+	nd	+	nd	nd	nd	nd	nd	nd	–	nd	+	nd	nd	+[j]	nd	nd	nd	nd	+	+[j]	nd	nd	nd	nd	+[j]	nd
Cephaloridine hydrochloride (1.0)	+	+	nd	+	nd	d	nd	nd	nd	–	–	–	nd	nd	d	–	nd	+	+	–	nd	–	nd	nd	nd	nd	+	nd
Cephaloridine hydrochloride (2.0)	–	+	nd	+	nd	d	nd	nd	nd	–	–	–	nd	nd	d	–	nd	–	–	–	nd	–	nd	nd	nd	nd	–	nd
Chloramphenicol (25.0)	+	+	nd	–	nd	nd	nd	nd	nd	nd	–	nd	+	nd	nd	+	nd	nd	nd	nd	+	+	nd	nd	nd	nd	+	nd
Chlortetracycline hydrochloride (1.0)	+	–	nd	–	nd	d	nd	nd	nd	–	–	–	nd	nd	d	–	nd	–	+	–	nd	–	nd	nd	nd	nd	–	nd
Gentamicin sulfate (0.5)	d	+	nd	+	nd	d	nd	nd	nd	d	+	d	nd	nd	+	+	nd	+	+	–	nd	+	nd	nd	nd	nd	+	nd
Gentamicin sulfate (1.0)	d	–	nd	–	nd	–	nd	nd	nd	–	+	–	nd	nd	–	–	nd	–	+	–	nd	–	nd	nd	nd	nd	–	nd
Gentamicin sulfate (4.0)	–	–	nd	–	nd	–	nd	nd	nd	–	+	–	nd	nd	–	–	nd	–	–	–	nd	–	nd	nd	nd	nd	–	nd
Kanamycin (25.0)	d	+	nd	+	nd	nd	nd	nd	nd	nd	–	nd	+	nd	nd	+	nd	nd	nd	nd	+	+	nd	nd	nd	nd	+	nd
Lincomycin hydrochloride (2.0)	+	+	nd	+	nd	d	nd	nd	nd	d	+	d	nd	nd	+	+	nd	+	+	d	nd	+	nd	nd	nd	nd	+	nd
Lincomycin hydrochloride (8.5)	+	+	nd	–	nd	d	nd	nd	nd	d	–	–	nd	nd	d	+	nd	+	+	d	nd	–	nd	nd	nd	nd	–	nd
Metacycline hydrochloride (0.25)	d	+	nd	+	nd	+	nd	nd	nd	d	+	d	nd	nd	+	+	nd	+	+	–	nd	+	nd	nd	nd	nd	–	nd
Metacycline hydrochloride (2.0)	–	–	nd	–	nd	–	nd	nd	nd	–	–	–	nd	nd	d	–	nd	–	+	–	nd	–	nd	nd	nd	nd	–	nd
Neomycin sulfate (0.5)	+	+	nd	–	nd	–	nd	nd	nd	d	+	d	nd	nd	d	–	nd	–	+	–	nd	+	nd	nd	nd	nd	+	nd
Neomycin sulfate (2.0)	–	–	nd	–	nd	–	nd	nd	nd	–	+	–	nd	nd	–	–	nd	–	+	–	nd	–	nd	nd	nd	nd	–	nd
Rifampin (0.25)	+	+	nd	–	nd	+	nd	nd	nd	d	+	+	nd	nd	+	–	nd	+	+	+	nd	+	nd	nd	nd	nd	+	nd
Rifampin (2.0)	+	–	nd	–	nd	–	nd	nd	nd	–	–	–	nd	nd	–	–	nd	+	+	–	nd	+	nd	nd	nd	nd	–	nd
Streptomycin (50.0)	nd	nd	nd	nd	–	nd	nd	nd	nd	nd	nd	nd	nd	nd	nd	nd	nd	nd	nd	nd	nd	nd	+	nd	nd	nd	nd	+

Tobramycin sulfate (0.5)	+	−	nd	−	nd	d	nd	nd	nd	d	+	+	nd	nd	d	−	nd	+	+	−	nd	−[b]	nd	nd	nd	nd	+	nd
Tobramycin sulfate (1.5)	+	−	nd	−	nd	d	nd	nd	nd	−	+	d	nd	nd	−	−	nd	+	+	−	nd	+[b]	nd	nd	nd	nd	−	nd
Antimicrobial activity against:																												
Aspergillus niger	−	−	nd	−	nd	d	nd	nd	nd	−	−	−	nd	nd	−	−	−	−	−	−	−	−	nd	−	nd	nd	−	nd
Bacillus subtilis	−	−	d	−	nd	nd	nd	nd	nd	d	−	d	nd	nd	d	−	+	−	−	−[b], +[h]	−	−[b], +[h]	nd	+	nd	nd	−	nd
Candida albicans	nd	nd	nd	nd	nd	nd	nd	nd	nd	nd	nd	nd	nd	nd	nd	nd	−	nd	nd	−	−	−	nd	−	nd	nd	nd	nd
Escherichia coli	nd	nd	nd	nd	nd	nd	nd	nd	nd	nd	nd	nd	nd	nd	nd	nd	−	nd	nd	−	−	−	nd	−	nd	nd	nd	nd
Micrococcus luteus	nd	nd	nd	nd	nd	nd	nd	nd	nd	nd	nd	nd	nd	nd	nd	nd	−	nd	nd	+	−	+	nd	−	nd	nd	nd	nd
Pseudomonas aeruginosa	nd	nd	nd	nd	nd	nd	nd	nd	nd	nd	nd	nd	nd	nd	nd	nd	−	nd	nd	−	−	−	nd	−	nd	nd	nd	nd
Saccharomyces cerevisiae	nd	nd	nd	nd	nd	nd	nd	nd	nd	nd	nd	nd	nd	nd	nd	nd	−	nd	nd	−	−	−	nd	−	nd	nd	nd	nd
Staphylococcus aureus	−	−	nd	−	nd	d	nd	nd	nd	d	−	+	nd	nd	d	−	+	−	−	+[b], −[h]	−	−[b], +[h]	nd	+	nd	nd	−	nd
Streptomyces murinus	d	−	nd	−	nd	−	nd	nd	nd	d	−	−	nd	nd	d	−	−	−	+	−	−	+	nd	+	nd	nd	−	nd
Major fatty acids:[o]	c	n	m	n	n	b	c		m	b		m	m	m	b	c	h	n	b	b	c	b	n	h	m	m	b	n
Saturated straight-chain																												
$C_{14:0}$	0.7	0.7		3.7	1.7	1.2	2.8			2.1			1.0		4.8	2.0	1.8	1.4	1.8	0.8	6.5	0.7	6.2				3.2	3.7
$C_{15:0}$	19.2	10.5	3.0	4.5	8.0	8.3	11.0			6.0		1.2	2.3	1.0	4.2	4.5	1.5	4.9	2.1	9.9	15.3	2.3	4.2	1.1		2.3	4.7	6.7
$C_{16:0}$	1.7	5.6	1.7	2.5	1.3	18.3	14.5		1.6	24.3		4.6	15.5	3.6	17.7	11.4	6.5	1.1	6.0	4.1	11.9	4.9	7.7	2.5	1.5	4.4	15.5	6.5
$C_{17:0}$	6.8	2.1	5.1	3.4	4.6	2.5	5.9		1.0	5.7		4.7	7.2	3.0	15.5	5.1	1.2	4.3	7.9	12.0	9.6	9.2	4.4	3.5	1.2	5.1	28.8	4.2
$C_{18:0}$	0.1		1.4	0.9	0.8	9.6	0.3		3.1	10.2		8.9	9.3	3.8	12.0	3.3	1.5	0.9	9.0	12.2	2.9	4.3		1.8	1.5	4.4	14.0	
$C_{19:0}$						1.3				0.5					9.0				26.4	0.5	0.5	2.4					5.6	
Saturated branched																												
$C_{14:0}$ iso	1.3	0.9	1.5	2.8	3.0		0.4					2.0	1.6			0.5	1.7	3.4						1.0		1.9		0.9
$C_{15:0}$ iso	9.5	15.7	10.1	17.6	16.7	12.0	4.1		12.6	5.2		7.3	8.4	17.6	6.6	19.8	26.1	22.1	6.5	15.8	7.0	10.6	22.2	15.1	12.6	10.6	4.1	10.8
$C_{15:0}$ anteiso	1.2	3.2	24.5	6.9	4.7	27.2	1.5		10.6	1.6		12.6	22.9	3.4	8.8	1.8	13.1	4.9	7.0	29.1	0.4	13.3	1.4	8.8	11.7	11.8	3.8	3.6
$C_{16:0}$ iso	25.1	19.2	17.1	14.9	14.1	7.2	10.5		20.5	22.2		19.8	7.7	20.1	10.4	13.1	12.1	14.6	18.3	6.2	1.9	15.0	15.3	20.7	21.6	18.6	8.5	16.0
$C_{17:0}$ iso	1.1	2.9	2.1	5.0	6.0	0.3	4.7		7.3	1.2		2.1	1.1	9.7	1.1	4.0	3.0	9.0	3.5	0.8	1.3	7.0	4.2	3.8	5.0	2.4	3.8	4.1
$C_{17:0}$ anteiso	1.8		18.8		5.8	11.8	8.6		34.0	20.5		20.4	13.4	13.4	7.2	3.8	1.0	5.1	7.8	5.9		20.8	3.1	19.8	24.9	15.3	6.3	4.7
$C_{17:0}$ 10-methyl	0.6			0.9	0.8											0.2												
Unsaturated																												
$C_{15:1}$ ω8c	0.3																											
$C_{16:1}$ iso	1.6															0.6								1.6				
$C_{16:1}$ ω9c		5.6		10.6	2.3							1.2	1.9					3.4			0.3		3.3	3.4				5.9
$C_{17:1}$ iso	0.7			6.9			1.7									1.9	1.9							2.6				
$C_{17:1}$ anteiso	0.5								1.1							0.3								2.0	1.7			
$C_{17:1}$ ω10c																	3.3											
$C_{17:1}$ ω9c		13.4	1.9	5.3	18.4							1.4	1.3	1.3			2.2	8.9					11.1	6.7	3.1	1.9		21.5
$C_{17:1}$ ω8c	19.2						9.9									9.7												
$C_{17:1}$ ω6c	1.9						1.5																					
$C_{18:1}$ ω7c	0.4															1.1												
$C_{18:1}$ ω9c	0.4	6.2	5.0	5.3	7.1				1.0			4.6	4.6	3.9		6.5	2.9	4.6					7.5	4.3	3.6	5.0		3.9
$C_{16:1}$ ω7c and/or $C_{15:0}$ iso 2-OH	2.2						17.5									9.4												
$C_{19:1}$ ω11c and/or $C_{19:1}$ ω9c																												
Menaquinones:[p]																												
MK-7(H_4)	−					MI				−		−			−	−		−	−	−		−					−	

(continued)

TABLE 194. (continued)

Characteristic	1. *A. philippinensis*	2. *A. auranticolor*	3. *A. brasiliensis*	4. *A. campanulatus*	5. *A. capillaceus*	6. *A. consettensis*	7. *A. couchii*	8. *A. cyaneus*	9. *A. deccanensis*	10. *A. derwentensis*	11. *A. digitatis*	12. *A. durhamensis*	13. *A. ferrugineus*	14. *A. globisporus*	15. *A. humidus*	16. *A. italicus*	17. *A. liguriensis*	18. *A. lobatus*	19. *A. missouriensis*	20. *A. palleronii*	21. *A. rectilineatus*	22. *A. regularis*	23. *A. sichuanensis*	24. *A. teichomyceticus*	25. *A. toevensis*	26. *A. tereljensis*	27. *A. utahensis*	28. *A. xinjiangensis*
MK-7(H_6)	MI					–				–		–			MI	–		MI	MI	–		MI					–	
MK-7(H_8)	–					–				–		–			MI	–		–	–	–		–					–	
MK-8(H_2)	–					MI				MI		–			–	–		–	–	MI		–					–	
MK-8(H_4)	–					–				–		–			–	–		MI	–	–		–					–	
MK-8(H_6)	–					–				MI		–			MI	MI		MI	MI	MI		MI					MI	
MK-8(H_8)	–					MI				–		–			–	–		–	–	–		–					–	
MK-9(H_2)	–[b], MO[k]	MI				–	MI			MI		–			MI	MI		MI	MI	MI		MI[b], MO[k]					MI	
MK-9(H_4)	PR[b] MO[k]	PR		PR	PR	PR	PR			PR	PR	PR			PR	PR[b] MO[k]	PR	PR	PR	PR		PR	PR	PR			PR	PR
MK-9(H_6)	–	MI			MI	–	MO			MO	MI	MI			MO	MO		–	MI	MO		–	MI		PR	PR	MO	
MK-9(H_8)	–					MI	MI			–		MI			–	–		–	–	MI		–			MI	MO	–	
MK-10(H_2)	–					–	MI			–		–			–	–		–	–	–		–					–	
MK-10(H_4)	–[b], MI[k]				MI	–				–		–			–	–[b], MI[k]	PR	–[b], MI[k]	–	–		–		PR			–[b], MI[k]	
MK-10(H_6)	–					–	tr			–		–			–	–		–	–	–		–	MI				–	MI
MK-10(H_8)	–					–				–		–			–	–		–	–	–		–					–	
Polar lipids[q]																												
DPG	+	+				+	+			+		+			+	+			+	+	+	+			+	+		
PG	–	+				+	+			+		+			+	+			+[k], (+)[l]	+	(+)	+			–	–		
PE	+	+	+[m]		+	+	+		+[m]	+		+	+[m]	+[m]	+	+	+		+	+	+	–	+	+	+	+		+
PI	+	+				+	(+)			+		+			+	+			+	+	+	+	+		+	+		
PIM	+	+					(+)									+			+		+	+						
PC	–	–			–		–									–	–		–		–	–		–	–	–		
AGP	–	–			–		–									–			–			–						
HL							(+)																					
PL							(+)																					
L							+																					
GL															+													

[a]Symbols: +, >85% positive; d, different strains give different reactions (16–84% positive); –, 0–15% positive; (+), weakly positive, moderate; D, different reactions occur in different taxa (species of a genus or genera of a family); w, weak reaction; nd, not determined; nr, not reported.

[b]Data compiled from Goodfellow et al. (1990a)

[c]Kämpfer et al. (2007)

[d]Matsumoto et al. (2000)

[e]Palleroni (1989)

[f]Thiemann (1967)

[g]Vobis and Kothe (Vobis and Kothe, 1989)

[h]Wink et al. (2006)

[i]Schäfer (1973)

[j]Kothe (1987)

[k]Stackebrandt and Kroppenstedt (1987)

[l]Lechevalier et al. (1977)

[m]Ara et al. (2010)

[n]Sun et al. (2009); deviations or incoherent data are indicated by superscript letters.

[o]Percentages of fatty acids are shown; for unsaturated fatty acids, the position of the double bond is located by counting from the methyl (ω) end of the carbon chain; *cis* isomers are indicated by the suffix *c*.

[p]Menaquinones: PR, predominant component (>50%); MO, moderate component (11–50%); MI, minor component (1–10%); tr, traces (<1%).

[q]Polar lipids: DPG, diphosphatidylglycerol; PG, phosphatidylglycerol; PE, phosphatidylethanolamine; PI, phosphatidylinositol; PIM, phosphatidylinositol mannoside; PC, phosphatidylcholine; APG, amino-containing glycophospholipid; HL, highly hydrophilic lipid; PL, unknown phospholipids; L, unknown polar lipid; GL, uncharacterized glycolipid.

[r]Occurs in spirals (G. Vobis, unpublished observation).

system (Kroppenstedt, 1985; Wink et al., 2006), but usually in minor amounts (Matsumoto et al., 2000; Stackebrandt and Kroppenstedt, 1987). Smaller amounts of MK-10(H_2) and traces of MK-10(H_6) are characteristic for *Actinoplanes couchii* (Kämpfer et al., 2007). The additional presence of MK-7(H_4, H_6, H_8) and MK-8(H_2, H_4, H_6, H_8) can be observed in some species, where they may occur in minor amounts (Goodfellow et al., 1990b). The distribution patterns of polar lipids and menaquinones for the species are given in Table 194.

Colonial characteristics. *Actinoplanes* strains grow well on various complex media, forming compact colonies up to about 3 cm in diameter after 4–6 weeks of incubation. They are characteristically elevated and convoluted, and frequently have protuberances in the center. The marginal areas can be ridged or flat. The colonies can be divided into sectors. Their consistency is generally hard, but may occasionally be soft, especially for "*Ampullariella*" strains. Colonies can be covered with a slight glittering whitish bloom if abundant sporangia are produced on the surface of the substrate mycelium. Aerial mycelium is rarely developed and generally only in a rudimentary state, giving the colonies a whitish gray surface layer.

Colonies are usually brightly colored, with orange being the basic color component. Szaniszlo (1968) has identified spectrophotometrically the orange pigment of various strains as being associated with carotenoids. Besides all shades of orange, a great variety of additional colors exists, depending on the individual strains: cream to yellow, brown, rusty brown, red, purple, violet, green, or black (Palleroni, 1989; Parenti and Coronelli, 1979; Vobis, 1987). Segments with different colors may even occur in a single colony (Vobis, 1992). Palleroni (1989) reported a red pigment analyzed from two strains of *Actinoplanes* that had the structure of an anthraquinone derivative. Methanolic extracts of *Actinoplanes ferrugineus* lack a characteristic carotenoid spectrum (Palleroni, 1979).

Soluble pigments are also produced in several species, ranging from yellowish, greenish, and auburn to dark brown (Couch, 1963). "*Ampullariella violaceochromogenes*" produces a violet to dark purple soluble pigment on various agar media (Nonomura et al., 1979). A cherry red pigment is the distinctive mark of *Actinoplanes italicus*, although this red pigment is not produced and substrate mycelium appears deep orange on exposure to light during growth (Beretta, 1973). The soluble blue pigment of *Actinoplanes cyaneus* has been shown to belong to the celocomycin-actinorodine group (Terekhova et al., 1977). Melanoid pigments are produced in *Actinoplanes digitatis* (Couch, 1963).

Production of secondary metabolites. *Actinoplanes* strains have shown a high capacity to produce antibiotics and other useful secondary metabolites. First reviews revealed about 20 antibiotics (Palleroni, 1983; Parenti and Coronelli, 1979). The number increased to more than 120, covering a wide range of chemical diversity including peptides, glycopeptides, anthracyclines, nucleosides, polyenes, and quinones (Okami and Hotta, 1988; Vobis, 1992). The amino acid derivatives are predominant and some are of clinical relevance (Lazzarini et al., 2000), e.g. teicoplanin, a glycopeptide from *Actinoplanes teichomyceticus* ATCC 31121^T (Bardone et al., 1978), actaplanin, a glycopeptide from *Actinoplanes missouriensis* ATCC 23342 (Debono et al., 1984), and ramoplanin, a glycolipodepsidpeptide from *Actinoplanes* sp. ATCC 33076 (Ciabatti and Cavalleri, 1989). Further examples of other interesting antibiotics are the friulimicins, lipopetides from "*Actinoplanes friuliensis*" DSM 7358 (Aretz et al., 2000), purpuromycin, a naphthoquinone from "*Actinoplanes ianthinogenes*" ATCC 21884 (Coronelli et al., 1974), the chloride-containing lipiarmycin from *Actinoplanes deccanensis* ATCC 21983^T (Parenti et al., 1975), and actagardine, an oligopeptide from *Actinoplanes liguriensis* ATCC 31048^T (Parenti et al., 1976; Vértesy et al., 1999). The list of "Practically used Antibiotics and Their Related Substances" (Sezaki and Miyadoh, 2001) also includes acarbose, a pseudotetrasaccharide produced by *Actinoplanes* sp. SE-50 (Truscheit et al., 1981), a very effective inhibitor of α-glucosidase, successfully applied in cases of diabetes mellitus (Creutzfeld, 1988).

Life cycle. The life cycle of *Actinoplanes* is characterized by an alternation between terrestrial and aquatic habitats (Figure 213H) (Vobis, 1987, 1992). The growth of vegetative mycelium on plant or animal debris or other adequate substrates (Figure 213F) culminates in the differentiation into sporangia (Figure 213H, I), which are generally produced on the surface of the substrate, directly in contact with the air (Bland and Couch, 1981). The sporangia can easily lose their connection to the degenerating mycelium and are disseminated as diaspores by the wind (Figure 213H, II), or by soil fauna such as mites, collembola, or arthropods. The sporangial envelope and the intrasporangial matrix hold together the mass of spores and shield the thin-walled spores (Figure 212C, D) from mechanical and physiological stress. The sporangia withstand prolonged desiccation (Figure 213H, III) and survive for many years (Makkar and Cross, 1982).

The sporangial envelope is usually hydrophobic. It can be rehydrated by sufficient moisture, e.g. during periods of fog or rain, and flagellated spores are released from the sporangia. Under laboratory conditions (Figure 213A, B, C), this process takes 10–60 min (Higgins, 1967; Vobis, 1987). The process can be facilitated by the addition of wetting agents like Tween 80 (Higgins, 1967). During immersion in water, the spores swell, becoming more clearly visible within the sporangium. In general, spores show motility inside the sporangium before they are released by disruption of the sporangial envelope (Lechevalier and Holbert, 1965). As Higgins (1967) has shown, both swelling and motility of the spores contribute to their release, thus contradicting the suggestion made by Couch (1963) that this may be a consequence of the swelling of the intersporal material. Of the two factors, swelling of the spores is the more important, as sometimes sporangium dehiscence gives nonmotile spores (Palleroni, 1989). In some instances, motility begins after spore release. Spores of *Actinoplanes brasiliensis* retain their motility in water or in diluted buffers for more than 1 d at room temperature (Palleroni, 1983). In baiting experiments simulating a natural aquatic microhabitat, pollen or other natural substrates are exposed to the surface of water. The zoospores, once released from the submerged sporangia, are able to swim to the surface (Figure 213H, IV), attach to the natural substrates, germinate (Figure 213E, H, V), and colonize them within several days (Figure 213F) (Couch, 1963; Vobis, 1987). This may be a result of aerotactic and chemotactic behavior of the spores (Cross, 1986). Although the chemotactic response is used effectively in the isolation method of Palleroni (1980), up to now the exact physiological explanation is not known. If the liquid contains nutrients, germination may ensue after a few hours and the emergence of the filamentous stage completes the cycle of morphogenetic changes (Palleroni, 1989).

The state of swarming sporangiospores (Cross, 1986) may be skipped over during the life cycle. In nearly all cases, under culture conditions, it is possible to induce the spores to germinate, if they are still enclosed within the sporangia. Exposed on the surface of agar media, the spores germinate synchronously, and can develop directly into new mycelium (Figure 213G) (Vobis, 1987).

Metabolism and physiology. *Actinoplanes* strains are aerobic and mesophilic, growing between 10 and 35°C, with an optimal growth temperature of about 23–28°C. Some species, e.g. *Actinoplanes deccanensis*, show moderate thermotolerance, growing at a maximum temperature of 45°C (Parenti and Coronelli, 1979). No growth occurs at 50°C (Kothe, 1987; Parenti et al., 1978). The minimum growth temperature observed for many species is 4°C (Goodfellow et al., 1990b).

Species of *Actinoplanes* can be characterized as neutrophilic micro-organisms, growing well at pH 6.0–8.0. Goodfellow et al. (1990b) could not register growth at either pH 5.6 and lower or at pH 9.0 and higher for 13 species cultivated on modified Bennett's agar. In contrast, Kothe (1987) observed growth in acid conditions at pH 4.0 for four species and in alkaline conditions at pH 10.0 for nine species grown on yeast extract-starch agar (Table 195). These differences can probably be explained by the use of the two distinct agar media, perhaps having diverse buffer effects.

Strains of *Actinoplanes* can be cultured on most complex media generally used for actinomycetes, including various ISP media (Wink et al., 2006). Further recommendations and/or recipes of media suitable for isolation, vegetative growth, sporulation, and many kinds of physiological tests are given by Goodfellow et al. (1990b), Palleroni (1989), and Vobis (1992).

Isolates of *Actinoplanes* are able to grow on chemically defined agar media prepared with inorganic salts and a single compound as the sole source of carbon. In general, the test method follows the recommendations of Shirling and Gottlieb (1966) using about ten organic compounds (Schäfer, 1973; Thiemann et al., 1969; Wink et al., 2006). Other authors used different basal media and/or amplified the number of tested carbohydrates (Goodfellow et al., 1990b; Kämpfer et al., 2007; Kothe, 1987; Palleroni, 1989). D-Glucose and, to a lesser extent, L-arabinose, D-mannose, and maltose are used as carbon sources by all species (Table 195).

As typical saprophytic inhabitants of soil, *Actinoplanes* strains are able to degrade organic compounds (Table 195). Casein, chitin, gelatin, DNA, RNA, and lecithin are decomposed by most species, as well as starch as a representative polymer of plant origin. Pectin is decomposed by many species. Different results are reported for cellulose degradation. Goodfellow et al. (1990b) found that none of the 55 strains investigated were able to break down cellulose, whereas tests of Schäfer (1973) and Solans and Vobis (2003) presented positive results for 21 strains. Hemicellulose can be degraded and lignocellulose is used as a preferred substrate by various strains (Solans and Vobis, 2003). The compound guanine and the polymer keratin, both of typical animal origin, are not degraded. Further enzymic activities have been studied recently (Wink et al., 2006), specifically hydrolysis of chromogenic substances (Kämpfer et al., 2007) (Table 195).

Studies on other physiological features, including the ability to grow in media supplemented with sodium chloride, have been tested for 15 species with different results (Table 195): concentrations of up to 2.0% (w/v) were tolerated by six species; *Actinoplanes campanulatus* and one strain of *Actinoplanes palleronii* could grow at 3.0% (w/v) NaCl (Goodfellow et al., 1990b); and a concentration of 5% (w/v) sodium chloride was not tolerated by eight species including *Actinoplanes campanulatus* (Kothe, 1987).

Many species have been tested by Goodfellow et al. (1990b) for their capacity to grow in the presence of various other organic and inorganic chemical compounds at different concentrations, as well as their resistance to several antibiotics (Table 194). Aside from well-studied antibiotics produced by *Actinoplanes*, antimicrobial activities against *Aspergillus niger* and *Candida albicans*, as representatives of eukaryotic micro-organisms, and against Gram-stain-negative and Gram-stain-positive bacteria have been carried out to characterize various species (Goodfellow et al., 1990b; Wink et al., 2006) (Table 194). Further biochemical tests like reduction of nitrate, production of urease and hydrogen sulfide, and coagulation and peptonization of milk are additional useful taxonomic markers (Table 195).

The production of sporangia occurs preferably on certain agar media, depending on the strain. In general, starch-casein and a minimal agar medium with certain carbon sources promotes sporangia formation (Palleroni, 1989). Humic acid and fulvic acid are possible factors to stimulate sporangiogenesis (Willoughby et al., 1968; Willoughby and Baker, 1969), as well as tea infusion (Parenti et al., 1978). Good results have been obtained when sections of mycelia, originally grown on nutrient-rich complex media, were transferred onto nutrient-poor media like artificial soil agar (Vobis, 1992; Vobis and Kothe, 1985).

The motile sporangiospores of *Actinoplanes* are, in general, vigorous swimmers, moving in a straight line. They interrupt the movement for moments, tumbling and "re-arranging" their flagella, and continue to swim in a different direction (Palleroni, 1983). Couch and Koch (1962) have observed that 1% Casamino acid induces the motility of sporangiospores of an "*Ampullariella*" strain that is a rather poor swimmer and that L-arginine hydrochloride (6 mM) and urea (0.01 M) greatly increase the motility. In spore-releasing studies on *Actinoplanes rectilineatus*, Higgins (1967) could demonstrate that flagellation and motility decreased in old spores, but motility can be regained if glucose is supplied as exogenous carbon source, which must be applied within 180 min of the initial wetting. The addition of amino acids or phosphate buffer permitted flagellation, but not motility. Deflagellation/reflagellation experiments indicated that functional flagella can be regenerated only in the presence of both amino acids and glucose. Inoperative flagella were formed in the presence of inhibitors of nucleic acid synthesis, such as 6-azauracil, but inhibitors of protein synthesis, such as chloramphenicol, did not interfere with reflagellation. Inhibitors such as sodium *p*-chlormercuribenzoate, sodium iodoacetate, 2-iodoacetamide, sodium azide, and 2,4-dinitrophenol inhibit the formation of flagella and also immobilized fully motile spores (Higgins, 1967). Carbonylcyanide-*m*-chlorophenylhydrazone, which is effective against oxidative phosphorylation, has the same physiological effect (Palleroni, 1983).

TABLE 195. Physiological and biochemical tests differentiating *Actinoplanes* species[a]

Characteristic	1. *A. philippinensis*	2. *A. auranticolor*	3. *A. brasiliensis*	4. *A. campanulatus*	5. *A. capillaceus*	6. *A. consettensis*	7. *A. couchii*	8. *A. cyaneus*	9. *A. deccanensis*	10. *A. derwentensis*	11. *A. digitatis*	12. *A. durhamensis*	13. *A. ferrugineus*	14. *A. globisporus*	15. *A. humidus*	16. *A. italicus*	17. *A. liguriensis*	18. *A. lobatus*	19. *A. missouriensis*	20. *A. palleronii*	21. *A. rectilineatus*	22. *A. regularis*	23. *A. sichuanensis*	24. *A. teichomyceticus*	25. *A. tereljensis*	26. *A. toevensis*	27. *A. utahensis*	28. *A. xinjiangensis*
Utilization of sole carbon sources:																												
Acetate	nd	nd	nd	nd	nd	nd	–	nd	nd	nd	nd	nd	nd	nd	nd	–	nd	nd	nd	nd	–	nd	+	nd	nd	nd	nd	+
N-Acetyl-D-galactosamine	nd	nd	nd	nd	nd	nd	–	nd	nd	nd	nd	nd	nd	nd	nd	–	nd	nd	nd	nd	–	nd	nd	nd	nd	nd	nd	nd
N-Acetyl-D-glucosamine	nd	nd	nd	nd	nd	nd	–	nd	nd	nd	nd	nd	nd	nd	nd	–	nd	nd	nd	nd	–	nd	nd	nd	nd	nd	nd	nd
trans-Aconitate	nd	nd	nd	nd	nd	nd	–	nd	nd	nd	nd	nd	nd	nd	nd	–	nd	nd	nd	nd	–	nd	nd	nd	nd	nd	nd	nd
Adipate	+	nd	–	nd	nd	nd	–	nd	nd	nd	nd	nd	–	nd	nd	+[e], –[c]	nd	nd	–	nd	–	nd	nd	nd	nd	nd	+	nd
Adonitol	–	–	nd	–	nd	–	–	nd	nd	–	–	–	nd	nd	–	–	nd	–	–	–	–	–	+	nd	nd	nd	–	+
β-Alanine	+	nd	–	nd	nd	nd	–	nd	nd	nd	nd	nd	+	nd	nd	+	nd	nd	–	nd	–	nd	nd	nd	nd	nd	+	nd
4-Aminobutyrate	nd	nd	nd	nd	nd	nd	–	nd	nd	nd	nd	nd	nd	nd	nd	(+)	nd	nd	nd	nd	–	nd	nd	nd	nd	nd	nd	nd
γ-Aminobutyrate	+	nd	+	nd	nd	nd	nd	nd	nd	nd	nd	nd	–	nd	nd	+	nd	nd	–	nd	nd	nd	nd	nd	nd	nd	–	nd
D-Arabinose	–	+	+	–	nd	d	nd	nd	nd	d	–	d	+	nd	+	–	nd	–	–	–	nd	–	+	nd	nd	nd	–	+
L-Arabinose	+	+	+	+	+	+	+	nd	+	+	+, d[g]	+	+	+	+	+	–[h], +[e]	+	+	d[b], –[h]	+	+	nd	+	nd	nd	+	nd
Arabitol	–	nd	+	nd	nd	nd	nd	nd	nd	nd	nd	nd	+	nd	nd	–	nd	nd	d	nd	nd	nd	nd	nd	nd	nd	–	nd
p-Arbutin	nd	nd	nd	nd	nd	nd	+	nd	nd	nd	nd	nd	nd	nd	nd	–	nd	nd	nd	nd	+	nd	nd	nd	nd	nd	nd	nd
L-Arginine	+	nd	–	nd	nd	nd	nd	nd	nd	nd	nd	nd	–	nd	nd	+	nd	nd	–	nd	nd	nd	nd	nd	nd	nd	+	nd
L-Asparagine	(+)	nd	+	nd	nd	nd	nd	nd	nd	nd	nd	nd	(+)	nd	nd	–	nd	nd	–	nd	nd	nd	nd	nd	nd	nd	(+)	nd
L-Aspartate	nd	nd	nd	nd	nd	nd	+	nd	nd	nd	nd	nd	nd	nd	nd	(+)	nd	nd	nd	nd	(+)	nd	nd	nd	nd	nd	nd	nd
Azelate	nd	nd	nd	nd	nd	nd	–	nd	nd	nd	nd	nd	nd	nd	nd	–	nd	nd	nd	nd	–	nd	nd	nd	nd	nd	nd	nd
Catechol	–	–	nd	–	nd	–	nd	nd	nd	–	–	–	nd	nd	–	–	nd	–	–	–	nd	–	nd	nd	nd	nd	–	nd
Cellobiose	–	–	nd	–	nd	–	nd	nd	nd	–	–	–	nd	nd	–	–	nd	–	–	–	nd	–	–	nd	nd	nd	–	–
Cellulose	+	–	+	nd	nd	nd	nd	nd	–	nd	nd	nd	–	–	nd	–	–	nd	–	nd	–	nd	nd	–	nd	nd	–	nd
Citrate	–	nd	–	nd	nd	nd	–	nd	nd	nd	nd	nd	–	nd	nd	+	nd	nd	–	nd	–	nd	+	nd	nd	nd	–	+
L-Citrulline	(+)	nd	–	nd	nd	nd	nd	nd	nd	nd	nd	nd	–	nd	nd	(+)	nd	nd	d	nd	nd	nd	nd	nd	nd	nd	+	nd
Coumarin	–	–	nd	–	nd	–	nd	nd	nd	–	–	–	nd	nd	–	–	nd	–	–	–	nd	–	nd	nd	nd	nd	–	nd
Dextrin	+	+	nd	+	nd	+	nd	nd	nd	+	+	+	nd	–	+	+	nd	+	+	d	nd	+	–	nd	nd	nd	+	–
Dulcitol	nd	nd	nd	nd	nd	nd	nd	nd	nd	nd	nd	nd	nd	–	nd	nd	nd	nd	nd	nd	nd	nd	nd	nd	nd	nd	nd	nd
Erythritol	–	–	nd	–	nd	–	nd	nd	nd	–	–	–	nd	nd	–	–	nd	–	–	–	nd	–	nd	nd	nd	nd	–	nd
D-Fructose	+	+	+	+	+	+	+	nd	–	+	+	+	(+)	+	+	+	–[h], +[e]	+	+	+	+	+	+	+	nd	nd	+	+
L-Fucose	–	nd	–	nd	nd	nd	nd	nd	nd	nd	nd	nd	+	nd	nd	–	nd	nd	d	nd	nd	nd	nd	nd	nd	nd	+	nd
Fumarate	nd	nd	nd	nd	nd	nd	–	nd	nd	nd	nd	nd	nd	nd	nd	+	nd	nd	nd	nd	+	nd	nd	nd	nd	nd	nd	nd
D-Galactose	+	+	nd	+	+	+	+	nd	nd	+	+	+	nd	+	+	+	nd	+	+	d	+	–	+	nd	nd	nd	+	–
Gluconate	nd	nd	nd	nd	nd	nd	+	nd	nd	nd	nd	nd	nd	nd	nd	+	nd	nd	nd	nd	+	nd	+	nd	nd	nd	nd	+
D-Glucose	+	+	+	+	+	+	+	nd	+	+	+	+	+	+	+	+	+	+	+	+	+	+	+	+	+	+	+	+
L-Glutamine	–	nd	+	nd	nd	nd	nd	nd	nd	nd	nd	nd	–	nd	nd	+	nd	nd	+	nd	nd	nd	nd	nd	nd	nd	+	nd
Glutarate	nd	nd	nd	nd	nd	nd	(+)	nd	nd	nd	nd	nd	nd	nd	nd	(+)	nd	nd	nd	nd	(+)	nd	nd	nd	nd	nd	nd	nd
Glycerol	nd	nd	nd	+	+	nd	nd	nd	nd	nd	+	nd	–	nd	nd	nd	nd	+	nd	nd	nd	+	nd	nd	nd	nd	nd	nd

L-Histidine	+	nd	–	nd	nd	nd	–	nd	nd	nd	nd	nd	–	nd	nd	+	nd	nd	d	nd	–	nd	nd	nd	nd	nd	–	nd
3-Hydroxybenzoate	nd	nd	nd	nd	nd	nd	–	nd	nd	nd	nd	nd	nd	nd	nd	–	nd	nd	nd	nd	–	nd	nd	nd	nd	nd	nd	nd
4-Hydroxybenzoate	nd	nd	nd	nd	nd	nd	–	nd	nd	nd	nd	nd	nd	nd	nd	–	nd	nd	nd	nd	–	nd	nd	nd	nd	nd	nd	nd
m-Hydroxybenzoic acid	d	+	nd	–	nd	–	nd	nd	nd	–	–	d	nd	nd	–	–	nd	–	–	d	nd	+	nd	nd	nd	nd	–	nd
o-Hydroxybenzoic acid	–	–	nd	–	nd	–	nd	nd	nd	–	–	–	nd	nd	–	–	nd	–	–	–	nd	–	nd	nd	nd	nd	–	nd
p-Hydroxybenzoic acid	–	–	nd	–	nd	–	nd	nd	nd	d	–	+	nd	nd	+	–	nd	–	+	–	nd	+	nd	nd	nd	nd	–	nd
β-Hydroxybutyrate	+	nd	+	nd	nd	nd	nd	nd	nd	nd	nd	nd	–	nd	nd	+	nd	nd	+	nd	nd	nd	nd	nd	nd	nd	+	nd
DL-3-Hydroxybutyrate	nd	nd	nd	nd	nd	nd	–	nd	nd	nd	nd	nd	nd	nd	nd	(+)	nd	nd	nd	nd	–	nd	nd	nd	nd	nd	nd	nd
Inositol[k]	+	+	+	–	–	–	+	nd	nd	+	+	d	–	+	d	+	+[e], –[h]	–	–	d[b], –[h]	+[h], –[c]	–	–	–	nd	nd	–	–
Inulin	nd	nd	nd	nd	nd	nd	nd	nd	nd	nd	nd	nd	nd	+	nd	nd	nd	nd	nd	nd	nd	nd	nd	nd	nd	nd	nd	nd
Itaconate	nd	nd	nd	nd	nd	nd	–	nd	nd	nd	nd	nd	nd	nd	nd	–	nd	nd	nd	nd	–	nd	nd	nd	nd	nd	nd	nd
α-Ketoglurate	–	nd	–	nd	nd	nd	nd	nd	nd	nd	nd	nd	–	nd	nd	+	nd	nd	d	nd	nd	nd	nd	nd	nd	nd	–	nd
DL-Lactate	+	nd	+	nd	nd	nd	–	nd	nd	nd	nd	nd	–	nd	nd	+	nd	nd	+	nd	–	nd	nd	nd	nd	nd	–	nd
D-Lactose	+	+	nd	–	nd	+	nd	nd	+	+	+	+	+	+	+	+	–	+	+	d	+	–	–	(+)	(+)	+	–	–
D-Malate	–	nd	–	nd	nd	nd	nd	nd	nd	nd	nd	nd	–	nd	nd	(+)	nd	nd	–	nd	nd	nd	nd	nd	nd	nd	+	nd
L-Malate	nd	nd	nd	nd	nd	nd	+	nd	nd	nd	nd	nd	nd	nd	nd	(+)	nd	nd	nd	nd	(+)	nd	nd	nd	nd	nd	nd	nd
Maltitol	nd	nd	nd	nd	nd	nd	–	nd	nd	nd	nd	nd	nd	nd	nd	(+)	nd	nd	nd	nd	–	nd	nd	nd	nd	nd	nd	nd
D-Maltose	+	+	nd	+	+	+	+	nd	nd	+	+	+	nd	+	+	+	nd	+	+	+	+	+	+	nd	+	+	+	–
D-Mannitol	+	+	+	+	+	+	nd	nd	–	+	–	+	+	+	+	+	–	+[g], –[b]	+	+[h], d[b]	+	–	–	+	–	–	+	–
D-Mannose	+	+	nd	+	+	+	+	nd	+	+	+	+	+	+	+	+	+	+	+	+	+	+	–	+	nd	nd	+	+
Melezitose	+	+	–	–	nd	d	nd	nd	d	nd	–	+	–	nd	d	+	nd	–	–	d	nd	+	nd	nd	nd	nd	–	nd
Melibiose (α-D-)[c]	–	nd	+	nd	–	nd	–	nd	nd	nd	nd	nd	–	nd	nd	–	nd	nd	–	nd	–	nd	–	nd	nd	nd	–	–
Mesaconate	nd	nd	nd	nd	nd	nd	–	nd	nd	nd	nd	nd	nd	nd	nd	–	nd	nd	nd	nd	–	nd	nd	nd	nd	nd	nd	nd
Methyl-α-D-glucoside	–	+	nd	–	nd	–	nd	nd	nd	–	–	d	nd	nd	d	–	nd	–	–	–	nd	–	–	nd	nd	nd	–	–
Methyl-β-D-glucoside	+	+	nd	+	nd	+	nd	nd	nd	+	–	+	nd	nd	+	+	nd	+	+	–	nd	–	nd	nd	nd	nd	+	nd
L-Ornithine	–	nd	+	nd	nd	nd	–	nd	nd	nd	nd	nd	–	nd	nd	(+)	nd	nd	–	nd	–	nd	nd	nd	nd	nd	–	nd
Phenylacetate	–	nd	–	nd	nd	nd	–	nd	nd	nd	nd	nd	(+)	nd	nd	–	nd	nd	d	nd	–	nd	nd	nd	nd	nd	–	nd
L-Proline	+	nd	+	nd	nd	nd	+	nd	nd	nd	nd	nd	+	nd	nd	+[e], (+)[c]	nd	nd	d	nd	–	nd	nd	nd	nd	nd	+	nd
Propionate	nd	nd	nd	nd	nd	nd	–	nd	nd	nd	nd	nd	nd	nd	nd	–	nd	nd	nd	nd	–	nd	nd	nd	nd	nd	nd	nd
Putrescine	nd	nd	nd	nd	nd	nd	–	nd	nd	nd	nd	nd	nd	nd	nd	(+)	nd	nd	nd	nd	–	nd	nd	nd	nd	nd	nd	nd
Pyruvate	nd	nd	nd	nd	nd	nd	+	nd	nd	nd	nd	nd	nd	nd	nd	–	nd	nd	nd	nd	(+)	nd	nd	nd	nd	nd	nd	nd
Quinate	+	nd	nd	nd	nd	nd	nd	nd	nd	nd	nd	nd	–	nd	nd	–	nd	nd	–	nd	nd	nd	nd	nd	nd	nd	+	nd
Raffinose	+[e,f], –[b]	+	–	–	–	+	nd	nd	–	d	–[b], +[g]	+	–	–	d	–	–	–[b], +[g]	–	d[b], –[h]	–[e], +[h]	–	–	–	nd	nd	–	–
L-Rhamnose	+	+	+	+	+	+	+	nd	+	d	+	+	–	+	d	+	+	–[b], +[g]	+	d[b], +[h]	+	+	–	–	nd	nd	+	–
D-Ribose	+	+	–	+	–	–	–	nd	nd	–	–[b], +[g]	d	–	–	d	+	nd	+	–	–	–	+	+	nd	–	–	+	+
Salicin	nd	+	nd	nd	nd	nd	(+)	nd	nd	nd	nd	nd	+	nd	nd	–	–	nd	nd	nd	+	nd	–	(+)	nd	nd	nd	–
L-Serine	nd	nd	nd	nd	nd	nd	+	nd	nd	nd	nd	nd	nd	nd	nd	–	nd	nd	nd	nd	–	nd	nd	nd	nd	nd	nd	nd
Sodium acetate	+	–	nd	–	nd	d	nd	nd	nd	d	–	d	nd	nd	d	–	nd	–	+	d	nd	–	–	nd	nd	nd	–	–
Sodium fumarate	+	–	nd	–	nd	d	nd	nd	nd	+	–	–	nd	nd	d	–	nd	–	+	–	nd	–	nd	nd	nd	nd	–	nd
Sodium succinate	+	–	nd	–	nd	d	nd	nd	nd	d	–	–	nd	nd	d	–	nd	–	–	–	nd	–	nd	nd	nd	nd	–	nd
D-Sorbitol	(+)[e], +[b]	+	+	+	nd	d	–	nd	nd	+	–	d	–	–	+	+[b], (+)[e], –[c]	nd	+	+[e], –[b]	+	–	+	+	nd	nd	nd	(+)[e], +[b]	+
Sorbose	nd	nd	nd	nd	nd	nd	nd	nd	nd	nd	nd	nd	nd	–	nd	nd	nd	nd	nd	nd	nd	nd	nd	nd	nd	nd	nd	nd
Spermine	–	nd	–	nd	nd	nd	nd	nd	nd	nd	nd	nd	+	nd	nd	–	nd	nd	–	nd	nd	nd	nd	nd	nd	nd	–	nd

(continued)

TABLE 195. (continued)

Characteristic	1. *A. philippinensis*	2. *A. auranticolor*	3. *A. brasiliensis*	4. *A. campanulatus*	5. *A. capillaceus*	6. *A. consettensis*	7. *A. couchii*	8. *A. cyaneus*	9. *A. deccanensis*	10. *A. derwentensis*	11. *A. digitatis*	12. *A. durhamensis*	13. *A. ferrugineus*	14. *A. globisporus*	15. *A. humidus*	16. *A. italicus*	17. *A. liguriensis*	18. *A. lobatus*	19. *A. missouriensis*	20. *A. palleronii*	21. *A. rectilineatus*	22. *A. regularis*	23. *A. sichuanensis*	24. *A. teichomyceticus*	25. *A. tereljensis*	26. *A. toevensis*	27. *A. utahensis*	28. *A. xinjiangensis*
Suberate	nd	nd	nd	nd	nd	nd	–	nd	nd	nd	nd	nd	nd	nd	nd	–	nd	nd	nd	nd	–	nd	nd	nd	nd	nd	nd	nd
Sucrose	+	+	+	+	(+)	nd	+	nd	+	nd	+	nd	–	+	nd	+	–	+	+	+	+[h], –[e,c]	+	+	+	nd	nd	+	+
Syringaldehyde	–	–	nd	–	nd	nd	nd	nd	nd	–	–	–	nd	nd	–	–	nd	–	–	–	–	–	nd	nd	nd	nd	–	nd
Syringic acid	–	–	nd	–	nd	d	nd	nd	nd	–	–	–	nd	nd	d	–	nd	–	–	–	–	nd	nd	nd	nd	nd	–	nd
Trehalose	nd	nd	nd	nd	nd	nd	+	nd	nd	nd	nd	nd	nd	nd	nd	+	nd	nd	nd	nd	+	nd	–	nd	nd	nd	nd	nd
Vanillin	+	–	nd	–	nd	–	nd	nd	nd	–	–	–	nd	nd	–	–	nd	+	–	nd	nd	–	nd	nd	nd	nd	–	nd
Xylitol	–	nd	+	nd	nd	nd	nd	nd	nd	nd	nd	nd	–	nd	nd	–	nd	nd	–	nd	nd	nd	nd	nd	nd	nd	–	nd
D-Xylose	+	+	+	+	+	+	+	nd	+	+	+	+	+	+	+	+	–	–[b], +[g]	+	+[b], –[h]	+	–[h], +[b,g]	+	–	+	+	+	+
Degradation of:																							nd		nd	nd		nd
Adenine	nd	–	nd	nd	–	nd	+	nd	nd	nd	nd	nd	nd	nd	nd	+	nd	–	nd	nd	+	nd	–	nd	–	–	nd	–
Calcium malate	+	nd	+	nd	nd	nd	nd	nd	+	nd	nd	nd	+	–	nd	–	–	nd	+	nd	nd	nd	nd	–	–	–	+	nd
Casein	+	+	+	+	+	+	+	nd	–	+	+[b], –[i]	+	+	+[e], (+)[f]	+	+	+	+	+	+	+	+	–	+	(+)	+	+	–
Cellulose	–[b], +[i]	–	nd	–	d	–	nd	nd	nd	–	–	–	nd	nd	–	nd	nd	–	–	–	nd	–[b], +[i]	nd	nd	nd	nd	–	nd
Chitin	+	–	nd	+	nd	d	nd	nd	nd	d	+	d	nd	nd	d	+	nd	+	+	d	nd	+	nd	nd	nd	nd	+	nd
DNA	d	+	nd	+	nd	d	nd	nd	nd	d	+	d	nd	nd	+	+	nd	+	+	–	nd	+	nd	nd	nd	nd	+	nd
Elastin	+	–	nd	–	nd	+	nd	nd	nd	+	+	d	nd	nd	d	–	nd	+	+	–	nd	–	–	nd	nd	nd	+	–
Esculin	nd	nd	nd	nd	+	nd	nd	nd	nd	nd	nd	nd	nd	nd	nd	nd	nd	nd	nd	nd	nd	nd	nd	nd	nd	nd	nd	nd
Gelatin	+	+[e], –[i]	+	+[b], –[g]	–	+	nd	nd	+	+	+[b], v[g]	+	+	–	+	+	+[h], –[e]	+	+[b], –[e]	+	+	+	+	+	–	+	+	+
Guanine	–	–	nd	–	–	–	nd	nd	nd	–	–	–	nd	nd	–	–	nd	–	–	–	nd	–	–	nd	nd	nd	–	–
Hypoxanthine	–	–	nd	–	–	–	+	nd	nd	–	–	–	nd	nd	–	+[c], –[b]	nd	–	–	–	+	–	–	nd	–	–	–	–
Lecithin	+	–	nd	+	nd	+	nd	nd	nd	d	+	+	nd	nd	+	+	nd	+	+	+	nd	–	nd	nd	nd	nd	+	nd
Pectin	+	+	nd	–	nd	d	nd	nd	nd	d	–	–	nd	nd	d	+	nd	+	+	d	nd	–	nd	nd	nd	nd	+	nd
RNA	+	+	nd	+	nd	+	nd	nd	nd	d	+	d	nd	nd	+	+	nd	+	+	d	nd	+	nd	nd	nd	nd	+	nd
Starch	+	+	+	+	+	+	+	nd	+	+	+	+	+	+	+	+	+	+	+	+	+	+	+	+	–	+	+	–
Tyrosine	+	–	–	–	nd	d	+	nd	nd	d	+[b], –[g]	d	+	nd	d	+	–	+[b], –[g]	+	d	+	–	+	–	(+)	(+)	–[b], +[e]	–
Xanthine	–	–	nd	–	–	–	+	nd	nd	–	–	–	nd	nd	–	–[b], +[c]	nd	–	–	–	+	–	–	nd	–	–	–	–
Xylan	nd	nd	nd	nd	nd	nd	+	nd	nd	nd	nd	nd	nd	nd	nd	+	nd	nd	nd	nd	+	nd	nd	nd	nd	nd	nd	nd
Hydrolysis of:[c]																												
Esculin	nd	nd	nd	nd	nd	nd	+	nd	nd	nd	nd	nd	nd	nd	nd	+	nd	nd	nd	nd	+	nd	+	nd	nd	nd	nd	+
L-Alanine pNA[l]	nd	nd	nd	nd	nd	nd	(+)	nd	nd	nd	nd	nd	nd	nd	nd	–	nd	nd	nd	nd	–	nd	nd	nd	nd	nd	nd	nd
Arbutin[b]	–	+	nd	+	nd	+	nd	nd	nd	+	+	+	nd	nd	+	+	nd	+	+	d	nd	+	nd	nd	–	–	+	nd
2-Deoxythymidine-5′-pNP phosphate[m]	nd	nd	nd	nd	nd	nd	(+)	nd	nd	nd	nd	nd	nd	nd	nd	–	nd	nd	nd	nd	–	nd	nd	nd	nd	nd	nd	nd
L-Glutamate-γ-3-carboxy pNA	nd	nd	nd	nd	nd	nd	–	nd	nd	nd	nd	nd	nd	nd	nd	–	nd	nd	nd	nd	–	nd	nd	nd	nd	nd	nd	nd
oNP β-D-glucopyranoside[n]	nd	nd	nd	nd	nd	nd	+	nd	nd	nd	nd	nd	nd	nd	nd	–	nd	nd	nd	nd	+	nd	nd	nd	nd	nd	nd	nd
pNP α-D-glucopyranoside	nd	nd	nd	nd	nd	nd	+	nd	nd	nd	nd	nd	nd	nd	nd	+	nd	nd	nd	nd	+	nd	nd	nd	nd	nd	nd	nd

pNP β-D-glucopyranoside	nd	nd	nd	nd	nd	nd	+	nd	nd	nd	nd	nd	nd	nd	nd	+	nd	nd	nd	nd	+	nd	nd	nd	nd	nd	nd	nd
pNP β-D-glucuronide	nd	nd	nd	nd	nd	nd	–	nd	nd	nd	nd	nd	nd	nd	nd	–	nd	nd	nd	nd	–	nd	nd	nd	nd	nd	nd	nd
pNP phenylphosphate	nd	nd	nd	nd	nd	nd	+	nd	nd	nd	nd	nd	nd	nd	nd	+	nd	nd	nd	nd	–	nd	nd	nd	nd	nd	nd	nd
bis-pNP phosphate	nd	nd	nd	nd	nd	nd	+	nd	nd	nd	nd	nd	nd	nd	nd	+	nd	nd	nd	nd	–	nd	nd	nd	nd	nd	nd	nd
pNP phosphorylcholine	nd	nd	nd	nd	nd	nd	+	nd	nd	nd	nd	nd	nd	nd	nd	(+)	nd	nd	nd	nd	–	nd	nd	nd	nd	nd	nd	nd
pNP β-D-xylopyranoside	nd	nd	nd	nd	nd	nd	+	nd	nd	nd	nd	nd	nd	nd	nd	+	nd	nd	nd	nd	+	nd	nd	nd	nd	nd	nd	nd
L-Proline pNA	nd	nd	nd	nd	nd	nd	–	nd	nd	nd	nd	nd	nd	nd	nd	–	nd	nd	nd	nd	–	nd	nd	nd	nd	nd	nd	nd
Enzymic activities (API test):[h]																												
N-Acetyl-β-glucosaminidase	nd	nd	nd	nd	nd	nd	nd	nd	nd	nd	nd	nd	nd	nd	nd	nd	+	nd	nd	+	+	–	+	–	+	+	nd	+
Acid phosphatase	nd	nd	nd	nd	nd	nd	nd	nd	nd	nd	nd	nd	nd	nd	nd	nd	+	nd	nd	+	+	+	+	+	+	+	nd	+
Alkaline phosphatase	nd	nd	nd	nd	nd	nd	nd	nd	nd	nd	nd	nd	nd	nd	nd	nd	+	nd	nd	+	+	–	+	+	(+)	+	nd	+
Arginine dihydrolase	nd	nd	nd	nd	nd	nd	nd	nd	nd	nd	nd	nd	nd	nd	nd	nd	–	nd	nd	–	+	+	nd	+	nd	nd	nd	nd
Chymotrypsin	nd	nd	nd	nd	nd	nd	nd	nd	nd	nd	nd	nd	nd	nd	nd	nd	+	nd	nd	–	+	+	nd	–	(+)	–	nd	nd
Citrate utilization	nd	nd	nd	nd	nd	nd	nd	nd	nd	nd	nd	nd	nd	nd	nd	nd	–	nd	nd	–	+	+	nd	–	nd	nd	nd	nd
Cystine arylamidase	nd	nd	nd	nd	nd	nd	nd	nd	nd	nd	nd	nd	nd	nd	nd	nd	–	nd	nd	–	+	–	nd	–	+	+	nd	nd
Esterase (C4)	nd	nd	nd	nd	nd	nd	nd	nd	nd	nd	nd	nd	nd	nd	nd	nd	+	nd	nd	+	+	–	nd	+	–	–	nd	nd
Esterase lipase (C8)	nd	nd	nd	nd	nd	nd	nd	nd	nd	nd	nd	nd	nd	nd	nd	nd	+	nd	nd	+	+	–	nd	+	(+)	(+)	nd	nd
α-Fucosidase	nd	nd	nd	nd	nd	nd	nd	nd	nd	nd	nd	nd	nd	nd	nd	nd	–	nd	nd	–	–	–	nd	–	+	+	nd	nd
α-Galactosidase	nd	nd	nd	nd	nd	nd	nd	nd	nd	nd	nd	nd	nd	nd	nd	nd	+	nd	nd	+	+	–	nd	–	–	–	nd	nd
β-Galactosidase	nd	nd	nd	nd	nd	nd	nd	nd	nd	nd	nd	nd	nd	nd	nd	nd	+	nd	nd	v	+	v	+	–	+	+	nd	+
α-Glucosidase	nd	nd	nd	nd	nd	nd	nd	nd	nd	nd	nd	nd	nd	nd	nd	nd	+	nd	nd	+	+	+	nd	–	–	–	nd	nd
β-Glucosidase	nd	nd	nd	nd	nd	nd	nd	nd	nd	nd	nd	nd	nd	nd	nd	nd	+	nd	nd	+	+	+	nd	–	+	+	nd	nd
β-Glucuronidase	nd	nd	nd	nd	nd	nd	nd	nd	nd	nd	nd	nd	nd	nd	nd	nd	+	nd	nd	–	–	–	nd	–	–	–	nd	nd
Indole production	nd	nd	nd	nd	nd	nd	nd	nd	nd	nd	nd	nd	nd	nd	nd	nd	–	nd	nd	–	–	–	nd	–	nd	nd	nd	nd
Leucine arylamidase	nd	nd	nd	nd	nd	nd	nd	nd	nd	nd	nd	nd	nd	nd	nd	nd	–	nd	nd	+	+	–	nd	–	–	–	nd	nd
Lipase (C14)	nd	nd	nd	nd	nd	nd	nd	nd	nd	nd	nd	nd	nd	nd	nd	nd	–	nd	nd	–	–	–	nd	–	–	–	nd	nd
Lysine decarboxylase	nd	nd	nd	nd	nd	nd	nd	nd	nd	nd	nd	nd	nd	nd	nd	nd	–	nd	nd	–	+	+	nd	+	nd	nd	nd	nd
α-Mannosidase	nd	nd	nd	nd	nd	nd	nd	nd	nd	nd	nd	nd	nd	nd	nd	nd	+	nd	nd	–	+	–	nd	–	–	+	nd	nd
Naphthol-AS-BI-phosphohydrolase	nd	nd	nd	nd	nd	nd	nd	nd	nd	nd	nd	nd	nd	nd	nd	nd	+	nd	nd	–	+	–	nd	+	(+)	+	nd	nd
Ornithine decarboxylase	nd	nd	nd	nd	nd	nd	nd	nd	nd	nd	nd	nd	nd	nd	nd	nd	–	nd	nd	–	–	+	nd	+	nd	nd	nd	nd
Trypsin	nd	nd	nd	nd	nd	nd	nd	nd	nd	nd	nd	nd	nd	nd	nd	nd	+	nd	nd	–	+	–	nd	–	–	–	nd	nd
Tryptophan deaminase	nd	nd	nd	nd	nd	nd	nd	nd	nd	nd	nd	nd	nd	nd	nd	nd	–	nd	nd	–	–	–	nd	–	nd	nd	nd	nd
Urease	+	–[b], +[j]	nd	+	nd	–	nd	nd	nd	d	–[b], +[j]	–	+	nd	d	+	+	–	+	–	+	+[h,j], –[b]	nd	+	nd	+	+	nd
Valine arylamidase	nd	nd	nd	nd	nd	nd	nd	nd	nd	nd	nd	nd	nd	nd	nd	nd	–	nd	nd	–	–	–	nd	–	–	–	nd	nd
Biochemical tests:																												
H_2S production	+	+	–	–	nd	d	nd	nd	–	d	–	–	–	–	d	+[e], –[b]	+[b], –[h,e]	nd	–	–	–	–	–	+[e], –[h]	nd	nd	+	+
Melanin formation	–	+	–	–	–	nd	nd	nd	+	nd	+	+	+	–	nd	+	–	–	–	nd	+	–	nd	+	nd	nd	+	+
Nitrate reduction	+	–	+	+	+	–	nd	nd	+	d	+	d	+	–	+	–[b], +[e,j]	–	+	+	d	–	+	+	+	+	–	+	+
Coagulation of milk	–	nd	–	+	d	nd	nd	nd	–	nd	+	nd	nd	–	nd	–	–	+	–	nd	+	+	nd	–	nd	nd	–	nd
Peptonization of milk	–	nd	–	+	d	nd	nd	nd	–	nd	+	nd	nd	–	nd	+	–	+	–	nd	+	–	nd	+	nd	nd	–	nd
Other physiological properties:																												
Growth at:																												
pH 4.0	–[b], +[k]	–	nd	–	nd	–	nd	nd	nd	–	–	–	+	nd	–	–[b], +[j]	nd	–	–	–	+	–	nd	nd	–	nd	–[b], +[j]	nd
pH 6.0	+	+	nd	+	nd	+	nd	nd	nd	d[b]	+	d[b]	+	nd	+	+	nd	+	+	–	+	+	–	nd	+	+	–[b], +[j]	–
pH 8.0	+	+	nd	+	nd	+	nd	nd	nd	d[b]	+	d[b]	+	nd	+	+	nd	+	+	+	+	+	+	nd	+	+	+	+
pH 10.0	–[b]	–[b], +[j]	nd	–[b], +[j]	nd	–	nd	nd	nd	–	–[b], +[j]	–	+	nd	–	–[b], +[j]	nd	–	–	–	+	–[b], +[j]	+	nd	–	+	–[b], +[j]	–

(continued)

TABLE 195. (continued)

Characteristic	1. *A. philippinensis*	2. *A. auranticolor*	3. *A. brasiliensis*	4. *A. campanulatus*	5. *A. capillaceus*	6. *A. consettensis*	7. *A. couchii*	8. *A. cyaneus*	9. *A. deccanensis*	10. *A. derventensis*	11. *A. digitatis*	12. *A. durhamensis*	13. *A. ferrugineus*	14. *A. globisporus*	15. *A. humidus*	16. *A. italicus*	17. *A. liguriensis*	18. *A. lobatus*	19. *A. missouriensis*	20. *A. palleronii*	21. *A. rectilineatus*	22. *A. regularis*	23. *A. sichuanensis*	24. *A. teichomyceticus*	25. *A. tereljensis*	26. *A. toevensis*	27. *A. utahensis*	28. *A. xinjiangensis*
Tolerance to (%, w/v):																												
Sodium chloride (1.0)	+	–	nd	+	nd	+	nd	nd	nd	d	+	+	nd	nd	+	+	nd	–	+	+	nd	–	+	nd	+	+	–	+
Sodium chloride (2.0)	+	–	nd	+	nd	d	nd	nd	nd	–	+	+	nd	nd	–	–	nd	–	+	+	nd	–	+	nd	+	+	–	+
Sodium chloride (3.0)	–	–	nd	+	–	–	nd	nd	nd	–	–	–	nd	nd	–	–	nd	–	–	d	nd	–	+	nd	+	–	–	+

[a]Symbols: +, >85% positive; d, different strains give different reactions (16–84% positive); –, 0–15% positive; (+), weakly positive; w, weak reaction; nd, not determined.

[b]Data from Goodfellow et al. (1990a).

[c]Kämpfer et al. (2007).

[d]Matsumoto et al. (2000).

[e]Palleroni (1989).

[f]Thiemann (1967).

[g]Vobis and Kothe (1989).

[h]Wink et al. (2006).

[i]Schäfer (1973).

[j]Kothe (1987); deviations are indicated by superscript letters.

[k]For isomers of inositol see[b,c,d,e].

[l]pNA, *p*-nitroanilide.

[m]pNP, *p*-nitrophenyl.

[n]oNP, *o*-nitrophenyl.

The zoospores of *Actinoplanes* exhibit chemotactic properties. In *Actinoplanes brasiliensis*, Palleroni (1976) found bromide and chloride ions acting as attractants at a relatively high concentration (0.1 M). Addition of methionine stimulated this chemotactic effect, suggesting that protein methylation may be involved (Palleroni, 1983). Spores of *Actinoplanes missouriensis* were attracted by fungal conidia and sclerotia and to their respective exudates (Arora, 1986). Phototactic effects have not been observed, but an apparent microaerophilic behavior was seen in *Actinoplanes brasiliensis* (Palleroni, 1976).

Sporangiospores of *Actinoplanes brasiliensis* are able to germinate in a minimal medium with glucose as sole carbon source. The germination rate is nearly 100% (Palleroni, 1989). In contrast, *Actinoplanes rectilineatus* gave a much lower proportion of spores able to germinate under minimal conditions and normal levels of germination depended on the addition of amino acids. Germination could be inhibited by chloramphenicol and actinomycin D and, even though it was stimulated by 6-azauracil, this compound inhibited further growth (Higgins, 1967).

Ecology. While studying the distribution and abundance of *Actinoplanes* in soils of Japan, Nonomura and Takagi (1977) introduced the convenient term "actinoplanetes", which was accepted for further ecological investigations (Makkar and Cross, 1982). In a more extended sense, the name "Actinoplanetes" was used for a long time for a suprageneric group within actinomycete systematics (Goodfellow, 1989) until the genera belonging to it could be harbored in the well-defined taxon *Micromonosporaceae* (Goodfellow et al., 1990b). Now, the term actinoplanetes can be used again in its original meaning, characterizing an ecological group of actinomycetes.

Actinoplanetes are widely distributed in soil throughout the world (Couch, 1963; Parenti and Coronelli, 1979; Schäfer, 1973). Strains of *Actinoplanes* occur in all types of soil, in arid desert areas (Couch, 1957; Garrity et al., 1996; Makkar and Cross, 1982), in sand dune systems close to seashores (Palleroni, 1976), and in subtropical and tropical regions. In a large-scale investigation of the distribution of the actinoplanetes in soil in Japan, Nonomura and Takagi (1977) demonstrated a correlation between their abundance, the type of soil, its pH value, and the content of organic matter. Relatively few actinoplanetes occurred in soils with pH 4.0–5.0 and abundant organic matter content. Their number increased with lower humus content and a pH value between 6.4 and 7.2. Soils with a permanent high content of water (e.g. paddy rice fields) have no advantage compared with cultivated fields, which are dry for longer periods.

Strains of *Actinoplanes* can also colonize plant or animal debris (Cross, 1981b; Makkar and Cross, 1982). A frequent drying and wetting of the substrates increases their occurrence. Favored habitats are edges of ponds, drainage ditches, and barnyards (Shearer, 1987). Sediments of rivers are also a good source for the isolation of *Actinoplanes* strains (Goodfellow et al., 1990b). They occur frequently on twigs submerged in streams (Willoughby, 1971), muddy dead leaves that are caught and dried on branches of overhanging trees (Cross, 1981b), and on allochthonous leaf litter cast up on the shores of lakes (Willoughby, 1969b). In general, the sporangiate actinoplanetes can be considered as normal inhabitants of soil and leaf litter (Cross, 1981b), although they can also be isolated directly from lake or river water (Willoughby, 1969a, 1971).

After 20 years of experience, Couch (1963) concluded that about 66% of the soil samples collected from all over the world contain sporangiate actinomycetes. This is in accordance with the observations of Schäfer (1973), who isolated *Actinoplanes* strains from 56% of soil samples investigated. Similar results were also obtained from geographically limited regions like Japan (strains isolated from 75% samples; Nonomura and Takagi, 1977) or Argentina (strains isolated from 65% samples; Vobis, 1987).

The function of actinoplanetes in soil ecosystems is poorly known. With a behavior of typical saprophytic micro-organisms, abilities to degrade any kind of biological material may be possible. Chitin has been used as a carbon source for the isolation of strains by Makkar and Cross (1982). However, chitin degradation seems to be extremely slow and the greatest advantage using chitin media is the inhibition/decrease in growth of other micro-organisms (Willoughby, 1968). Degradation tests using chitin from insects and fungi gave negative results (Schäfer, 1973), which is in contrast to the positive results of Goodfellow et al. (1990b), using the basal medium of Gordon (1967) supplemented with 0.5% (w/v) chitin.

Because *Actinoplanes* strains exhibit good growth on xylose and arabinose, it is conceivable that they play a role in decomposing pentosans of plant origin (Parenti and Coronelli, 1979). It was assumed that strains of *Actinoplanes* cannot decompose cellulose, with the exception of *Actinoplanes brasiliensis* (Palleroni, 1989; Parenti and Coronelli, 1979), but recently, a study of saprophytic actinomycete strains associated with the root system of the actinorhizal plant *Ochetophila* (*Discaria*) *trinervis*, revealed that all 27 isolated *Actinoplanes* strains can degrade starch, cellulose, and pectin. Furthermore, a third of them were also able to decompose hemicellulose and/or colonized preferably thin sections of dead wood (Solans and Vobis, 2003). It could be demonstrated in plant growth assays that one of the most active *Actinoplanes* strains, BCRU-ME 3, promotes *Frankia* symbiosis of *Ochetophila trinervis* as well as *Sinorhizobium meliloti*/*Medicago sativa* symbiosis (Solans, 2007; Solans et al., 2009). This helper effect could be increased by coinoculations with strains of *Streptomyces* and *Micromonospora*, respectively, which were isolated from the rhizosphere of the same host plant (Solans, 2007). The actinomycetes involved produce phytohormones, which seem to play an important role in those interactions with higher plants (Solans, 2008). Presumably, single strains of *Actinoplanes* have a multiple function in terrestrial ecosystems, as shown in this example.

Enrichment and isolation procedures

All isolation methods stated below, with exception of the incubation technique, are based on the knowledge of morphological or physiological properties of the organisms. Three main aspects of their life cycle are utilized: the sporangia are dry-resistant survival units that can release motile spores when sufficiently wet and the spores are attracted by organic or inorganic chemical substances. It is recommended that freshly collected samples are dried at about 35°C. Well enclosed in a glass vessel or even a plastic bag, the air-dried probes are suitable for later investigations and can be stored for a long time, preferably at 4°C (Makkar and Cross, 1982; Nonomura and Takagi, 1977).

Baiting technique. The baiting technique is the traditional isolation method, which was originally designed to isolate water molds and chytrids, but led to the discovery of actinomycetes with motile spores (Couch, 1949, 1950).

About 1 g soil is placed in a small sterilized Petri dish (3 or 4 cm in diameter), which is then covered with sterile water. After cautiously stirring, the particles settle to the bottom. Natural baits are placed singly or in combination on the surface of the water: e.g. pollen of *Pinus*, *Liquidambar*, or *Sparganium*, boiled *Paspalum* grass leaves or other material of biological origin (Bland and Couch, 1981; Couch, 1949, 1954; Makkar and Cross, 1982; Schäfer, 1973). A ring of Parafilm can be used to ensure the baits do not stick to the wall of the Petri dish (Hayakawa, 2003). Pollen grains of *Pinus* are the most recommended baits (Hayakawa, 2003; Palleroni, 1989). The baits must be presterilized, depending on their consistency, either chemically with ethanol or propylene dioxide or by autoclaving (Gaertner, 1955; Makkar and Cross, 1982; Schäfer, 1973). The enrichment cultures with floating baits are stored undisturbed at room temperature for several days or weeks. The water level can be regulated by addition of sterile distilled water. The examination for actinoplanetes is carried out with a dissecting microscope, preferably with ×100 magnification and horizontal lighting (Bland and Couch, 1981). Typical sporangia of *Actinoplanes* are recognizable as glistening beads on the air-exposed sides of the baits. Such baits are then removed carefully from the water and transferred to a 3% agar plate (Bland and Couch, 1981). Individual sporangia can be separated from the bait and rolled several centimeters over the surface of agar, using a micromanipulator or a thin-pointed tungsten needle, which has a tip curved like a hockey stick (Vobis, 1992). In this way, contaminants are removed from the sporangial surface. Cleaned sporangia are transferred onto a Petri dish with suitable agar media such as Czapek sucrose or peptone Czapek agars (Bland and Couch, 1981), half-concentrated Casamino acids-peptone Czapek agar (Schäfer, 1973), or Emerson's yeast extract-starch agar (Vobis, 1992). The colonies originating from the individual sporangia are visible to the naked eye after 1–4 weeks.

Makkar and Cross (1982) designed a special apparatus to apply an additional enrichment step. The colonized baits are transferred onto a Nucleopore polycarbonate membrane filter, washed with distilled water, and immediately dried at 28°C for 7 d. If this sporangia-containing material is resuspended and incubated in water, the fluid is highly enriched with motile spores, and is used for spreading onto agar plates.

Hayakawa et al. (1991c) enhanced the effect of releasing the spores by desiccating the sporangia-bearing baits for 2 h at 30°C in a mixture of fine soil particles and silica gel. These specially treated baits were then immersed in water and portions of the liquid, now enriched with zoospores, were plated on humic acid-vitamin agar.

Another convenient variation consists of making a wet mount with the colonized pollen grains in a drop of water on a slide and following the spore liberation microscopically. When this occurs, the liquid, enriched with swimming spores, can be used as an inoculum for isolation (Palleroni, 1989).

Dehydration-rehydration technique. This technique utilizes the ability of the sporangia to withstand desiccation and to release motile spores when they are subsequently in contact with water. Besides soil samples, it is also applicable to leaf litter, decaying plant material from aquatic habitats, organic debris, etc. (Makkar and Cross, 1982).

Samples are dried at 28–30°C for 7 d. For rehydration, 0.5 g soil or corresponding substrate is mixed with 50 ml sterile tap water in a 150 ml beaker or Erlenmeyer flask, which is covered with sterile aluminum foil (Shearer, 1987; Vettermann and Prauser, 1979). The suspension is incubated at 20–30°C for about 1 h. During the first 30 min, the vessel can be shaken at irregular intervals. After that, the particles should be permitted to settle. From the supernatant, 0.5–1.0 ml is removed with a sterile Pasteur pipette and spread onto agar plates (Shearer, 1987). If necessary, dilutions can be prepared from the inoculation suspension (Makkar and Cross, 1982). Soil extract agar or colloidal chitin agar containing cycloheximide and nystatin (Makkar and Cross, 1982), oatmeal-soil extract agar, or starch-casein-sulfate agar (Shearer, 1987) have been used in combination with this technique. Plates are incubated at 28°C for 2–4 weeks.

Rehydration and centrifugation method. Hayakawa et al. (2000) described an enrichment method incorporating differential centrifugation. Samples are rehydrated with 10 mM phosphate buffer containing 10% soil extract at 30°C for 90 min. The liquid enriched with zoospores is centrifuged at 1500 × *g* for 20 min. Portions of the supernatant are then plated on humic acid-vitamin agar supplemented with nalidixic acid and trimethoprim. The centrifugation procedure specifically eliminates strains of *Streptomyces* and other nonmotile actinomycetes (Hayakawa, 2003).

Chemotactic method. Investigations of the chemotactic behavior of the sporangiospores of *Actinoplanes brasiliensis* demonstrated, among other things, that they are attracted by chloride ions (Palleroni, 1976). This fact induced the same author to develop a new, very effective, and time-saving isolation method (Palleroni, 1980).

The essential tool of this technique is the isolation chamber, a sterilizable plastic block (80 × 40 × 12 mm) with two cylindrical wells (9 mm deep and 24 mm in diameter) with centers 32 mm apart. The wells are connected by a channel that is 2 mm wide and 3 mm deep (Palleroni, 1980).

The soil sample (1 g) is divided equally between the two compartments. Sterile water is added nearly to the rim. After incubation for approximately 1 h at 30°C, spores are released from the sporangia and move freely in the water. A sterile 1 µl glass capillary about 32 mm long, filled with 0.01 M phosphate buffer (pH 7.0), containing 0.01 M KCl as chemoattractant (Palleroni, 1980), is placed in the channel. Buffer and attractants may vary: phosphate buffer (5–10 mM, pH 6.8) and KCI (2 mM) (Palleroni, 1989), or colloidine (100 mM) (Hayakawa, 2003). The capillary must be submerged about 1 mm below the surface of the liquid. One hour after immersion of the capillary, sufficient spores are attracted and trapped in the lumen of the capillary, which is then removed and washed from the outside with sterile water. The contents of the capillary are blown into 1 ml sterile water or buffer. Portions of the spore suspension are taken with a sterile pipette and spread onto agar plates. The plates are then incubated at 28°C. Starch casein-sulfate agar is recommended as the isolation medium by Palleroni (1980) and humic acid-vitamin agar is recommended by Hayakawa (2003). Although colonies can be selected after 4 d, slowly growing actinoplanetes may only be detectable after 3 weeks.

Moist incubation procedure. This method is suitable for the direct detection of actinoplanetes on natural substrates.

Although the ability to produce motile spores obviously plays no role, *Actinoplanes* strains can be readily enriched. Willoughby (1968) emphasized the importance of selecting materials, such as plant residues found on the shores of lakes and streams, that are alternatively wetted and dried as the level of water rises and falls.

Portions of decaying leaves or other biological substrates, freshly collected from the field, are washed with sterile water to remove adhering detritus. They are placed in Petri dishes, the bottoms of which have been covered with moist filter paper or layers of cellulose before autoclaving. The Petri dishes, acting as humid chambers, are sealed and incubated for about 4 weeks at 25°C. Examination by both dissecting and light microscopy is necessary to identify the sporangia of the actinoplanetes (Willoughby, 1969b). The isolation media recommended by Willoughby (1968) include starch-casein agar and chitin agar.

Maintenance procedures

The traditional method for maintaining *Actinoplanes* strains is lyophilization. The experiments of Miller and Couch (1959) show that both the mycelium and the sporangia can survive the freeze-drying process. As pointed out by Palleroni (1989), actinoplanetes do not differ from other actinomycetes in their conservation requirements. Good results are also obtained by freezing concentrated spore suspensions in liquid media or in dilute phosphate buffer with 5% (v/v) glycerol and storing them at −20°C or in liquid nitrogen.

Differentiation of the genus *Actinoplanes* from other related genera and taxonomic comments

With the discovery of *Actinoplanes* strains, Couch (1949) demonstrated for the first time the existence of actinomycetes having a motile phase during their life cycle, represented by flagellated spores produced within sporangia. In addition, he confirmed the close affinity to the genus *Micromonospora*, correlated by colonial characters like production of substrate mycelium and absence of aerial mycelium (Couch, 1950). During the following decades, newly described or formerly known sporangia-forming genera with or without motility were classified either closer or more distantly related to *Actinoplanes* (Bland and Couch, 1981; Goodfellow and Cross, 1984; Palleroni, 1989; Vobis, 1989b). Considering chemotaxonomic properties, DNA pairing experiments, rRNA cistron similarity, and 16S rRNA oligonucleotide cataloging, the genera *Ampullariella* and *Amorphosporangium* were combined in a redefined genus *Actinoplanes* (Stackebrandt and Kroppenstedt, 1987). Supported by chemotaxonomic and numerical taxonomic studies, as well as 16S rRNA gene sequence-based phylogenetic clustering, other sporangiate genera were separated from the "*Micromonospora–Actinoplanes–*complex" and classified within other families (Goodfellow et al., 1990b; Miyadoh et al., 1997, 2001; Stackebrandt et al., 1997). This was the case for the genera *Acrocarpospora, Cryptosporangium, Kutzneria, Planobispora, Planomonospora, Planotetraspora, Sphaerisporangium, Spirillospora, Streptoalloteichus*, and *Streptosporangium*, and the genera with multilocular sporangia, *Dermatophilus, Frankia*, and *Geodermatophilus*.

The genera *Actinoplanes* and *Micromonospora* form the central part of the well-established family *Micromonosporaceae* (Goodfellow et al., 1990b), which can be distinguished from other suprageneric groups by 16S rRNA gene sequence similarity (Koch et al., 1996a). Each genus of the family *Micromonosporaceae* is characterized by distinctive morphological and/or chemotaxonomic features (Ara et al., 2008b). The morphological heterogeneity of spore-producing structures allows differentiation of the genus *Actinoplanes* from the other genera. Members of the *Micromonospora, Salinospora*, and *Verrucosispora* can be clearly separated morphologically by forming single, nonmotile spores on substrate hyphae. Species of *Actinocatenispora, Asanoa, Catellatospora, Catenuloplanes, Longispora, Polymorphospora*, and *Spirilliplanes* produce spores in chains on the surface of the substrate mycelium. The spores of *Catenuloplanes* and *Spirilliplanes* are motile, a feature in common with *Actinoplanes*, but they are not developed within sporangia. Morphologically more related may be the group comprising the genera *Couchioplanes, Krasilnikovia*, and *Pseudosporangium*, which form spore chains within pseudosporangium-like structures or pseudosporangia on substrate mycelium; spores are motile in the case of *Couchioplanes*, but not in *Pseudosporangium* or *Krasilnikovia*.

The genera *Dactylosporangium, Luedemannella, Pilimelia, Planosporangium*, and *Virgisporangium* appear morphologically much more related to *Actinoplanes* because they all develop real sporangia on the surface of the substrate mycelium. *Luedemanella* can be distinguished by the production of nonmotile sporangiospores. The genera *Dactylosporangium, Planosporangium*, and *Virgisporangium* form finger-like or rod-shaped, oligospore sporangia, with 2–6 or more motile spores per sporangium in a single row. *Pilimelia* is a genus characterized by multispored sporangia and motile sporangiospores, features in common with *Actinoplanes*. The spores of most *Actinoplanes* species are spherical, but in the case of the former "*Ampullariella*" species, they are also rod-like as in *Pilimelia* species. The distinguishing properties are the size of spores and the type of flagellation: the rod-like spores of *Actinoplanes* are 0.5–1.0 × 2.0–4.0 μm and have a polarly inserted tuft of flagella; and the rod-like spores of *Pilimelia* are 0.3–0.7 × 0.7–1.5 μm with a lateral tuft of flagella. Based on the type of flagellation, the mode of movement is completely different. The rod-like spores of *Actinoplanes* rotate around their longitudinal axis, whereas those of *Pilimelia* rotate around the lateral axis (Vobis, 1992).

Differentiation of species of the genus *Actinoplanes*

Morphological characters are easy to recognize, but the strains have to be cultured onto suitable agar media and investigated microscopically at the right time, e.g. on M3 agar of Rowbotham and Cross (1977), supplemented with 0.1% (w/v) fructose, after incubation for 14 d at 25°C (Goodfellow et al., 1990b). The artificial soil agar of Henssen and Schäfer (1971) is also recommended (Vobis, 1992). Sporangia are the joint characteristic structures, which can be used to limit groups of species or even species. The spores produced within spherical sporangia have in general a globose shape and are arranged in coiled chains. The observation "irregular arrangement" depends possibly on an advanced process of maturity. All species with globose spores belong to *Actinoplanes* in the original concept of Couch (1963). Sporangia shapes other than spherical are characterized by spore chains orientated in parallel and rod-like spores, which are characteristic for the former "*Ampullariella*" species (Vobis and Kothe, 1989). Within this group, *Actinoplanes capillaceus* stands out, with a hairy ornamentation on the surface of the sporangia, and *Actinoplanes rectilineatus*, with globose spores.

Another useful morphological marker is the presence of aerial mycelium, which is well developed in two species, *Actinoplanes rectilineatus* and *Actinoplanes regularis*, and observed in a rudimentary form in *Actinoplanes couchii, Actinoplanes ferrugineus*, and *Actinoplanes liguriensis* (Table 194). None of the other species produce aerial mycelium. Fragmentation of substrate mycelium is described only in *Actinoplanes couchii* (Kämpfer et al., 2007) and *Actinoplanes philippinensis* (Couch and Bland, 1974b).

In many cases, the different colors of the substrate mycelium maybe useful for the separation of species. Physiological conditions like the type of agar medium or the age of cultures are factors that can influence exact determinations. Additionally, zonal divisions of the colonies like radial sectors or central and marginal areas may cause differences. Using modified Bennett's agar, Goodfellow et al. (1990b) assigned a total of 163 isolates to eight color groups. In Table 194, the spectrum of colors of substrate mycelium is restricted to six main colors. Their combinations permit specification of a characteristic color for each species. The color white also comprises the variations ivory, creamish, or pale; yellow also represents buff, tawny, amber, and ochre. Orange is the most frequent color, with all shades like apricot or orange-red. Orange may lead to the variations of red, brick-like, cherry, coral, pink, scarlet, cinnabar, purple, and mahogany. The color brown includes cinnamon, chestnut, rufous or rusty. Black may be interpreted as a very dense concentration of many colors.

The soluble pigments are considered by Goodfellow et al. (1990b) as a character of possible diagnostic value to define *Actinoplanes* clusters at the 83% similarity level (S_{SM}). Because the production and/or the color variations depend on the culture media, RNA agar was recommended. The soluble pigments may influence the basic orange color of the substrate mycelium and then, in the case of overproduction, are diffused into the substrate (Parenti and Coronelli, 1979). In Table 194, the most common pigments are compiled for each species; e.g. the rare blue soluble pigment allows a rapid diagnosis for the species *Actinoplanes cyaneus*. Brown pigments are formed by many strains; however, it is not known if these pigments are melanoid. The production of melanin, regarded as a constant physiological feature for *Streptomyces* strains (Shirling and Gottlieb, 1966), is also used as a taxonomic character for *Actinoplanes* species. Nevertheless, Goodfellow et al. (1990b) could not obtain reproducible results. Additionally, Kothe (1987) demonstrated by absorption spectra that the type strains of *Actinoplanes philippinensis* and *Actinoplanes brasiliensis* characterized as melanin-negative organisms show the same peaks at 275 nm and 345 nm like in *Actinoplanes utahensis*, a typically melanin-positive species (Table 195). The feature of melanin production seems to have no diagnostic value.

The assimilation of different organic compounds as sole sources of carbon is used as a diagnostic feature to recognize species. A total of 86 compounds is listed in Table 195, some of which have a differentiating value (Goodfellow et al., 1990b; Kämpfer et al., 2007; Palleroni, 1989; Wink et al., 2006). However, more than 70% of the tests have not been determined. Enzymic activities, including degradation and hydrolysis of organic compounds, can be used to differentiate *Actinoplanes* species, but the list is also incomplete for many species (Table 195). Other physiological properties like pH and sodium chloride tolerances (Table 195), growth in the presence of organic and inorganic substances, and resistance to various antibiotics (Table 194) supply further diagnostic data. Tests for antimicrobial activities against micro-organisms may also be integrated into the large list of diagnostic characters (Table 194).

Among the chemotaxonomic features, the fatty acid profiles, menaquinone components, and polar lipid patterns show sufficient quantitative differences, which could be used in particular cases for separating species of *Actinoplanes* (Table 194).

Many species are described only on the basis of their corresponding type strain. In this case, the species description is identical with the characterization of a single strain in all phylogenetic and phenotypic aspects. The studies of Goodfellow et al. (1990b), Kothe (1987), Matsumoto et al. (2000), and Schäfer (1973) have shown that not all features that are characteristic for one strain have diagnostic value to differentiate between species, but allow intraspecific variations. In this sense, the concept of many species of the genus *Actinoplanes* seems to be still ill-defined (Palleroni, 1989).

List of species of the genus *Actinoplanes*

The following list is based upon J.P. Euzéby's "List of Prokaryotic names with Standing in Nomenclature – Genus *Actinoplanes*" (http://www.bacterio.cict.fr/a/actinoplanes.html).

1. **Actinoplanes philippinensis** Couch 1950, 89[AL]

phi.lip.pi.nen'sis. N.L. masc. adj. *philippinensis* of or pertaining to the Philippines.

Sporangia are spherical to subspherical, 8.8–25.0 µm in diameter (most are about 16 µm thick), produced on the surface of substrate mycelium by palisade hyphae. Spores are arranged in coils within the sporangia, globose, 1.0–1.2 µm in diameter, motile with a tuft of flagella. Substrate hyphae are 0.5–1.5 µm wide, branched, and sparingly septate. Conidia may be developed under certain circumstances.

Growth on Czapek agar is moderate; colonies are flat, or slightly elevated, light buff to tawny, occasionally changing to purplish brown with age. Hyphae in the upper layer are arranged in palisades with abundant production of sporangia. On peptone Czapek agar, growth is very good, surface of the colonies with concentric rings and radial grooves, formation of sporangia is very rare. Hyphae may fragment into spheres and rods. Color is apricot-orange to orange-chrome. Vigorous growth is observed on potato glucose agar, with the color at first near apricot-orange, then mahogany red, chestnut, or cinnamon-rufous. On nutrient agar, colonies are flat, slightly elevated in the center, color orange to cinnamon-rufous. A dark brown soluble pigment is produced on potato glucose and glycerine Czapek agars.

Produces macrocyclic lactone antibiotics.

Source: the type strain was isolated from rice paddy soil (the Philippines).

DNA G+C content (mol%): 72.1 (Bd).

Type strain: ATCC 12427, CBS 107.58, DSM 43019, NBRC 13878, HAMBI 1927, JCM 3001, NRRL 2506, RIA 468, UNCC P-15, VKM Ac-842.

Sequence accession no. (16S rRNA gene): D85474.

Additional comments: the type strain shows the highest sequence similarity to the following *Actinoplanes* species (based on ~1470 nt): *Actinoplanes sichuanensis* 03-723^T, 99.0%; *Actinoplanes capillaceus* NBRC 16408^T, 98.9%; *Actinoplanes campanulatus* NBRC 12511^T, 98.9%.

2. **Actinoplanes auranticolor** (Couch 1963) Stackebrandt and Kroppenstedt 1988, 328^{VP} (Effective publication: Stackebrandt and Kroppenstedt 1987, 113.) (*Amorphosporangium auranticolor* Couch 1963, 65)

au.ran'ti.co.lor. N.L. n. *aurantium* a bitter orange; L. n. *color* tint, hue; N.L. adj. *auranticolor* orange colored.

Sporangia are very irregular in shape, multilobed, 6.0–25.0 µm wide and 8.0–15.0 µm long. Spore chains are irregularly arranged within the sporangium. Sporangiophores may be branched. Spores are short rod-shaped, 0.5–0.7 × 1.0–1.5 µm. The original description by Couch (1963) considers the spores to be nonmotile, but Kane Hanton (1968) later reported motility by means of polar flagella. Nonmotile microconidia may be produced on tyrosine medium.

Growth on Czapek agar is very good and colonies are frequently elevated and convoluted with slippery surfaces and apricot-orange in color; colonies are scarlet on peptone Czapek agar. Good growth is also seen on oatmeal agar and on casein media, but poor growth is observed on tyrosine media. In all cases, colony colors are various shades of orange. A diffusible dark pigment is formed in tyrosine media whereas, in other media, a yellow diffusible pigment may be observed.

Source: the type strain and a further strain were isolated from soil collected from meadows (Nevada, USA).

DNA G+C content (mol%): 72 (T_m).

Type strain: ATCC 15330, DSM 43031, NBRC 12245, HAMBI 1975, JCM 3038, NRRL B-3343, UNCC 253, VKM Ac-648.

Sequence accession no. (16S rRNA gene): D85471.

Additional comments: the type strain shows highest sequence similarity to the following *Actinoplanes* species (based on ~1470 nt): *Actinoplanes missouriensis* NBRC 13243^T, 99.8%; *Actinoplanes cosettensis* NBRC 14913^T, 98.7%; *Actinoplanes digitatis* NBRC 12512^T, 98.7%; *Actinoplanes humidus* NBRC 14915^T, 98.7%.

3. **Actinoplanes brasiliensis** Thiemann, Beretta, Coronelli and Pagani 1969, 119^{AL}

bra.si.li.en'sis. N.L. masc. adj. *brasiliensis* of or pertaining to Brazil.

Sporangia are produced abundantly on soil extract agar, calcium malate agar, and starch-casein agar. They are umbraculiformis (umbrella shaped) to irregular or very occasionally globose, with very wrinkled surfaces, measuring at their widest from 3.5 to 11.5 µm. Spores vary from subspherical (1.2 µm) to rod-shaped (1.2 µm in width and 1.7–2.3 µm in length).

Good growth occurs on skim milk, glucose asparagine, and potato agars. Growth varies from very good to moderate on different ISP media (Shirling and Gottlieb, 1966). Colonies have smooth to wrinkled surfaces and vary in color from light pink, orange rose, and orange to deep orange. No aerial mycelium is produced.

Produces the acidic antibiotic A/672.

Source: the type strain was isolated from a soil sample (State Bahia, Brazil).

DNA G+C content (mol%): 70.5 (HPLC).

Type strain: ATCC 25844, DSM 43805, NBRC 13938, JCM 3196, NRRL B-16714, VKM Ac-1320.

Sequence accession no. (16S rRNA gene): D85470.

Additional comments: the type strain shows the highest sequence similarity to the following *Actinoplanes* species (based on ~1470 nt): *Actinoplanes deccanensis* NBRC 13994^T, 98.2%; *Actinoplanes digitatis* NBRC 12512^T, 98.0%.

4. **Actinoplanes campanulatus** (Couch 1963) Stackebrandt and Kroppenstedt 1988, 328^{VP} (Effective publication: Stackebrandt and Kroppenstedt 1987, 113.) ["*Ampullaria campanulata*" Couch 1963, 59; *Ampullariella campanulata* (Couch 1963) Couch 1964, 29]

cam.pa.nu.la'tus. N.L. dim. n. *campanella* small bell; L. masc. suff. *-atus* suffix denoting provided with; N.L. masc. adj. *campanulatus* bell-shaped.

Sporangia are characteristically bell shaped, frequently lobed or irregular, sometimes pyriform or cylindrical, and 5.0–15.0 × 6.0–12.0 µm. They often have a papillate appearance at the proximal end because the spore chains are of unequal length. Spore chains are arranged in parallel rows within the sporangia. The spores are rod-shaped (0.5–1.0 × 2.0–4.0 µm) and motile by a polarly inserted tuft of flagella.

Good growth occurs on Czapek, peptone Czapek, and casein agars and moderate growth occurs on tyrosine agar. Colonies are elevated and convoluted with ridged areas at the margin. The color of the substrate mycelium is coral-red, coral-pink, orange, brown, or black. Soluble yellowish, greenish, and brownish pigments may be produced on various media. Formation of sporangia occurs on Czapek agar. D-Mannitol is utilized by the type strain; inositol and raffinose are not utilized. Gelatin is not liquefied.

The iso- and anteiso- fatty acids are unsaturated.

Source: the type strain was isolated from soil (Douglas, Kansas, USA). Three further isolates were from Chapel Hill, North Carolina, USA, and four were from soils collected in Tahiti (Couch, 1963). Additionally, 17 strains were isolated from soil by D. Schäfer (Marburg, Germany) originating from: Kenya (2), Ceylon (4), Corsica (1), Mexico (2), Austria (1), Taiwan (1), Germany (1), South Africa (1), Italy (1), USA (2), and Turkey (1) (Kothe, 1987).

DNA G+C content (mol%): 71 (T_m).

Type strain: ATCC 15348, CBS 190.64, DSM 43148, NBRC 12511, JCM 3059, NRRL B-3344, UNCC 65, VKM Ac-1319.

Sequence accession no. (16S rRNA gene): AB036995.

Additional comments: the type strain shows the highest sequence similarity to the following *Actinoplanes* species (based on ~1470 nt): *Actinoplanes capillaceus* K95-5561^T, 99.9%; *Actinoplanes lobatus* NBRC 12513^T, 99.3%.

5. **Actinoplanes capillaceus** Matsumoto, Takahashi, Kudo, Seino, Iwai and Ōmura 2001, 793[VP] (Effective publication: Matsumoto, Takahashi, Kudo, Seino, Iwai and Ōmura 2000, 114.)

ca.pil.la′ce.us. L. masc. adj. *capillaceus* capillary, hairlike, referring to the hairy surface of the sporangium.

Sporangia are produced by substrate hyphae on short sporangiophores and are bell-shaped (5.0–10.0 × 5.0–15.0 µm). The surface of the sporangia is covered with short hair-like elements. Spores are oblong in shape (0.7–0.8 × 2.0–3.0 µm) and motile by means of a polar tuft of flagella.

Good growth occurs on inorganic salts-starch agar; colonies are brick-red in color. Colonies on yeast extract-starch agar are pink to yellow colored. No soluble or melanoid pigments are produced. No growth occurs in the presence of 3% (w/v) NaCl. Sensitive to streptomycin (50 µg/ml), but resistant to novobiocin (20 µg/ml).

Source: two strains, including the type strain, were isolated from a soil sample collected in Sayama City, Saitama Prefecture, Japan. Regarding the subjective synonym species, further three strains were isolated from soil (China) (Jiang and Ruan, 1982; Juan and Zhang, 1974).

DNA G+C content (mol%): 71.4 (HPLC).

Type strain: DSM 44859, NBRC 16408, JCM 10268, K95-5561, NBRC 16408.

Sequence accession no. (16S rRNA gene): AB013495.

Additional comments: the type strain shows the highest sequence similarity to the following *Actinoplanes* species (based on ~1470 nt): *Actinoplanes campanulatus*, NBRC 12511[T], 99.9%; *Actinoplanes lobatus*, NBRC 12513[T], 99.2%. Subjective synonyms (*sensu* Matsumoto et al., 2000) are: "*Ampullariella cylindrica*" Jiang and Ruan 1982 (type strains: JCM 3329, NBRC 14264); "*Ampullariella pekinesis*" Juan (Ruan) and Zhang 1974 (type strains: JCM 3174, NBRC 13662, DSM 46148); and "*Ampullariella pilifera*" Jiang and Ruan 1982 (type strains: JCM 3330, NBRC 14265).

6. **Actinoplanes consettensis** Goodfellow, Stanton, Simpson and Minnikin 1990a, 320[VP] (Effective publication: Goodfellow, Stanton, Simpson and Minnikin 1990b, 33.)

con.set.ten′sis. N.L. masc. adj. *consettensis* of or belonging to Consett, UK, a town near the site where the organism was isolated.

Sporangia are globose; spores are motile and arranged irregularly within sporangia.

Abundant growth occurs on modified Bennett's agar, developing light to yellow-brown substrate mycelium. Degrades arbutin, casein, DNA, elastin, lecithin, starch, and RNA. Adonitol, L-arabinose, dextrin, D-fructose, D-galactose, D-glucose, D-lactose, maltose, D-mannitol, D-mannose, methyl β-D-glucoside, raffinose, L-rhamnose, and D-xylose are utilized as sole carbon sources, but not cellobiose, erythritol, *myo*-inositol, methyl α-D-glucoside, D-ribose, catechol, *m*-, *o*- or *p*-hydroxybenzoic acid, syringaldehyde, or vanillin. Nitrate is not reduced and urea is not hydrolyzed. Tolerant to a number of antibiotics and chemical inhibitors. Grows between 4 and 30°C.

The peptidoglycan of the cell wall contains *meso*- and hydroxy-diaminopimelic acid. The principal wall sugars are arabinose, galactose, glucose, mannose, and xylose. The organism contains complex mixtures of straight- and branched-chain fatty acids, with hexadecanoic and 12-methyltetradecanoic acids predominating. Tetrahydrogenated menaquinones with nine isoprene units are the predominant isoprenologue, and major amounts of phosphatidylglycerol, phosphatidylethanolamine, phosphatidylglycerol, and phosphatidylinositol are present.

Source: the type strain and five further strains were isolated from sediments of the River Derwent at Allensford, County Durham, UK.

DNA G+C content (mol%): not known.

Type strain: ATCC 49799, DSM 43942, NBRC 14913, LA 97, NCIB (now NCIMB) 20027, NRRL B-16688.

Sequence accession no. (16S rRNA gene): AB036996.

Additional comments: the type strain shows the highest sequence similarity to the following *Actinoplanes* species (based on ~1470 nt): *Actinoplanes humidus* NBRC 14915[T], 100%; *Actinoplanes digitatis* NBRC 12512[T], 98.8%.

7. **Actinoplanes couchii** Kämpfer, Huber, Thummes, Grün-Wollny and Busse 2007, 722[VP]

cou′chi.i. N.L. gen. masc. n. *couchii* of Couch, named after J.N. Couch (1896–1986), who proposed the genus name *Actinoplanes* in 1950.

Sporangia are globose to oval. A rudimentary sterile aerial mycelium is formed. Substrate hyphae fragment easily in irregular rod-shaped cells.

Good growth occurs on nutrient agar and DSMZ medium 65 at 25–30°C. Color of the substrate mycelium is yellow-orange to orange-red. A red to brown soluble pigment develops on DSMZ medium 65. Starch, xylan, tyrosine, casein, hypoxanthine, adenine, and xanthine are degraded.

The quinone system consists of MK-9(H_4) (75%) as the predominant compound, with moderate amounts of MK-9(H_6) (11%), minor amounts of MK-9(H_2) (4%), MK-9(H_8) (2%), and MK-10(H_2) (2%), and traces of MK-10(H_6) (<0.1%). The polar lipid profile contains the major compounds diphosphatidylglycerol and phosphatidylethanolamine, but lacks phosphatidylcholine and aminoglycolipids. Additionally, moderate amounts of phosphatidylinositol, two unknown phospholipids, a highly hydrophilic glycolipid, and a polar lipid are detectable and traces of a single phosphatidylinositol mannoside are present.

Source: the type strain was isolated from soil close to the Marmore waterfalls, Terni, Italy.

DNA G+C content (mol%): not known.

Type strain: CIP 109316, DSM 45050, GW8-1761.

Sequence accession no. (16S rRNA gene): AM400230.

Additional comments: the type strain shows the highest sequence similarity to the following *Actinoplanes* species (based on ~1470 nt): *Actinoplanes italicus* NBRC 13911[T], 99.2%; *Actinoplanes rectilineatus* NBRC 13941[T], 98.9%.

8. **Actinoplanes cyaneus** Terekhova, Sadikova and Preobrazhenskaya 1987, 179[VP] (Effective publication: Terekhova, Sadikova and Preobrazhenskaya 1977, 1059.)

cy.a.ne′us. L. masc. adj. *cyaneus* dark blue.

Spherical sporangia with diameters of 30–60 µm are formed by substrate mycelium, sporangiospores are globose

(1.2–1.4 µm) and motile. Aerial mycelium is not developed. On synthetic media, a soluble blue pigment is produced, belonging to the chemical group of celicomycin-actinorodine.

Source: the type strain was isolated from soil (Siberia, Russia).

DNA G+C content (mol%): not known.

Type strain: DSM 46137, NBRC 14990, INA 1569, JCM 9082, VKM Ac-1095.

Sequence accession no. (16S rRNA gene): AB036997.

Additional comments: the type strain shows the highest sequence similarity to the following *Actinoplanes* species (based on ~1470 nt): *Actinoplanes rectilineatus* NBRC 13941^T, 98.4%; *Actinoplanes italicus* NBRC 13911^T, 98.3%.

9. **Actinoplanes deccanensis** Parenti, Pagani and Beretta 1975, 248[AL]

dec.ca.nen'sis. N.L. masc. adj. *deccanensis* of or pertaining to the Indian locality of Decca.

Sporangia are globose, small with irregular surfaces (4.0–7.0 µm in diameter), produced on hyphae of substrate mycelium, especially when grown on soil extract agar. Spores are subspherical, 1.0 × 1.5 µm in diameter, and motile. Soil extract agar promotes formation of sporangia. Hyphae of substrate mycelium are branched and about 1.0 µm in diameter. Aerial mycelium is not formed.

Growth is abundant on ISP agar media 2, 4, and 7 (Shirling and Gottlieb, 1966), and on other media like oatmeal, potato, Hickey–Tresner, Bennett's, or skim milk agars. Surfaces of the colonies are wrinkled, crusty, or rough. Color varies from light amber, light orange, light orange pinkish to orange. No growth on calcium malate agar. On tyrosine agar, a brown diffusible pigment is produced. Thermotolerant, grows between 26 and 42°C.

Produces the antibiotic lipiarmycin.

Source: the type strain was isolated from soil (Decca, India).

DNA G+C content (mol%): 72.0 (HPLC).

Type strain: A/10655, ATCC 21983, DSM 43806, NBRC 13994, JCM 3247, NRRL B-16715.

Sequence accession no. (16S rRNA gene): AB036998.

Additional comments: the type strain shows the highest sequence similarity to the following *Actinoplanes* species (based on ~1470 nt): *Actinoplanes missouriensis* NBRC 13243^T, 98.5%; *Actinoplanes digitatis* NBRC 12512^T, 98.3%.

10. **Actinoplanes derwentensis** Goodfellow, Stanton, Simpson and Minnikin 1990a, 320[VP] (Effective publication: Goodfellow, Stanton, Simpson and Minnikin 1990b, 33.)

der.wen.ten'sis. N.L. masc. adj. *derwentensis* of or pertaining to The Derwent, a river in County Durham, England, from which the organism was isolated.

Globose sporangia are readily produced; spores are motile and arranged irregularly within sporangia.

Color of substrate mycelium is orange to dark orange on modified Bennett's agar. Degrades arbutin, casein, chitin, DNA, elastin, lecithin, RNA, and starch; utilizes adonitol, L-arabinose, dextrin, D-fructose, D-galactose, D-glucose, *myo*-inositol, D-lactose, maltose, D-mannitol, D-mannose, methyl β-D-glucoside, D-xylose, and sodium fumarate as sole carbon sources but not methyl α-D-glucoside, D-ribose, catechol, coumarin, *m*-, *o*- or *p*-hydroxybenzoic acid, syringaldehyde, syringic acid, or vanillin. Grows between 4 and 30°C.

The peptidoglycan of the cell walls contains *meso*- and hydroxy-diaminopimelic acid. The principal wall sugars are arabinose, galactose, glucose, mannose, and xylose. The organism contains complex mixtures of straight- and branched-chain fatty acids, with hexadecanoic, 14-methylpentadecanoic, and 14-methylhexadecanoic acids predominating. Tetrahydrogenated menaquinones with nine isoprene units are the predominant isoprenologue, and major amounts of diphosphatidylglycerol, phosphatidylethanolamine, phosphatidylglycerol, and phosphatidylinositol are present.

Source: the type strain and seven further strains were isolated from sediments of the River Derwent, Allensford, Durham, UK.

DNA G+C content (mol%): not known.

Type strain: ATCC 49798, DSM 43941, NBRC 14935, JCM 7556, LA 107, NCIB (now NCIMB) 12875, NRRL B-16692.

Sequence accession no. (16S rRNA gene): AB036999.

Additional comments: the type strain shows the highest sequence similarity to the following *Actinoplanes* species (based on ~1470 nt): *Actinoplanes couchii* GW8-1761^T, 98.6%; *Actinoplanes rectilineatus* NBRC 13941^T, 98.5%; *Actinoplanes utahensis* NBRC 13244^T, 98.5%.

11. **Actinoplanes digitatis** (Couch 1963) Stackebrandt and Kroppenstedt 1988, 328[VP] (Effective publication: Stackebrandt and Kroppenstedt 1987, 113.) ["*Ampullaria digitata*" Couch 1963, 61; *Ampullariella digitata* (Couch 1963) Couch 1964, 29]

di.gi.ta'tis. L. masc. adj. *digitatis* (*sic*) having fingers.

Sporangia are usually digitate, sometimes subcylindrical or lobed, rarely bottle shaped, and 3.0–9.0 × 6.0–14.0 µm. Spore chains are arranged in parallel rows within the sporangia. Formation of sporangia occurs on Czapek agar. The motile spores are rod-shaped (0.5–1.0 × 2.0–4.0 µm).

Good growth occurs on Czapek, peptone Czapek, and tyrosine agars. Colonies are flat, wrinkled, or convoluted, and frequently ridged. The margin is sometimes fimbriate. The color of substrate mycelium is pinkish cinnamon or red-cinnamon to blackish brown, sometimes red with black sectors or very dark with red sectors, flesh-colored, or dirty buff. The marginal areas are usually lighter in color. Soluble yellowish, greenish, and brownish pigments may be produced on various media.

Inositol and raffinose are utilized, but D-mannitol is not utilized. Melanoid pigments are produced.

Source: the type strain was isolated from soil (Sheboygan, Michigan, USA); two further strains from soil were collected in Parfrey's Glen, Wisconsin, USA, and from garden soil (Highfield Cresent, Hindhead, UK). D. Schäfer (Marburg, Germany) isolated a total of 29 strains from soil samples originating from: Germany (11), Corsica (1), Portugal (3), Austria (3), Uruguay (1), England (2), Switzerland (3), Italy (3), Sardinia (1), and the Azores (1) (Kothe, 1987).

DNA G+C content (mol%): 73.0 (T_m).

Type strain: ATCC 15349, DSM 43149, NBRC 12512, JCM 3060, NRRL B-3345, VKM Ac-649.

Sequence accession no. (16S rRNA gene): AB037000.

Additional comments: the type strain shows the highest sequence similarity to the following *Actinoplanes* species (based on ~1470 nt): *Actinoplanes cosettensis* NBRC 14913^T, 98.8%; *Actinoplanes humidus* NBRC 14915^T, 98.8%; *Actinoplanes auranticolor* NBRC 12245^T, 98.7%.

12. **Actinoplanes durhamensis** Goodfellow, Stanton, Simpson and Minnikin 1990a, 320^{VP} (Effective publication: Goodfellow, Stanton, Simpson and Minnikin 1990b, 33.)

dur.ham.en'sis. N.L. masc. adj. *durhamensis* of or belonging to Durham, a city in the north-east of England.

Sporangia are globose; sporogenous hyphae are arranged irregularly within sporangia. Spores are motile.

Light to dark orange substrate mycelium is formed on modified Bennett's agar. Melanin is produced on ISP 7 medium.

Arbutin, casein, lecithin, and starch are degraded; L-arabinose, dextrin, D-fructose, D-galactose, D-glucose, *myo*-inositol, D-lactose, maltose, D-mannitol, D-mannose, melezitose, methyl β-D-glucoside, raffinose, L-rhamnose, D-sorbitol, D-xylose, and *p*-hydroxybenzoic acid are used as sole carbon sources, but not cellobiose, erythritol, catechol, coumarin, *o*-hydroxybenzoic acid, sodium fumarate, sodium succinate, syringaldehyde, syringic acid, or vanillin. Hydrogen sulfide and urease are not formed. Sensitive to a number of antibiotics and chemical inhibitors. Shows activity against *Staphylococcus aureus*. Grows between 4 and 30°C, but not at 37°C.

The peptidoglycan contains *meso*- and hydroxy-diaminopimelic acid. The principal cell-wall sugars are arabinose, galactose, glucose, mannose, and xylose. The organism contains a mixture of straight- and branched-chain fatty acids, with octadecanoic, 14-methylpentadecanoic, and 14-methylhexadecanoic acids predominating. Tetrahydrogenated menaquinones with nine isoprene units are the predominant isoprenologue, and major amounts of diphosphatidylglycerol, phosphatidylethanolamine, phosphatidylglycerol, and phosphatidylinositol are present.

Source: the type strain and four further strains were isolated from sediments of the River Derwent, Allensford, Durham, UK.

DNA G+C content (mol%): 70.8 (HPLC).

Type strain: ATCC 49800, DSM 43939, NBRC 14914, JCM 7625, LA 139, NCIB (NCIMB) 20041, NRRL B-16689.

Sequence accession no. (16S rRNA gene): AB037001.

Additional comments: the type strain shows the highest sequence similarity to the following *Actinoplanes* species (based on ~1470 nt): *Actinoplanes ferrugineus* NBRC 15555^T, 98.3%; *Actinoplanes digitatis* NBRC 12512^T, 97.7%.

13. **Actinoplanes ferrugineus** Palleroni 1979, 55^{AL}

fer.ru.gi'ne.us. L. masc. adj. *ferrugineus* of the color of iron rust.

Sporangia are globose to irregular, 4.0–12.0 µm in diameter, and develop on the surface of substrate mycelium. Abundant production of sporangia is observed on minimal media supplemented with single carbon sources like L-rhamnose, D-galactose, or D-fructose. Spores are spherical (0.9–1.0 µm), motile by means of a tuft of flagella. Rudimentary aerial mycelium may be formed on ISP medium 7.

Abundant growth occurs on various ISP agar media, Czapek-glucose, and potato-glucose agars. Colonies have flat to wrinkled surface, elevated in the center, and edges penetrating into the agar. Colors vary from amber to deep brown. On ISP 7 agar medium, a reddish-brown soluble pigment is produced.

Produces the proline analog L-azetidine-2-carboxylic acid.

Source: the type strain was isolated from a red soil sample collected at Dorrigo Mountain, Australia.

DNA G+C content (mol%): 70.5 (HPLC).

Type strain: ATCC 29868, DSM 43807, NBRC 15555, JCM 3277, NRRL B-16718, X-14695.

Sequence accession no. (16S rRNA gene): AB037002.

Additional comments: the type strain shows the highest sequence similarity to the following *Actinoplanes* species (based on ~1470 nt): *Actinoplanes tereljensis* MN07-A0371^T, 98.6; *Actinoplanes toevensis* MN07-A0368^T, 98.6%.

14. **Actinoplanes globisporus** (Thiemann 1967) Stackebrandt and Kroppenstedt 1988, 328^{VP} (Effectively published: Stackebrandt and Kroppenstedt 1987, 113.) (*Amorphosporangium globisporus* Thiemann 1967, 239)

glo.bi.spo'rus. L. n. *globus* a ball, sphere; Gr. n. *spora* a seed, and in biology a spore; N.L. masc. adj. *globisporus* round spored.

Sporangia (4.0–7.0 µm wide and 3.0–5.0 µm long) develop on short sporangiophores directly from the substrate mycelium, highly irregular in shape, resembling masses of spores not surrounded by a sporangial envelope. Spores are spherical, 0.8–1.0 µm, and motile (*fide* Palleroni, 1989). Production of chlamydospore-like structures is quite frequent.

Colors of the colonies vary from ivory, creamish, pale orange to orange.

Good growth occurs on Bennett's and Hickey–Tresner agars; moderate growth occurs on glucose asparagine, glycerol asparagine, oatmeal, starch, tyrosine, skim milk, and calcium malate agars. No growth is observed on Czapek, peptone iron, or cellulose agars. Grows on synthetic media only if supplemented with vitamin solution.

Source: the type strain was isolated from soil (Appiano Gentile, Como, Italy).

DNA G+C content (mol%): 70.2 (HPLC).

Type strain: ATCC 23056, DSM 43857, NBRC 13912, JCM 3186.

Sequence accession no. (16S rRNA gene): AB037003.

Additional comments: the type strain shows the highest sequence similarity to the following *Actinoplanes* species (based on ~1470 nt): *Actinoplanes deccanensis* NBRC 13994^T, 96.9%; *Actinoplanes ferrugineus* NBRC 15555^T, 96.5%.

15. **Actinoplanes humidus** Goodfellow, Stanton, Simpson and Minnikin 1990a, 320^{VP} (Effective publication: Goodfellow, Stanton, Simpson and Minnikin 1990b, 34.)

hu'mi.dus. L. masc. adj. *humidus* moist, damp, wet.

Abundant production of spherical sporangia; sporogenous hyphae are arranged irregularly within sporangia. Spores are motile.

Light to yellow-orange brown substrate mycelium is formed on modified Bennett's agar and a dark diffusible pigment is produced on Bacto Tryptic soy agar supplemented with RNA.

Degrades arbutin, casein, DNA, elastin, lecithin, pectin, RNA, starch, and tyrosine. Utilizes D- and L-arabinose, dextrin, D-fructose, D-galactose, D-glucose, D-lactose, maltose, D-mannitol, D-mannose, methyl β-D-glucoside, D-sorbitol, D-xylose, *p*-hydroxybenzoic acid, sodium fumarate, and sodium succinate as sole carbon sources, but not cellobiose, erythritol, *myo*-inositol, catechol, coumarin, *m*- and *o*-hydroxybenzoic acid, syringaldehyde, syringic acid, or vanillin. Nitrate is reduced. Tolerant to a range of antibiotics and chemical inhibitors. Grows at 4 and 30°C, but not at 37°C.

Peptidoglycan contains *meso*- and hydroxy-diaminopimelic acid. The main cell-wall sugars are arabinose, galactose, mannose, and xylose. Contains complex mixtures of straight- and branched-chain fatty acids, with straight-chain components predominating. Tetrahydrogenated menaquinones with nine isoprene units are the predominant isoprenologue, and major amounts of diphosphatidylglycerol, phosphatidylethanolamine, phosphatidylglycerol, phosphatidylinositol, and two uncharacterized glycolipids are present.

Source: the type strain and eight further strains were isolated from sediments of the River Derwent (Allensford, Durham, UK).

DNA G+C content (mol%): not known.

Type strain: ATCC 49801, DSM 43938, NBRC 14915, JCM 7555, LA 6, NCIB (now NCIMB) 20000, NRRL B-16690.

Sequence accession no. (16S rRNA gene): AB037004.

Additional comments: the type strain shows the highest sequence similarity to following *Actinoplanes* species (based on ~1470 nt): *Actinoplanes cosettensis* NBRC 14913^T, 100.0%; *Actinoplanes digitatis* NBRC 12512^T, 98.8%.

16. **Actinoplanes italicus** Beretta 1973, 42^{AL}

i.ta'li.cus. L. masc. adj. *italicus* of or pertaining to Italy.

Sporangia are produced abundantly on starch (ISP 4) and skim milk agars, supported by straight sporangiophores arising from substrate mycelium. They are irregular in shape, varying from globose to oval and pyriform, 6.0–11.0 µm in diameter. Sporangiospores are spherical to oval, with a diameter of 1.0–2.0 µm, and highly motile.

Colonies on starch agar have a smooth surface and a dome-shaped center. Cherry-red substrate mycelium is produced on most media, but on ISP 2 and ISP 7 media (Shirling and Gottlieb, 1966), the color is orange; on Hickey–Tresner, Bennett's and nutrient agars, colonies are amber-brown to brown.

Production of a cherry-red soluble pigment is seen on oatmeal and starch agars; colors vary on other media from yellowish pink, rose, and amber to deep orange yellow. No pigment is produced on Bennett's, nutrient, or calcium-malate agar media.

Source: the type strain was isolated from a soil sample collected in an orchard at Pontelongo, Italy.

DNA G+C content (mol%): not known.

Type strain: A 5221, ATCC 27366, DSM 43146, NBRC 13911, JCM 3165, NRRL B-16722.

Sequence accession no. (16S rRNA gene): AB037005.

Additional comments: the type strain shows the highest sequence similarity to the following *Actinoplanes* species (based on ~1470 nt): *Actinoplanes couchii* GW8-1761^T, 99.2%; *Actinoplanes rectilineatus* NBRC 13941^T, 98.6%.

17. **Actinoplanes liguriensis** Wink, Kroppenstedt, Schuhmann, Seibert and Stackebrandt 2006, 2128^{VP} ("*Actinoplanes liguriae*" Parenti, Pagani and Beretta 1976, 505)

li.gu.ri.en'sis. N.L. masc. adj. *liguriensis* of or pertaining to the Italian region of Liguria.

Sporangia are globose to oval, 15.0–25.0 µm in diameter. Spore chains are coiled within the sporangia. Spherical motile spores (1.5–2.0 µm) with smooth surfaces are found. A rudimentary aerial mycelium is formed on ISP 6 medium. Abundant growth occurs on various agar media. Color of substrate mycelium is yellow-orange on ISP media 2, 3, 4, 5, 6, and 7. A yellow soluble pigment is produced on ISP media 4 and 5, and a red one on ISP 7.

Peptidoglycan contains *meso*-diaminopimelic acid; diagnostic sugars are xylose and arabinose. MK-9(H_4) and MK-10(H_4) are the principal menaquinones. The major phospholipid is phosphatidylethanolamine.

Produces the antibiotics gardimycin (Parenti and Coronelli, 1979) and Ala (*O*)-actagardine (Vértesy et al., 1999).

Source: the type strain was isolated from garden soil, Liguria region, Italy (Parenti et al., 1976).

DNA G+C content (mol%): not known.

Type strain: A/6353, ATCC 31048, BCRC 12121, CBS 355.75, DSM 43865, FH 2244, JCM 3250, KCC A-0250, NRRL B-16723.

Sequence accession no. (16S rRNA gene): AB047497.

Additional comments: the type strain shows the highest sequence similarity to the following *Actinoplanes* species (based on ~1470 nt): *Actinoplanes rectilineatus* NBRC 13941^T, 98.7%; *Actinoplanes regularis* NBRC 12514^T, 98.7%; *Actinoplanes palleronii* NBRC 14916^T, 98.5%.

18. **Actinoplanes lobatus** (Couch 1963) Stackebrandt and Kroppenstedt 1988, 328^{VP} (Effective publication: Stackebrandt and Kroppenstedt 1987, 113.) ["*Ampullaria lobata*" Couch 1963, 59; *Ampullariella lobata* (Couch 1963) Couch 1964, 29]

lo'ba.tus. N.L. masc. adj. *lobatus* lobed.

Strains have sporangia that vary in size and shape, are typically lobed and/or irregular, sometimes cylindrical, and are 4.0–20.0 × 12.0–23.0 µm. They often have a papillate appearance at the proximal end because of the varying length of the spore chains, which are arranged in parallel rows within the sporangia. The lobed sporangia sometimes are divided in several parts, giving the appearance of fused sporangia. Formation of sporangia occurs only on Czapek agar. The motile spores are rod-shaped (0.5–1.0 × 2.0–4.0 µm). Oval moniliform conidia may also be produced.

Colonies are up to 25 mm in diameter on Czapek agar or peptone Czapek agar and up to 10 mm on casein or tyrosine agars after 6 weeks' incubation. They are flattish or convoluted, and frequently ridged. The color of the substrate mycelium varies from coral red to dragon's blood red, jasper red, coral red, and brick red. Albino cultures may occur. On casein agar, the color is light coral red to ferruginous. Soluble pale yellowish green pigments are produced on Czapek agar. Casein and tyrosine agars are slightly darkened. D-Mannitol and raffinose are utilized, but not inositol. Gelatin is liquefied.

Source: the type strain and a further strain were isolated from soil samples collected at Madison, Wisconsin, USA. Another isolate originated from pasture near Charleston, West Virginia, USA.

DNA G+C content (mol%): not known.

Type strain: ATCC 15350, DSM 43150, NBRC 12513, JCM 3061, NRRL B-3346, VKM Ac-676.

Sequence accession no. (16S rRNA gene): AB037006.

Additional comments: the type strain shows the highest sequence similarity to the following *Actinoplanes* species (based on ~1470 nt): *Actinoplanes sichuanensis* 03-723^T, 99.5%; *Actinoplanes campanulatus* NBRC 12511^T, 99.3%.

19. **Actinoplanes missouriensis** Couch 1963, 69^{AL}

mis.sou.ri.en'sis. N.L. masc. adj. *missouriensis* of or pertaining to the state of Missouri.

Sporangia are subglobose to globose, sometimes irregular, 6.0–14.0 µm in diameter. Spores are arranged in irregular coils within the sporangia, subspherical (1.0–1.2 µm), weakly motile by a tuft of polarly inserted flagella. Hyphae are less than 1.0 µm in diameter and branched. Production of palisade hyphae is rare.

Growth on Czapek agar is very good; colonies have an elevated center and flat border. Color is mostly ochraceous salmon, with whitish areas on the surface, where sporangia are abundant. Also, very good growth is observed on peptone Czapek agar (color zinc-orange to ochraceous orange). Production of sporangia is restricted to limited areas.

Produces the antibiotic 5-azacytidine.

Source: the type strain was isolated from soil collected near Hamilton, Missouri, USA. Members of the species also appeared in ten soil collections from the Mississippi Valley to the West Coast, USA.

DNA G+C content (mol%): not known.

Type strain: ATCC 14538, DSM 43046, NBRC 13243, JCM 3121, NBRC 102363, NRRL B-3342, UNCC 431.

Sequence accession no. (16S rRNA gene): AB037008.

Additional comments: the type strain shows the highest sequence similarity to the following *Actinoplanes* species (based on ~1470 nt): *Actinoplanes auranticolor* NBRC 12245^T, 99.2%; *Actinoplanes cosettensis* NBRC 14913^T, 98.6%; *Actinoplanes digitatis* NBRC 12512^T, 98.6%; *Actinoplanes humidus* NBRC 14915^T, 98.6%.

20. **Actinoplanes palleronii** Goodfellow, Stanton, Simpson and Minnikin 1990a, 320^{VP} (Effective publication: Goodfellow, Stanton, Simpson and Minnikin 1990b, 34.)

pal.le.ro'ni.i. N.L. gen. masc. n. *palleronii* of Palleroni, named after Noberto Palleroni, a distinguished taxonomist.

Sporangia are spherical; sporogenous hyphae are arranged irregularly within the sporangia. Spores are motile.

Light to yellow-brown substrate mycelium is formed on modified Bennett's agar. Melanoid pigments are produced on ISP 7 medium. Degrades arbutin, casein, lecithin, pectin, and starch; utilizes D-fructose, D-glucose, maltose, D-mannose, and D-xylose as sole carbon sources, but not D-arabinose, cellobiose, dextrin, erythritol, D-galactose, methyl β-D-glucoside, D-ribose, D-sorbitol, catechol, coumarin, *m*-, *o*- or *p*-hydroxybenzoic acid, sodium acetate, sodium fumarate, sodium succinate, syringaldehyde, syringic acid, or vanillin. Hydrogen sulfide and urease are not formed. Sensitive to potassium tellurite (0.005%, w/v), gentamicin sulfate (0.5%, w/v), and methacycline hypochloride (0.25%, w/v). Grows between 4 and 30°C, but not at 37°C.

The peptidoglycan contains *meso*- and hydroxy-diaminopimelic acid. The principal wall sugars are arabinose, galactose, glucose, rhamnose, ribose, and xylose. Contains complex mixtures of straight- and branched-chain fatty acids, with 12-methyltetradecanoic acid predominating. Tetrahydrogenated menaquinones with nine isoprene units are the predominant isoprenologue, and major amounts of diphosphatidylglycerol, phosphatidylethanolamine, phosphatidylglycerol, and phosphatidylinositol are present.

Source: the type strain and five further strains were isolated from sediments of the River Derwent at Allensford, Durham, UK.

DNA G+C content (mol%): not known.

Type strain: ATCC 49797, DSM 43940, NBRC 14916, JCM 7626, LA 83, NCIB (now NCIMB) 20021, NRRL B-16691.

Sequence accession no. (16S rRNA gene): AB037009.

Additional comments: the type strain shows the highest sequence similarity to the following *Actinoplanes* species (based on ~1470 nt): *Actinoplanes regularis* NBRC 12514^T, 99.4%; *Actinoplanes rectilineatus* NBRC 13941^T, 99.0%.

21. **Actinoplanes rectilineatus** Lechevalier and Lechevalier 1975, 371^{AL}

rec.ti.li.ne.a'tus. L. adj. *rectus* straight, upright; L. adj. *lineatus* striped, marked by fine parallel lines; N.L. masc. adj. *rectilineatus* marked with straight lines.

Sporangia are cylindrical, 8.0–15.0 µm, containing straight, longitudinally arranged rows of spores. Sporangiospores are globose, 1.5–2.0 µm in diameter, motile, and equipped with a polarly inserted tuft of about 30 flagella.

Color of the colonies vary from white, grayish tan, yellow-tan to yellow-brown (Lechevalier and Lechevalier, 1975), or pale, ochre, light-orange to orange (Kothe, 1987), depending on the agar medium used. White aerial mycelium is formed on some media. Grows between 10 and 37°C; abundant growth is observed at 23–28°C.

Source: the type strain was isolated from garden soil collected in Somerset, New Jersey, USA. The effective description (not validly published) of the species "*Actinoplanes penicillatus*" (Schäfer, 1973) was based on 15 strains, isolated from soil and characterized as having the same morphological features as *Actinoplanes rectilineatus*. The strains originated from Kenya (12), Australia (1, type strain), Corsica (1), and Spain (1).

DNA G+C content (mol%): 69 (T_m).

Type strain: ATCC 29234, DSM 43808, NBRC 13941, IMRU 3919, JCM 3194, NRRL B-16090, LL 7-10.

Sequence accession no. (16S rRNA gene): AB037010.

Additional comments: the type strain shows the highest sequence similarity to the following *Actinoplanes* species (based on ~1470 nt): *Actinoplanes couchii* GW8-1761^T, 98.9%; *Actinoplanes liguriensis* NBRC 13997^T, 98.7%.

22. **Actinoplanes regularis** (Couch 1963) Stackebrandt and Kroppenstedt 1988, 328[VP] (Effective publication: Stackebrandt and Kroppenstedt 1987, 113.) ["*Ampullaria regularis*" Couch 1963, 57; *Ampullariella regularis* (Couch 1963) Couch 1964, 29]

re.gu.la'ris. L. masc. adj. *regularis* regular.

Sporangia are mostly cylindrical, measuring 5.0–14.0 × 8.0–30.0 µm. The base of the sporangia is frequently mound shaped and resembles a corked bottle. Formation of sporangia occurs on Czapek and casein agars. Spores are arranged in parallel rows inside the sporangia, and are rod-shaped, 0.5–1.0 µm wide and 2.0–4.0 µm long, and motile by a polarly inserted tuft of flagella. Brushlike conidiophores may be produced on Czapek agar.

Good growth occurs on Czapek, peptone Czapek, casein, and tyrosine agars. Colonies are flat, frequently convoluted in the center with radial ridges on the margin. The color of the substrate mycelium is orange, red, brownish, ochre, or salmon to coral-pink, sometimes with white or gray areas on the surface, where the sporangia are produced. Soluble yellowish, greenish, or brownish (perhaps melanoid) pigments may be produced on various media.

Inositol, D-mannitol, and raffinose are not utilized by the type strain as sole carbon sources. Gelatin is liquefied.

Three strain variants are distinguishable on casein and tyrosine agars (Couch, 1963). The first variant utilizes casein and tyrosine and darkens both agars. The second variant utilizes casein and tyrosine, but darkens only the tyrosine agar. The third variant utilizes casein and tyrosine, but darkens only casein agar. The type strain of *Actinoplanes regularis* is from the first variant.

Source: the type strain was isolated from soil collected from a white pine grove, University of Wisconsin Arboretum, Madison, Wisconsin, USA. A further 28 strains were isolated, most from forest soils from North and South Carolina, Virginia, Mississippi, Wisconsin and Canada. D. Schäfer (Marburg, Germany) isolated a total of 26 strains from soil samples, originating from Germany (1), Kenya (3), Ceylon (1), Corsica (2), Tenerife (1), Austria (1), USA (3), Argentina (2), the Philippines (2), Taiwan (1), Uruguay (1), Cameroon (1), Mexico (1), France (2), the Azores (1), South Africa (1), and Yugoslavia (1) (Kothe, 1987).

DNA G+C content (mol%): 72.3 (Bd).

Type strain: CBS 193.64, DSM 43151, NBRC 12514, JCM 3062, NRRL B-3347, UNCC 65, VKM Ac-650.

Sequence accession no. (16S rRNA gene): AB037011.

Additional comments: the type strain shows the highest sequence similarity to the following *Actinoplanes* species (based on ~1470 nt): *Actinoplanes palleronii* NBRC 14916[T], 99.4%; *Actinoplanes rectilineatus* NBRC 13941[T], 98.9%.

23. **Actinoplanes sichuanensis** Sun, Dong, Zhang, Wei, Li, Yu, Klenk and Zhang 2009, 2766[VP]

si.chu.a.nen'sis. N.L. masc. adj. *sichuanensis* of or pertaining to Sichuan Province, southwest of China, soil of which was the source of the type strain.

Spherical sporangia are formed. The sporangiospores are motile and the surface is smooth. Aerial mycelium is absent. Color of substrate mycelium is cream to reddish orange on ISP 2, 3, 4, 5, 6, 7, yeast extract-starch agar, nutrient agar, Czapek solution agar, and modified Bennett's agar. Grows in the presence of 0–4% NaCl, but not in 5–10% NaCl. Growth occurs at initial pH of 6.5–10.5 and between 10 and 37°C, but not at pH 4.5–6.0 or at 40°C. Soluble pigments and melanin are not formed.

Esculin is hydrolyzed and nitrate is reduced to nitrite. Amylase and gelatinase are produced. H_2S is not produced. MK-9(H_4) (94.4%) is the predominant menaquinone, with MK-9(H_6) (3.7%) and MK-10(H_6) (1.9%) present as minor components. The major phospholipids are phosphatidylethanolamine and phosphatidylinositol. The predominant cellular fatty acids (>10%) are $C_{15:0}$ iso (22.2%), $C_{16:0}$ iso (15.3%) and *cis*-9-$C_{17:1}$ (11.1%); other fatty acids occurring in relatively small amounts (>5%) are $C_{14:0}$ (6.2%), $C_{16:0}$ (7.7%) and *cis*-9-$C_{18:1}$ (7.5%).

The organism is sensitive to filter-paper disks soaked in 1 ml of (µg/ml) erythromycin (15), gentamicin (10), and kanamycin (30), whereas it is resistant to ciprofloxacin (5), novobiocin (5), oxacillin (1), streptomycin (10), and tobramycin (10).

Source: the type strain was isolated from soil (Sichuan Province, China).

DNA G+C content (mol%): 70.4 (T_m).

Type strain: 03-723, CCM 7526, KCTC 19460.

Sequence accession no. (16S rRNA gene): EU531458.

Additional comments: the type strain shows the highest sequence similarity to the following *Actinoplanes* species (based on ~1470 nt): *Actinoplanes lobatus* NBRC 12513[T], 99.5%; *Actinoplanes philippinensis* NBRC 13878[T], 99.0%.

24. **Actinoplanes teichomyceticus** Wink, Kroppenstedt, Schuhmann, Seibert and Stackebrandt 2006, 2129[VP]

tei.cho.my.ce'ti.cus. Gr. n. *teichos* wall; Gr. n. *mukês -êtos* fungus; L. adj. suff. *-icus* suffix used with the sense of belonging to; N.L. masc. adj. *teichomyceticus* literally belonging to a fungus cell wall (referring to inhibition of cell-wall synthesis by teichomycin, produced by the type strain).

Spherical to oval sporangia (15.0–25.0 µm) are produced abundantly on many agar media, mainly on the dome of the colonies. Spores are highly motile, globose to oval, and 1.5–2.0 µm in diameter.

Abundant growth occurs on various agar media. Colonies may have a central protuberance or dome, and a smooth to wrinkled surface. Color of substrate mycelium may vary from beige on ISP media 5 and 6, pastel yellow on ISP media 2, 3, 4, and 7, and light orange on Bennett's and Czapek agars to deep orange on peptone glucose agar. On Hickey–Tresner, potato, and skim milk agars, it is light brownish to amber.

White aerial mycelium is produced on ISP 3 medium and Hickey–Tresner agar, rudimentary aerial mycelium with light pink or rose tinge is formed on Bennett's, Czapek glucose, and glycerol asparagine agars. A red-colored soluble pigment is formed on ISP medium 5, a yellow pigment on ISP media 4 and 5, and brown (melanoid) on ISP media 3 and 6.

Peptidoglycan contains *meso*-diaminopimelic acid; diagnostic sugars are xylose and arabinose; MK-9(H_4) and MK-10(H_4) are the principal menaquinones. Major phospholipid is phosphatidylethanolamine.

Produces the antibiotic teicoplanin RS-1 to RS-4, originally described as teichomycin A_1 and A_2 (Parenti et al., 1978).

Source: the type strain was isolated from a soil sample collected at Nimodi Village, Indore, India (Parenti et al., 1978).

DNA G+C content (mol%): not known.

Type strain: AB8327, ATCC 31121, BCRC 12106, DSM 43866, FH 2149, NBRC 13999, JCM 3252, KCC A-0252, NCIMB 12640, NRRL B-16726.

Sequence accession no. (16S rRNA gene): AB047513.

Additional comments: the type strain shows the highest sequence similarity to the following *Actinoplanes* species (based on ~1470 nt): *Actinoplanes liguriensis* NBRC 13997^T, 98.4%; *Actinoplanes palleronii* NBRC 14916^T, 98.3%. An earlier description of "*Actinoplanes teichomyceticus*" (Parenti, Beretta, Berti and Arioli 1978, 277) was not included in the Approved Lists of Bacterial Names (Skerman et al., 1980) and therefore treated in *Bergey's Manual of Systematic Bacteriology* (1st edition) under "*Species incertae sedis*" (Palleroni, 1989). Both species descriptions are based on the same type strain.

25. **Actinoplanes tereljensis** Ara, Yamamura, Tsetseg, Daram and Ando 2010, 925[VP]

te.rel.jen'sis. N.L. masc. adj. *tereljensis* of or pertaining to the Mongolian region of Terelj, where the type strain was isolated.

Numerous irregular sporangia are formed on the surface of water agar, ISP 3 and 4, and Bennett's agar media; spores are motile. Cells grow well on ISP 2, 3, and 7, Bennett's, and yeast extract-starch agars, forming well-developed, extensively branched, and non-fragmented substrate hyphae. Optimal growth is at 20–28°C; no growth occurs at 37°C. Growth pH is 6–9. Tolerates up to 3% NaCl. Color of substrate mycelium is moderate grayish brown to dark brownish red on different agar media. A pale pinkish brown soluble pigment produced on ISP 7 and aerial mycelium is not observed.

The menaquinone system consists of MK-9(H_6) (85%) with a minor amount of MK-9(H_8) (15%). The polar lipid profile contains diphosphatidylglycerol, phosphatidylethanolamine, phosphatidyl-*N*-methylethylethanolamine, and phosphatidylinositol; phosphatidylcholine and phosphatidylglycerol have not been detected (phospholipid type PII). Major fatty acid methyl esters are $C_{16:0}$ iso (19%), $C_{17:0}$ anteiso (15%), $C_{15:0}$ anteiso (12%), and $C_{15:0}$ iso (11%).

Source: the type strain was isolated from soil (Terelj, Mongolia).

DNA G+C content (mol%): 70.6 (HPLC).

Type strain: MN07-A0371, NBRC 105297, VTCC D9-010.

Sequence accession no. (16S rRNA gene): AB468944.

Additional comments: the type strain shows the highest sequence similarity to the following *Actinoplanes* species (based on ~1470 nt): *Actinoplanes toevensis* MN07-A0368^T, 99.7%; *Actinoplanes ferrugineus* NBRC 15555^T (AB037002), 98.6%.

26. **Actinoplanes toevensis** Ara, Yamamura, Tsetseg, Daram and Ando 2010, 924[VP]

to.e.ven'sis. N.L. masc. adj. *toevensis* of or pertaining to the Mongolian region of Töv Province, where the type strain was isolated.

Numerous globose to oval sporangia are formed on the surface of water agar, ISP 3 and 4, and Bennett's agar media. Spores formed in the sporangium are motile. Cells grow well on ISP 2, 3, and 4, Bennett's, and yeast extract-starch agar, forming well-developed, extensively branched, and non-fragmented substrate hyphae. Aerial mycelium is not observed. Optimal growth is at 25–30°C; no growth occurs at 40°C. Growth pH is 6–11. Tolerates up to 2% NaCl. Color of substrate mycelium is moderate yellow brown to orange brown on different agar media. A pale pinkish brown soluble pigment is produced on ISP 7 agar.

The menaquinone system consists of MK-9(H_6) (97%) with a minor amount of MK-9(H_8) (3%). The polar lipid profile contains diphosphatidylglycerol, phosphatidylethanolamine, phosphatidyl-*N*-methylethylethanolamine, and phosphatidylinositol; phosphatidylcholine and phosphatidylglycerol have not been detected (phospholipid type PII). Major fatty acid methyl esters are $C_{17:0}$ anteiso (25%), $C_{16:0}$ iso (22%), $C_{15:0}$ iso (13%), and $C_{15:0}$ anteiso (12%).

Source: the type strain was isolated from soil (Töv Province, Mongolia).

DNA G+C content (mol%): 70.6 (HPLC).

Type strain: MN07-A0368, NBRC 105298, VTCC D9-011.

Sequence accession no. (16S rRNA gene): AB468943.

Additional comments: the type strain shows the highest sequence similarity to the following *Actinoplanes* species (based on ~1470 nt): *Actinoplanes tereljensis* MN07-A0371^T, 99.7%; *Actinoplanes ferrugineus* NBRC 15555^T, 98.6%.

27. **Actinoplanes utahensis** Couch 1963, 67[AL]

u.tah.en'sis. N.L. masc. adj. *utahensis* of or pertaining to the state of Utah.

Sporangia are very irregular in size and shape, lobed, pyriform, club-shaped, or digitate, 5.0–18.0 µm in diameter. Spores are arranged in irregular coils within the sporangia, subglobose, 1.0–2.0 µm in diameter, motile by polarly inserted polytrichous flagella. Microspores may be produced on peptone Czapek agar.

Good growth occurs on Czapek, peptone Czapek, casein, and tyrosine agars. Colonies are convoluted or flat; surfaces are slippery moist with minute or conspicuous bumps. Sporangial development is sparse and only on Czapek, oatmeal, and starch-casein agars. Color of the substrate mycelium is apricot-orange to salmon-orange. On peptone Czapek agar, the color may become ferrugineus toward the center. Substrate mycelium on oatmeal agar is amber colored and on peptone Czapek agar is slightly darkened.

Produces cyclic peptide antibiotics.

Source: the type strain was isolated from soil taken from Liberty Park, Salt Lake City, Utah, USA, together with two isolates from distinct samples from the same site. A further isolate originated from a soil sample collected between Carlin and Dunphy, Nevada, USA.

DNA G+C content (mol%): 70 (T_m).

Type strain: ATCC 14539, DSM 43147, NBRC 13244, JCM 3122, NRRL B-16727, UNCC 260, VKM Ac-674.

Sequence accession no. (16S rRNA gene): AB037012.

Additional comments: the type strain shows the highest sequence similarity to the following *Actinoplanes* species (based on ~1470 nt): *Actinoplanes regularis* NBRC 12514^T, 98.7%; *Actinoplanes palleronii* NBRC 14916^T, 98.6%.

28. **Actinoplanes xinjiangensis** Sun, Dong, Zhang, Wei, Li, Yu, Klenk and Zhang 2009, 2767[VP]

xin.ji.an.gen'sis. N.L. masc. adj. *xinjiangensis* of or pertaining to Xinjiang Uyghur Autonomous Region, north-west China, soil of which was the source of the type strain.

Spherical to oval sporangia are formed. The sporangiospores are motile and the surface is almost smooth. Aerial mycelium is absent.

Grows well on ISP 2, 3, 4, 5, and 7, yeast extract-starch agar, nutrient agar, Czapek solution agar, and modified Bennett's agar; grows moderately on ISP 6 agar. Grows in the presence of 0–7% NaCl. Growth occurs at an initial pH value of 6.5–8.5 and between 10 and 37°C.

Esculin is hydrolyzed and nitrate is reduced to nitrite. Amylase is not produced but gelatinase is produced. H_2S production is positive. Melanin is produced on ISP 6 agar, but soluble pigment is not formed.

MK-9(H_4) (96.1%) is the predominant menaquinone and MK-10(H_6) (3.9%) is present as a minor component. The major phospholipid is phosphatidylethanolamine. The predominant cellular fatty acids (>10%) are $C_{15:0}$ iso (10.8%), $C_{16:0}$ iso (16.0%) and *cis*-9-$C_{17:1}$ (21.5%); other fatty acids occurring in relatively small amounts (>5%) are $C_{15:0}$ (6.7%), $C_{16:0}$ (6.5%) and *cis*-9-$C_{16:1}$ (5.9%).

The organism is sensitive to filter-paper disks soaked in 1 ml of (μg/ml) kanamycin (30), but resistant to (1 ml; μg/ml) ciprofloxacin (5), erythromycin (15), gentamicin (10), novobiocin (5), oxacillin (1), streptomycin (10), and tobramycin (10).

Source: the type strain was isolated from soil from Xinjiang Uyghur Autonomous Region, north-west China.

DNA G+C content (mol%): 71.0 (T_m).

Type strain: 03-8772, CCM 7527, KCTC 19461.

Sequence accession no. (16S rRNA gene): EU531457.

Additional comments: the type strain shows the highest sequence similarity to the following *Actinoplanes* species (based on ~1470 nt): *Actinoplanes sichuanensis* 03-723^T, 99.0%; *Actinoplanes lobatus* NBRC 12513^T, 98.9%.

Species and subspecies *incertae sedis*

The following list contains species and subspecies of *Actinoplanes*, *Ampullariella*, and *Amorphosporangium* that are effectively, but not validly published.

1. **"Actinoplanes arizonaensis"** Karwowski, Jackson, Theriault, Prokop, Maus, Hansen and Hensey 1988, 1210
   Produces arizonins.
   Type strain: AB660-122, ATCC 49796, NBRC 14837, JCM 9648.

2. **"Actinoplanes awaijnensis"** Torikata and Enokita 1978 *in* Torikata, Enokita, Imai, Itoh, Nakajima, Haneishi and Arai (1978)
   Produces 5-azacytidine.
   Type strain: ATCC 33917, JCM 9334, SANK 90277.

3. **"Actinoplanes awaijnensis subsp. mycoplanecinus"** Torikata, Enokita, Okazaki, Nakajima, Iwado, Haneishi and Arai 1983, 959
   Produces mycoplanecins.
   Type strain: ATCC 33919, NBRC 14279, JCM 6112.

4. **"Actinoplanes coloradoensis"** Jackson, Karwowski, Theriault, Fernandes, Semon and Kohl 1987, 1381
   Produces coloradocin.
   Type strain: AB 921J-26.

5. **"Actinoplanes deccanensis subsp. azaserinus"** Torikata, Enokita, Imai, Itoh, Nakajima, Haneishi and Arai (1978)
   Produces azaserine.
   Type strain: ATCC 33916, JCM 9333.

6. **"Actinoplanes friuliensis"** Aretz, Meiwes, Seibert, Vobis and Wink 2000, 813
   Produces friulimycins.
   Type strain: DSM 7358, HAG 010964.
   Additional comment: strain HAG 010964 (= DSM 7358) is a patent strain (EP 0 629 636, Hoechst AG, 1994). This strain can be obtained from the DSMZ for scientific research by a special agreement. According to Rules 27(3) and 30, "*Actinoplanes friuliensis*" is not validly published because, at the time of publication, the type strain was not deposited in two publicly accessible service collections in different countries.

7. **"Actinoplanes garbadinensis"** Parenti, Pagani and Beretta 1976, 502
   Produces gardimycin.
   Type strain: ATCC 31049, NBRC 13995, Lepetit A/10889.

8. **"Actinoplanes ianthinogenes"** Coronelli, Pagani, Bardone and Lancini 1974, 161
   Produces purpuromycin.
   Type strain: ATCC 21884, NBRC 13996, Lepetit A/1668.

9. **"Actinoplanes ianthinogenes subsp. octamycini"** Gause and Sveshnikova 1979 *in* Gause, Sveshnikova, Maksimova and Olkhovatova 1979
   Produces purpuromycin and octamycin.
   Type strain: ATCC 43632, NBRC 14524, JCM 9649.

10. **"Actinoplanes nipponensis"** Routien 1977 *in* Celmer, Moppett, Cullen, Routien, Jefferson, Shibakawa and Tone 1977a
    Produces antibiotic 41012.
    Type strain: ATCC 31145, DSM 43867, NBRC 14063, JCM 3264.

11. **"Actinoplanes nirasakinensis"** Torikata, Enokita, Imai, Itoh, Nakajima, Haneishi and Arai 1978, 92
    Produces actinomycins.
    Type strain: ATCC 33918, JCM 9335.

12. **"Actinoplanes pallidoaurantiacus"** Ruan, Zhang and Jiang 1976, 297
    Type strain: DSM 46145, NBRC 13968, JCM 3242.

13. **"Actinoplanes penicillatus"** Schäfer 1973, 185
    Type strain: CBS 558.75, DSM 46142, SE 2.
    Additional comments: the species "*Actinoplanes penicillatus*" (type strain: SE 2^T), was described by Schäfer (1973), describing

the same morphological features as those described for *Actinoplanes rectilineatus* (type strain: 7-10). A second investigation of both type strains (7-10 and SE 2) concluded in the description of a novel subspecies "*Actinoplanes rectilineatus* subsp. *penicillatus*" (*ex* Schäfer) nov. (sic!) rev. comb. nov. (Kothe, 1987).

14. **"Actinoplanes purpeobrunneus"** Ruan and Jiang 1979, 236
    Type strain: NBRC 14020, JCM 3253, A58.
15. **"Actinoplanes pyriformis"** Ruan and Jiang 1979, 236
    Type strain: NBRC 14030, JCM 3262, A68.
16. **"Actinoplanes rectilineatus subsp. penicillatus"** Kothe 1987, 70
    See: "*Actinoplanes penicillatus*".
17. **"Actinoplanes roseosporangius"** Ruan, Zhang and Jiang 1976, 292
    Type strain: DSM 46143, NBRC 13969, JCM 3243, 71-C29.
18. **"Actinoplanes rutilosporangius"** Ruan, Zhang and Jiang 1976, 298
    Type strain: DSM 46151, NBRC 13970, JCM 3244, 71-C6.
19. **"Actinoplanes sarveparensis"** Japanese Patent (Kokai) 53-2402, 1978
    Produces antibiotic L 13365.
    Type strain: Lepetit A/13826, DSM 43901, NBRC 13993.
20. **"Actinoplanes tuftoflagellus"** Ruan and Jiang 1979, 235
    Type strain: NBRC 14021, JCM 3254, A5.
21. **"Actinoplanes violaceus"** Jiang, Xu and Ruan 1983a
    Type strain: ATCC 43537, NBRC 14458, JCM 3353, Y80-610.
22. **"Actinoplanes yunnanensis"** Jiang, Xu and Ruan 1983b, 212
    Type strain: ATCC 43538, NBRC 14459, JCM 3354, Y79-21.
23. **"Ampullariella kinshanensis"** Ruan and Zhang 1974, 35
    Type strain: DSM 46147, NBRC 13661, JCM 3173, 71-C11.
24. **"Ampullariella kunmingensis"** Jiang, Xu and Ruan 1983b, 214
    Type strain: ATCC 43539, JCM 3355, Y79-15.
25. **"Ampullariella regularis subsp. intermedia"** Nonomura, Iino and Hayakawa 1979, 84
    Type strain: DSM 43898, NBRC 14065, JCM 3235.
26. **"Ampullariella regularis subsp. mannitophila"** Itoh, Enokita, Okazaki, Iwado, Torikata, Haneishi and Arai 1981, 930
    Produces candiplanecin.
    Type strain: ATCC 33986, JCM 9336.
27. **"Ampullariella violaceochromogenes"** Nonomura, Iino and Hayakawa 1979, 84
    Type strain: DSM 43899, NBRC 14066, JCM 3236.
28. **"Amorphosporangium castaneum"** Jiang and Yan 1984
    Type strain: ATCC 43631, DSM 43914, NBRC 14428, JCM 3341, B-133.

Genus IV. **Asanoa** Lee and Hah 2002, 970^{VP}

The Editorial Board

As.a.n'a. N.L. fem. n. *Asanoa* named after Kozo Asano, the Japanese microbiologist who made the original description of the genus *Catellatospora*.

Weakly sporulating branched vegetative hyphae (0.3–0.4 µm in diameter). **Gram-stain-positive**. Nonmotile. **Aerobic**. Mesophilic. Cell wall contains diamino acids *meso*-diaminopimelic acid and 3-hydroxydiaminopimelic acid. Orange colony mass. Sporulation only occurs on tap-water agar and glycerol/calcium malate agar. Aerial mycelium and globose bodies are not observed. Possess a glycolylated peptidoglycan and whole-cell sugars of arabinose, rhamnose, ribose, xylose, galactose, mannose and glucose. Catalase-positive. Urease-negative. Nitrate is not reduced to nitrite. H_2S is not produced. Mycolic acids are not present. MK-10(H_6, H_8) are the major menaquinones. Polar lipid profile comprises phosphatidylethanolamine (a phospholipid type PII pattern). The fatty acid pattern is 2d type.

DNA G+C content (mol%): 69–71.5.

Type species: **Asanoa ferruginea** Lee and Hah 2002, 970^{VP}.

Further descriptive information

Phylogenetic analysis of the 16S rRNA gene positions the genus within the family *Micromonosporaceae*. The closest phylogenetic neighbor is *Micromonospora auratinigra* strain TT1-11 (97.8% sequence similarity, accession no. NR028659), which was isolated from a peat swamp forest in Thailand (Thawai et al., 2004b). An environmental clone with high 16S rRNA gene sequence similarity has been detected in cellulosic waste from a simulated low-level-radioactive-waste site (98.4 %, accession no. GQ263486) (Field et al., 2010).

Enrichment and isolation procedures

Asanoa ferruginea strain 6257-C^T was isolated by a dilution method from woodland soil samples collected in Yamanashi, Japan (Asano and Kawamoto, 1986).

Maintenance procedures

Strains are cultivated on oatmeal agar (International *Streptomyces* Project, ISP, medium 3), inorganic salts/starch agar (ISP medium 4) and tap-water agar at 28°C for 21 d or on yeast extract/glucose broth for 3 d at 30°C. Stock cultures are maintained on yeast extract/malt extract agar (ISP medium 2) at 4°C or as a suspension in 20% (v/v) glycerol at −20°C (Asano and Kawamoto, 1986).

Differentiation of the genus *Asanoa* from closely related genera

Production of chains of nonmotile spores only from the vegetative mycelium differentiates *Asanoa* from phylogenetically related genera: *Micromonospora* produce single nonmotile spores on the vegetative hyphae (Luedemann, 1974); *Actinoplanes, Ampullariella, Amorphosporangiurn,* and *Pilimelia* form unique globose or cylindrical sporangia containing numerous motile spores (Couch, 1950, 1963, 1964); *Dactylosporangiurn* produce finger-shaped sporangia containing three to four motile spores arranged in a single row; and *Glycomyces* form chains of nonmotile spores on the aerial mycelium (Asano and Kawamoto, 1986). *Asanoa* possess major menaquinones MK-10(H_6, H_8), while members of the related genus *Catellatospora* have the major menaquinones MK-9(H_4, H_6) or MK-10(H_4).

List of species of the genus *Asanoa*

1. **Asanoa ferruginea** (Asano and Kawamoto 1986) Lee and Hah 2002, 970[VP] (Basonym: *Catellatospora ferruginea* Asano and Kawamoto 1986, 516.)

fer.ru.gi'ne.a. L. fem. adj. *ferruginea* rust-colored.

Growth occurs at mesophilic temperature. Does not grow on 3% NaCl or 0.01% lysozyme. Grows on 0.001% brilliant green and 0.0001% crystal violet. Antibiotic susceptibility (μg/ml): 50, gentamicin; 5, neomycin; and 100, streptomycin. Antibiotic resistance (μg/ml): 50, novobiocin; 20, vancomycin; and 10, tetracycline. Utilizes D-arabinose, L-arabinose, dextran, D-cellobiose, D-fructose, D-galactose, D-glucose, D-lactose, maltose, D-mannose, melibiose, methyl α-D-glucoside, D-raffinose, L-rhamnose, D-ribose, salicin, starch, sucrose, D-trehalose, D-xylose, adonitol, and D-mannitol. Does not utilize gluconate, inulin, D-melezitose, L-sorbose, dulcitol, butanol, *meso*-erythritol, ethanol, glycerol, *meso*-inositol, 2-propanol, 1-propanol, D-sorbitol, or D-xylitol. Hydrolyzes elastin and starch, but not casein, DNA, or gelatin. Does not decompose adenine, hippurate, hypoxanthine, DL-tyrosine, or xanthine. The phospholipid profile comprises diphosphatidylglycerol, phosphatidylinositol, phosphatidylinositol mannoside, phosphatidylethanolamine, phosphatidylglycerol, and an unknown phospholipid. The major fatty acids are $C_{15:0}$ anteiso (21%), $C_{17:0}$ (20%), $C_{16:0}$ iso (18%), $C_{17:1}$ (15%), and $C_{15:0}$ iso (10%) acids. Minor fatty acids are $C_{15:0}$ (5%), $C_{18:0}$ (4%), $C_{16:0}$ (3%), $C_{14:0}$ iso (2%), and $C_{17:0}$ iso (1%).

For all other characteristics, refer to the genus description.

DNA G+C content (mol%): 71.5 (HPLC).

Type strain: IMSNU 22009, NBRC 14496, DSM 44099.

Sequence accession no. (16S rRNA gene): AF152108.

2. **Asanoa iriomotensis** Tamura and Sakane 2005, 726[VP]

i.ri.o.mo.ten'sis. N.L. fem. adj. *iriomotensis* of or belonging to Iriomote Island, Okinawa, Japan, the origin of the soil sample from which the type strain was isolated.

Spore chains borne on the tip of short sporophores arising directly from the agar surface form on water agar and HV agar. Growth occurs optimally between 20 and 30°C; no growth at 15 or 37°C. Does not grow on 4% NaCl. Hydrolyzes starch. Does not reduce nitrate. Utilizes D-mannitol, D-melibiose, D-maltose, L-rhamnose, methyl α-D-glucoside, D-raffinose, D-galactose, D-mannose, and glucose. Does not utilize L-erythritol, adonitol, D-lactose, L-inositol, D-sorbitol, or dulcitol. Phosphatidylethanolamine is present as the diagnostic phospholipid. Unsaturated fatty acids and 10-methylated fatty acids are not detected. Major fatty acids are $C_{15:0}$ anteiso (22%), $C_{17:0}$ anteiso (20), $C_{15:0}$ iso (18%), $C_{17:0}$ (14%), and $C_{16:0}$ iso (12%). Minor fatty acids are $C_{17:0}$ iso, $C_{16:0}$, $C_{18:0}$, $C_{15:0}$, and $C_{14:0}$ iso.

For all other characteristics, refer to the genus description.

DNA G+C content (mol%): 69.0 (HPLC).

Type strain: TT 97-02, NBRC 100142, DSM 44745.

3. **Asanoa ishikariensis** Lee and Hah 2002, 971[VP] (Basonym: '*Catellatospora ishikariense*'.)

ish.i.ka.ri.en'sis. N.L. fem. adj. *ishikariensis* of or belonging to Ishikari-gun, Hokkaido, Japan, the origin of the soil sample from which the type stain was isolated.

Growth occurs at mesophilic temperatures below 37°C. Grows on 0.001% Brilliant green, but not on 0.0001% crystal violet, 3% NaCl or 0.01% lysozyme. Antibiotic susceptiblity (μg/ml): 50, gentamicin; 5, neomycin; 100, streptomycin; and 10, tetracycline. Antibiotic resistance (μg/ml): 50, novobiocin and 20, vancomycin. Utilizes D-arabinose, L-arabinose, dextran, D-cellobiose, D-fructose, D-galactose, gluconate, D-glucose, D-lactose, maltose, D-mannose, melibiose, methyl α-D-glucoside, D-raffinose, L-rhamnose, D-ribose, starch, sucrose, D-trehalose, D-xylose, adonitol, dulcitol, and D-mannitol. Does not utilize inulin, D-melezitose, salicin, L-sorbose, butanol, *meso*-erythritol, ethanol, glycerol, *meso*-inositol, 2-propanol, 1-propanol, D-sorbitol, or D-xylitol. Hydrolyzes casein, gelatin and starch but not DNA or elastin. Does not decompose adenine, hippurate, hypoxanthine, DL-tyrosine, or xanthine. The phospholipid profile consists of diphosphatidylglycerol, phosphatidylinositol, phosphatidylinositol mannoside, phosphatidylethanolamine, and an unknown phospholipid. A trace amount of phosphatidylglycerol is also detected. The major fatty acids are $C_{15:0}$ anteiso (25%), $C_{17:1}$ (24%), $C_{15:0}$ iso (16%) and $C_{17:0}$ (12%) acids. Minor fatty acids are $C_{17:0}$ iso (6%), $C_{16:0}$ iso (4%), $C_{15:0}$ (3%), $C_{18:0}$ (2%), $C_{16:0}$ (1%), and $C_{19:0}$ (1%).

For all other characteristics, refer to the genus description.

DNA G+C content (mol%): 71.1 (HPLC).

Type strain: IMSNU 22004, NBRC 14551.

Sequence accession no. (16S rRNA gene): AJ294715.

Genus V. Catellatospora Asano and Kawamoto 1986, 516^{VP} emend. Lee and Hah 2002, 971^{VP} emend. Ara, Bakir and Kudo 2008a, 1958^{VP}

Fred A. Rainey

Ca.tel.la.to.spo'ra. L. n. *catella* small chain; Gr. n. *spora* a seed and in biology a spore; N.L. fem. n. *Catellatospora* (organism forming) small chain of spores.

Gram-stain-positive, forming chains of nonmotile spores that arise singly or in tufts from vegetative hyphae. Vegetative hyphae are branched but not fragmented. Substrate mycelia are light yellow to bright yellow. Aerial mycelia are absent. **Aerobic** and **chemo-organotrophic**. **Mesophilic**. **Catalase-positive**. The diagnostic amino acids of the peptidoglycan are ***meso*-diaminopimelic acid** and 3-hydroxydiaminopimelic acids. The muramic acid *N*-acyl type is glycolyl. The cell-wall hydrolysates contain arabinose, galactose, ribose, mannose, glucose and xylose. Rhamnose is present in some species. Mycolic acids are not present. The major fatty acids are $C_{17:0}$, $C_{17:1}$ ω8*c*, $C_{15:0}$ iso and $C_{16:0}$ iso. The polar lipids can include phosphatidylethanolamine, diphosphatidylglycerol, phosphatidylinositol and phosphatidylinositol mannosides. **Menaquinone MK-9(H_4)** is the major menaquinone. Species of the genus have been isolated from sandy and woodland soils from Japan and Bangladesh. Based on 16S rRNA gene sequence phylogeny the species of the genus form a distinct lineage within the family *Micromonosporaceae.*

DNA G+C content (mol%): 70.7–71.4.

Type species: **Catellatospora citrea** Asano and Kawamoto 1986, 516^{VP}.

List of species of the genus *Catellatospora*

1. **Catellatospora citrea** Asano and Kawamoto 1986, 516^{VP}

cit're.a. L. fem. adj. *citrea* of or pertaining to the citrus-tree, intended to mean lemon yellow.

Aerobic, Gram-stain-positive staining, short chains of nonmotile spores arise singly or in tufts from vegetative hyphae. Sporulation observed on Tyrosine agar, oatmeal agar, 1/5 yeast extract-starch agar, oatmeal-nitrate agar, tap-water agar and sucrose-nitrate agar. Yellow vegetative hyphae on oatmeal and Hickey–Tresner agar medium. Soluble or melanin-like pigments are not produced.

The cell-walls contain *meso*-diaminopimelic and 3-hydroxydiaminopimelic acids. The muramic acid *N*-acyl type is glycolyl. The cell-wall sugars are arabinose, galactose, rhamnose, ribose and xylose. Whole-cell fatty acids when grown at 30°C on yeast extract-starch broth include: $C_{15:0}$ (3.9%), $C_{16:0}$ (1.5%), $C_{17:0}$ (8.1%), $C_{18:0}$ (2.2%), $C_{16:1}$ 2-OH (5.5%), $C_{17:1}$ ω8*c* (7.8%), $C_{18:1}$ ω9*c* (2.2%), $C_{14:0}$ iso (2.4%), $C_{15:0}$ iso (28.4%), $C_{15:0}$ anteiso (5.3%), $C_{15:1}$ iso (2.6%), $C_{16:1}$ iso(1.8%), $C_{16:0}$ iso (10.5%), $C_{17:1}$ iso ω9*c* (1.4%), $C_{17:0}$ iso (3.3%), $C_{17:0}$ anteiso (4.3%), 10-methyl $C_{18:0}$ (1.0%) and summed feature 6 (1.5%) (Ara and Kudo, 2006). The polar lipid profile consists of phosphatidylethanolamine, phosphatidylglycerol, phosphatidylinositol, phosphatidylinositol mannosides. MK-9(H_4) is the major menaquinone with minor amounts of MK-9(H_6). Growth occurs between 20 and 37°C, with optimal growth at 31°C. Growth occurs between pH 6 and 9. Growth does not occur in the presence of 1 % (w/v) NaCl. Negative for milk coagulation and peptonization. Gelatin is not liquefied. Substrates used as sole carbon sources include: L-arabinose, glycerol, lactose, D-galactose, D-glucose, maltose, D-mannose, α-D(+)-melibiose, methyl α-D-glucoside, L-rhamnose, salicin, starch, sucrose, trehalose, and D-xylose. Adonitol, erythritol, D-(+)-raffinose, are not used as sole carbon sources. Utilization of D- fructose, *myo*-inositol, D-mannitol and D- ribose has been reported as both positive and negative for the type strain. Sensitive to the following antibiotics: novobiocin (20 μg/ml), vancomycin (50 μg/ml), gentamicin (50 μg/ml), demethylchlortetracycline (500 μg/ml) and streptomycin (100 μg/ml). Growth occurs in the presence of crystal violet at 0.0001% (w/v) but not at 0.001% (w/v).

Isolated from woodland soil collected in Itsukaichi-shi Tokyo, Japan.

DNA G+C content (mol%): 71.5 (T_m).

Type strain: 6183-E, ATCC 49964, CIP 107011, DSM 44097, NBRC 14495, JCM 7542, NRRL B-16429, VKM Ac-1421.

Sequence accession no. (16S rRNA gene): X93197.

2. **Catellatospora bangladeshensis** Ara and Kudo 2006, 399^{VP}

ban.gla.desh.en'sis. N.L. fem. adj. *bangladeshensis* of or pertaining to Bangladesh, the origin of the soil from which the type strain was isolated.

Aerobic, Gram-stain-positive staining, short chains of nonmotile spores arise singly or in tufts from vegetative hyphae. Spores are spherical to cylindrical; spore surface is smooth. Forms well developed branched substrate mycelium. Sporulation observed on sucrose-nitrate agar and oatmeal-nitrate agar. Light yellow to bright yellow substrate mycelium form on yeast extract starch agar. Soluble or melanin-like pigments are not produced. The cell-walls contain *meso*-diaminopimelic and 3-hydroxydiaminopimelic acids. The muramic acid *N*-acyl type is glycolyl. The cell-wall hydrolysates contain arabinose, galactose, rhamnose, ribose, mannose, glucose and xylose. Mycolic acids are not present. Whole-cell fatty acids when grown at 30°C on yeast extract-starch broth include: $C_{15:0}$ (1.7%), $C_{17:0}$ (3.1%), $C_{17:1}$ ω8*c* (8.8%), $C_{18:1}$ ω9*c* (3.0%), $C_{14:0}$ iso (5.2%), $C_{15:0}$ iso (19.8%), $C_{15:0}$ anteiso (2.5%), $C_{16:1}$ iso (3.7%), $C_{16:0}$ iso (35.4%), $C_{17:1}$ iso ω9*c* (1.8%), $C_{17:0}$ iso (2.8%), $C_{17:0}$ anteiso (2.0%), $C_{18:0}$ iso (1.2%), 10-methyl $C_{18:0}$ (2.3%), and summed feature 3 (1.2%). The polar lipid profile consists of phosphatidylethanolamine, diphosphatidylglycerol, phosphatidylinositol, phosphatidylinositol mannosides. Glucosamine containing phospholipids and phosphatidylcholine are absent. MK-9(H_4) is the major menaquinone with minor amounts of MK-9(H_6) and MK-9(H_2). Growth occurs between 25 and 30°C and between pH 6.8 and 7.2. Growth does not occur in the presence of 1 % (w/v) NaCl. Substrates used as sole carbon sources include: L-arabinose, D-glucose, D-galactose, lactose, D-mannose, maltose, α-D(+)-melibiose, L-rhamnose, salicin, sucrose, trehalose, and D-xylose. Weakly positive utilization of adonitol, glycerol, and D-mannitol. Erythritol, D- fructose, *myo*-inositol, methyl α-D-glucoside, D-(+)-raffinose, and D- ribose are not used as sole carbon sources.

Isolated from sandy soil collected at Chokoria, Bandladesh.

DNA G+C content (mol%): 71.0 (HPLC).

Type strain: 2-70(23), JCM 12949, DSM 44899.

Sequence accession no. (16S rRNA gene): AB200233.

3. **Catellatospora chokoriensis** Ara and Kudo 2006, 397[VP]

cho.kor.i.en'sis. N.L. fem. adj. *chokoriensis* of or pertaining to Chokoria, Bangladesh, the origin of the soil from which the type strain was isolated.

Aerobic, Gram-stain-positive staining, short chains of spores are borne directly on substrate mycelium. Spores are spherical to cylindrical and nonmotile. Forms well developed branched substrate mycelium. Aerial mycelium is absent. Sporulation observed on glycerol-asparagine agar, tyrosine agar, yeast-malt extract agar, oatmeal agar, Bennett agar, water agar, sucrose-nitrate agar, 1/5 yeast-starch agar and oatmeal-nitrate agar. Light yellow to bright yellow substrate mycelium form on yeast extract starch agar. Soluble or melanin-like pigments are not produced. The cell-walls contain *meso*-diaminopimelic and 3-hydroxydiaminopimelic acids. The muramic acid *N*-acyl type is glycolyl. The cell-wall hydrolysates contain arabinose, galactose, rhamnose, ribose, mannose, glucose and xylose. Mycolic acids are not present. Whole-cell fatty acids when grown at 30°C on yeast extract-starch broth include: $C_{15:0}$ (2.2%), $C_{17:0}$ (10.9%), $C_{18:0}$ (4.3%), $C_{17:1}$ ω8*c* (3.5%), $C_{14:0}$ iso (3.6%), $C_{15:0}$ iso (30.3%), $C_{15:0}$ anteiso (6.3%), $C_{16:1}$ iso (2.5%), $C_{16:0}$ iso (22.9%), $C_{17:0}$ iso (5.5%), and $C_{17:0}$ anteiso (7.9%). The polar lipid profile consists of phosphatidylethanolamine, diphosphatidylglycerol, phosphatidylinositol, phosphatidylinositol mannosides. Glucosamine containing phospholipids and phosphatidylcholine are absent. MK-9(H_4) is the major menaquinone with minor amounts of MK-9(H_6) and MK-9(H_2). Growth occurs between 15 and 30°C, between pH 6.0 and 9.0, and in the presence of 1 % (w/v) NaCl. Substrates used as sole carbon sources include: L-arabinose, D-glucose, D-galactose, lactose, D-mannose, maltose, α-D(+)-melibiose, salicin, sucrose, trehalose, and D-xylose. Weakly positive utilization of adonitol. Erythritol, D- fructose, glycerol, *myo*-inositol, D-mannitol, methyl α- D-glucoside, D-(+)-raffinose, L-rhamnose, and D-ribose are not used as sole carbon sources.

Isolated from sandy soil collected at a forest-side waterfall, Chokoria, Bangladesh.

DNA G+C content (mol%): 71.0 (HPLC).

Type strain: 2-25(1), JCM 12950, DSM 44900.

Sequence accession no. (16S rRNA gene): AB200231.

4. **Catellatospora coxensis** Ara and Kudo 2006, 398[VP]

cox.en'sis. N.L. fem. adj. *coxensis* of or pertaining to Cox's Bazar, Bangladesh, the origin of the soil from which the type strain was isolated.

Aerobic, Gram-stain-positive staining, short chains of nonmotile spores arise singly or in tufts from vegetative hyphae. Spores are spherical to cylindrical and nonmotile. Forms well developed branched substrate mycelium. Aerial mycelium is absent. Sporulation observed on oatmeal agar, water agar, 1/5 yeast-starch agar and oatmeal-nitrate agar. Light yellow to bright yellow substrate mycelium form on yeast extract starch agar. Soluble or melanin-like pigments are not produced. The cell-walls contain *meso*-diaminopimelic and 3-hydroxydiaminopimelic acids. The muramic acid *N*-acyl type is glycolyl. The cell-wall hydrolysates contain arabinose, galactose, rhamnose, ribose, mannose, glucose and xylose. Mycolic acids are not present. Whole-cell fatty acids when grown at 30°C on yeast extract-starch broth include: $C_{15:0}$ (4.8%), $C_{16:0}$ (1.6%), $C_{17:0}$ (14.4%), $C_{18:0}$ (1.8%), $C_{19:0}$ (1.2%), $C_{17:1}$ ω8*c* (7.9%), %), $C_{18:1}$ ω9*c* (1.6%), $C_{11.0}$ iso (4.8%), $C_{15:0}$ iso (22.2%), $C_{15:0}$ anteiso (7.4%), $C_{16:0}$ iso (18.5%), $C_{17:0}$ iso (2.2%), and $C_{17:0}$ anteiso (4.5%). The polar lipid profile consists of phosphatidylethanolamine, diphosphatidylglycerol, phosphatidylinositol, phosphatidylinositol mannosides. Glucosamine containing phospholipids and phosphatidylcholine are absent. MK-9(H_4) is the major menaquinone with minor amounts of MK-9(H_6) and MK-9(H_2). Growth occurs between 20 and 30°C and between pH 6.0 and 9.0. Growth does not occur in the presence of 1 % (w/v) NaCl. Substrates used as sole carbon sources include: L-arabinose, D-glucose, D-galactose, glycerol, lactose, D-mannose, maltose, α-D(+)-melibiose, L-rhamnose, D-ribose, sucrose, trehalose, and D-xylose. Weakly positive utilization of D- fructose, and methyl α- D-glucoside. Adonitol, erythritol, *myo*-inositol, D-mannitol, D-(+)-raffinose, salicin, and are not used as sole carbon sources.

Isolated from sandy soil collected at a forest-side waterfall, Chokoria, Cox's Bazar, Bangladesh.

DNA G+C content (mol%): 71.0 (HPLC).

Type strain: 2-29(17), JCM 12951, DSM 44901.

Sequence accession no. (16S rRNA gene): AB200232.

5. **Catellatospora methionotrophica** (*ex* Asano and Kawamoto 1988) Ara and Kudo 2006, 399[VP] (*Catellatospora citrea* subsp. *methionotrophica* Asano and Kawamoto 1988)

me.thi.o.no.tro'phi.ca. N.L. n. *methioninum* methionine; Gr. adj. *trophikos* nursing, tending or feeding; N.L. fem. adj. *methionotrophica*, methionine auxotroph.

Aerobic, Gram-stain-positive staining, straight chains of smooth surfaced spores. Light yellow to bright yellow vegetative hyphae formed. Soluble or melanin-like pigments are not produced. Forms well developed, branched substrate mycelium. Aerial mycelium is absent. The cell-walls contain *meso*-diaminopimelic and 3-hydroxydiaminopimelic acids. The muramic acid *N*-acyl type is glycolyl. The cell-wall hydrolysates contain arabinose, xylose, galactose, ribose, mannose, and glucose. Mycolic acids are absent. Whole-cell fatty acids when grown at 30°C on yeast extract-starch broth include: $C_{15:0}$ (2.9%), $C_{16:0}$ (1.3%), $C_{17:0}$ (8.9%), $C_{18:0}$ (1.7%), $C_{16:1}$ 2-OH (1.4%), $C_{17:1}$ ω8*c* (9.3%), $C_{18:1}$ ω9*c* (2.9%), $C_{14:0}$ iso (1.4%), $C_{15:0}$ iso (37.6%), $C_{15:0}$ anteiso (7.5%), $C_{16:0}$ iso (5.7%), $C_{17:1}$ iso ω9*c* (3.2%), $C_{17:0}$ iso (4.5%), $C_{17:0}$ anteiso (5.0%) and summed feature 6 (1.4%). The polar lipid profile consists of phosphatidylethanolamine, diphosphatidylglycerol, phosphatidylglycerol, phosphatidylinositol,

phosphatidylinositol mannosides. Glucosamine containing phospholipids and phosphatidylcholine are absent. MK-9(H_4) is the major menaquinone with minor amounts of MK-9(H_6) and MK-9(H_2). Growth occurs between 20 and 34°C and between pH 6.8 and 7.2. Growth does not occur in the presence of 1 % (w/v) NaCl. Negative for milk coagulation and peptonization. Gelatin is not liquefied. Requires methionine which cannot be replaced by cysteine or homoserine. Substrates used as sole carbon sources include: D-arabinose, L-arabinose, cellobiose, lactose, D-galactose, D-glucose, maltose, D-mannose, L-rhamnose, salicin, starch, sucrose, trehalose, and D-xylose. Weakly positive utilization of erythritol, glycerol, D-fructose, methyl α-D-glucoside and D-(+)-raffinose. Adonitol, ethanol, gluconate, D-mannitol, melezitose, melibiose, α-D(+)-melibiose, methanol, methyl α-D-glucoside, D- ribose and sorbitol are not used as sole carbon sources. Sensitive to the following antibiotics: novobiocin (50 μg/ml), vancomycin (50 μg/ml), gentamicin (50 μg/ml), demethylchlortetracycline (500 μg/ml) and streptomycin (100 μg/ml). Growth occurs in the presence of crystal violet at 0.0001% (w/v) but not at 0.001% (w/v).

Isolated from a woodland soil sample collected in Yamanashi, Japan.

DNA G+C content (mol%): 71.0 (HPLC).

Type strain: 6257-B, ATCC 49965, CIP 107012, NBRC 14553, NRRL B-16431, VKM Ac-2008, IMSNU 22006, JCM 7543, DSM 44098.

Sequence accession no. (16S rRNA gene): AF152107.

Genus VI. **Catenuloplanes** Yokota, Tamura, Hasegawa and Huang 1993, 809[VP]

THE EDITORIAL BOARD

Ca.te.nul.o.plan′es. L. fem. n. *catenula* short chain; Gr. masc. n. *planes* a wanderer; N.L. masc. n. *Catenuloplanes* a short chain wanderer; intended to signify a motile short chain.

Gram-stain-positive. Not acid-fast. **Strictly aerobic**. **Forms branching and non-fragmenting vegetative hyphae**. Aerial mycelium is rudimentarily developed or absent. If produced, spores are arranged in chains, which arise from the vegetative hyphae or are formed on the rudimentary aerial hyphae. The spore chains are aggregated into clusters and may be enveloped by outer sheaths. The configuration of spore chains is curly or spiral with one or two turns and sometimes branched, The spores are rod-shaped, straight or curved (0.6–0.8 × 2–4 μm) with smooth surfaces, and motile by means of peritrichous flagella. **Cell-wall type is VI and the peptidoglycan contains D-glutamate, D-serine, L-serine, glycine, D-alanine, and L-lysine**. **Cell-wall sugars are mannose, xylose, ribose and glucose**. The muramic acid in the glycan moiety is N-glycolated. The Mycolic acids are absent. Major menaquinones are MK-10(H_4) and MK-11(H_4); some strains may also possess small amounts of MK-10, MK-10(H_2), MK-10(H_8), MK-10(H_6), MK-11(H_2) and MK-11(H_6). The major cellular fatty acids are $C_{18:1}$, $C_{16:0}$, and $C_{17:0}$ anteiso. Small amounts of $C_{17:0}$, $C_{16:1}$, $C_{16:0}$ iso, and $C_{18:0}$ may also be present. The diagnostic phospholipid is phosphatidylcholine.

DNA G+C content (mol%): 70.0–72.0.

Type species: **Catenuloplanes japonicas** Yokota, Tamura, Hasegawa and Huang 1993, 810[VP].

Further descriptive information

Phylogenetic analysis of the 16S rRNA gene classifies the genus within family *Micromonosporaceae*. The closest phylogenetic neighbor is *Asanoa iriomotensis* (97.3%, accession no. AB112081).

Enrichment and isolation procedures

Isolated from soil or leaf litter in Japan, India, and Nepal. Many strains were isolated on a medium containing 1.0% soluble starch, 0.1% casein, 0.05% K_2HPO_4, and 1.5% agar (pH 7.0–7.5) supplemented with (per ml) 25 μg nalidixic acid, 12.5 μg kanamycin, 5.0 μg cefsulodin, and 6.25 μg kabicidin (Yokota et al., 1993).

Maintenance procedures

Strains are cultivated on yeast extract/starch agar containing 2 g yeast extract (Difco), 10 g soluble starch and 15 g agar per 1 distilled water (pH 7.3) at 28°C for 14 d and maintained at 8°C.

Differentiation of the genus *Catenuloplanes* from closely related genera

Spore motility differentiates *Catenuloplanes* from closely related genera with nonmotile spores: *Actinocatenispora, Asanoa, Catellatospora, Krasilinkovia, Longispora, Luedemannella, Micromonospora, Polymorphospora, Salinispora,* and *Verrucosispora.* Sporangia or spore vesicles are absent in *Catenuloplanes*, but are present in *Actinoplanes, Dactylosporangium, Luedemannella, Pilimelia, Virgisporangium,* and *Planosporangium.* Within the family, only *Catenuloplanes* and *Couchioplanes* contain the diamino acid L-Lys in their cell walls. All other genera contain *m*-DAP. *Catenuloplanes* is differentiated from *Couchioplanes* by phospholipid type (III and II, respectively) and major menaquinones [MK-10(H_4) and MK-11(H_4) versus MK-9(H_4), respectively].

Taxonomic comments

Planopolyspora crispa was described nearly simultaneously with *Catenuloplanes japonicas* (Petrolini et al., 1993; Yokota et al., 1993). Because the name *Catenuloplanes japonicas* was validated first, it has priority, and *Planopolyspora crispa* was transferred to the genus *Catenuloplanes* as *Catenuloplanes crispus* by Kudo et al. (1999).

List of species of the genus *Catenuloplanes*

1. **Catenuloplanes japonicus** Yokota, Tamura, Hasegawa and Huang 1993, 810[VP]

ja.pon'i.cus. N.L. masc. adj. *japonicus* of or pertaining to Japan.

Vegetative mycelia of the strains are pale orange to orange yellow, and the aerial mycelia are white to pale yellow. They produce a pale yellowish soluble pigment on Czapek-sucrose agar, glucose-asparagine agar, and calcium malate agar. H_2S is produced. Nitrate is not reduced to nitrite. Gelatin liquefaction is positive. Hydrolyzes starch, but not hippurate. Decomposes calcium malate, tyrosine, esculin, and urea, but not adenine, xanthine, hypoxanthine, or cellulose. Susceptible to lysozyme. Milk is coagulated and peptonized. Utilizes acetate, lactate, malate, pyruvate, succinate, L-arabinose, cellobiose, fructose, galactose, glucose, glycerol, inositol, lactose, maltose, mannitol, melezitose, melibiose, α-methyl-D-glucoside, raffinose, L-ribose, salicin, starch, sucrose, trehalose, and D-xylose. Does not utilize benzoate, citrate, mucate, oxalate, ribitol, galactitol, L-erythritol, sorbitol, or L-sorbose. Growth occurs optimally at 21–28°C. Major fatty acids are $C_{18:1}$, $C_{16:0}$ and $C_{17:0}$ anteiso.

For all other characteristics, refer to the genus description.

DNA G+C content (mol%): 71.0 (HPLC).

Type strain: N381-16, NBRC 14176, ATCC 31637, DSM 44102, JCM 9106, VKM Ac-875.

Sequence accession no. (16S rRNA gene): X93201.

2. **Catenuloplanes crispus** Kudo, Nakajima and Suzuki 1999, 1858[VP]

cris'pus. L. masc. adj. *crispus* curly.

Yellow to brown vegetative mycelia on most media. Yellowish diffusible pigment is produced in glycerol/asparagine agar. Decomposes casein, elastin, esculin, testosterone, and urea, but not adenine, DNA, hypoxanthine, tyrosine, or xanthine. Utilizes L-arabinose, D-cellobiose, D-fructose, D-galactose, D-glucose, glycerol, *myo*-inositol, D-lactose, maltose, D-mannitol, D-mannose, L-rhamnose, salicin, starch, sucrose, D-trehalose, D-xylose, fumarate, L-malate, and succinate. Does not utilize adonitol, dulcitol, iso-erythritol, D-melezitose, methyl-α-D-glucoside, D-raffinose, citrate, mucate, benzoate, oxalate, or L-tartrate. Acids produced from L-arabinose, D-fructose, D-galactose, D-glucose, *myo*-inositol, L-rhamnose, sucrose and D-xylose, but not from D-melezitose, melibiose, methyl α-D-glucoside, D-raffinose, D-ribose or L-sorbose. No growth in the presence of 2% NaCl. Antibiotic susceptibility (20 μg/ml): ampicillin, benzylpenicillin, cephalexin, novobiocin, and kanamycin.

For all other characteristics, refer to the genus description.

DNA G+C content (mol%): 70.0 (HPLC).

Type strain: JCM 9312, ATCC 51431, DSM 44128, NBRC 15622, IPV 2867, NCB 1173, VKM Ac-1992.

Sequence accession no. (16S rRNA gene): AB024701.

3. **Catenuloplanes niger** Tamura, Yokota, Huang, Hasegawa and Hatano 1995, 860[VP]

ni'ger. L. masc. adj. *niger* black, referring to the production of a black soluble pigment.

The morphological, physiological, and chemotaxonomic characteristics are the same as those given previously for *Catenuloplanes japonicus*. Additionally, *Catenuloplanes niger* grows at 37°C and produces a black soluble pigment in peptone-yeast extract-iron agar (ISP medium 6) and nutrient agar, but does not produce any pigment in inorganic salts-starch agar (ISP medium 4). Resistant to penicillin, cephaloridine, cephalexin, and benzylpenicillin.

For all other characteristics, refer to the genus description.

DNA G+C content (mol%): 72.2 (HPLC).

Type strain: N406-14, NBRC 14177, ATCC 31638, DSM 44711, JCM 9533, VKM Ac-1964.

Sequence accession no. (16S rRNA gene): AB523881.

4. **Catenuloplanes indicus** Tamura, Yokota, Huang, Hasegawa and Hatano 1995, 860[VP]

in'di.cus. L. masc. adj. *indicus*, of or pertaining to India, where the organism was isolated.

The morphological, physiological, and chemotaxonomic characteristics are the same as those given previously for *Catenuloplanes japonicus*. Additionally, *Catenuloplanes indicus* grows at 37°C and does not produce any soluble pigment in inorganic salts-starch agar (ISP medium 4), tyrosine agar (ISP medium 7), Bennett agar, and Bennett agar containing maltose. Susceptible to penicillin, cephaloridine, cephalexin, and benzylpenicillin.

For all other characteristics, refer to the genus description.

DNA G+C content (mol%): 71.3–71.9 (HPLC).

Type strain: RA328[T], NBRC 15575, IMSNU 22099, ATCC 700014, DSM 44709, JCM 9534, VKM Ac-1999.

Sequence accession no. (16S rRNA gene): AJ294717.

5. **Catenuloplanes atrovinosus** Tamura, Yokota, Huang, Hasegawa and Hatano 1995, 860[VP]

at.ro.vi.no'sus. L. adj. *ater -tra -trum* dark; L. masc. adj. *vinosus* full of wine; N.L. masc. adj. *atrovinosus*, full of dark wine, dark wine color (red).

The morphological, physiological, and chemotaxonomic characteristics are the same as those given previously for *Catenuloplanes japonicus*. Additionly, *Catenuloplanes atrovinosus* grows at 37°C and produces a black soluble pigment in peptone-yeast extract-iron agar (ISP medium 6) and a pale brown soluble pigment in nutrient agar, but does not produce any soluble pigment in Bennett agar and Bennett agar containing maltose. It forms reddish colonies on inorganic salts-starch agar (ISP medium 4) and glycerol-asparagine agar (ISP medium 5). Galactose, mannose, xylose, and glucose are present as whole-cell sugars. Resistant to penicillin, cephaloridine, cephalexin, and benzylpenicillin.

For all other characteristics, refer to the genus description.

DNA G+C content (mol%): 72.2–72.7 (HPLC).

Type strain: RA332[T], NBRC 15579, IMSNU 22012, ATCC 700015, DSM 44707, JCM 9535, VKM Ac-1972.

Sequence accession no. (16S rRNA gene): AJ294716.

6. **Catenuloplanes castaneus** Tamura, Yokota, Huang, Hasegawa and Hatano 1995, 860[VP]

cas.ta'ne.us. L. masc. adj. *castaneus* chestnut-colored.

The morphological, physiological, and chemotaxonomic characteristics are the same as those given previously for *Catenuloplanes japonicus*. Additionally, *Catenuloplanes castaneus* grows at 37°C and produces a scarlet soluble pigment in glycerol-asparagine agar (ISP medium 5), but does not produce any soluble pigments in inorganic salts-starch agar (ISP medium 4), tyrosine agar (ISP medium 7), and Bennett agar. Resistant to penicillin and benzylpenicillin, but susceptible to cephaloridine and cephalexin.

For all other characteristics, refer to the genus description.

DNA G+C content (mol%): 72.0–72.4 (HPLC).

Type strain: RA344, NBRC 15584, ATCC 700016, DSM 44708, JCM 9537, VKM Ac-1973.

Sequence accession no. (16S rRNA gene): AB523883.

7. **Catenuloplanes nepalensis** Tamura, Yokota, Huang, Hasegawa and Hatano 1995, 860[VP]

ne.pal.en′sis. N.L. masc. adj. *nepalensis* of or pertaining to Nepal, where the organisms were isolated.

The morphological, physiological, and chemotaxonomic characteristics are the same as those given previously for *Catenuloplanes japonicus*. In addition, *Catenuloplanes nepalensis* does not produce a soluble pigment in inorganic salts-starch agar (ISP medium 4), tyrosine agar (ISP medium 7), Bennett agar, and Bennett agar containing maltose. Resistant to penicillin, cephaloridine, and benzylpenicillin, but susceptible to cephalexin.

For all other characteristics, refer to the genus description.

DNA G+C content (mol%): 71.1 (HPLC).

Type strain: RA343, NBRC 15583, ATCC 700017, DSM 44710, JCM 9536, VKM Ac-1996.

Sequence accession no. (16S rRNA gene): AB523882.

Genus VII. **Couchioplanes** Tamura, Nakagaito, Nishii, Hasegawa, Stackebrandt and Yokota 1994, 199[VP]

Tomohiko Tamura

Couch.i.o.pla′nes. N.L. masc. n. *Couchius* a personal name, referring to J.N. Couch (1896–1986), a mycologist who contributed to the taxonomy of the family *Actinoplanaceae*; Gr. masc. n. *planes* a wanderer; N.L. masc. n. *Couchioplanes* a wanderer organism of the family *Actinoplanaceae* named after J.N. Couch.

Gram-stain-positive bacterium producing fine, nonfragmenting, branching mycelia. Not acid-fast. Strictly aerobic. The spore chains and aerial mycelia often **aggregate into clusters resembling sporangia**, but true sporangia are not observed (Figure 214, Figure 215, Figure 216 and Figure 217). Aerial mycelia with short spore chains are **arranged in spirals** that have one to five turns and are hooked or rarely flexuous (Figure 214). Several spores are present per spore chain, and the spores are **oval to short rods** (0.5–0.9 × 1.0–1.5 µm) and smooth. Upon immersion in water or phosphate buffer including soil extract, **motile spores** are released from the spore chain, but in many instances motility begins over 30–60 min after spore release. **Polar flagella** are present in motile spores (Figure 217). Cell wall contains D-**glutamic acid**, D- **and** L-**serine, glycine**, L-**alanine, and** L-**lysine** (molar ratio approx. 1:1:1:1:1). Xylose, arabinose, and galactose are present in whole-cell hydrolysates. Phosphatidylglycerol and phosphatidylethanolamine are present, but phosphatidylcholine is absent. $C_{16:0}$ iso and $C_{17:0}$ anteiso are the major cellular fatty acids. The major menaquinone is MK-9(H_4); in addition, small amounts of MK-9(H_6), MK-9(H_8), and MK-9(H_2) are also present. The acyl type of the cell-wall polysaccharides is glycolyl. Mycolic acid is absent.

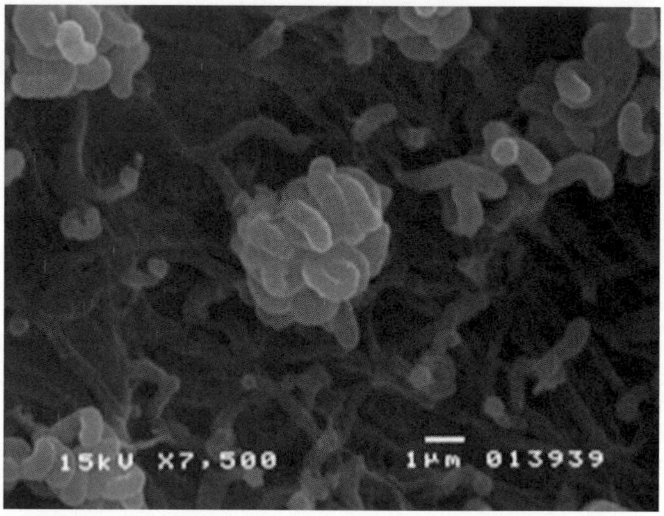

FIGURE 215. Scanning electron micrograph of *Couchioplanes caeruleus* subsp. *caeruleus* showing spore-chain clusters. Bar = 1 µm.

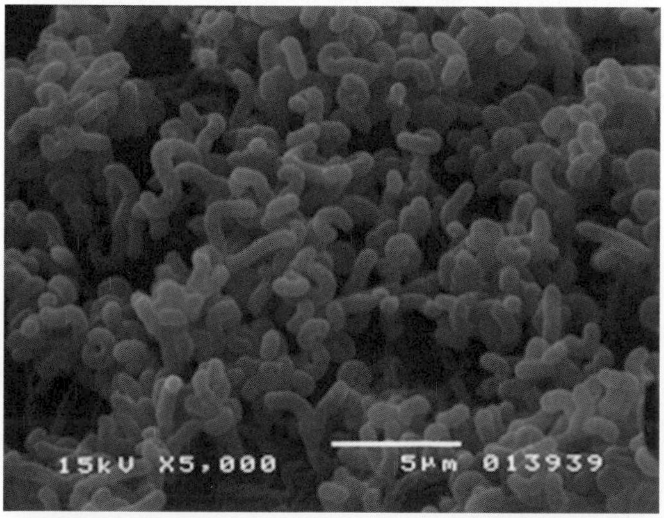

FIGURE 214. Scanning electron micrograph of *Couchioplanes caeruleus* subsp. *caeruleus* showing spiral, hook and/or rarely flexuous spore-chain forms. Bar = 5 µm.

DNA G+C content (mol%): 70–72 (HPLC).

Type species: **Couchioplanes caeruleus** (Horan and Brodsky 1986a) Tamura, Nakagaito, Nishii, Hasegawa, Stackebrandt and Yokota 1994, 200[VP] (*Actinoplanes caeruleus* Horan and Brodsky 1986a, 189).

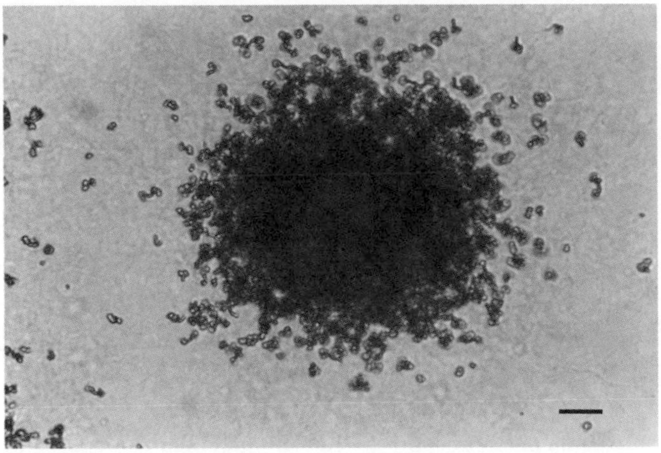

FIGURE 216. Light micrograph of the colony of *Couchioplanes caeruleus* subsp. *caeruleus*. Spore-chain cluster and aerial mycelium look like a sporangium. Bar = 10 µm.

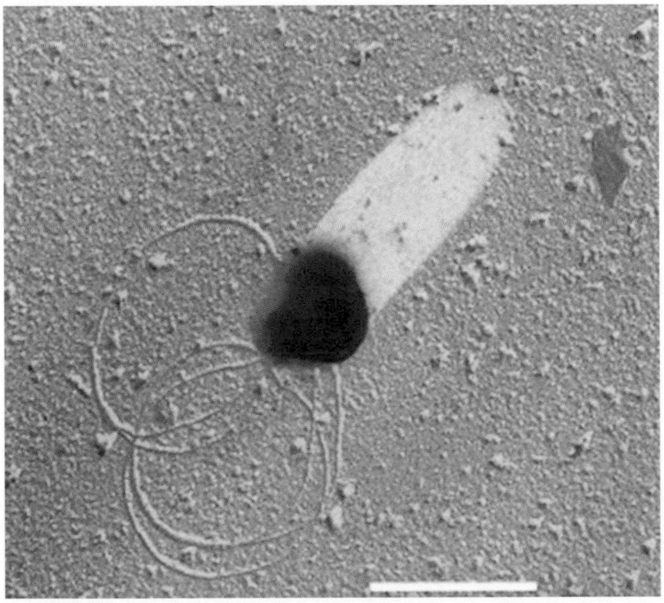

FIGURE 217. Transmission electron micrograph of a shadowing motile spore of *Couchioplanes caeruleus* subsp. *caeruleus* showing the polar flagellum. Bar = 1.0 µm.

Further descriptive information

In general, the vegetative mycelia of the strains are pale to yellowish orange in young cultures and change to a dark blue color in mature cultures. The aerial mycelia are white to gray. The zoospore exhibits active motility after incubation at 28°C for 1 h in 0.01 M phosphate buffer (pH 7.0) containing 10% soil extract. Using light microscopy (Figure 216), the physical appearance of spore chains and aerial mycelia may look like the sporangia of the genus *Actinoplanes*.

Good growth occurs at temperatures between 22 and 28°C. Grows well on yeast extract-malt extract agar, inorganic salts-starch agar, glycerol-asparagine agar, and Bennett's agar. Cultures grow well in yeast extract-glucose broth, consisting of yeast extract (1%) and D-glucose (1%) (pH 7.0), at 28°C on a rotary shaker for 4 d.

The wall chemotype is type VI according to the classification of Lechevalier and Lechevalier (1970a), and the peptidoglycan type is type A (most probably type A3α) according to the classification of Schleifer and Kandler (1972).

Enrichment and isolation procedures

The currently known strains of *Couchioplanes* have been isolated from soil. Generally, the enrichment and isolation methods for actinomycetes producing motile spores, such as centrifuge and capillary methods, may be applicable for strains of the genus *Couchioplanes*.

Maintenance procedures

Strains of the genus *Couchioplanes* are maintained by freezing in water containing 10–30% glycerol at −70°C. Lyophilization of suspensions in 10% skim milk +1% monosodium glutamate and L-drying in 0.01 M potassium phosphate buffer (pH 7.0) containing 3% monosodium glutamate (Sakane and Kuroshima, 1997) are also recommended for long-term preservation.

Differentiation of the genus *Couchioplanes* from other genera

The genera *Couchioplanes* and *Catenuloplanes* differ from other genera of the family *Micromonosporaceae* because their cell-wall peptidoglycan contains L-lysine instead of *meso*-diaminopimelate. These genera are also similar in the formation of motile arthrospores. However, the arthrospores of the genus *Couchioplanes* are oval to short rods with polar flagella, whereas the *Catenuloplanes* arthrospores are rods with peritrichous flagella. The genera *Couchioplanes* and *Catenuloplanes* also differ in their menaquinone systems [MK-9(H_4) vs MK-9(H_8), MK-10(H_8)], phospholipid types (PII vs PIII), and cellular fatty acids ($C_{16:0}$ iso and $C_{17:0}$ anteiso vs $C_{18:1}$, $C_{16:0}$ and $C_{17:0}$ anteiso) (Yokota et al., 1993).

Taxonomic comments

Although *Couchioplanes caeruleus* differs from other species of the genus *Actinoplanes* by forming a deep blue pigment in the vegetative mycelia, by the absence of diaminopimelic acid in its cell wall, by its ability to hydrolyze adenine and hypoxanthine, by its resistance to lysozyme, and by its inability to utilize L-arabinose, D-xylose, and succinate as sole carbon sources, Horan and Brodsky (1986a) included this organism in the genus *Actinoplanes* for the following reasons. It formed irregular to globose sporangia, which upon wetting released spherical to oval motile spores that were partially flagellated, and arabinose and xylose were present as diagnostic whole-cell sugars. However, Stackebrandt and Kroppenstedt (1987) reported that this organism should not be included in the genus *Actinoplanes* because of its peptidoglycan type. The results of numerical taxonomy studies of the genus *Actinoplanes* performed by Goodfellow et al. (1990b) also supported this conclusion.

Based on 16S rRNA gene sequence analysis, the closest neighbor of the genus *Couchioplanes* is the genus *Actinoplanes*, both of which belong to the family *Micromonosporaceae* of the order *Micromonosporales*.

List of species of the genus *Couchioplanes*

1. **Couchioplanes caeruleus** (Horan and Brodsky 1986a) Tamura, Nakagaito, Nishii, Hasegawa, Stackebrandt and Yokota 1994, 200[VP] (*Actinoplanes caeruleus* Horan and Brodsky 1986a, 189)

ca.e.ru'le.us. L. masc. adj. *caeruleus* dark blue, referring to the blue vegetative mycelial pigment.

A yellow to pale brownish soluble pigment is produced on peptone-yeast extract-iron agar. Hydrogen sulfide is produced. Reduces nitrate to nitrite. Gelatin liquefaction is positive. Hydrolyzes starch. Does not decompose calcium malate. Does not coagulate milk. Fructose, glucose, inositol, and sucrose are utilized as carbon sources, but arabinose, raffinose, and xylose are not.

Source: soil.

DNA G+C content (mol%): 70–72 (HPLC).

Type strain: SCC 1014, ATCC 33937, DSM 43634, NBRC 13939, JCM 3195, NRRL 5325, VKM Ac-1257.

Sequence accession no. (16S rRNA gene): D85479.

The species has subsequently been divided into subspecies.

1a. **Couchioplanes caeruleus subsp. caeruleus** (Horan and Brodsky 1986a) Tamura, Nakagaito, Nishii, Hasegawa, Stackebrandt and Yokota 1994, 201[VP] (*Actinoplanes caeruleus* Horan and Brodsky 1986a, 189)

A yellow to pale brownish soluble pigment is produced on glycerol-asparagine agar and glucose-asparagine agar, but this pigment is not produced on yeast extract-malt extract agar. Rhamnose and mannitol are utilized as carbon sources. A novel heptaene antifungal antibiotic is produced (Wagman et al., 1975), which has a broad spectrum of activity against pathogenic fungi (Wright et al., 1977). No growth occurs in the presence of 2% NaCl.

Source: soil.

DNA G+C content (mol%): 70 (HPLC).

Type strain: SCC 1014, ATCC 33937, DSM 43634, NBRC 13939, JCM 3195, NRRL 5325, VKM Ac-1257.

Sequence accession no. (16S rRNA gene): D85479.

1b. **Couchioplanes caeruleus subsp. azureus** Tamura, Nakagaito, Nishii, Hasegawa, Stackebrandt and Yokota 1994, 201[VP]

a.zu're.us. N.L. masc. adj. *azureus* azure blue, referring to the blue vegetative mycelial pigment.

A yellow to pale brownish soluble pigment is produced on yeast extract-malt extract agar, but this pigment is not produced on glycerol-asparagine and glucose-asparagine agar. Growth occurs in the presence of 2% NaCl. Rhamnose is not utilized as a carbon source, and mannitol is weakly utilized. A mixture of antibiotics, which includes a number of macrocyclic lactones and depsipeptides, is produced (Celmer et al., 1977b). The individual compounds exhibit significant antibiotic activity. The crude antibiotic mixture or combinations of a pure macrocyclic lactone and a pure depsipeptide demonstrate marked synergistic antibiotic activity. These antibiotics act as growth promotants in chicks and swine and are effective in the treatment of swine dysentery. "*Actinoplanes azureus*" (Celmer et al., 1977b) is the basonym of this subspecies.

Source: soil.

DNA G+C content (mol%): 72 (HPLC).

Type strain: ATCC 31157, DSM 43900, NBRC 13993, JCM 3246, VKM Ac-2019.

Sequence accession no. (16S rRNA gene): D85478, X93202.

Genus VIII. **Dactylosporangium** Thiemann, Pagani and Beretta 1967, 43[AL]

Gernot Vobis

Dac.ty.lo.spo.ran'gi.um. Gr. n. *daktylos* finger; Gr. n. *spora* a seed, and in biology a spore; Gr. neut. n. *angeion* (L. translit. *angium*) vessel; N.L. neut. n. *Dactylosporangium* an organism with finger-shaped, spore-containing vessels (sporangia).

Finger-shaped to claviform sporangia (0.6–1.4 × 2.5–6.0 µm) are **formed on short sporangiophores on the substrate mycelium**. They develop singly or in clusters above the surface of the substrate. **Each sporangium contains a single row of normally three to four spores. The spores are oblong, ellipsoidal, ovoid, or slightly pyriform** (0.4–1.3 × 0.5–1.8 µm) and **motile** by means of a polarly inserted tuft of flagella. True aerial mycelium is not formed. Hyphae of the substrate mycelium are 0.5–1.0 µm in diameter, branched, and rarely septate. Large single spores, **globose bodies** (1.7–2.8 µm in diameter) are formed on short branches on substrate mycelium. Organisms are Gram-stain-positive and not acid-fast. The peptidoglycan of the cell walls contains *meso*-**diaminopimelic acid** (*meso*-**DAP) and glycine, with xylose and arabinose** as characteristic sugars of whole-cell hydrolysates. Colonies grow on various agar media. They are compact, somewhat tough and leathery, and mostly flat or sometimes elevated with a smooth to slightly wrinkled surface. **The color of the substrate mycelium is pale orange to deep orange, rose or wine-colored to brown**. Aerobic, chemo-organotrophic, with optimum growth between 25 and 37°C and at pH 6.0–7.0.

DNA G+C content (mol%): 71–73 (T_m).

Type species: **Dactylosporangium aurantiacum** Thiemann, Pagani and Beretta 1967, 43[AL].

Further descriptive information

Phylogeny. The genus belongs to the family *Micromonosporaceae*. Based on 16S rRNA gene sequence analysis, it comprises six species with validly published names: *Dactylosporangium aurantiacum*, *Dactylosporangium fulvum*, *Dactylosporangium matsuzakiense*, *Dactylosporangium roseum*, *Dactylosporangium thailandense*,

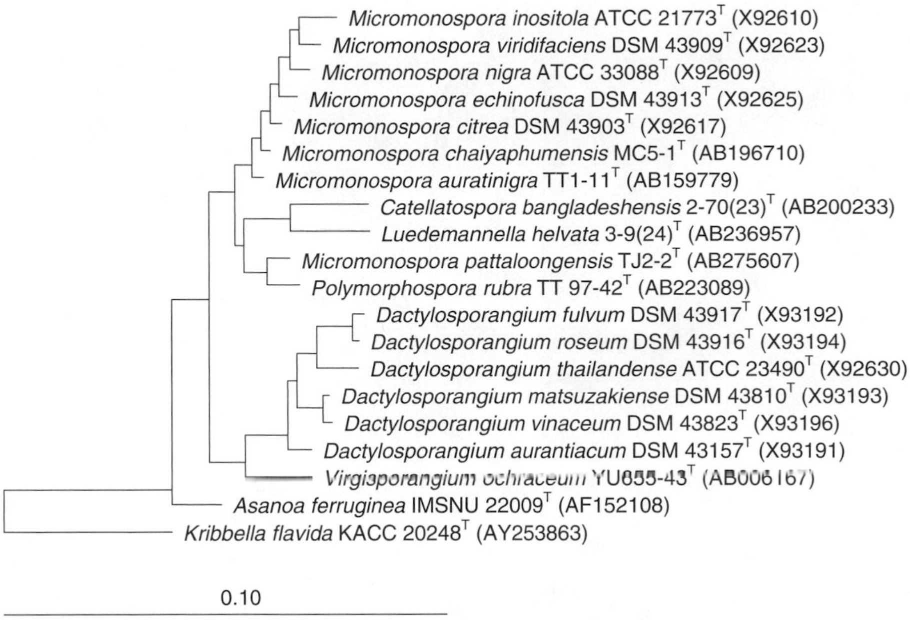

FIGURE 218. Phylogenetic tree showing the relationship of type strains of the genus *Dactylosporangium* and type strains of related genera of the family *Micromonosporaceae*. The tree was reconstructed with the maximum-likelihood method using the software environment ARB (Ludwig et al., 2004) and the corresponding SILVA SSURef 95 database (release July 2008; Pruesse et al., 2007). Bar = 0.10 substitutions per site.

and *Dactylosporangium vinaceum*. Sequence similarities between species range from 99.7 to 97.8%. The nearest genera of the family *Micromonosporaceae* are *Asanoa*, *Catellatospora*, *Luedemannella*, *Micromonospora*, *Polymorphospora*, and *Virgisporangium*, with sequence similarities between type strains of the genus *Dactylosporangium* and type strains of related genera ranging from 97.2 to 96.6% (genus *Polymorphospora*), 97.1 to 94.5% (genus *Micromonospora*), 97.1 to 95.7% (genus *Virgisporangium*), 96.8 to 96.1% (genus *Asanoa*), 96.5 to 94.4% (genus *Catellatospora*), and 96.5 to 95.4% (genus *Luedemannella*). A phylogenetic tree is shown in Figure 218.

Cell morphology and fine structure. The substrate hyphae are 0.5–1.0 µm in diameter and irregularly branched. They are rarely septate and do not separate into fragments either in agar or in liquid cultures. A true aerial mycelium is not formed; however, short hyphae in contact with the air are observed occasionally (Shomura et al., 1983b; Thiemann et al., 1967). *Dactylosporangium fulvum* develops rudimentary aerial mycelium (Shomura et al., 1986). The substrate hyphae may form coremia, also bearing sporangia and globose bodies (Shomura et al., 1986). The cell walls of the hyphae, the globose bodies, and the zoospores each consist of a single layer (Miyadoh et al., 1997; Vobis, 1987). The mesosomes are tubular-vesicular (Williams et al., 1973). Crystalline phage particles have been detected in the cytoplasm of the substrate hyphae of a strain of *Dactylosporangium thailandense* (Higgins and Lechevalier, 1969).

The oligosporous sporangia of the genus *Dactylosporangium* are formed on the surface of the colonies, singly or, more frequently, in tufts (Figure 219A). They are club- or finger-shaped, 0.6–1.4 µm in diameter, and 2.5–6.0 µm in length. The short sporangiophores are 0.5–1.5 µm long and usually branched (Figure 219B). Scanning and transmission electron micrographs have revealed a collar-like structure (Figure 219A, B) at the sporangiophore–sporangium juncture (Ensign, 1978; Sharples et al., 1974; Shomura et al., 1980; Vobis and Kothe, 1985). Each sporangium contains a single straight chain of three to four spores. A minimum of two spores and a maximum of five are produced (Shomura et al., 1980, 1983b, 1985; Thiemann, 1974; Thiemann et al., 1967). The formation of the sporangiospores corresponds to the scheme of spore formation as proposed for the sporangiate actinomycetes by Lechevalier and Holbert (1965). Inside a thin, expanding envelope, the unbranched sporogenous hypha grows up to the final length of the sporangium (Figure 219B). It is then divided into spore-shaped sections, which round off immediately (Miyadoh et al., 1997). The separating cross walls are double-layered (Vobis, 1989c; Vobis and Kothe, 1985). New sporangia are formed by lateral branches of the sporangiophore in a subterminal position at the base of an older sporangium (Figure 219B) (Ensign, 1978; Vobis and Kothe, 1985). Strains with abnormally long and branched sporangia are described by Thiemann (1970a).

The sporangiospores (zoospores) produced inside the sporangia have a smooth surface and are variable in shape, from oblong, ellipsoidal, and ovoid to slightly pyriform (Figure 219A, B, C). They measure 0.4–1.3 µm in diameter and 0.5–1.8 µm in length (Shomura et al., 1985; Shomura et al., 1980; Shomura et al., 1983b; Thiemann et al., 1967). Young spores are thin-walled and include various reserve substances in their cytoplasm (Vobis, 1987). The zoospores are motile by means of a polar tuft of flagella (Figure 219C) (Higgins et al., 1967; Lechevalier and Lechevalier, 1970; Shomura et al., 1985; Thiemann, 1974).

In addition to the zoospores, large globose bodies or aleuriospores of 1.7–2.8 µm in diameter are formed singly on substrate

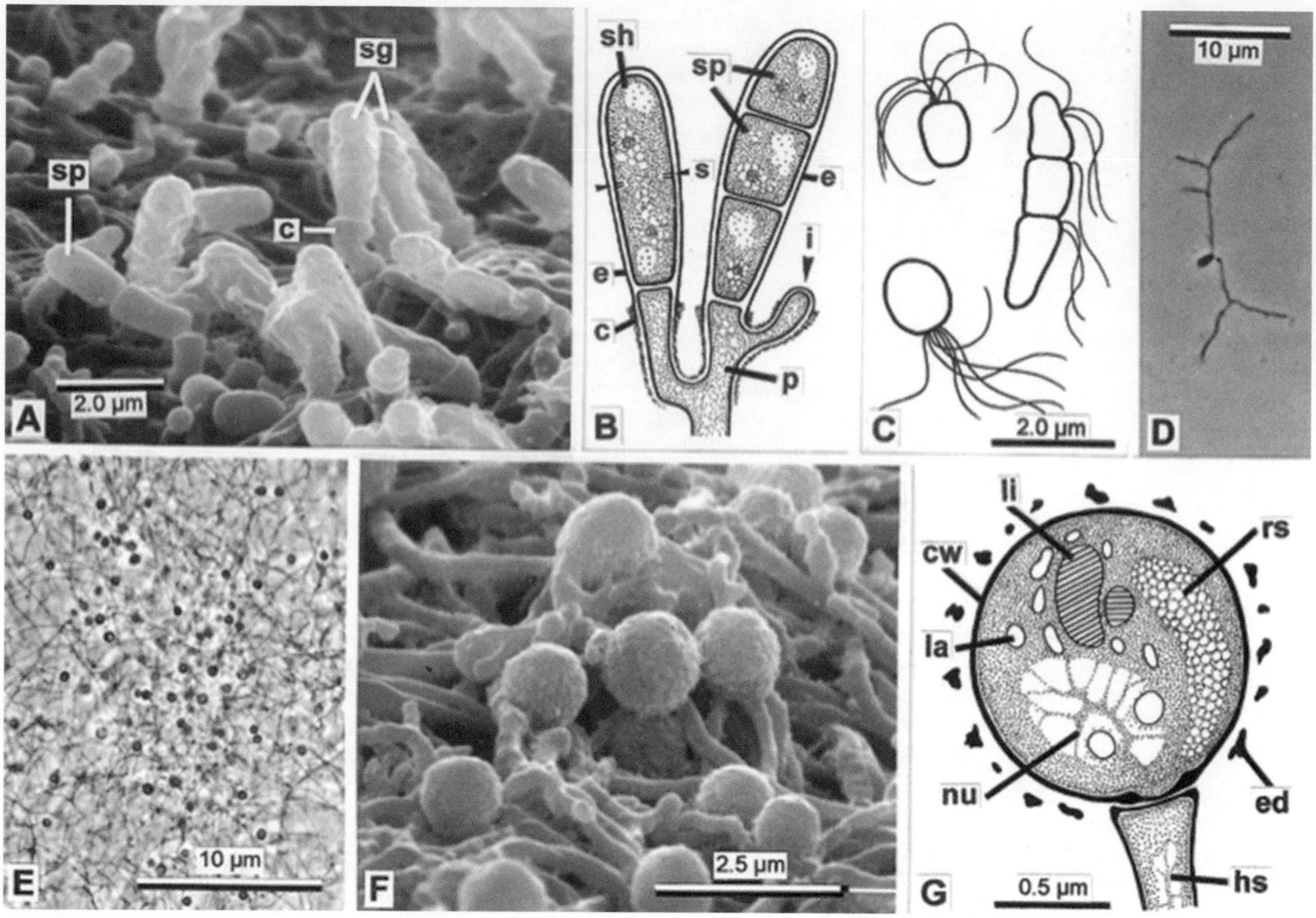

FIGURE 219. Morphological aspects of *Dactylosporangium*. A, Oligosporous sporangia (sg) with collars (c) and sporangiospores (sp) on the surface of the colony of *Dactylosporangium thailandense* (strain ATCC 23490^{T}; SEM); B, scheme of sporangial development (c, collar; e, envelope; i, initium; p, sporangiophore; s, septum; sh, sporogenous hypha; sp, spore); C, flagellated sporangiospores (zoospores); D, germinated zoospore with hyphae developed after 36 h in distilled water (strain MB-VS 704; PHACO); E, globose bodies produced by substrate mycelium (strain MB-VS 699; PHACO); F, substrate hyphae and globose bodies, partially covered with granular deposits (strain A 1486; SEM); G, scheme of the ultrastructure of a globose body (cw, cell wall; ed, extracellular deposit; hs, hyphal stalk; la, light area; li, lamellate inclusion; nu, area of DNA; rs, reserve substance). Abbreviations: PHACO, phase contrast microscopy; SEM, scanning electron microscopy.

hyphae. They can be embedded in the agar or freely exposed on the surface (Figure 219E, F), but are also produced in liquid cultures (Thiemann et al., 1967). When observed by light microscopy using the phase-contrast technique (Figure 219E), they appear as refractile spores (Ensign, 1978; Vobis, 1992). The globose bodies contain nuclear material, large diffuse electron-transparent areas, smaller defined light areas, lamellated, protein-containing paracrystalline inclusions, and possible phage particles (Sharples and Williams, 1974). The deposit of electron-dense material, irregularly distributed and closely attached to the wall surface (Figure 219F, G), is conspicuous (Miyadoh et al., 1997).

Cell-wall composition. The peptidoglycan of the cell wall contains 3-hydroxy-DAP and/or *meso*-DAP and glycine, with xylose and arabinose as diagnostic sugars in whole-cell hydrolysates (Hasegawa et al., 1983; Lechevalier and Lechevalier, 1970b; Shomura et al., 1980, 1983b, 1985, 1986). The chemical composition of cell walls therefore conforms to chemotype II and sugar pattern D in the classification scheme of Lechevalier and Lechevalier (1970). *Dactylosporangium aurantiacum*, *Dactylosporangium thailandense*, and "*Dactylosporangium salmoneum*" possess cell-wall peptidoglycans of the *N*-glycolylmuramic acid type, exhibiting amounts of 74.6, 60.0, and 92.0 nmol glycolyl residues per mg dried bacterial cells, respectively (Uchida and Seino, 1997). An unknown sugar other than 3-*O*-methylrhamnose, which can be related to that found in species of *Catellatospora*, was detected in the type strains of *Dactylosporangium matsuzakiense* and *Dactylosporangium thailandense* (Asano et al., 1989b). This unknown sugar is probably identical to the deoxyhexose reported by Szaniszlo and Gooder (1967).

The polar lipid profile is characterized by the presence of diphosphatidylglycerol, phosphatidylethanolamine (PE), phosphatidylglycerol (PG), phosphatidylinositol, and phosphatidylinositol mannosides (Goodfellow et al., 1990b; Lechevalier et al., 1977). The presence of PG can be variable (Lechevalier et al., 1977). Further uncharacterized glycolipids may be present and the type strain of *Dactylosporangium thailandense* is additionally characterized by two ninhydrin-positive, phosphate-positive lipids (Goodfellow et al., 1990b). Based upon the presence of PE and the absence of phosphatidylcholine and amino-containing phosphoglycolipid (GluNu) as diagnostic phospholipids, *Dactylosporangium* can be classified as having type II phospholipids of Lechevalier et al. (1981).

Fatty acids. Analysis of the fatty acids of the cytoplasmic membranes show that the principal components are branched iso- and anteiso-fatty acids (Kroppenstedt and Kutzner, 1978; Lechevalier et al., 1977). Cyclopropane and 10-methyl-branched fatty acids are not present (Kroppenstedt, 1979). In some strains, small amounts of unknown unsaturated fatty acids are found, which are assumed to have branched chains (Lechevalier et al., 1977). The type strains of *Dactylosporangium aurantiacum* and *Dactylosporangium thailandense* exhibit relatively high proportions of branched chain fatty acids, mainly iso-15 (4.1–15.8%), anteiso-15 (2.0–29.1%), iso-16 (6.2–18.3%), and anteiso-17 (4.4–14.2%), and smaller amounts of straight-chain constituents (13.7–32.9%) (Goodfellow et al., 1990b). The pattern of fatty acids corresponds to type 2d of the classification system of Kroppenstedt (1985).

The composition of the isoprenoid quinones is characterized by the possession of menaquinones with nine isoprene units (MK-9), whereas isoprenologues with ten units (MK-10) are absent. MK-9(H_2) and MK-9(H_4) are present in minor amounts, and MK-9(H_6) and MK-9(H_8) occupy predominant positions (Collins et al., 1984; Goodfellow et al., 1990b; Ruan et al., 1988). The presence of MK-9(H_4), MK-9(H_6), and MK-9(H_8) indicates that *Dactylosporangium* belongs to the menaquinone type 4b of the classification scheme of Kroppenstedt (1985).

Colonial characteristics. Colonies of *Dactylosporangium aurantiacum* are mostly flat with a smooth surface. The color of the substrate mycelium varies from whitish to pale orange to deep orange. The surface of the colonies of *Dactylosporangium thailandense* is usually smooth, but can be wrinkled on certain media. The substrate mycelium is light orange, amber, or brownish with a rose tinge (Thiemann et al., 1967). Colonies or *Dactylosporangium vinaceum* are compact, tough, and somewhat leathery; the mycelia are wine-colored to brown, depending on the medium (Shomura et al., 1983b). The colonial characteristics of *Dactylosporangium matsuzakiense* are similar to those of *Dactylosporangium vinaceum*; the substrate mycelium is orange (Shomura et al., 1983b). *Dactylosporangium roseum* has a rose-colored substrate mycelium on certain agar media (Shomura et al., 1985). Colonies of *Dactylosporangium fulvum* are yellowish brown (Shomura et al., 1986). Soluble pigments are produced by *Dactylosporangium vinaceum* (wine-colored to deep red), *Dactylosporangium thailandense* (amber to brown), and *Dactylosporangium matsuzakiense* (light brownish pink) (Shomura et al., 1985). Melanoid pigments are not produced (Table 196).

Antibiotic metabolites. The variety of the chemical structures of antibiotics isolated from *Dactylosporangium* strains indicates the great biosynthetic capability of this genus (Lancini and Lorenzetti, 1993). The aminoglycoside antibiotic dactimicin is produced by the type strains of *Dactylosporangium matsuzakiense* (SF-2052^T = ATCC 31570^T) and *Dactylosporangium vinaceum* (SF-2127^T = ATCC 35207^T) (Shomura et al., 1980; Shomura et al., 1983b). Dactimicin, a member of the pseudosaccharide group of antibiotics, is active against wide variety of Gram-stain-positive and Gram-stain-negative bacteria, including resistant strains with aminoglycoside-modifying enzymes (Omoto et al., 1987). *Dactylosporangium thailandense* strain G-367 produces the aminoglycoside antibiotics G-367-1 (2′-*N*-formylsisomicin) and G-367-2, a steric isomer of sisomicin, as well as sisomicin and gentamicin C1a and C2 (Fujii et al., 1982). Further aminoglycosides of the fortimicin antibiotic group are produced by *Dactylosporangium matsuzakiense* ATCC 31570^T (Dairi et al., 1992), and by the unidentified *Dactylosporangium* strain G 308. The antibiotic complex SF-2107 A-1, A_2, B and C, a member of the orthosomycin group, is produced by *Dactylosporangium roseum* strains SF-2107 (= ATCC 31744) and SF-2186^T (= NBRC 14352^T). The SF-2107 series substance is effective against Gram-stain-positive and Gram-stain-negative bacteria (Shomura et al., 1985).

Tiacumicins, a complex of 18-membered macrolide antibiotics, are isolated from strain AB718C-41 (= NRRL 18085), described as *Dactylosporangium aurantiacum* subsp. "*hamendensis*" (Theriault et al., 1987). The diarrhea-associated bacterium *Clostridium difficile* can be combated successfully *in vitro* and *in vivo* with tiacumicins B and C (Swanson et al., 1991).

The three strains FD 25647 (= ATCC 31222), FD 25712 (= ATCC 31223), and FD 25718 (= ATCC 718), determined as "*Dactylosporangium salmoneum*", produce the polycyclic ether antibiotic 44161, which is active against Gram-stain-positive bacteria, fungi, and protozoa, and promotes growth of animals. It exhibits potent anticoccidial activity (Celmer et al., 1978). The compound 44161 is identical to nigerimicin, which is also known as a herbicidal agent (Heisey and Putnam, 1990).

The tetracyclic antibiotic dactylocyclinone Sch 34164 (4α-hydroxy-8-methoxychlortetracycline) has been isolated from a culture of strain SCC 1695 (= ATCC 39499), determined as "*Dactylosporangium vescum*" (Patel et al., 1987). Sch 34164 (dactylocycline A) and dactylocycline B, isolated from strain SC 14051 (= ATCC 53693) have tested positive against tetracycline-resistant bacteria (Tymiak et al., 1993). Another polyketide-derived antibiotic, DK-7814-A (hydroxypurpuromycin), has been isolated from "*Dactylosporangium purpureum*" (Lancini and Lorenzetti, 1993). DK-7814-A, DK-7814-B, and DK-7814-CO are active against Gram-stain-positive bacteria and show antitumor activity.

The polypeptide compound capreomycin can also be obtained from strain D-409-5 (= ATCC 31203), originally described as "*Dactylosporangium variesporium*" (Tomita et al., 1977). Capreomycin is of primary interest for its use as an antituberculosis agent. Strain SF-2253 produces L-threo-β-hydroxyaspartic acid, an antibiotic useful against a wide spectrum of micro-organisms. This amino acid is also an inhibitor of glutamate uptake, frequently used in neurological studies (Alexander et al., 1997).

The type strain of *Dactylosporangium fulvum* (SF 2113^T = ATCC 43301^T) produces pyridomycin (Shomura et al., 1986), which is known as an antimycobacterial antibiotic (Maeda et al., 1953). The acidic substance SF 2185 is an antibiotic against plant pathogens, particularly the causal organisms of cucumber downy mildew and rice blast. The producing strain is SF-2185, determined as *Dactylosporangium aurantiacum* subsp. "*gifuense*" (Matsumoto et al., 1985). *Dactylosporangium aurantiacum* strain SANK 61299 produces the plant growth inhibitors streptol, A-79197-2 (disaccharide of streptol) and A-79197-3 (trisaccharide

TABLE 196. Diagnostic and physiological characteristics of the species of the genus *Dactylosporangium*[a]

Characteristic	*D. aurantiacum*	*D. fulvum*	*D. matsuzakiense*	*D. roseum*	*D. thailandense*	*D. vinaceum*
Morphological and colonial:						
Formation of globose bodies	+	+	$-^{c}$, v^{g}	$-^{d}$, p^{g}	+	$+^{b}$, v^{g}
Formation of coremia	–	+	–	–	–	–
Color of substrate mycelium:						
Yellowish brown	–	+	–	–	–	–
Orange	+	–	+	–	–	–
Orange to brown	–	–	–	–	+	–
Wine to brown	–	–	–	–	–	+
Rose	–	–	–	+	–	–
Formation of diffusible pigment:						
Light brownish pink	–	v	v	–	–	–
Amber or brown	–	–	–	–	v	–
Wine to deep red	–	–	–	–	–	+
None	+	–	–	+	–	–
Production of melanoid pigment	–	–	–	–	–	–
Utilization of carbon sources:						
L-Arabinose	+	+	+	v	(+)	+
Dextrin	(+)	nd	nd	nd	(+)	nd
D-Dulcitol	–	nd	nd	nd	–	nd
D-Fructose	+	+	+	+	+	+
D-Galactose	+	nd	nd	nd	(+)	nd
D-Glucose	+	+	+	+	(+)	+
Glycerol	–	–	–	nd	–	–
i-Inositol	–	–	–	–	–	–
Inulin	(+)	nd	nd	nd	(+)	nd
Lactose	(+)	nd	nd	nd	(+)	nd
Maltose	(+)	nd	nd	nd	+	nd
D-Mannitol*	+	+	+	–	(+)	+
D-Mannose	+	nd	nd	nd	(+)	nd
Melibiose*	+	nd	–	nd	–	nd
Raffinose	$(+)^{e}$, $-^{b,c}$	–	–	–	$+^{e}$, $-^{b,c}$	–
L-Rhamnose*	(+)	–	+	–	(+)	+
D-Ribose*	–	nd	nd	nd	+	nd
D-Sorbitol	–	nd	nd	nd	–	nd
Sorbose	–	nd	nd	nd	nd	nd
Sucrose	(+)	+	+	+	+	+
D-Xylose	+	+	+	v	+	+
Degradation of:						
Calcium malate	–	nd	nd	nd	–	nd
Casein	+	+	nd	nd	+	nd
Cellulose	–	nd	nd	nd	–	nd
Esculin	nd	+	nd	nd	nd	nd
Gelatin	$-^{e}$, $+^{b,g}$	–	–	+	$+^{e}$, $-^{b,g}$	+
Hypoxanthine	nd	–	nd	nd	nd	nd
Starch*	+	+	+	–	+	+
Tyrosine	–	nd	nd	nd	+	nd
Xanthine	nd	–	nd	nd	nd	nd
Peptonization of milk	$+^{f}$, $-^{g}$	–	–	–	$+^{f}$, $-^{g}$	+
Coagulation of milk*	–	–	–	–	–	+
Reduction of nitrate*	+	+	–	+	–	–
Production of H_2S	+	nd	nd	nd	+	nd
Other physiological properties:						
Optimum pH for growth	6.0–7.0	nd	nd	nd	6.0–7.0	nd
Optimum temperature for growth, °C	28–37	26–34	25–37	28–37	28–37	25–37
Tolerance to NaCl, % (w/v)	3.0	3.0	<1.5	1.5	1.5	3.0

[a]v, Strain instability; p, poor; nd, not determined; (+), moderate. Characters of diagnostic value are marked with an asterisk. Deviations or incoherent data are indicated by superscript letters:

[b]Shomura et al. (1983b).

[c]Shomura et al. (1980).

[d]Shomura et al. (1985).

[e]Thiemann et al. (1967).

[f]Thiemann (1974).

[g]Shomura et al. (1986).

of streptol), which inhibit the germination of *Brassica rapa*. The streptol moiety seems to be important for herbicidal activity (Kizuka et al., 2002).

Life cycle. Members of the genus *Dactylosporangium* produce two kinds of spores. The zoospores, which do not seem to be dormant, may function to ensure dissemination of the organisms. The nonmotile globose bodies or aleuriospores, which are constitutively dormant, may function to provide insurance for surviving during long periods (Ensign, 1978).

The few-spored sporangia are developed by substrate mycelium in contact with the air, standing out permanently from the substrate. The short spore chains inside the sporangia are held together by the thin sporangial envelope. They may function as an "aerophilic" survival unit, easily distributed in the soil environment or by the wind. If the sporangia are immersed in water, the zoospores are released after a period of 10–60 min (Thiemann et al., 1967). A solution of soil extract promotes this process (Hayakawa et al., 2000). Spore release is probably initiated by the swelling of an intrasporangial matrix (Thiemann et al., 1967), or by swelling of the substance forming the collar at the base of the sporangium (Figure 219A, B). The short chain of sporangiospores is pressed upward and pushed through the apex of the sporangial envelope. High motility starts after a time lag of about 30 min when the spores first separate from one another (Shomura et al., 1980, 1983b; Thiemann et al., 1967). The cytoplasm of the spores includes reserve material (Vobis, 1987, 1992). Spores are able to swim for up to 30 h. Germination was reported after 24 h on agar medium (Thiemann et al., 1967). It occurs also in distilled water (Figure 219D). By a consequent production of new mycelium and sporangia, the typical "aero-aquatic" live cycle of an actinoplanete is formed (Cross, 1986; Vobis, 1987, 1992).

The globose bodies, which are formed by substrate hyphae, are in general submersed in the substrate, but can also develop very closely attached to the surface (Figure 219E, F). Like the zoospores, they contain reserve material (Figure 219G) (Vobis, 1992). Ensign (1978) demonstrated that the globose bodies can germinate in a 10% (w/v) yeast extract solution and concluded that they have the same function as true, nonmotile spores. Compared with the flagellated sporangiospores, the globose bodies have a different physiological behavior when treated with the chemical germicide benzethonium chloride (BC). After being exposed for 30 min at 30°C in 0.01% solution of BC, 51% of the globose bodies survived, whereas the zoospores of *Dactylosporangium* and *Actinoplanes* did not (Hayakawa, 2003). This is compatible with the ecological observation of Johnston and Cross (1976), namely that *Dactylosporangium* species can survive in lake sediments that contain no viable *Actinoplanes*. Cross (1986) concluded that *Dactylosporangium* has two different types of survival stages.

Cultural characteristics and physiology. *Dactylosporangium* species are aerobic and mesophilic. The optimum temperature for growth is between 25 and 37°C (Table 196). The species *Dactylosporangium aurantiacum*, *Dactylosporangium matsuzakiense*, and *Dactylosporangium roseum* tolerate temperatures up to 42°C; no growth occurs at 45°C (Shomura et al., 1986; Thiemann et al., 1967). Good vegetative growth occurs on various media, e.g. oatmeal agar, Bennett agar, Hickey–Tresner agar, nutrient agar, glucose-asparagine agar, sucrose-nitrate agar, sucrose-yeast extract agar, yeast extract-malt extract agar, and inorganic salt-starch agar. However, individual species grow well on only a small spectrum of these media (Shomura et al., 1980, 1983b, 1985, 1986; Thiemann et al., 1967).

For carbon utilization tests, a different basal media was employed because individual strains do not grow on a common basal medium. A modified M/40 medium of Magni and von Borstel was used for *Dactylosporangium aurantiacum*, which had to be additionally supplemented with vitamins for *Dactylosporangium thailandense* (Thiemann et al., 1967). For other species, carbohydrate utilization was determined on Luedemann–Brodsky basal medium (Shomura et al., 1986). The results of the carbon utilization tests are given in Table 196. *Dactylosporangium* can be characterized as a neutrophilic micro-organism, growing well at pH 6.0–7.0 (Thiemann et al., 1967). Species can grow on Luedemann agar medium containing concentrations of NaCl up to 3%, but do not tolerate a concentration of 4% (Shomura et al., 1986). The ability to degrade polymers of biological origin and other physiological characters are not sufficiently studied. The physiological characteristics are summarized in Table 196.

The development of sporangia depends on the agar media used. They can be formed after 2–3 d under favorable conditions (Ensign, 1978), although normally they are only evident after 5–15 d of incubation (Shomura et al., 1983b). Sporangial formation can be promoted by soil agar, calcium malate agar, and inorganic salts-starch agar (Shomura et al., 1986; Thiemann et al., 1967). Globose bodies appear to be produced mainly on complex agar media that promote the growth of substrate mycelium, but not the formation of sporangia (Ensign, 1978). *Dactylosporangium fulvum* develops globose bodies abundantly on glucose-asparagine agar and tyrosine agar (Shomura et al., 1986).

The physiological behavior of the flagellated sporangiospores has been demonstrated by their chemotactic responses towards chemoattractants (Hayakawa et al., 1991d). The studies were carried out by microcapillary assays and included the type strains of *Dactylosporangium thailandense*, *Dactylosporangium aurantiacum*, *Dactylosporangium vinaceum*, *Dactylosporangium matsuzakiense*, *Dactylosporangium roseum*, and *Dactylosporangium fulcrum*. γ-Collidine at a concentration of 100 mM was the most effective chemical attractant, followed by vanillin (10 or 100 mM), xylose (10 mM), and KCl (10 mM) (Hayakawa, 2003).

Phages. A poorly lytic bacteriophage was discovered in a strain of *Dactylosporangium thailandense* that did not infect other actinomycetes (Higgins and Lechevalier, 1969). Cross-infections with phages of various actinomycetes were not successful (Prauser, 1984; Willoughby et al., 1972).

Ecology. Members of the genus *Dactylosporangium* are distributed worldwide. Thiemann (1970a) reported that out of 454 soil samples from various parts of the world, 140 isolates of *Dactylosporangium* were obtained. They were present in sandy as well as in loamy soils. No correlation could be established between the type of soil, its pH (4.0–9.0), and the incidence of *Dactylosporangium*. A total of 33 strains could be isolated from different soil samples from Thailand, Brazil, and Argentina (Thiemann et al., 1967). Further strains could be isolated from soil of an uncultivated field of grass in Colombia, South America (Shearer, 1987), and from soil samples collected in tropical and subtropical regions in Yunnan, China (Xu et al., 1996).

The dactylocycline producing strain SCC 1695 (=ATCC 39449) was isolated from a soil sample collected in Zambia, Africa. We isolated strains MB-VS 699 und MB-VS 704, presented in Figure 219, from soil samples collected at Lüneburger Heide (Germany). Strain A 1486, together with two further undetermined *Dactylosporangium* strains, were isolated from a soil sample from the Nationalpark Taman Negawa, Malaysia (G. Vobis and J.M. Wink, unpublished results).

The species *Dactylosporangium fulvum*, *Dactylosporangium matsuzakiense*, *Dactylosporangium roseum*, and *Dactylosporangium vinaceum* all originated from soil samples collected at various localities in Japan (see "List of Species"). Hayakawa and Nonomura (1987a) obtained isolates from soil samples from vegetable and corn fields with pH ranging from 5.4 to 6.1, collected in different Prefectures (Nagano, Mie, Gunma, and Iwate) in Japan. Field soils seem to be the most fruitful sources for isolating *Dactylosporangium* strains, together with other diverse rare actinomycete taxa, but they were also isolated frequently from mountainous forest soils. Field, mountain grass-land, and rice paddy with soil of pH 6.0–7.0, organic matter content <5%, and immature brown humic acid <0.8 (Δlog K) are the characteristic soil habitats of *Dactylosporangium* (Hayakawa, 2003).

Only a few sources other than soil were successfully tested as natural substrates inhabited by *Dactylosporangium* or simply utilized as intermediate locations by its resistant structures like the globose bodies. Johnston and Cross (1976) isolated strains from the surface muds of two lakes of the English Lake District in Great Britain. *Dactylosporangium* strains were also found on plant debris (Lechevalier, 1981). Leaf litter in marsh water in New Jersey (USA) was used as substrate to isolate the antibiotic dactylocycline-producing strain SC 14051 (= ATCC 53693). More recently, Okazaki (2003) reported an antibiotic-producing strain of *Dactylosporangium aurantiacum* isolated from fresh plant leaves of *Cucubalus* sp.

Enrichment and isolation procedures

The details of the very effective isolation technique used by Thiemann et al. (1967) were never published. Strains of *Dactylosporangium* can be detected by isolation methods used for other genera of actinoplanetes, including *Micromonospora* (Vobis, 1992). They are characterized as slow-growing organisms, recognizable on poor nutritive agar media (Shomura et al., 1980). In general, the soil samples are pretreated, e.g. with dry heat, and dilutions are spread onto specific selective isolation media, which can be supplemented with antibiotic agents (Hayakawa and Nonomura, 1987a; Nonomura, 1984; Shearer, 1987; Xu et al., 1996). It seems that all these abovementioned methods, including also the traditional chemotactic technique of Palleroni (1980), have only incidentally recovered *Dactylosporangium* species in soil.

More recently, Hayakawa (2003) summarized his own valuable experiences of isolating rare actinomycetes. To isolate selectively high numbers of strains of *Dactylosporangium* from soil, combinations of several techniques are recommended. At first, the soil samples are dried slowly at room temperature for a week, sieved, and ground slightly in a mortar (Nonomura and Ohara, 1969). After that, the samples can be pretreated physically with dry heat (120°C) for 1 h, followed by treatment with the chemical germicide BC (0.01 or 0.03%), exposed for 30 min at 30°C (Hayakawa et al., 1991a). Especially in the latter case, the globose bodies (aleuriospores) function as the surviving units.

The enrichment procedures profit by the release of zoospores from the sporangia in an aqueous environment. One possibility is an improved chemotactic method employing the capillary technique of Palleroni (1980), but using γ-collidine or vanillin (100 mM) as chemoattractant instead of the traditional 0.01 M KCl (Hayakawa et al., 1991d). A further enrichment method is named "rehydration and centrifugation" (RC) (Hayakawa et al., 2000). The samples are flooded with 10 mM phosphate buffer containing 1% soil extract at 30°C for 90 min. The fluid is centrifuged at 1500 × ***g*** for 20 min and the supernatant, containing actively swimming zoospores, is used for plating.

The preferable selective isolation medium is humic acid-vitamin (HV) agar, containing (per liter) 1.0 g humic acid, 0.02 g $CaCO_3$, 0.5 g NaH_2PO_4, 1.7 g KCl, 0.5 g $MgSO_4 \cdot 7H_2O$, 0.01 g $FeSO_4 \cdot 7H_2O$, 18.0 g agar, and B vitamins (0.5 mg each of thiamine·HCl, riboflavin, niacin, pyridoxin·HCl, inositol, Ca-pentothenate, *p*-aminobenzoic acid, and 0.25 mg biotin); pH 7.2 (Hayakawa and Nonomura, 1987b, 1987a). The HV agar is supplemented with nalidixic acid (20 mg/l) (Hayakawa et al., 1995; Hayakawa et al., 1991d; Nonomura and Hayakawa, 1988), and additionally with leucomycin or tunicamycin (20 mg/l or 30 mg/l) in the case of dry heat and BC pretreatment (Hayakawa et al., 1991a), or trimethoprim (20 mg/l) for the RC enrichment method (Hayakawa et al., 2000). Cycloheximide (50 mg/l) is also generally added as an antifungal compound. The plates are incubated at 30°C for 4–6 weeks.

An alternative source for isolation of *Dactylosporangium* was demonstrated by Okazaki (2003). Instead of soil, leaf samples can be investigated as a habitat for actinomycetes. Freshly picked leaves are cut into several pieces, rinsed with sterile water, and soaked in 70% ethanol for 1 min. They are then washed once more with sterile water, soaked in 1% NaClO for 3 min, and rinsed again with sterile water. After these treatments, the leaf pieces are incubated on 0.8% water agar for several weeks. Although the percentage of *Dactylosporangium* species recovered seems to be very low, a strain of particular interest based on its secondary metabolite production was isolated (Kizuka et al., 2002).

Maintenance procedures

Slant agar cultures can be stored for several weeks at room temperature or at 6°C. For long-term preservation, the strains can be lyophilized or maintained by other techniques recommended for the aerobic actinomycetes.

Differentiation of the genus *Dactylosporangium* from other related genera

The morphological characteristics of *Dactylosporangium*, namely the oligosporous sporangia with 2–5 flagellated sporangiospores and the typical globose bodies, both produced by substrate mycelium, are sufficient to differentiate the genus from other genera of the family *Micromonosporaceae* (Vobis, 1992). The newly described genera *Virgisporangium* (Tamura et al., 2001) and *Planosporangium* (Wiese et al., 2008) have similar sporangia, but small differences are observed (Table 197). Thiemann (1970a) observed *Dactylosporangium*-like strains with deviating morphologies of sporangia, resembling *Planosporangium* or *Virgisporangium* species, and also strains having short chains of nonmotile spores like *Catellatospora* species. He announced the description of new genera, which was never realized. The closely related genera *Dactylosporangium*, *Planosporangium*, *Virgisporangium*, *Catellatospora*,

TABLE 197. Characteristics differentiating *Dactylosporangium* from closely related genera forming either oligosporous sporangia with motile sporangiospores or single, nonmotile spores/globose bodies on substrate mycelium[a,b]

Characteristic	*Dactylosporangium*	*Planosporangium*	*Virgisporangium*	*Catellatospora*	*Asanoa*
Production of sporangia	+	+	+	–	–
Number of sporangiospores	2–5	3 or more	6 or more	–	–
Type of motility of sporangiospores	Tuft of flagella	Single polar flagella	nd	–	–
Production of single spores (0.7–1.5 µm in diameter)	–	+	–	–	–
Production of globose bodies (1.7–2.8 µm in diameter)	+[c]	–	–	+[b]	–
Production of short chains of nonmotile spores	–	–	–	+	+
Fatty acid type[d]	2d	3b	2d	nd	2d
Major menaquinones	MK-9(H_4, H_6, H_8)	MK-9(H_4), MK-10(H_4)	MK-10(H_4, H_6, H_8)	MK-9(H_4)	MK-10(H_6, H_8)

[a]Symbols: +, >85% positive; –, 0–15% positive; nd, not determined.

[b]Data compiled from Ara et al. (2008a), Asano and Kawamoto (1986), Lee and Hah (2002), Lee et al. (2000b), Tamura et al. (2001) and Wiese et al. (2008).

[c]Not confirmed with certainty for all species.

[d]According to the classification of Kroppenstedt (1985).

and *Asanoa* can be differentiated by a combination of morphological and chemical characteristics (Table 197).

Taxonomic comments

When the first two species of *Dactylosporangium* were described by Thiemann et al. (1967), the genus was classified in the family *Actinoplanaceae* (Couch, 1955a, 1955b), which comprised all sporangia-forming actinomycetes. Within this morphologically defined family, the three genera *Dactylosporangium*, *Planomonospora*, and *Planobispora* were characterized by having oligosporous sporangia and flagellated sporangiospores (Thiemann, 1970a). At the same time, it could be shown that the *Actinoplanaceae* could be divided into two groups, one characterized by cell-wall chemotype II and the other by chemotype III (Lechevalier and Lechevalier, 1970). By having cell-wall chemotype II, *Dactylosporangium* was closely related to the genus *Actinoplanes*, but could be plainly distinguished from other genera by morphological and chemotaxonomic characteristics (Vobis, 1989b, 1989c). The genus *Dactylosporangium* constantly kept its taxonomic position as the "good little brother" of *Actinoplanes* throughout all changes of classification concepts of generic and suprageneric ranks (Bland and Couch, 1981; Goodfellow, 1989; Goodfellow and Cross, 1984; Goodfellow et al., 1990b; Koch et al., 1996a; Miyadoh et al., 1997, 2001; Stackebrandt and Kroppenstedt, 1987; Stackebrandt et al., 1997; Zhi et al., 2009).

Differentiation of species of the genus *Dactylosporangium*

The species of *Dactylosporangium* have essentially the same morphological characteristics. They cannot be differentiated based on sporangia and zoospores, although globose bodies may or may not be present (Table 196). The formation of coremia is a special morphological characteristic for *Dactylosporangium fulvum* (Shomura et al., 1986).

Species differentiation is mainly based on the color of the substrate mycelium and on the production of soluble pigments. *Dactylosporangium aurantiacum* has orange substrate mycelium and does not produce pigment. *Dactylosporangium thailandense* has orange to brown mycelium and produces pigments on some media that are amber to brown with a reddish tinge (Thiemann, 1974; Thiemann et al., 1967). A wine-colored to deep red diffusible pigment characterizes *Dactylosporangium vinaceum* and the mycelium of this species is similarly colored. *Dactylosporangium matsuzakiense* has orange colonies and produces a light brownish-pink pigment on tyrosine agar (Shomura et al., 1983b). *Dactylosporangium roseum* is characterized by the rose color of its substrate mycelium (Shomura et al., 1985), and *Dactylosporangium fulvum* by a yellowish brown substrate mycelium and light brownish pink diffusible pigment (Shomura et al., 1986). *Dactylosporangium* species differ slightly in their patterns of carbon utilization, e.g. D-mannitol or L-rhamnose, and some physiological properties like degradation of starch and reduction of nitrate (Table 196).

List of species of the genus *Dactylosporangium*

1. **Dactylosporangium aurantiacum** Thiemann, Pagani and Beretta 1967, 43[AL]

au.ran.ti.a'cum. N.L. neut. adj. *aurantiacum* orange colored.

Sporangia (4.0–6.0 µm in length) develop on the surface of soil agar and on calcium malate agar. The zoospores, released 10–15 min after placing the sporangia in water, are oval, oblong, and slightly pyriform (1.5 µm long by 1.0 µm wide). They are extremely vigorous swimmers. Globose bodies are formed by substrate mycelium.

Colonies on agar media are mostly flat with a smooth surface. Abundant to good growth occurs on oatmeal agar and nutrient agar. The color of the substrate mycelium is pale orange to orange. On Hickey–Tresner agar, growth is moderate with very pale mycelium. On tyrosine, nutrient, and skim milk agars, the colonies are orange to deep orange. Whitish colonies occur on glucose-asparagine agar and on starch agar. On glycerol-asparagine agar and on peptone-iron agar, the colonies are hyaline.

D-Mannitol, L-rhamnose, and melibose are utilized for growth; D-ribose is not. NaCl is tolerated up to 3% (w/v). Nitrate is reduced to nitrite. Litmus milk is not coagulated. No soluble pigments are produced on any media. The type strain shows the highest sequence similarity to *Dactylosporangium matsuzakiense* (98.9%) and *Dactylosporangium vinaceum* (98.6%). All other species show sequence similarities below 98.5%.

Source: the type strain was isolated from soil; no further information is given regarding its origin.

DNA G+C content (mol%): 73 (T_m).

Type strain: ATCC 23491, D-748, DSM 43157, NBRC 12592, JCM 3083, NRRL B-8111.

Sequence accession no. (16S rRNA gene): X93191.

2. **Dactylosporangium fulvum** Shomura, Amano, Yoshida and Kojima 1986, 169[VP]

ful'vum. L. neut. adj. *fulvum* yellowish brown, referring to the color of the vegetative mycelium.

Sporangia (0.7–1.0 × 2.5–5.0 µm in size) with three to four spores in a single row are produced abundantly on sodium succinate agar, and moderately on calcium malate agar and inorganic salts-starch agar. Sporangiospores are motile. Globose bodies are formed abundantly on glucose-asparagine agar and tyrosine agar (ISP medium 5), and moderately on oatmeal agar, glycerol-asparagine agar and calcium malate agar. Coremia are formed abundantly on inorganic salts-starch agar and oatmeal agar, and moderately on yeast extract-malt extract agar and sucrose-yeast extract agar. Sporangia and globose bodies also develop directly from coremia.

The color of the substrate mycelium is yellowish brown and does not change when treated with alkaline or acid solutions. Except for a light brownish pink color on tyrosine agar, neither melanoid nor distinct diffusible pigments are not produced. Good growth occurs on sucrose-yeast extract agar, yeast extract-malt extract agar, and inorganic salts-starch (ISP medium 4) with light brown to cinnamon color. It grows moderately on Bennett agar, nutrient agar, tyrosine agar (ISP medium 7), oatmeal agar (ISP medium 3), and sucrose nitrate agar.

Decomposes casein and esculin; hydrolyzes starch. Nitrate is reduced to nitrite. Decomposition of xanthine and hypoxanthine, liquefaction of gelatin, and peptonization and coagulation of milk are negative. Optimum growth is between 26 and 34°C. Tolerates 3% NaCl, but no growth occurs on media containing more than 4% NaCl.

Produces pyridomycin. The type strain shows the highest sequence similarity to that of *Dactylosporangium roseum* (99.6%). The type strains of all other *Dactylosporangium* species show sequence similarities below 98.5%.

Source: the type strain was isolated from a soil sample collected in Chiba, Japan.

DNA G+C content (mol%): not determined.

Type strain: ATCC 43301, DSM 43917, NBRC 14381, JCM 5631, NRRL B-16292, SF-2113.

Sequence accession no. (16S rRNA gene): X93192.

3. **Dactylosporangium matsuzakiense** Shomura and Niida *in* Shomura, Kojima, Yoshida, Ito, Amano, Totsugawa, Niwa, Inouye, Ito and Niida 1983a, 672[VP] (Effective publication: Shomura, Kojima, Yoshida, Ito, Amano, Totsugawa, Niwa, Inouye, Ito and Niida 1980, 928.)

mat.su.za.ki.en'se. N.L. neut. adj. *matsuzakiense* of or pertaining to Matsuzaki-cho, Izu Peninsula, Japan.

Finger-shaped sporangia are 4.0–6.0 × 0.9–1.4 µm in size, formed on short sporangiophores (0.5–1.0 µm), and occurring singly or in clusters. Each sporangium contains usually three spores, arranged in a single row. Sporangiospores are cylindrical to oblong (1.1–1.6 × 0.8–1.3 µm), motile. Abundant production of sporangia occurs on inorganic salts-starch agar, but is rare on Czapek, oatmeal, and tyrosine agars. Formation of globose bodies is not observed.

Colonies grow well on inorganic salts-starch agar, with a russet-orange substrate mycelium. Moderate growth occurs on Czapek agar and on yeast extract-malt extract agar, with an amber to light brown mycelium. Colonies grow moderately on glucose-asparagine agar and on oatmeal agar; the substrate mycelium is russet to orange. Poor growth occurs on glycerol-asparagine, nutrient, and calcium malate agars, the color of substrate mycelium varies from light yellow to light orange. Moderate growth occurs on tyrosine agar; the color of colonies is dusty orange to light brown and a light brownish-pink soluble pigment is produced.

D-Mannitol and L-rhamnose are utilized as sole carbon sources; melibiose is not. Growth occurs between 15 and 42°C. NaCl tolerance is lower than 1.5% (w/v). Gelatin is not liquefied; nitrate is not reduced. Skim milk is neither peptonized nor coagulated.

Produces the aminoglycoside antibiotic dactimicin. The type strain shows the highest sequence similarity to the type strains of *Dactylosporangium vinaceum* (99.7%) and *Dactylosporangium aurantiacum* (98.8%). The type strains of all other *Dactylosporangium* species show sequence similarities below 98.5%.

Source: the type strain was isolated from a soil sample collected at Matsuzaki-cho, Izu Peninsula, Japan.

DNA G+C content (mol%): not determined.

Type strain: ATCC 31570, DSM 43810, FERM-P 4670, NBRC 14259, JCM 3311, NRRL B-16293, SF-2052.

Sequence accession no. (16S rRNA gene): X93193.

4. **Dactylosporangium roseum** Shomura, Amano, Tohyama, Yoshida, Ito and Niida 1985, 4[VP]

ro'se.um. L. neut. adj. *roseum* rose colored, pink.

Pod-shaped sporangia (2.5–5.5 × 0.8–1.1 µm), containing a single row of spores, are abundantly formed on chemically defined media such as sucrose-nitrate, glucose-asparagine, glycerol-asparagine, inorganic salts-starch, calcium malate, and tyrosine agars, but rarely on yeast extract-malt extract, nutrient, and Bennett agars. Sporangiospores are ellipsoidal (0.8–1.1 × 0.6–1.5 µm) and motile by means of polarly inserted flagella. The formation of globose bodies has not been observed.

Colonies grow well on sucrose-yeast extract agar, and moderately on yeast extract-malt extract, inorganic salts-starch, oatmeal, Bennett, and tyrosine agars. The typical rose color of the substrate mycelium occurs on sucrose-yeast extract, inorganic salts-starch, and yeast extract-malt extract agars. On Bennett agar, the color of mycelium is pastel orange and on tyrosine agar it is yellowish. Colonies on sucrose-nitrate, glucose-asparagaine, glycerol-asparagine, calcium malate, oatmeal, and nutrient agars are colorless. Soluble pigments are not produced.

D-Mannitol and L-rhamnose are not utilized as sole carbon sources. NaCl is tolerated up to 1.5% (w/v). Growth occurs between 20 and 40°C; the temperature optimum ranges from 28 to 37°C. Nitrate is reduced to nitrite and gelatin is liquefied. Starch is not hydrolyzed; skim milk is neither peptonized nor coagulated. Melanoid pigments are not produced.

Produces antibiotic complex SF-2107, a member of orthosomycin group. The type strain shows the highest sequence similarity to the type strain of *Dactylosporangium fulvum* (99.6%). The type strains of all other *Dactylosporangium* species show sequence similarities below 98.5%.

Source: the type strain was isolated from a sample collected at Shizuoka, Japan; a further strain (SF-2107) was from soil collected at Yokohama, Japan.

DNA G+C content (mol%): not determined.

Type strain: DSM 43916, NBRC 14352, JCM 3364, NRRL B-16295, SF-2186.

Sequence accession no. (16S rRNA gene): X93194.

5. **Dactylosporangium thailandense** Thiemann, Pagani and Beretta 1967, 49[AL]

thai.lan.den'se. N.L. neut. adj. *thailandense* of or pertaining to Thailand. The correct Latin epithet of the species was proposed by Thiemann (1970b).

Abundant sporangial formation occurs on soil agar, calcium malate agar, and starch agar. Colonies grow on various agar media with a wrinkled or smooth surface. Good growth occurs on oatmeal agar with light orange-brown substrate mycelium, producing a light amber diffusible pigment. On Hickey–Tresner agar, growth is also good; the mycelium is brown with a rose tinge and the pigment is brown to reddish pink. Growth on glycerol-asparagine, glucose-asparagine, and starch agars is moderate, with a light orange to orange substrate mycelium. On nutrient agar, growth is good and the mycelium is pale orange.

D-Ribose, D-mannitol, and L-rhamnose are utilized for growth; melibiose is not. Tolerates up to 1.5% (w/v) NaCl. Tyrosine is hydrolyzed. Nitrate is not reduced to nitrite. Milk is not coagulated. No growth occurs at 42°C. Production of brown soluble pigments occurs on some media. Sequence similarities between the type strain and those of all other species of the genus are below 98.5%.

Source: the type strain was isolated from a soil sample collected in Thailand.

DNA G+C content (mol%): 71 (T_m).

Type strain: ATCC 23490, D-449, DSM 43158, NBRC 12593, JCM 3084.

Sequence accession no. (16S rRNA gene): X92630.

6. **Dactylosporangium vinaceum** Shomura, Yoshida, Miyadoh, Ito and Niida 1983b, 312[VP]

vi.na'ce.um. L. neut. adj. *vinaceum* of or belonging to wine, intended to mean wine-colored.

Finger-shaped sporangia (3.0–5.5 × 0.8–1.1 µm), produced singly or in tufts. Sporangia are occasionally apparent on Czapek and oatmeal agars, but less apparent on inorganic salts-starch and on calcium malate agar. Each sporangium contains three sporangiospores arranged in a single row. Spores are cylindrical to oblong (0.9–1.8 × 0.6–0.9 µm) and motile. Globose bodies are produced on glucose-asparagine and nutrient agars.

Colonies grow well on Czapek, glucose-asparagine, inorganic salts-starch, oatmeal, Bennett, and Hickey–Tresner agars. The color of substrate mycelium or reverse color ranges from light to dark wine-colored or occasionally ebony brown. Poor growth occurs on glycerol-asparagine agar with apricot substrate mycelium and no or very light wine-colored pigment production. Production of a wine red-colored diffusible pigment is conspicuous on various agar media, shaded from rose-wine to dark red or cherry. The pigments of the substrate mycelium and the soluble pigment are stable with changes in pH.

D-Mannitol and L-rhamnose are utilized as carbon source for growth. Gelatin and casein are hydrolyzed; milk is coagulated. Nitrate is not reduced to nitrite. NaCl concentrations of up to 3% (w/v) are tolerated.

Produces the pseudodisaccharide antibiotic dactimicin. The type strain shows the highest sequence similarity to the type strains of *Dactylosporangium matsuzakiense* (99.7%) and *Dactylosporangium aurantiacum* (98.6%). The type strains of all other *Dactylosporangium* species show sequence similarities below 98.5%.

Source: the type strain was isolated from a soil sample collected at Sekigahara, Gifu Prefecture, Japan.

DNA G+C content (mol%): not determined.

Type strain: ATCC 35207, DSM 43823, NBRC 14181, JCM 3307, NRRL B-16297, SF-2127.

Sequence accession no. (16S rRNA gene): X93196.

Species *incertae sedis*

1. **"Dactylosporangium salmoneum"** Routien *in* Celmer, Cullen, Moppett, Routien, Jefferson, Shibakawa and Tone 1978, 3 (U.S. Patent 4081532)

sal.mon'e.um. L. n. *salmon -onis* salmon; L. adj. suff. *-eus -a -um* suffix used with various meanings; N.L. neut. adj. *salmoneum* salmon-colored.

The species is not included in the Approved Lists of Bacterial Names and the name is not validly published in the *International Journal of Systematic Bacteriology.*

Finger-like sporangia (5.5–8.0 × 1.6 µm) containing 3–4 ellipsoidal to tear-drop shaped sporangiospores, develop abundantly on calcium malate agar. The color of the substrate mycelium is salmon, pale pink, orange, pink orange, pinkish, or creamish. Soluble pigments are not produced. No growth occurs on starch agar. Three polycyclic ether antibiotic-producing strains have been described as "*Dactylosporangium salmoneum*".

Source: the type strain was isolated from a soil sample collected in Japan.

Type strain: ATCC 31222, JCM 3272, Pfizer FD 25647.

2. **"Dactylosporangium variesporum"** Tomita, Kobaru, Hanada and Tsukiara 1977, 3 (U.S. Patent 4026766)

The species is not included in the Approved Lists of Bacterial Names and the name is not validly published in the *International Journal of Systematic Bacteriology*.

Finger shaped sporangia containing spores of various shapes. The substrate mycelium is orange to light reddish brown, creamy to light yellowish orange. A yellow pale brownish-to reddish-orange diffusible pigment is produced on various agar media. Glycerol, inositol, and D-ribose are utilized as sole carbon sources; L-rhamnose is not utilized. Nitrate is reduced to nitrite.

"*Dactylosporangium variesporum*" was established for a strain producing the antibiotic capreomycin.

Source: the type strain was isolated from a soil sample collected in India.

Type strain: ATCC 31203, Bristol Labs. D409-5, JCM 3273, NRRL B-16296, DSM 43911, NBRC 14104, KCC A-0273, NBRC 14104.

Further comment: "*Dactylosporangium variesporum*" differs in *N*-acyl type of muramyl residues of peptidoglycan from all other *Dactylosporangium* species in having acetyl. The taxonomic position is furthermore doubtful because of the lack of sporangiospore formation (Uchida and Seino, 1997).

Genus IX. **Krasilnikovia** Ara and Kudo 2007, 2449[VP] (Effective publication: Ara and Kudo 2007a, 6.)

Susmitha Seshadri

Kra.sil.ni.kov'i.a. N.L. fem. n. *Krasilnikovia* referring to N.A. Krasil'nikov, a Russian actinomycetologist who contributed to the taxonomy of the family *Micromonosporaceae*.

Aerobic with **branching hyphae**. **Gram-type positive**. Non-acid-fast. Pseudosporangia on short sporangiophores contain **oval to reniform, nonmotile spores** with a smooth surface. Growth occurs between **20–37°C** and **pH 5–9**. Did not grow on 3% NaCl. Cell wall contains ***meso*-diaminopimelic acid**. Major whole-cell sugars are galactose, mannose, xylose, arabinose, ribose, and glucose. The major menaquinone is **MK-9(H_6), and** the predominant fatty acids are **$C_{16:0}$ iso, $C_{14:0}$ iso** and **$C_{18:1}$ (ω9*c*)**. Phosphatidylethanolamine is the diagnostic phospholipid.

DNA G+C content (mol%): 71.

Type species: **Krasilnikovia cinnamomea** Ara and Kudo 2007, 2449[VP] (Effective publication: Ara and Kudo 2007a, 8.).

Further descriptive information

Isolated from sandy soils of a forest-side waterfall at Chokoria, Cox's Bazar, Bangladesh. Light microscopy revealed large, single or clustered spherical to irregularly shaped structures (~2.0–5.0 μm) on the substrate mycelium. These structures are pseudosporangia, i.e. lacking a sporangial wall, containing highly aggregated and hooked spore chains and substrate mycelium. The non-fragmenting substrate mycelium is light yellow to cinnamon in color. The acyl type of the cell-wall peptidoglycan is glycolyl. Most 16S rRNA gene nucleotide signatures of the family *Micromonosporaceae* are present except that a C is at position 222 instead of a U, a C–G pair is at positions 445:489 and 446:488 instead of a G–C pair, a U–A is at 1011:1018 instead of a C–G and G–U is at position 1263:1272 instead of A–U or C–G.

Enrichment and isolation procedures

Isolation was performed after 21 d of growth at 30°C on humic acid-vitamin (HV) agar supplemented with nalidixic acid (20 mg/l), nystatin and cycloheximide (each 50 mg/l). Strain 3-54(41), the type strain, was isolated by standard dilution plating and was further purified on yeast extract-malt extract agar.

Maintenance procedures

Routine subculturing is done on yeast-starch agar containing soluble starch (15.0 g); yeast extract (4.0 g); K_2HPO_4 (0.5 g); $MgSO_4 \cdot 7H_2O$ (0.5 g); and agar (15.0 g) in 1 liter of distilled water (pH 7.2).

Differentiation of the genus *Krasilnikovia* from closely related genera

16S rRNA gene sequence analysis showed that *Krasilnikovia* was 94.2–97.5% similar to the other genera in the family *Micromonosporaceae*. Phylogenetically, the closest relatives are *Couchioplanes* and *Actinoplanes globisporus*. *Krasilnikovia* lacks a true sporangium like *Couchioplanes*. However, its spores are nonmotile, unlike the motile spores of *Couchioplanes* and *Actinoplanes globisporus*. Also, *Krasilnikovia* differs from *Actinoplanes globisporus* in its menaquinone and whole cell sugar hydrolysate content. 16S rRNA gene signature sequences also differentiate this genus from phylogenetically closely related genera. Hence, on the basis of phylogenetic, phenotypic and chemotaxonomic studies, *Krasilnikovia* is classified as a novel genus in the family *Micromonosporaceae*.

List of species of the genus *Krasilnikovia*

1. **Krasilnikovia cinnamomea** Ara and Kudo 2007, 2449[VP] (Effective publication: Ara and Kudo 2007a, 8.)

cin.na.mo'mea. L. n. *cinnamomum* cinnamon; L. fem. suff. *-ea* suffix used with various meanings; N.L. fem. adj. *cinnamomea* cinnamon-colored.

The characteristics are as described for the genus with the following additional information. Spores are oval to short-rods (0.2–0.4 × 0.8–1.0 μm) and nonmotile. Growth is exhibited on glucose-asparagine agar, inorganic salts-starch agar, oatmeal-nitrate agar, sucrose-nitrate agar, yeast extract-starch agar, oatmeal agar, ISP medium 2, Bennett agar, Hickey–Tresner agar and glucose-yeast extract agar. Good sporulation is observed on glucose-asparagine agar, sucrose-nitrate agar and HV agar media. D-glucose, glycerol, D-xylose, D-galactose, D-fructose,

D-mannose, salicin and maltose are utilized, while adonitol, L-rhamnose, D-mannitol, α-D(+)melibiose, D-raffinose and trehalose are poorly utilized, and erythritol, L-arabinose, D-ribose, *myo*-inositol, α-methyl-D-glucoside and lactose are not utilized. In addition to the MK-9(H_6), small quantities of MK-9(H_4) and MK-9(H_8) are present. Saturated fatty acid $C_{18:0}$, $C_{17:0}$ 10 methyl, $C_{15:0}$ iso, $C_{15:0}$ anteiso, $C_{17:0}$ anteiso, and saturated $C_{16:0}$ are cellular fatty acids present in low amounts. Phospholipids present include phosphatidylethanolamine, diphosphatidylglycerol, phosphatidylglycerol, and phosphatidylinositol and phosphatidylinositol mannosides.

DNA G+C content (mol%): 71 (HPLC).

Type strain: 3-54(41), JCM 13252, MTCC 8094.

Sequence accession no. (16S rRNA gene): AB236956.

Genus X. **Longispora** Matsumoto, Takahashi, Shinose, Seino, Iwai and Ōmura 2003, 1558[VP]

Yōko Takahashi and Atsuko Matsumoto

Lon.gi.spo′ra. L. adj. *longus* long; Gr. fem. n. *spora* seed and in biology a spore; N.L. fem. n. *Longispora* long spore.

Leathery colonies. **Aerial hyphae bear long spore-chains of more than 20 spores on the tips of short sporophores** that branch from the vegetative mycelia. No fragmentation of vegetative mycelia. **No sporangia, synnemata, or sclerotia formed**. Cells are Gram-stain-positive, aerobic, non-acid-fast and **nonmotile**. Cell-wall peptidoglycans contain ***meso*-diaminopimelic acid (DAP)**, glycine, and alanine. Arabinose, galactose, and xylose are detected in whole-cell hydrolysates. The acyl form of muramic acid in the peptidoglycans is **glycolyl**. Predominant menaquinones are **MK-10(H_4) and MK-10(H_6)**; MK-10(H_8) is a minor component. Mycolic acids are not detected. Diagnostic phospholipid is phosphatidylethanolamine (phospholipid type II). Mesophilic.

DNA G+C content (mol%): 70 (HPLC).

Type species: **Longispora albida** Matsumoto, Takahashi, Shinose, Seino, Iwai and Ōmura 2003, 1558[VP].

Further descriptive information

The almost-complete 16S rRNA gene sequence (1496 nt) [positions 10–1506, according to the *Escherichia coli* numbering system of Brosius et al. (1978)] of the type strain of *Longispora albida* has been determined. The phylogenetic tree based on 16S rRNA gene sequences revealed that the genus *Longispora* fell within the cluster of the family *Micromonosporaceae* and belonged to no other genera in the family (Figure 220). The pattern of the 16S rRNA gene signature nucleotides (Stackebrandt et al.,

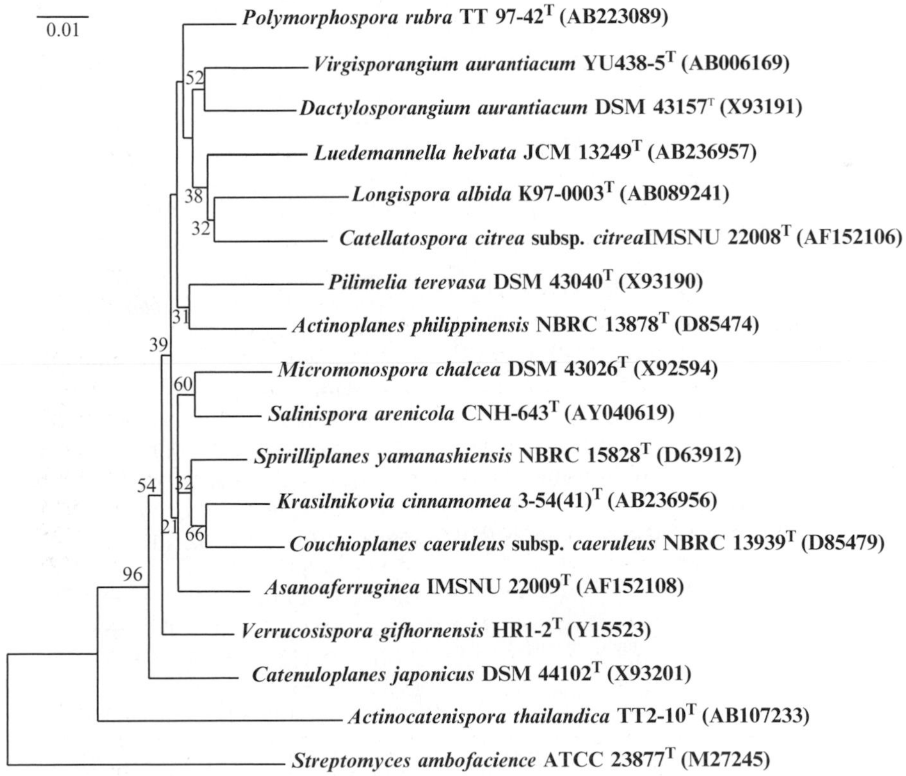

FIGURE 220. Phylogenetic tree based on 16S rRNA gene sequences of 17 genera in the family *Micromonosporaceae*. *Streptomyces ambofaciens* was used as an outgroup. Numbers at nodes indicate the level (%) of bootstrap support based on neighbor-joining analysis of 1000 resampled datasets. Only values >20% are shown. Bar, 1 nucleotide substitution per 100 nucleotides.

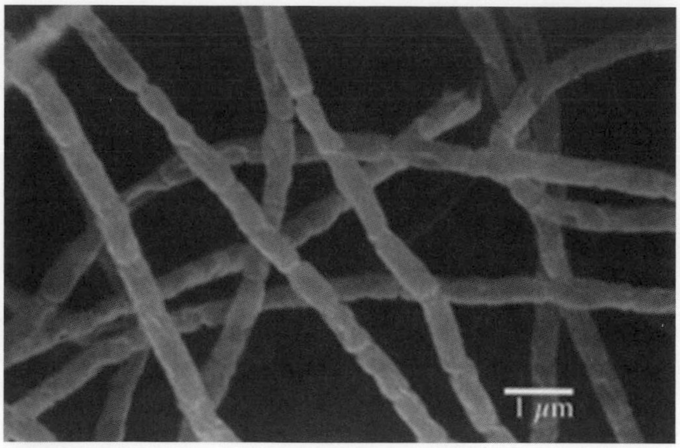

FIGURE 221. Scanning electron micrograph of spore-chains of *Longispora albida* K97-0003[T] grown on 1/10 V8+$CaCl_2$ gellan gum medium for 20 d at 27°C. Bar = 1 μm.

TABLE 198. Aerial mycelium formation of *Longispora albida* on various media

Medium	Solidifying agent			
	Agar		Gellan gum	
	Growth	Aerial mycelium[a]	Growth	Aerial mycelium[a]
Yeast extract-malt extract (ISP 2)	+	+[t]	++	+[t]
Oatmeal (ISP 3)	+	−	+	−
Tyrosine (ISP 7)	+	−	+	+
Glucose-peptone	+	−	++	−
Nutrient	+	−	+	−
Water/proline	+	−	++	−
1/10 V8	+	−	+	+
1/10 V8 + $CaCl_2$	+	+[t]	+	++

[a]Symbols: +[t], Trace aerial mycelium observed with a light microscope; +, poor aerial mycelium; ++, aerial mycelium produced (better than +).

1997) only differs from that of the family *Micromonosporaceae* (Koch et al., 1996a) at position 502 (A). The similarity values of the 16S rRNA gene with strains of other members of the family *Micromonosporaceae*, i.e. *Catellatospora citrea* subsp. *citrea*, *Virgisporangium aurantiacum*, *Asanoa ferruginea*, *Dactylosporangium aurantiacum*, *Pilimelia terevasa*, *Catenuloplanes japonicus*, *Verrucosispora gifhornensis*, *Spirilliplanes yamanashiensis*, *Micromonospora chalcea*, *Couchioplanes caeruleus* subsp. *caeruleus*, and *Actinoplanes philippinensis*, were 91.8–93.0%.

Short sporophores branch from the vegetative mycelia. Spore-chains from the sporophores have more than 20 spores per chain. Spores are cylindrical (Figure 221). Whirls, sclerotic granules, sporangia, and flagellated spores are not produced.

Longispora albida grows well on yeast extract/malt extract agar (ISP 2, Shirling and Gottlieb, 1966)*, oatmeal agar (ISP 3)†, peptone/yeast extract/iron agar (ISP 6)‡, and nutrient agar§, but aerial mycelia are not produced. When gellan gum is used as a solidifying agent, aerial mycelia grow slightly on some media. Spores are produced on 1/10 V8 + $CaCl_2$ gellan gum medium consisting of 2% V8 juice (Campbell's soup), 0.03% $CaCO_3$, 0.06% $CaCl_2 \cdot 2H_2O$ and 1.0% gellan gum in tap water (Table 198).

Enrichment and isolation procedures

Longispora albida was isolated from soil sample using water/proline/gellan gum medium consisting of 1% proline and 1% gellan gum in tap water. A feature of this medium is that gellan gum is used instead of agar as a solidifying agent. Samples (100 μl) from soil suspensions diluted with sterilized water were mixed with the water/proline/gellan gum medium kept at 55°C after sterilization at 121°C for 15 min in Petri dishes. The Petri dishes were incubated for 10 d at 27°C (Matsumoto et al., 2003). Although the recommended medium is water/proline/gellan gum medium, agar medium may be attempted for isolation of *Longispora* strains as *Longispora albida* grows on agar media.

Maintenance procedures

The organism can be maintained in the laboratory by transfer to the same media used for isolation. Viability of isolates during long-term storage is improved by addition of stabilizers. Recommended conditions are storage at −80°C or lyophilization in the presence of stabilizers such as 10% skim milk.

Differentiation of the genus *Longispora* from other genera

At the time of writing, the family *Micromonosporaceae* comprised 17 genera. Table 199 lists the characteristics that are useful for distinguishing *Longispora* from other genera in the family *Micromonosporaceae*. The phylogenetic tree revealed that *Longispora* branches within the family *Micromonosporaceae* (Figure 220). Although these genera have similar cultural, morphological, and chemotaxonomic characteristics, there are some differences. Genera that have spore-chains similar to those of *Longispora* are *Actinocatenispora* (Thawai et al., 2006a), *Asanoa* (Lee and Hah, 2002), *Catellatospora* (Asano and Kawamoto, 1986), *Polymorphospora* (Tamura et al., 2006), *Catenuloplanes* (Yokota et al., 1993), *Couchioplanes* (Tamura et al., 1994), and *Spirilliplanes* (Tamura et al., 1997). The latter three genera possess motile spores. The four genera that do not possess motile spores, *Actinocatenispora*, *Asanoa*, *Catellatospora*, and *Polymorphospora*, share this trait with *Longispora*. However, the genera *Asanoa* and *Catellatospora* form distinctive spore-chains that are borne directly from the vegetative hyphae growing on the surface of agar media and do not produce true aerial mycelia. *Actinocatenispora* and *Polymorphospora* contain MK-9 as major menaquinone.

*Yeast extract/malt extract agar: 0.4% yeast extract, 1.0% malt extract, 0.4% glucose, and 2.0% agar in distilled water, pH 7.3.

†Oatmeal agar. Cook 20 g oatmeal in 1.0 l distilled water for 20 min. Filter through gauze. Add distilled water to restore filtrate to 1.0 l, 1.0 ml of trace salts solution (consisting of 0.1 g $FeSO_4 \cdot 7H_2O$, 0.1 g $MnCl_2 \cdot 4H_2O$, 0.1 g, and $ZnSO_4 \cdot 7H_2O$ in 100 ml distilled water), and agar 18 g, pH 7.2.

‡Peptone/yeast extract/iron agar: 3.6% peptone iron agar, dehydrated (Difco), 0.1% yeast extract in distilled water, pH 7.0.

§Nutrient agar: 0.3% beef extract, 0.5% peptone, and 1.5% agar, pH 7.0.

TABLE 199. Differential characteristics of the genus *Longispora* and other genera in the family *Micromonosporaceae*[a,b]

Characteristic	*Longispora*	*Micromonospora*	*Actinocatenispora*	*Actinoplanes*	*Asanoa*	*Catellatospora*	*Catenuloplanes*	*Couchioplanes*	*Dactylosporangium*	*Krasilnikovia*	*Luedemannella*	*Pilimelia*	*Polymorphospora*	*Salinispora*	*Spirilliplanes*	*Verrucosispora*	*Virgisporangium*
Spore chain	+	–	+	–	+	+	+	+	–	–	–	–	+	–	+	–	–
Motile spore	–	–	–	+	–	–	+	+	+	–	–	–	–	–	+	–	+
Diamino acid	*meso*-DAP	*meso*-DAP	*meso*-DAP	*meso*-DAP	*meso*-DAP	*meso*-DAP	L-Lys	L-Lys	*meso*-DAP	*meso*-DAP	*meso*-DAP	*meso*-DAP	*meso*-DAP	*meso*-DAP	*meso*-DAP	*meso*-DAP	*meso*-DAP
Major menaquinones	MK-10($H_{4,6}$)	MK-10($H_{4,6}$), 9($H_{4,6}$)	9($H_{4,6}$)	MK-9(H_4)	MK-10($H_{6,8}$)	MK-9 ($H_{4,6}$) or 10(H_4)	MK-10(H_4), 11(H_4)	MK-9(H_4)	MK-9($H_{6,8}$)	MK-9 ($H_{6,4,8}$)	MK-9 ($H_{6,4,2,8}$)	MK-9($H_{4,2}$)	MK-10 ($H_{6,4}$), 9($H_{6,4}$)	MK-9(H_4)	MK-10(H_4)	MK-9(H_4)	MK-10($H_{4,6}$)
Phospholipid type[c]	PII	PII	PII	PII	PII	PII	PIII	PII	PII	PII	PII	PII	PII	PII	PII	PII	PII
Characteristic whole-cell sugars	Ara, Gal, Xyl	Ara, Xyl	Ara, Gal, Xyl	Ara, Xyl	Ara, Gal, Xyl	Ara, Gal, Xyl, or Xyl	Xyl	Ara, Gal, Xyl	Ara, Xyl	Gal, Man, Xyl, Ara, Rib,	Xyl, Gal, Man, Rham, Rib, Ara	Ara, Xyl	Xyl	Ara, Gal, Xyl	Ara, Xyl	Man, Xyl	Ara, Gal, Xyl
DNA G+C content (mol%)	70	71–72	72	72–73	71–72	71–73	70–72	69–73	71–73	71	71	nd	71	70–73	69	70	71

[a]Symbols: +, >85% positive; –, 0–15% positive; nd, not determined.

[b]Data from Matsumoto et al. (2003), Goodfellow et al. (1990b), Stackebrandt and Kroppenstedt (1987), Yokota et al. (1993), Vobis (1989b), Lee et al. (2000b), Lee and Hah (2002), Kudo et al. (1999), Tamura et al. (1994, 1997, 2001, 2006). Maldonado et al. (2005), Thawai et al. (2006a), Ara and Kudo (2007a, 2007d), and Rheims et al. (1998).

[c]According to the classification of Lechevalier et al. (1977).

List of species of the genus *Longispora*

1. **Longispora albida** Matsumoto, Takahashi, Shinose, Seino, Iwai and Ōmura 2003, 1558[VP]

al.bi′da. L. fem. adj. *albida* somewhat white.

Longispora albida grows on yeast extract/malt extract agar, oatmeal agar, peptone/yeast extract/iron agar, and nutrient agar, and the color of vegetative mycelium is light ivory to cream. Growth is better on gellan gum media than on agar media. Although aerial mycelia are not produced on these agar media, when gellan gum is used as a solidifying agent instead of agar, aerial mycelia grow on some media. Spores are well-produced on 1/10 V8 gellan gum medium containing $CaCl_2$ (Table 198). Spores are cylindrical (0.4–0.5 × 1.0–1.4 µm) and have a smooth surface (Figure 221). The temperature range for growth is 12–37°C (optimum range is 21–33°C). Growth occurs at pH 6–9. Melanoid pigment is not produced. Positive for nitrate reduction, gelatin liquefaction, milk coagulation, and milk peptonization. D-Glucose is utilized, but L-arabinose, D-fructose, *myo*-inositol, D-mannitol, melibiose, raffinose, L-rhamnose, sucrose, and D-xylose are not utilized. Cellulose is not decomposed. No growth occurs in the presence of 2% NaCl. The predominant components of the cellular fatty acids are $C_{16:0}$ iso, $C_{17:1}$, and $C_{18:1}$. Produces actinohivin, a novel antibiotic active against human immunodeficiency virus (Chiba et al., 2001).

Source: soil.

DNA G+C content (mol%): 70 (HPLC).

Type strain: K97-0003, NRRL B-24201, JCM 11711.

Sequence accession no. (16S rRNA gene): AB089241.

Acknowledgements

We thank Satoshi Ōmura, Akio Seino, and Yuzuru Iwai for their kind suggestions, and Ismet Ara for help with making phylogenetic tree.

Further reading

Suzuki, S., K. Takahashi, T. Okuda and S. Komatsubara. 1998. Selective isolation of *Actinobispora* on gellan gum plate. Can. J. Microbiol. *44*: 1–5.

Genus XI. **Luedemannella** Ara and Kudo 2007c, 1372[VP]

The Editorial Board

Lu.e.de.man.nel′la. N.L. fem. dim. n. *Luedemannella* here refering to George M. Luedemann, a Russian actinomycetologist who contributed to the taxonomy of family *Micromonosporaceae*.

Single or clustered spherical to irregular sporangia (variable in size, ~3.0–5.0 µm) with branched hyphae, forms a non-fragmenting substrate mycelium. Gram-stain-positive. Nonmotile. **Aerobic.** Forms nonmotile spherical to oval shaped spores (0.2–0.4 µm) with a smooth surface and loosely arranged in sporogenous hyphae. Temperature range for growth 20–37°C, pH range 5–9, NaCl tolerance <3%. Cell wall contains the diamino acid *meso*-diaminopimelic acid. The *N*-acyl group of muramic acid is glycolyl. Mycolic acids are not present. MK-9(H_6) and MK-9(H_4) are the major menaquinones, and small amounts of MK-9(H_2) and MK-9(H_8) are present. Polar lipid profile comprises phosphatidylethanolamine, diphosphatidylglycerol, phosphatidylglycerol, phosphatidylinositol and phosphatidylinositol mannosides.

DNA G+C content (mol%): 71.

Type species: **Luedemannella helvata** Ara and Kudo 2007c, 1372[VP].

Further descriptive information

Diagnostic whole-cell sugars are galactose, mannose, xylose and rhamnose as well as small amounts of ribose and arabinose (sugar pattern is D). Major fatty acids are $C_{17:0}$ anteiso (30.0–38.0%), $C_{15:0}$ anteiso (12.5–14.0%), $C_{16:0}$ iso (10.0–15.0%) and $C_{15:0}$ iso (10.1–12.0%) (fatty acid type 2d).

Phylogenetic analysis of the 16S rRNA gene positions the genus within the family *Micromonosporaceae*. The closest phylogenetic neighbor is *Micromonospora pattaloongensis* (96.7% sequence similarity) (Thawai et al., 2008). An environmental clone with high 16S rRNA gene sequence similarity has been detected in undisturbed tallgrass prairie soil in central Oklahoma (97.3%, accession no. FJ479324; Youssef et al., 2009).

Enrichment and isolation procedures

Strains 3-9(24)[T], 3-21(27) and 7-40(26)[T] were isolated from sandy soil collected in Chokoria and Cox's Bazar, Bangladesh. Serial diluents of the soil were transferred onto humic acid-vitamin agar (HV) (Hayakawa and Nonomura, 1987a) supplemented (per liter) with 50 mg cycloheximide, 50 mg nystatin and 20 mg nalidixic acid. Plates were incubated for 21 d at 30°C, then strains were transferred and purified on ISP medium 2 (yeast extract-malt extract agar, medium 2 of the International *Streptomyces* Project) (Shirling and Gottlieb, 1966).

Maintenance procedures

Working cultures are maintained on JCM medium 61 (per liter: 15.0 g soluble starch, 4.0 g yeast extract, 0.5 g K_2HPO_4, 0.5 g $MgSO_4 \cdot 7H_2O$, and 15.0 g agar; pH 7.2).

Differentiation of the genus *Luedemannella* from closely related genera

Possession of major menaquinones MK-9(H_6) and MK-9(H_4) as well small amounts of MK-9(H_2) and MK-9(H_8) distinguishes *Luedemannella* is from other phylogenetically related genera, which possess a variety of different menaquinone profiles: *Longispora*, MK-10(H_4) and MK-10(H_6); *Micromonospora*, MK-10(H_4), MK-10(H_6), MK-9(H_4) and MK-9(H_6); *Salinispora*, MK-9(H_4); *Actinocatenispora*, MK-9(H_4) and MK-9(H_6); *Polymorphospora*, MK-10(H_6), MK-10(H_4), MK-9(H_6) and MK-9(H_4); *Actinoplanes*,

MK-9(H_4) and MK-10(H_4); *Asanoa*, MK-10(H_6) and MK-10(H_8); *Catellatospora*, MK-9(H_4) and MK-9(H_6) or MK-10(H_4) and MK-10(H_6); *Catenuloplanes*, MK-9(H_8) and MK-10(H_8); *Couchioplanes*, MK-9(H_4); *Dactylosporangium*, MK-9(H_4), MK-9(H_6), and MK-9(H_8); *Pilimelia*, MK-9(H_4) and MK-9(H_2); *Spirilliplanes*, MK-10(H_4); *Verrucosispora*, MK-9(H_4); *Virgisporangium*, MK-10(H_4), MK-10(H_6), and MK-10(H_8); *Plantactinospora*, MK-10(H_6), MK-10(H_8), and MK-10(H_4); *Planosporangium*, MK-9(H_4) and MK-10(H_4); *Pseudosporangium*, MK-9(H_6). *Luedemannella* is also differentiated from other genera within the family by the presence of diagnostic whole-cell sugars including galactose, mannose, xylose and rhamnose along with small amounts of ribose and arabinose. *Luedemannella* possess all family-specific nucleotide signatures of 16S rRNA gene, except for the "U–A" pair at position 139:224, "A" at position 381, the "A/G–C" pair at position 656:750 and the C–U pair at position 999:1041.

List of species of the genus *Luedemannella*

1. **Luedemannella helvata** Ara and Kudo 2007c, 1372[VP]

hel.va'ta. N.L. fem. adj. *helvata* honey yellow, referring to the color of the substrate mycelium.

Vegetative mycelia are shell to melon yellow in color and aerial mycelia are not present. D-glucose, L-arabinose, maltose, and trehalose are utilized. D-Xylose, lactose, sucrose and D-raffinose are moderately utilized; erythritol, D-galactose and salicin are poorly utilized. Growth is not affected by glycerol, adonitol, D-ribose, D-fructose, D-mannose, L-rhamnose, *myo*-inositol, D-mannitol, α-methyl-D-glucoside, and α-D-melibiose. Temperature range for growth is 20–37°C, pH range is 5–9, NaCl tolerance is <3%. Grows well on yeast-extract-malt extract agar, Bennett agar, and glucose yeast extract agar. Moderate growth on ISP media 4 and 3, Hickey–Tresner agar, sucrose-nitrate agar, yeast extract-starch agar (JCM medium 61), oatmeal-nitrate agar, 1/5 yeast extract-starch agar (JCM medium 202), sucrose-beef extract agar, ISP medium 1, and 1/20 V8 juice agar. Melanin pigment production on ISP medium 7 is negative. Abundant sporulation occurs on nutrient agar, sucrose-nitrate agar, yeast extract-starch agar, 1/5 yeast extract-starch agar, and 1/20 V8 juice agar, and moderate sporulation occurs on glucose-asparagine agar, glycerol-asparagine agar, tyrosine agar, yeast extract-malt extract agar, oatmeal agar, Bennett agar, glucose-yeast extract agar, Hickey–Tresner agar, tap water agar, and oatmeal-nitrate agar.

For all other characteristics, refer to the genus description.

DNA G+C content (mol%): 71 (HPLC).

Type strain: 3-9(24), JCM 13249, MTCC 8091.

Sequence accession no. (16S rRNA gene): AB236957.

2. **Luedemannella flava** Ara and Kudo 2007c, 1372[VP]

fla'va. L. fem. adj. *flava* yellow, referring to the color of the substrate mycelium.

Vegetative mycelia are cream yellow to wheat yellow in color and aerial mycelia are not present. D-glucose, L-arabinose, maltose, trehalose, D-xylose, D-galactose, D-mannose, L-rhamnose, salicin, lactose, α-D-melibiose, and sucrose are utilized. Glycerol, D-ribose, D-fructose, *myo*-inositol, D-mannitol, and α-methyl-D-glucoside are poorly utilized. Growth is not affected by erythritol, adonitol, and D-raffinose. Temperature range for growth 20–30°C, pH range 6–9, NaCl tolerance <2%. Grows well on oatmeal agar (ISP medium 3), Bennett agar, glucose-yeast extract agar, Hickey–Tresner agar, yeast extract-starch agar (JCM medium 61) and 1/5 yeast extract-starch agar (JCM medium 202). Moderate growth on yeast extract-malt extract agar and oatmeal-nitrate agar. Melanin pigment production on ISP medium 7 is negative. Abundant sporulation occurs on glucose-asparagine agar, ISP media 5 and 4, tap water agar, and 1/5 yeast extract-starch agar; moderate sporulation on ISP medium 7, glucose-yeast extract agar, Hickey–Tresner agar, sucrose-nitrate agar, and oatmeal-nitrate agar; and no sporulation on nutrient agar, yeast extract-malt extract agar, ISP medium 3, Bennett agar, or yeast extract-starch agar.

For all other characteristics, refer to the genus description.

DNA G+C content (mol%): 71 (HPLC).

Type strain: 7-40(26), JCM 13250, MTCC 8095.

Sequence accession no. (16S rRNA gene): AB236959.

Genus XII. **Pilimelia** Kane 1966, 225[AL]

GERNOT VOBIS AND PETER KÄMPFER

Pi.li.mel'i.a. L. n. *pilus* a hair; Gr. fem. n. *melia* Melia, a nymph loved by the river god Inachus; N.L. fem. n. *Pilimelia* an aquatic organism growing on hair substrate.

Members of the genus produce substrate mycelium. Hyphae are Gram-stain-positive, 0.2–0.8 µm in diameter, branched, and septate. True aerial mycelium is not developed. **Sporangia** are produced **on the surface of the substrate** on sporangiophores. The shape of sporangia is **spherical, ovoid, pyriform, campanulate, or cylindrical**, approximately 10–15 µm in size. Sporangia contain **numerous spores in chains** that are arranged in parallel or irregularly in swirl-like rows. **Spores (zoospores)** are **rod-shaped** (0.4 × 1.2 µm) and **motile** by means of a **laterally inserted tuft of flagella**. Nonmotile spores develop in free chains arranged similarly to the zoospores. Colonies grow only on complex media and are small, compact, soft, pasty, or

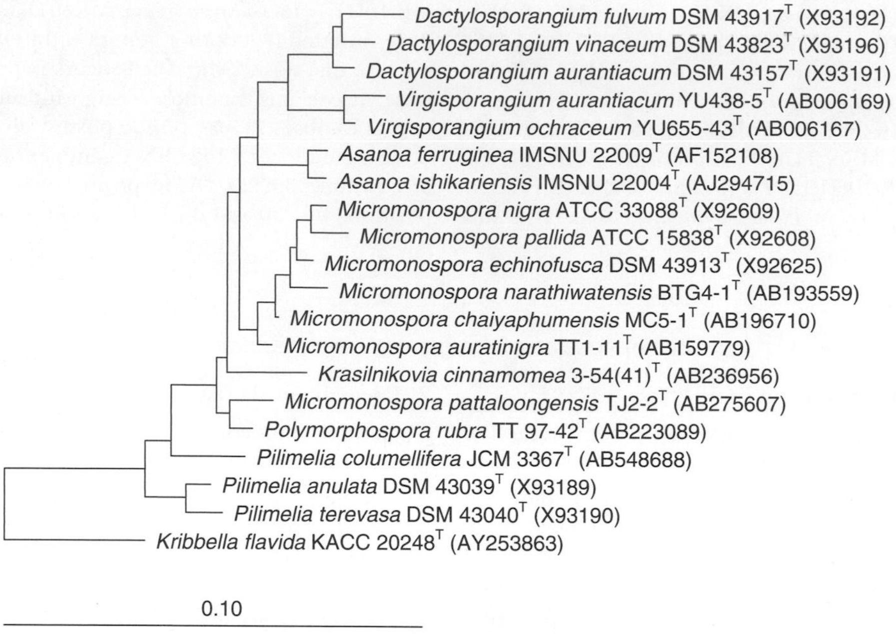

FIGURE 222. Phylogenetic tree showing the relationship of type strains of the genus *Pilimelia* and the related genus *Dactylosporangium*. The tree was reconstructed using the maximum-likelihood method using the software environment ARB (Ludwig et al., 2004) and the corresponding SILVA SSURef 95 database (release July 2008; Pruesse et al., 2007). Bar = 0.10 substitutions per site.

solid. **Color of substrate mycelium is pale lemon-yellow, golden yellow, orange, or pale**, turning brown to dark with age. Aerobic and chemo-organotrophic. Optimal growth is at pH 6.5–7.5 and 20–30°C (minimum 10°C, maximum 38°C). **Strains decompose keratinic substances** (e.g. hair of mammals). The peptidoglycan of the cell walls contains ***meso*-diaminopimelic acid** (***meso*-DAP**) and **glycine**, with **xylose** and **arabinose** as characteristic sugars of whole-cell hydrolysates.

DNA G+C content (mol%): 71 (Miyadoh et al., 2001).

Type species: **Pilimelia terevesa** Kane 1966, 225[AL].

Further descriptive information

Phylogeny. On the basis of 16S rRNA gene sequence analysis, the genus *Pilimelia* belongs to the family *Micromonosporaceae* and contains three species and two subspecies with validly published names. *Pilimelia* species form a distinct phylogenetic cluster within the family *Micromonosporaceae* although the genus does not appear to be monophyletic. The 16S rRNA gene sequence similarity between the three species is between 96.1 and 98.6%. Similarity values of *Pilimelia columellifera* subsp. *columellifera*, *Pilimelia anulata*, and *Pilimelia terevasa* and type strains of the closely related genera are 94.1–96.85% for *Micromonospora*, 96.2–96.7% for *Polymorphospora*, 95.5–96.0% for *Krasilnikovia*, and 95.5–96.0% for *Asanoa*. Interestingly, *Pilimelia columellifera* subsp. *columellifera* JCM 3367^T shares a similarity value of 97.8% with *Micromonospora pattaloongensis* TJ2-2^T; however, chemotaxonomic and other phenotypic data indicate that this strain does not belong to the genus *Micromonospora*. A phylogenetic tree is shown in Figure 222. At present, no sequence is available for the type strain of *Pilimelia columellifera* subsp. *pallida*.

Cell morphology. The fine structure of the walls of hyphae and spores of *Pilimelia* is typical for those of Gram-stain-positive cells. A single compact layer surrounds the cytoplasm, which contains irregularly elongated nucleoid regions, ribosomes, and vacuole-like structures. Mesosomes are connected to the nuclear material or involved in the formation of cross-walls (Bland, 1968; Vobis, 1984). The cross-walls of substrate hyphae have the typical solid structure of Gram-stain-positive bacteria (cross-wall "type 1" of Williams et al., 1973). Typical interspace or "split" septa are observable in sporangial development (Vobis, 1984; Vobis et al., 1986b). Hyphae in contact with the air are additionally covered either by a thin sheath or by a thick layer of perhaps fibrous and mucilaginous material. The sporangial envelopes originate from these outer layers (Bland, 1968; Vobis, 1984).

The sporangia have different shapes and sizes and each species can be characterized by a special morphological type of sporangia (Figure 223; Table 200). Typical cylindrical sporangia are produced by *Pilimelia anulata*, reaching lengths of 10–35 µm (Kane, 1966); *Pilimelia columellifera* has spherical or oval to pyriform sporangia with diameters of 7–15 µm (Vobis et al., 1986b). Globose sporangia are also developed by *Pilimelia terevasa*, their size ranging between 5 and 23 µm (Kane Hanton, 1974). Other strains of *Pilimelia* have campanulate, inverse conical, heart-shaped, or flabelliform sporangia (Figure 223E) (Gaertner, 1955; Karling, 1954; Rothwell, 1957; Vobis et al., 1986b). Presumably, they belong to *Pilimelia terevasa*. After repeated subculturing, the strains may lose the ability to produce sporangia (Kane Hanton, 1974). The polysporous sporangia are developed at the tip of thickened hyphae that penetrate the surface of the substrate. These hyphae may be called palisade hyphae (Bland, 1968) or sporangiophores (Vobis, 1997). When sporangia are mature, the cytoplasm of the sporangiophores is

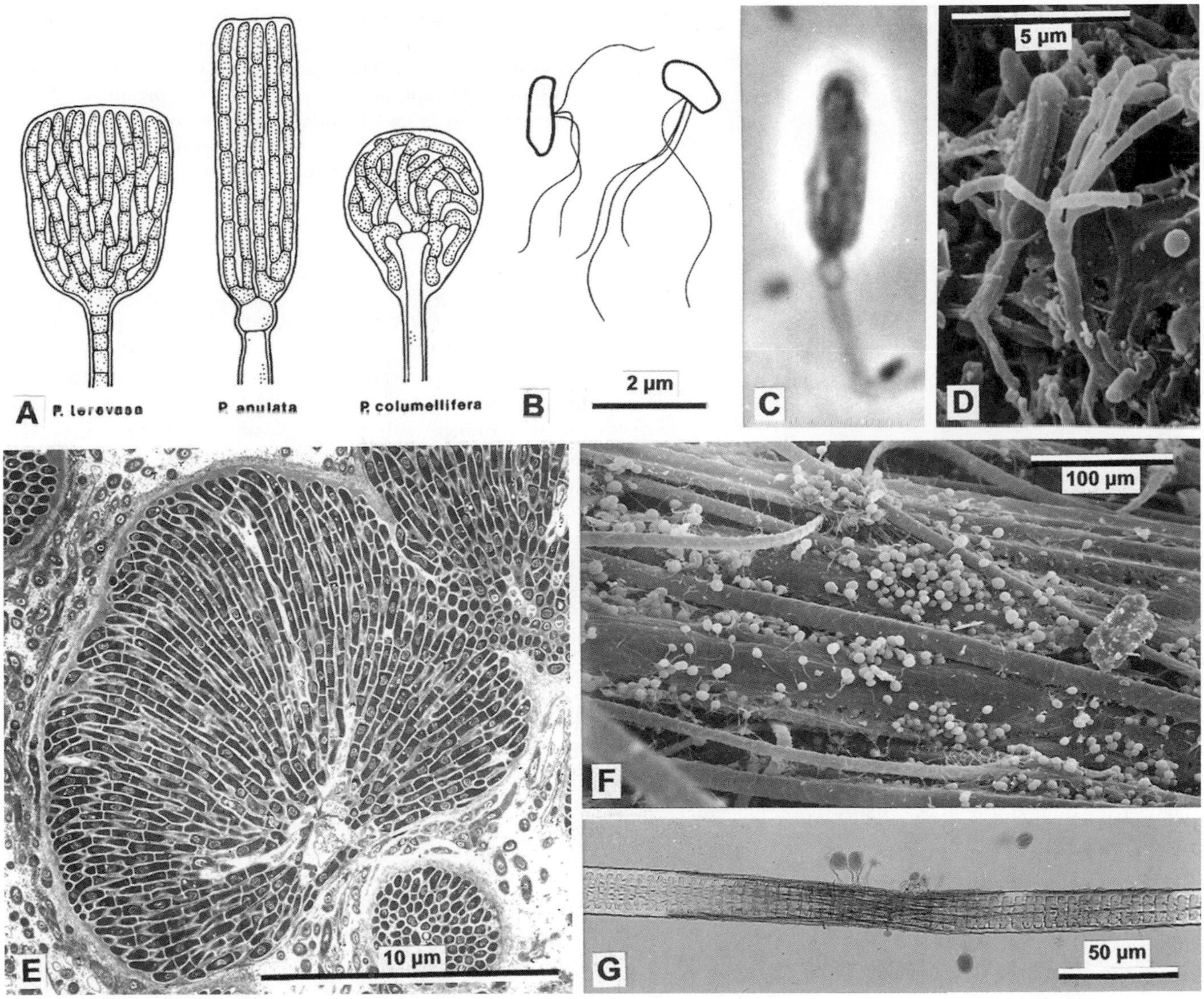

FIGURE 223. Morphological aspects of *Pilimelia*. A, Scheme of the different types of sporangia; B, rod-like sporangiosphores with laterally inserted tuft of flagella; C, cylindrical sporangium of *Pilimelia anulata* ATCC 25604^T with a basal ring-like structure (PHACO, magnification as in D); D, conidiophore of *Pilimelia anulata* ATCC 25604^T bearing chains of conidia (SEM); E, transverse section of a flabellate sporangium of strain MB-VK 122 with branched parallel rows of spores (TEM); F, globose sporangia on the surface of a colony of *Pilimelia columellifera* strain MB-SK6 cultivated on artificial soil agar and hair of white mice (SEM); G, hair colonized by a strain of *Pilimelia*, attacked section is darkly stained with cotton blue (LM). Abbreviations: LM, light microscopy; PHACO, phase contrast microscopy; SEM, scanning electron microscopy; TEM, transmission electron microscopy.

generally autolyzed. The different morphological characters of the sporangiophores are of diagnostic value (Table 200). In *Pilimelia terevasa* and *Pilimelia anulata*, they are fragmented by double-layered cross-walls (Vobis et al., 1986b). The uppermost fragment can be swollen to a ring-like structure or annulus, the distinctive structure of *Pilimelia anulata* (Figure 223C). In *Pilimelia columellifera*, the sporangiophores are not septate, but extend into the lumen of the sporangia to form clearly visible, nail-shaped columellae (Schäfer, 1973). The sporangiophores are arranged in either branched parallel (*Pilimelia terevasa* and *Pilimelia anulata*) or swirl-like (*Pilimelia columellifera*) rows (Figure 223A). It is estimated that one sporangium can contain from hundreds to several thousand spores (Vobis, 1984). The ultrastructure of sporangial development was described by Vobis (1984), confirming the scheme of sporangium formation as proposed by Lechevalier and Holbert (1965) for sporangiate actinomycetes.

The shape of the zoospores released from sporangia is rod-like. Occasional variations of slightly curved or reniform spores originate from branching segments of the sporogenous hyphae (Figure 223A, E). The spores can vary from 0.3 to 0.7 μm in diameter and 0.7 to 1.5 μm in length, and bear a laterally inserted tuft of flagella (Figure 223B) (Vobis et al., 1986b). A single flagellum might reach 5 μm in length with a diameter of about 11 nm. The flagella are frequently bundled and may function as a unit (Schäfer, 1973; Vobis, 1984).

TABLE 200. Diagnostic characters of the species of the genus *Pilimelia*[a]

Characteristic	*P. terevasa*	*P. anulata*	*P. columellifera*
Shape of sporangia:			
Spherical	+	–	+
Pyriform	–	–	+
Campanulate	+	–	–
Flabelliform	+	–	–
Cylindrical	–	+	–
Arrangement of spore chains:			
Parallel rows	+	+	–
Swirl-like	–	–	+
Sporangiophore:			
Septate	+	+	–
Annulate	–	+	–
Extended as columella	–	–	+
Consistency of colonies:			
Solid	–	–	+
Soft	+	+	–
Color of colonies:			
Lemon-yellow, yellow-gray	+	+	–
Golden-yellow, orange	–	–	+
Pale	–	–	+

[a]Symbols: +, >85% positive; –, 0–15% positive.

Nonmotile spores (conidia) may be produced (Figure 223D), mainly in the aqueous milieu of the substrate (Kane, 1966). The conidia develop in chains that are arranged in a pattern similar to the spore chains inside the corresponding sporangia (Kane, 1966; Schäfer, 1973; Vobis and Kothe, 1985). The formation of these unflagellated spores may be a variation of sporangial development in which the sporangial envelope is not formed (Gaertner, 1955).

Cell-wall composition. The peptidoglycan of the cell wall contains *meso*-DAP and glycine according to chemotype II of Lechevalier and Lechevalier (1970a), with xylose and arabinose as characteristic sugars in whole-cell hydrolysates (sugar pattern D) (Szaniszlo and Gooder, 1967; Vobis et al., 1986b). The acyl groups of the muramyl residues of the cell-wall peptidoglycans are of the glycolyl type, exhibiting high amounts of glycolic acid with 149 nmol/mg of cells for *Pilimelia anulata*, 141 nmol/mg for *Pilimelia terevasa*, and 74.0 nmol/mg for *Pilimelia columellifera* subsp. *columellifera* (Uchida and Seino, 1997).

The polar lipid profile is characterized by the presence of phosphatidylethanolamine and phosphatidylcholine and the absence of the unknown glucosamine-containing phospholipid (Vobis et al., 1986b). The phospholipid type of the cell membrane corresponds to type II of Lechevalier et al. (1981).

Fatty acids. *Pilimelia* strains have a high content (>50%) of C_{15} iso, $C_{15:1}$ iso, and $C_{17:1}$ fatty acids, and lack substantial amounts of C_{18} fatty acids (Stackebrandt and Kroppenstedt, 1987). The fatty acid pattern corresponds to type 2b of Kroppenstedt (1985).

The menaquinone composition in *Pilimelia* is predominantly MK-9(H_2) and MK-9(H_4), thus fitting into type 4a of the classification scheme of Kroppenstedt (1985).

Colonial characteristics. *Pilimelia* strains grow very slowly. Small colonies of about 5 mm in diameter develop after 4–6 weeks of incubation. These are compact and grow only with substrate mycelium, aerial mycelium is not produced. *Pilimelia terevasa* and *Pilimelia anulata* have soft and pasty colonies, colored bright lemon-yellow to yellow-gray. In both species, intra-mycelial pigments have been reported. The properties of absorption spectra are associated with carotenoids, with peaks at 479, 451, and 425 nm with an absorption maximum at 451 nm (Szaniszlo, 1968). In comparison with the spectral properties of extracts of *Actinoplanes* strains, the absorption peaks are shifted about 22 nm toward the UV region, indicating that the conjugated double bond system of the carotenoids of *Pilimelia terevasa* and *Pilimelia anulata* has one double bond less than that of *Actinoplanes* species (Parenti and Coronelli, 1979). The colonies of *Pilimelia columellifera* are solid and hard, either golden yellow to orange (subsp. *columellifera*) or pale (subsp. *pallida*) (Table 200). Growth occurs only on complex media such as 50% diluted skim milk agar (Gordon and Smith, 1955), Casamino acids-peptone Czapek agar (Henssen and Schäfer, 1971), nutrient-sugar agar (Henssen and Schäfer, 1971), peptone-yeast extract-iron agar (Shirling and Gottlieb, 1966), oatmeal-yeast extract agar (Vobis et al., 1986b), and yeast extract-starch agar (Emerson, 1958). On these nutrient-rich media, the colonies do not penetrate deeply into the agar but grow upward. Sporangial development rarely occurs. Nutrient-poor media with addition of natural keratinic substances promote the production of sporangia. Good examples are the artificial soil extract agar of Henssen and Schäfer (1971) combined with hair of white mice (Figure 223F), or highly diluted skim milk-mineral agar with the addition of cattle horn meal (Schäfer, 1973; Vobis, 1984, 1992). On these media, the colonies are more flat and the substrate hyphae penetrate deeply into the agar. The skim milk-cattle horn meal agar is highly recommended for starting a new culture (Vobis, 1992).

Life cycle. Sporangia with thick envelopes may function as a resistant form surviving in soil or disseminated as diaspores, possibly attached to soil particles and distributed by actively moving members of the edaphon like nematodes and arthropods. Transport over long distances by the wind is also easily imaginable. If sporangia are dipped into water, numerous flagellated spores are released, leaving behind the sporangial envelope (Vobis, 1984). The zoospores are also adapted to disseminate in water. If sporangia of *Pilimelia columellifera* are flooded with distilled water, motile spores are released after 45–60 min (Vobis et al., 1986b). The spores seem to swim randomly (Vobis, 1984). The laterally inserted tuft of flagella functions as a propelling flagellum and pushes the rod-shaped spore forwards. In microscopic water preparations, this type of movement in one direction, perpendicularly orientated to the longitudinal axis of the spore, is only hardly noticeable. Additionally, the spore moves like a slowly rotating screw. In addition to single spores, short chains of two or three spores can also be observed swimming actively in water (Vobis, 1984). On reaching natural keratinic substances, e.g. hair of mammals, they colonize the new substrate to produce mycelium and sporangia within 14 d (Vobis, 1989a). The life cycle of *Pilimelia* can be considered as "aeroaquatic" (Vobis, 1987), typical for all ecologically defined "actinoplanetes" (Makkar and Cross, 1982; Nonomura and Takagi, 1977).

TABLE 201. Other characteristics of species and subspecies of the genus *Pilimelia*[a,b]

Characteristic	*P. terevasa*	*P. anulata*	*P. columellifera* subsp. *columellifera*	*P. columellifera* subsp. *pallida*
Hydrolysis of starch	–	–	–	–
Degradation of tyrosine	+	+	–	–
Production of melanoid pigments[c]	v	v	v	+
Liquefaction of gelatin	–	–	+	+
Peptonization of casein	+	+	+	+
Reduction of nitrate	–	–	+	–
pH growth range	6.5–7.6	6.5–7.8	6.5–7.6	5.0–7.5
Temperature growth range (°C)	10–35	15–35	15–35	10–30

[a]Symbols: +, >85% positive; –, 0–15% positive; v, strain instability.

[b]Data from studies of type strains (Vobis et al., 1986b).

[c]Data compiled from Kane Hanton (1974), Schäfer (1973), and Vobis et al. (1986b).

Metabolism and physiology. Members of the genus *Pilimelia* are aerobic and mesophilic with growth optima between 20 and 30°C, although exceptional growth has been observed at 42°C (Schäfer, 1973). The pH range of growth is restricted to between 5.0 and 7.8. Colonies grow well at about pH 7.0. For more details and other physiological properties see Table 201. *Pilimelia* strains utilize neither the various carbon sources recommended by Shirling and Gottlieb (1966) (Vobis et al., 1986b) nor individual or combinations of purified amino acids as a sole source of nutrition (Kane Hanton, 1974). Four *Pilimelia* strains isolated from rhizosphere of the actinorhizal plant *Ochetophila* (= *Discaria*) *trinervis* did not show any enzymic activity to decompose substrates of vegetable origin like starch, cellulose, hemicellulose, pectin, or lignin (Solans and Vobis, 2003). The capacity to degrade natural keratinic substances, which can be observed easily by microscopic preparations of infected hairs, is highly conspicuous (Figure 223G). Further physiological investigations are needed to understand the high specialized enzymic activity of *Pilimelia*.

Ecology. Strains of *Pilimelia* must have a worldwide distribution (see also species descriptions). The discoverer of these rare actinomycetes, Karling (1954), first detected them in baited soil cultures from New York City, USA, in 1938 and later in soil samples from various parts of the Amazon Valley in Brazil, as well as from New Jersey, Virginia, Louisiana, Indiana, and Iowa (USA). Schäfer (1973) registered a total of 96 out of 427 soil samples, collected from different geographical regions of the world (22%), that produced of *Pilimelia* sporangia in enrichment cultures. Gaertner (1955) confirmed that 16% of his soil samples from the African continent were positive for the presence of *Pilimelia*. Strains presumptively identified as *Pilimelia terevasa* and *Pilimelia columellifera* have been reported to be widely distributed in diverse soils from England (Tribe and Abu El-Souod, 1979).

The investigations of Garrity et al. (1996) demonstrated that in the arid environment of the Mojave Desert along the California–Nevada border, 34 *Pilimelia*-like isolates could be recovered from eight out of 32 soil samples using the baiting technique. The total number of isolates was estimated as low, but the overall diversity was high. Accordingly, organisms belonging to *Pilimelia* do not occur so rarely in nature, but they are very difficult to isolate and cultivate (Schäfer, 1973; Vobis et al., 1986b).

Pilimelia strains are remarkable "nutrition specialists". In baiting experiments, they colonize and decompose natural keratinic material like hair of cows, horses, mice, rats, dogs, and deer (Gaertner, 1955; Makkar and Cross, 1982; Schäfer, 1973), and human hair (Kane, 1966; Rothwell, 1957). Infected parts of the hair lose their stability and external structure (Vobis, 1984). In addition to the degradation of mammalian hair, feathers of birds or snake skin can also be colonized and attacked by *Pilimelia* strains (Karling, 1954; Schäfer, 1973). These unusual keratinophilic micro-organisms may occupy an important ecological role in soil by decomposing the recalcitrant scleroproteins of vertebrates, especially hair of mammals (Gaertner, 1955). They have not yet been found to be dermatophytes.

Enrichment and isolation procedures

Strains of *Pilimelia* can be baited with natural keratinic substrates (Bland and Couch, 1981; Kane, 1966; Karling, 1954; Schäfer, 1973). Soil samples are placed in Petri dishes and stirred with double-distilled water. Sterilized hairs or bits of snake skin are laid upon the surface. After 3–4 weeks of incubation at room temperature, the hairs are examined microscopically. When they are covered with sporangia, they are transferred onto agar. With a thin, fine-pointed tungsten needle, slightly hooked at the tip, individual sporangia are removed and rolled in zig-zag curves over the surface of agar to remove bacteria, before transferring them to a suitable growth medium, such as 5% diluted skim milk-mineral agar containing cattle horn meal (Schäfer, 1973; Vobis, 1992) or Emerson's yeast extract-starch agar (Kane, 1966). Colonies of 1 mm in diameter can be transferred onto slants after about 3 weeks. Garrity et al. (1996) transferred the infected baits to sterile Whatman filter discs and washed them under vacuum on a mini-sieve. After a period of drying of 7–10 d, the filter discs with baits are suspended in 10 ml in a small Petri dish and incubated for 45–60 min at 22°C. The solutions are diluted (10^{-1} to 10^{-3}) and plated onto suitable or selective agar medium.

Maintenance procedures

Colonies growing on agar media must be subcultured after 4–5 weeks because autolysis of vegetative mycelium may occur after 2 weeks of incubation. For long-term preservation, *Pilimelia* strains must be lyophilized.

Procedures for testing special characters

The ability to decompose keratinic substances can be tested with hair of white mice. Sterilized hairs are incubated together with mycelium and checked microscopically after 2 weeks. Damaged hair segments are stainable with cotton blue (aniline blue) (Gams et al., 1980). Usually, the hairs are also deformed and splintered in the longitudinal axis (Figure 223G).

Differentiation of the genus *Pilimelia* from other related genera

The genus *Pilimelia* has a unique taxonomic position among the sporangiate genera belonging to the family *Micromonosporaceae.* It can be differentiated by morphological, physiological, and chemotaxonomic criteria. The genera *Dactylosporangium, Planosporangium,* and *Virgisporangium* can be separated from *Pilimelia* because they have oligospore sporangia with 2–6 or more spores per sporangium, organized in a single row. Among the polysporous sporangia producing genera of *Micromonosporaceae, Luedemanella* can be distinguished by the formation of nonmotile sporangiophores. *Actinoplanes,* which also forms polysporous sporangia with motile sporangiophores, produces globose or rod-like spores; *Pilimelia* spores are exclusively rod-like. The distinguishing properties are the size of the spores and the type of flagellation: the spores of *Pilimelia* are 0.3–0.7 × 0.7–1.5 µm with a laterally tuft of flagella, whereas the rod-like spores of *Actinoplanes,* characterizing the formerly "*Ampullariella*" species, are 0.5–1.0 × 2.0–4.0 µm and equipped with a polarly inserted tuft of flagella (Vobis, 1989b). In comparison to other members of *Micromonosporaceae,* the growth of *Pilimelia* strains is very slow; they need natural keratinic substances to produce well developed colonies (Vobis, 1989a, 1992). The menaquinone composition of MK-9(H_2) and MK-9(H_4), characteristic of type 4a of Kroppenstedt (1985), differentiates *Pilimelia* from other genera of the *Micromonosporaceae,* including *Micromonospora, Actinoplanes, Catellatospora, Catenuloplanes, Couchioplanes, Dactyloporangium, Spirilliplanes,* and *Verrucosispora* (Kudo, 2001). The pattern of fatty acids is also significantly different from those of other genera with high amounts of C15 and C17, and a lack of C18 fatty acids, corresponding to type 2b of Kroppenstedt (1985).

Taxonomic comments

Members of the genus *Pilimelia* were discovered by Karling (1954), Gaertner (1955), and Rothwell (1957), who were studying hair of mammals attacked by unusual, keratinophilic micro-organisms. Only preliminary diagnoses could be made, since pure cultures were not obtainable at this time. After successful culturing of single strains, Kane (1966) described the genus *Pilimelia* based on two species, *Pilimelia terevasa* and *Pilimelia anulata.* According to the original descriptions, the type strains appeared to be keratinophilic members of the genus *Ampullariella* (Cross and Goodfellow, 1973). This questionable taxonomic position could be clarified by re-examination of the type strains in connection with the description of a third species, *Pilimelia columellifera* (Schäfer, 1973; Vobis et al., 1986b). Finally, the recognition of the genus *Pilimelia* was supported by chemotaxonomic and phylogenetic investigations (Koch et al., 1996a; Kroppenstedt, 1985).

The proposals to transfer *Pilimelia terevasa* and *Pilimelia anulata* to "*Ampullariella*" (Juan and Zhang, 1974), and *Pilimelia columellifera* (Schäfer, 1973) to *Spirillospora* (Tribe and Abu El-Souod, 1979) were not considered in the Approved Lists of Bacterial Names (Skerman et al., 1980).

Differentiation of species of the genus *Pilimelia*

The species of *Pilimelia* are distinguishable by morphological and colonial criteria. Differential characteristics are listed in Table 200. Other physiological characters of the species and subspecies are indicated in Table 201.

List of species of the genus *Pilimelia*

1. **Pilimelia terevasa** Kane 1966, 225[AL]

ter.e.vas′a. L. adj. *teres* rounded (i.e. circular in transverse sections, tapering or narrow cylindric); L. pl. n. *vasa* vessels; N.L. n. *terevasa* rounded vessels, indicating "rounded", spherical sporangia.

Sporangia develop on hair of mammals floating on soil–water and on artificial soil extract agar. The shape of sporangia is spherical (type strain) or campanulate (Figure 223). The sporangia can reach diameters of up to 24 µm. The rod-shaped spores (0.3–0.6 × 0.7–1.5 µm) are motile by means of a tuft of laterally inserted flagella. Spores are released after 15–20 min from sporangia when flooded with water. Sporangiophores are septate, approximately 1 µm in diameter.

Colonies on agar media are bright lemon-yellow with rough lobed borders. The surface is tuberculate and warty with curled protrusions. The consistency is soft and pasty.

Source: the type strain was isolated from a soil sample collected in the area of St. Joseph Co., Indiana (USA).

DNA G+C content (mol%): unknown.

Type strain: ATCC 25603, CBS 570.75, DSM 43040, NBRC 15964, JCM 3091, KCC A-0091.

Sequence accession no. (16S rRNA gene): X93190.

2. **Pilimelia anulata** Kane 1966, 225[AL]

an.u.la′ta. L. fem adj. *anulata* having a ring.

Sporangia develop on mammalian hair and on agar media, supported by septate sporangiophores, where the uppermost fragment is swollen to a ring-like structure (Figure 223A, C). The sporangia are cylindrical, 2.8–11.2 µm wide and up to 35 µm long. Inside the sporangium, the spores are arranged in parallel chains. The spores are rod-shaped (0.3–0.7 × 0.8–1.3 µm) and equipped with a laterally inserted tuft of flagella. Conidiophores with conidia in chains (Figure 223D) are occasionally produced.

Colonies on agar media are bright lemon-yellow to yellow-gray and with lobed margins. The surface of the colonies is tuberculate with narrow and often branched, curled protrusions. The consistency is soft and pasty.

Source: the type strain was isolated from a soil sample collected in the area of St. Joseph Co., Indiana (USA).

DNA G+C content (mol%): unknown.

Type strain: ATCC 25604, DSM 43039, NBRC 16051, NBRC 15533, JCM 3090, KCC A-0090, NCIMB 12892.

Sequence accession no. (16S rRNA gene): X93189.

3. **Pilimelia columellifera** (*ex* Schäfer 1973) Vobis, Schäfer, Kothe and Renner 1986a, 573[VP] (Effective publication: Vobis, Schäfer, Kothe and Renner 1986b, 72.) (Schäfer 1973, 190.)

co.lu.mel.li′fe.ra. L. n. *columella* small column; L. suff. *-fer -fera -ferum* carrying; N.L. fem. adj. *columellifera* bearing a small column.

Sporangia are produced on the surface of hair and on agar media (Figure 223F). The shape of sporangia is spherical, ovoid, or pyriform and 7–16 µm in diameter. Sporangiophores develop in chains that are arranged like swirls. The nonseptate sporangiophore is extended into the lumen of the sporangium, thus forming a columella (Figure 223A). If sporangia are dipped into the water, spores start to swarm after 45 min. Occasionally, swarming occurs inside the sporangia until finally the envelope tears and the spores escape. The rod-shaped zoospores (0.35–0.45 µm wide and 0.8–1.5 µm long) are motile by means of two to four laterally inserted flagella.

Colonies on agar media are small, about 5 mm in diameter after 4 weeks of incubation. They are compact, irregular, and pulvinate. The surface is warty and squamous, and the consistency is solid. The color of mycelium is golden yellow to orange or colorless to pale brownish. The type strain shows the highest sequence similarity to *Micromonospora pattaloongensis* TJ2-2[T] (97.3%), *Micromonospora auratinigra* TT1-11[T] (97.2%), and *Polymorphospora rubra* (97%).

Source: Pilimelia columellifera was preliminarily described as "organism I" by Gaertner (1955), and could be registered by enrichment culture in 45 soil samples collected in Africa and in one sample from Germany. Tribe and Abu El-Souod (1979) reported the frequent presence of the species in soil samples collected in the Cambridge region (England).

DNA G+C content (mol%): unknown.

Type strain: ATCC 43728, CBS 569.75, DSM 43797, NBRC 16052, JCM 3367, MB-SK6.

Sequence accession no. (16S rRNA gene): AB548688.

3a. **Pilimelia columellifera subsp. columellifera** (*ex* Schäfer 1973) Vobis, Schäfer, Kothe and Renner 1986a, 573[VP] (Effective publication: Vobis, Schäfer, Kothe and Renner 1986b, 72) (Schäfer 1973, 190.)

The description is as for the species. It differs from the subspecies *Pilimelia columellifera* subsp. *pallida* by its golden yellow to orange substrate mycelium and by its ability to reduce nitrate.

Source: the type strain originated from a soil sample from Peru; a further 19 strains were isolated from soil samples collected in Germany, British Isles, Ceylon, Spain, Greece, Yugoslavia, Israel, Persia, and Austria (Schäfer, 1973).

DNA G+C content (mol%): unknown.

Type strain: ATCC 43728, CBS 569.75, DSM 43797, NBRC 16052, JCM 3367, MB-SK6.

Sequence accession no. (16S rRNA gene): AB548688.

3b. **Pilimelia columellifera subsp. pallida** (*ex* Schäfer 1973) Vobis, Schäfer, Kothe and Renner 1986a, 573[VP] (Effective publication: Vobis, Schäfer, Kothe and Renner 1986b, 72.) (Schäfer 1973, 192.)

pal′li.da. L. fem. adj. *pallida* pale.

The subspecies *pallida* is distinguishable from *Pilimelia columellifera* subsp. *columellifera* by a colorless to pale brownish substrate mycelium. The center of the colonies occasionally becomes blackish with increasing age. Colonies are flat or convex to slightly raised and umbonate. Sporangial development occurs occasionally on oatmeal-yeast extract agar or on artificial soil extract agar. Melanoid pigments are produced on peptone-yeast extract-iron agar and on tyrosine agar. Nitrate is not reduced to nitrite.

Source: the type strain was isolated from a soil sample collected in Germany; in addition, three further strains were isolated from soil in Germany and Portugal (Schäfer, 1973).

DNA G+C content (mol%): unknown.

Type strain: DSM 43799, MB-SK8.

Sequence accession no. (16S rRNA gene): not available.

Genus XIII. **Polymorphospora** Tamura, Hatano and Suzuki 2006, 1961[VP]

Tomohiko Tamura

Po.ly.mor.pho.spo′ra. Gr. adj. *polumorphos* multiform; Gr. fem. n. *spora* seed, and in biology, a spore; N.L. fem. n. *Polymorphospora* polymorphic spore.

Gram-stain-positive, not acid-fast, producing a fine, nonfragmenting, branching mycelium. Strictly aerobic. **Short spore chains** develop on short sporophores on the substrate mycelium. Immature spores are **oval or of various shapes**, and **short rods** are formed (0.6–0.9 × 0.8–1.5 µm wide) on maturation. Spores are **nonmotile**. Optimum temperature for growth generally ranges between 20 and 30°C. Cell wall contains D-glutamate, glycine, D-alanine, and *meso*-diaminopimelate. Mannose, 3-*O*-methylmannose, glucose, and galactose are detected as whole-cell sugars. The predominant cellular fatty acid is $C_{16:0}$ iso, followed by $C_{17:1}$ and $C_{17:0}$ anteiso. The major menaquinones are MK-10(H_6), MK-10(H_4), MK-9(H_6), and MK-9(H_4). Phosphatidylethanolamine is present as the diagnostic phospholipid, whereas phosphatidylcholine is absent (phospholipid pattern type PII). The acyl type of the cell wall is glycolyl. Mycolic acid is absent.

DNA G+C content (mol%): 70–71 (HPLC).

Type species: **Polymorphospora rubra** Tamura, Hatano and Suzuki 2006, 1963[VP].

Further descriptive information

The organism shows good growth on oatmeal agar, inorganic salts-starch agar, and peptone-yeast extract-iron agar. Cultures also grow well in yeast extract-glucose broth consisting of yeast extract (1%) and D-glucose (1%), pH adjusted to 7.0, upon incubation at 28°C on a rotary shaker for 4 d. The chemotaxonomy of the genus *Polymorphospora* is chemotype II according to Lechevalier and Lechevalier (1970a), and the peptidoglycan type is presumed to be A1γ according to Schleifer and Kandler (1972). A unique nucleotide signature includes a U at position 1244 of the 16S rRNA gene.

Enrichment and isolation procedures

Samples are inoculated on humic acid-vitamin (HV) agar (Hayakawa and Nonomura, 1987a) using the yeast extract–SDS method (Hayakawa and Nonomura, 1989). Incubation is at 28°C for 2 weeks. The described strains were isolated from soil samples collected near the roots of *Bruguiera gymnorrhiza* and *Sonneratia alba* at the mouth of the River Shiira on Iriomote Island, Okinawa, Japan (Tamura et al., 2006).

Maintenance procedures

Strains of the genus *Polymorphospora* are maintained by freezing in water containing 10–30% glycerol at −70°C. Lyophilization of suspensions in 10% skim milk +1% monosodium glutamate and liquid-drying (Sakane and Kuroshima, 1997) in 0.01 M potassium phosphate buffer (pH 7.0) containing 3% monosodium glutamate are also recommended for long-term preservation.

Differentiation of the genus *Polymorphospora* from other genera

The genus *Polymorphospora* contains *meso*-diaminopimelic acid and glycine in its peptidoglycan. Unlike members of the genera *Catenuloplanes* and *Spirilliplanes*, arabinose is not detected in whole cells (Tamura et al., 1997; Yokota et al., 1993). 3-*O*-Methylmannose is contained as a whole-cell sugar, as in the genera *Spirilliplanes* and *Virgisporangium* (Tamura et al., 2001). The genus *Polymorphospora* forms spore chains and differs in this respect from the genera *Dactylosporangium* and *Micromonospora*, which are the closest phylogenetic neighbors.

Taxonomic comments

Based on 16S rRNA gene sequence analysis, the genus *Polymorphospora* forms an independent monophyletic clade within the family *Micromonosporaceae* of the order *Micromonosporales*. In addition, the two strains isolated in the original study showed DNA relatedness higher than 70% and are accommodated in the same species (Tamura et al., 2006).

List of species of the genus *Polymorphospora*

1. **Polymorphospora rubra** Tamura, Hatano and Suzuki 2006, 1963[VP]

ru'bra. L. fem. adj. *rubra* red.

Colonies that develop on ISP media 2, 3, and 4 are red to reddish-orange in color. Utilizes D-mannitol, melibiose, maltose, L-rhamnose, methyl α-D-glucoside, D-galactose, D-mannose, and D-glucose. Positive in tests for starch hydrolysis and urea decomposition. No growth in the presence of 4% NaCl. Esculin is not hydrolyzed. The major cellular fatty acids are $C_{16:0}$ iso and $C_{16:1}$.

Source: the type strain was isolated from rhizosphere soil of mangrove.

DNA G+C content (mol%): 70–71 (HPLC).

Type strain: TT 97-42, DSM 44947, NBRC 101157.

Sequence accession no. (16S rRNA gene): AB223089.

Genus XIV. **Salinispora** Maldonado, Fenical, Jensen, Kauffman, Mincer, Ward, Bull and Goodfellow 2005, 1763[VP]

Paul R. Jensen, Luis A. Maldonado and Michael Goodfellow

Sa.li.ni.spo'ra. L. adj. *salinus* saline; Gr. fem. n. *spora* a seed and, in bacteriology, a spore; N.L. fem. n. *Salinispora* a spore-forming bacterium originating from a saline habitat, indicating the marine habitat of the organism.

Obligately aerobic, Gram-stain-positive, non-acid-fast, nonmotile actinomycetes that form extensively branched substrate hyphae carrying single or clusters of smooth-surfaced spores, which may be sessile or borne on short sporophores. Vegetative hyphae are finely branched but do not show any fragmentation. Strains grow well at 10–30°C and pH 7–12. Colonies range from bright to pale orange and dark brown changing to black as they sporulate with no aerial mycelia observed. Brown diffusible pigments that darken agar media are normally produced. **Whole-organism hydrolysates are rich in *meso*-diaminopimelic acid and contain major amounts of arabinose, galactose, and xylose. The muramic acid moiety of the peptidoglycan is glycolated. Cells contain: diphosphatidylglycerol, phosphatidylethanolamine, phosphatidylglycerol, and phosphatidylinositol as major polar lipids; tetrahydrogenated menaquinones with nine isoprene units as the predominant isoprenologue; and complex mixtures of saturated iso- and anteiso-fatty acids but lack mycolic acids. Do not grow when seawater is replaced with deionized water in standard complex growth media.** The phylogenetic position of *Salinispora*, as determined by 16S rRNA gene sequence analysis, is in the family *Micromonosporaceae.*

Source: isolated from marine sediments and benthic organisms.

DNA G+C content (mol%): 69.4–69.5 (from genome sequences).

Type species: **Salinispora arenicola** Maldonado, Fenical, Jensen, Kaufman, Mincer, Ward, Bull and Goodfellow 2005, 1764[VP].

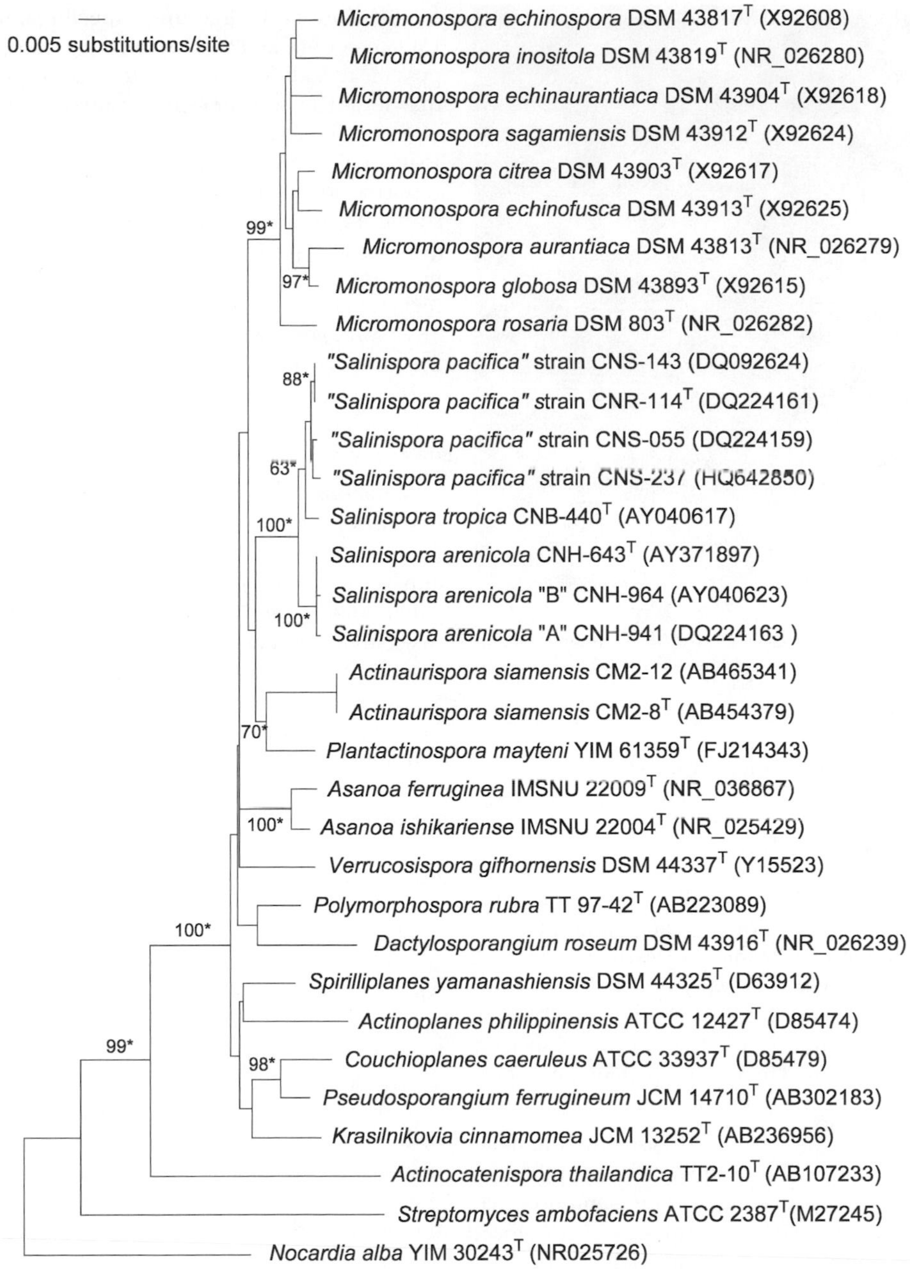

FIGURE 224. Neighbor-joining tree based on nearly complete 16S rRNA gene sequences showing relationships between *Salinispora* species and representatives of other genera classified in the family *Micromonosporaceae*. The numbers at the nodes indicate levels of bootstrap support based on an analysis of 1000 re-sampled datasets, only values above 60% are given. Asterisks indicate nodes that are also supported using the maximum-likelihood and maximum-parsimony methods. Bar = 0.005 substitutions per nucleotide position.

Further descriptive information

Phylogeny. The genus *Salinispora* forms a tight subclade in the 16S rRNA *Micromonosporaceae* gene tree (Figure 224). It encompasses two species with validly published names, *Salinispora arenicola* and *Salinispora tropica*, and a putative third species, "*Salinispora pacifica*" (Jensen and Mafnas, 2006). All *Salinispora* strains examined to date share 16S rRNA gene sequence identities greater than or equal to 99% and thereby constitute what has been described as a multidiverse ribotype cluster (Acinas et al., 2004). Initially, the genus was considered to be most closely related to the genus *Micromonospora* (Maldonado et al., 2005) but it is now known that its nearest phylogenetic neighbors are the genera *Actinaurispora* (Thawai et al., 2010) and *Plantactinospora* (Qin et al., 2009). Micromonosporae and salinisporae can also be distinguished by analysis of 16S–23S intergenic spacer regions (Maldonado et al., 2005) and by *gyrB* nucleotide sequence data (Jensen and Mafnas, 2006).

Cell morphology. *Salinispora* strains present extensively branched substrate hyphae which range from 0.25 to 0.5 µm in diameter. The substrate hyphae carry single or clusters of

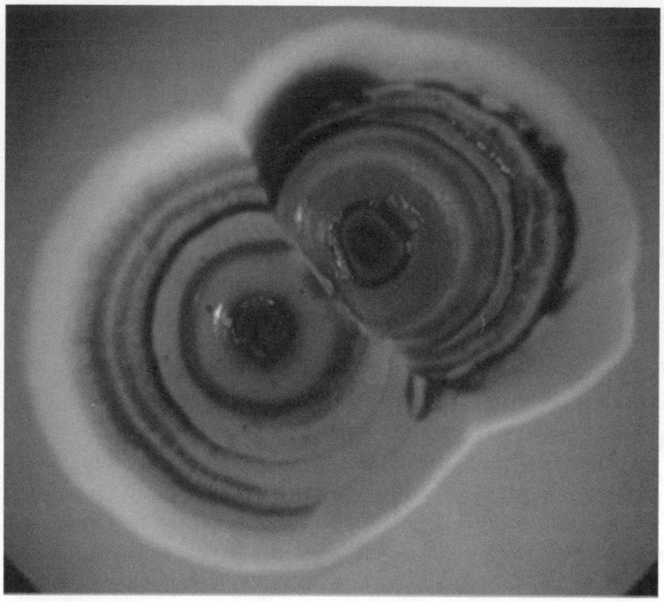

FIGURE 225. Colony of *Salinispora tropica* growing on seawater agar. Concentric rings of spores are visible on the colony surface. Photo credit: Erin Gontang.

smooth-surfaced spores. Spores range from 0.8 to 3.8 μm in diameter. Aerial hyphae have not been observed and spores may be sessile or borne on short sporophores.

Colony morphology. *Salinispora* strains grow slowly taking from 1 to 3 weeks to form small colonies on primary isolation plates (Jensen et al., 2005). Colonies are tough, leathery, and adhere to agar surfaces. On nutrient-rich media they exhibit a range of orange to brown pigments with spores blackening the colony surface (Figure 225), although on occasion poorly pigmented strains are encountered. In liquid culture, most strains are dispersed while some form dense clumps. The orange pigmentation produced in liquid culture generally fades to black over time.

Chemotaxonomy. Salinisporae contain *meso*-diaminopimelic acid as the diagnostic diamino acid of the cell-wall peptidoglycan, muramic acid residues that are *N*-glycolated, and produce whole-organism hydrolysates rich in arabinose, galactose, and xylose. They also contain diphosphatidylglycerol, phosphatidylethanolamine, phosphatidylglycerol, and phosphatidylinositol as diagnostic polar lipids, tetrahydrogenated menaquinones with nine isoprene units as the major isoprenologue, and complex mixtures of saturated iso- and anteiso-fatty acids, but lack mycolic acids (Maldonado et al., 2005). The predominant fatty acids (as percentages of total fatty acid content) are 13-methyltetradecanoic acid ($C_{15:0}$ iso; 3.7–13.3%), 14-methylpentadecanoic acid (C6:0; 36.4–53.5%), 16-methylheptadecanoic acid ($C_{18:0}$ iso; 3.5–10.2%), heptadecanoic acid ($C_{17:0}$; 4.1–7.8%) and 10-methyloctadecanoic acid ($C_{18:0}$ 10-methyl; 0.9–6.8%).

Nutrition and growth conditions. *Salinispora* strains were first recognized as being distinct from other actinomycetes based on their inability to grow when seawater was replaced with deionized water in the complex growth medium M4 (Jensen et al., 1991). Subsequently, it was reported that they failed to grow when sodium salts were replaced with equimolar concentrations of potassium salts suggesting a specific sodium ion requirement (Mincer et al., 2002). However, it is now clear that growth is possible at sodium concentrations as low as 5 mM (approx. 1% that found in seawater) if the medium is supplemented with sufficient concentrations of the appropriate non-sodium salts (Kim et al., 2005; Tsueng and Lam, 2008a). *Salinispora* cells are reported to lyze in the absence of a sufficiently high osmotic strength growth medium (Tsueng and Lam, 2010) and will only grow if the appropriate salts and high osmotic strength environment are provided (Tsueng and Lam, 2008b). *Salinispora* strains form well-developed colonies on media commonly employed to grow *Micromonospora* [e.g. yeast-malt extract agar (ISP medium 2); Shirling and Gottlieb, (1966)] when deionized water is replaced with seawater or an appropriate salts solution. Strains grow well on nutrient-rich agar media prepared with either natural or artificial seawater (Mincer et al., 2002).

Metabolism and metabolic pathways. *Salinispora* strains are heterotrophic and obligately aerobic. *Salinispora arenicola* and *Salinispora tropica* strains degrade arbutin, casein, elastin, gelatin, and starch, but not cellulose, chitin, tributyrin, or xylan (Maldonado et al., 2005). Similarly, cellobiose, α-lactose, melezitose, and starch are used as sole carbon sources, but not fructose, mannose, ribose, sorbose, or xylose. Salinisporae are best known metabolically for the production of secondary metabolites including the highly selective proteasome inhibitor salinosporamide A (Feling et al., 2003), which is undergoing clinical trials for the treatment of cancer (Fenical et al., 2009). Based on the analysis of two genome sequences, secondary metabolism is the major functionally annotated class of metabolic genes that differentiates the two species (Penn et al., 2009). This is supported by the observation that secondary metabolite production occurs in species-specific patterns with *Salinispora arenicola* strains producing rifamycins and staurosporines while *Salinispora tropica* strains produce salinosporamides and sporolides (Jensen et al., 2007). To date, at least 16 distinct structural types have been characterized from this genus including the cyanosporosides from "*Salinispora pacifica*" (Oh et al., 2006) and the sporolides from *Salinispora tropica* (Buchanan et al., 2005), both of which are proposed to be derived from enediyne intermediates. Other new structures from *Salinispora arenicola* include the polyketide-derived arenicolides (Williams et al., 2007) and the depsipeptide arenamides (Asolkar et al., 2009), while the antitumor antibiotic lomaivitacin (He et al., 2001) was isolated from what is now proposed to be "*Salinispora pacifica*".

Genetics. Representatives of all *Salinispora* 16S rRNA gene sequence types were originally aligned to *Escherichia coli* (accession no. J01695) and 71 type strains within the family *Micromonosporaceae*. This alignment revealed five genus-specific signature nucleotides consisting of A, C, T, T, and G (Mincer et al., 2002). This was subsequently revised to A, C, T, and G (Maldonado et al., 2005), while a new analysis including sequence data from the genera *Actinaurispora* (Thawai et al., 2010) and *Plantactinospora* (Qin et al., 2009) reveals three genus-specific nucleotides (C, T, T). Mean DNA:DNA hybridization values for *Salinispora tropica* CNB-440^{T} and *Salinispora arenicola* CNH-643^{T} are 68%. A PCR-targeted mutagenesis protocol has been developed to generate gene knock-outs in *Salinispora* species (Eustáquio et al., 2008) and used to characterize the function of enzymes involved in secondary metabolite biosynthesis in this taxon (Eustáquio et al., 2009).

Genomics. The complete genome sequences of *Salinispora tropica* strain CNB-440^T (accession no. NC_009380) and *Salinispora arenicola* strain CNS-205 (accession no. NC_009953) have been determined and analyzed (Penn et al., 2009; Udwary et al., 2007). Both chromosomes are circular with the *Salinispora arenicola* genome (5.8 Mb) being 600 kb larger in size. The *Salinispora arenicola* strain was chosen for sequencing because, unlike the type strain for this species (CNH-643^T), it produces the unusual cyclic peptide cyclomarin A (Renner et al., 1999). The two strains devote considerable proportions of their genomes (9–11%) to secondary metabolite production, which occurs in species-specific patterns (Jensen et al., 2007). Most genetic differences between the two strains are located in 21 major genomic islands, which house all of the species-specific secondary metabolic pathways. The mean nucleotide identity calculated for the two strains is 87.2%, far below the 94% value that has been suggested to delineate bacterial species (Konstantinidis and Tiedje, 2005). The genome sequences also provide evidence of marine adaptation in the form of a highly duplicated family of polymorphic membrane proteins (Penn et al., 2009). Intragenomic variability was not observed among the three copies of the 16S rRNA gene operon present in each genome.

Antibiotic sensitivity. Good growth is observed in the presence of (µg/ml) gentamicin sulfate (5 and 25), neomycin sulfate (5), streptomycin sulfate (5), novobiocin (5 and 25) penicillin G (5) and rifampin (5). *Salinispora arenicola* strains that produce compounds in the rifamycin class are also resistant to 25 µg/ml rifampin, whereas other *Salinispora* species are not.

Ecology. To date, *Salinispora* strains have only been cultured from tropical and subtropical marine environments. Most observations have been from sediments, where they have been shown to increase in abundance with depth (Jensen et al., 1991) and to occur at concentrations of 2–5 × 10^3 c.f.u./ml wet sediment (Mincer et al., 2005). Although the maximum depth limits are not known, strains have been successfully cultured from marine sediments collected at a depth of 1100 m (Mincer et al., 2005). Salinisporae have also been reported from a marine sponge (Kim et al., 2005) and are readily cultured from a variety of benthic marine plants and invertebrates (unpublished data, Scripps Institution of Oceanography), though there is not any evidence of specificity in these associations. The genus is broadly distributed in tropical and subtropical sediments (Jensen and Mafnas, 2006) but has yet to be cultured from more temperate environments despite culture-independent evidence that they occur in these regions (unpublished data, Scripps Institution of Oceanography). *Salinispora* species show distinct biogeographical patterns, with the only reports to date for *Salinispora tropica* coming from the Caribbean. *Salinispora arenicola* has the broadest distribution and has been recovered from all sites in which the genus has been reported, while "*Salinispora pacifica*" has an intermediate distribution, having yet to be cultured from the Caribbean (Jensen and Mafnas, 2006). Culture-independent studies have not revealed any new species-level diversity, while DNA extraction experiments have shown that most strains recovered from sediments were present as spores (Mincer et al., 2005).

Enrichment and isolation procedures

Salinispora strains are readily cultured from marine sediments using either of two approaches. The first involves drying approximately 1 g sediment at room temperature overnight in a laminar flow hood. A sterile foam plug is first stamped onto the dried sediment and then repeatedly stamped onto an appropriate agar growth medium creating a serial dilution effect. After 2–3 weeks of incubation at room temperature, *Salinispora* colonies can be seen on the agar surface. The second method involves serially diluting (1:10) 1 ml wet sediment in sterile seawater. The dilutions are then heated to 50°C for 6 min and 50 µl is spread-plated onto an appropriate growth medium using a sterile glass rod. Effective media formulations include seawater agar (1 liter natural or artificial seawater, 16 g agar) and A1 agar (10 g soluble starch, 4 g yeast extract, 2 g peptone, 16 g agar, 1 liter natural or artificial seawater). Cycloheximide (100 µg/ml final concentration) is used to supplement isolation media to reduce fungal growth, while rifampin (20 µg/ml) can be added to select for *Salinispora arenicola*, which produces related compounds in the rifamycin class and demonstrates a higher level of resistance to this antibiotic than the other taxa.

Maintenance procedures

Salinispora strains can be maintained as glycerol suspensions (20%, v/v) at −20 and −80°C for long-term preservation as suggested by Wellington and Williams (1978). The glycerol suspensions are prepared using 4–5 loopfuls of fresh biomass scraped using sterile disposable loops or sterile toothpicks from GYM or YEME agar plates that had been streaked for single colonies and incubated at 28–30°C for 10–14 d. Short-term storage can be in GYM or YEME agar slants which are inoculated from purified colonies grown at 28–30°C for 7–10 d.

Differentiation of the genus *Salinispora* from other genera

The most appropriate way of distinguishing salinisporae from other sporoactinomycetes is by PCR followed by sequencing. Suitable targets are 16S rRNA, *gyrB*, and 16S–23S rRNA intergenic spacer loci. Primers suitable for 16S rRNA gene amplification include FC27 (5′-AGAGTTTGATCCTGGCTCAG) and RC 1492 (5′-TACGGCTACCTTGTTACGACTT), as discussed by Mincer et al. (2005). *Salinispora* strains can be distinguished from well established genera classified in the family *Micromonosporaceae* using a combination of chemotaxonomic and morphological properties (see Table 190 in the treatment of the family *Micromonosporaceae*); they can also be separated from the genera *Actinaurispora* and *Plantactinospora* on this basis (Qin et al., 2009; Thawai et al., 2010). A quick and simple method is available for distinguishing between salinisporae and micromonosporae as only the latter grow on complex growth medium 4 prepared using deionized water (Jensen et al., 1991).

Taxonomic comments

The genus *Salinispora* was proposed by Maldonado et al. (2005) to accommodate representatives of large numbers of strains isolated from geographically diverse tropical and/or subtropical locations and designated MAR1 (Mincer et al., 2002). The MAR1 strains were distinguished by morphological characteristics, 16S rRNA gene signature nucleotides, and failure to grow when seawater was replaced with deionized water in the growth medium. Comparative 16S rRNA gene sequence analysis showed that seven of the isolates formed a monophyletic clade in the family *Micromonosporaceae*, which suggested novelty at the genus level. The MAR1 isolates were provisionally assigned to a taxon that was informally designated "*Salinospora*" (Feling et al., 2003; Mincer et al., 2002).

The single most distinguishing feature of the genus is that, to date, it has only been reported from marine environments. It also appears to be unique among actinomycete genera in that strains fail to grow when seawater is replaced with deionized water in a complex growth medium or in defined media that lack specific salt combinations and a sufficiently high osmolarity (Tsueng and Lam, 2008b). Evidence that strains clading with "*Salinispora pacifica*" (M101 and SW02; Kim et al., 2005) have distinct spore arrangements relative to *Salinispora arenicola* warrants further investigation into the possibility that this is a species-defining morphological trait.

Differentiation of species of *Salinispora*

Members of the two species with validly published names can be distinguished using analysis of 16S rRNA gene sequence, DNA:DNA relatedness and phenotypic data (Maldonado et al., 2005). *Salinispora arenicola* strains, unlike those of *Salinispora tropica*, use L-proline, salicin, L-threonine, and L-tyrosine, but not galactose or inulin as sole carbon sources, and grow in the presence of 20 µg/ml rifampin. In addition, members of these taxa synthesize core-sets of secondary metabolites, *Salinispora arenicola* strains produce compounds that belong to the rifamycin and staurosporine classes whereas representatives of *Salinispora tropica* produce salinosporamides and sporalides (Jensen et al., 2007). In contrast, some "*Salinispora pacifica*" strains synthesize the structurally novel metabolites cyanosporasides A and B, salinipyrones A and B, and the polyketides pacificanones A and B (Oh et al., 2008).

List of species of the genus *Salinispora*

1. **Salinispora arenicola** Maldonado, Fenical, Jensen, Kaufman, Mincer, Ward, Bull and Goodfellow 2005, 1764^{VP}

a.re.ni'co.la. L. n. *arena* sand; L. suff. *-cola* (from L. n. *incola*) inhabitant, dweller; N.L. n. *arenicola* sand-dweller, indicating isolation from marine sediments.

Aerobic, Gram-stain-positive actinomycetes, which form an extensively branched substrate mycelium that carries smooth-surfaced spores either singly or in clusters. Grows from 10–30°C; optimal growth occurs between 20 and 28°C.

Additional phenotypic properties are cited either in the genus description or in the text.

Source: isolated from coarse sand off the Bahamas.

DNA G+C content (mol%): 69.5 (from whole genome sequence).

Type strain: ATCC BAA-917, CNH-643, DSM 44817.

Sequence accession nos: AY040619 (16S rRNA gene); AY371897 (16S–23S rRNA intergenic spacer region); DQ228681 (*gyrB*); NC_009953 (genome of strain CNS-205).

2. **Salinispora tropica** Maldonado, Fenical, Jensen, Kaufman, Mincer, Ward, Bull and Goodfellow 2005, 1764^{VP}

tro'pi.ca. L. fem. adj. *tropica* tropical, pertaining to the tropics, the source of the isolates.

Aerobic, Gram-stain-positive actinomycetes that form an extensively branched substrate mycelium that carries smooth-surfaced spores either singly or in clusters. Grows from 10–30°C; optimal growth occurs between 15–28°C.

Additional phenotypic properties are cited either in the genus description or in the text.

Source: isolated from sediment samples collected in the Caribbean. The type strain (CNB-440^{T}) was isolated from a sediment sample collected at a depth of 20 m off the coast of the Bahamas.

DNA G+C content (mol%): 69.4 (from whole genome sequence).

Type strain: ATCC BAA-916, CNB-440, DSM 44818.

Sequence accession nos: AY040617 (16S rRNA gene); AY371895 (16S–23S rRNA spacer region); DQ2288684 (*gyrB*); NC_009380 (whole genome of strain CNB-440^{T}).

Species *incertae sedis*

1. **"Salinispora pacifica"**

This interesting actinomycete was initially mentioned by Jensen and Mafnas (2006) but has yet to be deposited in culture collections and the name has not been validly published. Nonetheless, it provides insight into the phenotypic and phylogenetic diversity of the genus *Salinispora*. A representative of the taxon forms a distinct lineage in the 16S rRNA *Salinispora* gene tree (Figure 224) and has a range of phenotypic properties and DNA:DNA relatedness values that distinguish it from the type strains of *Salinispora arenicola* and *Salinispora tropica*. Members of this taxon have been isolated from marine sediments collected from around the islands of Guam and Palau in the Pacific Ocean (Gontang et al., 2007).

Genus XV. **Spirilliplanes** Tamura, Hayakawa and Hatano 1997, 101[VP]

Tomohiko Tamura

Spi.ril.li.plan'es. N.L. dim. neut. n. *spirillum* a small spiral; Gr. masc. n. *planes* a wanderer; N.L. fem. (*sic*) n. *Spirilliplanes* an organism with wandering cells, in spirals.

Gram-stain-positive, not acid-fast, producing a fine, nonfragmenting, branching mycelium. Strictly aerobic. The aerial hyphae **aggregate into clusters resembling coils** (Figure 226), but true sporangia are not observed; 14-d-old cultures grown on inorganic salts-starch agar have hyphae arranged in spirals of 5–10 turns with several spores per chain. Spores are ovals or short rods (0.5–0.7 × 0.7–1.0 µm) with smooth surfaces. Upon immersion in water or phosphate buffer containing soil extract, **motile spores** are released from spore chains, but in many instances motility begins over 30–60 min after spore release. **Polar flagella** are present in motile spores. In general, the vegetative mycelia are yellow to orange and aerial hyphae are white. Cell wall contains D-glutamate, glycine, D-alanine, and *meso*-diaminopimelate. Mannose, 3-*O*-methylmannose, glucose, xylose, and galactose are detected in the whole-cell sugars. $C_{17:1}$, $C_{17:0}$, $C_{15:0}$ anteiso, $C_{15:0}$ iso, and $C_{16:0}$ iso are present as major cellular fatty acids. The major menaquinone is MK-10(H_4); small amounts of MK-10(H_6) and MK-10(H_8) are also present. Phosphatidylethanolamine and phosphatidylinositol are diagnostic phospholipids (phospholipid pattern type PII). The acyl type of the cell-wall polysaccharides is glycolyl. Mycolic acid is absent.

DNA G+C content (mol%): 69 (HPLC).

Type species: **Spirilliplanes yamanashiensis** Tamura, Hayakawa and Hatano 1997, 102[VP].

Further descriptive information

The chains of very narrow and coiled sporogenous hyphae with zoospores often appear to be sporangium-like structures under a light microscope (Figure 227). Under a scanning electron microscope, however, these structures are found to be aggregated hyphae and not true sporangia (Figure 226). After incubation at 28°C for 1 h in either water or 0.01 M phosphate buffer (pH 7.0) containing 10% (v/v) soil extract, spores showed active motility.

Good growth occurs at temperatures between 25 and 30°C and on oatmeal agar, inorganic salts-starch agar, and peptone-yeast extract-iron agar.

Cultures also grow well in yeast extract-glucose broth consisting of yeast extract (1%) and D-glucose (1%), pH adjusted to 7.0, and incubated at 28°C on a rotary shaker for 4 d.

The wall chemotype is type II according to the scheme of Lechevalier and Lechevalier (1970a) and the peptidoglycan type is presumed to be type A1γ according to the classification of Schleifer and Kandler (1972).

Enrichment and isolation procedures

The type strain was isolated from a soil sample from Kofu, Yamanashi, Japan. The sample was spread on humic acid-vitamin (HV) agar (Hayakawa and Nonomura, 1987a) for isolation after dry heating (120°C, 1 h) (Nonomura and Ohara, 1969). Incubation was at 28°C for 2 weeks.

Maintenance procedures

Strains of the genus *Spirilliplanes* are maintained by freezing in water containing 10–30% glycerol at −70°C. Lyophilization of suspensions in 10% skim milk + 1% monosodium glutamate and L-drying in 0.01 M potassium phosphate buffer (pH 7.0) containing 3% monosodium glutamate (Sakane and Kuroshima, 1997) are also recommended for long-term preservation.

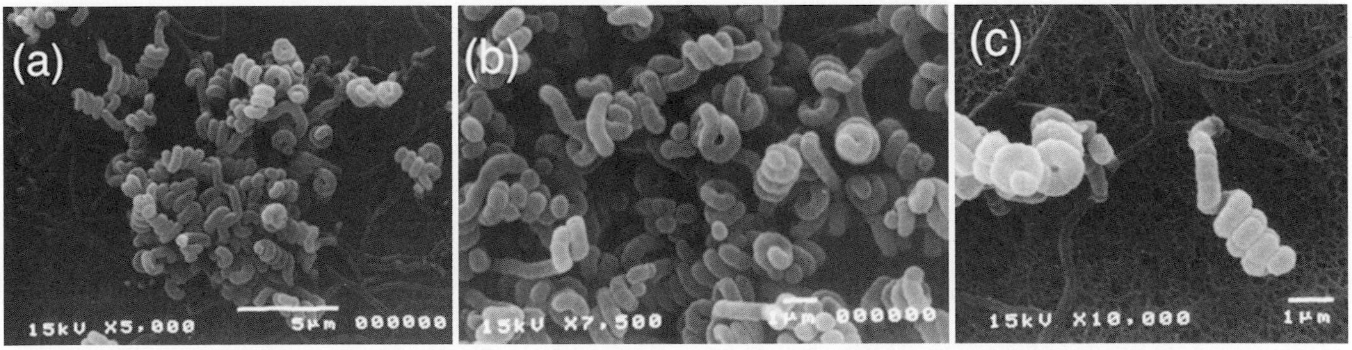

FIGURE 226. Scanning electron micrograph of a 14-d-old culture of *Spirilliplanes yamanashiensis* grown on HV agar revealed the presence of short hyphae arranged in spirals, which could arise from the substrate mycelia. The aerial mycelium at maturity formed short chains of spores (a–c), but no true sporangia were observed because the hyphae were not covered with a sheath.

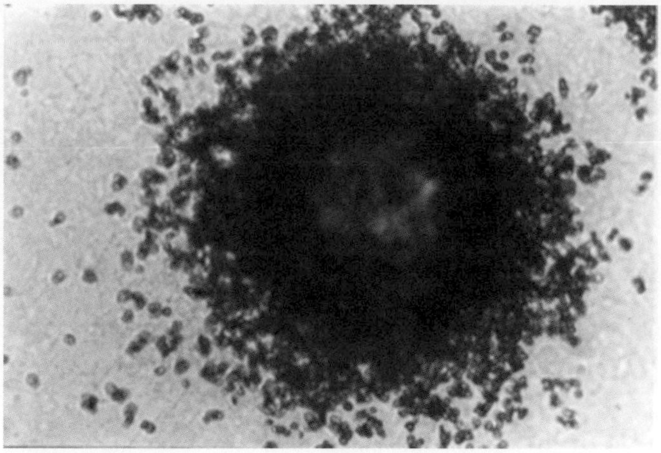

FIGURE 227. Light micrograph of a colony of a 14-d-old culture of *Spirilliplanes yamanashiensis* grown on HV agar. The sporogenous hyphae aggregate into clusters resembling sporangia under the light microscope, but are not true sporangia (see legend to Figure 226).

Differentiation of the genus from other genera

The genus *Spirilliplanes* possesses a chemotype II cell wall and chains of very narrow, coiled sporogenous hyphae with zoospores. Members of the order *Micromonosporales* with chemotype II cell walls are the genera *Actinoplanes, Dactylosporangium,* and *Pilimelia,* and they also have motile elements or spores. The genus *Spirilliplanes* can be distinguished from these genera because it does not form sporangia, lacks arabinose in whole cells, and possesses a different menaquinone pattern.

Taxonomic comments

Based on 16S rRNA gene sequence analysis, the genus *Spirilliplanes* clearly forms an independent lineage in the phylogenetic tree of the family *Micromonosporaceae* of the order *Micromonosporales.* The morphological and chemotaxonomic characteristics also clearly separate the genus *Spirilliplanes* from the other genera.

List of species of the genus *Spirilliplanes*

1. **Spirilliplanes yamanashiensis** Tamura, Hayakawa and Hatano 1997, 102[VP]

ya.ma.na.shi.en'sis. N.L. fem. adj. *yamanashiensis* of or pertaining to Yamanashi Prefecture, Japan, the source of soil from which the organism was isolated.

Brownish soluble pigment is produced on tyrosine agar (ISP medium 7). Liquefies gelatin. Hydrolyzes starch. Does not decompose calcium malate. Does not coagulate milk. Xylose, glucose, inositol, raffinose, rhamnose, mannitol, and sucrose are utilized, but fructose and inositol are not.

Source: soil sample from Kofu, Yamanashi, Japan.

DNA G+C content (mol%): 69 (HPLC).

Type strain: YU127-1, DSM 44325, NBRC 15828, JCM 10032, VKM Ac-1993.

Sequence accession no. (16S rRNA gene): D63912.

Genus XVI. **Verrucosispora** Rheims, Schumann, Rohde and Stackebrandt 1998, 1125[VP]

Erko Stackebrandt

Ver.ru.co.si.spo'ra. L. adj. *verrucosus -a -um* warty; Gr. fem. n. *spora* a seed and in biology a spore; N.L. fem. n. *Verrucosispora* an organism with warty spores.

Gram-stain-positive, non-acid-fast, **aerobic organism with branching hyphae**. **Well developed septate mycelium** averaging 0.4 µm in diameter. **Nonmotile spores are borne singly, sessile, or on short or long sporophores**. Warty spore surface changes to a hairy appearance with increased age. **Aerial mycelium is absent**. Strictly aerobic. Chemo-organotroph. Good growth occurs at temperatures between 30 and 40°C. The peptide side-chain of the peptidoglycan contains *meso*-diaminopimelic acid (*meso*-A_2pm), glycine, alanine, and glutamic acid. The peptide side-chains are directly cross-linked (type A1γ). The acyl type of the cell-wall polysaccharides is glycolyl. Mannose and xylose are present in whole cell hydrolysates; arabinose is absent. Characteristic phospholipids are phosphatidylethanolamine, diphosphatidylglycerol, phosphatidylinositol mannoside, and phosphatidylserine. Major menaquinone is MK-(H_4); minor amounts of MK-9(H_6), MK-10(H_4), and MK-9(H_2) are found. Major fatty acids (>15% of total) are $C_{15:0}$ iso, $C_{16:0}$ iso, and $C_{17:0}$ anteiso; $C_{17:0}$ iso, $C_{17:1}$ iso, and $C_{18:0}$ iso are found in minor amounts (>4% to <15%). Phylogenetically, a member of the *Micromonosporaceae.*

DNA G+C content (mol%): 70 (HPLC).

Type species: **Verrucosispora gifhornensis** Rheims, Schumann, Rohde and Stackebrandt 1998, 1126[VP].

Further descriptive information

The family *Micromonosporaceae* was phylogenetically placed in the suborder *Micromonosporineae* as one of several suborders of the order *Actinomycetales*, subclass *Actinobacteridae,* class *Actinobacteria* (Stackebrandt et al., 1997), but in the present volume the suborder has been elevated to order *Micromonosporales,* class *Actinobacteria.* The type strains of members of the most closely related genera are indicated in the phylogenetic tree

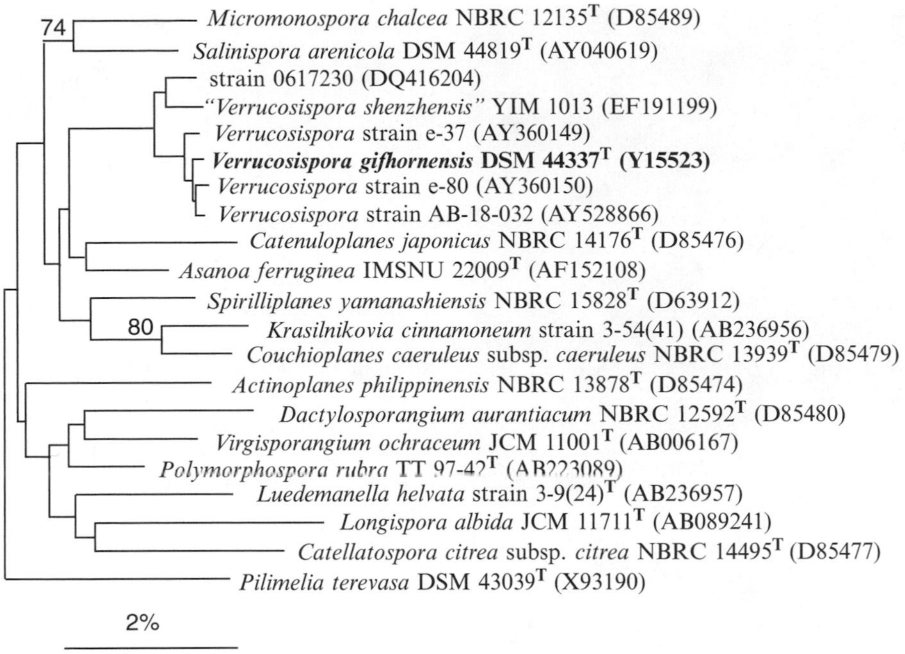

FIGURE 228. Phylogenetic tree showing the position of *Verrucosispora gifhornensis* and related strains among type strains of the family *Micromonosporaceae*, based upon 16S rRNA gene sequence analysis. The sequence of *Streptomyces ambofaciens* served as an outgroup sequence. Evolutionary distances were calculated by the method of Jukes and Cantor (1969). Phylogenetic dendrograms (DeSoete, 1983) were constructed using sequences from different sets of reference strains. Bootstrap analysis (>70% are shown) was used to evaluate the tree topology of the neighbor-joining data by performing 500 resamplings (Felsenstein, 1985). Bar = 2 nucleotide substitutions per 100 nucleotides.

based on 16S rRNA gene sequence similarities (Figure 228). All genera are closely related phylogenetically (Koch et al., 1996a; Maldonado et al., 2005; Matsumoto et al., 2003; Tamura et al., 1997, 2001, 2006; Thawai et al., 2006a), sharing >91% similarity among each other. *Verrucosispora gifhornensis* appears equally closely related (96.0–97.4%) to the majority of genera, sharing a slightly higher degree of relatedness with members of the *Catenuloplanes*, *Asanoa*, and *Spirilloplanes*. However, bootstrap values are low for most branching points indicating the low statistical significance of the branching order [see also dendrograms shown by Tamura et al. (2006) and Matsumoto et al. (2003)]. The close relatedness among type strains is also supported by DNA–DNA reassociation values. The experiments, carried between strain *Verrucosispora gifhornensis* and two close relatives, revealed 41.4% and 40.7% hybridization with the type strains of *Micromonospora olivasterospora* and *Spirilliplanes yamanashiensis*, respectively. The similarity value for the latter pair of organisms was 30.8%. A recently published 16S rRNA gene tree on the phylogenetic position of *Polymorphosporangium rubra* (Tamura et al., 2006) differs from other trees published for newly described type strains of novel genera of the *Micromonosporaceae* in that *Verrucosispora gifhornensis* represents the most deeply branching organism of the family.

Verrucosispora gifhornensis DSM 44337^T produces substrate mycelium on all media that promote growth, but does not produce aerial mycelium. The substrate mycelium is well developed. Hyphae are approximately 0.4 µm in diameter. Sporangia are not formed. On older parts, granular structures are observed on the mycelial surface. Spores are borne singly from the substrate mycelium and are 0.8 µm in diameter. The spore surface appears warty in younger states and changes to a hairy appearance with increased age (Figure 229). Spores are nonmotile.

Members of the genus are strictly aerobic. On all media tested, *Verrucosispora gifhornensis* develops orange colonies which became brownish with increased age. Best growth is observed on agar plates with Difco tryptic soy broth. Growth on the media recommended by the International *Streptomyces* Project (ISP) (Shirling and Gottlieb, 1966) is good on medium 1 (tryptone-yeast extract broth), medium 4 (inorganic salts-starch agar), and medium 5 (glycerol-asparagine agar); moderate growth is observed on medium 2 (yeast extract-malt extract agar), medium 6 (peptone-yeast extract iron agar), and medium 7 (tyrosine agar); no growth is observed on medium 3. Growth on HV-agar (Hayakawa and Nonomura, 1987a; Tamura et al., 1997) is slow. The type strain, DSM 44337^T, grows well in 0–2% NaCl, moderately in 2–4%, but fails to grow at concentrations above 4% NaCl.

No growth is observed below 20°C or above 40°C. Good growth is seen between 30 and 40°C, with optimum growth at 35°C. The pH range for growth is 6.5–8.2, with an optimum at 7.5.

Growth of *Verrucosispora gifhornensis* DSM 44337^T is inhibited by the following antibiotics. Strong inhibition (diameter of the inhibition zone: 31–50 mm): amikacin (concentration: 30 µg per disc), bacitracin (10 µg), cephalozin (30 µg), doxycycline (30 µg), gentamicin (10 µg), imipenem (10 µg), kanamycin (30 µg), neomycin (30 µg), polymyxin B (300 µg), and tetracycline

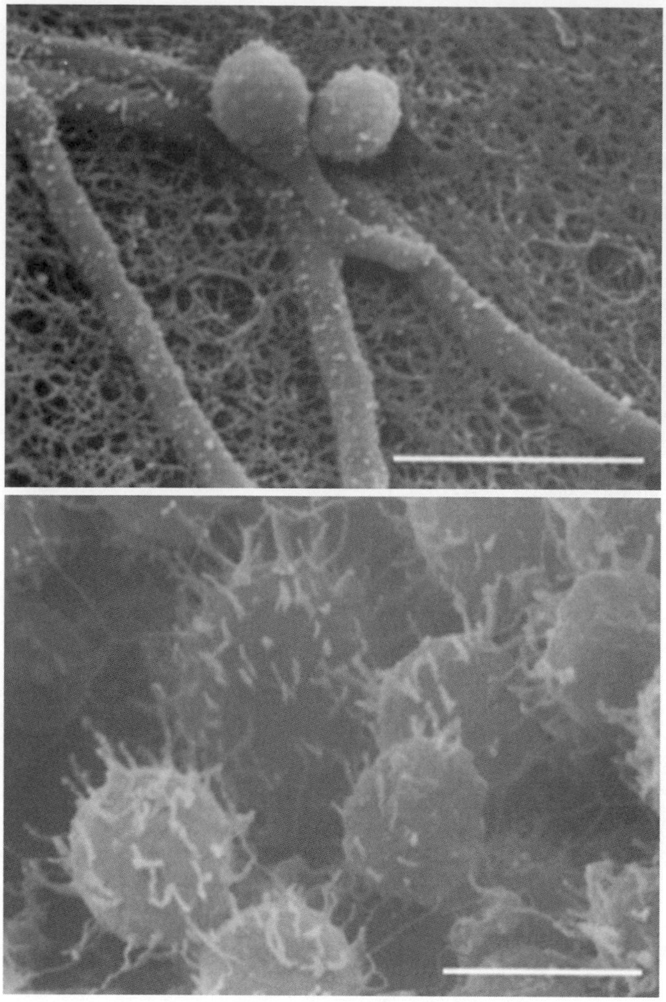

FIGURE 229. Scanning electron microscopic images of *Verrucosispora gifhornensis* DSM 44337^T. (Top) Young spores in detail; bar = 2 µm. (Bottom) Old spores in detail; the spore surface has become hairy; bar = 1 µm. (Reproduced with permission from Rheims et al., 1998. Int. J. Syst. Bacteriol. *48*: 1119–1127.)

(30 µg). Good inhibition (21–30 mm): aztreonam (30 µg), cefotaxime (30 µg), cephalothin (30 µg), norfloxacin (10 µg), penicillin G (6 µg), and ticarcillin (75 µg). Poor inhibition (10–20 mm): ampicillin (10 µg), azlocillin (30 µg), chloramphenicol (30 µg), colistin sulfate (10 µg), erythromycin (15 µg), mezlocillin (30 µg), and oflaxacin (5 µg). No inhibition is observed with lincomycin (15 µg), nitrofurantoin (100 µg), oxacillin (5 µg), or pipemidic acid (20 µg).

Strain AB-18-032, which is affiliated to the genus *Verrucosispora*, is the producer of the polycyclic polyketide Abyssomicin C. This antibiotic is an inhibitor of *para*-iminobenzoic acid biosynthesis and therefore inhibits folic acid biosynthesis at an early stage (Bister et al., 2004). Gram-stain-positive bacteria, including multiresistant and vancomycin-resistant *Staphylococcus aureus* strains, are strongly inhibited.

Enrichment and isolation procedures

Verrucosispora gifhornensis was isolated from a sample taken from a peat bog near Gifhorn, Lower Saxony, Germany (10°33′E, 52°30′N). Samples were taken from a depth of 20–40 cm after removal of the top peat layer. A dilution series of 10 g of an aged peat sample, stored at 4°C for 12 months, was set up in sterile tap water to a dilution of 10^{-6}. Aliquots of 100 µl were plated on Actinomycete isolation agar, pH 8.2 (Difco). After 11 d of aerobic incubation at 30°C, growth of strain DSM 44337^T was indicated by an expanding clear circular zone in the medium. After further incubation, an orange-red colony became visible which was transferred onto fresh Actinomycete isolation agar.

Several *Verrucosispora* strains have been isolated recently in studies exploring microbial diversity [Wang et al., (1999) (i.e. AF131629); Riedlinger et al., 2004; Fiedler et al., 2005; Rifaat et al., 2002; H Muramatsu and others, unpublished (i.e. AB123463)]. Several of these strains were isolated from the marine environment (L.A. Maldonado and others, unpublished, e.g. AY371894 and AY360149), mangrove (L.H. Liu and K. Hong, unpublished; DQ416204) and sponges (S. Jiang and others, unpublished, e.g. DQ994712). The name "*Verrucosispora shenzhensis*" has been proposed for one strain, the sequence of which is available under EF191199 (Z.L. Liao and others, unpublished). The phylogenetic position of a few strains, for which the almost complete 16S rRNA gene sequence has been determined, is shown in Figure 228. Their inter-strain relationship is >99.1%.

Maintenance procedures

Strains can be stored for some weeks as slants at 4°C and as 20% (w/v) glycerol suspensions at –20°C and –80°C. Long-term preservation methods include freeze-drying in skim milk and in liquid nitrogen at –196°C.

Differentiation of the genus *Verrucosispora* from other genera

Members of the genera can be distinguished mainly by their phylogenetic position, morphological features, e.g. presence or absence of sporangia, and motility of spores, and a combination of chemotaxonomic properties (see Table 191). The fatty acid profile of isolate DSM 44337^T is characterized by the predominance of $C_{15:0}$ iso, followed by $C_{16:0}$ iso and $C_{17:0}$ anteiso The predominance of $C_{15:0}$ iso and $C_{17:0}$ iso fatty acids over $C_{15:0}$ anteiso and $C_{17:0}$ anteiso fatty acids, the occurrence of significant amounts of unsaturated fatty acids, and the absence of 10-methyl- and 2-OH-fatty acids matches the diagnostic fatty acid type 2d (Kroppenstedt, 1985). This type has also been reported in members of the genera *Actinoplanes*, *Longispora*, *Virgisporangium*, *Actinocatenispora*, *Pilimelia*, and *Spirilliplanes*. The *meso*-A_2pm-containing, directly cross-linked peptidoglycan of the A1γ variation (cell-wall type II) is represented in all members of the family, except for members of the *Couchioplanes* and *Catenuloplanes* (type IV). The major isoprenoid quinone component, MK-9(H_4), is present in several members of the family, though often in combination with other menaquinones. The polar lipid pattern is PII (Lechevalier and Lechevalier, 1970a), consisting of the diagnostic phosphatidylethanolamine (phospholipid type PII), diphosphatidylglycerol, phosphatidylinositol mannoside, and phosphatidylserine. This combination is common among species of the family. *Verrucosispora gifhornensis* lacks arabinose among its whole cell sugars, a feature shared with members of the genera *Polymorphospora* (Tamura et al., 2006), *Catenuloplanes* (Yokota et al., 1993), and *Spirilliplanes* (Tamura et al., 1997).

List of species of the genus *Verrucosispora*

1. **Verrucosispora gifhornensis** Rheims, Schumann, Rohde and Stackebrandt 1998, 1126[VP]

gif.horn.en'sis. N.L. fem. n. *gifhornensis* of or belonging to the city of Gifhorn, adjacent to the peat bog from which the organism was isolated.

Morphological, chemotaxonomic, and general characteristics are as described for the genus. An orange pigment is produced on all ISP media tested. Gelatin liquefaction and peptonization of milk is positive. Hydrolyzes starch. Nitrite is not produced from nitrate. Cellulose decomposition is negative. D-Xylose, D-glucose, D-galactose, maltose, sucrose, D-arabinose, and α-trehalose are utilized; D-ribose, D-fructose, L-rhamnose, L-sorbose, lactose, α-melibiose, melezitose, raffinose, glycerol, dulcitol, *myo*-inositol, and salicin are not utilized. L-Serine, L-aspartic acid, L-glutamic acid, L-histidine, L-arginine, and L-phenylalanine are used as nitrogen source. Major fatty acids are (%) $C_{15:0}$ iso (31.1), $C_{16:0}$ iso (18.6), $C_{17:0}$ anteiso (16.0); minor components (>1.5% to <10%) are $C_{17:0}$ iso (8.3), $C_{17:1}$ iso (4.0), $C_{18:0}$ iso (4.4), $C_{15:0}$ anteiso (2.2), $C_{17:0}$ (2.9), and $C_{18:1}$ (1.8). The other chemotaxonomic markers are indicated in the genus description. Highly sensitive against a wide range of antibiotics.

Source: isolated from a peat bog near Gifhorn, Lower Saxony, Germany.

DNA G+C content (mol%): 70 (HPLC).

Type strain: HR1-2, DSM 44337.

Sequence accession no. (16S rRNA gene): Y15523.

Genus XVII. **Virgisporangium** corrig. Tamura, Hayakawa and Hatano 2001, 1814[VP]

Tomohiko Tamura

Vir.gi.spo.ran'gi.um. L. n. *virga* a slender green branch, rod; N.L. neut. n. *sporangium* (from Gr. n. *spora* a seed and, in biology, a spore; Gr. n. *angeion* vessel) sporangium (spore-containing vessel); N.L. neut. n. *Virgisporangium* an organism with rod-shaped sporangia (spore-containing vessels).

Gram-stain-positive, not acid-fast, produces a fine, nonfragmenting, branching mycelium. Strictly aerobic. **Slender sporangia** are formed on short sporangiophores on the substrate mycelium (Figure 230 and Figure 231). Each sporangium typically contains a single row of **six or more spores**. Spores are oval or short rods (0.6–0.9 × 0.8–1.5 µm) and exhibit motility. Cell wall contains D-glutamate, glycine, alanine, and 3-OH-diaminopimelate. Mannose, 3-*O*-methylmannose, rhamnose, glucose, arabinose, xylose, and galactose are present in whole-cell hydrolysates. The major cellular fatty acid is $C_{17:0}$ anteiso. The major menaquinones are MK-10(H_4) and MK-10(H_6). Phosphatidylethanolamine is present as the diagnostic phospholipid (phospholipid pattern type PII). The acyl type of the cell wall is glycolyl. Mycolic acid is not detected.

DNA G+C content (mol%): 71 (HPLC).

Type species: **Virgisporangium ochraceum** Tamura, Hayakawa and Hatano 2001, 1815[VP].

Further descriptive information

Good growth occurs between 20 and 30°C. In general, the vegetative mycelia are yellow to orange.

Cultures grow well in yeast extract-glucose broth, consisting of yeast extract (1%) and D-glucose (1%), pH adjusted to 7.0, and incubated at 28°C on a rotary shaker for 4 d.

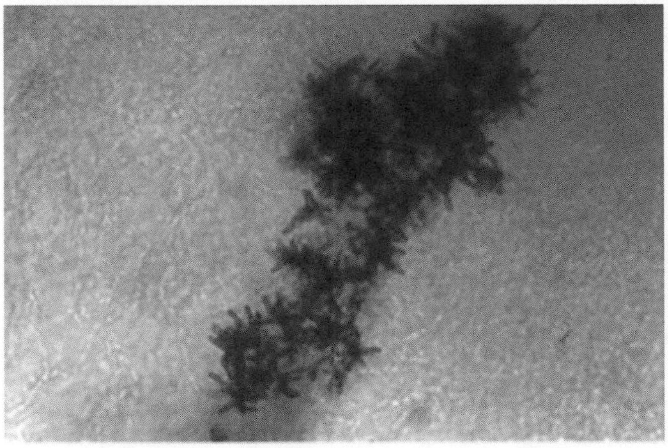

FIGURE 230. Light micrograph of 14-d-old cultures of *Virgisporangium ochraceum* isolates grown on HV agar showing rod-shaped sporangia on the substrate mycelium, which developed singly or in clusters above the surface of the substrate.

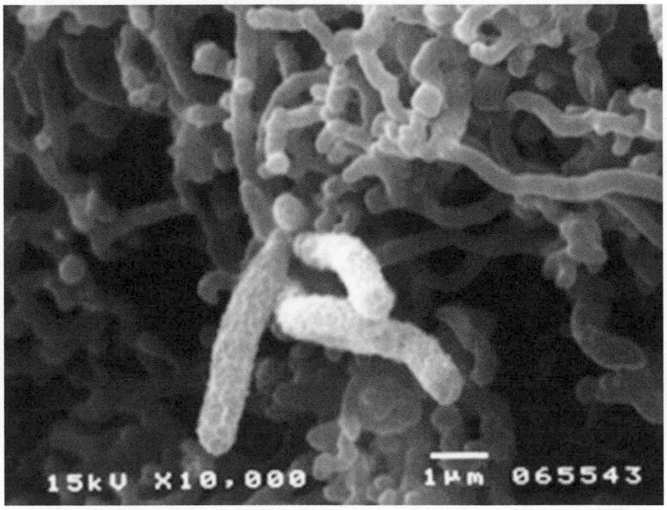

FIGURE 231. Scanning electron micrograph of 14-d-old cultures of *Virgisporangium ochraceum* isolates grown on HV agar showing that each sporangium typically contains a single row of six or more spores. Globose bodies were not observed. After incubation at 28°C for 1 h in distilled water, many spores exhibited active motility.

The cell-wall chemotype is II according to Lechevalier and Lechevalier (1970a) and the peptidoglycan type is presumed to be A1γ according to Schleifer and Kandler (1972). The cellular fatty acids are predominantly $C_{17:0}$ anteiso, with lesser amounts of $C_{18:1}$, $C_{16:0}$ iso, and $C_{15:0}$ iso, depending upon the species.

Enrichment and isolation procedures

For isolation of the strains of the genus *Virgisporangium*, samples are inoculated on humic acid-vitamin (HV) agar (Hayakawa and Nonomura, 1987a) using the capillary method (Hayakawa et al., 1991d; Palleroni, 1980) with vanillin as an attractant. Incubation is at 28°C for 2 weeks. The described species were isolated from a vegetable field soil in Yamanashi, from a mulberry field soil in Okinawa, and from a potato field soil in Okinawa, Japan (Tamura et al., 2001).

Maintenance procedures

Strains of the genus *Virgisporangium* are maintained by freezing in water containing 10–30% glycerol at −70°C. Lyophilization of suspensions in 10% skim milk + 1% monosodium glutamate and L-drying in 0.01 M potassium phosphate buffer (pH 7.0) containing 3% monosodium glutamate (Sakane and Kuroshima, 1997) are also recommended for long-term preservation.

Differentiation of the genus *Virgisporangium* from other genera

Strains of the genus *Virgisporangium* form rod-shaped sporangia on short sporangiophores, which resemble the extended sporangia of the genus *Dactylosporangium*. However, the genus *Virgisporangium* can be distinguished from the genus *Dactylosporangium* in the development of rod-shaped sporangia with six or more spores, the absence of 10-methylated fatty acids, and the presence of MK-10(H_4, H_6, H_8). Cell walls of *Virgisporangium* contain exclusively 3-OH diaminopimelic acid as the diamino acid. Some members of the genera *Micromonospora* and *Dactylosporangium* are known to have 3-OH-diaminopimelic acid as well as *meso*-diaminopimelic acid in their cell walls.

Taxonomic comments

The original spelling *Virgisporangium* (*sic*) proposed by Tamura et al. (2001), has been corrected by the List Editor, IJSEM (2001).

Based on 16S rRNA gene sequence analysis, this genus represents an independent lineage in the phylogenetic tree of the family *Micromonosporaceae* of the order *Micromonosporales*. *Virgisporangium aurantiacum* was proposed based upon a single strain. In contrast, the description of the type species *Virgisporangium ochraceum* was based upon three strains with relatively low values of DNA relatedness (40–60%), but high phenotypic and 16S rRNA gene sequence similarity (Tamura et al., 2001).

List of species of the genus *Virgisporangium*

1. **Virgisporangium ochraceum** corrig. Tamura, Hayakawa and Hatano 2001, 1815^{VP}

och.ra'ce.um. L. n. *ochra* ochre, yellow ochre; N.L. neut. adj. *ochraceum* of the color of ochre, rust-colored.

Morphological, chemotaxonomic, and general characteristics are as given above for the genus. Brownish soluble pigment is produced on tyrosine agar (ISP medium 7). Gelatin liquefaction is negative. Hydrolyzes starch. Does not decompose calcium malate. Coagulation and peptonization of milk are positive. Optimum temperature for growth is 15–30°C. Grows at 37°C. Does not grow on 7% NaCl. Glucose, D-fructose, D-xylose, L-arabinose, glycerol, and mannose are utilized, but inulin is not. The major cellular fatty acids are $C_{17:0}$ anteiso, $C_{18:1}$, and $C_{15:0}$ iso, with $C_{16:0}$ iso, $C_{17:0}$, and $C_{19:0}$ in smaller amounts.

Source: soil.

DNA G+C content (mol%): 71 (HPLC).

Type strain: YU655-43, CIP 107213, NBRC 16418, JCM 11001.

Sequence accession no. (16S rRNA gene): AB006167.

2. **Virgisporangium aurantiacum** corrig. Tamura, Hayakawa and Hatano 2001, 1815^{VP}

au.ran.ti.a'cum. N.L. neut. adj. *aurantiacum* orange-colored.

Morphological, chemotaxonomic, and general characteristics are as given above for the genus. Brownish soluble pigment is produced on tyrosine agar (ISP medium 7). Gelatin liquefaction is negative. Hydrolyzes starch. Does not decompose calcium malate. Coagulation and clearing of milk are positive. Optimum temperature for growth is 15–30°C. Does not grow at 37°C. Does not grow on 4% NaCl. Glucose, D-fructose, D-xylose, L-arabinose, glycerol, maltose, and mannose are utilized, but inositol, D-sorbitol, and inulin are not. The major cellular fatty acids are $C_{17:0}$ anteiso and $C_{16:0}$ iso.

Source: the type strain was isolated from soil.

DNA G+C content (mol%): 71 (HPLC).

Type strain: YU438-5, CIP 107212, NBRC 16421, JCM 11002.

Sequence accession no. (16S rRNA gene): AB006169.

References

Acinas, S.G., V. Klepac-Ceraj, D.E. Hunt, C. Pharino, I. Ceraj, D.L. Distel and M.F. Polz. 2004. Fine-scale phylogenetic architecture of a complex bacterial community. Nature *430*: 551–554.

Alexander, D.C., D.J. Devlin, D.D. Hewitt, A.C. Horan and T.J. Hosted. 2003. Development of the *Micromonospora carbonacea* var. *africana* ATCC 39149 bacteriophage pMLP1 integrase for site-specific integration in *Micromonospora* spp. Microbiology *149*: 2443–2453.

Alexander, G.M., J.R. Grothusen, S.W. Gordon and R.J. Schwartzman. 1997. Intracerebral microdialysis study of glutamate reuptake in awake, behaving rats. Brain Res *766*: 1–10.

Ara, I. and T. Kudo. 2006. Three novel species of the genus *Catellatospora*, *Catellatospora chokoriensis* sp. nov., *Catellatospora coxensis* sp. nov. and *Catellatospora bangladeshensis* sp. nov., and transfer of *Catellatospora citrea* subsp. *methionotrophica* Asano and Kawamoto 1988 to *Catellatospora methionotrophica* sp. nov., comb. nov. Int. J. Syst. Evol. Microbiol. *56*: 393–400.

Ara, I. and T. Kudo. 2007a. *Krasilnikovia* gen. nov., a new member of the family *Micromonosporaceae* and description of *Krasilnikovia cinnamonea* sp. nov. Actinomycetologica *21*: 1–10.

Ara, I. and T. Kudo. 2007b. Two new species of the genus *Micromonospora*: *Micromonospora chokoriensis* sp. nov. and *Micromonospora coxensis* sp. nov., isolated from sandy soil. J. Gen. Appl. Microbiol. *53*: 29–37.

Ara, I. and T. Kudo. 2007c. List of new names and new combinations previously effectively, but not validly, published. Int. J. Syst. Evol. Microbiol. *57*: 1371–1373.

Ara, I. and T. Kudo. 2007d. *Luedemannella* gen. nov., a new member of the family *Micromonosporaceae* and description of *Luedemannella helvata* sp. nov. and *Luedemannella flava* sp. nov. J. Gen. Appl. Microbiol. *53*: 39–51.

Ara, I. and T. Kudo. 2007e. *In* List of new names and new combinations previously effectively, but not validly, published. Validation List no. 116. Int. J. Syst. Evol. Microbiol. *57*: 1371–1373.

Ara, I. and Takuji Kudo. 2007f. *In* List of new names and new combinations previously effectively, but not validly, published. Validation List no. 118. Int. J. Syst. Evol. Microbiol. *57*: 2449–2450.

Ara, I., M.A. Bakir and T. Kudo. 2008a. Transfer of *Catellatospora koreensis* Lee *et al.* 2000 as *Catelliglobosispora koreensis* gen. nov., comb. nov. and *Catellatospora tsunoense* Asano *et al.* 1989 as *Hamadaea tsunoensis* gen. nov., comb. nov., and emended description of the genus *Catellatospora* Asano and Kawamoto 1986 emend. Lee and Hah 2002. Int. J. Syst. Evol. Microbiol. *58*: 1950–1960.

Ara, I., A. Matsumoto, M.A. Bakir, T. Kudo, S. Ōmura and Y. Takahashi. 2008b. *Pseudosporangium ferrugineum* gen. nov., sp. nov., a new member of the family *Micromonosporaceae.* Int. J. Syst. Evol. Microbiol. *58*: 1644–1652.

Ara, I., H. Yamamura, B. Tsetseg, D. Daram and K. Ando. 2010. *Actinoplanes toevensis* sp. nov. and *Actinoplanes tereljensis* sp. nov., isolated from Mongolian soil. Int. J. Syst. Evol. Microbiol. *60*: 919–927.

Aretz, W., J. Meiwes, G. Seibert, G. Vobis and J. Wink. 2000. Friulimicins: novel lipopeptide antibiotics with peptidoglycan synthesis inhibiting activity from *Actinoplanes friuliensis* sp. nov. I. Taxonomic studies of the producing microorganism and fermentation. J. Antibiot. (Tokyo) *53*: 807–815.

Arora, D.K. 1986. Chemotaxis of *Actinoplanes missouriensis* Zoospores to Fungal Conidia, Chlamydospores and Sclerotia. J. Gen. Microbiol. *132*: 1657–1663.

Asano, K. and I. Kawamoto. 1986. *Catellatospora*, a new genus of the *Actinomycetales.* Int. J. Syst. Bacteriol. *36*: 512–517.

Asano, K. and I. Kawamoto. 1988. *Catellatospora citrea* subsp. *methionotrophica* subsp. nov., a methionine-deficient auxotroph of the *Actinomycetales.* Int. J. Syst. Bacteriol. *38*: 326–327.

Asano, K., I. Masunaga and I. Kawamoto. 1989a. *Catellatospora matsumotoense* sp. nov. and *Catellatospora tsunoense* sp. nov., actinomycetes found in woodland soils. Int. J. Syst. Bacteriol. *39*: 309–313.

Asano, K., H. Sano, I. Masunaga and I. Kawamoto. 1989b. 3-O-Methylrhamnose: Identification and Distribution in *Catellatospora* Species and Related Actinomycetes. Int. J. Syst. Bacteriol. *39*: 56–60.

Asolkar, R.N., K.C. Freel, P.R. Jensen, W. Fenical, T.P. Kondratyuk, E.J. Park and J.M. Pezzuto. 2009. Arenamides A-C, cytotoxic NFkappaB inhibitors from the marine actinomycete *Salinispora arenicola.* J. Nat. Prod. *72*: 396–402.

Bardone, M.R., M. Paternoster and C. Coronelli. 1978. Teichomycins, new antibiotics from *Actinoplanes teichomyceticus* nov. sp. II. Extraction and chemical characterization. J. Antibiot. (Tokyo) *31*: 170–177.

Bérdy, J. 2005. Bioactive microbial metabolites. J. Antibiot. (Tokyo) *58*: 1–26.

Beretta, G. 1973. *Actinoplanes italicus*, a new red-pigmented species. Int. J. Syst. Bacteriol. *23*: 37–42.

Bister, B., D. Bischoff, M. Strobele, J. Riedlinger, A. Reicke, F. Wolter, A.T. Bull, H. Zahner, H.P. Fiedler and R.D. Sussmuth. 2004. Abyssomicin C-A polycyclic antibiotic from a marine *Verrucosispora* strain as an inhibitor of the p-aminobenzoic acid/tetrahydrofolate biosynthesis pathway. Angew Chem. Int. Ed. Engl. *43*: 2574–2576.

Bland, C.E. 1968. Ultrastructure of *Pilimelia anulata* (*Actinoplanaceae*). J. Elisha Mitchell Sci. Soc. *84*: 8–15.

Bland, C.E. 1970. Fine structure of the motile cells and flagella in a member of the *Actinoplanaceae* (*Actinomycetales*). Proc. Natl. Acad. Sci. U S A *67*: 1550–1557.

Bland, C.E. and J.N. Couch. 1981. The family *Actinoplanaceae. In* The Prokaryotes, A Handbook on Habitats, Isolation and Identification of Bacteria (edited by Starr, Stolp, Trüper, Balows and Schlegel). Springer-Verlag, New York, pp. 2004–2010.

Brosius, J., M.L. Palmer, P.J. Kennedy and H.F. Noller. 1978. The complete nucleotide sequence of a 16S ribosomal RNA gene from *Escherichia coli.* Proc. Natl. Acad. Sci. U.S.A. *75*: 4801–4805.

Buchanan, G.O., P.G. Williams, R.H. Feling, C.A. Kauffman, P.R. Jensen and W. Fenical. 2005. Sporolides A and B: structurally unprecedented halogenated macrolides from the marine actinomycete *Salinispora tropica.* Org. Lett. *7*: 2731–2734.

Burman, N.P., C.P. Oliver and J.K. Stevens. 1969. Membrane filtration techniques for the isolation from water, of coli-aerogenes, *Escherichia coli*, faecal streptococci, *Clostridium perfringens*, actinomycetes and microfungi. *In* Isolation Methods for Microbiologists (edited by Shapton and Gould). Academic Press, London, pp. 127–134.

Burman, N.P. 1973. The occurrence and significance of actinomycetes in water supply. *In Actinomycetales:* Characteristics and Practical Importance (edited by Sykes and Skinner). Academic Press, London, pp. 219–230.

Caso, J.L., C. Hardisson and J.E. Suarez. 1990. Structure of the DNA of five bacteriophages infecting *Micromonospora.* Microbiologia *6*: 94–99.

Celmer, W.D., C.E. Moppett, W.P. Cullen, J.B. Routien, M.T. Jefferson, R. Shibakawa and J. Tone. 1977a. Antibiotic compound 41,012. U.S. Patent 4001397.

Celmer, W.D., W.P. Cullen, C.E. Moppet, J.B. Routien, R. Shbakawa and J. Tone. 1977b. Mixture of antibiotics produced by a species of *Actinoplanes.* US Patent 4,038,383.

Celmer, W.D., W.P. Cullen, E. Moppett, J.B. Routien, M.T. Jefferson, R. Shibakawa, J. Tone and Pfizer Inc. 1978. Polycyclic ether antibiotic produced by new species of *Dactylosporangium.* U.S. Patent 4081532 (March 28).

Chiba, H., J. Inokoshi, M. Okamoto, S. Asanuma, K. Matsuzaki, M. Iwama, K. Mizumoto, H. Tanaka, M. Oheda, K. Fujita, H. Nakashima, M. Shinose, Y. Takahashi and S. Ōmura. 2001. Actinohivin, a novel anti-HIV protein from an actinomycete that inhibits syncytium formation: isolation, characterization, and biological activities. Biochem. Biophys. Res. Commun. *282*: 595–601.

Ciabatti, R. and B. Cavalleri. 1989. Ramoplanin (A/16686): a new glycolipodepsipeptide antibiotic from *Actinoplanes.* Progr. Ind. Microbiol. *27*: 205–219.

Collins, M.D., M. Faulkner and R.M. Keddie. 1984. Menaquinone composition of some sporeforming actinomycetes. Syst. Appl. Microbiol. *5*: 20–29.

Conn, V.M. and C.M. Franco. 2004. Analysis of the endophytic actinobacterial population in the roots of wheat (*Triticum aestivum* L.) by terminal restriction fragment length polymorphism and sequencing of 16S rRNA clones. Appl. Environ. Microbiol. *70*: 1787–1794.

Coombs, J.T. and C.M. Franco. 2003. Isolation and identification of actinobacteria from surface-sterilized wheat roots. Appl. Environ. Microbiol. *69*: 5603–5608.

Coronelli, C., H. Pagani, M.R. Bardone and G.C. Lancini. 1974. Purpuromycin, a new antibiotic isolated from *Actinoplanes ianthinogenes* n. sp. J. Antibiot. (Tokyo) *27*: 161–168.

Couch, J. 1957. A new horizon in soil microbiology. Proc. Natl. Acad. Sci. India *27*: 69–73.

Couch, J.N. 1949. A new group of organisms related to actinomycetes. J. Elisha Mitchell Sci. Soc. *65*: 315–318.

Couch, J.N. 1950. *Actinoplanes*, a new genus of the *Actinomycetales.* J. Elisha Mitchell Sci. Soc. *66*: 87–92.

Couch, J.N. 1954. The genus *Actinoplanes* and its relatives. Trans. N.Y. Acad. Sci. *16*: 315–318.

Couch, J.N. 1955a. A new genus and family of the *Actinomycetales* with a revision of the genus *Actinoplanes.* J. Elisha Mitchell Sci. Soc. *71*: 148–155.

Couch, J.N. 1955b. *Actinosporangiaceae* should be *Actinoplanaceae.* J. Elisha Mitchell Sci. Soc. *71*: 269.

Couch, J.N. and W.J. Koch. 1962. Induction of motility in the spores of some *Actinoplanaceae.* Science *138*: 987.

Couch, J.N. 1963. Some new genera and species of the *Actinoplanaceae*. J. Elisha Mitchell Sci. Soc. *79*: 53–70.

Couch, J.N. 1964. A proposal to replace the name *Ampullaria* Couch with *Ampullariella*. J. Elisha Mitchell Sci. Soc. *80*: 29.

Couch, J.N. and C.E. Bland. 1974a. Genus V. *Ampullariella*. *In* Bergey's Manual of Determinative Bacteriology, 8th edn (edited by Buchanan and Gibbons). Williams and Wilkins, Baltimore, MD, pp. 717–718.

Couch, J.N. and C.E. Bland. 1974b. Genus 1. *Actinoplanes*. *In* Bergey's Manual of Determinative Bacteriology, 8th edn (edited by Buchanan and Gibbons). Williams & Wilkins, Baltimore, pp. 708–710.

Creutzfeld, W. 1988. Acarbose for the Treatment of Diabetes Mellitus (edited by Creutzfeld). Springer-Verlag, Berlin.

Cross, T. and M. Goodfellow. 1973. Taxonomy and classification of the actinomycetes. *In Actinomycetales:* Characteristics and Practical-Importance (edited by Sykes and Skinner). Academic Press, London, pp. 11–112.

Cross, T. and R.W. Attwell. 1974. Recovery of viable thermoactinomycete endospores from deep mud cores. *In* Spore Research 1973 (edited by Barker, Gould and Wolf). Academic Press, London, pp. 11–20.

Cross, T. 1981a. The monosporic actinomycetes. *In* The Prokaryotes: a Handbook on Habitats, Isolation, and Identification of Bacteria (edited by Starr, Stolp, Trüper, Balows and Schlegel). Springer, New York, pp. 2091–2102.

Cross, T. 1981b. Aquatic actinomycetes: a critical survey of the occurrence, growth and role of actinomycetes in aquatic habitats. J. Appl. Bacteriol. *50*: 397–423.

Cross, T. 1986. The occurrence and role of actinoplanetes and motile actinomycetes in natural ecosystems. *In* Perspectives in Microbial Ecology (edited by Megusar and Gantar). Slovene Society for Microbiology, Ljubljana, pp. 265–270.

Dairi, T., T. Ohta, E. Hashimoto and M. Hasegawa. 1992. Organization and nature of fortimicin A (astromicin) biosynthetic genes studied using a cosmid library of *Micromonospora olivasterospora* DNA. Mol. Gen. Genet. *236*: 39–48.

Dassain, M., G. Tiraby, M.A. Laneelle and J. Asselineau. 1983. [Comparative study of the lipid composition of seven species of "*Micromonospora*"]. Ann. Microbiol. (Paris) *134A*: 9–17.

Debono, M., K.E. Merkel, R.M. Molloy, M. Barnhart, E. Presti, A.H. Hunt and R.L. Hamill. 1984. Actaplanin, new glycopeptide antibiotics produced by *Actinoplanes missouriensis*. The isolation and preliminary chemical characterization of actaplanin. J. Antibiot. (Tokyo) *37*: 85–95.

DeSoete, G. 1983. A least square algorithm for fitting additive trees to proximity data. Psychometrika *48*: 621–626.

Embley, T.M. and E. Stackebrandt. 1994. The molecular phylogeny and systematics of the actinomycetes. Annu. Rev. Microbiol. *48*: 257–289.

Emerson, R. 1958. Mycological organization. Mycologia *50*: 589–621.

Ensign, J.C. 1978. Formation, properties, and germination of actinomycete spores. Annu. Rev. Microbiol. *32*: 185–219.

Ensign, J.C. 1982. Developmental biology of actinomycetes. *In* Over Production of Microbial Products (edited by Krumphanzl, Sikytha and Vaneck). Academic Press, London, pp. 127–140.

Erikson, D. 1935. The pathogenic aerobic organisms of the actinomyces group. Med. Res. Coun. Spec. Rep. Ser. No. *203*: 5–61.

Erikson, D. 1941. Studies on some lake-mud strains of *Micromonospora*. J. Bacteriol. *41*: 277–300.

Eustáquio, A.S., F. Pojer, J.P. Noel and B.S. Moore. 2008. Discovery and characterization of a marine bacterial SAM-dependent chlorinase. Nat. Chem. Biol. *4*: 69–74.

Eustáquio, A.S., R.P. McGlinchey, Y. Liu, C. Hazzard, L.L. Beer, G. Florova, M.M. Alhamadsheh, A. Lechner, A.J. Kale, Y. Kobayashi, K.A. Reynolds and B.S. Moore. 2009. Biosynthesis of the salinosporamide A polyketide synthase substrate chloroethylmalonyl-coenzyme A from S-adenosyl-L-methionine. Proc. Natl. Acad. Sci. U.S.A. *106*: 12295–12300.

Euzéby, J.P. and B.J. Tindall. 2004. A replacement name of the specific epithet aurantiaca in *Micromonospora aurantiaca* Sveshnikova *et al.* 1969 (Approved Lists 1980) and a proposal to treat the combination *Micromonospora aurantiaca* Sveshnikova *et al.* 1969 as a rejected name. Request for an Opinion. Int. J. Syst. Evol. Microbiol. *54*: 1905–1906.

Feling, R.H., G.O. Buchanan, T.J. Mincer, C.A. Kauffman, P.R. Jensen and W. Fenical. 2003. Salinosporamide A: a highly cytotoxic proteasome inhibitor from a novel microbial source, a marine bacterium of the new genus *Salinospora*. Angew Chem. Int. Ed. Engl. *42*: 355–357.

Felsenstein, J. 1985. Confidence limits on phylogenies: an approach using the bootstrap. Evolution *39*: 783–791.

Fenical, W., P.R. Jensen, M.A. Palladino, K.S. Lam, G.K. Lloyd and B.C. Potts. 2009. Discovery and development of the anticancer agent salinosporamide A (NPI-0052). Bioorg. Med. Chem. *17*: 2175–2180.

Fiedler, H.P., C. Bruntner, A.T. Bull, A.C. Ward, M. Goodfellow, O. Potterat, C. Puder and G. Mihm. 2005. Marine actinomycetes as a source of novel secondary metabolites. Antonie Van Leeuwenhoek *87*: 37–42.

Field, E.K., S. D'Imperio, A.R. Miller, M.R. VanEngelen, R. Gerlach, B.D. Lee, W.A. Apel and B.M. Peyton. 2010. Application of molecular techniques to elucidate the influence of cellulosic waste on the bacterial community structure at a simulated low-level-radioactive-waste site. Appl. Environ. Microbiol. *76*: 3106–3115.

Foulerton, A.G.R. 1905. New species of *Streptothrix* isolated from the air. Lancet *1*: 1199–1200.

Fujii, T., S. Satoi, N. Muto, M. Hayashi, A. Kodama, M. Otani and Toyo Jozo Kabushiki Kaisha. 1982. Aminoglycoside antibiotic G-367-2. U.S. Patent 4349667 (September 7).

Gaertner, A. 1955. [Two unusual keratinophilic organisms in the soil]. Arch. Mikrobiol. *23*: 28–37.

Gams, W., H.A. van der Aa, A.J. van der Plaats-Niterink, R.A. Samson and J.A. Stalpers. 1980. *In* CBS Course of Mycology, 2nd edn. Centraalbureau voor Schimmelcultures, Baarn.

Garcia, L.C., E. Martinez-Molina and M.E. Trujillo. 2010. *Micromonospora pisi* sp. nov., isolated from root nodules of *Pisum sativum*. Int. J. Syst. Evol. Microbiol. *60*: 331–337.

Garrity, G.M., B.K. Heimbuch and M. Gagliardi. 1996. Isolation of zoosporogenous actinomycetes from desert soils. J. Ind. Microbiol. Biotech. *17*: 260–267.

Gause, G.F., M.A. Sveshnikova, T.S. Maksimova and O.L. Olkhovatova. 1979. Production of antibiotic complex 4041 by *Actinoplanes ianthinogenes* subsp. *octamycini* subsp. nov. Antibiotiki *24*: 563–566.

Gontang, E.A., W. Fenical and P.R. Jensen. 2007. Phylogenetic diversity of gram-positive bacteria cultured from marine sediments. Appl. Environ. Microbiol. *73*: 3272–3282.

González, I., A. Ayuso-Sacido, A. Anderson and O. Genilloud. 2005. Actinomycetes isolated from lichens: Evaluation of their diversity and detection of biosynthetic gene sequences. FEMS Microbiol. Ecol. *54*: 401–415.

Goodfellow, M. and S.T. Williams. 1983. Ecology of actinomycetes. Annu. Rev. Microbiol. *37*: 189–216.

Goodfellow, M. and T. Cross. 1984. Classification. *In* The Biology of the Actinomycetes (edited by Goodfellow, Mordarski and Williams). Academic Press, London, pp. 7–164.

Goodfellow, M. and J.A. Haynes. 1984. Actinomycetes in marine sediments. *In* Biological, Biochemical and Biomedical Aspects of Actinomycetes (edited by Ortiz-Ortiz, Bojalil and Yakoleff). Academic Press, Orlando, pp. 453–472.

Goodfellow, M. 1989. Suprageneric classification of actinomycetes. *In* Bergey's Manual of Systematic Bacteriology, vol. 4 (edited by Williams, Sharpe and Holt). Williams & Wilkins, Baltimore, pp. 2333–2339.

Goodfellow, M., L. J. Stanton, K.E. Simpson and D.E. Minnikin. 1990a. *In* Validation of the publication of new names and new combinations previously effectively published outside the IJSB. List no. 34. Int. J. Syst. Bacteriol. *40*: 320–321.

Goodfellow, M., L.J. Stanton, K.E. Simpson and D.E. Minnikin. 1990b. Numerical and chemical classification of *Actinoplanes* and some related actinomycetes. J. Gen. Microbiol. *136*: 19–36.

Gordon, R.E. and M.M. Smith. 1955. Proposed group of characters for the separation of *Streptomyces* and *Nocardia*. J. Bacteriol. *69*: 147–150.

Gordon, R.E. 1967. The taxonomy of soil bacteria. *In* The Ecology of Soil Bacteria (edited by Gray and Parkinson). Liverpool University Press, Liverpool.

Hanton, W.K. 1968. *Amorphosporangium* (*Actinoplanaceae*): report of motility and additional characters. J. Gen. Microbiol. *53*: 317–320.

Hardisson, C. and J.E. Suarez. 1979. Fine structure of spore formation and germination in *Micromonospora chalcea*. J. Gen. Microbiol. *110*: 233–237.

Hasegawa, T., M. Takizawa and S. Tanida. 1983. A rapid analysis for chemical grouping of aerobic actinomycetes. J. Gen. Appl. Microbiol. *29*: 319–322.

Hayakawa, M. and H. Nonomura. 1987a. Humic acid-vitamin agar, a new medium for the selective isolation of soil actinomycetes. J. Ferment. Technol. *65*: 501–509.

Hayakawa, M. and H. Nonomura. 1987b. Efficacy of artificial humic acid as a selective nutrient in HV agar used for the isolation of soil actinomycetes. J. Ferment. Technol. *65*: 609–616.

Hayakawa, M. and H. Nonomura. 1989. A new method for the intensive isolation of actinomycetes from soil. Actinomycetologica *3*: 95–104.

Hayakawa, M., T. Kaihura and H. Nonomura. 1991a. New methods for the highly selective isolation of *Streptosporangium* and *Dactylosporangium* from soil. J. Ferment. Technol. Bioeng. *72*: 327–333.

Hayakawa, M., T. Sadakata, T. Kajiura and H. Nonomura. 1991b. New methods for the highly selective isolation of *Micromonospora* and *Microbispora* from soil. J. Ferment. Bioeng. *72*: 320–326.

Hayakawa, M., T. Tamura, H. Iino and H. Nonomura. 1991c. Pollen-baiting and drying method for the highly selective isolation of *Actinoplanes* spp. from soil. J. Ferment. Bioeng. *72*: 433–438.

Hayakawa, M., T. Tamura and H. Nonomura. 1991d. Selective isolation of *Actinoplanes* and *Dactylosporangium* from soil by using g-collidine as the chemoattractant. J. Ferment. Bioeng. *72*: 426–432.

Hayakawa, M., M. Ariizumi, T. Yamazaki and H. Nonomura. 1995. Chemotaxis in the zoosporic actinomycete *Catenuloplanes japonicus*. Actinomycetologica *9*: 152–163.

Hayakawa, M., M. Otoguro, T. Takeuchi, T. Yamazaki and Y. Iimura. 2000. Application of a method incorporating differential centrifugation for selective isolation of motile actinomycetes in soil and plant litter. Antonie van Leeuwenhoek *78*: 171–185.

Hayakawa, M. 2003. Selective isolation of rare actinomycete genera using pretreatment techniques. *In* Selective Isolation of Rare Actinomycetes (edited by Kurtböke). University of Sunshine Coast, Queensland, pp. 55–81.

He, H., W.D. Ding, V.S. Bernan, A.D. Richardson, C.M. Ireland, M. Greenstein, G.A. Ellestad and G.T. Carter. 2001. Lomaiviticins A and B, potent antitumor antibiotics from *Micromonospora lomaivitiensis*. J. Am. Chem. Soc. *123*: 5362–5363.

Heisey, R.M. and A.R. Putnam. 1990. Herbicidal activity of the antibiotics geldanamycin and nigericin. J. Plant Growth Reg. *9*: 19–25.

Henssen, A. and D. Schäfer. 1971. Emended description of the genus *Pseudonocardia* Henssen and description of the new species *Pseudonocardia spinosa*. Int. J. Syst. Bacteriol. *21*: 29–34.

Higgins, M.L. 1967. Release of sporangiospores by a strain of *Actinoplanes*. J. Bacteriol. *94*: 495–498.

Higgins, M.L., M.P. Lechevalier and H.A. Lechevalier. 1967. Flagellated actinomycetes. J. Bacteriol. *93*: 1446–1451.

Higgins, M.L. and M.P. Lechevalier. 1969. Poorly lytic bacteriophage from *Dactylosporangium thailandensis* (*Actinomycetales*). J. Virol. *3*: 210–216.

Hirsch, P., U. Mevs, R.M. Kroppenstedt, P. Schumann and E. Stackebrandt. 2004a. Cryptoenclolithic actinomycetes from antarctic sandstone rock samples: *Micromonospora endolithica* sp. nov. and two isolates related to *Micromonospora coerulea* Jensen 1932. Syst. Appl. Microbiol. *27*: 166–174.

Hirsch, P., U. Mevs, R.M. Kroppenstedt, P. Schumann and E. Stackebrandt. 2004b. *In* Validation of new names and new combinations previously effectively, but not validly, published outside the IJSEM. List no. 97. Int. J. Syst. Evol. Microbiol. *54*: 631–632.

Horan, A.C. and B. Brodsky. 1986a. *Actinoplanes caeruleus* sp. nov., a blue-pigmented species of the genus *Actinoplanes*. Int. J. Syst. Bacteriol. *36*: 187–191.

Horan, A.C. and B.C. Brodsky. 1986b. *Micromonospora rosaria* sp. nov., nom. rev., the rosaramicin producer. Int. J. Syst. Bacteriol. *36*: 478–480.

Horan, A.C. 1999. Secondary metabolite production, actinomycetes, other than *Streptomyces*. *In* Bioprocess Technology: Fermentation, Biocatalysis and Bioseparation (edited by Flickinger and Drew). Wiley, New York, pp. 2333–2348.

Hoskisson, P.A., G. Hobbs and G.P. Sharples. 2000. Response of *Micromonospora echinospora* (NCIMB 12744) spores to heat treatment with evidence of a heat activation phenomenon. Lett. Appl. Microbiol. *30*: 114–117.

Hosted, T.J. Jr, T. Wang and A.C. Horan. 2005. Characterization of the *Micromonospora rosaria* pMR2 plasmid and development of a high G+C codon optimized integrase for site-specific integration. Plasmid *54*: 249–258.

Hsu, S.C. and J.L. Lockwood. 1975. Powdered chitin agar as a selective medium for enumeration of actinomycetes in water and soil. Appl. Microbiol. *29*: 422–426.

Huang, H., J. Lv, Y. Hu, Z. Fang, K. Zhang and S. Bao. 2008. *Micromonospora rifamycinica* sp. nov., a novel actinomycete from mangrove sediment. Int. J. Syst. Evol. Microbiol. *58*: 17–20.

Hunter, J.C., D.E. Eveleigh and G. Casella. 1981. Actinomycetes of a salt march. Zentralbl. Bakteriol. Mikrobiol. Hyg. Abt. 1 Suppl. *11*: 195–200.

Hunter, J.C., M. Fonda, L. Sotos, B. Toso and A. Belt. 1984. Ecological approaches to isolation. Dev. Ind. Microbiol. *25*: 247–266.

Inahashi, Y., A. Matsumoto, H. Danbara, S. Ōmura and Y. Takahashi. 2010. *Phytohabitans suffuscus* gen. nov., sp. nov., an actinomycete of the family *Micromonosporaceae* isolated from plant roots. Int. J. Syst. Evol. Microbiol. *60*: 2652–2658.

Inouye, M., Y. Takada, N. Muto, T. Beppu and S. Horinouchi. 1994. Characterization and expression of a P-450-like mycinamicin biosynthesis gene using a novel *Micromonospora-Escherichia coli* shuttle cosmid vector. Mol. Gen. Genet. *245*: 456–464.

Itoh, Y., R. Enokita, T. Okazaki, S. Iwado, A. Torikata, T. Haneishi and M. Arai. 1981. Candiplanecin, a new antibiotic from *Ampullariella regularis* subsp. *mannitophila* subsp. nov. I. Taxonomy of producing organism and fermentation. J. Antibiot. (Tokyo) *34*: 929–933.

Ivanitskaia, L.P., E.M. Singal, M.V. Bibikova and S.N. Vostrov. 1978. [Directed isolation of *Micromonospora* generic cultures on a selective medium with gentamycin]. Antibiotiki *23*: 690–692.

Jackson, M., J.P. Karwowski, R.J. Theriault, P.B. Fernandes, R.C. Semon and W.L. Kohl. 1987. Coloradocin, an antibiotic from a new *Actinoplanes*. I. Taxonomy, fermentation and biological properties. J. Antibiot. (Tokyo) *40*: 1375–1382.

Jensen, H.L. 1930. The genus *Micromonospora* Ørskov, a little known group of soil microorganisms. Proc. Linnean Soc. N.S.W. *55*: 231–248.

Jensen, H.L. 1932. Contribution to our knowledge of *Actinomycetales*. III. Further observations on the genus *Micromonospora*. Proc. Linnean Soc. N.S.W. *57*: 173–180.

Jensen, P.R., R. Dwight and W. Fenical. 1991. Distribution of actinomycetes in near-shore tropical marine sediments. Appl. Environ. Microbiol. *57*: 1102–1108.

Jensen, P.R., E. Gontang, C. Mafnas, T.J. Mincer and W. Fenical. 2005. Culturable marine actinomycete diversity from tropical Pacific Ocean sediments. Environ. Microbiol. *7*: 1039–1048.

Jensen, P.R. and C. Mafnas. 2006. Biogeography of the marine actinomycete *Salinispora*. Environ. Microbiol. *8*: 1881–1888.

Jensen, P.R., P.G. Williams, D.C. Oh, L. Zeigler and W. Fenical. 2007. Species-specific secondary metabolite production in marine actinomycetes of the genus *Salinispora*. Appl. Environ. Microbiol. *73*: 1146–1152.

Jiang, C. and J. Ruan. 1982. Two new species and a new variety of *Ampullariella*. Acta Microbiol. Sin. *22*: 207–211.

Jiang, C., L. Xu and J. Ruan. 1983a. A new species of *Actinoplanes*. Acta Microbiol. Sin. *23*: 295–297.

Jiang, C., L. Xu and J. Ruan. 1983b. New species of *Actinoplanes* and *Ampullariella*. Acta Microbiol. Sin. *23*: 210–215.

Jiang, Z. and X. Yan. 1984. A new species of *Amorphosporangium*. Acta Microbiol. Sin. *24*: 129–133.

Johnston, D.W. and T. Cross. 1976. The occurrence and distribution of actinomycetes in lakes of the English Lake District. Freshwater Biol. *6*: 457–463.

Jongrungruangchok, S., S. Tanasupawat and T. Kudo. 2008a. *Micromonospora krabiensis* sp. nov., isolated from marine soil in Thailand. J. Gen. Appl. Microbiol. *54*: 127–133.

Jongrungruangchok, S., S. Tanasupawat and T. Kudo. 2008b. *Micromonospora chaiyaphumensis* sp. nov., isolated from Thai soils. Int. J. Syst. Evol. Microbiol. *58*: 924–928.

Juan, C.S. and Y. Zhang. 1974. A taxonomic study of *Actinoplanaceae*. I. Classification of *Ampullariella*. Acta Microbiol. Sin. *14*: 31–41.

Jukes, T.H. and C. Cantor. 1969. Evolution of protein molecules. *In* Mammalian Protein Metabolism (edited by Murano). Academic Press, New York pp. 21–132.

Kämpfer, P., B. Huber, K. Thummes, I. Grun-Wollny and H.-J. Busse. 2007. *Actinoplanes couchii* sp. nov. Int. J. Syst. Evol. Microbiol. *57*: 721–724.

Kane Hanton, W. 1974. Genus *Pilimelia*. *In* Bergey's Manual of Determinative Bacteriology, 8th edn. The Williams and Wilkins Co., Baltimore, pp. 718–719.

Kane, W.D. 1966. A new genus of *Actinoplanaceae*, *Pilimelia*, with a description of two species, *Pilimelia terevasa* and *Pilimelia anulata*. J. Elisha Mitchell Sci. Soc. *82*: 220–230.

Karling, J.S. 1954. An unusual keratinophilic microorganism. Proc. Indianapolis Acad. Sci. *63*: 83–86.

Karwowski, J.P., M. Jackson, R.J. Theriault, J.F. Prokop, M.L. Maus, C.F. Hansen and D.M. Hensey. 1988. Arizonins, a new complex of antibiotics related to kalafungin. I. Taxonomy of the producing culture, fermentation and biological activity. J. Antibiot. (Tokyo) *41*: 1205–1211.

Kasai, H., T. Tamura and S. Harayama. 2000. Intrageneric relationships among *Micromonospora* species deduced from *gyrB*-based phylogeny and DNA relatedness. Int. J. Syst. Evol. Microbiol. *50*: 127–134.

Kawamoto, I., R. Okachi, H. Kato, S. Yamamoto and I. Takahashi. 1974. The antibiotic XK-41 complex. I. Production, isolation and characterization. J. Antibiot. (Tokyo) *27*: 492–501.

Kawamoto, I., T. Oka and T. Nara. 1981. Cell wall composition of *Micromonospora olivoasterospora*, *Micromonospora sagamiensis*, and related organisms. J. Bacteriol. *146*: 527–534.

Kawamoto, I., T. Oka and T. Nara. 1982. Spore resistance of *Micromonospora olivasterospora*, *Micromonospora sagamiensis* and related organisms. Agric. Biol. Chem. *43*: 221–231.

Kawamoto, I., M. Yamamoto and T. Nara. 1983. *Micromonospora olivasterospora* sp. nov. Int. J. Syst. Bacteriol. *33*: 107–112.

Kawamoto, I. 1989. Genus *Micromonospora* Ørskov. *In* Bergey's Manual of Systematic Bacteriology, vol. 4 (edited by Williams, Sharpe and Holt). Williams and Wilkins, Baltimore, pp. 2442–2450.

Kikuchi, M. and D. Perlman. 1977. Bacteriophages infecting *Micromonospora purpurea*. J. Antibiot. (Tokyo) *30*: 423–424.

Kikuchi, M. and D. Perlman. 1978. Characteristics of bacteriophages for *Micromonospora purpurea*. Appl. Environ. Microbiol. *36*: 52–55.

Kim, T.K., M.J. Garson and J.A. Fuerst. 2005. Marine actinomycetes related to the "*Salinospora*" group from the Great Barrier Reef sponge *Pseudoceratina clavata*. Environ. Microbiol. *7*: 509–518.

Kimura, M. 1980. A simple method for estimating evolutionary rates of base substitutions through comparative studies of nucleotide sequences. J. Mol. Evol. *16*: 111–120.

Kirby, B.M. and P.R. Meyers. 2010. *Micromonospora tulbaghiae* sp. nov., isolated from the leaves of wild garlic, *Tulbaghia violacea*. Int. J. Syst. Evol. Microbiol. *60*: 1328–1333.

Kizuka, M., R. Enokita, K. Shibata, Y. Okamoto, Y. Inoue and T. Okazaki. 2002. Kizuka, M., R. Enokita, K. Shibata, Y. Okamoto, Y. Inoue and T Okazaki. 2002. Studies on actinomycetes from plant leaves - New plant growth inhibitors A-79197-2 and -3 from *Dacthylosporangium* (sic) *aurantiacum* SANK 61299. Actinomycetologist *16*: 14–16.

Koch, C., R.M. Kroppenstedt, F.A. Rainey and E. Stackebrandt. 1996a. 16S ribosomal DNA analysis of the genera *Micromonospora*, *Actinoplanes*, *Catellatospora*, *Catenuloplanes*, *Couchioplanes*, *Dactylosporangium*, and *Pilimelia* and emendation of the family *Micromonosporaceae*. Int. J. Syst. Bacteriol. *46*: 765–768.

Koch, C., R.M. Kroppenstedt and E. Stackebrandt. 1996b. Intrageneric relationships of the actinomycete genus *Micromonospora*. Int. J. Syst. Bacteriol. *46*: 383–387.

Konstantinidis, K.T. and J.M. Tiedje. 2005. Genomic insights that advance the species definition for prokaryotes. Proc. Natl. Acad. Sci. U.S.A. *102*: 2567–2572.

Kothe, H.-W. 1987. Die Gattung *Actinoplanes* und ihre Stellung innerhalb der *Actinomycetales*. Dissertation, Marburg.

Krasil'nikov, N.A. 1938. Ray Fungi and Related Organisms – *Actinomycetales*. Akad. Nauk. S.S.S.R. Moscow.

Kroppenstedt, R.M. and H.J. Kutzner. 1978. Biochemical taxonomy of some problem actinomycetes. Zentralbl. Bakteriol. Parasitenkd. Infektionskr. Hyg. Abt. I *Suppl. 6*: 125–133.

Kroppenstedt, R.M. 1979. Chromatographische Indifizierung von Mikroorganismen, dargestellt am Beispiel der Actinomyceten Kontakte (Merk) *2*: 12–21.

Kroppenstedt, R.M. 1985. Fatty acid and menaquinone analysis of actinomycetes and related organisms. *In* Chemical Methods in Bacterial Systematics (edited by Goodfellow and Minnikin). Academic Press, London, pp. 173–199.

Kroppenstedt, R.M., S. Mayilraj, J.M. Wink, W. Kallow, P. Schumann, C. Secondini and E. Stackebrandt. 2005a. Eight new species of the genus *Micromonospora*, *Micromonospora citrea* sp. nov., *Micromonospora echinaurantiaca* sp. nov., *Micromonospora echinofusca* sp. nov. *Micromonospora fulviviridis* sp. nov., *Micromonospora inyonensis* sp. nov., *Micromonospora peucetia* sp. nov., *Micromonospora sagamiensis* sp. nov., and *Micromonospora viridifaciens* sp. nov. Syst. Appl. Microbiol. *28*: 328–339.

Kroppenstedt, R.M., S. Mayilraj, J.M. Wink, W. Kallow, P. Schumann, C. Secondini and E. Stackebrandt. 2005b. *In* Validation of publication of new names and new combinations previously effectively, but not validly, published outside the IJSEM. List no. 105. Int. J. Syst. Evol. Microbiol. *55*: 1743–1745.

Kudo, T., Y. Nakajima and K.-i. Suzuki. 1999. *Catenuloplanes crispus* (Petrolini *et al.* 1993) comb. nov.: incorporation of the genus *Planopolyspora* Petrolini 1993 into the genus *Catenuloplanes* Yokota *et al.* 1993 with an amended description of the genus *Catenuloplanes*. Int. J. Syst. Bacteriol. *49*: 1853–1860.

Kudo, T. 2001. Methods for chemotaxonomy. *In* Identification Manual of Actinomycetes (edited by Miyadoh, Hamada, Hotta, Seino, Suzuki and Yokota), Tokyo, Japan, pp. 49–82.

Lancini, G. and R. Lorenzetti. 1993. *In* Biotechnology of antibiotics and other bioactive microbial metabolites. Plenum Press, New York and London, pp. 49–57.

Lazzarini, A., L. Cavaletii, G. Toppo and F. Marinelli. 2000. Rare genera of actonomycetes as potential producers of new antibiotics. Antonie van Leeuwenhoek *78*: 399–405.

Lechevalier, H. and P.E. Holbert. 1965. Electron microscopic observation of the sporangial structure of a strain of *Actinoplanes*. J. Bacteriol. *89*: 217–222.

Lechevalier, H.A., M.P. Lechevalier and P.E. Holbert. 1966. Electron microscopic observation of the sporangial structure of strains of *Actinoplanaceae*. J. Bacteriol. *92*: 1228–1235.

Lechevalier, H.A. and M.P. Lechevalier. 1970. A critical evaluation of the genera of aerobic actinomycetes. *In* The *Actinomycetales* (edited by Prauser). Gustav Fischer Verlag, Jena, pp. 395–405.

Lechevalier, M.P. and H.A. Lechevalier. 1970a. Chemical composition as a criterion in the classification of aerobic actinomycetes. Int. J. Syst. Bacteriol. *20*: 435–443.

Lechevalier, M.P. and H.A. Lechevalier. 1970b. Composition of whole-cell hydrolysates as a criterion in the classification of aerobic actinomycetes. *In* The *Actinomycetales* (edited by Prauser). Gustav Fischer Verlag, Jena, pp. 311–316.

Lechevalier, M.P. and H.A. Lechevalier. 1975. Actinoplanete with cylindrical sporangia, *Actinoplanes rectilineatus* sp. nov. Int. J. Syst. Bacteriol. *25*: 371–376.

Lechevalier, M.P., C. De Bièvre and H. Lechevalier. 1977. Chemotaxonomy of aerobic actinomycetes: phospholipid composition. Biochem. Syst. Ecol. *5*: 249–260.

Lechevalier, M.P. 1981. Ecological associations involving actinomycetes. Zentralbl. Bakteriol. Mikrobiol. Hyg. I Abt. Suppl. *11*: 159–166.

Lechevalier, M.P., A.E. Stern and H.A. Lechevalier. 1981. Phospholipids in the taxonomy of actinomycetes. Zentralbl. Bakteriol. Parasitenkd. Infektionskr. Hyg. I Abt. Orig. *Suppl. 11*: 111–116.

Lee, S.D., M. Goodfellow and Y.C. Hah. 1999. A phylogenetic analysis of the genus *Catellatospora* based on 16S ribosomal DNA sequences, including transfer of *Catellatospora matsumotoense* to the genus *Micromonospora* as *Micromonospora matsumotoense* comb. nov. FEMS Microbiol. Lett. *178*: 349–354.

Lee, S.D., M. Goodfellow and Y.C. Hah. 2000a. *In* Validation of publication of new names and new combinations previously effectively published outside the IJSEM. Validation List no. 72. Int. J. Syst. Evol. Microbiol. *50*: 3–4.

Lee, S.D., S.O. Kang and Y.C. Hah. 2000b. *Catellatospora koreensis* sp. nov., a novel actinomycete isolated from a gold-mine cave. Int. J. Syst. Evol. Microbiol. *50*: 1103–1111.

Lee, S.D. and Y.C. Hah. 2002. Proposal to transfer *Catellatospora ferruginea* and '*Catellatospora ishikariense*' to *Asanoa* gen. nov. as *Asanoa ferruginea* comb. nov. and *Asanoa ishikariensis* sp. nov., with emended description of the genus *Catellatospora*. Int. J. Syst. Evol. Microbiol. *52*: 967–972.

Li, X., X. Zhou and Z. Deng. 2003. Vector systems allowing efficient autonomous or integrative gene cloning in *Micromonospora* sp. strain 40027. Appl. Environ. Microbiol. *69*: 3144–3151.

Li, X., X. Zhou and Z. Deng. 2004. Isolation and characterization of *Micromonospora* phage PhiHAU8 and development into a phasmid. Appl. Environ. Microbiol. *70*: 3893–3897.

List Editor. 2001. Notification that new names and new combinations have appeared in volume 51, part 5, of the IJSEM. Int. J. Syst. Evol. Microbiol. *51*: 1947–1948.

Ludwig, W., O. Strunk, R. Westram, L. Richter, H. Meier, Yadhukumar, A. Buchner, T. Lai, S. Steppi, G. Jobb, W. Forster, I. Brettske, S. Gerber, A.W. Ginhart, O. Gross, S. Grumann, S. Hermann, R. Jost, A. Konig, T. Liss, R. Lussmann, M. May, B. Nonhoff, B. Reichel, R. Strehlow, A. Stamatakis, N. Stuckmann, A. Vilbig, M. Lenke, T. Ludwig, A. Bode and K.H. Schleifer. 2004. ARB: a software environment for sequence data. Nucleic Acids Res. *32*: 1363–1371.

Luedemann, G.M. and B.C. Brodsky. 1964. Taxonomy of gentamicin-producing *Micromonospora*. Antimicrob. Agents Chemother. (Bethesda) *1963*: 116–124.

Luedemann, G.M. and B. Brodsky. 1965. *Micromonospora Carbonacea* sp. n., an everninomicin-producing organism. Antimicrob. Agents Chemother. (Bethesda) *1964*: 47–52.

Luedemann, G.M. 1971. *Micromonospora purpureochromogenes* (Waksman and Curtis 1916) comb. nov. (subjective synonym: *Micromonospora fusca* Jensen 1932). Int. J. Syst. Bacteriol. *21*: 240–247.

Luedemann, G.M. and C.J. Casmer. 1973. Electron microscope study of whole mounts and thin sections of *Micromonospora chalcea* ATCC 12452. Int. J. Syst. Bacteriol. *23*: 243–255.

Luedemann, G.M. 1974. Genus *Micromonospora*. *In* Bergey's Manual of Determinative Bacteriology, 8th edn (edited by Buchanan and Gibbons). Williams & Wilkins, Baltimore, pp. 846–855.

Maeda, K., H. Kosaka, Y. Okami and H. Umezawa. 1953. A new antibiotic, pyridomycin. J. Antibiot. (Tokyo) *6*: 140.

Magarvey, N.A., J.M. Keller, V. Bernan, M. Dworkin and D.H. Sherman. 2004. Isolation and characterization of novel marine-derived actinomycete taxa rich in bioactive metabolites. Appl. Environ. Microbiol. *70*: 7520–7529.

Makkar, N.S. and T. Cross. 1982. Actinoplanetes in soil and on plant litter from freshwater habitats. J. Appl. Bacteriol. *52*: 209–218.

Maldonado, L.A., W. Fenical, P.R. Jensen, C.A. Kauffman, T.J. Mincer, A.C. Ward, A.T. Bull and M. Goodfellow. 2005. *Salinispora arenicola* gen. nov., sp. nov. and *Salinispora tropica* sp. nov., obligate marine actinomycetes belonging to the family *Micromonosporaceae*. Int. J. Syst. Evol. Microbiol. *55*: 1759–1766.

Matsumoto, A., Y. Takahashi, T. Kudo, A. Seino, Y. Iwai and S. Ōmura. 2000. *Actinoplanes capillaceus* sp. nov., a new species of the genus *Actinoplanes*. Antonie Van Leeuwenhoek *78*: 107–115.

Matsumoto, A., Y. Takahashi, T. Kudo, A. Seino, Y. Iwai and S. Ōmura. 2001. *In* Validation of publication of new names and new combinations previously effectively published outside the IJSEM. List no. 80. Int. J. Syst. Evol. Microbiol. *51*: 793–794.

Matsumoto, A., Y. Takahashi, M. Shinose, A. Seino, Y. Iwai and S. Ōmura. 2003. *Longispora albida* gen. nov., sp. nov., a novel genus of the family *Micromonosporaceae*. Int. J. Syst. Evol. Microbiol. *53*: 1553–1559.

Matsumoto, A., Y. Takahashi, M. Fukumoto and S. Ōmura. 2007. *Actinocatenispora sera* sp. nov., isolated by long-term culturing. Int. J. Syst. Evol. Microbiol. *57*: 2651–2654.

Matsumoto, K., T. Shomura, M. Shimura, J. Yoshida, M. Ito, T. Watanabe and T. Ito. 1985. A new antibiotic SF-2185 produced by *Dactylosporangium*. I. Taxonomy, fermentation and biological properties. J. Antibiot. (Tokyo) *38*: 1487–1493.

Miller, C.E. and J.N. Couch. 1959. Lyophilization of the *Actinoplanaceae*. Mycologia *51*: 146–150.

Mincer, T.J., P.R. Jensen, C.A. Kauffman and W. Fenical. 2002. Widespread and persistent populations of a major new marine actinomycete taxon in ocean sediments. Appl. Environ. Microbiol. *68*: 5005–5011.

Mincer, T.J., W. Fenical and P.R. Jensen. 2005. Culture-dependent and culture-independent diversity within the obligate marine actinomycete genus *Salinispora*. Appl. Environ. Microbiol. *71*: 7019–7028.

Miwa, T.K., K.L. Mikolajczak, F.R. Earle and I.A. Wolff. 1960. Gas chromatographic characterization of fatty acids. Identification constants for mono- and dicarboxylic methyl esters. Anal. Chem. *32*: 1739–1742.

Miyadoh, S., M. Hamada, K. Hotta, T. Kudo, A. Seino, G. Vobis and A. Yokota. 1997. Atlas of Actinomycetes. Asakura Publishing, Tokyo.

Miyadoh, S., M. Hamada, K. Hotta, T. Kudo, A. Seino, K. Suzuki and A. Yokota. 2001. Identification Manual of Actinomycetes. Business Center for Academic Societies, Japan.

Nara, T., S. Takasawa, R. Okashi, I. Kawaoto, M. Yamamoto, S. Sato, T. Sato and A. Morikawa. 1977. Antibiotic XK-62-2 and process for production thereof. US Patent 4,045,298.

Nonomura, H. and Y. Ohara. 1969. Distribution of actinomycetes in soil. VI. A culture method effective for both preferential isolation and enumeration of *Microbispora* and *Streptosproangium* strains in Soil. Part I. J. Ferment. Technol. *47*: 463–469.

Nonomura, H. and S. Takagi. 1977. Distribution of actinoplanetes in soils of Japan. J. Ferment. Technol. *55*: 423–428

Nonomura, H., S. Iino and M. Hayakawa. 1979. Classification of actinomycetes of genus *Ampullariella* from soils of Japan. Hakkokogaku Kaishi *57*: 79–85.

Nonomura, H. 1984. Design of a new medium for the isolation of soil actinomycetes. The Actinomycetes *18*: 206–209.

Nonomura, H. and M. Hayakawa. 1988. New methods for the selective isolation of soil actinomycetes. *In* Biology of Actinomycetes (edited by Okami, Beppu and Ogawara). Japan Scientific Societies Press, Tokyo, pp. 288–293.

Oh, D.C., P.G. Williams, C.A. Kauffman, P.R. Jensen and W. Fenical. 2006. Cyanosporasides A and B, chloro- and cyano-cyclopenta[a] indene glycosides from the marine actinomycete "*Salinispora pacifica*". Org. Lett. *8*: 1021–1024.

Oh, D.C., E.A. Gontang, C.A. Kauffman, P.R. Jensen and W. Fenical. 2008. Salinipyrones and pacificanones, mixed-precursor polyketides from the marine actinomycete *Salinispora pacifica*. J. Nat. Prod. *71*: 570–575.

Okami, Y. and T. Okazaki. 1978. Actinomycetes in marine environments. Zentralbl. Bakteriol. Parasitenkd. Infektionskr. Hyg. Abt. 1 Suppl. *6*: 145–152.

Okami, Y. and K. Hotta. 1988. Search and discovery of new antibiotics. *In* Actinomycetes in Biotechnology (edited by Goodfellow, Williams and Mordarski). Academic Press, San Diego, pp. 33–67.

Okazaki, T. 2003. Studies on actinomycetes isolated from plant leaves. *In* Selective Isolation of Rare Actinomycetes (edited by Kurtböke). University of the Sunshine Coast, Queensland, pp. 102–122.

Omoto, S., T. Yoshida, M. Kurebe and S. Inouye. 1987. Dactimicin, a new, less toxic aminoglycoside antibiotic active against resistant bacteria. Drugs Exp. Clin. Res. *13*: 719–725.

Ørskov, J. 1923. Investigations into the Morphology of the Ray Fungi. Levin and Munksgaard, Copenhagen, Denmark.

Palleroni, N.J. 1976. Chemotaxis in *Actinoplanes*. Arch. Microbiol. *110*: 13–18.

Palleroni, N.J. 1979. New species of the genus *Actinoplanes*, *Actinoplanes ferrugineus*. Int. J. Syst. Bacteriol. *29*: 51–55.

Palleroni, N.J. 1980. A chemotactic method for the isolation of *Actinoplanaceae*. Arch. Microbiol. *128*: 53–55.

Palleroni, N.J. 1983. Biology of *Actinoplanes*. Actinomycetes *17*: 46–65.

Palleroni, N.J. 1989. Genus *Actinoplanes*. *In* Bergey's Manual of Systematic Bacteriology, 1st edn, vol. 4 (edited by Williams, Sharpe and Holt). Williams and Wilkins, Baltimore, MD, pp. 2419–2428.

Parag, Y. and M.E. Goedeke. 1984. A plasmid of the sisomicin producer *Micromonospora inyoensis*. J. Antibiot. (Tokyo) *37*: 1082–1084.

Parenti, F., H. Pagani and G. Beretta. 1975. Lipiarmycin, a new antibiotic from *Actinoplanes*. I. Description of the producer strain and fermentation studies. J. Antibiot. (Tokyo) *28*: 247–252.

Parenti, F., H. Pagani and G. Beretta. 1976. Gardimycin, a new antibiotic from *Actinoplanes*. I. Description of the producer strain and fermentation studies. J. Antibiot. (Tokyo) *29*: 501–506.

Parenti, F., G. Beretta, M. Berti and V. Arioli. 1978. Teichomycins, new antibiotics from *Actinoplanes teichomyceticus* Nov. Sp. I. Description of the producer strain, fermentation studies and biological properties. J. Antibiot. (Tokyo) *31*: 276–283.

Parenti, F. and C. Coronelli. 1979. Members of the genus *Actinoplanes* and their antibiotics. Annu. Rev. Microbiol. *33*: 389–411.

Patel, M., V.P. Gullo, V.R. Hegde, A.C. Horan, J.A. Marquez, R. Vaughan, M.S. Puar and G.H. Miller. 1987. A new tetracycline antibiotic from a *Dactylosporangium* species. Fermentation, isolation and structure elucidation. J. Antibiot. (Tokyo) *40*: 1414–1418.

Penn, K., C. Jenkins, M. Nett, D.W. Udwary, E.A. Gontang, R.P. McGlinchey, B. Foster, A. Lapidus, S. Podell, E.E. Allen, B.S. Moore and P.R. Jensen. 2009. Genomic islands link secondary metabolism to functional adaptation in marine *Actinobacteria*. ISME J. *3*: 1193–1203.

Petrolini, B., S. Quaroni, M. Saracchi and P. Sardi. 1993. A new genus of the maduromycetes: *Planopolyspora* gen. nov. Actinomycetes *4*: 8–16.

Prauser, H. 1984. Phage host ranges in the classification and identification of Gram-positive branched and related bacteria. *In* Biological, Biochemical, and Biomedical Aspects of Actinomycetes (edited by Ortiz-Ortiz, Bojalil and Yakoleff). Academic Press, Orlando, pp. 617–633.

Pruesse, E., C. Quast, K. Knittel, B. Fuchs, W. Ludwig, J. Peplies and F.O. Glöckner. 2007. SILVA: a comprehensive online resource for quality checked and aligned rRNA sequence data compatible with ARB. Nucleic Acids Res. *35*: 7188–7196.

Qin, S., J. Li, Y.Q. Zhang, W.Y. Zhu, G.Z. Zhao, L.H. Xu and W.J. Li. 2009. *Plantactinospora mayteni* gen. nov., sp. nov., a member of the family *Micromonosporaceae*. Int. J. Syst. Evol. Microbiol. *59*: 2527–2533.

Renner, M.K., Y.-C. Shen, X.-C. Cheng, P.R. Jensen, W. Frankmoelle, C.A. Kauffman, W. Fenical, E. Lobkovsky and J. Clardy. 1999. Cyclomarins A–C, new antiinflammatory cyclic peptides produced by a marine bacterium (*Streptomyces* sp.). J. Am. Chem. Soc. *121*: 11273–11276.

Rheims, H., P. Schumann, M. Rohde and E. Stackebrandt. 1998. *Verrucosispora gifhornensis* gen. nov., sp. nov., a new member of the actinobacterial family *Micromonosporaceae*. Int. J. Syst. Bacteriol. *48*: 1119–1127.

Riedlinger, J., A. Reicke, H. Zahner, B. Krismer, A.T. Bull, L.A. Maldonado, A.C. Ward, M. Goodfellow, B. Bister, D. Bischoff, R.D. Sussmuth and H.P. Fiedler. 2004. Abyssomicins, inhibitors of the *para*-aminobenzoic acid pathway produced by the marine *Verrucosispora* strain AB-18-032. J. Antibiot. (Tokyo) *57*: 271–279.

Rifaat, H.M., K. Marialigeti and G. Kovacs. 2002. Investigations on rhizoplane *Actinobacteria* communities of papyrus (*Cyperus papyrus*) from an Egyptian wetland. Acta Microbiol. Immunol. Hung. *49*: 423–432.

Rothwell, F.M. 1957. A further study of Karling's keratinophilic organisms. Mycologia *49*: 68–72.

Rowbotham, T.J. and T. Cross. 1977. Ecology of *Rhodococcus coprophilus* and associated actinomycetes in fresh water and agricultural habitats. J. Gen. Microbiol. *100*: 231–240.

Ruan, J. and Y. Zhang. 1974. A taxonomic study of *Actinoplanaceae*. I. Classification of *Ampullariella*. Acta Microbiol. Sin. *14*: 31–41.

Ruan, J., Y. Zhang and C. Jiang. 1976. A taxonomic study of *Actinoplanaceae*. II. Four new species of *Actinoplanes*. Acta Microbiol. Sin. *16*: 291–300.

Ruan, J. and C. Jiang. 1979. A taxonomic study of *Actinoplanaceae*. III. Three new species of *Actinoplanes*. Acta Microbiol. Sin. *19*: 235–242.

Ruan, J.S., L. Xiaotao, Z. Yamei and S. Yanlin. 1988. Numerical classification and chemotaxonomy of actinoplanetes and nocardiae. *In* Biology of Actinomycetes '88 (edited by Okami, Beppu and Ogaware). Japan Scientific Societies Press, Tokyo, pp. 221–226.

Saitou, N. and M. Nei. 1987. The neighbor-joining method: a new method for reconstructing phylogenetic trees. Mol. Biol. Evol. *4*: 406–425.

Sakane, T. and K. Kuroshima. 1997. Viabilities of dried cultures of various bacteria after preservation for 20 years and their production by the accelerated storage test. Microbiol. Cult. Coll. *13*: 1–7.

Sandrak, N.A. 1977. [Degradation of cellulose by micromonospores]. Mikrobiologiia *46*: 478–481.

Sasser, M. 1990. Identification of bacteria by gas chromatography of cellular fatty acids. MIDI Technical Note 101, Newark, Delaware, MIDI Inc.

Schäfer, D. 1973. Beiträge zur Klassifizierung und Taxonomie der Actinoplanaceen. PhD dissertation. Marburg, Germany.

Schleifer, K.H. and O. Kandler. 1972. Peptidoglycan types of bacterial cell walls and their taxonomic implications. Bacteriol. Rev. *36*: 407–477.

Seino, A. 1983. Surface ornamentation of sporangia of *Actinoplanaceae*. Hakko to Kogyo (Fermentation and Industry) *41*: 3–4.

Seo, S.H. and S.D. Lee. 2009. *Actinocatenispora rupis* sp. nov., isolated from cliff soil, and emended description of the genus *Actinocatenispora*. Int. J. Syst. Evol. Microbiol. *59*: 3078–3082.

Sezaki, M. and S. Miyadoh. 2001. Practically used antibiotics and their related substances. *In* Identification Manual of Actinomycetes (edited by Miyadoh, Hamada, Hotta, Seino, Suzuki and Yokota). Business Center for Academic Societies Japan, pp. 349–389.

Sharples, G.P. and S.T. Williams. 1974. Fine structure of the globose bodies of *Dactylosporangium thailandense* (*Actinomycetales*). J. Gen. Microbiol. *84*: 219–222.

Sharples, G.P., S.T. Williams and R.M. Bradshaw. 1974. Spore formation in the *Actinoplanaceae* (*Actinomycetales*). Arch. Microbiol. *101*: 9–20.

Shearer, M.C. 1987. Methods for the isolation of non-streptomycete actinomycetes. Dev. Indust. Microbiol. *28*: 91–97.

Shirling, E.B. and D. Gottlieb. 1966. Methods for characterization of *Streptomyces* species. Int. J. Syst. Bacteriol. *16*: 313–340.

Shomura, T., M. Kojima, J. Yoshida, M. Ito, S. Amano, K. Totsugawa, T. Niwa, S. Inouye, T. Ito and T. Niida. 1980. Studies on a new aminoglycoside antibiotic, dactimicin. I. Producing organism and fermentation. J. Antibiot. (Tokyo) *33*: 924–930.

Shomura, T., M. Kojima, J. Yoshida, M. Ito, S. Amano, K. Totsugawa, T. Niwa, S. Inouye, T. Ito and T. Niida. 1983a. *In* Validation of the publication of new names and new combinations previously effectively published outside the IJSB. List no. 11. Int. J. Syst. Bacteriol. *33*: 672–674.

Shomura, T., J. Yoshida, S. Miyadoh, T. Ito and T. Niida. 1983b. *Dactylosporangium vinaceum* sp. nov. Int. J. Syst. Bacteriol. *33*: 309–313.

Shomura, T., S. Amano, H. Tohyama, J. Yoshida, T. Ito and T. Niida. 1985. *Dactylosporangium roseum* sp. nov. Int. J. Syst. Bacteriol. *35*: 1–4.

Shomura, T., S. Amano, J. Yoshida and M. Kojima. 1986. *Dactylosporangium fulvum* sp. nov. Int. J. Syst. Bacteriol. *36*: 166–169.

Skerman, V.B.D., V. McGowan and P.H.A. Sneath. 1980. Approved Lists of Bacterial Names. Int. J. Syst. Bacteriol. *30*: 225–420.

Solans, M. and G. Vobis. 2003. Actinomycetes saprofíticos asociados a la rizósfera y rizoplano de Discaria trinervis. Ecol. Austral. *13*: 97–107.

Solans, M. 2007. *Discaria trinervis – Frankia* symbiosis promotion by saprophytic actinomycetes. J. Basic Microbiol. *47*: 243–250.

Solans, M. 2008. Influencia de rizoactinomycetes nativos sobre el desarrollo de la planta actinorrícica *Ochetophila trinervis*. PhD thesis, Bariloche.

Solans, M., G. Vobis and L.G. Wall. 2009. Saprophytic actinomycetes promote nodulation in *Medicago sativa–Sinorhizobium* symbiosis in the presence of high N. J. Plant Growth Regulation *28*: 106–114.

Stackebrandt, E. and R.M. Kroppenstedt. 1987. Union of the genera *Actinoplanes* Couch, *Ampullariella* Couch, and *Amorphosporangium* Couch in a redefined genus *Actinoplanes*. Syst. Appl. Microbiol. *9*: 110–114.

Stackebrandt, E. and R.M. Kroppenstedt. 1988. *In* Validation of the publication of new names and new combinations previously effectively published outside the IJSB. List no. 26. Int. J. Syst. Bacteriol. *38*: 328–329.

Stackebrandt, E., F.A. Rainey and N.L. Ward-Rainey. 1997. Proposal for a new hierarchic classification system, *Actinobacteria* classis nov. Int. J. Syst. Bacteriol. *47*: 479–491.

Stevens, R.T. 1975. Fine structure of sporogenesis and septum formation in *Micromonospora globosa* Kriss and *M. fusca* Jensen. Can. J. Microbiol. *21*: 1081–1088.

Suárez, J.E., C. Barbes and C. Hardisson. 1980. Germination of spores of *Micromonospora chalcea*: physiological and biochemical changes. J. Gen. Microbiol. *121*: 159–167.

Sun, W., G.X. Dong, Y.Q. Zhang, Y.Z. Wei, Q.P. Li, L.Y. Yu, H.P. Klenk and Y.Q. Zhang. 2009. *Actinoplanes sichuanensis* sp. nov. and *Actinoplanes xinjiangensis* sp. nov. Int. J. Syst. Evol. Microbiol. *59*: 2763–2768.

Sveshnikova, M., T. Maximova and E. Kudrina. 1969. The species belonging to the genus *Micromonospora* Ørskov 1923, and their taxonomy. Mikrobiologiya *38*: 883–893.

Sveshnikova, M.A., N.T. Chormonova, N.V. Lavrova, L.P. Terekhova and T.P. Preobrazhenskaia. 1976. [Isolation of soil actinomycetes on selective media with novobiocin]. Antibiotiki *21*: 784–787.

Swanson, R.N., D.J. Hardy, N.L. Shipkowitz, C.W. Hanson, N.C. Ramer, P.B. Fernandes and J.J. Clement. 1991. In vitro and in vivo evaluation of tiacumicins B and C against *Clostridium difficile*. Antimicrob. Agents Chemother. *35*: 1108–1111.

Szabó, Z. and C. Fernandez. 1984. *Micromonospora brunnea* Sveshnikova, Maksimova, and Kudrina 1969 is a Junior Subjective Synonym of *Micromonospora purpureochromogenes* (Waksman and Curtis 1916) Luedemann 1971. Int. J. Syst. Bacteriol. *34*: 463–464.

Szaniszlo, P.J. and H. Gooder. 1967. Cell wall composition in relation to the taxonomy of some *Actinoplanaceae*. J. Bacteriol. *94*: 2037–2047.

Szaniszlo, P.J. 1968. The nature of the intramycelial pigmentation of some *Actinoplanaceae*. J. Elisha Mitchell Sci. Soc. *84*: 24–26.

Szvoboda, G., T. Lang., I. Gado, G. Ambrus, C. Kari, K. Fodor and L. Alfoldi. 1980. Fusion of *Micromonospora* protoplasts. *In* Advances in Protoplast Research (edited by Ferenczy and Farkas). Pergamon Press, Oxford, UK, pp. 235–240.

Takada, Y., M. Inouye, T. Morohoshi, N. Muto, F. Kato, M. Kizuka, M. Tanaka and Y. Koyama. 1994. Establishment of the host-vector system for *Micromonospora griseorubida*. J. Antibiot. (Tokyo) *47*: 1167–1170.

Takizawa, M., R.R. Colwell and R.T. Hill. 1993. Isolation and diversity of actinomycetes in the chesapeake bay. Appl. Environ. Microbiol. *59*: 997–1002.

Tamura, K., J. Dudley, M. Nei and S. Kumar. 2007. MEGA4: Molecular Evolutionary Genetics Analysis (MEGA) Software Version 4.0. Mol. Biol. Evol. *24*: 1596–1599.

Tamura, T., Y. Nakagaito, T. Nishii, T. Hasegawa, E. Stackebrandt and A. Yokota. 1994. A new genus of the order *Actinomycetales*, *Couchioplanes* gen. nov., with description of *Couchioplanes caeruleus* (Horan and Brodsky 1986) comb. nov. and *Couchioplanes caeruleus* subsp. *Azureus* subsp. nov. Int. J. Syst. Bacteriol. *44*: 193–203.

Tamura, T., A. Yokota, L.H. Huang, T. Hasegawa and K. Hatano. 1995. Five new species of the genus *Catenuloplanes: Catenuloplanes niger* sp. nov., *Catenuloplanes indicus* sp. nov., *Catenuloplanes atrovinosus* sp. nov., *Catenuloplanescastaneus* sp. nov., and *Catenuloplanes nepalensis* sp. nov. Int. J. Syst. Bacteriol. *45*: 858–860.

Tamura, T., M. Hayakawa and K. Hatano. 1997. A new genus of the order *Actinomycetales*, *Spirilliplanes* gen. nov., with description of *Spirilliplanes yamanashiensis* sp. nov. Int. J. Syst. Bacteriol. *47*: 97–102.

Tamura, T. and K. Hatano. 2001. Phylogenetic analysis of the genus *Actinoplanes* and transfer of *Actinoplanes minutisporangius* Ruan *et al.* 1986 and '*Actinoplanes aurantiacus*' to *Cryptosporangium minutisporangium* comb. nov. and *Cryptosporangium aurantiacum* sp. nov. Int. J. Syst. Evol. Microbiol. *51*: 2119–2125.

Tamura, T., M. Hayakawa and K. Hatano. 2001. A new genus of the order *Actinomycetales*, *Virgosporangium* gen. nov., with descriptions of *Virgosporangium ochraceum* sp. nov. and *Virgosporangium aurantiacum* sp. nov. Int. J. Syst. Evol. Microbiol. *51*: 1809–1816.

Tamura, T. and T. Sakane. 2005. *Asanoa iriomotensis* sp. nov., isolated from mangrove soil. Int. J. Syst. Evol. Microbiol. *55*: 725–727.

Tamura, T., K. Hatano and K. Suzuki. 2006. A new genus of the family *Micromonosporaceae*, *Polymorphospora* gen. nov., with description of *Polymorphospora rubra* sp. nov. Int. J. Syst. Evol. Microbiol. *56*: 1959–1964.

Tanasupawat, S., S. Jongrungruangchok and T. Kudo. 2010. *Micromonospora marina* sp. nov., isolated from sea sand. Int. J. Syst. Evol. Microbiol. *60*: 648–652.

Terekhova, L.P., O.A. Sadikova and T.P. Preobrazhenskaia. 1977. [New species of *Actinoplanes cyanea* sp. nov. and its antagonistic properties]. Antibiotiki *22*: 1059–1063.

Terekhova, L.P., O.A. Galatenko and T.P. Preobrazhenskaya. 1987. *In* Validation of the publication of new names and new combinations previously effectively published outside the IJSB. List no. 23. Int. J. Syst. Bacteriol. *37*: 179–180.

Thawai, C., S. Tanasupawat, T. Itoh, K. Suwanborirux and T. Kudo. 2004a. *In* Validation of publication of new names and new combinations previously effectively published outside the IJSEM. List no. 99. Int. J. Syst. Evol. Microbiol. *54*: 1425–1426.

Thawai, C., S. Tanasupawat, T. Itoh, K. Suwanborirux and T. Kudo. 2004b. *Micromonospora aurantionigra* sp. nov., isolated from a peat swamp forest in Thailand. Actinomycetologica *18*: 8–14.

Thawai, C., S. Tanasupawat, T. Itoh, K. Suwanborirux and T. Kudo. 2005a. *Micromonospora siamensis* sp. nov., isolated from Thai peat swamp forest. J. Gen. Appl. Microbiol. *51*: 229–234.

Thawai, C., S. Tanasupawat, T. Itoh, K. Suwanborirux, K. Suzuki and T. Kudo. 2005b. *Micromonospora eburnea* sp. nov., isolated from a Thai peat swamp forest. Int. J. Syst. Evol. Microbiol. *55*: 417–422.

Thawai, C., S. Tanasupawat, T. Itoh and T. Kudo. 2006a. *Actinocatenispora thailandica* gen. nov., sp. nov., a new member of the family *Micromonosporaceae*. Int. J. Syst. Evol. Microbiol. *56*: 1789–1794.

Thawai, C., S. Tanasupawat, T. Itoh, K. Suwanborirux and T. Kudo. 2006b. *In* List of new names and new combinations previously effectively, but not validly, published. Validation List no. 107. Int. J. Syst. Evol. Microbiol. *56*: 1–6.

Thawai, C., S. Tanasupawat, K. Suwanborirux, T. Itoh and T. Kudo. 2007. *Micromonospora narathiwatensis* sp. nov., from Thai peat swamp forest soils. J. Gen. Appl. Microbiol. *53*: 287–293.

Thawai, C., S. Tanasupawat and T. Kudo. 2008. *Micromonospora pattaloongensis* sp. nov., isolated from a Thai mangrove forest. Int. J. Syst. Evol. Microbiol. *58*: 1516–1521.

Thawai, C., S. Tanasupawat, K. Suwanborirux and T. Kudo. 2010. *Actinaurispora siamensis* gen. nov., sp. nov., a new member of the family *Micromonosporaceae*. Int. J. Syst. Evol. Microbiol. *60*: 1660–1666.

Theriault, R.J., J.P. Karwowski, M. Jackson, R.L. Girolami, G.N. Sunga, C.M. Vojtko and L.J. Coen. 1987. Tiacumicins, a novel complex of 18-membered macrolide antibiotics. I. Taxonomy, fermentation and antibacterial activity. J. Antibiot. (Tokyo) *40*: 567–574.

Thiemann, J.E. 1967. A new species of the genus *Amorphosporangium* isolated from Italian soil. Mycopathol. *33*: 233–240–240.

Thiemann, J.E., H. Pagani and G. Beretta. 1967. A new genus of the *Actinoplanaceae*: *Dactylosporanguim*, gen. nov. Arch. Mikrobiol. *58*: 42–52.

Thiemann, J.E., G. Beretta, C. Coronelli and H. Panani. 1969. Antibiotic production by new form-genera of the *Actinomycetales*. II. Antibiotic A-672 isolated from a new species of *Actinoplanes*: A/ Thiemann JE, Beretta G, Coronelli C, Pagani H: Antibiotic production by new form-genera of the *Actinomycetales*. II. Antibiotic A-672 isolated from a new species of *Actinoplanes*: *Actinoplanes brasiliensis* nov. sp. J. Antibiot. (Tokyo) *22*: 119–125.

Thiemann, J.E. 1970a. Study of some new genera and species of the *Actinoplanaceae*. *In* The *Actinomycetales* (edited by Prauser). VEB Gustav Fischer Verlag, Jena, pp. 245–257.

Thiemann, J.E. 1970b. *Dactylosporangium thailandensis* should be *D. thailandense*. Int. J Syst. Bacteriol.: 59.

Thiemann, J.E. 1974. Genus *Dactylosporangium*. *In* Bergey's Manual of Determinative Bacteriology, 8th edn (edited by Buchanan and Gibbons). The Williams and Wilkins Co., Baltimore, pp. 721–722.

Tilley, B.C., J.L. Meyertons and M.P. Lechevalier. 1990. Characterization of a temperate actinophage, MPphiWR-1, capable of infecting *Micromonospora purpurea* ATCC 15835. J. Ind. Microbiol. *5*: 167–182.

Tomita, K., S. Kobaru, M. Hanada, H. Tsukiara and B.-M. Company. 1977. Fermentation process. U.S. Patent 4026766 (May 31).

Tomita, K., Y. Hoshino, N. Ohkusa, T. Tsuno and T. Miyaki. 1992a. *In* Validation of the publication of new names and new combinations previously effectively published outside the IJSB. List no. 43. Int. J. Syst. Bacteriol. *42*: 656–657.

Tomita, K., Y. Hoshino, N. Ohkusa, T. Tsuno and T. Miyaki. 1992b. *Micromonospora chersina* sp. nov. Actinomycetologica *6*: 21–28.

Torikata, A., R. Enokita, H. Imai, Y. Itoh, M. Nakajima, T. Haneishi and M. Arai. 1978. Studies on the antibiotics from genus *Actinoplanes*. I. Taxonomy of the producers of three antibiotics and their isolation and identification with azaserine, 5-azacytidine and actinomycins. Ann. Rep. Sankyo Res. Lab. *30*: 84–97.

Torikata, A., R. Enokita, T. Okazaki, M. Nakajima, S. Iwado, T. Haneishi and M. Arai. 1983. Mycoplanecins, novel antimycobacterial antibiotics from *Actinoplanes awajiensis* subsp. mycoplanecinus subsp. nov. J. Antibiot. *36*: 957–960.

Tribe, H.T. and S.M. Abu El-Souod. 1979. Colonization of hair in soil-water cultures, with especial reference to the genera *Pilimelia* and *Spirillospora* (*Actinomyceteales*). Nova Hedwigia *31*: 789–805.

Trujillo, M.E., C. Fernandez-Molinero, E. Velazquez, R.M. Kroppenstedt, P. Schumann, P.F. Mateos and E. Martinez-Molina. 2005. *Micromonospora mirobrigensis* sp. nov. Int. J. Syst. Evol. Microbiol. *55*: 877–880.

Trujillo, M.E., R.M. Kroppenstedt, P. Schumann, L. Carro and E. Martinez-Molina. 2006. *Micromonospora coriariae* sp. nov., isolated from root nodules of *Coriaria myrtifolia*. Int. J. Syst. Evol. Microbiol. *56*: 2381–2385.

Trujillo, M.E., R.M. Kroppenstedt, C. Fernandez-Molinero, P. Schumann and E. Martinez-Molina. 2007. *Micromonospora lupini* sp. nov. and *Micromonospora saelicesensis* sp. nov., isolated from root nodules of *Lupinus angustifolius*. Int. J. Syst. Evol. Microbiol. *57*: 2799–2804.

Truscheit, E., W. Frommer, B. Junge, L. Müller, D.D. Schmidt and W. Wingender. 1981. Chemie und Biochemie mikrobieller α-Glucosidasen-Inhibitoren. Angew. Chem. *93*: 738–755.

Tsueng, G. and K.S. Lam. 2008a. A low-sodium-salt formulation for the fermentation of salinosporamides by *Salinispora tropica* strain NPS21184. Appl. Microbiol. Biotechnol. *78*: 821–826.

Tsueng, G. and K.S. Lam. 2008b. Growth of *Salinispora tropica* strains CNB440, CNB476, and NPS21184 in nonsaline, low-sodium media. Appl. Microbiol. Biotechnol. *80*: 873–880.

Tsueng, G. and K.S. Lam. 2010. A preliminary investigation on the growth requirement for monovalent cations, divalent cations and medium ionic strength of marine actinomycete *Salinispora*. Appl. Microbiol. Biotechnol. *86*: 1525–1534.

Tymiak, A.A., C. Aklonis, M.S. Bolgar, A.D. Kahle, D.R. Kirsch, J. O'Sullivan, M.A. Porubcan, P. Principe and W.H. Trejo. 1993. Dactylocyclines: novel tetracycline glycosides active against tetracycline-resistant bacteria. J Org. Chem. *58*: 535–537.

Uchida, K. and A. Seino. 1997. Intra- and intergeneric relationships of various actinomycete strains based on the acyl types of the muramyl residue in cell-wall peptidoglycans examined in a glycolate test. Int. J. Syst. Bacteriol. *47*: 182–190.

Udwary, D.W., L. Zeigler, R.N. Asolkar, V. Singan, A. Lapidus, W. Fenical, P.R. Jensen and B.S. Moore. 2007. Genome sequencing reveals complex secondary metabolome in the marine actinomycete *Salinispora tropica*. Proc. Natl. Acad. Sci. U.S.A. *104*: 10376–10381.

Valdés, M., N.O. Perez, P. Estrada de los Santos, J. Caballero-Mellado, J.J. Pena-Cabriales, P. Normand and A.M. Hirsch. 2005. Non-*Frankia* actinomycetes isolated from surface-sterilized roots of *Casuarina equisetifolia* fix nitrogen. Appl. Environ. Microbiol. *71*: 460–466.

Vértesy, L., W. Aretz, A. Bonnefoy, E. Ehlers, M. Kurz, A. Markus, M. Schiell, M. Vogel, J. Wink and H. Kogler. 1999. Ala(0)-actagardine, a new lantibiotic from cultures of *Actinoplanes liguriae* ATCC 31048. J. Antibiot. (Tokyo) *52*: 730–741.

Vettermann, R. and H. Prauser. 1979. Comparative studies on the isolation of actinoplanetes. Poster presentation, Fourth International Symposium on Actinomycete Biology, Cologne.

Vobis, G. 1984. Sporogenesis in the *Pilimelia* species. *In* Biological, Biochemical, and Biomedical Aspects of Actinomycetes (edited by Ortiz-Ortiz, Bojalil and Yakoleff). Academic Press, Orlando, pp. 423–439.

Vobis, G. and H.-W. Kothe. 1985. Sporogenesis in sporangiate actinomycetes. *In* Frontiers in Applied Microbiology, vol. 1 (edited by Mukerji, Pathak and Singh). Print House, Lucknow, India, pp. 25–47.

Vobis, G., D. Schäfer, H.W. Kothe and B. Renner. 1986a. *In* Validation of the publication of new names and new combinations previously effectively published outside the IJSB. List no. 22. Int. J. Syst. Bacteriol. *36*: 573–576.

Vobis, G., D. Schäfer, H.W. Kothe and B. Renner. 1986b. Descriptions of *Pilimelia columellifera* (*ex* Schäfer 1973) nom. rev. and *Pilimelia columellifera* subsp. *pallida* (*ex* Schäfer 1973) nom. rev. Syst. Appl. Microbiol. *8*: 67–74.

Vobis, G. 1987. Sporangiate Actinoplaneten, Actinomycetales mit aero-aquatischem Lebenszyklus. Forum Mikrobiol. *10*: 416–424.

Vobis, G. 1989a. Genus *Pilimelia*. *In* Bergey's Manual of Systematic Bacteriology, 1st edn, vol. 4 (edited by Williams, Sharpe and Holt). Williams & Wilkins, Baltimore, pp. 2433–2437.

Vobis, G. 1989b. Actinoplanetes. *In* Bergey's Manual of Systematic Bacteriology, vol. 4 (edited by Williams, Sharpe and Holt). Williams and Wilkins, Baltimore, pp. 2418–2450.

Vobis, G. 1989c. Genus *Dactylosporangium*. *In* Bergey's Manual of Systematic Bacteriology, 1st edn, vol. 4 (edited by Williams, Sharpe and Holt). The Williams & Wilkins Co., Baltimore, pp. 2437–2442.

Vobis, G. and H.-W. Kothe. 1989. Genus *Ampullariella*. *In* Bergey's Manual of Systematic Bacteriology, 1st edn, vol. 4 (edited by Williams, Sharpe and Holt). Williams and Wilkins, Baltimore, MD, pp. 2429–2433.

Vobis, G. 1992. The genus *Actinoplanes* and related genera. *In* The Prokaryotes, a Handbook on the Biology of Bacteria: Ecophysiology, Isolation, Identification, Application, 2nd edn, vol. 2 (edited by Balows, Trüper, Dworkin, Harder and Schleifer). Springer Verlag, New York, pp. 1029–1060.

Vobis, G. 1997. Morphology of actinomycetes. *In* Atlas of Actinomycetes (edited by Miyadoh, Hamada, Hotta, Kudo, Seino, Vobis and Yokota). Asakura Publishing, Tokyo pp. 180–191.

Vukov, N. and B. Vasiljevic. 1998. Analysis of plasmid pMZ1 from *Micromonospora zionensis*. FEMS Microbiol. Lett. *162*: 317–323.

Wagman, G.H., J.A. Waitz, J. Marquez, A. Murawaski, E.M. Oden, R.T. Testa and M.J. Weinstein. 1972. A new *Micromonospora*-produced macrolide antibiotic, rosamicin. J. Antibiot. (Tokyo) *25*: 641–646.

Wagman, G.H., R.T. Testa, J.A. Marquez and M.J. Weinstein. 1974. Antibiotic G-418, a new *Micromonospora*-produced aminoglycoside with activity against protozoa and helminths: fermentation, isolation, and preliminary characterization. Antimicrob. Agents Chemother. *6*: 144–149.

Wagman, G.H., R.T. Testa, M. Patel, J.A. Marquez, E.M. Oden, J.A. Waitz and M.J. Weinstein. 1975. New polyene antifungal antibiotic produced by a species of *Actinoplanes*. Antimicrob. Agents Chemother. *7*: 457–461.

Wagman, G.H. and M.J. Weinstein. 1980. Antibiotic from *Micromonospora*. Annu. Rev. Microbiol. *34*: 537–557.
Wakisaka, Y., Y. Kawamura, Y. Yasuda, K. Koizumi and Y. Nishimoto. 1982. A selective isolation procedure for *Micromonospora*. J. Antibiot. (Tokyo) *35*: 822–836.
Waksman, S.A. and R.E. Curtis. 1916. The *Actinomyces* of the soil. Soil Sci. *1*: 99–134.
Waksman, S.A. 1961. The Actinomycetes, vol. 2. Classification, Identification and Descriptions of Genera and Species. Williams & Wilkins, Baltimore.
Wang, Y., Z. Zhang, J.S. Ruan and S. Ali. 1999. Investigations of actinomycete diversity in the tropical rainforests of Singapore. J. Clin. Microbiol. *23*: 178–187.
Watson, E.T. and S.T. Williams. 1974. Studies of the ecology of actinomycetes in soil. VII. Actinomycetes in a coastal sand belt. Soil. Biol. Biochem. *6*: 43–52.
Weinstein, M.J., G.M. Luedemann, E.M. Oden and G.H. Wagman. 1968. Halomicin, a new *Micromonospora*-produced antibiotic. Antimicrob. Agents Chemother. (Bethesda) *1967*: 435–441.
Weinstein, M.J., J.A. Marquez, R.T. Testa, G.H. Wagman, E.M. Oden and J.A. Waitz. 1970. Antibiotic 6640, a new *Micromonospora*-produced aminoglycoside antibiotic. J. Antibiot. (Tokyo) *23*: 551–554.
Wellington, E.M.H. and S.T. Williams. 1978. Preservation of actinomycete inoculum in frozen glycerol. Microbiol. Lett. *6*: 151–159.
Weyland, H. 1969. Actinomycetes in North Sea and Atlantic Ocean sediments. Nature *223*: 858.
Weyland, H. 1981. Distribution of actinomycetes on the sea floor. Zentrabl. Bakteriol. Mikrobiol. Hyg. I. Abt. Orig. Suppl. *11*: 185–193.
Wiese, J., Y. Jiang, S.K. Tang, V. Thiel, R. Schmaljohann, L.H. Xu, C.L. Jiang and J.F. Imhoff. 2008. A new member of the family *Micromonosporaceae, Planosporangium flavigriseum* gen. nov., sp. nov. Int. J. Syst. Evol. Microbiol. *58*: 1324–1331.
Williams, P.G., E.D. Miller, R.N. Asolkar, P.R. Jensen and W. Fenical. 2007. Arenicolides A-C, 26-membered ring macrolides from the marine actinomycete *Salinispora arenicola*. J. Org. Chem. *72*: 5025–5034.
Williams, S.T., G.P. Sharples and R.M. Bradshaw. 1973. The fine structure of the *Actinomycetales*. *In Actinomycetales*: Characteristics and Practical Importance (edited by Sykes and Skinner). Academic Press, London, pp. 113–130.
Willoughby, L.G. 1966. A conidial *Actinoplanes* isolate from Blelham Tarn. J. Gen. Microbiol. *44*: 69–72.
Willoughby, L.G. 1968. Aquatic *Actinomycetales* with particular reference to the *Actinoplanaceae*. Veröff. Inst. Meeresforschung Bremerhaven, Sonderband *3*: 19–26.
Willoughby, L.G., C.D. Baker and S.E. Foster. 1968. Sporangium formation in the *Actinoplanaceae* induced by humic acid. Cell. Mol. Life Sci. *24*: 730–731.
Willoughby, L.G. 1969a. A study of aquatic actinomycetes of Blelham Tarn. Hydrobiologija *34*: 465–483.
Willoughby, L.G. 1969b. A study of aquatic actinomycetes, the allochthonous leaf component. Nova Hedwigia *18*: 45–113.
Willoughby, L.G. and C.D. Baker. 1969. Humic and fulvic acids and their derivatives as growth and sporulation media for aquatic actinomycetes. Verh. Int. Verein. Limnol. *17*: 795–801.
Willoughby, L.G. 1971. Observations on some aquatic Actinomycetes of streams and rivers. Freshwater Biol. *1*: 23–27.
Willoughby, L.G., S.M. Smith and R.M. Bradshaw. 1972. Actinomycete virus in fresh water. Freshwater Biol. *2*: 19–26.
Wink, J.M., R.M. Kroppenstedt, P. Schumann, G. Seibert and E. Stackebrandt. 2006. *Actinoplanes liguriensis* sp. nov. and *Actinoplanes teichomyceticus* sp. nov. Int. J. Syst. Evol. Microbiol. *56*: 2125–2130.
Wright, J.J., D. Greeves, A.K. Mallams and D.H. Picker. 1977. Structural elucidation of heptaene macrolide antibiotics 67-121-A and 67-121.C. J. Chem. Soc. Chem. Commun. *20*: 710–712.
Xu, L., Q. Li and C. Jiang. 1996. Diversity of soil actinomycetes in Yunnan, China. Appl. Environ. Microbiol. *62*: 244–248.
Yokota, A., T. Tamura, T. Hasegawa and L.H. Huang. 1993. *Catenuloplanes japonicus* gen. nov., sp. nov., nom. rev., a new genus of the order *Actinomycetales*. Int. J. Syst. Bacteriol. *43*: 805–812.
Youssef, N., C.S. Sheik, L.R. Krumholz, F.Z. Najar, B.A. Roe and M.S. Elshahed. 2009. Comparison of species richness estimates obtained using nearly complete fragments and simulated pyrosequencing-generated fragments in 16S rRNA gene-based environmental surveys. Appl. Environ. Microbiol. *75*: 5227–5236.
Zenova, G.M., T.A. Gracheva and A.A. Likhacheva. 1994. Actinomycetes of the genus *Micromonospora* in terrestrial ecosystems. Microbiology *63*: 313–317.
Zenova, G.M. and D.G. Zviagintsev. 2002. [Actinomycetes of the genus *Micromonospora* in meadow ecosystems]. Mikrobiologiia *71*: 662–666.
Zenova, G.M., Y.V. Zakalyukina, V.V. Selyanin and D.G. Zvyagintsev. 2004. Isolation and growth of acidophilic soil actinomycetes from the *Micromonospora* genus. Eurasian Soil Sci. *37*: 737–742.
Zhi, X.-Y., W.-J. Li and E. Stackebrandt. 2009. An update of the structure and 16S rRNA gene sequence-based definition of higher ranks of the class *Actinobacteria*, with the proposal of two new suborders and four new families and emended descriptions of the existing higher taxa. Int. J. Syst. Evol. Microbiol. *59*: 589–608.

Order XII. **Propionibacteriales** ord. nov.

Sheila Patrick and Andrew McDowell

Pro.pio.ni.bac.te.ri.a′les. N.L. neut. n. *Propionibacterium* type genus of the order; suff. *-ales* ending to denote an order; N.L. fem. pl. n. *Propionibacteriales* the *Propionibacterium* order.

This order is formed by elevation of the suborder *Propionibacterineae* Zhi et al. 2009. The pattern of 16S rRNA signatures defining this order consists of nucleotides at positions 127:234 (A–U), 598:640 (U–A), 657:749 (G–C), 828 (U), 829:851 (A–C), 832:854 (U–C), 833:853 (G–U), 952:1229 (C–G), and 986:1219 (U–A).

The order contains the families *Propionibacteriaceae* and *Nocardioidaceae*.

Type genus: **Propionibacterium** Orla-Jensen 1909, 337[AL] emend. Charfreitag, Collins and Stackebrandt 1988, 356.

References

Charfreitag, O., M.D. Collins and E. Stackebrandt. 1988. Reclassification of *Arachnia propionica* as *Propionibacterium propionicus* comb. nov. Int. J. Syst. Bacteriol. *38*: 354–357.
Orla-Jensen, S. 1909. Die Hauptlinien des natürlichen Bakteriensystems. Zentralbl. Bakteriol. Parasitenkd. Infektionskr. Hyg. Abt. 2 *22*: 305–346.
Zhi, X.-Y., W.-J. Li and E. Stackebrandt. 2009. An update of the structure and 16S rRNA gene sequence-based definition of higher ranks of the class *Actinobacteria*, with the proposal of two new suborders and four new families and emended descriptions of the existing higher taxa. Int. J. Syst. Evol. Microbiol. *59*: 589–608.

Family I. **Propionibacteriaceae** Delwiche 1957[AL] emend. Rainey, Ward-Rainey and Stackebrandt 1997, 484 emend. Zhi, Li and Stackebrandt 2009, 599

Sheila Patrick and Andrew McDowell

Pro.pio.ni.bac.te.ri.a.ce'a.e. N.L. neut. n. *Propionibacterium* type genus of the family; suff. *-aceae* ending denoting family; N.L. fem. pl. n. *Propionibacteriaceae* the *Propionibacterium* family.

The pattern of 16S rRNA signatures consists of nucleotides at positions 328 (U), 407:435 (C–G), 451 (A), 453 (G), 819 (G), 825:875 (A–U), 827 (C), 828 (U), 832:854 (U–C), 833:853 (G–U), and 844 (U).

Type genus: **Propionibacterium** Orla-Jensen 1909, 337[AL] emend. Charfreitag, Collins and Stackebrandt 1988, 356.

Other genera of the family *Propionibacteriaceae* are: *Aestuariimicrobium* Jung et al. 2007; *Brooklawnia* Rainey et al. 2006 (*in* Bae et al. 2006b); *Friedmanniella* Schumann et al. 1997; *Granulicoccus* Maszenan et al. 2007; *Luteococcus* Tamura et al. 1994; *Microlunatus* Nakamura et al. 1995; *Micropruina* Shintani et al. 2000; *Propionicicella* Bae et al. 2006a; *Propionicimonas* Akasaka et al. 2003b; *Propioniferax* Yokota et al. 1994; *Propionimicrobium* Stackebrandt et al. 2002; and *Tessaracoccus* Maszenan et al. 1999b.

Genus I. **Propionibacterium** Orla-Jensen 1909, 337[AL] emend. Charfreitag, Collins and Stackebrandt 1988, 356

Sheila Patrick and Andrew McDowell*

Pro.pio.ni.bac.te'ri.um. N.L. n. *acidum propionicum* propionic aid; L. neut. n. *bacterium* a small rod; N.L. neut. n. *Propionibacterium* propionic (acid) bacterium.

Pleomorphic rods, 0.2–1.5 μm ×, 1–5 μm, often diphtheroid or club-shaped with one end rounded and the other tapered or pointed; **however, cells may be coccoid, bifid, branched, or filamentous with lengths up to 20 μm**. Cells may occur singly, in pairs or short chains, in V or Y configurations, or in clumps with "Chinese character" arrangement. Swollen spherical cells resembling sphaeroplasts, up to 5–20 μm in diameter, are formed by some strains. Highly filamentous microcolonies of long, branched, septate or non-septate filaments may be formed. **Gram-stain-positive, nonmotile**, non-acid-fast, nonsporeforming. **Anaerobic-to-aerotolerant-to-microaerophilic, although some strains can grow aerobically, most grow better anaerobically**. **Chemo-organotrophic mostly with complex nutritional requirements, grow in standard complex media:** fermentation of sugars, **polyhydroxy alcohols**, or lactate produced by the fermentative activities of other bacteria (secondary fermentation) and **produce large amounts of propionic and acetic acids** which is a distinguishing feature of the genus. Lesser amounts of iso-valeric, formic, succinic, or lactic acids and carbon dioxide are also formed. **Generally catalase-positive**. **Optimum growth temperature 30–37°C**. Colonies on solid media are smooth, convex, or rough. Floccular or granular masses of variable size may be observed in liquid media. White, gray, pink, red, yellow, or orange in color. Cell-wall peptidoglycan contains either *meso-* or LL-diaminopimelic acid (DAP) as the dibasic amino acid. Tetrahydrogenated menaquinones with nine isoprene units [**MK-9(H_4)**] are the major respiratory quinones. The long-chain **cellular fatty acids are of the straight-chain saturated, iso- and anteiso-methyl branched types, principally 12 and 13-methyltetradecanoic acids ($C_{15:0}$ iso; $C_{15:0}$ anteiso)**; monounsaturated acids may also be present in small amounts.

DNA G+C content (mol%): 57–70.

Type species: **Propionibacterium freudenreichii** van Niel 1928, 162[AL].

Further descriptive information

Two principal groups of organisms are traditionally described in the genus *Propionibacterium*; namely the "classical" or "dairy" (Table 202) and the "cutaneous" (Table 203) propionibacteria (Table 204). Although the phylogenetic relationship between *Propionibacterium* species based on 16S rRNA gene sequences does not mirror exactly their classification based on habitat (Figure 232), the latter is still useful for categorizing some species.

1. The "classical" or "dairy" propionibacteria: typified by strains from dairy products and used, for example, in the manufacture of hard rennet Swiss-type cheese. The genus was first described as a result of the study of propionic acid-producing bacteria isolated from cheese by a number of authors in the early 1900s (Orla-Jensen, 1909; Von Freudenreich and Orla-Jensen, 1906). They are used as starter cultures in the dairy industry and have been studied in relation to the commercial production of propionic acid. They have also been isolated from fermenting food and plant materials such as silage and fermenting olives (Cancho et al., 1980; Cancho et al., 1970; Plastourgos and Vaughn, 1957), and from soil (van Niel, 1928), including soil from rice paddy fields (Hayashi and Furusaka, 1979). Related bacteria have been found associated with nematodes in the Zebra gut (Krecek et al., 1992) and anaerobic digestors (Riedel and Britz, 1993; Sarada and Joseph, 1994). The described species *Propionibacterium microaerophilum* and *Propionibacterium cyclohexanicum*, isolated from olive mill wastewater (Koussémon et al., 2001) and spoiled off-flavor orange juice (Kusano et al., 1997) respectively, do not fit the "classical" or "dairy" propionibacteria groups based on habitat, but they are more closely related to these species based on 16S rRNA gene sequence (97.5%

*This section is an updated version of the text written by Cecil S. Cummins and John L. Johnson in the previous edition of this *Manual*, 1986.

identity between *Propionibacterium microaerophilum* and *Propionibacterium acidipropionici*; 97.1% identity between *Propionibacterium cyclohexanicum* and *Propionibacterium freudenreichii*) compared with the "cutaneous" species. Similarly, based on 16S rRNA gene sequence, *Propionibacterium australiense*, isolated from granulomatous bovine lesions (Bernard et al., 2002b) clusters with *Propionibacterium cyclohexanicum* and is more closely related to *Propionibacterium freudenreichii* than "cutaneous" species. The recently described novel species *Propionibacterium acidifaciens*, isolated from human carious dentine, clusters with *Propionibacterium australiense* on the basis of 16S rRNA gene analysis (97.7% identity; Downes and Wade, 2009).

2. The "cutaneous" propionibacteria: strains isolated from human skin, although they also colonize the mouth, genitourinary tract, and the large intestine. *Propionibacterium propionicum*, which also causes human infection, clusters with "cutaneous" species based on 16S rRNA gene sequence.

On the basis of morphology, the human "cutaneous" propionibacteria were originally assigned to the genus *Corynebacterium* (up to seventh edition of the *Manual of Determinative Bacteriology*). They were transferred from that genus to *Propionibacterium* in the eighth edition because (a) they were anaerobic, (b) propionic acid is the main metabolic product, (c) LL-DAP is present in the peptidoglycan of most species, (d) they produce C_{15} iso- and anteiso- acids as the principal fatty acids of cell lipids, and (e) they lack mycolic acids and the arabinogalactan characteristic of *Corynebacterium* sensuo stricto. Prévot (1976), however, proposed that this group of organisms, because of their pathogenic and immunomodulatory properties, should not be transferred to *Propionibacterium* but should instead be accommodated in a separate subgenus *Coryneformis* in the family *Corynebacteriaceae*. Analysis of 16S rRNA gene sequence data confirms the distinction from the *Corynebacteriaceae* and relatedness to the other propionibacteria (Charfreitag and

TABLE 202. Characteristics differentiating the "classical" and "dairy" species of the genus *Propionibacterium*[a,b,c]

Characteristic	*P. freudenreichii*	*P. acidipropionici*	*P. jensenii*	*P. thoenii*
Acid from:				
Maltose	–	+	d+	d+
Sorbitol	–	+	–	d+
Sucrose	–	+	+	d+
Nitrate reduction	d	+	–	–
β-Hemolysis	–	–[d]	–	+[e]
Color/pigment	Cream	Cream-orange-yellow	Cream	Red-brown
DAP isomer	*meso*-	LL-	LL-	LL-

[a]Symbols: +, positive reaction in 90–100% of isolates; –, negative reaction in 90–100% of isolates; d, positive reaction in 11–89% of isolates; d+, reaction positive in 40–90% of isolates.

[b]Reactions determined by procedures given in Holdeman et al., (1977).

[c]DAP, diaminopimelic acid.

[d]May show slight β-hemolysis under area of confluent growth.

[e]Hemolytic on blood agar containing human, bovine, equine, sheep, rabbit, and pig blood (Vedamuthu et al., 1971).

TABLE 203. Principal characteristics of "cutaneous" species of the genus *Propionibacterium*[a]

Characteristic	*P. acnes* phylotype			*P. avidum* biovar		*P. granulosum*
	I	II	III[b]	I	II	
Fermentation of:						
L-Arabinose	–	–	–	d	d	–
Glycerol	d	d	+	+	+	+
Maltose	–	–	–	+	+	d+
Sorbitol	d+	–	–	–	–	–
Sucrose	–	–	nd	+	+	+
Esculin hydrolysis	–	–	nd	+	+	–
Biochemical tests:						
Gelatin liquefaction	+	+	nd	+	+	–
β-Hemolysis (rabbit blood) (5 d/37°C)	d+	–	–	+	+	–[c]
Indole production	+	+	+	–	–	–
Nitrate reduction	+	+	+	–	–	–
CAMP reaction (sheep blood) (2 d/37°C)	+	+	nd	nd	nd	–
Cell-wall composition:						
A_2pm isomer	LL-	LL- (*meso*-)	nd	LL-	*meso*-(LL-)	LL-
Amino acids	Ala, Gly, Glu	Ala, Gly[d], Glu	nd	Ala, Gly, Glu	Ala, Gly[d], Glu	Ala, Gly, Glu
Sugars	Galactose, glucose, mannose	Glucose, mannose	nd	Galactose, glucose, mannose	Glucose, mannose	Galactose, mannose

[a]Symbols: +, positive reaction in 90% or more of isolates; –, 90% or more of isolates are negative; d, 11–89% of isolates are positive; d+, 40–90% of isolates are positive; parentheses, present in some isolates.

[b]Data based on the study of five isolates (McDowell et al., 2008).

[c]Generally nonhemolytic on sheep, horse, or rabbit blood agar, but reported to be β-hemolytic on rabbit blood by Hoeffler, (1977).

[d]Gly not present in isolates with *meso*-DAP.

TABLE 204. Characteristics differentiating the species of the genus *Propionibacterium*[a]

Characteristic	"Classical" propionibacteria						"Cutaneous" propionibacteria					
	P. freudenreichii	*P. acidipropionici*	*P. jensenii*	*P. thoenii*	*P. cyclohexanicum*[b]	*P. microaerophilum*[c]	*P. acnes*	*P. avidum*	*P. granulosum*	*P. australiense*	*P. acidifaciens*[d]	*P. propionicum*
Acid from:												
Maltose	–	+	d+	d+	+	nd	–	+	d+	–	+	+
Sucrose	–	+	+	d+	nd	nd	–	+	+	–	+	+
Esculin hydrolysis	+	+	+	+	+	–	–	+	–	–	–	–
Indole production	d	–	–	–	–	nd	d+	–	–	–	–	–
Nitrate reduction	–	+	–	–	–	+[e]	d+	–	–	+	–	+
Gelatin hydrolysis	–	–	–	–	–	nd	+	+	d–	–	–	d
β-Hemolysis	–	–[f]	–	+[g]	nd	nd	d+	d+	–[h]	–	nd	+[i]
DAP isomer	*meso-*	LL-	LL-	LL-	*meso-*	nd	LL- (*meso*)	LL- (*meso*)	LL-	*meso-*	nd	LL-
DNA G+C content (mol%)	64–67	66–68	65–68	66–67	66.8	67.7	57–60	62–63	61–63	nd	70	63–65

[a]Symbols: +, positive reaction in 90% or more of isolates; –, 90% or more of isolates are negative; d, 11–89% of isolates are positive; d+, 40–90% of isolates are positive; d–, 10–40% isolates are positive; parentheses, present in some isolates.

[b]Data taken from Kusano et al. (1997).

[c]Data taken from Koussémon et al. (2001).

[d]Data taken from Downes and Wade (2009).

[e]Nitrate reduced to nitrogen rather than nitrite.

[f]May show slight β-hemolysis under area of confluent growth.

[g]Hemolytic on blood agar containing human, bovine, equine, sheep, rabbit, and pig blood (Vedamuthu et al., 1971).

[h]Generally nonhemolytic on sheep, horse, or rabbit blood agar, but reported to be β-hemolytic on rabbit blood by Hoeffler, (1977).

[i]+, for β-hemolysis of human blood; –, for sheep blood; d, for horse blood.

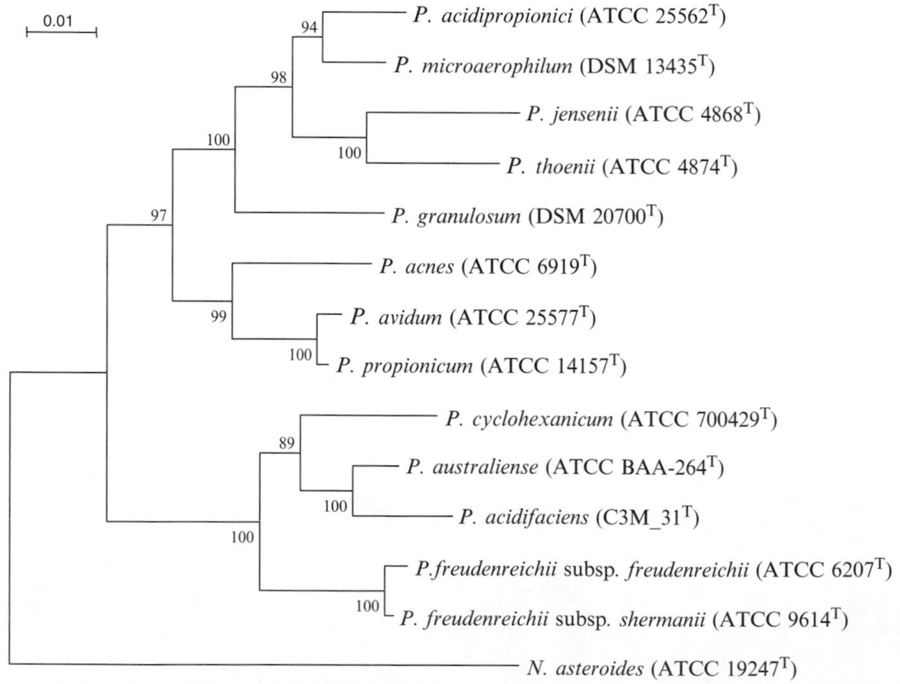

FIGURE 232. Phylogenetic tree of 16S rRNA gene sequences for the genus *Propionibacterium*. The neighbor joining tree was constructed using the Jukes–Cantor-based algorithm. The sequence input order was randomized, and bootstrapping resampling statistics were performed using 500 data sets. The 16S rRNA gene sequence from *Nocardia asteroides* was used as a distant outgroup to root the tree as it is also belongs to the class *Actinobacteria*. Bootstrap values are shown at each node of the tree. Bar = 1 substitution per 100 nucleotide positions.

Stackebrandt, 1989), and there is now general acceptance that *Propionibacterium* is a more fitting genus than *Corynebacterium* (Greenman, 1995) (see below for further discussion). This group has also been referred to as "anaerobic coryneforms", "anaerobic diphtheroids", "acnes group strains", or "cutaneous propionbacters".

In the early literature, anaerobic coryneform bacteria associated with infection were referred to as either *Corynebacterium acnes* (Douglas and Gunter, 1946) or *Corynebacterium parvum* (Cummins and Johnson, 1974). Douglas and Gunter (1946) proposed that these anaerobic coryneforms should be placed in the genus *Propionibacterium* due to the production of propionic acid, catabolic processes, and preference for anaerobic conditions, while others favored the retention of *Corynebacterium* (Puhvel, 1968). Furthermore, it is now clear that strains identified as *Corynebacterium acnes* included the two species *Propionibacterium acnes* and *Propionibacterium granulosum* (Cummins and Johnson, 1974). Cummins and Johnson (1974) also demonstrated that of 59 strains labeled *Corynebacterium parvum*, 90% were in fact *Propionibacterium acnes* by serological and physiological tests as well as DNA similarity, and almost all of the remainder were *Propionibacterium avidum*. The name *Corynebacterium parvum* has been used in immunological and oncological literature to describe the organism which has been extensively used as a reticulostimulant or immunomodulator but should be considered a synonym for *Propionibacterium acnes*. It should also be noted that the National Collection of Type Cultures (NCTC) strain 10387, still designated by the NCTC as *Corynebacterium parvum*, is identical with *Propionibacterium granulosum* ATCC 11829 (Cummins, 1980) and that NCTC 10390 (ATCC 12930), deposited as *Corynebacterium parvum*, is a strain of *Propionibacterium acnes* (phylotype II; see Table 203). We have kept much of the genus and species descriptions of Johnson and Cummins as given in the previous edition of the *Manual*, but have added to them in the light of recent molecular phylogenetic analyses. It should be noted that *Propionibacterium lymphophilum* (Torrey, 1916) has now been reclassified as *Propionimicrobium lymphophilum* (Stackebrandt et al., 2002) and *Propionibacterium innocuum* (Pitcher and Collins, 1991) as *Propioniferax innocua* (Yokota et al., 1994).

Morphologically, the "classical" propionibacteria tend to be shorter and rather thicker than the "cutaneous" species, although all strains may be very variable in morphology, especially in early exponential phase cultures. Strains of *Propionibacterium acnes* in particular give longer more slender irregular rods in young cultures, much resembling the "classical" description of *Corynebacterium diptheriae mitis*, making it easy to understand why the organism was called *Corynebacterium acnes*. In stationary phase cultures, all strains tend to be more coccal. The species *Propionibacterium propionicum* (Buchanan and Pine, 1962) and strains of *Propionibacterium acnes* phylotype III (McDowell et al., 2008) can form filaments up to at least 20 μm in length. Some strains of propionibacteria (e.g., *Propionibacterium avidum* and *Propionibacterium thoenii*) can be observed as capsulated (Skogen et al., 1974; Stackebrandt et al., 2006), and it seems that a considerable number of strains of all species may produce extracellular slime not organized in the form of clear-cut individual capsules. The extracellular material is carbohydrate in nature (Skogen et al., 1974). Purple/red polyphosphate granules may be observed by light microscopy with methylene blue stain or as electron-dense inclusions by electron microscopy with osmium staining, particularly in older cultures (Vorobjeva, 1999). Variations in the peptidoglycan composition and other sugar moieties of the cell envelopes of *Propionibacterium* species are detailed in Table 205. It has been suggested that peptidoglycan containing *meso*-DAP is related to a more coccoid morphology and LL-DAP with a coryneform morphology (Vorobjeva, 1999).

The cell lipids of both "classical" and "cutaneous" strains are characterized by the presence of large amounts of C_{15} branched-chain fatty acids (Moss et al., 1969). Mannose-containing phospholipids have been described in "*Propionibacterium shermanii*" strains (Prottey and Ballou, 1968). The mycolic acids, characteristic of the *Corynebacterium–Mycobacterium–Nocardia* group, are not found. Spermidine and spermine are the predominant compounds within the polyamine profile of the propionibacteria (Hamana, 1995).

With respect to nutritional requirements, the two groups of propionibacteria seem similar. Early studies indicated a requirement for pantothenate in all strains examined and many also require biotin, with growth generally much improved by thiamine and nicotinamide. The addition of oleate may stimulate growth and some strains require *p*-aminobenzoic acid (Delwiche, 1949; Ferguson and Cummins, 1978; Holland et al., 1979). In contrast to the auxotrophic behavior of most propionibacteria species, *Propionibacterium microaerophilum* does not require such growth factors and is, therefore, prototrophic. Amino acid requirements are complex for the "cutaneous" group (Ferguson and Cummins, 1978), but at least some of the "classical" propionibacteria can grow with ammonium sulfate as a nitrogen source (Wood et al., 1938). Most strains of propionibacteria produce vitamin B_{12}, and some, especially strains of *Propionibacterium freudenreichii*, produce large amounts (Hettinga and Reinbold, 1972; Janicka et al., 1976; Vorobjeva, 1999). Fermentation of glucose, pyruvate, dioxyacetate, and glycerol all result in propionic acid, acetic acid, and carbon dioxide production.

Bacteriophages have been reported for *Propionibacterium freundenreichii* subsp. *shermanii*; isometric and filamentous bacteriophage has been described and lysogeny demonstrated (Herve et al., 2001; Vorobjeva, 1999). Bacteriophages specific to *Propionibacterium acnes* have been used to distinguish types I and II (Webster and Cummins, 1978). The complete genome sequence of *Propionibacterium acnes* KPA171202 contains a putative pro-phage sequence (Brüggemann et al., 2004). The *Propionibacterium acnes* specific bacteriophage PA6, which has an icosahedral head and non-contractile tail characteristic of the *Siphoviridae*, has a genome sequence organization similar to the temperate mycobacteriophages, but does not appear to contain genes related to lysogeny (Farrar et al., 2007).

Isolation and maintenance procedures

Most methods for the primary isolation of propionibacteria from dairy products have relied on yeast extract-sodium lactate media (Malik et al., 1968). More complex media such as brain heart infusion broth or agar, however, have often been used for the "cutaneous" strains. Growth of all propionibacteria can be obtained in a trypticase-yeast extract-glucose medium containing 0.05% (v/v) Tween 80 (Cummins and Johnson, 1981). Defined media for the growth of propionibacteria have been devised (Ferguson and Cummins, 1978; Holland et al., 1979; Kurman, 1960; Reddy et al., 1973). A medium containing

TABLE 205. Summary of metabolic reactions and other biochemical characteristics of species of the genus *Propionibacterium*[a,b]

	"Classical" propionibacteria						"Cutaneous" propionibacteria					
Reaction	*P. freudenreichii*	*P. acidipropionici*	*P. jensenii*	*P. thoenii*	*P. cyclohexanicum*	*P. microaerophilium*	*P. acnes*	*P. avidum*	*P. granulosum*	*P. acidifaciens*	*P. australiense*	*P. propionicum*
Acid from:												
Adonitol	d+	+	d+	d+	–	nd	d+	d+	–	nd	+	d
Amygdalin	–	–	d+	d+	+	–	–	–	d–	nd	–	d
D-Arabinose	+	+	–	–	–	–	–	d+	–	nd	–	–
Cellobiose	–	+	d–	–	+	nd	–	–	–	–	–	–
Dulcitol	–	–	–	–	–	nd	–	–	–	nd	nd	–
Erythritol	+	+	+	d+	–	nd	d+	+	–	nd	–	d
Esculin	–	d+	–	+	+	nd	–	–	–	nd	–	nd
D-Fructose	+	+	+	+	+	nd	d+	+	+	+	+	+
Galactose	+	+	+	+	+	nd	d+	+	d–	nd	d–	d
D-Glucose	+	+	+	+	+	nd	d+	+	+	+	+	+
Glycerol	+	+	+	+	+	nd	d+	+	+	nd	nd	d
Glycogen	–	–	–	d+	–	+	–	–	–	nd	–	–
Inositol	d+	+	d+	d+	–	+	d–	d+	–	nd	+	d
Inulin	–	–	–	–	–	nd	–	–	–	nd	nd	–
Lactose	d–	+	d+	d–	+	–	–	d+	–	+	d–	d
Maltose	–	+	d+	d+	+	nd	–	+	d+	+	–	+
Mannitol	–	+	+	–	–	+	d–	d–	d+	+	d+	+
D-Mannose	+	+	+	+	+	nd	d+	+	+	+	d+	d
Melezitose	–	+	d+	d+	+	+	–	d+	d–	–	–	–
Melibiose	d–	d+	+	d+	nd	–	–	d+	d–	+	nd	d
D-Raffinose	–	d–	+	d+	–	–	–	d+	d+	+	d+	+
Rhamnose	–	+	–	–	–	+	–	–	–	+	–	–
Ribose	d+	+	+	+	–	nd	d+	d+	d–	+	d+	d
Salicin	–	+	+	d+	+	nd	–	d+	–	–	d+	d
Sorbitol	–	+	–	d+	–	+	d+	–	–	nd	–	d
L-Sorbose	–	d+	–	–	–	+	–	–	–	nd	nd	–
Starch	–	+	–	+	–	+	–	–	–	nd	–	d
Sucrose	–	+	+	d+	nd	nd	–	+	+	+	–	+
Trehalose	–	+	+	+	+	nd	–	+	d+	nd	–	d
D-Xylose	–	d+	d+	d+	–	–	–	d–	–	nd	–	–
Esculin hydrolysis	+	+	+	+	+	–	–	+	–	–	+	–
Gelatin hydrolysis	–	–	–	–	–	nd	+	+	d–	–	–	d
Starch hydrolysis	–	–	–	d+	–	nd	–	–	–	nd	–	d
Milk:												
Curd	d+	d–	d–	d–	nd	nd	d+	+	+	nd	d–	d
Digestion	–	–	d–	–	nd	nd	d+	+	–	nd	d–	d

(continued)

TABLE 205. (continued)

	"Classical" propionibacteria						"Cutaneous" propionibacteria					
Reaction	*P. freudenreichii*	*P. acidipropionici*	*P. jensenii*	*P. thoenii*	*P. cyclohexanicum*	*P. microaerophilium*	*P. acnes*	*P. avidum*	*P. granulosum*	*P. acidifaciens*	*P. australiense*	*P. propionicum*
Indole-produced	–	–	–	–	–	nd	d+	–	–	–	–	–
Nitrate-reduced	–	+	–	–	–	+[c]	d+	–	–	–	+	+
Catalase	+	d+	d+	+	–	–	d+	+	+	–	–	–
Gas[d]	–, 1	1, 2	–, 1	–, 1	nd	nd	–, 2	–, 1	–, 1	nd	–	–
Acetoin	–	d+	–	–	nd	nd	–	–	–	nd	nd	–
Growth in 20% bile[e]	+	+	+	–	nd	nd	+	+	–	–	–	–
Optimum temp for growth (°C)	30–32	30–32	30–32	30–32	35	30	36–37	36–37	36–37	37	35–37	35–37
Major respiratory quinone	MK-9(H_4)	MK-9(H_4)	MK-9(H_4)	MK-9(H_4)	MK-9(H_4)	nd	MK-9(H_4)	MK-9(H_4)	MK-9(H_4)	nd	nd	MK-9(H_4)
G+C content (mol%)	64–67	66–68	65–68	66–67	66.8	67.7	57–60	62–63	61–63	70	nd	63–65
Major fatty acid	Branched	Branched	nd	nd	ω-Cyclohexane	nd	Branched	Branched	Branched	nd	Branched	Straight
A_2pm isomer	*meso-*	LL-	LL-	LL-	*meso-*	nd	LL-(*meso*)	LL-(*meso*)	LL-	nd	*meso-*	LL-
Amino acids in peptidoglycan	Ala, Glu	Ala, Glu, Gly	Ala, Glu, Gly	Ala, Glu, Gly	Ala, Glu	nd	Ala, Glu, (Gly)	Ala, Glu, (Gly)	Ala, Glu, Gly	nd	nd	Ala, Gly, Glu
Sugars	Galactose, mannose, rhamnose	Glucose, (galactose), (mannose)	Glucose, galactose, mannose	Glucose, galactose,	Glucose, galactose, mannose, rhamnose, ribose	nd	Glucose, mannose, (galactose)	Glucose, mannose, (galactose)	Galactose, mannose	nd	nd	(Glucose), galactose, (mannose)

[a]Symbols: +, positive reaction in 90–100% of isolates, or pH below 5.7 in 90–100% of isolates; –, negative reaction in 90–100% of isolates, or pH 5.7 or above in 90–100% of isolates; d–, reaction positive in 10–40% of isolates; d+ reaction positive in 40–90% of isolates; d, reaction positive in 11–89% of isolates; parentheses, present in some isolates; nd, not determined.

[b]Reactions determined by procedures given in Holdeman et al. (1977).

[c]Nitrate reduced to nitrogen rather than nitrite.

[d]Numbers refer to amount of gas produced on a 1–4 scale.

[e]Data taken from Bernard et al., 2002b; –, unable to grow or grow poorly in 20% bile; +, able to grow well or luxuriantly in bile.

sodium oleinate for the growth of propionibacteria and inhibition of *Actinomyces* strains from human skin swabs has been described (Kishishita et al., 1980). The addition of bromocresol purple to the medium permits identification of *Propionibacterium acnes* and other cutaneous propionibacteria which appear as yellow or slightly yellow opaque colonies with yellow zones of lysis due to the fermentation of glycerol and the production of acid. *Staphylococcus epidermidis* appears as white translucent colonies which therefore enables visual differentiation from propionibacteria. Propionibacteria can also be recovered from skin swabs using TYEG agar (2% w/v tryptone, 1% w/v yeast extract, 0.5% w/v glucose) containing furazolidone (6 µg/ml) to inhibit staphylococcal growth (Ross et al., 2003). Although some strains of propionibacteria may be aerotolerant, a culture methodology used for the primary isolation for strict anaerobes should still be adopted, such as is detailed in the Wadsworth KTL Anaerobic Bacteriology Manual (Jousimies-Somer et al., 2002). Growth media and diluents should be pre-reduced and pre-equilibrated under anaerobic conditions. The incorporation of L-cysteine hydrochloride (0.05% w/v) into liquid media and diluents provides a suitable reducing agent. An anaerobic gas atmosphere within an anaerobic jar or cabinet, from which oxygen has been scavenged by palladium catalists, should be used for incubation. A lack of adherence to strict anaerobiosis when handling and processing clinical specimens will result in reduced detection of pathogenic *Propionibacterium* species (Martin-Rabadan et al., 2008; Tunney et al., 1998). The Hungate roll tube technique (and modifications thereof), where pre-reduced anaerobically sterilized (PRAS) medium is inoculated under N_2 gas (Holdeman et al., 1977), has also been used successfully (Sarada and Joseph, 1994). The "classical" propionibacteria grow best at 30–32°C, and the "cutaneous" group strains at 36–37°C. In general, maximum growth in complex media of pure cultures is attained in 48 h; primary isolation from clinical samples may, however, require incubation for 7 d or more for the "cutaneous" group and up to 14 d for the "classical" group.

For maintenance, chopped meat medium in stoppered tubes under N_2 or another anaerobic gas atmosphere is excellent. After 48 h at the optimum temperature for growth, cultures are kept at room temperature and will maintain viability better than at +4°C (Stackebrandt et al., 2006). For maintenance it is better to omit glucose from the medium to avoid formation of excess acid. The bacterium can also be maintained successfully at −70°C on commercially available microbeads stored in a cryopreservative solution.

Differentiation of the genus *Propionibacterium* from other genera

If molecular identification methods are not used, members of the genus *Propionibacterium* may need to be distinguished from other Gram-stain-positive, nonsporeforming, nonmotile, mainly anaerobic organisms with a rather irregular morphology. Although the production of large amounts of propionic and acetic acids is generally distinctive, the following may cause confusion in some cases.

Some members of the genus *Clostridium* produce considerable amounts of propionic acid, and not all will spore freely. This refers especially to strains of *Clostridium haemolyticum, Clostridium novyi, Clostridium botulinum* types C and D, *Clostridium propionicum, Clostridium quericolum,* and "*Clostridium arcticum*". However, all of these, except "*Clostridium arcticum*", produce large amounts of hydrogen during growth, unlike propionibacteria which produce largely CO_2.

The other genus whose members are most likely to cause confusion is *Corynebacterium.* Here the range of G+C contents (53–67 mol%) is very similar to that in the propionibacteria, and propionic acid may be an end product of metabolism. The corynebacteria generally grow much better aerobically than anaerobically and are characterized by the presence of arabinogalactan and mycolic acids in the cell wall, neither of which is found in propionibacteria. Some corynebacteria species, however, are lacking mycolic acids; these include *Corynebacterium amycolatum* (Collins et al., 1988), *Corynebacterium atypicum* (Hall et al., 2003), *Corynebacterium caspium* (Collins et al., 2004), *Corynebacterium ciconiae* (Fernández-Garayzabal et al., 2004), and *Corynebacterium kroppenstedtii* (Collins et al., 1988). On the other hand, corynebacteria do not have the large amount of C_{15} branched-chain fatty acids in the membrane lipids which are found in propionibacteria. Three types of menaquinones have been recognized in the genus; MK-8(H_4), MK-9, and MK-9(H_4) (Yamada et al., 1976). Corynebacteria have relatively low polyamine contents, with spermidine normally the major compound (Altenburger et al., 1997).

Taxonomic comments

In relation to the "classical" propionibacteria, the species "*Propionibacterium coccoides*" shows low reassociation values in DNA–DNA hybridizations with the propionibacteria, and 16S rRNA gene sequencing indicates that it is related to *Luteococcus* (Tamura et al., 1994; Vorobjeva, 1999).

List of species of the genus *Propionibacterium*

1. **Propionibacterium freudenreichii** van Neil 1928, 162[AL] ("*Bacterium acidi propionici a*", "*Bacterium acidi propionici d*" Von Freudenreich and Orla-Jensen 1906)

freu.den.reich′i.i. N.L. gen. masc. n. *freudenreichii* of Freudenreich; named for Edouard von Freudenreich, the Swiss bacteriologist who first isolated this species.

Description based on literature descriptions including those of Sakaguchi et al. (1941), Janoschek (1944), and Werkman and Brown (1933) and on a study of two strains of *Propionibacterium freudenreichii* (Williams E1.51 and ATCC 6207) and six strains of "*Propionibacterium shermanii*" (van Niel 1.11. IAM 1714 and ATCC 8262, 9615, 9617, 13673).

Surface colonies on horse blood agar (2 d) are 0.2–0.5 µm, circular, entire, convex, semi-opaque, glistening, gray-to-white (may become cream, tan, or pink). Colonies in deep agar are lenticular-to-4 mm, white, tan, or pink. Glucose broth cultures are turbid with a smooth or granular sediment, or clear with a granular sediment, and terminal pH of 4.5–4.9. Nonhemolytic based on Vedamuthu et al., (1971).

Rarely grows on the surface of agar incubated aerobically; grows in deep broth incubated aerobically, but more slowly than anaerobically. Strongly catalase-positive.

Strains require pantothenic acid and some also require biotin and thiamine. For others, thiamine is not essential but is stimulatory, while for all strains, *p*-aminobenzoic acid is not required (Delwiche, 1949).

The major long-chain fatty acids produced in thioglycollate cultures (Moss et al., 1969) are 12-methyltetradecanoic (~43%) and a 17-carbon branched-chain acid (~12%). Will produce large amounts of free proline in peptide-containing media (Langsrud et al., 1977, 1978). Peptidoglycan contains alanine, glutamic acid, and *meso*-DAP; cell-wall sugar components are galactose and moderate amounts of mannose and rhamnose but no glucose.

Source: raw milk, Swiss cheese and other dairy products; particularly associated with the ripening, flavor and aroma of Swiss cheese.

DNA G+C content (mol%): 64–67 (T_m).

Type strain: ATCC 6207, CCUG 7433, CIP 103026, DSM 20271, HAMBI 274, LMG 16412, NBRC 12424, NCTC 10470, NRRL B-3523.

Sequence accession no. (16S rRNA gene): X53217.

Further comments: member of the "classical" or "dairy" group of propionibacteria. On the basis of 16S rRNA gene sequence analysis, *Propionibacterium freudenreichii* clusters with *Propionibacterium australiense* (isolated from bovine lesions) and *Propionibacterium cycohexanicum*, (isolated from spoiled pasteurized orange juice (Figure 232). Lipolysis generated by esterase activity ripens the cheese and the free fatty acids that are produced contribute to flavor. Carbon dioxide produced from fermentation generates the characteristic holes in, for example, Emmental cheese (Dherbecourt et al., 2008). Used commercially as a starter culture in cheese production (Benjelloun et al., 2007). Strains in this species differ broadly from the other "classical" propionibacteria in the following ways:

1. They are generally very short rods, often almost coccal in shape.
2. They are more heat resistant (Malik et al., 1968).
3. They ferment a restricted range of carbohydrates: in particular the inability to ferment sucrose and maltose seems a reliable distinction from other species.
4. Their peptidoglycan contains *meso*-DAP instead of the LL-isomer and the cell wall contains rhamnose.

1a. **Propionibacterium freudenreichii subsp. freundereichii** van Niel 1928, 162[AL] emend. Moore and Holdeman 1970

freu.den.reich'i.i. N.L. gen. masc. n. *freudenreichii* of Freudenreich; named for Edouard von Freudenreich, the Swiss bacteriologist who first isolated this species.

Description as for *Propionibacterium freudenreichii*; distinguished from subsp. *shermanii* by nitrate reduction and by a lack of lactose fermentation.

Source: raw milk, Swiss cheese and other dairy products; particularly associated with the ripening, flavor and aroma of Swiss cheese.

DNA G+C content (mol%): 64–67 (T_m).

Type strain: ATCC 6207, CCUG 7433, CIP 103026, DSM 20271, HAMBI 274, LMG 16412, NBRC 12424, NCTC 10470, NRRL B-3523.

Sequence accession no. (16S rRNA gene): X53217.

1b. **Propionibacterium freudenreichii subsp. shermanii** van Niel 1928, 162[AL] emend. Moore and Holdeman 1970 ("*Propionibacterium shermanii*" van Niel 1928, 162)

sher.man'i.i. N.L. gen. masc. n. *shermanii* of Sherman; named for James M. Sherman, an American bacteriologist.

Description as for *Propionibacterium freudenreichii*; distinguished from subsp. *freudenreichii* by lactose fermentation and a lack of nitrate reduction.

DNA G+C content (mol%): 67 (T_m).

Type strain: ATCC 9614, CCUG 36819, CIP 103027, DSM 4902, LMG 16424.

Sequence accession no. (16S rRNA gene): Y10819.

Further comments: member of the "classical" or "dairy" group of propionibacteria. Strains of *Propionibacterium freudenreichii* subsp. *freudenreichii* and *Propionibacterium freudenreichii* subsp. *shermanii* have high DNA similarity with each other (Johnson and Cummins, 1972), and there seems to be little justification for separation at the species level. Of the species described by Sakaguchi et al. (1941), it is possible that "*Propionibacterium globosum*", "*Propionibacterium orientum*", and "*Propionibacterium coloratum*" were in fact other variants of *Propionibacterium freudenreichii*, but no labeled strains appear to be extant and available for testing. Complete genome sequencing of the *Propionibacterium freudenreichii* subsp. *shermanii* strain ATCC9614 (~2.6 Mb; G+C content 67%) has been undertaken (Meurice et al., 2004).

2. **Propionibacterium acidifaciens** Downes and Wade 2009, 2780[VP]

a.ci.di.fa'ci.ens. N.L. n. *acidum* an acid; L. v. *facio* to produce; N.L. part. adj. *acidifaciens* acid-producing.

Description based on three strains isolated from the human oral cavity (Downes and Wade, 2009).

Microscopically, cultures are pleomorphic rods; straight, slightly curved, or club-shaped cells are arranged singly, in pairs, or short chains with some branched diphtheroids (0.7–0.8 × 1.2–4 µm). Surface colonies are 0.7–1.1 mm in diameter, circular, entire, high-convex-to-dome-shaped, white-to-pale cream, solid, and non-translucent after 7 d growth at 37°C on fastidious anaerobe agar (LabM) supplemented with 5% (v/v) horse blood under an atmosphere of 80% N_2, 10% H_2, and 10% CO_2.

Growth is evident in peptone yeast extract broth, with moderate turbidity that is enhanced by fermentable carbohydrates (1%). In peptone-yeast extract-glucose broth, large amounts of acetic and propionic acid are produced and, in moderate amounts, succinic acid is produced.

Growth was not detectable in air enriched with 5% CO_2. Catalase-negative.

Bile tolerant, although colonies are smaller on tryptone soy agar containing 20% bile compared to tryptone soy agar only.

Source: carious lesions in the human mouth.

DNA G+C content (mol%): 70 (T_m).

Type strain: C3M_31, CCUG 57100, DSM 21887, JCM 16571.

Sequence accession no. (16S rRNA gene): AB565481, EU979537.

Further comments: based on 16S rRNA gene sequences (1442 bp), the three strains differed between one and two bases. *Propionibacterium australiense* is the most closely related species based on 16S rRNA gene sequence (97.7% identity), although the species only share 8% DNA–DNA relatedness. On the basis of RpoB amino acid sequence, *Propionibacterium acnes* is the most closely related species. The strains examined represent a homogeneous group and are representative of the dominant *Propionibacterium* taxon associated with dentinal caries (Downes and Wade, 2009). Sensitive to vancomycin and kanamycin, but resistant to colistin.

3. **Propionibacterium acidipropionici** corrig. Orla-Jensen 1909, 337[AL] (*"Bacillus acidi propionici"* von Freudenreich and Orla-Jensen 1906; *"Propionibacterium pentosaceum"* van Niel 1928; *"Propionibacterium arabinosum"* Hitchner 1932)

a.ci.di.pro.pio'ni.ci. N.L. n. *acidum propionicum* propionic acid; N.L. gen. n. *acidipropionici* of propionic acid.

Description based on the study of the type strain ATCC 25562, ATCC 4875, *Propionibacterium pentosaceum*" (van Niel 4), ATCC 4965 (*"Propionibacterium arabinosum"*, Hitchner), and IAM 1725. Phenotypic characteristics of these strains are similar to original descriptions.

Surface colonies on horse blood agar after 2 d anaerobic incubation are punctiform–1 mm, circular to slightly irregular, convex, entire or slightly scalloped, gray or white, and semiopaque. Colonies in deep agar are white, becoming pink with continued incubation.

Glucose broth cultures are turbid with smooth cream-colored sediment. Terminal pH 4.1–4.9.

Usually nonhemolytic, but may show slight β-hemolysis under area of confluent growth. May grow under both aerobic and anaerobic conditions. Weak or negative for catalase (van Niel, 1957; Werkman and Brown, 1933).

Requires pantothenate and biotin for growth. Thiamine is not essential but is stimulatory (Delwiche, 1949).

The major long-chain fatty acids produced in thioglycolate cultures are 13-methyltetradecanoic acid (17–40%) and 12-methyltetradecanoic acid (12–23%) (Moss et al., 1969). Peptidoglycan contains alanine, glutamic acid, glycine, and LL-DAP as diamino acid; cell-wall sugars are glucose with galactose and/or mannose (some strain variation: see Table 5 in Johnson and Cummins, 1972).

Source: dairy products.

DNA G+C content (mol%): 66–68 (T_m).

Type strain: ATCC 25562, CIP 103025, DSM 4900, VPI 399.

Sequence accession no. (16S rRNA gene): AJ704569.

Further comments: member of the "classical" or "dairy" group of propionibacteria. On the basis of 16*S* rRNA gene sequence analysis, *Propionibacterium acidipropionici* clusters with *Propionibacterium microaerophilum* (Figure 232). The strains originally described by van Niel (1928) and Hitchner (1932) as *"Propionibacterium pentosaceum"* and *"Propionibacterium arabinosum"*, respectively, show high DNA similarity with each other and are also characterized by a weak or negative catalase reaction (Cummins and Johnson, 1986). Strains of *"Propionibacterium arabinosum"* were described as being unable to ferment xylose and rhamnose, while those of *"Propionibacterium pentsaceum"* could ferment these sugars. However, as in the case of *Propionibacterium freudenreichii* and *"Propionibacterium shermanii"* with lactose fermentation, this is not considered sufficient to warrant speciation. *Propionibacterium acidipropionici* produces a high yield of propionic acid, with low acetic acid production, from glycerol fermentation and has been studied as an alternative to petrochemicals as a commercial source of propionic acid (Himmi et al., 2000).

4. **Propionibacterium acnes** (Gilchrist 1900) Douglas and Gunter 1946, 22[AL] (*"Bacillus acnes"* Gilchrist 1900; *"Corynebacterium acnes"* Eberson 1918)

ac'nes. Gr. n. *acme* a point; incorrectly transliterated as N.L. n. *acne* acne; N.L. gen. n. *acnes* of acne.

Description based on a study of several hundred strains by Holdeman et al. (1977), Cummins and Johnson (1986), including the type strain ATCC 6919, and on the results of Kishishita et al. (1979), McGinley et al. (1978), and McDowell et al. (2008, 2005).

Four distinct genetic groups have been described within *Propionibacterium acnes*. These clades, known as types IA, IB, II, and III, share over 99.9% identity based on the analysis of 16S rRNA gene sequences (1484 bp section) but can be differentiated based on polymorphisms in *recA* and a putative hemolysin/cytotoxin gene (*tly*) (McDowell et al., 2005, 2008) (Figure 233).

Colonies in deep agar appear lenticular, 4 mm or less; colonies of some strains become tan, pink or orange in 3 weeks. Surface colonies on blood (horse or rabbit) agar (2–3 d) are punctiform-to-0.5 mm, circular, entire-to-pulvinate, translucent-to-opaque, white-to-gray, and glistening. On brain heart infusion (BHI) agar, type III strains may form very small, flat, dry colonies, which are difficult to recover from the plate, in contrast to types I and II which form circular, convex, glistening, opaque colonies. Strains within the type III clade have the capacity to form long slender filaments in addition to the "classical" coryneform shape (Figure 234 and Figure 235), a characteristic not been previously described for *Propionibacterium acnes*, but known for other propionibacteria such as *Propionibacterium propionicum* (Figure 236 and Figure 237) (Pine and Georg, 1969) and *Propionibacterium australiense* (Bernard et al., 2002b). Strains of phylotype II may appear more coccoid compared to those of phylotype I and are most similar to previous descriptions of *"Clostridium parvum"* and *"Clostridium adamsoni"* which are synonyms of *Propionibacterium acnes*.

In defined medium broth culture, *Propionibacterium acnes* type IA will form a turbid suspension, while in proteose peptone yeast (PPY) or BHI broth, the organism forms a settled granular sediment (due to aggregation) with a clear solution. With increasing concentrations of glucose greater than 4%, however, aggregation is significantly reduced and the broth becomes more turbid (Patrick and Glenn unpublished). Studies with types IB and II have shown the formation of a slight fine sediment upon culture in BHI broth and a turbid solution that contains suspended cells (Cohen et al., 2005). In suitable media with good growth, the final pH is 4.5–5.0. The auto-aggregating properties of type IA strains in broth culture may reflect, in part, a hydrophobic cell surface (Cohen et al., 2005).

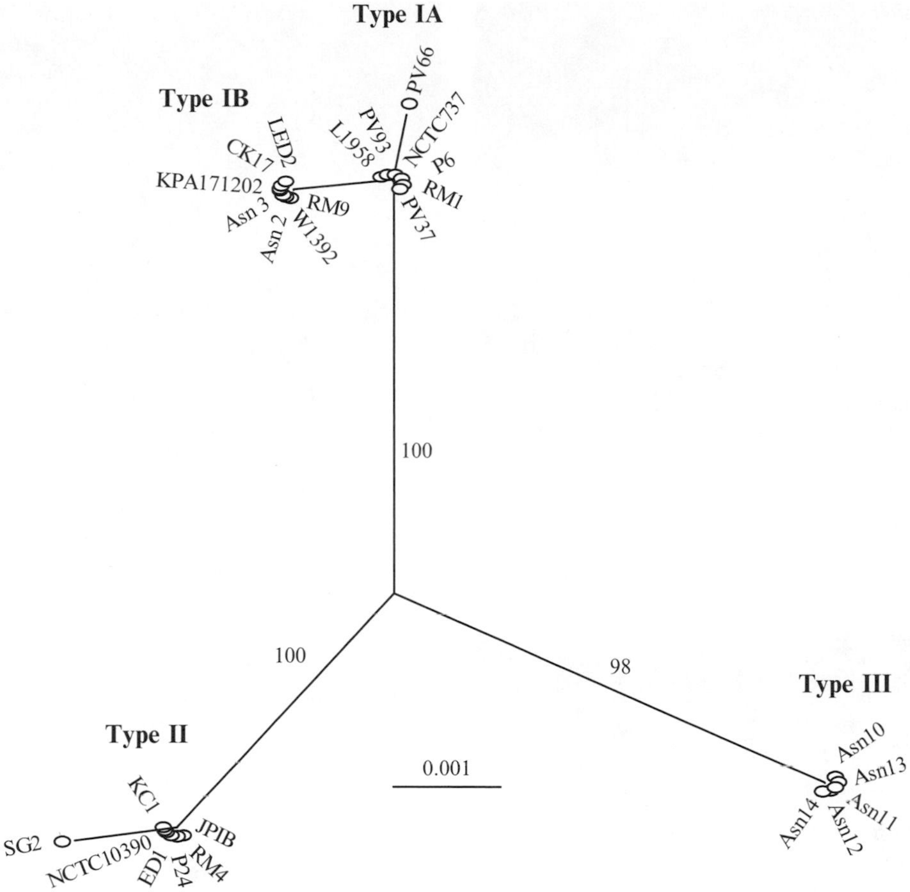

FIGURE 233. Unrooted phylogenetic tree of *Propionibacterium acnes* based on the complete *recA* gene sequence. The neighbor-joining tree was constructed using the Jukes–Cantor-based algorithm. The sequence input order was randomized, and bootstrapping resampling statistics were performed using 100 data sets. Bootstrap values are shown on the arms of the tree. The phylotype status for the different strains analyzed is shown. (Printed with permission from McDowell et al., 2008. J. Med. Microbiol. *57*: 218–224.)

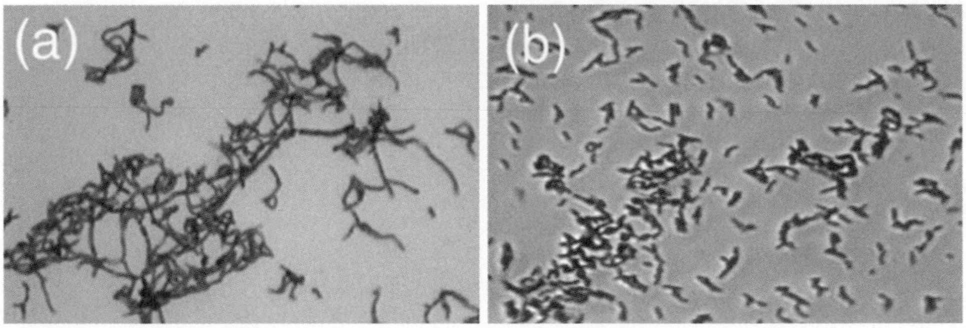

FIGURE 234. Gram stain of *Propionibacterium acnes* prepared from a mature colony (anaerobic blood agar; 7 d; 37°C); micrograph (obj. ×100). (a) Type III (Asn 10); (b) type IA (NCTC 737).

Approximately 65% of type I isolates (n=85) show β-hemolysis on horse blood, but this has not been observed with type II isolates (n=31) or any type III isolates examined to date (n=5). A gene family consisting of five homologs of the *Streptococcus agalactiae* co-hemolysin or CAMP (Christie–Atkins–Munch–Peterson)-factor has been identified within the genome of all *Propionibacterium acnes* phylotypes, and the co-hemolysin reaction similar to that originally described by Christie et al. (1944).has also been demonstrated among 50 isolates representing types IA, IB, and II (Valanne et al., 2005). Differential production of the different CAMP factor proteins has been observed among the phylotypes. In partic-

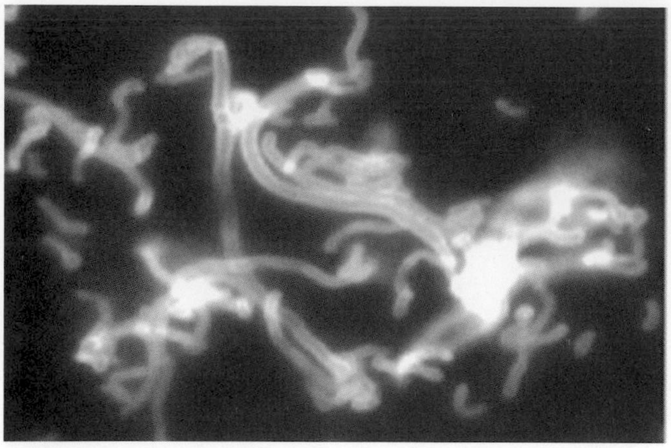

FIGURE 235. Immunolabelling of *Propionibacterium acnes* type III prepared from a mature colony (anaerobic blood agar; 7 d; 37°C); micrograph (obj × 100). Bacterium was labeled with a mouse IgG monoclonal antibody QUBPa3 (reacts with all *Propionibacterium acnes* strains) and a fluorescein isothiocyanate-conjugated goat anti-mouse IgG antibody.

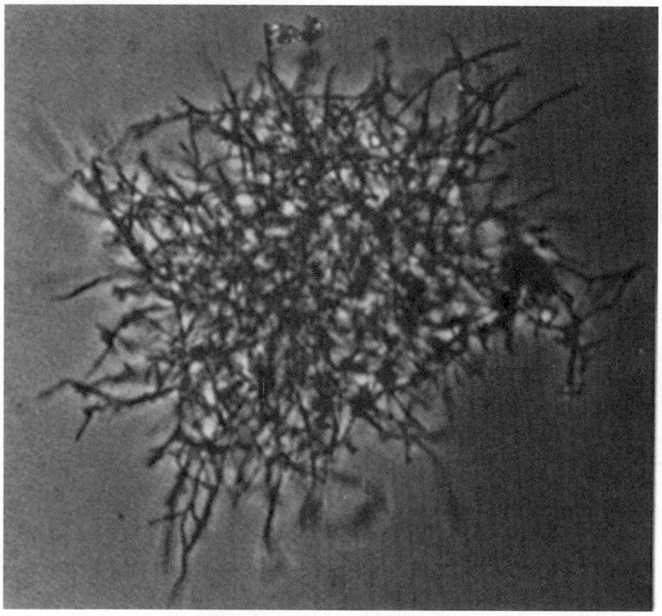

FIGURE 236. *Propionibacterium propionicum,* serovar 1. Wet mount in lactophenol cotton blue mounting fluid prepared from a 24 h culture in Tarozzi broth; phase contrast micrograph (640×).

ular, strains of type II and a commonly isolated type IB strain (ST-10 by MLST) produce large quantities of CAMP factor 1 compared to type IA organisms (Valanne et al., 2005).

The presence of oxygen is not lethal to *Propionibacterium acnes,* but anaerobiosis is usually required for isolation and growth (Cove et al., 1983). A small number of strains grow on the surface of blood agar plates incubated aerobically, while others (~30%) display variable growth on the surface of blood agar plates incubated in a candle jar for microaerophilic growth. Many strains grow in deep glucose broth without special provision for anaerobiosis. Strains are catalase-positive, although cultures need to be exposed

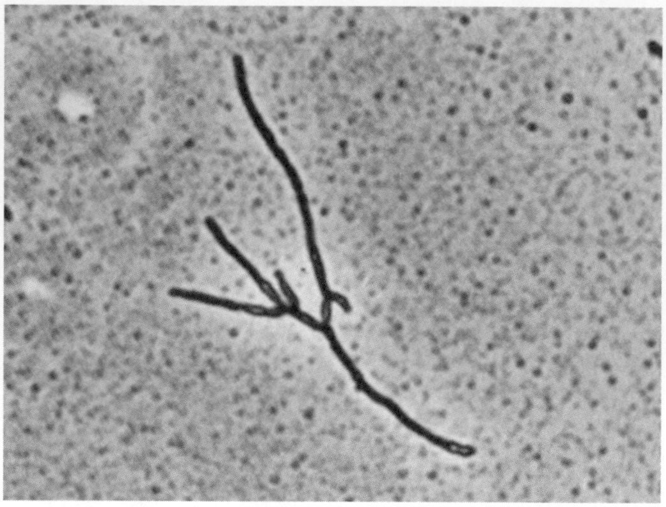

FIGURE 237. *Propionibacterium propionicum,* serovar 1. Young (20-h) microcolony on CC medium (slide culture); phase-contrast micrograph in situ (1600×).

to air for a time before testing. McGinley et al. (1978), who exposed colonies on BHI agar to air for 1 h before testing, found all 231 strains to be-positive for catalase. A study with one strain of *Propionibacterium acnes* has shown that increases in oxygen tension will alter the production of extracellular enzymes due to a decrease in growth rate (Cove et al., 1983).

The ability to ferment sorbitol is a characteristic of some type I strains, but not type II or type III strains and can, therefore, be used as a simple method of identification. On the collective basis of sorbitol, ribose, and erythritol fermentation, Kishishita et al. (1979) identified five fermentation biotypes (B1–B5) among 128 isolates of *Propionibacterium acnes* recovered from healthy facial skin. Biotypes B1-B5 were all identified among type I strains, but only B2 (ribose and erythritol fermenters) was characteristic of the type II clade. All type III strains examined to date (n=5) belong to biotype 4 (ribose fermenters) (McDowell et al., 2008).

Studies have shown that different strains produce many enzymes responsible for the degradation of host derived molecules. These include ribonuclease (Smith, 1969), neuraminidase (von Nicolai et al., 1980), hyaluronidase (Hoeffler, 1980; Ingham et al., 1980), acid phosphatase (Ingham et al., 1980) lecithinase (Werner, 1967), lipase (Ingham et al., 1981; Smith and Willett, 1968), and proteinase (Ingram et al., 1983). Analysis of the *Propionibacterium acnes* genome sequence has now identified the specific gene loci encoding these various extracellular products (Brüggemann et al., 2004). Free fatty acids released from skin lipids (e.g., sebum) by the action of lipase may (a) act as tissue irritants and (b) promote the growth of *Propionibacterium acnes* (e.g., oleate, see below). They may also assist bacterial adherence and colonization of the sebaceous follicle (Brüggemann, 2005).

Some strains also produce bacteriocin-like substances inhibitory for Gram-stain-positive and -negative anaerobes

(Fujimura and Nakamura, 1978; Ko et al., 1978; Paul and Booth, 1988).

All strains tested to date have an absolute requirement for pantothenate, while thiamine, biotin, and nicotinamide are stimulatory. Some strains of phylotype II require heme and vitamin K to grow (Ferguson and Cummins, 1978). The organic acids lactate, pyruvate, and 2-oxoglutarate also stimulate growth, as does oleate (usually used in the form of Tween 80). Amino acid requirements are complex (Ferguson and Cummins, 1978; Holland et al., 1979; Nielsen, 1983).

The major long-chain fatty acid produced in thioglycolate cultures (Moss et al., 1969) is 13-methyltetradecanoic acid (32–62%).

Peptidoglycan contains alanine, glutamic acid, glycine, LL-DAP and, in certain strains of type II, occasionally *meso*-DAP. Cell-wall sugars are glucose, mannose, and galactose. In contrast to *Propionibacterium acnes* type I, galactose is not found in the cell wall of type II strains. On the basis of differences in cell-wall composition, types I and II can be identified by serology and monoclonal antibody labeling methods (Johnson and Cummins, 1972; McDowell et al., 2005). Cell-wall sugar content of type III strains has not been determined.

Source: the bacterium primarily colonizes the sebaceous gland-rich areas of human skin, but the organism is also found in the mouth, as well as the genito-urinary tract and large intestine. In relation to infections, it has been recovered from comedones of acne vulgaris, wounds, blood, pus, soft tissue abscesses, the surface of indwelling medical devices, and eye infections. Levels of *Propionibacterium acnes* colonization on the skin vary from person-to-person and from the area of the body sampled (McLorinan et al., 2005). Levels can be as high as 10^6 organisms/cm^2, with the neck, forehead, and shoulder showing some of the highest concentrations of the bacterium compared to other sites such as the abdomen, hip, knee, and chest where levels are lower (Patel et al., 2009).

DNA G+C content (mol%): 57–60 (T_m).

Type strain: ATCC 6919, CCUG 1794, CIP 53.117, DSM 1897, JCM 6425, LMG 16711, NCTC 737, NRRL B-4224, VKM Ac-1450.

Sequence accession no. (16S rRNA gene): AB042288.

Further comments: member of the "cutaneous" group of propionibacteria. On the basis of 16S rRNA gene sequence analysis, *Propionibacterium acnes* clusters with *Propionibacterium avidum* and *Propionibacterium propionicum.* Multilocus Sequence Typing (MLST) of *Propionibacterium acnes* isolates based on the analysis of 7 core housekeeping genes (*aroE, atpD, gmk, guaA, lepA, recA,* and *sodA*) supports the IA, IB, II and III phylogenetic divisions (http://pubmlst.org/pacnes/; McDowell, Gao, Dowson and Patrick, unpublished). Strains of phylotypes I and II share DNA–DNA hybridization values of 88–99%. DNA–DNA hybridization values for phylotype III versus type I and II have not been determined. The *Propionibacterium acnes* genome sequence (type IB strain KPA171202; ST-10 by MLST) has revealed the presence of genes for oxidative phosphorylation, the Embden-Meyerhof and pentose phosphate pathways, and the tricarboxylic acid cycle (Brüggemann et al., 2004). Systems for anaerobic respiration, such as nitrate reductase, are also present. Reconstruction of the metabolic pathways in the bacterium demonstrates its capacity to deal with changing oxygen tensions and confirms laboratory observations (Cove et al., 1983).

Although the role of *Propionibacterium acnes* in the inflammatory condition acne has been debated, its involvement is generally accepted since acne which is refractory to antibiotic treatment is associated with strains resistant to the same antibiotic (Eady et al., 1989). Other conditions associated with *Propionibacterium acnes* include opportunistic biofilm infections of medical implants such as prosthetic joints (Tunney et al., 1998), native and prosthetic heart valves (Clayton et al., 2006), intravascular catheters (Martin-Rabadan et al., 2008), central nervous system shunts (Brook and Frazier, 1991), and also endophthalmitis post-eye surgery (Aldave et al., 1999). It has been also linked to synovitis-acne-pustulosis-hyperostosis-osteitis (SAPHO) syndrome (Kotilainen et al., 1996), sarcoidosis (Eishi et al., 2002), and chronic infection of the prostate gland, potentially leading to prostate cancer (Cohen et al., 2005). The possibility of wide variation in the potential for virulence among different *Propionibacterium acnes* isolates should not be discounted. The heterogeneity of *Propionibacterium acnes* in relation to putative virulence determinants is well documented (Lodes et al., 2006; McDowell et al., 2008; Valanne et al., 2005). Variation in expression of immune-reactive surface and secreted proteins has been described. For example, variation of proteins that bind to dermatan sulfate with carboxy-terminal repeats and the amino acid motif LPXTG, characteristic of MSCRAMMs (microbial surface components recognizing adhesive molecules), is generated by a variable number of C residues in a C_nTC_n motif upstream of the putative signal peptide for the proteins. In addition, there are variable numbers of repeats towards the mid-region and the carboxy-terminus that alter molecular mass and expression of the proteins (Lodes et al., 2006). This variation is probably generated by slipped strand mispairing during replication leading to phenotypic variation.

The strong pro-inflammatory nature of *Propionibacterium acnes* has been known for some time and homologs of GroEL, DnaK, and several other heat-shock proteins (Graham et al., 2004), which are major targets for the immune system, are present in the genome (Brüggemann et al., 2004). Three gene clusters putatively involved in extracellular polysaccharide biosynthesis have been identified in the genome sequence and may also be important in modulating immunogenicity. The production of porphyrins may lead to inflammation and generate toxic reduced oxygen species that cause cell damage. The immunostimulatory activity and adjuvant properties of isolates previously, and currently by some authors, referred to as "*Corynebacterium parvum*" are exploited in studies of the inflammatory response (Fink et al., 2008). *Propionibacterium acnes* has also been used as a non-specific immune stimulant in the equine industry (EqStim; Flaminio et al., 1998).

Propionibacterium acnes is often dismissed as a skin contaminant when present in clinical samples, especially when present in low numbers. This has undoubtedly led to an under-reporting and, consequently, under-recognition of its association with different infections and conditions in the literature. Furthermore, many samples are not routinely cultured under anaerobic conditions. The use of protocols for anaerobic handling and isolation from clinical material increases the isolation of *Propionibacterium acnes* from clinical samples (Tunney et al., 1998). It has been demonstrated that aerobic incubation of agar plates inoculated with clinical specimens for up to 96 h prior to anaerobic reincubation will result in reduced *Propionibacterium acnes* colony numbers (Martin-Rabadan et al., 2008). This may have ramifications for the detection of *Propionibacterium acnes* associated with clinical samples that contain lower levels of the bacterium. In addition, the use of prophylactic antibiotics may also hinder growth of the organism from tissue samples. For phylotype III strains, the filamentous-like cellular morphology observed upon routine Gram-staining (Figure 234) and colony appearance (see above) may also confuse identification. When anaerobic growth from samples is performed, agar plates should be incubated long enough to observe growth (7–14 d). Infection of indwelling medical devices due to *Propionibacterium acnes* may also be under-described due to inappropriate sampling of the device and failure to detect the bacterium growing as an adherent biofilm (Tunney et al., 1998, 1999, 2007).

Analysis of isolates causing surgical, foreign body, and septicemia infections in Europe found that 3% were resistant to tetracycline, 15% to clindamycin, and 17% to erythromycin, while none were resistant to linezolid, benzylpenicillin, or vancomycin (Oprica and Nord, 2005). A study of acne patients in Europe found that the widespread use of topical formulations of erythromycin and clindamycin to treat this skin condition has resulted in a significant dissemination of cross-resistant propionibacteria strains. With the exception of Sweden and the UK, resistance rates to the orally administered tetracycline group of antibiotics are low (Ross et al., 2003). Similar results have also been described among acne patients being treated in Asia (Ishida et al., 2008; Tan et al., 2001). Resistance to erythromycin was most commonly encountered, and erythromycin-resistant strains were frequently cross-resistant to clindamycin. Among the tetracycline group of drugs, the mean MICs were higher than that for doxycycline and minocycline (Tan et al., 2001). Strains of *Propionibacterium acnes* are resistant to metronidazole (Chow et al., 1975), sulfonamides (Pochi and Strauss, 1961), and aminoglycosides (Wang et al., 1977). Studies have also shown that susceptibility of *Propionibacterium acnes* to common antimicrobials decreases as the bacterium forms a biofilm (Coenye et al., 2007; Ramage et al., 2003).

5. **Propionibacterium australiense** Bernard, Shuttleworth, Munro, Forbes-Faulkner, Pitt, Norton and Thomas 2002a, 1915[VP] (Effective publication: Bernard, Shuttleworth, Munro, Forbes-Faulkner, Pitt, Norton and Thomas 2002b, 45.)

aus.tra.li.en'se. N.L neut. adj. *australiense* of or belonging to Australia, the country where the bacterium was first isolated.

Description based on the literature description of the type strain 98A072 and the study of six strains in total (Bernard et al., 2002b; Forbes-Faulkner et al., 2000).

Microscopically, cultures are pleomorphic rods with filamentous, branching, and curved forms sometimes evident. Surface colonies grown on pre-reduced brain heart infusion agar supplemented with 5% sheep blood (BHI) under anaerobic conditions at 35°C are off-white or cream, rounded, convex, and do not adhere to the agar surface. Growth is evident in peptone-yeast broth with and without glucose. Nonhemolytic on sheep blood agar and negative for the CAMP and reverse CAMP reactions.

Growth is optimal under strict anaerobiosis and poor or non-existent in air enriched with 5% CO_2. Primary isolation from material derived from bovine granulomatous lesions required 5 d incubation on pre-reduced BHI agar. Good growth is, however, obtained within 24–48 h with repeated subculture. Catalase-negative. Peptidoglycan contains *meso*-DAP.

Source: granulomatous bovine lesions located throughout the animals (0.5–15 cm in diameter), from cattle in Queensland, Australia.

DNA G+C content (mol%): not available.

Type strain: 98A072, NML 98A072, ATCC BAA-264, CCUG 46075.

Sequence accession no. (16S rRNA gene): AF225962.

Further comments: all six strains examined had identical 16S rRNA gene sequences (>1460 bp) and cluster with *Propionibacterium acidifaciens* and *Propionibacterium cyclohexanicum* (Figure 232). These granulomatous lesions were distinguishable from those caused by *Mycobacterium bovis, Rhodococcus equi, Actinomyces bovis*, and fungi. The lesions consisted of a characteristic fibrous outer capsule surrounding thick yellow viscous pus with caseating granulomas. The lesions were present in large numbers throughout the animal, most commonly on the external surface of the rumen and reticulum, tongue, peritoneal surface of the gastrointestinal tract, and, less commonly, on the omemtum, flank, and flank muscles. Lesions were also more rarely present on the skin over the ribs, the cheek, or under the jaw. Affected carcasses were condemned. Cases were observed in cattle from different geographical sites in northern and central Queensland, Australia, on seven occasions over a period of more than two years. In one case, up to 30% of a group of 324 cows, between 10–12 yrs old, had lesions. Cases were also reported in three-year-old steers, bulls, and breeding cattle that were in calf. Initially observed on abattoir post-mortem inspection, multiple external lesions were subsequently found on live cattle at breeding properties. Post-mortem, these animals also had multiple internal lesions. The source of the *Propionibacterium australiense* has yet to be determined, although animal-to-animal transmission is thought to be most likely (Forbes-Faulkner et al., 2000).

6. **Propionibacterium avidum** (Eggerth 1935) Moore and Holdeman 1969, 7[AL] ("*Bacteroides avidus*" Eggerth 1935; "*Corynebacterium avidum*" Prévot 1938; "*Mycobacterium avidum*" Krasil'nikov 1949)

a'vi.dum. L. neut. adj. *avidum* greedy, voracious.

Description based on a study of 20 strains including ATCC 25577 by Holdeman et al. (1977) and 23 strains by Cummins and Johnson (1986).

Surface colonies on suitable media are 0.5–1.0 mm at 2–3 d, smooth, entire, circular, and white to light cream color. In general, grows better and gives larger colonies than either *Propionibacterium acnes* or *Propionibacterium granulosum.*

Glucose broth is generally turbid with a smooth abundant sediment which resuspends readily. Generally β-hemolytic on sheep, horse, or rabbit blood agar (19/23 strains; Cummins and Johnson, 1986) which agrees well with the results of Hoeffler (1977) on sheep, human, and rabbit blood agar. A thiol-activated extracellular hemolysin has been partially isolated (Fujimura et al., 1982).

Grows anaerobically or as a microaerophile when freshly isolated, but will often grow well aerobically after a few transfers. Appears to be the best adapted species of the "cutaneous" group for growth in aerobic environments (Cove et al., 1983). Catalase-positive.

Nutritional requirements are less exacting than either *Propionibacterium acnes* or *Propionibacterium granulosum* and, after several transfers, will grow in simple medium consisting of salts, glucose, and vitamins (Ferguson and Cummins, 1978). Pantothenic acid is an absolute requirement for growth.

Peptidoglycan contains alanine, glutamic acid, and glycine amino acid residues. Type I peptidoglycan contains LL-DAP, whereas type II contains *meso*-DAP, but a few strains have LL-DAP and no glycine (thus resembling similar strains of *Propionibacterium acnes* type II). Type I strains contain glucose, galactose, and mannose sugars in their cell wall, while type II contains glucose and mannose only, again resembling type II *Propionibacterium acnes* (Johnson and Cummins, 1972).

Strains are normally positive for gelatinase and deoxyribonuclease; most are negative for lecithinase, hyaluronidase, and chondroitin sulfatase (Hoeffler, 1977).

Source: principally, the moister areas of the skin: e.g., vestibule of the nose, axilla, perineum, and from chronically infected areas such as sinuses and occasionally from abscesses.

DNA G+C content (mol%): 62–63 (T_m).

Type strain: ATCC 25577, CCUG 36754, CIP 103261, DSM 4901, NBRC 15671, NCTC 11864, VPI 179.

Sequence accession no. (16S rRNA gene): AJ003055.

Further comments: member of the "cutaneous" group of propionibacteria. *Propionibacterium avidum* can be divided into two serological groups designated type I and II (Johnson and Cummins, 1972). By DNA–DNA hybridization, sequences within each group were at least 90% similar, but between the two groups the mean similarity was 80% (Goodsell et al., 1991). On the basis of 16S rRNA gene sequence analysis, *Propionibacterium avidum* is most closely related to *Propionibacterium propionicus* but also clusters with *Propionibacterium acnes* (Figure 232). Only two out of 22 type I strains (including the type strain) fermented inositol, in contrast to type II strains which all ferment inositol. *Propionibacterium avidum* has been rarely associated with severe infection after invasive procedures. These include skeletal infection such as sacroiliitis and septic arthritis of the hip (Million et al., 2008) and soft tissue infection such as splenic (Vohra et al., 1998) and breast abscesses (Panagea et al., 2005). The immunostimulatory properties of *Propionibacterium avidum* strain KP-40 have been studied and anti-metastatic activity demonstrated in mice (Isenberg et al., 1994; Isenberg et al., 1995). Previous studies have shown sensitivity to benzylpenicillin, ampicillin, cephalothin, rifampin, clindamycin, erythromycin, and minocycline (Eady et al., 1989). Little information is available on the current susceptibility of *Propionibacterium avidum* strains, but analysis of an isolate responsible for a post-surgical breast abscess revealed sensitivity to penicillin G, ampicillin, erythromycin, tetracycline, and vancomycin and resistance to metronidazole (Panagea et al., 2005). Erythromycin-resistant strains have been isolated from acne patients treated with oral erythromycin and topical clindamycin (Eady et al., 1989).

7. **Propionibacterium cyclohexanicum** Kusano, Yamada, Niwa, Yamasato 1997, 830[VP]

cy.clo.hex.a'ni.cum. Gr. n. *kuklos* circle; Gr. n. *hex* six; L. neut. suff. *-icum* suffix used with the sense of pertaining to; N.L. neut. adj. *cyclohexanicum* relating to ω-cyclohexyl fatty acid, the characteristic cellular fatty acid of the organism.

Description based on the literature description of the type strain TA-12[T] (Kusano et al., 1997; Walker and Phillips, 2007).

After 3 d growth, surface colonies on peptone-yeast extract-glucose (PYG) agar at 35°C in a gas atmosphere of 20% CO_2: less than 0.1% O_2 are circular, white-to-creamy, translucent, and 0.2–0.5 mm in diameter. Optimal growth temperature of 35°C.

Lactic acid is a major product of fermentation in addition to propionic and acetic acids.

Colony formation on plates incubated aerobically. Catalase-negative.

Growth at pH 3.2–7.5; optimum pH 5.5–6.5; can survive up to 90°C in PYG broth and 95°C in orange juice at pH 3.9 for 10 min.

ω-Cyclohexyl undecanoic acid accounts for 52.7% of the total fatty acids. Other fatty acids determined include the straight-chain and anteiso-branched fatty acids n-C_{15} (16.8%), C_{15}, (6.4%), n-C_{16} (2.8%), and n-C_{17} (5.3%).

Peptidoglycan contains alanine, glutamic acid, and *meso*-DAP; whole-cell sugars are galactose, glucose, mannose, ribose, and rhamnose.

Source: spoiled off-flavor pasteurized orange juice.

DNA G+C content (mol%): 66.8 (T_m).

Type strain: TA-12, ATCC 700429, CCUG 48885, CIP 105414, IAM 14535, JCM 21245, NBRC 103082, NRIC 0247.

Sequence accession no. (16S rRNA gene): D82046.

Further comments: on the basis of 16S rRNA gene sequence analysis, *Propionibacterium cyclohexanicum* is most closely related to *Propionibacterium australiense* but also clusters with *Propionibacterium freundenreichii*, the closest of the "classical" propionibacteria (Figure 232). The ω-cyclohexane fatty acids are also found in the cell membranes of *Alicyclobacillus* which is acidophilic and thermotolerant but does not belong to the *Actinobacteria*, and *Curtobacterium pusillum*. It is the most thermotolerant of the propionibacteria as it can grow and survive in a number of fruit juices; it may therefore survive pasteurization processes used in the fruit juice industry (Kusano et al., 1997; Walker and Phillips, 2007).

8. **Propionibacterium granulosum** (Prévot 1938) Moore and Holdeman 1970, 15[AL] ("*Corynebacterium granulosum*" Prévot 1938)

gra.nu.lo'sum. L. n. *granulum* a small grain; L. neut. suff. *-osum* suffix meaning full of; N.L. neut. adj. *granulosum* full of granules.

Description based largely on a study of the reference strains and 36 other strains by Holdeman et al. and of 30 additional strains originally isolated by Charles Evans, University of Washington, Seattle, and examined by Cummins and Johnson (1986). Surface colonies on suitable media up to 1 mm at 2–3 d, generally white or grayish, smooth, circular, entire, usually larger and more whitish than colonies of *Propionibacterium acnes*. Glucose broth cultures are generally turbid with a rather coarsely granular deposit; often rather viscid and difficult to centrifuge cleanly. Generally nonhemolytic on sheep, horse, or rabbit blood, although reported β-hemolytic on rabbit blood by Hoeffler (1977). Very poor or no growth under aerobic conditions. Catalase-positive. Peptidoglycan amino acid residues are alanine, glutamic acid, glycine, and LL-DAP. Cell-wall sugar components are galactose, mannose, and glucosamine (Johnson and Cummins, 1972). Nutritional requirements are generally the same as *Propionibacterium acnes* requiring pantothenate, although some strains appear to require additional unidentified factors (Ferguson and Cummins, 1978).

Most strains had an active deoxyribonuclease andlecithinase; chondroitin sulfatase, hyaluronidase, and phosphates activity was detected in a few strains, but none produced gelatinase (Hoeffler, 1977). *Propionibacterium granulosum* has a lipase which is considerably more active than that of *Propionibacterium acnes* (Greenman et al., 1981).

Source: sebum-rich oily areas of the skin but in significantly smaller numbers than *Propionibacterium acnes* (McGinley et al., 1978). *Propionibacterium granulosum* is found along with *Propionibacterium acnes* in acne comedones and may play some part in the pathogenesis of acne. It is especially common in the region of the alae nasi (McGinley et al., 1978). The type strains was isolated as a contaminant from a culture labeled *Staphylococcus aerogenes*.

DNA G+C content (mol%): 63 (T_m).

Type strain: ATCC 25564, CCUG 32987, CIP 103262, DSM 20700, LMG 16726, NCTC 11865, VPI 507.

Sequence accession no. (16S rRNA gene): AJ003057.

Further comments: member of the "cutaneous" group of propionibacteria. *Propionibacterium granulosum* seems to have been first separated from *Propionibacterium acnes* (although those names were not used) by Brzin (1964) who recognized a group of strains that did not produce indole or reduce nitrate but could ferment sucrose and maltose. This distinction was later confirmed by Voss (1970) who showed that the indole-negative, nitrate-negative strains were serologically distinct from typical "*acnes*" strains. On the basis of acid extracts containing cell-wall polysaccharide, all strains belong to a single type, with little or no cross-reaction against polysaccharides from *Propionibacterium acnes* or *Propionibacterium avidum* (Cummins, 1975). However, Hoeffler et al. (1977) have found at least three types by tube or slide agglutination of intact suspensions which points to the existence of several surface antigens. Strains of *Propionibacterium granulosum* show low DNA similarity (~12–15%) to both *Propionibacterium acnes* and *Propionibacterium avidum* which is similar to the level of relationship that is found between *Propionibacterium acnes* and the "classical" propionibacteria (Johnson and Cummins, 1972). This is further illustrated by 16S rRNA gene sequence analysis which shows *Propionibacterium granulosum* more closely related to the "classical" propionibacteria than to *Propionibacterium acnes* or *Propionibacterium avidum* (Figure 232). Adjuvent therapy using *Propionibacterium granulosum* strain KP-45 resulted in a significant improvement in the survival of patients with stage I and stage II colorectal cancer (Isenberg et al., 1994). Antibiotic susceptibility studies with a septicemia isolate demonstrated sensitivity to penicillin G, ampicillin, erythromycin, cefotaxime, clindamycin, rifampin, and imipenem, but resistance to metronidazole, sulfonamide, fosfomycin, and tobramycin (Branger et al., 1987). Studies with an isolate responsible for endocarditis revealed susceptibility to penicillin, clindamycin, augmentin, piperacillin, chloramphenicol, cefotaxime, and cefazolin, but resistance to metronidazole (Chaudhry et al., 2000). As with other "cutaneous" propionibacteria, erythromycin-resistant strains of *Propionibacterium granulosum* have been isolated from the skin of acne patients treated with oral erythromycin and topical clindamycin (Eady et al., 1989).

9. **Propionibacterium jensenii** van Niel 1928, 163[AL] ("*Bacterium acidi propionici b*" von Freudenreich and Orla-Jensen 1906; "*Propionibacterium jensenii* var. *raffinosaceum*" van Neil 1928; "*Propionibacterium peterssonii*" van Niel 1928; "*Propionibacterium technicum*" van Niel 1928; "*Propionibacterium raffinosaceum*" Werkman and Kendall 1931; "*Propionibacterium zeae*" Hitchner 1932)

jen.se'ni.i. N.L. gen. masc. n. *jensenii* of Jensen; named for Sigurd Orla-Jensen (1870–1949), the Danish bacteriologist who first isolated this organism.

Description based on literature descriptions, including those of Werkman and Brown (1933), Sakaguchi et al. (1941), and Janoschek (1944) and on the study of 13 strains including 3 of *Propionibacterium jensenii* [ATCC 4867 (van Neil 24), ATCC 4868 (van Neil 29), and ATCC 4869 (van Neil 1)]; 2 of "*Propionibacterium technicum*" [ATCC 14073 (van Neil E.6.1) and ISL 106 (van Neil 22)]; 1 of "*Propionibacterium raffinosaceu*" [ISL 103 (van Neil 29)]; 1 of "*Propionibacterium peterssoni*" [ATCC 4870 (van Neil 20)]; 1 of "*Propionibacterium zeae*" [(ATCC 4964 (Hitchner)]; and ATCC 4871. Phenotypic characteristics of these strains are similar to original descriptions of *Propionibacterium jensenii*, "*Propionibacterium raffinosaceum*", "*Propionibacterium technicum*", "*Propionibacterium peterssonii*", or "*Propionibacterium zeae*".

Surface colonies on horse blood agar (2 d) are punctiform, circular, entire, convex, glistening, semiopaque, and white, cream, or pink. Colonies in deep agar are 4 mm or less, lenticular and white, pink or red-brown.

Glucose broth cultures are turbid or clear with smooth, granular or ropy sediment and terminal pH of 4.4–4.9.

Only strains producing the red polyene pigment granadaene have been shown to be hemolytic (Vedamuthu et al., 1971). Twenty pigmented strains produced β-hemolytic activity on horse blood agar. The hemolytic system of *Propionibacterium jensii* appears similar to that of *Streptococcus agalactiae* (Vanberg et al., 2007).

Some strains grow as well aerobically as anaerobically. Variable catalase activity.

Requires pantothenate and biotin for growth. Some require *p*-aminobenzoic acid while for others it is stimulatory. Thiamine is also stimulatory but not essential (Delwiche, 1949).

Peptidoglycan contains alanine, glutamic acid, glycine, and LL-DAP as diamino acid; cell-wall sugars are glucose and galactose, usually with small amounts of mannose.

Source: dairy products and silage.

DNA G+C content (mol%): 65–68 (T_m).

Type strain: ATCC 4868, CCUG 48883, CIP 103028, DSM 20535.

Sequence accession no. (16S rRNA gene): AJ704571, X53219.

Further comments: member of the "classical" or "dairy" group of propionibacteria. On the basis of 16S rRNA gene sequence analysis, *Propionibacterium jensenii* clusters with *Propionibacterium thoenii* (Figure 232). Some preliminary data suggest that the *Propionibacterium jensenii* strain PJ702 may have application as a living vaccine vector for tuberculosis and other mucosally transmitted diseases (Adams et al., 2005).

10. **Propionibacterium microaerophilum** Koussémon, Combet-Blanc, Patel, Cayol, Thomas, Garcia and Ollivier 2001, 1380[VP]

mi.cro.a.e.ro'phi.lum. Gr. adj. *mikros* small, Gr. n. *aer* air; Gr. adj. *philos -ê -on* loving; N.L. neut. adj. *microaerophilum* slightly air-loving.

Description based on the literature description of the type strain M5[T] (Koussémon et al., 2001).

Surface colonies on yeast extract minimal salts basal medium (YEM; prepared and stored using oxygen-free nitrogen) agar prepared under microaerophilic conditions (95:5% N_2/O_2) at 30°C are white after 1 week and lens-shaped with smooth edges and 2–3 mm in diameter.

Growth pH range 4.5–9.5; optimum 7.0. Mesophilic, with an optimum growth temperature of 30°C; range 20–45°C. Killing was observed at 80°C.

Growth occurs under anaerobic and aerobic conditions with an optimum O_2 concentration of 5% in the gas phase of the culture. Catalase-negative. Nitrate is reduced to nitrogen rather than nitrite.

The type strain M5 does not require growth factors such as pantothenate, biotin, or thiamin and is, therefore, prototrophic. This is in contrast to other propionibacteria which display auxotrophic behavior.

Source: decantation reservoir of olive mill wastewater.

DNA G+C content (mol%): 67.7 (T_m).

Type strain: M5, CNCM I-2360, DSM13435.

Sequence accession no. (16S rRNA gene): AJ234623.

Further comments: on the basis of 16S rRNA gene sequence analysis, *Propionibacterium microaerophilum* clusters with the "classical" propionibacteria and is most closely related to *Propionibacterium acidipropionici* (97.5% similarity) (Figure 232). Despite this, DNA–DNA hybridization experiments with *Propionibacterium acidipropionici* DSM 4900 only showed a reassociation value of 56.2%. Originally isolated from a decantation reservoir of olive mill wastewater that is stored to enable natural bioremediation prior to river disposal. Isolated by enrichment in deep minimal salts glucose basal medium agar (0.6 g KH_2PO_4, 0.2 g $MgCl_2·6H_2O$, 1 g NII_2%Cl, 1 g NaCl, and 10 ml trace element solution supplemented with 0.1 g yeast extract L <MIN>1 and 10 mM glucose) prepared under anaerobic conditions (Koussémon et al., 2001). This species is unlike other members of the genus as growth is sustained on a minimal salts glucose medium with ammonium salts as the only nitrogen source, although yeast extract enhances growth.

11. **Propionibacterium propionicum** corrig. (Buchanan and Pine 1962) Charfreitag, Collins and Stackebrandt 1988, 357[VP] ("*Actinomyces propionicus*" Buchanan and Pine 1962)

pro.pio'ni.cum. N.L. n. *acidum propionicum* propionic acid; N.L. neut. adj. *propionicum* pertaining to propionic acid.

Previously *Actinomyces propionicus* (Buchanan and Pine, 1962) and *Arachnia propionica* (Pine and Georg, 1969; Schaal, 1986). Original spelling of the specific epithet, "propionicus" was corrected by Moore and Moore (1992).

Short irregular rods, 0.2–0.3 µm in diameter and 3–5 µm long, which may or may not be branched, as well as slender branching filaments 5–20 µm in length or longer (Figure 236). These filaments are especially evident in young liquid media cultures but commonly break up in older cultures into short rods (Figure 237). Rods are of variable length, often with clubbed ends and are commonly arranged in pairs, in Y or V configurations, or in parallel rows forming palisades. Gram stain may be uneven, with a beaded appearance. Swollen spherical cells resembling sphaeroplasts, up to 5–20 µm in diameter, are formed by some strains.

Cellular morphology varies with sugar substrate and growth conditions. Colony appearance may be variable; off-white to buff, breadcrumb, gritty, pitting or smooth, convex, entire edged. Young microcolonies are commonly composed of tangled filaments; older colonies may be heaped or convoluted with an undulant edge. Both types (rough and smooth) may occur together. Microscopically, colonies are mycelial with long hyphal branched elements but with no aerial filaments and can resemble *Actinomyces israelii*. Red fluorescence observed under ultraviolet light (365 nm) on blood agar.

Hemolytic on human blood agar, variable hemolysis on horse blood agar, and non-hemolytic on sheep blood agar.

Grows in pure culture both aerobically and anaerobically. Increased CO_2 is reported as a requirement for growth anaerobically and to enhance growth aerobically by Schofield and Schaal (1981) whereas Buchanan and Pine (1962) report that CO_2 does not stimulate growth in broth culture. Obligately anaerobic and slow-growing on primary isolation from clinical samples, requiring up to 14 d incubation (Hall, 2006). Catalase-negative.

Peptidoglycan contains LL-DAP as the dibasic amino acid. Two distinct cell-wall sugar patterns have been described. Strains contain essentially only galactose or galactose and glucose with trace amounts of mannose, but not rhamnose or 6-deoxylatose.

Little or no proteoytic activity. API tests for hyaluronidase, esterase (C_4), esterase lipase (C_8), leucine arylamidase, and α-glucosidase are positive for all or the majority of isolates. Alkaline and acid phophatases, lipase (C_{14}), valine and cystine arylamidase, phosphoamidase, β-glucouronidase, β-glucosidase, α-mannosidase, and α-fucosidase are negative. Chymotrypsin, α-galactosidase, β-galactosidase, and *N*-acetyl-β-glucosamine are variable with ~10–50% isolates positive (Kilian, 1978; Schofield and Schaal, 1981).

Source: isolated from the oral cavity of humans including the mucosa, tonsils, and dental plaque, occasionally isolated from the gastrointestinal tract and female genital tract. It seems unlikely to represent part of the indigeous microbiotia. It may also be present in the oral cavity of other animals (Bowden, 1991). Isolated from lacrimal canaliculitis (inflammation of the short passage which drains tears form the lacrimal lake to the lacrimal sac).

DNA G+C content (mol%): 63–65 mol% (T_m).

Type strain: ATCC 14157, CCUG 4939, CIP 101941, DSM 43307, JCM 5830, NBRC 14587, NCTC 12967, VKM Ac-1449.

Sequence accession no. (16S rRNA gene): AJ003058, AJ315953, X53216.

Further comments: description based largely on the reference strain and eight other strains (Buchanan and Pine, 1962; Charfreitag et al., 1988; Cummins and Moss, 1990; Johnson and Cummins, 1972; Schofield and Schaal, 1981) and numerous clinical isolates (Hall, 2006). Two distinct serovars, based on fluorescent antibody labeling, have been described: serovar 1 (type strain ATCC 14157) and 2 (ATCC 29326, WVU† 346, F. Lentze strain "Fleischmann": VPI5067, CDCW904; isolated from a typical case of human actinomycosis in Germany), (Gerencser and Slack, 1967; Holmberg and Forsum, 1973). These serotypes also form separate clusters on numerical phenetic analysis (Schofield and Schaal, 1981). To date, there is no published information relating to any phylogenetic differences.

On the basis of 16S rRNA gene sequence analysis, *Propionibacterium propionicum* clusters with *Propionibacterium avidum* (Figure 232). Microscopically, the bacterium can be mistaken for *Streptococcus* species in clinical material as it can form chains of cells with coccoid appearance. Recognized cause of actinomycosis, a chronic granulomatous disease which commonly affects the cervicofacial area but may also cause infection of, for example, the thorax, abdomen, or pelvis. The organism is sometimes associated with the use of intra-uterine contraceptive devices. (Hall, 2006). Most commonly associated with infections of the lacrimal apparatus such as lacrimal canaliculitis (Brazier and Hall, 1993). It is not possible to clinically distinguish between *Actinomyces* species and *Propionibacterium propionicum* as the underlying cause of these diseases. *Propionibacterium propionicum* is morphologically similar to *Actinomyces israelii* but a distinguishing feature is propionic acid production from glucose metabolism. Long-term, high dose antibiotic therapy is frequently necessary; late diagnosis and insufficient treatment can lead to fatality (Hall, 2006). *Propionibacterium propionicum* is also associated with primary and persistent ondodontic infection (Siqueira and Rocas, 2003). Johnson and Cummins (1972) describe VPI 5077 as an unidentifiable isolate with 1% DNA similarity to ATCC 14157 (Type 1 *Propionibacterium propionicum*) and no antigenic cross-reactivity, received in their laboratory as "*Arachnia propoionica*". In the results, this isolate is listed as VPI 5067, a synonym for ATCC 29326: WVU 346 (F Lentz strain "Fleishman") reported by Gerencser and Slack (1967) as Type 2 "*Actinomyces propionicus*". Subsequent publications have quoted the results obtained by Johnson and Cummins (1972) as relating to VPI 5067. It is not known if the tabular designation VPI5067 in Johnson and Cummins (1972) is an error. Most isolates are sensitive to a wide selection of antibiotics including penicillins, erythromycin, clindamycin, imipenem cephalosporins, chloramphenicol, and tetracyclines. Resistant to metronidazole, fluoroquinolones, aztreonam, and aminoglycosides.

12. **Propionibacterium thoenii** van Niel 1928, 164[AL] ("*Bacterium acidi propionici* var. *rubrum*" Thöni and Allemann 1910; "*Propionibacterium rubrum*" van Niel 1928)

tho.e'ni.i. N.L. gen. masc. n. *thoenii* of Thöni; named for J. Thöni, the Swedish bacteriologist who first isolated this organism.

Description based on a study of the type strain and ATCC 4871 ("*Propionibacterium rubrum*" van Niel 23) and ATCC 4872 ("*Propionibacterium rubrum*" van Niel 19).

Surface colonies at 4 d are circular, entire, smooth, and generally orange or red-brown.

Broth cultures show generalized turbidity with brownish-red or orange-red deposit; terminal pH in glucose broth is 4.7–4.9.

Shows β-hemolysis on blood agar containing human, bovine, equine, sheep, rabbit or pig blood.

Less strictly anaerobic than the type species (van Niel, 1957). Catalase-positive.

Pantothenate and biotin are essential for growth. Thiamine is required by some strains and is stimulatory to others, while *p*-aminobenzoic acid is not required (Delwiche, 1949).

Peptidoglycan contains alanine, glutamic acid, glycine, and LL-DAP as diamino acid, and cell-wall sugars are glucose and galactose.

Source: originally isolated from red spots in Emmentaler cheese (Thöni and Alleman, 1910), also found in other dairy products.

DNA G+C content (mol%): 66–67 (T_m).

Type strain: ATCC 4874, CCUG 28149, CIP 103029, DSM 20276, HAMBI 247, JCM 6437, LMG 16455.

Sequence accession no. (16S rRNA gene): AJ704572, X53220.

Further comments: member of the "classical" or "dairy" group of propionibacteria. On the basis of 16S rRNA gene sequence analysis, *Propionibacterium thoenii* clusters with *Propionibacterium jensenii* (Figure 232). It has been recognized for some time that the strains described under the names *Propionibacterium thoenii* and "*Propionibacterium rubrum*" have many features in common, not least the production of an intense red or reddish brown pigment. The nature of the pigment is not known. Classically, (van Niel, 1957) "*Propionibacterium rubrum*" ferments raffinose and mannitol but not sorbitol, while *Propionibacterium thoenii* ferments sorbitol, but not raffinose or mannitol. However, strains of *Propionibacterium thoenii* and "*Propionibacterium rubrum*" show high DNA similarity (Johnson and Cummins, 1972) and, therefore, it was recommended that they be combined in a single species. Sequence comparison of the complete 16S rRNA genes revealed only 12 bp differences thus confirming the similarity (de Carvalho et al., 1995). Some strains of *Propionibacterium thoenii* have been shown to produce bacteriocins under different growth conditions (Ben-Shushan et al., 2003; Van der Merwe et al., 2004).

Genus II. **Aestuariimicrobium** Jung, Kim, Song, Lee, Oh and Yoon 2007, 2117[VP]

Susmitha Seshadri

A.es.tu.a.ri.i.mi.cro'bi.u.m. L. n. *aesturarium* part of the sea coast which, during the flood-tide, is overflowed, but at the ebb-tide is left covered with mud or slime, a tidal flat; N.L. neut. n. *microbium* microbe; N.L. neut. n. *Aestuariimicrobium* a microbe isolated from a tidal flat.

Short to coccoid nonsporeforming rods. Cells are nonflagellated and stain **Gram-positive**. **Aerobic**. Growth occurs between 4–40°C with **30°C as the optimal temperature**. The optimal pH range for growth is 7.5–8.5. Colonies are pigmented yellow. Nitrate is reduced to nitrite. **Catalase-positive**. The major **menaquinone is MK-9(H_4)** and the **predominant fatty acid is $C_{15:0}$ antesio**.

DNA G+C content (mol%): 68.8–69.2.

Type species: **Aestuariimicrobium kwangyangense** Jung, Kim, Song, Lee, Oh and Yoon 2007, 2117[VP].

Further descriptive information

The only species in this genus was isolated from an oil contaminated tidal flat sediment by enrichment with diesel oil and exhibits diesel oil degradation activity. Phylogenetic studies based upon the 16S rRNA indicate that *Aestuariimicrobium* is closely affiliated to the family *Propionibacteriaceae*. On the basis of phylogenetic, phenotypic, chemotaxonomic and genetic studies, *Aestuariimicrobium* is classified as a novel genus. Cell wall contains LL-diaminopimelic acid.

Enrichment and isolation procedures. Isolation was performed by inoculating 0.5 mg of the oil-contaminated sediment samples in 100 ml of Bushnell–Haas broth with 2% (w/v) diesel oil. Cultures were incubated at 30°C on a horizontal shaker at 150 r.p.m. Four strains R27[T], R44, R45, and R47 with diesel oil degradation activity were isolated by dilution plating on R2A agar.

Maintenance procedures

Routine subculturing is done on R2A agar at 30°C. Can be cultivated in R2A broth without the agar, supplemented with 2% (v/v) Hutner's mineral base and 0.1% (v/v) trace element solution.

Differentiation of the genus *Aestuariimicrobium* from closely related genera

The almost-complete 1472 nucleotide long 16S rRNA gene sequence of the four isolated strains showed low sequence similarity to other genera in the family *Propionibacteriaceae* and form a distinct lineage. *Aestuarimicrobium* is aerobic in comparison to phylogenetically closely related genera such as *Propionibacterium*, *Propioniferax*, *Tessaracoccus* and *Luteococcus*, which are all facultative anaerobes. The four above-mentioned genera differ from *Aestuariimicrobium* in one or more of phenotypic properties such as their shape, temperature and pH requirements for growth, catalase production, and nitrate reduction. *Aestuariimicrobium* and *Tessaracoccus* both contain only $C_{15:0}$ antesio as the major fatty acid while the other genera contain more than one major fatty acid. *Tessaracoccus* contains both MK-9(H_4) and MK-7(H_4) while *Aestuariimicrobium* contains only MK-9(H_4) as the predominant menaquinone.

List of species of the genus *Aestuariimicrobium*

1. **Aestuariimicrobium kwangyangense** Jung, Kim, Song, Lee, Oh and Yoon 2007, 2117[VP]

   kwang.yang.en'se. N.L. neut. adj. *kwangyangense* of or belonging to Kwangyang, Korea, the source of isolation.

   Cells are short rods or cocci (0.6–1.2 × 1.2–2.0 μm). Colonies measure 0.8–1.0 mm after 3 d of growth at 30°C on R2A agar. Colonies are circular, convex, smooth and yellow in color. No growth occurs anaerobically with or without nitrate supplementation on R2A agar. Esculin, casein and Tweens 20, 40, 60, and 80 are hydrolyzed, while starch, hypoxanthine, xanthine and tyrosine are not. D-Glucose, D-fructose, D-galactose, D-cellobiose, D-mannose, D-xylose, sucrose, maltose and salicin, are utilized as sole carbon and energy sources, but L-arabinose, acetate, citrate, succinate, L-malate, formate and L-glutamate are not utilized. Utilization of trehalose, benzoate, and pyruvate are variable. Susceptible to penicillin G, chloramphenicol, ampicillin, cephalothin, novobiocin, tetracycline and carbenicillin, but not to polymyxin B, streptomycin, gentamicin, kanamycin, lincomycin, oleandomycin or neomycin. The API ZYM system detects the presence of esterase (C4), esterase lipase

(C8), leucine, arylamidase, acid phosphatase, α-galactosidase, β-galactosidase, α-glucosidase and β-glucosidase but not alkaline phosphatase, lipase (C14), valine arylamidase, cystine arylamidase, trypsin, α-chymotrypsin, *N*-acetyl-β-glucosaminidase, α-mannosidase and α-fucosidase. Naphthol-AS-BI-phosphohydrolase and β glucuronidase activities are negative for the type strain, but other strains are positive.

DNA G+C content (mol%): 68.8–69.2 (HPLC).

Type strain: R27, KCTC 19182, JCM 14204.

Sequence accession no. (16S rRNA gene): DQ830982.1.

Genus III. **Brooklawnia** Rainey, da Costa and Moe 2006, 1981[VP] (*in* Bae, Moe, Yan, Tiago, da Costa and Rainey 2006b)

Milton S. da Costa, Fred A. Rainey and William M. Moe

Brook.law'ni.a N.L. fem. n. *Brooklawnia* named after Brooklawn, the contaminated site from which members of the genus were first isolated.

Pleomorphic rods. Colonies are white. Nonsporeforming. Gram-stain-positive. Strains are **nonmotile**. **Mesophilic** with a temperature optimum for growth of about 37°C; neutrophilic with an optimum pH for growth of about 6.5. **Menaquinone 9(H_4)** is the predominant respiratory quinone. Fatty acids are primarily **iso-** and **anteiso-branched**; straight-chain fatty acids are also present. Cell-wall peptidoglycan contains *meso*-DAP (A1γ). Cells are catalase-positive and oxidase-negative, and nitrate is not reduced. **Facultatively anaerobic**. **Chemoheterotrophic**. Propionate and acetate are the predominant products of glucose fermentation. Strains of this genus were isolated from groundwater contaminated with chlorosolvents.

DNA G+C content (mol%): 67.5–67.9.

Type species: **Brooklawnia cerclae** Rainey, da Costa and Moe 2006, 1981[VP] (*in* Bae, Moe, Yan, Tiago, da Costa and Rainey 2006b).

Further descriptive information

One species, *Brooklawnia cerclae,* is currently classified in this genus (Bae et al., 2006b). The isolates of this species form colorless colonies that are about 1–3 mm in diameter under anaerobic growth conditions. The organisms are able to grow under aerobic and anaerobic conditions, but growth is better under an atmosphere of H_2 (10%), CO_2 (10%), and N_2 (80%) or under CO_2 (5%) and N_2 (95%).

Strains BL-34[T] and BL-35 are chemoheterotrophic, growing on a variety of complex media, namely Plate count Agar, Colombia Anaerobic Sheep Blood, and Peptone/Yeast Extract/Glucose (PYG) medium. Several carbon sources are assimilated, among them carbohydrates, organic acids, and methanol. The major fermentation products of glucose fermentation are propionate and acetate.

Iso-and anteiso-branched fatty acids are the major fatty acids; $C_{15:0}$ anteiso, and C_{15} iso account for about 84–86% of the total, which is much higher than in other closely related species classified in other genera (Bae et al., 2006b). The major respiratory quinone is MK-9(H_4) like those of other closely related organisms. Strains BL-34[T] and BL-35 are not able to assimilate or transform 1,2-dichloroethane (1,2-DCA) or 1,1,2-trichloroethane (1,1,2-TCA) despite being isolated from water containing high levels of these chlorosolvents. However, they are extremely resistant to these solvents and able to grow in media with solvent concentrations of up to 9.8 mM. The resistance to chlorosolvents may explain their ability to colonize and their subsequent isolation from the Brooklawn site where petrochemicals were disposed and where 1,1,2,2-tetrachloroethane, 1,1,2-trichloroethane, 1,2-dichloroethane, 1,2-dichloropropane, hexachloro-1,3-butadiene, hexachlorobenzene, vinyl chloride, and polycyclic aromatic hydrocarbons (PAHs) are present in ground water in large quantities (Clement et al., 2002; US EPA, 2005). Another organism isolated from this site, *Propionicicella superfundia* (Bae et al., 2006c), also tolerated high levels of chlorosolvents, but was not able to transform them.

Enrichment, isolation, and growth conditions. The isolates of *Brooklawnia cerclae* were recovered from ground water at the Brooklawn Site (Petro-Processors of Louisiana, Inc. Superfund Site) located near Baton Rouge, LA (USA). Strain BL-34[T] was isolated on Nutrient Agar (Difco) supplemented with 0.5 g/l, L-cysteine, 1.0 mg/l, and resazurin, adjusted to pH 7.0 prior to autoclaving. Incubation was at 30°C in an anaerobic chamber containing a gas mixture of 10% H_2, 10% CO_2, and 80% N_2 (by vol.). Strain BL-35 was isolated on Plate Count Agar (Difco) supplemented with the same amendments and incubated under the same conditions. The strains are maintained on Colombia Anaerobic Sheep Blood agar plates (BBL) or PYG agar plates (Akasaka et al., 2003b).

Maintenance procedures

The strains are maintained on Colombia Anaerobic Sheep Blood agar plates (BBL) or PYG agar plates (Akasaka et al., 2003b). Strains have been successfully maintained long-term in nutrient broth containing 15% (v/v) glycerol at −80°C.

Taxonomic comments

The genus *Brooklawnia* belongs to the family *Propionibacteriaceae* (Delwiche 1957; Stackebrandt et al. 1997) within the order *Propionibacteriales* (formerly suborder *Propionibacterineae*; Stackebrandt et al., 1997), sharing several characteristics with the species of the most closely related genera. The genera that fall within the radiation of the family *Propionibacteriaceae* include *Propionibacterium* Orla-Jensen 1909, *Aestuarimicrobium* Jung et al. 2007, *Friedmanniella* Schumann et al. 1997, *Granulicoccus* Maszenan et al. 2007, *Luteococcus* Tamura et al. 1994, *Microlunatus* Nakamura et al. 1995, *Micropruina* Shintani et al. 2000, *Propioniferax* Yokota et al. 1994, *Propionicicella* Bae et al. 2006c, *Propionicimonas* Akasaka et al. 2003b, *Propionimicrobium* Stackebrandt et al. 2002, and *Tessaracoccus* Maszenan et al. 1999b. The characteristics that can be used to differentiate *Brooklawnia* from the other genera of the family *Propionibacteriaceae* are shown in Table 206. *Brooklawnia* can be differentiated from *Propionimicrobium* (its closest phylogenetic relative based on branching – see

TABLE 206. Characteristics differentiating the genus *Brooklawnia* from other genera of the family *Propionibacteriaceae*[a]

Characteristic	*Brooklawnia*[b]	*Aestuariimicobium*[c]	*Friedmanniella*[d]	*Granulicoccus*[e]	*Luteococcus*[f]	*Microlunatus*[g]	*Micropuina*[h]	*Propionibacterium*[i]	*Propionicicella*[j]	*Propionicimonas*[k]	*Propioniferax*[l]	*Propionimicrobium*[m]	*Tessaracoccus*[n]
Origin	Chlorosolvent contaminated groundwater	Tidal flat sediment	Antarctic sandstone, activated sludge	Phenol degrading aerobic granules	Soil and water, human blood, human peritoneum	Activated sludge, soil	Activated sludge reactor	Dairy products, human sources, bovine lesions, wastewater, spolied orange juice	Chlorosolvent contaminated groundwater	Plant residue in rice field soil	Human epidermis	Human lymph nodes	Activated sludge, marine sediment
Cell morphology	Pleomorphic rods	Short rods, cocci	Cocci, in packets	Cocci	Coccoi, arranged in pairs and tetrads	Cocci	Cocci	Pleomorphic rods, cocci	Rods	Pleomorphic rods	Pleomorphic rods	Pleomor-phic rods	Coccoid, arranged in tetrads
O_2 metabolism	Facultatively anaerobic	Aerobic	Aerobic	Facultatively anaerobic	Facultatively anaerobic	Aerobic	Aerobic	Anaerobic, facultative anaero-bic, aerotolerant	Facultatively anaerobic	Facultatively anaerobic	Facultatively anaerobic	Anaerobic	Facultatively anaerobic
Catalase/ oxidase	+/−	+/−	+/v	−/+	+/+	+/v	+/+	v/v	−/−	−/−	+/+	v/−	+/−
Nitrate reduction	−	+	−	−	−	v	+	v	−	−	+	v	+
DNA G+C content (mol%)	67.5	68.8–69.2	69–74	69	66–68	67.9–70.9	70.5	57–68	69.9	68.7	59–63	53–56	74
Diamino acid in peptidoglycan	*meso*-DAP	LL-DAP	LL-DAP	LL-DAP	LL-DAP	LL-DAP	*meso*-DAP	*meso*-DAP; LL-DAP; *meso*- and LL-DAP	*meso*-DAP	*meso*-DAP	LL-DAP	Lys–Asp	LL-DAP
Murein type	A1γ	nd	A3γ′	A3γ	A3γ′	A3γ′	A1γ	A1γ; A3γ, A3γ′; A1γ and A3γ	nd	nd	A3γ′	A4-α	A3γ′
Major quinone	MK-9(H_4)	MK-9(H_4)	MK-9(H_4) or MK-9(H_2), MK-9(H_2) and MK-7(H_2) or MK-9(H_2) and MK-9(H_2)	MK-9(H_4)	MK-9(H_4)	MK-9(H_4)	MK-9(H_4)	MK-9(H_4)	MK-9	MK-9(H_4); MK-10(H_4)	MK-9(H_4)	MK-9(H_4)	MK-9(H_4), MK-7(H_4)

[a]Symbols: +, >85% positive; −, 0–15% positive; v, variable; nd, not determined.

[b]Data from Bae et al., (2006b).

[c]Data from Jung et al., (2007).

[d]Data from Schumann et al. (1997); Maszenan et al. (1999a); Lawson et al. (2000c).

[e]Data from Maszenan et al. (2007).

[f]Data from Tamura et al. (1994); Collins et al. (2000); Collins et al. (2003).

[g]Data from Nakamura et al. (1995); Cui et al. (2007); Wang et al. (2008).

[h]Data from Shintani et al. (2000).

[i]Data from Stackebrandt and Schaal (2006a).

[j]Data from Bae et al. (2006c).

[k]Data from Akasaka et al. (2003b).

[l]Data from Yokota et al. (1994).

[m]Data from Stackebrandt et al. (2002).

[n]Data from Maszenan et al. (1999b); Lee et al. (2008).

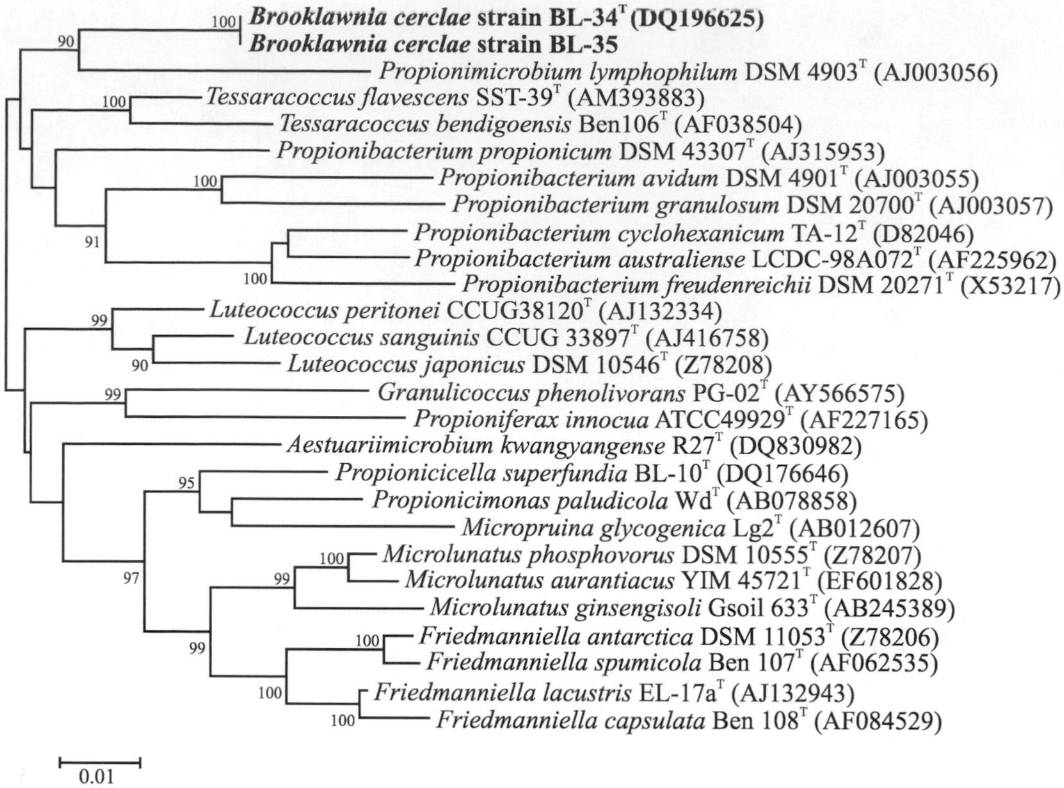

FIGURE 238. 16S rRNA gene sequence based phylogeny indicating the relationship of the genus *Brooklawnia* to other taxa of the family *Propionibacteriaceae*. The scale bar = 1 inferred nucleotide substitution per 100 nucleotides.

Figure 238) on the basis of its cell-wall type and G+C content of the genomic DNA. Differentiation from other genera is, in many cases, based on a combination of chemotaxonomic, physiological, and genomic characteristics (Table 206).

The species *Brooklawnia cerclae* is shown by 16S rRNA gene sequence comparisons to fall within the radiation of the species of the genera of the family *Propionibacteriaceae* (Figure 238). Pairwise sequence similarity values between the 16S rRNA gene sequence of *Brooklawnia cerclae* (DQ196625) and species within the family *Propionibacteriaceae* are in the range 93.5–95.6%. The highest pairwise similarities are to species of the genera *Tessaracoccus* and *Luteococcus* which have short branches as shown in Figure 238. *Brooklawnia cerclae* actually branches with *Propionimicrobium lymphophilum* although the branching is only supported by a bootstrap value of 90%.

List of species of the genus *Brooklawnia*

1. **Brooklawnia cerclae** Rainey, da Costa and Moe 2006, 1981[VP] (*in* Bae, Moe, Yan, Tiago, da Costa and Rainey 2006b)

cer'cla.e. N.L. gen. fem. n. *cerclae* of CERCLA, arbitrary name formed from CERCLA, acronym for Comprehensive Environmental Response, Compensation, and Liability Act, which has mandated cleanup of many hazardous waste sites in the United States.

Cells are Gram-stain-positive, nonmotile, nonspore-forming, pleomorphic rods. Growth occurs at 10°C–40°C; the optimum growth temperature is about 37°C. Growth occurs at pH 4.5–8.0; the optimum pH for growth is 6.5. Growth is not stimulated by addition of NaCl, but is sustained in the presence of up to 3% NaCl (v/v). Facultative anaerobic growth is supported by fermentation. Propionate and acetate are the main products of glucose fermentation. Chemotaxonomic features are the same as the genus description. Growth occurs on arabinose, fructose, glucose, galactose, maltose, rhamnose, xylose, ribose, mannose, starch, glycogen, glycerol, mannitol, lactate, and pyruvate, but not on acetate, ribitol, cellobiose, cellulose, galactitol, erythritol, ethanol, fucose, fumarate, inositol, lactose, malate, methanol, raffinose, sorbitol, succinate, sucrose, or xylan. Strains of this species have been isolated from chlorosolvent-contaminated ground water. Strain BL-35 (LMG 23249, NRRL B-41419) also belongs to this species.

DNA G+C content (mol%) of the type strain: 67.5 (HPLC).

Type strain: BL-34, JCM 14918, LMG 23248, NRRL B-41418.

Sequence accession no. (16S rRNA gene): DQ196625.

Genus IV. **Friedmanniella** Schumann, Prauser, Rainey, Stackebrandt and Hirsch 1997, 282[VP]

Peter Schumann and Rüdiger Pukall

Fri.ed.man.ni.el'la. N.L. fem. dim. n. *Friedmanniella* named after E. Imre Friedmann (1921–2007), an American microbiologist, in recognition of his contributions to Antarctic microbiology.

Spherical to ellipsoidal cells that occur mostly in more or less regular packets which develop through cell division in three perpendicular planes. The packets aggregate, forming clusters. **Gram-stain-positive**. **Nonmotile**. Nonsporeforming. Non-acid-fast. **Strictly aerobic**. **Oxidase-negative**. **Catalase-positive**. Chemo-organotrophic. Only a few carbohydrates, organic acids, and other carbon sources are metabolized. The **peptidoglycan type** is **A3γ** containing **LL-diaminopimelic acid**, glycine in position 1 of the peptide subunit, and a single glycine residue as interpeptide bridge. The major **menaquinone** is **MK-9(H$_4$)**, and the main **fatty acids** are 12- and 13-methyltetradecanoic acid (**$C_{15:0}$ anteiso and $C_{15:0}$ iso**). Mycolic acids are absent. The **phospholipid pattern** includes **phosphatidylglycerol, diphosphatidylglycerol, phosphatidylinositol**, and one unknown phospholipid.

Phylogenetically, this genus is affiliated to the family *Propionibacteriaceae* Delwiche 1957, emend. Rainey, Ward-Rainey and Stackebrandt 1997 of the order *Propionibacteriales* Zhi et al. 2009.

DNA G+C content (mol%): 69–74.

Type species: **Friedmanniella antarctica** Schumann, Prauser, Rainey, Stackebrandt and Hirsch 1997, 282[VP].

Further descriptive information

The four species of the genus *Friedmanniella*, *Friedmanniella antarctica* (Schumann et al., 1997), *Friedmanniella capsulata* (Maszenan et al., 1999a), *Friedmanniella lacustris* (Lawson et al., 2000a, 2000b) and *Friedmanniella spumicola* (Maszenan et al., 1999a), form a distinct cluster together with the members of the genera *Microlunatus* (Nakamura et al., 1995), *Micropruina* (Shintani et al., 2000), *Propionicicella* (Bae et al., 2006c), and *Propionicimonas* (Akasaka et al., 2003b) within the phylogenetic tree based on 16S rRNA gene sequences of the suborder *Propionibacterinae* (Figure 239). However, the affiliation of these genera to a family is the subject of conflicting taxonomic opinions (Garrity et al., 2007; Stackebrandt and Schaal, 2006b). For

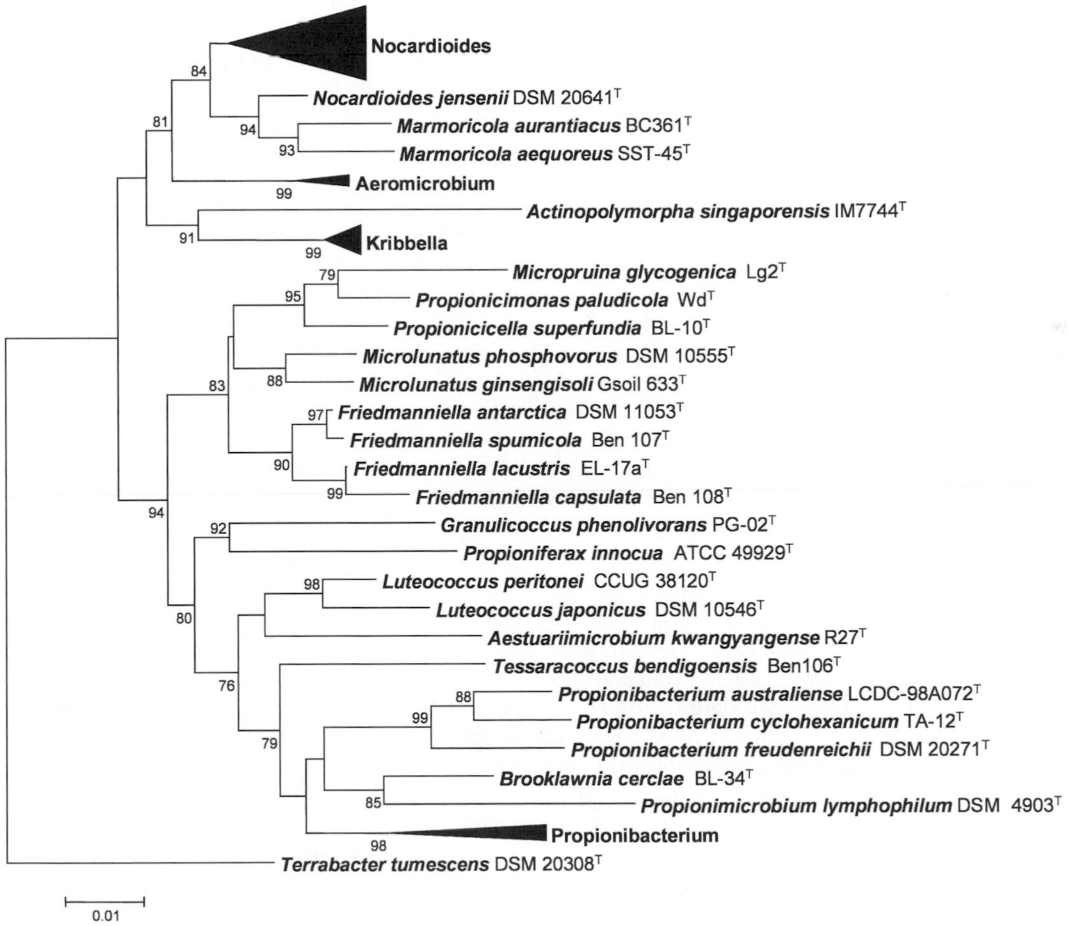

FIGURE 239. Neighbor joining analysis of almost complete 16S rRNA gene sequences from type strains of the genus *Friedmanniella* and other representatives of the *Actinobacteria*. The phylogenetic tree was conducted using Mega 3.1 (Kumar et al., 2004). The Kimura-2-parameter method was used for correction (Kimura, 1980). Bootstrap values were calculated from 1000 resamplings, but values >70% are indicated at branching points only. Bar = 1 subsitution per 100 nucleotides.

reasons given in *Taxonomic comments*, the genus *Friedmanniella* is considered a member of the family *Propionibacteriaceae* in this chapter. The type species with the highest 16S rRNA gene sequence similarity to *Friedmanniella antarctica* are *Microlunatus phosphovorus* (Nakamura et al., 1995), *Propionicicella superfundia* (Bae et al., 2006c), *Propionicimonas paludicola* (Akasaka et al., 2003b), *Aestuariimicrobium kwangyangense* (Jung et al., 2007), *Micropruina glycogenica* (Shintani et al., 2000), and *Brooklawnia cerclae* (Bae et al., 2006b) with values of 94.6%, 94.1%, 94.1%, 93.9%, 93.5%, and 93.6%, respectively. The type species *Friedmanniella antarctica* shows 16S rRNA gene sequence similarity values of 98.8%, 97.3%, and 96.5% to *Friedmanniella spumicola*, *Friedmanniella lacustris*, and *Friedmanniella capsulata*, respectively (all binary 16S rRNA gene sequence similarity values were calculated by the EzTaxon server, Chun et al., 2007). The DNA–DNA similarity values of type strains of *Friedmanniella spumicola* and *Friedmanniella capsulata* to *Friedmanniella antarctica* are 50% and 27%, respectively. *Friedmanniella spumicola* and *Friedmanniella capsulata* are related by more than 97% 16S rRNA gene sequence similarity but show only 29% DNA–DNA similarity (Maszenan et al., 1999a).

Members of the genus *Friedmanniella* are Gram-stain-positive, nonmotile cocci (usually 1.2–1.5 µm in diameter) or cells of nearly spherical shape which show a pronounced tendency to aggregate in clusters of packets which result from cell divisions in three perpendicular planes. The clusters are surrounded and internally subdivided by an extracellular capsular polymer (Maszenan et al., 1999a). Well-developed colonies are pigmented in shades of orange which may become more intense under the influence of light (Schumann et al., 1997). Ageing colonies change in color from orange to faint yellow.

The development of colonies of *Friedmanniella* strains on all tested agar media is usually slow and takes several days or up to 3 weeks. Growth in submerged shaking and standing cultures is often poor. *Friedmanniella lacustris* requires the vitamins biotin, thiamine, and nicotinic acid and grows optimally with 4% (w/v) NaCl. Medium PYGV (DSMZ medium 621; Staley, 1968) supplemented with artificial sea water and vitamins is recommended for cultivation of *Friedmanniella lacustris.* The details for preparation of this medium are given by Stackebrandt and Schaal (2006b). R-Medium (Yamada and Komagata, 1972) or modified organic medium 79 [containing per liter distilled water 10 g glucose, 10 g bacto peptone (Difco), 2 g Casamino acids (Difco), 2 g yeast extract (Serva), 6 g NaCl, and 15 g agar, pH 7.5] can be recommended for the cultivation of *Friedmanniella antarctica.* R2A agar [DSMZ medium 830, containing per liter distilled water 0.50 g yeast extract, 0.50 g proteose peptone (Difco no. 3), 0.50 g Casamino acids (Difco), 0.50 g glucose, 0.50 g soluble starch, 0.30 g Na pyruvate, 0.30 g K_2HPO_4, 0.05 g $MgSO_4 \cdot 7H_2O$, 15.00 g agar – adjust to pH 7.2 with crystalline K_2HPO_4 or KH_2PO_4 before adding agar] is suited for the cultivation of *Friedmanniella capsulata* and *Friedmanniella spumicola.*

Biochemical and physiological characteristics are given in the species descriptions and in Table 207. *Friedmanniella* strains are capable of polyphosphate accumulation (*Friedmanniella lacustris* has not been tested). *Friedmanniella lacustris, Friedmanniella capsulata,* and *Friedmanniella spumicola* are sensitive to penicillin; the latter two species were found also to be susceptible to chloramphenicol and vancomycin (Lawson et al., 2000a; Maszenan et al., 1999a). No data are available on the antibiotic sensitivity of *Friedmanniella antarctica.*

All members of the genus *Friedmanniella* exhibit the peptidoglycan type A3γ′ (A42.1, http://www.peptidoglycan-types.info) based on LL-diaminopimelic acid (LL-A_2pm), in which one glycine residue is included in the interpeptide bridge and another one is found at position 1 of the peptide subunit. The same peptidoglycan structure is found in the following members of the family *Propionibacteriaceae*: *Microlunatus* (Schumann et al., 1997), *Tessaracoccus* (Maszenan et al., 1999b), and *Propionibacterium propionicus* (Weiss et al., 1981). The predominating menaquinone is MK-9(H_4), while MK-7(H_4), MK-8(H_4), MK-7(H_2), and MK-9(H_2) may occur as minor components. $C_{15:0}$ anteiso exceeds the value of 50% in the cellular fatty acid profiles of all *Friedmanniella* strains, followed by $C_{15:0}$ iso (13.0–35.8%) and $C_{14:0}$ iso as minor component (1.8–3.9%). The polar lipids are phosphatidylglycerol, diphosphatidylglycerol, phosphatidylinositol, and one unidentified phospholipid. An unknown glycolipid additionally occurs in *Friedmanniella capsulata* and *Friedmanniella spumicola. Friedmanniella antartica* contains spermidine and spermine and corresponds in this polyamine pattern to *Microlunatus, Luteococcus,* and *Propioniferax* as members of the family *Propionibacteriacea* but differs from all tested species of the genera *Nocardioides* and *Aeromicrobium* of the family *Nocardioidaceae* which contains mainly cadaverine (Busse and Schumann, 1999).

Friedmanniella antarctica and *Friedmanniella lacustris* originate from Antarctica. *Friedmanniella antarctica* was isolated from a sandstone sample containing a cryptoendolithic microbial community from the Linnaeus Terrace (1600 m above ocean level), McMurdo Dry Valleys, Asgard Range of the Transarctic Mountains. *Friedmanniella lacustris* was isolated from a water sample taken at 1 m depth from the hypersaline and meromictic Ekho Lake in the ice-free area of the Vestfold Hills in East Antarctica. *Friedmanniella spumicola* was isolated from a stable surface foam of the aerobic reactor in an activated sludge plant treating mainly wastewater from an orange-juice-processing plant in Mildura, Victoria, Australia. The type strain of *Friedmanniella capsulata* was obtained from an activated sludge biomass sample from Haman Island, Queensland, Australia. Only a small number of additional strains related to *Friedmanniella spumicola* or *Friedmanniella antarctica* has been isolated from environmental habitats. *Friedmanniella* sp. Ellin 163 was isolated from pasture soil by using the liquid serial dilution culture method as described by Schoenborn et al. (2004). The 16S rRNA gene sequence of the isolate shows 99% similarity with *Friedmanniella spumicola,* but is also related to not yet cultivatable bacteria whose sequences have been submitted to EMBL. For instance, *Friedmanniella* sp. Ellin 163 shares 98–99% 16S rRNA gene sequence similarity with uncultured bacteria detected by analysis of the bacterial diversity in indoor dust (Rintala et al. not yet published; e.g., accession number AM697104) and also to sequences derived from analysis of clone libraries originating from metagenomes of endophytes and symbionts enriched from stem bark of the spurge *Trewia pudiflora.* Sequences related to *Friedmanniella spumicola* were also found in metagenomic analyses from soil as reported by Fierer et al. (2007). A *Friedmanniella spumicola*-like strain was isolated from pasture soil of perennial ryegrass (*Lolium perene*) and white clover (*Trifolium repens*) (Joseph et al., 2003), but also from 12,000 years old glacial ice, Sojana, Bolivia (Christner, 2002). Partial sequence analysis from airborne bacterial populations (Fierer et al., 2008) led to the detection of a *Friedmanniella antarctica*-related strain (97% sequence similarity). Further 16S rRNA gene sequences from uncultured *Friedmanniella*-like bacteria were detected in soil under eucalyptus trees (Silveira et al., 2006) and bronchoalveolar lavage fluid from children with cystic fibrosis (Harris et al., 2007) showing similarity values of 98% to *Friedmanniella capsulata* and 97% to *Friedmanniella antarctica*.

Isolation and maintenance procedures

Friedmanniella antartica was isolated by sprinkling of loosened sandstone material onto PYGV agar (pH 6.9). Colonies developed around sand grains after incubation for 5 months at 9°C in dim light were picked and streaked onto the same medium. Resulting single colonies were subcultured after

TABLE 207. Characteristics differentiating *Friedmanniella* and *Microlunatus* species[a]

Characteristic	*F. antarctica*[b]	*F. capsulata*[c]	*F. lacustris*[d]	*F. spumicola*[c]	*M. phosphovorus*[e]	*M. ginsengisoli*[f]
Storage products	Polyphosphate	Polyphosphate	No intracellular granules	Polyphosphate	Polyphosphate	nt
NaCl tolerance range (%)	0–2	Inhibition	0–6	Inhibition	0–6	0–5
H_2S production	+	+	–	+	nt	–
Hydrolysis of gelatin	–	–	+	–	nt	+
Nitrate reduction to NO_2^-	–	–	+	–	+ (anaerobically)	+
Carbon source:						
Acetate	–	–	+	–	+	–
N-Acetyl-L-glutamate	nt	–	+	–	nt	nt
Alaninamide	–	–	+	+	nt	nt
Alanine	–	+	–	–	–	+
L-Alanylglycine, 2,3-butanediol, inosine, glucose-6-phosphate	–	+	–	–		
Asparagine,	–	+	–	–	+	+
β-Cyclodextrin	–	+	–	–	nt	nt
Dextrin	–	+	–	+	nt	nt
L-Fructose	–	+	+	–	nt	nt
Galactose	–	–	+	+	+	+
Gentiobiose	–	–	+	–	nt	nt
Glucose, mannose, raffinose	–	–	+	–	+	+
Glycogen	–	+	–	+	–	+
γ-Hydroxybutyrate	nt	–	+	–	nt	–
Inositol	–	–	+	–	+	–
Lactamide/succinamate	–	+	–	–	nt	nt
D-Lactate methylester	nt	+	–	+	nt	nt
Lactose	–	–	+	–	–	+
Malate	–	+	+	–	–	+
Maltose	–	+	+	–/+[g]	+	+
Maltotriose	nt	+	+	–	nt	nt
D-Mannitol	–	+	+	+	+	–
D-Melibiose	nt	+	+	–	+	+
Melicitose	nt	–	+	–	nt	nt
α-Methyl-D-galactoside, β-methyl-D-glucoside, palatinose, D-psicose	–	+	+	+	nt	nt
β-Methyl-D-galactoside	nt	–	+	–	nt	nt
3-Methyl-glucose	+	+	+	–	nt	nt
α-Methyl-D-mannoside	nt	–	+	–	nt	nt
Methylpyruvate	–	–	–	+	nt	nt
2-Oxoglutarate	nt	+	–	–	nt	nt
Propionate	–	+	+	–	–	+
Pyruvate	–	+	–	+	+	+
L-Rhamnose	–	+	+	–	nt	+
Succinate	–	+	–	+	–	–
Sucrose	–	+	+	–/+[g]	+	+
Thymidine	–	–	+	+	nt	nt
Thymidine-5′-monophosphate	–	+	–	+	nt	nt
Trehalose	–	–	+	–/+[g]	+	+
Turanose	–	–	+	+	nt	nt

(continued)

TABLE 207. (continued)

Characteristic	*F. antarctica*[b]	*F. capsulata*[c]	*F. lacustris*[d]	*F. spumicola*[c]	*M. phosphovorus*[e]	*M. ginsengisoli*[f]
Menaquinones (molar ratio)	MK-9(H_4), MK-9(H_2) (63:8)	MK-9(H_4), MK-9(H_2), MK-7(H_4) (72:10:8)	MK-9(H_4), MK-8(H_4) (83:4)	MK-9(H_4), MK-7(H_2), MK-8(H_2), MK-9(H_2), MK-8(H_4) (62:11:7:6:4)	MK-9(H_4)	MK-9(H_4)
Origin	Sandstone, Antarctica	Activated sludge, Australia	Lake water, Antarctica	Activated sludge, Australia	Activated sludge, Japan	Soil of a ginseng field, South Korea

[a]Symbols: +, positive reaction; –, negative reaction; nt, not tested.

[b]Data from Schumann et al. (1997); Maszenan et al. (1999a).

[c]Data from Maszenan et al. (1999a).

[d]Data from Lawson et al. (2000a).

[e]Data from Nakamura et al. (1995); Maszenan et al. (1999a).

[f]Data from Cui et al. (2007).

[g]Reactions variable in different test systems.

5 months on PYGV agar slants at 4–6°C. Serial transfers at 2-week intervals on R-agar at 22°C under diffuse daylight kept the type strain viable. Strain AA-1042[T] could be recultivated successfully from lyophilized conserves stored at 4°C for 27 months (Schumann et al., 1997). *Friedmanniella antarctica* DSM 11053[T] and *Friedmanniella capsulata* DSM 12936[T] could also be recovered successfully from freeze-dried cultures after a 10 year period of storage. Viability testing revealed 10^3–10^5 c.f.u./ml depending on the growth behavior of the strain (strength of aggregating packets and cluster formation).

The type strains of *Friedmanniella spumicola* and *Friedmanniella capsulata* were isolated from a foam sample and activated sludge biomass, respectively, by micromanipulation (Skerman, 1968). Colonies of both strains developed on standard methods agar (SMA, Difco) supplemented with 1% sterile horse serum at 25°C after 7–10 d. Subsequent purification was done on R2A agar. Short-term storage is possible in R2A or trypticase soy yeast extract medium containing 20% glycerol at –80°C. At the DSMZ, cultures are freeze-dried for long-term conservation according to described procedures which have been summarized in the Cabri guidelines (www.Cabri.org). In addition, preservation by freezing and low temperature storage in glass capillary tubes can be applied. *Friedmanniella lacustris* was isolated from a 20 ml water sample after enrichment by inoculation into 100 ml autoclaved PYGV medium (pH 8.0) prepared with filtered Ekho Lake water for 12 d at 15°C. Single colonies of strain EL17A[T] were obtained after streaking the enrichment culture onto PYGV agar. Cultures of the strain were freeze-dried and stored at 8°C in the dark.

Procedures for testing special characteristics

Because the peptidoglycan type A3γ′, variation A42.1, is a significant feature of members of the genera *Friedmanniella*, *Microlunatus*, *Tessaracoccus*, and of *Propionibacterium propionicus*, refer to the chapter on the genus *Terracoccus* (this volume) where the elucidation of the peptidoglycan structure is described. The structural variation A42.1 of the peptidoglycan based on LL-A_2pm can be concluded from the occurrence of the peptide Gly→D-Glu instead of the dipeptide L-Ala→D-Glu which is commonly found in A-type peptidoglycans (Schleifer and Kandler, 1972). The structure is confirmed by the molar ratio of the amino acids of ca. 1 LL-A_2pm:1 Glu:2 Gly:1 Ala, determined by gas chromatography as described by MacKenzie (1987).

Differentiation of the genus *Friedmanniella* from closely related genera

The genera with the highest phylogenetic relationship to the genus *Friedmanniella* are *Microlunatus* (Nakamura et al., 1995), *Micropruina* (Shintani et al., 2000), *Propionicicella* (Bae et al., 2006c), and *Propionicimonas* (Akasaka et al., 2003b). The genera *Micropruina*, *Propionicicella*, and *Propionicimonas* differ from the genus *Friedmanniella* in displaying the diagnostic diamino acid *meso*-A_2pm (Table 208). The latter two species can also be differentiated by their morphology. The unsaturated menaquinone MK-9 was reported for the genus *Propionicicella* (Bae et al., 2006c) which is unique within the families *Propionibacteriaceae* and *Nocardioidaceae*. The menaquinone profile of *Propionicimonas* differs from that of the genus *Friedmanniella* by the additional occurrence of MK-10(H_4) (Akasaka et al., 2003b). The genus *Microlunatus* shares not only the highest 16S rRNA gene sequence similarity with *Friedmanniella antarctica* but also peptidoglycan structure, major cellular fatty acids, menaquinone, and polar lipids (Table 208). As distinguishing chemotaxonomic characteristics are lacking, the differentiation of the genera *Friedmanniella* and *Microlunatus* is based on the lower tendency to aggregate and to form capsular polymers, better growth in submerged cultures, lack of orange pigments, and metabolism of a broader range of organic compounds of strains of the lat-

TABLE 208. Characteristics differentiating the genus *Friedmanniella* from the phylogenetically related genera *Microlunatus*, *Micropruina*, *Propionicicella*, and *Propionicimonas*[a]

Characteristic	*Friedmanniella*[b]	*Microlunatus*[c]	*Micropruina*[d]	*Propionicicella*[e]	*Propionicimonas*[f]
Morphology	Cocci in packets	Cocci	Coccoid	Rods	Irregular rods
Diamino acid of the peptidoglycan (peptidoglycan type)	LL-A_2pm (A3γ′, A42.1)	LL-A_2pm (A3γ′, A42.1)	*meso*-A_2pm (A1γ, A31)	*meso*-A_2pm (A1γ, A31)	*meso*-A_2pm (A1γ, A31)
Major menaquinones	MK-9(H_4)	MK-9(H_4)	MK-9(H_4)	MK-9	MK-9(H_4), MK-10(H_4)
Major fatty acids	$C_{15:0}$ anteiso, $C_{15:0}$ iso, $C_{14:0}$ iso	$C_{15:0}$ anteiso, $C_{15:0}$ iso, $C_{16:0}$ iso	$C_{15:0}$ anteiso, $C_{14:0}$ iso $C_{16:0}$, $C_{16:0}$ iso	$C_{15:0}$ anteiso, $C_{15:0}$, $C_{16:0}$ iso	$C_{15:0}$, $C_{15:0}$ anteiso, $C_{14:0}$ iso
Polar lipids	PI, PG, DPG, PL, (GL)	PI, PG, DPG, PL	nd	nd	nd
DNA G+C content (mol%)	69–74	67.9–69.8	70.5	69.9	67.4–68.7

[a]Abbreviations: A_2pm, diaminopimelic acid; peptidoglycan types according to http://www.peptidoglycan-types.info; MK-9(H_4), partially saturated menaquinone with two of 9 isoprene units hydrogenated; MK-10(H_4), partially saturated menaquinone with two of 10 isoprene units hydrogenated; MK-9, unsaturated menaquinone with 9 isoprene units; DPG, diphosphatidylglycerol; PG, phosphatidylglycerol; PI, phosphatidylinositol; PL, unidentified phospholipid; GL, unidentified glycolipid; nd, no data available.

[b]Data from Schumann et al. (1997); Maszenan et al. (1999a); Lawson et al. (2000a).

[c]Data from Nakamura et al. (1995); Maszenan et al. (1999a); Cui et al. (2007).

[d]Data from Shintani et al. (2000).

[e]Data from Bae et al. (2006c).

[f]Data from Akasaka et al. (2003b); Bae et al. (2006c).

ter genus (Schumann et al., 1997). The differentiation of both genera by the oxidase test (Maszenan et al., 1999a; Schumann et al., 1997) lost its importance as Lawson et al. (2000a) reports a weak oxidase reaction also for *Friedmanniella lacustris* and as the second *Microlunatus* species, *Microlunatus ginsengisoli* (Cui et al., 2007), differs from the type species of *Microlunatus* in its negative oxidase reaction. Members of the genus *Mirolunatus* differ from *Friedmanniella* species (except for *Friedmanniella lacustris*) in their capability to reduce nitrate. Several physiological and biochemical traits contribute to the differentiation of *Friedmanniella* and *Microlunatus* species (see Table 207).

Taxonomic comments

Stackebrandt and Schaal (2006b) considered the genus *Friedmanniella* as well as the genus *Micropruina* members of the family *Propionibacteriaceae* Delwiche (1957), emend. Rainey, Ward-Rainey and Stackebrandt (1997), while both genera were listed as members of the family *Nocardioidaceae* Nesterenko et al., (1985), emend. Rainey, Ward-Rainey and Stackebrandt (1997) in the Taxonomic Outline of the *Bacteria* and *Archaea* (Garrity et al., 2007). Because members of the genus *Friedmanniella* represent a phylogenetic subcluster within the clade of the family *Propionibacteriaceae* as shown in Figure 239, and because the peptidoglycan type A3γ′ in combination with the major menaquinone MK-9(H_4) and the polyamines spermidine and spermine can only be found in members of the family *Propionibacteriaceae*, the authors of this chapter take the view of Stackebrandt and Schaal (2006b) that the genus *Friedmanniella* shares the membership in the family *Propionibacteriaceae* with the genera *Propionibacterium*, *Luteococcus*, *Microlunatus*, *Micropruina*, *Propioniferax*, *Propionimicrobium*, and *Tessaracoccus* but also with *Propionicimonas* (Akasaka et al., 2003b), *Propionicicella* (Bae et al., 2006c), *Aestuariimicrobium* (Jung et al., 2007), *Granulicoccus* (Maszenan et al., 2007), and *Brooklawnia* (Bae et al., 2006b). The species *Jiangella gansuensis* YIM 002[T] (Song et al., 2005) was excluded from the phylogenetic analysis because all sequences available at present from EMBL and NCBI database are assigned to the family *Pseudonocardiaceae*. The EzTazon (Chun et al., 2007) also does not place the strain into the family *Nocardioidaceae* as stated by Song et al. (2005).

The species of the genus *Friedmanniella* are clearly differentiated from one another by their metabolic properties (Table 207). Both *Microlunatus* species were included in Table 207, as biochemical and physiological properties are useful for the differentiation of *Friedmanniella* and *Microlunatus* species. *Friedmanniella lacustris* and *Friedmanniella capsulata* are the metabolically most versatile organisms of the genus while *Friedmanniella antarctica* shows the lowest amount of positive reactions. *Friedmanniella lacustris* differs from the other members of the genus by lacking H_2S production, tolerance of up to 6% NaCl, growth below 9°C, and its ability to hydrolyze gelatin and reduce nitrate. The growth of *Friedmanniella capsulata* and *Friedmanniella spumicola* is inhibited by addition of NaCl to the media.

Acknowledgements

This chapter is dedicated to Dr. Helmut Prauser on the occasion of his 80th birthday in honor of his contribution to pioneering work on the taxonomy of actinomycetes and of the genera *Nocardioides* and *Friedmanniella* in particular.

List of species of the genus *Friedmanniella*

1. **Friedmanniella antarctica** Schumann, Prauser, Rainey, Stackebrandt and Hirsch 1997, 282[VP]

an.tarc'ti.ca. L. fem. adj. *antarctica* southern, isolated from Antarctica.

The spherical cells are 0.5–2.2 mm in diameter. The cells are arranged in more or less regular packets which adhere to one another to form clusters. Growth in submerged standing and shaking cultures is slow and poor. Colonies on R agar are up to 2 mm in diameter and raised and have irregular edges and shapes. The colony surface is crumbly and dull to smooth and shiny depending on the growth medium. The orange color of colonies may become more intense when the organism is cultured in diffuse daylight. The optimum growth temperature is 22°C. The temperature range for growth is approximately 9–25°C; no growth occurs at 6°C and 28°C. The optimal pH range for growth is 6.0–7.2. Catalase-positive. Oxidase-negative. Urease-positive. Does not reduce nitrate to nitrite. Acid is produced from D-ribose, is produced weakly from L-arabinose, and is produced very weakly from D-xylose. No acid is produced from L-rhamnose, D-glucose, D-fructose, D-mannose, D-galactose, maltose, lactose, sucrose, D-cellobiose, trehalose, D-raffinose, glycerol, D-mannitol, and *myo*-inositol. Starch and esculin are hydrolyzed, and Tween 80 and DNA are only weakly hydrolyzed. Casein and gelatin are not hydrolyzed. Sodium formate is utilized as a carbon source. Sodium acetate, sodium aconitate, sodium benzoate, sodium citrate, disodium succinate, and potassium hydrogentartrate are not utilized. Hypoxanthine, xanthine, adenine, DL-tyrosine, and sodium hippurate are not decomposed. H_2S is produced.

Source: a cryptoendolithic microbial community in sandstone on Linnaeus Terrace, McMurdo Dry Valleys, Antarctica.

DNA G+C content (mol%): 73 (HPLC).

Type strain: AA-1042, DSM 11053, JCM 11651, NBRC 16127.

Sequence accession no. (16S rRNA gene): Z78206.

2. **Friedmanniella capsulata** Maszenan, Seviour, Patel, Schumann, Burghardt, Webb, Soddell and Rees 1999a, 1678[VP]

cap.su.la'ta. L. n. *capsula* a small box or chest; L. fem. suff. *-ata* suffix denoting provided with; N.L. fem. adj. *capsulata* with a chest, capsuled.

This species is characterized by a bright orange color when grown in both solid and liquid media. Cell diameter is 0.6–1.2 µm. Can store polyphosphate aerobically. The optimum growth temperature is 20–25°C. The pH range for growth is 5.5–7.5, with an optimal growth pH of 6.5–7.0. Utilizes L-rhamnose, maltose, sucrose, β-cyclodextrin, L-fucose, 2-aminoethanol, 2,3-butanediol, glucose 6-phosphate, maltotriose, sedoheptulosan, stachyose, D-tagatose, lactamide, 2′-deoxyadenosine, α-ketoglutaric acid, propionic acid, succinamic acid, 3-methylglucose, bromosuccinic acid, glucuronamide, D-alanine, L-alanyl glycine, L-asparagine, L-aspartic acid, glycyl-L-aspartic acid, L-histidine, hydroxy-L-proline, L-threonine, DL-carnitine, γ-aminobutyric acid, urocnic acid, inosine, D-melezitose, malate, and trypsin. The following substrates are not utilized: *i*-erythritol, D-saccharic acid, formic acid, adonitol, D-galactose, D-psicose, turanose, methyl-pyruvate, α-ketobutyric acid, alaninamide, and thymidine. Cellular fatty acid profile is characterized by the presence of $C_{16:0}$ iso, $C_{17:0}$ iso, and $C_{17:0}$ anteiso. Contains the menaquinones MK-9(H_4) and MK-9(H_2) and characterized by the presence of MK-7(H_4). Mycolic acids are absent.

Source: activated sludge.

DNA G+C content (mol%): 74 (HPLC).

Type strain: Ben 108, ACM 5120, CCUG 43143, DSM 12936, JCM 13522.

Sequence accession no. (16S rRNA gene): AF084529.

3. **Friedmanniella lacustris** Lawson, Collins, Schumann, Tindall, Hirsch and Labrenz 2000b, 1953[VP] (Effective publication: Lawson, Collins, Schumann, Tindall, Hirsch and Labrenz 2000a, 226.)

la.cus'tris. N.L. fem. adj. *lacustris* (from L. n. *lacus*, a lake) belonging to the lake.

Cocci with some extracellular polymer, in older cultures also short rods. Cells nonmotile, 0.9–1.3 µm, nonsporeforming, aggregating in short chains, tetrads, or even packets. Grow well on medium PYGV + ASW containing vitamins. Agar colonies flat, watery, and slimy, with a brownish to pink color; older colonies more orange. Aerobic heterotrophs; carbon sources utilized for growth are acetate (weakly), pyruvate, α-D-glucose, glutamate, and citrate, but does not grow on succinate, malate, butyrate, or methanol. A large number of sugar compounds and several organic acids offered by the Biolog GP test system are metabolized. Polymers hydrolyzed: gelatin, starch, and DNA (weakly); does not hydrolyze alginate, casein, or Tween 80. Requires biotin, thiamine, and nicotinic acid for growth. Nitrate is weakly reduced aerobically. NH_3 is formed from peptone; produces acids from glucose. Voges–Proskauer, indole formation, and H_2S production-negative. Temperature optimum is 26°C; the range for growth is 3–3.5°C. The pH tolerance range for growth is 5.5–9.5, with an optimum at pH 7.5. Tolerates up to 6% (w/v) of NaCl with an optimum at 4%. Sensitive to chloramphenicol, penicillin G, and vancomycin. Peptidoglycan type is A3γ′ based on LL-diaminopimelic acid, a single glycine residue as interpeptide bridge, and a glycine residue at position 1 of the peptide subunit. Cell-wall sugars: glucose, mannose, ribose, rhamnose, and galactose. The major respiratory lipoquinone is MK-9(H_4) (83%), with MK-8(H_4) (4%) as minor component. Major fatty acids: $C_{15:0}$ anteiso and $C_{15:0}$ iso. Polar lipids include phosphatidylinositol, phosphatidylglycerol, diphosphatidylglycerol, and an unidentified phospholipid. Mycolic acids absent.

Source: a 1 m water sample from hypersaline Ekho Lake, East Antarctica.

DNA G+C content (mol%): 73 (HPLC).

Type strain: EL-17A, ATCC BAA-165, CIP 106992, DSM 11465, JCM 11951, NCFB 3066.

Sequence accession no. (16S rRNA gene): AJ132943.

4. **Friedmanniella spumicola** Maszenan, Seviour, Patel, Schumann, Burghardt, Webb, Soddell and Rees 1999a, 1678[VP]

spu.mi'co.la. L. fem. n. *spuma* foam; L. suffix *-cola* inhabitant; N.L. masc./fem. n. *spumicola* inhabitant of foam.

Cells adhere to one another and form aggregates of four and eight. The cell diameter is 0.5–1.4 μm. Growth in liquid and solid media is slow and poor requiring up to 2 weeks. Colony color when grown on solid and liquid media is dark yellow to pale orange. The temperature range for growth is 15–37°C with an optimum temperature of 25°C. The pH range for growth is 5.5–8.0, with an optimum pH of 7.0–7.5. Cells store PolyP granules aerobically. Can utilize the following substrates: i-erythritol, D-saccharic acid, formic acid, D-galactose, turanose, methylpyruvate, α-ketobutyric acid, alaninamide, and thymidine. The following substrates were not utilized: L-rhamnose, maltose, sucrose, β-cyclodextrin, L-fucose, 2-aminoethanol, 2,3-butanediol, glucose 6-phosphate, maltotriose, sedoheptulosan, stachyose, D-tagatose, lactamide, 2′-deoxyadenosine, α-ketoglutaric acid, propionic acid, succinamic acid, glucuronamide, D-alanine, L-alanyl-glycine, L-asparagine, L-aspartic acid, glycyl-L-aspartic acid, L-histidine, hydroxy-L-proline, L-threonine, DL-carnitine, γ-aminobutyric acid, uroconic acid, inosine, D-melezitose, malate, 3-methylglucose, bromosuccinic acid, and trypsin. Characterized by the presence of $C_{18:1}$ and $C_{15:0}$ iso. Unbranched fatty acids with 14, 15, and 16 carbons were found in trace amounts. Its major menaquinones are MK-9(H_4) and MK-7(H_2), with MK-9(H_2), MK-8(H_2), and MK-8(H_4) present in trace amounts. Mycolic acids are absent.

Source: activated sludge.

DNA G+C content (mol%): 69 (HPLC).

Type strain: Ben 107, ACM 5121.

Sequence accession no. (16S rRNA gene): AF062535.

Genus V. **Granulicoccus** Maszenan, Jiang, Tay, Schumann, Kroppenstedt and Tay 2007, 733[VP]

The Editorial Board

Gra.nu.li.coc′cus. L. neut. n. *granulum* a small grain; N.L. masc. n. *coccus* (from Gr. masc. n. *kokkos* grain, seed) coccus; N.L. masc. n. *Granulicoccus* a coccus from (sludge) granules, here referring to the isolation source.

Non-spore-forming coccus (0.3–1.4 μm diameter). **Gram-stain-positive**. Nonmotile. **Facultative anaerobe**. Temperature range for growth 15–37°C, pH range 5–8.5. Cell wall contains type A3γ peptidoglycan (LL-A_2pm←Gly with alanine at position 1 of the peptide subunit). MK-9(H_4) is the major menaquinone, MK-8(H_4) is present as well in a 42:1 compositional ratio. Polar lipid profile comprises diphosphatidylglycerol and phosphatidylglycerol.

DNA G+C content (mol%): 69.

Type species: **Granuliococcus phenolivorans** Maszenan, Jiang, Tay, Schumann, Kroppenstedt and Tay 2007, 733[VP].

Further descriptive information

Major fatty acids are 13-methyltetradecanoic acid ($C_{15:0}$ iso, 50.5%) and 1,1-dimethoxy-iso-pentadecane ($C_{15:0}$ iso DMA, 37.4%). Stains contain polyphosphate granules but not poly-β-hydroxyalkanoates. Produces capsular material and can autoaggregate.

Phylogenetic analysis of the 16S rRNA gene affiliates the genus with the family *Propionibacteriaceae*. The closest phylogenetic neighbors are *Luteococcus peritonei* (93.9% sequence similarity) (Collins et al., 2000), *Microlunatus panaciterrae* (93.6%) (An et al., 2008), and *Propioniferax innocua* (93.6%) (Pitcher and Collins, 1991; Yokota et al., 1994).

Enrichment and isolation procedures

Strain PG-02[T] was isolated from phenol-degrading aerobic granules cultivated in a laboratory-scale sequencing batch reactor fed with synthetic wastewater containing phenol as the sole carbon source (Jiang et al., 2004). Activated sludge was the initial seed for the reactor. Granules (2.5 g) were added to 15 ml MP medium, and serial diluents of the supernatant were spread onto agar plates containing MP medium with 1.2% Bacto agar (Difco). Plates were incubated for 28 d at 25°C (2007). Colony morphology is visible after 10 d of incubation on MP agar plates.

Maintenance procedures

Strain PG-02[T] can be grown on synthetic wastewater medium with phenol as the sole carbon source with the following composition (per liter): phenol, 0.5 g; NH_4Cl, 0.20; $MgSO_4 \cdot 7H_2O$, 0.13 g; K_2HPO_4, 1.65 g; KH_2PO_4, 1.35 g; and 1 ml of micronutrient solution. Stock cultures are maintained as a 20% glycerol suspension at −80°C.

Differentiation of the genus *Granuliococcus* from closely related genera

Presence of $C_{15:0}$ iso DMA differentiates *Granulicoccus* from the related genera *Luteococcus, Friedmanniella, Tessaracoccus, Propioniferax, Micropruina* and *Microlunatus*. *Granulicoccus* differs from *Microlunatus* in the possession of MK-8(H_4) and lack of phosphatidylinositol. The pleomorphic rod morphology and the polar lipid phosphatidylethanolamine of *Propioniferax* distinguish it from *Granulicoccus*.

List of species of the genus *Granuliococcus*

1. **Granulicoccus phenolivorans** Maszenan, Jiang, Tay, Schumann, Kroppenstedt and Tay 2007, 733[VP]

phe.no.li.vo'rans. N.L. neut. n. *phenolum* phenol; L. part. adj. *vorans* devouring, consuming; N.L. part. adj. *phenolivorans* consuming phenol.

Growth occurs at mesophilic temperature and pH 5–8.5, with optima at 30°C and pH 7. Arginine dihydrolase, lysine decarboxylase, ornithine decarboxylase and tryptophan deaminase tests are negative. H_2S and indole are not produced. Voges–Proskauer-negative and does not produce acetoin or reduce nitrate to nitrite. Catalase-positive and oxidase-negative. Utilizes phenol, Tweens 40 and 80, L-arabinose, α-D-glucose, α-D-lactose, lactulose, maltose, maltotriose, D-mannose, D-melezitose, D-melibiose, methyl α-D-galactoside, methyl β-D-galactoside, 3-methyl glucose, methyl α-D-glucoside, methyl β-D-glucoside, psicose, D-raffinose, L-rhamnose, D-ribose, salicin, sedoheptulosan, stachyose, sucrose, D-tagatose, D-trehalose, turanose, D-xylose, *myo*-inositol, D-mannitol, D-sorbitol, xylitol, 2,3-butanediol, glycerol, DL-α-glycerol phosphate, glucose 1-phosphate, glucose 6-phosphate, adenosine, AMP, TMP, UMP and fructose 6-phosphate (Biolog GN and GP systems and API 20E). Acids and their derivatives utilized include methyl pyruvate, monomethyl succinate, acetic acid, citric acid, D-galactonic acid lactone, D-gluconic acid, D-glucuronic acid, α-, β- and γ-hydroxybutyric acids, *p*-hydroxyphenylacetic acid, itaconic acid, α-ketobutyric acid, α-ketoglutaric acid, α-ketovaleric acid, lactamide, D-lactic acid methyl ester, L- and DL-lactic acid, D- and L-malic acid, propionic acid, pyruvic acid, quinic acid, D-saccharic acid, sebacic acid, succinic acid, bromosuccinic acid, succinamic acid, *N*-acetylglutamic acid, L-glutamic acid, glycyl L-glutamic acid and L-pyroglutamic acid. Amino acid compounds utilized glucuronamide, alaninamide, D-alanine, L-alanine, L-alanyl glycine, L-asparagine, L-phenylalanine, L-proline, L-serine, inosine, uridine, thymidine and putrescine. Gentiobiose is weakly utilized. Growth is not affected by α-cyclodextrin, β-cyclodextrin, dextrin, glycogen, inulin, mannan, amygdalin, adonitol, D-arabitol, arbutin, cellobiose, i-erythritol, D-fructose, L-fucose, D-galactose, 2-aminoethanol, *N*-acetyl-D-galactosamine, *N*-acetyl-D-glucosamine, *N*-acetylmannosamine, phenyl ethylamine, deoxyadenosine, L-histidine, hydroxy-L-proline, L-leucine, L-ornithine, D-serine, L-threonine, DL-carnitine, D-galacturonic acid, formic acid, D-glucosaminic acid, malonic acid, L-aspartic acid, γ-aminobutyric acid and urocanic acid. Both the API ZYM and API 20E detect alkaline phosphatase, esterase, lipase, leucine arylamidase, valine arylamidase, naphthol-AS-BI-phosphohydrolase, α-galactosidase, β-galactosidase, β-glucuronidase, α-glucosidase and β-glucosidase. API 20E also detects β-galactosidase, urease and gelatinase. API ZYM did not detect acid phosphatase, esterase lipase, cystine arylamidase, trypsin, chymotrypsin, *N*-acetyl-β-glucosaminidase, α-mannosidase and α-fucosidase.

For all other characteristics, refer to the genus description.

DNA G+C content (mol%): 69 (HPLC).

Type strain: PG-02[T] ATCC BAA-1292, DSM 17626.

Sequence accession no. (16S rRNA gene): AY566575.

Genus VI. **Luteococcus** Tamura, Takeuchi and Yokota 1994, 355[VP] emend. Collins, Lawson, Nikolaitchouk and Falsen 2000, 181

Tomohiko Tamura

Lu.te.o.coc'cus. L. adj. *luteus* yellow; N.L. masc. n. *coccus* (from Gr. masc. n. *kokkos* grain, seed) coccus; N.L. masc. n. *Luteococcus* yellow coccus.

Gram-stain-positive coccus or pleomorphic rod. Spherical cells are 0.7–1.0 µm in diameter and occur **singly, in pairs, or in tetrads**. Nonsporeforming. Colonies are circular and smooth and may be cream colored to yellow. **Facultatively anaerobic**. Catalase- and oxidase-positive. Urease-negative. The cells may or may not reduce nitrate to nitrite. Starch is hydrolyzed. Tween 20, 40, 60, and 80 are not hydrolyzed. Acid is produced from glucose and some other sugars. Propionic acid is the major product formed from glucose. Optimum growth temperature is 26–28°C.

Cell-wall peptidoglycan contains **LL-diaminopimelic acid**, alanine, glycine, and glutamate (approximately 1:2:1:1). The major menaquinone is MK-9(H_4). Mycolic acid is not present. The major cellular fatty acid is $C_{16:1}$, and among the minor components a small amount of $C_{18:0}$ iso 2OH is also present. Arabinose is present as a diagnostic sugar in the cell wall. The polar lipids phosphatidylinositol, diphosphatidyiglycerol, and phosphatidyiglycerol, are present.

DNA G+C content (mol%): 64–67.

Type species: **Luteococcus japonicus** Tamura, Takeuchi and Yokota 1994, 355[VP].

Further descriptive information

The description of the genus *Luteococcus* has been emended by Collins et al. (2000) to include pleomorphic rods as well as cocci. The type species grows well at temperatures of 25–30°C on oatmeal agar, inorganic salts-starch agar, and peptone-yeast extract-iron agar. For chemotaxonomic studies, these strains were grown in shake cultures (nutrient broth; Difco) at 28°C, and cells were harvested in the stationary phase, washed twice with water, and then, if necessary, freeze-dried.

Cellular fatty acids of the genus *Luteococcus* comprise predominantly $C_{15:1}$, $C_{17:1}$, and $C_{18:1}$ in addition to $C_{16:1}$. The composition differs among species. *Luteococcus peritonei* and *Luteococcus sanguinis* contains predominantly $C_{17:1}$ and followed by $C_{16:1}$ and $C_{15:1}$.

Other properties of the species of *Luteococcus* are presented in Table 209.

Enrichment and isolation procedures

The two strains of *Luteococcus japonicus* were isolated from soil on Tokara Island, Japan, and from spring water for brewing

TABLE 209. Phenotypic characteristics of *Luteococcus* species[a]

Characteristic	*L. japonicus*[b]	*L. peritonei*[c]	*L. sanguinis*[d]
Cell morphology	Cocci	Rods	Cocci
Source	Soil, water	Human peritoneum	Human blood
Acid from:			
L-Arabinose	+	–	nd
D-Glucose	+	+	+
Glycogen	nd	–	+
Mannitol	+	+	+
Maltose	+	v	+
Raffinose	+	–	nd
Ribose	+	–	–
Sucrose	+	+	+
Trehalose	+	–	nd
D-Xylose	–	–	–
Hydrolysis of:			
Esculin	+	+	+
Gelatin	v	–	+
Hippurate	+	–	nd
Acid phosphatase	nd	v	w
Alkaline phosphatase	nd	v	+
Chymotrypsin	nd	–	+
β-Glucuronidase	nd	+	–
Trypsin	nd	–	+
Urease	–	–	–
Reduction of nitrate to nitrite	–	+	+

[a]Symbols: +, >85% positive; –, 0–15% positive; nd, no data; v, variable. The data of each species are those of the type strains.

[b]Tamura et al. (1994).

[c]Collins et al. (2000).

[d]Collins et al. (2003).

named "miyamizu" in Hyogo prefecture, Japan, respectively (Oda, 1935). *Luteococcus peritonei* was isolated from human peritoneum during a fetal autopsy. *Luteococcus sanguinis* was isolated from a blood sample (one out of four bottles) of a 32-year-old man. Although the latter two species were isolated from human samples, the pathogenicity is not known.

Maintenance procedures

The strains of the genus *Luteococcus* are maintained by freezing in water containing 10–30% glycerol at –70°C. Lyophilization of suspensions in 10% skim milk + 1% monosodium glutamate and l-drying in 0.01 M potassium phosphate buffer (pH 7.0) containing 3% monosodium glutamate are also recommended for long-term preservation.

Differentiation of the genus *Luteococcus* from other genera

Luteococcus species show the same chemotaxonomic characteristics as most other genera of the family *Propionibacteriaceae* in possessing a cell-wall peptidoglycan containing LL-diaminopimelic acid (LL-A_2pm) and menaquinone MK-9(H_4). However, the cellular fatty acid composition containing predominantly straight monounsaturated acids (approximately 90% of total acids) is unusual and characteristic for the genus *Luteococcus* (Collins et al., 2000, 2003; Tamura et al., 1994). In contrast, the other members of the family *Propionibacteriaceae* possess mostly iso- and anteiso-branched fatty acids.

Taxonomic comments

Based on 16S rRNA gene sequence analysis, the species of the genus *Luteococcus* form an independent clade in the family *Propionibacteriaceae* of the order *Propionibacteriales.* Within the genus, phylogenetic distances based on 16S rRNA gene sequences between *Luteococcus peritonei* and the other two species are both 94%, indicating the distinctness of this species. In contrast, the sequence difference between *Luteococcus japonicus* and *Luteococcus sanguinis* is 96.9%, indicative of a closer relationship (Collins et al., 2003). However, species status was confirmed by the low DNA relatedness of 49% between these species (Collins et al., 2003). These relationships are confirmed by a comparison of the other phenotypic properties of these species, where *Luteococcus japonicus* and *Luteococcus sanguinis* share similarities in morphologies and carbon sources not seen in *Luteococcus peritonei* (Table 209).

Differentiation of species of the genus *Lutoecoccus*

Luteococcus peritonei can be distinguished from *Luteococcus japonicus* and *Luteococcus sanguinis* by its pleomorphic rod-shaped morphology. Nitrate is reduced to nitrite by *Luteococcus peritonei* and *Luteococcus sanguinis* but not by *Luteococcus japonicus.*

List of species of the genus *Luteococcus*

1. **Luteococcus japonicus** Tamura, Takeuchi and Yokota 1994, 355[VP]

ja.po'ni.cus. N.L. masc. adj. *japonicus* of or pertaining to Japan, where the organisms were isolated.

Cells are spherical and 0.7–1.0 µm in diameter and occur singly, in pairs, or in tetrads. Nonsporeforming. Gram-stain-positive. Facultatively anaerobic. Colonies are circular and smooth and may be cream colored to yellow. Catalase- and oxidase-positive. Urease-negative. Nitrate is not reduced to nitrite. Oxidation-fermentation is fermentative. Acid is produced from D-glucose, D-ribose, D-galactose, D-mannose, D-fructose, sucrose, maltose, trehalose, raffinose, glycerol, mannitol, inositol, and L-arabinose but not from D-xylose, D-arabinose, or L-rhamnose. Propionic acid is produced from glucose. Starch is hydrolyzed. Gelatin is weakly or not hydrolyzed. Optimum growth temperature is 26–28°C.

Source: soil of Tokara Islands, Japan.

DNA G+C content (mol%): 67 (HPLC).

Type strain: ATCC 51526, CCUG 38731, CIP 104067, DSM 10546, JCM 9415, NBRC 12422, VKM Ac-1951.

Sequence accession no. (16S rRNA gene): D21245, D85487, Z78208.

2. **Luteococcus peritonei** Collins, Lawson, Nikolaitchouk and Falsen 2000, 181[VP]

pe.ri.to.ne'i. L. n. *peritoneum* peritoneum; L. gen. neut. n. *peritonei* of the peritoneum.

Cells consist of pleomorphic rods that are Gram-stain-positive. Pigment is not produced. Facultatively anaerobic and catalase-positive. Acid is produced from glucose, lactose, sucrose, mannitol, and methyl β-D-glucopyranoside. Acid may or may not be produced from maltose. Acid is not produced from L-arabinose, D-arabitol, cyclodextrin, glycogen, melibiose, melezitose, pullulan, ribose, raffinose, sorbitol, trehalose, tagatose, or D-xylose. Esculin is hydrolyzed but gelatin and hippurate are not. α-Galactosidase, β-galactosidase, β-galacturonidase, α-glucosidase, β-glucosidase, β-glucuronidase, leucine arylamidase, and pyrazinamidase are produced. Arginine dihydrolase, lipase C14, chymotrypsin, α-fucosidase, glycyl-tryptophan arylamidase, *N*-acetylglucosaminidase, α-mannosidase, β-mannosidase, pyroglutamic acid arylamidase, trypsin, valine arylamidase, and urease are not produced. Activity for alkaline phosphatase and acid phosphatase may or may not be detected. Voges–Proskauer reaction is negative. Nitrate is reduced to nitrite.

Source: human peritoneum.

DNA G+C content (mol%): 65 (method of determination not reported).

Type strain: ATCC BAA-60 = CCUG 38120 = CIP 106441 = JCM 11685.

Sequence accession no. (16S rRNA gene): AJ132334.

3. **Luteococcus sanguinis** Collins, Hutson, Nikolaitchouk, Nyberg and Falsen 2003, 1891[VP]

san'gui.nis. L. gen. n. *sanguinis* of blood.

Nonmotile coccus; Gram-stain-positive. Facultatively anaerobic and catalase-positive.

Acid is produced from glucose, glycogen, mannitol, maltose, lactose, and sucrose, but not from ribose or D-xylose. Esculin and gelatin are hydrolyzed. When tested by using commercial API Coryne and API ZYM systems, acid phosphatase (weak reaction), alkaline phosphatase, chymotrypsin, ester lipase C8 (weak), cystine arylamidase (weak), α-galactosidase, β-galactosidase, α-glucosidase, β-glucosidase, leucine arylamidase, phosphoamidase (weak), pyrrolidonyl arylamidase, pyrazinamidase, valine arylamidase (weak), and trypsin are detected. Esterase C4, lipase C14, α-fucosidase, β-glucuronidase, *N*-acetyl-β-glucosaminidase, α-mannosidase, pyroglutamic acid arylamidase, and urease are not detected. Voges–Proskauer reaction is negative. Nitrate is reduced to nitrite.

Source: human blood.

DNA G+C content (mol%): 64 (method of determination not reported).

Type strain: CCUG 33897 = CIP 107216 = JCM 12371.

Sequence accession no. (16S rRNA gene): AJ416758.

Genus VII. **Microlunatus** Nakamura, Hiraishi, Yoshimi, Kawaharasaki, Masuda and Kamagata 1995, 21[VP]

Satoshi Hanada and Kazunori Nakamura

Mi.cro.lu.na'tus. Gr. adj. *mikros* small; L. masc. adj. *lunatus* half moon-shaped; N.L. masc. n. *Microlunatus* small moon-like microorganism.

Coccoid or spherical cells, 0.3–2.0 μm in diameter. Occurs singly and in pairs. Occasionally forms clusters. Some species show rod-shaped morphology. **Gram-stain-positive**. Nonmotile. Nonsporeforming. Catalase-positive. **Mesophilic**. Good growth occurs at 20–30°C. **Aerobic and chemo-organotrophic**. The following sugars and sugar alcohols can support good growth: glucose, arabinose, mannose, maltose, melibiose, rhamnose, and sorbitol. Some species reduce nitrate under anaerobic conditions and accumulate phosphate inside cells. The major quinone is MK-9(H_4). The cell-wall peptidoglycan contains LL-diaminopimeric acid. The major fatty acids are $C_{15:0}$ anteiso, $C_{15:0}$ iso, and $C_{16:0}$ iso. Isolated from activated sludge in wastewater treatment system, soil, spawn of a mushroom, and an indoor wall.

DNA G+C content (mol%): 65.1–70.9.

Type species: **Microlunatus phosphovorus** Nakamura, Hiraishi, Yoshimi, Kawaharasaki, Masuda and Kamagata 1995, 21[VP].

Further descriptive information

Microlunatus phosphovorus, the type species of the genus, was isolated from activated sludge in the wastewater treatment process (Nakamura et al., 1995). The process, consisting of alternating anaerobic and aerobic conditions, was designed to exhibit high phosphate removal activity. Because *Microlunatus phosphovorus* can accumulate a large amount of phosphate inside cells as a polyphosphate, it is considered to significantly contribute to phosphate removal from wastewater. The uptake and accumulation of phosphate are observed under aerobic conditions. Conversely, the accumulated polyphosphate is degraded and released outside under anaerobic conditions. Such phosphate uptake under aerobic conditions is also found in *Microlunatus aurantiacus*, but the amount of accumulation is obviously smaller than that of *Microlunatus phosphovorus* (Wang et al., 2008).

The cells of *Microlunatus phosphovorus* are 0.8–2.0 μm in diameter and occurred singly or in pairs (Figure 240). Small irregular clusters of cells are occasionally formed. A cell-wall structure (segmentation) is frequently observed in the middle of the spherical cells (Figure 241).

The species of the genus *Microlunatus* are mesophilic and neutrophilic and show good growth at 25–30°C and around pH 7.0. Nitrate reduction under anaerobic condition was included in the description of the genus based on the characteristic of *Microlunatus phosphovorus*. However, *Microlunatus ginsengsoli* and *Microlunatus panaciterrae* are negative for nitrate reduction. Although some species can reduce nitrate to nitrite under anaerobic conditions, the genus *Microlunatus* is basically a chemo-organotroph which grows by oxygen respiration using sugars and sugar alcohols as substrates.

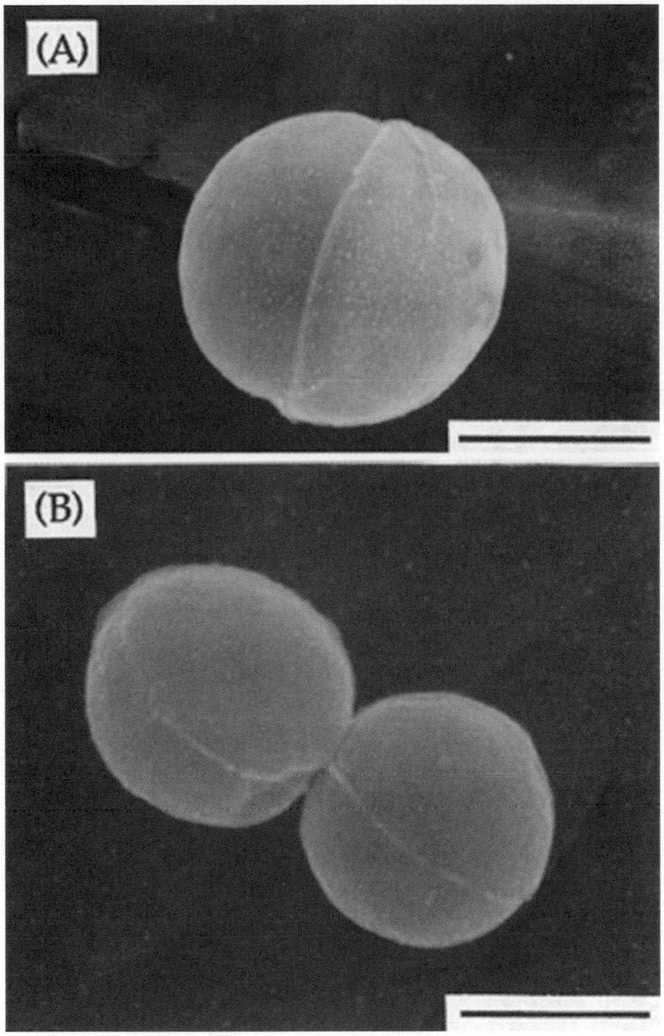

FIGURE 240. Scanning electron micrographs the type strain of *Microlunatus phosphovorus.* (A) Single cell. (B) Cells in pair. Bars = 1 µm. (Reproduced with permission from Nakamura et al., 1995. Int. J. Syst. Bacteriol. *45*: 17–22.)

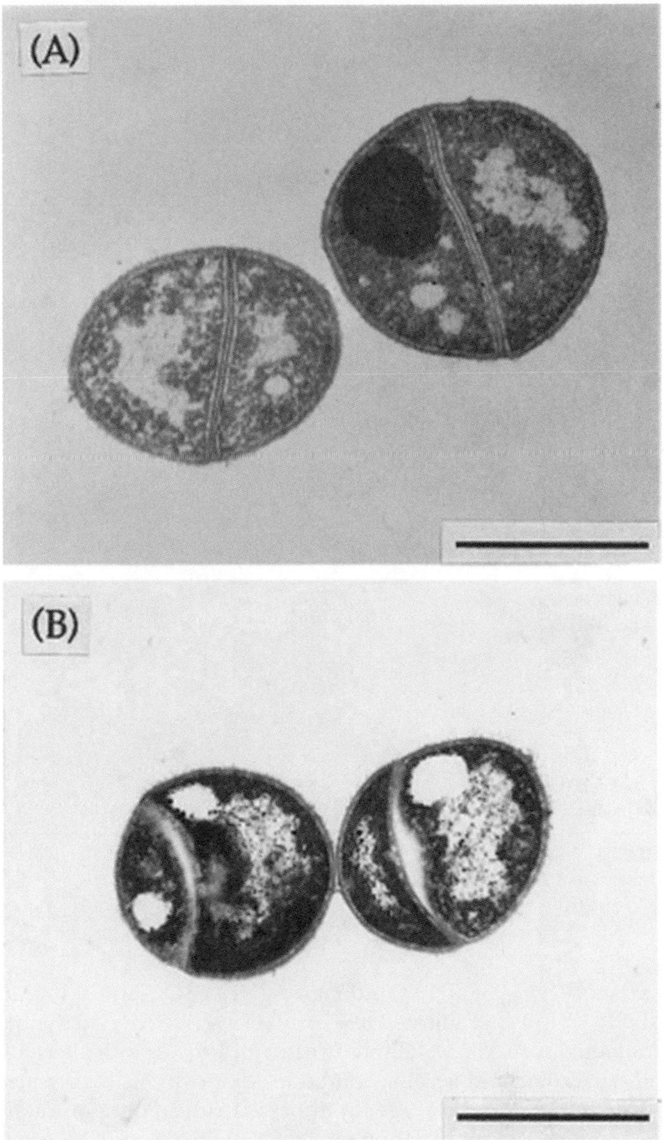

FIGURE 241. Transmission electron micrographs of thin sections of *Microlunatus phosphovorus* showing cell-wall structure in the middle of spherical cells and accumulation of polyphoisphate. (A) Cells harvested at the exponential phase. (B) Cells harvested at the stationary phase showing intracellular storage of polyphosphate (clear areas). Bars = 1 µm. (Reproduced with permission from Nakamura et al., 1995. Int. J. Syst. Bacteriol. *45*: 17–22.)

In addition to *Microlunatus phosphovorus* as the type species, the genus *Microlunatus* contains five species. *Microlunatus ginsengisoli, Microlunatus aurantiacus,* and *Microlunatus panaciterrae* were all isolated from soil (An et al., 2008; Cui et al., 2007; Wang et al., 2008). While *Microlunatus phosphovorus* is slow growing with a doubling time of 13 h, two species of these isolates from soil, i.e., *Microlunatus ginsengisoli* and *Microlunatus panaciterrae,* can grow quickly on R2A medium. *Microlunatus soli* and *Microlunatus parietis* were isolated from the spawn of the mushroom *Agaricus brasiliensis* and an indoor wall, respectively. They are both oxidase-positive species like *Microlunatus panaciterrae. Microlunatus soli* contains unsaturated fatty acids, e.g., $C_{18:1}$ ω7*c*, that are rarely found in any other *Microlunatus* species.

The peptidoglycan structure of the genus *Microlunatus* was determined for *Microlunatus phosphovorus* (Schumann et al., 1997) and *Microlunatus soli* (Kämpfer et al., 2010b) and reveals that the interpeptide bridge is a single glycine and that position 1 of the peptide subunit is substituted by glycine. The structure is designated A3γ′ according to the classification of Schleifer and Kandler (1972).

Although these six species of the genus *Microlunatus* have almost similar phenotypic characteristics as mentioned in the genus description above, they can be differentiated by comparing oxidase activity, nitrate reduction under anaerobic conditions, phosphate accumulation, and/or nutritional profiles (Table 210).

Enrichment and isolation procedures

For isolation of *Microlunatus phosphovorus,* a slow-growing bacterium inhabiting activated sludge, a relatively oligotrophic medium is used. Such an oligotrophic medium inhibits fast-growing bacteria from predominating. The isolation medium contains the following ingredients (per liter): 0.5 g of glucose, 0.5 g of peptone, 0.5 g of monosodium glutamate, 0.5 g of yeast

TABLE 210. Characteristics differentiating species of the genus *Microlunatus*[a,b]

Characteristics	*M. phosphovorus*	*M. aurantiacus*	*M. ginsengisoli*	*M. panaciterrae*	*M. parietis*	*M. soli*
Morphology	Cocci, occasionally. clusters	Cocci	Cocci	Cocci	Cocci to rods	Cocci
Cell diameter (μm)	0.8–2.0	0.9–1.3	0.5–0.8	0.3–0.7	nd	1.0–1.5
Habitat	Activated sludge	Soil	Soil	Soil	Indoor wall	Spawn of a mushroom
DNA G+C content (mol%)	67.9	70.9	69.8	65.1	nt	nt
Oxidase	w	−	−	+	+	+
Nitrate reduction under anaerobic conditions	+	−	−	nt	nt	nt
Phosphate uptake	+	w	−	−	nt	nt
Utilization of:						
Acetate	w	w	w	−	−	+
N-Acetyl-D-glucosamine	+	+	+	−	w	+
Adonitol	+	−	+	−	+	+
p-Arbutin	+	−	+	−	−	+
D-Cellobiose	+	+	+	−	+	+
D-Fructose	+	+	+	−	+	+
D-Galactose	+	w	+	−	+	+
myo-Inositol	+	+	+	−	+	+
L-Malate	+	+	w	−	−	w
D-Mannitol	+	+	+	−	+	+
Propionate	−	−	−	−	−	+
Sucrose	−	+	+	+	+	+
Salicin	+	−	+	+	−	+
D-Trehalose	+	+	+	−	+	+
D-Xylose	+	+	+	−	+	+

[a]Symbols: +, positive; −, negative; w, weakly positive, nt, not tested.

[b]Data from An et al. (2008) and Kämpfer et al. (2010a, 2010b).

extract, 0.44 g of KH_2PO_4, 0.1 g of $(NH_4)_2SO_4$, and 0.1 g of $MgSO_4 \cdot 7H_2O$ (pH adjusted to 7.0 with NaOH). Activated sludge (obtained from the wastewater treatment process with alternating anaerobic and aerobic conditions that exhibits high phosphate removal activity) is gently dispersed with an ultrasonicator and streaked on the isolation medium solidified with 1.5% agar. Circular, smooth, convex, and buff colonies of *Microlunatus phosphovorus* emerge on an agar plate after a few weeks of incubation at 25°C. The same liquid medium is used to enrich *Microlunatus phosphovorus* and is incubated at 25°C with gentle shaking.

The isolation procedure of *Microlunatus aurantiacus* from a soil sample includes an additional pre-incubation. Prior to spreading on agar plates, a soil sample dried at room temperature is suspended in a phosphate buffer solution (pH 7.0) containing 0.1% sodium cholate and incubated for 1 h at 45°C with vigorous shaking in order to eliminate fast-growing bacteria and disperse soil aggregates. The glycerol-asparagine based agar medium (ISP5 medium; Shirling and Gottlieb, 1966) is used as the isolation medium, and inoculated agar plates are incubated at 28°C for 21–30 d. To enrich this species, the modified ISP2 agar medium containing the following ingredients (per liter) is used: 4 g of glucose, 4 g of yeast extract, 5 g of malt extract, and a vitamin/amino acid mixture (1 mg of vitamin B_1, 1 mg of vitamin B_2, 1 mg of vitamin B_6, 1 mg of biotin, 1 mg of nicotinic acid, 1 mg of phenylalanine, 0.3 g of alanine); pH 7.2.

To isolate *Microlunatus ginsengisoli* and *Microlunatus panaciterrae*, MR2A agar medium is used. The MR2A agar medium is composed of (per liter): 0.25 g of tryptone, 0.25 g of peptone, 0.25 g of yeast extract, 0.125 g of malt extract, 0.125 g of beef extract, 0.25 g of Casamino acids, 0.25 g of soytone, 0.5 g of glucose, 0.3 g of soluble starch, 0.2 g of xylan, 0.3 g of sodium pyruvate, 0.3 g of K_2HPO_4, 0.05 g of $MgSO_4$, 0.05 g of $CaCl_2$, and 15 g of agar. These species can be maintained using standard R2A agar and nutrient agar as well as the modified R2A agar at 30°C.

Microlunatus soli and *Microlunatus parietis* can be cultured on R2A agar and nutrient agar at 30°C.

Maintenance procedures

A liquid culture of the type species, *Microlunatus phosphovorus*, retains viability for several months at room temperature in the dark. Long-term preservation at −80°C is possible in the presence of 10% (w/v) glycerol. Freeze-drying is also available with a suitable protective matrix.

Microlunatus ginsengisoli, *Microlunatus aurantiacus*, and *Microlunatus panaciterrae* can be preserved at −70°C in a 20% (w/v) glycerol suspension. Long-term preservation of *Microlunatus soli* and *Microlunatus parietis* is feasible by storage in 20% (v/v) glycerol stock and by lyophilization.

Differentiation of the genus *Microlunatus* from other genera

The genus *Microlunatus* can be differentiated clearly from other genera within the family *Propionibacteriaceae* by 16S rRNA gene sequence analysis (Figure 242). All species belonging to the genus *Microlunatus* are closely related to each other (the

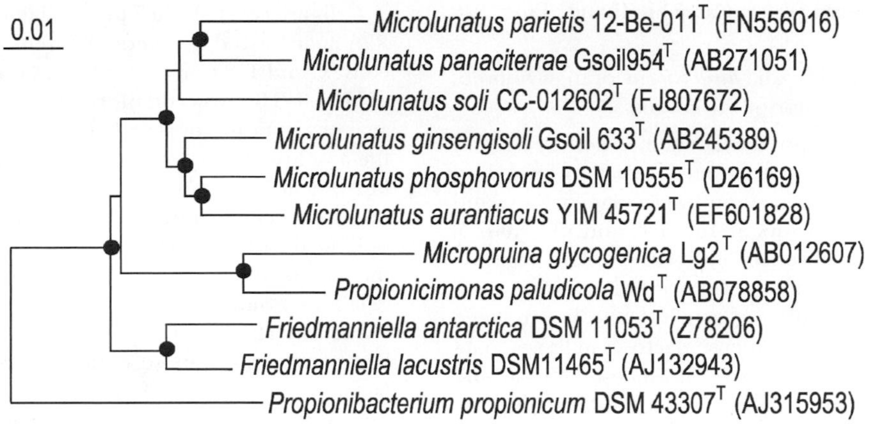

FIGURE 242. Phylogenetic relationships of the species of the genus *Microlunatus* based on 16S rRNA gene sequences constructed using the neighbor joining method. The EMBL/GenBank accession number of each sequence is indicated in parentheses. A bootstrap value (1000 replications) greater than 60% is shown as a closed circle at each branch point. The scale bar shows substitutions per 100 nucleotide positions.

sequence similarities among them are more than ~96%) and form a coherent cluster in the phylogenetic tree.

The related genera are *Micropruina*, *Propionicimonas*, and *Friedmanniella*, but all the members in the genus *Microlunatus* are phylogenetically distant from any species belonging to these related genera. All species in the genus *Microlunatus* and the related genera share the following common features in their phenotype: all are Gram-stain-positive, nonsporeforming, nonmotile cocci; contain $C_{15:0}$ anteiso as a main fatty acid component; and have menaquinone-9(H_4) as the major quinone. However, they can be differentiated from each other by comprehensive comparison of phenotypic characteristics such as their peptidoglycan types, nutritional profiles, and genomic G+C content.

Taxonomic comments

The type species of the genus, *Microlunatus phosphovorus*, was first described in 1995 with a proposal of the genus *Microlunatus* (Nakamura et al., 1995). The species was found in activated sludge in the wastewater treatment process. Since the first description, no species was newly proposed in this genus for more than ten years. From 2007 to 2008, three new bacteria belonging to the genus *Microlunatus* were found and proposed as new species. These new species, *Microlunatus ginsengisoli*, *Microlunatus aurantiacus*, and *Microlunatus panaciterrae*, were all isolated from soil, revealing that *Microlunatus* species inhabit not only the hydrosphere but also the lithosphere (An et al., 2008; Cui et al., 2007; Wang et al., 2008). In addition to these species, *Microlunatus soli* and *Microlunatus parietis* were newly isolated from the spawn of an edible mushroom, *Agaricus brasiliensis*, and an indoor wall, respectively. They were classified in the genus *Microlunatus* in 2010 (Kämpfer et al., 2010a, 2010b). At present, the genus *Microlunatus* consists of the following six species: *Microlunatus phosphovorus*, *Microlunatus aurantiacus*, *Microlunatus ginsengisoli*, *Microlunatus panaciterrae*, *Microlunatus parietis*, and *Microlunatus soli*.

List of species of the genus *Microlunatus*

1. **Microlunatus phosphovorus** Nakamura, Hiraishi, Yoshimi, Kawaharazaki, Masuda and Kamagata 1995, 21[VP]

phos.pho'vo.rus. L. n. *phosphorus* (from Gr. n. *phōsphoros* the light-bringer), the morning-star and, in chemistry, phosphorus; N.L. adj. *vorus* devouring; N.L. masc. adj. *phosphovorus* intended to mean phosphorus-accumulating microorganism.

The cells have a diameter of 0.8–2.0 µm and occur singly, in pairs, and occasionally in clusters. Often have a cell-wall structure in the middle of the cell and are hemispherical, especially at the stationary growth phase. The doubling time is about 13 h in a liquid medium. The colonies are circular (diameters, 0.5–1 mm), smooth, convex, and cream colored at the early stage of growth. After 10–14 d of incubation, the colonies are 1–2 mm in diameter and yellowish. Oxidase is positive but weak. The cells reduce nitrate to nitrite but do not reduce nitrite to nitrogen. The species has high phosphate-accumulating activity in the absence of any carbon substrate in the medium when it is exposed to exogenous phosphate under aerobic conditions (an intracellular phosphorus content sometimes increases more than 10% on a dry cell weight). The cells utilize glucose, mannose, galactose, xylose, arabinose, sucrose, maltose, cellobiose, trehalose, and melibiose, but not lactose. Starch is utilized, but glycogen is not utilized. Sugar alcohols such as inositol, dulcitol, and mannitol are utilized. Alcohols like methanol, ethanol, propanol, and glycerol are not utilized. Pyruvate is utilized, and acetate is utilized slowly. Growth occurs at 5–35°C with an optimum growth temperature of 25°C. The optimum pH is 7.0.

Source: sludge operating under alternating anaerobic and aerobic conditions.

DNA G+C content (mol%): 67.9 (HPLC).

Type strain: NM-1, ATCC 700054, CIP 104466, DSM 10555, HAMBI 2303, JCM 9379, NBRC 101784, VKM Ac-1990.

Sequence accession no. (16S rRNA gene): D26169, Z78207.

2. **Microlunatus aurantiacus** Wang, Cai, Zhi, Zhang, Tang, Xu, Cui and Li 2008, 1875[VP]

au.ran.ti.a'cus. N.L. masc. adj. *aurantiacus* orange-colored, referring to the orange color of the colonies.

Cells are cocci, 0.9–1.3 μm in diameter. Colonies are very small (~0.5–2.0 mm in diameter after incubation for 4 d on ISP2 agar medium at 28°C), smooth, circular, convex, and orange–yellow. Growth occurs at 15–37°C and pH 7.0–7.5. Growth occurs in the absence of NaCl. Oxidase-negative. Nitrate is reduced under anaerobic conditions. Cells show a weak phosphate-accumulating activity. Acid is produced from amygdalin, fructose, D-glucose, lactose, maltose, mannose, melibiose, raffinose, sucrose, trehalose, xylitol, and L-xylose. Utilizes arbutin, fructose, D-adonitol, esculin, alanine, amygdalin, fumarate, inulin, D-mannose, methyl α-D-galactoside, methyl β-D-galactoside, methyl α-D-glucoside, raffinose, D-ribose, salicin, D-sorbitol, starch, sucrose, trehalose, Tween 20, urea, xylitol, DL-xylose, adenine, asparagine, arginine, glutamate, histidine, hypoxanthine, threonine, tyrosine, and xanthine as sole carbon sources. The phospholipid pattern consists of diphosphatidylglycerol, phosphatidylglycerol, and phosphatidylinositol.

Source: a soil sample collected from the rhizosphere of *Taxus chinensis* in Yunnan Province, China.

DNA G+C content (mol%): 70.9 (HPLC).

Type strain: YIM 45721, CCTCC AB 206067, DSM 18424.

Sequence accession no. (16S rRNA gene): EF601828.

3. **Microlunatus ginsengisoli** Cui, Im, Yin, Yang and Lee 2007, 715[VP]

gin.seng.i.so'li. N.L. n. *ginsengum* ginseng; L. n. *solum* soil; N.L. gen. n. *ginsengisoli* of soil of a ginseng field, the source of the type strain.

Cells are 0.5–0.8 μm in diameter. Colonies are very small, smooth, circular, non-glossy, yellowish, and convex. Grows well at 20–30°C and at pH 5.5–8.5. Grows on nutrient agar but not MacConkey agar. Growth on MR2A agar occurs both in the absence of NaCl and in the presence of 4.0% (w/v) NaCl. No growth at 6% NaCl and higher. Oxidase-negative. Nitrate, as a nitrogen source, is reduced under aerobic conditions, but not under anaerobic conditions. β-Galactosidase and gelatinase activities and the Voges–Proskauer reaction are positive (API 20E). The following compounds are utilized as sole carbon sources: D-glucose, L-rhamnose, D-fructose, D-lyxose, D-ribose, L-xylose, propionate, valerate, fumarate, salicin, lactate, malate, tartrate, sucrose, D-trehalose, D-raffinose, gluconate, D-adonitol, D-sorbitol, xylitol, amygdalin, inulin, dextran, alanine, asparagine, aspartate, histidine, phenylalanine, praline, and tyrosine.

Source: soil from a ginseng field in Pocheon province, South Korea.

DNA G+C content (mol%): 69.8 (HPLC).

Type strain: Gsoil 633, DSM 17942, JCM15306, KCTC 13940.

Sequence accession no. (16S rRNA gene): AB245389.

4. **Microlunatus panaciterrae** An, Im and Yoon 2008, 2736[VP]

pa.na.ci.ter'ra.e. N.L. n. *Panax -acis* scientific name for ginseng; L. n. *terra* soil; N.L. gen. n. *panaciterrae* of soil of a ginseng field.

Cells are cocci, 0.3–0.7 μm in diameter. Colonies grown on R2A agar for 5 d are smooth, circular, non-glossy, yellowish, convex, and 1–2 mm in diameter. Grows well at 20–30°C and at pH 5.0–9.0. Grows on nutrient agar, but not on MacConkey agar or trypticase soy agar. Growth on R2A agar occurs both in the absence of NaCl and in the presence of 5.0% (w/v) NaCl but not at 6.0% NaCl. Oxidase-positive. Anaerobic growth does not occur. Nitrate is not reduced under anaerobic condition. Acid is produced from amygdalin, L-arabinose, D-glucose, melibiose, and L-rhamnose. β-Galactosidase, β-glucosidase, gelatinase, and the Voges–Proskauer reaction are positive (API20E). The following compounds are utilized as sole carbon sources: D-arabinose, citrate, D-fucose, L-fucose, D-glucose, lactate, maltose, D-mannose, melibiose, L-rhamnose, D-ribose, salicin, D-sorbitol, and sucrose. Starch is degraded. The polar lipids detected are phosphatidylethanolamine, diphosphatidylglycerol, and phosphatidylglycerol.

Source: soil from a ginseng field in Pocheon Province, South Korea.

DNA G+C content (mol%): 65.1 (HPLC).

Type strain: Gsoil 954, DSM 18662, KCTC 13058.

Sequence accession no. (16S rRNA gene): AB271051.

5. **Microlunatus parietis** Kämpfer, Schäfer, Lodders and Martin 2010a, 2422[VP]

pa.ri'e.tis. L. gen. n. *parietis* of the wall of a house.

Cells are rod-shaped and cocci. Good growth occurs on R2A agar, tryptone soy agar, M79 (modified Letheen Broth) agar, and nutrient agar at 25–30°C. Utilizes arbutin, cellobiose, D-fructose, D-galactose, sucrose, salicin, trehalose, D-xylose, adonitol, *myo*-inositol, maltitol, and D-mannitol. Weakly utilizes *N*-acetyl-D-glucosamine. The polar lipid profile mainly consists of diphosphatidylglycerol, phosphatidylglycerol, phosphatidylinositol, two unknown phospholipids, and one unknown glycolipid.

Source: an indoor wall in Berlin, Germany.

DNA G+C content (mol%): not determined.

Type strain: 12-Be-011, CCM 7636, DSM 22083.

Sequence accession no. (16S rRNA gene): FN556016.

6. **Microlunatus soli** Kämpfer, Young, Busse, Chu, Schumann, Arun, Shen and Rekha 2010b, 827[VP]

so'li. L. gen. n. *soli* of soil, the source of the type strain.

Cocci, 1.0–1.5 μm in diameter. Good growth occurs on R2A agar and nutrient agar at 25–30°C. Oxidase-positive. Utilizes *N*-acetyl-D-glucosamine, arbutin, cellobiose, D-fructose, D-galactose, sucrose, salicin, trehalose, D-xylose, adonitol, *myo*-inositol, maltitol, D-mannitol, acetate, and propionate. Weakly utilizes L-malate, L-histidine, L-proline, and L-serine. Cell-wall peptidoglycan is A3γ′ containing LL-diaminopimelic acid and glycine. The polyamine pattern is composed of spermidine and spermine as major compounds. Major polar lipids are phosphatidylglycerol and an unknown phospholipid followed by diphosphatidylglycerol, some unknown phospholipids, and an unknown glycolipid. A significant amount of an unsaturated fatty acid, $C_{18:1}$ ω7*c*, is the major cellular fatty acid. A small amount of $C_{14:0}$ 2-OH is also found.

Source: spawn of the edible mushroom *Agaricus brasiliensis*.

DNA G+C content (mol%): not determined.

Type strain: CC-12602, CCM7685, DSM21800.

Sequence accession no. (16S rRNA gene): FJ807672.

Genus VIII. **Micropruina** Shintani, Liu, Hanada, Kamagata, Miyaoka, Suzuki and Nakamura, 2000, 205[VP]

Lyudmila I. Evtushenko

Mi.cro.prui′na. Gr. adj. *mikros* small, fine; L. fem. n. *pruina* hoarfrost; N.L. fem. n. *Micropruina* fine hoarfrost.

Spherical cells, 0.5–2.2 μm in diameter (mostly about 1.0 μm) that are arranged in pairs, short chains, or clusters. Nonmotile. Nonsporeforming. **Gram-stain-positive**. Capsules are produced. **Chemo-organotroph, with a respiratory type of metabolism. Catalase and oxidase activities are positive.** Nitrate is used as an electron acceptor under anaerobic conditions; no production of nitrogen gas. The only known species, *Micropruina glycogenica*, is a slowly growing organism. Carbohydrates, sugar alcohols, carbonic acids, and amino acids are utilized as carbon sources; oxidative acid production occurs. **Capable of accumulating cellular glycogen**. Mesophilic and neutrophilic; growth occurs at 20–35°C and at pH 6–8. The optimum temperature and pH are 30°C and 7.0.

The cell-wall peptidoglycan contains *meso*-diaminopimelic acid (*meso*-A_2pm). Menaquinones are the only detected respiratory quinines; the tetrahydrogenated menaquinone with nine isoprene units, **MK-9(H_4), is the predominant** component. The major cellular fatty acids are $C_{15:0}$ anteiso, $C_{14:0}$ iso, $C_{16:0}$ iso, and $C_{16:0}$. No tuberculostearic or other 10-methyl branched acids are detected. Mycolic acids are absent.

The type species was isolated from an anaerobic-aerobic sequential batch biofilter reactor. No evidence of pathogenic properties.

DNA G+C content (mol%): 70.5.

Type species: **Micropruina glycogenica** Shintani, Liu, Hanada, Kamagata, Miyaoka, Suzuki and Nakamura 2000, 206[VP].

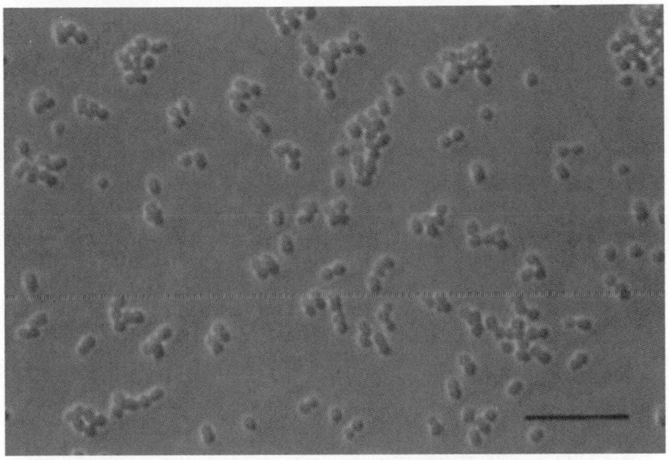

FIGURE 243. Phase-contrast photomicrograph of *Micropruina glycogenica* cells grown in NM-1 medium at 30°C. Bar = 10 μm. (Reproduced with permission from Shintani et al., 2000; Int. J. Syst. Evol. Microbiol. *50*: 201–207).

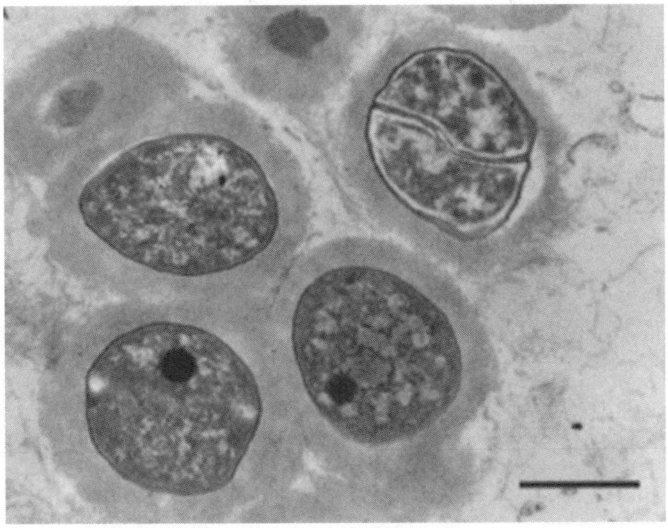

FIGURE 244. Transmission electron micrograph of *Micropruina glycogenica* cells with surrounding capsules. Cells were grown in NM-1 medium with a gentle shaking at 30°C. Bar = 1 μm. (Reproduced with permission from Shintani et al., 2000; Int. J. Syst. Evol. Microbiol. *50*: 201–207).

Further descriptive information

The growing cells divide at the middle by a septum to form pairs of cells which may be arranged in short chains, packets, or clusters (Figure 243 and Figure 244). Cells are individually surrounded by a capsule. The cell-wall structure is typical of Gram-stain-positive bacteria; Gram-stain-positive.

Little information is available on the habitats and ecology of organisms of the genus *Micropruina*. The type strain of *Micropruina glycogenica* was isolated from an anaerobic-aerobic sequential batch biofilter reactor exhibiting a biological phosphorus removal activity. However, neither orthophosphate uptake activity nor accumulation of intracellular polyphosphate has been detected in this bacterium by Neisser staining. Two organisms closely related to *Micropruina glycogenica* (>98% 16S rRNA gene sequence similarity) have been revealed in activated sludge and in sea water (GenBank accession numbers EU104262 and FJ545597, respectively).

Enrichment and isolation procedures

The *Microlunatus* agar medium (NM-1) containing glucose (0.5 g), peptone (0.5 g), monosodium glutamate (0.5 g), yeast extract (0.5 g), KH_2PO_4 (0.44 g), $(NH_4)_2SO_4$ (0.1 g), $MgSO_4 \cdot 7H_2O$ (0.1 g), agar (15 g), and 1 liter of distilled water (final pH, 7.0) (Nakamura et al., 1995) was used to isolate the type strain as described by Liu et al. (1997).

Maintenance procedures

Long-term conservation is achieved by freeze-drying and in liquid nitrogen.

Differentiation of the genus *Micropruina* from related genera

Tables 206, 208, and 212 lists characteristics differentiating *Micropruina* from the other genera of the family *Propionibacteriaceae*. Coccoid cells, along with the positive catalase and oxidase reactions and the presence of the predominant menaquinone MK-9(H_4), are the most salient characteristics that differentiate *Micropruina* from the phylogenetically closest genera *Propionicicella* and *Propionicimonas*. The extent of growth with (or without) oxygen may also be used for the purposes of delineating *Micropruina* from the facultativelly anaerobic *Propionicicella* and *Propionicimonas*. The *meso*-A_2pm isomer in the cell-wall peptidoglycan readily differentiates *Micropruina* from the phylogenetically close genera of coccoid bacteria which have LL-A_2pm as the diagnostic diamino acid.

Taxonomic comments

On the basis of 16S rRNA gene analysis, *Micropruina glycogenica* was originally shown to form a separate phylogenetic cluster with the genera *Microlunatus* and *Friedmanniella*, which was loosely associated with a cluster encompassing *Nocardioides* and closely related genera of the family *Nocardioidaceae* (Shintani et al., 2000). Further phylogenetic analysis with the extended selection of related sequences led Stackebrandt and Schaal (2006a) to the conclusion that the genus *Micropruina*, along with *Microlunatus* and *Friedmanniella*, should be placed in the family *Propionibacteriaceae*.

Acknowledgements

The author was supported by the program MCB of the Russian Academy of Sciences.

List of species of the genus *Micropruina*

1. **Micropruina glycogenica** Shintani, Liu, Hanada, Kamagata, Miyaoka, Suzuki and Nakamura 2000, 206[VP]

gly.co.ge.ni'ca. N.L. n. *glycogenum* glycogen; L. fem. suff. *-ica* suffix used with the sense of pertaining to; N.L. fem. adj. *glycogenica* referring to the ability to accumulate glycogen.

Morphological and general phenotypic characteristics of the species are the same as described for the genus. Punctuated nonpigmented colonies are formed after 2–3 weeks on NM-1 agar medium. Large amount (up to 8.4% dry wt) of intracellular glycogen is accumulated as a storage material under aerobic and anaerobic conditions.

Aerobic growth in the liquid NM-1 medium is very slow. The doubling time is approximately 12.6 h at 30°C (the optimum growth temperature). The temperature range for growth is 20–35°C; growth is not observed below 16°C or above 37°C. Tolerates NaCl up to 3.0% (w/v). The pH range for growth was 6–8, with no growth at pH 5 and 9.

Reduces nitrate to nitrite under anaerobic conditions, but formation of nitrogen gas is not observable. The following carbon sources are utilized aerobically as determined using the Biolog system: glucose, arabinose, cellobiose, galactose, lactose, maltose, mannose, melibiose, sucrose, xylose, inositol, mannitol, acetate, propionate, pyruvate, succinate, alanine, arginine, and histidine. Trehalose, dulcitol, ethanol, glycerol, methanol, propanol, malate, asparagine, glutamate, and glutamine are not utilized. Starch and glycogen are hydrolyzed.

The cellular fatty acids of the type strain include $C_{15:0}$ anteiso (37.1%), $C_{16:0}$ (12.9%), $C_{14:0}$ iso (14.1%), $C_{16:0}$ iso (10.8%), $C_{15:0}$ iso (5.9%), $C_{18:0}$ (4.6%), $C_{14:0}$ (3.7%), and $C_{15:0}$ iso (2.2%).

Source: an anaerobic-aerobic sequential batch biofilter reactor.

DNA G+C content (mol%): 70.5 (HPLC).

Type strain: strain Lg2, DSM 15918, JCM 10248.

Sequence accession no. (16S rRNA gene): AB012607.

Genus IX. **Propionicicella** Bae, Moe, Yan, Tiago, da Costa and Rainey 2006a, 2026[VP]

THE EDITORIAL BOARD

Pro.pi.o.ni.ci.cel'la. N.L. n. *acidum propionicum* propionic acid; L. fem. n. *cella* a storeroom, chamber, and in biology, a cell; N.L. fem. n. *Propionicicella* propionic-acid-producing cells.

Nonsporeforming rods, (0.5 × 1.0–2.5 μm). **Gram-stain-positive**. Nonmotile. **Facultative anaerobe**. Temperature range for growth 15–37°C, pH range 4.5–8.5, NaCl tolerance 0–4%. The peptidoglycan type is A1γ with *meso*-diaminopimelic (*meso*-DAP) in the cell wall. MK-9 is the major menaquinone. Propionate and acetate are produced when grown anaerobically with glucose.

DNA G+C content (mol%): 69.9.

Type species: **Propionicicella superfundia** Bae, Moe, Yan, Tiago, da Costa and Rainey 2006a, 2026[VP].

Further descriptive information

Major fatty acids are $C_{15:0}$ (54.3%), $C_{15:0}$ (16.3%), $C_{16:0}$ iso (12.8%), $C_{17:0}$ anteiso (5.4%), $C_{14:0}$ iso (5.3%), $C_{17:0}$ (2.9%) and $C_{16:0}$ (1.1%).

Phylogenetic analysis of the 16S rRNA gene suggests that the genus is affilitiated with the family *Propionibacteriaceae*. The closest phylogenetic neighbor is *Propionicimonas* sp. F6 (97.3% sequence similarity, accession AY570689), which was isolated from production waters of a low-temperature, biodegraded oil reservoir (Grabowski et al., 2005). Nearest type strains are *Propionicimonas paludicola* (accession no. AB078858; Akasaka et al., 2003b) and *Micropruina glycogenica* (accession no. AB012607; Shintani et al., 2000), 97.5% and 95.4%, respectively. An environmental clone with high 16S rRNA gene sequence similarity have been detected colonizing the gut epithelium (97.5%, accession no. AB198497) (Nakajima et al., 2005).

Enrichment and isolation procedures

Strain BL-10^T was isolated from groundwater samples collected from well W-1024-1 located in the dense non-aqueous-phase liquid (DNAPL) source zone at the Brooklawn portion of the Petro-Processors of Louisiana, Inc. (PPI) EPA Superfund Site, located approximately 10 miles north of Baton Rouge, LA (USA). Sterile, 1-l glass sample collection bottles were filled with groundwater, leaving little or no headspace, and placed on ice during transport to the laboratory (approx. 1 h). Serial diluents (anaerobically in 100 mM potassium phosphate buffer, pH 7.0) were transferred onto Columbia Anaerobic Sheep Blood agar plates (CASB, BBL) and incubated anaerobically under an atmosphere of 90% N_2, 5% CO_2, and 5% H_2 at 30°C for up to 5 weeks.

Maintenance procedures

Strains are maintained anaerobically on CASB or PYG media (Akasaka et al., 2003b). Stock cultures are maintained as a 15% (v/v) glycerol and 5% (v/v) dimethylsulfoxide suspension at −80°C.

Differentiation of the genus *Propionicicella* from closely related genera

Facultative anaerobic growth, rod-shaped morphology, and predominant menaquinone MK-9 differentiate *Propionicicella* from phylogenetically related genera *Ponticoccus*, *Friedmanniella*, *Microlunatus*, and *Micropruina*, which are all aerobic cocci with MK-9(H_4) as the predominant menaquinone. Predominant menaquinones MK-9(H_4) and MK-10(H_4) distinguishes *Propionicimonas* from *Propionicicella*. *Ponticoccus*, *Friedmanniella*, *Microlunatus*, and *Micropruina* are catalase-positive, and *Propionicicella* is catalase-negative. *Ponticoccus* and *Micropruina* reduce nitrate and *Propionicicella* does not.

List of species of the genus *Propionicicella*

1. **Propionicicella superfundia** Bae, Moe, Yan, Tiago, da Costa and Rainey 2006a, 2026[VP]

su.per.fun′di.a. L. prep. *super* above/on top; L. masc. n. *fundus* land owned by someone; L. adjectival ending *-ius -ia -ium* indicating the meaning of "belonging to"; N.L. fem. adj. *superfundia* referring to land designated as a US Environmental Protection Agency Superfund Site.

Colonies are white, circular, convex, smooth and 2–3 mm in diameter on PYG agar. Growth occurs at mesophilic temperatures and pH 4.5 and 8.5, with optima at 30°C and pH 6.5. NaCl tolerance ≤ 4.0% (w/v). Catalase- and oxidase-negative. Nitrate is not reduced. Glucose fermentation products are formic acid, acetic acid, propionic acid, and succinic. Utilizes adonitol, erythritol, fructose, glucose, glycerin, lactate, maltose, mannitol, mannose, pyruvate, sorbitol, sucrose, and xylose. Does not utilize acetate, arabinose crystalline cellulose, dulcitol, ethanol, fucose, galactose, lactose, malate, meliobiose, methanol, raffinose, rhamnose, starch, succinate, and fumarate. Rapid ID 32A kits (bioMérieux) detects α-glucosidase, β-glucosidase, glycine arylamidase, histidine arylamidase, proline arylamidase, leucyl glycine arylamidase, phenylalanine arylamidase, leucine arylamidase, and alanine arylamidase, but not glutamyl glutamic acid arylamidase, alkaline phosphatase, α-fucosidase, glutaminc acid decarboxylase, arginine dehydrogenase, urease, α-galactosidase, and β-galactosidase. Fermentation occurs in the presence of 0–9.8 mM and 0–5.9 mM 1,2-dichloroethane and 1,1,2-trichloroethane, respectively.

For all other characteristics, refer to the genus description.

DNA G+C content (mol%): 69.9 (HPLC).

Type strain: BL-10, ATCC BAA-1218, LMG 23096.

Sequence accession no. (16S rRNA gene): DQ176646.

Genus X. **Propionicimonas** Akasaka, Ueki, Hanada, Kamagata and Ueki 2003b, 1996[VP]

Atsuko Ueki, Hiroshi Akasaka and Katuji Ueki

Pro.pio.ni.ci.mo′nas. N.L. n. *acidum propionicum* propionic acid; L. fem. n. *monas* a unit, monad; N.L. fem. n. *Propionicimonas* propionic acid-producing monad.

Irregular, often slightly curved rods. Gram-stain-positive. Endospores are not produced. Nonmotile. **Facultatively anaerobic. Chemo-organotrophic.** Negative for oxidase, catalase, and nitrate reduction. Mesophilic. **In the presence of an excess amount of cobalamin, acetate and propionate are produced anaerobically as major fermentation products from glucose.** Cell-wall peptidoglycan contains ***meso*-diaminopimelic acid (DAP)**, and major cellular fatty acids are **$C_{13:0}$, $C_{15:0}$ anteiso, and $C_{15:0}$**. The major respiratory quinones are **MK-9(H_4) and MK-10(H_4)**. Based on 16S rRNA gene sequences, species in the genera *Micropruina*, *Microlunatus*, and *Friedmanniella* in the *Nocardioidaceae* are the most closely related to those of *Propionicimonas*. Isolated from rice plant residue (straw and roots) in irrigated rice field soil in Japan.

DNA G+C content (mol%): 68.7 (HPLC).

Type species: **Propionicimonas paludicola** Akasaka, Ueki, Hanada, Kamagata and Ueki 2003b, 1996[VP].

Further descriptive information

The following information is based on the description of strains of *Propionicimonas paludicola*. Grows weakly under aerobic conditions; much better growth occurs anaerobically. Colonies are white and 2–3 mm in diameter after 2–3 d of anaerobic cultivation. Growth rates in PYG medium (Holdeman et al., 1977) are very slow for most strains, but growth, as well as propionate production, is stimulated significantly by the addition of cobalamin to the medium. Some strains do grow well in PYG medium without cobalamin and produce substantial amounts of acetate and propionate; however, their other phenotypic and phylogenetic characteristics are almost identical to those of strains that do require cobalamin. The cell-wall structure is typical of Gram-stain-positive bacteria. Intracellular storage compounds in the cells are shown as electron-translucent regions by electron microscopy.

Isolation procedures

The plant residue samples collected from irrigated rice field soil are washed with anoxic diluent, cut to pieces, and homogenized using a Waring blender under N_2 gas. The homogenized samples are diluted consecutively with anoxic diluent and inoculated into the medium by the anaerobic roll-tube method (Holdeman et al., 1977). Colonies formed in the roll tubes during incubation for 2 weeks are picked up and strains can be purified by repeating the roll-tube method (Akasaka et al., 2003a). Strains can also often be isolated from roots of living rice plants of the same rice field.

TABLE 211. Differential characteristics of the genus *Propionicimonas* and closely related genera[a,b]

Characteristic	*Propionicimonas*	*Micropruina*	*Microlunatus*	*Friedmanniella*
Source	Plant residue in paddy soil	Activated sludge reactor	Activated sludge reactor	Antarctic sandstone
Cell shape	Irregular rods	Cocci (single, pair, or packet)	Cocci (single or pair)	Cocci (arranged in packet)
Cell size (μm)	0.4–0.5 × 1.4–2.2	0.5–2.2	0.8–2.0	0.5–2.2
Color of colony	White	White	Cream	Orange
Optimum growth temperature (°C)	35	30	25–30	20–26
O_2 requirement	Facultative anaerobe	Aerobe	Aerobe	Aerobe
Oxidase	–	+	+	–
Catalase	–	+	+	+
Nitrate reduction	–	+	+	–
Acid production from glucose	+	+	+	–
DNA G+C content (mol%)	67.4–68.7	70.5	67.9	69–74
Major quinone	MK-9(H_4), MK-10(H_4)	MK-9(H_4)	MK-9(H_4)	MK-9(H_4)
Major cellular fatty acids	$C_{13:0}$, $C_{15:0}$ anteiso, $C_{15:0}$	$C_{14:0}$ iso, $C_{15:0}$ anteiso, $C_{16:0}$ iso, $C_{16:0}$	$C_{15:0}$ iso, $C_{15:0}$ anteiso	$C_{15:0}$ iso, $C_{15:0}$ anteiso
Peptidoglycan	*meso*-DAP	*meso*-DAP	LL-DAP	LL-DAP

[a]Symbols: +, >85% positive; -, 0–15% positive.

[b]Data from Shintani et al. (2000); Nakamura et al. (1995); Schumann et al. (1997); Maszenan et al. (1999a); Lawson et al. (2000c).

Maintenance procedures

The strains are cultivated anaerobically at 30°C using PY medium (Holdeman et al., 1977) as a basal medium with oxygen-free N_2/CO_2 (95:5) mixed gas as headspace closed with butyl-rubber stoppers. PY medium supplemented with (per litre) 0.25 g each of glucose, cellobiose, maltose, and soluble starch, as well as 50 ml plant residue extract (RE) and 15 g agar (pH 7.3) is used for maintenance in agar slants. Rice plant residue collected from irrigated rice field soil during the flooding period is autoclaved (120°C for 30 min) with a fivefold amount (wet weight basis) of deionized H_2O, and the supernatant obtained after centrifugation is used as RE for the provision of growth factors (Akasaka et al., 2003a). RE prepared as above contains 3–5 μg/l cobalamin (Akasaka et al., 2004). Cobalamin itself is not usually added to the medium for slant cultures to avoid excess growth. Anaerobic slant cultures kept at 4°C can remain viable for at least several months, or the slant cultures can be maintained at −70°C without transfer for more than 5 years.

Taxonomic comments

Based on 16S rRNA gene sequences, the species forms a cluster close to the genera *Micropruina* (Shintani et al., 2000), *Microlunatus* (Nakamura et al., 1995), and *Friedmanniella* (Schumann et al., 1997) in the *Actinobacteria*. The closest relative is *Micropruina glycogenica* (Shintani et al., 2000), with sequence similarity of 95.8%. The closest relative to the type strain of *Propionicimonas paludicola* is an environmental clone, SJA-181, which is derived from an anaerobic microbial consortium in a trichlorobenzene-transforming bioreactor (16S rRNA gene sequence similarity of 98.3%). *Propionicimonas paludicola* is only distantly related to the propionate-producing species *Propionibacterium propionicus* and *Propioniferax innocua* in the *Propionibacteriaceae*, with sequence similarities of 91.5–92.0% and 90.0%, respectively.

Differentiation of the genus *Propionicimonas* from other genera

Unlike *Propionicimonas paludicola*, all species in the closely related genera *Micropruina*, *Microlunatus*, and *Friedmanniella* have spherical cells, often arranged in packets. Related species are strictly aerobic and catalase-positive. Both *Micropruina* and *Microlunatus* species have oxidase and nitrate-reducing activities. Major cellular fatty acids of these relatives are $C_{14:0}$ iso, $C_{15:0}$ iso, $C_{16:0}$ iso, and $C_{16:0}$ in addition to $C_{15:0}$ anteiso. $C_{13:0}$ and $C_{15:0}$ fatty acids are absent or only present as minor components. All relatives also have MK-9(H_4) as a major respiratory quinone, but do not contain MK-10(H_4), which is found in *Propionicimonas paludicola* (Lawson et al., 2000c; Maszenan et al., 1999a; Nakamura et al., 1995; Schumann et al., 1997; Shintani et al., 2000). Among the close relatives, only *Micropruina* species has *meso*-DAP in the peptidoglycan of the cell wall (Shintani et al., 2000); species in other two genera have LL-DAP (Table 211).

List of species of the genus *Propionicimonas*

1. **Propionicimonas paludicola** Akasaka, Ueki, Hanada, Kamagata and Ueki 2003b, 1996

pa.lu.di'co.la. L. n. *palus -udis* a swamp, marsh; L. suff. *-cola* (from L. n. *incola*) inhabitant, dweller; N.L. masc. n. *paludicola* an inhabitant of swamps.

Cells are pleomorphic rods, frequently arranged in irregular V- or crescent-shapes, 1.8–2.0 × 0.4–0.5 μm. In the presence of excess amounts of cobalamin in the medium, all strains produce acetate and propionate in the molar ratio of 2:1, with small amounts of lactate and succinate, from glucose; however, lactate production is predominant under cobalamin limitation for cobalamin-requiring strains. Optimal growth occurs at 35°C and pH 6.5. The species grows in the presence of up to 2.0% (w/v) NaCl. Acids are produced from arabinose, xylose, fructose, galactose, glucose, mannose, cellobiose, maltose, sucrose, trehalose, glycerol, and mannitol. Ribose and lactose are poorly utilized. Grows on pyruvate and lactate, and very weakly on malate, fumarate, and succinate. Does not utilize fucose, rhamnose, sorbose, melibiose, melezitose, raffinose, cellulose, glycogen, soluble starch, xylan, adonitol, dulcitol, erythritol, inositol, sorbitol, ethanol, methanol, or propanol. Other characteristics are as described for the genus.

Source: isolated from rice plant residue (straw and roots) in irrigated rice field soil in Japan.

DNA G+C content (mol%): 68.7 (HPLC).

Type strain: JCM 11033, DSM 15597.

Sequence accession no. (16S rRNA gene): AB078858.

Genus XI. **Propioniferax** Yokota, Tamura, Takeuchi, Weiss and Stackebrandt 1994, 581[VP]

Akira Yokota

Pro.pio.ni.fe'rax. N.L. n. *acidum propionicum* propionic acid; L. adj. *ferax* fertile; N.L. fem. n. *propioniferax* propionic acid-producing.

Cells are **Gram-stain-positive**, non-acid-fast, **nonmotile**, nonsporing, **pleomorphic rods**, appearing in clusters and V forms. **Facultatively anaerobic** but luxuriant growth occurs aerobically. Oxidase- and catalase-positive. Cell wall contains **LL-diaminopimelic acid**, arabinose, and mannose, but not galactose. Glycine forms the interpeptide bridge of peptidoglycan. Mycolic acids are not present. Major fatty acids are $C_{15:0}$ anteiso and $C_{15:0}$ iso. The major respiratory quinone is **MK-9(H_4)**.

DNA G+C content (mol%): 59–63 (T_m).

Type species: **Propioniferax innocua** Yokota, Tamura, Takeuchi, Weiss and Stackebrandt 1994, 581[VP].

Further descriptive information

Propionibacterium innocuum was proposed by Pitcher and Collins (1991) for a coryneform bacterium species containing LL-diaminopimelic acid (LL-A_2pm) in the cell wall. However, in contrast to other species of the genus *Propionibacterium* (Charfreitag et al., 1988; Cummins and Johnson, 1986), strains of *Propionibacterium innocuum* showed aerobic growth, did not require blood, serum, or Tween 80 for growth, and contained arabinose in cell-wall hydrolyzates. In addition, 16S rRNA analysis indicated that *Propionibacterium innocuum* was only remotely related to other members of the genus *Propionibacterium.* On the basis of physiological and chemotaxonomic characteristics, and a 16S rRNA gene sequence comparison, the species *Propionibacterium innocuum* was transferred from the genus *Propionibacterium* to a new genus, *Propioniferax*, as *Propioniferax innocua* by Yokota et al. (1994). The genus *Propioniferax* is a single species genus. Some characteristics in the original description of the genus by Yokota et al. (1994) were transferred to the species description. The genus *Propioniferax* is a member of the family *Propionibacteriaceae* within the order *Propionibacteriales* represented by the genus *Propionibacterium* Orla-Jensen 1909 (Figure 245).

Isolation and maintenance procedures

Strains of *Propioniferax innocua* originated from human skin, isolated from laboratories in Philadelphia (USA), Leiden (The Netherlands), and London (UK). Strains were routinely cultured aerobically on nutrient agar (Oxoid) at 37°C. Anaerobic growth on nutrient agar under an atmosphere of 95% H_2 and 5% CO_2 was much reduced. Cells grow well in CPY broth (casein peptone, 10 g; yeast extract, 5 g; and H_2O, 1 liter; pH 7.2) and tryptone soy broth (Difco) supplemented with 1% yeast extract.

Strains of the genus *Propioniferax* can be preserved by lyophilization suspended in 10% skim milk containing 1% monosodium glutamate, as well as by deep-freezing suspended in 10–20% glycerol solution at temperatures below −80°C.

Differentiation of the genus *Propioniferax* from other genera

The genus *Propioniferax* can be differentiated from closely related genera in the family *Propionibacteriaceae* based on chemotaxonomic and phylogenetic data, as shown in Table 212.

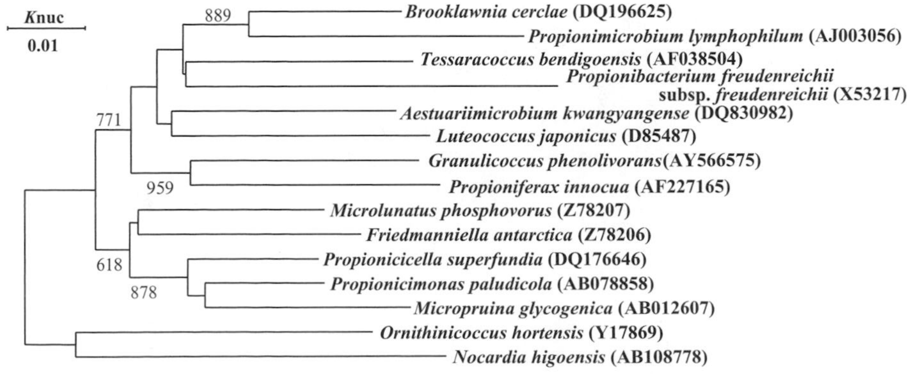

FIGURE 245. Phylogenetic relationships of the type strains of *Propioniferax* and named genera of the family *Propionibacteriaceae.* The neighbor-joining method was used for tree construction, 1000 bootstrap trees were generated and bootstrap values (>600; out of 1000 reiterations) are indicated at branch points.

TABLE 212. Differential characteristics of the genus *Propioniferax* and other genera of the family *Propionibacteriaceae*[a]

Characteristic	*Propioniferax*	*Propionibacterium*	*Aestuariimicrobium*	*Brooklawnia*	*Friedmanniella*	*Granulicoccus*	*Luteococcus*	*Microlunatus*	*Micropruina*	*Propionicicella*	*Propionicimonas*	*Propionimicrobium*	*Tessaracoccus*
O_2 requirement	Facultative anaerobes	Facultative anaerobes	Aerobes	Facultative anaerobes	Aerobes	Facultative anaerobes	Facultative anaerobes	Aerobes	Aerobes	Facultative anaerobes	Facultative anaerobes	Anaerobes	Facultative anaerobes
Cell morphology	Pleomorphic rods	Pleomorphic rods	Short rods or cocci	Pleomorphic rods	Cocci in packets	Cocci, singly and in pairs	Cocci, singly and in pairs	Cocci, singly and in pairs	Cocci	Rods	Irregular rods	Pleomorphic rods	Cocci, in tetrads
Cell size, μm	0.5–1.2	0.2–0.8	nk	nk	0.5–2.2	0.3–1.4	0.7–1.0	0.8–2.0	0.5–2.2	0.5–1.7	0.4–0.5 × 1.8–2.0	1.0–2.5	0.8–2.0
Isolation source(s)	Human epidermal surface	Human oral cavity, cervicovaginal secretion	Tidal flat	Chlorosolvent-contaminated groundwater	Sandstone of Antarctica	Phenol-degrading aerobic granules	Soil, water	Sewage treatment plant	Activated sludge reactor	Groundwater	Plant residue in paddy soil	Lymph nodes of patients	Sewage treatment plant
Growth temperature (°C):													
Optimum	37	35–37	30	37	9–25	30	26–28	25–30	nd	30	35	36–37	25
Range	10–40	30–37	4–40	10–40	22	15–37	12–38	5–35	20–30	15–37	10–40	nd	20–37
pH for growth:													
Optimum	7.0	nd	7.5–8.5	6.5	6.0–7.2	7.0	nd	7.0	nd	6.5	6.5	nd	7.5
Range	nd	nd	nd	4.5–8.0	nd	5.0–8.5	nd	5.0–9.0	6–8	4.5–8.5	4.5–7.5	nd	5.5–9.3
Presence of metachromatic granules	+, NK	nd	nd	nd	nd	+, Poly P	nd	+, Poly P	–	nd	nd	nd	+, Poly P
Oxidase	+	nd	–	–	–	–	+	+w	+	–	–	–	–
Catalase	+	–	+	+	+	+	+	+	+	–	–	v	+
Production of indole	–	–	nd	nd	nd	–	–	+	nd	nd	nd	–	–
Production of H_2S	–	+	nd	nd	+	–	–	nd	nd	nd	nd	nd	–
Major menaquinone(s)	MK-9(H_4)	MK-9(H_4)	MK-9(H_4)	MK-9(H_4)	MK-9(H_4)	MK-9(H_4), MK-8(H_4)	MK-9(H_4)	MK-9(H_4)	MK-9(H_4)	MK-9	MK-9(H_4), MK-10(H_4)	MK-9(H_4)	MK-9(H_4), MK-7(H_4)
A_2pm/murein type	LL-A_2pm/A3-γ	LL-A_2pm/A3-γ′	LL-A_2pm	*meso*-A_2pm/A1-γ	LL-A_2pm/A3-γ′	LL-A_2pm/A3-γ	LL-A_2pm/A3-γ	LL-A_2pm/A3-γ′	*meso*-A_2pm	*meso*-A_2pm	*meso*-A_2pm	Lysine/A4-α	LL-A_2pm/A3-γ′
Polar lipids[b]	PE, PG, PL, GL	nd	nd	nd	PG, DPG, PI, PL	PG, DPG	PG, DPG, PI, GL	PG, DPG, PI, GL	nd	nd	nd	nd	PG, DPG, PI, GL
Urease	+	–	–	nd	nd	+	–	+	nd	nd	nd	nd	–
Nitrate reduction	+	+	+	–	nd	–	–	+	+	–	–	v	+
DNA G+C content (mol%)	59–63	63–65	69	68	73	69	66–68	68	71	70	69	56	74

[a]Data for the various genera are from the following sources: *Aestuariimicrobium*, Jung et al. (2007); *Brooklawnia*, Bae et al. (2006b); *Friedmanniella*, Schumann et al. (1997); *Granulicoccus*, Maszenan et al. (2007); *Luteococcus*, Tamura et al. (1994); *Microlunatus*, Nakamura et al. (1995); *Micropruina*, Shintani et al. (2000); *Propionicicella*, Bae et al. (2006c); *Propionicimonas*, Akasaka et al. (2003b); *Propionimicrobium*, Stackebrandt et al. (2002); *Tessaracoccus*, Maszenan et al. (1999b). Symbols: +, positive; –, negative; +w, weak positive; v, variable; nd, not determined; NK, not known.

[b]DPG, Diphosphatidylglycerol; GL, unknown glycolipid; PE, phosphatidylethanolamine; PG, phosphatidylglycerol; PI, phosphatidylinositol; PL, unknown phospholipid.

List of species of the genus *Propioniferax*

1. **Propioniferax innocua** (Pitcher and Collins 1992) Yokota, Tamura, Takeuchi, Weiss and Stackebrandt 1994, 581^VP (*Propionibacterium innocuum* Pitcher and Collins 1992, 327)

in.noc'u.a. L. fem. adj. *innocua* harmless.

Cells are Gram-stain-positive, non-acid-fast, nonmotile, nonsporing, pleomorphic rods appearing in clusters and V forms. Colonies are white, shining, and convex to domed. The optimum growth temperature is approximately 37°C. Grows at 10 and 40°C. Facultatively anaerobic, but substantial growth occurs aerobically (colonies are approximately 0.5–3 mm in diameter after growth on nutrient or horse blood agar at 37°C for 2 d). Catalase- and oxidase-positive. The principal carboxylic acid produced from glucose is propionic acid. Glucose, sucrose, maltose, trehalose, fructose, mannose, and glycerol are fermented. Inositol, arabinose, adonitol, salicin, erythritol, mannitol, cellobiose, dulcitol, raffinose, lactose, sorbose, sorbitol, and rhamnose are not utilized, and galactose utilization is variable. Gelatin is hydrolyzed. Growth occurs in the presence of 7.5% NaCl, but not in 10% NaCl. Reduces nitrate. Hydrolyzes urea and starch. Growth is not stimulated by Tween 80, but Tween 80 is hydrolyzed. The following tests are negative: arginine dihydrolase, esculin hydrolysis, glutamate oxidation, arginine, lysine and ornithine decarboxylase, phenylalanine deaminase, citrate and malonate utilization, indole, H_2S, and decomposition of tyrosine and xanthine.

The cell wall contains LL-A_2pm, arabinose, and mannose, but not galactose. Major fatty acids are $C_{15:0}$ anteiso and $C_{15:0}$ iso. The major respiratory quinone is MK-9(H_4).

Source: human epidermal surface.

DNA G+C content (mol%): 59 to 63 (T_m).

Type strain: strain L60, ATCC 49929, CCUG 33480, DSM 8251, JCM 13395, LMG 16732, NCTC 11082.

Sequence accession no. (16S rRNA gene): AF227165.

Genus XII. **Propionimicrobium** Stackebrandt, Schumann, Schaal and Weiss 2002, 1926^VP

Erko Stackebrandt

Pro.pio.ni.mi.cro'bi.um. N.L. *acidum propionicum* propionic acid; N.L. neut. n. *microbium* (from Gr. adj. *mikros* small and Gr. n. *bios* life) a microbe; N.L. neut. n. *Propionimicrobium* propionic acid-producing microbe.

Pleomorphic rods, 0.5–0.8 × 1–2.5 μ, often diphtheroid or club-shaped. **Cells may be coccoid**. They may occur singly, in pairs or short chains, in V or Y configurations, or in clumps. Gram-stain-positive, nonmotile, nonsporeforming chemo-organotroph. **Anaerobic**. **Fermentation end products include propionic acid, acetic acid, succinic acid, iso-valeric acid, and lesser amounts of formic acid**. One out of two strains is catalase-positive. The major menaquinone is MK-9(H_4). **The peptidoglycan contains lysine and aspartic acid (Lys–Asp type)**. Major fatty acids are $C_{18:1}$ ω9*c*, $C_{15:0}$ anteiso, and $C_{16:0}$. The genus is a member of the family *Propionibacteriaceae* Delwiche (1957), emend. Stackebrandt, Rainey and Ward-Rainey (1997) *in* Rainey, Ward-Rainey and Stackebrandt (1997).

DNA G+C content (mol%): 53–56.

Type species: **Propionimicrobium lymphophilum** (Torrey 1916) Stackebrandt, Schumann, Schaal and Weiss 2002, 1926^VP ("*Bacillus lymphophilus*" Torrey 1916; *Propionibacterium lymphophilum* Johnson and Cummins 1972, 1057).

Further descriptive information

The exclusion of *Propionimicrobium lymphophilum* species from the genus *Propionibacterium* (Stackebrandt et al., 2002) was based upon phylogenetic evidence (Dasen et al., 1998) and information given by Cummins and Johnson (1986) based on the characterization of four strains (Holdeman et al., 1977; Johnson and Cummins, 1972). The description of the species is based on two strains, VIP 0202 and VIP 0383, which show 75% DNA similarity to one another. *Propionimicrobium lymphophilum* DSM 4309^T shares less than 91.8% sequence similarity with the *Propionibacterium* species (Dasen et al., 1998) which themselves show more than 93% similarity among each other. Depending upon the algorithm used, analysis of the 16S rRNA gene indicates that *Propionimicrobium lymphophilum* forms either the deepest branch of the genus *Propionibacterium* (Koussémon et al., 2001) (neighbor-joining program; Felsenstein, 1993), or branches among the other genera of the family *Propionibacteriaceae* (maximum-likelihood; Felsenstein, 1993) and distance matrix analyses (De Soete, 1983).

Enrichment and isolation procedures

The organism was first isolated by Torrey (1916) from lymph glands of a patient with Hodgkin's disease. Later, strains were isolated from urinary tract infections and from a mesenteric ganglion of a monkey inoculated with an unidentified "actinobacterium" isolate. Strains of the species have been reported to thrive in soil of rice paddy fields (Hayashi and Furusaka, 1980) and have been found in preparations of green olives (Cancho et al., 1980), but the identification of these strains did not include verification of the distinct cell-wall composition or the base composition of DNA (Cummins and Johnson, 1986).

Maintenance procedures

The organism can be maintained on modified Peptone-Yeast-Glucose agar (DSMZ medium 104; DSMZ (2001) or Columbia agar supplemented with 5% defibrinated sheep blood (Becton and Dickinson). Plates are sealed in plastic bags containing an Anerocult A (Merck) bag. Storage is in 20% (v/v) glycerol at −80°C or as lyophilized cultures.

Taxonomic comments

Strain VIP 0202 was originally described as "*Bacillus lymphophilus*" Torrey 1916, then as "*Corynebacterium lymphophilum*" (Torrey 1916) Eberson 1918 and as "*Mycobacterium lymphophilum*" (Torrey 1916) Krasil'nikov 1949, before it was included in a taxonomic study on coryneforms and propionibacteria (Johnson and Cummins, 1972). While membership of strain VIP 0202 in

Corynebacterium was excluded because of low DNA reassociation with members of this genus, it was tentatively classified as *Propionibacterium lymphophilum* based on its anaerobic growth and its ability to form propionic acid. *Propionibacterium lymphophilum* was included in the *Approved Lists of Bacterial Names* (Skerman et al., 1980), with reference to the publication of Johnson and Cummins (1972) and to the non-formal description of physiological and morphological properties by Holdeman et al. (1977) in the *Anaerobe Laboratory Manual of the Virginia Polytechnic Institute and State University*, Blacksburg, VA (VIP). The formal description of the species was made by Cummins and Johnson (1986) in their coverage of the genus *Propionibacterium* in *Bergey's Manual of Systematic Bacteriology*. Interestingly, both Holdeman et al. (1977) and Skerman et al. (1980) indicated strain VIP 7625B^T (=ATCC 27520^T) as the type strain, but neither Johnson and Cummins (1972), Holdeman et al. (1977), nor Cummins and Johnson (1986) refer to strain VIP 7625B^T but to strain VIP 0202. According to information obtained from the Technical Services of the ATCC in August 2001, strain ATCC 27520^T had originally been received as strain VIP 7625B^T.

In order to determine, whether strain VIP 7625B^T (=ATCC 27520^T = DSM 4309^T) was a subculture of strain VIP 0202, Stackebrandt et al. (2002) reinvestigated the base composition of DNA and the composition of peptidoglycan of strain DSM 4309^T, originally obtained for strain VIP 0202. The presence of lysine and the A4α (Lys–Asp) peptidoglycan type were confirmed as was the rather low base composition of DNA of 56 mol% G+C, which was slightly higher than the 53–54 mol% reported for strain VIP 0202.

Differentiation of the genus *Propionibacterium* from other genera

16S rRNA gene sequence analysis reveals the moderate 16S rRNA gene sequence relationship of strain DSM 4903^T to members of the genus *Propionibacterium* (>92%). The position of *Propionibacterium lymphophilum* outside the *Propionibacterium* cluster is not supported by high bootstrap values and its position changes with the algorithm applied. While it branches with *Tessarococcus bendigoensis* ACM 5119^T and *Propionibacterium propionicum* DSM 43307^T in the tree of Stackebrandt et al. (2002), it branches with *Luteococcus japonicus* IFO 12422^T in the tree of Dasen et al. (1998).

Physiologically the species can be differentiated from members of the genus *Propionibacterium* by a combination of results from esculin hydrolysis, indole production, and nitrate reduction, as well as by a combination of acid production from adonitol, erythritol, maltose, ribose, and L-sorbose, (Koussémon et al., 2001; Kusano et al., 1997). The composition of major fatty acids consisting of $C_{18:1}$ ω9*c*, $C_{15:0}$ anteiso, and $C_{16:0}$ differs significantly from those reported for *Propionibacterium* strains in which branched fatty acids (Moss et al., 1969) or ω-cyclohexane (Kusano et al., 1997) dominate.

List of species in the genus *Propionimicrobium*

1. **Propionimicrobium lymphophilum** (Torrey 1916) Stackebrandt, Schumann, Schaal and Weiss 2002, 1926^{VP} ("*Bacillus lymphophilus*" Torrey 1916; *Propionibacterium lymphophilum* Johnson and Cummins 1972, 1057)

lym.pho'phi.lum L. fem. n. *lympha* clear water and, in biology, lymph; N.L. neut. adj. *philum* (from Gr. neut. adj. *philon*), friend, loving; N.L. neut. adj. *lymphophilum*, lymph-loving.

Surface colonies on horse blood in 4 d are punctiform to 0.5 mm, circular, entire, convex to pulvinate, white, glistening, and smooth. Glucose cultures (24 h) are turbid, becoming clear, with a ropy sediment and a terminal pH of 5.4–5.7. Anaerobic, producing no growth on the agar surface when incubated aerobically, but growth develops in deep broth incubated aerobically. Acid production from adonitol, erythritol, fructose, glucose, maltose, and ribose, as well as from starch and inositol (reaction positive in 40–90% of strains). Acid is not formed from amygdalin, arabinose, cellobiose, dulcitol, esculin, galactose, glycerol, glycogen, inulin, lactose, mannitol, mannose, melezitose, raffinose, rhamnose, salicin, sorbitol, or sorbose. Nitrate reduction is positive for some strains. Esculin and gelatin are not hydrolyzed; indole, and acetoin are not produced. Hemolysis-negative. Weak production of gas and weak growth in 20% bile.

Cell walls contain alanine, glutamic acid, lysine, and aspartic acid. Glucose, galactose, and mannose are the principal cell-wall sugars. Major fatty acids are $C_{18:1}$ ω9*c*, $C_{15:0}$ anteiso, and $C_{16:0}$; smaller amounts (>2 < 5%) of $C_{14:0}$, $C_{15:0}$ iso, $C_{17:0}$ anteiso, and $C_{18:1}$ ω7*c* are present.

Source: urinary tract infections and a mesenteric ganglion of a monkey. Strains originally described by Torrey (1916) were from lymph nodes of patients suffering from Hodgkin's disease.

DNA G+C content (mol%): 53–54 mol% (T_m), 56 (HPLC).

Type strain: ATCC 27520, CCUG 27816, CIP 103263, DSM 4903, JCM 5829, LMG 16728, NCTC 11866, VPI 7625B.

Sequence accession no. (16S rRNA gene): AJ003056.

Genus XIII. **Tessaracoccus** Maszenan, Seviour, Patel, Schumann and Rees 1999b, 466^{VP}

Robert J. Seviour and Abdul M. Maszenan

Tes.sa.ra.coc'cus. Gr. adj. num. *tessares* four; N.L. masc. n. *coccus* (from Gr. masc. n. *kokkos* grain, seed) coccus; N.L. masc. n. *Tessaracoccus*, four round cells.

Gram-stain-positive, nonsporeforming, facultatively anaerobic nonmotile cocci 0.5–1.1 μm in diameter, arranged in regular tetrads. **Contains polyphosphate granules. Oxidase-negative, catalase-positive, and nitrate reduced.** Cell wall contains LL-diaminopimelic acid with an **A3γ′ peptidoglycan type and a glycine residue at position 1 of the tetrapeptide. Major menaquinones are MK-9(H_4) and MK-7(H_4), while the polar lipids are phosphatidylinositol, phosphatidylglycerol, and diphosphatidylglycerol,**

together with three unidentified glycolipids. The major fatty acid is anteiso hexadecanoic acid. Growth at 20–37°C, with the optimal temperature at 25°C. Growth at pH 6.0–9.0 with optimum pH at 7.5.

DNA G+C content (mol%): 74.

Type species: **Tessaracoccus bendigoensis** Maszenan, Seviour, Patel, Schumann and Rees 1999b, 466[VP].

Further descriptive information

Tessaracoccus is currently placed as a member of the family *Propionibacteriaceae* emend. Rainey, Ward-Rainey and Stackebrandt despite different signature nucleotides in some positions (e.g. G–T instead of A–T and A instead of G at positions 602:636 and 686, respectively) from those proposed by Stackebrandt et al. (1997) to circumscribe members of this family (Maszenan et al., 1999b). Only one species, *Tessaracoccus bendigoensis*, is currently recognized. It is related to members of the genus *Propionibacterium* with *Propionibacterium cyclohexanicum*, *Propionibacterium freudenreichii*, *Propionimicrobium lymphophilum*, and *Propionibacterium propionicum* as its closest relatives (Figure 246). It can utilize a wide range of sugars and sugar derivatives, as well as organic acids, although it does not utilize most amino acids and amines (Maszenan et al., 1999b). *Tessaracoccus* was isolated from activated sludge and synthesizes polyphosphate under aerobic conditions, thereby raising the possibility that it might play some role in enhanced biological phosphate removal (EBPR). However, there are not any studies which show this organism occurs in high numbers in such systems, and 16S rRNA targeted probes have not been generated for its *in situ* identification. There is not any evidence that the organism can accumulate poly β-hydroxyalkanoates (Maszenan et al., 1999b), a feature generally considered necessary in any putative polyphosphate accumulating organism in such systems (Seviour et al., 2003). It has also been detected in, but not cultured from, a Cr (IV) degrading community by denaturing gradient gel electrophoresis profiling of its 16S rRNA gene fragments (Arias and Tebo, 2003).

Enrichment and isolation procedures

Tessaracoccus bendigoensis was isolated with difficulty by micromanipulation (Skerman, 1968) from activated sludge biomass taken from an anaerobic:aerobic laboratory scale EBPR sequencing batch reactor seeded from a full scale EBPR activated sludge plant (Maszenan et al., 1999b). The only medium successful for its isolation and subsequent growth was the GS medium described by Williams and Unz (1985).

Maintenance procedures

Tessaracoccus bendigoensis was cultivated on GS medium and incubated at 25°C. After purity was confirmed by microscopic examination, cultures were stored successfully on GS medium in 20% glycerol (v/v) at −80°C.

Taxonomic comments

Only a single strain of a single species of *Tessaracoccus* has been characterized and described so far, hence it seems likely that the genus description may require emending as more species are isolated and characterized. Based collectively on its G+C mol% composition, 16S rRNA gene sequence, menaquinone and fatty acid composition, and its peptidoglycan type (A3γ′), with LL-diaminopimelic acid (LL-A_2pm), and only a single glycine residue in its interpeptide bridge, and glycine at position 1 in the tetrapeptide, *Tessaracoccus* seems to differ markedly from all other genera in the family *Propionibacteriaceae* (Table 213). *Propionimicrobium lymphophilum*, for example, has aspartate as the interpeptide bridge and an A4α cell-wall type (Stackebrandt et al., 2002). In addition to the MK-9(H_4) common to all members of the family *Propionibacteriaceae*, it contains MK-7(H_4), while its predominant fatty acid is $C_{15:0}$ anteiso.

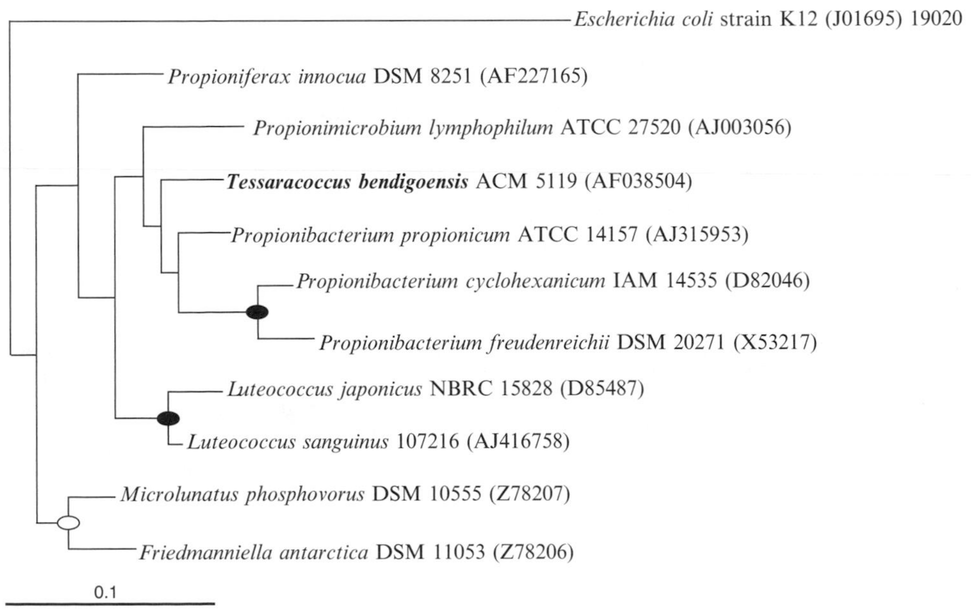

FIGURE 246. Phylogenetic tree based on 16S rRNA sequence data indicating the placement of *Tessaracoccus bendigoensis* within the family *Propionibacteriaceae.* Nodes receiving ≥75% and <97% bootstrap support (1000 replicates) with parsimony and neighbor-joining algorithms (○). Those receiving ≥97% are indicated (●). Scale bar = 0.1 nucleotide substitutions per site.

TABLE 213. Characteristics differentiating *Tessaracoccus bendigoensis* and related taxa[a]

Characteristics	*Tessaracoccus bendigoniensis*	*Luteococcus japonicus*	*Propionibacterium freudenreichii*	*Propionibacterium propionicum*	*Propioniferax innocua*	*Propionimicrobium lymphophilum*
Origin	Activated sludge	Soil and water	Cheese and dairy produce	Human oral cavity and cervicovaginal secretion	Human skin	Lymph nodes of patients
Cell shape and arrangement	Coccoid cells in tetrads	Coccoid cells; arranged in pairs and tetrads	Pleomorphic rods	Pleomorphic rods	Pleomorphic rods	Pleomorphic rods
Optimal temperature for growth	25	26–28	30–32	35–37	37	36–37
Optimal pH for growth (optimum)	6.0–9.0 (7.5)	nd	4.5–8.5	nd	7.0	nd
O_2 requirement	Facultatively anaerobic	Facultatively anaerobic	Anaerobic, aerotolerant	Facultatively anaerobic	Facultatively anaerobic	Anaerobic
Catalase	+	+	+	−	+	v
Oxidase	−	+	nd	nd	+	−
Nitrate reduction	+	−	v	+	+	v
DNA G+C content (mol%)	74	66–68	64–67	63–65	59–63	53–56
Major diamino acid	LL-A_2pm	LL-A_2pm	*meso*-A_2pm	LL-A_2pm	LL-A_2pm	Lysine
Peptidoglycan type	A3γ′	A3γ	nd	A3γ′	A3γ′	A4-α
Major menaquinone	MK-9(H_4), MK-7(H_4)	MK-9(H_4)	MK-9(H_4)	MK-9(H_4)	MK-9(H_4)	MK-9(H_4)

[a]Symbols: +, positive; −, negative; v, variable; nd, not determined.

[b]References: *Tessaracoccus bendigoniensis* (Maszenan et al., 1999b); *Luteococcus japonicus* (Maszenan et al., 1999b; Tamura et al., 1994); *Propionibacterium freudenreichii* (Bae et al., 2006b; Cummins and Johnson, 1986; Kusano et al., 1997); *Propionibacterium propionicum* (Cummins and Moss, 1990; Maszenan et al., 1999b); *Propioniferax innocua* (Maszenan et al., 1999b; Pitcher and Collins, 1991; Schumann et al., 1997; Yokota et al., 1994); *Propionimicrobium lymphophilum* (Stackebrandt et al., 2002).

List of species of the genus *Tessaracoccus*

1. **Tessaracoccus bendigoensis** Maszenan, Seviour, Patel, Schumann and Rees 1999b, 466[VP]

ben.di.go′en.sis. N.L. masc. adj. *bendigoensis* of or belonging to Bendigo, Australia, the place of origin of the isolate.

The following enzymes are detected by APIzyme: *N*-acetyl-β-glucosaminidase, acid phosphatase, alkaline phosphatase, cystine arylamidase, esterase, esterase lipase, α-galactosidase, β-galactosidase, α-glucosidase, β-glucosidase, gelatinase, leucine aryl amidase, lipase, α-mannosidase, naphthol-AS-B1-phosphohydrolase, and valine aryl amidase. Produces acetoin, but not H_2S or indole. It can liquefy gelatin, and grows at 20–37°C and pH 6.0–9.0. Optimal growth is at 25°C and optimal pH is 7.5.

The following substrates are utilized as detected with Biolog™: acetate, *N*-acetylglucosamine, amygdalin, L-arabinose, arbutin, α-cyclodextrin, β-cyclodextrin, dextrin, D-fructose, D-galactose, gentiobiose, D-gluconic acid, α-D-glucose, glucose 1-phosphate, glucose 6-phosphate, glycerol, DL-glycerol phosphate, glycogen, α-hydroxybutyrate, DL-lactate, L-lactate, α-D-lactose, lactulose, maltose, maltotriose, D-mannitol, D-mannose, D-melibiose, methyl-α-D-glucoside, methyl-β-D-glucoside, methyl pyruvate, palatinose, pyruvate, D-raffinose, D-ribose, salicin, stachyose, sucrose, turanose, D-xylose, uridine, and UMP.

Source: activated sludge.

DNA G+C content (mol%): 74 (T_m).

Type strain: Ben 106, ACM 5119, DSM 12906, JCM 13525.

Sequence accession no. (16S rRNA gene): AF038504.

References

Adams, M.C., M.L. Lean, N.C. Hitchick and K.W. Beagley. 2005. The efficacy of *Propionibacterium jensenii* 702 to stimulate a cell-mediated response to orally administered soluble *Mycobacterium tuberculosis* antigens using a mouse model. Lait *85*: 75–84.

Akasaka, H., T. Izawa, K. Ueki and A. Ueki. 2003a. Phylogeny of numerically abundant culturable anaerobic bacteria associated with degradation of rice plant residue in Japanese paddy field soil. FEMS Microbiol. Ecol. *43*: 149–161.

Akasaka, H., A. Ueki, S. Hanada, Y. Kamagata and K. Ueki. 2003b. *Propionicimonas paludicola* gen. nov., sp. nov., a novel facultatively anaerobic, Gram-positive, propionate-producing bacterium isolated from plant residue in irrigated rice-field soil. Int. J. Syst. Evol. Microbiol. *53*: 1991–1998.

Akasaka, H., K. Ueki and A. Ueki. 2004. Effects of plant residue extract and cobalamin on growth and propionate production of *Propionicimonas paludicola* isolated from plant residue in irrigated rice field soil. Microbes Environ. *19*: 112–119.

Aldave, A.J., J.D. Stein, V.A. Deramo, G.K. Shah, D.H. Fischer and J.I. Maguire. 1999. Treatment strategies for postoperative *Propionibacterium acnes* endophthalmitis. Ophthalmology *106*: 2395–2401.

Altenburger, P., P. Kämpfer, V.N. Akimov, W. Lubitz and H.-J. Busse. 1997. Polyamine distribution in actinomycetes with group B peptidoglycan and species of the genera *Brevibacterium*, *Corynebacterium*, and *Tsukamurella*. Int. J. Syst. Bacteriol. *47*: 270–277.

An, D.S., W.T. Im and M.H. Yoon. 2008. *Microlunatus panaciterrae* sp. nov., a beta-glucosidase-producing bacterium isolated from soil in a ginseng field. Int. J. Syst. Evol. Microbiol. *58*: 2734–2738.

Arias, M.Y. and B.M. Tebo. 2003. Cr(VI) reduction by sulfidogenic and nonsulfidogenic microbial consortia. Appl. Environ. Microbiol. *69*: 1847–1853.

Bae, H.-S., W.M. Moe, J. Yan, I. Tiago, M.S. da Costa and F.A. Rainey. 2006a. *In* List of new names and new combinations previously effectively, but not validly, published. Validation List no. 111. Int. J. Syst. Evol. Microbiol. *56*: 2025–2027.

Bae, H.S., W.M. Moe, J. Yan, I. Tiago, M.S. da Costa and F.A. Rainey. 2006b. *Brooklawnia cerclae* gen. nov., sp. nov., a propionate-forming bacterium isolated from chlorosolvent-contaminated groundwater. Int. J. Syst. Evol. Microbiol. *56*: 1977–1983.

Bae, H.S., W.M. Moe, J. Yan, I. Tiago, M.S. da Costa and F.A. Rainey. 2006c. *Propionicicella superfundia* gen. nov., sp. nov., a chlorosolvent-tolerant propionate-forming, facultative anaerobic bacterium isolated from contaminated groundwater. Syst. Appl. Microbiol. *29*: 404–413.

Ben-Shushan, G., V. Zakin and N. Gollop. 2003. Two different propionicins produced by *Propionibacterium thoenii* P-127. Peptides *24*: 1733–1740.

Benjelloun, H., M.R. Ravelona and J.M. Lebeault. 2007. Characterization of growth and metabolism of commercial strains of propionic acid bacteria by pressure measurement. Eng. Life Sci. *7*: 143–148.

Bernard, K.A., L. Shuttleworth, C. Munro, J.C. Forbes-Faulkner, D. Pitt, J.H. Norton and A.D. Thomas. 2002a. *In* Validation of the publication of new names and new combinations previously effectively published outside the IJSEM. List no. 88. Int. J. Syst. Evol. Microbiol. *52*: 1919–1916.

Bernard, K.A., L. Shuttleworth, C. Munro, J.C. Forbes-Faulkner, D. Pitt, J.H. Norton and A.D. Thomas. 2002b. *Propionibacterium australiense* sp. nov. derived from granulomatous bovine lesions. Anaerobe *8*: 41–47.

Bowden, G.H. 1991. *Actinomyces* and *Arachnia*. *In* Anaerobes in Human Disease (edited by Duerden and Drasar). Edward Arnold, London, pp. 132–161.

Branger, C., B. Bruneau and P. Goullet. 1987. Septicemia caused by *Propionibacterium granulosum* in a compromised patient. J. Clin. Microbiol. *25*: 2405–2406.

Brazier, J.S. and V. Hall. 1993. *Propionibacterium propionicum* and infections of the lacrimal apparatus. Clin. Infect. Dis. *17*: 892–893.

Brook, I. and E.H. Frazier. 1991. Infections caused by *Propionibacterium* species. Rev. Infect. Dis. *13*: 819–822.

Brüggemann, H., A. Henne, F. Hoster, H. Liesegang, A. Wiezer, A. Strittmatter, S. Hujer, P. Durre and G. Gottschalk. 2004. The complete genome sequence of *Propionibacterium acnes*, a commensal of human skin. Science *305*: 671–673.

Brüggemann, H. 2005. Insights in the pathogenic potential of *Propionibacterium acnes* from its complete genome. Semin. Cutan. Med. Surg. *24*: 67–72.

Brzin, B. 1964. Studies on the *Corynebacterium acnes*. Acta Pathol. Microbiol. Scand. *60*: 599–608.

Buchanan, B.B. and L. Pine. 1962. Characterization of a propionic acid producing actinomycete, *Actinomyces propionicus*, sp. nov. J. Gen. Microbiol. *28*: 305–323.

Busse, H.-J. and P. Schumann. 1999. Polyamine profiles within genera of the class *Actinobacteria* with LL-diaminopimelic acid in the peptidoglycan. Int. J. Syst. Bacteriol. *49*: 179–184.

Cancho, F.G., M. Nosti Vega, M. Fernandez Diaz and N.J.Y. Buzcu. 1970. Especies de *Propionibacterium* relacionades con la zapateria. Factores que influyen en su desarrollo. Microbiol. Esp. *23*: 233–252.

Cancho, F.G., L.R. Navarro and R. de la Borbolla y Alcala. 1980. La formacion de acido propionico durante la conservacion de las aceitunas verdes de mesa. III. Microorganismos responsables. Grasas Aceites *31*: 245–250.

Charfreitag, O., M.D. Collins and E. Stackebrandt. 1988. Reclassification of *Arachnia propionica* as *Propionibacterium propionicus* comb. nov. Int. J. Syst. Bacteriol. *38*: 354–357.

Charfreitag, O. and E. Stackebrandt. 1989. Inter- and intrageneric relationships of the genus *Propionibacterium* as determined by 16S rRNA sequences. J. Gen. Microbiol. *135*: 2065–2070.

Chaudhry, R., B. Dhawan, A. Pandey, S.K. Choudhary and A.S. Kumar. 2000. *Propionibacterium granulosum*: a rare cause of endocarditis. J. Infect. *41*: 284.

Chow, A.W., V. Patten and L.B. Guze. 1975. Susceptibility of anaerobic bacteria to metronidazole: relative resistance of non-spore-forming Gram-positive baccilli. J. Infect. Dis. *131*: 182–185.

Christie, R., N.E. Atkins and E. Munch-Peterson. 1944. A note on a lytic phenomenon shown by group B streptococci. Aust. J. Exp. Biol. Med. Sci. *22*: 197–200.

Christner, B.C. 2002. Recovery of bacteria from glacial and subglacial environments. PhD thesis, Ohio State University.

Chun, J., J.H. Lee, Y. Jung, M. Kim, S. Kim, B.K. Kim and Y.W. Lim. 2007. EzTaxon: a web-based tool for the identification of prokaryotes based on 16S ribosomal RNA gene sequences. Int. J. Syst. Evol. Microbiol. *57*: 2259–2261.

Clayton, J.J., W. Baig, G.W. Reynolds and J.A. Sandoe. 2006. Endocarditis caused by *Propionibacterium species*: a report of three cases and a review of clinical features and diagnostic difficulties. J. Med. Microbiol. *55*: 981–987.

Clement, T.P., M.J. Truex and P. Lee. 2002. A case study for demonstrating the application of U.S. EPA's monitored natural attenuation screening protocol at a hazardous waste site. J. Contam. Hydrol. *59*: 133–162.

Coenye, T., E. Peeters and H.J. Nelis. 2007. Biofilm formation by *Propionibacterium acnes* is associated with increased resistance to antimicrobial agents and increased production of putative virulence factors. Res. Microbiol. *158*: 386–392.

Cohen, R.J., B.A. Shannon, J.E. McNeal, T. Shannon and K.L. Garrett. 2005. *Propionibacterium acnes* associated with inflammation in radical prostatectomy specimens: a possible link to cancer evolution? J. Urol. *173*: 1969–1974.

Collins, M.D., R.A. Burton and D. Jones. 1988. *Corynebacterium-Amycolatum* Sp-Nov a New Mycolic Acid-Less *Corynebacterium* Species from Human-Skin. FEMS Microbiol. Lett. *49*: 349–352.

Collins, M.D., P.A. Lawson, N. Nikolaitchouk and E. Falsen. 2000. *Luteococcus peritonei* sp. nov., isolated from the human peritoneum. Int. J. Syst. Evol. Microbiol. *50*: 179–181.

Collins, M.D., R.A. Hutson, N. Nikolaitchouk, A. Nyberg and E. Falsen. 2003. *Luteococcus sanguinis* sp. nov., isolated from human blood. Int. J. Syst. Evol. Microbiol. *53*: 1889–1891.

Collins, M.D., L. Hoyles, G. Foster and E. Falsen. 2004. *Corynebacterium caspium* sp. nov., from a Caspian seal (Phoca caspica). Int. J. Syst. Evol. Microbiol. *54*: 925–928.

Cove, J.H., K.T. Holland and W.J. Cunliffe. 1983. Effects of oxygen concentration on biomass production, maximum specific growth rate and extracellular enzyme production by three species of cutaneous propionibacteria grown in continuous culture. J. Gen. Microbiol. *129*: 3327–3334.

Cui, Y.S., W.T. Im, C.R. Yin, D.C. Yang and S.T. Lee. 2007. *Microlunatus ginsengisoli* sp. nov., isolated from soil of a ginseng field. Int. J. Syst. Evol. Microbiol. *57*: 713–716.

Cummins, C.S. and J.L. Johnson. 1974. *Corynebacterium parvum*: a synonym for *Propionibacterium acnes*? J. Gen. Microbiol. *80*: 433–442.

Cummins, C.S. 1975. Identification of *Propionibacterium acnes* and related organisms by precipitin tests with trichloracetic acid extracts. J. Clin. Microbiol. *2*: 104–110.

Cummins, C.S. 1980. Serology of propionibacteria. *In* Anaerobic bacteria - selected topics (edited by Lambe, Genco and Mayberry-Carson). Plenum Press, New York, pp. 205–221.

Cummins, C.S. and J.L. Johnson. 1981. The genus *Propionibacterium*. *In* The Prokaryotes: a Handbook on Habitats, Isolation and Identification of Bacteria (edited by Starr, Stolp, Trüper, Balows and Schlegel). Springer, New York, pp. 1894–1902.

Cummins, C.S. and J.L. Johnson. 1986. Genus I. *Propionibacterium*. *In* Bergey's Manual of Systematic Bacteriology, vol. 2 (edited by Sneath, Mair, Sharpe and Holt). Williams & Wilkins, Baltimore, pp. 1346–1353.

Cummins, C.S. and C.W. Moss. 1990. Fatty acid composition of *Propionibacterium propionicum* (*Arachnia propionica*). Int. J. Syst. Bacteriol. *40*: 307–308.

Dasen, G., J. Smutny, M. Teuber and L. Meile. 1998. Classification and identification of Propionibacteria based on ribosomal RNA genes and PCR. Syst. Appl. Microbiol. *21*: 251–259.

de Carvalho, A.F., S. Guezenec, M. Gautier and P.A. Grimont. 1995. Reclassification of "*Propionibacterium rubrum*" as *P. jensenii*. Res. Microbiol. *146*: 51–58.

De Soete, G. 1983. A least squares alogorithm for fitting additive trees to proximity data. Psychometrika *48*: 621–626.

Delwiche, E.A. 1949. Vitamin requirements of the genus *Propionibacterium*. J. Bacteriol. *58*: 395–398.

Delwiche, E.A. 1957. Family *Propionibacteriaceae*. *In* Bergey's Manual of Determinative Bacteriology, 7th edn (edited by Breed, Murray and Smith). Williams & Wilkins, Baltimore, p. 569.

Dherbecourt, J., H. Falentin, S. Canaan and A. Thierry. 2008. A genomic search approach to identify esterases in *Propionibacterium freudenreichii* involved in the formation of flavour in Emmental cheese. Microb. Cell. Fact. *7*: 16.

Douglas, H.C. and S.E. Gunter. 1946. The taxonomic position of *Corynebacterium acnes*. J. Bacteriol. *52*: 15–23.

Downes, J. and W.G. Wade. 2009. *Propionibacterium acidifaciens* sp. nov., isolated from the human mouth. Int. J. Syst. Evol. Microbiol. *59*: 2778–2781.

DSMZ. 2001. Catalogue of Strains. German Collection of Microorganisms and Cell Cultures, 7th edn. DSMZ - Deutsche Sammlung von Mikroorganismen und Zellkulturen, Braunschweig, Germany.

Eady, E.A., J.H. Cove, K.T. Holland and W.J. Cunliffe. 1989. Erythromycin resistant propionibacteria in antibiotic treated acne patients: association with therapeutic failure. Br. J. Dermatol. *121*: 51–57.

Eberson, F. 1918. A bacteriologic study of the diphtheroid organisms with special reference to Hodgkin's disease. J. Infect. Dis. *23*: 1–42.

Eggerth, A.H. 1935. The gram-positive non-spore-bearing anaerobic bacilli of human feces. J. Bacteriol. *30*: 277–290.

Eishi, Y., M. Suga, I. Ishige, D. Kobayashi, T. Yamada, T. Takemura, T. Takizawa, M. Koike, S. Kudoh, U. Costabel, J. Guzman, G. Rizzato, M. Gambacorta, R. du Bois, A.G. Nicholson, O.P. Sharma and M. Ando. 2002. Quantitative analysis of mycobacterial and propionibacterial DNA in lymph nodes of Japanese and European patients with sarcoidosis. J. Clin. Microbiol. *40*: 198–204.

Farrar, M.D., K.M. Howson, R.A. Bojar, D. West, J.C. Towler, J. Parry, K. Pelton and K.T. Holland. 2007. Genome sequence and analysis of a *Propionibacterium acnes* bacteriophage. J. Bacteriol. *189*: 4161–4167.

Felsenstein, D. 1993. PHYLIP (Phylogeny Inference Package) 3.57 edn. Department of Genetics, University of Washington, Seattle.

Ferguson, D.A. and C.S. Cummins. 1978. Nutritional requirements of anaerobic coryneforms. J. Bacteriol. *135*: 858–867.

Fernández-Garayzábal, J.F., A.I. Vela, R. Egido, R.A. Hutson, M.P. Lanzarot, M. Fernandez-Garcia and M.D. Collins. 2004. *Corynebacterium ciconiae* sp. nov., isolated from the trachea of black storks (Ciconia nigra). Int. J. Syst. Evol. Microbiol. *54*: 2191–2195.

Fierer, N., M. Breitbart, J. Nulton, P. Salamon, C. Lozupone, R. Jones, M. Robeson, R.A. Edwards, B. Felts, S. Rayhawk, R. Knight, F. Rohwer and R.B. Jackson. 2007. Metagenomic and small-subunit rRNA analyses reveal the genetic diversity of bacteria, archaea, fungi, and viruses in soil. Appl. Environ. Microbiol. *73*: 7059–7066.

Fierer, N., Z. Liu, M. Rodriguez-Hernandez, R. Knight, M. Henn and M.T. Hernandez. 2008. Short-term temporal variability in airborne bacterial and fungal populations. Appl. Environ. Microbiol. *74*: 200–207.

Fink, H., M. Helming, C. Unterbuchner, A. Lenz, F. Neff, J.A. Martyn and M. Blobner. 2008. Systemic inflammatory response syndrome increases immobility-induced neuromuscular weakness. Crit. Care. Med. *36*: 910–916.

Flaminio, M.J., B.R. Rush and W. Shuman. 1998. Immunologic function in horses after non-specific immunostimulant administration. Vet. Immunol. Immunopathol. *63*: 303–315.

Forbes-Faulkner, J.C., D. Pitt, J.H. Norton, A.D. Thomas and K. Bernard. 2000. Novel *Propionibacterium* infection in cattle. Aust. Vet. J. *78*: 175–178.

Fujimura, S. and T. Nakamura. 1978. Purification and properties of a bacteriocin-like substance (acnecin) of oral *Propionibacterium acnes*. Antimicrob. Agents Chemother. *14*: 893–898.

Fujimura, S., H.L. Ko, G. Pulverer and J. Jeljaszewicz. 1982. Hemolysin of *Propionibacterium avidum*. Zentralbl. Bakteriol. Mikrobiol. Hyg. A. *252*: 108–115.

Garrity, G.M., T.G. Lilburn, J.R. Cole, S.H. Harrison, J. Euzéby and B.J. Tindall. 2007. Part 10 - The *Bacteria*: phylum "*Actinobacteria*": class *Actinobacteria*. *In* Taxonomic Outline of the Bacteria and Archaea, Release 7.7 (edited by Board of Trustees 2001–2007). Michigan State University, pp. 399–539.

Gerencser, M.A. and J.M. Slack. 1967. Isolation and characterization of *Actinomyces propionicus*. J. Bacteriol. *94*: 109–115.

Gilchrist, T.C. 1900. A bacteriological and microscopical study of over three hundred vesicular and pustular lesions of the skin, with a research upon the etiology of *Acne vulgaris*. Johns Hopkins Hospital Report *9*: 409–430.

Goodsell, M.E., J. Toth, J.L. Johnson and C.S. Cummins. 1991. Two types of *Propionibacterium avidum* with different isomers of diaminopimelic acid. Curr. Microbiol. *22*: 225–230.

Grabowski, A., O. Nercessian, F. Fayolle, D. Blanchet and C. Jeanthon. 2005. Microbial diversity in production waters of a low-temperature biodegraded oil reservoir. FEMS Microbiol. Ecol. *54*: 427–443.

Graham, G.M., M.D. Farrar, J.E. Cruse-Sawyer, K.T. Holland and E. Ingham. 2004. Proinflammatory cytokine production by human keratinocytes stimulated with *Propionibacterium acnes* and *P. acnes* GroEL. Br. J. Dermatol. *150*: 421–428.

Greenman, J., K.T. Holland and W.J. Cunliffe. 1981. Effects of glucose concentration n biomass, maximum specific growth rate and extracellular enzyme production by three species of cutaneous Propionibacteria grown in continuous culture. J. Gen. Microbiol. *127*: 371–376.

Greenman, J. 1995. *Propionibacterium* species. *In* Medical and Dental Aspects of Anaerobes (edited by Duerden, Wade, Brazier, Eley, Wren and Hudson). Science Reviews, Middlesex, pp. 9–18.

Hall, V., M.D. Collins, R.A. Hutson, P.A. Lawson, E. Falsen and B.I. Duerden. 2003. *Corynebacterium atypicum* sp. nov., from a human clinical source, does not contain corynomycolic acids. Int. J. Syst. Evol. Microbiol. *53*: 1065–1068.

Hall, V. 2006. Anaerobic *Actinomyces* and related organisms. *In* Principles and Practice of Clinical Bacteriology, 2nd edn. John Wiley & Sons, Chichester, pp. 575–586.

Hamana, K. 1995. Polyamine distribution patterns in coryneform bacteria and related Gram-positive eubacteria. Annu. Rep. Coll. Med. Care. Technol. Gunma. Univ. *16*: 69–77.

Harris, J.K., M.A. De Groote, S.D. Sagel, E.T. Zemanick, R. Kapsner, C. Penvari, H. Kaess, R.R. Deterding, F.J. Accurso and N.R. Pace. 2007. Molecular identification of bacteria in bronchoalveolar lavage fluid from children with cystic fibrosis. Proc. Natl. Acad. Sci. U.S.A. *104*: 20529–20533.

Hayashi, S. and C. Furusaka. 1979. Studies on *Propionibacterium* isolated from paddy soils. Antonie van Leeuwenhoek *45*: 565–574.

Hayashi, S. and C. Furusaka. 1980. Enrichment of *Propionibacterium* in paddy soil by addition of various organic substances. Antonie van Leeuwenhoek *46*: 313–320.

Herve, C., A. Coste, A. Rouault, J.M. Fraslin and M. Gautier. 2001. First evidence of lysogeny in *Propionibacterium freudenreichii* subsp. *shermanii*. Appl. Environ. Microbiol. *67*: 231–238.

Hettinga, D.H. and G.W. Reinbold. 1972. The propionic acid bacteria-a review. III. Miscellaneous metabolic activities. J. Milk Food Technol. *35*: 436–447.

Himmi, E.H., A. Bories, A. Boussaid and L. Hassani. 2000. Propionic acid fermentation of glycerol and glucose by *Propionibacterium acidipropionici* and *Propionibacterium freudenreichii* ssp. *shermanii*. Appl. Microbiol. Biotechnol. *53*: 435–440.

Hitchner, E.R. 1932. A cultural study of the propionic acid bacteria. J. Bacteriol. *23*: 40–41.

Hoeffler, U. 1977. Enzymatic and hemolytic properties of *Propionibacterium acnes* and related bacteria. J. Clin. Microbiol. *6*: 555–558.

Hoeffler, U. 1980. Production of hyaluronidase E.C.3.2.1.36 by propionibacteria from different origins. Zentralbl. Bakteriol. Parasitenkd. Infektionskr. Hyg. I. Abt. Orig. A *245*: 1–2.

Holdeman, L.V., E.P. Cato and W.E.C. Moore (editors). 1977. Anaerobe Laboratory Manual, 4th edn. Anaerobe Laboratory, Virginia Polytechnic Institute and State University, Blacksburg, VA.

Holland, K.T., J. Greenman and W.J. Cunliffe. 1979. Growth of cutaneous propionibacteria on synthetic medium; growth yields and exoenzyme production. J. Appl. Bacteriol. *47*: 383–394.

Holmberg, K. and U. Forsum. 1973. Identification of *Actinomyces, Arachnia, Bacterionema, Rothia,* and *Propionibacterium* species by defined immunofluorescence. Appl. Microbiol. *25*: 834–843.

Ingham, E., K.T. Holland, G. Gowland and W.J. Cunliffe. 1980. Purification and partial characterization of an acid phosphatase (EC 3.1.3.2) produced by *Propionibacterium acnes*. J. Gen. Microbiol. *118*: 59–65.

Ingham, E., K.T. Holland, G. Gowland and W.J. Cunliffe. 1981. Partial purification and characterization of lipase (EC 3.1.1.3) from *Propionibacterium acnes*. J. Gen. Microbiol. *124*: 393–401.

Ingram, E., K.T. Holland, G. Gowland and W.J. Cunliffe. 1983. Studies of the extracellular proteolytic activity produced by *Propionibacterium acnes*. J. Appl. Bacteriol. *54*: 263–271.

Isenberg, J., H. Ko, G. Pulverer, R. Grundmann, H. Stutzer and H. Pichlmaier. 1994. Preoperative immunostimulation by *Propionibacterium granulosum* KP-45 in colorectal cancer. Anticancer Res. *14*: 1399–1404.

Isenberg, J., B. Stoffel, U. Wolters, J. Beuth, H. Stutzer, H.L. Ko and H. Pichlmaier. 1995. Immunostimulation by propionibacteria – effects on immune status and antineoplastic treatment. Anticancer Res. *15*: 2363–2368.

Ishida, N., H. Nakaminami, N. Noguchi, I. Kurokawa, S. Nishijima and M. Sasatsu. 2008. Antimicrobial susceptibilities of *Propionibacterium acnes* isolated from patients with acne vulgaris. Microbiol. Immunol. *52*: 621–624.

Janicka, I., M. Maliszewska and F. Pedziwilk. 1976. Utilization of lactose and production of corrinoids in selected strains of propionic acid bacteria in cheese-whey and casein media. Acta Microbiol. Pol. *25*: 205–210.

Janoschek, A. 1944. Zur Systematik der Propionsäurebakterien. Zentralbl. Bakteriol. Parasitenkd. Infektionskr. Hyg. Abt. *2*: 321–337.

Jiang, H.-L., J.-H. Tay, A.M. Maszenan and S.T.-L. Tay. 2004. Bacterial diversity and function of aerobic granules engineered in a sequencing batch reactor for phenol degradation. Appl. Environ. Microbiol. *70*: 6767–6775.

Johnson, J.L. and C.S. Cummins. 1972. Cell wall composition and deoxyribonucleic acid similarities among the anaerobic coryneforms, classical propionibacteria, and strains of *Arachnia propionica*. J. Bacteriol. *109*: 1047–1066.

Joseph, S., J.P. Hugenholtz, P. Sangwan, C.A. Osborne and P.H. Janssen. 2003. Laboratory cultivation of widespread and previously uncultured soil bacteria. Appl. Environ. Microbiol. *69*: 7210–7215.

Jousimies-Somer, H.R., P. Summanen, D.M. Citron, E.J. Baron, H.M. Wexler and S.M. Finegold. 2002. Wadsworth-KTL Anaerobic Bacteriology Manual, 6th edn (edited by Finegold and Jousimies-Somer). Star Publishing Company, Belmont, CA.

Jung, S.Y., H.S. Kim, J.J. Song, S.G. Lee, T.K. Oh and J.H. Yoon. 2007. *Aestuariimicrobium kwangyangense* gen. nov., sp. nov., an LL-diaminopimelic acid-containing bacterium isolated from tidal flat sediment. Int. J. Syst. Evol. Microbiol. *57*: 2114–2118.

Kämpfer, P., J. Schäfer, N. Lodders and K. Martin. 2010a. *Microlunatus parietis* sp. nov., isolated from an indoor wall. Int. J. Syst. Evol. Microbiol. *60*: 2420–2423.

Kämpfer, P., C.C. Young, H.-J. Busse, J.N. Chu, P. Schumann, A.B. Arun, F.T. Shen and P.D. Rekha. 2010b. *Microlunatus soli* sp. nov., isolated from soil. Int. J. Syst. Evol. Microbiol. *60*: 824–827.

Kilian, M. 1978. Rapid identification of *Actinomycetaceae* and related bacteria. J. Clin. Microbiol. *8*: 127–133.

Kimura, M. 1980. A simple method for estimating evolutionary rates of base substitutions through comparative studies of nucleotide sequences. J. Mol. Evol. *16*: 111–120.

Kishishita, M., T. Ushijima, Y. Ozaki and Y. Ito. 1979. Biotyping of *Propionibacterium acnes* isolated from normal human facial skin. Appl. Environ. Microbiol. *38*: 585–589.

Kishishita, M., T. Ushijima, Y. Ozaki and Y. Ito. 1980. New medium for isolating propionibacteria and its application to assay of normal flora of human facial skin. Appl. Environ. Microbiol. *40*: 1100–1105.

Ko, H.L., G. Pulverer and J. Jeljaszewicz. 1978. Propionicins, bacteriocins produced by *Propionibacterium avidum*. Zentralbl. Bakteriol. Orig. A. *241*: 325–328.

Kotilainen, P., R. Merilahti-Palo, O.P. Lehtonen, I. Manner, I. Helander, T. Mottonen and E. Rintala. 1996. *Propionibacterium acnes* isolated from sternal osteitis in a patient with SAPHO syndrome. J. Rheumatol. *23*: 1302–1304.

Koussémon, M., Y. Combet-Blanc, B.K. Patel, J.L. Cayol, P. Thomas, J.L. Garcia and B. Ollivier. 2001. *Propionibacterium microaerophilum* sp. nov., a microaerophilic bacterium isolated from olive mill wastewater. Int. J. Syst. Evol. Microbiol. *51*: 1373–1382.

Krasil'nikov, N.A. 1949. Guide to the bacteria and actinomycetes. Akad. Nauk. S.S.S.R., Moscow.

Krecek, R.C., H.J. Els, S.C. de Wet and M.M. Henton. 1992. Studies on ultrastructure and cultivation of microorganisms associated with Zebra nematodes. Microb. Ecol. *23*: 87–95.

Kumar, S., K. Tamura and M. Nei. 2004. MEGA3: Integrated software for Molecular Evolutionary Genetics Analysis and sequence alignment. Brief Bioinform. *5*: 150–163.

Kurman, J. 1960. Ein vollsynthetischer Nährboden für Propionsäurebakterien. Pathol. Microbiol. *23*: 700–711.

Kusano, K., H. Yamada, M. Niwa and K. Yamasato. 1997. *Propionibacterium cyclohexanicum* sp. nov., a new acid-tolerant omega-cyclohexyl fatty acid-containing propionibacterium isolated from spoiled orange juice. Int. J. Syst. Bacteriol. *47*: 825–831.

Langsrud, T., G.W. Reinbold and E.G. Hammond. 1977. Proline production by *Propionibacterium shermanii* P59. J. Dairy Sci. *60*: 16–23.

Langsrud, T., G.W. Reinbold and E.G. Hammond. 1978. Free proline production by strains of propionibacteria. J. Dairy Sci. *61*: 303–308.

Lawson, P.A., M.D. Collins, P. Schumann, B.J. Tindall, P. Hirsch and M. Labrenz. 2000a. New LL-diaminopimelic acid-containing actinomycetes from hypersaline, heliothermal and meromictic Antarctic Ekho Lake: *Nocardioides aquaticus* sp. nov. and *Friedmanniella* [correction of Friedmannielly] lacustris sp. nov. Syst. Appl. Microbiol. *23*: 219–229.

Lawson, P.A., M.D. Collins, P. Schumann, B.J. Tindall, P. Hirsch and M. Labrenz. 2000b. *In* Validation of publication of new names and new combinations previously effectively published outside the IJSEM. List no. 77. Int. J. Syst. Evol. Microbiol. *51*: 1953.

Lawson, P.A., M.D. Collins, P. Schumann, B.J. Tindall, P. Hirsch and M. Labrenz. 2000c. New LL-diaminopimelic acid-containing actinomycetes from hypersaline, heliothermal and meromictic Antarctic Ekho Lake: *Nocardioides aquaticus* sp. nov. and *Friedmanniella lacustris* sp. nov. Syst. Appl. Microbiol. *23*: 219–229.

Lee, D.W. and S.D. Lee. 2008. *Tessaracoccus flavescens* sp. nov., isolated from marine sediment. Int. J. Syst. Evol. Microbiol. *58*: 785–789.

Liu, W.T., K. Nakamura, T. Matsuo and T. Mino. 1997. Internal energy-based competition between polyphosphate- and glycogen-accumulating bacteria oin biological phosporus removal reactor-effect of the P/C feeding ratio. Water Res. *31*: 1430–1438.

Lodes, M.J., H. Secrist, D.R. Benson, S. Jen, K.D. Shanebeck, J. Guderian, J.F. Maisonneuve, A. Bhatia, D. Persing, S. Patrick and Y.A. Skeiky. 2006. Variable expression of immunoreactive surface proteins of *Propionibacterium acnes*. Microbiol. *152*: 3667–3681.

MacKenzie, S.L. 1987. Gas chromatographic analysis of amino acids as the *N*-heptafluorobutyryl isobutyl esters. J. Assoc. Off. Anal. Chem. *70*: 151–160.

Malik, A.C., G.W. Reinbold and E.R. Vedamuthu. 1968. An evaluation of the taxonomy of *Propionibacterium*. Can. J. Microbiol. *14*: 1185–1191.

Martin-Rabadan, P., P. Gijon, L. Alcala, M. Rodriguez-Creixems, N. Alvarado and E. Bouza. 2008. *Propionibacterium acnes* is a common colonizer of intravascular catheters. J. Infect. *56*: 257–260.

Maszenan, A.M., R.J. Seviour, B.K. Patel, P. Schumann, J. Burghardt, R.I. Webb, J.A. Soddell and G.N. Rees. 1999a. *Friedmanniella spumicola* sp. nov. and *Friedmanniella capsulata* sp. nov. from activated sludge foam: Gram-positive cocci that grow in aggregates of repeating groups of cocci. Int. J. Syst. Bacteriol. *49*: 1667–1680.

Maszenan, A.M., R.J. Seviour, B.K. Patel, P. Schumann and G.N. Rees. 1999b. *Tessaracoccus bendigoensis* gen. nov., sp. nov., a Gram-positive coccus occurring in regular packages or tetrads, isolated from activated sludge biomass. Int. J. Syst. Bacteriol. *49*: 459–468.

Maszenan, A.M., H.L. Jiang, J.-H. Tay, P. Schumann, R.M. Kroppenstedt and S.T.-L. Tay. 2007. *Granulicoccus phenolivorans* gen. nov., sp. nov., a Gram-positive, phenol-degrading coccus isolated from phenol-degrading aerobic granules. Int. J. Syst. Evol. Microbiol. *57*: 730–737.

McDowell, A., S. Valanne, G. Ramage, M.M. Tunney, J.V. Glenn, G.C. McLorinan, A. Bhatia, J.F. Maisonneuve, M. Lodes, D.H. Persing and S. Patrick. 2005. *Propionibacterium acnes* types I and II represent phylogenetically distinct groups. J. Clin. Microbiol. *43*: 326–334.

McDowell, A., A.L. Perry, P.A. Lambert and S. Patrick. 2008. A new phylogenetic group of *Propionibacterium acnes*. J. Med. Microbiol. *57*: 218–224.

McGinley, K.J., G.F. Webster and J.J. Leyden. 1978. Regional variations of cutaneous propionibacteria. Appl. Environ. Microbiol. *35*: 62–66.

McLorinan, G.C., J.V. Glenn, M.G. McMullan and S. Patrick. 2005. *Propionibacterium acnes* wound contamination at the time of spinal surgery. Clin. Orthop. Relat. Res. *437*: 67–73.

Meurice, G., D. Jacob, C. Deborde, S. Chaillou, A. Rouault, P. Leverrier, G. Jan, A. Thierry, M.B. Maillard, P. Amet, M. Lalande, M. Zagorec, P. Boyaval and D. Dimova. 2004. Whole genome sequencing project of a dairy *Propionibacterium freudenreichii* subsp. *shermanii* genome: progress and first bioinformatic analysis. Lait *84*: 15–24.

Million, M., F. Roux, J. Cohen Solal, P. Breville, N. Desplaces, J. Barthas, J.C. Nguyen Van and G. Rajzbaum. 2008. Septic arthritis of the hip with *Propionibacterium avidum* bacteremia after intraarticular treatment for hip osteoarthritis. Joint Bone Spine *75*: 356–358.

Moore, W.E.C. and L.V. Holdeman. 1969. Outline of Clinical Methods in Anaerobic Bacteriology (edited by Cato, Cummins, Holdeman, Johnson, Moore, Smibert and Smith). Virginia Polytechnic Institute, Anaerobe Laboratory, Blacksburg, Virginia.

Moore, W.E.C. and L.V. Holdeman. 1970. *Propionibacterium*, *Arachnia*, *Actinomyces*, *Lactobacillus* and *Bifidobacterium*. *In* Outline of Clinical Methods in Anaerobic Bacteriology, 2nd edn (edited by Cato, Cummings, Holdeman, Johnson, Moore, Smibert and Smith). Virginia Polytechnic Institute Anaerobe Laboratory, Blacksburg, VA, pp. 15–22.

Moore, W.E.C., Moore, L.V.H. 1992. Index of the Bacterial and Yeast Nomenclatural Changes Published in the International Journal of Systematic Bacteriology since the 1980 Approved Lists of Bacterial Names (1st January 1980 to 1st January 1992). American Society for Microbiology, Washington, D.C.

Moss, C.W., V.R. Dowell, Jr, D. Farshtchi, L.J. Raines and W.B. Cherry. 1969. Cultural characteristics and fatty acid composition of propionibacteria. J. Bacteriol. *97*: 561–570.

Nakajima, H., Y. Hongoh, R. Usami, T. Kudo and M. Ohkuma. 2005. Spatial distribution of bacterial phylotypes in the gut of the termite *Reticulitermes speratus* and the bacterial community colonizing the gut epithelium. FEMS Microbiol. Ecol. *54*: 247–255.

Nakamura, K., A. Hiraishi, Y. Yoshimi, M. Kawaharasaki, K. Masuda and Y. Kamagata. 1995. *Microlunatus phosphovorus* gen. nov., sp. nov., a new Gram-positive polyphosphate-accumulating bacterium isolated from activated-sludge. Int. J. Syst. Bacteriol. *45*: 17–22.

Nesterenko, O.A., E.I. Kvasnikov and T.M. Nogina. 1985. *Nocardioidaceae* fam. nov., a new family of the order *Actinomycetales* Buchanan 1917. Mikrobiol. Zhurnal *47*: 3–12.

Nielsen, P.A. 1983. Role of reduced sulfur compounds in nutrition of *Propionibacterium acnes*. J. Clin. Microbiol. *17*: 276–279.

Oda, M. 1935. Bacteriological studies on water used for brewing sake (part 6). I. Bacteriological studies on "miyamizu" (8) and (9). *Micrococcus* and *Actinomyces* isolated from "miyamizu". (In Japanese) Jozogaku Zasshi *13*: 1202–1228.

Oprica, C. and C.E. Nord. 2005. European surveillance study on the antibiotic susceptibility of *Propionibacterium acnes*. Clin. Microbiol. Infect. *11*: 204–213.

Orla-Jensen, S. 1909. Die Hauptlinien des natürlichen Bakterien-systems. Zentralbl. Bakteriol. Parasitenkd. Infektionskr. Hyg. Abt. 2 *22*: 305–346.

Panagea, S., J.E. Corkill, M.J. Hershman and C.M. Parry. 2005. Breast abscess caused by *Propionibacterium avidum* following breast reduction surgery: case report and review of the literature. J. Infect. *51*: e253–255.

Patel, A., R.P. Calfee, M. Plante, S.A. Fischer and A. Green. 2009. *Propionibacterium acnes* colonization of the human shoulder. J. Shoulder Elbow Surg. *18*: 897–902.

Paul, G.E. and S.J. Booth. 1988. Properties and characteristics of a bacteriocin-like substance produced by *Propionibacterium acnes* isolated from dental plaque. Can. J. Microbiol. *34*: 1344–1347.

Pine, L. and L.K. Georg. 1969. Reclassification of *Actinomyces propionicus*. Int. J. Syst. Bacteriol. *19*: 267–272.

Pitcher, D.G. and M.D. Collins. 1991. Phylogenetic analysis of some LL-diaminopimelic acid-containing coryneform bacteria from human skin: description of *Propionibacterium innocuum* sp. nov. FEMS Microbiol. Lett. *84*: 295–300.

Pitcher, D.G. and M.D. Collins. 1992. *In* Validation of the publication of new names and new combinations previously effectively published outside the IJSB. List no. 41. Int. J. Syst. Bacteriol. *42*: 327–329.

Plastourgos, S. and R.H. Vaughn. 1957. Species of *Propionibacterium* associated with zapatera spoilage of olives. Appl. Microbiol. *5*: 267–271.

Pochi, P.E. and J.S. Strauss. 1961. Antibiotic sensitivity of *Corynebacterium acnes* (*Propionibacterium acnes*). J. Invest. Dermatol. *36*: 423–429.

Prévot, A.R. 1938. Études de systématique bactérienne. III. Invalidité du genre *Bacteroides* Castellani et Chalmers démembrement et reclassification. Ann. Inst. Pasteur *20*: 285–307.

Prévot, A.R. 1976. New concept of the taxonomic position of anaerobic corynebacteria. C.R. Acad. Sci. Hebd. Seances Acad. Sci. D. *282*: 1079–1081.

Prottey, C. and C.E. Ballou. 1968. Diacyl myoinositol monomannoside from *Propionibacterium shermanii*. J. Biol. Chem. *243*: 6196–6201.

Puhvel, S.M. 1968. Characterization of *Corynebacterium acnes*. J. Gen. Microbiol. *50*: 313–320.

Rainey, F.A., N.L. Ward-Rainey and E. Stackebrandt. 1997. Proposal for a new hierarchic classification system. *Actinobacteria* classis nov. Int. J. Syst. Bacteriol. *47*: 479–491.

Ramage, G., M.M. Tunney, S. Patrick, S.P. Gorman and J.R. Nixon. 2003. Formation of *Propionibacterium acnes* biofilms on orthopaedic biomaterials and their susceptibility to antimicrobials. Biomaterials *24*: 3221–3227.

Reddy, M.S., F.D. Williams and G.W. Reinbold. 1973. Sulfonamide resistance of propionibacteria: nutrition and transport. Antimicrob. Agents Chemother. *4*: 254–258.

Riedel, K.H.J. and T.J. Britz. 1993. *Propionibacterium* species diversity in anaerobic digestors. Biodivers. Conserv. *2*: 400–411.

Ross, J.I., A.M. Snelling, E. Carnegie, P. Coates, W.J. Cunliffe, V. Bettoli, G. Tosti, A. Katsambas, J.I. Galvan Perez Del Pulgar, O. Rollman, L. Torok, E.A. Eady and J.H. Cove. 2003. Antibiotic-resistant acne: lessons from Europe. Br. J. Dermatol. *148*: 467–478.

Sakaguchi, K., M. Iwasaki and S. Yamada. 1941. Studies on the propionic acid fermentation. J. Agric. Chem. Soc. Jpn I *1*: 127–158.

Sarada, R. and R. Joseph. 1994. Characterization and enumeration of microorganisms associated with anaerobic digestion of tomato-processing waste. Bioresour. Technol. *49*: 261–265.

Schaal, K.P. 1986. Genus *Arachnia*. *In* Bergey's Manual of Systematic Bacteriology, vol. 2 (edited by Sneath, Mair, Sharpe and Holt). Williams & Wilkins, Baltimore, pp. 1332–1342.

Schleifer, K.H. and O. Kandler. 1972. Peptidoglycan types of bacterial cell walls and their taxonomic implications. Bacteriol. Rev. *36*: 407–477.

Schoenborn, L., P.S. Yates, B.E. Grinton, P. Hugenholtz and P.H. Janssen. 2004. Liquid serial dilution is inferior to solid media for isolation of cultures representative of the phylum-level diversity of soil bacteria. Appl. Environ. Microbiol. *70*: 4363–4366.

Schofield, G.M. and K.P. Schaal. 1981. A numerical taxonomic study of members of the *Actinomycetaceae* and related taxa. J. Gen. Microbiol. *127*: 237–259.

Schumann, P., H. Prauser, F.A. Rainey, E. Stackebrandt and P. Hirsch. 1997. *Friedmanniella antarctica* gen. nov., sp. nov., an LL-diaminopimelic acid-containing actinomycete from antarctic sandstone. Int. J. Syst. Bacteriol. *47*: 278–283.

Seviour, R.J., T. Mino and M. Onuki. 2003. The microbiology of biological phosphorus removal in activated sludge systems. FEMS Microbiol. Rev. *27*: 99–127.

Shintani, T., W.T. Liu, S. Hanada, Y. Kamagata, S. Miyaoka, T. Suzuki and K. Nakamura. 2000. *Micropruina glycogenica* gen. nov., sp. nov., a new Gram-positive glycogen-accumulating bacterium isolated from activated sludge. Int. J. Syst. Evol. Microbiol. *50*: 201–207.

Shirling, E.B. and D. Gottlieb. 1966. Methods for characterization of *Streptomyces* species. Int. J. Syst. Bacteriol. *16*: 313–340.

Silveira, E.L., R.M. Pereira, D.C. Scaquitto, E.A. Pedrinho, S.P. Val-Moares, E. Wickert, L.M. Carareto-Alves and E.G.M. Lemos. 2006. Bacterial diversity of soil under eucalyptus assessed by 16S rDNA sequencing analysis. Pesqui. Agropecu. Bras. *10*: 1507–1516.

Siqueira, J.F., Jr and I.N. Rocas. 2003. Polymerase chain reaction detection of *Propionibacterium propionicus* and *Actinomyces radicidentis* in primary and persistent endodontic infections. Oral Surg. Oral Med. Oral Pathol. Oral Radiol. Endod. *96*: 215–222.

Skerman, V.B. 1968. A new type of micromanipulator and microforge. J. Gen. Microbiol. *54*: 287–297.

Skerman, V.B.D., V. McGowan and P.H.A. Sneath. 1980. Approved Lists of Bacterial Names. Int. J. Syst. Bacteriol. *30*: 225–420.

Skogen, L.O., G.W. Reinbold and E.R. Vedamuthu. 1974. Capsulation of *Propionibacterium*. J. Milk Food Technol. *37*: 314–321.

Smith, R.F. and N.P. Willett. 1968. Lipolytic activity of human cutaneous bacteria. J. Gen. Microbiol. *52*: 441–445.

Smith, R.F. 1969. Role of extracellular ribonuclease in growth of *Corynebacterium acnes*. Can. J. Microbiol. *15*: 749–752.

Song, L., W.J. Li, Q.L. Wang, G.Z. Chen, Y.S. Zhang and L.H. Xu. 2005. *Jiangella gansuensis* gen. nov., sp. nov., a novel actinomycete from a desert soil in north-west China. Int. J. Syst. Evol. Microbiol. *55*: 881–884.

Stackebrandt, E., F.A. Rainey and N.L. Ward-Rainey. 1997. Proposal for a new hierarchic classification system, *Actinobacteria* classis nov. Int. J. Syst. Bacteriol. *47*: 479–491.

Stackebrandt, E., P. Schumann, K.P. Schaal and N. Weiss. 2002. *Propionimicrobium* gen. nov., a new genus to accommodate *Propionibacterium lymphophilum* (Torrey 1916) Johnson and Cummins 1972, 1057[AL] as *Propionimicrobium lymphophilum* comb. nov. Int. J. Syst. Evol. Microbiol. *52*: 1925–1927.

Stackebrandt, E., C.S. Cummins and J.L. Johnson. 2006. Family *Propionibacteriaceae:* The Genus *Propionibacterium*. *In* The Prokaryotes: a Handbook on the Biology of Bacteria, 3rd edn (edited by Dworkin, Falkow, Rosenberg, Schleifer and Stackebrandt). Springer, New York, pp. 400–418.

Stackebrandt, E. and K. Schaal. 2006a. The family *Propionibacteriaceae*: the genera *Friedmanniella, Luteococcus, Microlunatus, Micropruina, Propioniferax, Propionimicrobium* and *Tessarococcus*. *In* The Prokaryotes: a Handbook on the Biology of Bacteria, 3rd edn, vol. 3, *Archaea, Bacteria, Firmicutes*, Actinomycetes (edited by Dworkin, Falkow, Rosenberg, Schleifer and Stackebrandt). Springer, New York, pp. 383–399.

Stackebrandt, E. and K.P. Schaal. 2006b. The family *Propionibacteriaceae*: the genera *Friedmanniella, Luteococcus, Microlunatus, Micropruina, Propioniferax, Propionimicrobium* and *Tessarococcus*. *In* The Prokaryotes: a Handbook on the Biology of Bacteria, 3rd edn, vol. 3, *Archaea, Bacteria, Firmicutes*, Actinomycetes (edited by Dworkin, Falkow, Rosenberg, Schleifer and Stackebrandt). Springer, New York, pp. 383–399.

Staley, J.T. 1968. *Prosthecomicrobium* and *Ancalomicrobium*: new prosthecate freshwater bacteria. J. Bacteriol. *95*: 1921–1942.

Tamura, T., M. Takeuchi and A. Yokota. 1994. *Luteococcus japonicus* gen. nov., sp. nov., a new gram-positive coccus with LL-diaminopimelic acid in the cell wall. Int. J. Syst. Bacteriol. *44*: 348–356.

Tan, H.H., C.L. Goh, M.G. Yeo and M.L. Tan. 2001. Antibiotic sensitivity of *Propionibacterium acnes* isolates from patients with acne vulgaris in a tertiary dermatological referral centre in Singapore. Ann. Acad. Med. Singapore *30*: 22–25.

Thöni, J. and O. Alleman. 1910. Über das Vorkommen von gefärbten, makroskopischen Bakterienkolonien in Emmentalerkäsen. Zentralbl. Bakteriol. Parasitenkd. Infektionskr. Hyg. Abt. *2*: 8–30.

Torrey, J.C. 1916. Bacteria associated with certain types of abnormal lymph glands. J. Med. Res. *34*: 65–80 61.

Tunney, M.M., S. Patrick, S.P. Gorman, J.R. Nixon, N. Anderson, R.I. Davis, D. Hanna and G. Ramage. 1998. Improved detection of infection in hip replacements. A currently underestimated problem. J. Bone Joint Surg. Br. *80*: 568–572.

Tunney, M.M., S. Patrick, M.D. Curran, G. Ramage, D. Hanna, J.R. Nixon, S.P. Gorman, R.I. Davis and N. Anderson. 1999. Detection of prosthetic hip infection at revision arthroplasty by immunofluorescence microscopy and PCR amplification of the bacterial 16S rRNA gene. J. Clin. Microbiol. *37*: 3281–3290.

Tunney, M.M., N. Dunne, G. Einarsson, A. McDowell, A. Kerr and S. Patrick. 2007. Biofilm formation by bacteria isolated from retrieved failed prosthetic hip implants in an in vitro model of hip arthroplasty antibiotic prophylaxis. J. Orthop. Res. *25*: 2–10.

US Environmental Protection Agency. 2005. Petro-Processors of Louisiana, Inc. Fact Sheet. US Environmental Protection Agency, Washington, DC, pp. 1–6.

Valanne, S., A. McDowell, G. Ramage, M.M. Tunney, G.G. Einarsson, S. O'Hagan, G.B. Wisdom, D. Fairley, A. Bhatia, J.F. Maisonneuve, M. Lodes, D.H. Persing and S. Patrick. 2005. CAMP factor homologues in *Propionibacterium acnes*: a new protein family differentially expressed by types I and II. Microbiology *151*: 1369–1379.

Van der Merwe, I.R., R. Bauer, T.J. Britz and L.M. Dicks. 2004. Characterization of thoeniicin 447, a bacteriocin isolated from *Propionibacterium thoenii* strain 447. Int. J. Food Microbiol. *92*: 153–160.

van Niel, C.B. 1928. The Propionic Acid Bacteria. J. W. Boissevain & Co., Haarlem, The Netherlands.

van Niel, C.B. 1957. Genus *Propionibacterium*. *In* Bergey's Manual of Determinative Bacteriology, 7th edn (edited by Breed, Murray and Smith). Williams & Wilkins, Baltimore, pp. 569–576.

Vanberg, C., B.F. Lutnaes, T. Langsrud, I.F. Nes and H. Holo. 2007. *Propionibacterium jensenii* produces the polyene pigment granadaene and has hemolytic properties similar to those of *Streptococcus agalactiae*. Appl. Environ. Microbiol. *73*: 5501–5506.

Vedamuthu, E.R., C.J. Washam and G.W. Reinbold. 1971. Isolation of inhibitory factor in raw milk whey active against propionibacteria. Appl. Microbiol. *22*: 552–556.

Vohra, A., E. Saiz, J. Chan, J. Castro, R. Amaro and J. Barkin. 1998. Splenic abscess caused by *Propionibacterium avidum* as a complication of cardiac catheterization. Clin. Infect. Dis. *26*: 770–771.

Von Freudenreich, E. and S. Orla-Jensen. 1906. über die in Emmentalerkäse stafffindene Propionsaüre-gärung. Zentralbl. Bakteriol. Parasitenkd. Infektionskr. Hyg. Abt. *2*: 529–546.

von Nicolai, H., U. Hoffler and F. Zilliken. 1980. Isolation, purification, and properties of neuraminidase from *Propionibacterium acnes*. Zentralbl. Bakteriol. A. *247*: 84–94.

Vorobjeva, L.I. 1999. Propionibacteria. Kluwer Academic Publishers, The Netherlands.

Voss, J.G. 1970. Differentiation of two groups of *Corynebacterium acnes*. J. Bacteriol. *101*: 392–397.

Walker, M. and C.A. Phillips. 2007. The growth of *Propionibacterium cyclohexanicum* in fruit juices and its survival following elevated temperature treatments. Food Microbiol. *24*: 313–318.

Wang, W.L., E.D. Everett, M. Johnson and E. Dean. 1977. Susceptibility of *Propionibacterium acnes* to seventeen antibiotics. Antimicrob. Agents Chemother. *11*: 171–173.

Wang, Y.X., M. Cai, X.Y. Zhi, Y.Q. Zhang, S.K. Tang, L.H. Xu, X.L. Cui and W.J. Li. 2008. *Microlunatus aurantiacus* sp. nov., a novel actinobacterium isolated from a rhizosphere soil sample. Int. J. Syst. Evol. Microbiol. *58*: 1873–1877.

Webster, G.F. and C.S. Cummins. 1978. Use of bacteriophage typing to distinguish *Propionibacterium acne* types I and II. J. Clin. Microbiol. *7*: 84–90.

Weiss, N., K.H. Schleifer and O. Kandler. 1981. The peptidoglycan types of Gram positive anaerobic bacteria and their taxonomic implications. Rev. Inst. Pasteur Lyon *14*: 3–12.

Werkman, C.H. and S.E. Kendall. 1931. The propionic acid bacteria. I: Classification and nomenclature. Iowa State J. Sci. *6*: 17–32.

Werkman, C.H. and R.W. Brown. 1933. The propionic acid bacteria. II. Classification. J. Bacteriol. *26*: 393–417.

Werner, H. 1967. Lipase and lecithinase activities of aerobic and anaerobic *Corynebacterium* and *Propionibacterium* species. Zentralbl. Bakteriol. Orig. *204*: 127–138.

Williams, M.W. and R.F. Unz. 1985. Isolation and characterization of filamentous bacteria present in bulking activated sludge. Appl. Microbiol. Biotechnol. *22*: 273–280.

Wood, H.G., A.A. Andersen and C.H. Werkman. 1938. Nutrition of the propionic acid bacteria. J. Bacteriol. *36*: 201–214.

Yamada, K. and K. Komagata. 1972. Taxonomic studies on coryneform bacteria. IV. Morphological, cultural, biochemical and physiological characteristics. J. Gen. Appl. Microbiol. *18*: 399–416.

Yamada, Y., G. Inouye, Y. Tahara and K. Kondo. 1976. The menaquinone system in the classification of coryneform and nocardioform bacteria and related organisms. J. Gen. Appl. Microbiol. *22*: 203–214.

Yokota, A., T. Tamura, M. Takeuchi, N. Weiss and E. Stackebrandt. 1994. Transfer of *Propionibacterium innocuum* Pitcher and Collins 1991 to *Propioniferax* gen. nov. as *Propioniferax innocua* comb. nov. Int. J. Syst. Bacteriol. *44*: 579–582.

Zhi, X.-Y., W.-J. Li and E. Stackebrandt. 2009. An update of the structure and 16S rRNA gene sequence-based definition of higher ranks of the class *Actinobacteria*, with the proposal of two new suborders and four new families and emended descriptions of the existing higher taxa. Int. J. Syst. Evol. Microbiol. *59*: 589–608.

Family II. **Nocardioidaceae** Nesterenko, Kvasnikov and Nogina 1990, 320[VP] (Effective publication: Nesterenko, Kvasnikov and Nogina 1985a, 9.) emend. Rainey, Ward-Rainey and Stackebrandt 1997, 484 emend. Zhi, Li and Stackebrandt 2009, 599

Lyudmila I. Evtushenko and Elena V. Ariskina

No.car.di.o.i.da.ce.a.e. N.L. masc. n. *Nocardioides* type genus of the family; suff. *-aceae* ending to denote a family; N.L. fem. pl. n. *Nocardioidaceae* the *Nocardioides* family.

Young cultures exhibit different morphologies, ranging from extensively branching vegetative hyphae, growing on and penetrating the surface of agar media, **to irregular rods** and **spherical cells. Aerial hyphae are produced** by organisms of many taxa and may be abundant or scant, at times discernible only microscopically. **The vegetative and aerial hyphae**, depending on organisms and growth conditions, **undergo varying degrees of fragmentation and differentiation eventually resulting in rod-like and coccoid cells (arthrospores)** often arranged in chains or small aggregates. **Cell division and hyphal elongation by pronounced budding** are characteristic of some taxa. **Clusters of tightly packed irregularly shaped cells** can be observed. The non-mycelial bacteria of the genus *Aeromicrobium* and the majority of *Nocardioides* species usually exhibit **a rod–coccoid morphogenetic cycle** and **can be motile. Spherical cells during all growth phases** are characteristic mostly of the genus *Marmoricola.* **Gram-stain-positive type of cell wall.** Non-acid-fast. **Colonies are non-pigmented** or of different intensity and shades of **cream, yellow, or orange.** Aerial mycelium, if any, is typically white, but may be cream or light yellow in color.

Chemo-organotrophs, having a respiratory type of metabolism, with a potential for metabolic flexibility. **Grow under aerobic conditions on standard laboratory media**, including chemically defined (synthetic) media. Some organisms are nutritionally fastidious. Some show weak growth in media rich in organics under anaerobic conditions. Mostly catalase-positive. Oxidase activity varies with the species. Utilize a wide range of carbon and nitrogen sources, including unusual organic compounds and toxic environmental pollutants, and possess a wide spectrum of enzymatic activities. **Mostly mesophiles, non-halophiles. Prefer a neutral to mildly alkaline pH.** However, thermophilic, salt-requiring, and alkaliphilic species occur.

The cell-wall peptidoglycan contains LL-diaminopimelic acid and glycine as the diagnostic amino acid **(the peptidoglycan type A3γ)** The muramic acid residue of the peptidoglycan is *N*-acetylated. **Anionic (acidic) polysaccharides are usually present** in the cell wall, which can be either phosphorous-containing or phosphorous-free (**teichoic, teichuronic, teichulosonic acids** or some not yet identified polymers). Menaquinones are the sole respiratory quinines detected; **the predominant component is menaquinone with partially saturated side chain consisting of either 8, 9, or 10 isoprene units – MK-8(H_4), MK-9(H_4), MK-9(H_6), or MK-10(H_4).** Cellular fatty acids are **different combinations of branched-chain saturated (iso-, anteiso-, 10-methyl- and 9-methyl-branched), straight-chain saturated and monounsaturated, and hydroxylated components.** Mycolic acids are absent. Phospholipids of the three types are present: **type I (with no nitrogenous phospholipids), type III (phosphatidylcholine as diagnostic component), and rarely type II (with diagnostic phosphatidylethanolamine).** Occur in various terrestrial and aquatic environments, including polluted sites, and can be found in microbial communities associated with plants, algae, lichens, animals, or humans. No medically relevant strains or species have been described within the family so far.

DNA G+C content (mol%): 65.5–74.8 (T_m, HPLC).

Type genus: **Nocardioides** Prauser 1976, 61[AL].

Further descriptive information

The family *Nocardioidaceae* (Nesterenko et al., 1985a, 1990; Zhi et al., 2009) belongs to the order *Propionibacteriales*, class *Actinobacteria* (Stackebrandt et al., 1997). Genera currently assigned to this family form a 16S rRNA-based phylogenetic group which includes two main clusters (Figure 247). The first one comprises the genera *Nocardioides* (type genus of the family), *Marmoricola* and *Aeromicrobium.* The second cluster includes the genera *Actinopolymorpha, Flindersiella**, *Kribbella,* and *Thermasporomyces**. Delineation of genera within the *Nocardioidaceae,* like in many other actinomycete families, is primarily based on 16S rRNA gene sequence phylogenetic clustering and on morphological and chemotaxonomic characteristics of species included (Table 214).

Morphology. Three main morphological groups can be distinguished among organisms of this family: (1) organisms producing rod-shaped cells and typically displaying a rod-to-coccus morphogenetic cycle, (2) organisms with mostly spherical cells during all growth phases, and (3) organisms forming a mycelium during the morphogenetic cycle. The first morphological group consists of the majority of *Nocardioides* species and all members of the genus *Aeromicrobium.* Cells display irregular rods in young cultures and divide by transverse septum formation, often showing V-forms. While growth proceeds, the cells typically become shorter, and a significant proportion of cells can be coccoid in older cultures. The distinct rod–coccoid growth cycle resembling that of *Arthrobacter* (Keddie et al., 1986) is often observed, especially in synchronized cultures on rich agar media. The second morphological group is mostly represented by coccoid organisms of the genus *Marmoricola.* To this group might also be assigned some *Nocardioides* species producing very short rods and cocci, and showing outward morphological resemblance with *Marmoricola,* e.g. *Nocardioides aquaticus* (Lawson et al., 2000a). The third morphological group encompasses organisms (genera *Actinopolymorpha, Kribbella, Flindersiella, Thermasporomyces,* and two *Nocardioides* species, *Nocardioides albus* and *Nocardioides luteus*) with a common feature of forming branched hyphae but varying significantly in developmental events and cell differentiation (Table 214). Both vegetative (substrate) and, if any, aerial hyphae vary in length and branching intensity, depending on the genus or species and the composition and consistency of the growth medium. The substrate hyphae

*After this chapter was accepted for publication, the descriptions of genera *Flindersiella* (Kaewkla and Franco, 2010a) and *Thermasporomyces* (Yabe et al., 2011) were published. The main characteristics of these genera are outlined only in this chapter and mentioned in the chapters "Genus *Actinopolymorpha*" and "Genus *Nocardioides*" (for further details, see the original descriptions).

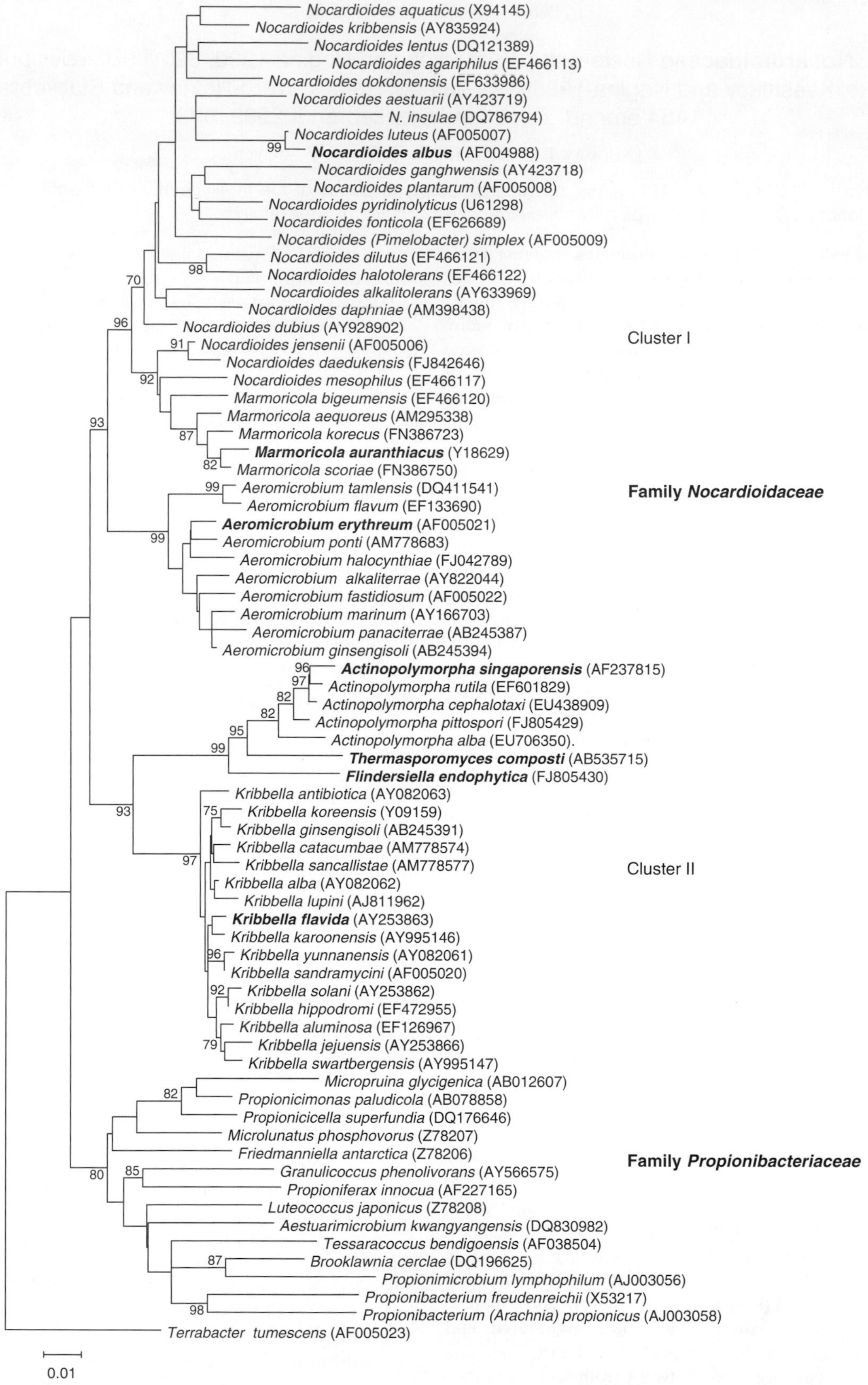

FIGURE 247. Phylogenetic dendogram based on 16S rRNA comparison of type strains of the species comprising the family *Nocardioidaceae* (genera *Aeromicrobium, Actinopolymorpha, Flindersiella, Kribbella, Marmoricola,* and *Thermasporomyces,* and representative species of the genus *Nocardioides*). The type species of the genera are given in bold. For the extended tree of the genus *Nocardioides,* see Figure 248. Bar = 0.01 inferred substitutions per nucleotide. Values at nodes indicate bootstrap values for 1000 replicates.

can be of uneven thickness and/or have local swellings, often located at the top of hyphae. The hyphae usually break up into elongated, Y-shaped, or rod-like fragments which can undergo further division by transverse septum formation. The fragments may become thicker and rounder, becoming spore-like cells (arthrospores) after maturation, which can be strung together into short, straight, or zig-zag shaped chains. In some cultures (e.g. in the genus *Kribbella*), the substrate and aerial hyphae or significant hyphal area may remain stable, and no distinct regular fragmentation into rod-like elements can be observed *in situ*. In addition, short chains of poorly to well-differentiated cells (conidia) may occasionally occur on terminal parts of aerial hyphae (*Flindersiella, Kribbella*). A characteristic feature of some representatives of the family, especially of the genus *Actinopolymorpha*, is the production of novel cells through marked apical and lateral budding and hyphal elongation by budding (similar to that in *Pseudonocardia*; Henssen, 1989; Henssen et al., 1983), along with the hyphal growth in an ordinary way. Cells produced by budding, and also by septum formation, can form aggregates or compact clusters (most pronounced in *Actinopolymorpha* and some *Kribbella*). The cells in clusters, as recently revealed in kribbellae, are usually tightly packed and irregular in size and shape (often angular) and are produced via transverse and/or differently directed septa, like that reported, e.g. for *Geodermatophilus* (Eppard et al., 1996; Ishiguro and Wolfe, 1970). Both the hyphal fragments and products of other modes of cell fission usually give rise to new hyphae (or buds) when transferred to a fresh medium. Notably, all organisms of the first and the second morphological groups with rod-shaped and coccoid cells fall into the phylogenetic cluster I (Figure 247). The third morphological group, except two nocardioform *Nocardioides* species (*Nocardioides albus* and *Nocardioides luteus*), corresponds to the phylogenetic cluster II.

Chemotaxonomy. All species comprising the genera of the family *Nocardioidaceae* are characterized by the cell-wall peptidoglycan containing LL-diaminopimelic (LL-A_2pm) acid (cell-wall chemotype I *sensu* Lechevalier and Lechevalier, 1970). The peptidoglycan type is A3γ *sensu* Schleifer and Kandler (1972) (type A41.1 according to DSMZ Catalog of Strains, 2001), which was determined for several species (Fiedler et al., 1970; Lawson et al., 2000a; Prauser, 1986; Schippers et al., 2005; Schleifer and Kandler, 1972; Schumann et al., 1997; Trujillo et al., 2006; Urzì et al., 2000) or inferred from available data on the peptidoglycan amino acid composition (e.g. Lee et al., 2000; Miller et al., 1991; Park et al., 1999; Tamura and Yokota, 1994). The subunit of this polymer is characterized by the tetrapeptide L-Ala–D-Glu–LL-A_2pm–D-Ala and a glycine residue as an interpeptide bridge linking the amino group located on the D-carbon of LL-A_2pm and the C-terminal D-alanine of an adjacent subunit. It should be emphasized that this polymer structure differs from those reported for some other LL-A_2pm-containing actinobacteria, belonging, e.g. to the phylogenetically neighboring family *Propionibacteriaceae* or to more distant *Intrasporangiaceae*. Peptidoglycans of organisms of these families may have additional glycine molecules in the interpeptide bridge, along with glycine or glycine amide linked to the α-carboxyl group of D-glutamic acid at position 2 of the peptide subunit (type A3γ; A41.2) or contain glycine instead of alanine in position 1 of the peptide subunit (type A3γ′; A42.1) (DSMZ, 2001; Schleifer and Kandler, 1972; Schumann et al., 1997, 2009; Stackebrandt and Schaal, 2006; Stackebrandt and Schumann, 2006; Weon et al., 2007). Thus, the peptidoglycans of members of the family *Nocardioidaceae* analyzed so far contain one glycine in the peptide subunit, in contrast to that of some *Propionibacteriaceae* or *Intrasporangiaceae*, which may contain up to 4 glycine molecules per peptide subunit. An exception might be the genus *Thermasporomyces* reported to possess peptidoglycan with glycine, glutamic acid, alanine, and LL-A_2pm in a molar ratio of 3.9:1.0:0.6:0.5, although the peptidoglycan structure was not yet studied (Yabe et al., 2011). For a few organisms of the family *Nocardioidaceae*, the acyl type of muramyl residues of the peptidoglycans was determined and found to be acetyl type (Kaewkla and Franco, 2010a; Kubota et al., 2005a; Lee et al., 2000; Uchida and Seino, 1997; Urzì et al., 2000).

The cell-wall monosugars of genera of the family *Nocardioidaceae* routinely recorded during chemotaxonomic studies lack coherence. The sugars revealed in cell walls or whole cells of different species include various combinations of glucose, galactose, mannose, rhamnose, and less frequently 2-*O*-methyl-D-galactose (madurose), 2,3-*O*-dimethyl-D-galactose, ribose, xylose, as well as some unidentified sugars. On the other hand, the neutral cell-wall monosugars may be absent (*Nocardioides plantarum*), while aminosugars, polyols, carbonic acids, and some unusual compounds originating from different types of peptidoglycan-attached polysaccharides may be present (see the chapters on *Aeromicrobium, Kribbella*, and *Nocardioides* in this volume for more details). The data available suggest that the monosugar patterns, individual sugars, and/or other components originating from the cell-wall polysaccharides linked to the peptidoglycan are indicative of individual species or species group within a genus and may also predict membership of a bacterium to a certain genus.

The peptidoglycan-linked polysaccharides from representatives of this family were found to possess mostly anionic (acidic) polymers of different types (Table 214). One such acidic polymer type is represented by teichoic acids which are poly(polyol phosphate) polymers often including glycosyl moieties in the basic chain and bearing lateral branches (substituents) (Baddiley, 1970; Ward, 1981). The teichoic acids were found in almost all organisms of the genera *Nocardioides* and *Aeromicrobium* so far analyzed (Naumova et al., 2001; Shashkov et al., 2000b; Shashkov et al., 1999; Takeuchi and Yokota, 1989; Tul'skaya, 2009). The exception is *Nocardioides plantarum* (which lacks teichoic acids and other phosphate-containing polysaccharides, and most likely has a different kind of the peptidoglycan-linked polymer). *Aeromicrobium fastidiosum*, in addition to a ribitol teichoic acid, has another type of phosphate-containing polymer, presumably a phosphorhamnan. Two types of phosphorous-free acidic polymers (along with minor amounts of a neutral glycopolymer, mannan) have been found in the cell walls of 15 *Kribbella* strains representing different phylogenetic groups within the genus (Shashkov et al., 2009; A.S. Shashkov, E.M. Tulskaya, and L.I. Evtushenko, unpublished). One type is represented by teichuronic acids with a rare diaminosugar, 2,3-diacetamido-2,3-dideoxyglucose, in the basic chain. The second type comprises unusual glycopolymers with the backbone containing a nine-carbon sialic-acid-like keto sugar, a nonulosonic (mostly pseudaminic) acid, and is characterized by an unusual linkage in the polymeric chain (Figure 275). By analogy with teichuronic acids, the name for nonulosonic acid-containing polymers was

TABLE 214. Differential characteristics of the genera in the family *Nocardioidaceae*[a]

Characteristic	*Nocardioides*	*Aeromicrobium*	*Marmoricola*	*Kribbella*	*Actinopolymorpha*	*Flindersiella*	*Thermasporomyces*
Phylogenetic group	I	I	I	II	II	II	II
Morphology	Hyphae[b], rods, coccoid cells[c,d]	Rods, coccoid cells[d]	Coccoid cells	Hyphae, rods, coccoid cells[c], conidia, hyphal swellings, cell clusters[e]	Hyphae, pleomorphism, marked budding, cell aggregates (clusters)[e]	Hyphae, rods[c], conidia	Hyphae, rods, coccoid cells[c]
Aerial mycelium	D +	–	–	+	D + (w)	+	–
Motility	D–	D–	D–	–	–	–	–
Optimal temperature (°C)	20–30	25–35	28–30	25–30	27–28	27–35	50–55
Major menaquinones[f]	8/4[g]	9/4	8/4[h]	9/4[i]	9/4 or 9/6 (9/4–8, 10/4–8)	10/6 (10/8,4,2)	9/4, 10/4, 11/4
Predominant fatty acids (>10%)[j]	$C_{16:0}$ iso, ($C_{18:1}$ ω9*c*, $C_{17:1}$ ω6/8c, 10-Me-$C_{18:0}$, 10-Me-$C_{17:0}$, $C_{17:0}$ anteiso, $C_{16:0}$, $C_{16:1}$ iso, $C_{17:0}$, $C_{17:0}$ iso, $C_{15:0}$ iso)	$C_{18:1}$ ω9*c*, 10-Me-$C_{18:0}$, $C_{16:0}$, ($C_{16:0}$ 2-OH, $C_{18:0}$, 10-Me-$C_{16:0}$)	$C_{16:0}$, $C_{18:1}$ ω9*c*, (10-Me $C_{18:0}$, $C_{17:1}$ ω8*c*, $C_{16:1}$) or $C_{16:0}$ iso[k]	$C_{15:0}$ anteiso, $C_{16:0}$ iso, ($C_{15:0}$ iso, $C_{17:1}$ iso, $C_{14:0}$ iso, $C_{17:0}$ anteiso, 9-Me-$C_{16:0}$)	$C_{16:0}$ iso, $C_{15:0}$ iso, ($C_{17:0}$ iso, $C_{17:0}$ anteiso, $C_{16:1}$ iso, $C_{15:0}$ anteiso)	$C_{16:0}$ iso, $C_{17:0}$ anteiso, $C_{15:0}$ anteiso	$C_{17:0}$ anteiso, $C_{15:0}$ anteiso, $C_{17:0}$ iso, $C_{15:0}$ iso
Polar lipids[l]	DPG, PG, PL (PG-OH, PI, PI-OH, PIM, acyl-PG, PC[m], PE[m])	DPG, PG (PE, PI, PL)	PI, DPG, PG, (PC, PL)	PC, PI, DPG, PG, (PI-OH, PL, GL)	PG, PIM, (DPG, PI)	DPG, PG	PG, DPG, PGL, GL
Major polyamines[n] (>20%)	CAD (PUT, SPM)	CAD (SPD, SPM)	nd	nd	nd	nd	nd
Major murein-linked glycopolymers[o]	Teichoic acids[p]	Teichoic acids[q]	nd	Teichuronic or teichulosonic acids	nd	nd	nd
DNA G+C content (mol%)	67.5–74.8	65.5–74.0	71.0–72.9	66.3–71.3	66.6–69.6	68.8	69.2

Symbols and abbreviations: +, present; –, absent; D, different between species within a genus (character for the type species is indicated); anteiso, iso-, 9-Me-, and 10-Me indicate anteiso-, iso-, 9-, and 10-methyl-branched acids, respectively; 2-OH, 2-hydroxylated acids; DPG, diphosphatidylglycerol; PC, phosphatidylcholine; PE, phosphatidylethanolamine; PG, phosphatidylglycerol; PG-OH, phosphatidylglycerol containing 2-hydroxy fatty acids; PGL, ninhydrin-positive phosphoglycolipid; PI, phosphatidylinositol; PI-OH, phosphatidylinositol containing 2-hydroxy fatty acids; PIM, phosphatidylinositol mannosides; PL, unidentified phospholipid(s); GL, glycolipid(s); CAD, cadaverine; PUT, putrescine; SPD spermidine; SPM, spermine.

[a]Data compiled from the original descriptions of the genera, the species composing the genera and publications cited in the respective genus chapters.

[b]Vegetative and aerial hyphae develop in cultures of *Nocardioides albus* and *Nocardioides luteus* only.

[c]Rod-shaped and coccoid cells are usually produced owing to fragmentation of vegetative and/or aerial hyphae, with different genera and species showing varied degrees of hyphal fragmentation and additional cell differentiation.

[d]Coccoid cells are typically produced in later stages of the morphogenetic cycle.

[e]Cells in clusters are often tightly packed and irregular in shape and size (can be produced internally by differently oriented septa, as shown for kribbellae and supposed for *Actinopolymorpha*).

[f]Numerals indicate the numbers of isoprene units and the number of hydrogen atoms in the partially saturated side chain, e.g. 8/4 is a menaquinone with 8 isoprene units and 4 hydrogen atoms in the side chain. Components irregularly detectable among species or reported to occur in lesser amounts are given in parentheses.

[g]Menaquinones 6/4, 7/4, 8/0, 8/2, 8/6, and 9/4 may be produced as minor components or in trace amounts (see the chapter on *Nocardioides* for more details and references).

[h]Menaquinone system including 8/4, 7/4, 8/2, and 6/4 (peak area ratio, 73:4:1:1) has been reported for *Marmoricola aurantiacus* (Urzì et al., 2000).

[i]Menaquinone 9/4 constitutes 93% (Carlsohn et al., 2007) or more of the total.

[j]Numbers before and after colons represent chain lengths and numbers of double bonds of fatty acids; ω indicates the double bond position. Compounds are listed in the order of decreasing amounts and frequency among species within a genus. Compounds found to irregularly contribute more than 10% among species of multi-species genera or in different experiments are given in parenthesis (see the respective genus chapters in this volume for details). Relative amounts of the predominant components in the fatty acid profiles may vary depending on growth conditions and growth phase.

[k]Found only in *Marmoricola bigeumensis.*

[l]PL usually stands for principal unidentified or incompletely identified compounds which may be the same or different. Parenthesis indicates that a compound was present in only some species within a genus. Minor or trace amounts of other unidentified polar lipids may also be present (see respective genus chapters in this volume).

[m]The key components of the phospholipid types II and III (according to Lechevalier et al., 1977) were reported for *Nocardioides daedukensis* and *Nocardioides dubius* (phosphatidylethanolamine) and *Nocardioides furvisabuli* (phosphatidylcholine).

[n]Polyamines found in some representatives of a genus are given in parentheses.

[o]See the text and relevant genus chapters for the composition of polymers indicated. Minor or trace amounts of neutral polysaccharides may occur in organisms of some taxa so far tested.

[p]An exception is *Nocardioides plantarum,* which most likely contains a peptidylglycan-linked polymer of a different type.

[q]Available for the only species, *Aeromicrobium fastidiosum;* this species also contains another polysaccharide, presumably phosphorhamnan.

suggested to be teichulosonic acids (Knirel, 2009). Notably, each *Kribbella* strain contained as the major polysaccharide either teichuronic or teichulosonic acid, or, occasionally both, while a neutral polymer (mannan) was present in varying amounts in all 15 strains investigated. The teichulosonic acids of a different type were found previously (along with some other acidic polymers) in the cell wall of some streptomycetes, including plant pathogenic strains (Shashkov et al., 2002a; Shashkov et al., 2000a; Shashkov et al., 2002b; Tul'skaya et al., 2007; Tul'skaya, 2009). Pseudaminic acid (or its derivatives) are a rather common component of polysaccharides of Gram-negative bacteria (Knirel et al., 2003; Schoenhofen et al., 2006; Vimr et al., 2004); in the cell walls of Gram-positives, the pseudaminic acid and pseudaminic acid-containing polymers have been found so far solely in kribbellae. Lipoteichoic acids (which, in contrast to the cell-wall teichoic acids, are membrane-anchored) were found in abundance in all representatives of the family investigated so far, i.e. in several mycelium-forming *Nocardioides* strains (E.M. Tulskaya and L.I. Evtushenko, unpublished).

Members of the *Nocardioidaceae* are characterized by different types and the predominant components of cellular fatty acids (Table 214). The complex fatty acid type (Suzuki and Komagata, 1983a; Suzuki et al., 1993), which includes both straight- and branched-chain fatty acids (iso-, anteiso, and 10-methyl-branched) and also hydroxylated acids is characteristic of the genus *Nocardioides*. The fatty acid profile of the majority *Nocardioides* species are dominated by $C_{16:0}$ iso which may contribute up to 65–70% (Choi et al., 2007; Kim et al., 2008a; Schumann et al., 1997). Species of the genera *Aeromicrobium* and *Marmoricola*, except *Marmoricola bigeumensis*, contain large proportions of straight-chain fatty acids as well as 10-methyl-branched and hydroxylated acids, whereas iso- and anteiso-branched components are usually present in minor amounts. In contrast, the four remaining (mycelium-forming) genera of the family are characterized by the predominance of iso- and/or anteiso-branched acids (Table 214). The fatty acid profiles in members of the family may vary considerably with culture age, growth conditions and analytical procedure. Nevertheless, the fatty acid type and, in general, the patterns of major fatty acids, together with some rarely encountered components, e.g. 9-methyl branched chain acids (*Kribbella*) appear to be indicative of certain genera (Table 214).

Three phospholipid types *sensu* Lechevalier et al. (1977) can be identified: type I (no nitrogenous phospholipids), type II (phosphatidylethanolamine as diagnostic phospholipid), and type III (phosphatidylcholine as diagnostic phospholipid). Other polar lipids include phosphatidylglycerol, diphosphatidylglycerol, phosphatidylinositol, phosphatidylinositol mannosides, hydroxy-phosphatidylglycerol, as well as unidentified phospholipids and glycolipids (Table 214). The type of phospolipids and the pattern of principal polar lipids are uniform in some genera (e.g. *Kribbella*, all species of which contain phosphatidylcholine) and heterogenous particularly in the genera *Nocardioides*, *Aeromicrobium*, and *Marmoricola*. As with the fatty acids, the polar lipid patterns may be affected by growth conditions and analytical procedure, resulting in conflicting data reported by different authors (see, e.g. Collins et al., 1989, 1983; Komura et al., 1975b; Lechevalier et al., 1981, 1977; Lee and Kim, 2007; O'Donnell et al., 1982; Prauser, 1989; Tamura and Yokota, 1994). Besides, many publications report "unknown" polar lipid components, which might be the same as or different from those identified by other authors. Detailed analysis of the polar lipids in members of the family will indeed increase their discriminative power.

Polyamines have so far been detected in only a few representatives of the genus *Nocardioides* and two strains of *Aeromicrobium* (Busse and Schumann, 1999). In most species, cadaverine (which distinguishes these two genera from the other LL-A_2pm-containing actinobacteria analyzed) was present as the primary polyamine with either putrescine, spermine, or spermidine as a secondary one. Others such as 1,3-diaminopropane, tyramine, and sym-homospermidine were detected in some strains in minor or trace amounts (Busse and Schumann, 1999).

Physiology. Bacteria of the *Nocardioidaceae* are considered to be heterotrophic aerobes growing in media containing peptone and/or yeast extract as well as in chemically defined (synthetic) media with glucose and other sugars or organic acids as sole carbon sources; some require vitamins and other growth factors. Representatives of the family, especially *Nocardioides* strains, also exhibit the capacity to degrade and metabolize recalcitrant molecules, including diverse hydrocarbons and numerous substituted aromatic compounds, many of which are toxic (see respective genera chapters in this volume for references and details). Some species of the genus *Kribbella* show weak or moderate anaerobic growth on agar media rich in organics in an atmosphere of $H_2/CO_2/N_2$ (5:10:85), but not on ISP 9 with glucose as the sole carbon source (Kirby et al., 2006). The ability of some kribbellae to grow aerobically on tap-water agar (Lee et al., 2000; L.M. Baryshnikova and L.I. Evtushenko, recent observations) suggests that they may possess an oligotrophic or even autotrophic lifestyle under certain conditions, as reported earlier for representatives of the genus *Pseudonocardia* (Goodfellow and Lechevalier, 1986; Lechevalier et al., 1986; Mahendra and Alvarez-Cohen, 2005; Parales et al., 1994; Takamiya and Tubaki, 1956). There is some indirect evidence of possible chemolithotrophic growth with CO and hydrogen (King and Weber, 2007; Mattes et al., 2005; Osborne et al., 2010). Bacteria of the family are mostly mesophiles, exhibiting optimal growth within the temperature range 25–37°C; some grow best at 16–26°C (Lawson et al., 2000a). So far, only one thermophilic species, *Thermasporomyces composti*, growing at 35–62°C (optimum growth at 50–55°C) has been reported (Yabe et al., 2011). Some mesophilic species can grow at temperatures up to 45°C (Cao et al., 2009; Kaewkla and Franco, 2010a; Kirby et al., 2006; Song et al., 2011). Members of the family usually prefer a neutral to mildly alkaline pH, but some (representatives of *Marmoricola*, *Kribbella*, and *Nocardioides*) resist alkaline conditions and can grow at initial pH values up to pH 12 (e.g. Dastager et al., 2008b; Lee, 2007a; Lee and Lee, 2010; Lee et al., 2010; Yoon et al., 2005a). Moreover, some species appear to be alkaliphilic, in particular, *Marmoricola scoriae* (optimal growth at initial pH 8–11; Lee et al., 2010). Some species show weak growth at pH 4.5 or slightly below (Everest and Meyers, 2008). Bacteria of the family are largely non-halophiles, but some species of *Actinopolymorpha*, *Aeromicrobium*, and *Nocardioides* require salt, grow best in the presence of NaCl up to 6–8% (w/v), and/or tolerate NaCl concentrations up to 15% (w/v) (e.g. Bruns et al., 2003; Cao et al., 2009; Choi et al., 2007; Kim et al., 2008a; Lawson et al., 2000a; Lee and Lee, 2008; Wang et al., 2001).

Genomic characteristics. The G+C content of the DNA are largely about 70 mol%, varying between 65.5 mol% (HPLC) determined for *Aeromicrobium panaciterrae* (Cui et al., 2007a) and 74.8 mol% (HPLC) reported for *Nocardioides lentus* (Yoon et al., 2006a). The sequences of the internal transcribed spacer (ITS) region of the 16S–23S rRNA gene (Yoon and Park, 2000) and the ribonuclease P (RNase P) RNA gene (Yoon et al., 1998a) were analyzed in representatives of *Nocardioides*, *Aeromicrobium*, and *Kribbella*. It is worth noting that the 16S–23S ITS and RNase P RNA genes do not necessarily show higher sequence differences between genera than between some *Nocardioides* species. In particular, differences were greater between *Nocardioides albus* and some rod-shaped *Nocardioides* species than between representatives of this and the other genera of the family included in the study. The RNase P RNA gene transfer between organisms of this group can probably take place, as follows from the nearly identical (>99% similarity) sequences of this gene in *Nocardioides jensenii* and *Luteococcus japonicus* (*Propionibacteriaceae*) (Yoon and Park, 2000, 2006). The whole genome has been sequenced for only two representatives of the family, a rod-shaped *Nocardioides* sp. JS614 (Copeland et al., 2006; GenBank accession no. NC_008699) capable of assimilating vinyl chloride and ethene as carbon and energy sources (Coleman et al., 2002; Mattes et al., 2005) and the type strain of *Kribbella flavida*, DSM 17836 (Pukall et al., 2010); GenBank accession no. NC_013729). The genome of *Nocardioides* sp. JS614 is represented by a 4.99-Mb circular chromosome, 91% of which possesses a protein-coding capacity (4645 protein coding genes), and harbors 55 predicted pseudogenes. The strain also contains a circular plasmid with DNA G+C content of 68 mol%, which is lower than that calculated for the genome (71 mol%) of this strain. The *Kribbella flavida* genome, a circular chromosome of 7.58 Mb, has 7086 protein-coding genes, 60 RNA genes, 2 rRNA operons, and 143 predicted pseudogenes.

Ecology and habitats. Bacteria of the *Nocardioidaceae* are widespread in soil and also in other terrestrial and aquatic environments, including sub-zero-temperature, deep subsurface, and nutrient-limited ecosystems, as well as sites polluted by toxic organic compounds (for references and details, see respective generic chapters in this volume). They have also been detected in uranium and nuclear waste-contaminated sites (Desantis et al., 2006; Fredrickson et al., 2004). It has been reported that representatives of the *Nocardioidaceae* from desert top soils, along with other bacteria and fungi, survive long-distance transport through the atmosphere by dust storms (Griffin, 2007; Polymenakou et al., 2008) and occur in urban aerosols (Brodie et al., 2007). It is worth mentioning that the dust flux only from the Saharan-Sahel region to the atmosphere approximates 1 billion tons per year (Moulin et al., 1997) and, accordingly, huge bacterial and fungal mass are transferred through the atmosphere. Cells or spores of some *Kribbella* remained alive in a soil sample after exposure at 120°C (dry heating) for 1 h (Kirby et al., 2006). Thus, presumably members of this family, like many other actinobacteria, can survive environmental hazards, including desiccation, low and high temperatures, oxygen radicals, UV damage, toxic compounds, etc.

Bacteria of the family, like other actinomycetes, are considered to be consumers of organic material in ecosystems. Many organisms, especially of the genus *Nocardioides*, show biodegradative activities, secreting a range of extracellular enzymes and exhibiting the capacity to metabolize recalcitrant and toxic environmental pollutants. They may also use traces of organics or engage in chemolithotrophic metabolism with input from some atmospheric gases and minerals. Organisms of the genus *Kribbella*, along with some other soil bacteria, are suggested to utilize hydrogen at low concentrations (in the plant rhizosphere) and might contribute to the function of soil as a sink in the global hydrogen cycle (Osborne et al., 2010). Members of the *Nocardioidaceae* can be associated with plants and may exist as mutualistic plant endophytes (e.g. Collins et al., 1994; Coombs and Franco, 2003; Coombs et al., 2003; Kaewkla and Franco, 2010b, 2010a; Song et al., 2004; Trujillo et al., 2006). They can also occur in association with algae, lichens, fungi, and other eukaryotic organisms, including warm-blooded animals (e.g. Fall et al., 2007; Gill et al., 2006; Harris et al., 2007; Lauer et al., 2007; Lee and Kim, 2007; Li et al., 2007b; Sfanos et al., 2005; Tóth et al., 2008). They have also been found in the human microbiome (Grice et al., 2008, 2009). All members of the family are considered nonpathogenic to humans, vertebrate animals, and plants, although some representatives can be occasionally detected among members of bacterial populations found in diseased humans and plants (e.g. Filion et al., 2004; Harris et al., 2007; Song et al., 2004).

Taxonomic comments

The family *Nocardioidaceae* (Nesterenko et al., 1985a, 1990), as initially described on the basis of data accumulated by that time, included the genus *Nocardioides* (Prauser, 1976) with the species *Nocardioides albus* (Prauser, 1976), *Nocardioides luteus* (Prauser, 1984b, 1985), *Nocardioides simplex* (O'Donnell et al., 1982, 1983), and related organisms containing LL-A_2pm in the cell wall, i.e. *Pimelobacter jensenii* (Suzuki and Komagata, 1983b, 1983c) and *Arthrobacter tumescens* (Conn and Dimmick, 1947; Jensen, 1934). In the same publication, the authors also proposed to reclassify *Pimelobacter jensenii* Suzuki and Komagata (1983b, 1983c) as *Nocardioides jensenii*, but the name was not validated, and treated *Arthrobacter tumescens* (*Pimelobacter tumescens*, according to Suzuki and Komagata, 1983b, 1983c) as a separate taxon outside the genus *Nocardioides*. In this context, the genus *Pimelobacter* with the species *Pimelobacter simplex*, *Pimelobacter jensenii*, and *Pimelobacter tumescens* was validly described by Suzuki and Komagata (1983b, 1983c) almost simultaneously with the proposal of O'Donnell et al. (1982, 1983) to reclassify *Arthrobacter simplex* (Lochhead, 1957) as *Nocardioides simplex*, and shortly before the publication of Nesterenko and colleagues (Nesterenko et al., 1985a) (see the *Nocardioides* chapter, below, for more details on the taxonomic and nomenclatural history of the organisms).

Subsequent taxonomic re-evaluation of this group, involving the use of 16S rRNA analysis, showed that *Nocardioides albus*, *Nocardioides luteus*, *Pimelobacter jensenii*, and *Pimelobacter simplex* formed a common phylogenetic group, whereas *Pimelobacter tumescens* represented a separate line of descent (Collins et al., 1989). Based on the sequence data which were mainly in accordance with the data on polar lipid and fatty acid composition (Collins et al., 1983; O'Donnell et al., 1982), and the results of phage typing (Prauser, 1976), the authors transferred *Pimelobacter jensenii* to the genus *Nocardioides* and reclassified *Pimelobacter tumescens* in a newly established genus *Terrabacter* (currently within the family *Intrasporangiaceae*; Stackebrandt et al., 1997). Simultaneously, Collins and Stackebrandt (1989a, 1989b) proposed the species *Nocardioides fastidiosus* which was later transferred (Tamura and Yokota, 1994) to the genus *Aeromicrobium* (Miller et al., 1991). In 1994, another species, *Nocardioides plantarum*, was added to the genus *Nocardioides* (Collins et al., 1994). Thus, the four *Nocardioides* species (*Nocardioides albus, Nocardioides luteus, Nocardioides simplex, Nocardioides jensenii*, and *Nocardioides plantarum*) and two species of *Aeromicrobium* (*Aeromicrobium erythromycini* and *Aeromicrobium fastidiosum*) had been recognized by 1997, when E. Stackebrandt and colleagues (1997) proposed the emended description of the family *Nocardioidaceae* in a paper introducing a novel hierarchic classification scheme of actinobacteria, in which delineation of higher taxa was based on the 16S rRNA gene sequence-based phylogenetic clustering and distribution of signature nucleotides.

The remaining genera, i.e. *Actinopolymorpha, Flindersiella, Kribbella, Hongia, Marmoricola*, and *Thermasporomyces* were added to this family during the years 1999–2011 as a result of using a taxonomic strategy based on the polyphasic approach (Collwell, 1970; Stackebrandt, 2006; Vandamme et al., 1996) which integrates genomic and phenotypic, including chemotaxonomic, characteristics and assumes a certain level of their consensus while establishing or revising the genera and delineating their boundaries. Among them, the genera *Kribbella* (Park et al., 1999) and *Hongia* (Lee et al., 2000) were independently published by different research teams to accommodate the phenotypically and phylogenetically very similar mycelial organisms. Later, Sohn et al. (2003) reclassified the only species of the genus *Hongia, Hongia koreensis*, as *Kribbella koreensis*, providing strong evidence that the name *Hongia* Lee et al. (2000) is a junior heterotypic synonym of *Kribbella* Park et al. (1999).

The genus *Jiangella* was also assigned at its original description to the family *Nocardioidaceae* (Song et al., 2005). The family affiliation of this genus was accepted by Zhi et al. (2009) who suggested an updated structure and 16S rRNA gene sequence-based definition of higher ranks of the class *Actinobacteria*. The authors, along with other proposals concerned with establishment and emendation of higher taxa, provided the emended description of the family *Nocardioidaceae* (comprising the genera *Nocardioides, Actinopolymorpha, Aeromicrobium, Jiangella, Kribbella*, and *Marmoricola*). However, the genus *Jiangella* has recently been transferred from the *Nocardioidaceae* to the newly proposed family *Jiangellaceae*, suborder *Jiangellineae* (Tang et al., 2010), which has been elevated to order *Jiangellales* in the present volume. In addition, two novel genera, *Flindersiella* and *Thermasporomyces*, have been added to the *Nocardioidaceae*. All the above data suggest that the current family *Nocardioidaceae* Nesterenko et al. (1990) emend. Zhi et al. (2009) is in need of re-evaluation and further emendation.

Our recent analyses of 16S rRNA gene sequences using different clustering algorithms and different selections of strains, including type strains of *Flindersiella, Thermasporomyces*, and representatives of the family *Propionibacteriaceae*, showed, as already mentioned, that the genera comprising the current family *Nocardioidaceae* form at least two separate phylogenetic clusters (Figure 247) which can be equated with families. The two clusters are approximately equidistant from each other and from the cluster of the family *Propionibacteriaceae*; the grouping appears to be consistent with morphological and chemotaxonomic features of the genera encompassed (Table 214). The first group, the *bona fide Nocardioidaceae* (in general, corresponding to the *Nocardioidaceae* Nesterenko et al. (1990) emend. Rainey et. al. (1997), includes the genera *Nocardioides, Aeromicrobium*, and *Marmoricola*. The second group, here provisionally named "*Kribellaceae*" comprises the genera *Kribbella, Actinopolymorpha, Flindersiella*, and *Thermasporomyces*. Properties shared primarily or exclusively by organisms of the bona fide *Nocardioidaceae* include: (a) relatively simple morphology and developmental cycles (mostly irregular rods to coccoid cells, or occasionally fragmenting substrate hyphae giving rise to scant aerial hyphae); (b) the presence of motile cells; (c) significant proportions of straight-chain saturated and unsaturated fatty acids (and their 10-methyl-branched derivatives); (d) one predominating menaquinone (containing a tetra-hydrogenated side chain with 8 or 9 isoprene units) in the respiratory chain; (e) teichoic acids or other phosphorous-containing cell-wall polymers (*Nocardioides, Aeromicrobium*).

In contrast to the *bona fide Nocardioidaceae*, the "*Kribellaceae*" comprises organisms which are characterized by: (a) generally more complex cell morphology reflecting more complex developmental events, manifested in the course of their reproductive cycles; (b) the absence of motile cells; (c) fatty acid profile dominated by iso- and anteiso-branched acids with minor quantity of straight-chain components; (d) more complex menaquinone system (and, therefore the respiratory system as a whole) with several major isoprenologues tending to have longer and more saturated side chains; (e) phosphorous-free acidic polysaccharides in the cell wall, i.e. teichulosonic and teichuronic acids (found in *Kribbella*). In addition, representatives of these two groups have different-sized genomes, i.e. 4.99 Mb (*Nocardioides* sp. JS614) and 7.58 Mb (*Kribbella flavida* DSM 17836), with different numbers of protein-coding genes, 4645 and 7086, respectively (Copeland et al., 2006; Pukall et al., 2010). Thus, it appears reasonable to divide the current family *Nocardioidaceae* into at least two families, to reflect more robust phylogenetic grouping of the genera encompassed and to achieve more focused and practicable family definitions. In this way, the bona fide *Nocardioidaceae* would be restricted to the genera *Nocardioides, Aeromicrobium*, and *Marmoricola*, while the family provisionally named "*Kribellaceae*" is suggested to include *Kribbella, Actinopolymorpha, Flindersiella*, and *Thermasporomyces*. Along with the 16S rRNA gene sequence data used to define actinobacterial families, the data on this and the other actinobacterial families tend to suggest the feasibility of the phenotypic circumscription (at least of selected families at the beginning) within the framework of contemporary taxonomic structure of the class *Actinobacteria* and the taxonomic methods presently available.

In general, the current multi-species genera of the group under consideration are clearly defined by 16S rRNA gene sequence clustering (Figure 247), and the allocation of a novel strain to a certain genus is achievable by determination of full or partial 16S rRNA gene sequences. Exceptions are the genera *Marmoricola* and *Nocardioides*, members of which form a common clade at a periphery of the *Nocardioides* radiation. Additionally, *Nocardioides dubius* and some other species rather occupy a phylogenetic position intermediate between this group and the remaining *Nocardioides* species. Correspondingly, phylogenetic delineation of organisms of the genera *Marmoricola* and *Nocardioides* might be problematic owing to this situation and also to uncertainty with the chemotaxonomic differentiation between these two genera (Table 214). Allocation of novel strains to the genus *Marmoricola* or related *Nocardioides* species can be achieved via the determination of both the phylogenetic position and chemotaxonomic characteristics (mostly fatty acid composition), as well as by comparison with individual species of these two genera.

It is quite likely that the taxonomic structure of the genus *Marmoricola* will be revised to reflect more focused definition of the inter-generic boundaries between *Marmoricola* and *Nocardioides* with removal of *Marmoricola bigeumensis* from the genus *Marmoricola*. Notably, the establishment of the genus *Marmoricola* was based on the priority of differences in morphology, lipid composition, and several secondary-structure-forming nucleotides between the novel strain and the phylogenetically closest *Nocardioides* species (Urzì et al., 2000). The proposal of *Marmoricola bigeumensis*, in contrast, relied mostly on the result of the 16S rRNA gene-clustering (Dastager et al., 2008b), although the organism has a higher (97%) binary 16S rRNA gene sequence similarity to *Nocardioides jensenii* than to the type species *Marmoricola aurantiacus* and markedly differs from the two *Marmoricola* species described by that time (Lee and Kim, 2007; Urzì et al., 2000) at least in fatty acid type (which is commonly considered to be a chemotaxonomic marker differentiating actinobacterial taxa above species level; Kroppenstedt, 1985; Suzuki and Komagata, 1983a). Recent descriptions of two additional *Marmoricola* species (Lee and Lee, 2010; Lee et al., 2010) and a few related *Nocardioides* species provided additional evidence for the *insertae sedis* status of the species *Marmoricola bigeumensis*.

The data accumulated since the original description of the genus *Nocardioides* (Prauser, 1976) tend to suggest that the taxonomic structure of this genus will be re-evaluated to achieve a more coherent phylogenetic and phenotypic circumscription. The original description of the genus *Nocardioides* (created for mycelium-producing actinomycetes with the peptidoglycan type similar to that of the genus *Streptomyces*, and showing susceptibility to specific actinophages) does not reflect characteristics of organisms subsequently described under the generic name *Nocardioides*. At present the genus comprises phylogenetically distant and phenotypically dissimilar species, including both the mycelium-forming and rod-shaped organisms. The most unrelated species show 16S rRNA gene sequence similarity of ~92–93%, which is equal to or lower than cut-off values separating many well-defined genera in this and other actinomycete families. In addition, the differences between some *Nocardioides* species in the 16S–23S ITS and RNase P gene sequences exceed those found between representatives of different genera (Yoon et al., 1998a; Yoon and Park, 2000). As for chemotaxonomic heterogeneity, some *Nocardioides* species differ from the type species in the phospholipid types (*sensu* Lechevalier et al., 1977) which is usually believed to differentiate genera (Kämpfer, 2006; Kroppenstedt and Evtushenko, 2006; Lechevalier et al., 1977, 1981), and in the polar lipid patterns as a whole (Table 214). Strains of several *Nocardioides* species so far tested differ in the second major polyamine (Busse and Schumann, 1999) and in the nature of the peptidoglycan-bound cell-wall polysaccharides (Shashkov et al., 1999, 2000b; Tul'skaya, 2009), although the taxonomic value of such characteristics in the family *Nocardioidaceae* is in need of further evaluation. On the other hand, the cell-wall peptidoglycan structures, which are well recognized taxonomic markers differentiating actinomycete genera, have not yet been determined for the majority of *Nocardioides* species, and might turn out to be dissimilar in some organisms.

It can be expected that further comprehensive study of phenotypic and genotypic characteristics of organisms currently encompassed by the genus *Nocardioides*, including new relevant isolates, and involving the genomic and proteomic data for focused circumscription, will provide stronger grounds for dividing the current genus *Nocardioides* into several genera. Dissection of the taxonomic structure of this genus is consistent with the recent way that classification schemes have been improved for other phylogenetically heterogeneous actinomycete genera. Phenotypic (chemotaxonomic) characteristics in particular were used to discriminate between incoherent species and establish separate genera (Behrendt et al., 2011; Tamura et al., 2009, and other studies published in IJSEM during the last decade). Dissection is also in line with the recent tendencies to describe species with more homogenous sets of strains and very high 16S rRNA gene sequence similarity (99% and higher) to recognized species. The tendencies, in turn, reflect genome variation among strains within current bacterial species evolving in different ecological settings (Konstantinidis et al., 2006; Konstantinidis and Tiedje, 2005, 2007).

Acknowledgements

This work was supported by the program MCB of the Russian Academy of Sciences.

Genus I. **Nocardioides** Prauser 1976, 61[AL]

Lyudmila I. Evtushenko, Valentina I. Krausova and Jung-Hoon Yoon

No.car.di.o.i'des. N.L. fem. n. *Nocardia* name of a genus; L. suff. *-oides* (from Gr. suff. *-eides* from Gr. n. *eidos*, that which is seen, form, shape, figure) ressembling, similar; N.L. masc. n. *Nocardioides Nocardia*-like, referring to the similarity of life cycles of the type species of this genus and *Nocardia*.

Abundantly branched vegetative hyphae or irregular rods may be formed in young cultures. The morphogenetic cycle is usually observable, with **different organisms showing more or less complex succession of morphological stages**. The morphogenetic cycle usually starts with the coccoid cells or short rods which may simply germinate into rods or longer filaments, show elementary branching or form extensively branched hyphae on and below the surface of agar media, and may give rise to **aerial mycelium**. The latter consists of irregular, sparsely branching or unbranched hyphae and may totally or partially cover the primary mycelium, or be discernible only microscopically. Both the **vegetative and aerial hyphae and the rod-shaped cells undergo various degrees of fragmentation** (division via septa formation). Fragmentation finally results in the next generation of short rod-like and coccoid cells. No endospores are formed. **Rod-shaped bacteria may be motile. Gram-stain-positive** type of cell wall. Non-acid-fast. The colony color is mainly **whitish, creamy, or yellow** of different tint and intensity and rarely **orange**. Diffusible pigments are not usually produced. Colonies not covered by aerial mycelium are **mostly pasty**, with **smooth to wrinkled surface**. Chemo-organotrophic, with an oxidative type of metabolism. Predominantly catalase-positive. The level of oxidase activity varies among species. Grows under aerobic conditions on standard laboratory media, including chemically defined (synthetic) media or media with low nutrient concentrations. Certain vitamins or other growth factors may be required. Utilizes a wide range of carbon and nitrogen sources, including unusual organic compounds and toxic environmental pollutants. Mostly mesophilic and neutrophilic; some grow at initial pH values of 5–5.5 and/or 11–12. Mostly non-halophilic, but salt-requiring organisms occasionally occur.

The cell-wall peptidoglycan type is A3γ, with LL-diaminopimelic acid and glycine as the diagnostic amino acids. **Muramic acid is of the acetyl type**. The **cell-wall teichoic acids are present** in most organisms examined. **Menaquinones** are the sole respiratory quinones, the predominant component is **MK-8(H_4)** containing a tetra-hydrogenated side chain with eight isoprene units. **Cellular fatty acids are complex mixtures** of saturated and monounsaturated, straight-chain and iso-, anteiso- and 10-methyl-branched components, including 10-methyl octadecanoic acid (tuberculostearic acid, TBSA), among which **14-methyl pentadecanoic acid ($C_{16:0}$ iso) usually predominates.** Mycolic acids are not present. **The principal phospholipids are typically composed of non-nitrogenous components**, but phosphatidylethanolamine and phosphatidylcholine may occasionally occur. The polyamine patterns usually include **cadaverine as the predominant component**, with putrescine, spermine, or spermidine representing the second major polyamine.

Natural habitats include various terrestrial and aquatic environments. Can be associated with plants, animals, and humans. Some are occasionally found among bacteria associated with human diseases, but considered to be of no relevance to the disease agents.

DNA G+C content (mol%): 67.5 (T_m)–74.8 (HPLC).

Type species: **Nocardioides albus** Prauser 1976, 61[AL].

Further descriptive information

The genus *Nocardioides* belongs to the family *Nocardioidaceae*, order *Propionibacteriales*. Based on the 16S rRNA gene sequence analysis, the species currently comprising this genus form a phylogenetic radiation which is clearly separated from the other genera of the family *Nocardioidaceae*, except for *Marmoricola* (Figure 247 and Figure 248). Several *Nocardioides* species rather occupy intermediate phylogenetic position between the majority of *Nocardioides* species and the genus *Marmoricola* or are intermixed with members of the genus *Marmoricola*. The 16S rRNA gene sequence similarity levels between *Nocardioides* species are quite dissimilar and range from ~92–93 to 99.6%.

Morphology and colony appearance. Bacteria of the genus *Nocardioides* usually display life cycles that most closely resemble those of *Nocardia* or *Arthrobacter*. Information on the mycelium-forming organisms presented in this section is based to a large extent on the works of H. Prauser (Prauser, 1976; Prauser, 1984a; Prauser, 1989).

Visible growth of *Nocardioides albus* and *Nocardioides luteus* and related mycelium-forming organisms is observed within 1–2 d on standard nutrient media. On media rich in organic nitrogen, and in submerged shaken culture, the extent of mycelium development and its persistence are reduced. The hyphae of the primary mycelium are 0.5–0.8 μm in diameter and irregularly septate. Preceding fragmentation, additional septa are formed. Fragmentation of the vegetative hyphae begins in the older parts of the colonies (Figure 249). Depending on the growth media, the elements resulting from fragmentation may be irregular, rod-like, or coccoid (Figure 249 and Figure 250; Prauser, 1989; Suzuki and Komagata, 1983c). The fragments may give rise to new mycelia by extruding one, two, or more hyphae. The irregularly shaped and branched hyphae of the aerial mycelium (Figure 251 and Figure 252) are slightly thicker (0.6–1.0 μm). The aerial hyphae become septate and break up more regularly than those of the primary mycelium. They usually fragment completely into rod- or coccus-like elements that resemble at maturity the surface-smooth conidia (arthrospores) of some streptomycetes (Figure 253 and Figure 254). The spore-like elements germinate by producing one or two germ tubes when transferred to a fresh medium. Motility does not occur at any stage of the life cycle of *Nocardioides albus* and *Nocardioides luteus*. Indeed, the spore-like elements may be unrecognizable *in situ* but observable after mechanical disruption of the mycelium threads. Notably strains may occasionally lose the ability to form aerial mycelium on continued subculture. Both hyphae and fragments show the ultrastructure typical for Gram-stain-positive bacteria (Figure 250 and Figure 254).

The colony appearance of *Nocardioides albus*, *Nocardioides luteus*, and related filamentous organisms is influenced to a large extent by the culture conditions. The colony surface on agar media may be smooth, rough or wrinkled, or covered by aerial mycelium. Depending on the growth medium, individual strain, and the culture age, the aerial mycelium may totally cover the primary mycelium, may be formed only in patches or at the margins of the colonies, may be visible only microscopically or absent. Colonies not covered by aerial mycelium may be dull to bright, but usually are faintly glistening. The colony consistency (mass of the primary mycelium) is mostly pasty, particularly in the center of colonies where hyphal fragmentation begins and in older agar cultures on rich media.

The majority of species currently attributed to the genus *Nocardioides* produce neither primary nor aerial mycelium. Cells of such species are mostly slender irregular rods in young cultures (Table 215; Figure 255, Figure 256, Figure 257, Figure 258, Figure 259A, Figure 261, and Figure 262) that may occur singly, in short chains, in palisades or other side-by-side formations (Figure 258) probably due to adhesive properties. V-forms may be produced. Older cultures of the majority of species predominantly or exclusively consist of coco-bacillary or coccoid cells (Figure 259B). A marked rod-to-coccus morphogenetic cycle resembling that of *Arthrobacter* was reported or suggested for nearly all rod-shaped *Nocardioides* species. Larger club-shaped or spherical forms significantly exceeding the remaining cells in diameter (Figure 257) may be occasionally observable both in young and older cultures (Dastager et al., 2009d; Dastager et al., 2008f, c, 2009e; Jensen, 1934). The cell division proceeds by means of irregular or binary

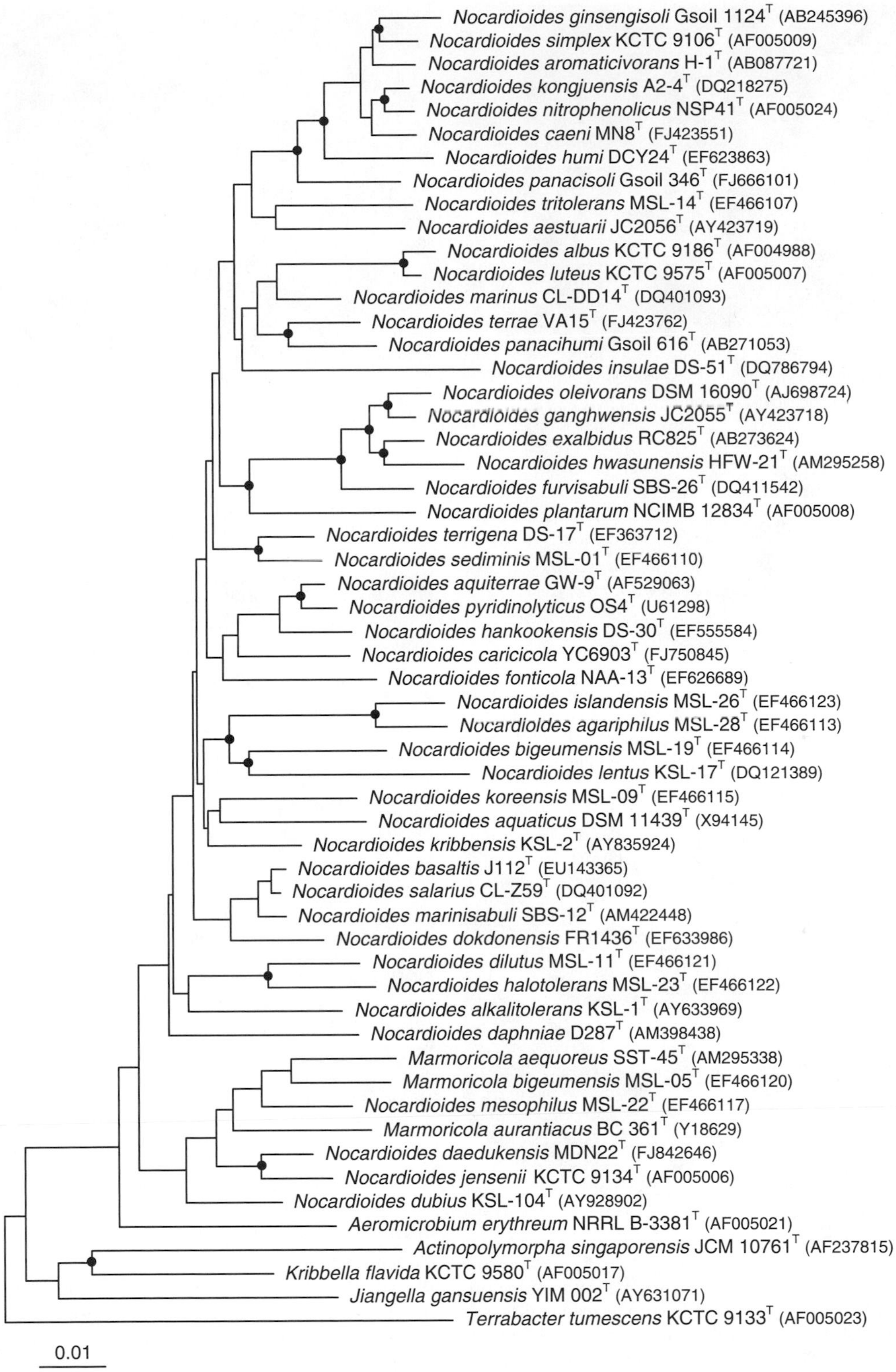

FIGURE 248. Neighbor-joining phylogenetic tree based on 16S rRNA gene sequences showing the positions of *Nocardioides* species and some related taxa. Filled circles indicate that the corresponding nodes were also recovered in the trees generated with the maximum-likelihood and maximum-parsimony algorithms. The bar = 0.01 substitutions per nucleotide position.

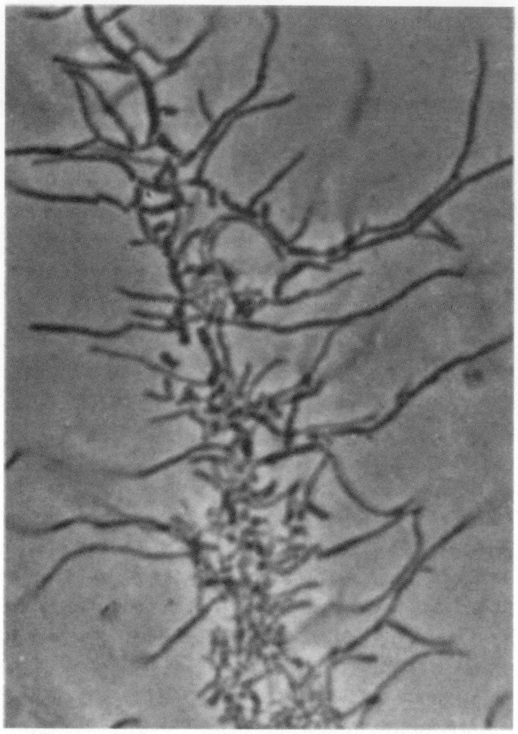

FIGURE 249. *Nocardioides albus* IMET 7807. Fragmentation of hyphae of the primary mycelium *in situ*; 7-d-old culture on glycerol asparagine agar. Phase-contrast micrograph (1600×). (Reprinted from Prauser, 1986. *Bergey's Manual of Systematic Bacteriology*, vol. 2, Williams & Wilkins, Baltimore, pp. 1481–1485.)

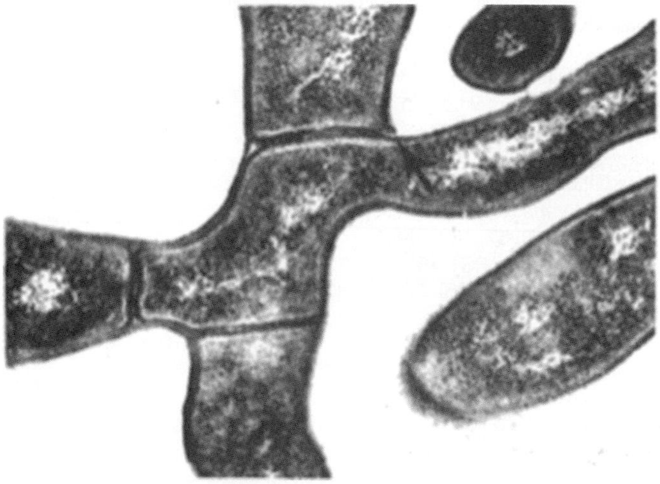

FIGURE 250. *Nocardioides albus* IMET 7807. Part of a hypha of the primary mycelium with branches originating from one segment. Beginning of fragmentation of the hypha at the upper right angle (arrow). Electron micrograph (~40,000×). (Reprinted from Prauser, 1986). *Bergey's Manual of Systematic Bacteriology*, vol. 2, Williams & Wilkins, Baltimore, pp. 1481–1485.)

fission via septa production, so as very short rods or coccoid cells are eventually formed. At the same time, a bud-like mode of cell division may occur. The cell reproduction through constricting was also reported (Kvasnikov et al., 1974). Interestingly, the rod-shaped strain *Nocardioides* sp. JS614, which is the only member of the genus whose complete genome is sequenced, contains a single *ssg* gene that is most likely functionally related to *ssgB*

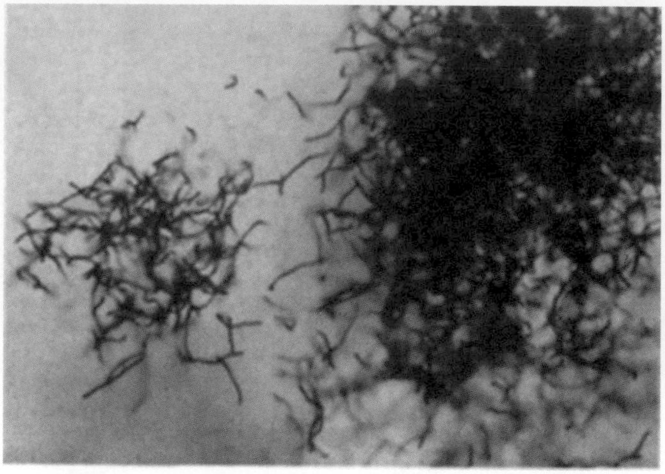

FIGURE 251. *Nocardioides albus* IMET 7807. Aerial mycelium of 11-d-old culture on chitin agar. Phase-contrast micrograph (~400×). (Reprinted from Prauser, 1986. *Bergey's Manual of Systematic Bacteriology*, vol. 2, Williams & Wilkins, Baltimore, pp. 1481–1485.)

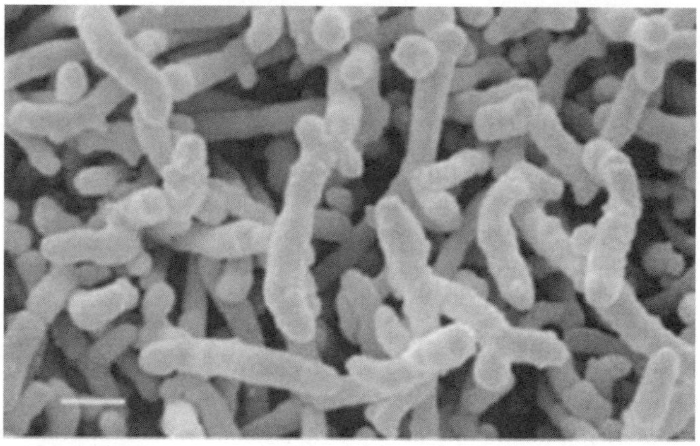

FIGURE 252. Aerial mycelium of *Nocardioides albus* KCTC 9186 on glucose-asparagine agar. Scanning electron micrograph. Bar = 1 μm. (Reprinted from Yoon and Park, 2006. *The Prokaryotes*, 3rd edn, vol. 3, Springer, New York, pp. 1099–1113.)

(hypothesized to be essential for septa formation in actinomycetes) (Traag and van Wezel, 2008). The cells may be nonmotile or motile (Table 215) either with single polar, subpolar, or peritrichous flagella (Figure 260, Figure 261, and Figure 262). Colonies of rod-shaped organisms are usually small, circular, convex, or rarely flat, smooth and glistening, with entire margins. They typically are indistinct in color (whitish, light creamy, or yellowish white), with some organisms producing distinct yellow pigments of different tint and intensity; orange-pigmented colonies may occur on some media (Table 215).

Chemotaxonomy. The cell-wall peptidoglycan contains 2,6-LL-diaminopimelic acid (LL-A_2pm), along with alanine, glutamic acid, and glycine. As available for several species (Lawson et al., 2000a; Prauser, 1976; Schippers et al., 2005; Schleifer and Kandler, 1972), the peptidoglycan type is A3γ type *sensu* Schleifer and Kandler (1972); variation A41.1 (http://www.peptidoglycan-types.info). It contains LL-A_2pm in position 3 and L-alanine in position 1 of the tetrapeptide subunit, with a glycine residue forming the interpeptide bridge (like that in the majority of LL-A_2pm-containing actinomycetes). This polymer differs from the

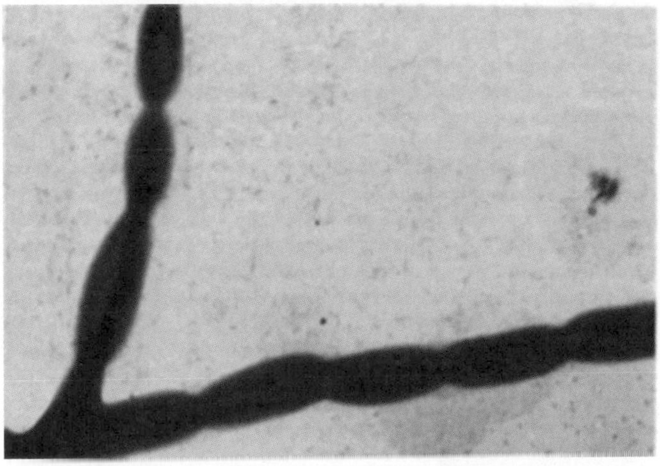

FIGURE 253. *Nocardioides albus* IMET 7807. Aerial hyphae fragmented into spore-like elements; 14-d-old culture on yeast extract-malt extract agar. Electron micrograph (~17,000×). (Reprinted from Prauser, 1986. *Bergey's Manual of Systematic Bacteriology*, vol. 2, Williams & Wilkins, Baltimore, pp. 1481–1485.)

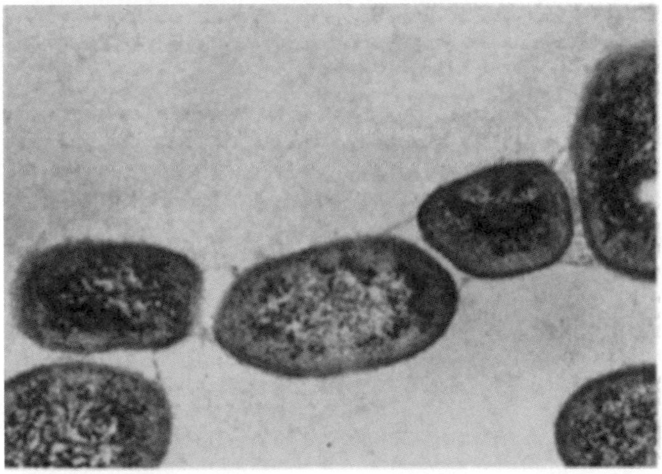

FIGURE 254. *Nocardioides albus.* Developing spore-like elements (arthrospores) still connected by the surface sheath of the aerial hypha. Electron micrograph (~50,000×). (Reprinted from Prauser, 1986. *Bergey's Manual of Systematic Bacteriology*, vol. 2, Williams & Wilkins, Baltimore, pp. 1481–1485.)

LL-A$_2$pm-based peptidoglycans of some other representatives of the suborder *Propionibacterineae* or the family *Intrasporangiaceae*, which may contain 3 glycine molecules in the interpeptide bridge and glycine or glycine amide at D-glutamic acid of the peptide subunit (type A3γ; variation A41.2), or contain glycine instead of alanine in position 1 of the peptide subunit (type A3γ′; variation A42.1) (Schleifer and Kandler, 1972; Schumann et al., 2009; Stackebrandt and Schaal, 2006; Weon et al., 2007; and the respective chapters in this volume). No isomer of A$_2$pm was reported for *Nocardioides dilutus.* The acetyl type of the muramyl residues in the cell-wall peptidoglycans was reported for a few *Nocardioides* species so far studied with respect to this characteristic (Kubota et al., 2005a; Uchida and Seino, 1997). The data on cell-wall sugars routinely examined in actinomycetes during taxonomic studies are also available for some species only, namely, *Nocardioides albus, Nocardioides luteus, Nocardioides simplex, Nocardioides jensenii,* and *Nocardioides aquaticus.* These typically include galactose and different combinations of glucose, mannose, and rhamnose (Cummins and Harris, 1959; Keddie and Cure, 1977; Lawson et al., 2000a; Prauser, 1989; Sadikov et al., 1983; Shashkov et al., 2000b; Shashkov et al., 1999; Takeuchi and Yokota, 1989; Tul'skaya, 2009). On the other hand, no monosugars have been revealed in the cell wall of the type strain of *Nocardioides plantarum* (Tul'skaya, 2009).

The cell-wall teichoic acids of different structure were found in strains of *Nocardioides albus, Nocardioides luteus, Nocardioides jensenii, Nocardioides simplex,* and some mycelium-producing organisms of this genus (Evtushenko et al., 1984; Sadikov et al., 1983; Shashkov et al., 2000b; Shashkov et al., 1999; Takeuchi and Yokota, 1989; Tul'skaya, 2009). No phosphorous-containing polymers was revealed in the cell wall of *Nocardioides plantarum* (Tul'skaya, 2009), which, along with the absence of monosugars, is rather indicative of a different kind of the peptidoglycan-linked polymer in this species. The type strain of *Nocardioides albus* possesses a galactosylglycerol phosphate polymer, with repeating units joined by phosphodiester links involving the glycerol C3 and the β-D-galactopyranose C3 atoms. The β-D-galactopyranosyl residues are substituted at C2 with acetate groups and at C4 with β-D-glucopyranose carrying a 4,6 pyruvate ketal group (Shashkov et al., 1999). The type strain of *Nocardioides simplex* was reported to contain a poly(glycerol phosphate) teichoic acid (Sadikov et al., 1983; Takeuchi and Yokota, 1989). Additional sugars (galactose, mannose, glucose, and/or rhamnose) detected by these authors might be involved in the side chains of the polymer(s). The teichoic acid of the type strain of *Nocardioides jensenii* was not identified, but reported to include glycerol, galactose, *N*-acetylglycosamine, and pyruvic acid (Tul'skaya, 2009). Takeuchi and Yokota (1989) revealed galactose and glycerol as predominant components, together with lesser amounts of *N*-acetylglycosamine, glucose, and rhamnose, in a phosphorus-containing polysaccharide fraction of the cell wall of *Nocardioides jensenii.* The type strain of *Nocardioides luteus* and several very closely related strains, including *Nocardioides luteus* ("*Nocardioides flavus*") VKM Ac-2525 (= DSM 46114 = IMET 7844 = J.-S. Ruan, 71-N54) and *Nocardioides* sp. ("*Nocardioides fulvus*") VKM Ac-2526 (= DSM 46115 = IMET 7846 = J.-S. Ruan, 71-N86) have identical 1,5-poly(ribitol phosphate) teichoic acids (Shashkov et al., 1999; Tul'skaya, 2009). All ribitol molecules of these strains are substituted at C4 with α-D-galactopyranosyl residues carrying a 4,6 pyruvate ketal group (Shashkov et al., 2000b; Tul'skaya, 2009). It is noteworthy that regardless of the polymer structure, galactose and pyruvic acid are present in the teichoic acids of all *Nocardioides* strains so far characterized (no precise data are available for *Nocardioides simplex* from early work; Sadikov et al., 1983). High abundance of lipoteichoic acids that, in contrast to the cell-wall teichoic acids, are the membrane-anchored molecules in the cell envelopes, were found in all mycelium-forming *Nocardioides* strains investigated to date (E.M. Tul'skaya and L.I. Evtushenko, unpublished).

The predominant menaquinone in *Nocardioides* species is MK-8(H$_4$). The data available for some species show that other menaquinones, i.e. MK-6(H$_4$), MK-7(H$_4$), MK-8, MK-8(H$_2$), MK-8(H$_6$), and MK-9(H$_4$) may be produced as minor components or in trace amounts (Collins et al., 1994; Collins et al., 1979; Collins et al., 1983; Lawson et al., 2000a; Schippers et al., 2005; Suzuki and Komagata, 1983c; Yamada et al., 1976; Yoon et al., 1997b).

TABLE 215. Descriptive and differential phenotypic characteristics of *Nocardioides* species[a,b]

Characteristic	1. *N. albus*	2. *N. aestuarii*	3. *N. agariphilus*	4. *N. alkalitolerans*	5. *N. aquaticus*	6. *N. aquiterrae*	7. *N. aromaticivorans*	8. *N. basaltis*	9. *N. bigeumensis*	10. *N. caeni*	11. *N. caricicola*	12. *N. daedukensis*	13. *N. daphniae*	14. *N. dilutus*	15. *N. dokdonensis*	16. *N. dubius*
Colony color[c]	White	Ivory	White to cream	White	Cream to dull orange	Cream	White	Cream	Cream	Grayish yellow	White	Yellowish	Yellowish	White to cream	Cream	Yellowish white
Mycelium (M) or rods (R)	M	R	R	R	R	R	R	R	R	R	R	R	R	R	R	R
Motility	–	–	+	–	–	+	–	–	+	–	–	–	–	+	–	+
Cell size (μm)[d]	0.5–0.8 (width)	0.3–0.4 × 0.9–2.1	0.4–0.5 × 1.3–3.0	0.8–1.0 × 1.5–2.0	0.9–1.0 × 0.9–1.4	0.8–1.0 × 1.7–2.0	0.5–0.7 × 1.0–2.0	0.7–1.0 × 1.2–2.0	0.3–0.8 × 0.8–4.0	0.3–0.7 × 0.7–2.5	0.4–0.6 × 2.0–5.0	0.4–0.8 × 0.8–3.0	0.8–1.0 × 1.2–2.2	0.4–0.8 × 1.9–4.0	0.6–0.9 × 1.2–1.8	0.8–1.0 × 1.5–2.5
Temperature optimum (°C)	28	30	28	25–30	16–26	30	30	25–30	28	30	30	30	28	26–28	25	30
Temperature range (°C)[e]	15–42	20–35	25–37	4–34	16–26	15–42	22–40	10–37	20–35	10–35	10–45	4–37	4–38	nd	4–30	10–37
NaCl requirement, optimum (%)[f]	–	– (0–2)	–	–	– (1–6)	–	–	+ (1–2)	–	– (0–0.5)	–	– (0–0.5)	–	–	– (0–3)	– (0)
NaCl maximum (%)[g]	8	8	<2	5	15	nd	nd	10	<1	1.0	0.5	9	5	<1	7	5
pH optimum	~7	7	7.5	7–9	7–8	6–7	7	6–7	7.5–9	6.5–7.5	8	7–8	7.5–8.5	7–8	7	7–8
pH range[h]	6–9	6–10	nd	5.0–12	5.5–9.5	nd	5–8	5.5–8	nd	6.0–9.5	7–9	6–10	5.5–10.5	nd	5–10	6–10.5
Catalase	+	+	–	+	+	+	+	+	–	+	+	+	+	nd	+	+
Oxidase	–	–	–	+	–	+	–	–	–	+	–	+	–	nd	–	+
Nitrate reduction	–	–	+	+	+	+	–	–	–	–	+	+	+	+	+	–
Decomposition of:																
Esculin	w	w	–	–	–	v	+	–	–	–	+	+	+	–	–	+
Casein	+	+	nd	+	+	+	+	nd	–	+	+	+	w	+	–	+
DNA	w	+	–	nd	+	+	nd	nd	–	nd	nd	nd	nd	–	–	nd
Gelatin	+	+	–	v	+	+	+	+	–	+	–	+	+	–	–	v
Starch	+	–	–	–	v	v	–	–	–	+	–	–	+	–	–	–
Tween 80	+	+	+	+	+	v	v	–	+	+	+	–	v	+	+	–
Tyrosine	+	–	–	+	+	–	–	–	+	–	–	–	+	+	–	–
Urea	–	–	nd	–	–	–	+	–	–	–	–	–	–	–	+	–
Hypoxanthine	+	–	nd	–	–	–	+	–	nd	–	nd	–	–	–	nd	–
Xanthine	+	–	nd	–	–	v	–	–	nd	–	nd	–	–	–	–	–
Enzymes (API ZYM):																
N-Acetyl-glucosaminidase	–	–	–	.–	.–	–	.–	–	–	–	–	–	.–	–	–	–
Acid phosphatase	–	w	nd	+	+	+	v	–	nd	+	+	–	w	+	+	+
Alkaline phosphatase	+	w	+	+	v	–	+	+	+	+	+	–	w	–	+	+
α-Chymotrypsin	–	+	nd	–	–	–	–	–	nd	–	–	–	–	–	+	–
Cystine arylamidase	–	–	+	–	v	w	v	w	–	w	–	–	–	–	–	–
Esterase (C4)	+	+	+	+	w	–	v	+	+	–	+	–	+	+	+	+
Lipase (C14)	–	–	nd	v	–	–	–	–	nd	–	–	–	w	+	–	–
α-Fucosidase	–	–	–	–	–	–	–	–	–	–	–	–	–	–	–	–
α-Galactosidase	–	–	nd	–	–	–	–	–	nd	–	–	–	–	–	–	–
β-Galactosidase	v	+	–	–	–	v	–	–	–	–	+	–	–	v	–	–
α-Glucosidase	+	+	nd	v	+	+	+	+	nd	+	+	–	–	v	+	+
β-Glucosidase	w	–	+	–	–	w	v	–	+	–	+	–	–	+	–	+

Leucine arylamidase	+	+	nd	v	+	+	+	+	nd	+	+	+	+	*v*	+	+
α-Mannosidase	v	–	nd	–	–	–	–	–	nd	–	–	–	–	–	–	–
Naphthol-AS-BI-phosphohydrolase	w	w	nd	w	–	+	v	+	nd	w	–	w	w	+	+	+
Trypsin	+	+	–	–	w	v	v	w	–	–	+	–	–	–	+	–
Valine arylamidase	–	+	–	–	w	–	v	–	–	w	–	–	w	nd	w	–
Utilization of carbon source:[i]	*ISP9, B-YC*	*B-YC*	*MS1+ GF1*	*ISP9*[j]	*API CH50, B-YC*	*ISP9, B-YC*	*API CH50, MS2+VE*	*API 50 CH*	*MS1+GF1*	*API 50 CH*	*API 20NE*	*ISP9*	*ISP9*[j]	*ISP9*	*YNB+CA*	*ISP9*[j]
L-Arabinose	+	–	nd	v	–	–	+	–	nd	+	–	–	–	–	–	–
Cellobiose	+	+	nd	+	–	+	+	–	nd	+	nd	–	+	+	–	v
D-Fructose	+	+	–	–	+	+	+	–	–	–	nd	–	+	–	+	v
D-Galactose	+	+	–	–	+	+	–	–	–	–	nd	–	–	+	–	–
D-Glucose	+	+	nd	–	+	+	+	+	nd	–	–	–	+	+	–	+
Lactose	v	w	–	–	–	v	–	–	+	–	nd	nd	+	+	–	–
Maltose	+	nd	–	v	nd	+	+	–	–	–	–	–	–	–	–	–
D-Mannitol	+	+	–	–	+	+	+	+	–	–	–	nd	–	+	+	+
D-Mannose	+	–	–	–	–	v	v	–	+	–	–	–	+	+	+	–
Melezitose	–	nd	nd	+	nd	–	–	+	nd	–	nd	nd	nd	nd	+	nd
Melibiose	–	nd	+	–	nd	–	v	–	+	–	nd	nd	–	+	+	–
D-Raffinose	–	w	+	+	–	–	–	–	–	–	nd	nd	–	+	–	–
L-Rhamnose	+	–	–	–	nd	v	+	–	–	+	nd	nd	–	+	–	–
D-Ribose	v	–	–	–	–	v	+	–	–	–	nd	nd	–	–	+	–
Sucrose	+	+	nd	v	+	+	+	+	nd	+	nd	–	+	+	–	v
D-Trehalose	–	nd	nd	+	nd	+	+	+	nd	+	nd	+	nd	nd	+	–
D-Xylose	+	+	nd	v	+	+	+	–	nd	–	nd	–	+	+	w	v
Glycerol	+	w	nd	nd	+	–	v	–	nd	–	nd	nd	nd	nd	+	nd
Inositol	–	–	–	–	–	–	–	–	–	–	nd	nd	+	+	–	–
Inulin	–	–	nd	nd	–	–	–	–	nd	–	nd	nd	nd	nd	–	nd
Salicin	w	+	nd	–	–	–	–	+	nd	–	nd	–	nd	nd	–	w
N-Acetylglucosamine	w	–	nd	–	–	–	–	–	nd	–	–	nd	nd	nd	–	nd
Principal phospholipids[k]	PG, DPG, PI, APG, [PIMs, PI-OH][l,m,n]	nd	DPG, PG	nd	PI, PG, DPG, PL	nd	nd	nd	DPG, PG	nd	nd	DPG, PL, L [PG, PE, PL][n]	DPG, PG PL	DPG, PG	nd	DPG, PG, PE, PI
DNA G+C content (mol%)[p]	68.6 (T_m)	70	69.4	72.4–73.6	69	73	72.0–72.4	68 (T_m)	69.3	71.5	71.7	68.7	69.9	71.8	69.1	70.6
Isolation source	Soil	Tidal flat sediment	Soil	Alkaline soil	Saline lake, Antarctica	Ground water	Polluted environments	Black sand	Soil	Sludge of wastewater	Halophilic plant, *Carex* sp.	Soil	Water flea *Daphnia cucullata*	Soil	Beach sand sediment	Alkaline soil

(continued)

TABLE 215. (continued)

Characteristic	17. *N. exalbidus*	18. *N. fonticola*	19. *N. furvisabuli*	20. *N. ganghwensis*	21. *N. ginsengisoli*	22. *N. halotolerans*	23. *N. hankookensis*	24. *N. humi*	25. *N. hwasunensis*	26. *N. insulae*	27. *N. islandensis*	28. *N. jensenii*	29. *N. kongjuensis*	30. *N. koreensis*	31. *N. kribbensis*	32. *N. lentus*
Colony color[c]	White	Yellowish	Yellow	Ivory	Yellow-white	Cream-white	White	Pale yellow	Yellowish	Ivory	White to cream	Yellowish white	Yellowish white	Cream white	Cream	Yellow
Mycelium (M) or rods (R)	R	R	R	R	R	R	R	R	R	R	R	R	R	R	R	R
Motility	–	–	+	–	–	nd	–	+	–	–	–	–	–	+	–	–
Cell size (μm)[d]	~ 0.5 × 2.0	~ 0.8 × 2.0–9.0	0.4–0.5 × 0.6–1.2	0.4–0.5 × 0.9–4.5	0.2–0.4 × 0.8–1.2	~ 0.6 × 1.2–3.4	0.4–0.8 × 1.5–10.0	0.3–0.5 × 0.8–1.0	0.4–0.7 × 1.0–1.7	0.6–1.0 × 1.3–6.0	0.5–0.6 × 1.0–3.7	0.6–0.8 × 3.0–7.0	0.4–0.7 × 0.8–3.0	0.4–0.6 × 0.8–2.0	0.8–1.0 × 1.5–2.0	0.4–0.7 × 1.0–4.5
Temperature optimum (°C)	30	30	30	30	30	28	25–28	30–37	30	30	28	28	30–37	30	30	28
Temperature range (°C)[e]	15–35	25–37	4–37	10–40	15–37	nd	10–34	25–42	4–37	10–34	nd	18–37	10–40	27–37	4–35	4–34
NaCl requirement, optimum (%)[f]	–	+ (0.5–1)	–	– (0–1)	–	– (3)	– (0–0.5)	nd	–	– (0)	nd	–	– (0)	–	–	– (0.5)
NaCl maximum (%)[g]	nd	1	6	8	5	10	2	nd	4	3	7	7	5	5	3	5
pH optimum	~7	7–8	~7	7	7	7–8	6–7	~7	~7	8	7	7–9	7–8	7–8	9	8
pH range[h]	~6–9	5–9	5.1–10.1	6–10	5–8.5	nd	5.5–8.0	5–11	5–9	6.5 – >8	5–12	nd	5.5– >8	nd	6–11	6.5–9.5
Catalase	+	+	+	+	+	–	+	–	+	+	–	+	+	–	+	+
Oxidase	–	–	–	–	–	–	+	+	–	+	–	–	+	–	+	W
Nitrate reduction	–	–	+	+	–	+	–	–	–	+	–	+	–	–	+	+
Decomposition of:																
Esculin	–	+	–	w	nd	–	+	+	–	–	–	–	–	–	+	–
Casein	nd	+	+	+	nd	–	+	nd	–	+	c	+	+	nd	+	+
DNA	nd	+	+	+	+	nd	nd	nd	nd	nd	nd	+	nd	–	nd	nd
Gelatin	nd	+	–	+	+	nd	+	–	v	+	–	+	+	–	+	+
Starch	–	+	+	+	nd	v	+	nd	+	–	+	–	–	+	–	–
Tween 80	nd	+	–	+	nd	v	w	+	nd	+	+	+	+	+	+	+
Tyrosine	nd	nd	–	+	nd	+	v	nd	–	–	–	+	–	+	+	v
Urea	–	–	–	–	v	+	–	–	–	–	–	+	–	nd	–	–
Hypoxanthine	nd	nd	–	–	nd	nd	–	nd	–	–	–	+	–	nd	–	–
Xanthine	nd	nd	–	w	nd	nd	–	nd	–	–	–	+	–	nd	–	–
Enzymes (API ZYM):																
N-Acetyl-glucosaminidase	nd	–	–	–	nd	–	–	–	–	–	–	–	–	–	–	–
Acid phosphatase	w	+	–	w	nd	+	+	+	w	+	+	v	+	nd	+	+
Alkaline phosphatase	+	+	+	+	nd	+	+	+	+	+	+	w	+	+	+	+
α-Chymotrypsin	nd	–	–	–	nd	–	–	+	–	–	–	–	–	nd	+	–
Cystine arylamidase	w	–	–	w	nd	c	v	+	–	–	–	–	–	–	–	–
Esterase (C4)	w	+	–	v	nd	+	+	+	w	+	–	w	v	+	+	+
Lipase (C14)	–	–	–	–	nd	–	–	–	–	–	–	–	–	nd	–	–
α-Fucosidase	–	–	–	–	nd	–	–	–	–	–	–	–	–	–	–	–
α-Galactosidase	+	–	w	+	nd	–	–	–	nd	–	–	–	–	nd	–	–
β-Galactosidase	+	+	+	+	–	+	v	+	–	–	–	–	–	–	w	–
α-Glucosidase	+	+	–	+	nd	+	–	+	w	–	v	v	+	nd	+	–

β-Glucosidase	w	+	–	–	+	+	–	+	nd	–	v	–	v	+	–	+
Leucine arylamidase	+	+	+	+	nd	nd	+	+	+	–	–	+	+	nd	+	+
α-Mannosidase	–	–	–	–	nd	–	–	–	–	–	–	–	–	nd	–	–
Naphthol-AS-BI-phosphohydrolase	+	+	–	–	nd	+	+	+	–	–	–	v	+	nd	+	+
Trypsin	–	–	–	–	nd	–	v	+	–	–	–	v	v	–	–	–
Valine arylamidase	+	–	w	+	nd	–	–	+	+	–	–	v	v	–	–	–
Utilization of carbon source[i]:	*API CH50*	*API CH50*	*ISP9*	*B-YC, ISP9*	*MS2+VE*	*ISP9*	*API CH50*	*API 20NE, API 32GN*	*ISP9*	*API 20E*	*ISP9*	*ISP9, B-YC*[j]	*ISP9, API 50 CH, MS2+VE*	*MS1+ GF1*	*ISP9*	*ISP9*
L-Arabinose	–	+	+	+	c	–	+	–	v	–	–	–	v	nd	–	w
Cellobiose	–	+	+	+	nd	–	+	nd	+	+	v	–	–	nd	+	+
D-Fructose	+	+	+	+	+	–	+	nd	+	–	–	v	v	–	–	–
D-Galactose	–	+	+	+	nd	+	+	nd	+	+	–	–	–	–	+	+
D-Glucose	+	+	+	+	+	+	+	+	+	+	+	+	+	nd	+	+
Lactose	–	nd	+	+	nd	+	–	nd	–	nd	+	–	–	+	–	v
Maltose	–	–	+	+	+	–	+	–	+	w	–	v	–	–	+	+
D-Mannitol	+	nd	+	+	+	+	+	+	+	nd	–	–	–	–	+	+
D-Mannose	–	+	+	+	–	+	–	–	+	w	v	–	v	+	–	–
Melezitose	–	nd	–	nd	nd	nd	–	nd	–	nd	nc	nd	–	nd	–	+
Melibiose	+	nd	–	+	–	nd	+	+	–	nd	+	–	–	+	–	–
D-Raffinose	–	nd	+	+	nd	+	–	nd	–	nd	–	–	–	–	+	+
L-Rhamnose	+	nd	–	v	–	+	+	+	+	nd	–	+	v	–	+	+
D-Ribose	–	nd	–	–	–	–	–	+	–	nd	–	–	–	–	+	–
Sucrose	+	–	–	+	+	nd	+	+	+	–	+	+	+	nd	+	+
D-Trehalose	–	–	+	+	nd	+	+	nd	+	+	+	+	+	nd	+	+
D-Xylose	–	+	+	+	+	+	+	nd	+	w	+	v	–	nd	–	v
Glycerol	–	nd	+	+	nd	nd	–	nd	–	nd	nd	w	–	nd	nd	nd
Inositol	–	nd	w	–	–	+	–	–	–	–	–	–	–	–	–	–
Inulin	–	nd	+	–	–	nd	–	nd	+	nd	nd	–	–	nd	nd	nd
Salicin	–	nd	–	+	–	nd	–	–	+	w	nd	–	–	nd	nd	nd
N-Acetylglucosamine	–	nd	nd	+	v	nd	–	–	nd	nd	nd	–	–	nd	nd	nd
Principal phospholipids[k]	DPG, PI	nd	PG, PC, PI, PL	nd	nd	DPG, PG, PL	nd	DPG, PG, PL	DPG, PI PG, PL	nd	DPG, ▪G, PL	DPG, PG, PI, OH-PG	nd	DPG, PG, PL	nd	nd
DNA G+C content (mol%)[p]	74	71.8	69.1	72	70.2	69.7	71.3	71.0	71.1–72.2	71.1	71.4	68.8 (T_m)	72.1	69.9	73–74	74.6–74.8
Isolation source	Lichen	Freshwater from spring	Beach sand	Tidal flat	Soil	Agricultural soil	Soil	Soil	Sea water	Soil	Agricultural soil	Soil	Soil	Soil	Alkaline soil	Alkaline soil

(continued)

TABLE 215. (continued)

Characteristic	33. *N. luteus*	34. *N. marinisabuli*	2.515 pt	36. *N. mesophilus*	37. *N. nitrophenolicus*	38. *N. oleivorans*	39. *N. panacihumi*	40. *N. panacisoli*	41. *N. plantarum*	42. *N. pyridinolyticus*	43. *N. salarius*	44. *N. sediminis*	45. *N. simplex*	46. *N. terrae*	47. *N. terrigena*	48. *N. tritolerans*
Colony color[c]	Yellow to orange	Pale yellow	Creamy white	Cream whitish	Yellowish white	Orange	None or grayish	Yellowish white	White	Cream	Creamy white	Cream whitish	Yellowish white	Cream	Ivory	Cream
Mycelium (M) or rods (R)	M	R	R	R	R	R	R	R	R	R	R	R	R	R	R	R
Motility	–	–	–	+	+	–	–	–	–	+	–	+	+	–	–	+
Cell size (μm)[d]	0.5–1.0, width	0.6–0.8 × 1.4–2.1	0.4–0.6 × 1.0–1.8	0.3–0.8 × 0.9–1.4	0.5–0.8 × 1.0–3.0	~ 0.3 × 1.1	0.3–0.5 × 0.7–1.2	0.2–0.4 × 0.8–1.2	0.4–0.7 × 0.8–2.4	0.5–0.6 × 1.2–1.6	0.3–0.6 × 0.6–1.6	0.3–0.4 × 0.9–1.4	0.5–0.7 × 1.5–3.0	0.2–0.5 × ~ 1 or more	0.4–0.7 × 0.7–2.0	0.3–0.5 × 0.8–4.0
Temperature optimum (°C)	28	30	25–30	28	30	30	30	30	25	35	25–30	28	26–30	29	30	28–30
Temperature range (°C)[e]	20–45	4–40	10–40	20–37	15–40	25–30	15–30	10–42	<37	20–40	10–35	20–37	10–37	16–34	4–35	20–40
NaCl requirement, optimum (%)[f]	–	–	+ (1–3)	nd	–	–	–	+(2)	–	–	+(3)	nd	–	0–0.25	–(0.5–1)	–
NaCl maximum (%)[g]	8	8	8	nd	nd	>2	1	<3	<5	nd	10	7–7.5	4–5	1	3	7
pH optimum	~7	7–8	7–8	7–7.5	8	7	7	7	~7	8	6–7	nd	~7	6.2–6.5	8	7–7.5
pH range[h]	5–9	5–12	6–9	nd	6–10	nd	5–8	5.5–8.5	nd	5–9	6–10	nd	nd	5.5–8.5	nd	6–12
Catalase	+	+	w	+	+	+	nd	+	+	+	+	+	+	nd	+	–
Oxidase	v	–	–	–	v	–	nd	+	–	–	–	–	+	–	+	–
Nitrate reduction	–	–	–	–	–	–	w	+	–	+	–	+	–	+	+	c
Decomposition of:																
Esculin	+	+	+	–	+	+	+	nd	w	+	+	–	+	+	–	–
Casein	+	+	+	+	+	+	+	–	v	+	+	+	+	+	+	+
DNA	+	nd	nd	nd	+	+	+	nd	–	+	nd	nd	+	nd	nd	–
Gelatin	+	–	–	–	+	–	+	+	+	+	+	+	+	–	+	–
Starch	+	w	–	+	v	–	–	–	–	+	w	+	v	–	w	+
Tween 80	+	–	+	–	+	–	nd	nd	+	–	+	+	+	–	w	+
Tyrosine	+	–	–	+	+	–	nd	nd	–	+	+	+	+	nd	–	+
Urea	–	–	–	–	v	–	–	–	–	–	–	–	–	–	–	–
Hypoxanthine	+	–	–	–	–	–	nd	nd	–	–	–	–	–	nd	–	–
Xanthine	v	–	–	–	–	–	nd	nd	–	+	–	–	–	nd	–	+
Enzymes (API ZYM):																
N-Acetyl-glucosaminidase	–	–	–	–	–	–	+	–	–	–	–	–	–	+	–	–
Acid phosphatase	–	–	w	+	+	–	+	+	w	+	+	–	w	+	–	+
Alkaline phosphatase	v	+	+	+	+	+	+	+	–	+	+	+	+	+	+	+
α-Chymotrypsin	–	–	–	–	–	–	–	+	–	–	+	–	–	–	–	–
Cystine arylamidase	–	–	–	–	w	w	w	–	+	–	+	–	w	–	–	+
Esterase (C4)	+	–	w	+	–	w	+	–	+	–	+	+	–	–	+	+
Lipase (C14)	–	–	–	–	–	–	–	–	w	–	–	–	–	–	–	–
α-Fucosidase	–	–	–	–	–	–	+	–	–	–	–	–	–	–	–	–
α-Galactosidase	–	–	–	–	–	–	–	–	–	–	–	–	–	–	–	–

β-Galactosidase	v	–	–	–	–	w	–	–	–	–	–	–	–	+	–	–
α-Glucosidase	v	+	+	–	+	+	+	+	+	+	+	–	+	+	+	+
β-Glucosidase	–	–	–	–	v	–	+	–	+	–	–	–	v	–	–	–
Leucine arylamidase	+	+	+	+	+	+	+	–	+	+	+	+	+	+	+	+
α-Mannosidase	+	–	–	–	–	–	+	–	–	–	–	–	–	–	–	–
Naphthol-AS-BI-phosphohydrolase	v	–	–	+	+	v	–	+	+	+	+	–	v	+	+	nd
Trypsin	+	–	+	–	+	–	+	+	–	+	+	–	v	+	–	–
Valine arylamidase	–	–	w	–	+	+	w	–	w	w	+	–	w	–	–	+
Utilization of carbon source:[j]	*ISP9, B-YC*	*ISP9*	*API 20 NE*	*ISP9*	*ISP9, B-YC, API 50 CH, MS2+VE*	*MS1+ GF2*	*API 20NE*	*MS2+VE*	*MS4+GF3, B-YC*	*YNB+CA, B-YC*	*MS3+YE*	*ISP9*	*ISP9, B-YC, API 50 CH, MS2+VE*	*API 20 NE*	*API CH 50*	*LBM*
L-Arabinose	+	–	–	–	v	–	–	–	–	–	–	–	–	–	–	–
Cellobiose	+	–	nd	+	–	+	–	–	+	+	+	+	–	nd	+	+
D-Fructose	v	+	+	–	+	+	+	+	+	+	–	+	–	nd	+	–
D-Galactose	–	+	+	–	–	+	–	nd	–	v	–	+	–	nd	+	+
D-Glucose	+	–	+	+	+	+	+	+	+	+	+	+	+	+	+	+
Lactose	–	–	–	–	–	+	nd	nd	–	v	–	+	–	nd	+	+
Maltose	+	v	nd	–	v	+	+	+	+	+	nd	+	v	+	+	–
D-Mannitol	+	–	+	+	v	+	–	–	–	v	+	+	v	+	+	+
D-Mannose	+	–	–	–	v	+	–	–	–	v	–	v	v	–	–	+
Melezitose	nd	nd	nd	nd	–	nd	nd	nd	+	+	nd	–	–	nd	–	nd
Melibiose	nd	nd	nd	–	v	+	–	–	nd	–	nd	+	–	nd	+	+
D-Raffinose	–	–	–	+	–	nd	+	nd	–	–	–	–	v	nd	–	+
L-Rhamnose	v	–	+	–	+	+	+	–	+	+	–	+	–	nd	+	+
D-Ribose	–	–	–	–	+	–	–	–	–	+	+	–	v	nd	–	–
Sucrose	w	–	+	+	+	+	+	+	+	+	+	+	+	nd	+	+
D-Trehalose	nd	+	+	nd	+	+	–	nd	–	+	+	+	+	nd	+	nd
D-Xylose	+	+	+	+	+	–	+	–	–	+	–	+	–	nd	+	+
Glycerol	w	–	–	nd	–	–	nd	–	+	–	–	–	–	nd	–	nd
Inositol	–	–	–	+	–	–	nd	–	–	+	–	+	–	nd	–	+
Inulin	–	+	nd	nd	–	–	nd	nd	–	–	nd	–	–	nd	–	nd
Salicin	–	–	nd	nd	v	–	+	nd	+	–	–	–	–	nd	–	nd
N-Acetylglucosamine	+	–	nd	nd	–	+	nd	–	–	–	nd	nd	–	–	nd	nd
Principal phospholipids[k]	PG, DPG, PI, APG, PL [PIMs][m,n,o]	PG, PI	nd	nd	nd	nd	nd	PG, PI	nd	nd	nd	nd	DPG, PG, PI, PL, [OH-PG][n]	nd	nd	PG, DPG
DNA G+C content (mol%)[p]	68 (T_m)	73.1	72.9	68.7	71.4	nd	73	73	69 (T_m)	72.5	73.3	71.5	71.7–73.5	71.6 (T_m)	71.5	67.6
Isolation source	Soil	Beach sand	Sea water	Soil	Industrial wastewater	Crude oil	Soil	Soil	Herbage	Oil shale column	Zooplankton enriched sea water	Sediment	Soil	Soil	Soil	Soil

(continued)

TABLE 215. (continued)

[a]Symbols and abbreviations: +, positive; –, negative; w, weakly positive; v, variable results (between strains, different experiments or the test methods), or conflicting data; nd, not determined. DPG, diphosphatidylglycerol; PG, phosphatidylglycerol; PC, phosphatidylcholine; PE, phosphatidylethanolamine; PI, phosphatidylinositol; PI-OH, phosphatidylinositol apparently containing hydroxylated fatty acids; OH–PG, hydroxy phosphatidylglycerol; PL, unidentified phospholipids; L, unidentified lipids; APG, acylphosphatidylglycerol; PIMs, phosphatidylinositol mannosides. Data are from An et al. (2007), Cho et al. (2010), Choi et al. (2007), Chou et al. (2008), Cui et al. (2009), Collins et al. (1994, 1989, 1979, 1983), Dastager et al. (2009a, 2008d, 2008a, 2008e, 2008f, 2008c, 2009e, 2010), Jones and Collins (1986), Kim et al. (2008a), Kim et al. (2009a, 2009b), Kubota et al. (2005a), Lawson et al. (2000), Lee (2007b), Lee et al. (2007), Lee et al. (2008), Li et al. (2007b), O'Donnell et al. (1982), Park et al. (1999), Park et al. (2008), Prauser (1976, 1984a, 1989), Schippers et al. (2005), Song et al. (2011), Suzuki and Komagata (1983c), Tóth et al. (2008), Yano et al. (1970, 1971), Yi and Chun (2004b, 2004a), Yoon et al. (2008, 2007a, 1999, 2007b, 2009, 2004, 2005a, 2005b, 2005d, 2006a, 2006b, 2010, 1997b), Zhang et al. (2009a) and recent observations.

[b]See also the tree (Figure 248) for phylogenetic position of species and Table 216, Table 217, Table 218, and Table 221 for characteristics of some species groups.

[c]Differences in the color intensity and shade may be influenced by growth conditions and culture age (see also the species description section for more details). For *Nocardioides albus* and *Nocardioides luteus*, the color of vegetative mycelium is indicated.

[d]Cell morphology and size as usually observable in young cultures; some differences in the cell size may be caused by different growth conditions. In older cultures, cells as a rule become shorter and often consist predominantly or exclusively of coccoid or coccobacillary forms (see the species descriptions for details).

[e]The temperature range in which the growth was registered; the actual growth temperature range may be broader for some species (see the species descriptions for further details).

[f]Optimal NaCl concentrations (%, w/v) where available are indicated in parentheses. Zero indicates that addition of NaCl to culture medium is not required for growth, but minor or trace amounts of NaCl may be present. The salinity level may slightly differ depending on the test conditions (see the species descriptions for further details).

[g]Maximal NaCl concentration for growth as registered in the test media used (see the species descriptions for further details); the values may be influenced by the culture medium composition.

[h]Initial pH of test media, with exception of data for *Nocardioides fonticola*; the pH range values may slightly vary depending on the test conditions.

[i]Data on utilization of carbon sources were obtained using different methods, and, therefore, the results presented may not be comparable in some cases. Basal media used in conventional tests to monitor the growth on carbon sources are as follows: *ISP9* (Shirling and Gottlieb, 1966); *B–YC*, basal mineral salts medium (Baumann et al., 1972), supplemented with 2% (v/v) Hutner's mineral base (Cohen-Bazire et al., 1957) and modified by reducing the concentration of sea salts to half-strength (Yi and Chun, 2004a); *LBM*, Leifson's basal medium (Leifson, 1963); *MS1+GF1*, mineral salts medium supplemented with the following growth factors (GF): yeast extract, Oxoid (0.02 g/l), bio-Lactysat, bioMérieux (0.02 g/l), a vitamin solution and a trace element solution (Kämpfer et al., 1990); *MS1+GF2*, the same test medium, but bio-Lactysat is replaced by peptone (Merck) (0.02 g/l) (Kämpfer et al., 1991; Schippers et al., 2005); *MS2+VE*, mineral salts medium supplemented with vitamins and element solution (Cui et al., 2009); *MS3+YE*, mineral salts medium with 0.05 (g/l) of yeast extract (Kim et al., 2008a); *MS4+GF3*, basal test medium containing (per liter) mineral base E (Owens and Keddie, 1969), 0.2 g of Bacto Yeast Extract (Difco), 12 g of agar, 2 μg of vitamin B_{12}, 10 mg of sodium glutamate, and 10 mg of methionine (Collins et al., 1994); *YNB+CA*, Bacto Yeast Nitrogen Base without amino acids (Difco), modified by the addition of 10 mg/l Casamino acids (Difco), and agar (Stevenson, 1967).

[j]Data obtained with the Biolog GP2 test system are displayed in Table 217.

[k]Some differences in quantitative and qualitative composition of phospholipids may be influenced by growth conditions, the culture age and analytical procedure (see the *Further descriptive information* section for more details).

[l]Data from Collins at al. (1983), Lechevalier at al. (1981, 1977), and O'Donnell et al. (1982). Lechevalier et al. (1981, 1977) identified PG, PI, APG, and traces of PIMs (see also the *Further descriptive information* section and Figure 263).

[m]Glycolipids may be detected among principal polar lipids (see the *Further descriptive information* section and Figure 263).

[n]Polar lipids detected in minor amounts are given in square brackets.

[o]*Nocardioides luteus* is generally similar to *Nocardioides albus* in the principal polar lipids (see the *Further descriptive information* section and Figure 263).

[p]DNA base composition as determined by direct quantification of nucleosides by the HPLC method, unless indicated; data for *Nocardioides simplex* were estimated by the thermal denaturation method, with the HPLC-based method for the type strain.

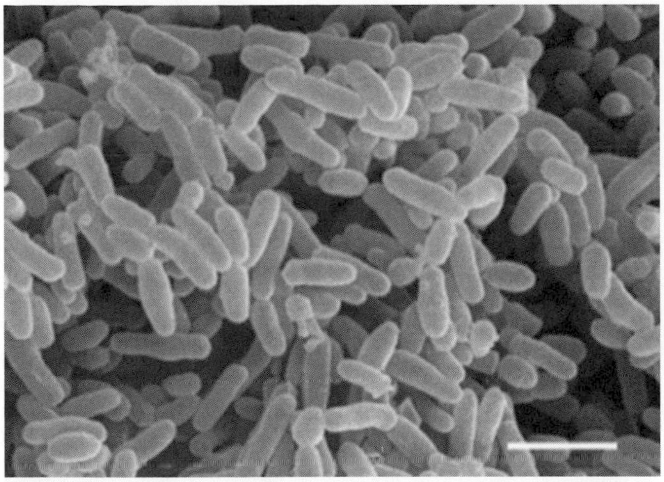

FIGURE 255. *Nocardioides koreensis* MSL-09, 7-d-old culture on R2A agar (28°C). Scanning electron micrograph. Bar = 2 µm. (Reproduced with permission from Dastager et al., 2008d. Int. J. Syst. Evol. Microbiol. *58*: 2292–2296.)

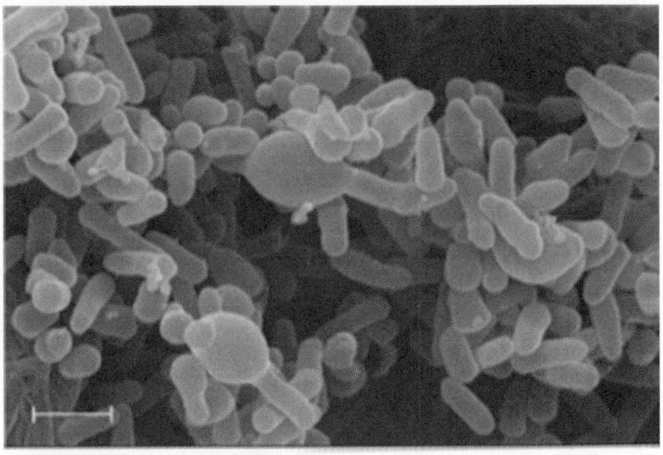

FIGURE 257. *Nocardioides tritolerans* MSL-14; 7-d-old culture grown on R2A agar at 28°C. Scanning electron micrograph. Bar = 1.0 µm. (Reproduced with permission from Dastager et al., 2008c. J. Microbiol. Biotechnol. *18*: 1203–1206.)

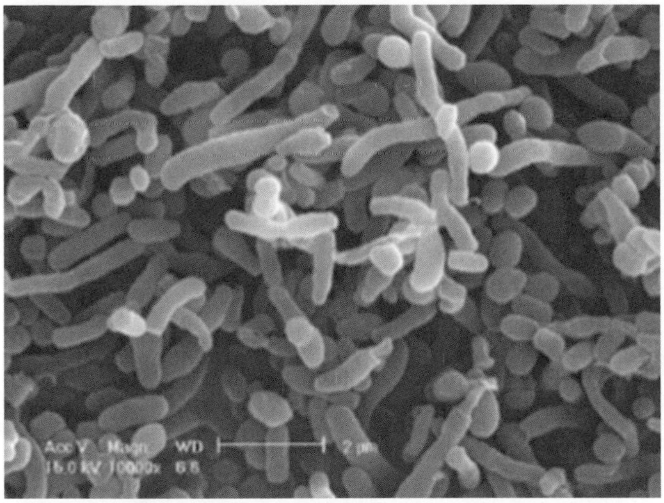

FIGURE 256. *Nocardioides caricicola* YC6903; 5-d-old culture grown in half-strength R2A broth at 30°C. Scanning electron micrograph. Bar = 2 µm. (Reproduced with permission from Song et al., 2011. Int. J. Syst. Evol. Microbiol. *61*: 105–109.)

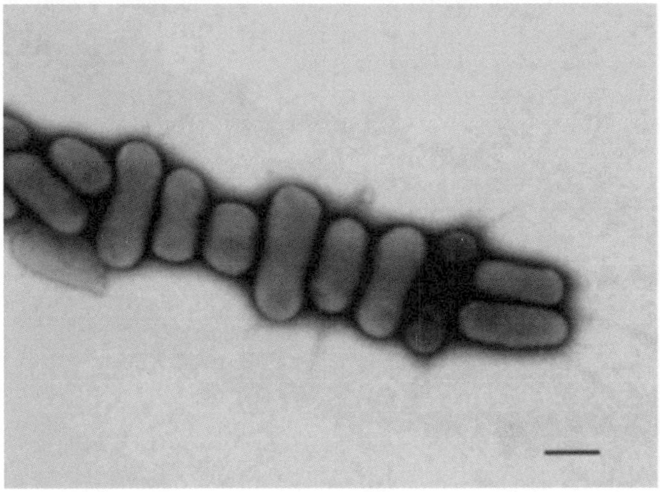

FIGURE 258. Palisade-arranged cells of *Nocardioides furvisabuli* SBS-26; 3-d-old culture grown at 30°C on ISP 2 medium prepared with 60% (v/v) natural seawater. Negative stain; transmission electron microscopy. Bar = 0.5 µm. (Reproduced with permission from Lee, 2007b. Int. J. Syst. Evol. Microbiol. *57*: 35–39.)

The cellular fatty acid profiles of *Nocardioides* species contain complex mixtures of saturated and monounsaturated, straight-chain and iso-, anteiso- and 10-methyl-branched acids, including 10-methyl-octadecanoic acid (tuberculostearic acid, TBSA), as well as 2-OH and 3-OH hydroxylated acids. In addition, minor quantities of ω-alicyclic fatty acids were reported (Song et al., 2010). The growth (developmental) phase, growth conditions, e.g. temperature, pH, oxygen supply, and growth medium composition (carbon sources, an excessive supply of primer sources) and consistency (solid, liquid), along with analytical procedure, may considerably influence the proportions of principal components or the components considered to be diagnostic for this genus (see Kaneda, 1991; Kroppenstedt, 1985; Suzuki and Komagata, 1983a; Suzuki et al., 1993; Yano et al., 1970, 1971); and also the species descriptions in this chapter). The majority of the data on fatty acids for *Nocardioides* species were obtained using the procedure of Sasser (1990) to extract methyl esters and then gas chromatography analysis. Many recent data were obtained using an automated MIDI system consisting of a Hewlett Packard model 5890 capillary gas chromatograph and a computer with specific software (Microbial ID, Inc., Newark, DE) to automatically identify and quantify the fatty acids.

Comparison of data on fatty acids determined in some *Nocardioides* cultures of different age grown on Nutrient agar (Difco) at 30°C for 4 and 7 d (Yoon et al., 1997a; Yoon et al., 2010) and in some corynebacteria (Suzuki and Komagata, 1983a) demonstrated that the content of TBSA and its homologs increases as the cells aged. On the other hand, the content of 10-methyl acids plus their biosynthetic precursors (respective monoenoic acids) in some analyzed bacteria is rather constant, at least throughout 120 h of cultivation (Suzuki and Komagata, 1983a). Correspondingly, the total contents of these fatty acids

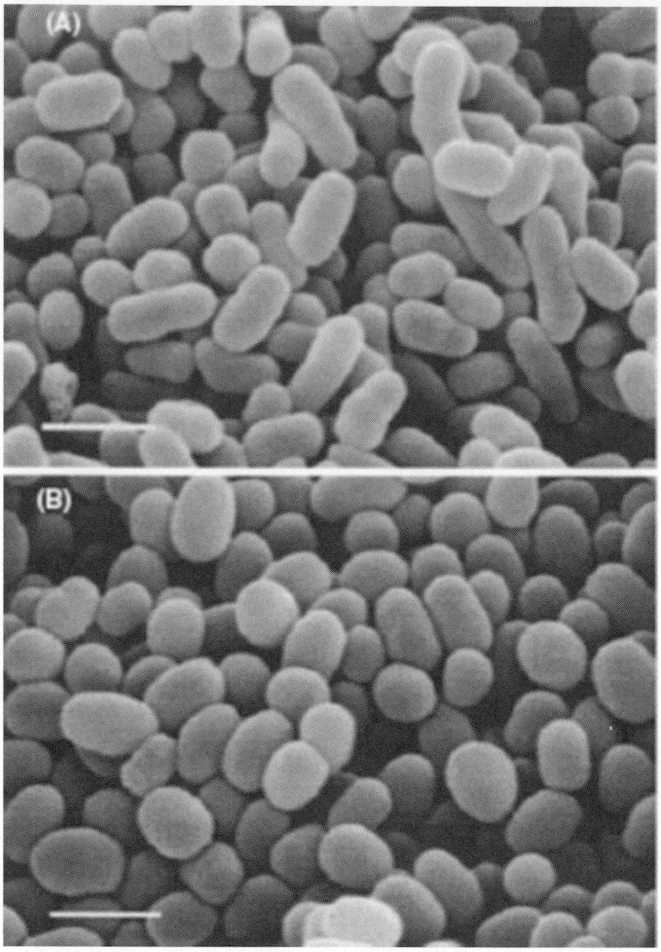

FIGURE 259. *Nocardioides pyridinolyticus* KCTC 0074BP; 3-d-old (A) and 7-d-old (B) cultures on nutrient agar. Scanning electron micrographs. Bar = 1 µm. (Reprinted from Yoon and Park, 2006. *The Prokaryotes*, 3rd edn, vol. 3, Springer, New York, pp. 1099–1113.)

(and other biosynthetically related acids) should be taken into account when comparing the cellular fatty acid compositions of organisms and drawing conclusions about the fatty acid type. In addition, some *Nocardioides* strains grown at higher temperatures usually exhibit a larger proportion of saturated acids (Song et al., 2011; Suzuki and Komagata, 1983a; Yoon et al., 2008). The content of saturated and unsaturated fatty acids may be influenced by the consistency of the medium (liquid, solid) and the extent of aeration. Nevertheless, irrespective of the culture conditions and analytical procedure, the 14-methylpentadecanoic acid ($C_{16:0}$ iso) predominate in most cases and may reach up to 65–82% (Choi et al., 2007; Kim et al., 2008a; Kim et al., 2009a; Schumann et al., 1997; Yoon et al., 2006a). A substantially larger proportion of $C_{16:0}$ iso (65–70%) might be a chemotaxonomic marker of some species, as demonstrated by the data on the phylogenetically closest relatives *Nocardioides salarius* (Kim et al., 2008a) and *Nocardioides basaltis* (Kim et al., 2009b). Similar profiles were obtained independently by different research teams for cultures of different age (1 and 3 d) grown on marine agar (MA; Difco) at 30°C. Very high content of $C_{16:0}$ iso (71.5%) was also reported for *Nocardioides marinus* (MA agar, 30°C, 1 d), also isolated from sea water but phylogenetically distant from the above species (Choi et al., 2007).

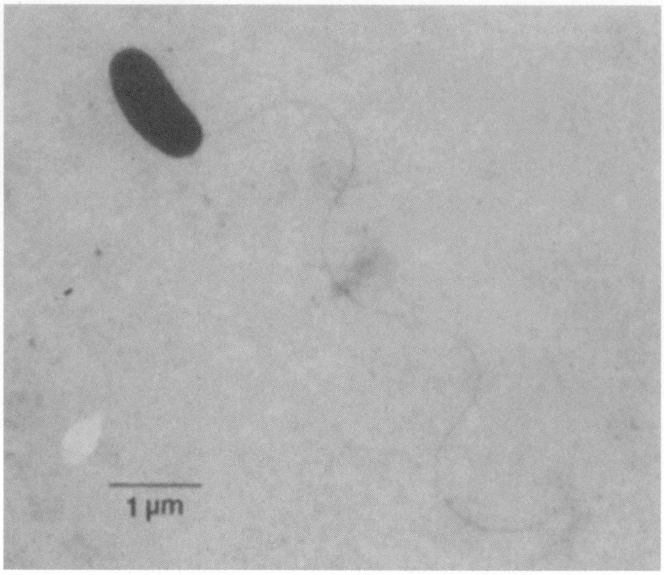

FIGURE 260. *Nocardioides pyridinolyticus* KCTC 0074BP. Flagellated cell from 3-d-old culture on nutrient agar. Negative stain; transmission electron microscopy. Bar = 1 µm. (Reprinted from Yoon. and Park, 2006. *The Prokaryotes*, 3rd edn, vol. 3, Springer, New York, pp. 1099–1113.)

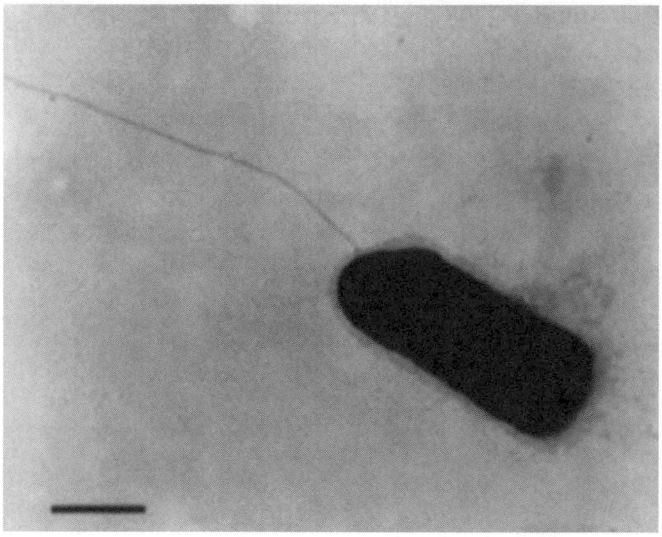

FIGURE 261. Flagellated cell of *Nocardioides nitrophenolicus* KCTC 0457BP; 2-d-old culture grown on nutrient agar at 30°C. Negative stain; transmission electron microscopy. Bar = 0.5 µm. (Reprinted from Yoon et al., 1999. Int. J. Syst. Bacteriol. *49*: 675–680.)

Different predominant fatty acids were also reported, which may be characteristic of certain species or species groups, or reflect the particular culture conditions, or both. For example, a straight-chain unsaturated acid $C_{18:1}$ ω9*c* (27.5%), and $C_{16:0}$ iso (18.6%) were the main components for *Nocardioides oleivorans* (Schippers et al., 2005). In one report (Dastager et al., 2009e), $C_{18:1}$ ω7*c* (50.3%) predominates with a substantially smaller content of $C_{16:0}$ iso (4.2%) for *Nocardioides islandensis* cultured on R2A agar (Difco) for 10 d at 28°C. *Nocardioides aquaticus* grown at 16–20°C also contained an abundance of straight-chain unsaturated acid, $C_{18:1}$ (30.6%), as well as $C_{16:0}$ (14.1%), and $C_{17:0}$ anteiso (10.6%) (Lawson et al., 2000a). *Nocardioides fonticola* had

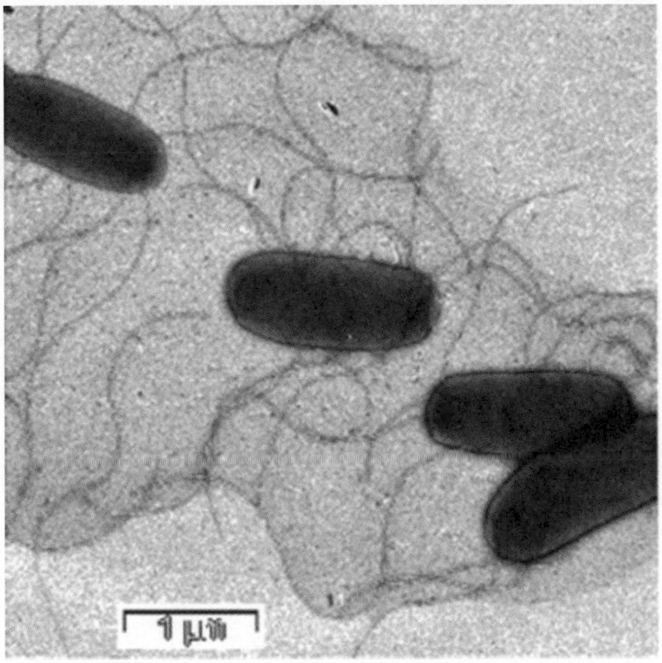

FIGURE 262. Cells of *Nocardioides humi* DCY24 with peritrichous flagella at different developmental stages; cells grown on Luria–Bertani agar at 30°C for 18 h. Transmission electron micrographs of negatively stained cells. Bar = 1 μm. (Reproduced with permission from Kim et al., 2009b. Int. J. Syst. Evol. Microbiol. *59*: 2724–2728.)

$C_{16:0}$ iso and $C_{17:0}$ in equal proportions, nearly 20% when grown on R2A for 3 d at 25°C (Chou et al., 2008). *Nocardioides hankookensis*, on the other hand, contained nearly similar proportions of three acids, $C_{14:0}$, $C_{16:0}$ iso, and $C_{18:1}$ ω5*c* (9–11%) when grown on twofold-diluted R2A agar for 3 d at 30°C (Song et al., 2011), while $C_{16:0}$ iso (42.7%) was found to be the major component in cells mass-harvested from the standard R2A agar after 7 d incubation at 25°C (Yoon et al., 2008). Cells of *Nocardioides furvisabuli*, harvested from MA after incubation at 30°C contained almost equal proportions of $C_{18:0}$ and $C_{16:0}$ iso (19.5%), whereas only $C_{16:0}$ iso (34.1%) was predominating in cells of the same species grown on ISP 2 medium prepared with natural sea water (Lee, 2007b). Strains of *Nocardioides simplex* may exhibit (13–18%) $C_{17:0}$ anteiso larger proportions of $C_{17:0}$ anteiso (13–18%) and $C_{18:1}$ (13–28%) in some experiments (Suzuki and Komagata, 1983c) or have $C_{16:0}$ iso as predominating (reaching up to 41.8%) (Schumann et al., 1997). However, when analyses are performed under identical experimental conditions, strains of this and some other species typically exhibit very similar fatty acid profiles (see, e.g. Lee et al., 2008; Yoon et al., 2005a, 2005b, 2009) and this characteristic may be useful to discriminate *Nocardioides* species, including closely related ones (Cui et al., 2009; Song et al., 2011; Yi and Chun, 2004a; Yoon et al., 1997a, 2010). On the other hand, species (type strains) with rather undistinguishable profiles also occur.

With regard to polar lipids, phosphatidylglycerol (PG), diphosphatidylglycerol (DPG) as well as several incompletely characterized or "unknown" phosphorus-containing or other lipids are typically reported among principal components in *Nocardioides* species (Table 215). The majority of species do not contain nitrogenous phospholipids (i.e. possess phospholipids type I *sensu* Lechevalier et al., 1981, 1977). On rare occasions, nitrogenous phospholipids, such as phosphatidylethanolamine (PE) (Yoon et al., 2005d, 2010) and phosphatidylcholine (PC) (Lee, 2007b) were reported, that are diagnostic of phospholipid types II and III (Lechevalier et al., 1977, 1981). The other phospholipids revealed for some organisms of this genus included phosphatidylinositol (PI), hydroxy-phosphatidylglycerol (PG-OH), acylphosphatidylglycerol (APG), and phosphatidylinositol mannosides (PIMs) (Collins at al., 1983, 1989; Lechevalier et al., 1981, 1977; O'Donnell et al., 1982; see Table 215). It should be emphasized that certain "unknown" phospholipids reported by different authors for some *Nocardioides* species may correspond to those identified (named) in other studies, or only slightly differ from such phospholipids in R_f value owing to some differences in chemical structure of the components involved, mostly fatty acids. Indeed, different patterns of principal phospholipids were reported for some strains in this bacterial group (e.g. Collins et al., 1989, 1983; Komura et al., 1975a; Lechevalier et al., 1977, 1981; O'Donnell et al., 1982; Tamura and Yokota, 1994; Yano et al., 1970, 1971). Glycolipids are usually not mentioned among major polar lipids for *Nocardioides* species. However, several distinct glycolipids can be detected in *Nocardioides albus*, *Nocardioides luteus* (Figure 263), and some other *Nocardioides* strains.

There is now substantial evidence for the effect of culture conditions on the quantitative and qualitative composition of polar lipids in *Nocardioides* strains (which is reminiscent of the phenomenon known for the cellular fatty acids). A prominent example is the polar lipid profile of *Nocardioides albus*. Lechevalier et al. (1981, 1977) reported the presence of PG, PI, APG, and traces of PIMs for *Nocardioides albus*, emphasizing the lack of DPG. According to O'Donnell et al. (1982) and Collins et al. (1983), DPG still occurs among the major phospholipids of this species, along with PG and several incompletely characterized phospholipids, two of which supposedly contain a 2-hydroxy acid. In our recent experiments (N.G. Vinokurova and L.I. Evtushenko, unpublished), both young (8 h, mycelial stage) and older (24 h, fragmentation stage) cultures contained PG, DPG, PL_1 (supposedly APG, run above the DPG spot, just after neutral lipids) and two (or three) characteristic very closely related phospholipids. One of them is most likely PI (contained inositol) and the second was tentatively identified as hydroxy-acid-containing phosphatidylinositol (PI-OH). The older cultures *Nocardioides albus* and *Nocardioides luteus* differed from young cultures mainly in that they contained additional glycolipids (varying amounts) and had an unknown phosphorous-involving yellowish-pigmented compound PL_x in some experiments (Figure 263C, D). The latter, which stained well with Mo-blue (Dittmer and Lester, 1964; Komagata and Suzuki, 1987), shows very weak coloration when sprayed with 5% molybdophosphoric acid in ethanol, but has no distinct specific staining either with ninhydrin, α-naphthol (Jacin and Mishkin, 1965; Komagata and Suzuki, 1987), periodate Schiff reagent (Komagata and Suzuki, 1987; Minnikin et al., 1977) or 5% $AgNO_3$ in aqueous ammonia solution. Certain differences between the polar lipid patterns of *Nocardioides albus* and *Nocardioides luteus* also occur, especially in older cultures (Figure 263C, D). Notably, PL_1 (supposedly APG and/or another closely related compound) characteristic of *Nocardioides albus* and *Nocardioides luteus*, was not revealed in *Nocardioides simplex*. Thus, the polar lipid pattern, even with some components incompletely identified and varying to some extent between experiments, may serve to differentiate species

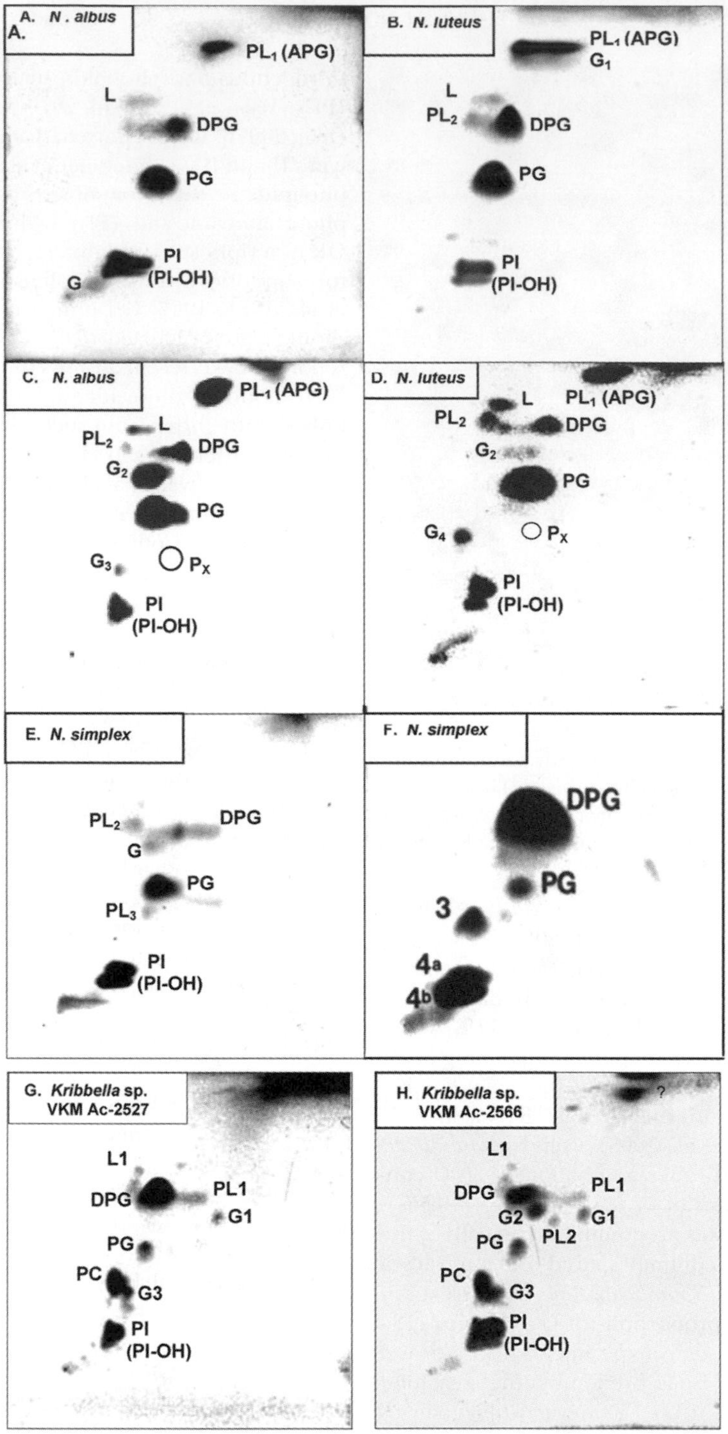

FIGURE 263. Two-dimensional thin-layer chromatograms of polar lipids from the type strains of *Nocardioides albus*, *Nocardioides luteus*, *Nocardioides simplex*, and representatives of the genus *Kribbella* (A–D) Polar lipids of *Nocardioides albus* VKM Ac-805 and *Nocardioides luteus* VKM Ac-1246 in young (nonfragmenting mycelium) (A, C) and older (fragmentation stage) (B, D) cultures grown in complex liquid medium, PYGP (peptone, 0.5%; yeast extract, 0.3%; glucose, 0.5%; K_2HPO_4, 0.02%; pH 7.2). Chloroform-methanol-water (65:25:4, by vol.) was used in the first direction and chloroform-acetic acid-methanol-water (80:15:12:5, v/v) in the second direction (modified from Minnikin et al., 1977). (E and F) Polar lipids patterns from subcultures of the type strain of *Nocardioides simplex*, VKM Ac-1118 (E) and *Nocardioides simplex* NCIB 8929 (F) grown under different conditions. (G and H) Examples of PC-containing polar lipids patterns from nocardioform strains of *Kribbella* (PYGP, 16 h, 28°C; mycelial stage). Abbreviations: DPG, diphosphatidylglycerol; PG, phosphatidylglycerol; PI, phosphatidylinositol; PI-OH, a compound (supposedly phosphatidylinositol) that contains hydroxylated fatty acid(s); PL, unknown phospholipids; PL_1, supposedly acylphosphatidylglycerol (APG); G, glycolipids; L, unknown non-phosphorylated polar lipid; P_x, unknown phosphorus-containing pigment (see the text for details). (A–E, G, and H are courtesy of Natalia G. Vinikurova; F is reprinted from O'Donnell et al., 1982. Arch. Microbiol. *133*: 323–329.)

within this genus and from members of other genera. For differential purposes, the polar lipid patterns have to be analyzed under standardized and controlled experimental conditions and in the same laboratory. Further detailed studies of polar lipids in members of the genus *Nocardioides* will elucidate the spectra of the polar lipid components characteristic of particular species (species groups) and hopefully increase the discriminative power of this characteristic.

Polyamine patterns have been analyzed for 10 strains comprising five *Nocardioides* species (Busse and Schumann, 1999). The majority of strains grown in liquid R medium (Yamada and Komagata, 1972) contained cadaverine in maximum proportion (34–77%). The second major component (nearly 20% or more) for most strains was putrescine (*Nocardioides albus*, *Nocardioides luteus*, and *Nocardioides jensenii*), spermidine (the type strain of *Nocardioides simplex*), or spermine (*Nocardioides simplex* IMET 10283). For *Nocardioides plantarum*, spermine was found to predominate (42%), followed by cadaverine. This strain was also distinguished by a significantly lower concentration of total polyamines. Other compounds, such as 1,3-diaminopropane, tyramine, and sym-homospermidine, were revealed in lesser or trace quantities, with larger proportions (>10%) of 1,3-diaminopropane and tyramine found in a few strains.

In combination with cadaverine, putrescine (in high concentration and proportion in the majority of *Nocardioides* strains) appears to be useful in distinguishing many species of this genus from *Aeromicrobium*. Both *Nocardioides* and *Aeromicrobium* differed from all other analyzed LL-A$_2$pm-containing coryneform and nocardioform bacteria used in the comparative study (families *Propionibacteriaceae*, *Intrasporangiaceae*, and *Sporichthyaceae*) by the presence of cadaverine among major components (Busse and Schumann, 1999; Yoon et al., 2007b). The distinguishing polyamine patterns, along with the overall polyamine concentrations, indicates the potential of polyamines to differentiate individual species or species groups within the genus, as demonstrated by the data on major and additional polyamines for the type strains: *Nocardioides albus* DSM 43109 (cadaverine, 49%; putrescine, 50%); *Nocardioides luteus* DSM 43366 (cadaverine, 57%; putrescine, 28%; spermidine, 11%); *Nocardioides jensenii* DSM 20641 (cadaverine, 54%; putrescine, 22%; spermidine, 12%; spermine, 9.3%); *Nocardioides plantarum* DSM 11054 (spermine, 42%; cadaverine, 34%; tyramine, 12%; 1,3-diaminopropane, 6.8%; spermidine, 5.1%); *Nocardioides simplex* DSM 20130 (cadaverine, 57%; spermidine, 19%; putrescine, 15%; spermine, 5.6%; 1,3-diaminopropane, 4.2%). Notably, the polyamine profiles of strains within the species *Nocardioides albus* and *Nocardioides simplex* (for which more than one strain were studied) were not uniform. A prominent example is the pair *Nocardioides albus* DSM 43874 (53% cadaverine, 30% putrescine, 15% 1,3-diaminopropane, with the total polyamine concentration of 1.5 μmol/g, dry wt) and the type strain *Nocardioides albus* DSM 43109 (49% cadaverine and 50% putrescine, and the total polyamine content of 10.5 μmol/g). The heterogeneity of the polyamine patterns, at least for strains referred to as *Nocardioides albus*, is most likely evidence for the presence of different species within this group (see *Taxonomic comments* for more information). Further studies of polyamines in *Nocardioides* strains strictly proven to be members of one species will elucidate the polyamine set of diagnostic importance for individual species or species groups within the genus.

Nutrition and growth conditions. Strains of *Nocardioides* species usually grow well in standard nutrient media with glucose, peptone, and yeast extract under aerobic conditions. Some organisms of this genus prefer the nutrient-poor media, including a tenfold-diluted nutrient agar (NA; Difco) (Yoon et al., 2005a, 2005b, 2005d), tenfold-diluted R2A agar (Difco) (Dastager et al., 2008f), and some other diluted complex media. Knowledge of minimal nutritional requirements of *Nocardioides* species is incomplete. Most species are nutritionally non-exacting and grow well on a suitable minimal salts medium, e.g. ISP 9 (Shirling and Gottlieb, 1966)* with glucose or other compounds as a sole carbon and energy source and an ammonium salt as a sole nitrogen source (Table 215).

Some organisms, including *Nocardioides albus*, *Nocardioides luteus*, *Nocardioides dokdonensis*, *Nocardioides exalbidus*, *Nocardioides jensenii*, and *Nocardioides simplex* (Collins et al., 1989; Keddie et al., 1986; Li et al., 2007b; Park et al., 2008) and probably many others can use nitrate as a sole nitrogen source. Traces of complex organic substrates (e.g. yeast extract or casein peptone) in minimal salts media commonly enhance growth rates and biomass production of *Nocardioides* species. The good growth of *Nocardioides plantarum* was reported to require thiamine (Collins et al., 1994), while that of *Nocardioides aquaticus* needed both thiamine and biotin (Lawson et al., 2000a). However, both these organisms can grow in the absence of thiamine and biotin (Yi and Chun, 2004a), e.g. in a mineral salts medium consisting of the basal medium (BM; Baumann et al., 1972, 1971) supplemented with 2% (v/v) Hutner's mineral base (Cohen-Bazire et al., 1957), and modified by reducing the concentration of sea salts to half-strength. The BM medium (Baumann et al., 1971) contained 50 mM tris(hydroxymethyl) aminomethane [(Tris)-hydrochloride pH 7.5], 190 mM NH_4Cl, 0.33 mM $K_2HPO_4 \cdot H_2O$, 0.1 mM $FeSO_4 \cdot 7H_2O$, and half-strength artificial sea water (MacLeod, 1968).

Some mycelium-forming non-pigmented *Nocardioides* strains (capable of growing on glucose or sucrose with ISP 9 as a basal medium) can utilize other sugars (L-rhamnose, D-xylose, and D-mannitol) as carbon and energy sources only in the presence of thiamine or a vitamin solution (thiamine, riboflavin, nicotinic acid, pyridoxine, and *p*-aminobenzoic acid) (Suzuki and Komagata, 1983c). A similar situation probably takes place with some other species, carbon sources, and growth factors. Some components (or their concentrations) of the above vitamin solution probably inhibit growth or growth rate (Suzuki and Komagata, 1983c). The ability to grow in mineral salts medium with certain carbon sources may also depend on other specific test conditions (e.g. composition of salts in the test medium, carbon source concentrations, the growth medium used to obtain cells for testing, amount of cell mass used as inoculate, etc.).

Bacteria of this genus are largely non-halophilic and neutrophilic mesophiles. A few species, particularly isolated from marine and marine-related environments are salt-requiring and/or grow better in the presence of 0.5–6% NaCl and show resistance to NaCl concentration up to 10–15% (w/v) (Table 215). Some organisms (*Nocardioides bigeumensis* and *Nocardioides caricicola*), in contrast, were reported to be highly sensitive to salt and did not grow even in the presence of 1% (w/v) NaCl (Dastager et al., 2008d; Song et al., 2011). A number of species

*See Shirling and Gottlieb, (1966) (Int. J. Syst. Bacteriol. *16*: 313–340) for the composition of ISP media cited here and in other sections of this chapter.

may start growth at pH 5–5.5 and/or pH 11–12, while others grow in a narrow pH range (pH 7–9) (Table 215). The reported optimal growth temperatures vary insignificantly among species (most organisms grow well between 25 and 30°C; some show best growth at lower temperatures, 16–26°C. Among the recognized *Nocardioides* species, *Nocardioides caricicola* (Song et al., 2011) had the highest growth temperature (45°C) and weak growth was observable for *Nocardioides luteus* (recent experiments). Representatives of several species, i.e. *Nocardioides albus*, *Nocardioides aquiterrae*, *Nocardioides humi*, and *Nocardioides panacisoli* exhibit growth at 42°C (Cho et al., 2010; Kim et al., 2009b; Prauser, 1984b; Yoon, 2004). However, not all organisms have been tested for growth at 42 and/or 45°C.

Metabolism, enzymic and degradative activities. Bacteria of this genus are commonly considered to be chemo-organotrophic aerobes that are able to utilize and degrade a wide range of carbon and energy sources, including various unusual organic compounds. The ability to grow in nutrient-poor media, as well as the finding of representatives of *Nocardioides* in environments of nutrient scarcity, e.g. in cavities (Barton et al., 2004; Groth and Saiz-Jimenez, 1999; Groth et al., 2001), suggests that they may be sufficiently adapted to the oligotrophic lifestyle, be involved in a complex metabolic network of oligotrophic microbial communities, or even engage in chemolithotrophic metabolism with possible input from some atmospheric gases and minerals. In particular, strain *Nocardioides* sp. JS614 is probably able to grow chemolithotrophically on CO, as it might be suggested both from the finding of form I *Cox* genes in its genome (Mattes et al., 2005) and the experiments demonstrating CO-oxidizing activity and chemolithotrophic growth on CO by some actinomycetes containing this gene (King and Weber, 2007). The ability of strains of the recognized species to grow anaerobically on heterotrophic media under different gas atmospheres was not revealed (An et al., 2007; Chou et al., 2008; Cui et al., 2009; Dastager et al., 2008c, 2009e, 2010; Kim et al., 2008a; Lawson et al., 2000a; Schippers et al., 2005; Tóth et al., 2008; Yi and Chun, 2004a, 2004b).

All or the majority of *Nocardioides* strains studied taxonomically possessed activity for esterase lipase (C 8) (conflicting results were reported for a few species by Cho et al., 2010), but were negative in tests for β-glucuronidase, α-fucosidase, and α-mannosidase (API ZYM, bioMérieux). A limited number of strains tested to date (mostly by the API 20NE system; bioMérieux) were negative in the tests for arginine dihydrolase, lysine decarboxylase, ornithine decarboxylase, and tryptophan deaminase. Negative results were also typically registered in the tests for production of H_2S from peptone or cysteine. The only species, *Nocardioides panacisoli*, was reported to produce H_2S from thiosulfate (Cho et al., 2010). For all strains studied so far, tests showed no utilization of thiamine, L-ascorbate, salicylate, 2-propanol, dulcitol, erythritol, sorbitol, xylitol), and polyethylene glycol as carbon sources (Kubota et al., 2005a; Yi and Chun, 2004a, 2004b; Yoon et al., 2009). The results of Voges–Proskauer test vary with species and experiments (Cui et al., 2009; Lee, 2007b; Yoon et al., 2009). The ability to degrade adenine is uncommon (Prauser, 1989; Yi and Chun, 2004a, 2004b), although many species degrade DNA (Table 215) and some utilize thymine (Collins et al., 1994). Although conventional tests have not shown pronounced cellulolytic activity or utilization of cellulose (carboxymethylcellulose) as a sole or principal carbon source in recognized *Nocardioides* species, some unnamed strains of this genus were found to degrade rice straw pieces in minimal salts medium (Abdulla and El-Shatoury, 2007) and many species possess some relevant enzymatic activities. A few organisms can degrade chitin (Prauser, 1976; Tóth et al., 2008) or xylan (Park et al., 2008). Several bacteria were unable to decompose alginate (Lawson et al., 2000a; Yi and Chun, 2004b, a). Varying or conflicting test results (Table 215) for some characteristics of some species (type strains) were reported by different or the same authors. In most cases (disregarding obvious technical mistakes), such data indicate that a particular activity or ability exists, but it is weak or indistinct under the experimental conditions employed.

A prominent feature of members of the genus *Nocardioides* is their capability of performing transformation and degradation of complex and unusual compounds, including very common persistent and toxic environmental pollutants. Examples of such compounds and pollutants include alkanes of various lengths, petroleum chemicals, crude oil and its derivatives (Hamamura and Arp, 2000; Hamamura et al., 2001; Iizuka and Komagata, 1964; Jung et al., 2002; Purswani et al., 2008; Schippers et al., 2005), including carbazole, a recalcitrant *N*-heterocyclic aromatic compound derived from crude oil, creosote, and shale oil (Inoue et al., 2007; Inoue et al., 2006), as well as a sulfur heterocyclic compound dibenzothiophene (Sandhya et al., 1997; Sandhya et al., 1995). There are reports on the ability of members of this genus to degrade and utilize phenols and nitrophenols (Collins et al., 1994; Cui et al., 2009; Gundersen and Jensen, 1956; Keddie et al., 1986; Yoon et al., 2009), including 2,4-dinitrophenol and 2,4,6-trinitrophenol (picric acid) (Behrendt and Heesche-Wagner, 1999; Cho et al., 1998, 2000; Ebert et al., 1999, 2001; Rajan et al., 1996; Yoon et al., 1999), as well as pyridine (Lee et al., 1994; Lee et al., 1991; Rhee et al., 1997; Yoon et al., 1997b). They are capable of degrading a three aromatic ring compound, phenanthrene (Iwabuchi and Harayama, 1997, 1998a, 1998b; Saito et al., 1999, 2000), heterocyclic compounds such as dibenzofurans and chloroaromatics, including dibenzo-*p*-dioxins (Futamata et al., 2004; Inoue et al., 2007; Inoue et al., 2006; Kubota et al., 2005a; Sukda et al., 2009) and 2,4,5-trichlorophenoxyacetic acid which is one of the most persistent herbicides (Ebert et al., 1999, 2001; Golovleva et al., 1990; Männisto et al., 2001; Männisto et al., 1999; Traag and van Wezel, 2008; Travkin et al., 1999). There are reports on the ability to degrade different other herbicides, including *s*-triazine herbicides (Mulbry et al., 2002; Topp et al., 2000; Vibber et al., 2007; Yamazaki et al., 2008). *Nocardioides* sp. PD653 is the first naturally occurring aerobic bacterium reported to be capable of mineralizing hexachlorobenzene, a recalcitrant environmental pollutant, and degrading pentachloronitrobenzene (Takagi et al., 2009).

Nocardioides sp. JS614 is one of the most studied representatives of *Nocardioides* with respect to degradative activities, genetics, and metabolism (Chuang and Mattes, 2007; Mattes et al., 2003, 2005; Owens et al., 2009) and the only *Nocardioides* strain for which the complete genome (CP000509) has been sequenced so far. It is capable of aerobic growth with ethene (a plant hormone and greenhouse gas) and vinyl chloride (a known human carcinogen and common groundwater contaminant), which is often generated in groundwater by the incomplete reduction of chlorinated solvents (Coleman et al., 2002). This strain also transforms other short-chain alkenes, including propene, 1-butene, and *trans*-2-butene to their corresponding epoxyalkanes, and produces highly enantio-enriched epoxyalkanes via stereoselective monooxygenase-mediated alkene epoxidation (Owens et al., 2009).

Nocardioides simplex and other *Nocardioides* strains possess various enzymes transforming sterols (Arima et al., 1969; Fokina et al., 2003a, 2003b; Nagasawa et al., 1969; Yu et al., 2007). An extracellular adenosine deaminase was described for *Nocardioides* sp. J-326TK (Jun et al., 1994). Histamine dehydrogenase, catalyzing the oxidative deamination of histamine to produce imidazole acetaldehyde and an ammonium ion, and also oxidizing agmatine and putrescine had been reported for *Nocardioides simplex*; the gene encoding this enzyme had been characterized and overexpressed in *Escherichia coli* (Fujieda et al., 2004, 2005; Limburg et al., 2005; Reed et al., 2008; Tsutsumi et al., 2008). Two mycelium-forming strains, *Nocardioides albus* SC13912 (ATCC 55426) and *Nocardioides luteus* SC13911, produce enzymes facilitating the production of a 10-deacetylbaccatin III, a precursor used for semisynthesis of paclitaxel and analogs. An extracellular enzyme from *Nocardioides albus* is a polypeptide (47 kDa) that specifically removes the C-13 side chain from paclitaxel, cephalomannine, and other analogs. An intracellular 10-deacetylase (40 kDa) from *Nocardioides luteus* SC13912 removes the 10-acetate from baccatin III and paclitaxel (Hanson et al., 1994, 2004). *Nocardioides kongjuensis* is able to degrade *N*-hexanoyl-L-homoserine lactone, a signaling molecule in quorum-sensing system of many bacteria (Yoon et al., 2006b). Extremely thermostable deoxycytidine deaminase (52 kDa) with the activity maximum at 99°C was revealed in *Nocardioides* sp., strain CT16 (Sakai et al., 2002). A chitosanase catalyzing the hydrolysis of glycosidic bonds in chitosan and its encoding gene were described for a soil isolate NIO6 identified as *Nocardioides* sp. on chemotaxonomic grounds (Masson et al., 1995). A number of other medically and industrially useful enzymes and degradation activities of *Nocardioides* strains were reported (e.g. Hanson et al., 1994; Jun et al., 1994; Masson et al., 1995; Nishimoto et al., 1996; Nobile and Belleville, 1958; Patel et al., 2000; Siddiqui et al., 2000). Numerous cloning, sequencing, and three-dimensional structure studies of enzymes and relevant genes have been published and the literature and sequences are available from public databases.

Antibiotic sensitivity and antibiotic production. The antibiotic sensitivity and production were determined for some *Nocardioides* species (*Nocardioides aquiterrae, Nocardioides aquaticus, Nocardioides caricicola, Nocardioides daedukensis, Nocardioides dokdonensis, Nocardioides fonticola, Nocardioides hankookensis, Nocardioides insulae, Nocardioides pyridinolyticus*, and *Nocardioides terrigena*). All strains, except for *Nocardioides pyridinolyticus* (Song et al., 2011) were sensitive to chloramphenicol and streptomycin in different concentrations. Various other antibiotics (ampicillin, carbenicillin, cephalothin, gentamicin, kanamycin, lincomycin, neomycin, novobiocin, oleandomycin, penicillin G, polymyxin B, and tetracycline) suppress growth of individual strains (species) as given in the species descriptions in this chapter. The type strain of *Nocardioides albus* was reported to be highly susceptible to antifungal drugs, such as bifonazole, econazole, miconazole, and clotrimazole, but not to the triazoles fluconazole and voriconazole (100 µg/ml) (Dabbs et al., 2003). No production of β-lactamases associated with resistance to β-lactam antibiotics was revealed for type strains of *Nocardioides albus, Nocardioides luteus, Nocardioides plantarum, Nocardioides pyridinolyticus*, and *Nocardioides simplex* (Ogawara et al., 1999).

The data about antibiotic production of *Nocardioides* strains are relatively sparse as compared with members of some other actinomycete genera. Representatives of *Nocardioides* inhabiting marine shellfish were reported to exhibit wide-spectrum antimicrobial activities and also show antitumor activities (El-Shatoury et al., 2009). Mycelium-forming strains of this genus from wheat roots were capable of suppressing *in vitro* some fungal pathogens of wheat, including *Rhizoctonia solani, Pythium* spp., and *Gaeumannomyces graminis* var. *tritici* (Coombs et al., 2004). A *Nocardioides* strain phylogenetically close to *Nocardioides oleivorans* inhibited growth of filamentous fungi (representatives of *Aspergillus* and *Fusarium*) but was not active towards the yeasts or bacteria tested (Romanenko et al., 2008). A strain, identified as *Nocardioides* sp., produces along with known piericidins, a new biologically active 4-pyridinol compound that inhibits NADH dehydrogenase and electron transfer; the new compound inhibited cell division of fertilized starfish (*Asterina pectinifera*) eggs at the minimum inhibitory concentration of 0.09 mg/ml (Kubota et al., 2003). Strains DSM 3176 and DSM 3177 identified as *Nocardioides albus* on morphological and chemotaxonomic grounds have been reported to synthesize leucylblasticidin S, a precursor in the biosynthesis of blasticidin S, as well as to produce a peptidylnucleoside (Dellweg et al., 1988), named rodaplutin (Gullo et al., 1988). They show insecticidal and acaricidal activities, along with weak antimycotic and antibacterial ones (Dellweg et al., 1988). Some strains described under the generic name *Nocardioides* produce macrolide antibiotics, such as luminamicin with activity against anaerobic bacteria (Ōmura et al., 1985), or an antibiotic with antifungal activities against *Candida albicans* and *Candida tropicalis* (Loppinet et al., 1997). Indeed, data on phylogenetic position of some antibiotic-producing strains described in early works under the name *Nocardioides* are not available and some of them (e.g. *Nocardioides* ATCC 39419, a producer of antitumor antibiotic, sandramycin) might not belong to this genus (Matson. and Bush, 1989). This strain had been reclassified as a member of the genus *Kribbella* (Park et al., 1999).

Bacteriophages. A large collection of bacteriophages that multiply in *Nocardioides albus, Nocardioides luteus, Nocardioides simplex*, and *Nocardioides jensenii* strains but not in strains of other actinomycete genera, including *Aeromicrobium*, have been described (Miller et al., 1991; Prauser, 1976, 1984b; Prauser, 1984a; Prauser, 1989; Prauser and Falta, 1968; Wellington and Williams, 1981; Williams et al., 1980). Some *Streptomyces* phages were reported to cause clearing effects, i.e. phage-dependent lysis without phage propagation, on strains of the species *Nocardioides albus, Nocardioides luteus*, and related mycelium-forming organisms of this genus (Prauser, 1976, 1984b, 1989). The sensitivity to specific *Nocardioides* bacteriophages was used to differentiate the *Nocardioides* species from other mycelium-forming actinomycetes (Kurtboke and Williams, 1991; Prauser, 1976, 1984b) and was an additional criterion that justified the transfer of *Arthrobacter simplex* and *Arthrobacter jensenii* to the genus *Nocardioides* (Collins et al., 1989; O'Donnell et al., 1982). The phage susceptibility is considered to rely to a certain degree on adsorption to specific bacterial cell-wall receptors present in a limited number of closely related bacterial strains (Kurtboke and Williams, 1991; Young, 1967). In the case of the aforementioned *Nocardioides* species and the specific *Nocardioides* phages (Prauser, 1976, 1984b; Prauser, 1984a; Prauser and Falta, 1968), the critical determinants in the initial phage binding site are most likely located on teichoic acids, as reported for representatives of *Bacillus*, *Listeria* and other bacteria (Archibald, 1976; Monteville et al., 1994; Schleifer and Steber, 1974; Wendlinger et al., 1996). Such specific binding sites in strains of *Nocardioides albus*, *Nocardioides luteus*, and related mycelium-forming *Nocardioides* strains may include D-galactose and/or D-glucose linked to pyruvate, which are common structural elements of teichoic acids in these organisms (Shashkov et al., 2000b; Shashkov et al., 1999; Tul'skaya, 2009).

Genomic characteristics and plasmids. The mol% G+C of the DNA of members of this genus is around 70 (Table 215), with a lower value of 66.5 (T_m) reported for strain *Nocardioides albus* IMET 7801 (Prauser, 1976) and a higher value of 74.8 (HPLC) for a strain of *Nocardioides lentus* (Yoon et al., 2006a). Analysis of the 16S–23S rRNA gene's internally transcribed spacer (ITS) region found no tRNA sequences in more than 30 strains of the genus *Nocardioides* and related organisms (Yoon et al., 1998a) in contrast to some other actinomycetes (Baylis and Bibb, 1988; Kim et al., 1993; Normand et al., 1992; Pernodet et al., 1989; Yoon et al., 1997b). Moreover, two types of 16S–23S ITS sequences (differing in size from 1 to 12 bp) were found in a few strains of *Nocardioides albus* (DSM 43874, JCM 5851, and JCM 5862) and the strain *Nocardioides simplex* NCIMB 12919, which suggests that at least two rRNA operons might occur in genomes of these organisms. The 16S–23S ITSs in *Nocardioides* strains vary in size from 328 (*Nocardioides nitrophenolicus*) to 539 bp (the *Nocardioides albus* group). The levels of nucleotide similarity between the type strains ranged from 48.4% (for the pair *Nocardioides jensenii* – *Nocardioides nitrophenolicus*) or 51.3% (for the pair *Nocardioides jensenii* – *Nocardioides simplex*) to 84.8% (for the type strains of *Nocardioides albus* and *Nocardioides luteus*). The sequences of *Nocardioides simplex* strains used in the study were identical or nearly identical (386–388 bp) and exhibited a high nucleotide sequence similarity (97.7–100%). The 16S–23S ITSs of the type strain of *Nocardioides luteus* and "*Nocardioides fulvus*" 71-N86 (= IMET 7846 = DSM 46115 = JCM 3335), "*Nocardioides flavus*" 71-N54 (= IMET 7844 = DSM 46114 = NBRC 14396), and 71-N82 (= KCTC 9579 = NBRC 14397) were also identical in size (473 bp) and invariant or almost similar (1 bp difference) in the sequences. The data indicate that these three strains (isolated from soils from different parts of worldwide) are *Nocardioides luteus* (Prauser, 1984a; Ruan and Zhang, 1979; Tille et al., 1978). In contrast, the 18 strains referred to as *Nocardioides albus* substantially differed in the 16S–23S ITS nucleotide sequences, sharing 84.2–100% sequence similarity (the former figure is lower than the value for type strains of this species and *Nocardioides luteus*). They also differed in size of 16S–23S ITS sequences (468–539 bp), with the type strain containing 514 bp and some strains containing 473 bp (identical to that in *Nocardioides luteus*).

The sequences of RNase P RNA genes were analyzed for seven *Nocardioides* species and representatives of other genera (Yoon and Park, 2000). The sequence similarity between the type strains of *Nocardioides* species was 77.6–94.7 (approx. 64–89% if gaps are included). Like its 16S–23S ITS sequences, the RNase P RNA gene sequence of the type strain of *Nocardioides luteus* is identical or almost identical (differing by 1 bp) to the sequences in "*Nocardioides flavus*" 71-N54 and "*Nocardioides fulvus*" 71-N86. Phylogenetic analysis based on RNase P RNA gene sequences showed the clustering of the type strain of *Nocardioides albus* with *Nocardioides luteus* (94% sequence similarity), *Nocardioides simplex* with *Nocardioides nitrophenolicus* (94.7% similarity), and the formation of distinct lineages by other species (*Nocardioides jensenii*, *Nocardioides plantarum* and *Nocardioides pyridinolyticus*). Importantly, the sequence similarity values (77.6–78.6%) between more distant *Nocardioides* species were lower than, or close to, the values between *Nocardioides* species and strains of *Kribbella* (79.3–83.3%) or *Aeromicrobium* (77.4–82.0%). Accordingly, the *Nocardioides* species did not form a coherent cluster in the extended tree including members of these and other genera used in the study. Furthermore, *Luteococcus japonicus* (*Propionibacteriaceae*) possessed a gene with a nucleotide sequence that was nearly identical (>99% similarity) to that of *Nocardioides jensenii* (Yoon and Park, 2000), which suggests that the RNase P RNA gene transfer can take place between these groups.

As mentioned before, *Nocardioides* sp. JS614 is the only completely sequenced member of the genus capable of assimilating vinyl chloride and ethene, and degrading a number of short-chain alkenes (GenBank accession no. NC_008699; Copeland et al., 2006). The genome is a circular molecule (4.99 Mb), which harbors 4546 protein-coding genes and 55 predicted pseudogenes. The circular plasmid pNOCA01 (307,8 bp) of this strain (GenBank accession number CP000508) has the DNA G+C content of 68 mol%, which is lower than that calculated for the genome (71 mol%) and carries genes encoding key enzymes participating in alkene oxidation and assimilation (Chuang and Mattes, 2007; Mattes et al., 2007; Mattes et al., 2003; Mattes et al., 2005). A few other plasmids were also found in *Nocardioides* strains possessing degradative activities. A plasmid with the *dfdBC* gene cluster encoding a ring-cleavage dioxygenase that degrades dibenzofuran (an oxygen-containing heterocyclic compound) was reported for strain *Nocardioides* sp. DF412 (Miyauchi et al., 2008). A plasmid (34.2 kb) bearing genes encoding the sulfur oxidizing enzyme system involved in degradation of dibenzothiophene (an environmentally persistent heterocyclic sulfur compound) was found in *Nocardioides* sp. PKSP 12 (Sandhya et al., 1995, 1997).

Habitat and ecology. Bacteria of the recognized species of the genus *Nocardioides* have been isolated from various environments (Table 215). Numerous ecological studies indicate the presence of these organisms in various terrestrial and aquatic ecosystems worldwide, including low-temperature and deep subsurface ecosystems as well as nutrient-limited environments (e.g. Barton et al., 2007; Boivin-Jahns, et al., 1995; Gontang et al., 2007; Groth and Saiz-Jimenez, 1999; Groth et al., 1999; Katayama et al., 2006; Rintala et al., 2008; Vishnivetskaya et al., 2006; Ward and Bora, 2006; Zhang et al., 2009a). Mycelial *Nocardioides* strains were reported to predominate in kaolin prepared for the ceramic industry (Prauser, 1976, 1986). There is now substantial evidence that representatives of the genus *Nocardioides* also occur in soils and industrial wastewater polluted by different organic and inorganic compounds, including toxic polyaromatics, heavy metals, and nuclear waste (e.g. Coleman et al., 2002; Desantis et al., 2006; Fredrickson et al., 2004;

Futamata et al., 2004; Golovleva et al., 1990; Hamamura and Arp, 2000; Iizuka and Komagata, 1964; Kubota et al., 2005a; Lee et al., 1994, 1991); Maltseva and Oriel, 1997; Männisto et al., 1999; Mattes et al., 2003; Miyauchi et al., 2008; Rajan et al., 1996; Sandhya et al., 1995; Schippers et al., 2005; Suzuki and Komagata, 1983c; Yoon and Park, 2006; Yoon et al., 1999).

Members of the genus *Nocardioides*, like many other actinobacteria, are most likely involved in the turnover of organic matter in ecosystems because they can degrade and metabolize a wide range of natural organic compounds. The ability to degrade aromatic compounds, including polyaromatics and other chemicals, implies a significant role in the natural degradation of such compounds. In organics-limited environments, organisms of this genus, as mentioned before, may use traces of organics dissolved in water or derived from the decomposition of other micro-organisms, or interact mutualistically with other micro-organisms in such environments (Barton et al., 2004; Barton et al., 2007; Groth and Saiz-Jimenez, 1999). Bacteria of the genus *Nocardioides* can also be detected in the water phase of tropospheric clouds (Amato et al., 2006) where they may be active and take part (as described for other bacteria) in the biodegradation of organic compounds dissolved in cloud water, reaching up to 20 mg per liter (Vaitilington et al., 2010). In desert dust clouds, they can migrate great distances through the atmosphere during dust storms (Griffin, 2007; Polymenakou et al., 2008).

All the above data indicate that the resting and/or vegetative cells of organisms of the genus *Nocardioides* can survive and maintain a suitable metabolic activity under extreme conditions such as increased UV and nuclear radiation, and heavy metal or other pollutant contaminations. The data also indicate the high adaptive potential of organisms of this genus. At the genomic level, this potential is often associated with increased numbers of genes in the hypervariable regions, often called "gene islands", and the ability to move horizontally (Dobrindt et al., 2004; Wilmes et al., 2009). In the latter context, it might be relevant to note that thin channel-like structures connecting cells of *Nocardioides* are occasionally seen on electron micrographs.

Bacteria of this genus occur in the rhizosphere, roots, and the above-ground parts of plants; some have been reported to be mutualistic plant endophytes (Coombs and Franco, 2003; Coombs et al., 2004; Qin et al., 2009; Song et al., 2011; Tian, 2007; Ulrich et al., 2008). Similar to other endophytic bacteria, they may obtain nutrients from plants, obtain protection from abiotic stress like desiccation or freezing, promote plant growth via different mechanisms, and prevent plant disease caused by plant-pathogenic bacteria, fungi, and insects by secreting secondary metabolites (Coombs and Franco, 2003; Dimock et al., 1988; Firakova et al., 2007; Hallmann et al., 1997; Kloepper et al., 1991; Reiter et al., 2002; Schrey et al., 2005; Tian et al., 2007; Tian et al., 2004). Members of this genus can also be found in association with lichens, daphnia, termites and other eukaryotic organisms, including vertebrates and humans (e.g. Delbes et al., 2007; El-Shatoury et al., 2009; Fall et al., 2007; Grice et al., 2009; Li et al., 2007b; Tóth et al., 2008). Although bacteria of the genus *Nocardioides* have been occasionally found in bacterial populations associated with human diseases (Cox et al., 2003; Frank et al., 2007; Harris et al., 2007), they are considered nonpathogenic either to humans or other warm-blooded animals.

Enrichment and isolation procedures

Most *Nocardioides* strains have been isolated at random from terrestrial or marine environments during studies of bacterial diversity in certain biotopes or during screening for strains possessing biodegradative activities. Organic media with low nutrient concentrations are generally advantageous for recovering diverse *Nocardioides* species from natural habitats. *Nocardioides* strains are successfully isolated by the standard dilution plating technique on suitable nonselective ("total count") agar media including the standard R2A agar (Difco), twofold or tenfold-diluted R2A, tenfold-diluted nutrient agar (NA, Difco), oligotrophic medium PYGV (Staley, 1968) and other media (see, e.g. Prauser, 1976; Yoon and Park, 2006), and the descriptions of *Nocardioides* species in this chapter). Oligotrophic medium PYGV (Staley, 1968) with sea water(Lawson et al., 2000a), marine agar (Difco), and a number of other nutrient media supplemented with sea water or salts in a proper concentration (Choi et al., 2007; Lee, 2007a; Park et al., 2008; Yi and Chun, 2004b, a) can be used for isolation of *Nocardioides* strains from marine ecosystems or other hypersaline environments. For isolation of *Nocardioides* strains from alkaline soil, a diluted nutrient agar (NA; Difco) with the pH adjusted to 9.0–10.0 using Na_2CO_3 can be applied (Yoon et al., 2005a; Yoon et al., 2005b; Yoon et al., 2005d, 2006a). *Nocardioides* can also be isolated using media and procedures recommended for arthrobacters (Keddie et al., 1986) and some rhodococci (Goodfellow and Maldonado, 2006).

Since the cells or mycelium fragments are usually closely associated with the mineral and organic particles of the soil, a special procedure for their suspending may be useful, e.g. vigorous shaking of the sample with diluent or mechanical or chemical desorption methods [see Futamata et al. (2004), Herron and Wellington (1990), and Kämpfer (2006), for details and references]. Soil samples pretreated with electric pulses (at a field intensity of 16 kV/cm^2) can result in a highly selective outgrowth of filamentous *Nocardioides*-like strains (Bulina et al., 1998). A subsequent treatment of sample suspension and plating procedures are generally similar to those reported for other mycelial and coryneform actinomycetes [see, e.g. Kämpfer (2006), and Keddie et al. (1986), for details and references]. Isolation of endophytic representatives of *Nocardioides* is carried out from the thoroughly washed and surface-sterilized roots or other parts of plants as described (e.g. Coombs and Franco, 2003; Chung et al., 2008; Song et al., 2011; Tian et al., 2004; Zinniel et al., 2002; or in other works dealing with isolation of plant endophytes). Nutrient-poor media such as humic acid-vitamin B agar (HV; Hayakawa and Nonomura, 1987), tap water-yeast extract agar (TWYE, containing 0.25 g of yeast extract, 0.5 g of K_2HPO_4, and 18 g of agar per liter of tap water) (Crawford et al., 1993), yeast extract-casein hydrolysate agar (YECD, containing 0.3 g of yeast extract, 0.3 g of D-glucose, 2 g of K_2HPO_4, and 18 g of agar per liter of distilled water) (Coombs and Franco, 2003) were successfully applied to isolate endophytic *Nocardioides* from wheat roots. To control fungal growth, the above media can be supplemented with 50 μg/ml benomyl (DuPont). Other antibacterial and antifungal agents might be helpful, but some, including antifungal agents (Dabbs et al., 2003) may inhibit the growth of *Nocardioides* organisms.

Selective enrichment procedures and special media containing certain synthetic organic chemicals (pollutants) or their derivatives are advantageous for isolation of *Nocardioides* strains from polluted environments. Such procedures and media were reported while isolating strains of *Nocardioides pyridinolyticus* (Yoon et al., 1997b), *Nocardioides nitrophenolicus* (Yoon et al., 1999), *Nocardioides aromaticivorans* (Kubota et al., 2005a), *Nocardioides simplex* ("*Brevibacterium lipolyticum*") (Iizuka and Komagata, 1964), *Nocardioides oleivorans* (Schippers et al., 2005), and a number of other not yet named representatives of the genus (e.g. Maltseva and Oriel, 1997; Männisto et al., 1999; Mattes et al., 2003; Miyauchi et al., 2008; Rajan et al., 1996). Because organisms of this genus may be sufficiently adapted to the environmental conditions, isolation conditions may be employed that mimic natural environmental conditions. While intended for isolation of strains belonging to or closely related to the described *Nocardioides* species, the selective (semi-selective) media may be prepared taking into account isolation sources and the specific characteristics listed in Table 215 and mentioned in the species description.

Incubation of isolation plates or enrichment cultures is usually carried out at 25–30°C for an appropriate time period, but may also be carried out at a lower (10–15°C) or higher (40°C) temperature (Iwabuchi et al., 1998; Kim et al., 2008a; Lawson et al., 2000a). Colonies of mycelium-producing *Nocardioides* strains formed on many isolation media can be distinguished from those of other actinomycetes (mostly streptomycetes that are abundantly present in the same soil sample) based on the absence of mycelium or scant aerial mycelium, pasty or sand-pasty consistency of colonies, and whitish, light creamy-white, yellowish-white, or yellow colony color. These features, however, are also characteristic of representatives of some other nocardioform actinomycetes, including *Agromyces*, *Kribbella*, or *Promicromonospora* often present in the same soil sample. The *Nocardioides* species with rod-shaped cell morphologies usually occur among coryneform soil isolates that have slender cells and white, yellowish, or indistinctly pigmented colonies.

Maintenance procedures

Cultures may be maintained as 20% glycerol suspensions at −20 and −80°C. Long-term conservation is achieved by freeze-drying or in liquid nitrogen by standard procedures.

Differentiation of the genus *Nocardioides* from other genera

The characteristics essential for phenotypic delineation of the currently recognized genus *Nocardioides* are listed in Table 214 and Table 216. The major menaquinone MK-8(H_4) is the most prominent chemotaxonomic marker that differentiates members of the genus *Nocardioides* from all other genera of the family, except for *Marmoricola*. The presence of well-developed branching hyphae or relatively long rods (in young cultures), along with a marked rod-to-coccoid morphogenetic life cycle, allows primary separation of the majority of *Nocardioides* species from members of the genus *Marmoricola*, which typically produce spherical cells. Genus level identification of novel strains that form short rods to coccoid cells in young cultures or exhibit to a greater or lesser extent phylogenetic relatedness to the genus *Marmoricola* and related *Nocardioides* species is achieved (in the current classification system) via comparison of such strains with respective individual species of these genera (Table 216). The data on cellular fatty acids are of a practical importance, along with other phenotypic characteristics (Table 214 and Table 216).

Taxonomic comments

The genus *Nocardioides*, with the type species *Nocardioides albus*, originally encompassed the actinomycetes that produced branched fragmenting hyphae, possessed the peptidoglycan type similar to that of the genus *Streptomyces*, and showed susceptibility to specific actinophages (Prauser, 1976). The second mycelium-forming species, *Nocardioides luteus*, was added to the genus in 1984 (Prauser, 1984b, 1985). Shortly before, O'Donnell et al. (1982, 1983) reclassified *Arthrobacter simplex*, which was first described as *Corynebacterium simplex* (Jensen, 1934) and then affiliated to the genus *Arthrobacter* (Lochhead, 1957), as *Nocardioides simplex*. There was general agreement by 1982, mainly on chemotaxonomic grounds, that atypical LL-A_2pm-containing arthrobacters should be removed from this genus (for further details and references, see, e.g. Keddie and Jones, 1981; Keddie et al., 1986; Suzuki and Komagata, 1983c). The distant relatedness of *Arthrobacter simplex* to *Arthrobacter globiformis* was also shown by the rRNA cataloguing studies (Stackebrandt et al., 1980). However, the LL-A_2pm-containing arthrobacters remained species "in search of a genus" at that time.

The proposal of O'Donnell et al. (1982, 1983) to affiliate *Arthrobacter simplex* to *Nocardioides* was based on similarity of the type strain of this species to the mycelium-forming *Nocardioides* in lipid composition (fatty acids, phospholipids, and menaquinones), as well as on other supportive data including, in particular, data on the presence the peptidoglycan type A3γ (Prauser, 1978; Schleifer and Kandler, 1972), the sensitivity to some of specific *Nocardioides* actinophages (Prauser, 1976, 1981; Wellington and Williams, 1981), the DNA G+C content (Tille et al., 1978; Yamada and Komagata, 1970), and the DNA–DNA similarity level of 15–20% (Prauser, 1981). While discussing the generic affiliation of coryneform *Arthrobacter simplex* to *Nocardioides*, O'Donnell et al. (1982) also referred to some actinomycete genera that accommodate morphologically diverse strains and to the evidence from early 16S rRNA sequencing studies (Stackebrandt et al., 1980; Stackebrandt and Woese, 1981) that morphological properties are not always reliable indicators of natural relationships.

Owing to the inclusion of *Arthrobacter simplex* into *Nocardioides* (the transfer was validated in 1983), the genus became heterogeneous with respect to morphology. The authors proposed the emended description of the genus *Nocardioides* with inclusion of the extended range of morphological and chemotaxonomic characters (O'Donnell et al., 1982). This proposal, however, was not validated. There were conflicting opinions with regard to the assignment of organisms with markedly different morphology to one genus, mainly for practical reasons prevailing at that time. The aggregation of a typical nocardioform organism with abundant primary and aerial mycelium (*Nocardioides albus*) with a motile, single-celled bacterium (*Arthrobacter simplex*) was questioned, in particular, by the author of the genus *Nocardioides*, H. Prauser (Prauser, 1986). In line with this view, Suzuki and Komagata (1983b, 1983c) validly published the new genus *Pimelobacter* to accommodate some LL-A_2pm-containing

TABLE 216. Selected characteristics for preliminary differentiation of species of the *Nocardioides jensenii* assemblage, some phylogenetically related *Nocardioides* species, and species of the genus *Marmoricola*[a,b,c]

Characteristic	*N. jensenii*	*N. daedukensis*	*N. dubius*	*N. mesophilus*	*N. daphnia*	*N. alkalitolerans*	*N. dilutus*	*N. halotolerans*	*M. aurantiacus*	*M. aequoreus*	*M. bigeumensis*	*M. korecus*	*M. scoriae*
Colony color[d]	Yellowish white	Yellowish	Yellowish white	Cream-white	Yellowish	Yellowish	Cream-white	Cream-white	Orange	Yellow	Lemon yellow	Yellow	Vivid yellow
Cell appearanced[e]	Irregular rods to cocci	Irregular rods to cocci	Irregular rods to cocci	Short rods to cocci	Irregular rods to cocci	Irregular rods to cocci	Irregular rods to cocci	Irregular rods to cocci	Cocci	Cocci	Cocci	Cocci	Cocci
Cell diameter (μm)	0.6–0.8	0.4–0.8	0.8–1.0	0.3–0.8	0.8–1.0	0.8–1.0	0.4–0.8	0.6–0.9	0.5–0.7	0.5–0.7	0.3–0.5	1.1–1.2	0.6–1.0
Motility	–	–	+	+	–	–	+	nd	–	–	+	–	–
Growth at 37°C	+	+	+	+	+	–	nd	nd	–	+	+	+	+
Catalase activity	+	+	+	+	+	+	nd	–	+	+	–	+	+
Oxidase activity	–	+	+	–	–	+	nd	–	–	–	–	–	–
Nitrate reduction	+	+	–	–	+	+	+	+	–	+	+	–	–
Decomposition of:													
Esculin	–	+	+	–	+	–	–	–	+	+	+	+	+
Starch	–	–	–	+	+	–	–	v	–	–	+	–	–
Xanthine	+	–	–	–	–	–	–	–	–	–	+	–	–
Maximal NaCl (%)[f]	>5	9	5	nd	5	5	<1	10	2	5 (7 w)	7	2.0	3.0
Optimum and (max.) pH[f]	7	7–8 (10)	7–8 (10.5)	7–7.5	7.5–8.5 (10.5)	7–9 (12)	7–8	7–8	5.1–8.7	7.1 (12)	7.2 (12)	6.1–10.1 (5.1–12.1)	8.1–11.1 (6.1–12.1)
Major fatty acids (>20%) (*growth conditions*)[g]	$C_{16:0}$ iso (*NA, 30°C, 7d*)	$C_{16:0}$ iso (*NA, 30°C, 7d*)	$C_{16:0}$ iso (*NA, 30°C, 7d*)	$C_{16:0}$ iso (*R2A, 28°C, 7d*)	$C_{16:0}$ iso (*TSA, 28°C*)	$C_{16:0}$ iso, 10-Me-$C_{18:0}$ (*NA, 30°C, 7d*)	$C_{16:0}$ iso (*1/2 R2A, 28°C, 7d*)	$C_{16:0}$ iso (*1/2 R2A, 28°C, 7d*)	$C_{16:0}$, $C_{18:1}$ (*R2A, 30°C, 5 d; TSB, 28°C*)	$C_{18:1}$, $C_{16:0}$ (*R2A, 30°C, 5 d; TSB, 30°C, 3d*)	$C_{16:0}$ iso (*R2A, 30°C, 5 d; TSA*)	$C_{16:0}$, $C_{17:1}$ ω8*c* (*R2A, 30°C, 5 d*)	$C_{16:0}$, $C_{18:1}$ ω9 (*R2A, 30°C, 5 d*)
Phospholipids	DPG, PG, PI, OH-PG, PL	DPG, PL, UL [PG, PE, PL]	DPG, PG, PE, PI	nd	DPG, PG, PL	nd	DPG, PG	nd	DPG, PG, PI, PL	DPG, PG, PI, UL	DPG, PG, PI, UL	DPG, PC, PG, PI, PL	DPG, PC, PG, PI, PL

(continued)

TABLE 216. (continued)

Characteristic	*N. jensenii*	*N. daedukensis*	*N. dubius*	*N. mesophilus*	*N. daphnia*	*N. alkalitolerans*	*N. dilutus*	*N. halotolerans*	*M. aurantiacus*	*M. aequoreus*	*M. bigeumensis*	*M. korecus*	*M. scoriae*
DNA G+C content (mol%)[h]	68.8 (T_m)	68.7	70.6	68.7	69.9	69.9	71.8	67.9	72.0	72.4	72.9	71.0	72.0
Source of isolation	Soil	Soil	Alkaline soil	Soil	Water flea	Alkaline soil	Soil	Agricultural soil	Marble statue	Beach sand	Soil	Volcanic ash	Volcanic ash

[a]Data are for the type strains. Symbols and abbreviations: +, positive; −, negative; w, weak; v, variable or conflicting data; nd, no data available; UL, unknown polar lipid(s); PL, unidentified phospholipids(s); other abbreviations, see also Table 215.

[b]Data are from Suzuki and Komagata (1983c), Yi and Chun (2004a), Dastager et al. (2008d, 2008f, 2010), Lee (2007a), Urzì et al. (2000), Yoon et al. (2005a, 2005d, 2010), Tóth et al. (2008), Lee and Lee (2010), and Lee et al. (2010).

[c]See Figure 248 and Figure 247 for phylogenetic position of the organisms.

[d]Differences in the color intensity and shade and the cell appearance may be influenced by the growth conditions and culture age.

[e]Older cultures of *Nocardioides* species as a rule consist of short rods, coco-bacillary and/or coccoid forms.

[f]Tolerance to NaCl and pH and optimal pH values for growth may vary depending on the test medium composition (mineral, rich in organics, liquid and/or agar-containing).

[g]*NA*, nutrient agar (Difco); *R2A*, R2A agar (Difco); *1/2 R2A*, half-strength R2A agar (1.6 g of R2A broth powder supplemented with 1.5% agar in 1 l of distilled water); *TSA*, trypticase soy agar (Difco); *TSB*, trypticase soy broth (Difco).

[h]DNA base composition as determined by the HPLC method; data for *Nocardioides jensenii* were estimated by the thermal denaturation method.

arthrobacters. Pointing out a rather close relatedness of the studied atypical arthrobacters to *Nocardioides albus* in chemotaxonomic characteristics along with a dissimilarity in the cell morphology and colony appearance, the authors noted some difference between these organisms in the fatty acid and menaquinone composition, a higher DNA G+C content in strains assigned to *Pimelobacter*, and low DNA–DNA similarity values clearly showing distant relationship of *Nocardioides albus* and *Pimelobacter* (Suzuki and Komagata, 1983c). The following species were included in the genus: the type species *Pimelobacter simplex* (*Arthrobacter simplex*), *Pimelobacter jensenii* (created for a single strain NCMB 9770 = JCM 1364, formerly a strain of *Arthrobacter simplex*; Jensen and Gundersen, 1956), and *Pimelobacter tumescens* (the former *Arthrobacter tumescens*; Jensen, 1934; Conn and Dimmick, 1947) currently belonging to the genus *Terrabacter* (see below).

Thus, the valid descriptions of *Arthrobacter simplex* (Lochhead, 1957), *Nocardioides simplex* (O'Donnell et al., 1982, 1983), and *Pimelobacter simplex* (Suzuki and Komagata, 1983a, 1983c) are based on the same type strain and their names are considered to be objective synonyms, with *Arthrobacter simplex* being basonym. In the first edition of *Bergey's Manual of Systematic Bacteriology*, this species was treated under *Arthrobacter*, in Addendum II to "The list of species of the genus *Arthrobacter*", mentioning *Nocardioides simplex* and *Pimelobacter simplex* in the list of synonyms and emphasizing that "the taxonomic position of *Arthrobacter simplex* (and *Arthrobacter tumescens*) remains unresolved" (Keddie et al., 1986). The species *Pimelobacter jensenii* was briefly discussed in the *Further comments* to the description of *Arthrobacter simplex* (Keddie et al., 1986).

Subsequent taxonomic study of the above group using the 16S rRNA-based phylogenetic analysis showed that *Nocardioides albus*, *Nocardioides luteus*, *Nocardioides* (*Pimelobacter*) *simplex*, and *Pimelobacter jensenii* formed a coherent phylogenetic cluster among the strains included in the study, whereas *Pimelobacter tumescens* was phylogenetically distinct (Collins et al., 1989). Accordingly, *Pimelobacter jensenii* was transferred to the genus *Nocardioides*, while *Pimelobacter tumescens* was reclassified as a representative of the newly established genus *Terrabacter* (in the family *Intrasporangiaceae*). Shortly thereafter, Collins and Stackebrandt (1989a, 1989b) proposed the new species *Nocardioides fastidiosus* that seemed at that time to be phylogenetically affiliated to the genus *Nocardioides*. This species was later transferred by Tamura and Yokota (1994) to the genus *Aeromicrobium* (Miller et al., 1991).

Proposals of all other *Nocardioides* species published within the period 1994–2010 and described in this volume rely on the 16S RNA-based phylogenetic analysis, key chemotaxonomic characteristics as well as differences from related species in phenotypic traits and DNA–DNA hybridization values (currently accepted for species delineation) (Stackebrandt et al., 2002; Wayne et al., 1987). The DNA–DNA hybridization experiments (mostly by the method of Ezaki et al., 1989) were performed mainly when particular strains and closely related organisms had high 16S rRNA sequence similarities. However, DNA–DNA hybridization values or other data justifying delineation at the genomic level from the closest related species are absent in a few cases. These mostly include species proposed independently and almost at the same time by different research groups.

At present the genus *Nocardioides* displays an assemblage of phylogenetically rather distant and phenotypically dissimilar species whose features are not reflected by the original genus description created for mycelium-forming actinomycetes (Prauser, 1976). The most unrelated species share 92–93% sequence similarities which are equal to or even less than values separating the well-defined genera in the family *Nocardioidaceae* and other families of the suborder *Actinomycetales*. In addition, as stated before, the differences in both 16S–23S ITS and RNase P gene sequence similarities between some *Nocardioides* species are less than those between some *Nocardioides* species and representatives of other actinomycete genera used in the study (Yoon et al., 1998a; Yoon and Park, 2000).

At the chemotaxonomic level, there are only two characteristics identical for all *Nocardioides* species, i.e. the presence of LL-A_2pm in the cell-wall peptidoglycan and the predominance of the menaquinone MK-8(H_4). Both these characteristics are also typical of some other actinomycete genera (e.g. in the family *Intrasporangiaceae*). In general, *Nocardioides* species seem to display a common fatty acid type, the complex fatty acid type *sensu* Suzuki et al. (1993) and type 3 according to Kroppenstedt (1985). However, it appears that several different subtypes may be distinguished. The data on other chemotaxonomic features are absent for most *Nocardioides* species; if reported, do not appear to be uniform. In particular, the phospholipid types, I, II, and III (with no nitrogenous phospholipids or with phosphatidylethanolamine and phosphatidylcholine as diagnostic components) are usually considered to represent chemotaxonomic markers useful in distinguishing actinomycete genera (Kämpfer, 2006; Kroppenstedt and Evtushenko, 2006; Kroppenstedt and Goodfellow, 2006; Lechevalier et al., 1977, 1981). On the other hand, some specific component(s), e.g. an acylphosphatidylglycerol-like phospholipid (PL_1 in Figure 263) are variable among *Nocardioides* species with the phospholipid type I. Some representatives of this genus also differ in the composition of peptidoglycan-bound cell-polymers (Shashkov et al., 1999, 2000b; Tul'skaya, 2009), and in polyamine patterns (Busse and Schumann, 1999). Furthermore, the cell-wall peptidoglycan structure (undetermined for the majority of *Nocardioides* species) might turn out to be dissimilar in some phylogenetically distant organisms of this genus. This has been shown, in particular, for some LL-A_2pm-containing organisms of the family *Propionibacteriaceae* (the sister family of the *Nocardioidaceae*) and for members of the family *Intrasporangiaceae* (see Schumann et al., 1997, 2009; Stackebrandt and Schaal, 2006; and the respective chapters in this volume). The acyl type of muramic acid, a characteristic feature of the cell envelope differentiating some genera within a family or a suborder (Uchida and Seino, 1997), might be dissimilar for distantly related *Nocardioides* species as well. Organisms of the genus *Nocardioides*, as mentioned before, also differ in cell morphology, catalase activity (a possible indication that different factors are utilized to prevent oxygen damage or cope with certain environmental conditions), optimal and maximal salinity for growth, pH growth range, and other properties affected by ecological niche and in turn affecting speciation.

The data available suggest that further taxonomic study of *Nocardioides*, involving the data on genomics and proteomics, will result in the dissection of this genus into several genera and the establishment of phenotypically and phylogenetically coherent taxa. Dissection of the genus *Nocardioides* is consistent with the current trend toward improving the classification schemes of heterogeneous actinomycete genera, e.g. by finding of additional phenotypic (chemotaxonomic) differentiating characteristics and establishing novel genera (see recent publications in IJSEM). Dissection is also in line with a tendency to describe novel species (often originating from the same or similar ecological niches) that show very high 16S rRNA gene sequence (99% and higher) and include strains exhibiting greater genomic and phenotypic homogeneity. Finally, the classification system must take into account ecological studies revealing diverse novel organisms associated with, or belonging to, the recent genus *Nocardioides* and showing high 16S rRNA gene sequence similarity to species of this genus.

At the moment, some recognized *Nocardioides* species need to be revised taking into account their more focused circumscriptions and current taxonomic concepts. Prominent examples of such species requiring further revision and emendation of the species description are *Nocardioides albus* Prauser 1976 and *Nocardioides luteus* Prauser 1984b. According to the original description of *Nocardioides albus*, based on the study of 17 strains from a variety of soils and related sources from different sites of the world (Prauser, 1976), the species encompassed strains showing a certain phenotypic similarity, mostly in formation of indistinctly colored (whitish to faintly yellowish or faintly brownish) primary mycelium and white aerial mycelium. The species *Nocardioides luteus* was proposed to accommodate strains distinguished by distinct yellow to orange-yellow primary mycelium and cream aerial mycelium (when well developed) (Prauser, 1984b). The separate species status of these organisms was justified by DNA–DNA hybridization experiments with the type strains of the two above species. However, relatedness to *Nocardioides albus* of the remaining strains studied in the original work (Prauser, 1976) and also many other mycelial organisms later described in the literature under the species names *Nocardioides albus* (Collins et al., 1994; Evtushenko and Zelenkova, 1989; Prauser, 1976; Suzuki and Komagata, 1983c; Yoon et al., 1998a; Yoon et al., 1998b) has not been verified in by DNA–DNA hybridization experiments or other adequate genomic analysis. An exception is *Nocardioides albus* IMET 7832 with a high DNA–DNA relatedness to the type strain (Prauser, 1984b) and probably some strains that harbor 16S rRNA gene and 16S–23S ITS sequences identical or almost identical to those of the type strain (Yoon et al., 1998a; Yoon et al., 1998b).

Meanwhile, much evidence has accumulated showing that strains referred to as *Nocardioides albus* are heterogeneous group. As mentioned before, some strains of this assemblage are affiliated with *Nocardioides luteus* or distinct from *Nocardioides albus* or *Nocardioides luteus* on the basis of 16S rRNA gene and 16S–23S rDNA ITS sequence analyses (Yoon et al., 1998a, 1998b). Some were found to contain teichoic acids identical to those of the type strain of *Nocardioides luteus* or differ from those of both the *Nocardioides luteus* and *Nocardioides albus* type strains (Tul'skaya, 2009). Differences in susceptibility to individual actinophages (Prauser, 1976) might indicate differences in cell-wall chemistry between strains of the *Nocardioides albus* complex. Representatives of this group tend also to differ from the *Nocardioides albus* type strain in the proportions of the predominant polyamines (Busse and Schumann, 1999). There is also a report of one *Nocardioides albus* strain (IMET 7819) with a fatty acid composition anomalous to that of the type and other strains studied under the same experimental conditions (O'Donnell et al., 1982). In addition, there are some differences in physiological characteristics, including the range of pH and NaCl concentration for growth (Prauser, 1976; Suzuki and Komagata, 1983c; L. Dorofeeva and L. Evtushenko, unpublished). Some strains of the *Nocardioides albus* assemblage might be eventually be proven to be *Nocardioides luteus* strains and described as non-pigmented subspecies (biovars), while other strains most likely will be assigned to novel species.

Another group of *Nocardioides* species in need of further attention includes a few recent species proposed shortly after the description of their phylogenetic relatives. For such strains, the DNA–DNA hybridization or other relevant experiments justifying their separation into species at the genomic level were not performed. As an example, *Nocardioides basaltis* (Kim et al., 2009a) is closely related to *Nocardioides salarius* (Kim et al., 2008a) (Figure 248). Both these species have very similar fatty acid profiles and other phenotypic traits (Table 215) and originate from marine-related environments in the same geographic region. In this context it is worth noting that many bacteria including actinomycetes have more than one rRNA operon and multiple 16S rRNA gene variants (Acinas et al., 2004). The difference between 16S rRNA genes in one genome may range from 0 to several percent (Conville and Witebsky, 2007; Wang et al., 1997; Yap et al., 1999). In the case of *Nocardioides* species, at least two rRNA operons with two types of 16S–23S ITS sequences (differing in size by up to 12 bp) may occur in some strains (Yoon et al., 1998a). Analogously, other organisms of this genus may contain more than one rRNA operon, which may include 16S rRNA genes with substantially different sequences. Accordingly, the species status of organisms, which is mostly based on direct 16S rRNA sequencing of a single isolate, might be ambiguous (particularly taking into account the limitations of measuring DNA–DNA relatedness by the hybridization technique and of delineating phenotypic features with certainty). The *Nocardioides* strains in assemblages in need of further attention include those very similar in phenotypic traits and isolated from the same or similar ecological niches located at the same or nearby geographical sites. Further comparative taxonomic studies of such species and newly isolated strains, utilizing 16S rRNA gene sequence analysis and reanalysis, with cloning and evaluation for the presence of multiple copies of this gene as well as multilocus sequence analysis (MLSA; Gevers et al., 2005; Rong and Huang, 2010) would help clarify the taxonomy of this genus.

Differentiation of the species of the genus *Nocardioides*

The phenotypic properties differentiating species of the genus *Nocardioides* are listed in Table 215, Table 216, Table 217, Table 218, Table 219, and Table 220. Additional details are given in the *List of species of the genus Nocardioides*.

Assignment of a previously unknown or a dubious isolate to a certain species normally is a two-stage process. In the first stage, the strain in question is assigned to a certain genus; in the second stage, the species level is identified. During stage one the data from phylogenetic analyses are matched to a spectrum

TABLE 217. Oxidization of carbon sources by the type strains *Nocardioides jensemi* and some related species tested by the Biolog GP2[a,b]

Carbon source	*N. akalitolerans*	*N. daphniae*	*N. dubius*	*N. jensemi*
D-Cellobiose	+	−	−	−
D-Fructose	+	+	−	−
D-Glucose	+	+	−	−
Maltose	−	−	+	−
Maltotriose	+	−	+	−
D-Mannose	+	+	−	−
Palatinose	+	+	−	−
D-Psicose	+	+	−	−
D-Ribose	+	+	−	−
Sucrose	+	−	−	−
Trehalose	+	+	−	−
3-Methyl glucose	+	−	−	−
Mannan	+	−	−	−
L-Lactic acid	+	+	−	+
Succinic acid monomethyl ester	+	−	+	−
Propionic acid	+	−	−	+
Pyruvic acid methyl ester	−	+	+	+
L-Alaninamide	−	−	+	+
L-Alanyl glycine	−	−	+	−
Inosine	−	+	+	+
Tween 40	+	−	+	+
Tween 80	−	−	+	+

[a]Symbols: +, positive; −, negative.

[b]Data from Tóth et al. (2008).

TABLE 218. Differential characteristics of organisms of the *Nocardioides simplex* assemblage[a,b]

Characteristic	*N. aromaticivorans*	*N. caeni*	*N. ginsengisoli*	*N. kongjuensis*	*N. nitrophenolicus*	*N. simplex*
Colony color[c]	MW	GY	YW	YW	YW	W, YW
Motility	−	−	−	−	+	+
Growth at 37°C	+	−	+	+	+	+
Growth at 4% (w/v) NaCl	nd	−	+	+	nd	+
Oxidase test	−	+	−	+	+[d]	+
Carbon source utilization:[e]						
L-Arabinose	+	+	−	−	−	−
D-Cellobiose	+	+	nd	−	−	−
D-Fructose	+	−	+	+	+	−
D-Glucose	+	−	+	+	+	+
D-Maltose, D-mannitol	+	−	+	−	−[f]	−[f]
L-Rhamnose	+	+	−	−[f]	+	−
D-Ribose	+	−	−	−	+	−[f]
D-Xylose	+	−	+	−	+	−
Utilization of aromatic chemicals:[g]						
Dibenzofuran (100 p.p.m.)	+	nd	+	+	+	−
Dibenzofuran (200 p.p.m.)	+	nd	−	−	(+)	−
p-Nitrophenol (50 p.p.m.)	+	nd	(+)	+	+	−
p-Nitrophenol (100 p.p.m.)	(+)	nd	−	−	+	−
Decomposition of:						
Esculin	+	−	nd	−	+	+
Starch	−	+	nd	−	+	−[h]
Tyrosine	−	−	nd	−	+	+
Hypoxanthine	+	−	nd	−	−	−
DNA G+C content (mol%) (HPLC)	72.1	71.5	70.2	72.1	71.4	71.7

[a]Data presented are for the type strains. Symbols and abbreviations: +, positive; −, negative; (+), weakly positive; nd, no data available; MW, milky white; GY, grayish yellow; YW, yellowish white; W, white.

[b]Data from Yoon et al. (2006b, 2009), Kubota et al. (2005a), Suzuki and Komagata (1983c), Yi and Chun (2004a), Cui et al. (2009) and unpublished data of the authors of this chapter.

[c]Some variation in the color shade and intensity may be influenced by the growth conditions and culture age.

[d]Negative reaction was reported by Yi and Chun (2004a).

[e]Data on carbon source utilization for all species, except for *Nocardioides ginsengisoli* according to the API 50 CH test system (bioMérieux), with the AUX (bioMérieux) used as the inoculation medium (Yoon et al., 2009). Data for *Nocardioides ginsengisoli* and some data for other species (except for *Nocardioides caeni*) were also obtained using a mineral salts medium supplemented with vitamins and trace element solution as the basal medium (Cui et al., 2009).

[f]Positive results were reported by Cui et al. (2009).

[g]Data from Cui et al. (2009).

[h]Weak activity was reported by Yi and Chun (2004b).

TABLE 219. Percentage cellular fatty acid compositions of the type strains of the *Nocardioides simplex* assemblage[a]

Fatty acid	*N. aromaticiborans*[b]	*N. ginsengisoli*[b]	*N. kongjuensis*[b]	*N. nitrophenolicus*[b]	*N. simplex*[b]	*N. caeni*[c]
Saturated, straight-chain:						
$C_{16:0}$	2.5	1.5	6.4	6.4	1.4	4.6
$C_{17:0}$	2.4		2.9	2.8		1.3
$C_{18:0}$					5.1	1.2
Saturated, iso-branched:						
$C_{14:0}$ iso	2.2	2.9	1.7		1.4	1.0
$C_{15:0}$ iso	2.1	1.9	3.2	4.6	1.1	2.9
$C_{16:0}$ iso	52.0	57.5	52.9	39.8	35.8	40.3
$C_{17:0}$ iso	2.8	2.4	5.4	9.3	3.5	6.3
$C_{18:0}$ iso	5.5	2.3	3.7	8.2		1.9
Saturated anteiso-branched:						
$C_{17:0}$ anteiso			1.7	3.4		0.8
$C_{19:0}$ anteiso	1.8					
Unsaturated straight-chain:						
$C_{17:1}$ ω6*c*	7.5	8.1	2.3	5.6	13.3	8.0
$C_{17:1}$ ω8*c*	5.4	9.0	2.4	1.2	1.0	8.6
$C_{18:1}$ ω7*c*						1.7
$C_{18:1}$ ω9*c*	8.4	5.1	6.7	8.0	8.1	16.4
Unsaturated, iso-branched:						
$C_{16:1}$ iso		1.1			2.1	
10-Methyl-branched:						
$C_{16:0}$ 10-methyl					4.6	
$C_{17:0}$ 10-methyl	3.7	3.2	3.1	1.3	6.2	0.6
$C_{18:0}$ 10-methyl (TBSA)	1.1	1.7	2.0	4.3	11.0	
Hydroxylated acids:						
$C_{15:0}$ 3-OH	3.2		4.2			
$C_{15:0}$ iso 2-OH and/or $C_{16:1}$ ω7*c*	1.3	3.3	1.5		2.2	2.2

[a]Fatty acids that represented <1% in all strains are omitted (except for $C_{17:0}$ anteiso.and $C_{17:0}$ 10-methyl for *Nocardioides caeni*).

[b]Data from Cui et al. (2009). Fatty acids as determined in cell mass harvested from trypticase soy agar (TSA; Difco) after incubation at 30°C for 2 d.

[c]Data from Yoon et al. (2009). Fatty acids as determined in cell mass harvested from nutrient agar (Difco) after incubation at 30°C for 7 d (different proportions of fatty acids will be obtained for this strain if grown on TSA for 2 d).

of phenotypic characters identified with a supposedly coherent group bearing the genus name. During stage two, characters said to be associated with the genus taxon but in a less consistent way may be the focus of attention. Sometimes the results of such a preliminary screening provides useful hints to help further the species level differentiation. In particular, the morphological features (extensively branched hyphae or rod-shaped cells) and chemotaxonomic traits (fatty acid composition, polar lipids, cell-wall polysaccharides, and polyamine patterns) are helpful. Strains phylogenetically related to the *Marmoricola* species require special attention (Figure 247 and Figure 248). The morphological, chemotaxonomic, and other phenotypic characteristics distinguishing *Nocardioides* and *Marmoricola*-related organisms is of practical significance and some key characteristics are given in Table 216.

Strains can be preliminarily assigned to *Nocardioides* species or species group using the multiplex PCR assay and specific primers generated by aligning 16S rRNA gene sequences of *Nocardioides* and one universal reverse primer (Park et al., 1998). Most *Nocardioides* species were described on the basis of a single isolate designated the type strain. Therefore, the other members of such species might differ in some growth and physiological features from the type strains. In addition, some enzymes are inducible, so the test results for enzymic activities (and some other phenotypic characteristics) may be influenced by the medium used to obtain the cells used in a particular experiment. Moreover, the type and reference strains maintained for many years in culture collections might have characteristics altered to some extent since their isolation. A prominent example is the strain IMET 7801 that lost the ability to produce aerial mycelium (Prauser, 1976; Vandamme et al., 1996). In light of very close 16S rRNA gene relatedness and phenotypic similarity of some species, the precise affiliation of a novel strain with a recognized *Nocardioides* species or its classification as a representative of a novel species requires thorough comparison of phenotypic characteristics and DNA–DNA hybridization results or analyses of other adequate genomic characteristics capable of taxonomic resolution at the strain-species level (Gevers et al., 2005; Rong and Huang, 2010; Rosselló-Mora and Amann, 2001; Schumann et al., 2009). Analysis of the 16S–23S rRNA gene internal transcribed spacer region (Yoon et al., 1998a), RNase P RNA gene (Yoon and Park, 2000), or *gyrB* gene (encoding the β-subunit of DNA gyrase, a type II DNA topoisomerase) for novel strains and related species may facilitate the affiliation strains to species. Correlation between the DNA–DNA similarity level and the *gyrB*-based genetic distance was demonstrated for species of the related genus *Kribbella* (Kirby et al., 2010).

Acknowledgements

This work was supported by the program MCB of the Russian Academy of Sciences. The author is grateful to Drs L.V. Dorofeeva, N.G. Vinokurova, E.M. Tul'skaya, and A.M. Shashkov for helpful cooperation.

TABLE 220. Differential characteristics of species of the *Nocardioides ganghwensis* assemblage[a,b]

Characteristic	*N. exalbidus*	*N. furvisabuli*	*N. ganghwensis*	*N. hwasunensis*	*N. oleivorans*
Colony color[c]	W	Y	I	YW	O
Maximal NaCl for growth (%)	nd	6	8	4	(2.34)[d]
Motility	−	+	−	−	−
Nitrate reduction	−	+	+	−	−
Utilization of carbon sources:[e]					
Adonitol	−	+	−	−	−
Glycerol, D-raffinose	−	+	+	−	−
L-Arabinose, D-xylose	−	+	+	+	−
Cellobiose, D-galactose, D-mannose	−	+	+	+	+
D-Salicin	−	−	+	+	−
Sucrose	+	−	+	+	+
Melibiose	+	−	+	−	+
L-Rhamnose	+	−	−[f]	+	+
Inulin	−	+	−	+	−
Enzymes (API ZYM):					
α-Glucosidase	+	−	+	+	+
β-Galactosidase	+	+	+	−	+[g]
Naphthol-AS-BI-phosphohydrolase	+	−	−	−	−[h]
Degradation of:					
Esculin	−	−	(+)	−	+
Casein	nd	+	+	−	+
Starch	−	+	+	+	−
Tyrosine	nd	−	+	−	−
DNA G+C content (mol%) (HPLC)	74	69.1	72.0	71.1	nd

[a]Data are for the type strains. Symbols and abbreviations: +, positive; −, negative; (+), weakly positive; nd, no data available; I, ivory; O, orange; YW, yellowish white; W, white.

[b]Data from Yi and Chun (2004a), Schippers et al. (2005), Li et al. (2007b), Lee (2007b), and Lee et al. (2008).

[c]Some variation in the color shade and intensity may be influenced by the growth conditions and culture age.

[d]No data are available on the ability to grow at a higher concentration of NaCl.

[e]Data on utilization of carbon sources for *Nocardioides exalbidus* are according to the *API CH50* test system (Li et al., 2007a). Data for the remaining species are according to conventional methods (see Table 215 and the original species description for the basal media employed, as well as for the other methods).

[f]Positive result was reported by Lee (2007b).

[g]Activity may be weak (Lee et al., 2008).

[h]Weak activity may be observed (Li et al., 2007b).

List of species of the genus *Nocardioides*

1. **Nocardioides albus** Prauser 1976, 61[AL]

al'bus. L. masc. adj. *albus* white, referring to the white aerial mycelium.

Characteristics are as described for the genus and listed in Table 215. Additional information given below is taken from the works of Prauser (1976, 1989) and Yi and Chun (2004a), unless indicated.

Primary mycelium is white or may be whitish to faintly cream-colored in aged cultures on some ISP media, including glycerol-nitrate agar. White aerial mycelium is typically well formed on yeast extract-malt extract agar, oatmeal agar, chitin agar, and glucose-asparagine agar (Figure 251 and Figure 252); usually fairly well formed on inorganic salts-starch agar and glycerol-asparagine agar, and absent on glycerol-nitrate agar and many complex media rich in organics. No distinct soluble pigments are usually produced. Colonies lacking aerial mycelium are mostly pasty, with smooth to wrinkled and dull to bright surfaces. Growth is optimum at about 28°C, good at 10°C and 37°C, but not at 50°C; weak growth may occur at 42°C. Grows in up to 8% (w/v) NaCl and at initial pH 6–9; weak growth may be observable at initial pH 5 and 10, as assessed in PYGP medium containing 0.5% peptone, 0.3% yeast extract, 0.5% glucose, and 0.02% K_2HPO_4 (L. Dorofeeva and L. Evtushenko, unpublished data). Nutritionally non-exacting; grows well on a suitable mineral salts medium with glucose as sole carbon-plus-energy source and an ammonium salt or nitrate as sole nitrogen source. Growth and biomass production are enhanced in the presence of vitamins (Lawson et al., 2000a) and traces of organics. Utilization of some sugars may depend on the presence of vitamins; thiamine supports growth with D-mannitol, while a solution of vitamins (thiamine, riboflavin, nicotinic acid, pyridoxine and *p*-aminobenzoic acid) supports growth on L-rhamnose or D-xylose (Suzuki and Komagata, 1983c). The type strain utilizes succinate, L-asparagine (weak), L-ornithine (weak), and *N*-acetylglucosamine (weak), but not acetamide, L-arginine, L-lysine, and D-sorbitol as sole carbon sources. The ability to use azelate, malonate, suberate, L-proline, histamine, and tetradecane as sole carbon and energy sources was reported (Collins et al., 1994). Acetate, citrate, and benzoate are utilized, but the test results may vary between test methods (Park et al., 1999; Yi and Chun, 2004a). Acid is produced by oxidation of D-glucose, L-arabinose, sucrose, D-xylose, D-mannitol, D-fructose and L-rhamnose in conventional tests, but not from inositol and raffinose. However, acid was not produced from glucose using the API 20NE test system. No distinct reaction is usually observable in the test for oxidase activity (assessed with tetramethyl-*p*-phenylenediamine). The type strain degrades chitin and grows on chitin agar (Prauser, 1976) as a sole source of carbon and nitrogen. The type strain is highly susceptible to antifungal drugs such as the imidazoles (bifonazole, econazole, miconazole, and clotrimazole), while resistant to triazoles fluconazole and voriconazole (100 µg/ml) (Dabbs et al., 2003).

Peptidoglycan type is A3γ (LL-A_2pm-glycine). The cell-wall sugars are galactose and glucose. The cell wall of the type strain contains poly(galactosylglycerol phosphate) teichoic acid, each monomeric unit containing glycerol, galactose, glucose, pyruvate, and phosphate (Shashkov et al., 1999). The presence of glycerol in the cell wall is the most salient characteristic differentiating this species from *Nocardioides luteus* with cell wall containing ribitol-based teichoic acid (Shashkov et al., 1999, 2000b). The polyamine pattern of the type strain of *Nocardioides albus* grown in rich medium (Yamada and Komagata, 1972) contains almost equal amounts of diamines cadaverine and putrescine (49.0 and 49.6%, respectively), with a minor quantities (<1%) of 1,3-diaminopropane and tyramine, and traces of spermidine and spermine (Busse and Schumann, 1999). The predominant isoprenoid quinone is MK-8(H_4); minor amounts of MK-8(H_2) and MK-8 are detected (Collins et al., 1983; O'Donnell et al., 1982; Suzuki and Komagata, 1983c). The cellular fatty acid profile to a large extent depends on the growth conditions and analytical procedures, but $C_{16:0}$ iso is typically predominating (25–71%) and hydroxylated acids are not detected (Collins et al., 1983; Lee et al., 2000; Miller et al., 1991; Schumann et al., 1997; Suzuki and Komagata, 1983c; Yoon et al., 1997a). The major fatty acids of cells harvested from Nutrient agar (Difco) after 4 d incubation at 30°C were $C_{16:0}$ iso (53.6%), $C_{17:1}$ ω6*c* (15.6%), and $C_{17:0}$ 10-methyl (12.2%); other acids included tuberculostearic acid (TBSA; 2.3%) and homologous components, as well as saturated, unsaturated, iso-, and anteiso- branched acids (Yoon et al., 1997b). The fatty acids of the type strain grown in trypticase soy broth at 28°C for 24–48 h contained mainly $C_{16:0}$ iso (70.8%), along with small proportions of other fatty acids, including TBSA (2.6%) (Schumann et al., 1997). A larger proportion of TBSA (18.3%) was detected in cells from modified Sauton's medium (30°C; Mordarska et al., 1972); $C_{16:0}$ iso contributed 26.6% (O'Donnell et al., 1982). A higher level of TBSA, 34%, was reported by Lee et al. (2000). The qualitative and quantitative composition of polar lipids is also influenced by culture conditions, but phosphatidyl glycerol (PG), diphosphatidyl glycerol (DPG), phosphatidyl inositol (PI) as well as phospholipid (PL_1) [corresponding to acyl phosphatidyl glycerol (APG) in chromatographic behavior] and PI-OH (supposedly a hydroxylated PI) are usually among the principal phosphorous-containing components. Glycolipids may be detected in significant amounts (see the section *Further descriptive information* and Figure 263). Susceptible to specific *Nocardioides* actinophages. The DNA–DNA similarity of the type strain of *Nocardioides albus* to the type strain of *Nocardioides luteus* was 49% and 38% in different experiments (Prauser, 1984b, 1989). Typically occur in soil. The type strain of *Nocardioides albus* was isolated on asparagine agar, containing asparagine 1.0 g, glucose 10.0 g, K_2HPO_4 (anhyd.) 1.0 g, $MgSO_4$ (anhyd.) 0.5 g, NaCl 0.5 g, agar 20.0 g and distilled water 1 liter; pH 7.0–7.2 (Prauser, 1976).

Source (type strain): soil of a lavender field, Tihany peninsula, Balaton region, Hungary.

DNA G+C content (mol%): 68.6 (T_m) (for the type strain; Suzuki and Komagata, 1983c).

Type strain: IMET 7807, ATCC 27980, DSM 43109, KCTC 9186, CCUG 37987, CIP 103451, IFO (now NBRC) 13917, JCM 3185, LMG 16326, NRRL B-5389, VKM Ac-805.

Sequence accession no. (16S rRNA gene): AF004988, X53211.

Additional remarks: the mycelium-forming strains mentioned in the literature under the species name *Nocardioides albus* compose a rather phylogenetically and chemotaxonomically heterogeneous group and some strains most likely will be described as novel species in future (see Further descriptive information for more information).

2. **Nocardioides aestuarii** Yi and Chun 2004b, 2152[VP]

a.es.tu.a'ri.i. L. gen. n. *aestuarii* of the tidal flat.

Characteristics are as described for the genus and listed in Table 215. Information presented below is taken from the original species description (Yi and Chun, 2004b).

Cells are nonmotile rods (ca. 0.3–0.4 × 0.9–2.1 µm) in 3-d-old culture on marine agar 2216 (MA; Difco). Colonies on MA are ivory, approximately 0.5–1 mm in the diameter after 3 d at 30°C, and reach the maximum diameter of 1–2 mm after 5 d. Substrate or aerial mycelium is not observed. Grows at 20–35°C, 0–8% (w/v) NaCl, and at initial pH 6–10; optimal growth was observed at 30°C, 0–2% (w/v) NaCl, and pH 7 as tested on ZoBell medium (Yi and Chun, 2004a; ZoBell, 1941) containing Bacto agar (Difco) 15 g, Bacto peptone (Difco) 5 g, yeast extract (Difco) 1 g, ferric citrate 0.1 g, sea salts 40 g, and 1 liter of distilled water (sea salts were not added when the relation to NaCl was tested). No growth occurs at 15 or 40°°C, at 9% (w/v) NaCl, and at pH 5 and 11 under the same experimental conditions. Utilizes acetate, L-lysine (weak), succinate, and other substrates (Table 215) in basal medium described by Baumann et al. (1972) supplemented with 2% (v/v) Hutner's mineral base (Cohen-Bazire et al., 1957) and modified by reducing the concentration of sea salts to half-strength. L-Arginine, L-asparagine, L-ornithine, D-sorbitol, and tartrate are not used as sole carbon sources on the same test medium. Acid was not produced from glucose using the API 20NE test system. Major fatty acids determined for the cells grown on MA at 30°C for 3 d were $C_{16:0}$ iso (52.0%), $C_{16:1}$ iso H (14.5%), $C_{17:1}$ ω8*c* (7.0%), $C_{17:0}$ 10-methyl (3.1%), and a small quantity (0.3%) of TBSA. The type strain (and the only strain described) was isolated using R2A agar (Difco) supplemented with artificial sea salts (Sigma) by the dilution plating method.

Source (type strain): sediment of tidal flat in Ganghwa Island, Korea (37°36′22.3 N, 126°22′59.4″ E).

DNA G+C content (mol%): 70 (HPLC).

Type strain: JC2056, IMSNU 14029, KCTC 9921, JCM 12125.

Sequence accession no. (16S rRNA gene): AY423719.

3. **Nocardioides agariphilus** Dastager, Lee, Ju, Park and Kim 2008f, 2295[VP]

a.ga.ri.phi'lus. N.L. n. *agarum* agar; N.L. adj. *philus -a -um* (from Gr. adj. *philos -ê -on*) friend, loving; N.L. masc. adj. *agariphilus* agar-loving.

Characteristics are as described for the genus and listed in Table 215. Information presented below is taken from the original species description (Dastager et al., 2008f).

Cells are irregular rods (0.1–0.5 × 1.3–3.0 µm) or cocci. Club-shaped and larger spherical forms reaching up to 2 µm in diameter occur in 7-d-old culture (28°C) on R2A agar. No aerial or substrate mycelium is formed. Motility is reported. Colonies on R2A agar are white to cream in color, and 1.5–2.7 mm in diameter after 7 d of incubation at 28°C. Will grow at 25–35°C, with optimum at 28°C; no growth occurs below 20°C or above 37°C. Optimum pH for growth is 7.5. Prefers a low salinity level; prefers not to grow in twofold-diluted R2A medium (Difco) in the presence of 2% (w/v) NaCl. The carbon sources for growth (according to Kämpfer, 1991) using mineral salts medium supplemented with yeast extract, Oxoid (0.02 g/l), bio-Lactysat, bioMérieux (0.02 g/l), a vitamin solution, and trace element solution (Kämpfer et al., 1990) are listed in Table 215. The organism was also reported to show growth on agar medium without any addition of carbon and nitrogen sources. Major fatty acids of cells harvested from twofold-diluted R2A broth (pH 7.5) after incubation for 10 d at 28°C were $C_{16:0}$ iso (37.6%), $C_{17:1}$ ω8*c* (9.8%), $C_{17:0}$ (7.1%), $C_{17:0}$ iso (5.8%), $C_{17:1}$ iso ω9*c* (5.1%), $C_{15:0}$ iso (4.9%), $C_{18:1}$ ω9c (4.3%), $C_{17:0}$ anteiso (3.9%), and $C_{17:0}$ 10-methyl (3.0%). Neither TBSA nor hydroxylated fatty acids were reported among components exceeding 2.4% of the total fatty acids. The type strain (and the only strain described) was isolated using a tenfold-diluted R2A medium (Difco).

Source (type strain): soil, Bigeum Island, Korea.

DNA G+C content (mol%): 69.4 (HPLC).

Type strain: MSL-28, DSM 19323, JCM 16020, KCTC 19276.

Sequence accession no. (16S rRNA gene): EF466113.

4. **Nocardioides alkalitolerans** Yoon, Kim, Lee, Lee and Oh 2005a, 813[VP]

al.ka.li.to'le.rans. Arabic article *al* the; Arabic n. *qaliy* ashes of saltwort; L. part. adj. *tolerans* tolerating; N.L. masc. part. adj. *alkalitolerans* referring to the ability to tolerate high pH.

Characteristics are as described for the genus and listed in Table 215, Table 216, and Table 217. Information presented below based on the original species description Yoon et al. (2005a), unless indicated.

Cells are rods (0.8–1.0 × 1.5–2.0 µm) in the exponential phase of growth and show rod-to-coccus morphogenesis from the early exponential phase to the stationary phase. Neither substrate nor aerial mycelium is formed. Gram-stain-positive, but Gram staining is variable in old cultures. Colonies on twofold-diluted nutrient agar (NA; Difco) adjusted to pH 9.0 are milky-white in color and 0.7–1.0 mm in diameter after 7 d incubation at 30°C. Growth occurs at 4 and 34°C, but not at or above 35°C. Optimal pH for growth is 7.0–9.0, with the pH growth range of 5.5–12.0 [assessed in two-fold diluted nutrient broth, adjusted to alkaline pH with Na_2CO_3 (below pH 10.5) or KOH (above pH 10.5)]. Growth at pH 5.0 is variable, with no growth for type strain. Growth occurs at in up to 5% (w/v) NaCl (in trypticase soy broth). Nutritionally non-exacting; grows on a mineral salts medium (ISP 9) with glucose as carbon-plus-energy source and an ammonium salt as a sole nitrogen source. Adonitol and D-sorbitol are not utilized on the basal medium ISP 9; utilization of maltose varies among strains (the type strain is negative). Tweens 20, 40, and 60 are hydrolyzed; some strains, including the type strain, are able to hydrolyze gelatin. Some test results may vary between experiments (Dastager et al., 2010; Yoon et al., 2005a). The ability to oxidize some carbon substrates (Tóth et al., 2008) in the Biolog GP2 system are shown in Table 216. The major fatty acids determined for 4 strains grown on two-fold diluted NA agar (pH 9.0) at 30°C for 7 d, included TBSA (21.1–23.2%), $C_{16:0}$ iso (19.9–22.9%), $C_{18:1}$ ω9*c* (10.2–13.4%), $C_{17:1}$ ω6*c* (8.2–10.4%), $C_{16:0}$ (5.9–8.1%), $C_{18:0}$ (4.8–5.9%), and $C_{18:0}$ iso (4.0–5.2%). The DNA–DNA similarity levels between four strains of this species, including the type strain and strains KSL-9, KSL-10, and KSL-12) are 85–91%. The four characterized strains of this species were isolated using the dilution plating technique at 30°C on tenfold-diluted nutrient agar (NA; Difco) adjusted to pH 10.0 with Na_2CO_3.

Source (type strain): alkaline serpentinite soil, Korea.

DNA G+C content (mol%): 72.4–73.6 (HPLC); 71.7 for the type strain.

Type strain: KSL-1, KCTC 19037, DSM 16699, JCM 13365.

Sequence accession no. (16S rRNA gene): AY633969 for the type strain, and AY633970–AY633972 for strains KSL-9, KSL-10, and KSL-12.

5. **Nocardioides aquaticus** Lawson, Collins, Schumann, Tindall, Hirsch and Labrenz 2000b, 1953[VP] (Effective publication: Lawson, Collins, Schumann, Tindall, Hirsch and Labrenz 2000a, 226.)

a.qua'ti.cus. L. masc. adj. *aquaticus* living in water.

Characteristics are as described for the genus and listed in Table 215. Information presented below is based on the original species description (Lawson et al., 2000a), unless indicated.

Cells are predominantly coccoid (around 1.0 µm or more in diameter) or very short rods (0.9 × 1.0 µm), arranged in pairs or small clusters, as observable in 10–14 d cultures grown at 26°C on oligotrophic medium PYGV (DSM No. 621) (DSMZ, 2001; Staley, 1968) prepared with artificial sea water. Neither substrate nor aerial mycelium is formed. Colonies are mostly creamy or dull orange, and older cultures may be orange-colored. Grows at 16–26°C, in the presence of 1–6% NaCl (w/v) and at pH 7–8. The ranges of salinity and pH growth are 0–15% (w/v) NaCl and pH 5.5–9.5 (the maximal pH value examined); prefers not to grow at

3°C or 33.5°C. Growth occurs on marine agar (Difco) (Yi and Chun, 2004a) and is enhanced in the presence of thiamine and biotin (Lawson et al., 2000a). Acetate, pyruvate, glutamate, succinate, malate, and butyrate are utilized as carbon sources for growth in mineral salts medium supplemented with vitamins and traces of yeast extract (Labrenz et al., 1999). No growth with methanol as a carbon source. According to Yi and Chun (2004a), the type strain is able to grow with glucose, acetate, succinate, and other substrates as sole carbon sources on mineral salts medium (Baumann et al., 1971) supplemented with 2% (v/v) Hutner's mineral base (Cohen-Bazire et al., 1957) and modified by reducing the concentration of sea salts; does not utilize benzoate, tartrate, L-arginine L-asparagine, L-lysine, L-ornithine, sorbitol, and acetamide as sole carbon sources for growth. Weak (Lawson et al., 2000a) or no growth (Yi and Chun, 2004a) occurs with citrate. Tween 40 is hydrolyzed. NH_3 is produced from peptone. Methyl red test is negative. No acid is produced from glucose using the API 20NE strip (Yi and Chun, 2004a). More than 20 substrates are oxidized using the Biolog GP test system, including D-glucose, D-fructose, maltose, D-melibiose, D-psicose, L-rhamnose, stachyose, sucrose, D-trehalose, turanose, D-xylose, adonitol, glycerol, inositol, D-mannitol, acetate, succinamate, thymidine, and Tweens 40 and 80. Sensitive to chloramphenicol, penicillin G, and streptomycin as determined with bioDiscs (bioMérieux) after 4 d of growth. Peptidoglycan type is A3γ (LL-A_2pm-glycine). The cell-wall sugars are galactose and glucose. The major respiratory menaquinone is MK-8(H_4) (94% of the total), with MK-6(H_4) and MK-7(H_4) as minor components. The fatty acids analyzed according to Groth et al. (1996) in relatively young cells grown in trypticase soy broth (Difco) at 20°C included the following components: $C_{18:1}$ (30.6%), $C_{16:0}$ (14.1%), $C_{17:0}$ anteiso (10.6%), $C_{16:0}$ iso (8.1%), $C_{18:0}$ (6.9%), $C_{17:0}$ (5.9%), as well as TBSA (3.6%). The type strain (and the only strain characterized) was isolated using PYGV medium prepared with the natural water from Antarctic Ekho Lake, both for enrichment (15°C, 12 d) and subsequent isolation.

Source (type strain): water sample from Antarctic Ekho Lake (depth 1 m, salinity of 9.5%, temperature of 2.8°C, and pH of 8.04) located in the Vestfold Hills in East Antarctica.

DNA G+C content (mol%): 69 (HPLC).

Type strain: EL-17K, ATCC BAA-164, CIP 106993, DSM 11439, JCM 11266, NBRC 100371, NCIMB 703076 (formerly NCFB 3076).

Sequence accession no. (16S rRNA gene): X94145

6. **Nocardioides aquiterrae** Yoon, Kim, Kang, Oh and Park 2004, 74[VP]

a.qui.ter'ra.e. L. n. *aqua* water; L. gen. fem. n. *terrae* of earth or ground; N.L. gen. fem. n. *aquiterrae* pertaining to groundwater.

Characteristics are as described for the genus and listed in Table 215. Information presented below is based on the publications of Yoon et al. (2008, 2007a, 2004), unless indicated.

Cells are irregular rods (0.8–1.0 × 1.7–2.0 µm) in 7 d culture on nutrient agar (NA; Difco) at 30°C, forming small (0.5–1.0 mm in diameter) cream-colored colonies and show rod-to-coccus morphogenesis from the early exponential phase to the stationary phase. Motile, a single lateral flagellum was observed. Neither substrate nor aerial mycelium is formed. Stained Gram-stain-positive or may be Gram-stain-variable in old cultures. Grows at 15 and 42°C (optimum, 30°C), but not at 10°C or temperatures above 43°C. Optimal pH for growth is 6.0–7.0; no growth occurs at pH 5.0. Nutritionally non-exacting; grows on a suitable mineral salts medium (ISP 9 medium) with glucose as the carbon-plus-energy source and an ammonium salt as the sole nitrogen source. Utilizes gentiobiose, turanose, gluconate, and malate as sole carbon and energy sources but not D-arabinose, D- and L-fucose, D-lyxose, D-tagatose, sorbose, stachyose, L-xylose, adonitol, L-arabitol, D-sorbitol, xylitol, glycogen, methyl β-D-xyloside, methyl α-D-mannoside, methyl α-D-glucoside, amygdalin, adipate, 2- and 5-ketogluconate, caprate, and phenylacetate. According to Yi and Chun (2004a), the type strain also uses acetate, succinate, tartrate (weak), acetamide, L-asparagine, and L-lysine as carbon sources for growth and energy, but not benzoate, L-arginine, and L-ornithine. Tributylin is hydrolyzed (Song et al., 2011). Utilization of citrate, lactose, L-rhamnose, D-ribose, and esculin, hydrolysis of Tween 80 and starch, activities for β-galactosidase and trypsin may vary with experiments or the test method (Song et al., 2011; Yi and Chun, 2004a; Yoon et al., 2008; Yoon et al., 2004). Acid not produced from glucose using the API 20NE test system (Yi and Chun, 2004a). Sensitive to chloramphenicol (100 µg), streptomycin (50 µg) and ampicillin (10 µg), but tolerates tetracycline (30 µg) and rifampin (30 µg), as assessed using filter-paper discs (Song et al., 2011).

The fatty acids of cells harvested from NA after 6 d incubation at 30°C included (>4%): $C_{16:0}$ iso (57.6%), $C_{17:0}$ anteiso (9.0%), $C_{16:1}$ iso H (6.8%), $C_{15:0}$ iso (4.9%), and $C_{17:0}$ 10-methyl (4.1%) as well as TBSA (0.8%). The fatty acids determined for cells grown on half-strength R2A agar plates for 3 d at 30°C (Song et al., 2011) were a smaller proportion of $C_{16:0}$ iso (29.9%) and larger proportions of straight-chain unsaturated (>15%) and saturated (7.7%) acids. Song et al. (2011) compared the fatty acid profiles of *Nocardioides aquiterrae* and phylogenetically related species (*Nocardioides pyridinolyticus*, *Nocardioides hankookensis*, and *Nocardioides caricicola*). The levels of DNA–DNA hybridization between the type strains of *Nocardioides aquiterrae* and the phylogenetically closest species *Nocardioides pyridinolyticus* (99.2% 16S rRNA gene similarity) were 32.5 and 28.7% in reciprocal experiments using the procedure of Ezaki et al. (1989). A lower DNA similarity value (19%) was determined with respect to *Nocardioides hankookensis* (98.1% 16S rRNA gene sequence similarity) (Yoon et al., 2008). The following characteristics are helpful in distinguishing *Nocardioides aquiterrae* from the above-mentioned close species: the cell size and motility, optimal and maximal growth temperatures, activity for cytochrome oxidase, nitrate reduction, alkaline phosphatase, esterase (C4) and α-glucosidase, utilization of D-melezitose, D-melibiose, gentiobiose, D-ribose, D-arabitol, and inositol, as well as sensitivity to antibiotics (ampicillin, chloramphenicol, streptomycin, rifampin, and tetracycline; Song et al., 2011; Yoon et al., 2004, 2008). The type strain (and the only strain described) was isolated by the dilution plating technique using NA at 30°C.

Source (type strain): groundwater, Korea.
DNA G+C content (mol%): 73 (HPLC).
Type strain: GW-9, KCCM 41647, JCM 11813.
Sequence accession no. (16S rRNA gene): AF529063.

7. **Nocardioides aromaticivorans** Kubota, Kawahara, Sekiya, Uchida, Hattori, Futamata and Hiraishi 2005b, 984[VP] (Effective publication: Kubota, Kawahara, Sekiya, Uchida, Hattori, Futamata and Hiraishi 2005a, 172.)

a.ro.ma.ti.ci.vo'rans. L. adj. *aromaticus* aromatic, fragrant; L. part. adj. *vorans* devouring; N.L. part. adj. *aromaticivorans* devouring aromatic (compounds).

Characteristics are as described for the genus and listed in Table 215, Table 218, Table 219, and Table 221. Information presented below is based on the original species description (Kubota et al., 2005a), unless indicated.

Cells are nonmotile rods (0.5–0.7 × 1.0–2.0 µm). Transmission electron microscopy with negatively stained cells demonstrated the presence of septa in the middle of some rods, which suggests that coccoid cells may occur in old cultures on some media. Neither substrate nor aerial mycelium is formed. Colonies are milky-white on complex agar media, e.g. on a peptone-beef extract–yeast extract agar medium (Futamata et al., 2004). Growth occurs in the temperature range of 22–40°C (optimum, 30°C) and pH range 5–8 (optimum, pH 7). Dibenzofuran, biphenyl, and dibenzo-*p*-dioxin (weak) are utilized as sole carbon and energy sources, with formation of yellow-orange pigments. Cui et al. (2009) reported the ability of the type strain to utilize *p*-nitrophenol. Cellobiose, maltose, D-mannose, sucrose, trehalose, and D-mannitol are utilized as carbon and energy sources (API CH50), but growth may be weak with these sugars. The carbon sources not utilized include D-arabinose, L-sorbose, L-xylose, fucose, D-tagatose, adonitol, arabitol, dulcitol, erythritol, sorbitol, xylitol, gluconate, 2- and 5-ketogluconate, glycogen, amygdalin, α-methyl-D-glucoside, and α-methyl-D-mannoside (API CH50). Yoon et al. (2009) additionally reported positive reaction of the type strain in the tests for utilization of D- and L-fucose, D- and L-arabitol, adipate and malate, as well as negative results in the tests for utilization of gentiobiose, arbutin, esculin, starch, caprate, and phenylacetate (API 50 CH, API 20NE). Weak or no growth is observable with D-lyxose (Kubota et al., 2005a; Yoon et al., 2009). The type strain grows with L-alanine, L-histidine, L-proline, L-serine, acetate, caprate, 2-ketogluconate, malic acid, propionic acid, suberic acid, and valeric acid as sole carbon and energy sources (conventional tests with mineral salts medium supplemented with vitamins and element solution) but not with adipate, citrate, itaconate, 3- and 4-hydroxybenzoic acids, 3-hydroxybutyrare, 5-ketogluconate, lactate, malonate under the same test conditions (Cui et al., 2009). Production of acetoin was reported (Cui et al., 2009). The major cellular fatty acids determined using the method of Suzuki and Komagata (1983a, 1983c) for seven strains grown at 25°C in a complex liquid medium included $C_{16:0}$ iso (32.8–49.0%), $C_{18:1}$ (11.8–22.1%), $C_{17:0}$ iso (11.9–13.5%), $C_{18:0}$ iso (4.9–7.0%), $C_{17:1}$ (3.0–5.6%), $C_{19:0}$ iso (2.6–3.7%), $C_{17:0}$ (2.0–5.0%), and $C_{17:0}$ anteiso (2.4–4.2%). Neither 10-methyl-branched nor hydroxylated fatty acids were detected. In contrast, the cells harvested from trypticase soy agar (Difco) after incubation at 30°C for 2 d produced $C_{17:0}$ 10-methyl and TBSA, as well as $C_{15:0}$ 3-OH acid (Cui et al., 2009) as shown in Table 219 (also contains data on fatty acids of the type strains of the phylogenetically closest species). The levels of DNA–DNA hybridization between seven strains (H-1, H-2, A-1, A-2, A-3, NSA1-1, and NSA1-2) assigned to *Nocardioides aromaticivorans* exceed 78%. The two strains (H-1 and NSA1-2) affiliated with this species showed DNA–DNA hybridization levels of 17–55% with the type strains of the phylogenetically closest *Nocardioides* species Table 221. The DNA–DNA hybridization values for type strains of more distant species obtained by the method of Ezaki et al. (1989) were 25–29% (*Nocardioides plantarum*) to 32–40% (*Nocardioides albus*, *Nocardioides luteus*, and *Nocardioides jensenii*). Selected phenotypic characteristics differentiating *Nocardioides aromaticivorans* from the phylogenetically closest species are listed in Table 218. Occur in various environments (river water, river sediments, and soil) polluted with polychlorinated dibenzo-*p*-dioxins and dibenzofurans. A dibenzofuran-containing mineral salts medium supplemented with a vitamin solution can be used for enrichment and subsequent isolation. Colonies showing soluble yellow pigment production on the dibenzofuran-containing agar are picked up and subjected to the standard purification procedure, culture on a suitable medium, and strain identification. For details on the enrichment and

TABLE 221. Mean levels of DNA–DNA hybridization (%) between the type strains of *Nocardioides simplex* and closely related species[a,b]

Species	*N. simplex*	*N. aromaticivorans*	*N. nitrophenolicus*	*N. kongjuensis*
1. *N. aromaticivorans*	48–55			
2. *N. nitrophenolicus*	41	48–49		
3. *N. kongjuensis*	21–28	22–36	38–46	
4. *N. ginsengisoli*[c]	33	28	25	22
5. *N. caeni*	13	17	22	31

[a]Data from Yoon et al. (1999, 2006b, 2009), Kubota et al. (2005a), Kim et al. (2009b) and Cui et al. (2009) obtained by the method of Ezaki et al. (1989).

[b]There are no data on the DNA–DNA hybridization level between *Nocardioides ginsengisoli* and *Nocardioides caeni*. These species were described independently by different research groups almost simultaneously (Cui et al., 2009; Yoon et al., 2009).

[c]Close binding values (18–36%) in the reciprocal experiments were obtained for the four species (*Nocardioides aromaticivorans*, *Nocardioides kongjuensis*, *Nocardioides nitrophenolicus*, and *Nocardioides simplex*).

the isolation procedure, see Futamata et al. (2004), Hiraishi and Kitamura (1984), and Kubota et al. (2005b).

Source (type strain): water from the Hikichi river polluted with polychlorinated dibenzo-*p*-dioxins and dibenzofurans, Kanagawa, Japan.

DNA G+C content (mol%): 72.0–72.4 (HPLC).

Type strain: H-1, CIP 108782, DSM 15131, IAM 14992, JCM 11674.

Sequence accession no. (16S rRNA gene; type strain): AB087721.

8. **Nocardioides basaltis** Kim, Roh, Chang, Nam, Yoon, Jeon, Oh and Bae 2009a, 46[VP]

ba.sal'tis. L. gen. n. *basaltis* of basalt, pertaining to the composition of sand, a source of isolation.

Characteristics are as described for the genus and listed in Table 215. Characteristics presented below are taken from the original species description (Kim et al., 2009a).

Cells are nonmotile rods (0.7–1.0 × 1.2–2.0 µm), forming creamy-colored colonies (0.5–1.5 mm in diameter) after 3 d growth on marine agar (MA; Difco) at 30°C. Grows at 10–37°C (optimum about 25–30°C) but not at 4 or 41°C. Grows in marine broth at initial pH 5.5–8.0 (optimally at pH 6–7). Salt-dependent; grows in marine broth containing 1–10% (w/v) NaCl (optimally at 1–2% NaCl), but not in the absence of NaCl. No growth also occurs on the standard R2A agar (Difco) or tripticase soy agar (TSA; Difco). Positive for assimilation of D-arabitol, gluconate and turanose according to the API 50 CH test system examined using inoculation medium AUX supplemented with 1.5% (w/v) NaCl. Negative for assimilation of D-adonitol, amygdalin, D-arabinose, L-arabitol, arbutin, dulcitol, erythritol, D-fucose, L-fucose, gentiobiose, glycogen, inulin, 2- and 5-ketogluconate, D-lyxose, methyl α-D-glucopyranoside, methyl α-D-mannopyranoside, methyl D-xylose, D-sorbitol, L-sorbose, D-tagatose, L-xylose, and xylitol under the same test conditions. No acid is produced from glucose using the API 20NE test system. Fatty acids (>2%) determined for cells grown on MA at 30°C for 3 d included $C_{16:0}$ iso as the predominant component (70.3%), along with $C_{17:1}$ ω8*c* (4.3%), $C_{16:1}$ iso H (3.7%), $C_{14:0}$ iso (3.5%), $C_{17:0}$ 10-methyl (3.2%), $C_{18:1}$ ω9*c* (2.8%), and $C_{18:0}$ iso (2.7%). No TBSA was revealed among fatty acids exceeding 1%. Generally similar profile, with $C_{16:0}$ iso predominating (65.3%), was reported for the phylogenetically closest species, *Nocardioides salarius*, obtained for a younger (1 d) culture grown under the same conditions (Kim et al., 2008a). The fatty acid profile of another phylogenetically close species, *Nocardioides marinisabuli*, grown for 3 d and analyzed under the same experimental conditions as *Nocardioides basaltis*, differed from the latter in that it contained a smaller proportion of $C_{16:0}$ iso (35.9%), and larger proportions of $C_{18:1}$ ω9*c* (15.4%) and $C_{17:0}$ iso (12.3%). The DNA–DNA relatedness between the type strains of this species and phylogenetically very close species *Nocardioides marinisabuli* (99.2% 16S rRNA similarity) was 15.8±1.5%. Mean DNA–DNA similarities with the type strains of more distant species, *Nocardioides terrigena* and *Nocardioides kribbensis*, were 7.0 and 28.7%, respectively. There are no available data on DNA–DNA relatedness to the phylogenetically closest species, *Nocardioides salarius* (Kim et al., 2008a). Sharing many phenotypic features in common with *Nocardioides salarius*, including the fatty acid profile and physiological and biochemical traits (Table 215), the type strain of *Nocardioides basaltis* still appears to differ from that of *Nocardioides salarius* by absence of acid phosphatase, α-chymotrypsin, and valine arylamidase (API ZYM), inability to decompose esculin, Tween 80, and tyrosine, ability to use salicin as a carbon source, having larger-sized cells, and a lower DNA G+C content. The type strain (and the only strain described) was isolated by the dilution plating technique using NA at 30°C.

Source (type strain): black sand, beach, Soesoggak, Jeju Island, Korea.

DNA G+C content (mol%): 68 (T_m).

Type strain: J112, JCM 14945, KCTC 19365.

Sequence accession no. (16S rRNA gene): EU143365.

Additional remarks: the species *Nocardioides basaltis* Kim et al. 2009a described shortly after the establishment of the species *Nocardioides salarius* Kim et al. 2008a has very similar 16S rRNA gene sequence similarity (99.6%) to *Nocardioides salarius*, while the data on the DNA–DNA relatedness between these species are absent. Further comparative taxonomic studies of these two species seem to be needed to justify the separate species status of *Nocardioides basaltis*.

9. **Nocardioides bigeumensis** Dastager, Lee, Ju, Park and Kim 2008f, 2295[VP]

bi.ge.um.en'sis. N.L. masc. adj. *bigeumensis* of or pertaining to Bigeum Island, Korea, the geographical origin of the type strain.

Characteristics are as described for the genus and listed in Table 215. Information presented below based on the original species description (Dastager et al., 2008f).

Cells are cocci or irregular rods (0.3–0.8 × 0.8–4.0 µm). Larger spherical forms, up to 1.5 µm in diameter are reported. Motile. Colonies are cream in color, flat and 1.0–2.5 mm in diameter after 4–5 d incubation on R2A at 28°C. No aerial or substrate mycelium is produced. Grows at 20–35°C (optimum 28°C) and at pH 7.5–9.0. Prefers low salinity; no growth is observed in twofold-diluted R2A in the presence of 1% (w/v) NaCl or more. The carbon sources for growth (Kämpfer, 1991) using mineral salts medium supplemented with yeast extract, Oxoid (0.02 g/l), bio-Lactysat, bioMérieux (0.02 g/l), a vitamin solution, and trace element solution (Kämpfer et al., 1990) are listed in Table 215. The major fatty acids determined in cell mass harvested from twofold-diluted R2A broth (pH 7.5) after incubation for 10 d at 28°C were reported to include $C_{16:0}$ iso (38.3), $C_{15:0}$ iso (13.1), $C_{14:0}$ iso (9.0), $C_{18:1}$ ω9*c* (6.3), $C_{16:0}$ 10-methyl (4.8), $C_{16:0}$ (4.4), and $C_{17:1}$ ω8*c* (3.9). Neither TBSA nor hydroxylated fatty acids exceeded 1%. The type strain (and the only strain described) was isolated using a tenfold-diluted R2A medium by the standard dilution plating method.

Source (type strain): soil, Bigeum Island, Korea.

DNA G+C content (mol%): 69.3 (HPLC).

Type strain: MSL-19, DSM 19320, JCM 16021, KCTC 19290.

Sequence accession no. (16S rRNA gene): EF466114.

10. **Nocardioides caeni** Yoon, Kang, Park, Kim and Oh 2009, 2796[VP]

ca.e'ni. L. gen. n. *caeni* of sludge, isolated from wastewater.

Characteristics are as described for the genus and listed in Table 215, Table 218, Table 219, and Table 221. Information presented below is taken from the original species description (Yoon et al., 2009), unless indicated.

Cells are nonmotile rods or cocci (0.3–0.7 × 0.7–2.5 μm) and can undergo a rod-to-coccus morphogenetic cycle from the early exponential phase to the stationary phase. Gram-stain-positive but results of Gram-staining vary in old cultures. Colonies on nutrient agar (NA; Difco) are grayish yellow and 1.5–2.5 mm in diameter after incubation for 7 d at 30°C. Neither substrate nor aerial mycelium is formed. Optimal temperature for growth is 30°C. The organism grows at 10 and 35°C, but not at 4 or 37°C. Growth occurs in the presence of 0–1.0% (w/v) NaCl, with optimum growth at 0–0.5% (w/v) NaCl and no growth at 2% (w/v) NaCl (assessed in trypticase soy broth prepared according to the formula of the Difco medium without NaCl). Optimal pH for growth is 6.5–7.5; growth occurs at pH 6.0 and 9.5, but not at pH 5.5 or 10.0 [initial pH of the test medium, nutrient broth (NB; Difco) adjusted to various pH by adding HCl or Na_2CO_3]. Positive for assimilation of gluconate, adipate, and malate, but negative for assimilation of 2- and 5-ketogluconate, caprate, and phenylacetate (API 20NE), as well as D-arabinose, sorbose, adonitol, D- and L-arabitol, dulcitol, erythritol, sorbitol, xylitol, methyl α-D-mannoside, methyl α-D-glucoside, amygdalin, arbutin, glycogen, gentiobiose, D-tagatose, D-fucose, and L-fucose (API 50 CH, AUX as suspending medium). Tweens 20, 40, and 60 are hydrolyzed. Susceptible to the following antibiotics (amounts per disc): carbenicillin (100 μg), cephalothin (30 μg), chloramphenicol (100 μg), gentamicin (30 μg), kanamycin (30 μg), neomycin (30 μg), novobiocin (5 μg), oleandomycin (15 μg), penicillin G (20 U), polymyxin B (100 U), streptomycin (50 μg), and tetracycline (30 μg). Resistant to the following antibiotics (amounts per disc): ampicillin (10 μg) and lincomycin (15 μg). Major fatty acids (>5.0% of the total) determined in cell mass harvested from NA plates after incubation for 7 d at 30°C included $C_{16:0}$ iso, $C_{18:1}$ ω9*c*, $C_{17:1}$ ω8*c*, $C_{17:1}$ ω6*c*, and $C_{17:0}$ iso (Table 219); no TBSA was not detected among fatty acids exceeding 0.5%. The type strain exhibited the DNA–DNA similarity values of 13–31% to type strains of the most closely related species (Table 221). Selected phenotypic characteristics differentiating *Nocardioides caeni* from phylogenetically closely related species are listed in Table 218. The type strain (and the only strain described) was isolated by means of the standard dilution plating technique on NA at 30°C.

Source (type strain): sludge of domestic wastewater, Korea.

DNA G+C content (mol%): 71.5 (HPLC).

Type strain: MN8, KCTC 19600, CCUG 57506.

Sequence accession no. (16S rRNA gene): FJ423551.

11. **Nocardioides caricicola** Song, Yasir, Bibi, Chung, Jeon and Chung 2011, 108[VP]

ca.ri.ci'co.la. L. n. *carex -icis* reed-grass, rush or sedge, and also a botanical genus name (*Carex*); L. suff. *-cola* (from L. n. *incola*), inhabitant, dweller; N.L. n. *caricicola*, *Carex*-dweller, isolated from a halophytic plant *Carex scabrifolia* Steud.

Characteristics are as described for the genus and listed in Table 215. Information presented below is taken from the original species description (Song et al., 2011).

Cells are irregular rods (0.4–0.6 × 2.0–5.0 μm) to cocci (Figure 256) in 5-d-old culture grown in half-strength R2A broth (Difco). No flagella were observed for cells grown in the same medium for 2 d at 30°C. Colonies are convex, glistening, white in color and 0.5–0.7 mm in diameter after 5 d of incubation at 30°C on half-strength R2A agar (1.6 g R2A broth powder supplemented with 1.5% agar in 1 L of distilled water). The organism grows in half-strength R2A broth with 0.5% (w/v) NaCl and on marine agar 2216 (MA, Difco). No growth is observable in half-strength R2A broth supplemented with 1% (w/v) NaCl. The pH range for growth reported is also narrow, 7.0–9.0 (optimum pH 8.0); no growth at pH 6.5 and 9.5 (assessed on half-strength R2A agar). Grows at 10 and 45°C, where the latter is the maximal growth temperature reported for the recognized *Nocardioides* species, and not at 50°C (half-strength R2A agar plates). Acid is produced from D-glucose (API 20E). No assimilation of any carbon source was observed using API 20NE test system (no inoculation medium is reported). Tween 20 is decomposed. Tributylin or xylan is not decomposed. Nitrate is reduced to nitrogen gas. Susceptible to the following antibiotics (μg per filter-paper disc): chloramphenicol (100), gentamicin (10), kanamycin (30), penicillin (10), rifampin (30), streptomycin (50), and vancomycin (30); resistant to ampicillin (10) and tetracycline (30). The major cellular fatty acids (>5.0%) determined for cells harvested from half-strength R2A agar plates after incubation for 3 d at 30°C comprised $C_{16:0}$ iso (28.9%), $C_{18:1}$ ω5*c* (7.0%), and $C_{14:0}$ (6.6%), as well as $C_{18:2}$ ω6*c*, $C_{18:2}$ ω9*c*, and/or $C_{18:0}$ anteiso (8.1%). Minor contents of TBSA (1.7%) and $C_{17:0}$ 10-methyl (3.3%) were also detected. Song et al. (2011) compared the fatty acid profiles of the type strains of *Nocardioides caricicola* and related species (*Nocardioides aquiterrae, Nocardioides hankookensis,* and *Nocardioides pyridinolyticus*) obtained under the same experimental conditions. DNA–DNA relatedness between the type strains of this species and *Nocardioides pyridinolyticus* was determined to be 53.5 ± 5.5%. The type strain (and the only strain described) was isolated from the surface-sterilized plant roots. Following sterilization (70% ethanol, 1.0% NaOCl and again 70% ethanol), washing and checking for the surface sterility, a root sample was dried, grounded in sterile sea water and plated on 1/10-strength R2A agar, followed by incubation at 25°C for 1–2 weeks (for details on the isolation procedure see Chung et al., 2008; Song et al., 2011).

Source (type strain): roots of a halophytic plant, *Carex scabrifolia* Steud, growing on sand dunes, Namhae Island, Korea.

DNA G+C content (mol%): 71.7 (HPLC).

Type strain: YC6903, KACC 13778, DSM 22177.

Sequence accession no. (16S rRNA gene): FJ750845.

12. **Nocardioides daedukensis** Yoon, Park, Kang, Lee, Lee and Oh 2010, 1337[VP]

da.e.duk.en'sis. N.L. masc. adj. *daedukensis* of or pertaining to Daeduk Science Park, where the Korea Research Institute of Bioscience and Biotechnology is located.

Characteristics are as described for the genus and listed in Table 215 and Table 216. Information presented below is taken from the original species description (Yoon et al., 2010).

Cells are nonmotile irregular rods or cocci (0.4–0.8 × 0.8–3.0 μm) and show rod-to-coccus morphogenesis from the early exponential phase to the stationary phase. Gram-stain-positive but Gram-staining is variable in old cultures. Colonies on nutrient agar (NA; Difco) are pale yellow in color, glistening, raised, and 1.0–1.5 mm in diameter after incubation for 7 d at 30°C. Neither substrate nor aerial mycelium is formed. Grows at 4 and 37°C (optimally at 30°C), but not at 40°C. Growth is observable at pH 6.0 and 10.0 (optimum at pH 7.0–8.0), but not at pH 5.5 and 10.5, as assessed in nutrient broth (NB; Difco) adjusted to different pH values with HCl or Na_2CO_3. Tolerates salinity up to 9.0% (w/v) NaCl (tested in trypticase soy broth prepared according to the Difco formula, with a definite NaCl concentration), however prefers low NaCl concentrations for growth, 0–0.5% (w/v). Nutritionally non-exacting; grows on a mineral salts medium (ISP 9 medium) with glucose as carbon-plus-energy source and an ammonium salt as sole nitrogen source. Glutamate, acetate, L-malate and pyruvate are utilized as a sole carbon and energy sources on the same basal medium, while negative test results are observable for benzoate, citrate, formate, and succinate. Tweens 20, 40, and 60 are hydrolyzed. Susceptible to the following antibiotics (amounts per disc): carbenicillin (100 μg), cephalothin (30 μg), chloramphenicol (100 μg), gentamicin (30 μg), kanamycin (30 μg), lincomycin (15 μg), neomycin (30 μg), oleandomycin (15 μg), penicillin G (20 U), polymyxin B (100 U), streptomycin (50 μg), and tetracycline (30 μg). Resistant to ampicillin (10 μg) and novobiocin (5 μg).The major fatty acids determined for cells harvested from NA plates after cultivation for 7 d at 30°C were $C_{16:0}$ iso (40.9%), $C_{17:1}$ (15.5%), $C_{17:0}$ 10-methyl (12.2%), $C_{18:1}$ (7.6%), TBSA (6.3%), and $C_{16:1}$ iso (3.5%), as well as minor amounts of other acids, including $C_{17:0}$ 3-OH (1.2%). The fatty acid profile differed from that of the type strain of phylogenetically close *Nocardioides jensenii* and *Nocardioides dubius* (analyzed under the same experimental conditions). In particular, *Nocardioides jensenii* contained a greater proportion of $C_{16:1}$ iso (10.9%), whereas *Nocardioides dubius* had a smaller proportion of $C_{17:1}$ (1.4%) and a significantly larger total iso-branched acids. The major polar lipids of cells grown in NB for 7 d at 30°C are diphosphatidylglycerol, an unidentified phospholipid, and two unidentified lipids. Minor components detected included phosphatidylglycerol, phosphatidylethanolamine, and an unidentified phospholipid. The organism did not contain a compound identical to phosphatidylinositol that was reported for *Nocardioides jensenii* (Collins et al., 1989), *Nocardioides dubius* (Yoon et al., 2005d), and *Marmoricola aurantiacus* (Urzì et al., 2000).

The type strain exhibited mean DNA–DNA relatedness values of 19% to the type strain of the phylogenetically closest species, *Nocardioides jensenii* (98.3% 16S rRNA similarity) and of 10% to *Nocardioides dubius.* The type strain (and the only strain described) was isolated by the standard dilution plating technique on ten-diluted NA at 30°C.

Source (type strain): soil at Taejon, South Korea.

DNA G+C content (mol%): 68.7 (HPLC).

Type strain: MDN22, KCTC 19601, CCUG 57505.

Sequence accession no. (16S rRNA gene): FJ842646.

13. **Nocardioides daphniae** Tóth, Kéki, Homonnay, Borsodi, Márialigeti and Schumann 2008, 81[VP]

daph.ni'a.e. N.L. gen. n. *daphniae* of *Daphnia*, generic name of water flea (*Daphnia cucullata*) from which the type strain was isolated.

Characteristics are as described for the genus and listed in Table 215, Table 216, and Table 217. Information presented below is based on the original description (Tóth et al., 2008), unless indicated.

Cells are short rods or coccoids (0.8–1.0 × 1.2–2.2 μm). Nonmotile. Colonies on King B agar medium (King et al., 1954) are of yellowish color. Neither substrate nor aerial mycelium is produced.

The organism grows at 4–38°C, at initial pH values of 5.5–10.5, and at NaCl concentrations up to 5% (w/v), with optimum growth at 28°C and pH 7–9; no growth occurs at 45°C, at pH 11, and in the presence of 10% (w/v) NaCl (King B agar or broth as basal media). Good growth also occurs on standard nutrient agar (NA, Difco) and trypticase soy agar. The organism grows on mineral salts medium (ISP 9) with glucose and other substrates (Table 215) as carbon-plus-energy sources and an ammonium salt as a sole nitrogen source (Dastager et al., 2010). Produces chitinase (tested using a modified version of the method of Holding and Collee, 1971). No acid produced from glucose in the O/F test (Hugh and Leifson, 1953). Hemolysin is not detected. Data on the oxidation of carbon substrates assessed with the Biolog GP2 test system are listed in Table 216. In addition, the type strain oxidizes methyl-β-D-glucoside, sorbitol, stachyose, acetic acid, β-hydroxybutyric acid, γ-hydroxybutyric acid, α-ketovaleric acid, L-lactic acid, D-malic acid, methyl-pyruvate, glycyl L-glutamic acid, pyruvic acid, 2,3-butanediol, adenosine, 2-deoxyadenosine, thymidine, uridine, adenosine 5′-monophosphate, and uridine 5′-monophosphate but not the remaining Biolog GP2 substrates (about 50 substrates). The major fatty acids determined for cells grown on trypticase soy agar (Difco) at 28°C included $C_{16:0}$ iso (42.7%), $C_{18:1}$ ω9*c* (9.9%), $C_{17:0}$ 10-methyl (9.7%), $C_{17:0}$ iso (8.2%), and $C_{17:1}$ iso ω9*c* (8.1%). Components detected in smaller proportions were $C_{17:1}$ ω8*c* (3.6), $C_{16:1}$ iso (2.6), $C_{17:0}$ anteiso (2.4), $C_{18:0}$ iso (2.4), $C_{16:1}$ ω7*c* (2.2), and some others typical of the genus. The type strain (and the only strain described) was isolated using King B agar medium (King et al., 1954).

Source (type strain): whole-body homogenate sample of *Daphnia cucullata* (Crustacea: Cladocera) adult individuals, Lake Balaton at Tihany (46°55′20″N; 17°55′39″E), Hungary.

DNA G+C content (mol%): 69.9 (HPLC).

Type strain: D287, DSM 18664, CCM 7403, JCM 16608.

Sequence accession no. (16S rRNA gene): AM398438.

14. **Nocardioides dilutus** Dastager, Lee, Ju, Park and Kim 2009b, 1555[VP] (Effective publication: Dastager, Lee, Ju, Park and Kim 2008d, 572.)

di.lu'tus. L. masc. adj. *dilutus* weak, diluted, intended to mean that the organism is able to grow in 100 times diluted R2A medium.

Characteristics are as described for the genus and listed in Table 215 and Table 216. Information presented below is taken from Dastager et al. (2008d, 2010).

Cells are irregular rods (0.4–0.8 μm in diameter; up to 4.0 μm in length) or cocci. The 5-d-old culture grown on R2A agar (Difco) at 28°C is composed of shorter cells (mostly 0.4–0.6 × 0.8–1.6 μm) with a small proportion of coccoid cells. Motile. Colonies on R2A are irregular, smooth, flat, and white to cream in color; 1.5–2.7 mm in diameter after 5 d incubation at 28°C. Neither substrate nor aerial mycelium is formed. Nutritionally non-exacting; grows on a mineral salts medium (e.g. ISP 9) with glucose and other substrates as carbon-plus-energy sources and an ammonium salt as a sole nitrogen source. Growth occurs on solid media with low nutrient concentrations, including 100-fold diluted R2A. Prefers not to grow in R2A broth supplemented with 1% (w/v) NaCl or more. Cell-wall peptidoglycan contains A_2pm as the diagnostic diamino acid (no A_2pm isomer is reported). Fatty acids determined in cell mass harvested from R2A broth after incubation for 7 d at 28°C included (% of the total): $C_{16:0}$ iso (46.7%), $C_{18:1}$ ω9*c* (12.4%), $C_{17:1}$ ω8*c* (5.5%), $C_{14:0}$ iso (4.3%), $C_{17:0}$ anteiso (4.2%), $C_{16:0}$ (3.8%), $C_{17:0}$ (2.9%), $C_{18:0}$ (2.7%), $C_{15:0}$ anteiso (2.5%), $C_{15:0}$ iso (2.4%), $C_{16:1}$ iso (2.0%), and $C_{17:0}$ (1.2%). The type strain (and the only strain described) was isolated by serial dilution plating on tenfold-diluted R2A agar at 30°C after 7 d incubation.

Source (type strain): soil, Bigeum Island, Korea.

DNA G+C content (mol%): 71.8 (HPLC).

Type strain: MSL-11, KCTC 19288, DSM 19318.

Sequence accession no. (16S rRNA gene): EF466121.

15. **Nocardioides dokdonensis** Park, Baik, Kim, Chun and Seong 2008, 2622[VP]

dok.do.nen′sis. N.L. masc. adj. *dokdonensis* of or pertaining to Dokdo, the Korean island from where the type strain was isolated.

Characteristics are as described for the genus and listed in Table 215. Information presented below based on the original description (Park et al., 2008).

Cells are relatively short rods (0.6–0.9 × 1.2–1.8 μm) in the exponential phase of growth. Nonmotile. Colonies on trypticase soy agar (TSA; Difco) are cream in color and approximately 1.0–2.0 mm in diameter after 3 d at 25°C, reaching a maximum diameter of 3 mm after 7 d. Substrate and aerial mycelia are not observed. Grows at 4–30°C (optimum, 25°C), at pH 5–10 (optimum, pH 7), and in up to 7% (w/v) NaCl, with the optimum growth in 0–3% (w/v) NaCl, as assessed on TSA. The following substrates are utilized as sole carbon and energy sources for growth: adonitol, sodium acetate, sodium citrate, sodium propionate, sodium pyruvate, and other carbon sources (Table 215) as tested on Stevenson's basal medium (Stevenson, 1967).* Acetate, adipate, caprate, phenyl acetate, malate, gentiobiose, dextran, DL-xylitol are not utilized on the same basal medium. The following substrates are utilized as sole nitrogen sources: L-cysteine, L-hydroxyproline, L-phenylalanine, L-threonine, L-valine, and nitrate, but not DL-α-amino-*n*-butyric acid or L-histidine (examined using Tsukamura's medium; Tsukamura, 1975). Allantoin, guanine, Tween 20, and xylan are hydrolyzed, but arbutin, elastin, and hippurate are not. Acids are not produced from D-glucose and maltose using the API 20 NE system. In the Biolog GP2 system, the type strain oxidizes 29 of 97 substrates including acetate, adenosine, adenosine 5′-monophosphate, β-cyclodextrin, 2′-deoxyadenosine, D-fructose, glycerol, β-hydroxybutyric acid, γ-hydroxybutyric acid, inosine, α-ketovaleric acid, maltotriose, D-mannitol, D-mannose, melezitose, melibiose, propionic acid, D-psicose, pyruvic acid, D-ribose, sedoheptulosan, D-sorbitol, succinic acid, thymidine, trehalose, Tween 40, Tween 80, uridine, and D-xylose. The type strain is sensitive to the following antibiotics (μg per disc, unless indicated): amikacin (30), ampicillin (10), chloramphenicol (30), erythromycin (15), gentamicin (10), kanamycin (30), nalidixic acid (30), streptomycin (10), tetracycline (30), vancomycin (30), penicillin (10 U), and polymyxin B (300 U). The major cellular fatty acids determined for the type strain grown on TSA for 2 d at 25°C were $C_{16:0}$ iso (40.4%), $C_{18:1}$ ω9*c* (11.2%), $C_{16:0}$ (7.7%), $C_{18:0}$ (7.2%), $C_{17:0}$ iso (4.8%), and $C_{17:1}$ ω8*c* (4.6%). TBSA was not detected; the only 10-methyl-branched acid, i.e. $C_{17:0}$ 10-methyl contributed 1.4%. The DNA–DNA similarity between the type strains of *Nocardioides dokdonensis* and *Nocardioides marinisabuli* was 23.1±5.6%. The type strain (and the only strain described) was isolated by the standard dilution plating technique using R2A agar (Difco) supplemented with 3.5% artificial sea salts (Sigma).

Source (type strain): sand sediment, a beach on Dokdo Island, Korea (37°05′N; 131°13′E).

DNA G+C content (mol%): 69.1 (HPLC).

Type strain: FR1436, KCTC 19309, JCM 14815.

Sequence accession no. (16S rRNA gene): EF633986.

16. **Nocardioides dubius** Yoon, Lee and Oh 2005d, 2211[VP]

du′bi.us. L. masc. adj. *dubius* doubtful of the taxonomic position.

Characteristics are as described for the genus and listed in Table 215, Table 216, and Table 217. Information presented below is based on the original description (Yoon et al., 2005d; Yoon et al., 2010), unless indicated.

Cells are irregular rods (0.8–1.0 × 1.5–2.5 μm) in the exponential phase of growth and show rod-to-coccus morphogenesis from the early exponential phase to the stationary phase. Motile; a single flagellum was observed. Neither substrate nor aerial mycelia is formed. Colonies are irregular, raised with erose margins, yellowish-white in color and 2.0–3.0 mm in diameter after 3 d incubation on nutrient agar (NA; Difco) at 30°C. Grows at 10 and 37°C, but not at 4 or 38°C. Grows in up to 5% (w/v) NaCl, but better without addition of NaCl in trypticase soy broth (TSB; Difco). Optimal pH for growth is 7.0–8.0; grows at pH 6.0 and 10.5, but not at pH 5.5 (initial pH values; examined on NA as a basal medium after 10 d incubation at 30°C). Nutritionally non-exacting; grows on a mineral salts medium (ISP 9) with glucose and some other substrates (Table 215) as carbon-plus-energy source and an ammonium salt as sole

*Bacto Yeast Nitrogen Base medium without amino acids (Difco) supplemented with Casamino acids (10 mg per liter) and agar (15 g per liter), and neutralized by K_2HPO_4.

nitrogen source. Does not utilize D-cellobiose, D-fructose, sucrose, and D-xylose (Yoon et al., 2010); conflicting data were reported by Dastager et al. (2010) using the same basal medium (ISP 9). Utilizes glutamate and pyruvate as a sole carbon and energy sources in the same basal medium, but not benzoate, citrate, formate, and L-malate. Table 216 shows the carbon sources in the Biolog GP2 test system that are oxidized. Tweens 20, 40, and 60 are not hydrolyzed. The composition of the major polar lipids is unique: phosphatidylethanolamine, along with diphosphatidylglycerol, phosphatidylglycerol, and phosphatidylinositol [cells for analysis were grown in NB (Difco) at 30°C]. The major fatty acids (more than 3% of the total in at least one experiment) determined in cells harvested from NA after 6 and 7 d incubation at 30°C were $C_{16:0}$ iso (58.5 and 57.8%), $C_{17:0}$ 10-methyl (7.0 and 8.7%), $C_{16:0}$ 10-methyl (4.2 and 5.3%), $C_{18:0}$ iso (4.3 and 3.0%), $C_{16:1}$ iso (3.7 and 5.1%), $C_{17:0}$ iso (3.2 and 2.4%), and $C_{14:0}$ iso (2.5 and 3.0%). TBSA (2.7%) was detected only in the 6-d-old culture (among fatty acids exceeding 0.5%). The fatty acid profile of 7-d-old culture of *Nocardioides dubius* differed from those of *Nocardioides jensenii* and *Nocardioides daedukensis* (analyzed under the same experimental conditions) in the proportion of iso-branched acids, which was larger contributing nearly 75%, and in the proportions of some individual components (for details, see the descriptions of *Nocardioides jensenii* and *Nocardioides daedukensis* in this chapter and the paper of Yoon et al., 2010). The mean DNA–DNA relatedness of the type strain to the type strain of *Nocardioides daedukensis* was 19% (Yoon et al., 2010). The type strain (and the only strain described) was isolated by the standard dilution plating technique on a tenfold-diluted NA with the pH 10.0 adjusted using Na_2CO_3 (30°C).

Source (type strain): alkaline soil, Kwangchun, Korea.

DNA G+C content (mol%): 70.6 (HPLC).

Type strain: KSL-104, KCTC 9992, JCM 13008.

Sequence accession no. (16S rRNA gene): AY928902.

17. **Nocardioides exalbidus** Li, Xie and Yokota 2007a, 2449[VP] (Effective publication: Li, Xie and Yokota 2007b, 24.)

ex.al.bi'dus. L. masc. adj. *exalbidus* whitish, referring to the color of the colonies.

Characteristics are as described for the genus and listed in Table 215 and Table 220. Additional information presented below is based on the original description (Li et al., 2007b).

Cells are nonmotile, coccoid- to rod-shaped (~ 0.5 μm wide, up to 2.0 μm long), predominantly coccoid or very short rods (~0.6–0.8 μm in diameter) when grown 2 d on ISP 2 agar at 30°C, and show rod-to-coccus morphogenesis from the early exponential phase to the stationary phase. Colony color is white, as observed on trypticase soy agar (TSA; BBL) and on IAM-A1 agar (that is a modified Detmer's medium; see IAM Catalog of Strains, 2004). Grows at 15–35°C, optimally at 30°C, and over a pH range of 6.0–9.0, with the optimal pH around 7.0. The organism grows in a mineral salts medium (IAM-A1 agar) with glucose as carbon-plus-energy source and nitrate as sole nitrogen source. The type strain utilizes amygdalin, 2-keto-gluconate, and 5-keto-gluconate as sole carbon and energy sources but not gluconate, gentiobiose, D-lyxose, sorbose, D-tagatose, D-turanose, adonitol, D- and L-arabitol, dulcitol, erythritol, sorbitol, xylitol, methyl β-D-xyloside, methyl α-D-mannoside, methyl α-D-glucoside, arbutin, starch, and glycogen.

Diphosphatidylglycerol and phosphatidylinositol were reported as major polar lipids; no phosphatidylcholine was detected. The main fatty acids determined in culture grown on TSA for 3 d at 27°C, included $C_{16:0}$ iso (28.0%), TBSA (18.2%), $C_{17:0}$ 10-methyl (7.7%), $C_{17:0}$ iso (9.1%), $C_{17:0}$ anteiso (6.9%), and $C_{18:0}$ (5.8%), and $C_{17:1}$ ω6*c* (4.7%). The DNA–DNA relatedness of the type strain of *Nocardioides exalbidus* to the type strains of phylogenetically neighboring species was 44% (*Nocardioides oleivorans*) and 39% (*Nocardioides ganghwensis*). Selected phenotypic characteristics useful in distinguishing *Nocardioides exalbidus* from the above and other phylogenetically closely related species are listed in Table 220. The type strain (the only strain characterized) was isolated from a lichen sample that was washed and crushed in sterilized water, followed by streaking the effluent on IAM-A1 agar medium, after incubation for 2 weeks and purification on TSA.

Source (type strain): lichen, Izu-Oshima Island, Japan.

DNA G+C content (mol%): 74 (HPLC).

Type strain: RC825, IAM 15416, JCM 23199, CCTCC AA206016.

Sequence accession no. (16S rRNA gene): AB273624.

18. **Nocardioides fonticola** Chou, Cho, Arun, Young and Chen 2008, 1866[VP]

fon.ti'co.la. L. masc. n. *fons, fontis* a spring, fountain; L. suff. *-cola* (from L. masc. or fem. n. *incola*) an inhabitant of a place; N.L. n. *fonticola* an inhabitant of a fountain or spring.

Characteristics are as described for the genus and listed in Table 215. Information presented below is based on the original species description (Chou et al., 2008).

Cells are nonmotile rods (~ 0.8 × 2.0–9.0 μm). Colonies on R2A (Difco) are pale yellowish, approximately 0.9–1.0 mm in diameter after 3 d incubation at 25°C, and reach a maximum of 2 mm after 5 d of incubation. Grows at 25–37°C, with optimum at 30°C. Grows at low salinity (0.5–1.0%, w/v, NaCl; optimum at 0.5%), but not in the absence of NaCl or at 2% (w/v) NaCl as assessed in nutrient broth prepared according to the Difco formula without NaCl. The pH range for growth is 5–9 (optimum pH 7–8), as examined in nutrient broth (Difco) with appropriate buffers; no growth is observed at pH 4.5 or 9.5 under the same experimental conditions. No acid is produced by oxidation of D-glucose, D-xylose, D-mannitol, maltose, D-lactose, sucrose, D-ribose, and glycogen (API Coryne test system). Negative for fermentation of glucose, and for assimilation of gluconate, malate, caprate, adipate, citrate, and phenylacetate (API 20NE). Positive for pyrazinamidase and pyrrolidonyl arylamidase. Tweens 20, 40, and 60 are hydrolyzed. Corn oil is not degraded. Sensitive to the following antibiotics (μg per disk): ampicillin (10), chloramphenicol (30), gentamicin (10), kanamycin (30), novobiocin (30), rifampin (5), penicillin G (10), streptomycin (10), and tetracycline (30). Resistant to nalidixic acid (30), ceftizoxime (30), or sulfamethoxazole (23.75) plus trimethoprim (1.25). Major fatty acids detected for cells harvested from

R2A after 3 d incubation at 25°C included $C_{16:0}$ iso (19.9%), $C_{17:0}$ (19.9%), $C_{17:1}$ ω8*c* (13.3%), $C_{18:1}$ ω9*c* (11.7%), $C_{18:0}$ (4.8%), $C_{18:0}$ iso (4.5%), as well as $C_{19:1}$ ω11*c* and/or $C_{19:1}$ ω9*c* (8.3%), 10-methyl-branched fatty acids [10-Me-$C_{17:0}$ (2.6%) and TBSA (1.7%)], and other acids [$C_{16:0}$, $C_{19:0}$, and $C_{17:0}$ iso (each 2.6–3.6%)]. The type strain (and the only strain described) was isolated by plating a water sample on R2A agar medium and incubation at 25°C for 3 d.

Source (type strain): freshwater of spring, Kaoshiung County, Taiwan.

DNA G+C content (mol%): 71.8 (HPLC).

Type strain: NAA-13, BCRC 16874, JCM 16703, LMG 24213.

Sequence accession no. (16S rRNA gene): EF626689.

19. **Nocardioides furvisabuli** Lee 2007b, 37[VP]

fur.vi.sa'bu.li. L. neut. adj. *furvum* black-colored; L. neut. n. *sabulum* gravel, sand; N.L. gen. n. *furvisabuli* of black-colored sand, the source of isolation of the type strain.

Characteristics are as described for the genus and listed in Table 215 and Table 220. Information presented below is based on the original publication (Lee, 2007b).

Cells are short motile rods (0.4–0.5 × 0.6–1.2 µm) after 3 d in culture at 30°C on YE/SW agar (ISP 2 medium prepared in 60%, v/v, natural sea water), and tend to adhere and form palisades (Figure 258). Colonies are yellow and 0.6–0.8 mm in diameter after 5 d incubation under the same conditions. The organism shows good growth on ISP 2 medium, marine agar (MA; Difco), and nutrient agar (Difco) with or without the addition of natural sea water. Prefers not to grow on standard trypticase soy agar (Difco). Grows at 4–37°C, with optimum at 30°C. The reported pH range for growth is pH 5.1–10.1 (initial pH of media), and the optimum is about 7.0. Grows in up to 6% (w/v) NaCl, as determined in ISP 2 as the basal medium. Nutritionally non-exacting; grows in a suitable mineral salts medium with glucose as carbon-plus-energy source and an ammonium salt as sole nitrogen source (e.g. ISP 9 medium). Citrate, formate, malate, succinate, tartrate, adonitol, D-dulcitol, and methyl α-D-mannoside but not benzoate, D-arabinose, L-sorbose, dextran, methyl α-D-glucoside, *meso*-erythritol, D-xylitol, and 2,3-butanediol are utilized as sole carbon and energy sources. Weakly positive for utilization of acetate, D-sorbitol, and 1,2-propanediol. Negative for acid production (API 20NE) from glucose. Elastin is degraded. The fatty acid profiles were reported for cells grown for 5 d at 30°C on YE/SW agar and MA agar. In the YE/SW culture, $C_{16:0}$ iso (34.1%), $C_{18:1}$ ω9*c* (27.2%), $C_{18:0}$ (7.9%), and $C_{16:0}$ (7.5%) were predominating; the straight-chain saturated acids comprised 17.1% of the total, and straight-chain unsaturated acids comprised almost 33%. Cells grown on MA agar had a significantly larger proportion (44%) of straight-chain saturated components, including $C_{18:0}$ (19.3%) and $C_{16:0}$ (14.6%) and a smaller proportion of straight-chain unsaturated acids (10.5%). The other major acids were $C_{16:0}$ iso (19.6%), $C_{18:1}$ ω9*c* (6.5%), $C_{16:1}$ iso (7.2%), and $C_{15:0}$ anteiso (7.5%). Notably, cells grown on MA agar had anteiso-branched acids ($C_{15:0}$ and $C_{17:0}$; 9.6% in sum), whereas cells grown on YE/SW had one 10-methyl-branched acid (10-Me-$C_{17:0}$; 1.7%). The polar lipid pattern (determined for cells harvested from YE/SW broth after 3 d incubation at 30°C) for *Nocardioides* is unique: phosphatidylcholine, phosphatidylglycerol, phosphatidylinositol, and an unidentified phospholipid. Selected phenotypic characteristics differentiating *Nocardioides furvisabuli* from the phylogenetically closely related species are listed in Table 220. The type strain (the only strain described) was isolated on ISP 4 medium prepared with 60% (v/v) natural sea water.

Source (type strain): black sand, Samyang Beach on Jeju Island, Korea.

DNA G+C content (mol%): 69.1 (HPLC).

Type strain: SBS-26, JCM 13813, NRRL B-24465.

Sequence accession no. (16S rRNA gene): DQ411542.

20. **Nocardioides ganghwensis** Yi and Chun 2004, 1298[VP]

gang.hwen'sis. N.L. masc. adj. *ganghwensis* of or belonging to Ganghwa island in Korea, the geographical origin of the type strain.

Characteristics are as described for the genus and listed in Table 215 and Table 220. Information presented below is based on the original species description (Yi and Chun, 2004a), unless indicated.

Cells are nonmotile irregular rods (~0.4–0.5 × 0.9–4.5 µm) in 3-d-old culture on marine agar (Difco). Colonies on marine agar (MA; Difco) are ivory, butyraceous, and approximately 1–2 mm in diameter after 3 d at 30°C and reach a maximum diameter of 3–4 mm after 5 d. Substrate or aerial mycelium is not observed. Grows at 10–40°C (with no growth at 5 and 50°C), at initial pH 6–10 (no growth at pH 5 and 11), and in the presence of NaCl up to 8% (w/v), as assayed on synthetic ZoBell medium (ZoBell, 1941).* The best growth is observed at 30°C, pH 7, and at a low NaCl concentration (1% or less). Nutritionally non-exacting; grows on a suitable mineral salts medium (e.g. ISP 9) with glucose as sole carbon-plus-energy source and an ammonium salt as sole nitrogen source (Lee, 2007b). The type strain utilizes acetate, citrate, succinate, L-asparagine, L-ornithine, and *N*-acetylglucosamine and other substrates (Table 215) as sole carbon and energy sources in mineral salts medium (Baumann et al., 1971) supplemented with 2% (v/v) Hutner's mineral base (Cohen-Bazire et al., 1957) and modified by reducing the concentration of sea salts to half-strength. Benzoate, tartrate, L-arginine, L-lysine, acetamide, and D-sorbitol are not used under the same test conditions. According to Lee (2007b), the organism can also utilize maltose, melibiose, and trehalose. L-Rhamnose is utilized as a carbon source for growth and energy (Schippers et al., 2005; Yi and Chun, 2004a); a conflicting result was reported by Lee (2007b) with ISP 9 as the basal medium. Crude oil is not degraded (Schippers et al., 2005). No acid production from glucose (API 20NE kit). Major fatty acids determined for cells grown at 30°C on MA for 3 d included $C_{16:0}$ iso (30.4%), $C_{17:1}$ ω8*c* (25.7%), $C_{18:1}$ ω9*c* (5.7%), $C_{16:1}$ iso H (4.4%), and $C_{17:0}$ 10-methyl (3.9%). Other fatty acids (various saturated, unsaturated, iso- and anteiso-branched and hydroxylated fatty acids) were present in smaller quantities.

*ZoBell medium: Bacto agar (Difco) 15 g; Bacto peptone (Difco), 5 g; yeast extract (Difco), 1 g; ferric citrate, 0.1 g; sea salts, 40 g; distilled water, 1 liter; sea salts were not added when the relation to NaCl was tested.

TBSA was not detected under the test conditions employed. The DNA–DNA similarity to the phylogenetically closest species, *Nocardioides oleivorans* and *Nocardioides exalbidus*, was respectively 32% (Schippers et al., 2005) and 39% (Li et al., 2007b). Selected phenotypic characteristics differentiating *Nocardioides ganghwensis* from these and other phylogenetically closely related species are listed in Table 220. The type strain (the only strain described) was isolated using R2A (Difco) supplemented with artificial sea salts (Sigma).

Source (type strain): sediment sample of getbol (tidal flat), Ganghwa Island, South Korea (37°35′31.9″N; 126°27′ 24.5″E).

DNA G+C content (mol%): 72 (HPLC).

Type strain: JC2055, IMSNU 14028, KCTC 9920, JCM 12124.

Sequence accession no. (16S rRNA gene): AY423718.

21. **Nocardioides ginsengisoli** Cui, Lee and Im 2009, 3048[VP]

gin.sen.gi.so′li. N.L. n. *ginsengum* ginseng; L. n. *solum* soil; N.L. gen. n. *ginsengisoli* of soil of a ginseng field, the source of the organism.

Characteristics are as described for the genus and listed in Table 215, Table 218, Table 219, and Table 221. Information presented below is taken from the original species description (Cui et al., 2009).

Cells are nonmotile, slender, short rods (~ 0.2–0.4 × 0.8–1.2 μm) after culture on R2A agar (Difco) for 3 d at 30°C. Colonies developed under the same growth conditions are light yellow–white and convex. Grows at 15–37°C (optimum, 30°C), but not at 4 or 42°C after 5 d incubation. Growth occurs at pH 5.0–8.5, with optimum at pH 7.0, and in up to 5.0% (w/v) NaCl (assessed on R2A agar as basal medium after culture for 5 d). The organism grows on nutrient agar (Difco) and mineral salt medium* supplemented with glucose as a sole carbon source for growth-plus-energy, vitamins (Widdel and Bak, 1992), trace element solution SL-10 (Widdel et al., 1983), and selenite/tungstate solution (Tschech and Pfennig, 1984), but not on MacConkey agar. Utilizes n-nitrophenol (50 p.p.m.) and dibenzofuran (50–150 p.p.m.) as well as the following compounds as sole carbon and energy sources in the above basal medium: D-lyxose, pyruvic acid, acetate, propionate, 3-hydroxybutyrate, valerate, phenylacetate, benzoic acid, 3- and 4-hydroxybenzoate, malate, oxalate, gluconate, L-alanine, arginine, asparagine, aspartic acid, glutamic acid, glutamine, glycine, histidine, isoleucine, leucine, lysine, phenylalanine, L-proline, L-serine, threonine, tryptophan, tyrosine, and valine. The organism is unable to use the following substrates as sole carbon sources under the same test conditions: L-fucose, L-sorbose, citrate, formate, malonate, glutaric acid, tartaric acid, adipate, caprate, itaconate, maleinate, suberate, ethanol, methanol, dulcitol, D-sorbitol, xylitol, amygdalin, glycogen, dextran, cysteine, and methionine. Voges–Proskauer reaction is observable. Tests for degradation of chitin and xylan are negative. No acid or gas is produced from glucose. Nitrate is not used as a terminal electron acceptor in R2A broth supplemented with KNO_3 (10 mM) and thioglycolate with incubation under a nitrogen gas atmosphere. The fatty acid profiles determined for the type strain of this and closely related species are displayed in Table 219. Other selected phenotypic characteristics useful in preliminary discriminating *Nocardioides ginsengisoli* from the closely related species are listed in Table 218. DNA–DNA relatedness of the *Nocardioides ginsengisoli* type strain to the type strains of the phylogenetically closest species were 22–33% (Table 221). The type strain (the only strain described) was isolated from a soil sample suspended in 50 mM phosphate buffer (pH 7.0). After serial dilution with the same phosphate buffer, the suspension was spread on plates with one-fifth strength modified-R2A agar†, and the plates were incubated at 30°C for 1 month.

Source (type strain): soil of a ginseng field, Pocheon province, South Korea.

DNA G+C content (mol%): 70.2 (HPLC).

Type strain: Gsoil 1124, KCTC 19135, CCUG 52478, DSM 17921, JCM 16930.

Sequence accession no. (16S rRNA gene): AB245396.

22. **Nocardioides halotolerans** Dastager, Lee, Ju, Park and Kim 2009c, 1555[VP] (Effective publication: Dastager, Lee, Ju, Park and Kim 2008a, 26.)

ha.lo.to′le.rans. Gr. n. *hals halos* salt; L. part. adj. *tolerans* tolerating; N.L. part. adj. *halotolerans*, salt-tolerating, referring to the ability of the organism to tolerate high salt concentrations.

Characteristics are as described for the genus and listed in Table 215 and Table 216. Information presented below is taken from the original species description (Dastager et al., 2008a).

Cells are irregular rods (normally about 0.6 μm but may be 0.9 μm in diameter, up to 3.4 μm in length) on twofold diluted R2A agar plates at 28°C. A 7-d-old culture is mostly composed of coccoid cells and very short rods, up to 1 μm long. Data on motility are conflicting. Colonies are cream-white and 1–2 mm in diameter after 3 d under the same growth conditions. Nutritionally non-exacting; grows in a suitable mineral salts medium (e.g. ISP 9) with glucose as a sole carbon-plus-energy source and an ammonium salt as a sole nitrogen source. Also grows on twofold dilute R2A medium with up to 10% (w/v), optimally 3%, NaCl. Acid production from carbohydrates is rather uncommon or weak; no acid formed from any of the carbon sources tested. The major fatty acids determined for cells harvested from twofold-diluted R2A after incubation at 28°C for 7 d included $C_{16:0}$ iso (59.29%), $C_{17:0}$ anteiso (8.2%), $C_{18:1}$ ω9*c* (6.8%) and 10 methyl-$C_{16:0}$ (5.6%). The type strain (and the only strain described) was isolated by diluting plating method on tenfold-diluted R2A.

Source (type strain): farming soil, Bigeum Island, South Korea.

DNA G+C content (mol%): 69.7 (HPLC).

Type strain: MSL-23, KCTC 19274, DSM 19273.

Sequence accession no. (16S rRNA gene): EF466122.

---

*Mineral salt medium: 1.8 g of K_2HPO_4, 1.08 g of KH_2PO_4, 0.5 g of $NaNO_3$, 0.5 g of NH_4Cl, 0.1 g of KCl, 0.1 g of $MgSO_4$, and 0.05 g of $CaCl_2$ per 1 liter.

† One-fifth strength modified-R2A agar (per liter of distilled water): 0.25 g of tryptone, 0.25 g of peptone, 0.25 g of yeast extract, 0.125 g of malt extract, 0.125 g of beef extract, 0.25 g of Casamino acids, 0.25 g of Soytone, 0.5 g of glucose, 0.3 g of soluble starch, 0.2 g of xylan, 0.3 g of sodium pyruvate, 0.3 g of K_2HPO_4, 0.05 g of $MgSO_4$, 0.05 g of $CaCl_2$, and 15 g of agar.

23. **Nocardioides hankookensis** Yoon, Kang, Lee and Oh 2008, 437[VP]

han.ko.o.ken'sis. N.L. masc. adj. *hankookensis* of or belonging to Hankook, the Korean name of South Korea from where the type strain was isolated.

Characteristics are as described for the genus and listed in Table 215. Information presented below is based on the original species description (Yoon et al., 2008), unless indicated.

Cells are nonmotile irregular rods (0.4–0.8 μm in width, 1.5–10 μm in length). No coccoid cells are observable if grown on R2A (Difco) or trypticase soy agar (Difco) at 25°C. Neither substrate nor aerial mycelium is formed. Gram-stain-positive, but Gram-stain variable in old cultures. Colonies on R2A agar are white in color and 0.7–1.0 mm in diameter after 7 d incubation at 25°C. Grow at 10 and 34°C (optimum, 25°C), but not at 4 or 35°C, at pH 5.5–8.0, but not at pH 5.0 or 8.5 [in trypticase soy broth (Difco) with the pH adjusted by HCl or Na_2CO_3], and in the presence of 0–2% (w/v) NaCl (optimum 0–0.5%), but not 3% NaCl (in a complex medium prepared according to the Difco trypticase soy broth formula, except that NaCl was omitted). Utilizes gentiobiose (weak), turanose, adipate, gluconate (weak), and malate as carbon sources, but not D-arabinose, D-lyxose, D- and L-fucose, sorbose, D-tagatose, L-xylose, adonitol, D-arabitol, dulcitol, erythritol, sorbitol, xylitol, caprate, citrate, 2- and 5-ketogluconate, phenylacetate, methyl β-D-xyloside, methyl α-D-mannoside, methyl α-D-glucoside, glycogen, and amygdalin (API 20NE and API 50 CH; AUX suspending medium). Tweens 20, 40, and 60 are decomposed. Tributylin is hydrolyzed (Song et al., 2011). Weak activity for cystine arylamidase, β-galactosidase, and trypsin, as well as tyrosine hydrolysis, but the test results vary between experiments or test methods (Song et al., 2011; Yoon et al., 2008). The type strain is susceptible to the following antibiotics (μg per disc, unless indicated otherwise): carbenicillin (100), chloramphenicol (100), gentamicin (30), kanamycin (30), lincomycin (15), neomycin (30), oleandomycin (15), polymyxin B (100 U), and streptomycin (50), but resistant to ampicillin (10), cephalothin (30), novobiocin (5), and penicillin G (20 U). The type strain is also sensitive to rifampin (30 μg) but reports on its tetracycline (30 μg) sensitivity conflict (Song et al., 2011). Major fatty acids (>5% of the total) of cells grown on R2A agar for 7 d at 25°C were $C_{16:0}$ iso (42.7%), $C_{17:1}$ ω8*c* (14.8%), $C_{18:1}$ ω9*c* (13.1%), $C_{17:0}$ 10-methyl (5.5%), and TBSA (2.3%). A different fatty acid profile was reported for cells grown on double-strength R2A agar plates for 3 d at 30°C (Song et al., 2011). The most pronounced differences were in a smaller proportion of $C_{16:0}$ iso (9.1%) and a larger proportion of total straight-chain saturated acids (26.3%) with $C_{14:0}$ predominating (10.2%). Song et al. (2011) reported the fatty acid profiles for the type strains of *Nocardioides hankookensis* and phylogenetically related species (*Nocardioides aquiterrae, Nocardioides pyridinolyticus,* and *Nocardioides caricicola*) obtained under the same experimental conditions.

The type strain exhibited DNA–DNA hybridization levels of 19 and 22% with the type strains of phylogenetically closest species, *Nocardioides aquiterrae* and *Nocardioides pyridinolyticus,* respectively. Phenotypic characteristics useful in distinguishing *Nocardioides hankookensis* from both above species include the cell appearance, size and non-motility, a lower growth temperature optimum and maximum, a lower DNA G+C content, the ability to produce esterase (C4), growth on L-arabinose and D-melibiose, sensitivity to rifampin (30 μg per disc), as well inability to reduce nitrate and produce α-glucosidase (Song et al., 2011; Yoon et al., 2004, 2008; Table 215). The type strain (and the only strain described) was isolated by the standard dilution plating technique at 25°C with tenfold-diluted nutrient agar (Difco).

Source (type strain): soil, Dokdo Island, Korea (37°14′12″N, 131°52′07″E).

DNA G+C content (mol%): 71.3 (HPLC).

Type strain: DS-30, JCM 15302, KCTC 19216, CCUG 54522.

Sequence accession no. (16S rRNA gene): EF555584.

24. **Nocardioides humi** Kim, Srinivasan, Park, Sathiyaraj, Kim and Yang 2009b, 2725[VP]

hu'mi. L. gen. n. *humi* of/from soil, pertaining to the source of isolation of the type strain.

Characteristics are as described for the genus and listed in Table 215. Information presented below is taken from the original species description (Kim et al., 2009b).

Cells incubated on Luria–Bertani (LB) agar at 30°C for 18 h are small rods (0.3–0.5 × 0.8–1.0 μm) and motile with peritrichous flagella (Figure 262). Colonies developed on R2A agar (Difco) after 3 d at 30°C are irregular, flat, and pale yellow in color. Grows at 25 and 42°C (the maximal temperature tested) optimally at 30–37°C, but not at 4°C. Optimal pH for growth is 6.5–7.5, the pH growth range is 5–11. L-Alanine, L-proline, L-serine, glycogen, 2-ketogluconate, 3-hydroxybutyrate, acetate, gluconate, itaconate, malate, propionate, and valerate (API 20NE, API 32GN), but not D-sorbitol, L-histidine, 3-hydroxybenzoate, 4-hydroxybenzoate, 5-ketogluconate, adipate, caprate, citrate, lactate, malonate, phenylacetate, and suberate are utilized as carbon sources. No acid produced from glucose (API 20NE). The major fatty acids (>5%) determined in cell mass harvested from TSA after 2 d incubation at 30°C included $C_{16:0}$ iso (23.1%), $C_{17:0}$ iso (16.2%), $C_{18:1}$ ω9*c* (11.8%), TBSA (9.7%), $C_{17:1}$ ω6*c* (7.7%), $C_{18:0}$ (6.0%), and $C_{16:0}$ (5.2%). The type strain (the only strain described) was isolated using the standard dilution-plating method on a tenfold-diluted R2A agar.

Source (type strain): surface soil of a ginseng field, South Korea.

DNA G+C content (mol%): 71.0 (HPLC).

Type strain: DCY24, JCM 14942, JCM 17026, KCTC 19265, LMG 24128

Sequence accession no. (16S rRNA gene): EF623863.

25. **Nocardioides hwasunensis** Lee, Lee and Kim 2008, 280[VP]

hwa.sun.en'sis. N.L. masc. adj. *hwasunensis* of or belonging to Hwasun, the place where the type strain was isolated.

Characteristics are as described for the genus and listed in Table 215 and Table 220. Information presented below is based on the original species description (Lee et al., 2008).

Nonmotile short rods (0.4–0.7 × 1.0–1.7 μm), occurring singly, in pairs, or in aggregates, are observed in young cultures grown in marine broth (MB; Difco) or in trypticase soy broth at 30°C. Shorter rods, and oval or coccoid cells are observed in older cultures. Colonies are yellowish cream in color on Marine agar (MA; Difco) and light yellow on ISP 2 agar or trypticase soy agar (TSA; Difco). Grows at 4–37°C (optimum, 30°C) but not at 42°C. Grows at pH 5.1 and 9.1, but not at pH 10.1 (on MA with pH adjusted using HC1 or NaOH) and in 4% (w/v) NaCl, but not in 5% (w/v) NaCl (on ISP 2 as the basal medium). Nutritionally non-exacting; grows on a suitable mineral salts medium with glucose as sole carbon-plus-energy source and an ammonium salt as sole nitrogen source (e.g. ISP 9 medium). Acid is produced from glucose by oxidation (API 20NE system), but glucose is not fermented. Acetate, citrate, formate (weak), malate, succinate, tartrate, D-sorbitol, and dextran are used as sole carbon and energy sources, but not D-arabinose, L-sorbose, adonitol, D-dulcitol, *meso*-erythritol, D-xylitol, 2,3-butanediol, 1,2-propanediol, benzoate, methyl-β-D-glucoside, methyl-β-D-mannoside. Utilization of L-arabinose is variable between strains, with the type strain showing positive reaction. The ability to hydrolyze gelatin varies among strains. The fatty acid profiles determined for cells grown on TSA at 30°C for 3 d were almost identical for two strains studied. The major fatty acids (>5%) for the type strains were as follows: $C_{16:0}$ iso (27.8%), $C_{16:0}$ (10.8%), $C_{18:0}$ (8.8%), $C_{18:1}$ ω9*c* (8.2%), $C_{15:0}$ iso (7.5%). 10-Methyl-branched acids were represented by TBSA (4.5%), 10-Me-$C_{17:0}$ (3.5%), and 10-Me-$C_{16:0}$ (2.9%). Selected characteristics differentiating *Nocardioides hwasunensis* from the phylogenetically closely related species are listed in Table 220. The two described strains (HFW-18 and HFW-21) were isolated from a water sample using ISP 2 medium prepared with 60% (v/v) natural sea water.

Source (type strain): water, the area, where running water from a valley merges into sea water on Hwasun beach, Jeju Island, Korea.

DNA G+C content (mol%): 71.1–72.2 (HPLC); 71.1 for the type strain.

Type strain: HFW-21, DSM 18584, JCM 15307, KCTC 19197.

Sequence accession no. (16S rRNA gene): AM295258.

26. **Nocardioides insulae** Yoon, Kang, Lee and Oh 2007b, 138[VP]

in.su'la.e. L. fem. gen. n. *insulae* of an island, referring to the source of isolation of the type strain.

Characteristics are as described for the genus and listed in Table 215. Information presented below is based on the original species description (Yoon et al., 2007b).

Cells are irregular rods (0.6–1.0 × 1.3–6.0 μm) or cocci in the exponential phase of growth and show rod-to-coccus morphogenesis from the early exponential phase to the stationary phase. Neither substrate nor aerial mycelium is formed. Colonies are ivory in color and 1.0–1.5 mm in diameter after 7 d incubation on nutrient agar (NA; Difco) at 30°C.

Grows at 10 and 34°C (optimally at 30°C), but not at 4 or 35°C, and at low NaCl concentration [up to 3%, w/v, in trypticase soy broth (Difco formula without NaCl)]. Media without the addition of NaCl are preferable for growth. Growth is optimal at pH 8.0, observed at pH 6.5, but not observed at pH 6.0 [tested in Nutrient broth (Difco) adjusted to various pH values by the addition of HCl or Na_2CO_3]. In the API 20E test, acetate, L-malate, pyruvate, L-glutamate, and salicin (weakly), but not citrate, succinate, benzoate or formate are utilized as sole carbon and energy sources. Tweens 20, 40, and 60 are hydrolyzed. The type strain is susceptible to the following antibiotics (μg per disk, unless indicated): carbenicillin (100), cephalothin (30), chloramphenicol (100), gentamicin (30), kanamycin (30), lincomycin (15), neomycin (30), oleandomycin (15), polymyxin B (100 U), streptomycin (50), tetracycline (30), but not ampicillin (10), novobiocin (5), or penicillin G (20 U). The predominant fatty acid, as determined in cells harvested from NA plates after incubation at 30°C for 7 d, was $C_{16:0}$ iso (49.7%). The amounts of TBSA (8.3%), $C_{17:0}$ 10-methyl (2.9%), $C_{18:1}$ ω9*c* (7.4%), $C_{17:1}$ ω6*c* (5.4%), $C_{17:0}$ anteiso (4.9%), and other acids typical of the genus were smaller. The type strain (and the only strain described) was isolated on tenfold-diluted NA at 30°C using the standard dilution plating technique.

Source (type strain): soil, Dokdo Island, Korea.

DNA G+C content (mol%): 71.1 (HPLC).

Type strain: DS-51, JCM 15308, KCTC 19180, DSM 17944.

Sequence accession no. (16S rRNA gene): DQ786794.

27. **Nocardioides islandensis** Dastager, Lee, Ju, Park and Kim 2009d, 1555[VP] (Effective publication: Dastager, Lee, Ju, Park and Kim 2008e, 405.)

is.lan.den'sis. N.L. masc. adj. *islandensis* from or pertaining to Bigeum Island in Korea, from where the type strain was isolated.

Characteristics are as described for the genus and listed in Table 215. Information presented below is taken from the original species description (Dastager et al., 2008e).

Cells (cultured 10 d on R2A agar or twofold-diluted R2A at 28°C) are mostly slender irregular rods (ca. 0.5–0.6 μm in diameter, up to 3.7 μm in length) with nearly flat (just after division) and rounded ends. Short chains and branching forms may occur. Coccoid cells and larger spherical or club-shaped forms may be up to 1.6 μm in diameter. Nonmotile. Colonies are white to cream in color, moist and approximately 1–2 mm in diameter after 3 d incubation R2A agar at 28°C. Aerial mycelium is not formed. Nutritionally non-exacting; grows on ISP 9 with glucose as the carbon-plus-energy source and an ammonium salt as the sole nitrogen source. Acid production from carbohydrates is rather uncommon or weak; no acid formed from any of the carbon sources examined (the test method, not reported). Acetoin not produced. The major fatty acids (cells from R2A agar plates after incubation for 10 d at 28°C) included (% of the total): $C_{18:1}$ ω7*c* (50.3), $C_{16:0}$ (11.5), $C_{14:0}$ (7.5), $C_{16:0}$ iso (4.2), and $C_{15:0}$ anteiso (4.2); other acids ($C_{15:0}$ iso, $C_{17:1}$ ω8*c*, $C_{18:0}$, $C_{18:1}$ ω5*c*, $C_{17:0}$ iso, $C_{17:0}$ anteiso, and $C_{17:0}$) each contributed 1–1.9%. No TBSA was detected (limit, 1%).

The type strain (and the only strain described) was isolated by the standard dilution plating on tenfold-diluted R2A agar at 28°C.

Source (type strain): farming soil, Bigeum Island, Korea.

DNA G+C content (mol%): 71.4 (HPLC).

Type strain: MSL-26, KCTC 19275, DSM 19321.
Sequence accession no. (16S rRNA gene): EF466123.

28. **Nocardioides jensenii** (Suzuki and Komagata 1983b) Collins, Dorsch and Stackebrandt 1989, 3[VP] (*Pimelobacter jensenii* Suzuki and Komagata 1983b, 673[VP]; effective publication: Suzuki and Komagata 1983c, 69.)

jen.se'ni.i. N.L. gen. masc. n. *jensenii* of Jensen; named after H.L. Jensen, the Danish bacteriologist who contributed to the taxonomy of coryneform bacteria.

Characteristics are as described for the genus and listed in Table 215, Table 216, and Table 217. Information presented below based on the descriptions of Suzuki and Komagata (1983b), Collins et al. (1989), Yi and Chun (2004a, 2004b), and Yoon et al. (2010) unless otherwise indicated.

Cells are nonmotile irregular rods [mostly 0.6–0.8 × 2.0–4.0 µm or longer, up to 7.0 µm, after culture 1 d on YM agar (0.5% Bacto-peptone, 0.3% yeast extract, 0.3% malt extract, 1% glucose, and 2% agar)] can undergo a distinct rod–coccus morphogenetic cycle. Branching cells and V-forms occasionally occur, but neither substrate nor aerial mycelium is formed. Week-old cultures on YM agar are mostly coccoid cells and very short rods (0.6–1.0 × 0.8–1.0 µm) arranged singly, in pairs, short chains, or small clusters. Colonies are white, yellowish white, and may be light yellowish gray in older cultures on some media. A yellowish soluble pigment may be occasionally produced in older cultures, e.g. on Chapek's agar with glucose instead of sucrose. Grows at 18 and 37°C, but not at 42°C; in 5% (w/v) NaCl, but not in 10% NaCl. The ability to grow in 7% (w/v) NaCl was reported (Nesterenko et al., 1985b). Nutritionally non-exacting; grows in a suitable mineral salts medium with glucose as sole carbon-plus-energy source and an ammonium salt or nitrogen as sole nitrogen source. The type strain of *Nocardioides jensenii* can degrade dinitro-*o*-cresol herbicides (Gundersen and Jensen, 1956). Assimilates acetate, but shows no alkaline reaction with citrate and succinate (in tests described by Yamada and Komagata, 1972). The type strain utilizes acetate, pyruvate, L-asparagine, L-lysine (weak), but not succinate, acetamide, L-arginine, L-ornithine, and D-sorbitol as sole carbon and energy sources for growth (Yi and Chun, 2004a). Uses azelate, suberate, thymine, and uracil, but not tetradecane as sole carbon and energy sources (Collins et al., 1994). Some test results displayed in Table 215, including citrate and benzoate utilization, may vary among experiments or test methods (Dastager et al., 2010; Yi and Chun, 2004a; Yoon et al., 2010). No acid produced during glucose oxidation in the API 20NE test (Yi and Chun, 2004a) or in conventional tests in peptone-based media. The test results for oxidation of carbon substrates (Tóth et al., 2008) obtained with the Biolog GP2 system are shown in Table 217.

The cell-wall peptidoglycan contains LL-A_2pm, glycine, alanine, and glutamic acid. The peptidoglycan type is A3γ (A41.1; http://www.peptidoglycan-types.info), as reported by Schleifer and Kandler (1972) for the type strain *Nocardioides jensenii* (*Arthrobacter simplex* NCIB 9770). Conflicting data on the composition of peptidoglycan (with three-glycine interpeptide bridge) was reported by Collins et al. (1989). Re-examination of the data for the type strain of *Nocardioides jensenii* (L. Evtushenko and L. Dorofeeva, unpublished) showed the molar ratio of alanine, glutamate, LL-A_2pm, and glycine to be 2.1:1.0:1.0:1.2, which is similar to the peptidoglycan composition reported by Schleifer and Kandler (1972). The sugars detectable in the cell wall include galactose, glucose, and rhamnose, along with *N*-acetylglucosamine and glycerol (Takeuchi and Yokota, 1989; Tul'skaya, 2009) and minor amounts of mannose (Tul'skaya, 2009). The cell wall contains teichoic acid of unidentified structure, which involves glycerol, galactose, *N*-acetylglycosamine, and pyruvate as components, along with phosphate (Takeuchi and Yokota, 1989; Tul'skaya, 2009). The polyamine pattern of the type strain grown in rich (R) medium (Yamada and Komagata, 1972) includes cadaverine as the predominant component (54%), followed by putrescine (22%), spermidine (12%), spermine (9%), and 1,3-diaminopropane (2.5%) (Busse and Schumann, 1999). The major isoprenoid quinone is MK-8(H_4), with isoprenyl units saturated at sites II and III.

Fatty acid composition depends on the culture conditions and analytical procedures (Miller et al., 1991; Schumann et al., 1997; Suzuki and Komagata, 1983a, 1983c; Yoon et al., 2010; Yoon et al., 1997b). The fatty acids [for cells harvested from nutrient agar (NA, Difco) after 4 and 7 d cultivation at 30°C] included $C_{16:0}$ iso (45.5 and 33.4%), $C_{16:1}$ iso (14.0 and 10.9%), $C_{17:1}$ (8.7 and 10.1%), $C_{18:1}$ (3.2 and 4.6%), $C_{15:0}$ iso (3.7 and 4.0%), $C_{17:0}$ iso (1.7 and 2.1%), $C_{18:0}$ iso (2.7 and 1.2%), $C_{18:0}$ (2.0 and 2.9%), $C_{17:0}$ 10-methyl (5.4 and 5.5%), $C_{16:0}$ 10-methyl (2.5 and 6.4%), TBSA (2.4 and 9.5%), and other fatty acids characteristic of the genus, each contributing 0.3–1.4% (Yoon et al., 2007b; Yoon et al., 2010). The major fatty acids (for cells grown in TSB medium for 24–48 h at 28°C) included $C_{16:0}$ iso (28.6%), $C_{15:0}$ iso (16.7%), $C_{18:1}$ (10.3%), $C_{17:0}$ iso (9.6%) and $C_{17:1}$ (6.5%), with TBSA constituting 2.4% (Schumann et al., 1997). A larger proportion of TBSA (19.1%), along with similar amounts of $C_{17:0}$ 10-methyl (5.3%), $C_{16:0}$ 10-methyl (3.6), and $C_{16:0}$ iso (34%), was reported by Miller et al. (1991). In some experiments, 2-hydroxy fatty acids may be detected (O'Donnell et al., 1982; Suzuki and Komagata, 1983c; Suzuki and Komagata, 1983a). Diphosphatidylglycerol, phosphatidylglycerol, phosphatidylinositol and supposedly hydroxy-phosphatidylglycerol were the principal polar lipids (Collins et al., 1989). Susceptible to some actinophages (Collins et al., 1989) that multiply in mycelium-forming *Nocardioides* strains (Prauser, 1976; Prauser, 1984a; Prauser, 1989). The mean DNA–DNA relatedness was 16% to the type strains of *Nocardioides albus* and *Nocardioides simplex*, 13% to the type strain of *Terrabacter tumescens* (membrane filter method with tritium-labeled DNA; Suzuki and Komagata, 1983c), and 10% to *Nocardioides daedukensis* (Yoon et al., 2010; procedure of Ezaki et al., 1989). The type strain (the only strain described) was isolated as a herbicide-decomposing bacterium (Jensen and Gundersen, 1956).

Source (type strain): soil.
DNA G+C content (mol%): 68.8 (T_m).
Type strain: ATCC 49810, CCUG 37988, CIP 102404, CNF 091, DSM 20641, IAM 12581, IFO (now NBRC) 14755, KCTC 9134, NBRC 14755, JCM 1364, LMG 16325, NCIB (now NCIMB) 9770, VKM Ac-1878.
Sequence accession no. (16S rRNA gene): AF005006, X53214.

Additional remarks: the type strain of this species was originally identified as a strain of *Arthrobacter simplex* (Jensen and Gundersen, 1956). Suzuki and Komagata (1983c) proposed to reclassify it as *Pimelobacter jensenii.* Collins et al. (1989) transferred this species to *Nocardioides* as *Nocardioides jensenii* (see *Further descriptive information* for more details).

29. **Nocardioides kongjuensis** Yoon, Lee, Jung, Kim, Kim and Oh 2006b, 1786[VP]

kong.ju.en'sis. N.L. masc. adj. *kongjuensis* of or belonging to Kongju, Korea, pertaining to the geographical origin of the type strain.

Characteristics are as described for the genus and listed in Table 215, Table 218, Table 219, and Table 221. Information presented below is from Yoon et al. (2009, 2006b), unless indicated.

Cells are irregular rods (0.4–0.7 × 0.8–3.0 μm) in the exponential phase of growth and exhibit rod-to-coccus morphogenesis from the early exponential phase to the stationary phase. Neither substrate nor aerial mycelium is formed. Colonies are yellowish white in color, reaching 3.0–5.0 mm diameter after 3 d incubation on nutrient agar (NA; Difco) at 30°C. Grows at 10 and 40°C, but not at 4 or 41°C, at pH 5.5, but not at pH 5.0 [optimum, 7.0–8.0; nutrient broth (NB; Difco) with the pH adjusted with HCl or Na_2CO_3 prior to sterilization], and in up to 5% (w/v) NaCl, with optimal growth in the absence of NaCl (trypticase soy broth prepared according to the Difco formula without NaCl).

Nutritionally non-exacting; grows on a suitable mineral salts medium (e.g. ISP 9 medium) with glucose as sole carbon-plus-energy source and an ammonium salt as sole nitrogen source. The organism is capable of degrading *N*-hexanoyl-L-homoserine lactone and using this compound as a sole source of carbon and energy for growth. L-Arabinose, D-mannose (weak), malate, succinate, and pyruvate are utilized as carbon sources, but not D-fructose, acetate, benzoate, citrate, and formate (basal medium ISP 9). In the API 20NE or API 50 CH (AUX as suspending medium) test, D-fructose, adipate, citrate (weak), gluconate, and malate were used, but not D- and L-arabinose, D- and L-fucose, gentiobiose, D-lyxose, D-mannose, D-melibiose, L-rhamnose, L-sorbose, D-tagatose, D-turanose, L-xylose, D-adonitol, D-and L-arabitol, dulcitol, erythritol, D-sorbitol, xylitol, methyl α-D-mannoside, methyl α-D-glucoside, methyl β-D-xyloside, *N*-acetylglucosamine, amygdalin, arbutin, esculin, starch, glycogen, 2- and 5-ketogluconate, caprate, and phenylacetate. In conventional tests with mineral salts medium supplemented with vitamins and element solution, this organism used L-alanine, L-histidine, L-proline, L-serine, acetate, 3- and 4-hydroxybenzoic acids, 3-hydroxybutyrate, malic acid, propionic acid, suberic acid, and valeric acid as sole carbon sources for growth, but not lactate, itaconate, and malonate (Cui et al., 2009). Tweens 20, 40, and 60 are hydrolyzed. Acetoin is produced (Cui et al., 2009). For cells harvested from NB agar after incubation at 30°C for 6 d, the fatty acids (5% or more of the total) included $C_{16:0}$ iso (27.1%), TBSA (14.4%), $C_{17:0}$ 10-methyl (6.0%), $C_{17:1}$ ω6*c* (11.2%), $C_{18:1}$ ω9*c* (5.0%), and $C_{17:0}$ iso (6.3%). A different fatty acid profile was reported for cells grown on trypticase soy agar (Difco) at 30°C for 2 d (Table 219). The fatty acids composition of the type strain of *Nocardioides kongjuensis* and type strains of related species are compared in Table 219. A 21–46% DNA–DNA similarity was found between the type strain of *Nocardioides kongjuensis* and the type strains of phylogenetically closest species (Table 221). Selected phenotypic characteristics differentiating *Nocardioides ginsengisoli, Nocardioides caeni,* and other phylogenetically related species are listed in Table 218. The type strain (the only strain described) was isolated using minimal medium (Leadbetter and Greenberg, 2000) supplemented with *N*-hexanoyl-L-homoserine lactone as the carbon source and NH_4Cl as the nitrogen source; pH 6.5 (for further details on the isolation procedure see Yoon et al., 2006b).

Source (type strain): soil, Kongju, South Korea.

DNA G+C content (mol%): 72.1 (HPLC).

Type strain: A2-4, KCTC 19054, JCM 12609.

Sequence accession no. (16S rRNA gene): DQ218275.

30. **Nocardioides koreensis** Dastager, Lee, Ju, Park and Kim 2008f, 2294[VP]

ko.re.en'sis. N.L. masc. adj. *koreensis* of or pertaining to Korea, the geographical origin of the type strain.

Characteristics are as described for the genus and listed in Table 215. Information presented below is based on the original species description (Dastager et al., 2008f).

Cells are typically irregular rods (mostly 0.4–0.7 μm wide, up to 3.2 μm long). The irregular rods do not exceed 2 μm in length and coccoid cells occasionally occur after 7 d of incubation in R2A medium (Difco) at 28°C (Figure 255). Branching cells may be observed. Motility is reported. Colonies are cream to whitish in color, flat, and 0.9–1.4 mm in diameter after 4–5 d incubation in R2A medium at 30°C. No aerial or substrate mycelium is formed. Optimum growth is at pH 7.0–8.0 and 30°C. Grows at 27–37°C, but not at 25°C or above 37°C, and in up to 5% (w/v) NaCl (in twofold-diluted R2A medium). The carbon sources for growth tested according to Kämpfer (1991) by using mineral salts medium supplemented with yeast extract, Oxoid (0.02 g/l), bio-Lactysat, bioMérieux (0.02 g/l), a vitamin solution, and trace element solution (Kämpfer et al., 1990) are listed in Table 215. The major fatty acids reported for cells harvested from twofold-diluted R2A broth (pH 7.5) after incubation for 10 d at 28°C included (% of the total): $C_{16:0}$ iso (62.9%), $C_{18:1}$ ω9*c* (5.9%), $C_{14:0}$ iso (4.0%), $C_{17:0}$ 10-methyl (3.4%), $C_{16:0}$ (3.3%), $C_{16:1}$ iso (3.1%). Neither TBSA nor hydroxylated fatty acids were reported among components equal or exceeding 1% of the total fatty acids. The type strain (and the only strain described) was isolated using a tenfold-diluted R2A.

Source (type strain): soil, Bigeum Island, South Korea.

DNA G+C content (mol%): 69.9 (HPLC).

Type strain: MSL-09, DSM 19266, JCM 16022, KCTC 19272.

Sequence accession no. (16S rRNA gene): EF466115.

31. **Nocardioides kribbensis** Yoon, Kim, Lee and Oh 2005b, 1614[VP]

krib.ben'sis. N.L. masc. adj. *kribbensis* pertaining to KRIBB, arbitrary adjective formed from the acronym of the Korea Research Institute of Bioscience and Biotechnology, KRIBB, where taxonomic studies on this species were performed.

Characteristics are as described for the genus and listed in Table 215. Information presented below from the original publication (Yoon et al., 2005b).

Cells are irregular rods (0.8–1.0 × 1.5–2.0 µm) in the exponential phase of growth and can undergo a rod-to-coccus morphogenetic cycle from the early exponential phase to the stationary phase. Neither substrate nor aerial mycelium is formed. Gram-stain-positive; Gram-staining is variable in old cultures. Colonies are cream in color and 1.0–1.5 mm in diameter after 6 d incubation on twofold diluted NA (pH 9.0). Grows at 4 and 35°C but not at 36°C, and in up to 3% (w/v) NaCl. Growth occurs at pH 6.0 and 11.0, with optimum growth at pH 9, but not at pH 5.5 or pH 12 [in twofold diluted NA adjusted to alkaline pH with Na_2CO_3 (below pH 10.5) or KOH (above pH 10.5)]. Nutritionally non-exacting; grows on a suitable mineral salts medium (e.g. on ISP 9) with glucose as sole carbon-plus-energy source and an ammonium salt as sole nitrogen source. D-Sorbitol but not adonitol is used as a carbon source. Tweens 20, 40, and 60 are hydrolyzed. The predominant fatty acids of three strains of this species [harvested from twofold-diluted NA (pH 9.0) after incubation for 5 d at 30°C] were almost identical and that of the type strain (% of the total) were: $C_{16:0}$ iso (42.5), $C_{15:0}$ iso (9.7), $C_{18:1}$ ω9*c* (7.2), $C_{17:0}$ iso (5.7), $C_{14:0}$ iso (5.4), $C_{17:0}$ anteiso (4.2), $C_{16:0}$ (3.6), and TBSA (1.2%). DNA–DNA relatedness between the three strains of this species were 88–93%. Mean DNA–DNA hybridization values of 28.7% and 18% were reported between the type strain of *Nocardioides kribbensis* and type strains of *Nocardioides basaltis* and *Nocardioides salarius*, respectively (Kim et al., 2008a; Kim et al., 2009a). DNA–DNA relatedness of strains of *Nocardioides kribbensis* to the type strain of *Nocardioides aquiterrae* was 8–15% (Yoon et al., 2005b). The three described strains of this species were isolated on tenfold-diluted NA with pH adjusted to 10.0 using Na_2CO_3.

Source (type strain): alkaline soil, Korea.

DNA G+C content (mol%): 73–74 (HPLC); 74 for type strain.

Type strain: KSL-2, DSM 16314, JCM 13594, KCTC 19038.

Sequence accession no. (16S rRNA gene): AY835924.

32. **Nocardioides lentus** Yoon, Lee and Oh 2006a, 274[VP]

len'tus. L. masc. adj. *lentus* slow, delayed, referring to slow growth.

Characteristics are as described for the genus and listed in Table 215. Characteristics presented below taken from the original description (Yoon et al., 2006a).

Cells are irregular rods (0.4–0.7 × 1.0–4.5 µm) in the exponential phase of growth and can undergo rod-to-coccus morphogenesis from the early exponential phase to the stationary phase. Neither substrate nor aerial mycelium is formed. Gram-stain-positive; staining is variable in old cultures. Colonies are yellow in color and 0.5–1.0 mm in diameter after 10 d incubation on twofold diluted nutrient agar (NA; Difco) at 28°C. Grows at 4 and 34°C, but not at 35°C, at pH 6.5 and 9.5, but not at pH 6.0 or 10.0 [initial pH of the test medium, twofold diluted nutrient broth (Difco) adjusted prior to sterilization with HCl or Na_2CO_3], and optimally at pH 8.0. Growth occurs in up to 5% (w/v) NaCl, and is optimal in 0.5% [w/v; as assessed using trypticase soy broth (Difco)]. Nutritionally non-exacting; grows on a suitable mineral salts medium (e.g. ISP 9 medium) with glucose as sole carbon-plus-energy source and an ammonium salt as sole nitrogen source. Lactose and D-xylose are also used as carbon sources by some strains, including the type strain. Utilizes D-sorbitol but not adonitol as a carbon source. Positive for hydrolysis of Tweens 20, 40, and 60; hydrolysis of tyrosine is variable among strains and positive for the type strain. The fatty acids of the type strain (cells harvested from twofold diluted NA [pH 8.0] after incubation for 10 d at 28°C) included $C_{16:0}$ iso (60.9%), $C_{17:1}$ ω8*c* (6.6%), and $C_{17:0}$ 10-methyl (5.8%) as well as TBSA (1.6%), a hydroxylated fatty acid ($C_{17:0}$ iso 3-OH; 2.7%) and minor amounts of other fatty acids characteristic of *Nocardioides*. One strain of this species, KSL-18, was found to contain $C_{16:0}$ iso in the highest proportion (81.9%) reported for the genus *Nocardioides* so far. Strains KSL-17, KSL-18, and KSL-19 are identical in their 16S rRNA gene sequences and exhibit 85–90% DNA–DNA hybridization to each other. The type strain and both reference strains described were isolated by the standard dilution plating technique on tenfold-diluted NA with the pH adjusted to 9.0 using Na_2CO_3 at 30°C.

Source (type strain): alkaline soil, Kwangchun, Korea.

DNA G+C content (mol%): 74.6–74.8 (HPLC).

Type strain: KSL-17, DSM 16315, JCM 14046, KCTC 19039.

Sequence accession no. (16S rRNA gene): DQ121389 for the type strain and DQ121390 and DQ121391 for the reference strains KSL-18 and KSL-19, respectively.

33. **Nocardioides luteus** Prauser 1985, 223[VP] (Effective publication: Prauser 1984b, 647.) ("*Nocardioides flavus*" Ruan and Zhang 1979, 347; "*Nocardioides fulvus*" Ruan and Zhang 1979, 350.)

lu'te.us. L. masc. adj. *luteus* yellow, referring to the yellow primary mycelium of the type strain.

Characteristics are as described for the genus and listed in Table 215. Information presented below is based on the author's description (Prauser, 1984b, 1989) and recent observations, unless indicated.

The branching hyphae of the primary mycelium are usually yellow on most ISP media to orange-yellow (oatmeal agar, glycerol-nitrate agar) or cream to faintly yellowish-brown on some peptone-containing media, particularly in aged cultures. The aerial mycelium is white or cream-colored if well developed. No pronounced soluble pigments are usually produced, but a light reddish-brown or light yellowish-brown pigment may be observable in older cultures on some media. Colonies lacking aerial mycelium are usually pasty on most media, with smooth to wrinkled and dull to bright surfaces. Growth is optimum at about 28°C, observable at 10°C and 42°C, and may occur at 45°C but not at 50°C. The organism grows well in up to 8% (w/v) NaCl (weak growth at 9%) and at initial pH values of 5–9, but not at pH 10, as assessed using a complex medium, PYGP (peptone, 0.5%; yeast extract, 0.3%; glucose, 0.5%; K_2HPO_4, 0.02%). Nutritionally non-exacting; grows aerobically on a suitable mineral salts medium (e.g. ISP 9) with glucose or sucrose as carbon-plus-energy source and an ammonium salt or nitrate as sole nitrogen source. Growth and biomass production is enhanced in the presence of vitamins (Lawson et al., 2000a) and traces of yeast extract or peptone. Acids are produced from D-glucose, L-arabinose, D-xylose, D-mannitol, and D-fructose under aerobic conditions in conventional tests; some strains may produce acids from L-rhamnose and sucrose. However, no acid production from glucose was observed with the API 20NE system (Yi and Chun, 2004a). Citrate, succinate, and benzoate but not tartrate are used as carbon sources. The type strain also utilizes acetate but not acetamide, L-arginine, L-asparagine, L-lysine, L-ornithine, and D-sorbitol as a sole carbon sources (Yi and Chun, 2004a), uses azelate, malonate, suberate, L-proline, tetradecane, and phenol as sole carbon and energy sources (Collins et al., 1994). Uses D-fructose, L-rhamnose and sucrose as carbon sources, and can hydrolyze xanthine, but the test results may vary between experiments or test conditions (Prauser, 1984b; Yi and Chun, 2004a). Relatively weak but distinct oxidase activity is observable (assessed with tetramethyl-*p*-phenylenediamine in young culture). Production of α-mannosidase was reported (Yi and Chun, 2004a).

Galactose is the only cell-wall monosugar which is clearly detectable in routine chemotaxonomic studies; traces of glucose and/or rhamnose may also be found (Shashkov et al., 2000b; Tul'skaya, 2009). The cell wall contains a 1,5-poly(ribitol phosphate) teichoic acid (Shashkov et al., 2000b), which is the most essential chemotaxonomic marker differentiating *Nocardioides luteus* from *Nocardioides albus* that possesses a poly (galactosylglycerol phosphate) polymer (see Further descriptive information for details). The major isoprenoid quinone is MK-8(H_4); minor amounts of MK-8(H_2) and MK-8 may be detected (O'Donnell et al., 1982). The type strain of *Nocardioides luteus* harvested from rich (R) medium (Yamada and Komagata, 1972) had nearly 6 μmol/g (dry wt) of polyamines, with the predominant compounds cadaverine (53.7%) and putrescine (28.2%), and lesser amounts of spermidine (11.3%), 1,3-diaminopropane (2.6%), spermine (2.4%), and tyramine (1.7%) (Busse and Schumann, 1999). The cellular fatty acid profile of the type strain of *Nocardioides luteus* is generally similar to that obtained for *Nocardioides albus* under the same experimental conditions (Miller et al., 1991; O'Donnell et al., 1982; Yoon et al., 1997b). In some experiments, a small quantity of a 3-hydroxy fatty acid may additionally occur (Yoon et al., 1997b) and a significantly larger proportion of TBSA may be detected (Miller et al., 1991). The qualitative polar lipid composition of the type strain of *Nocardioides luteus* is also generally reminiscent of that detected in *Nocardioides albus* (Figure 263). Susceptible to taxon-specific *Nocardioides* actinophages. The DNA–DNA hybridization levels between the type strains of *Nocardioides luteus* and *Nocardioides albus* were 49% and 38% in different experiments (Prauser. 1984a, 1989). The DNA–DNA relatedness of *Nocardioides albus* type strain to "*Nocardioides flavus*" 71-N54 (= DSM 46114) and "*Nocardioides fulvus*" 71-N86 (= DSM 46115) was 33% and 41%, respectively (Prauser, 1989).

Source (type strain): soli.

DNA G+C content (mol%): 67.5 (T_m).

Type strain: KCTC 9575, ATCC 43052, CCUG 37986, CIP 103450, DSM 43366, IMET 7830, JCM 3358, LMG 16209, IFO (now NBRC) 14491, VKM Ac-1246.

Sequence accession no. (16S rRNA gene): AF005007, X53212.

Additional remarks: some strains without distinct yellow pigmentation cited in the literature under the species name *Nocardioides albus* might also belong to *Nocardioides luteus* (see the section *Further descriptive information* for more details).

34. **Nocardioides marinisabuli** Lee, Hyun and Lee 2007, 2961[VP]

ma.ri.ni.sa'bu.li. L. adj. *marinus -a -um* of the sea, marine; L. neut. n. *sabulum* gravel, sand; N.L. gen. n. *marinisabuli* of sea sand, referring to the sand sample from which the type strain was isolated.

Characteristics are as described for the genus and listed in Table 215. Data presented below based on information published by Lee et al. (2007), unless indicated.

Cells are irregular rods (0.6–0.8 × 1.4–2.1 μm). Colonies are pale yellow in color approximately 1.0–1.5 mm in diameter after growth on trypticase soy agar (TSA; Difco) for 7 d.

Grows at pH 5.1–12.1, in up to 8% NaCl, as determined in the basal medium ISP 2, at 4–40°C but not at 42°C. Nutritionally non-exacting; grows on a suitable mineral salts medium with glucose as the carbon-plus-energy source and an ammonium salt as the sole nitrogen source (e.g. ISP 9 medium). Citrate and gluconate are assimilated, while D-arabinose, D-sorbitol, acetate, adipate, caproate, citrate, malate, and phenylacetate are not sole carbon and energy sources. Glucose is not fermented (API 20NE). The major fatty acids (in cells harvested from TSA and MA after 3 d incubation at 30°C) were mainly the same components but had slightly different proportions (% of the total): $C_{16:0}$ iso (48.7 and 35.9), $C_{17:0}$ iso (10.8 and 12.3), $C_{18:1}$ ω9*c* (5.6 and 15.4), $C_{17:1}$ ω8*c* (2.8 and 5.6), $C_{17:0}$ anteiso (4.3 and 4.1). Less than 1% TBSA was found in both experiments; the other 10-methyl-branched acid were $C_{17:0}$ 10-methyl (1.7%) and $C_{16:0}$ 10-methyl (<1%) for the MA culture and $C_{17:0}$ 10-methyl (3.2%) for the TSA culture. The fatty acid profile of the type strain of a phylogenetically close species, *Nocardioides basaltis* (grown on MA at 30°C for 3 d and analyzed using the same procedure) differed from that of *Nocardioides marinisabuli* mainly by having a larger proportion of $C_{16:0}$ iso (70%), smaller proportions of $C_{18:1}$ ω9*c* (2.8%), and traces of $C_{17:0}$ iso and $C_{17:0}$ anteiso (<1%). A larger proportion of $C_{16:0}$ iso (65.3%) was also detected in a younger culture of *Nocardioides salarius* grown on MA at 30°C (Kim et al., 2008a). The type strain of *Nocardioides marinisabuli* had DNA–DNA hybridization values of 33±9% and 15.8±1.5%, respectively, with type strains of the most closely related species (99.2% 16S rRNA gene sequence similarity) *Nocardioides salarius* and *Nocardioides basaltis* (Kim et al., 2008a, 2009a) and 23.1±5.6% (Park et al., 2008) and 18 ± 6% (Lee et al., 2007) with type strains of more distant species, *Nocardioides dokdonensis* and *Nocardioides kribbensis* (<97.7% sequence similarity). The type strain (the only strain described) was isolated using ISP 4 medium (Shirling and Gottlieb, 1966) prepared with 60% (v/v) natural sea water.

Source (type strain): sand, Samyang Beach, Jeju Island, Korea.

DNA G+C content (mol%): 73.1 (HPLC).

Type strain: SBS-12, DSM 18965, JBRI 2003, KCCM 42681

Sequence accession no. (16S rRNA gene): AM422448.

35. **Nocardioides marinus** Choi, Kim, Noh and Cho 2007, 778[VP]

ma.ri'nus. L. masc. adj. *marinus* of or belonging to the sea, from where the type strain was isolated.

Characteristics are as described for the genus and listed in Table 215. The data presented below are from Choi et al. (2007), unless indicated.

Cells are nonmotile small rods (0.4–0.6 × 1.0–1.8 µm) in the exponential phase and coccoid forms in the stationary phase. Colonies on marine agar (MA; Difco) are creamy white.

Grows at 10–40°C (optimum, 25–30°C) but not at 5 or 45°C, and in 0.5–8% (w/v) NaCl (optimum, 1–3%), but not 0% NaCl (test medium, ZoBell broth*; ZoBell, 1941). Utilizes L-asparagine, but not benzoate, citrate, succinate, tartrate, acetamide, L-arginine, L-lysine, and L-ornithine as a sole or principal carbon source for growth and energy. Glucose is not fermented (API 20NE). The major fatty acids in cells grown on MA agar plates at 30°C for 1 d included $C_{16:0}$ iso (71.5%), $C_{17:0}$ 10-methyl (5.1%), $C_{18:1}$ ω9*c* (4.4%), and $C_{17:1}$ ω8*c* (3.8%). Other components of the fatty acid profile were various saturated, monounsaturated, iso-, anteiso- and 10-methyl-branched fatty acids (0.3–2.1%), including TBSA (1.7%) and $C_{15:0}$ iso 2-OH and/or $C_{16:1}$ ω7*c* (1.3%). Level of DNA–DNA similarity (thermal denaturation and renaturation method) between the type strain of this species and the type strain of *Nocardioides terrae* was 16% (Zhang et al., 2009a). The type strain (the only strain described) was isolated using a selective S medium† (Fialho et al., 1999).

Source (type strain): sea water around Dokdo Island, the East Sea, Korea.

DNA G+C content (mol%): 72.9 (HPLC).

Type strain: CL-DD14, JCM 15615, KCCM 42321, DSM 18248.

Sequence accession no. (16S rRNA gene): DQ401093.

36. **Nocardioides mesophilus** Dastager, Lee, Pandey and Kim 2010, 2291[VP]

me.so.phi'lus. Gr. adj. *mesos* middle; Gr. adj. *philos* loving; N.L. masc. adj. *mesophilus* middle (temperature)-loving, i.e. mesophilic.

Characteristics are as described for the genus and listed in Table 215 and Table 216. Information presented below is taken from the original species description (Dastager et al., 2010).

Cells are slender short rods, 0.3–0.8 × 0.9–1.4 µm on R2A agar. Motility but no flagella were reported. Colonies are translucent, cream-whitish, and slightly raised after 2 d incubation at 28°C on twofold-diluted R2A agar. Optimum growth temperature and pH are 28°C and 7.0–7.5, respectively. Grows at 20–37°C. Acid is not produced from any of the carbon sources tested. Nutritionally non-exacting; grows on a mineral salts medium (ISP 9) with glucose as the carbon-plus-energy source and an ammonium salt as the sole nitrogen source. The major cellular fatty acids in cells harvested from R2A plates after incubation at 28°C for 7 d included $C_{16:0}$ iso (44.6%), $C_{16:1}$ (12.1%), $C_{14:0}$ iso (11.0%), $C_{16:0}$ (6.5%), $C_{18:1}$ ω9*c* (3.1%), $C_{16:0}$ 10-methyl (2.8%), $C_{15:0}$ iso (2.3%), and $C_{15:0}$ (2.2%). No TBSA was reported. DNA–DNA relatedness between the type strains of *Nocardioides mesophilus* and *Nocardioides jensenii* was 34±2.0% (Dastager et al., 2010). The type strain (and the only strain described) was isolated by serial dilution plating on tenfold-diluted R2A agar after 7 d incubation at 28°C.

Source (type strain): soil, Bigeum Island, Jeollanam-do Province, Korea.

DNA G+C content (mol%): 68.7 (HPLC).

Type strain: MSL-22, DSM 19432, KCTC 19310.

Sequence accession no. (16S rRNA gene): EF466117.

37. **Nocardioides nitrophenolicus** Yoon, Cho, Lee, Suzuki, Nakase and Park 1999, 679[VP]

ni.tro.phe.no'li.cus. N.L. n. *nitrophenol* nitrophenol; L. masc. suff. *-icus* suffix used with the sense of pertaining to; N.L. masc. adj. *nitrophenolicus* relating to nitrophenols.

Characteristics are as described for the genus and listed in Table 215, Table 218, Table 219, and Table 221. Information presented below is from Yoon et al. (1999, 2009, 2006b), unless indicated.

Cells are irregular rods (0.5–0.8 × 1.0–3.0 µm) in 2 d culture on nutrient agar and can undergo rod-to-coccus morphogenesis from the early exponential phase to the stationary phase. Motile, a single flagellum was observed (Figure 261). Neither substrate nor aerial mycelium is produced. Gram-stain-positive but the Gram staining test result is variable in old cultures. Colonies are yellowish white on nutrient agar (NA; Difco). Grows at 15 and 40°C (optimum, 30°C) but not at 45 or 10°C, as assessed after one-month incubation on nutrient agar (NA, Difco), and at initial pH 6 and 10 (optimum, pH 8; tested on NA).

Nutritionally non-exacting; grows on a suitable mineral salts medium (e.g. ISP 9) with glucose and an ammonium salt as sole carbon and nitrogen sources. The organism is capable of utilizing phenol, *p*-nitrophenol, and dibenzofuran (at suitable concentrations in the test medium) as sole carbon sources for growth (Cui et al., 2009; Kubota et al., 2005a; Yoon et al., 1999; Table 218). D-Mannose (weak), acetate, fumarate, succinate are used as sole carbon and energy sources for growth with ISP9 as the basal medium, but not L-arabinose, lactose, maltose, D-melibiose, adonitol, D-mannitol, D-sorbitol, citrate, and benzoate. The organism can also grow on a mineral salts medium with L-arginine, but not with tartrate, L-asparagine, L-lysine, L-ornithine, acetamide, D-salicin, and *N*-acetylglucosamine as sole carbon sources (Yi and Chun, 2004a). In API 20NE or API 50 CH tests with AUX as suspending medium, adipate, gluconate, and malate were used as sole carbon sources but not D- and L-arabinose, D- and L-fucose, gentiobiose, D-lactose, D-melibiose, L-xylose, D-sorbose, D-tagatose, adonitol, D- and L-arabitol, dulcitol, erythritol, sorbitol, xylitol, methyl α-D-mannoside, methyl

*Zobell Broth: 5 g of Bacto peptone, 1 g of yeast extract, 0.1 g of ferric citrate, and 1 liter of distilled water.

†S medium: 10 g of Na_2HPO_4, 3 g of KH_2PO_4, 1 g of K_2SO_4, 30 g of NaCl, 0.2 g of $MgSO_4 \cdot 7H_2O$, 0.01 g of $CaCl_2$, 0.001 g of $FeSO_4 \cdot 7H_2O$, 1 g of Casamino acids, 1 g of yeast extract, 20 g of glucose, 20 g of Bacto agar, and 1 liter of distilled water.

α-D-glucoside, *N*-acetylglucosamine, amygdalin, arbutin, esculin, starch, glycogen, citrate, 2-and 5-ketogluconate, caprate, and phenylacetate. According to Cui et al. (2009), the type strain grows with L-alanine, L-histidine, L-proline, L-serine, lactate, caprate, 3- and 4-hydroxybenzoic acids, 3-hydroxybutyrate, malic acid, propionate, suberate, valerate, as well as with maltose, D-mannose, D-mannitol, melibiose, and salicin as carbon and energy sources but not with adipate, itaconate, and malonate (tested on mineral salts medium supplemented with vitamins and an element solution). Acetoin is produced (Cui et al., 2009). According to Yi and Chun (2004a), the type strain cannot produce acid from glucose (API 20NE kit). Arbutin is weakly hydrolyzed. Elastin is not degraded.

Major fatty acids in cells harvested from NA after 4 d incubation at 30°C were $C_{16:0}$ iso (28.1%), TBSA (13.9%), $C_{17:1}$ ω6*c* (10.8%), $C_{17:0}$ iso (12.3%), $C_{17:0}$ anteiso (5.8%), $C_{17:0}$ 10-methyl (5.1%), $C_{16:0}$ 10-methyl (3.6%), $C_{18:1}$ ω9*c* (4.5%), and $C_{15:0}$ iso (3.1%). The major fatty acids for cells grown on trypticase soy agar (Difco) at 30°C for 2 d (Table 219) differed in having larger proportions of saturated iso-branched acids (61.7% in total, with $C_{16:0}$ iso contributing 39.8%) and saturated straight-chain acids (14.4%), smaller proportions of total 10-methyl-branched acids (5.7%) and the absence of 2-hydroxylated acids. *Nocardioides nitrophenolicus* is substantially distinguished from *Nocardioides simplex* and other species of this genus by the length of the 16S–23S ITS region (328 bp, the shortest among the analyzed strains) and the nucleotide sequences (Yoon et al., 1998a). Distinct differences from other species, including *Nocardioides simplex*, was also revealed in the RNase P RNA gene sequences (Yoon and Park, 2000). The DNA–DNA similarity values (Table 221) between the type strain of *Nocardioides nitrophenolicus* and the type strains of the closest species ranged from 22 to 49% (Cui et al., 2009; Kubota et al., 2005a; Yoon et al., 1999; Yoon et al., 2009; Yoon et al., 2006b). Selected phenotypic characteristics useful in distinguishing *Nocardioides nitrophenolicus* from phylogenetically close species are listed in Table 218. The type strain (the only strain described) was isolated using minimal salts medium supplemented with a trace element solution and *p*-nitrophenol for enrichment and subsequent isolation (for further details on the isolation procedure see Yoon et al., 1999).

Source (type strain): industrial wastewater, Cheong-ju, South Korea.

DNA G+C content (mol%): 71.4 (HPLC).

Type strain: NSP41, CIP 107017, DSM 15529, KCTC 0457BP, JCM 10703.

Sequence accession no. (16S rRNA gene): AF005024.

38. **Nocardioides oleivorans** Schippers, Schumann and Spröer 2005, 1502[VP]

o.le.i.vo'rans. L. n. *oleum* oil; L. v. *vorare* to devour; N.L. part. adj. *oleivorans* capable of utilizing oil (hydrocarbons).

Characteristics are as described for the genus and listed in Table 215 and Table 220. Information presented below is from the original description (Schippers et al., 2005), unless indicated.

Nonmotile, irregular rods (about 0.3 × 1.1 μm). Colonies (maximum diameter after 2 weeks, 2 mm) are translucent and orange-pigmented. Grows at 30°C and in 2% NaCl. According to the original species description, *N*-acetyl-D-glucosamine, acetate, propionate, fumarate, DL-3-hydroxybutyrate, DL-lactate, L-malate, pyruvate, L-aspartate, L-histidine, L-proline, putrescine, phenylacetate, L-ornithine, and other substrates but not aconitate, adipate, citrate, α-D-galacturonate, 3-hydroxybenzoate, 4-hydroxybenzoate, suberate, adonitol, sorbitol, L-alanine, L-hydroxyproline, L-serine, glycogen, and *N*-acetyl-D-galactosamine are utilized as sole carbon and energy sources (tested using a basal medium containing vitamins and trace elements (Kämpfer et al., 1991). Crude oil is also used as a growth substrate. Acid is not produced from D-glucose, rhamnose, sucrose, adonitol, inositol, xylose, or sorbitol under aerobic conditions in conventional tests (Kämpfer et al., 1991). Glucose is not fermented (API 20NE; Lee, 2007b). The following substrates are hydrolyzed: *p*-nitrophenyl (pNP) *N*-acetyl-β-D-glucosaminide, pNP β-D-galactopyranoside, pNP α-D-glucopyranoside, pNP β-D-glucopyranoside, pNP α-D-maltoside, bis-pNP phosphate, benzolphosphonacid pNP-ester, pNP phosphocholine, 2-deoxythymidine 5′-pNP phosphate, L-alanine *p*-nitroanilide (pNA), L-glutamate pNA, L-glutamate-γ-3-carboxy pNA, L-leucine pNA, and L-lysine pNA, whereas pNP *N*-acetyl-β-D-galactosaminide, pNP α-L-arabinopyranoside, pNP β-D-cellobioside, pNP β-D-glucuronide, pNP β-D-lactoside, pNP α-D-mannoside, pNP β-D-xyloside, glycine pNA, L-proline pNA, and L-valine pNA are not hydrolyzed. Some additional characteristics of this organism were reported by Lee (2007b). The cell-wall peptidoglycan is of A3γ-type (LL-A_2pm-glycine). MK-8(H_4) is the predominant menaquinone and MK-8(H_2) is a minor menaquinone component. The major fatty acids determined for cells grown in basal medium* included $C_{18:1}$ ω9*c* (27.5%), $C_{16:0}$ iso (18.6%), TBSA (9.0%), $C_{18:0}$ (7.5%), $C_{17:0}$ iso (6.6%), and $C_{17:1}$ ω6*c* (5.7%).

The DNA–DNA relatedness (using the spectrophotometric method) of the type strain of *Nocardioides oleivorans* to the type strain of the phylogenetically closest species, *Nocardioides ganghwensis* (99% 16S rRNA gene sequence similarity) was 32% whereas its relatedness to *Nocardioides exalbidus* (using the method of Ezaki et al., 1989) was 44% (Li et al., 2007b). Selected phenotypic characteristics useful in distinguishing *Nocardioides oleivorans* from phylogenetically closely related species are listed in Table 220. The type strain (the only strain described) was isolated by enrichment in mineral salts medium (Fedorak and Westlake, 1981) supplemented with crude oil (30°C; several weeks) and subsequent isolation on the complex oil-free medium (Bosecker et al., 1991; Schippers et al., 2005).

Source (type strain): crude oil sample, the oilfield Oerrel of the Gifhorn Trough, Germany.

DNA G+C content (mol%): not determined.

Type strain: BAS3, DSM 16090, JCM 14342, NCIMB 14004.

Sequence accession no. (16S rRNA gene): AJ698724.

*Basal medium: 23.4 g of NaCl, 0.75 g of KCl, 7.0 g of $MgSO_4 \cdot 7H_2O$, 0.5 g of peptone from meat, 0.5 g of peptone from casein, 1.0 g of yeast extract, and 18 g of agar per 1 liter of water; pH 7.3.

39. **Nocardioides panacihumi** An, Im, Lee and Yoon 2007, 2145[VP]

pa.na.ci.hu'mi. N.L. n. *Panax* scientific name of ginseng; N.L. gen. n. *panacis* of ginseng; L. n. *humus* soil; N.L. gen. n. *panacihumi* of soil of a ginseng field.

Characteristics are as described for the genus and listed in Table 215. Other characteristics presented below are from the original publication (An et al., 2007), unless indicated.

Cells are nonmotile, small irregular rods (0.3–0.5 × 0.7–1.2 μm), forming white colonies, after 3 d incubation on R2A agar at 30°C. Undergo rod-to-coccus morphogenesis from the early exponential phase to the stationary phase. No growth occurs on MacConkey agar.

Grows well at 30°C and at pH 7.0; growth occurs at 15 and 30°C and at pH 5.0–8.0, but not at 4 and 37°C and at pH 8.5. Prefers low salinity, growing on R2A agar in 1% (w/v) NaCl, but not in 2%. Acetate, gluconate, glycogen, 3-hydroxybenzoate, 4-hydroxybenzoate, 3-hydroxybutyrate, propionate, valerate, phenylacetate, L-alanine, and L-proline but not L-fucose, D-sorbitol, adipate, caprate, citrate, itaconate, 2- and 5-ketogluconate, L-lactate, malate, malonate, suberate, L-histidine and L-serine are utilized as sole carbon sources. Acids are produced from D-glucose, D-fructose, D-lyxose, D-raffinose, L-rhamnose, sucrose, D-turanose, and *N*-acetylglucosamine (API 50 CH), but not from D- and L-fucose, gentiobiose, D-lactose, D-lyxose, D-mannose, melezitose, melibiose, L-sorbose, D-tagatose, L-xylose, adonitol, arabitol, dulcitol, erythritol, glycerol, inositol, D-mannitol, D-sorbitol, xylitol, gluconate, 2- or 5-ketogluconate, starch, glycogen, arbutin, inulin, salicin, methyl α-glucopyranoside, methyl α-D-mannoside, or methyl β-D-xyloside. In API 50 CH tests, acids are also not produced from cellobiose, D-galactose, maltose, D-ribose, and trehalose (characteristics that differentiate this species from the phylogenetically close species *Nocardioides terrae*; Zhang et al., 2009a). Glucose is not fermented (API 20NE). Nitrate reduction is weak under aerobic conditions. Catalase-positive (Cho et al., 2010), oxidase negative (Zhang et al., 2009a). Xylan and chitin are not degraded. The major fatty acids of cells harvested from R2A agar after incubation for 6 d at 30°C were $C_{16:0}$ iso (45.5%), $C_{17:1}$ ω6*c* (7.5%), $C_{16:0}$ (6.4%), and $C_{17:0}$ (5.9%); the minor fatty acids were TBSA and its homologs, as well as $C_{15:0}$ iso 3-OH. The DNA–DNA relatedness (thermal denaturation and renaturation method) of the type strains of *Nocardioides panacihumi* and *Nocardioides terrae* was 21% (Zhang et al., 2009a). To isolate the type strain (the only strain described), the sample was suspended in 50 mM phosphate buffer (pH 7.0) and seeded on one-fifth-strength R2A agar plates after serial dilution in the same buffer.

Source (type strain): soil of a ginseng field, Pocheon Province, South Korea.

DNA G+C content (mol%): 73 (HPLC).

Type strain: Gsoil 616, DSM 18660, JCM 15309, KCTC 19187.

Sequence accession no. (16S rRNA gene): AB271053.

40. **Nocardioides panacisoli** Cho, Lee, An, Whon and Kim 2010, 390[VP]

pa.na.ci.so'li. N.L. n. *Panax -acis* scientific name of ginseng; L. n. *solum* soil; N.L. gen. n *panacisoli* of soil of a ginseng field, the source of isolation of the organism.

Characteristics are as described for the genus and listed in Table 215. Information presented below is taken from the original species description (Cho et al., 2010).

Cells are slender short rods (0.2–0.4 × 0.8–1.2 μm) after 3 d in culture on R2A agar (Difco) at 30°C, forming light yellowish and convex colonies. Will grow within the temperature range of 10–42°C but not at 4 or 45°C, with the best growth at 18–37°C. Grows at pH 5.5–8.5 (on R2A agar prepared with suitable buffers). Grows on R2A agar with 2.0% (w/v) but not 3.0% NaCl. Grows in mineral salts medium (1.8 g of K_2HPO_4, 1.08 g of KH_2PO_4, 0.5 g of $NaNO_3$, 0.5 g of NH_4Cl, 0.1 g of KCl, 0.1 g of $MgSO_4$, and 0.05 g of $CaCl_2$ per 1 liter), supplemented with glucose, vitamins (Widdel and Bak, 1992), trace element solution SL-10 (Widdel et al., 1983), and selenite/tungstate solution (Tschech and Pfennig, 1984). Acetate, adipate, citrate, 3-hydroxybutyrate, gluconate, L-malate, L-proline, propionate, and valerate are also used as a sole carbon source for growth-plus-energy as tested on the above basal medium, but not the following compounds: L-alanine, caprate, L-fucose, glycogen, L-histidine, 3-hydroxybenzoic acid, itaconate, 5-ketogluconate, L-lactic acid, malonate, phenylacetate, L-serine, D-sorbitol, and suberate. The organism produces H_2S from sodium thiosulfate. Xylan or chitin is not degraded. Neither acid nor gas is produced from glucose (API 20E). Xylan or chitin is not degraded. Major fatty acids (>5% of the total) determined in cell mass scraped from trypticase soy agar Difco) after incubation at 30°C for 2 d included $C_{16:0}$ iso (28.0%), $C_{18:1}$ ω9*c* (10.7%), $C_{17:1}$ ω8*c* (9.0%), $C_{17:1}$ ω6*c* (5.8%), TBSA (6.7%), and $C_{16:0}$ (6.1%). The type strain (and the only strain described) was isolated according to the procedure and medium described by Cui et al. (2007b) for isolation of *Nocardioides ginsengisoli*; the plates were incubated at 30°C for 1 month under aerobic conditions.

Source (type strain): soil of a ginseng field in Pocheon Province, South Korea (37°58′N; 127°15′E).

DNA G+C content (mol%): 73.0 (HPLC).

Type strain: Gsoil 346, JCM 16953, KCTC 19470, DSM 21348.

Sequence accession no. (16S rRNA gene): FJ666101.

41. **Nocardioides plantarum** Collins, Cockcroft and Wallbanks 1994, 525[VP]

plan.ta'rum. N.L. gen. pl. n. *plantarum* of plants.

Characteristics are as described for the genus and listed in Table 215. Data presented below based on information published by Collins et al. (1994), unless indicated.

Cells are irregular rods (mean 0.4–0.7 × 0.8–2.4 μm or longer) in young culture and may produce V-forms. In older cultures, coccoid cells usually predominate. Nonmotile. Colonies are typically non-pigmented, but may be light brownish-gray in older cultures on some media. Growth occurs at 5 and 30°C, but not at 37°C (see Table 215) or in the presence of 5% NaCl. The organism grows well on nutrient agar (Difco) (Yi and Chun (2004a) and some other organic media based on yeast extract and peptone. Growth occurs with glucose and other sources (including fumarate, propionate, succinate, and isobutyrate) as sole or principal carbon sources on the basal mineral salts medium containing (per liter) mineral base E (Owens and Keddie, 1969), 0.2 g of Bacto Yeast Extract (Difco), 12 g of agar, 2 μg of vitamin B_{12}, 10 mg of sodium glutamate, and 10 mg of methionine (Collins et al., 1994). Some strains utilize L-proline and tetradecane. Histidine, histamine, malonate, phenol, azelate, suberate, crotonate, pentoate, thymine, and uracil are not utilized. Tween 40 and Tween 60 are hydrolyzed. Hippurate is not decomposed. According to Lawson et al. (2000a), thiamine is requited as the only growth factor in a glucose-mineral salts medium (Grainger, 1963). However, the type strain can also grow with glucose or other substrates as sole carbon sources in minimal media without thiamine (Yi and Chun, 2004a), for instance on mineral salts medium (Baumann et al., 1971) supplemented with 2% (v/v) Hutner's mineral base (Cohen-Bazire et al., 1957) and modified by reducing the concentration of sea salts to half-strength. On the above basal medium, according to Yi and Chun (2004a), the type strain is able to grow with acetate, citrate, L-arginine, and L-ornithine as sole carbon sources, but not with benzoate, tartrate, D-sorbitol, L- and D-asparagine, L-lysine, and acetamide. No acid production from glucose (API 20NE kit) (Yi and Chun (2004a). The main component of the quinone system is MK-8(H_4); a minor amount of MK-7(H_4) is detectable. Neither cell-wall neutral monosugars nor cell-wall teichoic acids were revealed (Tul'skaya, 2009).

Polyamine pattern of the type strain grown in rich (R) medium (Yamada and Komagata, 1972) included cadaverine (33.9%) and spermine (42.4%) as the predominant components, with minor amounts of tyramine (11.9%), 1,3-diaminopropane (6.8%), and spermidine (5.1%) (Busse and Schumann, 1999). The major fatty acids reported for cells harvested after 1-d incubation on marine agar (Difco) at 30°C included $C_{16:0}$ (10.5%), TBSA (8.4%), $C_{18:0}$ (6.8%), $C_{16:0}$ iso (6.2%), $C_{17:0}$ 10-methyl (6.0%), $C_{10:0}$ (5.7%), and $C_{18:1}$ ω9*c* (5.0%) (Yi and Chun, 2004a). The main cellular fatty acids determined for the type strain grown in TSB medium at 28°C for 24–48 h included $C_{16:0}$ iso (26.0%), $C_{17:0}$ anteiso (16.8%), $C_{18:1}$ (15.7%), $C_{17:0}$ (6.1%), and $C_{17:0}$ iso (5.4%) (Schumann et al., 1997). The two strains described (Grainger J70 and Grainger J6) were isolated from herbage as reported by Grainger (1963).

Source (type strain): herbage.

DNA G+C content (mol%): 69 (T_m).

Type strain: Grainger J70, ATCC 51889, CIP 104157, DSM 11054, JCM 9626, LMG 16210, NCIMB 12834, VKM Ac-1998.

Sequence accession no. (16S rRNA gene): AF005008, X69973.

42. **Nocardioides pyridinolyticus** Yoon, Rhee, Lee, Park and Lee 1997b, 935[VP]

py.ri.di.no.ly'ti.cus. N.L. n. *pyridinum* pyridine; N.L. masc. adj. *lyticus* (from Gr. masc. adj. *lutikos*) able to loosen, able to dissolve; N.L. masc. adj. *pyridinolyticus* pyridine-dissolving.

Characteristics are as described for the genus and listed in Table 215. Data presented below based on information published by Yoon et al. (2008, 2004, 1997b), unless indicated.

Cells are small rods (0.5–0.6 × 1.2–1.6 µm) in young culture on minimal salts medium supplemented with 0.05% (w/v) yeast extract. Show rod-to-coccus morphogenesis from the early exponential phase to the stationary phase (Figure 259). Gram-stain-positive, but Gram straining is indistinct in old cultures. Motile, a single flagellum was observed (Figure 260). Neither substrate nor aerial mycelium is produced. Colonies are cream colored on nutrient agar (NA; Difco) and are approximately 1.0–1.5 mm in diameter after 5 d at 35°C. The organism grows well on minimal salts media supplemented with additional growth factors, yeast extract and tryptone, at pH 5 and 9 (optimum, pH 8) and at 20 and 40°C (optimum, 35°C). Utilizes pyridine and phenol as carbon and energy sources for growth, and also acetate, adipate, L-arabitol, turanose and other carbon sources (Table 215) as tested on basal Stevenson's medium* (Stevenson, 1967). The following substrates are not utilized on the same basal medium: D- and L-fucose, gentiobiose, D-lyxose, D-tagatose, sorbose, stachyose, adonitol, dulcitol, erythritol, D-sorbitol, xylitol, caprate, 2-ketogluconate, 5-ketogluconate, malate, phenylacetate, methyl α-D-glucoside, methyl α-D-mannoside, methyl β-D-xyloside, amygdalin, glycogen, esculin, and arbutin. According to Yi and Chun (2004b, 2004a), L-asparagine and succinate, but not benzoate, tartrate, L-arginine, L-lysine, L-ornithine, and acetamide are utilized as carbon and energy sources. Tributylin is hydrolyzed (Song et al., 2011). Data on utilization of D-galactose, lactose, D-mannose and D-mannitol as sole sources of carbon and energy vary between experiments or the test methods (Yi and Chun, 2004a, 2004b; Yoon et al. 1997b, 2004, 2008). Similarly, hydrolysis of tyrosine and production of cysteine arylamidase vary between experiments or test methods (Song et al., 2011; Yoon et al., 2008). Acid production from glucose is not observable with the API 20NE test system (Yi and Chun, 2004a, 2004b). Sensitive to tetracycline (30 µg) and shows resistance to ampicillin (10 µg), chloramphenicol (100 µg), streptomycin (50 µg), and rifampin (30 µg), as assessed using filter-paper discs (Song et al., 2011). The predominant menaquinone is MK-8(H_4); minor or trace amounts of MK-7(H_4) and MK-8(H_2) may occur. The major cellular fatty acids determined for cells harvested after 4 d incubation on nutrient agar (NA; Difco) at 30°C included $C_{16:0}$ iso (47.3%), $C_{17:0}$ anteiso (14.2%), $C_{15:0}$ iso (5.1%), $C_{17:0}$ 10-methyl (5.5%), $C_{16:0}$ 10-methyl (5.0%), TBSA (3.4%), and $C_{16:1}$ iso H (4.2%). A different fatty acid profile was recently reported for cells grown on 1/2-strength R2A agar (1/2-R2A) for 3 d at 30°C (Song et al., 2011). The most remarkable differences were in $C_{14:0}$ iso 3-OH (8.8%), $C_{18:1}$ ω5*c* (8.8%) and minor amounts of cyclic fatty acids, which were not mentioned among fatty acids exceeding 0.5% in cells grown on NA (Yoon et al. 1997a, 2004), as well as a smaller proportion of $C_{16:0}$ iso (22%). The fatty acid profiles of *Nocardioides pyridinolyticus* and phylogenetically close species (*Nocardioides aquiterrae, Nocardioides hankookensis*, and *Nocardioides caricicola*) grown on 1/S-R2A agar are compared by Song et al. (2011). The DNA–DNA relatedness of the type strains of *Nocardioides pyridinolyticus* to it closest phylogenetic relative, *Nocardioides aquaterrae* (99.2% 16S rRNA gene sequence similarity) were

*Stevenson's Medium: Bacto yeast nitrogen base medium without amino acids (Difco) supplemented with Casamino acids (10 mg per liter) and agar (15 g per liter), and neutralized by K_2HPO_4.

found to be 28.7 and 32.5% in reciprocal experiments using the procedure of Ezaki et al. (1989). Mean similarity to *Nocardioides hankookensis* using the same method was 22%. A higher DNA–DNA relatedness (53.5±5.5%) determined by the same method was reported for the pair *Nocardioides pyridinolyticus*–*Nocardioides caricicola* (Song et al., 2011). The characteristics useful in distinguishing *Nocardioides pyridinolyticus* from the phylogenetically closest species, *Nocardioides aquiterrae* and *Nocardioides hankookensis,* include the growth temperature range (higher optimal temperature for growth), cell motility, negative reaction in the test for cytochrome oxidase activity, the ability to grow with inositol, D-melezitose and D-ribose as sole carbon sources, and resistance to chloramphenicol (100 μg per disc), streptomycin (50 μg per disc), and probably the ability to utilize pyridine. The type strain (and the only strain taxonomically characterized) was isolated as described by Lee et al. (1991).

Source (type strain): oxic zone of an oil shale column.

DNA G+C content (mol%): 72.5 (HPLC).

Type strain: OS4, CIP 106800, DSM 15530, KCTC 0074BP, JCM 10369.

Sequence accession no. (16S rRNA gene): U61298.

43. **Nocardioides salarius** Kim, Choi, Hwang and Cho 2008a, 2062[VP]

sa.la'ri.us. L. masc. adj. *salarius* of salt, referring to the salt resistance of this micro-organism.

Characteristics are as described for the genus and listed in Table 215. Additional data presented below are based on information published by Kim et al. (2008a).

Cells are nonmotile rods (0.3–0.6 × 0.6–1.6 μm), forming creamy white colonies on marine agar (MA; Difco) MA at 30°C. Grows at 10–35°C, optimally at 25–30°C; no growth occurs at pH 5 or at temperatures 5 or 40°C. Salt-requiring; grows at NaCl concentrations of 1–10% (w/v), with optimum growth at 3% NaCl, as tested in ZoBell broth (ZoBell, 1941), containing 5 g of Bacto peptone, 1 g of yeast extract, 0.1 g of ferric citrate, and 1 liter of distilled water; no growth is observable at 0.5% (v/w) or without NaCl in the same test medium. Growth occurs in marine broth at initial pH values of 6 and 10 (the highest value examined); the optimum pH for growth is 6–7. A number of carbon sources (Table 215), as well as L-arginine and L-ornithine enhance growth in a basal mineral salts medium* supplemented with 0.05 g of yeast extract. Citrate, succinate, L-asparagine, and L-lysine are not used as sole or principal carbon sources. Glucose is not fermented (API 20NE). The predominant fatty acid detected in cells harvested from MA after 1 d incubation at 30°C was $C_{16:0}$ iso (65.3%), followed by $C_{16:0}$ 10-methyl (5.2%) and other iso-branched acids, such as $C_{16:1}$ iso H (4.6%), $C_{15:0}$ iso (3.4%), $C_{14:0}$ iso (3.2%), and $C_{15:0}$ iso2-OH (and/or $C_{16:1}$ ω7*c*) (3.2%). Components detected in lesser amounts (up to 2.5%) included mostly straight-chain monounsaturated and iso- and anteiso-branched fatty acids, as well as TBSA (0.5%) and $C_{17:0}$ 2-OH (0.7%). A mean DNA–DNA hybridization value of 33±9% was obtained to the type strain of *Nocardioides marinisabuli* (99.2% 16S rRNA gene sequence identity), with a lower DNA–DNA similarity (18 ± 6%) to more distant *Nocardioides kribbensis.* There are no data on the DNA–DNA similarity between *Nocardioides salarius* and the phylogenetically closest (99.6% 16S rRNA gene sequence similarity) species *Nocardioides basaltis* described by Kim et al. (2009a) shortly after the establishment of the species *Nocardioides salarius*). The type strains of these two species show many phenotypic features in common, including the fatty acid composition and physiological and biochemical traits. See the description of the species *Nocardioides basaltis* in this section, Table 215 and the original publications (Kim et al., 2008a, 2009a) for characteristics that might be helpful in distinguishing these two species. The type strain (and the only strain described) was isolated by enrichment culture (sea water supplemented with zooplankton incubated at 10–15°C for about 1 year) on low-nutrient heterotrophic medium (for details on the isolation procedure and the medium composition see Kim et al., 2008a; Cho and Giovannoni, 2004).

Source (type strain): sea water enriched with zooplankton, the South Sea, Korea.

DNA G+C content (mol%): 73.3 (HPLC).

Type strain: CL-Z59, KCCM 42320, DSM 18239.

Sequence accession no. (16S rRNA gene): DQ401092.

Additional remarks: the absence of data on the DNA–DNA similarity between *Nocardioides salarius* and the phylogenetically closest species *Nocardioides basaltis,* along with their very similar phenotypic traits suggest that further comparative taxonomic studies of these two species, are needed to support distinctions between these two species.

44. **Nocardioides sediminis** Dastager, Lee, Ju, Park and Kim 2009e, 281[VP]

se.di.mi'nis. L. gen. n. *sediminis* of a sediment, referring to the source from which the type strain was isolated.

Characteristics are as described for the genus and listed in Table 215. Information presented below is taken from the original species description (Dastager et al., 2009e).

Cells are mostly slender irregular rods (0.3–0.4 × 0.9–1.4 μm) after 3–5 d incubation on R2A agar (Difco) at 28°C. Non-separated coccoid cells and larger club-shaped and spherical forms up to 0.8 μm in diameter may occur. The cells from two-fold diluted R2A agar appeared to be motile under light microscopy but no flagella were detected. Colonies are translucent, cream–whitish, and slightly raised on R2A agar after 2 d growth at 28°C. Nutritionally non-exacting; grows on a mineral salts medium (ISP 9) with glucose as the carbon-plus-energy source and an ammonium salt as the sole nitrogen source. Growth also occurs with turanose but not with adonitol, adipate, D-arabitol, esculin, arbutin, dulcitol, erythritol, D-fucose, gentiobiose, gluconate, glycogen, D-lyxose, malate, sorbose, sorbitol, xylitol, or L-xylose tested under the same experimental conditions. Acid production from carbohydrates is rather uncommon or weak; no acid was formed from any of the tested carbon sources. The predominant fatty acids determined for cell mass harvested from R2A plates after incubation at 28°C for 7 d included $C_{16:0}$ iso (48.3%), $C_{17:1}$ ω8*c* (17.1%), $C_{17:0}$ iso (7%), $C_{17:0}$ 10-methyl (4.8%), and $C_{18:1}$ ω9*c* (4.3%). DNA–DNA

*Mineral salts medium: 11.8 g of NaCl, 0.32 g of KCl, 2.26 g of MgCl2·6H_2O, 2.97 g of $MgSO_4$·7H_2O, 0.65 g of $CaCl_2$·2H_2O, 0.2 g of $NaNO_3$, 0.2 g of NH_4Cl in 1 liter of distilled water; pH 6.7±0.5.

similarity of the two type strains of the species *Nocardioides sedimentis* and *Nocardioides terrigena* was reported to be 34%. The type strain (the only strain described) was isolated using standard dilution techniques on tenfold-diluted R2A at 28°C.

Source (type strain): sediment sample taken at a depth of 60–70 m, Bigeum Island, Jeollanam-do Province, South Korea.

DNA G+C content (mol%): 71.5 (HPLC).

Type strain: MSL-01, DSM 19263, KCTC 19271.

Sequence accession no. (16S rRNA gene): EF466110.

45. **Nocardioides simplex** (Jensen 1934) O'Donnell, Goodfellow and Minnikin 1983, 896[VP] (Effective publication: O'Donnell, Goodfellow and Minnikin 1982, 327.) [*Pimelobacter simplex* (Jensen 1934) Suzuki and Komagata 1983b, 673[VP]; effective publication: Suzuki and Komagata 1983c, 69; *Arthrobacter simplex* (Jensen) Lochhead 1957, 608[AL]; "*Corynebacterium simplex*" Jensen, 1934, 43.]

sim′plex. L. masc. adj. *simplex* simple.

Characteristics are as described for the genus and listed in Table 215, Table 218, Table 219, and Table 221. Additional data are from O'Donnell et al. (1982), Suzuki and Komagata (1983c), and Keddie et al. (1986), unless indicated.

Cells are typically slender irregular rods, exhibiting a rod-to-coccus morphogenesis during the growth cycle. Irregular rods (~0.4–0.5 × 1.0–3.0 µm or longer) in late lag and exponential phase cultures and coccoid cells or very short rods (~0.4–0.5 µm in diameter) in older cultures were reported for the type strain on complex media by O'Donnell et al. (1982). Slightly wider cells (~0.5–0.9 µm in diameter), both in young and older cultures, were reported for this species by Keddie et al. (1986). Rods are typically motile with single- or peritrichous flagella. Colonies on yeast-peptone agar and most ISP media usually show no distinctive pigmentation. On YM agar and some other rich media, they may be slightly yellowish white or ivory in color, particularly in older cultures. A light brownish soluble pigment may occur in older cultures on some peptone-based media. Optimum growth temperature is ~26–30°C; grows at 10°C, and may or may not grow at 37°C. The type strain grows well in 5% NaCl but not in 7% NaCl (Nesterenko et al., 1985b).

Nutritionally non-exacting; grows in a suitable mineral salts medium with glucose as sole carbon-plus-energy source and an ammonium salt or nitrate as sole nitrogen source (Keddie et al., 1966; Owens and Keddie, 1969). The organism utilized about 60 out of the 180 compounds tested as sole or principal sources of carbon and energy. These include a very narrow range of carbohydrates and sugar derivatives, a wide range of fatty acids, simple alcohols, and amino acids, including some hydroxy-acids, oxo-acids, amines, pyrimidines, and other compounds (Keddie, 1974). The type strain utilizes acetate, citrate, succinate, fumarate, acetamide, L-asparagine, L-arginine, L-ornithine (Yi and Chun, 2004b, a), azelate, malonate, L-proline, histamine, thymine, uracil, and tetradecane (Collins et al., 1994) as sole carbon and energy sources, and tartrate in conventional tests. Negative results were recorded for utilization of L-lysine, benzoate, and hippurate(Yi and Chun, 2004b, a). Among carbohydrates tested by Yi and Chun (2004a), only D-glucose and sucrose supported growth. In the API 20NE or API 50 CH test system (AUX as suspending medium), the type strain utilized a few carbon sources for growth and energy, such as D-glucose, sucrose, trehalose, gluconate, adipate, and malate but not many other substrates, including D- and L-arabinose, D-cellobiose, D-fructose, D- and L-fucose, D-galactose, gentiobiose, D-lactose, D-melibiose, D-lyxose, D-maltose, D-mannose, D-melezitose, D-raffinose, L-rhamnose, D-ribose, D-tagatose, D-turanose, D- and L-xylose, D-sorbose, D- and L-adonitol, D- and L-arabitol, dulcitol, erythritol, glycerol, inositol, D-mannitol, D-sorbitol, xylitol, methyl α-D-mannoside, methyl α-D-glucoside, methyl β-D-xyloside, *N*-acetylglucosamine, amygdalin, arbutin, esculin, D-salicin, inulin, starch, glycogen, 2- and 5-ketogluconate, caprate, and phenylacetate (Kubota et al., 2005b; Yoon et al., 2009). Cui et al. (2009) reported the ability of the type strain to grow with L-alanine, L-histidine, L-serine, lactate, caprate, 3- and 4-hydroxybenzoic acids, 3-hydroxybutyrate, malic acid, propionate, suberate, and valerate (on mineral salts medium supplemented with vitamins and element solution); the strain in addition was reported to grow with maltose, D-mannose, melibiose, D-mannitol, salicin, *N*-acetylglucosamine as carbon and energy sources under the same experimental condition.

No acid is produced from glucose or other sugars by oxidation in peptone-based media in conventional tests (Keddie et al., 1986). No acid is produced from glucose and other sugars by the type strain using the API 20NE and API 50 CH test systems (Kubota et al., 2005a; Yi and Chun, 2004a). As mentioned before, the *Nocardioides simplex* strains transform sterols and their derivates (Arima et al., 1969; Nagasawa et al., 1969). They also were reported to utilize phenol as a sole carbon and energy source (Collins et al., 1994; Keddie et al., 1986), but the type strain could not grow in the presence of phenol at a concentration of 125 p.p.m. (Cui et al., 2009). The authors also found that this strain utilized dibenzofuran at 50 p.p.m. in the test medium (higher concentrations were growth-inhibiting) (Cui et al., 2009). According to Kubota et al. (2005a), the type strain *Nocardioides simplex* JCM 1363 did not use biphenyl and dibenzofuran as a carbon and energy source, in contrast to *Nocardioides aromaticivorans*. Five tested strains of *Nocardioides simplex*, in contrast to closely related species, were also unable to utilize *p*-nitrophenol as the sole carbon and energy source (Cui et al., 2009; Yoon et al., 1999) (Table 218).

The cell-wall peptidoglycan is of the A3γ type (LL-A_2pm-glycine) (Schleifer and Kandler, 1972). The well-documented cell-wall sugar is galactose (Keddie and Cure, 1977; Sadikov et al., 1983; Takeuchi and Yokota, 1989). In addition, the presence of glucose (Takeuchi and Yokota, 1989) or rhamnose (Sadikov et al., 1983) was reported for the type strain, and mannose for strain NBRC (IFO) 12679 = JCM 1366 (Takeuchi and Yokota, 1989). (Relationship to *Nocardioides simplex* was supported by DNA–DNA hybridization experiments; Suzuki and Komagata, 1983c; Yoon et al., 1999.) The wall polysaccharides of both the above strains were also found to contain a small or trace amount of glucosamine (Takeuchi and Yokota, 1989). Both strains had cell walls containing glycerol teichoic acids (Sadikov et al., 1983; Takeuchi and Yokota, 1989). The polyamine composition of the type strain, DSM 43109, harvested from rich (R) medium (Yamada and Komagata, 1972) was cadaverine

(56.6%), spermidine (18.6%), putrescine (15.0%), spermine (5.6%), and 1,3-diaminopropane (4.2%), with a trace amount of *sym*-homospermidine (Busse and Schumann, 1999). The predominant isoprenoid quinone is MK-8(H_4); minor quantities of MK-7(H_4), MK-8, MK-8(H_6), and a trace amount of MK-9(H_4) may be detected (Collins et al., 1979; Collins et al., 1983; O'Donnell et al., 1982; Suzuki and Komagata, 1983c; Yamada et al., 1976).

The fatty acids of *Nocardioides simplex* to a large extent depends on the culture conditions and analytical procedures (Collins et al., 1983; Cui et al., 2009; Miller et al., 1991; O'Donnell et al., 1982; Schumann et al., 1997; Suzuki and Komagata, 1983c; Yoon et al., 1999; Yoon et al., 1997a). When strains of this and other species of this genus are analyzed under similar experimental conditions, the fatty acid profiles of *Nocardioides simplex* are very similar (Yoon et al., 1999) and differ from those of other species (Cui et al., 2009; Yoon et al., 1999), including closely related ones (Table 219). The cellular fatty acids (4% or more), determined for 5 strains grown on nutrient agar (NA; Difco) for 4 d at 30°C were $C_{16:0}$ iso (29.0–36.6%), $C_{18:1}$ ω9*c* (7.3–12.0%), $C_{17:1}$ ω6*c* (11.9–14.8%), TBSA (9.3–15.3%), $C_{17:0}$ 10-methyl (2.4–6.3%), $C_{16:0}$ 10-methyl (1.1–4.2%), $C_{16:1}$ iso H (2.8–6.6%), and $C_{17:0}$ iso (2.2–6.4%). The cells of the type strain grown in trypticase soy broth (Difco) for 24–48 h at 28°C contained different proportions of predominant fatty acids: $C_{16:0}$ iso (41.8%), $C_{18:1}$ (13.1%), $C_{17:1}$ (5.2%), TBSA (5.2%), $C_{16:1}$ (5.0%), $C_{14:0}$ iso (4.6%), and also had a lower level of 10-methyl-branched components (Schumann et al., 1997). The presence of 2-hydroxylated acids was also detected (Collins et al., 1983; Suzuki and Komagata, 1983c; Yoon et al., 1999), which may contribute up to 10% in some experiments (Suzuki and Komagata, 1983c). The phospholipids also varied between different studies and included diphosphatidylglycerol, phosphatidylglycerol, and a few incompletely characterized components (Collins et al., 1989; Collins et al., 1983; O'Donnell et al., 1982). The presence of phosphatidylglycerol with a significant amount of 2-hydroxy fatty acids at the 2-position was reported by Yano et al. (1970, 1971). There is also a report about tentative identification of substantial proportions of phosphatidylinositol mannosides and phosphatidylethanolamine in *Arthrobacter simplex* ("*Corynebacterium*" *simplex*) grown on *n*-alkanes (Yanagawa et al., 1972). In our recent experiment, only three clearly dominating phospholipids were detected for the type strain of *Nocardioides simplex*, namely, phosphatidylglycerol and two similar unidentified phospholipids, one of which was found to include inositol and is most likely phosphatidylinositol, while the second is supposedly phosphatidylinositol containing a hydroxy acid (Figure 263).

Nocardioides simplex is susceptible to some actinophages among those multiplying on mycelium-forming *Nocardioides* strains (Prauser, 1976, 1981, 1989).

Strains affiliated with this species had 60–100% DNA–DNA similarity (using the membrane filter method with tritium-labeled DNA; Suzuki and Komagata, 1983c) and more than 85% (Yoon et al., 1999, using the method of Ezaki et al., 1989). The levels of DNA–DNA hybridization between the type strain of *Nocardioides simplex* and type strains of the species composing the tight *Nocardioides simplex* phylogenetic group were reported to be 13–55% (Cui et al., 2009; Kubota et al., 2005a; Yoon et al., 1999, 2006b, 2009; Table 221). The DNA–DNA hybridization values of the type strain of *Nocardioides simplex* with those of the type strains of more distant species (*Nocardioides plantarum*, *Nocardioides albus*, *Nocardioides luteus*, and *Nocardioides jensenii*) obtained by the method of Ezaki et al. (1989) were reported to be 28–33% (Kubota et al., 2005a). A lower DNA–DNA binding level (16–22%) was found between members of *Nocardioides simplex* and the type strains of *Nocardioides albus* and *Nocardioides jensenii* (using the membrane filter method; Suzuki and Komagata, 1983c). Similar values (15–20%) were reported for *Nocardioides simplex* and mycelium-forming strains of *Nocardioides* by Prauser (1981). Strains of *Nocardioides simplex* are very similar in the length and nucleotide sequences of the 16S–23S ITS region and clearly distinguished from *Nocardioides nitrophenolicus* and other members of this genus by these characteristics (Yoon et al., 1998a). The type strains of *Nocardioides simplex* and other species included in the study are also different in the RNase P RNA gene sequences (Yoon and Park, 2000).

Selected phenotypic characteristics useful in distinguishing *Nocardioides simplex* from phylogenetically closely related species are listed in Table 218. Occur mostly in soil. The type strain was isolated as described by Jensen (1934).

Source (type strain): rice soli.

DNA G+C content (mol%): 71.7–73.5 (HPLC, T_m); 71.7 (HPLC) for the type strain.

Type strain: AJ 1420, ATCC 6946, CCM 1652, CCUG 23611, CIP 82.106, DSM 20130, HAMBI 90, HAMBI 1861, IAM 1660, IFO (now NBRC) 12069, JCM 1363, KCTC 9106, LMG 16261, NRRL B-14051, NRRL B-3157, VKM Ac-1118.

Sequence accession no. (16S rRNA gene): X53213, AF005009.

Additional remarks: first described as "*Corynebacterium simplex*" (Jensen, 1934), this species was then affiliated to the genus *Arthrobacter* as *Arthrobacter simplex* (Lochhead, 1957) and later reclassified, mainly on chemotaxonomic grounds, as *Nocardioides simplex* (O'Donnell et al., 1982, 1983). Independently, Suzuki and Komagata (1983b, 1983c) established the genus *Pimelobacter* to accommodate *Arthrobacter simplex* and some related organisms. Those, in particular, included two strains (IAM 1398 = IFO 1366 and IAM 1413 = IFO 1367) isolated from soils of an oilfield in Japan and originally described as "*Brevibacterium lipolyticum*" (Iizuka and Komagata, 1964). Strain NCIB 9770, originally identified as *Arthrobacter simplex* (Jensen and Gundersen, 1956) was reclassified as a representative of a novel species of the genus *Pimelobacter*, *Pimelobacter jensenii* (Suzuki and Komagata, 1983b, 1983c), and subsequently transferred to *Nocardioides* as *Nocardioides jensenii* (Collins et al., 1989). Some other strains (ATCC 13260, 19565, and 19566) originally described under the name *Nocardioides simplex* were reidentified as members of the genus *Rhodococcus* (Yoon et al., 1997a). Notably, other strains cited in the literature under the species name *Nocardioides simplex* might also not belong to this species, particularly in light of recent proposals regarding some species phylogenetically closely related to *Nocardioides simplex* (Table 218). Although some of these strains (e.g. *Nocardioides simplex* VKM Ac-2033D involved in transformation of sterols (Fokina et al., 2003a, 2003b) or *Nocardioides simplex* 3E involved in degradation of 2,4,5-trichlorophenoxyacetic acid (Golovleva et al., 1990) exhibit a very high 16S rRNA gene sequence similarity to the type strain of *Nocardioides simplex*, their relationship to this species was not firmly supported by DNA–DNA hybridization or other relevant genotypic studies.

46. **Nocardioides terrae** Zhang, Liu and Liu 2009a, 2447[VP]

ter'ra.e. L. gen. n. *terrae* of or from the earth.

Characteristics are as described for the genus and listed in Table 215. Information presented below is based on the original species description (Zhang et al., 2009a).

Cells are slender irregular rods or coccoid. Data on the cell size are conflicting. In the species description, the cells are 0.2–0.3 µm in diameter and 0.3–1.0 µm long. As seen from the micrograph provided, most cells after 3 d in culture on R2A agar (30°C) appear as short rods, mostly about 0.3–0.5 µm in diameter and up to 1.0 µm long; longer cells, up to 2–2.3 µm may also occur. Coccoid cells of the same diameter, slightly smaller or increased are observable. The general appearance of the cells suggests the occurrence of a rod-to-coccoid growth cycle. Nonmotile, no flagella are observed. Colonies are cream-colored on R2A agar. Grows at 16–34°C (optimum, 29°C), but prefers not to grow at 14 or 36°C. Grows at pH 5.5–8.5 (optimum pH 6.2–6.5) and in 0–1% (w/v) NaCl (optimum 0–0.25%), as assessed in R2A broth. Along with substrates listed in Table 215, the organism assimilates gluconate as a carbon source, but not adipate, caprate, and citrate (API 20 NE). Acids are produced from D-glucose, L-rhamnose, and weakly from starch (API 50 CH). Acids are also formed from D-cellobiose, D-galactose, maltose, D-ribose, and trehalose (characteristics differentiating this species from the phylogenetically close species *Nocardioides panacihumi*). The type strain contains *N*-acetyl-β-glucosaminidase. Tween 20 is not hydrolyzed. Glucose fermentation (API 20NE) occurs. The major cellular fatty acids (>5% of the total) determined for cell mass scraped from R2A after incubation at 30°C for 3 d were $C_{16:0}$ iso (18.1%), TBSA (11.8%), $C_{17:0}$ 10-methyl (10.7%), $C_{17:1}$ ω8*c* (7.0%), $C_{15:0}$ iso (6.1%), $C_{18:1}$ ω9*c* (5.3%), and $C_{16:0}$ 10-methyl (5.1%). Levels of DNA–DNA similarity between the type strain of this species and the type strains of *Nocardioides panacihumi* and *Nocardioides marinus* were reported to be 21 and 16%, respectively (thermal denaturation method). The type strain (the only strain described) was isolated on VL55 medium (Sait et al., 2002) modified by adding 10 ml per liter of an amino acid mixture (Davis et al., 2005) and using gellan as a solidifying agent. The strain was isolated from a plate inoculated with the 10^{-7} dilution after 1 week of incubation at 30°C.

Source (type strain): forest soil, the Changbai Mountains, Heilongjiang Province, China.

DNA G+C content (mol%): 71.6 (T_m).

Type strain: VA15,CGMCC 1.7056, JCM 16799, NBRC 104259.

Sequence accession no. (16S rRNA gene): FJ423762.

47. **Nocardioides terrigena** Yoon, Kang, Lee and Oh 2007a, 2474[VP]

ter.ri.ge'na. L. masc. or fem. n. *terrigena* child of the earth, referring to the isolation of the type strain from soil.

Characteristics are as described for the genus and listed in Table 215. Additional information is from the original species description (Yoon et al., 2007a).

Cells are nonmotile rods or cocci (0.4–0.7 × 0.7–2.0 µm) in the exponential phase of growth. A rod–coccus morphogenetic growth cycle occurs during growth on complex media. Neither substrate nor aerial mycelium is formed. Gram-staining is variable in old cultures. Colonies are ivory in color and 1.0–1.2 mm in diameter after 7 d incubation on R2A agar at 30°C. Grows at 4 and 35°C, but prefers not to grow at 36°C. Grows in 0–3% but not at 4% (w/v) NaCl, with optimum growth at 0.5–1.0% (w/v) NaCl (assessed in R2A broth prepared according to the Difco formula without agar). Optimal pH for growth is around 8.0. Gentiobiose, D-lyxose (weak), D-turanose, adipate, gluconate and L-malate, but not D-arabinose, D- and L-fucose, sorbose, D-tagatose, L-xylose, D- and L-arabitol, erythritol, dulcitol, sorbitol, xylitol, caprate, 2- and 5-ketogluconate, phenylacetate, methyl β-D-xyloside, methyl α-D-mannoside, methyl α-D-glucoside, glycogen, esculin, amygdalin, and arbutin are utilized as sole carbon and energy sources. Tweens 20, 40, and 60 are hydrolyzed. The type strain is susceptible to the following antibiotics (µg per disk, unless indicated): ampicillin (10), carbenicillin (100), cephalothin (30), chloramphenicol (100), kanamycin (30), lincomycin (15), neomycin (30), novobiocin (5), oleandomycin (15), penicillin G (20 U), and streptomycin (50) and weakly susceptible to tetracycline (30); resistant to gentamicin (30) or polymyxin B (100 U). The main fatty acids included $C_{16:0}$ iso (35.5%), $C_{17:1}$ ω8*c* (30.2%), and $C_{17:0}$ (11.5%). Minor compounds reported were represented by various saturated, monounsaturated, iso-, anteiso- and 10-methyl-branched fatty acids characteristic of the genus *Nicardioides*, each contributing 0.8–2.7%. TBSA was not detected among fatty acids that exceeded 0.5%. The DNA–DNA hybridization value was 34% for the type strain *Nocardioides terrigena* and type strain of the most closely related species, *Nocardioides sediminis* (Dastager et al., 2009e), and 7% for the type strains of *Nocardioides terrigena* and *Nocardioides basaltis* (Kim et al., 2009a). The type strain (the only strain described) was isolated using the standard dilution plating technique on tenfold-diluted nutrient agar (Difco) at 25°C.

Source (type strain): soil, Dokdo, Korea (37°14′12″N; 131°52′07″E).

DNA G+C content (mol%): 71.5 (HPLC).

Type strain: DS-17, KCTC 19217, JCM 14582.

Sequence accession no. (16S rRNA gene): EF363712

48. **Nocardioides tritolerans** Dastager, Lee, Ju, Park and Kim 2009a, 1555[VP] (Effective publication: Dastager, Lee, Ju, Park and Kim 2008c, 1205.)

tri.to'le.rans L. pref. *tri-* (from L. num. adj. *tris*), three; L. part. adj. *tolerans* tolerating; N.L. part. adj. *tritolerans* referring to the ability of the organism referring to the ability of the organism to tolerate relatively high salinity, alkalinity and temperature.

Characteristics are as described for the genus and listed in Table 215. Information presented below is taken from the original work on species description (Dastager et al., 2008c).

Cells are typically motile irregular rods, mostly about 0.3–0.5 µm wide and up to 4.0 µm long on R2A medium. A week-old culture on the same medium is mostly composed of short rods, not exceeding 1 µm in length; a small fraction of coccoid cells is present. Larger club-shaped or spherical forms, up to 1.4 μm in diameter may occur (Figure 257). Colonies are translucent, cream in color, slightly raised, and 0.9–1.3 mm in diameter after 4–5 d incubation on R2A medium at 30°C. Grows at 20–40°C, but not at 19°C or 41°C. Grows well on R2A medium at initial pH 6.0–11.0 (optimum, pH 7.0–7.5), with weak growth at pH 12, and on R2A medium containing up to 7.0% (w/v) NaCl.

Acids are produced from many sugars listed in Table 215, as tested by conventional methods. The fatty acids of cells harvested from R2A after incubation for 5 d at 28°C included $C_{16:0}$ iso (40.6%), $C_{17:1}$ ω8*c* (7.3%), $C_{18:1}$ ω9*c* (6.1%), $C_{17:1}$ (4.5%), $C_{15:0}$ iso (4.5%), $C_{17:0}$ anteiso (4.4%), $C_{17:0}$ iso (4.1%), $C_{17:0}$ 10-methyl (2.9%), $C_{17:1}$ ω6*c* (2.9%), $C_{17:1}$ iso ω9*c* (2.8%), $C_{16:1}$ iso (2.6%), and $C_{16:0}$ (2.4%); other acids reported were $C_{14:0}$ iso, $C_{18:0}$ iso, $C_{18:0}$, and $C_{15:1}$ anteiso, each contributing 1.4–1.8%. The type strain (and the only strain described) was isolated by serial dilution plating on tenfold-diluted R2A agar at 30°C after 7 d incubation.

Source (type strain): soil, Bigeum Island, Korea.

DNA G+C content (mol%): 67.6 (HPLC).

Type strain: MSL-14, KCTC 19289, DSM 19319.

Sequence accession no. (16S rRNA gene): EF466107.

Genus II. **Actinopolymorpha** Wang, Zhang, Xu, Ruan and Wang 2001, 472[VP]

Lyudmila I. Evtushenko

Ac.ti.no.po.ly.mor'pha. Gr.n. *actis, actinos* a ray; Gr. adj. *polumorphos* multiform, manifold; N.L. fem. n. (N.L. fem. adj. used as a substantive) *Actinopolymorpha* actinomycete of many shapes.

Branched fragmenting vegetative hyphae of uneven thickness to highly pleomorphic cells of different sizes, both usually **growing on the agar surface**. Marked **apical and lateral budding** is often observable. Cells may remain attached after division, forming **short chains and small aggregates**. Colonies are white, yellow, or orange in color, mostly of a pasty consistency, smooth, or wrinkled. Scant or moderate aerial mycelium occasionally occurs in older cultures on some media. Endospores are not formed. Nonmotile. Gram-stain-positive. Non-acid-fast. **Chemo-organotrophic with a respiratory type of metabolism.** Catalase- and oxidase positive. Grow aerobically on standard laboratory media, including the chemically defined (synthetic) media. Oxidative acid production from glucose and some other carbohydrates. Mesophilic (optimum temperature ~28°C), non-salt-requiring. Diagnostic diamino acid of the cell wall is **LL-diaminopimelic acid**. Menaquinones are the sole respiratory quinones detected; the major component is tetrahydrogenated or hexahydrogenated menaquinone with nine isoprene units, **MK-9(H_4) or MK-9(H_6)**. The fatty acid profile dominated by **iso-branched acids ($C_{16:0}$ iso, $C_{15:0}$ iso, $C_{17:0}$ iso, and $C_{16:1}$)**; anteiso-branched, hydroxylated, and 10-methyl branched fatty acids usually contribute lesser or minor amounts. Polar lipid pattern includes **phosphatidylinositol mannosides, phosphatidylinositol, diphosphatidylglycerol, and phosphatidylglycerol.** The described *Actinopolymorpha* species have been isolated from soil or plant tissues.

DNA G+C content (mol%): 66.6–69.6.

Type species: **Actinopolymorpha singaporensis** Wang, Zhang, Xu, Ruan and Wang 2001, 472[VP].

Further descriptive information

Based on the 16S rRNA gene sequence analysis, the genus *Actinopolymorpha* represents, together with the recently described genera *Thermasporomyces* (Yabe et al., 2011) and *Flindersiella* (Kaewkla and Franco, 2010a), a separate subcluster within the family *Nocardioidaceae*, order *Propionibacteriales* (formerly suborder *Propionibacterineae* Stackebrandt et al. 1997) (see Figure 247). The genus currently includes five species: the type species *Actinopolymorpha singaporensis* (Wang et al., 2001), *Actinopolymorpha rutila* (Wang et al., 2008b), *Actinopolymorpha alba* (Cao et al., 2009), *Actinopolymorpha cephalotaxi* (Yuan et al., 2010), and *Actinopolymorpha pittospori* (Kaewkla and Franco, 2010a). The information given below is from the original species descriptions, unless indicated.

All organisms, when cultured on agar media, grow on the agar surface, and occasionally weakly penetrate it (reported for *Actinopolymorpha singaporensis*). The colonies are pasty, usually smooth at first and later wrinkled. A marked morphological differentiation within colonies may be observable (Figure 264). The colony color varies depending on species and growth medium, and usually is white or yellow to orange of different intensity and shade (see Table 222 and the species description for more details). Aerial mycelia white in color may develop in older cultures on certain agar media (Table 222), which is mostly scant but may be moderate to good, e.g. mycelia of *Actinopolymorpha*

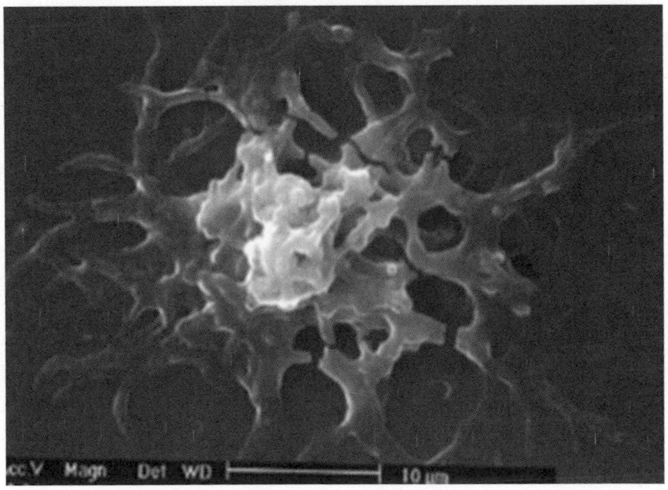

FIGURE 264. Scanning electron micrograph of a microcolony of *Actinopolymorpha rutila* YIM 45725 grown on ISP 2 agar at 28°C for 12 d. (Reproduced with permission from Wang et al., 2008b. Int. J. Syst. Evol. Microbiol. *58*: 2443–2446.)

TABLE 222. Descriptive and differentiating characteristics of *Actinopolymorpha* species[a]

Characteristic	*A. singaporensis*	*A. alba*	*A. cephalotaxi*	*A. pittospori*	*A. rutila*
Colony color on ISP 2	Brilliant orange	Milk white	Brilliant orange	Pale yellow	Deep orange-yellow
Colony color on ISP 3	Yellow	Light grey-white	Yellow or buff	Yellowish white	Brilliant orange-yellow
Aerial hyphae (media)	w (ISP 3)	w (ISP 2)	–	+ (HPDA), w (BA, ISP 2)	w (ISP 2)
Diffusible pigment (media)	–	+ (ISP 7)	+ (ISP, NA)	–	–
Growth at:[b]					
15°C	–	+[c]	+	+	+
37°C	+	+[c]	– or w	–	+
45°C	– or w	+[c]	–	–	–
Maximal % of NaCl (medium)[d]	8, 15 w (TSB)	7 (ISP 2)	5 (TSB)	3 w (ISP 2)	5 (ISP 2)
Nitrate reduction	+	–	+	nd	–
Utilization of C-sources:[e]					
D-Glucose	+	+	+	+	+
D-Arabinose	–	+	+	w	+
Fructose	+	–	+	+	+
D-Galactose	+	–	+	+	+
myo-Inositol	+	–	+	–	+
D-Mannitol	+	–	+	–	+
Raffinose	–	+	+	w	+
L-Rhamnose	+	v	+	nd	+
Sorbitol	+	–	+	+	+
Sucrose	+	–	+	+	v
D-Xylose	+	v	+	nd	+
Acid production from:[f]					
D-Glucose	+	+	+	+	+
Fructose	+	+	+	–	+
Dulcitol	w	–	–	–	w
D-Galactose	+	+	+	–	v
Maltose	v	+	+	–	+
D-Mannose	+	+	+	w	+
myo-Inositol	+	w	+	–	– or w
D-Mannitol	+	w	+	–	+
Salicin	+	+	+	w	+
Sucrose	+	+	+	–	+
Trehalose	+	+	+	–	+
D-Xylose	+	+	+	+	+
Adonitol	+	+	–	w	+
Sorbitol	+	–	+	–	v
Cell-wall diamino acid	LL-A_2pm	LL-A_2pm	LL-A_2pm	LL-A_2pm	LL-A_2pm
Whole-cell sugars	Glc, Rha, Rib	Glc, Rha, Rib	Glc	Glc, Rib, Gal	Glc, Rib, Gal
Menaquinone pattern[g]	9/6 (9/4, 9/8, 10/4)	9/6 (9/8, 10/6, 10/8)	9/4 (9/6, 9/8, 10/4)	9/4 (9/6, 10/4, 10/8)	9/4 (9/6, 9/8, 10/6)
Phospholipids	DPG, PG, PIM, PI	DPG, PG, PIM, PI	PI, PG, PIM	nd	DPG, PG, PIM, PI
Major acids (>5%):[h]	$C_{16:0}$ iso (**25.3**; 19.4), $C_{15:0}$ iso (**33.7**; 33.3), $C_{17:0}$ iso (**12.1**; 7.0), $C_{17:0}$ anteiso (**11.1**; 7.9), $C_{15:0}$ anteiso (**<0.5**; 4.0), $C_{16:1}$ iso (**<0.5**; 21.6)	$C_{16:0}$ iso (**25.3**; 7.1), $C_{15:0}$ iso (**33.7**; 28.1), $C_{17:0}$ iso (**12.1**; n.d), $C_{17:0}$ anteiso (**11.1**; n.d.), $C_{15:0}$ anteiso (**<0.5**; 14.5), $C_{16:1}$ iso (**<0.5**; 6.6)	$C_{16:0}$ iso (**40.5**; 21.4), $C_{15:0}$ iso (**17.5**; 12.0), $C_{17:0}$ iso (**10.6**; 3.0), $C_{17:0}$ anteiso (**9.4**; 6.0), $C_{15:0}$ anteiso (**1.2**; 1.5), $C_{16:1}$ iso (**1.6**; 32.9)	$C_{16:0}$ iso (**29.2**), $C_{15:0}$ iso (**17.0**), $C_{17:0}$ iso (**16.4**), $C_{17:0}$ anteiso (**20.7**), $C_{15:0}$ anteiso (**7.5**), $C_{16:1}$ iso (**<0.5**)	$C_{16:0}$ iso (**35.8**; 22.4), $C_{15:0}$ iso (**15.3**; 22.1), $C_{17:0}$ iso (**10.0**; 8.9), $C_{17:0}$ anteiso (**9.8**; 8.3), $C_{15:0}$ anteiso (**3.5**; 3.7), $C_{16:1}$ iso (**1.9**; 21.6)
10-Methyl-branched[i]	$C_{16:0}$, $C_{17:0}$	$C_{16:0}$, $C_{17:0}$	$C_{16:0}$, $C_{17:0}$	$C_{16:0}$, $C_{17:0}$	$C_{17:0}$
Hydroxylated[i]	$C_{9:0}$ 3-OH, $C_{15:0}$ 3-OH, $C_{14:0}$ iso 3-OH	$C_{14:0}$ iso 3-OH (11.8), $C_{9:0}$ 3-OH, $C_{15:0}$ 3-OH	$C_{9:0}$ 3-OH, $C_{14:0}$ 3-OH, $C_{15:0}$ 3-OH, $C_{16:1}$ 2-OH	$C_{9:0}$ 3-OH	$C_{14:0}$ 3-OH, $C_{15:0}$ 3-OH, $C_{18:0}$ 2-OH
DNA G+C content (mol%)	69.5 (HPLC)	66.6 (HPLC)	69.3 (T_m)	69.6 (HPLC)	67.7 (HPLC)

(continued)

TABLE 222. (continued)

Characteristic	*A. singaporensis*	*A. alba*	*A. cephalotaxi*	*A. pittospori*	*A. rutila*
Source of type strain	Soil, primary rain-forest	Soil, plant rhizo-sphere	Soil, rhizosphere of Chinese cowtail pine (*Cephalotaxus fortunei*)	Leaves of apricot tree (*Pittosporum phylliaeoides*)	Soil, forest

[a]Based on characteristics of type strains. Data from Wang et al. (2008b, 2001), Cao et al. (2009), Yuan et al. (2010), and Kaewkla and Franco (2010b, 2010a). Symbols and abbreviations: +, positive; –, negative; v, variable between experiments or test methods; w, weak; LL-A_2pm, 2,6-diaminopimelic acid; Glc, glucose; Gal, galactose; Rib, ribose; Rha, rhamnose; HPDA, half-strength potato dextrose agar (Atlas, 1993); BA, Bennett's agar (Atlas, 1993); NA, nutrient agar (Atlas, 1993); TSB, tryptic soy broth (Difco); nd, not determined.

[b]Tested on ISP 2 agar, except *Actinopolymorpha alba* examined on TSA [3% (w/v) trypticase soy broth (BBL); 1.5% (w/v) Bacto agar (Difco)] (data from the original species descriptions and Kaewkla and Franco, (2010a, 2010b).

[c]Conflicting test results were reported by Kaewkla and Franco (2010b, 2010a).

[d]Lower NaCl concentrations allowing growth of *Actinopolymorpha singaporensis* (3%), *Actinopolymorpha alba* (1%), and *Actinopolymorpha rutila* (3%) were obtained using basal medium ISP 2 at 27°C (Kaewkla and Franco, 2010a, 2010b).

[e]Tested on basal medium ISP 9 (data from the original species descriptions, and from Kaewkla and Franco, 2010b).

[f]According to the method of Gordon et al. (1974).

[g]Predominant menaquinone and components detected in lesser amounts (in parenthesis); numerals indicate the numbers of isoprene units and the number of hydrogen atoms on the side chain (e.g. 9/4 is partially saturated chain with 4 hydrogen atoms on the side chain containing 9 isoprene units).

[h]Obtained using the standard protocol of the MIDI System in cells cultured under different conditions. Data for cells grown in liquid Tryptic Soya Broth medium, TSB (Oxoid) at 27°C for 10 d (Kaewkla and Franco, 2010b) are given in bold; data for cells grown at 28°C for 6 d in liquid TSB (Difco) (Yuan et al., 2010) are underlined; data for *Actinopolymorpha rutila* (Wang et al., 2008b) and *Actinopolymorpha alba* (Cao et al., 2009) cells cultured for 5 d at 28°C on TSA agar [3% trypticase soy broth (BBL) and 1.5% Bacto agar (Difco)] are presented neither in bold nor underlined.

[i]Other diagnostic fatty acids detectable irregularly in minor (< 5%) or trace amounts, except $C_{14:0}$ iso 3-OH (which is reported to constitute 11.8% in 5-d cultures of *Actinopolymorpha alba* grown at 28°C on TSA agar (Cao et al., 2009).

pittospori grown on half-strength potato dextrose agar (HPDA; Atlas, 1993).

Various morphological forms, extending from well developed branched hyphae with swollen hyphal segments to highly pleomorphic cells (irregularly-sized globular, coccoid, rod-shaped, and angular), which are often produced by apical and lateral budding, is a characteristic feature of organisms of this genus (Figure 265, Figure 266, and Figure 267). The observable morphologies depend on species, growth medium, the culture age, and location in colonies. The cells produced by budding often remain attached after division, forming aggregates and short chains (pseudomycelium). The cells in chains resemble at maturation exospores (arthrospores) of spore-forming actinomycetes (Figure 265c). The cells may give rise to new buds which further develop into cells of different shapes or hyphae as growth proceeds or when a culture is transferred to fresh medium. The marked hyphal elongation by budding was emphasized for *Actinopolymorpha singaporensis* (which has been described in more morphological detail) at a later growth stage (Figure 265c), but probably takes place in some other species of this genus. No data are available on the fine structure of thickened hyphae, hyphal swellings or globular forms (presumably cell clusters) examined by electron microscopy of thin sections. There are certain analogues of images of such morphological structures with those observed, e.g. in strains of *Kribbella* (chapter in this volume), *Friedmanniella* (Schumann et al., 1997), and, especially, *Pseudonocardia* which, in addition, are characterized by the growth of hyphae by budding (Agre et al., 1984; Evtushenko et al., 1989; Henssen et al., 1983; Kuimova and Malishkaite, 1984). Likewise with these genera, it might be suggested that the cells composing clusters or thickened hyphal masses may divide internally by differently oriented septa. Members of *Actinopolymorpha* were reported not to produce "classical" sporangia typical of sporangial actinomycetes (e.g. *Actinoplanes*). However, it cannot be excluded that the aforementioned cell clusters (subclusters) may have a common envelope (to be designated as "sporangia"?).

Gram-stain-positive. Non-acid and non-alcohol-fast (examined for *Actinopolymorpha pittospori*). The cell wall contains LL-diaminopimelic acid (LL-A_2pm), as reported for all species; no data on the other peptidoglycan amino acids are available. The whole-cell sugars detected include glucose, galactose, rhamnose, and ribose in different combinations (Table 222). The menaquinone systems contain MK-9(H_6) as the predominant compound for *Actinopolymorpha singaporensis* and *Actinopolymorpha alba*, or MK-9(H_4) for the remaining species. Additional menaquinones, i.e. MK-9(H_8), MK-10(H_4), MK-10(H_6), and MK-10(H_8) in different combinations have been found in lesser or minor amounts (Table 222). Some shifts in the predominant menaquinone and the menaquinone patterns in this genus may be influenced by the medium composition and the age (developmental stage) of cultures, as suggested from some examples with other actinomycetes (Li et al., 2003; Saddler et al., 1986; Tanaka et al., 1996). All species contain similar patterns of principal polar lipids, i.e. phosphatidylinositol mannosides, phosphatidylinositol, diphosphatidylglycerol, and phosphatidylglycerol. An exception is *Actinopolymorpha cephalotaxi* which has been reported to lack diphosphatidylglycerol; no data are available for *Actinopolymorpha pittospori.* The fatty acid profiles are dominated by iso-branched components, mostly $C_{16:0}$ iso, $C_{15:0}$ iso, $C_{17:0}$ iso, and $C_{16:1}$ iso (Table 222). In addition, anteiso-branched and hydroxylated acids may be detected in significant proportions for certain species in some experiments. 10-Methyl-branched acids ($C_{16:0}$ 10-methyl and $C_{17:0}$ 10-methyl) are usually

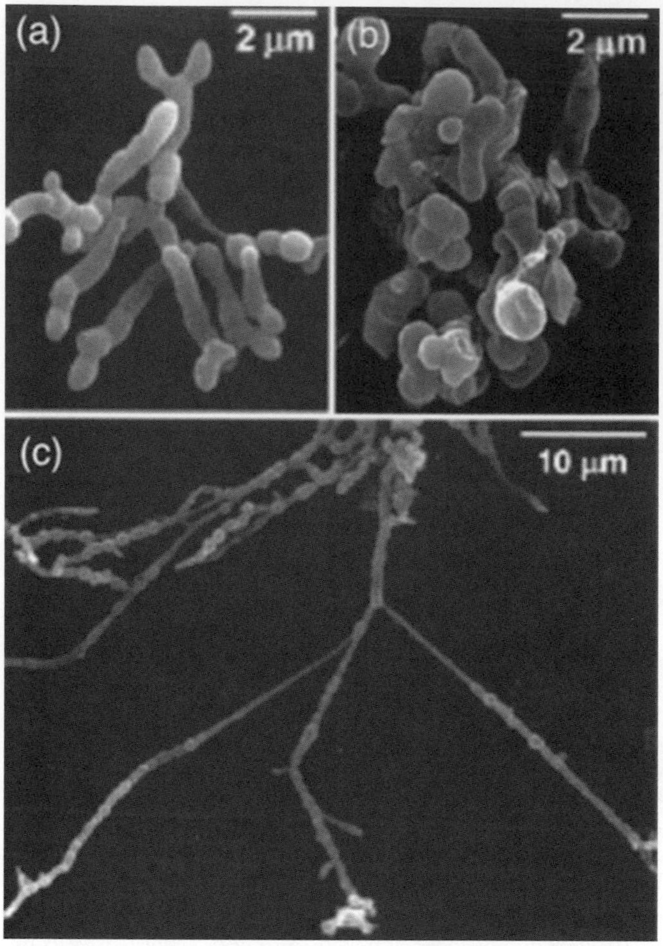

FIGURE 265. Cellular morphology of *Actinopolymorpha singaporensis*. Scanning electron micrographs of cells of strain *Actinopolymorpha singaporensis* IM 7744 grown on ISP 4 (a) and ISP 3 (b) media for 8 d; hyphae at the periphery of a colony on ISP 3 medium in 35-d-old culture (c). (Reproduced with permission from Wang et al., 2001. Int. J. Syst. Evol. Microbiol. *51*: 467–473.)

present but not in excess of 5%. The patterns of predominant fatty acids (and the fatty acid profiles as a whole) vary with species and the culture age (developmental stage). The most common change for cultures grown on trypticase soy broth (agar) is that the proportions of saturated iso-branched acids increase in older (10 d) cultures compared with younger (5–6 d) cultures, while that of unsaturated ($C_{16:1}$ iso) decreases (Table 222).

All species grow aerobically on complex agar media at nearly neutral pH at ~28°C and also are capable of growing on chemically defined ISP* media, including minimal salts medium ISP 9 supplemented with glucose or some other carbon sources. Some species were reported to grow on inorganic salt-starch agar (ISP 4) and Czapek's agar (Pridham and Lyons, 1980). The species differ in the rate and abundance of growth, but all form colonies within 8–10 d at 27–28°C on yeast extract-malt extract agar (ISP 2) and some other media.

*See Shirling and Gottlieb (1966) for the composition of ISP media cited here and in other sections of this chapter.

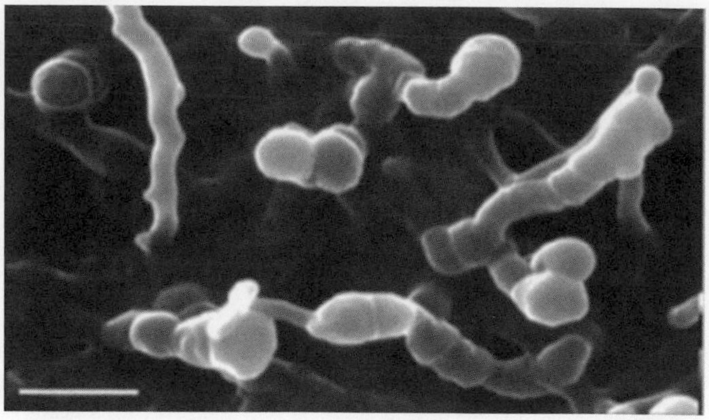

FIGURE 266. Cellular morphology of *Actinopolymorpha alba*. Scanning electron micrograph of cells of strain *Actinopolymorpha alba* YIM 48868 grown on ISP 2 agar for 12 d at 28°C. Bar = 2 μm. (Reproduced with permission from Cao et al., 2009. Int. J. Syst. Evol. Microbiol. *59*: 2200–2203.)

All species produce acid during aerobic growth from glucose and other carbohydrates (Table 222). All tested strains exhibit positive reactions for catalase and oxidase (no data on oxidase for *Actinopolymorpha pittospori*). Nitrate reduction varies with species (Table 222). All tested strains hydrolyze gelatin, but show negative results in the classical tests for hydrolysis of cellulose or starch (no data on *Actinopolymorpha pittospori*). *Actinopolymorpha singaporensis* and *Actinopolymorpha rutila* so far examined test positive for hydrolysis of casein, Tween 20, but negative for decomposition of Tween 80, and adenine. The four species (*Actinopolymorpha alba*, *Actinopolymorpha cephalotaxi*, *Actinopolymorpha rutila*, and *Actinopolymorpha singaporensis*) so far tested with the API ZYM system exhibit various enzymatic activities (see the species descriptions for details). The optimal growth temperature is ~28°C, some species can grow at 37°C or somewhat higher temperatures in appropriate conditions (Table 222). The recognized *Actinopolymorpha* species are non-salt-requiring, but some may tolerate up to 7% NaCl (e.g. *Actinopolymorpha alba* when tested on ISP 2 medium at 28°C; Cao et al., 2009), or even 15% (weak growth was reported for *Actinopolymorpha singaporensis* in liquid TSB as basal medium; Wang et al., 2001). A lower salt resistance of the above species and also *Actinopolymorpha rutila* was observed by Kaewkla and Franco (2010a, 2010b) on ISP 2 as basal medium at a slightly lower temperature (27°C). Organisms of this genus are considered to be neutrophilic, grow best at pH 6–8. They also may show growth at initial pH 6 and 10 on ISP 2 medium; some can grow at pH 5 (Kaewkla and Franco, 2010a, 2010b).

The DNA base ratio is 66.6–69.6 mol% (Table 222). The 16S rRNA gene sequence similarity ranges from 97.5% (the pair *Actinopolymorpha alba*–*Actinopolymorpha pittospori*) to 99.5% (*Actinopolymorpha cephalotaxi*–*Actinopolymorpha rutila*). The DNA–DNA similarity between the phylogenetically closest species obtained by the thermal denaturation method was 33.8% (Yuan et al., 2010). The highest value of DNA–DNA similarity (53.1%) was reported (Cao et al., 2009) for the pair *Actinopolymorpha alba*–*Actinopolymorpha rutila*, obtained by using the modified (Christensen et al., 2000) fluorometric micro-well method of Ezaki et al. (1989).

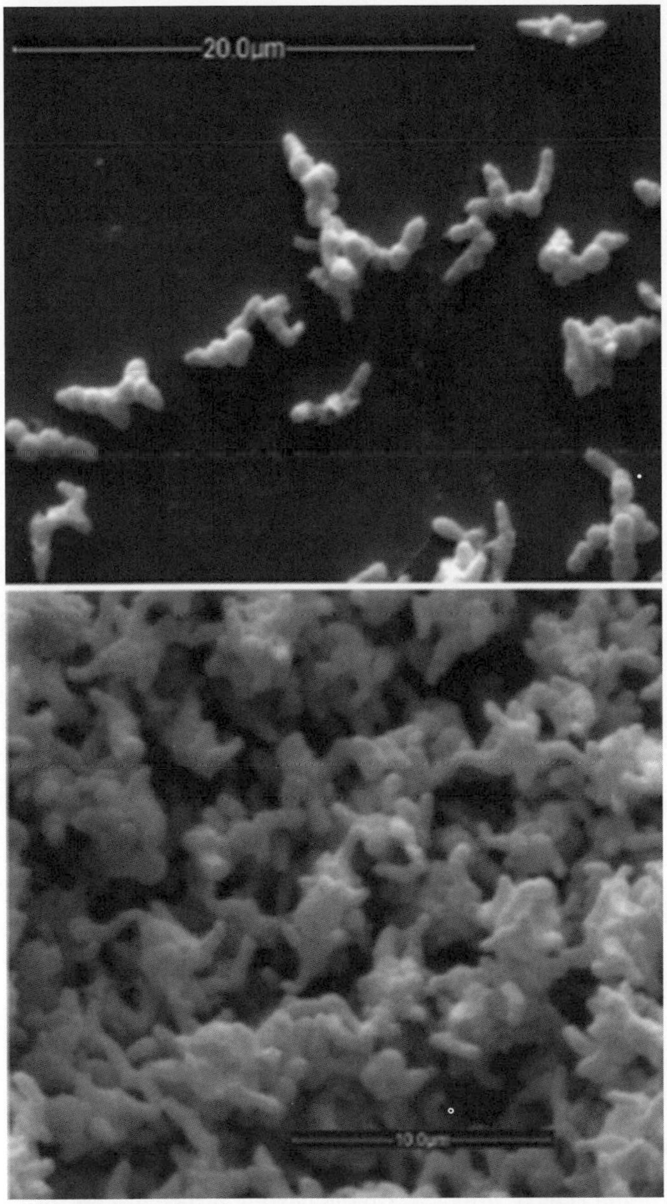

FIGURE 267. Cellular morphology of *Actinopolymorpha cephalotaxi*. Scanning electron micrographs of strain *Actinopolymorpha cephalotaxi* I06-2230 grown on ISP 2 agar for 14 d at 28°C. (Reproduced with permission from Yuan et al., 2010. Int. J. Syst. Evol. Microbiol. *60:* 51–54.)

All species are sensitive to 0.1% phenol. The sensitivity to antibiotics has not been tested directly, but it may be assumed from the composition of the isolation media that *Actinopolymorpha cephalotaxi* resists aztreonam (25 mg/l), while *Actinopolymorpha pittospori* tolerates nalidixic acid (20 µg/ml). *Actinopolymorpha singaporensis* may be resistant to penicillin or streptomycin, or both (the isolation procedure included pre-incubation of a soil sample in LB medium containing both these antibiotics, each 10 µg/ml). *Actinopolymorpha rutilus* YIM 45725 produces an estrogenic ligand, actinopolymorphol A, that preferentially induces heterodimerization of estrogen receptors ERα and ERβ and modulate estrogen receptor function (Huang et al., 2010; Powell et al., 2010).

Knowledge of the natural habitat of *Actinopolymorpha* species is fragmentary. A few available strains of this genus were isolated from forest soils or plant rhizosphere. An endophytic species, *Actinopolymorpha pittospori*, inhabited leaves of the Australian native apricot tree, *Pittosporum phylliraeoides*. Notably, members of *Actinopolymorpha* have not been reported so far in numerous ecological studies performed by direct non-culture (metagenomic) methods, in contrast to vast diversity of other organisms (clones) of the family *Nocardioidaceae*. No data have been reported on the possible pathogenic properties.

Enrichment and isolation procedures

Several strains of this genus have been isolated by plating on suitable agar media and incubation at 27–30°C for 3–11 weeks. The isolation media were yeast extract–malt extract agar (ISP 2), oatmeal agar (ISP 3) used for isolation of *Actinopolymorpha singaporensis*, inorganic salt-starch agar (Gauze 1; Gauze et al., 1983) for *Actinopolymorpha alba*, and starch-glycerol-proline agar with aztreonam (25 mg/l) for *Actinopolymorpha cephalotaxi*. A complex medium VL70 containing a mixture of 17 amino acids and solidified with 0.8% gellan gum (Hudson et al., 1989; Schoenborn et al., 2004) and supplemented with nalidixic acid and nystatin (each 20 µg/ml) was used to isolate *Actinopolymorpha pittospori* from the surface-sterilized and crushed leaves of *Pittosporum phylliraeoides*. The sterilization procedure included treatment with 70% ethanol and 6% hypochlorite (for 5 min each), followed by repeated rinsing with sterile water, treatment with 10% $NaHCO_3$ (10 min), and final rinsing with sterile water. Seeded plates were kept for 11 weeks until isolation in plastic sealed boxes, which contained wet paper towels to maintain humidity. To isolate *Actinopolymorpha singaporensis*, a soil sample was pre-incubated in Luria–Bertani (LB) medium containing penicillin and streptomycin (each 10 µg/ml), which eliminated many fast-growing bacteria. The soil sample was first dried in a chemical fume hood for 3 d and ground in a mortar. One milligram of the sample was suspended in 10 ml of LB medium containing antibiotics and incubated for a 3 h at 37°C with vigorous shaking. Then, the culture was centrifuged, washed three times with 10 ml of sterile water, resuspended in 1 ml of sterile water, spread onto an appropriate agar media, and incubated at 28–30°C for a month.

Maintenance procedures

Cultures may be maintained as 20% glycerol suspensions at −20 and −80°C. Long-term conservation is achieved by freeze-drying or in liquid nitrogen by standard procedures.

Differentiation of the genus *Actinopolymorpha* from other genera

Characteristics useful for phenotypic delineation of *Actinopolymorpha* from the other genera of the family *Nocardioidaceae* are listed in Table 222. The menaquinone systems dominated by MK-9(H_6) or MK-9(H_4), and the growth temperature range and optima are the most notable phenotypic characteristics that, along with morphological features, differentiate species of *Actinopolymorpha* from the phylogenetically neighboring genera *Flindersiella* and *Thermasporomyces*. The phosphatidylcholine-lacking polar lipid profile sharply differentiates this genus from *Kribbella*. *Actinopolymorpha* is readily distinguished by the morphological features from the phylogenetically more distant non-mycelium-forming *Aeromicrobium*, *Marmoricola* and the majority of *Nocardioides* species. In addition, *Actinopolymorpha* can by separated by the menaquinone pattern both from the rod-shaped and mycelium-forming *Nocardioides* species and *Marmoricola*, which possess the predominant menaquinone MK-8(H_4).

Taxonomic comments

After the establishment of the genus *Actinopolymorpha* based on the study of a single strain (Wang et al., 2001), additional four species have been isolated and described on the basis of polyphasic taxonomic approach within 2008–2011. The data accumulated since 2001 show that some characteristics in the original genus description do not reflect features of all species currently composing the genus, or the detailed data provided conflict with those of individual species. The genus description therefore needs harmonization while preparing the emended description of this genus.

Acknowledgements

The author was supported by the program MCB of the Russian Academy of Sciences.

Differentiation of species of the genus *Actinopolymorpha*

Primary differentiation of the currently recognized species of this genus is achievable taking into consideration the colony color and morphological features, differences in salt tolerance, the composition of whole-cell sugars, fatty acids, menaquinones, and some other characteristics listed in Table 222 and outlined in the species descriptions below. When using the chemotaxonomic characteristics for differentiation purposes, the possible age- and medium-dependent shifts in the fatty acid compositions and probably in the menaquinone and whole cell sugar patterns, should be taken into account. In comparative taxonomic studies, cells must be grown and analyzed (tested) under similar conditions.

List of species of the genus *Actinopolymorpha*

1. **Actinopolymorpha singaporensis** Wang, Zhang, Xu, Ruan and Wang 2001, 472[VP]

sin.ga.po.ren'sis. N.L. fem. adj. *singaporensis* of or belonging to Singapore, signifying the country where the type strain was isolated.

Characteristics are as described for the genus and listed in Table 222. Information given below is based on the data and microphotographs provided in the original paper (Wang et al., 2001), unless indicated.

Grows usually on the agar surface; poor penetration of vegetative hyphae into the agar medium, e.g. ISP 3, may occur after prolonged cultivation. The colony color is brilliant orange on ISP 2 medium, and yellow on ISP 3, ISP 4, TSA (Difco), and Bennett's agar (Atlas, 1993). Aerial mycelium is absent or scarcely formed in old cultures on some solid media. No diffusible pigment is produced. The cells of varied size and shape (irregular rods, squarish, and coccoid) are usually observed. The cells can be arranged in small aggregates and short chains. Budding cell division and hyphal elongation by budding take place. Grows at 25–37°C; weak growth may be observable at 45°C on ISP 2 medium (Kaewkla and Franco, 2010a). Grows well in liquid TSB medium containing NaCl up to 8% and slowly with 10–15% NaCl. A lower salt resistance may be observed on other growth media (Kaewkla and Franco, 2010b, a). Utilizes maltose, dextrin, glycerol, and salicin, as sole carbon sources, but not cellobiose or lactose (Wang et al., 2008b; Wang et al., 2001). An alkaline color of the phenol red in the tests (Gordon et al., 1974) for utilization of citrate or malonate, but not succinate (Wang et al., 2008b). Hydrolyzes hypoxanthine and xanthine. Positive in the API ZYM test for esterase lipase (C8), but negative for esterase (C4) and urease (Yuan et al., 2010).

Source (type strain): soil of the primary rainforest of the Bukit Timah nature reserve, Singapore.

DNA G+C content (mol%): 69.5 (HPLC).

Type strain: IM 7744, JCM 10761, NBRC 100040, NRRL B-24113.

Sequence accession no. (16S rRNA gene): AF237815.

2. **Actinopolymorpha alba** Cao, Jiang, Wu, Xu and Jiang 2009, 2202[VP]

al'ba. L. fem. adj. *alba* white, referring to the white substrate mycelium.

Characteristics are as described for the genus and listed in Table 222. Information presented below is taken from the original paper (Cao et al., 2009), unless indicated.

Good growth occurs at 28°C on ISP 2, ISP 2, ISP 5, Czapek's agar, and potato agar, but not nutrient agar and ISP 4. The colony color is usually milk–white to gray-white on the all media tested. Sparse aerial hyphae develop on ISP 2 medium but not on the other test media. No diffusible pigments are observable, except brown melanoid pigment on tyrosine agar, ISP 7 (Kaewkla and Franco, 2010b). Branched vegetative hyphae of uneven thickness, with marked budding and swellings (Figure 266). The ranges of temperatures and salinity allowing growth are 0–7% NaCl and 10–45°C (on ISP 2) according to the original description; lower threshold values were reported (Kaewkla and Franco, 2010b, a). Utilizes dextrin, fucose, maltose, mannose, and ribose as carbon sources, but not cellobiose, lactose, glycerol, or xylitol. Negative for milk coagulation and peptonization. Hydrolyzes L-tyrosine, but not hypoxanthine, xanthine, or urea. Positive in the API ZYM tests for α-chymotrypsin, esterase (C4), β-galactosidase, α-glucosidase, *N*-acetyl-β-glucosaminidase, leucine arylamidase, α-mannosidase, naphthol-AS-BI-phosphohydrolase, but negative for acid phosphatase, alkaline phosphatase, cystine arylamidase, esterase lipase (C8), α-fucosidase, α-galactosidase, β-glucosidase, β-glucuronidase, lipase (C14), trypsin, and valine arylamidase.

Source (type strain): rhizosphere soil, Yunnan Province, China.

DNA G+C content (mol%): 66.6 (HPLC).

Type strain: YIM 48868, CCTCC AA 208030, DSM 45243, JCM 16897.

Sequence accession no. (16S rRNA gene): EF601829.

3. **Actinopolymorpha cephalotaxi** Yuan, Zhang,Yu, Sun, Wei, Liu, Li and Zhang 2010, 53[VP]

ce.pha.lo.ta'xi. N.L. gen. n. cephalotaxi of *Cephalotaxus*, of *Cephalotaxus fortunei*, a plant from which the rhizosphere soil sample was collected for isolation.

Characteristics are as described for the genus and listed in Table 222. Information presented below is taken from the original paper (Yuan et al., 2010), unless indicated.

Grows at 28°C, forming brilliant orange colonies on ISP 2, and buff or yellow on ISP 3, ISP 4, ISP 5, Czapek's agar, nutrient agar (Difco), and tomato-paste-oatmeal agar (Waksman, 1961). Aerial mycelium is absent. Buff diffusible pigment is produced on ISP 3 agar, nutrient agar and tomato-paste-oatmeal agar. Cells usually of irregular shapes, divide through apical and lateral budding, and remain attached after division, forming small aggregates and short chains (Figure 267). Grows well in the temperature range of 20–28°C, will grow at 15°C (Kaewkla and Franco, 2010b) but not at 4°C or above 37°C. Tolerates up to 5% NaCl both on ISP 2 agar (Kaewkla and Franco, 2010b) and in liquid TSB medium. Utilizes glucosamine as sole carbon sources, but not acetate, citrate, gluconate, or tartrate. Positive for milk coagulation and peptonization. H_2S is not produced. Positive in the API ZYM tests for acid and alkaline phosphatases, esterase, esterase lipase, urease, trypsin, chymotrypsin, naphthol-AS-BI-phosphohydrolase, α- and β-galactosidases, α- and β-glucosidases, *N*-acetyl-β-glucosaminidase, and α-mannosidase.

Source (type strain): rhizosphere soil of *Cephalotaxus fortunei*; Yunnan Province, China.

DNA G+C content (mol%): 69.3 (T_m).

Type strain: I06-2230, CCM 7466T, DSM 45117, KCTC 19293.

Sequence accession no. (16S rRNA gene): EU438909.

4. **Actinopolymorpha pittospori** Kaewkla and Franco 2010b, 000[VP]

pit.to.spo'ri. N.L. gen. n. *pittospori* of *Pittosporum*, isolated from *Pittosporum phylliraeoides*

Characteristics are as described for the genus and listed in Table 222. Information presented below is taken from the original paper (Kaewkla and Franco, 2010b).

Good growth on ISP 2, Bennett's agar, and half-strength potato dextrose agar (HPDA; Atlas, 1993); poor growth is observed on ISP 3, ISP 4, ISP 5, ISP 7, and nutrient agar (Atlas, 1993). Colony color is white on ISP 4, and pale yellow or yellowish white on the all other above media. Aerial mycelium is moderate to good on HPDA, scant on ISP 2 and Bennett's agar, and absent on the other test media. Diffusible pigments are not produced. Branched substrate mycelium of irregular thickness is well developed on most media. In the later stage of growth, hyphae fragment to rod-like elements or V-shaped forms. Short chains and aggregates of cells occur. Will grow between 15–27°C, but not at 4 or 37°C. Good growth is observed at a low (1%) NaCl concentration, with weak growth at 3% NaCl, as assessed on ISP 2 agar.

Source (type strain): surface-sterilized leaves of *Pittosporum phylliraeoides*; the campus of Flinders University, Adelaide, South Australia.

DNA G+C content (mol%): 69.6 (HPLC).

Type strain: PIP 143, DSM 45354, ACM 5288.

Sequence accession no. (16S rRNA gene): FJ805429.

5. **Actinopolymorpha rutila** Wang, Zhang, Xu and Li 2008b, 2445[VP]

ru.ti'la. L. fem. adj. *rutila* red inclining to golden-yellow, referring to the color of colonies.

Characteristics are as described for the genus and listed in Table 222. Information presented below is taken from the original paper (Wang et al., 2008b), unless indicated.

Colonies develop well within 3–4 d at 28°C on suitable agar media. Good growth occurs on ISP 2, ISP 5, and potato agar; moderate growth on ISP 3, ISP 4, Czapek's agar, and nutrient agar. The color of colonies is deep orange-yellow on ISP 2 and potato agar; brilliant orange-yellow on ISP 3, ISP 5, and Czapek's agar; pink orange-yellow on ISP 4 and nutrient agar. Scant aerial hyphae are formed on ISP 2 in the later stages of growth. Diffusible pigments are not produced. Branched substrate mycelia of uneven thickness are formed, and fragment in the early stages of growth. Hyphae exhibit varied degrees of fragmentation and differentiation as growth proceeds, with short to elongated rod-like or V-shaped elements typically observable in older cultures. Will grow at 15–37°C, and in the presence of 3% NaCl on ISP 2 medium; data on higher values of the above growth parameters are conflicting (Kaewkla and Franco, 2010b, a; Wang et al., 2008b). Positive oxidase reaction with 1% *p*-aminodimethylalanine oxalate. Assimilates cellobiose, lactose, maltose, mannose, melibiose, melezitose, ribose, dextrin, salicin, adonitol, glycerol, and xylitol as sole carbon sources but not arabitol or dulcitol. Acid is produced from fucose, raffinose, D-xylose, and erythritol. Utilizes malonate or succinate, but not acetate, citrate, or oxalate (tested according to Gordon et al., 1974). Negative test results for hydrolysis of DNA, Tween 60, or L-tyrosine, and for H_2S production. Hydrolyzes hypoxanthine and xanthine and decomposes urea in conventional tests (but negative for urease in the API ZYM test system; Yuan et al., 2010). Positive in the API ZYM for alkaline phosphatase, α-glucosidase, β-glucosidase, α-galactosidase, β-galactosidase, and α-mannosidase, but negative for *N*-acetyl-β-glucosaminidase, lipase (C14), β-glucuronidase, acid phosphatase, naphthol-AS-BI-phosphohydrolase, leucine arylamidase, cystine arylamidase, valine arylamidase, α-fucosidase, chymotrypsin, and trypsin. Variable results in the API ZYM for esterase (C4) and esterase lipase (C8) (Wang et al., 2008b; Yuan et al., 2010).

Source (type strain): forest soil collected in Yunnan Province, south-west China.

DNA G+C content (mol%): 67.7 (HPLC).

Type strain: YIM 45725, CCTCC AA 206004, DSM 18448, JCM 16537.

Sequence accession no. (16S rRNA gene): EF601829.

Genus III. **Aeromicrobium** Miller, Woese and Brenner 1991, 367[VP] emend. Yoon, Lee and Oh 2005c, 2174

Lyudmila I. Evtushenko and Valentina I. Krausova

A.e.ro.mi.cro'bi.um. Gr. n. *aer, aeros* air; N.L. neut. n. *microbium* microbe; N.L. neut. n. *aeromicrobium* aerobic microbe.

Small irregular rods (mostly 0.3–0.6 µm in diameter) **to coccoid forms.** Elongated rods, branched filamentous elements, and V-forms rarely occur. A marked rod-to-coccoid growth cycle can be observed. Endospores are not formed. Nonmotile or motile. Extracellular diffuse haloes or capsules can be produced. **Gram-positive** cell-wall chemotype, non-acid-fast. Colonies are non-pigmented, yellow or beige to amber-beige in color. **Chemo-organotrophs, metabolism primarily respiratory** with oxygen as the terminal electron acceptor. **Mostly catalase-positive;** negative catalase reaction may occasionally occur. Nitrate can be reduced by some species. Acids are produced oxidatively from some carbohydrates. Generally nutritionally exacting; grows aerobically on complex media based on peptone, yeast extract, and similar sources of nutrients. Some species can grow in chemically defined media. Mesophilic, optimal growth at 25–37°C; growth range is ~ 4–42°C. Mostly non-halophilic or slightly halophilic; some species are salt-requiring. Prefer a neutral to mildly alkaline pH. **The cell-wall peptidoglycan contains LL-diaminopimelic acid, along with alanine, glutamic acid, and glycine.** Menaquinones are the sole respiratory quinones detected; the tetrahydrogenated menaquinone with nine isoprene units **[MK-9(H_4)] is predominant. The major cellular fatty acids are octadecenoic ($C_{18:1}$ ω9*c*), hexadecanoic ($C_{16:0}$), tuberculostearic ($C_{18:0}$ 10-methyl), and 2-hydroxy hexadecanoic ($C_{16:0}$ 2-OH) acids.** Mycolates are absent. Principal polar lipids reported are diphosphatidylglycerol, phosphatidylglycerol, and phosphatidylethanolamine, or diphosphatidylglycerol, phosphatidylglycerol, and phosphatidylinositol. Occur in terrestrial and aquatic environments and can be associated with plants, animals, and humans.

DNA G+C content (mol%): 65.5–75.9.

Type species: **Aeromicrobium erythreum** Miller, Woese and Brenner 1991, 367[VP].

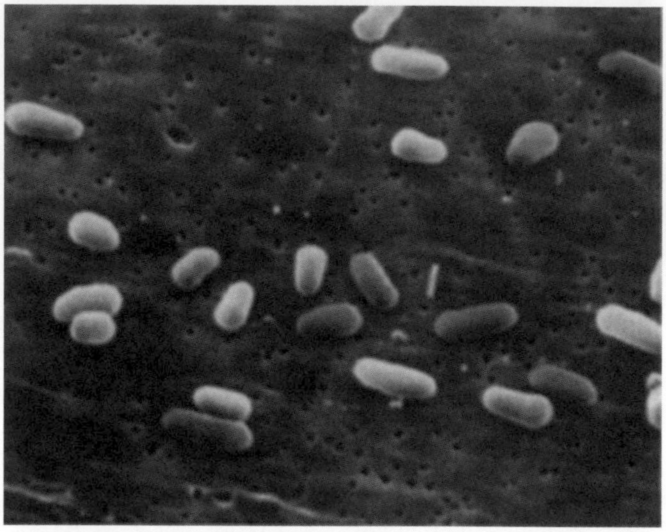

FIGURE 268. Cells of *Aeromicrobium erythreum*, 36-h culture grown in Marine broth (Difco). Scanning electron micrograph. Bar = 1 µm. (Reproduced with permission from Miller et al., 1991. Int. J. Syst. Bacteriol. *41*: 363–368.)

Further descriptive information

The genus *Aeromicrobium* belongs to the family *Nocardioidaceae* and the order *Propionibacteriales* (formerly suborder *Propionibacterineae* Stackebrandt et al. 1997). The current *Aeromicrobium* species form a coherent 16S rRNA-based phylogenetic cluster which is clearly separated from other genera and species comprising the family *Nocardioidaceae* (Figure 247).

Morphology and colony appearance. The cells of *Aeromicrobium* species are typically small irregular rods (mostly 0.3–0.6 µm in width and up to 2.4 µm in length) and can be coccoid (Figure 268 and Figure 269a, b; Table 223). Longer cells can be produced by *Aeromicrobium tamlense* (up to 4.8 µm; Lee and Kim, 2007) and *Aeromicrobium halocynthiae* (up to 6.0 µm or more; Kim et al., 2010). Cells of *Aeromicrobium halocynthiae* can also form straight or curved filaments with rudimentary branching (in A1+C liquid medium containing 10 g of starch, 4 g of peptone, 2 g of yeast extract, and 1 g of calcium carbonate in 1 liter of filtered seawater; pH 7.0; Kim et al., 2010). Where reported, the cells of *Aeromicrobium* species stain Gram-positive.

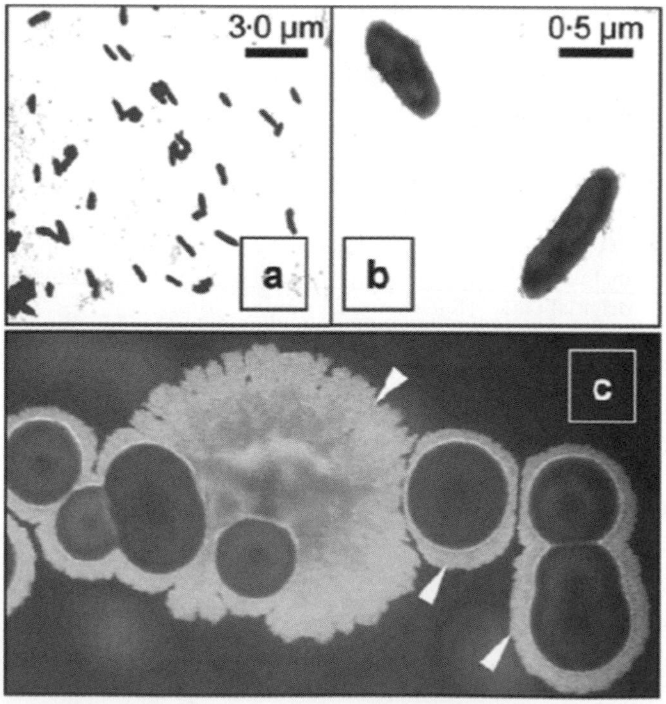

FIGURE 269. Morphological features of *Aeromicrobium marinum.* (a, b) Transmission electron micrographs of negatively stained cells from a 10-d-old culture in Marine broth; (c) colony morphology on Marine agar observed by light and epifluorescence microscopy; haloes are indicated by arrowheads. (Reproduced with permission from Bruns et al., 2003. Int. J. Syst. Evol. Microbiol. *53*: 1917–1923.)

TABLE 223. Descriptive and differential characteristics of *Aeromicrobium* species[a,b,c]

Characteristic	1. *A. erythreum*	2. *A. alkaliterrae*	3. *A. fastidiosum*	4. *A. flavum*	5. *A. ginsengisoli*	6. *A. halocynthiae*	7. *A. marinum*	8. *A. panaciterrae*	9. *A. ponti*	10. *A. tamlense*
Colony color	Beige, yellow	Cream	Whitish	Yellow	White	Light yellowish	Ivory	Light yellow	Yellow	Yellow
Cell width (μm)[d]	0.5	0.3–0.5	0.4–0.5	0.2–0.4	0.3–0.4	0.4–0.5	0.3–0.5	0.2–0.4	0.7	0.4–0.6
Cell length (μm)[d]	0.5–1.2	0.8–1.4	1.5–2.2	0.6–1.2	0.5–1.2	4.1–6.0	0.7–1.3	1.0–1.5	2.4	0.8–4.8
Motility	–	–	+	–	–	–	–	–	–	–
Optimal temperature (°C)	35	25	25	30	30	25	25	~30	30–37	30
Temperature range (°C)[e]	21–40	4–35	5–30	25–37	4–30	10–42	4–35	15–30	4–42	10–42
NaCl growth range (%, w/v)	0–4	0–8	0–4	0–4	0–4(w)	0–6	0.1–10.7	0–3	0–10	0–5
Optimal pH[f]	~7.0	7.0–7.5	7.0–7.4	~7.0	~7.0	~7.0	7.0–8.5	5.0–8.5	5.1–7.1	~7.0
pH growth range[f]	5.0–9.0	6.0–11.0	5.0–8.0	5.0–10.0	5.0–8.5	5.0–10.0	5.5–9.5	5.0–8.5	4.1–12.1	5.1–10.1
Catalase activity	+	+	+	+	–	+	+	+	+	+
Oxidase activity	+	–	+	+	+	–	–	–	–	–
Nitrate reduction[g]	–	–[h]	–[h]	+	+		–[h]	–[h]	–	–
Growth on ISP 9 with glucose[i]	–	+	–				–		+	+
Utilization of carbon sources:[j]										
L-Arabinose	+	v	+	–	+	+	–	–	+	–
D-Cellobiose	–	+	+	–	+	–	+	+	+	+
D-Fructose	+	–	+	+	+	+	–	–	+	+
D-Galactose	+	+	+		+	–	+	+	+	+
D-Glucose	+	+	+	+	+	+	v	+	+	+
Lactose	–	–	–		+	–	–	–	–	+
Maltose	–	+	v	+	+	+	v	+	+	+
D-Mannose	–	–	+	–	+	+	v	+	+	+
Melibiose	–				–	–		+		
D-Ribose	+	–		+	–	+		–		
L-Rhamnose	–	v	v	–	+	–	–	–	–	v
L-Sorbose	–	–	–	+	–	–	–	–	–	–
Sucrose	+	+	+	+	+	+	–	v	+	+
D-Trehalose	+	+	+	+	+	+	+	+	+	+
D-Xylose	+	–	+	–	+	+	–	–	+	–
Dextran	+	+	–	+	–		w		+	+
Glycerol	+	+	+	+	+	+	v	+	+	+
Inositol	+	+	+		+	+	–	–	–	–
D-Mannitol	–	v	v		–	+	+	v	+	–
Inulin	–			–	+	–			–	–
Salicin	–	+		–	–	–	v	+	–/w	–
Alanine	+	–	+		+		–	+		
Acetate	+	v	+	+	+	+	+	v	+	+
Adipate		+	+		–		+	+		
Citrate	v	v	v		–	–	v	v	v	–
Caprate		+	+		–		+	+		
Gluconate		+	+		+	–	+	+		
3-Hydroxybutyrate		+	+	–	+		+	–		
DL-Lactate		–	+		–		–	–		
Malate	+	+	+		+		+	+	+	+
Malonate		+	–		–		–	–		
Phenyl aceatate		+	+		–		+	+		
Propionate		+	+		+		–	–		
Pyruvate		–	+	+	+		+			–
Suberate		+	+		–		–	–		
Succinate	+	+	–		+		+	+	+	+
Valerate		+	+		+		–	–		
Hydrolysis/decomposition:[k]										
Esculin	+	–		+		–				+
Casein	+	+	+	–			–	–	+	–
Cellulose	+	–	–		–		–	–	+	
DNA	w	+	+	–	–		w	–	+	
Elastin	+	+	–				–	–	+	
Gelatin	+	+	v	+	+		–	+	+	+

(continued)

TABLE 223. (continued)

Characteristic	1. *A. erythreum*	2. *A. alkaliterrae*	3. *A. fastidiosum*	4. *A. flavum*	5. *A. ginsengisoli*	6. *A. halocynthiae*	7. *A. marinum*	8. *A. panaciterrae*	9. *A. ponti*	10. *A. tamlense*
Hypoxanthine	–	–	–				–	–	–	+
Starch	+	–	+			–	–	–	–	
Tween 80		+		–						+
Tyrosine	w	–	–		+		–	–	+	–
Enzyme activities (*API*, bioMérieux):										
Acid phosphatase	+	+	+	–	+	–	–	+	+	+
Alkaline phosphatase	v	–	+	–	–	–	–	+	+	+
Arginine dihydrolase	–	–	–	–	–	–	–	v	–	–
Esterase (C 4)	+	+	+	+	+	+	+	w	+	–
Esterase lipase (C8)	+	+	+	+	+	+	+	+	+	w
Lipase (C14)	–	–	–	w	–	–	–	–	–	–
β-Galactosidase	–	v	v	–	+	–	v	v	–	–
α-Glucosidase	+	+	+	+	v	w	–	–	v	+
β-Glucosidase	v	v	v	+	+	–	v	v	–	v
Leucine arylamidase	+	+	+	+		+	+	+	–	+
Naphthol-AS-BI-phosphohydrolase	–/w	+	w	–	+	w	w	+	–	w
N-Acetyl-β-glucosaminidase	–	–/w	–	–	+	–	–/w	–/w	–	–
Valine arylamidase	–	–	–	–		w	–	–	w	–
Urease	–	–	–	w	–	–	–	–	–	–
Trypsin	–	–	–	–	–	–	–	–	–	–/w
Voges–Proskauer reaction	+	–	–	w		+	+	–	+	w
Major menaquinone	$9(H_4)$	$9(H_4)$	$9(H_4)$	$9(H_4)$, $8(H_4)$	$9(H_4)$	$9(H_4)$	$9(H_4)$	$9(H_4)$	$9(H_4)$	
Principal phospholipids	DPG, PG, PE		DPG, PG, PE						DPG, PG, PI, PL	DPG, PG, PI
DNA G+C content (mol%)	71–73	71.5	71–72	73.3	66.8	75.9	70.6	65.5	74.0	72.7
Isolation source	Soil	Soil	Herb-age	Air	Soil	Ascidian	Sea water	Soil	Sea water	Sea-weed

[a]Based on characteristics of type strains. Symbols and abbreviations: –, negative; +, positive; v, variable between different experiments or test methods; w, weak or slow; –/w, either negative, or weak and slow; $9(H_4)$ and $8(H_4)$, menaquinone having nine and eight isoprene units, two of which are saturated; PE, phosphatidylethanolamine, PG, phosphatidylglycerol; DPG, diphosphatidylglycerol; PI, phosphatidylinositol, PL, unidentified phospholipids.

[b]Data compiled from: Miller et al. (1991), Tamura and Yokota (1994), Collins and Stackebrandt (1989b), Bruns et al. (2003), Yoon et al. (2005c), Cui et al. (2007a), Lee and Kim (2007), Kim et al. (2008b), Lee and Lee (2008), Tang et al. (2008), and Kim et al. (2010).

[c]The following characteristics were reported to be negative for all species (type strains) tested: Activities for lysine decarboxylase, ornithine decarboxylase, tryptophan deaminase; decomposition of chitin or chitinase activity (no data for *Aeromicrobium halocynthiae*); H_2S production and activities for cystine arylamidase, α-chymotrypsin, α-fucosidase, α-galactosidase, β-glucuronidase and α-mannosidase (no data for *Aeromicrobium ginsengisoli*); decomposition of xanthine (no data for *Aeromicrobium flavum*, *Aeromicrobium ginsengisoli* and *Aeromicrobium halocynthiae*).

The following characteristics were tested only for a few species (type strains) and were reported to be negative: Utilization of raffinose, dulcitol, and D-sorbitol (*Aeromicrobium ginsengisoli*, *Aeromicrobium halocynthiae*, *Aeromicrobium ponti*, and *Aeromicrobium tamlense*); utilization of tartrate (*Aeromicrobium ginsengisoli*, *Aeromicrobium ponti*, and *Aeromicrobium tamlense*); utilization of adonitol (*Aeromicrobium erythreum*, *Aeromicrobium ginsengisoli*, *Aeromicrobium halocynthiae*, and *Aeromicrobium tamlense*); utilization of benzoate (*Aeromicrobium alkaliterrae*, *Aeromicrobium ginsengisoli*, *Aeromicrobium ponti*, and *Aeromicrobium tamlense*); utilization of formate (*Aeromicrobium alkaliterrae*, *Aeromicrobium ginsengisoli*, *Aeromicrobium panaciterrae*, *Aeromicrobium ponti*, and *Aeromicrobium tamlense*); utilization of methyl-α-D-mannoside (*Aeromicrobium flavum*, *Aeromicrobium ponti*, and *Aeromicrobium tamlense*); indole production (*Aeromicrobium alkaliterrae*, *Aeromicrobium flavum*, *Aeromicrobium halocynthiae*, and *Aeromicrobium tamlense*), and xylanase activity/xylan degradation (*Aeromicrobium ginsengisoli* and *Aeromicrobium panaciterrae*).

[d]Cell size as usually observable in young cultures. Some differences in the cell size and morphology can be caused by different growth conditions; *Aeromicrobium halocynthiae* can produce filamentous elements with primary branching. In older (synchronized) cultures of some species, cells may become shorter and may consist predominantly or exclusively of shorter rods, coco-bacillary or coccoid forms (see the species descriptions for details).

[e]The lower and higher growth temperatures reported in the original species descriptions are indicated; actual growth temperature range can be slightly broader for some species (for more information, see the species description).

[f]Initial pH values of media.

[g]According to the data reported in the original species descriptions.

[h]Conflicting results obtained with the API 20NE (bioMérieux) test have been reported (Kim et al., 2008b).

[i]Basal mineral medium (Shirling and Gottlieb, 1966) without vitamins.

[j]Determined using different conventional methods, the API 20E, API 20NE, API ID32 GN (bioMérieux), and Biolog GP2 test systems, therefore the data for some species can be incomparable. See Table 224 and the original species descriptions for methods.

[k]See Lee and Lee (2008), Cui et al. (2007a), Kim et al. (2008b), and Kim et al. (2010) for methods.

Although cocci and irregular rods are characteristic of many species, little information is available on the rod–coccus growth cycle, which is characteristic of arthrobacters and many other actinobacteria. In the cycle, irregular rods in young cultures are replaced by coccoid forms in older cultures, and these coccoid forms, when transferred to fresh medium, produce outgrowths to give irregular rods again (Jones and Keddie, 2006). No correlation was observed between the age of the culture and prevalence of rod-shaped or coccoid cells for *Aeromicrobium erythreum* in TYE broth (1.6% tryptone, 1% yeast extract, 0.5% NaCl, and 0.4% D-glucose; Miller at al., 1991). The cell morphology (irregular rod-shaped cells) of *Aeromicrobium marinum* in Marine broth (MB; Difco) was also reported to remain constant irrespective of the culture age (Bruns et al., 2003). On the other hand, irregular spherical forms were reported in a 5-d-old culture of *Aeromicrobium ginsengsoli* on R2A agar (Difco; Kim et al., 2008b). Our recent observations showed that cultures of *Aeromicrobium fastidiosum* can exhibit a distinct rod–coccoid growth cycle on some organic media, e.g. on modified Corynebacterium agar (DSMZ medium No. 53) containing glucose (5 g), yeast extract (5 g), Casamino acids (10 g), NaCl (5 g), agar (15 g), and distilled water (1 liter); pH 7.0–7.2. When resting cells of old (2–3 weeks) or lyophilized cultures of this species are transferred to fresh agar medium, the cells grow as rods (mean, 0.3–0.5 × 1.8–2.2 µm) in young cultures (1–2 d) and become shorter (~3–4 fold) rods or coccoid cells in 5-d-old cultures. Most likely the irregular rods of *Aeromicrobium flavum* and elongated cells or branched filamentous elements of *Aeromicrobium halocynthiae* are also replaced by short rods and/or coccoid cells in older cultures, as can be suggested from the data and micrographs provided in the original species descriptions (Kim et al., 2010; Tang et al., 2008). The marked change of morphology during the growth cycle can probably be shown by some other *Aeromicrobium* species in synchronized cultures on suitable agar media, at least those species reportedly having both irregular rods and coccoid forms. As in arthrobacters and many other coryneform bacteria (Cure and Keddie, 1973; Jones and Keddie, 2006), the composition of the nutrient medium probably influences the extent to which the morphology of *Aeromicrobium* species changes during the growth cycle. The replacement (if any) of the (irregular) rods in cultures by coccoid or coco-bacillary forms varies with individual species and culture conditions and can take up to 5–7 d or even more.

Colonies are usually 0.5–1.5 mm in diameter; larger colonies (up to 3 mm) are occasionally formed (reported for 5 d-old culture R2A agar; Cui et al., 2007a). Colonies are typically circular, smooth, convex, and may have thin, translucent edges, probably due to production of extracellular slime as demonstrated for *Aeromicrobium marinum* (Bruns et al., 2003) and *Aeromicrobium flavum* (Tang et al., 2008). After treatments aimed at detection of proteins or acidic polysaccharides, the diffuse haloes around the *Aeromicrobium marinum* colonies stain weakly with Alcian blue, indicating that the haloes may contain acidic polysaccharides (Bruns et al., 2003). The color of colonies is white or whitish to yellow and amber beige. The colony pigment color and intensity may vary depending on growth medium and culture age, as reported, e.g. for *Aeromicrobium erythreum* (Miller et al., 1991). Diffusible brown or purple pigments can be produced under certain conditions (Miller et al., 1991).

Chemotaxonomy. All species contain LL-diaminopimelic acid (LL-A_2pm) as the diamino acid in the cell-wall peptidoglycan. The amino acid composition (LL-A_2pm, alanine, glutamic acid, and glycine) reported for *Aeromicrobium erythreum* and *Aeromicrobium fastidiosum* (Busse and Schumann, 1999; Miller et al., 1991; Tamura and Yokota, 1994) is consistent with the peptidoglycan type A3γ *sensu* Schleifer and Kandler (1972). The cell wall of the type strain of *Aeromicrobium fastidiosum* was found to have glucose, rhamnose, ribitol, and amino sugars, which are derived from at least two cell-wall anionic polymers. One of these polymers is a ribitol teichoic acid that contains an amino sugar, and another is supposedly phosphorhamnan (Tul'skaya, 2009). The major component of the respiratory quinine system is MK-9(H_4); other menaquinones, MK-8(H_4), MK-7(H_4), and MK-9(H_6), may occur in minor or trace amounts (Tamura and Yokota, 1994). A larger proportion (25.4%) of MK-8(H_4) has been reported for *Aeromicrobium flavum* (Tang et al., 2008).

The fatty acids of *Aeromicrobium* species are of complex type (Suzuki and Komagata, 1983a). The predominant components are mostly $C_{18:1}$ $\omega 9c$ (reaching up to 68%; Tang et al., 2008), $C_{16:0}$ (up to 47%; Schumann et al., 1997), $C_{18:0}$ 10-methyl (up to 42%; Tamura and Yokota, 1994), and $C_{16:0}$ 2-OH (up to 23%; and Lee, 2008). Their proportions (and the fatty acid profiles as a whole) vary with species, and usually are influenced by the composition of media, the growth temperature, the availability of oxygen, the culture age, and analytical procedure (Lee and Lee, 2008; Lee et al., 2000; Miller et al., 1991; Park et al., 1999; Tamura and Yokota, 1994; Yoon et al., 2005c). The saturated iso- and anteiso-branched fatty acids that dominate the fatty acid profiles of many other coryneform bacteria, including members of the family *Nocardioidaceae*, occur in minor or trace amounts. Principal polar lipids reported for *Aeromicrobium erythreum* and *Aeromicrobium fastidiosum* are phosphatidylethanolamine and phosphatidylglycerol (Tamura and Yokota, 1994); in addition, Lee and Kim (2007) reported the presence of diphosphatidylglycerol in these species. A different phospholipid pattern, including phosphatidylinositol, diphosphatidylglycerol, and phosphatidylglycerol, has been reported for *Aeromicrobium tamlense* (Lee and Kim, 2007) and *Aeromicrobium ponti* (Lee and Lee, 2008); additionally three unidentified phospholipids have been indicated for *Aeromicrobium ponti* (Lee and Lee, 2008).

Analysis of polyamines in the type strains of *Aeromicrobium erythreum* and *Aeromicrobium fastidiosum* (Busse and Schumann, 1999) revealed that both strains contained a large amount of cadaverine (48.6% and 24.8%, respectively, of the polyamine content). The principal polyamine in *Aeromicrobium fastidiosum* was spermine (58.4%), whereas the second most abundant polyamine in *Aeromicrobium erythreum* was spermidine (37.2%). Putrescine and 1,3-diaminopropane occurred in minor amounts in both strains. Notably, the total polyamine concentration in *Aeromicrobium fastidiosum* was rather low.

Nutrition and growth conditions. Bacteria of this genus will grow aerobically in nutritionally complex media containing peptone and yeast extract and utilize a wide range of carbohydrates and other compounds as carbon sources in appropriate growth media (Table 223). The data on minimal nutritional requirements of *Aeromicrobium* species are incomplete. *Aeromicrobium fastidiosum* requires thiamine and biotin (Tamura and Yokota, 1994). *Aeromicrobium erythreum* additionally needs nicotinic acid (Miller et al., 1991). Bruns et al. (2003) reported that *Aeromicrobium marinum* will grow on a minimal, artificial seawater medium with defined carbon sources (e.g. mannitol and amino acids), a 10-vitamin solution (Balch et al., 1979) and 1 ml of trace element solution (Widdel and Bak, 1992). *Aeromicrobium panaciterrae* grows in a minimal salts medium with glucose (or some other carbon sources), vitamins, and trace elements (Cui et al., 2007a). Vitamins (biotin, thiamine, and nicotinic acid) are not necessary for growth of *Aeromicrobium alkaliterrae* (Yoon et al., 2005c). Most likely *Aeromicrobium ponti* and *Aeromicrobium tamlense* do not require vitamins for growth on basal medium ISP 9 (Shirling and Gottlieb, 1966) containing defined carbon sources (Lee and Lee, 2008; Lee and Kim, 2007).

The optimal growth temperatures (25–37°C) and the temperature ranges (4–42°C or slightly higher) vary with species (Table 223). The recognized *Aeromicrobium* species are considered to be non-salt-requiring, except for *Aeromicrobium marinum*, which is reported to be obligately salt-dependent and exhibits optimum growth at salt concentrations similar to that of sea water (Bruns et al., 2003). Some other species have also been reported to prefer salt-containing media. *Aeromicrobium halocynthiae*, which grows on A1+C agar (prepared with seawater supplemented with 0–6% NaCl), shows robust growth in the presence of NaCl in this medium, but does not grow on standard TSA medium (Difco) (Kim et al., 2010). *Aeromicrobium flavum* displays the optimal growth rate in LB broth with 1–2% NaCl (Tang et al., 2008). Bacteria of this genus prefer a neutral to mildly alkaline pH. However, they may also grow in test media with acidic (up to 4.1) or alkaline (up to 11–12.1) initial pH values (Table 223).

Genomic characteristics. The DNA G+C content of most *Aeromicrobium* species are higher than 70 mol% (up to 75.9 mol%; Kim et al., 2010); lower values (65.5 and 66.8 mol%; HPLC) have been reported for *Aeromicrobium panaciterrae* (Cui et al., 2007a) and *Aeromicrobium ginsengisoli* (Kim et al., 2008b). The current *Aeromicrobium* species form a coherent phylogenetic group based on 16S rRNA gene sequence analysis (similarity, >96%; Figure 247). The similarity based on 16S–23S ITS sequence analysis for the type strains of *Aeromicrobium erythreum* and *Aeromicrobium fastidiosum* was found to be 71.2% (Yoon et al., 1998a). The authors emphasized that the 16S–23S ITS similarity of *Aeromicrobium erythreum* was higher to the type strain of *Nocardioides jensenii* (73.1%) than to *Aeromicrobium fastidiosum*, and higher to three *Streptomyces* species (62.9–70.0%) and one *Frankia* strain (68.8%) than to some organisms of the family *Nocardioidaceae*. Likewise, *Aeromicrobium fastidiosum* had high, but slightly lower, similarity to *Nocardioides jensenii* (70.1%) and higher similarity to representatives of *Streptomyces* and *Frankia* than to most members of the family *Nocardioidaceae* used in the analysis. Similarity of the ribonuclease P RNA gene sequences between the type strains of *Aeromicrobium erythreum* and *Aeromicrobium fastidiosum*, in contrast, was highest (88% or 78.6% when nucleotide gaps were included) than between *Aeromicrobium* strains and organisms of other genera used in the study (Yoon and Park, 2000).

Bacteriophages. Four bacteriophages infectious for *Nocardioides albus*, *Nocardioides simplex*, and/or *Terrabacter tumescens* were not infectious for *Aeromicrobium erythreum* NRRL B-3381; the complete absence of plaques suggests the existence of barriers to phage adsorption or phage replication (Miller et al., 1991).

Antibiotic sensitivity and antibiotic production. The growth of the type strains of two *Aeromicrobium* species (*Aeromicrobium erythreum* and *Aeromicrobium flavum*) are inhibited by tetracycline, chloramphenicol, kanamycin, and vancomycin, as well as other antibiotics as indicated in the species descriptions (Miller et al., 1991; Tang et al., 2008). *Aeromicrobium erythreum* JCM 8359 but not *Aeromicrobium fastidiosum* JCM 8088 produces β-lactamase (Ogawara et al., 1999). The type strain of *Aeromicrobium erythreum* produces the macrolide antibiotic erythromycin A (French et al., 1970; Miller et al., 1991). The erythromycin-biosynthetic (*ery*) gene cluster of *Aeromicrobium erythreum* was cloned and characterized (Brikun et al., 2004). The 55.4-kb cluster contained 25 *ery* genes, and homologs were found for each gene in the previously characterized *ery* gene cluster from *Saccharopolyspora erythraea*. In addition, four new predicted *ery* genes were revealed (Brikun et al., 2004). The characteristics of the erythromycin resistance gene *ermA* of *Aeromicrobium erythreum* were described by Roberts et al. (1985).

Habitats and ecology. The recognized *Aeromicrobium* species have been isolated from soils of different origin, plants, air, and aquatic environments (Table 223). *Aeromicrobium halocynthiae* was recovered from the siphon tissue of a marine ascidian, *Halocynthia roretzi* (Kim et al., 2010).

The species *Aeromicrobium tamlense* and *Aeromicrobium marinum* were or could be associated with algae. *Aeromicrobium marinum* was found to occur in large numbers in the German Wadden Sea and to assimilate mannitol, which is present in seaweed in large amount as well as in algal exudates (Bruns et al., 2003; Budavari, 1989; Spencer, 1990). This bacterium, in addition, produces abundant exopolysaccharides that might serve to attach to algal surfaces (Bruns et al., 2003). The ability of another seawater species, *Aeromicrobium ponti*, to use mannitol as a carbon source and to hydrolyze cellulose (Lee and Lee, 2008) might be indicative of its association with algae as well. Recent ecological studies show that bacteria of the genus *Aeromicrobium* are distributed in various other terrestrial and aquatic environments, including low-temperature and deep subseafloor ecosystems (Glöckner et al., 2000; Hansen et al., 2007; Imazaki and Kobori, 2010; Katayama et al., 2007; Kobayashi et al., 2008; Lesaulnier et al., 2008; Rintala et al., 2008; Stevens et al., 2007; Zhang et al., 2007; Zhang et al., 2009b). Some unnamed members of this genus have been found in polluted environments, including high-level nuclear waste-contaminated sediments (Fredrickson et al., 2004), hydrocarbon-contaminated soils (Militon et al., 2010), and copper-contaminated substrates (Sun et al., 2010). They can occur on the surface of copper coins (Santo et al., 2010) and are associated with plant tissues (Sun et al., 2010; Wang et al., 2008a) and deep-water marine sponge (Sfanos et al., 2005). Representatives of the genus also can be isolated from the skin microbiota of salamanders (Lauer et al., 2008) and humans (e.g. GenBank numbers HM336148 and HQ616221) and are also among viable microorganisms in retropharyngeal lymph nodes (organs once believed to be largely amicrobic in the absence of overt disease) of healthy wild ungulates (mule deer, *Odocoileus hemionus*; Wittekindt et al., 2010). No species or strains pathogenic for humans, other warm-blooded animals, or plants have yet been identified.

Enrichment and isolation procedures

No selective medium for the isolation of strains of *Aeromicrobium* has been described. Dilution plating and different media, e.g. R2A agar (Difco), tenfold-diluted nutrient agar (NA, Difco), and other agar media containing peptone and yeast extract may be used for isolation. Incubation is usually carried out at 25–30°C from a week to a month. Adjustment of pH to 10.0 with Na_2CO_2, as reported for *Aeromicrobium alkaliterrae* (Yoon et al., 2005c), facilitates isolation of alkalitolerant (alkaliphilic) species. Marine agar (MA, Difco) and SC-SW medium (1% soluble starch, 0.03% casein, 0.2% KNO_3, 0.2% NaCl, 0.002% $CaCO_3$, 1.8% agar, 0.005% $MgSO_4 \cdot 7H_2O$, and 0.001% $FeSO_4 \cdot 7H_2O$ in a 60:40 mixture of natural seawater and distilled water; Lee and Kim, 2007) can be used to isolate *Aeromicrobium* species from marine environments. Pre-incubation of a water sample in Marine broth (Difco) as described for isolation of *Aeromicrobium marinum* (Bruns et al., 2003) can be useful. When isolating strains of *Aeromicrobium* species, selective (semi-selective) media may be developed on the basis of characteristics listed in Table 223, the minimal nutritional requirements of *Aeromicrobium* species, and resistance to certain antibiotics. Characteristics of environmental samples should be taken into account. Alkaline pre-treatment before serial dilution and plating, suggested for isolation of some bacteria (Påhlson et al., 1986; Wakisaka et al., 1982), facilitates reducing numerous Gram-negative bacteria. The filtration of (suspended) environmental samples through membrane filters with a pore size of 0.45 or 0.65 µm or other approaches may be useful to separate small cells of *Aeromicrobium* (which may be even smaller in environmental samples) from large-sized microorganisms occurring in the same sample.

Maintenance procedures

Cultures may be maintained as 20% glycerol suspensions at −20 and −70°C. Long-term conservation is achieved by freeze-drying by standard procedures.

Differentiation of the genus *Aeromicrobium* from other genera

Table 214 lists the phenotypic characteristics differentiating *Aeromicrobium* from other genera of the family *Nocardioidaceae*. *Aeromicrobium* can be easily distinguished from the genera and species comprising mycelium-forming organisms on the basis of cell morphology. The predominant menaquinone MK-9(H_4) is the most salient characteristic which differentiates *Aeromicrobium* from all members of the genera *Nocardioides* and *Marmoricola*. Other chemotaxonomic characteristics which may serve to distinguish *Aeromicrobium* from related genera with LL-A_2pm in the cell wall include the profiles of cellular fatty acids, polar lipids, polyamines, and probably the composition of cell-wall polysaccharides.

Taxonomic comments

The genus *Aeromicrobium* with the type species *Aeromicrobium erythreum* was described by Miller et al. (1991) to accommodate the strain "*Arthrobacter*" sp. NRRL B-3381, an unusual non-mycelium-forming producer of the macrolide antibiotic erythromycin A, that differed in many ways from *Arthrobacter sensu stricto*. The second species, *Aeromicrobium fastidiosum*, was added to the genus by Tamura and Yokota (1994) as a result of the reclassification of *Nocardioides fastidiosa* (Collins and Stackebrandt, 1989a, 1989b).

Eight remaining *Aeromicrobium* species included in this volume were described within the 2003–2010 period on the basis of taxonomic study of single environmental isolates. The descriptions relied on 16S rRNA gene sequence analysis, DNA-DNA hybridization studies (examined in cases of high 16S rRNA sequence similarity), and phenotypic traits. The DNA-DNA similarities between *Aeromicrobium* species obtained by different methods ranged from very low (3–11.5%; Cui et al., 2007a; Tang et al., 2008; Yoon et al., 2005c) to high (up to 63.5% for the *Aeromicrobium halocynthiae-Aeromicrobium ponti* pair; Kim et al., 2010). The original genus description (Miller et al., 1991), which was based on one strain and included many features specific for this strain, was emended by Yoon et al. (2005c) to take into account other relevant data accumulated by 2005.

Differentiation of species of the genus *Aeromicrobium*

Phenotypic characteristics useful in distinguishing the currently recognized ten species of the genus *Aeromicrobium* are listed in Table 224. Additional characteristics useful for differentiation are given in Table 223 and in the species descriptions below. Since many test results can be influenced by the test method, a comparative experimental study of phenotypic characteristics of a novel isolate and type strains of recognized species rather than comparisons with data presented is recommended (Schumann et al., 2009; Tindall et al., 2010).

Acknowledgements

The authors were supported by the program MCB RAS of the Russian Academy of Sciences.

TABLE 224. Differential characters useful for identification of *Aeromicrobium* species[a,b]

Characteristic	1. *A. erythreum*	2. *A. alkaliterrae*	3. *A. fastidiosum*	4. *A. flavum*	5. *A. ginsengisoli*	6. *A. halocynthiae*	7. *A. marinum*	8. *A. panaciterrae*	9. *A. ponti*	10. *A. tamlense*
Colony color[c]	Beige, yellow	Cream	White	Yellow	White	Light yellowish	Ivory	Light yellow	Yellow	Yellow
Motility	–	–	+	–	–	–	–	–	–	–
Growth at:										
10°C	–	+	+	–	+	+	+	–	+	+
42°C	–	–	–	–	–	+	–	–	+	+
6% NaCl	–	+	–	–	–	+	+	–	+	–
8% NaCl	–	+	–	–	–	–	+	–	+	–
Catalase activity	+	+	+	+	–	+	+	+	+	+
Oxidase activity	+	–	+	+	+	–	–	–	–	–
Utilization of carbon sources:[d]										
L-Arabinose	+	+[e]	+	–	+	+	–	–	+	–
D-Cellobiose	–	+	+	–	+	–	+	+	+	+
D-Fructose	+	–	+	+	+	+	–	–	+	+
D-Glucose	+	+	+	+	+	+	–[e]	+	+	+
D-Mannose	–	–	+	–	+	+	–[e]	+	+	+
D-Xylose	+	–	+	–	+	+	–	–	+	–
Succinate	+	+	–		+		+	+	+	+
Hydrolysis of casein	+	+	+	–			–	–	+	–
Hydrolysis of tyrosine	w	–	–		+		–	–	+	–
Acid phosphatase	+	+	+	–	–	–	–	+	+	+
Alkaline phosphatase	v	–	+	–	–	–	–	+	+	+
DNA G+C content (mol%)	71–73 (T_m)	71.5 (HPLC)	71–72 (T_m)	73.3 (T_m)	66.8 (HPLC)	75.9 (HPLC)	70.6 (HPLC)	65.5 (HPLC)	74.0 (HPLC)	72.7 (HPLC)

[a]Based on characteristics of type strains. Symbols: –, negative; +, positive; v, variable between experiments; w, weakly positive.

[b]See Table 223 for references.

[c]Color intensity and shade may vary depending on growth conditions; for details concerning *Aeromicrobium erythreum*, see the species description.

[d]Identical test results (unless indicated) were obtained by using both conventional methods and the API test systems (data for *Aeromicrobium erythreum*, *Aeromicrobium alkaliterrae*, *Aeromicrobium fastidiosum*, *Aeromicrobium marinum*, and *Aeromicrobium panaciterrae*). The results for *Aeromicrobium ponti* and *Aeromicrobium tamlense* were obtained using ISP 9 (Shirling and Gottlieb, 1966) as the basal medium (Lee and Lee, 2008; Lee and Kim, 2007). API test systems (bioMérieux) were used to obtain the data for *Aeromicrobium ginsengisoli* (Kim et al., 2008b) and *Aeromicrobium halocynthiae* (Kim et al., 2010), and the Biolog GP2 test system was used to obtain the data for *Aeromicrobium flavum* (Tang et al., 2008).

[e]The opposite result was obtained using the API test system (Kim et al., 2008b).

List of species of the genus *Aeromicrobium*

1. **Aeromicrobium erythreum** Miller, Woese and Brenner 1991, 367[VP]

e.ryth're.um. N.L. neut. adj. *erythreum* intended to mean erythromycin-producing.

Characteristics are as described for the genus and listed in Table 223. Additional information given below is taken from the original paper (Miller et al., 1991) unless indicated. Predominantly irregular rods to coccoid forms (0.5 × 0.5–1.2 µm; Figure 268) in TEY broth (1.6% tryptone, 1% yeast extract, 0.5% NaCl, and 0.4% D-glucose) at all growth phases (24, 48, and 120 h). Cells are non-motile. Colonies are beige to amber beige and 1 mm in diameter after 3 d of cultivation on TYE agar medium (1% tryptone, 0.5% yeast extract, 0.8% NaCl, and 1.5% agar) at 32°C. Colonies sometimes have thin, translucent edges. The colony color changes to a shade of yellow on media allowing even slight acid production and to distinct yellow (e.g. on R2YE medium containing sucrose; Hopwood et al., 1985) or other media (Miller et al., 1991; Tamura and Yokota, 1994). Diffusible brown or purple pigments may be produced on R2YE plates (but are not observed on TYE agar) after extended incubation at 32°C. The optimum growth temperature is 35 ± 2°C; no growth occurs at 18°C or 43°C. Grows in nutritionally complex media. Biotin, nicotinic acid, and thiamine are required for growth in chemically defined media. A wide range of organic compounds are utilized as sole or principal carbon plus energy sources for growth (Table 223). Acid production from carbohydrates is generally very weak, as determined using the phenol red indicator medium (Difco), but growth on fructose gives readily detectable acidification. The type strain produces the macrolide antibiotic erythromycin A (which can be detected by standard growth inhibition assays with susceptible bacteria on various media). The strain tested using a Gram-positive minimal inhibitory concentration (MIC) panel (Baxter, Sacramento, CA) is sensitive to amikacin, cefotaxime, ceftriaxone, chloramphenicol, ciprofloxacin, gentamycin, hygromycin, kanamycin, neomycin, oxacillin, streptomycin, sulfamethoxazole, tetracycline, thiostrepton, vancomycin, and viomycin. Resistant to amoxicillin, ampicillin, cefamandole, clindamycin, erythromycin, imipenem, naladixic acid, nitrofurantoin, norfloxacin, penicillin, rifampin, spiramycin, and tylosin. Ogawara et al. (1999) reported that benzylpenicillin potassium inhibited growth at a concentration of 50 µg/ml, but not at 10 µg/ml.

The major compounds of polyamine pattern include cadaverine (48.6%) and spermidine (37.2%); other compounds, i.e. spermine, 1,3-diaminopropane, and putrescine, occur in minor amounts (Busse and Schumann, 1999). The predominant menaquinone is MK-9(H_4), with minor amounts of MK-7(H_4) and MK-8(H_4) (Tamura and Yokota, 1994). The major phospholipids are phosphatidylethanolamine, phosphatidylglycerol, and diphosphatidylglycerol (Lee and Kim, 2007; Tamura and Yokota, 1994). The major (>5%) fatty acids, as determined in cells grown under various conditions and harvested at different growth phase were reported to be $C_{18:0}$ 10-methyl (16.2–39%), $C_{18:1}$ ω9*c* (15.1–29.1%), $C_{16:0}$ (16.2–25.2%), $C_{16:0}$ 2-OH (11.3–14.4%), $C_{17:0}$ 10-methyl (0.3–9.5%), and $C_{18:0}$ (4.3–10.9%) (Lee and Lee, 2008; Lee et al., 2000; Miller et al., 1991; Park et al., 1999; Tamura and Yokota, 1994). A higher content of $C_{16:0}$ (47%) and lower proportions of other acids were detected for cells harvested from trypticase soy broth (Difco) after incubation at 28°C for 24–48 h (Schumann et al., 1997).

Source (type strain): tropical soil in Puerto Rico.

DNA G+C content (mol%): 71–73 (T_m).

Type strain: NRRL B-3381, ATCC 51598, DSM 8599, JCM 8359, LMG 16472, NBRC 15406, NRRL B-3381.

Sequence accession no. (16S rRNA gene): AF005021

2. **Aeromicrobium alkaliterrae** Yoon, Lee and Oh 2005c, 2174[VP]

al.ka.li.ter'ra.e. N.L. n. *alkali* (from Arabic *al-qalyi*, the ashes of saltwort), soda ash; L. gen. n. *terrae* of the soil or earth; N.L. gen. n. *alkaliterrae* of alkaline soil.

Characteristics are as described for the genus and listed in Table 223. Additional information given below is taken from the original paper (Yoon et al., 2005c) unless indicated. Small rods (0.3–0.5 × 0.8–1.4 µm) or cocci. No motility by flagella. Colonies are yellow and 1.0–1.5 mm in diameter after 7 d of cultivation at 25°C on nutrient agar (Difco). Capable of growth on simple chemically defined media, including ISP 9 medium (Shirling and Gottlieb, 1966) supplemented with glucose or other carbon sources. No growth occurs under anaerobic conditions on trypticase soy agar (TSA; Difco) or on TSA supplemented with nitrate. Grows at 35°C, but not at 37°C. Tolerates NaCl up to 8% in trypticase soy broth (TSB; Difco), but optimal growth is observed when no salt is added. Growth occurs at pH 6.0 and pH 11.0, but not at pH 5.5 or 11.5 (initial pH values of medium, TSB). Displays the ability to hydrolyze Tweens 20, 40, and 60. The major fatty acids (5% or more), as determined in cultures grown on TSA and MA at 25°C for 6 d, are $C_{16:0}$ (34.1 and 33.8%), $C_{18:0}$ 10-methyl (16.3 and 20.9%), $C_{16:0}$ 2-OH (17.6 and 15.5%), $C_{18:1}$ ω9*c* (5.7 and 9.9%), and $C_{17:0}$ 10-methyl (5.0 and 1.0%), respectively. Generally similar proportions of the above acids, except for slightly decreased $C_{18:1}$ ω9*c* (2.3%) were reported by Lee and Lee (2008) for cells grown on TSA agar at 30°C for 5 d.

Source (type strain): an alkaline soil in Kwangchun, Korea.

DNA G+C content (mol%): 71.5 (HPLC).

Type strain: strain KSL-107, KCTC 19073, DSM 16824, JCM 13518.

Sequence accession no. (16S rRNA gene): AY822044.

3. **Aeromicrobium fastidiosum** (Collins and Stackebrandt 1989b) Tamura and Yokota 1994, 610[VP] (*Nocardioides fastidiosa* Collins and Stackebrandt 1989b, 293)

fas.ti.di.o'sum. L. neut. adj. *fastidiosum* fastidious, referring to the nutritionally fastidious nature of the organism when it is first isolated.

Characteristics are as described for the genus and listed in Table 223. Additional information given below is taken from the original descriptions (Collins and Stackebrandt, 1989b; Tamura and Yokota, 1994), unless indicated. Irregular rods (average, 0.4–0.5 × 1.5–2.2 µm) in young culture (1–2 d) on modified CB agar at 26°C (recent observation). The rods fragment and become shorter as growth proceeds (usually after 3 d) and can be short cell chains at this stage. After 5–6 d of culture, coco-bacillary to coccoid cells, usually single or in pairs (recent observation), predominate. In contrast to all other *Aeromicrobium* species, *Aeromicrobium fastidiosum* is motile. Colonies are white or grayish-white. Grows in nutritionally complex media. Biotin and thiamine are required for growth in a suitable minimal salts medium with glucose. Growth occurs at 30°C, but not at 40°C. A wide range of organic compounds are utilized as sole or principal carbon plus energy sources for growth (Table 223), including L-phenylalanine, L-threonine, tyrosine, butanol, crotonate, dodecane, hexadecane, oxalacetate, and pentoate. Acid is produced from glucose. The cell-wall polysaccharides contain glucose, rhamnose, an aminosugar, and ribitol, which originate from at least two anionic polymers, i.e. the ribitol-based teichoic acid and supposedly phosphorhamnan (Tul'skaya, 2009). *Aeromicrobium fastidiosum* is characterized by a rather low concentration of polyamines; the polyamime pattern includes spermine (58.4%), cadaverine (24.8%), and minor amounts of putrescine, spermidine, and 1,3-diaminopropane (Busse and Schumann, 1999). The major menaquinone is MK-9(H_4), with minor amounts of MK-9(H_6) and MK-8(H_4). Principal phospholipids are phosphatidylethanolamine, phosphatidylglycerol, and diphosphatidylglycerol (Lee and Kim, 2007; Tamura and Yokota, 1994). The major fatty acids (>5%) determined in cells grown under different conditions are $C_{18:1}$ ω9*c* (12.3–48.2%), $C_{18:0}$ 10-methyl (9.1–42%), $C_{16:0}$ 2-OH (8.9–23.2%), $C_{16:0}$ (10–18.3%), $C_{18:0}$ (6.1–16.1%), and $C_{17:0}$ 10-methyl (<1–8.7%) as reported by Lee and Lee (2008), Lee et al. (2000), Park et al. (1999), and Tamura and Yokota (1994).

Source (type strain): herbage.

DNA G+C content (mol%): 71–72 (T_m).

Type strain: strain J41, ATCC 49363, DSM 10552, IFO (now NBRC) 14897, JCM 8088, LMG 16205, NCIB (now NCIMB) 12713, VKM Ac-1324.

Sequence accession no. (16S rRNA gene): AF005022; X53189, X76862, Z78209.

4. **Aeromicrobium flavum** Tang, Zhou, Zhang, Mao, Luo, Wang and Fang 2008, 1862[VP]

fla'vum. L. neut. adj. *flavum* yellow, referring to the colony color.

Characteristics are as described for the genus and listed in Table 223. Additional information given below is taken from the original description (Tang et al., 2008). Cells are mostly irregular rods (ca. 0.2–0.4 in width and up to 1.2 µm in length) as observed after 16 h of culture on LB agar, and become shorter in a 2-d-old culture. The cells are surrounded by an amorphous polysaccharide-like layer and form conglomerates. No motility by flagella. Colonies are yellow pigmented. Grows readily in nutritionally complex media at 25–37°C but not at 10 and 42°C. The optimum NaCl concentration for growth is 1–2% (tested in LB broth). In addition to characteristics listed in Table 223, the type strain shows positive test reactions in the Biolog GP2 test system for D-psicose, turanose, α-ketovaleric acid, γ-hydroxybutyric acid, and D-fructose-6-phosphate. Negative responses have been recorded for D-arabitol, L-fucose, α-cyclodextrin, glycogen, mannan, *N*-acetyl-β-D-mannosamine, amygdalin, L-alanyl glycine, arbutin, L-asparagine, 2′-deoxyadenosine, putrescine, sedoheptulosan, glycyl-L-glutamic, α-hydroxybutyric, and L-lactic acids. Displays also negative responses in the Biolog GP2 plate for assimilation of dextrin, thymidine, and uridine, which along with other features listed in Table 223 and Table 224 differentiate this species from its phylogenetically closest relative *Aeromicrobium tamlense* (98.4% 16S rRNA gene sequence similarity). The DNA-DNA relatedness value between the type strains of these two species was reported to be 35%. Susceptible to tetracycline, chloramphenicol, kanamycin, vancomycin, erythromycin, rifampicin, and tobramycin; resistant to penicillin and nalidixic acid (determined by the agar-diffusion method using antibiotic-impregnated discs as described by Buczolits et al., 2002). The major menaquinones are MK-9(H_4) (74.6%) and MK-8(H_4) (25.4%). The fatty acids (1% or more) recorded for cells grown for 24 h at 30°C in trypticase soy broth (Difco) include $C_{18:1}$ ω9*c* (68.4%), $C_{18:0}$ (11.7%), $C_{16:0}$ (8.7%), $C_{16:0}$ 2-OH (5.3%), $C_{18:0}$ 10-methyl (2.3%), and $C_{16:1}$ ω6*c* (1.0%).

Source (type strain): air (on the campus of Wuhan University, China).

DNA G+C content (mol%): 73.3 (T_m).

Type strain: TYLN1, CCTCC AB 206046, DSM 19355.

Sequence accession no. (16S rRNA gene): EF133690.

5. **Aeromicrobium ginsengisoli** Kim, Park, Im and Yang 2008b, 2028[VP]

gin.sen.gi.so'li. N.L. n. *ginsengum* ginseng; L. n. *solum* soil; N.L. gen. n. *ginsengisoli* of/from ginseng soil.

Characteristics are as described for the genus and listed in Table 223. Additional information given below is taken from the original description (Kim et al., 2008b). Cells (typically 0.3–0.4 × 0.5–1.2 µm) display coccoid forms after 5 d in culture at 30°C on R2A agar (Difco). No motility was observed with a light microscope. Colonies are white on R2A after 5 d. Will grow in nutritionally complex media at temperatures up to 30°C or slightly higher, but not at 37°C. Grows on R2A agar with 3% NaCl and weakly with 4% NaCl. No growth on MacConkey agar. In contrast to all other recognized *Aeromicrobium* species, this species was reported to display negative catalase reaction. Oxidase activity (evaluated by using 1% *p*-aminodimethylaniline oxalate) is positive. Reduces nitrate to nitrite, but not to nitrogen gas. Positive for assimilation of D-fucose, D-lyxose, ethanol, fumarate, L-asparagine, glutamate, glutamine, L-leucine, L-phenylalanine, L-proline, L-serine, L-tryptophan, and L-tyrosine in the API 20E and API 20NE tests (bioMérieux). Negative for assimilation of L-fucose, melibiose, L-xylose, *myo*-inositol, methanol, glycogen, 3- and 4-hydroxybenzoate, glutarate, 2- and 5-ketogluconate, itaconate, maleic acid, oxalate, L-arginine, L-aspartate, L-cysteine, L-glycine, L-histidine, L-isoleucine, L-lysine, L-methionine, L-threonine, and L-valine. Fatty acids determined in cells grown on trypticase soy agar (Difco) at 30°C for 48 h include $C_{16:0}$ (32.7%), $C_{18:0}$ 10-methyl (20.6%), $C_{18:0}$ (12.9%), $C_{17:0}$ (3.2%), $C_{19:0}$ 10-methyl (3.1%), and $C_{17:0}$ 10-methyl (2.4%).

Source (type strain): soil of a ginseng field in Daejeon, South Korea.

DNA G+C content (mol%): 66.8 (HPLC).

Type strain: Gsoil 098, GBS 39, JCM 14732, KCTC 19207.

Sequence accession no. (16S rRNA gene): AB245394.

6. **Aeromicrobium halocynthiae** Kim, Yang, Sohn and Kwon 2010, 2797[VP]

ha.lo.cyn.thi'a.e. N.L. gen. n. *halocynthiae* of *Halocynthia roretzi*, the ascidian from which the type strain was isolated.

Characteristics are as described for the genus and listed in Table 223. Additional information presented below is taken from the original paper (Kim et al., 2010). Cells are typically elongated rods (approximately 0.4–0.5 × 4.1–6.0 µm); straight, curved and rudimentarily branched filaments can be produced (e.g. in culture growing in A1+C liquid medium at 25°C). Colonies are light yellowish and 0.6–1 mm in diameter after incubation on A1+C agar plates at 25°C for 5 d. Grows on marine agar (Difco) and some other complex media but not on standard TSA medium (Difco). Will grow on A1+C agar with 0–6% NaCl, but NaCl is required for robust growth. Grows at 10–42°C (but not at 7 and 45°C), with optimal growth at 25°C. Will grow in liquid A1+C medium over a pH range of 5.0–10.0. The following carbon sources are utilized (API 50 CHE system): turanose, D-tagatose and 5-ketogluconate. Erythritol, D-arabinose, L-xylose, adonitol, methyl-β-xyloside, dulcitol, sorbitol, methyl α-D-mannoside, methyl-α-D-glucoside, *N*-acetylglucosamine, amygdalin, arbutin, esculin, salicin, melezitose, raffinose, starch, glycogen, xylitol, β-gentiobiose, D-lyxose, D-fucose, L-fucose, D-arabitol, L-arabitol, and 2-ketogluconate are not utilized as sole carbon and energy sources. Oxidizes arabinose, does not oxidize glucose, mannitol, inositol, sorbitol, rhamnose, sucrose, melibiose or amygdalin. α-Galactosidase and β-glucuronidase activities are not detected. The type strain produces taurocholic acid, a bile acid, as a major secondary metabolite. Predominant fatty acids (>5%) are $C_{18:1}$ ω9*c* (35.2%), $C_{16:0}$ (31.5%), $C_{18:0}$ 10-methyl (21.1%), and $C_{18:0}$ (5.5%). The hydroxylated fatty acid ($C_{16:0}$ 2-OH), which typically occurs in significant amounts in all other recognized *Aeromicrobium* species, has been not detected in *Aeromicrobium halocynthiae*. This species is phylogenetically closest to *Aeromicrobium ginsengisoli*, *Aeromicrobium erythreum* and *Aeromicrobium ponti* (97.7, 97.6, and 97.5% 16S rRNA gene sequence similarities, respectively). The DNA–DNA similarity values between the type strain of *Aeromicrobium halocynthiae* and the type strains of the aforementioned species are 49.6–63.5%.

Source (type strain): siphon tissue of a marine ascidian, *Halocynthia roretzi*, collected at a depth of 15 m near Kyung-Po beach, coast of Gangneung, Korea.

DNA G+C content (mol%): 75.9 (HPLC).

Type strain: KME 001, JCM 15749, KCCM 90079.

Sequence accession no. (16S rRNA gene): FJ042789.

7. **Aeromicrobium marinum** Bruns, Philipp, Cypionka and Brinkhoff 2003, 1922[VP]

ma.ri′num. L. neut. adj. *marinum* of the sea, marine.

Characteristics are as described for the genus and listed in Table 223. Additional information presented below is taken from the original description (Bruns et al., 2003) unless indicated. Cells are irregular rods (0.3–0.5 × 0.7–1.3 µm; Figure 269a,b). The morphology is reported to remain constant in Marine broth (Difco), irrespective of the culture age (up to 10 d). No flagella, pili, or other appendages are observed. Colonies are ivory-colored and 0.5–1.0 mm in diameter; diffuse haloes of different size (Figure 269c) are observed around the colonies. It is the sole species within the genus proven to be salt-requiring so far. Grows well in ASW/YPG medium (1 liter of artificial sea water, 0.3% yeast extract, 0.6% peptone, and 0.3% glucose) with salt concentrations in the range 0.63–10.7%. A maximal growth rate is observable at 5.35%; growth is weak at 0.08% salinity and absent when no salts are added. The type strain grows with carbon sources such as crotonic acid, fumarate, 2-oxoglutarate, arginine, and glutamic acid but not with pimelic acid, nicotinic acid, salicylic acid, ethanol, propanol, butanol, asparagine, aspartic acid, cysteine, glutamine, glycine, or histidine. Cannot grow with D-glucose as a carbon source in minimal salts media containing vitamins and microelements (Bruns et al., 2003; Cui et al., 2007a). Unable to hydrolyze vegetable oil. Bruns et al. (2003) and Lee and Lee (2008) reported the following major fatty acids (>5%) for the type strain: $C_{18:1}$ ω9*c* (28.5 and 21.7%), $C_{16:0}$ (15.0 and 16.6%), $C_{16:0}$ 2-OH (14.0 and 7.9%), and $C_{18:0}$ 10-methyl (12.5 and 8.2%). Lee and Lee (2008) reported additionally significant amounts of $C_{17:0}$ (9.9%) and $C_{18:0}$ (11.5%) for cells harvested from TSA agar (Difco) after incubation at 30°C for 5 d.

Source (type strain): surface waters of the German Wadden Sea.

DNA G+C content (mol%): 70.6 (HPLC).

Type strain: T2, DSM 15272, JCM 13314, LMG 21768.

Sequence accession no. (16S rRNA gene): AY166703.

8. **Aeromicrobium panaciterrae** Cui, Im, Yin, Lee, Lee and Lee 2007a, 690[VP]

pa.na.ci.ter′ra.e. N.L. n. *Panax, -acis* scientific name of ginseng; L. n. *terra* soil; N.L. gen. n. *panaciterrae* of soil of a ginseng field.

Characteristics are as described for the genus and listed in Table 223. Information presented below is taken from the original paper (Cui et al., 2007a). Cells are short rods (0.2–0.4 × 1.0–1.5 µm) after culture for 3 d in modified R2A broth. No motility was observed with a light microscope. Colonies grown on R2A agar (Difco) for 5 d are light yellowish white and reach 1–3 mm in diameter. Grows well at 15–30°C (but not at 4 or 35°C), in the pH range of 5.0–8.5, and in the presence of up to 3% NaCl on R2A agar. Will grow readily on nutrient agar (0.5% peptone, 0.3% beef extract, and 1.5% agar). Growth on TSA agar is slow and looks like very thin layer on the agar surface after 5 d at 30°C. No growth on MacConkey agar. Capable of growth in suitable minimal salts media with glucose, vitamins, and trace elements. Utilizes melibiose, L-lactate, D-glycogen, and glycine as carbon sources but not D-fucose, D-lyxose, L-lysine, and DL-serine. Does not produce acid or gas from glucose oxidatively or fermentatively as determined using O-F basal medium (Atlas, 1993) with bromothymol blue. The major components (>5%) of the fatty acid profile determined in cells grown in modified R2A broth for 6 d are $C_{18:0}$ 10-methyl (14.5%), $C_{16:0}$ (13.0%), $C_{16:0}$ 2-OH (12.8%), $C_{17:0}$ 10-methyl (11.9%), $C_{16:0}$ 10-methyl (11.7%), $C_{16:1}$ ω7*c*, $C_{15:0}$ iso 2-OH (6.6%), and $C_{18:1}$ ω9*c* (5.3%).

Source (type strain): soil of a ginseng field in Pocheon Province, South Korea.

DNA G+C content (mol%): 65.5 (HPLC).

Type strain: Gsoil 161, KCTC 19131, DSM 17939, CCUG 52476.

Sequence accession no. (16S rRNA gene): AB245387.

9. **Aeromicrobium ponti** Lee and Lee 2008, 990[VP]

pon′ti. L. gen. n. *ponti* of the sea, belonging to the sea, referring to the isolation of the type strain from seawater.

Characteristics are as described for the genus and listed in Table 223. Information presented below is taken from the original paper (Lee and Lee, 2008). Cells are irregular rods (0.7 × 2.4 µm) as observed in 24-h culture on TSB at 30°C. Colonies are yellow in color and reach 0.8–1.2 mm in diameter after 5 d on TSA (Difco). Neither motility nor flagella are observed. Will grow on simple chemically defined media, e.g. ISP 9 (Shirling and Gottlieb, 1966) with glucose or other carbohydrates, sugar alcohols, or organic acids (Table 223). Does not grow on ISP 9 medium with D-arabinose, melezitose, methyl-α-D-glucoside, or *meso*-erythritol. Grows well at 30–37°C; can grow at 42°C or even higher. Displays growth in the pH range of 4.1–12.1 (initial pH of medium, TSA), but the optimal pH for growth is 5.1–7.1. Tolerates 10% NaCl as tested on ISP 2 agar medium. Able to hydrolyze carboxymethylcellulose. The polar lipids are phosphatidylinositol, diphosphatidylglycerol, phosphatidylglycerol, and three unidentified phospholipids. Predominant fatty acids determined in cells grown on TSA agar at 30°C for 5 d are $C_{18:1}$ ω9*c* (34.7%), $C_{16:0}$ (19.7%), $C_{16:0}$ 2-OH (14.8%), and $C_{18:0}$ 10-methyl (11.0%).

Source (type strain): seawater from the coast of Jeju Island, Republic of Korea.

DNA G+C content (mol%): 74.0 (HPLC).

Type strain: HSW-1, DSM 19178, KACC 20565.

Sequence accession no. (16S rRNA gene): AM778683.

10. **Aeromicrobium tamlense** Lee and Kim 2007, 339[VP]

tam.len′se. N.L. neut. adj. *tamlense* of or belonging to Tamla, the old name of Jeju, Republic of Korea, the site of isolation of the type strain.

Characteristics are as described for the genus and listed in Table 223. Additional information presented below is taken from the original paper (Lee and Kim, 2007) unless indicated. Cells are irregular rods ranging in size from 0.4–0.6 × 0.8–1.2 μm to 0.5 × 3.8–4.8 μm, as observed in 3-d-old cultures on YE-SW agar at 30°C. Colonies are yellow and 0.6–0.8 mm in diameter after incubation for 5 d on YE-SW at 30°C. No growth occurs at 4 or 45°C. Growth is observed at pH 10.1 (initial pH of medium, TSA) and in the presence of up to 5% NaCl (tested on ISP 2 medium; Shirling and Gottlieb, 1966). Grows in simple chemically defined media, e.g. ISP 9 (Shirling and Gottlieb, 1966), with glucose or other carbohydrates, sugar alcohols, or organic acids (Table 223) and also with L-ribose, or methyl-α-glucoside. The following carbon sources are not utilized as sole carbon and energy sources: D-arabinose, D-melezitose, 2,3-butanediol, *meso*-erythritol, and 1,2-propanediol. The type strain displays positive responses in the Biolog GP2 plate for assimilation of dextrin, thymidine, and uridine (Tang et al., 2008), which, along with other characteristics listed in Table 223 and Table 224, may serve to differentiate *Aeromicrobium tamlense* from the phylogenetically closest species *Aeromicrobium flavum*. Principal phospholipids include phosphatidylinositol, diphosphatidylglycerol, and phosphatidylglycerol. The major fatty acids in cells harvested from TSA agar (Difco) after incubation at 30°C for 5 d are $C_{18:1}$ ω9*c* (47.2%), $C_{16:0}$ (13.8%), $C_{18:0}$ (13.6), $C_{18:0}$ 10-methyl (8.7%), and $C_{16:0}$ 2-OH (6.8%).

Source (type strain): dried seaweed collected from Samyang beach in Jeju Island, Korea.

DNA G+C content (mol%): 72.7 (HPLC).

Type strain: SSW1-57, JCM 13811, CIP 09549, DSM 19087, NRRL B-24466.

Sequence accession no. (16S rRNA gene): DQ411541.

Genus IV. **Kribbella** Park, Yoon, Shin, Suzuki, Kudo, Seino, Kim, Lee and Lee 1999, 750^VP^ emend. Sohn, Hong, Bae and Chun 2003, 1007

Lyudmila I. Evtushenko and Valentina I. Krausova

Krib.bel'la. N.L. fem. dim. n. *Kribbella* arbitrary name formed from the acronym of the Korea Research Institute of Bioscience and Biotechnology, KRIBB, where taxonomic studies of this taxon were performed.

Extensively branched vegetative (substrate) hyphae, mostly ~0.4–0.7 μm in diameter, growing on the surface of, and penetrating, agar media. **Aerial hyphae are typically produced** on suitable agar media. Both the substrate and aerial **hyphae usually undergo various degrees of fragmentation** eventually **resulting in irregular rod-shaped to coccoid, nonmotile, elements**. Short chains of well-to-poorly differentiated conidia may occur on aerial hyphae. Bud-like cells, apical swellings, and cell conglomerates may be produced by the substrate mycelium. No endospores are formed. **Colonies are pasty to soft-leathery**, often with lichenous shape and usually no distinct coloration. Diffusible pigments are usually not produced, except for melanoid pigments characteristic of some species. **Gram-stain-positive**, but may be variable. Non-acid-fast. **Chemo-organotrophs, having a respiratory type of metabolism**, with a potential for mixotrophy and metabolic flexibility. **Grow well under aerobic conditions** on standard laboratory media, including the chemically defined (synthetic) media. Some species show also weak growth under anaerobic conditions. Catalase-positive. Oxidase test reaction intensity varies with species. Utilize a wide range of carbon and nitrogen sources and possess a significant spectrum of hydrolytic activities. Mesophilic, optimal growth temperature is ~28–30°C; the lower and higher growth temperatures reported for different species are 8°C and 45°C. Usually non-halophilic and prefer a neutral to mildly alkaline pH; some species grow at initial pH values of 10 and slightly below 5.0. **The cell-wall peptidoglycan contains LL-diaminopimelic acid**, along with alanine, glutamic acid, and glycine. The muramic acid residue of the peptidoglycan is ***N*-acetylated.** The **main secondary cell-wall polymer is teichuronic or teichulosonic acid.** Menaquinones are the sole respiratory quinones detected; the tetrahydrogenated menaquinone with nine isoprene units **[MK-9(H_4)] is the major component. The saturated iso- and anteiso branched acids ($C_{15:0}$ anteiso, $C_{16:0}$ iso, and $C_{15:0}$ iso) are the main** cellular fatty acids, with **$C_{15:0}$ anteiso usually contributing a maximal proportion**. Monounsaturated iso-branched, saturated 9-methyl branched fatty acids typically occur and may constitute significant amounts. Other fatty acids, including 2-hydroxy and 10-methyl branched acids may occur in minor or trace quantities. Mycolates are absent. **The diagnostic component of the polar lipid patterns is phosphatidylcholine** (PC); additional principal polar lipids reported are diphosphatidylglycerol (DPG), phosphatidylglycerol (PG), and phosphatidylinositol (PI).

Natural habitats are various terrestrial environments; some organisms are associated with plants and fungi, and occur in the human skin microbiome.

DNA G+C content (mol%): 67–71.3 (T_m; HPLC).

Type species: **Kribbella flavida** Park, Yoon, Shin, Suzuki, Kudo, Seino, Kim, Lee and Lee 1999, 750^VP^.*

Further descriptive information

The genus *Kribbella* belongs to the family *Nocardioidaceae*, order *Propionibacteriales*. Based on the 16S rRNA gene sequence analysis, the current 16 *Kribbella* species form a tight, clearly separated phylogenetic cluster (Figure 247) with a high 16S rRNA gene sequence similarity (>97.5% and up to 99.6%).

Morphology and colony appearance. Aerial mycelium varies in abundance, depending on growth media and the individual strain or species, and at times is discernible only microscopically or lacking, e.g. on some agar media rich in organics.

* *Hongia* Lee, Kang and Hah 2000, 197^VP^ is a later heterotypic synonym of *Kribbella* Park, Yoon, Shin, Suzuki, Kudo, Seino, Kim, Lee and Lee 1999, 750^VP^

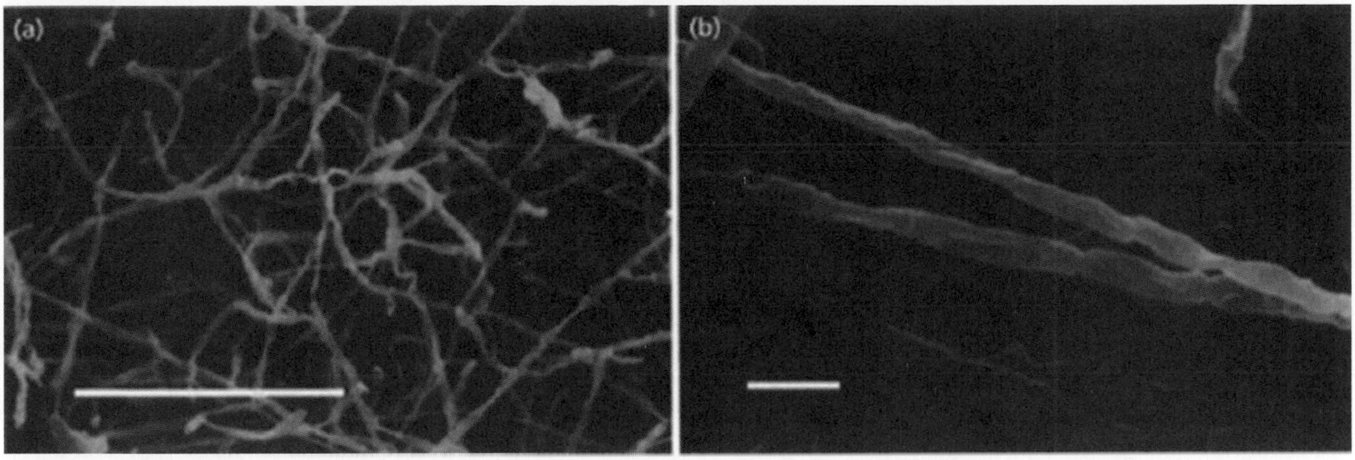

FIGURE 270. Aerial mycelium of *Kribbella flavida* on oatmeal agar (ISP 3 medium). Scanning electron micrographs. Bars = 10 µm (a) and 1 µm (b). (Reproduced with permission from Park et al., 1999. Int. J. Syst. Bacteriol. *49*: 743–752.)

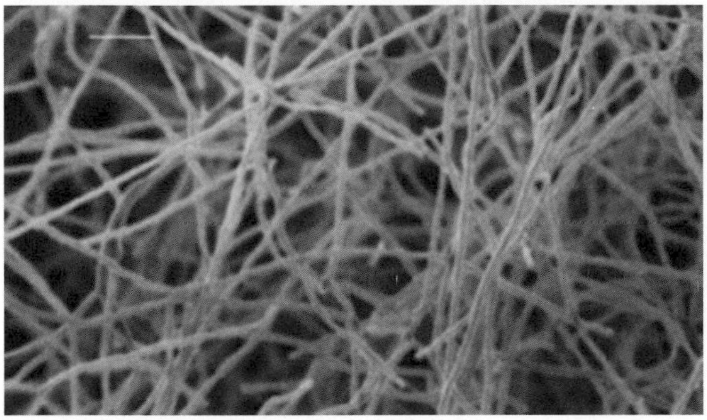

FIGURE 271. Aerial mycelium of *Kribbella swartbergensis*, 14-d-old culture on oatmeal agar (ISP 3). Scanning electron micrograph. Bar = 3 µm. (Reproduced with permission from Kirby et al., 2006. Int. J. Syst. Evol. Microbiol. *56*: 1097–1101.)

The aerial mycelium, if present, is mostly white or pale yellow and consists of straight or curved branched hyphae (Figure 270 and Figure 271). Well-developed and extensively branching substrate mycelium is typically white, off-white, pale yellow, cream, or rarely orange (Table 225). Both the aerial and substrate hyphae, depending on the individual organism and the consistency and composition of growth medium, undergo various degrees of fragmentation to yield nonmotile elongated, Y-shaped, and irregular rod-like elements with square ends (Figure 272, Figure 273, and Figure 274). Shorter fragments often become rounded-off and have the appearance of arthrospores at maturation. Hyphae in many areas may remain stable and not fragmented *in situ* on agar or in liquid media. Short chains of well-to-poorly differentiated conidia can be observed on aerial hyphae. Some organisms may produce bud-like cells on substrate hyphae (Figure 274). Hyphae of many organisms have a tendency to form apical swellings (Lee et al., 2000; recent observation of the authors of this chapter). Spherical or irregularly-shaped structures (up to 3–5 µm in diameter) can be produced on substrate hyphae of some organisms, most abundantly on agar media rich in organics when aerial mycelium is absent (recent observations). These consist of tightly packed, irregularly sized polymorphic cells remaining closely associated with each other after cell division and embedded in a common matrix of extracellular polysaccharide-like material. Both the hyphal fragments (arthrospores) and the cells produced by different division modes usually give rise to new vegetative hyphae when transferred to a fresh medium.

Growth may be pasty, sand-pasty, soft-leathery, or leathery, depending on species, growth media, and growth phase. Diffusible pigments are usually absent, except for melanoid pigments, which are produced on tyrosine agar (ISP 7 medium*) (Table 225) and occasionally on peptone-yeast extract agar (ISP 6 medium). The cell-wall architecture of *Kribbella* species is typical of Gram-positive bacteria (recent observation), but the Gram-staining test may show variable results (Park et al., 1999).

Chemotaxonomy. All *Kribbella* species are characterized by the presence of LL-diaminopimelic acid (LL-A_2pm) as the diamino acid in the cell wall. The peptidoglycan type is A3γ according to classification of Schleifer and Kandler (1972), as follows from the peptidoglycan amino acid composition [alanine, LL-2,6-diaminopimelic acid (LL-A_2pm), glutamic acid, and glycine] of the total cell-wall hydrolysate, and specific dipeptides (L-Ala–D-Glu, LL-A_2pm–D-Ala and LL-A_2pm–Gly) found in the partial cell-wall hydrolysate of *Kribbella lupini* (Trujillo et al., 2006). The monomeric unit of the A3γ polymer type contains L-alanine in the first position and LL-A_2pm in the third position of the tetrapeptide subunit, with a single glycine residue as an interpeptide bridge. The same peptidoglycan amino acids were reported for *Kribbella flavida*, *Kribbella sandramycini* (Park et al., 1999), *Kribbella koreensis* (Lee et al., 2000), *Kribbella solani*, *Kribbella jejuensis* (Song et al., 2004), and *Kribbella ginsengisoli* (Cui et al., 2010); Park et al. (1999) indicated the presence of both D- and L-alanine for *Kribbella flavida* and *Kribbella sandramycini*.

The sugar patterns determined in the cell walls or whole organisms of the recognized *Kribbella* species include combinations of galactose, glucose, mannose, ribose, xylose, and

*See Shirling and Gottlieb (1966) for the composition of ISP media cited here and in other sections of this chapter.

TABLE 225. Differential characters useful for preliminary identification of *Kribbella* species[a,b]

Characteristic	1. *K. flavida*	2. *K. alba*	3. *K. aluminosa*	4. *K. antibiotica*	5. *K. catacumbae*	6. *K. ginsengsoli*	7. *K. hippodromi*	8. *K. jejuensis*	9. *K. karoonensis*	10. *K. koreensis*	11. *K. lupini*	12. *K. sancticallisti*	13. *K. sandramycini*	14. *K. solani*	15. *K. swartbergensis*	16. *K. yunnanensis*
Aerial mycelium[c]	White	White to soft yellow	White	Yellow white	White	White	White	White	White, pale cream	White	White	White	White	White	White	White to moderate yellow
Substrate mycelium[c]	Pale yellow	Pale yellow to soft yellow	Beige to pale yellow	Yellow white to yellow	Yellow to yellow-orange[d]	Cream	Cream	Cream	Cream to yellow	Cream	Cream	Cream	Pale yellow	Cream	Cream	Yellow white to yellow
Melanin production	+	–	–	+	+	nr	–	–	–	+	nr	w	–	–	–	w
Anaerobic growth	w[e]	nr	nr	–[e]	–[f]	nr	–[e]	w[e]	–[e]	w[e]	nr	–[f]	w[e]	+[e,f]	–[e]	nr
Maximal NaCl (%)[g]	3[h]	nr	2	<2	3	4	6	1–2[i]	4	2	7	5	4	5	4	nr
Growth at 37°C	v	nr	+	–	–	+	+	+	+	v	+	+	v	–	+	–
Growth at 45°C	–	nr	–	–	–	–	–	–	–	–	–	–	–	–	+	–
Growth at initial pH 5[g]	v	nr	+	–	+	–	+	–	+	–	–	+	+	–	v	–
Growth at initial pH 9[g]	+	–	–	–	+	–	v	–	+	+	+	+	+	–	+	–
Oxidase	+	nr	–	+	+	+	nr	nr	nr	–	+	+	+	nr	nr	nr
Nitrate reduction	+	–	–	–	–	–	+	–	+	v	–	+	–	v	+	–
H_2S production	–	–	+	–	nr	–	+	nr	+	+	nr	nr	–	nr	+	–
Growth on sole carbon sources:[j]																
L-Arabinose	–	+	+	+	+	+	+	+	–	+	+	+	w	+	w	+
D-Galactose	–	+	nr	+	+	+	+	+	nr	+	+	+	+	–	nr	–
D-Xylose	–	+	+	+	+	+	–	+	+	+	nr	+	+	+	w	+
Inositol	+	+	+	+	–	+	+	–	+	+	nr	–	+	v	w	+
Mannitol	+	+	+	+	+	+	+	–	+	+	+	+	+	v	w	+
Degradation of:[k]																
Casein	+	–	+	+	+	+	+	v	+	+	+	+	+	v	+	–
Gelatin	–	+	nr	+	+	+	+	+	+	+	+	+	+	nr	+	–
Hypoxanthine	–	nr	+	+	w	nr	+	v	+	+	nr	+	–	v	+	nr
Starch	–	+	+	+	–	w	w	v	+	v	–	–	v	v	+	+
Tween 80	+	nr	+	nr	+	nr	+	–	w	–	nr	+	+	+	+	nr
Urea	+	nr	+	nr	–	–	–	+	+	+	–	–	+	–	–	nr
DNA G+C content (mol%)	70	67.9	nr	67	nr	66.3	nr	68	nr	71.3	68	nr	68.3	69	nr	68.6

Sugar composition[l]	Man, Glu, Gal	Glu, Gal, Rib	Glu, Man, Rib	Glu, Xyl, Rib	nr	Gal, Rib, Xyl	Glu, US1	Man, Glu, Gal Rib	Man, Rib	Man, Glu, Gal, Rib	Gal, US2	nr	Man, Glu, Gal	Man, Glu, Gal, Rib	Man, Rib	Man, Glu, Gal, Rib
Principal phospholipids	PC, DPG, PG, PI	PC, DPG, PG, PI	PC, DPG, PG, PI [m]	PC, DPG, PG, PI	PC, DPG, PG, PI[m]	PC, DPG, PG, PI	nr	PC, DPG, PI	nr	PC, DPG, PG, PI	PC, DPG, PG, PI	PC, DPG, PG, PI [m]	PC, PG, DPG, PI	PC, DPG, PI	nr	PC, PG, DPG
Source of isolation	Soil	Soil	Medieval alum slate mine	Soil	Catacomb, deteriorated surface	Soil	Soil	Soil	Soil	Gold-mine cave	Root of lupine	Catacomb, tufacean surface	Soil	Potato tuber	Soil	Soil

[a]Based on characteristics of the type strains, except for *Kribbella aluminosa*, *Kribbella catacumbae*, and *Kribbella sancticallisti*. Symbols and abbreviations: +, positive; –, negative; w, weakly positive or delayed; v, variable among experiments or methods used (sometimes because of weak growth or reaction); Glc, glucose; Gal, galactose; Man, mannose; Rib, ribose; Xyl, xylose, US1 and US2, unidentified sugars; PC, phosphatidylcholine; PG, phosphatidylglycerol; DPG, diphosphatidylglycerol; PI, phosphatidylinositol; nr, Not reported.

[b]Data from: Park et al. (1999), Lee et al. (2000), Sohn et al. (2003), Song et al. (2004), Li et al. (2004, 2006), Kirby et al. (2006), Trujillo et al. (2006), Carlsohn et al. (2007), Urzì et al. (2008), Everest and Meyers (2008), and Cui et al. (2010).

[c]According to the original descriptions; color may slightly change depending on the growth medium and the culture age.

[d]Dark orange in old colonies on Luedemann medium (Luedemann, 1968).

[e]Obtained on ATCC Medium No. 172 (Cote et al., 1984) according to Kirby et al. (2006) and Everest and Meyers (2008).

[f]Tested on VL agar (Tiecco, 1975) according to Urzì et al. (2008).

[g]Tested on solidified or/and in liquid media of different composition (see the text for details).

[h]Obtained on nutrient agar as basal medium (Trujillo et al., 2006); no growth on ISP 2 as basal medium (Kirby et al., 2006).

[i]Different results reported (Carlsohn et al., 2007; Kirby et al., 2006; Song et al., 2004) are likely influenced by different test conditions.

[j]ISP 9 (Shirling and Gottlieb, 1966) as basal medium, except for *Kribbella ginsengisoli* tested on a minimal salts medium with vitamins and microelements, or using the API API 20NE and ID 32GN test systems (Cui et al., 2010).

[k]See the original species description for test methods.

[l]Determined in whole cells or purified cell walls. Cui et al. (2010) reported glucosamine instead of glucose for *Kribbella flavida*, *Kribbella koreensis*, *Kribbella sandramycini* and *Kribbella yunnanensis*, and mannose instead of glucose for *Kribbella alba* and *Kribbella antibiotica*.

[m]Unidentified phospholipids and glycolipids reported in addition (Carlsohn et al., 2007; Urzì et al., 2008).

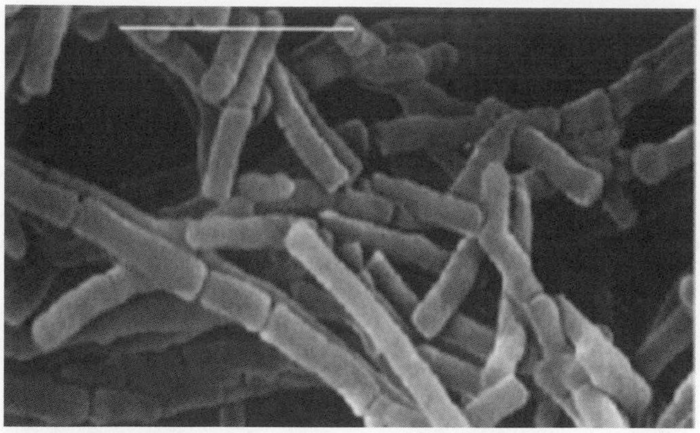

FIGURE 272. Fragmented aerial mycelium of *Kribbella koreensis*, 14-d-old culture on inorganic salts/starch agar (ISP medium 4). Scanning electron micrograph. Bar = 2 µm. (Reproduced with permission from Lee et al., 2000. Int. J. Syst. Evol. Microbiol. *50*: 191–199.)

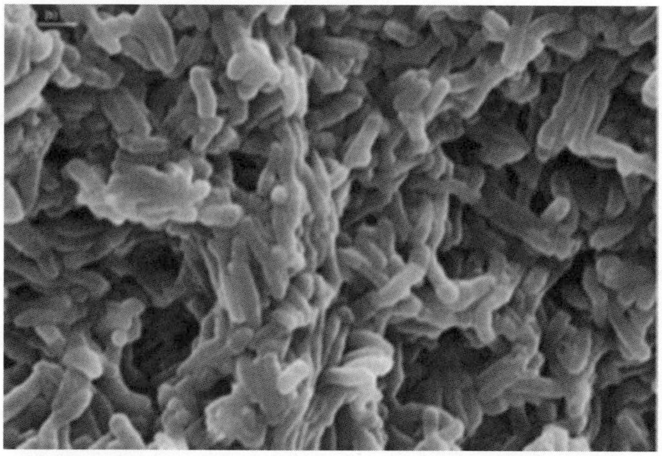

FIGURE 273. Fragmented substrate mycelium of *Kribbella karoonensis*, 14-d-old culture on oatmeal agar (ISP 3 medium). Samples were prepared by placing agar blocks containing substrate mycelium in liquid nitrogen and freezing prior to viewing by cryo-SEM. Bar = 3 µm. (Reproduced with permission from Kirby et al., 2006. Int. J. Syst. Evol. Microbiol. *56*: 1097–1101.)

unidentified sugars (Table 225). In addition, 3-*O*-methyl-D-galactose (madurose), 2,3-di-*O*-methyl-D-galactose, and rhamnose were found in several recently analyzed *Kribbella* strains which supposedly (based on 16S rRNA gene sequence analysis and phenotypic traits) represent at least four novel species (Shashkov et al., 2009; E.M. Tul'skaya and L.I. Evtushenko, unpublished results). In traditional terms (Lechevalier and Lechevalier, 1970), neither individual cell-wall sugars nor sugar pattern are characteristic of all organisms of the genus *Kribbella*. However, the sugars, along with other compounds derived from the cell-wall polysaccharides linked to the peptidoglycan, can serve as chemical markers of individual species or species groups within the genus. Recent studies performed with six of the aforementioned novel *Kribbella* strains revealed that each strain contained two cell-wall polysaccharides, i.e. an acidic phosphate-less polymer as the predominant one and a neutral polysaccharide, a mannan, found in minor or trace amounts (Shashkov et al., 2009; Tul'skaya, 2009). Notably, the mannan

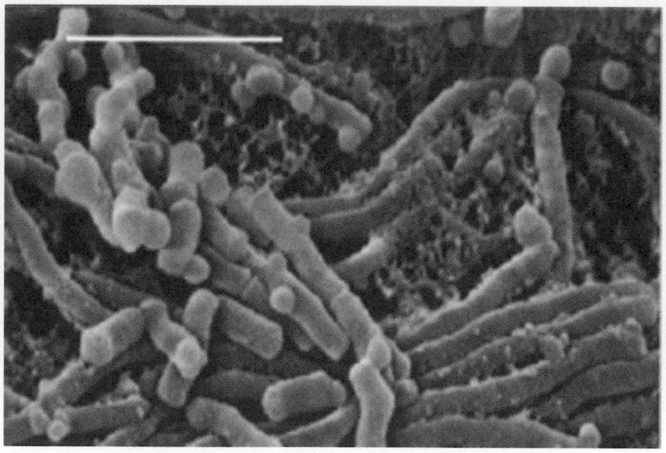

FIGURE 274. Irregularly fragmented substrate mycelium of *Kribbella koreensis* with bud-like cells, 14-d-old culture on inorganic salts/starch agar (ISP medium 4). Scanning electron micrograph. Bar = 2 µm. (Reproduced with permission from Lee et al., 2000. Int. J. Syst. Evol. Microbiol. *50*: 191–199.)

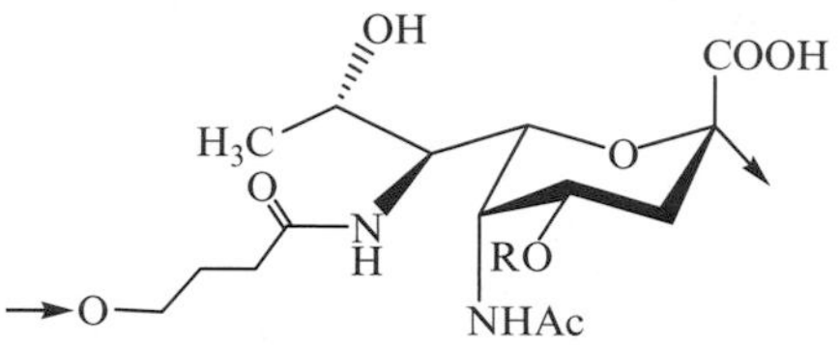

FIGURE 275. Repeating unit of teichulosonic acid of strain *Kribbella* sp. VKM Ac-2541, where R = H (45%), α-D-Gal*p*3OMe (37%) or α-D-Gal*p*2,3OMe (18%).

(the basic chain built of 1–6-linked α-mannopyranose substituted at O-2 with α-mannopyranose) was common to all six strains studied. The predominant acidic polymers, in contrast, differed to a greater or lesser extent, and were represented either by different teichuronic acids (Ward, 1981) or unusual polysaccharides, the so-called teichulosonic acids (polymers with nonulosonic acids; Knirel, 2009; Knirel et al., 1987, 2003). The disaccharide subunits of the two identified teichuronic acids consisted of aminomannuronic acid and a rare sugar, 2,3-diacetamido-2,3-dideoxy-α-glucopyranose (strain VKM Ac-2539), or of the same sugar and diaminomannuronic acid (strains VKM Ac-2538 and VKM Ac-2540). The subunit of teichulosonic acid of strain VKM Ac-2541 contained in the basic chain a residue of oxybutyrate and pseudaminic acid, a sialic acid-like nine-carbon compound (5,7-diacylamido-3,5,7,9-tetradeoxy-L-glycero-β-L-manno-nonulosonic acid; Knirel et al., 2003; Vimr et al., 2004) with madurose and 2,3-di-*O*-methyl-D-galactose as irregular substituents at O-4 (Figure 275). The teichulosonic acids of the two other strains had the same structure as in strain VKM Ac-2541 plus rhamnose as an additional substituent, or the polymer subunit contained additional galactose (unknown position). Remarkably, neither teichoic acids nor other related phosphorous-containing polysaccharides have been revealed in the cell walls of more than 15 *Kribbella* strains examined up to now (Shashkov et al., 2009; E.M. Tul'skaya, A.S. Shashkov, and L.I. Evtushenko, unpublished), in contrast to organisms of the genera *Nocardioides* and *Aeromicrobium* (Naumova et al., 2001; Takeuchi and Yokota, 1989; Tul'skaya, 2009). Also, although pseudaminic acid (or its derivatives) is common in capsular polysaccharides, lipopolysaccarides, pili, and flagella of many Gram-negative bacteria (see, e.g. Knirel, 2009; Knirel et al., 1987, 2003; Schoenhofen et al., 2006; Vimr et al., 2004), it has not been found until recently in Gram-positive bacteria except kribbellae.

The predominant menaquinone is MK-9(H_4) in all bacteria of the genus *Kribbella* (Table 226) and may represent up to 93% of the total quinone content as reported for *Kribella aluminosa* (Carlsohn et al., 2007), or even more. The cellular fatty acids of *Kribbella* species are dominated by the saturated iso- and anteiso branched acids, mainly $C_{15:0}$ anteiso, $C_{16:0}$ iso, and $C_{15:0}$ iso. Among these, $C_{15:0}$ anteiso was the predominant component in all species (usually contributing about 30%), except *Kribbella sancticallisti* and *Kribbella swartbergensis* which were reported to contain $C_{16:0}$ iso in maximal proportions (~24–27%) (Carlsohn et al., 2007; Urzì et al., 2008). Monounsaturated *iso*-branched (mostly $C_{17:1}$ iso) as well as saturated 9-methyl branched (mostly 9-methyl-$C_{16:0}$) fatty acids typically occur and may reach 10% or more of the total fatty acids. Other components, including 2-hydroxy (mostly $C_{16:0}$ iso 2-OH) and 10-methyl branched acids ($C_{18:0}$ 10-methyl and $C_{17:0}$ 10-methyl) were reported for some species in minor amounts. Notably, the qualitative and quantitative compositions of fatty acids in strains of this genus may depend on the growth conditions, the culture age, and analytical procedure employed, as demonstrated, e.g. for *Kribbella flavida* and *Kribbella sandramycini* (Park et al., 1999; Sohn et al., 2003; Song et al., 2004; Trujillo et al., 2006).

The principal phospholipids of most *Kribbella* species include PC, DPG, PG, and PI. PG was not reported in *Kribbella jejuensis* and *Kribbella solani* (Song et al., 2004), while PI was not mentioned among main polar lipids of *Kribbella yunnanensis* (Li et al., 2006). On the other hand, an additional phospholipid (tentatively identified as phosphatidylinositol containing 2-hydroxy fatty acids) and some unidentified minor phospholipids occur in all 15 yet-unnamed *Kribbella* strains analyzed (N.G. Vinokurova and L.I. Evtushenko, unpublished data). In addition, one to three glycolipids may present in lesser but significant amounts in *Kribbella* strains, together with a variety of minor or trace amounts of unidentified phospho-, glyco-, and other kinds of polar lipids (Figure 263 in the chapter Genus *Nocardioides*).The clearly detectable glycolipids have chromatographic mobility close to that of PC and DPG. Several unknown minor phospho- and glycolipids, along with PC, DPG, PG, and PI, were also reported for *Kribbella aluminosa* (Carlsohn et al., 2007), *Kribbella catacumbae*, and *Kribbella sancticallisti* (Urzì et al., 2008).

Nutrition and growth conditions. The *Kribbella* strains usually grow well under aerobic conditions at neutral pH or mildly alkaline pH on standard laboratory media, including the chemically defined media, e.g. ISP media used to culture streptomycetes. Kribbellae also utilize a wide range of other organic compounds as carbon sources in ISP9 or other minimal salts media, and possess various enzymatic activities towards complex macromolecules resulting in production of low molecular weight compounds which can be used as carbon (and/or nitrogen) sources (Table 225 and Table 226; see the section *List of species of the genus Kribbella*). *Kribbella ginsengsoli* was reported (Cui et al., 2010) to grow on a minimal medium containing basal salts, a vitamin solution (Widdel and Bak, 1992), trace elements (SL-10; Widdel et al., 1983), selenite/tungstate solution (Tschech and Pfennig, 1984), and a definite carbon source (0.1%). Acids are rather weakly produced by kribbellae under aerobic conditions from carbohydrates as shown for *Kribbella koreensis* (Lee et al., 2000) and some other species. Optimal temperatures for growth vary insignificantly among species, but all organisms of this genus grow well between 25 and 30°C. The higher and lower growth temperatures reported are 45°C (*Kribbella swartbergensis*; Everest and Meyers, 2008) and 8°C (*Kribbella ginsengisoli*; Cui et al., 2010). Members of the genus *Kribbella* usually prefer neutral or slightly alkaline pH for growth. The pH growth range varies with species (Table 225 and Table 226). A few species can grow at initial pH of 5; *Kribbella hippodromi* showed weak growth even at pH 4.3 (Everest and Meyers, 2008). Some species grow at initial pH 10 (*Kribbella catacumbae*, *Kribbella koreensis*, and *Kribbella sancticallisti*; Lee et al., 2000; Urzì et al., 2008). Bacteria of this genus are typically non-halophilic and tolerate low NaCl concentrations; *Kribbella antibiotica* and *Kribbella jejuensis* are the most sensitive to salts (cells grow in 2% NaCl; Carlsohn et al., 2007; Kirby et al., 2006; Song et al., 2004).

It should be noted that actinomycetes of the genus *Kribbella* were originally described as chemo-organotrophs with strictly aerobic metabolism (Lee et al., 2000; Park et al., 1999). However, some species were subsequently shown to grow moderately well (*Kribbella solani*) or weakly (*Kribbella flavida*, *Kribbella jejuensis*, *Kribbella koreensis*, and *Kribbella sandramycini*) on ATCC Medium No. 172* anaerobically in an atmosphere of $H_2/CO_2/N_2$ (5:10:85) (Kirby et al., 2006). According to Urzì et al. (2008), *Kribbella solani* also grew anaerobically on VL agar (Tiecco, 1975). Some yet-unnamed *Kribbella* strains exhibit weak but distinct growth in organic semi-solid medium (0.5% glucose, 0.5% peptone, 0.3% yeast extract, 0.02% K_2HPO_4, 0.3% agar; pH 7.2) under anaerobic and/or microaerophilic conditions; L.M. Baryshnikova and L.I Evtushenko, recent observation). None of the tested kribbellae grew anaerobically on ISP 9 agar with glucose (Kirby et al., 2006). On the other hand, weak aerobic growth was observed for some yet-unnamed *Kribbella* strains on ISP 9 medium without addition of a carbon source (recent observation). *Kribbella koreensis*, when seeded with a diluted soil suspension (during the isolation procedure), showed the ability to grow aerobically on a tap-water agar with antifungal antibiotics (Lee et al., 2000). The data suggest that members of the genus *Kribbella* can adapt to an oligotrophic life style and grow in the presence of traces of carbon and/or nitrogen sources, and probably may scavenge nutrient substances from air, e.g. ammonium. Perhaps, some organisms may even have an autotrophic life style under certain conditions, like that reported or assumed, e.g. for some *Pseudonocardia* and *Streptomyces* (Goodfellow and Lechevalier, 1986; O'Donnell et al., 1993; Selesi et al., 2005; Takamiya and Tubaki, 1956). Recent finding showing that kribbellae increase in numbers in a soil microcosm after hydrogen exposure is indicative of their possible ability to utilize hydrogen as a source of energy (Osborne et al., 2010). All the above data, along with data showing that *Kribbella* often occur in nutrient-limited environments, including acidic and heavy-metal-contaminated ones (see below the section Habitats and ecology), suggest that at least certain organisms of this genus are rather aerobic chemo-organotrophs with a potential for mixotrophy and metabolic flexibility.

*Contains 10 g of glucose, 20 g of soluble starch, 5 g of yeast extract, 5 g of N-Z amine type A, 1 g of $CaCO_3$, 15 g of agar, and 1 liter of distilled water (Cote et al., 1984).

TABLE 226. Additional differentiating and descriptive characteristics of *Kribbella* species[a]

Characteristic	1. *K. flavida*	2. *K. alba*	3. *K. aluminosa*	4. *K. antibiotica*	5. *K. catacumbae*	6. *K. ginsengsoli*	7. *K. hippodromi*	8. *K. jejuensis*	9. *K. karoonensis*	10. *K. koreensis*	11. *K. lupini*	12. *K. sancticallisti*	13. *K. sandramycini*	14. *K. solani*	15. *K. swartbergensis*	16. *K. yunnanensis*
Predominant menaquinone[b]	$9(H_4)$	$9(H_4)$	$9(H_4)$	$9(H_4)$	$9(H_4)$	$9(H_4)$	nr	$9(H_4)$	nr	$9(H_4)$	$9(H_4)$	$9(H_4)$	$9(H_4)$	$9(H_4)$	nr	$9(H_4)$
pH growth range	(5.0)–9.0	5.5–8.5	5.0–9.0	nr	5.0–(10.0)	5.5–8.5	(4.3)–(9.0)	5.5–8.5	5.0–9.9	7.0–10.0	6.0–9.0	5.0–10.0	5.0–9.0	5.5–8.5	(5.0)–9.0	5.5–8.5
Catalase reaction	+	+	+	nr	+	+	nr	+	+	+	+	+	+	+	+	+
Growth on sole carbon sources:[c]																
D-Cellobiose	nr	nr	nr	nr	nr	+	+	+	nr	+	+	nr	nr	+	nr	nr
D-Fructose	v	+	+	+	+	+	+	nr	+	+	nr	+	+	nr	+	+
D-Glucose	+	+	+	+	+	+	+	+	+	+	+	+	+	+	+	+
Lactose	v	+	nr	+	+	+	w/–	+	+	+	nr	+	v	+	w	+
Maltose	+	+	nr	+	+	+	+	nr	w/–	+	–	+	+	nr	+	+
D-Mannose	v	+	nr	+	+	+	+	–	+	+	+	+	v	+	w	+
D-Melezitose	+	nr	nr	nr	nr	nr	+	+	nr	+	–	nr	+	+	nr	nr
Melibiose	+	nr	nr	nr	nr	+	+	+	–	+	+	nr	+	+	+	nr
Raffinose	v	+	+	+	w	+	+	+	+	+	+	+	+	+	w	+
L-Rhamnose	v	+	+	+	+	nr	+	+	+	+	+	+	+	+	w	+
Sucrose	+	+	+	+	+	+	+	+	+	+	+	+	+	+	+	+
Trehalose	nr	nr	nr	nr	nr	+	+	+	nr	+	+	nr	nr	+	nr	nr
Adonitol	+	nr	nr	nr	nr	–	+	–	w/–	+	nr	nr	+	v	nr	nr
Glycerol	–	+	nr	+	+	+	+	nr	+	+	nr	+	+	+	w	+
Sorbitol	v	+	nr	+	w	–	nr	nr	w	v	+	–	v	nr	+	+
Xylitol	–	nr	nr	nr	nr	nr	–	–	nr	–	+	nr	+	–	nr	nr
Inulin	+	nr	nr	nr	nr	–	+	+	–	+	nr	nr	+	–	–	nr
N-Acetyl-β-glucosamine	nr	nr	nr	nr	+	+	nr	nr	nr	v	+	+	nr	nr	nr	nr
Utilization of organic acids:[d]																
Acetate	–	+	+	+	nr	+	w	nr	w/–	+[e]	nr	nr	–	nr	w	+
Citrate	v	–	+	+	–	+	w	nr	w/–	+[e]	–	–	–	nr	w	+
Lactate	nr	nr	nr	nr	nr	+	nr	nr	w/–	+[e]	nr	nr	nr	nr	w	nr
Malate	nr	nr	+	nr	nr	+	nr	nr	nr	+[e]	+	+	nr	nr	nr	nr
Oxalate	+	+	nr	+	nr	+	nr	nr	nr	+[e]	nr	nr	–	nr	w	+
Succinate	+	nr	+	nr	nr	+	w	nr	+	+[e]	nr	nr	+	nr	w	nr
Degradation of:																
Adenine	nr	nr	+	nr	nr	nr	+	–	+	–	nr	nr	nr	v	+	nr
Esculin	+	nr	+	nr	+	nr	+	+	+	+	+	+	+	+	+	nr
Arbutin	+	nr	nr	nr	+	nr	+	+	+	+	+	w	+	+	+	nr
Cellulose	–	–	–	–	–	–	–	nr	nr	–	nr	nr	–	nr	nr	–
Elastin	+	nr	nr	nr	w	nr	nr	–	nr	+	nr	+	+	–	nr	nr

Tyrosine	nr	nr	+	nr	+	nr	+	−	+	+	+	+	nr	v	+	nr
Xanthine	+	nr	+	nr	nr	nr	−	−	nr	+	nr	w	nr	−	−	nr
Xylan	nr	nr	nr	nr	nr	−	−	−	nr	nr	−	nr	nr	−	nr	nr
API ZYM (bioMérieux):																
Cystine arylamidase	+	nr	−	nr	nr	nr	nr	w	nr	+	w	nr	+	w	+	nr
Esterase (C4)	+	nr	+	nr	nr	nr	nr	nr	nr	+	−	nr	+	nr	nr	nr
α-Fucosidase	nr	nr	+	nr	nr	nr	nr	+	nr	nr	+	nr	nr	−	−	nr
β-Galactosidase	+	nr	+	nr	nr	nr	nr	nr	nr	+	w	nr	+	nr	nr	nr
Lipase (C14)	nr	nr	−	nr	nr	nr	nr	−	nr	nr	+	nr	nr	+	+	nr
Naphthol-AS-BI-phosphohydrolase	+	nr	+	nr	nr	nr	nr	nr	nr	−	−	nr	+	nr	nr	nr

[a]See Table 225 for symbols, for references and other details. Data in parentheses (for the pH growth range) indicate variable, weak or delayed growth; w/−, doubtful growth.

[b]9(H_4), menaquinone having nine isoprene units two of which are saturated.

[c]ISP 9 (Shirling and Gottlieb, 1966) as basal medium. See Table 225 for *Kribbella ginsengisoli*; positive results were also obtained for *Kribbella koreensis* in the same study (Cui et al., 2010) with D-glucose, lactose, maltose, D-mannose, D-melibiose, L-rhamnose, and sucrose.

[d]Growth with the listed organic acids as sole carbon sources on basal medium ISP 9 (Shirling and Gottlieb, 1966), unless other methods indicated.

[e]An alkaline reaction, according to the method of Gordon et al. (1974); growth also occurs with acetate, lactate, or malate, but not with citrate (API 20NE; Cui et al., 2010).

Genomic characteristics. The DNA base ratio of species of the genus *Kribbella* ranges from 66.3 mol% (Cui et al., 2010) to 71.3 mol% (Lee et al., 2000), with 70 mol% reported for the type strains of the type species *Kribbella flavida* (Park et al., 1999). As mentioned before, the current 16 *Kribbella* species form a coherent phylogenetic cluster, with some species showing more than 99% 16S rRNA gene sequence similarity. The DNA–DNA similarities between *Kribbella* species obtained by different methods ranged from 0% for the pair *Kribbella koreensis–Kribbella sandramycini* (Sohn et al., 2003), which looks rather ambiguous, to 61.2% for *Kribbella yunnanensis–Kribbella sandramycini* (Li et al., 2006). Recently, the sequences of *gyrB* gene (encoding the β-subunit of DNA gyrase, a type II DNA topoisomerase) have been analyzed for strains of the 15 recognized *Kribbella* species (Kirby et al., 2010). The type strains of *Kribbella* species had partial *gyrB* gene sequence (1108 nt, 56% of the full-length *gyrB*) similarity values in the range 89.91–94.95%, except for the pair *Kribbella solani–Kribbella hippodromi* (98.22%).

The 16S–23S rDNA internally transcribed spacer (ITS) sequences and the ribonuclease P RNA gene sequences were analyzed for *Kribbella flavida* and *Kribbella sandramycini* (Yoon et al., 1998a; Yoon and Park, 2000). The type strain of *Kribbella sandramycini* was reported to contain two types of 16S–23S ITS sequences (*rrn* 1, 429 bp; *rrn* 2, 439 bp) showing high sequence divergence, with the nucleotide similarity level of 80.8% (and 78.4% when gaps are included). The type strain of *Kribbella flavida* exhibits a higher degree of ITS sequence similarity to that of *Kribbella sandramycini* (75.4%, *rrn* 1; 73.6%, *rrn* 2) than to the ITS sequences of representatives of *Nocardioides* and *Aeromicrobium* used in the study. The same *Kribbella* strains shared 92±2% similarity of the ribonuclease P RNA gene sequences and formed a cluster distinct from the strains of other genera analyzed (Yoon and Park, 2000). Gao and Gupta (2005) revealed in *Kribbella sandramycini*, along with an insertion in the 23S rRNA gene, three conserved indels in three widely distributed proteins, i.e. a deletion in cytochrome *c* oxidase subunit 1, an insertion in CTP synthetase, and an insertion in glutamyl-tRNA synthetase that are distinctive characteristics of most *Actinobacteria* and not found in any other groups of bacteria. The phylogenetic analysis based on CTP synthetase sequences showed that the overall topology of the tree of actinobacteria studied was very similar to that seen in the 16S rRNA gene trees. One of the interesting exceptions relevant to this chapter is *Kribbella sandramycini*, which is grouped with *Saccharopolyspora erythrea* DSM 40517 (not with *Nocardioides simplex*) and has an almost identical CTP synthetase sequence to that of *Saccharopolyspora erythrea* DSM 40517 (Gao and Gupta, 2005). Recently, the sequence of the complete genome for *Kribbella flavida* DSM 17836 (a circular chromosome of 7,579,488 nt, 7086 genes with a protein-coding capacity, 60 RNA genes, and 143 predicted pseudogenes) was determined (Pukall et al. 2010, GenBank accession no. NC_013729). The majority of the protein-coding genes (70.7%) were assigned with a putative function while those remaining were annotated as hypothetical proteins).

Antibiotic sensitivity and antibiotic production. Strains of *Kribbella flavida, Kribbella antibiotica, Kribbella jejuensis, Kribbella karoonensis, Kribbella koreensis, Kribbella sandramycini, Kribbella solani,* and *Kribbella swartbergensis* were reported to be sensitive to ampicillin sodium salt (100 μg/ml) and also (except *Kribbella jejuensis*) to cefotaxime (10 μg/ml) (Kirby et al., 2006). Except for *Kribbella solani*, all the above strains, as well as *Kribbella catacumbae* and *Kribbella sancticallisti*, are inhibited by tobramycin (10 μg/ml) (Kirby et al., 2006; Urzì et al., 2008). All the above-mentioned species are resistant to carbenicillin (100 μg/ml) and, except *Kribbella sancticalisti* and *Kribbella ginsengisoli*, and to streptomycin (10 μg/ml) (Cui et al., 2010; Kirby et al., 2006; Urzì et al., 2008). The listed and other members of this genus are sensitive to a wide range of other antibiotics (see the *List of species of the genus Kribbella*). A few *Kribbella* strains were reported to produce biologically active compounds. *Kribbella sandramycini* ATCC 39419 produces an antitumor antibiotic, sandramycin (a cyclic decadepsipeptide with a two-fold axis of symmetry and 3-hydroxyquinaldic acid as an appended chromophore; Matson and Bush, 1989; Matson et al., 1993). *Kribbella koreensis*is produces novel neuropilin/growth factor complexes (Alitalo et al., 2003). *Kribbella antibiotica* possesses antifungal activity (Li et al., 2004) and *Kribbella jejuensis* inhibits growth of *Streptomyces scabiei* (Song et al., 2004).

Habitats and ecology. Kribbellae are widely distributed in terrestrial ecosystems worldwide, mostly in soils of different origin, including acidic ones (Ding et al., GenBank accession no. FJ938352; Zheng and Huang, GenBank accession no. EU697199) and polluted sites (Cho et al., 2006; Leigh et al., 2007). Representatives of *Kribbella* were found in the root zone of an Austrian pine (*Pinus nigra* L.) growing naturally in soil contaminated with polychlorinated biphenyls (PCBs) (Leigh et al., 2007). Furthermore, kribbellae along with *Pseudonocardia, Nocardiodes,* and *Sphingomonas* were the predominating representatives of 75 genera which utilized carbon from PCBs (Leigh et al., 2007). Organisms of this genus often occur in sub-subsurface and nutrient-limited environments, including gold-mine caves (Lee et al., 2000), an alum slate mine (Carlsohn et al., 2007), and catacombs (Urzì et al., 2008). Heterotrophic lifestyle and enzymatic activity towards complex macromolecules assume that kribbellae function as consumers of organic material in soil. In environments of nutrient scarcity, they probably use traces of organics dissolved in water or derived from the decomposition of other microorganisms, are involved in mutualistic interactions with other microorganisms, or engage in chemolithotrophic metabolism with input from some atmospheric gases and minerals, like those suggested for other actinomycetes from such environments (Barton et al., 2007; Groth and Saiz-Jimenez, 1999). Organisms of this genus and some other soil bacteria are assumed to be important utilizers of hydrogen at low concentrations in soil, and could be important contributors to the function of soil as a sink in the global hydrogen cycle (Osborne et al., 2010). Kribbellae also occur among microbial endophytes in plant tissues (Conn and Franco, 2004; Trujillo et al., 2006) and in bacterial communities associated with spores of the endomycorrhizal fungus *Gigaspora margarita* (L. Long, H. Zhu, and Q. Yao, GenBank accession no. EU589433). *Kribbella solani* was isolated from a potato tuber with scab lesions (Song et al., 2004), although there is no evidence of its plant pathogenic properties. Representatives of this genus were also found in the bacterial community of human skin (Kong et al., GenBank accession no. HM277981). To the authors' knowledge, kribbellae have not been discovered up to now in marine ecosystems or in clinical specimens. The spores or cells of *Kribbella* are thought to survive or maintain a low metabolic activity under extreme conditions in nature, since these remain alive in a soil sample (from which some species have been isolated) after exposure to temperature as high as 120°C for 1 h (dry heating) (Kirby et al., 2006).

Enrichment and isolation procedures

The available strains of the genus *Kribbella* are mostly random isolates obtained during studies on (actinobacterial) microbiota of certain biotopes or screenings for biologically active compounds. Strains of this genus are normally isolated by plating on suitable "total count" agar media, e.g. R2A agar (Difco), as well as those used for isolation of certain actinomycete groups. Incubation is usually carried out at 25–30°C for a week to a month. Generally, kribbellae relatively frequently occur among the mycelial soil isolates having sand-pasty or soft-leathery, agar-penetrating and non-pigmented, yellowish or light cream colonies developing on R2A agar. Other media, reported in the original species descriptions, e.g. glycerol-asparagine agar (ISP 5 medium), glucose-yeast extract-malt extract (ISP 2 medium), SM1 agar (proposed for isolation of *Amycolatopsis*; Tan et al., 2006), yeast extract-mannitol agar (used for isolation of *Rhizobia*; Vincent, 1970), tap-water agar, etc. were successfully employed for isolation of kribbellae. Various isolation strategies used to isolate mycelial actinomycetes, including air-drying and heating (up to 120°C for 1 h; Kirby et al., 2006) of soil samples, adding antifungal antibiotics, exposition in alkaline solutions, etc., may be used to reduce the numbers or inhibit the growth of accompanying microbes. When intending isolation of strains belonging or closely related to the described *Kribbella* species, some selective (semi-selective) media may be formulated on the basis of characteristics listed in Table 225 and Table 226, and a consideration of the resistance to certain antibiotics and heavy metal salts and the isolation sources.

Maintenance procedures

Cultures may be maintained as 20% glycerol suspensions at −20 and −80°C. Long-term conservation is achieved by freeze-drying or in liquid nitrogen by standard procedures.

Differentiation of the genus *Kribbella* from other genera

The phenotypic characteristics essential for delineation of *Kribbella* from other genera of the family are listed in Table 214. The formation of well-developed branching hyphae allows primary separation of *Kribbella* from the genera *Aeromicrobium*, *Marmoricola*, and the vast majority of *Nocardioides* species. The predominant menaquinone MK-9(H_4) and the presence of phosphatidylcholine among principal polar lipids clearly distinguish kribbellae from all species of the genus *Nocardioides*. The polar lipid pattern containing phosphatidylcholine is also the most distinctive feature that differentiates kribbellae both from the genera of the family *Nocardioidaceae* and other actinomycete genera comprising mycelium-forming organisms with LL-A_2pm in the cell wall. A rapid method based on single-digestion restriction analysis of the PCR-amplified 16S rRNA gene with endonucleases (*Mbo*I, *Vsp*I, *Sph*I, *Sna*BI, *Sal*I, and *Age*I), useful for identification of *Kribbella* strains, is described (Cook and Meyers, 2003).

Taxonomic comments

The genus *Kribbella* was established by Park et al. (1999) with *Kribbella flavida* (type species) and *Kribbella sandramycini* to accommodate two strains, which were previously ascribed to the genus *Nocardioides*, i.e. "*Nocardioides fulvus*" (Ruan and Zhang, 1979) and *Nocardioides* sp. ATCC 39419, a producer of the antibiotic sandramycin (Matson and Bush, 1989). Almost simultaneously, Lee et al. (2000) independently published the description of *Hongia koreensis* for an original isolate which was very close to *Kribbella* at the phylogenetic level and similar in many phenotypic characteristics. Subsequent taxonomic analysis of the above species resulted in the reclassification of *Hongia koreensis* as *Kribbella koreensis* (Sohn et al., 2003). The remaining 13 *Kribbella* species were described within the 2004–2010 period on the basis of taxonomic studies mostly of single environmental isolates. As mentioned before, some species of this genus are very close phylogenetically and share up to 99% or more 16S rRNA gene sequence similarity, with the highest value being between *Kribbella solani* and *Kribbella hippodromi* (99.64%) by local alignment (Everest and Meyers, 2008). The type strains of *Kribbella solani* and *Kribbella hippodromi* also display the closest identity (98.2%) by *gyrB* gene sequence similarity (Kirby et al., 2010). Nevertheless, these two species had a DNA–DNA similarity of approximately 40%, as revealed by hybridization experiments, and differed in a number of phenotypic properties (Everest and Meyers, 2008). Analogously, the DNA–DNA similarity values between the type strains of other phylogenetically very close species (see Cui et al., 2010, and Kirby et al., 2010, for details and references) are below the threshold value of 70% recommended for the definition of bacterial species (Wayne et al., 1987).

Differentiation of species of the genus *Kribbella*

The phenotypic properties useful in preliminarily distinguishing of *Kribbella* species are listed in Table 225. Additional differentiating and descriptive characteristics are given in Table 226 and outlined in the descriptions of species. It is worth noting that the available differential physiological and biochemical characteristics are based, in most cases, on the data obtained for a single representative of a species. Moreover, conflicting results have been reported for strains which give rather weak or doubtful growth on some carbon sources and showed very weak activities in degradation tests. Some differences in salt resistance and the pH growth range of a strain can be influenced by the test medium composition (Carlsohn et al., 2007; Cui et al., 2010; Kirby et al., 2006; Song et al., 2004; Trujillo et al., 2006). Chemotaxonomic features such as the composition of cell-wall sugars, polar lipids, and fatty acid profiles are important for differentiation purposes. It should be emphasized that whole cells but not cell walls were analyzed in most *Kribbella* species or the purified cell walls were obtained (and analyzed) by different methods, and this should be taken into account when comparing the sugar composition. Also the qualitative and quantitative compositions of fatty acids (and polar lipids) in strains of this genus may be influenced by growth conditions and analytical procedures. For the comparison of fatty acid profiles, the data should therefore be obtained solely by growing cells under standardized cultivation conditions and performing analyses in the same laboratory rather than from the literature. The distance dendrogram generated on the basis of fatty acid profiles by the software of the Sherlock Microbial Identification System may be useful to separate strains of phylogenetically related species (Urzì et al., 2008). Matrix Assisted Laser Desorption Ionization Time-of-Flight (MALDI/TOF) mass spectrometry of proteins and the RiboPrint analyses (with *Pvu*II as the restriction enzyme) are also quite helpful for clustering of phylogenetically close *Kribbella* strains at the strain-species level (Carlsohn et al., 2007; Urzì et al., 2008).

DNA–DNA hybridization studies may be necessary to support the species affiliation of a strain or strain group which is closely related to known species on the basis of 16S rRNA gene sequences or other characteristics. Analysis of *gyrB* gene sequences is advantageous to assess membership of a strain in question to an established species. According to Kirby et al. (2010), strains with *gyrB*-based genetic distance exceeding 0.04 to a known species (calculated from partial *gyrB* gene sequences, 1108 nt) most likely represent a distinct species. This value therefore is suggested as a cut-off point to determine whether DNA-DNA hybridization is required. A 390-nucleotide sequence of the *gyrB* gene (a variable region from 1010 to 1400 bp according to *Streptomyces avermitilis* MA-4680 *gyrB* numbering) is sufficient to preliminarily assess whether a strain is likely to represent a new species. The GenBank numbers for the *gyrB* gene sequences of *Kribbella* strains, PCR primers for amplification of the *Kribbella gyrB* gene, and the details of the method are available from Kirby et al. (2010).

Acknowledgements

The authors were supported by the program MCB of the Russian Academy of Sciences.

List of species of the genus *Kribbella*

1. **Kribbella flavida** Park, Yoon, Shin, Suzuki, Kudo, Seino, Kim, Lee and Lee 1999, 750[VP]

fla'vi.da. L. fem. adj. *flavida* yellowish, pale yellow.

Characteristics are as described for the genus and listed in Table 225 and Table 226. Additional information given below is taken from the original paper (Park et al., 1999), unless indicated.

Substrate mycelium is usually pale yellow and consists of extensively branched hyphae penetrating into agar media and often fragmenting into rod-shaped to coccoid elements. Aerial hyphae appear straight or curved and break up into short to elongated rod-like fragments (Figure 270). Colonies are usually pasty and have lichenous shapes. Melanin is produced on ISP 7 medium. Gram-staining reaction is variable. Grows at 12–30°C and at pH 9 (initial pH value of test media); growth at 37°C and at pH 5 varies with experiments (Kirby et al., 2006; Park et al., 1999; Trujillo et al., 2006). Grows at 3% NaCl on nutrient agar (Trujillo et al., 2006) but not on ISP 2 medium supplemented with the same salt concentration (Kirby et al., 2006). Produces pyrrolidonyl arylamidase (API ZYM; Trujillo et al., 2006).

Grows in the presence (μg/ml) of carbenicillin disodium salt (100), cephaloridine (10), chloramphenicol (50), cycloheximide (10), erythromycin (10), gentamycin sulfate (10, weak), and streptomycin sulfate (10). Growth is inhibited by ampicillin sodium salt (100 mg/ml), cefotaxime (10), neomycin sulfate (10), tobramycin sulfate (10), and vancomycin hydrochloride (10) (Kirby et al., 2006). The major fatty acids (>5% of the total registered at least in one experiment) reported for cultures grown under different conditions were $C_{15:0}$ anteiso (24.9–32), $C_{16:0}$ iso (14.0–21.3), $C_{16:1}$ iso (2–13.5), $C_{17:0}$ anteiso (3.2–10.0), $C_{14:0}$ iso (3–9.1), $C_{15:0}$ iso (5.0–5.4), $C_{17:1}$ iso ω9*c* (< 1–6.7), 9-methyl-$C_{16:0}$ (< 1–8) (Park et al., 1999; Sohn et al., 2003; Song et al., 2004), $C_{17:1}$ ω8*c* (10.6) (Song et al., 2004), and $C_{17:1}$ ω9*c* (7.0) (Sohn et al., 2003).

Source (type strain): soil near Beijing, China.

DNA G+C content (mol%): 70 (HPLC); 70.6% (calculated for the genome; Pukall et al., 2010).

Type strain: DSM 17836, KACC 20248, KCTC 9580, IFO (now NBRC) 14399, JCM 10339.

Sequence accession no. (16S rRNA gene): AY253863, AF005017.

Sequence accession no. (complete genome): NC_013729.

2. **Kribbella alba** Li, Wang, Zhang, Xu and Jiang 2006a, 1459[VP] (Effective publication: Li, Wang, Zhang, Xu and Jiang 2006b, 34.)

al'ba. L. fem. adj. *alba* white.

Characteristics are as described for the genus and listed in Table 225 and Table 226. Additional information given below is taken from the original paper (Li et al., 2006).

Substrate mycelium ranges from pale yellow on ISP 3, ISP 4, Czapek's agar, and potato agar to soft yellow on ISP 2 and nutrient agar. The hyphae are extensively branched and often fragment into irregular rod-shaped elements. Aerial mycelium is developed on all the media tested, and it is white (on ISP 3, 4, and 5 media and on potato agar), yellow-white (on ISP 2 and Czapek's agar) or pale yellow (on nutrient agar). Some areas of aerial hyphae fragment into short to elongated rod-like elements which can be arranged in chains. The major fatty acids (> 5% of the total) determined in cells grown on TSA at 28°C include $C_{15:0}$ anteiso (27.9), $C_{16:0}$ iso (15.3), $C_{15:0}$ iso (9.1), $C_{17:1}$ iso ω9*c* (10.4), and $C_{17:0}$ anteiso (6.1).

Source: soil in Yunnan Province, the west of China.

DNA G+C content (mol%): 67.9 (T_m).

Type strain: YIM 31075, CCTCC AA 001020, DSM 15500, JCM 14306, NBRC 103561.

Sequence accession no. (16S rRNA gene): AY082062.

3. **Kribbella aluminosa** Carlsohn, Groth, Spröer, Schütze, Saluz, Munder and Stackebrandt, 2007, 1946[VP]

a.lu.mi.no'sa. L. fem. adj. *aluminosa* aluminous, foil of alum, alum-containing, referring to the source of isolation of the first strains.

Characteristics are as described for the genus and listed in Table 225 and Table 226. Additional information given below is taken from the original paper (Carlsohn et al., 2007).

Extensively branched, beige to pale yellow substrate mycelium is produced on all media tested. White aerial mycelium varies in abundance and typically forms well on ISP 5 medium and less abundantly on ISP media 2, 3, and 4, but not on organic medium 79 (Prauser and Falta, 1968). Both vegetative and aerial hyphae fragment into irregular rod-shaped elements. Colonies on organic medium 79 are wrinkled and of a pasty consistency. Grows at 20 and 37°C; no growth occurs at 6 or 42°C. No growth is also observed at pH 4.5 or 9.0 (initial pH values of liquid organic medium 79). Grows in the same medium with 2% NaCl but not with 4% NaCl. Growth occurs on minimal agar medium (Amoroso et al., 2000) in the presence of $NiCl_2$ (5.0 mM) or $CuSO_4$ (0.5 mM). Aconitate (delayed) is utilized as a sole carbon source, but benzoate and DL-tartrate are not. In API ZYM tests, positive reactions are observed for leucine arylamidase, valine arylamidase, α-galactosidase, α-glucosidase, β-glucosidase, esterase lipase (C8), α-mannosidase, *N*-acetyl-β-glucosaminidase, acid phosphatase, and alkaline phosphatase, but not β-glucuronidase or trypsin; weak or negative reaction for α-chymotrypsin. Hippurate is hydrolyzed. Indole is not produced. Susceptible to the following antibiotics (μg per disc): ampicillin (10), chloramphenicol (30), ciprofloxacin (5, weakly), imipenem (10), novobiocin (5, weakly), ofloxacin (10, weakly), oxytetracycline hydrochloride (30, weakly), as well as polymyxin B (300 IU per disc, weakly). Resistant to (μg per disc) lincomycin hydrochloride (2), methicillin (5), nalidixic acid (30), norfloxacin (10), and penicillin G (10 IU per disc). The major fatty acid (>5% of the total registered at least for one strain) as determined in three strains grown at 28°C in tryptic soy broth (TSB; Sigma-Aldrich) and harvested at exponential growth phase include $C_{15:0}$ anteiso (36.0–44.6), $C_{16:0}$ iso (7.5–17.6), $C_{15:0}$ iso (6.1–9.7), $C_{14:0}$ iso (2.7–6.1), $C_{17:0}$ anteiso (4.6–5.4), and 9-methyl-$C_{16:0}$ (7.0–8.7).

Source: acidic and heavy-metal-containing rock surfaces, a medieval alum slate mine in Thuringia, Germany.

DNA G+C content (mol%): not determined.

Type strain: HKI 0478, DSM 18824, JCM 14599.

Sequence accession no. (16S rRNA gene): EF126967.

4. **Kribbella antibiotica** Li, Wang, Zhang, Schumann, Stackebrandt, Xu and Jiang 2004a, 1425[VP] (Effective publication: Li, Wang, Zhang, Schumann, Stackebrandt, Xu and Jiang 2004b, 164.)

an.ti.bio'ti.ca. Gr. prep. *anti* against, in opposition to; Gr. n. *bios* life; L. suff. *-icus -a -um*, suffix used in adjectives with the sense of belonging to, related to; N.L. fem. adj. *antibiotica* related to antibiotic (the type strain of *Kribbella antibiotica* is an antifungal strain).

Characteristics are as described for the genus and listed in Table 225 and Table 226. Additional information given below is taken from the original paper (Li et al., 2004b), unless indicated.

Substrate mycelium is yellow white (on ISP 3, 4, and 5 media) or yellow (on potato agar), extensively branched, and often fragments into irregular rod-shaped elements. Aerial mycelium is developed on all the media tested, particularly on Czapek's agar and ISP 5 medium, and fragments into elongated to short rod-like elements. The color of aerial mycelium is yellow white on ISP 3, ISP 4, ISP 5, Czapek's agar, and nutrient agar, soft yellow on ISP 2, and gray yellow on potato agar. Melanin is produced on ISP 7 medium. Grows at 22–30°C. No growth occurs at 12 or 37°C (Trujillo et al., 2006). This species does not grow on ISP 2 medium supplemented with 2% NaCl (Kirby et al., 2006). Grows in the presence (μg/ml) of carbenicillin disodium salt (100), cephaloridine (10), chloramphenicol (20, weak), cycloheximide (10), streptomycin sulfate (10), and vancomycin hydrochloride (10) but not in the presence of ampicillin sodium salt (100), cefotaxime (10), chloramphenicol (50), erythromycin (10), gentamycin sulfate (10), neomycin sulfate (10), and tobramycin sulfate (10) (Kirby et al., 2006). Produces a substance with antifungal activity. The predominant fatty acids determined in cells grown in liquid ISP 2 medium at 28°C were $C_{15:0}$ anteiso and $C_{15:0}$ iso.

Source (type strain): soil in Yunnan Province, China.

DNA G+C content (mol%): 67 (T_m).

Type strain: YIM 31530, CCTCC AA 001021, DSM 15501, JCM 13523, NBRC 101882.

Sequence accession no. (16S rRNA gene): AY082063.

5. **Kribbella catacumbae** Urzì, De Leo and Schumann 2008, 2095[VP]

ca.ta.cum'ba.e. L. gen. n. *catacumbae* of a catacomb, isolated from a Roman catacomb.

Characteristics are as described for the genus and listed in Table 225 and Table 226. Additional information given below is taken from the original paper (Urzì et al., 2008).

Substrate hyphae (diameter 0.36 μm) are branched and tend to penetrate into the agar. They are pale yellow to orange-yellow on the test media, becoming dark orange in old cultures on Luedemann medium [0.5% yeast extract (Difco), 0.5% NZ Amine type A (Sheffield Chemical Co., Norwich, NY), 1% glucose, 2% soluble starch (Difco), 0.1% $CaCO_3$, and 1.5% agar (Luedemann, 1968)]. Colonies are irregular (diameter 5–10 mm), with lobate margin, crateriform profile and rough surface, that are covered with white aerial hyphae. No distinct formation of rod-shaped elements due to fragmentation has been observed in cultures on Luedemann medium, but morphological structures resembling short chains of non-separated arthrospores are observable by electron microscopy. Melanin is produced on ISP 7 medium. Grows at 15–30°C; weak growth is observed at 10°C. Grows in the presence of 2% NaCl and weakly at 3%, but not at 5% as determined on Luedeman agar without $CaCO_3$ addition. Grows at pH 5–9, with weak growth at pH 10 (initial pH values of the same medium). Weak growth on D-raffinose and sorbitol after 21 d. Tweens 20, 40, and 60 are hydrolyzed, and DNA is weakly hydrolyzed. Resistant to the following antibiotics (μg/ml): cephaloridine (10), neomycin sulfate (1.5), streptomycin sulfate (10), and vancomycin hydrochloride (10). Growth is inhibited by chloramphenicol (50), gentamycin sulfate (3), and tobramycin sulfate (1.5). Sensitive to 0.01% lysozyme. Major fatty acids (>10% of the total) as determined for 6 strains grown on TSA at 28°C for 24 h were $C_{15:0}$ anteiso (22.4–28.3), $C_{15:0}$ iso (20.5–28.7), and $C_{17:1}$ iso ω9*c* (9.9–10.6).

Source: deteriorated surfaces in the St. Callistus catacombs, Rome, Italy.

DNA G+C content (mol%): not determined.

Type strain: BC631, DSM 19601, JCM 14968.

Sequence accession no. (16S rRNA gene): AM778575.

6. **Kribbella ginsengisoli** Cui, Lee, Lee and Im 2010, 366[VP]

gin.sen.gi.so'li. N.L. n. *ginsengum* ginseng; L. n. *solum* soil; N.L. gen. n. *ginsengisoli* of soil of a ginseng field.

Characteristics are as described for the genus and listed in Table 225 and Table 226. Additional information given below is taken from the original paper (Cui et al., 2010).

Extensively branched substrate mycelium penetrating into the agar and aerial mycelium are observed on R2A agar plates (Difco) and most other media tested. Cells are not lyzed in the KOH test according to the method of Buck (1982). Grows at 8–37°C but not at 6 or 42°C. Will grow on R2A agar with up to 4.0% NaCl, and in the pH range of 5.5–8.5. Grows on nutrient agar (Difco) but not on MacConkey agar. Unable to reduce nitrate or nitrite under either aerobic or anaerobic conditions. Utilizes glucose as a sole carbon source but does not produce acid or gas from it (API 20NE and ID 32GN tests systems). In addition to compounds listed in Table 225 and Table 226, the following carbon sources are utilized: amygdalin, asparagine, aspartate, fumarate, gluconate, glutamate, glutamine, glycogen, L-histidine, oxalate, phenylalanine, pyruvate, threonine, and valerate; the following compounds are not utilized as sole carbon sources: adipate, L-alanine, arginine, benzoic acid, 3- and 4-hydroxybenzoates, caprate, cysteine, dextran, dulcitol, ethanol, formic acid, D-fucose, glutarate, glycine, isoleucine, itaconate, leucine, lysine, D-lyxose, maleic acid, malonate, methanol, methionine, phenyl acetate, proline, propionate, D-rhamnose, D-ribose, salicin, L-serine, L-sorbose, suberate, tartarate, tryptophan, tyrosine, valine, or xylitol (conventional method with minimal salts medium containing vitamins and microelements, and/or the API 20NE and ID 32GN test systems). Negative for production of arginine dihydrolase, lysine decarboxylase, ornithine decarboxylase, and tryptophan deaminase. Degrades chitin, but not DNA. Indole is not produced from tryptophan. Voges–Proskauer test is positive. Growth is inhibited by the following antibiotics (μg per disc): streptomycin sulfate (6), gentamycin sulfate (4), tetracycline hydrochloride (10), chloramphenicol (15), kanamycin (6), penicillin (18), or erythromycin (6). The major fatty acids (5% or more), as determined for cells grown in tryptic soy broth for 5 d at 30°C include $C_{15:0}$ anteiso (20.6), $C_{16:0}$ iso (11.5), $C_{16:0}$ (8.8), $C_{16:0}$ 2-OH (8.1), $C_{15:0}$ iso (6.4), $C_{17:0}$ iso (5.0), and $C_{18:0}$ (6.3).

Source (type strain): soil from of a ginseng field, Pocheon Province, South Korea.

DNA G+C content (mol%): 66.3 (HPLC).

Type strain: Gsoil 001, KCTC 19134, DSM 17941.

Sequence accession no. (16S rRNA gene): AB245391.

7. **Kribbella hippodromi** Everest and Meyers, 2008, 444[VP]

hip.po.dro'mi. Gr. masc. n. *hippodromos* horse racecourse, N.L. gen. masc. n. *hippodromi* of/from a horse racecourse, referring to the source of isolation of the type strain, Kenilworth Racecourse, Cape Town, Western Cape, South Africa.

Characteristics are as described for the genus and listed in Table 225 and Table 226. Additional information given below is taken from the original paper (Everest and Meyers, 2008).

Substrate mycelium is usually cream-colored, with extensively branched hyphae, which fragment in both liquid and agar cultures. Aerial mycelium is white on ISP 4 medium. Colonies appear convoluted with irregular edges on most media. Grows at 20–37°C, but not at 45°C. Growth occurs in the presence of 5% NaCl, with very weak growth at 6% NaCl, and no growth at 7% NaCl (tested on ISP 2 medium). Grows weakly at pH 4.3 (initial pH value of Bennett's agar). Utilizes salicin as a sole carbon source. Utilizes L-arginine, L-asparagine, L-histidine, L-threonine, and nitrate as sole nitrogen sources, with weak growth on L-cysteine and L-serine. Gives also weak growth on DL-α-amino-*n*-butyric acid, L-4-hydroxyproline, and L-valine as sole nitrogen sources and no growth on L-methionine and L-phenylalanine (characteristics distinguishing this species from *Kribbella solani* which shows clear growth on these amino acids). Allantoin, guanine, and pectin are not degraded or hydrolyzed. Resistant to the following antibiotics (μg/ml): lincomycin hydrochloride (100), neomycin sulfate (50), oleandomycin phosphate (100), rifampicin (50), streptomycin sulfate (100), and penicillin G (10 IU /ml). Sensitive to (μg/ ml) cephaloridine (100), tobramycin sulfate (50), and vancomycin hydrochloride (50). No antibacterial activity against *Mycobacterium aurum.*

Source (type strain): soil sample from a fynbos-rich site, Kenilworth Racecourse, Cape Town, South Africa.

DNA G+C content (mol%): not determined.

Type strain: S1.4, DSM 19227, JCM 15572, NRRL B-24553.

Sequence accession no. (16S rRNA gene): EF472955.

8. **Kribbella jejuensis** Song, Kim, Hong, Cho, Sohn, Chun, Sun, 2004, 1347[VP]

je.ju.en'sis. N.L. fem. adj. *jejuensis* of or belonging to Jeju, Korea.

Characteristics are as described for the genus and listed in Table 225 and Table 226. Additional information given below is taken from the original paper (Song et al., 2004), unless indicated.

Substrate mycelium is usually cream-colored and consists of broadly branched hyphae. Aerial hyphae are white and fragmenting into rod-shaped elements. Colonies are usually sand-pasty and have lichenous shapes. Grows at 28–30 and 42°C (Carlsohn et al., 2007; Song et al., 2004) and in the presence of 2% NaCl on ISP 2 medium (Kirby et al., 2006); no growth is observed (Carlsohn et al., 2007) in liquid organic medium 79 (Prauser and Falta, 1968) at the same salt concentration. No growth occurs at initial pH 5 or pH 9 as assessed both on Bennet's agar and in liquid organic medium 79 (Carlsohn et al., 2007; Kirby et al., 2006). Utilizes DL-arginine, L-histidine, DL-homoserine, DL-α-amino-*n*-butyric acid, *trans*-4-hydroxy-L-proline, ammonium, and nitrate as nitrogen sources. Does not utilize salicin as a sole carbon source. Hydrolyzes tributyrin. Does not decompose guanine, allantoin, and chitin. Produces trypsin (Carlsohn et al., 2007). Tolerant to 0.01% (w/v) lysozyme, 0.1% (w/v) phenyl ethanol, and 0.001% (w/v) potassium tellurite, but not to 0.01% (w/v) sodium azide, 0.01% (w/v) thallous acetate, or 0.1% (w/v) phenol. Grows in the presence (μg/ ml) of carbenicillin disodium salt (100), cefotaxime (10), chloramphenicol (50), cycloheximide (10), erythromycin (10), gentamycin (0.4), neomycin sulfate (10, weak), rifampicin (4), streptomycin sulfate (10), tetracycline (1), tobramycin (4, weak), and vancomycin hydrochloride (10) but not in the presence of ampicillin sodium salt (100), cephaloridine (10), gentamycin sulfate (10), streptomycin (20), tobramycin sulfate (10), and troleandomycin (10) (Kirby et al., 2006; Song et al., 2004). Weak growth occurs in the presence of penicillin G (10 U/ml). Carlsohn et al. (2007) reported that the type strain of this species resisted ampicillin (10 μg per disk) and imipenem (10 μg per disk). Shows weak antimicrobial activity towards *Streptomyces scabiei* but not against *Bacillus subtilis, Pseudomonas aeruginosa, Micrococcus luteus, Candida albicans, Aspergillus niger, Trichoderma harzianum, Fusarium acuminatum,* and *Colletotrichum gloeosporioides*. The major fatty acids (>5% of the total) determined for the cultures grown on TSA (BBL) at 30°C (Song et al., 2004) and in TSB at 28°C and harvested at the logarithmic growth phase (Carlsohn et al., 2007) were similar: $C_{15:0}$ anteiso (31.2 and 27.1), $C_{16:0}$ iso (16.2 and 17.3), $C_{15:0}$ iso (10.8 and 11.3), $C_{14:0}$ iso (7.1 and 10.4), 9-methyl-$C_{16:0}$ (9.3 and 7.9), $C_{16:0}$ iso 2-OH (6.5 and 6.3).

Source (type strain): soil from Jeju, Korea.

DNA G+C content (mol%): 68 (T_m).

Type strain: HD9, JCM 12204, KACC 20266, NBRC 101070.

Sequence accession no. (16S rRNA gene): AY253866.

9. **Kribbella karoonensis** Kirby, Le Roes and Meyers, 2006, 1100[VP]

ka.ro.o.nen'sis. N.L. fem. adj. *karoonensis* of or pertaining to the Karoo Desert National Botanical Garden in Worcester, South Africa.

Characteristics are as described for the genus and listed in Table 225 and Table 226. Additional information given below is taken from the original paper (Kirby et al., 2006).

Substrate mycelium is cream/yellow on inorganic salts starch agar (ISP 4) and fragment into rod-shaped elements in broth and on agar media (Figure 273). Aerial mycelium is pale cream/white and fragments into rod-shaped elements. Grows at 20°C (minimal temperature tested) and 37°C, but not at 45°C. Grows well on ISP 2 medium supplemented with 2% NaCl, with weak growth in the presence of 3 or 4% NaCl. Survives in a soil sample heated (dry heating) at 120°C for 1 h. Grows in the presence of (μg/ml) carbenicillin disodium salt (100), chloramphenicol (20), cycloheximide (10), erythromycin (10), and streptomycin sulfate (10). Growth is inhibited by ampicillin sodium salt (100), cefotaxime (10), cephaloridine (10), chloramphenicol (50), gentamycin sulfate (10), neomycin sulfate (10), tobramycin sulfate (10), and vancomycin hydrochloride (10).

Source (type strain): soil collected from the base of a giant quiver tree, *Aloe pillansii*, in the Karoo Desert National Botanical Garden, Worcester, Western Cape Province, South Africa.

DNA G+C content (mol%): not determined.

Type strain: Q41, DSM 17344, JCM 14304, NBRC 101884, NRRL B-24425.

Sequence accession no. (16S rRNA gene): AY995146.

10. **Kribbella koreensis** (Lee, Kang and Hah 2000) Sohn, Ohn, Hong, Bae and Chun 2003, 1007[VP] (*Hongia koreensis* Lee, Kang and Hah, 2000, 197)

ko.re.en'sis. N.L. fem. adj. *koreensis* of or pertaining to Korea, the location of the soil sample from which the type strain was isolated.

Characteristics are as described for the genus and listed in Table 225 and Table 226. Additional information given below is taken from the papers of Lee et al. (2000) and Sohn et al. (2003), unless indicated.

Substrate mycelium is usually cream on the media tested and consists of irregularly branched hyphae tending to form apical swellings (observed, e.g. on ISP 3 medium). The hyphae fragment into elongated or short rod-shaped elements; bud-like cells are formed on some media, for instance, on ISP 3 medium (Figure 274). Aerial mycelium is white and fragmenting (Figure 272). Melanin is produced on ISP 7 medium. Grows at 10 and 30°C; growth at 37°C varies with experiments (Kirby et al., 2006; Lee et al., 2000; Trujillo et al., 2006). Grows well on ISP 2 medium in the presence of 2% NaCl (Kirby et al., 2006); no growth occurs at 3% NaCl (Kirby et al., 2006; Lee et al., 2000; Trujillo et al., 2006). Growth is observable at initial pH 7–10. Grows on D-arabinose, D-lyxose, methyl-α-D-glucoside, methyl-α-D-mannoside, and salicin as sole carbon sources; no growth is observed with dextran, D-glucosamine, L-sorbose, and dulcitol. Growth occurs on a tap-water agar (supplemented with cycloheximide and nystatin). Produces pyrrolidonyl arylamidase (API ZYM; Trujillo et al., 2006). Positive response in the API 20NE and ID 32GN systems for utilization of acetate, adipate, L-fucose, gluconate, 3-hydroxybutyrate, itaconate, malate, D-ribose, and suberate as sole carbon sources, and production of β-glucosidase; negative for utilization of caprate, citrate, glycogen, 3- and 4-hydroxybenzoate, 2- and 5-ketogluconate, L-alanine, malonate, phenyl acetate, L-proline, propionate, valerate, L-histidine, and L-serine, as well as reduction of nitrate to nitrogen, production of arginine dihydrolase and indole, and acidification of glucose (Cui et al., 2010). Acid is produced from D-raffinose in the Bacto OF basal medium (Difco); no acid production was found with all other compounds tested: D-glucose, D-arabinose, L-arabinose, D-cellobiose, D-fructose, D-galactose, inulin, D-lactose, maltose, D-mannose, D-melezitose, melibiose, methyl-α-D-glucoside, methyl-α-D-mannoside, L-rhamnose, L-ribose, salicin, sucrose, D-trehalose, D-xylose, adonitol, 2, 3-butanediol, *meso*-erythritol, glycerol, *meso*-inositol, D-mannitol, or 1, 2-propanediol. Gives an alkaline reaction in the tests for utilization of organic acids, such as α-ketoglutarate, malonate, propionate, pyruvate, and *trans*-aconitate with phenol red indicator; negative responses with *cis*-aconitate, benzoate, formate, maleate, salicylate, sebacate, sorbate, or tartarate (according to the method of Gordon et al. (1974). DNA and pectin are not hydrolyzed. Sodium hippurate is decomposed. Resistant to lysozyme, 0.1% (w/v) phenol, 0.01% (w/v) potassium tellurite, 0.00005% (w/v) crystal violet, and 0.001% (w/v) brilliant green, but not to 0.01% (w/v) sodium azide, 0.3% (w/v) phenylethanol, or 0.01% (w/v) thallous acetate. Grows in the presence of (μg/ml) carbenicillin disodium salt (100), cephaloridine (10), chloramphenicol (20, weak), cycloheximide (10), erythromycin (10), neomycin sulfate (10, weak), streptomycin sulfate (10), and vancomycin hydrochloride (10); growth is inhibited by (μg/ml) ampicillin sodium salt (100), cefotaxime (10), chloramphenicol (50), gentamycin sulfate (10), and tobramycin sulfate (10) (Kirby et al., 2006). No antimicrobial activity was observed against *Bacillus subtilis, Micrococcus luteus, Staphylococcus aureus, Streptomyces murinus, Escherichia coli, Enterobacter aerogenes, Saccharomyces cerevisiae, Candida albicans,* and *Aspergillus niger*. Produces neuropilin/growth factor complexes (Alitalo et al., 2003). The major cellular fatty acids (representing >5% of the total recorded in at least one experiment) reported for cultures grown under different conditions were: $C_{15:0}$ anteiso (24.7–36.4), $C_{16:0}$ iso (8.4–24.0), $C_{15:0}$ iso (12.3–20.1), $C_{17:0}$ iso (4.5–9.0), $C_{17:0}$ anteiso (3.1–5.0), 9-methyl-$C_{16:0}$ (<1–12.7), $C_{17:1}$ ω9*c* (<1–5.0), $C_{14:0}$ iso (2.0–5.9), $C_{16:0}$ (<1–6.6), and 10-methy-$C_{16:0}$ (<1–5.1) (Lee et al., 2000; Sohn et al., 2003; Song et al., 2004).

Source (type strain): a gold-mine cave in Kongju, Republic of Korea.

DNA G+C content (mol%): 71.3 (T_m).

Type strain: LM 161, IMSNU 50530, JCM 10977, NBRC 101069.

Sequence accession no. (16S rRNA gene): Y09159.

11. **Kribbella lupini** Trujillo, Kroppenstedt, Schumann and Martinez-Molina 2006, 410[VP]

lu.pi'ni. L. gen. n. *lupini* of lupin, isolated from *Lupinus angustifolius.*

Characteristics are as described for the genus and listed in Table 225 and Table 226. Additional information given below is taken from the original paper (Trujillo et al., 2006) unless indicated.

Substrate mycelium is white to cream, extensively branching and fragmenting into rod-shaped and coccoid elements. Abundant white aerial mycelia are produced that break up into rod-shaped fragments. Grows at 12 and 37°C. Tolerates up to 7% NaCl, as tested on nutrient agar. Growth is observed at pH 6–9, but not at pH 5 (initial pH values of nutrient agar). Grows on L-sorbose, adipate, gluconate, L-alanine, L-arginine, L-histidine, L-lysine, L-proline, and DL-valine but not caprate and L-serine as sole carbon sources. Acid is produced from glucose. Produces arginine dihydrolase, esterase lipase (C8), acid and alkaline phosphatases, leucine arylamidase, valine arylamidase, α-chymotrypsin, trypsin, α-galactosidase, α-glucosidase, β-glucosidase, *N*-acetyl-β-glucosaminidase, and α-mannosidase but not β-glucuronidase or pyrrolidonyl arylamidase (API ZYM). Sensitive (µg per disk) to amoxicillin (30), gentamycin (10), neomycin (5), novobiocin (5), oxytetracycline (30), streptomycin (300), tobramycin (10), vancomycin (30), and tetracycline (30). Resistant to ampicillin (2 µg/disk), penicillin G (10 U/disk), and rifampicin (2 µg/disk). The purified cell wall contains galactose and an unidentified sugar which is neither arabinose, glucose, mannose, rhamnose, ribose nor xylose. The major fatty acids (>5% of the total), as determined for cells grown in tryptic soy broth, were $C_{15:0}$ anteiso (28.2), $C_{16:0}$ iso (21.3), $C_{17:1}$ iso (11.0), $C_{14:0}$ iso (9.1), $C_{15:0}$ iso (5.3), $C_{16:1}$ iso (8.1), and $C_{17:0}$ anteiso (5.4).

Source (type strain): root nodules of *Lupinus angustifolius.*

DNA G+C content (mol%): 68 (T_m).

Type strain: LU14, DSM 16683, JCM 14303, LMG 22957, NBRC 101883.

Sequence accession no. (16S rRNA gene): AJ811962.

12. **Kribbella sancticallisti** Urzì, De Leo and Schumann 2008, 2095[VP]

sanc.ti.cal.li'sti. L. n. *Sanctus Callistus* Saint Callistus; N.L. gen. n. *sancticallisti* of Saint Callistus, isolated from the Saint Callistus Roman catacombs.

Characteristics are as described for the genus and listed in Table 225 and Table 226. Additional information given below is taken from the original paper (Urzì et al., 2008).

Substrate hyphae (diameter, 0.51 µm) are cream on Luedemann medium (Luedemann, 1968), branched, and tend to penetrate into the agar. Colonies are irregular (diameter 5–10 mm), with lobate margin, crateriform profile, and rough surface that is covered with white aerial hyphae. Produces melanin (weakly) on ISP 7 medium. No distinct formation of rod-shaped elements due to fragmentation has been observed in cultures on Luedemann medium, but morphological structures resembling short chains of non-separated or separated arthrospores are observable by electron microscopy. Grows at 15 and 37°C with weak growth at 10°C. Grows in the presence of 4% NaCl and weakly in the presence of 5% on Luedeman agar (Luedemann, 1968) without $CaCO_3$ addition. Growth occurs at pH 5 and 10 (initial pH values of the above medium). Utilizes gluconate; hydrolyzes DNA and Tweens 20, 40, and 60. Resistant to the following antibiotics (µg/ml) cephaloridine (10), gentamycin sulfate (3), oleandomycin (15), rifampicin (4) tobramycin (1.5), and penicillin G (10 U /ml); growth is inhibited by chloramphenicol (50), streptomycin sulfate (10), and vancomycin chloride (10). Sensitive to 0.01% lysozyme. The major fatty acids (>10% of the total) determined for three strains grown on TSA at 28°C for 24 h were $C_{16:0}$ iso (24.0–27.6), $C_{15:0}$ anteiso (19.7–20.8), and $C_{15:0}$ iso (9.1–10.8).

Source: deteriorated tufacean surfaces with grey-whitish alterations in the Roman catacombs of St. Callistus (Rome, Italy).

DNA G+C content (mol%): not determined.

Type strain: BC633, DSM 19602, JCM 14969.

Sequence accession no. (16S rRNA gene): AM778577.

13. **Kribbella sandramycini** Park, Yoon, Shin, Suzuki, Kudo, Seino, Kim, Lee and Lee 1999, 750[VP]

san.dra.my.ci'ni. N.L. n. *sandramycinum* sandramycin (an antibiotic); N.L. gen. n. *sandramycini* of sandramycin, intended to mean producing sandramycin.

Characteristics are as described for the genus and listed in Table 225 and Table 226. Additional information given below is taken from the original paper (Park et al., 1999), unless indicated.

Substrate mycelium consists of extensively branched hyphae which penetrate into agar media; they often fragment into rod- to coccoid-shaped and rod elements. Aerial hyphae appear straight or curved and break up into short to elongated rod-like elements. Colonies are pasty and have lichenous shapes with irregular edges. Gram-stain variable. Grows at 20 and 28–30°C; growth at 37°C varies with experiments (Kirby et al., 2006; Park et al., 1999; Trujillo et al., 2006). Weak growth may occur at 40°C. Grows well on ISP 2 medium in the presence of 2% NaCl, with moderate growth at 3% NaCl and weak growth in the presence of 4% NaCl (Kirby et al., 2006). Produces pyrrolidonyl arylamidase (Trujillo et al., 2006). Growth is observed at initial pH 5 and pH 9 (Kirby et al., 2006; Park et al., 1999; Trujillo et al., 2006). Grows in the presence of (µg/ml) carbenicillin disodium salt (100), chloramphenicol (50), cycloheximide (10), erythromycin (10), neomycin sulfate (10), streptomycin sulfate (10), and vancomycin hydrochloride (10) but not in the presence of (μg/ml) cefotaxime (10), cephaloridine (10), gentamycin sulfate (10), and tobramycin sulfate (10) (Kirby et al., 2006). Produces an anti-tumor antibiotic, sandramycin (Matson et al., 1993). The major fatty acids (>5% of the total at least in one experiment) reported for cultures grown under different conditions were $C_{15:0}$ anteiso (32.2–42.6), $C_{14:0}$ iso (3.0–14.8), $C_{15:0}$ iso (11.2–15.0), $C_{16:0}$ iso (5.6–8.1), $C_{17:0}$ iso (1.6–6.0), $C_{17:1}$ iso ω8*c* (<1–7.5), $C_{17:1}$ ω8*c* (3.9–6.1), and 9-methyl-$C_{16:0}$ (<1–7.0) (Park et al., 1999; Sohn et al., 2003; Song et al., 2004; Trujillo et al., 2006).

Source (type strain): soil in Mexico.

DNA G+C content (mol%): 68.3 (HPLC).

Type strain: ATCC 39419, KCTC 9609, JCM 10340, NBRC 100341.

Sequence accession no. (16S rRNA gene): AF005020, AY253864.

14. **Kribbella solani** Song, Kim, Hong, Cho, Sohn, Chun and Sun 2004, 1347[VP]

so.la'ni. N.L. gen. n. *solani* of *Solanum*, the genus of the potato, *Solanum tuberosum*, from which the type strain was isolated.

Characteristics are as described for the genus and listed in Table 225 and Table 226. Additional information given below is taken from the original paper (Song et al., 2004), unless indicated. The species description is based on the type strain only.

Substrate mycelium is usually cream, the hyphae are broadly branched. Aerial mycelium is white and consists of hyphae fragmenting into rod-shaped elements. Colonies are typically sand-pasty, with irregular edges, and have lichenous shapes. Among the strains tested, the best (but moderate) anaerobic growth of the type strain of this species is observed on ATCC Medium No. 172 (Kirby et al., 2006). According to Urzì et al. (2008), the strain also grows anaerobically on VL agar (Tiecco, 1975). Grows well in liquid organic medium 79 (Prauser and Falta, 1968) and on ISP 2 agar supplemented with 4% NaCl under aerobic conditions; moderate or weak growth is observed in the presence of 5% NaCl on ISP 2 medium, with no growth at 6% (Carlsohn et al., 2007; Everest and Meyers, 2008; Kirby et al., 2006). No growth is observed at initial pH 5 or pH 9 as assessed in liquid organic medium 79 (Carlsohn et al., 2007) and on Bennett's agar (Kirby et al., 2006). Grows on L-histidine, DL-arginine, and nitrate as nitrogen sources. Gives also clear growth on DL-α-amino-*n*-butyric acid, L-4-hydroxyproline, L-valine, L-methionine, and L-phenylalanine as sole nitrogen sources, in contrast to the phylogenetically closest relative *Kribbella hippodromi*, which shows weak or doubtful growth on these amino acids (Everest and Meyers, 2008; Song et al., 2004). Utilizes salicin as a sole carbon source. Produces (weakly) trypsin (Carlsohn et al., 2007). DNA and tributyrin are hydrolyzed but guanine, allantoin, and chitin are not. Tolerant to 0.01% lysozyme, 0.1% phenyl ethanol, and 0.01% potassium tellurite but not to 0.01% sodium azide, 0.01% thallous acetate, and 0.1% phenol. Grows in the presence (μg/ml) of carbenicillin disodium salt (100), chloramphenicol (50), cycloheximide (10), erythromycin (10), neomycin (80), streptomycin (20), tetracycline (4), rifampicin (4), tobramycin sulfate (10, weak), and vancomycin hydrochloride (10), with weak growth in the presence of penicillin G (10 IU/ml); growth is inhibited (μg/ml) by ampicillin sodium salt (100), cefotaxime (10), cephaloridine (10), and treoleandomycin (10) (Kirby et al., 2006; Song et al., 2004). Resists ampicillin (10 μg per disk) and shows weak growth in the presence of imipenem (10 μg per disk) (Carlsohn et al., 2007). No antimicrobial activity against *Bacillus subtilis, Micrococcus luteus, Streptomyces scabiei, Pseudomonas aeruginosa, Candida albicans, Aspergillus niger, Trichoderma harzianum, Fusarium acuminatum,* and *Colletotrichum gloeosporioides.* The major fatty acids (>5%) as determined for cultures grown on TSA agar (BBL) (Song et al., 2004) and in tryptic soy broth (Carlsohn et al., 2007) were $C_{15:0}$ anteiso (21.6 and 31.2), $C_{16:0}$ iso (19.5 and 16.9), $C_{15:0}$ iso (5.8 and 10.0), $C_{14:0}$ iso (12.6 and 10.4), and $C_{16:1}$ iso (6.2 and 1.5).

Source (type strain): a potato tuber with scab lesions, Jeju, Korea.

DNA G+C content (mol%): 69 (T_m).

Type strain: DSA1, JCM 12205, KACC 20196, NBRC 101071.

Sequence accession no. (16S rRNA gene): AY253862.

15. **Kribbella swartbergensis** Kirby, Le Roes and Meyers 2006, 1100[VP]

swart.berg.en'sis. N.L. fem. adj. *swartbergensis* of or pertaining to the Groot Swartberg mountain range, South Africa.

Characteristics are as described for the genus and listed in Table 225 and Table 226. Additional information given below is taken from the original paper (Kirby et al., 2006), unless indicated.

Substrate hyphae are cream-colored on inorganic salts/starch agar (ISP 4 medium) and fragment into rod-shaped elements in broth and on agar media. Aerial hyphae appear straight (Figure 271) and white in color on ISP 4 agar. Optimal temperature for growth is 30°C; grows occurs at 20 and 45°C (minimal and maximal temperatures tested). The ability to grow at 45°C readily distinguishes this species from all other kribbellae. It also grows in the presence of 3% NaCl, with weak growth at 4% NaCl as tested on ISP 2 medium. Grows at initial pH 5 (weak) and pH 9 both on Bennett's agar (Kirby et al., 2006) and in liquid organic medium 79 (Carlsohn et al., 2007). Produces trypsin (Carlsohn et al., 2007). Grows in the presence (μg/ml) of carbenicillin disodium salt (100), cephaloridine (10), chloramphenicol (20), cycloheximide (10), erythromycin (10), gentamycin sulfate (10), neomycin sulfate (10, weak), and streptomycin sulfate (10) but not in the presence of ampicillin sodium salt (100), cefotaxime (10), chloramphenicol (50), tobramycin sulfate (10), and vancomycin hydrochloride (10). Carlsohn et al. (2007) reported that the type strain of this species resisted ampicillin (10 μg per disk) and imipenem (10 μg per disk). The major fatty acids (>5%) determined for the culture grown in TSB were $C_{16:0}$ iso (27.4), $C_{15:0}$ anteiso (16.4), 9-methyl $C_{16:0}$ (13.8), $C_{17:0}$ anteiso (11.7), and $C_{15:0}$ iso (5.7) (Carlsohn et al., 2007).

Source (type strain): soil from the banks of the River Gamka, Die Hel, Groot Swartberg mountain range, Western Cape Province, South Africa.

DNA G+C content (mol%): not determined.

Type strain: HMC25, DSM 17345, JCM 14305, NBRC 101885, NRRL B-24426.

Sequence accession no. (16S rRNA gene): AY995147.

16. **Kribbella yunnanensis** Li, Wang, Zhang, Xu and Jiang, 2006a, 1459[VP] (Effective publication: Li, Wang, Zhang, Xu and Jiang 2006b, 33.)

yun.nan.en'sis. N.L. fem. adj. *yunnanensis* of or pertaining to Yunnan, a province of south-west China in which the sample was collected.

Characteristics are as described for the genus and listed in Table 225 and Table 226. Additional information given below is taken from the original paper (Li et al., 2006) unless indicated.

Substrate mycelium is moderate yellow/soft yellow on ISP 2, potato agar, and nutrient agar, and pale yellow or yellow white on ISP 3, ISP 4, ISP 5, and Czapek's agar. The hyphae are extensively branched and often fragmenting into irregular rod-shaped elements. Aerial mycelia develop on all media tested, especially on Czapek's agar; it is typically white on ISP 3, ISP 4, and ISP 5 media and potato agar, yellow-white on ISP 2 medium and Czapek's agar, and pale yellow on nutrient agar. Aerial hyphae fragment into short to elongated rod-like elements which can be arranged in short chains and have an arthrospore appearance on some media, e.g. on ISP 2 agar. Melanin is weakly produced on ISP 7 medium (Urzì et al., 2008). The major fatty acids (>5% of the total) determined in cells grown on TSA at 28°C include $C_{15:0}$ anteiso (31.6), $C_{15:0}$ iso (14.8), $C_{16:0}$ iso (10.2), $C_{14:0}$ iso (7.8), $C_{17:1}$ iso ω9*c* (6.7), and $C_{17:0}$ iso (6.5).

Source (type strain): soil in Yunnan Province, the west of China.

DNA G+C content (mol%): 68.6 (T_m).

Type strain: YIM 30006, CCTCC AA 001019, DSM 15499, JCM 14307, NBRC 103562.

Sequence accession no. (16S rRNA gene): AY082061.

Genus V. **Marmoricola** Urzì, Salamone, Schumann and Stackebrandt 2000, 534[VP] emend. Lee and Lee 2010, 2138

Lyudmila I. Evtushenko

Mar.mo.ri'co.la. L. neutr. n. *marmor* marble; L. masc. suff. *-cola* inhabitant of; *Marmoricola* inhabitant of marble.

Spherical cells, mostly about 0.5–1.0 µm in diameter, that occur singly, in pairs, short chains, or small clusters. Nonmotile or motile. Endospores are not formed. **Gram-positive** cell-wall type. Non-acid-fast. **Chemo-organotrophic with a respiratory type of metabolism. Catalase-positive, test for oxidase is usually negative.** Grow under aerobic conditions on a range of standard nutrient media, including the chemically defined (synthetic) media, forming circular, pasty, and **yellow or orange colonies.** Acids are produced oxidatively rather weakly from carbohydrates. Nitrate reduction varies among species. **Mesophilic**, grow best at 28–30°C; no growth occurs at 42°C. Prefer a neutral to mildly alkaline pH; some species are alkalitolerant or display alkaliphilic properties. Usually non-halophilic. **The cell-wall peptidoglycan contains LL-diaminopimelic acid and glycine** as the diagnostic amino acid **(the peptidoglycan type A3γ).** The acyl type of muramic acid is acetyl. **Menaquinones are the only respiratory quinones; the major component is the tetrahydrogenated menaquinone with eight isoprene units, MK-8(H_4).** The cellular fatty acid pattern is dominated by **straight-chain saturated and unsaturated ($C_{16:0}$ and $C_{18:1}$) components. Tuberculostearic ($C_{18:0}$ 10-methyl) acid is usually detected.** Mycolic acids are absent. The principal identified polar lipids are **phosphatidylinositol, phosphatidylglycerol, and diphosphatidylglycerol; phosphatidylcholine may be present.** Occur in various terrestrial and marine environments and can be found in bacterial communities associated with plants and humans.

DNA G+C content (mol%): 72–73 (HPLC).

Type species: **Marmoricola aurantiacus** Urzì, Salamone, Schumann and Stackebrandt 2000, 534[VP].

Further descriptive information

The genus *Marmoricola* belongs to the family *Nocardioidaceae*, order *Propionibacteriales* (formerly suborder *Propionibacterineae* Stackebrandt et al. 1997), and currently harbors five species, with *Marmoricola aurantiacus* as the type species (Table 227). Based on the 16S rRNA gene analysis, the *Marmoricola* species together with *Nocardioides jensenii* and a few related species of this genus form a group at the periphery of the genus *Nocardioides* radiation (Figure 247). The 16S rRNA gene sequence similarities between the recognized *Marmoricola* species range from 95.3 to 98.7%.

The information given below originates from the papers on the descriptions of the genus *Marmoricola* (Urzì et al., 2000) and the species currently assigned to the genus (Dastager et al., 2008b; Lee and Lee, 2010; Lee, 2007a; Lee et al., 2010; Urzì et al., 2000), unless indicated.

Cells of most species are non-motile cocci, ranging from 0.5 to 1.0 µm in diameter (Figure 276) or a bit more, while the coccoid cells of *Marmoricola bigeumensis* are motile and smaller (~0.3–0.5 µm). A rod–coccus morphogenetic cycle does not occur (data for *Marmoricola aurantiacus* and *Marmoricola aequoreus*). Colonies are typically circular, entire, convex, shiny, pasty, and yellow- or orange-pigmented. They may become rough, crater-shaped, or irregularly shaped in old (1–1.5 months) cultures.

Marmoricola aurantiacus contains the cell-wall peptidoglycan A3γ type *sensu* Schleifer and Kandler (1972), with LL-diaminopimelic acid (LL-A_2pm) at the third position of the peptide chain and a single glycine as an interpeptide bridge. The same diamino acid, LL-A_2pm, has been reported for the remaining

TABLE 227. Descriptive and differential characteristics of *Marmoricola* species[a]

Characteristic	*M. aurantiacus*	*M. aequoreus*	*M. bigeumensis*	*M. korecus*	*M. scoriae*
Colony color[b]	Orange	Yellow	Lemon yellow	Yellow	Vivid yellow
Cell shape (size, μm)	Cocci (0.5–0.7)	Cocci (0.5–0.7)	Cocci (~0.3–0.5)	Cocci (1.1–1.2)	Cocci (0.6–1.0)
Motility	–	–	+	–	–
Temperature range (°C)[c]	18–28	4(w)–37	20–37	4–37	10–37
Growth at pH[d]	5.1–8.7	5.1–12.1	6.0–12.0	5.1–12.1	6.1–12.1
Optimal pH[d]	~7.0	~7.1	~7.2	6.1–10.1	8.1–11.1
Max. NaCl (%, w/v)	2.0	7.0 (w)	7.0	2.0	3.0
Catalase activity	+	+	+	+	+
Oxidase test	–	–	–	–	–
Nitrate reduction	–	+	+	–	–
Hydrolysis of:					
Casein	–	+	–	+	+
DNA	–		–	–	+
Gelatin	–	+	–	+	+
Starch	–	–	+	–	–
Tyrosine	–	–	+	–	–
Hypoxanthine	+	–	nd	–	–
Xanthine	–	–	+	–	–
Utilization of:[e]					
L-Arabinose	+	–	+	nd	+
D-Arabinose	+	–	–	–	+
D-Cellobiose	+	+	+	+	–
Dextran	+	+	+	–	–
D-Fructose	+	+	–	–	+
D-Galactose	+	+	–	–	+
D-Lactose	+	+	+	–	–
Maltose	+	+	+	–	+
D-Mannose	+	+	+[f]	+	v
D-Melezitose	+	+	–	–	+
Raffinose	+	–	–	–	+
L-Rhamnose	+	–	+	–	+
L-Ribose	+	+	–	–	+
Salicin	+	–	–	–	+
Sucrose	+	+	–	–	+
D-Trehalose	+	+	–	–	+
D-Xylose	+	+	+	+	+
Adonitol	+	–	–	+	–
Dulcitol	–	–	–	+	–
meso-Erythritol	–	–	–	+	–
Glycerol	+	+	–	–	–
D-Mannitol	+	+	+[f]	+	+
D-Sorbitol	+	+	+	–	–
D-Xylitol	+	–	+	+	–
Acetate	+	+	+	–	+
Citrate	+	+	–	–	+
DL-Malate	+	+	–	–	+
Succinate	+	+	–	+	–
DL-Tartrate	+	–	–	–	–
Enzymatic activity (API ZYM):					
Acid phosphatase	+	–	+	+	+
Cystine arylamidase	–	w	+	w	w
Valine arylamidase	w	+	–	+	+
Esterase (C4)	w	–	+	–	–
β-Galactosidase	w	+	v	–	+
β-Glucosidase	+	w	+	w	+
Trypsin	–	–	+	w	w
α-Chymotrypsin	–	–	+	–	–
Naphthol-AS-BI-phosphohydrolase	+	–	+	+	+
Cell-wall diamino acid	LL-A_2pm	LL-A_2pm	LL-A_2pm	LL-A_2pm	LL-A_2pm
Predominant fatty acid type	Straight-chain	Straight-chain	iso-Branched	Straight-chain	Straight-chain

(continued)

TABLE 227. (continued)

Characteristic	*M. aurantiacus*	*M. aequoreus*	*M. bigeumensis*	*M. korecus*	*M. scoriae*
Major fatty acids (>10% of the total)[g]	$C_{16:0}$, $C_{18:1}$ ω9*c*, ($C_{16:1}$)[h]	$C_{16:0}$, $C_{18:1}$ ω9*c*	$C_{16:0}$ iso	$C_{16:0}$, $C_{17:1}$ ω8*c*, $C_{18:1}$ ω9*c*	$C_{16:0}$, $C_{18:1}$ ω9*c*, $C_{18:0}$ 10-methyl
Phospholipids	DPG, PG, PI, PL	DPG, PG, PI, PL	DPG, PG, PI, PL	DPG, PC, PG, PI, PL	DPG, PC, PG, PI, PL
Major menaquinone	MK-8(H_4)	MK-8(H_4)	MK-8(H_4)	MK-8(H_4)	MK-8(H_4)
DNA G+C content (mol%)	72.0	72.4	72.9	71.0	72.0
Source of type strain	Marble statue	Beach sandy sediment	Agricultural soil	Volcanic ash	Volcanic ash

[a]Based on characteristics of type strains; see the text for other descriptive characteristics. Data from Urzì et al. (2000), Lee (2007a), Dastager et al. (2008b), Lee and Lee (2010), and Lee et al. (2010). Abbreviations: LL-A_2pm, LL-2,6-diaminopimelic acid; PI, phosphatidylinositol; PG, phosphatidylglycerol; DPG, diphosphatidylglycerol; PC, phosphatidylcholine; PL, unknown phospholipid(s); MK-8(H_4), a menaquinone with eight isoprenoic units in the side chain, two of which are saturated. Symbols: –, negative; +, positive; w, weak or slow; v, variable between experiments; nd, no data available; see the text for other symbols.

[b]The pigmentation intensity and shade may vary depending on the growth medium and the culture age.

[c]The recorded lower and higher growth temperatures are indicated; the actual growth temperature range can be slightly broader for some species (see the text).

[d]Initial pH values of media.

[e]Data from Lee and Lee (2010) and Lee et al. (2010) obtained using ISP 9 (Shirling and Gottlieb, 1966) as basal medium.

[f]The opposite result has been reported by Dastager et al. (2008b) obtained using different test medium.

[g]Either $C_{16:0}$ or $C_{18:1}$ ω9*c* may represent a larger proportion depending on culture conditions (see the text and original papers for the culture conditions and the methods).

[h]$C_{16:1}$ may be undetectable or comprising <1% of the total fatty acids (Lee and Lee, 2010).

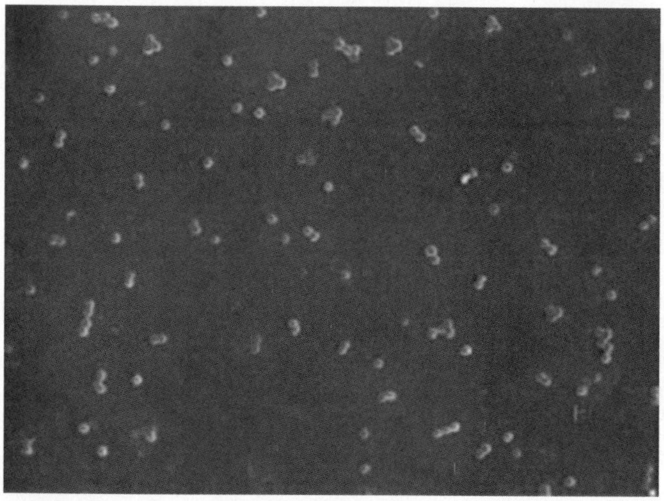

FIGURE 276. Cellular morphology of *Marmoricola aurantiacus*. Cells from a 14-d-old culture on Luedemann medium (Luedemann, 1968). Scanning electron micrograph. Bar = 1 μm. (Reproduced with permission Urzì et al., 2000. Int. J. Syst. Evol. Microbiol. *50*: 529–536.)

species. The acyl type of peptidoglycan is acetyl, as determined for *Marmoricola aurantiacus* (using the glycolate test of Uchida and Aida, 1984). The whole-cell sugars were also reported only for this species and included glucose and traces of ribose. The major component of the respiratory quinone system is MK-8(H_4); minor amounts of MK-7(H_4), MK-8(H_2), and MK-6(H_4) have been reported for *Marmoricola aurantiacus* (peak area ratio, 73:4:1:1). The cellular fatty acids of most species are dominated by straight-chain saturated and straight-chain unsaturated acids. A significant proportion (up to 10%) of tuberculostearic acid ($C_{18:0}$ 10-methyl) can be detected, along with other branched-chain and hydroxylated components occurring in minor or trace amounts (see Table 227 and *List of species of the genus Marmoricola* for more details). In *Marmoricola bigeumensis*, in contrast, branched-chain fatty acids predominate, with $C_{16:0}$ iso comprising 27.6 or almost 60% depending on experiment. All species are characterized by similar principal phospholipids in the polar lipid fraction, including phosphatidylinositol (PI), phosphatidylglycerol (PG), diphosphatidylglycerol (DPG), and an additional unidentified component (unidentified phospholipids in all cases except for *Marmoricola bigeumensis*). There is no evidence, however, that this unidentified component is the same for all species or different. Recently, phosphatidylcholine has been reported in *Marmoricola scoriae* and *Marmoricola korecus* along with PI, PG, DPG, and unidentified phospholipids (Lee and Lee, 2010; Lee et al., 2010).

Bacteria of this genus grow aerobically on a range of standard laboratory media based on peptone and yeast extract, and also display growth on mineral media, e.g. ISP 9* (contains essential salts, trace microelements, and ammonium as a source of nitrogen) supplemented with particular carbon sources, including various carbohydrates, sugar alcohols, organic acids, and other compounds. According to Lee and Lee (2010) and Lee et al. (2010), all species of this genus can grow on ISP 9 with D-glucose, D-mannose, D-mannitol, xylose, or α-methyl-D-glucoside, but not with formate, *meso*-inositol, α-methyl-D-mannoside, or L-sorbose, and display specificity with regard to a variety of other carbon sources (Table 227). Generally, species of this genus are rather non-exacting in their growth requirements, but some may be nutritionally fastidious. *Marmoricola aurantiacus* is unable to grow on certain media rich in organics, including *Corynebacterium* agar (DSMZ medium No. 53). *Marmoricola bigeumensis* has been reported (Dastager et al., 2008b) not to grow on yeast extract-malt extract agar (ISP 2), oatmeal agar (ISP 3), and inorganic salt-starch agar (ISP 4). Bacteria of this genus decompose and hydrolyze a range of

*See Shirling and Gottlieb (1966) for the composition of ISP media cited here and in other sections of this chapter.

complex organic compounds and possess various enzymatic activities (Table 227). In addition, all species show activities for alkaline phosphatase, esterase lipase (C8), leucine arylamidase, α-glucosidase, β-glucosidase, but are negative in tests for lipase (C14), α-galactosidase, β-glucuronidase, *N*-acetyl-β-glucosaminidase, α-mannosidase, and α-fucosidase (the API ZYM assay). All give negative responses in the API 20NE tests for arginine dihydrolase, lysine decarboxylase, ornithine decarboxylase, urease, tryptophan deaminase, and indole production, but are positive in the Voges–Proskauer test. Only two species (*Marmoricola aequoreus* and *Marmoricola bigeumensis*) display positive reaction in the test for nitrate reduction. H_2S production is not observed (examined for *Marmoricola aequoreus* and *Marmoricola scoriae*). All species hydrolyze esculin. Do not decompose carboxymethyl cellulose (data for *Marmoricola korecus* and *Marmoricola scoriae*). *Marmoricola* species so far described are mesophilic and generally considered to be non-halophilic, but some have been reported to grow best at a low salinity (1.5–2% NaCl; *Marmoricola bigeumensis*) or tolerate up to 7% NaCl (*Marmoricola aequoreus* and *Marmoricola bigeumensis*). All prefer a neutral to mildly alkaline pH; some species can grow in test media at high initial pH values (up to pH 12); some exhibit alkaliphilic properties (*Marmoricola scoriae*).

Bacteria of this genus occur in various environments, including nutrient-limited ones. The recognized *Marmoricola* species have been isolated from agricultural soil, sandy beach sediments, volcanic ash, and a marble statue (Table 227). Ecological studies show that representatives of *Marmoricola* or phylogenetically very close organisms occur in soils of different origins, including pastures (Schoenborn et al., 2004) and the mound of a soil-feeding termite (*Cubitermes niokoloensis*; Fall et al., 2007). Bacteria of this genus have also been associated with dust particles transported from desert top soils by Saharan dust storms (Polymenakou et al., 2008) and with urban aerosols (Brodie et al., 2007), and found in indoor environments (Rintala et al., 2008; GenBank no. FM872516 and no. AM697095). They also occur in marine ecosystems (Gontang et al., 2007) and among bacteria cultured from granular activated carbon filters in water treatment installations (Magic-Knezev et al., 2009). Bacteria of the genus *Marmoricola* are detected in human skin microbiome (Grice et al., 2009, 2008; see, e.g. GenBank numbers HM267307, HM326534, etc.) and have been found in microbial populations of bronchoalveolar lavage fluid from children with cystic fibrosis (Harris et al., 2007). A bacterium remotely related to *Marmoricola aurantiacus* was also detected in the rhizosphere-associated bacterial and fungal communities of diseased seedlings (Filion et al., 2004). However, no species or strains pathogenic for humans or warm-blooded animals or plants have yet been identified within the genus *Marmoricola*.

Enrichment and isolation procedures

No selective media or enrichment procedures have been devised for isolating organisms of the genus *Marmoricola*. The available strains of this genus are isolates obtained during studies on microbiota of certain biotopes. *Marmoricola aurantiacus* was isolated from a marble statue on Bunt and Rovira medium (Bunt and Rovira, 1955), modified by the addition of 0.5% glucose, 0.5% NaCl, and 0.03% Na_2CO_3, pH 8.6 (BRII) and further cultivated in Luedemann medium (Luedemann, 1968). *Marmoricola korecus* and *Marmoricola scoriae* were isolated on starch–casein agar (1% soluble starch, 0.03% casein, 0.2% KNO_3, 0.2% NaCl, 0.2% KH_2PO_4, 0.002% $CaCO_3$, 0.005% $MgSO_4 \cdot 7H_2O$, 0.001% $FeSO_4 \cdot 7H_2O$, and 1.8% agar, pH 7.2). SC-SW agar (the same composition but without addition of KH_2PO_4 and prepared in a 60:40 mixture of natural sea water and distilled water; Lee, 2006) was used for isolation of *Marmoricola aequoreus*, and one-tenth-strength R2A agar (Difco) was used for isolation of *Marmoricola bigeumensis*. Isolations were carried out using the serial dilution plating method, followed by incubation at 28–30°C for ~2–4 weeks.

Maintenance procedures

Cultures may be maintained as 20% glycerol suspensions at −20 and −80°C. Long-term conservation is achieved by freeze-drying or in liquid nitrogen by standard procedures.

Differentiation of the genus *Marmoricola* from other genera

The phenotypic characteristics differentiating *Marmoricola* from other genera of the family are listed in Table 214 and Table 216. The genus *Marmoricola* is readily distinguished from most genera of the family by the presence of coccoid cells and predominant menaquinone MK-8(H_4). The same major menaquinone is also characteristic of the genus *Nocardioides*, but members of the latter have mycelial or rod-shaped cell morphologies in young cultures and usually form coccoid cells (along with short rods) at later growth stages. Further, the *Marmoricola* species readily differ from the genus *Nocardioides* in the fatty acid profile dominated by straight-chain compounds (exception is *Marmoricola bigeumensis*). It should be emphasized that delineation of the current genera *Marmoricola* and *Nocardioides*, as a whole, remains somewhat uncertain (See *Taxonomic comments*, below). Therefore, the side-by-side comparison of novel strains with individual species of the genus *Marmoricola* and closely related *Nocardioides* species is advisable during the identification process. Also, (short) rods in some organisms of this group, owing to their possible fast fragmentation into cocci on certain media, may be difficult to observe.

Taxonomic comments

The genus *Marmoricola* was described by Urzì et al. (2000) with one species, *Marmoricola aurantiacus*, represented by a single nonmotile coccoid strain having high (96.4%) 16S rRNA gene sequence similarity with the phylogenetically neighboring species, *Nocardioides jensenii*. The establishment of the genus *Marmoricola* was based on the assumed priority of unusual phenotypic, including chemotaxonomic, features of the strain and its difference from other members of the family *Nocardioidaceae* described by that time in several secondary-structure-forming nucleotides. The key chemotaxonomic marker separating the genus *Marmoricola* from *Nocardioides* included the cellular fatty acid profile (mostly straight-chain components with minor amounts of $C_{18:0}$ 10-methyl). The species *Marmoricola aequoreus*, *Marmoricola scoriae*, and *Marmoricola korecus*, described subsequently (Lee and Lee, 2010; Lee, 2007a; Lee et al., 2010) have generally similar morphological and chemotaxonomic properties to those of *Marmoricola aurantiacus*, except the presence of phosphatidylcholine (detected in *Marmoricola scoriae* and *Marmoricola korecus*). This finding rendered the genus heterogeneous with respect to principal polar lipids, and the genus description has been emended (Lee and Lee, 2010) to include this characteristic and also reflect some features of *Marmoricola bigeumensis* described in 2008 (Dastager et al., 2008b). The latter was assigned to the genus mainly on the basis 16S rRNA-based phylogenetic clustering (Dastager et al., 2008b).

However, *Marmoricola bigeumensis*, as mentioned before, markedly differs from the other *Marmoricola* species by possessing a large proportion of saturated iso-branched acids and resembling *Nocardioides* by the fatty acid profile. Notably, iso- and anteiso-branched acids are synthesized in a different way than the straight-chain saturated acids (for details and references see Suzuki et al., 1993), and the respective difference in fatty acid composition is typically used as a criterion to differentiate the actinomycete genera (Kroppenstedt, 1985; Suzuki and Komagata, 1983a; Urzì et al., 2000). *Marmoricola bigeumensis* is also distinguished from the other *Marmoricola* species by motile and smaller cells, and exceeds those in NaCl resistance. In addition, it is not quite clear whether the cells of *Marmoricola bigeumensis* are cocci at all growth phases, or arise from (short) rods at later stages of the developmental cycle as in *Nocardioides* species. Furthermore, with the recent descriptions of additional *Marmoricola* species and closely related organisms of the genus *Nocardioides*, it becomes clear that *Marmoricola bigeumensis* is branching outside the group of the remaining *Marmoricola* species. This species is also closest to members of the genus *Nocardioides* (*Nocardioides jensenii* and *Nocardioides mesophilus*) in terms of 16S rRNA gene sequence similarity (97%; calculated from 1420 bp and 1396 bp, respectively), while the sequence similarity to the other *Marmoricola* species is 96.6–95.3% (1398–1435 bp compared). Taken together, all the above data indicate that the current taxonomic structure of the genus *Marmoricola* is in need of re-evaluation to improve coherence of its phylogenetic and phenotypic circumscription and delimitation from the genus *Nocardioides*. The available data suggest that further detailed taxonomic studies of *Marmoricola bigeumensis* at the genomic and phenotypic level, involving available and newly isolated relevant strains, may provide strong grounds for movement of *Marmoricola bigeumensis* to the genus *Nocardioides* or to a new genus.

Differentiation of species of the genus *Marmoricola*

Phenotypic characteristics useful in distinguishing the current *Marmoricola* species are discussed in the above sections and listed in Table 227 and in the *List of species of the genus Marmoricola*, below. When using the compositions of fatty acids and polar lipids for differentiation, note that these characteristics may depend on growth conditions, age of the culture, and analytical procedure. In comparative taxonomic studies, cells must be grown and analyzed under the same conditions.

Acknowledgements

The author was supported by the MCB program of the Russian Academy of Sciences.

List of species of the genus *Marmoricola*

1. **Marmoricola aurantiacus** Urzì, Salamone, Schumann and Stackebrandt 2000, 534[VP]

au.ran.ti'a.cus. N.L. masc. adj. *aurantiacus* orange-colored.

Characteristics are as described for the genus and listed in Table 227. Additional information presented below is taken from the original paper (Urzì et al., 2000), unless indicated. Spherical cells (0.5–0.7 μm in diameter) occurring singly, in pairs, as short chains, or small clusters (Figure 276). Nonmotile. No rod–coccus life cycle. Cells are Gram-stain-positive and not lysed in the KOH test. Non-acid-fast. Colonies are orange-pigmented and 2–5 mm diameter after 30-d incubation on Luedemann medium (Luedemann, 1968) at 28°C. Grows at 18°C; no growth occurs at 6 or 37°C. Growth is restricted on Luedemann medium supplemented with 2% NaCl and absent in the presence of 4% NaCl. The type strain will grow on brain heart infusion agar (BHIA; Oxoid), potato glucose agar (PDA; Oxoid), ISP 2 agar, and in Bacto tryptic soy broth (Difco). According to Lee and Lee (2010), growth is also supported by R2A agar (Difco). No growth is observed in several other tested complex media. Will grow on inorganic medium ISP 9 with various carbon sources (Lee and Lee, 2010) but not on Czapek–Dox modified agar (CZ; Oxoid) or on water agar. Acid is not produced from D-glucose, D-ribose, L-arabinose, D-galactose, D-cellobiose, lactose, maltose, D-mannose, D-raffinose, L-rhamnose, D-trehalose, D-xylose, glycerol, D-mannitol, or *myo*-inositol, as registered on the medium containing 0.5% tryptone (Oxoid), 0.4% Bacto Casamino acids (Difco), 0.07% $(NH_4)_2HPO_4$, 0.5% NaCl, 0.003% bromocresol purple, and the substrate at a concentration of 1%. Hydrolyzes Tween 80 (weak). The whole-cell sugars are glucose and traces of ribose. The major cellular fatty acids (about 5% or more at least in one experiment) determined in cells grown in Bacto tryptic soy broth at 28°C (Urzì et al., 2000) and in cells cultured on R2A agar at 30°C for 5 d (Lee and Lee, 2010) are $C_{16:0}$ (41.4 and 37%, respectively), $C_{18:1}/C_{18:1}$ ω9*c* (33.7 and 27), $C_{16:1}/C_{16:1}$ ω9*c* (14.9 and <1), $C_{18:0}$ 10-methyl (2.8 and 5.4), and $C_{16:0}$ 2-OH (1.4 and 4.7). In addition, Lee and Lee (2010) detected a significant amount of $C_{16:1}$ ω7*c* and/or $C_{15:0}$ iso 2-OH (9.9%) in cells from R2A agar.

Source (type strain): a Carrara marble statue (Wagmüller's monument) located in the Nordfriedhof Cemetery in Munich, Germany.

DNA G+C content (mol%): 72.1 (HPLC).

Type strain: BC 361, DSM 12652, CIP 106770, JCM 10917.

Sequence accession no. (16S rRNA gene): Y18629.

2. **Marmoricola aequoreus** Lee 2007a, 1392[VP]

a.e.qu.o.re'us. L. masc. adj. *aequreus* belonging to the sea, referring to the isolation site of the type strain.

Characteristics are as described for the genus and listed in Table 227. Additional information presented below is taken from the papers of Lee (2007a) and Lee and Lee (2010).

Spherical cells (0.5–0.7 μm in diameter) occurring singly, in pairs, as short chains, or small clusters. Nonmotile. No rod–coccus life cycle is observed. Colonies are bright yellow after incubation for 5 d at 30°C on YE-SW agar (ISP 2 agar prepared in a 60:40 mixture of natural sea water and distilled water). Growth is good at 10–37°C (with an optimum at 30°C), poor at 4°C, and absent at 42°C. Will grow on ISP medium 2 with 5% NaCl, but poorly at 6–7% NaCl. Grows at pH 5.1–12.1 (initial pH of medium, YE-SW agar) and optimally at pH 7.1. The type strain produces acid from dextran, maltose, and salicin but not from D-glucose and almost 30 other carbon sources tested using Bacto OF basal medium (Difco) supplemented with a filter-sterilized carbon source at a final concentration of 1%. The cellular fatty acids (>5% at least in one experiment), determined in cells grown on TSA for 3 d at 30°C and on R2A agar for 5 d at the same temperature, include $C_{18:1}$ $\omega 9c$ (40.1 and 26.4), $C_{16:0}$ (35 and 30.2), $C_{18:0}$ (7.6 and 2.3%), $C_{16:1}$ $\omega 9c$ (6.6 and <1). In cells harvested from R2A agar, 10-methyl-branched acids were detected (7.1% $C_{18:0}$ 10-methyl and 1.7% $C_{17:0}$ 10-methyl), as well as $C_{19:0}$ anteiso (6.3%), $C_{16:1}$ 2-OH (4.3%), and $C_{16:1}$ $\omega 7c$ and/or $C_{15:0}$ iso 2-OH (8.2%).

Source (type strain): sandy sediment 1 m below the surface of Samyang Beach on Jeju Island, Republic of Korea.

DNA G+C content (mol%): 72.4 (HPLC).

Type strain: SST-45, JCM 13812, NRRL B-24464.

Sequence accession no. (16S rRNA gene): AM295338.

3. **Marmoricola bigeumensis** Dastager, Lee, Ju, Park and Kim 2008b, 1062[VP]

bi.ge.um.en'sis. N.L. masc. adj. *bigeumensis* of or pertaining to Bigeum Island, Korea, from where the type strain was isolated.

Characteristics are as described for the genus and listed in Table 227. Additional information presented below is taken from the original paper (Dastager et al., 2008b), unless indicated.

Coccoid and motile cells (0.3–0.5 μm in diameter). Colonies are lemon-yellow and 1–3 mm in diameter on R2A agar. Grows in the presence of up to 7% NaCl and at pH 6.0–12.0 (initial pH of medium), but best at 1.5–2% NaCl and pH ~7.2 (all tested on twofold diluted in R2A agar). Will grow in the temperature range of 20–37°C. No growth is observed on yeast extract-malt extract agar (ISP 2), oatmeal agar (ISP 3), and inorganic salt-starch agar (ISP 4). D-Glucose, L-arabinose, L-rhamnose, D-ribose, and D-xylose are utilized as carbon sources for growth and energy, as determined both in the original study (Dastager et al., 2008b) and using ISP 9 as basal medium (Lee and Lee, 2010); growth with D-mannose and D-mannitol is variable (Dastager et al., 2008b; Lee and Lee, 2010). No acid is produced (the Hugh–Leifson test; Hugh and Leifson, 1953) from the aforementioned carbohydrates and also from cellobiose, D-galactose, lactose, maltose, and raffinose. Tween 80 is hydrolyzed. The reported cellular fatty acids [determined in cells grown on Trypticase soy agar plates (BBL) using the procedures described by Miller (1982)] include $C_{16:0}$ iso (59.1%), $C_{17:0}$ anteiso (6.9%), $C_{16:0}$ (4.0%), and $C_{14:0}$ iso (3.3%). A wider spectrum of fatty acids was detected by Lee and Lee (2010) for cells grown on R2A agar at 30°C for 5 d, with $C_{16:0}$ iso predominating (27.6%). The other components (3% or more) were $C_{16:0}$ (6.8), $C_{17:0}$ (3.8), $C_{17:1}$ $\omega 8c$ (6.9), $C_{18:1}$ $\omega 9c$ (4.2), $C_{14:0}$ iso (4.3), $C_{15:0}$ iso (5.5), $C_{17:0}$ iso (3.5), $C_{15:0}$ anteiso (3.3), $C_{17:0}$ anteiso (5.9), $C_{16:0}$ iso 3-OH (3.0), and $C_{16:1}$ 2-OH (3.5); a 10-methyl-branched acid ($C_{17:0}$ 10-methyl) was found as a minor component (2.2%).

Source (type strain): a soil sample collected from an agricultural area of Bigeum Island, Korea.

DNA G+C content (mol%): 72.9 (HPLC).

Type strain: MSL-05, KCTC 19287, JCM 15624, DSM 19426.

Sequence accession no. (16S rRNA gene): EF466120.

4. **Marmoricola korecus** Lee, Lee and Ko 2010, 000[VP]

ko.re'cus. N.L. masc. adj. *korecus* pertaining to Korea, where the type strain was isolated.

Characteristics are as described for the genus and listed in Table 227. Additional information presented below is taken from the original paper (Lee et al., 2010).

Spherical (1.1–1.2 μm diameter) and nonmotile cells. Colonies are yellow-pigmented after incubation for 5 d on ISP 2 agar at 30°C (optimal temperature). Growth occurs at 4–37°C but not at 42°C. Grows on ISP 2 with 2% NaCl (but not with 3%), and in the pH range of pH 5.1–12.1; optimal pH is 6.0–10.1 (initial pH of the same test medium). The major fatty acids determined in cells grown on R2A agar at 30°C for 5 d are $C_{16:0}$, $C_{17:1}$ $\omega 8c$, $C_{18:1}$ $\omega 9c$, as well as $C_{16:1}$ $\omega 7c$ and/or $C_{15:0}$ iso 2-OH.

Source (type strain): a red-colored layer of volcanic ash, Jeju, Republic of Korea.

DNA G+C content (mol%): 71.0 (HPLC).

Type strain: Sco-A36, KCTC 19596, DSM 22128.

Sequence accession no. (16S rRNA gene): FN386723.

5. **Marmoricola scoriae** Lee and Lee 2010, 2138[VP]

sco.ri'a.e. L. gen. n. *scoriae* of scoria (volcanic ash), referring to the site from which the type strain was isolated.

Characteristics are as described for the genus and listed in Table 227. Additional information presented below is taken from the original paper (Lee and Lee, 2010).

Spherical cells (0.6–1.0 μm in diameter), occur singly, in pairs or in clusters. Nonmotile. Colonies are vivid yellow, 0.1–0.2 mm in diameter after incubation on ISP 2 agar for 5 d at 30°C (optimal growth temperature). Will grow at 10–37°C but not at 4 or 42°C. Growth occurs at pH 6.1–12.1; optimal pH is 8.1–11.1 (initial pH values of test medium, ISP 2). Grows on the same medium at 1–3% NaCl but better in the absence of NaCl. The major fatty acids (>5% of the total) determined in cells grown on R2A agar at 30°C for 5 d are $C_{16:0}$ (27.7), $C_{18:1}$ $\omega 9$ (25.9), $C_{18:0}$ 10-methyl (10.2), $C_{16:0}$ 2-OH (8), as well as $C_{16:1}$ $\omega 7c$ and/or $C_{15:0}$ iso 2-OH (6).

Source (type strain): volcanic ash of Oreum (a parasitic volcanic cone), Jeju, Republic of Korea.

DNA G+C content (mol%): 72.0 (HPLC).

Type strain: Sco-D01, KCTC 19597, DSM 22127.

Sequence accession no. (16S rRNA gene): FN386750.

References

Abdulla, H.M. and S.A. El-Shatoury. 2007. Actinomycetes in rice straw decomposition. Waste Manag. *27*: 850–853.

Acinas, S.G., L.A. Marcelino, V. Klepac-Ceraj and M.F. Polz. 2004. Divergence and redundancy of 16S rRNA sequences in genomes with multiple *rrn* operons. J. Bacteriol. *186*: 2629–2635.

Agre, N.S., V.P. Shekhovtsev, T.F. Kuimova, Y.B. Malishkaite and L.S. Sharaya. 1984. Micromorphology and fine structure of the 3LS isolate. Actinomycetes *18*: 54–66.

Alitalo, K., U. Eriksson, B. Olofsson and T. Makinen. 2003. Novel neuropilin/growth factor binding and uses thereof. Patent: JP 2003508009-A 2 04-MAR.

Amato, P., M. Parazols, M. Sancelme, P. Laj, G. Mailhot and A.M. Delort. 2006. Microorganisms isolated from the water phase of tropospheric clouds at the Puy de Dôme: major groups and growth abilities at low temperatures. FEMS Microbiol. Ecol. *59*: 242–254.

Amoroso, M.J., D. Schubert, P. Mitscherlich, P. Schumann and E. Kothe. 2000. Evidence for high affinity nickel transporter genes in heavy metal resistant *Streptomyces* spec. J. Basic. Microbiol. *40*: 295–301.

An, D.S., W.T. Im, S.T. Lee and M.H. Yoon. 2007. *Nocardioides panacihumi* sp. nov., isolated from soil of a ginseng field. Int. J. Syst. Evol. Microbiol. *57*: 2143–2146.

Archibald, A.R. 1976. Cell wall assembly in *Bacillus subtilis*: development of bacteriophage-binding properties as a result of the pulsed incorporation of teichoic acid. J. Bacteriol. *127*: 956–960.

Arima, K., M. Nagasawa, M. Bae and G. Tamura. 1969. Microbial transformation of sterols. Part 1: Decomposition of cholesterol by microorganisms. Agric. Biol. Chem. *33*: 1636–1643.

Atlas, R.M. 1993. Handbook of Microbiological Media.. CRC Press, Boca Raton, FL.

Baddiley, J. 1970. Structure, biosynthesis and function of teichoic acids. Account. Chem. Res. *3*: 98–105.

Balch, W.E., G.E. Fox, L.J. Magrum, C.R. Woese and R.S. Wolfe. 1979. Methanogens: reevaluation of a unique biological group. Microbiol. Rev. *43*: 260–296.

Barton, H.A., M.R. Taylor and N.R. Pace. 2004. Molecular phylogenetic analysis of a bacterial community in an oligotrophic cave environment. Geomicrobiol. *21*: 11–20.

Barton, H.A., N.M. Taylor, M.P. Kreate, A.C. Springer, S.A. Oehrle and J.L. Bertog. 2007. The impact of host rock geochemistry on bacterial community structure in oligotrophic cave environments. Int. J. Speleol. *36*: 93–104.

Baumann, L., P. Baumann, M. Mandel and R.D. Allen. 1972. Taxonomy of aerobic marine eubacteria. J. Bacteriol. *110*: 402–429.

Baumann, P., L. Baumann and M. Mandel. 1971. Taxonomy of marine bacteria: the genus *Beneckea*. J. Bacteriol. *107*: 268–294.

Baylis, H.A. and M.J. Bibb. 1988. Transcriptional analysis of the 16S rRNA gene of the *rrnD* gene set of *Stveptomyces coelicolov* A3(2). Mol. Microbiol. *2*: 569–579.

Behrendt, U. and K. Heesche-Wagner. 1999. Formation of hydride-Meisenheimer complexes of picric acid (2,4,6-trinitrophenol) and 2,4-dinitrophenol during mineralization of picric acid by *Nocardioides* sp. strain CB 22-2. Appl. Environ. Microbiol. *65*: 1372–1377.

Behrendt, U., P. Schumann, M. Hamada, K. Suzuki, C. Spröer and A. Ulrich. 2011. Reclassification of *Leifsonia ginsengi* (Qiu *et al.* 2007) as *Herbiconiux ginsengi* gen. nov., comb. nov. and description of *Herbiconiux solani* sp. nov., an actinobacterium associated with the phyllosphere of *Solanum tuberosum* L. Int. J. Syst. Evol. Microbiol. *61*: 1039–1047.

Boivin-Jahns, V., A. Bianchi, R. Ruimy, J. Garcin, S. Daumas and R. Christen. 1995. Comparison of phenotypical and molecular methods for the identification of bacterial strains isolated from a deep subsurface environment. Appl. Environ. Microbiol. *61*: 3400–3406.

Bosecker, K., M. Teschner and H. Wehner. 1991. Biodegradation of crude oils. *In* Developments in Geochemistry 6: Diversity of Environmental Biogeochemistry (edited by Berthelin). Elsevier, Amsterdam, pp. 195–204.

Brikun, I.A., A.R. Reeves, W.H. Cernota, M.B. Luu and J.M. Weber. 2004. The erythromycin biosynthetic gene cluster of *Aeromicrobium erythreum*. J. Ind. Microbiol. Biotechnol. *31*: 335–344.

Brodie, E.L., T.Z. DeSantis, J.P. Parker, I.X. Zubietta, Y.M. Piceno and G.L. Andersen. 2007. Urban aerosols harbor diverse and dynamic bacterial populations. Proc. Natl. Acad. Sci. U.S.A. *104*: 299–304.

Bruns, A., H. Philipp, H. Cypionka and T. Brinkhoff. 2003. *Aeromicrobium marinum* sp. nov., an abundant pelagic bacterium isolated from the German Wadden Sea. Int. J. Syst. Evol. Microbiol. *53*: 1917–1923.

Buck, J.D. 1982. Nonstaining (KOH) method for determination of gram reactions of marine bacteria. Appl. Environ. Microbiol. *44*: 992–993.

Buczolits, S., E.B. Denner, D. Vybiral, M. Wieser, P. Kämpfer and H.-J. Busse. 2002. Classification of three airborne bacteria and proposal of *Hymenobacter aerophilus* sp. nov. Int. J. Syst. Evol. Microbiol. *52*: 445–456.

Budavari, S. 1989. The Merck Index. Merck, Rathway, NJ.

Bulina, T.I., L.P. Terekhova and M.V. Tiurin. 1998. Use of electric impulses for selective isolation of actinomycetes from soil. Mikrobiologiya *67*: 556–560.

Bunt, J.S. and A.D. Rovira. 1955. Microbiological studies of some subantarctic soils. J. Soil Sci. *6*: 119–128.

Busse, H.-J. and P. Schumann. 1999. Polyamine profiles within genera of the class *Actinobacteria* with LL-diaminopimelic acid in the peptidoglycan. Int. J. Syst. Bacteriol. *49*: 179–184.

Cao, Y.R., Y. Jiang, J.Y. Wu, L.H. Xu and C.L. Jiang. 2009. *Actinopolymorpha alba* sp. nov., isolated from a rhizosphere soil. Int. J. Syst. Evol. Microbiol. *59*: 2200–2203.

Carlsohn, M.R., I. Groth, C. Spröer, B. Schütze, H.P. Saluz, T. Munder and E. Stackebrandt. 2007. *Kribbella aluminosa* sp. nov., isolated from a medieval alum slate mine. Int. J. Syst. Evol. Microbiol. *57*: 1943–1947.

Cho, C.H., J.S. Lee, D.S. An, T.W. Whon and S.G. Kim. 2010. *Nocardioides panacisoli* sp. nov., isolated from the soil of a ginseng field. Int. J. Syst. Bacteriol. *60*: 387–392.

Cho, J.C. and S.J. Giovannoni. 2004. Cultivation and growth characteristics of a diverse group of oligotrophic marine *Gammaproteobacteria*. Appl. Environ. Microbiol. *70*: 432–440.

Cho, M.H., K.S. Whang, S.M. Han and H.J. Baek. 2006. Ecological characteristics of actinomycetes from mercury and chrome polluted soil. Korean J. Environ. Biol. *24*: 38–45.

Cho, Y.G., J.H. Yoon, Y.H. Park and S.T. Lee. 1998. Simultaneous degradation of p-nitrophenol and phenol by a newly isolated *Nocardioides* sp. J. Gen. Appl. Microbiol. *44*: 303–309.

Cho, Y.G., S.K. Rhee and S.T. Lee. 2000. Influence of phenol on biodegradation of p-nitrophenol by freely suspended and immobilized *Nocardioides* sp. NSP41. Biodegradation *11*: 21–28.

Choi, D.H., H.M. Kim, J.H. Noh and B.C. Cho. 2007. *Nocardioides marinus* sp. nov. Int. J. Syst. Evol. Microbiol. *57*: 775–779.

Chou, J.H., N.T. Cho, A.B. Arun, C.C. Young and W.M. Chen. 2008. *Nocardioides fonticola* sp. nov., a novel actinomycete isolated from spring water. Int. J. Syst. Evol. Microbiol. *58*: 1864–1868.

Christensen, H., O. Angen, R. Mutters, J.E. Olsen and M. Bisgaard. 2000. DNA-DNA hybridization determined in micro-wells using covalent attachment of DNA. Int. J. Syst. Evol. Microbiol. *50*: 1095–1102.

Chuang, A.S. and T.E. Mattes. 2007. Identification of polypeptides expressed in response to vinyl chloride, ethene, and epoxyethane in *Nocardioides* sp. strain JS614 by using peptide mass fingerprinting. Appl. Environ. Microbiol. *73*: 4368–4372.

Chung, B.S., A. Zubair, G.G. Kim, S.K. Kang, J.W. Ahn and Y.R. Chung. 2008. A bacterial endophyte, *Pseudomonas brassicacearum* YC5480, isolated from the root of artemisia sp. producing antifungal and phytotoxic compound. Plant Pathol. *24*: 461–468.

Cohen-Bazire, G., W.R. Sistrom and R.Y. Stanier. 1957. Kinetic studies of pigment synthesis by non-sulfur purple bacteria. J. Cell Phys. *49*: 25–68.

Coleman, N.V., T.E. Mattes, J.M. Gossett and J.C. Spain. 2002. Phylogenetic and kinetic diversity of aerobic vinyl chloride-assimilating bacteria from contaminated sites. Appl. Environ. Microbiol. *68*: 6162–6171.

Collins, M.D., M. Goodfellow and D.E. Minnikin. 1979. Isoprenoid quinones in the classification of coryneform and related bacteria. J. Gen. Microbiol. *110*: 127–136.

Collins, M.D., R.M. Keddie and R.M. Kroppenstedt. 1983. Lipid composition of *Arthrobacter simplex*, *Arthrobacter tumescens* and possibly related taxa.. Syst. Appl. Microbiol. *4*: 18–26.

Collins, M.D., M. Dorsch and E. Stackebrandt. 1989. Transfer of *Pimelobacter tumescens* to *Terrabacter* gen. nov. as *Terrabacter tumescens* comb. nov. and of *Pimelobacter jensenii* to *Nocardioides* as *Nocardioides jensenii* comb. nov. Int. J. Syst. Bacteriol. *39*: 1–6.

Collins, M.D. and E. Stackebrandt. 1989a. *In* Validation of the publication of new names and new combinations previously effectively published outside the IJSB. Int. J. Syst. Bacteriol. *39*: 371.

Collins, M.D. and E. Stackebrandt. 1989b. Molecular taxonomic studies on some LL-diaminopimelic acid-containing coryneforms from herbage: description of *Nocardioides fastidiosa* sp. nov. FEMS Microbiol. Lett. *48*: 289–293.

Collins, M.D., S. Cockcroft and S. Wallbanks. 1994. Phylogenetic analysis of a new LL-diaminopimelic acid-containing coryneform bacterium from herbage, *Nocardioides plantarum* sp. nov. Int. J. Syst. Bacteriol. *44*: 523–526.

Collwell, R.R. 1970. Polyphasic taxonomy of bacteria. *In* Culture Collections of Microorganisms (edited by Iisuka and Hasegava). University of Tokyo Press, Tokyo, pp. 421–436.

Conn, H.J. and I. Dimmick. 1947. Soil bacteria similar in morphology to *Mycobacterium* and *Corynebacterium*. J. Bacteriol. *54*: 291–303.

Conn, V.M. and C.M. Franco. 2004. Effect of microbial inoculants on the indigenous actinobacterial endophyte population in the roots of wheat as determined by terminal restriction fragment length polymorphism. Appl. Environ. Microbiol. *70*: 6407–6413.

Conville, P.S. and F.G. Witebsky. 2007. Analysis of multiple differing copies of the 16S rRNA gene in five clinical isolates and three type strains of *Nocardia* species and implications for species assignment. J. Clin. Microbiol. *45*: 1146–1151.

Cook, A.E. and P.R. Meyers. 2003. Rapid identification of filamentous actinomycetes to the genus level using genus-specific 16S rRNA gene restriction fragment patterns. Int. J. Syst. Evol. Microbiol. *53*: 1907–1915.

Coombs, J.T. and C.M. Franco. 2003. Isolation and identification of actinobacteria from surface-sterilized wheat roots. Appl. Environ. Microbiol. *69*: 5603–5608.

Coombs, J.T., P.P. Michelsen and C.M.M. Franco. 2003. Evaluation of endophytic actinobacteria as antagonists of *Gaeumannomyces graminis* var. *tritici* in wheat. Biol. Control *29*: 359–366.

Coombs, J.T., P.P. Michelson and C.M. Franco. 2004. Evaluation of endophytic actinobacteria as antagonists of *Gaeumannomyces graminis* var. *tritici* in wheat. Biol. Control *29*: 359–366.

Copeland, A., S. Lucas, A. Lapidus, K. Barry, J.C. Detter, T. Glavina del Rio, N. Hammon, S. Israni, E. Dalin, H. Tice, S. Pitluck, L.S. Thompson, T. Brettin, D. Bruce, C. Han, R. Tapia, J. Schmutz, F. Larimer, M. Land, L. Hauser, N. Kyrpides, E. Kim, T. Mattes, J. Gossett and P. Richardson. 2006. Complete sequence of Chromosome1 of *Nocardioides* sp. JS614. GenBank, CP000509.

Cote, R., P.-M. Daggett, M.J. Gantt, R. Hay, S.-C. Jong and P. Pienta. 1984. ATCC Media Handbook, 1st edn. American Type Culture Collection. Rockville, MD.

Cox, C.J., K.E. Kempsell and J.S. Gaston. 2003. Investigation of infectious agents associated with arthritis by reverse transcription PCR of bacterial rRNA. Arthritis Res Ther *5*: R1–8.

Crawford, D.L., J.M. Lynch, J.M. Whipps and M.A. Ousley. 1993. Isolation and characterization of actinomycete antagonists of a fungal root pathogen. Appl. Environ. Microbiol. *59*: 3899–3905.

Cui, Y.-S., J.-S. Lee, S.-T. Lee and W.-T. Im. 2010. *Kribbella ginsengisoli* sp. nov., isolated from soil of a ginseng field. Int. J. Syst. Evol. Microbiol. *60*: 364–368.

Cui, Y.S., W.T. Im, C.R. Yin, J.S. Lee, K.C. Lee and S.T. Lee. 2007a. *Aeromicrobium panaciterrae* sp. nov., isolated from soil of a ginseng field in South Korea. Int. J. Syst. Evol. Microbiol. *57*: 687–691.

Cui, Y.S., W.T. Im, C.R. Yin, D.C. Yang and S.T. Lee. 2007b. *Microlunatus ginsengisoli* sp. nov., isolated from soil of a ginseng field. Int. J. Syst. Evol. Microbiol. *57*: 713–716.

Cui, Y.S., S.T. Lee and W.T. Im. 2009. *Nocardioides ginsengisoli* sp. nov., isolated from soil of a ginseng field. Int. J. Syst. Evol. Microbiol. *59*: 3045–3050.

Cummins, C.S. and H. Harris. 1959. Taxonomic position of *Arthrobacter*. Nature *184*: 831–832.

Cure, G.L. and R.M. Keddie. 1973. Methods for the morphological examination of aerobic coryneforms bacteria. *In* Sampling – Microbiological Monitoring of Environments (edited by Board and Lovelock). Academic Press, London, pp. 123–135.

Dabbs, E.R., S. Naidoo, C. Lephoto and N. Nikitina. 2003. Pathogenic *Nocardia*, *Rhodococcus*, and related organisms are highly susceptible to imidazole antifungals. Antimicrob. Agents Chemother. *47*: 1476–1478.

Dastager, S.G., J.C. Lee, Y.J. Ju, D.J. Park and C.J. Kim. 2008a. *Nocardioides halotolerans* sp. nov., isolated from soil on Bigeum Island, Korea. Syst. Appl. Microbiol. *31*: 24–29.

Dastager, S.G., J.C. Lee, Y.J. Ju, D.J. Park and C.J. Kim. 2008b. *Marmoricola bigeumensis* sp. nov., a member of the family *Nocardioidaceae*. Int. J. Syst. Evol. Microbiol. *58*: 1060–1063.

Dastager, S.G., J.C. Lee, Y.J. Ju, D.J. Park and C.J. Kim. 2008c. *Nocardioides tritolerans* sp. nov., Isolated from soil in Bigeum Island, Korea. J. Microbiol. Biotechnol. *18*: 1203–1206.

Dastager, S.G., J.C. Lee, Y.J. Ju, D.J. Park and C.J. Kim. 2008d. *Nocardioides dilutes* sp. nov. isolated from soil in Bigeum Island, Korea. Curr. Microbiol. *56*: 569–573.

Dastager, S.G., J.C. Lee, Y.J. Ju, D.J. Park and C.J. Kim. 2008e. *Nocardioides islandiensis* sp. nov., isolated from soil in Bigeum Island Korea. Antonie van Leeuwenhoek *93*: 401–406.

Dastager, S.G., J.C. Lee, Y.J. Ju, D.J. Park and C.J. Kim. 2008f. *Nocardioides koreensis* sp. nov., *Nocardioides bigeumensis* sp. nov. and *Nocardioides agariphilus* sp. nov., isolated from soil from Bigeum Island, Korea. Int. J. Syst. Evol. Microbiol. *58*: 2292–2296.

Dastager, S.G., J.C. Lee, Y.J. Ju, D.J. Park and C.J. Kim. 2009a. *Nocardioides tritolerans* sp. nov. List of new names and new combinations previously effectively, but not validly, published. Validation List no. 128. Int. J. Syst. Evol. Microbiol. *59*: 1555–1556.

Dastager, S.G., J.C. Lee, Y.J. Ju, D.J. Park and C.J. Kim. 2009b. *In* List of new names and new combinations previously effectively, but not validly, published. Validation List no. 128. Int. J. Syst. Evol. Microbiol. *59*: 1555–1556.

Dastager, S.G., J.C. Lee, Y.J. Ju, D.J. Park and C.J. Kim. 2009c. *In* List of new names and new combinations previously effectively, but not validly, published. Validation List no. 128. Int. J. Syst. Evol. Microbiol. *59*: 1555–1556.

Dastager, S.G., J.C. Lee, Y.J. Ju, D.J. Park and C.J. Kim. 2009d. *In* List of new names and new combinations previously effectively, but not validly, published. Validation List no. 128. Int. J. Syst. Evol. Microbiol. *59*: 1555–1556..

Dastager, S.G., J.C. Lee, Y.J. Ju, D.J. Park and C.J. Kim. 2009e. *Nocardioides sediminis* sp. nov., isolated from a sediment sample. Int. J. Syst. Evol. Microbiol. *59*: 280–284.

Dastager, S.G., J.-C. Lee, A. Pandey and C.-J. Kim. 2010. *Nocardioides mesophilus* sp. nov., isolated from soil. Int. J. Syst. Evol. Microbiol. *60*: 2288–2292.

Davis, K.E., S.J. Joseph and P.H. Janssen. 2005. Effects of growth medium, inoculum size, and incubation time on culturability and isolation of soil bacteria. Appl. Environ. Microbiol. *71*: 826–834.

Delbes, C., L. Ali-Mandjee and M.C. Montel. 2007. Monitoring bacterial communities in raw milk and cheese by culture-dependent and -independent 16S rRNA gene-based analyses. Appl. Environ. Microbiol. *73*: 1882–1891.

Dellweg, H., J. Kurz, W. Pfluger, M. Schedel, G. Vobis and C. Wunsche. 1988. Rodaplutin, a new peptidylnucleoside from *Nocardioides albus*. J. Antibiot. (Tokyo) *41*: 1145–1147.

Desantis, T.Z., D.C. Joyner, S.M. Baek, J.T. Larsen, G.L. Andersen, T.C. Hazen, P.M. Richardson, D.J. Herman, T.K. Tokunaga, J.M. Wan and M.K. Firestone. 2006. Application of a high-density oligonucleotide microarray approach to study bacterial population dynamics during uranium reduction and reoxidation. Appl. Environ. Microbiol. *72*: 6288–6298.

Dimock, M.B., R.M. Beach and P.S. Carlson. 1988. Endophytic bacteria for the delivery of crop protection agents. Biotechnol. Biol. Pestic. *1*: 88–92.

Dittmer, J.C. and R.L. Lester. 1964. A simple, specific spray for the detection of phospholipids on thin-layer chromatograms. J. Lipid Res *15*: 126–127.

Dobrindt, U., B. Hochhut, U. Hentschel and J. Hacker. 2004. Genomic islands in pathogenic and environmental microorganisms. Nat. Rev. Microbiol. *2*: 414–424.

DSMZ. 2001. Catalogue of Strains. German Collection of Microorganisms and Cell Cultures, 7th edn. DSMZ - Deutsche Sammlung von Mikroorganismen und Zellkulturen, Braunschweig, Germany.

Ebert, S., P.G. Rieger and H.J. Knackmuss. 1999. Function of coenzyme F420 in aerobic catabolism of 2,4, 6-trinitrophenol and 2,4-dinitrophenol by *Nocardioides simplex* FJ2–1A. J. Bacteriol. *181*: 2669–2674.

Ebert, S., P. Fischer and H.J. Knackmuss. 2001. Converging catabolism of 2,4,6-trinitrophenol (picric acid) and 2,4-dinitrophenol by *Nocardioides simplex* FJ2–1A. Biodegradation *12*: 367–376.

El-Shatoury, S.A., N.S. El-Shenawy and I.M. Abd El-Salam. 2009. Antimicrobial, antitumor and in vivo cytotoxicity of actinomycetes inhabiting marine shellfish. World J. Microbiol. Biotechnol. *25*: 1547–1555.

Eppard, M., W.E. Krumbein, C. Koch, E. Rhiel, J.T. Staley and E. Stackebrandt. 1996. Morphological, physiological, and molecular characterization of actinomycetes isolated from dry soil, rocks, and monument surfaces. Arch. Microbiol. *166*: 12–22.

Everest, G.J. and P.R. Meyers. 2008. *Kribbella hippodromi* sp. nov., isolated from soil from a racecourse in South Africa. Int. J. Syst. Evol. Microbiol. *58*: 443–446.

Evtushenko, L.I., N.A. Ianushkene, G.M. Streshinskaia, I.B. Naumova and N.S. Agre. 1984. [Distribution of teichoic acids in representatives of the order *Actinomycetales*]. Dokl. Akad. Nauk SSSR *278*: 237–239.

Evtushenko, L.I., V.N. Akimov, S.V. Dobritsa and S.D. Taptykova. 1989. A new species of actinomycete, *Amycolata alni*. Int. J. Syst. Bacteriol. *39*: 72–77.

Evtushenko, L.I. and N.F. Zelenkova. 1989. The taxonomic position of *Proactinomyces farineus*. Mikrobiologiya *58*: 498–500.

Ezaki, T., Y. Hashimoto and E. Yabuuchi. 1989. Fluorometric DNA-DNA hybridization in microdilution wells as an alternative to membrane filter hybridization in which radioisotopes are used to determine genetic relatedness among bacterial strains. Int. J. Syst. Bacteriol. *39*: 224–229.

Fall, S., J. Hamelin, F. Ndiaye, K. Assigbetse, M. Aragno, J.L. Chotte and A. Brauman. 2007. Differences between bacterial communities in the gut of a soil-feeding termite (*Cubitermes niokoloensis*) and its mounds. Appl. Environ. Microbiol. *73*: 5199–5208.

Fedorak, P.M. and D.W. Westlake. 1981. Microbial degradation of aromatics and saturates in Prudhoe Bay crude oil as determined by glass capillary gas chromatography. Can J. Microbiol. *27*: 432–443.

Fialho, A.M., L.O. Martins, M.L. Donval, J.H. Leitao, M.J. Ridout, A.J. Jay, V.J. Morris and I.I. Sá-Correia. 1999. Structures and properties of gellan polymers produced by *Sphingomonas paucimobilis* ATCC 31461 from lactose compared with those produced from glucose and from cheese whey. Appl. Environ. Microbiol. *65*: 2485–2491.

Fiedler, F., K.H. Schleifer, B. Cziharz, E. Interschick and O. Kandler. 1970. Murein types in *Arthrobacter*, brevibacteria, corynebacteria and microbacteria. Publ. Fak. Sci. Univ. J. E. Purkyne, Brno *47*: 111–122.

Filion, M., R.C. Hamelin, L. Bernier and M. St-Arnaud. 2004. Molecular profiling of rhizosphere microbial communities associated with healthy and diseased black spruce (*Picea mariana*) seedlings grown in a nursery. Appl. Environ. Microbiol. *70*: 3541–3551.

Firakova, S., B. Proksa and M. Sturdikova. 2007. Biosynthesis and biological activity of enniatins. Pharmazie *62*: 563–568.

Fokina, V.V., G.V. Sukhodol'skaia, S.A. Gulevskaia, E. Gavrish, L.I. Evtushenko and M.V. Donova. 2003a. [The 1(2)-dehydrogenation of steroid substrates by *Nocardioides simplex* VKM Ac-2033D]. Mikrobiologiia *72*: 24–29.

Fokina, V.V., G.V. Sukhodolskaya, B.P. Baskunov, K.F. Turchin, G.S. Grinenko and M.V. Donova. 2003b. Microbial conversion of pregna-4,9(11)-diene-17alpha,21-diol-3,20-dione acetates by *Nocardioides simplex* VKM Ac-2033D. Steroids *68*: 415–421.

Frank, D.N., A.L. St Amand, R.A. Feldman, E.C. Boedeker, N. Harpaz and N.R. Pace. 2007. Molecular-phylogenetic characterization of microbial community imbalances in human inflammatory bowel diseases. Proc. Natl. Acad. Sci. U.S.A. *104*: 13780–13785.

Fredrickson, J.K., J.M. Zachara, D.L. Balkwill, D. Kennedy, S.M. Li, H.M. Kostandarithes, M.J. Daly, M.F. Romine and F.J. Brockman. 2004. Geomicrobiology of high-level nuclear waste-contaminated vadose sediments at the Hanford Site, Washington State. Appl. Environ. Microbiol. *70*: 4230–4241.

French, J.C., J.D. Howells and L.E. Anderson. 1970. Erythromycin process. USA patent 3,551,294.

Fujieda, N., A. Satoh, N. Tsuse, K. Kano and T. Ikeda. 2004. 6-S-cysteinyl flavin mononucleotide-containing histamine dehydrogenase from *Nocardioides simplex*: molecular cloning, sequencing, overexpression, and characterization of redox centers of enzyme. Biochemistry *43*: 10800–10808.

Fujieda, N., N. Tsuse, A. Satoh, T. Ikeda and K. Kano. 2005. Production of completely flavinylated histamine dehydrogenase, unique covalently bound flavin, and iron-sulfur cluster-containing enzyme of nocardioides simplex in *Escherichia coli*, and its properties. Biosci. Biotechnol. Biochem. *69*: 2459–2462.

Futamata, H., T. Uchida, N. Yoshida, Y. Yonemitsu and A. Hiraishi. 2004. Distribution of dibenzofuran-degrading bacteria in soils polluted with different levels of polychlorinated dioxins. Microbes Environ. *19, No. 2*: 172–176.

Gao, B. and R.S. Gupta. 2005. Conserved indels in protein sequences that are characteristic of the phylum *Actinobacteria*. Int. J. Syst. Evol. Microbiol. *55*: 2401–2412.

Gauze, G.F., T.P. Preobrazhenskaya, M.A. Sveshnikova, L.P. Terekhova and T.S. Maximova. 1983. A guide for the determination of actinomycetes. *In* Genera *Streptomyces, Streptoverticillium, and Chainia*. Nauka, Moscow.

Gevers, D., F.M. Cohan, J.G. Lawrence, B.G. Spratt, T. Coenye, E.J. Feil, E. Stackebrandt, Y. Van de Peer, P. Vandamme, F.L. Thompson and J. Swings. 2005. Opinion: Re-evaluating prokaryotic species. Nat. Rev. Microbiol. *3*: 733–739.

Gill, J.J., P.M. Sabour, J. Gong, H. Yu, K.E. Leslie and M.W. Griffiths. 2006. Characterization of bacterial populations recovered from the teat canals of lactating dairy and beef cattle by 16S rRNA gene sequence analysis. FEMS Microbiol. Ecol. *56*: 471–481.

Glöckner, F.O., E. Zaichikov, N. Belkova, L. Denissova, J. Pernthaler, A. Pernthaler and R. Amann. 2000. Comparative 16S rRNA analysis of lake bacterioplankton reveals globally distributed phylogenetic clusters including an abundant group of actinobacteria. Appl. Environ. Microbiol. *66*: 5053–5065.

Golovleva, L.A., R.N. Pertsova, L.I. Evtushenko and B.P. Baskunov. 1990. Degradation of 2,4,5-trichlorophenoxyacetic acid by a *Nocardioides simplex* culture. Biodegradation *1*: 263–271.

Gontang, E.A., W. Fenical and P.R. Jensen. 2007. Phylogenetic diversity of gram-positive bacteria cultured from marine sediments. Appl. Environ. Microbiol. *73*: 3272–3282.

Goodfellow, M. and M.P. Lechevalier. 1986. Genus *Nocardia*. *In* Bergey's Manual of Systematic Bacteriology, vol. 2 (edited by Sneath, Mair, Sharpe and Holt). Williams & Wilkins, Baltimore, pp. 1459–1471.

Goodfellow, M. and L.A. Maldonado. 2006. The families *Dietziaceae*, *Gordoniaceae*, *Nocardiaceae* and *Tsukamurellaceae*. *In* The Prokaryotes: a Handbook on the Biology of Bacteria, 3rd edn, vol. 3, *Archaea, Bacteria, Firmicutes*, Actinomycetes (edited by Dworkin, Falkow, Rosenberg, Schleifer and Stackebrandt). Springer, New York, pp. 843–888.

Gordon, R.E., D.A. Barnett, J.E. Handerhan and C.H.-N. Pang. 1974. *Nocardia coeliaca*, *Nocardia autotrophica*, and the Nocardin Strain. Int. J. Syst. Bacteriol. *24*: 54–63.

Grainger, J.M. 1963. Studies on coryneform bacteria from soil and herbage. PhD thesis, University of Reading, Reading.

Grice, E.A., H.H. Kong, G. Renaud, A.C. Young, G.G. Bouffard, R.W. Blakesley, T.G. Wolfsberg, M.L. Turner and J.A. Segre. 2008. A diversity profile of the human skin microbiota. Genome Res. *18*: 1043–1050.

Grice, E.A., H.H. Kong, S. Conlan, C.B. Deming, J. Davis, A.C. Young, G.G. Bouffard, R.W. Blakesley, P.R. Murray, E.D. Green, M.L. Turner and J.A. Segre. 2009. Topographical and temporal diversity of the human skin microbiome. Science *324*: 1190–1192.

Griffin, D.W. 2007. Atmospheric movement of micro-organisms in clouds of desert dust and implications for human health. Clin. Microbiol. *20*: 459–477.

Groth, I., P. Schumann, N. Weiss, K. Martin and F.A. Rainey. 1996. *Agrococcus jenensis* gen. nov., sp. nov., a new genus of actinomycetes with diaminobutyric acid in the cell wall. Int. J. Syst. Bacteriol. *46*: 234–239.

Groth, I. and C. Saiz-Jimenez. 1999. Actinomycetes in hypogean environments. Geomicrobiology *16*: 1–8.

Groth, I., R. Vettermann, B. Schuetze, P. Schumann and C. Saiz-Jimenez. 1999. Actinomycetes in Karstic caves of northern Spain (Altamira and Tito Bustillo). J. Microbiol. Methods *36*: 115–122.

Groth, I., P. Schumann, L. Laiz, S. Moral-Sanchez, J.C. Canaveras and C. Saiz-Jimenez. 2001. Geomicrobiological study of the grotta dei Cervi. Geomicrobiology *18*: 241–258.

Gullo, V., M. Conover, R. Cooper, C. Federbush, A.C. Horan, T. Kung, J. Marquez, M. Patel and A. Watnick. 1988. Sch 36605, a novel antiinflammatory compound. Taxonomy, fermentation, isolation and biological properties. J. Antibiot. (Tokyo) *41*: 20–24.

Gundersen, K. and L. Jensen. 1956. A soil bacterium decomposing organic nitrophenols. Agric. Scand. *6*: 1.

Hallmann, J., A. QuadtHallmann, W.F. Mahaffee and J.W. Kloepper. 1997. Bacterial endophytes in agricultural crops. Can. J. Microbiol. *43*: 895–914.

Hamamura, N. and D.J. Arp. 2000. Isolation and characterization of alkane-utilizing *Nocardioides* sp. strain CF8. FEMS Microbiol. Lett. *186*: 21–26.

Hamamura, N., C.M. Yeager and D.J. Arp. 2001. Two distinct monooxygenases for alkane oxidation in *Nocardioides* sp. strain CF8. Appl. Environ. Microbiol. *67*: 4992–4998.

Hansen, A.A., R.A. Herbert, K. Mikkelsen, L.L. Jensen, T. Kristoffersen, J.M. Tiedje, B.A. Lomstein and K.W. Finster. 2007. Viability, diversity and composition of the bacterial community in a high Arctic permafrost soil from Spitsbergen, Northern Norway. Environ. Microbiol. *9*: 2870–2884.

Hanson, R.L., J.M. Wasylyk, V.B. Nanduri, D.L. Cazzulino, R.N. Patel and L.J. Szarka. 1994. Site-specific enzymatic hydrolysis of taxanes at C-10 and C-13. J. Biol. Chem. *269*: 22145–22149.

Hanson, R.L., J. Kant and R.N. Patel. 2004. Conversion of 7-deoxy-10-deacetylbaccatin-III into 6-alpha-hydroxy-7-deoxy-10-deacetylbaccatin-III by *Nocardioides luteus*. Biotechnol. Appl. Biochem. *39*: 209–214.

Harris, J.K., M.A. De Groote, S.D. Sagel, E.T. Zemanick, R. Kapsner, C. Penvari, H. Kaess, R.R. Deterding, F.J. Accurso and N.R. Pace. 2007. Molecular identification of bacteria in bronchoalveolar lavage fluid from children with cystic fibrosis. Proc. Natl. Acad. Sci. U.S.A. *104*: 20529–20533.

Hayakawa, M. and H. Nonomura. 1987. Humic acid-vitamine agar, a new medium for the selective isolation of soil actinomycetes. J. Ferment. Technol. *65*: 501–509.

Henssen, A., C. Happachkasan, B. Renner and G. Vobis. 1983. *Pseudonocardia compacta* sp. nov. Int. J. Syst. Bacteriol. *33*: 829–836.

Henssen, A. 1989. Genus *Pseudonocardia* Henssen 1957. *In* Bergey's Manual of Systematic Bacteriology, vol. 2 (edited by Sneath, Mair, Sharpe and Holt). Williams & Wilkins, Baltimore, pp. 1485–1488.

Herron, P.R. and E.M. Wellington. 1990. New method for extraction of streptomycete spores from soil and application to the study of lysogeny in sterile amended and nonsterile soil. Appl. Environ. Microbiol. *56*: 1406–1412.

Hiraishi, A. and H. Kitamura. 1984. Distribution of phototrophic nonsulfur bacteria in activated sludge systems and other aquatic environments. Bull. Jpn. Sci. Soc. Fish. *50*: 1929–1937.

Holding, A.J. and J.G. Collee. 1971. Routine biochemical tests. Methods Microbiol. *6A*: 1–5.

Hopwood, D.A., M.J. Bibb, K.F. Chater, T. Kieser, C.J. Bruton, H.M. Kieser, D.J. Lydiate, C.P. Smith, J.M. Ward and S. H. 1985. Genetic Manipulation of *Streptomyces*. A Laboratory Manual. John Innes Foundation, Norwich, UK.

Huang, S.X., E. Powell, S.R. Rajski, L.X. Zhao, C.L. Jiang, Y. Duan, W. Xu and B. Shen. 2010. Discovery and total synthesis of a new estrogen receptor heterodimerizing actinopolymorphol A from *Actinopolymorpha rutilus*. Org. Lett. *12*: 3525–3527.

Hudson, J.A., K.M. Schofield, H.W. Morgan and R.M. Daniel. 1989. *Thermonema lapsum* gen. nov., sp. nov., a thermophilic gliding bacterium. Int. J. Syst. Bacteriol. *39*: 485–487.

Hugh, R. and E. Leifson. 1953. The taxonomic significance of fermentative versus oxidative metabolism of carbohydrates by various Gram-negative bacteria. J. Bacteriol. *66*: 24–26.

IAM. 2004. IAM Catalogue of Strains, 3rd edn. Institute of Molecular and Cellular Biosciences, The University of Tokyo, Tokyo.

Iizuka, H. and K. Komagata. 1964. Microbiological studies on petroleum and natural gas. 1. Determination of hydrocarbon-utilizing bacteria. J. Gen. Appl. Microbiol. *10*: 207–221.

Imazaki, I. and Y. Kobori. 2010. Improving the culturability of freshwater bacteria using FW70, a low-nutrient solid medium amended with sodium pyruvate. Can. J. Microbiol. *56*: 333–341.

Inoue, K., H. Habe, H. Yamane and H. Nojiri. 2006. Characterization of novel carbazole catabolism genes from gram-positive carbazole degrader *Nocardioides aromaticivorans* IC177. Appl. Environ. Microbiol. *72*: 3321–3329.

Inoue, K., Y. Ashikawa, Y. Usami, H. Noguchi, Z. Fujimoto, H. Yamane and H. Nojiri. 2007. Crystallization and preliminary crystallographic analysis of the ferredoxin component of carbazole 1,9*a*-dioxygenase from *Nocardioides aromaticivorans* IC177. Acta Crystallogr. Sect. F Struct. Biol. Cryst. Commun. *63*: 855–857.

Ishiguro, E.E. and R.S. Wolfe. 1970. Control of morphogenesis in *Geodermatophilus*: ultrastructural studies. J. Bacteriol. *104*: 566–580.

Iwabuchi, T. and S. Harayama. 1997. Biochemical and genetic characterization of 2-carboxybenzaldehyde dehydrogenase, an enzyme involved in phenanthrene degradation by *Nocardioides* sp. strain KP7. J. Bacteriol. *179*: 6488–6494.

Iwabuchi, T. and S. Harayama. 1998a. Biochemical and genetic characterization of trans-2′-carboxybenzalpyruvate hydratase-aldolase from a phenanthrene-degrading *Nocardioides* strain. J. Bacteriol. *180*: 945–949.

Iwabuchi, T. and S. Harayama. 1998b. Biochemical and molecular characterization of 1-hydroxy-2-naphthoate dioxygenase from *Nocardioides* sp. KP7. J. Bacteriol. *273*: 8332–8336.

Iwabuchi, T., Y. Inomata-Yamauchi, A. Katsuta and S. Harayama. 1998. Isolation and characterization of marine *Nocardioides* capable of growing and degrading phenanthrene at 42°C. J. Mar. Biotechnol. *6*: 86–90.

Jacin, H. and A.R. Mishkin. 1965. Separation of carbohydrates on borate-impregnated silica gel G Plates. J. Chromatogr *18*: 170–173.

Jensen, H.L. 1934. Studies on saprophytic mycobacteria and corynebacteria.. Proc. Linn. Soc. N.S.W. *59*: 19–61.

Jensen, H.L. and K. Gundersen. 1956. A soil bacterium decomposing organic nitro-compounds. Acta Agric. Scand. *6*: 100–114.

Jones, D. and M.D. Collins. 1986. Irregular, non-sporing Gram-positive rods. *In* Bergey's Manual of Systematic Bacteriology, vol. 2 (edited by Sneath, Mair, Sharpe and Holt). Williams & Wilkins, Baltimore, pp. 1261–1434.

Jones, D. and R.M. Keddie. 2006. The Genus *Arthrobacter*. *In* The Prokaryotes, 3rd edn, vol. 3 (edited by Dworkin, Falkow, Rosenberg, Schleifer and Stackebrandt). Springer, New York, pp. 945–960.

Jun, H.K., T.S. Kim and Y. Yeeh. 1994. Purification and characterization of an extracellular adenosine deaminase from *Nocardioides* sp. J-326TK. Biotechnol. Appl. Biochem. *20*: 265–277.

Jung, C.M., C. Broberg, J. Giuliani, L.L. Kirk and L.F. Hanne. 2002. Characterization of JP-7 jet fuel degradation by the bacterium *Nocardioides luteus* strain BAFB. J. Basic Microbiol. *42*: 127–131.

Kaewkla, O. and C.M. Franco. 2010a. *Flindersiella endophytica* gen. nov., sp. nov., an endophytic actinobacterium isolated from the root of Grey Box, an endemic eucalyptus tree. Int. J. Syst. Evol. Microbiol., first published on 1 October 2010 as doi: doi:10.1099/ijs.0.026757-0.

Kaewkla, O. and C.M. Franco. 2010b. *Actinopolymorpha pittospori* sp. nov., an endophytic actinobacterium isolated from surface-sterilized leaves of an Australian native apricot tree. Int. J. Syst. Evol. Microbiol., first published on 10 December 2010 as doi: doi:10.1099/ijs.0.029579-0.

Kämpfer, P., W. Dott and R.M. Kroppenstedt. 1990. Numerical classification and identification of some nocardioform bacteria. J. Gen. Appl. Microbiol. *36*: 309–331.

Kämpfer, P. 1991. Application of miniaturized physiological tests in numerical classification and identification of some bacilli. J. Gen. Appl. Microbiol. *37*: 225–247.

Kämpfer, P., M. Steiof and W. Dott. 1991. Microbiological characterization of a fuel-oil contaminated site including numerical identification of heterotrophic water and soil bacteria. Microbial Ecology *21*: 227–251.

Kämpfer, P. 2006. The family *Streptomycetaceae*, Part I: Taxonomy. *In* The Prokaryotes: a Handbook on the Biology of Bacteria, 3rd edn, vol. 3, *Archaea, Bacteria, Firmicutes*, Actinomycetes (edited by Dworkin, Falkow, Rosenberg, Schleifer and Stackebrandt). Springer, New York, pp. 538–604.

Kaneda, T. 1991. Iso- and anteiso-fatty acids in bacteria: biosynthesis, function, and taxonomic significance. Microbiol. Rev *55*: 288–302.

Katayama, T., M. Fukuda, J. Moriizumi, T. Nakamura, A. Brouchkov, K. Asano, M. Tanaka, J. Beget and F. Tomita. 2006. A late quaternary ice wedge from the Fox Permafrost Tunnel in central Alaska is a time capsule for gas and bacteria. 10–15.

Katayama, T., M. Tanaka, J. Moriizumi, T. Nakamura, A. Brouchkov, T.A. Douglas, M. Fukuda, F. Tomita and K. Asano. 2007. Phylogenetic analysis of bacteria preserved in a permafrost ice wedge for 25,000 years. Appl. Environ. Microbiol. *73*: 2360–2363.

Keddie, R.M., B.G.S. leask and J.M. Grainger. 1966. A comparison of coryneforms bacteria from soil and herbage: cell-wall composition and nutrition. J. Appl. Bacteriol. *29*: 17–43.

Keddie, R.M. 1974. Genus *Arthrobacter*. *In* Bergey's Manual of Determinative Bacteriology, 8th edn (edited by Buchanan and Gibbons). Williams & Wilkins, Baltimore.

Keddie, R.M. and G.L. Cure. 1977. The cell-wall composition and distribution of free mycolic acids in named strains of coryneform bacteria and in isolates from various natural sources. J. Appl. Bacteriol. *42*: 229–252.

Keddie, R.M. and D. Jones. 1981. Saprophytic, aerobic coryneform bacteria. *In* The Prokaryotes: a Handbook on Habitats, Isolation, and Identification of Bacteria (edited by Starr, Stolp, Trüper, Balows and Schlegel). Springer, New York, pp. 1838–1878.

Keddie, R.M., M.D. Collins and D. Jones. 1986. Genus *Arthrobacter*. *In* Bergey's Manual of Systematic Bacteriology, vol. 2 (edited by Sneath, Mair, Sharpe and Holt). Williams & Wilkins, Baltimore, pp. 1288–1301.

Kim, E., H. Kim, S.P. Hong, K.H. Kang, Y.H. Kho and Y.H. Park. 1993. Gene organization and primary structure of a ribosomal RNA gene cluster from *Streptomyces griseus* subsp. *griseus*. Gene *132*: 21–31.

Kim, H.M., D.H. Choi, C.Y. Hwang and B.C. Cho. 2008a. *Nocardioides salarius* sp. nov., isolated from seawater enriched with zooplankton. Int. J. Syst. Evol. Microbiol. *58*: 2056–2064.

Kim, K.-H., S.W. Roh, H.-W. Chang, Y.-D. Nam, J.-H. Yoon, C.O. Jeon, H.-M. Oh and J.-W. Bae. 2009a. *Nocardioides basaltis* sp. nov., isolated from black beach sand. Int. J. Syst. Evol. Microbiol. *59*: 42–47.

Kim, M.K., M.J. Park, W.T. Im and D.C. Yang. 2008b. *Aeromicrobium ginsengisoli* sp. nov., isolated from a ginseng field. Int. J. Syst. Evol. Microbiol. *58*: 2025–2030.

Kim, M.K., S. Srinivasan, M.J. Park, G. Sathiyaraj, Y.J. Kim and D.C. Yang. 2009b. *Nocardioides humi* sp. nov., a β-glucosidase-producing bacterium isolated from soil of a ginseng field. Int. J. Syst. Evol. Microbiol. *59*: 2724–2728.

Kim, S.H., H.O. Yang, Y.C. Sohn and H.C. Kwon. 2010. *Aeromicrobium halocynthiae* sp. nov., a taurocholic acid-producing bacterium isolated from the marine ascidian *Halocynthia roretzi*. Int. J. Syst. Evol. Microbiol. *60*: 2793–2798.

King, E.O., M.K. Ward and D.E. Raney. 1954. Two simple media for the demonstration of pyocyanin and fluorescin. J. Lab. Clin. Med. *44*: 301–307.

King, G.M. and C.F. Weber. 2007. Distribution, diversity and ecology of aerobic CO-oxidizing bacteria. Nat. Rev. Microbiol. *5*: 107–118.

Kirby, B.M., M. Le Roes and P.R. Meyers. 2006. *Kribbella karoonensis* sp. nov. and *Kribbella swartbergensis* sp. nov., isolated from soil from the Western Cape, South Africa. Int. J. Syst. Evol. Microbiol. *56*: 1097–1101.

Kirby, B.M., G.J. Everest and P.R. Meyers. 2010. Phylogenetic analysis of the genus *Kribbella* based on the *gyrB* gene: proposal of a *gyrB*-sequence threshold for species delineation in the genus *Kribbella*. Antonie van Leeuwenhoek *97*: 131–142.

Kloepper, J.W., R.M. Zablotowitz, E.M. Tipping and R. Lifshitz. 1991. Plant growth promotion mediated by bacterial rhizosphere colonozers. *In* The Rhizosphere and Plant Growth (edited by Keister and Cregan). Kluwer Academic Publishers, Dordrecht, pp. 315–326.

Knirel, Y.A., N.A. Kocharova, A.S. Shashkov, B.A. Dmitriev, N.K. Kochetkov, E.S. Stanislavsky and G.M. Mashilova. 1987. Somatic antigens of *Pseudomonas aeruginosa*. The structure of O-specific polysaccharide chains of the lipopolysaccharides from *P. aeruginosa* O5 (Lanyi) and immunotype 6 (Fisher). Eur. J. Biochem. *163*: 639–652.

Knirel, Y.A., A.S. Shashkov, Y.E. Tsvetkov, P.E. Jansson and U. Zahringer. 2003. 5,7-diamino-3,5,7,9-tetradeoxynon-2-ulosonic acids in bacterial glycopolymers: chemistry and biochemistry. Adv. Carbohydr. Chem. Biochem. *58*: 371–417.

Knirel, Y.A. 2009. Structures of bacterial polysaccharides. *In* Progress in the Synthesis of Complex Carbohydrate Chains of Plant and Microbial Polysaccharides (edited by Nifantiev). Transworld Research Network, Kerala, India, pp. 181–198.

Kobayashi, T., O. Koide, K. Mori, S. Shimamura, T. Matsuura, T. Miura, Y. Takaki, Y. Morono, T. Nunoura, H. Imachi, F. Inagaki, K. Takai and K. Horikoshi. 2008. Phylogenetic and enzymatic diversity of deep subseafloor aerobic microorganisms in organics- and methane-rich sediments off Shimokita Peninsula. Extremophiles *12*: 519–527.

Komagata, K. and K. Suzuki. 1987. Lipid and cell-wall analysis in bacterial systematic. *In* Methods in Microbiology, vol. 19 (edited by Colwell and Grigorova). Academic Press, London, pp. 161–207.

Komura, I., K. Yamada and K. Komagata. 1975a. Taxonomic significance of phospholipid composition in aerobic Gram positive cocci. J. Gen. Appl. Microbiol. *21*: 97–107.

Komura, I., K. Yamada, S. Otsuka and K. Komagata. 1975b. Taxonomic significance of phospholipids in coryneform and nocardioform bacteria. J. Gen. Appl. Microbiol. *21*: 251–261.

Konstantinidis, K.T. and J.M. Tiedje. 2005. Genomic insights that advance the species definition for prokaryotes. Proc. Natl. Acad. Sci. U.S.A. *102*: 2567–2572.

Konstantinidis, K.T., A. Ramette and J.M. Tiedje. 2006. The bacterial species definition in the genomic era. Philos. Trans. R. Soc. Lond. B Biol. Sci. *361*: 1929–1940.

Konstantinidis, K.T. and J.M. Tiedje. 2007. Prokaryotic taxonomy and phylogeny in the genomic era: advancements and challenges ahead. Curr. Opin. Microbiol. *10*: 504–509.

Kroppenstedt, R.M. 1985. Fatty acid and menaquinone analysis of actinomycetes and related organisms. *In* Chemical Methods in Bacterial Systematics (edited by Goodfellow and Minnikin). Academic Press, London, pp. 173–199.

Kroppenstedt, R.M. and L.I. Evtushenko. 2006. The family *Nocardiopsaceae*. *In* The Prokaryotes: a Handbook on the Biology of Bacteria, 3rd edn (edited by Dworkin, Falkow, Rosenberg, Schleifer and Stackebrandt). Springer, New York, pp. 745–795.

Kroppenstedt, R.M. and M. Goodfellow. 2006. The family *Thermomonosporaceae*: *Actinocorallia*, *Actinomadura*, *Spirillospora* and *Thermomonospora*. *In* The Prokaryotes: a Handbook on the Biology of Bacteria, 3rd edn, vol. 3, *Archaea*, *Bacteria*, *Firmicutes*, Actinomycetes (edited by Dworkin, Falkow, Rosenberg, Schleifer and Stackebrandt). Springer, New York, pp. 682–724.

Kubota, M., K. Kawahara, K. Sekiya, T. Uchida, Y. Hattori, H. Futamata and A. Hiraishi. 2005a. *Nocardioides aromaticivorans* sp. nov., a dibenzofuran-degrading bacterium isolated from dioxin-polluted environments. Syst. Appl. Microbiol. *28*: 165–174.

Kubota, M., K. Kawahara, K. Sekiya, T. Uchida, Y. Hattori, H. Futamata and A. Hiraishi. 2005b. *In* Validation of publication of new names and new combinations previously effectively published outside the IJSEM. List no. 103. Int. J. Syst. Evol. Microbiol. 55: 983–985.

Kubota, N.K., E. Ohta, S. Ohta, F. Koizumi, M. Suzuki, M. Ichimura and S. Ikegami. 2003. Piericidins C5 and C6: new 4-pyridinol compounds produced by *Streptomyces* sp. and *Nocardioides* sp. Bioorg. Med. Chem. *11*: 4569–4575.

Kuimova, T.F. and Y.B. Malishkaite. 1984. Fine structure characteristics of *Nocardia autotrophica*. Microbiologiya *53*: 342–345.

Kurtboke, D.I. and S.T. Williams. 1991. Use of actinophage for selective isolation purposes: current problems. Actinomycetes *2*: 31–34.

Kvasnikov, E.I., E.N. Pisarchuk, V.V. Stepaniuk and O.A. Nesterenko. 1974. [Characteristics of the biology of *Arthrobacter simplex* (Jensen) Lochhead]. Izv Akad. Nauk. SSSR Biol. *4*: 587–590.

Labrenz, M., M.D. Collins, P.A. Lawson, B.J. Tindall, P. Schumann and P. Hirsch. 1999. *Roseovarius tolerans* gen. nov., sp. nov., a budding bacterium with variable bacteriochlorophyll *a* production from hypersaline Ekho Lake. Int. J. Syst. Bacteriol. *49*: 137–147.

Lauer, A., M.A. Simon, J.L. Banning, E. André, K. Duncan and R.N. Harris. 2007. Common cutaneous bacteria from the Eastern Red-backed Salamander can inhibit pathogenic fungi. Copeia *2007*: 630–640.

Lauer, A., M.A. Simon, J.L. Banning, B.A. Lam and R.N. Harris. 2008. Diversity of cutaneous bacteria with antifungal activity isolated from female four-toed salamanders. ISME J. *2*: 145–157.

Lawson, P.A., M.D. Collins, P. Schumann, B.J. Tindall, P. Hirsch and M. Labrenz. 2000a. New LL-diaminopimelic acid-containing actinomycetes from hypersaline, heliothermal and meromictic Antarctic Ekho Lake: *Nocardioides aquaticus* sp. nov. and *Friedmanniella lacustris* sp. nov. Syst. Appl. Microbiol. *23*: 219–229.

Lawson, P.A., M.D. Collins, P. Schumann, B.J. Tindall, P. Hirsch and M. Labrenz. 2000b. *In* Validation of the publication of new names and new combinations previously effectively published outside the IJSEM. List no. 77. Int. J. Syst. Evol. Microbiol. *50*: 1953.

Leadbetter, J.R. and E.P. Greenberg. 2000. Metabolism of acylhomoserine lactone quorum-sensing signals by *Variovorax paradoxus*. J. Bacteriol. *182, No. 24*: 6921–6926.

Lechevalier, M.P. and H.A. Lechevalier. 1970. Chemical composition as a criterion in the classification of aerobic actinomycetes. Int. J. Syst. Bacteriol. *20*: 435–443.

Lechevalier, M.P., C. De Bièvre and H. Lechevalier. 1977. Chemotaxonomy of aerobic actinomycetes: phospholipid composition. Biochem. Syst. Ecol. *5*: 249–260.

Lechevalier, M.P., A.E. Stern and H.A. Lechevalier. 1981. Phospholipids in the taxonomy of actinomycetes. *In* Actinomycetes: Proceedings of the 4th International Symposium on Actinomycete Biology, Cologne, 1979 (edited by Schaal and Pulverer). Gustav Fischer, Stuttgart, pp. 111–116.

Lechevalier, M.P., H. Prauser, D.P. Labeda and J.S. Ruan. 1986. Two new genera of nocardioform actinomycetes, *Amycolata* gen. nov. and *Amycolatopsis* gen. nov. Int. J. Syst. Bacteriol. *36*: 29–37.

Lee, D.W., C.G. Hyun and S.D. Lee. 2007. *Nocardioides marinisabuli* sp. nov., a novel actinobacterium isolated from beach sand. Int. J. Syst. Evol. Microbiol. *57*: 2960–2963.

Lee, D.W. and S.D. Lee. 2008. *Aeromicrobium ponti* sp. nov., isolated from seawater. Int. J. Syst. Evol. Microbiol. *58*: 987–991.

Lee, D.W. and S.D. Lee. 2010. *Marmoricola scoriae* sp. nov., isolated from volcanic ash. Int. J. Syst. Evol. Microbiol. *60*: 2135–2139.

Lee, S.-T., S.-K. Rhee and G.M. Lee. 1994. Biodegradation of pyridine by freely suspended and immobilized *Pimelobacter* sp. Appl. Environ. Microbiol. *41*: 652–657.

Lee, S.D., S.O. Kang and Y.C. Hah. 2000. *Hongia* gen. nov., a new genus of the order *Actinomycetales*. Int. J. Syst. Evol. Microbiol. *50*: 191–199.

Lee, S.D. 2006. *Kineococcus marinus* sp. nov., isolated from marine sediment of the coast of Jeju, Korea. Int. J. Syst. Evol. Microbiol. *56*: 1279–1283.

Lee, S.D. 2007a. *Marmoricola aequoreus* sp. nov., a novel actinobacterium isolated from marine sediment. Int. J. Syst. Evol. Microbiol. *57*: 1391–1395.

Lee, S.D. 2007b. *Nocardioides furvisabuli* sp. nov., isolated from black sand. Int. J. Syst. Evol. Microbiol. *57*: 35–39.

Lee, S.D. and S.J. Kim. 2007. *Aeromicrobium tamlense* sp. nov., isolated from dried seaweed. Int. J. Syst. Evol. Microbiol. *57*: 337–341.

Lee, S.D., D.W. Lee and J.S. Kim. 2008. *Nocardioides hwasunensis* sp. nov. Int. J. Syst. Evol. Microbiol. *58*: 278–281.

Lee, S.D., D.W. Lee and Y.H. Ko. 2010. *Marmoricola korecus* sp. nov. Int J Syst Evol Microbiol., first published on 6 August 2010 as doi: 10.1099/ijs.0.025460-0.

Lee, S.T., S.B. Lee and Y.H. Park. 1991. Characterization of a pyridine-degrading branched Gram-positive bacterium isolated from the anoxic zone of an oil shale column. Appl. Environ. Microbiol. *35*: 824–829.

Leifson, E. 1963. Determination of carbohydrate metabolism of marine bacteria. J. Bacteriol. *85*: 1183–1184.

Leigh, M.B., V.H. Pellizari, O. Uhlik, R. Sutka, J. Rodrigues, N.E. Ostrom, J. Zhou and J.M. Tiedje. 2007. Biphenyl-utilizing bacteria and their functional genes in a pine root zone contaminated with polychlorinated biphenyls (PCBs). ISME J *1*: 134–148.

Lesaulnier, C., D. Papamichail, S. McCorkle, B. Ollivier, S. Skiena, S. Taghavi, D. Zak and D. van der Lelie. 2008. Elevated atmospheric CO_2 affects soil microbial diversity associated with trembling aspen. Environ. Microbiol. *10*: 926–941.

Li, B., C.H. Xie and A. Yokota. 2007a. *In* List of new names and new combinations previously effectively, but not validly, published. Validation List no. 118. Int. J. Syst. Evol. Microbiol. *57*: 2449–2450.

Li, B., C.H. Xie and A. Yokota. 2007b. *Nocardioides exalbidus* sp. nov., a novel actinomycete isolated from lichen in Izu-Oshima Island, Japan. Actinomycetologica *21*: 22–26.

Li, W.J., P. Xu, L.P. Zhang, S.K. Tang, X.L. Cui, P.H. Mao, L.H. Xu, P. Schumann, E. Stackebrandt and C.L. Jiang. 2003. *Streptomonospora alba* sp. nov., a novel halophilic actinomycete, and emended description of the genus *Streptomonospora* Cui *et al.* 2001. Int. J. Syst. Evol. Microbiol. *53*: 1421–1425.

Li, W.J., D. Wang, Y.Q. Zhang, P. Schumann, E. Stackebrandt, L.H. Xu and C.L. Jiang. 2004a. *In* Validation of publication of new names and new combinations previously effectively published outside the IJSEM. Validation List no. 99. Int. J. Syst. Bacteriol. *54*: 1425–1426.

Li, W.J., D. Wang, Y.Q. Zhang, P. Schumann, E. Stackebrandt, L.H. Xu and C.L. Jiang. 2004b. *Kribbella antibiotica* sp. nov., a novel nocardioform actinomycete strain isolated from soil in Yunnan, China. Syst. Appl. Microbiol. *27*: 160–165.

Li, W.J., D. Wang, Y.Q. Zhang, L.H. Xu and C.L. Jiang. 2006a. *In* List of new names and new combinations previously effectively, but not validly, published. Validation of List no. 110. Int. J. Syst. Evol. Microbiol. *56*: 1459–1460.

Li, W.J., D. Wang, Y.Q. Zhang, L.H. Xu and C.L. Jiang. 2006b. *Kribbella yunnanensis* sp. nov., *Kribbella alba* sp. nov., two novel species of genus *Kribbella* isolated from soils in Yunnan, China. Syst. Appl. Microbiol. *29*: 29–35.

Limburg, J., M. Mure and J.P. Klinman. 2005. Cloning and characterization of histamine dehydrogenase from *Nocardioides simplex*. Arch. Biochem. Biophys. *436*: 8–22.

Lochhead, A.G. 1957. Genus IV. *Arthrobacter*. *In* Bergey's Manual of Determinative Bacteriology, 7th edn (edited by Breed, Murray and Smith). Williams & Wilkins, Baltimore, pp. 605–612.

Loppinet, V., L. Hilali, N. Youssef, R. Bonaly and C. Finance. 1997. Isolation of antifungal macrolide from soil sample *Nocardioides* strain: production and structure elucidation. *In* Expanding Indications for the New Macrolides, Azalides and Streptogramins (edited by Zinner, Young and Neu). Marcel Dekker, New York, pp. 286–292.

Luedemann, G.M. 1968. *Geodermatophilus*, a new genus of the *Dermatophilaceae* (*Actinomycetales*). J. Bacteriol. *96*: 1848–1858.

MacLeod, R.A. 1968. On the role of inorganic ions in the physiology of marine bacteria. Adv. Microbiol. Sea *1*: 95–126.

Magic-Knezev, A., B. Wullings and D. Van der Kooij. 2009. *Polaromonas* and *Hydrogenophaga* species are the predominant bacteria cultured from granular activated carbon filters in water treatment. J. Appl. Microbiol. *107*: 1457–1467.

Mahendra, S. and L. Alvarez-Cohen. 2005. *Pseudonocardia dioxanivorans* sp. nov., a novel actinomycete that grows on 1,4-dioxane. Int. J. Syst. Evol. Microbiol. *55*: 593–598.

Maltseva, O. and P. Oriel. 1997. Monitoring of an alkaline 2,4,6-trichlorophenol-degrading enrichment culture by DNA fingerprinting methods and isolation of the responsible organism, haloalkaliphilic *Nocardioides* sp. strain M6. Appl. Environ. Microbiol. *63*: 4145–4149.

Männisto, M.K., M.A. Tiirola, M.S. Salkinoja-Salonen, M.S. Kulomaa and J.A. Puhakka. 1999. Diversity of chlorophenol-degrading bacteria isolated from contaminated boreal groundwater. Arch. Microbiol. *171*: 189–197.

Männisto, M.K., M.S. Salkinoja-Salonen and J.A. Puhakka. 2001. *In situ* polychlorophenol bioremediation potential of the indigenous bacterial community of boreal groundwater. Water Res *35*: 2496–2504.

Masson, J.Y., I. Boucher, W.A. Neugebauer, D. Ramotar and R. Brzezinski. 1995. A new chitosanase gene from a *Nocardioides* sp. is a third member of glycosyl hydrolase family 46. Microbiology *141*: 2629–2635.

Matson, J.A. and J.A. Bush. 1989. Sandramycin, a novel antitumor antibiotic produced by a *Nocardioides* sp. Production, isolation, characterization and biological properties. J. Antibiot. (Tokyo) *42*: 1763–1767.

Matson, J.A., K.L. Colson, G.N. Belofsky and B.B. Bleiberg. 1993. Sandramycin, a novel antitumor antibiotic produced by a *Nocardioides* sp. II. Structure determination. J. Antibiot. (Tokyo) *46*: 162–166.

Mattes, T.E., N.V. Coleman, J.C. Spain and J.M. Gossett. 2003. Evidence that vinyl chloride monooxygenase genes are encoded by a megaplasmid in *Nocardioides* strain JS614. Abstracts of the 103rd General Meeting, American Society for Microbiology: 525.

Mattes, T.E., N.V. Coleman, J.C. Spain and J.M. Gossett. 2005. Physiological and molecular genetic analyses of vinyl chloride and ethene biodegradation in *Nocardioides* sp. strain JS614. Arch. Microbiol. *183*: 95–106.

Mattes, T.E., N.V. Coleman, A.S. Chuang, A.J. Rogers, J.C. Spain and J.M. Gossett. 2007. Mechanism controlling the extended lag period associated with vinyl chloride starvation in *Nocardioides* sp. strain JS614. Arch. Microbiol. *187*: 217–226.

Militon, C., D. Boucher, C. Vachelard, G. Perchet, V. Barra, J. Troquet, E. Peyretaillade and P. Peyret. 2010. Bacterial community changes during bioremediation of aliphatic hydrocarbon-contaminated soil. FEMS Microbiol. Ecol. *74*: 669–681.

Miller, E.S., C.R. Woese and S. Brenner. 1991. Description of the erythromycin producing bacterium *Arthrobacter* sp. strain NRRLB-3381 as *Aeromicrobium erythreum* gen. nov., sp. nov. Int. J. Syst. Bacteriol. *41*: 363–368.

Miller, L.T. 1982. Single derivatization method for routine analysis of bacterial whole-cell fatty acid methyl esters, including hydroxy acids. J. Clin. Microbiol. *16*: 584–586.

Minnikin, D.E., P.V. Patel, L. Alshamaony and M. Goodfellow. 1977. Polar lipid composition in the classification of *Nocardia* and related bacteria. Int. J. Syst. Bacteriol. *27*: 104–117.

Miyauchi, K., P. Sukda, T. Nishida, E. Ito, Y. Matsumoto, E. Masai and M. Fukuda. 2008. Isolation of dibenzofuran-degrading bacterium, *Nocardioides* sp. DF412, and characterization of its dibenzofuran degradation genes. J. Biosci. Bioeng. *105*: 628–635.

Monteville, M.R., B. Ardestani and B.L. Geller. 1994. Lactococcal bacteriophages require a host cell wall carbohydrate and a plasma membrane protein for adsorption and ejection of DNA. Appl. Environ. Microbiol. *60*: 3204–3211.

Mordarska, H., M. Mordarski and M. Goodfellow. 1972. Chemotaxonomic characters and classification of some nocardioform bacteria. J. Gen Microbiol. *71*: 77–86.

Moulin, C., C.E. Lambert, F. Dulac and U. Dayan. 1997. Control of atmospheric export of dust from North Africa by the North Atlantic oscillation. Nature *387*: 691–694.

Mulbry, W.W., H. Zhu, S.M. Nour and E. Topp. 2002. The triazine hydrolase gene trzN from *Nocardioides* sp. strain C190: cloning and construction of gene-specific primers. FEMS Microbiol. Lett. *206*: 75–79.

Nagasawa, M., M. Bae, G. Tamura and K. Arima. 1969. Microbial transformation of sterols. Part II: Cleavage of sterol side chains by microorganisms. Agric. Biol. Chem. *33*: 1644–1650.

Naumova, I.B., A.S. Shashkov, E.M. Tul'skaya, G.M. Streshinskaya, Y.I. Kozlova, N.V. Potekhina, L.I. Evtushenko and E. Stackebrandt. 2001. Cell wall teichoic acids: structural diversity, species specificity in the genus *Nocardiopsis*, and chemotaxonomic perspective. FEMS Microbiol. Rev. *25*: 269–284.

Nesterenko, O.A., E.I. Kvasnikov and T.M. Nogina. 1985a. *Nocardioidaceae* fam. nov., a new family of the order *Actinomycetales* Buchanan 1917. Mikrobiol. Zhurnal *47*: 3–12.

Nesterenko, O.A., E.I. Kvasnikov and T.M. Nogina. 1985b. Nocardiaform and coryneform bacteria. Naukova Dumka, Kiev, Ukraine (In Russian).

Nesterenko, O.A., E.I. Kvasnikov and T.M. Nogina. 1990. *Nocardioidaceae* fam. nov. Int. J. Syst. Bacteriol. *40*: 320–321.

Nishimoto, T., M. Nakano, T. Nakada, H. Chaen, S. Fukuda, T. Sugimoto, M. Kurimoto and Y. Tsujisaka. 1996. Purification and properties of a novel enzyme, trehalose synthase, from *Pimelobacter* sp. R48. Biosci. Biotechnol. Biochem. *60*: 640–644.

Nobile, A. and N.J. Belleville. 1958. Process for production of dienes by corynebacteria. US Patent 2,837,464.

Normand, P., B. Cournoyer, P. Simonet and S. Nazaret. 1992. Analysis of a ribosomal RNA operon in the actinomycete *Frankia*. Gene *111*: 119–124.

O'Donnell, A.G., M. Goodfellow and D.E. Minnikin. 1982. Lipids in the classification of *Nocardioides*: reclassification of *Arthrobacter simplex* (Jensen) Lochhead in the genus *Nocardioides* (Prauser) emend. O'Donnell *et al.* as *Nocardioides simplex* comb. nov. Arch. Microbiol. *133*: 323–329.

O'Donnell, A.G., M. Goodfellow and D.E. Minnikin. 1983. *Nocardioides simplex* comb. nov. Int. J. Syst. Bacteriol. *33*: 896–897.

O'Donnell, A.G., C. Falconer, M. Goodfellow, A.C. Ward and E. Williams. 1993. Biosystematics and diversity amongst novel carboxydotrophic actinomycetes. Antonie van Leeuwenhoek *64*: 325–340.

Ogawara, H., N. Kawamura, T. Kudo, K.I. Suzuki and T. Nakase. 1999. Distribution of β-lactamases in actinomycetes. Antimicrob. Agents Chemother. *43*: 3014–3017.

Ōmura, S., R. Iwata, Y. Iwai, S. Taga, Y. Tanaka and H. Tomoda. 1985. Luminamicin, a new antibiotic. Production, isolation and physico-chemical and biological properties. J. Antibiot. (Tokyo) *38*: 1322–1326.

Osborne, C.A., M.B. Peoples and P.H. Janssen. 2010. Detection of a reproducible, single-member shift in soil bacterial communities exposed to low levels of hydrogen. Appl. Environ. Microbiol. *76*: 1471–1479.

Owens, C.R., J.K. Karceski and T.E. Mattes. 2009. Gaseous alkene biotransformation and enantioselective epoxyalkane formation by *Nocardioides* sp. strain JS614. Appl. Microbiol. Biotechnol. *84*: 685–692.

Owens, J.D. and R.M. Keddie. 1969. The nitrogen nutrition of soil and herbage coryneform bacteria. J. Appl. Bacteriol. *32*: 338–347.

Påhlson, C., A. Hallen and U. Forsum. 1986. Improved yield of *Mobiluncus* species from clinical specimens after alkaline treatment. Acta Pathol. Microbiol. Immunol. Scand. B *94*: 113–116.

Parales, R.E., J.E. Adamus, N. White and H.D. May. 1994. Degradation of 1,4-dioxane by an actinomycete in pure culture. Appl. Environ. Microbiol. *60*: 4527–4530.

Park, S.C., K.S. Baik, M.S. Kim, J. Chun and C.N. Seong. 2008. *Nocardioides dokdonensis* sp. nov., an actinomycete isolated from sand sediment. Int. J. Syst. Evol. Microbiol. *58*: 2619–2623.

Park, Y.H., J.H. Yoon and S.T. Lee. 1998. Application of multiplex PCR using species specific primers within the 16S rRNA gene for rapid identification of *Nocardioides* strains. Int. J. Syst. Bacteriol. *48*: 895–900.

Park, Y.H., J.H. Yoon, Y.K. Shin, K. Suzuki, T. Kudo, A. Seino, H.J. Kim, J.S. Lee and S.T. Lee. 1999. Classification of '*Nocardioides fulvus*' IFO 14399 and *Nocardioides* sp. ATCC 39419 in *Kribbella* gen. nov., as *Kribbella flavida* sp. nov. and *Kribbella sandramycini* sp. nov. Int. J. Syst. Bacteriol. *49*: 743–752.

Patel, R.N., A. Banerjee and V.V. Nanduri. 2000. Enzymatic acetylation of 10-deacetylbaccatin III to baccatin III by C-10 deacetylase from *Nocardioides luteus* SC 13913. Enzyme Microb. Technol. *27*: 371–375.

Pernodet, J.L., F. Boccard, M.T. Alegre, J. Gagnat and M. Guerineau. 1989. Organization and nucleotide sequence analysis of a ribosomal RNA gene cluster from *Streptomyces ambofaciens*. Gene *79*: 33–46.

Polymenakou, P.N., M. Mandalakis, E.G. Stephanou and A. Tselepides. 2008. Particle size distribution of airborne microorganisms and pathogens during an Intense African dust event in the eastern Mediterranean. Environ. Health Perspect. *116*: 292–296.

Powell, E., S.X. Huang, Y. Xu, S.R. Rajski, Y. Wang, N. Peters, S. Guo, H.E. Xu, F.M. Hoffmann, B. Shen and W. Xu. 2010. Identification and characterization of a novel estrogenic ligand actinopolymorphol A. Biochem. Pharmacol. *80*: 1221–1229.

Prauser, H. and R. Falta. 1968. [Phage sensitivity, cell-wall composition and taxonomy of actinomyctes]. Z. Allg. Mikrobiol. *8*: 39–46.

Prauser, H. 1976. *Nocardioides*, a new genus of order *Actinomycetales*. Int. J. Syst. Bacteriol. *26*: 58–65.

Prauser, H. 1978. Considerations on taxonomic relations among Gram-positive, branching bacteria. *In Nocardia* and *Streptomyces* (edited by Mordarski, Kurylowicz and Jeliaszewicz). Gustav Fischer Verlag, Stuttgart, pp. 3–12.

Prauser, H. 1981. Nocardioform organisms: General characterisation and taxonomic relationships. Zentralbl. Bakteriol. Mikrobiol. Hyg. *11*: 17–24.

Prauser, H. 1984a. Phage host ranges in the classification and identification of Gram-positive branched and related bacteria. *In* Biological, Biochemical, and Biomedical Aspects of Actinomycetes (edited by Ortiz-Ortiz, Bojalil and Yakoleff). Academic Press, Orlando, pp. 617–633.

Prauser, H. 1984b. *Nocardioides luteus* spec. nov. Z. Allg. Micr0biol. *24*: 647–648.

Prauser, H. 1985. *In* Validation of the publication of new names and new combinations previously effectively published outside the IJSB, List no. 34. Int. J. Syst. Bacteriol. *35*: 223–225.

Prauser, H. 1986. Genus *Nocardioides*. *In* Bergey's Manual of Systematic Bacteriology, vol. 2 (edited by Sneath, Mair, Sharpe and Holt). Williams & Wilkins, Baltimore, pp. 1481–1485.

Prauser, H. 1989. Genus *Nocardioides* Prauser 1976. *In* Bergey's Manual of Systematic Bacteriology, vol. 4 (edited by Williams, Sharpe and Holt). Williams & Wilkins, Baltimore, pp. 2371–2375.

Pridham, T.G. and A. J. Lyons. 1980. Methodologies for *Actinomycetales* with special reference to streptomycetes and streptoverticillia. *In* Actinomycete Taxonomy, Special Publication no. 6. Society for Industrial Microbiology, Arlington, VA, pp. 153–224.

Pukall, R., A. Lapidus, T. Glavina Del Rio, A. Copeland, H. Tice, J.F. Cheng, S. Lucas, F. Chen, M. Nolan, K. Labutti, A. Pati, N. Ivanova, K. Mavromatis, N. Mikhailova, S. Pitluck, D. Bruce, L. Goodwin, M. Land, L. Hauser, Y.J. Chang, C.D. Jeffries, A. Chen, K. Palaniappan, P. Chain, M. Rohde, M. Goker, J. Bristow, J.A. Eisen, V. Markowitz, P. Hugenholtz, N.C. Kyrpides, H.P. Klenk and T. Brettin. 2010. Complete genome sequence of *Kribbella flavida* type strain (IFO 14399). Stand. Genomic Sci. *2*: 186–193.

Purswani, J., C. Pozo, M. Rodriguez-Diaz and J. Gonzalez-Lopez. 2008. Selection and identification of bacterial strains with methyl-tert-butyl ether, ethyl-tert-butyl ether, and tert-amyl methyl ether degrading capacities. Environ. Toxicol. Chem. *27*: 2296–2303.

Qin, S., J. Li, H.H. Chen, G.Z. Zhao, W.Y. Zhu, C.L. Jiang, L.H. Xu and W.J. Li. 2009. Isolation, diversity, and antimicrobial activity of rare actinobacteria from medicinal plants of tropical rain forests in Xishuangbanna, China. Appl. Environ. Microbiol. *75*: 6176–6186.

Rainey, F.A., N.L. Ward-Rainey and E. Stackebrandt. 1997. Proposal for a new hierarchic classification system. *Actinobacteria* classis nov.. Int. J. Syst. Bacteriol. *47*: 479–491.

Rajan, J., K. Valli, R.E. Perkins, F.S. Sariaslani, S.M. Barns, A.L. Reysenbach, S. Rehm, M. Ehringer and N.R. Pace. 1996. Mineralization of 2,4,6-trinitrophenol (picric acid): characterization and phylogenetic identification of microbial strains. J. Ind. Microbiol. *16*: 319–324.

Reed, T.M., H. Hirakawa, M. Mure, E.E. Scott and J. Limburg. 2008. Expression, purification, crystallization and preliminary X-ray studies of histamine dehydrogenase from *Nocardioides simplex*. Acta Crystallogr. Sect. F Struct. Biol. Cryst. Commun. *64*: 785–787.

Reiter, B., U. Pfeifer, H. Schwab and A. Sessitsch. 2002. Response of endophytic bacterial communities in potato plants to infection with *Erwinia carotovora* subsp. *atroseptica*. Appl. Environ. Microbiol. *68*: 2261–2268.

Rhee, S.K., G.M. Lee, J.H. Yoon, Y.H. Park, H.S. Bae and S.T. Lee. 1997. Anaerobic and aerobic degradation of pyridine by a newly isolated denitrifying bacterium. Appl. Environ. Microbiol. *63*: 2578–2585.

Rintala, H., M. Pitkaranta, M. Toivola, L. Paulin and A. Nevalainen. 2008. Diversity and seasonal dynamics of bacterial community in indoor environment. BMC Microbiol. *8*: 56.

Roberts, A.N., G.S. Hudson and S. Brenner. 1985. An erythromycin-resistance gene from an erythromycin-producing strain of *Arthrobacter* sp. Gene *35*: 259–270.

Romanenko, L.A., N. Tanaka, M. Uchino, N.I. Kalinovskaya and V.V. Mikhailov. 2008. Diversity and antagonistic activity of sea ice bacteria isolated from the Sea of Japan. Microbes Environ. *23*: 209–214.

Rong, X. and Y. Huang. 2010. Taxonomic evaluation of the *Streptomyces griseus* clade using multilocus sequence analysis and DNA–DNA hybridization, with proposal to combine 29 species and three subspecies as 11 genomic species. Int. J. Syst. Evol. Microbiol. *60*: 696–703.

Rosselló-Mora, R. and R. Amann. 2001. The species concept for prokaryotes. FEMS Microbiol. Rev *25*: 39–67.

Ruan, J.-S. and Y.-M. Zhang. 1979. Two new species of *Nocardioides*. Acta Microbiol. Sinica *19*: 347–352.

Saddler, G.S., M. Goodfellow, D.E. Minnikin and A.G. O'Donnell. 1986. Influence of the growth cycle on the fatty acid and menaquinone composition of Streptomyces cyaneus NCIB 9616. J. Appl. Microbiol. *60*: 51–56.

Sadikov, B.M., N.V. Potekhina, V.D. Kuznetsov and I.B. Naumova. 1983. [Detection of teichoic acids in cells of bacteria of the genus *Arthrobacter*]. Dokl. Akad. Nauk. SSSR *271*: 459–461.

Sait, M., P. Hugenholtz and P.H. Janssen. 2002. Cultivation of globally distributed soil bacteria from phylogenetic lineages previously only detected in cultivation-independent surveys. Environ. Microbiol. *4*: 654–666.

Saito, A., T. Iwabuchi and S. Harayama. 1999. Characterization of genes for enzymes involved in the phenanthrene degradation in *Nocardioides* sp. KP7. Chemosphere *38*: 1331–1337.

Saito, A., T. Iwabuchi and S. Harayama. 2000. A novel phenanthrene dioxygenase from *Nocardioides* sp. strain KP7: expression in *Escherichia coli*. J. Bacteriol. *182*: 2134–2141.

Sakai, T., T. Daikai, H. Monma and H. Maeda. 2002. Isolation from *Nocardioides* sp. strain CT16, purification, and characterization of a deoxycytidine deaminase extremely thermostable in the presence of DL-dithiothreitol. Biosci. Biotechnol. Biochem. *66*: 1646–1651.

Sandhya, S., S.K. Prabu and R.B.T. Sundari. 1995. Microbial degradation of dibenzothiophene by *Nocardioides*. J. Environ. Sci. Health *A30*: 1995–2006.

Sandhya, S., S.K. Prabu and R. Bala. 1997. Transformation and expression of plasmid from *Nocardioides* sp. to *Pseudomonas putida*. Lett. Appl. Microbiol. *24*: 240–242.

Santo, C.E., P.V. Morais and G. Grass. 2010. Isolation and characterization of bacteria resistant to metallic copper surfaces. Appl. Environ. Microbiol. *76*: 1341–1348.

Sasser, M. 1990. Identification of bacteria by gas chromatography of cellular fatty acids. MIDI Technical Note 101, Newark, Delaware, MIDI Inc.

Schippers, A., P. Schumann and C. Spröer. 2005. *Nocardioides oleivorans* sp. nov., a novel crude-oil-degrading bacterium. Int. J. Syst. Evol. Microbiol. *55*: 1501–1504.

Schleifer, K.H. and O. Kandler. 1972. Peptidoglycan types of bacterial cell walls and their taxonomic implications. Bacteriol. Rev. *36*: 407–477.

Schleifer, K.H. and J. Steber. 1974. [Chemical studies on the phage receptor of *Staphylococcus epidermidis* (author's transl.)]. Arch. Microbiol. *98*: 251–270.

Schoenborn, L., P.S. Yates, B.E. Grinton, P. Hugenholtz and P.H. Janssen. 2004. Liquid serial dilution is inferior to solid media for isolation of cultures representative of the phylum-level diversity of soil bacteria. Appl. Environ. Microbiol. *70*: 4363–4366.

Schoenhofen, I.C., D.J. McNally, J.R. Brisson and S.M. Logan. 2006. Elucidation of the CMP-pseudaminic acid pathway in *Helicobacter pylori*: synthesis from UDP-N-acetylglucosamine by a single enzymatic reaction. Glycobiology *16*: 8C-14C.

Schrey, S.D., M. Schellhammer, M. Ecke, R. Hampp and M.T. Tarkka. 2005. Mycorrhiza helper bacterium *Streptomyces* AcH 505 induces differential gene expression in the ectomycorrhizal fungus *Amanita muscaria*. New Phytol. *168*: 205–216.

Schumann, P., H. Prauser, F.A. Rainey, E. Stackebrandt and P. Hirsch. 1997. *Friedmanniella antarctica* gen. nov., sp. nov., an LL-diaminopimelic acid-containing actinomycete from antarctic sandstone. Int. J. Syst. Bacteriol. *47*: 278–283.

Schumann, P., P. Kämpfer, H.-J. Busse and L.I. Evtushenko. 2009. Proposed minimal standards for describing new genera and species of the suborder Micrococcineae. Int. J. Syst. Evol. Microbiol. *59*: 1823–1849.

Selesi, D., M. Schmid and A. Hartmann. 2005. Diversity of green-like and red-like ribulose-1,5-bisphosphate carboxylase/oxygenase large-subunit genes (*cbbL*) in differently managed agricultural soils. Appl. Environ. Microbiol. *71*: 175–184.

Sfanos, K., D. Harmody, P. Dang, A. Ledger, S. Pomponi, P. McCarthy and J. Lopez. 2005. A molecular systematic survey of cultured microbial associates of deep-water marine invertebrates. Syst. Appl. Microbiol. *28*: 242–264.

Shashkov, A.S., E.M. Tul'skaya, L.I. Evtushenko and I.B. Naumova. 1999. A teichoic acid of *Nocardioides albus* VKM Ac-805^T cell walls. Biochemistry (Mosc.) *64*: 1305–1309.

Shashkov, A.S., G.M. Streshinskaya, L.N. Kosmachevskaya, L.I. Evtushenko and I.B. Naumova. 2000a. A polymer of 8-*O*-glucosylated 2-keto-3-deoxy-D-*glycero*-D-*galacto*-nonulosonic acid (Kdn) in the cell wall of *Streptomyces* sp. VKM Ac-2090. Mendeleev Commun. *10*: 167–168.

Shashkov, A.S., E.M. Tul'skaya, L.I. Evtushenko, A.A. Gratchev and I.B. Naumova. 2000b. Structure of a teichoic acid from *Nocardioides luteus* VKM Ac-1246^T cell wall. Biochemistry (Mosc) *65*: 509–514.

Shashkov, A.S., L.N. Kosmachevskaya, G.M. Streshinskaya, L.I. Evtushenko, O.V. Bueva, V.A. Denisenko, I.B. Naumova and E. Stackebrandt. 2002a. A polymer with a backbone of 3-deoxy-D-*glycero*-D-*galacto*-non-2-ulopyranosonic acid, a teichuronic acid, and a β-glucosylated ribitol teichoic acid in the cell wall of plant pathogenic *Streptomyces* sp. VKM Ac-2124. Eur. J. Biochem. *269*: 6020–6025.

Shashkov, A.S., E.M. Tul'skaya, L.I. Evtushenko, V.A. Denisenko, V.G. Ivanyuk, A.A. Stomakhin, I.B. Naumova and E. Stackebrandt. 2002b. Cell wall anionic polymers of *Streptomyces* sp. MB-8, the causative agent of potato scab. Carbohydr. Res. *337*: 2255–2261.

Shashkov, A.S., E.M. Tul'skaya, G.M. Streshinskaya, S.N. Senchenkova, A.N. Avtukh and L.I. Evtushenko. 2009. New cell wall glycopolymers of the representatives of the genus *Kribbella*. Carbohydr. Res. *344*: 2255–2262.

Shirling, E.B. and D. Gottlieb. 1966. Methods for characterization of *Streptomyces* species. Int. J. Syst. Bacteriol. *16*: 313–340.

Siddiqui, J.A., S.M. Shoeb, S. Takayama, E. Shimizu and T. Yorifuji. 2000. Purification and characterization of histamine dehydrogenase from Nocardioides simplex IFO 12069. FEMS Microbiol. Lett. *189*: 183–187.

Sohn, K., S.G. Hong, K.S. Bae and J. Chun. 2003. Transfer of *Hongia koreensis* Lee *et al.* 2000 to the genus *Kribbella* Park *et al.* 1999 as *Kribbella koreensis* comb. nov. Int. J. Syst. Evol. Microbiol. *53*: 1005–1007.

Song, G.C., M. Yasir, F. Bibi, E.J. Chung, C.O. Jeon and Y.R. Chung. 2011. *Nocardioides caricicola* sp. nov., an endophytic bacterium isolated from a halophyte, *Carex scabrifolia* Steud. Int. J. Syst. Evol. Microbiol. *61*: 105–109.

Song, J., B.Y. Kim, S.B. Hong, H.S. Cho, K. Sohn, J. Chun and J.W. Suh. 2004. *Kribbella solani* sp. nov. and *Kribbella jejuensis* sp. nov., isolated from potato tuber and soil in Jeju, Korea. Int. J. Syst. Evol. Microbiol. *54*: 1345–1348.

Song, L., W.J. Li, Q.L. Wang, G.Z. Chen, Y.S. Zhang and L.H. Xu. 2005. *Jiangella gansuensis* gen. nov., sp. nov., a novel actinomycete from a desert soil in north-west China. Int. J. Syst. Evol. Microbiol. *55*: 881–884.

Spencer, K.G. 1990. Lipids and polyols from microalgae. *In* Algae and Human Affairs (edited by Lembi). Cambridge University Press, Cambridge, pp. 248–249.

Stackebrandt, E., B.J. Lewis and C.R. Woese. 1980. The phylogenetic structure of the coryneform group of bacteria. Zentralbl. Bakteriol. Mikrobiol. Hyg. Abt. II Orig. C. *1*: 137–149.

Stackebrandt, E. and C.R. Woese. 1981. The evolution of prokaryotes. *In* Molecular and Cellular Aspects of Microbial Evolution (edited by Carlile, Collins and Moseley). Cambridge University Press, Cambridge, pp. 1–31.

Stackebrandt, E., F.A. Rainey and N.L. Ward-Rainey. 1997. Proposal for a new hierarchic classification system, *Actinobacteria* classis nov. Int. J. Syst. Bacteriol. *47*: 471–491.

Stackebrandt, E., W. Frederiksen, G.M. Garrity, P.A. Grimont, P. Kämpfer, M.C. Maiden, X. Nesme, R. RossellüMora, J. Swings, H.G. Trüper, L. Vauterin, A.C. Ward and W.B. Whitman. 2002. Report of the *ad hoc* committee for the re-evaluation of the species definition in bacteriology. Int. J. Syst. Evol. Microbiol. *52*: 1043–1047.

Stackebrandt, E. 2006. Defining taxonomic ranks. *In* The Prokaryotes: a Handbook on the Biology of Bacteria, 3rd edn, vol. 1 (edited by Dworkin, Falkow, Rosenberg, Schleifer and Stackebrandt). Springer, New York, pp. 29–57.

Stackebrandt, E. and K.P. Schaal. 2006. The family *Propionibacteriaceae*: the genera *Friedmanniella, Luteococcus, Microlunatus, Micropruina, Propioniferax, Propionimicrobium* and *Tessarococcus*. *In* The Prokaryotes: a Handbook on the Biology of Bacteria, 3rd edn, vol. 3, *Archaea, Bacteria, Firmicutes*, Actinomycetes (edited by Dworkin, Falkow, Rosenberg, Schleifer and Stackebrandt). Springer, New York, pp. 383–399.

Stackebrandt, E. and P. Schumann. 2006. Introduction to the taxonomy of actinobacteria. *In* The Prokaryotes: a Handbook on the Biology of Bacteria, 3rd edn, vol. 3, *Archaea, Bacteria, Firmicutes*, Actinomycetes (edited by Dworkin, Falkow, Rosenberg, Schleifer and Stackebrandt). Springer, New York, pp. 297–321.

Staley, J.T. 1968. *Prosthecomicrobium* and *Ancalomicrobium*: new prosthecate freshwater bacteria. J. Bacteriol. *95*: 1921–1942.

Stevens, H., T. Brinkhoff, B. Rink, J. Vollmers and M. Simon. 2007. Diversity and abundance of Gram-stain-positive bacteria in a tidal flat ecosystem. Environ. Microbiol. *9*: 1810–1822.

Stevenson, I.L. 1967. Utilisation of aromatic hydrocarbons by *Arthrobacter* spp. Can. J. Microbiol. *13*: 205–212.

Sukda, P., N. Gouda, E. Ito, K. Miyauchi, E. Masai and M. Fukuda. 2009. Characterization of a transcriptional regulatory gene involved in dibenzofuran degradation by *Nocardioides* sp. strain DF412. Biosci. Biotechnol. Biochem. *73*: 508–516.

Sun, L.N., Y.F. Zhang, L.Y. He, Z.J. Chen, Q.Y. Wang, M. Qian and X.F. Sheng. 2010. Genetic diversity and characterization of heavy metal-resistant-endophytic bacteria from two copper-tolerant plant species on copper mine wasteland. Bioresour. Technol. *101*: 501–509.

Suzuki, K. and K. Komagata. 1983a. Taxonomic significance of cellular fatty acid composition in some coryneform bacteria. Int. J. Syst. Bacteriol. *33*: 188–200.

Suzuki, K. and K. Komagata. 1983b. *In* Validation of the publication of new names and new combinations previously effectively published outside the IJSB: List no. 11. Int. J. Syst. Bacteriol. *33*: 672–674.

Suzuki, K. and K. Komagata. 1983c. *Pimelobacter* gen. nov., a new genus of coryneform bacteria with LL-diaminopimelic acid in the cell wall. J. Gen. Appl. Microbiol. *29*: 59–71.

Suzuki, K., M. Goodfellow and A.G. O'Donnell. 1993. Cell envelopes and classification. *In* Handbook of New Bacterial Systematics (edited by Goodfellow and O'Donnell). Academic Press, London, pp. 195–250.

Takagi, K., A. Iwasaki, I. Kamei, K. Satsuma, Y. Yoshioka and N. Harada. 2009. Aerobic mineralization of hexachlorobenzene by newly isolated pentachloronitrobenzene-degrading *Nocardioides* sp. strain PD653. Appl. Environ. Microbiol. *75*: 4452–4458.

Takamiya, A. and K. Tubaki. 1956. A new form of *Streptomyces* capable of growing autotrophically. Arch. Mikrobiol. *25*: 58–64.

Takeuchi, M. and A. Yokota. 1989. Cell-wall polysaccharides in coryneform bacteria. J. Gen. Appl. Microbiol. *35*: 233–252.

Tamura, T. and A. Yokota. 1994. Transfer of *Nocardioides fastidiosa* Collins and Stackebrandt 1989 to the genus *Aeromicrobium* as *Aeromicrobium fastidiosum* comb. nov. Int. J. Syst. Bacteriol. *44*: 608–611.

Tamura, T., Y. Ishida, Y. Nozawa, M. Otoguro and K. Suzuki. 2009. Transfer of *Actinomadura spadix* Nonomura and Ohara 1971 to *Actinoallomurus spadix* gen. nov., comb. nov., and description of *Actinoallomurus amamiensis* sp. nov., *Actinoallomurus caesius* sp. nov., *Actinoallomurus coprocola* sp. nov., *Actinoallomurus fulvus* sp. nov., *Actinoallomurus iriomotensis* sp. nov., *Actinoallomurus luridus* sp. nov., *Actinoallomurus purpureus* sp. nov. and *Actinoallomurus yoronensis* sp. nov. Int. J. Syst. Evol. Microbiol. *59*: 1867–1874.

Tan, G.Y., A.C. Ward and M. Goodfellow. 2006. Exploration of *Amycolatopsis* diversity in soil using genus-specific primers and novel selective media. Syst. Appl. Microbiol. *29*: 557–569.

Tanaka, Y., K. Yazawa, Y. Mikami, T. Kudo, K-i. Suzuki, H. Komaki, T. Tojo and K. Kadowaki. 1996. Changes in menaquinone composition associated with growth phase and medium composition in *Amycolatopsis* species. Microbiol. Cult. Coll. *12*: 11–16.

Tang, S.-K., X.-Y. Zhi, Y. Wang, R. Shi, K. Lou, L.-H. Xu and W.-J. Li. 2010. *Haloactinopolyspora alba* gen. nov., sp. nov., a novel halophilic filamentous actinomycete isolated from a salt lake in China, with proposal of *Jiangellaceae* fam. nov. and *Jiangellineae* subord. nov. Int. J. Syst. Evol. Microbiol. *61*: 194–200.

Tang, Y., G. Zhou, L. Zhang, J. Mao, X. Luo, M. Wang and C. Fang. 2008. *Aeromicrobium flavum* sp. nov., isolated from air. Int. J. Syst. Evol. Microbiol. *58*: 1860–1863.

Tian, X., L. Cao, H. Tan, W. Han, M. Chen, Y. Liu and S. Zhou. 2007. Diversity of cultivated and uncultivated actinobacterial endophytes in the stems and roots of rice. Microb. Ecol. *53*: 700–707.

Tian, X.L., L.X. Cao, H.M. Tan, Q.G. Zeng, Y.Y. Jia and S.N. Zhou. 2004. Study on the communities of endophytic fungi and endophytic actinomycetes from rice and their antipathogenic activities *in vitro*. World J. Microbiol. Biotechnol. *20*: 303–309.

Tiecco, G. 1975. Microbiologia Degli Alimenti Di Origine Animale. Bologna: Edizioni Agricole.

Tille, D., H. Prauser, K. Szyba and M. Mordarski. 1978. On the taxonomic position of *Nocardioides albus* Prauser by DNA/DNA-hybridization. Z. Allg. Microbiol. *18*: 459–462.

Tindall, B.J., R. Rossello-Mora, H.-J. Busse, W. Ludwig and P. Kämpfer. 2010. Notes on the characterization of prokaryote strains for taxonomic purposes. Int. J. Syst. Evol. Microbiol. *60*: 249–266.

Topp, E., W.M. Mulbry, H. Zhu, S.M. Nour and D. Cuppels. 2000. Characterization of s-triazine herbicide metabolism by a *Nocardioides* sp. isolated from agricultural soils. Appl. Environ. Microbiol. *66*: 3134–3141.

Tóth, E.M., Z. Keki, Z.G. Homonnay, A.K. Borsodi, K. Marialigeti and P. Schumann. 2008. *Nocardioides daphniae* sp. nov., isolated from *Daphnia cucullata* (Crustacea: Cladocera). Int. J. Syst. Evol. Microbiol. *58*: 78–83.

Traag, B.A. and G.P. van Wezel. 2008. The SsgA-like proteins in actinomycetes: small proteins up to a big task. Antonie van Leeuwenhoek *94*: 85–97.

Travkin, V.M., E.V. Linko and L.A. Golovleva. 1999. Purification and characterization of maleylacetate reductase from *Nocardioides simplex* 3E utilizing phenoxyalcanoic herbicides 2,4-D and 2,4,5-T. Biochemistry (Mosc.) *64*: 625–630.

Trujillo, M.E., R.M. Kroppenstedt, P. Schumann and E. Martinez-Molina. 2006. *Kribbella lupini* sp. nov., isolated from the roots of *Lupinus angustifolius*. Int. J. Syst. Evol. Microbiol. *56*: 407–411.

Tschech, A. and N. Pfennig. 1984. Growth yield increase linked to caffeate reduction in *Acetobacterium woodii*. Arch. Microbiol. *137*: 163–167.

Tsukamura, M. 1975. Numerical-analysis of relationship between *Mycobacterium, Rhodochrous* group, and *Nocardia* by use of hypothetical median organisms. Int. J. Syst. Bacteriol. *25*: 329–335.

Tsutsumi, M., N. Fujieda, S. Tsujimura, O. Shirai and K. Kano. 2008. Thermodynamic redox properties governing the half-reduction characteristics of histamine dehydrogenase from *Nocardioides simplex*. Biosci. Biotechnol. Biochem. *72*: 786–796.

Tul'skaya, E., A. Shashkov, O. Buyeva and L. Evtushenko. 2007. Anionic carbohydrate-containing cell wall polymers of *Streptomyces melanosporofaciens* and related species. Microbiology *76*: 39–44.

Tul'skaya, E.M. 2009. Teichoic acids and glycopolymers of actinomycetes: structural diversity, taxonomic and ecological aspects. Moscow State University, Moscow.

Uchida, K. and K. Aida. 1984. An improved method for the glycolate test for simple identification of the acyl type of bacterial cell walls. J. Gen. Appl. Microbiol. *30*: 131–134.

Uchida, K. and A. Seino. 1997. Intra- and intergeneric telationships of various actinomycete strains based on the acyl types of the muramyl residue in cell-wall peptidoglycans examined in a glycolate test. Int. J. Syst. Bacteriol. *47*: 182–190.

Ulrich, K., A. Ulrich and D. Ewald. 2008. Diversity of endophytic bacterial communities in poplar grown under field conditions. FEMS Microbiol. Ecol. *63*: 169–180.

Urzì, C., P. Salamone, P. Schumann and E. Stackebrandt. 2000. *Marmoricola aurantiacus* gen. nov., sp. nov., a coccoid member of the family *Nocardioidaceae* isolated from a marble statue. Int. J. Syst. Evol. Microbiol. *50*: 529–536.

Urzì, C., F. De Leo and P. Schumann. 2008. *Kribbella catacumbae* sp. nov. and *Kribbella sancticallisti* sp. nov., isolated from whitish-grey patinas in the catacombs of St Callistus in Rome, Italy. Int. J. Syst. Evol. Microbiol. *58*: 2090–2097.

Vaitilington, M., P. Amato, M. Sancelme, P. Laj, M. Leriche and A.-M. Delort. 2010. Contribution of microbial activity to carbon chemistry in clouds. Appl. Environ. Microbiol. *76*: 23–29.

Vandamme, P., B. Pot, M. Gillis, P. De Vos, K. Kersters and J. Swings. 1996. Polyphasic taxonomy, a consensus approach to bacterial systematics. Microbiol. Rev. *60*: 407–438.

Vibber, L.L., M.J. Pressler and G.M. Colores. 2007. Isolation and characterization of novel atrazine-degrading microorganisms from an agricultural soil. Appl. Microbiol. Biotechnol. *75*: 921–928.

Vimr, E.R., K.A. Kalivoda, E.L. Deszo and S.M. Steenbergen. 2004. Diversity of microbial sialic acid metabolism. Microbiol. Mol. Biol. Rev *68*: 132–153.

Vincent, J.M. 1970. The cultivation, isolation and maintenance of rhizobia. *In* A Manual for the Practical Study of the Root-Nodule Bacteria (edited by Vincent). Blackwell Scientific Publications, Oxford, pp. 1–13.

Vishnivetskaya, T.A., M.A. Petrova, J. Urbance, M. Ponder, C.L. Moyer, D.A. Gilichinsky and J.M. Tiedje. 2006. Bacterial community in ancient Siberian permafrost as characterized by culture and culture-independent methods. Astrobiology *6*: 400–414.

Wakisaka, Y., Y. Kawamura, Y. Yasuda, K. Koizumi and Y. Nishimoto. 1982. A selective isolation procedure for *Micromonospora*. J. Antibiot. (Tokyo) *35*: 822–836.

Waksman, S.A. 1961. The Actinomycetes, vol. 2. Classification, Identification and Descriptions of Genera and Species. Williams & Wilkins, Baltimore.

Wang, H.X., Z.L. Geng, Y. Zeng and Y.M. Shen. 2008a. Enriching plant microbiota for a metagenomic library construction. Environ. Microbiol. *10*: 2684–2691.

Wang, Y.-X., Y.-Q. Zhang, L.-H. Xu and W.-J. Li. 2008b. *Actinopolymorpha rutila* sp. nov., isolated from a forest soil. Int. J. Syst. Evol. Microbiol. *58*: 2443–2446.

Wang, Y., Z. Zhang and N. Ramanan. 1997. The actinomycete *Thermobispora bispora* contains two distinct types of transcriptionally active 16S rRNA genes. J. Bacteriol. *179*: 3270–3276.

Wang, Y.M., Z.S. Zhang, X.L. Xu, J.S. Ruan and Y. Wang. 2001. *Actinopolymorpha singaporensis* gen. nov., sp. nov., a novel actinomycete from the tropical rainforest of Singapore. Int. J. Syst. Evol. Microbiol. *51*: 467–473.

Ward, A.C. and N. Bora. 2006. Diversity and biogeography of marine actinobacteria. Curr. Opin. Microbiol. *9*: 279–286.

Ward, J.B. 1981. Teichoic and teichuronic acids: biosynthesis, assembly, and location. Microbiol. Rev. *45*: 211–243.

Wayne, L.G., D.J. Brenner, R.R. Colwell, P.A.D. Grimont, O. Kandler, M.I. Krichevsky, L.H. Moore, W.E.C. Moore, R.G.E. Murray, E. Stackebrandt, M.P. Starr and H.G. Trüper. 1987. International Committee on Systematic Bacteriology. Report of the *ad hoc* committee on reconciliation of approaches to bacterial systematics. Int. J. Syst. Bacteriol. *37*: 463–464.

Wellington, E.M.H. and S.T. Williams. 1981. Host ranges of phages isolated to *Streptomyces* and other genera. Zentralbl. Bakteriol. Mikrobiol. Hyg. I. Abt. *Suppl. 11*: 93–98.

Wendlinger, G., M.J. Loessner and S. Scherer. 1996. Bacteriophage receptors on *Listeria monocytogenes* cells are the *N*-acetylglucosamine and rhamnose substituents of teichoic acids or the peptidoglycan itself. Microbiology *142*: 985–992.

Weon, H.Y., P. Schumann, R.M. Kroppenstedt, B.Y. Kim, J. Song, S.W. Kwon, S.J. Go and E. Stackebrandt. 2007. *Terrabacter aerolatus* sp. nov., isolated from an air sample. Int. J. Syst. Evol. Microbiol. *57*: 2106–2109.

Widdel, F., G.W. Kohring and F. Mayer. 1983. Studies on dissimilatory sulfate-reducing bacteria that decompose fatty acids. 3. Characterization of the filamentous gliding *Desulfonema limicola* gen. nov. sp. nov., and *Desulfonema magnum* sp. nov. Arch. Microbiol. *134*: 286–294.

Widdel, F. and F. Bak. 1992. Gram-negative mesophilic sulfate-reducing bacteria. *In* The Prokaryotes: a Handbook on the Biology of Bacteria: Ecophysiology, Isolation, Identification, Applications, 2nd edn, vol. 4 (edited by Balows, Trüper, Dworkin, Harder and Schleifer). Springer, New York, pp. 3352–3378.

Williams, S.T., E.M.H. Wellington and L.S. Tipler. 1980. The taxonomic implications of the reactions of representative *Nocardia* strains to actinophage. J. Gen. Microbiol. *119*: 173–178.

Wilmes, P., S.L. Simmons, V.J. Denef and J.F. Banfield. 2009. The dynamic genetic repertoire of microbial communities. FEMS Microbiol. Rev. *33*: 109–132.

Wittekindt, N.E., A. Padhi, S.C. Schuster, J. Qi, F. Zhao, L.P. Tomsho, L.R. Kasson, M. Packard, P. Cross and M. Poss. 2010. Nodeomics: pathogen detection in vertebrate lymph nodes using meta-transcriptomics. PLoS One *5*: e13432.

Yabe, S., Y. Aiba, Y. Sakai, M. Hazaka and A. Yokota. 2011. *Thermasporomyces composti* gen. nov., sp. nov., a thermophilic actinomycete isolated from compost. Int. J. Syst. Evol. Microbiol. *61*: 86–90.

Yamada, K. and K. Komagata. 1970. Taxonomic studies on coryneform bacteria. III. DNA base composition of coryneform bacteria. J. Gen. Appl. Microbiol. *16*: 215–224.

Yamada, K. and K. Komagata. 1972. Taxonomic studies on coryneform bacteria. IV. Morphological, cultural, biochemical and physiological characteristics. J. Gen. Appl. Microbiol. *18*: 399–416.

Yamada, Y., G. Inouye, Y. Tahara and K. Kondo. 1976. The menaquinone system in the classification of coryneform and nocardioform bacteria and related organisms. J. Gen. Appl. Microbiol. *22*: 203–214.

Yamazaki, K., K. Fujii, A. Iwasaki, K. Takagi, K. Satsuma, N. Harada and T. Uchimura. 2008. Different substrate specificities of two triazine hydrolases (TrzNs) from *Nocardioides* species. FEMS Microbiol. Lett *286*: 171–177.

Yanagawa, S., K. Fujii, A. Tanaka and S. Fukui. 1972. Lipid composition and localization of 10-methyl branched-chain fatty acids in *Corynebacterium simplex* grown on n-alkanes. Agric. Biol. Chem. *36*: 2123–2128.

Yano, I., Y. Furukawa and M. Kusunose. 1970. 2-Hydroxy fatty acid-containing phospholipid of *Arthrobacter simplex*. Biochim. Biophys. Acta *210*: 105–115.

Yano, I., Y. Furukawa and M. Kusunose. 1971. Fatty acid composition of *Arthrobacter simplex* grown on hydrocarbons. Occurrence of α-hydroxy-fatty acids. Eur J. Biochem *23*: 220–228.

Yap, W.H., Z. Zhang and Y. Wang. 1999. Distinct types of rRNA operons exist in the genome of the actinomycete *Thermomonospora chromogena* and evidence for horizontal transfer of an entire rRNA operon. J. Bacteriol. *181*: 5201–5209.

Yi, H. and J. Chun. 2004a. *Nocardioides ganghwensis* sp. nov., isolated from tidal flat sediment. Int. J. Syst. Evol. Microbiol. *54*: 1295–1299.

Yi, H. and J. Chun. 2004b. *Nocardioides aestuarii* sp. nov., isolated from tidal flat sediment. Int. J. Syst. Evol. Microbiol. *54*: 2151–2154.

Yoon, J.-H., S.-J. Kang, S.-Y. Lee and T.-K. Oh. 2007a. *Nocardioides terrigena* sp. nov., isolated from soil. Int. J. Syst. Evol. Microbiol. *57*: 2472–2475.

Yoon, J.-H., S.-J. Kang, M.-H. Lee and T.-K. Oh. 2008. *Nocardioides hankookensis* sp. nov., isolated from soil. Int. J. Syst. Evol. Microbiol. *58*: 434–437.

Yoon, J.H., J.S. Lee, Y.K. Shin, Y.H. Park and S.T. Lee. 1997a. Reclassification of *Nocaridioides simplex* ATCC 13260, ATCC 19565, and ATCC 19566 as *Rhodococcus erythropolis*. Int. J. Syst. Bacteriol. *47*: 904–907.

Yoon, J.H., S.K. Rhee, J.S. Lee, Y.H. Park and S.T. Lee. 1997b. *Nocardioides pyridinolyticus* sp. nov., a pyridine-degrading bacterium isolated from the oxic zone of an oil shale column. Int. J. Syst. Bacteriol. *47*: 933–938.

Yoon, J.H., S.T. Lee and Y.H. Park. 1998a. Genetic analyses of the genus *Nocardioides* and related taxa based on 16S–23S rDNA internally transcribed spacer sequences. Int. J. Syst. Bacteriol. *48*: 641–650.

Yoon, J.H., S.T. Lee and Y.H. Park. 1998b. Inter- and intraspecific phylogenetic analysis of the genus *Nocardioides* and related taxa based on 16S rDNA sequences. Int. J. Syst. Bacteriol. *48*: 187–194.
Yoon, J.H., Y.G. Cho, S.T. Lee, K. Suzuki, T. Nakase and Y.H. Park. 1999. *Nocardioides nitrophenolicus* sp. nov., a p-nitrophenol-degrading bacterium. Int. J. Syst. Bacteriol. *49*: 675–680.
Yoon, J.H. and Y.F. Park. 2000. Comparative sequence analyses of the ribonuclease P (RNase P) RNA genes from LL-2,6-diaminopimelic acid-containing actinomycetes. Int. J. Syst. Evol. Microbiol. *50*: 2021–2029.
Yoon, J.H., I.G. Kim, K.H. Kang, T.K. Oh and Y.H. Park. 2004. *Nocardioides aquiterrae* sp. nov., isolated from groundwater in Korea. Int. J. Syst. Evol. Microbiol. *54*: 71–75.
Yoon, J.H., I.G. Kim, M.H. Lee, C.H. Lee and T.K. Oh. 2005a. *Nocardioides alkalitolerans* sp. nov., isolated from an alkaline serpentinite soil in Korea. Int. J. Syst. Evol. Microbiol. *55*: 809–814.
Yoon, J.H., I.G. Kim, M.H. Lee and T.K. Oh. 2005b. *Nocardioides kribbensis* sp. nov., isolated from an alkaline soil. Int. J. Syst. Evol. Microbiol. *55*: 1611–1614.
Yoon, J.H., C.H. Lee and T.K. Oh. 2005c. *Aeromicrobium alkaliterrae* sp. nov., isolated from an alkaline soil, and emended description of the genus *Aeromicrobium*. Int. J. Syst. Evol. Microbiol. *55*: 2171–2175.
Yoon, J.H., C.H. Lee and T.K. Oh. 2005d. *Nocardioides dubius* sp. nov., isolated from an alkaline soil. Int. J. Syst. Evol. Microbiol. *55*: 2209–2212.
Yoon, J.H., C.H. Lee and T.K. Oh. 2006a. *Nocardioides lentus* sp. nov., isolated from an alkaline soil. Int. J. Syst. Evol. Microbiol. *56*: 271–275.
Yoon, J.H., J.K. Lee, S.Y. Jung, J.A. Kim, H.K. Kim and T.K. Oh. 2006b. *Nocardioides kongjuensis* sp. nov., an *N*-acylhomoserine lactone-degrading bacterium. Int. J. Syst. Evol. Microbiol. *56*: 1783–1787.
Yoon, J.H. and Y.H. Park. 2006. The genus *Nocardioides*. *In* The Prokaryotes: a Handbook on the Biology of Bacteria, 3rd edn, vol. 3, *Archaea, Bacteria, Firmicutes*, Actinomycetes (edited by Dworkin, Falkow, Rosenberg, Schleifer and Stackebrandt). Springer, New York, pp. 1099–1113.
Yoon, J.H., S.J. Kang, C.H. Lee and T.K. Oh. 2007b. *Nocardioides insulae* sp. nov., isolated from soil. Int. J. Syst. Evol. Microbiol. *57*: 136–140.
Yoon, J.H., S.J. Kang, S. Park, W. Kim and T.K. Oh. 2009. *Nocardioides caeni* sp. nov., isolated from wastewater. Int. J. Syst. Evol. Microbiol. *59*: 2794–2797.
Yoon, J.H., S. Park, S.J. Kang, J.S. Lee, K.C. Lee and T.K. Oh. 2010. *Nocardioides daedukensis* sp. nov., a halotolerant bacterium isolated from soil. Int. J. Syst. Bacteriol. *60*: 1334–1338.
Young, F.E. 1967. Requirement of glucosylated teichoic acid for adsorption of phage in *Bacillus subtilis* 168. Proc. Natl. Acad. Sci. U.S.A. *58*: 2377–2384.
Yu, C.P., H. Roh and K.H. Chu. 2007. 17beta-estradiol-degrading bacteria isolated from activated sludge. Environ. Sci. Technol. *41*: 486–492.
Yuan, L.J., Y.Q. Zhang, L.Y. Yu, C.H. Sun, Y.Z. Wei, H.Y. Liu, W.J. Li and Y.Q. Zhang. 2010. *Actinopolymorpha cephalotaxi* sp. nov., a novel actinomycete isolated from rhizosphere soil of the plant *Cephalotaxus fortunei*. Int. J. Syst. Evol. Microbiol. *60*: 51–54.
Zhang, G., F. Niu, X. Ma, W. Liu, M. Dong, H. Feng, L. An and G. Cheng. 2007. Phylogenetic diversity of bacteria isolates from the Qinghai-Tibet Plateau permafrost region. Can. J. Microbiol. *53*: 1000–1010.
Zhang, J.Y., X.Y. Liu and S.J. Liu. 2009a. *Nocardioides terrae* sp. nov., isolated from forest soil. Int. J. Syst. Evol. Microbiol. *59*: 2444–2448.
Zhang, X., X. Ma, N. Wang and T. Yao. 2009b. New subgroup of *Bacteroidetes* and diverse microorganisms in Tibetan plateau glacial ice provide a biological record of environmental conditions. FEMS Microbiol. Ecol. *67*: 21–29.
Zhi, X.-Y., W.-J. Li and E. Stackebrandt. 2009. An update of the structure and 16S rRNA gene sequence-based definition of higher ranks of the class *Actinobacteria*, with the proposal of two new suborders and four new families and emended descriptions of the existing higher taxa. Int. J. Syst. Evol. Microbiol. *59*: 589–608.
Zinniel, D.K., P. Lambrecht, N.B. Harris, Z. Feng, D. Kuczmarski, P. Higley, C.A. Ishimaru, A. Arunakumari, R.G. Barletta and A.K. Vidaver. 2002. Isolation and characterization of endophytic colonizing bacteria from agronomic crops and prairie plants. Appl. Environ. Microbiol. *68*: 2198–2208.
ZoBell, C.E. 1941. Studies on marine bacteria. I. The cultural requirements of heterotrophic aerobes. J. Mar. Res. *4*: 42–75.

Order XIII. **Pseudonocardiales** ord. nov.

David P. Labeda and Michael Goodfellow

Pseu.do.no.car.di′a.les. N.L. fem. n. *Pseudonocardia* type genus of the order; suff. *-ales* ending to denote an order; N.L. fem. pl. n. *Pseudonocardiales* the *Pseudonocardia* order.

The order was formed by elevation of the suborder *Pseudonocardineae* Stackebrandt, Rainey, Ward-Rainey 1997, 486[VP] emend. Zhi, Li and Stackebrandt 2009, 599. Labeda et al. (2011) proposed the elimination of the family *Actinosynnemataceae* Labeda and Kroppenstedt 2000 and emended the description of the family *Pseudonocardiaceae* to include the genera formerly found within this family. As a result, the order *Pseudonocardiales* contains a single family and **the description and signature nucleotides of the 16S rRNA gene are that of the family *Pseudonocardiaceae*.**

Type genus: **Pseudonocardia** Henssen 1957, 408[AL].

References

Henssen, A. 1957. Beiträge zur Morphologie und Systematik der thermophilen Actinomyceten. Arch. Mikrobiol. *26*: 373–414.
Labeda, D.P. and R.M. Kroppenstedt. 2000. Phylogenetic analysis of *Saccharothrix* and related taxa: proposal for *Actinosynnemataceae* fam. nov. Int. J. Syst. Evol. Microbiol. *50*: 331–336.
Labeda, D.P., M. Goodfellow, J. Chun, X.-Y. Zhi and W.-J. Li. 2011. Reassessment of the systematics within the suborder *Pseudonocardineae*: transfer of the genera within the family *Actinosynnemataceae* Labeda and Kroppenstedt 2000 emend. Zhi *et al.* 2009 into an emended family *Pseudonocardiaceae* Embley *et al.* 1989 emend. Zhi *et al.* 2009. Int. J. Syst. Evol. Microbiol. *61*: 1259–1264.
Stackebrandt, E., F.A. Rainey and N.L. Ward-Rainey. 1997. Proposal for a new hierarchic classification system, *Actinobacteria* classis nov. Int. J. Syst. Bacteriol. *47*: 479–491.
Zhi, X.-Y., W.-J. Li and E. Stackebrandt. 2009. An update of the structure and 16S rRNA gene sequence-based definition of higher ranks of the class *Actinobacteria*, with the proposal of two new suborders and four new families and emended descriptions of the existing higher taxa. Int. J. Syst. Evol. Microbiol. *59*: 589–608.

Family I. **Pseudonocardiaceae** Embley, Smida and Stackebrandt 1989, 205^{VP} emend. Labeda, Goodfellow, Chun, Zhi and Li 2010a

David P. Labeda and Michael Goodfellow

Pseu.do.no.car.di.a.ce′a.e. N.L. fem. n. *Pseudonocardia* the type genus of the family; suff. *-aceae* ending to denote a family; N.L. fem. pl. n. *Pseudonocardiaceae* the *Pseudonocardia* family.

Aerobic, mesophilic, or thermophilic, Gram-stain-positive, non-acid-fast, catalase-positive, lysozyme resistant, actinomycetes comprising the type genus *Pseudonocardia* (Henssen 1957) emend. Park et al. 2008, as well as the genera *Actinoalloteichus* Tamura et al. 2000, *Actinokineospora* (Hasegawa 1988b) Labeda et al. 2010b, *Actinomycetospora* Jiang et al. 2008, *Actinosynnema* Hasegawa et al. 1978, *Allokutzneria* Labeda and Kroppenstedt 2008, *Amycolatopsis* (Lechevalier et al. 1986) Lee 2009, *Crossiella* Labeda 2001, *Goodfellowiella* Labeda et al. 2008 (previous illegitimate name *Goodfellowia* Labeda and Kroppenstedt 2006), *Kibdelosporangium* Shearer et al. 1986a, *Kutzneria* Stackebrandt et al. 1994, *Lechevalieria* Labeda et al. 2001, *Lentzea* (Yassin et al. 1995) Labeda et al. 2001, *Prauserella* (Kim and Goodfellow 1999) Li et al. 2003a, *Saccharomonospora* Nonomura and Ohara 1971, *Saccharopolyspora* (Lacey and Goodfellow 1975) Korn-Wendisch et al. 1989, *Saccharothrix* (Labeda et al. 1984) Labeda and Lechevalier 1989, *Sciscionella* Tian et al. 2009, *Streptoalloteichus* (Tomita et al. 1987) Tamura et al. 2008b, *Thermocrispum* Korn-Wendisch et al. 1995, and *Umezawaea* Labeda and Kroppenstedt 2007. Figure 277 shows a phylogenetic tree for the genera of the family *Pseudonocardiaceae* calculated from almost-complete 16S rRNA gene sequences.

Morphologically heterogeneous; single or short chains of spores may be present on aerial mycelium and the substrate mycelium. Vegetative mycelium branches, diameter approximately 0.5–0.7 µm; aerial mycelium is produced and fragments into single or chains of smooth-surfaced, rod-shaped elements in some genera. Other structures such as synnemata or dome-like bodies, sporangia or pseudosporangia may be produced by some genera. Some taxa may fail to produce aerial mycelia. Motile spores may be produced by some genera. Marked fragmentation of hyphae occurs in some taxa, but is absent in others. Most taxa are chemo-organotropic, although some are autotrophic. A few taxa are halophilic. Members of the family are found in a diversity of environments, including soils, plant material, manure, and clinical or veterinary samples.

All genera contain *meso*-diaminopimelic acid as the diamino acid in their peptidoglycan and galactose as one of many diagnostic whole-cell sugars. Mycolic acids are not present in any of the genera. Tetrahydrogenated menaquinones of nine isoprene units are characteristic components, although menaquinones containing eight isoprene units predominate in the genus *Pseudonocardia*. The phospholipid profile generally includes phosphatidylethanolamine, sometimes containing hydroxylated fatty acids, as a major constituent, although representatives of one or more genera may also contain phosphatidylcholine. Chemotaxonomic properties of genera of the family are shown in Table 228. The pattern of 16S rRNA signatures consists of nucleotides at positions 127:234 (G–C), 564 (U), 672:734 (U–G), 831:855 (U–G), 832:854 (G–Y), 833:853 (U–G), 952:1229 (U–A), and 986: 1219 (U–A).

DNA G+C content (mol%): 66–76.

Type genus: **Pseudonocardia** Henssen 1957, 408^{AL} emend. Park, Park, Lee and Kim 2008, 2477.

Further descriptive information

The family *Pseudonocardiaceae* Embley, Smida and Stackebrandt 1989, 205^{VP} emend. Zhi, Li and Stackebrandt 2009, 599 was proposed by Embley et al. (1988a, 1989) and Warwick et al. (1994) on the basis of 16S rRNA gene sequence phylogeny for the genera. The family was emended by Stackebrandt et al. (1997) and, subsequently, Labeda and Kroppenstedt (2000) proposed that the genera *Actinosynnema*, *Actinokineospora*, *Lentzea*, and *Saccharothrix* be placed in the new family *Actinosynnemataceae* based on phylogenetic analysis of 16S rRNA gene sequences for a subset of all taxa within the family. The description of the family *Actinosynnemataceae* was emended by Zhi et al. (2009) to include additional genera described since 2000, namely *Lechevalieria* Labeda et al. 2001 and *Umezawaea* Labeda and Kroppenstedt 2007. A recent study of the taxonomic status of the families *Actinosynnemataceae* and *Pseudonocardiaceae* (Labeda et al., 2010a), based on the 16S rRNA gene sequence data available for the 142 validly named taxa and the chemotaxonomic and morphological properties available from the literature, concluded that the retention of the family *Actinosynnemataceae* could no longer be supported, nor could adequate support be found for any other subdivision of the order *Pseudonocardiales* at this time. It was proposed that the genera within the family *Actinosynnemataceae* be included in the family *Pseudonocardiaceae* and the family description emended accordingly.

Key to the genera of the family *Pseudonocardiaceae*

I. Produces sporangia on the colony surface observable with light microscopy.

A. Polar lipid pattern includes phosphatidylethanolamine containing hydroxylated fatty acids and *lyso*-phosphatidylethanolamine.

→Genus *Allokutzneria*

B. Polar lipid pattern does not include phosphatidylethanolamine containing hydroxylated fatty acids or *lyso*-phosphatidylethanolamine; whole-cell sugar pattern includes arabinose and galactose.

→Genus IX. *Kibdelosporangium*

C. Polar lipid pattern does not include phosphatidylethanolamine containing hydroxylated fatty acids or *lyso*-phosphatidylethanolamine; whole-cell sugar pattern includes galactose and rhamnose.

→Genus X. *Kutzneria*

II. Produces pseudosporangia on the colony surface.

A. Polar lipid pattern includes phosphatidylethanolamine containing hydroxylated fatty acids.

→Genus XVIII. *Thermocrispum*

B. Polar lipid pattern does not include phosphatidylethanolamine containing hydroxylated fatty acids.

1. Polar lipid pattern includes phosphatidylinositol and phosphatidylinositol mannosides; whole-cell sugar pattern includes rhamnose.

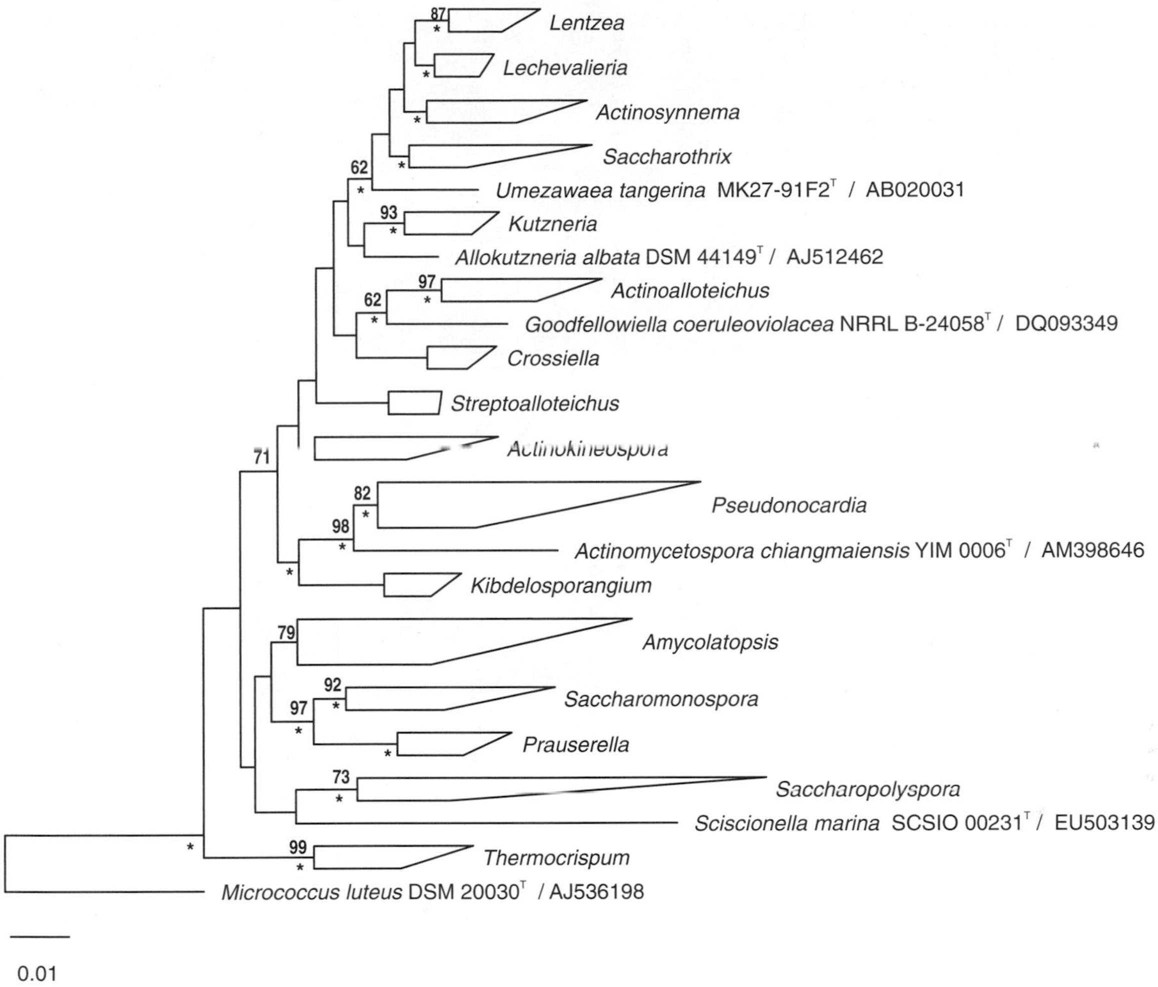

FIGURE 277. Phylogenetic tree for the genera of the family *Pseudonocardiaceae* calculated from almost-complete 16S rRNA gene sequences using the Kimura evolutionary distance method (Kimura, 1980) and the neighbor-joining algorithm of Saitou and Nei (1987). Percentages at the nodes represent levels of bootstrap support from 100 resampled datasets; values less than 60% are not shown. Branches also conserved in maximum-parsimony (Felsenstein, 1993) and maximum-likelihood (Stamatakis et al., 2002) trees are marked with an asterisk. Bar = 0.01 nucleotide substitutions per site.

→Genus VII. *Crossiella*

2. Polar lipid pattern does not include phosphatidylinositol or phosphatidylinositol mannosides; whole-cell sugar pattern does not include rhamnose.

→Genus XVII. *Streptoalloteichus*

III. Does not produce sporangia or pseudosporangia.

A. Produces motile spores.

1. Produces chains of motile spores borne on synnemata or raised dome-like structures on the vegetative mycelium.

→Genus V. *Actinosynnema*

2. Produces chains of motile spores on the aerial mycelia; synnemata not produced

→Genus III. *Actinokineospora*

B. Does not produce motile spores.

1. Does not contain arabinose as a diagnostic whole-cell sugar.

a. Polar lipid pattern includes phosphatidylethanolamine containing hydroxylated fatty acids and *lyso*-phosphatidylethanolamine.

→Genus XIX. *Umezawaea*

b. Polar lipid pattern includes phosphatidylethanolamine containing hydroxylated fatty acids but not *lyso*-phosphatidylethanolamine.

i. Diagnostic whole-cell sugars include mannose and rhamnose.

→Genus XVI. *Saccharothrix*

ii. Diagnostic whole-cell sugars do not include either mannose or rhamnose.

→Genus VIII. *Goodfellowiella*

c. Polar lipid pattern does not include phosphatidylethanolamine containing hydroxylated fatty acids.

i. 16S rRNA gene contains diagnostic nucleotide signature pattern of TT (844–845) and GGT (1107–1109).

→Genus XI. *Lechevalieria*

ii. 16S rRNA gene contains diagnostic nucleotide signature patterns of TCAA (617–620) and GCC (843–845).

→Genus XII. *Lentzea*

2. Contains arabinose as diagnostic whole-cell sugar.

a. Polar lipid pattern contains phosphatidylcholine and/or phosphatidylethanolamine.

TABLE 228. Comparison of chemotaxonomic profiles of genera within the family *Pseudonocardiaceae*

Character	*Pseudonocardia*	*Actinoalloteichus*	*Actinokineospora*	*Actinomycetospora*	*Actinosynnema*	*Allokutzneria*	*Amycolatopsis*	*Crossiella*	*Goodfellowiella*	*Kibdelosporangium*	*Kutzneria*
Sporangia produced	None	None	None	None	Synnemata	Yes, no spores	None	None	None	Yes	Yes
Motile spores	No	No	Variable	No	Yes	No	No	No	No	No	No
Whole-cell sugar pattern[b]	Ara, Gal	Glu, Gal, Man, Rib	Ara, Gal, Rha, Man	Ara, Gal	Gal, Man	Ara, Gal, Man	Ara, Gal	Gal, Man, Rha, Rib	Gal, Rib	Ara, Gal, Glu (v), Mad (v), Rha (v)	Gal, Rha
Phospholipids[a]	PC, PE, PME, PI, PIM, OH-PE	PE, PIM, PI, PG, DPG PME	PE, DPG, PI	PC, PI, PG	PE, OH-PE, DPG	PE, PME, OH-PE, PI, lyso-PME, DPG, PG, *lyso*-PE	PE, DPG, PG, PI	PE, PME, PI, PIM	PE, DPG, OH-PE, PME	PE, PI, PME, PG, DPG, PIM	PE, DPG, PI, PG, PME
Predominant menaquinone(s)	MK-8 (H_4)	MK-9 (H_4)	MK-9 (H_4)	MK-9(H_4)	MK-9(H_4), some MK-9(H_6)	MK-9 (H_4)	MK-9(H_4)	MK-9 (H_4)	MK-9(H_4), MK-10(H_4)	MK-9(H_4), MK-9(H_6), MK-9 (H_{10})	MK-9(H_4)
DNA G+C content, mol%	68–69	72–72.5	72.0	69.0	73.0	71.6	66.0–69.0	74.1	69.2	66	70.3–70.7

Character	*Lechevalieria*	*Lentzea*	*Prauserella*	*Saccharomonospora*	*Saccharopolyspora*	*Saccharothrix*	*Sciscionella*	*Streptoalloteichus*	*Thermocrispum*	*Umezawaea*
Sporangia produced	None	None	None	None	None	None	None	Pseudosporangia	Pseudosporangia	None
Motile spores	No	No	No	No	No	No	No	Variable	No	No
Whole-cell sugar pattern[b]	Gal, Man, Rha (trace)	Gal, Man, Rib	Ara, Gal	Ara, Gal	Ara, Gal	Gal, Rha, Man (trace)	Ara, Gal, Glu	Gal, Man, Rha,	Ara, Glu, Gal (trace), Man	Gal, Man, Rib, Rha (trace)
Phospholipids[a]	PE, DPG, PG, PI	PE, DPG, DPG, PI	DPG, PE	PE, DPG, PG, PI	PC, PE, DPG, PG, PI	PE, OH-PE, DPG, PG, PI, PIM	DPG, PC, PE, PI, PL, PME	PE, DPG, PI, PIM, DPG, PME	PE, PI, OH-PE	PE, PI, OH-PE, *lyso*-PE
Predominant menaquinone(s)	MK-9(H_4)	MK-9 (H_4)	MK-9 (H_2), MK-9 (H_4)	MK-9(H_4)	MK-9 (H_4)	MK-9(H_4), MK-10(H_4)	MK-9 (H_4)	MK-9(H_6), MK-10(H_6)	MK-9 (H_4)	MK-9(H_4), MK-10(H_4)
DNA G+C content, mol%	68.0–71.4	71.4	67–68.9	66.0–70.0	66.0–74.0	71.4	69.0	71.6	69.0–73.0	74.0

[a]Ara, Arabinose; Gal, galactose; Glu, glucose; Mad, madurose; Man, mannose; Rha, rhamnose; Rib, ribose. v, Variable.

[b]DPG, Diphosphatidylglycerol; OH-PE, phosphatidylethanolamine with hydroxy fatty acids; *lyso*-PE, phosphatidylethanolamine where one fatty acid chain is missing from the glycerol backbone; *lyso*-PME, phosphatidylmethylethanolamine where one fatty acid chain is missing from the glycerol backbone; PC, phosphatidylcholine; PE, phosphatidylethanolamine; PG, phosphatidylglycerol; PI, phosphatidylinositol; PIM, phosphatidylinositol mannosides; PL, unknown phospholipids; PME, phosphatidylmethylethanolamine.

i. Polar lipid pattern includes phosphatidylethanolamine containing hydroxylated fatty acids; predominant menaquinone is MK-8(H_4); chains of spores not produced on aerial mycelia, but vegetative mycelium fragments.
→Genus I. *Pseudonocardia*

ii. Polar lipid pattern does not include phosphatidylethanolamine containing hydroxylated fatty acids; predominant menaquinone is MK-9(H_4); chains of spores not produced on aerial mycelia, but vegetative mycelium fragments.
→Genus *Sciscionella*

b. Polar lipid pattern contains phosphatidylcholine; predominant menaquinone is MK-9(H_4).

i. Short chains of spores, sometimes with spiny to hairy ornamentation, are produced on the aerial mycelia.
→Genus XV. *Saccharopolyspora*

ii. Short chains of smooth-surfaced spores are produced from the vegetative mycelium; phosphatidylinositol and phosphatidylglycerol are only other polar lipids.
→Genus III. *Actinomycetospora*

c. Polar lipid pattern contains phosphatidylethanolamine but not phosphatidylcholine.

i. Single spores, often densely packed, are borne along the aerial mycelium.
→Genus XIV. *Saccharomonospora*

ii. Substrate mycelium fragments into irregular rods; aerial mycelium may be produced which fragments into spore chains; nucleotide signature regions in the 16S rRNA gene sequence include 51 (A/C), 208 (U), 210 (G), 213 (A), 280 (C), 1004 (A), and 1257 (U).
→Genus VI. *Amycolatopsis*

iii. Substrate mycelium fragments into irregular rods; aerial mycelium may be produced which fragments into spore chains; nucleotide signature regions in the 16S rRNA gene sequence include 51 (A), 72 (A), 183:194 (G–C), 208 (G), 210 (A), 213 (C), 280 (U), 1004 (G), and 1257 (G).
→Genus XIII. *Prauserella*

Genus I. **Pseudonocardia** Henssen 1957, 408[AL] emend. Park, Park, Lee and Kim 2008, 2477

Ying Huang and Michael Goodfellow

Pseu.do.no.car′di.a. Gr. adj. *pseudês* false; N.L. fem. n. *Nocardia* a bacterial genus name; N.L. fem. n. *Pseudonocardia* the false *Nocardia*.

Aerobic, Gram-stain-positive, non-acid-fast, nonmotile, catalase-positive, actinomycetes which form branched substrate hyphae that may fragment into rod-shaped elements. Hyphae vary in thickness and in the degree of branching. Aerial hyphae, when formed, may be sterile, fragmented into chains of oval or squarish elements, or into chains of two or more spores. **Substrate and aerial hyphae may exhibit cell division in different directions with a tendency to form apical or intercalary swellings. Spores are usually smooth and are produced on the substrate and/or aerial mycelium by acropetal budding and/or by basipetal septation (fragmentation), or else are formed in longitudinal pairs on vegetative hyphae and singly or in longitudinal pairs on aerial hyphae. In some species, the mycelium is covered by an electron-dense outer layer.** Grows on a variety of organic and synthetic media. Some species are facultative autotrophs. **Cell wall contains *meso*-diaminopimelic acid (*meso*-A2pm), arabinose, and galactose. The major menaquinone is MK-8(H_4) and the predominant fatty acid is iso-branched hexadecanoic acid. Mycolic acids are absent.** The polar lipid profile, which varies between species, may include any of the following: phosphatidylcholine, phosphatidylethanolamine, phosphatidylglycerol, phosphatidylmethylethanolamine, and glucosamine-containing phospholipids. **The 16S rRNA gene sequence contains two unique regions which correspond to nucleotide positions 179–219 and 813–833 of the *Streptomyces ambofaciens* 16S rRNA gene** (Pernodet et al., 1989). The phylogenetic position of *Pseudonocardia*, as determined by 16S rRNA gene sequence analysis, is in the family *Pseudonocardiaceae*.

DNA G+C content (mol%): 68–79 (HPLC, T_m).

Type species: **Pseudonocardia thermophila** Henssen 1957, 408[AL].

Further descriptive information

Phylogeny. The genus *Pseudonocardia* forms a distinct clade in the *Pseudonocardiaceae* 16S rRNA gene tree (Bowen et al., 1989; Embley et al., 1988b, 1988c; Warwick et al., 1994). It encompasses 29 species with validly published names, half of which have been proposed in the past 10 years on the basis of polyphasic taxonomic studies (Chen et al., 2009; Huang et al., 2002; Kämpfer and Kroppenstedt, 2004; Lee et al., 2000a; Park et al., 2008). *Pseudonocardia* species can be phylogenetically assigned to four multimembered clades and one single-membered clade (Figure 278). The earliest described species in the three multimembered subclades, which are supported by bootstrap values that range from 70 to 100%, are *Pseudonocardia halophobica*, *Pseudonocardia hydrocarbonoxydans*, and *Pseudonocardia nitrificans*. The earliest described species in the least well circumscribed multi-membered subclade is *Pseudonocardia thermophila*. *Pseudonocardia spinosispora* forms a single-membered clade that is loosely associated with the *Pseudonocardia nitrificans* phyletic line. The most distant phylogenetic relationship in the genus is between the type strains of *Pseudonocardia ammonioxydans* and *Pseudonocardia thermophila* and these organisms share a 16S rRNA gene sequence similarity of 93%. In contrast, the type strains of *Pseudonocardia chloroethenivorans* and *Pseudonocardia dioxanivorans* have identical 16S rRNA gene sequences.

Cell morphology. *Pseudonocardia* strains form extensive substrate and aerial hyphae that vary in the degree of branching and in diameter (0.3–2 µm). Some species, such as *Pseudonocardia asaccharolytica*, *Pseudonocardia halophobica*, and *Pseudonocardia sulfidoxydans*, form swollen hyphal segments up to 5 µm in diameter, zig-zag mycelial growth, and longitudinal and transverse septa in the mycelium (Akimov et al., 1989; Reichert et al., 1998). A characteristic feature of many *Pseudonocardia* species, such as *Pseudonocardia ammonioxydans*, *Pseudonocardia spinosa*, and *Pseudonocardia thermophila*, is acropetal budding (Figure 279, Figure 280, and Figure 281); a constriction is formed behind the tip of the terminal segment, the tip then enlarges to form a new segment, another constriction is formed near the tip, and the process continues (Figure 279, Figure 280, Figure 281, Figure 282, Figure 283, and Figure 284; Henssen and Schäfer,

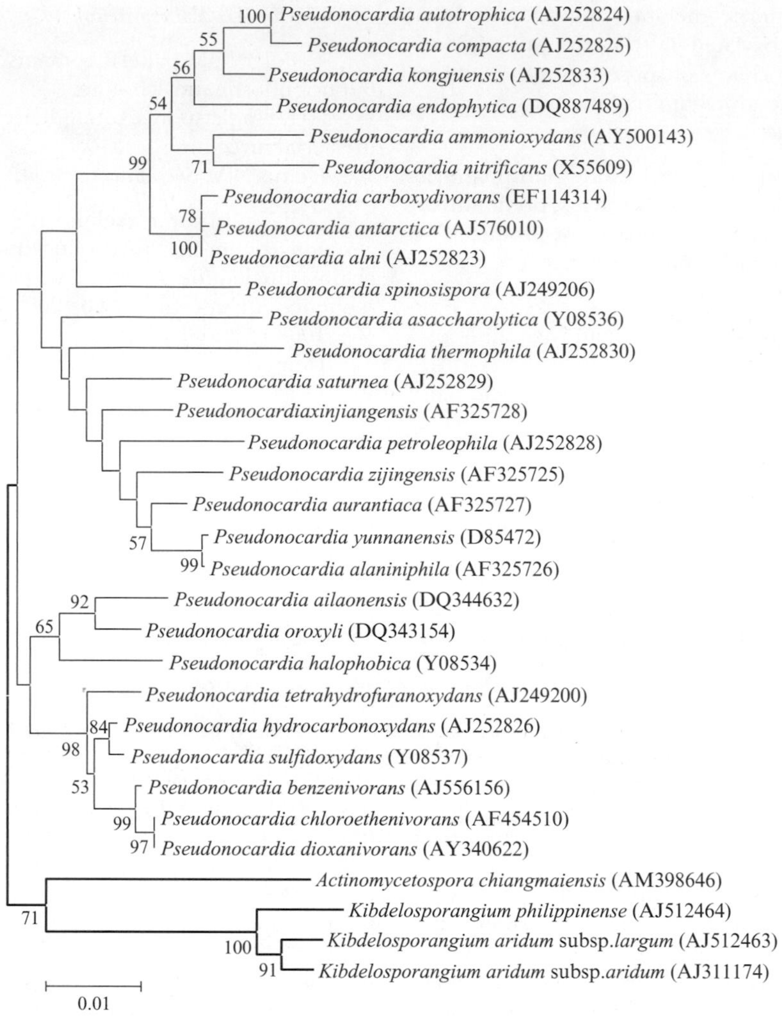

FIGURE 278. Neighbor-joining tree based on nearly complete 16S rRNA gene sequences showing relationships between type strains belonging to the genus *Pseudonocardia*. The numbers at the nodes indicate levels of bootstrap support based on a neighbor-joining analysis of 1000 resampled datasets; only values above 50% are given. Bar = 0.01 substitutions per nucleotide position.

1971). The constrictions may be secondarily separated by septa (Figure 280), which may be formed some distance behind the constriction, as in *Pseudonocardia compacta*. Side branches are usually formed below a septatum, more rarely from the center of a septation (Figure 281 and Figure 282). Aerial hyphae can be sterile, may fragment into chains of squarish to oval fragments, or differentiate into chains of spores (Lechevalier et al., 1986).

Spores may be formed on substrate or aerial hyphae by successive acropetal budding, by basipetal septation (fragmentation), or by irregular spore formation along senescing hyphae (Henssen, 1989; Henssen and Schnepf, 1967). *Pseudonocardia alaniniphila*, *Pseudonocardia aurantiaca*, *Pseudonocardia xinjiangensis*, and *Pseudonocardia yunnanensis* bear spores either in longitudinal pairs on substrate hyphae or singly on aerial hyphae (Xu et al., 1999). In *Pseudonocardia chloroethenivorans*, the aerial and substrate hyphae carry single spores at the ends of hyphae (Lee et al., 2004). Single spores are also formed at the ends of substrate hyphae in *Pseudonocardia tetrahydrofuranoxydans* (Kämpfer et al., 2006). Spores are smooth-walled or spiny and vary greatly in size, but are usually 0.5–1 μm wide by 1.5–4.5 μm long. Spore walls are of uniform thickness and intersporal pads are not formed. In *Pseudonocardia spinosa*, the ornamentation of the spiny spores is formed by folds in the fibrous sheath.

The fine structure of some *Pseudonocardia* species, notably *Pseudonocardia autotrophica*, *Pseudonocardia spinosa*, and *Pseudonocardia thermophila*, has been investigated extensively by Henssen and coworkers (Henssen and Schäfer, 1971; Henssen et al., 1983; Kothe et al., 1989; Kuimova and Malishkaite, 1984). The substrate and aerial mycelial wall of these organisms is composed of two layers: an inner electron-transparent, uniformly thick layer; and an outer electron-dense irregular layer to which the fibrous sheath is attached (Figure 281, Figure 282, and Figure 283). The cross walls are interspace septa (Henssen et al., 1981, 1983) of type 2 *sensu* Williams et al. (1973). Hyphal fragments may become subdivided by septa growing inwards at different angles. Spores of *Pseudonocardia thermophila* show resistance to dry and wet heat at 100°C (Fergus, 1967).

Colony morphology. *Pseudonocardia* strains form well developed colonies on most standard media used to cultivate filamentous actinomycetes (Shirling and Gottlieb, 1966). The substrate

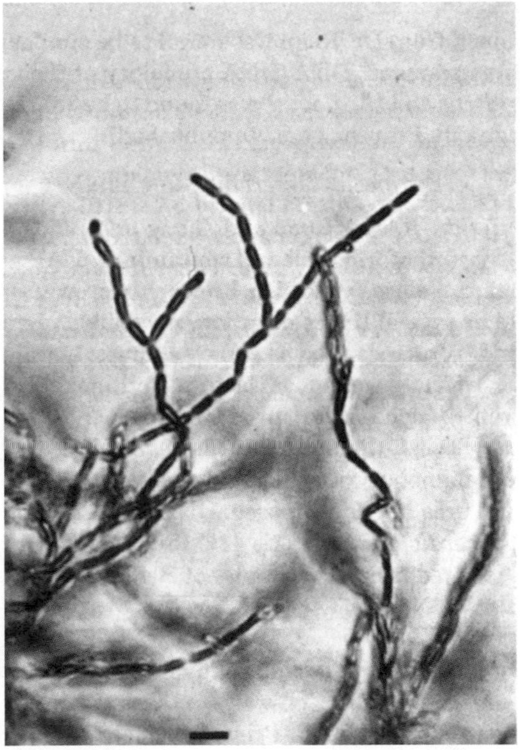

FIGURE 279. Budding, zig-zag-shaped hyphae of *Pseudonocardia spinosa* strain MB SF-1 (light microscopy; bar = 5 µm). (Reproduced with permission from Henssen and Schäfer, 1971. Int. J. Syst. Bacteriol. *21*: 29–34.)

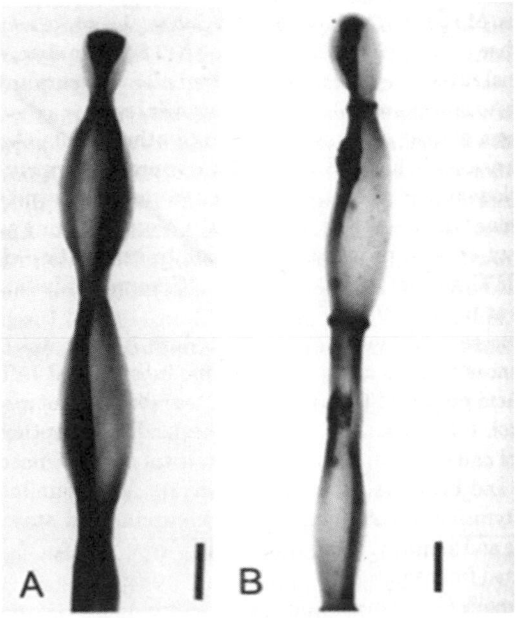

FIGURE 280. A and B. Budding aerial hyphae of *Pseudonocardia thermophila* strain MB-A18.B. Constrictions separated by septae (whole mount silhouettes, transmission electron microscopy; bar = 0.5 µm). (Reproduced with permission from Henssen and Schäfer, 1971. Int. J. Syst. Bacteriol. *21*: 29–34.)

mycelium is typically yellowish to brown and the aerial mycelium, if produced, is whitish. The colors of substrate and aerial mycelia, however, can vary between species and are influenced by the cultivation medium. *Pseudonocardia compacta* strains form

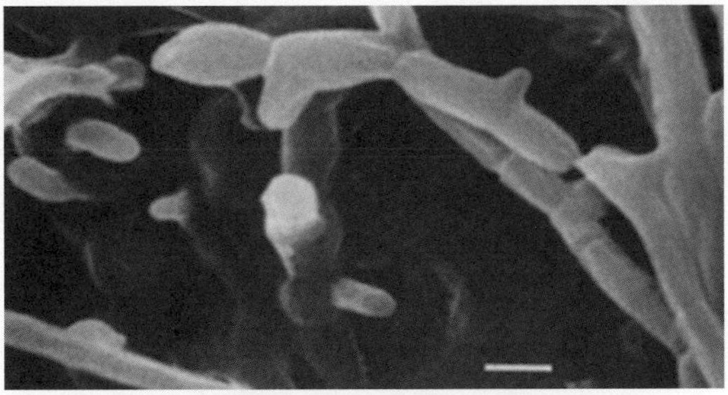

FIGURE 281. Budding substrate and aerial hyphae of *Pseudonocardia compacta* strain MB H-146 (scanning electron microscopy; bar = 1 µm). (Reproduced with permission from Henssen et al., 1983. Int. J. Syst. Bacteriol. *33*: 829–836.)

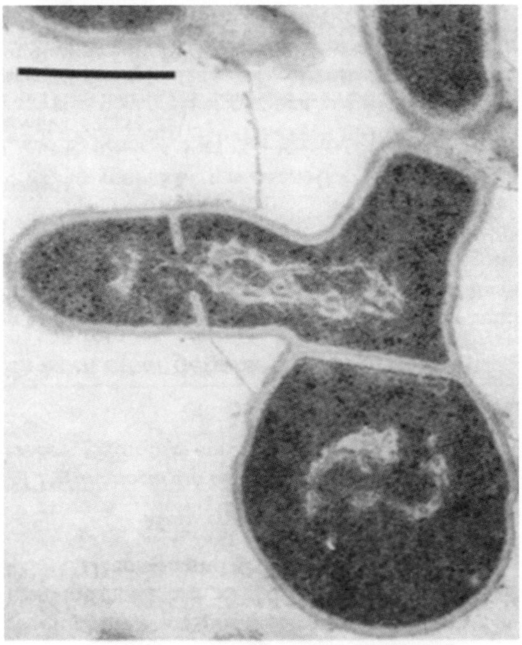

FIGURE 282. *Pseudonocardia thermophila* MB-A18: tip of aerial hyphae with apical swelling and side branch showing septate formation (TEM, glutaraldehyde/osmium tetroxide fixation; bar = 0.5 µm).

lumpy colonies composed of densely aggregated aerial hyphae, which frequently bear apical and intercalary swellings (Henssen et al., 1983). The straight or occasionally helically twisted hyphae of these organisms are more or less regularly segmented. The substrate mycelium of *Pseudonocardia spinosa* forms a compact mass composed of non-septate hyphae of various thicknesses constricted at intervals (Henssen and Schäfer, 1971).

Cell-wall composition. The peptidoglycan of *Pseudonocardia* strains contains *meso*-A_2pm, arabinose, and galactose; these components, which are observed in whole-organism hydrolysates (Embley, 1992; Huang et al., 2002; Reichert et al., 1998; Takeuchi et al., 1992; Warwick et al., 1994), are typical of wall chemotype IV *sensu* Lechevalier and Lechevalier (1970). They typically contain tetrahydrogenated menaquinones with eight isoprene units [MK-8(H_4)] as the predominant isoprenologue

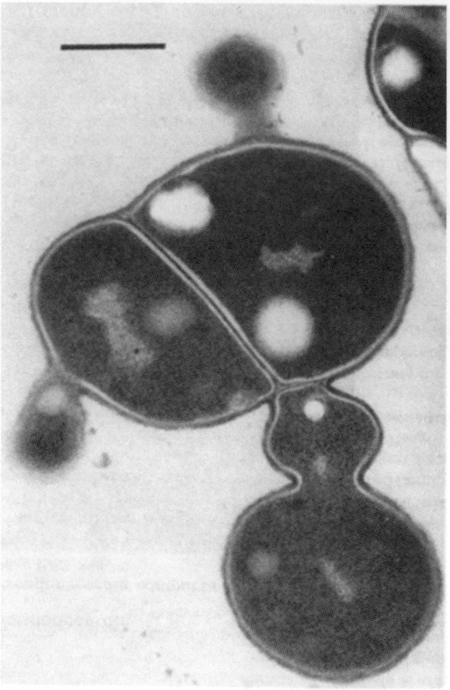

FIGURE 283. Budding cells of *Pseudonocardia spinosa* MB SF-1, the upper bud with constriction (TEM, glutaraldehyde/osmium tetroxide fixation; bar = 0.5 µm).

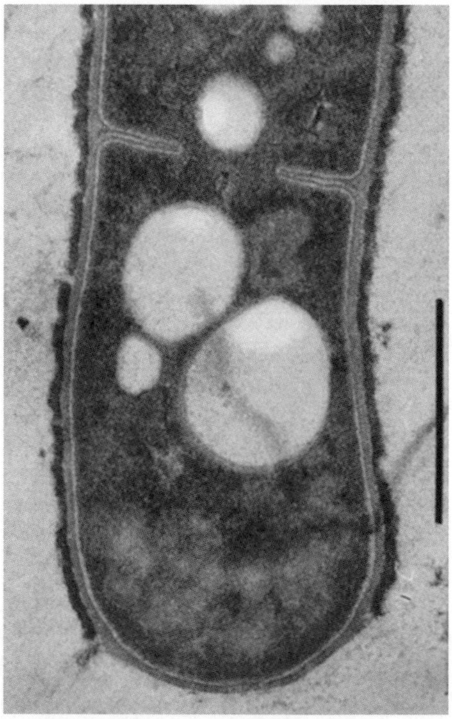

FIGURE 284. *Pseudonocardia compacta* MB H-146: tip of spore wall formed by inward growth of double annulus (TEM; bar = 0.5 µm). (Reproduced with permission from Henssen et al., 1983. Int. J. Syst. Bacteriol. *33*: 829–836.)

(Chen et al., 2009; Collins et al., 1985; Huang et al., 2002; Kroppenstedt, 1985; Liu et al., 2006; Qin et al., 2008b) and iso-branched hexadecanoic acid ($C_{16:0}$ iso) as the major fatty acid (Gu et al., 2006; Huang et al., 2002; Kroppenstedt, 1985; Park et al., 2008), but lack mycolic acids (Goodfellow and Minnikin, 1981; Lechevalier et al., 1986; Lee et al., 2004). *Pseudonocardia* species may exhibit either a phospholipid type II (phosphatidylethanolamine or its derivatives as major diagnostic components) or type III (phosphatidylcholine) pattern *sensu* Lechevalier et al. (1981, 1977a). Variations in fatty acid profiles and polar lipid patterns are cited in the species descriptions. *Pseudonocardia carboxydivorans* is unusual as it contains MK-9 as the predominant menaquinone (Park et al., 2008).

Nutrition and growth conditions. Many pseudonocardiae grow on standard media, such as Czapek (Waksman, 1967), glucose-asparagine (ISP medium 5, Shirling and Gottlieb, 1966), and yeast extract (Henssen, 1989) agars. Some species such as *Pseudonocardia petroleophila*, however, grow poorly on most media (Hirsch and Engel, 1956). *Pseudonocardia spinosispora* produces sparse, if any, substrate mycelium on most agar media, but does grow in stationary, but not shaken, broth cultures (Lee et al., 2002). *Pseudonocardia halophobica* does not grow in the presence of 3% NaCl (Akimov et al., 1989). Most pseudonocardiae grow well between 15 and 37°C, although *Pseudonocardia thermophila* grows optimally between 40 and 50°C (Henssen, 1957) and *Pseudonocardia antarctica* is psychrotolerant (Prabahar et al., 2004a).

Metabolism. *Pseudonocardia* strains have an oxidative metabolism and exhibit considerable physiological versatility. Some species, such as *Pseudonocardia ammonioxydans*, *Pseudonocardia autotrophica*, *Pseudonocardia dioxanivorans*, *Pseudonocardia petroleophila*, and *Pseudonocardia saturnea*, are facultative autotrophs, whereas most *Pseudonocardia* species are heterotrophs. *Pseudonocardia autotrophica*, *Pseudonocardia petroleophila*, and *Pseudonocardia saturnea* grow in the presence of CO_2, H_2, and O_2 (Hirsch and Engel, 1956; Lechevalier et al., 1986; Takamiya and Tubaki, 1956); *Pseudonocardia carboxydivorans* uses CO as a sole carbon source (Park et al., 2008); *Pseudonocardia ammonioxydans* oxidizes ammonia to nitrate as an energy source, and grows autotrophically on modified nitrifying medium and heterotrophically on Luria–Bertani medium (Liu et al., 2006).

Pseudonocardia (previously *Amycolata*) *autotrophica* strains can degrade lignin and related phenolic compounds (Ball et al., 1989; Haider et al., 1978; Malarczyk et al., 1987; McCarthy and Broda, 1984). When grown in a medium containing carbohydrates and supplemented with a radiolabeled dehydropolymer of coniferyl alcohol they show a wide variety of activities, releasing $^{14}CO_2$ from the methoxyl group, the 2-carbon of the side chain, and from the benzene ring (Haider et al., 1978). In contrast, only small amounts of $^{14}CO_2$ were released from the labeled lignin component of corn stalks and *Pseudonocardia autotrophica* strains also do not significantly degrade ^{14}C-lignin labeled wheat lignocellulose (McCarthy and Broda, 1984). Malarczyk et al. (1987) studied the metabolism of radiolabeled methoxyphenolic acids supplied to *Pseudonocardia autotrophica* strains as sole carbon sources and found that they could release $^{14}CO_2$ from vanillic acid irrespective of the position of the label, and from ring-labeled benzylovanillic, isovanillic and veratric acids. They proposed a pathway for the degradation of vanillic acid that proceeded via guaiacol, isovanillic acid, and protocatechuic acid.

Pseudonocardia hydrocarbonoxydans and *Pseudonocardia petroleophila* grow well in an atmosphere containing simple hydrocarbons (Hirsch and Engel, 1956; Nolof and Hirsch, 1962); *Pseudonocardia hydrocarbonoxydans* is able to partially oxidize gaseous aldehydes, hydrocarbons, monocarboxylic acids (C_6–C_{14}), and two dicarboxylic acids, namely sebacic and succinic acids (Lacey, 1988; Nolof, 1962). *Pseudonocardia petroleophila* oxidizes the cyclohexane ring of methylcyclohexane (Tonge and Higgins, 1974) and *Pseudonocardia nitrificans* is unusual in its ability to convert ethyl-, ethyl-N-ethyl-, *n*-butyl-, and *n*-propyl-carbamates to nitrite and use these compounds as sole carbon, energy, and nitrogen sources (Isenberg et al., 1954). *Pseudonocardia* sp. strain TY-7 oxidizes propane to 1-propanol and 2-propanol through both terminal and subterminal oxidations (Kotani et al., 2006).

Pseudonocardiae have physiological properties which suggest that they may play a role in the degradation of hazardous compounds and in biogeochemical cycles: *Pseudonocardia asaccharolytica* and *Pseudonocardia sulfidoxydans* utilize dimethyl disulfide as an energy source (Reichert et al., 1998); *Pseudonocardia benzenivorans* utilizes 1,2,3,5-tetrachlorobenzene as a sole carbon source (Kämpfer and Kroppenstedt, 2004); *Pseudonocardia chloroethenivorans* degrades chloroethene, *cis*-1,2-dichloroethene, and trichloroethene, and metabolizes phenol as a source of carbon and energy by a *meta*-cleavage pathway (Lee et al., 2004); *Pseudonocardia dioxanivorans* degrades 1,4-dioxane (Mahendra and Alvarez-Cohen, 2005); and *Pseudonocardia hydrocarbonoxydans* oxidizes hydrocarbons (Lechevalier et al., 1986; Nolof and Hirsch, 1962). Similarly, *Pseudonocardia nitrificans* oxidizes urethane to nitrite as a sole source of carbon, nitrogen, and energy (Schatz et al., 1954). *Pseudonocardia* sp. strain ENV478 degrades 1,4-dioxane and other potentially important ether pollutants and, hence, may be a useful biocatalyst for *in situ* and *ex situ* systems designed for treating recalcitrant pollutants (Vainberg et al., 2006).

Pseudonocardia thermophila exhibits an inducible exo- and intracellular carboxymethylcellulase and β-D-glucosidase activity, although these attributes are suppressed in the presence of free glucose (Malfait et al., 1984). Isolates of *Pseudonocardia thermophila* from cattle compost metabolize different types of native and crystalline cellulose under laboratory conditions (Goddon and Penninckx, 1984). Ca^{2+}-alginate-immobilized cells of *Pseudonocardia thermophila* IFO (now NBRC) 12133 have been used to produce optically active α-hydroxy-esters (Ishihara et al., 1997).

It is evident from the phylogenetic tree calculated from 16S rRNA gene sequences that the type strains of *Pseudonocardia benzenivorans*, *Pseudonocardia hydrocarbonoxydans*, *Pseudonocardia sulfidoxydans*, and *Pseudonocardia tetrahydrofuranoxydans* form a distinct subclade with high bootstrap support (Figure 278). All of these organisms grow on 4-aminobutyrate, 4-hydroxybutyrate, serine, and succinate (each at 10 mM) although the growth yields differ (Kämpfer et al., 2006). Moreover, these species, as well as *Pseudonocardia dioxanivorans*, tolerate a high concentration of tetrahydrofuran (THF, 60 mM), but exhibit differences in their ability to form visible cell aggregations in liquid media containing this compound (Kämpfer et al., 2006; Mahendra and Alvarez-Cohen, 2005). *Pseudonocardia benzenivorans* and *Pseudonocardia sulfidoxydans* form cell aggregates at a low THF concentration (10 mM), whereas *Pseudonocardia hydrocarbonoxydans* and *Pseudonocardia tetrahydrofuranoxydans* exhibit dispersed cell suspensions at this concentration. THF degradation in *Pseudonocardia tetrahydrofuranoxydans* is initiated by a three-component binuclear iron-containing monooxygenase which contains an NADH-dependent reductase component with an unusual covalently bound flavin (Thiemer et al., 2001, 2003).

In general, *Pseudonocardia* species produce acid from a range of sugars and use diverse compounds as sole carbon and nitrogen sources, but tend to vary in their ability to degrade adenine, casein, gelatin, hypoxanthine, starch, tyrosine, and xanthine (Chen et al., 2009; Goodfellow, 1971; Lechevalier et al., 1986; Lee et al., 2004; Liu et al., 2006; Orchard and Goodfellow, 1980; Prabahar et al., 2004a). They are also a potentially rich source of useful enzymes, as exemplified by *Pseudonocardia thermophila*, which produces cobalt-containing nitrile hydratases (Miyanaga et al., 2001; Peplowski et al., 2007) and a constitutively expressed thermostable amidase (Egorova et al., 2004), as well as exhibiting cellulase and xylanase activity (Li et al., 1984; Malfait et al., 1984; Zimmermann et al., 1988). *Pseudonocardia autotrophica* strains can convert vitamin D_3 into hydroxylated active forms, such as 1α and 2α,25-dihydroxyvitamin D_3 (Kim et al., 2002b; Takeda et al., 2006). Kang et al. (2006) optimized the culture conditions for the bioconversion of vitamin D_3 to 1α,25-dihydroxyvitamin D_3 using *Pseudonocardia autotrophica* ID 9302. *Pseudonocardia* strains are also known to metabolize 4-ethylpyridine and 4-methylpyridine (Lee et al., 2006a), and polyethylene glycol (Yamashita et al., 2004).

Pseudonocardia strains have been reported to produce antimicrobial compounds such as the novel glycopeptides helvecardins A and B (Takeuchi et al., 1991), phenazostatin D (Maskey et al., 2003), and eight new quinolone compounds with selective and potent anti-*Helicobacter pylori* activity (Dekker et al., 1998). There has also been a report that *Pseudonocardia autotrophica* produces a broad spectrum antifungal compound (Antoun et al., 1978). *Pseudonocardia* strains isolated from coastal soils have also been found to produce metabolites inhibitory against *Micrococcus luteus* (Srivibool and Sukchotiratana, 2006).

Genetics. There have been relatively few studies on the genetics of *Pseudonocardia* strains, although a nitrile hydratase gene from *Pseudonocardia thermophila* has been cloned and sequenced (Yamaki et al., 1997), as has the gene for cytochrome P-450 hydroxylase in *Pseudonocardia autotrophica* (Kim et al., 2002b). There have been studies on gene clusters involved in fatty acid metabolism in *Pseudonocardia autotrophica* BCRC 12444 (Chen et al., 2005), THF degradation in *Pseudonocardia tetrahydrofuranoxydans* (Thiemer et al., 2003), and on the cloning and expression of a gene that encodes an ether-bond-cleaving enzyme, diglycolic acid dehydrogenase, in polyethylene glycol-utilizing *Pseudonocardia tetrahydrofuranoxydans* $K1^T$ (Yamashita et al., 2004). Two gene clusters encoding putative propane monooxygenases have been cloned from *Pseudonocardia* sp. strain TY-7 and shown to be induced by *n*-alkanes, suggesting that the products of these genes are involved in gaseous *n*-alkane oxidation (Kotani et al., 2006).

A cryptic polyene gene cluster has been isolated and partially characterized from a strain of *Pseudonocardia autotrophica*, an organism that does not exhibit antifungal activity (Lee et al., 2006b), and this strain was shown to contain a novel polyene-specific cytochrome P-450 hydroxylase (CYP) genes. Genomic DNA library screening using the probe for the polyene-specific CYP gene identified a positive cosmid clone containing a DNA fragment of approximately 34.5 kb. Complete sequencing of this DNA fragment showed the presence of seven complete and two incomplete open reading frames that were found to be unique, but comparatively similar to previously known polyene biosynthetic genes. The nine open reading frames were thought to be involved in the biosynthesis of a novel cryptic *Pseudonocardia* polyene metabolite. Polyene-specific CYP genes were not detected, however, in representatives of several other

Pseudonocardia species (Hwang et al., 2007). Two large cryptic plasmids, about 80 and 120 MDa, have also been detected in *Pseudonocardia alni* strains (Dobritsa, 1984).

Pathogenicity. There is no definitive evidence that *Pseudonocardia* strains have a role as clinical or veterinary pathogens. *Pseudonocardia autotrophica* has been occasionally isolated from clinical materials and may be an opportunistic pathogen of immunocompromised hosts (Mishra et al., 1980; Schaal and Beaman, 1984). In addition, unidentified *Pseudonocardia* strains have been associated with allergic diseases in Kuwait (Diab and Al-Gunaim, 1982).

Antibiotic sensitivity. Little is known about the antibiotic sensitivity patterns of *Pseudonocardia* species, although certain closely related species can be distinguished on the basis of their susceptibility to a range of antibiotics (Evtushenko et al., 1989; Prabahar et al., 2004a; Xu et al., 1999). The antibiotic sensitivity profiles of individual taxa, where known, are provided in the species descriptions.

Ecology. Pseudonocardiae have been primarily isolated from enrichment cultures, plant materials, and soils, but have also been recovered from various other habitats, including air, clinical materials, coastal and marine sediments, composts, and sludge, although population numbers, distribution, and roles of *Pseudonocardia* species in natural habitats is virtually unknown. Several species, including *Pseudonocardia ailaonensis*, *Pseudonocardia alaniniphila*, *Pseudonocardia aurantiaca*, *Pseudonocardia xinjiangensis*, *Pseudonocardia yunnanensis*, and *Pseudonocardia zijingensis*, were isolated from soils collected in Yunnan and Xinjiang Provinces in China (Huang et al., 2002; Jiang et al., 1991; Qin et al., 2008b; Xu et al., 1999). Similarly, *Pseudonocardia ammonioxydans* was isolated from a coastal sediment (Liu et al., 2006), *Pseudonocardia antarctica* from a moraine sample from the McMurdo Dry Valleys region of Antarctica (Prabahar et al., 2004a), *Pseudonocardia benzenivorans* from contaminated soil (Kämpfer and Kroppenstedt, 2004), and *Pseudonocardia kongjuensis* and *Pseudonocardia spinosispora* from gold mine cave soil (Lee, 1996; Lee et al., 2001, 2002).

Several (Aragno and Schlegel, 1981)species have been isolated from plant materials, including *Pseudonocardia alni*, which is thought to be associated with the nitrogen-fixing root nodules and rhizospheres of the alder species *Alnus glutinosa* and *Alnus incana* (Evtushenko et al., 1989; Sharaya et al., 1982) where it might utilize the excess hydrogen produced by nitrogen-fixing bacteria (Aragno and Schlegel, 1981). *Pseudonocardia asaccharolytica* and *Pseudonocardia sulfidoxydans* were isolated from bark compost biofilters in an animal rendering plant (Reichert et al., 1998), *Pseudonocardia endophytica* from the Chinese medicinal plant, *Lobelia clavata* (Chen et al., 2009), and *Pseudonocardia oroxyli* from the surface-sterilized elongation root zone of *Oroxylum indicum*, another traditional Chinese medicinal plant (Gu et al., 2006). Enrichment cultures from contaminated industrial sludge and waste water plants led to the isolation of strains of *Pseudonocardia benzenivorans* (Kämpfer and Kroppenstedt, 2004) and *Pseudonocardia tetrahydrofuranoxydans* (Kämpfer et al., 2006), respectively. *Pseudonocardia* strains have been isolated from marine sediments following their detection using culture-independent procedures (Maldonado et al., 2005).

It seems likely that *Pseudonocardia* species with an ability to oxidize hydrogen and fix carbon dioxide (Aragno and Schlegel, 1981; Hirsch, 1960) are widely distributed in hydrogen-rich habitats that are abundant in nature (Aragno and Schlegel, 1981). *Amycolata autotrophica* is able to colonize the surfaces of substrates and actively oxidizes hydrogen when continuously supplied with a mixture of H_2:O_2:CO_2 (7:2:1) in the presence of a small amount of mineral medium (Kriukov, 1981). This organism has been isolated from decomposing vegetable matter, marshland, and soil (Kuznetsov et al., 1978; Lechevalier et al., 1986; Okazaki et al., 1983), and from degraded polyester polyurethane (Pommer and Lorenz, 1986). *Pseudonocardia autotrophica* has been shown to have the capacity to degrade methoxyphenolic compounds (Haider et al., 1978; Malarczyk et al., 1987) and thus may have a role in the turnover of phenolic compounds derived from lignin degradation.

The relative biodiversity of toluene-degrading bacteria isolated from a compost biofilter treating toluene vapor was studied by Juteau et al. (1999). The toluene-degrading community was composed of *Pseudonocardia* and *Rhodococcus* strains that dominated usually faster-growing bacteria, such as members of the genera *Acinetobacter* and *Pseudomonas*. These authors considered that the actinomycetes might be *K*-strategists (adapted to a resource-restricted and crowded environment) and that the compost biofilter was a *K*-environment. These observations would explain why the actinomycetes were not outnumbered by the faster-growing acinetobacters and pseudomonads, which can be considered as *r*-strategists (adapted to a resource-abundant and undercrowded environment).

Pseudonocardia thermophila has been isolated from cattle manure and may play a part in the turnover of plant materials (Goddon and Penninckx, 1984; Henssen, 1989; Malfait et al., 1984). Goddon and Penninckx (1984) studied the succession of the microflora in composting cattle manure using a bench-scale reactor and observed that *Pseudonocardia thermophila* was one of the most abundant actinomycetes [approx. 10^6–10^7 colony forming units (c.f.u.)/g dry weight compost] when the temperature exceeded 40°C, although the numbers stabilized around 10^5–10^6 c.f.u./g after 15 d when the temperature fell to 10°C.

Morón et al. (1999) generated genus-specific primers for PCR identification of *Pseudonocardia* species. The primers were used to identify 106 strains presumptively assigned to the family *Pseudonocardiaceae* that had been isolated from geographically diverse soil and decomposing plant material samples. Nearly half of the isolates produced the genus-specific amplification product and partial 16S rRNA gene sequencing of representative isolates was used to validate their genus assignment. Fatty acid analyses of all of the isolates showed that they formed a heterogeneous group indicating that the genus *Pseudonocardia* is underspeciated.

Mutualistic associations. Attine leaf-cutting ants maintain two highly specialized, vertically transmitted, mutualistic symbionts, basidiomycete fungi (*Agaricales*: mostly *Lepiotaceae*: *Leucocopineae*) that are cultivated for food in underground gardens (Chapela et al., 1994; Martin, 1970; Weber, 1966, 1972), and pseudonocardiae, which produce antibiotics that specifically inhibit the growth of *Escovopsis* parasites (*Ascomycota*: anamorphic *Hypocreales*) of the fungal gardens (Cafaro and Currie, 2005; Currie et al., 1999a, 1999b, 2003a; Kost et al., 2007). The ants provide the fungus with fresh substrate and protection against competition and pathogens (Bass and Cherett, 1994; Mueller et al., 2004 ; North et al., 1997), and virgin ant queens carry their mother's symbiont when leaving their colony to mate. In return, the fungus produces nutrient-rich bodies (gongylidia), which workers harvest as a sole source of food for their larvae and the queen. The relationships between attine ants, their fungal cultivars, pseudonocardial mutualists, and cultivar antagonists is considered to be one that has evolved over 50 million years (Currie, 2001; Currie et al., 2003b, 2006; Mueller et al., 2001; Schultz and Brady, 2008).

The fungal gardens are propagated asexually as clonal monocultures and hence are susceptible to parasitism by micro-organisms that are competitively superior to the fungus cultivated by the ants (Weber, 1966), notably to a highly specialized parasite belonging to the genus *Escovopsis* (Currie et al., 1999a). Genetically diverse strains of the horizontally transmitted pathogen (Gerardo et al., 2006; Taerum et al., 2007) are capable of rapidly devastating fungal gardens and thereby threaten the survival of ant colonies (Currie et al., 1999a; Currie and Stuart, 2001). Bioassay and *in vivo* infection experiments indicate differences in the inhibitory capabilities of ant-associated *Pseudonocardia* strains, as well as variation in *Escovopsis* strain susceptibility to different antibiotics (Poulsen et al., 2007).

Pseudonocardial symbionts typically reside on the cuticle of ants in elaborate cuticular crypts linked to endocrine glands, which apparently produce nutrients to support bacterial growth (Currie et al., 2006). A degree of broad-scale co-evolution is considered to be apparent between *Pseudonocardia* and fungus-growing ants (Cafaro and Currie, 2005), though this relationship is not clear-cut due to the ability of *Pseudonocardia* strains to cross attine species boundaries, both within and between genera (Poulsen et al., 2005). These workers detected frequent *Pseudonocardia* exchanges between sympatric ant species within the genus *Acromyrineae*, thereby indicating that horizontal transmission had occurred. However, symbiotic recognition and behavioral choice may play a crucial role in the fungus-growing, ant-pseudonocardial mutualism (Bot et al., 2001; Mueller et al., 2004; Viana et al., 2001), possibly by recognizing clone-specific chemical signatures (Richard et al., 2007). The operation of such factors might allow ants to retain the ecological flexibility they need to defend their gardens from *Escovopsis* strains and, at the same time, resolve potential conflict that can arise from rearing competing symbiotic *Pseudonocardia* strains (Zhang et al., 2007b).

Symbiotic pairing assay experiments have revealed the presence and extent of antagonistic interactions between *Pseudonocardia* symbionts of fungus-growing ants (Poulsen et al., 2007). The widespread ability of ant-associated pseudonocardiae to inhibit one another suggests that competition will often arise when *Pseudonocardia* strains mix and that such interactions may have a direct impact on host/symbiont dynamics. Indeed, the detection of antagonistic interactions across the phylogenetic diversity of pseudonocardial symbionts indicates that such interactions may have shaped the ant/pseudonocardial symbioses from its inception. Antagonism between *Pseudonocardia* strains can be expected to prevent new strains from gaining a foothold in ant colonies, thereby enforcing single strain rearing within individual ant colonies. However, while ant recognition of *Pseudonocardia* strains may prevent conflict between them, it may also allow ants to replace their resident *Pseudonocardia* strain with an organism that produces novel antibiotics against the garden parasite, *Escovopsis*. The results of the experiments conducted by Poulsen and his colleagues support the view that the evolution of sociality required dramatic increases in anti-microbial defenses and that micro-organisms have been powerful selective agents (Stow and Beattie, 2008).

The coevolution of attine ants and pseudonocardiae has been reassessed by Mueller et al. (2008). They found that phylogenetic data from culture-dependent and -independent microbial surveys indicated close relationships between free-living and ant-associated pseudonocardiae, and a lack of topological correspondence between ant and *Pseudonocardia* phylogenies, indicating frequent pseudonocardial acquisition from environmental sources. Identity of ant-associated pseudonocardiae and isolates from plants and soil implicated them as sources from which attine ants acquire *Pseudonocardia* strains. These data are at variance with the prevailing views of specific coevolution between attine ants, pseudonocardiae, and garden diseases. Consequently, the effectiveness of pseudonocardial antibiotics may not be due to advantages in the coevolutionary arms race with specialized garden diseases, but from the frequent recruitment of effective *Pseudonocardia* strains from the environment. Current models of *Pseudonocardia*/disease coevolution need to be re-examined in light of these new findings.

Multiple lines of evidence shown that a black yeast, closely related to the genus *Phialophora* (Ascomycota), is the fifth symbiont to be recognized in the fungus-growing attine ant/microbe symbiosis (Little and Currie, 2007). The black yeasts, which form a monophyletic group, grow on the ants' cuticle and derive nutrients from the pseudonocardial symbiont (Little and Currie, 2008). Ants infected with black yeasts are significantly less effective at defending their fungal gardens from *Escovopsis*. It is possible that the reduction of mutualistic actinobacterial biomass in ants, probably caused by the black yeast symbionts, reduces the quantity of antibiotics available to inhibit the garden symbiont. However, the reduction in the ability of ants to inhibit *Escovopsis* indicates that it is an integral component of a complex symbiotic association which is not fully understood. Nevertheless, it seems clear that each member of the fungus-growing ant community is directly and/or indirectly influenced by the other partners in both positive and negative ways (Little and Currie, 2008; Poulsen and Currie, 2006).

Enrichment and isolation procedures

Most *Pseudonocardia* species have been isolated on nutrient-poor media under aerobic conditions at 20–30°C. Several species, including *Pseudonocardia xinjiangensis* and *Pseudonocardia yunnanensis*, have been recovered from soil samples collected from Xinjiang and Yunnan Provinces in China following inoculation of soil suspensions onto AV (Nonomura and Ohara, 1971) and HV (Hayakawa, 1990) agars and incubation at 28°C for 21–28 d (Xu et al., 1999). *Pseudonocardia ailaonensis* and *Pseudonocardia zijingensis* were also isolated from Yunnan soil samples, albeit on starch-casein and yeast extract-starch agars, respectively (Huang et al., 2002; Qin et al., 2008b). Similarly, *Pseudonocardia antarctica* was isolated on yeast extract-peptone agar (Prabahar et al., 2004a), and *Pseudonocardia kongjuensis* and *Pseudonocardia spinosispora* were isolated on tap water agar (Lee et al., 2001, 2002). *Pseudonocardia ammonioxydans* was isolated on a modified nitrifying medium inoculated with a coastal sediment suspension prior to incubation at 30°C for a month (Liu et al., 2006), and *Pseudonocardia oroxyli* was recovered from sterile root samples inoculated onto BL-2 agar plates supplemented with penicillin and incubated at 27°C for up to 4 weeks (Gu et al., 2006).

The remaining *Pseudonocardia* species that oxidize hydrocarbons or deleterious organic compounds can be selectively isolated either by enrichment culture or by plating suspensions onto media containing selected compounds as sole carbon sources. *Pseudonocardia asaccharolytica* and *Pseudonocardia sulfidoxydans* were isolated from samples enriched with methyl sulfide-containing gas (Reichert et al., 1998), *Pseudonocardia carboxydivorans* from soil enriched with low concentrations of CO (Park et al., 2008), and *Pseudonocardia chloroethenivorans* by using trichloroethene-contaminated air with phenol as a source of carbon for energy and growth (Lee et al., 2004). Similarly, *Pseudonocardia benzenivorans* was isolated on a medium containing 1,2,3,5-tetrachlorobenzene as a sole carbon source (Kämpfer and Kroppenstedt, 2004), *Pseudonocardia dioxanivorans* on a basal medium supplemented with THF and yeast extract (Parales et al., 1994), *Pseudonocardia nitrificans* on tap-water agar supplemented with urethan and trace metal salts (Schatz et al., 1954), and *Pseudonocardia tetrahydrofuranoxydans* on a medium enriched with THF (Kohlweyer et al., 2000). *Pseudonocardia petroleophila* and *Pseudonocardia saturnea* strains were isolated using CO_2 enrichments of mineral salts solutions seeded with compost (Hirsch, 1960), and *Pseudonocardia carboxydivorans* was isolated from soil enriched with low concentrations of CO (Park et al., 2008). *Pseudonocardia hydrocarbonoxydans* was isolated as an aerial contaminant on a silica gel plate (Nolof and Hirsch, 1962).

The type strain of *Pseudonocardia autotrophica* was isolated as an airborne contaminant of phosphate buffer that had been left standing next to a coal gas leak (Hirsch, 1960; Takamiya and Tubaki, 1956). *Pseudonocardia autotrophica*, in common with other autotrophic *Pseudonocardia* species, grows faster under heterotrophic conditions. Strains of *Pseudonocardia autotrophica*, "*Amycolata* (i.e. *Pseudonocardia*) *autotrophica* subsp. *amethystina*" and "*Amycolata* (i.e. *Pseudonocardia*) *autotrophica* subsp. *canberrica*" were isolated from suspensions of eucalyptus forest soil, lake side sand, and lake bottom mud using heterotrophic media supplemented with selective agents (Okazaki et al., 1983).

Henssen (1957) isolated the type strain of *Pseudonocardia thermophila* from fresh and rotten manure using a nutrient medium supplemented with cellulose dextrin as a carbon source (Fuller and Norman, 1942). The exact composition of this nutrient medium is not entirely clear from the original paper, however, as a number of variations were used. Subsequently, large numbers of *Pseudonocardia thermophila* strains were isolated from the heating phase of composting cattle manure after 20 d at 45°C on a cellulose medium supplemented with streptomycin sulfate (Goddon and Penninckx, 1984). *Pseudonocardia composta* and *Pseudonocardia spinosa* have been isolated from soil using nutrient-poor media, such as soil extract agar, and incubated at 20–30°C (Henssen, 1989; Henssen and Schäfer, 1971).

Maintenance procedures

Working cultures of *Pseudonocardia* can be maintained at 4°C on appropriate standard media. Long-term preservation of strains is best achieved as frozen stocks in 20% (v/v) aqueous glycerol at −20°C, −80°C (mechanical freezer), or −172°C (liquid nitrogen vapor phase), or by traditional lyophilization procedures.

Procedures for testing special characters

Two pairs of genus specific oligonucleotide primers, AMP3 (5′-GCGGCACAGAGACCGTGGAAT-3′)/AMP2 (5′-GTGGAAAGTTTTTTCGGCTGGGG-3′) and AMP4 (5′-GCGGCACAGAAACCGTGGAAT-3′)/AMP5 (5′-GTGGAAAGTTTTTTCGGTGGGGG-3′), can be used to identify members of the genus *Pseudonocardia* at annealing temperatures of 53 and 60°C respectively (Morón et al., 1999). The amplification conditions are: initial denaturation at 95°C for 4 min; 40 cycles of 30 s at 93°C, 30 s at 53°C for primer set AMP3/AMP2 or 30 s at 60°C for primer set AMP4/AMP5, and 2 min at 72°C; followed by 10 min at 72°C. Both primers sets yield a 640 bp genus-specific amplification product. The AMP3/AMP2 primer set is particularly effective for the detection and identification of *Pseudonocardia* strains at a lower annealing temperature.

Differentiation of the genus *Pseudonocardia* from other genera

The genus *Pseudonocardia* can be distinguished from other genera classified in the family *Pseudonocardiaceae* by using a combination of chemotaxonomic and morphological criteria (see Table 228 in the treatment of the family *Pseudonocardiaceae*). It also forms a distinct clade in the *Pseudonocardiaceae* 16S rRNA gene tree and can be distinguished from the phylogenetically close genera *Actinomycetospora* and *Kibdelosporangium* (Figure 278) as they contain MK-9(H_4) as the predominant isoprenologue and form pseudosporangia (*Kibdelosporangium*) or no aerial mycelium (*Actinomycetospora*). *Pseudonocardia* strains can also be recognized using genus-specific oligonucleotide primer sets (Morón et al., 1999) and can be distinguished from other genera classified in the family *Pseudonocardiaceae* based on partial amino acid sequencing of ribosomal AT-L30 proteins (Ochi, 1995).

Taxonomic comments

The genus *Pseudonocardia* was proposed to accommodate *Pseudonocardia thermophila*, a moderately thermophilic organism that showed distinctive morphological properties (Henssen, 1957). Morphologically similar strains were designated *Pseudonocardia compacta* (Henssen, 1957) and *Pseudonocardia spinosa* (Henssen and D. Schäfer, 1971; Schäfer, 1969). Another thermophilic organism, "*Pseudonocardia thermospinosa*", was isolated from soil samples from Hanoi, Vietnam (Lu and Yan, 1978), but the name was not validly published. The three species with validly published names could be distinguished from members of other genera of actinomycetes which formed aerial spores in chains by exhibiting acropetal budding of substrate and aerial hyphae (Henssen, 1970; Henssen and Schäfer, 1971; Henssen and Schnepf, 1967), and by forming a two-layered wall and interspace septa (Henssen et al., 1981). However, most of the 29 species with validly published names assigned to the genus have been circumscribed using a combination of genotypic and phenotypic properties. These taxa include species that were previously classified in the genera *Actinobispora* (Xu et al., 1999), *Amycolata* (Lechevalier et al., 1986), and *Pseudoamycolata* (Akimov et al., 1989).

The genus *Actinobispora* was proposed to encompass members of a single species, *Actinobispora yunnanensis*, which formed spores in longitudinal pairs on vegetative hyphae, in longitudinal pairs or singly on aerial hyphae, and which contained *meso*-A_2pm as the wall diamino acid, arabinose, galactose and xylose in whole-organism hydrolysates (an unusual wall chemotype IV), phosphatidylethanolamine and glucosamine-containing phospholipids (phospholipid type 1), and MK-7(H_2) and MK-9(H_2) as predominant isoprenologues, but lacked mycolic acids (Jiang et al., 1991). Subsequently, three additional species, *Actinobispora alaniniphila*, *Actinobispora aurantiaca*, and *Actinobispora xinjiangensis*, were described (Xu et al., 1999). Suzuki et al. (1998) selectively isolated *Actinobispora* strains on gellan gum plates. Huang et al. (2002) found that *Actinobispora* and *Pseudonocardia* species were intermixed in a distinct clade in the *Pseudonocardiaceae* 16S rRNA gene tree, shared key chemical markers, and gave a taxon-specific amplification product using PCR primers designed to differentiate the genus *Pseudonocardia* from related taxa. They also found that the type strains of the original *Actinobispora* species contained MK-8(H_4) as the major isoprenologue. Huang and her colleagues proposed that the genus *Actinobispora* should become a junior synonym of the genus *Pseudonocardia* and that *Actinobispora alaniniphila*, *Actinobispora aurantiaca*, *Actinobispora xinjiangensis*, and *Actinobispora yunnanensis* be transferred to this genus.

The genus *Amycolata* was proposed by Lechevalier et al. (1986) for a group of actinomycetes that formed substrate hyphae that tended to fragment into squarish subunits, and aerial hyphae which, when formed, were either sterile or fragmented into chains of squarish to oval elements or spore-like structures. The organisms had a wall chemotype IV, a type III phospholipid pattern, and MK-8(H_2, H_4) as predominant isoprenologues, but lacked mycolic acids. The genus provided a taxonomic home for organisms that had previously been misclassified in the genera *Nocardia* and *Streptomyces*. The type species, *Amycolata autotrophica*, included strains previously classified as *Nocardia autotrophica* (Hirsch, 1960) or "*Streptomyces autotrophicus*" (Takamiya and Tubaki, 1956), and also encompassed some strains of *Nocardia coeliaca* (Gordon et al., 1974). *Amycolata autotrophica* strains isolated from root nodules and rhizospheres of *Alnus* species were reclassified as *Amycolata alni* following DNA–DNA pairing studies (Evtushenko et al., 1989). Similarly, *Amycolata hydrocarbonoxydans* and *Amycolata saturnea* were initially described as *Nocardia* species by Nolof and Hirsch (1962) and Hirsch (1960), respectively. "*Streptomyces nitrificans*" and "*Nocardia petroleophila*" were also considered to be species of *Amycolata* based on 16S rRNA gene sequence data (Embley, 1992). Two subspecies of *Amycolata autotrophica* have been described but their names have not been validly published, namely "*Amycolata autotrophica* subsp. *amethystine*" and "*Amycolata autotrophica* subsp. *canberrica*".

In an extensive phylogenetic study of the family *Pseudonocardiaceae*, Warwick et al. (1994) found that the 16S rRNA gene sequences of the type strains of *Amycolata* and *Pseudonocardia* species formed a mixed, but distinct, phyletic branch. These results were in line with those from previous studies which showed that members of these genera contained phosphatidylcholine as a diagnostic phospholipid (Embley et al., 1988b; Lechevalier et al., 1977a, 1986), MK-8(H_4) as the predominant isoprenologue (Embley et al., 1988b; Lechevalier et al., 1986), formed a characteristic electron-dense layer in the substrate and aerial mycelia (Kothe et al., 1989), and contained ribosomal AT-130 proteins that had very similar electrophoretic mobilities (Ochi and Yoshida, 1991) and partial amino acid sequences (Ochi, 1995). Consequently, they proposed that the genus *Amycolata* Lechevalier et al. (1986) be recognized as a junior synonym of the genus *Pseudonocardia* Henssen (1957); they also provided an emended description of the genus, and proposed that *Amycolata alni*, *Amycolata autotrophica*, *Amycolata hydrocarbonoxydans*, *Amycolata saturnea*, "*Nocardia petroleophila*" and "*Streptomyces nitrificans*" be transferred to the genus *Pseudonocardia*.

The monospecific genus *Pseudoamycolata* was proposed for two strains that had previously been classified as *Amycolata autotrophica* (Akimov et al., 1989). The strains resembled *Amycolata* species in overall morphology and phenotype, and contained MK-8(H_4) as the major isoprenologue, but lacked phosphatidylcholine, the absence of which was judged to be sufficient to justify the proposal for a new genus. McVeigh et al. (1994) found that apart from the absence of phosphatidylcholine, *Pseudoamycolata halophobica* resembled *Pseudonocardia* species with respect to cultural, morphological, and physiological properties. They also noted that bacteria synthesized phosphatidylcholine by sequential methylation of phosphatidylethanolamine and speculated that minor mutations might prevent this transformation. These considerations led them to propose that *Pseudoamycolata halophobica* be reclassified in the genus *Pseudonocardia* as *Pseudonocardia halophobica*.

Nearly all of the *Pseudonocardia* species with validly published names form distinct phyletic lines in the *Pseudonocardia* 16S rRNA gene tree (Figure 278). However, comparative taxonomic studies, including DNA–DNA relatedness experiments, are needed to determine whether *Pseudonocardia chloroethenivorans* and *Pseudonocardia dioxanivorans* are distinct species as the type strains of these taxa have identical 16S rRNA gene sequences. The position of *Pseudonocardia spinosa* in the *Pseudonocardia* 16S rRNA tree remains to be established. It has been reported that the type strain of this organism is no longer extant (Embley, 1992), but this is not so as the strain is held in the JCM and NBRC collections.

Improvements in the classification of wall chemotype IV actinomycetes that lack mycolic acids have helped to clarify the taxonomy of the genus *Pseudonocardia* (Embley et al., 1988a, 1988b; Embley, 1992; Goodfellow et al., 1989b; Warwick et al., 1994) and have thereby facilitated the recognition and description of novel species (Gu et al., 2006; Kämpfer et al., 2006; Liu et al., 2006). The improved taxonomy also led to the reclassification of some species previously assigned to the genus. Ōmura et al. (1983) proposed *Pseudonocardia azurea* for an organism that produced two water-soluble antibiotics, azureamycin A and B (Ōmura et al., 1979). This organism was considered to show acropetal budding, as was "*Pseudonocardia fastidiosa*", which produced a macrobicyclic peptide antibiotic (Celmer et al., 1977). Henssen et al. (1987) were unable to confirm that these organisms exhibited acropetal budding, but did find that they had chemotaxonomic and morphological properties consistent with their reclassification in the genus *Amycolatopsis*, as *Amycolatopsis azurea* and *Amycolatopsis fastidiosa*.

It is clear that additional *Pseudonocardia* species remain to be described as it is evident that the genus is underspeciated (Bredholdt et al., 2007; Morón et al., 1999; Schabereiter-Gurtner et al., 2001; Song et al., 2001; Zhang et al., 2008a). It seems likely that the isolation of additional novel pseudonocardiae will be facilitated by the application of innovative selective isolation procedures (Li et al., 2002).

Differentiation of the species of the genus *Pseudonocardia*

The chemotaxonomic, morphological, and physiological properties shown in Table 229 can be used to distinguish between species of *Pseudonocardia*, but members of these taxa have yet to be examined using a common set of tests. The type strains of all but two *Pseudonocardia* species can be separated based on their 16S rRNA gene sequences and phylogeny, as seen in Figure 278.

Acknowledgements

The authors thank Dr T. Kudo (JCM) for providing the type strain of *Pseudonocardia spinosa*, and Dr J.P. Euzéby for clarifying the etymology of *Pseudonocardia nitrificans*. We are particularly indebted to Dr D.P. Labeda (USDA, Peoria) for critically reading our initial manuscript and thereby making many helpful suggestions on how it could be improved.

TABLE 229. Diagnostic characteristics for *Pseudonocardia* species[a]

Characteristic	1. *P. thermophila*	2. *P. ailaonensis*	3. *P. alaniniphila*	4. *P. alni*	5. *P. ammonioxydans*	6. *P. antarctica*	7. *P. asaccharolytica*	8. *P. aurantiaca*	9. *P. autotrophica*	10. *P. benzenivorans*	11. *P. carboxydivorans*	12. *P. chloroethenivorans*	13. *P. compacta*	14. *P. dioxanivorans*	15. *P. endophytica*	16. *P. halophobica*	17. *P. hydrocarbonoxydans*	18. *P. kongjuensis*	19. *P. nitrificans*	20. *P. oroxyli*	21. *P. petroleophila*	22. *P. saturnea*	23. *P. spinosa*	24. *P. spinosispora*	25. *P. sulfidoxydans*	26. *P. tetrahydrofuranoxydans*	27. *P. xinjiangensis*	28. *P. yunnanensis*	29. *P. zijingensis*
Fragmentation of:																													
Substrate mycelium	+	+	−	+	+	nd	+	−	+	+	nd	−	+	+	+	+	+	nd	nd	+	+	+	+	nd	+	nd	−	−	+
Aerial mycelium	+	+	−	+	+	+	+	−	+	+	+	−	+	+	−	+	+	+	+	+	+	+	+	+	+	+	−	−	+
Single spores	−	−	+	−	−	−	−	+	−	−	−	+	−	−	−	−	−	−	−	−	−	−	−	−	−	−	+	+	−
Pairs of spores	−	−	+	−	−	−	−	+	−	−	−	−	−	−	−	−	−	−	−	−	−	−	−	−	−	−	+	+	−
Acid produced from:																													
Adonitol	nd	+	−	+	+	−	−	−	+	nd	nd	−	−	+	−	−	−	+	+	nd	nd	−	nd	+	−	−	+	+	+
L-Arabinose	+	+	−	+	+	−	−	(+)	+	−	nd	+	−	−	−	+	+	−	nd	−	nd	+	−	+	+	nd	+	−	(+)
Cellobiose	+	+	+	−	−	−	−	+	+	−	−	+	+	−	−	−	+	+	nd	−	nd	+	−	−	−	nd	+	+	+
Erythritol	+	+	−	+	nd	−	+	−	+	−	nd	−	−	−	+	+	−	−	nd	−	nd	−	−	+	+	nd	−	−	−
Fructose	+	+	−	+	+	−	−	+	+	+	−	+	+	+	(+)	−	−	(+)	nd	+	nd	+	−	+	+	nd	+	+	−
Glucose	+	+	(+)	+	+	nd	−	−	+	+	nd	+	+	+	+	+	+	(+)	nd	+	nd	+	−	+	+	nd	+	−	+
Lactose	+	−	−	−	−	nd	−	−	−	−	−	+	−	−	−	−	+	−	nd	−	nd	−	−	−	−	nd	+	−	+
Maltose	+	+	−	+	+	+	−	+	+	−	−	+	−	−	+	+	−	+	nd	+	nd	+	−	−	−	nd	(+)	−	(+)
Mannitol	+	+	+	−	nd	(+)	−	+	+	−	−	+	+	−	−	+	+	+	nd	+	nd	+	−	−	−	nd	+	+	(+)
Rhamnose	+	+	+	−	−	+	−	+	−	nd	nd	+	−		−	+	+	nd	−	nd	nd	−	−	+	−	nd	+	+	+
Salicin	+	+	−	+	nd	nd	−	(+)	−	−	nd	+	−	−	+	−	+	−	nd	−	nd	−	−	−	−	nd	(+)	(+)	−
Sorbitol	+	−	−	+	−	−	−	−	+	−	−	−	−	−	+	−	−	−	nd	−	nd	−	−	−	−	nd	−	−	+
Trehalose	+	+	−	+	+	−	−	−	+	−	−	+	−	+	+	+	+	+	nd	+	nd	+	−	−	+	nd	−	−	+
D-Xylose	+	+	+	+	(+)	+	−	−	+	+	−	+	−	(+)	−	(+)	+	+	nd	−	nd	+	−	+	+	nd	+	+	+
Decomposition of:																													
Adenine		−	−	+	nd	nd	−	−	+	−	−	−	−	−	nd	(+)	−	nd	nd	+	nd	+		−	−		−	−	−
Hypoxanthine	nd	+	+	+	−	+	−	+	−	+	nd	+	−	+	+	+	−	+	nd	+	nd	+	−	−	+	nd	+	+	+
L-Tyrosine	−	+	−	+	+	+	−	−	+	−	nd	+	−	−	+	−	−	+	nd	+	nd	+	−	−	+	nd	−	−	+
Xanthine	nd	+	−	+	−	−	−	−	−	−	nd	+	−	+	nd	+	−	−	nd	+	nd	+	−	−	+	nd	−	(+)	+
Growth in NaCl (w/v):																													
3%	+	+	−	+	+	+	−	−	+	+	+	(+)	−	+	+	+	−	+	nd	+	+	+	−	−	+	+	−	−	+
4%	nd	+	−	+	+	+	−	−	+	+	+	−	−	+	+	+	−	+	nd	+	+	nd	−	−	−	+	−	−	+
5%	nd	+	−	+	+	+	−	−	+	−	−	−	−	(+)	+	(+)	−	+	nd	+	+	−	−	−	−	−	−	−	(+)
7%	nd	−	−	+	+	+	−	−	+	−	−	−	−	−	−	(+)	−	+	nd	(−)	+	−	−	−	−	−	−	−	−
Urease production	+	+	−	+	+	+	−	+	+	nd	+	−	−	nd	−	−	−	+	nd	−	nd	+	−	+	−	nd	+	−	−

[a]+, Positive; −, negative; (+), weakly positive; nd, not determined.

List of species of the genus *Pseudonocardia*

1. **Pseudonocardia thermophila** Henssen 1957, 408[VP]

ther.mo'phi.la. Gr. n. *thermê* heat; Gr. adj. *philus* loving; N.L. fem. adj. *thermophila* heat loving.

Substrate hyphae are septate, often zig-zag shaped; swellings are usually present and become divided by transverse and longitudinal septa. Aerial hyphae are often zig-zag shaped, young stages with constrictions; later on they are septate throughout; swellings are rarely present. Spores on substrate and aerial mycelium are formed by budding or secondary septation of hyphal segments. The inner wall layer in hyphae and spores is uniformly thin and is not thickened in mature spores. Good growth occurs on nutrient and yeast agars. Colonies are yellow and covered by thick white aerial hyphae. Good growth also occurs on asparagine-glycerol and yeast-glucose agars; colonies are yellow and covered by sparse aerial hyphae. Does not grow on oatmeal agar. Grows between at 28 and 60°C (optimally between 40 and 50°C).

Casein and starch are degraded, but not gelatin. H_2S is produced. Acid is produced from galactose, inositol, maltose, melezitose, and 1,2-propanediol, but not from mannose, methyl α-D-glucoside, or sucrose. Additional phenotypic features are shown in Table 229.

Source: isolated from fresh and rotten manure.

DNA G+C content (mol%): not determined.

Type strain: strain MB A-18, ATCC 19285, CBS 277.66, DSM 43832, JCM 3095, NBRC 15559, VKM Ac-896.

Sequence accession no. (16S rRNA gene): X53195.

2. **Pseudonocardia ailaonensis** Qin, Su, Zhang, Wang, Jiang, Xu and Li 2008b, 2088[VP]

ai.la.o.nen'sis. N.L. fem. adj. *ailaonensis* pertaining to Ailao Mountain, Yunnan province, China, the source of the soil sample from which the type strain was isolated.

Forms substrate hyphae that fragment into rod-shaped elements and aerial hyphae that differentiate into long chains of smooth-walled rod-shaped spores. Cream-white aerial hyphae are formed on Czapek, glucose-asparagine, potato, and yeast extract-malt extract agars. The color of the substrate mycelium is orange-yellow on Czapek agar, yellow-brown on glucose-asparagine, and deep orange-yellow on yeast extract-malt extract and potato agars; does not form diffusible pigments on any of these media. Melanin pigments are produced on peptone-yeast extract-iron agar. Grows between 15 and 37°C (optimum at 28°C). The pH range for growth is 6.0–8.0.

Does not liquefy gelatin, produce H_2S, lecithinase, or degrade either cellulose or starch. Milk is coagulated and peptonized. Reduces nitrate. Acid is produced from dulcitol (weak), galactose, glycerol, mannose, and sucrose, but not from inositol, D-lactulose, melezitose, methyl-D-glucoside, or raffinose. Grows on cellobiose, inositol, lactose, mannitol, melezitose (weak), oxalate, ribose, sorbitol, and sucrose as sole carbon sources, but not on acetate, galactose, or xylose. L-Arginine, L-cysteine, L-ornithine, L-threonine, and L-tyrosine are used as sole carbon sources.

The predominant fatty acids are $C_{16:0}$ iso (35.5%), $C_{16:0}$ iso 2-OH (10.8%), and $C_{16:0}$ 10-methyl (9.0%). The cellular polar lipid pattern contains diphosphatidylglycerol, phosphatidylglycerol, and phosphatidylinositol.

Source: the type strain was isolated from a soil sample collected from Ailao Mountain in Yunnan Province, south-west China.

DNA G+C content (mol%): 74.1 (HPLC).

Type strain: strain YIM 45505, DSM 44979, KCTC 19315.

Sequence accession no. (16S rRNA gene): DQ344632.

3. **Pseudonocardia alaniniphila** (Xu, Jin, Mao, Lu, Cui and Jiang 1999) Huang, Wang, Lu, Hong, Liu, Tan and Goodfellow 2002, 981[VP] (Basonym: *Actinobispora alaniniphila* Xu, Jin, Mao, Lu, Cui and Jiang 1999, 885)

a.la.ni.ni'phi.la. N.L. n. *alaninum* alanine; Gr. adj. *philus* loving; N.L. fem. adj. *alaniniphila* alanine loving.

Forms substrate and aerial hyphae that are branched, but which do not fragment. Spores are borne in longitudinal pairs on vegetative hyphae and either singly or in longitudinal pairs on aerial hyphae at 45°C. Substrate hyphae are orange-yellowish and aerial hyphae are sparse and pink-white. Neither diffusible nor melanin pigments are produced. Does not grow at 45°C.

Pectin is degraded, but not cellulose or starch. Nitrate is reduced. Does not produce H_2S. Cells utilize D-fructose, L-rhamnose, D-xylose, D-mannitol, alanine, L-histidine, and proline, but not adonitol cellobiose, raffinose, or inulin. Acid is not produced from these carbon sources. Alanine, L-histidine, and proline are used as sources of nitrogen. Resistant to neomycin and rifampicin (each at 50 μg/ml). Weak antifungal activity is shown against *Aspergillus niger*. Additional phenotypic features are shown in Table 229.

The cellular polar lipid pattern contains phosphatidylcholine, phosphatidylethanolamine, phosphatidylglycerol, phosphatidylmethylethanolamine, and glucosamine-containing phospholipids.

Source: the type strain was isolated from a soil sample collected at Xichou, Yunnan Province, China.

DNA G+C content (mol%): 69.3 (T_m).

Type strain: strain Y-16303, CCTCC AA 97001, CGMCC 4.1536, DSM 44660, JCM 11837.

Sequence accession no. (16S rRNA gene): EU722519.

4. **Pseudonocardia alni** (Evtushenko, Akimov, Dobritsa and Taptykova 1989) Warwick, Bowen, McVeigh and Embley 1994, 298[VP] (Basonym: *Amycolata alni* Evtushenko, Akimov, Dobritsa and Taptykova 1989, 76)

al'ni. L. gen. n. *alni* of alder, referring to the isolation of the type strain and some other strains from alder associations.

Substrate and aerial mycelia fragment into rod-shaped and oval elements. Abundant chains of spores (up to 1.5 μm in diameter) are formed on aerial hyphae. Substrate hyphae are characterized by the presence of swellings, polygonal shaped cells, and their conglomerates. Swollen hyphal segments (up to 2.5 μm in diameter and 10 μm long) are present and become divided by transverse and longitudinal septa. The aerial mycelium is white to cream colored and the substrate mycelium is orange to brown. Moderate growth occurs on water agar.

Degrades gelatin, starch, and Tweens 40, 60 and 80, but not casein or cellulose. Esculin is hydrolyzed, but not allantoin. Acid is produced from D-arabinose, fructose, galactose, glycerol, methyl α-D-glucoside, 1,2-propanediol, ribose, and sucrose, but not from dulcitol, galactose, inositol, inulin, mannose, melezitose, rhamnose, ribose, or sucrose. Acetate, aconitate, benzoate, citrate, formate, fumarate, 2-oxoglutarate, lactate, malate, maleate, propionate, pyruvate, sebacate, succinate, and *trans*-aconitate are assimilated. Arginine, asparagine, glutamine, histidine, leucine, ornithine, and proline are used as nitrogen sources. Grows in the presence of phenol (0.001%, w/v) and thymol (0.01%, w/v). Susceptible (μg/ml) to chloramphenicol (32), fucidin (10), gentamicin (4), kanamycin (16), lysozyme (50), monomycin (8), nalidixic acid (64), neomycin (4), novobiocin (0.25), rifampin (0.125), roseofungin (10), streptomycin (4), tetracycline (1), and vancomycin (0.5), but resistant to ampicillin (1–16), benzylpenicillin (0.5–10), carbenicillin (1–10), ristomycin (0.05), tobramycin (5), and vancomycin (0.25). Additional phenotypic features are shown in Table 229.

The major fatty acids are $C_{16:0}$ iso (36–40%) and $C_{17:0}$ anteiso acid (14–19%); also contains smaller amounts of $C_{16:1}$ iso (5.8–8.6%), $C_{16:0}$ (3.8–8.1%), $C_{17:0}$ iso (4.4–14.4%), $C_{17:0}$ (4.5–9.6%), $C_{18:0}$ methyl (3.5–6.9%), and $C_{18:0}$ anteiso (1.2–7.5). The cellular polar lipid pattern contains phosphatidycholine and phosphatidylethanolamine.

Source: isolated from root nodules and rhizospheres of alders [*Alnus glutionosa* (L.) Gaerth. and *Alnus incana* (L.) Moench.], and from marine sediments.

DNA G+C content (mol%): 72–74 (T_m).

Type strain: strain 3LS, DSM 44104, IFO (now NBRC) 14991, JCM 9103, NBRC 14991, VKM Ac-901.

Sequence accession no. (16S rRNA gene): AJ252823.

5. **Pseudonocardia ammonioxydans** Liu, Wu, Liu and Liu 2006, 556[VP]

am.mo.ni.o.xy'dans. N.L. n. *ammonia* ammonia; N.L. part. adj. *oxydans* oxidizing; N.L. part. adj. *ammonioxydans* oxidizing ammonia.

Forms aerial and substrate mycelia which fragment into rod-shaped elements. Short chains of smooth-surfaced spores are formed on the substrate mycelium by acropetal budding. A brown substrate mycelium and a white aerial mycelium are formed on Luria–Bertani and trypticase soy broth agars. Grows between 4 and 40°C.

Ammonia is oxidized to nitrate in a modified nitrifying medium (nitrification in MNM broth) and in blends of MNM and Luria–Bertani media (dissimilatory, heterotrophic ammonia oxidation). Degrades casein, gelatin (weak), and starch (weak). Produces H_2S. Acid is produced from *N*-acetyl-D-glucosamine, D-arabitol, D-galacturonic acid, gluconic acid, glycerol, 1,2-propanediol, and ribose, but not from *N*-acetyl-β-D-mannosamine, arbutin, L-fucose, D-tagatose, gentiobiose, inositol, α-D-lactose, lactulose, maltotriose, mannose, methyl α-D-galactoside, methyl β-D-galactoside, methyl α-D-glucoside, methyl α-D-mannoside, melezitose, palatinose, D-psicose, raffinose, sucrose, turanose, or xylitol.

Growth occurs in the presence of 0–8% NaCl, with optimal growth at 3.5% NaCl. Additional phenotypic features are shown in Table 229.

The major fatty acids are $C_{16:0}$ iso (41.1%), $C_{16:1}$ iso (15.7%), and $C_{17:1}$ ω8*c* (12.1%). The cellular polar lipid pattern contains diphosphatidylglycerol, phosphatidylcholine, phosphatidylglycerol, phosphatidylinositol mannosides, and phosphatidylmethylethanolamine, but not glucosamine-containing phospholipids.

Source: the type strain was isolated from coastal sediment collected from the Jiao-Dong peninsula, near Qingdao, Shandong Province, China.

DNA G+C content (mol%): 69.6 (T_m).

Type strain: strain H9, CGMCC 4.1877, JCM 12462.

Sequence accession no. (16S rRNA gene): AY500143.

Further comments: metabolizes diverse carbon compounds on Biolog GP2 microplates (Liu et al., 2006).

6. **Pseudonocardia antarctica** Prabahar, Dube, Reddy and Shivaji 2004b, 1005[VP] (Effective publication: Prabahar, Dube, Reddy and Shivaji 2004a, 69.)

an.tarc'ti.ca. L. fem. adj. *antarctica* southern, and by extension, pertaining to the continent Antarctica.

Filamentous brown substrate and aerial mycelia form a white conglomerate. Spores are formed by fragmentation of aerial hyphae. Grows at 7–38°C (optimum at 25°C) and pH 4–10. β-Galactosidase-, oxidase-, and phosphatase-positive, but negative for arginine dihydrolase, arginine decarboxylase, esculin, lipase, and lysine. Acid is produced from galactose, mannose, and ribose, but not from fructose, inositol, or melezitose.

Resistant to co-trimoxazole, nalidixic acid, and norfloxacin. Additional phenotypic features are shown in Table 229.

The major fatty acids are $C_{16:0}$ iso (30%) and $C_{16:0}$ (15%). The cellular polar lipid pattern contains diphosphatidylglycerol, phosphatidylcholine, phosphatidylethanolamine, phosphatidylglycerol, phosphatidylinositol, and phosphatidylmethylethanolamine.

Source: isolated from a moraine sample collected from the McMurdo Dry Valleys region of Antarctica.

DNA G+C content (mol%): 71 (T_m).

Type strain: strain DVS 5a1, DSM 44749, JCM 12172, MTCC 4297.

Sequence accession no. (16S rRNA gene): AJ576010.

Further comments: the specific epithet *antarctica* is a L. adj. not a N.L. adj. as cited in the paper by Prabahar et al. (2004a).

7. **Pseudonocardia asaccharolytica** Reichert, Lipski, Pradella, Stackebrandt and Altendorf 1998, 447[VP]

a.sac.cha.ro.ly'ti.ca. Gr. pref. *a* not; Gr. n. *sakchâr* sugar; Gr. fem. adj. *lutikê* able to dissolve, able to loosen; N.L. fem. adj. *asaccharolytica* referring to the failure to produce acid from carbohydrates.

Substrate hyphae are yellow on trypticase soy agar and fragment into coccoid and rod-shaped elements. Swollen hyphal segments up to 2 μm in diameter are formed, as are transverse hyphal septa. White, aerial hyphae carry long chains of oval spores. Spore surfaces are smooth. Dimethyl disulfide is oxidized. Acid is produced from 1,2-propanediol, but not from galactose, inositol, mannose, melezitose, methyl α-D-glucoside, sucrose, or xylitol. Additional phenotypic features are shown in Table 229.

The major fatty acids are $C_{16:0}$ iso (21.5–28.1%), $C_{16:0}$ (19.3–23.3%), $C_{17:0}$ iso (11.6–12.7%), $C_{15:0}$ iso (8.2–11.4%), and $C_{17:0}$ anteiso (4.8–6.1%); also contains small amounts of 10-methyl fatty acids. Mycolic acids are not formed. The cellular polar lipid pattern contains diphosphatidylglycerol, hydroxyphosphatidylethanolamine, phosphatidylglycerol, phosphatidylinositol, phosphatidylinositol mannosides, phosphatidylmethylethanolamine, and a ninhydrin-positive component.

Source: isolated from tree-bark compost biofilters from an animal-rendering plant.

DNA G+C content (mol%): not determined.

Type strain: strain 580, CIP 105685, DSM 44247, IFO (now NRBC) 16224, JCM 10410.

Sequence accession no. (16S rRNA gene): Y08536.

8. **Pseudonocardia aurantiaca** (Xu, Jin, Mao, Lu, Cui and Jiang 1999) Huang, Wang, Lu, Hong, Liu, Tan and Goodfellow 2002, 981[VP] (Basonym: *Actinobispora aurantiaca* Xu, Jin, Mao, Lu, Cui and Jiang 1999, 885.)

au.ran.ti'a.ca. N.L. fem. adj. *aurantiaca* orange-colored.

Forms substrate and aerial hyphae that are branched, but do not fragment. Spores are borne in longitudinal pairs on vegetative hyphae and either singly or in longitudinal pairs on aerial hyphae. The substrate mycelium is orange to orange-yellow and is covered by sparse, pink-white aerial hyphae. A brilliant yellow diffusible pigment is produced on oatmeal agar. Melanin pigments are formed. Does not grow at 45°C.

Pectin is degraded, but not cellulose or starch. Gelatin is liquefied. Milk is neither coagulated nor peptonized. Does not produce H_2S, lecithinase, or reduce nitrate. Cellulose, fructose, raffinose, and rhamnose are used as sole carbon sources, but not adonitol, inositol, mannitol, or xylose. Acid is not produced from these sugars. Alanine, L-histidine, and proline are used as sole nitrogen sources.

Shows weak antimicrobial activity against *Bacillus subtilis*. Resistant to neomycin, but not to rifampin (each at 50 µg/ml). Additional phenotypic features are shown in Table 229.

The cellular polar lipid pattern contains phosphatidylcholine, phosphatidylethanolamine, phosphatidylglycerol, and glucosamine-containing phospholipids.

Source: the type strain was isolated from a soil sample collected at Jianchuan, Yunnan Province, China.

DNA G+C content (mol%): 71.5 (T_m).

Type strain: strain Y-14860, CCTCC AA 97002, CGMCC 4.1537, JCM 11838.

Sequence accession no. (16S rRNA gene): AF056707.

9. **Pseudonocardia autotrophica** (Takamiya and Tubaki 1956) Warwick, Bowen, McVeigh and Embley 1994, 298[VP] [Basonym: *Nocardia autotrophica* Hirsch 1961, 362; "*Streptomyces autotrophicus*" Takamiya and Tubaki 1956, 59; *Amycolata autotrophica* (Takamiya and Tubaki 1956) Lechevalier, Prauser, Labeda and Ruan 1986, 34]

au.to.tro'phi.ca. Gr. pron. *autos* himself; N.L. fem. adj. *trophica* (from Gr. fem. adj. *trophikê*) nursing, tending or feeding; N.L. fem. adj. *autotrophica* self-nourishing, referring to the ability to grow at the expense of H_2 and CO_2.

Yellow to brown substrate hyphae fragment into squarish spores of unequal size, especially in liquid media. White to cream aerial hyphae differentiate into long chains of cylindrical to oval spores. Zig-zag mycelial growth and swollen hyphal segments are formed. Longitudinal and transverse septa are formed in the mycelium. Grows at 10–37°C, but not at 42°C. Autotrophic. Hydrogen is utilized in the presence of oxygen and carbon dioxide in a mineral salts medium.

Decarboxylates acetate, citrate, lactate, malate, propionate, pyruvate, and succinate, but not benzoate, mucate, oxalate, or tartrate. Produces phosphatase and H_2S. Does not hydrolyze esculin. Degrades testosterone, and Tweens 20, 40, 60 and 80, but not casein, gelatin, or starch. Acid is produced from galactose, inositol, mannose, melezitose, methyl α-D-glucoside, 1,2-propanediol, and sucrose.

Does not grow in lysozyme, or salicylate broth. Susceptible to (µg/ml) gentamicin (50), minocycline (50), rifampin (50), streptomycin (50), and vancomycin (50), but is resistant to erythromycin (50), fusidic acid (50), penicillin (10 IU). Additional phenotypic features are shown in Table 229.

The major fatty acids are $C_{16:0}$ iso (33.5%), $C_{17:0}$ iso (7.4%), and $C_{17:0}$ ω-methyl (10%). The cellular polar lipid pattern contains diphosphatidylglycerol, phosphatidylcholine, phosphatidylglycerol, phosphatidylinositol, phosphatidylglycerol, phosphatidylinositol, phosphatidylinositol mannoside, and phosphatidylmethylethanolamine, but not phosphatidylethanolamine.

Source: isolated from phosphate buffer solution, aluminum hydroxide gel, vegetable matter, soil, and clinical specimens.

DNA G+C content (mol%): 70 (T_m).

Type strain: ATCC 19727, CIP 107114, DSM 535, IFO (now NRBC) 12743, JCM 4348, NRRL B-11275, VKM Ac-941.

Sequence accession no. (16S rRNA gene): AJ252824.

10. **Pseudonocardia benzenivorans** Kämpfer and Kroppenstedt 2004, 751[VP]

ben.ze.ni.vo'rans. N.L. neut. n. *benzenum* benzene; L. part. adj. *vorans* devouring, digesting; N.L. fem. part. adj. *benzenivorans* digesting benzene.

Forms pale substrate hyphae that fragment into rod-shaped and coccoid elements. Aerial mycelium is white. Good growth on R2A and nutrient agars at 25–30°C after incubation for 3 d. Does not grow at 4–20 or 40–55°C. L-Alanine-*p*-nitroanilide (NA), *p*-nitrophenyl-(NP)-α-D-glucopyranoside, *p*NP-phenylphosphonate (weak), *bis*-*p*NP-phosphate, and L-proline-*p*NA are hydrolyzed, but not esculin, 2-deoxythymidine-5′-*p*NP-phosphate, *o*NP-β-D-galactopyranoside, *p*NP-β-D-glucopyranoside, L-glutamate-3-carboxy-*p*NA, *p*NP-phosphorylcholine, proline-*p*NA, or *p*NP-α-D-xylopyranoside. Acid is produced from galactose, glycerol, D-lactulose, and mannose, but not from dulcitol, inositol, melezitose, methyl α-D-glucoside, raffinose, sorbose, or sucrose. Grows on 4-aminobutyrate and 4-hydroxybutyrate (each at 10 mM). Additional phenotypic features are shown in Table 229.

The predominant fatty acids are $C_{16:0}$ iso (52.1%), $C_{16:1}$ iso *cis*9 (6.0%), $C_{17:0}$ iso (6.9%), $C_{17:1}$ *cis*9 (5.2%), and $C_{17:0}$

10-methyl (5.5%); small amounts of methyl-branched fatty acids are also present. The cellular polar lipid pattern contains diphosphatidylglycerol, phosphatidylethanolamine, and phosphatidylinositol, but not phosphatidylcholine.

Source: the type strain, which originated from a soil sample collected in Bitterfeld, Germany, was isolated from an enrichment culture containing 1,2,3,5-tetrachlorobenzene as the sole source of carbon.

DNA G+C content (mol%): not determined.

Type strain: strain B5, CCUG 49018, CIP 107928, DSM 44703, JCM 12694.

Sequence accession no. (16S rRNA gene): AJ556156.

Further comments: metabolizes a diverse range of carbon compounds on Biolog GP2 microplates (Kämpfer and Kroppenstedt, 2004)

11. **Pseudonocardia carboxydivorans** Park, Park, Lee and Kim 2008, 2477[VP]

car.bo.xy.di.vo'rans. N.L. neut. n. *carboxydum* carbon monoxide; L. part. adj. *vorans* devouring, digesting; N.L. part. adj. *carboxydivorans* digesting carbon monoxide.

Forms brown substrate mycelium and a white aerial mycelium. The aerial hyphae fragment into rod-shaped elements. Grows optimally at 25°C. Oxidizes carbon monoxide. Does not produce oxidase. Does not degrade gelatin or starch. Acid is produced from inulin, but not from cellobiose, inositol, maltose, mannitol, mannose, melezitose, ribose, trehalose, or xylose. *N*-Acetyl-D-glucosamine, inulin, and mannan are used as sole carbon sources, but not cellobiose, fructose, galactose, *myo*-inositol, maltose, mannitol, mannose, melezitose, raffinose, L-rhamnose, D-ribose, sorbitol, sucrose, trehalose, xylitol, or D-xylose. Additional phenotypic features are shown in Table 229.

The predominant fatty acids are $C_{16:0}$ iso (47.2%) and $C_{16:1}$ iso (22.8%); also contains smaller amounts of $C_{15:0}$ iso (5.0%), $C_{16:1}$ *cis*9 (5.9%), $C_{16:0}$ 10-methyl (4.6%), $C_{17:0}$ iso (3.0%), $C_{17:0}$ anteiso (2.6%), and $C_{17:1}$ *cis*9 (2.4%). The major menaquinone is MK-9.

Source: the type strain was isolated from soil collected at a roadside in Seoul, Korea.

DNA G+C content (mol%): 77 (HPLC).

Type strain: strain Y8, JCM 14827, KCCM 42678.

Sequence accession no. (16S rRNA gene): EF114314.

12. **Pseudonocardia chloroethenivorans** Lee, Strand, Stensel and Herwig 2004, 138[VP]

chlo.ro.e.the.ni.vo'rans. N.L. neut. n. *chloroethenum* chloroethene; L. part. adj. *vorans* devouring; N.L. fem. part. fem. adj. *chloroethenivorans* chloroethene-devouring.

Forms substrate and aerial hyphae that are branched, but do not fragment. Spores are formed at the ends of hyphae. Substrate hyphae are white. Diffusible pigments not formed. Grows on phenol as a source of carbon and energy. Chloroethane, *cis*1,2-dichloroethene, and trichloroethane are degraded. Additional phenotypic features are shown in Table 229.

The major fatty acids are $C_{16:0}$ iso (37.3%), $C_{15:0}$ iso (4.5%), $C_{16:0}$ (5.6%), $C_{16:0}$ 10-methyl (6.1%), $C_{17:0}$ iso (5.9%), $C_{17:0}$ anteiso (6.0%).

Source: the type strain was isolated from an aerobic laboratory enrichment in the Department of Civil and Environmental Engineering, University of Washington, Seattle, WA, USA.

DNA G+C content (mol%): not determined.

Type strain: strain SL-1, ATCC BAA-742, DSM 44698, JCM 12679.

Sequence accession no. (16S rRNA gene): AF454510.

Further comments: metabolizes a wide range of substrates on Biolog GP2 microplates (Lee et al., 2004).

13. **Pseudonocardia compacta** Henssen, Happach-Kasan, Renner and Vobis 1983, 834[VP]

com.pac'ta. L. fem. adj. *compacta* compact.

Substrate hyphae are septate and densely branched; swellings are common and, in part, multidivided. Aerial hyphae are compact, constricted or septate, and show apical and intercalary swellings. Spores of varying shapes and lengths are formed on substrate and aerial mycelia. The inner layer of hyphal walls vary in thickness. The inner layer of spore walls are thick in mature spores. Moderate to good growth occurs on artificial soil agar, with scanty yellow substrate hyphae and abundant white aerial hyphae. Optimum temperature for growth is 20–30°C.

Does not degrade casein, gelatin, or starch. Acid is not formed from galactose, mannose, melezitose, methyl α-D-glucoside, inositol, 1,2-propanediol, sucrose, or xylitol. Additional phenotypic features are shown in Table 229.

The cellular polar lipid pattern contains diphosphatidylglycerol, phosphatidylcholine, phosphatidylethanolamine, phosphatidylinositol, phosphatidylinositol mannosides, and phosphatidylmethylethanolamine. The major fatty acids are $C_{16:0}$ iso (34.5%), $C_{15:0}$ (10.0%), and $C_{17:0}$ (4.8%).

Source: isolated from garden soil collected in Wohra, near Marburg, Germany.

DNA G+C content (mol%): not determined.

Type strain: strain MB H-146, ATCC 35407, CBS 160.82, DSM 43592, IFO (now NRBC) 14343, JCM 7438, NRRL B-16170, VKM Ac-897.

Sequence accession no. (16S rRNA gene): AJ252825.

14. **Pseudonocardia dioxanivorans** Mahendra and Alvarez-Cohen 2005, 597[VP]

di.o.xa.ni.vo'rans. N.L. n. *dioxanum* dioxane; L. part. adj. *vorans* devouring; N.L. part. adj. *dioxanivorans* dioxane-devouring.

Forms white aerial hyphae and yellowish branched substrate hyphae which fragment into rod-shaped elements. Hyphal swellings and budding are sometimes observed. Does not form diffusible pigments. Optimal growth temperature is 30°C.

Aerobic autotrophic growth occurs on CO_2 and H_2. Fixes dinitrogen. Grows on 1,3- and 1,4-dioxane, butyl methyl ether, 2-methyl-1,3-dioxolane, THF, and tetrahydropyran. Acid is produced from glycerol, but not from dulcitol, inositol, lactulose, mannose, melezitose, or sucrose. Additional phenotypic features are shown in Table 229.

The major fatty acids are $C_{16:0}$ iso (27.5%), $C_{16:1}$ iso *cis*9 (9.3%), and $C_{17:1}$ iso *cis*9 (23.6%). The predominant menaquinone is MK-8(H_4); also contains a small amount of MK-7(H_4).

Source: isolated from a 1,4-dioxane-contaminated sludge sample collected at Darlington, South Carolina, USA (Parales et al., 1994).

DNA G+C content (mol%): 74 (HPLC).

Type strain: strain CB1190, ATCC 55486, DSM 44775, JCM 13855.

Sequence accession no. (16S rRNA gene): AY340622.

Further comments: metabolizes a diverse range of carbon compounds on Biolog GP2 and SEP2 microplates, and nitrogen sources on Biolog PM3 plates (Mahendra and Alvarez-Cohen, 2005).

15. **Pseudonocardia endophytica** Chen, Qin, Li, Zhang, Xu, Jiang, Kim and Li 2009, 559[VP]

en.do.phy'ti.ca. Gr. *endo* within; Gr. n. *phyton* plant; L. fem. suff. *-ica* adjectival suffix used with the sense of belonging to; N.L. fem. adj. *endophytica* within plant, endophytic, pertaining to the original isolation from plant tissues.

Forms yellowish-brown substrate mycelium and white aerial mycelium on tryptic soy and yeast extract-malt extract agars. The substrate mycelium fragments into rod-shaped elements. Spore chains form on aerial mycelium. Does not form diffusible pigments. Growth occurs at 15–37°C and pH 6.0–8.0.

Catalase-positive and oxidase-negative. Malonate and ornithine decarboxylase tests are positive, but negative for the gluconate test, H_2S production, milk coagulation and peptonization, and nitrate reduction. Does not degrade casein, cellulose, gelatin, or starch. Acid is produced from arbutin, esculin, galactose, glycerol, melezitose, D-ribose, sucrose, trehalose, and turanose, but not from amygdalin, arabitol, dulcitol, fucose, gentiobiose, gluconate, glycogen, inositol, inulin, 2-ketogluconate, lyxose, mannose, melibiose, methyl α-D-glucoside, methyl α-D-mannoside, methyl β-D-xyloside, *N*-acetylglucosamine, raffinose, sorbose, starch, tagatose, or xylitol. Utilizes L-asparagine, L-glutamic acid, *p*-hydroxyphenylacetic acid, α-ketoglutaric acid, α-ketovaleric acid, D-malic acid, *N*-acetyl-L-glutamic acid, propionic acid, putrescine, L-pyroglutamic acid, pyruvic acid, succinamic acid, and succinic acid, but not acetic acid, L-alaninamide, D- or L-alanine, L-alanyl glycine, glycyl L-glutamic acid, α-, β- or γ-hydroxybutyric acid, lactamide, L-lactic acid, L-malic acid, methyl pyruvate, monomethyl succinate, pyruvic acid, or L-serine. Additional phenotypic features are shown in Table 229.

The cellular polar lipid pattern contains diphosphatidylglycerol, phosphatidylcholine, phosphatidylethanolamine, and phosphatidylmethylethanolamine. The major fatty acids are $C_{16:0}$ iso (34.0%), $C_{17:1}$ *cis*9 (14.5%), and $C_{15:0}$ iso (9.6%). The predominant menaquinone is MK-8 (H_4); also contains traces of MK-7(H_6) and MK-9.

Source: isolated from a plant sample collected from Xishuangbanna, Yunnan Province, south-western China.

DNA G+C content (mol%): 70.3 (HPLC).

Type strain: strain YIM 56035, CCTCC AA 206026, DSM 44969, KCTC 19150.

Sequence accession no. (16S rRNA gene): DQ887489.

16. **Pseudonocardia halophobica** (Akimov, Evtushenko and Dobritsa 1989) McVeigh, Munro and Embley 1994, 302[VP] (Basonym: *Pseudoamycolata halophobica* Akimov, Evtushenko and Dobritsa 1989, 460)

ha.lo.pho'bi.ca. Gr. n. *hals* halos salt, the sea; Gr. n. *phobos* fear, dread; L. fem. suff. *-ica* suffix used with the sense of pertaining to; N.L. fem. adj. *halophobica* salt-fearing, referring to the inability to grow in the presence of 3% NaCl.

Forms substrate and aerial hyphae which tend to fragment into rod-shaped and oval elements. Shows zig-zag mycelial growth and longitudinal and transverse hyphal septa. Short chains of spores may be produced on aerial hyphae. Swollen hyphal segments (up to 3 μm long) are formed. The aerial mycelium is white to cream colored and the substrate mycelium ranges from yellowish orange to brown, depending on the medium. An orange substrate mycelium is formed on peptone-yeast extract agar.

Degrades Tweens 40 and 60. Arbutin and esculin are hydrolyzed, but not allantoin. Decarboxylates aconitate, formate, lactate, mannose, malate, propionate, pyruvate, succinate, sebacate, and *trans*-aconitate, but not maleate or salicylate. Gaseous aliphatic hydrocarbons (C6–C14) are used for growth. Grows on 4-aminobutyrate and 4-hydroxybutyrate (each at 10 mM). Acid is produced from inositol and sucrose, but not from dulcitol, glycerol, D-lactulose, mannose, melezitose, methyl α-D-glucoside, raffinose, or sorbose. Alanine, arginine, hydroxyproline, ornithine, phenylalanine, proline, threonine, tryptophan, and valine are used as sole nitrogen sources, but not glycine, histidine, leucine, lysine, or tyrosine.

Susceptible to (μg/ml) fucidin (10), lincomycin (1), methacycline (1), novobiocin (50), penicillin (10), polymyxin B (20), roseofungin (10), and tetracycline (10). Does not grow in the presence of crystal violet (0.000001, w/v); azide (0.001%, w/v) or sodium chloride (3%, w/v). Additional phenotypic features are shown in Table 229.

Major fatty acids are $C_{16:0}$ iso (19–26%), $C_{17:0}$ anteiso (17–21%), and $C_{15:0}$ iso (5.5–9%); also contains smaller amounts of $C_{15:0}$, $C_{16:0}$, $C_{16:1}$, $C_{17:0}$ 10-methyl, and $C_{18:0}$ 10-methyl. The cellular polar lipid pattern contains diphosphatidylglycerol, hydroxyphosphatidylethanolamine, phosphatidylethanolamine, phosphatidylglycerol, phosphatidylinositol, and phosphatidylinositol mannoside; also contains small amounts of phosphatidylmethylethanolamine and a ninhydrin-positive component, but does not contain either phosphatidylcholine or glucosamine-containing phospholipids.

Source: isolated from soil.

DNA G+C content (mol%): 72 (T_m).

Type strain: strain SS1/1, ATCC 51535, DSM 43089, IFO (now NRBC) 15408, IMRU 1300, JCM 9421, NRRL B-16514, VKM Ac-1069.

Sequence accession no. (16S rRNA gene): AJ252827.

17. **Pseudonocardia hydrocarbonoxydans** (Nolof and Hirsch 1962) Warwick, Bowen, McVeigh and Embley 1994, 298[VP] [Basonym: *Nocardia hydrocarbonoxydans* Nolof and Hirsch 1962, 267; *Amycolata hydrocarbonoxydans* (Nolof and Hirsch 1962) Lechevalier, Prauser, Labeda and Ruan 1986, 34]

hy.dro.car.bo.no'xy.dans. Gr. n. *hudôr* water; L. n. *carbo -onis* coal, charcoal; N.L. part. adj. *oxydans* oxidizing; N.L. pres. part. *hydrocarbonoxydans* oxidizing hydrocarbons.

Zig-zag mycelial growth and swollen hyphal segments are evident. Longitudinal and transverse hyphal septa are formed. Aerial and substrate hyphae may fragment into squarish elements. Sparse to moderate white aerial hyphae are formed. Aerial hyphae are white and sparse to moderate; substrate hyphae are off-white, yellowish white, gold, or brown depending on the growth medium. Grows at 10–37°C, but not at 45°C.

Gaseous aliphatic hydrocarbons (C6–C14) are used for growth. Grows on 4-aminobutyrate and 4-hydroxybutyrate (each at 10 mM). Positive for catalase, phosphatase, and esculin. Decarboxylates lactate, propionate, and pyruvate, but not benzoate, mucate, or tartrate. Does not degrade casein, or grow in lysozyme or salicylate broths. Acid is produced from inositol and sucrose, but not from dulcitol, glycerol, D-lactulose, mannose, melezitose, methyl α-D-glucoside, raffinose, or sorbose. Additional phenotypic features are shown in Table 229.

L-Alanine-*p*NA, *o*NP-β-galactopyranoside (weak), *p*NP-α- and *p*NP-β-D-glucopyranoside, *p*NP-phenylphosphonate, *bis*-*p*NP-phosphate, L-proline, *p*NA-glucopyranoside, and *p*NP-β-D-xylopyranoside are hydrolyzed, but not 2-deoxythymidine-5′-*p*NP-phosphate, esculin, L-glutamate-γ-3-carboxy-*p*-NA, or *p*NP-phosphorylcholine.

The predominant fatty acids are $C_{15:0}$ iso (10.9%), $C_{16:0}$ iso (30.8%), and $C_{16:0}$ (25.8%). The cellular polar lipid pattern contains diphosphatidylglycerol, phosphatidycholine, phosphatidylglycerol, phosphatidylinositol, phosphatidylinositol mannosides, and phosphatidylmethylethanolamine, but not phosphatidylethanolamine. The predominant menaquinones are MK-8(H_2) and MK-8(H_4).

Source: air contaminant; isolated from a silica gel plate.

DNA G+C content (mol%): 69 (T_m).

Type strain: ATCC 15104, DSM 43281, IFO (now NRBC) 14498, JCM 3392, NRRL B-16171, VKM Ac-799.

Sequence accession no. (16S rRNA gene): AJ252826.

Further comments: metabolizes a diverse range of substrates on Biolog GP2 and SFP2 microplates (Mahendra and Alvarez-Cohen, 2005).

18. **Pseudonocardia kongjuensis** Lee, Kim, Min, Lee, Kang and Hah 2001, 1509[VP]

kong.ju.en′sis. N.L. fem. adj. *kongjuensis* of Kongju, Republic of Korea.

Forms an abundant yellowish brown substrate mycelium and a white aerial hyphae mycelium that fragments into rod-shaped spores. The spore surface is smooth. Growth occurs between 4 and 37°C.

Catalase-positive. Produces H_2S. Casein is degraded, but not gelatin or starch. Acid is produced from galactose, glycerol (weak), inositol, mannose, melezitose, and sucrose, but not from 2,3-butanediol, dulcitol, melibiose, methyl α-D-glucoside, methyl α-D-mannoside, 1,2-propanediol, raffinose, sorbose, or D-xylitol. Additional phenotypic features are shown in Table 229.

The major fatty acids are $C_{16:0}$ iso (31.4%), $C_{17:1}$ (13.8%), $C_{18:0}$ (9.6%), $C_{16:1}$ (7.4%), $C_{16:0}$ (6.0%), $C_{16:1}$ iso (6.7%), and $C_{17:1}$ iso (6.2%). The cellular polar lipid pattern contains diphosphatidylglycerol, phosphatidylcholine, phosphatidylethanolamine, phosphatidylglycerol, and phosphatidylinositol. The predominant menaquinone is MK-8(H_4).

Source: isolated from a gold mine cave in Kongju, Republic of Korea.

DNA G+C content (mol%): 71 (HPLC).

Type strain: strain LM 157, DSM 44525, IMSNU 50583, JCM 11896, KCTC 9990, NBRC 100380.

Sequence accession no. (16S rRNA gene): AJ252833.

19. **Pseudonocardia nitrificans** (*ex* Schatz, Isenberg, Angrist and Schatz 1954) Warwick, Bowen, McVeigh and Embley 1994, 298[VP] ("*Streptomyces nitrificans*" Schatz, Isenberg, Angrist and Schatz 1954, 2)

ni.tri.fi′cans. N.L. part. adj. *nitrificans* nitrifying.

Produces straight, branched, sporulating aerial hyphae. Grows on various simple and complex agar media. Grows better at 25–30°C than at 37°C. Develops a brick-red mycelium without hemolysis on blood agar, and a gray sporulated surface with a pink to buff reverse on other media; does not produces diffusible pigments. Forms a well sporulating pellicle in nutrient broth and other liquid media. Growth is gnarled and wrinkled on beet, carrot, and potato slants.

Urethane is used as a sole source of carbon, energy, and nitrogen. Produces nitrite from carbamate, but does not oxidize urethane nitrogen beyond nitrite. Slowly alkalinizes milk without curdling. Does not degrade cellulose or produce indole. Ethanol, fumarate, lactate, and *n*-propanol are used as sole carbon sources. Ammonia, guanidine, nitrate, nitrite, purines, urea, and several amino acids are used as sole nitrogen sources. Additional phenotypic features are shown in Table 229.

Source: isolated from soil enriched with urethane as a sole carbon, nitrogen, and energy source.

DNA G+C content (mol%): not determined.

Type strain: IFAM 379.

Sequence accession no. (16S rRNA gene): X55609.

20. **Pseudonocardia oroxyli** Gu, Luo, Zheng, Liu and Huang 2006, 2194[VP]

o.ro.xy′li. N.L. gen. n. *oroxyli* of the plant genus *Oroxylum*.

Forms aerial and substrate hyphae which fragment into smooth rod-shaped elements. Swellings are produced at hyphal tips. The cinnamon-buff substrate mycelium carries pinkish cinnamon aerial hyphae. Grows well on glucose-yeast extract-malt extract agar after 3 d at 25–30°C. Does not produce diffusible pigments.

Acid is produced from galactose, glycerol, mannose, melezitose, and sucrose, but not from dulcitol, inositol, D-lactulose, or D-sorbose. Acetate, L-alanine, L-arabinose, L-arginine, cellobiose, citrate, fructose, galactose, glucose, glutamic acid (weak), glycerol, glycogen, inulin, inositol, lactose, D-lactulose, L-leucine, malate, mannitol, maltose, mannose, melezitose, methyl α-D-glucoside, L-methionine, L-ornithine, oxalate (weak), L-phenylalanine, L-proline, raffinose, rhamnose, ribose, salicin (weak), sorbitol, succinate, sucrose, trehalose, L-valine, and D-xylose are used as sole carbon sources, but not L-arginine, L-cysteine, L-leucinamide, L-tyrosine, or malonate. L-Ornithine is used as a sole nitrogen source, but not L-cysteine, L-threonine, or L-tyrosine. Additional phenotypic features are shown in Table 229.

The predominant fatty acid is $C_{16:0}$ iso (45.1%); smaller amounts of $C_{16:0}$ 10-methyl (11.1%), $C_{17:1}$ ω6*c* (7.4%), iso-H-$C_{16:1}$ (7.1%), $C_{18:1}$ ω9*c* (5.1%), $C_{15:0}$ iso (4.4%), $C_{16:1}$ ω7*c*

(4.3%), $C_{14:0}$ iso (3.5%), and $C_{16:0}$ (3.0%) are also formed. The cellular polar lipid pattern contains phosphatidylethanolamine, phosphatidylinositol, phosphatidylmethylethanolamine, two glucosamine-containing phospholipids of unknown structure, and two glycolipids.

Source: the type strain was isolated from a surface-sterilized root of *Oroxylum indicum* collected in the rain forest around Liusha River, southwest of Jinghong City, Yunnan Province, China.

DNA G+C content (mol%): 70.6 (T_m).

Type strain: strain D10, CGMCC 4.3143, DSM 44984, JCM 13909.

Sequence accession no. (16S rRNA gene): DQ343154.

Further comments: metabolizes a diverse range of carbon compounds on Biolog GP2 and SEP2 microplates, and nitrogen sources on Biolog PM3 plates (Mahendra and Alvarez-Cohen, 2005).

21. **Pseudonocardia petroleophila** (Hirsch and Engel 1956) Warwick, Bowen, McVeigh and Embley 1994, 298[VP] (Basonym: *Nocardia petroleophila* Hirsch and Engel 1956, 445.)

pe.tro.le.oph'il.a. Gr. n. *petra* stone; L. n. *oleum* oil; Gr. adj. *philus* loving; N.L. fem. adj. *petroleophila* loving stone oil, petroleum.

Substrate mycelium is yellowish, citron, or ochre-yellow depending on the carbon or nitrogen source. It fragments into bacillary elements under autotrophic and heterotrophic growth conditions, but much more rapidly in the latter case. Abundant snow-white aerial hyphae, which fragment into spore-like fragments with age, are produced on mineral media. Optimal temperature for growth is 25–28°C; slight growth is evident at 37°C, but does not grow at 50°C. Grows at pH 6.8–8.3. Grows on certain organic media, but without aerial mycelium production and with rapid fragmentation of substrate mycelium. Grows slowly but abundantly on all mineral salts media. Growth rate increases in a petroleum atmosphere. Atmospheric carbon dioxide is used as a carbon and energy source. Does not coagulate or peptonize milk. Grows in the presence of 10% NaCl. Additional phenotypic features are shown in Table 229.

Source: isolated from soil collected in Germany.

DNA G+C content (mol%): not determined.

Type strain: ATCC 15777, CIP 104515, DSM 43193, IFAM 78, IFO (now NRBC) 14406, JCM 3378, JCM 3394, NRRL B-16301, VKM Ac-865.

Sequence accession no. (16S rRNA gene): AJ252828.

22. **Pseudonocardia saturnea** (Hirsch 1960) Warwick, Bowen, McVeigh and Embley 1994, 298[VP] [Basonym: *Nocardia saturnea* Hirsch 1960, 401; *Amycolata saturnea* (Hirsch 1960) Lechevalier, Prauser, Labeda and Ruan 1986, 34.]

sa.tur'ne.a. L. n. *Saturnus* Saturn, Roman god of seed sowing; N.L. fem. adj. *saturnea* pertaining to Saturn, referring to the colonies which have a Saturnian shape.

Yellowish white to bright butter yellow substrate hyphae fragment into rod-shaped elements, especially on organic media. Aerial hyphae are white to yellowish and fragment into long rectangular elements on nutritionally poor media. Grows at 10–37°C (optimum at 28–30°C), but does not grow at 42°C. CO_2 is used as a sole source of carbon for energy and growth. Positive for catalase, esculin, and phosphatase. Does not degrade casein. Decarboxylates benzoate, lactate, propionate, and pyruvate, but not mucate or tartrate. Acid is produced from inositol, mannose, melezitose, methyl α-D-glucoside, 1,2-propanediol, and sucrose, but not from galactose or xylitol. Additional phenotypic features are shown in Table 229.

The cellular polar lipid pattern contains phosphatidycholine and phosphatidylmethylethanolamine, but not phosphatidylethanolamine. The predominant menaquinones are MK-8(H_2) and MK-8(H_4).

Source: isolated from air and compost in Germany.

DNA G+C content (mol%): 72 (T_m).

Type strain: ATCC 15809, DSM 43195, IFO (now NRBC) 14499, IMRU 1181, JCM 3187, NRRL B 16172, VKM Ac 781.

Sequence accession no. (16S rRNA gene): AJ252829.

23. **Pseudonocardia spinosa** Schäfer *in* Henssen and Schäfer 1971, 31[AL]

spi.no'sa. L. fem. adj. *spinosa* spiny.

Substrate hyphae are compact, irregularly branched, constricted, and septate; swellings are common. Aerial hyphae are constricted or septate. Spores are formed on substrate and aerial mycelia either by budding or secondary septation of hyphal segments. Inner wall layers of hyphae and spores vary in thickness. Grows extremely slowly. Moderate to good growth occurs on oatmeal and yeast-starch agars. Yellow colonies covered by abundant white aerial mycelium are formed on asparagine-glycerol agar. Optimal temperature for growth is between 20 and 30°C. Additional phenotypic features are shown in Table 229.

Source: isolated from soil collected in Turkey.

DNA G+C content (mol%): not determined.

Type strain: strain MB SH-1, ATCC 25924, CBS 818.70, IFO (now NRBC) 16002, JCM 3136.

Sequence accession no. (16S rRNA gene): not determined.

24. **Pseudonocardia spinosispora** Lee, Kim, Kang and Hah 2002, 1607[VP]

spi.no.si.spo'ra. L. adj. *spinosus* thorny; Gr. fem. n. *spora* seed; N.L. fem. n. *spora* spore; N.L. fem. n. *spinosispora* thorny spore.

Substrate hyphae are sparse or absent on most cultivation media; the reverse color is deep brown in older cultures. A well developed white aerial mycelium fragments into rod-shaped elements (approx. 0.54 × 2.25 µm), which are covered with numerous spines. Growth occurs in standing, but not in shaken, cultures. Grows at temperatures between 4 and 30°C.

Produces H_2S, but does not degrade casein, gelatin, or starch. Acid is produced from glycerol, inositol, mannose, and D-xylitol, but not from 2,3-butanediol, dulcitol, galactose, melezitose, melibiose, methyl α-D-glucoside, methyl α-D-mannoside, 1,2-propanediol, raffinose, sorbose, or sucrose. Additional phenotypic features are shown in Table 229.

The major fatty acids contain $C_{16:0}$ iso (33.8%), $C_{15:0}$ iso (13.8%), $C_{17:0}$ iso (10.0%), $C_{17:1}$ (9.5%), $C_{16:1}$ iso (7.4%), $C_{17:1}$ iso (7.2%), and $C_{16:1}$ iso (6.0%); also contains small amounts of 10-methyl-branched fatty acids, but not tuberculostearic acid or hydroxy fatty acids. The cellular polar lipid pattern contains diphosphatidylglycerol, phosphatidylcholine, phosphatidylethanolamine, phosphatidylglycerol, phosphatidylinositol, phosphatidylinositol mannosides, phosphatidylmethylethanolamine, hydroxyphosphatidylethanolamine, and an unknown phospholipid.

Source: isolated from soil from a gold mine cave near Kongju City, Korea.

DNA G+C content (mol%): 70.4 (HPLC).

Type strain: strain LM 141, IMSNU 50581, KCTC 9991, NRRL B-24156.

Sequence accession no. (16S rRNA gene): AJ249206.

25. **Pseudonocardia sulfidoxydans** Reichert, Lipski, Pradella, Stackebrandt and Altendorf 1998, 448[VP]

sul.fi.do'xy.dans. N.L. n. *sulfidum* sulfide; N.L. part. adj. *oxydans* oxidizing; N.L. part. adj. *sulfidoxydans* oxidizing sulfides.

Aerial and substrate mycelia fragment into rod-shaped elements. Hyphal segments are sometimes swollen up to 5 µm. Shows zig-zag mycelial growth and longitudinal and transverse hyphal septa. Yellow substrate hyphae and white aerial hyphae, which carry long chains of oval spores, are formed on trypticase soy agar.

Acid is produced from galactose, glycerol, mannose, and sucrose, but not from dulcitol, inositol, D-lactulose, melezitose, methyl α-D-glucoside, D-sorbose, or raffinose. L-Aniline-*p*NA, *o*NP-β-D-galactopyranoside, *p*NP-α-D-glucopyranoside, *bis*-*p*NP-phosphate, *p*NP-phenyl-phosphate, and L-proline-*p*NA are hydrolyzed, but not 2-deoxythymidine-5′-*p*NP-phosphate, esculin, *p*NP-β-D-glucopyranoside, L-glutamate-γ-3-carboxy-*p*NA, or *p*NP-phosphorylcholine. Dimethyl disulfide and dimethyl sulfide are oxidized. Grows on 4-hydroxybutyrate and 4-aminobutyrate (10 mM each). Additional phenotypic features are shown in Table 229.

The major fatty acids are $C_{16:0}$ iso (24.8–32.3%), $C_{16:0}$ (15.0–30.5%), and $C_{15:0}$ iso (11.2–16.4%). Mycolic acids are not formed. The cellular polar lipid pattern contains diphosphatidylglycerol, hydroxyphosphatidylethanolamine, phosphatidylethanolamine, phosphatidylglycerol, phosphatidylinositol, phosphatidylinositol mannosides, phosphatidylmethylethanolamine, and a ninhydrin-positive component.

Source: isolated from tree-bark compost biofilters from an animal-rendering plant.

DNA G+C content (mol%): not determined.

Type strain: strain 592, CIP 105686, DSM 44248, IFO (now NRBC) 16205, JCM 10411.

Sequence accession no. (16S rRNA gene): Y08537.

Further comments: metabolizes a diverse range of substrates on Biolog GP2 and SFP2 microplates and nitrogen sources on Biolog PM3 plates (Mahendra and Alvarez-Cohen, 2005).

26. **Pseudonocardia tetrahydrofuranoxydans** Kämpfer, Kohlweyer, Thiemer and Andreesen 2006, 1536[VP]

te.tra.hy.dro.fu.ra.no'xy.dans. N.L. n. *tetrahydrofuranum* tetrahydrofuran; N.L. v. *oxydare* to make acid, to oxidize; N.L. part. adj. *tetrahydrofuranoxydans* oxidizing tetrahydrofuran.

Forms branched filaments (e.g. 1.3 µm in width) that produce cell aggregates in THF-containing medium. Single spore-like bodies are formed at the end of cells. Aerial mycelium on agar is white, branched, and becomes fragmented. Good growth occurs on nutrient and R3A agars after 3 d at 25–30°C. Growth on THF occurs between 11 and 36°C.

L-Aniline-*p*NA and L-proline-*p*NA are hydrolyzed, but not 2-deoxythymidine-5′-*p*NP-phosphate, esculin, *o*NP-β-D-galactopyranoside, *p*NP-β-D-glucuronide, *p*NP-α-glucopyranoside, *p*NP-β-D-glucopyranoside, L-glutamate-γ-3-carboxyl-*p*NA, *bis-p*NP-phosphate, *p*NP-phenylphosphonate, *p*NP-phosphorylcholine, or *p*NP-α-D-xylopyranoside. Grows on 4-aminobutyrate, 1,4-butanediol, 2,4-diaminobutyrate, 4-hydroxybutyrate, DL-3-hydroxybutyrate, and some purines and ethers, but not on 4-aminobutyrate, chlorinated benzenes, dimethylsulfide, or hydrocarbons (C6 or petroleum). Additional phenotypic features are shown in Table 229.

Major fatty acids are $C_{16:0}$ iso (24.8–32.3%), $C_{15:0}$ iso (11.2–16.4%), and $C_{16:0}$ (15.0–30.5%); small to moderate amounts of methyl-branched fatty acids are also present.

Source: the type strain was isolated from an enrichment culture containing THF as the sole source of carbon, and originated from sludge of a wastewater treatment plant in Göttingen, Germany.

DNA G+C content (mol%): 71.3 (HPLC).

Type strain: strain K1, CIP 109050, CCUG 52126, DSM 44239, JCM 14745.

Sequence accession no. (16S rRNA gene): AJ249200.

Further comments: metabolizes a broad range of carbon compounds on Biolog GP2 plates (Kämpfer et al., 2006).

27. **Pseudonocardia xinjiangensis** (Xu, Jin, Mao, Lu, Cui and Jiang 1999) Huang, Wang, Lu, Hong, Liu, Tan and Goodfellow 2002, 981[VP] (Basonym: *Actinobispora xinjiangensis* Xu, Jin, Mao, Lu, Cui and Jiang 1999, 885.)

xin.ji.ang.en'sis. N.L. fem. adj. *xinjiangensis* pertaining to Xinjiang, an autonomous region in northwest China.

Forms substrate and aerial hyphae that are branched, but do not fragment. Spores are borne in longitudinal pairs on vegetative hyphae and either singly or in longitudinal pairs on aerial hyphae. An orange-yellowish substrate mycelium carries sparse, white aerial hyphae. Neither diffusible nor melanin pigments are produced. Does not grow at 45°C. Does not degrade cellulose, pectin, or starch, produce lecithinase or H_2S, or reduce nitrate. Adonitol, cellobiose, fructose, inositol, inulin, mannitol, raffinose, and xylose are used as sole carbon sources. Acid is not produced from these carbon sources. Alanine, L-histidine, and proline are used as nitrogen sources. Additional phenotypic features are shown in Table 229.

The cellular polar lipid pattern contains phosphatidycholine, phosphatidylethanolamine, phosphatidylglycerol, phosphatidylinositol, and glucosamine-containing phospholipids.

Source: the type strain was isolated from a soil sample collected from Tolufan, Xinjiang Province, China.

DNA G+C content (mol%): 72.1 (T_m).

Type strain: strain XJ-45, CCTCC AA 97020, CGMCC 4.1538, DSM 44661, JCM 11839.

Sequence accession no. (16S rRNA gene): EU722520.

28. **Pseudonocardia yunnanensis** (Jiang, Xu, Yang, Guo, Ma and Liu 1991) Huang, Wang, Lu, Hong, Liu, Tan and Goodfellow 2002, 981[VP] (Basonym: *Actinobispora yunnanensis* Jiang, Xu, Yang, Guo, Ma and Liu 1991, 527)

yun.nan.en'sis. N.L. fem. adj. *yunnanensis* pertaining to Yunnan, a province of south China.

Substrate and aerial hyphae are branched and do not fragment. A yellow substrate mycelium carries sparse, white aerial hyphae. Spores are borne in longitudinal pairs on vegetative hyphae and either singly or in longitudinal pairs on aerial hyphae. Neither diffusible nor melanin pigments are produced. Grows at 45°C. Does not degrade cellulose, gelatin, starch, or pectin. Milk is coagulated and peptonized. Nitrate is not reduced; lecithinase and H_2S are not produced. Adonitol, cellobiose, fructose, inositol, mannitol, raffinose, and xylose are used as sole carbon sources.

Susceptible to neomycin and rifampin (each at 50 μg/ml). Additional phenotypic features are shown in Table 229.

The cellular polar lipid pattern contains phosphatidylethanolamine, phosphatidylmethylethanolamine, and glucosamine-containing phospholipids

Source: isolated from a soil sample collected in Weixin, Yunnan Province, China.

DNA G+C content (mol%): 73.4 (T_m).

Type strain: strain Y-11981, CCTCC M 90959, CGMCC 4.1542, DSM 44253, NBRC 15681, JCM 9330, VKM Ac-1991.

Sequence accession no. (16S rRNA gene): AJ252822.

29. **Pseudonocardia zijingensis** Huang, Wang, Lu, Hong, Liu, Tan and Goodfellow 2002, 981[VP]

zi.jing.en'sis. N.L. fem. adj. *zijingensis* pertaining to Zijing, the source of the soil from which the organism was isolated.

The aerial and substrate mycelia fragment into rod-shaped elements. Chains of smooth spores are produced by acropetal budding on the substrate mycelium. Branched yellow substrate hyphae and white aerial hyphae are formed on trypticase soy broth agar. Does not produce diffusible pigments. Growth occurs at 15–45°C. Produces lipase. Additional phenotypic features are shown in Table 229.

The cellular polar lipid pattern contains diphosphatidylglycerol, phosphatidycholine, phosphatidylglycerol, phosphatidylinositolmannosides, and phosphatidylmethylethanolamine, but not glucosamine-containing phospholipids.

Source: isolated from a soil sample of Zijing Mountain, Yunnan Province, China.

DNA G+C content (mol%): 70.9 (T_m).

Type strain: strain 6330, CGMCC 4.1545, DSM 44774, JCM 11117.

Sequence accession no. (16S rRNA gene): AF325725.

Genus II. **Actinoalloteichus** Tamura, Zhiheng, Yamei and Hatano 2000, 1439[VP]

Tomohiko Tamura

Ac.ti.no.al.lo.tei'chus. Gr. n. *actis*, *actinos* ray (used to refer to actinomycetes); Gr. adj. *allos* another, the other; Gr. masc. n. *teichos* wall; N.L. masc. n. *Actinoalloteichus* actinomycete with a different wall.

Stains Gram-positive, not-acid-fast. A fine, non-fragmenting, branching mycelium is produced. Strictly aerobic. Non-fragmentary substrate mycelia are present. Aerial mycelia are formed with an **aggregate of straight spore chains**. The aerial and substrate mycelia tend to fragment. Good growth occurs at 20–37°C. The organism shows good growth on yeast extract-malt extract agar and oatmeal agar. Cell walls contain glutamate, glucosamine, alanine, and *meso*-diaminopimelate. $C_{17:0}$ anteiso, $C_{15:0}$ iso, $C_{16:0}$ iso, and $C_{15:0}$ anteiso are the major cellular fatty acids present. The major menaquinone is MK-9(H_4); small amounts of MK-8(H_4) and MK-9(H_2) are also present. Phosphatidylethanolamine and phosphatidylmonomethylethanolamine are present as diagnostic phospholipids. The acyl type of the cell wall is acetyl.

DNA G+C content (mol%): 72–73.

Type species: **Actinoalloteichus cyanogriseus** Tamura, Zhiheng, Yamei and Hatano 2000, 1439[VP].

Further descriptive information

Ribose, mannose, galactose, and glucose are detected in whole-cell sugars, but arabinose is not. The wall corresponds to chemotype III according to Lechevalier and Lechevalier (1970). The fatty acid and menaquinone compositions of the species of *Actinoalloteichus* are given in Table 230. *Actinoalloteichus hymeniacidonis* can be distinguished from the other species of the genus by its higher concentrations of $C_{17:1}$ ω8*c*, $C_{17:0}$, and $C_{15:0}$ anteiso. Species of *Actinoalloteichus* grow well on a variety of sugars. Other physiological characteristics are summarized in Table 231.

Enrichment and isolation procedures

The type strain of *Actinoalloteichus cyanogriseus* was isolated from a soil sample collected from Yunnan, China (Liu et al., 1984). The type strain of *Actinoalloteichus hymeniacidonis* was isolated

TABLE 230. Fatty acids and menaquinones of type strains of *Actinoalloteichus* species

Characteristic	*A. cyanogriseus*[a]	*A. hymeniacidonis*[b]	*A. spitiensis*[c]
Cellular fatty acids (%):			
$C_{14:0}$ iso	5	4	6
$C_{15:0}$ iso	15	6	17
$C_{15:0}$ anteiso	10	20	7
$C_{15:0}$		6	
$C_{16:1}$ iso	8		6
$C_{16:1}$	2		2
$C_{16:0}$ iso G		6	
$C_{16:0}$ iso	19	16	33
$C_{16:0}$	2		2
$C_{17:0}$ iso	3		3
$C_{17:0}$ anteiso	20	4	8
$C_{17:1}$ ω8*c*		19	
$C_{17:0}$		11	
Others	16	8	16
Menaquinones (%):			
MK-8(H_4)	10		5
MK-9(H_2)	9		
MK-9(H_4)	75	64	82
MK-9(H_6)	2	23	2
MK-9(H_8)		12	
MK-10(H_4)	3		9

[a]Tamura et al. (2000).
[b]Zhang et al. (2006).
[c]Singla et al. (2005).

TABLE 231. Physiological characteristics of *Actinoalloteichus* species[a]

Characteristic	*A. cyanogriseus*[b]	*A. hymeniacidonis*[c]	*A. spitiensis*[d]
Utilization of:			
Arabinose	w	–	–
Fructose	–	+	–
Glucose	+	+	–
Maltose	+	+	–
Mannitol	+	+	+
Mannose	+	+	–
Raffinose	–	–	+
Rhamnose	+	+	–
Sucrose	w	+	+
Sorbitol	+	+	–
Sodium citrate	–	+	+
Sodium succinate	–	–	+
Xylose	+	+	–
Decomposition of casein	–	+	+
Hydrolysis of starch	w	+	+
Resistance to methyl violet	–	+	–
Pigmentation on:			
ISP 1	Absent	Black	Absent
ISP 6	Black	Black	Absent

[a]Data are for the type strain of each species. +, Positive; w, weakly positive; –, negative.
[b]Tamura et al. (2000).
[c]Zhang et al. (2006).
[d]Singla et al. (2005).

from a marine sponge collected at the inter-tidal beach of Dalian, on the Chinese Yellow Sea, located in northern China. Freshly collected sponge specimens were rinsed five times in sterile seawater to remove transient and loosely attached bacteria and then thoroughly homogenized in a sterile mortar. A 10-fold dilution series of sponge homogenate was plated in triplicate on modified arginine-glycerol agar (ISP medium 5) plates, consisting of glycerol (0.6%, v/v), arginine (0.1%, w/v), K_2HPO_4 (0.1%, w/v), $MgSO_4$ (0.05%, w/v), and agar (1.8%, w/v) in natural seawater. The type strain of *Actinoalloteichus spitiensis* was isolated from a soil sample collected from the Lahaul–Spiti Valley, a cold desert of the Indian Himalayas, by using a dilution plating technique on actinomycetes isolation agar, containing (in w/v): sodium caseinate (0.2%), asparagine (0.01%), sodium propionate (0.4%), dipotassium phosphate (0.05%), magnesium sulfate (0.01%), ferrous sulfate (0.0001%), and agar (1.5%), pH 8.1. Incubation was at 28°C for 2 to 4 weeks.

Maintenance procedures

Strains of the genus *Actinoalloteichus* are maintained by freezing in water containing 10–30% glycerol at −70°C. Lyophilization of suspensions in 10% skim milk+1% monosodium glutamate and L-drying in 0.01 M potassium phosphate buffer (pH 7.0) containing 3% monosodium glutamate (Sakane and Kuroshima, 1997) are also recommended for long-term preservation.

Differentiation of the genus *Actinoalloteichus* from other genera

The genus *Actinoalloteichus* develops aerial mycelium with straight chains of spores, contains galactose as the characteristic whole-cell sugar, and *meso*-diaminopimelate, glutamate, and alanine as the cell-wall amino acids. Phylogenetic analysis indicates that the genus *Actinoalloteichus* is closely related to the genera *Goodfellowiella* (Labeda and Kroppenstedt, 2006) and *Streptoalloteichus* (Tomita et al., 1987). The genus *Goodfellowiella* has $C_{17:0}$ 10-methyl and anteiso-branched 2-hydroxy fatty acids. *Streptoalloteichus* develops motile spores in sporangia. On the other hand, the genus *Actinoalloteichus* lacks 10-methyl fatty acids and does not develop sporangia or motile cells. The genera *Saccharothrix* (Labeda et al., 1984), *Actinosynnema* (Hasegawa et al., 1978), and *Thermocrispum* (Korn-Wendisch et al., 1995) have *meso*-diaminopimelic acid in the cell wall (cell-wall type III) and long chains of spores. However, the genus *Actinoalloteichus* can be distinguished from these genera by its morphological characteristics, fatty acid components, and the absence of motility.

Taxonomic comments

The species *Streptomyces caeruleus* has been reclassified as *Actinoalloteichus cyanogriseus* based on 16S rRNA gene sequence analysis and DNA hybridization (Tamura et al., 2008a). The genus *Actinoalloteichus* belongs to the family *Pseudonocardiaceae* of the order *Pseudonocardiales*.

Differentiation of the species of the genus *Actinoalloteichus*

Characteristics of the three *Actinoalloteichus* species are given in Table 230 and Table 231.

List of species of the genus *Actinoalloteichus*

1. **Actinoalloteichus cyanogriseus** Tamura, Zhiheng, Yamei and Hatano 2000, 1439[VP]

cya.no.gri'se.us. L. adj. *cyaneus* deep blue; N.L. adj. *griseus* gray; N.L. masc. adj. *cyanogriseus* blue-gray.

The vegetative mycelia are brown to gray in color and the aerial mycelia are blue to gray. A black soluble pigment is produced on yeast extract-malt extract agar. Decomposition of urea, growth in Sabouraud glucose broth and MacConkey agar, hydrolysis of esculin and hippurate, utilization of sodium citrate, sodium succinate and calcium malate, gelatin liquefaction and reduction of nitrate are all negative. Glucose, maltose, xylose, mannitol, rhamnose, mannose, and sorbitol are utilized, but inositol, fructose, and raffinose are not. No growth at 15 or 45°C.

Source: the type strain was isolated from cultivated soil.

DNA G+C content (mol%): 73 (HPLC).

Type strain: AS 4.1159, CIP 106755, DSM 43889, NBRC 14455, JCM 6095, NRRL B-16252.

Sequence accession no. (16S rRNA gene): AB006178.

2. **Actinoalloteichus hymeniacidonis** Zhang, Zheng, Huang, Luo, Jin, Zhang, Liu and Huang 2006, 2311[VP]

hy.me.ni.a.ci'do.nis. N.L. gen. n. *hymeniacidonis* of *Hymeniacidon*, the generic name of the marine sponge *Hymeniacidon perleve*, the source of the type strain.

Aerial mycelia are produced with aggregates of straight chains of spores (0.6–0.8 µm). Grows well on yeast extract-malt extract agar and oatmeal agar at 20–37°C. A black soluble pigment is produced on yeast extract-malt extract agar and peptone-yeast extract-iron agar. Decomposition of urea, growth in Sabouraud glucose broth and MacConkey agar, hydrolysis of esculin and hippurate, utilization of calcium malate, sodium oxalate and sodium succinate, and reduction of nitrate are all negative. Fructose, glucose, maltose, mannose, mannitol, rhamnose, sucrose, sorbitol, and xylose are utilized as sole carbon sources, but arabinose, inositol, and raffinose are not. Grows weakly at 15°C and does not grow at 45°C. The cell-wall chemotype is III. The major menaquinone is MK-9(H_4); small amounts of MK-9(H_6) and MK-9(H_8) are also present. The phospholipid profile comprises mainly phosphatidylethanolamine, phosphatidylglycerol, phosphatidylinositol, and phosphatidylinositol mannoside. Major cellular fatty acids are $C_{15:0}$ anteiso (20%), $C_{17:1}$ ω8*c* (19%), $C_{16:0}$ iso (16%), and $C_{17:0}$ (11%).

Source: the type strain was isolated from a marine sponge, *Hymeniacidon perleve*, in Dalian, China (type strain).

DNA G+C content (mol%): not reported.

Type strain: HPA177, CGMCC 4.2500, JCM 13436.

Sequence accession no. (16S rRNA gene): DQ144222.

3. **Actinoalloteichus spitiensis** Singla, Mayilraj, Kudo, Krishnamurthi, Prasad and Vohra 2005, 2563[VP]

spi.ti.en'sis. N.L. masc. adj. *spitiensis* pertaining to Spiti Valley, located in the Indian Himalayas, where the type strain was isolated.

No aerial mycelium or spores are produced. Positive for utilization of D-mannitol, raffinose, sucrose, salicin, sodium citrate, and sodium succinate as sole carbon sources. Negative for utilization of D-glucose, maltose, *myo*-inositol, D-fructose, D-arabinose, D-xylose, L-rhamnose, D-mannose, and D-sorbitol as sole carbon sources. Positive for decomposition of casein and negative for decomposition of urea. No growth occurs on MacConkey agar or in Sabouraud glucose broth. Positive for hydrolysis of starch and negative for hydrolysis of hippurate and esculin. Tolerates up to 2% NaCl. Grows at temperatures between 20 and 37°C, with an optimum temperature of 25°C; it cannot grow at 15 or 42°C. Growth occurs at initial pH values between 6 and 11, the optimum being pH 8.0. Major fatty acids are $C_{16:0}$ iso (33.0%) and $C_{15:0}$ iso (17.0%), with lesser amounts of $C_{15:0}$ anteiso (7.0%) and $C_{17:0}$ anteiso (8.0%). The major menaquinone is MK-9(H_4) (82%). MK-9(H_2) is absent.

Source: the type strain was isolated from soil 3600 m above sea level, at Rangrik Village in Spiti Valley, Himachal Pradesh, India.

DNA G+C content (mol%): 72 (HPLC).

Type strain: RMV-1378, DSM 44848, JCM 12472, MTCC 6194.

Sequence accession no. (16S rRNA gene): AY426714.

Genus III. **Actinokineospora** Hasegawa 1988a, 449[VP] (Effective publication: Hasegawa 1988b, 33.)

David P. Labeda

Ac.ti.no.ki.ne.o.spo'ra. Gr. n. *aktis -inos* ray; Gr. v. *kineo* to set in motion; Gr. fem. n. *spora* seed and in biology a spore; N.L. fem. n. *Actinokineospora* actinomycete bearing zoospores.

Forms hyphae (approx. 0.5 µm in diameter) that differentiate into a vegetative mycelium that penetrates the agar medium and forms colonies on the surface; aerial mycelium arises from the colony. **Aerial hyphae bear chains of conidia capable of forming flagella in an aqueous environment.** Gram-stain-positive. Catalase-positive. Aerobic. **The cell wall contains *meso*-diaminopimelic acid as the diamino acid along with glycine, D-glutamic acid, and L-alanine, properties characteristic of type A1γ peptidoglycan. The characteristic whole-cell sugars are arabinose and galactose, but very little arabinose is found in purified cell walls. The phospholipid pattern consists of significant amounts of phosphatidylethanolamine along with diphosphatidylglycerol and phosphatidylinositol. The principal menaquinone is MK-9(H_4).** Phylogenetically, within the the order *Pseudonocardiales* based on 16S rRNA gene sequences.

DNA G+C content (mol%): 72.0 (T_m).

Type species: **Actinokineospora riparia** Hasegawa 1988a, 449[VP] (Effective publication: Hasegawa 1988b, 33.).

Further descriptive information

The genus *Actinokineospora* was described by Hasegawa (1988b, 1988a) for a novel actinomycete strain isolated from a soil sample collected on the side of the Ado River in Shiga Prefecture, Japan. This strain was unique in that it produced motile zoospores (Figure 285) from aerial hyphae (Figure 286) and was found to have both arabinose and galactose as diagnostic sugars in whole-cell hydrolysates, thereby distinguishing it from all described actinomycete genera. Subsequently, members of an additional six species of this genus that exhibited similar morphology and chemotaxonomy were isolated and described. Phylogenetic analysis of *Actinokineospora* species based on 16S rRNA gene sequences places the genus within the suborder *Pseudonocardineae* (Figure 287), elevated to the order *Pseudonocardiales* in this volume, along with other genera which produce motile zoospores, such as *Actinosynnema* and *Streptoalloteichus*, but members of these taxa produce structures such as synnemata and sporangia, respectively, which are not present in *Actinokineospora*. All *Actinokineospora* species described to date have been found associated with plant leaves and soil samples.

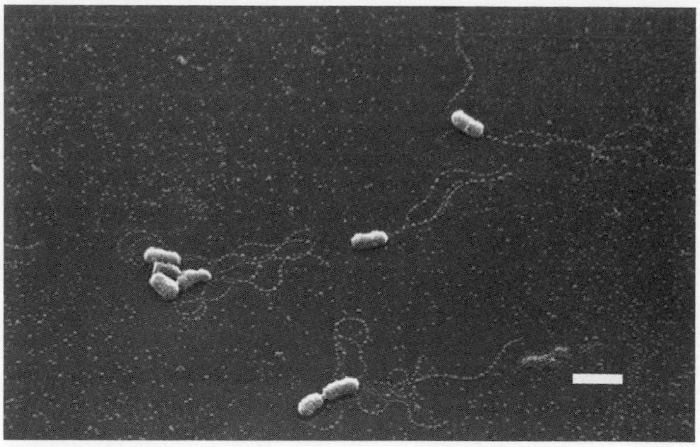

FIGURE 286. Electron micrograph of zoospores of *Actinokineospora enzanensis* NBRC 16517^T. Bar = 1.0 µm. (Electron micrograph courtesy of Misa Otogura, NBRC, Japan.)

Enrichment and isolation procedures

Strains of *Actinokineospora* have been routinely isolated on humic acid-vitamin agar (HV agar; Hayakawa and Nonomura, 1987), using a modification of the procedure of Makkar and Cross (1982). Soil or leaf litter samples are air-dried at 28°C for 7 d, 0.5 g sample is mixed with 50 ml sterile tap water, and the mixture is incubated at 20°C for 55 min with occasional shaking. Aliquots (0.1 ml) of dilutions of the supernatant are spread on the surface of plates and incubated at 28°C for 2–3 weeks.

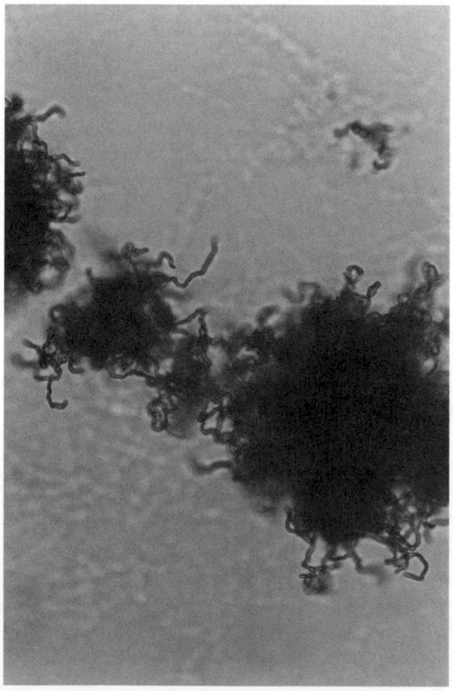

FIGURE 285. Light micrograph of the aerial mycelium on a culture of *Actinokineospora riparia* on sucrose-nitrate agar.

Hayakawa et al. (2000) described an isolation method for strains of *Actinokineospora* and other actinomycetes that have motile zoospores from soils and plant material. They placed a 0.5 g sample of air-dried soil or leaf litter in a beaker and gently flooded it with 50 ml sterile 10 mM phosphate buffer (pH 7.0) containing 10% soil extract. After incubating the sample for 90 min at 30°C, the motile zoospores that might be present are released. Following low speed (i.e. 1500 × *g*) centrifugation for 20 min and a subsequent 30 min incubation period, the samples were serially diluted and 200 µl aliquots plated onto an appropriate isolation medium, such as HV medium (Hayakawa and Nonomura, 1987) containing 50 µg/ml cycloheximide and with/without trimethoprim (20 µg/ml) and nalidixic acid (10 µg/ml). After a 2–3 week incubation period at 30°C, the plates are observed using a microscope fitted with a long working distance objective for the tentative identification of putative *Actinokineospora* strains based on morphological criteria.

Maintenance procedures

Working cultures of *Actinokineospora* can be maintained as refrigerated (4°C) agar slants on an appropriate medium such as NZamine medium (DSMZ medium 554; DSMZ, 2001). The slants are generally incubated for 5–7 d at 28–30°C prior to refrigeration and then should be subcultured at monthly intervals. Long-term preservation of strains is best accomplished as frozen stocks in 20% (v/v) aqueous glycerol at −80°C (mechanical freezer) to −172°C (liquid nitrogen vapor phase) or by using traditional lyophilization techniques.

Procedures for testing special characters

Strains are routinely cultivated on NZamine medium (DSMZ medium 554) at 28°C. Morphological observations are made on the media of Shirling and Gottlieb (1966) and NZamine medium.

Chemotaxonomic analysis of strains for fatty acids, menaquinones, and polar lipids are performed using methods described previously by Grund and Kroppenstedt (1989), Minnikin et al. (1984), Saddler et al. (1991), and Sasser (1990).

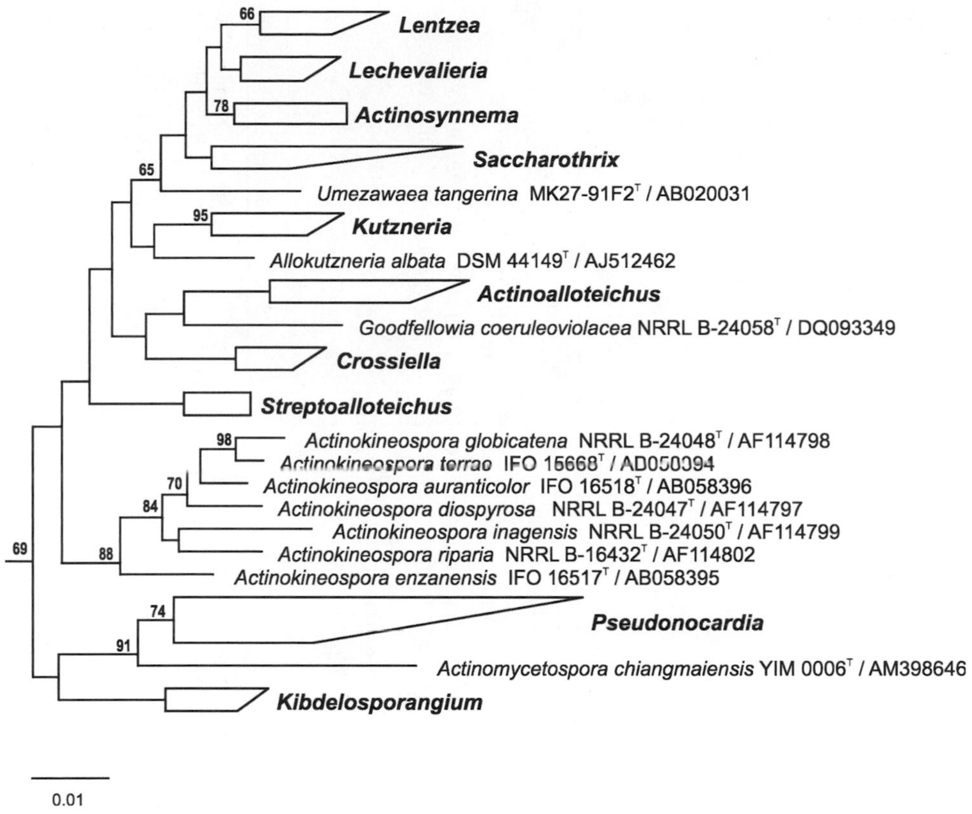

FIGURE 287. Phylogenetic tree for the genus *Actinokineospora* and related genera within the suborder *Pseudonocardineae* (now the order *Pseudonocardiales*) calculated from almost-complete 16S rRNA gene sequences using the Kimura's evolutionary distance method (Kimura, 1980) and the neighbor-joining algorithm of Saitou and Nei (1987). Numbers at the nodes represent levels of bootstrap support (%; Felsenstein, 1989) from 100 resampled datasets; values less than 60% are not shown. Bar = 0.01 nucleotide substitutions per site.

Physiological tests, including production of acid from carbohydrates, utilization of organic acids, and hydrolysis and decomposition of adenine, casein, esculin, guanine, hippurate, hypoxanthine, tyrosine, xanthine, and urea are typically determined using the media of Gordon et al. (1974). Phosphatase activity is evaluated by using the method of Kurup and Schmitt (1973) substituting NZamine agar as the basal medium. Temperature range for growth is determined on slants of NZamine medium and salt tolerance on slants of the same medium supplemented with 4 and 5% (w/v) NaCl.

Differentiation of the genus *Actinokineospora* from other genera

Actinokineospora strains can be easily differentiated from other genera morphologically based on their production of motile zoospores from aerial mycelia in the absence of elaborate structures such as sporangia or synnemata. They are chemotaxonomically distinct from other related actinomycete genera producing motile zoospores in the presence of arabinose in the whole-cell sugar hydrolysate pattern (Table 232) and are phylogenetically distinct based on 16S rRNA gene sequences (Figure 287).

Differentiation of the species of the genus *Actinokineospora*

The physiological characteristics of *Actinokineospora* species are summarized in Table 233 and can be used to differentiate between species, although a common set of physiological tests has not been used in the description of all species. Gross colonial morphology provides some additional information of the species based on the color of the substrate mycelium, production and color of aerial mycelium, and on the formation and color of soluble pigments when growing on agar media.

TABLE 232. Chemotaxonomic profiles of *Actinokineospora* and phylogenetically closely related genera

Character	*Actinokineospora*	*Actinoalloteichus*	*Actinosynnema*	*Allokutzneria*	*Crossiella*	*Goodfellowiella*	*Kibdelosporangium*	*Kutzneria*	*Streptoalloteichus*
Motile spores	+	–	+	–	–	–	–	–	+
Whole-cell sugar pattern	Galactose, arabinose, rhamnose	Glucose, galactose, mannose, ribose	Galactose, mannose	Arabinose, galactose, mannose	Galactose, mannose, rhamnose, ribose	Galactose, ribose	Arabinose, galactose, glucose, rhamnose	Galactose, rhamnose	Galactose, mannose, rhamnose, ribose
Phospholipids[a]	PE, OH-PE	PIM, PI, PG, DPG, PME	PE, OH-PE, PG	PE, PME, OH-PE, PI, *lyso*-PME, DPG,PG, *lyso*-PE	PE, DPG, PI, PIM, PME	PE, DPG, OH-PE, PME	PE, PME, PG, PI	PE, DPG, PI, PG, PME	PE, DPG, PI, PIM, DPG, PME
Predominant menaquinones	MK-9(H_4), MK-7(H_4), MK-8(H_4)	MK-9(H_4)	MK-9(H_4), MK-9(H_6)	MK-9(H_4)	MK-9(H_4)	MK-9(H_4), MK-10(H_4)	MK-9(H_4)	MK-9(H_4)	MK-9(H_6), MK-10(H_6)
DNA G+C content (mol%)	69.1–72.0	72–72.5	71	71.6	71.4	69.2	66	70.3–70.7	71.6

[a]Abbreviations: DPG, diphosphatidylglycerol; OH-PE, phosphatidylethanolamine with hydroxy fatty acids; *lyso*-PE, phosphatidylethanolamine where one fatty acid chain is missing from the glycerol backbone; *lyso*-PME, phosphatidylmethylethanolamine where one fatty acid chain is missing from the glycerol backbone; PE, phosphatidylethanolamine; PG, phosphatidylglycerol; PI, phosphatidylinositol; PIM, phosphatidylinositol mannosides; PME, phosphatidylmethylethanolamine.

TABLE 233. Physiological properties of *Actinokineospora* species[a,b]

Character	*A. riparia*	*A. auranticolor*	*A. diospyrosa*	*A. enzanensis*	*A. globicatena*	*A. inagensis*	*A. terrae*
Colony reverse color	Yellow/brown	Orange	Yellow/brown	Gray	Yellow/brown	Yellow/brown	Yellow/brown
Hydrolysis of:							
Calcium malate	–	+	–	+	+	–	v
Elastin	–	+	+	+	+	+	+
Gelatin	–	+	+	+	+	–	+
Milk (peptonization)	–	nd	+	nd	–	–	+
Starch	–	nd	+	nd	+	–	+
Testosterone	+	+	+	+	v	–	+
Production of:							
Nitrate reductase	+	–	–	+	–	+	–
Hydrogen sulfide	–	+	+	+	+	+	+
Growth on sole carbon source (1.0%, w/v):							
Arabinose	–	–	–	–	–	–	w
D-Fructose	w	–	+	–	+	–	+
Galactose	–	nd	–	nd	–	–	–
Glucitol	–	nd	–	nd	–	–	–
Glycerol	+	+	+	+	+	–	v
Maltose	–	+	+	–	+	w	+
D-Mannose	w	–	w	–	+	–	+
Rhamnose	–	–	w	–	+	w	w
Sucrose	–	+	+	–	+	–	+
Trehalose	nd	+	nd	+	nd	nd	nd
Growth on sole carbon source (0.1%, w/v):							
D-Alanine	–	v	v	+	v	–	+
L-Proline	–	+	+	+	–	+	+
Sodium acetate	–	v	–	v	–	+	v
Growth in the presence of (w/v):							
Bismuth citrate, 0.001%	+	+	v	+	–	–	v
Brilliant green, 0.001%	–	–	v	–	–	–	+
Furazolidone, 0.004%	–	–	+	–	+	+	+
Potassium tellurite 0.01%	–	+	+	+	+	–	+
Sodium chloride, 2.0%	+	+	+	+	+	–	+
Sodium chloride, 3.0%	–	v	+	–	+	–	+
Vanillin, 0.05%	–	v	–	v	v	+	+
Resistance to antibiotics:							
Ampicillin, 10 µg/ml	–	+	–	+	–	–	–
Benzyl penicillin, 10 µg/ml	–	–	–	+	–	+	–
Cephaloridine, 10 µ/ml	–	–	–	+	+	+	v
Chloramphenicol, 10 µg/ml	–	+	v	–	–	–	v
Lincomycin, 20 µg/ml	–	+	v	–	v	–	v
Norfloxacin, 40 µg/ml	–	+	v	v	–	–	–
Oleandomycin, 5 µg/ml	–	+	+	–	v	–	v
Rifampin, 20 µg/ml	+	–	–	+	–	–	–
Growth at:							
10°C	–	+	+	–	+	–	+
37°C	+	+	v	–	–	–	v

[a]Symbols: v, variable reaction; w, weak positive reaction; nd, not determined.

[b]All species hydrolyzed hippurate but not xylan. None grew on erythritol, inositol, D-lactose, mannitol, raffinose, sorbitol, or xylose as sole carbon sources or in the presence of 0.2% *p*-hydroxybenzaldehyde or 0.1% phenylethanol. All species grew at 25°C and 30°C and were resistant to bekanamycin (20 and 40 µg/ml), kanamycin (40 and 80 µg/ml), oxytetracycline (20 µg/ml), and vancomycin (0.25 µg/ml).

List of species of the genus *Actinokineospora*

1. **Actinokineospora riparia** Hasegawa 1988a, 449[VP] (Effective publication: Hasegawa 1988b, 33.)

ri.pa'ri.a. L. fem. adj. *riparia* that frequents the banks of rivers, riverside, referring to the collection site of the soil sample from which the type strain was isolated, along the Ado River, Japan.

Vegetative mycelium varies in color from colorless on most media to yellowish-white on some media and tan to brown on yeast extract-malt extract agar. Aerial mycelium is white in color when produced. Tyrosine is decomposed, but not adenine, hypoxanthine, or xanthine. Growth is observed on D-glucose, soluble starch, and trehalose, with weak growth on cellulose, inulin, and melibiose; does not grow on adonitol, erythritol, ribose, salicin, L-sorbose, or dulcitol. Temperature for growth is 28°C.

Source: isolated from a soil sample collected on the side of the Ado River, Shiga Prefecture, Japan.

DNA G+C content (mol%): 72 (T_m).

Type strain: C-39162, ATCC 49499, DSM 44259, NBRC 14541, JCM 7471, NRRL B-16432, VKM Ac-1980.

Sequence accession no. (16S rRNA gene): AF114802.

2. **Actinokineospora auranticolor** Otoguro, Hayakawa, Yamazaki, Tamura, Hatano and Iimura 2003, 1[VP] (Effective publication: Otoguro, Hayakawa, Yamazaki, Tamura, Hatano and Iimura 2001, 38.)

au.ran.ti.co'lor. N.L. n. *Aurantium* generic name of the orange; L. n. *color* tint, hue; N.L. adj. *auranticolor* orange colored.

Yellowish-orange substrate mycelium. Aerial mycelium is white to gray. Temperature for growth is 28°C.

Source: isolated from leaves and soil.

DNA G+C content (mol%): 71.3 (HPLC).

Type strain: YU 961-1, DSM 44650, NBRC 16518, JCM 11646.

Sequence accession no. (16S rRNA gene): AB058396.

3. **Actinokineospora diospyrosa** Tamura, Hayakawa, Nonomura, Yokota and Hatano 1995, 378[VP]

di.o.spy'ro.sa. N.L. fem. adj. *diospyrosa* pertaining to the fruit tree *Diospyros kaki*.

Tan substrate mycelium. Aerial mycelia are white to gray. Temperature for growth is 28°C.

Source: isolated from fallen leaves of the persimmon tree, *Diospyros kaki*.

DNA G+C content (mol%): 69.3 (HPLC).

Type strain: YU8-1, DSM 44255, NBRC 15665, JCM 9921, NRRL B-24047, VKM Ac-1984.

Sequence accession no. (16S rRNA gene): AF114797.

4. **Actinokineospora enzanensis** Otoguro, Hayakawa, Yamazaki, Tamura, Hatano and Iimura 2003, 1[VP] (Effective publication: Otoguro, Hayakawa, Yamazaki, Tamura, Hatano and Iimura 2001, 38.)

en.za.nen'sis. N.L. fem. adj. *enzanensis* pertaining to Enzan City, Yamanashi, Japan, where the organism was isolated.

Vegetative mycelium is greenish-gray. Aerial mycelium is sparse and white, when formed. Temperature for growth is 28°C.

Source: isolated from soil.

DNA G+C content (mol%): 70.0 (HPLC).

Type strain: YU 924-101, DSM 44649, NBRC 16517, JCM 11647.

Sequence accession no. (16S rRNA gene): AB058395.

5. **Actinokineospora globicatena** Tamura, Hayakawa, Nonomura, Yokota and Hatano 1995, 377[VP]

glo.bi.ca.te'na. L. n. *globus* a ball, sphere; L. n. *catena* chain; N.L. n. *globicatena* ball of chains, referring to the mass of spore chains that aggregate into balls.

Vegetative mycelium is yellow to golden brown. Aerial mycelium is white to gray. Temperature for growth is 28°C.

Source: isolated from soil and fallen leaves.

DNA G+C content (mol%): 69.5–69.8 (HPLC).

Type strain: YU6-1, DSM 44256, NBRC 15664, JCM 9922, NRRL B-24048, VKM Ac-1981.

Sequence accession no. (16S rRNA gene): AF114798.

6. **Actinokineospora inagensis** Tamura, Hayakawa, Nonomura, Yokota and Hatano 1995, 377[VP]

in.ag.en'sis. N.L. fem. adj. *inagensis* pertaining to Lake Inaga, the lake from which the organism was first isolated.

Yellow to tan substrate mycelium is produced. Aerial mycelium is white to gray in color. Temperature for growth is 28°C.

Source: isolated from fallen leaves.

DNA G+C content (mol%): 69.1 (HPLC).

Type strain: YU4-1, DSM 44258, NBRC 15663, JCM 9923, NRRL B-24050, VKM Ac-1982.

Sequence accession no. (16S rRNA gene): AF114799.

7. **Actinokineospora terrae** Tamura, Hayakawa, Nonomura, Yokota and Hatano 1995, 377[VP]

ter'ra.e. L. gen. n. *terrae* of the earth.

Yellow to golden brown substrate mycelium is produced. Aerial mycelium is white to gray. Temperature for growth is 28°C.

Source: isolated from soil and fallen leaves.

DNA G+C content (mol%): 70.0 (HPLC).

Type strain: YU6-3, DSM 44260, NBRC 15668, JCM 9924, NRRL B-24049, VKM Ac-1983.

Sequence accession no. (16S rRNA gene): AB058394.

Genus IV. **Actinosynnema** Hasegawa, Lechevalier and Lechevalier 1978, 305[AL]

David P. Labeda

Ac.ti.no.syn′ne.ma. Gr. n. *actis, actinos* ray; Gr. adv. *syn* together; Gr. n. *nema, nematos* thread; N.Gr. n. *synnema*, threads wrapping together, synnema; N.L. neut. n. *Actinosynnema* indicates a synnema-forming actinomycete.

Fine hyphae (about 0.5 µm in diameter) are **differentiated into substrate mycelium**, with long branching hyphae that penetrate the agar and also grow into and **form synnemata, dome-like bodies, or flat colonies on the agar surface, and aerial hyphae** (0.5–1.0 µm in diameter) **that arise from synnemata, dome-like bodies, or flat colonies. The aerial hyphae bear chains of spores capable of forming flagella in an aqueous environment. The cell walls contain major amounts of *meso*-diaminopimelic acid (*meso*-DAP)**, glutamic acid, alanine, glucosamine, and muramic acid. **The whole-cell sugar pattern consists of galactose and mannose. Principal phospholipids include phosphatidylethanolamine, phosphatidylethanolamine containing 2-hydroxy fatty acids, and diphosphatidylglycerol. Menaquinones are predominantly MK-9(H_4), with some MK-9(H_6)**. Gram-stain-positive. Non acid-fast. Catalase-positive. Aerobic. Mesophilic. Most strains isolated directly from plant tissue. Phylogenetically nearest neighbors are the genera *Lechevalieria* and *Lentzea.*

DNA G+C content (mol%): 71–73 (T_m).

Type species: **Actinosynnema mirum** Hasegawa, Lechevalier and Lechevalier 1978, 305[AL].

Further descriptive information

The genus *Actinosynnema* was described by Hasegawa et al. (1978) to accommodate actinomycetes that produce unique morphological structures called synnemata (Figure 288) or dome-like bodies on most media. Aerial mycelia are produced on these synnemata or dome-like bodies and are initially whitish in color and become yellow to yellowish-orange in color. Regular septation occurs in mature aerial hyphae making it look bamboo-like when observed microscopically and then the hyphae become chains of spores. Suspension of the aerial mycelia in liquid media under a coverslip permits the observation of peritrichously motile zoospores (Figure 289) within 30 min to 1 h.

Phylogenetic analyses based on 16S rRNA gene sequencing demonstrates that the genus *Actinosynnema* is related to the genera *Lechevalieria, Lentzea,* and *Saccharothrix,* being intermediate between the first two genera and *Saccharothrix* (Figure 290) and is the type genus for the family *Actinosynnemataceae* Labeda and Kroppenstedt (2000), now part of the family *Pseudonocardiaceae.*

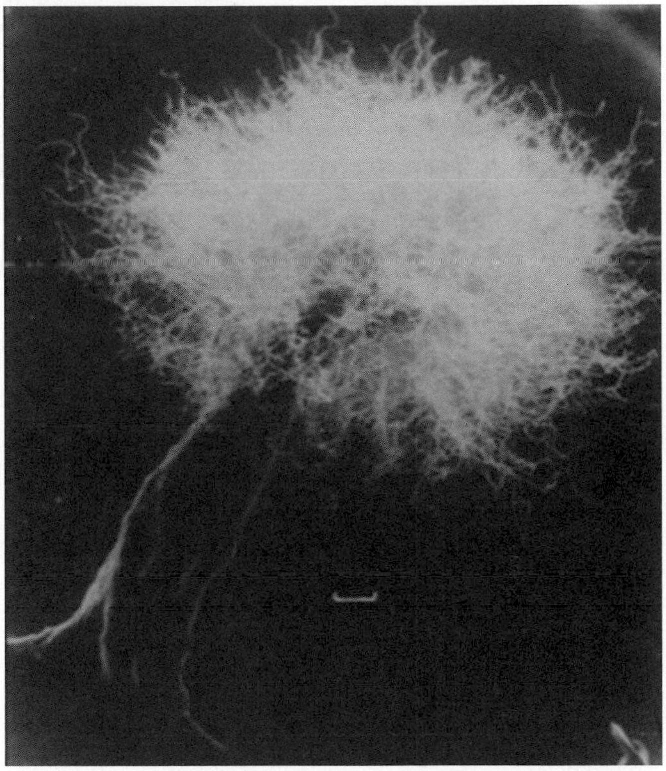

FIGURE 288. Scanning electron micrograph of a synnemata from *Actinosynnema mirum.* Bar = 10 µm.

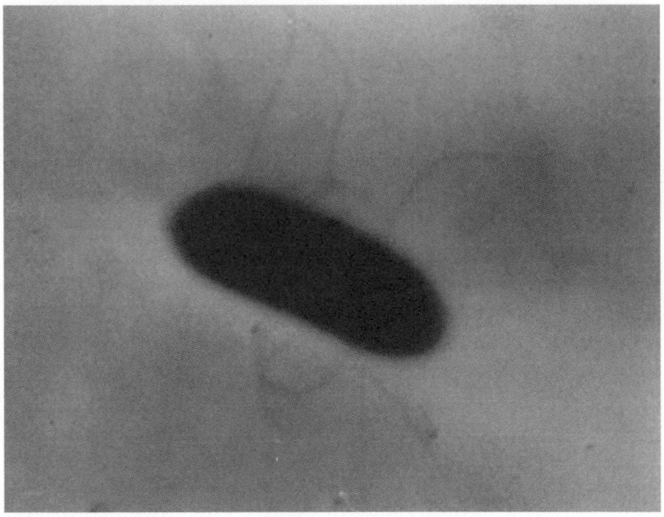

FIGURE 289. Electron micrograph of a zoospore of *Actinosynnema mirum.*

Enrichment and isolation procedures

Strains of the genus *Actinosynnema* can be isolated from grass blades by placing them on the surface of yeast extract agar plates (0.02% yeast extract and 1.5% agar in distilled water) and incubating for 3 weeks at 28°C. The agar surface will most likely be covered with various types of microbial growth, but small synnemata can be observed on the grass blade itself using a stereoscopic microscope. These synnemata can be carefully removed with a sterile loop and transferred to fresh growth media. *Actinosynnema mirum* has been reported not to be inhibited by the presence of 100 µg/ml nystatin or 50 µg/ml candicidin and *Actinosynnema pretiosum* has been found to be resistant to 100 µg/ml amphotericin B so it might be possible to use these antibiotics to suppress the growth of fungal competitors. Hayakawa et al. (2000) described an isolation method for strains of *Actinokineospora* and other actinomycetes that have motile zoospores which has also been reported to permit the isolation of *Actinosynnema* strains from soils and plant material. They placed a 0.5 g sample of air-dried soil or leaf litter in a beaker and gently flooded it with 50 ml sterile 10 mM phosphate buffer (pH 7.0) containing 10% soil extract. After incubating the sample for 90 min at 30°C, the motile zoospores that might be present are released. Following

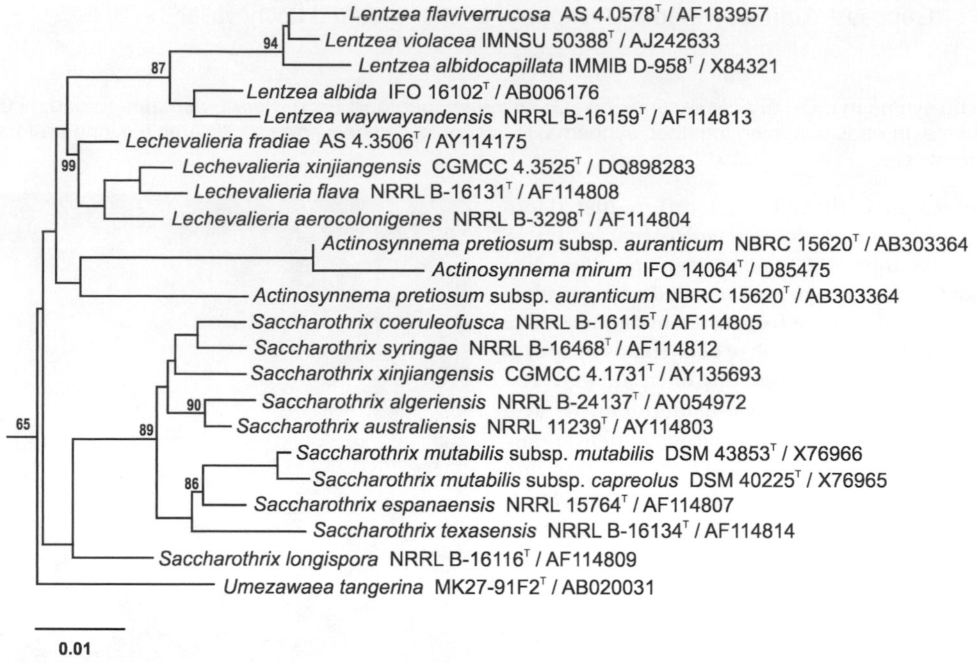

FIGURE 290. Phylogenetic tree for the genera *Actinosynnema, Lechevalieria, Lentzea, Saccharothrix,* and *Umezawaea* of the suborder *Pseudonocardineae* (now order *Pseudonocardiales*) calculated from almost-complete 16S rRNA gene sequences using the Kimura's evolutionary distance method (Kimura, 1980) and the neighbor-joining algorithm of Saitou and Nei (1987). Percentages at the nodes represent levels of bootstrap support (Felsenstein, 1989) from 100 resampled datasets; values less than 60% are not shown. Bar = 0.01 nucleotide substitutions per site.

low speed (i.e. 1,500 × *g*) centrifugation for 20 min and a subsequent 30 min incubation period, the samples are serially diluted and 200 µl aliquots plated onto an appropriate isolation medium such as HV medium (Hayakawa and Nonomura, 1987) containing 50 µg/ml cycloheximide with and without trimethoprim (20 µg/ml) and nalidixic acid (10 µg/ml). After a 2–3 week incubation period at 30°C, the plates are observed for the presence of synnemata using a microscope fitted with a long working distance objective and those observed can be transferred to fresh media.

Maintenance procedures

Working cultures of *Actinosynnema* can be maintained as plate or agar slant cultures on an appropriate medium such as NZamine agar (DSMZ medium 554; DSMZ, 2001) with bi-weekly subculturing. Survival of the cultures is often better at room temperature than at 4°C. Longer term preservation of strains is best accomplished as frozen stocks in 40% (v/v) aqueous glycerol at −80°C (mechanical freezer) to −172°C (liquid nitrogen vapor phase) or by using traditional lyophilization techniques.

Procedures for testing special characters

Strains are routinely cultivated on NZamine medium (DSMZ medium 554) at 28°C. Morphological observations are made on the media of Shirling and Gottlieb (1966) and NZamine medium, but for production of synnemata, strains should be grown on plates of 1.5% agar in tap water or Bennett's agar (DSMZ medium 548). Chemotaxonomic analysis of strains for fatty acids, polar lipids, and menaquinones are performed using methods described previously by Grund and Kroppenstedt (1989), Minnikin et al. (1984), Saddler et al. (1991), and Sasser (1990).

Physiological tests, including production of acid from carbohydrates, utilization of organic acids, and hydrolysis and decomposition of adenine, casein, esculin, guanine, hippurate, hypoxanthine, tyrosine, xanthine, and urea are typically determined using the media of Gordon et al. (1974). Phosphatase activity is evaluated by using the method of Kurup and Schmitt (1973) substituting NZamine agar as the basal medium. Temperature range for growth is determined on slants of NZamine medium and salt tolerance is evaluated on slants of the same medium supplemented with 2% to 5% NaCl.

Differentiation of the genus *Actinosynnema* from other genera

Strains of *Actinosynnema* can be differentiated easily from other actinomycetes by observation of the very characteristic synnemata produced on most growth media. Numerous other actinomycete genera produce motile zoospores, notably *Actinokineospora, Actinoplanes, Planobispora, Planomonospora, Spirillospora,* and several others, but none produce synnemata. The chemotaxonomic profile of *Actinosynnema* species is different from those of members of the other genera in the *Actinosynnemataceae*, particularly the whole-cell sugar pattern consisting of only galactose and mannose as the diagnostic sugars. The phospholipid pattern of *Actinosynnema* strains is quite similar to those of *Saccharothrix* species, but the lack of rhamnose in the whole-cell sugar pattern and the presence of MK-9(H_6) and lack of MK-10(H_4) menaquinones differentiates them from members of this genus. Characteristics of *Actinosynnema* and some related genera are given in Table 234.

Differentiation of the species of the genus *Actinosynnema*

The physiological characteristics of *Actinosynnema* species are summarized in Table 235 and can be used to differentiate between species, although a common set of physiological tests has not been used in the description of all species. Gross colonial morphology provides some additional information of the species based on the color of the substrate mycelium, production and color of aerial mycelium, and production and color of soluble pigments when growing on agar media.

TABLE 234. Chemotaxonomic and morphological characteristics of *Actinosynnema* and related genera[a,b]

Characteristic	*Actinosynnema*	*Lechevalieria*	*Lentzea*	*Saccharothrix*	*Umezawaea*
Production of synnemata and motile zoospores	+	–	–	–	–
Whole-cell sugar pattern	Galactose, mannose	Galactose, mannose, rhamnose	Galactose, mannose, ribose	Galactose, rhamnose, mannose (trace)	Galactose, rhamnose
Phospholipids	PE, OH-PE, PI, PIM, DPG	PE	PE, DPG, PG, PI	PE, OH-PE, PI, PIM, DPG, PG (v)	PE
Menaquinone(s)	MK-9(H_4), MK-9(H_6)	MK-9(H_4)	MK-9(H_4)	MK-10(H_4), MK-9(H_4)	MK-9(H_4), trace MK-10(H_4)

[a]Symbols: DPG, diphosphatidylglycerol; OH-PE, phosphatidylethanolamine containing hydroxylated fatty acids; PE, phosphatidylethanolamine; PG, phosphatidylglycerol; PI, phosphatidylinositol; PIM, phosphatidylinositol mannosides; PME, phosphatidylmethylethanolamine.

[b]All genera have *meso*-DAP as the cell-wall diamino acid, are of cell-wall chemotype III, and contain straight-chain, mono-unsaturated, iso, and anteiso fatty acids.

TABLE 235. Physiological properties of *Actinosynnema* species[a]

Character	*A. mirum*	*A. pretiosum* subsp. *pretiosum*	*A. pretiosum* subsp. *auranticum*
Decomposition of:			
Adenine	–	nd	nd
Casein	+	+	+
Esculin	nd	+	+
Gelatin	+	+/–	+/–
Hypoxanthine	–	–	–
Starch	+	+	+
Tyrosine	+	+	+
Urea	–	–	–
Xanthine	–	–	–
Growth on:			
Arabinose	+/–	+	–
Galactose	+	+	+
Glucose	+	+	+
Glycerol	+	–	+
Inositol	–	–	–
Lactose	–	–	+/–
Mannitol	+	+	+
Maltose	+	+/–	+
Mannose	+	+	+
Melibiose	–	+	+/–
Raffinose	–	+/–	+/–
Rhamnose	+	+	+
Soluble starch	+/–	+	+
Sorbitol	–	–	–
Sucrose	+	+	+
Trehalose	+	+	+
Xylose	+	+	+
Production of:			
Catalase	+	+	+
Hydrogen sulfide	nd	–	–
Nitrate reductase	+	+	+
Phosphatase	+	+	+
Growth in the presence of:			
NaCl (2%, w/v)	nd	+	+
Lysozyme	+	+	+
Amphotericin B (100 µg/ml)	nd	+	+
Candicidin (50 µg/ml)	+	nd	nd
Chloramphenicol (20 µg/ml)	–	–	–
Dihydrostreptomycin (10 µg/ml)	–	nd	nd
Nystatin (100 µg/ml)	+	nd	nd
Streptomycin (20 µg/ml)	nd	–	–
Sulbenicillin (100 µg/ml)	nd	+	+
Tetracyline (1 µg/ml)	–	nd	nd
Tetracyline (20 µg/ml)	nd	–	–
Growth at:			
10°C	+	–	–
37°C	–	+	+

[a]Symbols: +, positive; –, negative; +/–, doubtful response; nd, not determined.

List of species of the genus *Actinosynnema*

1. **Actinosynnema mirum** Hasegawa, Lechevalier and Lechevalier 1978, 305[AL]

mi'rum. L. neut. adj. *mirum* marvellous.

Substrate mycelium is yellow to orange-yellow in color, producing whitish synnemata that become yellow to orange-yellow in color on some media, notably thin potato-carrot agar of Higgins et al. (1967). White to yellowish-white aerial mycelia are produced on most media. A pale yellow-brown soluble pigment is produced on tyrosine agar and a pale greenish pigment on oatmeal agar.

Lactate, malate, pyruvate, and tartrate are assimilated for growth, but not acetate, benzoate, citrate, or succinate. Acid is produced from adonitol, L-arabinose, cellobiose, dextrin, fructose, galactose, glucose, glycogen, lactose, maltose, D-mannitol, D-mannose, D-melibiose, raffinose, L-rhamnose, soluble starch, sucrose, trehalose, and xylose, but not from dulcitol, erythritol, *i*-inositol, inulin, glycerol, methyl α-D-glucoside, methyl α-D-mannoside, D-ribose, salicin, or sorbitol. Temperature for growth is 28–30°C.

Source: isolated from a sedge (grass) blade.

DNA G+C content (mol%): 73 ± 1 (T_m).

Type strain: ATCC 29888, DSM 43827, NBRC 14064, IMRU 3971, JCM 3225, NRRL B-12336, VKM Ac-843.

Sequence accession no. (16S rRNA gene): D85475.

2. **Actinosynnema pretiosum** Hasegawa, Tanida, Hatano, Higashide and Yoneda 1983b, 314[VP]

pre.ti.o'sum. L. neut. adj. *pretiosum* precious.

2a. **Actinosynnema pretiosum subsp. pretiosum** Hasegawa, Tanida, Hatano, Higashide and Yoneda 1983b, 317[VP]

pre.ti.o'sum. L. neut. adj. *pretiosum* precious.

Substrate mycelium is pale yellow to pale orange-yellow in color. Branched hyphae in liquid culture fragment into motile elements with peritrichous flagella. The aerial mycelium consists of long, straight, helical, or (rarely) branching hyphae that are white to pale yellow in color. The aerial hyphae form in tufts at the tips of synnemata or on the surface of dome-like or irregular structures on the surface of the colonies. The aerial hyphae have a bamboo-like appearance and transform into chains of peritrichously flagellated spores. Yellow soluble pigments are produced on glucose-asparagine and peptone-yeast extract agars. The type strain produces antibiotics of the ansamitocin complex and tomaymycin. Temperature for growth is 28–30°C.

Source: isolated from a blade of *Carex* grass.

DNA G+C content (mol%): 71 ± 1 (T_m).

Type strain: C-15003(N-1), ATCC 31281, DSM 44132, FERM-P 3992, NBRC 15621, JCM 7344, NRRL B-16060, VKM Ac-1963.

Sequence accession no. (16S rRNA gene): AF114800.

2b. **Actinosynnema pretiosum subsp. auranticum** Hasegawa, Tanida, Hatano, Higashide and Yoneda 1983b, 320[VP]

au.ran'ti.cum. N.L. neut. adj. *auranticum* orange.

Substrate mycelium is yellowish-orange or orange on various media. Aerial hyphae are white to yellowish white in color. Pale yellowish-brown soluble pigments are produced on several media. The type strain produces the antibiotics dnacins and ansamitocins. Temperature for growth is 28–30°C.

Source: isolated from a blade of *Carex* species grass.

DNA G+C content (mol%): 71 ± 1 (T_m).

Type strain: C-14482(N-1001), ATCC 31309, DSM 44131, FERM-P 4130, NBRC 15620, JCM 7343, NRRL B-16078, VKM Ac-1961.

Sequence accession no. (16S rRNA gene): AB303364.

Genus V. **Amycolatopsis** Lechevalier, Prauser, Labeda and Ruan 1986, 34[VP] emend. Lee 2009, 1403

Geok Yuan Annie Tan and Michael Goodfellow

A.my.co.la.top'sis. N.L. fem. n. *Amycolata* genus belonging to the order *Actinomycetales*; Gr. fem. n. *opsis* aspect, appearance; N.L. fem. n. *Amycolatopsis* that which appears similar to *Amycolata*.

Aerobic to facultatively anaerobic, Gram-stain-positive, non-acid-fast, nonmotile, catalase-positive actinomycetes that form branching substrate hyphae which fragment into squarish and rod-shaped elements. When formed, aerial hyphae may be sterile or differentiate into chains of smooth-walled, squarish to ellipsoidal spore-like structures. Chemo-organotrophic to facultatively autotrophic. Grows on a broad range of organic and synthetic media. Mesophilic or thermophilic. **Whole-organism hydrolysates are rich in *meso*-2,6-diaminopimelic acid, arabinose, and galactose.** The peptidoglycan is of the A1γ type. Muramic acid moieties are *N*-acetylated. **The diagnostic phospholipid is phosphatidylethanolamine (with or without phosphatidylmethylethanolamine) or phosphatidylmethylethanolamine** with variable occurrence of diphosphatidylglycerol, phosphatidylglycerol, phosphatidylinositol, and phosphatidylinositol mannosides; contains **complex mixtures of saturated and branched chain fatty acids**. Does not contain mycolic acids. The phylogenetic position of *Amycolatopsis*, as determined by 16S rRNA gene sequence analysis, is in the family *Pseudonocardiaceae*.

Common in arid soils but has also been isolated from activated sludge, equine placentas, and from clinical and plant material.

DNA G+C content (mol%): 66–75 (T_m; HPLC).

Type species: **Amycolatopsis orientalis** (Pittenger and Brigham 1956) Lechevalier, Prauser, Labeda and Ruan 1986, 35[VP].

Further descriptive information

Phylogeny. The genus *Amycolatopsis* forms a clade in the *Pseudonocardiaceae* 16S rRNA gene tree. It encompasses 39 species with validly published names, most of which have been proposed in the past five years using polyphasic taxonomic approaches (Carlsohn et al., 2007; Labeda et al., 2003; Lee, 2009; Tan et al., 2006a, 2007). These species can be assigned to five multimembered and two single-membered phyletic lines in the *Amycolatopsis* 16S rRNA gene tree (Figure 291). The taxonomic integrity of the multimembered taxa, the *Amycolatopsis methanolica*, *Amycolatopsis nigrescens*, *Amycolatopsis orientalis*, *Amycolatopsis palatopharyngis*, and *Amycolatopsis sulphurea* subclades, are supported by all of the tree-making algorithms and by high bootstrap values.

Amycolatopsis methanolica subclade strains grow well at temperatures up to 60°C (Chun et al., 1999; De Boer et al., 1990; Henssen et al., 1987; Kim et al., 2002a) and can thereby be considered to be thermophilic actinomycetes (Brock, 1986; Cross, 1968). Further comparative taxonomic studies between members of this and the other multimembered subclades are needed to determine whether the *Amycolatopsis methanolica* subclade merits generic status. The type strain of *Amycolatopsis fastidiosa*, now reclassified as *Actinokineospora fastidiosa*, which grows between 10 and 60°C, forms a distinct phyletic line at the foot of the *Amycolatopsis* 16S rRNA gene tree (Goodfellow et al., 2001; Huang et al., 2001, 2004; Labeda et al., 2010b).

Everest and Meyers (2009) found good congruence between 16S rRNA and partial *gyrB* phylogenies of representative *Amycolatopsis* strains. Differences were found between the overall topologies of the trees, though groups of closely related strains clustered together in both trees, a result consistent with corresponding studies on the genera *Gordonia* (Shen et al., 2006) and *Micromonospora* (Kasai et al., 2000). A concatenated tree based on *gyrB*/16S rRNA gene sequences closely resembled the *gyrB* tree and contained conserved phyletic lines evident in the separate trees. It was apparent that the number of phyletic lines supported by high bootstrap values was much greater in the concatenated gene tree than in the corresponding 16S rRNA and partial *gyrB* trees. The type strain of *Amycolatopsis fastidiosa* formed a distinct phyletic line outside the evolutionary radiation occupied by the other *Amycolatopsis* strains in both the partial *gyrB* and concatenated trees.

DNA–DNA relatedness studies show that the type strain of *Amycolatopsis fastidiosa* has little, if any, DNA in common with the other *Amycolatopsis* strains to which it has been compared (Table 236). This is also the case with the type strain of *Amycolatopsis sulphurea*, an organism which forms a distinct phyletic line with *Amycolatopsis jejuensis* NRRL B-24427[T] in the 16S rRNA/*gyrB* concatenated tree (Everest and Meyers, 2009) though its nearest neighbor in the 16S rRNA gene tree, *Amycolatopsis sulphurea*, was not included in this study. In general, good congruence exists between DNA–DNA pairing and corresponding 16S rRNA gene sequence data.

Cell morphology. *Amycolatopsis* strains show regular to occasional fragmentation of either the substrate mycelium or the aerial mycelium or both (Figure 292). They do not form sclerotia, spore vesicles, or synnemata. Kothe et al. (1989) examined the morphology and fine structure of members of the family *Pseudonocardiaceae* and found that *Amycolatopsis* strains produced long chains of smooth, squarish to ellipsoidal spore-like structures on substrate and aerial mycelia. They also noted that the fine structure of the hyphae varied as some *Amycolatopsis* strains produced an electron-dense layer that covered the hyphal wall but others did not. Some strains show an unusual degree of morphological variation, as exemplified by the formation of short chains of oval, smooth-surfaced spores on the substrate mycelium of *Amycolatopsis taiwanensis* KCTC 19116[T] (Figure 293) and the production of globose, smooth-surfaced pseudosporangia by *Amycolatopsis decaplanina* DSM 44594[T] (Figure 294). However, spores were not detected either inside or outside the pseudosporangiae (Wink et al., 2004).

Chemotaxonomy. *Amycolatopsis* strains have cell walls which contain *meso*-diaminopimelic acid (*meso*-A_2pm), arabinose, and galactose (Carlsohn et al., 2007; Lechevalier et al., 1986; Takeuchi et al., 1992), that is, they have a wall chemotype IV *sensu* Lechevalier and Lechevalier (1970); an A1γ type peptidoglycan (Schleifer and Kandler, 1972); muramic acid in the *N*-acetylated form (Lee and Hah, 2001; Tan et al., 2006a, 2007), and phosphatidylethanolamine (PE; taxonomically significant phospholipid) and phosphatidylglycerol (PG) as major polar lipids (phospholipid type II *sensu* Lechevalier et al., (1981, 1977a) with a variable occurrence of diphosphatidylglycerol (DPG), hydroxyphosphatidylethanolamine (HPE), phosphatidylinositol (PI), phosphatidylinositol mannosides (PIMs), phosphatidylserine (PS), and phosphatidylmethylethanolamine (PME). The type strains of *Amycolatopsis nigrescens* and *Amycolatopsis saalfeldensis*, for instance, contain DPG, PE, HPE, PG, PI, PS, and uncharacterized glycolipids (Carlsohn et al., 2007; Groth et al., 2007). Details of the polar lipid composition of individual species are given in the species descriptions.

In general, most *Amycolatopsis* species have di-, tetra-, or hexahydrogenated menaquinones with nine isoprene units [MK-9(H_2,H_4,H_6)] as the predominant isoprenologue (Alderson et al., 1981; Lechevalier et al., 1986; Tan et al., 2006a, 2006b, 2007; Wink et al., 2003b; Yassin et al., 1991). In contrast, *Amycolatopsis decaplanina* contains a mixture of tetrahydrogenated menaquinones with eight and nine isoprene units (Wink et al., 2004), whereas *Amycolatopsis nigrescens* is characterized by the presence of major amounts of tetrahydrogenated menaquinones with eleven isoprene units and corresponding components with nine, ten, and twelve isoprene units (Groth et al., 2007). Variations in the proportions of di-, tetra-, and hexahydrogenated components with nine isoprene units amongst several *Amycolatopsis* species were attributed to the collection of biomass from different stages of the growth cycle (Yassin et al., 1991).

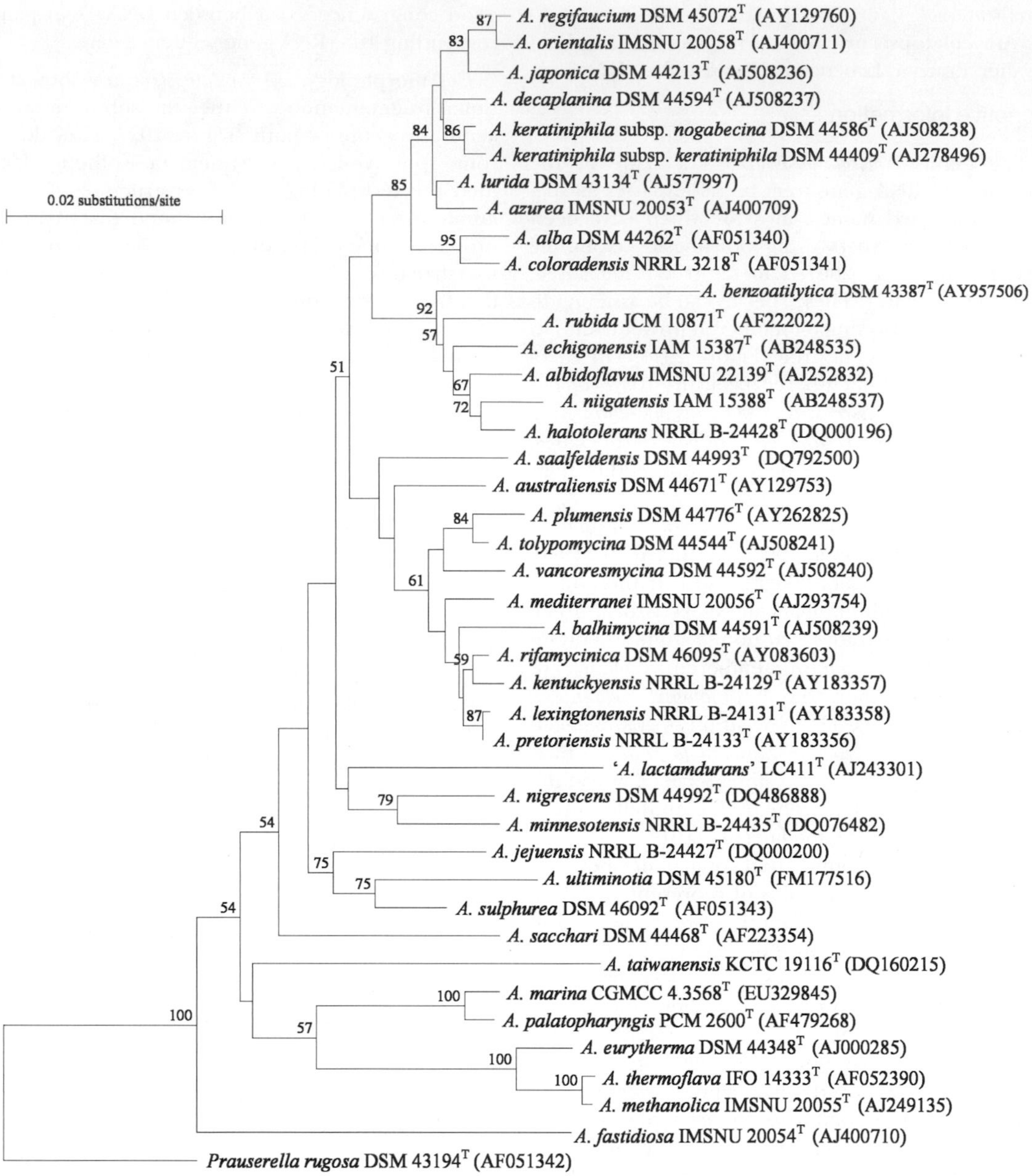

FIGURE 291. Neighbor-joining tree (Saitou and Nei, 1987) based on nearly complete 16S rRNA gene sequences showing relationships between the type strains of *Amycolatopsis* species. The numbers at the nodes indicate the levels of bootstrap support based on 1000 resampled datasets; only values above 50% are given. Bar = 0.02 substitutions per nucleotide position.

Amycolatopsis strains contain complex mixtures of iso-branched, straight chain saturated, and unsaturated components (Alderson et al., 1981; Kroppenstedt, 1985; Wink et al., 2003b; Yassin et al., 1993), but lack mycolic acids (Carlsohn et al., 2007; Huang et al., 2004; Lechevalier et al., 1986). Most strains have similar qualitative fatty acid profiles in which 14-methylpentadecanoic acid ($C_{16:0}$ iso) is the major component (Carlsohn et al., 2007; Kothe et al., 1989; Mertz and Yao, 1993; Wink et al., 2003b) though considerable amounts of hexadecanoic ($C_{16:0}$), 12-methyltridecanoic ($C_{14:0}$ iso), 13-methyltetradecanoic ($C_{15:0}$ iso), heptadecanoic ($C_{17:0}$), and octadecanoic ($C_{18:0}$) acids may be present (Huang et al., 2004; Lee, 2006; Lee and Hah, 2001). Quantitative differences in fatty acid profiles have been used to distinguish between *Amycolatopsis* species (Groth et al., 2007; Mertz and Yao, 1993).

Colony morphology. *Amycolatopsis* strains form well developed colonies on most standard media used to cultivate filamen-

TABLE 236. Percentage DNA–DNA relatedness values between *Amycolatopsis* species[a]

	A. orientalis	*A. alba*	*A. albidoflavus*	*A. azurea*	*A. balhimycina*	*A. benzoatilytica*	*A. coloradensis*	*A. decaplanina*	*A. echigonensis*	*A. eurytherma*	*A. fastidiosa*	*A. japonica*	*A. kentuckyensis*	*A. keratiniphila*	*A. lexingtonensis*	*A. lurida*	*A. marina*	*A. mediterranei*	*A. methanolica*	*A. niigatensis*	*A. regifaucium*	*A. rifamycinica*	*A. rubida*	*A. sulphurea*	*A. thermoflava*	*A. tolypomycina*	*A. tucumanensis*	*A. ultiminotia*	*A. vancoresmycina*
A. orientalis		24–30	18–19			24	0–8	51			0–7					48–59		0–7			43			0					
A. alba			56				27				0					24		25						0					
A. albidoflavus						46														33–34									
A. azurea		56					33	56	45–48		0					37		0						0					
A. balhimycina																		46								61			
A. benzoatilytica																													
A. coloradensis		27	33								0					37		0						0					53
A. decaplanina																32													
A. eurytherma																			60								40		
A. fastidiosa		0	0				0									0		11						7					
A. japonica														60															
A. kentuckyensis																		14				67							
A. lexingtonensis													43		42			22				47							
A. lurida	46	24–45	37				37	32			0							6						0					
A. marina																													
A. mediterranei		25	0				0				0											40		0		55			54
A. methanolica										58–60															21				
A. niigatensis								58–62																					
A. rubida								39–41												28–35									
A. palatopharyngis																	49												
A. pretoriensis													26		54			31				48							
A. sulphurea	0	0	0	–							7					0		0										29	
A. tolypomycina																													54
A. tucumanensis																													
A. ultiminotia																								31					

[a]Data taken from Al-Musallam et al. (2003), Bala et al. (2004), Chun et al. (1999), Kim et al. (2002a), Labeda (1995), Lechevalier et al. (1986), Everest and Meyers (2009), Stackebrandt et al. (2004), Tan et al. (2007), and Wink et al. (2004).

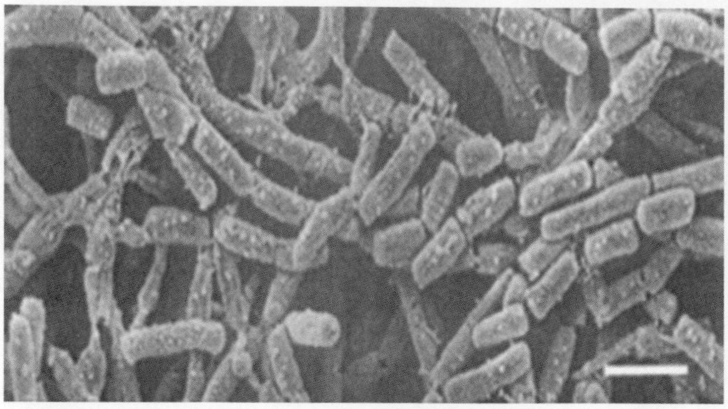

FIGURE 292. Scanning electron micrograph of *Amycolatopsis orientalis* KCTC 9412^{T} showing the formation of squarish fragments from the mycelium after growth on modified Bennett's agar at 28°C for 14 d. Bar = 1.0 μm.

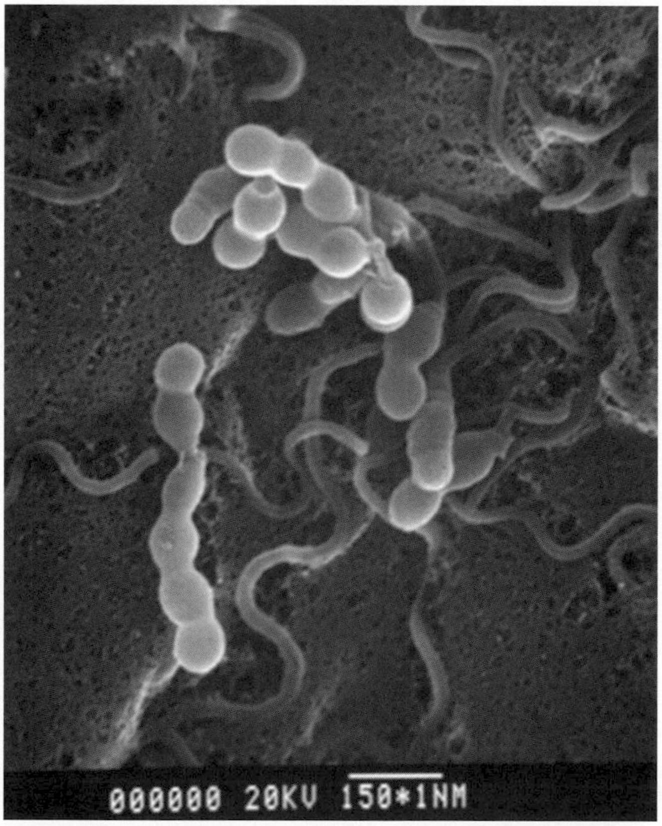

FIGURE 293. Scanning electron micrograph of *Amycolatopsis taiwanensis* KCTC 19116^{T} showing the formation of spore chains following growth on HV agar at 28°C for 14 d. Bar = 1.5 μm (Reproduced with permission of Tseng et al., 2006. Int. J. Syst. Evol. Microbiol. *56*: 1811–1815.)

tous actinomycetes and may carry abundant aerial hyphae (Mertz and Yao, 1993; Wink et al., 2003b, 2004). They grow particularly well on modified Bennett's agar supplemented with mannitol (0.5%, w/v) and soybean flour (0.5%, w/v) following incubation at 28°C for 14 d (Tan et al., 2006b). Tan and her colleagues were able to assign strains growing under these conditions to 35 color-groups based on aerial hyphal color, substrate mycelial pigmentation, and diffusible pigment color. Strains taken to represent many of the color groups formed distinct phyletic lines in the 16S rRNA *Amycolatopsis* gene tree indicating that color groupings can be predictive from a taxonomic perspective. Isolates assigned to two of these color groups were subsequently classified as novel *Amycolatopsis* species, namely *Amycolatopsis australiensis* and *Amycolatopsis regifaucium* (Tan et al., 2006a, 2007).

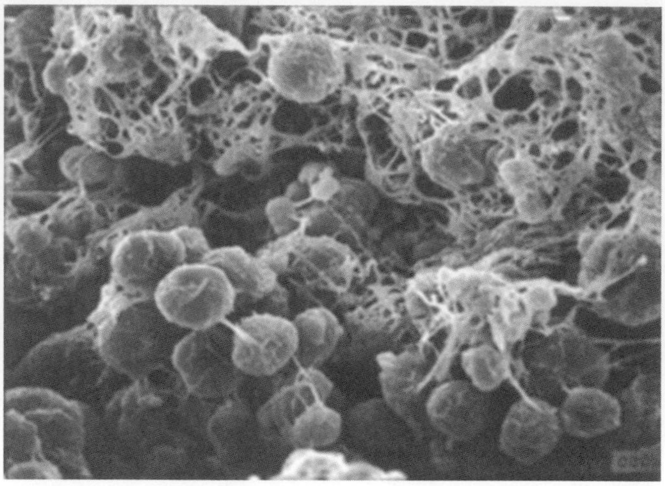

FIGURE 294. Scanning electron micrograph (×1000) of *Amycolatopsis decaplanina* DSM 44594^{T} showing the formation of pseudosporangiae following growth on oatmeal agar at 28°C for 14 d (Reproduced with permission from Wink et al., 2004. Int. J. Syst. Evol. Microbiol. *54*: 235–239.)

Nutrition and growth conditions. *Amycolatopsis* strains grow well on standard media such as glycerol-asparagine, inorganic-salts starch, oatmeal, peptone-yeast extract-iron, tyrosine, tryptone-yeast extract and yeast extract-malt extract agars (ISP media 2–7; Shirling and Gottlieb, 1966), and on modified Bennett's agar supplemented with mannitol and soybean flour (Tan et al., 2006b). *Amycolatopsis eurytherma*, *Amycolatopsis fastidiosa*, *Amycolatopsis methanolica*, and *Amycolatopsis thermoflava* strains grow well at 50°C or above (Chun et al., 1999; De Boer et al., 1990; Henssen et al., 1987; Kim et al., 2002a) and can be considered to be thermophilic actinomycetes (Brock, 1986; Cross, 1968). A fourth species, *Amycolatopsis sacchari*, contains organisms that are moderate thermophiles (Goodfellow et al., 2001). In contrast, organisms assigned to the remaining *Amycolatopsis* species are mesophiles (Lee and Hah, 2001; Tan et al., 2006a, 2007). *Amycolatopsis* strains grow well from pH 6.0 to 9.0.

Metabolism. *Amycolatopsis* strains are aerobic to facultatively anaerobic, chemo-organotrophic to facultatively autotrophic actinomycetes that degrade several organic substrates, produce acid from a range of sugars, and use diverse compounds as sole sources of carbon for energy and growth (Chun et al., 1999; De Boer et al., 1990; Gordon et al., 1978; Groth et al., 2007). The

TABLE 237. Bioactive compounds produced by members of the genus *Amycolatopsis*

Strain/taxon	Product	Feature	Reference
A. alba	Glycopeptide antibiotic	Inhibits peptidoglycan synthesis	Mertz & Yao (1993)
A. azurea	Azureomycins A and B	Glycopeptide antibiotics	Ōmura et al. (1979)
A azurea	Octacosamicins	Antifungal activity	Dobashi et al. (1988)
A. balhimycina	Balhimycin	Glycopeptide antibiotic	Nadkarni et al. (1994)
A. coloradensis	Avoparcin	Glycopeptide antibiotic	Kunstmann et al. (1968)
A. fastidiosa	Antibiotics 41034 and 41494	Macrobicyclic peptides	Celmer et al. (1977)
A. japonica	*(S,S)-N,N′*-ethylene-diaminedisuccinic acid	Inhibits phospholipase C	Nishikori et al. (1984)
	Dethymicin	Immunosuppressant	Ueno et al. (1992)
A. keratiniphila subsp. *nogabecina*	Nogabecin	Glycopeptide antibiotic	Shorin et al. (1957)
"*A. lactamdurans*"	Cephamycin C	β-Lactam antibiotic	Stapley et al. (1972)
	Efrotomycin	β-Isomer	Wax et al. (1976)
	3-Methylpseudouridine	Polyether	Nielsen and Arison (1989)
A. mediterranei	Rifamycins	Clinically useful ansamycins, active against *Mycobacterium* spp.	Margalith and Beretta (1960); Lancini et al. (1967); Birner et al. (1972); Lancini and Sartori (1976)
	31-Homorifamycin W		Wang et al. (1994)
	Kanglemycin A	Ansamycin-type antibiotic	Wang et al. (1988)
	Dethymicin	Immunosuppressant	Ueno et al. (1992)
	3-Hydroxyrifamycin S	Ansamycin antibiotic	Traxler et al. (1981)
	Protorifamycins	Ansamycin antibiotics	Ghisalba et al. (1978, 1979, 1980)
	Aromatic amino acids	Suitable for strain improvement	de Boer et al. (1990)
A. orientalis	Vancomycin	Glycopeptide antibiotic, active against severe bacterial infections	Pittenger and Brigham (1956)
	Glycopeptide compounds	Glycosyltransferase gene, *gtfA*	Baltz (2000)
	Muraceins	Muramyl peptide inhibitors	Bush et al. (1984)
	N-Demethylvancomycin	Vancomycin analog	Boeck et al. (1984)
	D-Amino acid specific peptidase		Sugie et al. (1988)
	MM47761, MM49721, MM55266, MM55268	Glycopeptide antibiotics	Box et al. (1991, 1990)
	Orienticin	Glycopeptide antibiotic	Tsuji et al. (1988)
	Quartromicin	Antiviral antibiotics	Tsunakawa et al. (1992)
	Antibiotic UK-69753	Efrotomycin-like antibiotic	Ruddock et al. (1987); Pacey et al. (1989)
A. lurida	Benzathrins	Quinone antibiotic with antitumor activity	Philip et al. (1957); Theriault et al. (1986); Rasmussen et al. (1986)
	Ristocetin	Glycopeptide antibiotic	Grundy et al. (1957)
A. regifaucium	Kigamicins	Antitumor antibiotics	Kunimoto et al. (2003)
A. sulphurea	Cetocycline	Tetracycline derivative	Proctor et al. (1978)
A. tolypomycina	Tolypomycin	Ansamycin-type antibiotic	Hasegawa et al. (1971); Kishi et al. (1972)
A. vancoresmycina	Homorifamycin; vancoresmycin	Ansamycin antibiotic; polyketide antibiotic	Hopmann et al. (2002).

type strain of *Amycolatopsis benzoatilytica* metabolizes *m*-hydroxybenzoate through a central protocatechuate intermediate, unlike other *Amycolatopsis* species which are unable to use this compound as growth substrate (Grund et al., 1990).

Members of the genus are a rich source of antibiotics and secondary metabolites (Table 237). Medically important antibiotics produced by *Amycolatopsis* strains include balhimycin (Nadkarni et al., 1994), dethymicin (Ueno et al., 1992), rifamycin (Sensi et al., 1959), and vancomycin (Barna and Williams, 1984). Balhimycin and vancomycin show antibiotic activity against methicillin-resistant *Stapylococcus aureus* strains, whereas dethymicin has a mode of action which distinguishes it from other immunosuppresants, such as cyclosporine, FK506, and rapamycin. Rifamycin is used to treat leprosy and tuberculosis, and to control infections in organ implants and AIDS patients.

Much of the emphasis in studies on the metabolism of *Amycolatopsis* strains has been focused on the application of biochemical and genetic approaches to the synthesis of antibiotics, as exemplified by studies on the biosynthesis of rifampin (Floss and Yu, 2005; Ghisalba and Nuesch, 1981; Hu et al., 1999; Schupp et al., 1998; Xu et al., 2005). The polyketide framework of rifamycin B is assembled from 3-amino-5-hydroxybenzoic acid (AHBA), a product of the aminoshikimate pathway (Kim et al., 1998; Yu et al., 2001), two molecules of acetate and eight of propionate. Five multifunctional proteins (RifA–RifE) and an amide synthase (RifF) catalyze the synthesis of the core structure of rifampin (August et al., 1998; Schupp et al., 1998; Xu et al., 2005).

A number of non-antibiotic bioactive metabolites are formed by *Amycolatopsis* strains, including (*S,S*)-*N,N'*-ethylenediamine disuccinic acid from *Amycolatopsis japonica* (Nishikori et al., 1984), a D-amino acid-specific peptidase from *Amycolatopsis orientalis* (Sugie et al., 1988), and a polylactic acid depolymerase from *Amycolatopsis* sp. strain K104-1 (Nakamura et al., 2001). *Amycolatopsis orientalis* produces chitinases of potential value in the production of chitin hexamers (Tominaga and Tsujisaka, 1976; Usui et al., 1987); this organism converts tetra-*N*-acetyl-chitotelraose to a mixture of hexa-*N*-acetyl-chitohexaose and the corresponding dimer by a transglycosylation reaction (Usui et al., 1987). *Amycolatopsis* sp. strain HR167 is used to produce vanillic acid from ferulic acid (Rabenhorst and Hopp, 1997), and an *Amycolatopsis* isolate for the bioconversion of lovastatin to the novel compound wuxistatin (Zhuge et al., 2008).

Amycolatopsis methanolica NCIB 11946^T is a facultative methylotrophic actinomycete which was previously assigned to the genera *Nocardia* and *Streptomyces* (De Boer et al., 1990). The organism assimilates formaldehyde by the ribulose monophosphate shunt (Hazeu et al., 1983), a finding that made it a candidate for fermentative overproduction of aromatic amino acids (Dijkhuizen et al., 1985; Morinaga and Hirose, 1984). Regulation of aromatic amino acid biosynthesis in the strain has been studied in detail (Abou-Zeid et al., 1995; De Boer et al., 1989; Euverink et al., 1992, 1995) and stable mutants have been isolated and grown under diverse conditions in batch and continuous culture (De Boer et al., 1988; Hazeu et al., 1983). Prephenate dehydratase (PDT) is a key regulatory enzyme in L-phenylalanine biosynthesis, the PDT protein of which has been purified and characterized as a homotetrameric enzyme with 34 kDa subunits (Euverink et al., 1995). It has been shown that *Amycolatopsis methanolica* metabolizes glucose via the Embden–Meyerhof–Parnas pathway (Alves et al., 1994), and produces a PPi-dependent phosphofructokinase with biochemical characteristics similar to both ATP- and PPi-dependent enzymes during growth on glucose (Alves et al., 1996).

A copper-resistant *Amycolatopsis* strain was isolated from copper-polluted sediment by Albarracin et al. (2008). The organism accumulates high levels of copper and is considered to have potential as an agent of bioremediation. It was subsequently characterized and described as a novel *Amycolatopsis* species, *Amycolatopsis tucumanensis* (Albarracin et al., 2010). These workers also found that the type strain of *Amycolatopsis eurytherma*, a close relative of *Amycolatopsis tucumanensis*, has a moderate copper resistance profile.

Genetics. Genetic analyses show that the antibiotic potential of members of the genus *Amycolatopsis* is much greater than previously recognized (Banskota et al., 2006; Wood et al., 2007). Improved genetic tools, including new cloning, expression, and shuttle vectors are being developed to manipulate and regulate genes involved in secondary metabolite production and to gain an insight into the origin, evolution, and functional properties of extrachromosomal elements (Malhotra and Lal, 2007; te Poele et al., 2007, 2008).

Amycolatopsis methanolica NCIB 11946 contains a 13.3 kb conjugative element, pMEA300, which is maintained as an integrated element, but may also be present as a free circular plasmid (Malhotra and Lal, 2007; te Poele et al., 2007, 2008). Various cloning vectors based on these plasmids have been constructed (Madón and Hutter, 1991; Vrijbloed et al., 1995a), and used to clone genes involved in glucose and methanol metabolism and in aromatic amino acid biosynthesis into *Amycolatopsis methanolica* (Alves et al., 1996; Hektor, 1997; Vrijbloed et al., 1995c, 1995d). The conjugative element pMEA100 (23.7 kb) from *Amycolatopsis mediterranei* LBG A3136 also exists in an integrated form and in the free state, but in *Amycolatopsis mediterranei* ATCC 13685 it is only found in the free form (Moretti et al., 1985).

The integrated elements of *Amycolatopsis mediterranei* and *Amycolatopsis methanolica* have been sequenced (te Poele et al., 2006; te Poele et al., 2008; Vrijbloed, 1996), and on the basis of structural and functional similarities shown to belong to a class of integrative and conjugative elements (ICEs; Moretti et al., 1985; Burrus et al., 2002) that have been found in several other filamentous actinomycetes, such as pSE211 and pSE101 from *Saccharopolyspora erythraea* (Brown et al., 1988, 1994), pSAM2 from *Streptomyces ambofaciens* (Boccard et al., 1989; Pernodet et al., 1984; Smokvina et al., 1991), SLP1 from "*Streptomyces coelicolor*" A3(2) (Bibb et al., 1981) and pSG1 from *Streptomyces griseus* (Cohen et al., 1985). This class of elements integrate site-specifically in a tRNA gene of the host genome, but unlike other ICEs, replicate autonomously like a plasmid. Most are self-transmissible and several can mobilize chromosomal markers (Bibb et al., 1981; Brown et al., 1988; Hopwood et al., 1984; Moretti et al., 1985; Smokvina et al., 1988; Vrijbloed, 1996). The integrase (Int) directs site-specific DNA recombination between the attP site and a chromosomal attB site (Boccard et al., 1989).

Characterization of deletion derivatives of pMEA300 led to the identification of genes needed for replication, regulation, integration, and conjugation (Vrijbloed et al., 1994, 1995a, 1995b, 1995c, 1995d). The organization of genes involved in replication, excision, and integration (attp, int, repAM, xis) in pMEA300 is conserved (te Poele et al., 2007, 2008). It has been shown that most of these elements can mediate the pock-formation phenotype, reflecting growth retardation of the recipient which occurs during conjugation (Vrijbloed et al., 1995b). These workers found that the *traJ* gene was required for the transfer of pMEA300 into recipient strains lacking this element. TraJ of pMEA300 shows 33% identity to TraJ of pMEA100 (te Poele et al., 2008) and 27% identity to a cell division FtsK/SpoIII protein of *Frankia* strain EAN1 pec, and to TraB of *Streptomyces ghanaenis* plasmid pS65 (te Poele et al., 2007). This latter protein is a septal DNA translocator which mediates a unique conjugation mechanism that translocates unprocessed double-stranded DNA molecules to recipient strains (Reutler et al., 2006).

te Poele et al. (2007, 2008) determined the presence and distribution of a new class of replication initiator and transfer proteins, namely RepAM and TraJ. They screened a collection of over 100 *Amycolatopsis* strains that had been isolated from different geographical locations and found pMEA sequences to be widely distributed. They identified two geographically distinct pMEA-like elements. Phylogenetic analysis of their deduced RepAM and TraJ protein sequences revealed clusters with protein sequences of either pMEA100 or pMEA300. The sequences which clustered with pMEA100 were from European strains and those from pMEA300 encompassed Australasian strains. The *repAM* and *traJ* genes were linked with the 16S rRNA genes of the host strains suggesting that the pMEA-elements co-evolved with their hosts rather than being dispersed by horizontal transfer of the free replicating form.

Amycolatopsis benzoatilytica (previously *Amycolatopsis orientalis*) DSM 43387^T (Majumdar et al., 2006) contains a 29.6 kb non-integrative, freely replicating plasmid, pA387 (Lal et al., 1991). This plasmid has been completely sequenced and used to construct cloning and conjugative shuttle vectors (Lal et al., 1991; Malhotra et al., 2008) and to generate *Escherichia coli–Amycolatopsis* shuttle vectors (Ding et al., 2003; Khanna et al., 1998; Kumar et al., 1994; Lal et al., 1998; Priefert et al., 2002; Tuteja et al., 2000). These vectors replicate and are maintained in members of several *Amycolatopsis* species, including *Amycolatopsis japonica* (Stegmann et al., 2001), "*Amycolatopsis lactamdurans*" (Kumar et al., 1994), *Amycolatopsis mediterranei* (Ding et al., 2003; Lal et al., 1991, 1998; Tuteja et al., 2000), *Amycolatopsis orientalis* (Lal et al., 1991), and *Amycolatopsis rifamycinica* (Khanna et al., 1998; Lal et al., 1998; Tuteja et al., 2000). The value of pA387-derived shuttle vectors has been demonstrated in several studies (Kumar et al., 1996; Xu et al., 2005), notably by increased production of cephamycin C by "*Amycolatopsis lactamdurans*" (Chary et al., 1997, 2000).

The direct mycelium transformation system developed for *Amycolatopsis mediterranei* by Madón and Hütter (1991) was applied and optimized to enhance the conversion of ferulic acid into vanillin using *Amycolatopsis* sp. strain HR 167 (Priefert et al., 2002). The density of the mycelial suspensions in the transformation mixture and the methylation state of the plasmid DNA used for the transformation were critical factors. The transformation rates achieved with plasmid DNA isolated from *Escherichia coli* ET 12567 was 3,500-fold higher than those obtained with DNA isolated from *Escherichia coli* XLI-Blue. Other interesting developments include the isolation of balhimycin genes from *Amycolatopsis mediterranei* using a reverse cloning procedure (Pelzer et al., 1999), the discovery that the *pfp* gene of *Amycolatopsis methanolica*, which encodes PPi-dependent phosphofructokinase, is located on a 2.3 kb *Pvu*II fragment (Alves et al., 1996), and the detection of genes involved in the modification of the polyketide backbone of rifamycin B by *Amycolatopsis mediterranei* S699 (Xu et al., 2005).

Antibiotic sensitivity. Comprehensive comparative studies on the antibiotic sensitivity patterns of all *Amycolatopsis* species are needed now that there is evidence that members of this genus may be considered as emerging pathogens. Preliminary studies show that closely related *Amycolatopsis* species can be distinguished using antibiotic sensitivity data (De Boer et al., 1990).

Pathogenicity. There is no conclusive evidence that *Amycolatopsis* strains can cause disease in humans, although the type strain of *Amycolatopsis benzoatilytica* has been implicated as an agent of submandibular mycetoma (Majumdar et al., 2006; Scharfen, 1971). In addition, members of the genus are occasionally encountered in clinical material (Mishra et al., 1980; Schaal and Beaman, 1984). *Amycolatopsis orientalis* has been isolated from cerebrospinal fluid (Gordon et al., 1978) and *Amycolatopsis palatopharyngis* has been isolated from infected palatopharyngeal mucosa of a 70-year-old male (Huang et al., 2004). There is evidence that actinomycetes can cause placentitis and abortion in horses (Donahue and Williams, 2000; Giles et al., 1993; Hong et al., 1993). The distinctive type of nocardiosis in horses, nocardioform placentitis, is recognized by the formation of lesions on the chorionic surface of the placenta and the isolation of Gram-stain-positive, branching bacteria. Three *Amycolatopsis* species, *Amycolatopsis kentuckyensis*, *Amycolatopsis lexingtonensis*, and *Amycolatopsis pretoriensis*, were proposed to accommodate strains isolated from equine placentas (Labeda et al., 2003).

Ecology. Most members of the genus *Amycolatopsis* have been isolated from soil and vegetable matter though their roles in these environments have not been studied. It seems likely that the primary habitat is soil, as exemplified by the isolation of *Amycolatopsis australiensis* from an arid composite soil (Tan et al., 2006a), *Amycolatopsis eurytherma* from scrubland soil (Kim et al., 2002a), *Amycolatopsis keratiniphila* from marsh soil (Al-Musallam et al., 2003), *Amycolatopsis plumensis* from a brown hypermagnesian ultramafic soil (Saintpierre-Bonaccio et al., 2005), *Amycolatopsis rubida* from a coniferous forest soil (Huang et al., 2001), and *Amycolatopsis ultiminotia* from rhizosphere soil (Lee, 2009). Additional *Amycolatopsis* species have been recovered from more unusual sources, such as the isolation of *Amycolatopsis echigonensis* from a filtrate substrate (Ding et al., 2007), *Amycolatopsis jejuensis* from a natural cave (Lee, 2006), *Amycolatopsis nigrescens* from a Roman catacomb (Groth et al., 2007), *Amycolatopsis saalfeldensis* from a medieval alum slate mine (Carlsohn et al., 2007), and *Amycolatopsis tucumanensis* from copper-polluted sediments (Albarracin et al., 2010). *Amycolatopsis* strains have also been isolated from hyper-arid soils of the Atacama Desert (Okoro et al., 2009).

Large numbers of *Amycolatopsis* strains have been recovered from Australian arid soils, and from an environmental sample collected from under a sycamore tree in Newcastle-upon-Tyne, UK, using a dilution plate procedure and three selective media (Tan et al., 2006b). Many of the isolates were related to either *Amycolatopsis mediterranei* or *Amycolatopsis orientalis* on the basis of 16S rRNA gene sequencing, but others formed new phyletic lines. Some of the novel isolates were classified as *Amycolatopsis australiensis* sp. nov. (Tan et al., 2006a) and *Amycolatopsis regifaucium* sp. nov. (Tan et al., 2007).

Enrichment and isolation procedures

Members of most *Amycolatopsis* species have been isolated by plating dilutions of soil suspensions onto nonselective media, supplemented with antifungal antibiotics, followed by colony selection and characterization. This hit and miss approach, for instance, led to the isolation of *Amycolatopsis halotolerans* and *Amycolatopsis jejuensis* on starch-casein agar (Lee, 2006), *A mediterranei* on Bennett's agar (Margalith and Beretta, 1960), *Amycolatopsis nigrescens* on nutrient agar (Groth et al., 2007), *A rubida* on glucose-asparagine agar (Huang et al., 2001), *Amycolatopsis saalfeldensis* on casein mineral agar (Carlsohn et al., 2007), and *Amycolatopsis taiwanensis* on HV agar (Tseng et al., 2006); in the main, isolation plates were incubated at either 28°C or 30°C for 14 d.

Several *Amycolatopsis* species have been isolated from environmental samples using less subjective isolation procedures. *Amycolatopsis keratiniphila* was isolated from soil enriched with sterilized and defatted wool (Al-Musallam et al., 2003), *Amycolatopsis palatopharyngis* was isolated on brain-heart infusion agar at 37°C for 5 d under microaerophilic conditions (Huang et al., 2004), and *Amycolatopsis plumensis* was isolated on yeast extract-malt extract agar supplemented with streptomycin (Saintpierre-Bonaccio et al., 2005). In addition, *Amycolatopsis sacchari* strains were recovered from diverse environmental samples using a wind tunnel technique (Lacey, 1971; Lacey and Dutkiewicz, 1976a); the Andersen sampler was loaded with Petri dishes containing half-strength nutrient agar (Goodfellow et al., 2001).

Tan et al. (2006b) designed a number of agar media to isolate *Amycolatopsis* strains from soil using putative selective agents drawn from phenotypic databases generated in numerical taxonomic studies on members of the family *Pseudonocardiaceae*. Three of the resultant media formulations proved to be particularly effective in recovering *Amycolatopsis* strains from soil samples, namely media SM1 (Stevenson's basal medium supplemented with D-sorbitol and neomycin sulfate), SM2 (Stevenson's basal medium supplemented with D-melezitose and neomycin sulfate), and SM3 (Gauze's medium supplemented with nalidixic acid and novobiocin); all three media contained antifungal antibiotics. Inoculated plates were incubated at 28°C for up to 21 d; the appearance of crusty or leathery colonies covered with white or whitish-yellow aerial hyphae proved to be characteristic of *Amycolatopsis* strains. Representative isolates were found to have morphological and chemotaxonomic features consistent with their assignment to the genus *Amycolatopsis*. Tan and her colleagues generated a set of genus-specific oligonucleotides which were used for the rapid identification of putative *Amycolatopsis* strains recovered from the selective isolation plates.

Maintenance procedures

Amycolatopsis strains can be cultivated on glucose-yeast extract agar or modified Bennett's agar. After incubation at 28°C to allow abundant growth, cultures may be maintained at 4°C or room temperature for up to 6 months. Long-term preservation can be achieved by lyophilization or in 20% glycerol frozen at −20°C or −80°C.

Differentiation of the genus *Amycolatopsis* from other genera

The genus *Amycolatopsis* can be distinguished from other genera in the *Pseudonocardiaceae* family by using a combination of chemotaxonomic and morphological markers (see Table 228 in the treatment of the family *Pseudonocardiaceae*). *Amycolatopsis* strains can also be distinguished from related taxa by using the genus-specific primers AMY2 and ATOP (Tan et al., 2006b).

Taxonomic comments

The genus *Amycolatopsis* was proposed by Lechevalier et al. (1986) for actinomycetes that formed substrate hyphae that tended to fragment into squarish elements, and aerial hyphae which, when formed, were either sterile or fragmented into chains of squarish to oval bodies or spore-like structures. The organisms had a wall chemotype IV (*meso*-diaminopimelic acid, arabinose, and galactose), a type II phospholipid pattern (PE with or without PME), and MK-9(H_2,H_4) as predominant isoprenologues, but lacked mycolic acids.

The founder members of the genus, *Amycolatopsis orientalis* (comprising *Amycolatopsis orientalis* subsp. *orientalis* and *Amycolatopsis orientalis* subsp. *lurida*), *Amycolatopsis mediterranei*, *Amycolatopsis rugosa*, and *Amycolatopsis sulphurea* were formally members of other taxa, notably the genera *Nocardia* and *Streptomyces*. The type species, *Amycolatopsis orientalis*, had been classified as *Actinomyces orientalis* (Krasil'nikov, 1981), *Nocardia orientalis* (Pridham, 1970), *Proactinomyces orientalis* (Rautenstein et al., 1975), and *Streptomyces orientalis* (Pittenger and Brigham, 1956). Similarly, *Amycolatopsis mediterranei* had previously been known as *Nocardia mediterranei* (Thiemann et al., 1969) and "*Streptomyces mediterranei*" (Margalith and Beretta, 1960). "*Nocardia lurida*" (Grundy et al., 1957), *Nocardia rugosa* (di Marco and Spalla, 1957), and "*Nocardia sulphurea*" (Oliver and Sinclair, 1964) were transferred to the genus *Amycolatopsis* as *Amycolatopsis orientalis* subsp. *lurida*, *Amycolatopsis rugosa*, and *Amycolatopsis sulphurea*, respectively.

The valid publication of *Amycolatopsis orientalis* subsp. *lurida* (*ex* Grundy et al. (1957) Lechevalier et al. (1986) automatically created the name *Amycolatopsis orientalis* subsp. *orientalis* (Pittenger and Brigham, 1956) Lechevalier et al. (1986) according to Rule 46 of the Bacteriological Code (Howey et al., 1990). However, the elevation of *Amycolatopsis orientalis* subsp. *lurida* to full species status automatically meant that the name *Amycolatopsis orientalis* subsp. *orientalis* reverted to the earlier name, *Amycolatopsis orientalis* (Pittenger and Brigham, 1956) Lechevalier et al. (1986), according to Rule 40d (formerly Rule 46) of the Bacteriological Code.

The taxonomic positions of *Amycolatopsis mediterranei*, *Amycolatopsis orientalis*, and *Amycolatopsis sulphurea* were underpinned by molecular systematic data (Embley et al., 1988a; Warwick et al., 1994), but the remaining species, *Amycolatopsis rugosa*, was subsequently reclassified as *Prauserella rugosa* gen. nov., sp. nov. (Kim and Goodfellow, 1999). Similarly, *Amycolatopsis orientalis* subsp. *lurida* was found to merit species status as *Amycolatopsis lurida* (Stackebrandt et al., 2004). Several other taxa were soon assigned to the genus, notably *Amycolatopsis coloradensis* Labeda (1995), *Amycolatopsis azurea* Henssen et al. (1987), and *Amycolatopsis fastidiosa* Henssen et al. (1987); these taxa previously had been classified as *Streptomyces candidis* (Kunstmann et al., 1968), *Pseudonocardia azurea* (Ōmura et al., 1979), and *Amycolatopsis fastidiosa* (Henssen et al., 1987), respectively. Another organism, *Amycolatopsis methanolica* de Boer et al. (1990) had formally been a member of the genus *Nocardia* (Hazeu et al., 1983) and, prior to that, of the genus *Streptomyces* (Kato et al., 1977).

The species *Amycolatopsis fastidiosa* was proposed by Henssen et al. (1987), based on chemotaxonomic and morphological properties, for a strain initially described as "*Pseudonocardia fastidiosa*" by Celmer et al. (1977) in a US patent. However, it has become increasingly clear from 16S rRNA gene sequence and *gyrB* sequence analysis, and DNA–DNA relatedness data mentioned earlier, that the strain is misplaced in the genus *Amycolatopsis*. It has also been noted that the strain did not give an amplification product using *Amycolatopsis*-specific oligonucleotide primers (Tan et al., 2006b). The organism has recently been transferred to the genus *Actinokineospora* as *Actinokineospora fastidiosa* comb. nov. following a careful evaluation of its chemotaxonomic, morphological, and physiological properties and of its position in the order *Pseudonocardiales* (formerly suborder *Pseudonocardineae*) 16S rRNA gene tree (Labeda et al., 2010b).

Differentiation of the species of the genus *Amycolatopsis*

Amycolatopsis species can be distinguished using a combination of 16S rRNA gene sequence (Figure 291), DNA: DNA relatedness (Table 236) and phenotypic (Table 237) data, although a common set of phenotypic tests has not been used in the description of all species. Some *Amycolatopsis* species have been differentiated using MALDI-TOF MS spectra (Groth et al., 2007) and on the basis of fatty acid (Ding et al., 2007; Lee, 2006; Mertz and Yao, 1993) and ribotype (Wink et al., 2003b, 2004) patterns.

List of species of the genus *Amycolatopsis*

1. **Amycolatopsis orientalis** (Pittenger and Brigham 1956) Lechevalier, Prauser, Labeda and Ruan 1986, 35[VP] (Basonym: *Nocardia orientalis* Pridham and Lyons 1969, 183; "*Streptomyces orientalis*" Pittenger and Brigham 1956, 642.)

or.i.en.tal'is. L. fem. adj. *orientalis* of the orient.

Forms an extensive branched substrate mycelium which appears slightly zig-zag in places. When produced, the aerial mycelium differentiates into cylindrical, occasionally ovoid spores which are carried in straight to flexuous chains. Blue colored colonies and a light brown diffusible pigment are formed on modified Bennett's agar. Grows at 10–42°C, but not at 45°C.

Positive for *N*-acetyl-β-glucosamidase, acid and alkaline phosphatases, α-chymotrypsin, esterase (C4), and esterase lipase (C8), but negative for cystine arylamidase, α-fucosidase, α- and β-galactosidases, α- and β-glucosidases, β-glucuronidase, α-mannosidase, trypsin, and valine arylamidase (API ZYM tests). Positive for β-galactosidase, but not for arginine dehydrolase or lysine and ornithine decarboxylases (API 20E tests). Acetoin-positive, but indole-negative. Arbutin is hydrolyzed, but does not produce H_2S.

Elastin, Tween 40, uric acid, and xylan are degraded, but not adenine. Acid is not produced from either dulcitol or methyl β-D-xyloside. Acetate, citrate, malate, propionate, pyruvate, and succinate are used as sole carbon sources, but not benzoate, mucate, oxalate, or tartrate.

Resistant (μg/ml) to gentamicin sulfate (5), neomycin sulfate (8), novobiocin (10), penicillin G (20), polymyxin B (50), rifampin (10), streptomycin sulfate (16), tobramycin sulfate (8), and vancomycin hydrochloride (0.25), but sensitive to lysozyme. Additional phenotypic properties are shown in Table 238.

Whole-organism hydrolysates are rich in arabinose and galactose. Cellular fatty acid profiles contain major proportions of $C_{15:0}$ iso (15.2%), $C_{16:0}$ iso (23.8%), $C_{17:0}$ anteiso (13.7%), and $C_{17:0}$ (12.6%), and smaller proportions (<10%) of $C_{14:0}$ iso (4.3%), $C_{15:0}$ anteiso (11%), $C_{15:1}$ (1.3%), $C_{15:0}$ (7.2%), $C_{16:1}$ (3.5%), $C_{16:0}$ (8.6%), $C_{17:0}$ iso (1.5%), $C_{17:1}$ (1.2%), $C_{18:1}$ iso (1.5%), and $C_{18:0}$ (3.0%). Mycolic acids are absent. The polar lipid pattern contains PE as the diagnostic component. The predominant menaquinones are MK-9(H_2,H_4).

Source: isolated from soil, vegetable matter, and clinical specimens.

DNA G+C content (mol%): 66 (T_m).

Type strain: ATCC 19795, CIP 107113, DSM 40040, IFO (now NRBC) 12806, IMSNU 20058, JCM 4235, JCM 4600, NRRL 2450, UNIQEM 181, VKM Ac-866.

Sequence accession nos: AJ400711 (16S rRNA gene); EU822906 (*gyrB*).

2. **Amycolatopsis alba** Mertz and Yao 1993, 719[VP]

al'ba. L. fem. adj. *alba* white, referring to the white aerial hyphae.

Forms a yellowish-brown substrate mycelium and a white aerial mycelium on Bennett's, glycerol-asparagine, tomato paste-oatmeal, and yeast extract-glucose agars. A light brown soluble pigment is produced on Emerson's and yeast extract-malt extract agars. Cylindrical, smooth spores are formed with a typical cobweb morphology. Grows between 15 and 37°C, but not at 10 or 45°C.

Arbutin is hydrolyzed and H_2S is produced. Elastin, Tween 40, uric acid, and xylan are degraded.

Positive for *N*-acetyl-β-glucosamidase, acid and alkaline phosphatases, α-chymotrypsin, cystine arylamidase, esterase (C4), esterase lipase (C8), β-galactosidase, α-glucosidase, β-glucuronidase, leucine arylamidase, α-mannosidase, naphthol-AS-BI-phosphohydrolase, trypsin, and valine phosphatase, but negative for α-fucosidase, α-galactosidase, and lipase C14 (API ZYM tests). Positive for arginine dihydrolase, β-galactosidase, and lysine and ornithine decarboxylases (API 20E tests).

Acid is produced from ribose, but not from cellulose, dulcitol, ethanol, glycogen, inulin, or xylitol. Acetate, butyrate, formate, lactate, malate, propionate, pyruvate, and succinate are used as sole carbon sources, but not oxalate, mucate, or tartrate.

Resistant (μg/ml, unless otherwise indicated) to bacitracin (10 U), kanamycin (30), nalidixic acid (30), neomycin (30), oleandomycin (15), rifampin (5), streptomycin (10), sulfonamide (200), and tetracycline (30), but sensitive to cephalothin (30), gentamicin (10), lincomycin (2), penicillin (10 U), streptomycin (10), tobramycin (10), and vancomycin (30). Grows in the presence of lysozyme. Additional phenotypic properties are shown in Table 238.

Whole-organism hydrolysates are rich in arabinose and galactose. The cellular fatty acid profile includes major proportions of $C_{15:0}$ iso (30.0%) and $C_{17:0}$ (18.6%), and smaller proportions (<10%) of $C_{14:0}$ iso (2.9%), $C_{15:0}$ anteiso (6.6%), $C_{15:0}$ (6.0%), $C_{16:0}$ iso (9.3%), $C_{16:0}$ iso (7.8%), $C_{17:0}$ iso (2.8%), $C_{17:0}$ anteiso (3.8%), and $C_{18:0}$ (1.5%). Mycolic acids are absent. PE is the diagnostic polar lipid. The predominant menaquinone is MK-9(H_4); also contains minor amounts of MK-8(H_4).

Source: isolated from soil.

TABLE 238. Characteristics differentiating the species of the genus *Amycolatopsis*[a]

Characteristic	1. *A. orientalis*	2. *A. alba*	3. *A. albidoflavus*	4. *A. australiensis*	5. *A. azurea*	6. *A. balhimycina*	7. *A. benzoatilytica*	8. *A. coloradensis*	9. *A. decaplanina*	10. *A. echigonensis*	11. *A. eurytherma*	12. *A. halotolerans*	13. *A. japonica*	14. *A. jejuensis*	15. *A. kentuckyensis*	16a. *A. keratiniphila* subsp. *keratiniphila*	16b. *A. keratiniphila* subsp. *nogabecina*	17. *A. lexingtonensis*	18. *A. lurida*	19. *A. marina*	20. *A. mediterranei*	21. *A. methanolica*	22. *A. minnesotensis*	23. *A. nigrescens*	24. *A. niigatensis*	25. *A. palatopharyngis*	26. *A. plumensis*	27. *A. pretoriensis*	28. *A. regifaucium*	29. *A. rifamycinica*	30. *A. rubida*	31. *A. saalfeldensis*	32. *A. sacchari*	33. *A. sulphurea*	34. *A. taiwanensis*	35. *A. thermoflava*	36. *A. tolypomycina*	37. *A. tucumanensis*	38. *A. ultiminotia*	39. *A. vancoresmycina*
Acid production from:																																								
Adonitol	+	+	+	−	+	nr	+	−	nr	$+^{w}$	+	+	+	+	+	+	nr	+	+	nr	−	+	+	nr	$+^{w}$	+	+	−	−	+	+	nr	+	−	nr	+	−	nr	−	−
Arabinose	+	+	+	+	+	+	+	−	+	nr	+	−	+	+	+	+	+	+	+	−	+	−	+	+	nr	+	+	+	+	+	+	+	+	−	+	+	+	nr	−	+
Cellobiose	+	+	+	+	+	nr	+	+	nr	nr	+	+	+	−	+	+	nr	+	+	+	+	+	+	nr	nr	−	+	+	+	+	+	nr	+	−	nr	+	+	nr	+	+
Dextrin	+	−	nr	v	nr	nr	+	+	nr	nr	$+^{w}$	nr	+	nr	+	nr	nr	+	+	nr	+	−	−	nr	nr	−	nr	+	−	nr	−	nr	+	+	nr	−	nr	nr	nr	nr
meso-Erythritol	+	+	+	nr	+	nr	+	−	nr	$+^{w}$	+	−	+	−	−	nr	nr	$+^{w}$	+	nr	−	+	+	nr	$+^{w}$	+	nr	+	−	nr	+	nr	+	−	nr	+	nr	nr	−	nr
Fructose	+	+	+	+	+	+	+	+	+	nr	+	+	+	+	+	+	+	+	+	+	+	+	+	+	nr	+	+	+	nr	+	+	+	+	nr	−	+	+	$+^{w}$	−	+
Galactose	+	+	+	+	+	nr	+	+	nr	+	+	+	+	+	+	+	nr	+	+	+	+	+	+	nr	+	+	−	+	+	+	+	nr	+	+	nr	+	+	nr	+	+
Glucose	+	+	−	+	nr	nr	+	nr	+	$+^{w}$	nr	−	nr	−	+	nr	nr	+	nr	+	+	nr	+	+	$+^{w}$	nr	nr	+	nr	nr	nr	+	nr	nr	+	nr	nr	nr	+	nr
Glycerol	+	+	nr	+	nr	nr	nr	nr	nr	nr	nr	+	nr	−	+	+	nr	+	+	nr	+	nr	+	nr	nr	nr	nr	+	nr	nr	nr	nr	nr	nr	nr	nr	nr	nr	+	nr
myo-Inositol	+	+	+	+	+	+	+	+	−	+	+	+	+	+	+	+	+	+	+	+	+	−	+	+	−	+	nr	+	+	+	+	+	−	+	nr	−	+	−	+	+
Lactose	+	+	+	+	+	nr	+	−	nr	+	$+^{w}$	−	+	−	+	+	nr	+	+	−	+	−	+	nr	−	−	−	+	−	−	−	nr	+	−	nr	+	−	nr	+	−
Maltose	+	+	−	v	nr	nr	−	+	nr	nr	−	−	+	−	+	nr	nr	+	+	+	+	nr	+	nr	nr	−	+	+	nr	−	−	nr	+	+	nr	−	nr	nr	+	nr
Mannitol	+	+	−	+	+	+	+	+	+	−	+	+	+	+	−	+	+	$+^{w}$	+	+	+	+	+	+	+	−	+	−	nr	−	+	+	+	+	nr	+	+	nr	+	+
Mannose	−	+	+	+	nr	nr	+	nr	nr	−	nr	+	nr	+	+	nr	nr	+	nr	nr	+	nr	+	nr	−	nr	nr	+	+	nr	nr	nr	nr	nr	nr	nr	nr	nr	+	nr
Melezitose	+	−	−	−	+	nr	nr	−	nr	nr	−	−	−	−	−	nr	nr	−	nr	nr	−	+	−	nr	nr	−	nr	−	−	−	−	−	−	nr	−	−	nr	nr	−	nr
Melibiose	+	+	−	−	nr	+	nr	−	nr	nr	−	+	+	−	+	+	nr	+	−	nr	+	−	−	nr	nr	−	−	+	−	nr	−	nr	−	−	nr	+	−	nr	−	−
Methyl α-D-glucoside	+	+	−	nr	nr	nr	nr	+	nr	nr	−	−	+	−	+	nr	nr	+	+	nr	+	−	−	nr	nr	−	nr	+	nr	nr	−	nr	+	−	nr	+	nr	nr	−	nr
Raffinose	−	+	−	v	+	+	nr	−	−	nr	−	−	+	−	+	+	−	+	−	−	+	−	−	+	nr	+	−	+	−	−	−	−	nr	−	nr	+	−	−	−	+
Rhamnose	+	−	nr	nr	nr	+	+	−	−	+	+	−	−	−	+	+	+	+	−	+	+	+	+	$+^{w}$	+	−	−	+	nr	−	+	$+^{w}$	+	−	nr	−	+	$+^{w}$	+	+
Salicin	+	+	−	nr	nr	nr	nr	+	nr	nr	−	−	+	−	+	nr	nr	+	+	nr	+	$+^{w}$	+	nr	nr	−	nr	+	nr	nr	+	nr	+	−	−	+	nr	nr	−	nr
D(−)-Sorbitol	−	−	−	−	nr	−	−	−	nr	+	+	−	−	−	+	nr	nr	−	−	−	$+^{w}$	+	+	nr	$+^{w}$	−	−	$+^{w}$	−	−	−	nr	−	−	nr	+	−	nr	−	−
Sucrose	+	−	+	v	nr	+	−	+	−	nr	−	+	+	+	+	+	+	+	+	−	+	+	+	−	nr	−	−	+	−	nr	+	+	+	+	+	−	+	−	−	+
Trehalose	+	+	+	v	nr	nr	+	+	nr	nr	+	−	+	−	+	nr	nr	−	+	+	+	+	+	nr	nr	+	+	+	nr	nr	+	nr	+	+	nr	+	nr	$+^{w}$	−	nr
Xylose	+	+	+	+	+	−	+	+	−	+	+	−	−	−	+	+	+	+	+	+	+	+	+	+	−	+	−	+	+	+	+	+	+	−	nr	+	−	$+^{w}$	+	−
Decomposition of:																																								
Allantoin	−	+	nr	v	−	nr	+	−	nr	nr	+	nr	+	nr	−	nr	nr	−	nr	nr	nr	−	nr	nr	nr	−	nr	−	−	−	−	nr	+	nr	nr	+	nr	nr	nr	nr
Casein	+	+	+	+	+	nr	−	+	nr	nr	+	+	+	+	+	nr	nr	+	+	−	+	+	+	+	nr	+	+	+	nr	+	+	+	+	+	+	+	+	+	+	−
Esculin	+	+	+	+	+	nr	+	+	nr	nr	−	nr	+	nr	+	nr	nr	+	+	nr	+	$+^{w}$	nr	+	nr	−	nr	+	+	+	+	+	+	+	+	+	nr	+	+	nr
Gelatin	+	+	+	+	+	nr	−	+	+	nr	+	+	+	+	+	nr	nr	+	+	+	+	+	+	nr	nr	+	+	+	nr	+	$+^{w}$	+	+	+	+	−	+	nr	+	−
Hypoxanthine	+	+	+	v	−	nr	−	+	nr	nr	$+^{w}$	+	+	+	+	nr	nr	+	+	+	+	+	+	+	nr	+	−	+	+	+	+	−	−	+	$+^{w}$	+	nr	nr	+	nr
Tyrosine	+	nr	+	nr	nr	nr	nr	+	nr	nr	nr	+	+	+	+	nr	nr	+	+	−	+	+	+	+	nr	+	+	$+^{w}$	+	nr	nr	+	+	+	nr	nr	nr	nr	+	nr
Xanthine	+	+	+	−	−	−	−	−	nr	nr	+	+	+	−	−	nr	nr	−	+	nr	−	−	+	+	nr	+	−	−	+	−	+	+	+	−	−	+	+	nr	+	−

	1	2	3	4	5	6	7	8	9	10	11	12	13	14	15	16	17	18	19	20
Production of:																				
Amylase	−	+	+	+	−	−	−	+	nr	nr	−	−	+	−	−	nr	nr	−	−	$+^{w}$
Nitrate reductase	+	−	+	+	+	−	nr	+	nr	+	+	+	−	+	−	+	−	+	+	−
Urease	+	+	$+^{w}$	−	+	+	+	−	+	+	+	+	−	$+^{w}$	+	+	+	+	+	−
Growth in/ at:																				
5% NaCl	+	+	+	−	+	nr	nr	+	nr	nr	+	+	+	−	+	+	nr	+	−	+
10°C	+	−	+	v	+	nr	nr	+	nr	+	−	+	+	+	−	+	nr	nr	+	+
45°C	−	−	−	+	−	−	nr	−	nr	+	+	−	+	−	+	−	nr	+	−	+

	21	22	23	24	25	26	27	28	29	30	31	32	33	34	35	36	37	38	39	40
Production of:																				
Amylase	−	−	−	−	nr	−	+	−	+	−	−	−	−	−	−	−	−	nr	−	−
Nitrate reductase	+	+	+	+	−	+	nr	−	+	nr	+	−	+	+	+	+	+	−	−	−
Urease	+	−	+	+	+	$+^{w}$	+	$+^{w}$	−	+	+	+	+	−	−	+	+	+	−	+
Growth in/ at:																				
5% NaCl	−	nr	+	+	nr	+	−	+	+	$+^{w}$	+	−	+	+	−	+	+	nr	+	+
10°C	+	−	+	−	+	+	−	nr	+	+	+	$+^{w}$	−	−	−	−	nr	nr	+	nr
45°C	−	+	−	−	+	−	−	−	−	−	−	−	+	−	−	+	+	nr	−	−

[a]+, Positive; $+^{w}$, weakly positive; −, negative; v, variable; nr, not reported.

DNA G+C content (mol%): not determined.

Type strain: A83850, ATCC 51368, DSM 44262, NBRC 15602, JCM 10030, NRRL 18532.

Sequence accession nos: AF051340 (16S rRNA gene); EU822885 (*gyrB*).

3. **Amycolatopsis albidoflavus** Lee and Hah 2001, 649[VP]

al.bi.do.fla'vus. L. adj. *albidus* white; L. adj. *flavus* yellow; N.L. adj. *albidoflavus* whitish yellow.

Forms well developed aerial and vegetative mycelia that fragment into rod-shaped elements. The aerial mycelium is white and the substrate mycelium is orange yellow; a yellow soluble pigment is formed on modified Bennett's agar. Grows between 10 and 37°C, but not at 45°C.

Phosphatase-positive. Does not produce acid from L-sorbose or xylitol. Benzoate, citrate, lactate, oxalate, and propionate are used as sole carbon sources, but not mucate or tartrate. Grows in the presence of 7% (w/v) NaCl.

Resistant (μg/ml) to gentamicin sulfate (5), neomycin sulfate (8), novobiocin (10), penicillin G (20), polymyxin B (50), rifampin (10), streptomycin sulfate (16), tobramycin sulfate (8), and vancomycin hydrochloride (0.25). Additional phenotypic properties are shown in Table 238.

Whole-organism hydrolysates are rich in arabinose and galactose. Muramic acid is in the *N*-acetylated form. The cellular fatty acid profile includes major proportions of $C_{16:0}$ iso (26.1%), $C_{14:0}$ iso (9.8%), and $C_{16:0}$ (10.2%), and smaller proportions (<10%) of $C_{14:0}$ (1.1%), $C_{15:0}$ iso (8.7%), $C_{15:1}$ (2.3%), $C_{15:0}$ (5.8%), $C_{16:1}$ (4.0%), $C_{17:0}$ iso (1.6%), $C_{17:1}$ (4.4%), $C_{17:0}$ (7.8%), and $C_{18:0}$ (6.2%). The cellular polar lipid pattern includes PE and PI. The predominant menaquinone is MK-9(H_4).

Source: isolated from soil.

DNA G+C content (mol%): 68.5 (HPLC).

Type strain: ATCC 53205, DSM 44639, IMSNU 22139, JCM 11300, KCTC 9471, NBRC 100337, NRRL B-24149.

Sequence accession nos: AJ252832 (16S rRNA gene); EU822886 (*gyrB*).

4. **Amycolatopsis australiensis** Tan, Robinson, Lacey and Goodfellow 2006a, 2299[VP]

aus.tra.li.en'sis. N.L. fem. adj. *australiensis* pertaining to Australia, the source of the soil from which the first strains were isolated.

Forms an extensively branched substrate mycelium which fragments into squarish rod-shaped elements. Abundant, white aerial hyphae, a pale yellow substrate mycelium, and a medium yellow diffusible pigment are produced on modified Bennett's agar supplemented with mannitol and soybean flour. Melanin pigments are not formed on peptone-yeast extract-iron or tyrosine agars. Grows between 28 and 45°C, and from pH 5 to 7.

Tween 40 and uric acid are degraded. Acid is produced from arabitol, glycogen, and turanose, but not from dulcitol, D-ribose, or xylitol. Citrate, oxalate, and propionate are used as sole carbon sources, but not benzoate, lactate, or mucate (all at 0.1%, w/v).

Resistant (μg/ml) to gentamicin sulfate (5), neomycin sulfate (8), polymyxin B (50) and tobramycin sulfate (8), but sensitive to novobiocin (10), penicillin G (20), rifampin (10), and streptomycin sulfate (16). Additional phenotypic properties are shown in Table 238.

Whole-organism hydrolysates are rich in arabinose and galactose. Muramic acid is *N*-acetylated. Mycolic acids are not present. The polar lipid pattern includes DPG, PE, PG, PI, and PME. The predominant menaquinone is MK-9(H_4); also contains MK-8(H_2) (26%) and MK-9(H_8) (8%).

Source: isolated from arid soil collected from Western Australia.

DNA G+C content (mol%): not determined.

Type strain: GY048, DSM 44671, JCM 15587, NCIMB 14142.

Sequence accession nos: AY129753 (16S rRNA gene); EU822887 (*gyrB*).

5. **Amycolatopsis azurea** (Ōmura, Tanaka, Tanaka, Spiri-Nakagawa, Oiwa, Takahashi, Matsuyama and Iwai 1983) Henssen, Kothe and Kroppenstedt 1987, 294[VP] (Basonym: *Pseudonocardia azurea* Ōmura, Tanaka, Tanaka, Spiri-Nakagawa, Oiwa, Takahashi, Matsuyama and Iwai 1983, 673.)

a.zu're.a. N.L. fem. adj. *azurea* azure blue, referring to the color of aerial mycelium.

Forms irregularly branched substrate hyphae which tend to be zig-zag-shaped. The aerial mycelium is usually white, but is blue on sucrose-nitrate and tyrosine agars, and pink on glucose-peptone agar. The substrate mycelium is brownish black and the aerial mycelium is light gray on modified Bennett's agar. Blue soluble pigments are produced on glucose-nitrate and starch-yeast extract agars. Aerial spores are smooth and cylindrical (0.4–1.1 × 3.7 μm). Grows between 10 and 36°C, but not at 37 or 45°C.

Positive for *N*-acetyl-β-glucosamidase, α-chymotrypsin, esterase lipase (C8), β-galactosidase, α- and β-glucosidases, leucine arylamidase, naphthol-AS-BI-phosphohydrolase, and trypsin, but negative for alkaline phosphatase, cystine arylamidase, esterase (C4), α-fucosidase, α-galactosidase, β-glucuronidase, lipase (C14), α-mannosidase, and valine arylamidase (API ZYM tests). Positive for arginine dihydrolase, β-galactosidase, and lysine and ornithine decarboxylases (API 20E tests). Acetoin-positive, but indole-negative. Arbutin is hydrolyzed, but does not produce H_2S. Elastin, Tween 40, uric acid, and xylan are degraded. Citrate, oxalate, and propionate are used as sole carbon sources, but not benzoate, lactate, or mucate (all at 0.1%, w/v). Grows in the presence of 7% (w/v) NaCl.

Resistant (μg/ml) to gentamicin sulfate (5), kanamycin (30), nalidixic acid (30), neomycin sulfate (8), penicillin G (20), polymyxin B (50), rifampin (10), streptomycin sulfate (16), sulfonamide (200), tobramycin sulfate (8), and vancomycin hydrochloride (30). Additional phenotypic properties are shown in Table 238.

Whole-organism hydrolysates are rich in arabinose and galactose. Cellular fatty acid profiles contain major proportions of $C_{15:0}$ iso (30.0%), $C_{17:0}$ (11.8%), and $C_{17:0}$ (25.7%), and smaller proportions (>10) of $C_{15:0}$ anteiso (3.9%), $C_{15:0}$ (6.5%), $C_{16:0}$ iso (2.9%), $C_{16:1}$ (8.8%), $C_{16:0}$ (5.7%), $C_{17:0}$ iso (2.6%), and $C_{18:0}$ iso (2.1%). Mycolic acids are not present. The polar lipid pattern contains DPG, PE, PG, PI, and PIMs. The predominant menaquinone is MK-9(H_4); also contains MK-8(H_4) and MK-9(H_4).

Source: isolated from soil.

DNA G+C content (mol%): 66 (T_m).

Type strain: AM-3696, ATCC 51273, DSM 43854, FERM-P 4738, IFO (now NRBC) 14573, IMSNU 20053, JCM 3275, NRRL 11412, VKM Ac-1418.

Sequence accession nos: AJ400709 (16S rRNA gene); EU822888 (*gyrB*).

6. **Amycolatopsis balhimycina** Wink, Kroppenstedt, Ganguli, Nadkarni, Schumann, Seibert and Stackebrandt 2003a, 935[VP] (Effective publication: Wink, Kroppenstedt, Ganguli, Nadkarni, Schumann, Seibert and Stackebrandt 2003b, 44.)

bal.hi.my'ci.na. N.L. n. *balhimycinum* balhimycin (an antibiotic produced by the organism); L. fem. suff. *-ina* suffix used with the sense of belonging to; N.L. fem. adj. *balhimycina* pertaining to balhimycin.

Forms a chrome yellow substrate mycelium on glycerol-asparagine, inorganic salts-starch, oatmeal, peptone-yeast extract, tyrosine, and yeast-extract-malt extract agars (ISP media 2–7); soluble pigments are not formed on any of these media. Fragmentation of the substrate mycelium and white aerial hyphae into more or less irregular arthrospores is only seen on oatmeal agar.

Positive for alkaline phosphatase, *N*-acetyl-β-glucosamidase, α-chymotrypsin, cystine arylamidase, α-fucosidase, α- and β-galactosidases, α-glucosidase, leucine arylamidase, lipase (C14), naphthol-AS-BI-phosphohydrolase, trypsin, and valine arylamidase, but negative for β-glucuronidase and α-mannosidase (API ZYM tests). Produces H_2S. Additional phenotypic properties are shown in Table 238.

Whole-organism hydrolysates are rich in arabinose and galactose. Cellular fatty acid profiles contain major proportions of $C_{16:0}$ iso (36.0%), and smaller proportions (<10%) of $C_{15:0}$ (5.3%), $C_{16:0}$ (2.0%), $C_{17:0}$ (2.2%), $C_{15:1}$ (5.1%), $C_{14:0}$ iso (3.8%), $C_{15:0}$ iso (5.7%), C_{161} (7.3%), $C_{17:0}$ iso (1.2%), $C_{15:0}$ anteiso (1.5%), $C_{17:0}$ anteiso (4.1%), $C_{16:1}$ *cis*9 (4.0%), $C_{17:1}$ *cis*9 (7.6%), $C_{16:0}$ 10-methyl (2.4%), $C_{17:0}$ 10-methyl (3.2%), 2-hydroxy-$C_{16:0}$ (7.7%), and $C_{17:0}$ anteiso 2-OH (0.7%). Mycolic acids are absent. The polar lipid pattern contains DPG, HPE, PE, PG, and PI. The predominant menaquinone is MK-9(H_4).

Source: isolated from a soil sample collected in India.

DNA G+C content (mol%): not determined.

Type strain: FH 1894, DSM 44591, JCM 12668, NRRL B-24207.

Sequence accession nos: AJ508239 (16S rRNA gene); EU822889 (*gyrB*).

7. **Amycolatopsis benzoatilytica** Majumdar, Prabhagaran, Shivaji and Lal 2006, 202[VP]

ben.zo.a.ti.ly'ti.ca. N.L. n. *benzoas -atis* benzoate; N.L. adj. *lyticus-a-um* (from Gr. adj. *lutikos-ê-on*) dissolving; N.L. fem. adj. *benzoatilytica* benzoate-degrading.

Forms cream-colored, irregularly shaped, rough, flat colonies with undulating margins. The substrate mycelium tends to fragment in liquid medium. Grows between 10 and 37°C, and from pH 5 to 10.

Tween 80 is degraded and *m*-hydroxybenzoate is metabolized. Additional phenotypic properties are shown in Table 238.

Whole-organism hydrolysates are rich in arabinose and galactose. Cellular fatty acid profiles contain major proportions of $C_{14:0}$ iso (11.3%), $C_{15:0}$ (14.7%), $C_{15:1}$ ω6*c* (12.8%), $C_{16:0}$ iso (12.8%), $C_{17:0}$ (11.3%), and $C_{17:1}$ ω8*c* (11.3%), and smaller proportions (<10%) of $C_{15:0}$ iso (6.8%), $C_{15:0}$ anteiso (3%), $C_{16:0}$ (5.2%), and $C_{15:0}$ 2-OH (1%). The polar lipid pattern contains DPG, PE, PG, PI, and PME. The predominant menaquinone is MK-9(H_4).

Source: isolated from a patient with submandibular mycetoma.

DNA G+C content (mol%): not determined.

Type strain: AK 16/65, ATCC 55165, DSM 43387, IMRU 1389, JCM 13851.

Sequence accession no. (16S rRNA gene): AY957506.

Additional comments: the strain was initially identified as *Nocardia brasiliensis* (Scharfen, 1971) and reclassified as *Nocardia orientalis* by Gordon et al. (1978).

8. **Amycolatopsis coloradensis** Labeda 1995, 126[VP]

co.lo.rad.en'sis. N.L. fem. adj. *coloradensis* of Colorado, the source of the soil sample from which the type strain was isolated.

Forms a yellow to orange substrate mycelium. When formed, the aerial mycelium is sparse and white to olive buff. Aerial hyphae differentiate into chains of straight to flexuous cylindrical to ovoid spores. An orange yellow soluble pigment is produced on most growth media, including modified Bennett's agar. Grows at 10 to 37°C, but not at 45°C.

Arbutin is not hydrolyzed nor is H_2S produced. Elastin, Tween 40, and xylan are degraded, but not uric acid. Positive for *N*-acetyl-β-glucosamidase, acid and alkaline phosphatases, α-chymotrypsin, esterase (C8), leucine arylamidase, lipase (C14), and trypsin, but negative for cystine arylamidase, esterase (C4), α-fucosidase, α- and β-galactosidases, α- and β-glucosidases, β-glucuronidase, α-mannosidase, naphthol-AS-BI-phosphohydrolase, and valine arylamidase (API ZYM tests). Positive for arginine dihydrolase, β-galactosidase, and lysine and ornithine decarboxylases, but not for β-galactosidase (API 20E tests). Acetate, malate, propionate, pyruvate, and succinate are used as sole carbon sources, but not benzoate, citrate, mucate, oxalate, or tartrate.

Resistant (μg/ml) to gentamicin sulfate (5), neomycin sulfate (8), polymyxin B (50), streptomycin sulfate (16), tobramycin sulfate (8), and vancomycin hydrochloride (0.25). Additional phenotypic properties are shown in Table 238.

Whole-organism hydrolysates are rich in arabinose and galactose. Cellular fatty acid profiles contain major proportions of $C_{15:0}$ (14.6%), $C_{15:1}$ ω6*c* (12.8%), $C_{16:0}$ iso (25) and $C_{17:0}$ (11.3%), and smaller proportions of $C_{14:0}$ iso (0.5%), $C_{15:0}$ iso (6.8%), $C_{15:0}$ anteiso (3.0%), $C_{16:0}$ (5.2%), $C_{16:1}$ iso H (0.2%), $C_{17:0}$ iso (0.4%), $C_{17:0}$ anteiso (0.8%), and $C_{18:0}$ (0.4%). Mycolic acids are absent. The cellular polar lipid pattern contains PE as the diagnostic nitrogen-containing component. The predominant menaquinone is MK-9(H_2,H_4).

Source: isolated from a soil sample collected in Colorado.

DNA G+C content (mol%): 66 (T_m).

Type strain: ATCC 53629, DSM 44225, IFO (now NRBC) 15804, JCM 9869, NRRL 3218, VKM Ac-1732.

Sequence accession nos: AF051341 (16S rRNA gene); EU822890 (*gyrB*).

9. **Amycolatopsis decaplanina** Wink, Gandhi, Kroppenstedt, Seibert, Sträubler, Schumann and Stackebrandt 2004, 237[VP]

de.ca.pla'ni.na. N.L. neut. n. *decaplaninum* decaplanin; L. fem. suff. *-ina* related to; N.L. fem. adj. *decaplanina* related to decaplanin, an antibiotic produced by the organism.

Honey yellow substrate mycelium is formed on glycerol-asparagine, inorganic salts-starch, oatmeal, peptone-yeast extract-iron, tyrosine, tryptone-yeast extract, and yeast extract-malt extract agars (ISP media 1–7). Only produces an aerial mycelium on oatmeal agar and a soluble red pigment on tyrosine agar. Melanin pigments are not formed. Regular shaped to globose and smooth-surfaced pseudosporangia are produced, but spores are not seen either inside or outside these bodies.

Positive for *N*-acetyl-β-glucosamidase, acid and alkaline phosphatases, α-chymotrypsin, β-galactosidase, α- and β-glucosidases, leucine arylamidase, α-mannosidase, naphthol-AS-BI-phosphohydrolase, trypsin, and valine arylamidase, but negative for cystine arylamidase, esterase (4), esterase lipase (C8), α-fucosidase, α-galactosidase, β-glucuronidase, and lipase (C14) (API ZYM tests). Positive for β-galactosidase, but not for arginine dihydrolase or lysine and ornithine decarboxylases (API 20E tests). Produces acetone, but not indole. Arbutin is hydrolyzed, but H_2S is not produced. Elastin and Tween 40 are degraded, but not uric acid or xylan. Additional phenotypic properties are shown in Table 238.

Whole-organism hydrolysates are rich in arabinose and galactose. Cellular fatty acid profiles contain major proportions of $C_{15:0}$ iso (22.5%), $C_{17:0}$ (11.4%), and $C_{16:0}$ iso (10.3%), and smaller proportions (<10%) of $C_{14:0}$ iso (6.4%), $C_{15:0}$ anteiso (9.4%), $C_{15:0}$ iso 2-OH (8.7%), $C_{15:0}$ (7.6%), $C_{17:1}$ (5.8%), $C_{16:0}$ (3.3%), $C_{15:1}$ (2.1%), $C_{17:0}$ iso (1.7%), $C_{17:0}$ anteiso (3.4%), $C_{16:0}$ iso 2-OH (2.8%), $C_{15:0}$ anteiso 2-OH (1.9%), $C_{17:0}$ 2-OH (1.5%) and $C_{17:0}$ 10-methyl iso (1.2%). The polar lipid pattern contains DPG, HPE, PE, PG, PI, and PIMs. The predominant menaquinones are MK-8(H_4) and MK-9(H_4).

Source: isolated from a soil sample collected in India.

DNA G+C content (mol%): not determined.

Type strain: FH 1845, DSM 44594, JCM 12669, NRRL B-24209.

Sequence accession nos: AJ508237 (16S rRNA gene); EU822891 (*gyrB*).

Additional comments: the type strain of *Amycolatopsis decaplanina* has a ribotype pattern which distinguishes it from phylogenetically related *Amycolatopsis* strains (Wink et al., 2004).

10. **Amycolatopsis echigonensis** Ding, Hirose and Yokota 2007, 1750[VP]

e.chi.go.nen'sis. N.L. fem. adj. *echigonensis* referring to Echigo, the old name of Niigata Prefecture, Japan, the source of the type strain.

The substrate mycelium is light yellow and the aerial mycelium white. Grows between 5 and 45°C, and optimally at 30°C, and from pH 6 to 11, optimally at pH 9.0.

Positive for *N*-acetyl-β-glucosamidase, acid and alkaline phosphatases, α-chymotrypsin, cystine arylamidase, esterase (C4), esterase lipase (C8), α- and β-galactosidases, α-glucosidase, β-glucuronidase, leucine arylamidase, α-mannosidase, trypsin, and valine arylamidase, but negative for α-fucosidase, β-glucosidase, and lipase (C14) (API ZYM tests). Grows in the presence of 7% (w/v) NaCl. Additional phenotypic properties are shown in Table 238.

Whole-organism hydrolysates are rich in arabinose and galactose. Cellular fatty acid profiles contain major proportions of $C_{15:0}$ anteiso (16.7%) and $C_{16:0}$ iso (31.9%), and smaller proportions (<10%) of $C_{14:0}$ iso, (6.5%), $C_{14:0}$ (0.4%), $C_{15:0}$ iso (7.0%), $C_{15:0}$ (5.2%), $C_{15:1}$ ω6*c* (3.6%), $C_{16:0}$ (2.5%), $C_{16:1}$ iso OH (1.3%), $C_{17:0}$ iso (1.1%), $C_{17:}$ (6.9%), $C_{17:0}$ (3.4%), $C_{17:1}$ ω6*c* (5.6%), and $C_{17:1}$ ω8*c* (2.4%). Mycolic acids are absent. The predominant menaquinone is MK-9(H_4).

Source: isolated from filtration material made from volcanic soil, Niigata, Japan.

DNA G+C content (mol%): 72.4 (HPLC).

Type strain: LC2, CCTCC AB206019, IAM 15387, JCM 21831.

Sequence accession nos: AB248535 (16S rRNA gene); EU822892 (*gyrB*).

11. **Amycolatopsis eurytherma** Kim, Sahin, Tan, Zakrzewska-Czerwinska and Goodfellow 2002a, 893[VP]

eur.y.ther'ma. Gr. adj. *eurus* wide, broad; Gr. adj. *thermos* hot; N.L. fem. adj. *eurytherma* grows over a wide temperature range.

Extensively branched substrate hyphae fragment into squarish, rod-like elements (approx. 0.4–0.5 × 0.7–1.6 µm). Forms a yellow substrate mycelium on modified Bennett's and glucose-yeast extract agars, but distinct pigments are not evident on Czapek-Dox, glycerol-asparagine, or oatmeal agars; diffusible pigments are not formed on any of these media. Melanin pigments are not produced on peptone-yeast extract-iron or tyrosine agars. Aerial mycelium is white, sparse, and sterile. Grows from 25 to 55°C, but not at 10 or 60°C and from pH 6 to 9.

Negative for β-galactosidase and *N*-acetyl-β-glucosamidase. Elastin and xylan are decomposed, but not chitin or guanine. Grows on adonitol, L-arabinose, cellobiose, fructose, galactose, *myo*-inositol, lactose, mannitol, mannose, melezitose, melibiose, rhamnose, ribose, sorbitol, starch, sucrose, trehalose, turanose, xylitol, and xylose as sole carbon sources, but not on raffinose (all at 1%, w/v).

Resistant (µg/ml) to ampicillin (32), bacitracin (16), carbenicillin (12), cefoxitin (32), cephaloridine (128), cephradine (32), doxycyline hydrochloride (4), ethionamide (16), fusidic acid (4), gentamicin sulfate (32), gramicidin (8), isoniazid (16), lincomycin hydrochloride (128), lividomycin A (8), nalidixic acid (32), neomycin sulfate (8), novobiocin (4), oleandomycin phosphate (128), penicillin G (15 IU), polymyxin B (50), rifampin (64), spiramycin (10), streptomycin sulfate (64), tetracycline hydrochloride (32), tobramycin sulfate (8), vancomycin hydrochloride (64), and viomycin sulfate (20), but sensitive to amikacin (4), bacitracin (32), doxycyclin (6), fusidic acid (8), gentamicin (64), lividomycin A (16), neomycin sulfate (32), novobiocin (16), streptomycin sulfate (64), ticarcillin (16), and tyrothricin (16). Additional phenotypic properties are shown in Table 238.

Whole-organism hydrolysates are rich in arabinose and galactose. Cellular fatty acid profiles contain major proportions of $C_{16:0}$ iso (46%) and $C_{16:0}$ (23%), and smaller proportions (<10%) of $C_{15:0}$ iso, (4%), $C_{17:0}$ iso (6%), $C_{17:0}$ anteiso (9%), $C_{17:0}$ (3%), and $C_{16:0}$ (3%). Mycolic acids are absent. The polar lipid pattern contains DPG, PE, PG, PI, PI dimannoside, and PME. The predominant isoprenologue is MK-9(H_4).

Source: isolated from arid and scrubland soil collected in Madurai, India, and Van, Turkey, respectively.

DNA G+C content (mol%): 72.2–74.0 (HPLC).

Type strain: NT202, DSM 44348, JCM 12071, NBRC 100338, NCIMB 13795.

Sequence accession nos: AJ000285 (16S rRNA gene); EU822893 (*gyrB*).

12. **Amycolatopsis halotolerans** Lee 2006, 552[VP]

ha.lo.to'le.rans. Gr. n. *hals halos* salt; L. part. adj. *tolerans* tolerating, enduring; N.L. part. adj. *halotolerans* salt-tolerating.

Forms well-developed, branched aerial and substrate mycelia that fragment into rod-shaped elements. The substrate mycelium is brown and the aerial mycelium is grayish white on yeast extract-malt extract agar (ISP medium 2). Grows between 10 and 37°C, but not at or above 45°C.

Does not produce acid from D-arabinose, dulcitol, methyl α-D-mannoside, sorbose, or xylitol. Growth occurs in the presence of 7%, but not 10% (w/v) NaCl. Additional phenotypic properties are shown in Table 238.

Whole-organism methanolyzates are rich in arabinose and galactose. Muramic acid moieties are *N*-acetylated. Cellular fatty acids contain major proportions of $C_{15:0}$ iso (13.0%), $C_{16:0}$ (16.2%), $C_{16:0}$ iso (17.6%), $C_{17:0}$ (14.5%), and $C_{18:0}$ (13.2%), and smaller proportions (<10%) of $C_{15:0}$ (3.7%), $C_{17:0}$ iso, (5.0%), $C_{15:0}$ anteiso (3.2%), $C_{17:0}$ anteiso, (3.1%), $C_{14:0}$ 2-OH (0.8%), and $C_{15:0}$ iso 3-OH (5.2%). Mycolic acids are absent. The polar lipid pattern contains DPG, PE, PG, and PI. The predominant menaquinone is MK-9(H_4); minor amounts of MK-9(H_6) and MK-9(H_8) are also present.

Source: isolated from soil collected inside a natural cave on Jeju Island, Republic of Korea.

DNA G+C content (mol%): 72.5 (HPLC).

Type strain: N4-6, DSM 45041, JCM 13279, NRRL B-24428.

Sequence accession nos: DQ000196 (16S rRNA gene); EU822895 (*gyrB*).

13. **Amycolatopsis japonica** corrig. Goodfellow, Brown, Cai, Chun and Collins 1997b, 915[VP] (Effective publication: Goodfellow, Brown, Cai, Chun, Collins 1997a, 81.)

ja.po'ni.ca. N.L. fem. adj. *japonica* pertaining to Japan.

Forms substrate and aerial mycelia which fragment into squarish elements. A yellow brown substrate mycelium, a white aerial mycelium and a dark olive brown diffusible pigment are formed on modified Bennett's agar. Grows between 10 and 45°C.

Positive for *N*-acetyl-β-glucosamidase, acid and alkaline phosphatases, α-chymotrypsin, cystine arylamidase, esterase (C4), esterase lipase (C8), α-fucosidase, α- and β-galactosidases, α- and β-glucosidases, leucine arylamidase, lipase (C14), α-mannosidase, naphthol-AS-BI-phosphohydrolase, trypsin, and valine arylamidase, but negative for β-glucuronidase (API ZYM tests). Positive for β-galactosidase, arginine dihydrolase, and lysine and ornithine decarboxylases (API 20E tests). Acetoin-positive, but indole- and lecithanase-negative. Produces lipolytic enzymes. Arbutin is hydrolyzed, but not hippurate. H_2S is not produced.

DNA, elastin, RNA, Tweens 20, 40, 60 and 80, uric acid, and xylan are degraded, but not adenine, chitin, or pectin. Adonitol, L-arabinose, fructose, galactose, *myo*-inositol, lactose, mannitol, melibiose, raffinose, rhamnose, salicin, sucrose, and xylose are used as sole carbon sources, but not dextran, melezitose, or xylitol (all at 0.1%, w/v). Similarly, citrate, propionate, and pyruvate are used as sole carbon sources (all at 1.0%, w/v). DL-α-Amino-*n*-butyric acid, L-cysteine, L-histidine, L-hydroxyproline, L-phenylalanine, potassium nitrate, and L-threonine are used as sole carbon and nitrogen sources, but not L-valine. Grows in the presence of crystal violet (0.0001%, w/v) and potassium tellurite (0.001%, w/v), but is inhibited by phenol (0.1%, w/v), sodium azide (0.01%, w/v), sodium chloride (7%, w/v), and thallous acetate (0.001%, w/v).

Resistant (μg/ml) to gentamicin sulfate (5), neomycin sulfate (8), novobiocin (10), penicillin G (20), polymyxin B (50), rifampin (10), streptomycin sulfate (16), tobramycin sulfate (8), and vancomycin hydrochloride (0.25), but sensitive to neomycin sulfate (50), triacyl oleandomycin (100) and rifampin (50). Additional phenotypic properties are shown in Table 238.

Whole-organism methanolyzates are rich in arabinose and galactose. Cellular fatty acid profiles contain major proportions of $C_{16:0}$ iso (41.4%), $C_{16:0}$ (13.2%), and $C_{15:0}$ iso (13.5%). Mycolic acids are absent. The polar lipid pattern contains DPG, PE, PG, PI, and PIMs. The predominant menaquinone is MK-9(H_4).

Source: isolated from soil.

DNA G+C content (mol%): 69.5 (T_m).

Type strain: MG417-CF17, CIP 106801, DSM 44213, JCM 10140, NRRL B-24138.

Sequence accession nos: AJ508236 (16S rRNA gene); EU822896 (*gyrB*).

Additional comments: the original spelling, *Amycolatopsis japonicum* (*sic*) was corrected at validation in line with Rule 61 of the Bacteriological Code (Associate Editor, 1997).

14. **Amycolatopsis jejuensis** Lee 2006, 552[VP]

je.ju.en'sis. N.L. fem. adj. *jejuensis* of Jeju Island, Republic of Korea.

Forms well-developed, branched aerial and substrate mycelia that fragment into rod-shaped elements. The substrate mycelium is yellowish brown and the aerial mycelium is white on yeast extract-malt extract agar (ISP medium 2); does not produce soluble pigments. Grows between 10 and 30°C.

Acid is produced from D-arabinose, but not from dextran, dulcitol, inulin, methyl α-D-mannoside, ribose, sorbose, L-xylose, or D-xylitol. Grows in the presence of 2%, but not 3% (w/v) NaCl. Additional phenotypic properties are shown in Table 238.

Whole-organism hydrolysates are rich in arabinose and galactose. Muramic acid moieties are *N*-acetylated. Cellular fatty acids profiles contain major proportions of $C_{15:0}$ iso (14.3%), $C_{16:0}$ (20.2%), $C_{16:0}$ iso (14.6%), and $C_{18:0}$ (17.5%), and minor proportions (<10%) of $C_{14:0}$ (1.2%), $C_{16:1}$ (0.9%), $C_{14:0}$ iso (2.2%), $C_{15:0}$ (2.6%), $C_{17:0}$ (4.8%), $C_{17:0}$ iso (5.0%), $C_{15:0}$ anteiso (3.2%), $C_{17:0}$ anteiso (3.5%), $C_{20:0}$ (1.1%), $C_{14:0}$ 2-OH (0.6%), and $C_{15:0}$ iso 3-OH (7.6%). Mycolic acids are absent. The polar lipid pattern contains DPG, PE, PG, PI, and PIMs. The predominant menaquinone is MK-9(H_4).

Source: isolated from dried bat dung from a natural cave on Jeju Island, Republic of Korea.

DNA G+C content (mol%): 71.7 (HPLC).

Type strain: N7-3, DSM 45042, JCM 13280, NRRL B-24427.

Sequence accession nos: DQ000200 (16S rRNA gene); EU822897 (*gyrB*).

15. **Amycolatopsis kentuckyensis** Labeda, Donahue, Williams, Sells and Henton 2003, 1602[VP]

ken.tuc.ky.en'sis. N.L. fem. adj. *kentuckyensis* from Kentucky, named after the place of origin, the state of Kentucky, USA.

Forms a well-developed, yellow-orange to brownish-orange substrate mycelium and a light orangish-white to grayish orange-white aerial mycelium on most media. A faint brownish soluble pigment is produced on some media. Grows from 15 to 42°C.

Hippurate is hydrolyzed. Does not degrade adenine. Acid is produced from dulcitol, but not from methyl β-xyloside. Acetate and citrate are used as sole carbon sources; benzoate, lactate, malate, mucate, oxalate, propionate, succinate, and DL-tartrate are used weakly, if at all. Additional phenotypic properties are shown in Table 238.

Whole-organism hydrolysates are rich in arabinose and galactose. Cellular fatty acids profiles contain major proportions of $C_{15:0}$ iso (14.2%), $C_{16:0}$ iso (17.2%) and $C_{17:1}$ *cis*9 (10.1%), and smaller proportions (<10%) of $C_{14:0}$ iso (2.2%), $C_{15:0}$ anteiso (3.3%), $C_{15:1}$ (1.0%), $C_{15:0}$ (5.1%), $C_{16:1}$ *cis*9 (2.3%), $C_{16:0}$ (3.0%), $C_{16:0}$ 9-methyl (3.3%), $C_{17:0}$ iso (2.4%), $C_{17:0}$ anteiso (6.6%), $C_{16:0}$ 2-OH (2.9%), $C_{17:0}$ (6.2%), and $C_{17:0}$ 10-methyl (5.3%). The major phospholipid is PE; smaller amounts of PME are also present. The predominant menaquinones are MK-9(H_2) and MK-9(H_4).

Source: isolated from equine placentas.

DNA G+C content (mol%): not determined.

Type strain: LDDC 9447-99, DSM 44652, JCM 12570, NRRL B-24129.

Sequence accession no. (16S rRNA gene): AY183357.

16. **Amycolatopsis keratiniphila** Al-Musallam, Al-Zarban, Fasasi, Kroppenstedt and Stackebrandt 2003, 872[VP]

ke.ra.ti.ni'phi.la. N.L. n. *keratinum* keratin; N.L. adj. *philus -a -um* (from Gr. adj. *philos -ê -on*) friend, loving; N.L. fem. adj. *keratiniphila* keratin-loving, referring to the ability of the species to degrade keratin.

Additional comments: the type strain NRRL B-24117 was mistakenly cited as NRRL B24117 in Al-Musallam et al. (2003). The taxon was subsequently divided into two subspecies.

16a. **Amycolatopsis keratiniphila subsp. keratiniphila** Wink, Kroppenstedt, Ganguli, Nadkarni, Schumann, Seibert and Stackebrandt 2003a, 935[VP] (Effective publication: Wink, Kroppenstedt, Ganguli, Nadkarni, Schumann, Seibert and Stackebrandt 2003b, 44.) (*Amycolatopsis keratiniphila* Al-Musallam, Al-Zarban, Fasasi, Kroppenstedt and Stackebrandt 2003, 872)

Light gray aerial mycelium is formed; does not form a soluble pigment. Grows between 10 and 28°C.

Positive for *N*-acetyl-β-glucosamidase, acid and alkaline phosphatases, α-chymotrypsin, cystine arylamidase, esterase (C4), esterase lipase (C8), α- and β-galactosidases, α- and β-glucosidases, leucine arylamidase, lipase (C14), trypsin, and valine arylamidase, but negative for α-fucosidase, β-glucuronidase, α-mannosidase and naphthol-AS-BI-phosphohydrolase (API ZYM tests). Positive for arginine dehydrolase, β-galactosidase, and lysine and ornithine decarboxylases (API 20E tests). Acetoin-positive, but indole-negative. Arbutin is hydrolyzed, but does not produce H_2S. Elastin and Tween 40 are degraded, but not uric acid or xylan.

Resistant (filter paper discs soaked in µg/ml) to nalidixic acid (30), streptomycin (10), sulfonamide (200), and vancomycin (30), but sensitive to kanamycin (30). Additional phenotypic properties are shown in Table 238.

Whole-organism hydrolysates are rich in arabinose and galactose. The cellular fatty acid profile contains major proportions of $C_{16:0}$ iso (33.5%) and $C_{14:0}$ iso (13.9%), and smaller proportions (<10%) $C_{17:1}$ (9.9%), $C_{15:0}$ iso (8.0%), $C_{17:0}$ (7.6%), 2-hydroxy-$C_{16:0}$ iso (7.4%), $C_{15:0}$ (7.1%), $C_{16:1}$ (4.9%), $C_{16:0}$ (2.0%), $C_{15:0}$ anteiso, (1.9%), $C_{17:0}$ anteiso (1.3%), $C_{17:0}$ 10-methyl iso (1.0%), 2-hydroxy-$C_{15:0}$ anteiso (0.9%) and $C_{17:0}$ iso (0.8%). Mycolic acids are not formed. The polar lipid pattern contains DPG, HPE, PE, and PI. The predominant menaquinone is MK-9(H_4).

Source: isolated from agricultural soil in Kuwait using animal wool in bait.

DNA G+C content (mol%): not determined.

Type strain: D2, DSM 44409, JCM 12683, NRRL B-24117.

Sequence accession nos: AJ278496 (16S rRNA gene); EU822898 (*gyrB*).

16b. **Amycolatopsis keratiniphila subsp. nogabecina** Wink, Kroppenstedt, Ganguli, Nadkarni, Schumann, Seibert and Stackebrandt 2003a, 935[VP] (Effective publication: Wink, Kroppenstedt, Ganguli, Nadkarni Schumann, Seibert and Stackebrandt 2003b, 44.)

no.ga.be'ci.na. N.L. n. *nogabecinum* nogabecin (an antibiotic produced by the organism); L. fem. suff. *-ina* suffix used with sense of belonging to; N.L. fem. adj. *nogabecina* pertaining to nogabecin, an antibiotic produced by the organism.

The substrate mycelium is sand yellow on oatmeal, peptone-yeast extract-iron, and yeast extract-malt extract agars and beige on glycerol-asparagine, inorganic salts, and tyrosine agars; soluble pigments are not formed on any of these media. Aerial hyphae are not produced.

Positive for *N*-acetyl-β-glucosamidase, acid and alkaline phosphatases, α-chymotrypsin, cystine arylamidase, esterase

(C8), α- and β-galactosidases, α-glucosidase, leucine arylamidase, lipase (C14), α-mannosidase, naphthol-AS-BI-phosphohydrolase, trypsin, and valine arylamidase, but negative for esterase (C4), α-fucosidase, and β-glucuronidase (API ZYM tests). Positive for arginine dihydrolase, β-galactosidase, and lysine and ornithine decarboxylases (API 20E tests) Arbutin is hydrolyzed and H_2S produced. Tween 40 is degraded, but not elastin, uric acid, or xylan.

Resistant (filter paper discs soaked in μg/ml) to nalidixic acid (30), streptomycin (10) and vancomycin, but is susceptible to kanamycin (30) and sulfonamide (200). Additional phenotypic properties are shown in Table 238.

Whole-organism hydrolysates are rich in arabinose and galactose. The fatty acid profile contains major proportions of $C_{16:0}$ iso (31.9%), $C_{14:0}$ iso (12.3%), and $C_{17:0}$ (10.9%), and smaller proportions of $C_{15:0}$ (9.1%), $C_{16:0}$ (2.2%), $C_{17:1}$ *cis*9 (3.8%), $C_{15:0}$ iso (9.8%), $C_{17:0}$ iso (0.9%), $C_{15:0}$ anteiso (2.0%), $C_{17:0}$ anteiso (1.8%), $C_{18:0}$ iso (0.4%), $C_{17:0}$ 10-methyl iso (2.1%), $C_{17:0}$ 10-methyl (0.3%), 2-hydroxy-$C_{15:0}$ iso (3.9%), 2-hydroxy-$C_{15:0}$ iso (0.9%), 2-hydroxy-$C_{16:0}$ iso (7.7%), and 2-hydroxy-$C_{17:0}$ anteiso (0.9%). Mycolic acids are absent. The polar lipid pattern contains DPG, HPE, PE, and PI. The predominant menaquinone is MK-9(H_4).

Source: isolated from a soil sample collected in India.

DNA G+C content (mol%): not determined.

Type strain: FH 1893, DSM 44586, JCM 12671, NRRL B-24206.

Sequence accession nos: AJ508238 (16S rRNA gene); EU822899 (*gyrB*).

17. **Amycolatopsis lexingtonensis** Labeda, Donahue, Williams, Sells and Henton 2003, 1603[VP]

lex.ing.ton.en'sis. N.L. fem. adj. *lexingtonensis* from Lexington, named after the place of origin, Lexington, Kentucky, USA.

Forms an abundant dark orange-brown to dark reddish-brown substrate mycelium, a light yellow to purplish-tan aerial mycelium, and a dark red to reddish-brown soluble pigment on most media. Grows between 15 to 42°C.

Hippurate is hydrolyzed, but does not degrade adenine. Acid is produced from dulcitol (weak), but not from methyl α-D-glucoside. Acetate, citrate, oxalate, and propionate are used as sole carbon sources, but not benzoate, lactate, malate, mucate, succinate, or DL-tartrate. Additional phenotypic properties are shown in Table 238.

Whole-organism hydrolysates are rich in arabinose and galactose. Cellular fatty acid profiles contain major proportions of $C_{15:0}$ iso (10.2%) and $C_{16:0}$ iso (42.9%), and smaller proportions (<10%) of $C_{14:0}$ iso (4.1%), $C_{15:0}$ anteiso (1.2%), $C_{15:1}$ (1.7%), $C_{15:0}$ (4.2%), $C_{16:1}$ *cis*9 (4.0%), $C_{16:0}$ (2.2%), 9-methyl-$C_{16:0}$ (1.5%), $C_{17:0}$ iso (2.1%), $C_{17:0}$ anteiso (1.7%), $C_{17:1}$ *cis*9 (7.8%), $C_{16:0}$ 2-OH (7.8%), $C_{17:0}$ (3.3%), and $C_{17:0}$ 10-methyl (1.8%). The major phospholipid is PE. The predominant menaquinone is MK-9(H_4).

Source: isolated from an equine placenta.

DNA G+C content (mol%): not determined.

Type strain: LDDC 12275-99, DSM 44653, JCM 12672, NRRL B-24131.

Sequence accession no. (16S rRNA gene): AY183358.

18. **Amycolatopsis lurida** (Lechevalier, Prauser, Labeda and Ruan 1986) Stackebrandt, Kroppenstedt, Wink and Schumann 2004, 267[VP] [Basonym: *Amycolatopsis orientalis* subsp. *lurida* (*ex* Grundy, Sinclair, Theriault, Goldstein, Rickher, Warren, Oliver and Sylvester 1957) Lechevalier, Prauser, Labeda and Ruan 1986, 35; "*Nocardia lurida*" Grundy, Sinclair, Theriault, Goldstein, Rickher, Warren, Oliver and Sylvester 1957, 687.]

lu'ri.da. L. fem. adj. *lurida* pale yellow, sallow.

White aerial mycelium carries cylindrical, occasionally ovoid, smooth spores in straight to flexuous chains. Yellow to beige substrate mycelium branches frequently and appears to be slightly zig-zag. Grows at 10°C, but not at 45°C.

Positive for *N*-acetyl-β-glucosamidase, acid and alkaline phosphatases, α-chymotrypsin, esterase lipase, (C8), β-galactosidase, β-glucosidase, leucine arylamidase, lipase (C14), naphthol-AS-BI-phosphohydrolase, and trypsin, but negative for esterase (C4), cystine arylamidase, α-fucosidase, α-galactosidase, α-glucosidase, β-glucuronidase, α-mannosidase, and valine arylamidase (API ZYM tests). Positive for β-galactosidase, but not for arginine dihydrolase, or lysine and ornithine decarboxylases (API 20E tests). Acetoin positive, but indole-negative. Arbutin is hydrolyzed, but does not produce H_2S. Elastin and xylan are degraded, but not Tween 40 or uric acid.

Resistant (filter paper discs soaked in μg/ml) to nalidixic acid (30), streptomycin (10) and vancomycin (30), but sensitive to kanamycin (30) and sulfonamide (200). Grows in the presence of lysozyme. Additional phenotypic properties are shown in Table 238.

Whole-organism hydrolysates are rich in arabinose and galactose. Mycolic acids are present. The polar lipid pattern contains DPG as the diagnostic component. The predominant menaquinones are MK-9(H_2,H_4).

Source: isolated from soil.

DNA G+C content (mol%): 67 (T_m).

Type strain: ATCC 14930, DSM 43134, IFO (now NRBC) 14500, JCM 3141, LMG 4064, NRRL 2430, NRRL WC-3860, VKM Ac-1242.

Sequence accession nos: AJ577997 (16S rRNA gene); EU822900 (*gyrB*).

19. **Amycolatopsis marina** Bian, Li, Wang, Song, Liu, Dai, Ren, Gao, Hu, Liu, Li and Zhang 2009, 480[VP]

ma.ri'na. L. fem. adj. *marina* of the sea, marine.

Sparse, white aerial mycelium and branched yellow to yellow-brown substrate mycelium fragment into rod-like elements on yeast extract-malt extract agar medium (ISP medium 2). Diffusible pigments are not produced. Grows between 10 and 45°C (optimally at 28°C) and at pH 6–9 (optimally between pH 7 and 8).

H_2S is not produced. Tweens 20, 40, 60, and 80 are degraded, but not adenine or elastin. Cellobiose, fructose, galactose, maltose, mannitol, *myo*-inositol, rhamnose, ribose, trehalose, xylitol, and sodium acetate are used as sole carbon sources, but not DL-arabinose, methyl-D-lactose, raffinose, sorbitol, sucrose, or sodium citrate dehydrate (all at 1%, w/v). Grows in the presence of 12% (w/v) NaCl.

Resistant (filter paper discs soaked in 30 μg/ml antibiotic) to amikacin, carbenicillin, clarithromycin, kanamycin, penicillin G, rifampin, sulfamethoxazole and tobramycin, but sensitive to acetylspiramycin, carbenicillin, cephalothin, chloramphenicol, doxycycline, erythromycin, gentamicin, midecamycin, minocycline, novobiocin, and streptomycin. Additional phenotypic properties are shown in Table 238.

Whole-organism hydrolysates are rich in arabinose and galactose. Cellular fatty acid profiles contain major proportions of $C_{16:0}$ iso (40.4%) and $C_{16:0}$ iso 2-OH (11.4%), and smaller proportions (<10%) of $C_{17:1}$ *cis*9 (8.6%), $C_{16:1}$ *cis*9, (7.8%), $C_{16:0}$ (7.1%), $C_{17:0}$, (4.9%), $C_{17:0}$ iso (3.5%), $C_{15:0}$ (2.9%), $C_{16:1}$ iso H (2.7%), $C_{17:0}$ anteiso (2.3%), $C_{15:1}$ B (1.6%), $C_{15:0}$ iso (1.5%), $C_{16:0}$ 10-methyl (1.4%), $C_{18:0}$ (1.4%), $C_{18:1}$ *cis*9 (1%), $C_{14:0}$ iso (0.9%), and $C_{18:0}$ iso (0.8%). The polar lipid pattern contains DPG, PE, PG, PI, PIMs, and PME. The predominant menaquinone is MK-9(H_4) (79%); also contains minor amounts of MK-8(H_4).

Source: isolated from a sediment sample collected from the South China Sea.

DNA G+C content (mol%): 70.1 (HPLC).

Type strain: Ms392A, CGMCC 4.3568, NBRC 104263.

Sequence accession no. (16S rRNA gene): EU329845.

20. **Amycolatopsis mediterranei** (Margalith and Beretta 1960) Lechevalier, Prauser, Labeda and Ruan 1986, 35[VP] [Basonym: *Nocardia mediterranei* (Margalith and Beretta 1960) Thiemann, Zucco, Pelizza 1969, 106; "*Streptomyces mediterranei*" Margalith and Beretta 1960, 321.]

med.i.ter.ra'ne.i. L. neut. gen. n. *mediterranei* of the interior of the land, from the Mediterranean area.

Substrate hyphae may show a zig-zag appearance. Aerial hyphae, when formed, differentiate into long, straight to flexuous chains of ellipsoid to oblong spores. Yellow colored colonies are formed on modified Bennett's agar. Grows from 10 to 42°C, but not at 45°C.

Acid and alkaline phosphatase-positive, but is negative for *N*-acetyl-β-glucosamidase, α-chymotrypsin, cystine arylamidase, esterase (C4), esterase lipase C8), α-fucosidase, α- and β-galactosidases, α- and β-glucosidases, leucine arylamidase, lipase (C14), α-mannosidase, naphthol-AS-BI-phosphohydrolase, trypsin, and valine arylamidase. Acetate, citrate, lactate, malate, oxalate, propionate, pyruvate, and succinate are decarboxylated, but not benzoate, mucate, or tartrate.

Resistant (μg/ml) to gentamicin sulfate (5), neomycin sulfate (8), tobramycin sulfate (8), novobiocin (10), polymyxin B (50), rifampin (10) and vancomycin hydrochloride (0.25). Additional phenotypic properties are shown in Table 238.

Whole-organism hydrolysates are rich in arabinose and galactose. Muramic acid moieties are *N*-acetylated. The cellular fatty acid profile contains major proportions of $C_{16:0}$ iso (44.9%), and smaller proportions (<10%) of $C_{14:0}$ iso (1.2%), $C_{15:0}$ iso (9.3%), $C_{15:0}$ anteiso (1.4%), $C_{15:1}$ (2.0%), $C_{15:0}$ (3.1%), $C_{16:1}$ iso H (2.3%), $C_{16:0}$ (1.0%), $C_{16:1}$ *cis*9 (1.3%), $C_{17:1}$ iso G (1.5%), $C_{17:0}$ iso (2.0%), $C_{17:0}$ anteiso (4.0%), $C_{17:1}$B (6.1%), $C_{17:1}$ C (9.0%), $C_{17:0}$ C (2.9%), and $C_{18:1}$ (5.6%). Does not contain mycolic acids. The predominant menaquinones are MK-9(H_4) and MK-9(H_6).

Source: isolated from a soil sample collected in a *Pinus* arboretum near St Raphael in France.

DNA G+C content (mol%): 67–69 (T_m).

Type strain: ATCC 13685, CCUG 43144, CIP 107074, DSM 43304, NBRC 13415, IMET 7651, ISP 5501, JCM 4789, KCTC 1739, NRRL B-3240, VKM Ac-798.

Sequence accession nos: AJ293754 (16S rRNA gene); EU822901 (*gyrB*).

Additional comments: Amycolatopsis mediterranei DSM 46095 and DSM 40696 have been reclassified as *Amycolatopsis rifamycinica* (Bala et al., 2004).

21. **Amycolatopsis methanolica** de Boer, Dijkhuizen, Grobben, Goodfellow, Stackebrandt, Parlett, Whitehead and Witt 1990, 203[VP]

me.tha.no'li.ca. N.L. n. *methanol* methanol; L. fem. suff. *-ica* suffix used with the sense of pertaining to; N.L. fem. adj. *methanolica* relating to methanol.

Forms a yellow substrate mycelium which bears white aerial hyphae that differentiate into smooth, squarish to oval spores (0.4 × 0.6–0.8 μm) on long, straight to flexuous chains on Czapek Dox agar. Grows from 10 to 50°C.

Decomposes DNA and elastin, but not adenine, arbutin, or tributyrin. Acetate, benzoate, fumarate, 2-oxoglutarate, lactate, malate, propionate, pyruvate, and succinate are used as sole carbon sources, but not citrate, formate, oxalate or gluconate. Acid is produced from ribose, but not from dulcitol.

Resistant to (μg/ml) cephaloridine hydrochloride (10), gentamicin sulfate (5), lincomycin hydrochloride (10), neomycin sulfate (10), oleandomycin phosphate (2), penicillin G (20), novobiocin (10), polymyxin B (50), and vancomycin hydrochloride (0.25), but sensitive to lysozyme. Additional phenotypic properties are shown in Table 238.

Whole-organism hydrolysates are rich in arabinose and galactose. The cellular fatty acid profile contains major proportions of $C_{16:0}$ iso (32.0%), $C_{16:0}$ (24.3%), and $C_{17:0}$ anteiso (21.8%). Mycolic acids are absent. The polar lipid pattern contains DPG, PE, PME, and PIMs. The predominant menaquinones are MK-9(H_2,H_4).

Source: isolated from a soil sample collected in New Guinea.

DNA G+C content (mol%): not determined.

Type strain: 239, DSM 44096, NBRC 15065, IMSNU 20055, JCM 8087, LMD 80.32, NCIB (now NCIMB) 11946, NRRL B-24139.

Sequence accession nos: AJ249135 (16S rRNA gene); EU822902 (*gyrB*).

Additional comments: the type strain of *Amycolatopsis methanolica* cleaves a broad range of 7-amino-methylcoumarin and 4-methylumbelliferone conjugated substrates and grows in a mineral medium broth containing diverse sole carbon compounds (De Boer et al., 1990).

22. **Amycolatopsis minnesotensis** Lee, Kinkel and Samac 2006c, 268[VP]

min.ne.sot.en'sis. N.L. fem. adj. *minnesotensis* pertaining to Minnesota, the origin of the soil sample from which the type strain was isolated.

Forms a well-developed, white aerial mycelium and a yellow vegetative mycelium that fragment into rod-shaped elements. Grows from 10 and 30°C, but not at 37°C.

Positive for acid and alkaline phosphatase, esterase (C4), and naphthol-AS-BI-phosphohydrolase, but negative for α-chymotrypsin, cystine arylamidase, α- and β-galactosidases, α-glucosidase, lipase (C14), α-mannosidase, trypsin, and valine arylamidase. Acid is produced from D-xylitol, but not from 2,3-butanediol, dextran, dulcitol, or 1,2-propanediol. Additional phenotypic properties are shown in Table 238.

Resistant (filter paper discs soaked in µg/ml) to kanamycin (30), streptomycin (10) and vancomycin (30), but sensitive to nalidixic acid (30) and sulfonamide (200).

Predominant menaquinones are MK-9(H_4); major fatty acids are $C_{16:0}$ iso, $C_{16:0}$, $C_{15:0}$ iso, and $C_{17:0}$.

Source: isolated from a prairie soil.

DNA G+C content (mol%): 69.5.

Type strain: 32U-2, DSM 44988, JCM 14545, KCCM 42246, NRRL B-24435.

Sequence accession nos: DQ076482 (16S rRNA gene); EU822903 (*gyrB*).

23. **Amycolatopsis nigrescens** Groth, Tan, González, Laiz, Carlsohn, Schütze, Wink and Goodfellow 2007, 517[VP]

ni.gres'cens. L. part. adj. *nigrescens* becoming black.

Forms extensively branched hyphae (0.7–0.9 µm in diameter) that fragment into squarish rod-like elements. The color of the substrate mycelium changes from orange to black with the production of a dark reddish-black soluble pigment. The substrate mycelium carries sparse to moderate white or pale-orange aerial hyphae. Grows between 20 and 40°C, but not at 10 or 42°C, and between pH 5 and 9 but not at pH 4.5 or 10.

Tween 80 is degraded, but not adenine. Positive for *N*-acetyl-β-glucosamidase, acid and alkaline phosphatases, α-chymotrypsin, cystine arylamidase, esterase (C4), esterase lipase (C8), leucine arylamidase, lipase (C14), naphthol-AS-BI-phosphohydrolase, and valine arylamidase, but negative for α-galactosidase, α-glucosidase, α-mannosidase, and trypsin (API ZYM tests).

Hippurate is hydrolyzed. L-arabinose, fructose, glucose, mannitol, *myo*-inositol, raffinose, and xylose are used as sole carbon sources, but not cellulose, or sucrose. Similarly, acetate, aconitate, citrate, malate, and succinate are used as sole carbon sources, but not benzoate or DL-tartrate. Grows in the presence of 6% (w/v) NaCl.

Resistant (filter paper discs soaked in µg/ml) to lincomycin hydrochloride (2), kanamycin (30), meticillin (5), nalidixic acid (30), norfloxacin (10), streptomycin (10), sulfonamide (200), and vancomycin (30), but sensitive to chloramphenicol (30), imipenem (10), ofloxacin (10), oxytetracycline hydrochloride (30), and rifampin (30). Additional phenotypic properties are shown in Table 238.

Whole-organism hydrolysates are rich in arabinose and galactose. Muramic acid moieties are *N*-acetylated. The cellular fatty acid profile contains major proportions of $C_{16:0}$ iso (27–28%), and smaller proportions (<10%) of $C_{17:0}$ (7.3–9.3%), $C_{17:1}$ *cis*9 (8.4–9.8%), $C_{16:0}$ (7.2–7.4%), $C_{16:0}$ iso 2-OH (6.4–7.9%), $C_{15:0}$ iso (5.2–5.7%), $C_{18:1}$ *cis*9 (4.8–4.9%), $C_{16:1}$ iso (4.2–4.4%), $C_{17:0}$ anteiso (3.7–3.8%), $C_{16:1}$ *cis*9 (3.1–3.8%), $C_{18:0}$ (3.3%), and an unidentified component (3.0–3.7%). Mycolic acids are absent. The polar lipid pattern includes DPG, HPE, PG, PI, and PS. The predominant menaquinone is MK-11(H_4) (52–54%); also contains smaller proportions of MK-12(H_4) (18%), MK-9(H_4) (8–9%), and MK-10(H_4) (9%).

Source: isolated from the wall of St Callistus hypogean Roman catacomb.

DNA G+C content (mol%): not determined.

Type strain: CSC17Ta-84, DSM 44992, HKI 0330, JCM 14717, NRRL B-24473.

Sequence accession nos: DQ486888 (16S rRNA gene); EU822904 (*gyrB*).

24. **Amycolatopsis niigatensis** Ding, Hirose and Yokota 2007, 1750[VP]

ni.i.ga.ten'sis. N.L. fem. adj. *niigatensis* referring to Niigata Prefecture, Japan, the source of the type strain.

The substrate mycelium is purple brown and the aerial mycelium white to light yellow. Grows between 5 to 45°C, optimally at 30°C, and from pH 6 to 11, optimally at pH 9.

Positive for *N*-acetyl-β-glucosamidase, alkaline phosphatase, α-chymotrypsin, cystine arylamidase, esterase (4), esterase lipase (C8), α- and β-galactosidases, α-glucosidase, leucine arylamidase, α-mannosidase, naphthol-AS-BI-phosphohydrolase, and valine arylamidase, but negative for α-fucosidase, β-glucosidase, and lipase (C14). Grows in the presence of 7% (w/v) NaCl. Additional phenotypic properties are shown in Table 238.

Whole-organism hydrolysates are rich in arabinose and galactose. Cellular fatty acids contain a major proportion of $C_{16:0}$ iso (40%), and smaller proportions (<10%) of $C_{14:0}$ iso (7.1%), $C_{14:0}$ (0.4%), $C_{15:0}$ iso (9.3%), $C_{15:0}$ anteiso (7.7%), $C_{15:0}$ (4.5%), $C_{16:0}$ (2.1%), $C_{16:1}$ iso OH (2.6%), $C_{17:0}$ iso (1.6%), $C_{17:0}$ anteiso (6.9%), $C_{17:0}$ (2.6%), $C_{17:1}$ ω6*c* (6.6%), and $C_{17:1}$ ω8*c* (1.7%). Mycolic acids are absent. The predominant menaquinone is MK-9(H_4).

Source: isolated from filtration material made from volcanic soil, Niigata, Japan.

DNA G+C content (mol%): 72.4 (HPLC).

Type strain: LC11, CCTCC AB206020, IAM 15388, JCM 21832.

Sequence accession nos: AB248537 (16S rRNA gene); EU822905 (*gyrB*).

25. **Amycolatopsis palatopharyngis** Huang, Paściak, Liu, Xie and Gamian 2004, 361[VP]

pa.la.to.pha.ryn'gis. N.L. n. *palatopharynx* palatopharynx; N.L. gen. n. *palatopharyngis* of the palatopharynx.

Forms a branched yellow to yellow-brown substrate mycelium that fragments into rod-like elements. White aerial hyphae are produced sparsely on Bennett's, glucose-yeast extract-malt extract, and brain-heart infusion agars and moderately on inorganic salts-starch agars. Aerial hyphae differentiate into long chains of spore-like structures. Diffusible pigments are not produced. Grows between 10 and 40°C, but not at 45°C, and between pH 6 and 10. Grows in the presence of 10% (w/v) NaCl.

Resistant (µg/ml) to ampicillin (10), carbenicillin (100), and cephalothin (30), but sensitive to acetylspiramycin (15), chloramphenicol (30), clarithromycin (10), doxycycline (30), gentamicin (10), kanamycin (30), midecamycin (15), minocycline (30), novobiocin (5), penicillin (30), rifampin (5), streptomycin (10), and tobramycin (10). Additional phenotypic properties are shown in Table 238.

Whole-organism hydrolysates are rich in arabinose and galactose. The cellular fatty acid profile contains major proportions (<10%) of $C_{16:0}$ iso (50.3%), $C_{16:0}$ (21.7%), and $C_{17:0}$ anteiso (12.4%), and smaller proportions of $C_{16:0}$ anteiso (3.6%), $C_{17:0}$ (8.5%), and $C_{18:0}$ (3.7%). Mycolic acids are absent. The polar lipid pattern includes PE, DPG, and PI. The predominant menaquinone is MK-9(H_4).

Source: isolated from an infected palatopharyngeal mucosa of an elderly human patient.

DNA G+C content (mol%): 65.8 (T_m).

Type strain: 1BDZ, AS 4.1729, DSM 44832, JCM 12460, PCM 2600.

Sequence accession nos: AF479268 (16S rRNA gene); EU822907 (*gyrB*).

26. **Amycolatopsis plumensis** Saintpierre-Bonaccio, Amir, Pineau, Tan and Goodfellow 2005, 2060[VP]

plum.en'sis. N.L. fem. adj. *plumensis* referring to the Plum region of the main island of New Caledonia, the source of the soil from which the type strain was isolated.

Forms extensively branched substrate hyphae that fragment into squarish, rod-like elements. A pale-orange substrate mycelium is formed on modified Bennett's agar, but diffusible pigments are absent. The substrate mycelium carries abundant pale-orange aerial hyphae. Grows between 20 and 37°C, but not at 10 or 45°C, and between pH 4 to12.

H_2S is formed. Elastin, Tween 80, and xylan are degraded.

Resistant (μg/ml) to erythromycin (4), gentamicin sulfate (10), rifampin (6), and streptomycin sulfate (5). Grows in the presence of crystal violet (0.0002%, w/v), phenol (0.01%, w/v), and potassium tellurite (0.005%, w/v). Additional phenotypic properties are shown in Table 238.

Whole-organism hydrolysates are rich in arabinose and galactose. The predominant menaquinone is MK-9(H_4).

Source: isolated from brown hypermagnesian ultramafic soil at the southern end of the main island of New Caledonia.

DNA G+C content (mol%): not determined.

Type strain: SBHS Strp1, DSM 44776, JCM 13852, NBRC 102106, NRRL B-24324.

Sequence accession nos: AY262825 (16S rRNA gene); EU822908 (*gyrB*).

27. **Amycolatopsis pretoriensis** Labeda, Donahue, Williams, Sells and Henton 2003, 1604[VP]

pre.tor.i.en'sis. N.L. fem. adj. *pretoriensis* from Pretoria, named after the place of origin, Pretoria, South Africa.

Forms a well-developed grayish-yellow to orange-brown substrate mycelium and an abundant white to orange-white aerial mycelium on most media. Faint soluble pigments are produced on some media, such as yeast extract-malt extract agar. Grows from 15 to 37°C.

Positive for cystine arylamidase and β-galactosidase, but negative for α-fucosidase, β-glucosidase, and valine arylamidase (API ZYM tests). Hippurate is hydrolyzed. Acid is produced from dulcitol, but not from methyl β-xyloside or melezitose. Acetate, benzoate (weak), citrate (weak), and lactate (weak) are used as a sole carbon sources, but not malate, mucate, or tartrate. Additional phenotypic properties are shown in Table 238.

Whole-organism hydrolysates are rich in arabinose and galactose. Cellular fatty acid profiles contain major proportions of $C_{16:0}$ iso (45.5%) and $C_{16:0}$ 2-OH (12.3%), and smaller proportions (<10%) of $C_{14:0}$ iso (3.4%), $C_{15:0}$ iso (7.9%), $C_{15:1}$ (3.3%), $C_{15:0}$ (3.1%), $C_{17:0}$ anteiso (1.3%), $C_{17:1}$ *cis*9 (8.9%), $C_{17:0}$ (2.7%), and $C_{17:0}$ 10-methyl (2.6%). The major polar lipids are PE and PME. The predominant menaquinones are MK-9(H_2) and MK-9(H_4).

Source: isolated from an equine placenta in Pretoria, South Africa.

DNA G+C content (mol%): not determined.

Type strain: ARC OVI 0181, DSM 44654, JCM 12673, NRRL B-24133.

Sequence accession no. (16S rRNA gene): AY183356.

28. **Amycolatopsis regifaucium** Tan, Robinson, Lacey, Brown, Kim and Goodfellow 2007, 2566[VP]

re.gi.fau'ci.um. L. n. *rex regis* king; L. gen. pl. n. *faucium* of a defile; N.L. gen. pl. n. *regifaucium* of King's Canyon, Australia, the source of the soil from which the first strains were isolated.

Forms an extensively branched substrate mycelium which fragments into squarish, rod-like elements. An abundant, light-gray aerial mycelium and a dark yellow–brown substrate mycelium with filamentous edges are formed on modified Bennett's agar supplemented with mannitol and soybean flour; a dark gray–brown diffusible pigment is also produced. Grows between 10 and 37°C, and between pH 5 and 10.

Arbutin is hydrolyzed. Elastin, Tween 40, uric acid, and xylan are degraded. L-Arabinose, arabitol, cellobiose, dextrin, galactose, glucose, glycerol, glycogen, *myo*-inositol, maltose, mannitol, methyl α-D-glucoside, ribose, sucrose, glycogen, and xylitol are used as sole carbon sources, but not adonitol, *meso*-erythritol, melezitose, melibiose, raffinose, or sorbitol.

Resistant (μg/ml) to gentamicin sulfate (5), neomycin sulfate (8), novobiocin (10), penicillin G (20), polymyxin B sulfate (50), rifampin (10), streptomycin sulfate (16), and tobramycin sulfate (8). Additional phenotypic properties are shown in Table 238.

Whole-organism hydrolysates are rich in arabinose and galactose. Does not contain mycolic acids. The polar lipid pattern includes DPG, PE, PG, PI, PIMs, and PME. The predominant menaquinone is MK-9(H_4) (71%); also contains MK-9(H_6) (29%).

Source: isolated from an arid soil collected from King's Canyon, Australia.

DNA G+C content (mol%): not determined.

Type strain: GY080, DSM 45072, JCM 15588, NCIMB 14277.

Sequence accession nos: AY129760 (16S rRNA gene); EU822909 (*gyrB*).

29. **Amycolatopsis rifamycinica** Bala, Khanna, Dadhwal, Prabagaran, Shivaji, Cullum and Lal 2004, 1148[VP]

rif.a.my.ci'na. N.L. n. *rifamycinum* rifamycin; L. suff. *-icus -a -um* related to; N.L. fem. adj. *rifamycinica* referring to the ability to produce rifamycin.

Orange-colored substrate mycelium is produced on yeast extract and glucose-asparagine agars and a white to very pale pink aerial mycelium is produced on oatmeal and yeast extract-molasses agars. A light pale to brown yellow pigment is formed on tyrosine, yeast extract-glucose, and oatmeal agars. Grows from 10 to 37°C, but not at 45°C.

Positive for cystine arylamidase, β-galactosidase, β-glucosidase, and valine arylamidase, but negative for α-fucosidase (API ZYM tests). Additional phenotypic properties are shown in Table 238.

The cellular fatty acid profile contains major proportions of $C_{16:0}$ iso (24%), $C_{17:0}$ anteiso (11%), and $C_{18:1}$ (25%), and smaller proportions (<10%) of $C_{15:0}$ iso (3%), $C_{16:0}$ (4%), iso $C_{17:0}$ (9%), $C_{17:1}$ (4%), and $C_{18:4}$ (4%). The polar lipid pattern includes cardiolipin, PE, PG, and PI.

Source: isolated from a soil sample in an arid region near Alice Springs, Northern Territory, Australia.

DNA G+C content (mol%): not determined.

Type strain: nt 19, ATCC 27643, DSM 46095, JCM 12674.

Sequence accession nos: AY083603 (16S rRNA gene); EU822910 (*gyrB*).

Additional comments: DNA restriction profiles of seven *Amycolatopsis mediterranei* strains provided further evidence that strain DSM 46095 was misclassified in this species (Bala et al., 2004), a result in line with the results of an earlier study (Kaur et al., 2001).

30. **Amycolatopsis rubida** Huang, Qi, Lu, Liu and Goodfellow 2001, 1096[VP]

ru.bi'da. L. fem. adj. *rubida* reddish.

Forms branching white to yellowish substrate mycelium which fragments into squarish elements (0.4–0.5 × 1.2–3.0 μm). An abundant white aerial mycelium which fragments into squarish elements is produced on modified Bennett's agar. A reddish diffusible pigment is formed on glucose-asparagine agar. Grows between 10 and 40°C. Grows in the presence of 7% (w/v) NaCl.

Resistant (μg/ml) to gentamicin sulfate (5), neomycin sulfate (8), novobiocin (10), penicillin G (20), polymyxin B (50), streptomycin sulfate (30), tobramycin sulfate (8), and vancomycin hydrochloride (0.25), but sensitive to chloramphenicol (30) and erythromycin (15). Also resistant to lysozyme (0.005%, w/v). Additional phenotypic properties are shown in Table 238.

Whole-organism hydrolysates are rich in arabinose and galactose. The cellular fatty acid profile contains major proportions of $C_{14:0}$ iso (11.3%), $C_{15:0}$ (14.6%), $C_{15:1}$ ω6*c* (12.8%), $C_{16:0}$ iso (12.8%) and $C_{17:0}$ (11.3%), and smaller proportions (<10%) of $C_{14:0}$ (0.5%), $C_{15:0}$ iso (6.8%), $C_{15:0}$ anteiso (3%), $C_{16:0}$ (5.2%), $C_{16:1}$ iso H (0.2%), $C_{17:0}$ iso (0.4%), $C_{17:0}$ anteiso (0.8%), and $C_{18:0}$ (0.4%). Mycolic acids are absent. The polar lipid pattern includes DPG, PE, PIMs, and PME. The predominant menaquinone is MK-9(H_4).

Source: isolated from coniferous forest soil collected in Guangxi Province, China.

DNA G+C content (mol%): 67.4 (T_m).

Type strain: 13.4, AS4. 1541, CIP 107102, DSM 44637, JCM 10871 NBRC 100041, NRRL B-24150.

Sequence accession nos: AF222022 (16S rRNA gene); EU822911 (*gyrB*).

31. **Amycolatopsis saalfeldensis** Carlsohn, Groth, Tan, Schütze, Saluz, Munder, Yang, Wink and Goodfellow 2007, 1644[VP]

sa.al.fel.den'sis. N.L. fem. adj. *saalfeldensis* from Saalfeld, named after the place of origin, a town in Thuringia, Germany.

Forms extensively branched substrate mycelium (hyphal diameter 0.5–0.6 μm) that fragments into squarish rod-like elements on oatmeal and yeast-extract-malt extract agars. The substrate mycelium carries moderate amounts of white aerial hyphae on oatmeal agar which fragment into squarish rod-like elements. Diffusible pigments are not produced. Grows between 20 and 35°C, and between pH 4.5 and 8, but not at pH 9.

Positive for *N*-acetyl-β-glucosamidase, acid and alkaline phosphatases, α-chymotrypsin, esterase (C4), esterase lipase (C8), α-glucosidase, naphthol-AS-BI-phosphohydrolase, and valine arylamidase, but negative for cystine arylamidase, α-fucosidase, α- and β-galactosidases, lipase (C14), α-mannosidase, and trypsin (API ZYM tests). Hippurate is hydrolyzed, and H_2S and oxidase are produced, but does not form indole. Tween 80 is degraded, but not adenine. L-Arabinose, fructose, glucose, *myo*-inositol, mannitol, raffinose, rhamnose (weakly), sucrose, and xylose are used as sole carbon sources, but not cellulose (all at 1%, w/v). Similarly, acetate, aconitate, benzoate (weakly), citrate, malate, and succinate are used as sole carbon sources, but not DL-tartrate (all at 0.2%, w/v).

Resistant (filter paper discs soak in μg/ml) to ampicillin (10), methicillin (5), norfloxacin (10), novobiocin (5), penicillin G (10 IU), and polymyxin B (300 IU), but sensitive to chloramphenicol (30), ciprofloxacin (5), imipenem (10), kanamycin sulfate (30), lincomycin hydrochloride (2), ofloxacin (10), oxytetracycline hydrochloride (30), rifampin (30), streptomycin sulfate (10), and vancomycin hydrochloride (30). Additional phenotypic properties are shown in Table 238.

Whole-organism hydrolysates are rich in arabinose and galactose. Muramic acid moieties are *N*-acetylated. Cellular fatty acid profiles contain major proportions of $C_{16:0}$ iso (41–42%), and smaller proportions of $C_{17:0}$ iso (8–10%), $C_{14:0}$ iso (7–9%), $C_{15:0}$ iso (8–9%), $C_{16:0}$ iso 2-OH (4–6%), $C_{15:0}$ (6%), $C_{17:0}$ (5–7%), and $C_{17:0}$ iso (8–10%). Mycolic acids are absent. The polar lipid pattern includes DPG, hydroxyphosphatidylethanolamine, PE, PG, and PS. The predominant menaquinone is MK-9(H_4) (86.9%); also contains minor amounts of MK-8(H_4) (4.5%), MK-9(H_6) (1.0–3.0%), and MK-10(H_4) (2.0%).

Source: isolated from the surface of rocks in a medieval alum slate mine.

DNA G+C content (mol%): not determined.

Type strain: HKI 0457, DSM 44993, JCM 14909, NRRL B-24474.

Sequence accession nos: DQ792500 (16S rRNA gene); EU822912 (*gyrB*).

Additional comments: grows on minimal medium supplemented with $CuSO_4$ (2 mM) and $NiCl_2$ (5 mM) (Carlsohn et al., 2007). The type strain DSM 44993 was mistakenly cited as DSM 44493 in the protologue of the valid publication.

32. **Amycolatopsis sacchari** Goodfellow, Kim, Minnikin, Whitehead, Zhou and Mattinson-Rose 2001, 191[VP]

sac'char.i. N.L. n. *Saccharum* generic name of sugar cane; N.L. gen. n. *sacchari* of sugar cane.

Forms branched substrate mycelium that fragments into rod-like elements when grown on modified Bennett's and Czapek Dox-yeast extract-casein (CYC) agars. The substrate mycelium carries moderate to abundant white aerial hyphae which differentiate into straight chains of spore-like structures on CYC agar. Diffusible pigments are not produced. Grows between 20 and 45°C, but not at 55°C, and between pH 5 and 8.

Positive for alkaline phosphatase, α-chymotrypsin, cystine arylamidase, esterase (C4), α-glucosidase, lysine (C14), naphthol-AS-BI-phosphohydrolase, and valine arylamidase, but negative for α- and β-galactosidases, α-mannosidase, and trypsin. Testosterone, and Tweens 20, 40, 60 and 80 are degraded. Adonitol, D- and L-arabinose, cellobiose, dextrin, *meso*-erythritol, fructose, D- and L-fucose, galactose, gentiobiose, glycerol, glycogen, lactose, maltose, mannitol, mannose, methyl α-D-glucoside, methyl β-D-glucoside, rhamnose, ribose, salicin, sorbose, sucrose, trehalose, xylitol, and xylose are used as sole carbon sources, but not amygdalin, arabitol, dulcitol, *myo*-inositol, melibiose, sorbitol, or tyrosine (all at 1%, w/v). Similarly, L-alanine, androsterone, arbutin, butan-1-ol, butyrate, ergosterol, ethanol, hippurate, propan-1-ol, propan-2-ol, propionate, propylase glycol, protocatechuic acid, pyruvate, quinic acid, L-serine, shikimic acid, succinate, testosterone, and tyrosine are used as sole carbon sources, but not acetamide, benzamide, catechol, L-citrulline, *p*-cresol, fumarate, *m*- or *p*- hydroxybenzoic acid, β-hydroxybutyric acid, 15-mandelic acid, squalene, syringealdehyde, tartrate, trimethylamine, or vanillin (at 0.1% w/v or v/v).

Grows in the presence of (μg/ml) bismuth citrate (10), crystal violet (1), phenol (100), phenyl ethanol (4000), potassium tellurite (10), sodium azide (10), sodium chloride (7%, w/v), teepol (100), tetrazolium (100), and thallous acetate (10), but sensitive to bismuth citrate (100), crystal violet (100), sodium chloride (10), tetrazolium (1000), and thallous acetate (100).

Resistant to (μg/ml) cephaloridine hydrochloride (250), chloramphenicol (50), demeclocycline hydrochloride(8), gentamicin sulfate (5), kanamycin sulfate (5), lincomycin hydrochloride (10), neomycin sulfate (10), nalidixic acid (30), oleandomycin phosphate (2), penicillin G (20), polymyxin B sulfate (50), rifampin (2), streptomycin sulfate (16), tobramycin sulfate (8), and vancomycin hydrochloride (0.25), but sensitive to kanamycin sulfate (50) and tetracycline hydrochloride (5). Additional phenotypic properties are shown in Table 238.

Whole-organism hydrolysates are rich in arabinose and galactose. The cellular fatty acid profile contains major proportions of $C_{17:0}$ anteiso (30–58%), and smaller proportions of $C_{16:0}$ (4–14%), $C_{17:0}$ (2–5%), $C_{17:1}$ (1–3%), $C_{17:0}$ anteiso (7–11%), and $C_{17:0}$ iso (7–11%). Mycolic acids are absent. The polar lipid pattern includes DPG, PE, PG, and PI. The predominant menaquinone is MK-9(H_4) (96%).

Source: isolated from sugar cane bagasse and from floor dust of a hemp factory.

DNA G+C content (mol%): not determined.

Type strain: K24, CIP 107029, DSM 44468, JCM 11272, KCTC 9863, NBRC 100339.

Sequence accession nos: AF223354 (16S rRNA gene); EU822913 (*gyrB*).

Additional comments: cleaves a broad range of 7-amino-4-methyl coumarin and 4-methylumbelliferone conjugated substrates (Goodfellow et al., 2001).

33. **Amycolatopsis sulphurea** Lechevalier, Prauser, Labeda and Ruan 1986, 35[VP] ("*Nocardia sulphurea*" Oliver and Sinclair 1964)

sul.phu're.a. L. n. *sulphur* sulfur; L. masc. suff. *-eus* suffix used with various meanings; N.L. fem. adj. *sulphurea* of sulfur, referring to the yellow color of the substrate hyphae.

Forms a white to yellowish to olive substrate mycelium, which tends to break down into fragments. The aerial mycelium is light yellow. A dark brown diffusible pigment is produced on modified Bennett's agar. Grows between 10 to 37°C.

Does not degrade adenine. Acetate, citrate, lactate, malate, propionate, pyruvate, and succinate are used as sole carbon sources, but not benzoate, mucate, oxalate, or tartrate.

Resistant (μg/ml) to lysozyme, neomycin sulfate (4), tobramycin sulfate (8), novobiocin (10), polymyxin B (50), and vancomycin hydrochloride (0.25). Additional phenotypic properties are shown in Table 238.

Whole-organism hydrolysates are rich in arabinose and galactose. Mycolic acids are absent. The cellular fatty acid profile contains major proportions of $C_{16:0}$ iso (21.0%) and $C_{16:0}$ (17.0%), and smaller proportions (<10%) of $C_{14:0}$ iso (3–7%), $C_{15:0}$ iso (9–3%), $C_{15:0}$ anteiso 5.0%), $C_{15:0}$ (3–4%), *cis*9 $C_{16:0}$ (3–9%), $C_{17:0}$ iso (2.0%), $C_{17:0}$ anteiso (5–6%), $C_{17:1}$ (8–9%), $C_{17:0}$ iso (9.3%), and $C_{18:0}$ (3.2%). The polar lipid pattern includes PE and PME. The predominant menaquinones are MK-9(H_2,H_4).

Source: isolated from soil.

DNA G+C content (mol%): 67 (T_m).

Type strain: ATCC 27624, DSM 46092, NBRC 13270, IMET 7649, JCM 3142, VKM Ac-1244.

Sequence accession nos: AF051343 (16S rRNA gene); EU822914 (*gyrB*).

Additional comments: the name "*Nocardia sulphurea*" Oliver and Sinclair (1964) was described in a patent which means that the name has not been effectively published [Rule 25b (5) Bacteriological Code]. Consequently, this nomenclatural name cannot be revived (Rule 28a) and the citation of *Nocardia sulphurea* cannot refer to the original authors.

34. **Amycolatopsis taiwanensis** Tseng, Yang, Li and Jiang 2006, 1814[VP]

tai.wan.en'sis. N.L. fem. adj. *taiwanensis* of Taiwan, where the type strain was isolated.

The substrate mycelium is light yellow on Bennett's, peptone-yeast extract-iron, tyrosine and yeast extract-malt extract agars; yellowish-white on glucose-asparagine, inorganic salts-starch and oatmeal agars, and purple-yellow on tryptone-yeast extract agars. Short chains of oval spores are formed, albeit poorly, on glycerol-asparagine, inorganic salts, starch, and tyrosine agars. Soluble pigments are not formed. Does not produce aerial hyphae. Grows at 20–40°C. Additional phenotypic properties are shown in Table 238.

Whole-organism hydrolysates are rich in arabinose and galactose. Muramic acid moieties are *N*-acetylated. The cellular fatty acid profile contains major proportions of $C_{16:0}$ iso (38.1%), and $C_{17:1}$ (25.4%), and smaller proportions (<10%) of $C_{17:0}$ anteiso (5.2%), $C_{16:0}$ (5.1%), and $C_{17:0}$ (4.0%). Mycolic acids are not present. The diagnostic polar lipid is PE. The predominant menaquinone is MK-9(H_4).

Source: isolated from soil collected from Yilan county, Taiwan.

DNA G+C content (mol%): 68.9 (HPLC).

Type strain: 0345M-7, BCRC 16802, JCM 14925, KCTC 19116, NBRC 102103.

Sequence accession nos: DQ160215 (16S rRNA gene); EU822915 (*gyrB*).

35. **Amycolatopsis thermoflava** Chun, Kim, Oh, Seong, Lee, Bae, Lee, Kang, Hah and Goodfellow 1999, 1372[VP]

ther.mo.fla'va. Gr. n. *thermê* heat; L. adj. *flavus* yellow; N.L. fem. adj. *thermoflava* thermophilic, yellow.

Forms an extensively branched substrate mycelium which fragments into squarish elements (0.6–0.7 × 6.5–14.0 µm). Aerial hyphae are sterile, sparse, and white; the substrate mycelium is brown and a brown diffusible pigment is produced on modified Bennett's agar. Grows at 28–55°C, but not at 10 or 60°C.

Resistant (µg/ml) to gentamicin sulfate (5), neomycin sulfate (8), novobiocin (10), rifampin (10), tobramycin sulfate (8), and vancomycin hydrochloride (0.25). Grows in the presence of lysozyme. Additional phenotypic properties are shown in Table 238.

Whole-organism hydrolysates are rich in arabinose and galactose. The cellular fatty acid profile contains major proportions of $C_{16:0}$ iso (29%), $C_{16:0}$ iso-α OH (16%), and $C_{17:0}$ anteiso (14%). Mycolic acids are not present. The predominant menaquinone is MK-9(H_4).

Source: isolated from a soil sample collected from Hainan Island, China.

DNA G+C content (mol%): 75.0 (T_m).

Type strain: N1165, CIP 106795, DSM 44574, NBRC 14333, JCM 10669, NRRL B-24140.

Sequence accession nos: AF052390 (16S rRNA gene); EU822916 (*gyrB*).

36. **Amycolatopsis tolypomycina** Wink, Kroppenstedt, Ganguli, Nadkarni, Schumann, Seibert and Stackebrandt 2003a, 935[VP] (Effective publication: Wink, Kroppenstedt, Ganguli, Nadkarni, Schumann, Seibert and Stackebrandt 2003b, 44.)

to.ly.po.my'ci.na. N.L. n. *tolypomycinum* tolypomycin (an antibiotic produced by the organism); L. fem. suff. *-ina* suffix used with the sense of belonging to; N.L. fem. adj. *tolypomycina* pertaining to tolypomycin.

The substrate mycelium is pure yellow on yeast extract-malt extract agar, melon yellow on oatmeal agar, yellow orange on inorganic salts-starch and tyrosine agars, and colorless on glycerol-asparagine agar; soluble pigments are not formed on any of these media. White, fragmenting aerial hyphae are produced on glycerol-asparagine, oatmeal, and tyrosine agars. Fragmentation of the substrate mycelium is seen on glycerol-asparagine, oatmeal, and tyrosine agars.

Positive for *N*-acetyl-β-glucosamidase, acid and alkaline phosphatases, cystine arylamidase, esterase (C8), α- and β-galactosidases, α-glucosidase, β-glucuronidase, leucine arylamidase, lipase (C14), α-mannosidase, trypsin, and valine arylamidase, but negative for esterase (C4), α-fucosidase, and naphthol-AS-BI-phosphohydrolase (API ZYM tests). Additional phenotypic properties are shown in Table 238.

Whole-organism hydrolysates are rich in arabinose and galactose. Cellular fatty acids contain major proportions of $C_{16:0}$ iso (14.5%), $C_{15:0}$ iso (10.3%), and $C_{17:0}$ anteiso (15.2%), and smaller proportions of $C_{14:0}$ (0.6%), $C_{15:0}$ (3.3%), $C_{16:0}$ (7.7%), $C_{17:0}$ (7.5%), $C_{15:1}$ (1.0%), $C_{16:1}$ (6.1%), $C_{17:1}$ (8.9%), $C_{18:1}$ *cis9* (1.0%), $C_{14:0}$ iso (0.8%), $C_{15:0}$ anteiso (3.6%), $C_{16:1}$ (0.8%), $C_{17:0}$ iso (4.1%), $C_{15:1}$ anteiso (0.5%), $C_{16:0}$ 10-methyl iso (0.4%), $C_{16:0}$ 10-methyl (2.6%), $C_{17:0}$ 10-methyl (3.1%), 2-hydroxyl-$C_{15:0}$ anteiso (0.5%), 2-hydroxy-$C_{16:0}$ iso (3.1%), 2-hydroxyl-$C_{17:0}$ anteiso (0.5%), and 2-hydroxyl-$C_{17:0}$ anteiso (3.9%). Mycolic acids are absent. The polar lipid pattern includes DPG, HPE, PE, PG, and PI. The predominant menaquinone is MK-9(H_4).

Source: isolated from soil in Tokyo, Japan.

DNA G+C content (mol%): not determined.

Type strain: ATCC 21177, DSM 44544, NBRC 14664, NRRL B-24205.

Sequence accession nos: AJ293757 (16S rRNA gene); EU822917 (*gyrB*).

37. **Amycolatopsis tucumanensis** Albarracin, Alonso-Vega, Trujillo, Amoroso and Abate 2010, 400[VP]

tu.cu.ma.nen'sis. N.L. fem adj. *tucumanensis* of or pertaining to Tucumán, Argentina, the origin of the soil sample from which the type strain was isolated.

Forms an extensively branched honey-yellow substrate mycelium that fragments into squarish elements on glycerol-asparagine, inorganic salts-starch, peptone-yeast extract, tyrosine, and yeast extract-malt extract agars; diffusible pigments are not formed on these media. A white aerial mycelium is produced on glycerol-asparagine, tyrosine, and yeast extract-malt extract agars. Aerial hyphae differentiate into straight to flexuous chains of spore-like elements (0.3–0.8 × 1.5 µm). Grows between 15 and 55°C and from pH 5.0 to 10.

Produces β-galactosidase, *N*-acetyl-β-glucosamidase, and phosphatase. Raffinose is used as a sole source of carbon, but not cellobiose, rhamnose, or xylose.

Resistant to lysozyme (100 µg/ml) and high concentrations of copper (up to 3 mM). Additional phenotypic properties are shown in Table 238.

Whole-organism hydrolysates are rich in arabinose and galactose. The fatty acid profile contains major proportions of $C_{16:0}$ iso (23%), $C_{16:0}$ (12%), and $C_{17:0}$ anteiso (11%). Mycolic acids are absent. The polar lipid pattern includes DPG, PI, and HPE. The predominant menaquinone is MK-9(H_4); also contains minor amounts of MK-9(H_2), MK-9(H_6), and MK-10(H_2).

Source: isolated from a sediment sample polluted with copper in Tucumán, Argentina.

DNA G+C content (mol%): not determined.

Type strain: ABO, BCCM/LMG 24814, DSM 45259.
Sequence accession no. (16S rRNA gene): DQ886938.

38. **Amycolatopsis ultiminotia** Lee 2009, 1403[VP]

ul.ti.mi.no'ti.a. L. sup. adj. *ultimus* farthest, extreme; L. fem. adj. *notia* southern; N.L. fem. adj. *ultiminotia* farthest southern, implying that the type strain was isolated from the southernmost part of the Republic of Korea.

Forms a white aerial mycelium and a cream to yellow substrate mycelium which fragment into rod-shaped elements. Grows from 10–37°C, but not at 42°C, and from pH 5.1 and 12.1.

DNA and elastin are degraded, but not chitin or cellulose. Acid is produced from dextran and dulcitol, but not from D-arabinose, inulin, methyl α-D-mannoside, ribose, sorbose, or xylitol.

Whole-organism hydrolysates are rich in arabinose and galactose. The cellular fatty acid profile contains major proportions of $C_{17:0}$ (23.7%), $C_{15:0}$ (19.8%), $C_{15:0}$ iso (10.0%), and $C_{16:0}$ iso (13.2%), and smaller proportions of $C_{13:0}$ (0.8%), $C_{14:0}$ (0.8%), $C_{14:0}$ iso (7.9%), $C_{14:0}$ 2-OH (3.3%), $C_{15:1}$ (1.7%), $C_{15:0}$ anteiso (2.7%), $C_{15:0}$ iso 3-OH (2.1%), $C_{17:0}$ anteiso (2.0%), $C_{17:0}$ iso (0.7%), $C_{18:0}$ (4.1%), and $C_{18:1}$ 0.7%). Does not contain mycolic acids. The polar lipid pattern consists of PME and an unknown ninhydrin-positive phospholipid. The predominant menaquinone is MK-9(H_4).

Source: isolated from the rhizosphere of a cliff associated plant (*Peucedanmum japonicium* Thunb).
DNA G+C content (mol%): 67.5.
Type strain: DSM 45180, NRRL B-24662.
Sequence accession no. (16S rRNA gene): FM177516.

39. **Amycolatopsis vancoresmycina** Wink, Kroppenstedt, Ganguli, Nadkarni, Schumann, Seibert and Stackebrandt 2003a, 935[VP] (Effective publication: Wink, Kroppenstedt, Ganguli, Nadharni, Schumann, Seibert and Stackebrandt 2003b, 44.)

van.co.res.my'ci.na. N.L. n. *vancoresmycinum* an antibiotic, vancoresmycin, produced by the organism; L. fem. suff. *-ina* suffix used with the sense of belonging to: N.L. fem. adj. *vancoresmycina* pertaining to vancoresmycin.

Substrate mycelium is brown-beige on peptone-yeast extract-iron agar, saffron yellow on yeast extract-malt extract agar, and maize yellow on glycerol asparagine, inorganic salts-starch, oatmeal, and tyrosine agars; a brown soluble pigment is formed on yeast extract-malt extract agar. White fragmenting aerial hyphae are produced on inorganic salts-starch, oatmeal, and yeast extract-malt extract agars. Fragmentation of the substrate mycelium is seen on oatmeal and yeast extract-malt extract agars.

Positive for *N*-acetyl-β-glucosamidase, alkaline phosphatase, α-chymotrypsin, cystine arylamidase, esterase (C8), leucine arylamidase, lipase (C14), naphthol-AS-BI-phosphohydrolase, trypsin, and valine arylamidase, but negative for esterase (4), α-fucosidase, α- and β-galactosidases, α-glucosidase, β-glucuronidase, and α-mannosidase (API ZYM tests).

Whole-organism hydrolysates are rich in arabinose and galactose. The fatty acids profile contains major proportions of $C_{16:0}$ iso (41.9%) and smaller proportions (<10%) of $C_{15:0}$ (1.6%), $C_{16:0}$ (1.9%), $C_{17:0}$ (1.8%), $C_{15:1}$ (1.9%), $C_{14:0}$ iso (1.6%), $C_{15:0}$ iso (9.9%), $C_{16:1}$ iso (1.9%), $C_{17:0}$ iso (4.0%), $C_{15:0}$ anteiso (1.1%), $C_{17:0}$ anteiso (4.2%), *cis*9-$C_{16:1}$ (3.0%), $C_{17:1}$ *cis*9 (3.8%), $C_{16:0}$ 10-methyl (2.8%), $C_{16:1}$ 10-methyl (1.4%), $C_{17:0}$ 10-methyl iso (3.3%), $C_{17:0}$ 10-methyl (2.4%), 2-OH-$C_{15:0}$ iso (2.4%), and 2-hydroxy-$C_{16:0}$ iso (8.4%). Additional phenotypic properties are shown in Table 238.

Source: isolated from Indian soil.
DNA G+C content (mol%): not determined.
Type strain: ST 101170, DSM 44592, JCM 12675, NRRL B-24208.
Sequence accession nos: AJ508240 (16S rRNA gene); EU822918 (*gyrB*).

Species *incertae sedis*

1. "**Amycolatopsis lactamdurans**" Barreiro, Pisabarro and Martin 2000, 22

Forms a slight yellow substrate mycelium, a white aerial mycelium and a bright yellow substrate mycelium on modified Bennett's agar supplemented with mannitol and soybean flour. Grows from 25–37°C, but not at 10 or 45°C, and from pH 6.0–9.0.

Allantoin, esculin, and urea are hydrolyzed, but not arbutin. Does not produce H_2S or reduce nitrate. Elastin, hypoxanthine, L-tyrosine, Tween 40, uric acid, xanthine, and xylan are degraded, but not starch. Acid is produced from arabitol, cellobiose, galactose, glucose, glycerol, *myo*-inositol, mannitol, mannose, raffinose, salicin, sorbitol, trehalose, and xylose, but not from adonitol, cellobiose, dulcitol, *meso*-erythritol, glycogen, lactose, maltose, melezitose, melibiose, methyl α-D-glucoside, ribose, sucrose, turanose, or xylitol.

L-Arabinose, D-arabitol, cellobiose, dextrin, glycerol, *myo*-inositol, mannitol, sorbitol, trehalose, xylitol, and xylose are used as sole carbon sources, but not adonitol, *meso*-erythritol, glycogen, lactose, maltose, melezitose, melibiose, methyl α-D-glucoside, raffinose, ribose, or sucrose (all at 1%, w/v). Similarly, citrate, propionate, oxalate, and tartrate are used as sole carbon sources, but not benzoate, lactate, or mucate (all at 0.1%, w/v).

Resistant (μg/ml) to gentamicin sulfate (5), neomycin sulfate (8), streptomycin sulfate (16), novobiocin (10), polymixin B (50), and vancomycin hydrochloride (0.25), but sensitive to penicillin (20) and rifampin (10).

Source: soil.
DNA G+C content (mol%): not determined.
Type strain: ATCC 27382, NRRL 3802.
Sequence accession no. (16S rRNA gene): AJ243301.

Genus VI. **Crossiella** Labeda 2001, 1578[VP]

David P. Labeda

Cros.si.el'la. N.L. fem. dim. n. *Crossiella* named for Thomas Cross, a microbiologist at the University of Bradford who made many contributions to actinomycete biology and systematics.

Aerobic. Gram-stain-positive, non-acid fast, nonmotile actinomycetes. Forms branched substrate mycelium (approx. 0.5 µm in diameter) and, on some media, aerial mycelia are produced. **Vegetative mycelium may fragment into rod-shaped elements and sclerotia-like pseudosporangia may be produced on the substrate mycelium. Swellings may be produced at or near the tip of aerial hyphae.** Mycolic acids are absent. Catalase-positive. **Contains *meso*-diaminopimelic acid as the diamino acid and acetylated peptidoglycan. The whole-cell sugar pattern consists of galactose, mannose, rhamnose, and ribose. The phospholipid pattern consists of phosphatidylethanolamine, phosphatidylmethylethanolamine, phosphatidylinositol, and phosphatidylinositol mannosides. The predominant menaquinone is MK-9(H_4).** Has a fatty acid profile rich in branched-chain and saturated components. **Phylogenetically, the nearest neighbor is the genus *Kutzneria*.**

DNA G+C content (mol%): 71.4 (T_m).

Type species: **Crossiella cryophila** Labeda 2001, 1579[VP].

Further descriptive information

The type species of *Crossiella* was originally described by Takahashi et al. (1986) as *Nocardiopsis mutabilis* subsp. *cryophilis* for a novel soil isolate that produced the antibiotic dopsisamine. The authors observed that the morphological characteristics of the strain were somewhat different from those of *Nocardiopsis* but, based on their chemotaxonomic data, considered that the genus *Nocardiopsis* was the closest fit for their strain. The transfer of *Nocardiopsis mutabilis* and its subspecies to the genus *Saccharothrix* was proposed by two independent studies (Grund and Kroppenstedt, 1989; Labeda and Lechevalier, 1989) based on more detailed chemotaxonomic analyses. Evaluation of DNA relatedness of the type strain of *Nocardiopsis mutabilis* subsp. *cryophilis* with other species within the genus *Saccharothrix* (Labeda and Lechevalier, 1989) demonstrated that it was distinct and therefore constituted a novel species for which the name *Saccharothrix cryophila* was proposed. The DNA relatedness between this strain and strains of the other species was low (2–11%) and there was some question regarding its generic identity. Subsequent phylogenetic analysis of the type strains of many species of the genera within the suborder *Pseudonocardineae*, elevated in this volume to order *Pseudonocardiales*, based on 16S rRNA gene sequences (Labeda and Kroppenstedt, 2000) supported the existence of at least two families, *Actinosynnemataceae* and *Pseudonocardiaceae*, and also demonstrated that a number of *Saccharothrix* species were misclassified. *Saccharothrix cryophilus* is phylogenetically located outside of *Actinosynnemataceae sensu stricto*, far from *Saccharothrix* and closest to the genera *Kutzneria*, *Actinoalloteichus*, and *Streptoalloteichus* as seen in Figure 295. Consideration of the morphological, chemotaxonomic, and phenotypic properties of this strain resulted in the proposal for the creation of the new genus, *Crossiella*, for this species (Labeda, 2001).

Actinomycetes were reported as a significant emergent agent of placentitis and abortion in horses in Kentucky after first being observed in 1986 at the Livestock Disease Diagnostic Center at the University of Kentucky (Donahue and Williams, 2000; Giles et al., 1993; Hong et al., 1993). Nocardioform placentitis is the term used to describe this distinct type of placentitis in horses and the infection is diagnosed based on the location of the lesion on the chorionic surface of the placenta and the recovery of Gram-stain-positive branching micro-organisms upon culture. The actinomycete biomass may infiltrate up to 30% of the surface area of the placenta (light gray area in Figure 296), but does not invade fetal tissue and probably contributes to death or weakening of the unborn foal through competition for nutrients. Infections can result in spontaneous abortions, still birth, full-term deliveries of weak foals, or no apparent effect and healthy foals. The 16S rRNA gene sequences from strains isolated from placental tissues exhibit high similarity (98.1%) to that of the type strain of *Crossiella cryophila* and morphological and chemotaxonomic characteristics are also typical for the genus. Physiological characteristics for all strains are quite similar and 16S rRNA gene sequences are identical. A novel species, *Crossiella equi*, was therefore proposed for these equine isolates (Donahue et al., 2002). The manner in which *Crossiella equi* strains become introduced into the equine uterus and infect the placenta is still unknown, as are possible environmental reservoirs for this species. The incidence of *Crossiella equi* equine placentitis is quite variable and numbers of infected placentas observed are relatively low in most years.

Enrichment and isolation procedures

Crossiella grows well on typical actinomycete growth media such as yeast extract-malt extract agar (Shirling and Gottlieb, 1966) or NZamine medium [DSMZ medium 554 (DSMZ, 2001)]. Strains have been isolated from lesions on equine placentas on tryptic soy agar containing 5% blood, but other media without blood can be used. The type strain of *Crossiella cryophilus* was isolated from dilutions of a soil sample from Shosenkyo, Yamanashi Prefecture, Japan, but details of the isolation procedure are lacking and there have been no other reports of isolation of strains of *Crossiella* from environmental samples.

Maintenance procedures

Working cultures of *Crossiella* can be maintained as refrigerated (4°C) agar slants on an appropriate medium such as yeast extract-malt extract medium (Shirling and Gottlieb, 1966) or NZamine medium (DSMZ medium 554). The slants are generally incubated for 5–7 d at 28–30°C prior to refrigeration and then should be subcultured at monthly intervals. Longer term preservation of strains is best accomplished as frozen stocks in 20% aqueous glycerol at −80°C (mechanical freezer) to −172°C (liquid nitrogen vapor phase) or by using traditional lyophilization techniques.

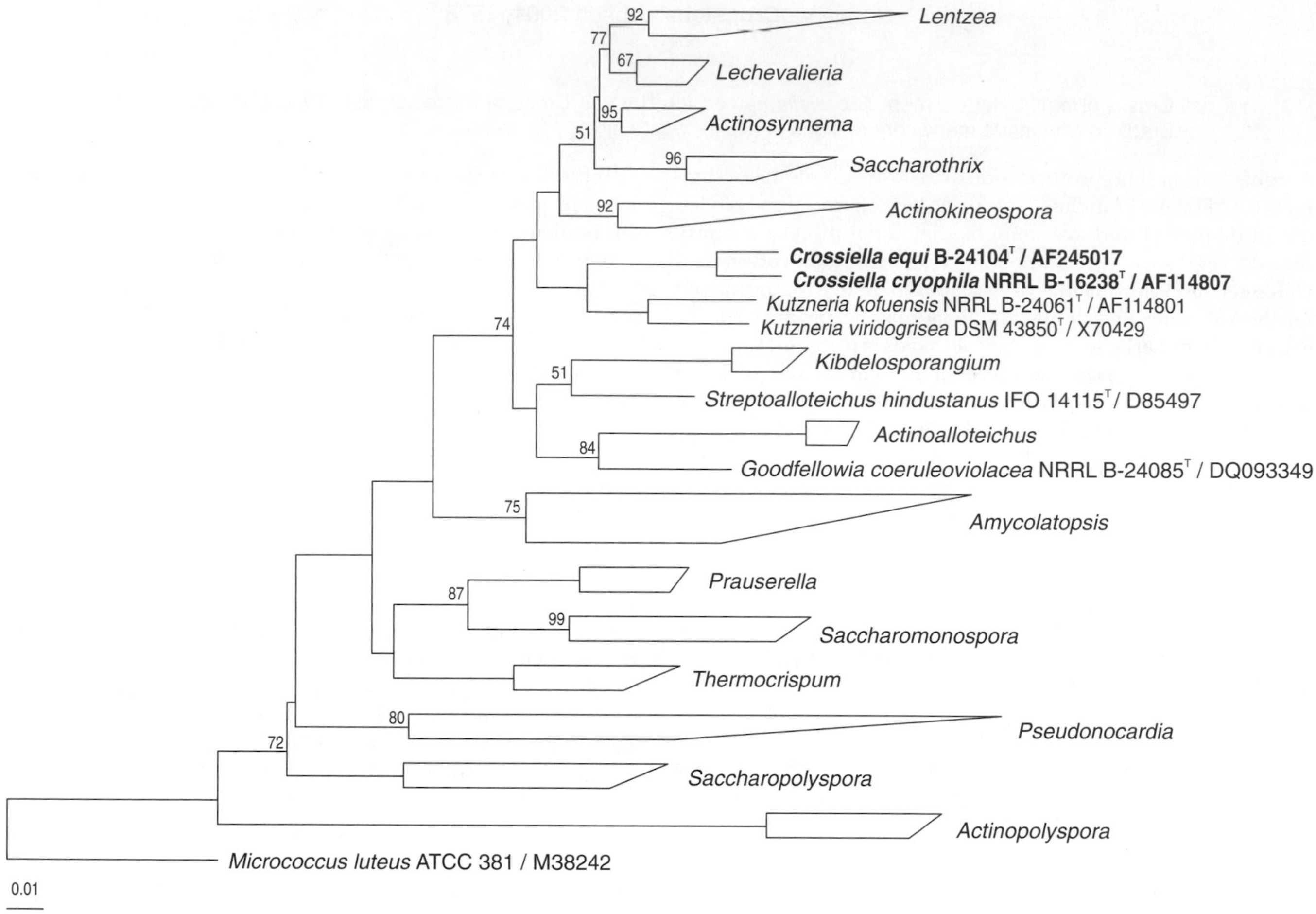

FIGURE 295. Phylogenetic tree for families of the order *Pseudonocardiales*, calculated from almost-complete 16S rRNA gene sequences using Kimura's evolutionary distance methods (Kimura, 1980) and the neighbor-joining method of Saitou and Nei (1987) illustrating the taxonomic position of *Crossiella cryophila* and *Crossiella equi* and the other taxa in the order. Bar = 0.01 nucleotide substitutions per site.

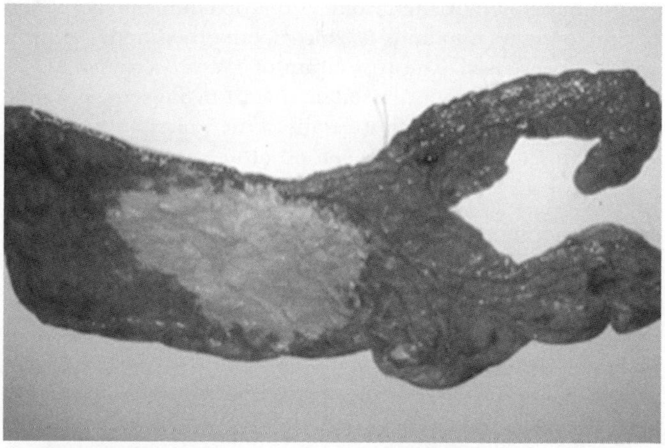

FIGURE 296. Photograph of an equine placenta exhibiting symptoms of *Crossiella equi* nocardioform placentitis. The light gray colored area is infiltrated with *Crossiella* mycelium. (Printed with permission of N.M. Williams, Livestock Disease Diagnostic Center, University of Kentucky, Lexington, KY, USA.)

Procedures for testing special characters

Strains are routinely cultivated on NZamine medium at 28°C. Morphological observations are made on the media of Shirling and Gottlieb (1966) and on NZamine medium.

Chemotaxonomic analyses of strains for polar lipids, menaquinones, and fatty acids are performed using methods previously described by Grund and Kroppenstedt (1989), Minnikin et al. (1984), and Sasser (1990).

Physiological tests, including production of acid from carbohydrates, utilization of organic acids, and hydrolysis and decomposition of adenine, casein, guanine, esculin, hypoxanthine, tyrosine, urea, and xanthine, are typically evaluated using the media of Gordon et al. (1974). Phosphatase activity is evaluated by using the method of Kurup and Schmitt (1973) substituting NZamine agar as the basal medium. Temperature range for growth is determined on slants of DSMZ medium 554 and salt tolerance is determined on slants of the same medium supplemented with 4% and 5% NaCl.

Differentiation of the genus *Crossiella* from other genera

The genus *Crossiella* is phylogenetically distinct from neighboring genera based on analyses of 16S rRNA gene sequences (Figure 295). Moreover, the micromorphology of *Crossiella* is distinct from that exhibited by neighboring taxa. *Crossiella* strains exhibit substrate mycelium fragmenting into rod-shaped elements, the presence of sclerotia or pseudosporangium-like bodies on the colony surface (Figure 297), and swellings near the tips of mycelium (Figure 297); the latter two properties have not been observed in members of the genus *Goodfellowiella.* Motile spores have not been observed and true sporangia such as those found in species of the genera *Kutzneria, Kibdelosporangium,* and *Streptoalloteichus* are not produced. Chains of spores typical of *Actinoalloteichus* species have not been observed in *Crossiella* species. The chemotaxonomic profile of *Crossiella* species is most similar to that of *Streptoalloteichus* (Table 239), but MK-10(H_4) menaquinones, while present in *Streptoalloteichus,* are not found in *Crossiella* strains.

Differentiation of the species of the genus *Crossiella*

The physiological characteristics of *Crossiella* species are summarized in Table 240. The physiological properties of six strains of *Crossiella equi* were evaluated for the original description of the species and there was some variation observed. Determination of acid production from the various sugars and utilization of organic acids in this species was complicated by the fact that all of the strains tested absorbed the pH indicator dye, bromothymol blue, into their substrate mycelium making it difficult to evaluate test results. The two species can be differentiated by the many physiological differences indicated in Table 240, but most obviously by the lack of growth of *Crossiella cryophila* at temperatures of 37°C or greater. The growth rate of *Crossiella equi* strains is also very much greater than that of *Crossiella cryophila* as well as most other actinomycetes. There are also differences between the fatty acid profiles of the species as can be seen in Table 241.

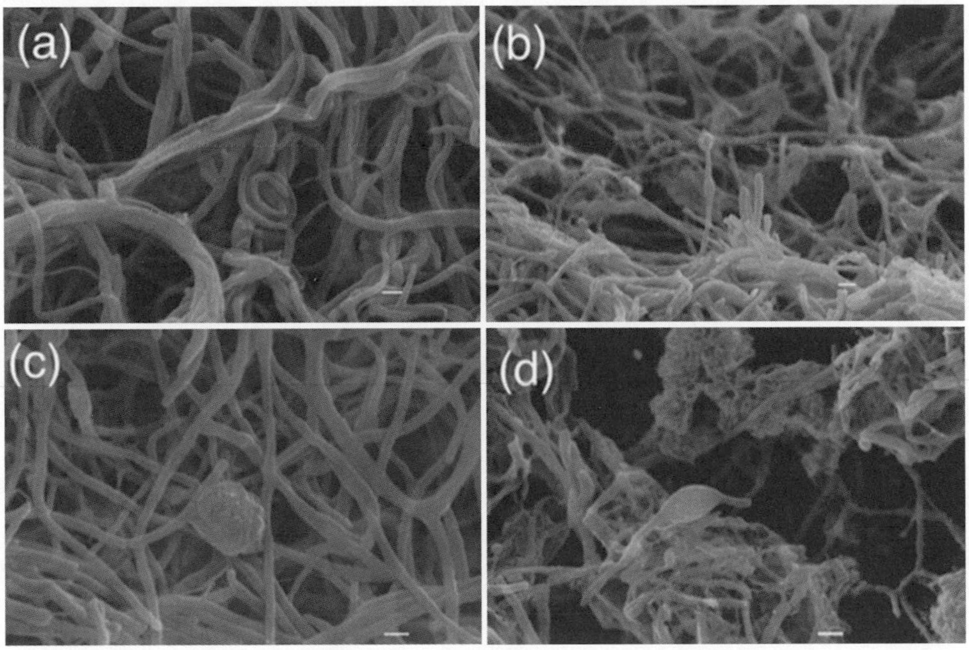

FIGURE 297. Comparison of the micromorphological properties of *Crossiella cryophilus* NRRL B-16238[T] (a, c) and *Crossiella equi* NRRL B-24104[T] (b, d). Note the pseudosporangia on the substrate mycelium in (a) and (b) and the swollen mycelial tips in (c) and (d). Bars = 1 µm.

TABLE 239. Comparison of chemotaxonomic profile of *Crossiella* with phylogenetically nearest genera

Character	*Crossiella*	*Actinoalloteichus*	*Goodfellowiella*	*Kibdelosporangium*	*Kutzneria*	*Streptoalloteichus*
Whole-cell sugar pattern[a]	Gal, Man, Rha, Rib	Glu, Gal, Man, Rib	Gal, Rib	Ara, Gal, Glu, Rha	Gal, Rha	Gal, Man, Rha, Rib
Phospholipids[b]	PE, DPG, PI, PIM, Methyl-PE	PIM, PI, PG, DPG, Methyl-PE	PE, DPG, OH-PE, Methyl-PE	PE, PME, PG, PI	PE, DPG, PI, PG, Methyl-PE	PE, DPG, PI, PIM, DPG, Methyl-PE
Predominant menaquinones	MK-9(H_4)	MK-9(H_4)	MK-9(H_4), MK-10(H_4)	MK-9(H_4)	MK-9(H_4)	MK-9(H_4), MK-10(H_4)

[a]Ara, Arabinose; Gal, galactose; Glu, glucose; Man, mannose; Rha, rhamnose; Rib, ribose.

[b]DPG, Diphosphatidylglycerol; PE, phosphatidylethanolamine; PG, phosphatidylglycerol; PI, phosphatidylinositol; PIM, phosphatidylinositol mannosides.

TABLE 240. Physiological characteristics of species of the genus *Crossiella*[a]

Character	*Crossiella cryophila*	*Crossiella equi* (6 strains)
Decomposition of:		
Adenine	–	–
Allantoin	–	–
Casein	–	+
Esculin	+	+
Gelatin	+	+(4/6)
Hippurate	–	+ (5/6)
Hypoxanthine	–	– (4/6)
Starch	+	+
Tyrosine	+	+
Urea	+	–
Xanthine	–	–
Acid production from:		
Adonitol	–	–
Arabinose	–	+ (3/6)
Cellobiose	w	+
Dextrin	w	+
Dulcitol	–	–
Erythritol	–	–
Fructose	+	+
Galactose	+	+
Glucose	+	+
Glycerol	+	+
Inositol	+	+
Lactose	w	– (4/6)
Maltose	+	+
Mannose	+	+
Mannitol	–	– (4/6)
Melibiose	w	+
Methyl α-D-glucoside	–	–
Methyl β-D-xyloside	–	–
Raffinose	w	+
Rhamnose	–	+ (5/6)
Salicin	–	+ (5/6)
Sorbitol	–	– (5/6)
Sucrose	–	– (5/6)
Trehalose	+	+ (5/6)
Xylose	–	+ (5/6)
Utilization of:		
Acetate	+	w (5/6)
Benzoate	–	w (4/6)
Citrate	–	w (3/6)
Lactate	+	– (4/6)
Malate	+	–
Mucate	–	–
Oxalate	+	–
Propionate	+	w (3/6)
Succinate	+	w (4/6)
Tartrate	–	–
Production of:		
Nitrate reductase	+	+
Phosphatase	+	+
Growth in the presence of:		
4% NaCl	+	+
5% NaCl	–	+
Growth at:		
10°C	+	+
37°C	–	+
42°C	–	+
45°C	–	–

[a]Symbol: w, weak positive.

TABLE 241. Fatty acid profiles of *Crossiella* species[a]

Fatty acid	*C. cryophila*	*C. equi* (6 strains)
$C_{14:0}$ iso	1.91	1.02
$C_{15:0}$ iso	44.28	57.51
$C_{15:0}$ anteiso	2.29	3.01
$C_{15:1}$ B	3.06	3.34
$C_{16:1}$ iso-H	3.16	1.98
$C_{16:0}$ iso	9.77	7.25
$C_{16:1}$ ω9*c*	3.11	0.80
$C_{16:0}$	1.76	0.66
$C_{16:0}$ iso 10-methyl	1.09	0.00
$C_{16:0}$ 9?-methyl	11.58	4.20
$C_{17:0}$ iso	11.36	4.54
$C_{17:0}$ anteiso	1.86	1.05
$C_{17:1}$ ω9*c*	1.86	0.82
$C_{17:0}$	1.21	0.50
$C_{17:0}$ iso 3-OH	0.00	4.68

[a]Fatty acids are listed as percentages of total fatty acids as determined by the Microbial Identification System software (MIDI Inc.) peak naming table.

List of species of the genus *Crossiella*

1. **Crossiella cryophila** (Labeda and Lechevalier 1989) Labeda 2001, 1579[VP] (*Saccharothrix cryophilus* Labeda and Lechevalier 1989, 420; "*Nocardiopsis mutabilis* subsp. *cryophilus*" Takahashi, Hotta, Saito, Morioka, Okami and Umezawa 1986, 179)

cry.o′phi.la. Gr. n. *kruos* icy cold, frost; N.L. adj. *philus* -*a* -*um* (from Gr. adj. *philos* -*ê* -*on*) friend, loving; N.L. fem. adj. *cryophila* cold loving, referring to the low permissive temperature range for growth.

Pale-yellow to light-brown substrate mycelium is produced on most media. White to yellowish-white aerial mycelium is formed, particularly on inorganic salts-starch agar or glycerol-asparagine agar. Soluble pigments are not produced. Physiological properties are shown in Table 240. Temperature range for growth is 10–33°C.

Source: isolated from soil.

DNA G+C content (mol%): 74.1 (T_m).

Type strain: NRRL B-16238, ATCC 51143, DSM 44230, IFO 14475, NBRC 14475, Okami TS-1980.

Sequence accession no. (16S rRNA gene): AF114806.

2. **Crossiella equi** Donahue, Williams, Sells and Labeda 2002, 2172[VP]

e′qui. L. gen. n. *equi* of the horse, referring to the source of isolation of this micro-organism, equine placentas.

Pale orange to light-brown substrate mycelium is formed on most media. Copious white aerial mycelium is produced on most media. Physiological properties are shown in Table 240. Temperature for growth is 10–42°C.

Source: isolated from equine placentas.

DNA G+C content (mol%): 74.1 (T_m).

Type strain: NRRL B-24104, CIP 107800, DSM 44580, LDDC 22291-98.

Sequence accession no. (16S rRNA gene): AF245017.

Genus VII. **Goodfellowiella** Labeda, Kroppenstedt, Euzéby and Tindall 2008, 1048[VP]

David P. Labeda

Good.fel.low.i.el′la. N.L. fem. dim. n. *Goodfellowiella* named for Michael Goodfellow, a microbiologist at the University of Newcastle-upon-Tyne, in recognition of his contributions to microbial systematics.

Aerobic. Gram stain positive, non acid-fast, nonmotile actinomycetes. Branched substrate mycelium (approx. 0.5 µm in diameter) and, on some media, aerial mycelia are produced. **Ovoid conidia are produced by fragmentation of substrate mycelium.** Catalase-positive. **Contains *meso*-diaminopimelic acid as the diamino acid. The whole-cell sugar pattern consists of galactose and ribose. The phospholipid pattern consists of diphosphatidylglycerol, phosphatidylethanolamine, phosphatidylethanolamine with hydroxylated fatty acids, and traces of phosphatidylinositol and phosphatidylinositol mannosides. The predominant menaquinones are MK-9(H_4) and MK-10(H_4). Has a fatty acid profile rich in branched chain and saturated components including 10-methyl-branched heptadecanoic acid and anteiso-branched 2-hydroxy fatty acids.** Phylogenetically, its nearest neighbor is the genus *Actinoalloteichus*.

DNA G+C content (mol%): 68.2 (T_m).

Type species: **Goodfellowiella coeruleoviolacea** (Preobrazhenskaya and Terekhova 1987) Labeda, Kroppenstedt, Euzéby and Tindall 2008, 1048[VP].

Further descriptive information

The type species was originally described by Preobrazhenskaya et al. (1987, 1976) as *Actinomadura coeruleoviolacea* and was subsequently transferred by Kroppenstedt et al. (1990, 1991) to the genus *Saccharothrix* as *Saccharothrix coeruleoviolacea.* During a phylogenetic evaluation of species of the genus *Saccharothrix* with validly published names based on almost complete 16S rRNA gene sequences (Labeda and Kroppenstedt, 2000), it was noted that the type strain of *Saccharothrix coeruleoviolacea* was not related to other species of the genus and represented a new genus within the suborder *Pseudonocardineae,* elevated in this volume to order *Pseudonocardiales.* A polyphasic investigation of the characteristics of this strain demonstrated that it was also chemotaxonomically distinct from *Saccharothrix* and the other genera within the suborder and the new genus *Goodfellowia* was proposed by Labeda and Kroppenstedt (2006). It was later discovered that the genus name *Goodfellowia* had been used previously in as the name of an avian genus and thus *Goodfellowia* Labeda and Kroppenstedt 2006 was taxonomically illegitimate. A proposal was subsequently published to correct this situation by emending the genus name to *Goodfellowiella* (Preobrazhenskaya and Terekhova 1987) Labeda, Kroppenstedt, Euzéby and Tindall 2008.

Maintenance procedures

Working cultures of *Goodfellowiella* can be maintained as refrigerated (4°C) agar slants on appropriate media such as yeast extract-malt extract medium (Shirling and Gottlieb, 1966) or NZamine medium (DSMZ medium 554; DSMZ, 2001). The slants are generally incubated for 5–7 d at 28–30°C prior to refrigeration and then should be subcultured at monthly intervals. Longer term preservation of strains is best accomplished as frozen stocks in 20% (v/v) aqueous glycerol at −80°C (mechanical freezer) to −172°C (liquid nitrogen vapor phase) or by using traditional lyophilization techniques.

Procedures for testing special characters

Strains are routinely cultivated on NZamine medium at 28°C. Morphological observations are made on the media of Shirling and Gottlieb (1966) and NZamine medium.

Chemotaxonomic analyses of strains for fatty acids, menaquinones, and polar lipids are performed using methods described previously by Grund and Kroppenstedt (1989), Minnikin et al. (1984), Saddler et al. (1991), and Sasser (1990).

Physiological tests, including production of acid from carbohydrates, utilization of organic acids, and hydrolysis and decomposition of adenine, guanine, hypoxanthine, tyrosine, xanthine, casein, esculin, urea, and hippurate, are typically evaluated using the media of Gordon et al. (1974). Phosphatase activity is evaluated by using the method of Kurup and Schmitt (1973) substituting NZamine agar as the basal medium. Temperature range for growth is determined on slants of DSMZ medium 554 and salt tolerance is determined on slants of the same medium supplemented with 4% and 5% NaCl.

Differentiation of the genus *Goodfellowiella* from other genera

Goodfellowiella coeruleoviolacea is phylogenetically separate from the genus *Saccharothrix,* the genus in which this micro-organism was originally classified, and appears to be most closely related to the genus *Actinoalloteichus,* as can be seen in Figure 298. The chemotaxonomic properties of *Goodfellowiella* distinguish it from *Actinoalloteichus* and the other related taxa within the order *Pseudonocardiales,* as can be seen in Table 242. The whole-cell sugar pattern consisting of only galactose and ribose differs from those of other genera, as does the phospholipid pattern, i.e. a lack of phosphatidylinositol and the presence of both phosphatidyl-

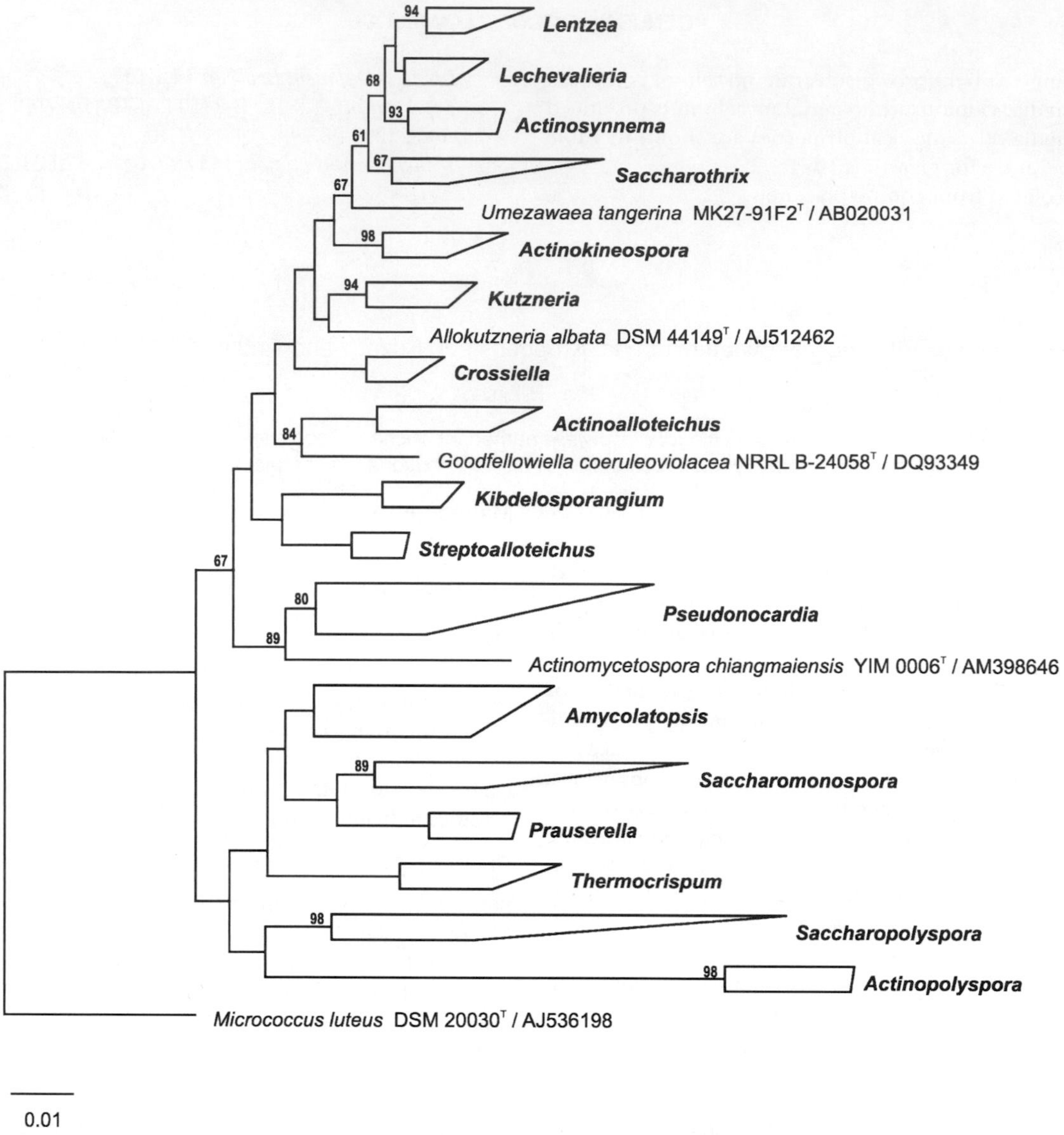

FIGURE 298. Phylogenetic tree for the genera of the order *Pseudonocardiales* calculated from almost-complete 16S rRNA gene sequences using the Kimura's evolutionary distance method (Kimura, 1980) and the neighbor-joining algorithm of Saitou and Nei (1987). Numbers at the nodes represent levels (%) of bootstrap support from 100 resampled datasets; values less than 60% are not shown. Bar = 0.01 nucleotide substitutions per site.

TABLE 242. Comparison of chemotaxonomic profile of *Goodfellowiella* with phylogenetically nearest genera and *Kibdelosporangium*

Character	*Goodfellowiella*	*Crossiella*	*Actinoalloteichus*	*Allokutzneria*	*Kibdelosporangium*	*Kutzneria*	*Streptoalloteichus*
Whole-cell sugar pattern[a]	Gal, Rib	Gal, Man, Rha, Rib	Glu, Gal, Man, Rib	Ara, Gal, Man	Ara, Gal, Glu, Rha	Gal, Rha	Gal, Man, Rha, Rib
Phospholipids[b]	PE, DPG, OH-PE, PME	PE, DPG, PI, PIM, PME	PIM, PI, PG, DPG, PME	PE, PME, OH-PE, PI, *lyso*-PME, DPG, PG, *lyso*-PE	PE, PME, PG, PI	PE, DPG, PI, PG, PME	PE, DPG, PI, PIM, DPG, PME
Predominant menaquinones	MK-9(H_4), MK-10(H_4)	MK-9(H_4)	MK-9(H_4)	MK-9(H_4)	MK-9(H_4)	MK-9(H_4)	MK-9(H_6), MK-10(H_6)
DNA G+C content (mol%)	69.2	71.4	72–72.5	71.6	66	70.3–70.7	nd

[a]Ara, Arabinose; Gal, galactose; Glu, glucose; Man, mannose; Rha, rhamnose; Rib, ribose.

[b]DPG, Diphosphatidylglycerol; PE, phosphatidylethanolamine; PG, phosphatidylglycerol; PI, phosphatidylinositol; PIM, phosphatidylinositol mannosides; OH-PE, phosphatidylethanolamine with hydroxy fatty acids; *lyso*-PE, phosphatidylethanolamine where one fatty acid chain is missing from the glycerol backbone; *lyso*-PME, phosphatidylmethylethanolamine where one fatty acid chain is missing from the glycerol backbone; PME, phosphatidylmethylethanolamine.

TABLE 243. Fatty acid content of *Goodfellowiella coeruleoviolacea* and representative species of phylogenetically related genera

Fatty acid	*Goodfellowiella coeruleoviolacea* DSM 43935[T]	*Actinoalloteichus cyanogriseus* DSM 43889[T]	*Allokutzneria albata* DSM 44149[T]	*Crossiella cryophila* DSM 44230[T]	*Kibdelosporangium aridum* subsp. *aridum* DSM 43828[T]	*Kutzneria viridogrisea* DSM 43850[T]	*Streptoalloteichus hindustanus* DSM 44523[T]
$C_{14:0}$ iso	–	5.0	6.4	0.7	5.2	–	–
$C_{14:0}$	–	–	–	–	1.8	–	–
$C_{15:0}$ iso	6.7	15.0	6.0	40.0	8.7	2.7	34.2
$C_{15:0}$ anteiso	5.0	10.0	0.5	2.2	12.7	–	10.7
$C_{15:1}$(*cis*9)	–	–	3.2	3.8	–	–	–
$C_{15:0}$	6.2	trace	2.0	2.6	0.7	1.6	–
$C_{16:1}$ iso	4.5	8.0	10.5	3.2	6.0	0.6	–
$C_{16:0}$ iso	19.8	19.0	42.5	10.7	34.5	30.3	2.6
$C_{16:1}$ (*cis*9)	–	2.0	1.5	3.9	0.3	3.8	0.6
$C_{15:0}$ iso 2-OH	1.7	–	–	–	4.2	–	–
$C_{15:0}$ anteiso 2-OH	–	–	–	–	2.5	–	–
$C_{16:0}$	–	2.0	1.6	2.2	1.5	–	–
$C_{16:1}$ 10-methyl?	1.7	–	–	–	–	–	–
$C_{16:0}$ iso 10-methyl	–	–	–	0.9	–	–	–
$C_{16:0}$ 10-methyl	–	–	–	–	0.6	8.6	–
$C_{16:0}$ 2-OH	5.8	–	–	–	–	–	–
$C_{17:1}$ iso	–	–	–	–	–	–	14.5
$C_{17:1}$ anteiso	–	–	–	–	0.4	–	1.2
$C_{17:0}$ iso	1.7	3.0	2.3	10.3	1.0	6.2	10.6
$C_{17:0}$ anteiso	19.39	20.0	0.8	1.2	10.8	7.5	22.8
$C_{17:1}$(*cis*9)	–	–	8.0	2.3	–	2.1	1.0
$C_{16:0}$ iso 2-OH	–	–	–	–	5.4	11.7	–
$C_{17:0}$	3.1	trace	3.8	1.8	0.4	4.3	–
$C_{17:0}$ 10-methyl	9.4	–	–	–	–	5.9	–
$C_{16:0}$ iso 3-OH	–	–	3.0	–	–	–	–
$C_{18:0}$ iso	–	–	1.0	–	0.3	0.9	–
$C_{18:1}$(*cis*9)	–	–	1.0	–	–	0.8	–
$C_{17:0}$ anteiso 2-OH	6.5	–	–	–	–	4.5	–
$C_{17:0}$ iso 3-OH	–	–	3.8	–	0.9	–	–
$C_{17:0}$ anteiso 3-OH	–	–	–	4.1	–	–	0.3
$C_{17:0}$ 3-OH	–	–	0.7	–	2.1	–	–
$C_{18:0}$ 10-methyl	–	–	–	–	–	1.5	–
$C_{19:0}$ anteiso	–	–	–	–	–	–	0.5

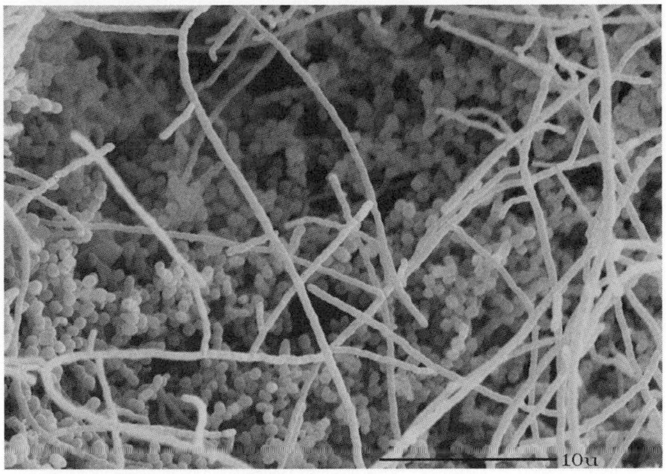

FIGURE 299. Scanning electron micrograph of 21-d growth of *Goodfellowiella coeruleoviolacea* NRRL B-24058[T] on yeast extract-malt extract agar. Note that spores are produced by fragmentation of the vegetative mycelium. Bar = 10 μm.

monomethylethanolamine and phosphatidylethanolamine with 2-hydroxy fatty acids. Substantial quantities of menaquinone MK-10(H_4) are present and this is distinct from other taxa within the suborder. The fatty acid profile of *Goodfellowiella coeruleoviolacea* (Table 243) clearly distinguishes it from its nearest phylogenetic neighboring genera by the presence of significant quantities of $C_{17:0}$ 10-methyl fatty acid and various hydroxylated fatty acids. Scanning electron microscopic observations of colony growth on several different media have not revealed the presence of sporangia and the substrate mycelium appears to fragment into coccoidal rod elements (Figure 299). Spore chains typical of those observed in *Actinoalloteichus* species were not observed.

List of species of the genus *Goodfellowiella*

1. **Goodfellowiella coeruleoviolacea** (Preobrazhenskaya and Terekhova 1987) Labeda, Kroppenstedt, Euzeby and Tindall 2008, 1048[VP] [*Actinomadura coeruleoviolacea* Preobrazhenskaya and Terekhova 1987, 179; *Saccharothrix coeruleoviolacea* (Preobrazhenskaya and Terekhova 1987) Kroppenstedt, Stackebrandt and Goodfellow 1990, 179; illegitimate synonym *Goodfellowia coeruleoviolacea* (Preobrazhenskaya and Terekhova 1987) Labeda and Kroppenstedt 2006, 1206]

co.e.ru.le.o.vi.o.la′ce.a. L. adj. *coeruleus* dark-colored, dark blue; L. adj. *violaceus* violet-colored, violet; N.L. fem. adj. *coeruleoviolacea* dark violet-colored.

Vegetative mycelium is pale yellow to dark brownish-yellow, depending on medium; white aerial hyphae are produced on most media becoming blue in color on several media including inorganic salts-starch (ISP4) agar and yeast extract-malt extract (ISP2) agar. Pale violet soluble pigment is produced on inorganic salts-starch agar and blue-green soluble pigment is produced on yeast extract-malt extract agar. Degrades or hydrolyzes casein, esculin, gelatin, hypoxanthine, starch, tyrosine, and urea. Does not degrade adenine, allantoin, or xanthine. Weakly reduces nitrates. Assimilates acetate, citrate, malate, oxalate, propionate, and succinate; does not assimilate benzoate, lactate, mucate, or tartrate. Acid is produced from adonitol, arabinose, cellobiose, dextrin, erythritol, fructose, galactose, glucose, glycerol, inositol, lactose, maltose, mannitol, mannose, melibiose, methyl α-D-glucoside, methyl β-xyloside, raffinose, rhamnose, salicin, sorbitol, sucrose, trehalose, and xylose; no acid is produced from dulcitol or melezitose. Grows weakly in the presence of 4% NaCl and not at all at higher salt concentrations.

Temperature range for growth is 15–45°C, with optimum growth around 30°C.

Source: isolated from soil.

DNA G+C content (mol%): 68.2 (T_m).

Type strain: NRRL B-24058, DSM 43935, INA 3564, JCM 9110, NBRC 14988, VKM Ac-1083.

Sequence accession no. (16S rRNA gene): DQ093349.

Genus VIII. **Kibdelosporangium** Shearer, Colman, Ferrin, Nisbet and Nash 1986a, 48[VP]

David P. Labeda

Kib.del.o.spo.ran′gi.um. Gr. adj. *kibdelos* false, ambiguous; Gr. n. *spora* seed, and in biology a spore; Gr. n. *angeion* a vessel; N.L. neut. n. *Kibdelosporangium* false or ambiguous sporangium.

Aerobic, catalase-positive, Gram-stain-positive, non-acid-fast, filamentous organisms that produce a substrate mycelium that penetrates the agar and forms a compact layer on top of the agar; aerial mycelium originates from the substrate mycelium. **Substrate mycelium may exhibit varying degrees of fragmentation and usually bears specialized structures that appear to be dichotomously branched, septate hyphae radiating from a common stalk. The aerial mycelia bear long chains of spores and sporangium-like structures. The sporangium-like structures are surrounded by well-defined walls, but do not contain spores and germinate directly, producing one or more germ tubes when placed on solid growth media. Motile elements or spores are not present. Contains *meso*-diaminopimelic acid as the diamino acid. Whole-cell sugar pattern consists of arabinose and D-galactose, with traces of madurose usually present.** The phospholipid pattern consists of diphosphatidylglycerol, phosphatidylethanolamine, phosphatidylinositol, phosphatidylinositol mannosides, and phosphatidylmethylethanolamine. **Mycolic acids are not present.**

DNA G+C content (mol%): 66 (T_m).

Type species: **Kibdelosporangium aridum** Shearer, Colman, Ferrin, Nisbet and Nash 1986a, 48[VP].

Further descriptive information

The genus *Kibdelosporangium* is placed phylogenetically within the family *Pseudonocardiaceae* in the order *Pseudonocardiales* (Figure 298) based on 16S rRNA gene sequences and shares similar chemotaxonomic characteristics with many of the genera within the family. The reproductive structures of members of the genus are quite distinct, however, and *Kibdelosporangium* colonies are frequently covered with straight to flexuous chains of rod-shaped spores (Figure 300) when viewed microscopically from above. Sporangium-like structures are observed beneath the spore chains, closer to the substrate mycelium (Figure 301). The sporangium-like structures and the spore chains can arise from the same aerial hyphae and may be borne apically on branched or unbranched hyphae, as well as terminally on short lateral hyphal branches.

The sporangium-like structures originate as small round swellings at the tips of the hyphae that continue to enlarge and, at maturity, are usually 9 to 35 μm in diameter (Figure 302). They are surrounded by a well-defined sporangial wall and contain septate, branched hyphae in an amorphous matrix (Figure 303). When placed on a suitable growth medium, the sporangium-like bodies germinate, usually within 24 to 48 h, and produce one or more germ tubes.

A third species, *Kibdelosporangium albatum*, was described by Tomita and others in 1993 because it appeared morphologically and chemotaxonomically most similar to the described members of the genus *Kibdelosporangium* (Tomita et al., 1993). This species has now been transferred by Labeda and Kroppenstedt (2008) to the new genus *Allokutzneria* as *Allokutzneria albata* on the basis of the phylogenetic position of this species, nearest to the genus *Kutzneria*, and its unique chemotaxonomic characteristics.

Enrichment and isolation procedures

Kibdelosporangium species can be isolated from soil by plating serial dilutions of the samples on starch-casein-nitrate agar (Küster and Williams, 1964) or arginine-glycerol-salt agar (El-Nakeeb and Lechevalier, 1963) media supplemented with

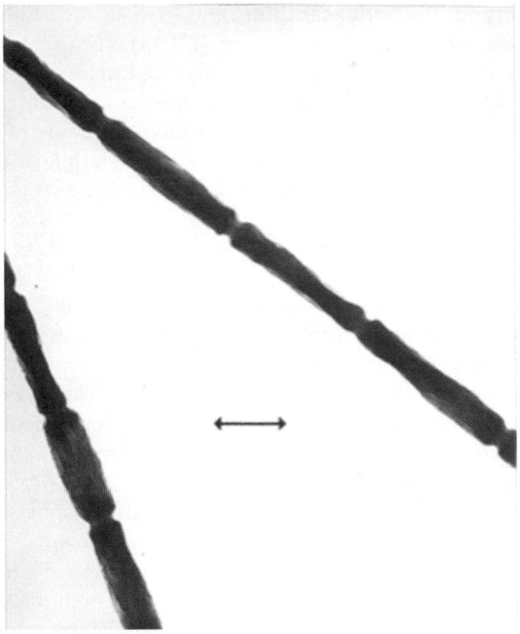

FIGURE 300. Transmission electron micrograph of *Kibdelosporangium aridum* spore chains (19-d-old culture on Bennett's agar). Bar = 1 μm. (Reproduced with permission from Shearer et al., 1986a. Int. J. Syst. Bacteriol. *36*: 47–54.)

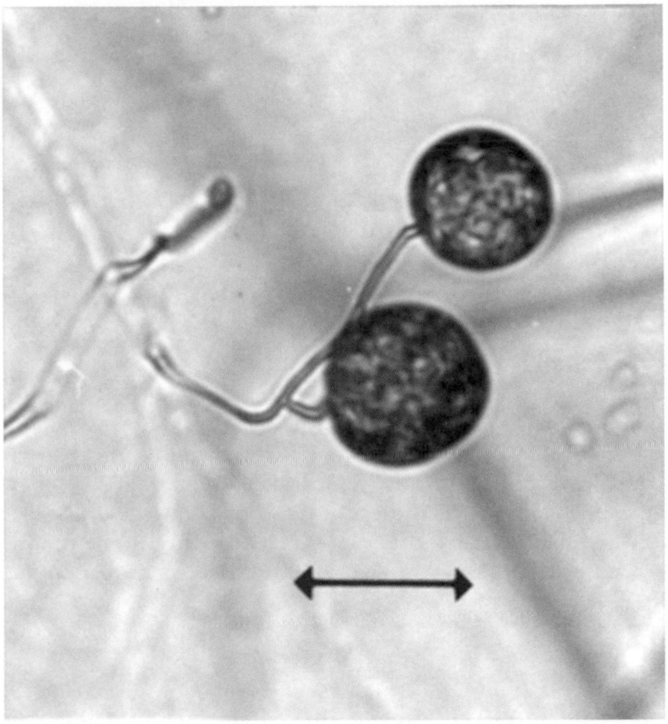

FIGURE 302. Micrograph of sporangium-like structures of *Kibdelosporangium aridum* (8-week-old culture on water agar). Bar = 14 μm. (Reproduced with permission from Shearer et al., 1986a. Int. J. Syst. Bacteriol. *36*: 47–54.)

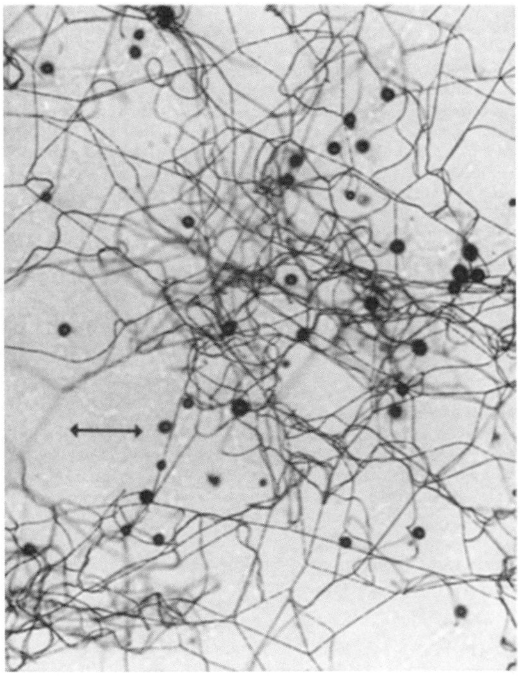

FIGURE 301. Micrograph of aerial mycelium with long, irregularly curved chains of spores and, nearer the agar surface, sporangium-like structures (*Kibdelosporangium aridum*; 26-d culture on water agar). Bar = 55 μm.

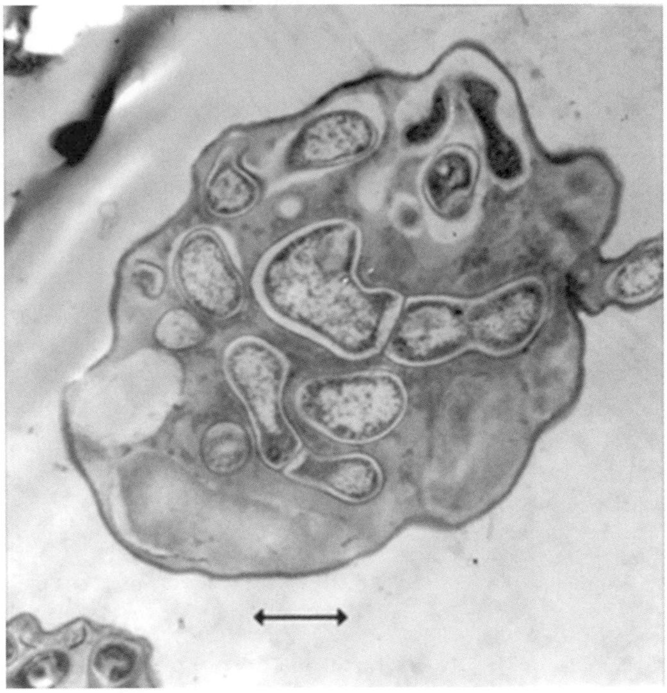

FIGURE 303. Transmission electron micrograph of thin section of sporangium-like structure of *Kibdelosporangium aridum* (2.5-week-old culture on thin potato-carrot agar). Bar = 1.2 μm. (Reproduced with permission from Shearer et al., 1986a. Int. J. Syst. Bacteriol. *36*: 47–54.)

either vancomycin (1–25 µg/ml) or gentamicin (2.5–5.0 µg/ml) and antifungal antibiotics (e.g. cycloheximide, 25 µg/ml) (Shearer, 1987). Pre-treating air-dried soil samples by heating at 120°C for 60 min may also prove helpful by eliminating many competing soil micro-organisms. Plates are routinely incubated at 28°C for 4 weeks.

Kibdelosporangium colonies resemble those of many other actinomycetes and are therefore not easily identified on isolation plates. Observation of sporulating colonies on isolation plates microscopically (400×) is the best method to identify typical *Kibdelosporangium* colonies on these plates.

Maintenance procedures

Serial transfer of strains is not an acceptable method of maintenance for *Kibdelosporangium* species because they tend to lose the ability to produce sporangium-like structures after several transfers. Moderate to long-term preservation of strains is best accomplished as frozen stocks in 10% (v/v) aqueous glycerol at −80°C (mechanical freezer) to −172°C (liquid nitrogen vapor phase) or by using traditional lyophilization techniques.

Procedures for testing special characters

Strains are routinely cultivated on yeast extract-malt extract (ISP2) agar at 30°C. Morphological observations are made on the media of Shirling and Gottlieb (1966), thin potato-carrot agar (Higgins et al., 1967), soil extract agar (Shearer et al., 1983), or tap water agar.

Chemotaxonomic analysis of strains for fatty acids, menaquinones, and polar lipids are performed using the previously described methods of Grund and Kroppenstedt (1989), Minnikin et al. (1984), Sasser (1990), and Saddler et al. (1991).

Physiological tests, including production of acid from carbohydrates, utilization of organic acids, and hydrolysis and decomposition of adenine, casein, esculin, guanine, hippurate, hypoxanthine, tyrosine, xanthine, and urea, are typically evaluated using the media of Gordon et al. (1974). Phosphatase activity is determined by using the method of Kurup and Schmitt (1973) substituting NZamine agar as the basal medium. Temperature range for growth is determined on slants of DSMZ medium 554 (DSMZ, 2001) and salt tolerance is determined on slants of the same medium supplemented with 2 to 8% NaCl.

Differentiation of the genus *Kibdelosporangium* from other genera

Species of the genus *Kibdelosporangium* can be easily differentiated from the other sporangium- and pseudosporangium-forming genera within the order *Pseudonocardiales*, where they are firmly placed on the basis of 16S rRNA gene phylogeny (Figure 298), as well as on morphological and chemotaxonomic criteria. All *Kibdelosporangium* species produce sporangium-like bodies that do not appear to contain spores, similar to *Allokutzneria*, but are different from *Kutzneria* species whose sporangia do contain spores, see Table 244. Both *Kibdelosporangium* and *Allokutzneria* produce chains of spores on their aerial hyphae, but this is not observed in *Kutzneria* species. The chemotaxonomic profile for *Kibdelosporangium* species is also different from the other sporangiate or pseudosporangiate genera within the order, as can be seen in Table 244.

Differentiation of the species of the genus *Kibdelosporangium*

The physiological characteristics of *Kibdelosporangium* species and subspecies are summarized in Table 245 and can be used to differentiate between them. The strains are phylogenetically distinct based on 16S rRNA gene sequences (Figure 304) and gross colonial morphology provides some additional information for differentiation between the *Kibdelosporangium aridum* subspecies and *Kibdelosporangium philippinense*.

TABLE 244. Comparison of chemotaxonomic profiles of *Kibdelosporangium* with related genera classified in the order *Pseudonocardiales*

Character	*Kibdelosporangium*	*Allokutzneria*	*Kutzneria*	*Streptoalloteichus*	*Thermocrispum*
Sporangia/spores	Sporangia, 1–32 µm in diameter; spore chains formed on aerial hyphae	Sporangium-like bodies, 8–20 µm in diameter; spore chains on aerial hyphae	Sporangia, 10–40 µm in diameter; no spore chains.	Sporangium-like vessel (1 species), 1.5–4.5 × 2.7–7.0 µm containing 1–4 oval to rod-shaped spores; spore chains formed on aerial hyphae	Pseudosporangial fragment into rod-shaped spores
Motile spores	No	No	No	Variable (1 species)	No
Whole-cell sugar pattern[a]	Ara, Gal, Mad (v), Glu (v), Rha (v)	Ara, Gal, Man	Gal, Rha	Gal, Man, Rha (v)	Ara, Man, Glu, Gal (trace)
Phospholipids[b]	PE, PI, PME, PG, DPG, PIM	PE, PME, OH-PE, PI, *lyso*-PME, DPG, PG, *lyso*-PE	PE, DPG, PI, PG, PME	PE, DPG, PI, PIM, DPG, PME	PE, PI, OH-PE
Predominant menaquinones	MK-9(H_4), MK-9(H_6), MK-9(H_{10})	MK-9(H_4)	MK-9(H_4)	MK-9(H_6), MK-10(H_6)	MK-9(H_4)
DNA G+C content (mol%)	66	71.6	70.3–70.7	71.6	69.0–73.0

[a]Abbreviations: Ara, Arabinose; Gal, galactose; Glu, glucose; Mad, madurose; Man, mannose; Rha, rhamnose. v, Variable.

[b]Abbreviations: DPG, diphosphatidylglycerol; OH-PE, phosphatidylethanolamine with hydroxy fatty acids; *lyso*-PE, phosphatidylethanolamine where one fatty acid chain is missing from the glycerol backbone; *lyso*-PME, phosphatidylmethylethanolamine where one fatty acid chain is missing from the glycerol backbone; PE, phosphatidylethanolamine; PG, phosphatidylglycerol; PI, phosphatidylinositol; PIM, phosphatidylinositol mannosides; PME, phosphatidylmethylethanolamine.

TABLE 245. Physiological characteristics of species of the genus *Kibdelosporangium*[a]

Character	*K. aridum* subsp. *aridum*	*K. aridum* subsp. *largum*	*K. philippinense*
Decomposition/hydrolysis of:			
Adenine	–	–	–
Allantoin	+	w	–
Casein	+	–	+
Cellulose	–	–	nd
Esculin	+	+	+
Gelatin	+	+	+
Hippurate	+	+	+
Hypoxanthine	+	+	+
Starch	–	–	–
Tyrosine	+	+	+
Urea	+	+	+
Xanthine	–	–	–
Acid from:			
Adonitol	–	nd	–
L-Arabinose	+	+	–
Cellobiose	+	+	+
Dextrin	+	+	–
Dulcitol	–	–	–
iso-Erythritol	–	–	–
D-Fructose	+	+	+
D-Galactose	+	+	+
Glucose	+	+	+
Glycerol	+	+	+
iso-Inositol	+	+	+
Inulin	–	–	–
Lactose	+	–	+
Maltose	+	+	+
D-Mannitol	+	+	+
D-Mannose	+	+	+
Melezitose	–	+	+
Melibiose	+	+	+
Methyl α-D-glucoside	+	+	+
Methyl β-D-xyloside	+	nd	nd
Raffinose	+	+	–
Rhamnose	+	+	+
D-Ribose	+	+	+
Salicin	v	v	–
D-Sorbitol	–	–	–
L-Sorbose	–	–	–
Sucrose	+	+	–
Trehalose	+	+	+
D-Xylose	+	+	+
Utilization of:			
Acetate	+	+	+
Benzoate	–	–	–
Citrate	+	+	+
Formate	+	+	+
Lactate	+	+	+
Malate	+	+	+
Oxalate	+	+	+
Propionate	+	+	+
Pyruvate	+	+	+
Succinate	+	+	+
Tartrate	–	–	–
Production of:			
Nitrate reductase	–	–	+
Phosphatase	+	+	+
Hydrogen sulfide	+	+	+
Melanin	+	+	+

(continued)

TABLE 245. (continued)

Character	*K. aridum* subsp. *aridum*	*K. aridum* subsp. *largum*	*K. philippinense*
Growth in:			
Lysozyme broth	–	–	–
2% NaCl	+	+	+
4% NaCl	+	v	–
5–7% NaCl	v	v	–
8% NaCl	–	–	–
Survival at 50°C/8 h	+	nd	nd
Growth at:			
10°C	v	–	–
15°C	+	+	–
42°C	+	+	–
45°C	tr	tr	–

[a]Symbols: w, weak positive; nd, not determined; v, variable; tr, trace amounts.

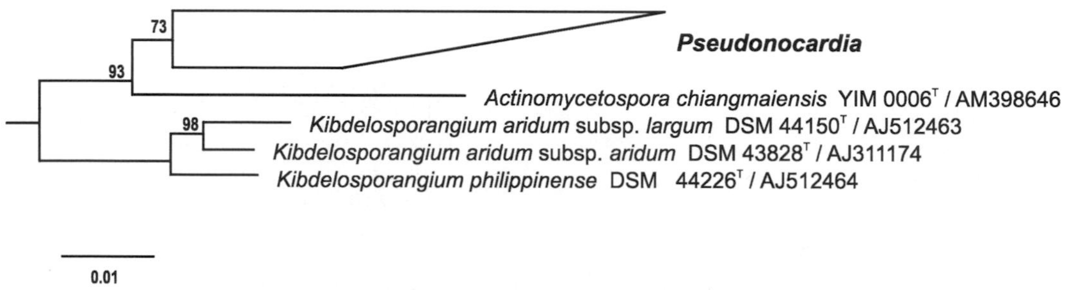

FIGURE 304. Phylogenetic tree for the genera *Kibdelosporangium*, *Actinomycetospora*, and *Pseudonocardia* calculated from almost-complete 16S rRNA gene sequences using the Kimura's evolutionary distance method (Kimura, 1980) and the neighbor-joining algorithm of Saitou and Nei (1987). Percentages at the nodes represent levels of bootstrap support (Felsenstein, 1993) from 100 resampled datasets; values less than 60% are not shown. Bar = 0.01 nucleotide substitutions per site.

List of species of the genus *Kibdelosporangium*

1. **Kibdelosporangium aridum** Shearer, Colman, Ferrin, Nisbet and Nash 1986a, 48[VP]

a'ri.dum. L. neut. adj. *aridum* dry, arid.

The substrate mycelium is off-white to grayish yellow-brown in color and is well developed with moderately branching, septate hyphae that are 0.4–1.0 µm in diameter. Fragmentation of substrate hyphae without hyphal displacement frequently occurs in plate cultures. Characteristic crystals are produced in the agar on many media. No pigments other than melanin or yellow-brown soluble pigments are produced. Produces the glycopeptides antibiotics aridicins A, B, and C (Shearer et al., 1985; Sitrin et al., 1985). Temperature for growth is 30°C.

Source: isolated from soil.

DNA G+C content (mol%): 66 (T_m).

Type strain: ATCC 39323, DSM 43828, JCM 7912, NBRC 14493, NRRL B-16436, SK&F AAD-216, VKM Ac-1316.

Sequence accession no. (16S rRNA gene): AJ311174.

Subsequently this species has been divided into subspecies.

1a. **Kibdelosporangium aridum subsp. aridum** Shearer, Colman, Ferrin, Nisbet and Nash 1986a, 48[VP]

The description is as for the species.

Type strain: ATCC 39323, DSM 43828, JCM 7912, NBRC 14493, NRRL B-16436, SK&F AAD-216, VKM Ac-1316.

Sequence accession no. (16S rRNA gene): AJ311174.

1b. **Kibdelosporangium aridum subsp. largum** Shearer, Giovenella, Grappel, Hedde, Mehta, Oh, Pan, Pitkin and Nisbet, 1988, 136[VP] (Effective publication: Shearer, Giovenella, Grappel, Hedde, Mehta, Oh, Pan, Pitkin and Nisbet 1986b, 1391.)

lar'gum. L. neut. adj. *largum* abundant, plentiful, numerous.

The substrate mycelium is off-white to yellow-brown in color and shows branching. The aerial mycelium is white to light gray in color, when produced. Pale yellow-brown to yellow-brown soluble pigments are produced on some media. Characteristic crystals are formed in the agar on some media. Temperature for growth is 30°C. Produces the glycopeptides antibiotics aridicins A, B, and C (Shearer et al., 1985; Sitrin et al., 1985) and the kibdelins A, B, C, and D (Folena-Wasserman et al., 1986; Shearer et al., 1986b).

Source: isolated from soil.

DNA G+C content (mol%): not determined.

Type strain: ATCC 39922, DSM 44150, JCM 9107, NBRC 15152, SK&F AAD-609.

Sequence accession no. (16S rRNA gene): AJ512463.

2. **Kibdelosporangium philippinense** Mertz and Yao 1988, 286[VP]

phil.ip.pi.nen′se. N.L. neut. adj. *philippinense* pertaining to the Philippines.

The branching substrate mycelium is generally pale yellow to orange-yellow in color. Aerial mycelium is white in color, when produced. Light brown to light reddish-brown soluble pigments are formed on some growth media. Production of crystals is not observed on any growth medium. Produces a ristocetin-like glycopeptide antibiotic. Temperature for growth is 30°C.

Source: isolated from soil.

DNA G+C content (mol%): not determined.

Type strain: A80407, ATCC 49844, DSM 44226, JCM 9918, NBRC 15154, NRRL 18198.

Sequence accession no. (16S rRNA gene): AJ512464.

Genus IX. **Kutzneria** Stackebrandt, Kroppenstedt, Jahnke, Kemmerling and Gürtler 1994, 267[VP]

David P. Labeda

Kutz.ne′ri.a. N.L. fem. n. *Kutzneria* named after Hans-Jürgen Kutzner, a German microbiologist.

Aerobic, Gram-stain-positive, mesophilic actinomycete which forms a stable, branched, cottony aerial mycelium. **Globose sporangia are large (diameter of 10–48 µm).** Sporangial walls are thick and strong. **Sporangiophores more than 50 µm long are formed by septation of coiled, unbranched hyphae within sporangiophores. Spores are spherical, rod-shaped, or ovoid and nonmotile.** Some species are thermotolerant. Chemoorganotrophic. **Cell walls contain *N*-acetylated muramic acid, *meso*-diaminopimelic acid, and generally galactose as the characteristic sugar.** Rhamnose may also be present as a whole-cell diagnostic sugar. **MK-9(II, III-H_4) are the major menaquinones. Major phospholipids include diphosphatidylglycerol, hydroxyphosphatidylethanolamine, phosphatidylethanolamine, and phosphatidylinositol. Major fatty acids are $C_{16:0}$ iso, $C_{16:0}$ iso 2-OH, $C_{16:0}$ 10-methyl, $C_{17:0}$ anteiso, and $C_{17:0}$ 2-OH anteiso; $C_{10:0}$, $C_{14:0}$ iso, and $C_{14:0}$ fatty acids are absent.** Natural habitat is soil.

DNA G+C content (mol%): 70.3–70.7 (T_m).

Type species: **Kutzneria viridogrisea** (Okuda, Furumai, Watanabe, Okugawa and Kimura 1966) Stackebrandt, Kroppenstedt, Jahnke, Kemmerling and Gürtler 1994, 268[VP].

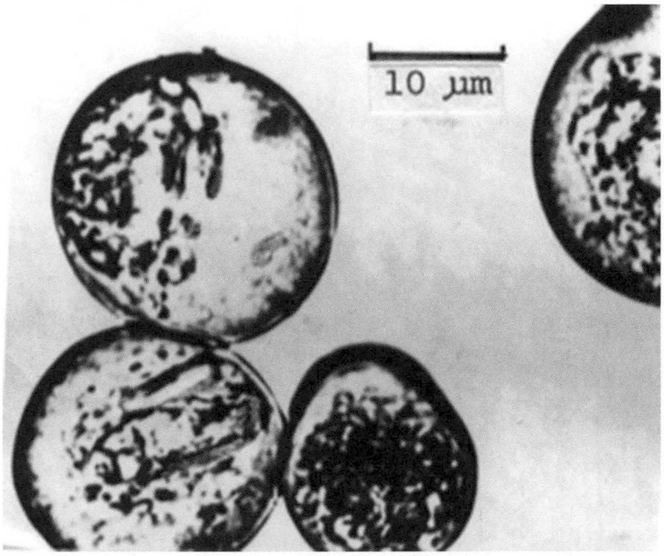

FIGURE 305. *Kutzneria kofuensis* sporangia. Note that sporangial walls are thick. Light micrograph stained with methylene blue. (Reproduced with permission from Nonomura and Ohara, 1969b. J. Ferment. Technol. *47*: 648.)

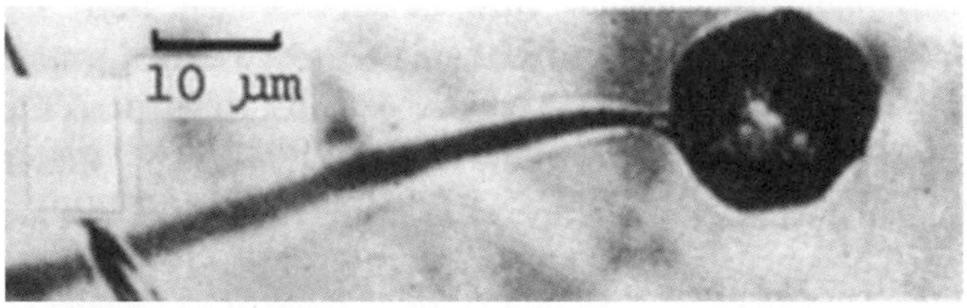

FIGURE 306. Light micrograph showing the long sporangiophores of *Kutzneria kofuensis.* (Reproduced with permission from Nonomura and Ohara, 1969b. J. Ferment. Technol. *47*: 648.)

Further descriptive information

All of the species currently classified within the genus *Kutzneria* were originally described as *Streptosporangium* species with validly published names based on their production of sporangia (Figure 305). The sporangia are produced on long sporophores (Figure 306) containing chains of nonmotile spores (Figure 307). Stackebrandt et al. (1994) proposed that these species be transferred to the new genus *Kutzneria* because phylogenetic analysis based on 16S rRNA genes (Kemmerling et al., 1993), as well as previous phylogenetic studies based on 5S rRNA genes (Kudo et al., 1993) and electrophoretic mobility of ribosomal protein AT-L30 (Ochi and Miyadoh, 1992), demonstrated that they were phylogenetically distinct from species of the genus *Streptosporangium*, as can be seen by their phylogenetic position within the suborder *Pseudonocardineae*, now the order *Pseudonocardiales*, shown in Figure 277 in the treatment of the family *Pseudonocardiaceae*. The chemotaxonomic characteristics of the species transferred to *Kutzneria* were also observed to be distinct from those of *Streptosporangium sensu stricto*, as can be seen in Table 246, particularly with regard to the lack of madurose in the whole-cell sugar pattern, the lack of ninhydrin-positive and sugar-positive phospholipids in their polar lipid patterns, and major differences in fatty acid and menaquinone profiles.

Enrichment and isolation procedures

Nonomura and Ohara (1969a) described a novel method for the isolation of *Microbispora* and *Streptosporangium* from soil

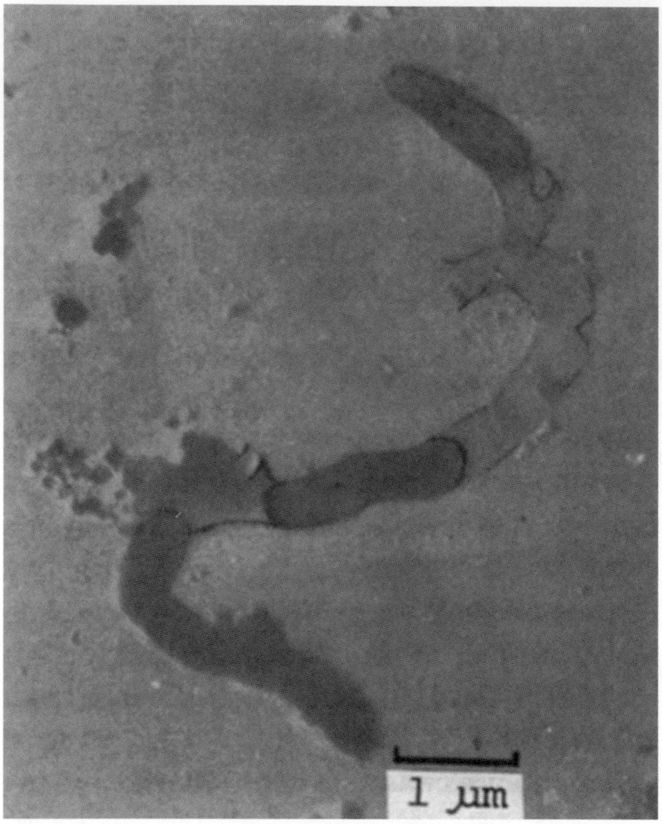

FIGURE 307. Electron micrograph showing *Kutzneria kofuensis* sheathed spores from within the sporangium. (Reproduced with permission from Nonomura and Ohara, 1969b. J. Ferment. Technol. *47*: 648.)

samples; the type strain of *Kutzneria kofuensis* was isolated using this procedure. Soil samples are pretreated by air drying, grinding in a mortar, and then heating at 120°C for 1 h. Aliquots of serial dilutions of the soil samples are spread onto the surface of arginine-vitamin (AV) agar typically containing antifungal antibiotics (e.g. cycloheximide and/or nystatin at 50 µg/ml) and plates are incubated for 40 d at 30°C. AV agar consists of (per l): L-arginine hydrochloride, 0.3 g; glucose, 1.0 g; glycerol, 1.0 g; K_2HPO_4, 0.3 g; $MgSO_4 \cdot 7H_2O$, 0.2 g; NaCl, 0.3 g; thiamine hydrochloride, 0.5 mg; riboflavin, 0.5 mg; niacin, 0.5 mg; pyridoxine hydrochloride, 0.5 mg; inositol, 0.5 mg; calcium pantothenate, 0.5 mg; *p*-aminobenzoic acid, 0.5 mg; biotin, 0.25 mg; $Fe_2(SO_4)_3$, 10 mg; $CuSO_4 \cdot 5H_2O$, 1 mg; $MnSO_4 \cdot H_2O$, 1 mg; $ZnSO_4 \cdot 7H_2O$, 1 mg; agar, 15 g. The pH should be adjusted to 6.4 prior to sterilization. The vitamins should be filter-sterilized and added to autoclaved and tempered medium prior to dispensing.

Maintenance procedures

Working cultures of *Kutzneria* can be maintained as refrigerated (4°C) agar slants on appropriate medium such as NZamine agar (DSMZ medium 554; DSMZ, 2001). The slants are generally incubated for 5–7 d at 28–30°C prior to refrigeration and then should be subcultured at monthly intervals. Longer term preservation of strains is best accomplished as frozen stocks in 20% (v/v) aqueous glycerol at −80°C (mechanical freezer) to −172°C (liquid nitrogen vapor phase) or by using traditional lyophilization techniques. Strains have also been successfully stored for shorter periods as quick-frozen stationary-phase broth cultures or mycelium suspensions in 20% aqueous glycerol at −20 to −72°C.

Procedures for testing special characters

Strains are routinely cultivated on NZamine agar at 28°C. Morphological observations are made on the media of Shirling and Gottlieb (1966) and NZamine agar.

Chemotaxonomic analysis of strains for fatty acids, menaquinones, and polar lipids are performed using methods previously described by Grund and Kroppenstedt (1989), Minnikin et al. (1984), Sasser (1990), and Saddler et al. (1991). Physiological tests, including production of acid from carbohydrates, utilization of organic acids, and hydrolysis and decomposition of adenine, casein, esculin, guanine, hippurate, hypoxanthine, tyrosine, xanthine, and urea, are typically evaluated using the media of Gordon et al. (1974). Phosphatase activity is determined by using the method of Kurup and Schmitt (1973), but substituting NZamine agar as the basal medium. Temperature range for growth is determined on slants of DSMZ medium 554 and salt tolerance is determined on slants of the same medium supplemented with 4% and 5% NaCl.

Differentiation of the genus *Kutzneria* from other genera

The genus *Kutzneria* is phylogenetically distinct from all of the other sporangium- or pseudosporangium-forming genera within the suborder *Pseudonocardineae*, now the order *Pseudonocardiales*, but is phylogenetically most closely related to the genus *Allokutzneria*, as can be seen in Figure 308. The morphology of the sporangia produced by *Kutzneria* species and their chemotaxonomic characteristics (Table 246) differentiate them from similar sporangium-producing genera. The chemotaxonomy of the genus *Kutzneria* is distinct from that of *Allokutzneria* in containing galactose and mannose as whole-cell sugars, i.e. arabinose, galactose, and mannose versus galactose and rhamnose that are found in *Allokutzneria*. The phospholipid profiles of these two genera are also quite different: *Allokutzneria* contains *lyso*-phosphatidylethanolamine, *lyso*-phosphatidylmethylethanolamine, and phosphatidylethanolamine containing hydroxylated fatty acids, whereas these are absent in *Kutzneria*. The sporangial morphology and chemotaxonomic profile, particularly in regard to the whole-cell sugar pattern and predominant menaquinones, between *Kutzneria* and *Kibdelosporangium*, are also quite different, as can be seen in Table 246.

Differentiation of the species of the genus *Kutzneria*

The morphological and physiological characteristics of *Kutzneria* species are summarized in Table 247 and can be used to differentiate between species, although a common set of physiological tests has not been used in the description of all species. The whole-cell sugar patterns of each species also appear to be distinct. The strains can also be distinguished based on their 16S rRNA gene sequences and phylogeny (Figure 308).

TABLE 246. Comparison of chemotaxonomic profile of *Kutzneria* with phylogenetically nearest genera and *Streptosporangium*

Character	*Kutzneria*	*Streptosporangium*	*Allokutzneria*	*Actinoalloteichus*	*Crossiella*	*Goodfellowiella*	*Kibdelosporangium*	*Saccharothrix*	*Streptoalloteichus*	*Umezawaea*
Production of sporangia and size	Diameter 11–50 μm, contain chains of sheathed spores	Diameter up to 10 μm, contains coiled chains of oval to rod-shaped spores	Diameter 8–20 μm, contains coiled hyphae, no spores	None	None	None	Diameter 9–35 μm, contains hyphae in an amorphous matrix	None	Pseudosporangia	None
Whole-cell sugar pattern[a]	Gal, Rha	Gal (v), Mad (v)	Ara, Gal, Man	Glu, Gal, Man, Rib	Gal, Man, Rha, Rib	Gal, Rib	Ara, Gal, Glu (v), Mad (v), Rha (v)	Gal, Rha, Man (tr)	Gal, Man, Rha, Rib	Gal, Man, Rib, Rha (tr)
Phospholipids[b]	PE, DPG, PI, PG, PME	PE, OH-PE, PI, DPG, GlN	PE, PME, OH-PE, PI, *lyso*-PME, DPG, PG, *lyso*-PE	PIM, PI, PG, DPG PME	PE, DPG, PI, PIM, PME	PE, DPG, OH-PE, PME	PE, PI, PME, PG, DPG, PIM	PE, OH-PE, DPG, PG, PI, PIM	PE, DPG, PI, PIM, DPG, PME	PE, PI, OH-PE, *lyso*-PE
Predominant menaquinones	MK-9(H_4)	MK-9(H_2), MK-9(II,VIII-H_4), and/or MK-9(H_6)	MK-9(H_4)	MK-9(H_4)	MK-9(H_4)	MK-9(H_4), MK-10(H_4)	MK-9(H_4), MK-9(H_6), MK-9(H_{10})	MK-9(H_4), MK-10(H_4)	MK-9(H_6), MK-10(H_6)	MK-9(H_4), MK-10(H_4)
DNA G+C content (mol%)	70.3–70.7	69–71	71.6	72–72.5	71.4	69.2	66	72.2–74.0	ndND	74.0

[a]Ara, Arabinose; Gal, galactose; Glu, glucose; Mad, madurose; Man, mannose; Rha, rhamnose; Rib, ribose. v, Variable; tr, trace.

[b]Abbreviations: DPG, diphosphatidylglycerol; GlN, *N*-acetylglucosamine-containing phospholipid; OH-PE, phosphatidylethanolamine with hydroxy fatty acids; *lyso*-PE, phosphatidylethanolamine where one fatty acid chain is missing from the glycerol backbone; *lyso*-PME, phosphatidylmethylethanolamine where one fatty acid chain is missing from the glycerol backbone; PE, phosphatidylethanolamine; PG, phosphatidylglycerol; PI, phosphatidylinositol; PIM, phosphatidylinositol mannosides; PME, phosphatidylmethylethanolamine.

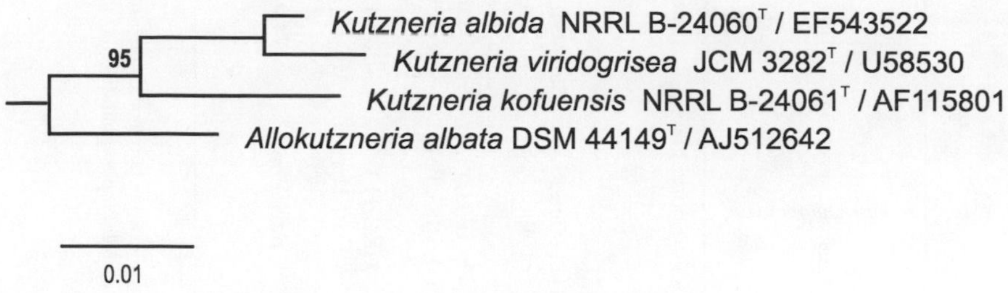

FIGURE 308. Phylogenetic tree for the genus *Kutzneria* and *Allokutzneria* calculated from almost-complete 16S rRNA gene sequences using the Kimura's evolutionary distance method (Kimura, 1980) and the neighbor-joining algorithm of Saitou and Nei (1987). Percentages at the nodes represent levels of bootstrap support (Felsenstein, 1989) from 100 resampled datasets; values less than 60% are not shown. Bar = 0.01 nucleotide substitutions per site.

TABLE 247. Morphological and physiological characteristics of *Kutzneria* species[a]

Characteristic	*K. viridogrisea*	*K. albida*	*K. kofuensis*
pH for growth	4–9	nd	nd
Sporangium:			
11–15 µm	–	(+)	+
16–20 µm	–	(+)	+
21–30 µm	+	+	–
31–50 µm	+	–	–
Sporangiospores:			
Short (10 µm)	–	–	+
Long (50 µm)	+	+	+
Spores:			
Spherical to ovoid	+	+	–
Rods	–	–	+
Color of spore mass	Greenish-gray	White	Greenish-gray
Color of substrate mycelium	Yellowish-brown to brown	Yellowish-brown to brown	Yellowish-brown to brown
Production of:			
Soluble pigments	–	–	–
Nitrate reductase	+	+	–
Hydrolysis of:			
Cellulose	–	–	nd
Gelatin	+	–	+
Milk (peptonization)	+	+	w
Starch	+	–	+
Growth as sole carbon source:			
Arabinose	w	+	+
Dextrin	nd	–	nd
Dulcitol	–	nd	nd
Fructose	+	+	nd
Galactose	+	nd	nd
Glucose	+	+	+
Glycerol	+	+	+
Inositol	w	+	+
Inulin	–	nd	nd
Lactose	w	+	nd
Maltose	+	nd	nd
Mannitol	+	+	nd
Mannose	+	+	nd
Raffinose	+	+	nd
Rhamnose	w	+	w
Salicin	–	–	nd
Sorbitol	+	nd	nd
Starch	+	nd	nd
Sucrose	+	+	nd
Xylose	w	+	nd

(continued)

TABLE 247. (continued)

Characteristic	*K. viridogrisea*	*K. albida*	*K. kofuensis*
Assimilation of:			
Malate	+	+	nd
Growth at:			
10°C	−	nd	nd
27°C	+	+	+
37°C	+	nd	+
42°C	nd	nd	+
50°C	v	nd	v

[a]Symbols: (+), sometimes observed; w, weak positive reaction; v, variable; nd, not determined.

List of species of the genus *Kutzneria*

1. **Kutzneria viridogrisea** (Okuda, Furumai, Watanabe, Okugawa and Kimura 1966) Stackebrandt, Kroppenstedt, Jahnke, Kemmerling and Gürtler 1994, 268[VP] (Basonym: *Streptosporangium viridogriseum* Okuda, Furumai, Watanabe, Okugawa and Kimura 1966, 126.)

vi.ri.do.gri'se.a. L. adj. *viridis* green; N.L. adj. *griseus* gray; N.L. fem. adj. *viridogrisea* greenish gray.

Substrate mycelium is yellowish-brown in color on most growth media. Cottony aerial mycelium is produced, initially white in color, then becoming greenish-gray to olive-gray, depending upon the growth medium. Soluble pigments are not produced on any medium. Temperature range for growth is 17–45°C, with an optimum at 37°C. Whole-cell sugars are galactose and rhamnose. Produces sporoviridin.

Source: isolated from soil.

DNA G+C content (mol%): 70.3 (HPLC).

Type strain: ATCC 25242, DSM 43850, JCM 3282, NBRC 15561, NRRL B-24059, VKM Ac-1297.

Sequence accession no. (16S rRNA gene): U58530.

2. **Kutzneria albida** (Furumai, Ogawa and Okuda 1968) Stackebrandt, Kroppenstedt, Jahnke, Kemmerling and Gürtler 1994, 268[VP] (Basonym: *Streptosporangium albidum* Furumai, Ogawa and Okuda 1968, 174)

al'bi.da. L. fem. adj. *albida* white.

Substrate mycelium is colorless to yellowish-brown on synthetic media with production of thin white aerial mycelia which become cottony or floccose and brownish-white. Sporangia (10–30 µm) can be easily observed microscopically. Substrate mycelium is pale yellow to reddish-brown on organic media. Soluble pigments are not produced on any medium. Temperature range for growth is 17°C to 45°C with an optimum at 37°C. Whole-cell sugar is rhamnose. Produces the antibiotic sporoviridin.

Source: isolated from soil.

DNA G+C content (mol%): 70.3 (HPLC).

Type strain: ATCC 25243, DSM 43870, JCM 3240, NBRC 13901, NRRL B-24060.

Sequence accession no. (16S rRNA gene): EF543522.

3. **Kutzneria kofuensis** (Nonomura and Ohara 1969b) Stackebrandt Kroppenstedt, Jahnke, Kemmerling and Gürtler 1994, 268[VP] (Basonym: *Streptosporangium viridogriseum* subsp. *kofuense* Nonomura and Ohara 1969b, 708)

ko.fu.en'sis. N.L. fem. adj. *kofuensis* belonging to Kofu, a district in Japan, where the organism was isolated.

Grows well on oatmeal-yeast extract-glucose agar and yeast-malt agar. Aerial mycelium is greenish-gray. Soluble pigments are not produced on any medium. Temperature range for growth is 17–45°C, with an optimum at 37°C. Whole-cell sugars are rhamnose and galactose. Produces chloramphenicol.

Source: isolated from soil.

DNA G+C content (mol%): 70.3 (HPLC).

Type strain: ATCC 27102, DSM 43851, JCM 3157, NBRC 13989, NRRL B-24061.

Sequence accession no. (16S rRNA gene): AF114801.

Genus X. **Lechevalieria** Labeda, Hatano, Kroppenstedt and Tamura 2001, 1049[VP]

David P. Labeda

Le.che.va.li.e'ri.a. N.L. fem. n. *Lechevalieria* named after the American microbiologists Hubert and Mary Lechevalier, who contributed substantially to the field of actinomycete biology during their careers at the Waksman Institute of Microbiology.

Branching vegetative mycelium (approx. 0.5 µm in diameter) is produced. **Very scant aerial mycelium is formed on some media.** Gram-stain-positive. Lysozyme resistant. Catalase-positive and aerobic. **The cell wall contains *meso*-diaminopimelic acid as the diamino acid. Whole-cell sugar pattern consists of galactose, mannose, and traces of rhamnose. Phospholipid pattern consists of significant quantities of phosphatidylethanolamine lacking hydroxylated fatty acids, as well as diphosphatidylglycerol, phosphatidylglycerol, and**

phosphatidylinositol. The major menaquinone is MK-9(H_4). Fatty acid profile consists of saturated and mono-unsaturated iso and anteiso fatty acids. **Phylogenetically represents a line of descent in the family *Pseudonocardiaceae*; adjacent to the genus *Saccharothrix* and close to the genera *Actinosynnema* and *Lentzea*. 16S rRNA gene sequence contains genus-specific diagnostic nucleotide signature pattern of TT (844–845) and GGT (1107–1109).**

DNA G+C content (mol%): 68–71.4 (T_m).

Type species: **Lechevalieria aerocolonigenes** (Shinobu and Kawato 1960) Labeda, Hatano, Kroppenstedt, and Tamura 2001, 1050[VP].

Further descriptive information

The genus *Lechevalieria* was described by Labeda et al. (2001) to contain strains formerly placed in the genus *Saccharothrix* that were found to be distinct by phylogenetic analysis of 16S rRNA gene sequences. Both of the constituent species, *Lechevalieria aerocolonigenes* and *Lechevalieria flava*, had been transferred previously into *Saccharothrix* from the genera *Streptomyces* and *Nocardiopsis*, respectively, although the latter taxon had been previously transferred first from the genus *Actinomadura* to the genus *Nocardiopsis*. Previously, the classification of these organisms had been based largely on morphology and then on chemotaxonomy. Subsequent phylogenetic analyses based on 16S rRNA gene sequences elucidated the significance of new chemotaxonomic markers which are essential for distinguishing *Lechevalieria* species from neighboring taxa.

Enrichment and isolation procedures

Lechevalieria strains, as well as those of the genera *Lentzea* and *Saccharothrix*, can be isolated from soil and plant residue samples by spreading soil dilutions onto the surface of routine media used for the general isolation of actinomycetes, such as 1.5% crude agar and 0.4% casein hydrolysate in tap water. Antibiotics can be used to more selectively isolate members of these genera as has been described by Shearer (1987). The recently described species *Lechevalieria fradiae* and *Lechevalieria xinjiangensis* were isolated from soil samples by serial plating on yeast extract-malt extract agar (Shirling and Gottlieb, 1966) and modified Bennett's agar (Jones, 1949), respectively.

Maintenance procedures

Working cultures of *Lechevalieria* can be maintained as refrigerated (4°C) agar slants on appropriate medium such as NZamine agar (DSMZ medium 554; DSMZ, 2001). The slants are generally incubated for 5 to 7 d at 28–30°C prior to refrigeration and then should be subcultured at monthly intervals. Longer term preservation of strains is best accomplished as frozen stocks in 20% (v/v) aqueous glycerol at −80°C (mechanical freezer) to −172°C (liquid nitrogen vapor phase) or by using traditional lyophilization techniques. Strains have also been successfully stored for shorter periods as quick-frozen stationary-phase broth cultures or mycelium suspensions in 20% aqueous glycerol at −20 to −72°C.

Procedures for testing special characters

Strains are routinely cultivated on NZamine agar (DSMZ medium 554) at 28°C. Morphological observations are made on the media of Shirling and Gottlieb (1966) and NZamine agar.

Chemotaxonomic analysis of strains for fatty acids, menaquinones, and polar lipids are performed using methods described previously by Grund and Kroppenstedt (1989), Minnikin et al. (1984), Sasser (1990), and Saddler et al. (1991).

Physiological tests, including production of acid from carbohydrates, utilization of organic acids, and hydrolysis and decomposition of adenine, casein, esculin, guanine, hippurate, hypoxanthine, tyrosine, xanthine, and urea, are typically evaluated using the media of Gordon et al. (1974). Phosphatase activity is determined by using the method of Kurup and Schmitt (1973) substituting NZamine agar as the basal medium. Temperature range for growth is determined on slants of DSMZ medium 554 and salt tolerance on slants of the same medium supplemented with 4% and 5% NaCl.

Differentiation of the genus *Lechevalieria* from other genera

The type strains of *Lechevalieria aerocolonigenes*, *Lechevalieria flava*, *Lechevalieria fradiae*, and *Lechevalieria xinjiangensis* form a monophyletic lineage distinct from the genus *Lentzea* (mean nucleotide similarity of 96.7%), and intermediary between *Lentzea* and the genus *Actinosynnema* (mean nucleotide similarity of 97.3%) as can be seen in Figure 290. The aligned sequences of the 16S rRNA genes for *Lechevalieria*, *Actinosynnema*, *Lentzea*, and *Saccharothrix* (Figure 309) illustrate that *Lechevalieria* strains can be clearly distinguished from the other genera on the basis of the diagnostic nucleotide signatures TT (844–845) and GGT (1107–1109). Note that the published 16S rRNA gene sequence for *Lechevalieria fradiae* contains the signature GGC (1107–1109) and this has not yet been confirmed to be accurate.

Species of the genus *Lechevalieria* can be distinguished from the genera *Actinosynnema* and *Saccharothrix* by the lack of hydroxy-substituted fatty acids in the phosphatidylethanolamine component of the diagnostic phospholipids (see Table 248), similar to species of the genus *Lentzea*. *Lechevalieria*, *Saccharothrix*, and *Umezawaea* strains tend to contain varying amounts of rhamnose in whole-cell hydrolysates, whereas *Lentzea* strains are observed to totally lack or contain only trace amounts of rhamnose and may also contain ribose. *Actinosynnema* species have only galactose and mannose as their diagnostic whole-cell sugar pattern.

Lechevalieria strains produce very sparse aerial mycelium on agar media, their substrate mycelium fragments into coccoid to coccoidal-rod-shaped elements, and they have not been observed to produce sporangia, coremia, or motile spores, which differentiates them from the genera *Actinokineospora* and *Actinosynnema*.

Differentiation of the species of the genus *Lechevalieria*

The physiological characteristics of *Lechevalieria* species are summarized in Table 249 and can be used to differentiate between species, although a common set of physiological tests has not been used in the description of all species. Gross colonial morphology provides some additional information for differentiation

	601	611	621	841
Lechevalieria aerocolonigenes NRRL B-3298^T	**AAACTTGGGG**	**CTTAACCCCG**	**AGCCTGCGGT**	**ACGTTCTCCG**
Lechevalieria flava NRRL B-16131^T	....	....	....	....
Lechevalieria fradiae CGMCC 4.3506^T	....	....	....	....**C**.....
Lechevalieria xinjiangensis CGMCC 4.3525^T	....	....	....	....
Lentzea albida IFO 16102^T	....	......**T**..**A**	...**T**......	...**CC**...**T**.
Lentzea albidocapillata DSM 44073^T	....	......**T**..**A**	...**T**......	...**CC**...**T**.
Lentzea californiensis NRRL B-16137^T	....	......**T**..**A**	...**T**...**C**..	...**CC**...**T**.
Lentzea flaviverrucosa AS4.0578^T	....	......**T**..**A**	...**T**......	...**CC**...**T**.
Lentzea kentuckyensis NRRL B-24416^T	....	......**T**..**A**	...**T**.......	...**CC**...**T**.
Lentzea violacea IMSNU 50388^T	....	......**T**..**A**	...**T**......	...**CC**...**T**.
Lentzea waywayandensis NRRL B-16159^T	....	......**T**..**A**	...**T**......	...**CC**...**T**.
Actinosynnema mirum DSM 43827^T	....	....	....	....**C**.....
Actinosynnema pretiosum subsp. *pretiosum* NRRL B-16060^T	....	....	...**T**......	....**C**.....
Saccharothrix australiensis NRRL 11239^T	......**CAC**.	......**GTG**.	....	....**C**.....
Saccharothrix algeriensis NRRL B-24137^T	......**CAC**.	......**GTG**.	....	....**C**.....
Saccharothrix coeruleofusca NRRL B-16115^T	......**CAC**.	......**GTG**.	....	....**C**.....
Saccharothrix espanaensis NRRL 15764^T	......**CAC**.	......**GTG**.	....	....**C**.....
Saccharothrix longispora NRRL B-16116^T	......**CAC**.	......**GTG**.	.......**A**..	....**C**.....
Saccharothrix mutabilis subsp. *capreolus* DSM 40225^T	......**CAC**.	...**N**..**GTG**.	....	....**C**.....
Saccharothrix mutabilis subsp. *mutabilis* DSM 43853^T	......**CAC**.	..**N**...**GTG**.	....	....**C**.....
Saccharothrix syringae NRRL B-16468^T	......**CAC**.	......**GTG**.	....	....**C**.....
Saccharothrix texasensis NRRL B-16134^T	......**CAC**.	......**GTG**.	.......**A**..	....**C**.....
Saccharothrix xinjinagensis AS.4.1731^T	......**CAC**.	......**GTG**.	....	....**C**.....
Umezawaea tangerinus MK27-91F2^T	....**C**.**ACA**.	......**TGT**.	**G**......**A**..	....

FIGURE 309. Nucleotide signatures in the 16S rRNA gene for the genera *Saccharothrix*, *Lechevalieria*, *Lentzea*, and *Umezawaea*.

TABLE 248. Chemotaxonomic characteristics of *Lechevalieria* and related genera[a]

Characteristic	*Lechevalieria*	*Actinosynnema*	*Lentzea*	*Saccharothrix*	*Umezawaea*
Whole-cell sugar pattern[b]	Gal, Man, Rha	Gal, Man	Gal, Man, Rib	Gal, Rha, Man (tr)	Gal, Rha
Phospholipids[c]	PE	PE, OH-PE, PI, PIM, DPG	PE, DPG, PG, PI	PE, OH-PE, PI, PIM, DPG, PG (v)	PE
Predominant menaquinone(s)	MK-9(H_4)	MK-9(H_4), MK-9(H_6)	MK-9(H_4)	MK-10(H_4), MK-9(H_4)	MK-9(H_4), MK-10(H_4) (tr)

[a]All genera have *meso*-diaminopimelic acid (A_2pm) as the cell-wall diamino acid, are of cell-wall chemotype III, and contain straight chain, mono-unsaturated, iso, and anteiso fatty acids. tr, Trace; v, variable.

[b]Gal, Galactose; Man, mannose; Rha, rhamnose; Rib, ribose.

[c]DPG, Diphosphatidylglycerol; OH-PE, phosphatidylethanolamine containing hydroxylated fatty acids; PE, phosphatidylethanolamine; PG, phosphatidylglycerol; PI, phosphatidylinositol; PIM, phosphatidylinositolmannosides; PME, phosphatidylmethylethanolamine.

TABLE 249. Physiological characteristics of species of the genus *Lechevalieria*[a]

Characteristic	*L. aerocolonigenes*	*L. flava*	*L. fradiae*	*L. xinjiangensis*
Decomposition of:				
Adenine	–	–	–	nd
Allantoin	–	nd	nd	nd
Casein	+	–	+	+
Esculin	+	+	+	nd
Gelatin	+	nd	+	nd
Hippurate	–	nd	nd	nd
Hypoxanthine	+	+	–	–
Starch	+	nd	+	nd
Tyrosine	+	+	+	–
Urea	w	nd	+	nd
Xanthine	–	–	nd	nd
Acid from:				
Adonitol	–	–	–	nd
Arabinose	+	+	+	nd
Cellobiose	+	+	–	nd

(continued)

TABLE 249. (continued)

Characteristic	*L. aerocolonigenes*	*L. flava*	*L. fradiae*	*L. xinjiangensis*
Dextrin	+	+	–	nd
Dulcitol	–	nd	nd	nd
Erythritol	–	nd	+	nd
Fructose	+	+	–	nd
Galactose	+	nd	–	nd
Glucose	+	+	–	nd
Glycerol	+	nd	–	nd
Inositol	+	+	–	–
Lactose	+	+	–	–
Maltose	+	nd	+	nd
Mannose	+	nd	–	nd
Melibiose	+	+	–	nd
Methyl α-D-glucoside	w	nd	+	nd
Methyl β-D-xyloside	w	nd	nd	nd
Raffinose	+	+	+	nd
Rhamnose	+	+	–	nd
Salicin	+	nd	–	–
Sorbitol	–	–	–	nd
Sucrose	+	+	–	nd
Trehalose	+	nd	+	nd
Xylose	+	nd	+	nd
Utilization of:				
Acetate	+	–	+	+
Benzoate	–	–	–	nd
Citrate	+	+	–	+
Lactate	+	–	+	nd
Malate	+	nd	+	nd
Mucate	–	nd	nd	nd
Oxalate	+	nd	–	+
Propionate	+	nd	nd	+
Succinate	+	nd	+	nd
Tartrate	+	nd	–	nd
Production of:				
Nitrate reductase	+	+	+	+
Phosphatase	+	nd	nd	nd
Growth in the presence of:				
4% NaCl	+	+	+	+
5% NaCl	w	–	+	+
Growth at:				
10°C	+	+	–	+
37°C	+	+	+	+
42°C	w	+	+	+
45°C	–	–	+	+

[a]Symbols: w, weak positive; nd, not determined.

of the species based on the color of the substrate mycelium and production and color of soluble pigments, but some of the species appear quite similar when grown on agar media. The DNA relatedness between species, as summarized in Table 250, was determined during the characterization of the most recently described species (Wang et al., 2007; Zhang et al., 2007a) and provides evidence that these species are distinct.

TABLE 250. DNA relatedness (%) between *Lechevalieria* species[a]

	L. aerocolonigenes	*L. flava*	*L. fradiae*
L. flava	nd		
L. fradiae	45	37	
L. xinjiangensis	44	28	34

[a]Data are from Zhang et al. (2007a) and Wang et al. (2007).

List of species of the genus *Lechevalieria*

1. **Lechevalieria aerocolonigenes** (Shinobu and Kawato 1960) Labeda, Hatano, Kroppenstedt and Tamura 2001, 1050^{VP} [*Saccharothrix aerocolonigenes* (Shinobu and Kawato 1960) Labeda 1986, 109; *Streptomyces aerocolonigenes* Shinobu and Kawato 1960, 215]

ae.ro.co.lo.ni'ge.nes. Gr. n. *aer* air; L. n. *colonia* a colony; N.L. suff. *-genes* (from Gr. v. *gennaô* to produce) producing; N.L. fem. adj. *aerocolonigenes* producing aerial colonies.

Branching substrate mycelium is produced (diameter approx. 0.5 µm); on some media extremely sparse aerial hyphae are formed. Some strains may appear to form clumps of interwoven hyphae or "colonies" in the aerial mycelium. Substrate and aerial mycelium fragment. Strains usually lose their capacity to form aerial mycelium during subcultivation.

Yellowish to brownish substrate mycelium is produced along with extremely sparse white aerial hyphae. Yellowish to brownish soluble pigment is produced on several media. Temperature for growth is 28°C.

Source: isolated from soil.

DNA G+C content (mol%): 70–71 (T_m).

Type strain: NRRL B-3298, ATCC 23870, CCRC 13661, DSM 40034, IFO 13195, JCM 4150.

Sequence accession no. (16S rRNA gene): AF114804.

2. **Lechevalieria flava** (Gauze, Maksimova, Ollkhovatova, Sveshnikova, Kochetkova and Ilchenko 1974) Labeda, Hatano, Kroppenstedt and Tamura 2001, 1050^{VP} [*Saccharothrix flava* (Gauze, Maksimova, Ollkhovatova, Sveshnikova, Kochetkova and Ilchenko 1974) Grund and Kroppenstedt 1990a, 320 (Effective publication: Grund and Kroppenstedt 1989, 271.); *Nocardiopsis flava* (Gauze, Maksimova, Ollkhovatova, Sveshnikova, Kochetkova and Ilchenko 1974) Gauze and Sveshnikova 1985, 224; (Effective publication: Preobrazhenskaya, Sveshnikova and Gauze 1982, 111.); *Actinomadura flava* Gauze, Maksimova, Ollkhovatova, Sveshnikova, Kochetkova and Ilchenko 1974^{AL}]

fla'va. L. fem adj. *flava* yellow (referring to the color of the substrate mycelium).

Substrate mycelium is usually yellowish in color and aerial mycelium, if present, is white. Special spore forming hyphae are absent. Long, straight aerial mycelium fragments completely into spores. Temperature for growth is 28°C.

Source: isolated from soil.

DNA G+C content (mol%): not determined.

Type strain: NRRL B-16131, ATCC 29533, CCRC 13328, DSM 43885, NBRC 14521, INA 2171, JCM 3296.

Sequence accession no. (16S rRNA gene): AF114808.

3. **Lechevalieria fradiae** Zhang, Xie, Liu and Goodfellow 2007a, 834^{VP}

fra'di.ae. N.L. gen. n. *fradiae* of Fradia, a patronymic.

Substrate mycelium is yellow to orange in color. Aerial mycelium is not produced. Substrate mycelium fragments into coccoid to coccoidal–rod-shaped elements. No soluble pigments are produced. Temperature for growth is 28°C.

Source: isolated from soil.

DNA G+C content (mol%): 68.0 (T_m).

Type strain: Z6, CGMCC 4.3506, JCM 14205, NRRL B-24612.

Sequence accession no. (16S rRNA gene): AY114175.

4. **Lechevalieria xinjiangensis** Wang, Zhang, Tang, Mao, Wei, Huang, Liu, Shi and Goodfellow 2007, 2821^{VP}

xin.ji.ang.en'sis. N.L. fem. adj. *xinjiangensis* referring to Xinjiang, north-western China, the source of the isolate.

Substrate mycelium is yellow to orange in color. Moderate amounts of white to yellow aerial mycelium produced on Bennett's or Gauze No. 1 agar, whereas brownish aerial mycelium is formed on other media. Substrate mycelium fragments into coccoid to coccoidal–rod-shaped elements. No soluble pigments are produced. Temperature for growth is 28°C.

Source: isolated from soil.

DNA G+C content (mol%): 68.6 (T_m).

Type strain: R24, CGMCC 4.3525, DSM 45081, NRRL B-24613.

Sequence accession no. (16S rRNA gene): DQ898283.

Genus XI. **Lentzea** Yassin, Rainey, Brzezinka, Jahnke, Weissbrodt, Budzikiewicz, Stackebrandt and Schaal 1995, 362^{VP} emend. Labeda, Hatano, Kroppenstedt and Tamura 2001, 1049

David P. Labeda

Lent'ze.a. N.L. fem. n. *Lentzea* named after Friedrich A. Lentze, a German microbiologist who devoted a considerable part of his life to studying pathogenic actinomycetes.

Branched vegetative mycelia (diameter approx. 0.5 to 0.7 µm); aerial mycelium is produced and fragments into rod-shaped elements. Gram-stain-positive. Resistant to lysozyme. Catalase-positive. Aerobic. **The cell wall contains *meso*-diaminopimelic acid as the diamino acid. The characteristic whole-cell sugars are galactose, mannose, and ribose. The phospholipid pattern consists of significant amounts of phosphatidylethanolamine along with diphosphatidylglycerol, phophatidylglycerol, and phosphatidylinositol. The principal menaquinone is MK-9(H_4).** The fatty acid profile consists of straight-chain saturated, unsaturated, and branched-chain saturated fatty acids of the iso and anteiso types, in addition to tuberculostearic acid. **Phylogenetically, the genus *Lentzea* represents a line of descent adjacent to the genus *Actinosynnema* and close**

to the genera *Lechevalieria*, *Saccharothrix*, and *Umezawaea*. The 16S rRNA gene sequence contains genus-specific diagnostic nucleotide signature patterns of TCAA (617–620) and GCC (843–845).

DNA G+C content (mol%): 68.6–79.6 (HPLC, T_m)

Type species: **Lentzea albidocapillata** Yassin, Rainey, Brzezinka, Jahnke, Weissbrodt, Budzikiewicz, Stackebrandt and Schaal 1995, 362[VP].

Further descriptive information

The genus *Lentzea* was proposed by Yassin et al. (1995) for a single strain, which was isolated from a tissue sample taken from an abdominal mass of a patient with peritoneal carcinomatosis, on the basis of 16S rRNA gene phylogeny, and chemotaxonomic properties that distinguished it from the closely related genus *Saccharothrix*. Lee et al. (2000b) subsequently proposed, based on their chemotaxonomic and phylogenetic studies, that the genus *Lentzea* should be considered a later synonym of *Saccharothrix*. A subsequent study by Labeda et al. (2001), however, demonstrated that *Lentzea* was indeed a valid genus and could be differentiated from the genera *Saccharothrix* and *Lechevalieria* on the basis of phylogenetic position (see Figure 290 in the treatment of the genus *Lechevalieria*), diagnostic signature nucleotides TCAA (617–620) and GCC (843–845) in 16S rRNA gene sequences, and the presence of distinct chemotaxonomic characteristics. *Lentzea* species lack phosphatidylethanolamine containing 2-hydroxy-fatty acids in their polar lipid profiles, differentiating them from members of the genus *Saccharothrix*, and their whole-cell sugar pattern, consisting of galactose, mannose, and small quantities of ribose, differentiates them from both *Lechevalieria* and *Saccharothrix*.

Enrichment and isolation procedures

The first human clinical isolate of *Lentzea albidocapillata* was from a tissue sample; it was isolated from a streak culture on a Columbia blood agar plate. Subsequently, described species have generally been isolated from soil samples from China, Korea, and the United States, indicating that they most likely have a global distribution. Most recently, *Lentzea kentuckyensis* was isolated from an equine placenta, although this species is probably not pathogenic. Media typically used to isolate actinomycetes from environmental samples (i.e. soil and water samples), such as tap water agar or starch-casein agar (Küster and Williams, 1964) supplemented with antifungal antibiotics (cycloheximide, 50 µg/ml), can be used to obtain novel isolates of the genus *Lentzea*.

Maintenance procedures

Working cultures of *Lentzea* can be maintained as refrigerated (4°C) agar slants on an appropriate medium such as yeast extract-malt extract medium (Shirling and Gottlieb, 1966) or NZamine medium (DSMZ medium 554; DSMZ, 2001). The slants are generally incubated for 5 to 7 d at 28–30°C prior to refrigeration and then should be subcultured at monthly intervals. Longer term preservation of strains is best accomplished as frozen stocks in 20% (v/v) aqueous glycerol at −80°C (mechanical freezer) to −172°C (liquid nitrogen vapor phase) or by using traditional lyophilization techniques. Strains have also been successfully stored for shorter periods as quick-frozen stationary-phase broth cultures or mycelial suspensions in 20% aqueous glycerol at −20 to −72°C.

Procedures for testing special characters

Strains are routinely cultivated on NZamine medium (DSMZ medium 554) at 28°C. Morphological observations are made on the media of Shirling and Gottlieb (1966) and NZamine medium.

Chemotaxonomic analysis of strains for cell-wall diamino acid isomer and diagnostic whole-cell sugar content are determined by the methods of Staneck and Roberts (1974) and Saddler et al. (1991). Determination of cellular fatty acids, menaquinones, and polar lipids are performed using methods described previously by Grund and Kroppenstedt (1989), Minnikin et al. (1984), and Sasser (1990).

Physiological tests, including production of acid from carbohydrates, utilization of organic acids, and hydrolysis and decomposition of adenine, casein, esculin, guanine, hippurate, hypoxanthine, tyrosine, urea, and xanthine, are typically evaluated using the media of Gordon et al. (1974). Phosphatase activity is evaluated by using the method of Kurup and Schmitt (1973) substituting NZamine agar as the basal medium. The temperature range for growth is determined on slants of DSMZ medium 554 and salt tolerance is determined on slants of the same medium supplemented with 4% and 5% NaCl.

Differentiation of the genus *Lentzea* from other genera

Morphologically, *Lentzea* species are quite similar to species within the genera *Lechevalieria* and *Saccharothrix* producing aerial mycelium which may exhibit a "zig-zag" morphology (Figure 310a) and fragments into coccoidal-rod-shaped elements (Figure 310b). The chemotaxonomic characteristics of members of this genus (see Table 248 in the treatment of the genus *Lechevalieria*) include a polar lipid profile which lacks 2-hydroxy-fatty acid containing phosphatidylethanolamine and contains phosphatidylinositol and phosphatidylglycerol, differentiating them from *Saccharothrix* species. The sugars observed in their whole-cell sugar profiles include galactose, mannose, and small quantities of ribose, but no rhamnose, which differentiates them from both *Lechevalieria* and *Saccharothrix* species. The genus *Lentzea* can also be differentiated from the genera *Lechevalieria* and *Saccharothrix* phylogenetically and based on the genus-specific nucleotide signature patterns TCAA (617–620) and GCC (843–845) in their 16S rRNA gene sequence (see Figure 290 and Figure 309).

Differentiation of the species of the genus *Lentzea*

The physiological characteristics of *Lentzea* species are summarized in Table 251 and can be used to differentiate between species, although a common set of physiological tests has not been used in the description of all species. Gross colonial morphology provides some additional information of the species based on color of the substrate mycelium, production and color of aerial mycelium, and production and color of soluble pigments when grown on agar media. The distinct fatty acid profiles of each *Lentzea* species, as shown in Table 252, can also be used to differentiate between species.

TABLE 251. Physiological properties of *Lentzea* species[a]

Character	*L. albidocapillata*	*L. albida*	*L. californiensis*	*L. flaviverrucosa*	*L. kentuckyensis*	*L. violacea*	*L. waywayandensis*
Hydrolysis of:							
Adenine	–	–	–	–	–	–	–
Allantoin	nd	–	–	–	–	nd	–
Casein	+	+	+	–	+	+	+
Esculin	+	+	+	–	+	+	+
Gelatin	+	nd	nd	+	+	–	+
Hypoxanthine	+	+	+	+	+	+	+
Starch	+	+	+	+	+	+	+
Tyrosine	+	+	+	+	+	+	+
Urea	v	v	+	–	+	+	+
Xanthine	–	–	–	–	+	–	–
Production of:							
Nitrate reductase	–	–	+	+	–	–	+
Phosphatase	+	nd	nd	+	–	nd	+
Acid from:							
Adonitol	+	+	–	–	–	–	+
Arabinose	+	+	+	nd	+	+	+
Cellobiose	+	+	+	+	+	–	+
Dextrin	nd	+	+	+	nd	nd	+
Dulcitol	nd	–	–	nd	–	nd	–
Erythritol	nd	–	–	+	–	nd	+
Fructose	+	+	+	nd	+	+	+
Galactose	+	+	+	nd	+	+	+
Glucose	+	+	+	nd	+	+	+
Glycerol	nd	+	+	+	+	nd	+
Inositol	+	+	+	+	+	–	+
Lactose	+	w	+	–	+	+	+
Maltose	+	+	+	+	+	–	+
Mannitol	+	+	+	+	+	–	+
Mannose	nd	+	+	+	+	–	+
Melibiose	nd	+	+	+	+	–	+
Methyl β-xyloside	nd	–	–	nd	v	nd	–
Raffinose	v	–	+	+	+	+	+
Rhamnose	+	+	+	–	+	–	+
Salicin				+	+	–	+
Sucrose	+	+	+	+	+	–	+
Trehalose	+	+	+	–	+	–	+
Xylose	+	+	+	–	+	–	+
Assimilation of:							
Acetate	–	+	+	+	–	+	+
Benzoate	–	–	–	–	–	–	–
Citrate	–	+	+	–	v	–	+
Lactate	–	–	–	–	v	+	+
Malate	+	+	+	+	+	–	+
Mucate	nd	–	–	nd	–	nd	–
Oxalate	nd	–	–	–	+	+	+
Propionate	nd	+	+	+	–	+	+
Succinate	nd	+	+	–	+	+	+
Tartrate	nd	–	–	–	–	–	–
Growth in the presence of:							
4% (w/v) NaCl	+	+	+	–	+	+	+
5% (w/v) NaCl	nd	+	+	–	+	nd	+
Growth at:							
10°C	+	–	+	nd	+	+	+
37°C	+	+	+	+	+	+	w
42°C	–	+	–	+	–	–	–
45°C	–	+	–	–	–	–	–

[a]Symbols: nd, not determined; v, variable; w, weakly positive.

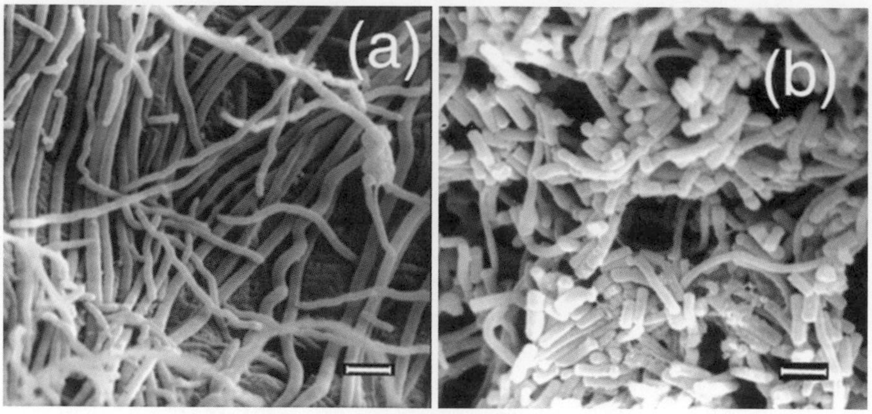

FIGURE 310. Scanning electron micrographs of *Lentzea albidocapillata* IMMIB D-958^{T} grown for 14 d on yeast extract-malt extract agar illustrating "zig-zag" aerial mycelium (a) and coccoidal-rod-shaped fragmentation of the aerial mycelium (b). Bars = 2 μm. (Micrographs courtesy of A.F. Yassin, University of Bonn.)

TABLE 252. Fatty acid profiles of *Lentzea* species

Fatty acid (%)	*L. albidocapillata*	*L. albida*	*L. californiensis*	*L. flaviverrucosa*	*L. kentuckyensis*	*L. violacea*	*L. waywayandensis*
$C_{13:0}$ iso	–	0.46	–	–	0.19	–	–
$C_{13:0}$ anteiso	–	–	–	–	0.13	–	–
$C_{13:0}$	–	0.44	–	–	–	–	–
$C_{14:0}$ iso	8.00	3.28	10.28	5.87	8.88	8.90	12.70
$C_{14:1}$ *cis*9	–	–	–	–	–	0.21	–
$C_{14:0}$	0.56	1.91	1.85	1.55	0.52	1.27	–
$C_{15:1}$ iso	–	–	–	–	0.23	–	–
$C_{15:0}$ iso	3.58	14.45	9.90	5.63	10.07	6.15	6.55
$C_{15:0}$ anteiso	2.57	11.33	16.21	6.25	8.20	3.79	4.26
$C_{15:1}$ *cis*9	1.24	1.38	–	0.84	0.79	1.47	0.93
$C_{15:0}$	0.75	7.00	2.21	2.53	1.01	1.00	0.58
$C_{16:1}$ iso	9.27	0.62	–	1.85	3.83	4.05	10.07
$C_{16:0}$ iso	55.63	19.29	23.47	31.97	47.40	45.03	45.60
$C_{16:1}$ *cis*9	9.66	8.42	8.70	16.26	4.64	15.52	8.87
$C_{15:0}$ anteiso 2-OH	–	1.12	–	–	–	–	–
$C_{16:0}$	2.21	9.91	17.08	12.01	1.9	4.01	2.07
$C_{16:0}$ 10-methyl	3.04	–	4.14	5.48	–	1.40	4.10
$C_{17:1}$ iso	–	1.12	–	–	1.39	–	–
$C_{17:1}$ anteiso	–	–	–	–	0.68	–	–
$C_{17:0}$ iso	–	2.22	–	0.77	1.21	0.58	–
$C_{17:0}$ anteiso	0.97	5.64	4.01	2.32	4.61	1.18	1.42
$C_{17:1}$ *cis*9	1.57	4.14	–	2.09	1.95	1.87	1.48
$C_{17:1}$ *cis*11	–	–	–	–	1.03	2.49	–
$C_{16:0}$ iso 2-OH	0.95	0.63	–	1.08	–	–	1.35
$C_{17:0}$	–	4.75	2.16	1.62	0.65	0.39	–
$C_{17:0}$ 10-methyl	–	–	–	–	0.18	–	–
$C_{18:0}$ iso	–	–	–	–	0.34	–	–
$C_{18:1}$ *cis*9	–	0.75	–	1.03	0.17	0.69	–
$C_{18:0}$	–	1.05	–	0.84	–	–	–

List of species of the genus *Lentzea*

1. **Lentzea albidocapillata** Yassin, Rainey, Brzezinka, Jahnke, Weissbrodt, Budzikiewicz, Stackebrandt and Schaal 1995, 362[VP]

al.bi.do.ca.pil.la'ta. L. adj. *albidus* white; L. adj. *capillatus* hairy; N.L. fem. adj. *albidocapillata* white haired, referring to the abundant whitish aerial hyphae.

Substrate mycelium is yellow to yellowish-brown. Aerial mycelium is white to whitish-yellow. Soluble pigments are not produced. Temperature for growth is 30°C.

Source: isolated originally from the peritoneal cavity of a cancer patient.

DNA G+C content (mol%): 68.6 (HPLC).

Type strain: DSM 44073, AS 4.1519, CIP 107111, NBRC 15855, JCM 9732, NBRC 100372.

Sequence accession no. (16S rRNA gene): X84321.

Further comment: Lee et al. (2000b) suggested that *Lentzea albidocapillata* be transferred to the genus *Saccharothrix* Labeda et al. 1989, but phylogenetic and chemotaxonomic data do not support this proposal according to Labeda et al. (2001). *Lentzea* is considered to be the current correct name of a genus in which the type strain is *Lentzea albidocapillata*.

2. **Lentzea albida** Labeda, Hatano, Kroppenstedt and Tamura 2001, 1049[VP]

al.bi'da. L. fem. adj. *albida* whitish, referring to the color of the aerial mycelium.

Substrate mycelium is yellowish-orange in color on most media. Copious white aerial mycelium is produced. Soluble pigments are not produced. Temperature for growth is 28°C.

Source: isolated from soil.

DNA G+C content (mol%): not determined.

Type strain: NBRC 16102, AS 4.1727, DSM 44437, JCM 9734, KCTC 19911, NRRL B-24073.

Sequence accession no. (16S rRNA gene): AB006167.

3. **Lentzea californiensis** Labeda, Hatano, Kroppenstedt and Tamura 2001, 1049[VP]

cal.i.forn.i.en'sis. N.L. fem. adj. *californiensis* pertaining to California, referring to the source of this isolate, soil from California.

Substrate mycelium is yellow to orange-brown. White aerial mycelium is produced. An orange soluble pigment may be produced on Czapek's agar. Temperature for growth is 28°C.

Source: isolated from soil.

DNA G+C content (mol%): not determined.

Type strain: NRRL B-16137, DSM 43393, IMRU 550, JCM 11305, KCTC 19912.

Sequence accession no. (16S rRNA gene): AF174435.

4. **Lentzea flaviverrucosa** Xie, Wang, Huang, Wu, Ba and Liu 2002, 1818[VP]

fla.vi.ver.ru.co'sa. L. adj. *flavus* yellow; L. adj. *verrucosus* rough, rugged, verrucose; N.L. fem. adj. *flaviverrucosa* yellowish and verrucose, referring to the yellowish, verrucose colony morphology observed on the agar surface.

Substrate mycelium is pale yellow to yellowish-brown on most media. Sparse white to yellowish-white aerial mycelium is produced on some media (e.g. oatmeal agar). Neither melanin nor soluble pigments are produced. Temperature for growth is 28°C.

Source: isolated from soil.

DNA G+C content (mol%): 64.1 (T_m).

Type strain: AS 4.0578, CIP 107743, DSM 44664, JCM 11373, NBRC 100042.

Sequence accession no. (16S rRNA gene): AF183957.

5. **Lentzea kentuckyensis** Labeda, Donahue, Sells and Kroppenstedt 2007, 1782[VP]

ken.tuck.y.en'sis. N.L. fem. adj. *kentuckyensis* from Kentucky, named after the place of origin of the type strain, the state of Kentucky, USA.

Yellow to strong yellow substrate mycelium is produced on most media. Aerial mycelium ranging in color from white to yellowish-white is produced on most media. A faint brown soluble pigment is produced on some media. Temperature for growth is 28°C.

Source: isolated from an equine placenta.

DNA G+C content (mol%): not determined.

Type strain: NRRL B-24416, LDDC 2876-05, DSM 44909.

Sequence accession no. (16S rRNA gene): DQ291145.

6. **Lentzea violacea** (Lee, Kim, Roe, Kim, Kang and Hah 2000b) Labeda, Hatano, Kroppenstedt and Tamura 2001, 1049[VP] (*Saccharothrix violacea* Lee, Kim, Roe, Kim, Kang and Hah 2000b, 1320)

vi.o.la.ce'a. L. fem. adj. *violacea* violet-colored, violet.

Violet substrate mycelium and a white aerial mycelium produced on most media. Reddish-brown soluble pigment is produced. Melanin pigment is not produced. Temperature for growth is 28°C.

Source: isolated from soil.

DNA G+C content (mol%): 79.6 (HPLC).

Type strain: IMSNU 50388, DSM 44796, JCM 10975, KCTC 9948.

Sequence accession no. (16S rRNA gene): AJ242633.

7. **Lentzea waywayandensis** (Labeda and Lyons 1989) Labeda, Hatano, Kroppenstedt and Tamura 2001, 1049[VP] (*Saccharothrix waywayandensis* Labeda and Lyons 1989, 357)

way.way.an.den'sis. N.L. fem. adj. *waywayandensis* of or belonging to Lake Waywayanda, NJ, of the soil samples from which the organism was first isolated.

Substrate mycelium is pale yellow to dark yellow. Aerial mycelium is white when produced, particularly on inorganic salts-starch agar (ISP medium 4). Soluble pigments are not produced. Temperature for growth is 28°C.

Source: isolated from soil.

DNA G+C content (mol%): 71 (T_m).

Type strain: NRRL B-16159, AS 4.1646, ATCC 51594, DSM 44232, NBRC 14970, JCM 9114, NCIMB 13164, VKM Ac-1970.

Sequence accession no. (16S rRNA gene): AF114813.

Genus XII. **Prauserella** Kim and Goodfellow 1999, 510[VP] emend. Li, Xu, Tang, Xu, Kroppenstedt, Stackebrandt and Jiang 2003c, 1547

Seung Bum Kim and Michael Goodfellow

Prau.se.rel′la. N.L. fem. dim. n. *Prauserella* named after Helmut Prauser, a German microbiologist who made many contributions to actinomycete systematics.

Aerobic, Gram-stain-positive, non-acid–alcohol-fast, nonmotile actinobacteria which form an extensively branched substrate mycelium (0.6–0.8 µm in diameter) that **fragments into irregular rod-shaped elements within 24–48 h on rich media**. Aerial hyphae, when formed, may differentiate into branched short or, at maturity, long chains which have a straight to flexuous arrangement. Spores are nonmotile. **Most strains grow optimally in the presence of 10% or between 10 and 15% NaCl at 28°C and pH 7.0.** Optimal growth occurs between pH 6.8 and 7.2. The temperature range for growth is 10–45°C, with optimal growth between 28 and 34°C. **Contains *meso*-diaminopimelic acid as the major diamino acid, an acetylated peptidoglycan, major amounts of arabinose and galactose, MK-9(H_4) as the predominant menaquinone, either phosphatidylcholine or phosphatidylethanolamine as the diagnostic phospholipid, and major amounts of branched chains and saturated fatty acids, but not mycolic acids.**

The genus *Prauserella*, as determined by 16S rRNA gene sequence analysis, is classified in the family *Pseudonocardiaceae*.

DNA G+C content (mol%): 65.8–70.1.

Type species: **Prauserella rugosa** (Lechevalier, Prauser, Labeda and Ruan 1986) Kim and Goodfellow 1999, 510[VP].

Further descriptive information

Phylogeny. The genus *Prauserella* forms a distinct branch within the evolutionary radiation occupied by the family *Pseudonocardiaceae* and is most closely related to the genus *Saccharomonospora* (Labeda et al., 2010a). The nine species assigned to the genus (Figure 311) share 16S rRNA gene sequence similarities which range from 95.8% between *Prauserella halophila* and *Prauserella muralis* to 100% between *Prauserella flava* and *Prauserella salsuginis*. The level of DNA–DNA relatedness between the latter two species is 56.9% (Li et al., 2009b).

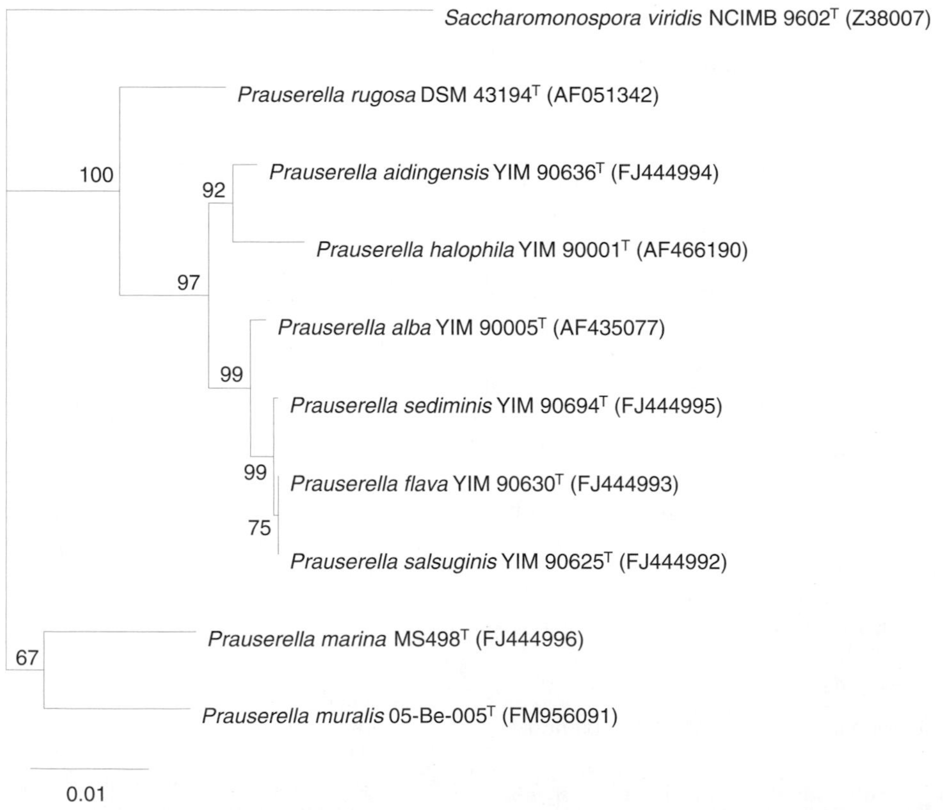

FIGURE 311. Neighbor-joining tree (Saitou and Nei, 1987) based on nearly complete 16S rRNA gene sequences showing the relationships between *Prauserella* species and representatives of some other genera classified in the family *Pseudonocardiaceae*. Numbers at the nodes indicate levels of bootstrap support (Felsenstein, 1985) based on analysis of 1000 datasets. Bar = 0.01 substitutions per nucleotide position.

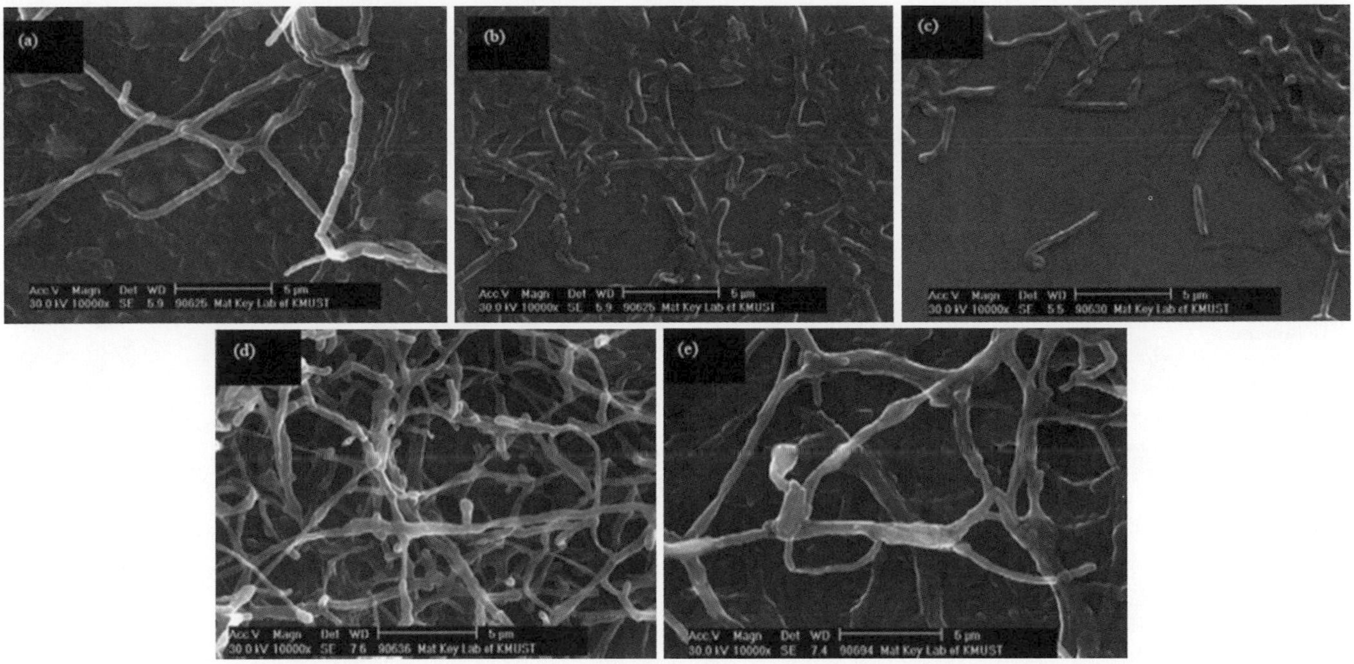

FIGURE 312. Scanning electron micrographs showing the morphology of *Prauserella aidingensis* YIM 90636^{T}, *Prauserella flava* YIM 90630^{T}, *Prauserella salsuginis* YIM 90625^{T}, and *Prauserella sediminis* YIM 90694^{T} grown on yeast extract-malt extract agar (ISP medium 2; Shirling and Gottlieb, 1966) at 28°C for 23 d (bars = 5 µm). (a) Spore chains of strain YIM 90625^{T}; (b) substrate mycelium of strain YIM 90625^{T}; (c) substrate mycelium of strain YIM 90630^{T}; (d) spore chains of strain YIM 90636^{T}; and (e) spore chains of strain YIM 90694^{T}. (Reproduced with permission from Li et al., 2009b. Int. J. Syst. Evol. Microbiol. *59*: 2928–2932.)

Cell morphology. *Prauserella* strains form a substrate mycelium which usually carries aerial hyphae, although *Prauserella flava* and *Prauserella rugosa* do not form an aerial mycelium (Lechevalier et al., 1986; Li et al., 2009b). Substrate mycelia fragment into rod-shaped elements and aerial hyphae may differentiate into branched short or, at maturity, long chains, which are straight to flexuous (Figure 312).

Chemotaxonomy. *Prauserella* strains contain *meso*-diaminopimelic acid as the diagnostic wall diamino acid and arabinose and galactose as major whole-cell sugars (Lechevalier et al., 1986; Li et al., 2009b; Mertz and Yao, 1993), i.e. they have wall chemotype IV *sensu* Lechevalier and Lechevalier (1970). Other major sugars may be detected in whole-cell hydrolysates, notably ribose (Table 253). In addition, all *Prauserella* strains have tetrahydrogenated menaquinones, MK-9(H_4), as the predominant isoprenologue, although *Prauserella rugosa* also contains a large amount of MK-9(H_2); small amounts of other menaquinones (<7%) may be detected (Lechevalier et al., 1986; Li et al., 2009b; Schäfer et al., 2010). *Prauserella rugosa* has *N*-acetylated muramic acid (Henssen et al., 1987).

Prauserellae can be assigned to two groups based on the discontinuous distribution of diagnostic polar lipids, although they all have patterns which include diphosphatidylglycerol, phosphatidylglycerol, phosphatidylinositol, phosphatidylethanolamine, and, in some instances, unknown phospholipids (Li et al., 2009b; Schäfer et al., 2010; Wang et al., 2010). *Prauserella alba*, *Prauserella halophila*, *Prauserella muralis*, and *Prauserella rugosa* contain phosphatidylethanolamine as the diagnostic polar lipid, whereas phosphatidylcholine is the diagnostic marker in *Prauserella aidingensis*, *Prauserella flava*, *Prauserella marina*, *Prauserella sediminis*, and *Prauserella salsuginis*; hence, these two groups have phospholipid patterns II and III, respectively (Lechevalier et al., 1977b, 1981). Hydroxyphosphatidylethanolamine and hydroxymethylphosphatidylethanolamine have been detected in *Prauserella rugosa* (Lechevalier et al., 1986; Yassin et al., 1993) and phosphatidylserine has been found in *Prauserella muralis* (Schäfer et al., 2010).

Prauserella strains contain complex mixtures of branched-chain and saturated fatty acids (Henssen et al., 1987; Li et al., 2003c; Mertz and Yao, 1993), but lack mycolic acids (Henssen et al., 1987; Lechevalier et al., 1986). The principal fatty acid, 14-methylpentadecanoic acid ($C_{16:0}$ iso), accounts for between 20 and 42% of total fatty acid composition; other major components (<10%) may include $C_{16:0}$, $C_{17:1}$ anteiso, and $C_{16:1}$ ω9*c* (Li et al., 2009b; Schäfer et al., 2010; Wang et al., 2010).

Nutritional and growth conditions. Prauserellae tend to grow well on standard media used to cultivate filamentous actinobacteria, such as Czapek's agar, glycerol-asparagine agar (ISP medium 5), inorganic salts-starch agar (ISP medium 4), nutrient agar, oatmeal agar (ISP medium 3) and yeast extract-malt extract agar (ISP medium 2) supplemented with 10% (w/v) NaCl and incubated at 28–37°C; the composition of the ISP media is given by Shirling and Gottlieb (1966). Most strains are either halophilic or halotolerant. However, *Prauserella marina* grows optimally on yeast extract-malt extract agar without NaCl at 28–37°C and pH 7.0 (Wang et al., 2010).

Metabolism and ecology. Little is known about the biological properties of *Prauserella* strains or of their distribution and activities in natural habitats. They are aerobic, have an oxida-

TABLE 253. Differential characteristics of *Prauserella* species[a,b]

Characteristic	*P. rugosa*	*P. aidingensis*	*P. alba*	*P. flava*	*P. halophila*	*P. marina*	*P. muralis*	*P. salsuginis*	*P. sediminis*
Aerial mycelium	–	+	+	–	+	+	+	+	+
Growth in NaCl:									
Range (%)	0–20	5–15	0–25	5–15	5–25	0–10	nd	5–15	5–20
Optimum (%)	5–10	8–10	10–15	8–10	10–15	0–5	nd	8–10	10
Degradation of:									
Gelatin	–	+	+	+	+	+	nd	+	+
Urea	+	–	–	–	+	–	nd	–	–
Carbon source utilization:									
L-Arabinose	+	–	+	–	–	+	+	+	–
Cellobiose	+	–	+	–	+	–	+	+	–
D-Fructose	+	+	+	+	+	–	+	+	–
D-Galactose	+	+	+	–	–	+	+	+	–
myo-Inositol	–	–	+	+	+	–	–	+	–
Lactose	+	+	nd	+	nd	–	nd	+	+
Maltose	+	–	+	–	–	+	+	–	–
D-Mannitol	+	–	+	–	+	+	+	+	–
D-Mannose	+	–	nd	–	nd	+	+	+	–
Raffinose	+	–	nd	–	nd	–	nd	–	–
L-Rhamnose	+	+	+	–	+	+	+	+	+
D-Ribose	+	+	+	+	+	+	+	+	–
Trehalose	+	+	–	–	+	+	+	+	–
D-Xylitol	+	–	+	–	+	–	nd	+	–
D-Xylose	+	+	+	+	+	+	+	+	–
Nitrogen source utilization:									
L-Arginine	–	–	+	–	+	+	nd	–	–
L-Glycine	+	–	+	–	+	nd	nd	–	–
L-Hydroxyproline	–	+	+	+	+	+	nd	+	+
L-Lysine	–	–	+	+	+	–	nd	+	+
L-Serine	–	–	+	–	+	+	nd	–	–
L-Threonine	–	+	+	+	+	+	nd	+	+
Cell-wall sugars[c]	nd	Ribose	Ribose	Ribose	Ribose	nd	Glucose	Ribose	Ribose
Phospholipids[d]	PE	PC	PE	PC	PC	PE, PC	PE, PS	PC	PC
DNA G+C content (mol%)	67.0–68.9	70.1	66.7	69.9	65.8	66.1	nd	69.1	69.1

[a]Symbols: +, positive; –, negative; nd, not determined.

[b]Data from Kim and Goodfellow (1999), Li et al. (2003c, 2009b), Schäfer et al. (2010), and Wang et al. (2010).

[c]In addition, all species contain arabinose and galactose.

[d]PC, Phosphatidylcholine; PE, phosphatidylethanolamine; PS, phosphatidylserine.

tive metabolism, grow on a broad range of sole carbon and sole nitrogen compounds, and tend to grow optimally in the presence of 8–15% (w/v) NaCl (Li et al., 2003c, 2009b; Schäfer et al., 2010; Wang et al., 2010). The type strain of *Prauserella rugosa* hydrolyzes arbutin, allantoin, and esculin, degrades several organic compounds, and is active against a broad range of 4-methylumbelliferone and 7-amino-4-methylcoumarin-conjugated fluorogenic compounds (De Boer et al., 1990); it also produces alkaline hydrolases when grown on *n*-dodecane (Smits et al., 1999). Most *Prauserella* strains have been isolated from saline habitats (Li et al., 2003c, 2009b; Wang et al., 2010), although *Prauserella muralis* was obtained from a wall colonized by fungi (Schäfer et al., 2010) and *Prauserella rugosa* was from the rumen of a cow (di Marco and Spalla, 1957).

Isolation procedures

Prauserella species have been isolated by plating serial dilutions of saline samples onto starch-casein agar supplemented with 20% (w/v) NaCl and incubating for about 4 weeks at 28°C (Li et al., 2003c) and by plating onto cellulose-casein multi-salt medium and incubating for 3 weeks at 37°C (Li et al., 2009b; Tang et al., 2008). *Prauserella marina* was obtained from a marine sediment sample following incubation at 22°C for 4 weeks on MOPS-proline agar (Wang et al., 2010) and *Prauserella muralis* was isolated by shaking a 1 g sample of plaster colonized by molds in 10 ml NaCl (0.9%, w/v) containing Tween 80 (0.01%, v/v) then plating aliquots of the suspension over starch-mineral agar plates (Gauze, 1985), which were then incubated for 2 weeks at 28°C (Schäfer et al., 2010).

Maintenance procedures

Working cultures of *Prauserella* can be maintained at 4°C on standard media such as modified Bennett's (Jones, 1949), organic medium M79 (Schäfer et al., 2010), and yeast extract-malt extract (Shirling and Gottlieb, 1966) agars supplemented with 10% (w/v) NaCl for halophilic and halotolerant strains. Moderate to long-term preservation can be achieved by storing mycelial spore suspensions in aqueous glycerol (10 or 20%, v/v) at −20 or −80°C (mechanical freezers), at −172°C (liquid nitrogen vapor phase), or by standard lyophilization techniques.

Procedures for testing special characters

Reliable and well established procedures are available for analysis of the isomers of diaminopimelic acid and whole-cell sugars (Hasegawa et al., 1983a; Staneck and Roberts, 1974), fatty acids (Sasser, 1990), and menaquinones and polar lipids (Minnikin et al., 1984). Physiological features, such as the production of acids from sugars, growth on sole carbon and nitrogen compounds, and degradation of complex organic substrates, can be determined using standard methods (De Boer et al., 1990; Shirling and Gottlieb, 1966; Williams et al., 1983).

Differentiation of the genus *Prauserella* from other genera

Prauserella strains can be distinguished from other actinobacteria classified in the family *Pseudonocardiaceae* by using a combination of chemotaxonomic and morphological features and by comparative 16S rRNA gene sequence analyses (Kim and Goodfellow, 1999; Labeda et al., 2010a), as shown in the treatment of the family *Pseudonocardiaceae* in this volume. The genus can be separated from other filamentous actinobacteria using genus-specific 16S rRNA gene restriction fragment patterns (Cook and Meyers, 2003).

Taxonomic comments

The monospecific genus *Prauserella*, with *Prauserella rugosa* as the type species, was proposed by Kim and Goodfellow (1999) for an organism which had been isolated from the rumen of a cow and designated "*Nocardia rugosa*" (di Marco and Spalla, 1957). The taxon was well described, although a type species was not designated until the species was transferred to the genus *Amycolatopsis* as *Amycolatopsis rugosa*, a move based on chemotaxonomic and morphological criteria (Lechevalier et al., 1986). This transfer proved to be a temporary one as the organism was shown to differ from *bona fide* members of the genus *Amycolatopsis* based on fatty acid (Henssen et al., 1987; Mertz and Yao, 1993), DNA–DNA relatedness (Labeda, 1995), and extensive phenotypic (De Boer et al., 1990) data. In addition, the organism did not give the characteristic amplification product with *Amycolatopsis* genus-specific primers (Tan et al., 2006b) and formed a distinct single-membered cluster in a numerical taxonomic study which included strains that were subsequently classified in the genus *Amycolatopsis* (Goodfellow, 1971). The genus *Prauserella* currently contains nine species, most of which were circumscribed in polyphasic studies which included DNA–DNA relatedness data (Li et al., 2003c, 2009b). The original description of the genus was emended by Li et al. (2003c).

Differentiation of the species of the genus *Prauserella*

Prauserella species can be distinguished using a combination of phenotypic properties (Table 253) and by DNA–DNA relatedness data (Li et al., 2003c, 2009b). Qualitative and quantitative differences have been found in the fatty acid profiles of some species (Li et al., 2009b; Schäfer et al., 2010; Wang et al., 2010).

List of species of the genus *Prauserella*

1. **Prauserella rugosa** (Lechevalier, Prauser, Labeda and Ruan 1986) Kim and Goodfellow 1999, 510[VP] (Basonym: *Amycolatopsis rugosa* Lechevalier, Prauser, Labeda and Ruan 1986, 35 *ex* di Marco and Spalla 1957; "*Nocardia rugosa*" di Marco and Spalla 1957, 28.)

ru.go'sa. L. fem. adj. *rugosa* wrinkled.

Forms an extensively branched substrate mycelium that fragments into irregular rods. Aerial hyphae are not formed. Cream colored, irregular, flat, and veined colonies (10 × 6 mm) are formed on Czapek Dox-yeast extract agar. Good growth occurs on Bennett's, modified Bennett's, and nutrient agars. Produces cream to yellowish glistening colonies which may be wrinkled or folded. Brownish soluble pigments are formed on some media. The temperature range for growth is 10–45°C, with an optimum at 34°C. Grows optimally between pH 6.8 and 7.2.

Allantoin and esculin are hydrolyzed. Produces phosphatase, but does not reduce nitrate or produce hydrogen sulfide. Acetate, benzoate, lactate, malate, propionate, pyruvate, and succinate are decarboxylated, but not citrate, mucate, oxalate, or tartrate. 2-Deoxythymidine-5-*para*-nitrophenyl (*p*NP) phosphate, *p*NP-α-D-glucopyranoside, and *p*NP-β-D-glucopyranoside are hydrolyzed, but not L-alanine-D-nitroanilide (*p*NA), *bis*-*p*NP phosphate, *o*NP-β-D-galactopyranoside, *p*NP-glucopyranoside, *p*NP-β-D-glucuronide, L-glutamate-γ-3-carbonyl-*p*NP, *p*NP-phenylphosphonate, *p*NP-phosphorylcholine, L-proline-*p*NA, or *p*NP-β-D-xylopyranoside.

Arbutin, casein, hypoxanthine, L-tyrosine, tributyrin, Tweens 20, 40, 60 and 80, and xanthine are degraded, but not adenine, elastin, starch, or testosterone.

Acid is formed from adonitol, L-arabinose, erythritol, fructose, galactose, glucose, glycerol, mannitol, mannose, rhamnose, salicin, trehalose, and xylose, but not from D-arabinose, cellobiose, dextrin, inositol, lactose, maltose, melibiose, methyl α-D-glucoside, raffinose, sorbitol, sucrose, or methyl β-D-xyloside.

Adonitol, amygdalin, arbutin, dextrin, erythritol, glycerol, glycogen, salicin, and sucrose are used as sole carbon sources for energy and growth, but not arabitol, dulcitol, D- or L-fucose, gentiobiose, glucose, sorbitol, or turanose (at 1%, w/v). L-Alanine, androsterone, cholesterol, ergosterol, L-glycine, L-proline, protocatechuic acid, shikimic acid, butyrate, propionate, pyruvate, succinate, and L-tyrosine are also used as sole carbon sources, but not α-D-alanine, L-arginine, L-asparagine, catechol, L-citrulline, creatine, L-cysteine, ferulic acid, gluconate, glucuronate, histamine, L-hydroxyproline, phthallic acid, progesterone, quinic acid, L-serine, fumarate, tartrate, syringealdehyde, L-threonine, or vanillin (all at 0.1%, w/v). Does not use L-histidine, L-ornithine, or L-phenylalanine as sole nitrogen sources.

Grows in the presence of (μg/ml) cephaloridine (2), demeclocycline (2), lincomycin (10), neomycin (3), oleandomycin (2), penicillin G (10), streptomycin (16), and vancomycin (0.25), but is sensitive to cephaloridine (25), chloramphenicol (25), gentamicin (0.5), kanamycin (5), rifampin (2), tetracycline (5), and tobramycin (8). Similarly, growth occurs in the presence of (μg/ml) bismuth citrate (1), crystal violet (1), phenol (100), phenyl ethanol (0.1%, v/v), potassium tellurite (10), sodium azide (1), teepol (100), tetrazolium (100), thallous acetate (10), and sodium chloride (13%, w/v), but is sensitive to adenine (0.4%, w/v), crystal violet (10), phenyl ethanol (0.4%, v/v), potassium tellurite (100), sodium azide (100), and thallous acetate (100). Sensitive to lysozyme.

The type strain produces vitamin B_{12}.

Additional phenotypic properties are cited in the text and in Table 253.

Cellular fatty acids consist of major proportions of $C_{16:1}$ iso OH, $C_{16:0}$ iso, and $C_{17:1}$ ω6*c*; smaller proportions (<10%) of $C_{15:0}$ iso, $C_{15:1}$ ω6*c*, $C_{16:1}$ ω7*c* and/or $C_{15:0}$ iso 2-OH, $C_{15:0}$, $C_{16:0}$, $C_{16:0}$ 10-methyl, $C_{17:0}$ iso, $C_{17:1}$ ω8*c*, $C_{17:0}$, $C_{18:1}$ iso, and $C_{18:0}$ are found.

Source: isolated from the rumen of a cow.

DNA G+C content (mol%): 67.0–68.9 (T_m).

Type strain: ATCC 43014, CIP 106520, DSM 43194, NBRC 14506, IMRU 3760, JCM 9736, NCIMB 8926, NRRL B-2295, VKM Ac-1243.

Sequence accession no. (16S rRNA gene): AF051342.

2. **Prauserella aidingensis** Li, Tang, Chen, Wu, Zhi, Zhang and Li 2009b, 2926[VP]

ai.ding.en'sis. N.L. fem. adj. *aidingensis* of or belonging to Aiding Lake, where the type strain was isolated.

Forms a substrate mycelium that fragments and which carries a white aerial mycelium. Colonies are light gray-white on glycerol-asparagine agar (ISP medium 5), slightly gray-white on oatmeal agar (ISP medium 3), and brilliant yellow on potato and yeast extract-malt extract (ISP medium 2) agars. Grows at 15–45°C, pH 6.0–9.0, and 5–15% (w/v) NaCl on ISP 2 medium; optimal conditions are 28–37°C, pH 7, and 8–10% NaCl. Catalase-positive, but oxidase-negative. Coagulates milk, but is negative for H_2S production. L-Alanine, L-histidine, L-hydroxyproline, L-phenylalanine, L-proline, DL-tryptophan, L-tyrosine, L-xanthine, and L-valine are used as sole nitrogen sources, but not adenine or DL-methionine.

The DNA–DNA relatedness value between the type strain and that of *Prauserella sediminis* is 47.9%.

Additional phenotypic properties are cited in the text and in Table 253.

Cellular fatty acids contain major proportions of $C_{16:0}$ iso and $C_{16:1}$ ω9*c*; smaller proportions (<10%) of $C_{14:0}$, $C_{15:0}$, $C_{15:0}$ iso, $C_{15:1}$ B, $C_{16:0}$, $C_{16:0}$ iso 2-OH, $C_{16:1}$ iso H, $C_{17:0}$, $C_{17:0}$ iso 2-OH, $C_{17:0}$ anteiso, $C_{17:1}$ anteiso, and $C_{18:1}$ 2-OH, and traces of $C_{17:0}$ anteiso 2-OH are found.

Source: isolated from a salt lake.

DNA G+C content (mol%): 70.1 (HPLC).

Type strain: CCTCC AA 208053, DSM 45266, YIM 90636.

Sequence accession no. (16S rRNA gene): FJ444994.

3. **Prauserella alba** Li, Xu, Tang, Xu, Kroppenstedt, Stackebrandt and Jiang 2003c, 1548[VP]

al'ba. L. fem. adj. *alba* white, referring to the white aerial mycelium.

Forms a branched substrate mycelium which undergoes fragmentation. Substrate mycelia may be light orange-yellow (Czapek's and nutrient agars), gray-white (oatmeal agar; ISP medium 3), orange-yellow (yeast extract-malt extract agar; ISP medium 2), and ranges from yellow-white to light yellow on glucose-asparagine agar (ISP medium 5), inorganic salts-starch agar (ISP medium 4), and nutrient agar; all of these media were supplemented with 10% (w/v) NaCl. Diffusible pigments are not formed. White aerial hyphae differentiate into branched short or, at maturity, long, straight to flexuous spore chains on Czapek's agar. Optimal growth occurs on Czapek's agar supplemented with 10% (w/v) NaCl at 28°C and pH 7.0.

Does not hydrolyze esculin. *p*NP-α-D-Glucopyranoside is hydrolyzed, but not L-alanine-*p*NA, *bis-p*NP phosphate, 2-deoxythymidine-5-*p*NP phosphate, *o*NP-β-D-galactopyranoside, *p*NP-glucopyranoside, *p*NP-β-D-glucuronide, L-glutamate-γ-carbonyl-*p*NP, *p*NP-phenylphosphonate, *p*NP-phosphorylcholine, L-proline-*p*NA, or *p*NP-β-D-xylopyranoside. Utilizes acetate, *N*-acetyl-D-glucosamine, *cis-* and *trans*-aconitate, adipate, L-alanine, L-arabinose, citrate, L-histidine, 3-hydroxy-DL-butyrate, DL-lactate, L-leucine, L-malate, maltose, D-mannose, oxoglutarate, phenylacetate, L-phenylalanine, L-proline, pyruvate, L-serine, D-sorbitol, suberate, and trehalose, but not 4-aminobutyrate, L-aspartate, azelate, glutarate, itaconate, 4-hydroxybenzoate, or L-ornithine. DNA–DNA relatedness values between the type strain and those of other *Prauserella* species are as follows: *Prauserella aidingensis*, 41.4%; *Prauserella flava*, 43.6%; *Prauserella salsuginis*, 47.2%; and *Prauserella sediminis*, 51.6%.

Additional phenotypic features are cited in the text and in Table 253.

Cellular fatty acids contain major proportions of $C_{16:0}$ and $C_{16:0}$ 2-OH; smaller proportions (<10%) of $C_{14:0}$ iso, $C_{15:0}$ iso, $C_{15:1}$ B, $C_{16:0}$, $C_{16:1}$ *cis*9, $C_{16:1}$ iso H, $C_{17:0}$ iso, $C_{17:0}$ anteiso, $C_{17:0}$ anteiso 2-OH, and $C_{17:1}$ ω9*c*, and traces of $C_{15:0}$, $C_{17:0}$, and $C_{17:0}$ iso 2-OH are found.

Source: isolated from soil in hypersaline habitats.

DNA G+C content (mol%): 66.7 (T_m).

Type strain: CCTCC AA 001016, DSM 44590, YIM 90005.

Sequence accession no. (16S rRNA gene): AF435077.

Additional comments: in the original paper by Li et al. (2003c), strain CCTCC AA 001016 is wrongly cited as CCTCC AA001016.

4. **Prauserella flava** Li, Tang, Chen, Wu, Zhi, Zhang and Li 2009b, 2926[VP]

fla'va. L. fem. adj. *flava* yellow, referring to the color of the substrate mycelium.

Forms a substrate mycelium that undergoes fragmentation. Colonies are light yellow on glycerol-asparagine agar (ISP medium 5), brown on oatmeal agar (ISP medium 3), gray-yellow on potato agar, and brilliant yellow on yeast extract-malt extract agar (ISP medium 2). Does not form aerial hyphae on any of these media. Grows at 15–45°C, pH 6.0–9.0, and 5–15% (w/v) NaCl on ISP medium 2; optimal conditions are 28–37°C, pH 7, and 8–10% NaCl. Catalase-positive, but oxidase-negative. Coagulates milk, but is negative for H_2S production, nitrate reduction, and starch hydrolysis. L-Alanine, L-histidine, L-phenylalanine, L-proline, DL-tryptophan, L-tyrosine, xanthine, and L-valine are used as sole nitrogen sources, but not adenine, L-arginine, or DL-methionine.

DNA–DNA relatedness values of the type strain of *Prauserella flava* with those of *Prauserella aidingensis* and *Prauserella sediminis* are 55.3 and 40.9%, respectively.

Additional phenotypic properties are cited in the text and in Table 253.

Cellular fatty acids consist of major proportions of $C_{16:0}$ iso and $C_{17:1}$ anteiso; smaller proportions (<10%) of $C_{14:0}$, $C_{15:0}$, $C_{15:0}$ iso, $C_{15:1}$ B, $C_{16:0}$, $C_{16:0}$ iso 2-OH, $C_{16:1}$ ω9*c*, $C_{16:1}$ iso H, $C_{17:0}$, $C_{17:0}$ anteiso, and $C_{17:0}$ anteiso 2-OH, and traces of $C_{17:1}$ ω9*c* and $C_{18:1}$ 2-OH are found.

Source: isolated from a salt lake.

DNA G+C content (mol%): 69.9 (HPLC).

Type strain: CCTCC AA 208052, DSM 45265, YIM 90630.

Sequence accession no. (16S rRNA gene): FJ444993.

5. **Prauserella halophila** Li, Xu, Tang, Xu, Kroppenstedt, Stackebrandt and Jiang 2003c, 1548[VP]

ha.lo'phi.la. Gr. n. *hals halos* salt; N.L. adj. *philus -a -um* (from Gr. adj. *philos -ê -on*) friend, loving; N.L. fem. adj. *halophila* salt-loving, referring to the ability to grow at high NaCl concentrations.

Forms a branched substrate mycelium which undergoes fragmentation. Substrate mycelium is light orange-brown on Czapek's agar, light gray-white on arginine-glycerol agar (ISP medium 5), light yellow on inorganic salts-starch and yeast extract-malt extract agars (ISP media 4 and 2, respectively), deep gray-white on oatmeal agar (ISP medium 3), and deep yellow on potato agar [all media supplemented with 10% (w/v) NaCl]. Diffusible pigments are not formed. A white to yellow aerial mycelium is well developed on all of these media, apart from ISP medium 2. Aerial hyphae differentiate into branched short or, at maturity, long straight to flexuous spore chains. Optimal growth occurs on Czapek's agar supplemented with 10–15% (w/v) NaCl at 28°C and pH 7.0. Does not hydrolyze esculin, L-alanine-*p*NA, *bis-p*NP phosphate, 2-deoxythymidine-5-*p*NP phosphate, *o*NP-β-D-galactopyranoside, *p*NP-α-D-glucopyranoside, *p*NP-glucopyranoside, *p*NP-β-D-glucuronide, L-glutamate-γ-3-cabonyl-*p*NP, *p*NP-phenylphosphonate, *p*NP-phosphorylcholine, L-proline-*p*NP, or *p*NP-β-D-xylopyranoside. Acetate, adipate, L-alanine, azelate, fumarate, L-histidine, 3-hydroxy-DL-butyrate, L-malate, L-phenylalanine, propionate, L-proline, pyruvate, D-sorbitol, and suberate are used as sole carbon sources, but not *N*-acetyl-D-glucosamine, *cis-* or *trans*-aconitate, 4-aminobutyrate, L-arabinose, citrate, glutarate, 4-hydroxybenzoate, itaconate, DL-lactate, L-leucine, maltose, D-mannose, L-ornithine, oxoglutarate, L-phenylacetate, L-serine, or trehalose.

Additional phenotypic properties are cited in the text and in Table 253.

Cellular fatty acids contain major proportions of $C_{16:0}$ iso, $C_{17:1}$ ω6*c*, $C_{16:1}$ ω7*c* and/or $C_{15:0}$ iso 2-OH, and $C_{17:0}$ anteiso, and smaller proportions (<10%) of $C_{14:0}$ iso, $C_{15:0}$ iso, $C_{15:1}$ ω6*c*, $C_{16:1}$ iso H, $C_{16:0}$, $C_{17:0}$ 10-methyl, $C_{17:1}$ ω8*c*, $C_{17:0}$, and $C_{18:1}$ ω9*c*.

DNA–DNA relatedness values between the type strain and those of other *Prauserella* species are as follows: *Prauserella aidingensis*, 53.7%; *Prauserella alba*, 30.2%; *Prauserella flava*, 41.3%; *Prauserella salsuginis*, 20.8%; and *Prauserella sediminis*, 45.2%.

Source: isolated from soil in hypersaline habitats.

DNA G+C content (mol%): 65.8 (T_m).

Type strain: CCTCC AA 001015, DSM 44617, YIM 90001.

Sequence accession no. (16S rRNA gene): AF466190.

Additional comments: the etymology of the first compound of the specific epithet should be Gr. n. *halo halos* salt, not Gr. n. *halo* salt as cited by Li et al. (2003c). Similarly, the specific epithet *halophila* is a "N.L. fem. adj." not a "N.L. gen. adj." as cited by Li and his colleagues who also cited strain CCTCC AA 001015 as CCTCC AA001015.

6. **Prauserella marina** Wang, Li, Bian, Tang, Ren, Chen, Li and Zhang 2010, 988[VP]

ma.ri'na. L. fem. adj. *marina* of the sea, marine.

Forms a substrate mycelium which undergoes fragmentation. Substrate mycelium is light gray-white on glycerol-asparagine agar (ISP medium 5), moderate reddish-brown on yeast extract-malt extract agar (ISP medium 2), and pale pink on oatmeal (ISP medium 3) and potato agars. Optimal growth occurs on ISP 2 agar prepared without NaCl at 28–37°C and pH 7.0. Growth occurs at 15–45°C, pH 6.0–9.0 and 0–10% (w/v) NaCl. Catalase-positive, but oxidase-negative. Coagulates milk, but is negative for hydrogen sulfide production, nitrate reduction, and starch hydrolysis. L-Alanine, L-histidine, hypoxanthine, L-phenylalanine, L-proline, DL-tryptophan, L-tyrosine, L-valine, and xanthine are used as nitrogen sources, but not adenine or DL-methionine.

Additional phenotypic properties are cited in the text and in Table 253.

Cellular fatty acids contain major proportions of $C_{16:0}$ iso, $C_{16:0}$, and $C_{15:0}$ iso, and smaller proportions (<10%) of $C_{16:1}$ iso H and $C_{17:0}$.

Source: isolated from ocean sediment.

DNA G+C content (mol%): 66.1 (HPLC).

Type strain: CCTCC AA 208056, DSM 45268, MS498.

Sequence accession no. (16S rRNA gene): FJ444996.

7. **Prauserella muralis** Schäfer, Martin and Kämpfer 2010, 289[VP]

mu.ra'lis. L. fem. adj. *muralis* pertaining or belonging to walls.

Forms mycelial-like filaments (1.5 μm), and a white aerial mycelium which undergoes fragmentation. The substrate mycelium is gray to light orange on M79 agar. Grows well on nutrient and tryptone soy agars after 3 d at 25–30°C. Weakly oxidase-positive, but negative for esculin hydrolysis. *p*NP-α-D-Glucopyranoside and *p*NP-β-D-xylopyranoside are hydrolyzed, but not L-alanine-*p*NA, *bis-p*NP phosphate, 2-deoxythymidine-5-*p*NP phosphate, *o*NP-β-D-galactopyranoside, *p*NP-glucopyranoside, *p*NP-β-D-glucuronide, L-glutamate-γ-3-carbonyl-*p*NP, *p*NP-phenylphosphate, *p*NP-phosphorylcholine, or L-proline *p*NP.

Additional phenotypic properties are cited in the text or in Table 253.

Cellular fatty acids contain major proportions of $C_{16:0}$ iso; smaller proportions (<10%) of $C_{15:0}$ iso, $C_{15:1}$ ω6*c*, $C_{15:0}$, $C_{16:1}$ iso H, $C_{16:1}$ ω7*c* and/or $C_{15:0}$ iso 2-OH, $C_{16:0}$, $C_{16:0}$ 10-methyl, $C_{17:0}$ iso, $C_{17:0}$ anteiso, $C_{17:1}$ ω8*c*, $C_{17:1}$ ω6*c*, $C_{17:0}$, $C_{17:0}$ 10-methyl, and traces of $C_{14:0}$ iso and $C_{14:0}$ are found.

Source: isolated from the wall of a house colonized with molds.

DNA G+C content (mol%): not determined.

Type strain: 05-Be-005, CCM 7635, CCUG 57426, DSM 45305, NRRL B-24780.

Sequence accession no. (16S rRNA gene): FM956091.

8. **Prauserella salsuginis** Li, Tang, Chen, Wu, Zhi, Zhang and Li 2009b, 2926[VP]

sal.su'gi.nis. L. n. *salsugo -inis* brine, salt water, L. gen. n. *salsuginis* of salt water, from which the type strain was isolated.

Aerobic, Gram-stain-positive actinomycete which forms a substrate mycelium that undergoes fragmentation and produces a white aerial mycelium. Colonies are deep yellow on Czapek's agar, light gray-white on glycerol-asparagine agar (ISP medium 5), pale orange-yellow on nutrient and potato agars, and brilliant yellow on yeast extract-malt extract agar (ISP medium 2). Grows at 15–45°C, pH 6.0–9.0, and 5–15% (w/v) NaCl on ISP 2 medium; optimal conditions are 28–37°C, pH 7.0, and 8–10% NaCl.

Catalase-positive, but oxidase-negative. Coagulates milk, but is negative for melanin and H_2S production.

L-Alanine, L-histidine, L-phenylalanine, L-proline, DL-tryptophan, L-tyrosine, xanthine, and L-valine are used as sole nitrogen sources, but not adenine or DL-methionine.

DNA–DNA relatedness values between the type strain and those of other *Prauserella* species are: *Prauserella aidingensis*, 44.0%; *Prauserella flava*, 56.9%; and *Prauserella sediminis*, 60.3%.

Additional phenotypic properties are cited in the text and in Table 253.

Cellular fatty acids contain major proportions of $C_{16:0}$ iso; smaller proportions (<10%) of $C_{15:0}$, $C_{15:0}$ iso, $C_{15:1}$ B, $C_{16:0}$, $C_{16:0}$ iso 2-OH, $C_{16:1}$ ω9*c*, $C_{16:1}$ iso H, $C_{17:0}$ anteiso, $C_{17:0}$ anteiso 2-OH, $C_{17:1}$ anteiso, and $C_{18:1}$ 2-OH, and traces of $C_{17:1}$ ω9*c* are found.

Source: isolated from a salt lake.

DNA G+C content (mol%): 69.1 (HPLC).

Type strain: CCTCC AA 208051, DSM 45264, YIM 90625.

Sequence accession no. (16S rRNA gene): FJ444992.

9. **Prauserella sediminis** Li, Tang, Chen, Wu, Zhi, Zhang and Li 2009b, 2927[VP]

se.di'mi.nis. L. n. *sedimen -inis* sediment; L. gen. n. *sediminis* of sediment.

Forms a substrate mycelium that undergoes fragmentation and produces a white aerial mycelium on most media. Colonies are deep gray-white on glycerol-asparagine agar (ISP medium 5), pale yellow on oatmeal agar (ISP medium 3), gray-reddish orange on potato agar and orange-yellow on yeast extract-malt extract agar (ISP medium 2). Grows at 15–45°C, pH 6.0–9.0, and 5–20% (w/v) NaCl on ISP 2 medium; optimal conditions are 28–37°C, pH 7.0, and 10% NaCl.

Catalase-positive, but oxidase-negative. Coagulates milk, but is negative for H_2S production, nitrate reduction, and starch hydrolysis.

L-Alanine, L-histidine, L-phenylalanine, L-proline, DL-tryptophan, L-tyrosine, xanthine, and L-valine are used as sole nitrogen sources, but not adenine or DL-methionine.

Additional phenotypic properties are cited in the text and in Table 253.

Cellular fatty acids contain major proportions of $C_{16:0}$ iso; smaller proportions (<10%) of $C_{14:0}$ iso, $C_{15:0}$, $C_{15:0}$ iso, $C_{15:1}$ B, $C_{16:0}$, $C_{16:0}$ 2-OH, $C_{16:1}$ ω9*c*, $C_{16:1}$ iso H, $C_{17:0}$, $C_{17:0}$ iso 2-OH, $C_{17:0}$ anteiso, and $C_{17:1}$ anteiso, and traces of $C_{14:0}$, $C_{17:1}$ ω9*c*, and $C_{17:1}$ 2-OH are found.

Source: isolated from a salt lake.

DNA G+C content (mol%): 69.1 (HPLC).

Type strain: CCTCC AA 208054, DSM 45267, YIM 90694.

Sequence accession no. (16S rRNA gene): FJ444995.

Genus XIII. **Saccharomonospora** Nonomura and Ohara 1971, 899[AL]

Seung Bum Kim

Sac.cha.ro.mon.o.spo'ra. Gr. n. *sakchâr* sugar; Gr. adj. *monos* single, solitary; Gr. fem. n. *spora* seed; N.L. fem. n. *spora* a spore; N.L. fem. n. *Saccharomonospora* the sugar (-containing) single-spored (organism).

Gram-stain-positive, aerobic, and chemo-organotrophic. **Produces single or paired spores on aerial hyphae.** Spores may be formed on substrate mycelium. The aerial mycelium can be white, yellow-white, green, or light to dark blue; green pigmentation may also occur on the vegetative mycelium and diffuse into the surrounding medium. **Substrate mycelia are rarely fragmented**. Spores in pairs or short chains on vegetative or aerial hyphae are occasionally present. **The cell wall contains *meso*-diaminopimelic acid (*meso*-DAP), and the sugars arabinose and galactose (wall chemotype IV).** Mycolic acids are absent. Major amounts of iso- and anteiso- fatty acids are found; the main **menaquinone is MK-9(H_4)**. The diagnostic phospholipid is **phosphatidylethanolamine** (**phospholipid type II**), but some species may also contain glucosamine-containing phospholipids (type IV). **Mesophilic or thermophilic**; growth occurs between 24 and 60°C, and at neutral pH. **NaCl may be required for growth.** Isolated from soil, lake sediments, marsh soil, peat, manure, compost, and overheated fodder.

DNA G+C content (mol%): 68–74.

Type species: **Saccharomonospora viridis** (Schuurmans, Olson and San Clemente 1956) Nonomura and Ohara 1971, 899[AL] [*Thermoactinomyces viridis* Schuurmans, Olson and San Clemente 1956, 61; *Thermomonospora viridis* (Schuurmans, Olson and San Clemente 1956) Küster and Locci 1963].

Further descriptive information

Phylogeny. *Saccharomonospora* is phylogenetically related to the genera *Prauserella* and *Thermocrispum* (Kim and Goodfellow, 1999; Labeda and Kroppenstedt, 2006). In 16S rRNA gene sequence analyses, the species are divided into two main cluster groups, one including *Saccharomonospora azurea*, *Saccharomonospora cyanea*,

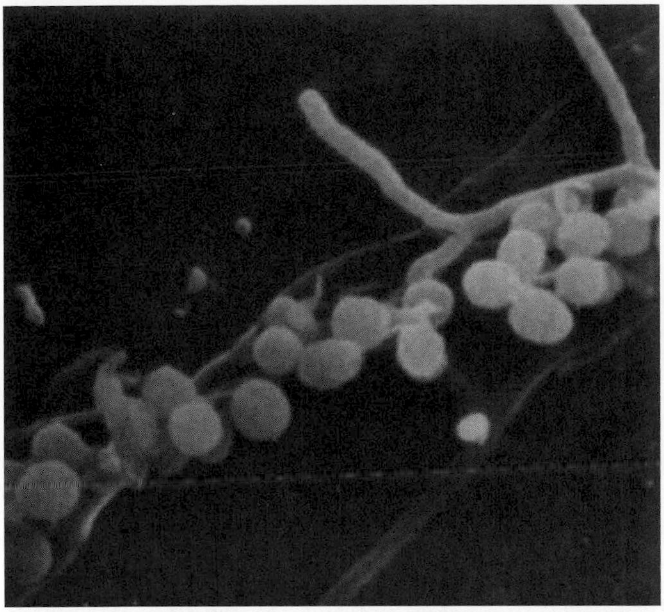

FIGURE 313. Aerial spores of *Saccharomonospora azurea* strain NA-128^T. Magnification 4800×. (Reproduced with permission from Hu, 1997. Int. J. Syst. Bacteriol. *47*: 60–61.)

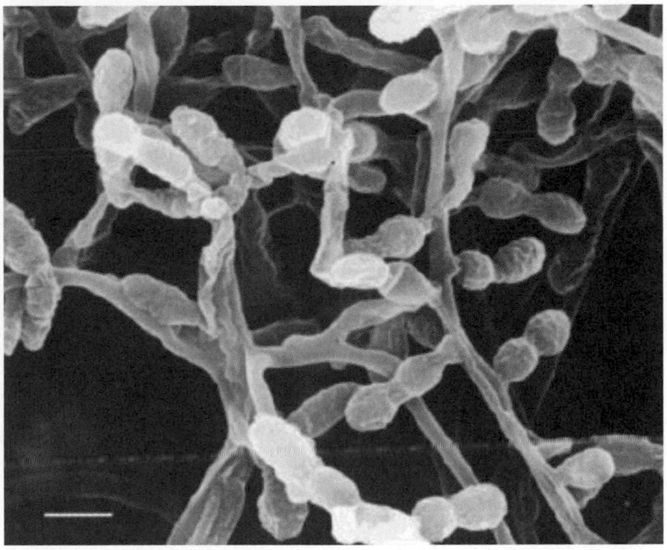

FIGURE 314. Aerial spore of *Saccharomonospora halophila* strain 8^T. Bar = 1 µm. (Reproduced with permission from Al-Zarban et al., 2002. Int. J. Syst. Evol. Microbiol. *52*: 555–558.)

Saccharomonospora glauca, *Saccharomonospora viridis*, and *Saccharomonospora xinjiangensis*, and the other including *Saccharomonospora halophila*, *Saccharomonospora marina*, *Saccharomonospora paurometabolica*, and *Saccharomonospora saliphila*, respectively (Al-Zarban et al., 2002; Li et al., 2003b; Syed et al., 2008; Liu et al., 2010). The 16S rRNA gene sequence similarity among the type strains ranges from 95.2 to 98.5%. The levels of DNA–DNA relatedness between the type strain of *Saccharomonospora saliphila* and those of *Saccharomonospora azurea*, *Saccharomonospora halophila*, and *Saccharomonospora paurometabolica* were reported as 46.0, 41.0, and 42.5%, respectively (Syed et al., 2008). The value between *Saccharomonospora paurometabolica* and *Saccharomonospora halophila* was 53.8% (Li et al., 2003b).

Cellular morphology. The aerial and vegetative mycelia are well developed and irregularly branched. The substrate mycelium is, in most cases, non-fragmented. Aerial mycelium is generally abundant, but can be absent in some strains. The color of aerial mycelia is green to blue, except for *Saccharomonospora paurometabolica* and *Saccharomonospora xinjiangensis* which is white, orange-white, or yellow-white. Single spores are borne at the tip of aerial hyphae in most species (Figure 313), but spores in longitudinal pairs are also produced in *Saccharomonospora marina*, *Saccharomonospora saliphila* and *Saccharomonospora xinjiangensis* (Figure 314).The spores are ovoid, ellipsoidal, or round (0.7–1.1 × 1.0–1.8 µm); the surface of individual spores is smooth for *Saccharomonospora azurea*, *Saccharomonospora marina*, *Saccharomonospora paurometabolica*, *Saccharomonospora saliphila*, and *Saccharomonospora xinjiangensis*, and warty for the remaining four species. Wrinkled spores may be observed for *Saccharomonospora marina*, *Saccharomonospora paurometabolica* and *Saccharomonospora saliphila*. The spores of *Saccharomonospora viridis* are not viable at temperatures above 70°C.

Nutrition and growth conditions. Strains of *Saccharomonospora glauca*, *Saccharomonospora viridis*, and *Saccharomonospora xinjiangensis* are thermotolerant, preferring temperatures of 45–50°C, whereas the remaining species are mesophilic. *Saccharomonospora halophila* and *Saccharomonospora paurometabolica* require NaCl for growth, but *Saccharomonospora cyanea*, *Saccharomonospora glauca*, *Saccharomonospora viridis*, and *Saccharomonospora xinjiangensis* cannot grow in the presence of 5% (w/v) NaCl. *Saccharomonospora azurea* and *Saccharomonospora saliphila* do not require NaCl for growth, but the former can grow in the presence of up to 7%, and the latter in up to 20% (w/v) concentration.

Cell-wall composition. Phosphatidylethanolamine is the main phospholipid for all species and, in addition, hydroxyphosphatidylethanolamine, lysophosphatidylethanolamine, phosphatidylinositol, phosphatidylglycerol, and diphosphatidylglycerol are variably found (corresponding to the type II phospholipid pattern *sensu* Lechevalier et al., 1981). However, *Saccharomonospora xinjiangensis* contains phosphatidylcholine and glucosamine-containing phospholipid in addition to phosphatidylethanolamine, exhibiting a type IV phospholipid pattern. The fatty acid profile is highly variable among species, but is generally a mixture of branched and straight chain saturated or unsaturated fatty acids, with smaller amounts of hydroxylated fatty acids. The main menaquinone is MK-9(H_4), whereas MK-8(H_4), MK-9(H_2), and MK-7(H_4) are variably present as minor components.

Ecology. Saccharomonosporae occur in a wide range of natural or synthetic habitats including soil (Al-Zarban et al., 2002; Jin et al., 1998; Li et al., 2003b; Syed et al., 2008), compost (Amner et al., 1988; Dees and Ghiorse, 2001; Song et al., 2001; Steger et al., 2007), plant materials (Abdulla and El-Shatoury, 2007; Gangwar et al., 1989; Roussel et al., 2005; Unaogu et al., 1994; Liu et al., 2010), marine sediment (Maldonado et al., 2009), and marine sponge (Selvin et al., 2009). Compost is clearly a preferred habitat for thermophilic *Saccharomonospora* (Dees and Ghiorse, 2001; Khan et al., 1995; Song et al., 2001; Unaogu et al., 1994).

Other properties. *Saccharomonospora viridis* is strongly implicated as one of the causative agents of hypersensitivity pneumonitis including farmer's lung disease (Greene et al., 1981; Harvey et al., 2001; Roberts et al., 1976; Treuhaft et al., 1980; Wenzel et al., 1974). Strains of *Saccharomonospora viridis*, *Saccharomonospora glauca*, "*Saccharomonospora caesia*", and "*Saccharomonospora internatus*" display antibiotic activities against Gram-stain-positive bacteria (Greiner-Mai et al., 1988). *Saccharomonospora viridis* is known to produce thermoviridin (Schuurmans et al., 1956). Many strains of *Saccharomonospora* produce degradative enzymes for natural or anthropogenic compounds, such as polyester, rice straw, mushroom compost, synthetic food waste compost, proteins, and starch (Abdulla and El-Shatoury, 2007; Collins et al., 1992; Dolashka et al., 1998; Song et al., 2001; Tseng et al., 2007).

Enrichment and isolation procedures

Mesophilic saccharomonosporae are present in soils, leaf litter, and manure; thermophilic strains can be found in compost, while marsh soils are a good source for halophilic strains. Conventional methods for the isolation of actinobacteria can be used for the isolation of saccharomonosporae, but improved recovery has been reported using selective isolation methods, such as dry heat treatment, addition of antibiotics to the isolation medium, use of sedimentation chamber with an Andersen air sampler, and a combination of the above methods (Amner et al., 1989; Andersen, 1958; Athalye et al., 1981; Kim et al., 1995; Nonomura and Ohara, 1971).

The isolation media include modified glycerol-asparagine agar [ISP medium 5 supplemented with 20% (w/v) NaCl] half-strength tryptone-soy agar, HV agar (Hayakawa, 1990), and R8 agar (Amner et al., 1989). For the prevention of fungal growth, antibiotics such as cycloheximide (50 μg/ml) may be added to the media.

Maintenance procedures

The following media can be used for the cultivation and maintenance of *Saccharmonospora* isolates: yeast extract-malt extract (ISP 2) agar, inorganic salts-starch (ISP 4) agar, Czapek–Dox yeast extract-Casamino acids (CYC) agar (Cross and Attwell, 1974), starch-nitrate agar with 10% (w/v) NaCl (Al-Zarban et al., 2002), and tryptic soy agar (TSA). *Saccharomonospora halophila* and *Saccharomonospora paurometabolica* require NaCl for growth at an optimal concentration of 10% (w/v). For long-term storage, spores and mycelial fragments are suspended in 10–20% (v/v) aqueous glycerol and stored at −20°C, or lyophilized.

Differentiation of the genus *Saccharomonospora* from other genera

Saccharomonospora can be distinguished from other members of the family *Pseudonocardiaceae* by the production of single spores on aerial hyphae. Fragmentation of the substrate mycelium is rarely observed, which is an important differential feature of the genus from closely related genera, including *Prauserella*, *Saccharopolyspora*, and *Thermoscrispum*. Sporangia-like structures have not been observed in *Saccharomonospora*, in contrast to *Crossiella*, *Kibdelosporangium*, *Kutzneria*, *Streptoalloteichus*, and *Thermocrispum*. The phospholipid pattern of *Saccharomonospora* is type II, with the exception of *Saccharomonospora xinjiangensis* (type IV), whereas that of *Saccharopolyspora* is type III.

Taxonomic comments

The genus *Saccharomonospora* currently contains nine species with validly published names (Figure 315). The type species *Saccharomonospora viridis* Nonomura and Ohara 1971 was first

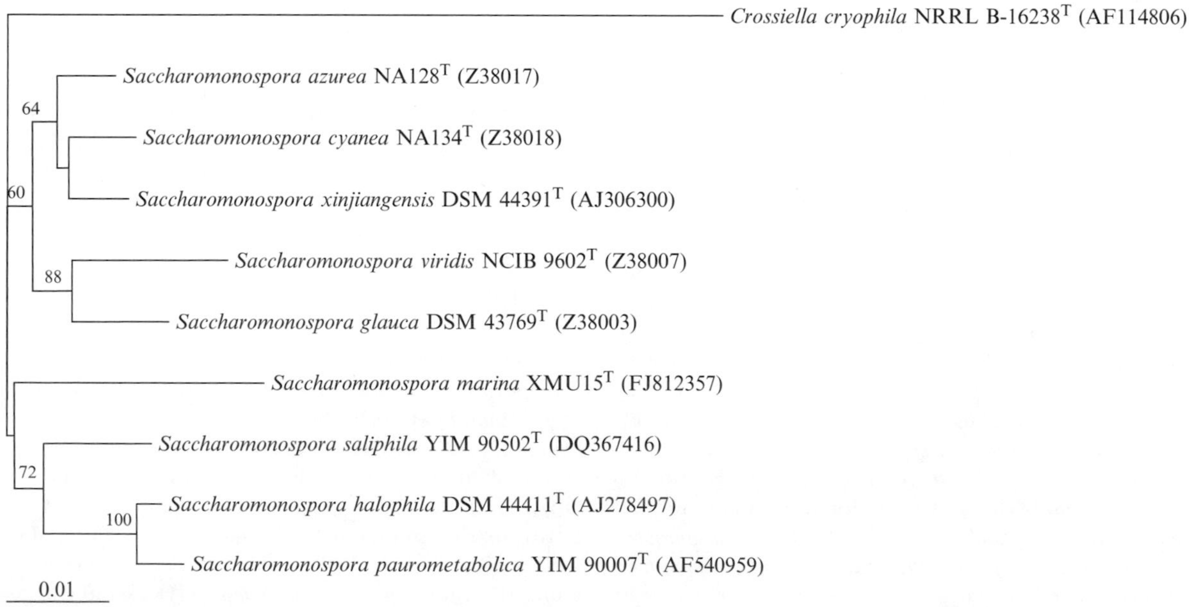

FIGURE 315. Phylogenetic tree of the genus *Saccharomonospora* based on 16S rRNA gene sequences. The Jukes–Cantor model was used in the estimation of evolutionary distances and neighbor-joining method was used for tree construction. Numbers at nodes indicate levels of bootstrap support (%). Bar = 0.01 substitutions per nucleotide position.

TABLE 254. Characteristics that differentiate the type strains of the genus *Saccharomonospora*[a]

Character	*S. viridis* NCIB 9602[T]	*S. azurea* NA-128[T]	*S. cyanea* NA-134[T]	*S. glauca* DSM 43769[T]	*S. halophila* DSM 44411[T]	*S. marina* KCTC 19701	*S. paurometabolica* YIM 90007[T]	*S. saliphila* YIM 90502[T]	*S. xinjiangensis* DSM 44391[T]
Aerial mycelium color	Green	Azure	Light to dark blue	Light blue to greenish	Light blue to greenish	White, gray to orange-white	White	Gray-red	Yellow-white
Spores in pairs	–	–	–	–	+	+	–	+	+
Spore ornamentation	Warty	Smooth	Warty	Warty	Warty	Smooth/ wrinkled	Smooth/ wrinkled	Smooth/ wrinkled	Smooth
Growth on sole carbon source (1%, w/v):									
L-Arabinose	–	–	nd	+	+	+	–	–	nd
Galactose	–	–	+	nd	+	+	–	+	nd
Glucose	Dbt	+	–	+	+	+	–	–	nd
Mannitol	Dbt	–	–	+	+	+	–	+	nd
Mannose	–	+	+	nd	+	+	nd	–	+
Melibiose	nd	+	–	nd	+	+	–	+	nd
Rhamnose	–	–	nd	–	+	+	–	+	+
Ribose	nd	+	+	nd	Dbt	–	–	nd	nd
Xylose	–	+	Dbt	+	–	+	–	+	+
Growth in NaCl (%, w/v):									
0	+	+	+	+	–	+	–	+	+
5	–	+	–	–	–	+	+	+	–
7	–	+	–	–	–	–	+	+	–
10	–	–	–	–	+	–	+	+	–
20	–	–	–	–	+	–	+	+	–
30	–	–	–	–	+	–	+	–	–
Growth temperature (°C[b])	35–50	24–40	24–40 (28–37)	37–60 (50)	(28–30)	(28–37)	(35–37)	(28–30)	45–50
Menaquinone	9(H_4)	nd	nd	9(H_4) (60%), 8(H_4) (20–30%)	9(H_4) (88%), 8(H_4) (12%)	9(H_4) (90%), 8(H_4) (10%)	9(H_4) (90%), 9(H_2) (10%)	9(H_4) (90%), 8(H_4) (10%)	9(H_2), 9(H_4), 7(H_4)
Phospholipid[c]	PI, PIM, DPG acyl-PG	nd	nd	PE-OH, lyso-PE	DPG, PI, PE-OH, lyso-PE	PI, PG, DPG	PI, PG, DPG, PE-OH	PI, PG, DPG	PC, GluNU
DNA G+C content (mol%)	69	nd	nd	nd	nd	68.1	71	71.8	nd

[a]+, Positive; –, negative; nd, not determined; Dbt, doubtful.

[b]Optimal temperatures are given in parentheses.

[c]PE-OH, hydroxyphosphatidylethanolamine; PI, phosphatidylinositol; PIM, phosphatidylinositol mannoside; PG, phosphatidylglycerol; DPG, diphosphatidylglycerol; PC, phosphatidylcholine; GluNU, glucosamine-containing phospholipid. All species contain phosphatidylethanolamine (PE).

proposed as "*Thermoactinomyces viridis*" by Schuurmans et al. (1956), and later reclassified as "*Thermomonospora viridis*" Küster and Locci 1963. *Saccharomonospora* was then proposed based on morphological and chemotaxonomic criteria (Nonomura and Ohara, 1971), which was further supported by numerical phenetic data (Goodfellow and Pirouz, 1982; McCarthy and Cross, 1984).

Phylogenetic studies on saccharomonosporae have been conducted using analysis of 16S rRNA gene sequences (Kim et al., 1995), internally transcribed spacer sequences of 16S–23S and 23S–5S rRNA genes (Yoon et al., 1997), and ribonuclease P (RNase P) RNA gene sequences (Cho et al., 1998). These data constantly suggested that "*Saccharomonospora caesia*" was a synonym of *Saccharomonospora azurea*. Additionally, DNA–DNA relatedness studies among the representative species of *Saccharomonospora* were also carried out (Yoon et al., 1999), confirming the former proposal. Ruan et al. (1994) used partial sequences of the 23S rRNA gene for the classification of *Saccharomonospora* and related taxa.

DNA–DNA hybridization data also confirmed that "*Saccharomonospora internatus*" (Greiner-Mai et al. 1987; Kurup 1981), formerly "*Micropolyspora internatus*" Agre et al. 1974, is a synonym of *Saccharomonospora viridis* Greiner-Mai et al. 1988, as the two taxa exhibited 90% DNA–DNA relatedness (Yoon et al., 1999).

A PCR-based method for the rapid detection of *Saccharomonospora* isolates has been developed using a set of genus-specific primers (Salazar et al., 2000). The sequence data indicate that these oligonucleotide primers are also applicable to the species that have been described subsequently, namely *Saccharomonospora halophila*, *Saccharomonospora paurometabolica*, and *Saccharomonospora xinjiangensis*. In another study, a rapid detection method based on fluorescence *in situ* hybridization (FISH) using 16S rRNA-targeted oligonucleotide probes was developed for the specific detection of *Saccharomonospora* spp. (Neef et al., 2003).

The primycin-producing actinobacterium, initially described as "*Thermomonospora galeriensis*" (Szabo et al., 1976), was reported to produce single spores and dark green aerial mycelium. This species contains a type IV cell wall and was considered to be a member of the genus *Saccharomonospora* by McCarthy and Cross (1984), but its correct taxonomic position needs to be clarified.

The strains known as "*Thermoactinomyces glaucus*" IFO 12530 (Henssen, 1957) and "*Thermoactinomyces monosporus*" IFO 14050 (Waksman and Cork, 1953) have been assigned to *Saccharomonospora glauca* based on chemotaxonomic and genotypic characterization (Yoon et al., 2000; Yoon and Park, 2000).

Further reading

Goodfellow, M. and T. Cross. 1984. Classification. *In* Goodfellow, Mordarski and Williams (Editors), The Biology of the Actinomycetes, Academic Press, London, pp. 7–164.

Differentiation of the species of the genus *Saccharomonospora*

The species of *Saccharomonospora* can be separated from one another by a combination of physiological and chemotaxonomic properties. *Saccharomonospora paurometabolica* produces white aerial mycelium, *Saccharomonospora xinjiangensis* yellow-white and *Saccharomonospora saliphila* gray-red, whereas most others produce blue to green aerial mycelia. The warty spore surface separates *Saccharomonospora cyanea*, *Saccharomonospora glauca*, *Saccharomonospora halophila*, and *Saccharomonospora viridis* from the remaining species. Unlike other species, *Saccharomonospora halophila* and *Saccharomonospora paurometabolica* can grow in the presence of up to 30% NaCl, but cannot grow without NaCl. Other differential properties are listed in Table 254.

List of species of the genus *Saccharomonospora*

1. **Saccharomonospora viridis** (Schuurmans, Olson and San Clemente 1956) Nonomura and Ohara 1971, 899[AL] [*Thermoactinomyces viridis* Schuurmans, Olson and San Clemente 1956, 61; *Thermomonospora viridis* (Schuurmans, Olson and San Clemente 1956) Küster and Locci 1963]. Subjective synonym: "*Saccharomonospora internatus*" (Agre et al. 1974) Greiner-Mai, Korn-Wendisch and Kutzner 1988.

vir′i.dis. L. fem. adj. *viridis* green.

Single spores are mainly produced on aerial mycelium, but short chains of spores may also be formed. The spores are heat-sensitive. Vegetative mycelia are branched and may produce spores. Leathery colonies are formed on agar media, with aerial hyphae covered by densely packed spores. Aerial mycelium is initially white, becoming gray-green to dark green. The aerial spore mass may be non-pigmented or lilac-colored. Vegetative mycelium may be green. Production of a green diffusible pigment is also observed. Amino acid and vitamin supplements (e.g. yeast extract) are required for good growth. Catalase, deaminase, and phosphatase are produced. Casein, gelatin, starch, xylan, and tyrosine are degraded, but not cellulose. A number of organic compounds can serve as sole sources of carbon, and the utilization of glycerol is characteristic. Optimal growth occurs at 35–50°C and pH 7.0–10. Growth occurs in the presence of up to 3% (w/v) NaCl, but not at concentrations of 5% or higher.

Source: manure, compost, overheated fodder, soil, lake sediments, peat.

DNA G+C content (mol%): 69–74 (T_m).

Type strain: ATCC 15386, CCUG 5913, DSM 43017, JCM 3036, NBRC 12207, NCIB 9602, NRRL B-3044, VKM Ac-681.

Sequence accession no. (16S rRNA gene): Z38007.

2. **Saccharomonospora azurea** Hu 1987, 61[VP]

Subjective synonym: "*Saccharomonospora caesia*" (Kalakoutskii 1964) Greiner-Mai, Kroppenstedt, Korn-Wendisch and Kutzner 1987.

a.zu.re′a. N.L. fem. adj. *azurea* azure, referring to the color of the aerial mycelium.

Substrate mycelium is non-fragmenting. Sporangia are not produced. Single spores are borne mainly on the aerial mycelium. Spores are oval or round, 0.8 to 1.0 µm. Sporophores are very short or sessile. Spore surface is smooth. The color of the aerial mycelium is azure on oatmeal and Czapek–Dox sucrose agars. No distinct soluble pigment is formed. D-Fructose, D-glucose, glycerol, lactose, maltose, mannose, melibiose, raffinose, L-rhamnose, ribose, sucrose, trehalose, and D-xylose are utilized as sole carbon sources, but not L-arabinose, galactose, inositol, or D-mannitol. Growth occurs between 24 and 40°C.

Source: soil.

DNA G+C content (mol%): not determined.

Type strain: NA-128, ATCC 43670, DSM 44631, JCM 7551, NBRC 14651, SIIA 86128.

Sequence accession no. (16S rRNA gene): Z38017.

3. **Saccharomonospora cyanea** Hu, Cheng and Wei 1988, 445[VP]

cy.a'ne.a. L. fem. adj. *cyanea* dark blue, referring to the color of aerial mycelium.

Substrate mycelium is non-fragmenting. No sporangium is observed. Single spores are borne mainly on aerial mycelium. Spores are oval to ellipsoidal (0.8–1.0 ×1.0–1.8 µm). Sporophores are very short or sessile. The color of the aerial mycelium is light to dark blue on various agar media. No distinct soluble pigment is formed. D-Fructose, glycerol, lactose, maltose, mannose, raffinose, L-rhamnose, ribose, sucrose, trehalose, and D-xylose are utilized as sole carbon sources, but not L-arabinose, D-glucose, *i*-inositol, D-mannitol, or melibiose. The temperature range for growth is between 24 and 40°C; optimal growth occurs between 28 and 37°C.

Source: soil.

DNA G+C content (mol%): not determined.

Type strain: NA-134, ATCC 43724, DSM 44106, JCM 7552, NBRC 14841, SIIA 86134.

Sequence accession no. (16S rRNA gene): Z38018.

4. **Saccharomonospora glauca** Greiner-Mai, Korn-Wendisch and Kutzner 1988, 403[VP]

glau'ca. L. fem. adj. *glauca* bluish, greenish, grayish blue, referring to the color of the aerial mycelium.

Branching, non-fragmenting aerial and substrate mycelia are formed. Single spores are produced tightly packed on the aerial hyphae. Spores are smooth or slightly roughened, round to oval, and 0.8 to 1.0 µm in diameter. Colonies produce light green to bluish green (turquoise) aerial mycelium, dark green substrate mycelium, and soluble pigment on glycerol-cornsteep (GC) and glycerol-yeast extract-malt extract (GYM) agars. The temperature range for growth is 37–60°C and the optimum temperature is 50°C. Strains are sensitive to lysozyme (200 U/ml) and tolerant to 7% (w/v) NaCl. No melanin is produced. Tyrosine, starch, triglycerides, blood cells (hemolysis), casein, collagen, and esculin are degraded. Arabinose, dextrin, D-glucose, and mannitol are used as sole carbon sources. Antibiotic activity is displayed against Gram-stain-positive bacteria.

All strains of *Saccharomonospora glauca* are sensitive to phage ϕ771 and, in addition, several strains are sensitive to phage ϕL1g. All strains are resistant to the Tm_1 family of phages, which is specific for *Saccharomonospora viridis.* All strains show identical total protein and DNA restriction patterns; esterase pattern III (four main bands) is found. The R_f value of malate dehydrogenase is 0.50.

Source: moldy hay, soil, compost, and manure.

DNA G+C content (mol%): not determined.

Type strain: K62, DSM 43769, JCM 7444, NBRC 14831.

Sequence accession no. (16S rRNA gene): Z38003.

5. **Saccharomonospora halophila** Al-Zarban, Al-Musallam, Abbas, Stackebrandt and Kroppenstedt 2002, 557[VP]

ha.lo.phi'la. L. n. *hals, halos* salt; N.L. adj. *philus -a -um* (from Gr. adj. *philos -ê -on*) friend, loving; N.L. fem. adj. *halophila* salt-loving, referring to the ability to grow at high NaCl concentration.

Light blue aerial mycelium is produced. Specific endo- or exo-pigments are not observed. Optimal growth is obtained on starch-nitrate agar supplemented with 10% (w/v) NaCl at 28°C. Grows in the presence of 10–30% NaCl. Feathers can be utilized as sole C and N source in the presence of 10% (w/v) NaCl. L-Arabinose, D-galactose, D-glucose, mannitol, mannose, melibiose, and L-rhamnose are utilized as sole carbon sources, but not D-xylose. The utilization of D-ribose is doubtful. Major cellular fatty acids are $C_{16:0}$ iso (22.5%), $C_{16:0}$ (15.8%), and $C_{16:1}$ (14.1%).

Source: salt marsh soil.

DNA G+C content (mol%): not determined.

Type strain: strain 8, DSM 44411, JCM 11761, NRRL B-24125.

Sequence accession no. (16S rRNA gene): AJ278497.

6. **Saccharomonospora marina** Liu, Li, Zheng, Huang and Li 2010, 1856[VP]

ma.ri'na. L. fem. adj. *marina* of the sea.

Nonmotile smooth or wrinkled spores are produced on the branched aerial mycelium singly, in pairs and occasionally in short chains. Optimal growth occurs at 28–37°C and at pH 7.0 on ISP medium 2. Growth occurs in the presence of up to 5% NaCl (w/v), with an optimum concentration of 0–3% (w/v). D-Arabinose, cellobiose, D-galactose, D-glucose, *myo*-inositol, maltose, D-mannitol, D-mannose, melibiose, L-rhamnose, sucrose and D-xylose are utilized as carbon sources, but not lactose, sorbitol, L-sorbose, raffinose and ribose. L-Alanine, L-arginine, L-cystine, L-glutamate, glycine, L-leucine, L-lysine, L-methionine, L-phenylalanine, L-proline, L-serine, L-threonine, L-tyrosine and L-valine are used as nitrogen sources, but not DL-asparagine, L-histidine, L-hydroxyproline and DL-tryptophan. Gelatin liquefaction, milk coagulation and nitrate reduction are positive, but cellulose and starch hydrolysis, hydrogen sulfide and melanin production are negative. Catalase, urease and oxidase activities are negative. Resistant to ampicillin (10 mg), carbenicillin (10 mg), cefuroxime (30 mg), ceftriaxone (30 mg), cephalothin (V) (30 mg), cephalothin (VI) (30 mg), chloramphenicol (30 mg), furoxone (30 mg), penicillin (10 mg), piperacillin (10 mg) and oxacillin (1 mg), but not to cephalothin (IV) (300 mg) and fortum (30 mg).

Major cellular fatty acids are $C_{16:0}$ iso (26.3%), $C_{17:1}$ ω6*c* (16.8%), $C_{15:0}$ (15.2%), $C_{16:0}$ (8.9%), $C_{17:1}$ ω8*c* (7.7%), and $C_{16:1}$ iso H (6.0%).

Habitat: ocean sediment.

DNA G+C content (mol%): 68.1 (HPLC).

Type strain: XMU15, KCTC 19701, CCTCC AA 209048.

Sequence accession no. (16S rRNA gene): FJ812357.

7. **Saccharomonospora paurometabolica** Li, Tang, Stackebrandt, Kroppenstedt, Schumann, Xu and Jiang 2003b, 1593[VP]

pau.ro.me.ta.bo'li.ca. Gr. adj. *pauros* little; Gr. adj. *metabolikos* changeable; N.L. fem. adj. *paurometabolica* little changeable, referring to the poor utilization of carbon sources.

Aerial mycelium is well developed on yeast extract-malt extract agar, glycerol-asparagine agar, nutrient agar, and Czapek's agar; moderate on oatmeal agar and poor on inorganic salts-starch agar and potato agar. White aerial mycelium is produced on all media except on nutrient agar where it is green-yellow. Sporulation is good on ISP 2, ISP 5, nutrient agar, and Czapek's agar, and moderate on ISP 3 agar, but poor on ISP 4 agar. Substrate mycelium is well developed on most media tested. The color of the substrate mycelium is deep orange-yellow (ISP 2), light yellow-brown (nutrient agar), light yellow-orange (potato agar), or white (ISP 4, ISP 5, and Czapek's agar). Nonmotile, single spores with smooth or wrinkled surface are borne on either the aerial or substrate mycelium. Optimum growth temperature is 35–37°C. Optimum NaCl concentration for growth is 10% (w/v). Nitrate is reduced. Milk peptonization and coagulation, gelatin liquefaction, growth in cellulose, H_2S and melanin production, starch hydrolysis, and urease production are not observed. Major cellular fatty acids are $C_{18:1}$ (44.3%), $C_{16:0}$ (20.7%), and $C_{16:0}$ iso (11.2%).

Source: saline soil.

DNA G+C content (mol%): 71 (T_m).

Type strain: BCRC (formerly CCRC) 16315, CCTCC AA 001018, DSM 44619, JCM 13241, YIM 90007.

Sequence accession no. (16S rRNA gene): AF540959.

8. **Saccharomonospora saliphila** Syed, Tang, Cai, Zhi, Agasar, Lee, Kim, Jiang, Xu and Li 2008, 572[VP]

sa.li'phi.la. L. n. *sal, salis* salt; N.L. adj. *philus -a -um* (from Gr. adj. *philos -ê -on*) friend, loving; N.L. fem. adj. *saliphila* salt-loving.

Aerial mycelium is well developed on yeast extract-malt extract agar (ISP 2), inorganic salts-starch agar (ISP 4), glycerol-asparagine agar (ISP 5), potato agar, and Czapek's agar; no growth is observed on oatmeal agar (ISP 3) or nutrient agar. Grayish to reddish-gray aerial mycelium is produced on the above media. Good sporulation is observed on ISP 2, ISP 4, ISP 5, potato agar, and Czapek's agar. Substrate mycelium is well developed on most media tested. The color of substrate mycelium is grayish red on Czapek's agar, dark red on ISP 2 and ISP 5, and blackish red on ISP 4 and potato agar. Nonmotile, single or pairs of spores with smooth or wrinkled surfaces are borne on aerial mycelium. Optimum growth temperature is 28°C; grows well at temperatures up to 40°C. Optimum growth is observed in 10% (w/v) NaCl, although it is not essential for growth. H_2S is produced. Milk peptonization and coagulation, gelatin liquefaction, growth in cellulose, melanin production, starch hydrolysis, and urease production are not observed. Cellobiose, fructose, D-galactose, D-glucose, maltose, raffinose, sorbitol, sucrose, and D-xylose are utilized as sole carbon sources, but not L-arabinose, *myo*-inositol, lactose, mannitol, rhamnose, trehalose, or xylitol. Major cellular fatty acids are $C_{16:0}$ iso (49.2%), $C_{17:1}$ ω6*c* (9.1%), $C_{15:0}$ (5.4%), and $C_{16:1}$ iso OH (5.0%).

Source: muddy soil.

DNA G+C content (mol%): 71.8 (T_m).

Type strain: DSM 45087, KCTC 19234, YIM 90502.

Sequence accession no. (16S rRNA gene): DQ367416.

9. **Saccharomonospora xinjiangensis** Jin, Xu, Mao, Hseu and Jiang 1998, 1097[VP]

xin.ji.ang.en'sis. N.L. fem. adj. *xinjiangensis* pertaining to Xinjiang, a province of north-west China.

Yellow-white aerial mycelium is formed on ISP 2, ISP 3, and nutrient agar (light green-gray on Czapek's agar), and the vegetative mycelium is light yellowish. The sporulation of both aerial and vegetative mycelia is good on most media tested. Spores are borne in longitudinal pairs on vegetative hyphae, and in longitudinal pairs (or in a few cases singly) on aerial hyphae. Light yellow-brown diffusible pigment is produced on potato extract-glucose agar, but melanin pigment is not produced on tyrosine agar. Adonitol, cellobiose, fructose, inositol, inulin, mannitol, raffinose, rhamnose, and xylose are utilized, but no acid is produced from these carbon sources. Alanine, histidine, and proline are utilized. Starch, cellulose, and lecithin are degraded. Hydrogen sulfide is produced. Autolysis of aerial hyphae is observed on yeast extract-malt extract agar and nutrient agar. Growth occurs between 45 and 50°C. Unlike other species of the genus, the phospholipid pattern is type IV, containing phosphatidylcholine and glucosamine-containing phospholipids in addition to phosphatidylethanolamine.

Source: soil.

DNA G+C content (mol%): not determined.

Type strain: XJ-54, CCTCC AA 97021, DSM 44391, JCM 11270.

Sequence accession no. (16S rRNA gene): AJ306300.

Genus XIV. **Saccharopolyspora** Lacey and Goodfellow 1975, 76[AL] emend. Korn-Wendisch, Kempf, Grund, Kroppenstedt and Kutzner 1989, 438

Seung Bum Kim and Michael Goodfellow

Sac.cha.ro.po.ly.spo'ra. N.L. n. *Saccharum* generic name of sugar cane; Gr. adj. *polus* many; Gr. n. *spora* a seed, and in biology a spore; N.L. fem. n. *Saccharopolyspora* the many spored (organism) from sugar cane.

Aerobic, Gram-stain-positive, non-acid-fast, nonmotile, catalase-positive actinobacteria which form an extensively branched substrate mycelium **that typically fragments into coccoid and/or rod-shaped elements**. In some species, the substrate hyphae remain intact or are partially transformed into chains of spores. **Aerial hyphae, when present, generally differentiate into bead-like chains of spores contained within a smooth sheath.** Spores are borne in straight, flexuous, hooked, looped, or spiral chains.

Spore surfaces can be hairy, smooth, spiny, rough, or warty. Substrate mycelia may be buff, brownish red, orange, or yellow and aerial mycelia are white to gray or pinkish white. Diverse compounds are used as sole carbon sources for energy and growth. **Whole-organism hydrolysates contain *meso*-diaminopimelic acid, arabinose, and galactose. Muramic acid moieties are *N*-acetylated. Cells contain tetrahydrogenated menaquinones with nine isoprene units as the predominant menaquinone, phosphatidylcholine, phosphatidylethanolamine, and phosphatidylmethylethanolamine as major polar lipids, and fatty acid profiles rich in iso- and anteiso-branched chain components, but lack mycolic acids.** The phylogenetic position of *Saccharopolyspora*, as determined by 16S rRNA gene sequencing, is in the family *Pseudonocardiaceae*.

DNA G+C content (mol%): 66–77.

Type species: **Saccharopolyspora hirsuta** Lacey and Goodfellow 1975, 78[AL].

Further descriptive information

Phylogeny. The genus *Saccharopolyspora* forms a distinct line of descent in the 16S rRNA *Pseudonocadiaceae* tree (Labeda et al., 2010a; Lu et al., 2001). The relationships between *Saccharopolyspora* species with validly published names are shown in Figure 316. The two most closely related taxa, *Saccharopolyspora hirsuta* subsp. *kobensis* and *Saccharopolyspora jiangxiensis*, share a 16S rRNA gene sequence similarity of 99% and the two most distantly related species, *Saccharopolyspora erythraea* and *Saccharopolyspora thermoflava*, share a similarity of 92%.

Cell morphology. Most saccharopolysporae, like the well-studied type strains of *Saccharopolyspora hirsuta* and *Saccharopolyspora rectivirgula* (formerly *Faenia rectivirgula*), form an extensively branched substrate mycelium, carrying aerial hyphae (Lacey, 1989a, 1989c; Locci, 1976). In *Saccharopolyspora hirsuta*, some substrate hyphae, like those of *Nocardia* strains, fragment into chains of cells in angular opposition (Figure 317), whereas others remain stable (Figure 318). Fragmented hyphae are most abundant in older parts of cultures, although they usually occur together with sterile hyphae. Substrate hyphae remain intact in some species such as *Saccharopolyspora halophila* (Tang et al., 2009a) and *Saccharopolyspora tripterygii* (Li et al., 2009a). In *Saccharopolyspora rectivirgula*, branching of the substrate hyphae is almost at right angles with chains of spores, mostly on short unbranched lateral and terminal sporophores (Figure 319). The developmental micromorphology of the organism has been studied by Locci (1976).

Aerial hyphae which differentiate into chains of spores are typical of *Saccharopolyspora* strains, as illustrated in Figure 320 and Figure 321. The spores of some species, such as *Saccharopolyspora hirsuta*, *Saccharopolyspora hordei*, and *Saccharopolyspor aspinosa*, are separated by lengths of apparently empty hyphae giving a characteristic bead-like appearance (Goodfellow et al., 1989b; Lacey and Goodfellow, 1975; Mertz and Yao, 1990). The spore chains of *Saccharopolyspora hirsuta* may be straight but are usually in loops or spirals (Korn-Wendisch et al., 1989; Lacey and Goodfellow, 1975). In contrast, straight to flexuous chains of 6–10 spores are seen in *Saccharopolyspora jiangxiensis*

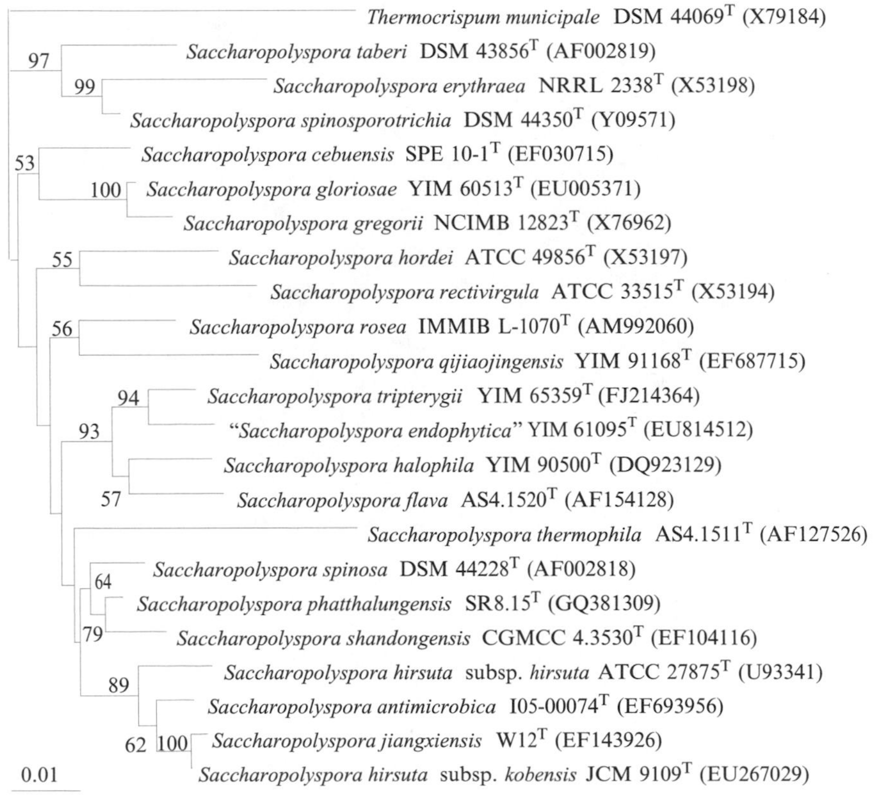

FIGURE 316. Neighbor-joining tree (Saitou and Nei, 1987) based on nearly complete 16S rRNA gene sequences showing relationships between *Saccharopolyspora* species. Numbers at the nodes indicate levels of bootstrap support (Felsenstein, 1985) based on an analysis of 1000 resampled datasets. Only values over 50% are given. Bar = 0.01 substitutions per nucleotide position.

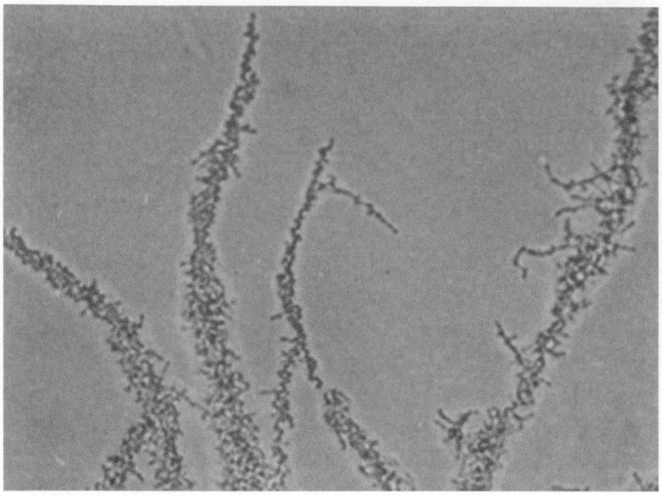

FIGURE 317. Fragmentation of substrate mycelium of *Saccharopolyspora hirsuta*. Glycerol-asparagine agar, incubation 40°C (500×). (Reproduced with permission from Lacey and Goodfellow, 1975. J. Gen. Microbiol. *88*: 75–85.)

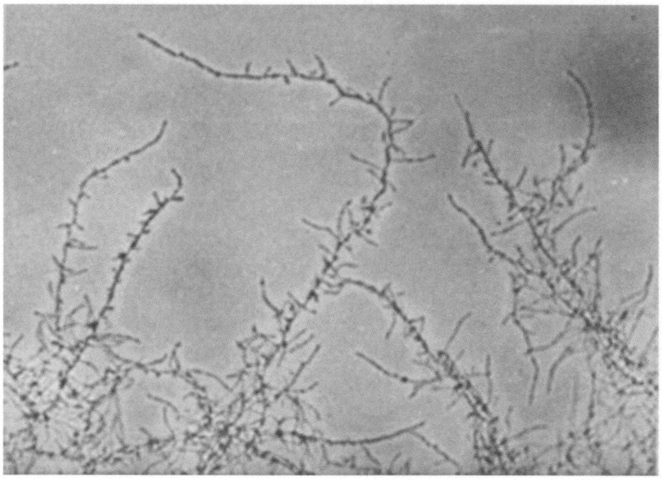

FIGURE 318. Morphology of substrate mycelium of *Saccharopolyspora hirsuta*. Glycerol-asparagine agar, incubation 40°C (550×). (Reproduced with permission from Lacey and Goodfellow, 1975. J. Gen. Microbiol. *88*: 75–85.)

(Zhang et al., 2009), hooked or flexuous chains of 4–6 spores in *Saccharopolyspora thermoflava* (Lu et al., 2001), spiral chains in *Saccharopolyspora spinosporotrichia* (Zhou et al., 1998), and short, straight chains in *Saccharopolyspora rectivirgula* (Figure 322).

Most *Saccharopolyspora* species have spores with smooth surfaces (Table 255). However, ornamented spores are not uncommon, as exemplified by the presence of hairy spores in *Saccharopolyspora hirsuta* (Lacey and Goodfellow, 1975), spiny spores in *Saccharopolyspora shandongensis* (Zhang et al., 2008b), warty spores in *Saccharopolyspora spinosporotrichia* (Zhou et al., 1998), and rough spores in *Saccharopolyspora rectivirgula*, as shown in Figure 322. Spores may be spherical as in *Saccharopolyspora spinosporotrichia* (Zhou et al., 1998), round to oval as in *Saccharopolyspora hirsuta* (Lacey and Goodfellow, 1975), or vesicular as shown by *Saccharopolyspora thermophila* (Lu et al., 2001). The spores of some species are covered by a sheath, as in *Saccharopolyspora gregorii*, *Saccharopolyspora hirsuta*, *Saccharopolyspora hordei*, and *Saccharopolyspora spinosa* (Goodfellow et al., 1989b; Lacey and Goodfellow, 1975; Mertz and Yao, 1990).

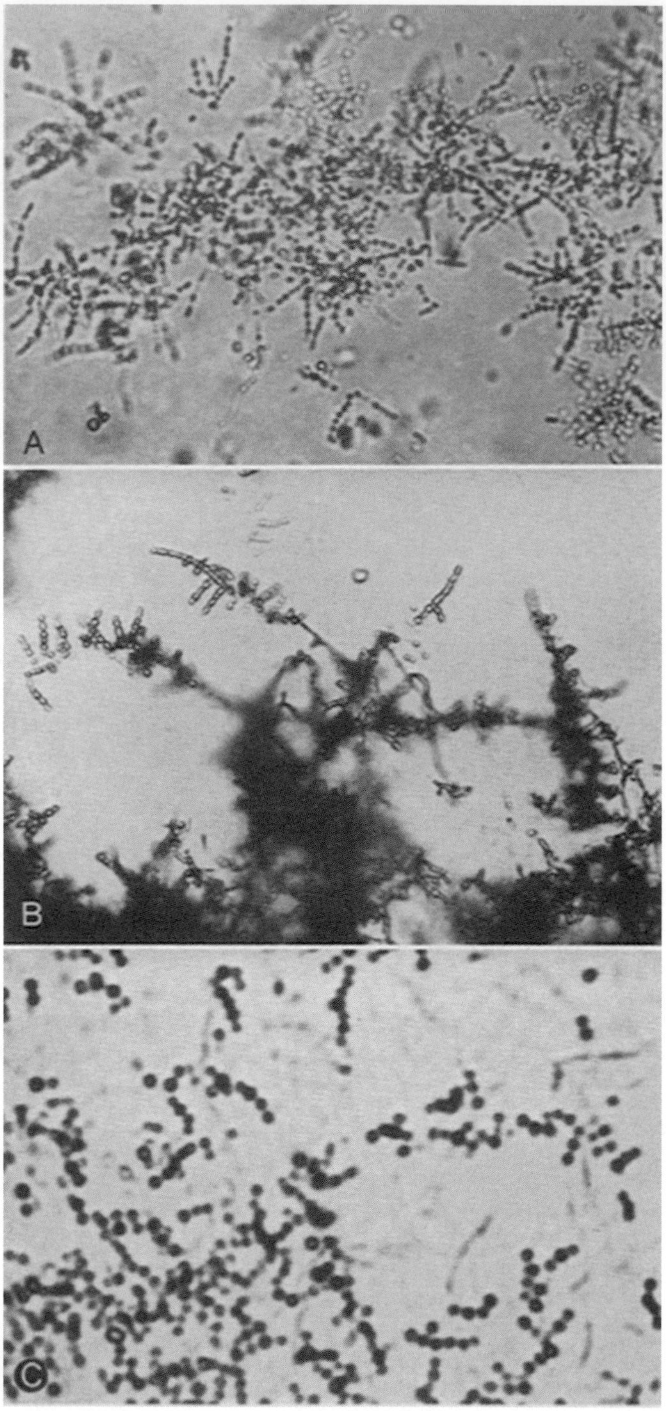

FIGURE 319. Morphology of substrate mycelium of *Saccharopolyspora rectivirgula*. (A) Appearance near growing margin (×650). (B) Typical right angle branching (650×). (C) Spore chains in older part of colony (1300×). Half-strength nutrient agar, 55°C.

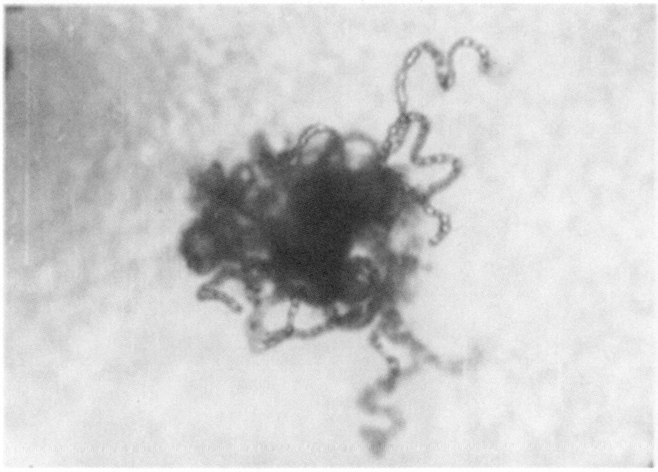

FIGURE 320. Spore chains on aerial mycelium of *Saccharopolyspora hirsuta* showing tufted appearance and typical curved chains. Half-strength nutrient agar, incubation 40°C (800×). (Reproduced with permission from Lacey and Goodfellow, 1975. J. Gen. Microbiol. *88*: 75–85.)

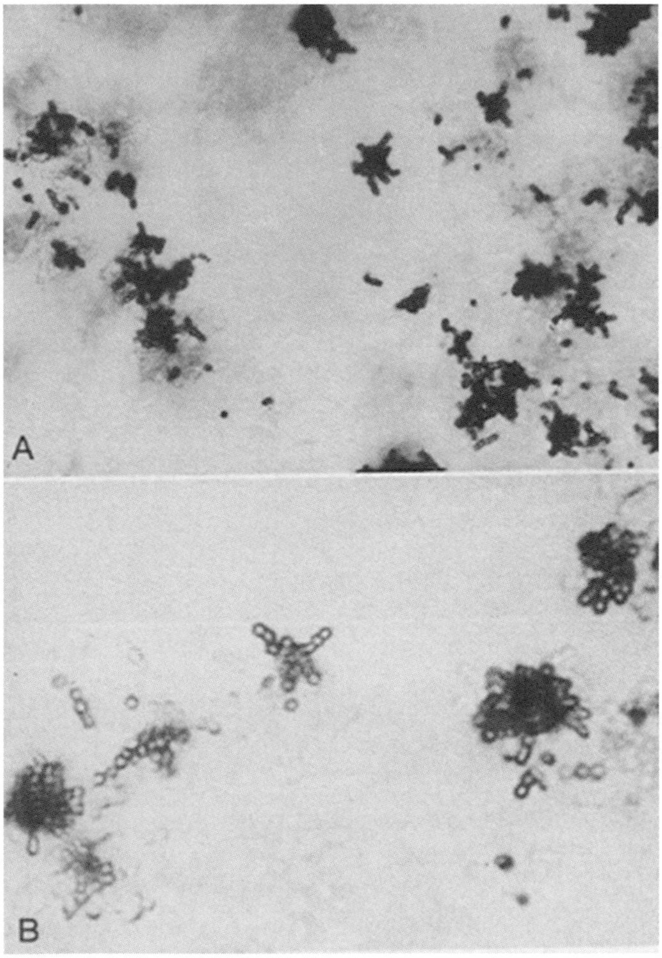

FIGURE 321. Aerial mycelium of *Saccharopolyspora rectivirgula* showing (A) sparse, tufted appearance (×300) and (B) formation of spore chains (780×). Half-strength nutrient agar, 55°C.

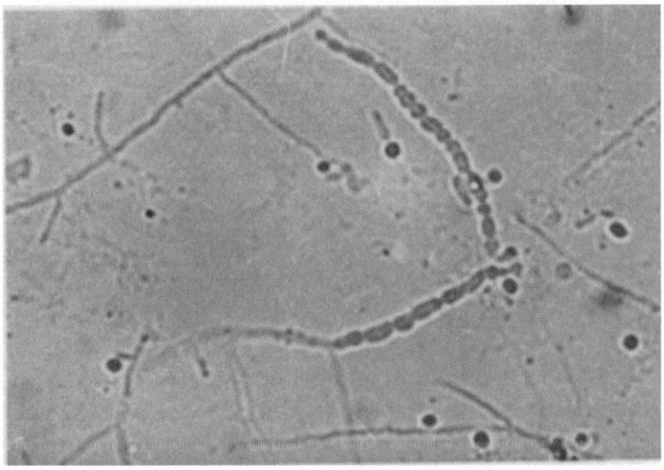

FIGURE 322. Scanning electron micrographs of (A) sporulating hyphae (3000×) and (B) spores (13,000×) of *Saccharopolyspora rectivirgula*.

Fine structure. In general, studies on *Saccharopolyspora hirsuta* and *Saccharopolyspora rectivirgula* show that their cell-wall structure resembles that of other actinobacteria (Lacey, 1989c; Lacey and Goodfellow, 1975). In thin section, *Saccharopolyspora hirsuta* hyphae (Figure 323) are bound by a wall 22–30 nm thick. Within this, a typical unit membrane encloses granular cytoplasm with axial diffuse nuclear material. Electron-transparent vacuoles, up to 0.3 µm in diameter and resembling lipid accumulations in other filamentous actinobacteria which undergo fragmentation (Williams et al., 1976), are sometimes abundant. Also occasionally present are electron-dense granules, up to 0.1 µm in diameter, resembling metachromatic or polyphosphate granules. Septation occurs by double ingrowth of the wall leading to fragmentation (type II; Williams et al., 1973). This may be associated with lamellar mesosomes up to 0.25 µm in diameter.

The sheath surrounding the spores of *Saccharopolyspora hirsuta* is 18–36 nm thick. It carries tufts of structureless hairs, triangular and 0.2–0.3 µm across at the base, which extend into apical filaments about 20 nm in diameter. The hairs are long, straight, or curved, and brittle (Figure 324). The morphology of the hairs is best seen on lengths of empty sheath (Figure 325) or by scanning electron microscopy (Figure 326). Spore walls are thickened uniformly to 50–60 nm, but their internal structure resembles that of hyphae, though they contain few vacuoles (Figure 327).

Two types of hyphae have been distinguished in thin sections of *Saccharopolyspora rectivirgula*, one having walls 19–25 nm thick and the other with walls 11–15 nm thick (Dorokhova et al., 1970). The cytoplasm in the thicker-walled cells is uniformly fine grained with a large nuclear zone extending the full-length of the cell. In thinner walled cells, the cytoplasm is less compact and homogeneous and the nuclear zone appears as small areas of low density. Mesosomes are less developed than in the thicker wall cells. Hyphae tend to autolyze during prolonged incubation at 55°C or at room temperature.

The spore chains of *Saccharopolyspora rectivirgula* are surrounded by a multilayered sheath, although this is less evident on spore chains formed on the substrate mycelium than on those produced on the aerial mycelium (Dorokhova et al., 1969; Williams et al., 1976). The spores are covered by a wall

TABLE 255. Differential properties of the type strains of species belonging to the genus *Saccharopolyspora*[a,b]

Property	1. *S. hirsuta*	2. *S. antimicrobia*	3. *S. cebuensis*	4. "*S. endophytica*"	5. *S. erythraea*	6. *S. flava*	7. *S. gloriosae*	8. *S. gregorii*	9. *S. halophila*	10. *S. hordei*	11. *S. jiangxiensis*	12. *S. phatthalungensis*	13. *S. qijiaojingensis*	14. *S. rectivirgula*	15. *S. rosea*	16. *S. shandongensis*	17. *S. spinosa*	18. *S. spinosporotrichia*	19. *S. taberi*	20. *S. thermo phila*	21. *S. tripterygii*
Spore chains	Straight to loose spirals	Straight	Straight	Straight to loose spirals	Open spirals	Straight	Hooks/ curved	Hooks/ flexuous	Straight	Hooks/ spirals	Straight to flexuous	Hooks/ open loops	Straight	Straight	Straight	Spiral	Hooks/ open loops	Spiral	–	Hooked/ flexuous	Straight
Spore surfaces	Hairy	Rough	Smooth	Smooth	Spiny	Smooth	Smooth	Smooth	Smooth	Smooth	Smooth or irregularly rough	Spiny	Smooth	Smooth or irregularly rough	Smooth	Spiny	Spiny	Warty	–	Smooth	Smooth
Degradation of:																					
Adenine	+	+	–	+	+	+	+	–	–	+	+	–	–	–	+	+	–	–	+	+	nd
Casein	+	+	–	–	–	–	–	+	+	+	–	+	–	–	+	+	–	+	+	–	nd
Chitin	–	–	–	+	+	–	–	–	–	+	–	–	–	–	nd	–	–	–	+	–	nd
Esculin	+	+	+	–	+	+	+	+	+	+	+	–	–	+	+	+	+	+	+	+	+
Elastin	+	–	nd	nd	+	–	nd	+	nd	+	+		nd	–	–	+	–	+	+	–	nd
Hypoxanthine	+	+	–	+	+	+	+	+	+	–	–	+	+	+	+	+	+	nd	+	+	nd
Starch	+	+	+	+	+	+	+	+	–	+	+	–	–	+	nd	+	–	+	+	+	–
Tyrosine	+	+	+	+	+	–	+	+	+	+	+	+	–	+	+	+	+	–	+	+	nd
Urea	+	+	nd	–	nd	+	nd	–	+	–	+	+	–	+	+	+	+	+			+
Xanthine	+	–	nd	+	+	+	+	+	nd	+	+	–	+	+	–	+	–	–	+	–	nd
Nitrate reduction	–	+	–	+	+	+	–	–	+	–	+	–	–	+	–	+	+	–	+	–	–
NaCl tolerance (w/v)	<7	≤7	2.5–12.5	≤15	<5	7	≤11	13	3–20	<13	<11	<7	6–22	<10	nd	<7	<11	2–3	7	7	≤12
Temperature range (°C)	25–50	20–45	15–37	20–45	20–42	28–37	10–32	10–35	10–45	20–60	15–45	18–42	20–40	37–63	22–42	15–38	15–37	28–37	20–45	45–55	10–37
Growth on carbon sources:																					
L-Arabinose	–	+	+	+	+	–	+	+	+	+	+	–[c]	–	–	+	+	+	–	–	–	+
D-Galactose	+	+	+	+	+	+	–	+	+	+	+	+[c]	+	+	+	+	–	+	+	+	+
D-Lactose	+	+	+	+	–	+	–	–	+	+	+	nd[c]	+	+	–	–	–	–	+	+	+
Maltose	+	+	+	+	+	+	+	+	+	+	+	–[c]	+	+	+	+	–	–	+	+	+
Mannitol	+	+	–	+	+	+	+	+	+	+	+	+[c]	+	+	+	+	+	+	+	+	+
Raffinose	+	+	+	+	+	+	–	+	+	+	+	–[c]	nd	+	–	+	–	+	+	+	+
L-Rhamnose	+	+	+	+	+	+	+	+	+	+	+	–[c]	+	+	–	+	–	+	+	+	–
Sucrose	+	+	+	+	+	+	–	+	+	+	+	–[c]	–	+	+	+	–	+	+	+	–
D-Xylose	+	+	+	+	+	+	+	+	+	+	+	+[c]	–	+	+	+	–	+	+	–	+

[a]Data for type strains. All strains were positive for the utilization of fructose, glucose, and mannose as the sole carbon sources for energy and growth.

[b]+, Positive; –, negative; nd, not determined.

[c]Determined by acid production from substrate.

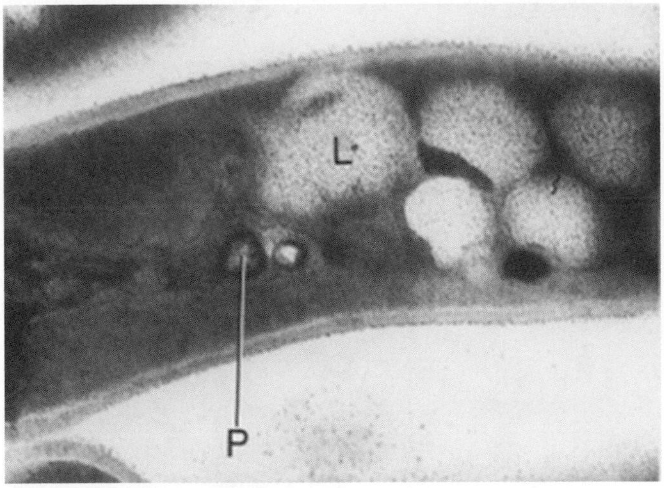

FIGURE 323. Longitudinal section of hyphae of *Saccharopolyspora hirsuta* showing possible lipid accumulation (L) and polyphosphate granules (P) (70,000×). (Reproduced with permission from Lacey and Goodfellow, 1975. J. Gen. Microbiol. *88*: 75–85.)

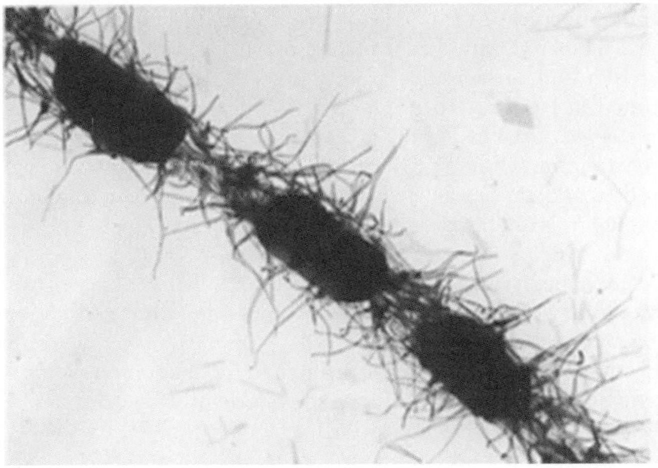

FIGURE 324. Electron micrograph of spore chain of *Saccharopolyspora hirsuta* (18,400×). (Reproduced with permission from Lacey and Goodfellow, 1975. J. Gen. Microbiol. *88*: 75–85.)

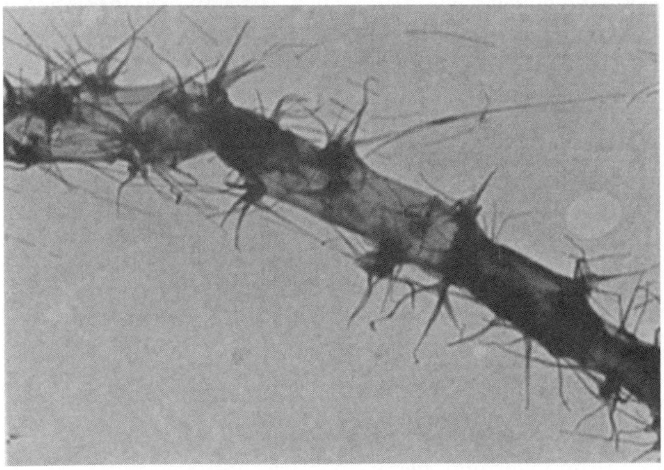

FIGURE 325. Electron micrograph of spore sheath of *Saccharopolyspora hirsuta* showing tufted production of hairs (17,600×). (Reproduced with permission from Lacey and Goodfellow, 1975. J. Gen. Microbiol. *88*: 75–85.)

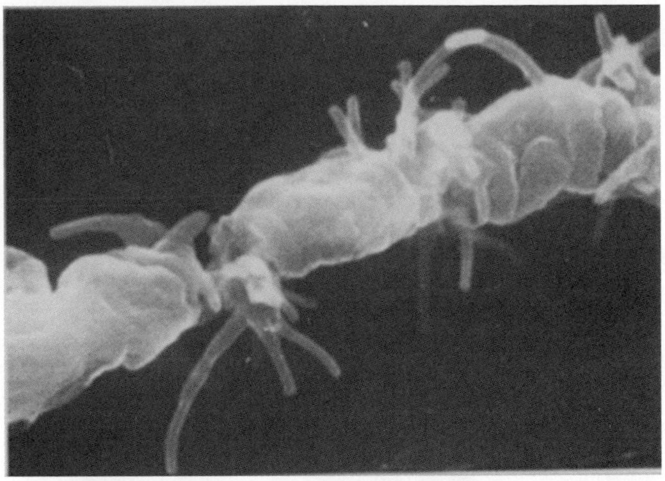

FIGURE 326. Scanning electron micrograph of spores of *Saccharopolyspora hirsuta* (15,000×). (Reproduced with permission from Lacey and Goodfellow, 1975. J. Gen. Microbiol. *88*: 75–85.)

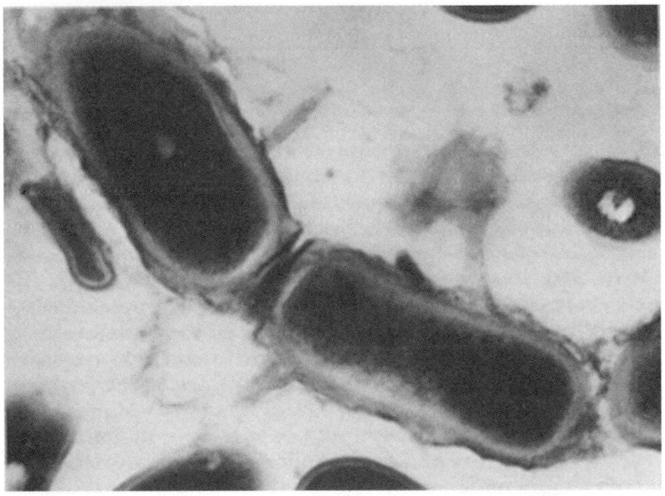

FIGURE 327. Longitudinal section of mature spore chain of *Saccharopolyspora hirsuta* showing sheath and hair bases (34,000×). (Reproduced with permission from Lacey and Goodfellow, 1975. J. Gen. Microbiol. *88*: 75–85.)

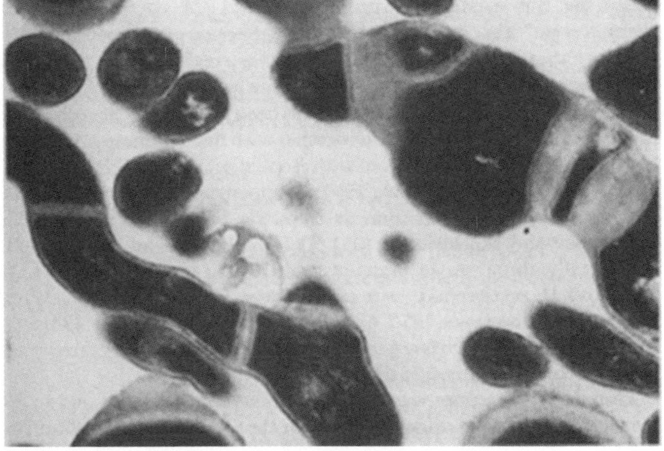

FIGURE 328. Sections of mycelium of *Saccharopolyspora rectivirgula* showing double septa in normal hyphae and irregular thickened septa in enlarged hyphae or aberrant spore chains (25,000×).

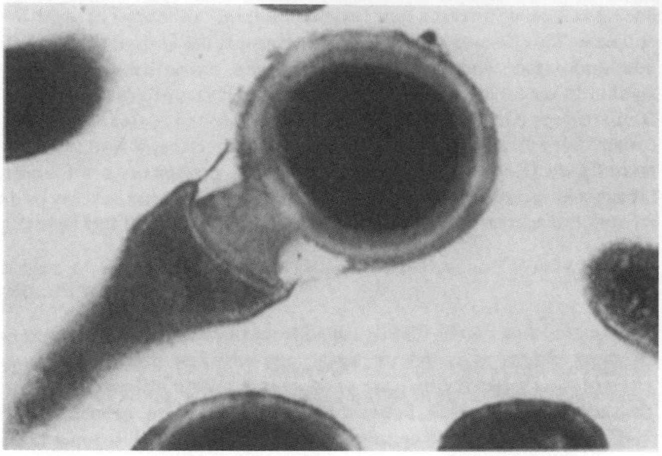

FIGURE 329. Longitudinal section through a developing spore of *Saccharopolyspora hirsuta* (40,000×).

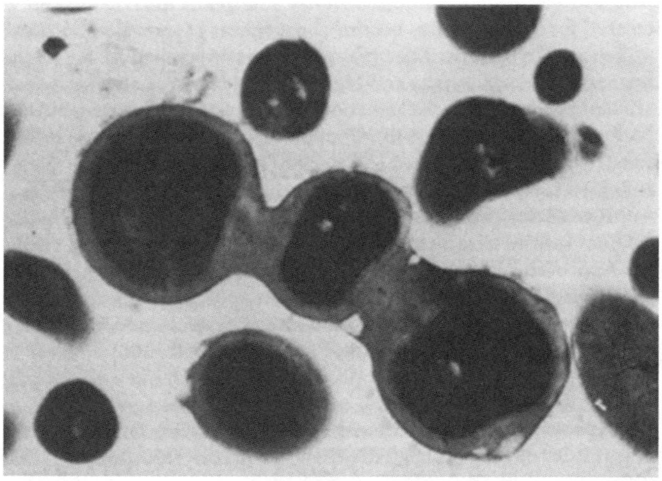

FIGURE 330. Longitudinal section through a developing spore of *Saccharopolyspora rectivirgula* showing interspore pads (25,000×).

70–100 nm thick in which two layers may be distinguished (Figure 328 and Figure 329), differing in thickness and electron density. Additional thickening of the cross-walls usually occurs giving characteristic interspore pads (Dorokhova et al., 1969) (Figure 330). These may sometimes be seen by light microscopy of stained preparations as conspicuous non-staining areas (Cross et al., 1968), but they may break down as the spore matures (Dorokhova et al., 1969). Plasmodesmata have been observed within the interspore pads. The protoplast is separated from the wall by a membrane and contains small, dark, densely packed ribosomes. Mesosomes are well developed and often adjoin the nuclear material. Although the spores are characteristically round or oval, spores of irregular shape are often seen in sections.

Colony morphology. Saccharopolysporae grow well on most standard media used to cultivate filamentous actinomycetes, such as Czapek's, glucose-asparagine (ISP medium 5), inorganic salts-starch (ISP medium 4), Sauton's, and V-8 vegetable juice agars. Colonies vary in size, in the extent of aerial hyphae production, and in aerial and substrate mycelia pigments. Those of *Saccharopolyspora hirsuta*, for instance, are thin,

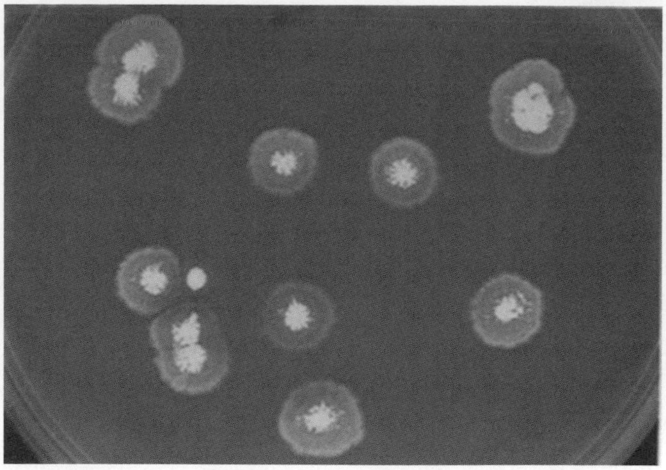

FIGURE 331. Colonies of *Saccharopolyspora hirsuta*. Half-strength nutrient agar, incubation 37°C (×1). (Reproduced with permission from Lacey and Goodfellow, 1975. J. Gen. Microbiol. *88*: 75–85.)

round or convex, wrinkled, and grow to about 1 cm in diameter in 7 d at 40°C, with a central area of white aerial mycelium on an almost colorless substrate mycelium (Lacey, 1975), as illustrated in Figure 331. In contrast, colonies of *Saccharopolyspora rectivirgula* grow to 5 mm in diameter in 7 d at 40–50°C; the substrate mycelium may be colorless, brown-yellow, or orange-yellow and the aerial mycelium may be white to light pink, though it is often sparse or absent (Korn-Wendisch et al., 1989; Lacey, 1989c). The aerial mycelium of *Saccharopolyspora gregorii* is also at best sparse (Goodfellow et al., 1989b), whereas macroscopically visible aerial hyphae have not been observed in *Saccharopolyspora taberi* (Labeda, 1987). *Saccharopolyspora flava* only produces aerial hyphae after prolonged incubation on oatmeal agar (Lu et al., 2001). Additional details on other species can be found in the species descriptions.

Chemotaxonomy. *Saccharopolyspora* species have a wall peptidoglycan which is characterized by the presence of *meso*-diaminopimelic acid, arabinose, and galactose (wall chemotype IV *sensu* Lechevalier and Lechevalier, 1970), an A1γ peptidoglycan (Schleifer and Kandler, 1972), and *N*-acetylated muramic acid (Duangmal et al., 2010). They have fatty acid profiles rich in iso- and anteiso-branched components in which $C_{15:0}$ iso, $C_{16:0}$ iso, and $C_{17:0}$ anteiso tend to predominate (Embley et al., 1987; Goodfellow et al., 1989b; Tang et al., 2009a, 2009b) and lack mycolic acids (Mertz and Yao, 1990; Minnikin et al., 1975; Pimentel-Elardo et al., 2008; Zhou et al., 1998), but typically contain major amounts of phosphatidylcholine, phosphatidylethanolamine, and phosphatidylmethylethanolamine (Embley et al., 1988b; Korn-Wendisch et al., 1989; Pimentel-Elardo et al., 2008) with a variable distribution of diphosphatidylglycerol, phosphatidylglycerol, phosphatidylinositol, and phosphatidylinositol mannosides (Li et al., 2009a; Yassin, 2009; Yuan et al., 2008) and, hence, have polar lipid pattern type III after Lechevalier et al. (1981, 1977b). A major glycolipid has been detected in *Saccharopolyspora erythraea*, *Saccharopolyspora hirsuta*, and *Saccharopolyspora rectivirgula* (Gamian et al., 1996). The predominant menaquinones are tetrahydrogenated with nine isoprene units [MK-9(H_4)] (Embley et al., 1988b; Korn-Wendisch et al., 1989; Tang et al., 2009a, 2009b), though smaller amounts of MK-9(H_2, H_6, H_8) may be present (Alderson et al., 1985; Collins et al., 1977; Lu et al., 2001; Qin et al., 2008a). The type strain of *Saccharopolyspora thermophila* contains MK-9(H_6, H_8) as the predominant isoprenologs (Lu et al., 2001), though this apparently anomalous result may reflect the time when the culture was sampled as there is evidence that the menaquinone composition of *Amycolatopsis* and *Streptomyces* strains can be age-dependent (Saddler et al., 1986; Yassin et al., 1991). There is preliminary evidence that the esterase patterns of *Saccharopolyspora* strains may be species-specific (Korn-Wendisch et al., 1989).

Nutrition and growth conditions. In general, saccharopolysporae do not have any specific growth requirements and can grow on a diverse range of compounds as sole sources of carbon and nitrogen (Lacey and Goodfellow, 1975; Li et al., 2009a; Zhang et al., 2009). They grow well in media supplemented with salt, notably *Saccharopolyspora halophila* and *Saccharopolyspora qijiaojingensis*, which grow in the presence of 20% (w/v) NaCl (Tang et al., 2009a, 2009b). The type strain of *Saccharopolyspora cebuensis* has a strict requirement for salt which suggests that it is an obligate marine actinomycete (Pimentel-Elardo et al., 2008). Saccharopolysporae show a range of temperature requirements; some, such as *Saccharopolyspora hirsuta*, *Saccharopolyspora hordei*, *Saccharopolyspora rectivirgula*, and *Saccharopolyspora thermophila*, are moderate thermophiles (Goodfellow et al., 1989b; Lacey and Goodfellow, 1975; Lu et al., 2001). Most strains grow within the pH range 5.0–9.0 (Qin et al., 2008a; Tang et al., 2009b; Zhang et al., 2009).

Metabolism. Saccharopolysporae are aerobic, catalase-positive, chemo-organotrophic actinomycetes which have an oxidative metabolism. They degrade a broad range of organic substrates and use diverse compounds as sole carbon sources (Goodfellow et al., 1989b; Lacey and Goodfellow, 1975; Yuan et al., 2008). Isolates from fodder and other plant material typically degrade hypoxanthine, starch, Tweens, tyrosine, and xanthine, but not xylan (Goodfellow et al., 1989b; Lacey, 1989c).

Members of the genus are best known as a source of secondary metabolites, notably *Saccharopolyspora erythraea*, which is used for the industrial scale production of the clinically important macrolide antibiotic, erythromycin A, and *Saccharopolyspora spinosa*, the source of spinosyns A and D, glycosylated polyketide-derived macrolides which are commercially marketed as the insecticide spinosad (Hong et al., 2006, 2008; Huang et al., 2009). However, despite the commercial importance of these organisms, relatively little is known about their metabolic properties, apart from studies on erythromycin (Chen et al., 2008; Katz and Donadio, 1995; Reeves et al., 2006, 2007; Staunton and Weissman, 2001; Weissman et al., 2004) and spinosyn (Gaisser et al., 2001; Hong et al., 2008; Huang et al., 2009; Kim et al., 2007; Kirst et al., 1991, 1992; Waldron et al., 2001) biosynthesis and on host–vector systems developed for *Saccharopolyspora erythraea* (Gaisser et al., 2000). "*Saccharopolyspora pogona*" (Lewer et al., 2002) synthesizes pogonins (butylene-spinosyns), pericidal macrolides which are similar to the spinosyns produced by *Saccharopolyspora spinosa* (Hahn et al., 2006). Another taxon that does not have a validly published name, "*Saccharopolyspora aurantica*", forms a complex of pesticidal compounds designated CL307-24 (Etienne et al., 1993).

The erythromycins are broad-spectrum antibiotics active against Gram-stain-positive bacteria (Labeda, 1987; Staunton and Weissman, 2001). Erythromycin A, the most widely used and clinically effective member of the family, contains a characteristic 14-membered macrolide to which are attached two unusual deoxysugars, desosamine and mycarose (Chen et al., 2008). The entire gene cluster governing the biosynthesis of erythromycin A has been cloned and each open reading frame has been characterized by targeted gene inactivation. Erythromycins B and C, which are biologically less active and cause greater side effects, are intermediates in erythromycin A biosynthesis.

Spinosyns are unique macrolides with a tetracyclic core, which consists of a 12-membered macrocyclic lactone fused to a 5,6,5-*cis*-anti-*trans* tricyclic ring system, to which the deoxysugars forosamine and tri-*O*-methylated rhamnose are attached (Huang et al., 2009). The biosynthetic pathway of spinosyns have been determined, notably in studies based on precursor-labeling, identification of intermediate metabolites using blocked mutants, and by *in vitro* analysis of enzymes involved in spinosyn biosynthesis (Kirst et al., 1993). To date, more than 25 spinosyns have been isolated and identified from *Saccharopolyspora spinosa* (Crouse et al., 2001). The most abundant spinosyns from fermentation broths of this organism are spinosyn A (approx. 85% of spinosad) and spinosyn D (approx. 15% of spinosad). Spinosad kills susceptible insects by causing rapid excitation of the insect nervous system, probably through the interaction and binding of the δ-aminobutyric acid and nicotinacetylcholine receptor sites (Millar and Denholm, 2007).

Bioactive compounds are also produced by other *Saccharopolyspora* species. *Saccharopolyspora hirsuta* synthesizes a macrolide, nargenicin A (Ikeda et al., 1985), an aminoglycoside complex, apramycin and derivatives (Kamiya et al., 1983; O'Connor et al., 1976), and a cyclic polyketide, nodusmicin (Whaley et al., 1980). *Saccharopolyspora hirsuta* subsp. *kobensis* produces sporaricin and related aminoglycoside antibiotics (Deushi et al., 1979; Umezawa et al., 1987) and *Saccharopolyspora* sp. strain AC 3440 produces 4-deamino-4-hydroxyapromycin (Awata et al., 1983). *Saccharopolyspora erythraea* is a rich source of bioactive compounds other than erythromycin, including erythronolide B (Martin and Rosenbrook, 1967), an *N*-acetylmuramidase (Morita et al., 1978), a trypsin-like protease (Yoshida et al., 1971), and a rennin-like enzyme (Sternberg, 1976).

The genome of *Saccharopolyspora erythraea* NRRL 2338^T has been sequenced (Oliynyk et al., 2007), a development which is promoting interest in the metabolism and biotechnological exploitation of this and related actinobacteria (Katz and Khosla, 2007; Peano et al., 2007). The genome of this organism shows considerable divergence from those of streptomycetes in gene organization and function thereby confirming previous taxonomic insights (Labeda, 1987). Oliynyk and his colleagues reported that the *Saccharopolyspora erythraea* genome contained at least 25 gene clusters for the production of known and predicted secondary metabolites, at least 72 genes predicted to confer resistance to a range of common antibiotic classes and many sets of duplicated genes to support the saprophytic way of life of the organism. The latter included genes involved in defense and stress responses, in ensuring or preserving correct protein folding, and those encoding a wide range of degradative enzymes, including seven chitinases and multiple glucanases and proteinases. The genome sequence of *Saccharopolyspora erythraea* sets the stage for understanding the biology of this organism both in nature and in the fermentation process and for explaining the higher levels of erythromycin production in industrially significant strains (Katz and Khosla, 2007).

Genetics. Most studies have been directed towards mapping and cloning of the erythromycin biosynthesis and resistance genes of *Saccharopolyspora erythraea* (Baltz et al., 1986; Bibb et al., 1986; Stanzak et al., 1986; Tuan et al., 1986; Vanden Boom, 2000; Weber et al., 1985) and on the cloning and analysis of the *Saccharopolyspora spinosa* spinosad biosynthetic gene cluster (Hong et al., 2008; Matsushima and Baltz, 1994; Matsushima et al., 1994; Waldron et al., 2001). The erythromycin genes are located on the chromosome, together with resistance and regulatory elements (Oliynyk et al., 2007; Stanzak et al., 1986; Weber et al., 1985). Most of the *Saccharopolyspora spinosa* genes involved in spinosyn biosynthesis are present in a 74 kb cluster with characterization studies suggesting that spinosyns are synthesized by mechanisms similar to those used to assemble complex macrolides from primary metabolic precursors as in other actinobacteria (Waldron et al., 2001). These workers noted that several unusual genes in the spinosyn gene cluster might encode enzymes which generate the tetracyclic polyketide aglycone nucleus, the most striking structural feature of spinosyns.

Integrated and conjugative elements (AICEs) have been detected in *Saccharopolyspora erythraea* (Brown et al., 1988, Brown et al., 1994; te Poele et al., 2008). te Poele and her colleagues found two novel putative AICEs in the *Saccharopolyspora erythraea* genome, one of which (PSE102) encoded a putative aminoglycoside phosphotransferase which may confer antibiotic resistance to the host. They also found that the AICEs of *Saccharopolyspora erythraea*, *Amycolatopsis mediterranei*, and *Amycolatopsis methanolica* have a highly conserved structural organization which consists of four functional modules (conjugative transfer, excision/integration, regulation, and replication). Identification and characterization of mobile genetic elements is important for the manipulation of commercially significant *Saccharopolyspora* strains.

The genes responsible for the biosynthesis of erythromycin were cloned into but not well expressed in "*Streptomyces lividans*" (Stanzak et al., 1986; Thompson et al., 1982), a problem reflecting the phylogenetic gulf between *Saccharopolyspora* and *Streptomyces* strains (Embley et al., 1988a) and, thereby, emphasizing the need to develop cloning systems based on indigenous *Saccharopolyspora* vectors (Gayer-Herkert et al., 1989; Katz et al., 1988). There is evidence that *Streptomyces* phages are unstable in *Saccharopolyspora* strains (Gayer-Herkert et al., 1989; Yamamoto et al., 1986). Plasmids have been detected in the type strain of *Saccharopolyspora erythraea* (Brown et al., 1986; Chiang et al., 1985) and methods for the transformation of *Saccharopolyspora erythraea* (Yamamoto et al., 1986) and *Saccharopolyspora rectivirgula* (Gayer-Herkert et al., 1989) have been described.

The *Saccharopolyspora spinosa* spinosad biosynthetic gene cluster has been cloned, analysed, and shown to contain five large open reading frames which encode a multifunctional type 1 polyketide synthase (Waldron et al., 2001). The enzymes involved in D-forosamine [(4-dimethylamino)-2,3,4,6-tetradeoxy-β-D-*threo*-hexopyranose] biosynthesis in the spinosyn pathway have been cloned and expressed heterologically and the corresponding proteins have been purified and their activities examined *in vitro*, developments which have elucidated the mechanisms for the synthesis of this highly deoxygenated sugar (Hong et al., 2008). Indeed, these studies have provided the basis for future work on the biosynthesis of 2,3,6-trideoxy- and 2,3,4,6-tetradeoxyhexoses present in many bioactive natural products.

The sequenced chromosome of the type strain of *Saccharopolyspora erythraea* (8.2 kb) is circular like those of *Amycolatopsis mediterranei*, *Corynebacterium diphtheriae*, and *Mycobacterium tuberculosis*, but unlike the linear chromosomes of "*Streptomyces coelicolor*" A(3)2 and *Streptomyces avermitilis* MA-4680 (Oliynyk et al., 2007; Peano et al., 2007; Zhao et al., 2010). Approximately, half of the chromosome, which contains more than 7200 genes, consists of a core region which includes most of the genes required for primary metabolism, cell division, information transfer, and sporulation (Katz and Khosla, 2007). Erythromycin biosynthesis is encoded in the core region, which suggests that this natural product has been integral to the evolutionary success of this organism. The remaining half of the chromosome contains most of the remaining gene clusters involved in secondary metabolism, including genes for the biosynthesis of unknown polyketides, ribosomal peptides, and terpenoids.

It is clear that *Saccharopolyspora erythraea* has a remarkable capacity for the production of secondary metabolites, a trait which seems likely to confer substantial advantages to the organism in the soil milieu. Indeed, it has been suggested that the ability to deploy a differential chemical arsenal may be a general evolutionary strategy used by actinobacteria which grow as filamentous mycelia in highly competitive environments (Challis and Hopwood, 2003; Jenke-Kodama et al., 2006). A detailed understanding of the transcriptional organization of the *Saccharopolyspora erythraea* chromosome has been attained by using GeneChip DNA microarrays derived from the complete gene sequence of this organism (Peano et al., 2007). The use of the *Saccharopolyspora erythraea* DNA microarray improved the specificity and sensitivity of gene expression analysis, allowing a global and, at the same time, detailed understanding of how *Saccharopolyspora erythraea* genes are modulated. The results confirmed that the erythromycin gene cluster is upregulated during the initial growth phase of the organism, and identified six additional clusters – for non-ribosomal peptides and terpenes – that are regulated in later growth phases.

Bacteriophages. Until recently, considerable interest was shown in *Saccharopolyspora* phages both as cloning vectors (Katz et al., 1988; Schneider and Kutzner, 1989) and taxonomic markers (Korn-Wendisch et al., 1989; Labeda, 1987; Prauser and Momirova, 1970). Most attention was focused on phages isolated from *Saccharopolyspora erythraea* and *Saccharopolyspora rectivirgula* (Donadio et al., 1986; Grund and Hutchinson, 1987; Katz et al., 1988; Kempf et al., 1987; Schneider et al., 1987) using methods similar to those developed for *Streptomyces* species (Hopwood et al., 1985; Lanning and Williams, 1982), albeit with minor modifications and the use of appropriate baiting and indicator strains (Greiner-Mai et al., 1987; Grund and Hutchinson, 1987; Kurup and Heinzen, 1978).

Some phages isolated from *Saccharopolyspora erythraea* were found to infect and multiply in *Saccharopolyspora rectivirgula* and *vice versa* (Korn-Wendisch et al., 1989; Smorawinska et al., 1988). Some *Saccharopolyspora erythraea* phages have been shown to infect *Saccharopolyspora hirsuta* and *Saccharopolyspora taberi* (Grund and Hutchinson, 1987; Korn-Wendisch et al., 1989). Phage life cycle and plaque morphology are also host-dependent, whereas phages lysogenic for one host may lyse another (Brzezinski et al., 1986; Kempf et al., 1987; Schneider and Kutzner, 1989).

Saccharopolyspora phages have a tail of variable length attached to an icosahedron capsid which encloses a double-stranded genome with *cos* termini present in some strains (Grund and Hutchinson, 1987; Katz et al., 1988; Kurup and Heinzen, 1978; Schneider et al., 1987; Schneider and Kutzner, 1989). The central regions of representative phages from *Saccharopolyspora erythraea* (Brzezinski et al., 1986) and *Saccharopolyspora rectivirgula* (Schneider et al., 1987) are variable due to the deletion or insertion of large gene fragments (Brzezinski et al., 1986; Schneider et al., 1987; Schneider and Kutzner, 1989; Smorawinska et al., 1988). Hybridization studies (Schneider and Kutzner, 1989) have shown that the same or related elements occur in the genomes of *Saccharopolyspora rectivirgula* phages representing different compatability groups and in a phage isolated from *Saccharopolyspora erythraea* (Brzezinski et al., 1986).

Antibiotic sensitivity patterns. In general, saccharopolysporae are resistant to a broad range of antibiotics though studies have been limited to species such as *Saccharopolyspora antimicrobica, Saccharopolyspora gregorii, Saccharopolyspora hirsuta, Saccharopolyspora hordei,* and *Saccharopolyspora spinosa.* The antibiotic sensitivity profiles of these and other *Saccharopolyspora* species are given in the species descriptions.

Pathogenicity. Saccharopolysporae are not known to cause infections though *Saccharopolyspora rectivirgula* is the chief agent of the extrinsic allergic alveolitis condition known as farmer's lung (Campbell, 1932; Pepys et al., 1963). This disease, which occurs widely in China, Europe, Japan, and the USA, is caused by the inhalation of large numbers of spores released when moldy substrates are disturbed in the presence of sensitized individuals (Lacey, 1981). In western Scotland and the Orkneys, it has been reported to affect up to 8.6% of farm workers (Grant et al., 1972), whereas in the US, 8.4% of Wisconsin dairy farmers had precipitins to actinobacteria, nearly 90% of them to *Saccharopolyspora rectivirgula* (Roberts et al., 1976). Farmer's lung is normally a chronic rather than an acute disease, although fatalities have been recorded (Lacey, 1988). *Saccharopolyspora hordei* spores are also found in high numbers in grain stores and may be involved as an agent of extrinsic allergic alveolitis (Lacey and Crook, 1988). Respiratory diseases resembling farmer's lung have been reported in cattle and horses exposed to moldy hay (Lacey, 1988; Pirie et al., 1971) and an acute outbreak resulting in several fatalities has been reported in Canadian cattle (Wilkie, 1978).

The criteria used for the diagnosis of extrinsic allergic alveolitis, including farmer's lung, have been considered by Lacey (1988). Antigens from *Saccharopolyspora rectivirgula* have been purified, which can be used for the detection of circulating antibodies in the sera of patients (Brummund et al., 1988; Mäntyjärvi and Kurup, 1988). An enzyme-linked immunosorbent assay has also been developed for this purpose (Ramasamy et al., 1987).

Antigenicity. The antigenicity of *Saccharopolyspora rectivirgula* has received a lot of attention since the organism, first known as *Thermopolyspora polyspora*, then as *Micropolyspora faeni* and *Faenia rectivirgula*, was implicated in farmer's lung (Pepys et al., 1963). Initially, three precipitin arcs were recognized in gel diffusion and immunoelectrophoresis tests using extracts of *Saccharopolyspora rectivirgula* and sera from farmer's lung patients. Using gel filtration, adsorption on DEAE columns and elution with crossed immunoelectrophoresis and immunodiffusion, the precipitins have been resolved into up to 75 individual antigenic components, as outlined by Lacey (1989a).

Ecology. Members of the genus *Saccharopolyspora* have been isolated from a broad range of habitats, notably plant material which has been stored and allowed to decay (Lacey, 1971, 1974, 1978). *Saccharopolyspora rectivirgula* is common in spontaneously heating vegetable material. It was first isolated from moldy hay which had been baled containing more than 35% water content and heated to 50–65°C (Gregory et al., 1963). Subsequently, it has been found in cereal grain, cotton bales, mushroom compost, straw, sugar corn bagasse, and the air over pastures (Lacey, 1978). Heating of the substrate is initiated by plant cells and mesophilic bacteria and fungi. The organism begins to grow at 30 to 35°C, but larger numbers are found in hays that heat to around 60°C after baling at about 39% water content. The change in pH from 5.5–6.0 to 7.0–8.0 caused by fungal proteolysis probably favors actinobacterial colonization though *Saccharopolyspora rectivirgula* can sometimes grow in hay without pretreatment (Gregory et al., 1963). It can survive in moist grains stored anaerobically in silos and its spores can withstand up to 20 min at 70°C.

Other species isolated from plant material include *Saccharopolyspora gregorii, Saccharopolyspora hirsuta,* and *Saccharopolyspora hordei. Saccharopolyspora hirsuta* is known mostly from moldy sugar cane bagasse that has heated spontaneously during storage. It was found in 12% of samples originating in India, Jamaica, Puerto Rico, and Trinidad (Lacey, 1974). Strains of *Saccharopolyspora hordei* are common in stored barley and hay and are found less frequently in sugar cane bagasse. This organism, together with *Saccharopolyspora rectivirgula,* was present in 15% of freshly harvested grain samples, but not in grass used to seal moist barley silos (Lacey, 1971). However, after storage they occurred in 33% of grass and straw samples and in 57% and 32% of grain samples in two seasons. *Saccharopolyspora rectivirgula* has been isolated from a range of vegetable material collected from different sites in the Nigerian states of Anambra and Enugu (Unaogu et al., 1994). *Saccharopolyspora gregorii* has been infrequently isolated from barley, stored hay, and straw (Goodfellow et al., 1989b). Saccharopolysporae may also be present on lichens (Gonzalez et al., 2005).

Developments in the systematics of *Saccharopolyspora* provide good grounds for the recognition of novel species isolated from diverse man-made and natural habitats. Novel species from terrestrial habitats include *Saccharopolyspora flava* and *Saccharopolyspora thermophila* from garden soil (Lu et al., 2001), *Saccharopolyspora phatthalungensis* from the rhizosphere of *Hevea brasiliensis* (Duangmal et al., 2010), *Saccharopolyspora shandongensis* from wheat field soil (Zhang et al., 2008b), *Saccharopolyspora tripterygii* from a surface-sterilized stem sample from *Trypterygium hypoglaucum* (Li et al., 2009a), and "*Saccharopolyspora endophytica*" from a root of *Mytenus austroyunnanensis* (Qin et al., 2008a). In addition, *Saccharopolyspora halophila* and *Saccharopolyspora qijiaojingensis* were isolated from saline salt lakes (Tang et al., 2009a, 2009b), *Saccharopolyspora cebuensis* was from a sponge (Pimentel-Elardo et al., 2008), *Saccharopolyspora rosea* was from a patient with bronchial carcinoma (Yassin, 2009), and *Saccharopolyspora spinosa* was from soil collected from a sugar mill rum still (Mertz and Yao, 1990).

Isolation procedures

The isolation of *Saccharopolyspora* from airborne dust, grain, fodder, and other vegetable material is best achieved by using an Andersen sampler (Andersen, 1958) to examine spore clouds generated using a wind tunnel or sedimentation chamber (Gregory and Lacey, 1963; Lacey and Dutkiewicz, 1976a, 1976b). Sedimentation of spore clouds should be allowed to proceed for 10–30 min in order to allow larger particles to settle as the lighter actinomycete spores remain airborne. The Andersen sampler should be loaded with appropriate isolation media supplemented with cycloheximide (50 μg/ml) to inhibit the growth of fungi (Cross et al., 1968).

Colonies of *Saccharopolyspora hirsuta* were first isolated from airborne sugar cane bagasse dust (Lacey, 1974) using a wind tunnel and an Andersen sampler (Lacey and Goodfellow, 1975). The medium, half-strength nutrient agar supplemented with cycloheximide, was incubated at 40°C following inoculation. The resultant colonies were colorless to pale brown, thin, and raised or convex and carried small tufts of white aerial mycelia in the centers. The same procedure was used to isolate *Saccharopolyspora gregorii* and *Saccharopolyspora hordei* from cereal grain and hay (Goodfellow et al., 1989b; Lacey, 1971). Isolation of *Saccharopolyspora gregorii* is enhanced at 30°C.

Dilution and direct plating procedures have been used to isolate *Saccharopolyspora rectivirgula*, but enumeration in hay and other vegetable matter is best achieved by using an Andersen sampler to impact airborne spores onto appropriate media. A range of media can be used, including half-strength nutrient agar (Lacey, 1971, 1974), half-strength tryptone soy agar supplemented with casein hydrolysates (Lacey, 1989a), R8 agar (Amner et al., 1989), and starch-casein-arginine agar (Iwasaki et al., 1979). Hippurate has been used as a selective carbon source for the isolation of *Saccharopolyspora rectivirgula* (Mattinson-Rose, 1986). This investigator used the membrane-filter method of Hirsch and Christensen (1983) to isolate *Saccharopolyspora hirsuta* from dilutions of substrate suspensions; only mycelial actinomycetes can grow through the pores in membrane filters and thereby reach and grow on the surface of isolation media.

Small numbers of taxonomically diverse saccharopolysporae have been isolated from aquatic and terrestrial habitats by plating suspensions of substrates onto non-selective media, as exemplified by the isolation of *Saccharopolyspora antimicrobica* on yeast extract-malt extract agar (Yuan et al., 2008) and *Saccharopolyspora flava* on oatmeal agar (Lu et al., 2001). Additional strains have been isolated using more selective, but nevertheless empirical procedures, as illustrated by the isolation of *Saccharopolyspora halophila* on cellulose-casein multi-salts agar (Tang et al., 2009a) and *Saccharopolyspora phatthalungensis* on starch-casein agar supplemented with ketokamazole and nalidixic acid (Duangmal et al., 2010). In general, inoculated plates were incubated at 28°C for between 1 and 4 weeks.

Maintenance procedures

Working cultures of *Saccharopolyspora* strains can be maintained using refrigerated (4°C) agar slopes on media such as glucose-yeast extract agar (Gordon and Mihm, 1962), V8 vegetable juice agar (Corbaz et al., 1963), and yeast extract-malt extract agar (ISP medium 2; Shirling and Gottlieb, 1966). Similarly, halophilic strains can be maintained on inorganic salts-starch agar (ISP medium 4; Shirling and Gottlieb, 1966). Longer term preservation of strains is best achieved as frozen suspensions of mycelia and spores at −80°C (mechanical freezer) to −172°C (liquid nitrogen vapor phase) or by using standard lyophilization techniques.

Differentiation of the genus *Saccharopolyspora* from other genera

Saccharopolyspora strains can be distinguished from members of other genera classified in the family *Pseudonocardiaceae* by using a combination of chemical and morphological markers and by comparative 16S rRNA gene sequence studies (see the family *Pseudonocardiaceae*, above). It can be separated from other genera in the family on the basis of PAGE and sequence analysis of ribosomal AT-L30 proteins (Ochi, 1995; Ochi and Yoshida, 1991) and by the use of a genus-specific primer (Morón et al., 1999).

Taxonomic comments

The genus *Saccharopolyspora* was proposed by Lacey and Goodfellow (1975) to encompass actinobacteria from sugar cane bagasse that produced aerial mycelia with bead-like chains of spores enclosed in a characteristic hairy sheath. Subsequently, Labeda (1987) transferred the type species of *Streptomyces erythraeus* to the genus *Saccharopolyspora* as *Saccharopolyspora erythraea* because its cell walls contained *meso*-diaminopimelic acid, arabinose, and galactose. He also proposed that "*Nocardia taberi*" be reclassified as *Saccharopolyspora hirsuta* subsp. *taberi*, an organism later given species status as *Saccharopolyspora taberi* (Korn-Wendisch et al., 1989). Another taxon, "*Saccharopolyspora hirsuta* subsp. *kobensis*" (Iwasaki et al., 1979) was formally recognized by these investigators. Two other taxa, "*Saccharopolyspora aurantiaca*" and "*Saccharopolyspora hirsuta* subsp. *kunmingensis*" were described for strains isolated from soil taken from a rice field (Jiang and Xu, 1986). Three other species, *Saccharopolyspora gregorii*, *Saccharopolyspora hordei*, and *Saccharopolyspora spinosa*, were early additions to the genus (Goodfellow et al., 1989b; Mertz and Yao, 1990).

There have been numerous twists and turns in the classification and nomenclature of *Saccharopolyspora rectivirgula*, an organism designated *Faenia rectivirgula* in the last edition of *Bergey's Manual of Systematic Bacteriology* (Lacey, 1989a). The name *Faenia rectivirgula* was proposed by Kurup and Agre (1983) for the thermophilic actinobacterium *Micropolyspora faeni* (Cross et al., 1968) which had been previously named "*Thermopolyspora polyspora*" (Corbaz et al., 1963) and "*Thermopolyspora rectivirgula*" (Krasil'nikov and Agre, 1964). The genus name *Micropolyspora* lost its standing in nomenclature when its type strain, *Micropolyspora brevicatena*, was reclassified as *Nocardia brevicatena* (Goodfellow and Pirouz, 1982). The subsequent reclassification of *Faenia rectivirgula* as *Saccharopolyspora rectivirgula* (Korn-Wendisch et al., 1989) was supported by a wealth of taxonomic information, including data derived from chemotaxonomic (Collins et al., 1977; Embley et al., 1988b; Kroppenstedt, 1985; Mordarskaia et al., 1973), phage host range (Schneider and Kutzner, 1989; Smorawinska et al., 1988), and 16S rRNA gene sequencing (Bowen et al., 1989; Embley et al., 1988a, 1988c) studies. The name of the mesophilic species "*Faenia rectivirgula*", proposed by Okazaki et al. (1987), has not been validly published.

The genus *Saccharopolyspora* currently contains 20 species with validly published names, most of which have been delineated using a broad range of genotypic and phenotypic features, in some instances including DNA–DNA relatedness data

(Li et al., 2009a; Qin et al., 2008a; Yuan et al., 2008; Zhang et al., 2008b, 2009). However, despite the recent rise in the number of *Saccharopolyspora* species, mainly based on the description of single isolates, there is evidence that the genus is underspeciated (Goodfellow et al., 1989b).

Differentiation of the species of the genus *Saccharopolyspora*

Saccharopolyspora species can be distinguished by using a range of phenotypic markers (Table 255), though results need to be interpreted with care as a common set of tests are not yet available for the delineation of all species and because many of the latter are based on the description of single isolates. In practice, closely related *Saccharopolyspora* species are delineated using a combination of 16S rRNA gene sequence, DNA–DNA relatedness, and phenotypic data (Duangmal et al., 2010; Qin et al., 2008a). Some *Saccharopolyspora* species have been distinguished by differences in quantitative fatty acid composition (Embley et al., 1988b; Mertz and Yao, 1990) and on the basis of esterase patterns, plasmid profiles, and phage host range studies (Korn-Wendisch et al., 1989; Schneider and Kutzner, 1989).

List of species of the genus *Saccharopolyspora*

1. **Saccharopolyspora hirsuta** Lacey and Goodfellow 1975, 78[VP]

hir.su'ta. L. fem. adj. *hirsuta* hairy, rough, shaggy, bristly, referring to the hairy spore chains produced by the organism.

Forms branched, substrate hyphae (0.4–0.6 µm diameter) which tend to fragment into rod-shaped elements towards the edges and in older parts of colonies. Aerial hyphae (0.5–0.7 µm diameter) differentiate into bead-like spores which tend to be separated by short lengths of apparently "empty" hyphae. Looped or loosely spiral chains of spores are produced, though long straight chains of spores are sometimes formed between tufts of aerial hyphae. Spores are round to oval (0.7–1.3 × 0.5–0.7 µm) and are covered by a sheath bearing tufts of long straight or curved hairs. Thin, raised or convex, and usually slightly wrinkled colonies are formed in 7 d at 40°C. Sparse white aerial hyphae are produced in tufts at the center of colonies. The substrate mycelium is colorless to cartridge buff and usually gelatinous or mucoid. Good growth occurs on V-8 vegetable juice and yeast extract-malt extract (ISP medium 2) agars; a yellow diffusible pigment is formed on the latter. Grows on agar media between 25 and 50°C, with an optimal temperature of about 37 to 40°C; does not grow at 10°C. Aerial hyphae are only formed at the optimal temperature.

Degrades adenine, but not xylan. Adonitol, cellobiose, erythritol, glycerol, inositol, lactose, methyl α-D-glucoside, methyl β-D-glucoside, sorbitol, and trehalose are used as sole carbon sources, but not arabinose, melezitose, or salicin (all at 1%, w/v). Similarly, acetate, benzoate, butyrate, citrate, fumarate, H-malate, succinate, and sebacate acid are used as sole carbon sources, but not adipate or tartrate (all at 0.1%, w/v).

Resistant (filter paper discs soaked in µg/ml antibiotics) to erythromycin (50), gentamicin (100), kanamycin (10), streptomycin (100), neomycin (50), tobramycin (100), and rifampin (50), but susceptible to dapsone (500) and septrin (500). Sensitive to lysozyme. Additional phenotypic properties are shown in Table 255.

Whole-organism hydrolysates contain arabinose and galactose as characteristic sugars and *meso*-diaminopimelic acid as the principal diamino acid. Cellular fatty acids contain major proportions of iso- and anteiso-branched chain components. The predominant menaquinone is MK-9(H_4).

Source: isolated from moldy sugarcane bagasse spontaneously heated during storage.

DNA G+C content (mol%): not determined.

Type strain: ATCC 27875, CBS 420.74, DSM 43463, IFO (now NBRC) 13919, JCM 3170, NCIB 11079, NRRL B-5792.

Sequence accession no. (16S rRNA gene): U93341.

Taxonomic comments:. subsequently, this species has been divided into subspecies.

1a. **Saccharopolyspora hirsuta subsp. hirsuta** Lacey and Goodfellow 1975, 78[VP]

Description is as for the species.

Additional comments: this subspecies was automatically created by the valid publication of *Saccharopolyspora hirsuta* subsp. *taberi* Labeda (1987) (Howey et al., 1990), a taxon subsequently raised to species status (Korn-Wendisch et al., 1989).

1b. **Saccharopolyspora hirsuta subsp. kobensis** (*ex* Iwasaki, Itoh and Mori 1979) Lacey 1989b, 496[VP] (Effective publication: Lacey 1989c, 2385.) (*Saccharopolyspora hirsuta* subsp. *kobensis* Iwasaki, Itoh and Mori 1979, 185)

ko.ben'sis. N.L. fem. adj. *kobensis* belonging to Kobe, a city in Japan where the organism was isolated.

Differs from *Saccharopolyspora hirsuta* subsp. *hirsuta* in having a yellow to pink substrate mycelium and a yellow to red soluble pigment, in reducing nitrate to nitrite, and by using inositol, rhamnose, sorbitol, and xylose. Source of the antibiotic sporacin.

Source: isolated from soil.

DNA G+C content (mol%): not determined.

Type strain: ATCC 20501, FERM-P 3912, NBRC 15151, JCM 9109, KC 6606.

Sequence accession no. (16S rRNA gene): DQ381814.

Additional comments: a revised name (Lacey, 1989b), cited in Validation List 31.

2. **Saccharopolyspora antimicrobica** Yuan, Zhang, Guan, Wei, Li, Yu, Li and Zhang 2008, 1184[VP]

an.ti.mi.cro'bi.ca. Gr. prep. *anti* against; N.L. n. *microbium* microbe; L. adj. suff. *-cus -a -um* suffix used with various meanings; N.L. fem. adj. *antimicrobica* antimicrobial.

Forms extensively branched, white, buff to pink substrate mycelia that later fragment into rod-shaped elements. Aerial hyphae differentiate into long straight chains of rough-surfaced spores. Grows well on glycerol-asparagine (ISP medium 5), inorganic salts-starch (ISP medium 4), oatmeal (ISP medium 3), and yeast extract-malt extract (ISP medium 2) agars. White aerial hyphae are produced on ISP 2 agar. Diffusible pigments, buff, pink to brown, are produced on some agar media. Grows in the presence of 0–7% (w/v) NaCl. Temperature and pH ranges for growth are 20–45°C and pH 6.0–8.5, respectively. Optimal temperature and pH for growth are 28–37°C and pH 7.0–7.5, respectively.

Gelatin and xylan are degraded, but not cellulose, chitin, or elastin. Milk is coagulated and peptonized, but does not produce H_2S. Positive for acid phosphatase, alkaline phosphatase, β-galactosidase, α-glucosaccharase, β-glucosaccharase, and *N*-acetylglucosaminidase. Adonitol, cellobiose, dutcitol, inositol, melezitose, melibiose, D-ribose, salicin, sorbitol, turanose, acetate, citrate, malonate, phenylalanine, glyconate, glucosamine, and tartrate are used as sole carbon sources, but not erythritol, methyl α-D-glucoside, or trehalose.

Resistant (μg/ml) to amikacin (30), ampicillin (10), aztreonam (30), cefotaxime (30), ceftazidime (30), chloromycetin (30), erythromycin (15), furadantin (300), gentamicin (10), oxacillin (1), penicillin G (10 U), streptomycin (10), and tobramycin (10). Inhibited by lysozyme. Shows antimicrobial activity against *Escherichia coli* and *Staphylococcus aureus*. Additional phenotypic features are shown in Table 255.

Whole-organism hydrolysates contain arabinose and galactose as characteristic sugars and *meso*-diaminopimelic acid as the principal diamino acid. Cellular fatty acids contain major proportions of $C_{15:0}$ iso, $C_{16:0}$ iso, $C_{17:0}$ iso, and $C_{17:0}$ anteiso. The predominant menaquinone is MK-9(H_4) and the major phospholipids are diphosphatidylglycerol, phosphatidylcholine, phosphatidylglycerol, and phosphatidylinositol.

Source: isolated from soil.

DNA G+C content (mol%): 69.3 (T_m).

Type strain: I05-00074, CCM 7463, DSM 45119, KCTC 19303.

Sequence accession no. (16S rRNA gene): EF693956.

3. **Saccharopolyspora cebuensis** Pimentel-Elardo, Tiro, Grozdanov and Hentschel 2008, 631[VP]

ce.bu.en'sis. N.L. fem. adj. *cebuensis* pertaining to the province of Cebu in the Philippines where the type strain was collected.

Forms extensively branched white substrate hyphae which fragment into rod-shaped elements. Aerial hyphae bear short chains of round to oval spores with smooth surfaces. Brown diffusible pigment is observed. Grows at 15–37°C on yeast extract-malt extract (ISP medium 2) or M1 agars which contain 2.5–12.5% (w/v) NaCl or 25–100% artificial seawater. Does not grow anaerobically.

Positive for acid phosphatase, alkaline phosphatase, esterase (C4), esterase lipase (C8), α-glucosidase, leucine arylamidase, lipase (C14), α-mannosidase, naphthol-A-BI-phosphohydratase, *N*-acetyl-β-glucosaminidase, and valine arylamidase (API ZYM tests). Catalase-positive and oxidase-negative.

Amygdalin, D-arabinose, arabitol, cellobiose, dulcitol, erythritol, esculin, fucose, gentiobiose, glycerol, glycogen, inulin, *N*-acetylglucosamine, potassium gluconate, ribose, starch, and trehalose are used as sole carbon sources. Additional phenotypic features are shown in Table 255.

Whole-organism hydrolysates contain arabinose and galactose as characteristic sugars and *meso*-diaminopimelic acid as the principal diamino acid. Cellular fatty acids contain major proportions of iso-, anteiso-, and 10-methyl-branched fatty acids. The predominant menaquinone is MK-9(H_4) and the major polar lipids are diphosphatidylglycerol, phosphatidylcholine, phosphatidylethanolamine, phosphatidylglycerol, phosphatidylinositol, and phosphatidylmethylethanolamine.

Source: isolated from a marine sponge (*Haliclona* sp.).

DNA G+C content (mol%): 72.6 (HPLC).

Type strain: DSM 45019, CIP 109355, SPE 10-1.

Sequence accession no. (16S rRNA gene): EF030715.

4. **Saccharopolyspora erythraea** (Waksman 1923) Labeda 1987, 21[VP] [*Actinomyces erythreus* Waksman 1923, 370; *Streptomyces erythreus* (Waksman 1923) Waksman and Henrici 1948, 938]

e.ry.thra'e.a. L. fem. adj. *erythraea* reddish, referring to colony color.

Forms an extensively branched substrate mycelium. Spore chains are usually short so that imperfect spirals and short to flexuous chains are common. Spore surfaces are spiny. Substrate mycelium ranges from pale orangish yellow to reddish brown, depending on the medium. Pale pink to moderately brownish gray aerial hyphae are formed on many media, including glycerol-asparagine agar (ISP medium 5). White aerial mycelia are seen. A faint yellow to pinkish orangish brown soluble pigment is produced on most media. Melanin pigments are not produced. Grows between 20 and 42°C.

Degrades gelatin, but does not produce phosphatase or hydrolyze allantoin. Acid is produced from adonitol, arabinose, cellobiose, dextrin, erythritol, fructose, galactose, glucose, glycerol, inositol, maltose, mannitol, mannose, melezitose, melibiose, raffinose, rhamnose, salicin, sucrose, and xylose, but not from dutcitol, lactose, methyl α-D-glucoside, methyl β-D-xyloside, or sorbitol. Acetate, citrate, lactate, malate, propionate, pyruvate, and succinate are used as sole carbon sources, but not benzoate, mucate, oxalate, or tartrate. Many strains produce erythromycins A or B. Additional phenotypic properties are shown in Table 255.

Whole-organism hydrolysates contain arabinose and galactose as characteristic sugars and *meso*-diaminopimelic acid as the predominant principal diamino acid. The predominant menaquinone is MK-9(H_4).

Source: isolated from soil.

DNA G+C content (mol%): 76.9 (T_m).

Type strain: ATCC 11635, DSM 40517, ISP 5517, NRRL 2338.

Sequence accession no. (16S rRNA gene): X53198.

Additional comments: Saccharopolyspora erythraea (Waksman 1923) Labeda 1987 comb. nov. was initially described as *Saccharopolyspora erythraea* sp. nov. However, since *Saccharopolyspora erythraea* Labeda 1987 contains the type strain of *Streptomyces erythraeus* (Waksman 1923) Waksman and Henrici 1948 (Approved Lists, Skerman et al., 1980), the taxon has to be considered as a new combination, *Saccharopolyspora erythraea* (Waksman 1923) Labeda 1987.

5. **Saccharopolyspora flava** Lu, Liu, Wang, Zhang, Qi and Goodfellow 2001, 322[VP]

fla'va. L. fem. adj. *flava* yellow, referring to the color of the substrate mycelium.

Forms an extensively branched, yellow substrate mycelium which fragments into rod-shaped elements after 3–4 d at 28°C. Aerial hyphae are produced after prolonged cultivation on oatmeal agar. The aerial mycelium carries abundant straight chains of 3–5 smooth surfaced spores (0.4–0.5 × 0.6 μm). The organism grows well on glucose-asparagine, modified Sauton's, and yeast extract-malt extract (ISP medium 2) agars. Does not produce diffusible pigments. Grows between 28 and 37°C. Weak growth occurs in the presence of 7% (w/v) NaCl.

Cellulose and xylan are degraded, but not guanine. Utilizes adonitol, cellobiose, erythritol, glycerol, inositol, methyl α-D-glucoside, salicin, sorbitol, acetate, adipate, benzoate, butyrate, citrate, fumarate, H-malate, propionate, pyruvate, and succinate as sole carbon sources, but not melezitose, sebacic acid, or tartrate.

Sensitive (discs soaked in 100 μg/ml antibiotic) to gentamicin, streptomycin, and rifampin. Sensitive to lysozyme. Additional phenotypic features are shown in Table 255.

Whole-organism hydrolysates contain arabinose and galactose as characteristic sugars and *meso*-diaminopimelic acid as the principal diamino acid. The predominant menaquinone is MK-9(H_4) and the polar lipid pattern includes phosphatidylcholine.

Source: isolated from garden soil.

DNA G+C content (mol%): 67 (T_m).

Type strain: 07, AS4.1520, DSM 44771, NBRC 16345, JCM 10665.

Sequence accession no. (16S rRNA gene): AF154128.

6. **Saccharopolyspora gloriosae** Qin, Chen, Klenk, Kim, Xu and Li 2010, 1150[VP]

glo.ri.o'sa.e. N.L. fem. gen. n. *gloriosae* of the plant genus *Gloriosa*, referring to the isolation of the type strain from a stem of *Gloriosa superba*.

Forms an extensively branched substrate mycelium. Curved and hooked chains of smooth-surfaced spores are formed on the aerial mycelium. Grows well on glycerol-asparagine (ISP medium 5), oatmeal (ISP medium 3), and yeast extract-malt extract (ISP medium 2) agars and moderately well on Czapek's and inorganic salts-starch (ISP medium 4) agars. Grows between 10 and 32°C (optimally at 28°C), pH 6.0 and 8.0, and in the presence of 0–11% NaCl (optimally with 0–5% salt).

Oxidase-negative. Adonitol, D-arabinose, arabitol, cellobiose, erythritol, esculin, fructose, gentiobiose, glycerol, glycogen, inositol, mannose, melezitose, starch, and trehalose are used as sole carbon sources. L-Alanine, D-arginine, L-asparagine, L-leucine, L-histidine, L-ornithine, L-proline, L-serine, and L-threonine are used as sole nitrogen sources. Additional phenotypic features are shown in Table 255.

Whole-organism hydrolysates contain arabinose and galactose as characteristic sugars and *meso*-diaminopimelic acid as the principal diamino acid. Cellular fatty acids contain major proportions of $C_{16:0}$ iso, $C_{17:0}$ anteiso, and $C_{17:1}$ *cis*9. The predominant menaquinone is MK-9(H_4) and the major polar lipids are diphosphatidylglycerol, phosphatidylcholine, phosphatidylethanolamine, phosphatidylglycerol, phosphatidylinositol, phosphatidylinositol mannosides, and several unknown phospholipids.

Source: isolated from a surface-sterilized stem of *Gloriosa superba* L.

DNA G+C content (mol%): 71.6 (HPLC).

Type strain: YIM 60513, KCTC 19243, CCTCC AA 207006.

Sequence accession no. (16S rRNA gene): EU005371.

7. **Saccharopolyspora gregorii** Goodfellow, Lacey, Athalye, Embley and Bowen 1989a, 496[VP] (Effective publication: Goodfellow, Lacey, Athalye, Embley and Bowen 1989b, 2137.)

gre.gor'i.i. N.L. gen. masc. n. *gregorii* of Gregory, named after P.H. Gregory, a British mycologist and aerobiologist.

Forms an extensively branched substrate mycelium, which fragments into coccoid elements. Sporulation or fragmentation of the aerial mycelium may sometimes be seen, forming hooks or flexuous chains of spores with smooth surfaces. Transmission electron micrographs show mainly vegetative rods and cocci, but occasionally spores with a smooth sheath are seen. Colonies are pale yellowish to buff with sparse white aerial mycelium at growth temperatures of about 20°C. Temperature range for growth is 10–35°C.

Degrades arbutin, gelatin, and Tweens 20 and 80. Adonitol, glycerol, D- and L-alanine, L-proline, propionate, and pyruvate are used as sole carbon sources, but not glycogen, trehalose, starch, or L-serine.

Resistant (μg/ml) to cephaloridine (2), lincomycin (16), oleandomycin (2), and benzylpenicillin (10), but sensitive to demecylocyline (2), gentamicin (4), neomycin (3), rifampin (2), streptomycin (4), tobramycin (8), and vancomycin (2). Grows in the presence of (%, w/v) bismuth citrate (0.001), crystal violet (0.0001), phenol (0.01), phenyl ethanol (0.1%, v/v), potassium tellurite (0.001), sodium azide (0.001), sodium chloride (13), tetrazolium (0.01), teepol (0.01), and thallous acetate (0.0001), but is sensitive to bismuth citrate (0.01), phenyl ethanol (0.4%, v/v), potassium tellurite (0.01), and thallous acetate (0.01). Additional phenotypic features are shown in Table 255.

Whole-organism hydrolysates contain arabinose and galactose as characteristic sugars and *meso*-diaminopimelic acid as the principal diamino acid. Cellular fatty acids contain major proportions of $C_{16:0}$ iso, $C_{17:0}$ anteiso, and $C_{17:0}$ iso. The predominant menaquinone is MK-9(H_4) and the major polar lipids are diphosphatidylglycerol, phosphatidylcholine, phosphatidylglycerol, phosphatidylinositol, and uncharacterized phospho- and glycolipids.

Source: isolated from grass, hay, straw, barley grain, and soil.

DNA G+C content (mol%): 74 (T_m).

Type strain: A85, ATCC 51265, DSM 44324, NBRC 15045, NCIMB 12823, JCM 9687, NRRL B-16506.

Sequence accession no. (16S rRNA gene): X76962.

8. **Saccharopolyspora halophila** Tang, Wang, Cai, Zhi, Lou, Xu, Jiang and Li 2009a, 557[VP]

ha.lo'phi.la. Gr. n. *hals halos* salt; N.L. adj. *philus -a -um* (from Gr. adj. *philos -ê -on*) friend, loving; N.L. fem. adj. *halophila* salt-loving, referring to the ability to grow at high NaCl concentrations.

Moderately halophilic. Forms a well-developed substrate mycelium that shows no evidence of fragmentation. Aerial hyphae differentiate into long straight chains of oval or spherical spores (0.6–0.7 × 0.6–1.1 μm) which have smooth surfaces. The aerial mycelium is white-yellow in color and the substrate mycelium is yellow to moderate orange-yellow. Does not produce diffusible pigments. Grows well on Czapek's, glycerol-asparagine (ISP medium 5), inorganic salts-starch (ISP medium 4), potato, and yeast extract-malt extract (ISP medium 2) agars. Moderate growth occurs on nutrient and oatmeal (ISP medium 3) agars. Temperature, pH, and NaCl tolerance ranges are 10–45°C, pH 6–8.5, and 3–20% (w/v), respectively. Good growth occurs at 28–37°C, pH 7–8, and with 10–15% (w/v) NaCl.

Does not degrade cellulose. Liquefies gelatin. Milk peptonization and coagulation are positive. Does not produce H_2S. Acid is produced on L-arabinose, glucose, *myo*-inositol, rhamnose, and xylose. Cellobiose, *myo*-inositol, xylitol, and acetate are used as sole carbon sources, but not sorbitol or trehalose. Alanine, asparagine, arginine, cystine, glycine, histidine, homocysteine, hypoxanthine, lysine, proline, threonine, tyrosine, valine, and urea are utilized as sole nitrogen sources, but not adenine, hydroxyproline, or glutamate. Additional phenotypic features are shown in Table 255.

Whole-organism hydrolysates contain arabinose and galactose as characteristic sugars and *meso*-diaminopimelic acid as the principal diamino acid. Cellular fatty acids contain major proportions of $C_{15:0}$ iso, $C_{16:0}$ iso, and $C_{17:0}$ anteiso. The predominant menaquinone is MK-9(H_4) and the major polar lipids are diphosphatidylglycerol and phosphatidylcholine; large amounts of phosphatidylinositol are also present.

Source: isolated from a saline lake.

DNA G+C content (mol%): 66.3 (HPLC).

Type strain: DSM 45007, KCTC 19162, YIM 90500.

Sequence accession no. (16S rRNA gene): DQ923129.

9. **Saccharopolyspora hordei** Goodfellow, Lacey, Athalye, Embley and Bowen 1989a, 496[VP] (Effective publication: Goodfellow, Lacey, Athalye, Embley and Bowen 1989b, 2137.)

hor'de.i. L. n. *hordeum* barley, and also the generic name of barley (*Hordeum*); L. gen. n. *hordei* from barley, referring to the isolation source of the strains.

Forms an extensively branched substrate mycelium which sometimes fragments into coccoid elements. Aerial mycelium is either not produced or is sparse and is white with short flexuous, hooked, or spiral spore chains often with a beaded appearance. Transmission electron micrographs show that the spores are covered by a smooth sheath with adjacent spores often separated by a short length of empty hypha. Colonies are colorless to light buff, conical, rounded, or wrinkled. Grows at 20–60°C.

Resistant (μg/ml) to lincomycin (16), neomycin (3), oleandomycin (16), benzylpenicillin (10), and tobramycin (4), but sensitive to cephaloridine (10), demeclocycline (2), gentamicin (16), rifampin (2), streptomycin (4), and vancomycin (2). Arbutin, DNA, guanine, RNA, testosterone, and Tweens 20 and 80 are degraded. Does not hydrolyze allantoin. Cellobiose, glycogen, glycerol, and starch are used as sole carbon sources. Grows well in the presence of (%, w/v) bismuth citrate (0.0001), crystal violet (0.0001), phenol (0.1), phenyl ethanol (0.2%, v/v), potassium tellurite (0.001), sodium azide (0.001), sodium chloride (10), teepol (0.01), tetrazolium (0.01), and thallous acetate (0.0001), but is sensitive to bismuth citrate (0.01), phenyl ethanol (0.4%, v/v), potassium tellurite (0.01), sodium azide (0.02), and thallous acetate (0.01). Additional phenotypic features are shown in Table 255.

Whole-organism hydrolysates contain arabinose and galactose as characteristic sugars and *meso*-diaminopimelic acid as the principal diamino acid. Cellular fatty acids contain major proportions of $C_{16:0}$ iso, $C_{17:0}$ anteiso, and $C_{17:0}$ iso. The predominant menaquinone is MK-9(H_4) and the major polar lipids are diphosphatidylglycerol, phosphatidylcholine, phosphatidylglycerol, phosphatidylinositol, and uncharacterized phospho- and glycolipids.

Source: isolated from barley and oat grains, grass hay, straw, and sugar cane bagasse.

DNA G+C content (mol%): 72 (T_m).

Type strain: A54, ATCC 49856, DSM 44065, NBRC 15046, JCM 8090, NCIMB 12824, NRRL B-16507.

Sequence accession no. (16S rRNA gene): X53197.

10. **Saccharopolyspora jiangxiensis** Zhang, Wu and Liu 2009, 1078[VP]

ji.ang.xi.en'sis. N.L. fem. adj. *jiangxiensis* pertaining to Jiangxi Province, China, the source of the grass-field soil from which the organism was isolated.

Produces an extensively branched (0.4–0.7 μm diameter), colorless to buff substrate mycelium which fragments *in situ* into coccoid- and rod-shaped elements after 3–4 d at 28°C. White to buff aerial hyphae (0.6–1.0 μm in diameter) are produced upon prolonged cultivation on GYM agar. The aerial mycelium carries long straight to flexuous chains of 6–10 spores (spore size 0.8–1.1 × 1.0–1.3 μm). The spore surface is smooth or irregularly rough. Colony elevation is convex to irregular and colony margins are filamentous. Does not form diffusible pigments. Grows well on GYM, inorganic salts-starch (ISP medium 4), oatmeal (ISP medium 3), and modified Sauton's agars, and at 15–45°C and pH 5.5–9.5. Weak growth occurs in the presence of NaCl at 11%, but not at 12% (w/v).

Does not degrade cellulose, guanine, or xylan. Catalase- and arbutin-positive. Acid is formed from adonitol, D-arabinose, dextrin, D-fructose, D-galactose, D-glucose, glycerol, inulin, maltose, D-mannitol, D-mannose, D-ribose, sucrose, and trehalose, but not from L-arabinose, cellobiose, ethanol, *meso*-erythritol, glycogen, *myo*-inositol, D-lactose, melezitose, melibiose, methyl α-D-glucoside, raffinose, α-L-rhamnose, D-salicin, D-sorbitol, or D-xylose. Adonitol, D-arabinose, *meso*-erythritol, methyl α-D-glucoside, *myo*-inositol, melezitose, sorbitol, trehalose, acetate, benzoate, citrate, fumarate, malate, succinate, and tartrate are used as sole carbon sources, but not salicin, malonate, or oxalate.

Resistant (μg per disc) to amoxycillin plus clavulanic acid (10), ampicillin (10), aztreonam (30), clindamycin (2), erythromycin (15), kanamycin (30), penicillin G (10 U), and tobramycin (10), but susceptible to amikacin (30), cefotaxime (30), chloramphenicol (30), ciprofloxacin (5), gentamicin (10), mezlocillin (75), ofloxacin (5), rifampin (5), streptomycin (10), and tetracycline (30). Grows in the presence of lysozyme (0.005%, w/v). Additional phenotypic properties are shown in Table 255.

Whole-organism hydrolysates contain arabinose and galactose as characteristic sugars and *meso*-diaminopimelic acid as the major diamino acid. Cellular fatty acids contain major proportions of $C_{15:0}$ iso, $C_{16:0}$ iso, $C_{17:0}$ iso, and $C_{17:0}$ anteiso. The predominant menaquinone is MK-9(H_4) and the major polar lipids are phosphatidylcholine, phosphatidylethanolamine, and phosphatidylglycerol.

DNA relatedness values between the type strain and corresponding strains of closely related species are as follows: *Saccharopolyspora antimicrobica*, 56.3%; *Saccharopolyspora hirsuta*, 47.8%; *Saccharopolyspora shangdongensis*, 25.5%; and *Saccharopolyspora spinosa*, 21.7%.

Source: isolated from a soil sample from a grass field.

DNA G+C content (mol%): 70.3 (T_m).

Type strain: CGMCC 4.3529, JCM 14613, W12.

Sequence accession no. (16S rRNA gene): EF143926.

11. **Saccharopolyspora phatthalungensis** Duangmal, Mingma, Thamchaipenet, Matsumoto and Takahashi 2010, 1907[VP]

phat.tha.lun.gen'sis. N.L. fem. adj. *phatthalungensis* referring to Phatthalung Province, Thailand, the source of the rhizospheric soil from which the organism was isolated.

Forms an extensively branched substrate mycelium which is yellowish to yellowish brown on glycerol-asparagine (ISP medium 5), glucose-yeast extract, inorganic salts-starch (ISP medium 4), peptone-yeast extract iron (ISP medium 5), tyrosine (ISP medium 7), tryptone-yeast extract (ISP medium 1), and yeast extract-malt extract (ISP medium 2) agars. A white aerial mycelium is produced on these media and a brownish-black soluble pigment is produced on ISP medium 2. Spores with spiny surfaces are borne in hooks/open loops. Grows between 18 and 42°C (optimally at 28–34°C) and between pH 5.0 and 9.0. Grows well on ISP medium 2 supplemented with 1–5% (w/v) NaCl, moderately well with 6% (w/v) NaCl, and poorly in the presence of 7% (w/v).

Gelatin is degraded, but not arbutin, cellulose, or guanine. Catalase-positive, but does not hydrolyze allantoin. H_2S is produced. Acid is produced from adonitol, fructose, galactose, glucose, glycerol, *myo*-inositol, mannitol, mannose, ribose, trehalose, and xylose (weakly), but not from arabinose, cellobiose, fucose, β-lactose, maltose, melibiose, rhamnose, raffinose, sorbitol, sorbose, sucrose, or xylitol. Positive for alkaline phosphatase and leucine aminopeptidase, but negative for acid phosphatase, α-chymotrypsin, cystine aminopeptidase, esterase (C4), α-fucosidase, α-galactosidase, β-galactosidase, α-glucosidase, β-glucosidase, β-glucuronidase, lipase (C8), lipase (C14), α-mannosidase, *N*-acetyl-β-glucosaminidase, trypsin phosphoamidase, and valine aminopeptidase (API ZYM tests). Additional phenotypic properties are shown in Table 255.

Whole-organism hydrolysates contain arabinose and galactose as characteristic sugars and *meso*-diaminopimelic acid as the principal diamino acid. The glycan moiety of the peptidoglycan is acetylated. Cellular fatty acids contain major proportions of $C_{16:0}$ and 10-methyl $C_{17:0}$. The predominant menaquinone is MK-9(H_4) and the major polar lipids are phosphatidylcholine, phosphatidylglycerol, and phosphatidylinositol.

Source: isolated from rhizospheric soil.

DNA G+C content (mol%): 70.3 (HPLC).

Type strain: BCC 35844, NRRL B-24798, SR8.15, TISTR 1921.

Sequence accession no. (16S rRNA gene): GQ381309.

12. **Saccharopolyspora qijiaojingensis** Tang, Wang, Wu, Cao, Lou, Xu, Jiang and Li 2009b, 2168[VP]

qi.ji.a.o.jing.en'sis. N.L. fem. adj. *qijiaojingensis* pertaining to Qijiaojing Lake, Xinjiang Province, north-west China, where the sample from which the type strain was isolated was collected.

Moderately halophilic. Forms a well-developed substrate mycelium that fragments into rod-shaped elements. Bead-like straight chains of smooth surfaced spores (0.6–0.7 × 0.7–1.1 µm in size) are formed on the aerial mycelium. Temperature, pH, and NaCl ranges for growth are 20–40°C, pH 5.0–8.0, and 6–22% (w/v) NaCl; optimal growth occurs at 28°C, pH 7.0, and 10–15% (w/v) NaCl.

Gelatin and Tween 20 are degraded, but not cellulose or Tween 80. Does not peptonize or coagulate milk or produce H_2S. Trehalose is used as a sole carbon source, but not cellobiose, glycerol, glycogen, or starch. DL-Alanine, L-hydroxyproline, hypoxanthine, L-histidine, L-asparagine, and xanthine are utilized as nitrogen sources, but not L-lysine, L-serine, adenine, L-tyrosine, or L-phenylalanine.

Resistant (µg per disc) to gentamicin (10), streptomycin (10), sulfamethoxazole/trimethoprin (23.75/1.25), and tobramycin, but sensitive to amoxycillin (10), ampicillin (10), chloroamphenicol (30), ciprofloxacin (5), erythromycin (15), rifampin (5), tetracycline (30), and vancomycin (30). Additional phenotypic properties are shown in Table 255.

Whole-organism hydrolysates contain arabinose and galactose as characteristic sugars and *meso*-diaminopimelic acid as the principal diamino acid. Cellular fatty acids contain major proportions of $C_{15:0}$ iso, $C_{16:0}$ iso, and $C_{17:0}$ iso. The predominant menaquinone is MK-9(H_4) and the major polar lipids are diphosphatidylglycerol, phosphatidylcholine, phosphatidylethanolamine, phosphatidylinositol, phosphatidylinositol mannosides, and an unknown phospholipid.

Source: isolated from a salt lake.

DNA G+C content (mol%): 70.1 (HPLC).

Type strain: DSM 45088, KCTC 19235, YIM 91168.

Sequence accession no. (16S rRNA gene): EF687715.

13. **Saccharopolyspora rectivirgula** (Krasil'nikov and Agre 1964) Korn-Wendisch, Kempf, Grund, Kroppenstedt and Kutzner 1989, 439[VP] [*Faenia rectivirgula* (Cross, Maciver and Lacey 1968) Kurup and Agre 1983, 664; *Micropolyspora faeni* (Corbaz, Gregory and Lacey 1963) Cross, Maciver and Lacey 1968, 354; Basonym: *Micropolyspora rectivirgula* (Krasil'nikov and Agre 1964) Prauser and Momirova 1970, 220; *Thermopolyspora polyspora* Corbaz, Gregory and Lacey 1963, 450; *Thermopolyspora polyspora* Henssen 1957, 396; *Thermopolyspora rectivirgula* Krasil'nikov and Agre 1964, 106]

rec.ti.vir'gu.la. L. adj. *rectus* straight; L. n. *virgula* twig; N.L. n. *rectivirgula* straight twig.

Forms chains of spores on both substrate and aerial hyphae. The spores are borne on short, unbranched, lateral, or terminal sporophores. Spore formation is basipetal and spore surfaces are rough. Substrate hyphae are branched, septate, 0.5–0.8 µm in diameter, and yellow to orange on GYM and TS agars. Similarly, a white to pink aerial mycelium is formed on these media: aerial hyphae are especially well developed in the presence of 5% (w/v) NaCl. Grows at 37–63°C (optimally at 50°C) and in the presence of 10% (w/v) NaCl.

Degrades gelatin, but not xylan. Allantoin and uric acid are hydrolyzed, but hemolysis and egg yolk reactions are negative. Sensitive to lysozyme. Fructose and inositol are used as sole carbon sources. Lysed by the five lytic *Saccharopolyspora rectivirgula* phages (e.g. P113); nonlysogenic strains are lysed by all temperate *Saccharopolyspora rectivirgula* phages (e.g. ϕFR114 and ϕFR-C) and three *Saccharopolyspora erythraea* phages (e.g. P517). Additional phenotypic properties are shown in Table 255.

Whole-organism hydrolysates contain arabinose and galactose as characteristic sugars and *meso*-diaminopimelic acid as the principal diamino acid. The predominant menaquinone is MK-9(H_4) and the major polar lipids are phosphatidylcholine, phosphatidylethanolamine, and phosphatidylmethylethanolamine.

Causative agent of farmer's lung disease.

Source: isolated from moldy hay, soil, compost, and manure.

DNA G+C content (mol%): 70.4 (HPLC).

Type strain: ATCC 33515, BKM A-810, DSM 43747, NBRC 12464, INMI 683, JCM 3057, NRRL B-16280, VKM Ac-810.

Sequence accession no. (16S rRNA gene): X53194.

14. **Saccharopolyspora rosea** Yassin 2009, 1151[VP]

ro.se'a. L. fem. adj. *rosea* rose-colored, pink, referring to the color of the diffusible pigment produced by the organism.

Forms extensively branched, brownish-yellow substrate hyphae (0.4–0.5 µm in diameter) which fragment into rod-shaped elements. Yellowish to white aerial hyphae (0.5–0.7 µm in diameter) differentiate into long straight chains of smooth-surfaced spores. Pink colored diffusible pigments are produced on oatmeal (ISP medium 3), inorganic salts-starch (ISP medium 4), and yeast extract-malt extract (ISP medium 2) agars, but does not form melanin pigments on peptone-yeast extract iron (ISP medium 6) or tyrosine (ISP medium 7) agars. Grows at 22–42°C and pH 6.0–9.0.

Gelatin is degraded, but not guanine, keratin, or testosterone. Catalase-positive and oxidase-negative. Acetate, adipate, adonitol, isoamyl alcohol, 2,3-butanediol, citrate, *meso*-erythritol, D-gluconate, *myo*-inositol, L-lactate, sorbitol, and trehalose are used as sole carbon sources, but not cellobiose, *m*-hydroxybenzoate, *p*-hydroxybenzoate, lactose, melezitose, or 1,2-propanediol. Acetamide, L-alanine, arginine, gelatin, ornithine, L-proline, and L-serine are used as simultaneous sources of carbon and nitrogen. Additional phenotypic properties are shown in Table 255.

Whole-organism hydrolysates contain arabinose and galactose as characteristic sugars and *meso*-diaminopimelic acid as the principal diamino acid. Cellular fatty acids contain major proportions of $C_{16:0}$ iso, $C_{17:0}$ iso, and $C_{17:0}$ anteiso. The predominant menaquinone is MK-9(H_4) and the major polar lipids are diphosphatidylglycerol, phosphatidylcholine, phosphatidylglycerol, and phosphatidylinositol.

Source: isolated from a bronchial lavage of patient with bronchial carcinoma.

DNA G+C content (mol%): unknown.

Type strain: DSM 45226, CCUG 56401, IMMIB L-1070.

Sequence accession no. (16S rRNA gene): AM992060.

15. **Saccharopolyspora shandongensis** Zhang, Wu, Zhang, Liu and Song 2008b, 1096[VP]

shan.dong.en'sis. N.L. fem. adj. *shandongensis* referring to Shandong Province, China, the source of the wheat-field soil from which the type strain was isolated.

Forms an extensively branched substrate mycelium (0.5–0.8 µm in diameter) which fragments *in situ* into coccoid- and rod-shaped elements after 3–4 d at 28°C. White aerial hyphae (0.7–1.0 µm in diameter) are produced upon prolonged cultivation on glucose-yeast extract-malt agar. The aerial mycelium carries abundant spiral chains of ten or more spiny-surfaced spores (0.8–1.0 × 1.0–1.2 µm) in a spiral arrangement. Grows well on oatmeal (ISP medium 3) and Sauton's agars. A brown diffusible pigment is formed on standard media. Grows at 15–38°C and pH 5.5–9.5. Weak growth occurs in the presence of 7% (w/v) NaCl, but no growth occurs in the presence of 8% (w/v) NaCl.

Does not degrade cellulose, guanine, or xylan. Catalase- and arbutin-positive. Acid is formed from adonitol, D-arabinose, cellobiose, dextrin, D-fructose, D-glucose, glycerol, maltose, D-mannitol, D-mannose, melibiose, D-ribose, and trehalose, but not from L-arabinose, ethanol, *meso*-erythritol, D-galactose, glycogen, *myo*-inositol, inulin, D-lactose, melezitose, methyl α-D-glucoside, raffinose, α-L-rhamnose, D-salicin, D-sorbitol, sucrose, or D-xylose. Adonitol, D-arabinose, cellobiose, *meso*-erythritol, D-fructose, D-glucose, glycerol, D-mannitol, D-mannose, methyl α-D-glucoside, melibiose, D-sorbitol, trehalose, acetate, benzoate, citrate, succinate, and tartrate are used as sole carbon sources, but not *myo*-inositol, melezitose, D-salicin, fumarate, malate, malonate, or oxalate.

Resistant (µg per filter paper disc) to aztreonam (30), clindamycin (2), and penicillin G (10 IU), but susceptible to amikacin (30), amoxycillin plus clavulanic acid (10), ampicillin (10), cefotaxime (30), chloramphenicol (30), ciprofloxacin (5), erythromycin (15), gentamicin (10), kanamycin (30), mezlocillin (75), ofloxacin (5), rifampin (5), streptomycin (10), tetracycline (30), and tobramycin (10). Grows in the presence of lysozyme (0.0005%, w/v). Additional phenotypic properties are shown in Table 255.

Whole-organism hydrolysates contain arabinose and galactose as characteristic sugars and *meso*-diaminopimelic acid as the major diamino acid. Cellular fatty acids contain major proportions of $C_{16:0}$ iso, $C_{17:0}$ anteiso, $C_{18:0}$ iso, $C_{17:1}$ ω8*c*, and $C_{17:0}$ iso. The predominant menaquinone is MK-9(H_4) and the diagnostic polar lipid is phosphatidylcholine.

The DNA–DNA relatedness value between the type strain and the corresponding strain of *Saccharopolyspora spinosa*, a closely related species, is 32.8%.

Source: isolated from wheat-field soil.

DNA G+C content (mol%): 70.1 (T_m).

Type strain: 88, CGMCC 4.3530, JCM 14614.

Sequence accession no. (16S rRNA gene): EF104116.

16. **Saccharopolyspora spinosa** Mertz and Yao 1990, 38[VP]

spi.no'sa. L. fem. adj. *spinosa* spiny, referring to the spiny spore sheath surface.

Forms an extensive substrate mycelium which fragments in liquid media. Well-formed aerial hyphae segment into long chains of spores arranged in hooks and open loops. Short and incomplete spiral spore chains are also seen. The spores (1.1 × 1.5 μm) have spiny surfaces. Aerial hyphae have a distinctive bead-like appearance with many empty spaces in the spore chains. The aerial mycelium is yellowish pink and the vegetative mycelium is yellow to yellowish brown. A soluble brown pigment is produced on some media. The temperature range for growth is 15 to 37°C. Does not survive a temperature of 50°C for 8 h. Growth occurs in the presence of 11% NaCl.

Allantoin, malate, and testosterone are degraded, but not guanine. Catalase- and phosphatase-positive. Acid is produced from adonitol, D-arabinose, *meso*-erythritol, D-fructose, D-glucose, glycerol, D-mannitol, D-mannose, D-ribose, and trehalose, but not from L-arabinose, cellobiose, cellulose, dextrin, dulcitol, ethanol, D-galactose, glycogen, inositol, inulin, lactose, maltose, melezitose, melibiose, methyl α-D-glucoside, raffinose, L-rhamnose, salicin, D-sorbitol, L-sorbose, sucrose, xylitol, or D-xylose. Acetate, butyrate, citrate, formate, lactate, malate, propionate, pyruvate, and succinate are used as sole carbon sources, but not benzoate, mucate, oxalate, or tartrate.

Resistant to (μg per disc) cephalothin (20), kasugamycin (500), nalidixic acid (30), novobiocin (20), oligomycin (100), oxytetracycline (10), penicillin (10 U), polymixin B (300 U), rifampin (5), and trimethoprim (30), but sensitive to bacitracin (10 U), chloromycetin (30), erythromycin (15), gramicidin S (10), gentamicin (10), kanamycin (10), lincomycin (2), mandelamine (3), mikamycin A and B (10), neomycin (30), oleandomycin (15), spiramycin (10), streptomycin (10), tetracycline (30), thiostreptin (10), tobramycin (10), timicamycin (5), valinomycin (10), and vancomycin (30). Additional phenotypic properties are shown in Table 255.

Whole-organism hydrolysates contain arabinose and galactose as characteristic sugars and *meso*-diaminopimelic acid as the principal diamino acid. The predominant menaquinone is MK-9(H_4) and the major polar lipids are cardiolipin and phosphatidylcholine.

Source: isolated from soil collected from a sugar mill rum still.

DNA G+C content (mol%): not known.

Type strain: A83543.1, ATCC 49460, DSM 44228, JCM 9375, NBRC 15153, NRRL 18395.

Sequence accession no. (16S rRNA gene): AF002818.

17. **Saccharopolyspora spinosporotrichia** Zhou, Liu, Qian, Kim and Goodfellow 1998, 56[VP]

spi.no.spo.ro.tri'chi.a. L. adj. *spinosus* thorny; N.L. n. *spora* a spore; N.L. n. *trichia* trichite; N.L. n. *spinosporotrichia* spores bearing needle-like spines.

Forms an extensively branched, reddish-brown substrate mycelium that fragments into rod-shaped elements. An abundant white to gray aerial mycelium bears long spiral chains of spherical spores with warty surfaces. Grows well on Bennett's, Czapek's, Gause's synthetic, inorganic salts-starch (ISP medium 4), oatmeal (ISP medium 3), and sucrose-yeast extract agars. A brown diffusible pigment is formed on most standard media. Grows between 25 and 37°C and in the presence of 2–3% (w/v) sodium chloride.

Glycerol, *myo*-inositol, melibiose, ribose, and sorbitol are used as sole carbon sources, but not sorbose. Additional phenotypic properties are shown in Table 255.

Whole-organism hydrolysates contain arabinose and galactose as characteristic sugars and *meso*-diaminopimelic acid as the predominant diamino acid. The predominant menaquinone is MK-9(H_4); the polar lipid profile includes major amounts of phosphatidylcholine.

Source: isolated from soil.

DNA G+C content (mol%): 70.4 (T_m).

Type strain: AS4.198, DSM 44350, NBRC 16190, JCM 10303.

Sequence accession no. (16S rRNA gene): Y09571.

18. **Saccharopolyspora taberi** (Labeda 1987) Korn-Wendisch, Kempf, Grund, Kroppenstedt and Kutzner 1989, 439[VP] (Basonym: *Saccharopolyspora hirsuta* subsp. *taberi* Labeda 1987, 21; *Saccharopolyspora hirsuta* subsp. *kobensis* Iwasaki, Itoh and Mori 1979, 185.)

ta'ber.i. N.L. gen. masc. n. *taberi* of Taber, named after Willard A. Taber, the American microbiologist who first isolated the organism.

Forms a substrate mycelium that ranges from yellow to orange and from reddish to brownish red on glucose-yeast extract and tryptic soy agars. Neither aerial hyphae nor spores have been detected. Grows from 20 to 45°C, optimally at 37°C. Grows poorly in the presence of 7% NaCl.

Degrades gelatin, but not DNA. Allantoin, arbutin, and uric acid are hydrolyzed. Negative for hemolysis and for the egg yolk reaction. Inositol is used as a sole carbon source.

The dark red metabolite texazone [2-(*N*-methylamino)-3H-phenoxazin-3-one-8-carboxylic acid] is produced. Lysed by three *Saccharopolyspora erythraea* phages (e.g. P517) and four temperate *Saccharopolyspora rectivirgula* phages (e.g. ϕFR-C). Additional phenotypic properties are shown in Table 255.

Whole-organism hydrolysates contain arabinose and galactose as characteristic sugars and *meso*-diaminopimelic acid as the principal diamino acid. Cellular fatty acids contain major proportions of $C_{16:0}$ iso, $C_{17:0}$ iso, $C_{15:0}$ iso, and $C_{17:0}$ anteiso. The predominant menaquinone is MK-9(H_4) and the major polar lipids are diphosphatidylglycerol, phosphatidylcholine, phosphatidylethanolamine, phosphatidylglycerol, phosphatidylmethylethanolamine, and hydroxyphosphatidylethanolamine.

Source: isolated from soil.

DNA G+C content (mol%): 70.8 (HPLC).

Type strain: ATCC 49842, DSM 43856, NBRC 15061, JCM 9383, LL-WRAT-210, NRRL B-16173.

Sequence accession no. (16S rRNA gene): AF002819.

19. **Saccharopolyspora thermophila** Lu, Liu, Wang, Zhang, Qi and Goodfellow 2001, 322[VP]

ther.mo'phi.la. Gr. n. *thermê* heat; N.L. adj. *philus -a -um* (from Gr. adj. *philos -ê -on*) friend, loving; N.L. fem. adj. *thermophila* heat-loving.

Forms an extensively branched colorless to buff substrate mycelium that fragments into rod-shaped elements after 4–5 d at 45°C and an abundant aerial mycelium which carries long hooked to flexuous chains of 4–6 smooth-surfaced, vesicular spores (0.7–1.1 × 0.85–1.5 μm). Grows well on oatmeal, modified Sauton's, and yeast extract-malt extract (ISP medium 2) agars. Does not produce diffusible pigments. The temperature growth range is 45–55°C. Growth occurs in the presence of 7% (w/v) NaCl.

Degrades guanine and xylan, but not cellulose. Adonitol, cellobiose, erythritol, glycerol, inositol, melezitose, methyl α-D-glucoside, salicin, sorbitol, trehalose, acetate, adipate, benzoate, butyrate, citrate, fumarate, H-malate, propionate, pyruvate, sebacic acid, and succinate are used as sole carbon sources, but not tartrate.

Inhibited by lysozyme, gentamicin, streptomycin, and rifampin. Additional phenotypic properties are shown in Table 255.

Whole-organism hydrolysates contain arabinose and galactose as characteristic sugars and *meso*-diaminopimelic acid as the principal diamino acid. The predominant menaquinones are MK-9(H_6) and MK-9(H_8); the polar lipid pattern contains phosphatidylcholine.

Source: isolated from grassland soil.

DNA G+C content (mol%): 73.1 (T_m).

Type strain: 216, AS4.1511, DSM 44575, NBRC 16346, JCM 10664.

Sequence accession no. (16S rRNA gene): AF127526.

20. **Saccharopolyspora tripterygii** Li, Zhao, Qin, Huang, Zhu, Xu and Li 2009a, 3042[VP]

trip.te.ry'gi.i. N.L. n. *Tripterygium* a botanical genus name; N.L. gen. n. *tripterygii* of *Tripterygium*, the plant genus from which this species was isolated.

Forms an extensively branched substrate mycelium and an aerial mycelium which carries long spore chains. Spores are elliptical or short rods (0.5–0.8 × 1.0–1.5 µm) with smooth surfaces. Grows well on glycerol-asparagine (ISP medium 5), inorganic salts-starch (ISP medium 4), potato, and yeast extract-malt extract (ISP medium 2) agars, and moderately well on Czapek's, nutrient, and oatmeal (ISP medium 3) agars. Soluble pigments are not produced. The substrate mycelium is orange-yellow and the aerial mycelium is white on most of the media cited above. Grows at 10–37°C and pH 7.0–8.0. The NaCl tolerance range is up to 12% (w/v).

Degrades Tweens 20 and 40, but not cellulose or gelatin. Catalase-positive, but does not coagulate or peptonize milk or produce H_2S. Acid is produced from amygdalin and esculin, but not from cellobiose, dulcitol, inositol, mannose, oxalate, or sorbitol. D-Arabinose, ribose, acetate, and citrate are used as sole carbon sources. Adenine, L-arginine, L-hydroxyproline, hypoxanthine, L-lysine, L-phenylalanine, L-serine, L-tyrosine, and L-valine are used as sole nitrogen sources, but not L-alanine, glycine, or xanthine. Additional phenotypic properties are shown in Table 255.

Whole-organism hydrolysates contain arabinose and galactose as characteristic sugars and *meso*-diaminopimelic acid as the principal acid. Cellular fatty acids contain major proportions (>10%) of $C_{15:0}$ iso, $C_{17:0}$ anteiso, $C_{18:0}$ iso, and $C_{17:0}$ iso. The predominant menaquinone is MK-9(H_4) and the major polar lipids are phosphatidylcholine, phosphatidylethanolamine, phosphatidylglycerol, phosphatidylinositol, phosphatidylinositol mannosides, phosphatidylmethylethanolamine, and one unknown phospholipid.

DNA–DNA relatedness values (± SD) between the type strain and corresponding strains of closely related species are as follows: "*Saccharopolyspora endophytica*", 57.5 ± 4.5%; *Saccharopolyspora flava*, 44.9 ± 3.5%; and *Saccharopolyspora spinosa*, 48.5 ± 3.0%.

Source: isolated from a surface-sterilized stem sample of *Tripterygium hypoglaucum*.

DNA G+C content (mol%): 70.5 (HPLC).

Type strain: CCTCC AA 208062, DSM 45269, YIM 65359.

Sequence accession no. (16S rRNA gene): FJ214364.

Species *incertae sedis*

1. **"Saccharopolyspora endophytica"** Qin, Li, Zhao, Chen, Xu and Li 2008a, 354

en.do.phy'ti.ca. Gr. adj. *endo* inside; Gr. *phyton* plant; L. fem. suff. *-ica*, adjectival suffix used with the sense of belonging to; N.L. fem. adj. *endophytica* within plant, pertaining to the original isolation from plant tissues.

Forms an extensively branched substrate mycelium. Long, smooth-surfaced spores are borne in straight chains or loose spirals on the aerial mycelium. Grows well on Czapek's, glycerol-asparagine (ISP medium 5), potato, and yeast extract-malt extract (ISP medium 2) agars. White aerial hyphae are formed on most of these media and pink diffusible pigments are formed on some agar media. NaCl tolerance range is 0–15% (w/v). Temperature and pH ranges for growth are 20–45°C and pH 5.0–9.0, respectively. Optimal temperature and pH for growth are 37°C and pH 7.0–7.5, respectively.

Cellulose, chitin, hypoxanthine, starch, tyrosine, Tweens 20 and 80, and xanthine are degraded, but not casein. Milk is not coagulated or peptonized. Does not hydrolyze esculin or urea, or produce H_2S. Cellobiose, dextrin, erythritol, inositol, malate, ribose, sorbitol, starch, trehalose, and xylitol are used as sole carbon sources, but not dulcitol, glycine, oxalate, propionate, sorbose, or succinate. Additional phenotypic features are shown in Table 255.

Whole-organism hydrolysates contain arabinose and galactose as characteristic sugars and *meso*-diaminopimelic acid as the principal diamino acid. Cellular fatty acids contain major proportions of $C_{15:0}$ iso, $C_{16:0}$ iso, $C_{17:0}$ iso, and $C_{17:0}$ anteiso. The predominant menaquinone is MK-9(H_4) and the major polar lipds are diphphatidylglycerol, phosphatidylcholine, phosphatidylethanolamine, phosphatidylglycerol, phosphatidylinositol, and phosphatidylinositol mannosides.

The DNA–DNA relatedness value between the type strain and the corresponding strain of *Saccharopolyspora flava*, a closely related species, is 31%.

Source: isolated from surface-sterilized root of *Maytenus austroyunnanensis*.

DNA G+C content (mol%): 66.2 (HPLC).

Type strain: YIM 61095, KCTC 19397, CCTCC AA 208003.

Sequence accession no. (16S rRNA gene): EU814512.

Genus XV. **Saccharothrix** Labeda, Testa, Lechevalier and Lechevalier 1984, 429[VP] emend. Labeda and Lechevalier 1989, 422

David P. Labeda

Sac.cha.ro'thrix. Gr. n. *sakchâr* sugar; Gr. fem. n. *thrix* hair; N.L. fem. n. *Saccharothrix* sugar-containing hair.

Branched vegetative mycelia (approximately 0.5–0.7 μm diameter); aerial mycelium is produced on some growth media. Both the vegetative and aerial hyphae fragment into coccoid to coccoid-rod, nonmotile elements. A "zig-zag" morphology of the aerial hyphae is typically observed during sporulation of most species. Gram-stain-positive. Resistant to lysozyme. Catalase-positive and aerobic. **Cell wall contains *meso*-diaminopimelic acid. Characteristic whole-cell sugars consist of galactose, rhamnose, and a trace of mannose. The phospholipid pattern contains significant amounts of phosphatidylethanolamine and phosphatidylethanolamine containing 2-hydroxy fatty acids, as well as diphosphatidylglycerol, phosphatidylglycerol, phosphatidylinositol, and phosphatidylinositol mannosides. The principal menaquinones are MK-9(H_4) and MK-10(H_4)**. The fatty acid profile consists predominantly of iso- and anteiso-pentadecanoic, hexadecanoic, and heptadecanoic saturated fatty acids along with smaller quantities of unsaturated, straight-chain pentadecanoic, hexadecanoic, and heptadecanoic fatty acids. Phylogenetically, the genus *Saccharothrix* represents a line of descent closest to the genera *Lentzea* and *Lechevalieria*. **The 16S rRNA gene sequence contains the genus-specific diagnostic nucleotide signature pattern of CAC (607–609) and GTG (617–619).**

DNA G+C content (mol%): 67–76 (T_m).

Type species: **Saccharothrix australiensis** Labeda, Testa, Lechevalier and Lechevalier 1984, 430[VP].

Further descriptive information

After the initial description of the genus *Saccharothrix* by Labeda et al. (1984), numerous species were transferred to this genus from other taxa, such as *Actinomadura* and *Nocardiopsis*, based on chemotaxonomic and morphological criteria. The number of species in this genus with validly published names has decreased in recent years as a result of taxonomic reclassifications based on molecular phylogenetic studies and reassessment of chemotaxonomic data. In this regard, *Crossiella cryophila*, *Goodfellowiella coeruleoviolacea*, *Lechevalieria aerocolonigenes*, *Lechevalieria flava*, *Lentzea waywayandensis*, and *Umezawaea tangerina* were all valid species within the genus *Saccharothrix* but were subsequently reclassified (Labeda, 2001; Labeda et al., 2001; Labeda and Kroppenstedt, 2006, 2007) as phylogenetic relationships based on analyses of the 16S rRNA gene were understood. In all of these taxa, a careful re-examination of the chemotaxonomic profiles found differences from that of *Saccharothrix sensu stricto* which were supportive of description as members of new or other existing genera.

Members of the genus *Saccharothrix* have been of great interest as sources of novel secondary metabolites and many of the species were isolated and described as a function of natural products discovery programs in the pharmaceutical industry. The antibiotics produced and the biotechnological activities shown by the type strains of several species of *Saccharothrix* can be seen in Table 256.

TABLE 256. Secondary metabolites produced by the type strains of *Saccharothrix* species

Species	Compound or activity
Saccharothrix algeriensis	Dithiolopyrrolone antibiotics
Saccharothrix australiensis	Antibiotic LL-BM-782-β
Saccharothrix espanaensis	Antibiotic LL-C19004-α
Saccharothrix mutabilis subsp. *capreolus*	Antibiotic capreomycin
Saccharothrix mutabilis subsp. *mutabilis*	Antibiotic polynitoxin
Saccharothrix syringae	Antibiotic nocamycin
Saccharothrix xinjiangensis	Degrades pyrene

Enrichment and isolation procedures

Strains of *Lechevalieria*, *Lentzea*, and *Saccharothrix* have been isolated from soil samples by spread-plating serial soil dilutions onto routine selective media (such as 1.5% crude agar and 0.4% casein hydrolysate in tap water) used for the general isolation of actinomycetes. Shearer (1987) reported the use of typical actinomycete isolation media, such as Gauze mineral medium no. 1 (Gauze et al., 1957) or starch-casein agar (Küster and Williams, 1964), supplemented with 5–10 μg/ml penicillin G and 15 μg/ml nalidixic acid antibiotics to selectively isolate *Saccharothrix* strains. *Saccharothrix algeriensis* was isolated from a Saharan soil sample by plating soil dilutions on the humic acid-vitamin medium of Hayakawa and Nonomura (1987) supplemented with streptomycin sulfate (10 μg/ml) and cycloheximide (50 μg/ml) (Zitouni et al., 2004). More recently, the pyrene-degrading species *Saccharothrix xinjiangensis* was isolated from filtered water samples (0.22 μm pore size) from Tianchi Lake by plating serial dilutions onto agar plates with anthracene, benzene, phenathrene, or pyrene as the sole carbon source (Hu et al., 2004).

Maintenance procedures

Working cultures of *Saccharothrix* can be maintained as refrigerated (4°C) agar slants on an appropriate medium such as NZamine medium (DSMZ medium 554; DSMZ, 2001). The slants are generally incubated for 5–7 d at 28–30°C prior to refrigeration and then should be subcultured at monthly intervals. Longer term preservation of strains is best accomplished as frozen stocks in 20% (v/v) aqueous glycerol at −80°C (mechanical freezer) to −172°C (liquid nitrogen vapor phase) or by using traditional lyophilization techniques.

Procedures for testing special characters

Strains are routinely cultivated on NZamine medium (DSMZ medium 554) at 28°C. Morphological observations are made on the media of Shirling and Gottlieb (1966) and NZamine medium.

Chemotaxonomic analyses of strains for fatty acids, polar lipids, and menaquinones are performed using methods previously described by Grund and Kroppenstedt (1989), Minnikin et al. (1984), Saddler et al. (1991), and Sasser (1990).

Physiological tests, including production of acid from carbohydrates, utilization of organic acids, and hydrolysis and decomposition of adenine, casein, esculin, guanine, hippurate, hypoxanthine, tyrosine, xanthine, and urea, are typically determined using the media of Gordon et al. (1974). Phosphatase activity is evaluated by using the method of Kurup and Schmitt

(1973) substituting NZamine agar as the basal medium. Temperature range for growth is determined on slants of NZamine medium and salt tolerance is determined on slants of the same medium supplemented with 4% and 5% NaCl.

Differentiation of the genus *Saccharothrix* from other genera

The type strains of the species of the genus *Saccharothrix* form a monophyletic lineage that is distinct from the genera *Actinosynnema, Lechevalieria,* and *Lentzea* and closest to, but distinct from, the genus *Umezawaea* (see Figure 290). The aligned sequences of the 16S rRNA genes for *Saccharothrix, Actinosynnema, Lechevalieria, Lentzea,* and *Umezawaea* (see Figure 309 in the treatment of the genus *Lechevalieria*) illustrate that *Saccharothrix* strains can easily be distinguished from those of other genera on the basis of the diagnostic nucleotide signatures CAC (607–609) and GTG (617–619). Salazar et al. (2002) also reported a set of specific PCR primers for the genus *Saccharothrix,* Stx2 (5′-AAGGCCCTTCGGGGTACACGAG-3′) and Stx1 (5′-TCGACCGCAGGCTCCACG-3′), that with a PCR annealing temperature of 66°C permitted the rapid detection of all species except *Saccharothrix texasensis.*

The species of the genus *Saccharothrix* can be distinguished chemotaxonomically from those of the genera *Lechevalieria, Lentzea,* and *Umezawaea* by the presence of phosphatidylethanolamine containing 2-hydroxy fatty acids in their phospholipid profiles (see Table 257), similar to the genus *Actinosynnema* from which they can be easily distinguished. *Saccharothrix* species can be differentiated from other phylogenetically related genera by the presence of galactose, rhamnose, and a trace of mannose as their whole-cell sugar pattern and the presence of both MK-9(H_4) and MK-10(H_4) as their predominant menaquinones (see Table 257).

Saccharothrix strains produce aerial mycelium on some growth media and both the substrate and aerial hyphae fragment into coccoid to coccoidal-rod-shaped elements. Sporangia, coremia, or motile spores are not produced, which differentiates *Saccharothrix* strains from members of the genera *Actinosynnema* and *Actinokineospora.*

Differentiation of the species of the genus *Saccharothrix*

The physiological characteristics of *Saccharothrix* species are summarized in Table 258 and can be used to differentiate between species, although a common set of physiological tests has not been used in the description of all species. Gross colonial morphology provides some additional information on the species based on the color of the substrate mycelium, production and color of aerial mycelium, and production and color of soluble pigments when grown on agar media.

TABLE 257. Chemotaxonomic characteristics of *Saccharothrix* and related genera[a]

Character	*Saccharothrix*	*Actinosynnema*	*Lechevalieria*	*Lentzea*	*Umezawaea*
Whole-cell sugar pattern[b]	Gal, Rha, Man (trace)	Gal, Man	Gal, Man, Rha	Gal, Man, Rib	Gal, Rha
Phospholipids[c]	PE, OH-PE, PI, PIM, DPG, PG (v)	PE, OH-PE, PI, PIM, DPG	PE	PE, DPG, PG, PI	PE
Predominant menaquinones	MK-9(H_4), MK-10(H_4)	MK-9(H_4), MK-9(H_6)	MK-9(H_4)	MK-9(H_4)	MK-9(H_4), trace MK-10(H_4)

[a]All genera have *meso*-diaminopimelic acid (A_2pm) as the cell-wall diamino acid, are of cell-wall chemotype III, and contain straight-chain, mono-unsaturated, iso, and anteiso fatty acids.

[b]Gal, Galactose; Man, mannose; Rib, ribose; Rha, rhamnose.

[c]DPG, Diphosphatidylglycerol; PE, phosphatidylethanolamine; PG, phosphatidylglycerol; PI, phosphatidylinositol; PIM, phosphatidylinositol mannosides; OH-PE, phosphatidylethanolamine containing hydroxylated fatty acids. v, Variable.

TABLE 258. Physiological properties of *Saccharothrix* species[a]

Character	*S. australiensis*	*S. algeriensis*	*S. coeruleofusca*	*S. espanaensis*	*S. longispora*	*S. mutabilis* subsp. *capreolus*	*S. mutabilis* subsp. *mutabilis*	*S. syringae*	*S. texasensis*	*S. xinjiangensis*
Hydrolysis of:										
Adenine	−	−	−	−	+	−	−	−	−	+
Casein	+	+	+	+	+	+	+	+	+	nd
Esculin	+	+	+	+	+	+	+	+	+	nd
Gelatin	+	+	+	+	+	+	+	+	+	nd
Hippurate	−	nd	−	+	−	+	+	+	+	nd
Hypoxanthine	−	−	−	+	−	+	+	−	−	nd

(continued)

TABLE 258. (continued)

Character	*S. australiensis*	*S. algeriensis*	*S. coeruleofusca*	*S. espanaensis*	*S. longispora*	*S. mutabilis* subsp. *capreolus*	*S. mutabilis* subsp. *mutabilis*	*S. syringae*	*S. texasensis*	*S. xinjiangensis*
Starch	–	–	+	–	+	+	+	+	+	–
Tyrosine	+	+	–	–	+	+	+	+	+	–
Urea	–	nd	–	–	+	–	–	–	–	nd
Xanthine	–	–	–	–	–	–	–	–	–	nd
Production of:										
Soluble pigments	+	+	–	–	–	–	–	+	–	+
Nitrate reductase	+	+	–	w	+	–	+	–	–	–
Phosphatase	–	nd	nd	+	nd	nd	+	nd	+	nd
Acid from:										
Adonitol	–	–	–	–	–	–	–	–	–	nd
Arabinose	–	–	+	–	+	+	+	+	+	+
Cellobiose	+	–	+	+	+	+	+	+	+	nd
Dextrin	+	–	+	–	+	+	+	+	+	nd
Dulcitol	–	–	nd	–	nd	nd	–	nd	–	nd
Erythritol	+	–	nd	–	nd	nd	–	nd	–	nd
Fructose	+	+	+	+	+	+	+	+	v	nd
Galactose	+	+	nd	nd	nd	nd	+	nd	+	nd
Glucose	+	+	+	+	+	+	+	+	+	nd
Glycerol	+	+	nd	+	+	nd	+	nd	+	nd
Inositol	–	–	–	–	–	+	+	–	+	nd
Lactose	–	–	+	–	+	–	+	+	+	+
Maltose	+	+	nd	v	nd	nd	nd	nd	nd	nd
Mannitol	nd	–	nd	–	nd	nd	nd	nd	+	nd
Mannose	+	–	nd	+	nd	nd	+	nd	+	nd
Melibiose	–	–	–	–	–	+	+	+	+	+
Methyl α-D-glucoside	–	–	+	–	–	–	+	–	+	+
Methyl β-xyloside	–	nd	nd	–	nd	nd	nd	nd	v	nd
Raffinose	–	–	+	–	–	–	+	+	–	+
Rhamnose	–	–	+	–	+	–	–	+	+	+
Salicin	–	nd	nd	–	nd	nd	+	nd	+	nd
Sorbitol	+	–	+	–	–	–	–	–	–	–
Sucrose	–	–	+	+	+	–	+	+	+	+
Trehalose	+	nd	nd	+	nd	nd	+	nd	+	nd
Xylose	–	–	+	v	+	+	+	+	+	+
Assimilation of:										
Acetate	+	+	nd	+	nd	nd	+	nd	+	nd
Benzoate	–	–	nd	–	–	nd	–	nd	–	–
Citrate	–	+	–	v	+	–	+	–	–	–
Lactate	v	nd	–	+	+	–	+	+	+	–
Malate	+	nd	–	+	+	+	+	+	+	nd
Mucate	–	nd	nd	–	nd	nd	–	–	–	nd
Oxalate	–	–	nd	–	nd	nd	–	nd	–	nd
Propionate	+	–	nd	v	nd	nd	+	nd	v	nd
Succinate	+	+	nd	+	nd	nd	+	nd	+	nd
Tartrate	–	–	nd	–	nd	nd	–	nd	–	–
Growth in the presence of:										
4% NaCl	+	+	+	+	+	+	–	+	–	–
5% NaCl	–	–	+	–	+	+	–	+	–	–
Growth at:										
10°C	+	+	+	+	+	+	+	+	+	+
37°C	+	+	+	+	+	+	+	+	+	+
45°C	+	+	+	–	–	+	+	+	–	+
50°C	nd	nd	nd	nd	nd	nd	nd	nd	–	+

[a]Symbols: nd, not determined; v, variable positive reaction; w, weak reaction.

List of species of the genus *Saccharothrix*

1. **Saccharothrix australiensis** Labeda, Testa, Lechevalier and Lechevalier 1984, 430[VP]

aus.tra.li.en'sis. N.L. fem. adj. *australiensis* of or belonging to Australia, referring to the location of the soil sample from which the organism was first isolated.

Substrate mycelium is brownish to grayish-yellow in color. Sparse to abundant white to yellowish-gray aerial mycelium produced on many media. Brownish soluble pigments are formed on several media. Temperature for growth is 28°C.

Source: isolated from soil.

DNA G+C content (mol%): 76 (T_m).

Type strain: NRRL 11239, ATCC 31497, DSM 43800, IAM 14291, NBRC 14444, IMSNU 21246, JCM 3370, KCTC 9193, KCTC 9388, NCIMB 13188, VKM Ac-894, LL-BM782Ce82.

Sequence accession nos (16S rRNA gene): X53193.1, AF114803.

2. **Saccharothrix algeriensis** Zitouni, Lamari, Boudjella, Badji, Sabaou, Gaouar, Matthieu, Lebrihi and Labeda 2004, 1380[VP]

al.ge.ri.en'sis. N.L. fem. adj. *algeriensis* of Algeria, where the type strain originated.

Substrate mycelium is vivid yellow, orange-yellow, or yellowish-brown. Copious yellow-orange aerial mycelium is produced. A bright yellow soluble pigment is produced on some media. Temperature for growth is 28°C.

Source: isolated from soil.

DNA G+C content (mol%): not determined.

Type strain: DSM 44581, JCM 13242, NRRL B-24137, SA 233.

Sequence accession no. (16S rRNA gene): AY054972.

3. **Saccharothrix coeruleofusca** (Preobrazhenskaya and Sveshnikova 1974) Grund and Kroppenstedt 1990a, 320[VP] (Effective publication: Grund and Kroppenstedt 1989, 270.) [*Actinomadura coeruleofusca* Preobrazhenskaya and Sveshnikova 1974, 864; *Nocardiopsis coeruleofusca* (Preobrazhenskaya and Sveshnikova 1974) Preobrazhenskaya and Sveshnikova 1985, 224]

co.e.ru.le.o.fus'ca. L. adj. *coeruleus* dark-colored, dark blue; L. adj. *fuscus* dark, swarthy, dusky, tawny; N.L. fem. adj. *coeruleofusca* blue-brown (referring to the color of aerial and substrate mycelium).

Substrate mycelium is yellowish in color. Only sparse aerial mycelia, if any, are produced on complex growth media. On glycerol-nitrate or starch-nitrate media, a blue to dark blue aerial mycelium is produced. Temperature for growth is 28°C.

Source: isolated from soil.

DNA G+C content (mol%): 67.0 (T_m) (Poschner et al., 1985).

Type strain: ATCC 35108, BCRC 13313, DSM 43679, IFO 14520, IMET 9602, IMSNU 21357, INA 1335, JCM 3313, KCTC 9389, NRRL B-16115, VKM Ac-855.

Sequence accession no. (16S rRNA gene): AF114805.

4. **Saccharothrix espanaensis** Labeda and Lechevalier 1989, 422[VP]

es.pa.na.en'sis. N.L. fem. adj. *espanaensis* of or belonging to Spain, referring to the source of the soil sample, Puerto Llano (Spain) from which the micro-organism was first isolated.

Substrate mycelium is grayish-yellow to yellowish-brown. Sparse white aerial mycelium is produced on some media, particularly glycerol-asparagine agar and inorganic salts-starch agar (ISP 3) and a yellow soluble pigment is also produced on these media. A brown to reddish-brown soluble pigment is formed on rich complex media. Temperature for growth is 28°C.

Source: isolated from soil.

DNA G+C content (mol%): 72.2 (T_m).

Type strain: ATCC 51144, CGMCC 4.1714, DSM 44229, IFO 15066, IMSNU 21342, JCM 9112, KCTC 9392, NBRC 15066, NRRL 15764, VKM Ac-1969.

Sequence accession no. (16S rRNA gene): AF114807.

5. **Saccharothrix longispora** (Preobrazhenskaya and Sveshnikova 1974) Grund and Kroppenstedt 1990a, 320[VP] (Effective publication: Grund and Kroppenstedt 1989, 272.) [*Actinomadura longispora* Preobrazhenskaya and Sveshnikova 1974, 866; *Nocardiopsis longispora* (Preobrazhenskaya and Sveshnikova 1974) Preobrazhenskaya and Sveshnikova 1985, 224]

lon.gi.spo'ra. L. adj. longus long; Gr. n. spora a seed and in biology a spore; N.L. n. *longispora* (nominative in apposition) the long spore, referring to the oblong shape of the spores.

Substrate mycelium has a yellow or red color and aerial mycelium is normally absent. On glycerol-nitrate agar a blue aerial mycelium is formed. Temperature for growth is 28°C.

Source: isolated from soil.

DNA G+C content (mol%): 68.0 (T_m) (Poschner et al., 1985).

Type strain: INA 10222, ATCC 35109, BCRC 13395, DSM 43749, HUT-6594, NBRC 14522, IMET 9603, IMSNU 21359, JCM 3314, KCTC 9394, NRRL B-16116, VKM Ac-907.

Sequence accession no. (16S rRNA gene): AF114809.

6. **Saccharothrix mutabilis** (Shearer, Colman and Nash 1983) Labeda and Lechevalier 1989, 422[VP] (*Nocardiopsis mutabilis* Shearer, Colman and Nash 1983, 374)

mu.ta'bi.lis. L. fem. adj. *mutabilis* changeable, variable, inconstant, referring to the variety of colony morphologies observed, particularly on rich organic media.

Substrate mycelium is yellow to yellowish-brown. White aerial mycelium is produced on most media. Upon exposure to light, the aerial mycelium may turn orange-yellow. Light yellow to yellowish-brown soluble pigments are produced on some media. Temperature for growth is 28°C.

Source: isolated from soil.

DNA G+C content (mol%): 73.1 (T_m).

Type strain: ATCC 31520, BCRC 12528, DSM 43853, IFM 240, NBRC 14310, IMSNU 21336, JCM 3380, KCTC 9397, NRRL B-16077, SKF AAA-025, VKM Ac-2023.

Sequence accession no. (16S rRNA gene): X76966.

This species has subsequently been divided into subspecies as follows.

6a. **Saccharothrix mutabilis subsp. mutabilis** (Shearer, Colman and Nash 1983) Labeda and Lechevalier 1989, 422[VP] (*Nocardiopsis mutabilis* Shearer, Colman and Nash 1983, 374)

Characteristics as above for *Saccharothrix mutabilis.*

6b. **Saccharothrix mutabilis subsp. capreolus** (*ex* Stark, Higgens, Wulfe, Hoehn and McGuire 1963) Grund and Kroppenstedt 1990a, 320[VP] (Effective publication: Grund and Kroppenstedt 1989, 273.) (*Streptomyces capreolus* Stark, Higgens, Wulfe, Hoehn and McGuire 1963, 596)

ca.pre'o.lus. L. masc. n. *capreolus* roebuck or chamois, two-pronged like the chamois, bifurcate.

Substrate mycelium is yellowish to brownish in color. Sparse, white aerial mycelium is produced on some media. Temperature for growth is 28°C.

Source: isolated from soil.

DNA G+C content (mol%): not determined.

Type strain: ATCC 23892, BCRC 13692, CBS 678.68, DSM 40225, NBRC 12847, IMSNU 21351, ISP 5225, JCM 4248, JCM 4630, KCTC 9395, NCIMB 9611, NCIMB 9801, NRRL 2773, RIA 1167, VKM Ac 1848.

Sequence accession no. (16S rRNA gene): X76865.

7. **Saccharothrix syringae** (Gauze and Sveshnikova 1985) Grund and Kroppenstedt 1990a, 320[VP] (Effective publication: Grund and Kroppenstedt 1989, 273.) (*Nocardiopsis syringae* Gauze, Sveshnikova, Ukholina, Komarova and Bazhanov 1977, 483)

sy.rin'ga.e. N.L. fem. n. *Syringa* generic name of lilac; N.L. fem. gen. n. *syringae* of the lilac (referring to the color of aerial mycelium).

Substrate mycelium is yellowish to brownish in color. Aerial mycelium is sparse and white in color on most media. The aerial mycelium is reported (Gauze et al., 1977) to be pale to dark lilac in color on Gauze no. 1 mineral agar (Gauze et al., 1957) or glycerol-nitrate agar (Lindenbein, 1952), hence the species name. Temperature for growth is 28°C.

Source: isolated from soil.

DNA G+C content (mol%): not determined.

Type strain: INA 2240, AS 4.1716, ATCC 51364, DSM 43886, NBRC 14523, IMET 9675, JCM 6844, KCTC 9398, NRRL B-16468, VKM Ac-1858.

Sequence accession no. (16S rRNA gene): AF114812.

8. **Saccharothrix texasensis** Labeda and Lyons 1989, 357[VP]

tex.as.en'sis. N.L. fem. adj. *texasensis* of or belonging to the state of Texas, referring to the source of the soil samples from which the species was first isolated.

Substrate mycelium is dark yellow to brownish-yellow. Sparse, white aerial mycelium is produced on some media. A brown to reddish-brown soluble pigment is formed on some media. Temperature for growth is 28°C.

Source: isolated from soil.

DNA G+C content (mol%): 71 (T_m).

Type strain: NRRL B-16134, AS 4.1713, ATCC 51593, DSM 44231, NBRC 14971, IMSNU 21343, JCM 9113, KCTC 9399, NCIMB 13186, VKM Ac-1968.

Sequence accession no. (16S rRNA gene): AF114814.

9. **Saccharothrix xinjiangensis** Hu, Zhou, Zhou, Liu and Liu 2004, 2094[VP]

xin.ji.ang.en'sis. N.L. fem. adj. *xinjiangensis* pertaining to Xinjiang, where the type strain was isolated.

The substrate mycelium is pinkish-buff in color on yeast extract-malt extract agar and pale orange to pale brown on other media. The aerial mycelium is grayish-white in color. Pale brown soluble pigments are produced on some media. Temperature for growth is 28°C.

Source: isolated from lake water.

DNA G+C content (mol%): 70.4 (T_m).

Type strain: AS 4.1731, DSM 44896, JCM 12329, NRRL B-24321, PYX-6.

Genus XVI. **Streptoalloteichus** Tomita, Nakakita, Hoshino, Numata and Kawaguchi 1987, 211[VP] emend. Tamura, Ishida, Otoguro, Hatano and Suzuki 2008b, 689

Seung Bum Kim and Tomohiko Tamura

Strep.to.al.lo.tei'chus. Gr. adj. *streptos* bent; Gr. adj. *allos* different; Gr. n. *teichos* wall; N.L. masc. n. *Streptoalloteichus* intended to mean streptomycete with different wall.

Gram-stain-positive, non-acid-fast, aerobic actinomycetes that show well-branched and non-fragmenting substrate hyphae. **Aerial mycelium bears chains of five to 50 spores** (0.5–1.2 µm wide) at the tip of well-branched hyphae. **Sporangia-like structures are formed**. The predominant menaquinones are **MK-9(H_6) and MK-10(H_6)**. The cell-wall peptidoglycan contains a major amount of ***meso*-diaminopimelic acid**. The predominant cellular fatty acids are **$C_{15:0}$ iso and $C_{17:0}$ anteiso**. Whole-cell hydrolysates contain **D-galactose and D-mannose**, whereas L-rhamnose and D-glucose are variably present. **Phosphatidylethanolamine** is present as the diagnostic phospholipid (phospholipid pattern type PII). Habitat: soil.

Type species: **Streptoalloteichus hindustanus** Tomita, Nakakita, Hoshino, Numata and Kawaguchi 1987, 211[VP].

Further descriptive information

The genus *Streptoalloteichus* currently belongs to the family *Pseudonocardiaceae* (Stackebrandt et al., 1997) and contains two species, *Streptoalloteichus hindustanus* and *Streptoalloteichus tenebrarius* (Table 259). However, the taxonomic affiliation of the genus at the family level looks rather unclear. *Streptoalloteichus*, together with *Crossiella* and *Kutzneria*, have been consistently placed outside both the *Pseudonocardiaceae* and *Actinosynnemataceae* clades, still residing within the former suborder *Pseudonocardineae* (Labeda, 2001; Labeda and Kroppenstedt, 2006), elevated in this volume to order *Pseudonocardiales*. Further phylogenetic studies on these genera and related taxa might be necessary to clarify their taxonomic positions within the order.

TABLE 259. Differential characteristics that distinguish the two species of the genus *Streptoalloteichus*[a]

Characteristic	*S. hindustanus* C677-91^T	*S. tenebrarius* NBRC 16177^T
Morphology:		
Sporangia-like vessels	+	–
Motile spores	+	–
Cell chemistry:		
Whole-cell sugars[b]	Gal, Man, Rha	Gal, Man, Glc
Menaquinones (MK-)	9(H_6), 10(H_6)	10(H_6), 10(H_4), 9(H_6), 9(H_4)
DNA G+C content (mol%)	nd	71.6
Cultural characteristics on inorganic salts-starch agar (ISP 4):		
Vegetative mycelium	Thin, colorless to grayish yellow	Grayish pink
Aerial mycelium	Pale pinkish yellow	Light grayish-yellow with white areas
Diffusible pigment	None	Grayish pink
Cultural characteristics on glycerol-asparagine agar (ISP 5):		
Growth	Restricted	Moderate
Vegetative mycelium	Thin, colorless to grayish yellow	Pale yellow
Aerial mycelium	Patches, white, turning yellowish gray later	Pale yellow with white areas
Cultural characteristics on Bennett's agar:		
Vegetative mycelium	Pale olivaceous yellow to light brown	Pale yellow
Aerial mycelium	Velvety, light yellowish beige	White
Utilization of:		
Maltose	–	+
L-Arabinose	–	+
D-Xylose	–	+
Salicin	–	+
myo-Inositol	–	+
Lactose	Weak	–
Antibiotic production	Tallysomycins A, B, and C; nebramycin factors II, IV′, and V′	Nebramycin factors I to XIII
Tolerance to 7% (w/v) NaCl[c]	–	+
Light sensitivity for formation of aerial mycelium	–	+

[a]Modified from Table 1 of Tamura et al. (2008b); nd, not determined.

[b]Gal, galactose; Man, mannose; Rha, rhamnose; Glc, glucose.

[c]Both grow at 5%, neither grows at 10%.

Streptoalloteichus hindustanus and *Streptoalloteichus tenebrarius* share 99.5% 16S rRNA gene sequence similarity, but only 24.3–37.9% DNA–DNA relatedness (Tamura et al., 2008b). The two species are clearly distinguishable by their morphological properties: *Streptoalloteichus hindustanus* produces oval or spherical sporangia-like vessels on the vegetative hyphae (Figure 332) and motile spores with a single polar flagellum. Sporangia-like structures can also be found, albeit with lower frequency, in *Streptoalloteichus tenebrarius* (Figure 333). However, the spores of *Streptoalloteichus tenebrarius* are nonmotile.

Purified cell wall preparations contain *meso*-diaminopimelic acid, but not the L-isomer or the hydroxy analogs. Alanine and glutamic acid are found in *Streptoalloteichus tenebrarius.* Whole-cell hydrolysates contain galactose and mannose; in addition, rhamnose and glucose are present in *Streptoalloteichus hindustanus* and *Streptoalloteichus tenebrarius*, respectively. The diagnostic phospholipid is phosphatidylethanolamine (phospholipid type PII). The predominant fatty acids of *Streptoalloteichus* are $C_{15:0}$ iso and $C_{17:0}$ anteiso.

Both species of *Streptoalloteichus* are known to produce nebramycins and *Streptoalloteichus hindustanus* is also known to produce tallysomycins.

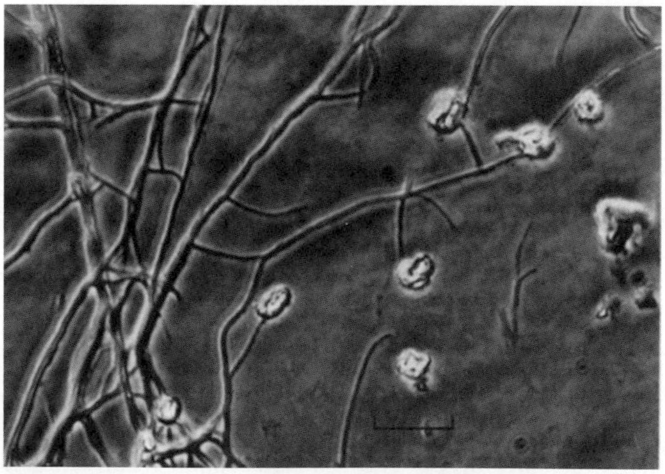

FIGURE 332. Sporangia-like vessels enveloping a single spore or a row of two to four spores of *Streptoalloteichus hindustanus*, which are formed on the sporangiophore of the substrate mycelium. Bar = 10 μm.

Enrichment and isolation procedures

Streptoalloteichus hindustanus was isolated from dry soil samples collected in Gujarat or adjacent states where the natural vegetation is dry tropical forest and scrub using nutrient agar supplemented with butirosin (50 μg/ml). Yeast extract-malt extract agar (ISP 2) or Bennett's agar, supplemented with kanamycin or gentamicin (10 μg/ml) and nystatin (100 μg/ml), can also be used. The agar plates are incubated for 1–2 weeks at 43°C. Colonies of *Streptoalloteichus hindustanus*

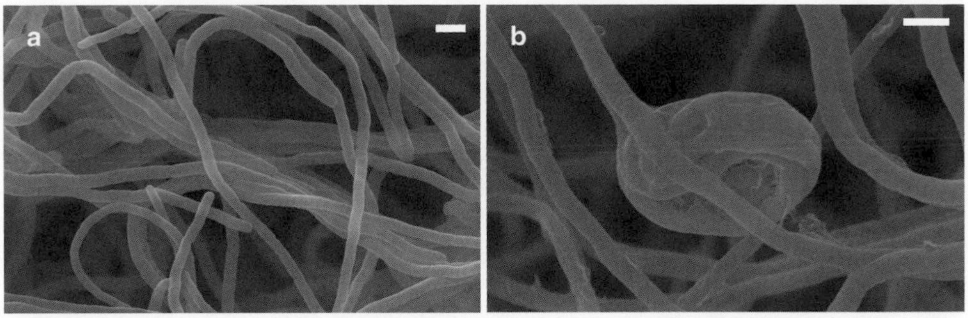

FIGURE 333. Scanning electron micrographic images of *Streptoalloteichus tenebrarius* showing the formation of arthrospore chains (a) and sporangia-like structure (b). Bars = 1 µm.

are distinguishable from those of most species of the genus *Streptomyces* by pale pinkish-yellow aerial mycelium, and the lack of a distinct reverse-side color.

Streptoalloteichus tenebrarius can be isolated from soil and grows on various media such as ISP 2, 4, and 5, and Bennett's agar medium.

Maintenance procedures

Streptoalloteichus strains can be maintained on media that is used universally for actinobacteria. The mixture of spores and mycelial fragments can be stored in 20% (v/v) glycerol solution at temperatures below −20°C or as lyophilized powder. Alternatively, the mixture is suspended in sterilized skim milk (10%) and stored at −80°C in screw-capped tubes. Lyophilization of the spore suspension is carried out by standard procedures for actinobacteria and the lyophilized culture is preserved under a vacuum of between 0.05 and 0.005 mmHg.

Procedures for testing special characters

Sporangia formation may be induced by growing the strains at 28°C for 3–4 weeks on ISP 2 or ISP 5 agar. Microscopic observation of sporangia and other structures in the vegetative mycelium can be achieved by preparing the cultures using the coverslip technique (Kawato and Shinobu, 1959).

Alternatively, ISP 5 broth or soil extract broth (Tomita, 1989) may be used. Media in metal-capped tubes (18×180 mm in size; 3 ml medium per tube) are autoclaved, inoculated heavily with a spore suspension, and incubated as slants for 4–6 weeks at 28°C. The sporangia can be observed at hyphal tips of mycelial masses.

Differentiation from closely related taxa

Streptoalloteichus is phylogenetically mostly related to members of the genera *Goodfellowiella*, *Crossiella*, and *Kibdelosporangium*, with 16S rRNA gene sequence similarities between strains of 95 to 96% (Figure 334). However, *Streptoalloteichus* can be differentiated from the other three genera based on chemotaxonomic properties. The major isoprenoid quinones of *Streptoalloteichus* are MK-9(H_6) and MK-10(H_6), whereas those of *Goodfellowiella* are MK-9(H_4) and MK-10(H_4), and that of *Crossiella* and *Kibdelosporangium* is MK-9(H_4). Mannose is one of the diagnostic sugars for *Streptoalloteichus*, but not for *Goodfellowiella* or *Kibdelosporangium*.

Streptoalloteichus can be distinguished from other genera of the family *Pseudonocardiaceae* by a combination of morphological and chemotaxonomic properties, such as spore chains, sporangia-like structures, phospholipid type, menaquinone profiles, and diagnostic sugars. In particular, the menaquinone profiles, consisting of MK-9(H_6) and MK-10(H_6), distinguish the genus from the rest of the family *Pseudonocardiaceae*. The presence of galactose and mannose in the cell wall also separates the genus from the genera *Actinoalloteichus*, *Actinopolyspora*, *Amycolatopsis*, *Goodfellowiella*, *Kibdelosporangium*, *Pseudonocardia*, *Prauserella*, *Saccharomonospora*, and *Saccharopolyspora*. The phospholipid pattern of *Streptoalloteichus* is type II *sensu* Lechevalier et al. (1977a), which is different from those of *Actinopolyspora* (type III), *Saccharopolyspora* (type III), and *Thermobispora* (type IV).

Formation of sporangia-like structures separates the genus from *Actinoalloteichus*, *Actinopolyspora*, *Goodfellowiella*, *Prauserella*, *Saccharopolyspora*, and *Thermobispora*. Mycelial fragmentation is not observed in *Streptoalloteichus*, which is different from *Actinoalloteichus*, *Amycolatopsis*, *Crossiella*, *Goodfellowiella*, *Kibdelosporangium*, *Prauserella*, *Saccharopolyspora*, and *Thermoscrispum*.

Streptoalloteichus can also be differentiated from related genera using 16S rRNA gene restriction fragment length polymorphism (RFLP) patterns, as suggested by Cook and Meyers (2003). All actinobacterial taxa can be divided into three distinct groups based on the RFLP patterns of 16S rRNA gene with *Sau*3A1. *Streptoalloteichus* belongs to Group 3; additional RFLP patterns with restriction enzymes *Asn*I and *Sph*I separate the genus from other genera of Group 3.

Taxonomic comments

The genus *Streptoalloteichus* with *Streptoalloteichus hindustanus* as the type and only species was formally published by Tomita et al. (1987), although the genus was originally proposed in 1978 (Tomita et al., 1978). *Streptoalloteichus tenebrarius* was originally described as "*Streptomyces tenebrarius*" (Higgins and Kastner, 1967). *Streptoalloteichus hindustanus* and *Streptoalloteichus tenebrarius* share a number of properties in common, e.g. cell-wall type, production of short spore chain clusters, smooth-surfaced spores, sclerotium formation, and pale pinkish-yellow aerial mycelium. On the other hand, the two species can be differentiated from each other based on various phenotypic characters. For example, the sporangia-like vessels enveloping motile spores are rarely observed in *Streptoalloteichus tenebrarius* and light sensitivity in the formation of aerial mycelium is only observed in *Streptoalloteichus tenebrarius*. Growth at 45°C has been reported for *Streptoalloteichus hindustanus*, but not for *Streptoalloteichus tenebrarius*. Similarly, *Streptoalloteichus tenebrarius* can grow in the presence of 7% NaCl, but *Streptoalloteichus hindustanus* cannot. The latter produces tallysomycins A, B, and C, and nebramycin factors II, IV′, and V′, whereas *Streptoalloteichus tenebrarius* produces nebramycin factors I to XIII.

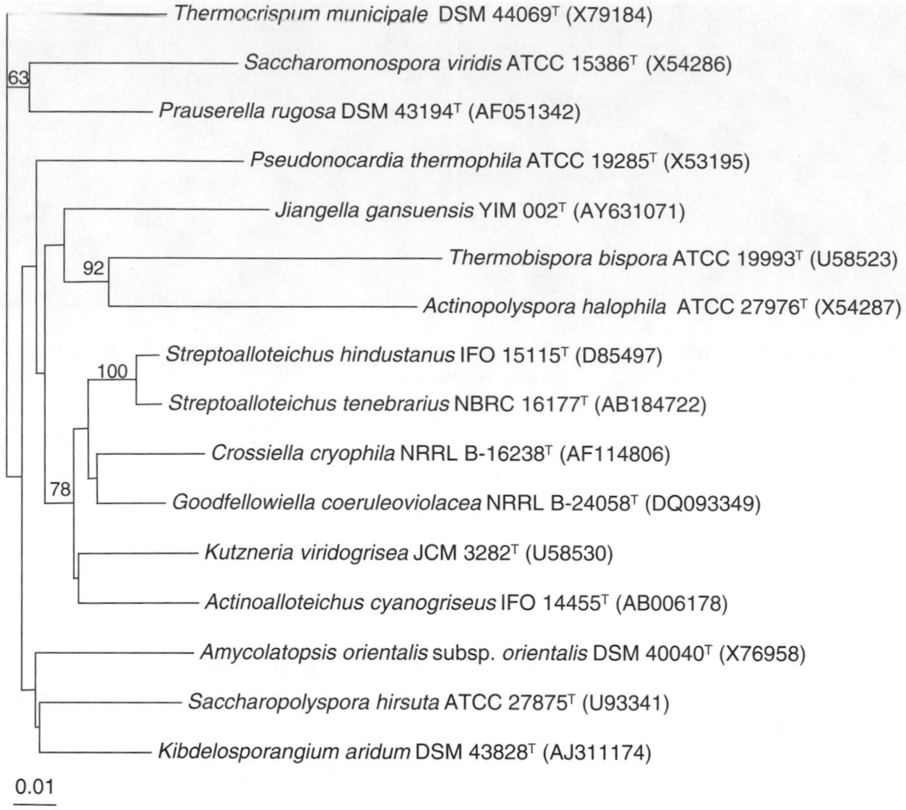

FIGURE 334. Phylogenetic tree of the genus *Streptoalloteichus* and related taxa based on 16S rRNA gene sequences. The Jukes–Cantor model was used in the estimation of evolutionary distances and the neighbor-joining method was used for tree construction. Numbers at nodes indicate levels of bootstrap support (%). Bar = 0.01 substitutions per nucleotide position.

Miscellaneous comments

Streptoalloteichus hindustanus has the ability to bear both non-motile conidiospores and motile sporangiospores. This morphology may be derived from adaptation to the original environments of this micro-organism, which have dry and rainy seasons in a year. *Streptoalloteichus hindustanus* was reported to produce tallysomycins A, B, and C, antitumor compounds, and BMY-28190, an antiviral agent (Ohkuma et al., 1988). *Streptoalloteichus* sp. 1454-19 was reported to produce siderophores IC202A, B, and C (Iijima et al., 1999a, 1999b). The bleomycin resistance (*ble*) gene from *Streptoalloteichus hindustanus* has been used as a dominant selection marker to construct transformation systems in animal, plant, fungal, and bacterial cells (Drocourt et al., 1990).

List of species of the genus *Streptoalloteichus*

1. **Streptoalloteichus hindustanus** Tomita, Nakakita, Hoshino, Numata and Kawaguchi 1987, 211[VP]

hin.du.stan'us. N.L. masc. adj. *hindustanus* of Hindustan, northwest region of India.

The aerial mycelium is abundant on most culture media. After sporulation, the aerial mycelium is pale in color and yellow, pink, or gray depending on the medium. The vegetative mycelium is colorless to light yellowish brown. Both substrate and aerial mycelia are well developed and branched, averaging 0.5 µm in diameter. Chains of spores are formed only at the tip of the aerial mycelium. These may be either long (10–50 spores in a chain) or short, hooked, branching spore chains. The individual arthrospores in the long chains are oval to cylindrical (0.5–2.0 µm in diameter) and the conidiospores in the short chains are barrel shaped. The spore surface is smooth in both cases. Spores are motile with a single polar flagellum. The spore chain cluster consists of curved or L-shaped conidiophore chains with many branches and often develops into a thick mass. The substrate mycelium does not fragment. The substrate mycelium penetrating the agar is thin, especially in chemically defined media; it is covered with a thick mass of white aerial mycelium that turns to pale yellow after sporulation. Single, oval, or spherical sporangia-like vessels enveloping one spore or a single row of two to four spores are randomly formed among the vegetative hyphae. Sporangia are 1.5–4.5 × 2.7–7.0 µm in size and sporangiospores are 0.9–1.5 × 1.2–4.0 µm. Globose dense bodies consisting of coalesced vegetative mycelium and sclerotia may be formed on the aerial mycelium. The temperature range for growth is 20–54°C; no growth is observed at 56°C. The optimal temperature for growth is 45°C. Additional cultural and physiological characteristics are shown in Table 260.

TABLE 260. Cultural characteristics of *Streptoalloteichus hindustanus*

Medium	Aerial mycelium	Vegetative mycelium	Diffusible pigments
Yeast extract-malt extract (ISP 2)	Thick pale yellowish pink	Light yellowish brown	None
Inorganic salts-starch (ISP 4)	Pale pinkish yellow	Thin, colorless to grayish yellow	None
Glycerol-asparagine (ISP 5)	Patches, white, turning yellowish gray later	Thin, colorless to grayish yellow	None
Peptone-yeast extract-iron (ISP 6)	Scant, white	Moderate brown	Light yellowish brown

Gelatin, casein, and starch are hydrolyzed; skim milk is coagulated and slightly peptonized. Melanoid pigments are not formed on ISP 1, 6, or 7. Catalase-positive, but tyrosinase-negative. Nitrate reduction occurs. Abundant growth is observed at 30–50°C, but no growth is seen at 56°C. Growth occurs with up to 5% (w/v) NaCl, but not with 7% (w/v). Utilizes D-fructose, D-galactose, D-glucose, and D-mannose as sole carbon sources, but not L-arabinose, inositol, D-mannitol, D-raffinose, L-rhamnose, D-sorbitol, or D-xylose. Weak growth occurs on lactose and sucrose. Resistant to ampicillin, cephalothin, erythromycin, gentamicin, kanamycin, and tetracycline at 100 μg/ml, and to novobiocin at 25 μg/ml. Less resistant to chloramphenicol and rifampin. Major menaquinones are MK-9(H_6) and MK-10(H_6).

Source: soil.

DNA G+C content (mol%): Not known.

Type strain: ATCC 31217, IFO 15115.

Sequence accession no. (16S rRNA gene): D85497.

2. **Streptoalloteichus tenebrarius** (*ex* Higgins and Kastner 1967) Tamura, Ishida, Otoguro, Hatano and Suzuki 2008b, 689[VP] (Subjective synonym: "*Streptomyces tenebrarius*" Higgins and Kastner 1967.)

te.ne.bra'ri.us. L. masc. adj. *tenebrarius* of or belonging to darkness, reflecting the sensitivity of the culture to light.

Straight spore chains are formed on short sporophores on the substrate mycelium. Spores have a smooth surface and are short, nonmotile rods (0.6–0.9 × 1.0–1.5 μm). Sporangia-like structures, though not common, can be formed. No soluble pigment is produced on ISP 7 agar medium.

Starch is hydrolyzed weakly and gelatin is hydrolyzed. Calcium malate is not decomposed and milk is not coagulated or peptonized. Optimum temperature for growth is 20–30°C; grows at 37°C, but not at 45°C. Grows in the presence of up to 8% (w/v) NaCl, but not in 10% (w/v) NaCl.

N-Acetyl-D-galactosamine, *N*-acetyl-D-glucosamine, arabonic acid, *p*-arbutin, D-fructose, gluconate, D-glucosamine, D-glucosaminic acid, D-glucose, glycogen, maltose, D-mannose, D-ribose, L-arabinose, D-xylose, sucrose, raffinose, salicin, starch, trehalose, adonitol, glycerol, *myo*-inositol, acetate, butyrate, caprate, isobutyrate, propionate, *cis*-aconitate, azelate, citrate, fumarate, DL-3-hydroxybutyrate, D-lactate, DL-lactate, DL-malate, L-malate, oxaloacetate, 2-oxoglutarate, pyruvate, acetamidocaprate, acetyl L-glutamine, acetyl glycine, D-alanine, β-alanine, L-arginine, L-asparagine, L-aspartate, casein, L-citrulline, L-glutamate, L-glutamine, L-histidine, L-isoleucine, L-leucine, L-ornithine, L-proline, L-serine, L-tryptophan, L-valine, and ethanolamine are utilized as sole carbon sources (as sodium salts where applicable). Major menaquinones are MK-10(H_6), MK-10(H_4), MK-9(H_6), and MK-9(H_4).

Source: soil.

DNA G+C content (mol%): 71.6 (HPLC)

Type strain: NBRC 16177, ATCC 17920, DSM 40477, JCM 4838, NRRL B-12390, ISP 5477.

Sequence accession no. (16S rRNA gene): AB184722.

Genus XVII. **Thermocrispum** Korn-Wendisch, Rainey, Kroppenstedt, Kempf, Majazza, Kutzner and Stackebrandt 1995, 73[VP]

Seung Bum Kim

Ther.mo.cris'pum. Gr. adj. *thermos* warm, hot; L. neut. adj. *crispum* tightly curled; N.L. neut. n. *Thermocrispum* a heat-loving, tightly curled organism.

Thermophilic, Gram-stain-positive, aerobic, catalase-positive, and non-acid-fast. Produces filamentous, branched hyphae. The **aerial mycelium is white**, and the **vegetative mycelium is yellow to light brown**. Soluble pigments are not produced. **Aerial hyphae are straight to flexuous and often aggregate into clusters (pseudosporangia) that fragment into rod-like structures**. Good growth occurs on Czapek Dox-yeast extract-Casamino acids (CYC) agar, glucose-yeast extract-malt extract (GYM) agar, Hickey–Tresner agar, oatmeal agar, potato-carrot agar, peptone-maize extract (PM) agar, R2A agar, and R8 agar, as well as on tryptic soy agar (TSA). **The temperature range for growth is 20–62.5°C**; the optimum range is 45–55°C. Growth occurs between pH 6.0 and 11.0, and also in the presence of NaCl (5%, w/v), novobiocin (25 μg/ml), and crystal violet (0.2 μg/ml). Nitrate is reduced to nitrite under aerobic conditions. The **cell wall contains *meso*-diaminopimelic acid and the whole-cell sugar pattern is type C; major amounts of arabinose, mannose, and glucose are present**, but only traces of galactose are detected. **Phospholipid pattern is type PII**, containing phosphatidylethanolamine as the diagnostic phospholipid. Mycolic acids are not present. The **predominant menaquinone is MK-9(H_4)**. Resistant to a set of phages that infect the genera *Amycolatopsis*, *Pseudonocardia*, *Saccharomonospora*, *Saccharopolyspora*, and *Saccharothrix*. Sensitive to genus-specific phages isolated from different habitats.

DNA G+C content (mol%): 69–73 (HPLC).

Type species: **Thermocrispum municipale** Korn-Wendisch, Rainey, Kroppenstedt, Kempf, Majazza, Kutzner and Stackebrandt 1995, 73[VP].

Further descriptive information

The genus *Thermocrispum* belongs to the family *Pseudonocardiaceae* and currently contains two species, *Thermocrispum agreste* and *Thermocrispum municipale* (Figure 335). The 16S rRNA gene sequence similarity between the type strains of the two species was reported as 98.2% (Korn-Wendisch et al., 1995), but BLAST (http://blast.ncbi.nlm.nih.gov/blast.cgi) results indicate a lower similarity value (96.0%).

Good growth of all *Thermocrispum* strains can be observed on the media mentioned in the genus description. Macroscopically, the isolates are similar to members of the genus *Saccharopolyspora*. All strains produce abundant white aerial mycelium and yellow to light brown substrate mycelium on CYC agar (Cross and Attwell, 1974) and potato-carrot agar (Cross et al., 1963), whereas sparse to moderate aerial mycelium is formed on TSA, GYM agar (ISP medium 4), oatmeal agar (ISP medium 3), PM agar (Agre, 1964), and R2A agar (Reasoner and Geldreich, 1985). Growth is poor on glycerol-arginine agar (El-Nakeeb and Lechevalier, 1963). The surfaces of colonies are covered with straight to flexuous aerial hyphae and both substrate and aerial mycelia often form aggregates (Figure 336). The aerial mycelium fragments into rod-like structures (Figure 337). Pseudosporangia similar to the structures of *Kibdelosporangium* can be observed in all strains. However, these structures are not surrounded by a well-defined wall and contain septate hyphae which fragment into rod-like structures (Figure 338).

Thermocrispum strains can degrade or hydrolyze casein, tyrosine, gelatin, DNA, esculin, Tween 80, and tributyrin, but not adenine, allantoin, cellulose, chitin, poly-β-hydroxybutyric acid, hypoxanthine, starch, urea, xanthine, or xylan. Pectin is weakly degraded and arbutin is hydrolyzed slowly. Melanoid pigments are not produced. Cellobiose, dextrin, galactose, glucose, inositol, mannose, sucrose, trehalose, sodium malonate, and sodium pyruvate are utilized as sole carbon sources, but adonitol, inulin, lactose, melibiose, raffinose, rhamnose, xylitol, xylose, and sodium propionate are not. Mannitol, melezitose, and sorbitol are utilized poorly.

All strains grow well at temperatures ranging from 28 to 60°C and between pH 6.0 and 11.0. Growth is weak at 20°C and inhibited at 65°C. All strains are resistant to novobiocin and crystal violet, and also to 5% (w/v) NaCl when the basal medium is GYM agar or TSA. All strains are susceptible to lysozyme degradation. *Thermocrispum* strains do not exhibit antibiotic activity against *Escherichia coli*, *Corynebacterium glutamicum*, *Bacillus subtilis*, or *Staphylococcus aureus*. The type strain of *Thermocrispum agreste* is slightly active against *Rhodococcus rhodochrous* and *Micrococcus luteus*, and also inhibits *Azotobacter chroococcum*, *Candida albicans*,

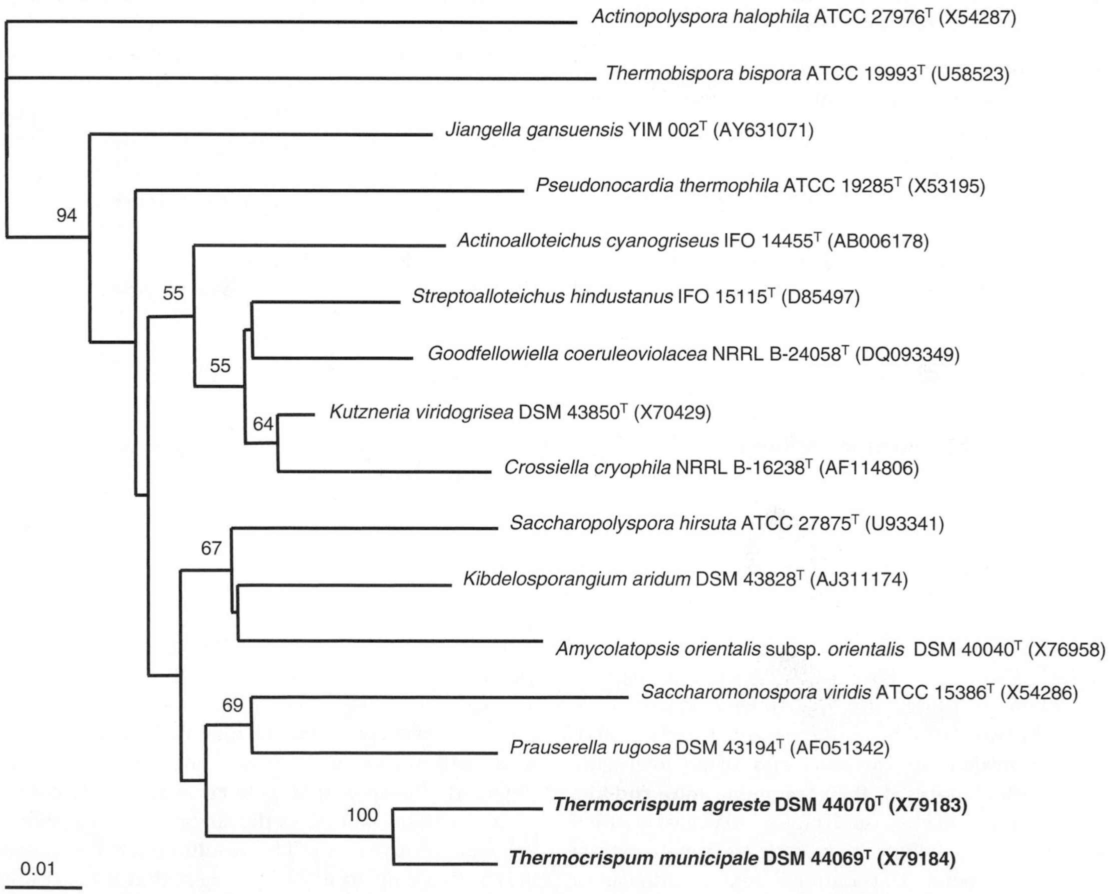

FIGURE 335. Phylogenetic tree of the genus *Thermocrispum* and related taxa based on 16S rRNA gene sequences. The Jukes–Cantor model was used in the estimation of evolutionary distances and the neighbor-joining method was used for tree construction. Numbers at nodes indicate levels of bootstrap support (%). Bar = 0.01 substitutions per nucleotide position.

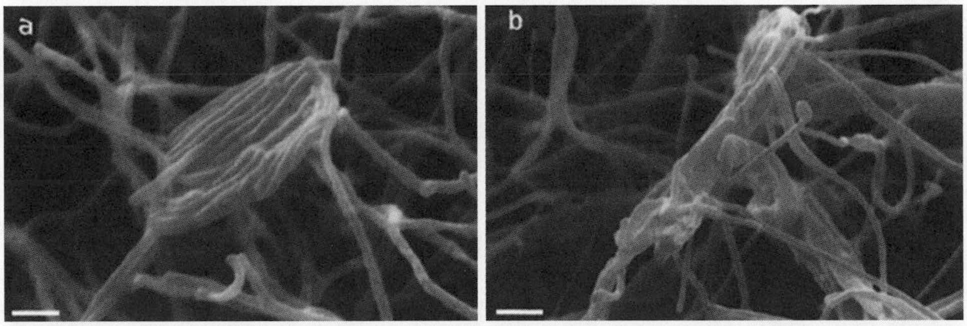

FIGURE 336. Scanning electron micrographic images of aerial mycelia of *Thermocrispum municipale* MKD35^{T} (a; bar = 1 µm) and *Thermocrispum agreste* CHB77^{T} (b; bar = 2 µm). (Reproduced with permission from Korn-Wendisch et al., 1995. Int. J. Syst. Bacteriol. *45*: 67–77.)

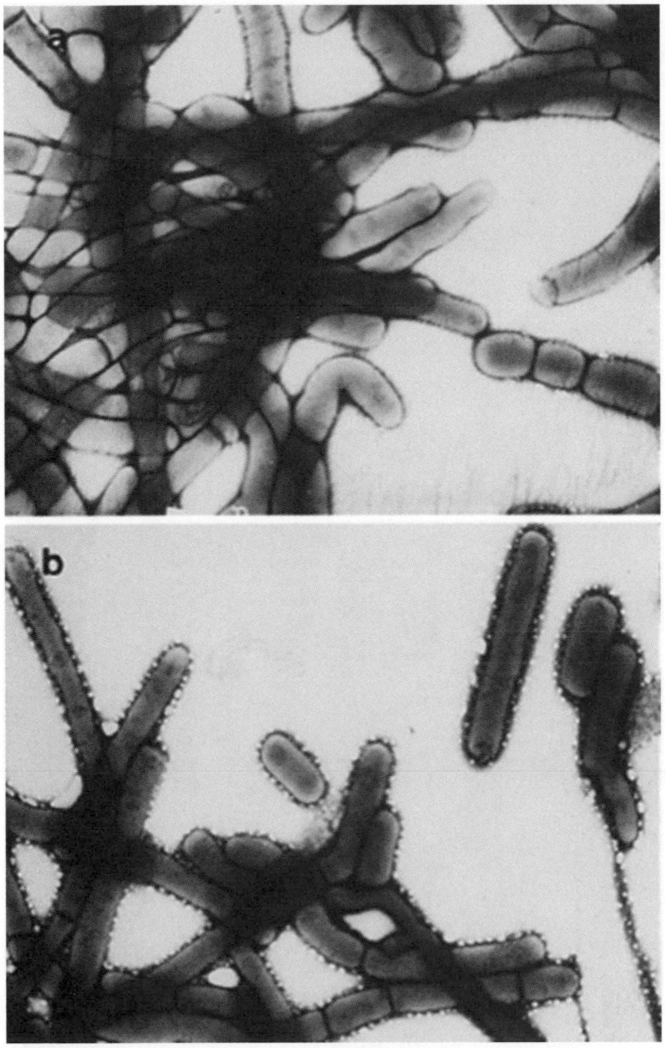

FIGURE 337. Transmission electron micrographic images of aerial mycelia of *Thermocrispum municipale* MKD35^{T} (a) and *Thermocrispum agreste* CHB77^{T} (b). Magnification 25,000×. (Reproduced with permission from Korn-Wendisch et al., 1995. Int. J. Syst. Bacteriol. *45*: 67–77.)

Saccharomyces cerevisiae, and *Geotrichum candidum*. A plasmid of about 35 kb was detected in *Thermocrispum municipale* MKD19.

All strains have similar phospholipid patterns and contain phosphatidylethanolamine, phosphatidylinositol, hydroxyphosphatidylethanolamine, and an unknown ninhydrin-positive lipid (phospholipid type PII). This phospholipid pattern is also found in the genera *Saccharothrix* and *Kutzneria* and in some other genera of the family *Pseudonocardiaceae*, including the genera *Amycolatopsis*, *Kibdelosporangium*, and *Saccharomonospora*. MK-9(H_4) is the predominant menaquinone (95–100%) in all strains of *Thermocrispum* studied. The DNA G+C contents of *Thermocrispum* strains range from 69 to 70 mol%, except for *Thermocrispum municipale* MKD19 (73 mol%). All strains exhibit qualitatively similar fatty acid profiles consisting mainly of 14-methylpentadecanoic acid ($C_{16:0}$ iso) and 2-hydroxy fatty acids (fatty acid type 3f); iso-, anteiso-, and 10-methyl branched fatty acids with 17 carbons are found in smaller amounts.

On the basis of the typing results using actinophages, the two species of *Thermocrispum* can be separated from each other. *Thermocrispum agreste* CHB77^{T} can be lyzed by its corresponding phage, f77, but is sensitive to only two of the five phages isolated for *Thermocrispum municipale* strains MKD8, MKD10, MKD19, MKD35^{T}, and MKD38, even if high phage titers are used. In contrast, only four of the nine *Thermocrispum municipale* strains can be lyzed by a high titer of f77. The strains of *Thermocrispum municipale* form three subgroups; group 1, strains MKD8, TMK2, and TMD78, can be lyzed by all five phages isolated for strains MKD8, MKD10, MKD19, MKD35^{T}, and MKD38; group 2, strains TMS14 and MKD38, are not lyzed by f10 even if high phage titers are used, and strain MKD35^{T} can be lyzed by this phage only if a high phage titer is used; group 3, strains MKD10, MKD19, and MKD57, are sensitive to phages f35 and f38 only if high phage titers are used.

The band patterns of esterases were shown to differentiate *Thermocrispum agreste* from strains of *Thermocrispum municipale* (Korn-Wendisch et al., 1995).

Enrichment and isolation procedures

Strains of *Thermocrispum municipale* can be isolated from municipal waste compost, and air of compost plants and refuse incineration plants. *Thermocrispum agreste* can be isolated from mushroom compost. Isolation methods include the dilution plate technique for isolation from various environments and use of an Andersen sampler for collecting isolates from aerosols at waste-composting plants. Since the organisms are thermophilic, high temperatures such as 50°C are used for isolation. *Thermocrispum* strains have been found to grow on the following six media combinations: R8 agar and TSA without antibiotics, TSA supplemented with rifampin (10 mg/ml), TSA supplemented with erythromycin and oleandomycin (100 mg/ml each), TSA supplemented with novobiocin (25 mg/ml), and PM agar supplemented with

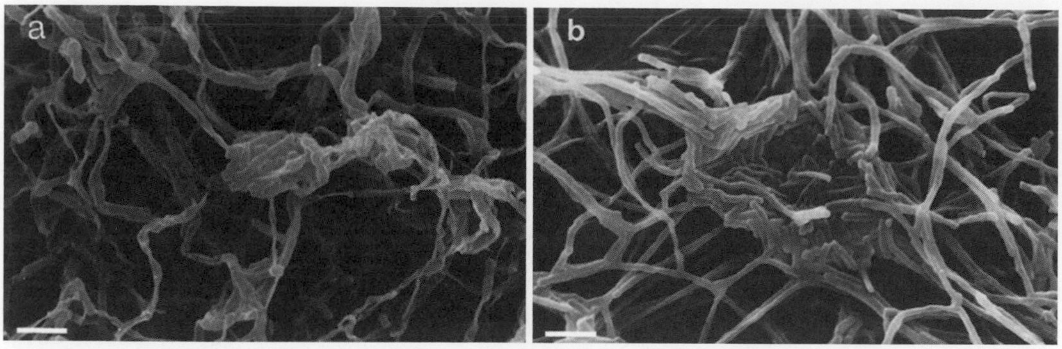

FIGURE 338. Scanning electron micrographic images of *Thermocrispum municipale* MKD35[T] showing pseudosporangia (a) and a pseudosporangium containing septate hyphae which fragmented into rod-like structures (b). Bars = 1 µm. (Reproduced with permission from Korn-Wendisch et al., 1995. Int. J. Syst. Bacteriol. *45*: 67–77.)

novobiocin (25 mg/ml). To inhibit fungal growth, cycloheximide and nystatin (50 mg/ml each) are added to the media.

Maintenance procedures

Strains of *Thermocrispum* can be maintained on CYC agar, GYM agar, Hickey–Tresner agar (Hickey and Tresner, 1952), oatmeal agar, potato-carrot agar, PM agar, R2A agar, and R8 agar (Amner et al., 1989), and TSA. The optimum temperature for growth is between 45 and 55°C at neutral pH. For long-term preservation, storage of mycelial or spore suspensions in 20% (v/v) glycerol at −20°C, or lyophilization is recommended.

Differentiation of the genus *Thermocrispum* from closely related taxa

Thermocrispum can be differentiated from related genera of the family *Pseudonocardiaceae* based on chemotaxonomic profiles (Korn-Wendisch et al., 1995). The cell-wall chemotype of *Thermocrispum* (type III) is different from that of *Actinopolyspora, Amycolatopsis, Kibdelosporangium, Pseudonocardia, Saccharomonospora,* and *Saccharopolyspora* (type IV). The phospholipid pattern of *Thermocrispum* (PII) is different from that of *Actinopolyspora, Pseudonocardia,* and *Saccharopolyspora*. MK-8(H_4), found among *Pseudonocardia* and *Saccharomonospora,* is not present in *Thermocrispum.*

From the phylogenetic analysis, *Amycolatopsis, Kibdelosporangium, Prauserella, Saccharomonospora,* and *Saccharopolyspora* are shown as the neighboring genera (Figure 335), but the levels of 16S rRNA gene sequence similarity between the species of *Thermocrispum* and the type species of these taxa are below 94.4% based on BLAST results (http://blast.ncbi.nlm.nih.gov/blast.cgi). *Thermocrispum* can also be differentiated from related genera using 16S rRNA gene restriction fragment patterns, as suggested by Cook and Meyers (2003).

Taxonomic comments

A recent report suggested that *Sciscionella marina* is loosely associated with *Thermocrispum municipale* sharing 93% sequence similarity (Tian et al., 2009); however, BLAST results indicate that the similarity level should be lower (91.9%). Since the initial description of the genus in 1995, no additional species have been described. The differences in phage sensitivity and fatty acid profiles among the strains of *Thermocrispum municipale* imply possible heterogeneity within the species (Korn-Wendisch et al., 1995). However, the 16S rRNA gene sequences of only two strains, including the type strain, are publicly available to date, and thus the phylogenetic relationship among the isolates is not clear. Further work on the isolates may be able to clarify the taxonomic positions of the *Thermocrispum municipale* strains.

TABLE 261. Differential characteristics of the two species of the genus *Thermocrispum*[a]

Properties	*T. municipale*	*T. agreste*
Growth at:		
20°C	w	tr
28°C	+++	++
62.5°C	(+)	++
65°C	−	−
Growth in the presence of NaCl at:		
7%	+++	++
10%	++	−
13%	−	−
Resistance to 25 µg/ml kanamycin	(+)	+
Degradation of:		
Elastin	−	+
Guanine	(+)	−
Utilization of:		
Fructose	+	tr
Maltose	+	(+)
Melezitose	v	tr
Mannitol	(+)	tr
Sorbitol (=glucitol)	v	tr
Salicin	(+)	+
Sodium acetate	+	tr
Sodium citrate	v	−
Hemolysis	−	+
Egg yolk reaction	−	+

[a]Data taken from Korn-Wendisch et al. (1995). Symbols: +++, good growth with abundant aerial mycelium; ++, good growth with moderate aerial mycelium; +, positive growth or reaction; (+), poor growth or moderate reaction; −, negative; w, weak; v, variable; tr, traces of growth.

Differentiation of the species of the genus *Thermocrispum*

Thermocrispum species can be distinguished from one another by using a combination of nutritional and physiological characteristics (Table 261).

List of species of the genus *Thermocrispum*

1. **Thermocrispum municipale** Korn-Wendisch, Rainey, Kroppenstedt, Kempf, Majazza, Kutzner and Stackebrandt 1995, 73^{VP}

mu.ni.ci.pa'le. L. neut. adj. *municipale* municipal, referring to the environment from which strains were isolated.

White aerial mycelium with long, straight, or flexuous chains of spores are formed. Aerial mycelia fragment into rod-like structures and both aerial and substrate mycelia often form aggregates. Sporangia-like structures are formed. Good growth occurs between 28 and 60°C, but weak growth is observed at 20 and 62.5°C. No growth occurs at 65°C. Good growth is also observed in the presence of up to 10% (w/v) NaCl, but not at 13%. Moderately resistant to kanamycin (25 µg/ml). Casein, DNA, esculin, gelatin, tributyrin, Tween 80, and tyrosine are degraded or hydrolyzed. Guanine is moderately degraded, pectin is weakly degraded, and arbutin is slowly hydrolyzed. In contrast, adenine, allantoin, cellulose, chitin, elastin, hypoxanthine, poly-β-hydroxybutyric acid, starch, urea, xanthine, and xylan are not degraded or hydrolyzed. Melanin is not produced. Dextrin, fructose, glucose, galactose, inositol, maltose, mannose, cellobiose, sucrose, trehalose, sodium acetate, sodium malonate, and sodium pyruvate are utilized as sole carbon sources, and mannitol and salicin are utilized weakly. In contrast, adonitol, inulin, lactose, melibiose, oxalate, raffinose, rhamnose, xylitol, xylose, and sodium propionate are not utilized as sole carbon sources. Melezitose, sorbitol, and sodium citrate may or may not be utilized as sole carbon sources. Hemolysis and egg yolk reaction are negative. Nitrate is reduced to form nitrite under aerobic conditions. Resistant to novobiocin and crystal violet.

Source: municipal waste compost, air of compost plants, and air of a refuse incineration plant.

DNA G+C content (mol%): 69–73 (HPLC).

Type strain: MKD35, DSM 44069.

Sequence accession no. (16S rRNA gene): X79184.

2. **Thermocrispum agreste** Korn-Wendisch, Rainey, Kroppenstedt, Kempf, Majazza, Kutzner and Stackebrandt 1995, 73^{VP}

a.gre'ste. L. neut. adj. *agreste* rural, referring to the origin of the compost from which the organism was isolated.

White aerial mycelium with long, straight, or flexuous chains of spores are formed. Aerial mycelia fragment into rod-like structures and both aerial and substrate mycelia often form aggregates. Sporangia-like structures are formed. Good growth occurs between 28 and 62.5°C; traces of growth are observed at 20°C, but not at 65°C. Good growth is also observed in the presence of up to 7% (w/v) NaCl, but not at 10%. Resistant to kanamycin (25 µg/ml). Casein, DNA, elastin, esculin, gelatin, tributyrin, Tween 80, and tyrosine are degraded or hydrolyzed. In contrast, adenine, allantoin, cellulose, chitin, guanine, hypoxanthine, poly-β-hydroxybutyric acid, starch, urea, xanthine, and xylan are not degraded or hydrolyzed. Melanin is not produced. Cellobiose, dextrin, glucose, galactose, inositol, mannose, salicin, sucrose, trehalose, sodium malonate, and sodium pyruvate are utilized as sole carbon sources, and maltose is utilized weakly. In contrast, adonitol, inulin, lactose, melibiose, oxalate, raffinose, rhamnose, xylitol, xylose, sodium citrate, and sodium propionate are not utilized as sole carbon sources. Only traces of growth occur in the presence of fructose, melezitose, mannitol, sorbitol, and sodium acetate as sole carbon sources. Hemolytic activity and egg yolk reactions are positive. Nitrate is reduced to form nitrite under aerobic conditions. Resistant to novobiocin and crystal violet. Inhibits growth of *Azotobacter chroococcum*, *Candida albicans*, *Saccharomyces cerevisiae*, and *Geotrichum candidum*, whereas slightly antagonistic against the Gram-stain-positive bacteria *Rhodococcus rhodochrous*, and *Micrococcus luteus*. No antagonistic activity is observed against *Escherichia coli*, *Corynebacterium glutamicum*, *Bacillus subtilis*, or *Staphylococcus aureus*.

Source: mushroom compost.

DNA G+C content (mol%): unknown.

Type strain: CHB77, DSM 44070.

Sequence accession no. (16S rRNA gene): X79183.

Genus XVIII. **Umezawaea** Labeda and Kroppenstedt 2007, 2761^{VP}

David P. Labeda

Um.e.za.wa'e.a. N.L. fem. n. *Umezawaea* named for Hamao Umezawa (1914–1986), of the Institute of Microbial Chemistry, Tokyo, in recognition of his leadership and contributions to the study of the biology and natural products of actinomycetes.

Aerobic. Gram-stain-positive, non-acidfast, nonmotile actinomycetes. Branched substrate mycelium (approx. 0.3–0.5 µm in diameter) is produced; aerial mycelia are formed on some media. **Ovoid or cylindrical conidia (0.3–0.5 × 1.1–1.9 µm) are produced by fragmentation of substrate mycelium. Pseudosporangia are produced on some media. Contains *meso*-diaminopimelic acid as the diamino acid. The muramic acid in the cell-wall peptidoglycan is acetylated. The whole-cell sugar pattern consists of galactose, mannose, and ribose, with a trace of rhamnose. The phospholipid pattern consists predominantly of phosphatidylethanolamine, phosphatidylinositol, phosphatidylethanolamine containing hydoxylated fatty acids, and *lyso*-phosphatidylethanolamine. The predominant menaquinone is MK-9(H_4), with a trace of MK-10(H_4)**. Mycolic acids are absent. Has a fatty acid profile consisting predominantly of $C_{16:0}$ iso fatty acids, with $C_{14:0}$ iso, $C_{15:0}$ iso, $C_{16:0}$, $C_{16:1}$, $C_{17:1}$, and $C_{16:1}$ iso as minor components. Phylogenetically, nearest neighbor is the genus *Saccharothrix*.

DNA G+C content (mol%): 74 (HPLC).

Type species: **Umezawaea tangerina** (Kinoshita, Igarashi, Ikeno, Hori and Hamada 2000) Labeda and Kroppenstedt 2007, 2761^{VP}.

Further descriptive information

Umezawaea tangerina was originally described as a species of the genus *Saccharothrix* (Kinoshita et al., 1999), although the production of pseudosporangia on the aerial mycelium, a feature not detected in other species in this genus, was noted. Phylogenetic analysis of species of the genus *Saccharothrix* and related genera revealed that *Saccharothrix tangerinus* consistently appeared distant from *Saccharothrix sensu stricto* (see Figure 290). A subsequent re-evaluation of the chemotaxonomic characteristics of this strain determined that it was different from the other taxa in the family *Pseudonocardiaceae*, particularly in regard to the presence of significant quantities of *lyso*-phosphatidylethanolamine in the phospholipid profile (Table 262). The nucleotide signatures in the 16S rRNA gene of *Umezawaea tangerina* are also different from those of *Saccharothrix* and related taxa (see Figure 309).

Enrichment and isolation procedures

Strains of *Lechevalieria*, *Lentzea*, and *Saccharothrix* have been isolated from soil samples by spread-plating serial soil dilutions onto routine selective media (such as 1.5% crude agar and 0.4% casein hydrolysate in tap water) used for the general isolation of actinomycetes. Shearer (1987) reported the use of typical actinomycete isolation media, such as Gauze mineral medium no. 1 (Gauze et al., 1957) or starch-casein agar (Küster and Williams, 1964), supplemented with 5–10 µg/ml penicillin G and 15 µg/ml nalidixic acid antibiotics to selectively isolate *Saccharothrix* strains that are phylogenetically close to *Umezawaea*.

Maintenance procedures

Working cultures of *Umezawaea* can be maintained as refrigerated (4°C) agar slants on an appropriate medium such as yeast extract-malt extract medium (Shirling and Gottlieb, 1966) or NZamine medium (DSMZ medium 554; DSMZ, 2001). The slants are generally incubated for 5–7 d at 28–30°C prior to refrigeration and then should be subcultured at monthly intervals. Longer term preservation of strains is best accomplished as frozen stocks in 20% aqueous glycerol at −80°C (mechanical freezer) to −172°C (liquid nitrogen vapor phase) or by using traditional lyophilization techniques.

Procedures for testing special characters

Strains are routinely cultivated on NZamine medium (DSMZ medium 554) at 28°C. Morphological observations are made on the media of Shirling and Gottlieb (1966) and NZamine medium.

Chemotaxonomic analyses of strains for whole-cell sugars, polar lipids, menaquinones, and fatty acids are performed using methods described previously by Saddler et al. (1991), Grund and Kroppenstedt (1990b), Minnikin et al. (1984), and Sasser (1990).

Physiological tests, including production of acid from carbohydrates, utilization of organic acids, and hydrolysis and decomposition of adenine, casein, esculin, hippurate, guanine, hypoxanthine, tyrosine, urea, and xanthine, are typically evaluated using the media of Gordon et al. (1974). Phosphatase activity is evaluated by using the method of Kurup and Schmitt (1973) substituting NZamine agar as the basal medium. Temperature range for growth is determined on slants of DSMZ medium 554 and salt tolerance on slants of the same medium supplemented with 4% and 5% NaCl.

Differentiation of the genus *Umezawaea* from other genera

Umezawaea strains can easily be differentiated from morphologically similar members of the suborder *Pseudonocardineae* by the presence of *lyso*-phosphatidylethanolamine as a predominant component in the polar lipid profiles in addition to phosphatidylethanolamine containing 2-hydroxy fatty acids (Table 262). The nucleotide signatures in the 16S rRNA gene sequences can be used to differentiate *Umezawaea* strains from *Saccharothrix* and other closely related genera in the family *Pseudonocardiaceae* (see Figure 309).

List of species of the genus *Umezawaea*

1. **Umezawaea tangerina** (Kinoshita, Igarashi, Ikeno, Hori and Hamada 2000) Labeda and Kroppenstedt 2007, 2761[VP] (*Saccharothrix tangerinus* Kinoshita, Igarashi, Ikeno, Hori and Hamada 2000, 949; effective publication: Kinoshita, Igarashi, Ikeno, Hori and Hamada 1999, 27.)

tan.ger.i'na. N.L. fem. adj. *tangerina* tangerine-colored, referring to the color of the vegetative growth.

The pale yellow, pale yellow-orange, or pale yellowish-brown substrate mycelium fragments into coccoidal-rod elements. White to brownish-white aerial mycelia are produced on some media and fragment into coccoid to coccoidal-rod elements. Pseudosporangia are sometimes produced on the aerial mycelia. Soluble pigments are faint brown or not produced. Melanin pigments are not produced. L-Arabinose, D-xylose, D-glucose, D-fructose, and inositol are utilized as sole carbon sources. Casein, hypoxanthine, starch, and tyrosine are hydrolyzed; adenine and urea are not hydrolyzed. Acid is produced from L-arabinose, cellobiose, dextrin, D-glucose, glycerol, and inositol, but not from adonitol or *meso*-erythritol. Citrate and oxalate are assimilated, but not DL-tartrate. Nitrate is not reduced. Phosphatase is produced. Grows in the presence of 4% (w/v) NaCl. Temperature range for growth is 20–30°C. Produces the antibiotic formamicin. Grows at 28°C.

Source: isolated from soil.

DNA G+C content (mol%): 74 (HPLC).

Type strain: MK27-91F2, DSM 44720, FERM P-16053, JCM 10302, NBRC 16184, NRRL B-24463.

Sequence accession no. (16S rRNA gene): AB020031.

TABLE 262. Chemotaxonomic characteristics of *Umezawaea* compared to other similar members of the family *Pseudonocardiaceae*[a]

Characteristic[b]	*Umezawaea*	*Actinokineospora*	*Actinosynnema*	*Crossiella*	*Goodfellowiella*	*Lechevalieria*	*Lentzea*	*Saccharothrix*
Whole-cell sugar pattern[c]	Gal, Man, Rib, Rha (tr)	Gal, Man, Rha	Gal, Man	Gal, Rha, Rib	Gal, Rib	Gal, Man, Rha	Gal, Man, Rib	Gal, Rha, Man (tr)
Phospholipids[d]	PE, PI, OH-PE, *lyso*-PE, DPG, DPG, PIM	PE	PE, OH-PE, PI, PIM, DPG	PE, DPG, PI, PIM, PME	PE, OH-PE, DPG	PE	PE, DPG, PG, PI	PE, OH-PE, PI, PIM, DPG, PG (v)
Predominant menaquinones	MK-9(H_4), (tr), MK-10(H_4)	MK-9(H_4)	MK-9(H_4), MK-9(H_6)	MK-9(H_4)	MK-10(H_4), MK-9(H_4)	MK-9(H_4)	MK-9(H_4)	MK-9(H_4), MK-10(H_4)

[a]tr, Trace; v, variable.

[b]All genera have *meso*-diaminopimelic acid as the cell-wall diamino acid and are of cell-wall chemotype III.

[c]Gal, Galactose; Man, mannose; Rha, rhamnose; Rib, ribose.

[d]DPG, diphosphatidylglycerol; OH-PE, phosphatidylethanolamine with hydroxy fatty acids; *lyso*-PE, phosphatidylethanolamine where one fatty acid chain is missing from the glycerol backbone; PE, phosphatidylethanolamine; PG, phosphatidylglycerol; PI, phosphatidylinositol; PIM, phosphatidylinositol mannosides; PME, phosphatidylmethylethanolamine.

References

Abdulla, H.M. and S.A. El-Shatoury. 2007. Actinomycetes in rice straw decomposition. Waste Manag. *27*: 850–853.

Abou-Zeid, A., G. Euverink, G.I. Hessels, R.A. Jensen and L. Dijkhuizen. 1995. Biosynthesis of L-phenylalanine and L-tyrosine in the actinomycete *Amycolatopsis methanolica*. Appl. Environ. Microbiol. *61*: 1298–1302.

Agre, N.S. 1964. A contribution to the technique of isolation and cultivation of thermophilic actinomycetes (in Russian). Mikrobiologiya *33*: 913–917.

Agre, N.S., L.N. Guzeva and L.A. Dorokhova. 1974. [New species of the genus *Micropolyspora–Micropolyspora internatus*]. Mikrobiologiia *43*: 679–685.

Akimov, V.N., L.I. Evtushenko and S.V. Dobritsa. 1989. *Pseudoamycolata halophobica* gen. nov., sp. nov. Int. J. Syst. Bacteriol. *39*: 457–461.

Al-Musallam, A.A., S.S. Al-Zarban, Y.A. Fasasi, R.M. Kroppenstedt and E. Stackebrandt. 2003. *Amycolatopsis keratiniphila* sp. nov., a novel keratinolytic soil actinomycete from Kuwait. Int. J. Syst. Evol. Microbiol. *53*: 871–874.

Al-Zarban, S.S., A.A. Al-Musallam, I. Abbas, E. Stackebrandt and R.M. Kroppenstedt. 2002. *Saccharomonospora halophila* sp. nov., a novel halophilic actinomycete isolated from marsh soil in Kuwait. Int. J. Syst. Evol. Microbiol. *52*: 555–558.

Albarracin, V.H., B. Winik, E. Kothe, M.J. Amoroso and C.M. Abate. 2008. Copper bioaccumulation by the actinobacterium *Amycolatopsis* sp. AB0. J. Basic Microbiol. *48*: 323–330.

Albarracin, V.H., P. Alonso-Vega, M.E. Trujillo, M.J. Amoroso and C.M. Abate. 2010. *Amycolatopsis tucumanensis* sp. nov., a copper-resistant actinobacterium isolated from polluted sediments. Int. J. Syst. Evol. Microbiol. *60*: 397–401.

Alderson, G., M. Goodfellow, E.M.H. Wellington, S.T. Williams, S.M. Minnikin and D.E. Minnikin. 1981. Chemical and numerical taxonomy of *Nocardia mediterranei*. Zentralbl. Bakteriol. Suppl. *11*: 39–46.

Alderson, G., M. Goodfellow and D.E. Minnikin. 1985. Menaquinone composition in the classification of *Streptomyces* and other sporoactinomycetes. J. Gen. Microbiol. *131*: 1671–1679.

Alves, A.M., G.J. Euverink, H.J. Hektor, G.I. Hessels, J. van der Vlag, J.W. Vrijbloed, D. Hondmann, J. Visser and L. Dijkhuizen. 1994. Enzymes of glucose and methanol metabolism in the actinomycete *Amycolatopsis methanolica*. J. Bacteriol. *176*: 6827–6835.

Alves, A.M., W.G. Meijer, J.W. Vrijbloed and L. Dijkhuizen. 1996. Characterization and phylogeny of the *pfp* gene of *Amycolatopsis methanolica* encoding PPi-dependent phosphofructokinase. J. Bacteriol. *178*: 149–155.

Amner, W., A.J. McCarthy and C. Edwards. 1988. Quantitative assessment of factors affecting the recovery of indigenous and released thermophilic bacteria from compost. Appl. Environ. Microbiol. *54*: 3107–3112.

Amner, W., C. Edwards and A.J. McCarthy. 1989. Improved medium for recovery and enumeration of the farmer's lung organism, *Saccharomonospora viridis*. Appl. Environ. Microbiol. *55*: 2669–2674.

Andersen, A.A. 1958. New sampler for the collection, sizing, and enumeration of viable airborne particles. J. Bacteriol. *76*: 471–484.

Antoun, H., L.M. Bordeleau, C. Gagnon and R.A. Lachance. 1978. [Identification of actinomycetes with antifungal activity which do not affect *Rhizobium meliloti*]. Can. J. Microbiol. *24*: 1073–1075.

Aragno, M. and H.G. Schlegel. 1981. The hydrogen oxidising bacteria. *In* The Prokaryotes: a Handbook on Habitats, Isolation, and Identification of Bacteria (edited by Starr, Stolp, Trüper, Balows and Schlegel). Springer, New York, pp. 865–893.

Associate Editor, IJSB. 1997. *In* Validation of the publication of new names and new combinations previously effectively published outside the IJSB, footnote (c). List no. 62. Int. J. Syst. Bacteriol. *47*: 915–916.

Athalye, M., J. Lacey and M. Goodfellow. 1981. Selective isolation and enumeration of actinomycetes using rifampicin. J. Appl. Bacteriol. *51*: 289–297.

August, P.R., L. Tang, Y.J. Yoon, S. Ning, R. Muller, T.W. Yu, M. Taylor, D. Hoffmann, C.G. Kim, X. Zhang, C.R. Hutchinson and H.G. Floss. 1998. Biosynthesis of the ansamycin antibiotic rifamycin: deductions from the molecular analysis of the *rif* biosynthetic gene cluster of *Amycolatopsis mediterranei* S699. Chem. Biol. *5*: 69–79.

Awata, M., S. Satoi, N. Muto, M. Hayashi, H. Sagai and H. Sakakibara. 1983. Saccharocin, a new aminoglycoside antibiotic. Fermentation, isolation, characterization and structural study. J. Antibiot. (Tokyo) *36*: 651–655.

Bala, S., R. Khanna, M. Dadhwal, S.R. Prabagaran, S. Shivaji, J. Cullum and R. Lal. 2004. Reclassification of *Amycolatopsis mediterranei* DSM 46095 as *Amycolatopsis rifamycinica* sp. nov. Int. J. Syst. Evol. Microbiol. *54*: 1145–1149.

Ball, A.S., W.B. Betts and A.J. McCarthy. 1989. Degradation of lignin-related compounds by actinomycetes. Appl. Environ. Microbiol. *55*: 1642–1644.

Baltz, R. 2000. Sweet home actinomycetes: The 1999 MDS Panlabs Lecture. J. Ind. Microbiol. Biotechnol. *24*: 79–88.

Baltz, R.H., P. Matsushima, R. Stanzak, B.E. Schoner and R.N. Rao. 1986. Efficient transformation in *Streptomyces* and cloning of the erythromycin biosynthesis genes. *In* Biological, Biochemical and Biomedical Aspects of Actinomycetes (edited by Szabó, Biró and Goodfellow). Akadémiai Kiadó, Budapest, pp. 55–66.

Banskota, A.H., J.B. McAlpine, D. Sorensen, A. Ibrahim, M. Aouidate, M. Piraee, A.M. Alarco, C.M. Farnet and E. Zazopoulos. 2006. Genomic analyses lead to novel secondary metabolites. Part 3. ECO-0501, a novel antibacterial of a new class. J. Antibiot. (Tokyo) *59*: 533–542.

Barna, J.C. and D.H. Williams. 1984. The structure and mode of action of glycopeptide antibiotics of the vancomycin group. Annu. Rev. Microbiol. *38*: 339–357.

Barreiro, C., A. Pisabarro and J.F. Martin. 2000. Characterization of the ribosomal *rrnD* operon of the cephamycin-producer '*Nocardia lactamdurans*' shows that this actinomycete belongs to the genus *Amycolatopsis*. Syst. Appl. Microbiol. *23*: 15–24.

Bass, M. and J.M. Cherett. 1994. The role of leaf-cutting ant workers (*Hymenoptera, Formicidae*) in fungal garden maintenance. Ecol. Entomol. *19*: 215–220.

Bian, J., Y. Li, J. Wang, F.H. Song, M. Liu, H.Q. Dai, B. Ren, H. Gao, X. Hu, Z.H. Liu, W.J. Li and L.X. Zhang. 2009. *Amycolatopsis marina* sp. nov., an actinomycete isolated from an ocean sediment. Int. J. Syst. Evol. Microbiol. *59*: 477–481.

Bibb, M.J., J.M. Ward, T. Kieser, S.N. Cohen and D.A. Hopwood. 1981. Excision of chromosomal DNA sequences from *Streptomyces coelicolor* forms a novel family of plasmids detectable in *Streptomyces lividans*. Mol. Gen. Genet. *184*: 230–240.

Bibb, M.J., G.R. Janssen and J.M. Ward. 1986. Cloning and analysis of the promoter region of the erythromycin resistance gene (*ermE*) of *Streptomyces erythreus*. Gene *41*: E357–E368.

Birner, J., P.R. Hodgson, W.R. Lane and E.H. Baxter. 1972. An Australian isolate of *Nocardia mediterranea* producing rifamycin SV. J. Antibiot. *25*: 356–359.

Boccard, F., T. Smokvina, J.L. Pernodet, A. Friedmann and M. Guerineau. 1989. The integrated conjugative plasmid pSAM2 of *Streptomyces ambofaciens* is related to temperate bacteriophages. EMBO J *8*: 973–980.

Boeck, L.D., F.P. Mertz, R.K. Wolter and C.E. Higgens. 1984. N-demethylvancomycin, a novel antibiotic produced by a strain of *Nocardia orientalis*. Taxonomy and fermentation. J. Antibiot. *37*: 446–453.

Bot, A.N.M., S.A. Rehner and J.J. Boomsma. 2001. Partial incompatibility between ants and symbiotic fungi in two sympatric species of *Acromyrmex* leaf-cutting ants. Evolution *55*: 1980–1991.

Bowen, T., E. Stackebrandt, M. Dorsch and T.M. Embley. 1989. The phylogeny of *Amycolata autotrophica, Kibdelosporangium aridum* and *Saccharothrix australiensis.* J. Gen. Microbiol. *135*: 2529–2536.

Box, S.J., A.L. Elson, M.L. Gilpin and D.J. Winstanley. 1990. MM 47761 and MM 49721, glycopeptide antibiotics produced by a new strain of *Amycolatopsis orientalis.* Isolation, purification and structure determination. J. Antibiot. (Tokyo) *43*: 931–937.

Box, S.J., N.J. Coates, C.J. Davis, M.L. Gilpin, C.S. Houge-Frydrych and P.H. Milner. 1991. MM 55266 and MM 55268, glycopeptide antibiotics produced by a new strain of *Amycolatopsis.* Isolation, purification and structure determination. J. Antibiot. (Tokyo) *44*: 807–813.

Bredholdt, H., O.A. Galatenko, K. Engelhardt, E. Fjaervik, L.P. Terekhova and S.B. Zotchev. 2007. Rare actinomycete bacteria from the shallow water sediments of the Trondheim fjord, Norway: isolation, diversity and biological activity. Environ. Microbiol. *9*: 2756–2764.

Brock, T.D. 1986. Introduction: an overview of the thermophiles. *In* Thermophiles: General, Molecular and Applied Microbiology (edited by Brock). Wiley, New York, pp. 1–17.

Brown, D., J. Dewitt and L. Katz. 1986. Integrated and autonomous forms of a pock-forming plasmid of *Streptomyces erythraeus.* H156:156. Presented at the Annual Meeting of the American Society for Microbiology., Anchorage.

Brown, D.P., S.J. Chiang, J.S. Tuan and L. Katz. 1988. Site-specific integration in *Saccharopolyspora erythraea* and multisite integration in *Streptomyces lividans* of actinomycete plasmid pSE101. J. Bacteriol. *170*: 2287–2295.

Brown, D.P., K.B. Idler, D.M. Backer, S. Donadio and L. Katz. 1994. Characterization of the genes and attachment sites for site-specific integration of plasmid pSE101 in *Saccharopolyspora erythraea* and *Streptomyces lividans.* Mol. Gen. Genet. *242*: 185–193.

Brummund, W., V.P. Kurup, A. Resnick, T.J. Milson, Jr and J.N. Fink. 1988. Immunologic response to *Faenia rectivirgula* (*Micropolyspora faeni*) in a dairy farm family. J. Allergy Clin. Immunol. *82*: 190–195.

Brzezinski, R., E. Surmacz, M. Kutner and A. Piekarowicz. 1986. Restriction mapping and close relationship of the DNA of *Streptomyces erythraeus* phages 121 and SE-5. J. Gen. Microbiol. *132*: 2937–2943.

Burrus, V., G. Pavlovic, B. Decaris and G. Guedon. 2002. Conjugative transposons: the tip of the iceberg. Mol. Microbiol. *46*: 601–610.

Bush, K., P.R. Henry and D.S. Slusarchyk. 1984. Muraceins–muramyl peptides produced by *Nocardia orientalis* as angiotensin-converting enzyme inhibitors. I. Taxonomy, fermentation and biological properties. J. Antibiot. *37*: 330–335.

Cafaro, M.J. and C.R. Currie. 2005. Phylogenetic analysis of mutualistic filamentous bacteria associated with fungus-growing ants. Can J. Microbiol. *51*: 441–446.

Campbell, J.M. 1932. Acute symptoms following work with hay. Br. Med. J. *2*: 1141–1144.

Carlsohn, M.R., I. Groth, G.Y.A. Tan, B. Schütze, H.P. Saluz, T. Munder, J. Yang, J. Wink and M. Goodfellow. 2007. *Amycolatopsis saalfeldensis* sp. nov., a novel actinomycete isolated from a medieval alum slate mine. Int. J. Syst. Evol. Microbiol. *57*: 1640–1646; erratum *57*: 2188.

Celmer, W.D., W.P. Cullen, C.E. Moppett, J.B. Routien, R. Shibakawa and J. Tone. 1977. Antibiotics produced by species of *Pseudonocardia.* US Patent 4031206. United States.

Challis, G.L. and D.A. Hopwood. 2003. Synergy and contingency as driving forces for the evolution of multiple secondary metabolite production by *Streptomyces* species. Proc. Natl. Acad. Sci. U.S.A. *100* 14555–14561.

Chapela, I.H., S.A. Rehner, T.R. Schultz and U.G. Mueller. 1994. Evolutionary history of the symbiosis between fungus-growing ants and their fungi. Science *266*: 1691–1694.

Chary, V.K., J.L. delaFuente, P. Liras and J.F. Martin. 1997. *amy* as a reporter gene for promoter activity in *Nocardia lactamdurans*: Comparison of promoters of the cephamycin cluster. Appl. Environ. Microbiol. *63*: 2977–2982.

Chary, V.K., J.L. de la Fuente, A.L. Leitao, P. Liras and J.F. Martin. 2000. Overexpression of the lat gene in *Nocardia lactamdurans* from strong heterologous promoters results in very high levels of lysine-6-aminotransferase and up to two-fold increase in cephamycin C production. Appl. Microbiol. Biotechnol. *53*: 282–288.

Chen, C.H., J.C. Cheng, Y.C. Cho and W.H. Hsu. 2005. A gene cluster for the fatty acid catabolism from *Pseudonocardia autotrophica* BCRC12444. Biochem. Biophys. Res. Commun. *329*: 863–868.

Chen, H.H., S. Qin, J. Li, Y.Q. Zhang, L.H. Xu, C.L. Jiang, C.J. Kim and W.J. Li. 2009. *Pseudonocardia endophytica* sp. nov., isolated from the pharmaceutical plant *Lobelia clavata.* Int. J. Syst. Evol. Microbiol. *59*: 559–563.

Chen, Y., W. Deng, J. Wu, J. Qian, J. Chu, Y. Zhuang, S. Zhang and W. Liu. 2008. Genetic modulation of the overexpression of tailoring genes *eryK* and *eryG* leading to the improvement of erythromycin A purity and production in *Saccharopolyspora erythraea* fermentation. Appl. Environ. Microbiol. *74*: 1820–1828.

Chiang, S.D., J. Tuan, D. Brown and L. Katz. 1985. Genetic instability of *Streptomyces erythraeus* NRRL 2338 plasmid pSE1. H193:140. Presented at the American Society for Microbiology.

Cho, M., J.H. Yoon, S.B. Kim and Y.H. Park. 1998. Application of the ribonuclease P (RNase P) RNA gene sequence for phylogenetic analysis of the genus *Saccharomonospora.* Int. J. Syst. Bacteriol. *48*: 1223–1230.

Chun, J.S., S.B. Kim, Y.K. Oh, C.N. Seong, D.H. Lee, K.S. Bae, K.J. Lee, S.O. Kang, Y.C. Hah and M. Goodfellow. 1999. *Amycolatopsis thermoflava* sp. nov., a novel soil actinomycete from Hainan Island, China. Int. J. Syst. Bacteriol. *49*: 1369–1373.

Cohen, A., D. Bar-Nir, M.E. Goedeke and Y. Parag. 1985. The integrated and free state of *Streptomyces griseus* plasmid pSG1. Plasmid *13*: 41–50.

Collins, B.S., C.T. Kelly and W.M. Fogarty. 1992. Maltogenic alpha-amylase of *Saccharomonospora viridis.* Biochem. Soc. Trans. *20*: 81S.

Collins, M.D., T. Pirouz, M. Goodfellow and D.E. Minnikin. 1977. Distribution of menaquinones in actinomycetes and corynebacteria. J. Gen. Microbiol. *100*: 221–230.

Collins, M.D., M. Goodfellow, D.E. Minnikin and G. Alderson. 1985. Menaquinone composition of mycolic acid-containing actinomycetes and some sporoactinomycetes. J. Appl. Bacteriol. *58*: 77–86.

Cook, A.E. and P.R. Meyers. 2003. Rapid identification of filamentous actinomycetes to the genus level using genus-specific 16S rRNA gene restriction fragment patterns. Int. J. Syst. Evol. Microbiol. *53*: 1907–1915.

Corbaz, R., P.H. Gregory and M.E. Lacey. 1963. Thermophilic and mesophilic actinomycetes in mouldy hay. J. Gen Microbiol. *32*: 449–454.

Cross, T., M.P. Lechevalier and H. Lechevalier. 1963. A new genus of the *Actinomycetales*: *Microellobosporia* gen. nov. J. Gen. Microbiol. *31*: 421–429.

Cross, T. 1968. Thermophilic actinomycetes. J. Appl. Bacteriol. *31*: 36–53.

Cross, T., A. Maciver and J. Lacey. 1968. The thermophilic actinomycetes in mouldy hay: *Micropolyspora faeni* sp. nov. J. Gen. Microbiol. *50*: 351–359.

Cross, T. and R.W. Attwell. 1974. Recovery of viable thermoactinomycete endospores from deep mud cores. *In* Spore Research 1973 (edited by Barker, Gould and Wolf). Academic Press, London, pp. 11–20.

Crouse, G.D., T.C. Sparks, J. Schoonover, J. Gifford, J. Dripps, T. Bruce, L.L. Larson, J. Garlich, C. Hatton, R.L. Hill, T.V. Worden and J.G. Martynow. 2001. Recent advances in the chemistry of spinosyns. Pest Manag. Sci. *57*: 177–185.

Currie, C.R., U.G. Mueller and D. Malloch. 1999a. The agricultural pathology of ant fungus gardens. Proc. Natl. Acad. Sci. U.S.A. *96*: 7998–8002.

Currie, C.R., J.A. Scott, R.C. Summerbell and D. Malloch. 1999b. Fungus-growing ants use antibiotic-producing bacteria to control garden parasites. Nature *398*: 701–704.

Currie, C.R. 2001. A community of ants, fungi, and bacteria: a multilateral approach to studying symbiosis. Annu. Rev. Microbiol. *55*: 357–380.

Currie, C.R. and A.E. Stuart. 2001. Weeding and grooming of pathogens in agriculture by ants. Proc. Biol. Sci. *268*: 1033–1039.

Currie, C.R., A.N.M. Bot and J.J. Boomsma. 2003a. Experimental evidence of a tripartite mutualism: bacteria protect ant fungal gardens from specialized parasites. Oikos *101*: 91–102.

Currie, C.R., B. Wong, A.E. Stuart, T.R. Schultz, S.A. Rehner, U.G. Mueller, G.H. Sung, J.W. Spatafora and N.A. Straus. 2003b. Ancient tripartite coevolution in the attine ant-microbe symbiosis. Science *299*: 386–388.

Currie, C.R., M. Poulsen, J. Mendenhall, J.J. Boomsma and J. Billen. 2006. Coevolved cryts and exocrine glands support mutualistic bacteria in fungus-growing ants. Science *311*: 81–83.

De Boer, L., W. Harder and L. Dijkhuizen. 1988. Phenylalanine and tyrosine metabolism in the facultative methylotroph *Nocardia* sp. 239. Arch. Microbiol. *149*: 459–465.

De Boer, L., J.W. Vrjbloed, G. Grobben and L. Dijkhuizen. 1989. Regulation of aromatic amino acid biosynthesis in the ribulose monophosphate cycle methylotroph *Nocardia* sp. 239. Arch. Microbiol. *152*: 319–325.

De Boer, L., L. Dijkhuizen, G. Grobben, M. Goodfellow, E. Stackebrandt, J.H. Parlett, D. Whitehead and D. Witt. 1990. *Amycolatopsis methanolica* sp. nov., a facultatively methylotrophic actinomycete. Int. J. Syst. Bacteriol. *40*: 194–204.

Dees, P.M. and W.C. Ghiorse. 2001. Microbial diversity in hot synthetic compost as revealed by PCR-amplified rRNA sequences from cultivated isolates and extracted DNA. FEMS Microbiol. Ecol. *35*: 207–216.

Dekker, K.A., T. Inagaki, T.D. Gootz, L.H. Huang, Y. Kojima, W.E. Kohlbrenner, Y. Matsunaga, P.R. McGuirk, E. Nomura, T. Sakakibara, S. Sakemi, Y. Suzuki, Y. Yamauchi and N. Kojima. 1998. New quinolone compounds from *Pseudonocardia* sp. with selective and potent anti-*Helicobacter pylori* activity: taxonomy of producing strain, fermentation, isolation, structural elucidation and biological activities. J. Antibiot. *51*: 145–152.

Deushi, T., A. Iwasaki, K. Kamiya, T. Kunieda, T. Mizoguchi, M. Nakayama, H. Itoh, T. Mori and T. Oda. 1979. A new broad-spectrum aminoglycoside antibiotic complex, sporaricin. I. Fermentation, isolation and characterization. J. Antibiot. (Tokyo) *32*: 173–179.

di Marco, C. and C. Spalla. 1957. La produzione di cobalamine de fermentazione con una nuova specie di *Nocardia*: *Nocardia rugosa*. G. Microbiol. *4*: 24–30.

Diab, A. and A.Y. Al-Gunaim. 1982. Species of thermophilic actinomycetes in the atmosphere of Kuwait associated with allergic diseases. J. Univ. Kuwait *9*: 119–128.

Dijkhuizen, L., T.A. Hansen and W. Harder. 1985. Methanol, a potential feedstock for biotechnological processes. Trends Biotechnol. *3*: 262–267.

Ding, L., T. Hirose and A. Yokota. 2007. *Amycolatopsis echigonensis* sp. nov. and *Amycolatopsis niigatensis* sp. nov., novel actinomycetes isolated from a filtration substrate. Int. J. Syst. Evol. Microbiol. *57*: 1747–1751.

Ding, X.M., Y.Q. Tian, J.S. Chiao, G.P. Zhao and W.H. Jiang. 2003. Stability of plasmid pA387 derivatives in *Amylcolatopsis mediterranei* producing rifamycin. Biotechnol. Lett. *25*: 1647–1652.

Dobashi, K., N. Matsuda, M. Hamada, H. Naganawa, T. Takita and T. Takeuchi. 1988. Novel antifungal antibiotics octacosamicins A and B. I. Taxonomy, fermentation and isolation, physico-chemical properties and biological activities. J. Antibiot. *41*: 1525–1532.

Dobritsa, S.V. 1984. Large Plasmids in an Actinomycete. FEMS Microbiol. Lett. *23*: 35–39.

Dolashka, P., D.N. Georgieva, S. Stoeva, N. Genov, R. Rachev, A. Gusterova and W. Voelter. 1998. A novel thermostable neutral proteinase from *Saccharomonospora canescens*. Biochim. Biophys. Acta *1382*: 207–216.

Donadio, S., R. Paladino, I. Costanzi, P. Sparapani, W. Schreil and M. Iaccarino. 1986. Characterization of bacteriophages infecting *Streptomyces erythreus* and properties of phage-resistant mutants. J. Bacteriol. *166*: 1055–1060.

Donahue, J.M. and N.M. Williams. 2000. Emergent causes of placentitis and abortion. Vet. Clin. North. Am. Equine. Pract. *16*: 443–456.

Donahue, J.M., N.M. Williams, S.F. Sells and D.P. Labeda. 2002. *Crossiella equi* sp. nov., isolated from equine placentas. Int. J. Syst. Evol. Microbiol. *52*: 2169–2173.

Dorokhova, L.A., N.S. Agre, L.V. Kalakoutskii and N.A. Krasil'nikov. 1969. Fine structure of sporulating hyphae and spores in a thermophilic actinomycete. *Micropolyspora rectivirgula*. J. Microsc. Biol. Cell *8*: 845–854.

Dorokhova, L.A., N.S. Agre, L.V. Kalahoutskii and N.A. Krasil'nikov. 1970. A study of the morphology of two cultures belonging to the genus *Micropolyspora*. Mikrobiologiya *39*: 95–100.

Drocourt, D., T. Calmels, J.P. Reynes, M. Baron and G. Tiraby. 1990. Cassettes of the *Streptoalloteichus hindustanus ble* gene for transformation of lower and higher eukaryotes to phleomycin resistance. Nucleic Acids Res. *18*: 4009.

DSMZ. 2001. Catalogue of Strains. German Collection of Microorganisms and Cell Cultures, 7th edn. DSMZ – Deutsche Sammlung von Mikroorganismen und Zellkulturen, Braunschweig, Germany.

Duangmal, K., R. Mingma, A. Thamchaipenet, A. Matsumoto and Y. Takahashi. 2010. *Saccharopolyspora phatthalungensis* sp. nov., isolated from rhizosphere soil of *Hevea brasiliensis*. Int. J. Syst. Evol. Microbiol. *60*: 1904–1908.

Egorova, K., H. Trauthwein, S. Verseck and G. Antranikian. 2004. Purification and properties of an enantioselective and thermoactive amidase from the thermophilic actinomycete *Pseudonocardia thermophila*. Appl. Microbiol. Biotechnol. *65*: 38–45.

El-Nakeeb, M.A. and H.A. Lechevalier. 1963. Selective isolation of aerobic Actinomycetes. Appl. Microbiol. *11*: 75–77.

Embley, M.T., J. Smida and E. Stackebrandt. 1988a. The phylogeny of mycolateless wall chemotype-IV actinomycetes and description of *Pseudonocardiaceae* fam. nov. Syst. Appl. Microbiol. *11*: 44–52.

Embley, M.T., J. Smida and E. Stackebrandt. 1989. *In* Validation of the publication of new names and new combinations previously effectively published outside the IJSB. List no. 29. Int. J. Syst. Bacteriol. *39*: 205–206.

Embley, T.M., R. Wait, G. Dobson and M. Goodfellow. 1987. Fatty-acid composition in the classification of *Saccharopolyspora hirsuta*. FEMS Microbiol. Lett. *41*: 131–135.

Embley, T.M., A.G. O'Donnell, J. Rostron and M. Goodfellow. 1988b. Chemotaxonomy of wall type IV actinomycetes which lack mycolic acids. J. Gen. Microbiol. *134*: 953–960.

Embley, T.M., J. Smida and E. Stackebrandt. 1988c. Reverse transcriptase sequencing of 16S ribosomal RNA from *Faenia rectivirgula*, *Pseudonocardia thermophila* and *Saccharopolyspora hirsuta*, three wall type IV actinomycetes which lack mycolic acids. J. Gen. Microbiol. *134*: 961–966.

Embley, T.M. 1992. The family *Pseudonocardiaceae*. *In* The Prokaryotes: a Handbook on the Biology of Bacteria: Ecophysiology, Isolation, Identification, Applications, 2nd edn (edited by Balows, Trüper, Dworkin, Harder and Schleifer). Springer, New York, pp. 996–1027.

Etienne, G., B. Fabre, E. Armau, F. Legendre, M. Ardourel and G. Tiraby. 1993. CL307-24, a new antibiotic complex from *Saccharopolyspora aurantiaca* sp. nov. I. Taxonomy, fermentation and purification. J. Antibiot. (Tokyo) *46*: 770–776.

Euverink, G.J., D.J. Wolters and L. Dijkhuizen. 1995. Prephenate dehydratase of the actinomycete *Amycolatopsis methanolica*: purification and characterization of wild-type and deregulated mutant proteins. Biochem. J. *308*: 313–320.

Euverink, G.J.W., G.I. Hessels, J.W. Vrijbloed, J.R. Coggins and L. Dijkhuizen. 1992. Purification and characterization of a dual function 3-dehydroquinate dehydratase from *Amycolatopsis methanolica*. J. Gen. Microbiol. *138*: 2449–2457.

Everest, G.J. and P.R. Meyers. 2009. The use of *gyrB* sequence analysis in the phylogeny of the genus *Amycolatopsis*. Int. J. Syst. Evol. Microbiol. *95*: 1–11.

Evtushenko, L.I., V.N. Akimov, S.V. Dobritsa and S.D. Taptykova. 1989. A new species of actinomycete, *Amycolata alni*. Int. J. Syst. Bacteriol. *39*: 72–77.

Felsenstein, D. 1993. PHYLIP (Phylogeny Inference Package) 3.57 edn. Department of Genetics, University of Washington, Seattle.

Felsenstein, J. 1985. Confidence limits on phylogenies: an approach using the bootstrap. Evolution *39*: 783–791.

Felsenstein, J. 1989. PHYLIP (Phylogeny Inference Package) version 3.5.1. Department of Genetics, University of Washington, Seattle.

Fergus, C.L. 1967. Resistance of spores of some thermophilic actinomycetes to high temperature. Mycopathol. Mycol. Appl. *32*: 205–208.

Floss, H.G. and T.W. Yu. 2005. Rifamycin-mode of action, resistance, and biosynthesis. Chem. Rev. *105*: 621–632.

Folena-Wasserman, G., B.L. Poehland, E.W.-K. Yeung, D. Staiger, L.B. Killmer, K.M. Snader, J.J. Dingerdissen and P.W. Jeffs. 1986. Kibdelins (AAD-609), novel glycopeptide antibiotics. II. Isolation, purification and structure. J. Antibiot. *39*: 1395–1406.

Fuller, W.H. and A.G. Norman. 1942. A cellulose dextrin medium for identifying cellulose organisms in soil. Proc. Soil Sci. Soc. Am. *7*: 243.

Furumai, T., H. Ogawa and T. Okuda. 1968. Taxonomic study on *Streptosporangium albidum* nov. sp. J. Antibiot. (Tokyo) *21*: 179–181.

Gaisser, S., J. Reather, G. Wirtz, L. Kellenberger, J. Staunton and P.F. Leadlay. 2000. A defined system for hybrid macrolide biosynthesis in *Saccharopolyspora erythraea*. Mol. Microbiol. *36*: 391–401.

Gaisser, S., R. Lill, G. Wirtz, F. Grolle, J. Staunton and P.F. Leadlay. 2001. New erythromycin derivatives from *Saccharopolyspora erythraea* using sugar *O*-methyltransferases from the spinosyn biosynthetic gene cluster. Mol. Microbiol. *41*: 1223–1231.

Gamian, A., H. Mordarska, I. Ekiel, J. Ulrich, B. Szponar and J. Defaye. 1996. Structural studies of the major glycolipid from *Saccharopolyspora* genus. Carbohydr. Res. *296*: 55–67.

Gangwar, M., Z.U. Khan, H.S. Randhawa and J. Lacey. 1989. Distribution of clinically important thermophilic actinomycetes in vegetable substrates and soil in north-western India. Antonie van Leeuwenhoek *56*: 201–209.

Gauze, G.F., T.P. Preobrazhenskaya, E.S. Kudrina, N.O. Blinov, I.D. Ryabova and M.A. Sveshnikova. 1957. Problems in the classification of antagonistic actinomycetes. State Publishing House for Medical Literature (in Russian). Medzig, Moscow.

Gauze, G.F., Maksimov, T.S., Olkhovat, O.L., Sveshnik, M.A., Kochetko, G.V. and G.B. Ilchenko. 1974. Production of madumycin, an antibacterial antibiotic, by *Actinomadura flava* sp. nov. Antibiotiki *19*: 771–775.

Gauze, G.F., M.A. Sveshnikova, R.S. Ukholina, G.N. Komarova and V.S. Bazhanov. 1977. [Formation of a new antibiotic, nocamycin, by a culture of *Nocardiopsis syringae* sp. nov.]. Antibiotiki *22*: 483–486.

Gauze, G.F. and M.A. Sveshnikova. 1985. *In* Validation of the publication of new names and new combinations previously effectively published outside the IJSB. List no. 17. Int. J. Syst. Bacteriol. *35*: 223–225.

Gauze, G.F., T.P. Preobrazhenskaya, M.A. Sveshnikova, L.P. Terekhova and T.S. Maksimova. 1985. *Opredelitei' Aktinomycetov. Rody Streptomyces, Streptoverticillium, Chainia*. Izol. Nauk (in Russian), Moscow.

Gayer-Herkert, G., J. Schneider and H.J. Kutzner. 1989. Transfection and transformation of protoplasts of the thermophilic actinomycete *Faenia rectivirgula*. Appl. Microbiol. Biotechnol. *31*: 371–375.

Gerardo, N.M., S.R. Jacobs, C.R. Currie and U.G. Mueller. 2006. Ancient host-pathogen associations maintained by specificity of chemotaxis and antibiosis. PLoS Biol. *4*: 1358–1363.

Ghisalba, O., P. Traxler and J. Nuesch. 1978. Early intermediates in the biosynthesis of ansamycins. I. Isolation and identification of protorifamycin I. J. Antibiot. *31*: 1124–1131.

Ghisalba, O., P. Traxler, H. Fuhrer and W.J. Richter. 1979. Early intermediates in the biosynthesis of ansamycins. II. Isolation and identification of proansamycin B-M1 and protorifamycin i-M1. J. Antibiot. *32*: 1267–1272.

Ghisalba, O., P. Traxler, H. Fuhrer and W.J. Richter. 1980. Early intermediates in the biosynthesis of ansamycins. III. Isolation and identification of further 8-deoxyansamycins of the rifamycin-type. J. Antibiot. *33*: 847–856.

Ghisalba, O. and J. Nuesch. 1981. A genetic approach to the biosynthesis of the rifamycin-chromophore in *Nocardia mediterranei*. IV. Identification of 3-amino-5-hydroxybenzoic acid as a direct precursor of the seven-carbon amino starter-unit. J. Antibiot. *34*: 64–71.

Giles, R.C., J.M. Donahue, C.B. Hong, P.A. Tuttle, M.B. Petrites-Murphy, K.B. Poonacha, A.W. Roberts, R.R. Tramontin, B. Smith and T.W. Swerczek. 1993. Causes of abortion, stillbirth, and perinatal death in horses: 3,527 cases (1986–1991). J. Am. Vet. Med. Assoc. *203*: 1170–1175.

Goddon, B. and M.J. Penninckx. 1984. Identification and evolution of the cellulolytic microflora present during composting of cattle manure: the role of actinomycetes. Ann. Mikrobiol. *135*: 69–78.

Gonzalez, I., A. Ayuso-Sacido, A. Anderson and O. Genilloud. 2005. Actinomycetes isolated from lichens: evaluation of their diversity and detection of biosynthetic gene sequences. FEMS Microbiol. Ecol. *54*: 401–415.

Goodfellow, M. 1971. Numerical taxonomy of some nocardioform bacteria. J. Gen. Microbiol. *69*: 33–80.

Goodfellow, M. and D.E. Minnikin. 1981. The genera *Nocardia* and *Rhodococcus*. *In* The Prokaryotes: a Handbook on Habitats, Isolation, and Identification of Bacteria (edited by Starr, Stolp, Trüper, Balows and Schlegel). Springer, New York, pp. 2016–2027.

Goodfellow, M. and T. Pirouz. 1982. Numerical classification of sporoactinomycetes containing *meso*-diaminopimelic acid in the cell wall. J. Gen. Microbiol. *128*: 503–527.

Goodfellow, M., J. Lacey, M. Athalye, T.M. Embley and T. Bowen. 1989a. *In* Validation of the publication of new names and new combinations previously effectively published outside the IJSB. List no. 31. Int. J. Syst. Bacteriol. *39*: 495–497.

Goodfellow, M., J. Lacey, M. Athalye, T.M. Embley and T. Bowen. 1989b. *Saccharopolyspora gregorii* and *Saccharopolyspora hordei* – two new actinomycete species from fodder. J. Gen. Microbiol. *135*: 2125–2139.

Goodfellow, M., A.B. Brown, J.P. Cai, J.S. Chun and M.D. Collins. 1997a. *Amycolatopsis japonicum* sp. nov., an actinomycete producing (S,S)-N,N′-ethylenediaminedisuccinic acid. Syst. Appl. Microbiol. *20*: 78–84.

Goodfellow, M., A.B. Brown, J.P. Cai, J.S. Chun and M.D. Collins. 1997b. *In* Validation of the publication of new names and new combinations previously effectively published outside the IJSB. List no. 62. Int. J. Syst. Bacteriol. *47*: 915–916.

Goodfellow, M., S.B. Kim, D.E. Minnikin, D. Whitehead, Z.H. Zhou and A.D. Mattinson-Rose. 2001. *Amycolatopsis sacchari* sp. nov., a moderately thermophilic actinomycete isolated from vegetable matter. Int. J. Syst. Evol. Microbiol. *51*: 187–193.

Gordon, R.E. and J.E. Mihm. 1962. Identification of *Nocardia caviae* (Erikson) comb. nov. Ann. N.Y. Acad. Sci. *98*: 628–636.

Gordon, R.E., D.A. Barnett, J.E. Handerhan and C.H.-N. Pang. 1974. *Nocardia coeliaca*, *Nocardia autotrophica*, and the Nocardin Strain. Int. J. Syst. Bacteriol. *24*: 54–63.

Gordon, R.E., S.K. Mishra and D.A. Barnett. 1978. Some bits and pieces of genus *Nocardia*: *Nocardia carnea*, *Nocardia vaccinii*, *Nocardia transvalensis*, *Nocardia orientalis* and *Nocardia aerocolonigenes*. J. Gen. Microbiol. *109*: 69–78.

Grant, I.W., W. Blyth, V.E. Wardrop, R.M. Gordon, J.C. Pearson and A. Mair. 1972. Prevalence of farmer's lung in Scotland: a pilot survey. Br. Med. J. *1*: 530–534.

Greene, J.G., M.W. Treuhaft and R.M. Arusell. 1981. Hypersensitivity pneumonitis due to *Saccharomonospora viridis* diagnosed by inhalation challenge. Ann. Allergy *47*: 449–452.

Gregory, P.H. and M.E. Lacey. 1963. Mycological examination of dust from mouldy hay associated with farmer's lung disease. J. Gen. Microbiol. *30*: 75–88.

Gregory, P.H., M.E. Lacey, G.N. Festerstein and F.A. Skinner. 1963. Microbial and biochemical changes during the moulding of key. J. Gen. Microbiol. *33*: 147–174.

Greiner-Mai, E., R.M. Kroppenstedt, F. Kornwendisch and H.J. Kutzner. 1987. Morphological and biochemical characterization and emended descriptions of thermophilic actinomycetes species. Syst. Appl. Microbiol. *9*: 97–109.

Greiner-Mai, E., F. Korn-Wendisch and H.J. Kutzner. 1988. Taxonomic revision of the genus *Saccharomonospora* and description of *Saccharomonospora glauca* sp. nov. Int. J. Syst. Bacteriol. *38*: 398–405.

Groth, I., G.Y.A. Tan, J.M. Gonzalez, L. Laiz, M.R. Carlsohn, B. Schütze, J. Wink and M. Goodfellow. 2007. *Amycolatopsis nigrescens* sp. nov., an actinomycete isolated from a Roman catacomb. Int. J. Syst. Evol. Microbiol. *57*: 513–519.

Grund, A.D. and C.R. Hutchinson. 1987. Bacteriophages of *Saccharopolyspora erythraea*. J. Bacteriol. *169*: 3013–3022.

Grund, E. and R.M. Kroppenstedt. 1989. Transfer of five *Nocardiopsis* species to the genus *Saccharothrix* Labeda *et al.* 1984. Syst. Appl. Microbiol. *12*: 267–274.

Grund, E., C. Knorr and R. Eichenlaub. 1990. Catabolism of benzoate and monohydroxylated benzoates by *Amycolatopsis* and *Streptomyces* spp. Appl. Environ. Microbiol. *56*: 1459–1464.

Grund, E. and R.M. Kroppenstedt. 1990a. *In* Validation of the publication of new names and new combinations previously effectively published outside the IJSB. List no. 34. Int. J. Syst. Bacteriol. *40*: 320–321.

Grund, E. and R.M. Kroppenstedt. 1990b. Chemotaxonomy and numerical taxonomy of the genus *Nocardiopsis*. Int. J. Syst. Bacteriol. *40*: 5–11.

Grundy, W.E., A.C. Sinclair, R.J. Theriault, A.W. Goldstein, C.J. Rickher, H.B. Warren, Jr, T.J. Oliver and S.J. C. 1957. Ristocetin, microbiologic properties. Antibiot. Annu. *1956–1957*: 687–792.

Gu, Q., H. Luo, W. Zheng, Z. Liu and Y. Huang. 2006. *Pseudonocardia oroxyli* sp. nov., a novel actinomycete isolated from surface-sterilized *Oroxylum indicum* root. Int. J. Syst. Evol. Microbiol. *56*: 2193–2197.

Hahn, D.R., G. Gustafson, C. Waldron, B. Bullard, J.D. Jackson and J. Mitchell. 2006. Butenyl-spinosyns, a natural example of genetic engineering of antibiotic biosynthetic genes. J. Ind. Microbiol. Biotechnol. *33*: 94–104.

Haider, K., J. Trojanowski and V. Sundman. 1978. Screening for lignin degrading bacteria by means of 14C-labelled lignins. Arch. Microbiol. *119*: 103–106.

Harvey, I., Y. Cormier, C. Beaulieu, V.N. Akimov, A. Meriaux and C. Duchaine. 2001. Random amplified ribosomal DNA restriction analysis for rapid identification of thermophilic Actinomycete-like bacteria involved in hypersensitivity pneumonitis. Syst. Appl. Microbiol. *24*: 277–284.

Hasegawa, T., E. Higashide and M. Shibata. 1971. Tolypomycin, a new antibiotic. II. Production and preliminary identification of tolypomycin Y. J. Antibiot. *24*: 817–822.

Hasegawa, T., M.P. Lechevalier and H.A. Lechevalier. 1978. New genus of *Actinomycetales*: *Actinosynnema* gen. nov. Int. J. Syst. Bacteriol. *28*: 304–310.

Hasegawa, T., M. Takizawa and S. Tanida. 1983a. A rapid analysis for chemical grouping of aerobic actinomycetes. J. Gen. Appl. Microbiol. *29*: 319–322.

Hasegawa, T., S. Tanida, K. Hatano, E. Higashide and M. Yoneda. 1983b. Motile actinomycetes: *Actinosynnema pretiosum* subsp. *pretiosum* sp. nov., subsp. nov., and *Actinosynnema pretiosum* subsp. Auranticum subsp. nov. Int. J. Syst. Bacteriol. *33*: 314–320.

Hasegawa, T. 1988a. *In* Validation of the publication of new names and new combinations previously effectively published outside the IJSB. List no. 27. Int. J. Syst. Bacteriol. *38*: 449.

Hasegawa, T. 1988b. *Actinokineospora*: a new genus of the *Actinomycetales*. Actinomycetologica *2*: 31–45.

Hayakawa, M. and H. Nonomura. 1987. Humic acid-vitamine agar, a new medium for the selective isolation of soil actinomycetes. J. Ferment. Technol. *65*: 501–509.

Hayakawa, M. 1990. Selective isolation methods and distribution of soil actinomycetes. Actinomycetologica *4*: 103–112.

Hayakawa, M., M. Otoguro, T. Takeuchi, T. Yamazaki and Y. Iimura. 2000. Application of a method incorporating differential centrifugation for selective isolation of motile actinomycetes in soil and plant litter. Antonie van Leeuwenhoek *78*: 171–185.

Hazeu, W., J.C. de Bruyn and J.P. van Dijken. 1983. *Nocardia* sp. 239, a facultative methanol utilizer with the ribulose monophosphate pathway of formaldehyde fixation. Arch. Microbiol. *135*: 205–210.

Hektor, H. 1997. Physiology and biochemistry of primary alcohol oxidation in Gram-positive bacteria. PhD thesis, University of Groningen, Netherlands.

Henssen, A. 1957. Beiträge zur Morphologie und Systematik der thermophilen Actinomyceten. Arch. Mikrobiol. *26*: 373–414.

Henssen, A. and E. Schnepf. 1967. [On the knowledge of thermophilic actinomycetes]. Arch. Mikrobiol. *57*: 214–231.

Henssen, A. 1970. Spore formation in thermophilic actinomycetes. *In* The *Actinomycetales* (edited by Prauser). Gustav Fischer-Verlag, Jena, pp. 205–210.

Henssen, A. and D. Schäfer. 1971. Emended description of the genus *Pseudonocardia* Henssen and description of the new species *Pseudonocardia spinosa*. Int. J. Syst. Bacteriol. *21*: 29–34.

Henssen, A., E. Weise, G. Vobis and B. Renner. 1981. Ultrastructure of sporogenesis in actinomycetes forming spores in chains. *In* Actinomycetes (edited by Schaal and Pulverer). Fischer-Verlag, Stuttgart, pp. 137–146.

Henssen, A., C. Happachkasan, B. Renner and G. Vobis. 1983. *Pseudonocardia compacta* sp. nov. Int. J. Syst. Bacteriol. *33*: 829–836.

Henssen, A., H.W. Kothe and R.M. Kroppenstedt. 1987. Transfer of *Pseudonocardia azurea* and "*Pseudonocardia fastidiosa*" to the genus *Amycolatopsis* with emended species descriptions. Int. J. Syst. Bacteriol. *37*: 292–295.

Henssen, A. 1989. Genus *Pseudonocardia*. *In* Bergey's Manual of Systematic Bacteriology, vol. 4 (edited by Williams, Sharpe and Holt). Williams & Wilkins, Baltimore, pp. 2371–2379.

Hickey, R.J. and H.D. Tresner. 1952. A cobalt-containing medium for sporulation of *Streptomyces* species. J. Bacteriol. *64*: 891–892.

Higgins, C.E. and R.E. Kastner. 1967. Nebramycin, a new broad-spectrum antibiotic complex. II. Description of *Streptomyces tenebrarius*. Antimicrob. Agents Chemother. *7*: 324–331.

Higgins, M.L., M.P. Lechevalier and H.A. Lechevalier. 1967. Flagellated actinomycetes. J. Bacteriol. *93*: 1446–1451.

Hirsch, C.F. and D.L. Christensen. 1983. Novel method for selective isolation of actinomycetes. Appl. Environ. Microbiol. *46*: 925–929.

Hirsch, P. and H. Engel. 1956. Über oligocarbophile Actinomyceten. Bericht Deutsch. Bot. Gesellschaft *69*: 441–454.

Hirsch, P. 1960. Einige weitere von Luftverunreinigungen lebende Actinomyceten und ihre Klassifizierung. Arch. Mikrobiol. *35*: 391–414.

Hirsch, P. 1961. Wasserstoffaktivierung und Chemoautotrophie bei Actinomyceten. Arch. Mikrobiol. *39*: 360–373.

Hong, C.B., J.M. Donahue, R.C. Giles, Jr, M.B. Petrites-Murphy, K.B. Poonacha, A.W. Roberts, B.J. Smith, R.R. Tramontin, P.A. Tuttle and T.W. Swerczek. 1993. Etiology and pathology of equine placentitis. J. Vet. Diagn. Invent. *5*: 56–63.

Hong, L., Z. Zhao and H.W. Liu. 2006. Characterization of SpnQ from the spinosyn biosynthetic pathway of *Saccharopolyspora spinosa*:

mechanistic and evolutionary implications for C-3 deoxygenation in deoxysugar biosynthesis. J. Am. Chem. Soc. *128*: 14262–14263.

Hong, L., Z. Zhao, C.E. Melancon, 3rd, H. Zhang and H.W. Liu. 2008. In vitro characterization of the enzymes involved in TDP-D-forosamine biosynthesis in the spinosyn pathway of *Saccharopolyspora spinosa*. J. Am. Chem. Soc. *130*: 4954–4967.

Hopmann, C., M. Kurz, M. Bronstrup, J. Wink and D. LeBeller. 2002. Isolation and structure elucidation of vancoresmycin - a new antibiotic from *Amycolatopsis* sp. ST 101170. Tetrahedron Lett. *43*: 435–438.

Hopwood, D.A., G. Hintermann, T. Kieser and H.M. Wright. 1984. Integrated DNA sequences in three streptomycetes form related autonomous plasmids after transfer to *Streptomyces lividans*. Plasmid *11*: 1–16.

Hopwood, D.A., M.J. Bibb, K.F. Chater, T. Kieser, C.J. Bruton, H.M. Kieser, D.J. Lydiate, C.P. Smith, J.M. Ward and S. H. 1985. Genetic manipulation of *Streptomyces*. A Laboratory Manual. John Innes Foundation, Norwich, UK.

Howey, R.T., C.M. Lock and L.V.H. Moore. 1990. Subspecies names automatically created by Rule 46. Int. J. Syst. Bacteriol. *40*: 317–319.

Hu, R.M. 1987. *Saccharomonospora azurea* sp. nov., a new species from soil. Int. J. Syst. Bacteriol. *37*: 60–61.

Hu, R.M., L. Cheng and G.Z. Wei. 1988. *Saccharomonospora cyanea* sp. nov. Int. J. Syst. Bacteriol. *38*: 444–446.

Hu, Y.T., P.J. Zhou, Y.G. Zhou, Z.H. Liu and S.J. Liu. 2004. *Saccharothrix xinjiangensis* sp. nov., a pyrene-degrading actinomycete isolated from Tianchi Lake, Xinjiang, China. Int. J. Syst. Evol. Microbiol. *54*: 2091–2094.

Hu, Z., D. Hunziker, C.R. Hutchinson and C. Khosla. 1999. A host-vector system for analysis and manipulation of rifamycin polyketide biosynthesis in *Amycolatopsis mediterranei*. Microbiology *145*: 2335–2341.

Huang, K.-x., L. Xia, Y. Zhang, X. Ding and J. Zahn. 2009. Recent advances in the biochemistry of spinosyns. Appl. Microbiol. Biotechnol. *82*: 13–23.

Huang, Y., W. Qi, Z. Lu, Z. Liu and M. Goodfellow. 2001. *Amycolatopsis rubida* sp. nov., a new *Amycolatopsis* species from soil. Int. J. Syst. Evol. Microbiol. *51*: 1093–1097.

Huang, Y., L. Wang, Z. Lu, L. Hong, Z. Liu, G.Y.A. Tan and M. Goodfellow. 2002. Proposal to combine the genera *Actinobispora* and *Pseudonocardia* in an emended genus *Pseudonocardia*, and description of *Pseudonocardia zijingensis* sp. nov. Int. J. Syst. Evol. Microbiol. *52*: 977–982.

Huang, Y., M. Paściak, Z. Liu, Q. Xie and A. Gamian. 2004. *Amycolatopsis palatopharyngis* sp. nov., a potentially pathogenic actinomycete isolated from a human clinical source. Int. J. Syst. Evol. Microbiol. *54*: 359–363.

Hwang, Y.-B., M.-Y. Lee, H.-J. Park, K. Han and E.-S. Kim. 2007. Isolation of putative polyene-producing actinomycetes strains via PCR-based genome screening for polyene-specific hydroxylase genes. Process Biochemistry *42*: 102–107.

Iijima, M., T. Someno, C. Imada, Y. Okami, M. Ishizuka and T. Takeuchi. 1999a. IC202A, a new siderophore with immunosuppressive activity produced by *Streptoalloteichus* sp. 1454-19. I. Taxonomy, fermentation, isolation and biological activity. J. Antibiot. *52*: 20–24.

Iijima, M., T. Someno, M. Ishizuka, R. Sawa, H. Naganawa and T. Takeuchi. 1999b. IC202B and C, new siderophores with immunosuppressive activity produced by *Streptoalloteichus* sp. 1454-19. J. Antibiot. *52*: 775–780.

Ikeda, Y., S. Kondo, F. Kanai, T. Sawa, M. Hamada, T. Takeuchi and H. Umezawa. 1985. A new destomycin-family antibiotic produced by *Saccharopolyspora hirsuta*. J. Antibiot. (Tokyo) *38*: 436–438.

Isenberg, H.D., A. Schatz, A.A. Angrist, V. Schatz and G.S. Trelawny. 1954. Microbial metabolism of carbamates. II. Nitrification of urethan by *Streptomyces nitrificans*. J. Bacteriol. *68*: 5–9.

Ishihara, K., M. Nishitani, H. Yamaguichi, N. Nakajima, T. Ohshima and K. Nakamura. 1997. Preparation of optimally active a-hydroxy esters: stereoselective reduction of a-keto esters using thermophilic actinomycetes. J. Ferment. Bioeng. *84*: 268–270.

Iwasaki, A., H. Itoh and T. Mori. 1979. A new broad-spectrum aminoglycoside antibiotic complex, sporaricin. II. Taxonomic studies on the sporaricin producing strain *Saccharopolyspora hirsuta* subsp. *Kobensis* nov. subsp. J. Antibiot. (Tokyo) *32*: 180–186.

Jenke-Kodama, H., T. Borner and E. Dittmann. 2006. Natural biocombinatorics in the polyketide synthase genes of the actinobacterium *Streptomyces avermitilis*. PLoS Comput. Biol. *2*: e132.

Jiang, C. and L. Xu. 1986. Identification of a new species of the genus *Saccharopolyspora*. Acta Microbiol. Sin. *26*: 17–20.

Jiang, C., L. Xu, Y.R. Yang, G.Y. Guo, J. Ma and Y. Liu. 1991. *Actinobispora*, a new genus of the order *Actinomycetales*. Int. J. Syst. Bacteriol. *41*: 526–528.

Jiang, Y., J. Wiese, S.K. Tang, L.H. Xu, J.F. Imhoff and C.L. Jiang. 2008. *Actinomycetospora chiangmaiensis* gen. nov., sp. nov., a new member of the family *Pseudonocardiaceae*. Int. J. Syst. Evol. Microbiol. *58*: 408–413.

Jin, X., L.H. Xu, P.H. Mao, T.H. Hseu and C.L. Jiang. 1998. Description of *Saccharomonospora xinjiangensis* sp. nov. based on chemical and molecular classification. Int. J. Syst. Bacteriol. *48*: 1095–1099.

Jones, K.L. 1949. Fresh isolates of actinomycetes in which the presence of sporogenous aerial mycelia is a fluctuating characteristic. J. Bacteriol. *57*: 141–145.

Juteau, P., R. Larocque, D. Rho and A. LeDuy. 1999. Analysis of the relative abundance of different types of bacteria capable of toluene degradation in a compost biofilter. Appl. Microbiol. Biotechnol. *52*: 863–868.

Kalakoutskii, L.V. 1964. A new species of *Micropolyspora – Micropolyspora caesia* n. sp. Mikrobiologiya *33*: 858–862.

Kamiya, K., T. Deushi, A. Iwasaki, I. Watanabe, H. Itoh and T. Mori. 1983. A new aminoglycoside antibiotic, KA-5685. J. Antibiot. (Tokyo) *36*: 738–741.

Kämpfer, P. and R.M. Kroppenstedt. 2004. *Pseudonocardia benzenivorans* sp. nov. Int. J. Syst. Evol. Microbiol. *54*: 749–751.

Kämpfer, P., U. Kohlweyer, B. Thiemer and J.R. Andreesen. 2006. *Pseudonocardia tetrahydrofuranoxydans* sp. nov. Int. J. Syst. Evol. Microbiol. *56*: 1535–1538.

Kang, D.-J., H.S. Lee, J.-T. Park, J.S. Bang, S.-K. Hong and T.Y. Kim. 2006. Optimization of culture conditions for the bioconversion of vitamin D_3 to 1a, 25-dihydroxyvitamin D_3 using *Pseudonocardia autotrophica* ID9302. Biotechnol. Bioprocess. *11*: 408–413.

Kasai, H., T. Tamura and S. Harayama. 2000. Intrageneric relationships among *Micromonospora* species deduced from *gyrB*-based phylogeny and DNA relatedness. Int. J. Syst. Evol. Microbiol. *50*: 127–134.

Kato, N., K. Tsuji, H. Ohashi, Y. Tani and K. Ogata. 1977. Two assimilation pathways of C1-compound in *Streptomyces* sp. No. 239 featuring growth on methanol. Agric. Biol. Chem. *41*: 29–34.

Katz, L., S.J. Chiang, J.S. Tuan and L.B. Zablen. 1988. Characterization of bacteriophage phi C69 of *Saccharopolyspora erythraea* and demonstration of heterologous actinophage propagation by transfection of *Streptomyces* and *Saccharopolyspora*. J. Gen. Microbiol. *134*: 1765–1771.

Katz, L. and S. Donadio. 1995. Macrolides. Biotechnology *28*: 385–420.

Katz, L. and C. Khosla. 2007. Antibiotic production from the ground up. Nat. Biotechnol. *25*: 428–429.

Kaur, H., J. Cortes, P. Leadlay and R. Lal. 2001. Cloning and partial characterization of the putative rifamycin biosynthetic gene cluster from the actinomycete *Amycolatopsis mediterranei*. DSM 46095. Microbiol. Res. *156*: 239–246.

Kawato, M. and R. Shinobu. 1959. On *Streptomyces herbaricolor* sp. nov., supplement: a simple technique for microscopical observation. Osaka Unit. Lib. Arts Educ. B Nat. Sci. *8*: 114–119.

Kemmerling, C., H. Gürtler, R.M. Kroppenstedt, R. Toalster and E. Stackebrandt. 1993. Evidence for the phylogenetic heterogeneity of the genus *Streptosporangium*. Syst. Appl. Microbiol. *16*: 369–372.

Kempf, A., E. Greiner-Mai, J. Schneider, F. Korn-Wendisch and H.J. Kutzner. 1987. A group of actinophages of *Faenia rectivirgula* Curr. Microbiol. *15*: 283–285.

Khan, Z.U., M. Gangwar, S.N. Gaur and H.S. Randhawa. 1995. Thermophilic actinomycetes in cane sugar mills: an aeromicrobiologic and seroepidemiologic study. Antonie van Leeuwenhoek *67*: 339–344.

Khanna, M., M. Dua and R. Lal. 1998. Selection of suitable marker genes for the development of cloning vectors and electroporation in different strains of *Amycolatopsis mediterranei*. Microbiol. Res. *153*: 205–211.

Kim, B., N. Sahin, G.Y.A. Tan, J. Zakrzewska-Czerwinska and M. Goodfellow. 2002a. *Amycolatopsis eurytherma* sp. nov., a thermophilic actinomycete isolated from soil. Int. J. Syst. Evol. Microbiol. *52*: 889–894.

Kim, C.G., T.W. Yu, C.B. Fryhle, S. Handa and H.G. Floss. 1998. 3-Amino-5-hydroxybenzoic acid synthase, the terminal enzyme in the formation of the precursor of mC7N units in rifamycin and related antibiotics. J. Biol. Chem. *273*: 6030–6040.

Kim, H.J., R. Pongdee, Q. Wu, L. Hong and H.W. Liu. 2007. The biosynthesis of spinosyn in *Saccharopolyspora spinosa*: synthesis of the cross-bridging precursor and identification of the function of SpnJ. J. Am. Chem. Soc. *129*: 14582–14584.

Kim, J.M., Y. Jin, C.-G. Hyun, J.-H. Kim, H.-S. Lee, D.-K. Kang, D.-J. Kang, T.-Y. Kim, J.-W. Suh, S.-S. Kang and S.-K. Hong. 2002b. Molecular cloning and analysis of the gene for P-450 hydroxylase from *Pseudonocardia autotrophica*. J. Microbiol. *40*: 211–218.

Kim, S.B., J.H. Yoon, H. Kim, S.T. Lee, Y.H. Park and M. Goodfellow. 1995. A phylogenetic analysis of the genus *Saccharomonospora* conducted with 16S rRNA gene sequences. Int. J. Syst. Bacteriol. *45*: 351–356.

Kim, S.B. and M. Goodfellow. 1999. Reclassification of *Amycolatopsis rugosa* Lechevalier *et al.* 1986 as *Prauserella rugosa* gen. nov., comb. nov. Int. J. Syst. Bacteriol. *49*: 507–512.

Kimura, M. 1980. A simple method for estimating evolutionary rates of base substitutions through comparative studies of nucleotide sequences. J. Mol. Evol. *16*: 111–120.

Kinoshita, N., M. Igarashi, S. Ikeno, M. Hori and M. Hamada. 1999. *Saccharothrix tangerinus* sp. nov., the producer of the new antibiotic formamicin: taxonomic studies. Actinomycetologica *13*: 20–31.

Kinoshita, N., M. Igarashi, S. Ikeno, M. Hori and M. Hamada. 2000. *In* Validation of publication of new names and combinations previously effectively published outside the IJSEM. List no. 74. Int. J. Syst. Evol. Microbiol. *50*: 949–950.

Kirst, H.A., K.H. Michel, J.W. Martin, L.C. Creemer, E.H. Chio, R.C. Yao, W.M. Nakatsukasa, L.D. Boeck, J.L. Occolowitz, J.W. Paschal, J.B. Deeter, N.D. Jones and G.D. Thompson. 1991. A83543A-D, unique fermentation-derived tetracyclic macrolides. Tetrahed. Lett. *32*: 4839–4842.

Kirst, H.A., K.H. Michel, J.S. Mynderse, E.H. Chio, R.C. Yao, W.H. Nakatsukasa, L. Boeck, J.L. Occolowitz, J.W. Paschal, J.B. Deeter and G.D. Thompson. 1992. Discovery, isolation and structure elucidation of a family of structurally unique fermentation derived tetracyclic macrolides. *In* Synthesis and Chemistry of Agrochemicals, vol. 3 (edited by Baker, Fenyes and Stellens). American Chemical Society, Washington, D.C., pp. 3214–3225.

Kirst, H.A., K.H. Michel, J.S. Mynderse, E.H. Chio, R.C. Yao, W.M. Nakatsukasa, L. Boeck, J.L. Occolowitz, J.W. Paschal, J.B. Deeter and G.D. Thomspon. 1993. Discovery and identification of a novel fermentation derived insecticide. *In* Development in Industrial Microbiology Series: Microbial Metabolites, vol. 32 (edited by Brown). Society for Industrial Microbiology, Washington, D.C., pp. 109–116.

Kishi, T., H. Yamana, M. Muroi, S. Harada and M. Asai. 1972. Tolypomycin, a new antibiotic. III. Isolation and characterization of tolypomycin Y. J. Antibiot. *25*: 11–15.

Kohlweyer, U., B. Thiemer, T. Schrader and J.R. Andreesen. 2000. Tetrahydrofuran degradation by a newly isolated culture of *Pseudonocardia* sp. strain K1. FEMS Microbiol. Lett. *186*: 301–306.

Korn-Wendisch, F., A. Kempf, E. Grund, R.M. Kroppenstedt and H.J. Kutzner. 1989. Transfer of *Faenia rectivirgula* Kurup and Agre 1983 to the genus *Saccharopolyspora* Lacey and Goodfellow 1975, elevation of *Saccharopolyspora hirsuta* subsp. *taberi* Labeda 1987 to species level, and emended description of the genus *Saccharopolyspora*. Int. J. Syst. Bacteriol. *39*: 430–441.

Korn-Wendisch, F., F. Rainey, R.M. Kroppenstedt, A. Kempf, A. Majazza, H.J. Kutzner and E. Stackebrandt. 1995. *Thermocrispum* gen. nov., a new genus of the order *Actinomycetales*, and description of *Thermocrispum municipale* sp. nov. and *Thermocrispum agreste* sp. nov. Int. J. Syst. Bacteriol. *45*: 67–77.

Kost, C., T. Lakatos, I. Bottcher, W.R. Arendholz, M. Redenbach and R. Wirth. 2007. Non-specific association between filamentous bacteria and fungus-growing ants. Naturwissenschaften *94*: 821–828.

Kotani, T., Y. Kawashima, H. Yurimoto, N. Kato and Y. Sakai. 2006. Gene structure and regulation of alkane monooxygenases in propane-utilizing *Mycobacterium* sp. TY-6 and *Pseudonocardia* sp. TY-7. J. Biosci. Bioeng. *102*: 184–192.

Kothe, H.W., G. Vobis, R.M. Kroppenstedt and A. Henssen. 1989. A taxonomic study of mycolateless, wall chemotype IV actinomycetes. Syst. Appl. Microbiol. *12*: 61–69.

Krasil'nikov, N.A. and N.S. Agre. 1964. On two new species of *Thermopolyspora*. Hindustan Antibiot. Bull. *6*: 97–107.

Krasil'nikov, N.A. 1981. Ray Fungi, Higher Forms. Amerind Publishing, New Delhi, India.

Kriukov, V.R. 1981. [Development of hydrogen bacteria on hard surfaces]. Mikrobiologiia *50*: 299–304.

Kroppenstedt, R.M. 1985. Fatty acid and menaquinone analysis of actinomycetes and related organisms. *In* Chemical Methods in Bacterial Systematics (edited by Goodfellow and Minnikin). Academic Press, London, pp. 173–199.

Kroppenstedt, R.M., E. Stackebrandt and M. Goodfellow. 1990. Taxonomic revision of the actinomycete genera *Actinomadura* and *Microtetraspora* Syst. Appl. Microbiol. *13*: 148–160.

Kroppenstedt, R.M., E. Stackebrandt and M. Goodfellow. 1991. *In* Validation of the publication of new names and new combinations previously effectively published outside the IJSB. List no. 36. Int. J. Syst. Bacteriol. *41*: 178–179.

Kudo, T., T. Itoh, S. Miyadoh, T. Shomura and A. Seino. 1993. *Herbidospora* gen. nov., a new genus of the family *Streptosporangiaceae* Goodfellow *et al.* 1990. Int. J. Syst. Bacteriol. *43*: 319–328.

Kuimova, T.F. and Y.B. Malishkaite. 1984. Fine structure characteristics of *Nocardia autotrophica*. Microbiologiya *53*: 342–345.

Kumar, C.V., J.J. Coque and J.F. Martin. 1994. Efficient transformation of the cephamycin C producer *Nocardia lactamdurans* and development of shuttle and promoter-probe cloning vectors. Appl. Environ. Microbiol. *60*: 4086–4093.

Kumar, V., J.L. de la Fuente, A.L. Leitao, P. Liras and J.F. Martin. 1996. Effect of amplification or targeted disruption of the beta-lactamase gene of *Nocardia* lactamdurans on cephamycin biosynthesis. Appl. Microbiol. Biotechnol. *45*: 621–628.

Kunimoto, S., J. Lu, H. Esumi, Y. Yamazaki, N. Kinoshita, Y. Honma, M. Hamada, M. Ohsono, M. Ishizuka and T. Takeuchi. 2003. Kigamicins, novel antitumor antibiotics. I. Taxonomy, isolation, physico-chemical properties and biological activities. J. Antibiot. (Tokyo) *56*: 1004–1011.

Kunstmann, M.P., L.A. Mitscher, J.N. Porter, A.J. Shay and M.A. Darken. 1968. LL-AV290, a new antibiotic. Fermentation, isolation and characterization. Antimicrob. Agents Chemother. *1968*: 242–245.

Kurup, P.V. and J.A. Schmitt. 1973. Numerical taxonomy of *Nocardia*. Can. J. Microbiol. *19*: 1035–1048.

Kurup, V.P. and R.J. Heinzen. 1978. Isolation and characterization of actinophages of *Thermoactinomyces* and *Micropolyspora*. Can. J. Microbiol. *24*: 794–797.

Kurup, V.P. 1981. Taxonomic study of some members of *Micropolyspora* and *Saccharomonospora*. Microbiologica *4*: 249–259.

Kurup, V.P. and N.S. Agre. 1983. Transfer of *Micropolyspora rectivirgula* (Krassilnikov and Agre 1964) Lechevalier, Lechevalier, and Becker 1966 to *Faenia* gen. nov. Int. J. Syst. Bacteriol. *33*: 663–665.

Küster, E. and R. Locci. 1963. Transfer of *Thermoactinomyces viridis* Schuurmans *et al.*, 1956 to the genus *Thermomonospora* as *Thermomonospora viridis* comb. nov. Int. Bull. Bacteriol. Nomencl. Taxon. *13*: 214–216.

Küster, E. and S.T. Williams. 1964. Selection of Media for Isolation of *Streptomycetes*. Nature *202*: 928–929.

Kuznetsov, V.D., V.R. Kruyov, E.G. Rodionova and S.N. Fiippova. 1978. Taxonomy of an autotrophic actinomycete isolated from a floodplain marsh in the Moscow region USSR. Mikrobiologiya *47*: 107–111.

Labeda, D.P., R.T. Testa, M.P. Lechevalier and H.A. Lechevalier. 1984. *Saccharothrix*: a new genus of the *Actinomycetales* related to *Nocardiopsis*. Int. J. Syst. Bacteriol. *34*: 426–431.

Labeda, D.P. 1986. Transfer of *Nocardia* aerocolonigenes (Shinobu and Kawato 1960) Pridham 1970 into the genus *Saccharothrix* Labeda, Testa, Lechevalier, and Lechevalier 1984 as *Saccharothrix aerocolonigenes* sp. nov. Int. J. Syst. Bacteriol. *36*: 109–110.

Labeda, D.P. 1987. Transfer of the type strain of *Streptomyces erythraeus* (Waksman 1923) Waksman and Henrici 1948 to the genus *Saccharopolyspora Lacey* and Goodfellow 1975 as *Saccharopolyspora erythraea* sp. nov., and designation of a neotype strain for *Streptomyces erythraeus*. Int. J. Syst. Bacteriol. *37*: 19–22.

Labeda, D.P. and M.P. Lechevalier. 1989. Amendment of the genus *Saccharothrix* Labeda *et al.* 1984 and descriptions of *Saccharothrix espanaensis* sp. nov., *Saccharothrix cryophilis* sp. nov., and *Saccharothrix mutabilis* comb. nov. Int. J. Syst. Bacteriol. *39*: 420–423.

Labeda, D.P. and A.J. Lyons. 1989. *Saccharothrix texasensis* sp. nov. and *Saccharothrix waywayandensis* sp. nov. Int. J. Syst. Bacteriol. *39*: 355–358.

Labeda, D.P. 1995. *Amycolatopsis coloradensis* sp. nov., the avoparcin (LL-AV290)-producing strain. Int. J. Syst. Bacteriol. *45*: 124–127.

Labeda, D.P. and R.M. Kroppenstedt. 2000. Phylogenetic analysis of *Saccharothrix* and related taxa: proposal for *Actinosynnemataceae* fam. nov. Int. J. Syst. Evol. Microbiol. *50*: 331–336.

Labeda, D.P. 2001. *Crossiella* gen. nov., a new genus related to *Streptoalloteichus*. Int. J. Syst. Evol. Microbiol. *51*: 1575–1579.

Labeda, D.P., K. Hatano, R.M. Kroppenstedt and T. Tamura. 2001. Revival of the genus *Lentzea* and proposal far *Lechevalieria* gen. nov. Int. J. Syst. Evol. Microbiol. *51*: 1045–1050.

Labeda, D.P., J.M. Donahue, N.M. Williams, S.F. Sells and M.M. Henton. 2003. *Amycolatopsis kentuckyensis* sp. nov., *Amycolatopsis lexingtonensis* sp. nov. and *Amycolatopsis pretoriensis* sp. nov., isolated from equine placentas. Int. J. Syst. Evol. Microbiol. *53*: 1601–1605.

Labeda, D.P. and R.M. Kroppenstedt. 2006. *Goodfellowia* gen. nov., a new genus of the *Pseudonocardineae* related to *Actinoalloteichus*, containing *Goodfellowia coeruleoviolacea* gen. nov., comb. nov. Int. J. Syst. Evol. Microbiol. *56*: 1203–1207.

Labeda, D.P., J.M. Donahue, S.F. Sells and R.M. Kroppenstedt. 2007. *Lentzea kentuckyensis* sp. nov., of equine origin. Int. J. Syst. Evol. Microbiol. *57*: 1780–1783.

Labeda, D.P. and R.M. Kroppenstedt. 2007. Proposal of *Umezawaea* gen. nov., a new genus of the *Actinosynnemataceae* related to *Saccharothrix*, and transfer of *Saccharothrix tangerinus* Kinoshita *et al.* 2000 as *Umezawaea tangerina* gen. nov., comb. nov. Int. J. Syst. Evol. Microbiol. *57*: 2758–2761.

Labeda, D.P. and R.M. Kroppenstedt. 2008. Proposal for the new genus *Allokutzneria* gen. nov. within the suborder *Pseudonocardineae* and transfer of *Kibdelosporangium albatum* Tomita *et al.* 1993 as *Allokutzneria albata* comb. nov. Int. J. Syst. Evol. Microbiol. *58*: 1472–1475.

Labeda, D.P., M. Goodfellow, J. Chun, W.-J. Li and X.-Y. Zhi. 2010a. Reassessment of the systematics within the suborder *Pseudonocardineae*: Elimination of the family *Actinosynnemataceae* (Labeda and Kroppenstedt 2000) Zhi *et al.* 2009 and emendation of the family *Pseudonocardiaceae* (Embley *et al.* 1989) Zhi *et al.* 2009. Int. J. Syst. Evol. Microbiol., first published on 2 July 2010 as doi: doi:10.1099/ijs.0.024984-0.

Labeda, D.P., N.R. Price, G.Y.A. Tan, M. Goodfellow and H.P. Klenk. 2010b. Emended description of the genus *Actinokineospora* Hasegawa 1988 and transfer of *Amycolatopsis fastidiosa* Henssen *et al.* 1987 as *Actinokineospora fastidiosa* comb. nov. Int. J. Syst. Evol. Microbiol. *60*: 1444–1449.

Lacey, J. 1971. The microbiology of moist barky storage in unsealed silos. Ann. Appl. Biol. *69*: 187–212.

Lacey, J. 1974. Moulding of sugar-cane bagasse and its prevention. Ann. Appl. Biol. *76*: 63–76.

Lacey, J. 1975. Airborne spores in pastures. Transactions of the British Mycological Society *64*: 265–281.

Lacey, J. and M. Goodfellow. 1975. Novel actinomycete from sugarcane Bagasse *Saccharopolyspora hirsuta* gen. et sp. nov. J. Gen. Microbiol. *88*: 75–85.

Lacey, J. and J. Dutkiewicz. 1976a. Isolation of actinomycetes and fungi from mouldy hay using a sedimentation chamber. J. Appl. Bacteriol. *41*: 315–319.

Lacey, J. and J. Dutkiewicz. 1976b. Methods for examining the microflora of mouldy hay. J. Appl. Bacteriol. *41*: 13–27.

Lacey, J. 1978. Ecology of actinomycetes in fodders and related substances. Zentralbl. Bakteriol. Parasitenkd. Infektionskr. Hyg. Abt. I *Suppl. 6*: 161–170.

Lacey, J. 1981. Airborne actinomycete spores as respiratory allergens. Zentralbl. Bakteriol. Mikrobiol. Hyg. I Abt *Suppl. 11*: 243–250.

Lacey, J. 1988. Actinomycetes as biodeteriogens and pollutants of the environment. *In* Actinomycetes in Biotechnology (edited by Goodfellow, Williams and Mordarski). Academic Press, London, pp. 359–432.

Lacey, J. and B. Crook. 1988. Fungal and actinomycete spores as pollutants of the workplace and occupational allergens. Ann. Occup. Hyg. *32*: 515–533.

Lacey, J. 1989a. Genus *Faenia*. *In* Bergey's Manual of Systematic Bacteriology, 2nd edn, vol. 4 (edited by Williams, Sharpe and Holt). Springer, New York, pp. 2387–2392.

Lacey, J. 1989b. *In* Validation of the publication of new names and new combinations previously effectively published outside the IJSB. List no. 31. Int. J. Syst. Bacteriol. *39*: 495–497.

Lacey, J. 1989c. Genus *Saccharopolyspora*. *In* Bergey's Manual of Systematic Bacteriology, 2nd edn, vol. 4 (edited by Williams, Sharpe and Holt). Springer, New York, pp. 2382–2386.

Lal, R., S. Lal, E. Grund and R. Eichenlaub. 1991. Construction of a hybrid plasmid capable of replication in *Amycolatopsis mediterranei*. Appl. Environ. Microbiol. *57*: 665–671.

Lal, R., R. Khanna, N. Dhingra, M. Khanna and S. Lal. 1998. Development of an improved cloning vector and transformation system in *Amycolatopsis mediterranei*. J. Antibiot. *51*: 161–169.

Lancini, G. and G. Sartori. 1976. Rifamycin G, a further product of *Nocardia mediterranei* metabolism. J. Antibiot. *29*: 466–468.

Lancini, G.C., J.E. Thiemann, G. Sartori and P. Sensi. 1967. Biogenesis of rifamycins. The conversion of rifamycin B into rifamycin Y. Experientia *23*: 899–900.

Lanning, S. and S.T. Williams. 1982. Methods for the direct isolation and enumeration of actinophages in soil. J. Gen. Microbiol. *128*: 2063–2071.

Lechevalier, M.P. and H.A. Lechevalier. 1970. Chemical composition as a criterion in the classification of aerobic actinomycetes. Int. J. Syst. Bacteriol. *20*: 435–443.

Lechevalier, M.P., C. De Bièvre and H. Lechevalier. 1977a. Chemotaxonomy of aerobic actinomycetes: phospholipid composition. Biochem. Syst. Ecol. *5*: 249–260.

Lechevalier, M.P., C. de Biévre and H.A. Lechevalier. 1977b. Chemotaxonomy of aerobic actinomycetes: Phospholipid composition. Biochemistry and Ecological Systems *5*: 249–260.

Lechevalier, M.P., A.E. Stern and H.A. Lechevalier. 1981. Phospholipids in the taxonomy of actinomycetes. Zentralbl. Bakteriol. Parasitenkd. Infektionskr. Hyg. I Abt. Orig. *Suppl. 11*: 111–116.

Lechevalier, M.P., H. Prauser, D.P. Labeda and J.S. Ruan. 1986. Two new genera of nocardioform actinomycetes, *Amycolata* gen. nov. and *Amycolatopsis* gen. nov. Int. J. Syst. Bacteriol. *36*: 29–37.

Lee, J.J., J.H. Yoon, S.Y. Yang and S.T. Lee. 2006a. Aerobic biodegradation of 4-methylpyridine and 4-ethylpyridine by newly isolated *Pseudonocardia* sp. strain M43. FEMS Microbiol. Lett. *254*: 95–100.

Lee, M.Y., J.S. Myeong, H.J. Park, K. Han and E.S. Kim. 2006b. Isolation and partial characterization of a cryptic polyene gene cluster in *Pseudonocardia autotrophica*. J. Ind. Microbiol. Biotechnol. *33*: 84–87.

Lee, S.-D. 1996. Classification of novel actinomycetes from a gold mine cave in Kongju, Korea. PhD thesis, Seoul National University, Seoul.

Lee, S.B., S.E. Strand, H.D. Stensel and R.P. Herwig. 2004. *Pseudonocardia chloroethenivorans* sp. nov., a chloroethene-degrading actinomycete. Int. J. Syst. Evol. Microbiol. *54*: 131–139.

Lee, S.D., E.S. Kim and Y.C. Hah. 2000a. Phylogenetic analysis of the genera *Pseudonocardia* and *Actinobispora* based on 16S ribosomal DNA sequences. FEMS Microbiol. Lett. *182*: 125–129.

Lee, S.D., E.S. Kim, J.H. Roe, J. Kim, S.O. Kang and Y.C. Hah. 2000b. *Saccharothrix violacea* sp. nov., isolated from a gold mine cave, and *Saccharothrix albidocapillata* comb. nov. Int. J. Syst. Evol. Microbiol. *50*: 1315–1323.

Lee, S.D. and Y.C. Hah. 2001. *Amycolatopsis albidoflavus* sp. nov. Int. J. Syst. Evol. Microbiol. *51*: 645–650.

Lee, S.D., E.S. Kim, K.L. Min, W.Y. Lee, S.O. Kang and Y.C. Hah. 2001. *Pseudonocardia kongjuensis* sp. nov., isolated from a gold mine cave. Int. J. Syst. Evol. Microbiol. *51*: 1505–1510.

Lee, S.D., E.S. Kim, S.O. Kang and Y.C. Hah. 2002. *Pseudonocardia spinosispora* sp. nov., isolated from Korean soil. Int. J. Syst. Evol. Microbiol. *52*: 1603–1608.

Lee, S.D. 2006. *Amycolatopsis jejuensis* sp. nov. and *Amycolatopsis halotolerans* sp. nov., novel actinomycetes isolated from a natural cave. Int. J. Syst. Evol. Microbiol. *56*: 549–553.

Lee, S.D., L.L. Kinkel and D.A. Samac. 2006c. *Amycolatopsis minnesotensis* sp. nov., isolated from a prairie soil. Int. J. Syst. Evol. Microbiol. *56*: 265–269.

Lee, S.D. 2009. *Amycolatopsis ultiminotia* sp. nov., isolated from rhizosphere soil, and emended description of the genus *Amycolatopsis*. Int. J. Syst. Bacteriol. *59*: 1401–1404.

Lewer, P., D.R. Hahn, L.L. Karr, P.R. Graupner, J.R. Gilbert, T. Worden, R. Yao and D.W. Norton. 2002. Pesticidal macrolides. U.S. Patent 6455504 (September 24).

Li, J., G.Z. Zhao, S. Qin, H.Y. Huang, W.Y. Zhu, L.-H. Xu and W.-J. Li. 2009a. *Saccharopolyspora tripterygii* sp. nov., an endophytic actinomycete isolated from the stem of *Tripterygium hypoglaucum*. Int. J. Syst. Evol. Microbiol. *59*: 3040–3044.

Li, W.-J., P. Xu, S.-K. Tang, L.-H. Xu, R.M. Kroppenstedt, E. Stackebrandt and C.-L. Jiang. 2003a. *Prauserella halophila* sp. nov. and *Prauserella alba* sp. nov., moderately halophilic actinomycetes from saline soil. Int. J. Syst. Evol. Microbiol. *53*: 1545–1549.

Li, W.-J., S.-K. Tang, E. Stackebrandt, R.M. Kroppenstedt, P. Schumann, L.-H. Xu and C.-L. Jiang. 2003b. *Saccharomonospora paurometabolica* sp. nov., a moderately halophilic actinomycete isolated from soil in China. Int. J. Syst. Evol. Microbiol. *53*: 1591–1594.

Li, W.-J., P. Xu, S.-K. Tang, L.-H. Xu, R.M. Kroppenstedt, E. Stackebrandt and C.L. Jiang. 2003c. *Prauserella halophila* sp. nov. and *Prauserella alba* sp. nov., moderately halophilic actinomycetes from saline soil. Int. J. Syst. Evol. Microbiol. *53*: 1545–1549.

Li, X., X. Zhou and Z. Deng. 1984. Growth and cellulose production of *Micromonospora chalcea* and *Pseudonocardia thermophila*. Ann. Microbiol. *135B*: 79–89.

Li, Y., S.K. Tang, Y.G. Chen, J.Y. Wu, X.Y. Zhi, Y.Q. Zhang and W.J. Li. 2009b. *Prauserella salsuginis* sp. nov., *Prauserella flava* sp. nov., *Prauserella aidingensis* sp. nov. and *Prauserella sediminis* sp. nov., isolated from a salt lake. Int. J. Syst. Evol. Microbiol. *59*: 2923–2928.

Li, Y.V., L.P. Terekhova and M.G. Gapochka. 2002. Isolation of actinomycetes from soil using extremely high frequency radiation. Microbiology *71*: 105–108.

Lindenbein, W. 1952. Über einige chemisch interessante Actinomyceten – stämme und ihre Klassifizierung. Arch. Mikrobiol. *17*: 361–383.

Little, A.E. and C.R. Currie. 2007. Symbiotic complexity: discovery of a fifth symbiont in the attine ant-microbe symbiosis. Biol. Lett. *3*: 501–504.

Little, A.E. and C.R. Currie. 2008. Black yeast symbionts compromise the efficiency of antibiotic defenses in fungus-growing ants. Ecology *89*: 1216–1222.

Liu, Z., Y. Zhang and X. Yan. 1984. A new genus of the order *Actinomycetales*. Acta Microbiol. Sinica *24*: 295–298.

Liu, Z.P., J.F. Wu, Z.H. Liu and S.J. Liu. 2006. *Pseudonocardia ammonioxydans* sp. nov., isolated from coastal sediment. Int. J. Syst. Evol. Microbiol. *56*: 555–558.

Liu, Z., Y. Li, L.Q. Zheng, Y.J. Huang and W.J. Li. 2010. *Saccharomonospora marina* sp. nov., isolated from an ocean sediment of the East China Sea. Int. J. Syst. Evol. Microbiol. *60*: 1854–1857.

Locci, R. 1976. Developmental morphology of actinomycetes. *In* Actinomycetes: The Boundary Microorganisms (edited by Arai). Toppan Co. Ltd, Tokyo, pp. 249–297.

Lu, Y. and X. Yan. 1978. Studies on the classification of thermophilic actinomycetes. Part 2: Determination of *Pseudonocardia*. Acta Microbiol. Sin. *18*: 8–10.

Lu, Z.T., Z.H. Liu, L.M. Wang, Y.M. Zhang, W.H. Qi and M. Goodfellow. 2001. *Saccharopolyspora flava* sp. nov. and *Saccharopolyspora thermophila* sp. nov., novel actinomycetes from soil. Int. J. Syst. Evol. Microbiol. *51*: 319–325.

Madón, J. and R. Hutter. 1991. Transformation system for *Amycolatopsis (Nocardia) mediterranei*: direct transformation of mycelium with plasmid DNA. J. Bacteriol. *173*: 6325–6331.

Mahendra, S. and L. Alvarez-Cohen. 2005. *Pseudonocardia dioxanivorans* sp. nov., a novel actinomycete that grows on 1,4-dioxane. Int. J. Syst. Evol. Microbiol. *55*: 593–598.

Majumdar, S., S.R. Prabhagaran, S. Shivaji and R. Lal. 2006. Reclassification of *Amycolatopsis orientalis* DSM 43387 as *Amycolatopsis benzoatilytica* sp. nov. Int. J. Syst. Evol. Microbiol. *56*: 199–204.

Makkar, N.S. and T. Cross. 1982. Actinoplanetes in soil and on plant litter from freshwater habitats. J. Appl. Bacteriol. *52*: 209–218.

Malarczyk, E., I. Korszen-Pilecka, J. Rogalski and A. Leonowicz. 1987. Guaicol and isovanillic acid as metabolites in the transformation of methoxyphenolic acids by *Nocardia autotrophica*. Phytochem *26*: 1321–1324.

Maldonado, L.A., J.E. Stach, W. Pathom-aree, A.C. Ward, A.T. Bull and M. Goodfellow. 2005. Diversity of cultivable actinobacteria in geographically widespread marine sediments. Antonie van Leeuwenhoek *87*: 11–18.

Maldonado, L.A., D. Fragoso-Yanez, A. Perez-Garcia, J. Rosellon-Druker and E.T. Quintana. 2009. Actinobacterial diversity from marine sediments collected in Mexico. Antonie van Leeuwenhoek *95*: 111–120.

Malfait, M., B. Godden and M.J. Penninckx. 1984. Growth and cellulose production of *Micromonospora chalcea* and *Pseudonocardia thermophila*. Ann. Microbiol. *135B*: 1321–1324.

Malhotra, S. and R. Lal. 2007. The genus *Amycolatopsis*: indigenous plasmids,cloning vectors and gene transfer systems. Ind. J. Microbiol. *47*: 3–14.

Malhotra, S., S. Majumdar, M. Kumar, V.K. Bhasin, K.H. Gartemann and R. Lal. 2008. Nucleotide sequence of plasmid pA387 of *Amycolatopsis benzoatilytica* and construction of a conjugative shuttle vector. J. Basic Microbiol. *48*: 177–185.

Mäntyjärvi, R.M. and V.P. Kurup. 1988. Dot-immunobinding assay in the detection of IgG antibodies against farmer's lung antigens. Mycopathologia *103*: 49–54.

Margalith, P. and G. Beretta. 1960. Rifomycin. XI. Taxonomic study on *Streptomyces mediterranei* nov. sp. Mycopathol. Mycol. Appl. *13*: 321–330.

Martin, J.R. and W. Rosenbrook. 1967. Studies on the biosynthesis of the erythromycins. II. Isolation and structure of a biosynthetic intermediate, 6-deoxyerythronolide B. Biochemistry *6*: 435–440.

Martin, M.M. 1970. The biochemical basis of the fungus-attine ant symbiosis. Science *169*: 16–20.

Maskey, R.P., I. Kock, E. Helmke and H. Laatsch. 2003. Isolation and structure determination of phenazostatin D, a new phenazine from a marine actinomycete isolate *Pseudonocardia* sp. B6273. Z. Naturforsch. *58B*: 692–694.

Matsushima, P. and R.H. Baltz. 1994. Transformation of *Saccharopolyspora spinosa* protoplasts with plasmid DNA modified *in vitro* to avoid host restriction. Microbiology *140*: 139–143.

Matsushima, P., M.C. Broughton, J.R. Turner and R.H. Baltz. 1994. Conjugal transfer of cosmid DNA from *Escherichia coli* to *Saccharopolyspora spinosa*: effects of chromosomal insertions on macrolide A83543 production. Gene *146*: 39–45.

Mattinson-Rose, D.M. 1986. Classification of amycolate wall chemotype IV actinomycetes. PhD thesis, Newcastle University, Newcastle upon Tyne.

McCarthy, A.J. and P. Broda. 1984. Screening for lignin-degrading actinomycetes and characterisation of their activity against ^{14}C-lignin-labelled wheat lignocellulose. J. Gen. Microbiol. *130*: 2905–2913.

McCarthy, A.J. and T. Cross. 1984. A taxonomic study of *Thermomonospora* and other monosporic actinomycetes. J. Gen. Microbiol. *130*: 5–25.

McVeigh, H.P., J. Munro and T.M. Embley. 1994. The phylogenetic position of *Pseudoamycolata halophobica* (Akimov et al. 1989) and a proposal to reclassify It as *Pseudonocardia halophobica*. Int. J. Syst. Bacteriol. *44*: 300–302.

Mertz, F.P. and R.C. Yao. 1988. *Kibdelosporangium philippinense* sp. nov. isolated from soil. Int. J. Syst. Bacteriol. *38*: 282–286.

Mertz, F.P. and R.C. Yao. 1990. *Saccharopolyspora spinosa* sp. nov. Isolated from soil collected in a sugar mill rum still. Int. J. Syst. Bacteriol. *40*: 34–39.

Mertz, F.P. and R.C. Yao. 1993. *Amycolatopsis alba* sp. nov., isolated from soil. Int. J. Syst. Bacteriol. *43*: 715–720.

Millar, N.S. and I. Denholm. 2007. Nicotinic acetylcholine receptors: targets for commercially important insecticides. Invert. Neurosci. *7*: 53–66.

Minnikin, D.E., L. Alshamaony and M. Goodfellow. 1975. Differentiation of *Mycobacterium*, *Nocardia*, and related taxa by thin-layer chromatographic analysis of whole-organism methanolysates. J. Gen. Microbiol. *88*: 200–204.

Minnikin, D.E., A. G. O'Donnell, M. Goodfellow, G. A. Alderson, M. Athalye, A. Schaal and J.H. Parlett. 1984. An integrated procedure for the extraction of isoprenoid quinones and polar lipids. J. Microbiol. Methods *2*: 233–241.

Mishra, S.K., R.E. Gordon and D.A. Barnett. 1980. Identification of nocardiae and streptomycetes of medical importance. J. Clin. Microbiol. *11*: 728–736.

Miyanaga, A., S. Fushinobu, K. Ito and T. Wakagi. 2001. Crystal structure of cobalt-containing nitrile hydratase. Biochem. Biophys. Res. Commun. *288*: 1169–1174.

Mordarskaia, G., L.N. Guzeva and N.S. Agre. 1973. [Mycelial lipids of thermophilic actinomycetes]. Mikrobiologiia *42*: 165–166.

Moretti, P., G. Hintermann and R. Hutter. 1985. Isolation and characterization of an extrachromosomal element from *Nocardia mediterranei*. Plasmid *14*: 126–133.

Morinaga, Y. and Y. Hirose. 1984. Production of metabolites by methylotrophs. *In* Methylotrophs: Microbiology, Biochemistry, and Genetics (edited by Hou). CRC Press, Boca Raton, pp. 107–118.

Morita, T., S. Hara and Y. Matsushima. 1978. Purification and characterization of lysozyme produced by *Streptomyces erythraeus*. J. Biochem. *83*: 893–903.

Morón, R., I. Gonzalez and O. Genilloud. 1999. New genus-specific primers for the PCR identification of members of the genera *Pseudonocardia* and *Saccharopolyspora*. Int. J. Syst. Bacteriol. *49*: 149–162.

Mueller, U.G., T.R. Schultz, C.R. Currie, R.M. Adams and D. Malloch. 2001. The origin of the attine ant-fungus mutualism. Q. Rev. Biol. *76*: 169–197.

Mueller, U.G., J. Poulin and R.M. Adams. 2004. Symbiotic choice in a fungus-growing ant (Attini, Formicidae). Behav. Ecol. *15*: 357–364.

Mueller, U.G., D. Dash, C. Rabelong and A. Rodrigues. 2008. Coevolution between antine ants and actinomycete bacteria: a re-evaluation. Evolution *62*: 2894–2912.

Nadkarni, S.R., M.V. Patel, S. Chatterjee, E.K. Vijayakumar, K.R. Desikan, J. Blumbach, B.N. Ganguli and M. Limbert. 1994. Balhimycin, a new glycopeptide antibiotic produced by *Amycolatopsis* sp. Y-86,21022. Taxonomy, production, isolation and biological activity. J. Antibiot. *47*: 334–341.

Nakamura, K., T. Tomita, N. Abe and Y. Kamio. 2001. Purification and characterization of an extracellular poly(L-lactic acid) depolymerase from a soil isolate, *Amycolatopsis* sp. strain K104-1. Appl. Environ. Microbiol. *67*: 345–353.

Neef, A., R. Schafer, C. Beimfohr and P. Kampfer. 2003. Fluorescence based rRNA sensor systems for detection of whole cells of *Saccharomonospora* spp. and *Thermoactinomyces* spp. Biosens. Bioelectron. *18*: 565–569.

Nielsen, J.B. and B.H. Arison. 1989. 3-Methylpseudouridine as a fermentation product. J. Antibiot. *42*: 1248–1252.

Nishikori, T., A. Okuyama, H. Naganawa, T. Takita, M. Hamada, T. Takeuchi, T. Aoyagi and H. Umezawa. 1984. Production by actinomycetes of (S,S)-N,N′-ethylenediaminedisuccinic acid. J. Antibiot. *XXXVII*: 426–427.

Nolof, G. 1962. Beiträge zur Kenntnis des Stoffwechsels von *Nocardia hydrocarbonoxydans* n. spec. Arch. Mikrobiol. *44*: 278–297.

Nolof, G. and P. Hirsch. 1962. *Nocardia hydrocarbonoxydans* n. spec., ein oligocarbophiler Actinomycet. Arch. Mikrobiol. *44*: 266–277.

Nonomura, H. and Y. Ohara. 1969a. Distribution of actinomycetes in soil. VII. A culture method effective for both preferential isolation and enumeration of *Microbispora* and *Streptosporangium* strains in soil. Part 2. Classification of the isolates. J. Ferment. Technol. *47*: 701–709.

Nonomura, H. and Y. Ohara. 1969b. Distribution of actionomycetes in soil. VI. A culture method effective for both preferential isolation and enumeration of *Microbispora* and *Streptosproangium* strains in soil. Part I. J. Ferment. Technol. *47*: 463–469.

Nonomura, H. and Y. Ohara. 1971. Distribution of actinomycetes in soil. X. New genus and species of monosporic actinomycetes in soil. J. Ferment. Technol. *49*: 895–903.

North, R.D., C.W. Jackson and P.E. Howse. 1997. Evolutionary aspects of ant-fungus interactions in leaf-cutting ants. Trends Ecol. Evol. *12*: 386–389.

O'Connor, S., L.K.T. Lam, N.D. Jones and M.O. Chaney. 1976. Apramycin, a unique aminocyclitol antibiotic. J. Org. Chem. *41*: 2087–2092.

Ochi, K. and M. Yoshida. 1991. Polyacrylamide gel electrophoresis analysis of mycolateless wall chemotype IV actinomycetes. Int. J. Syst. Bacteriol. *41*: 402–405.

Ochi, K. and S. Miyadoh. 1992. Polyacrylamide gel electrophoresis analysis of ribosomal protein AT-L30 from an actinomycete genus, *Streptosporangium*. Int. J. Syst. Bacteriol. *42*: 151–155.

Ochi, K. 1995. Amino acid sequence analysis of ribosomal protein at L30 from members of the family *Pseudonocardiaceae*. Int. J. Syst. Bacteriol. *45*: 110–115.

Ohkuma, H., O. Tenmyo, M. Konishi, T. Oki and H. Kawaguchi. 1988. BMY-28190, a novel antiviral antibiotic complex. J. Antibiot. *41*: 849–854.

Okazaki, T., N. Serizawa, R. Enokita, A. Torikata and A. Terahara. 1983. Taxonomy of actinomycetes capable of hydroxylation of ML-236B. J. Antibiot. *36*: 1176–1183.

Okazaki, T., R. Enokita, H. Miyaoka, T. Takatsu and A. Torikata. 1987. Chloropolysporins A, B and C, novel glycopeptide antibiotics from *Faenia interjecta* sp. nov. I. Taxonomy of producing organism. J. Antibiot. (Tokyo) *40*: 917–923.

Okoro, C.K., R. Brown, A.L. Jones, B.A. Andrews, J.A. Asenjo, M. Goodfellow and A.T. Bull. 2009. Diversity and cultivable actinomycetes in hyper-arid soils of the Atacama Desert, Chile. Antonie van Leeuwenhoek *95*: 121–133.

Okuda, T., T. Furumai, E. Watanabe, Y. Okugawa and S. Kimura. 1966. *Actinoplanaceae* antibiotics. II. Studies on sporaviridin. Taxonomic study on the sporaviridin-producing microorganism, *Streptosporangium viridogriseum* nov. sp. J. Antibiot. *19*: 121–127.

Oliver, T.J. and A.C. Sinclair. 1964. Antibiotic M-319. US Patent #3155582. United States Patent and Trademark Office.

Oliynyk, M., M. Samborskyy, J.B. Lester, T. Mironenko, N. Scott, S. Dickens, S.F. Haydock and P.F. Leadlay. 2007. Complete genome sequence of the erythromycin-producing bacterium *Saccharopolyspora erythraea* NRRL 23338. Nat. Biotechnol. *25*: 447–453.

Ōmura, S., H. Tanaka, Y. Tanaka, P. Spiri-Nakagawa, R. Oiwa, Y. Takahashi, K. Matsuyama and Y. Iwai. 1979. Studies on bacterial cell wall inhibitors. VII. Azureomycins A and B, new antibiotics produced by *Pseudonocardia azurea* nov. sp. Taxonomy of the producing organism, isolation, characterization and biological properties. J. Antibiot. *32*: 985–994.

Ōmura, S., H. Tanaka, Y. Tanaka, P. Spiri-Nakagawa, R. Oliva, Y. Takahashi, K. Matsuyama and Y. Iwai. 1983. *In* Validation of the publication of new names and new combinations previously effectively published outside the IJSB. List no. 11. Int. J. Syst. Bacteriol. *33*: 673.

Orchard, V.A. and M. Goodfellow. 1980. Numerical classification of some named strains of *Nocardia asteroides* and related isolates from soil. J. Gen. Microbiol. *118*: 295–312.

Otoguro, M., M. Hayakawa, T. Yamazaki, T. Tamura, K. Hatano and Y. Iimura. 2001. Numerical phenetic and phylogenetic analyses of *Actinokineospora* isolates, with a description of *Actinokineospora aurantticolor* sp. nov. and *Actinokineospora enzanensis* sp. nov. Actinomycetologica *15*: 30–39.

Otoguro, M., M. Hayakawa, T. Yamazaki, K. Hatano and Y. Iimura. 2003. *In* Validation of publication of new names and new combinations previously effectively published outside the IJSEM. List no. 89. Int. J. Syst. Evol. Microbiol. *53*: 1–2.

Pacey, M.S., M.R. Jefson, L.H. Huang, W.P. Cullen, H. Maeda, J. Tone, S. Nishiyama, K. Kaneda and M. Ishiguro. 1989. UK-69,753, a novel member of the efrotomycin family of antibiotics. I. Taxonomy of the producing organism, fermentation and isolation. J. Antibiot. *42*: 1453–1459.

Parales, R.E., J.E. Adamus, N. White and H.D. May. 1994. Degradation of 1,4-dioxane by an actinomycete in pure culture. Appl. Environ. Microbiol. *60*: 4527–4530.

Park, S.W., S.T. Park, J.E. Lee and Y.M. Kim. 2008. *Pseudonocardia carboxydivorans* sp. nov., a carbon monoxide-oxidizing actinomycete, and an emended description of the genus *Pseudonocardia*. Int. J. Syst. Evol. Microbiol. *58*: 2475–2478.

Peano, C., S. Bicciato, G. Corti, F. Ferrari, E. Rizzi, R.J. Bonnal, R. Bordoni, A. Albertini, L.R. Bernardi, S. Donadio and G. De Bellis. 2007. Complete gene expression profiling of *Saccharopolyspora erythraea* using GeneChip DNA microarrays. Microb. Cell Fact. *6*: 37–51.

Pelzer, S., R. Süssmuth, D. Heckmann, J. Recktenwald, P. Huber, G. Jung and W. Wohlleben. 1999. Identification and analysis of the balkimycin biosynthetic gene cluster and its use for manipulating glycopeptide biosynthesis in *Amycolatopsis mediterranei* DSM 5908. Antimicrob. Agents Chemother. *43*: 1565–1573.

Peplowski, L., K. Kubiak and W. Nowak. 2007. Insights into catalytic activity of industrial enzyme Co-nitrile hydratase. Docking studies of nitriles and amides. J. Mol. Model. *11*: 725–730.

Pepys, J., P.A. Jenkins, G.N. Festenstein, P.H. Gregory, M.E. Lacey and F.A. Skinner. 1963. Farmer's Lung. Thermophilic actinomycetes as a source of "Farmer's Lung Hay" antigen. Lancet *2*: 607–611.

Pernodet, J.L., J.M. Simonet and M. Guerineau. 1984. Plasmids in different strains of *Streptomyces ambofaciens*: free and integrated form of plasmid pSAM2. Mol. Gen. Genet. *198*: 35–41.

Pernodet, J.L., F. Boccard, M.T. Alegre, J. Gagnat and M. Guerineau. 1989. Organization and nucleotide sequence analysis of a ribosomal RNA gene cluster from *Streptomyces ambofaciens*. Gene *79*: 33–46.

Philip, J.E., J.R. Schenck and M.P. Hargic. 1957. Ristocetins A and B, two new antibiotics. Antibiot. Annu. *1956–1957*: 699–705.

Pimentel-Elardo, S.M., L.P. Tiro, L. Grozdanov and U. Hentschel. 2008. *Saccharopolyspora cebuensis* sp. nov., a novel actinomycete isolated from a Philippine sponge (*Porifera*). Int. J. Syst. Evol. Microbiol. *58*: 628–632.

Pirie, H.M., C.O. Dawson, R.G. Breeze, A. Wiseman and J. Hamilton. 1971. A bovine disease similar to farmer's lung: extrinsic allergic alveolitis. Vet. Rec. *88*: 346–351.

Pittenger, R.C. and R.B. Brigham. 1956. *Streptomyces orientalis* n. sp., the source of vancomycin. Antibiot. Chemother. *6*: 642–647.

Pommer, E.H. and G. Lorenz. 1986. The behaviour of polyester and polyether polyurethanes towards microorganisms. *In* Biodeterioration Society Occasional Publications (edited by Seal). International Biodeterioration and Biodegradation Society, Manchester, UK, pp. 77–86.

Poschner, J., R.M. Kroppenstedt, A. Fischer and E. Stackebrandt. 1985. DNA–DNA reassociation and chemotaxonomic studies on *Actinomadura, Microbispora, Microtetraspora, Micropolyspora* and *Nocardiopsis*. Syst. Appl. Microbiol. *6*: 264–270.

Poulsen, M., M. Cafaro, J.J. Boomsma and C.R. Currie. 2005. Specificity of the mutualistic association between actinomycete bacteria and two sympatric species of *Acromyrmex* leaf-cutting ants. Mol. Ecol. *14*: 3597–3604.

Poulsen, M. and C.R. Currie. 2006. Complexity of insect-fungal associations. Exploring the influence of micro-organisms in the attine antifungus symbiosis. *In* Insect Symbiosis, vol. 11 (edited by Bourtzis and Miller). CRC Press, Boca Raton.

Poulsen, M., D.P. Erhardt, D.J. Molinaro, T.L. Lin and C.R. Currie. 2007. Antagonistic bacterial interactions help shape host-symbiont dynamics within the fungus-growing ant-microbe mutualism. PLoS One *2*: e960.

Prabahar, V., S. Dube, G.S.N. Reddy and S. Shivaji. 2004a. *Pseudonocardia antarctica* sp. nov. an Actinomycetes from McMurdo Dry Valleys, Antarctica. Syst. Appl. Microbiol. *27*: 66–71.

Prabahar, V., S. Dube, G.S.N. Reddy and S. Shivaji. 2004b. *In* Validation of publication of new names and new combinations previously effectively published outside the IJSEM. List no. 98. Int. J. Syst. Evol. Microbiol. *54*: 1005–1006.

Prauser, H. and S. Momirova. 1970. [Phage sensitivity, cell wall composition and taxonomy of various thermophilic actinomycetes]. Z. Allg. Mikrobiol. *10*: 219–222.

Preobrazhenskaya, T.P. and M.A. Sveshnikova. 1974. New species of the *Actinomadura* genus. Mikrobiologiya *43*: 864–868.

Preobrazhenskaya, T.P., L.P. Terekhova, A.V. Laiko, T.I. Selezneva, V.A. Zenkova and N.O. Blinov. 1976. *Actinomadura coeruleoviolacea* sp. nov. and its antagonistic properties. Antibiotiki *21*: 779–784.

Preobrazhenskaya, T.P., M.A. Sveshnikova and G.F. Gauze. 1982. On the transfer of certain species belonging to the genus *Actinomadura* Lechevalier and Lechevalier 1970 into the genus *Nocardiopsis* Meyer 1976. Mikrobiologiya *51*: 111–113.

Preobrazhenskaya, T.P. and M.A. Sveshnikova. 1985. *In* Validation of the publication of new names and new combinations previously effectively published outside the IJSB. List no. 17. Int. J. Syst. Bacteriol. *35*: 223–225.

Preobrazhenskaya, T.P. and L.P. Terekhova. 1987. *In* Validation of the publication of new names and new combinations previously effectively published outside the IJSB. List no. 23. Int. J. Syst. Bacteriol. *37*: 179–180.

Pridham, T.G. and A.J. Lyons. 1969. Progress in clarification of the taxonomic and nomenclatural status of some problem actinomycetes. Dev. Indust. Microbiol. *10*: 183–221.

Pridham, T.G. 1970. New names and new combinations in the order *Actinomycetales* Buchanan 1917. U.S. Dept. Agric. Tech. Bull. *1424*: 1–55.

Priefert, H., S. Achetrholt and A. Steinbuchel. 2002. Tranformation of the *Pseudonocardiaceae Amycolatopsis* sp. strain HR 167 is highly dependent on the physiological state of the cells. Appl. Microbiol. Biotechnol. *58*: 454–460.

Proctor, R., W. Craig and C. Kunin. 1978. Cetocycline, tetracycline analog: in vitro studies of antimicrobial activity, serum binding, lipid solubility, and uptake by bacteria. Antimicrob. Agents Chemother. *13*: 598–604.

Qin, S., J. Li, G.Z. Zhao, H.H. Chen, L.H. Xu and W.-J. Li. 2008a. *Saccharopolyspora endophytica* sp. nov., an endophytic actinomycete isolated from the root of *Maytenus austroyunnanensis*. Syst. Appl. Microbiol. *31*: 352–357.

Qin, S., Y.Y. Su, Y.Q. Zhang, H.B. Wang, C.-L. Jiang, L.H. Xu and W.-J. Li. 2008b. *Pseudonocardia ailaonensis* sp. nov., isolated from soil in China. Int. J. Syst. Evol. Microbiol. *58*: 2086–2089.

Qin, S., H.H. Chen, H.P. Klenk, C.J. Kim, L.H. Xu and W.-J. Li. 2010. *Saccharopolyspora gloriosae* sp. nov., an endophytic actinomycete isolated from the stem of *Gloriosa superba* L. Int. J. Syst. Evol. Microbiol. *60*: 1147–1151.

Rabenhorst, J. and R. Hopp. 1997. Process for the preparation of vanillin and suitable microorganisms. European Patent Office EP 0761817 (A2).

Ramasamy, M., Z.U. Khan and V.P. Kurup. 1987. A partially purified antigen from *Faenia rectivirgula* in the diagnosis of farmer's lung disease. Microbios *49*: 171–182.

Rasmussen, R.R., M.E. Nuss, M.H. Scherr, S.I. Mueller, J.B. McAlpine and L.A. Mitscher. 1986. Benzanthrin A and B, a new class of quinone antibiotics. Isolation, structure elucidation and potential antitumor activity. J. Antibiot. *39*: 1516–1526.

Rautenstein, Y.I., V.D. Kuznetsov, E.G. Rodionova, I.V. Yangulova, S.V. Dmitrieva and L.A. Deshchits. 1975. Revision of the taxonomic position of *Actinomyces oreintalis* and its renaming as *Proactinomyces orientalis* nov. comb. Mikrobiologiya *44*: 528–533.

Reasoner, D.J. and E.E. Geldreich. 1985. A new medium for the enumeration and subculture of bacteria from potable water. Appl. Environ. Microbiol. *49*: 1–7.

Reeves, A.R., I.A. Brikun, W.H. Cernota, B.I. Leach, M.C. Gonzalez and J.M. Weber. 2006. Effects of methylmalonyl-CoA mutase gene knockouts on erythromycin production in carbohydrate-based and oil-based fermentations of *Saccharopolyspora erythraea*. J. Ind. Microbiol. Biotechnol. *33*: 600–609.

Reeves, A.R., I.A. Brikun, W.H. Cernota, B.I. Leach, M.C. Gonzalez and J.M. Weber. 2007. Engineering of the methylmalonyl-CoA metabolite node of *Saccharopolyspora erythraea* for increased erythromycin production. Metab. Eng. *9*: 293–303.

Reichert, K., A. Lipski, S. Pradella, E. Stackebrandt and K. Altendorf. 1998. *Pseudonocardia asaccharolytica* sp. nov. and *Pseudonocardia sulfidoxydans* sp. nov., two new dimethyl disulfide-degrading actinomycetes and emended description of the genus *Pseudonocardia*. Int. J. Syst. Bacteriol. *48*: 441–449.

Reutler, J., W. Wohlleben and G. Muth. 2006. Modular architecture of the conjugative plasmid p SVH1 from *Streptomyces venezuelae*. Plasmid *55*: 201–209.

Richard, F.J., M. Poulsen, A. Hefetz, C. Errard, D.R. Nash and J.J. Boomsma. 2007. The origin of chemical profiles of fungal symbionts: their significance for nest-male recognition in *Acromyrmex* leaf-cutting ants. Behav. Ecol. Sociolbiol. *61*: 1637–1649.

Roberts, R.C., F.J. Wenzel and D.A. Emanuel. 1976. Precipitating antibodies in a midwest dairy farming population toward the antigens associated with farmer's lung disease. J. Allergy Clin. Immunol. *57*: 518–524.

Roussel, S., G. Reboux, J.C. Dalphin, D. Pernet, J.J. Laplante, L. Millon and R. Piarroux. 2005. Farmer's lung disease and microbiological composition of hay: a case-control study. Mycopathologia *160*: 273–279.

Ruan, J.S., Y. Lang, Y. Shi, L. Qu and X. Yu. 1994. Chemical and molecular classification of *Saccharomonospora* strains. Int. J. Syst. Bacteriol. *44*: 704–707.

Ruddock, J.C., K.S. Holdom, H. Maeda, J. Tone and M.R. Jefson. 1987. European Patent Office Application 88300056.4.

Saddler, G.S., M. Goodfellow, D.E. Minnikin and A.G. O'Donnell. 1986. Influence of the growth cycle on the fatty acid and menaquinone composition of *Streptomyces cyaneus* NCIB 9616. J. Appl. Microbiol. *60*: 51–56.

Saddler, G.S., P. Tavecchia, S. Lociuro, M. Zanol, E. Colombo and E. Selva. 1991. Analysis of madurose and other actinomycete whole cell sugars by gas chromatography. J. Microbiol. Methods *14*: 185–191.

Saintpierre-Bonaccio, D., H. Amir, R. Pineau, G.Y.A. Tan and M. Goodfellow. 2005. *Amycolatopsis plumensis* sp. nov., a novel bioactive actinomycete isolated from a New-Caledonian brown hypermagnesian ultramafic soil. Int. J. Syst. Evol. Microbiol. *55*: 2057–2061.

Saitou, N. and M. Nei. 1987. The neighbor-joining method: a new method for reconstructing phylogenetic trees. Mol. Biol. Evol. *4*: 406–425.

Sakane, T. and K. Kuroshima. 1997. Viabilities of dried cultures of various bacteria after preservation for 20 years and their production by the accelerated storage test. Microbiol. Cult. Coll. *13*: 1–7.

Salazar, O., R. Moron and O. Genilloud. 2000. New genus-specific primers for the PCR identification of members of the genus *Saccharomonospora* and evaluation of the microbial diversity of wild-type isolates of *Saccharomonospora* detected from soil DNAs. Int. J. Syst. Evol. Microbiol. *50*: 2043–2055.

Salazar, O., I. Gonzalez and O. Genilloud. 2002. New genus-specific primers for the PCR identification of novel isolates of the genera *Nocardiopsis* and *Saccharothrix*. Int. J. Syst. Evol. Microbiol. *52*: 1411–1421.

Sasser, M. 1990. Identification of bacteria by gas chromatography of cellular fatty acids. USFCC Newsl. *20*: 1–6.

Schaal, K.P. and B.L. Beaman. 1984. Clinical significance of actinomycetes. *In* The Biology of the Actinomycetes (edited by Goodfellow, Mordarski and Williams). Academic Press, London, pp. 389–424.

Schabereiter-Gurtner, C., G. Pinar, W. Lubitz and S. Rolleke. 2001. An advanced molecular strategy to identify bacterial communities on art objects. J. Microbiol. Methods *45*: 77–87.

Schäfer, D. 1969. Eine neue *Streptosporangium* Art aus türkischer Steppenerde. Arch. Mikrobiol. *66*: 365–373.

Schäfer, J., K. Martin and P. Kampfer. 2010. *Prauserella muralis* sp. nov., from the indoor environment. Int. J. Syst. Evol. Microbiol. *60*: 287–290.

Scharfen, J. 1971. Trutnov 139–66. An unusual actinomycetes combining the contradictory properties of the genera *Nocardia* and *Actinomyces* – the causative agent of submandibular mycetoma. I. Introduction. J. Hyg. Epidemiol. Microbiol. Immunol. *15*: 43–51.

Schatz, A., H.D. Isenberg, A.A. Angrist and V. Schatz. 1954. Microbial metabolism of carbamates. I. Isolation of *Streptomyces nitrificans* spec. nov., and other organisms which grow on urethane. J. Bacteriol. *68*: 1–4.

Schleifer, K.H. and O. Kandler. 1972. Peptidoglycan types of bacterial cell walls and their taxonomic implications. Bacteriol. Rev. *36*: 407–477.

Schneider, J., I.A. Garcia and H.J. Kutzner. 1987. Characterization of a family of temperate actinophages of *Faenia rectivirgula*. J. Gen. Microbiol. *133*: 2263–2268.

Schneider, J. and H.J. Kutzner. 1989. Distribution of modules among the central regions of the genomes of several actinophages of *Faenia* and *Saccharopolyspora*. J. Gen. Microbiol. *135*: 1671–1678.

Schultz, T.R. and S.G. Brady. 2008. Major evolutionary transitions in ant agriculture. Proc. Natl. Acad. Sci. U.S.A. *105*: 5435–5440.

Schupp, T., C. Toupet, N. Engel and S. Goff. 1998. Cloning and sequence analysis of the putative rifamycin polyketide synthase gene cluster from *Amycolatopsis mediterranei*. FEMS Microbiol. Lett. *159*: 201–207.

Schuurmans, D.M., B.H. Olson and C.L.S. Clemente. 1956. Production and isolation of thermoviridin an antibiotic produced by *Thermoactinomyces viridis* n. sp. Appl. Microbiol. *4*: 61–66.

Selvin, J., S. Shanmugha Priya, G. Seghal Kiran, T. Thangavelu and N. Sapna Bai. 2009. Sponge-associated marine bacteria as indicators of heavy metal pollution. Microbiol. Res. *164*: 352–363.

Sensi, P., A.M. Greco and R. Ballotta. 1959. Rifomycin. I. Isolation and properties of rifomycin B and rifomycin complex. Antibiot. Annu. *7*: 262–270.

Sharaya, L.S., D. Taptykova, A.N. Parijskaya and L.V. Kalakoutskii. 1982. The characteristics of the life cycle of an actinomycete isolated from *Alnus incana* root nodules. Mikrobiologiya *51*: 657–663.

Shearer, M.C., P.M. Colman and C.H. Nash. 1983. *Nocardiopsis mutabilis*, a new species of nocardioform bacteria Isolated from soil. Int. J. Syst. Bacteriol. *33*: 369–374.

Shearer, M.C., P. Actor, B.A. Bowie, S.F. Grappel, C.H. Nash, D.J. Newman, Y.K. Oh, C.H. Pan and L.J. Nisbet. 1985. Aridicins, novel glycopeptide antibiotics. I. Taxonomy, production and biological activity. J. Antibiot. *38*: 555–560.

Shearer, M.C., P.M. Colman, R.M. Ferrin, L.J. Nisbet and C.H. Nash. 1986a. New genus of the *Actinomycetales*: *Kibdelosporangium aridum* gen. nov., sp. nov. Int. J. Syst. Bacteriol. *36*: 47–54.

Shearer, M.C., A.J. Giovenella, S.F. Grappel, R.D. Hedde, R.J. Mehta, Y.K. Oh, C.H. Pan, D.H. Pitkin and L.J. Nisbet. 1986b. Kibdelins, novel glycopeptide antibiotics. I. Discovery, production, and biological evaluation. J. Antibiot. *39*: 1386–1394.

Shearer, M.C. 1987. Methods for the isolation of non-streptomycete actinomycetes. Dev. Indust. Microbiol. *28*: 91–97.

Shearer, M.C., A.J. Giovenella, S.F. Grappel, R.D. Hedde, R.J. Mehta, Y.K. Oh, C.H. Pan, D.H. Pitkin and L.J. Nisbet. 1988. *In* Validation of the publication of new names and new combinations previously effectively published outside the IJSB. List no. 24. Int. J. Syst. Bacteriol. *38*: 136–137.

Shen, F.T., H.L. Lu, J.L. Lin, W.S. Huang, A.B. Arun and C.C. Young. 2006. Phylogenetic analysis of members of the metabolically diverse genus *Gordonia* based on proteins encoding the *gyrB* gene. Res. Microbiol. *157*: 367–375.

Shinobu, R. and M. Kawato. 1960. On *Streptomyces aerocolonigenes*, nov. sp., forming the secondary colonies on the aerial mycelia. Bot. Mag. *71*: 212–216.

Shirling, E.B. and D. Gottlieb. 1966. Methods for characterization of *Streptomyces* species. Int. J. Syst. Bacteriol. *16*: 313–340.

Shorin, V.A., S.D. Yudinstev, I.A. Kunrat, L. Goldberg, N.S. Pevzner, M.G. Brashnikova, N.N. Lomakina and E.F. Oparysheva. 1957. New antibiotics, actinoidin. Antibiotiki *2*: 44–49.

Singla, A.K., S. Mayilraj, T. Kudo, S. Krishnamurthi, G.S. Prasad and R.M. Vohra. 2005. *Actinoalloteichus spitiensis* sp. nov., a novel actinobacterium isolated from a cold desert of the Indian Himalayas. Int. J. Syst. Evol. Microbiol. *55*: 2561–2564.

Sitrin, R.D., G.W. Chan, J.J. Dingerdissen, W. Holl, J.R. Hoover, J.R. Valenta, L. Webb and K.M. Snader. 1985. Aridicins, novel glycopeptide antibiotics. II. Isolation and characterization. J. Antibiot. *38*: 561–571.

Skerman, V.B.D., V. McGowan and P.H.A. Sneath. 1980. Approved Lists of Bacterial Names. Int. J. Syst. Bacteriol. *30*: 225–420.

Smits, T.H.M., M. Röthlisberger, B. Witholt and J.B. Van Beilen. 1999. Molecular screening for alkane hydroxylase genes in Gram-negative and Gram-positive strains. Environ. Microbiol. *1*: 307–317.

Smokvina, T., F. Francou and M. Luzzati. 1988. Genetic analysis in *Streptomyces ambofaciens* element pSAM2. Plasmid *25*: 40–52.

Smokvina, T., F. Boccard, J.-L. Pernodet, A. Friedmann and M. Guerineau. 1991. Functional analysis in *Streptomyces ambofaciens* element pSAM2. Plasmid *25*: 40–52.

Smorawinska, M., F. Denis, C.V. dery, P. Magny and R. Brzezinski. 1988. Characterization of SE-3, a virulent bacteriophage of *Saccharopolyspora erythraea*. J. Gen. Microbiol. *134*: 1773–1778.

Song, J., H.Y. Weon, S.H. Yoon, D.S. Park, S.J. Go and J.W. Suh. 2001. Phylogenetic diversity of thermophilic actinomycetes and *Thermoactinomyces* spp. isolated from mushroom composts in Korea based on 16S rRNA gene sequence analysis. FEMS Microbiol. Lett. *202*: 97–102.

Srivibool, R. and M. Sukchotiratana. 2006. Bioperspective of actinomycetes isolates from coastal soils: a new source of antibiotic producers. J. Sci. Technol. *28*: 493–499.

Stackebrandt, E., R.M. Kroppenstedt, K.D. Jahnke, C. Kemmerling and H. Gurtler. 1994. Transfer of *Streptosporangium viridogriseum* (Okuda *et al.* 1966), *Streptosporangium viridogriseum* subsp. *kofuense* (Nonomura and Ohara 1969), and *Streptosporangium albidum* (Furumai *et al.* 1968) to *Kutzneria* gen. nov. as *Kutzneria viridogrisea* comb. nov., *Kutzneria kofuensis* comb. nov., and *Kutzneria albida* comb. nov., respectively, and emendation of the genus *Streptosporangium*. Int. J. Syst. Bacteriol. *44*: 265–269.

Stackebrandt, E., F.A. Rainey and N.L. Ward-Rainey. 1997. Proposal for a new hierarchic classification system, *Actinobacteria* classis nov. Int. J. Syst. Bacteriol. *47*: 479–491.

Stackebrandt, E., R.M. Kroppenstedt, J. Wink and P. Schumann. 2004. Reclassification of *Amycolatopsis orientalis* subsp. *lurida* Lechevalier *et al.* 1986 as *Amycolatopsis lurida* sp. nov., comb. nov. Int. J. Syst. Evol. Microbiol. *54*: 267–268.

Stamatakis, A.P., T. Ludwig, H. Meier and M.J. Wolf. 2002. AxML: a fast program for sequential and parallel phylogenetic tree calculations based on the maximum likelihood method. Proc. IEEE. Comput. Soc. Bioinform. Conf. *1*: 21–28.

Staneck, J.L. and G.D. Roberts. 1974. Simplified approach to identification of aerobic actinomycetes by thin-layer chromatography. Appl. Microbiol. *28*: 226–231.

Stanzak, R., P. Matsushima, R.H. Baltz and R.N. Rao. 1986. Cloning and expression in *Streptomyces livdans* of clustered erythromycin biosynthesis genes from *Streptomyces erythraeus*. Biotechnology *4*: 229–232.

Stapley, E.O., M. Jackson, S. Hernandez, S.B. Zimmerman, S.A. Currie, S. Mochales, J.M. Mata, H.B. Woodruff and D. Hendlin. 1972. Cephamycins, a new family of beta-lactam antibiotics. Antimicrob. Agents Chemother. *2*: 122–131.

Stark, W.M., C.E. Higgens, R.N. Wulfe, M.M. Hoehn and J.M. McGuire. 1963. Capreomycin, a new antimycobacterial agent produced by *Streptomyces capreolus* sp. n. Antimicrob. Agents Chemother. *1962*: 596–606.

Staunton, J. and K.J. Weissman. 2001. Polyketide biosynthesis: a millennium review. Nat. Prod. Rep. *18*: 380–416.

Steger, K., A. Jarvis, T. Vasara, M. Romantschuk and I. Sundh. 2007. Effects of differing temperature management on development of *Actinobacteria* populations during composting. Res. Microbiol. *158*: 617–624.

Stegmann, E., S. Pelzer, K. Wilken and W. Wohlleben. 2001. Development of three different gene cloning systems for genetic investigation of the new species *Amycolatopsis japonicum* MG417-CF17, the ethylenediaminedisuccinic acid producer. J. Biotechnol. *92*: 195–204.

Sternberg, M. 1976. Microbial rennets. Adv. Appl. Microbiol. *20*: 135–137.

Stow, A. and A. Beattie. 2008. Chemical and genetic defenses against disease in insect societies. Brain Behav. Immun. *22*: 1009–1013.

Sugie, M., H. Suzuki and N. Tomiyaka. 1988. Purification and some properties of D-amino acid specific peptidase from *Nocardia orientalis*. Rep. Ferment. Inst. *69*: 1–14.

Suzuki, S., Y. Takahashi, T. Okuda and S. Komatsubara. 1998. Selective isolation of *Actinobispora* on gellan gum plates. Can. J. Microbiol. *44*: 1–5.

Syed, D.G., S.K. Tang, M. Cai, X.Y. Zhi, D. Agasar, J.C. Lee, C.J. Kim, C.L. Jiang, L.H. Xu and W.J. Li. 2008. *Saccharomonospora saliphila* sp. nov., a halophilic actinomycete from an Indian soil. Int. J. Syst. Evol. Microbiol. *58*: 570–573.

Szabo, I.M., M. Marton, G. Kulcsar and I. Buti. 1976. Taxonomy of primycin producing actinomycetes. I. Description of the type strain of *Thermomonospora galeriensis*. Acta Microbiol. Acad. Sci. Hung. *23*: 371–376.

Taerum, S.J., M.J. Cafaro, A.E. Little, T.R. Schultz and C.R. Currie. 2007. Low host-pathogen specificity in the leaf-cutting ant-microbe symbiosis. Proc. Biol. Sci. *274*: 1971–1978.

Takahashi, A., K. Hotta, N. Saito, M. Morioka, Y. Okami and H. Umezawa. 1986. Production of novel antibiotic, dopsisamine, by a new subspecies of *Nocardiopsis mutabilis* with multiple antibiotic resistance. J. Antibiot. *39*: 175–183.

Takamiya, A. and K. Tubaki. 1956. A new form of *Streptomyces* capable of growing autotrophically. Arch. Mikrobiol. *25*: 58–64.

Takeda, K., K. Kominato, A. Sugita, Y. Iwasaki, M. Shimazaki and M. Shimizu. 2006. Isolation and identification of 2α,25-dihydroxyvitamin D3, a new metabolite from *Pseudonocardia autotrophica* 100U-19 cells incubated with Vitamin D3. Steroids *71*: 736–744.

Takeuchi, M., R. Enokita, T. Okazaki, T. Kagasaki and M. Inukai. 1991. Helvecardins A and B, novel glycopeptide antibiotics. I. Taxonomy, fermentation, isolation and physico-chemical properties. J. Antibiot. *44*: 263–270.

Takeuchi, M., T. Nishii and A. Yokata. 1992. Taxonomic significance of arabinose in the family *Pseudonocardiaceae*. Actinomycetologica *6*: 79–90.

Tamura, T., M. Hayakawa, H. Nonomura, A. Yokota and K. Hatano. 1995. Four new species of the genus *Actinokineospora*: *Actinokineospora inagensis* sp. nov., *Actinokineospora globicatena* sp. nov., *Actinokineospora terrae* sp. nov., and *Actinokineospora diospyrosa* sp. nov. Int. J. Syst. Bacteriol. *45*: 371–378.

Tamura, T., L. Zhiheng, Z. Yamei and K. Hatano. 2000. *Actinoalloteichus cyanogriseus* gen. nov., sp. nov. Int. J. Syst. Evol. Microbiol. *50*: 1435–1440.

Tamura, T., Y. Ishida, M. Otoguro, K. Hatano, D. Labeda, N.P. Price and K. Suzuki. 2008a. Reclassification of *Streptomyces caeruleus* as a synonym of *Actinoalloteichus cyanogriseus* and reclassification of *Streptomyces spheroides* and *Streptomyces laceyi* as later synonyms of *Streptomyces niveus*. Int. J. Syst. Evol. Microbiol. *58*: 2812–2814.

Tamura, T., Y. Ishida, M. Otoguro, K. Hatano and K. Suzuki. 2008b. Classification of '*Streptomyces tenebrarius*' Higgins and Kastner as *Streptoalloteichus tenebrarius* nom. rev., comb. nov., and emended description of the genus *Streptoalloteichus*. Int. J. Syst. Evol. Microbiol. *58*: 688–691.

Tan, G.Y.A., S. Robinson, E. Lacey and M. Goodfellow. 2006a. *Amycolatopsis australiensis* sp. nov., an actinomycete isolated from arid soils. Int. J. Syst. Evol. Microbiol. *56*: 2297–2301.

Tan, G.Y.A., A.C. Ward and M. Goodfellow. 2006b. Exploration of *Amycolatopsis* diversity in soil using genus-specific primers and novel selective media. Syst. Appl. Microbiol. *29*: 557–569.

Tan, G.Y.A., S. Robinson, E. Lacey, R. Brown, W. Kim and M. Goodfellow. 2007. *Amycolatopsis regifaucium* sp. nov., a novel actinomycete that produces kigamicins. Int. J. Syst. Evol. Microbiol. *57*: 2562–2567.

Tang, S.K., X.P. Tian, X.Y. Zhi, M. Cai, J.Y. Wu, L.L. Yang, L.H. Xu and W.J. Li. 2008. *Haloactinospora alba* gen. nov., sp. nov., a halophilic filamentous actinomycete of the family *Nocardiopsaceae*. Int. J. Syst. Evol. Microbiol. *58*: 2075–2080.

Tang, S.K., Y. Wang, M. Cai, X.Y. Zhi, K. Lou, L.H. Xu, C.L. Jiang and W.J. Li. 2009a. *Saccharopolyspora halophila* sp. nov., a novel halophilic actinomycete isolated from a saline lake in China. Int. J. Syst. Evol. Microbiol. *59*: 555–558.

Tang, S.K., Y. Wang, J.Y. Wu, L.L. Cao, K. Lou, L.H. Xu, C.L. Jiang and W.J. Li. 2009b. *Saccharopolyspora qijiaojingensis* sp. nov., a halophilic actinomycete isolated from a salt lake. Int. J. Syst. Evol. Microbiol. *59*: 2166–2170.

te Poele, E.M., H. Kloosterman, G.I. Hessels, H. Bolhuis and L. Dijkhuizen. 2006. RepAM of the *Amycolatopsis methanolica* integrative element pMEA300 belongs to a novel class of replication initiator proteins. Microbiology *152*: 2943–2950.

te Poele, E.M., M.N. Habets, G.Y.A. Tan, A.C. Ward, M. Goodfellow, H. Bolhuis and L. Dijkhuizen. 2007. Prevalence and distribution of nucleotide sequences typical for pMEA-like accessory genetic elements in the genus *Amycolatopsis*. FEMS Microbiol. *61*: 285–294.

te Poele, E.M., M. Samborskyy, M. Oliynyk, P.F. Leadlay, H. Bolhuis and L. Dijkhuizen. 2008. Actinomycete integrative and conjugative pMEA-like elements of *Amycolatopsis* and *Saccharopolyspora* decoded. Plasmid *59*: 202–216.

Theriault, R.J., R.R. Rasmussen, W.L. Kohl, J.F. Prokop, T.B. Hutch and G.J. Barlow. 1986. Benzanthrins A and B, a new class of quinone antibiotics. I. Discovery, fermentation and antibacterial activity. J. Antibiot. (Tokyo) *39*: 1509–1514.

Thiemann, J.E., G. Zucco and G. Pelizza. 1969. A proposal for the transfer of *Streptomyces mediterranei* Margalith and Beretta 1960 to the genus *Nocardia* as *Nocardia mediterranea* (Margalith and Beretta) comb. nov. Arch. Mikrobiol. *67*: 147–155.

Thiemer, B., J.R. Andreesen and T. Schrader. 2001. The NADH-dependent reductase of a putative multicomponent tetrahydrofuran monooxygenase contains a covalently bound FAD. Eur. J. Biochem. *268*: 3774–3782.

Thiemer, B., J.R. Andreesen and T. Schrader. 2003. Cloning and characterization of a gene cluster involved in tetrahydrofuran degradation in *Pseudonocardia* sp. strain K1. Arch. Microbiol. *179*: 266–277.

Thompson, C.J., R.H. Skinner, J. Thompson, J.M. Ward, D.A. Hopwood and E. Cundliffe. 1982. Biochemical characterization of resistance determinants cloned from antibiotic-producing streptomycetes. J. Bacteriol. *151*: 678–685.

Tian, X.P., X.Y. Zhi, Y.Q. Qiu, Y.Q. Zhang, S.K. Tang, L.H. Xu, S. Zhang and W.-J. Li. 2009. *Sciscionella marina* gen. nov., sp. nov., a marine actinomycete isolated from a sediment in the northern South China Sea. Int. J. Syst. Evol. Microbiol. *59*: 222–228.

Tominaga, Y. and Y. Tsujisaka. 1976. Purification and properties of two chitinases from *Streptomyces orientalis* which lyse *Rhizopus* cell wall. Agric. Biol. Chem. *40*: 2325–2333.

Tomita, K., Y. Uenoyama, E.I. Numata, T. Sasahira, Y. Hoshino, K.I. Fujisawa, H. Tsukiura and H. Kawaguchi. 1978. *Streptoalloteichus*, a new genus of the family *Actinoplanaceae*. J. Antibiot. *31*: 497–510.

Tomita, K., Y. Nakakita, Y. Hoshino, K. Numata and H. Kawaguchi. 1987. New genus of the *Actinomycetales*: *Streptoalloteichus hindustanus* gen. nov., nom. rev., sp. nov., nom. rev. Int. J. Syst. Bacteriol. *37*: 211–213.

Tomita, K. 1989. Genus *Streptoalloteichus*. *In* Bergey's Manual of Systematic Bacteriology, vol. 4 (edited by Williams, Sharpe and Holt). Williams & Wilkins, Baltimore, pp. 2569–2572.

Tomita, K., Y. Hoshino and T. Miyaki. 1993. *Kibdelosporangium albatum* sp. nov., producer of the antiviral antibiotics cycloviracins. Int. J. Syst. Bacteriol. *43*: 297–301.

Tonge, G.M. and I.J. Higgins. 1974. Microbial metabolism of alicyclic hydrocarbons. Growth of *Nocardia petroleophila* (NCIB 9438) on methylcyclohexane. J. Gen. Microbiol. *81*: 521–524.

Traxler, P., T. Schupp, H. Fuhrer and W.J. Richter. 1981. 3-Hydroxyrifamycin S and further novel ansamycins from a recombinant strain R-21 of *Nocardia mediterranei*. J. Antibiot. *34*: 971–979.

Treuhaft, M.W., J.G. Green, R. Arusel and A. Borge. 1980. Role of *Saccharomonospora viridis* in hypersensitivity pneumonitis. Am. Rev. Respir. Dis. *121*: 100.

Tseng, M., S.F. Yang, W.J. Li and C.L. Jiang. 2006. *Amycolatopsis taiwanensis* sp. nov., from soil. Int. J. Syst. Evol. Microbiol. *56*: 1811–1815.

Tseng, M., K.C. Hoang, M.K. Yang, S.F. Yang and W.S. Chu. 2007. Polyester-degrading thermophilic actinomycetes isolated from different environment in Taiwan. Biodegradation *18*: 579–583.

Tsuji, K., M. Kobayashi, T. Kamigauchi, Y. Yoshimura and T. Terui. 1988. New glycopeptides antibiotics. The structure of orienticins. J. Antibiot. *41*: 819–822.

Tsunakawa, M., O. Tenmyo, K. Tomita, N. Naruse, C. Kotake, T. Miyaki, M. Konishi and T. Oki. 1992. Quartromicin, a complex of novel antiviral antibiotics. I. Production, isolation, physico-chemical properties and antiviral activity. J. Antibiot. *45*: 180–188.

Tuan, J., J. Majer and L. Katz. 1986. Molecular cloning of a gene involved in the biosynthesis of erythromycin in *Streptomyces erythraeus*. H22:31. Presented at the Annual Meeting of the American Society for Microbiology.

Tuteja, D., M. Dua, R. Khanna, N. Dhingra, M. Khanna, H. Kaur, D.M. Saxena and R. Lal. 2000. The importance of homologous recombination in the generation of large deletions in hybrid plasmids in *Amycolatopsis mediterranei*. Plasmid *43*: 1–11.

Ueno, M., M. Iijima, T. Masuda, N. Kinoshita, H. Iinuma, H. Naganawa, M. Hamada, M. Ishizuka and T. Takeuchi. 1992. Dethymicin, a novel immunosupressant isolated from an *Amycolatopsis*. J. Antibiot. *45*: 1819–1826.

Umezawa, H., S. Gomi, Y. Yamagishi, T. Obata, T. Ikeda, M. Hamada and S. Kondo. 1987. 2″-*N*-Formimidoylsporaricin A produced by *Saccharopolyspora hirsuta* subsp. *kobensis*. J. Antibiot. (Tokyo) *40*: 91–93.

Unaogu, I.C., H.C. Gugnani and J. Lacey. 1994. Occurrence of thermophilic actinomycetes in natural substrates in Nigeria. Antonie van Leeuwenhoek *65*: 1–5.

Usui, T., Y. Hayashi, F. Nanjo, K. Sakai and Y. Ishido. 1987. Transglycosylation reaction of a chitinase purified from *Nocardia orientalis*. Biochim. Biophys. Acta *923*: 302–309.

Vainberg, S., K. McClay, H. Masuda, D. Root, C. Condee, G.J. Zylstra and R.J. Steffan. 2006. Biodegradation of ether pollutants by *Pseudonocardia* sp. strain ENV478. Appl. Environ. Microbiol. *72*: 5218–5224.

Vanden Boom, T.J. 2000. Recent developments in the molecular genetics of the erythromycin-producing organism *Saccharopolyspora erythraea*. *In* Adv. Appl. Microbiol., vol. 47. Academic Press, pp. 79–111.

Viana, A.M.M., A. Frezard, C. Malosse, T.M.C. Della Lucia, C. Errand and A. Lenoir. 2001. Colonial recognition of fungus in the fungus-growing ant *Acromyrmex subterranneus subterraneus* (Hypenoptera: Formicidae). Chemoecology *11*: 29–36.

Vrijbloed, J.W., J. Madon and L. Dijkhuizen. 1994. A plasmid from the methylotrophic actinomycete *Amycolatopsis methanolica* capable of site-specific integration. J. Bacteriol. *176*: 7087–7090.

Vrijbloed, J.W., M. Jelinkova, G.I. Hessels and L. Dijkhuizen. 1995a. Identification of the minimal replicon of plasmid pMEA300 of the methylotrophic actinomycete *Amycolatopsis methanolica*. Mol. Microbiol. *18*: 21–31.

Vrijbloed, J.W., J. Madon and L. Dijkhuizen. 1995b. Transformation of the methylotrophic actinomycete *Amycolatopis methanolica* with plasmid DNA: stimulatory effect of a pMEA300-encoded gene. Plasmid *34*: 96–104.

Vrijbloed, J.W., N.M. van der Put and L. Dijkhuizen. 1995c. Identification and functional analysis of the transfer region of plasmid pMEA300 of the methylotrophic actinomycete *Amycolatopsis methanolica*. J. Bacteriol. *177*: 6499–6505.

Vrijbloed, J.W., J. van Hylckama Vlieg, N.M. van der Put, G.I. Hessels and L. Dijkhuizen. 1995d. Molecular cloning with a pMEA300-derived shuttle vector and characterization of the *Amycolatopsis methanolica* prephenate dehydratase gene. J. Bacteriol. *177*: 6666–6669.

Vrijbloed, J.W. 1996. Functional analysis of the integrative plasmid pMEA 300 of the actinomycete *Amycolatopsis methanolica*. PhD thesis, University of Gröningen, Netherlands.

Waksman, S.A. 1923. Genus *Actinomyces*. *In* Bergey's Manual of Determinative Bacteriology, 1st edn (edited by Bergey, Harrison, Breed, Hammer and Huntoon). Williams & Wilkins, Baltimore, pp. 339–371.

Waksman, S.A. and A.T. Henrici. 1948. Family III. *Streptomycetaceae* Waksman and Henrici. *In* Bergey's Manual of Determinative Bacteriology, 6th edn (edited by Breed, Murray and Hitchens). Williams & Wilkins, Baltimore, pp. 929–980.

Waksman, S.A. and C.T. Cork. 1953. *Thermoactinomyces* Tsiklinksy, a genus of thermophilic actinomycetes. J. Bacteriol. *66*: 377–378.

Waksman, S.A. 1967. The Actinomycetes. A Summary of Current Knowledge. Ronald Press, New York.

Waldron, C., P. Matsushima, P.R. Rosteck, Jr, M.C. Broughton, J. Turner, K. Madduri, K.P. Crawford, D.J. Merlo and R.H. Baltz. 2001. Cloning and analysis of the spinosad biosynthetic gene cluster of *Saccharopolyspora spinosa*. Chem Biol. *8*: 487–499.

Wang, J., Y. Li, J. Bian, S.K. Tang, B. Ren, M. Chen, W.-J. Li and L.X. Zhang. 2010. *Prauserella marina* sp. nov., isolated from ocean sediment of the South China Sea. Int. J. Syst. Evol. Microbiol. *60*: 985–989.

Wang, N.J., Y. Fu, G.H. Yan, G.H. Bao, C.F. Xu and C.H. He. 1988. Isolation and structure of a new ansamycin antibiotic kanglemycin A from a *Nocardia* sp. J. Antibiot. *41*: 264–267.

Wang, N.J., B.L. Han, N. Yameshita and M. Sato. 1994. 31-Homorifamycin W, a novel metabolite from *Amycolatopsis mediterranei*. J. Antibiot. *47*: 613–615.

Wang, W., Z. Zhang, Q. Tang, J. Mao, D. Wei, Y. Huang, Z. Liu, Y. Shi and M. Goodfellow. 2007. *Lechevalieria xinjiangensis* sp. nov., a novel actinomycete isolated from radiation polluted soil in China. Int. J. Syst. Evol. Microbiol. *57*: 2819–2822.

Warwick, S., T. Bowen, H. McVeigh and T.M. Embley. 1994. A phylogenetic analysis of the family *Pseudonocardiaceae* and the genera *Actinokineospora* and *Saccharothrix* with 16S ribosomal RNA sequences and a proposal to combine the genera *Amycolata* and *Pseudonocardia* in an emended genus *Pseudonocardia*. Int. J. Syst. Bacteriol. *44*: 293–299.

Wax, R., W. Maises, R. Weston and J. Birnbaum. 1976. Efrotomycin, a new antibiotic from *Streptomyces lactamdurans*. J. Antibiot. *29*: 670–673.

Weber, J.M., C.K. Wierman and C.R. Hutchinson. 1985. Genetic analysis of erythromycin production in *Streptomyces erythreus*. J. Bacteriol. *164*: 425–433.

Weber, N.A. 1966. Fungus-growing ants. Science *153*: 587–604.

Weber, N.A. 1972. Gardening Ants: the Attines. American Philosophical Society, Philadelphia.

Weissman, K.J., H. Hong, M. Oliynyk, A.P. Siskos and P.F. Leadlay. 2004. Identification of a phosphopantetheinyl transferase for erythromycin biosynthesis in *Saccharopolyspora erythraea*. Chembiochem *5*: 116–125.

Wenzel, F.J., R.L. Gray, R.C. Roberts and D.A. Emanuel. 1974. Serologic studies in farmer's lung. Preciptins to the thermophilic actinomycetes. Am. Rev. Respir. Dis. *109*: 464–468.

Whaley, H.A., C.G. Chidester, S.A. Mizsak and R.J. Wnuk. 1980. Nodusmicin: the structure of a new antibiotic. Tetrahedr. Lett. *21*: 3659–3662.

Wilkie, B.N. 1978. Bovine allergic pneumonitis: an acute outbreak associated with mouldy hay. Can. J. Comp. Med. *42*: 10–15.

Williams, S.T., G.P. Sharples and R.M. Bradshaw. 1973. The fine structure of the *Actinomycetales*. *In Actinomycetales*: Characteristics and Practical Importance (edited by Sykes and Skinner). Academic Press, London, pp. 113–130.

Williams, S.T., G.P. Sharples, J.A. Serrano, A.A. Serrano and J. Lacey. 1976. The micromorphology and fine structure of nocardioform organisms. *In* The Biology of Nocardiae (edited by Goodfellow, Brownell and Serrano). Academic Press, London, pp. 102–104.

Williams, S.T., M. Goodfellow, G. Alderson, E.M. Wellington, P.H. Sneath and M.J. Sackin. 1983. Numerical classification of *Streptomyces* and related genera. J. Gen. Microbiol. *129*: 1743–1813.

Wink, J., J. Gandhi, R.M. Kroppenstedt, G. Seibert, B. Straubler, P. Schumann and E. Stackebrandt. 2004. *Amycolatopsis decaplanina* sp. nov., a novel member of the genus with unusual morphology. Int. J. Syst. Evol. Microbiol. *54*: 235–239.

Wink, J.M., R.M. Kroppenstedt, B.N. Ganguli, S.R. Nadkarni, P. Schumann, G. Seibert and E. Stackebrandt. 2003a. *In* Validation of publication of new names and new combinations previously effectively published outside the IJSEM. List no. 92. Int. J. Syst. Evol. Microbiol. *53*: 935–937.

Wink, J.M., R.M. Kroppenstedt, B.N. Ganguli, S.R. Nadkarni, P. Schumann, G. Seibert and E. Stackebrandt. 2003b. Three new antibiotic producing species of the genus *Amycolatopsis*, *Amycolatopsis balhimycina* sp. nov., *A. tolypomycina* sp. nov., *A. vancoresmycina* sp. nov., and description of *Amycolatopsis keratiniphila* subsp. *keratiniphila* subsp. nov. and *A. keratiniphila* subsp. *nogabecina* subsp. nov. Syst. Appl. Microbiol. *26*: 38–46.

Wood, S.A., B.M. Kirby, C.M. Goodwin, M. Le Roes and P.R. Meyers. 2007. PCR screening reveals unexpected antibiotic biosynthetic potential in *Amycolatopsis* sp. strain UM16. J. Appl. Microbiol. *102*: 245–253.

Xie, Q., Y. Wang, Y. Huang, Y. Wu, F. Ba and Z. Liu. 2002. Description of *Lentzea flaviverrucosa* sp. nov. and transfer of the type strain of *Saccharothrix aerocolonigenes* subsp. *staurosporea* to *Lentzea albida*. Int. J. Syst. Evol. Microbiol. *52*: 1815–1820.

Xu, J., E. Wan, C.J. Kim, H.G. Floss and T. Mahmud. 2005. Identification of tailoring genes involved in the modification of the polyketide backbone of rifamycin B by *Amycolatopsis mediterranei* S699. Microbiology *151*: 2515–2528.

Xu, L.H., X. Jin, P.H. Mao, Z.F. Lu, X.L. Cui and C.L. Jiang. 1999. Three new species of the genus *Actinobispora* of the family *Pseudonocardiaceae*, *Actinobispora alaniniphila* sp. nov., *Actinobispora aurantiaca* sp. nov. and *Actinobispora xinjiangensis* sp. nov. Int. J. Syst. Bacteriol. *49*: 881–886.

Yamaki, T., T. Oikawa, K. Ito and J. Nakamura. 1997. Cloning and sequencing of a nitrile hydratase gene from *Pseudonocardia thermophila* JCM 3095. J. Ferment. Bioeng. *5*: 474–477.

Yamamoto, H., K.H. Maurer and C.R. Hutchinson. 1986. Transformation of *Streptomyces erythraeus*. J. Antibiot. (Tokyo) *39*: 1304–1313.

Yamashita, M., A. Tani and F. Kawai. 2004. A new ether bond-splitting enzyme found in Gram-positive polyethylene glycol 6000-utilizing bacterium, *Pseudonocardia* sp. strain K1. Appl. Microbiol. Biotechnol. *66*: 174–179.

Yassin, A.F., K.P. Schaal, H. Brzezinka, M. Goodfellow and G. Pulverer. 1991. Menaquinone patterns of *Amycolatopsis* species. Zentralbl. Bakteriol. *274*: 465–470.

Yassin, A.F., B. Haggenei, H. Budzikiewicz and K.P. Schaal. 1993. Fatty-acid and polar lipid-composition of the genus *Amycolatopsis* - application of fast-atom-bombardment mass- spectrometry to structure-analysis of underivatized phospholipids. Int. J. Syst. Bacteriol. *43*: 414–420.

Yassin, A.F., F.A. Rainey, H. Brzezinka, K.D. Jahnke, H. Weissbrodt, H. Budzikiewicz, E. Stackebrandt and K.P. Schaal. 1995. *Lentzea* gen. nov., a new genus of the order *Actinomycetales*. Int. J. Syst. Bacteriol. *45*: 357–363.

Yassin, A.F. 2009. *Saccharopolyspora rosea* sp. nov., isolated from a patient with bronchial carcinoma. Int. J. Syst. Evol. Microbiol. *59*: 1148–1152.

Yoon, J.H., S.T. Lee, S.B. Kim, M. Goodfellow and Y.H. Park. 1997. Inter- and intraspecific genetic analysis of the genus *Saccharomonospora* with 16S to 23S ribosomal DNA (rDNA) and 23S to 5S rDNA internally transcribed spacer sequences. Int. J. Syst. Bacteriol. *47*: 661–669.

Yoon, J.H., S.B. Kim, S.T. Lee and Y.H. Park. 1999. DNA–DNA relatedness between *Saccharomonospora* species: '*Saccharomonospora caesia*' as a synonym of *Saccharomonospora azurea*. Int. J. Syst. Bacteriol. *49*: 671–673.

Yoon, J.H., K.C. Lee, Y.K. Shin and Y.H. Park. 2000. Transfer of '*Thermoactinomyces glaucus*' IFO 12530 and '*Thermoactinomyces monosporus*' IFO 14050 to the genus *Saccharomonospora* as members of *Saccharomonospora glauca*. J. Gen. Appl. Microbiol. *46*: 251–256.

Yoon, J.H. and Y.H. Park. 2000. Phylogenetic analysis of the genus *Thermoactinomyces* based on 16S rDNA sequences. Int. J. Syst. Evol. Microbiol. *50*: 1081–1086.

Yoshida, K., A. Sasaki and H. Inoue. 1971. An anionic trypsin-like enzyme from *Streptomyces erythraeus*. FEBS Lett *15*: 129–132.

Yu, T.W., R. Muller, M. Muller, X. Zhang, G. Draeger, C.G. Kim, E. Leistner and H.G. Floss. 2001. Mutational analysis and reconstituted expression of the biosynthetic genes involved in the formation of 3-amino-5-hydroxybenzoic acid, the starter unit of rifamycin biosynthesis in *Amycolatopsis mediterranei* S699. J. Biol. Chem. *276*: 12546–12555.

Yuan, L.J., Y.Q. Zhang, Y. Guan, Y.Z. Wei, Q.P. Li, L.Y. Yu, W.J. Li and Y.Q. Zhang. 2008. *Saccharopolyspora antimicrobica* sp. nov., an actinomycete from soil. Int. J. Syst. Evol. Microbiol. *58*: 1180–1185.

Zhang, H., W. Zheng, J. Huang, H. Luo, Y. Jin, W. Zhang, Z. Liu and Y. Huang. 2006. *Actinoalloteichus hymeniacidonis* sp. nov., an actinomycete isolated from the marine sponge *Hymeniacidon perleve*. Int. J. Syst. Evol. Microbiol. *56*: 2309–2312.

Zhang, H., W. Zhang, Y. Jin, M. Jin and X. Yu. 2008a. A comparative study on the phylogenetic diversity of culturable actinobacteria isolated from five marine sponge species. Antonie van Leeuwenhoek *93*: 241–248.

Zhang, J., Q. Xie, Z. Liu and M. Goodfellow. 2007a. *Lechevalieria fradiae* sp. nov., a novel actinomycete isolated from soil in China. Int. J. Syst. Evol. Microbiol. *57*: 832–836.

Zhang, J., D. Wu, J. Zhang, Z. Liu and F. Song. 2008b. *Saccharopolyspora shandongensis* sp. nov., isolated from wheat-field soil. Int. J. Syst. Evol. Microbiol. *58*: 1094–1099.

Zhang, J., D. Wu and Z. Liu. 2009. *Saccharopolyspora jiangxiensis* sp. nov., isolated from grass-field soil. Int. J. Syst. Evol. Microbiol. *59*: 1076–1081.

Zhang, M.M., M. Poulsen and C.R. Currie. 2007b. Symbiotic recognition of mutualistic bacteria by *Acromyrmex* leaf-cutting ants. ISME J. *1*: 313–320.

Zhao, W., Y. Zhong, H. Yuan, J. Wang, H. Zheng, Y. Wang, X. Cen, F. Xu, J. Bai, X. Han, G. Lu, Y. Zhu, Z. Shao, H. Yan, C. Li, N. Peng, Z. Zhang, Y. Zhang, W. Lin, Y. Fan, Z. Qin, Y. Hu, B. Zhu, S. Wang, X. Ding and G.P. Zhao. 2010. Complete genome sequence of the rifamycin SV-producing *Amycolatopsis mediterranei* U32 revealed its genetic characteristics in phylogeny and metabolism. Cell Res. *20*: 1096–1108.

Zhi, X.-Y., W.-J. Li and E. Stackebrandt. 2009. An update of the structure and 16S rRNA gene sequence-based definition of higher ranks of the class *Actinobacteria*, with the proposal of two new suborders and four new families and emended descriptions of the existing higher taxa. Int. J. Syst. Evol. Microbiol. *59*: 589–608.

Zhou, Z.H., Z.H. Liu, Y.D. Qian, S.B. Kim and M. Goodfellow. 1998. *Saccharopolyspora spinosporotrichia* sp. nov., a novel actinomycete from soil. Int. J. Syst. Bacteriol. *48*: 53–58.

Zhuge, B., H.Y. Fang, H. Yu, Z.M. Rao, W. Shen, J. Song and J. Zhuge. 2008. Bioconversion of lovastatin to a novel statin by *Amycolatopsis* sp. Appl. Microbiol. Biotechnol. *79*: 209–216.

Zimmermann, W., B. Winter and P. Broda. 1988. Xylanolytic enzyme-activities produced by mesophilic and thermophilic actinomycetes grown on graminaceous xylan and lignocellulose. FEMS Microbiol. Lett. *55*: 181–185.

Zitouni, A., L. Lamari, H. Boudjella, B. Badji, N. Sabaou, A. Gaouar, F. Mathieu, A. Lebrihi and D.P. Labeda. 2004. *Saccharothrix algeriensis* sp. nov., isolated from Saharan soil. Int. J. Syst. Evol. Microbiol. *54*: 1377–1381.

Order XIV. **Streptomycetales** ord. nov.

Peter Kämpfer

Strep.to.my.cet.al'es. N.L. masc. n. *Streptomyces* type genus of the order; suff. *-ales* ending to denote an order; N.L. fem. pl. n. *Streptomycetales* the *Streptomyces* order.

The order was created by elevation of suborder *Streptomycineae* Rainey et al. 1997.

The description is that of the type family *Streptomycetaceae.*

Type genus: **Streptomyces** Waksman and Henrici 1943, 339[AL].

References

Rainey, F.A., N.L. Ward-Rainey and E. Stackebrandt. 1997. *In* Stackebrandt, E., F.A. Rainey and N.L. Ward-Rainey. Proposal for a new hierarchic classification system, *Actinobacteria* classis nov. Int. J. Syst. Bacteriol. *47:* 479–491.

Waksman, S.A. and A.T. Henrici. 1943. The nomenclature and classification of the actinomycetes. J. Bacteriol. *46*: 337–341.

Family I. **Streptomycetaceae** Waksman and Henrici 1943, 339[AL] emend. Rainey, Ward-Rainey and Stackebrandt 1997, 486 emend. Kim, Lonsdale, Seong and Goodfellow 2003b, 113 emend. Zhi, Li and Stackebrandt 2009, 600

Peter Kämpfer

Strep.to.my.ce.ta.ce'a.e. N.L. masc. n. *Streptomyces* type genus of the family; L. suff. *-aceae* ending to denote a family; N.L. fem. pl. n. *Streptomycetaceae* the *Streptomyces* family.

Aerobic, Gram-stain-positive, non-acid–alcohol-fast actinomycetes that form an extensively branched substrate mycelium which rarely fragments. At maturity, the **aerial mycelium forms chains of three to many spores**. Members of a few species bear short chains of spores on the substrate mycelium. The organisms produce a **wide range of pigments**, which are responsible for the colors of the substrate and aerial mycelia. The organisms grow within different pH ranges and are chemo-organotrophic with an oxidative type of metabolism. **Walls of cells from the substrate mycelium contain either LL- or *meso*-diaminopimelic acid (A_2pm) as the predominant diamino acid; aerial or submerged spores contain LL-A_2pm (peptidoglycan type A3γ)**. Whole-organism **sugar profiles may contain major amounts of either galactose or galactose and rhamnose. Lipid profiles typically contain: hexa- and octahydrogenated menaquinones with nine isoprene units as the predominant isoprenologs; diphosphatidylglycerol, phosphatidylethanolamine, phosphatidylinositol, and phosphatidylinositol mannosides as major polar lipids; and complex mixtures of saturated, and iso- and anteiso-fatty acids. Mycolic acids are not present**. The pattern of 16S rRNA gene sequence signatures consists of nucleotides at positions: 127:234 (G–C), 449 (A), 672:734 (C–G), 950:1231 (U–G), 952:1229 (U–A), 955:1225 (C–G), 965 (C), 986:1219 (A–U), and 1362 (C). The family *Streptomycetaceae* belongs to the order *Streptomycetales*.

DNA G+C content (mol%): generally 66–74.

Type genus: **Streptomyces** Waksman and Henrici 1943, 339[AL].

Taxonomic comments

The family *Streptomycetaceae* was established in 1943 by Waksman and Henrici to accommodate actinomycetes with branched slender mycelia, rarely or not septate, spores on aerial hyphae, and not fragmenting into oidia (Waksman and Henrici, 1943). At that time, the description was mainly based on morphology. In the 8th edition of *Bergey's Manual of Determinative Bacteriology*, Pridham and Tresner (1974b) listed *Streptomyces*, *Streptoverticillium*, *Sporichthya*, and *Microellobosporia* as members of the family. Over the years, additional genera, like *Actinopycnidium*, *Actinosporangium*, *Chainia*, *Elytrosporangium*, *Kitasatoa*, and *Microellobosporia*, have been distinguished from *Streptomyces* by morphological criteria, but they have many phenotypic and genotypic characters in common with *Streptomyces* and have therefore been proposed as synonyms of this genus (Goodfellow et al., 1986a, 1986b, 1986c, 1986d, 1986e). The genus *Sporichthya* is now classified in the family *Sporichthyaceae* of the order *Frankiales* (Stackebrandt et al., 1997).

The genus *Streptoverticillium* was found to be distinguishable from *Streptomyces* by its verticillate sporophores, but also shared many characteristics with streptomycetes; numerical phenetic

(Kämpfer et al., 1991; Williams et al., 1983a) and rRNA–DNA similarities (Gladek et al., 1985; Witt and Stackebrandt, 1990) supported the proposal that the genus was a synonym of *Streptomyces* (Witt and Stackebrandt, 1990). Wellington et al. (1992) proposed the unification of *Kitasatospora* with *Streptomyces* on the basis of chemotaxonomic, biochemical, and 16S rRNA gene sequence similarities. However, Zhang et al. (1997) revived the genus mainly on the basis of the ratio of *meso*-A_2pm to LL-A_2pm in whole-cell hydrolysates. The *meso*-A_2pm content is 49–89% in *Kitasatospora* strains and 1–16% in *Streptomyces* strains. Furthermore, galactose is present in the whole-cell hydrolysates of *Kitasatospora* strains, but not in whole-cell hydrolysates of *Streptomyces* strains.

Kim et al. (2003b) proposed an additional member of the family, the genus *Streptacidiphilus*, which contained acidophilic actinomycetes isolated from acidic soils and litter. These actinomycetes grow over a pH range of 3.5–6.5, with optimum growth at pH 4.5–5.5.

The genera *Streptomyces*, *Kitasatospora*, and *Streptacidiphilus* are very difficult to differentiate on the basis of phenotypic features (including the few distinguishing chemotaxonomic markers given above). The few characteristic differentiating features are shown in Table 263. Although 16S rRNA gene sequence analyses have provided a framework for prokaryotic classification, the current classification based on this molecule has not clarified the taxonomic problems within the family *Streptomycetaceae* (Kämpfer, 2006). *Kitasatospora* and *Streptacidiphilus* form stable, separate sub-branches on the basis of phylogenetic analyses within the family *Streptomycetaceae*, but they are grouped within the large *Streptomyces* tree (Figure 339) and 16S rRNA gene sequence similarities are equally high to many *Streptomyces* species/groups and *Streptacidiphilus* species. On the basis of DNA–DNA microarray hybridizations, Hsiao and Kirby (2008) compared the genome content of *Streptomyces avermitilis*, "*Streptomyces cattleya*", "*Streptomyces maritimus*", and *Kitasatospora* (*Streptomyces*) *aureofaciens* with that of *Streptomyces coelicolor* A3(2) and found a very high agreement with the genome sequence data for *Streptomyces* and *Kitasatospora*.

For all these reasons, it is questionable whether a separate generic status for *Kitasatospora* and *Streptacidiphilus* is justified. In this edition of *Bergey's Manual*, these genera are cited as genera *incertae sedis*.

TABLE 263. Chemotaxonomic, morphological and physiological characteristics of *Kitasatospora*, *Streptacidiphilus*, and *Streptomyces* strains (according to Kim et al., 2003b)[a]

	Streptomyces	*Kitasatospora*	*Streptacidiphilus*
Long chains of spores formed on aerial hyphae	+	+	+
Optimal pH range	6.5–8.0[b]	nd	4.5–5.5
pH range for growth	5.0–11.5	5.5–9.0	3.5–6.0
A_2pm isomer(s) in whole-organism hydrolysates	LL-A_2pm	LL- and *meso*-A_2pm[c]	LL-A_2pm
Diagnostic sugars in whole-organism hydrolysates	None	Galactose[d]	Galactose, rhamnose
Predominant phospholipids[e]	DPG, PE, PI, PIMs	DPG, PE, PI, PIMs	DPG, PE, PI, PIMs
Major menaquiones[f]	MK-9(H_6,H_8)	MK-9(H_6,H_8)	MK-9(H_6,H_8)
Fatty acid pattern[g]	2c	2c	2c
DNA G+C content (mol%)	66–73	70–74	70–72

[a]Data taken from this and previous studies (Antony-Babu and Goodfellow, 2008; Lonsdale, 1985; Nakagaito et al., 1992a, 1992b; Ōmura et al., 1989; Shirling and Gottlieb, 1977; Williams et al., 1989).

[b]Alkaliphilic strains, which grow between pH 5.0 and 11.0, have an optimum at pH 9–9.5 (Mikami et al., 1982; Antony-Babu and Goodfellow, 2008).

[c]Aerial and submerged spores contain LL-A_2pm and vegetative mycelia *meso*-A_2pm.

[d]Rhamnose was detected in whole-organism hydrolysates of *Kitasatospora mediocidica* (Labeda, 1988).

[e]DPG, Diphosphatidylglycerol; PE, phosphatidylethanolamine; PI, phosphatidylinositol; PIMs, phosphatidylinositol mannosides.

[f]MK-9(H_6,H_8), hexa-and octa-hydrogenated menaquinones with nine isoprene units.

[g]Fatty acid group *sensu* Kroppenstedt (1985).

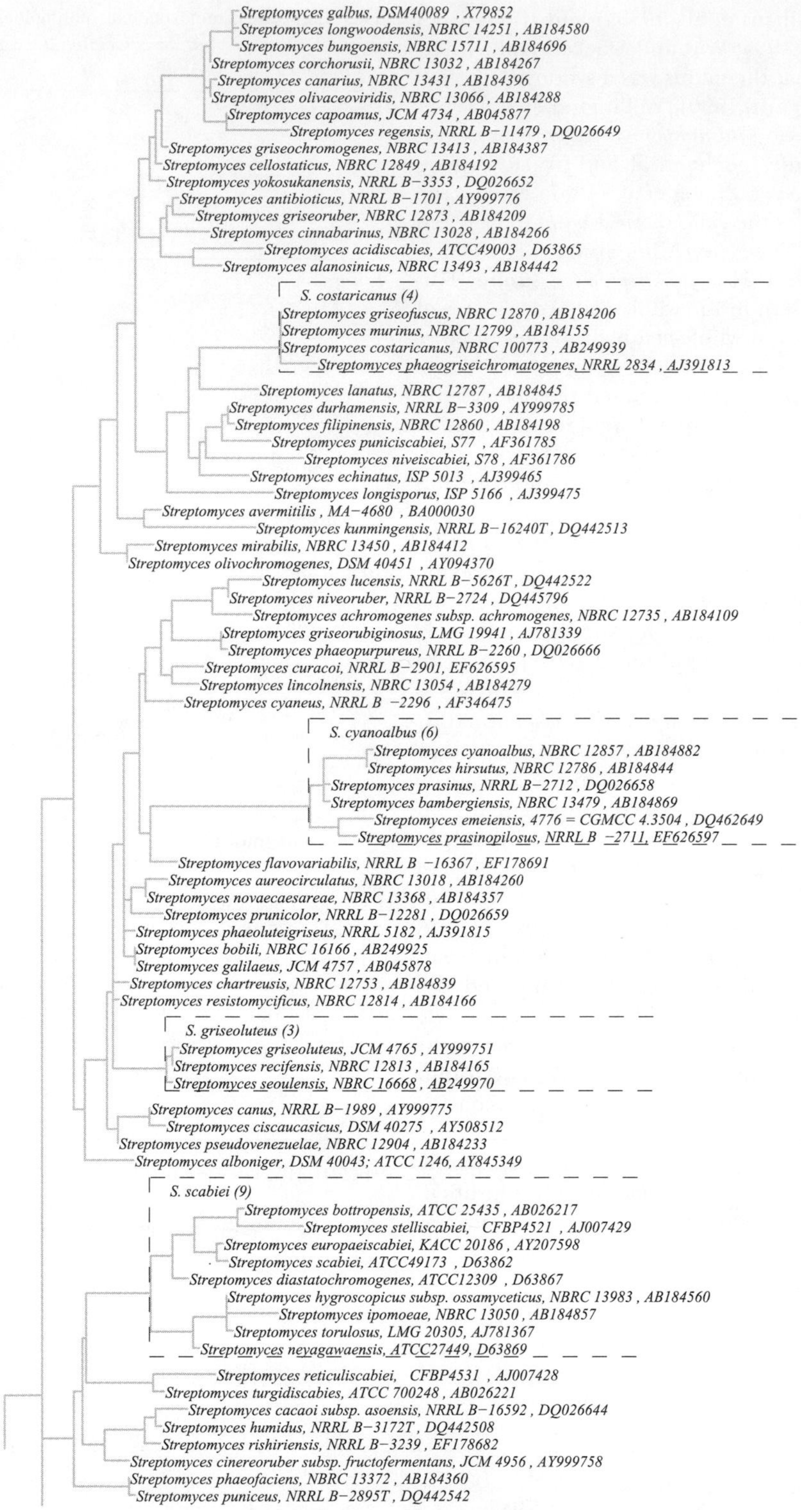

FIGURE 339. Phylogenetic analysis based on 16S rRNA gene sequences available from the EMBL database (accession nos are given in parentheses). The phylogenetic tree was constructed using the ARB software package (version December 2007; Ludwig et al., 2004) and the corresponding SILVA SSURef 95 database (version July 2008; Pruesse et al., 2007). Tree building was performed with all *Streptomycetaceae* sequences available in the SILVA database using the maximum-likelihood method with fastDNAml (Olsen et al., 1994) and an *Actionbacteria* conservatory filter. All strains in the figure are type strains.

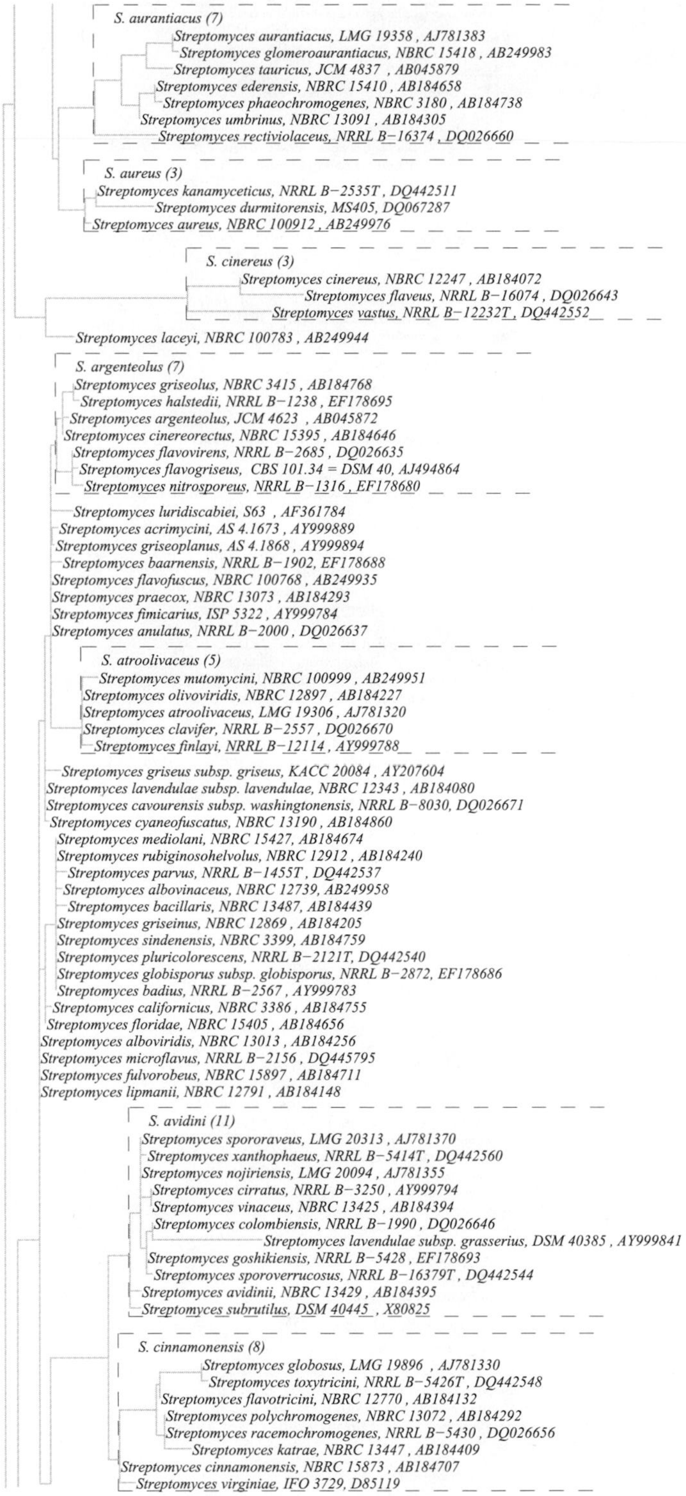

FIGURE 339. (continued)

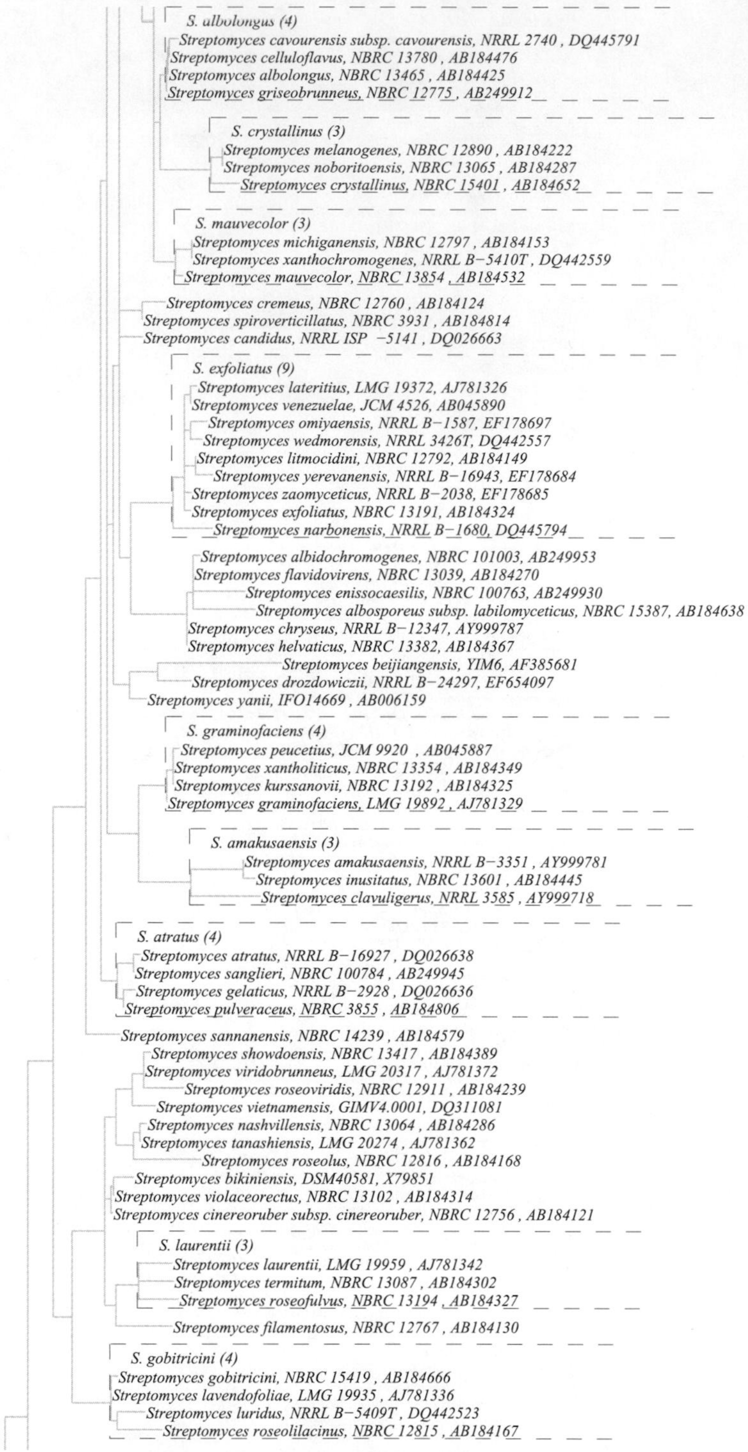
S. albolongus (4)
Streptomyces cavourensis subsp. cavourensis, NRRL 2740 , DQ445791
Streptomyces celluloflavus, NBRC 13780 , AB184476
Streptomyces albolongus, NBRC 13465 , AB184425
Streptomyces griseobrunneus, NBRC 12775 , AB249912
S. crystallinus (3)
Streptomyces melanogenes, NBRC 12890 , AB184222
Streptomyces noboritoensis, NBRC 13065 , AB184287
Streptomyces crystallinus, NBRC 15401 , AB184652
S. mauvecolor (3)
Streptomyces michiganensis, NBRC 12797 , AB184153
Streptomyces xanthochromogenes, NRRL B−5410T , DQ442559
Streptomyces mauvecolor, NBRC 13854 , AB184532
Streptomyces cremeus, NBRC 12760 , AB184124
Streptomyces spiroverticillatus, NBRC 3931 , AB184814
Streptomyces candidus, NRRL ISP −5141 , DQ026663
S. exfoliatus (9)
Streptomyces lateritius, LMG 19372, AJ781326
Streptomyces venezuelae, JCM 4526, AB045890
Streptomyces omiyaensis, NRRL B−1587, EF178697
Streptomyces wedmorensis, NRRL 3426T, DQ442557
Streptomyces litmocidini, NBRC 12792, AB184149
Streptomyces yerevanensis, NRRL B−16943, EF178684
Streptomyces zaomyceticus, NRRL B−2038, EF178685
Streptomyces exfoliatus, NBRC 13191, AB184324
Streptomyces narbonensis, NRRL B−1680, DQ445794
Streptomyces albidochromogenes, NBRC 101003, AB249953
Streptomyces flavidovirens, NBRC 13039, AB184270
Streptomyces enissocaesilis, NBRC 100763, AB249930
Streptomyces albosporeus subsp. labilomyceticus, NBRC 15387, AB184638
Streptomyces chryseus, NRRL B−12347, AY999787
Streptomyces helvaticus, NBRC 13382, AB184367
Streptomyces beijiangensis, YIM6, AF385681
Streptomyces drozdowiczii, NRRL B−24297, EF654097
Streptomyces yanii, IFO14669 , AB006159
S. graminofaciens (4)
Streptomyces peucetius, JCM 9920 , AB045887
Streptomyces xantholiticus, NBRC 13354 , AB184349
Streptomyces kurssanovii, NBRC 13192 , AB184325
Streptomyces graminofaciens, LMG 19892 , AJ781329
S. amakusaensis (3)
Streptomyces amakusaensis, NRRL B−3351 , AY999781
Streptomyces inusitatus, NBRC 13601 , AB184445
Streptomyces clavuligerus, NRRL 3585 , AY999718
S. atratus (4)
Streptomyces atratus, NRRL B−16927 , DQ026638
Streptomyces sanglieri, NBRC 100784 , AB249945
Streptomyces gelaticus, NRRL B−2928 , DQ026636
Streptomyces pulveraceus, NBRC 3855 , AB184806
Streptomyces sannanensis, NBRC 14239 , AB184579
Streptomyces showdoensis, NBRC 13417 , AB184389
Streptomyces viridobrunneus, LMG 20317 , AJ781372
Streptomyces roseoviridis, NBRC 12911 , AB184239
Streptomyces vietnamensis, GIMV4.0001, DQ311081
Streptomyces nashvillensis, NBRC 13064 , AB184286
Streptomyces tanashiensis, LMG 20274 , AJ781362
Streptomyces roseolus, NBRC 12816 , AB184168
Streptomyces bikiniensis, DSM40581, X79851
Streptomyces violaceorectus, NBRC 13102 , AB184314
Streptomyces cinereoruber subsp. cinereoruber, NBRC 12756 , AB184121
S. laurentii (3)
Streptomyces laurentii, LMG 19959 , AJ781342
Streptomyces termitum, NBRC 13087 , AB184302
Streptomyces roseofulvus, NBRC 13194 , AB184327
Streptomyces filamentosus, NBRC 12767 , AB184130
S. gobitricini (4)
Streptomyces gobitricini, NBRC 15419 , AB184666
Streptomyces lavendofoliae, LMG 19935 , AJ781336
Streptomyces luridus, NRRL B−5409T , DQ442523
Streptomyces roseolilacinus, NBRC 12815 , AB184167

FIGURE 339. (continued)

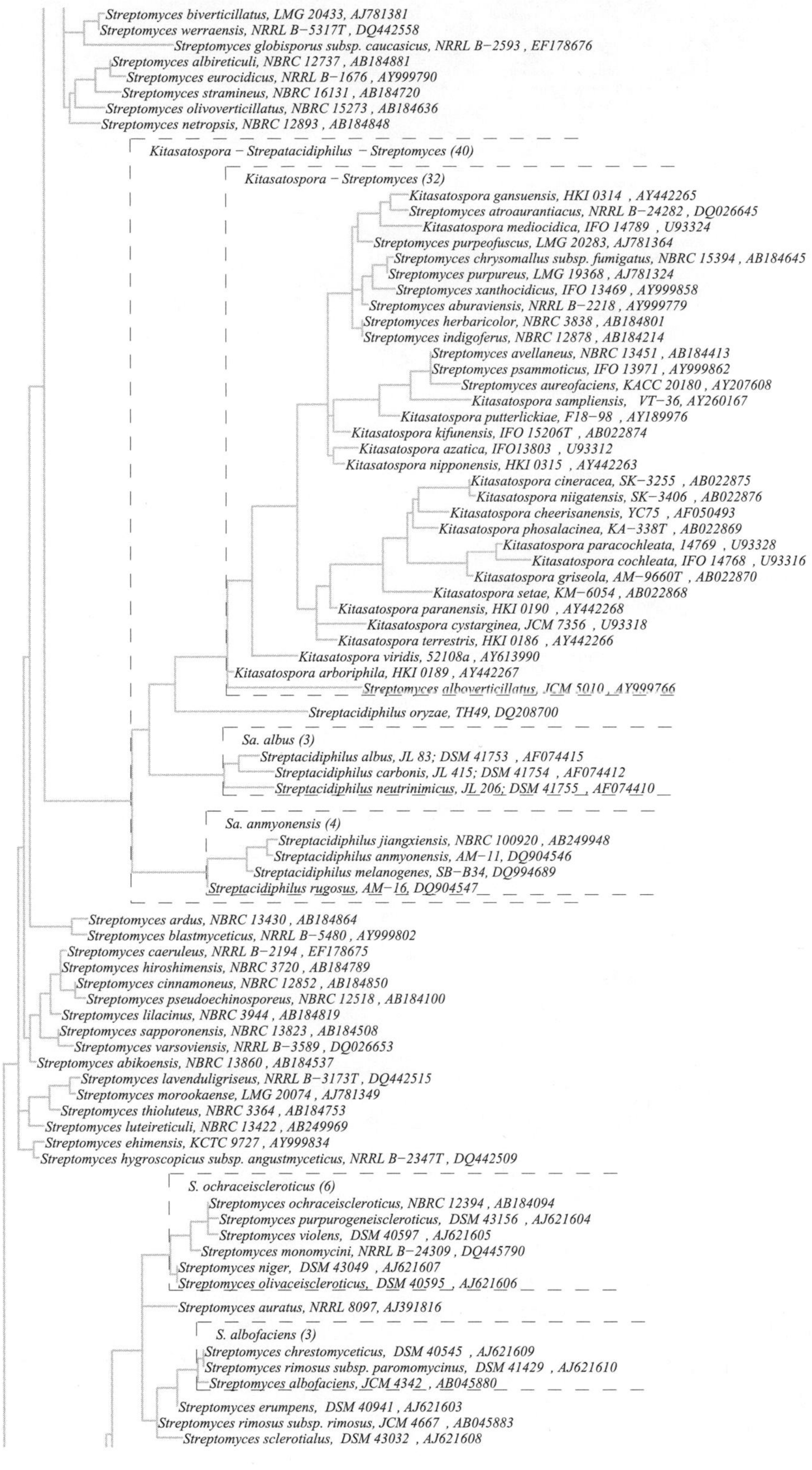

FIGURE 339. (continued)

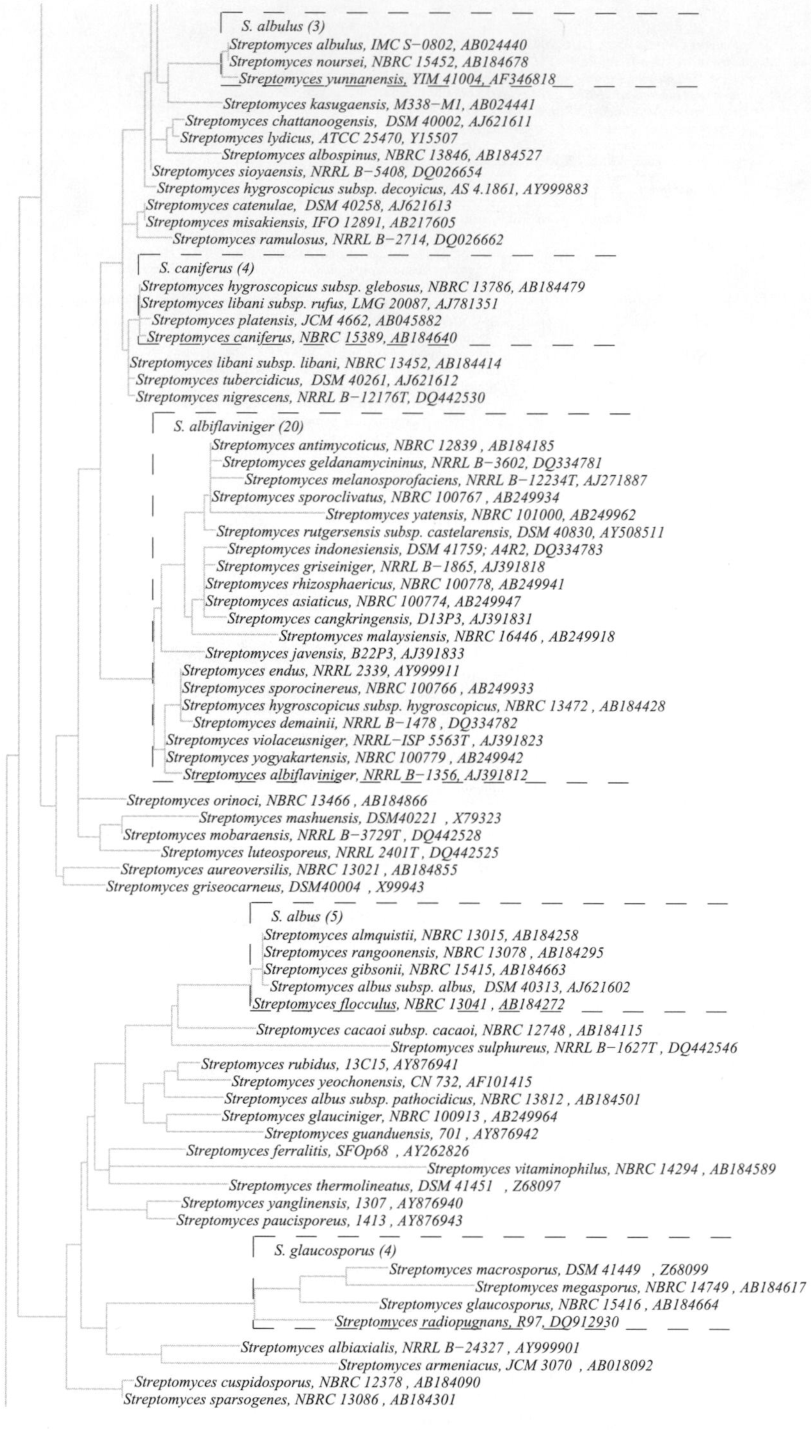

FIGURE 339. (continued)

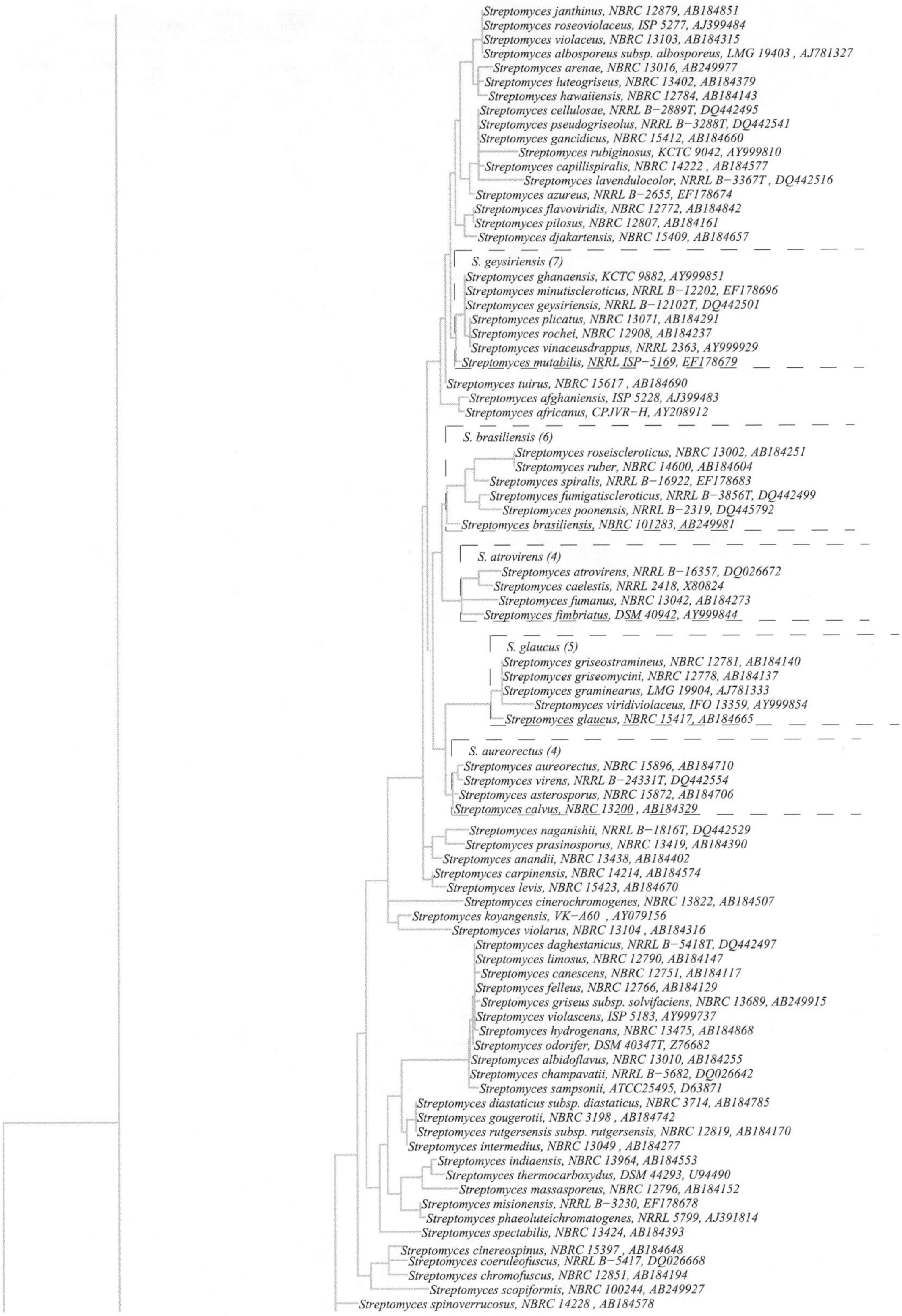

FIGURE 339. (continued)

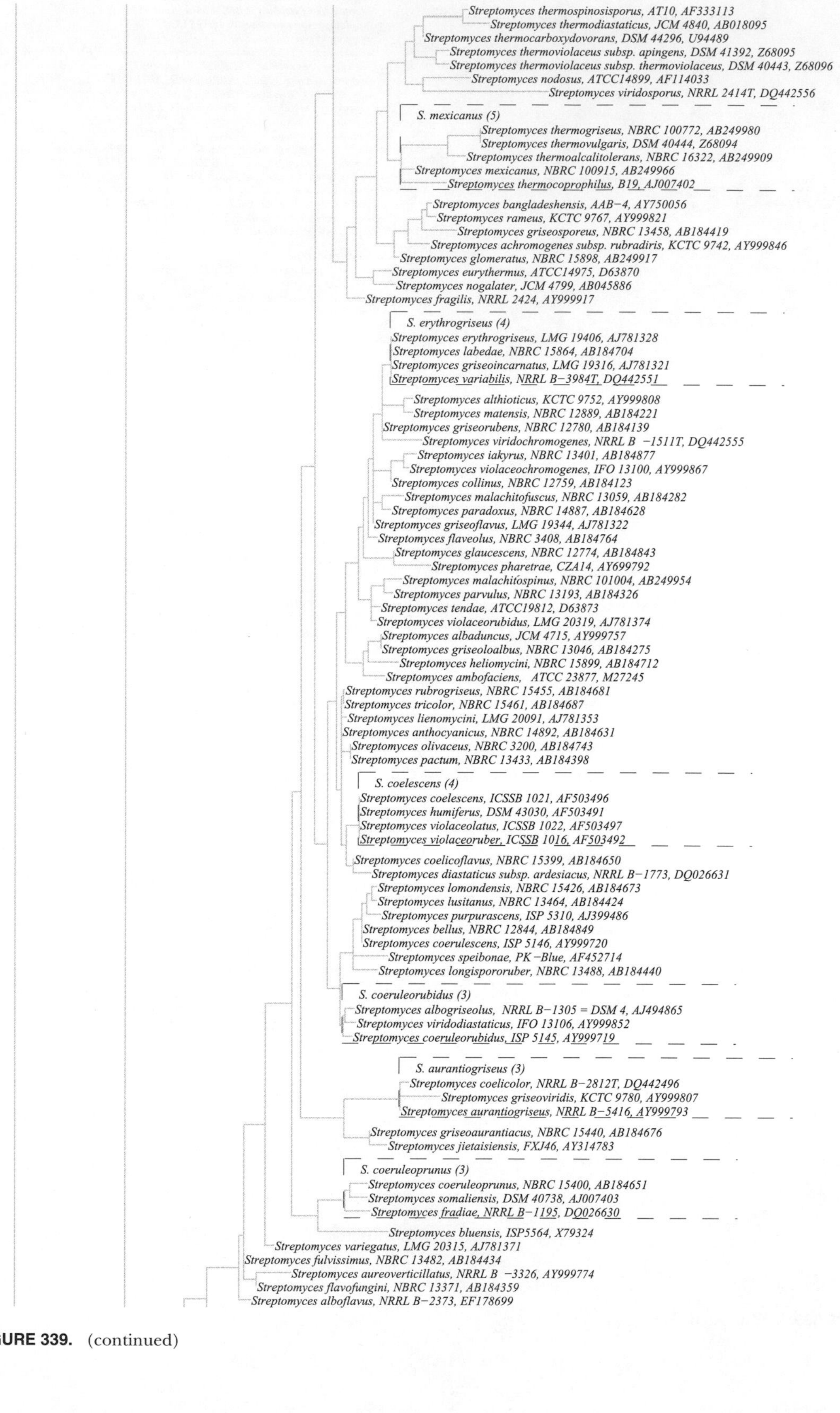

FIGURE 339. (continued)

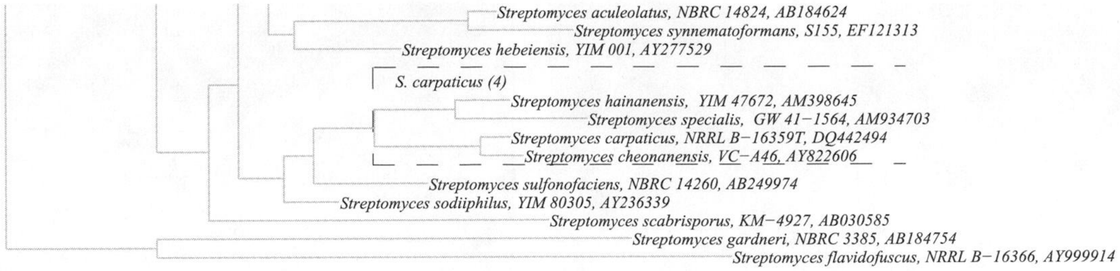

FIGURE 339. (continued)

Genus I. **Streptomyces** Waksman and Henrici 1943, 339[AL] emend. Witt and Stackebrandt 1990, 370 emend. Wellington, Stackebrandt, Sanders, Wolstrup and Jorgensen 1992, 159

Peter Kämpfer

Strep.to.my'ces. Gr. adj. *streptos* pliant, bent; Gr. masc. n. *mukês* fungus; N.L. masc. n. *Streptomyces* pliant or bent fungus.

Aerobic, Gram-stain-positive, non-acid-fast bacteria that form extensively branched substrate and aerial mycelia. Chemoorganotrophic, having an oxidative type of metabolism. The vegetative hyphae (0.5–2.0 µm in diameter) rarely fragment. **The aerial mycelium forms chains of three to many spores at maturity**. Some species show short chains of spores on the substrate mycelium and others form sclerotia, pycnidial-, sporangia-, and synnemata-like structures. The spores are nonmotile. **Colonies are discrete and lichenoid, leathery, or butyrous.** Often, colonies initially show a smooth surface, but later develop a weft of aerial mycelium that may appear floccose, granular, powdery, or velvety. **Can produce a wide variety of pigments responsible for the color of the vegetative and aerial mycelia. Colored diffusible pigments may also be formed.** Many strains are able to produce one or more antibiotic substances.

Catalase-positive. Generally reduce nitrates to nitrites and degrade polymeric substrates such as casein, gelatin, hypoxanthine, and starch, in addition to adenine and L-tyrosine. **Most species use a wide range of organic compounds as sole sources of carbon for energy and growth.** The temperature optimum for most species is in the range 25–35°C; some species, however, can grow at temperatures within the psychrophilic and thermophilic range; optimum pH range for growth is 6.5–8.0. The cell-wall peptidoglycan contains major amounts of **LL-A_2pm**. In some cases, low amounts of *meso*-A_2pm can be detected. **They lack mycolic acids, contain major amounts of saturated, iso- and anteiso-fatty acids, and typically possess either hexa- or octahydrogenated menaquinones with nine isoprene units as the predominant isoprenolog, although menaquinones with eight and ten isoprene units are also found. A complex polar lipid pattern is observed, containing typically diphosphatidylglycerol, phosphatidylethanolamine, phosphatidylinositol, and phosphatidylinositol mannosides. Widely distributed and abundant in soil, including composts.** A few species are pathogenic for animals and man, whereas some are phytopathogens.

DNA G+C content (mol%): 66–78 (T_m).

Type species: **Streptomyces albus** (Rossi Doria 1891) Waksman and Henrici 1943, 339[AL].

Further descriptive information

Morphology, fine structure, and life cycle. Early investigations of streptomycetes were dominated by a strong emphasis on morphology. The complex life cycle of streptomycetes (see below) offers three features for detailed microscopic characterization: (a) vegetative (substrate) mycelium (on solid and in liquid media); (b) aerial mycelium bearing chains of arthrospores (sometimes called "sporophores"); and (c) the arthrospores themselves (Kutzner, 1981). It is the last two categories that have provided most diagnostic information for taxonomists.

Early reports indicated that streptomycetes formed spore chains on the vegetative mycelium in both solid and liquid culture (e.g. Carvajal, 1947; Glauert and Hopwood, 1960; Tresner et al., 1967). However, it was never clarified whether these structures were analogous to the arthrospores formed on the aerial mycelium rather than deformations of the hyphae produced in staling cultures.

The fine structure and development of the aerial arthrospores have been studied extensively (Locci and Sharples, 1984). They are formed by septation and disarticulation of pre-existing hyphal elements within a thin fibrous sheath. The spore wall is formed, at least in part, from wall layers of the parent hypha; this is termed *holothallic development* (Locci and Sharples, 1984), and was found to be typical for many other sporo actinomycetes (Williams et al., 1973). The configuration of the spore chains (or sporophores) has played a very important role in species descriptions for many years. Often, the chains are long and contain more than 50 arthrospores. It is known that a number

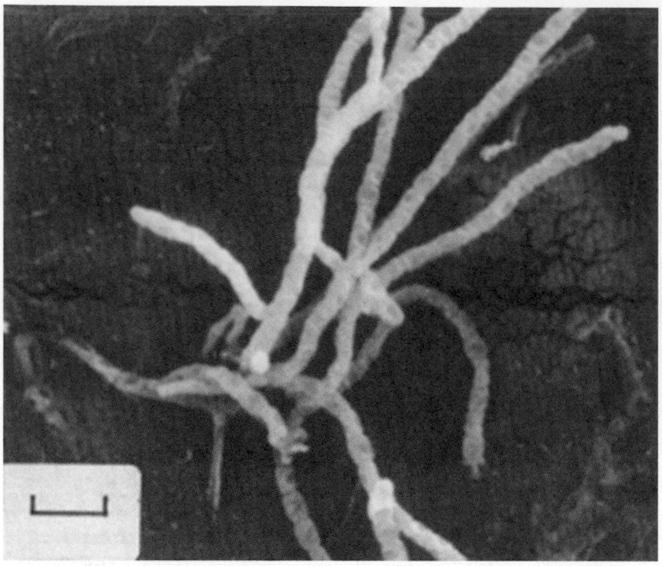

FIGURE 340. Straight to flexuous (*Rectiflexibles*) spore chains of *Streptomyces griseus*. Bar, 2 μm.

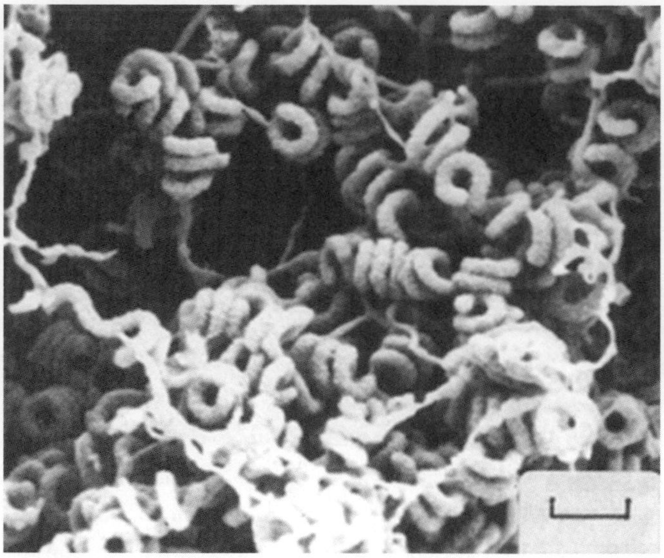

FIGURE 342. Spiral (*Spirales*) spore chains of *Streptomyces hygroscopicus*. Scanning electron microscopy. Bar, 5, μm.

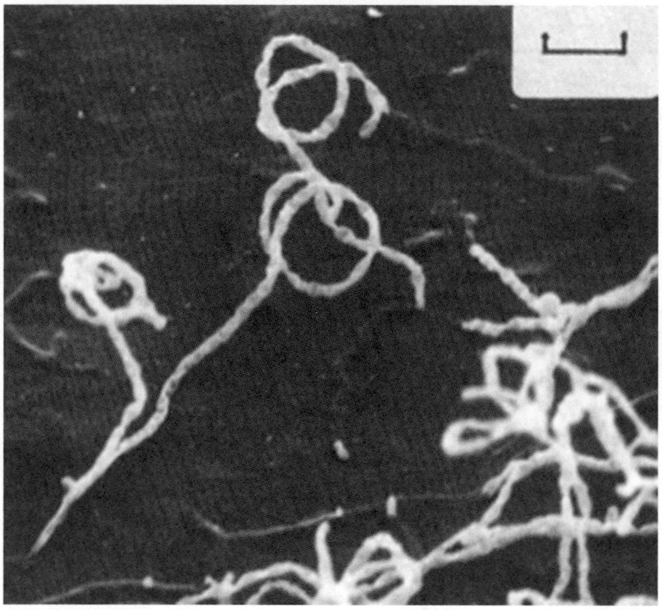

FIGURE 341. Looped (*Retinaculiaperti*) spore chains of *Streptomyces vinaceus*. Scanning electron microscopy. Bar, 5 μm.

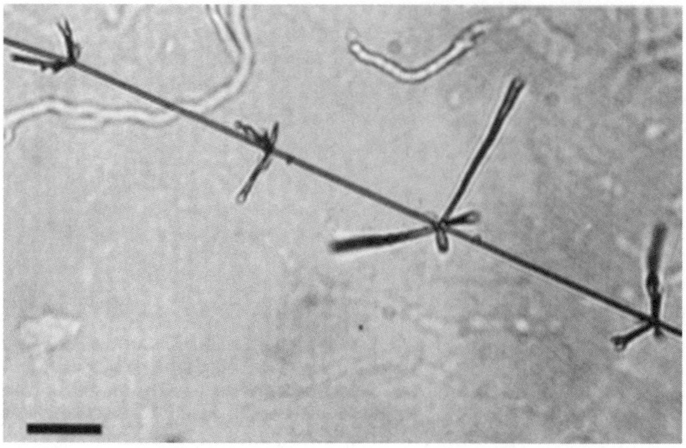

FIGURE 343. Aerial filament and initial verticils of *Streptoverticillium cinnamineum*. Light microscopy. Bar, 20 μm.

of different genes are involved in spore formation (Chater, 1979) and that different cultivation conditions can have an influence on spore formation. The range of spore chain morphologies is extensive and some workers have recognized many categories, for example, Ettlinger et al. (1958a) grouped strains into 15 morphological types. A simpler and practical scheme was proposed by Pridham et al. (1958) and adopted for the International *Streptomyces* Project (ISP; Shirling and Gottlieb, 1966). Three categories recognized were: (a) straight to flexuous (*Rectiflexibiles*) (Figure 340); (b) hooks, loops, or spirals with one to two turns (*Retinaculiaperti*) (Figure 341); and (c) spirals (*Spirales*) (Figure 342).

It should be noted, however, that even this simple system can pose problems, because it is not unusual that more than one category is seen in the same culture, and the distinction between *Retinaculiaperti* and *Spirales* is not always clear (Shirling and Gottlieb, 1977; Williams and Wellington, 1980). Aerial filaments can also differentiate into verticils (Figure 343, Figure 344, Figure 345, Figure 346, Figure 347, Figure 348, and Figure 349) as for most of species formerly grouped into the genus *Streptoverticillium* (Locci and Schofield, 1989).

Spore surface ornamentation has also been adopted as a taxonomic character. The ornaments, which are in fact borne on the spore sheath, can be grouped into the categories smooth, spiny, hairy, and warty. A further type, rugose, was proposed by Dietz and Mathews (1971) (Figure 350, Figure 351, Figure 352, Figure 353, and Figure 354). Spore surface ornamentation is a stable character, but the differences between smooth, warty, and rugose types can be difficult to observe. However, these problems can be resolved by using scanning electron microscopy.

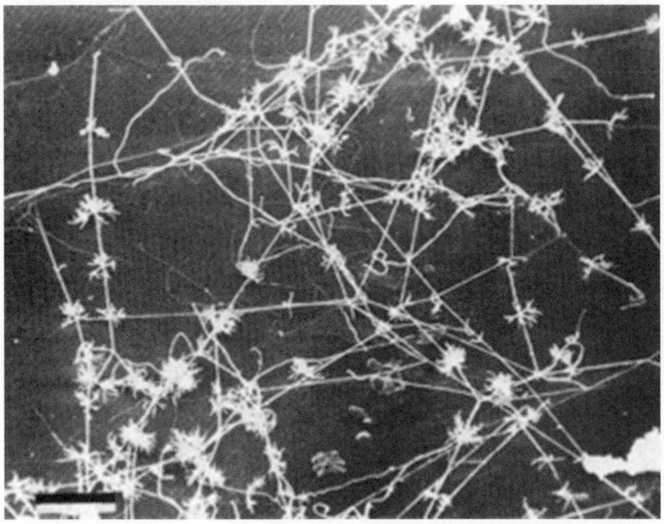

FIGURE 344. Aerial mycelium of *Streptoverticillium baldaccii*. Scanning electron microscopy. Bar, 50 μm.

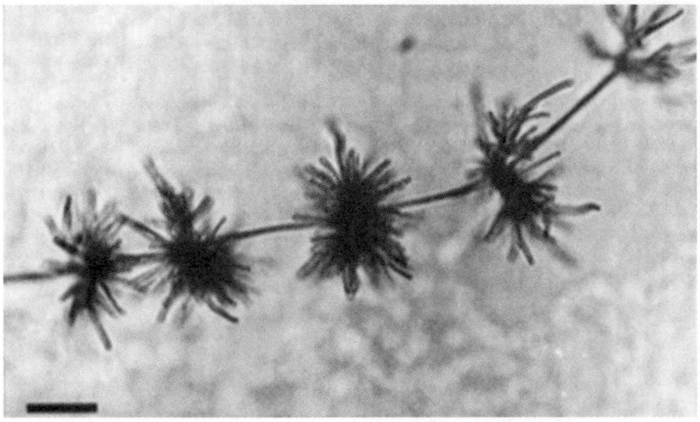

FIGURE 345. Sporulated aerial filament of *Streptoverticillium roseoverticillatum* subsp. *albosporum*. Light microscopy. Bar, 20 μm.

Normally, streptomycetes grow by tip extension as long, branching vegetative hyphae, which rarely have septae. Often, the compartments within the substrate hyphae contain numerous copies of the chromosomal DNA (Schrempf, 2006). The formation of aerial hyphae starts in response to nutrient depletion. The aerial structures can contain several different surface layers. The hydrophobic rodlet layer, which is one of them, comprises the proteins RdlA and RdlB. The corresponding genes have been studied in *Streptomyces coelicolor* M145 and "*Streptomyces lividans*" TK23. They are expressed within growing aerial hyphae, but not within spores (Claessen et al., 2002).

Comparisons of *Streptomyces coelicolor* A3(2) and *Streptomyces griseus* have shown that these strains share several orthologous genes that are important for the development of the aerial hyphae (Chater and Horinouchi, 2003). Some of these genes, e.g. *ramC* and *ramR*, are necessary for the production of the aerial hyphae, but are not essential for vegetative growth. The production of RamC requires additional developmental regulatory genes (*bldD*, *crpA*, and *ramR*, but not *bldN* and *bldM*; O'Connor et al., 2002). The expression of different *whi* (white)

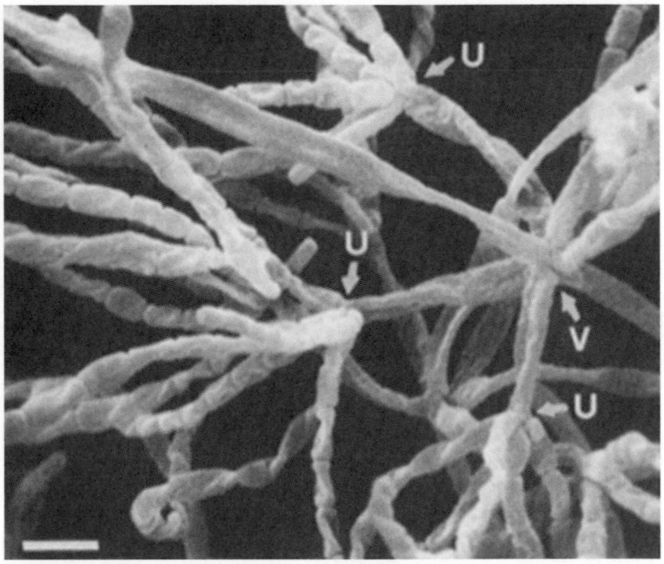

FIGURE 346. Verticil (V) and umbels (U) of spore chains of *Streptoverticillium kentuckense*. Scanning electron microscopy. Bar, 2 μm. (Reproduced with permission from Locci and Petrolini Baldan, 1971. Rivista di Patolgia Vegetale 7 *(Suppl.)*: 3–19.)

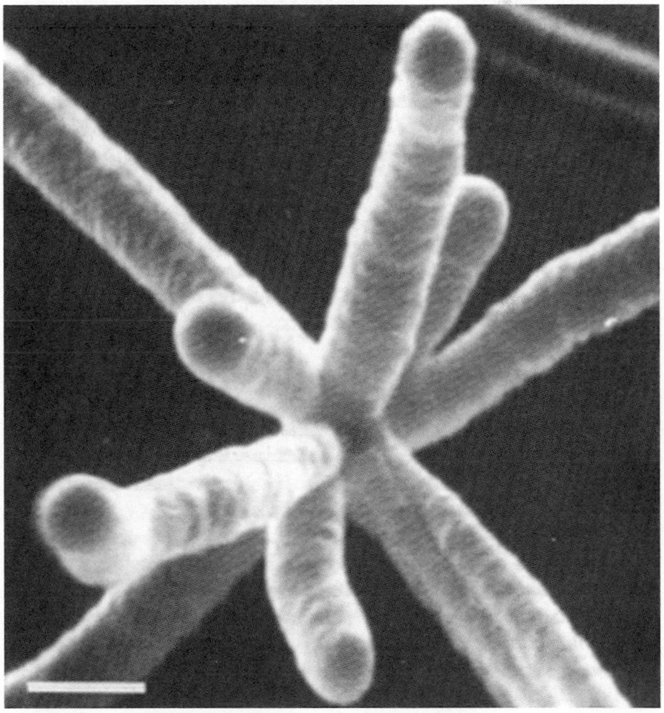

FIGURE 347. Verticil formation in *Streptoverticillium cinnamoneum*. Scanning electron microscopy. Bar, 1 μm. (Reproduced with permission from Locci and Petrolini Baldan, 1971. Rivista di Patolgia Vegetale 7 *(Suppl.)*: 3–19.)

genes is important for induction of curling of the aerial hyphae, their septation, and finally spore formation. In studies on *Streptomyces coelicolor* A3(2), it has been shown that some early regulatory *whi* genes (*A*, *B*, *G*, *H*, *I*, and *J*) are required for septation

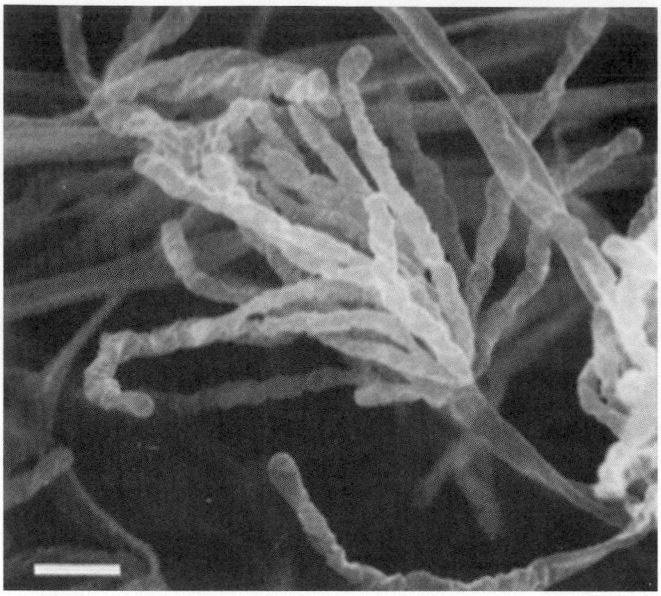

FIGURE 348. Mature umbel of spore chains of *Streptoverticillium baldaccii*. Scanning electron microscopy, Bar, 2 µm.

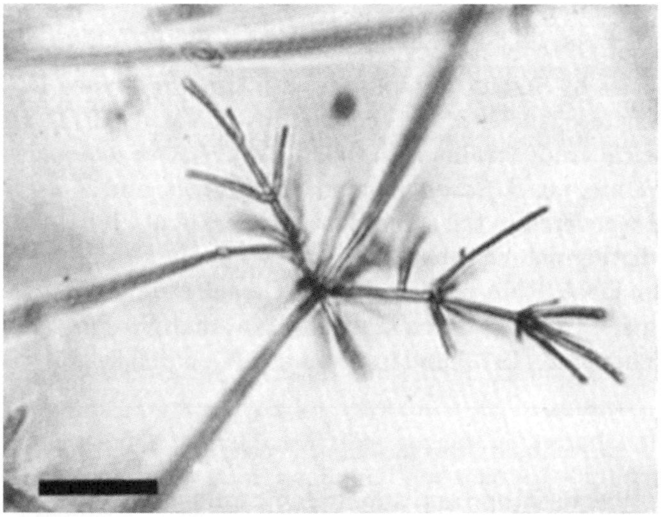

FIGURE 349. Biverticillate sporophore structure of *Streptoverticillium ehimense*. Light microscopy. Bar, 20 µm.

during sporulation. Furthermore, the *whiA* gene plays a major role in the extension process of the development of aerial hyphae towards septation (Ainsa et al., 2000).

Studies of the full genome of *Streptomyces coelicolor* A3(2) have shown that this strain encodes about 60 sigma (σ) factors. A subfamily of nine of these σ factors is involved in the late sporulation process and σ^H is involved in the development of aerial hyphae (in addition to responses to various stresses), whereas spore maturation is governed by σ^F (Schrempf, 2006). A recently published comparative genomic analysis (Chater and Chandra, 2006) gives a more detailed insight into this complex process. The partitioning of chromosomes is observed within the aerial hyphae; a comparatively synchronous septation leads

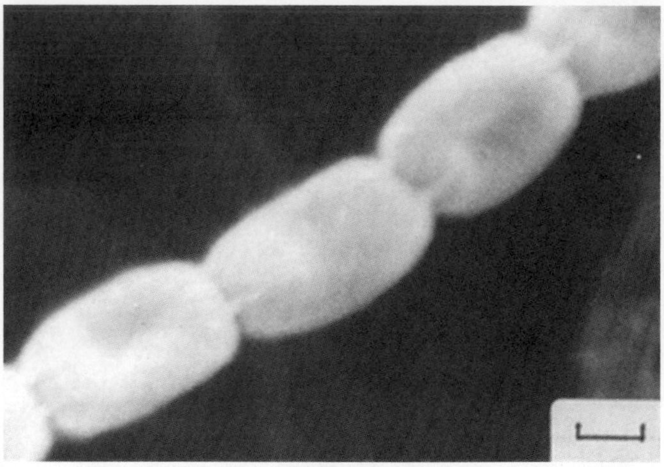

FIGURE 350. Smooth spores of *Streptomyces niveus*. Scanning electron microscopy. Bar, 0.25 µm.

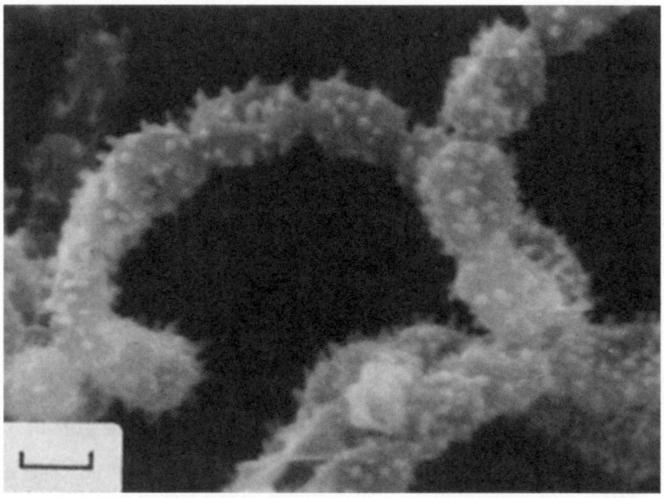

FIGURE 351. Spiny spores of *Streptomyces viridochromogenes*. Scanning electron microscopy. Bar, 0.5 µm.

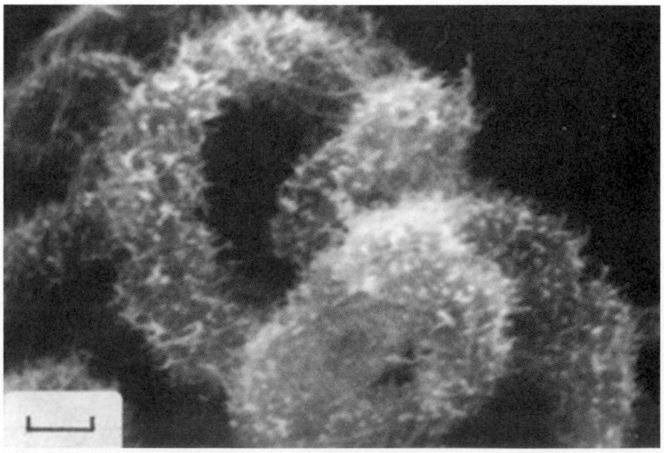

FIGURE 352. Hairy spores of *Streptomyces glaucescens*. Scanning electron microscopy. Bar, 0.5 µm.

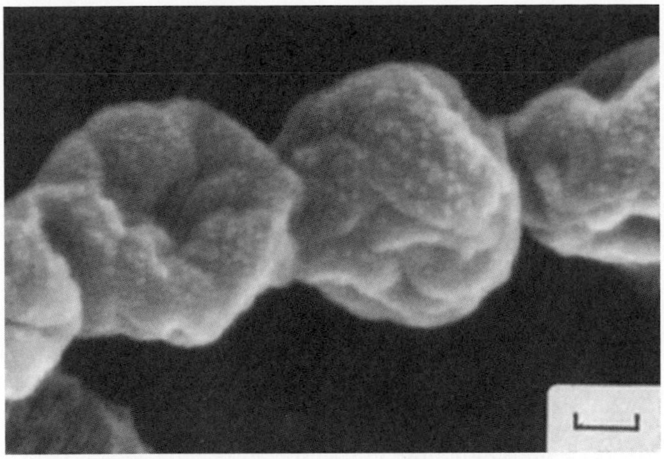

FIGURE 353. Warty spores of "*Streptomyces pulcher*". Scanning electron microscopy. Bar, 0.25 µm.

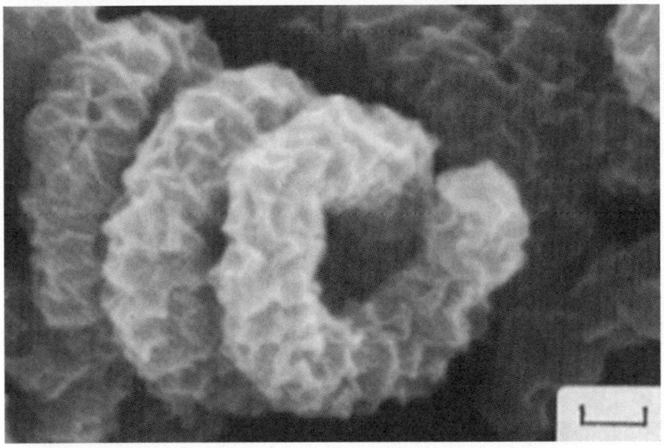

FIGURE 354. Rugose spores of *Streptomyces hygroscopicus*. Scanning electron microscopy. Bar, 0.5 µm.

to the compartment formation, which finally results in pore formation after maturation. Spores contain single chromosomes.

Cell division in *Streptomyces* is also a complex process. Normally, this is initiated by polymerization of the FtsZ protein on the inner surface of the cytoplasma membrane to form the Z-ring structure at the future division site (Lutkenhaus, 1997; Margolin, 2003). Studies on the transcription of genes in *Streptomyces griseus* suggest that *ftsZ* is expressed during both vegetative growth and sporulation (Dharmatilake and Kendrick, 1994). For *Streptomyces coelicolor* A3(2), it has been shown that FtsZ is required for septation within the vegetative substrate mycelium as well as for the synchronous formation of septae within the developing aerial hyphae prior to detectable partitioning of nucleoids (Grantcharova et al., 2003). The cycle of differentiation is usually observed on solid media, although for some strains [i.e. *Streptomyces griseus* (McCue et al., 1996); *Streptomyces coelicolor* A3(2) (Van Keulen et al., 2003)], sporulation in liquid culture is also reported. At the air interface of standing liquid cultures, gas vesicles may be present within the hyphae. For *Streptomyces coelicolor* A2(3), a gene cluster was reported that encoded proteins resembling gas vesicles of cyanobacteria and their homologs within halophilic archaea (Schrempf, 2006). For several *Streptomyces* strains, a transient slow down during growth in liquid culture is reported before entering the stationary phase. In this transition phase, an increase in ppGpp (guanosine 3′,5′-bispyrophosphate) and a decrease in GTP, as well as the activation of genes required for secondary metabolism, is observed. The synthesis of two ribosomal proteins is drastically reduced when the culture approaches the stationary phase (Blanco et al., 1994).

Cell-envelope composition.

Peptidoglycan. The ultrastructure and chemical composition of the cell walls of streptomycetes are typical for Gram-stain-positive bacteria (Schleifer and Kandler, 1972). Under the electron microscope, they appear as homogeneous electron-dense layers of about 16–35 nm. The cell walls are composed of peptidoglycan strands, which are multilayered. The heteropolymer peptidoglycan consists of heteropolysaccharide chains (the so-called "sugar backbone"), which are connected by peptide cross-links. The sugar backbone is composed of alternating β-1,4-linked units of the sugar derivatives *N*-acetylglucosamine and *N*-acetylmuramic acid. An oligopeptide of alternating D- and L-amino acids substitute the carboxyl group of the muramic acid (Schleifer and Kandler, 1972). In *Streptomyces*, the substitution is a tetrapeptide L-Ala–D-Glu–LL-A_2pm–D-Ala, which is cross-linked by a pentaglycine bridge extending from the C-terminal D-alanine of the peptide unit to the amino group located on the D carbon of LL-A_2pm. The resulting macromolecular structure forms the cell envelope. This LL-A_2pm–Gly_5 is also called the A3γ peptidoglycan type (Schleifer and Kandler, 1972) and is diagnostic for streptomycetes as well as some other combined-wall chemotype I actinomycetes (Lechevalier and Lechevalier, 1970a, 1970b, 1970c).

Lechevalier and coworkers used specific amino acids in purified cell walls to group aerobic actinomycetes into four so-called "wall chemotypes". Cell walls with *meso*-A_2pm and LL-A_2pm were the first to be detected. Cell-wall composition can vary with the developmental stage of streptomycetes. Takahashi et al. (1984) reported that submerged mycelium of strains having a cell wall with *meso*-A_2pm and LL-A_2pm consists of LL-A_2pm and glycine (wall chemotype I), whereas in the spores only *meso*-A_2pm could be detected (wall chemotype III according to Lechevalier and Lechevalier, 1970a, 1970b, 1970c). The quantitative distribution of cell-wall amino acids and cell-wall sugars differed in the cell-wall composition of aerial, substrate, and submerged mycelia of 11 streptomycetes. *N*-Acetylmuramic acid is found in the glycolyl type in the cell walls of *Streptomyces*, as in all other actinomycetes (Uchida and Aida, 1977).

Muramic acid phosphate residues are essential as attachment points to teichoic acids; the latter are polymeric substances containing repeating phosphodiester groups. They consist of polyols (i.e. the sugar alcohols glycerol and ribitol) or *N*-acetylamino sugars or both and are valuable for the identification of Gram-stain-positive bacteria. The structure of teichoic acids does not differ between streptomycetes and other Gram-stain-positive bacteria. The polymers consist either of ribitol phosphate or glycerol phosphate. Significant for the teichoic acids of actinomycetes is the absence of ester-bound D-alanine; instead, ester-linked acetic acid and sometimes succinic acid residues are present (Naumova et al., 1980).

The synthesis of either ribitol phosphate (e.g. *Streptomyces streptomycinii* and *Streptomyces violaceus*) or glycerol phosphate polymers (e.g. *Streptomyces antibioticus, Streptomyces levoris, Streptomyces rimosus,* and *Streptomyces thermovulgaris*) has been reported in streptomycetes (Naumova et al., 1980). In ribitol teichoic acids, positions 1 and 5 of ribitol are connected to the phosphates, but in glycerol teichoic acids, position 1 is commonly connected to 3, and, in other types, links to 2 (as in *Streptomyces antibioticus*) are uncommon. Polyol phosphates can be substituted with various combinations of sugars or amino sugars or both. The sugars or amino sugars are linked to glycerol or ribitol via glycosidic bonds. The role of teichoic acid in the taxonomy of *Streptomyces* is not clear as only a few strains have been analyzed in detail (Naumova et al., 1980).

Cell-wall polysaccharides. Cell-wall polysaccharides seem to be of no diagnostic value (Lechevalier et al., 1971) for strains that have LL-A_2pm in their cell wall. Occasionally, diagnostic sugars found in actinomycetes (e.g. arabinose, galactose, and xylose) have been reported in streptomycetes. The presence of diagnostic sugars in streptomycetes was extensively sought by Kroppenstedt (1977) who analyzed hundreds of strains. Glucose, mannose, and ribose are usually found in small amounts.

Phospho- and glycolipids. The lipids of streptomycetes consist mainly of diphosphatidylglycerol, phosphatidylethanolamine, phosphatidylinositol, and phosphatidylinositol mannosides. A summary of the lipid composition of actinomycetes can be found in Lechevalier et al. (1977). Glycolipids cannot be used for the identification of streptomycetes because they do not occur consistently and culture conditions largely determine their qualitative and quantitative lipid composition. Glycolipid content increases significantly under phosphate-limiting conditions.

Polar lipids have a significant taxonomic value in actinomycetes, as demonstrated by Lechevalier et al. (1977). The phospholipids of 97 actinomycete strains, representing 20 genera, were analyzed and assigned to five phospholipid types by Lechevalier et al. (1977). The phospholipid groups are characterized by the absence or presence of certain nitrogenous phospholipids. Members of the family *Streptomycetaceae* have phospholipid type II. The marker lipids of this type are phosphatidylethanolamine, methyl-phosphatidylethanolamine, hydroxy-phosphatidylethanolamine, and lyso-phosphatidylethanolamine, although differentiation can be made using additional lipids (e.g. phosphomonoester and hydroxy-phosphatidylethanolamine) and the presence or absence of phophatidylglycerol and phosphatidylinositol.

Menaquinones. Streptomycetes contain only menaquinones (Collins and Jones, 1981). The synthesized quinones have a partly saturated isoprenoid side chain at position 3 of the naphthoquinone ring. In this, streptomycetes resemble the majority of actinomycetes. Menaquinone composition has a great taxonomic value for the differentiation of actinomycetes. The following three variations are useful for classification and identification: 1) the different numbers of isoprene units; 2) the different degree of hydrogenation; 3) and the position of hydrogenated isoprene units (Figure 355, Table 264). The

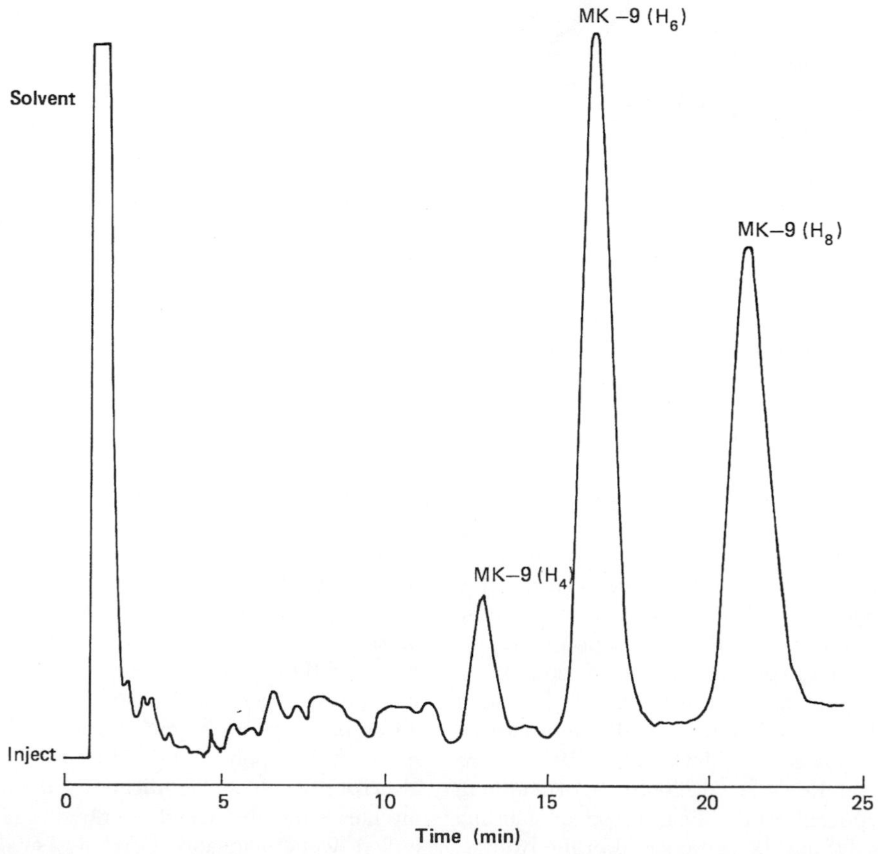

FIGURE 355. High-performance liquid chromatogram of menaquinones from *Streptomyces cyaneus* NCIB 9616.

TABLE 264. Key biochemical markers of genera assigned to the family *Streptomycetaceae* and to other families, which include strains producing an aerial mycelium (modified from Korn-Wendisch and Kutzner, 1992)[a]

Family and genus	A_2pm[b]	Glycine in IPB[b]	Peptidoglycan type[b]	Sugar type[b]	Phospholipid type[c]	Mycolic acids[b]	Fatty acid pattern[d]	Menaquinones[e]	DNA G+C content (mol%)
Streptomycetaceae:									
Streptomyces	LL	+	A3γ	–	PII	–	2c	$9(H_6)/(H_8)$	69–78
Kitasatospora	LL/*meso*	+	A3γ	C/E	PII	–	2c	$9(H_6)/(H_8)$	66–73
Streptacidiphilus	LL	+	A3γ	E	PII	–	2c	$9(H_6)/(H_8)$	70–72
Pseudonocardiaceae:									
Amycolatopsis	*meso*	–	A1γ	A	PII	–	3f	$9(H_4)(H_2)$	66–69
Kibdelosporangium[f]	*meso*	–	A1γ	A	PII	–	3f	nd	66
Pseudonocardia	*meso*	–	A1γ	A	PIII	–	2f	$8(H_4)$	79
Saccharopolyspora	*meso*	–	A1γ	A	PIII	–	2c/3e	$9(H_4)/10(H_4)/9(H_2)$	70–72
Saccharomonospora	*meso*	–	A1γ	A	PII	–	2a	$9(H_4)/8(H_4)$	69–74
Actinopolyspora	*meso*	–	A1γ	A	PIII	–	2c	$9(H_6)/9(H_4)$	64
Genera belonging to different families of the *Actinobacteria*[j]:									
Sporichthya	LL	+	A3γ	–	nd	–	3a	$9(H_6)/9(H_8)$	nd
Kineosporia[g]	LL/*meso*	(+)	nd	C	PIII	–	1	$9(H_4)$	69
Nocardioides	LL	+	A3γ	–	PI	–	3c	$8(H_4)$	66–73
Actinomadura[h]	*meso*	–	A1γ	B	PI	–	3a	$9(H_6)/(H_4)/(H_8)$	66–72
Microtetraspora[h]	*meso*	–	A1γ	B	PIV	–	3c	$9(H_4)/(H_2)/(H_0)$	66–69
Glycomyces	*meso*	+	nd	D	PI	–	2c	$9(H_4)/10(H_4)$	71–73
Saccharothrix	*meso*	–	nd	C/E	PII	–	3f	$9(H_4)/10(H_4)$	70–76
Nocardia[i]	*meso*	–	A1γ	A	PII	+	1b	cyclo $8(H_4)/9(H_2)$	64–72
Nocardiopsis	*meso*	–	nd	C	PIII	–	3d	$10(H_2)/(H_4)/(H_6)$	64–69
Streptoalloteichus	*meso*	–	nd	C	PII	–	nd	$9(H_6)/10(H_6)$	nd

[a]Abbreviations: IPB; interpeptide bridge. Symbols of sugar type: A, arabinose and galactose; B, madurose; C, no diagnostic sugars; D, arabinose and xylose; E, rhamnose and galactose; –, not applicable for LL-A_2pm. Phospholipid type: PI, phosphatidylglycerol (variable); PII, only phosphatidylethanolamine, PIII, phosphatidylcholine (with phosphatidylethanolamine, phosphatidylmethylethanolamine, and phosphatidylglycerol variable, no phospholipids containing glucosamine); PIV, phospholipids containing glucosamine (with phosphatidylethanolamine and phosphatidylmethylethanolamine variable). Menaquinones: number indicates number of isoprene units, H_x indicates presence of x hydrogenated menaquinones.

[b]Data from Goodfellow (1989).

[c]Data from Lechevalier et al. (1981, 1977).

[d]Data from Kroppenstedt(1985).

[e]Data from Kroppenstedt (1987) and R. Kroppenstedt (personal communication).

[f]Data from Bowen et al. (1989).

[g]Data from Itoh et al. (1989). The genus *Kineosporia* does not produce aerial mycelium.

[h]Data from R. Kroppenstedt (personal communication).

[i]Data from menaquinones from Howarth et al. (1986).

[j]For details see Stackebrandt and Schumann (2006).

menaquinones of streptomycetes have a highly hydrogenated isoprenoid chain and three to four (rarely five) saturated isoprene units. The actinomycetes, which belong to this type, can be differentiated by a different degree of saturation.

Fatty acids. *Streptomyces* species synthesize terminally branched fatty acids. 2-Methylbutyrate as a starting compound results in anteiso-branched fatty acids with an odd number of carbon atoms. In contrast, isovalerate and isobutyrate as starting compounds lead to the formation of iso-branched fatty acids with even and odd numbers of carbon atoms, respectively. For this reason, iso- and anteiso-branched fatty acids appear in pairs with odd numbers of carbon atoms only (Figure 356).

Colonial characteristics. Streptomycetes have many differential colonial features, such as pigmentation of spores, substrate mycelium, and diffusible exopigments, together with the morphology of colonies and the texture of the aerial mycelium. The production of different pigments has been widely used in classification and identification, but it is important to mention that colony morphology is too variable for use as a taxonomic character.

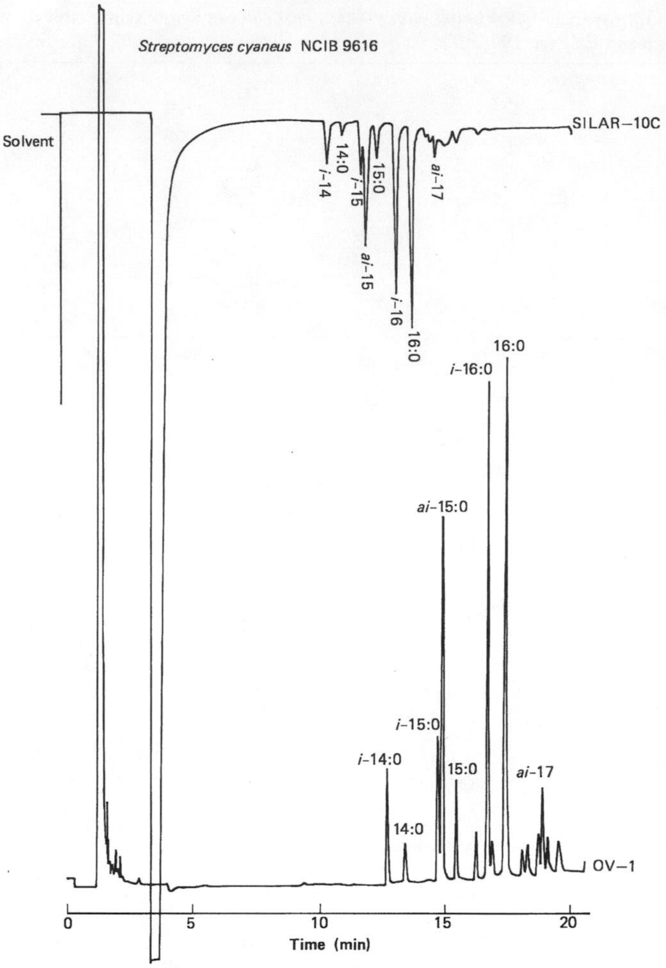

FIGURE 356. Gas chromatographic analysis of the fatty acid methyl esters of *Strepotomyces cyaneus* NCIB 9616.

One feature, widely used in streptomycete taxonomy, is spore mass color. Strains showing different spore colors have been assigned to "sections", "series", and "species-groups" (Burkholder et al., 1954; Ettlinger et al., 1958a; Flaig and Kutzner, 1954, 1960b; Gauze et al., 1957; Hesseltine et al., 1954; Krasil'nikov, 1960; Pridham et al., 1958). In the 8th edition of *Bergey's Manual* (Pridham and Tresner, 1974a), *Streptomyces* species were assigned to seven color series: Blue, Gray, Green, Red, Violet, White, and Yellow. In a later survey, the series were extended to accommodate additional colors (Kutzner, 1981). The color of the spore mass is still useful, but its determination may be difficult, because the color can be influenced by factors such as the medium, growth regime, and age of the culture. Sometimes, the color cannot clearly be attributed to any established category.

The color of the substrate mycelium and the soluble pigment are of high value when they are striking, e.g. blue, dark green, red, and violet. The color of the substrate mycelium has been used in a preliminary approach to group streptomycetes (Baldacci, 1958; Baldacci et al., 1954; Krasil'nikov et al., 1961), but again, the expression of the various pigments is often influenced by the medium composition, temperature, pH, and age of culture (Kutzner, 1981). Also, diffusible pigments and their pH sensitivity have been used as taxonomic characters (Jensen, 1930; Shirling and Gottlieb, 1970; Waksman and Curtis, 1916), but it may be that chemically different pigments exhibit the same color (Krasil'nikov, 1970a; Kutzner, 1981). Some *Streptomyces* strains produce anthracyclinglycoside, diazaindophenol, naphthoquinone, phenoxazinone, and prodigiosin pigments (Kutzner, 1981).

Despite the problems, determinations of aerial spore mass color, substrate mycelium color, and diffusible pigments were used for the descriptions of *Streptomyces* species in the ISP (Shirling and Gottlieb, 1966). It was important that the color determinations were based on standardized media and methods, but there was some disagreement on aerial spore color determination using the Tresner and Backus (1963) color chart. The diagnostic value of spore color was limited because over half of the test strains were placed in the Gray series. Good agreement was found for the determinations of pH sensitivity of the diffusible pigments, but not in the interpretation of substrate mycelium color (Shirling and Gottlieb, 1970).

Genetics and genomics. The genetics of streptomycetes is a rapidly developing topic and a wealth of information has been published within the last few years (for reviews and more information see, for example: Chen et al., 2002; Donadio et al., 2002; Paradkar et al., 2003; Schrempf, 2006; Ventura et al., 2007; Hopwood, 2003, 2007; Hsiao and Kirby, 2008; Dyson, 2010; and references therein). It is, of course, far beyond the scope of this short chapter to give a comprehensive summary of all the recent findings in this exciting area of research; however, despite this, it is not clear at present how this information can be used for taxonomic purposes.

The literature on *Streptomyces* genomes is extensive and information on the complex structure of genomes is increasing, mainly because streptomycetes are very abundant and important as soil inhabitants, where they are regarded as major agents in the cycling of organic carbon compounds. They are also able to produce many and diverse hydrolytic exoenzymes, like chitinases and cellulases. Summaries on their genomics and genetics are available (Paradkar et al., 2003; Schrempf, 2006; Ventura et al., 2007) and, hence, this paragraph is intended to give only a very short summary of existing knowledge on *Streptomyces* genetics.

In particular, one strain, *Streptomyces coelicolor* A3(2), is of major importance as it represents a major model organism for studying developmental complexity. In addition, streptomycetes are the most important natural source of antibiotic compounds and other bioactive metabolites. The whole genome sequences of strains of *Streptomyces* species have been studied and are published or available online: *Streptomyces coelicolor* A3(2) (representing the model streptomycete), *Streptomyces ambofaciens* strains ATCC 15154, DSM 40697, ETH 9247, and ETH 11317 (*Streptomyces ambofaciens* is known for its remarkable genetic instability), *Streptomyces avermitilis* MA-4680^T (the producer of avermectin), *Streptomyces griseus* subsp. *griseus* NBRC 13350 (producer of bioactive secondary metabolites), and *Streptomyces scabiei* 87.22 (causing potato scab).

All *Streptomyces* strains studied so far contain a large genome, which can be circular or linear. The linearity cannot be simply deduced from the extensive genetic linkage mapping of *Streptomyces coelicolor* and is not regarded to be linked with mycelial growth.

The *Streptomyces coelicolor* A3(2) genome was considered to be circular, but on the basis of studies on cosmid libraries in combination with comparisons of physical maps from the wild-type and mutant strains of *Streptomyces coelicolor* A3(2) and "*Streptomyces lividans*", it is quite likely that the chromosome occurs in circular and linear forms (Lin et al., 1993; Redenbach et al., 1996). Pulse-field gel electrophoresis (PFGE) studies have revealed the presence of a linear chromosome in other streptomycetes, including *Streptomyces ambofaciens* (Leblond et al., 1996), *Streptomyces antibioticus*, *Streptomyces moderatus*, *Streptomyces lipmanii*, *Streptomyces parvulus*, *Streptomyces rochei* (Lin et al., 1993), *Streptomyces griseus* (Lezhava et al., 1995), and *Streptomyces hygroscopicus* (Pang et al., 2002a, 2002b). A linear arrangement of the chromosome has also been shown for a *Streptoverticillium* sp. (Redenbach et al., 1998).

The *Streptomyces coelicolor* A3(2) chromosome contains about 8667 Mbp, which corresponds to 7825 genes. Twenty gene clusters encode known or predicted secondary metabolites (Bentley et al., 2002). It is noteworthy that the *Streptomyces coelicolor* chromosome was shown to carry more genes (i.e. 7825) than the eukaryote *Saccharomyces cerevisiae* (containing 6203 genes). The *Streptomyces avermitilis* MA-4680^T genome was shown to comprise about 9025 Mbp (mean G+C content 70.7 mol%), which corresponds to 7574 potential open reading frames, 35% of which constitute 721 paralogous families (Ōmura et al., 2001; Ventura et al., 2007). Thirty gene clusters encode secondary metabolites. It was found that one region of 6500 Mbp has been highly conserved with respect to gene order in the *Streptomyces avermitilis* MA-4680^T and *Streptomyces coelicolor* A3(2) genomes and, hence, may contain essential genes. The terminal regions are not conserved and contain "nonessential genes" (Ikeda et al., 2003). It is interesting to note that an ancient synteny (conservation of gene order) has been revealed between the central core of the *Streptomyces coelicolor* A3(2) chromosome and the whole chromosomes of *Corynebacterium diphtheriae* and *Mycobacterium tuberculosis* (Bentley et al., 2002).

Whole-genome synteny plots have shown a high conservation of the overall position and orientation of common genes in the chromosomes of *Streptomyces coelicolor* A3(2) and *Streptomyces avermitilis* MA-4680^T (Ventura et al., 2007). Ikeda et al. (2003) showed that about two-thirds (5283) of the genes of these strains represent conserved orthologs (estimated by reciprocal BLAST analysis). Similarly, 4837 genes were found to be orthologous between these two strains (Ventura et al., 2007). When *Streptomyces scabiei* ATCC 49173^T is added to this comparison, the number of genes conserved among the three strains drops to 4190 and, as pointed out by Ventura et al. (2007), a four-way analysis, including the hitherto unpublished sequence of *Streptomyces venezuelae* ATCC 10595, reduces the number to 3566. It can be expected that the number will fall further as more *Streptomyces* genomes become available. Ventura et al. (2007) pointed out that only about 17% of the 3566 genes common to the four *Streptomyces* genomes are present in *Escherichia coli* K-12 and *Bacillus subtilis* 168.

Hsiao and Kirby (2008) used DNA–DNA microarray hybridization to compare the genome content of *Streptomyces avermitilis* ATCC 31267^T, "*Streptomyces cattleya*" ATCC 35852, "*Streptomyces maritimus*" Yang-Ming and *Kitasatospora* (*Streptomyces*) *aureofaciens* ATCC 10762^T with that of *Streptomyces coelicolor* A3(2). About 93% agreement with the genome sequence data available for *Streptomyces avermitilis* ATCC 31267^T was shown and a number of trends in the genome structure for *Streptomyces* and closely related *Kitasatospora* species could be detected. The core central region was well conserved and a low degree of gene conservation in the terminal regions of the linear chromosome was observed across all four strains. Between these regions, two areas of intermediate gene conservation were detected by microarray analysis though some conserved genes were also identified within the terminal regions.

The replication process of the *Streptomyces* chromosome has been summarized by Schrempf (2006) and Ventura et al. (2007). The replication of linear *Streptomyces* chromosomes and plasmids is initiated from a fairly centrally located replication origin rich in DnaA box sequences and proceeds bidirectionally towards the telomeres. The chromosomal replication origin (*oriC*) region was found to be highly conserved in *Streptomyces coelicolor* A3(2) (Calcutt and Schmidt, 1992), "*Streptomyces lividans*" 66 (Zakrzewska-Czerwinska and Schrempf, 1992), "*Streptomyces lividus*" TK21, *Streptomyces antibioticus* ETH 7451, and *Streptomyces chrysomallus* ATCC 11523^T (Jakimowicz et al., 1998). Interestingly, it was found that contrary to the high overall G+C content (69–73 mol%) of *Streptomyces* DNA, the region of the origin (*oriC*) is rich in A+T (64 mol%). The chromosomal ends of some *Streptomyces* species have been shown to contain terminal inverted repeats (TIRs), which are covalently bound to proteins, most likely at their 5′ ends. TIR lengths among available sequenced *Streptomyces* chromosomes vary considerably: 174 bp for *Streptomyces avermitilis*; 18,488 bp for *Streptomyces scabiei*; 21,653 bp for *Streptomyces coelicolor* M145; and approximately 198 kb for *Streptomyces ambofaciens* (Ventura et al., 2007).

The telomeres are replicated by a special mechanism which is initiated by priming from a terminal protein covalently bound to the 5′ ends. Approximately 250–320 nucleotides at these ends are characteristic and possess a complex secondary structure (Ventura et al., 2007).

The origin of the linearity of the *Streptomyces* chromosome is thought to have occurred by single-crossover recombination between an initially circular chromosome and a linear plasmid. Several examples of exchange of ends between chromosomes and linear plasmids, resulting in hybrid molecules with different right and left ends, have been reported.

The genomes of streptomycetes contain several examples of apparent redundancy of metabolic genes, which is thought to be due to the complex morphological and physiological differentiation of streptomycetes. The metabolic genes comprise genes that encode functions involved in carbon storage transactions, genes specific for different hyphal cell types, genes comprising enzymes of the pentose phosphate pathway, and multiple *fabH*-like genes, which are important for the first step in fatty acid biosynthesis (some of these are linked with secondary metabolism gene sets) (Ventura et al., 2007). Interestingly, some genes are not present in streptomycetes. Two of the three subunits of exonuclease V (the *recB* and *recC* genes), for instance, are not found in streptomycetes, though they are present in other actinobacteria such as mycobacteria. The XerCD pathway, which is responsible for the resolution of circular chromosomes after replication, is absent from streptomycetes, a result in line with the linearity of *Streptomyces* chromosomes. The conserved *ftsA* gene, which is widely distributed in the domain *Bacteria* and involved in the complex cell division process, is generally absent from actinobacteria. In addition, the *minC* and *minE* genes, which are involved in the choice of division site in many unicellular bacteria are not found in *Streptomyces* strains (Ventura et al., 2007).

Streptomyces colonies often show a high spontaneous variability in antibiotic biosynthesis, pigmentation, and sporulation. The various antibiotic resistances, A-factor formation, and synthesis of tyrosinase or arginosuccinate are encoded by unstable genes. It is possible to stimulate this genetic instability by mutagens, such as ethidium bromide, mitomycin, ultraviolet light, and by gyrase- (topoisomerase II) inhibiting antibiotics. Often, these variations can be attributed to large chromosomal deletions, preferentially occurring at the telomeric and subtelomeric regions, and including up to 2 Mbp of DNA (Chen, 1995; Hütter and Eckhardt, 1988; Leblond and Decaris, 1994; Schrempf et al., 1989). More details about these processes are given by Schrempf (2006). The variability of the chromosomal DNA is further increased by its interaction with linear and circular plasmids, phages, transposons, and insertion elements. Streptomycetes inhabit quickly changing environments and, hence, the high plasticity of the genome is likely an effective prerequisite for quick adaptation.

***Streptomyces* extrachromosomal elements.** Many plasmids have been described (for a review, see Kieser et al., 2000). Linear plasmids, often very large, are widespread and diverse among streptomycetes, but circular plasmids are also found. The various types of circular plasmids differ in size. Schrempf et al. (1989) isolated the low-copy number circular plasmid SCP2 from a streptomycete and this is used as a cloning vector. The plasmid SCP2* (a variant of SCP2; Hopwood et al., 1985) has been sequenced and encodes 34 proteins, most of them of unknown function. The replication region was shown to contain the *repI* and the *repH* genes, both encoding small proteins. The *traA* gene, essential for DNA transfer and pock-formation, was identified in addition to 10 additional genes, which are thought to be involved in conjugation and DNA spreading. Plasmid pIJ101 (about 8.8 kb) is a high-copy-number natural conjugative *Streptomyces* plasmid. On this plasmid, regions for its replication, stability, transfer and distribution have been identified. Efficient conjugation among streptomycetes necessitates the plasmid-encoded *tra* gene and the *cis*-acting locus of transfer (Ducote et al., 2000). A further natural plasmid, pSG5 (initially found in *Streptomyces ghanaensis* DSM 2932), of about 12.3 kb, is naturally temperature-sensitive and cells may contain approximately 50 copies per chromosome. Additional multicopy plasmids pSN22 (Kataoka et al., 1991), pJV1 (Bailey et al., 1986), and pSMA2 (Pernodet et al., 1984) replicate via a rolling circle mechanism (Hagege et al., 1993; Servín-González, 1993).

Linear plasmids are also common in streptomycetes; however, knowledge of their encoding functions is still restricted to few functions. These include antibiotic production (Gravius et al., 1994; Kinashi et al., 1991) and mercury resistance (Ravel et al., 1998). Some of the linear plasmids are efficiently transferred during conjugation. The replication mechanism is best understood for pSLA2, initiated bidirectionally near the center and proceeding towards its telomeric ends, generating 3′ leading-strand overhangs. It should be mentioned that several other members of the order *Actinomycetales* harbor genes on linear plasmids, among them those required for isopropylbenzene and trichlorethylene catabolism (*Rhodococcus erythropolis*; Kebeler et al., 1996), biphenyl degradation (*Rhodococcus erythropolis* and *Rhodococcus globerulus*; Kosono et al., 1997), hydrogen autotrophy (*Rhodococcus opacus*, formerly "*Nocardia opaca*"; Kalkus et al., 1993), and fasciation in plants (*Rhodococcus fascians*; Crespi et al., 1992). Linear plasmids have also been discovered for mycobacteria (*Mycobacterium avium*, *Mycobacterium branderi*, *Mycobacterium celaturum*, and *Mycobacterium xenopi*). Their termini are similar to those of linear plasmids from *Streptomyces* and *Rhodococcus* species (Picardeau and Vincent, 1998); it is an open question whether circular and linear plasmids are exchanged during conjugation amongst actinomycetes. Relatively little is known about *Streptomyces* transposons; a summary of this topic is available (Schrempf, 2006). Phages with broad or narrow host ranges can be obtained from soil and several of them have been used for classifying strains (for review, see Kutzner, 1981). However, in current taxonomic studies, phage host range studies are not carried out.

DNA regions in mycelial actinobacterial genomes acquired by HGT. Despite the gross synteny between the central regions of *Streptomyces* genomes, there are hundreds of insertion-deletion (indel) differences between *Streptomyces coelicolor* A3(2) and *Streptomyces avermitilis* MA-4680^T, most of them involving one or a few genes. This often makes it difficult to recognize synteny at the level of small groups of genes. Streptomycetes also have numerous larger islands of species-specific DNA (Ventura et al., 2007). Prior to the publication of other *Streptomyces* genome sequences, 14 islands of likely laterally acquired DNA were found by Bentley et al. (2002) in the *Streptomyces coelicolor* A3(2) genome on the basis of gene content, atypical G+C content, and location next to a tRNA determinant. Around 50% of these islands were shared with *Streptomyces ambofaciens*. From this, it is clear that in pairwise synteny plots the genes in the "subtelomeric arms" of *Streptomyces* chromosomes are much less conserved between species than those in the central regions or "cores" and that the cores contain most of the genes conserved with other actinobacteria (Bentley et al., 2002; Choulet et al., 2006; Ikeda et al., 2003).

Production of extracellular enzymes. Streptomycetes are widely distributed in soil and play an important role in the recycling of organic matter. It is not surprising, therefore, that *Streptomyces* genomes encode high numbers of predicted secreted proteins; the approximately 800 proteins in *Streptomyces coelicolor* A3(2) have been found to comprise 147 hydrolases, of which seven are cellulases and five are chitinases (Bentley et al., 2002; Ventura et al., 2007). Almost all *Streptomyces* strains studied so far can use chitin not only as a carbon, but also as a nitrogen source (Blaak and Schrempf, 1995) and, often, several chitinases are produced (Miyashita et al., 1991; Robbins et al., 1988). For more details, see Schrempf (2006). Amylases and their inhibitors are also common in streptomycetes. An α-amylase gene (*aml*) of *Streptomyces limosus* ATCC 19778^T has been cloned by Virolle and Bibb (1988). It can be deduced from sequence information that the *Streptomyces coelicolor* A3(2) genome has many genes which code for glucosyltransferases. In addition, xylanases and their genes have been identified from several streptomycetes, i.e. "*Streptomyces lividans*" 10-164 (Pagé et al., 1996), *Streptomyces halstedii* JM8 (Ruiz-Arribas et al., 1998), and the thermophilic *Streptomyces thermoviolaceus* OPC-520 (Tsujibo et al., 1997). Laccases, including those produced by *Streptomyces cyaneus* CECT 3335, can be efficiently applied for biobleaching of kraft pulps (Arias et al., 2003).

Extracellular proteases are widely distributed among streptomycetes and several corresponding genes have been characterized (Kim and Lee, 1995). Streptomycetes also contain many genes for protease inhibitors (Taguchi et al., 1996), including leupeptin and subtilisin (Hiraga et al., 2000). In addition, keratinases are frequently found (for review, see Kutzner, 1981). A few extracellular lipases and their genes have been studied from different *Streptomyces* strains (Servín-González et al., 1997; Sommer et al., 1997), among them lipolytic enzymes expressed by *Streptomyces rimosus* R6-554W (Vujaklija et al., 2002).

Jendrossek et al. (1997) found out that *Streptomyces* strains are the predominant community members of latex rubber-degrading actinomycetes. Some streptomycetes can produce enzymes involved in the modification of pharmacologically relevant compounds and xenobiotics (Peczynska-Czoch and Mordarski, 1988).

Primary metabolism. In contrast to secondary metabolism, relatively few studies have been published on the primary metabolism of streptomycetes. Some genes, among them those encoding key enzymes, like fructose-1,6-bisphosphate aldolase and glucose-6-phosphate dehydrogenases, have been identified. Butler et al. (2002) identified two *zwf* genes determining isozymes of glucose-6-phosphate dehydrogenases [the first enzyme in the oxidative pentose phosphate pathway (PPP)] and one gene (*devB*) encoding 6-phosphogluconolactonase in "*Streptomyces lividans*" 66. The PPP and the tricarboxylic acid cycle relative to glucose uptake have been studied in *Streptomyces noursei* ATCC 11455^T (Jonsbu et al., 2001).

Similar to enteric bacteria, glutamine synthetase I (GSI) in *Streptomyces coelicolor* A3(2) is post-translationally controlled by adenylyltransferase (Hesketh et al., 2002a). A novel class of glutamate dehydrogenases (GDHs) has been detected in *Streptomyces clavuligerus* NRRL 3585^T (Minambres et al., 2000). In addition, *Streptomyces coelicolor* A3(2) can utilize fatty acids (C4 to C18) as sole carbon sources (Banchio and Gramajo, 1997) and the glyoxylate cycle also seems to be present, at least in *Streptomyces clavuligerus* NRRL 3585^T (Soh et al., 2001). Malonate is a well-known competitive inhibitor of succinate dehydrogenase. Kim and Goodfellow (2002) used the genes *matB* and *matC* in generating strain variants of *Streptomyces* used for the production of antibiotics.

The pathways and genes required for the biosynthesis of primary compounds, including their regulation pattern (Rodríguez-García et al., 1997), should be studied in more detail, in order to improve knowledge on metabolic fluxes (Obanye et al., 1996). This will certainly improve the biotechnological production of pharmacologically active compounds derived from primary metabolites. Detailed analyses of proteins involved in primary and secondary metabolism are available (Hesketh et al., 2002b; Huang et al., 2001). Proteomic and metabolomic data are currently being studied (e.g. Novotna et al., 2003).

Secondary metabolism. Streptomycetes have been the most important source of antibiotics since the discovery of actinomycin D, streptothricin, and streptomycin in the 1940s by Waksman and coworkers (for a review, see Hopwood, 2007). Streptomycetes synthesize a large variety of chemically different compounds, many of them acting as antibiotics, cytostatics, fungicides, or as modulators of immune responses (see e.g. Horinouchi, 2002; Bérdy, 2005; Challis and Hopwood, 2003; Van Wezel and Vijgenboom, 2004; and Hopwood, (2007) for more detailed information). Consequently, the study of *Streptomyces* genomes has been of great interest and has led to the discovery of 23 gene sets that code for these compounds in the *Streptomyces coelicolor* A3(2) chromosome and 30 in that of *Streptomyces avermitilis* MA-4680^T (Ventura et al., 2007). It is interesting that many of these gene sets are present in one genome, but not in others, and that the same position in different chromosomes can be occupied by different secondary metabolism clusters. For example, the *pks1* cluster of *Streptomyces avermitilis* MA-4680^T is replaced in *Streptomyces ambofaciens* ATCC 23877^T by a different secondary metabolism cluster of 28 genes and in *Streptomyces coelicolor* A3(2) by a 31-gene insertion (Choulet et al., 2006).

The subtelomeric chromosome arms often contain gene clusters for secondary metabolism, especially those that are species-specific. The more abundant genes for secondary metabolism, such as those for the production of pentalenolactone, different siderophores, and the odor compound geosmin, typically fall in syntenous locations within the central core region (Bentley et al., 2002; Ikeda et al., 2003). It is known that certain linear plasmids may also carry such clusters, thereby explaining the impact of lateral gene transfer between chromosomes present in different streptomycetes (Ventura et al., 2007).

The genes encoding different pharmacologically active substances of importance are located within DNA stretches of 20 kb to more than 100 kb. Successful cloning has been achieved by complementing mutants, by screening total genomic DNA or gene libraries with homologous or heterologous gene probes generated by cloning, or with the help of PCR, as well as by transposon mutagenesis (Schrempf, 2006). It is interesting that the genes for the biosynthesis of antibiotics are frequently located near one or more genes mediating resistance to the corresponding antibiotic. The following listing has been adapted from Schrempf (2006). The gene-cluster for the synthesis of the polyketide actinorhodin was achieved by complementation of mutants (Malpartida and Hopwood, 1984). Other gene clusters for polyketides were cloned using a gene-probe for the predicted key step for polyketide synthesis. These polyketides include: daunorubicin (Stutzman-Engwall and Hutchinson, 1989), frenolicin (Bibb et al., 1994), granaticin (Sherman et al., 1989), griseusin B (Yu et al., 1994), jadomycin B (Han et al., 1994), mithramycin (Lombo et al., 1996), tetracyclines (Binnie et al., 1989), tetracenomycin C (Motamedi and Hutchinson, 1987), tetrangomycin (Hong et al., 1997), and urdamycin A (Decker and Haag, 1995).

Genes for several clusters of macrolides have been identified, including the genes for carbomycin (Epp et al., 1987), tylosin (Fishman et al., 1987), oleandomycin (Swan et al., 1994), and rapamycin (Schwecke et al., 1995). Genes for peptide antibiotics [such as actinomycin (Hsieh and Jones, 1995; Stindl and Keller, 1994) and biolaphos (Murakami et al., 1986)], and cyclopentenoid antibiotics (such as methylenomycin; Chater and Bruton, 1985) have also been found. Additionally, genes have been detected for the synthesis of nikkomycin (a nucleoside-peptide; Bormann et al., 1996), nosiheptide (a thiopeptide; Dosch et al., 1988), undecylprodigiosin (a pyrrole; Feitelson and Hopwood, 1983; Malpartida et al., 1990), ansamycins [such as rubradirin (Sohng et al., 1997) and rifamycin (August et al., 1998)], aminoglycosides [such as puromycin (Lacalle et al., 1992) and streptomycin (Distler et al., 1987; Ohnuki et al., 1985)], carbapenems (Nakata et al., 1989), cephamycin (Aharonowitz et al., 1992; Paradkar et al., 1996), and cyclopilins (Pahl et al., 1997). Again, it is far beyond the scope of this chapter to give a comprehensive overview on the secondary metabolism of streptomycetes, which are covered in text books and/or reviews (e.g. Bérdy, 2005; Challis and Hopwood, 2003; Van Wezel and Vijgenboom, 2004; Hopwood, 2007; Dyson, 2010).

Ecology. Streptomycetes can be isolated in high numbers from soil, which is their primary natural habitat. As mentioned above, most streptomycetes can degrade complex and recalcitrant plant and animal materials, often polymeric residues including polysaccharides (e.g. cellulose, chitin, pectin, and starch), proteins (e.g. elastin and keratin), aromatic compounds, and lignocellulose. The biodegradative activities of actinomycetes have been the subject of several reviews (Crawford, 1988; Lechevalier, 1988; Peczynska-Czoch and Mordarski, 1988). Streptomycetes are able to degrade lignin, which occurs in nature together with cellulose and xylan (hemicellulose) in a lignocellulose complex. Experiments with ^{14}C-labeled lignin showed that streptomycetes (Antai and Crawford, 1981; Crawford, 1978), as well as other genera of actinomycetes, are involved in lignin decomposition (McCarthy et al., 1984, 1986; McCarthy and Broda, 1984), although fungi play a more important role in this process (Crawford, 1981; Janshekar and Fiechter, 1983; Kirk and Farrell, 1987). Ligniolytic streptomycetes can degrade the cellulose of the lignocellulose complex. For more details, see Ramachandra et al. (1988), Wang et al. (1990), Crawford et al. (1993), Chamberlain and Crawford (2000), Kormanec et al. (2001), Gottschalk et al. (2003), and Kaneko et al. (2003).

In addition, mesophilic and thermophilic streptomycetes have been reported to contain multicomponent cellulases, which consist of several endoglucanases and exoglucanases (Crawford and McCoy, 1972; Enger and Sleeper, 1965; Harchand and Singh, 1997; MacKenzie et al., 1984; Marri et al., 1997; Ulrich and Wirth, 1999; Wirth and Ulrich, 2002). Xylanases, which are involved in the decomposition of the lignocellulose complex, seem to be widespread among thermophilic actinomycetes, although they have also been found in mesophilic streptomycetes (Deobald and Crawford, 1987; Godden et al., 1989; Kluepfel and Ishaque, 1982; Kluepfel et al., 1986; McCarthy et al., 1985; Morosoli et al., 1999; Schäfer et al., 1996). Other polymeric compounds found in streptomycetes include pectinolytic complexes (Sato and Kaji, 1975, 1977, 1980a, 1980b) and chitinolytic complexes, which consist of chitinase and chitobiase, and have generally been isolated in full from *Streptomyces griseus* (Berger and Reynolds, 1958), *Streptomyces antibioticus* (Jeuniaux, 1966), and other streptomycetes (Beyer and Diekmann, 1985). More details are given by Schrempf (2006).

Starch, which is the primary material for the textile, paper, and food industries, can be degraded by a wide variety of fungi and bacteria. The enzymes involved are amylases, some of which have been found in several streptomycetes (Fairbairn et al., 1986; McKillop et al., 1986; Mordarski et al., 1970; Suganuma et al., 1980).

Next to degrading polymeric compounds, streptomycetes have the ability to degrade other organic materials, e.g. cotton and plant fibers (Khan et al., 1978; Lacey and Lacey, 1987), wool (Noval and Nickerson, 1959), hydrocarbons in jet fuel and emulsions (Genner and Hill, 1981), rubber (Cundell and Mulcock, 1975; Hutchinson et al., 1975), and plastics (Pommer and Lorenz, 1986). Lacey (1988) and Behal (2000) give a detailed review about the biodegradation of natural and synthetic substances. More details are given by Schrempf (2006).

In soil, streptomycetes can show pronounced mycelial growth. In this habitat, they are adapted to various, often quickly changing physical conditions (e.g. shifts in aeration, drought, frost, hydrostatic pressure and anaerobic conditions, moisture tension, and pH) by the formation of spores, which are semi-dormant stages in the life cycle and can survive in soil for long periods (Ensign, 1978; Mayfield et al., 1972). Viable cultures of cells have been reported by Morita (1985) from 70-year-old soil samples. Streptomycetes are almost always present as inactive spores in soil. One disadvantage of persisting as a spore is the very low germination efficiency, which may be caused by competition with other micro-organisms. Spores which pre-germinate can grow for a short time and then resporulate (Lloyd, 1969). Several factors may be responsible for the germination of spores. Next to special signaling factors, the presence of exogenous nutrients, water, and Ca^{2+} seems to be necessary (Ensign, 1978). In addition to germination, nutrients influence the extent of hyphal growth and the time of differentiation into aerial hyphae. Fodders and other organic material, freshwater, and marine habitats, as well as potable water systems, can come into contact with soil (Korn-Wendisch and Kutzner, 1992), e.g. through human and other activities. Natural substrates (e.g. grain, hay, fodder, and wood) and synthetic products (e.g. cotton textiles, fabric, paper, rubber, plastics, and plasticizers), which can be found in or transported to soil, can be degraded with the help of mesophilic and especially thermophilic streptomycetes. The contamination of creeks and rivers with soil streptomycetes is due to drainage, e.g. after heavy rainfalls. Streptomycetes find their way into the sediments of the lakes, rivers, and, after transport to the sea, into marine sediments. It may be possible that drinking water supplies become contaminated with streptomycetes. This is a problem because the compounds produced by some streptomycetes are odorous and lead to the spoilage of the water.

Streptomycetes as plant pathogens. Some of the many saprophytic *Streptomyces* species are plant pathogens which may cause economically important diseases, including potato scab. *Streptomyces scabiei* can still be regarded as the dominant pathogenic species worldwide, but is only one of many streptomycetes which cause very similar disease symptoms on plants. In addition to *Streptomyces scabiei* (Lambert and Loria, 1989b), *Streptomyces acidiscabies* (Lambert and Loria, 1989a), *Streptomyces turgidiscabies* (Miyajima et al., 1998), *Streptomyces europaeiscabiei*, *Streptomyces stelliscabiei* (Bouchek-Mechiche et al., 2000), *Streptomyces luridiscabiei*, *Streptomyces puniciscabiei*, and *Streptomyces niveiscabiei* (Park et al., 2003) have been shown to be plant pathogens that cause either common scab or netted scab, mostly in potatoes.

Streptomyces scabiei (previously know as "*Streptomyces scabies*"), which is the most important and oldest characterized potato scab pathogen, has been isolated from beets, carrot, peanut, and radish, among other crops (Loria et al., 2006). Strains of *Streptomyces scabiei* are phenotypically similar to *Streptomyces bottropensis*, *Streptomyces diastatochromogenes*, and *Streptomyces neyagawaensis*, a result confirmed by 16S rRNA gene sequence analyses. *Streptomyces acidiscabies* has been isolated from low pH soils in the north-eastern United States, amongst other locations. *Streptomyces turgidiscabies* has been isolated from cases of potato scab in Finland, but also from Japan and Korea (Loria et al., 2006). *Streptomyces europaeiscabiei*, the most closely related species to *Streptomyces scabiei* has been isolated from various locations in Europe. Three species, *Streptomyces luridiscabiei*, *Streptomyces niveiscabiei*, and *Streptomyces puniciscabiei*, are the causal agents of potato scab in Korea.

16S rRNA gene sequence analyses and DNA–DNA hybridization studies show that the documented pathogenic strains fall outside the described species listed above (Loria et al., 2006). This can be attributed to the polyphyletic nature of scab-causing species and the existence of a transmissible pathogenicity island, which seems to confer the pathogenic phenotype on some otherwise non-pathogenic species, as reviewed by Loria et al. (2006). The mechanisms used by plant-pathogenic species to manipulate their hosts have been studied in detail and summarized by Loria et al. (2008, 2006). The nitrated dipeptide phytotoxin, thaxtomin, plays an important role in inhibiting cellulose biosynthesis in expanding plant tissues, stimulating Ca^{2+} spiking, and causing cell death. In addition, a secreted necrogenic protein, Nec1, contributes to virulence on diverse plant species. A detailed genetic analysis revealed that the thaxtomin biosynthetic genes and *nec1* lie on a large mobilizable plasmid PAI, along with other putative virulence genes, including a cytokinin biosynthetic pathway and a saponinase homolog. The PAI is mobilized during conjugation and site-specifically inserts itself into the linear chromosome of recipient species, thereby accounting for the emergence of new pathogens in agricultural systems.

Streptomycetes as human pathogens. So far, only very few streptomycetes have been isolated from human pathological material. They include organisms that cause actinomycetoma, which is a localized chronic, destructive, and progressive infection of skin, subcutaneous tissue, and eventually bone (Develoux et al., 1999; McNeil and Brown, 1994). In certain tropical and subtropical regions, this disease is endemic and has a devastating effect on patients, as it frequently leads to deformities, disabilities, and eventually amputation of the affected organs. Although some of the main causal agents belong to other genera and species, i.e. *Actinomadura madurae*, *Actinomadura pelletieri*, *Nocardia brasiliensis*, *Nocardia otitidiscaviarum*, and *Nocardia transvalensis*, *Streptomyces somaliensis* is often implicated, notably in parts of the Sudan (Trujillo and Goodfellow, 2003). However, the recognition of a second species, *Streptomyces sudanensis*, by Quintana et al. (2008) suggests that some strains identified as *Streptomyces somaliensis* (Fahal, 2006; Gumaa, 1994; Gumaa and Mahgoub, 1975; Taha, 1983) may have been misclassified. Indeed, there is evidence that streptomycetes associated with cases of actinomycetoma in the Sudan (Fahal, 2004, 2006; Fahal and Hassan, 1992; Mahgoub, 1985) may be underspeciated (Quintana et al., 2008; Trujillo and Goodfellow, 2003).

Soil as a habitat. Abiotic and biotic factors, especially vegetation, content and kind of organic matter, soil type, season and climate, temperature, circulation of water and air, and pH influence the character and community composition of any habitat, including the microbial community. Streptomycetes are common in soils and have been the subject of several reviews: Goodfellow and Simpson (1987), Korn-Wendisch and Kutzner (1992), Lechevalier (1988), Williams (1982), Williams et al. (1982a). The most extensive studies have been carried out by Stan Williams and his colleagues (Flowers and Williams, 1977a, 1977b; Khan and Williams, 1975; Mayfield et al., 1972; Ruddick and Williams, 1972; Watson and Williams, 1974; Williams et al., 1971; Williams and Mayfield, 1971; Williams and Robinson, 1981). The typical life cycle of streptomycetes in soil, including its genetic control, can be found in Kieser et al. (2000). In most soils, streptomycetes constitute about 1–20% of the total viable count, that is 10^4–10^7 colony-forming units (c.f.u.) per g soil (Korn-Wendisch and Kutzner, 1992), and in some soils form the dominant population. Further details in the numbers and distribution of streptomycetes in soil can be found elsewhere (Flaig and Kutzner, 1960a; Küster, 1976; Misiek, 1955; Szabó and Marton, 1964).

The detection and localization of different *Streptomyces* species in their natural habitat are based mainly on cultivation-dependent techniques. Intrageneric classification of the genus *Streptomyces* is difficult and, hence, ecophysiological studies are often difficult to compare. Williams et al. (1969) assigned soil streptomycetes to color-groups, based on diffusible pigment colors formed on oatmeal agar, and on their capacity to produce melanin pigments on peptone-yeast extract-iron agar. This classification into color-groups was then used as a tool by other researchers to study the diversity of streptomycetes in natural habitats (e.g. Goodfellow and Haynes, 1984; Atalan et al., (2000) Sembiring et al., 2000). However, this color-grouping is a subjective method and comparison of data between different studies is difficult. Recently, a computer-assisted numerical analysis was carried out with 321 alkaliphilic streptomycetes that were assigned to color-groups (Antony-Babu et al., 2010). The authors argue that, with this method, distances between individual colors could be calculated more objectively and that the data can be compared with computer-assisted numerically defined color-groups in future investigations on streptomycete taxonomy in natural habitats.

Williams et al. (1972) have shown that streptomycetes resist desiccation because of their ability to form arthrospores. In addition, the water tension they need for growth can be much lower than for other bacteria, but on the other hand, they may be very sensitive to water-logged conditions.

Most attention has been focused on neutrophilic streptomycetes, which are common in neutral to alkaline soils (e.g. Flaig and Kutzner, 1960a), although acidotolerant and acidophilic streptomycetes are abundant in acidic soils and can be isolated using starch-casein agar adjusted to pH 5.0 supplemented with anti-fungal agents (Hagerdorn, 1976; Khan and Williams, 1975). Acidophilic streptomycetes produce specific and stable amylases (Williams and Flowers, 1978; Williams and Robinson, 1981). In contrast, alkalitolerant and alkaliphilic streptomycetes are common in alkaline soils (Antony-Babu and Goodfellow, 2008; Mikami et al., 1982, 1985; Taber, 1959, 1960).

Streptomycetes, as well as other soil bacteria, have been isolated from the intestinal tract of earthworms (Brüsewitz, 1959; Parle, 1963a, 1963b), the gut of arthropods (Bignell, 1984; Bignell et al., 1980, 1981; Szabó et al., 1967), and pellets produced by millipedes and woodlice (Márialigeti et al., 1984). Streptomycetes are also found in the rhizosphere (Goodfellow and Williams, 1983; Sembiring et al., 2000) where they may have an important role.

It has been suggested ever since the discovery of antibiotics from streptomycetes that antibiotic-producing organisms have a competitive advantage over nonproducing organisms. However, there is no clear evidence for the *in situ* production of antibiotics in soil (Williams, 1982). Antibiotics are difficult to detect in soil as they are found in low concentrations and may be unstable (Brian, 1957; Williams, 1982). In addition, they may be adsorbed onto soil colloids (Williams, 1982) and may also be produced at certain stages of the growth cycle (Williams, 1982; Williams and Khan, 1974).

However, antibiotic production in soil was reported by Rothrock and Gottlieb (1984) who supplemented sterilized soil with nutrients prior to adding a potent producer. In the control of fungal root pathogens, streptomycetes seem to play an important role (Goodfellow et al., 2007; Rothrock and Gottlieb, 1981; Sing and Mehrotra, 1980; Williams, 1978, 1982). It has also been observed that many streptomycetes are often successful in competition with other rhizosphere bacteria such as pseudomonads and bacilli, especially in relatively dry soils.

Thermophilic streptomycetes. The genus *Streptomyces* contains mainly mesophilic species, though some streptomycetes are thermotolerant (growing up to 45°C) and a few are thermophilic. So far, all described thermophilic streptomycetes grow at temperatures between 28–55°C and several grow at even higher temperatures. Kim et al. (1999) studied the taxonomy of thermophilic streptomycetes in detail. Additional thermophilic species (*Streptomyces thermocoprophilus* and *Streptomyces thermospinisporus*) were described by Kim et al. (2000) and Kim and Goodfellow (2002). The life cycle of thermophilic streptomycetes includes active growth at sites of high temperatures, e.g. compost, manure, and self-heating hay or grain. When the vegetative phase ends, the formation of large numbers of spores begins. The spores are returned with the compost or manure to the fields and pastures and can colonize plant material and hay directly or via soil dust (Korn-Wendisch and Kutzner, 1992). Therefore, the genus *Streptomyces* accounts for the majority of actinomycetes isolated from bioaerosols in the surroundings of composting facilities (P. Kämpfer and others, unpublished observation). Thermophilic actinomycetes are widespread and can be isolated from various sources like soils (Craveri and Pagani, 1962; Tendler and Burkholder, 1961), pig feces (Ohta and Ikeda, 1978), sewage-sludge compost (Millner, 1982), and freshwater habitats (Cross, 1981a, 1981b).

Freshwater environments, water supplies, and marine environments. Actinomycetes can easily be isolated from fresh water and especially from sediments of rivers and lakes. However, it is assumed that most of these organisms do not live naturally at these sites and are therefore inactive (Cross, 1981a, 1981b). Instead, they are wash-in forms ("aliens") from surrounding terrestrial environments. In particular, rivers carry vast amounts of various actinomycetes, including streptomycetes. Nevertheless, actinomycetes can survive as dormant spores in aquatic habitats for a long time (Al-Diwany and Cross, 1978). Burman (1973) found 59–200 streptomycetes and 10–20 micromonosporae per ml river water sampled from the River Thames, UK. The streptomycetes grew on decaying vegetation on riverbanks and mud flats at low water or on floating mats of decaying algae or other vegetation. They produce odorous substances which are washed into the water when river levels increase giving rise to "earthy tastes" in drinking water. Geosmin and methyl-isoborneol are the two most frequently detected odorous compounds (Gerber, 1979a, 1979b). Wood et al. (1983) noted that preventing the contamination of potable water with these compounds and, thus, the earthy tastes in reservoirs and water supply systems, depends on locating the production sites and determining the patterns of distribution of these compounds (Lechevalier et al., 1980; Silvey and Roach, 1975). Burman (1973) found that the number of streptomycetes in drinking water was reduced by filtration processes. He also detected a new, aquatic strain of *Streptomyces* in the distribution system (for details, see Burman, 1973).

The occurrence of streptomycetes in marine habitats, including sediments, has been considered by several workers (Cross, 1981b; Goodfellow and Haynes, 1984; Okazaki and Okami,

1976; Weyland, 1981a, 1981b; Weyland and Helmke, 1988). Streptomycetes have been found in the littoral and inshore zone and in deep-sea sediments. Although streptomycetes can be isolated from both localities, they are not necessarily part of the autochthonous microflora, but are probably derived from terrestrial habitats. Streptomycetes isolated from sediments (Roach and Silvey, 1959) and from decaying seaweed (Siebert and Schwartz, 1956) in littoral zones were able to grow on polymeric substances, such as agar and chitin (Humm and Shepard, 1946), alginate and laminarin (Chesters et al., 1956), and cellulose (Chandramohan et al., 1972), which are substances characteristic of these habitats.

In sediments, the ratio of different actinomycete taxa is dependent on the depth and the location of the sampling sites (Weyland, 1981b; Weyland and Helmke, 1988). In the open sea, only low numbers of actinomycetes are generally detected (viable counts about 100 c.f.u. per ml of wet sediment). It is assumed that the distribution of streptomycetes is correlated with barotolerance (Helmke, 1981), halotolerance, and psychrophilism (Weyland, 1981a) (horizontal as well as vertical) of streptomycetes, micromonosporae, and rhodococci. On the other hand, Goodfellow and Haynes (1984) did not find any correlation between salinity, pH, or depth and the number of actinomycetes recovered from marine sediments. These workers studied 732 isolates; 250 belonged to *Streptomyces*, 250 to *Micromonospora*, 140 to *Rhodococcus*, and 92 were assigned to the genus *Thermoactinomyces*. One of the streptomycetes was subsequently identified using a computer-assisted approach (Williams et al., 1983b) and about half of them were assigned to a cluster equated with *Streptomyces albidoflavus* (Williams et al., 1983a).

Streptomycetes are mainly found in sediments of shallow seas (70–520 m deep) with 300–1270 colonies per cm^3, whereas *Micromonospora* are dominant in samples 700–1600 m deep (Okami and Okazaki, 1978). However, these authors did not detect actinomycetes from depths of 2800 and 5000 m in the Pacific Ocean. On the other hand, Pathom-aree et al. (2006) isolated actinomycetes, including streptomycetes, from the Mariana Trench in the Pacific at a depth of 10,898 m. Marine streptomycetes showed a higher salt tolerance than their terrestrial counterparts, though salt tolerance among streptomycetes is widespread (Tresner et al., 1968). A few of the isolated marine streptomycetes were found to be obligate halophiles (Okazaki and Okami, 1976).

Many antibiotic-producing streptomycetes have been isolated from marine habitats (Goodfellow and Fiedler, 2010; Hotta et al., 1980; Okami and Okazaki, 1972; Okami et al., 1976) including seaweed (Nissen, 1963). Recently, Goodfellow and Fiedler (2010) provided a review on a bioprospecting strategy based upon the premise that new secondary metabolites can be found by screening relatively small numbers of dereplicated, novel actinomycetes isolated from marine sediments.

Enrichment, isolation, and cultivation

Korn-Wendisch and Kutzner (1992) summarized extensively the procedures used to isolate streptomycetes. These procedures are briefly described here. Further information about isolation for special purposes, growth, and preservation of streptomycetes can be found in the excellent textbook *Practical Streptomyces Genetics* (Kieser et al., 2000).

Generally, isolation procedures for micro-organisms are dependent on the nature of the micro-organism, as well as the number of individuals relative to the number of other microbes within the habitat (Stolp and Starr, 1981). Direct plating of a serial dilution on a nutrient agar medium can readily lead to a pure culture, if the chosen organism is best adapted to the selected isolation conditions. However, this procedure does not work well for isolation of streptomycetes. Instead enrichment cultures or selective media and/or specific isolation conditions are usually used.

Members of the family *Streptomycetaceae* can be isolated using the following selective criteria (Korn-Wendisch and Kutzner, 1992; Williams et al., 1984a; Williams and Wellington, 1982a, 1982b) (1) choice of the material containing the selected micro-organisms; (2) pretreatment of the sample, and in some cases, enrichment of the chosen microbial groups; (3) use of selective media or selective incubation conditions or both; and (4) colony selection on the basis of colony morphology.

Streptomycetes occur in and can be isolated from a wide variety of habitats. In most cases, the organisms are extracted from soil or another environmental sample, followed by dilution of cells (cell aggregates) to allow cultivation on solid media.

Isolation and enrichment from soil. Vegetative mycelia and spore chains are often closely associated with soil mineral and organic particles. For isolation, vigorous shaking of the sample with the diluent is often needed to suspend the spores or mycelial fragments. It can be helpful to use glass beads and agitate the sample on a shaker. Additional methods are described in the literature, including the use of mechanical devices such as the Ultrasonics sonicator-disrupter, Ultra-Turrax homogenizer, Turmix blender, Waring blender, or a mortar and pestle. However, the efficiency of these pretreatments has not been compared in detail. Other procedures described include the use of chemical disruption methods to separate mycelia from spores. Herron and Wellington (1990) gently shook soil samples with an ion-exchange resin Chelex-100 (Bio-Rad) followed by differential centrifugation and filtration. For increasing the yield and diversity of actinobacteria from natural habitats, the dispersion and differential centrifugation (DDC) technique can be used, which is a multistage procedure that combines several physicochemical treatments (Goodfellow and Fiedler, 2010). The DDC technique was introduced by Hopkins et al. (1991).

Subsequent treatment of samples (i.e. preparing dilutions and plating) differs little from usual bacteriological practice. Before dilutions are made, coarse particles of the soil suspensions should be allowed to settle. Another possibility is the use of the soil particles for the incubation of "soil plates" (Warcup, 1950); this method is also used to isolate fungi. For streptomycetes, the addition of lime to soil can be a helpful enrichment factor [see chapter 2 of Kieser et al. (2000) and references therein]. Surface-inoculation of isolation plates may be carried out with a sterile glass rod (or Drigalski spatula).

To avoid the spread of motile bacteria via water films, plates can be dried at 45°C before incubation (Vickers and Williams, 1987). Another highly recommended procedure is to mix the soil suspension with the molten agar (Korn-Wendisch and Kutzner, 1992). A 100-fold increase in streptomycete colonies on isolation plates can be achieved by the addition of $CaCO_3$ to air-dried soil samples (10:1, w/w) and subsequent incubation at

26°C for 7–9 d in a water-saturated atmosphere (El-Nakeeb and Lechevalier, 1963; Tsao et al., 1960).

Jensen (1930) enriched streptomycetes by amending soil with keratin; it is also known that adding chitin to soil enhances the growth of streptomycetes (Williams and Mayfield, 1971). In addition, enrichment of acidophilic and neutrophilic streptomycetes in acidic soil and litter can be achieved using fungal chitin (Williams and Robinson, 1981). Chitin in the form of insect wings has also been used as an isolation strategy (Jagnow, 1957; Okafor, 1966; Veldkamp, 1955). Other selective isolation methods studied by Porter and Wilhelm (1961) include the use of various other organic materials, such as salmon viscera meal, peanut meal, cottonseed meal, and dried blood flour (15 mg/g of soil), as described by Porter and Wilhelm (1961). Additionally, the authors recorded an increase in the number of streptomycetes (up to 1000-fold) when the enrichment cultures were incubated under moist conditions.

Arginine glycerol agar is frequently used for the selective isolation of streptomycetes (El-Nakeeb and Lechevalier, 1963), as are: HV agar (Hayakawa and Nonomura, 1987a, 1987b), colloidal chitin agar (Hsu and Lockwood, 1975), and reduced arginine starch salts agar.

To reduce or inhibit other microbes, several biological, chemical, and physical methods have been studied (see reviews by: Goodfellow and Williams, 1986; Goodfellow and Fiedler, 2010). The centrifugation of soil suspensions for 20 min at 1600 × *g* separates streptomycetes spores (in the supernatant) from other bacteria and fungal spores (in the sediment) as applied by Nüesch (1965), though this method has not been very successful. El-Nakeeb and Lechevalier (1963) used a similar approach, but obtained a significantly smaller number of streptomycete colonies as compared with the control. A simple sedimentation method was described by Voelskow (1988/89), who suspended 1 g soil in 15 ml salt solution and mixed the preparation by vigorously shaking followed by ultrasonic vibrations. Samples were taken from different levels of this solution after 1, 2, and 4 h of sedimentation, further diluted, and plated onto agar surfaces.

Arthrospores have a relatively high resistance to low moisture tension; hence, initial drying and heating procedures can be applied to environmental samples to reduce the number of unwanted bacteria. A relative increase in streptomycete concentrations can be obtained by drying samples, or by prolonged storage at ambient temperatures for mesophiles and at 50–60°C for thermophiles. A significant reduction in the vegetative bacterial proportion without affecting the colony counts of streptomycetes can be maintained by heat treatment of soil (40–50°C, 2–16 h), as reported by Williams et al. (1972).

Membrane filtration has been used for the enrichment of streptomycetes from water (Burman et al., 1969), and from seawater and mud (Okami and Okazaki, 1972) samples. This method was used as a first step in the isolation of streptomycetes from soil. Trolldenier (1966) filtered 1 ml of a series of 10-fold dilutions through membranes (0.3 µm pore size) prior to placing the latter upside down on a suitable agar medium supplemented with 10% compost soil. Streptomycetes were able to grow through the pores and developed colonies between the agar surface and the membrane filter, whereas other bacteria and fungi were unable to grow through the pores. The application of this procedure led to a three to fivefold increase in the number of streptomycete colonies compared with poured plates without soil in the medium.

The use of cellulose ester membrane filters (pore size 0.01–3.0 µm) was introduced by Hirsch and Christensen (1983). The membrane filters were placed onto nutrient agar containing anti-fungal antibiotics (cycloheximide and candicidin) and samples of soil, water, and vegetable material were used to inoculate the plates. The hyphae of actinomycetes were able to penetrate the pores in the membrane filters and grow on the underlying agar medium after 4 d, whereas the growth of other bacteria was restricted to the surface of the filters. Afterwards, the membrane filters were removed and the plates were reincubated to allow further development of actinomycete colonies. Filters (0.22–0.45 µm) can also be used for the exclusive recovery of actinomycetes, as described by Polsinelli and Mazza (1984) and Hanka et al. (1985).

Several authors have added chemicals to environmental samples to improve isolation efficiency. One recommended method to eliminate bacteria and fungi involved phenol treatment of a dense soil suspension (1.4% for 10 min) though El-Nakeeb and Lechevalier (1963) obtained less favorable results with this method. Burman et al. (1969) found that streptomycetes and other actinomycetes were slightly more resistant to ammonia, chloramine, and sodium hypochlorite than other bacteria. Hence, they used these agents for the treatment of water samples.

Isolation of airborne spores. *Streptomyces* spores from self-heating material such as hay or compost can be agitated in a wind tunnel (Lacey and Dutkiewicz, 1976b) or sedimentation chamber (see below; Lacey and Dutkiewicz, 1976a), prior to using an Andersen sampler to inoculate plates with the resultant aerosol (Goodfellow and Williams, 1986). This method is widely employed for the isolation of thermophilic actinomycetes, but can also be used for the isolation of mesophilic streptomycetes from soil. Other devices such as filtration samplers (e.g. Sartorius MD 8) can be used for the sampling of airborne streptomycetes.

Use of selective media. For the isolation of desired micro-organisms, selective media are widely used. By varying a number of factors, the media are favorable for some microbes but not for others. Factors that can be varied are as follows: (1) the nutrient composition and concentration of the isolation medium can be adjusted to the preferred micro-organisms, i.e. choice of carbon and nitrogen sources preferred by the organisms; (2) chemical substances can be added to the medium to inhibit selectively the accompanying flora of the natural habitat or to stimulate the desired organisms; (3) adequate pH values for acidophilic, neutrophilic, and alkaliphilic organisms can be chosen; (4) various temperatures can be used, depending on the temperature optima of the organisms, e.g. thermophiles or psychrophiles.

For the isolation of streptomycetes, many different media have been empirically formulated. In Table 265, selected carbon and nitrogen compounds are listed that are especially suitable for the isolation of these organisms. Table 266 and Table 267 list the most frequently used media with their formulae. On the other hand, streptomycetes can also be grown on very poor media such as water agar.

Different carbon and nitrogen sources for enrichment. It was recognized early on that streptomycetes can degrade chitin (Jagnow, 1957; Veldkamp, 1955). On this basis, Lingappa and Lockwood (1962) described a chitin medium for selective

TABLE 265. Nutrients and selective agents recommended for isolation of streptomycetes from soil (according to Korn-Wendisch and Kutzner, 1992)

Preferred C and N source	Selective agents in the medium: Antibiotic	Selective agents in the medium: Others	Reference
Starch, KNO_3			Flaig and Kutzner (1960a)
Starch, casein, KNO_3			Küster and Williams (1964a)
Chitin			Lingappa and Lockwood (1962)
Glycerol, arginine			El-Nakeeb and Lechevalier (1963)
Glycerol, casein, KNO_3			Küster and Williams (1964a)
Raffinose, histidine			Vickers et al. (1984)
Starch, casein, KNO_3	Rifampin		Vickers et al. (1984)
Starch, casein, KNO_3	Cycloheximide, nystatin, penicillin, polymyxin		Williams and Davies (1965)
Glycerol, arginine	Cycloheximide, pimaricin, nystatin		Porter et al. (1960)
Glucose, asparagine	Cycloheximide		Corke and Chase (1956)
Asparagine		Propionate	Crook et al. (1950)
Starch, casein, KNO_3	Cycloheximide	Rose Bengal	Ottow (1972)

TABLE 266. Some media recommended for the selective isolation of streptomycetes[a]

	1	2	3	4	5
Ingredients (g/l)	Starch-casein-KNO_3 agar	Glycerol-arginine agar	*Actinomyces* isolation agar	Chitin agar	Raffinose-histidine agar
Chitin (colloidal)	–	–	–	4.0	–
Starch	10.0[b]	–	–	–	–
Glycerol	–	12.5	5.0[c]	–	–
Raffinose	–	–	–	–	10.0
Sodium propionate	–	–	4.0	–	–
KNO_3	2.0	–	–	–	–
Casein	0.3	–	–	–	–
Sodium caseinate	–	–	2.0	–	–
Asparagine	–	–	0.1	–	–
Arginine	–	1.0	–	–	–
Histidine	–	–	–	–	1.0
NaCl	2.0	1.0	–	–	–
KH_2PO_4	–	–	–	0.3	–
K_2HPO_4	2.0	1.0	0.5	0.7	1.0
$MgSO_4 \cdot 7H_2O$	0.05	0.5	0.1	0.5	0.5
$CaCO_3$	0.02	–	–	–	–
$Fe_2(SO_4)_3 \cdot 6H_2O$	–	0.01	–	–	–
$FeSO_4 \cdot 7H_2O$	0.01	–	0.001	0.01	0.01
$CuSO_4 \cdot 5H_2O$	–	0.001	–	–	–
$ZnSO_4 \cdot 7H_2O$	–	0.001	–	0.001	–
$MnSO_4 \cdot H_2O$	–	0.001	–	–	–
$MnCl_2 \cdot 4H_2O$	–	–	–	0.001	–
Agar[d]	18.0	15.0	15.0	20.0	12.0
pH				Adjusted to 7.0–7.5 or lower or higher depending on the flora to be isolated.	

[a]References for media: 1, according to Küster and Williams (1964a); 2, El-Nakeeb and Lechevalier (1963); 3, Difco; 4, Hsu and Lockwood (1975); 5, Vickers et al. (1984).

[b]Alternatively, glycerol at 10 g/l can be used.

[c]Not contained in the dehydrated medium; added at the time of preparation.

[d]The different amounts of the agar are due to the varying quality used by the individual authors.

TABLE 267. Composition of some media suitable for the cultivation of streptomycetes (according to Korn-Wendisch and Kutzner, 1992)[a]

Medium	Ingredients	Comments
1. Glucose-yeast extract-malt extract (GYM) agar	Glucose, 4.0 g Yeast extract, 4.0 g Malt extract, 10.0 g Agar, 12.0 g Distilled water, 1 l	Addition of $CaCO_3$, 2.0 g/liter, is advantageous for streptomycetes. Adjust medium to pH 7.2
2. Oatmeal agar	Oatmeal, 20.0 g Agar, 12.0 g Trace salts solution (see no. 5), 1.0 ml Distilled water, 1 l	Cook 20.0 g oatmeal in 1 liter distilled water for 20 min. Filter through cheesecloth. Add distilled water to restore volume of filtrate to 1 liter, then add trace salts solution and agar. Adjust to pH 7.2.
3. Inorganic salts-starch agar	Starch (soluble), 10.0 g $(NH_4)_2SO_4$, 2.0 g K_2HPO_4 (anhydrous basis), 1.0 g $MgSO_4 \cdot 7H_2O$, 1.0 g NaCl, 1.0 g $CaCO_3$, 2.0 g Trace salts solution (see no. 5), 1.0 ml Agar, 12.0 g Distilled water, 1 l	Make a paste of the starch with a small amount of cold distilled water and bring to a volume of 1 liter; then add the other ingredients. The pH should be between 7.0 and 7.4. Do not adjust it if it is within this range.
4. Glycerol-asparagine agar	Glycerol, 10.0 g L-Asparagine (anhydrous basis), 1.0 g K_2HPO_4, 1.0 g Trace salts solution (see no. 5), 1.0 ml Agar, 12.0 g Distilled water, 1 l	The pH should be about 7.0–7.4. Do not adjust if it is within this range
5. Trace salts solution	$FeSO_4 \cdot 7H_2O$, 0.1 g $MnCl_2 \cdot 4H_2O$, 0.1 g $ZnSO_4 \cdot 7H_2O$, 0.1 g Distilled water, 100.0 ml	
6. Trace elements solution SPV-4	$CaCl_2 \cdot 2H_2O$, 4.0 g Fe (III) citrate, 1.0 g $MnSO_4$, 0.2 g $ZnCl_2$, 0.1 g $CuSO_4 \cdot 5H_2O$, 0.04 g $CoCl_2$, 0.022 g $Na_2MoO_4 \cdot 2H_2O$, 0.025 g $Na_2B_4O_7 \cdot 10H_2O$, 0.1 g Distilled water, 1 l	SPV-4 is used as an alternative to (5). 5 ml of this stock solution is added to 1 liter of medium

[a]Recipes 1–5 are from Shirling and Gottlieb (1966) and recipe 6 is from Voelskow (1988/89).

isolation. However, this chitin medium was only a little better than water agar, which was recognized by the authors and later also by El-Nakeeb and Lechevalier (1963). A useful medium for the isolation of actinomycetes (*Streptomyces*, *Nocardia*, and *Micromonospora*) from water samples was developed by Hsu and Lockwood (1975), who added mineral salts to the chitin medium (Table 266). Unfortunately, it had little effect when isolating actinomycetes from soil. Note that chitinolytic activity is not a genus-specific feature for *Streptomyces*. Only 25% of over 300 strains were strongly chitinolytic (Williams et al., 1983a). Consequently, this widely used medium selects chitinolytic streptomycete strains which may not be the most abundant in soil. Starch is a suitable selective carbon source for streptomycetes, as it is degraded by the vast majority of streptomycetes. The combination of starch with nitrate is used by many streptomycetes in contrast to other bacteria (Flaig and Kutzner, 1960a). Küster and Williams (1964a, 1964b), who improved this medium, stated: "The three best media, allowing good development of streptomycetes while suppressing bacterial growth, were those containing starch or glycerol as the carbon source with casein, arginine, or nitrate as the nitrogen source." Streptomycete isolation is also favored by the combination of glycerol and arginine (Benedict et al., 1955). Further studies by El-Nakeeb and Lechevalier (1963) revealed that this medium (Table 266 and Table 267) was superior to nine other media, resulting in higher numbers and proportions of streptomycete colonies.

Other compounds that have been used successfully for the selective isolation of streptomycetes are, e.g. cholesterol (Brown and Peterson, 1966), elemental sulfur (Wieringa, 1966), pectin (Wieringa, 1955), poly-β-hydroxybutyrate (Delafield et al., 1965), rubber (Nette et al., 1959), and natural and artificial humic acids (Hayakawa and Nonomura, 1987a, 1987b). Most of these compounds strongly select organisms which produce visible zones of clearing or other changes in the medium.

Compounds with anti-fungal activity (antibiotics) are generally used to supplement isolation media to suppress fungal growth (Table 265), notably, cycloheximide (actidione, 50–100 μg/ml), as described by Williams and Davies (1965). Additionally, these authors found that pimaricin and nystatin (each at 10–50 μg/ml) were even more effective.

Compounds with anti-bacterial activity need to be used with care as some actinomycetes may also be sensitive to them. Polymyxin (5 μg/ml) and penicillin (1 μg/ml), for instance, suppress the growth of bacteria, but also inhibit some streptomycetes (Williams and Davies, 1965). Actinobacterial genera differ significantly in their sensitivity to anti-bacterial compounds, notably streptomycetes (Preobrazhenskaya et al., 1978). Thus, it may be more helpful to use anti-bacterial compounds for the isolation of other actinobacterial genera (Cross, 1982).

However, the selective isolation of certain species or groups of *Streptomyces* can be facilitated by supplementary media with antibiotics, as exemplified by the use of starch-casein agar containing rifampin (50 μg/ml) for the selective isolation of members of the *Streptomyces diastaticus* cluster by Williams et al. (1983a) and Vickers et al. (1984). A similar effect was described by Wellington et al. (1987), who used several media containing different C and N sources, as well as media supplemented with inhibitors.

Streptoverticil-producing *Streptomyces* species were isolated by Hanka et al. (1985) using a selective isolation medium supplemented with cycloheximide and nystatin (each at 50 μg/ml), to control fungal growth, and oxytetracycline (25 μg/ml) to suppress that of other actinomycete genera, including *Streptomyces* groups. By adding lysozyme (1000 μg/ml), Hanka and Schaadt (1988) enhanced the selectivity of this medium. The selectivity of this medium was enhanced by the addition of sodium propionate to suppress the growth of fungi (Crook et al., 1950; Table 265). Starch-casein-nitrate agar (Ottow, 1972) containing Rose Bengal (35 mg/l) suppresses the growth of most bacteria and inhibits spreading of fungi across isolation plates.

pH of isolation media and incubation temperatures. Most streptomycetes grow optimally at neutral pH values, i.e. they are neutrophilic organisms. Consequently, most isolation media have pH values of 7.0–7.5. In contrast, acidophilic streptomycetes grow best on media with a pH of 4.5 (Khan and Williams, 1975) and alkaliphilic strains show optimum growth on media with pH values of 10–11 (Mikami et al., 1982). Most streptomycetes isolated from soils are mesophilic; hence, isolation plates are usually incubated at 22–37°C (mostly at 28°C). In contrast, psychrophilic strains (e.g. from marine environments) grow best at 15–20°C, while thermotolerant and thermophilic strains grow well at higher temperatures (40, 45, 50, or 55°C). Thermophilic actinomycetes often form colonies within 2–5 d of incubation, whereas their mesophilic counterparts tend to produce visible colonies within 7–14 d. Marine and other psychrophilic actinomycetes may need several weeks (up to 10) for visible colonies to appear on isolation media.

In most cases, colonies of *Streptomyces* can be readily recognized by their macroscopic and microscopic appearance. *Streptomyces* can usually be purified by transferring colonies from isolation plates into nonselective medium. Williams and Wellington (1982b) stated that purification is "undoubtedly the most time-consuming and often the most frustrating stage of the isolation procedure". Acidiphilic streptomycetes can be readily isolated on acidified starch-casein agar supplemented with cycloheximide and nystatin (Kim et al., 2003b).

Isolation of antibiotic-producing actinomycetes. Antibiotic-producing streptomycetes are isolated following the same procedures as given above.

The activity of streptomycetes is normally tested after the isolation of pure cultures, though procedures are available to detect them on isolation plates. Antibiotic exhibiting strains can, for instance, be detected on initial dilution plates by flooding or spraying them with appropriate indicator organisms then incubating plates until zones of inhibition are detected (Lindner and Wallhausser, 1955; Wilde, 1964). Alternatively, antibiotic activity of the colonies against selected sensitive organisms can be examined by using a simple replication procedure (Lechevalier and Corke, 1953). Further information about selective techniques that can be used for the isolation and screening of antibiotic producing actinomycetes can be found in Nolan and Cross (1988). Protocols for the selective isolation of streptomycetes for the generation of spore suspensions and for more sophisticated experimental procedures are described by Kieser et al. (2000).

Isolation of thermophilic streptomycetes. As stated earlier, most actinomycetes, including thermophilic streptomycetes, originate from samples taken from high temperature environments (e.g. compost materials, manure heaps, and fodders). Consequently, high temperatures (45–60°C) should be used for the selective isolation of such organisms (Festenstein et al., 1965).

It is also important to incubate thermophilic streptomycetes in a humid atmosphere (Greiner-Mai et al., 1987) by incubating plates in large jars with water at the bottom; another effective method is to seal Petri dishes with masking tape.

Interestingly, the media recommended for the isolation of thermophilic actinomycetes, including streptomycetes, contain higher nutrient concentrations than those used for mesophilic strains. Sometimes, anti-fungal and anti-bacterial agents are added as supplements to such media (Goodfellow et al., 1987b; Lacey and Dutkiewicz, 1976b). Special procedures recommended for the isolation of thermophilic actinomycetes are available (Cross, 1968; Fergus, 1964; Gregory and Lacey, 1963; B. Kim et al., 2000; D. Kim et al., 1996; S.B. Kim et al., 1998; Uridil and Tetrault, 1959).

Isolation from aquatic habitats. The media listed in Table 266 can be used to isolate streptomycetes from water. Hsu and Lockwood (1975) found that chitin-agar was more effective than egg albumin, glycerol-arginine, starch-casein, and *Actinomyces* isolation agars for the incubation of actinomycetes from aquatic habitats (see also Table 266).

After dilution, water samples can be streaked directly onto solid medium. When low numbers of actinomycete samples are expected, they can be concentrated by membrane filtration [for details, see Burman et al. (1969)].

The selective isolation of streptomycetes from marine habitats can be enhanced when media are prepared using seawater or an equivalent. Media containing 25 or 75% seawater (Weyland, 1981a, 1981b), artificial seawater (Goodfellow and Haynes, 1984), and deionized water supplemented with 3.0% NaCl (Okami and Okazaki, 1978) have all proved to be effective. See Weyland (1981b) and Goodfellow and Haynes (1984) for further details.

Isolation from diseased plants. For the isolation of streptomycetes from diseased plant tissues, e.g. scabby potatoes or beet surface layers, three general steps have been recommended (see also Korn-Wendisch and Kutzner, 1992): sterilization of the surfaces of tubers, beets, or roots; maceration of plant tissues; and use of appropriate isolation media. Several authors have described detailed methods for the isolation of *Streptomyces scabiei* from potatoes (Adams and Lapwood, 1978; Archuleta and Easton, 1981; KenKnight and Munzie, 1939; Menzies and Dade, 1959; Taylor, 1936).

Nutritional requirements and media for sporulation. The vast majority of streptomycetes are nonfastidious organisms with a chemo-organotrophic metabolism. The nutritional requirements of streptomycetes are appropriate to an organic carbon source, such as starch, glucose, glycerol, and lactate, and usually met by the provision of a suitable inorganic nitrogen source, like ammonium or nitrate (Kutzner, 1981). However, different isolates can vary considerably in their carbon and nitrogen source utilization patterns, which are often used as taxonomic characters (e.g. Shirling and Gottlieb, 1966; Pridham and Tresner, 1974a; Williams et al., 1983a; Kämpfer et al., 1991). Widely used carbon sources include cellobiose, glucose, glycerol, D-mannose, and trehalose; useful nitrogen sources are ammonium, L-arginine, L-asparagine, and nitrate. Relatively few strains use organic acids, inulin, L-methionine, nitrite, or xylitol, but most can degrade casein, esculin, gelatin, and hypoxanthine. Streptomycetes grow well on many different media, but spore production is usually most prolific on those with a high carbon:nitrogen ratio (Kutzner, 1981). Streptomycetes generally require a good supply of free water for growth, but are unable to develop at high osmotic or matric potentials. Survival of streptomycetes in dry conditions is aided by the high resistance of arthrospores, in contrast to vegetative mycelia, to desiccation (Williams et al., 1972).

The need to add specific trace elements to culture media has not been studied in detail. Spicher (1955) described the positive effect of trace elements in soil on the growth of streptomycetes. However, many of the early media used (even the "synthetic" media) were not supplemented with trace elements. Indeed, recipes of many authors (Table 266 and Table 267) contained only a selected number of metal ions. A rather complete mixture (SPV-4; Table 267) has been found to be optimal for the growth of actinomycetes and other bacteria (Voelskow, 1988/89).

"Synthetic media" can be used for the cultivation of streptomycetes, though the need for specific nutritional requirements with respect to vitamins and organic growth factors has not been addressed. Growth rates and biomass production can be enhanced by using complex organic substrates (e.g. malt extract, oatmeal, or yeast extract). A combination of a complex organic carbon source with a single amino acid as nitrogen source (e.g. glutamic acid, arginine, or asparagine) is also suitable.

"General media" have been proposed for the growth of streptomycetes as they allow the completion of the *Streptomyces* life cycle, i.e. germination of spores, growth of substrate and aerial mycelium, and visible formation of spores (visible, because of the typical color of the spores). Some of these media were used in the International *Streptomyces* Project (ISP), among them glucose-yeast extract-malt extract, oatmeal, inorganic salts-starch, and glycerol-asparagine agars (Shirling and Gottlieb, 1966). Innumerable general media have been recommended for the growth of streptomycetes, four of which are of considerable practical value (Table 267). For additional media formulations, see Waksman (1961) and Williams and Cross (1971). $CaCO_3$ is added to some media, as Ca^{2+} promotes growth and neutralizes acids produced by many streptomycetes; such media also allow good sporulation. Cultures should be checked microscopically to detect the extent of sporulation as macroscopically heavy aerial mycelia may contain very few spores while aerial mycelia which are hardly detectable by the naked eye may be a good source of spores.

A list of specialized media especially for genetic studies on streptomycetes is given by Kieser et al. (2000).

Media containing soil, clay, minerals, and calcium humate. Soil promotes growth, sporulation, and pigmentation of actinomycetes/streptomycetes (Trolldenier, 1966) and is therefore often added to isolation media to increase the number of colonies. The addition of montmorillonite or calcium humate to liquid media stimulates the growth and metabolic activity of some actinomycetes (Martin et al., 1976). A similar effect has been observed for clay in dialysis tubes after a short lag period, an observation which may be explained by the adsorption of one or more inhibitory substances produced during growth. Adsorbing materials have a positive effect on the genetic stability of other bacteria and on fungi (Martin et al., 1976).

Temperature, pH, and oxygen. The environmental requirements and tolerances of streptomycetes have been described in detail by Kutzner (1981). Most streptomycetes grow at temperatures between 10 and 37°C and, hence, are regarded as mesophiles. Nevertheless, several species can grow at temperatures above 37°C, although most of them are thermotolerant rather than thermophilic in their responses. A variety of type strains studied by Williams et al. (1983a) grew at 10, 37, and 45°C, although a few grew slowly at 4°C (1983a). However, in many instances, the optimal temperature for rapid growth or maximal yield may not be the best choice for studying the production of secondary metabolites (e.g. antibiotics and pigments). This means that culture conditions are influenced by the aims of the study.

Most *Streptomyces* strains behave as neutrophiles in culture, growing between pH 5.0 and 9.0, with an optimum close to pH 7.0. Only a few of the type strains studied by Williams et al. (1983a) were able to grow at pH 4.3, though large populations of acidophilic and acidoduric strains have been reported from acidic soils (Hagedorn, 1976; Khan and Williams, 1975; Williams et al., 1971). Acidophilic streptomycetes grow from about pH 3.5 to 6.5, and optimally between pH 4.5–5.5. However, a wide spectrum of pH requirements exists among streptomycetes from acidic environments (Flowers and Williams, 1977b). Acidophilic strains are able to produce diastases (Williams and Flowers, 1978) and chitinases which have optima pH below that of corresponding enzymes from neutrophilic counterparts.

Populations of alkaliphilic streptomycetes with optimal growth at pH 9.0–9.5 have been isolated from soils in Japan by Mikami et al. (1982), who found that six of the type strains tested were able to grow at pH 11.5.

In addition, large populations of alkaliphilic streptomycetes have been isolated from a beach and dune sand system at Ross Links in Northumberland, UK (Antony-Babu and Goodfellow, 2008). Streptomycetes are generally regarded as obligate aerobes with a limited capacity for microaerophilic growth in culture (Kutzner, 1981) and dissimilatory reduction of nitrate is common. Whether streptomycetes grow aerobically or microaerophilically depends on the nutritional status of the medium.

In a nutrient medium, streptomycetes grow aerobically at the surface of the semisolid agar column, but in poor media or in a medium with a non-utilizable carbon source, they grow microaerophilically in semisolid agar. In stationary liquid culture, streptomycetes grow as pellicles at the surface, while the medium itself remains completely clear.

Cultivation and preparation of inoculum. Solid media in dishes or slants should be used for the growth of streptomycetes for subcultivation and maintenance as well as for most diagnostic tests. Many strains produce aerial mycelia and spores on solid media when the entire surface is covered by confluent growth. In contrast to most molds, *Streptomyces* colonies spread over a limited distance and, hence, a point inoculation will not usually lead result in confluent growth. However, if streaked onto a plate, some strains need empty spaces between the streaks (cross-hatch inoculation) to sporulate. Dry conditions are generally more suitable for sporulation. A horizontal incubation of slants for the first 2 d allows liquid to soak into the surface of the agar (Hopwood et al., 1985). A suspension of inoculum in liquid should be used as starting material for sporulation (Kieser et al., 2000).

For the propagation of cultures, single colonies should be selected and streaked onto fresh media. Successive rounds of mass culture should be avoided, especially in genetic studies, because this technique reduces the accumulation of revertants or the gradual loss of selected plasmids or both (Kieser et al., 2000). When streptomycetes are cultivated on solid media, morphological heterogeneity is often observed. More details can be found in Kieser et al. (2000).

Precultivation of grown colonies in liquid media is required to obtain a homogeneous suspension for some diagnostic tests (Kämpfer et al., 1991). Streptomycetes should be cultivated in liquid medium without agitation. This precultivation is necessary for certain physiological studies (e.g. degradation tests), for the provision of cell material for biochemical analysis, and for the production of secondary metabolites (e.g. antibiotics) or enzymes. For many detailed studies, e.g. for preparation of protoplasts for fusion, transformation or transfection, liquid cultures should also be started from an inoculum of spores.

The multicellular lifestyle of streptomycetes causes some problems in the study of metabolic properties because not all cells of the initial suspension are in the same physiological condition. In general, streptomycetes grow by mycelial elongation and branching. However, physiological homogeneity cannot be sustained when central parts of the colony become nutrient limited. Therefore, spore germlings are used in physiological studies, although large numbers of spores are needed. To avoid this problem, liquid cultures can be supplemented with, for example, dispersants, like agar, carboxymethylcellulose, Junlon, polyethylene glycol, starch, and sucrose. A summary of the advantages and disadvantages of these methods can be found in chapter 2 of Kieser et al. (2000). Cultures need to be shaken during incubation due to the highly aerobic nature of streptomycetes. Recommended procedures include Erlenmeyer flasks with the use of indentations or stainless steel springs, but tubes in a slanted position on a shaker or roller also allow an excellent supply of oxygen for small quantities of broth, 3–5 ml being enough for some physiological tests. However, it should be noted that some secondary metabolites (e.g. antibiotics and pigments, which are produced on solid media) may not be synthesized under these conditions.

Two media recommended by Korn-Wendisch and Kutzner (1992) have been widely used for submerged cultivation of streptomycetes (g/l): 1) GPYB broth (glucose, 10.0; peptone from casein, 5.0; yeast extract, 5.0; beef extract, 5.0; $CaCl_2 \cdot 2H_2O$, 0.74; pH 7.2); and 2) soybean meal-mannitol nutrient medium (soybean meal, 20.0; mannitol, 20.0; pH 7.2). Arthrospores and vegetative mycelium can be used as inoculation material for subculturing streptomycetes; vegetative mycelium occasionally includes "submerged spores" (Wilkin and Rhodes, 1955). Similar procedures are recommended by Kieser et al. (2000).

Spore suspensions stored at 4°C can be used for several weeks. A few glass beads should be added to the screw-cap tubes, as spores tend to settle and clump. Glass beads help to resuspend spores before use. The preparation of mycelia for detailed DNA or RNA studies is described in chapters 8 and 9 of Kieser et al. (2000).

Maintenance procedures

Several different procedures have been employed (Kirsop and Snell, 1984) for the short- and long-term preservation of microorganisms. Korn-Wendisch and Kutzner (1992) described three short-term preservation methods: (1) agar slope cultures may be stored at 4°C for few months; (2) spore suspensions can be mixed with soft water agar and kept at 4°C (Kutzner, 1972); and (3) glycerol can be added to spore suspensions (final concentration, 10%, v/v) which are then stored at −20°C (Wellington and Williams, 1978). After thawing, these cultures can serve as inocula for most diagnostic tests, except carbon utilization (Williams et al., 1983a).

For long-term preservation, Kieser et al. (2000) recommended the preparation of spore suspensions in 20% glycerol which can be frozen and maintained at −20°C. In another method, strains are grown in complex media like trypticase soy broth (TSB) agar, 20% glycerol plus 10% lactose are added, and the samples are stored in the vapor phase of liquid nitrogen. A third method uses drying on unglazed porcelain beads (Lange and Boyd, 1968), followed by soil culture (Pridham et al., 1973), and lyophilization (Hopwood and Ferguson, 1969).

Spore suspensions or homogenized mycelia mixed with glycerol to give a final concentration of 25% can be kept at −25°C for longer term preservation (Wellington and Williams, 1978). Alternatively, spores and mycelia suspended in 10% skim milk can be lyophilized. Liquid nitrogen cryopreservation is a very simple, reliable, and time-saving method. In this method, living cells are stored in small polyvinyl chloride (PVC) tubes ("straws") at −196°C; this procedure has been tested for various actinomycetes. First, strains are harvested from well-sporulated cultures grown on suitable agar media in Petri dishes. A 2 × 25 mm piece of sterile PVC tubing is pressed into the mycelial mat and agar, and carefully raised to excise the agar plug. This procedure is repeated until the tube is filled with agar. The tube is then placed in a sterile cryovial (the screw cap marked with the strain accession number); up to 13 tubes can be placed in a 1.8 ml vial. Two vials prepared for each strain are then fixed to a metal clamp for freezing in the gas phase of a liquid nitrogen container. After 10–15 min, when the temperature falls below −130°C, the clamp can be immersed in the liquid phase at −196°C. A container with a capacity of 250 l will hold at least 8000 vials or 4000 strains. For viability testing, one tube is removed from the vial within the nitrogen gas atmosphere of the container and placed directly and thawed on a suitable agar medium. The mycelium will be visible after a few days of incubation. Plugs may be pushed out of the tubes with a sterile needle when strains do not produce abundant mycelium.

Differentiation of the genus *Streptomyces* from other genera

It is standard practice to assign unknown actinomycetes to genera based on 16S rRNA gene sequence analyses. However, it can be difficult to distinguish between species using this approach, especially in the case of streptomycetes (e.g. Stackebrandt et al., 1991a, 1991b, 1992); Kumar and Goodfellow, 2008).

Members of the genus *Streptomyces* can often be distinguished from other filamentous actinomycetes on the basis of colony morphology (Table 268 and Table 269), in particular by aerial spore mass substrate mycelium and soluble pigment colors.

TABLE 268. Spore colors for the grouping of streptomycetes and representatives of each color group (according to Korn-Wendisch and Kutzner, 1992)

Color of aerial mycelium	Representative species (DSM no.)[a]
Yellow-gray: "griseus"	*S. griseus* (40236); *S. coelicolor* (40233)
Pink/light violet	*S. fradiae* (40063); *S. toxytricini* (40178)
Gray-pink/lavender: "cinnamomeus"	*S. lavendulae* (40069); *flavotricini* (40152)
Brown (plus gray or red)	*S. eurythermus* (40014); *S. fragilis* (40044)
Blue: "azureus"	*S. viridochromogenes* (40110); *S. cyaneus* (40108)
Blue-green: "glaucus"	*S. glaucescens* (40155)
Green: "prasinus"	*S. prasinus* (40099); *S. hirsutus* (40095)
Gray: "cinereus"	*S. violaceoruber* (40049); *S. echinatus* (40013)
White: "niveus"	*S. albus* (40313); *S. longisporus* (40166)
Not definable: white plus various plus various light-colored shades	*S. alboniger* (40043); *S. rimosus* (40260)

[a]DSM no. 40XXX, ISP no. 5XXX; e.g. 40236, ISP 5236.

TABLE 269. Colors of substrate mycelium and soluble pigment occurring in streptomycetes (according to Korn-Wendisch and Kutzner, 1992)

Color	Representative species (DSM no.)[a]
Orange to dark red (mainly endopigment)	*S. aurantiacus* (40412); *S. griseoruber* (40275)
	S. longispororuber (40599); *S. spectabilis* (40512)
Red to blue/violet (mainly endopigment)	*S. californicus* (40058); *S. cinereoruber* (40012)
	S. violaceus (40082); *S. purpurascens* (40310)
Red-violet to blue (endo- and/or exopigment)	*S. coelicolor* (40233); *S. cyaneus* (40108)
	S. violaceoruber (40049); *S. lateritius* (40163)
Yellow-orange/greenish-yellow (endo- and exopigment)	*S. atroolivaceus* (40137); *S. canarius* (40528)
	S. galbus (40089); *S. tendae* (40101)
Green to gray-olive (endo- and exopigment)	*S. flavoviridis* (40210); *S. olivoviridis* (40211)
	S. viridochromogenes (40110); *S. nigrifaciens* (40071)
Green (endopigment)	"*S. malachiticus*" (40167); "*S. malachitorectus*" (40333)
Red-brown to dark-brown (endo- and exopigment)	*S. badius* (40139); *S. eurythermus* (40014)
	S. phaeochromogenes (40073); *S. ramulosus* (40100)
Gray-brown to black (mainly endopigment)	*S. alboniger* (40043); *S. hygroscopicus* (40578)
	S. purpeofuscus (40283); *S. mirabilis* (40553)

[a]DSM no. 40XXX = ISP no. 5 XXX.

Traditional methods highly recommended for this purpose are described by Korn-Wendisch and Kutzner (1992). Antony-Babu et al. (2010) used a computer-assisted numerical analysis based on the color-grouping procedure of Williams et al. (1969) to group 321 alkaliphilic streptomycetes grown on oatmeal agar (ISP 3) and peptone-yeast extract-iron agar (ISP 6). With this method, large numbers of streptomycetes can be assigned without using polyphasic taxonomic approaches.

In addition, streptomycetes can often be distinguished from other filamentous actinomycetes on the basis of morphological properties, notably aerial mycelium, arthrospores, and vegetative mycelium (Figure 357, Figure 358, and Figure 359). Details on the procedure used to detect such properties can be found in Korn-Wendisch and Kutzner (1992) and chapter 3 of Kieser et al. (2000).

Members of the genus *Streptomyces* can also be distinguished from related taxa using chemotaxonomic procedures (Lechevalier and Lechevalier, 1970b). Streptomycetes typically contain LL-A_2pm in cell wall or whole-cell hydrolysates (Lechevalier and Lechevalier, 1970b, 1970c), they lack mycolic acids, and produce major amounts of iso- and anteiso-methyl branched fatty acids (Kroppenstedt, 1985) (Table 263). Major menaquinones are hexa- and octa-hydrogenated menaquinones with nine isoprene units (Kim et al., 2003b). An important chemotaxonomic character for the differentiation of *Kitasatospora* from *Streptomyces*

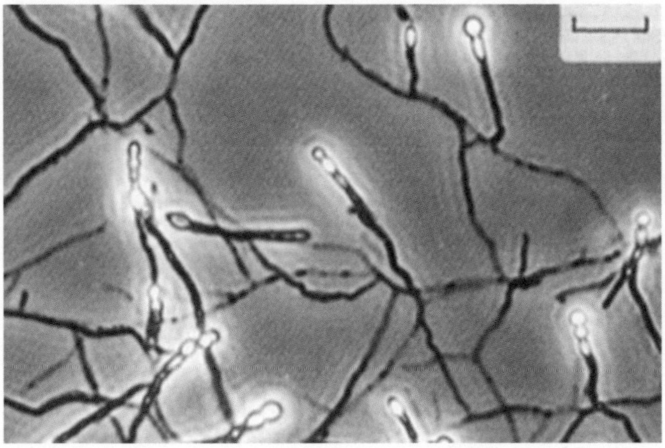

FIGURE 357. Spore chains *Streptomyces* (*Microellobosporia*) species on aerial mycelium. Light microscopy. Bar, 5 μm. (Courtesy of T. Cross, University of Bradford, Bradford, U.K.)

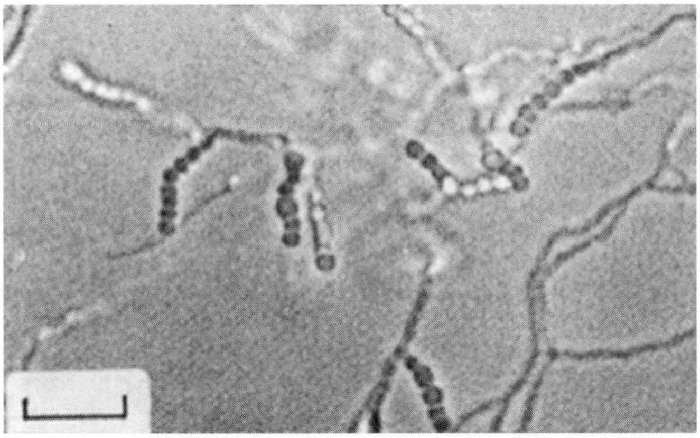

FIGURE 358. Spore chains of *Streptomyces carpinensis* (*Elytrosporangium carpinese*) on substrate mycelium. Light microscopy. Bar, 5 μm. (Courtesy of T. Cross, University of Bradford, Bradford, U.K.)

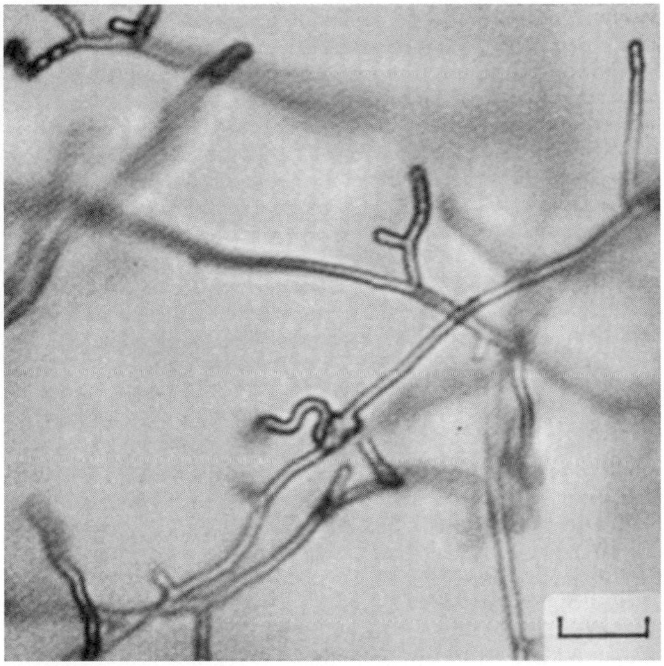

FIGURE 359. Spore chains of *Streptomyces carpinensis* (*Elytrosporangium carpinense*) on aerial mycelium. Light microscopy. Bar, 5 μm. (Courtesy of T. Cross, University of Bradford, Bradford, U.K.)

is the presence of *meso*-A_2pm in whole-cell hydrolysates (Table 263). In *Kitasatospora* strains, the *meso*-A_2pm content is 49–89%, whereas in *Streptomyces* strains it is 1–16% (Zhang et al., 1997). The predominant diamino acid of strains belonging to the genus *Streptacidiphilus* is (like *Streptomyces*) LL-A_2pm (Kim et al., 2003b). Further useful characters for species identification are shown in Table 274.

Taxonomic comments

Phenotypic methods comprise all procedures that are not directed towards analyses of DNA or RNA, and include chemotaxonomic techniques. Most early studies on streptomycetes (between 1916 and 1943) were carried out by soil microbiologists who were mainly interested in ecological questions; hence, only a few species were described at that time. Descriptions were mainly based on morphological criteria, pigmentation, and ecological requirements (Jensen, 1930; Waksman, 1919; Waksman and Curtis, 1916). However, the discovery of actinomycin from *Streptomyces antibioticus* (Waksman and Woodruff, 1940) promoted widespread interest in streptomycetes as a source of novel bioactive compounds. The focus on screening streptomycetes for novel bioactive compounds led to a widespread tendency for streptomycetes to be assigned to novel species on the basis of their ability to produce new natural products. This tendency led to an explosion of species descriptions and resulted in an overclassification of the genus with over 3000 species being recognized (Trejo, 1970).

The International *Streptomyces* Project (ISP) introduced 1964 standard criteria for determining species (Shirling and Gottlieb, 1968a, 1968b, 1969, 1972). These descriptions were based mainly on morphology (i.e. spore chain arrangement, spore surface ornamentation, color of spores, substrate mycelium, soluble pigments, and production of melanin pigments), and a few physiological properties, which were mainly restricted to utilization tests of different carbon sources. As a result, more than 450 *Streptomyces* species were redescribed and their type strains were deposited in internationally recognized culture collections. However, these studies did not directly lead to the generation of schemes for the identification of *Streptomyces* species.

Numerical taxonomic methods were developed in the 1960s for both the classification and identification of bacteria, including streptomycetes. Silvestri et al. (1962) carried out the first numerical taxonomic studies on streptomycetes and found considerable diversity within the genus, as well as groups which corresponded to initial morphological descriptions. However, these studies had little impact on *Streptomyces* systematics and did not

TABLE 270. Numerical classifications of streptomycetes (modified according to Korn-Wendisch and Kutzner, 1992)

Number of strains	No. characters (features)	Nature of material	No. clusters	No. unclustered strains	Reference
159	105	"Species"	24	16	Silvestri et al. (1962)
18	46	Isolates	5		Williams et al. (1969)
448	31	ISP "species"	14/21	168/37	Kurylowicz et al. (1967)[a]
618	24	ISP "species"	15	218	Gyllenberg (1976)
111	185	*Streptomyces* with verticals and pseudoverticils, formerly *Streptoverticillium*	24		Locci et al. (1981)
475	139	394 ISP species plus others	73	28	Williams et al. (1983a)
821	329	394 ISP species plus others	15 (major), 34 (minor)	40	Kämpfer et al. (1991)

[a]14 and 168 were obtained by the Wroclaw dendrite method. 21 and 37 were obtained by the centrifugal correlation method.

result in any nomenclatural changes, despite the development of other small databases for the identification of streptomycetes (Kurylowicz et al., 1975; Gyllenberg, 1976; Table 270).

A large-scale numerical taxonomic study was undertaken by Williams et al. (1983a) who analyzed 475 strains (including 394 *Streptomyces* type cultures from the ISP) for 139 unit characters using the simple matching (S_{SM}) and Jaccard coefficients (S_J), and the mean linkage algorithm UPGMA (unweighted pair group method with arithmetic means). The 394 type strains were assigned to 19 major (6–71 strains), 40 minor (2–5 strains), and 18 single clusters recovered at the 77±5% S_{SM} level. The minor and single-membered clusters were equated with species and the major clusters with species-groups.

The largest species group, *Streptomyces albidoflavus* (cluster 1), contained 71 strains, including 44 type strains, 15 representatives of species with names that were not validly published, and 12 unnamed strains. This taxon was subsequently subdivided into three subclusters: cluster 1a, *Streptomyces albidoflavus* subsp. *albidoflavus* (20 strains); cluster 1b, *Streptomyces albidoflavus* subsp. *anulatus* (38 strains); and cluster 1c, *Streptomyces albidoflavus* subsp. *halstedii* (13 strains; Williams et al., 1989). Cluster 1 included strains which showed considerable phenotypic diversity, though almost all of the strains formed yellow gray colonies, smooth spores in straight chains, did not produce melanin pigments, and were resistant to several antibiotics, including cephaloridine, lincomycin, and penicillin. Nearly 40% of the strains produced compounds with anti-fungal activity, 32% of the compounds were active against Gram-stain-positive micro-organisms, and 10% inhibited Gram-stain-negative micro-organisms (Williams et al., 1983b). This example highlights the extensive diversity found within a single cluster and exemplifies problems associated with streptomycete systematics (Anderson and Wellington, 2001).

The comprehensive survey of Williams et al. (1983a) led to a reduction in the number of described *Streptomyces* species (Williams et al., 1989), although the problem of overspeciation remained. Additionally, isolates from natural habitats did not match the reference strains used to construct identification schemes (Goodfellow and Dickenson, 1985). The problem was addressed by the generation of probability matrices for the identification of streptomycetes (Langham et al., 1989; Williams et al., 1983b), but these developments were not widely adopted by the scientific community.

The chapter on *Streptomyces* in the 1989 edition of *Bergey's Manual of Systematic Bacteriology* (Williams et al., 1989) was based on the numerical taxonomy studies of Williams and his colleagues. This chapter included descriptions of 142 *Streptomyces* species, in contrast to the 463 species described in the 1974 edition of *Bergey's Manual of Determinative Bacteriology* (Pridham and Tresner, 1974a).

Kämpfer et al. (1991) carried out an extensive numerical taxonomic analysis of the genus *Streptomyces* and, where possible, included more than one strain of each species. These workers examined 821 strains for 329 physiological properties and compared their data with those of Williams et al. (1983a). They also examined their strains for genetic and chemotaxonomic properties and compared them with the numerical data. They recognized many of the clusters defined by Williams et al. (1983a); for example, the *Streptomyces albidoflavus*, *Streptomyces anulatus*, *Streptomyces griseus*, and *Streptomyces halstedii* groups were recovered in both studies; 28 of the *Streptomyces griseus* strains were assigned to cluster 1. Interestingly, groups often contained strains which shared the same specific epithet indicating that some previous classifications were reliable. However, exceptions were also observed, for example, *Streptomyces hygroscopicus* strains were recovered in cluster 1 and in several other clusters and subclusters.

Data generated in this study were used to construct a probability matrix for the identification of streptomycetes (Kämpfer and Kroppenstedt, 1991), but this identification scheme was not widely used by other research groups.

Chemotaxonomic and molecular methods were used in parallel with the application of numerical taxonomic procedures for the classification of streptomycetes. The additional phenotypic methods used to study streptomycetes included cell-wall analysis (Lechevalier and Lechevalier, 1970a, 1970b, 1970c), fatty acid profiling (Hofheinz and Grisebach, 1965; Kroppenstedt, 1992; Lechevalier, 1977; Saddler et al., 1986, 1987), rapid biochemical tests based on the use of 4-methyl-umbelliferone-linked substrates (Goodfellow et al., 1987b), serological assays (Ridell et al., 1986), phage typing (Korn-Wendisch and Schneider, 1992; Wellington and Williams, 1981b), Curie-point pyrolysis MS of whole cells (Ferguson et al., 1997; Sanglier et al., 1992), whole organism protein profiling (Goodfellow and O'Donnell, 1993; Lanoot et al., 2002; Manchester et al.,

1990), and comparison of ribosomal protein patterns (Ochi, 1989, 1992, 1995).

Fatty acids. Hofheinz and Grisebach (1965) examined selected *Saccharopolyspora erythraeus* (formerly "*Streptomyces erythraeus*") and *Streptomyces halstedii* strains to clarify the biosynthetic pathway of branched-chain fatty acid synthesis. They found that *Streptomyces* strains synthesized terminally branched fatty acids; the starting compound, 2-methylbutyrate, led to the synthesis of anteiso-branched fatty acids with an odd number of carbon atoms. In contrast, isovalerate and isobutyrate were the starting compounds that lead to the formation of iso-branched fatty acids with even and odd numbers of carbon atoms, respectively. Consequently, iso- and anteiso-branched fatty acids appear in pairs with odd numbers of carbon atoms.

Hofheinz and Grisebach also identified individual fatty acids by first separating them as their methyl esters by GC on different stationary phases, then analyzing the results by comparing the equivalent chain-lengths of unknown fatty acids with those of standard mixtures. Preparative GC and physical methods, such as MS and NMR spectrometry, confirmed these results. Iso- and anteiso-branched fatty acids with chain lengths of 15 and 17 carbon atoms were detected in both *Saccharopolyspora erythraeus* and *Streptomyces halstedii*. In addition, high amounts of 14-methylpentadecanoic acid ($C_{16:0}$ iso) were found, whereas minor amounts of unbranched fatty acids, tuberculostearic acid and their homologs were detected in the *Saccharopolyspora erythraea* strains, but not in *Streptomyces halstedii*.

A limited number of streptomycetes synthesize small amounts of hydroxy fatty acids in the presence of optimal amounts of oxygen. These fatty acids are easily destroyed in a non-deactivated injection port of capillary GC systems. Hence, they are not always detected. However, some streptomycetes produce hydroxy fatty acids which are highly diagnostic when strains are grown under reproducible culture conditions. Hydroxy fatty acids have been detected in all strains of *Streptomyces coelicolor* (30), in 20 of 27 *Streptomyces hygroscopicus* strains, *Streptomyces rimosus* (14), and *Streptomyces violaceusniger* (18), but not in *Streptomyces albus* (33), *Streptomyces fradiae* (25), *Streptomyces glaucescens* (8), *Streptomyces griseus* (22), *Streptomyces lavendulae* (18), *Streptomyces violaceoruber* (16), or *Streptomyces viridochromogenes* (25; Kroppenstedt, 1992; R.M. Kroppenstedt, unpublished observations).

Standardized growth and cultivation conditions are prerequisites for generating fatty acid patterns for classification below the genus level (Saddler et al., 1986). These workers examined the fatty acid profiles of *Streptomyces cyaneus* strains and associated soil isolates, which produced a blue aerial spore mass; 13 of 19 blue-spored strains belonged to the *Streptomyces cyaneus* cluster (Hütter, 1962; Korn et al., 1978; Pridham and Tresner, 1974a). Saddler et al. (1987) showed that 8 of 10 blue-spored isolates were grouped together based on fatty acid data, whereas 17 of 34 *Streptomyces cyaneus* strains were assigned to a separate cluster. Saddler and his colleagues concluded that conventional features, like spore chain morphology, color, and ornamentation of spores, were not reliable for the classification of streptomycetes, but would be helpful for presumptive identification, a point also made by Williams et al. (1983a). However, Saddler and his colleagues demonstrated that the *Streptomyces cyaneus* taxon as defined by Williams et al. (1983a) was heterogeneous.

In general, fatty acid patterns cannot be used to delimit *Streptomyces* species (Phillips, 1992; R.M. Kroppenstedt, unpublished observation), but are still useful for the rapid characterization (independent of the taxonomic status) of large numbers of wild-type streptomycetes isolated from the environment when used under standardized conditions (Saddler et al., 1987). Identification and quantification can be done by using the automated commercially available MIDI system consisting of a Hewlett Packard model 5890 capillary GC and a computer with specific software (Microbial ID, Inc., Newark, DE). The MIDI system automatically identifies fatty acids using fatty acid standard mixtures for comparison. However, comparisons of different analytical methods show that numerical methods cannot be used for classification, because different methods give different groupings. Additionally, the resolution is not good; hence, too many strains are grouped together into some clusters (Kämpfer et al., 1991; Williams et al., 1983a).

Curie-point pyrolysis MS. Another method that has been applied to the classification and identification of actinomycetes is pyrolysis MS (PyMS; Sanglier et al., 1992; Ferguson et al., 1997). Whole cells exposed to high temperatures are degraded in a nonoxidative environment leading to the generation of pyrolysate, which can be analyzed by MS. This method needs to be rigorously standardized and results in the production of a fingerprint for each organism.

Sanglier et al. (1992) applied this method to strains belonging to the *Streptomyces albidoflavus* species-group defined by Williams et al. (1983a) and recovered *Streptomyces albidoflavus* and *Streptomyces anulatus* strains in distinct groups. The six *Streptomyces halstedii* strains (the third subgroup) were assigned to three groups. *Streptomyces albidoflavus* and *Streptomyces anulatus* strains were also recovered in different groups by Kämpfer et al. (1991). They also found that *Streptomyces anulatus* ISP 5361^{T}, the strain used to name the *Streptomyces anulatus* cluster, formed a single-membered cluster (Kämpfer et al., 1991).

Serology. Serological methods have rarely been used in *Streptomyces* systematics. Ridell et al. (1986) used antisera against the mycelia of streptomycetes, streptoverticillia, and *Nocardiopsis* species to confirm the high similarity between *Streptomyces lavendulae* and streptoverticillia (Kämpfer et al., 1991; Witt and Stackebrandt, 1990). The antisera of Kirby and Rybick (1986) were shown to be genus-specific and to a certain degree also group-specific when tested against *Streptomyces griseus* (*Streptomyces anulatus*, cluster 1B of Williams et al., 1983a) and "*Streptomyces cattleya*" (cluster 47 of Williams et al., 1983a). Wipat et al. (1994) generated a monoclonal antibody to "*Streptomyces lividans*" 1326 which was specific for "*Streptomyces lividans*" strain 1326 and for strains assigned to cluster 21 by Williams et al. (1983a).

Phage typing. Phage typing can be used for host identification at the genus and species levels (Korn et al., 1978; Kutzner, 1961a, 1961b; Wellington and Williams, 1981b; Welsch et al., 1957). Many actinophages, mainly virulent, have been used for phage typing. Two different groups of streptomycete phages exist, namely polyvalent phages (e.g. φC31; Chater et al., 1986) and species-specific phages (Anderson and Wellington, 2001; Table 271). Actinophages are specific at the genus level (e.g. Wellington and Williams, 1981b; Korn-Wendisch, 1982;

TABLE 271. Species-specific actinophages of the genus *Streptomyces* (modified according to Anderson and Wellington, 2001)

Phage	Host	Host species group	Host cluster no.[a]	Cluster no.[b]	Reference
S3	*S. albus* DSM 40313^T	*S. albus*	16	32	Korn-Wendisch and Schneider (1992)
SAt1	*S. azureus* ATCC 14921^T	*S. cyaneus*	18	9	Ogata et al. (1985)
100	"*S. caesius*" ATCC 19828	*S. griseoruber*	21	6	Wellington and Williams (1981b)
98	*S. coelicolor* Müller ATCC 23899^T	*S. albidoflavus*	1A	1-1	Wellington and Williams (1981b)
14, 24, 233	*S. coelicolor* Müller ATCC 23899^T	*S. albidoflavus*	1A	1-1	Korn-Wendisch and Schneider (1992)
90	*S. griseinus* ATCC 23915^T	*S. albidoflavus*	1B	1-3	Wellington and Williams (1981b)
89, DP 9	*S. griseus* ATCC 23345^T	*S. albidoflavus*	1B	1-3	Wellington and Williams (1981b)
41	*S. matensis* ATCC 23935^T	*S. rochei*	12	6	Wellington and Williams (1981b)
33	"*S. scabies*" ATCC 23962	*S. atroolivaceus*	3	1-3	Wellington and Williams (1981b)
SV1, SV2	*S. venezuelae* ATCC 10712^T	*S. violaceus*	6	2	Stuttard (1982)
4, 5a, 5b, 49	*S. violaceoruber* DSM 40049^T	*S. violaceoruber*	SMC	69	Korn-Wendisch and Schneider (1992)

[a]Clusters according to Williams et al. (1983a, 1983b); SMC, Single-member cluster (Williams et al., 1989).

[b]Clusters according to Kämpfer et al. (1991).

Prauser, 1984). Actinophage host range studies helped to justify the transfer of the genera *Actinopycnidium, Actinosporangium, Chainia, Elytrosporangium, Microellobosporia, Kitasatoa,* and *Streptoverticillium* to the genus *Streptomyces* (Goodfellow et al., 1986a, 1986c, 1986d, 1986e; Witt and Stackebrandt, 1990). Similar studies supported the transfer of *Actinoplanes armeniacus* to the genus *Streptomyces* (Kroppenstedt et al., 1981; Wellington and Williams, 1981a) and "*Streptomyces erythraeus*" to the genus *Saccharopolyspora* (Labeda, 1987). Phage typing has been shown to be less useful for species or group identification of *Streptomyces,* though there are a few exceptions (Table 271).

Phages are widely used in industrial microbiology (Carvajal, 1953; Ogata, 1980) and in genetic studies (see Chater, 1986), as exemplified in chapter 12 of Kieser et al. (2000). The temperate phage, φC31, is one of the best-investigated temperate actinophages and has a broad host range within the genus *Streptomyces* (Lomovskaya et al., 1980). This phage has been employed for many purposes (e.g. transfection, transduction, detection of transposon-like elements of host DNA, and cloning); details can be found in chapter 12 of Kieser et al. (2000).

Protein profiling. PAGE is used to analyze total protein extracts resulting in more or less complex banding patterns. These patterns have been used to clarify relationships between the species and subspecies levels of various bacterial genera. One-dimensional (1-D) and two-dimensional (2-D) protein electrophoresis can be used to determine protein patterns. Manchester et al. (1990) used one-dimensional protein electrophoresis to examine 37 *Streptomyces* strains, including five streptoverticillia, and found some taxonomic correlations between the resultant profiles and groups based on phenotypes (Kämpfer et al., 1991; Williams et al., 1983a) and DNA hybridization data (Table 272). However, only a few of these correlations were confirmed by Lanoot et al. (2002).

PAGE and DNA–DNA hybridizations were used to elucidate the taxonomy of *Streptomyces* isolates that caused common potato scab (Paradis et al., 1994). The isolates were the subject of an SDS-PAGE analysis and found to belong to two groups with a correlation coefficient of 0.75. The same groups were recovered in the DNA–DNA relatedness study, though not in the corresponding fatty acid analysis. This lack of correlation can be attributed to the influence of growth conditions on fatty acid profiles (Saddler et al., 1986, 1987). The protein profiling did not allow pathogenic and nonpathogenic strains to be differentiated. Lanoot et al. (2002) used SDS-PAGE of whole-cell proteins in an examination of 93 *Streptomyces* reference strains. Subsequent computer-assisted numerical analysis revealed 24 clusters, which included strains with very similar protein profiles. Several type strains were assigned to five clusters, which had visually identical patterns. DNA–DNA hybridizations of these type strains revealed similarities higher than 70%. On the basis of these results, *Streptomyces albosporeus* subsp. *albosporeus* LMG 19403^T was considered to be a subjective synonym of *Streptomyces aurantiacus* LMG 19358^T, *Streptomyces aminophilus* LMG 19319^T was a subjective synonym of *Streptomyces cacaoi* subsp. *cacaoi* LMG 19320^T, *Streptomyces niveus* LMG 19395^T and *Streptomyces spheroides* LMG 19392^T were subjective synonyms of *Streptomyces caeruleus* LMG 19399^T, and *Streptomyces violatus* LMG 19397^T was a subjective synonym of *Streptomyces violaceus* LMG 19360^T (Table 272).

Two-dimensional PAGE of the total cellular proteins gives greater resolution than one-dimensional studies. Very complex patterns can be obtained with 2-D PAGE, though this method seems to be too sensitive to differentiate between proteins with high rates of evolution (Hori and Osawa, 1987). Two-dimensional PAGE studies designed to distinguish between ribosomal proteins of streptomycetes were first described by Mikulik et al. (1982) and later by Ochi (1989). Subsequently, AT-L30 proteins were found to give genus-specific profiles (Ochi, 1992), while analyses of the N-terminal sequences of the ribosomal AT-L30 protein allowed streptomycete strains to be assigned to different taxonomic groups (Ochi, 1995); these groups were assigned to phylogenetic groupings which suggested that the genus *Streptomyces* was well described. However, no correlation was found between Ochi's groupings and numerical phenetic groups (Kämpfer et al., 1991; Williams et al., 1983a). Details of these groupings are in Table 272.

Multilocus enzyme electrophoresis (MLEE) results in more specific patterns than protein profiling. MLEE depends on the relative mobilities of cellular enzymes in a gel matrix. In a small study of 24 *Streptomyces* strains (Oh et al., 1996), both inter- and intraspecific characterization of the organisms was achieved, if the appropriate enzymes were used.

The isolation and sequencing of specific proteins led to more detailed taxonomic studies of some *Streptomyces*. The *Streptomyces* subtilisin inhibitor protein (SSI), for example, was used by Taguchi et al. (1996) to determine the taxonomic status of "*Streptomyces lividans*" 66, *Streptomyces coelicolor* Müller ISP 5233^T, and *Streptomyces coelicolor* A3(2); this protein plays unidentified role(s) in physiological or morphological regulation. Ribosomal sequence comparisons were supported by the alignments of the SSI indicating that *Streptomyces coelicolor* A3(2) is more closely related to "*Streptomyces lividans*" 66 [cluster 21 of Williams et al. (1983a)] than to the type strain, *Streptomyces coelicolor* Müller ISP 5233^T (cluster 1).

Genotypic methods. Genotypic methods encompass all procedures that are directed towards DNA and RNA molecules (Schleifer and Stackebrandt, 1983; Vandamme et al., 1996). The analysis of bacterial genomes by molecular methods has provided a new basis for studying bacterial taxonomy. In some cases, phylogenetic relationships of prokaryotes could be studied at the genus, species, and even subspecies level, leading to a wide use of these methods in modern taxonomic studies. Vandamme et al. (1996) described the general taxonomic values of different molecular techniques, while Anderson and Wellington (2001) considered their use in streptomycete systematics. In the following, the usefulness of these methods in the delineation of species within the genus *Streptomyces* is discussed briefly.

By comparative analysis of sequences of homologous and genetically stable semantides, it was shown that several classification systems based on morphology and physiology did not reflect the natural relationships among actinomycetes and related organisms (Stackebrandt and Schumann, 2006). In this respect, sequence analysis of the rRNA gene has revolutionized our insight into phylogenetic lineages of major taxonomic groups. Nevertheless, 16S rRNA gene sequencing does not always have the resolving power to delimit species. Additionally, while new strains should not be assigned to existing species solely on the basis of this molecule, rRNA gene sequence comparisons do play an important role in the taxonomy of *Streptomyces* and in studying horizontal gene transfer within the genus (Huddleston et al., 1997).

Three regions within 16S rRNA genes have been found to show enough variability to be useful as genus-specific (*a* and *b* regions) and species-specific (*c* regions) probes [see Stackebrandt et al. (1991a, 1991b, 1992) and Anderson and Wellington (2001)]. Next to 16S rRNA genes, 23S rRNA and 5S rRNA genes (Mehling et al., 1995), 16S–23S rRNA internally transcribed spacer (ITS) sequences (Song et al., 2003), and ribosomal protein sequences have been used to investigate relationships between *Streptomyces* (Liao and Dennis, 1994; Ochi, 1995). The ITS sequences of the six rDNA operons from two *Streptomyces ambofaciens* strains were analyzed by Wenner et al. (2002), who confirmed that a high degree of ITS variability was a common characteristic amongst *Streptomyces* species. They also showed that recombination frequently occurs between rDNA loci leading to the exchange of nucleotide blocks. Given such intraspecific variation and intragenomic heterogeneity rRNA gene sequences cannot solely be used for taxonomic studies.

DNA hybridization. The DNA–DNA hybridization values (%) and the decrease in thermal stability of hybrids are currently used as the "gold standards" for species delineation in bacteriology (Wayne et al., 1987), despite the recommendations of the Ad hoc committee for the re-evaluation of the species definition in light of the application of other methods (Stackebrandt et al., 2002), DNA–DNA similarity and changes in melting temperature (ΔT_m; Wayne et al., 1987) still remain the acknowledged standards for the definition of species.

DNA–DNA hybridizations of total chromosomal DNA have been widely used in the classification of *Streptomyces* species. In an initial study, strains of the *Streptomyces albidoflavus* cluster 1, defined by Williams et al. (1983a) using numerical phenetic methods, were the subject of DNA–DNA hybridization studies, performed by reassociation of labeled DNA on nitrocellulose filters (Mordarski et al., 1986). Good congruence was found between the results of the two approaches and the homogeneity of the *Streptomyces albidoflavus* subcluster *albidoflavus* was confirmed. However, two further subclusters obtained by DNA–DNA hybridization were not congruent with the *Streptomyces anulatus* or *Streptomyces halstedii* subclusters of Williams et al. (1983a), though some correlation was found with the groupings of Kämpfer et al. (1991). DNA–DNA pairing data supported the assignment of *Streptoverticillium* strains to the genus *Streptomyces* (reassociation of labeled DNA on filters; Witt and Stackebrandt, 1990), a result that was confirmed in the numerical phenetic study of Kämpfer et al. (1991).

The most extensive DNA–DNA hybridization (thermal renaturation method) studies on strains assigned to some of the major phenetic groups of Williams et al. (1983a) were carried out by Labeda and his colleagues (Labeda, 1993, 1996, 1998; Labeda and Lyons, 1991a, 1991b). They found little correlation between the DNA–DNA pairing and numerical phenetic data with respect to the *Streptomyces cyaneus* (Labeda and Lyons, 1991a), *Streptomyces violaceusniger* (Labeda and Lyons, 1991b), *Streptomyces lavendulae* (Labeda, 1993), the verticil-forming streptomycetes (formerly *Streptoverticillium* species; Labeda, 1996; Hatano et al., 2003), and *Streptomyces fulvissimus* and *Streptomyces griseoviridis* phenotypic clusters (Labeda, 1998), though some correlation was found with phenotypic groups circumscribed by Kämpfer et al. (1991) (Table 272). The fact that certain regions within the *Streptomyces* chromosome show considerable genetic instability supports the continued use of DNA–DNA pairing (Redenbach et al., 1993). However, DNA–DNA hybridization data can be influenced by the presence of large plasmids in *Streptomyces* strains. The general properties of *Streptomyces* plasmids and their use for gene cloning are considered in chapter 11 of Kieser et al. (2000).

TABLE 272. List of *Streptomyces* species (including *Kitasatospora* and *Streptacidiphilus* species) arranged according to the grouping given in Figure 339

Species names and groups[1]	No. in lists of species[2]	Type strain	Accession no.	Wil 83a[3]	Wil 89[4]	Käm 91[5]	Hat 03[6]	Lan 02[7]	Ful 95[8]	Och 95[9]	Kat 97[10]	Lab[11]	Sch[12]	Lan 04[13]	Lan 02[14]	Lan 04[15]	Lan 04[16]	Lan 04[17]	Lan 04[18]	Lan 04[19]	Lan 04[20]
Most closely related to group S. costaricanus et rel.:																					
S. galbus	195	DSM 40089, ATCC 23910, LMG 19879, ISP 5089	X79852	A 15	I 08	006 1-10							Sch00	Lan2-00	BENP	+		+			cl22
S. longwoodensis	295	DSM 41677, LMG 20096, NBRC 14251	AB184580										Sch00	Lan2-00	BENP	+		+			cl22
S. bungoensis	78	DSM 41781, LMG 20439, NBRC 15711	AB184696											Lan2-00	BENP	+		+			cl22
S. corchorusii	133	DSM 40340, ATCC 25444, LMG 20488, ISP 5340, NBRC 13032	AB184267	A 20	I 13	009 1-19							Sch04	Lan2-26	BENP	+	(a)	+			cl52
S. canarius	84	DSM 40528, ATCC 27423, LMG 20443, ISP 5528, NBRC 13431	AB184396	A 20	I 13	009 1-19							Sch00	Lan2-00	BENP	+		+			cl22
S. olivaceoviridis	351	DSM 40334, ATCC 23630, LMG 19324, ISP 5334, NBRC 13066	AB184288	A 20	I 13	009 1-19		La-21		OC-III			Sch04	Lan2-00	BENP +	+		+		+	cl53
S. capoamus	91	DSM 40494, ATCC 19006, LMG 20447, ISP 5494, JCM 4734	AB045877	C 45	II 13	1-7 1-15							Sch00		BENP	+		+		+	+
S. regensis	405	DSM 40551, ATCC 27461, LMG 20300, ISP 5551, NRRL B-11479	DQ026649	A 20	I 13	009 1-19							Sch00	Lan2-00	BENP	+		+			cl52
S. griseochromogenes	222	DSM 40499, ATCC 14511, LMG 19891, ISP 5499, NBRC 13413	AB184387	A 18	I 11	1-5 011						L2	Sch00	Lan2-00	BENP	+		+		+	cl22
S. cellostaticus	97	DSM 40189, ATCC 23894, LMG 20452, ISP 5189, NBRC 12849	AB184192	A 06	I 05	007 003							Sch15	Lan2-00	BENP	+		+			cl55
S. yokosukanensis	531	DSM 40224, ATCC 25520, LMG 21040, ISP 5224, NRRL B-3353	DQ026652	A 30	II 06	009 1-19							Sch00	Lan2-00	BENP	+		+			cl22
S. antibioticus	35	DSM 40234, ATCC 8663, LMG 20412, ISP 5234, NRRL B-1701	AY999776	A 31	I 21	1-7 1-15				OC-IV			Sch00	Lan2-00	BENP	+		+			cl13
S. griseoruber	232	DSM 40281, ATCC 23919, LMG 19325, ISP 5281, NBRC 12873	AB184209	A 21	I 14	018 023		La-00		OC-I			Sch00	Lan2-00	BENP +	+		+	cl15	+	+
S. cinnabarinus	115	DSM 40467, ATCC 23617, LMG 20467, ISP 5467, NBRC 13028	AB184266	A 18	I 11	009 1-19						L2	Sch00	Lan2-00	BENP	+		+		+	+
S. acidiscabies	5	DSM 41668, ATCC 49003, LMG 19856	D63865										Sch00	Lan2-00	BENP	+		+		+	+

(continued)

"Footnotes are given on pp. 1558-1559."

TABLE 272. (continued)

								Morphological characters[24]				Physiological tests[25]										
Guo 08[21] 16S rRNA	Guo 08[21] *atpD*	Guo 08[21] *gyrB*	Guo 08[21] *recA*	Guo 08[21] *rpoB*	Guo 08[21] *trpB*	Guo 08[21] concat.	Kim 04[23] *rpoB*	I	II	III	IV	I	II	III	IV	V	VI	VII	VIII	IX	X	XI
								Gy	S	C+	SM	+	+	+	–	+	n	–	+	+	n	–
								Gy	S	C–	SM	+	+	+	–	+	+	±	+	+	+	–
								Gy	S	C+	SPY	+	+	+	–	+	+	n	+	–	–	n
							+	Gy	S	C–	SM	+	+	+	+	+	n	+	+	+	n	+
								Y	S	C–	SM	+	+	+	+	+	+	+	+	+	–	+
							+	Gy	S	C–	SM	+	+	+	+	+	n	+	+	+	n	+
								R	RF	C+	SM	+	+	+	–	+	+	n	+	+	+	+
								Gy	S	C+	n	+	+	+	–	+	–	n	+	+	n	+
							+	Gy	S	C+	SPY	+	+	+	–	+	+	+	+	+	–	+
								Gy/R	S	C+	SPY	+	+	+	+	+	n	+	+	+	n	+
								R	S	C+	SPY	+	+	+	+	+	+	+	+	+	+	+
							Group A18	Gy	RF	C+	SM	+	+	+	+	+	+	–	+	+	–	–
								Gy	S	C+	SM	+	+	+	+	+	+	–	–	+	+	–
								R	RF	C+	SM	+	+	+	+	+	n	+	+	+	n	+
								R	RF	C–	SM	+	+	+	+	+	n	–	+	n	n	+

(continued)

TABLE 272. List of *Streptomyces* species continued

Species names and groups[1]	No. in lists of species[2]	Type strain	Accession no.	Wil 83a[3]	Wil 89[4]	Käm 91[5]	Hat 03[6]	Lan 02[7]	Ful 95[8]	Och 95[9]	Kat 97[10]	Lab[11]	Sch[12]	Lan 04[13]	Lan 02[14]	Lan 04[15]	Lan 04[16]	Lan 04[17]	Lan 04[18]	Lan 04[19]	Lan 04[20]
S. alanosinicus	10	DSM 40606, ATCC 15710, LMG 20391, ISP 5606, NBRC 13493	AB184442		IV 01 (gray series)	009 1-19							Sch34	Lan2-00	BENP	+		+			cl12
Group S. costaricanus et rel.:																					
S. griseofuscus	224	DSM 40191, ATCC 23916, LMG 19885, ISP 5191, NBRC 12870	AB184206	A 12	I 07	1-6 1-16			FU-6		KA-G		Sch00	Lan2-00	BENP	+		+		+	cl06
S. murinus	327	DSM 40091, ATCC 19788, LMG 10475, ISP 5091, NBRC 12799	AB184155	A 17	I 10	1-6 1-16						L3	Sch00	Lan2-00	BENP	+		+		+	cl59
S. costaricanus	134	DSM 41827, ATCC 55274, NBRC 100773	AB249939																		
S. phaeogriseichromatogenes	369	DSM 40710, NRRL 2834	AJ391813																		
Most closely related to group S. costaricanus et rel.:																					
S. lanatus	276	DSM 40090, ATCC 19775, LMG 19380, ISP 5090, NBRC 12787	AB184845	A 18	I 11	016 1-19		La-21				L2	Sch00	Lan2-00	BENP +	+		+	cl09		cl22
S. durhamensis	149	DSM 40539, ATCC 23194, LMG 20501, ISP 5539, NRRL B-3309	AY999785	A 30	II 06	009 1-19							Sch00	Lan2-00	BENP	+		+			cl22
S. filipinensis	169	DSM 40112, ATCC 23905, LMG 19333, ISP 5112, NBRC 12860	AB184198	A 30	II 06	009 1-19		La-10		OC-III			Sch00	Lan2-00	BENP +	+		+	cl09	+	cl23
S. puniciscabiei	392	KACC 20253, LMG 21391, S77	AF361785																	+	
S. niveiscabiei	339	LMG 21392, S78	AF361786																	+	
S. echinatus	151	DSM 40013, ATCC 19748, LMG 5972, ISP 5013	AJ399465	A 18	I 11	1-6 1-10			FU-1			L2	Sch00	Lan2-00	BENP	+		+			+
S. longisporus	294	DSM 40166, ATCC 23931, LMG 20053, ISP 5166	AJ399475	A 18	I 11	009 1-19						L2	Sch00	Lan2-00	BENP	+		+		+	cl22
S. avermitilis	60	ATCC 31267, MA-4680	BA000030																		
S. kunmingensis	271	DSM 41681, LMG 20521, NRRL B-16240	DQ442513										Sch00	Lan2-00	BENP	+		+			cl22
S. mirabilis	321	DSM 40553, ATCC 27447, LMG 20076, ISP 5553, NBRC 13450	AB184412	A 19	I 12	1-7 1-19							Sch00	Lan2-00	BENP	+		+		+	cl20
S. olivochromogenes	353	DSM 40451, ATCC 3336, LMG 20071, ISP 5451	AY094370	A 19	I 12	009 1-19							Sch00	Lan2-00	BENP	+		+			cl20
Most closely related to group S. cyanoalbus et rel.:																					
S. lucensis	296	DSM 40317, ATCC 17804, LMG 20065, ISP 5317, NRRL B-5626	DQ442522	A 31	I 21	1-5 1-16							Sch00	Lan2-00	BENP	+		+		+	+

(continued)

TABLE 272. (continued)

Guo 08[21] 16S rRNA	Guo 08[21] *atpD*	Guo 08[21] *gyrB*	Guo 08[21] *recA*	Guo 08[21] *rpoB*	Guo 08[21] *trpB*	Guo 08[21] concat.	Kim 04[23] *rpoB*	Morphological characters[24] I	II	III	IV	Physiological tests[25] I	II	III	IV	V	VI	VII	VIII	IX	X	XI
								Gy	S	C+	SPY	+	+	+	–	+	+	+	+	+	+	n
								Gy	S	C–	SM	+	+	+	–	+	n	–	+	–	n	–
								Gy	S	C–	SM	+	+	–	–	+	n	–	+	n	n	–
								Gy	S	C–	SM	+	+	–	–	+	+	–	n	n	+	–
								B	S	C+	SPY	+	+	+	+	+	n	+	+	+	n	+
								Gy	S	C+	SPY	+	+	+	–	+	+	+	+	+	–	n
								Gy	S	C+	SPY	+	+	+	–	+	+	+	+	+	–	+
									RF	C+/C–	SPY	+	+	+	+	+	n	+	+	+		+
								W/Gy	RF	C–	SM	+	+	+	+	+	n	+	+	+	n	+
								Gy	S	C+	SPY	+	+	+	+	+	+	+	+	+	–	–
								W	S	C+	SPY	+	+	+	+	+	n	+	+	+	n	+
								GY	S	C+	SM	+	+	+	+	+	n	n	n	+	n	–
								W	S	C–	n	n	+	+	+	n	n	+	+	–	n	–
								Gy	S	C+	SM	+	+	+	+	n	n	n	n	n	n	n
							+	Gy	S	C+	SM	+	+	+	–	+	+	+	+	+	+	n
								Gy	S	C+	SPY	+	+	+	–	+	n	–	+	–	n	+

(continued)

TABLE 272. List of *Streptomyces* species continued

Species names and groups[1]	No. in lists of species[2]	Type strain	Accession no.	Wil 83a[3]	Wil 89[4]	Käm 91[5]	Hat 03[6]	Lan 02[7]	Ful 95[8]	Och 95[9]	Kat 97[10]	Lab[11]	Sch[12]	Lan 04[13]	Lan 02[14]	Lan 04[15]	Lan 04[16]	Lan 04[17]	Lan 04[18]	Lan 04[19]	Lan 04[20]
S. niveoruber	340	DSM 40638, ATCC 14971, LMG 19379, NRRL B-2724	DQ445796		IV 08 (red series)	013 1-19		La-01					Sch00	Lan2-00	BENP +	+		+	cl04		+
S. achromogenes subsp. *achromogenes*	4a	DSM 40028, ATCC 12767, LMG 20387, ISP 5028, NBRC 12735	AB184109	A 19	I 12	1-5 009			FU-1		KA-B	L2	Sch00	Lan2-00	BENP	+		+		+	cl22
S. griseorubiginosus	233	DSM 40469, ATCC 23627, LMG 19941, ISP 5469	AJ781339	A 18	I 11	009 1-19						L2	Sch23	Lan2-00	BENP	+		+		+	cl19
S. phaeopurpureus	372	DSM 40125, ATCC 23946, LMG 20051, ISP 5125, NRRL B-2260	DQ026666	A 09	II 02	009 1-19							Sch23	Lan2-02	BENP	+	(b)	+			cl51
S. curacoi	137	DSM 40107, ATCC 13385, LMG 20491, ISP 5107, NRRL B-2901	EF626595	A 18	I 11	009 0-19						L2	Sch00	Lan2-00	BENP	+		+		+	cl55
S. lincolnensis	288	DSM 40355, ATCC 25466, LMG 20068, ISP 5355, NBRC 13054	AB184279	A 19	I 12	009 1-19							Sch00	Lan2-00	BENP	+		+		+	cl13
S. cyaneus	140	DSM 40108, ATCC 14923, LMG 20494, ISP 5108, NRRL B-2296	AF346475	A 18		009 1-19						L2	Sch00	Lan2-00	BENP	+	(g)	+		+	+
Group S. cyanoalbus et rel.:																					
S. cyanoalbus	141	DSM 40198, ATCC 15859, LMG 19343, ISP 5198, NBRC 12857	AB184882	A 37	I 17	007 003		La-17				Sch15	Lan2-00	MG-016	BEN-BOX +			+	cl14	+	+
S. hirsutus	249	DSM 40095, LMG 19927, ISP 5095, NBRC 12786	AB184844										Sch00	Lan2-00	BENP	+		+		+	cl49
S. prasinus	384	DSM 40099, ATCC 19800, LMG 20259, ISP 5099, NRRL B-2712	DQ026658	A 37	I 17	007 003							Sch00		BENP	+		+	cl14	+	cl49
S. bambergiensis	67	DSM 40590, ATCC 13879, LMG 19299, ISP 5590, NBRC 13479	AB184869	A Sm	III 10	075 1-25		La-20		OC-non			Sch00		BENP +	+		+	cl14		+
S. emeiensis	155	DSM 41884, CGMCC 4.3504	DQ462649																		
S. prasinopilosus	382	DSM 40098, ATCC 19799, LMG 19345, ISP 5098, NRRL B-2711	EF626597	A 37	I 17	007 003		La-20					Sch00	Lan2-00	BENP +	+		+	cl14	+	cl49
Most closely related to group S. cyanoalbus et rel.:																					
S. flavovariabilis	183	DSM 41479, LMG 19905, NRRL B-16367	EF178691										Sch00	Lan2-00	BENP	+		+		+	+
S. aureocirculatus	52	DSM 40386, ATCC 19823, LMG 21794, ISP 5386, NBRC 13018	AB184260	A 03	II 20	033 1-33							Sch00	Lan2-00	BENP	+		+		+	cl22

(continued)

TABLE 272. (continued)

								Morphological characters[24]				Physiological tests[25]										
Guo 08[21] 16S rRNA	Guo 08[21] *atpD*	Guo 08[21] *gyrB*	Guo 08[21] *recA*	Guo 08[21] *rpoB*	Guo 08[21] *trpB*	Guo 08[21] concat.	Kim 04[23] *rpoB*	I	II	III	IV	I	II	III	IV	V	VI	VII	VIII	IX	X	XI
								R	S	C–	SM	+	+	+	+	n	n	n	n	n	n	n
							+	Gy	RF	C+	SM	+	+	+	+	+	+	–	+	+	+	–
								Gy	RF	C+	SM	+	+	+	+	+	n	+	+	+	n	+
								Gy	RF	C+	SM	+	+	+	+	+	+	+	+	+	–	n
								B	S	C+	SPY	+	+	+	+	+	+	+	+	+	+	+
							+	R	RF	C+	SM	+	+	+	+	+	+	+	+	+	+	+
+	+	+	+	+	+	+	+	B	S	C+	SPY	+	+	+	+	n	n	n	n	n	n	n
								Gy	S	C–	H	+	+	+	+	+	+	+	+	–	–	+
								G	S	C–	SPY	+	+	+	+	+	n	+	+	+	n	+
								G	S	C–	SPY	+	+	+	+	+	+	–	+	+	–	+
							+	G	S	C–	H	n	n	n	n	n	n	n	n	n	n	n
								Gy	RF	C–	SPY	+	+	+	+	+	+	+	n	+	n	+
								G	S	C–	H	+	+	+	+	+	+	–	+	+	–	n
								R	S	C+	SPY	+	+	+	+	+	+	+	+	n	n	+
								W	RF	C–	SM	+	n	–	–	+	+	–	+	+	n	–

(continued)

TABLE 272. List of *Streptomyces* species continued

Species names and groups[1]	No. in lists of species[2]	Type strain	Accession no.	Wil 83a[3]	Wil 89[4]	Käm 91[5]	Hat 03[6]	Lan 02[7]	Ful 95[8]	Och 95[9]	Kat 97[10]	Lab[11]	Sch[12]	Lan 04[13]	Lan 02[14]	Lan 04[15]	Lan 04[16]	Lan 04[17]	Lan 04[18]	Lan 04[19]	Lan 04[20]
S. novaecaesareae	347	DSM 40358, ATCC 27452, LMG 20069, ISP 5358, NBRC 13368	AB184357	J Sm	III 25	004 006				OC-IV			Sch00	Lan2-00	BENP	+		+		+	cl22
S. prunicolor	385	DSM 40335, ATCC 25487, LMG 19311, ISP 5335, NRRL B-12281	DQ026659	A 11	III 01	1-1 1-1		La-00		OC-II			Sch09		BENP +	+		+		+	+
S. phaeoluteigriseus	371	DSM 41896, NRRL 5182, NRRL ISP-5182	AJ391815																	+	
S. bobili	75	DSM 40056, ATCC 3310, LMG 20436, ISP 5056, NBRC 16166	AB249925		IV 02 (white series)	1-7 1-15							Sch37	Lan2-00	BENP	+		+		+	cl22
S. galilaeus	196	DSM 40481, ATCC 14969, LMG 21790, ISP 5481, JCM 4757	AB045878	A 19	I 12	1-7 1-15							Sch37	Lan2-00	BENP	+		+		+	cl22
Most closely related to groups S. cyanoalbus et rel.: and S. griseoluteus et rel.:																					
S. chartreusis	102	DSM 40085, ATCC 14922, LMG 20455, ISP 5085, NBRC 12753	AB184839	A 18	I 11	009 1-19						L2	Sch30	Lan2-00	BENP	+		+		+	+
S. resistomycificus	406	DSM 40133, ATCC 19804, ISP 5133, NBRC 12814	AB184166	A 18	I 11	009 1-19			FU-12a			L2	Sch00	Lan2-00	BENP	+		+		+	+
Most closely related to group S. griseoluteus et rel.:																					
S. griseoluteus	228	DSM 40392, ATCC 12768, LMG 19356, ISP 5392, JCM 4765	AY999751	C 43	II 11	1-5 1-16		La-24		OC-III			Sch00		BENP +	+		+		+	cl41
S. recifensis	402	DSM 40115, ATCC 19803, LMG 20261, ISP 5115, NBRC 12813	AB184165	A 23	I 20	1-5 059							Sch00		BENP	+		+		+	cl41
S. seoulensis	437	NBRC 16255, NBRC 16668, JCM 10116	AB249970																		
Most closely related to groups S. cyanoalbus et rel. and S. griseoluteus et rel.:																					
S. canus	89	DSM 40017, ATCC 12237, LMG 19329, ISP 5017, NRRL B-1989	AY999775	A 25	III 02	009 1-19		La-21		OC-IV			Sch00	Lan2-00	BENP +	+		+	cl09	+	+
S. ciscaucasicus	119	DSM 40275, LMG 20474, ISP 5275	AY508512										Sch00	Lan2-05	BENP	+		+		+	cl19
S. pseudovenezuelae	389	NBRC 12904	AB184233																		
S. alboniger	21	DSM 40043, ATCC 12461, LMG 20397, ISP 5043	AY845349	A 1B	I 02	1-6 1-31							Sch00	Lan2-00	BENP	+		+		+	+
Most closely related to group S. scabiei et rel.:																					
S. bottropensis	76	DSM 40262, ATCC 25435, LMG 20437, ISP 5262	AB026217	A 19	I 12	009 1-19							Sch00	Lan2-00	BENP	+		+		+	+

(continued)

TABLE 272. (continued)

Guo 08[21] 16S rRNA	Guo 08[21] *atpD*	Guo 08[21] *gyrB*	Guo 08[21] *recA*	Guo 08[21] *rpoB*	Guo 08[21] *trpB*	Guo 08[21] concat.	Kim 04[23] *rpoB*	Morphological characters[24]				Physiological tests[25]										
								I	II	III	IV	I	II	III	IV	V	VI	VII	VIII	IX	X	XI
								n	n	C–	n	+	+	+	+	+	n	+	+	+	n	+
								R	RF	C–	SM	+	+	+	+	+	n	+	+	+	n	n
+	+	+	+	+	+	+		W	S	C+	SM	+	+	+	+	+	+	+	–	+	–	+
							Asn(AAC)[442]	Gy	S	C+	SM	+	+	+	+	n	n	n	n	n	n	n
								B	S	C+	SPY	+	+	+	+	+	+	+	+	+	+	+
								Gy	S	C+	SM	+	+	+	+	+	n	+	+	+	n	+
								Gy	RF	C+	SM	+	+	+	–	+	n	–	+	–	n	–
								Gy	S	C–	SM	+	+	+	–	+	+	+	+	–	+	+
								Gy	RF	C–	SM	+		–		+		+	+	–	+	+
								Gy	S	C–	SPY	+	+	+	+	+	+	±	+	+	n	+
								Gy	S	C–	SPY	+	+	+	+	+	n	+	n	n	n	n
							+	W	RF	C–	SM	+	+	+	–	+	+	–	+	+	+	–
								Gy	S	C+	SM	+	+	+	+	+	n	+	+	+	n	+

(continued)

TABLE 272. List of *Streptomyces* species continued

Species names and groups[1]	No. in lists of species[2]	Type strain	Accession no.	Wil 83a[3]	Wil 89[4]	Käm 91[5]	Hat 03[6]	Lan 02[7]	Ful 95[8]	Och 95[9]	Kat 97[10]	Lab[11]	Sch[12]	Lan 04[13]	Lan 02[14]	Lan 04[15]	Lan 04[16]	Lan 04[17]	Lan 04[18]	Lan 04[19]	Lan 04[20]
S. stelliscabiei	459	DSM 41803, NCPPB 4040, CFBP 4521	AJ007429																		
S. europaeiscabiei	162	DSM 41802, KACC 20186	AY207598																	+	
S. scabiei	433	DSM 41658, ATCC 49173, LMG 20323	D63862										Sch00	Lan2-00	BENP	+		+			cl25
S. diastatochromogenes	145	DSM 40449, ATCC 12309, LMG 20498, ISP 5449	D63867	A 19	I 12	009 1-19							Sch00	Lan2-00	BENP	+		+			cl25
S. hygroscopicus subsp. *ossamyceticus*	253e	DSM 40824, ATCC 15420, LMG 19951, NBRC 13983	AB184560		I 16	009 1-19							Sch20	Lan2-00	BENP	+		+			cl26
S. ipomoeae	260	DSM 40383, ATCC 25462, LMG 20520, ISP 5383, NBRC 13050	AB184857		IV 02 (blue series)	077 074						L2	Sch00	Lan2-00	BENP	+		+		+	cl26
S. torulosus	483	DSM 40894, NRRL B-3889, LMG 20305	AJ781367		IV 31 (gray series)	009 1-19							Sch00	Lan2-00	BENP	+		+		+	cl26
S. neyagawaensis	334	DSM 40588, ATCC 27449, LMG 20080, ISP 5588	D63869	A 18	I 11	009 1-19			FU-24			L2	Sch00	Lan2-00	BENP	+		+		+	cl26
Most closely related to group S. scabiei et rel.:																					
S. reticuliscabiei	407	DSM 41804, CIP 107061, CFBP 4531	AJ007428																		
S. turgidiscabies	488	ATCC 702348, ATCC 700248	AB026221																	+	
S. cacaoi subsp. *asoensis*	79b	DSM 41440, ATCC 19093, LMG 20440, NRRL B-16592	DQ026644										Sch00	Lan2-00	BENP	+		+		+	+
S. humidus	250	DSM 40263, ATCC 12760, LMG 19936, ISP 5263, NRRL B-3172	DQ442508	A 19	I 12	009 1-19							Sch00	Lan2-00	BENP	+		+			cl22
S. rishiriensis	410	DSM 40489, ATCC 14812, LMG 20297, ISP 5489, NRRL B-3239	EF178682	A 19	I 12	1-7 1-15			FU-12a				Sch00	Lan2-00	BENP	+		+			+
S. cinereoruber subsp. *fructofermentans*	111b	DSM 40692, NRRL 2588, LMG 20463, JCM 4956	AY999758		I 04	006 1-18							Sch00	Lan2-00	BENP	+		+		+	cl20
S. phaeofaciens	368	DSM 40367, LMG 20070, ISP 5367, NBRC 13372	AB184360										Sch00	Lan2-00	BENP	+		+		+	+
S. puniceus	391	DSM 40083, ATCC 19801, LMG 20258, ISP 5083, NRRL B-2895	DQ442542	A 09	II 02	005 029							Sch00	Lan2-00	BENP	+		+		+	cl22
Group S. aurantiacus et rel.:																					
S. aurantiacus	49	DSM 40412, ATCC 19822, LMG 19358, ISP 5412	AJ781383	C 45	II 13	012 019		La-01		OC-non			Sch00	Lan2-00	BENP cl01	+	(c)	+	cl05	+	cl24
S. glomeroaurantiacus	211	DSM 41782, NBRC 15418	AB249983																		

(continued)

TABLE 272. (continued)

								Morphological characters[24]				Physiological tests[25]										
Guo 08[21] 16S rRNA	Guo 08[21] *atpD*	Guo 08[21] *gyrB*	Guo 08[21] *recA*	Guo 08[21] *rpoB*	Guo 08[21] *trpB*	Guo 08[21] concat.	Kim 04[23] *rpoB*	I	II	III	IV	I	II	III	IV	V	VI	VII	VIII	IX	X	XI
								Gy	S	C+		+	+	+	+	+	n	+	+	+	n	+
								Gy	S	C+	n	+	+	+	+	+	n	+	+	+	n	+
								Gy	RF	C+/C–	SM	+	+	+	+	+	n	–	+	–	n	–
								Gy	S/RA	C+	SM	+	+	+	+	+	n	+	+	+	n	+
								Gy	S	C+	SM	+	+	+	+	+	+	+	+	+	–	+
								B	S	C–	SPY	+	+	+	+	+	n	+	+	+	n	+
								Gy	S	C+	WTY	+	+	+	+	+	+	+	+	+	–	n
								Gy	S	C+	SM	+	+	+	–	+	+	+	+	+	+	+
								Gy	RF	C–		+	+		+	+		+	+	+		+
								Gy	RF	C–	SM	+	+	+	+	+	n	+	+	+	n	+
								Gy	RF	C+	SM	+	+	+	n	+	+	+	n	+	n	+
								Gy	S	C–	SM	+	+	+	+	+	+	–	+	+	+	–
								Gy	S	C+	SM	+	+	+	+	+	+	+	–	+	+	+
							+	Gy	RF	C+	SM	+	+	+	+	+	+	–	–	–	+	n
								Gy	S	C+	SM	+	+	+	+	n	n	n	n	n	n	n
								Y	RF	C–	SM	+	+	+	–	+	+	–	+	–	+	±
								R	S	C–	SM	+	±	+	+	+	+	±	+	+	n	±
								R	S	C–	SM	+	+	–	–	+	+	–	+	+	–	–

(continued)

TABLE 272. List of *Streptomyces* species continued

Species names and groups[1]	No. in lists of species[2]	Type strain	Accession no.	Wil 83a[3]	Wil 89[4]	Käm 91[5]	Hat 03[6]	Lan 02[7]	Ful 95[8]	Och 95[9]	Kat 97[10]	Lab[11]	Sch[12]	Lan 04[13]	Lan 02[14]	Lan 04[15]	Lan 04[16]	Lan 04[17]	Lan 04[18]	Lan 04[19]	Lan 04[20]
S. tauricus	467	DSM 40560, ATCC 27470, LMG 20301, ISP 5560, JCM 4837	AB045879	A 19		012 019							Sch00	Lan2-00	BENP	+		+		+	cl24
S. ederensis	153	DSM 40741, ATCC 15304, LMG 20504, NBRC 15410	AB184658		IV 14 (gray series)	013 1-19							Sch00	Lan2-00	BENP	+		+			cl19
S. phaeochromogenes	367	DSM 40073, ATCC 3338, LMG 19348, ISP 5073, NBRC 3180	AB184738	A 40	I 18	009 0-19		La-01		OC-II			Sch00	Lan2-00	BENP +	+		+	cl06		-
S. umbrinus	489	DSM 40278, ATCC 19929, LMG 20280, ISP 5278, NBRC 13091	AB184305	A 05	I 04	1-6 1-16							Sch00	Lan2-15	BENP	+		+			cl54
S. rectiviolaceus	404	DSM 41459, LMG 20310, NRRL B-16374	DQ026660										Sch00	Lan2-00	BENP	+		+		+	-
Group S. aureus et rel.:																					
S. kanamyceticus	264	DSM 40500, LMG 19351, ISP 5500, NRRL B-2535	DQ442511					La-11					Sch00	Lan2-00	BENP +	+		+	cl02		cl23
S. durmitorensis	150	DSM 41863, MS405	DQ067287																		
S. aureus	57	DSM 41785, NCIMB 13927, NBRC 100912	AB249976																		
Group S. cinereus et rel.:																					
S. cinereus	113	DSM 43033, LMG 21310, NBRC 12247	AB184072										Sch00	Lan2-00	BENP	+		+		+	+
S. flaveus	174	DSM 43153, ATCC 15332, LMG 19323, NRRL B-16074	DQ026643					La-21							BENP +	+		+			-
S. vastus	493	DSM 40309, LMG 21043, NRRL B-12232	DQ442552										Sch21	Lan2-00	BENP	+		+		+	+
Most closely related to group S. cinereus et rel.:																					
S. laceyi	274	DSM 41788, NBRC 100783	AB249944																	+	
Group S. argenteolus et rel.:																					
S. griseolus	227	DSM 40067, ATCC 3325, LMG 19878, ISP 5067, NBRC 3415	AB184768	A 1C	I 03	1-2 015			FU-24		KA-B		Sch06		BENP	+		+		+	cl53
S. halstedii	242	DSM 40068, ATCC 10897, ISP 5068, NRRL B-1238	EF178695	A 1C	I 03	1-2 015			FU-24	OC-I	KA-B		Sch06					+			
S. argenteolus	41	DSM 40226, ATCC 11009, LMG 5967, ISP 5226, JCM 4623	AB045872	A 15	I 08	1-5 011					KA-B		Sch00	Lan2-00	BENP	+		+		+	cl23
S. cinereorectus	110	DSM 41469, LMG 20461, NBRC 15395	AB184646										Sch00	Lan2-30	BENP	+	(b)	+		+	cl28
S. flavovirens	184	DSM 40062, ATCC 3320, LMG 20516, ISP 5062, NRRL B-2685	DQ026635	A 1C		1-2 015							Sch05	Lan2-09	BENP	+	(a)	+		+	cl53

(continued)

TABLE 272. (continued)

Guo 08[21] 16S rRNA	Guo 08[21] *atpD*	Guo 08[21] *gyrB*	Guo 08[21] *recA*	Guo 08[21] *rpoB*	Guo 08[21] *trpB*	Guo 08[21] concat.	Kim 04[23] *rpoB*	Morphological characters[24] I	II	III	IV	Physiological tests[25] I	II	III	IV	V	VI	VII	VIII	IX	X	XI
								R	S	C–	SM	n	+	+	+	+	n	+	–	n	n	n
								Gy	RF	C+	SM	n	n	n	n	n	n	n	n	n	n	n
							+	R	RF	C+	SM	+	+	+	+	+	+	+	+	+	+	n
								R	RF	C+	SM	+	+	+	+	+	+	+	+	+	–	n
								V	RF	C–	SM	+	+	+	+	+	+	+	+	+	n	+
+	+	+	+	+	+	+		Y	RF	C–	SM	+	+	+	–	+	+	+	+	–	+	n
								Y	RF	C–	SM	+	+	–	–	+	+	+	+	–	n	+
+	+	+	+	+	+	+		Gy	S	C+	SM	n	n	n	n	n	n	n	+	+	n	n
								W	RF	C–	SM	+	+	+	+	+	+	+	+	+	+	+
								Gy	RF	C–	SM	+	+	+	+	+	+	+	+	+	+	+
								Gy	S	C–	SM	+	+	+	+	+	+	+	+	+	+	+
+	+	+	+	+	+	+		Gy/Y/R	S		SM								+	–		
+	+	+	+	+	+	+	+	Gy	RF	C–	SM	+	+	+	–	+	+	+	–	–	–	n
								Gy	RF	C–	SM	+	+	+	–	+	n	–	–	–	n	–
+	+	+	+	+	+	+	+	Gy	S	C–	SM	+	+	+	+	+	+	–	+	–	+	–
								Gy	RF	C–	SM	+	–	n	–	+	n	+	+	–	n	n
								Gy	RF	C–	SM	+	+	+	+	+	+	+	+	–	–	–

(continued)

TABLE 272. List of *Streptomyces* species continued

Species names and groups[1]	No. in lists of species[2]	Type strain	Accession no.	Wil 83a[3]	Wil 89[4]	Käm 91[5]	Hat 03[6]	Lan 02[7]	Ful 95[8]	Och 95[9]	Kat 97[10]	Lab[11]	Sch[12]	Lan 04[13]	Lan 02[14]	Lan 04[15]	Lan 04[16]	Lan 04[17]	Lan 04[18]	Lan 04[19]	Lan 04[20]
S. flavogriseus	180	DSM 40323, ATCC 25452, LMG 19887, ISP 5323, CBS 101.34	AJ494864	A 1C	I 03	1-2 015			FU-19b		KA-B		Sch00/0	Lan2-09	BENP	+		+			cl53
S. nitrosporeus	338	DSM 40023, LMG 20044, ISP 5023, NRRL B-1316	EF178680										Sch26	Lan2-00	BENP	+		+		+	+
Most closely related to groups S. argenteolus et rel.: and S. atroolivaceus et rel.:																					
S. luridiscabiei	297	KACC 20252, LMG 21390, S63	AF361784																	+	
S. acrimycini	6	DSM 40135, ATCC 19885, LMG 21798, ISP 5135, AS 4.1673	AY999889		IV 04 (green series)	1-3 010							Sch00	Lan2-00	BENP	+		+			cl05
S. griseoplanus	230	DSM 40009, ATCC 19766, LMG 19923, ISP 5009, AS 4.1868	AY999894	A 29	I 15	078 060							Sch00	Lan2-00	BENP	+		+		+	cl28
S. baarnensis	63	DSM 40232, ATCC 23885, LMG 20431, ISP 5232, NRRL B-1902	EF178688	A 1B	I 02	006 1-2					KA-B		Sch01		BENP	+		+		+	cl23
S. flavofuscus	179	DSM 41426, ATCC 19908, LMG 19900, NBRC 100768	AB249935										Sch00	Lan2-19	BENP	+		+		+	cl23
S. praecox	381	DSM 40393, ATCC 3374, LMG 20290, ISP 5393, NBRC 13073	AB184293		IV 08 (yellow series)	1-3 1-2							Sch01		BENP	+		+		+	cl23
S. fimicarius	171	DSM 40322, ATCC 25449, LMG 21044, ISP 5322	AY999784	A 1B	I 02	1-3 1-2			FU-9				Sch01		BENP	+		+		+	cl23
S. anulatus	37	DSM 40361, ATCC 27416, LMG 19301, ISP 5361, NRRL B-2000	DQ026637	A 1B	I 02	047 1-35		La-22		OC-I	KA-B		Sch01	Lan2-18	BENP +	+	(a) (e)	+	cl02	+	cl23
Group S. atroolivaceus et rel.:																					
S. mutomycini	329	DSM 41691, LMG 20098, NBRC 100999	AB249951										Sch00	Lan2-00	BENP	+		+		+	cl23
S. olivoviridis	357	DSM 40211, ATCC 15882, LMG 20057, ISP 5211, NBRC 12897	AB184227	A 03	II 20	1-3 010							Sch16		BENP	+		+		+	cl23
S. atroolivaceus	47	DSM 40137, ATCC 19725, LMG 19306, ISP 5137	AJ781320	A 03	II 20	006 0-10		La-23		OC-I			Sch16		BENP +	+		+	cl02		cl23
S. clavifer	121	DSM 40843, LMG 20476, NRRL B-2557	DQ026670										Sch00	Lan2-05	BENP	+		+			cl19
S. finlayi	172	DSM 40218, ATCC 23340, LMG 19373, ISP 5218, NRRL B-12114	AY999788	I Sm	III 24	22-4 043		La-00		OC-I			Sch00	Lan2-00	BENP +	+		+	cl02	+	cl23

(continued)

TABLE 272. (continued)

Guo 08[21] 16S rRNA	Guo 08[21] *atpD*	Guo 08[21] *gyrB*	Guo 08[21] *recA*	Guo 08[21] *rpoB*	Guo 08[21] *trpB*	Guo 08[21] concat.	Kim 04[23] *rpoB*	Morphological characters[24] I	II	III	IV	Physiological tests[25] I	II	III	IV	V	VI	VII	VIII	IX	X	XI
+	+	+	+	+	+	+		Gy	RF	C–	SM	+	+	+	+	+	n	–	+	–	n	–
								Gy	RF	C–	SM	+	+	+	+	–	+	–	–	–	–	–
+	+	+	+	+	+	+		Y/W	RF	C+	SM		+	+	+	+		+	+	+		+
Group I	Yes[22]	Yes	Yes	Yes	Yes	Yes		G	S	C–	H	+	+	–	+	+	n	–	+	+	n	–
Group I	No[22]	No	No	No	No	No		Gy	S	C–	WTY	+	+	+	–	+	+	+	–	–	–	–
								W	RF	C–	SM	+	+	+	+	+	+	–	+	+	n	n
								Y	RF	C–	SM											
Group I	No	No	No	No	No	No		Y	RF	C–	SM	+	+	+	+	+	+	+	+	–	+	n
Group I	Yes	Yes	Yes	Yes	Yes	Yes		Y	RF	C–	SM	+	+	+	+	+	+	–	+	–	–	n
Group I	No	No	No	No	No	No		Y	RF	C–	SM	+	+	+	+	+	n	–	+	–	n	–
+	+	+	+	+	+	+		Gy	S	C–	SPY	+	+	+		+			+			
								Gy	S	C–	SPY	+	+	+	+	+	+	–	+	–	–	–
+[2]	+	+	+	+	+	+		Gy	S	C–	WTY	+	+	+	+	+	n	n	n	n	n	n
								W	RF	C–	SM	+	+	–	+	+	+	–	+	–	–	n
+	+	+	+	+	+	+		Gy	S	C–	H	+	+	+	+	–	n	–	–	–	n	±

(continued)

TABLE 272. List of *Streptomyces* species continued

Species names and groups[1]	No. in lists of species[2]	Type strain	Accession no.	Wil 83a[3]	Wil 89[4]	Käm 91[5]	Hat 03[6]	Lan 02[7]	Ful 95[8]	Och 95[9]	Kat 97[10]	Lab[11]	Sch[12]	Lan 04[13]	Lan 02[14]	Lan 04[15]	Lan 04[16]	Lan 04[17]	Lan 04[18]	Lan 04[19]	Lan 04[20]
Most closely related to groups S. argenteolus et rel. and S. atroolivaceus et rel.:																					
S. griseus subsp. *griseus*	238a	DSM 40236, ATCC 23345, LMG 19302, ISP 5236, KACC 20084	AY207604	A 1B	I 02	1-3 1-2		La-22	FU-19		KA-B		Sch00	Lan2-00	BENP +	+		+	cl02	+	cl23
S. lavendulae subsp. *lavendulae*	280a	DSM 40069, ATCC 8664, LMG 19925, ISP 5069, NBRC 12343	AB184080	F 61	I 22	22-3 042			FU-12b	OC-I		L3/L	Sch00	Lan2-00	BENP	+	(e) (l)	+			-
S. cavourensis subsp. *washingtonensis*	96b	DSM 41423, LMG 20451, NRRL B-8030	DQ026671										Sch00	Lan2-00	BENP	+		+			cl23
S. cyaneofuscatus	139	DSM 40148, ATCC 23619, LMG 20493, ISP 5148, NBRC 13190	AB184860	A 1B	I 02	1-3 1-2							Sch00	Lan2-00	BENP	+		+		+	cl23
Not closely related to one of the groups:																					
S. mediolani	313	DSM 41058, DSM 41647, LMG 20093, NBRC 15427	AB184674										Sch20	Lan2-16	BENP	+		+			cl23
S. rubiginosohelvolus	424	DSM 40176, ATCC 19926, LMG 20267, ISP 5176, NBRC 12912	AB184240		IV 12 (red series)	006 1-2							Sch03		BENP	+		+			cl23
S. parvus	364	DSM 40348, ATCC 12433, LMG 20524, ISP 5348, NRRL B-1455	DQ442537	A 1B	I 02	1-3 1-2			FU-6		KA-B		Sch00	Lan2-00	BENP	+		+	cl02		cl23
S. albovinaceus	25	DSM 40136, ATCC 15823, LMG 20402, ISP 5136, NBRC 12739	AB249958	A 1B	I 02	1-3 008					KA-B		Sch03	Lan2-16	BENP	+		+		+	cl23
S. bacillaris	64	DSM 40598, ATCC 15855, LMG 8585, ISP 5598, NBRC 13487	AB184439	A 1B	I 02	1-3 1-2					KA-B		Sch00	Lan2-00	BENP	+		+		+	cl50
S. griseinus	218	DSM 40047, ATCC 23915, LMG 19875, ISP 5047, NBRC 12869	AB184205	A 1B	I 02	1-3 1-2			FU-6		KA-B		Sch03	Lan2-00	BENP	+		+			-
S. sindenensis	441	DSM 40255, ATCC 23963, LMG 21041, ISP 5255, NBRC 3399	AB184759	A 1B		1-3 1-2					KA-B		Sch00		BENP	+		+		+	cl23
S. pluricolorescens	378	DSM 40019, ATCC 19798, LMG 8576, ISP 5019, NRRL B-2121	DQ442540	A 1B	I 02	1-3 1-2					KA-B		Sch03		BENP	+		+		+	-
S. globisporus subsp. *globisporus*	208a	DSM 40199, ATCC 15864, LMG 8578, ISP 5199, NRRL B-2872	EF178686	A 1B	I 02	1-3 1-2					KA-B		Sch20	Lan2-00	BENP	+		+			cl23
S. badius	65	DSM 40139, ATCC 19888, LMG 19353, ISP 5139, NRRL B-2567	AY999783	C Sm	III 15	1-1 1-1		La-00		OC-I			Sch00		BENP +	+		+	cl02	+	cl23

(continued)

TABLE 272. (continued)

								Morphological characters[24]				Physiological tests[25]										
Guo 08[21] 16S rRNA	Guo 08[21] *atpD*	Guo 08[21] *gyrB*	Guo 08[21] *recA*	Guo 08[21] *rpoB*	Guo 08[21] *trpB*	Guo 08[21] concat.	Kim 04[23] *rpoB*	I	II	III	IV	I	II	III	IV	V	VI	VII	VIII	IX	X	XI
Group IV	Yes	Yes	Yes	Yes	No	Yes	Group A1B	Y	RF	C–	SM	+	+	–	–	+	+	–	+	–	+	–
								R	S	C+	SM	+	–	–	–	–	+	+	–	–	+	–
								Y	RF	C+	SM	+	+	+	–	+	n	–	+	–	n	–
+	+	+	+	+	+	+		Y	RF	C+	SM	+	+	–	+	+	+	–	+	–	+	+
Group II	Yes	Yes	Yes	Yes	Yes	Yes		Y	RF													
								R	RF	C–	SM	+	+	+	+	+	n	–	+	–	n	–
								Y	RF	C–	SM	+	+	+	+	+	+	–	+	–	–	n
Group II	Yes	Yes	Yes	Yes	Yes	Yes	+	W	RF	C–	SM	+	+	+	+	+	+	–	+	–	+	–
								Y	RF	C+	SM	+	+	–	–	+	+	–	+	+	+	n
Group II	Yes	Yes	Yes	Yes	Yes	Yes		Y	RF	C–	SM	+	+	+	+	+	+	–	+	–	–	–
Group II	No	No	Yes	Yes	Yes	No		Y	RF	C–	SM	+	+	+	–	+	+	–	+	–	+	n
								Y	RF	C–	SM	+	+	–	+	+	n	–	+	–	n	–
							Group A1B	Y	RF	C–	SM	+	+	+	+	+	n	–	+	–	n	–
Group II	No	No	Yes	Yes	Yes	No		Y	RF	C–	SM	+	+	+	–	+	n	–	+	–	n	–

(continued)

TABLE 272. List of *Streptomyces* species continued

Species names and groups[1]	No. in lists of species[2]	Type strain	Accession no.	Wil 83a[3]	Wil 89[4]	Käm 91[5]	Hat 03[6]	Lan 02[7]	Ful 95[8]	Och 95[9]	Kat 97[10]	Lab[11]	Sch[12]	Lan 04[13]	Lan 02[14]	Lan 04[15]	Lan 04[16]	Lan 04[17]	Lan 04[18]	Lan 04[19]	Lan 04[20]
S. californicus	82	DSM 40058, ATCC 3312/ ATCC 19734, LMG 19309, ISP 5058, NBRC 3386	AB184755	A 09	II 02	1-3 030		La-22	FU-6	OC-I			Sch32	Lan2-00	BENP +	+		+	cl02		cl23
S. floridae	187	DSM 40938, NCIB 9345, LMG 19899, NBRC 15405	AB184656		IV 04 (yellow series)	1-3 1-2							Sch32	Lan2-00	BENP	+		+		+	cl23
S. alboviridis	26	DSM 40326, ATCC 25425, LMG 20403, ISP 5326, NBRC 13013	AB184256	A 1B	I 02	1-3 1-2					KA-B		Sch02	Lan2-00	BENP	+		+		+	cl23
S. microflavus	319	DSM 40331, ATCC 13231, LMG 19327, ISP 5331, NRRL B-2156	DQ445795	A 23	I 20	1-3 1-2		La-22		OC-I			Sch02	Lan2-12	BENP +	+	(a)	+	cl02	+	cl23
S. fulvorobeus	192	DSM 41455, LMG 19901, NBRC 15897	AB184711										Sch00	Lan2-00	BENP	+		+		+	cl23
S. lipmanii	289	DSM 40070, ATCC 3331, LMG 20047, ISP 5070, NBRC 12791	AB184148	A 1B	I 02	1-3 1-2			FU-9		KA-B		Sch02	Lan2-12	BENP	+	(a)	+			cl23
Group S. avidini et rel.:																					
S. spororaveus	457	DSM 41462, LMG 20313	AJ781370										Sch00	Lan2-00	BENP	+		+		+	cl22
S. xanthophaeus	524	DSM 40134, ATCC 19819, LMG 21039, ISP 5134, NRRL B-5414	DQ442560	F 61	I 22	084 067						L5	Sch00	Lan2-00	BENP	+		+		+	cl22
S. nojiriensis	345	DSM 41655, LMG 20094	AJ781355										Sch00	Lan2-00	BENP	+		+		+	cl22
S. cirratus	118	DSM 40479, ATCC 14699, LMG 20473, ISP 5479, NRRL B-3250	AY999794	F 62	II 14	22-3 042							Sch00	Lan2-00	BENP	+		+			cl22
S. vinaceus	496	DSM 40515, ATCC 27476, LMG 20533, ISP 5515, NBRC 13425	AB184394	A 06	I 05	22-3 042					KA-A		Sch00	Lan2-23	BENP	+	(b)	+			cl08
S. colombiensis	132	DSM 40558, ATCC 27425, LMG 20487, ISP 5558, NRRL B-1990	DQ026646	F 61	I 22	22-3 042			FU-12b			L5	Sch00	Lan2-13	BENP	+	(b) (l)	+		+	-
S. lavendulae subsp. *grasserius*	280b	DSM 40385, LMG 19938	AY999841											Lan2-00	BENP	+		+			-
S. goshikiensis	213	DSM 40190, ATCC 23914, LMG 19884, ISP 5190, NRRL B-5428	EF178693	F 61	I 22	22-3 042						L3/L	Sch00		BENP	+		+			cl22
S. sporoverrucosus	458	DSM 41463, LMG 20314, NRRL B-16379	DQ442544										Sch00		BENP	+		+		+	cl22
S. avidinii	61	DSM 40526, ATCC 27419, LMG 20428, ISP 5526, NBRC 13429	AB184395	F 56		023 004							Sch00	Lan2-00	BENP	+		+			cl22
S. subrutilus	461	DSM 40445, ATCC 27467, LMG 20294, ISP 5445	X80825	F 61		22-3 042						L5	Sch00	Lan2-00	BENP	+		+			cl22

(continued)

TABLE 272. (continued)

Guo 08[21] 16S rRNA	Guo 08[21] *atpD*	Guo 08[21] *gyrB*	Guo 08[21] *recA*	Guo 08[21] *rpoB*	Guo 08[21] *trpB*	Guo 08[21] concat.	Kim 04[23] *rpoB*	Morphological characters[24] I	II	III	IV	Physiological tests[25] I	II	III	IV	V	VI	VII	VIII	IX	X	XI
+	+	+	+	+	+	+		Y	RF	C–	SM	+	+	–	–	+	+	–	+	–	–	–
+	+	+	+	+	+	+		Y	RF	C–	SM	+	+	–	–	+	+	–	+	–	–	n
Group III	Yes	Yes	Yes	Yes	Yes	Yes	Group A1B	Y	RF	C–	SM	+	+	–	+	+	n	–	+	–	n	–
Group III	Yes	Yes	Yes	Yes	Yes	Yes		Y	RF	C–	SM	+	+	–	+	+	+	–	+	–	+	–
Group III	No	No	No	No	No	No		R–Y	S	C–	SM	+	–	+	–	n	n	n	–	–	n	n
								Y	RF	C–	SM	+	+	–	+	+	+	+	+	–	+	±
+	+	+	+	+	+	+		Gy	S/RA	C+	WTY/ SM	+	–	–	–	–	–	–	n	n	n	n
							+	R	RF	C+	SM	+	–	–	–	–	n	–	–	–	n	–
+	+	+	+	+	+	+		Gy	S	C+	SM	+	–	–	–	–	–	–	–	–	+	–
+	+	+	+	+	+	+	Group F	Gy	S	C+	SM	+	+	+	–	+	+	–	–	–	–	+
								R	S	C+	SM	+	–	–	–	n	n	n	n	n	n	n
								R	S	C+	SM	+	–	–	–	n	n	n	n	n	n	n
								R	S	C+	SM	+	–	–	–	–	+	–	–	–	+	–
								R	S	C+	SM	+	–	–	–	+	n	–	–	–	n	–
								Gy/Y	S	C+	WTY											
							Asn(AAC)[442]	Gy	S	C+		n	–	–	n	n	n	–	n	n	n	n
+	+	+	+	+	+	+		R	RF	C+	SM	+	–	–	–	+	+	–	–	–	–	+

(continued)

TABLE 272. List of *Streptomyces* species continued

Species names and groups[1]	No. in lists of species[2]	Type strain	Accession no.	Wil 83a[3]	Wil 89[4]	Käm 91[5]	Hat 03[6]	Lan 02[7]	Ful 95[8]	Och 95[9]	Kat 97[10]	Lab[11]	Sch[12]	Lan 04[13]	Lan 02[14]	Lan 04[15]	Lan 04[16]	Lan 04[17]	Lan 04[18]	Lan 04[19]	Lan 04[20]
Group S. cinnamonensis et rel.:																					
S. globosus	209	DSM 40815, ATCC 14979, LMG 19896	AJ781330		IV 19 (gray series)	22-3 042							Sch00	Lan2-00	BENP	+		+		+	cl31
S. toxytricini	484	DSM 40178, ATCC 19813, LMG 20269, ISP 5178, NRRL B-5426	DQ442548	F 61		22-3 042						L3/L	Sch00	Lan2-00	BENP	+		+		+	cl31
S. flavotricini	182	DSM 40152, ATCC 23621, LMG 19880, ISP 5152, NBRC 12770	AB184132	F 61	I 22	22-3 042			FU-1			L3/L	Sch00	Lan2-00	BENP	+		+			cl22
S. polychromogenes	379	DSM 40316, ATCC 12595, LMG 20287, ISP 5316, NBRC 13072	AB184292	F 61	I 22	22-3 042						L3/L	Sch00	Lan2-00	BENP	+		+		+	cl22
S. racemochromogenes	397	DSM 40194, ATCC 23954, LMG 20273, ISP 5194, NRRL B-5430	DQ026656	F 61	I 22	22-3 042						L5	Sch00	Lan2-00	BENP	+		+			cl22
S. katrae	267	DSM 40550, ATCC 27440, LMG 19945, ISP 5550, NBRC 13447	AB184409	F 61	I 22	22-3 042						L5	Sch26	Lan2-00	BENP	+		+		+	cl22
S. cinnamonensis	116	DSM 40803, ATCC 12308, LMG 20468, NBRC 15873	AB184707		IV 02 (red series)	22-3 042							Sch00	Lan2-00	BENP	+		+			cl22
S. virginiae	510	DSM 40094, ATCC 19817, LMG 20534, ISP 5094, IFO 3729	D85119	F 61	I 22	22-3 042			FU-12b			L3/L	Sch00	Lan2-00	BENP	+		+			cl22
Group S. albolongus et rel.:																					
S. cavourensis subsp. *cavourensis*	96a	DSM 40300, ATCC 14889, LMG 20450, ISP 5300, NRRL 2740	DQ445791	A 1B	I 02	1-3 1-2			FU-6		KA-A		Sch00	Lan2-00	BENP	+		+		+	cl22
S. celluloflavus	98	DSM 40839, ATCC 29806, LMG 21796, NBRC 13780	AB184476		IV 01 (yellow series)	020 032							Sch00	Lan2-00	BENP	+		+		+	cl42
S. albolongus	20	DSM 40570, ATCC 27414, LMG 20396, ISP 5570, NBRC 13465	AB184425	F 63	II 15	22-4 043						L4	Sch00	Lan2-00	BENP	+		+			+
S. griseobrunneus	220	DSM 40066, ATCC 19762, LMG 19877, ISP 5066, NBRC 12775	AB249912	A 1B	I 02	1-3 1-2			FU-6		KA-A		Sch00	Lan2-00	BENP	+		+			cl50
Group S. crystallinus et rel.:																					
S. melanogenes	315	DSM 40192, ATCC 23937, LMG 20056, ISP 5192, NBRC 12890	AB184222	A 33	II 07	009 1-09							Sch00	Lan2-00	BENP	+		+		+	cl13
S. noboritoensis	342	DSM 40223, ATCC 25477, LMG 19337, ISP 5223, NBRC 13065	AB184287	A 33	II 07	009 1-09		La-19		OC-I			Sch00	Lan2-00	BENP +	+		+	cl06		cl34

(continued)

TABLE 272. (continued)

Guo 08[21] 16S rRNA	Guo 08[21] *atpD*	Guo 08[21] *gyrB*	Guo 08[21] *recA*	Guo 08[21] *rpoB*	Guo 08[21] *trpB*	Guo 08[21] concat.	Kim 04[23] *rpoB*	Morphological characters[24] I	II	III	IV	Physiological tests[25] I	II	III	IV	V	VI	VII	VIII	IX	X	XI
								Gy	RF	C+	SM	+	+	+	–	n	n	n	n	n	n	n
								R	S	C+	SM	+	–	–	–	±	n	–	–	–	n	–
								R	RF	C+	SM	+	–	–	–	±	n	–	–	–	n	–
								B	RF	C+	SM	+	+	+	–	+	+	–	–	–	+	n
								R	S	C+	SM	+	–	+	–	–	n	–	–	–	n	+
								R	S	C+	SM	+	–	–	–	+	+	+	–	–	–	n
								R	S	C+	SM	+	–	–	–	+	n	–	–	–	+	n
+	+	+	+	+	+	+	Group F	R	S	C+	SM	+	–	–	–	+	n	–	–	+	+	n
								Y/R	RF	C+	SM	+	+	–	–	+	n	–	+	–	n	n
								Y	RF	C–	SM	+	–	–	–	n	n	n	n	n	n	n
							+	W	RF	C–	SM	+	+	+	–	+	+	–	+	+	+	–
+	+	+	+	+	+	+		Y	RF	C+	SM	+	+	–	–	+	+	+	+	–	+	+
								R	RF	C+	SM	+	+	+	–	+	n	+	+	+	n	±
								Gy	RF	C+	SM	+	+	+	–	+	n	+	+	+	+	±

(continued)

TABLE 272. List of *Streptomyces* species continued

Species names and groups[1]	No. in lists of species[2]	Type strain	Accession no.	Wil 83a[3]	Wil 89[4]	Käm 91[5]	Hat 03[6]	Lan 02[7]	Ful 95[8]	Och 95[9]	Kat 97[10]	Lab[11]	Sch[12]	Lan 04[13]	Lan 02[14]	Lan 04[15]	Lan 04[16]	Lan 04[17]	Lan 04[18]	Lan 04[19]	Lan 04[20]
S. crystallinus	136	DSM 40945, LMG 20490, NBRC 15401	AB184652		IV 03 (red series)	009 1-09							Sch00	Lan2-00	BENP	+		+		+	cl34
Group S. mauvecolor et rel.:																					
S. michiganensis	318	DSM 40015, ATCC 14970, LMG 20042, ISP 5015, NBRC 12797	AB184153	A 06	I 05	005 029							Sch00	Lan2-00	BENP	+		+			cl14
S. xanthochromogenes	521	DSM 40111, ATCC 19818, LMG 19366, ISP 5111, NRRL B-5410	DQ442559	F 63	II 15	005 029		La-23		OC-I			Sch00	Lan2-00	BENP +	+		+	cl02	+	+
S. mauvecolor	312	DSM 41702, LMG 20100, NBRC 13854	AB184532										Sch00	Lan2-00	BENP	+		+		+	cl34
Not closely related to one of the groups:																					
S. cremeus	135	DSM 40147, ATCC 19897, LMG 20489, ISP 5147, NBRC 12760	AB184124	A 1B	I 02	002 1-7			FU-21				Sch00	Lan2-00	BENP	+		+			cl22
S. spiroverticillatus	453	DSM 40036, ATCC 19811, LMG 20254, ISP 5036, NBRC 3931	AB184814	A 06	I 05	002 1-7					KA-A		Sch00	Lan2-00	BENP	+		+		+	cl22
S. candidus	85	DSM 40141, ATCC 19891, ISP 5141	DQ026663	A 03		002 1-7							Sch00							+	
Group S. exfoliatus et rel.:																					
S. lateritius	277	DSM 40163, ATCC 19913, LMG 19372, ISP 5163	AJ781326	H Sm	III 23	22-3 1-08		La-00		OC-II			Sch00	Lan2-00	BENP +	+		+	cl07	+	cl22
S. venezuelae	494	DSM 40230, ATCC 10712, LMG 19308, ISP 5230, JCM 4526	AB045890	A 06	I 05	002 1-7		La-00			KA-C		Sch00	Lan2-00	BENP +	+		+	cl07		cl22
S. omiyaensis	358	DSM 40552, ATCC 27454, LMG 20075, ISP 5552, NRRL B-1587	EF178697	A 05	I 04	002 1-7					KA-C		Sch00	Lan2-00	BENP	+		+		+	cl23
S. wedmorensis	518	DSM 41676, ATCC 21239, LMG 21050, NRRL 3426	DQ442557										Sch00	Lan2-00	BENP	+		+			cl23
S. litmocidini	290	DSM 40164, ATCC 19914, LMG 20052, ISP 5164, NBRC 12792	AB184149	A 05	I 04	002 1-7					KA-C		Sch00		BENP	+		+			cl22
S. yerevanensis	529	DSM 43167, LMG 21053, NRRL B-16943	EF178684		III 18	080 066				OC-1					BENP	+		+			+
S. zaomyceticus	533	DSM 40196, ATCC 27482, LMG 19853, ISP 5196, NRRL B-2038	EF178685	A 05	I 04	002 1-7					KA-C		Sch00	Lan2-00	BENP	+		+			cl23
S. exfoliatus	164	DSM 40060, ATCC 12627, LMG 19307, ISP 5060, NBRC 13191	AB184324	A 05	I 04	002 1-7		La-00		OC-II	KA-C		Sch00	Lan2-00	BENP +	+		+	cl01	+	cl23

(continued)

TABLE 272. (continued)

Guo 08[21] 16S rRNA	Guo 08[21] *atpD*	Guo 08[21] *gyrB*	Guo 08[21] *recA*	Guo 08[21] *rpoB*	Guo 08[21] *trpB*	Guo 08[21] concat.	Kim 04[23] *rpoB*	Morphological characters[24] I	II	III	IV	Physiological tests[25] I	II	III	IV	V	VI	VII	VIII	IX	X	XI
								R	RF	C+	SM	+	n	n	n	n	n	n	n	n	n	–
								Y	RF	C+	SM	+	+	–	–	+	n	±	+	+	n	–
								Y	RF	C+	SM	+	+	±	±	+	n	±	+	±	n	±
+	+	+	+	+	+	+		Vi	S	C+	SPY	+	–	+	–	–	+	+	–	–	+	–
+	+	+	+	+	+	+		R	S	C–	SM	+	+	+	–	+	+	–	–	–	–	–
+	+	+	+	+	+	+		W	S	C–	SM	+	+	+	–	+	n	–	–	–	n	n
								W	RF	C–	SM	+	+	+	+	n	n	–	+	–	n	–
								R	S	C+	WTY	+	+	+	+	+	n	–	–	±	n	–
+	+	+	+	+	+	+		R	RF	C+	SM	+	–	–	–	–	n	–	–	–	n	–
								Gy	RF	C–	SM	+	+	–	+	–	+	–	–	–	–	n
								Gy	RF	C–	SM	+	+	+	+	+			+			
								Gy	RF	C+	SM	+	±	+	–	±	n	–	–	–	n	–
								Gy		C–	SM	+	+	+	+	+	–	+	+	+	+	+
								Gy	RF	C+	SM	+	+	+	–	–	+	–	–	–	+	+
+	+	+	+	+	+	+		R	RF	C–	SM	+	+	+	+	+	+	+	–	–	+	+

(continued)

TABLE 272. List of *Streptomyces* species continued

Species names and groups[1]	No. in lists of species[2]	Type strain	Accession no.	Wil 83a[3]	Wil 89[4]	Käm 91[5]	Hat 03[6]	Lan 02[7]	Ful 95[8]	Och 95[9]	Kat 97[10]	Lab[11]	Sch[12]	Lan 04[13]	Lan 02[14]	Lan 04[15]	Lan 04[16]	Lan 04[17]	Lan 04[18]	Lan 04[19]	Lan 04[20]
S. narbonensis	331	DSM 40016, ATCC 19790, LMG 20043, ISP 5016, NRRL B-1680	DQ445794	A 04	I 04	002 1-7					KA-C		Sch00	Lan2-00	BENP	+		+			cl44
Most closely related to group S. exfoliatus et rel.:																					
S. albidochromogenes	13	DSM 41800, NBRC 101003	AB249953																		
S. flavidovirens	176	DSM 40150, ATCC 19900, LMG 19387, ISP 5150, NBRC 13039	AB184270		IV 03 (yellow series)	026 033		La-22					Sch00	Lan2-00	BENP +	+		+		+	+
S. enissocaesilis	157	DSM 41454, LMG 20506, NBRC 100763	AB249930										Sch00	Lan2-06	BENP	+		+		+	cl58
S. albosporeus subsp. *labilomyceticus*	23b	DSM 41672, LMG 20400, NBRC 15387	AB184638										Sch00	Lan2-00	BENP	+		+		+	cl23
S. chryseus	108	DSM 40420, ATCC 19829, LMG 20458, ISP 5420, NRRL B-12347	AY999787	A 17	I 10	22-3 1-08						L3	Sch28	Lan2-00	BENP	+		+			cl23
S. helvaticus	246	DSM 40431, ATCC 19841, LMG 19940, ISP 5431, NBRC 13382	AB184367	F 62	II 14	22-3 043							Sch28	Lan2-00	BENP	+		+		+	cl23
Not closely related to one of the groups:																					
S. beijiangensis	69	DSM 41794, NBRC 100044, YIM6	AF385681																	+	
S. drozdowiczii	148	NRRL B-24297	EF654097																	+	
S. yanii	526	AS 4.1146, JCM 3331, IFO 14669	AB006159																		
Group S. graminofaciens et rel.:																					
S. peucetius	366	DSM 40754, NCIB 10972, LMG 20084, JCM 9920	AB045887		IV 09 (red series)	035 1-33							Sch00	Lan2-00	BENP	+		+			cl21
S. xantholiticus	523	DSM 40244, ATCC 27481, LMG 19402, ISP 5244, NBRC 13354	AB184349	C 24	II 05	062 024		La-21					Sch00	Lan2-00	BENP +	+		+	cl04	+	cl21
S. kurssanovii	272	DSM 40162, ATCC 15824, LMG 19933, ISP 5162, NBRC 13192	AB184325	F 60	IV 20 (gray series)	025 1-15							Sch00	Lan2-00	BENP	+		+		+	cl21
S. graminofaciens	216	DSM 40559, ATCC 12705, LMG 19892, ISP 5559	AJ781329	A 26	III 03	004 1-23				OC-I			Sch00	Lan2-00	BENP	+		+		+	+
Group S. amakusaensis et rel.:																					
S. amakusaensis	30	DSM 40219, ATCC 23876, LMG 19350, ISP 5219, NRRL B-3351	AY999781	B Sm	III 12	079 063		La-00		OC-I		L2	Sch00	Lan2-00	BENP +	+		+	cl14	+	+
S. inusitatus	259	DSM 41441, LMG 19955, NBRC 13601	AB184445										Sch00	Lan2-00	BENP	+		+		+	cl13

(continued)

TABLE 272. (continued)

Guo 08[21] 16S rRNA	Guo 08[21] *atpD*	Guo 08[21] *gyrB*	Guo 08[21] *recA*	Guo 08[21] *rpoB*	Guo 08[21] *trpB*	Guo 08[21] concat.	Kim 04[23] *rpoB*	Morphological characters[24]				Physiological tests[25]										
								I	II	III	IV	I	II	III	IV	V	VI	VII	VIII	IX	X	XI
								Gy	RF	C+	SM	+	+	+	+	+	n	+	–	–	n	+
								W	S	C+	SM	+	+	+	–	+	n	–	+	–	n	n
								Y/W	RF/ RA/S	C+	SM	+	+	+	+/–	+/–	n	–	+/–	+/–	n	+/–
								n	S	C–	SM	+	+	+	–	n	n	–	+	n	n	n
								W	RF	C–	SM	+	+	–	–	–	+	+	n	–	+	+
								Y	S	C–	SM	+	n	+	–	n	+	–	–	–	n	–
								Y	S	C–	SM	+	n	+	–	n	+	–	–	–	n	–
								n	RF–RA	C–	n	+	+	–	–	–	+	n	–	n	n	–
								Gy	S	C+												
+	+	+	+	+	+	+		Gy	RF	C–	SM	+	+	n	n	+	n	+	+	n	n	n
+	+	+	+	+	+	+		R	S	C–	SM	+	+	–	–	+	n	+	+	n	n	+
								W	S	C–	SM	+	–	–	–	n	+	–	n	n	n	–
								Gy	S	C+	SM	+	+	+	–	+	+	+	–	–	–	+
+	+	+	+	+	+	+		Gy	S	C–	WTY	+	+	+	+	n	n	n	n	n	n	n
								B	S	C+	SM	+	–	±	–	–	–	–	–	–	–	±
								B/Cy	S	C	SM	[illegible]					[illegible]					

(continued)

TABLE 272. List of *Streptomyces* species continued

Species names and groups[1]	No. in lists of species[2]	Type strain	Accession no.	Wil 83a[3]	Wil 89[4]	Käm 91[5]	Hat 03[6]	Lan 02[7]	Ful 95[8]	Och 95[9]	Kat 97[10]	Lab[11]	Sch[12]	Lan 04[13]	Lan 02[14]	Lan 04[15]	Lan 04[16]	Lan 04[17]	Lan 04[18]	Lan 04[19]	Lan 04[20]
S. clavuligerus	122	DSM 40751, ATCC 27064, LMG 20477, DSM 738, NRRL 3585	AY999718		IV 10 (gray series)	22-5 036							Sch00	Lan2-00	BENP	+		+		+	+
Group S. atratus et rel.:																					
S. atratus	45	DSM 41673, LMG 20420, NRRL B-16927	DQ026638										Sch00	Lan2-00	BENP	+		+		+	cl23
S. sanglieri	430	DSM 41791, NBRC 100784	AB249945																	+	
S. gelaticus	199	DSM 40065, ATCC 3323, LMG 19376, ISP 5065, NRRL B-2928	DQ026636	A Sm	III 11	003 1-3		La-00					Sch00	Lan2-00	BENP +	+		+		+	+
S. pulveraceus	390	DSM 41657, LMG 20322, NBRC 3855	AB184806										Sch00	Lan2-00	BENP	+		+			cl23
Not closely related to one of the groups:																					
S. sannanensis	431	DSM 41705, LMG 20329, NBRC 14239	AB184579										Sch07	Lan2-04	BENP	+		+			cl22
Most closely related to group S. laurentii et rel.:																					
S. showdoensis	440	DSM 40504, ATCC 15105, LMG 20298, ISP 5504, NBRC 13417	AB184389	A 06	I 05	22-2 037							Sch00	Lan2-00	BENP	+		+		+	cl23
S. viridobrunneus	513	DSM 41466, LMG 20317	AJ781372										Sch00	Lan2-00	BENP	+		+		+	cl22
S. roseoviridis	421	DSM 40175, ATCC 23959, LMG 20266, ISP 5175, NBRC 12911	AB184239	A 05	I 04	22-2 037							Sch00	Lan2-00	BENP	+		+		+	cl44
S. vietnamensis	495	CCTCC M 205143, JCM 21785, GIMV4.0001	DQ311081																		
S. nashvillensis	332	DSM 40314, ATCC 25476, LMG 20064, ISP 5314, NBRC 13064	AB184286	A 05	I 04	002 1-7							Sch00	Lan2-00	BENP	+		+		+	cl23
S. tanashiensis	466	DSM 40195, ATCC 23967, LMG 20274, ISP 5195	AJ781362		IV 30 (gray series)	002 1-7							Sch00	Lan2-00	BENP	+		+			cl22
S. roseolus	417	DSM 40174, ATCC 23210, LMG 20265, ISP 5174, NBRC 12816	AB184168	A 05	I 04	002 1-7							Sch00	Lan2-00	BENP	+		+		+	cl22
S. bikiniensis	71	DSM 40581, ATCC 11062, LMG 19367, ISP 5581	X79851	F 64	III 21	22-4 1-07		La-00		OC-II			Sch00	Lan2-00	BENP +	+		+	cl01		cl23
S. violaceorectus	500	DSM 40279, ATCC 25514, LMG 20281, ISP 5279, NBRC 13102	AB184314	A 05	I 04	002 1-7							Sch00		BENP	+		+		+	cl23
S. cinereoruber subsp. *cinereoruber*	111a	DSM 40012, ATCC 19740, LMG 20462, ISP 5012, NBRC 12756	AB184121	A 05	I 04	002 038			FU-6				Sch00		BENP	+		+		+	cl23

(continued)

TABLE 272. (continued)

Guo 08[21] 16S rRNA	Guo 08[21] *atpD*	Guo 08[21] *gyrB*	Guo 08[21] *recA*	Guo 08[21] *rpoB*	Guo 08[21] *trpB*	Guo 08[21] concat.	Kim 04[23] *rpoB*	Morphological characters[24]				Physiological tests[25]										
								I	II	III	IV	I	II	III	IV	V	VI	VII	VIII	IX	X	XI
								Gy	RF		SM											
								Gy	S	C–	SM	+	+	–	+	+	+	+	–	–	+	n
								Gy	S	C+	SM	n	+	+	+	+	+	+	n	n	n	+
								Gy	RF	C–	SM	+	+	–	+	–	+	+	–	–	+	+
+	+	+	+	+	+	+		Gy	S	C+	SM	n	+	–	+	+	+	+	–	–	+	–
								Gy	S	C–	SM	±	±	–	–	–	–	–	–	–	–	–
								Gy	RF	C+	SM	+	+	±	–	+	+	–	–	–	+	±
								Gy	RF	C+	SM	+		–	–			+	–	–		
								R	RF	C+	SM	+	+	+	–	–	+	–	–	–	–	n
								W	RF	C+	n	+	+	+	+	+	+	n	n	n	n	+
								Gy	RF	C+	SM	+	+	+	–	–	+	–	–	–	+	±
+	+	+	+	+	+	+		Gy	RF	C+	SM	+	+	+	–	–	+	–	–	–	+	–
								R	RF	C–	SM	+	+	+	+	±	n	–	–	–	n	–
							+	Gy	RF	C+	SM	+	+	–	–	–	+	–	–	–	±	±
								Gy	RF	C+	SM	+	+	+	–	+	n	–	–	–	n	+
								Gy	RF	C+	SM	+	+	+	–	–	+	–	–	–	+	–

(continued)

TABLE 272. List of *Streptomyces* species continued

Species names and groups[1]	No. in lists of species[2]	Type strain	Accession no.	Wil 83a[3]	Wil 89[4]	Käm 91[5]	Hat 03[6]	Lan 02[7]	Ful 95[8]	Och 95[9]	Kat 97[10]	Lab[11]	Sch[12]	Lan 04[13]	Lan 02[14]	Lan 04[15]	Lan 04[16]	Lan 04[17]	Lan 04[18]	Lan 04[19]	Lan 04[20]
Group S. laurentii et rel.:																					
S. laurentii	278	DSM 41684, LMG 19959	AJ781342										Sch00	Lan2-00	BENP	+		+		+	cl22
S. termitum	469	DSM 40329, ATCC 25499, LMG 20289, ISP 5329, NBRC 13087	AB184302	A 05	I 04	22-2 037							Sch00	Lan2-00	BENP	+		+			cl22
S. roseofulvus	415	DSM 40172, ATCC 19921, LMG 20263, ISP 5172, NBRC 13194	AB184327	A 14	II 04	002 1-7							Sch00	Lan2-00	BENP	+		+			cl22
Most closely related to group S. laurentii et rel.:																					
S. filamentosus	168	DSM 40022, ATCC 19753, LMG 20512, ISP 5022, NBRC 12767	AB184130	A 05	I 04	002 1-7							Sch00	Lan2-24	BENP	+	(b)	+		+	cl23
Group S. gobitricini et rel.:																					
S. gobitricini	212	DSM 41701, LMG 19910, NBRC 15419	AB184666										Sch00	Lan2-00	BENP	+		+			cl14
S. lavendofoliae	279	DSM 40217, ATCC 15872, LMG 19935, ISP 5217	AJ781336		IV 07 (red series)	22-3 1-08							Sch00	Lan2-00	BENP	+		+		+	cl14
S. luridus	298	DSM 40081, ATCC 19782, LMG 19365, ISP 5081, NRRL B-5409	DQ442523	F 62	II 14	22-3 1-08		La-17		OC-II			Sch00	Lan2-00	BENP +	+		+			-
S. roseolilacinus	416	DSM 40173, ATCC 19922, LMG 20264, ISP 5173, NBRC 12815	AB184167	G 68	II 18	22-5 039							Sch00	Lan2-00	BENP	+		+			cl12
Not closely related to one of the groups:																					
S. biverticillatus	72	DSM 40272, ATCC 23615, LMG 20433, ISP 5272	AJ781381		Sv. 01	22-1 040	Ha7					L4		Lan2-00	BENP	+	(j)	+		+	cl13
S. werraensis	519	DSM 40486, ATCC 14424, LMG 21047, ISP 5486, NRRL B-5317	DQ442558	A 12	I 07	006 1-18					KA-G		Sch00	Lan2-00	BENP	+		+		+	cl04
S. globisporus subsp. *caucasicus*	208b	DSM 40814, ATCC 19907, LMG 19895, NRRL B-2593	EF178676		I 02	1-1 1-1							Sch10		BENP	+		+		+	cl08
S. albireticuli	16	DSM 40051, ATCC 19721, LMG 20393, ISP 5051, NBRC 12737	AB184881	F SM	Sv. 11	076 069	Ha5						Sch00	Lan2-00	BENP	+	(j)	+			cl13
S. eurocidicus	161	DSM 40604, ATCC 27428, LMG 20509, ISP 5604, NRRL B-1676	AY999790	F 56	Sv. 02	22-1 040	Ha5					L4	Sch00	Lan2-00	BENP	+	(j)	+			cl13
S. stramineus	460	DSM 41783, NBRC 16131	AB184720				Ha16														

(continued)

TABLE 272. (continued)

								Morphological characters[24]				Physiological tests[25]										
Guo 08[21] 16S rRNA	Guo 08[21] *atpD*	Guo 08[21] *gyrB*	Guo 08[21] *recA*	Guo 08[21] *rpoB*	Guo 08[21] *trpB*	Guo 08[21] concat.	Kim 04[23] *rpoB*	I	II	III	IV	I	II	III	IV	V	VI	VII	VIII	IX	X	XI
									RF	C−	SM	+	+	−	−	−	+	−	−	−	n	+
								R	RF	C−	SM	+	+	−	n	−	n	−	−	−	n	n
								R	RF	C−	SM	+	+	+	+	+	+	+	−	−	+	+
								R	RF	C−	SM	+	+	+	−	−	+	−	−	−	−	+
								R	RF	C+	SM	+	+	+	+	+	+	−	−	+	−	n
								R	S	C+	SM	+	+	+	−	−	+	−	−	+	−	−
								R	S	C+	SM	+	+	+	−	±	n	−	−	+	n	−
								R	S	C−	SM	+	−	+	−	±	n	−	−	−	n	−
								Bi	VE	C+	SM	+	±	−	±	±	n	±	−	±	n	±
								Gy	S	C−	n	n	n	n	n	n	n	n	n	n	n	n
								Y	RF	C−	SM	+	+	+	−	n	n	n	n	n	n	n
							Group F	Ar	VE	C+	SM	+	±	±	±	±	n	±	±	+	n	+
								Ar	VE	C+	SM	+	−	−	−	±	n	−	−	±	n	+
								Y	VER	C+	SM	+	−	n	−	+	−	−	+	+	−	n

(continued)

TABLE 272. List of *Streptomyces* species continued

Species names and groups[1]	No. in lists of species[2]	Type strain	Accession no.	Wil 83a[3]	Wil 89[4]	Käm 91[5]	Hat 03[6]	Lan 02[7]	Ful 95[8]	Och 95[9]	Kat 97[10]	Lab[11]	Sch[12]	Lan 04[13]	Lan 02[14]	Lan 04[15]	Lan 04[16]	Lan 04[17]	Lan 04[18]	Lan 04[19]	Lan 04[20]
S. olivoverticillatus	356	DSM 40250, NRRL B-1994, LMG 20058, NBRC 15273	AB184636				Ha18					L4	Sch00	Lan2-00	BENP	+	(j)	+		+	cl13
S. netropsis	333	DSM 40259, ATCC 23940, LMG 5979, ISP 5259, NBRC 12893	AB184848	F 56	Sv. 01	22-1 040	Ha14		FU-21			L4	Sch00	Lan2-00	BENP	+	(j)	+		+	cl13
Group Kitasatospora–Streptacidiphilus–Streptomyces																					
Subgroup Kitasatospora–Streptomyces:																					
K. gansuensis	8*	DSM 44786, NBRC 101835, HKI 0314	AY442265																		
S. atroaurantiacus	46	DSM 41649, LMG 20421, NRRL B-24282	DQ026645										Sch00	Lan2-00	BENP	+		+		+	cl30
K. mediocidica	11*	DSM 43929, IFO 14789, IFO 14789	U93324																		
S. purpeofuscus	393	DSM 40283, ATCC 23952, LMG 20283, ISP 5283	AJ781364		IV 26 (gray series)	22-3 043							Sch00	Lan2-00	BENP	+		+			cl30
S. chrysomallus subsp. *fumigatus*	109b	DSM 41424, LMG 21793, NBRC 15394	AB184645										Sch00	Lan2-00	BENP	+		+		+	+
S. purpureus	395	DSM 43362, LMG 19368	AJ781324		I 23	22-3 1-05		La-18		OC-I			Sch00	Lan2-00						+	
S. xanthocidicus	522	DSM 40575, ATCC 27480, LMG 19370, ISP 5575, IFO 13469	AY999858	F 66	II 16	22-4 043		La-18					Sch00	Lan2-00	BENP +	+		+	cl03	+	cl29
S. aburaviensis	3	DSM 40033, ATCC 23869, LMG 19305, ISP 5033, NRRL B-2218	AY999779	A 02	II 01	22-3 043		La-00		OC-I			Sch00	Lan2-00	BENP +	+		+	cl03	+	+
S. herbaricolor	247	DSM 40123, ATCC 23922, LMG 19929, ISP 5123, NBRC 3838	AB184801	A 02	II 01	22-4 043							Sch00	Lan2-00	BENP	+		+		+	cl30
S. indigoferus	256	DSM 40124, LMG 19930, ISP 5124, NBRC 12878	AB184214										Sch00	Lan2-00	BENP	+		+			cl30
S. avellaneus	58	DSM 40554, ATCC 23730, LMG 20427, ISP 5554, NBRC 13451	AB184413		II 17	002 1-7							Sch00	Lan2-28	BENP	+		+			cl29
S. psammoticus	386	DSM 40341, ATCC 25488, LMG 20525, ISP 5341, IFO 13971	AY999862	F 67	II 17	011 1-21				OC-I			Sch09	Lan2-28	BENP	+		+		+	cl29
S. aureofaciens	53	DSM 40127, ATCC 10762, LMG 5968, ISP 5127, KACC 20180	AY207608	A 14	II 04	22-4 043				OC-I			Sch00	Lan2-28	BENP	+		+		+	cl29
K. sampliensis	18*	DSM 44898, NBRC 102069, VT-36	AY260167																		
K. putterlickiae	17*	DSM 44665, NBRC 100917, F18-98	AY189976																	+	

(continued)

TABLE 272. (continued)

Guo 08[21] 16S rRNA	Guo 08[21] *atpD*	Guo 08[21] *gyrB*	Guo 08[21] *recA*	Guo 08[21] *rpoB*	Guo 08[21] *trpB*	Guo 08[21] concat.	Kim 04[23] *rpoB*	Morphological characters[24] I	II	III	IV	Physiological tests[25] I	II	III	IV	V	VI	VII	VIII	IX	X	XI
								Ar	VE	C+	SM	+	–	±	–	–	n	±	–	±	n	+
							Group F	Ke	VE	C	SM	+	±	–	±	–	n	–	–	+	n	–
								W	RF	C+	SM	+	+	+	–	+	n	–	–	–	n	+
								Gy	RF	C+	SM	+	+	+	–	–	+	–	–	–	–	n
								Gy	RF	C–	SM	+	+	+	–	–	+	–	–	–	–	n
							+	Gy/R	RF	C+	SM	+	–	+	–	+	n	–	–	+	+	–
								Gy	RF	C–	SM	+	+	+	–	+	+	–	–	–	–	+
								Gy	RF	C–	SM	+	±	–	–	±	–	–	–	–	–	–
								Gy	RF	C+	SM	+	+	+	–	+	+	+	–	–	–	+
								Gy	RF	C+	SM	+	+	+	–	–	+	–	–	–	–	n
								Gy	S	C–	SM	+	±	–	–	+	n	–	n	–	n	+
								Gy	S	C–	SM	+	–	–	–	+	n	–	–	–	–	+
								Gy	S	C–	SM	+	±	+	–	+	+	–	–	–	–	+

(continued)

TABLE 272. List of *Streptomyces* species continued

Species names and groups[1]	No. in lists of species[2]	Type strain	Accession no.	Wil 83a[3]	Wil 89[4]	Käm 91[5]	Hat 03[6]	Lan 02[7]	Ful 95[8]	Och 95[9]	Kat 97[10]	Lab[11]	Sch[12]	Lan 04[13]	Lan 02[14]	Lan 04[15]	Lan 04[16]	Lan 04[17]	Lan 04[18]	Lan 04[19]	Lan 04[20]
K. kifunensis	10*	DSM 41654, IFO 15206	AB022874																		
K. azatica	3*	DSM 41650, LMG 20429, IFO 13803	U93312											Lan2-00	BENP	+		+		+	cl30
K. nipponensis	13*	DSM 44787, NBRC 101836, HKI 0315	AY442263																		
K. cineracea	5*	NRRL B-24134, SK-3255	AB022875																		
K. niigatensis	12*	IFO 16453, SK-3406	AB022876																		
K. cheerisanensis	4*	KCTC 2395, YC75	AF050493																		
K. phosalacinea	16*	DSM 43860, NRRL B-16230, LMG 20102, KA-338	AB022869											Lan2-00	BENP	+		+			cl27
K. paracochleata	14*	DSM 41656, NBRC 14769	U93328												BENP	+		+			cl28
K. cochleata	6*	DSM 41652, NBRC 14768	U93316											Lan2-30	BENP	+	(b)	+		+	cl28
K. griseola	9*	DSM 43859, NRRL B-16229, AM-9660	AB022870																		
K. setae	1*	DSM 43861, NBRC 14216, LMG 20529, KM-6054	AB022868											Lan2-00	BENP	+		+			cl27
K. paranensis	15*	DSM 44788, NBRC 101837, HKI 0190	AY442268																		
K. cystarginea	7*	DSM 41680, IFO 14836, JCM 7356	U93318																		
K. terrestris	19*	DSM 44789, NBRC 101838, HKI 0186	AY442266																		
K. viridis	20*	DSM 44826, 52108a	AY613990																		
K. arboriphila	2*	DSM 44785, NBRC 101834, HKI 0189	AY442267																		
S. alboverticillatus	24	DSM 41678, DSM 41500, LMG 20401, JCM 5010	AY999766				Ha6						Sch00	Lan2-00	BENP	+	(j)	+		+	cl17
Group Kitasatospora–Streptacidiphilus–Streptomyces:																					
Streptacidiphilus oryzae	7†	CGMCC 4.2012, JCM 13271, TH49	DQ208700																		
Subgroup Streptacidiphilus albus et rel.:																					
Streptacidiphilus albus	1†	DSM 41753, JL 83	AF074415																	+	
Streptacidiphilus carbonis	3†	DSM 41754, JL 415	AF074412																		
Streptacidiphilus neutrinimicus	6†	DSM 41755, NBRC 100921, JL 206	AF074410																		
Subgroup Streptacidiphilus anmyonensis et rel.:																					
Streptacidiphilus jiangxiensis	4†	NBRC 100920, JCM 12277	AB249948																		
Streptacidiphilus anmyonensis	2†	NBRC 103185, AM-11	DQ904546																		
Streptacidiphilus melanogenes	5†	NBRC 103184, SB-B34	DQ994689																		
Streptacidiphilus rugosus	8†	NBRC 103186, AM-16	DQ904547																		

(continued)

TABLE 272. (continued)

Guo 08[21] 16S rRNA	Guo 08[21] *atpD*	Guo 08[21] *gyrB*	Guo 08[21] *recA*	Guo 08[21] *rpoB*	Guo 08[21] *trpB*	Guo 08[21] concat.	Kim 04[23] *rpoB*	Morphological characters[24] I	II	III	IV	Physiological tests[25] I	II	III	IV	V	VI	VII	VIII	IX	X	XI
							Kitasatospora															
							Kitasatospora															
							Kitasatospora															
								Gy	S	C+	SM	+	–	+	–	–	n	n	n	–	n	–
							Kitasatospora															
							Kitasatospora															
								W		C–								–	–			

(continued)

TABLE 272. List of *Streptomyces* species continued

Species names and groups[1]	No. in lists of species[2]	Type strain	Accession no.	Wil 83a[3]	Wil 89[4]	Käm 91[5]	Hat 03[6]	Lan 02[7]	Ful 95[8]	Och 95[9]	Kat 97[10]	Lab[11]	Sch[12]	Lan 04[13]	Lan 02[14]	Lan 04[15]	Lan 04[16]	Lan 04[17]	Lan 04[18]	Lan 04[19]	Lan 04[20]
Not closely related to one of the groups:																					
S. ardus	39	DSM 40527, ATCC 27417, LMG 20415, ISP 5527, NBRC 13430	AB184864		Sv. 03	22-1 040	Ha2					L4	Sch00	Lan2-00	BENP	+	(j)	+			cl17
S. blastmyceticus	73	DSM 40029, ATCC 19731, LMG 20434, ISP 5029, NRRL B-5480	AY999802	F 58	Sv. 02	22-1 040	Ha3					L4	Sch00	Lan2-00	BENP	+	(j)	+		+	cl17
S. caeruleus	81	DSM 40103, ATCC 27421, LMG 19399, ISP 5103, NRRL B-2194	EF178675		IV 07 (gray series)	058 050		La-19					Sch00	Lan2-14	BENP cl19	+	(c)	+	cl09	+	cl47
S. hiroshimensis	248	DSM 40037, ATCC 19772, LMG 19924, ISP 5037, NBRC 3720	AB184789	F 57	Sv. 01	22-1 040	Ha7		FU-NC			L4	Sch00	Lan2-00	BENP	+	(j)	+			cl12
S. cinnamoneus	117	DSM 40005, ATCC 11874, LMG 8602, ISP 5005, NBRC 12852	AB184850	F 55	Sv. 02	22-1 040	Ha4					L4	Sch00		BENP	+	(j)	+			cl17
S. pseudoechinosporeus	387	DSM 43035, LMG 21052, NBRC 12518	AB184100										Sch00	Lan2-00	BENP	+		+		+	+
S. lilacinus	286	DSM 40254, ATCC 23930, LMG 20059, ISP 5254, NBRC 3944	AB184819		Sv. 16	22-1 040	Ha8						Sch00	Lan2-00	BENP	+	(j)	+		+	cl12
S. sapporonensis	432	DSM 41675, LMG 20324, NBRC 13823	AB184508				Ha4						Sch17	Lan2-00	BENP	+	(j)	+		+	cl17
S. varsoviensis	492	DSM 40346, ATCC 25505, LMG 20083, ISP 5346, NRRL B-3589	DQ026653	C 46	III 13	037 028		La-12		OC-II			Sch00	Lan2-00	BENP cl12	+	(c)	+		+	cl13
S. abikoensis	2	DSM 40831, NRRL B-2113, LMG 20386, NBRC 13860	AB184537				Ha1					L4	Sch00	Lan2-00	BENP	+	(j)	+		+	cl12
S. lavenduligriseus	281	DSM 40487, ATCC 13306, LMG 19943, ISP 5487, NRRL B-3173	DQ442515	A 34	Sv. 02	1-5 009						L4	Sch00	Lan2-00	BENP	+		+		+	cl59
S. morookaense	345	DSM 40503, ATCC 19166, LMG 20074, ISP 5503	AJ781349	F 59	Sv. 08	22-1 040	Ha13						Sch00	Lan2-00	BENP	+	(j)	+			+
S. thioluteus	482	DSM 40027, ATCC 12310, LMG 21037, ISP 5027, NBRC 3364	AB184753	F Sm	Sv. 21	22-1 040	Ha17						Sch24	Lan2-00	BENP	+	(j)	+		+	cl12
S. luteireticuli	300	DSM 40509, ATCC 27446, ISP 5509, NBRC 13422	AB249969		Sv.	1-8 1-17	Ha9					L4									
S. ehimensis	154	DSM 40253, ATCC 23903, LMG 20505, ISP 5253, KCTC 9727	AY999834		Sv. 09	22-1 040	Hal						Sch00	Lan2-00	BENP	+	(j)	+		+	cl12
S. hygroscopicus subsp. *angustmyceticus*	253b	DSM 41683, LMG 19958, NRRL B-2347	DQ442509										Sch00	Lan2-00	BENP	+		+			cl23

(continued)

TABLE 272. (continued)

								Morphological characters[24]				Physiological tests[25]										
Guo 08[21] 16S rRNA	Guo 08[21] *atpD*	Guo 08[21] *gyrB*	Guo 08[21] *recA*	Guo 08[21] *rpoB*	Guo 08[21] *trpB*	Guo 08[21] concat.	Kim 04[23] *rpoB*	I	II	III	IV	I	II	III	IV	V	VI	VII	VIII	IX	X	XI
								Ar	VE	C+	SM	+	–	–	–	+	n	±	–	+	n	+
								Mo	VE	C+	SM	+	–	–	±	–	n	±	–	±	n	+
								Gy	RF	C–	SM	+	–	–	–	n	n	n	n	n	n	n
							Group F	Hi	VE	C+	SM	+	–	–	–	–	n	–	–	+	n	±
								Ci	VE	C–	SM	+	±	–	±	±	n	±	±	+	n	+
								W/Gy		C+	SM	+	+	+	+	+	+	+	+			+
							Group F	Li	VE	C+	SM	+	–	–	–	–	n	–	–	±	n	±
								R		C–								–	–			
								W	S	C–	SM	+	–	–	–	+	+	–	+	+	+	–
							Group F	Ar	VE	C+	SM	–	–	–	–	–	n	–	–	–	n	–
								Mo	VE	C–	SM	+	±	±	±	+	n	+	±	+	n	+
								Y	VE	C–	SM	+	–	–	?	+	n	+	+	+	n	–
								Th	VE	C–	SM	+	–	–	–	–	n	–	–	±	n	±
								Y/Gy	VE	C+	SM	+	?	?	?	?	n	?	?	+	n	?
							Group F	Ke	VE	C+	SM	+	±	–	±	±	n	–	±	±	n	±
								Gy	S	C–	SM	+	–	–	–	n	±	±	+	–	–	+

(continued)

TABLE 272. List of *Streptomyces* species continued

Species names and groups[1]	No. in lists of species[2]	Type strain	Accession no.	Wil 83a[3]	Wil 89[4]	Käm 91[5]	Hat 03[6]	Lan 02[7]	Ful 95[8]	Och 95[9]	Kat 97[10]	Lab[11]	Sch[12]	Lan 04[13]	Lan 02[14]	Lan 04[15]	Lan 04[16]	Lan 04[17]	Lan 04[18]	Lan 04[19]	Lan 04[20]
Group S. ochraceiscleroticus et rel.:																					
S. ochraceiscleroticus	348	DSM 40594/ DSM 43155, ATCC 15814, LMG 19349, NBRC 12394	AB184094		III 08	069 1-26				OC-non			Sch00	Lan2-00	BENP +	+		+	cl08	+	+
S. purpurogeneiscleroticus	396	DSM 40271, DSM 43156, LMG 20331	AJ621604	A 40		069 1-26							Sch00	Lan2-00	BENP	+		+		+	+
S. violens	508	DSM 40597, ATCC 15898, LMG 20303, ISP 5597	AJ621605	A 40	I 18	069 1-26							Sch00	Lan2-00	BENP	+		+		+	+
S. monomycini	325	DSM 41801, NRRL B-24309	DQ445790																		
S. niger	335	DSM 40302, DSM 43049, LMG 20101	AJ621607	A 40	I 18	069 1-26							Sch00	Lan2-00	BENP	+		+		+	+
S. olivaceiscleroticus	350	DSM 40595, ATCC 15722, LMG 20081, ISP 5595	AJ621606		IV 24 (gray series)	069 1-26							Sch00	Lan2-00	BENP	+		+			+
Most closely related to groups S. ochraceiscleroticus et rel. and S. albofaciens et rel.:																					
S. auratus	51	DSM 41897, NRRL 8097	AJ391816																		
Group S. albofaciens et rel.:																					
S. chrestomyceticus	106	DSM 40545, ATCC 14947, LMG 20457, ISP 5545	AJ621609	B 42	I 19	035 1-33							Sch00	Lan2-00	BENP	+		+		+	cl23
S. rimosus subsp. *paromomycinus*	409b	DSM 41429, LMG 20308	AJ621610										Sch00	Lan2-00	BENP	+		+		+	cl23
S. albofaciens	17	DSM 40268, ATCC 25184, LMG 20394, ISP 5268, JCM 4342	AB045880	B 42	I 19	035 1-33							Sch00	Lan2-00	BENP	+		+		+	cl10
Most closely related to groups S. ochraceiscleroticus et rel. and S. albofaciens et rel.:																					
S. erumpens	158	DSM 40941, ATCC 23266, LMG 20507	AJ621603		IV 15 (gray series)	035 1-33							Sch21	Lan2-00	BENP	+		+		+	cl23
S. rimosus subsp. *rimosus*	409a	DSM 40260, ATCC 10970, LMG 19352, ISP 5260, JCM 4667	AB045883	B 42	I 19	035 1-33		La-09		OC-non			Sch00	Lan2-00	BENP +	+		+	cl13	+	cl10
S. sclerotialus	435	DSM 40269, DSM 43032, LMG 20528	AJ621608		I 18	069 1-26							Sch00	Lan2-00	BENP	+		+			cl02
Group S. albulus et rel.:																					
S. albulus	27	DSM 40492, ATCC 12757, LMG 20404, ISP 5492, IMC S-0802	AB024440	A 29	I 15	025 109							Sch00	Lan2-00	BENP	+		+		+	+
S. noursei	346	DSM 40635, ATCC 11455, LMG 5982, NBRC 15452	AB184678		IV 23 (gray series)	025 1-09								Lan2-00	BENP	+		+		+	+
S. yunnanensis	532	DSM 41793, CGMCC 4.1004, JCM 12115, YIM 41004	AF346818																	+	

(continued)

TABLE 272. (continued)

Guo 08[21] 16S rRNA	Guo 08[21] *atpD*	Guo 08[21] *gyrB*	Guo 08[21] *recA*	Guo 08[21] *rpoB*	Guo 08[21] *trpB*	Guo 08[21] concat.	Kim 04[23] *rpoB*	Morphological characters[24] I	II	III	IV	Physiological tests[25] I	II	III	IV	V	VI	VII	VIII	IX	X	XI
								W	S	C–	SM	+	+	+	+	+	+	+	+	+	+	n
								Gy	S	C–	SM	+	+	+	+	+	+	+	+	+	+	n
								n	n	C–	n	+	+	+	+	+	n	+	+	+	n	+
								W/Gy	S	C–	SM	+	+	–	–	+	n	–	+	–	n	n
								W/Gy	S	C–	SM	+	+	+	+	+	+	+	+	+	+	+
								Gy	S	C–	SM	+	+	+	+	n	n	n	n	n	n	n
								Gy	S	C+	SM	+	n	n	+	+	+	+	+	n	+	+
								W	S	C–	SM	+	–	–	–	+	+	–	+	–	–	n
								W	S	C–	SM	+	–	–	–	+	+	+	+	+	–	–
							+	W	S	C–	SM	+	±	+	–	+	n	+	+	+	n	±
Group IV	Yes	Yes	Yes	Yes	Yes	Yes		Gy	S	C–	SM	+	–	+	–	+	+	+	+	+	–	–
								W	S	C–	SM	+	–	+	–	+	+	+	+	+	–	n
								W	S	C–	SM	+	+	+	+	+	n	+	+	+	n	+
							+	Gy	S	C–	SPY	+	–	–	–	+	+	–	+	+	+	n
								Gy	S	C–	SPY	+	–	–	–	+	+	–	+	+	–	+
									S	C–	RU	+	–	+	+	+	n	+	+	+	n	–

(continued)

TABLE 272. List of *Streptomyces* species continued

Species names and groups[1]	No. in lists of species[2]	Type strain	Accession no.	Wil 83a[3]	Wil 89[4]	Käm 91[5]	Hat 03[6]	Lan 02[7]	Ful 95[8]	Och 95[9]	Kat 97[10]	Lab[11]	Sch[12]	Lan 04[13]	Lan 02[14]	Lan 04[15]	Lan 04[16]	Lan 04[17]	Lan 04[18]	Lan 04[19]	Lan 04[20]
Most closely related to groups S. ochraceiscleroticucs et rel., S. albofaciens et rel. and S. albulus et rel.:																					
S. kasugaensis	266	DSM 40819, LMG 19949, ISP 5819, M338-M1	AB024441										Sch00	Lan2-00	BENP	+		+		+	cl42
S. chattanoogensis	103	DSM 40002, ATCC 19739, LMG 19339, ISP 5002	AJ621611					La-00		OC-non			Sch00	Lan2-00	BENP +	+		+	cl12	+	cl23
S. lydicus	304	DSM 40461, ATCC 25470, LMG 19331, ISP 5461	Y15507	A 29	I 15	025 005		La-09	FU-21	OC-non			Sch00	Lan2-00	BENP +	+		+	cl12	+	cl23
S. albospinus	22	DSM 41674, LMG 20398, NBRC 13846	AB184527										Sch00	Lan2-00	BENP	+		+			+
S. sioyaensis	442	DSM 40032, ATCC 13989, LMG 20531, ISP 5032, NRRL B-5408	DQ026654	A 29	I 15	025 005							Sch00	Lan2-06	BENP	+		+		+	cl58
S. hygroscopicus subsp. *decoyicus*	253c	DSM 41427, LMG 19954, AS 4.1861	AY999883										Sch00	Lan2-00	BENP	+		+		+	cl23
Most closely related to groups S. ochraceiscleroticucs et rel., S. albofaciens et rel., S. albulus et rel. and S. caniferus et rel.:																					
S. catenulae	94	DSM 40258, ATCC 12476, LMG 20449, ISP 5258	AJ621613	C 43	II 11	035 041							Sch00	Lan2-00	BENP	+		+		+	cl23
S. misakiensis	322	DSM 40222, ATCC 23938, LMG 19369, ISP 5222, IFO 12891	AB217605	F 66	II 16	22-4 043		La-18		OC-non			Sch00	Lan2-00	BENP +	+		+		+	cl29
S. ramulosus	400	DSM 40100, ATCC 19802, LMG 19354, ISP 5100, NRRL B-2714	DQ026662	C Sm	III 16	035 041		La-00		OC-non			Sch00	Lan2-00	BENP +	+		+	cl12	+	cl23
Group S. caniferus et rel.:																					
S. hygroscopicus subsp. *glebosus*	253d	DSM 40823, LMG 19950, NBRC 13786	AB184479										Sch22	Lan2-00	BENP	+		+			cl23
S. libani subsp. *rufus*	284b	DSM 41230, LMG 20087	AJ781351										Sch22	Lan2-00	BENP	+		+		+	cl23
S. platensis	376	DSM 40041, ATCC 13865, LMG 20046, ISP 5041, JCM 4662	AB045882	A 29	I 15	025 005			FU-21				Sch22	Lan2-00	BENP	+		+		+	cl23
S. caniferus	88	DSM 41453, LMG 20446, NBRC 15389	AB184640										Sch00	Lan2-00	BENP	+		+			cl23
Most closely related to group S. caniferus et rel.:																					
S. libani subsp. *libani*	284a	DSM 40555, ATCC 23732, LMG 20077, ISP 5555, NBRC 13452	AB184414	A 29	I 15	025 005							Sch00		BENP	+		+			cl23

(continued)

TABLE 272. (continued)

Guo 08[21] 16S rRNA	Guo 08[21] *atpD*	Guo 08[21] *gyrB*	Guo 08[21] *recA*	Guo 08[21] *rpoB*	Guo 08[21] *trpB*	Guo 08[21] concat.	Kim 04[23] *rpoB*	Morphological characters[24] I	II	III	IV	Physiological tests[25] I	II	III	IV	V	VI	VII	VIII	IX	X	XI
								Gy	S	C–	SM	+	–	–	–	+		+	–	+		–
							+	Gy	S	C–	SPY	+	–	–	–	+	n	+	+	+	n	+
								Gy	S	C–	SM	+	+	+	–	+	+	+	+	–	n	+
								Gy	S	C–	SPY	+	±	–	–	+	+	+	+	+	+	–
							+	Gy	S	C–	SM	+	+	–	–	+	+	+	+	+	–	+
								Gy	S	C–	SM	+	+	–	–	+	+	–	+	+	–	n
							+	Gy	RF	C–	SM	+	–	–	–	+	+	–	+	–	–	n
								Gy	RF	C–	SM	+	–	–	–	+	n	+	+	+	n	+
								Gy	RF	C–	SM	+	–	–	–	±	+	+	+	–	–	–
								Gy	S	C–	SM	+	+	–	–	+	+	+	+	+	–	+
								Gy	S	C–	SM	+	+	+	–	+	n	+	n	+	n	+
								Gy	S	C–	SM	+	–	–	–	+	n	+	+	+	n	n
								Gy	S	C–	SM	+	–	–	+	–	–	–	+	+	n	n
							+	Gy	S	C–	SM	+	+	–	–	+	n	+	n	+	n	+
							+	Gy	S	C–	SM	+	–	–	–	+	n	+	+	+	n	+

(continued)

TABLE 272. List of *Streptomyces* species continued

Species names and groups[1]	No. in lists of species[2]	Type strain	Accession no.	Wil 83a[3]	Wil 89[4]	Käm 91[5]	Hat 03[6]	Lan 02[7]	Ful 95[8]	Och 95[9]	Kat 97[10]	Lab[11]	Sch[12]	Lan 04[13]	Lan 02[14]	Lan 04[15]	Lan 04[16]	Lan 04[17]	Lan 04[18]	Lan 04[19]	Lan 04[20]
S. tubercidicus	486	DSM 40261, ATCC 25502, LMG 19361, ISP 5261	AJ621612	C 47	III 14	025 005		La-02		OC-non			Sch00	Lan2-00	BENP +	+		+	cl12		cl23
S. nigrescens	336	DSM 40276, ATCC 23941, LMG 19332, ISP 5276, NRRL B-12176	DQ442530	A 29	I 15	025 005		La-02					Sch00		BENP +	+		+	cl12	+	cl23
Group S. albiflaviniger et rel.:																					
S. antimycoticus	36	DSM 40284, ATCC 23880, LMG 20413, ISP 5284, NBRC 12839	AB184185		IV 05 (gray series)	051 018							Sch27		BENP	+		+		+	cl16
S. geldanamycininus	200	DSM 41894, NRRL 3602, NRRL B-3602	DQ334781																	+	
S. melanosporofaciens	316	DSM 40318, ATCC 25473, LMG 20066, ISP 5318, NRRL B-12234	AJ271887	A 32	I 16	051 018						L1	Sch00	Lan2-00	BENP	+		+		+	cl16
S. sporoclivatus	456	DSM 41461, LMG 20312, NBRC 100767	AB249934										Sch27		BENP	+		+		+	cl16
S. yatensis	527	DSM 41771, NBRC 101000	AB249962																	+	
S. rutgersensis subsp. *castelarensis*	427b	DSM 40830, ATCC 15191, LMG 20304	AY508511		I 01	055 018							Sch00	Lan2-00	BENP	+		+			cl16
S. indonesiensis	257	DSM 41759, A4R2	DQ334783																	+	
S. griseiniger	217	DSM 41895, NRRL B-1865	AJ391818																	+	
S. rhizosphaericus	408	DSM 41760, NBRC 100778	AB249941																		
S. asiaticus	43	DSM 41761, NBRC 100774	AB249947																		
S. cangkringensis	87	DSM 41769, D13P3	AJ391831																	+	
S. malaysiensis	308	DSM 41697, LMG 20099, NBRC 16446	AB249918										Sch00	Lan2-00	BENP	+		+			l15
S. javensis	262	DSM 41764, B22P3	AJ391833																	+	
S. endus	156	DSM 40187, NRRL 2339, LMG 19393	AY999911					La-08				L1	Sch36	Lan2-29	BENP cl08	+	(f)	+	cl08	+	cl16
S. sporocinereus	455	DSM 41460, LMG 20311, NBRC 100766	AB249933										Sch36	Lan2-00	BENP	+		+			cl16
S. hygroscopicus subsp. *hygroscopicus*	253a	DSM 40578, ATCC 27438, LMG 19335, ISP 5578, NBRC 13472	AB184428	A 32	I 16	085 012		La-08	FU-6			L1	Sch00	Lan2-29	BENP cl08	+	(f)	+	cl08		cl16
S. demainii	143	DSM 41600, NRRL B-1478	DQ334782																		
S. violaceusniger	504	DSM 40563, ATCC 27477, LMG 19336, ISP 5563	AJ391823	A 32	I 16	051 018		La-07		OC-I		L1	Sch00	Lan2-00	BENP +	+	(f)	+	cl09		cl15
S. yogyakartensis	530	DSM 41766, NBRC 100779	AB249942																	+	
S. albiflaviniger	15	DSM 41598, NRRL B-1356	AJ391812																		

(continued)

TABLE 272. (continued)

Guo 08[21] 16S rRNA	Guo 08[21] *atpD*	Guo 08[21] *gyrB*	Guo 08[21] *recA*	Guo 08[21] *rpoB*	Guo 08[21] *trpB*	Guo 08[21] concat.	Kim 04[23] *rpoB*	Morphological characters[24] I	II	III	IV	Physiological tests[25] I	II	III	IV	V	VI	VII	VIII	IX	X	XI
								Gy	S	C–	SM	+	+	–	–	+	n	+	+	+	n	+
								Gy	S	C–	SM	+	+	+	+	+	n	–	+	+	+	+
								Gy	S	C–	RU	n	n	+	n	–	n	n	n	n	–	n
								Gy	S	C–	SM	+	+	+	+	+	n	+	+	+	n	–
								Gy	S	C–	WTY	+	+	–	–	+	+	+	+	n	n	n
								Gy	S	C–	RU	+	+	+	+	+	+	+	+	+	+	+
								Gy	S	C–	SM	+	+	+	+	+	+	+	+	–	+	–
								Gy	S	C+	RU	n	n	n	n	n	n	n	n	n	n	+
								Gy	S	C–	RU	n	n	–	n	+	n	n	n	n	–	n
								Gy	S	C–	RU											+
								Gy	S	C–	RU	n	n	n	n	n	n	n	n	n	n	+
								Gy	S	C–	RU	n	n	n	n	n	n	n	n	n	n	+
								W/Gy	S	C+	RU	+	+	+	+	+	+	+	+	+	n	–
								Gy	S	C–	RU											+
								Gy	S	C–	SM	+	+	+	+	+	+	–	+	–	+	–
								Gy	S	C–	WTY	+	n	n	n	–	+	n	n	n	n	n
								Gy	S	C–	SM	+	+	+	+	+	n	–	n	–	+	n
								Gy–Y	S	C–	RU	n	n	+	n	–	n	n	n	n	–	n
								Gy	S	C–	SM	+	+	+	+	+	n	+	+	+	+	n
								Gy	S	C–	RU	n	n	n	n	n	n	n	n	n	n	+
								W	S	C–	RU	n	n	+	n	+	n	n	n	n	–	n

(continued)

TABLE 272. List of *Streptomyces* species continued

Species names and groups[1]	No. in lists of species[2]	Type strain	Accession no.	Wil 83a[3]	Wil 89[4]	Käm 91[5]	Hat 03[6]	Lan 02[7]	Ful 95[8]	Och 95[9]	Kat 97[10]	Lab[11]	Sch[12]	Lan 04[13]	Lan 02[14]	Lan 04[15]	Lan 04[16]	Lan 04[17]	Lan 04[18]	Lan 04[19]	Lan 04[20]
Most closely related to groups S. ochraceiscleroticucs et rel., S. albofaciens et rel., S. albulus et rel., S. caniferus et rel. and S. albiflaviniger et rel.:																					
S. orinoci	359	DSM 40571, ATCC 23202, LMG 20079, ISP 5571, NBRC 13466	AB184866	F 58	Sv. 17	22-1 040	Ha15						Sch00	Lan2-00	BENP	+	(j)	+		+	cl15
S. mashuensis	309	DSM 40221, ATCC 23934, LMG 8603, ISP 5221	X79323	F 55	Sv. 03	22-1 040	Ha11					L4	Sch31		BENP	+	(j)	+			cl11
S. mobaraensis	324	DSM 40847, ATCC 29032, LMG 20086, NRRL B-3729	DQ442528		Sv. 07	22-1 040	Ha12		FU-12b			L4	Sch00		BENP	+	(j)	+			cl56
S. luteosporeus	302	DSM 40833, LMG 20085, NRRL 2401	DQ442525				Ha10						Sch00	Lan2-00	BENP	+	(j)	+			+
S. aureoversilis	55	DSM 40387, ATCC 15853, LMG 20425, ISP 5387, NBRC 13021	AB184855		Sv. 05	22-1 040	Ha7					L4	Sch00	Lan2-00	BENP	+	(j)	+		+	cl48
S. griseocarneus	221	DSM 40004, ATCC 12628, LMG 5973, ISP 5004	X99943	F 55	Sv. 03	22-1 040	Ha6	La-00	FU-12b			L4	Sch00	Lan2-00	BENP +	+	(j)	+		+	cl17
Group S. albus et rel.:																					
S. almquistii	28	DSM 40447, ATCC 618, LMG 21307, ISP 5447, NBRC 13015	AB184258	A 16	I 09	030 1-34							Sch24	Lan2-20	BENP	+		+		+	cl18
S. rangoonensis	401	DSM 40452, ATCC 6860, LMG 20295, ISP 5452, NBRC 13078	AB184295		IV 07 (white series)	030 1-34							Sch24	Lan2-20	BENP	+		+		+	cl18
S. gibsonii	203	DSM 43284, ATCC 6852, LMG 19912, NBRC 15415	AB184663		IV 05 (white series)	030 1-34							Sch24	Lan2-20	BENP	+		+		+	cl18
S. albus subsp. *albus*	1a	DSM 40313, ATCC 3004, ISP 5313	AJ621602	A 16	I 09	032 027			FU-6	OC-non			Sch24	Lan2-20						+	
S. flocculus	186	DSM 40327, ATCC 25453, LMG 19889, ISP 5327, NBRC 13041	AB184272	A 16	I 09	030 1-34							Sch24	Lan2-00	BENP	+		+			cl18
Most closely related to group S. albus et rel.:																					
S. cacaoi subsp. *cacaoi*	79a	DSM 40057, ATCC 3082, LMG 19320, ISP 5057, NBRC 12748	AB184115	A 16	I 09	031 1-34		La-05					Sch00	Lan2-17	BENP cl05	+	(c)	+	cl08	+	cl36
S. sulphureus	463	DSM 40104, ATCC 27468, LMG 19355, ISP 5104, NRRL B-1627	DQ442546	C Sm	III 17	068 002		La-00		OC-non			Sch00		BENP +	+		+		+	-
S. rubidus	423	CGMCC 4.2026, 13C15	AY876941																		

(continued)

TABLE 272. (continued)

Guo 08[21] 16S rRNA	Guo 08[21] *atpD*	Guo 08[21] *gyrB*	Guo 08[21] *recA*	Guo 08[21] *rpoB*	Guo 08[21] *trpB*	Guo 08[21] concat.	Kim 04[23] *rpoB*	Morphological characters[24] I	II	III	IV	Physiological tests[25] I	II	III	IV	V	VI	VII	VIII	IX	X	XI
								Ar	VE	C–	SM	+	–	–	–	±	n	±	–	–	n	±
								Ar	VE	C+	SM	+	–	–	–	+	n	±	–	+	n	+
								Mo	VE	C–	SM	+	±	±	–	+	n	–	±	±	n	+
								W	S	C–	SM	+	+	–	–	±	+	–	+	–	+	n
								Bi	VE	C+	SM	+	–	–	±	–	n	±	–	+	n	+
								Gr	VE	C+	SM	+	–	–	–	±	n	–	–	+	n	±
							Group A16	W	S	C–	SM	+	+	–	–	+	n	–	+	–	+	–
								W	S	C–	SM	+	+	±	–	+	n	–	+	–	n	–
								W	S	C–	SM	+	+	+	–	–	n	–	+	–	+	n
							Group A16	W	S	C–	SM	+	+	–	–	±	+	–	+	–	+	n
								W	S	C–	SM	+	+	+	–	+	+	+	+	+	+	n
								W	S	C–	SM	+	+	+	–	+	n	±	+	–	n	±
								Y	RF	C–	SM	+	+	+	–	+	n	+	n	–	+	n
									RF	C–	SM	+					+		+		+	

(continued)

TABLE 272. List of *Streptomyces* species continued

Species names and groups[1]	No. in lists of species[2]	Type strain	Accession no.	Wil 83a[3]	Wil 89[4]	Käm 91[5]	Hat 03[6]	Lan 02[7]	Ful 95[8]	Och 95[9]	Kat 97[10]	Lab[11]	Sch[12]	Lan 04[13]	Lan 02[14]	Lan 04[15]	Lan 04[16]	Lan 04[17]	Lan 04[18]	Lan 04[19]	Lan 04[20]
S. yeochonensis	528	NBRC 100782, JCM 12366, CN 732	AF101415																		
S. albus subsp. *pathocidicus*	1b	DSM 40799, LMG 20406, NBRC 13812	AB184501										Sch00	Lan2-00	BENP	+		+			cl07
S. glauciniger	205	LMG 22082, NBRC 100913	AB249964																		
S. guanduensis	239	CGMCC 4.2022, 701	AY876942																		
Most closely related to groups S. albus et rel. and S. glaucosporus et rel.:																					
S. ferralitis	166	DSM 41836, SFOp68	AY262826																	+	
S. vitaminophilus	517	DSM 41686, LMG 21051, NBRC 14294	AB184589										Sch00	Lan2-00	BENP	+		+		+	+
S. thermolineatus	477	DSM 41451, LMG 20309	Z68097										Sch00	Lan2-00	BENP	+		+		+	+
S. yanglinensis	525	CGMCC 4.2023, JCM 13275, 1307	AY876940																		
S. paucisporeus	365	CGMCC 4.2025, 1413	AY876943																		
Group S. glaucosporus et rel.:																					
S. macrosporus	305	DSM 41449	Z68099										Sch00	Lan2-00						+	
S. megasporus	314	DSM 41476, LMG 20092, NBRC 14749	AB184617										Sch00	Lan2-00	BENP	+		+			-
S. glaucosporus	206	DSM 41689, LMG 19907, NBRC 15416	AB184664										Sch00	Lan2-00	BENP	+		+			cl44
S. radiopugnans	398	DSM 41901, CGMCC 4.3519, R97	DQ912930																		
Most closely related to group S. glaucosporus et rel.:																					
S. albiaxialis	12	DSM 41799, NBRC 101002, NRRL B-24327	AY999901																	+	
S. armeniacus	42	DSM 43125, LMG 20418, JCM 3070	AB018092										Sch00	Lan2-00	BENP	+		+			+
Most closely related to groups S. albus et rel. and S. glaucosporus et rel.:																					
S. cuspidosporus	138	DSM 41425, LMG 20492, NBRC 12378	AB184090		IV 11 (gray series)	22-4 1-06							Sch08	Lan2-00	BENP	+		+		+	cl56
S. sparsogenes	445	DSM 40356, ATCC 25498, LMG 19378, ISP 5356, NBRC 13086	AB184301	A 32	I 16	010 1-19		La-07				L1	Sch00		BENP +	+		+			-
Most closely related to group S. geysiriensis et rel.:																					
S. janthinus	261	DSM 40206, ATCC 15870, LMG 8591, ISP 5206, NBRC 12879	AB184851	A 18	I 11	009 1-19						L2	Sch00		BENP	+		+			cl01
S. roseoviolaceus	420	DSM 40277, ATCC 25493, LMG 8594, ISP 5277	AJ399484	A 18	I 11	009 1-19			FU-1			L2	Sch00		BENP	+		+			cl01

(continued)

TABLE 272. (continued)

Guo 08[21] 16S rRNA	Guo 08[21] *atpD*	Guo 08[21] *gyrB*	Guo 08[21] *recA*	Guo 08[21] *rpoB*	Guo 08[21] *trpB*	Guo 08[21] concat.	Kim 04[23] *rpoB*	Morphological characters[24] I	II	III	IV	Physiological tests[25] I	II	III	IV	V	VI	VII	VIII	IX	X	XI
								Gy	RF	n	SM	n	n	n		n	n	n	n	n	+	n
								W	S	C–	SM	+	+	+	+	–	+	–	–	+	–	n
								Gy	S	C–	SM	+	+	+	+	+	+	+	+	+	n	+
								Gy/W	RF	C–	SM	+					+		+		+	
								W	S	C–	SM	+	n	–	n	n	+	n	+	n	n	n
										C–	SM	+	+	–	+	–	n	–	–	–	n	–
								G	RF	C–	SM	n	n	n	n	n	n	n	+	n	n	n
								W/Gy	RF	C–	SM	+	n	n	n	n	n	n	+	n	+	n
								W/Gy	RF	C+/C–	SM	+					+		+			
								Gy	S	C–	SPY	+	+	+	+	+	n	–	+	+	n	–
								G	S	C–	SPY/ WTY	+	+	+	+							
								G	S	C–	WTY	+	+	n	–	n	n	n	n	–	n	n
								W	S	C–	WTY	n	+	n	+	n	n	n	+	–	n	+
								W	S	C–	SM	+	+	+	+	n	n	+	–	–	n	+
								W	S	C–	n	+	+	+	+	+	+	+	–	n	+	+
								Gy	S	C–	SPY	+	+	+	+	+	+	+	+	+	+	+
								Gy	S	C–	SPY	+	+	+	+	+	±	+	+	±	–	+
								R	S	C+	SPY	+	+	+	+	+	+	+	+	+	+	+
								R	S	C+	SPY	+	+	+	+	+	n	+	+	+	n	+

(continued)

TABLE 272. List of *Streptomyces* species continued

Species names and groups[1]	No. in lists of species[2]	Type strain	Accession no.	Wil 83a[3]	Wil 89[4]	Käm 91[5]	Hat 03[6]	Lan 02[7]	Ful 95[8]	Och 95[9]	Kat 97[10]	Lab[11]	Sch[12]	Lan 04[13]	Lan 02[14]	Lan 04[15]	Lan 04[16]	Lan 04[17]	Lan 04[18]	Lan 04[19]	Lan 04[20]
S. violaceus	503	DSM 40082, ATCC 15888, LMG 20257, ISP 5082, NBRC 13103	AB184315	A 06	I 05	009 0-19				OC-III			Sch00	Lan2-00	BENP	+		+		+	cl01
S. albosporeus subsp. *albosporeus*	23a	DSM 40795, ATCC 15394, LMG 19403	AJ781327		IV 01 (red series)	063 049		La-01					Sch00	Lan2-00	BENP cl01	+	(c)	+	cl05	+	cl24
S. arenae	40	DSM 40293, ATCC 25428, LMG 20416, ISP 5293, NBRC 13016	AB249977	A 18	I 11	009 1-19						L2	Sch00	Lan2-00	BENP	+		+		+	cl01
S. luteogriseus	301	DSM 40483, ATCC 15072, LMG 20073, ISP 5483, NBRC 13402	AB184379	A 18	I 11	009 1-19						L2	Sch25	Lan2-00	BENP	+		+		+	cl01
S. hawaiiensis	243	DSM 40042, ATCC 12236, LMG 5975, ISP 5042, NBRC 12784	AB184143	A 18	I 11	009 1-19						L2	Sch34	Lan2-00	BENP	+		+			cl01
S. cellulosae	100	DSM 40362, ATCC 25439, LMG 19315, ISP 5362, NRRL B-2889	DQ442495	A 13	II 03	006 1-18		La-15		OC-non			Sch00	Lan2-00	BENP +	+		+	cl17		cl04
S. pseudogriseolus	388	DSM 40026, ATCC 12770, ISP 5026, NRRL B-3288	DQ442541	A 12	I 07	006 1-18					KA-G		Sch13	Lan2-00						+	
S. gancidicus	197	DSM 40935, NRRL B-1872, LMG 19898, NBRC 15412	AB184660		IV 17 (gray series)	006 1-18							Sch13	Lan2-00	BENP	+		+		+	cl04
S. rubiginosus	425	DSM 40177, ATCC 19927, LMG 20268, ISP 5177, KCTC 9042	AY999810	A 12	I 07	006 1-18							Sch13	Lan2-00	BENP	+		+		+	cl04
S. capillispiralis	90	DSM 41695, LMG 19909, NBRC 14222	AB184577										Sch00	Lan2-00	BENP	+		+		+	+
S. lavendulocolor	282	DSM 40216, ATCC 15871, LMG 19934, ISP 5216, NRRL B-3367	DQ442516	F 61	I 22	22-3 1-08						L5	Sch00	Lan2-00	BENP	+		+		+	cl12
S. azureus	62	DSM 40106, ATCC 14921, LMG 20430, ISP 5106, NRRL B-2655	EF178674	A 18	I 11	009 1-19			FU-1			L2	Sch00	Lan2-00	BENP	+		+		+	cl01
S. flavoviridis	185	DSM 40153, ATCC 19903, LMG 19881, ISP 5153, NBRC 12772	AB184842	A 28		006 1-10							Sch00		BENP	+		+		+	cl35
S. pilosus	375	DSM 40097, ATCC 19797, LMG 20049, ISP 5097, NBRC 12807	AB184161	A 37	I 17	006 1-10							Sch00		BENP	+		+		+	cl35
S. djakartensis	147	DSM 40743, ATCC 13441, LMG 21795, NBRC 15409	AB184657		IV 12 (gray series)	035 1-33							Sch00	Lan2-00	BENP	+		+		+	cl04
Group S. geysiriensis et rel.:																					
S. ghanaensis	202	DSM 40746, ATCC 14672, LMG 19894, KCTC 9882	AY999851		IV 05 (green series)	1-7 1-21							Sch00		BENP	+		+		+	+

(continued)

TABLE 272. (continued)

Guo 08[21] 16S rRNA	Guo 08[21] *atpD*	Guo 08[21] *gyrB*	Guo 08[21] *recA*	Guo 08[21] *rpoB*	Guo 08[21] *trpB*	Guo 08[21] concat.	Kim 04[23] *rpoB*	Morphological characters[24]				Physiological tests[25]										
								I	II	III	IV	I	II	III	IV	V	VI	VII	VIII	IX	X	XI
								R	S	C+	SY	+	+	+	+	+	+	+	n	+	n	+
								R	S	C–	SM	+	+	+	+	+	+	+	+	+	–	n
								Gy	S	C+	SPY	+	+	+	+	+	+	+	+	+	n	+
								Gy	S	C+	SM	+	+	+	+	+	+	+	+	+	+	+
								W	S/RA	C+	SPY	+	?	+	+	+	n	+	+	+	n	+
								Y	RF	C–	SM	+	+	+	+	+	n	–	+	+	n	n
								Gy	S	C–	SPY	+	+	+	+	+	+	–	+	+	+	n
								Gy	S	C–	SPY	+	+	+	+	+	+	–	+	+	–	–
								Gy	S	C–	SPY	+	+	±	+	+	n	–	+	+	n	+
								Gy	S	C–	H	n	n	–	n	n	n	–	n	n	–	–
								R	S	C+	SM	+	+	+	–	–	+	–	–	+	–	–
								B	S	C+	WTY	+	+	+	+	+	+	+	+	+	n	+
								Gy/G	S	C+	H	+	+	+	+	+	n	–	+	+	n	–
								Gy	S	C+	H	+	+	+	+	+	n	–	+	+	n	–
								Gy	S	C+	n	n	n	n	n	n	n	n	n	n	n	n
								G	S	C–	SPY	n	n	n	n	n	n	n	n	n	n	n

(continued)

TABLE 272. List of *Streptomyces* species continued

Species names and groups[1]	No. in lists of species[2]	Type strain	Accession no.	Wil 83a[3]	Wil 89[4]	Käm 91[5]	Hat 03[6]	Lan 02[7]	Ful 95[8]	Och 95[9]	Kat 97[10]	Lab[11]	Sch[12]	Lan 04[13]	Lan 02[14]	Lan 04[15]	Lan 04[16]	Lan 04[17]	Lan 04[18]	Lan 04[19]	Lan 04[20]
S. minutiscleroticus	320	DSM 40301, ATCC 17757, LMG 20062, ISP 5301, NRRL B-12202	EF178696	A 15	I 08	006 1-18					KA-G		Sch00	Lan2-03	BENP	+	(a)	+			cl46
S. geysiriensis	201	DSM 40742, ATCC 15303, LMG 19893, NRRL B-12102	DQ442501		IV 18 (gray series)	006 1-18							Sch14	Lan2-00	BENP	+		+			cl39
S. plicatus	377	DSM 40319, ATCC 25483, LMG 20288, ISP 5319, NBRC 13071	AB184291	A 12	I 07	006 1-18					A-E		Sch14	Lan2-00	BENP	+		+		+	cl39
S. rochei	411	DSM 40231, ATCC 10739, LMG 19313, ISP 5231, NBRC 12908	AB184237	A 12	I 07	006 1-18		La.13		OC-III	A-E		Sch00	Lan2-00	BENP +	+		+	cl17	+	cl39
S. vinaceusdrappus	497	DSM 40470, ATCC 25511, LMG 20296, ISP 5470, NRRL 2363	AY999929	A 12	I 07	006 1-18					A-E		Sch14	Lan2-00	BENP	+		+			cl39
S. mutabilis	328	DSM 40169, ATCC 19919, LMG 20054, ISP 5169	EF178679	A 12	I 07	006 1-18					A-E		Sch00	Lan2-00	BENP	+		+		+	+
Most closely related to group S. geysiriensis et rel.:																					
S. tuirus	487	DSM 40505, LMG 20299, NBRC 15617	AB184690	A 21	I 14	006 1-18							Sch00	Lan2-00	BENP	+		+			cl01
S. afghaniensis	8	DSM 40228, ATCC 23871, LMG 20390, ISP 5228	AJ399483	A 18	I 11	009 1-19			FU-1			L2	Sch00	Lan2-00	BENP	+		+		+	cl01
S. africanus	9	DSM 41829, NBRC 101005, CPJVR-H	AY208912																	+	
Group S. brasiliensis et rel.:																					
S. roseiscleroticus	412	DSM 40303, ATCC 17755, LMG 20284, ISP 5303, NBRC 13002	AB184251		II 19	049 022							Sch29	Lan2-01	BENP	+		+			cl38
S. ruber	422	DSM 40304, LMG 20285, NBRC 14600	AB184604		IV 11 (red series)	049 022							Sch29	Lan2-01	BENP	+		+			cl38
S. spiralis	452	DSM 43836, LMG 20332, NRRL B-16922	EF178683										Sch00	Lan2-00	BENP	+		+		+	+
S. fumigatiscleroticus	194	DSM 43154, LMG 19911, NRRL B-3856	DQ442499										Sch00	Lan2-00	BENP	+		+		+	+
S. poonensis	380	DSM 40596, ATCC 15723, LMG 19326, ISP 5596, NRRL B-2319	DQ445792	A 22	II 19	071 1-19		La-04		OC-III			Sch00	Lan2-00	BENP +	+		+	cl14	+	+
S. brasiliensis	70	DSM 43159, ATCC 23727, LMG 20438, NBRC 101283	AB249981										Sch38	Lan2-00	BENP	+		+		+	+
Group S. atrovirens et rel.:																					
S. atrovirens	48	DSM 41467, LMG 20422, NRRL B-16357	DQ026672										Sch00	Lan2-00	BENP	+		+		+	+

(continued)

TABLE 272. (continued)

								Morphological characters[24]				Physiological tests[25]										
Guo 08[21] 16S rRNA	Guo 08[21] *atpD*	Guo 08[21] *gyrB*	Guo 08[21] *recA*	Guo 08[21] *rpoB*	Guo 08[21] *trpB*	Guo 08[21] concat.	Kim 04[23] *rpoB*	I	II	III	IV	I	II	III	IV	V	VI	VII	VIII	IX	X	XI
								Gy	S	C–	n	+	n	+	+	+	+	–	+	n	+	–
								Gy	S	C–	H	n	n	n	n	n	n	n	n	n	n	n
								Gy	S	C–	SM	+	+	+	+	+	+	–	+	+	n	–
								Gy	S	C–	SM	+	+	+	+	+	+	–	+	+	+	–
								R	S	C–	SM	+	+	+	+	+	n	+	+	+	+	n
								Gy	S	C–	SM	+	+	+	+	+	+	–	+	+	–	±
								R	S	C+	SM	+	+	+	+	+	+	+	+	+	–	+
								Gy	S	C+	SPY	+	+	+	+	+	+	+	±	±	n	+
								B	S	C–	SPY	n	n	+	+	+	+	+	+	+	+	–
								R	S	C–	SM	+	+	+	+	+	+	–	+	–	–	n
								W/R	S	C–	SM	+	+	+	+	+		–	+			
								Y–Gy	S	C–	SM	+	–	+	+	+	+	+	+	+	+	+
									S	C–		+	+	+	–	n	–	n	+	–	–	?
								Gy	S	C–	SM	+	+	+	+	+	+	+	+	+	+	n
								Gy	S	C–	SM	+	n	+	+	+	+	+	+	+	+	+
								Gy	S	C–	H	n	+	–	+	+	n	+	+	+	n	n

(continued)

TABLE 272. List of *Streptomyces* species continued

Species names and groups[1]	No. in lists of species[2]	Type strain	Accession no.	Wil 83a[3]	Wil 89[4]	Käm 91[5]	Hat 03[6]	Lan 02[7]	Ful 95[8]	Och 95[9]	Kat 97[10]	Lab[11]	Sch[12]	Lan 04[13]	Lan 02[14]	Lan 04[15]	Lan 04[16]	Lan 04[17]	Lan 04[18]	Lan 04[19]	Lan 04[20]
S. caelestis	80	DSM 40084, ATCC 15084, LMG 5970, ISP 5084, NRRL 2418	X80824	A 18	I 11	009 1-19						L2	Sch00	Lan2-00	BENP	+		+		+	+
S. fumanus	193	DSM 40154, ATCC 19904, LMG 19882, ISP 5154, NBRC 13042	AB184273	A 18	I 11	1-7 1-19							Sch00	Lan2-00	BENP	+		+			cl12
S. fimbriatus	170	DSM 40942, ATCC 15051, LMG 20513	AY999844		IV 16 (gray series)	006 1-18							Sch00	Lan2-00	BENP	+		+		+	+
Group S. glaucus et rel.:																					
S. griseostramineus	235	DSM 40161, ATCC 23628, LMG 19932, ISP 5161, NBRC 12781	AB184140	F 60	IV 06 (green series)	006 1-10							Sch00	Lan2-10	BENP	+		+		+	cl04
S. griseomycini	229	DSM 40159, ATCC 23625, LMG 19883, ISP 5159, NBRC 12778	AB184137	A 12	I 07	006 1-10							Sch00	Lan2-10	BENP	+		+		+	cl04
S. graminearus	215	DSM 41747, LMG 19904	AJ781333										Sch00	Lan2-10	BENP	+		+		+	cl04
S. viridiviolaceus	512	DSM 40280, ATCC 27478, LMG 20282, ISP 5280, IFO 13359	AY999854		IV 35 (gray series)	006 1-18							Sch00	Lan2-00	BENP +	+		+			-
S. glaucus	207	DSM 41456, LMG 19902, NBRC 15417	AB184665										Sch00	Lan2-00	BENP	+		+		+	+
Group S. aureorectus et rel.:																					
S. aureorectus	54	DSM 41692, LMG 19908, NBRC 15896	AB184710										Sch19	Lan2-08	BENP	+		+			cl40
S. virens	509	DSM 41465, LMG 20316, NRRL B-24331	DQ442554										Sch00	Lan2-08	BENP	+		+			cl40
S. asterosporus	44	DSM 41452, LMG 20419, NBRC 15872	AB184706										Sch00	Lan2-08	BENP	+		+			cl40
S. calvus	83	DSM 40010, ATCC 13382, LMG 20442, ISP 5010, NBRC 13200	AB184329	A 12	I 07	006 1-18							Sch19	Lan2-08	BENP	+		+		+	cl40
Most closely related to groups S. geysiriensis et rel., S. brasiliensis et rel., S. atrovirens et rel., S. glaucus et rel. and S. aureorectus et rel.:																					
S. naganishii	330	DSM 40282, ATCC 23939, LMG 21042, ISP 5282, NRRL B-1816	DQ442529	A 31	I 21	1-6 1-15							Sch00	Lan2-00	BENP	+		+		+	+
S. prasinosporus	383	DSM 40506, ATCC 17918, LMG 19346, ISP 5506, NBRC 13419	AB184390	A 38	III 07	22-2 1-15		L.10		OC-III			Sch25	Lan2-00	BENP +	+		+	cl10	+	cl54
S. anandii	33	DSM 40535, ATCC 19388, LMG 8600, ISP 5535, NBRC 13438	AB184402	B 42	I 19	021 1-05							Sch00	Lan2-00	BENP	+		+		+	cl08

(continued)

TABLE 272. (continued)

Guo 08[21] 16S rRNA	Guo 08[21] *atpD*	Guo 08[21] *gyrB*	Guo 08[21] *recA*	Guo 08[21] *rpoB*	Guo 08[21] *trpB*	Guo 08[21] concat.	Kim 04[23] *rpoB*	Morphological characters[24]				Physiological tests[25]										
								I	II	III	IV	I	II	III	IV	V	VI	VII	VIII	IX	X	XI
								B	S	C+	SM	+	+	+	+	+	+	+	–	+	–	+
								R	S	C–	SM	+	+	+	+	+	n	+	+	–	n	–
								Gy	S	C+	SPY	+	+	+	+	+	+	+	+	+	–	n
								G	S	C+	H	+	+	+	+	+	n	n	+	+	n	–
								G	S/RA	C+	H/SM	+	+	+	+	+	n	–	+	+	n	–
								Gy	S	C–	SM	+		+	+	+		+	+	+		
								Gy	S	C–	SPY	n	n	n	n	n	n	n	n	n	n	n
								B/G	S	C–	H	+	+	+	+	n	+	n	+	+	n	n
								W	RF	C–	SM	n	+	n	n	n	+	+	n	n	n	n
								Gy	S	C–	SPY/ WTY	n	–	+	–	+	n	–	+	n	n	n
								Gy	S	C–	SPY	+	–	+	+	+	+	–	+	n	n	n
								Gy	S	C–	H	+	+	+	+	+	+	+	+	+	+	+
								Gy	S	C+	SM	+	+	+	+	+	+	+	+	+	+	–
								G	S	C+	H	+	+	+	+	+	n	n	+	+	+	–
								Gy	S	C+	SM	+	+	+	–	+	+	+	+	+	–	n

(continued)

TABLE 272. List of *Streptomyces* species continued

Species names and groups[1]	No. in lists of species[2]	Type strain	Accession no.	Wil 83a[3]	Wil 89[4]	Käm 91[5]	Hat 03[6]	Lan 02[7]	Ful 95[8]	Och 95[9]	Kat 97[10]	Lab[11]	Sch[12]	Lan 04[13]	Lan 02[14]	Lan 04[15]	Lan 04[16]	Lan 04[17]	Lan 04[18]	Lan 04[19]	Lan 04[20]
S. carpinensis	93	DSM 43835, LMG 19913, NBRC 14214	AB184574										Sch00	Lan2-00	BENP	+		+		+	+
S. levis	283	DSM 41458, LMG 20090, NBRC 15423	AB184670										Sch00	Lan2-00	BENP	+		+			cl01
S. cinerochromogenes	114	DSM 41651, LMG 20466, NBRC 13822	AB184507										Sch00	Lan2-00	BENP	+		+		+	cl57
S. koyangensis	270	NBRC 100598, VK-A60	AY079156																		
S. violarus	505	DSM 40205, ATCC 15891, LMG 20275, ISP 5205, NBRC 13104	AB184316	A 18	I 11	009 1-19						L2	Sch00		BENP	+		+			cl01
Not closely related to one of the groups:																					
S. daghestanicus	142	DSM 40149, ATCC 23620, LMG 20496, ISP 5149, NRRL B-5418	DQ442497	A 17		006 010						L3	Sch40	Lan2-11	BENP	+		+			cl37
S. limosus	287	DSM 40131, ATCC 19778, LMG 8570, ISP 5131, NBRC 12790	AB184147	A 1A	I 01	1-1 1-1			FU-1		KA-D		Sch00	Lan2-22	BENP	+	(k)	+		+	cl08
S. canescens	86	DSM 40001, ATCC 19736, LMG 20445, ISP 5001, NBRC 12751	AB184117	A 1A	I 01	1-1 1-1					KA-D		Sch10	Lan2-22	BENP	+	(k)	+			cl08
S. felleus	165	DSM 40130, ATCC 19752, LMG 20511, ISP 5130, NBRC 12766	AB184129	A 1A	I 01	1-1 1-1					KA-D		Sch00	Lan2-22	BENP	+	(k)	+			cl08
S. griseus subsp. *solvifaciens*	238d	DSM 40933, NRRL B-1561, LMG 19952, NBRC 13689	AB249915		I 02	1-1 1-1							Sch10	Lan2-00	BENP	+		+			cl08
S. violascens	506	DSM 40183, ATCC 23968, LMG 20272, ISP 5183	AY999737	A 06	I 05	002 1-7							Sch00	Lan2-00	BENP	+		+		+	cl23
S. hydrogenans	252	DSM 40586, ATCC 19631, LMG 19948, ISP 5586, NBRC 13475	AB184868	A 05	I 04	002 1-7							Sch00	Lan2-00	BENP	+		+		+	-
S. odorifer	349	DSM 40347, ATCC 6246, LMG 8572, ISP 5347	Z76682	A 1A	I 01	1-1 1-1					KA-D		Sch00		BENP	+	(k)	+			cl08
S. albidoflavus	14	DSM 40455, ATCC 25422, LMG 19300, ISP 5455, NBRC 13010	AB184255	A 1A	I 01	1-1 1-1		La-00	FU-1	OC-non	KA-D		Sch10		BENP +	+	(e) (k)	+	cl06		cl08
S. champavatii	101	DSM 40841, NRRL B-5682, LMG 20454	DQ026642		IV 02 (yellow series	)1-1 1-1							Sch00		BENP	+		+			cl08
S. sampsonii	429	DSM 40394, ATCC 25495, LMG 8574, ISP 5394	D63871	A 1A	I 01	1-1 1-1					KA-D		Sch00	Lan2-22	BENP	+	(k)	+		+	cl08
S. diastaticus subsp. *diastaticus*	144a	DSM 40496, ATCC 3315, LMG 19322, ISP 5496, NBRC 3714	AB184785	A 19	I 12	1-1 1-1		La-00	FU-1	OC-non			Sch00		BENP +	+		+		+	cl09

(continued)

TABLE 272. (continued)

								Morphological characters[24]				Physiological tests[25]										
Guo 08[21] 16S rRNA	Guo 08[21] *atpD*	Guo 08[21] *gyrB*	Guo 08[21] *recA*	Guo 08[21] *rpoB*	Guo 08[21] *trpB*	Guo 08[21] concat.	Kim 04[23] *rpoB*	I	II	III	IV	I	II	III	IV	V	VI	VII	VIII	IX	X	XI
								Gy	S	C–	SM	+	–	n	+	+	+	+	+	–	+	–
								V	S	C–	SM	+	+	+		+		+	+			
								Gy	S	C+	SM	n	–	+	+	n	n	–	–	–	+	+
								W/Gy	RF	C–	SM		+	+	–	+		–	+	–		–
								R	S	C+	SPY	+	n	+	+	+	+	+	n	+	n	+
								R	S	C–	SM	+	+	+	+	+	n	–	+	–	n	–
							+	Y	RF	C–	SM	+	+	+	–	+	n	–	+	–	n	
								Y	RF	C–	SM	+	–	+	–	+	n	–	–	–	–	–
							Asn (AAC)[442]	Y	RF	C–	SM	+	+	+	+	n	n	–	+	–	+	n
+	+	+	+	+	+	+		Y	RF	C–	SM	+	+	+	–	+	n	–	+	–	+	n
							+	V	S	C+	SPY	+	+	+	–	+	n	+	–	±	n	±
								W/Y/Gy	RF	C–	SM	+	+	+	+	–	n	–	–	–	n	–
								Y	RF	C–	SM	+	+	+	–	+	n	–	+	+	+	n
								Y	RF	C–	SM	+	+	+	–	+	n	–	+	–	n	–
								Y	RF	C–	SM	+	+	+	–	+	+	–	+	n	n	–
								Y	RF	C–	SM	+	+	+	–	+	n	–	+	–	n	–
								Gy/Y	RF	C–	SM	+	+	+	–	+	n	–	+	–	n	+

(continued)

TABLE 272. List of *Streptomyces* species continued

Species names and groups[1]	No. in lists of species[2]	Type strain	Accession no.	Wil 83a[3]	Wil 89[4]	Käm 91[5]	Hat 03[6]	Lan 02[7]	Ful 95[8]	Och 95[9]	Kat 97[10]	Lab[11]	Sch[12]	Lan 04[13]	Lan 02[14]	Lan 04[15]	Lan 04[16]	Lan 04[17]	Lan 04[18]	Lan 04[19]	Lan 04[20]
S. gougerotii	214	DSM 40324, ATCC 10975, LMG 19888, ISP 5324, NBRC 3198	AB184742	A 1A	I 01	1-1 1-1					KA-D		Sch00		BENP	+	(k)	+		+	cl09
S. rutgersensis subsp. *rutgersensis*	427a	DSM 40077, ATCC 3350, LMG 8568, ISP 5077, NBRC 12819	AB184170	A 1A	I 01	1-1 1-1					KA-D		Sch00		BENP	+	(k)	+		+	cl09
S. intermedius	258	DSM 40372, ATCC 3329, LMG 19304, ISP 5372, NBRC 13049	AB184277	A 1A	I 01	1-1 1-1		La-03			KA-D		Sch10	Lan2-00	BENP +	+		+	cl14		cl08
S. indiaensis	255	DSM 43803, LMG 19961, NBRC 13964	AB184553										Sch00	Lan2-00	BENP	+		+		+	cl04
S. thermocarboxydus	473	DSM 44293	U94490										Sch00							+	
S. massasporeus	310	DSM 40035, ATCC 19785, LMG 19362, ISP 5035, NBRC 12796	AB184152	D SM	III 19	015 1-19		La-12		OC-III			Sch00	Lan2-00	BENP +	+		+	cl17	+	cl01
S. misionensis	323	DSM 40306, ATCC 14991, LMG 20063, ISP 5306, NRRL B-3230	EF178678	A 31	I 21	1-6 1-16							Sch00	Lan2-15	BENP	+		+		+	cl54
S. phaeoluteichromatogenes	370	NRRL B-5799	AJ391814																		
S. spectabilis	447	DSM 40512, NRRL 2494, LMG 5986, ISP 5512, NBRC 13424	AB184393									L3	Sch00	Lan2-00	BENP	+		+		+	+
S. cinereospinus	112	DSM 41470, LMG 20464, NBRC 15397	AB184648										Sch00	Lan2-00	BENP	+		+		+	cl22
S. coeruleofuscus	127	DSM 40144, ATCC 23618, LMG 20482, ISP 5144, NRRL B-5417	DQ026668	A 18	I 11	009 1-19						L2	Sch00	Lan2-00	BENP	+		+			cl01
S. chromofuscus	107	DSM 40273, ATCC 23896, LMG 19317, ISP 5273, NBRC 12851	AB184194	A 15	I 08	006 1-18		La-06		OC-III			Sch00	Lan2-00	BENP +	+		+	cl10	+	cl09
S. scopiformis	436	DSM 41825, LMG 20251, NBRC 100244	AB249927																	+	
S. spinoverrucosus	451	DSM 41648, LMG 20321, NBRC 14228	AB184578										Sch00	Lan2-00	BENP	+		+		+	+
Most closely related to group S. mexicanus et rel.:																					
S. thermospinosisporus	479	DSM 41779, NBRC 100043, JCM 11756, AT10	AF333113																		
S. thermodiastaticus	475	DSM 40573, ATCC 27472, LMG 20302, ISP 5573, JCM 4840	AB018095	A 1C	I 03	006 1-18							Sch00	Lan2-00	BENP	+		+		+	cl04
S. thermocarboxydovorans	472	DSM 44296, LMG 19860	U94489										Sch00		BENP	+		+			cl04
S. thermoviolaceus subsp. *apingens*	480b	DSM 41392, LMG 20307	Z68095										Sch00		BENP	+		+		+	cl03
S. thermoviolaceus subsp. *thermoviolaceus*	480a	DSM 40443, ATCC 19283, LMG 19359, ISP 5443	Z68096	C 45	II 13	004 006		La-13					Sch00		BENP +	+		+	cl11	+	cl03

(continued)

TABLE 272. (continued)

Guo 08[21] 16S rRNA	Guo 08[21] *atpD*	Guo 08[21] *gyrB*	Guo 08[21] *recA*	Guo 08[21] *rpoB*	Guo 08[21] *trpB*	Guo 08[21] concat	Kim 04[23] *rpoB*	Morphological characters[24] I	II	III	IV	Physiological tests[25] I	II	III	IV	V	VI	VII	VIII	IX	X	XI
								Y	RF	C–	SM	+	–	+	–	n	n	n	n	n	n	n
								Y	RF	C–	SM	+	+	+	–	+	+	+	+	–	+	–
								Y	RF	C–	SM	+	+	+	–	+	+	+	+	–	+	+
								Gy	S	C+	SM	+	+	+	–	+	n	–	+	+	–	+
								Gy	RA	C–	WTY	+	–	n	n	+	n	–	+	+	n	–
								Gy	S	C+	SM	+	+	+	+	+	+	+	+	+	+	+
								Gy	S	C–	SM	+	+	+	+	+	+	+	+	+	–	–
							Asn (AAC)[442]	R	RF	C+	SM	+	+	–	–	+	+	+	+	+	–	–
								Gy	S	C–	SPY	+	–	–	+	+	n	+	n	–	n	n
								B	S	C+	SPY	+	+	+	+	+	n	+	+	+	n	+
								Gy	S	C+	SPY	+	+	+	+	+	n	+	+	+	n	–
								Gy	RF		SPY	+	+	+	+	+	+	–	–	+	n	+
								G	S	C+	SPY/ WTY	+	+	+	+	+	n	+	+	+	n	+
								Gy	RF	C–	SPY							+	+	+		+
								Gy	S	C–	WTY/ SPY	+	+	+	+	+	n	+	+	+	n	–
								Gy	RF	C–	SM	n	n	n	n	+	n	n	n	n	n	n
								Gy	S	??	WTY	n	n	n	n	n	n	n	n	n	n	n
								Gy	S	C+	SM	n	n	n	n	n	n	n	n	n	n	n

(continued)

TABLE 272. List of *Streptomyces* species continued

Species names and groups[1]	No. in lists of species[2]	Type strain	Accession no.	Wil 83a[3]	Wil 89[4]	Käm 91[5]	Hat 03[6]	Lan 02[7]	Ful 95[8]	Och 95[9]	Kat 97[10]	Lab[11]	Sch[12]	Lan 04[13]	Lan 02[14]	Lan 04[15]	Lan 04[16]	Lan 04[17]	Lan 04[18]	Lan 04[19]	Lan 04[20]
S. nodosus	343	DSM 40109, ATCC 14899, LMG 19430, ISP 5109	AF114033	A 35	II 08	006 1-11							Sch00	Lan2-00	BENP +	+		+	cl11	+	cl04
S. viridosporus	516	DSM 40243, ATCC 27479, LMG 20278, ISP 5243, NRRL 2414	DQ442556	A 15	I 08	006 1-18							Sch00		BENP	+		+			+
Group S. mexicanus et rel.:																					
S. thermogriseus	476	DSM 41756, LMG 20532, NBRC 100772	AB249980										Sch00	Lan2-07	BENP	+		+			cl32
S. thermovulgaris	481	DSM 40444, ATCC 19284, LMG 19342, ISP 5444	Z68094	A 36	II 09	021 002		La-00		OC-non			Sch00	Lan2-07	BENP +	+		+	cl10	+	cl32
S. thermoalcalitolerans	470	DSM 41741, LMG 19858, NBRC 16322	AB249909										Sch00	Lan2-00	BENP	+		+			+
S. mexicanus	317	DSM 41796, NBRC 100915	AB249966																		
S. thermocoprophilus	474	DSM 41700, LMG 19857, B19	AJ007402										Sch00	Lan2-00	BENP	+		+		+	+
Most closely related to group S. mexicanus et rel.:																					
S. bangladeshensis	68	NRRL B-24326, LMG 22738, AAB-4	AY750056																		
S. rameus	399	DSM 41685, LMG 20326, KCTC 9767	AY999821										Sch33		BENP	+		+		+	cl02
S. griseosporeus	234	DSM 40562, ATCC 27435, LMG 19947, ISP 5562, NBRC 13458	AB184419	A 23	I 20	1-7 1-19							Sch00	Lan2-00	BENP	+		+		+	+
S. achromogenes subsp. *rubradiris*	4b	DSM 40789, NRRL 3061, LMG 20388, KCTC 9742	AY999846		I 12	028 009							Sch00	Lan2-00	BENP	+		+		+	+
S. glomeratus	210	DSM 41457, LMG 19903, NBRC 15898	AB249917											Lan2-00	BENP	+		+		+	cl09
S. eurythermus	163	DSM 40014, ATCC 14975, LMG 20510, ISP 5014	D63870	A 23	I 20	1-5 009							Sch00	Lan2-00	BENP	+		+			cl08
S. nogalater	344	DSM 40546, ATCC 27451, LMG 19338, ISP 5546, JCM 4799	AB045886	A 34	III 06	1-5 009		La-14		OC-III			Sch00	Lan2-00	BENP +	+		+	cl14	+	cl04
S. fragilis	190	DSM 40044, ATCC 23908, LMG 19874, ISP 5044, NRRL 2424	AY999917	G SM	III 22	078 058				OC-III			Sch38	Lan2-00	BENP	+		+		+	+
Group S. erythrogriseus et rel.:																					
S. erythrogriseus	160	DSM 40116, ATCC 27427, LMG 19406, ISP 5116	AJ781328		IV 04 (red series)	074 1-27		La-15					Sch35	Lan2-25	BENP +	+		+	cl17	+	cl01
S. labedae	273	DSM 41446, LMG 19956, NBRC 15864	AB184704										Sch35	Lan2-25	BENP	+		+		+	cl01

(continued)

TABLE 272. (continued)

Guo 08[21] 16S rRNA	Guo 08[21] *atpD*	Guo 08[21] *gyrB*	Guo 08[21] *recA*	Guo 08[21] *rpoB*	Guo 08[21] *trpB*	Guo 08[21] concat.	Kim 04[23] *rpoB*	Morphological characters[24] I	II	III	IV	Physiological tests[25] I	II	III	IV	V	VI	VII	VIII	IX	X	XI
							+	Gy	S	C–	SM	+	+	–	+	+	n	–	+	+	n	–
								G	S	C–	SPY	+	+	+	+	+	+	–	+	+	–	±
										C–	SM	±	±		+	±		±	+	–		
								Gy	S	C–	SM	+	+	+	+	+	n	+	+	+	n	+
								Gy	n	C–	WTY	+	+	+	+	+	+	–	+	+	n	+
								Gy	RF	C–	SM	+	+	+	–	+	+	+	+	–	n	–
								Gy	RF	C+	SM	+	+	+	n	+	n	–	+	n	n	–
								Y–G	RF	C+	SM	+	–	+	+	+	n	+	+	+	+	n
							+	Gy	S	C+	SM	n	+	+	–	+	+	+	+	–	+	+
								Gy	S	C+	SM	+	+	+	+	+	+	+	+	+	+	+
								Gy	S	C+	SM	+	+	+	+	+	+	+	+	±	±	+
								Gy	S	C+	SM		+	+	+	+		+	+			
								Gy	S	C+	SM	+	+	+	–	+	+	+	+	–	–	+
								Gy	S	C–	SM	+	+	+	+	+	+	+	+	+	–	–
								R	S	C–	SM	+	+	+	–	–	+	–	–	–	–	±
								Gy/ R/W	S	C–	SPY/SM	+	+	+	+	+	n	–	+	+	n	–
								Gy	S	C–	SPY	+	+	–	+	+	+	–	+	+	–	–

(continued)

TABLE 272. List of *Streptomyces* species continued

Species names and groups[1]	No. in lists of species[2]	Type strain	Accession no.	Wil 83a[3]	Wil 89[4]	Käm 91[5]	Hat 03[6]	Lan 02[7]	Ful 95[8]	Och 95[9]	Kat 97[10]	Lab[11]	Sch[12]	Lan 04[13]	Lan 02[14]	Lan 04[15]	Lan 04[16]	Lan 04[17]	Lan 04[18]	Lan 04[19]	Lan 04[20]
S. griseoincarnatus	225	DSM 40274, ATCC 23623, LMG 19316, ISP 5274	AJ781321	A 13	II 03	006 1-18		La-15					Sch35	Lan2-25	BENP +	+		+	cl17	+	cl01
S. variabilis	490	DSM 40179, ATCC 19930, LMG 20270, ISP 5179, NRRL B-3984	DQ442551	A 12	I 07	006 1-18					KA-F		Sch35	Lan2-25	BENP	+		+			cl01
Most closely related to group S. erythrogriseus et rel.:																					
S. althioticus	29	DSM 40092, ATCC 19724, LMG 20408, ISP 5092, KCTC 9752	AY999808	A 12	I 07	006 1-18							Sch00	Lan2-00	BENP	+		+			+
S. matensis	311	DSM 40188, ATCC 23935, LMG 20055, ISP 5188, NBRC 12889	AB184221	A 12	I 07	006 1-18			FU-1				Sch00	Lan2-00	BENP	+		+		+	cl01
S. griseorubens	231	DSM 40160, ATCC 19909, LMG 19931, ISP 5160, NBRC 12780	AB184139	A 12	I 07	006 1-18					KA-F		Sch30	Lan2-00	BENP	+		+			cl01
S. viridochromogenes	514	DSM 40110, ATCC 14920, LMG 20260, ISP 5110, NRRL B-1511	DQ442555	A 27	III 04	009 1-19				OC-III		L2	Sch00	Lan2-00	BENP	+		+		+	cl54
S. iakyrus	254	DSM 40482, ATCC 15375, LMG 19942, ISP 5482, NBRC 13401	AB184877	A 18	I 11	009 1-19						L2	Sch00	Lan2-00	BENP	+		+		+	cl01
S. violaceochromogenes	498	DSM 40181, LMG 20271, IFO 13100	AY999867												BENP	+		+		+	cl01
S. collinus	131	DSM 40129, ATCC 19743, LMG 20486, ISP 5129, NBRC 12759	AB184123	A 18	I 11	009 1-19			FU-1			L2	Sch39	Lan2-00	BENP	+		+		+	+
S. malachitofuscus	306	DSM 40332, ATCC 25471, LMG 20067, ISP 5332, NBRC 13059	AB184282			006 1-18							Sch00	Lan2-00	BENP +	+		+		+	cl04
S. paradoxus	361	DSM 43350, LMG 20523, NBRC 14887	AB184628										Sch00	Lan2-00	BENP	+		+		+	cl01
S. griseoflavus	223	DSM 40456, ATCC 25456, LMG 19344, ISP 5456	AJ781322	A 37	I 17	006 1-18		La-04		OC-non			Sch00	Lan2-00	BENP +	+		+			cl01
S. flaveolus	173	DSM 40061, ATCC 3319, LMG 19328, ISP 5061, NBRC 3408	AB184764	A 24	II 05	1-6 1-13		La-12		OC-III			Sch00	Lan2-00	BENP +	+		+	cl17	+	cl59
S. glaucescens	204	DSM 40155, ATCC 23622, LMG 19330, ISP 5155, NBRC 12774	AB184843	A 28	III 05	006 1-10		La-16		OC-III		L2	Sch00	Lan2-00	BENP +	+		+		+	cl57
S. pharetrae	374	DSM 41856, NRRL B-24333, CZA14	AY699792																		
S. malachitospinus	307	IFO 101004, NBRC 101004	AB249954																		

(continued)

TABLE 272. (continued)

Guo 08[21] 16S rRNA	Guo 08[21] *atpD*	Guo 08[21] *gyrB*	Guo 08[21] *recA*	Guo 08[21] *rpoB*	Guo 08[21] *trpB*	Guo 08[21] concat.	Kim 04[23] *rpoB*	Morphological characters[24] I	II	III	IV	Physiological tests[25] I	II	III	IV	V	VI	VII	VIII	IX	X	XI
								Gy	S	C–	SPY	+	+	+	+	+	n	–	+	±	n	+
								Gy/R	S/RA	C–	SPY	+	+	+	+	+	n	–	+	+	n	–
								Gy	S	C–	SPY	+	+	+	+	+	+	–	+	+	n	±
								Gy	S	C–	SPY	+	+	+	+	+	+	–	+	+	n	–
								Gy	S	C–	SPY	+	+	±	+	+	n	–	+	±	n	–
								B	S	C+	SPY	+	+	+	+	+	+	+	+	+	+	+
								Gy	S	C+	SPY	+	+	+	+	+	+	+	+	+	+	n
								Gy	S	C+	SM	+	+	+	+	+	n	+	+	+	n	+
							Group A18	Gy	S	C+	SM	+	+	+	+	+	n	+	+	+	n	+
								Gy	S	C+	SPY	+	+	+	+	+	n	–	+	+	n	+
								Gy	RA	C+	SM	+	+	+	+	+	+	+	+	+	n	+
								Gy	S	C–	SPY	+	+	+	+	+	n	–	+	+	n	–
								Gy	S	C–	H	+	+	+	+	+	+	+	+	+	+	+
								B/G	S	C+	H	+	+	+	+	+	n	–	+	+	n	–
								Gy		C+	H			+	+				+			
								Gy	S	C	SPY	[illegible]		[illegible]			+					

(continued)

TABLE 272. List of *Streptomyces* species continued

Species names and groups[1]	No. in lists of species[2]	Type strain	Accession no.	Wil 83a[3]	Wil 89[4]	Käm 91[5]	Hat 03[6]	Lan 02[7]	Ful 95[8]	Och 95[9]	Kat 97[10]	Lab[11]	Sch[12]	Lan 04[13]	Lan 02[14]	Lan 04[15]	Lan 04[16]	Lan 04[17]	Lan 04[18]	Lan 04[19]	Lan 04[20]
S. parvulus	363	DSM 40048, ATCC 12434, ISP 5048, NBRC 13193	AB184326	A 12		006 1-18		La-24					Sch00	Lan2-00	BENP +	+		+	cl17		cl06
S. tendae	468	DSM 40101, ATCC 19812, LMG 19314, ISP 5101	D63873	A 12	I 07	006 1-18		La-14			KA-E		Sch00	Lan2-00	BENP +	+		+	+	+	cl06
S. violaceorubidus	502	DSM 41478, LMG 20319	AJ781374										Sch00	Lan2-00	BENP	+		+			cl08
S. albaduncus	11	DSM 40478, ATCC 14698, LMG 20392, ISP 5478, JCM 4715	AY999757		IV 02 (gray series)	006 1-10							Sch00	Lan2-00	BENP	+		+		+	cl05
S. griseoloalbus	226	DSM 40468, ATCC 23624, LMG 21308, ISP 5468, NBRC 13046	AB184275		IV 05 (yellow series	)017 007							Sch00	Lan2-00	BENP	+		+		+	cl22
S. heliomycini	245	DSM 41690, IFO 15899, LMG 19960, NBRC 15899	AB184712										Sch00	Lan2-00	BENP	+		+			cl05
S. ambofaciens	31	DSM 40053, ATCC 23877, LMG 20409, ISP 5053	M27245	A 23	I 20	006 1-18			FU-6				Sch00	Lan2-00	BENP	+		+		+	+
Most closely related to group S. coelescens et rel.:																					
S. rubrogriseus	426	DSM 41477, LMG 20318, NBRC 15455	AB184681										Sch00	Lan2-00	BENP	+		+			cl22
S. tricolor	485	DSM 41704, LMG 20328, NBRC 15461	AB184687										Sch33	Lan2-21	BENP	+	(b)	+			cl02
S. lienomycini	285	DSM 41475, LMG 20091	AJ781353										Sch00	Lan2-00	BENP	+		+		+	cl22
S. anthocyanicus	34	DSM 41422, LMG 20411, NBRC 14892	AB184631		IV 03 (gray series)	013 1-19							Sch00	Lan2-04	BENP	+		+		+	cl22
S. olivaceus	352	DSM 40072, ATCC 3335, LMG 19394, ISP 5072, NBRC 3200	AB184743	A 1C	I 03	042 014		La-23	FU-1				Sch00	Lan2-00	BENP +	+		+	cl16	+	cl13
S. pactum	360	DSM 40530, ATCC 27456, LMG 19357, ISP 5530, NBRC 13433	AB184398	C 44	II 12	22-4 035		La-11		OC-II			Sch00	Lan2-00	BENP +	+		+			+
Group S. coelescens et rel.:																					
S. coelescens	123	DSM 40421, ATCC 19830, LMG 20479, ISP 5421, ICSSB 1021	AF503496	A 21	I 14	006 1-18							Sch07		BENP	+	(k)	+		+	cl22
S. humiferus	251	DSM 43030, LMG 20519	AF503491										Sch08		BENP	+	(k)	+			cl22
S. violaceolatus	499	DSM 40438, ATCC 19847, LMG 20293, ISP 5438, ICSSB 1022	AF503497	A 21	I 14	006 1-18							Sch08		BENP	+	(k)	+		+	cl22
S. violaceoruber	501	DSM 40049, ATCC 14980, LMG 20256, ISP 5049, ICSSB 1016	AF503492		IV 34 (gray series)	069 1-26							Sch07		BENP	+	(d) (k)	+		+	cl22

(continued)

TABLE 272. (continued)

								Morphological characters[24]				Physiological tests[25]										
Guo 08[21] 16S rRNA	Guo 08[21] *atpD*	Guo 08[21] *gyrB*	Guo 08[21] *recA*	Guo 08[21] *rpoB*	Guo 08[21] *trpB*	Guo 08[21] concat.	Kim 04[23] *rpoB*	I	II	III	IV	I	II	III	IV	V	VI	VII	VIII	IX	X	XI
								Gy	S	C–	SM	+	+	+	+	+	n	–	+	+	+	+
								Gy	S	C–	SM	+	+	+	+	+	n	–	+	+	n	+
								Gy/W	S	C–	SM	+	+	+	+	+		+	+	+		
								Gy	S	C–	SPY	+	+	+	+	+	+	+	+	+	+	±
								Y	RF	C–	SM	+	+	+	+	+	n	n	+	+	n	+
								Gy	S	C–	WTY/ SPY/H	+	+		+	+			+	+		
								Gy	S	C	SM	+	+	+	+	+	n	–	+	+	n	+
								Gy/R	S	C–	SM											
								Gy	S	C–	SM	n	n	n	n	n	n	n	n	n	n	n
								Gy	S	C+	SM		+	+	+	+		+	+	+		
								Gy	S	C–	SM	+	n	+	+	+	+	–	–	+	n	–
								Gy	S	C–	SM	+	+	+	+	+	+	+	+	+	–	–
								Gy	S	C–	H	+	–	–	–	–	+	–	–	–	–	–
								Gy	S	C–	SM	+	n	+	n	n	n	n	–	n	n	n
								Gy	S	C–	SM	+	+	+	+	+	+	+	+	+	n	–
								Gy	S	C–	SM	+	+	+	+	+	n	+	+	+	n	+
								Gy	S	C–	SM	+	+	+	+	+	+	–	+	+	+	–

(continued)

TABLE 272. List of *Streptomyces* species continued

Species names and groups[1]	No. in lists of species[2]	Type strain	Accession no.	Wil 83a[3]	Wil 89[4]	Käm 91[5]	Hat 03[6]	Lan 02[7]	Ful 95[8]	Och 95[9]	Kat 97[10]	Lab[11]	Sch[12]	Lan 04[13]	Lan 02[14]	Lan 04[15]	Lan 04[16]	Lan 04[17]	Lan 04[18]	Lan 04[19]	Lan 04[20]
Most closely related to group S. coelescens et rel.:																					
S. coelicoflavus	124	DSM 41471, LMG 20480, NBRC 15399	AB184650										Sch00	Lan2-00	BENP	+		+			cl23
S. diastaticus subsp. *ardesiacus*	144b	DSM 40934, IFO 15402, LMG 20497, NRRL B-1773	DQ026631												BENP	+		+		+	+
Most closely related to group S. coeruleorubidus et rel.:																					
S. lomondensis	291	DSM 41428, LMG 20088, NBRC 15426	AB184673		IV 03 (blue series)	009 1-19							Sch00	Lan2-00	BENP	+		+			cl04
S. lusitanus	299	DSM 40568, ATCC 15842, LMG 20078, ISP 5568, NBRC 13464	AB184424	C 44	II 12	006 1-18							Sch00	Lan2-00	BENP	+		+		+	-
S. purpurascens	394	DSM 40310, ATCC 25489, LMG 20526, ISP 5310	AB045888	A 18	I 11	009 1-19						L2	Sch00	Lan2-00	BENP	+	(g)	+		+	cl01
S. bellus	70	DSM 40185, ATCC 14925, LMG 19401, ISP 5185, NBRC 12844	AB184849	A 18	I 11	061 1-22		La1-02				L2	Sch00		BENP +	+		+	cl17	+	cl01
S. coerulescens	130	DSM 40146, ATCC 19896, LMG 8590, ISP 5146	AY999720	A 18	I 11	009 1-19						L2	Sch00	Lan2-00	BENP	+		+		+	cl01
S. speibonae	448	DSM 41797, ATCC BAA-411, PK-Blue	AF452714																	+	
S. longispororuber	293	DSM 40599, ATCC 27443, LMG 20082, ISP 5599, NBRC 13488	AB184440	A 10	I 06	033 1-33					KA-F		Sch00	Lan2-00	BENP	+		+			cl01
Group S. coeruleorubidus et rel.:																					
S. albogriseolus	19	DSM 40003, ATCC 23875, LMG 20395, ISP 5003, NRRL B-1305	AJ494865	A 12	I 07	006 1-18					KA-F		Sch00	Lan2-00	BENP	+		+		+	+
S. viridodiastaticus	515	DSM 40249, ATCC 25518, LMG 20279, ISP 5249, NBRC 13106	AY999852		IV 36 (gray series)	006 1-18							Sch00	Lan2-00	BENP	+		+			+
S. coeruleorubidus	129	DSM 40145, ATCC 13740, LMG 20484, ISP 5145	AY999719	A 18	I 11	009 1-19						L2	Sch00		BENP	+	(g)	+			cl01
Group S. aurantiogriseus et rel.:																					
S. coelicolor	125	DSM 40233, ATCC 23899, LMG 8571, ISP 5233, NRRL B-2812	DQ442496	A 1A	I 01	1-1 1-1			FU-1		KA-D		Sch09	Lan2-00	BENP	+	(d) (k)	+		+	cl08
S. griseoviridis	237	DSM 40229, ATCC 23920, LMG 19321, ISP 5229, KCTC 9780	AY999807	A 17	I 10	006 010		La-06		OC-III		L3	Sch40	Lan2-11	BENP +	+	(h)	+		+	cl37

(continued)

TABLE 272. (continued)

Guo 08[21] 16S rRNA	Guo 08[21] *atpD*	Guo 08[21] *gyrB*	Guo 08[21] *recA*	Guo 08[21] *rpoB*	Guo 08[21] *trpB*	Guo 08[21] concat.	Kim 04[23] *rpoB*	Morphological characters[24] I	II	III	IV	Physiological tests[25] I	II	III	IV	V	VI	VII	VIII	IX	X	XI
								n	S	C–	SM	+	+	+	+	+	n	n	+	+	n	n
								Gy	S	C–	SM	+	+	+	+	n	n	n	n	n	n	n
								R/B	RF/S	C+	WTY/ SPY	+	+	+	+	+	+	+	+	+	n	+
								Gy	S	C–	SM	+	–	±	–	+	n	–	–	±	–	+
								R	S	C+	SPY	+	+	+	+	+	n	+	+	+	+	+
								B	S	C+	SPY	+	+	+	+	+	n	+	+	+	n	+
								B	S	C+	SPY	+	+	+	+	+	n	+	+	+	n	+
								Gy	S	C+	H		+	+	+	+	+	–	+	+	–	+
								W	S	C+	SM	n	n	n	n	n	n	n	n	n	n	n
							Asn (AAC)[442]	Gy	S	C–	SM	+	+	+	+	+	+	+	+	+	+	+
								Gy	S	C–	SPY	+	+	+	+	+	n	n	+	+	n	n
							+	B	S	C+	SPY	+	+	+	+	+	+	+	+	+	+	+
								Y	RF	C–	SM	+	+	+	–	+	n	–	+	–	n	n
							+	R	S	C–	SM	+	+	+	+	+	+	–	+	–	n	–

(continued)

TABLE 272. List of *Streptomyces* species continued

Species names and groups[1]	No. in lists of species[2]	Type strain	Accession no.	Wil 83a[3]	Wil 89[4]	Käm 91[5]	Hat 03[6]	Lan 02[7]	Ful 95[8]	Och 95[9]	Kat 97[10]	Lab[11]	Sch[12]	Lan 04[13]	Lan 02[14]	Lan 04[15]	Lan 04[16]	Lan 04[17]	Lan 04[18]	Lan 04[19]	Lan 04[20]
S. aurantiogriseus	50	DSM 40138, ATCC 23883, LMG 19298, NRRL-ISP 5138, NRRL B-5416	AY999793																		
Most closely related to group S. aurantiogriseus et rel.:																					
S. griseoaurantiacus	219	DSM 40430, ATCC 19840, LMG 21045, ISP 5430, NBRC 15440	AB184676	A 12	I 07	1-7 1-15							Sch00	Lan2-00	BENP	+		+		+	+
S. jietaisiensis	263	AS 4.1859, JCM 12279, FXJ46	AY314783																		
Group S. coeruleoprunus et rel.:																					
S. coeruleoprunus	128	DSM 41472, LMG 20483, NBRC 15400	AB184651										Sch00	Lan2-00	BENP	+		+		+	cl33
S. somaliensis	444	DSM 40738	AJ007403										Sch00	Lan2-00						+	
S. fradiae	189	DSM 40063, ATCC 10745, LMG 19371, ISP 5063, NRRL B-1195	DQ026630	G 68	II 18	22-5 039		La-00		OC-I			Sch12	Lan2-27	BENP +	+	(b)	+	cl09	+	cl45
Most closely related to group S. coeruleoprunus et rel.:																					
S. bluensis	74	DSM 40564, ATCC 27420, LMG 5969, ISP 5564	X79324	A 39	II 10	052 017						L2	Sch00	Lan2-00	BENP	+		+		+	cl07
Not closely related to one of the groups:																					
S. variegatus	491	DSM 41464, LMG 20315	AJ781371										Sch00	Lan2-00	BENP	+		+		+	+
S. fulvissimus	191	DSM 40593, ATCC 27431, LMG 19310, ISP 5593, NBRC 13482	AB184434	A 10		034 1-33		La-00		OC-IV		L3	Sch00	Lan2-00	BENP +	+	(h)	+	cl14		cl43
S. aureoverticillatus	56	DSM 40080, ATCC 15854, LMG 20426, ISP 5080, NRRL B-3326	AY999774	A 10	I 06	033 1-33						L3	Sch00	Lan2-00	BENP	+		+		+	+
S. flavofungini	178	DSM 40366, ATCC 27430, LMG 21799, ISP 5366, NBRC 13371	AB184359	B 42		033 1-33							Sch00	Lan2-00	BENP	+		+			+
S. alboflavus	18	DSM 40045, ATCC 12626, LMG 21038, ISP 5045, NRRL B-2373	EF178699	E 54	III 20	033 1-33		La-00		OC-IV			Sch00	Lan2-00	BENP +	+		+			cl43
S. aculeolatus	7	DSM 41644, LMG 19906, NBRC 14824	AB184624										Sch00	Lan2-00	BENP	+		+			+
S. synnematoformans	464	DSM 41902, CGMCC 4.2055, S155	EF121313																		
S. hebeiensis	244	DSM 41837, CCTCC AA-203005, YIM 001	AY277529																	+	

(continued)

TABLE 272. (continued)

Guo 08[21] 16S rRNA	Guo 08[21] *atpD*	Guo 08[21] *gyrB*	Guo 08[21] *recA*	Guo 08[21] *rpoB*	Guo 08[21] *trpB*	Guo 08[21] concat.	Kim 04[23] *rpoB*	Morphological characters[24] I	II	III	IV	Physiological tests[25] I	II	III	IV	V	VI	VII	VIII	IX	X	XI
								Gy	S	C+	SM	+	+	+	+	+	n	+	+	+	n	+
								Gy	S	C–	WTY	+	+	+	+	n	n	n	n	n	n	n
									RF	C–	SM	+	+				+		+		+	
								B	RF	C–	SM	+	+	+	+	+	–	–	n	+	n	n
							+	Y	RF	C–	SM	+	+	+	–	+	n	–	+	–	n	+
							+	R	S	C–	SM	+	+	+	–	–	+	+	–	–	–	–
								B	S	C–	SPY	+	+	+	+	+	+	+	+	+	+	+
								W/R	S	C–	SM		+	+	–	+		–				
								Y	RF	C+	SM	+	+	+	–	+	n	–	+	+	+	n
								R	S	C–	SM	+	+	+	–	+	+	+	+	+	+	n
								W/Y	RF	C–	SM	+	+	+	?	+	n	+	+	+	n	–
							+	Y	RF	C+	SM	+	+	+	–	+	+	+	+	+	–	+
								W,Y,R	S	n	WTY–SPY	+	+	+	+	+	n	+	+	–	n	–
								Gy–R	RF	C+	SM	+	n	n	n	+	+	n	n	n	–	–
									RF	C+	WTY		+				+					

(continued)

TABLE 272. List of *Streptomyces* species continued

Species names and groups[1]	No. in lists of species[2]	Type strain	Accession no.	Wil 83a[3]	Wil 89[4]	Käm 91[5]	Hat 03[6]	Lan 02[7]	Ful 95[8]	Och 95[9]	Kat 97[10]	Lab[11]	Sch[12]	Lan 04[13]	Lan 02[14]	Lan 04[15]	Lan 04[16]	Lan 04[17]	Lan 04[18]	Lan 04[19]	Lan 04[20]
Group S. carpaticus et rel.:																					
S. hainanensis	240	DSM 41900, CCTCC AA-205017, YIM 47672	AM398645																		
S. specialis	446	DSM 41924, CCM 7499, GW 41-1564	AM934703																		
S. carpaticus	92	DSM 41468, ATCC 43678, LMG 20448, NRRL B-16359	DQ442494										Sch00	Lan2-00	BENP	+		+			+
S. cheonanensis	104	NBRC 100940, VC-A46	AY822606																		
Most closely related to group S. carpaticus et rel.:																					
S. sulfonofaciens	462	DSM 41679, ATCC 31892, LMG 20325, NBRC 14260	AB249974										Sch00	Lan2-00	BENP	+		+		+	+
S. sodiiphilus	443	CCTCC AA-203015, JCM 13581, YIM 80305	AY236339																		
Not closely related to one of the groups:																					
S. scabrisporus	434	NBRC 100760, KM-4927	AB030585																		
S. gardneri	198	DSM 40064, ATCC 9604, LMG 19876, ISP 5064, NBRC 3385	AB184754	A 04		002 1-07			FU-23		KA-C		Sch00	Lan2-00	BENP	+		+		+	cl44
S. flavidofuscus	175	DSM 41473, ATCC 43683, NRRL B-16366	AY999914										Sch00	Lan2-00	BENP	+		+		+	-
Regarded as later heterotypic synonym of Streptomyces abikoensis (for references, see list of type strains):																					
S. luteoverticillatus	303	DSM 40038, ATCC 23933, LMG 20045, ISP 5038	AB184803	F 55	Sv. 03	22-1 040	Ha1		FU-12b				Sch00	Lan2-00	BENP	+	(j)	+		+	cl12
S. olivoreticuli subsp. *olivoreticuli*	355	DSM 40105, LMG 20050, ISP 5105	AB184853				Ha1						Sch00	Lan2-00	BENP	+	(j)	+		+	cl12
S. parvisporogenes	362	DSM 40473, ATCC 12568, LMG 20072, ISP 5473	AB249913		Sv. 02	22-1 040	Ha1						Sch00	Lan2-00	BENP	+	(j)	+			cl12
Regarded as later heterotypic synonym of Streptomyces anulatus (for references, see list of type strains):																					
S. chrysomallus subsp. *chrysomallus*	109a	DSM 40128, ATCC 11523, LMG 20459, ISP 5128	AB184644	A 1B		1-3 1-2			FU-22		KA-B		Sch01	Lan2-18	BENP	+	(a)	+			cl23
S. citreofluorescens	120	DSM 40265, ATCC 15858, LMG 20475, ISP 5265	AB184195	A 1B	I 02	1-3 1-2			FU-19b		KA-B		Sch01	Lan2-18	BENP	+	(a)	+		+	cl23
S. fluorescens	188	DSM 40203, ATCC 15860, LMG 8579, ISP 5203	AB184199	A 1B	I 02	1-3 1-2					KA-B		Sch01	Lab2-18	BENP	+	(a)	+		+	cl23

(continued)

TABLE 272. (continued)

Guo 08[21] 16S rRNA	Guo 08[21] *atpD*	Guo 08[21] *gyrB*	Guo 08[21] *recA*	Guo 08[21] *rpoB*	Guo 08[21] *trpB*	Guo 08[21] concat.	Kim 04[23] *rpoB*	Morphological characters[24] I	II	III	IV	Physiological tests[25] I	II	III	IV	V	VI	VII	VIII	IX	X	XI
								W	S	C+	SM	+	–	–	–	–	–	–	–	–	–	–
								W	S	C+	n	+	–	n	–	–	–	n	n	+	–	+
								Gy	S	C–	SM	+	+	+	+	+	n	+	+	+	n	n
								Gy	RF	C+	SM	n	+	+	+	+	n	+	+	+	–	+
								R/V	RF		SM	+	+	+	+	+	n	–	+	–	n	+
									RF	C–		–	–	–	+	–	–	–	–	–	n	–
								Gy	S	C–	RU	+	+	–	+	+	n	–	–	+	–	–
								Gy	RF	C+	SM	+	+	+	+	+	+	+	–	–	+	n
+	+	+	+	+	+	+		Y	S	C+	SM	+	n	+	+	+	n	+	+	+	n	n
								Lu	VE	C+	SM	+	–	–	–	+	n	±	+	–	n	+
								Lu	VE	C+	SM	+	–	–	–	–	n	±	–	±	n	+
								Ar	VE	C+	SM	+	–	–	–	±	n	–	–	±	n	±
								Y	RF	C–	SM	+	+	+	+	+	+	–	+	–	+	–
								Y	RF	C–	SM	+	+	+	+	+	+	–	+	–	–	–
								Y	RF	C–	SM	+	+	+	–	+	+	–	+	–	+	–

(continued)

TABLE 272. List of *Streptomyces* species continued

Species names and groups[1]	No. in lists of species[2]	Type strain	Accession no.	Wil 83a[3]	Wil 89[4]	Käm 91[5]	Hat 03[6]	Lan 02[7]	Ful 95[8]	Och 95[9]	Kat 97[10]	Lab[11]	Sch[12]	Lan 04[13]	Lan 02[14]	Lan 04[15]	Lan 04[16]	Lan 04[17]	Lan 04[18]	Lan 04[19]	Lan 04[20]
Regarded as later heterotypic synonym of Streptomyces avermitilis (for references, see list of type strains):																					
S. avermectinius	59																				
Regarded as later heterotypic synonym of Streptomyces cacaoi (for references, see list of type strains):																					
S. aminophilus	32	DSM 40186, ATCC 14961, LMG 19319, ISP 5186	AB184183	A 16	I 09	031 1-34		La-05					Sch00	Lan2-17	BENP cl05	+	(c)	+	cl08	+	cl36
Regarded as later heterotypic synonym of Streptomyces caeruleus (for references, see list of type strains):																					
S. niveus	341	DSM 40088, ATCC 19793, LMG 19395, ISP 5088	AB184160	A 1B	I 02	043 013		La-19					Sch11	Lan2-14	BENP cl19	+	(c)	+	cl09	+	cl47
S. spheroides	450	DSM 40292, ATCC 23965, LMG 19392, ISP 5292	EF178698	A 1B	I 02	040 048		La-19					Sch11	Lan2-00	BENP cl19	+	(c)	+	cl09		cl47
Regarded as later heterotypic synonym of Streptomyces cinnamoneus (for references, see list of type strains):																					
S. griseoverticillatus	236	DSM 40507, ATCC 27436, LMG 19944, ISP 5507	AB184862	F 58		22-1 040	Ha4						Sch17	Lan2-00	BENP	+	(j)	+		+	cl17
S. hachijoensis	240	DSM 40114, ATCC 19769, LMG 19928, ISP 5114	AB184141	F 55	Sv. 04	22-1 040	Ha4		FU-NC			L4	Sch00	Lan2-00	BENP	+	(j)	+		+	cl13
Regarded as later heterotypic synonym of Streptomyces chibaensis (for references, see list of type strains):																					
S. chibaensis	105	DSM 40220, ATCC 23895, LMG 20456, ISP 5220	AB184193	A 24	II 05	009 1-19							Sch04	Lan2-26	BENP	+	(a)	+		+	cl52
Regarded as later heterotypic synonym of Streptomyces filamentosus (for references, see list of type strains):																					
S. roseosporus	418	DSM 40122, ATCC 23958, LMG 20262, ISP 5122	AB184238	A 05	I 04	002 1-7							Sch00	Lan2-24	BENP	+	(b)	+			cl23
Regarded as later heterotypic synonym of Streptomyces flavofuscus (for references, see list of type strains):																					
S. globisporus subsp. *flavofuscus*	208c	ATCC 19908	DQ026648																	+	

(continued)

TABLE 272. (continued)

								Morphological characters[24]				Physiological tests[25]										
Guo 08[21] 16S rRNA	Guo 08[21] *atpD*	Guo 08[21] *gyrB*	Guo 08[21] *recA*	Guo 08[21] *rpoB*	Guo 08[21] *trpB*	Guo 08[21] concat.	Kim 04[23] *rpoB*	I	II	III	IV	I	II	III	IV	V	VI	VII	VIII	IX	X	XI
								W	S	C–	SM	+	+	+	–	+	+	–	+	–	+	n
								Y	S	C–	SM	+	+	+	+	+	+	–	+	–	–	n
								Y	S	C–	SM	+	+	–	+	+	n	–	+	–	n	n
								Ke	VE	C–	SM	+	–	–	–	–	n	±	–	+	n	+
							Group F	Ci	VE	C–	SM	+	±	–	±	±	n	±	±	+	n	±
								Gy	S	C–	SM	+	+	+	+	+	+	+	+	+	–	+
								R	RF	C–	SM	+	+	+	+	–	n	–	–	–	+	–
								Y	RF	C–	SM	+	+	+	+	n	n	n	n	n	n	n

(continued)

TABLE 272. List of *Streptomyces* species continued

Species names and groups[1]	No. in lists of species[2]	Type strain	Accession no.	Wil 83a[3]	Wil 89[4]	Käm 91[5]	Hat 03[6]	Lan 02[7]	Ful 95[8]	Och 95[9]	Kat 97[10]	Lab[11]	Sch[12]	Lan 04[13]	Lan 02[14]	Lan 04[15]	Lan 04[16]	Lan 04[17]	Lan 04[18]	Lan 04[19]	Lan 04[20]
Regarded as later heterotypic synonym of Streptomyces flavovirens (for references, see list of type strains):																					
S. nigrifaciens	337	DSM 40071, ATCC 19791, LMG 20048, ISP 5071	AB184158	A 1C	I 03	1-2 015					KA-B		Sch05	Lan2-09	BENP	+	(a)	+		+	cl53
Regarded as later heterotypic synonym of Streptomyces fradiae (for references, see list of type strains):																					
S. roseoflavus	414	DSM 40536, ATCC 13167, LMG 20535, ISP 5536			IV 10 (red series)	22-5 105							Sch12	Lan2-27	BENP	+	(b)	+			cl45
Regarded as later heterotypic synonym of Streptomyces griseocarneus (for references, see list of type strains):																					
S. septatus	438	DSM 40577, ATCC 27464, LMG 8604, ISP 5577	AB184883	F 55	Sv. 02	22-1 040	Ha6						Sch00	Lan2-00	BENP	+	(j)	+			+
Regarded as later heterotypic synonym of Streptomyces griseus (for references, see list of type strains):																					
S. setonii	439	DSM 40395, ATCC 25497, LMG 20291, ISP 5395	D63872	A 1B	I 02	1-3 1-2					KA-B		Sch00	Lan2-19	BENP	+		+		+	cl23
Regarded as later heterotypic synonym of Streptomyces hiroshimensis (for references, see list of type strains):																					
S. rectiverticillatus	403	DSM 40436, ATCC 19845, LMG 20292, ISP 5436	AB184296	F 57	Sv. 18	22-1 040	Ha7						Sch00	Lan2-00	BENP	+	(j)	+		+	cl48
S. roseoverticillatus	419	DSM 40039, ATCC 19807, LMG 20255, ISP 5039	AB184169		Sv. 01	22-1 040	Ha7					L4	Sch00	Lan2-00	BENP	+	(j)	+			cl13
S. salmonis	428	DSM 40895, NRRL B-1472, LMG 20306	X53169		Sv. 05	22-1 040	Ha7					L4	Sch00	Lan2-00	BENP	+	(j)	+		+	cl13
S. spitsbergensis	454	ATCC 51269, JCM 8881	AB184700				Ha7													+	
S. fervens	167	DSM 40086, ATCC 27429, ISP 5086	AB184871		Sv. 01	22-1 040						L4	Sch00								
Regarded as later heterotypic synonym of Streptomyces lilacinus (for references, see list of type strains):																					
S. kashmirensis	265	DSM 40336, LMG 19937, ISP 5336	AB184546				Ha8						Sch00	Lan2-00	BENP	+	(j)	+			cl12

(continued)

TABLE 272. (continued)

								Morphological characters[24]				Physiological tests[25]										
Guo 08[21] 16S rRNA	Guo 08[21] *atpD*	Guo 08[21] *gyrB*	Guo 08[21] *recA*	Guo 08[21] *rpoB*	Guo 08[21] *trpB*	Guo 08[21] concat.	Kim 04[23] *rpoB*	I	II	III	IV	I	II	III	IV	V	VI	VII	VIII	IX	X	XI
								Gy	RF	C–	SM	+	+	+	+	+	+	+	+	+	–	n
								R	S	C–	SM	+	+	+	–	–	–	–	–	–	n	–
								Y/R	VE	C+	SM	+	–	–	–	?	n	–	–	+	n	–
Group I	No	No	No	No	No	No	+	Y	RF	C–	SM	+	+	+	+	+	n	–	+	–	n	–
								Ke	VE	C+	SM	+	±	–	±	+	n	±	+	+	n	±
								Bi	VE	C+	SM	+	–	–	–	–	n	–	–	–	n	+
								Sa	VE	C+	SM	+	–	–	–	–	n	–	–	±	n	±
								R/V	RF		SM											
								Ba	VE	C+	SM	+	±	–	±	±	n	±	±	±	n	+
								Li	VE	C+	SM	–	–	–	–	–	n	–	–	–	n	–

(continued)

TABLE 272. List of *Streptomyces* species continued

Species names and groups[1]	No. in lists of species[2]	Type strain	Accession no.	Wil 83a[3]	Wil 89[4]	Käm 91[5]	Hat 03[6]	Lan 02[7]	Ful 95[8]	Och 95[9]	Kat 97[10]	Lab[11]	Sch[12]	Lan 04[13]	Lan 02[14]	Lan 04[15]	Lan 04[16]	Lan 04[17]	Lan 04[18]	Lan 04[19]	Lan 04[20]
Regarded as later heterotypic synonym of Streptomyces mashuensis (for references, see list of type strains):																					
S. kishiwadensis	269	DSM 40397, ATCC 25464, LMG 19939, ISP 5397	AB184858		Sv. 15	22-1 040	Hal1						Sch31		BENP	+	(j)	+		+	cl11
Regarded as later heterotypic synonym of Streptomyces microflavus (for references, see list of type strains):																					
S. griseus subsp. *alpha*	238b	DSM 40937, NRRL B-2249, LMG 19953	AB184668		I 02	1-3 1-2							Sch02	Lan2-12	BENP	+	(a)	+			cl23
S. griseus subsp. *cretosus*	238c	DSM 40561, ISP 5561	AB184418										Sch02	Lan2-12						+	
S. willmorei	520	DSM 40459, ATCC 6867, LMG 21046, ISP 5459	AB184374	A 1B	I 02	1-3 1-2					KA-B		Sch02	Lan2-12	BENP	+	(a)	+		+	cl23
Regarded as later heterotypic synonym of Streptomyces minutiscleroticus (for references, see list of type strains):																					
S. flaviscleroticus	177	DSM 40270, ATCC 19347, LMG 19886, ISP 5270	AB184634		I 08	017 007					KA-G		Sch00	Lan2-03	BENP	+	(a)	+		+	cl46
Regarded as later heterotypic synonym of Streptomyces mobaraensis (for references, see list of type strains):																					
S. ladakanum	275	DSM 40587, NRRL 3191	AB184430				Ha12					L4									
Regarded as later heterotypic synonym of Streptomyces netropsis (for references, see list of type strains):																					
S. distallicus	146	DSM 40846, NCIB 8936, LMG 20499	AB184703		Sv. 01	22-1 040	Hal4					L4		Lan2-13	BENP	+	(b) (j)	+		+	cl13
S. flavopersicus	181	DSM 40093, ATCC 19756, ISP 5093	AB249911	F 56		22-1 040	Ha14					L4	Sch00							+	
S. kentuckensis	268	DSM 40052, ATCC 12691, ISP 5052	AB184215	F SM	Sv. 11	22-1 040	Hal4					L4	Sch00							+	
S. syringium	465	DSM 41480, LMG 20320	AJ781375				Ha14						Sch00	Lan2-00	BENP	+	(j)	+		+	cl13
Regarded as later heterotypic synonym of Streptomyces phaeopurpureus (for references, see list of type strains):																					
S. phaeoviridis	373	DSM 40285, ATCC 23947, LMG 20061, ISP 5285	AB184230	A 19	I 12	009 1-19							Sch23	Lan2-02	BENP	+	(b)	+		+	cl51

(continued)

TABLE 272. (continued)

Guo 08[21] 16S rRNA	Guo 08[21] *atpD*	Guo 08[21] *gyrB*	Guo 08[21] *recA*	Guo 08[21] *rpoB*	Guo 08[21] *trpB*	Guo 08[21] concat.	Kim 04[23] *rpoB*	Morphological characters[24] I	II	III	IV	Physiological tests[25] I	II	III	IV	V	VI	VII	VIII	IX	X	XI
								Ar	VE	C+	SM	+	–	±	–	+	n	±	–	+	n	+
Group III	Yes	Yes	Yes	Yes	Yes	Yes		Y	RF	C–	SM	+	+	–	+	n	n	n	n	n	n	n
Group III	Yes	Yes	Yes	Yes	Yes	Yes		Y	RF	C–	SM	+	+	–	+	+	n	–	n	–	+	n
								Y	RF	C–	SM	+	+	–	+	n	n	n	n	n	n	n
								n	n	C–/C+	n	+	+	+	+	+	n	–	+	–	n	?
								W/Y	VE	C–	SM	+	–	–	–	+	n	–	–	–	n	–
								Ke	VE	C+	SM	+	±	–	±	±	n	±	–	+	n	±
								Ke	VE	C+	SM	+	–	–	–	±	n	–	–	+	n	±
								Ke	VE	C–	SM	+	±	–	–	±	n	±	±	+	n	+
								R	S	C+	SM	+	–	–	–	+	–	–	–	–	n	–
								R	S	C–	SM	+	+	+	+	+	n	+	+	n	n	+

(continued)

TABLE 272. List of *Streptomyces* species continued

Species names and groups[1]	No. in lists of species[2]	Type strain	Accession no.	Wil 83a[3]	Wil 89[4]	Käm 91[5]	Hat 03[6]	Lan 02[7]	Ful 95[8]	Och 95[9]	Kat 97[10]	Lab[11]	Sch[12]	Lan 04[13]	Lan 02[14]	Lan 04[15]	Lan 04[16]	Lan 04[17]	Lan 04[18]	Lan 04[19]	Lan 04[20]
Regarded as later heterotypic synonym of Streptomyces thermovulgaris (for references, see list of type strains):																					
S. thermonitrificans	478	DSM 40579, ATCC 23385, ISP 5579	Z68098	A 36	II 09	021 002														+	
Regarded as later heterotypic synonym of Streptomyces tricolor (for references, see list of type strains):																					
S. roseodiastaticus	413	DSM 41703, LMG 20327	AB184683										Sch33	Lan2-21	BENP	+	(b)	+			cl02
Regarded as later heterotypic synonym of Streptomyces olioverticillatus (for references, see list of type strains):																					
S. viridiflavus	511	LMG 20277	AB184702												BENP	+	(j)	+			cl13
Regarded as later heterotypic synonym of Streptomyces violaceus (for references, see list of type strains):																					
S. violatus	507	DSM 40209, ATCC 15892, LMG 19397, ISP 5209	AJ399480	A 18	I 11	050 019		La-12					Sch00		BENP cl12	+	(c)	+	cl14	+	cl01
No detailed sequence information available:																					
S. caviscabies	95	DSM 41811, ATCC 51928	AF112160																		
S. coeruleoflavus (no sequence available)	126																				
S. arabicus	38	DSM 40252, ATCC 23881, LMG 20414, ISP 5252	D44271	A 12	I 07	006 1-18							Sch00	Lan2-23	BENP	+	(b)	+		+	cl08
S. baldaccii	66	DSM 40845, ATCC 23654	X53164		Sv. 01	22-1 040	Ha7		FU-12b			L4								+	
S. cellulolyticus	99																			+	
S. echinoruber	152	DSM 41696, IFO 14238											Sch00							+	
S. erythraeus	159	DSM 40517, LMG 20508												Lan2-00	BENP	+		+		+	+
S. longisporoflavus	292	DSM 40165, ATCC 19915, LMG 19347, ISP 5165	AB184220	A 39	II 10	005 010		La-00		OC-non			Sch00	Lan2-00	BENP +	+		+	cl06	+	cl19
S. olivomycini	354																				
S. speleomycini	449																				
S. thermoautotrophicus	471	DSM 41605																		+	
Not in tree:																					
S. aldersoniae		DSM 41909, NRRL 18513	EU170123																		
S. alni		D65, CGMCC 4.3510, NRRL B-24611	DQ460470																		
S. angustmycinicus		DSM 41683, NRRL B-2347	EU170119																		
S. ascomycinicus		DSM 40822, NBRC 13981	EU170121																		

(continued)

TABLE 272. (continued)

								Morphological characters[24]				Physiological tests[25]										
Guo 08[21] 16S rRNA	Guo 08[21] *atpD*	Guo 08[21] *gyrB*	Guo 08[21] *recA*	Guo 08[21] *rpoB*	Guo 08[21] *trpB*	Guo 08[21] concat.	Kim 04[23] *rpoB*	I	II	III	IV	I	II	III	IV	V	VI	VII	VIII	IX	X	XI
								Gy	RF	C+	SM	+	–	–	–	–	+	–	+	+	–	–
								Gy	S	C–	SM	+	+	+	+	n	n	n	n	n	n	n
								W		C–								–	–			
							Group A18	R	S	C+	SPY	+	+	+	+	+	+	+	n	+	n	+
Group I	Yes	Yes	Yes	Yes	Yes	Yes		W	RF	C–	SM	n	n	n	n	n	n	+	n	n	n	n
								B	S	C–	SPY	n	+	+	+	+	n	+	n	+	n	n
								Gy	S	C–	SPY	+	+	+	+	+	n	–	+	+	n	n
								Ba	VE	C+	SM	+	–	–	–	–	n	±	±	±	n	+
								W Gy	RF S	C–	WTY SPY	+	+	+	–	+	n	n	n	+	n	+
								R	S	C–	SPY	+	+	+	+	+	+	+	+	+	+	n
								Y	S	C–	SM	+	+	+	+	+	+	–	+	–	+	–
									RF	C–	SM		+	+	+	+		+	+			
								Y/Gy Gy	RF	C–	SM		+	+	+	+	+			+		
								Gy/W	L	C–	SM	n	+	+	–	+	+	+	n	n	+	n
								W/Gy	RF	C–	SM	n	+	+	+	(+)	n	+	n	–	+	+
								Gy/W	S	C–	SM	n	–	–	–	+	+	+	n	n	–	n
								Gy/W	L	C–	SPY	n	+	+	+	+	+	+	n	n	+	n

(continued)

TABLE 272. List of *Streptomyces* species continued

Species names and groups[1]	No. in lists of species[2]	Type strain	Accession no.	Wil 83a[3]	Wil 89[4]	Käm 91[5]	Hat 03[6]	Lan 02[7]	Ful 95[8]	Och 95[9]	Kat 97[10]	Lab[11]	Sch[12]	Lan 04[13]	Lan 02[14]	Lan 04[15]	Lan 04[16]	Lan 04[17]	Lan 04[18]	Lan 04[19]	Lan 04[20]
S. atriruber		DSM 41860, LDDC6330-99, NRRL B-24165	EU812169																		
S. avicenniae		DSM 41943, MCCC 1A01535, CGMCC 4.5510	EU399234																		
S. axinellae		DSM 41948, Pol001, CIP 109838	EU683612																		
S. baliensis		ID03-0915, BTCC B-608, NBRC 104276	AB441718																		
S. castelarensis		DSM 40830, ATCC 15191	EF408732																		
S. deccanensis		DAS-139, KCTC 19241, CCTCC AA-207004	EF219459																		
S. decoyininicus		DSM 41427, NRRL 2666	EU170127																		
S. gulbargensis		DAS 131, KCTC 19179, CCTCC AA-206001	DQ317411																		
S. haliclonae		DSM 41970, Sp080513SC-31, NBRC 105049	AB473556																		
S. himastatinicus		DSM 41914, ATCC 53653	EF408736																		
S. hypolithicus		DSM 41950, HSM#10, NRRL B-24669	EU196762																		
S. iranensis		DSM 41954, HM 35, CCUG 57623	FJ472862																		
S. lunalinharesii		DSM 41876, ATCC BAA-1231, RCQ1071, CIP 108852	DQ094838																		
S. marinus		DSM 41968, Sp080513GE-26, NBRC 105047	AB473554																		
S. marokkonensis		DSM 41918, LMG 23016, R-22003, Ap1	AJ965470																		
S. mayteni		YIM 60475, KCTC 19383, CCTCC AA-207005	EU200683																		
S. milbemycinicus		DSM 41911, NRRL 5739	EU170126																		
S. modarskii		DSM 40771, NRRL B-1346	EF408735																		
S. nanshensis		SCSIO 01066, KCTC 19400, CCTCC AA-208005	EU589334																		
S. osmaniensis		OU-63, PCM 2690, CCTCC AA-209025	FJ613126																		
S. plumbiresistens		CCNWHX 13-160, ACCC 41207, HAMBI 2991	EU526954																		
S. polyantibioticus		DSM 44925, SPR, NRRL B-24448	DQ141528																		
S. rapamycinicus		ATCC 29253, NRRL 5491	EF408733																		
S. ruanii		DSM 40276, ISP 5276	EF408737																		
S. sedi		DSM 41942, YIM 65188, CCTCC AA-208020	EU925562																		

(continued)

TABLE 272. (continued)

Guo 08[21] 16S rRNA	Guo 08[21] *atpD*	Guo 08[21] *gyrB*	Guo 08[21] *recA*	Guo 08[21] *rpoB*	Guo 08[21] *trpB*	Guo 08[21] concat.	Kim 04[23] *rpoB*	Morphological characters[24]				Physiological tests[25]										
								I	II	III	IV	I	II	III	IV	V	VI	VII	VIII	IX	X	XI
								Gy	RF	C–	SM	+	+	+	+	+	+	+	+	+	+	±
									S		SM	n	+	+	+	n	+	+	+	–	n	+
									S		SM	+	+	–	+	+	+	–	+	n	–	–
									RF	C+	WTY/ SM	+	+	+	+	+	+	+	+	n	n	+
								Gy/Bl	S	C–	RU	+	+	n	+	+	+	+	+	n	n	–
								W	SC	C+	H	+	+	+	+	+	n	+	+	n	n	+
								Gy	S	C–	SM	n	+	–	–	+	+	–	n	n	–	n
									SC	C+	SM	+	+	+	+	n	+	+	+	n	n	+
								W	S	C–	SM	+	–	–	–	+	n	+	+	n	n	+
								Gy/Bl	S	C–	RU	+	+	n	+	+	+	+	+	n	n	+
									SC/RF	C–	SM	+	–	–	n	–	+	–	–	n	±	–
								Y–Gy/ Gy	S		RU	n	+	+	+	n	+	+	+	+	n	+
								Gy	S		SPY	n	–	+	+	n	n	–	n	n	n	+
									S	C–	SM	+	+	+	–	+	n	–	+	n	n	+
								Gy	S	C–	SM	+	+	±	+	+	+	–	+	±	+	+
									S/L		SM	–	–	–	–	+	+	–	n	–	n	+
								Gy	S/L	C–	WTY	n	+	+	+	–	–	–	n	n	–	n
								Gy/Bl	S	C–	RU	+	+	n	+	+	+	+	n	n	n	+
									S	C+	SM	+	+	n	+	+	+	+	+	+	n	+
								Gy–B	S	C+	SPY	n	+	+	+	+	+	+	+	n	n	+
								G–W	RF	C–	SM	+	+	+	+	+	+	+	n	n	n	+
									RF	C+	SM	+	+	–	–	+	+	+	–	n	+	–
								Gy/Bl	S	C–	RU	+	+	–	+	+	n	+	+	n	n	+
								Gy/Bl	S	C–	RU	n	+	–	+	+	+	n	n	n	n	+
									S		SM	n	–	n	–	+	–	–	–	–	n	n

(continued)

TABLE 272. List of *Streptomyces* species continued

Species names and groups[1]	No. in lists of species[2]	Type strain	Accession no.	Wil 83a[3]	Wil 89[4]	Käm 91[5]	Hat 03[6]	Lan 02[7]	Ful 95[8]	Och 95[9]	Kat 97[10]	Lab[11]	Sch[12]	Lan 04[13]	Lan 02[14]	Lan 04[15]	Lan 04[16]	Lan 04[17]	Lan 04[18]	Lan 04[19]	Lan 04[20]
S. silaceus		DSM 41861, LDDC 6638-99, NRRL B-24166	EU812170																		
S. tateyamensis		DSM 41969, Sp080513SC-30, NBRC 105048	AB473555																		
S. thinghirensis		DSM 41919, S10, CCMM B35	FM202482																		
S. tritolerans		DAS 165	DQ345779																		
S. wellingtoniae		DSM 40632, NRRL B-1503	EU170124																		
S. xiamenensis		DSM 41903, MCCC 1A01550, CGMCC 4.3534	EF012099																		
S. xinghaiensis		S187, KCTC 19546, CCTCC AA 208049, NRRL B-24674	EF577247																		
K. kazusensis		SK 60, KCTC 19565, JCM 14560	AB278569																		
K. saccharophila		SK 15, KCTC 19566, JCM 14559	AB278568																		

1. Species are grouped according to the maximum-likelihood tree in Figure 339.
2. Without symbols: list of type strains of *Streptomyces*; † list of type strains of *Streptacidiphilus*; * list of type strains of *Kitasatospora*.
3. Groups as described by Williams et al. (1983a), grouping on the basis of numerical identification.
4. Groups as described by Williams et al. (1989), grouping mainly on the basis of numerical identification according to Williams et al. (1983a).
5. Groups as described by Kämpfer et al. (1991), grouping on the basis of numerical identification.
6. Groups as described by Hatano et al. (2003), grouping on the basis of phenotypes, DNA–DNA hybridization and sequences of *gyrB*.
7. Groups as described by Lanoot et al. (2002), grouping on the basis of protein profiles.
8. Groups as described by Fulton et al. (1995), grouping on the basis of fingerprints of the rRNA operon.
9. Groups as described by Ochi (1995), grouping on the basis of the ribosomal AT-L30 protein.
10. Groups as described by Kataoka et al. (1997), grouping on the basis of partial 16S rRNA gene sequences containing a variable α region.
11. Groups as described by: Labeda and Lyons (1991b), L1; Labeda and Lyons (1991a), L2; Labeda (1998), L3; Labeda (1996), L4; Labeda (1993), L5 (grouping on the basis of DNA relatedness).
12. Groups as described by P. Schumann (unpublished), grouping on the basis of ribotyping.
13. Groups as described by Lanoot et al. (2004), grouping on the basis of Box-PCR.
14. Groups as described by Lanoot et al. (2002), grouping on the basis of protein profiles.
15. Groups as described by Lanoot (2004), grouping on the basis of DNA–DNA hybridization.
16. Groups as described by Lanoot (2004), grouping on the basis of DNA–DNA hybridization.
17. Groups as described by Lanoot (2004), grouping on the basis of ARDRA.
18. Groups as described by Lanoot (2004), grouping on the basis of ARDRA.
19. Groups as described by Lanoot (2004), grouping on the basis of analysis of the ITS region.
20. Groups as described by Lanoot (2004), grouping on the basis of 16S rRNA-ITS RFLP.
21. Groups as described by Guo et al. (2008), grouping on the basis of multilocus phylogeny calculated with the sequences of five housekeeping genes (*atpD, gyrB, recA, rpoB, trpB*) and the 16S rRNA gene.
22. +, Strains that were used in the study of Guo et al. (2008) but did not belong to one of the four detected groups; yes, strains that show the same grouping (I–IV) as in the 16S rRNA gene sequence tree; no, strains do not show the same grouping (I–IV) as in the 16S rRNA gene sequence tree.
23. Groups as described by Kim et al. (2004), grouping on the basis of *rpoB* gene sequences; +, strains that were used in this study; Group A16, group A18, group A1B, group F, *Kitasatospora* and Asn(AAC), names of species groups based on the *rpoB* gene according to Kim et al. (2004).
24. Morphological characters of species described before 1974 according to Pridham and Tresner (1974a, 1974b) and Baldacci and Locci (1974). n, Not determined.

I. Spore color en masse indicated as W (White), Gy (Gray), Y (Yellow), R (Red), B (Blue), G (Green), V (Violet), Bl (Black), Ba (substrate mycelium pink-red to orange-red, aerial mycelium pink, gray-pink, and violet-pink), Bi (substrate mycelium colorless, reddish and orange, yellow to brick red, aerial mycelium pinkish white), Hi (substrate mycelium brick red, aerial mycelium beige to pink-beige), Sa (substrate mycelium brick-red to orange, aerial mycelium white with pink and yellow shades), Lu (substrate mycelium yellow, yellowish to brown, aerial mycelium light yellow, yellowish to beige), Gr (substrate mycelium brownish yellow, aerial mycelium pinkish beige with lilac shades), Ci (substrate mycelium yellow to greenish, yellow and brown-yellow, aerial mycelium pinkish with beige and lilac shades), Ar (substrate mycelium light yellow to yellowish to pinkish yellow, aerial mycelium basically white with yellow, pink and gray), Ke (substrate mycelium yellow, yellowish to hazel-nut yellow, aerial mycelium beige

TABLE 272. (continued)

								Morphological characters[24]				Physiological tests[25]										
Guo 08[21] 16S rRNA	Guo 08[21] *atpD*	Guo 08[21] *gyrB*	Guo 08[21] *recA*	Guo 08[21] *rpoB*	Guo 08[21] *trpB*	Guo 08[21] concat.	Kim 04[23] *rpoB*	I	II	III	IV	I	II	III	IV	V	VI	VII	VIII	IX	X	XI
								W/Y–W	RF	C–	SM	+	+	–	–	+	+	+	+	+	+	–
								Gy	S	C–	SM	+	–	–	–	–	n	–	+	n	n	+
									S	C–	SM	+	–	n	+	+	+	–	+	n	n	±
								W/Gy	SC/RF	C+	SM	+	+	+	+	+	+	+	+	–	n	+
								R–Gy	S/L	C–	SM	n	+	+	+	+	+	+	n	n	–	n
									SC/RF	C–	SM	n	–	+	+	n	–	–	+	n	n	+
									SC/RF		SM	+	n	–	+	+	+	–	+	n	n	+
								W		C–	SM	+	+	+	–	–	+	–	–	–	n	–
								Gy	SC/RF	C–	SM	+	+	+	+	+	+	+	–	–	n	+

with shades toward yellow, pink and cinnamon), Mo (substrate mycelium yellow to greenish yellow, aerial mycelium grayish green), Li (substrate mycelium brown, aerial mycelium pinkish white), Th (substrate mycelium brown-yellow to greenish, aerial mycelium light yellow).

II. Spore chain morphology indicated as RF (Rectus Flexibilis), S (Spira), VE (Verticil), RA (Retinaculum-Apertum), SC (straight chains), L (loop).

III. C+, Melanoid pigments produced; C–, pigments not produced.

IV. Spore wall ornamentation indicated as SM (smooth), SPY (spiny), H (hairy), WTY (warty), RU (rugose).

25. Physiological tests of species described before 1974 according to Pridham and Tresner (1974a, 1974b) and Baldacci and Locci (1974). +, Positive for utilization of carbon compounds; –, negative for utilization of carbon compounds; n, not determined.

I. D-Glucose

II. D-Xylose

III. L-Arabinose

IV. L-Rhamnose

V. D-Fructose

VI. D-Galactose

VII. Raffinose

VIII. D-Mannitol

IX. i-Inositol

X. Salicin

XI. Sucrose

Fingerprinting techniques.

Randomly amplified polymorphic DNA (RAPD) PCR. In RAPD-PCR, single primers with arbitrary nucleotide sequences are used to amplify DNA at low annealing temperatures in order to detect polymorphisms. The method is used as a rapid screening method to detect similarities among streptomycetes. For meaningful results, a stringent standardization of reaction parameters is required; the latter include primer sequence, annealing temperature, buffer components, and concentration and quality of template DNA. RAPD-PCR results in a characteristic fingerprint of PCR products. The standardized procedure allows the detection of chromosomal differences between individual isolates without having any prior knowledge of chromosomal sequences.

Mehling et al. (1995) used this technique in a study of actinomycete species, but did not detect any characteristic banding patterns for closely related species unless they used a highly specific actinomycete primer. Even so, the resulting fingerprints contained only four bands. Similar results were reported by Huddleston et al. (1995), who used the method to try and determine interspecific relationships among members of the *Streptomyces albidoflavus* cluster of Williams et al. (1983a). Anzai et al. (1994) investigated 11 primers with various fragment patterns from zero to 20 and found that variations in fingerprint patterns can be obtained by substitution of a single base on the arbitrary primer; the most significant differences were observed when the sequence at the 3′ end was altered. Anzai and colleagues used an optimized procedure to study the relationship of *Streptomyces virginiae* and *Streptomyces lavendulae* strains; members of these taxa were assigned to the same numerically defined clusters by Williams et al. (1983a) and Kämpfer et al. (1991). Good correlation was found between the RAPD-PCR data and corresponding DNA–DNA hybridization, low-frequency restriction fragment analysis, and cultural and physiological tests, though the interspecific relationship of *Streptomyces lavendulae* and *Streptomyces virginiae* strains was not clarified.

Restriction digests of total chromosomal DNA. Low-frequency restriction fragment analysis is based on the digestion of the entire bacterial chromosome with restriction endonucleases that cut infrequently. Rare AT cutters are used for streptomycetes given their high DNA G+C content. PFGE is used to separate the resultant fragments. In the first study, Beyazova and Lechevalier (1993) examined 59 strains from eight species-groups and found that the method was useful as related strains clustered together. However, some discrepancies were found, for example, for strains grouped into the *Streptomyces cyaneus* cluster of Williams et al. (1983a). Like RAPD-PCR, the method seems to be useful for the detection of very closely related strains, but cannot be used to resolve interspecific relationships. A misinterpretation of banding patterns can result from large chromosomal amplifications or deletions (Rauland et al., 1995).

Nucleic acid sequence comparisons of 16S rRNA and other genes. In an early review of the application of 16S rRNA gene sequence analysis to the classification of streptomycetes, Stackebrandt et al. (1992) highlighted the importance of the region selected for comparison. They found that relationships between strains were influenced by which variable region (*a*, *b*, or *c*) was studied. Kataoka et al. (1997) were able to resolve inter- and intraspecies relationships between 89 streptomycete type strains, representing several clusters of Williams et al. (1983a) by sequencing the *c* region. Forty-two of the strains were found to have unique sequences; the remaining strains were assigned to 15 groups. In a more extensive study, Kataoka and his colleagues (1997) deposited the sequences of the *c* region of 485 *Streptomyces* strains in GenBank. At present, this is the largest publicly available set of streptomycete 16S rRNA gene sequence data.

A phylogenetic tree based on comparison of the *c* regions of representatives of the major cluster-groups defined by Williams et al. (1983a) was published by Anderson and Wellington (2001). They were able to confirm the taxonomic status of the phenotypic groups, apart from the *Streptomyces olivaceoviridis* and *Streptomyces griseoruber* strains which had identical *c* regions; these strains were found in clusters 20 and 21 of Williams et al. (1983a), respectively, but formed cluster 9 in the study of Kämpfer et al. (1991). The sequence data also showed that the 60 strains assigned to three species-groups in the *Streptomyces albidoflavus* group (Williams et al., 1983a) could be divided into six groups (Kataoka et al., 1997); the three phenotypic subgroups of Williams et al. (1983a) were maintained, but did not cluster together.

Hain et al. (1997) designed 16S rRNA oligonucleotide probes to determine intraspecific relationships within the *Streptomyces albidoflavus* group and found that the resultant sequences were useful for species delineation, but not for strain differentiation. The intergenic 16S–23S rRNA spacer regions were found to be more suitable for the delineation of intraspecific relationships within that cluster. Genus-specific probes have also been developed based on 23S rRNA gene (Mehling et al., 1995) and 5S rRNA gene (Park et al., 1991) sequences. The reclassification of the genera *Chainia*, *Elytrosporangium*, *Kitasatoa*, *Microellobosporia*, and *Streptoverticillium* into the genus *Streptomyces* was confirmed using 5S rRNA gene sequence data (Park et al., 1991).

At present, complete 16S rRNA gene sequences for almost all *Streptomyces* type strains are available from public databases. However, the variation in the 16S rRNA genes – even in the variable regions – is too limited to help resolve problems of species differentiation and to establish taxonomic structure within the genus (Anderson and Wellington, 2001; Stackebrandt et al., 1991a, 1991b, 1992; Witt and Stackebrandt, 1990). The situation is complicated by the fact that *Streptomyces* species may harbor different 16S rRNA gene sets, for example, *Streptomyces coelicolor* A3(2), "*Streptomyces lividans*", and several other *Streptomyces* species contain six ribosomal rRNA gene sets. Each set of rRNA genes comprises one gene copy for 16S, 23S, and 5S rRNA (van Wezel et al., 1991) and lacks tRNA genes.

Other genes can be used to establish inter- and intraspecies level relationships within the genus. Hatano et al. (2003) examined the partial sequences of the *gyrB* gene of 64 whorl-forming streptomycetes. This gene represents the structural gene of the B subunit of DNA gyrase. Most members of the 46 species, eight subspecies, and 13 species with names that have not been validly published (including 10 strains examined by the ISP) were assigned to two major groups. The larger group, which consisted of typical whorl-forming species (59 strains), was subdivided into six major clusters of three or more species, seven minor clusters of two species, and five single-member clusters at the 97% *gyrB* sequence similarity level. The major clusters contained *Streptomyces abikoensis*, *Streptomyces cinnamoneus*, *Streptomyces distallicus*, *Streptomyces griseocarneus*, *Streptomyces hiroshimensis*, and *Streptomyces netropsis* strains, results that were in line with phenotypic data.

With the exception of the *Streptomyces distallicus* cluster (which was divided phenotypically into the *Streptomyces distallicus* and *Streptomyces stramineus* subclusters) and the *Streptomyces netropsis* cluster (which was divided into the *Streptomyces netropsis* and *Streptomyces eurocidicus* subclusters), members of each of the clusters resembled one another closely, as did members of the minor clusters. Hatano and his colleagues classified 59 strains of typical whorl-forming *Streptomyces* species into the following 18 species [including subjective synonym(s)]: *Streptomyces abikoensis, Streptomyces ardus, Streptomyces blastmyceticus, Streptomyces cinnamoneus, Streptomyces eurocidicus, Streptomyces griseocarneus, Streptomyces hiroshimensis, Streptomyces lilacinus,* "*Streptomyces luteoreticuli*", *Streptomyces luteosporeus, Streptomyces mashuensis, Streptomyces mobaraensis, Streptomyces morookaense, Streptomyces netropsis, Streptomyces orinoci, Streptomyces stramineus, Streptomyces thioluteus,* and *Streptomyces viridiflavus* (Table 272). In addition, all of the strains, which showed 98.5–100% *gyrB* sequence similarities, had high DNA–DNA similarities (70–100%), indicating that *gyrB* sequences give a better resolution than corresponding 16S rRNA gene sequences.

Other conserved genes that are prime targets for taxonomic studies include the housekeeping genes (e.g. elongation factors and ATPase subunits; Ludwig and Schleifer, 1994) and tryptophan synthase genes (Huddleston et al., 1997), which were used to determine the phylogeny of streptomycin-producing streptomycetes and provide evidence for horizontal transfer of antibiotic resistance genes. Guo et al. (2008) sequenced five housekeeping genes (*atpD, gyrB, recA, rpoB, trpB*) and the corresponding 16S rRNA genes of 55 *Streptomyces* strains classified in the *Streptomyces griseus* clade. The strains were assigned to four clusters that contained strains with identical 16S rRNA gene sequences, but these results were not congruent with those based on the sequences of the individual or from corresponding concatenated gene sequences. Some of the strains with identical 16S rRNA gene sequences were assigned to different clusters in the housekeeping gene sequence trees. The trees based on individual gene sequences gave a higher resolution between strains than the concatenated tree. The authors concluded that phylogenetic trees generated on more than one gene sequence are more reliable and give a higher resolution power and topological stability.

Kim et al. (2004) analyzed the sequences of 16S rRNA and RNA polymerase β-subunit genes (*rpoB*) of 57 *Streptomyces* strains, five *Kitasatospora* strains, and a single *Micromonospora* strain and found that the resultant phylogenetic trees had similar topologies. They also found good congruence between the *rpoB* sequence and corresponding numerical phenetic data of Williams et al. (1983a). The five *Kitasatospora* strains were clearly separated from the *Streptomyces* strains in the *rpoB* gene tree. Such results show that sequence analysis of additional genes (i.e. other housekeeping genes) will help to give a better insight into the intraspecific structure of the genus *Streptomyces* (Stackebrandt et al., 2002).

Rapid methods for gene analysis in streptomycete taxonomy. Several alternative methods for gene analysis have been described which do not involve sequencing. These methods are based on either restriction analysis (Clarke et al., 1993; Fulton et al., 1995) or specialized gel electrophoresis techniques which are used to monitor the mobility of products (Hain et al., 1997; Heuer et al., 1997). Clarke et al. (1993) used a combination of *Bgl*I, *Eco*RI, *Pst*I, and *Pvu*II to obtain restriction fragment length polymorphism (RFLP) patterns of purified rRNA extracted from members of the *Streptomyces albidoflavus* cluster [subgroups 1A and 1B of Williams et al. (1983a)] and were able to differentiate between phenotypically similar strains, although profiles varied considerably between *Streptomyces albidoflavus* species-groups. Ribosomal restriction analysis of 98 named streptomycete strains, including members of cluster-groups A (comprising clusters 1–41) and F (comprising clusters 55–67) of Williams et al. (1983a) and some other strains, was performed by Fulton et al. (1995) using *Mse*I fingerprints of rRNA operons (RiDiTS) and 11 pattern types with varying degrees of similarity to the Williams subclusters were highlighted. Cluster-groups A and F were differentiated, albeit at a low resolution (70% similarity), but individual clusters could not be distinguished.

Further methods used to compare genotypic variation amongst streptomycetes include denaturing gradient gel electrophoresis (DGGE; Muyzer et al., 1993) with or without DNA-binding agents (Hain et al., 1997). Anderson and Wellington (2001) recommended DGGE in combination with other techniques. This method uses variable 16S rRNA regions and enables the delimitation of genus and species-groups. Isolates ASB33, ASB37, and ASSF22 were allocated to *Streptomyces albidoflavus, Streptomyces griseoruber,* and *Streptomyces albidoflavus,* respectively, by using a combination of techniques, including numerical taxonomy, PFGE, and sequence comparisons (Huddleston et al., 1997; Huddleston et al., 1995).

Additional comments. Nowadays, often the first step in characterization of an isolate is the determination of the 16S rRNA gene sequence. Those isolates that may represent novel species are then further characterized in order to find additional markers which are different from those reported for already established taxa. In several cases, only a very restricted set of phenotypic differences are found, especially within the genus *Streptomyces*, and hence the classification of novel taxa is based largely on the 16S rRNA gene sequence differences. More and more, additional (housekeeping) gene sequence differences are also reported, which are sometimes regarded as sufficient for the delineation of novel species. There is clearly a current trend to delineate species more and more on the basis of the genotype.

However, recent analyses have shown that, although the 16S rRNA gene sequence has been widely accepted as the "backbone" of bacterial systematics as part of the often called "tree of life", in the light of genome data the concept of a single universal tree of life appears increasingly obsolete, especially given the impact of lateral or horizontal gene transfer events (see, e.g. Bapteste et al., 2009; Bapteste and Boucher, 2008; Boucher and Bapteste, 2009; Dagan et al., 2008; Fournier and Gogarten, 2010; Fournier et al., 2009; Kreimer et al., 2008; Wolf et al., 2002; and others for a more detailed discussion). It is methodologically difficult to infer horizontal or lateral gene transfer unless one has an *a priori* hypothesis of relationship that indicates that the presumptive transferred genes are not homologs. But there is no doubt that horizontal gene transfer (HGT) is widespread in the microbial world. HGT can result in either acquisition of new genetic material or homologous replacement of existing genes. The evolutionary significance of homologous recombination in a population can be quantified by examining the relative rates at which polymorphisms are introduced from recombination (rho) and mutation [theta(w)]. In the study of Doroghazi and Buckley (2010), multilocus sequence analysis (MLSA) was used to quantify both intraspecies and interspecies homologous recombination among streptomycetes and some other Gram-stain-positive bacteria. These authors found that intraspecies recombination in *Streptomyces flavogriseus* isolated from soils at five locations spanning 1000 km showed a >99.8% nucleotide identity across the loci examined. The authors found remarkable levels of gene exchange within *Streptomyces flavogriseus* and that the population was in linkage equilibrium (standardized index of association=0.0018), providing evidence for a freely recombining sexual population structure. As an even more interesting result, extensive interspecies homologous recombination was found among different *Streptomyces* species summing up to 40% of housekeeping-genes that had been acquired through HGT. This recombination rate found for these named species exceeded by far that observed within many species of bacteria. Hence, it was concluded that this pattern of gene exchange and recombination clearly shaped the evolution of streptomycetes and this has also a tremendous effect on classification based on sequence data of housekeeping genes.

Molecular data can provide an enormous amount of information. However, at this point, we are far from able to interpret these data, especially the information behind them, well enough to draw decisive conclusions. There are numerous open questions, e.g. which genes belong to the conserved genome core and are considered probably useful to define a taxon and which belong to accessory dispensible genetic elements? The "overall" impact of processes such as lateral gene transfer, gene duplication, recombination, and rearrangements of genes in the genome is not clear and may vary considerably in different lineages (see Bapteste et al., 2009; Dagan et al., 2008; and other publications). In addition, the presence of genes and gene clusters (whether expressed or "silent") can have a totally different biological meaning and the roles of structural elements (some of them phenotypically recognizable by the so-called "chemotaxonomic" methods) and biochemical pathways (also recognized by studying the phenotype at different levels) should be consistent with the underlying genetic data, which is essentially the aim of a "polyphasic taxonomic" study.

Despite the advantages of molecular methods (including the generation and analyses of whole genome sequence data), it is often impossible to deduce phenotypic properties from the presence or absence of genes and gene clusters, because genes do not exist for their own sake. This is especially true for seemingly simple, but nevertheless "complex" phenotypic properties, like temperature, NaCl, or pH tolerances, and some complex chemotaxonomic features (just to name a few), which may be affected by very different and complex regulatory biochemical networks and are based on the underlying genetic potential and expression network. It should be emphasized again, as also pointed out by Tindall et al. (2010), that experience gained over the past six decades has continued to demonstrate the value of comparing different datasets and also of basing the description and delineation of taxa on as wide a dataset as possible. Only a combination of data acquired from DNA-based methods (DNA–DNA hybridization, gene sequences, genomic fingerprints) and phenotyping (chemotaxonomic, physiological, and morphological traits) provides a sound basis for the taxonomy of the prokaryotes in general, including streptomycetes (Tindall et al., 2010).

Differentiation and characteristics of the species of the genus *Streptomyces*

It is difficult to identify *Streptomyces* species, partly due to the high number of species with validly published names (Table 272). Most of these species are named on the basis of a single strain description and, hence, at present, it is not possible to recommend a single method or even a set of methods for identification at the species level. Species allocations on the basis of the results of one or few methods should be regarded with care. The ICSP Subcommittee on the Systematics of *Streptomycetaceae* (Kämpfer and Labeda, 2003) has recommended that more genomic information should be evaluated before a species concept is formulated for the genus *Streptomyces*. It was agreed "that the proposal of new species should only be accepted on the basis of very careful studies done with sufficient practice and considering all other species".

Sequence analysis of the rRNA gene has revolutionized our insight into phylogenetic lineages of major taxonomic groups, but this method does not have sufficient resolving power to delimit *Streptomyces* species. It is clearly not possible to assign new strains to existing species solely on the basis of this molecule, though trees are a useful visual aid for placing members of putative novel species next to their nearest relatives. Even if results from the application of different treeing algorithms differ, the judgment of what is a novel species and which are its closest relatives will be decided on the basis of individual sequence homologies, and on complementary genetic (e.g. DNA–DNA hybridization) data.

Given the difficulties of interpreting the 16S rRNA, the descriptions of all species with validly published names are presented in the species descriptions in alphabetical order. The species descriptions include details of 16S rRNA similarities of related species based on pairwise comparisons. A grouping of the species was performed on the basis of a phylogenetic analysis using the 16S rRNA gene sequences. Groups which were relatively stable on the basis of different treeing methods are given in Figure 339. A table with all *Streptomyces* species, grouped according to the clusters, lists further information to all *Streptomyces* species, including type strain numbers, their inclusion in the numerical taxonomic studies of Williams et al. (1983a) and Kämpfer et al. (1991), 16S rRNA gene sequence accession numbers, DNA–DNA hybridization, and numerous other studies (Table 272). Additional genotypic and phenotypic data can be obtained from these studies.

Acknowledgements

I thank Dr Jean Euzéby for his advice on numerous nomenclatural issues and for permission to include some information given on his website (http://www.bacterio.cict.fr/) in some of the species descriptions. I am indebted to Dr Wolfgang Ludwig for preparing the phylogenetic tree (Figure 339) and his help in interpreting sequence data, and to Professor Michael Goodfellow for his excellent and very helpful comments to this chapter. Many thanks go to my coworkers Dr Nicole Lodders, Dr Kathrin Thummes, Kerstin Fallschissel, and Corina Lang for support in writing and proofreading this chapter.

List of species of the genus *Streptomyces*

1a. **Streptomyces albus subsp. albus** (Rossi Doria 1891) Waksman and Henrici 1943, 339[AL] ["*Streptotrix alba*" (*sic*) Rossi Doria 1891, 421]

al'bus. L. masc. adj. *albus* white.

Spore chains in Section *Spirales*. Spirals are most abundant on oatmeal agar and may be poorly developed on yeast-malt agar, salts-starch agar, and glycerol-asparagine agar. Mature spore chains generally have 10–50 spores per chain, but shorter chains may be common on some media. Spore surface is smooth.

Color of colony: aerial mass color in the White color series on oatmeal agar and salts-starch agar; White or Yellow color series on yeast-malt agar and glycerol-asparagine agar. Reverse side of colony with no distinctive pigments (colorless to pale yellow) on yeast-malt agar, oatmeal agar, salts-starch agar, and glycerol-asparagine agar.

Color in medium: melanoid pigments are not formed in peptone-yeast-iron agar, tyrosine agar, or tryptone-yeast broth. No pigment is found in the medium in yeast-malt agar, oatmeal agar, salts-starch agar, or glycerol-asparagine agar.

D-Glucose, D-xylose, D-mannitol, and D-fructose are utilized for growth. Reports vary on utilization of L-arabinose and raffinose. No growth or only traces of growth with isoinositol, rhamnose, and sucrose.

Type strain shows the highest sequence similarity to: *S. almquistii*, AB184258, 100%; *S. rangoonensis*, AB184295, 100%; *S. gibsonii*, AB184663, 100%; *S. flocculus*, AB184272, 99.8%.

Source: not known.

DNA G+C content (mol%): not known.

Type strain: AS 4.164, ATCC 25426, ATCC 3004, IMRU 3004, CBS 410.63, CBS 924.69, BCRC (formerly CCRC) 10802, CCUG 33990, CECT 3077, CIP 104432, DSM 40313, HUT 6613, IFM 1119, IFO (now NBRC) 13014, NBRC 3710, IMET 40241, JCM 4450, JCM 4177, KCTC 1082, NCIMB 9558, NRRL B-1811, NRRL B-2208, NRRL-ISP 5313, RIA 1206, VKM Ac-35.

Sequence accession no. (16S rRNA gene): AJ621602.

1b. **Streptomyces albus subsp. pathocidicus** Nagatsu, Anzai and Suzuki 1962, 105[AL]

pa.tho.ci'di.cus. Gr. n. *pathos* disease; L. suff. *-cida* (from L. v. *caedo*, to cut or kill) murderer, killer; L. masc. suff. *-icus* suffix used with the sense of belonging to; N.L. masc. adj. *pathocidicus* belonging to pathocide, but obviously referring to the antibiotic pathocidin.

Aerial mycelium has a closed spiral or an open loop on glucose-asparagine agar and starch agar. Conidia are elliptical, 1.0 × 1.3–1.4 × 1.8 µm in size, and show a chain-like growth. Usually grows at 25–30°C on various media. At first, it has no color, then it changes from white to pale yellow, with white aerial mycelium. Hydrolyzes starch, does not liquefy gelatin, and forms small amounts of nitrite.

Type strain shows no sequence similarity over 99%.

Source: not known.

DNA G+C content (mol%): not known.

Type strain: AS 4.1633, ATCC 14510, BCRC 12331, CIP 104431, DSM 40799, NBRC 13812, JCM 4166, KCTC 9671, VKM Ac-598.

Sequence accession no. (16S rRNA gene): AB184501.

2. **Streptomyces abikoensis** (Umezawa, Tazaki and Fukuyama 1951) Witt and Stackebrandt 1991, 456[VP] (Effective publication: Witt and Stackebrandt 1990, 370.) ("*Streptomyces abokobensis*" Umezawa, Tazaki and Fukuyama 1951, 333; *Streptoverticillium abikoense* Locci, Baldacci and Petrolini Baldan 1969, 59; "*Verticillomyces abikobensis*" Shinobu 1965, 96)

a.bi.ko.en'sis. N.L. masc. adj. *abikoensis* of or belonging to Abiko (named after Abiko, Japan).

Spore chains are straight. Reverse colors tend to appear darker. Aerial mycelium can be beige, pale pink to pinkish white. Soluble pigments are present on Bacto Emerson agar. Utilizes starch, casein, and, more readily, gelatin. Growth at 37°C is the same as that at 27°C; however, aerial mycelium is less abundant and lighter in color. Brown soluble pigments are produced at 37°C on potato-glucose agar (Baldacci et al., 1954). The type strain produces abikoviromycin and exhibits polyenic anti-fungal activity.

Type strain shows the highest sequence similarity to: *S. lilacinus*, AB184819, 99.7%; *S. hygroscopius* subsp. *angustmyceticus*, DQ442509, 99.7%%; *S. ehimensis*, AY999834, 99.6%; *S. sapporonensis*, AB184508, 99.5%; *S. hiroshimensis*, AB184789, 99.5%; *S. caeruleus*, EF178675, 99.5%; *S. luteireticuli* AB249969, 99.3%; *S. thioluteus*, AB184753, 99.3%; *S. varsoviensis*, DQ026653, 99.2%; *S. morookaense* AJ781349, 99.1%; *S. lavenduligriseus*, DQ442515, 99.1%; *S. cinnamoneus*, AB184850, 99.1%; *S. blastmyceticus*, AY999802, 99%; *S. pseudoechinosporeus*, AB184100, 99%; *S. olivoverticillatus*, AB184636, 99%.

Source: soil.

DNA G+C content (mol%): not known.

Type strain: AS 4.1162, ATCC 12766, CBS 487.62, BCRC 12461, DSM 40831, NBRC 13860, JCM 4002, KCTC 9662, KCTC 9741, NRRL B-1518, NRRL B-2113, PCM 2364, RIA 497.

Sequence accession no. (16S rRNA gene): AB184537.

Further comments: in violation of Rule 33c of the *Bacteriological Code* (1990 Revision), in Validation List no. 38, *Streptomyces abikoensis* is proposed as a *nomen revictum* (basonym: "*Streptomyces abikoensis*" Umezawa, Tazaki and Fukuyama (1951).

According to Hatano et al. (2003), *Streptomyces abikoensis* (Umezawa et al. 1951) Witt and Stackebrandt 1991 is an earlier heterotypic synonym of *Streptomyces ehimensis* corrig. (Shibata et al. 1954) Witt and Stackebrandt 1991, of *Streptomyces luteoverticillatus* (Shinobu 1956) Witt and Stackebrandt 1991, of *Streptomyces olivoreticuli* (Arai et al. 1957) Witt and Stackebrandt 1991, and of *Streptomyces parvisporogenes* (Locci et al. 1969) Witt and Stackebrandt 1991. Hatano et al. (2003) also propose that *Streptomyces abikoensis* (Umezawa et al. 1951) Witt and Stackebrandt 1991 be a heterotypic synonym of "*Streptomyces olivoreticuli* subsp. *cellulophilus*" (NBRC 15929), of "*Streptomyces paucisporogenes*" (NBRC 13070), of "*Streptomyces takataensis*" (NBRC 13470), of "*Streptoverticillium rubrireticuli*" (NBRC 13082), and of "*Streptoverticillium rubroverticillatum*" (NBRC 15818). In the paper by Hatano et al. (2003), "*Streptomyces olivoreticuli* subsp. *cellulophilus*" is not in quotes. However, this name has no standing in bacterial nomenclature.

3. **Streptomyces aburaviensis** Nishimura, Kimura, Tawara, Sasaki, Nakajima, Shimaoka, Okamoto, Shimohira and Isono 1957, 206[AL].

a.bu.ra.vi.en'sis. N.L. masc. adj. *aburaviensis* of or belonging to Aburabi, Shiga Prefecture, Japan, the source of the soil from which the organism was isolated.

Spore chains in Section *Rectiflexibiles.* Mature spore chains are moderately long with 10–50, or sometimes more than 50, spores per chain. This morphology is seen on yeast-malt agar, oatmeal agar, salts-starch agar, and glycerol-asparagine agar. Spore surface is smooth.

Color of colony: aerial mass color in the Gray color series on yeast-malt agar, oatmeal agar, salts-starch agar, and glycerol-asparagine agar. Reverse side of colony is yellowish brown on yeast-malt agar, oatmeal agar, salts-starch agar, and glycerol-asparagine agar (this pigment is not a pH indicator).

Color in medium: melanoid pigments are not formed in peptone-yeast-iron agar, tyrosine agar, or tryptone-yeast broth. No pigment other than trace of yellow is found in medium in yeast-malt agar, oatmeal agar, salts-starch agar, or glycerol-asparagine agar.

Growth on carbon utilization media is generally poor. D-Glucose, D-fructose, and D-xylose are utilized for growth. Reports vary from no growth to traces of growth on L-arabinose, sucrose, iso-inositol, D-mannitol, rhamnose, and raffinose.

Type strain shows the highest sequence similarity to: *S. herbaricolor,* AB184801, 99.6%; *S. indigoferus,* AB184214, 99.6%; *S. purpureus,* AJ781324, 99.5%; *S. chrysomallus* subsp. *fumigatus,* AB184645, 99.5%; *S. xanthocidicus,* AY999858, 99.4%; *S. psammoticus,* AY999862, 99.1%; *S. purpeofuscus,* AJ781364, 99.1%; *S. avellaneus,* AB184413, 99%. Type strain shows the highest sequence similarity to following *Kitasatospora* species: *Kitasatospora kifunensis,* AB022874, 99.3%; *Kitasatospora gansuensis,* AY442265, 99%.

Source: soil from Japan.

DNA G+C content (mol%): not known.

Type strain: AS 4.1469, ATCC 23869, CBS 280.60, CBS 608.68, BCRC 11617, CECT 3315, DSM 40033, IFM 1083, NBRC 12830, IMET 43081, JCM 4613, JCM 4170, KCTC 9663, LMG 19305, NRRL B-2218, NRRL-ISP 5033, RIA 1107, RIA 732, VKM Ac-1868.

Sequence accession no. (16S rRNA gene): AY999779.

4a. **Streptomyces achromogenes subsp. achromogenes** (*sic*) Okami and Umezawa *in* Umezawa, Takeuchi, Okami and Tazaki 1953, 268[AL]

a.chro.mo'ge.nes. Gr. pref. *a-* not; Gr. n. *chroma* color; N.L. suff. *-genes* (from Gr. v. *gennaô* to produce), producing; N.L. adj. *achromogenes* not producing color.

Spore chains in Section *Rectiflexibiles.* Mature spore chains generally comprise 10–50 spores per chain. This morphology is seen on yeast-malt agar, oatmeal agar, salts-starch agar, and glycerol-asparagine agar. One observer recorded some loops or hooks on salts-starch agar. Spore surface is smooth.

Color of colony: aerial mass color in the Gray color series on yeast-malt agar, oatmeal agar, and glycerol-asparagine agar. Reverse side of colony with no distinctive pigments on yeast-malt agar, oatmeal agar, salts-starch agar, or glycerol-asparagine agar. Substrate pigment is not a pH indicator.

Color in medium: melanoid pigments formed in peptone-yeast-iron agar, tyrosine agar, and other organic media. Pigments other than melanoids are not formed in yeast-malt agar, oatmeal agar, salt-starch agar, or glycerol-asparagine agar.

D-Glucose, L-arabinose, D-mannitol, and D-fructose are utilized for growth. No growth or only traces of growth on sucrose, D-xylose, and raffinose. Variable reports of slight growth with rhamnose and iso-inositol.

Type strain shows the highest sequence similarity to: *S. cellostaticus,* AB184192, 99%; *S. niveoruber,* DQ445796, 99%.

Source: not known.

DNA G+C content (mol%): not known.

Type strain: ATCC 12767, ATCC 19719, CBS 458.68, BCRC 11618, CECT 3074, DSM 40028, HAMBI 1002, IFM 1173, NBRC 12735, IMET 43080, JCM 4561, JCM 4121, KCTC 1740, NRRL B-2120, NRRL-ISP 5028, PCM 2365, RIA 1000, RIA 756, UNIQEM 115, VKM Ac-1258.

Sequence accession no. (16S rRNA gene): AB184109.

4b. **Streptomyces achromogenes subsp. rubradiris** Bhuyan, Owen and Dietz 1965, 95[AL]

rub.ra.di'ris. L. adj. *ruber -bra -brum* red; L. n. *iris* the rainbow; N.L. n. *rubradiris* (*sic*) reddish rainbow.

S. achromogenes and its variants are melanin-positive and have gray aerial growth containing chains of oval and oblong spores. These chains contain more than 10 spores born in sporophores characterized as open loops. The cultures grow well at temperatures ranging from 18 to 37°C, but do not grow at 55°C. The effect of several nitrogen sources on antibiotic production was investigated in a medium containing starch as the carbohydrate source. A medium with Soludri, cornsteep liquor, $NaNO_3$, and starch gave the highest yields. Good growth and the proper pH level for antibiotic production were obtained in all cases, except when Wilson peptone liquor or Pharmamedia were used.

The effect of several carbohydrate sources was studied in a medium containing cornsteep liquor, Soludri, and $NaNO_3$. Cerelose or starch at a level of 1% were the preferred carbon sources. Equivalent growth was obtained in all cases, except for very heavy growth in glycerol-containing medium. Fermentations incubated at 25°C gave higher titers that those at 28 or 32°C. Incubation temperature during the first 2 d was critical for antibiotic production. Thus, the fermentation transferred to 28°C after 2 d of growth at 25°C gave the same yields as the fermentation at 25°C (21 μg/ml). However, when the fermentation was transferred to 25°C after 2 d growth at 28°C, lower yields (4 μg/ml) were obtained. Maximal yields were obtained when the pH was maintained near 8 during the production phase. Thus, when the pH was maintained at 7.2 until the third day by raising the starch concentration in the medium to 4% or by intermittent addition of acid, only 3 μg/ml antibiotic was produced as compared with 21 μg/ml when 1% starch was used.

During fermentation obtained with a medium containing starch, cornsteep liquor, Soludri, and $NaNO_3$, most of the carbohydrate was used during the first 24 h, after which little sugar was utilized. Free ammoniacal nitrogen utilized during the first 24 h resulted in a decrease in pH. Release of NH_3 after this period raised the pH on the second day to above 8, where it remained during the rest of the fermentation. Rubradirin was produced mainly during the later phase of the fermentation to give a maximal titer of 21 µg/ml after 4 d.

Rubradirin was active *in vitro* mainly against Gram-stain-positive organisms. It was also active *in vitro* against two clinical strains of *Staphylococcus aureus* which were resistant to several antibiotics. It was inactive at 100 µg/ml against *Escherichia coli, Salmonella paratyphi, Proteus vulgaris, Salmonella typhi, Klebsiella pneumoniae,* and *Pseudomonas aeruginosa.* The antibiotic inhibited *Nocardia asteroides* at 1 µg/ml, but was inactive against a group of fungi at 1 mg/ml. Rubradirin was not cross-resistant with penicillin, streptomycin, erythromycin, chloramphenicol, or albamycin.

In vivo, mice infected with *Staphylococcus aureus* and *Streptococcus pyogenes* were protected by non-toxic doses of rubradirin. The antibiotic was active both by the oral and subcutaneous routes. It was inactive against *Pasteurella multocida in vivo* and *in vitro.* Although the antibiotic was active against *Mycobacterium tuberculosis* H 37-RV *in vitro,* it was inactive against it *in vivo.*

The type strain shows the highest sequence similarities to: *S. bangladeshensis,* AY750056, 99%; *S. glomeratus,* AB249917, 99%.

Source: not known.

DNA G+C content (mol%): not known.

Type strain: AS 4.1601, CBS 566.70, CECT 3075, DSM 40789, NBRC 14000, JCM 4955, KCTC 9742, NCIMB 9516, NRRL 3061.

Sequence accession no. (16S rRNA gene): AY999846.

5. **Streptomyces acidiscabies** Lambert and Loria 1989a, 395[VP]

a.ci.di.sca'bi.es. L. adj. *acidus* sour, acid; L. n. *scabies* scab, mange; N.L. n. *acidiscabies* acid scab, referring to the ability of the organism to cause acid scab of potatoes.

Spores are 0.4–0.5 × 0.6 or 0.9–1.1 µm, smooth, and white (reddish on certain high pH media) and are borne in mature flexuous chains containing 20 or more spores. A diffusible pigment is produced which is red above pH 8.3 and golden-yellow below pH 8.3. Melanin is not produced. L-Arabinose, D-fructose, D-glucose, D-mannitol, rhamnose, sucrose, and D-xylose are used as carbon sources, but not raffinose. *S. acidiscabies* differs from all other streptomycetes that cause typical raised or pitted symptoms and have white spores, red pigment, and acid tolerance. It differs in at least one major characteristic from all other strains described in the ISP having red pigments or spores. *Streptomyces acidiscabies* is placed in the genus *Streptomyces* as it possesses typical morphology and cell walls with the LL-A_2pm isomer. To date, *Streptomyces acidiscabies* has been isolated only from potatoes (*Solanum tuberosum*), and its pathogenicity towards other species has not been determined. The ability of this species to persist in soil is relatively poor and it is usually transmitted by infected seed tubers.

The type strain shows no sequence similarity over 99%.

Source: potatoes.

DNA G+C content (mol%): 71.0.

Type strain: RL-110, ATCC 49003, ICMP 12536, DSM 41668, JCM 7913, KCTC 9736, LMG 19856, NRRL B-16524.

Sequence accession no. (16S rRNA gene): D63865.

6. **Streptomyces acrimycini** (Preobrazhenskaya, Blinov and Ryabova *in* Gauze, Preobrazhenskaya, Kudrina, Blinov, Ryabova and Sveshnikova 1957) Pridham, Hesseltine and Benedict 1958, 65[AL] ("*Actinomyces acrimycini*" Preobrazhenskaya, Blinov and Ryabova *in* Gauze, Preobrazhenskaya, Kudrina, Blinov, Ryabova and Sveshnikova 1957, 140).

a.cri.my.ci'ni. L. adj. *acer-cris-cre* sharp, keen, pungent; N.L. suff. *-mycinum,* -mycin (antibiotics produced by *Streptomyces* strains); N.L. gen. n. *acrimycini* of the sharp antibiotic.

Spore chains in Section *Spirales,* with many loose spirals, but spore chains representative of Section *Retinaculiaperti* and some flexuous chains are also observed. Mature spore chains generally have 10–50 spores per chain; longer chains are sometimes observed. This morphology is seen on yeast-malt agar, oatmeal agar, salts-starch agar, and glycerol-asparagine agar. Spore surface is hairy.

Color of colony: aerial mass color in the Green color series on yeast-malt agar and salts-starch agar; Green or Gray series on oatmeal agar and glycerol-asparagine agar. Reverse side of colony with no distinctive pigments on yeast-malt agar, oatmeal agar, salts-starch agar, or glycerol-asparagine agar. Substrate pigment is not a pH indicator (one observer reports a slight change from green to light green with addition of 0.05 M NaOH).

Color in medium: melanoid pigments are not formed in peptone-yeast-iron agar or tyrosine agar. No pigment is found in yeast-malt agar, oatmeal agar, salts-starch agar, or glycerol-asparagine agar.

D-Glucose, L-arabinose, D-xylose, iso-inositol, D-mannitol, D-fructose, and rhamnose are utilized for growth. No growth or only traces of growth on sucrose and raffinose.

Type strain shows the highest sequence similarity to: *S. griseoplanus,* AY999894, 100%; *S. anulatus,* DQ026637, 100%; *S. praecox,* AB184293, 100%; *S. flavofuscus,* AB249935, 100%; *S. fimicarius,* AY999784, 100%; *S. badius,* AY999783, 99.9%; *S. lavendulae* subsp. *lavendulae,* AB184080, 99.9%; *S. griseinus,* AB184205, 99.9%; *S. rubiginosohelvolus,* AB184240, 99.9%; *S. sindenensis,* AB184759, 99.9%; *S. mediolani,* AB184674, 99.9%; *S. cavourensis* subsp. *washingtonensis,* DQ026671, 99.9%; *S. pluricolorescens,* DQ442540, 99.9%; *S. fulvorobeus,* AB184711, 99.8%; *S. microflavus,* DQ445795, 99.8%; *S. albovinaceus,* AB249958, 99.8%; *S. cyaneofuscatus,* AB184860, 99.8%; *S. lipmanii,* AB184148, 99.8%; *S. baarnensis,* EF178688, 99.8%; *S. cinereorectus,* AB184646, 99.8%; *S. globisporus* subsp. *globisporus,* EF178686, 99.8%; *S. alboviridis,* AB184256, 99.8%; *S. californicus,* AB184755, 99.7%; *S. parvus,* DQ442537, 99.7%; *S. argenteolus,* AB045872, 99.7%; *S. floridae,* AB184656, 99.6%; *S. halstedii,* EF178695, 99.6%; *S. griseus* subsp. *griseus,* AY207604, 99.6%; *S. griseolus,* AB184768, 99.6%; *S. luridiscabiei,* AF361784, 99.6%; *S. flavovirens,* DQ026635, 99.6%; *S. flavogriseus,* AJ494864, 99.6%; *S. pulveraceus,* AB184806, 99.4%; *S. yanii,* AB006159, 99.4%; *S. olivoviridis,* AB184227, 99.4%; *S. bacillaris,* AB184439, 99.4%; *S. mutomycini,* AB249951, 99.4%; *S. nitrosporeus,* EF178680, 99.4%; *S. finlayi,* AY999788, 99.4%; *S. clavifer,*

DQ026670, 99.4%; *S. atroolivaceus*, AJ781320, 99.4%; *S. albolongus*, AB184425, 99.2%; *S. sanglieri*, AB249945, 99.2%; *S. griseobrunneus*, AB249912, 99.2%; *S. celluloflavus*, AB184476, 99.2%; *S. atratus*, DQ026638, 99.1%; *S. cavourensis* subsp. *cavourensis*, DQ445791, 99.1%; *S. spiroverticillatus*, AB184814, 99.1%; *S. gelaticus*, DQ026636, 99.1%; *S. candidus*, DQ026663, 99%.

Source: not known.

DNA G+C content (mol%): not known.

Type strain: AS 4.1673, ATCC 19720, ATCC 19885, CBS 459.68, BCRC 12220, CECT 3122, DSM 40135, NBRC 12736, INA 7699, JCM 4339, KCTC 9679, NRRL B-2565, NRRL-ISP 5135, RIA 1001, UNIQEM 117, VKM Ac-769.

Sequence accession no. (16S rRNA gene): AY999889.

7. **Streptomyces aculeolatus** Shomura, Gomi, Ito, Yoshida, Tanaka, Amano, Watabe, Ohuchi, Itoh, Sezaki, Takebe and Uotani 1988, 136[VP] (Effective publication: Shomura, Gomi, Ito, Yoshida, Tanaka, Amano, Watabe, Ohuchi, Itoh, Sezaki, Takebe and Uotani 1987, 738.)

acu.le.o.la'tus. N.L. masc. adj. *aculeolatus* (probably used in the place of the Latin adjective *aculeatus*) somewhat spiny, referring to the spore surfaces.

Mature spore chains have 10 or more spores per chain. This morphology is observed in sucrose-nitrate agar, glycerol-asparagine agar, and inorganic salts-starch agar. Spores are ellipsoidal in shape, 0.8–1.2 × 1.0–1.6 µm in size. Surface irregularities on spores are intermediate between warts and spines. Sporangia, flagellated spores, and sclerotic granules are not observed. Aerial mass color is in the White, Yellow, or Red color series. Vegetative mycelium is well developed and branched. The hyphae do not fragment into coccoid or bacillary elements. Aerial mycelium is simply branched and terminates in open or closed coils. The reverse side of colonies varies from pale yellow to orange depending on the medium. This orange color is somewhat pH-sensitive, changing from orange to reddish with addition of 0.05 M NaOH and from orange to yellowish with addition of 0.05 M HCl. Light brownish orange, water-soluble pigment is occasionally produced. On ISP 9 medium, utilizes D-glucose, D-fructose, D-xylose, L-arabinose, D-mannitol, raffinose, and rhamnose, but not iso-inositol or sucrose. Grows within the temperature range 15–37°C, with optimum growth at 26–30°C. Positive for hydrolysis of starch and liquefaction of gelatin. Reduction of nitrate, peptonization and coagulation of milk, and formation of melanoid pigment are all negative. Tolerates 3% NaCl, but no growth occurs on more than 4% NaCl. LL-A_2pm is detected in whole-cell hydrolysates of the culture.

The type strain shows no sequence similarity over 99%.

Source: not known.

DNA G+C content (mol%): not known.

Type strain: SF2415, DSM 41644, NBRC 14824, JCM 6055, KCTC 9680.

Sequence accession no. (16S rRNA gene): AB184624.

8. **Streptomyces afghaniensis** Shimo, Shiga, Tomosugi and Kamoi 1959, 1[AL]

af.gha.ni.en'sis. N.L. masc. adj. *afghaniensis* of or belonging to Afghanistan, the source of the soil from which the organism was isolated.

Spore chains in Section *Spirales*, but short spore chains usually result in hooks, incomplete spirals, or spirals with only one or two turns. Short, straight, and flexuous chains of only a few spores are common, but hooks and loops of wide diameter as found in *Retinaculum-Apertum* type cultures are not found. Spore chains are short, often with only 3–10 spores per chain, but more than 10 spores may be found on some chains. This morphology is best developed on oatmeal agar and salts-starch agar. Spore surface is spiny.

Color of colony: aerial mass color in the Blue color series on oatmeal agar. Immature aerial mycelium or mycelium producing few spores is in the Yellow color series. Observers have reported both Blue and Yellow color series for yeast-malt agar and salts-starch agar. Mature spores are not usually found on glycerol-asparagine agar. Reverse side of colony is grayed yellow on oatmeal agar, modified to orange or reddish brown on yeast-malt agar, salts-starch agar, and glycerol-asparagine agar; this pigment is not a pH indicator.

Color in medium: melanoid pigments are formed in peptone-yeast-iron agar, tyrosine agar, and peptone-yeast broth. Orange to red or brown pigments are found in the medium in yeast-malt agar, oatmeal agar, salts-starch agar, and glycerol-asparagine agar; these pigments are not pH-sensitive when tested with 0.05 M HCl or NaOH.

D-Glucose, L-arabinose, sucrose, D-xylose, iso-inositol, D-mannitol, D-fructose, rhamnose, and raffinose are utilized for growth.

Type strain shows the highest sequence similarity to: *S. africanus*, AY208912, 99.8%; *S. levis*, AB184670, 99.2%; *S. brasiliensis*, AB249981, 99.1%; *S. tuirus*, AB184690, 99.1%; *S. albosporeus* subsp. *albosporeus*, AJ781327, 99%; *S. cellulosae*, DQ442495, 99%; *S. azureus*, EF178674, 99%; *S. capillispiralis*, AB184577, 99%; *S. gancidicus*, AB184660, 99%; *S. pseudogriseolus*, DQ442541, 99%; *S. roseoviolaceus*, AJ399484, 99%; *S. violaceus*, AB184315, 99%; *S. janthinus*, AB184851, 99%; *S. hawaiiensis*, AB184143, 99%.

Source: soil from Afghanistan.

DNA G+C content (mol%): not known.

Type strain: ATCC 23871, CBS 610.68, DSM 40228, NBRC 12831, JCM 4340, NRRL B-5621, NRRL-ISP 5228, RIA 1169.

Sequence accession no. (16S rRNA gene): AJ399483.

9. **Streptomyces africanus** Meyers, Goodwin, Bennett, Aken, Price and Van Rooyen 2004, 1534[VP]

af.ri.ca'nus. L. masc. adj. *africanus* of Africa.

Spirales-type spore chains with spiny spore sheaths are produced. Aerial mycelium is blue and substrate mycelium is yellow. The color of the substrate mycelium is not pH-sensitive. Verticils are not present. The mycelium does not fragment. No diffusible pigments are produced on glycerol-asparagine agar or on any other medium tested. Melanin pigment is not produced on peptone-yeast extract-iron agar or on tyrosine agar. The cell wall contains LL-A_2pm (cell-wall type I); there are no diagnostic sugars. Excellent growth occurs on inorganic salts-starch agar. Very good growth occurs on yeast extract-malt extract agar, oatmeal agar, and Czapek's solution agar. Growth is good on Bennett's agar. Growth on glycerol-asparagine agar is moderate. Grows in the presence of 2-phenylethanol (0.3%), 7% NaCl (but not 10%), carbenicillin (100 µg/ml) cefataxime (100 µg/ml),

D-cycloserine (50 μg/ml), nalidixic acid (25 μg/ml), oleandomycin (100 μg/ml), and penicillin G (10 IU/ml). Grows at pH 4.3 and 45°C, but not at 4°C or in the presence of sodium azide, capreomycin (20 μg/ml), cephaloridine (100 μg/ml), chloramphenicol (50 μg/ml), erythromycin (50 μg/ml), gentamicin (100 μg/ml), kanamycin (10 μg/ml), lincomycin (100 μg/ml), neomycin (50 μg/ml), phenol (0.1%), rifampin (50 μg/ml), spectinomycin (20 μg/ml), streptomycin (100 μg/ml), tobramycin (50 μg/ml), or vancomycin (50 μg/ml). The organism degrades adenine, esculin, arbutin, casein, DNA, gelatin, hypoxanthine, starch, Tween 80, and L-tyrosine, but not allantoin, cellulose, guanine, urea, xanthine, or xylan. Utilizes adonitol, (+)-L-arabinose, (+)-D-cellobiose, (−)-D-fructose, (+)-D-galactose, glycerol, *myo*-inositol, inulin, lactose, maltose, D-mannitol, (+)-D-mannose, (+)-D-melibiose, methyl α-D-glucoside, raffinose, (+)-L-rhamnose, (−)-D-ribose, salicin, sodium acetate, sodium butyrate, sodium citrate, sodium DL-malate, sodium malonate, sodium propionate, sodium pyruvate, sodium salicylate, sodium succinate, sucrose, trehalose and (+)-D-xylose as sole carbon sources, but not *meso*-erythritol, (+)-D-melezitose, sodium benzoate, sodium formate, sodium maleate, sodium oxalate, sodium (+)-L-tartrate, (−)-L-sorbose, or xylitol. Utilizes 4-amino-n-butyric acid, DL-α-amino-n-butyric acid (weak growth), L-arginine, L-cysteine, L-histidine, L-hydroxyproline (weak growth), L-methionine, DL-ornithine, L-phenylalanine, potassium nitrate, L-serine, L-threonine, and L-valine as sole nitrogen sources. Tests for nitrate reductase and H_2S production are positive. Pectin is hydrolyzed, but hippurate is not. Protease, lipase, and lecithinase activities are produced on egg-yolk agar (the proteolytic reaction is weak). Weak antibiotic activity is exhibited against *Enterococcus faecium*, but no antibiotic activity is observed against *Escherichia coli* ATCC 25922 or *Pseudomonas aeruginosa* ATCC 27853.

Type strain shows the highest sequence similarity to: *S. afghaniensis*, AJ399483, 99.8%; *S. janthinus*, AB184851, 99.2%; *S. roseoviolaceus*, AJ399484, 99.2%; *S. albosporeus* subsp. *albosporeus*, AJ781327, 99.2%; *S. levis*, AB184670, 99.2%; *S. violaceus*, AB184315, 99.2%; *S. brasiliensis*, AB249981, 99.2%; *S. azureus*, EF178674, 99.1%; *S. tuirus*, AB184690, 99.1%; *S. mutabilis*, EF178679, 99%; *S. hawaiiensis*, AB184143, 99%; *S. pseudogriseolus*, DQ442541, 99%; *S. capillispiralis*, AB184577, 99%; *S. gancidicus*, AB184660, 99%; *S. cellulosae*, DQ442495, 99%.

Type strain shows DNA–DNA similarity to: *S. afghaniensis* NRRL B-5621^T, 46.2 ± 0.9%; "*S. steffisburgensis*" NRRL ISP 5547, 38.4 ± 0.5%.

Source: not known.

DNA G+C content (mol%): 73.2.

Type strain: CPJVR-H, DSM 41829, JCM 13243, NBRC 101005.

Sequence accession no. (16S rRNA gene): AY208912.

10. **Streptomyces alanosinicus** Thiemann and Beretta 1966, 158[AL]

a.la.no.si'ni.cus. N.L. n. *alanosinum* alanosine (name of an antibiotic); L. masc. suff. *-icus* suffix used with the sense of belonging to; N.L. masc. adj. *alanosinicus* belonging to alanosine.

Spore chains in Section *Spirales*. Mature spore chains are generally long, often with more than 50 spores per chain. This morphology is seen on yeast-malt agar, oatmeal agar, and salts-starch agar, but sporulating aerial mycelium is thin or absent on glycerol-asparagine agar. Spore surface is spiny.

Color of colony: aerial mass color in the Red color series (4ec, grayish yellowish pink, or 3ca, pale orange yellow) on yeast-malt agar, oatmeal agar, and salts-starch agar; sporulating aerial mycelium is thin or absent on glycerol-asparagine agar. Reverse side of colony has no distinctive pigments (moderate reddish brown to strong brown on yeast-malt agar; pale grayish yellow to light yellowish brown or olive brown on oatmeal agar, salts-starch agar, and glycerol-asparagine agar).

Color in medium: melanoid pigments are formed in peptone-yeast-iron agar, tyrosine agar, and tryptone-yeast broth, but reaction may be delayed or weak in tyrosine agar. No pigment is found in medium in salts-starch agar or glycerol-asparagine agar. Some yellow pigment may or may not be seen in yeast-malt agar and oatmeal agar.

D-Glucose, L-arabinose, D-xylose, iso-inositol, D-mannitol, D-fructose, and raffinose are utilized for growth. Reports vary on utilization of sucrose; no growth or only traces of growth with rhamnose.

Type strain shows no sequence similarity over 99%.

Source: not known.

DNA G+C content (mol%): not known.

Type strain: V119, AS 4.1634, ATCC 15710, CBS 348.69, CBS 794.72, DSM 40606, HAMBI 983, NBRC 13493, JCM 4714, KCTC 9683, NRRL B-3627, NRRL-ISP 5606, RIA 1454, VKM Ac-1752.

Sequence accession no. (16S rRNA gene): AB184442.

11. **Streptomyces albaduncus** Tsukiura, Okanishi, Ohmori, Koshiyama, Miyaki, Kitazima and Kawaguchi 1964b, 41[AL]

al.ba.dun'cus. L. adj. *albus* white; L. adj. *uncus* hooked, crooked; N.L. masc. adj. *albaduncus* (*sic*) white, hooked, probably referring to color of aerial mycelium and nature of spore chains of the organism.

Spore chains in Section *Retinaculiaperti* to *Spirales*. Short spore chains are flexuous or form hooks, loops, and incomplete or imperfect spirals. True spirals are not reported. Mature spore chains generally contain more than 10 spores per chain but are shorter than those found in *Retinaculum-Apertum* type cultures. This morphology is seen on yeast-malt agar, oatmeal agar, salts-starch agar, and glycerol-asparagine agar. Spore surface is spiny with numerous long spines.

Color of colony: aerial mass color in the White or Yellow or Gray color series on yeast-malt agar, oatmeal agar, salts-starch agar, and glycerol-asparagine agar. The nearest matching color tab in the Yellow series is 2ba, pale yellow, and in the Gray color series is d, light gray, 2fe, medium gray, or 3fe, light brownish gray. Reverse side of colony has no distinctive pigments (grayish yellow to yellowish brown, olive brown, or strong brown) on yeast-malt agar, oatmeal agar, salts-starch agar, and glycerol-asparagine agar.

Color in medium: melanoid pigments are not formed in peptone-yeast-iron agar, tyrosine agar, or tryptone-yeast broth. Yellow pigment is found in the medium in yeast-malt agar, oatmeal agar, salts-starch agar, and glycerol-asparagine agar. This pigment is not pH-sensitive when tested with 0.05 M NaOH or HCl.

D-Glucose, L-arabinose, D-xylose, iso-inositol, D-mannitol, D-fructose, and rhamnose are utilized for growth. No growth or only traces of growth with sucrose or raffinose.

Type strain shows the highest sequence similarity to: *S. griseoloalbus*, AB184275, 100%; *S. matensis*, AB184221, 99.3%; *S. paradoxus*, AB184628, 99.2%; *S. pseudogriseolus*, DQ442541, 99.2%; *S. gancidicus*, AB184660, 99.2%; *S. capillispiralis*, AB184577, 99.1%; *S. heliomycini*, AB184712, 99%; *S. malachitofuscus*, AB184282, 99%; *S. griseoflavus*, AJ781322, 99%; *S. ambofaciens*, M27245, 99%; *S. cellulosae*, DQ442495, 99%; *S. lusitanus*, AB184424, 99%.

Source: not known.

DNA G+C content (mol%): not known.

Type strain: AS 4.158, ATCC 14698, CBS 698.72, CECT 3226, DSM 40478, NBRC 13397, JCM 4715, KCTC 1741, NRRL B-3605, NRRL-ISP 5478, RIA 1358, VKM Ac-1753, ISP 5478.

Sequence accession no. (16S rRNA gene): AY999757.

12. **Streptomyces albiaxialis** Kuznetsov, Zajtseva, Vakulenko and Flippova 1993, 398[VP] (Effective publication: Kuznetsov, Zajtseva, Vakulenko and Flippova 1992, 90.)

al.bi.a.xi.a'lis. L. adj. *albus -a -um* white; L. n. *axis* axle; N.L. adj. *axialis -is -e* axial; N.L. masc. adj. *albiaxialis* white axial.

Forms spiral sporophores with 3–8 extended spirals; sporophores are distributed on a long axis as pseudowhorls. Spores are oblong with a smooth envelope. On solid growth media, the culture forms a white aerial mycelium and dark cream-colored substrate mycelium. The population of *S. albiaxialis* consists of three spontaneous variants: basic, faded, and oligosporous. Basic variant: colonies are slightly bulging, aerial mycelium is white, substrate mycelium is dark cream-colored. Faded variant: colonies are flattened, aerial mycelium is whitish, substrate mycelium is pale yellow. Oligosporous variant: colonies are slightly prominent, whitish aerial mycelium develops only in the center of the colony, substrate mycelium is beige-colored. Soluble pigment is not formed by any variant. It is interesting to note that a population of newly isolated culture originally was absolutely homogeneous and the above-mentioned variants were identified in the population of *Streptomyces albiaxialis* only after cultivation for 7–8 months under laboratory conditions. Probably, this is due to the specific and stable inhabitance conditions of the organism in highly mineralized abyssal (2000 m) stratal water.

The organism liquefies gelatin, coagulates and peptonizes milk, and weakly hydrolyzes starch. It assimilates glucose, sucrose, rhamnose, arabinose, and xylose as a sole carbon source, but rather weakly utilizes raffinose and does not assimilate mannitol or inositol. It assimilates oil hydrocarbons, wax, and Vaseline oil. It is halotolerant and thermotolerant and is able to develop in media containing from 3 to 30% NaCl. The temperature optimum for development is between 28 and 33°C; maximum at 48–50°C. The cell wall contains LL-A_2pm and no differentiating sugars (Type I cell wall). The organism inhibits growth of Gram-stain-positive bacteria, acid-fast mycobacteria, and actinomycetes. It has a slight effect on the growth of some mycelial fungi (*Helminthosporium sativum*), but does not inhibit growth of Gram-stain-negative bacteria and yeasts. The culture is sensitive to monomycin, streptomycin, kanamycin, neomycin, sisomycin, gentamicin, lincomycin, erythromycin, oleandomycin, ristomycin, levomycin, and fusidin, but is resistant to penicillin, carbenicillin, methicillin, polymyxin, tetracycline, and oxytetracycline.

Type strain shows no sequence similarity over 99%.

Source: not known.

DNA G+C content (mol%): not known.

Type strain: DSM 41799, NBRC 101002, VKM A-691, NRRL B-24327.

Sequence accession no. (16S rRNA gene): AY999901.

13. **Streptomyces albidochromogenes** Preobrazhenskaya 1986, 573[VP] (Effective publication: Preobrazhenskaya *in* Gause, Preobrazhenskaya, Sveshnikova, Terekhova and Maximova 1983.)

al.bi.do.chro.mo'ge.nes. L. adj. *albidus -a -um*, white; Gr. n. *chroma* color; N.L. suff. *-genes* (from Gr. v. *gennaô* to produce), producing; N.L. part. adj. *albidochromogenes* producing white color.

Spore chains are spiral. Spore surface is smooth. On mineral agar 1, glycerol-nitrate agar: aerial mycelium is whitish, creamy, yellow, grayish yellow; substrate mycelium is gray brownish yellow, yellow-gray-brown; no diffusible pigment. On starch-ammonia agar, glycerol-asparagine agar, oatmeal agar: aerial mycelium is white-yellow, yellow; substrate mycelium is colorless or yellow; no diffusible pigment. On organic agar 2: substrate mycelium is absent or whitish, grayish yellow; substrate mycelium and diffusible pigment are gray-brown. Melanoid pigment is found. Grows on glucose, fructose, xylose, arabinose, mannitol, and sucrose; no growth on rhamnose, raffinose, or inositol. Antibiotic is not produced.

Type strain shows the highest sequence similarity to: *S. flavidovirens*, AB184270, 100%; *S. chryseus*, AY999787, 99.9%; *S. helvaticus*, AB184367, 99.9%; *S. enissocaesilis*, AB249930, 99.1%.

Source: not known.

DNA G+C content (mol%): not known.

Type strain: DSM 41800, INA 11792, JCM 13858, NBRC 101003.

Sequence accession no. (16S rRNA gene): AB249953.

Further comments: the name of the individual to be credited for *Streptomyces albidochromogenes* proposed in reference Gause et al. (1983) was supplied by T.P. Preobrazhenskaya in a personal communication to the Associate Editor, IJSB.

Culture was originally described as *Actinomyces albidus invertens* Kudrina 1957.

14. **Streptomyces albidoflavus** (Rossi Doria 1891) Waksman and Henrici *in* Breed, Murray and Hitchens 1948, 949[AL] ["*Streptotrix albidoflava*" (*sic*) Rossi Doria 1891, 407; "*Streptothrix albidoflava*" Rossi Doria 1891, 407; "*Actinomyces albidoflavus*" Gasperini 1894, 87; "*Cladothrix albido-flava*" (*sic*) Macé 1901, 1095]

al.bi.do.fla'vus. L. adj. *albidus* white; L. adj. *flavus* yellow; N.L. masc. adj. *albidoflavus* whitish yellow.

Spore chains in Section *Rectiflexibiles*. Two of three observers were unable to find sporulating aerial mycelium. Mature spore chains, when formed, are generally short with 3–10 spores per chain. Sporulating aerial mycelium is sometimes found on yeast-malt agar and glycerol-asparagine agar, but is not seen on oatmeal agar or salts-starch agar. Spore surface is smooth. Special morphological characteristics are as follows: substrate mycelium fragments, forming conidia-like or irregular spores; and unusually large spores may sometimes be formed by fragmentation of aerial hyphae.

Color of colony: aerial mass color in the White or Gray color series on yeast-malt agar. Sporulating aerial mycelium is not produced on other ISP media. Reverse side of colony with no distinctive pigments (light yellow to grayish yellow or orange yellow) on yeast-malt agar, oatmeal agar, salts-starch agar, and glycerol-asparagine agar.

Color in medium: melanoid pigments are not formed in peptone-yeast-iron agar, tyrosine agar, or tryptone-yeast broth. No pigment is found in the medium in yeast-malt agar, oatmeal agar, salts-starch agar, or glycerol-asparagine agar.

D-Glucose, L-arabinose, D-xylose, D-mannitol, and D-fructose are utilized for growth. No growth or only traces of growth with sucrose, iso-inositol, rhamnose, and raffinose.

Type strain shows the highest sequence similarity to: *S. limosus*, AB184147, 100%; *S. felleus*, AB184129, 100%; *S. daghestanicus*, DQ442497, 100%; *S. odorifer*, Z76682, 100%; *S. violascens*, AY999737, 100%; *S. hydrogenans*, AB184868, 100%; *S. griseus* subsp. *solvifaciens*, AB249915, 100%; *S. champavatii*, DQ026642, 100%; *S. canescens*, AB184117, 100%; *S. sampsonii*, D63871, 99.9%; *S. koyangensis*, AY079156, 99.7%.

Source: not known.

DNA G+C content (mol%): not known.

Type strain: CBS 416.34, ATCC 25422, CBS 416.34, CBS 920.69, BCRC 13699, CIP 105122, DSM 40455, ICMP 12537, NBRC 13010, JCM 4446, KCTC 9202, LMG 19300, NCIMB 10043, NRRL B-1271, NRRL B-2663, NRRL-ISP 5455, RIA 1202, VKM Ac-746.

Sequence accession no. (16S rRNA gene): AB184255.

15. **Streptomyces albiflaviniger** Goodfellow, Kumar, Labeda and Sembiring 2008, 1[VP] (Effective publication: Goodfellow, Kumar, Labeda and Sembiring 2007, 197.)

al.bi.fla.vi.ni'ger. L. adj. *albus* white; L. adj. *flavus* yellow; L. adj. *niger* black; N.L. masc. adj. *albiflaviniger* white, yellow, and black colors.

Spore chains in Section *Spirales*; spore surface is rugose. On oatmeal agar, the aerial spore mass color is white, becoming black and moist when mature; the reverse side of colonial growth is yellow. Brown, orange, and yellow diffusible pigments are formed, but not melanin pigments.

Type strain shows the highest sequence similarity to: *S. violaceusniger*, AJ391823, 99.7%; *S. yogyakartensis*, AB249942, 99.7%; *S. demainii*, DQ334782, 99.3%; *S. endus*, AY999911, 99.3%; *S. sporocinereus*, AB249933, 99.3%; *S. hygroscopius* subsp. *hygroscopius*, AB184428, 99.3%. Type strain shows DNA–DNA similarity to: *S. geldanamycinus* NRRL 3602^T, 99.1%; *S. griseiniger* NRRL B1865^T, 99.1%.

Source: not known.

DNA G+C content (mol%): 70.5.

Type strain: DSM 41598, NRRL B-1356.

Sequence accession no. (16S rRNA gene): AJ391812.

16. **Streptomyces albireticuli** (Nakazawa 1955) Witt and Stackebrandt 1991, 456[VP] (Effective publication: Witt and Stackebrandt 1990, 370.) ("*Streptomyces albireticuli*" Nakazawa 1955, 248; "*Verticillomyces albireticuli*" Shinobu 1965; *Streptoverticillium albireticuli* Locci, Baldacci and Petrolini Baldan 1969, 59).

al.bi.re.ti'cu.li. L. adj. *albus* white; L. n. *reticulum* a small net; N.L. gen. n. *albireticuli* of a small white net.

Spore chains in Section *Verticillati*. Both monoverticillate and umbellate moverticillate (biverticillate) sporophores are found. Mature spore chains generally have 10–50 spores per chain on suitable media. This morphology is seen on yeast-malt agar, oatmeal agar, and salts-starch agar after 2–3 weeks; poor growth of aerial mycelium is seen on glycerol-asparagine agar and variable aerial growth is observed on oatmeal agar and salts-starch agar. Spore surface is smooth.

Color of colony: aerial mass color in the Yellow or White color series on yeast-malt agar, oatmeal agar, and salts-starch agar. Reverse side of colony with no distinctive pigments on yeast-malt agar, oatmeal agar, salts-starch agar, and glycerol-asparagine agar; substrate pigment is not a pH indicator.

Color in medium: melanoid pigments are formed in peptone-yeast-iron agar and tyrosine agar. Pigments other than melanoids not formed in yeast-malt agar, oatmeal agar, salts-starch agar, or glycerol-asparagine agar.

D-Glucose and iso-inositol are utilized for growth. No growth or only traces of growth on L-arabinose, sucrose, D-xylose, D-mannitol, rhamnose, and raffinose. Variable reports on growth with D-fructose.

Type strain shows the highest sequence similarity to: *S. eurocidicus*, AY999790, 99.8%; *S. biverticillatus*, AJ781381, 99.5%; *S. werraensis*, DQ442558, 99.5%; *S. blastmyceticus*, AY999802, 99.3%; *S. stramineus*, AB184720, 99%; *S. netropsis*, AB184848, 99%; *S. hiroshimensis*, AB184789, 99%.

Source: not known.

DNA G+C content (mol%): not known.

Type strain: AS 4.1649, ATCC 19721, CBS 460.68, BCRC 12427, CECT 3253, DSM 40051, HUT 6040, IFM 1068, NBRC 12737, NBRC 3400, JCM 4562, JCM 4116, KCTC 9685, NCIMB 9600, NRRL B-1670, NRRL B-5493, NRRL-ISP 5051, RIA 1002, UNIQEM 209.

Sequence accession no. (16S rRNA gene): AB184881.

Further comments: in violation of Rule 33c, in Validation List no. 38, *Streptomyces albireticuli* is proposed as a *nomen revictum* (basonym: "*Streptomyces albireticuli*" Nakazawa 1955).

According to Hatano et al. (2003), *Streptomyces albireticuli* (Nakazawa 1955) Witt and Stackebrandt 1991 is a later heterotypic synonym of *Streptomyces eurocidicus* (Okami et al. 1954) Witt and Stackebrandt 1991.

17. **Streptomyces albofaciens** Thirumalachar and Bhatt 1960, 63[AL]

al.bo.fa'ci.ens. L. adj. *albus* white; L. v. *facio* make; N.L. part adj. *albofaciens* making white.

Spore chain morphology in Section *Spirales*. Open irregular spirals sometimes appear to arise from an axial hypha, but true whorls typical of verticillate cultures are not formed. Mature spore chains generally have 10–50, or sometimes more than 50, spores per chain. This morphology is seen on yeast-malt agar, oatmeal agar, salts-starch agar, and glycerol-asparagine agar. Spore surface is smooth.

Color of colony: aerial mass color in the White color series on yeast-malt agar and glycerol-asparagine agar; White or Gray color series on oatmeal agar and inorganic salts-starch agar. Reverse side of colony with no distinctive pigments on yeast-malt agar, oatmeal agar, salts-starch agar, and glycerol-asparagine agar.

Color in medium: melanoid pigments are not formed in peptone-yeast-iron agar, tyrosine agar, or tryptone-yeast broth. No pigment or only trace of yellow pigment found in medium in yeast-malt agar, oatmeal agar, salts-starch agar, and glycerol-asparagine agar.

D-Glucose, L-arabinose, iso-inositol, D-mannitol, D-fructose, and raffinose are utilized for growth. No growth or only traces of growth on rhamnose. Utilization of sucrose and D-xylose is doubtful.

Type strain shows the highest sequence similarity to: *S. rimosus* subsp. *paromomycinus*, AJ621610, 99.7%; *S. chrestomyceticus*, AJ621609, 99.7%; *S. erumpens*, AJ621603, 99.1%.

Source: not known.

DNA G+C content (mol%): not known.

Type strain: AS 4.1655, ATCC 23873, ATCC 25184, CBS 612.68, BCRC 12072, CIP 104425, DSM 40268, NBRC 12833, IMET 43518, JCM 4342, KCTC 9686, NCIMB 10975, NRRL B-12172, NRRL-ISP 5268, RIA 1189, VKM Ac-724.

Sequence accession no. (16S rRNA gene): AB045880.

18. **Streptomyces alboflavus** (Waksman and Curtis 1916) Waksman and Henrici *in* Breed, Murray and Hitchens 1948, 954[AL] ("*Actinomyces alboflavus*" Waksman and Curtis 1916, 120)

al.bo.fla'vus. L. adj. *albus* white; L. adj. *flavus* yellow; N.L. masc. adj. *alboflavus* whitish yellow.

Typical aerial mycelium is not formed on yeast-malt agar, oatmeal agar, salts-starch agar, or glycerol-asparagine agar. Spore chain morphology, spore surface, and aerial mass color of colony cannot be observed on ISP media. Loss of ability to produce sporulating aerial mycelium was noted in an early description of this culture (Waksman 1919, 90). Early descriptions of the culture describe white or yellowish white aerial mycelium on synthetic agar or Czapek's agar only.

Special morphological characteristics: coremia formation on salts-starch agar and glycerol-asparagine agar was recorded by two observers. This same phenomenon was recorded in the original description on Czapek's agar (Waksman 1916, 120): "...aerial mycelium was found to have a tendency to produce...a mass of hyphae massed together into a rope, and from this rope fine filaments coming out in the shape of side branches. The structure looks like the root of a tree and fine rootlets coming out on the side". One ISP observer found straight spore chains within the coremia on glycerol-asparagine agar. Reverse side of colony with no distinctive pigment (grayed yellow or grayed greenish yellow) on yeast-malt agar, oatmeal agar, salts-starch agar, and glycerol-asparagine agar.

Color in medium: melanoid pigments form weakly in peptone-yeast-iron agar and tryptone-yeast broth, but not on tyrosine agar. Pigments other than melanoids are not formed in yeast-malt agar, oatmeal agar, salts-starch agar, or glycerol-asparagine agar.

D-Glucose, L-arabinose, D-xylose, iso-inositol, D-mannitol, D-fructose, and raffinose are utilized for growth. No growth or only traces of growth on rhamnose. Utilization of sucrose is doubtful.

Type strain shows the highest sequence similarity to: *S. fulvissimus*, AB184434, 99.3%; *S. flavofungini*, AB184359, 99%; *S. variegatus*, AJ781371, 99%.

Source: not known.

DNA G+C content (mol%): not known.

Type strain: IMRU 3008, AS 4.1461, ATCC 12626, ATCC 23874, CBS 613.68, BCRC 13664, CIP 104427, DSM 40045, NBRC 13196, IMET 42936, JCM 4615, KCTC 9674, NRRL B-1273, NRRL B-2373, NRRL-ISP 5045, RIA 1112, VKM Ac-972.

Sequence accession no. (16S rRNA gene): EF178699.

19. **Streptomyces albogriseolus** Benedict, Shotwell, Pridham, Lindenfelser and Haynes 1954, 653[AL]

al.bo.gri.se.o'lus. L. adj. *albus* white; N.L. adj. *griseus* gray; L. masc. suff. *-olus* diminutive ending; N.L. dim. masc. adj. *albogriseolus* white grayish.

Spore chains in Section *Spirales*. Spirals are open. Flexuous or *Retinaculum-Apertum* type spore chains are also common. Mature spore chains generally have 10–50 spores per chain. This morphology is seen on yeast-malt agar, oatmeal agar, and salts-starch agar, but not on glycerol-asparagine agar. Spore surface is warty. Warts are not prominent or regular and some smooth spores may be found.

Color of colony: aerial mass color in the Gray color series on yeast-malt agar, oatmeal agar, and salts-starch agar. Reverse side of colony with no distinctive pigments (colorless or yellowish gray on oatmeal agar, salts-starch agar, and glycerol-asparagine agar; grayed yellowish brown or olive brown on yeast-malt agar).

Color in medium: melanoid pigments are not formed in peptone-yeast-iron agar, tyrosine agar, or tryptone-yeast broth. No pigment is found in medium in yeast-malt agar, oatmeal agar, salts-starch agar, or glycerol-asparagine agar.

D-Glucose, sucrose, D-xylose, iso-inositol, D-mannitol, D-fructose, and rhamnose are utilized for growth, but growth on sucrose is less abundant than on other carbon sources. No growth or only traces of growth on raffinose.

Type strain shows the highest sequence similarity to: *S. viridodiastaticus*, AY999852, 99.8%; *S. coeruleorubidus*, AY999719, 99.4%; *S. bellus*, AB184849, 99.2%; *S. coerulescens*, AY999720, 99.2%; *S. griseorubens*, AB184139, 99.1%; *S. longispororuber*, AB184440, 99.1%; *S. atrovirens*, DQ026672, 99.1%; *S. griseoincarnatus*, AJ781328, 99%; *S. labedae*, AB184704, 99%; *S. ambofaciens*, M27245, 99%; *S. erythrogriseus*, AJ781328, 99%; *S. iakyrus*, AB184877, 99%; *S. lusitanus*, AB184424, 99%.

Source: not known.

DNA G+C content (mol%): not known.

Type strain: ATCC 23875, CBS 614.68, BCRC 12230, CIP 104424, CIP 104428, DSM 40003, HUT 6045, NBRC 12834, NBRC 3413, NBRC 3709, JCM 4616, JCM 4004, KCTC 9675, NCIMB 9604, NRRL B-1305, NRRL-ISP 5003, RIA 1101, VKM Ac-1200.

Sequence accession no. (16S rRNA gene): AJ494865.

Color of colony: aerial mass color in the White color series on yeast-malt agar, oatmeal agar, salts-starch agar, and glycerol-asparagine agar. Reverse side of colony with no distinctive pigments on yeast-malt agar, oatmeal agar, salts-starch agar, and glycerol-asparagine agar; substrate pigment is not a pH indicator.

Color in medium: melanoid pigments not formed in peptone-yeast-iron agar and tyrosine agar. No pigment found in medium in yeast-malt agar, oatmeal agar, salts-starch agar, or glycerol-asparagine agar.

D-Glucose, L-arabinose, D-xylose, D-mannitol, D-fructose, and rhamnose are utilized for growth. No growth or only traces of growth on sucrose, iso-inositol, and raffinose.

Type strain shows the highest sequence similarity to: *S. sindenensis*, AB184759, 100%; *S. pluricolorescens*, DQ442540, 100%; *S. mediolani*, AB184674, 100%; *S. badius*, AY999783, 100%; *S. griseinus*, AB184205, 100%; *S. rubiginosohelvolus*, AB184240, 100%; *S. fimicarius*, AY999784, 99.9%; *S. flavofuscus*, AB249935, 99.9%; *S. globisporus* subsp. *globisporus*, EF178686, 99.9%; *S. griseoplanus*, AY999894, 99.9%; *S. praecox*, AB184293, 99.9%; *S. anulatus*, DQ026637, 99.9%; *S. parvus*, DQ442537, 99.8%; *S. acrimycini*, AY999889, 99.8%; *S. californicus*, AB184755, 99.8%; *S. cavourensis* subsp. *washingtonensis*, DQ026671, 99.8%; *S. lavendulae* subsp. *lavendulae*, AB184080, 99.8%; *S. cinereorectus*, AB184646, 99.7%; *S. fulvorobeus*, AB184711, 99.7%; *S. microflavus*, DQ445795, 99.7%; *S. cyaneofuscatus*, AB184860, 99.7%; *S. luridiscabiei*, AF361784, 99.7%; *S. baarnensis*, EF178688, 99.7%; *S. lipmanii*, AB184148, 99.7%; *S. alboviridis*, AB184256, 99.7%; *S. argenteolus*, AB045872, 99.7%; *S. floridae*, AB184656, 99.7%; *S. griseus* subsp. *griseus*, AY207604, 99.6%; *S. flavogriseus*, AJ494864, 99.6%; *S. flavovirens*, DQ026635, 99.6%; *S. griseolus*, AB184768, 99.6%; *S. nitrosporeus*, EF178680, 99.5%; *S. halstedii*, EF178695, 99.5%; *S. pulveraceus*, AB184806, 99.5%; *S. bacillaris*, AB184439, 99.5%; *S. atroolivaceus*, AJ781320, 99.4%; *S. olivoviridis*, AB184227, 99.4%; *S. finlayi*, AY999788, 99.4%; *S. atratus*, DQ026638, 99.3%; *S. griseobrunneus*, AB249912, 99.3%; *S. celluloflavus*, AB184476, 99.3%; *S. yanii*, AB006159, 99.3%; *S. gelaticus*, DQ026636, 99.3%; *S. sanglieri*, AB249945, 99.3%; *S. albolongus*, AB184425, 99.3%; *S. clavifer*, DQ026670, 99.3%; *S. cavourensis* subsp. *cavourensis*, DQ445791, 99.2%; *S. spiroverticillatus*, AB184814, 99.1%; *S. mutomycini*, AB249951, 99.1%; *S. candidus*, DQ026663, 99.1%; *S. cremeus*, AB184124, 99%.

Source: not known.

DNA G+C content (mol%): not known.

Type strain: AS 4.1631, ATCC 15823, ATCC 19723, ATCC 23613, CBS 256.66, CBS 462.68, CCM 3005, BCRC 13757, DSM 40136, NBRC 12739, INA 273/53, JCM 4343, NCIMB 13010, NRRL B-2566, NRRL-ISP 5136, RIA 1004, UNIQEM 119, VKM Ac-572.

Sequence accession no. (16S rRNA gene): AB249958.

26. **Streptomyces alboviridis** (Duché 1934) Pridham, Hesseltine and Benedict 1958, 74[AL] ("*Actinomyces alboviridis*" Duché 1934, 317)

al.bo.vi'ri.dis. L. adj. *albus* white; L. adj. *viridis* green; N.L. masc. adj. *alboviridis* whitish green.

Spore chains in Section *Rectiflexibiles*. Mature spore chains are generally long with 10 to 50 or more spores per chain on yeast-malt agar, oatmeal agar, salts-starch agar, and glycerol-asparagine agar. Spore surface is smooth.

Color of colony: aerial mass color in the Yellow color series on yeast-malt agar and salts-starch agar; Yellow or white color series on oatmeal agar and glycerol-asparagine agar. Reverse side of colony with no distinctive pigments (pale yellow or grayish yellow) on yeast-malt agar, oatmeal agar, salts-starch agar, and glycerol-asparagine agar.

Color in medium: melanoid pigments are not formed in peptone-yeast-iron agar, tyrosine agar, or tryptone-yeast broth. No pigment is found in the medium in yeast-malt agar, oatmeal agar, salts-starch agar, or glycerol-asparagine agar.

D-Glucose, D-xylose, D-mannitol, D-fructose, and rhamnose are utilized for growth. Reports vary on utilization of L-arabinose. No growth or only traces of growth with sucrose, iso-inositol, and raffinose.

Type strain shows the highest sequence similarity to: *S. cavourensis* subsp. *washingtonensis*, DQ026671, 100%; *S. fulvorobeus*, AB184711, 100%; *S. microflavus*, DQ445795, 100%; *S. lipmanii*, AB184148, 100%; *S. lavendulae* subsp. *lavendulae*, AB184080, 100%; *S. luridiscabiei*, AF361784, 99.9%; *S. flavofuscus*, AB249935, 99.9%; *S. cinereorectus*, AB184646, 99.9%; *S. fimicarius*, AY999784, 99.9%; *S. griseoplanus*, AY999894, 99.9%; *S. cyaneofuscatus*, AB184860, 99.9%; *S. floridae*, AB184656, 99.9%; *S. praecox*, AB184293, 99.9%; *S. anulatus*, DQ026637, 99.9%; *S. pluricolorescens*, DQ442540, 99.8%; *S. acrimycini*, AY999889, 99.8%; *S. argenteolus*, AB045872, 99.8%; *S. griseinus*, AB184205, 99.8%; *S. badius*, AY999783, 99.8%; *S. rubiginosohelvolus*, AB184240, 99.8%; *S. sindenensis*, AB184759, 99.8%; *S. mediolani*, AB184674, 99.8%; *S. californicus*, AB184755, 99.8%; *S. parvus*, DQ442537, 99.7%; *S. halstedii*, EF178695, 99.7%; *S. baarnensis*, EF178688, 99.7%; *S. griseolus*, AB184768, 99.7%; *S. griseus* subsp. *griseus*, AY207604, 99.7%; *S. globisporus* subsp. *globisporus*, EF178686, 99.7%; *S. albovinaceus*, AB249958, 99.7%; *S. flavovirens*, DQ026635, 99.6%; *S. finlayi*, AY999788, 99.5%; *S. flavogriseus*, AJ494864, 99.5%; *S. yanii*, AB006159, 99.4%; *S. atroolivaceus*, AJ781320, 99.4%; *S. olivoviridis*, AB184227, 99.4%; *S. bacillaris*, AB184439, 99.4%; *S. pulveraceus*, AB184806, 99.4%; *S. celluloflavus*, AB184476, 99.3%; *S. griseobrunneus*, AB249912, 99.3%; *S. albolongus*, AB184425, 99.3%; *S. clavifer*, DQ026670, 99.3%; *S. nitrosporeus*, EF178680, 99.3%; *S. spiroverticillatus*, AB184814, 99.2%; *S. sanglieri*, AB249945, 99.2%; *S. cavourensis* subsp. *cavourensis*, DQ445791, 99.2%; *S. atratus*, DQ026638, 99.1%; *S. mutomycini*, AB249951, 99.1%; *S. candidus*, DQ026663, 99.1%; *S. gelaticus*, DQ026636, 99.1%; *S. cremeus*, AB184124, 99%.

Source: not known.

DNA G+C content (mol%): not known.

Type strain: AS 4.1627, ATCC 25425, CBS 923.69, BCRC 12054, DSM 40326, NBRC 13013, JCM 4449, KCTC 9667, NRRL B-3633, NRRL-ISP 5326, RIA 1205, VKM Ac-736.

Sequence accession no. (16S rRNA gene): AB184256.

27. **Streptomyces albulus** Routien *in* Pridham and Lyons 1969, 194[AL]

al.bu'lus. L. masc. adj. *albulus* whitish.

Spore chains in Section *Spirales*. Mature spore chains are moderately long with 10 to 50 or more spores per chain.

This morphology is seen on yeast-malt agar, oatmeal agar, salts-starch agar, and glycerol-asparagine agar. Spore surface is spiny.

Color of colony: aerial mass color in the Gray color series (5fe, light grayish reddish brown; 4li, brownish gray; 4ig, light grayish brown) on yeast-malt agar and oatmeal agar; Gray or Red color series on salts-starch agar and glycerol-asparagine agar (nearest tab in the Red color series is 5dc, grayish yellowish pink). Reverse side of colony with no distinctive pigments (pale or grayish yellow) on yeast-malt agar, oatmeal agar, salts-starch agar, and glycerol-asparagine agar.

Color in medium: melanoid pigments are not formed in peptone-yeast-iron agar, tyrosine agar, and tryptone-yeast broth. No pigment is found in medium in yeast-malt agar, oatmeal agar, salts-starch agar, or glycerol-asparagine agar.

D-Glucose, iso-inositol, D-mannitol, and D-fructose are utilized for growth. No growth or only traces of growth with L-arabinose, D-xylose, rhamnose, sucrose, and raffinose.

Type strain shows the highest sequence similarity to: *S. noursei*, AB184678, 100%; *S. yunnanensis*, AF346818, 99.6%.

Source: not known.

DNA G+C content (mol%): not known.

Type strain: AS 4.1585, ATCC 12757, CBS 711.72, BCRC 11819, DSM 40492, NBRC 13410, JCM 4718, KCTC 9668, NRRL B-5386, NRRL-ISP 5492, RIA 1371, IMC S-0802.

Sequence accession no. (16S rRNA gene): AB024440.

28. **Streptomyces almquistii** (Duché 1934) Pridham, Hesseltine and Benedict 1958, 74[AL] ["*Actinomyces almquisti*" (*sic*) Duché 1934, 278]

alm.quis'ti.i. N.L. gen. masc. n. *almquistii* of Ernst Bernhard Almquist (1852–1946), named for an early investigator of *Actinomycetales*.

Spore chains in Section *Spirales*; many straight to flexuous hyphae or immature spore chains may also be present. Mature spore chains generally have 10 to 50 or more spores per chain. This morphology is seen on yeast-malt agar, oatmeal agar, salts-starch agar, and glycerol-asparagine agar. Spore surface is smooth.

Color of colony: aerial mass color in the White color series on yeast-malt agar, oatmeal agar, salts-starch agar, and glycerol-asparagine agar. Reverse side of colony with no distinctive pigments (colorless to pale grayish yellow) on yeast-malt agar, oatmeal agar, salts-starch agar, and glycerol-asparagine agar.

Color in medium: melanoid pigments are not formed in peptone-yeast-iron agar, tyrosine agar, or tryptone-yeast broth. No pigment is found in the medium in yeast-malt agar, oatmeal agar, salts-starch agar, or glycerol-asparagine agar.

D-Glucose, D-xylose, D-mannitol, and D-fructose are utilized for growth. Reports vary on utilization of L-arabinose and iso-inositol. No growth or only trace of growth with rhamnose, sucrose, and raffinose.

Type strain shows the highest sequence similarity to: *S. rangoonensis*, AB184295, 100%; *S. gibsonii*, AB184663, 100%; *S. albus* subsp. *albus*, AJ621602, 100%; *S. flocculus*, AB184272, 99.9%.

Source: not known.

DNA G+C content (mol%): not known.

Type strain: AS 4.1685, ATCC 25427, ATCC 618, CBS 925.69, BCRC 12098, DSM 40447, HAMBI 50, HUT 6614, NBRC 13015, IMET 43380, JCM 4451, KCTC 9672, NRRL B-1685, NRRL-ISP 5447, RIA 1207.

Sequence accession no. (16S rRNA gene): AB184258.

29. **Streptomyces althioticus** Yamaguchi, Nakayama, Takeda, Tawara, Maeda, Takeuchi and Umezawa 1957, 196[AL]

al.thi.o'ti.cus. N.L. n. *althiomycinum* althiomycin, name of a sulfur-containing antibiotic; L. masc. suff. *-ticus* suffix used with the sense of belonging to; N.L. masc. adj. *althioticus* belonging to althiomycin.

Spore chains in Section *Spirales*, but flexuous sporophores are also common, especially on yeast-malt agar. Mature spore chains generally have 10–50 spores per chain. This morphology is observed on yeast-malt agar, oatmeal agar, salts-starch agar, and glycerol-asparagine agar. Spore surface is spiny, with only minor surface irregularities suggestive of spines or warts.

Color of colony: aerial mass color in the Gray color series on yeast-malt agar, oatmeal agar, salts-starch agar, and glycerol-asparagine agar. Reverse side of colony is grayed yellow, modified to blue-violet or red on yeast-malt agar, oatmeal agar, salts-starch agar, and glycerol-asparagine agar. Reverse pigment is pH indicator; changes from red or reddish brown to blue or blue-violet by addition of 0.05 M NaOH.

Color in medium: melanoid pigments are not formed or occur only in trace amounts in peptone-yeast-iron agar and tyrosine agar; blue-violet or red pigment (depending upon pH) occurs in yeast-malt agar, oatmeal agar, salts-starch agar, and glycerol-asparagine agar. This pigment is pH-sensitive, becoming blue or blue-violet when tested with 0.05 M NaOH and red when tested with 0.05 M HCl.

D-Glucose, L-arabinose, D-xylose, iso-inositol, D-mannitol, D-fructose, and rhamnose are utilized for growth. No growth or only traces of growth on raffinose. Variable reports on growth with sucrose.

Type strain shows the highest sequence similarity to: *S. matensis*, AB184221, 99.8%; *S. griseorubens*, AB184139, 99.4%; *S. labedae*, AB184704, 99.3%; *S. erythrogriseus*, AJ781328, 99.2%; *S. griseoflavus*, AJ781322, 99.2%; *S. variabilis*, DQ442551, 99.2%; *S. griseoincarnatus*, AJ781328, 99.2%; *S. paradoxus*, AB184628, 99.1%.

Source: not known.

DNA G+C content (mol%): not known.

Type strain: AS 4.1608, ATCC 19724, CBS 463.68, BCRC 13686, DSM 40092, NBRC 12740, NBRC 15956, JCM 4344, KCTC 9752, NRRL B-3981, NRRL-ISP 5092, RIA 1005, UNIQEM 120, VKM Ac-705.

Sequence accession no. (16S rRNA gene): AY999808.

30. **Streptomyces amakusaensis** Nagatsu, Anzai, Ohkuma and Suzuki 1963, 209[AL]

a.ma.ku.sa.en'sis. N.L. masc. adj. *amakusaensis* of or belonging to Amakusa Island, Japan, the source of the soil from which the organism was isolated.

Spore chains in Section *Spirales*. Mature spore chains have 10 to 50 or more spores per chain on yeast-malt agar and salts-starch agar. Sporulating aerial mycelium is poorly developed or absent on oatmeal agar and glycerol-asparagine agar. Spore surface is smooth.

Color of colony: aerial mass color of mature aerial mycelium is in the Blue color series on salts-starch agar and the Green or Blue color series on yeast-malt agar; younger mycelium may be in the Gray color series. Mature aerial mycelium is usually not formed on oatmeal agar or glycerol-asparagine agar. Reverse side of colony with distinctive pigments (colorless or pale yellow) on yeast-malt agar, oatmeal agar, salts-starch agar, or glycerol-asparagine agar.

This culture does not show good growth with any of the carbon sources tested on Pridham and Gottlieb carbon utilization medium. D-Glucose is utilized for growth. No significant growth occurs on D-xylose, iso-inositol, D-mannitol, D-fructose, rhamnose, or raffinose and only doubtful traces are seen on L-arabinose and sucrose.

Type strain shows the highest sequence similarity to: *S. inusitatus*, AB184445, 99.8%.

Source: not known.

DNA G+C content (mol%): not known.

Type strain: AS 4.1462, ATCC 23876, CBS 615.68, DSM 40219, NBRC 12835, JCM 4617, JCM 4167, KCTC 9753, LMG 19350, NRRL B-3351, NRRL-ISP 5219, RIA 1163, VKM Ac-995.

Sequence accession no. (16S rRNA gene): AY999781.

31. **Streptomyces ambofaciens** Pinnert-Sindico 1954, 702[AL]

am.bo.fa'ci.ens. L. adj. *ambo* both; L. part. adj. *faciens* producing; N.L. part. adj. *ambofaciens* producing both, referring to the production of two different antibiotics by the organism.

Spore chains in Section *Spirales*. Open terminal spirals on long spore chains also suggest *Retinaculum-Apertum* type morphology. Mature spore chains generally have 10–50 spores per chain. This morphology is seen on yeast-malt agar, oatmeal agar, salts-starch agar, and glycerol-asparagine agar, although sporulating aerial mycelium is not abundant on glycerol-asparagine agar. Spore surface is smooth to warty; surface irregularities suggesting warts are small.

Color of colony: aerial mass color in the Gray color series on yeast-malt agar, oatmeal agar, and salts-starch agar. Reverse side of colony with no distinctive pigments (colorless to grayed yellow) on yeast-malt agar, oatmeal agar, and glycerol-asparagine agar. Dark brown, dark blue, or almost black substrate mycelium pigments are found on salts-starch agar. This pigment is not pH-sensitive.

Color in medium: melanoid pigments are not formed in peptone-yeast-iron agar, tyrosine agar, or tryptone-yeast broth. Yellow pigments may be found in medium in salts-starch agar; no pigment is found in the medium in yeast-malt agar, oatmeal agar, or glycerol-asparagine agar.

D-Glucose, L-arabinose, D-xylose, iso-inositol, D-mannitol, D-fructose, and rhamnose are utilized for growth. No growth or only trace of growth on raffinose. Utilization of sucrose is doubtful.

Type strain shows the highest sequence similarity to: *S. collinus*, AB184123, 99.2%; *S. paradoxus*, AB184628, 99.2%; *S. griseoflavus*, AJ781322, 99.2%; *S. lienomycini*, AJ781353, 99.1%; *S. flaveolus*, AB184764, 99.1%; *S. heliomycini*, AB184712, 99.1%; *S. matensis*, AB184221, 99%; *S. griseorubens*, AB184139, 99%; *S. albaduncus*, AY999757, 99%; *S. albogriseolus*, AJ494865, 99%; *S. coelescens*, AF503496, 99%; *S. rubrogriseus*, AB184681, 99%; *S. violaceoruber*, AF503492, 99%.

Source: not known.

DNA G+C content (mol%): not known.

Type strain: ATCC 23877, CBS 616.68, BCRC 11857, CECT 3101, DSM 40053, NBRC 12836, JCM 4618, JCM 4204, KCTC 9111, NRRL 2420, NRRL B-2516, NRRL-ISP 5053, RIA 1115.

Sequence accession no. (16S rRNA gene): M27245.

32. **Streptomyces aminophilus** Foster *in* Oswald, Reedy and Randall *in* Hütter 1961, 370[AL]

a.mi.no'phi.lus. N.L. n. *aminum* amine; N.L. masc. adj. *philus* (from Gr. masc. adj. *philos*) loving; N.L. masc. adj. *aminophilus* amine-nitrogen loving.

Spore chains in Section *Spirales*. Mature spore chains have 3–10, or sometimes more than 10 spores per chain. This morphology is seen on yeast-malt agar, oatmeal agar, salts-starch agar, and glycerol-asparagine agar, but formation of sporulation aerial mycelium is poor on oatmeal agar. Spore surface is smooth.

Color of colony: aerial mass color in the Yellow color series on yeast-malt agar, salts-starch agar, and glycerol-asparagine agar; aerial mycelium poorly developed on oatmeal agar. Reverse side of colony with no distinctive pigments (yellow or grayed yellow) on yeast-malt agar, oatmeal agar, salts-starch agar, and glycerol-asparagine agar.

Color in medium: melanoid pigments not formed in peptone-yeast-iron agar, tyrosine agar, or tryptone-yeast broth. No pigments, or only traces of yellow pigment, are found in medium in yeast-malt agar, oatmeal agar, salts-starch agar, or glycerol-asparagine agar.

D-Glucose, D-xylose, D-mannitol, and D-fructose are utilized for growth. No growth or only traces of growth on iso-inositol and rhamnose. Reports vary on utilization of sucrose and raffinose.

For sequence similarity, see type strain of *S. cacaoi* subsp. *cacaoi*.

Source: not known.

DNA G+C content (mol%): not known.

Type strain: AS 4.1416, ATCC 13558, ATCC 14961, ATCC 23878, CBS 617.68, BCRC 11858, DSM 40186, NBRC 12837, JCM 4619, JCM 4275, KCTC 9673, LMG 19319, NCIMB 9827, NRRL 2390, NRRL-ISP 5186, RIA 1140, VKM Ac-706.

Sequence accession no. (16S rRNA gene): AB184183.

Further comments: according to Lanoot et al. (2002), *Streptomyces aminophilus* Foster 1961[AL] is a later heterotypic synonym of *Streptomyces cacaoi* subsp. *cacaoi* (Waksman 1932) Waksman and Henrici 1948[AL] emend. Lanoot et al. 2002.

33. **Streptomyces anandii** Batra and Bajaj 1965, 242[AL]

a.nan.di'i. N.L. gen. n. *anandii* of Anand, Gujarat, India, the source of the soil from which the organism was isolated.

Spore chains in Section *Spirales*, but open loops and flexuous chains are also seen. Mature spore chains are moderately long with 10–50 spores per chain. This morphology is seen on yeast-malt agar, oatmeal agar, salts-starch agar, and glycerol-asparagine agar. Spore surface is smooth. One of three observers found globose bodies in the aerial mycelium on yeast-malt agar, and coremia formation on oatmeal agar and salts-starch agar.

Color of colony: aerial mass color in the White or Gray (2dc, yellowish gray or 3ge, light grayish yellowish brown) color series on yeast-malt agar and glycerol-asparagine agar; Gray or Red color series (2dc, yellowish gray to 4ec or 5cb, grayish yellowish pink) on oatmeal agar and salts-starch agar. Reverse side of colony with no distinctive pigments (pale or grayish yellow to olive brown or strong brown) on yeast-malt agar, oatmeal agar, salts-starch agar, and glycerol-asparagine agar.

Color in medium: melanoid pigments are formed in peptone-yeast-iron agar, tyrosine agar, tryptone-yeast broth, and Gause's medium no. 2. No pigment or only a trace of yellow is found in the medium in yeast-malt agar, oatmeal agar, salts-starch agar, and glycerol-asparagine agar.

D-Glucose, L-arabinose, D-xylose, iso-inositol, D-mannitol, D-fructose, sucrose, and raffinose are utilized for growth. No growth or only traces of growth with rhamnose.

Type strain shows the highest sequence similarity to: *S. geysiriensis*, DQ442501, 99.3%; *S. minutiscleroticus*, EF178696, 99.3%; *S. rochei*, AB184237, 99.2%; *S. ghanaensis*, AY999851, 99.2%; *S. plicatus*, AB184291, 99.2%; *S. vinaceusdrappus*, AY999929, 99.2%; *S. mutabilis*, EF178679, 99.1%; *S. djakartensis*, AB184657, 99.1%; *S. calvus*, AB184329, 99.1%; *S. asterosporus*, AB184706, 99%; *S. tuirus*, AB184690, 99%; *S. virens*, DQ442554, 99%; *S. aureorectus*, AB184710, 99%.

Source: isolated from soil from Anand, Gujarat, India.

DNA G+C content (mol%): not known.

Type strain: ATCC 19388, CBS 739.72, BCRC 11825, DSM 40535, NBRC 13438, JCM 4720, KCTC 9687, NRRL B-12487, NRRL B-3590, NRRL-ISP 5535, RIA 1399, VKM Ac-1920.

Sequence accession no. (16S rRNA gene): AB184402.

34. **Streptomyces anthocyanicus** (Krasil'nikov et al. 1965) Pridham 1970, 7[AL] ("*Actinomyces anthocyanicus*" Krasil'nikov, Sorokina, Alferova and Bezzubenkova 1965, 118)

an.tho.cya'ni.cus. N.L. n. *anthocyaninum* anthocyanin (a water-soluble vacuolar pigments that may appear red, purple, or blue according to pH), L. masc. suff. *-icus* suffix used with the sense of belonging to; N.L. masc. adj. *anthocyanicus* belonging to anthocyanin, presumably referring to dark blue color.

Spore chains are of typical *Retinaculum-Apertum* type. Blue-colored vegetative mycelium and diffusible pigments are formed on some media. Exhibits slight anti-bacterial activity.

Type strain shows the highest sequence similarity to: *S. tricolor*, AB184687, 100%; *S. violaceoruber*, AF503492, 99.9%; *S. coelescens*, AF503496, 99.9%; *S. rubrogriseus*, AB184681, 99.9%; *S. violaceolatus* AF503497, 99.8%; *S. humiferus*, AF503491, 99.8%; *S. lienomycini*, AJ781353, 99.7%; *S. tendae*, D63873, 99.5%; *S. violaceorubidus*, AJ781374, 99.5%; *S. coelicoflavus*, AB184650, 99.5%; *S. olivaceus*, AB184743, 99.1%; *S. pactum*, AB184398, 99.1%.

Source: not known.

DNA G+C content (mol%): not known.

Type strain: AS 4.1683, ATCC 19821, DSM 41422, NBRC 14892, JCM 5058, KCTC 9755, NRRL B-12341.

Sequence accession no. (16S rRNA gene): AB184631.

35. **Streptomyces antibioticus** (Waksman and Woodruff 1941) Waksman and Henrici *in* Breed, Murray and Hitchens 1948, 942[AL] ("*Actinomyces antibioticus*" Waksman and Woodruff 1941, 246)

an.ti.bio'ti.cus. N.L. masc. adj. *antibioticus* (from Gr. prep. *anti* against; Gr. n. *bios* life; L. suff. *-ticus -a -um* suffix of various meanings, but signifying in general made of or belonging to) against life, antibiotic.

Spore chains in Section *Rectiflexibiles.* Mature spore chains are short, often with only 3–10 spores per chain; longer chains are also found. This morphology is observed on yeast-malt agar, oatmeal agar, salts-starch agar, and glycerol-asparagine agar. Spore surface is smooth.

Color of colony: aerial mass color in the Gray color series on yeast-malt agar, salts-starch agar, and glycerol-asparagine agar; mature aerial mycelium poorly developed on oatmeal agar. Reverse side of colony with no distinctive pigments (grayed yellow-brown on yeast-malt agar, grayed yellow or dark grayish greenish yellow on oatmeal agar, salts-starch agar, and glycerol-asparagine agar); substrate mycelium pigment is not a pH indicator.

Color in medium: melanoid pigments are produced in peptone-yeast-iron agar, tyrosine agar, and peptone-yeast broth. No pigment or only trace of yellow in medium in yeast-malt agar, oatmeal agar, salts-starch agar, and glycerol-asparagine agar.

D-Glucose, L-arabinose, iso-inositol, D-mannitol, D-fructose, and rhamnose are utilized for growth. No growth or only trace of growth on sucrose and raffinose. Utilization of D-xylose is doubtful.

Type strain shows the highest sequence similarity to: *S. griseoruber*, AB184209, 99.3%.

Source: not known.

DNA G+C content (mol%): not known.

Type strain: ATCC 23879, ATCC 8663, CBS 478.48, CBS 659.68, CCM 3159, BCRC 12164, CECT 3225, DSM 40234, NBRC 12838, IMET 40227, JCM 4620, KCTC 9688, LMG 5966, NCIMB 8504, NRRL B-1701, NRRL B-2770, NRRL-ISP 5234, RIA 1174, VKM Ac-964.

Sequence accession no. (16S rRNA gene): AY999776.

36. **Streptomyces antimycoticus** Waksman *in* Breed, Murray and Smith 1957, 799[AL]

an.ti.my.co'ti.cus. N.L. masc. adj. *antimycoticus* (from Gr. prep. *anti* against; Gr. n. *mukês -etis* fungus; L. suff. *-icus -a -um* suffix of various meanings, but signifying in general made of or belonging to), against fungal, antimycotic.

Spore chains in Section *Spirales*; compact clusters of closed spirals on yeast-malt agar, oatmeal agar, salts-starch agar, and glycerol-asparagine agar; mature spore chains of generally 10–50 spores per chain are found on these media. Spore surface is spiny to warty. Surface irregularities on spores are intermediate between very short, thick spines, and warts.

Color of colony: aerial mass color in the Gray color series on yeast-malt agar, oatmeal agar, salts-starch agar, and glycerol-asparagine agar. Reverse side of colony with no distinct pigment (pale grayed yellow or grayed greenish yellow) on yeast-malt agar, oatmeal agar, salts-starch agar, or glycerol-asparagine agar.

Color in medium: melanoid pigments are not produced

on peptone-yeast-iron agar, tyrosine agar, or tryptone-yeast broth. No pigment is found in medium in yeast-malt agar, oatmeal agar, salts-starch agar, or glycerol-asparagine agar.

D-Glucose, L-arabinose, sucrose, D-xylose, iso-inositol, D-mannitol, D-fructose, rhamnose, and raffinose are utilized for growth. Growth on L-arabinose and D-xylose is generally less abundant than on the other carbon sources.

Type strain shows the highest sequence similarity to: *S. sporoclivatus*, AB249934, 100%; *S. rutgersensis* subsp. *castelarensis*, AY508511, 99.8%; *S. geldanamycininus*, DQ334781, 99.8%; *S. melanosporofaciens*, AJ271887, 99.6%; *S. rhizosphaericus*, AB249941, 99.1%; *S. asiaticus*, AB249947, 99.1%.

Source: not known.

DNA G+C content (mol%): not known.

Type strain: AS 4.1591, ATCC 23880, CBS 660.68, DSM 40284, NBRC 12839, JCM 4621, JCM 4228, KCTC 9694, NRRL 2421, NRRL-ISP 5284, RIA 1198, VKM Ac-1824.

Sequence accession no. (16S rRNA gene): AB184185.

37. **Streptomyces anulatus** (Beijerinck 1912) Waksman *in* Waksman and Lechevalier 1953, 40[AL] [“*Actinomyces Streptothrix annulatus*” (*sic*) Beijerinck 1912, 7; “*Actinomyces annulatus*” (*sic*) Beijerinck 1912, 4; “*Streptomyces annulatus*” (*sic*) Waksman *in* Waksman and Lechevalier 1953, 40]

a.nu.la'tus. L. masc. adj. *anulatus* furnished with a ring.

Spore chains in Section *Rectiflexibiles* on oatmeal agar, salts-starch agar, and glycerol-asparagine agar, but open spirals of two to several turns, hooks and loops of small diameter, as well as straight and flexuous chains are found on yeast-malt agar. The strain as observed on yeast-malt agar can therefore be placed in Section *Spirales* or *Retinaculum-Apertum* type. Spore surface is smooth.

Color of colony: aerial mass color in the Yellow or White color series on yeast-malt agar, oatmeal agar, salts-starch agar, and glycerol-asparagine agar. Observers selected color tab 2ba (pale yellow) from the Yellow color series and tab a (white) from the White color series; both yellow and white may be observed on the same medium. Reverse side of colony with no distinctive pigments (colorless to pale yellow; grayish yellow or yellowish brown) on yeast-malt agar, oatmeal agar, salts-starch agar, and glycerol-asparagine agar.

Color in medium: melanoid pigments are not formed in peptone-yeast-iron agar, tyrosine agar, or tryptone-yeast broth. No pigment (or only a trace of yellow) is found in the medium in yeast-malt agar, oatmeal agar, salts-starch agar, or glycerol-asparagine agar.

D-Glucose, L-arabinose, D-xylose, D-mannitol, D-fructose, and rhamnose are utilized for growth. No growth or only traces of growth with iso-inositol, sucrose, and raffinose.

Type strain shows the highest sequence similarity to: *S. cavourensis* subsp. *washingtonensis*, DQ026671, 100%; *S. fimicarius*, AY999784, 100%; *S. badius*, AY999783, 100%; *S. sindenensis*, AB184759, 100%; *S. rubiginosohelvolus*, AB184240, 100%; *S. pluricolorescens*, DQ442540, 100%; *S. griseinus*, AB184205, 100%; *S. acrimycini*, AY999889, 100%; *S. flavofuscus*, AB249935, 100%; *S. mediolani*, AB184674, 100%; *S. praecox*, AB184293, 100%; *S. lavendulae* subsp. *lavendulae*, AB184080, 100%; *S. griseoplanus*, AY999894, 100%; *S. alboviridis*, AB184256, 99.9%; *S. lipmanii*, AB184148, 99.9%; *S. albovinaceus*, AB249958, 99.9%; *S. globisporus* subsp. *globisporus*, EF178686, 99.9%; *S. cyaneofuscatus*, AB184860, 99.9%; *S. cinereorectus*, AB184646, 99.9%; *S. fulvorobeus*, AB184711, 99.9%; *S. microflavus*, DQ445795, 99.8%; *S. argenteolus*, AB045872, 99.8%; *S. parvus*, DQ442537, 99.8%; *S. californicus*, AB184755, 99.8%; *S. baarnensis*, EF178688, 99.8%; *S. halstedii*, EF178695, 99.7%; *S. luridiscabiei*, AF361784, 99.7%; *S. flavovirens*, DQ026635, 99.7%; *S. floridae*, AB184656, 99.7%; *S. flavogriseus*, AJ494864, 99.7%; *S. griseolus*, AB184768, 99.7%; *S. griseus* subsp. *griseus*, AY207604, 99.7%; *S. atroolivaceus*, AJ781320, 99.5%; *S. nitrosporeus*, EF178680, 99.5%; *S. olivoviridis*, AB184227, 99.5%; *S. finlayi*, AY999788, 99.5%; *S. bacillaris*, AB184439, 99.5%; *S. clavifer*, DQ026670, 99.5%; *S. pulveraceus*, AB184806, 99.5%; *S. yanii*, AB006159, 99.4%; *S. gelaticus*, DQ026636, 99.3%; *S. mutomycini*, AB249951, 99.3%; *S. albolongus*, AB184425, 99.3%; *S. sanglieri*, AB249945, 99.3%; *S. atratus*, DQ026638, 99.3%; *S. griseobrunneus*, AB249912, 99.3%; *S. celluloflavus*, AB184476, 99.3%; *S. spiroverticillatus*, AB184814, 99.2%; *S. candidus*, DQ026663, 99.1%; *S. cavourensis* subsp. *cavourensis*, DQ445791, 99.1%; *S. mauvecolor*, AB184532, 99%; *S. cremeus*, AB184124, 99%.

Source: not known.

DNA G+C content (mol%): not known.

Type strain: AS 4.1421, ATCC 27416, CBS 100.18, CBS 670.72, DSM 40361, NBRC 13369, IMET 43334, JCM 4721, KCTC 9756, LMG 19301, NRRL B-2000, NRRL-ISP 5361, RIA 1330, VKM Ac-728.

Sequence accession no. (16S rRNA gene): DQ026637.

Further comments: according to Lanoot et al. (2005b), *Streptomyces anulatus* (Beijerinck 1912) Waksman 1953[AL] emend. Lanoot et al. 2005b is an earlier heterotypic synonym of *Streptomyces chrysomallus* subsp. *chrysomallus* Lindenbein 1952[AL], an earlier heterotypic synonym of *Streptomyces citreofluorescens* (Korenyako et al. 1960) Pridham 1970[AL], and an earlier heterotypic synonym of *Streptomyces fluorescens* (Krasil'nikov 1958) Pridham 1970[AL].

38. **Streptomyces arabicus** Shibata, Nakazawa, Miyake, Inoue and Okabori 1957, 36[AL]

a.ra'bi.cus. L. masc. adj. *arabicus* of or belonging to Arabia, the source of the soil from which the organism was isolated.

Spore chains in Section *Spirales* on oatmeal agar and salts-starch agar. Short spore chains on yeast-malt agar and glycerol-asparagine agar may appear to be *Rectiflexibiles* morphology or may show only incomplete spirals or hooks. Mature spore chains on oatmeal agar and salts-starch agar generally have 10 to 50 or more spores per chain; shorter spore chains are found on yeast-malt agar and glycerol-asparagine agar. Spore surface is warty to spiny. Surface irregularities are short and blunt; smooth spores may also be found.

Color of colony: aerial mass color in the Gray color series on yeast-malt agar, oatmeal agar, and salts-starch agar; Red color series on glycerol-asparagine agar. Observers selected color tabs representing neutral grays to light grayish reddish brown (5fe) for the Gray color series and grayish yellowish pink (5dc) from the Red color series for glycerol-asparagine agar. Reverse side of colony is grayish yellowish brown on yeast-malt agar; colorless to pale yellowish gray or brown on oatmeal agar, salts-starch agar, and glycerol-asparagine agar. Substrate mycelium pigment is not a pH indicator.

Color in medium: melanoid pigment is not produced in peptone-yeast-iron agar, tyrosine agar, or tryptone-yeast broth. Trace of yellow pigment may be found in the medium in oatmeal agar and salts-starch agar or may be absent. This pigment is not pH-sensitive.

D-Glucose, L-arabinose, D-xylose, iso-inositol, D-mannitol, D-fructose, and rhamnose are utilized for growth. No growth or only trace of growth on raffinose. Utilization of sucrose is doubtful.

No detailed sequence information is available (only partial sequence).

Source: soil from Arabia.

DNA G+C content (mol%): not known.

Type strain: ATCC 23881, CBS 661.68, DSM 40252, HAMBI 995, HUT 6041, IFM 1118, NBRC 12840, NBRC 14035, NBRC 3406, JCM 4622, JCM 4161, NRRL B-1733, NRRL-ISP 5252, RIA 1178, RIA 512, VKM Ac-1754.

Sequence accession no. (16S rRNA gene): D44271.

Further comments: according to Lanoot et al. (2004), *Streptomyces arabicus* Shibata et al. 1957[AL] is a later heterotypic synonym of *Streptomyces vinaceus* Jones 1952[AL].

39. **Streptomyces ardus** (DeBoer et al. 1961) Witt and Stackebrandt 1996, 836[VP] (Effective publication: Witt and Stackebrandt 1990, 370.) ("*Streptomyces ardus*" DeBoer, Dietz, Lummis and Savage 1961; *Streptoverticillium ardum* Locci, Baldacci and Petrolini Baldan 1969, 59)

ar'dus. L. masc. adj. *ardus* dry, withered, shrunk up, shrivelled, meagre.

Sporulating aerial mycelium is usually absent or very poorly developed on yeast-malt agar, oatmeal agar, salts-starch agar, and glycerol-asparagine agar. Sterile aerial hyphae are straight or flexuous. Coremia may be produced on yeast-malt agar and oatmeal agar in 7–14 d. The original description reported that this strain sporulates with difficulty, producing only a few monoverticillate and biverticillate sporophores.

Color of colony: aerial mass color of sporulating aerial mycelium not observed. Sterile hyphae or scant aerial mycelia appear to be white to light gray when produced on yeast-malt agar, oatmeal agar, salts-starch agar, and glycerol-asparagine agar. Reverse side of colony with no distinctive pigments (pale grayish yellow to light yellowish brown) on yeast-malt agar, oatmeal agar, salts-starch agar, and glycerol-asparagine agar.

Color in medium: melanoid pigments are formed in peptone-yeast-iron agar and tryptone-yeast broth, but not on tyrosine agar or Gause's medium no. 2. No pigment is found in the medium in oatmeal agar, salts-starch agar, or glycerol-asparagine agar. One observer reports traces of red pigment in yeast-malt agar. This pigment is not pH-sensitive when tested with 0.05 M NaOH or HCl.

D-Glucose, iso-Inositol, and D-fructose are utilized for growth. Reports vary on utilization of L-arabinose. No growth or only traces of growth with D-xylose, D-mannitol, rhamnose, sucrose, and raffinose.

Type strain shows the highest sequence similarity to: *S. blastmyceticus*, AY999802, 99.4%; *S. hiroshimensis*, AB184789, 99.2%; *S. cinnamoneus*, AB184850, 99.1%; *S. caeruleus*, EF178675, 99%.

Source: not known.

DNA G+C content (mol%): not known.

Type strain: AS 4.167, ATCC 27417, CBS 731.72, BCRC 12319, CECT 3254, DSM 40527, NBRC 13430, JCM 4722, JCM 4543, RIA 1391, NRRL 2817, NRRL-ISP 5527, VKM Ac-930.

Sequence accession no. (16S rRNA gene): AB184864.

Further comments: according to Hatano et al. (2003), *Streptomyces ardus* (DeBoer et al. 1961) Witt and Stackebrandt 1991 is a heterotypic synonym of "*Streptomyces caespitosus*" (NBRC 13490).

40. **Streptomyces arenae** Pridham, Hesseltine and Benedict 1958, 67[AL]

a.re'na.e. L. n. *arena* sand; L. gen. n. *arenae* of sand, referring to a sandy area near Zion, Illinois, USA, from which the organism was isolated.

Spore chains in Section *Spirales*. Mature spore chains are moderately long with 10 to 50 or more spores per chain. This morphology is seen on yeast-malt agar, salts-starch agar, and glycerol-asparagine agar. Spore surface is spiny.

Color of colony: aerial mass color in the Red or Gray color series on yeast-malt agar, oatmeal agar, salts-starch agar, and glycerol-asparagine agar. Observers selected color tabs 5fe (light grayish reddish brown) from the Gray color wheel and 5cb or 5dc (grayish yellowish pink) from the red color series. One observer selected tab 11ec from the Violet color series as nearest matching color for sporulating aerial growth on salts-starch agar, oatmeal agar, and glycerol-asparagine agar. Reverse side of colony with no distinctive pigments (pale yellow to yellowish brown) on yeast-malt agar, oatmeal agar, salts-starch agar, and glycerol-asparagine agar.

Color in medium: melanoid pigments are formed in peptone-yeast-iron agar, tyrosine agar, or tryptone-yeast broth. Yellow pigment may be found in yeast-malt agar, oatmeal agar, and salts-starch agar. This pigment is not pH-sensitive when tested with 0.05 M NaOH or HCl.

D-Glucose, L-arabinose, sucrose, D-xylose, iso-inositol, D-mannitol, D-fructose, rhamnose, and raffinose are utilized for growth.

Type strain shows the highest sequence similarity to: *S. hawaiiensis*, AB184143, 99.7%; *S. luteogriseus*, AB184379, 99.4%; *S. massasporeus*, AB184152, 99.4%; *S. purpurascens*, AJ399486, 99.3%; *S. violaceus*, AB184315, 99.2%; *S. albosporeus* subsp. *albosporeus*, AJ781327, 99.2%; *S. janthinus*, AB184851, 99.2%; *S. roseoviolaceus*, AJ399484, 99.2%; *S. flavoviridis*, AB184842, 99%; *S. bellus*, AB184849, 99%; *S. indiaensis*, AB184553, 99%; *S. pilosus*, AB184161, 99%.

Source: the type strain was isolated from a sandy area near Zion, Illinois, USA.

DNA G+C content (mol%): not known.

Type strain: AS 4.1610, ATCC 25428, CBS 926.69, BCRC 11827, DSM 40293, NBRC 13016, JCM 4452, NRRL 2377, NRRL-ISP 5293, RIA 1208, VKM Ac-1201.

Sequence accession no. (16S rRNA gene): AB249977.

41. **Streptomyces argenteolus** Tresner, Davies and Backus 1961, 74[AL]

ar.gen.te'o.lus. L. masc. adj. *argenteolus* of silver.

Spore chains in Section *Spirales*. This culture is described as "not forming loops or spirals" in the original descriptions appearing in the patents cited above. However, the type strain was observed to produce spirals in subsequent studies by Pridham et al. (1958) and Tresner et al. (1961). Mature spore chains generally have 10 to 50 or more spores per chain on yeast-malt agar, oatmeal agar, salts-starch agar, and glycerol-asparagine agar. Spore surface is smooth.

Color of colony: aerial mass color in the Gray color series on yeast-malt agar, oatmeal agar, salts-starch agar, and glycerol-asparagine agar. Reverse side of colony is colorless to grayish yellowish green or gray on yeast-malt agar, oatmeal agar, salts-starch agar, and glycerol-asparagine agar; substrate mycelium pigment is not a pH indicator.

Color in medium: melanoid pigments are not formed on peptone-yeast-iron agar, tyrosine agar, or tryptone-yeast broth. No pigment is found in medium in yeast-malt agar, oatmeal agar, salts-starch agar, or glycerol-asparagine agar.

D-Glucose, L-arabinose, D-xylose, D-mannitol, D-fructose, and rhamnose are utilized for growth. No growth or only trace of growth on sucrose, iso-inositol, and raffinose.

Type strain shows the highest sequence similarity to: *S. cinereorectus*, AB184646, 100%; *S. lavendulae* subsp. *lavendulae*, AB184080, 99.9%; *S. cavourensis* subsp. *washing tonensis*, DQ026671, 99.9%; *S. microflavus*, DQ445795, 99.8%; *S. griseoplanus*, AY999894, 99.8%; *S. cyaneofuscatus*, AB184860, 99.8%; *S. praecox*, AB184293, 99.8%; *S. anulatus*, DQ026637, 99.8%; *S. fimicarius*, AY999784, 99.8%; *S. flavofuscus*, AB249935, 99.8%; *S. griseolus*, AB184768, 99.8%; *S. alboviridis*, AB184256, 99.8%; *S. lipmanii*, AB184148, 99.8%; *S. fulvorobeus*, AB184711, 99.8%; *S. griseus* subsp. *griseus*, AY207604, 99.7%; *S. halstedii*, EF178695, 99.7%; *S. albovinaceus*, AB249958, 99.7%; *S. sindenensis*, AB184759, 99.7%; *S. floridae*, AB184656, 99.7%; *S. globisporus* subsp. *globisporus*, EF178686, 99.7%; *S. acrimycini*, AY999889, 99.7%; *S. pluricolorescens*, DQ442540, 99.7%; *S. mediolani*, AB184674, 99.7%; *S. luridiscabiei*, AF361784, 99.7%; *S. griseinus*, AB184205, 99.7%; *S. flavovirens*, DQ026635, 99.7%; *S. baarnensis*, EF178688, 99.7%; *S. badius*, AY999783, 99.7%; *S. rubiginosohelvolus*, AB184240, 99.7%; *S. californicus*, AB184755, 99.6%; *S. flavogriseus*, AJ494864, 99.6%; *S. parvus*, DQ442537, 99.6%; *S. pulveraceus*, AB184806, 99.5%; *S. yanii*, AB006159, 99.4%; *S. nitrosporeus*, EF178680, 99.4%; *S. olivoviridis*, AB184227, 99.3%; *S. sanglieri*, AB249945, 99.3%; *S. atroolivaceus*, AJ781320, 99.3%; *S. bacillaris*, AB184439, 99.3%; *S. finlayi*, AY999788, 99.3%; *S. clavifer*, DQ026670, 99.3%; *S. gelaticus*, DQ026636, 99.2%; *S. atratus*, DQ026638, 99.2%; *S. celluloflavus*, AB184476, 99.1%; *S. cremeus*, AB184124, 99.1%; *S. spiroverticillatus*, AB184814, 99.1%; *S. griseobrunneus*, AB249912, 99.1%; *S. albolongus*, AB184425, 99.1%; *S. cavourensis* subsp. *cavourensis*, DQ445791, 99%; *S. mutomycini*, AB249951, 99%.

Source: not known.

DNA G+C content (mol%): not known.

Type strain: AS 4.1693, ATCC 11009, ATCC 23882, CBS 662.68, BCRC 11815, DSM 40226, NBRC 12841, IMET 43659, JCM 4623, JCM 4229, KCTC 1742, LMG 5967, NCIMB 9625, NRRL B-1806, NRRL-ISP 5226, RIA 1168, VKM Ac-747.

Sequence accession no. (16S rRNA gene): AB045872.

Further comments: according to Liu et al. (2005b), *Streptomyces argenteolus* Tresner et al. 1961[AL] is a later heterotypic synonym of *Streptomyces griseus* (Krainsky 1914) Waksman and Henrici 1948[AL] emend. Liu et al. 2005b.

According to Guo et al. (2008), *Streptomyces argenteolus* Tresner et al. 1961[AL] is not a later heterotypic synonym of *Streptomyces griseus* (Krainsky 1914) Waksman and Henrici 1948[AL].

42. **Streptomyces armeniacus** (Kalakoutskii and Kusnetsov 1964) Wellington and Williams 1981a, 80[VP] (*Actinoplanes armeniacus* Kalakoutskii and Kusnetsov 1964, 613)

ar.me.ni'a.cus. L. fem. n. *armeniaca* apricot-tree, N.L. masc. adj. *armeniacus* apricot-colored.

Three-week-old cultures show white aerial mycelium and development of spiral spore chains on chitin medium, nutrient agar, oatmeal agar, and ISP 4. On all media, the substrate mycelium is stable and cream-colored to brown. No distinctive pigments are produced in the substrate mycelium or in the medium. According to the original description of *A. armeniacus*, milk is not coagulated or peptonized, gelatin is liquefied, and nitrate is not reduced. The following carbon sources are utilized for growth: L-rhamnose, sorbitol, galactose, D-fructose, D-arabinose, trehalose, cellobiose, esculin, glycerin, and inulin. Growth is weak on glucose, sucrose, lactose, maltose, raffinose, sorbose, xylose, salicin, and dulcitol. No growth occurs on mannitol. Starch is hydrolyzed. Whole-cell hydrolysates contain LL-A_2pm, which is characteristic of cell-wall chemotype I. Susceptible to penicillin, streptomycin, chlortetracycline, erythromycin, and polymyxin; resistant to kanamycin and neomycin.

Type strain shows no sequence similarity over 99%.

Source: not known.

DNA G+C content (mol%): not known.

Type strain: 26 A-32, RIA 807, AS 4.1684, ATCC 15676, CBS 559.75, DSM 43125, IFM 1166, IFM 1244, NBRC 12555, IMET 9250, JCM 3070, KCTC 9120, NCIMB 10179, VKM Ac-905.

Sequence accession no. (16S rRNA gene): AB018092.

43. **Streptomyces asiaticus** Sembiring, Ward and Goodfellow 2001, 1619[VP] (Effective publication: Sembiring, Ward and Goodfellow 2000, 365.)

a.si.a'ti.cus. L. masc. adj. *asiaticus* Asian.

Spore chains are *Spirales*; the spore surface is rugose. On oatmeal agar, the spore mass is gray, the substrate mycelium is grayish-yellow, and the diffusible pigment is yellow. Melanin pigments are not produced. The strain degrades pectin and grows at 45°C.

Type strain shows the highest sequence similarity to: *Streptomyces* sp., AJ391828, 99.5%; *Streptomyces* sp., AJ391831, 99.5%; *S. cangkringensis*, AJ391831, 99.5%; *Streptomyces* sp., AJ391825, 99.4%; *Streptomyces* sp., AJ391826, 99.4%; *Streptomyces* sp., AJ391832, 99.4%; *Streptomyces* sp., A33R1, DSM 41763, AJ391832, 99.4%; *Streptomyces* sp., B23P1, DSM 41765, AJ391825, 99.4%; *Streptomyces* sp., DSM 41768, AJ391826, 99.4%; *Streptomyces* sp., AJ391818, 99.3%; *Streptomyces* sp.,

AJ391827, 99.3%; *Streptomyces* sp., AJ391829, 99.3%; *Streptomyces* sp., AJ391836, 99.3%; *S. griseiniger*, AJ391818, 99.3%; *S. yogyakartensis*, DSM 41766^{T}, 99.3%; *Streptomyces* sp., AJ391836, 99.3%; *S. violaceusniger*, AJ391822, 99.1%; *S. violaceusniger*, AJ391823, 99.1%; *S. antimycoticus*, DSM 40284^{T}, 99.1%; *S. violaceusniger*, NRRL-ISP 5562^{T}, 99.1%; *Streptomyces* sp., AJ391834, 99%; *S. albiflaviniger*, AJ391812, 99%; *S. albiaxialis*, DSM 41799^{T}, 99%; *S. rhizosphaericus*, DSM 41760^{T}, 99%; *Streptomyces* sp., DSM 40602, 99%.

Type strain shows DNA–DNA similarity to: *S. rhizosphaericus*, AB249941, 100%; *S. cangkringensis*, AJ391831, 99.8%; *S. griseinger*, AJ391818, 99.8%; *S. indonesiensis*, DQ334783, 99.7%; *S. antimycoticus*, AB184185, 99.1%; *S. sporoclivatus*, AB249934, 99.1%; *S. rutgersensis* subsp. *castelarensis*, AY508511, 99.1%; *S. geldanamycininus*, DQ334781, 99%.

Source: isolated from the rhizosphere of the tropical legume *Paraserianthes falcataria*.

DNA G+C content (mol%): not known.

Type strain: A14P1, DSM 41761, JCM 11443, NBRC 100774, NCIMB 13675.

Sequence accession no. (16S rRNA gene): AB249947.

44. **Streptomyces asterosporus** (*ex* Krasil'nikov 1970b) Preobrazhenskaya 1986, 573^{VP} (Effective publication: Preobrazhenskaya *in* Gause, Preobrazhenskaya, Sveshnikova, Terekhova and Maximova 1983.) ("*Actinomyces asterosporus*" Krasil'nikov (1970b)

as.te.ro.spo'rus. L. n. *aster*-eris a star; N.L. n. *spora* (from. Gr. n. *spora* seed) spore; N.L. masc. adj. *asterosporus* star-shaped spore.

The type strain, INMI 16^{T}, easily loses its ability to construct aerial mycelium. Spore chains *Spirales*; spores are spiny, of medium-size, and with wide bodies. On mineral agar 1: aerial mycelium is gray, sometimes black, depending on moisture and release; substrate mycelium is colorless, sometimes gray; no diffusible pigment. On glycerol-nitrate agar and oatmeal agar: aerial mycelium is gray; substrate mycelium is colorless or gray; no diffusible pigment. On starch-ammonia agar: aerial mycelium is gray; substrate mycelium is colorless to creamy yellowish; no diffusible pigment. On glycerol-asparagine agar: aerial mycelium is poor, gray; substrate mycelium is colorless; no diffusible pigment. On organic agar 2: aerial mycelium is absent or gray; substrate mycelium is colorless or gray; no diffusible pigment. Melanoid pigments are not formed. Grows on glucose, fructose, sucrose, arabinose, galactose, rhamnose, and mannitol; no growth on raffinose or xylose.

Type strain shows the highest sequence similarity to: *S. aureorectus*, AB184710, 100%; *S. calvus*, AB184329, 100%; *S. virens*, DQ442554, 99.8%; *S. minutiscleroticus*, EF178696, 99.1%; *S. geysiriensis*, DQ442501, 99.1%; *S. djakartensis*, AB184657, 99.1%; *S. mutabilis*, EF178679, 99%; *S. rochei*, AB184237, 99%; *S. plicatus*, AB184291, 99%; *S. anandii*, AB184402, 99%; *S. vinaceusdrappus*, AY999929, 99%; *S. tuirus*, AB184690, 99%; *S. ghanaensis*, AY999851, 99%.

Source: not known.

DNA G+C content (mol%): not known.

Type strain: AS 4.1605, DSM 41452, NBRC 15872, INMI 16, JCM 6912, VKM Ac-40.

Sequence accession no. (16S rRNA gene): AB184706.

45. **Streptomyces atratus** Shibata, Higashide, Yamamoto and Nakazawa 1962, 232^{AL}

a.tra'tus. L. masc. adj. *atratus* clothed in black.

Spore chains of atypical *Retinaculum-Apertum* type. Forms gray to black vegetative mycelium on some media. Probably grows poorly on Czapek's solution agar. Produces rufomycin A, rufomycin B, and other anti-bacterial activity.

Type strain shows the highest sequence similarity to: *S. sanglieri*, AB249945, 100%; *S. yanii*, AB006159, 99.8%; *S. pulveraceus*, AB184806, 99.7%; *S. gelaticus*, DQ026636, 99.6%; *S. badius*, AY999783, 99.3%; *S. pluricolorescens*, DQ442540, 99.3%; *S. sindenensis*, AB184759, 99.3%; *S. albovinaceus*, AB249958, 99.3%; *S. globisporus* subsp. *globisporus*, EF178686, 99.3%; *S. griseinus*, AB184205, 99.3%; *S. flavofuscus*, AB249935, 99.3%; *S. mediolani*, AB184674, 99.3%; *S. fimicarius*, AY999784, 99.3%; *S. anulatus*, DQ026637, 99.3%; *S. rubiginosohelvolus*, AB184240, 99.3%; *S. praecox*, AB184293, 99.3%; *S. lavendulae* subsp. *lavendulae*, AB184080, 99.2%; *S. argenteolus*, AB045872, 99.2%; *S. parvus*, DQ442537, 99.2%; *S. cinereorectus*, AB184646, 99.2%; *S. griseoplanus*, AY999894, 99.2%; *S. californicus*, AB184755, 99.2%; *S. cavourensis* subsp. *washingtonensis*, DQ026671, 99.2%; *S. alboviridis*, AB184256, 99.1%; *S. griseolus*, AB184768, 99.1%; *S. baarnensis*, EF178688, 99.1%; *S. griseus* subsp. *griseus*, AY207604, 99.1%; *S. lipmanii*, AB184148, 99.1%; *S. cyaneofuscatus*, AB184860, 99.1%; *S. fulvorobeus*, AB184711, 99.1%; *S. halstedii*, EF178695, 99.1%; *S. acrimycini*, AY999889, 99.1%; *S. flavogriseus*, AJ494864, 99.1%; *S. flavovirens*, DQ026635, 99.1%; *S. microflavus*, DQ445795, 99.1%; *S. floridae*, AB184656, 99.1%; *S. luridiscabiei*, AF361784, 99%; *S. nitrosporeus*, EF178680, 99%.

Source: not known.

DNA G+C content (mol%): not known.

Type strain: AS 4.1632, ATCC 14046, DSM 41673, NBRC 3897, JCM 3386, NRRL B-16927.

Sequence accession no. (16S rRNA gene): DQ026638.

46. **Streptomyces atroaurantiacus** Nakagaito, Shimazu, Yokota and Hasegawa 1993a, 624^{VP} (Effective publication: Nakagaito, Shimazu, Yokota and Hasegawa 1992a, 632.)

a.tro.au.ran.ti.a'cus. L. adj. *ater -tra -trum* black, dark; N.L. adj. *aurantiacus -a -um* orange; N.L. masc. adj. *atroaurantiacus* dark orange.

Mature spore chains are long and straight to slightly flexuous. Spores are cylindrical with a smooth surface. Melanoid pigments are produced on tyrosine agar. Brown soluble pigments are produced on yeast extract-malt extract agar and yellow pigments are produced on oatmeal agar. The color of vegetative mycelia is dark orange to slightly orange. The color of aerial mycelia is white. Aerial mycelia are formed slightly on yeast extract-malt extract agar, oatmeal agar, inorganic salts-starch agar, tyrosine agar, Bennett's agar, and water agar. The temperature range for growth is 10–37°C. The concentration of NaCl at which growth occurs is less than 2%. Nitrate is not reduced, starch is hydrolyzed, gelatin is not liquefied, and milk is peptonized and coagulated. D-Glucose, D-arabinose, D-fructose, sucrose, and D-xylose are utilized, but D-mannitol, rhamnose, raffinose, and iso-inositol are not utilized or are poorly utilized. The cell wall contains both LL- and *meso*-A_2pm and glycine. Galactose and a trace of madurose are detected as whole-cell sugars. The phospholipid pattern is type II. MK-9(H_6) and MK-9(H_8) are detected.

Type strain shows no sequence similarity over 99% to other *Streptomyces* species. Type strain shows the highest sequence similarity to following *Kitasatospora*: *Kitasatospora gansuensis*, AY442265, 99.4%.

Source: not known.

DNA G+C content (mol%): 70.2.

Type strain: ATCC 51343, DSM 41649, NBRC 14327, JCM 3337, NRRL B-24282.

Sequence accession no. (16S rRNA gene): DQ026645.

47. **Streptomyces atroolivaceus** (Preobrazhenskaya, Blinov and Ryabova *in* Gauze, Preobrazhenskaya, Kudrina, Blinov, Ryabova and Sveshnikova 1957) Pridham, Hesseltine and Benedict 1958, 68[AL] ("*Actinomyces atroolivaceus*" Preobrazhenskaya, Blinov and Ryabova *in* Gauze, Preobrazhenskaya, Kudrina, Blinov, Ryabova and Sveshnikova 1957, 143)

at.ro.o.li.va'ce.us. L. adj. *ater -tra -trum* black, dark; N.L. adj. *olivaceus* olive colored; N.L. masc. adj. *atroolivaceus* of a dark olive color, referring to the pigment on an organic medium.

Spore chains in Section *Rectiflexibiles*. Mature spore chains generally have 10–50 spores per chain. This morphology is seen on yeast-malt agar, oatmeal agar, salts-starch agar, and glycerol-asparagine agar. Spore surface is smooth.

Color of colony: aerial mass color in the Gray color series on yeast-malt agar, oatmeal agar, salts-starch agar, and glycerol-asparagine agar. Reverse side of colony with no distinctive pigments on yeast-malt agar, oatmeal agar, salts-starch agar, or glycerol-asparagine agar; substrate pigment is not a pH indicator.

Color in medium: melanoid pigments not formed in peptone-yeast-iron agar or tyrosine agar. No pigment found in medium in yeast-malt agar, oatmeal agar, salts-starch agar, or glycerol-asparagine agar.

D-Glucose, L-arabinose, D-xylose, D-fructose, and rhamnose are utilized for growth. Variable reports on growth with sucrose, iso-inositol, D-mannitol, and raffinose. Two collaborators reported difficulty in observing results because of poor growth on all carbon sources with Pridham and Gottlieb basal medium.

Type strain shows the highest sequence similarity to: *S. clavifer*, DQ026670, 100%; *S. olivoviridis*, AB184227, 100%; *S. finlayi*, AY999788, 99.7%; *S. mutomycini*, AB249951, 99.7%; *S. pluricolorescens*, DQ442540, 99.5%; *S. sindenensis*, AB184759, 99.5%; *S. cavourensis* subsp. *washingtonensis*, DQ026671, 99.5%; *S. mediolani*, AB184674, 99.5%; *S. lavendulae* subsp. *lavendulae*, AB184080, 99.5%; *S. flavofuscus*, AB249935, 99.5%; *S. griseoplanus*, AY999894, 99.5%; *S. anulatus*, DQ026637, 99.5%; *S. praecox*, AB184293, 99.5%; *S. fimicarius*, AY999784, 99.5%; *S. rubiginosohelvolus*, AB184240, 99.5%; *S. badius*, AY999783, 99.5%; *S. griseinus*, AB184205, 99.5%; *S. acrimycini*, AY999889, 99.4%; *S. albovinaceus*, AB249958, 99.4%; *S. globisporus* subsp. *globisporus*, EF178686, 99.4%; *S. alboviridis*, AB184256, 99.4%; *S. cinereorectus*, AB184646, 99.4%; *S. baarnensis*, EF178688, 99.4%; *S. microflavus*, DQ445795, 99.4%; *S. cyaneofuscatus*, AB184860, 99.4%; *S. lipmanii*, AB184148, 99.4%; *S. fulvorobeus*, AB184711, 99.4%; *S. parvus*, DQ442537, 99.3%; *S. floridae*, AB184656, 99.3%; *S. luridiscabiei*, AF361784, 99.3%; *S. griseolus*, AB184768, 99.3%; *S. flavovirens*, DQ026635, 99.3%; *S. argenteolus*, AB045872, 99.3%; *S. californicus*, AB184755, 99.3%; *S. halstedii*, EF178695, 99.2%; *S. griseus* subsp. *griseus*, AY207604, 99.2%; *S. flavogriseus*, AJ494864, 99.2%; *S. candidus*, DQ026663, 99.1%; *S. pulveraceus*, AB184806, 99%; *S. nitrosporeus*, EF178680, 99%; *S. bacillaris*, AB184439, 99%.

Source: not known.

DNA G+C content (mol%): not known.

Type strain: AS 4.1405, ATCC 19725, CBS 464.68, BCRC 12073, CCUG 11112, CECT 3316, DSM 40137, NBRC 12741, IMET 43088, INA 4776, JCM 4345, KCTC 9017, LMG 19306, NRRL-ISP 5137, RIA 1006, UNIQEM 121, VKM Ac-970.

Sequence accession no. (16S rRNA gene): AJ781320.

48. **Streptomyces atrovirens** (*ex* Preobrazhenskaya et al. 1971) Preobrazhenskaya and Terekhova 1986, 573[VP] (Effective publication: Preobrazhenskaya and Terekhova *in* Gause, Preobrazhenskaya, Sveshnikova, Terekhova and Maximova 1983.) ("*Actinomyces atrovirens*" Preobrazhenskaya et al. 1971)

a.tro.vi'rens. L. adj. *ater -tra -trum* black, dark; L. part. adj. *virens* being green; N.L. part. adj. *atrovirens* dark green.

Spore chains in *Spirales*; spore surface is covered with long hair. On mineral agar 1, glycerol-nitrate agar, oatmeal agar, starch ammonia agar, and glycerol-asparagine agar: aerial mycelium is gray; substrate mycelium is grayish green, blue green to dark green; diffusible pigment is pale blue green, grayish green. For several strains, colorless substrate mycelium on some media is possible. On organic agar 2: aerial mycelium is gray; substrate mycelium and diffusible pigment are dark green blue to dark green, black green. The strain builds blue-green color at low pH and red-violet to gray-brownish color at high pH. Melanoid pigments are not formed. Sucrose, mannitol, rhamnose, xylose, and inositol are utilized for growth; weak growth is seen on fructose and raffinose; no growth on arabinose. Produces antibiotic no. 300 (antimetabolite Leycin).

Type strain shows the highest sequence similarity to: *S. heliomycini*, AB184712, 99.2%; *S. viridodiastaticus*, AY999852, 99.1%; *S. flavoviridis*, AB184842, 99.1%; *S. albogriseolus*, AJ494865, 99.1%; *S. coeruleorubidus*, AY999719, 99%; *S. pilosus*, AB184161, 99%.

Source: not known.

DNA G+C content (mol%): not known.

Type strain: AS 4.1595, DSM 41467, NBRC 15388, INA 1551, JCM 6913, NRRL B-16357, VKM Ac-1213.

Sequence accession no. (16S rRNA gene): DQ026672.

49. **Streptomyces aurantiacus** (Rossi Doria 1891) Waksman *in* Waksman and Lechevalier 1953, 53[AL] ["*Streptotrix aurantiaca*" (*sic*) Rossi Doria 1891, 417; "*Actinomyces aurantiacus*" (*sic*) Gasperini 1892, 222; "*Actinomyces aurantiacus*" Gasperini 1894, 84; "*Cladothrix aurantiaca*" Macé 1897, 1033; "*Nocardia aurantiaca*" Chalmers and Christopherson 1916, 268]

au.ran.ti.a'cus. N.L. n. *aurantium* generic name of the orange; N.L. masc. adj. *aurantiacus* orange colored.

Spore chains in Section *Spirales*. Short spore chains of 3 to 10 or more spores per chain form irregular hooks and loops of small diameter and imperfect spirals of one to three turns. Sporulating aerial mycelium is often absent or poorly developed on yeast-malt agar, oatmeal agar, salts-starch agar, and glycerol-asparagine agar. Spore surface is smooth. Fragmenting substrate mycelium may be seen after 12–14 d on yeast-malt agar, oatmeal agar, salts-starch agar, and glycerol-asparagine agar. One observer also notes irregular terminal swellings on some substrate hyphae as well as the presence of subglobose to clavate bodies 5–8 μm in diameter.

Color of colony: aerial mycelium is generally poorly developed or absent on ISP media. When adequate sporulating aerial mycelium is produced, it is in the Red color series (5cb or 6ec, grayish yellowish pink) in yeast-malt agar, oatmeal agar, and salts-starch agar. Reverse side of colony is yellow to yellow brown and modified by red to reddish brown on yeast-malt agar and salts-starch agar and to grayish yellowish pink or reddish orange on oatmeal agar and glycerol-asparagine agar. The substrate pigment is not a pH indicator.

Color in medium: melanoid pigments are not formed in peptone-yeast-iron agar, tyrosine agar, or tryptone-yeast broth. No pigment is found in medium in yeast-malt agar, oatmeal agar, salts-starch agar, or glycerol-asparagine agar.

D-Glucose, L-arabinose, iso-inositol, D-mannitol, D-fructose, and rhamnose are utilized for growth. Two of three observers also record utilization of D-xylose, sucrose, and raffinose.

Type strain shows the highest sequence similarity to: *S. glomeroaurantiacus*, AB249983, 100%; *S. tauricus*, AB045879, 99%.

Source: not known.

DNA G+C content (mol%): not known.

Type strain: AS 4.1429, ATCC 19822, ATCC 25429, CBS 927.69, DSM 40412, NBRC 13017, INMI 1373, JCM 4453, LMG 19358, NRRL-ISP 5412, RIA 1209, VKM Ac-44.

Sequence accession no. (16S rRNA gene): AJ781383.

Further comment: Streptomyces aurantiacus (Rossi Doria 1891) Waksman 1953[AL] emend. Lanoot et al. 2002 is an earlier heterotypic synonym of *Streptomyces albosporeus* subsp. *albosporeus* (Krainsky 1914) Waksman and Henrici 1948[AL].

50. **Streptomyces aurantiogriseus** (Preobrazhenskaya *in* Gauze, Preobrazhenskaya, Kudrina, Blinov, Ryabova and Sveshnikova 1957) Pridham, Hesseltine and Benedict 1958, 67[AL] ("*Actinomyces aurantiogriseus*" Preobrazhenskaya *in* Gauze, Preobrazhenskaya, Kudrina, Blinov, Ryabova and Sveshnikova 1957, 74)

au.ran.tio.gri'se.us. N.L. n. *Aurantium* generic name of the orange; N.L. adj. *griseus* gray; N.L. masc. adj. *aurantiogriseus* orange, gray.

Spore chains in Section *Retinaculiaperti* or *Spirales*. Spirals are open and often are irregular and poorly developed. Long spore chains of the *Retinaculum-Apertum* type and flexuous chains are common. Mature spore chains may be slow to develop; they are moderately long with more than 10 spores per chain on oatmeal agar, salts-starch agar, and glycerol-asparagine agar. Spore surface is smooth.

Color of colony: aerial mass color is in both the Red and the Gray color series on yeast-malt agar, salts-starch agar, and glycerol-asparagine agar; both colors may appear on the same medium. One observer noted an increase in gray aerial color between 14 and 21 d on glycerol-asparagine agar; this tendency to change from red to gray was also included in the original description. The color tabs most frequently were 2ec (yellowish gray) from the Gray color series and 5ge (light grayish reddish brown) from the Red color series. Reverse side of colony with no distinctive pigment (grayed yellow on oatmeal agar and glycerol-asparagine agar to orange-yellow or brown on yeast-malt and salts-starch agar). Substrate mycelium pigment is not a pH indicator.

Color in medium: melanoid pigments are produced in peptone-yeast-iron agar, tyrosine agar, and tryptone-yeast broth. Pigments other than melanoids are not found in yeast-malt agar, oatmeal agar, salts-starch agar, or glycerol-asparagine agar.

D-Glucose, L-arabinose, sucrose, D-xylose, iso-inositol, D-mannitol, D-fructose, rhamnose, and raffinose are utilized for growth.

Type strain shows the highest sequence similarity to: *S. coelicolor*, DQ442496, 99.9%; *S. griseoviridis*, AY999807, 99.3%.

Source: not known.

DNA G+C content (mol%): not known.

Type strain: AS 4.1450, ATCC 19887, ATCC 23883, CBS 663.68, BCRC 13758, DSM 40138, NBRC 12842, INA 10369/58, JCM 4346, LMG 19298, NCIMB 9849, NRRL B-5416, NRRL-ISP 5138, RIA 1130, VKM Ac-1093.

Sequence accession no. (16S rRNA gene): AY999793.

51. **Streptomyces auratus** Goodfellow, Kumar, Labeda and Sembiring 2008, 1[VP] (Effective publication: Goodfellow, Kumar, Labeda and Sembiring 2007, 198.)

au.ra'tus. L. masc. adj. *auratus* gold colored.

Spore chains are *Spirales*; spore surface is smooth. On oatmeal agar, the aerial spore mass color is gray and the substrate mycelium grayish-yellow; an orange diffusible pigment is produced, but not melanin pigments. The organism produces hydrogen sulfide and degrades adenine, arbutin, chitin, hypoxanthine, Tweens 40, 60 and 80, uric acid, and xanthine, but not casein, guanine, tyrosine, or xylan. It does not reduce nitrate or hydrolyze esculin or urea. Butan-1,4-diol, cellobiose, citric acid, dextrin, D-fructose, L-fucose, D-galactose, D-glucose, *myo*-inositol, maltose, D-mannitol, D-mannose, melezitose, methanol, propanol, pyruvic acid, raffinose, L-rhamnose, D-ribose, L-salicin, and sucrose are used as sole carbon sources, but not α-lactose. α- and L-Alanine, L-aminobutyric acid, L-glutamic acid, L-glycine, L-histidine, L-proline, L-serine, and L-threonine are used as sole carbon and nitrogen sources, but not aspartic acid, L-leucine, DL-methionine, DL-norleucine, L-phenylalanine, L-tryptophan, or L-valine. Grows from pH 5 to 10, at 25 and 37°C, but not at 10 or 40°C. Growth occurs in the presence of 13% (w/v) NaCl. Resistant to carbenicillin, cefoxitin, cephaloridine, chlortetracycline hydrochloride, doxycycline hydrochloride, rifampin, and novobiocin, but sensitive to cefoxitin, cephaloridine, doxycycline hydrochloride, fusidic acid, lincomycin hydrochloride, and oleandomycin phosphate.

Type strain shows the highest sequence similarity to: *S. sioyaensis*, DQ026654, 99%.

Source: not known.
DNA G+C content (mol%): 65.6.
Type strain: DSM 41897, NRRL 8097.
Sequence accession no. (16S rRNA gene): AJ391816.

52. **Streptomyces aureocirculatus** (Krasil'nikov and Yuan *in* Krasil'nikov 1965) Pridham 1970, 8[AL] ("*Actinomyces aureocirculatus*" Krasil'nikov and Yuan *in* Krasil'nikov 1965, 33)

au.re.o.cir.cu.la'tus. L. adj. *aureus* golden; L. part. adj. *circulatus* made circular or round; N.L. masc. adj. *aureocirculatus* golden-curled.

Spore chains in Section *Rectiflexibiles* with some irregular hooks and loops or imperfect spirals. Mature spore chains, when present, are generally long, sometimes with more than 50 spores per chain. Aerial mycelium is poorly developed on all ISP media. Mature spore chains may be found on yeast-malt agar, oatmeal agar, and salts-starch agar. One observer records fragmentation of substrate mycelium on glycerol-asparagine agar at 21 d. Another observer notes the presence of globular bodies and sclerotia on yeast-malt agar, oatmeal agar, and salts-starch agar and conidia-like spores on the substrate mycelium on salts-starch agar at 7 d. Spore surface is smooth.

Color of colony: sporulating aerial mycelium is inadequate for determination of aerial mass color on yeast-malt agar, oatmeal agar, salts-starch agar, and glycerol-asparagine agar. The scant aerial mycelium that develops on yeast-malt agar, oatmeal agar, and salts-starch agar is in the White color series. Reverse side of colony with no distinctive pigments (pale yellow to light grayish yellow) on yeast-malt agar, oatmeal agar, salts-starch agar, and glycerol-asparagine agar.

Color in medium: melanoid pigments are not formed in peptone-yeast-iron agar, tyrosine agar, or tryptone-yeast broth. Yellow pigment is found in the medium in yeast-malt agar, oatmeal agar, and salts-starch agar. This pigment is not pH-sensitive when tested with 0.05 M NaOH or HCl.

D-Glucose, D-xylose, iso-inositol, D-mannitol, and D-fructose are utilized for growth. Reports vary on utilization of L-arabinose, rhamnose, sucrose, and raffinose. Only one of three observers records utilization of rhamnose and raffinose.

Type strain shows the highest sequence similarity to: *S. novaecaesareae*, AB184357, 99.5%; *S. galilaeus*, AB045878, 99.5%; *S. bobili*, AB249925, 99.4%; *S. pseudovenezuelae*, AB184233, 99.1%; *S. resistomycificus*, AB184166, 99.1%; *S. phaeoluteigriseus*, AJ391815, 99%; *S. chartreusis*, AB184839, 99%.

Source: not known.
DNA G+C content (mol%): not known.
Type strain: AS 4.1609, ATCC 15851, ATCC 19823, ATCC 25430, CBS 928.69, DSM 40386, NBRC 13018, INMI 735, JCM 4454, NRRL B-3324, NRRL-ISP 5386, RIA 1210, RIA 682.
Sequence accession no. (16S rRNA gene): AB184260.

53. **Streptomyces aureofaciens** Duggar 1948, 177[AL]

au.re.o.fa'ci.ens. L. adj. *aureus* golden; L. part. adj. *faciens* producing; N.L. part adj. *aureofaciens* golden-producing, referring pigment produced.

Spore chains in Section *Retinaculiaperti* but chains representative of Section *Rectiflexibiles* are also common. Mature spore chains generally have 10–50 spores per chain. This morphology is seen on yeast-malt agar, oatmeal agar, salts-starch agar, and glycerol-asparagine agar. Spore surface is smooth.

Color of colony: aerial mass color in the Gray color series on yeast-malt agar, oatmeal agar, salts-starch agar, and glycerol-asparagine agar. Reverse side of colony with no distinctive pigments (grayed yellow, orange-yellow, or brown on yeast-malt agar, and grayed yellow or greenish yellow on oatmeal agar, salts-starch agar, and glycerol-asparagine agar). Substrate mycelium pigment is not a pH indicator.

Color in medium: melanoid pigments are not formed in peptone-yeast-iron agar, tyrosine agar, or tryptone-yeast broth. No pigment is found in medium in yeast-malt agar, oatmeal agar, salts-starch agar, or glycerol-asparagine agar.

D-Glucose, L-arabinose, sucrose, D-xylose, and D-fructose are utilized for growth. No growth or only trace of growth on iso-inositol, D-mannitol, rhamnose, and raffinose.

Type strain shows the highest sequence similarity to: *S. psammoticus*, AY999862, 99.6%; *S. avellaneus*, AB184413, 99.5%; *S. xanthocidicus*, AY999858, 99.1%.

Source: not known.
DNA G+C content (mol%): not known.
Type strain: ATCC 10762, ATCC 23884, CBS 434.51, CBS 664.68, CCM 3239, BCRC 11610, CECT 3206, CIP 57.11, DSM 40127, HAMBI 313, HAMBI 1072, HUT 6048, HUT 6097, ICMP 499, IFM 1042, IFM 1218, IFM 1219, NBRC 12594, NBRC 12843, IMET 43577, JCM 4624, JCM 4008, KACC 20180, LMG 5968, NCAIM B.01479, NCIMB 8234, NRRL 2209, NRRL B-2183, NRRL B-2657, NRRL B-5404, NRRL-ISP 5127, RIA 1129, RIA 57, VKM Ac-771.
Sequence accession no. (16S rRNA gene): AY207608.

54. **Streptomyces aureorectus** (*ex* Taig, Solovieva and Braginskaya 1969) Taig and Solovieva 1986, 573[VP] (Effective publication: Taig and Solovieva *in* Gause, Preobrazhenskaya, Sveshnikova, Terekhova and Maximova 1983.) ("*Actinomyces aureorectus*" Taig, Solovieva and Braginskaya 1969)

au.re.o.rec'tus. L. adj. *aureus* golden; L. adj. *rectus* straight; N.L. masc. adj. *aureorectus* golden, straight.

Spore chains are straight, spores are smooth. On mineral agar 1 and glycerol-nitrate agar: aerial mycelium is white gray, substrate mycelium is lemon yellow, yellow, dark yellow; diffusible pigment is light yellow, yellow. On starch-ammonia agar (ISP 4): no aerial mycelium; colorless substrate mycelium; no diffusible pigment. On organic agar 2: no aerial mycelium; colorless substrate mycelium; no diffusible pigment. On glycerol-asparagine agar (ISP 5): white aerial mycelium; substrate mycelium and diffusible pigment are yellow. Melanoid pigments are not formed. Good digestion of xylose, galactose, maltose, and raffinose. Antibiotic aurenin is formed.

Type strain shows the highest sequence similarity to: *S. asterosporus*, AB184706, 100%; *S. calvus*, AB184329, 99.9%; *S. virens*, DQ442554, 99.9%; *S. minutiscleroticus*, EF178696, 99%; *S. geysiriensis*, DQ442501, 99%; *S. djakartensis*, AB184657, 99%; *S. anandii*, AB184402, 99%; *S. tuirus*, AB184690, 99%.

Source: not known.
DNA G+C content (mol%): not known.

Type strain: 2843-10, DSM 41692, NBRC 15896, INA A-78, JCM 9947, RIA 553, VKM Ac-1828.

Sequence accession no. (16S rRNA gene): AB184710.

55. **Streptomyces aureoversilis** corrig. (Locci, Baldacci and Petrolini Baldan 1969) Witt and Stackebrandt 1996, 836[VP] (Effective publication: Witt and Stackebrandt 1990, 370.) (*Streptoverticillium aureoversile* corrig. Locci, Baldacci and Petrolini Baldan 1969, 59)

au.re.o.ver'si.lis. L. adj. *aureus* golden; L. adj. *versilis* that may be turned, movable; N.L. masc. adj. *aureoversilis* golden, movable.

On potato-glucose agar (Baldacci et al., 1954) and Oxoid nutrient agar, reverse colors are brown; traces of whitish aerial mycelium on potato-glucose agar (Baldacci et al., 1954), more abundant on Oxoid nutrient agar. Growth is fair on Bacto Czapek agar and Casamino acids Czapek agar (1 g/l Difco vitamin-free Casamino acids, replacing sodium nitrate), the reverse being dirty pinkish.

On Casamino acids Czapek agar (1 g/l Difco vitamin-free Casamino acids, replacing sodium nitrate), glucose-asparagine agar (ISP medium 5 with 1% glucose replacing glycerol), glycerol-asparagine agar (ISP medium 5), inorganic salts-starch agar (ISP medium 4), and yeast extract-malt extract agar (ISP medium 2), aerial mycelium is pinkish. On inorganic salts-starch agar (ISP medium 4) and glucose-asparagine agar (ISP medium 5 with 1% glucose replacing glycerol), reverse color is red.

Grows equally well at 27 and 37°C; aerial mycelium production slightly less at 37°C and slow in appearing. The reference strain produces tetraene 380 and pentaene 380 and exhibits anti-bacterial activity.

Type strain shows the highest sequence similarity to: *S. hiroshimensis*, AB184789, 99.5%; *S. cinnamoneus*, AB184850, 99.2%; *S. pseudoechinosporeus*, AB184100, 99.1%; *S. caeruleus*, EF178675, 99.1%.

Source: not known.

DNA G+C content (mol%): not known.

Type strain: AS 4.1641, ATCC 15853, ATCC 25433, CBS 664.69, BCRC 12451, DSM 40387, NBRC 13021, INMI 380, JCM 4457, NRRL B-3325, NRRL-ISP 5387, RIA 1213, RIA 681, VKM Ac-884.

Sequence accession no. (16S rRNA gene): AB184855.

Further comments: according to Hatano et al. (2003), *Streptomyces aureoversilis* corrig. (Locci et al. 1969) Witt and Stackebrandt 1991 is a later heterotypic synonym of *Streptomyces hiroshimensis* (Shinobu 1955) Witt and Stackebrandt 1991.

56. **Streptomyces aureoverticillatus** (Krasil'nikov and Yuan 1960) Pridham 1970, 54[AL] ("*Actinomyces aureoverticillatus*" Krasil'nikov and Yuan 1960, 487)

au.re.o.ver.ti.cil.la'tus. L. adj. *aureus* golden; N.L. adj. *verticillatus* whorled; N.L. adj. *aureoverticillatus* golden, whorled (referring to color of the vegetative mycelium of the organism and nature of its morphology).

Spore chains in Section *Retinaculiaperti* but spore chains representative of Section *Rectiflexibiles* are common. Pseudoverticillate sporophores (suggesting Section *Verticillati*) are found, especially on salts-starch agar. Verticils are not evenly spaced on an enlarged axial hypha as in true verticillate forms. Mature spore chains generally have10–50 spores per chain. This morphology is seen on yeast-malt agar, oatmeal agar, salts-starch agar, and glycerol-asparagine agar, but aerial mycelium is not abundant on yeast-malt agar or oatmeal agar. Spore surface is smooth.

Color of colony: aerial mass color in the Red color series on salts-starch agar and glycerol-asparagine agar (White is also found on these media and poorly sporulated surface on yeast-malt agar). Reverse side of colony is grayed yellow and modified to red on yeast-malt agar, oatmeal agar, salts-starch agar, and glycerol-asparagine agar. Substrate pigment is a pH indicator; it changes from orange to yellow by addition of 0.05 M NaOH, and from orange to red with HCl.

Color in medium: melanoid pigments not formed in peptone-yeast-iron agar and tyrosine agar. No pigment found in medium in yeast-malt agar, oatmeal agar, salts-starch agar, or glycerol-asparagine agar.

D-Glucose, L-arabinose, D-xylose, iso-inositol, D-mannitol, D-fructose, and raffinose are utilized for growth. No growth or only trace of growth on rhamnose. Variable reports on growth with sucrose.

Type strain shows the highest sequence similarity to: *S. diastaticus* subsp. *diastaticus*, AB184785, 99%; *S. gougerotii*, AB184742, 99%; *S. intermedius*, AB184277, 99%; *S. rutgersensis* subsp. *rutgersensis*, AB184170, 99%.

Source: not known.

DNA G+C content (mol%): not known.

Type strain: AS 4.1666, ATCC 15854, ATCC 19726, CBS 465.68, BCRC 12185, DSM 40080, NBRC 12742, INMI 1007, JCM 4347, NRRL B-3326, NRRL-ISP 5080, RIA 1007, RIA 679, VKM Ac-48.

Sequence accession no. (16S rRNA gene): AY999774.

57. **Streptomyces aureus** Manfio, Atalan, Zakrzewska-Czerwinska, Mordarski, Rodriguez, Collins and Goodfellow 2003b, 1219[VP] (Effective publication: Manfio, Atalan, Zakrzewska-Czerwinska, Mordarski, Rodriguez, Collins and Goodfellow 2003a, 254.)

au're.us. L. masc. adj. *aureus* golden, referring to the color of the diffusible pigment on oatmeal agar.

Forms extensively branched substrate and aerial hyphae. Open looped chains of smooth-surfaced spores are evident on aerial hyphae. Aerial spore mass is gray; a reddish-orange colored substrate mycelium is formed on Bennett's, glycerol-asparagine, inorganic salts-starch, oatmeal, and yeast extract-malt extract agars. A golden colored diffusible pigment is produced on oatmeal agar. Hippurate is hydrolyzed and hypoxanthine and xanthine are degraded. Cleaves L-citrulline-7-amino-methylcoumarin (7AMC), L-phenylalanine-7AMC (endopeptidase substrates), L-cysteine-7AMC, L-glutamine-7AMC, L-ornithine-7AMC, L-proline-7AMC (exopeptidase substrates), and 4-methylumbelliferone (4MU)-elaidate (organic ester), but not methoxysuccinyl-L-alanine (endopeptidase substrate), L-pyroglutamate-7AMC (exopeptidase substrate), 4MU-β-D-galactopyranoside, 4MU-*N*-acetyl-β-D-glucosaminide, 4MU-α-D-glucoside, 4MU-2-deoxy-β-D-glucopyranoside, 4MU-β-D-xyloside (glycosides), or 4MU-pyrophosphate (inorganic ester). *myo*-Inositol, inulin, D(+)mannitol, citrate, and pyruvate are used as sole carbon sources for energy and growth but not dextran. L-Hydroxyproline is used as a sole carbon and nitrogen source, but not histidine. Grows between 10 and 35°C and in the presence of thallous acetate (0.001%, w/v), but does not show antimicrobial activity against *Bacillus subtilis* NCIMB 3610, *Candida albicans* CBS 562, *Micrococcus luteus* NCIMB 196, or *Streptomyces murinus* ISP 5091.

Type strain shows the highest sequence similarity to: *S. durmitorensis*, DQ067287, 99.6%; *S. kanamyceticus*, DQ442511, 99.5%.

Source: not known.

DNA G+C content (mol%): 67.0–73.0.

Type strain: B7319, DSM 41785, JCM 12605, NBRC 100912, NCIMB 13927.

Sequence accession no. (16S rRNA gene): AB249976.

58. **Streptomyces avellaneus** Baldacci and Grein 1966, 195[AL]

a.vel.la'ne.us. N.L. masc. adj. *avellaneus* (from L. n. *avellana* hazel) hazel colored, referring to color of the aerial mycelium of the organism.

Spore chains in Section *Rectiflexibiles*. Mature spore chains are moderately long with 10 to 50 or more spores per chain. This morphology is seen on yeast-malt agar, oatmeal agar, salts-starch agar, and glycerol-asparagine agar. Spore surface is smooth.

Color of colony: aerial mass color in the Gray color series (2dc, yellowish gray; 3fe, light brownish gray; or 5fe, light grayish reddish brown) on yeast-malt agar, oatmeal agar, salts-starch agar, and glycerol-asparagine agar. Reverse side of colony with no distinctive pigment (strong yellowish brown to dark brown on yeast-malt agar; grayish yellow to orange yellow, light brown, or strong brown on oatmeal agar, salts-starch agar, and glycerol-asparagine agar).

Color in medium: melanoid pigments are not formed in peptone-yeast-iron agar, tyrosine agar, or tryptone-yeast broth. No pigment or only a trace of yellow is found in medium in yeast-malt agar, oatmeal agar, salts-starch agar, and glycerol-asparagine agar.

D-Glucose, D-fructose, and sucrose are utilized for growth. No growth or only traces of growth with L-arabinose, D-xylose, iso-inositol, D-mannitol, rhamnose, and raffinose.

Type strain shows the highest sequence similarity to: *S. psammoticus*, AY999862, 100%; *S. aureofaciens*, AY207608, 99.5%; *S. chrysomallus* subsp. *fumigatus*, AB184645, 99.2%; *S. purpureus*, AJ781324, 99.2%; *S. aburaviensis*, AY999779, 99%.

Source: not known.

DNA G+C content (mol%): not known.

Type strain: AS 4.1687, ATCC 23730, CBS 752.72, BCRC 12219, CMI 126840, DSM 40554, NBRC 13451, JCM 4725, JCM 4321, NCIMB 11000, NRRL B-3447, NRRL-ISP 5554, RIA 1412, VKM Ac-1720.

Sequence accession no. (16S rRNA gene): AB184413.

59. **Streptomyces avermectinius** Takahashi, Matsumoto, Seino, Ueno, Iwai and Ōmura 2002, 2167[VP]

a.ver.mec.ti'ni.us. N.L. adj. *avermectinius* pertaining to avermectin, an antibiotic produced by the organism.

Forms spiral spore chains. Spores are oval in shape (0.8 × 1.2 µm) and have a smooth surface. Soluble pigment is produced. Vegetative mycelia are brown. Aerial mass color is gray. Melanin and H_2S production, hydrolysis of starch and casein, liquefaction of gelatin, and peptonization of milk are positive. Adenine, casein, hypoxanthine and xanthine are decomposed but not cellulose. Grows in the presence of 5% (w/v) NaCl, but is sensitive to streptomycin (20 µg/ml) and novobiocin (20 µg/ml). Arabinose, fructose, glucose, inositol, lactose, maltose, mannitol, mannose, raffinose, rhamnose, sucrose, and xylose are decomposed.

For sequence similarity, see type strain of *Streptomyces avermitilis*. Type strain shows DNA–DNA similarity to: *S. bottropensis* DSM 40262^T, 15%; *S. cinnabarinus* DSM 40467^T, 21%; *S. galbus* DSM 40089^T, 33%; *S. griseochromogenes* DSM 40499^T, 7%; *S. luteogriseus* DSM 40483^T, 22%; *S. mirabilis* DSM 40553^T, 16%; *S. olivochromogenes* DSM 40451^T, 26%; *S. phaeochromogenes* DSM 40073^T, 20%.

Source: not known.

DNA G+C content (mol%): 70.3.

Type strain: MA-4680, ATCC 31267, JCM 5070, NCIMB 12804, NRRL 8165.

Sequence accession no. (16S rRNA gene): AB078897.

Further comments: Streptomyces avermectinius Takahashi et al. 2002 is a later homotypic synonym of *Streptomyces avermitilis* (*ex* Burg et al. 1979) Kim and Goodfellow 2002.

60. **Streptomyces avermitilis** (*ex* Burg, Miller, Baker, Birnbaum, Curri, Hartman, Kong, Monaghan, Olson, Putter, Tunac, Wallick, Stapley, Oiwa and Ōmura 1979) Kim and Goodfellow 2002, 2013[VP] ("*Streptomyces avermitilis*" Burg, Miller, Baker, Birnbaum, Curri, Hartman, Kong, Monaghan, Olson, Putter, Tunac, Wallick, Stapley, Oiwa and Ōmura 1979, 363)

a.ver.mi'ti.lis. N.L. masc. adj. *avermitilis* intended to mean avermectin producer.

The spore chains consist of 15 or more spherical to oval-shaped spores with smooth surfaces. Sporulation occurs on standard media such as egg albumin, glycerol-asparagine, inorganic salts-starch, and oatmeal agars. A gray aerial spore mass is formed on oatmeal agar; the colony reverse is dark brown to tan. Melanin pigments are produced on peptone-yeast extract-iron agar and brown diffusible pigments are produced on a range of standard media. Forms an extensively branched substrate mycelium and aerial hyphae that differentiate into long, compact spiral chains which become more open as the culture ages. The culture grows well at 28 and 37°C, but does not grow at 50°C. It metabolizes casein, but not tyrosine. It liquefies gelatin and uses glucose, maltose, mannose, and rhamnose, but not cellulose as sole carbon sources for energy and growth. Avermectins, a family of 16-membered antiparasitic macromolecules, are produced.

Type strain shows no sequence similarity over 99%. Type strain shows DNA–DNA similarity to: *S. cinnabarinus* DSM 40467^T, 98±7%; *S. griseochromogenes* DSM 40499^T, 99±0%.

Source: isolated from a soil sample collected at Kawana, Ito City, Shizuoka Prefecture, Japan.

DNA G+C content (mol%): not known.

Type strain: MA-4680, ATCC 31267, JCM 5070, NBRC 14893, NCIMB 12804, NRRL 8165.

Sequence accession no. (16S rRNA gene): BA000030.

Further comments: Streptomyces avermitilis (*ex* Burg et al. 1979) Kim and Goodfellow 2002 is an earlier homotypic synonym of *Streptomyces avermectinius* Takahashi et al. 2002.

61. **Streptomyces avidinii** Stapley, Mata, Miller, Demny and Woodruff 1964, 20[AL]

a.vi.di'ni.i. N.L. n. *avidinum* avidin the name of a biotin-binding protein; N.L. gen. n. *avidinii* of avidin.

Produces the antibiotic MSD-235 complex, streptavidin, and antibiotic MSD-235S, a synergistic anti-bacterial complex; poor growth on Czapek's solution agar.

Type strain shows the highest sequence similarity to: *S. nojiriensis*, AJ781355, 99.7%; *S. subrutilus*, X80825, 99.7%; *S. sporoраveus*, AJ781370, 99.7%; *S. xanthophaeus*, DQ442560, 99.7%; *S. goshikiensis*, EF178693, 99.6%; *S. cirratus*, AY999794, 99.6%; *S. vinaceus*, AB184394, 99.6%; *S. sporoverrucosus*, DQ442544, 99.5%; *S. cinnamonensis*, AB184707, 99.5%; *S. colombiensis*, DQ026646, 99.5%; *S. virginiae*, D85119, 99.3%.

Source: not known.

DNA G+C content (mol%): not known.

Type strain: MA-833, AS 4.1583, ATCC 27419, CBS 730.72, BCRC 13384, DSM 40526, NBRC 13429, JCM 4726, IMET 43538, KCTC 9757, NCIMB 11996, NRRL 3077, NRRL-ISP 5526, PCM 2342, RIA 1390, VKM Ac-1074.

Sequence accession no. (16S rRNA gene): AB184395.

62. **Streptomyces azureus** Kelly, Kutscher and Tuoti 1959, 1334[AL]

a.zur'e.us. N.L. masc. adj. *azureus* azure-blue, referring to the color of the aerial mycelium of the organism.

Spore chains in Section *Spirales*. Some sporophores supporting spiral spore chains appear to arise singly, in pairs, or in simple or branched whorls along an axial hypha (suggesting verticillate morphology) especially on salts-starch agar and glycerol-asparagine agar. Mature spore chains generally have 10–50 spores per chain. This morphology is seen on yeast-malt agar, oatmeal agar, salts-starch agar, and glycerol-asparagine agar. Spore surface is smooth.

Color of colony: aerial mass color in the Blue color series on yeast-malt agar, oatmeal agar, salts-starch agar, and glycerol-asparagine agar. Reverse side of colony with no distinctive pigments on yeast-malt agar, oatmeal agar, salts-starch agar, and glycerol-asparagine agar; substrate pigment is not a pH indicator.

Color in medium: melanoid pigments formed in peptone-yeast-iron agar, tyrosine agar and tryptone-yeast extract broth. Pigments (other than melanoids) not formed in yeast-malt agar, oatmeal agar, salts-starch agar, or glycerol-asparagine agar.

D-Glucose, L-arabinose, sucrose, D-xylose, iso-inositol, D-mannitol, D-fructose, rhamnose, and raffinose are utilized for growth.

Type strain shows the highest sequence similarity to: *S. capillispiralis*, AB184577, 99.5%; *S. gancidicus*, AB184660, 99.5%; *S. pseudogriseolus*, DQ442541, 99.5%; *S. caelestis*, X80824, 99.5%; *S. cellulosae*, DQ442495, 99.4%; *S. levis*, AB184670, 99.4%; *S. paradoxus*, AB184628, 99.3%; *S. djakartensis*, AB184657, 99.2%; *S. coerulescens*, AY999720, 99.2%; *S. tuirus*, AB184690, 99.2%; *S. bellus*, AB184849, 99.2%; *S. carpinensis*, AB184574, 99.1%; *S. minutiscleroticus*, EF178696, 99.1%; *S. africanus*, AY208912, 99.1%; *S. mutabilis*, EF178679, 99.1%; *S. geysiriensis*, DQ442501, 99.1%; *S. afghaniensis*, AJ399483, 99%; *S. rochei*, AB184237, 99%; *S. plicatus*, AB184291, 99%; *S. vinaceusdrappus*, AY999929, 99%; *S. matensis*, AB184221, 99%; *S. ghanaensis*, AY999851, 99%.

Source: not known.

DNA G+C content (mol%): not known.

Type strain: AS 4.1675, ATCC 14921, ATCC 19728, CBS 467.68, BCRC 12479, DSM 40106, NBRC 12744, IMET 43765, JCM 4564, NRRL B-2655, NRRL-ISP 5106, PCM 2313, RIA 1009, UNIQEM 122, VKM Ac-719.

Sequence accession no. (16S rRNA gene): EF178674.

63. **Streptomyces baarnensis** Pridham, Hesseltine and Benedict 1958, 74[AL]

ba.arn.en'sis. N.L. masc. adj. *baarnensis* of or belonging to Baarn, a community in the Netherlands province of Utrecht.

This strain has apparently lost the ability to produce good sporulating aerial mycelium. Spore chains, when found, are straight (Section *Rectiflexibiles*) and usually contain more than 10 spores per chain. Spore surface is smooth.

Color of colony: aerial mycelium on yeast-malt agar, oatmeal agar, salts-starch agar, or glycerol-asparagine agar is inadequate for determination of aerial mass color. The original description of Duché (op. cit) describes early appearance of abundant greenish aerial mycelium on comparable media. Reverse side of colony with no distinctive pigment; yellow-brown on yeast-malt agar, light grayish yellow on oatmeal agar, salts-starch agar, and glycerol-asparagine agar.

Color in medium: melanoid pigments not formed in peptone-yeast-iron agar, tyrosine agar, or tryptone-yeast broth. No pigment found in medium in yeast-malt agar, oatmeal agar, salts-starch agar, or glycerol-asparagine agar.

D-Glucose, L-arabinose, D-xylose, iso-inositol, D-mannitol, D-fructose, and rhamnose are utilized for growth. No growth or only trace of growth on raffinose. Utilization of sucrose is doubtful.

Type strain shows the highest sequence similarity to: *S. griseoplanus*, AY999894, 99.9%; *S. fimicarius*, AY999784, 99.9%; *S. flavofuscus*, AB249935, 99.9%; *S. praecox*, AB184293, 99.9%; *S. pluricolorescens*, DQ442540, 99.8%; *S. sindenensis*, AB184759, 99.8%; *S. anulatus*, DQ026637, 99.8%; *S. griseinus*, AB184205, 99.8%; *S. acrimycini*, AY999889, 99.8%; *S. badius*, AY999783, 99.8%; *S. mediolani*, AB184674, 99.8%; *S. lavendulae* subsp. *lavendulae*, AB184080, 99.8%; *S. rubiginosohelvolus*, AB184240, 99.8%; *S. lipmanii*, AB184148, 99.7%; *S. fulvorobeus*, AB184711, 99.7%; *S. albovinaceus*, AB249958, 99.7%; *S. microflavus*, DQ445795, 99.7%; *S. cavourensis* subsp. *washingtonensis*, DQ026671, 99.7%; *S. argenteolus*, AB045872, 99.7%; *S. parvus*, DQ442537, 99.7%; *S. globisporus* subsp. *globisporus*, EF178686, 99.7%; *S. cinereorectus*, AB184646, 99.7%; *S. cyaneofuscatus*, AB184860, 99.7%; *S. alboviridis*, AB184256, 99.7%; *S. californicus*, AB184755, 99.7%; *S. griseolus*, AB184768, 99.6%; *S. griseus* subsp. *griseus*, AY207604, 99.6%; *S. floridae*, AB184656, 99.6%; *S. flavovirens*, DQ026635, 99.5%; *S. halstedii*, EF178695, 99.5%; *S. atroolivaceus*, AJ781320, 99.4%; *S. finlayi*, AY999788, 99.4%; *S. flavogriseus*, AJ494864, 99.4%; *S. olivoviridis*, AB184227, 99.4%; *S. pulveraceus*, AB184806, 99.4%; *S. bacillaris*, AB184439, 99.4%; *S. clavifer*, DQ026670, 99.3%; *S. yanii*, AB006159, 99.3%; *S. nitrosporeus*, EF178680, 99.3%; *S. albolongus*,

AB184425, 99.2%; *S. griseobrunneus*, AB249912, 99.2%; *S. celluloflavus*, AB184476, 99.2%; *S. sanglieri*, AB249945, 99.2%; *S. atratus*, DQ026638, 99.1%; *S. cavourensis* subsp. *cavourensis*, DQ445791, 99.1%; *S. luridiscabiei*, AF361784, 99.1%; *S. spiroverticillatus*, AB184814, 99.1%; *S. mutomycini*, AB249951, 99.1%; *S. gelaticus*, DQ026636, 99.1%; *S. candidus*, DQ026663, 99%.

Source: not known.

DNA G+C content (mol%): not known.

Type strain: AS 4.1607, ATCC 23885, CBS 306.55, CBS 665.68, DSM 40232, HAMBI 1044, NBRC 14727, JCM 4349, IMET 43091, NRRL B-1902, NRRL-ISP 5232, RIA 1172, VKM Ac-1774.

Sequence accession no. (16S rRNA gene): EF178688.

64. **Streptomyces bacillaris** (Krasil'nikov 1958) Pridham 1970, 9[AL] (*"Actinomyces bacillaris"* Krasil'nikov 1958, 258)

ba.cil.lar'is.L. n. *bacillus* a rodlet; L. masc. suff. *-aris* suffix used with the sense of pertaining to; N.L. masc. adj. *bacillaris* pertaining to a rodlet.

Spore chains in Section *Rectiflexibiles*. Mature spore chains are generally long with 10 to more than 50 spores per chain. This morphology is seen on yeast-malt agar, oatmeal agar, salts-starch agar, and glycerol-asparagine agar. Spore surface is smooth.

Color of colony: aerial mass color in the Yellow color series (2ba, 1ba, pale yellow) on yeast-malt agar, oatmeal agar, salts-starch agar, and glycerol-asparagine agar; some parts of the aerial mycelium on yeast-malt agar may also be in the Gray color series (3ge, light grayish yellow brown) and white aerial mycelium is also observed. Reverse side of colony with no distinctive pigments (orange-yellow to yellowish brown on yeast-malt agar; light grayish yellow on oatmeal agar; yellowish or orange-yellow on salts-starch agar or glycerol-asparagine agar).

Color in medium: melanoid pigments are formed in peptone-yeast-iron agar and tryptone-yeast broth, but not on tyrosine agar. No pigment or only a trace of yellow is found in the medium in yeast-malt agar, oatmeal agar, salts-starch agar, and glycerol-asparagine agar.

D-Glucose, D-mannitol, and D-fructose are utilized for growth. Utilization of D-xylose is doubtful. Only traces of growth are seen with L-arabinose, iso-inositol, rhamnose, sucrose, and raffinose.

Type strain shows the highest sequence similarity to: *S. globisporus* subsp. *globisporus*, EF178686, 99.7%; *S. pluricolorescens*, DQ442540, 99.6%; *S. griseinus*, AB184205, 99.6%; *S. rubiginosohelvolus*, AB184240, 99.6%; *S. sindenensis*, AB184759, 99.6%; *S. badius*, AY999783, 99.6%; *S. mediolani*, AB184674, 99.6%; *S. californicus*, AB184755, 99.6%; *S. griseoplanus*, AY999894, 99.5%; *S. fulvorobeus*, AB184711, 99.5%; *S. albolongus*, AB184425, 99.5%; *S. albovinaceus*, AB249958, 99.5%; *S. parvus*, DQ442537, 99.5%; *S. floridae*, AB184656, 99.5%; *S. cavourensis* subsp. *washingtonensis*, DQ026671, 99.5%; *S. anulatus*, DQ026637, 99.5%; *S. flavofuscus*, AB249935, 99.5%; *S. fimicarius*, AY999784, 99.5%; *S. lavendulae* subsp. *lavendulae*, AB184080, 99.5%; *S. praecox*, AB184293, 99.5%; *S. baarnensis*, EF178688, 99.4%; *S. acrimycini*, AY999889, 99.4%; *S. cavourensis* subsp. *cavourensis*, DQ445791, 99.4%; *S. celluloflavus*, AB184476, 99.4%; *S. lipmanii*, AB184148, 99.4%; *S. griseobrunneus*, AB249912, 99.4%; *S. nitrosporeus*, EF178680, 99.4%; *S. cinereorectus*, AB184646, 99.4%; *S. cyaneofuscatus*, AB184860, 99.4%; *S. flavovirens*, DQ026635, 99.4%; *S. microflavus*, DQ445795, 99.4%; *S. alboviridis*, AB184256, 99.4%; *S. argenteolus*, AB045872, 99.3%; *S. flavogriseus*, AJ494864, 99.3%; *S. halstedii*, EF178695, 99.2%; *S. griseus* subsp. *griseus*, AY207604, 99.2%; *S. finlayi*, AY999788, 99.2%; *S. griseolus*, AB184768, 99.2%; *S. pulveraceus*, AB184806, 99.2%; *S. luridiscabiei*, AF361784, 99.2%; *S. clavifer*, DQ026670, 99.1%; *S. sanglieri*, AB249945, 99%; *S. olivoviridis*, AB184227, 99%; *S. atroolivaceus*, AJ781320, 99%; *S. mutomycini*, AB249951, 99%; *S. yanii*, AB006159, 99%.

Source: not known.

DNA G+C content (mol%): not known.

Type strain: AS 4.1548, ATCC 15855, CBS 788.72, DSM 40598, NBRC 13487, INMI 445, JCM 4727, KCTC 9018, NRRL B-3038, NRRL-ISP 5598, RIA 1448, RIA 336, VKM Ac-58.

Sequence accession no. (16S rRNA gene): AB184439.

65. **Streptomyces badius** (Kudrina *in* Gauze, Preobrazhenskaya, Kudrina, Blinov, Ryabova and Sveshnikova 1957) Pridham, Hesseltine and Benedict 1958, 58[AL] (*"Actinomyces badius"* Kudrina *in* Gauze, Preobrazhenskaya, Kudrina, Blinov, Ryabova and Sveshnikova 1957, 87)

ba'di.us. L. masc. adj. *badius* brown.

Spore chains in Section *Rectiflexibiles*. Mature spore chains generally have 10–50 spores per chain. This morphology is found on yeast-malt agar, oatmeal agar, salts-starch agar, and glycerol-asparagine agar. Spore surface is smooth.

Color of colony: aerial mass color in the Yellow color series on yeast-malt agar, oatmeal agar, salts-starch agar, and glycerol-asparagine agar (or intermediate between Gray series and Yellow series – one observer). Reverse side of colony with no distinctive pigments on yeast-malt agar, oatmeal agar, salts-starch agar, and glycerol-asparagine agar; substrate pigment is not a pH indicator.

Color in medium: melanoid pigments not formed in peptone-yeast-iron agar, but some variable darkening is observed in tyrosine agar and tryptone-yeast broth; other pigments are not formed.

D-Glucose, L-arabinose, D-xylose, D-mannitol, and D-fructose are utilized for growth. No growth or only trace of growth on sucrose, iso-inositol, rhamnose, and raffinose.

Type strain shows the highest sequence similarity to: *S. sindenensis*, AB184759, 100%; *S. albovinaceus*, AB249958, 100%; *S. griseoplanus*, AY999894, 100%; *S. anulatus*, DQ026637, 100%; *S. globisporus* subsp. *globisporus*, EF178686, 100%; *S. pluricolorescens*, DQ442540, 100%; *S. praecox*, AB184293, 100%; *S. flavofuscus*, AB249935, 100%; *S. fimicarius*, AY999784, 100%; *S. mediolani*, AB184674, 100%; *S. griseinus*, AB184205, 100%; *S. rubiginosohelvolus*, AB184240, 100%; *S. californicus*, AB184755, 99.9%; *S. lavendulae* subsp. *lavendulae*, AB184080, 99.9%; *S. cavourensis* subsp. *washingtonensis*, DQ026671, 99.9%; *S. acrimycini*, AY999889, 99.9%; *S. parvus*, DQ442537, 99.9%; *S. fulvorobeus*, AB184711, 99.8%; *S. lipmanii*, AB184148, 99.8%; *S. microflavus*, DQ445795, 99.8%; *S. cyaneofuscatus*, AB184860, 99.8%; *S. floridae*, AB184656, 99.8%; *S. alboviridis*, AB184256, 99.8%;

S. baarnensis, EF178688, 99.8%; *S. cinereorectus*, AB184646, 99.8%; *S. griseus* subsp. *griseus*, AY207604, 99.7%; *S. griseolus*, AB184768, 99.7%; *S. luridiscabiei*, AF361784, 99.7%; *S. flavovirens*, DQ026635, 99.7%; *S. argenteolus*, AB045872, 99.7%; *S. bacillaris*, AB184439, 99.6%; *S. pulveraceus*, AB184806, 99.6%; *S. halstedii*, EF178695, 99.6%; *S. flavogriseus*, AJ494864, 99.6%; *S. atroolivaceus*, AJ781320, 99.5%; *S. olivoviridis*, AB184227, 99.5%; *S. finlayi*, AY999788, 99.5%; *S. nitrosporeus*, EF178680, 99.5%; *S. albolongus*, AB184425, 99.4%; *S. clavifer*, DQ026670, 99.4%; *S. griseobrunneus*, AB249912, 99.4%; *S. sanglieri*, AB249945, 99.4%; *S. celluloflavus*, AB184476, 99.4%; *S. yanii*, AB006159, 99.4%; *S. gelaticus*, DQ026636, 99.3%; *S. atratus*, DQ026638, 99.3%; *S. cavourensis* subsp. *cavourensis*, DQ445791, 99.3%; *S. mutomycini*, AB249951, 99.2%; *S. candidus*, DQ026663, 99.2%; *S. spiroverticillatus*, AB184814, 99.1%; *S. cremeus*, AB184124, 99.1%.

Source: not known.

DNA G+C content (mol%): not known.

Type strain: AS 4.1406, ATCC 19729, ATCC 19888, CBS 468.68, BCRC 13759, DSM 40139, HAMBI 1008, NBRC 12745, IMET 43089, INA 1203/53, JCM 4350, LMG 19353, NCIMB 13011, NRRL B-2567, NRRL-ISP 5139, RIA 1010, VKM Ac-735.

Sequence accession no. (16S rRNA gene): AY999783.

66. **Streptomyces baldaccii** corrig. (Farina and Locci 1966) Witt and Stackebrandt 1991, 456[VP] (Effective publication: Witt and Stackebrandt 1990, 370.) (*Streptoverticillium baldaccii* Farina and Locci 1966, 48)

bal.dac'ci.i. N.L. gen. masc. n. *baldacii* of Baldacci (named for Elio Baldacci (1909–1987), who introduced the genus *Streptoverticillium*).

On potato-glucose agar (Baldacci et al., 1954): abundant growth; color, reverse is pink to salmon pink and aerial mycelium is pale pink to salmon pink with whitish tufts. On Bacto Czapek agar: limited growth; color, reverse is colorless to pinkish and aerial mycelium is traces of pink. On Casamino acids Czapek agar (1 g/l Difco vitamin-free Casamino acids replacing sodium nitrate): growth slightly better than on Bacto Czapek agar; colors similar. On glucose-asparagine agar (ISP medium 5 with 1% glucose replacing glycerol): good growth; color, reverse is red and aerial mycelium is pale pink with whitish overgrowth. On glycerol-asparagine agar (ISP medium 5): good growth; color, reverse is red and aerial mycelium is pale pink. On inorganic salts-starch agar (ISP medium 4): good growth; color, reverse is red to cherry red and aerial mycelium is pale pink. On yeast extract-malt extract agar (ISP medium 2): good growth; color, reverse is orange-red and aerial mycelium is pink with whitish patches. On Bacto Emerson agar and Bennett agar (1% glucose, 0.1% Bacto beef agar, 0.1% yeast extract, 0.2% peptone, 1.5% agar): good growth; color, reverse is orange-red and aerial mycelium is pink. On Oxoid nutrient agar: good growth; color, reverse is initially orange-red turning to brown-red and aerial mycelium (poor) is pink with violet shades; brown soluble pigment. Optimal growth at 27°C; only substrate mycelium is formed at 37°C; no growth at 45°C. Type strain exhibits anti-bacterial and anti-fungal activity.

No detailed sequence information is available (only partial sequence).

Source: not known.

DNA G+C content (mol%): not known.

Type strain: ATCC 23654, DSM 40845, HUT 6222, IPV 1339, IPV 174, NBRC 14693, JCM 4272, NRRL B-3500.

Sequence accession no. (16S rRNA gene): X53164.

Further comments: D.P. Labeda proposes *Streptomyces biverticillatus* (Preobrazhenskaya and Ryabova 1957) Witt and Stackebrandt 1991, *Streptomyces fervens* (DeBoer et al. 1959–1960) Witt and Stackebrandt 1991, and *Streptomyces roseoverticillatus* (Shinobu 1956) Witt and Stackebrandt 1991 as later synonyms of *Streptomyces baldaccii* corrig. (Farina and Locci 1966) Witt and Stackebrandt 1991. This proposal is in violation of Rule 24b(1) because the senior epithet is *roseoverticillatus*. *Streptomyces baldaccii*, *Streptomyces biverticillatus*, and *Streptomyces fervens* are therefore to be regarded as later synonyms of *Streptomyces roseoverticillatus*.

Hatano et al. (1997) propose *Streptomyces baldaccii* corrig. (Farina and Locci 1966) Witt and Stackebrandt 1991 as an earlier heterotypic synonym of *Streptomyces spitsbergensis* Wieczorek et al. (1993).

According to Hatano et al. (2003), *Streptomyces baldaccii* corrig. (Farina and Locci 1966) Witt and Stackebrandt 1991 is a later heterotypic synonym of *Streptomyces hiroshimensis* (Shinobu 1955) Witt and Stackebrandt 1991.

67. **Streptomyces bambergiensis** Wallhäusser, Nesemann, Präve and Steigler 1966, 734[AL]

bam.ber.gi.en'sis. N.L. masc. adj. *bambergiensis* of or belonging to Bamberg, Germany, the source of the soil from which the organism was isolated.

Spore chains in Section *Retinaculiaperti* to *Rectiflexibiles*. Strongly flexuous chains with open or primitive spirals and terminal loops or turns. Mature spore chains are moderately long, usually with more than 10 spores per chain. This morphology is seen on yeast-malt agar, oatmeal agar, salts-starch agar, and glycerol-asparagine agar. Spore surface is hairy.

Color of colony: aerial mass color in the Green color series (1½li-ig, light grayish olive or olive gray to 24½ih-li, dark greenish gray or grayish olive green) on yeast-malt agar, oatmeal agar, and salts-starch agar. White, Yellow or Green color series on glycerol-asparagine agar. Reverse side of colony is strong brown on yeast-malt agar; light yellow on oatmeal agar; light grayish olive to olive brown on salts-starch agar; and yellowish pink, orange-yellow, or reddish brown on glycerol-asparagine agar. Reverse mycelium pigment is not a pH indicator.

Color in medium: melanoid pigments are not formed in peptone-yeast-iron agar, tyrosine agar, or tryptone-yeast broth. Greenish to reddish pigment is reported in the medium in yeast-malt agar, salts-starch agar, and glycerol-asparagine agar. One observer reports no pigments in these media. The pigment, when present, is not pH-sensitive.

D-Glucose, D-xylose, iso-inositol, D-mannitol, D-fructose, and rhamnose are utilized for growth. Reports vary on the utilization of L-arabinose. No growth or only traces of growth with sucrose and raffinose.

Type strain shows the highest sequence similarity to: *S. prasinus*, DQ026658, 99.9%; *S. hirsutus*, AB184844, 99.1%; *S. cyanoalbus*, AB184882, 99%.

Source: soil, Bamberg, Germany.

DNA G+C content (mol%): not known.

Type strain: AS 4.1439, ATCC 13879, CBS 780.72, CECT 3211, DSM 40590, NBRC 13479, JCM 4728, KCTC 9019, LMG 19299, NRRL B-12101, NRRL B-12521, NRRL-ISP 5590, RIA 1440, VKM Ac-975.

Sequence accession no. (16S rRNA gene): AB184869.

68. **Streptomyces bangladeshensis** Al-Bari, Bhuiyan, Flores, Petrosyan, García-Varela and Ul Islam 2005, 1976[VP]

ban.gla.desh.en'sis. N.L. masc. adj. *bangladeshensis* of or belonging to Bangladesh, the source of the soil from which the organism was isolated.

Forms highly branched substrate mycelium and aerial hyphae that differentiate into long *Rectiflexibiles* chains of 8–10 spores. Aerial spore mass color is yellow-green. Substrate mycelium is beige on standard media. Yellowish diffusible pigments are formed on Czapek–Dox agar. Melanin pigments are not produced on peptone-iron or tyrosine agars. Positive for H_2S production. Utilizes glucose, sucrose, *myo*-inositol, mannitol, mannose, maltose, fructose, L-arabinose, rhamnose, glycerol, raffinose, and trehalose as sole carbon sources. Growth occurs at 20–50°C, at pH 6.0–11.0, and in the presence of 2% (w/v) NaCl, neomycin sulfate (50 μg/ml), and penicillin (10 IU/ml). Produces bis-(2-ethylhexyl)-phthalate, an antimicrobial agent.

Type strain shows the highest sequence similarity to: *S. rameus*, AY999821, 99.8%; *S. glomeratus*, AB249917, 99.2%; *S. achromogenes* subsp. *rubradiris*, AY999846, 99%.

Source: isolated from soil from Bangladesh.

DNA G+C content (mol%): not known.

Type strain: AAB-4, LMG 22738, NRRL B-24326.

Sequence accession no. (16S rRNA gene): AY750056.

69. **Streptomyces beijiangensis** Li, Zhang, Xu, Cui, Lu, Xu and Jiang 2002b, 1698[VP]

bei.ji.ang.en'sis. N.L. masc. adj. *beijiangensis* of or pertaining to Beijiang, a place in Yinjiang province in western China.

Aerial mycelium at maturity forms long chains of spores that are straight to flexuous or occasionally *Retinaculiaperti* and are nonmotile. Aerial mycelium and substrate mycelium are well developed. Good growth on most media. Optimum growth temperature is between 8 and 20°C. Diffusible pigment is not produced. The color of the colonies is medium-dependent. Glucose, galactose, and glycerol are utilized; lactose, mannose, inulin, acetate, and oxalate are not utilized. Positive for nitrate reduction and urease. Diagnostic amino acid of peptidoglycan is *meso*-A_2pm. Whole-cell hydrolysates contain glucose and small quantities of xylose, galactose, and arabinose. The predominant menaquinones are MK-9(H_6) and MK-9(H_8) and phosphatidylethanolamine is the diagnostic phospholipid. Predominant cellular fatty acids are $C_{15:0}$ anteiso, $C_{16:0}$ iso, and $C_{17:0}$ cyclo.

Type strain shows no sequence similarity over 99%.

Source: isolated from soil in low-temperature habitats collected from Beijiang, western China.

DNA G+C content (mol%): not known.

Type strain: YIM6, AS 4.1718, CCTCC 99005, DSM 41794, JCM 11882, NBRC 100044.

Sequence accession no. (16S rRNA gene): AF385681.

70. **Streptomyces bellus** Margalith and Beretta 1960, 193[AL]

bel'lus. L. masc. adj. *bellus* pretty, handsome.

Spore chains in Section *Spirales*. Short spore chains may form incomplete spirals, hooks, and loops of small diameter or flexuous chains. Spirals are best developed on salts-starch agar. Hooks and loops are of small diameter and therefore are not typical of *Retinaculum-Apertum* cultures. Mature spore chains with 3–10, or sometimes more than 10, spores per chain are produced on yeast-malt agar, oatmeal agar, salts-starch agar, and glycerol-asparagine agar. Spore surface is spiny.

Color of colony: aerial mass color in the Blue color series on oatmeal agar and salts-starch agar; White or Blue color series on yeast-malt agar and glycerol-asparagine agar. Reverse side of colony: substrate color is modified by red (orange) pigment on yeast-malt agar and glycerol-asparagine agar and by red or blue pigments on oatmeal agar and salts-starch agar. These pigments are not pH indicators.

Color in medium: melanoid pigments are formed in peptone-yeast-iron agar, tyrosine agar, and tryptone-yeast broth. Trace of red pigment may be found in the medium in glycerol-asparagine agar; it is not a pH indicator. Pigments are not found in the medium in yeast-malt agar, oatmeal agar, or salts-starch agar.

D-Glucose, L-arabinose, sucrose, D-xylose, iso-inositol, D-mannitol, D-fructose, rhamnose, and raffinose are utilized for growth.

Type strain shows the highest sequence similarity to: *S. coerulescens*, AY999720, 100%; *S. coeruleorubidus*, AY999719, 99.6%; *S. lomondensis*, AB184673, 99.4%; *S. purpurascens*, AJ399486, 99.4%; *S. lusitanus*, AB184424, 99.4%; *S. iakyrus*, AB184877, 99.3%; *S. paradoxus*, AB184628, 99.2%; *S. viridodiastaticus*, AY999852, 99.2%; *S. parvulus*, AB184326, 99.2%; *S. matensis*, AB184221, 99.2%; *S. azureus*, EF178674, 99.2%; *S. indiaensis*, AB184553, 99.2%; *S. albogriseolus*, AJ494865, 99.2%; *S. longispororuber*, AB184440, 99.2%; *S. hawaiiensis*, AB184143, 99.2%; *S. violaceus*, AB184315, 99.1%; *S. albosporeus* subsp. *albosporeus*, AJ781327, 99.1%; *S. janthinus*, AB184851, 99.1%; *S. thermocarboxydus*, U94490, 99.1%; *S. spinoverrucosus*, AB184578, 99.1%; *S. griseorubens*, AB184139, 99.1%; *S. erythrogriseus*, AJ781328, 99%; *S. griseoincarnatus*, AJ781328, 99%; *S. labedae*, AB184704, 99%; *S. variabilis*, DQ442551, 99%; *S. massasporeus*, AB184152, 99%; *S. luteogriseus*, AB184379, 99%; *S. roseoviolaceus*, AJ399484, 99%; *S. arenae*, AB249977, 99%.

Source: not known.

DNA G+C content (mol%): not known.

Type strain: A/870, ATCC 14925, ATCC 23886, CBS 666.68, DSM 40185, NBRC 12844, JCM 4292, JCM 4625, NRRL B-2575, NRRL-ISP 5185, RIA 1139.

Sequence accession no. (16S rRNA gene): AB184849.

71. **Streptomyces bikiniensis** Johnstone and Waksman 1947, 294[AL] ("*Actinomyces bikiniensis*" Krasil'nikov 1949, 100)

bi.ki.ni.en'sis. N.L. masc. adj. *bikiniensis* of or pertaining to Bikini Atoll.

Spore chains in Section *Rectiflexibiles* with long, straight to slightly flexuous spore chains of more than 50 spores per chain. This morphology is seen on yeast-malt agar, oatmeal agar, salts-starch agar, and glycerol-asparagine agar. Spore surface is smooth.

Color of colony: aerial mass color in the Gray color series on yeast-malt agar, oatmeal agar, salts-starch agar, and glycerol-asparagine agar. Nearest matching color tabs are 3ih, dark gray; 2fe, medium gray; 3fe, light brownish gray; 5fe, light grayish reddish brown; and 2dc, yellowish gray. Reverse side of colony with no distinctive pigments (grayish yellow to yellowish brown or olive brown on yeast-malt agar; pale to grayish yellow or light grayish yellowish brown on oatmeal agar, salts-starch agar, and glycerol-asparagine agar).

Color in medium: melanoid pigments are formed in peptone-yeast-iron agar, tyrosine agar, and tryptone-yeast broth, but the reaction is weak in tyrosine agar. No pigment or only a trace of yellow is found in the medium in yeast-malt agar, oatmeal agar, salts-starch agar, and glycerol-asparagine agar.

D-Glucose, D-xylose, and D-fructose are utilized for growth. Reports vary on utilization of L-arabinose; utilization of D-mannitol, rhamnose, and raffinose is doubtful and no significant growth is seen with iso-inositol or sucrose.

Type strain shows the highest sequence similarity to: *S. violaceorectus*, AB184314, 99.7%; *S. vietnamensis*, DQ311081, 99.5%; *S. cinereoruber* subsp. *cinereoruber*, AB184121, 99.3%; *S. tanashiensis*, AJ781362, 99.1%; *S. nashvillensis*, AB184286, 99%; *S. viridobrunneus*, AJ781372, 99%; *S. laurentii*, AJ781342, 99%; *S. showdoensis*, AB184389, 99%.

Source: not known.

DNA G+C content (mol%): not known.

Type strain: AS 4.569, ATCC 11062, CBS 412.54, DSM 40581, HUT 6084, IFM 1057, NBRC 14598, IMET 41362, JCM 4011, KCTC 9172, NRRL B-1049, NRRL B-2690, NRRL-ISP 5581, RIA 471, RIA 74, VKM Ac-999.

Sequence accession no. (16S rRNA gene): X79851.

72. **Streptomyces biverticillatus** (Preobrazhenskaya *in* Gauze, Preobrazhenskaya, Kudrina, Blinov, Ryabova and Sveshnikova 1957) Witt and Stackebrandt 1991, 456^{VP} (Effective publication: Witt and Stackebrandt 1990, 370.) ["*Actinomyces biverticillatus*" Preobrazhenskaya *in* Gauze, Preobrazhenskaya, Kudrina, Blinov, Ryabova and Sveshnikova 1957, 75; "*Streptomyces biverticillatus*" Pridham, Hesseltine and Benedict 1958, 72; "*Streptoverticillium biverticillatus*" (*sic*) Baldacci 1958, 25; *Streptoverticillium biverticillatum* Farina and Locci 1966, 49]

bi.ver.ti.cil'la.tus. L. adv. num. *bis* twice; L. masc. n. *verticillus* whorl; N.L. adj. *verticillatus* whorled; N.L. masc. adj. *biverticillatus* whorled twice.

Spore chains in Section *Verticillati* with umbellate-monoverticillate (biverticillate) spore chains. Mature spore chains generally contain 10 or more spores per chain. This morphology is seen on yeast-malt agar, oatmeal agar, salts-starch agar, and glycerol-asparagine agar. Spore surface is smooth.

Color of colony: aerial mass color in the Red color series on yeast-malt agar, oatmeal agar, salts-starch agar, and glycerol-asparagine agar (one observer, only, placed this culture in the Violet color series on yeast-malt agar and salts-starch agar). Reverse side of colony is red on yeast-malt agar, oatmeal agar, salts-starch agar, and glycerol-asparagine agar; this pigment is pH-sensitive, changing from red to orange-red with addition of 0.05 M NaOH and from red to violet-red or blue with addition of 0.05 M HCl.

Color in medium: melanoid pigments are formed in less than 2 d in peptone-yeast-iron agar and tryptone-yeast broth, but more slowly in tyrosine agar. Pigments other than melanoids are not formed in yeast-malt agar, oatmeal agar, salts-starch agar, or glycerol-asparagine agar.

D-Glucose and possibly iso-inositol and D-fructose are utilized for growth; growth on iso-inositol and D-fructose is much less abundant than on D-glucose. No growth or only trace of growth on L-arabinose, sucrose, D-xylose, D-mannitol, rhamnose, and raffinose. Type strain shows the highest sequence similarity to: *S. werraensis*, DQ442558, 100%; *S. albireticuli*, AB184881, 99.5%; *S. netropsis*, AB184848, 99.3%; *S. eurocidicus*, AY999790, 99.3%; *S. cinnamoneus*, AB184850, 99%; *S. hiroshimensis*, AB184789, 99%.

Source: not known.

DNA G+C content (mol%): not known.

Type strain: ATCC 23615, CBS 668.68, NBRC 12845, JCM 4431, NRRL-ISP 5272, RIA 1190.

Sequence accession no. (16S rRNA gene): AJ781381.

Further comments: in violation of Rule 33c of the *Bacteriological Code* (1990 Revision), in Validation List no. 38, *Streptomyces biverticillatus* is proposed as a *nomen revictum* [basonym: "*Streptomyces biverticillatus*" (Preobrazhenskaya and Ryabova 1957) Pridham et al. 1958].

Labeda (1996) proposes *Streptomyces biverticillatus* (Preobrazhenskaya and Ryabova 1957) Witt and Stackebrandt 1991, *Streptomyces fervens* (DeBoer et al. 1959–1960) Witt and Stackebrandt 1991, and *Streptomyces roseoverticillatus* (Shinobu 1956) Witt and Stackebrandt 1991 as later synonyms of *Streptomyces baldaccii* corrig. (Farina and Locci 1966) Witt and Stackebrandt 1991. This proposal is in violation of Rule 24b of the *Bacteriological Code* (1990 Revision) because the senior epithet is *roseoverticillatus. Streptomyces baldaccii, Streptomyces biverticillatus*, and *Streptomyces fervens* are therefore to be regarded as later synonyms of *Streptomyces roseoverticillatus.*

According to Hatano et al. (2003), *Streptomyces biverticillatus* (Preobrazhenskaya and Ryabova 1957) Witt and Stackebrandt 1991 is a later heterotypic synonym of *Streptomyces hiroshimensis* (Shinobu 1955) Witt and Stackebrandt 1991.

73. **Streptomyces blastmyceticus** (Watanabe et al. 1957) Witt and Stackebrandt 1991, 456^{VP} (Effective publication: Witt and Stackebrandt 1990, 370.) ("*Streptomyces blastmyceticus*" Watanabe, Tanaka, Fukuhara, Miyairi, Yonehara and Umezawa 1957, 39; *Streptoverticillium blastmyceticum* Locci, Baldacci and Petrolini Baldan 1969, 43)

blast.my.ce'ti.cus. N.L. n. *blastomycinum* blastomycin; L. masc. suff. *-icus* suffix used with the sense of belonging to; N.L. masc. adj. *blastmyceticus* belonging to blastomycin.

Spore chains in Section *Verticillati.* Both monoverticillate and umbellate monoverticillate (biverticillate) spore chains are found. Spore chains are short; usually only 3–10 spores per chain. This morphology is seen on oatmeal agar; mature spore chains may not develop on other ISP media. Spore surface is smooth.

Color of colony: aerial mass color in the Gray color series on oatmeal agar. Better sporulation and color is produced on tomato paste-oatmeal agar or on Hickey and Tresner's agar. On these media, an olive-gray color (Gray color series) is produced. Immature and poorly sporulated aerial growth on other media may appear to place this species in the Yellow or Red color series. Reverse side of colony is characteristic grayed yellow on yeast-malt agar and grayed greenish yellow to olive-brown on oatmeal agar, salts-starch agar, and glycerol-asparagine agar; substrate pigment is not a pH indicator.

Color in medium: melanoid pigments are formed in peptone-yeast-iron agar and tryptone-yeast-broth. Pigments other than melanoids are not formed in yeast-malt agar, oatmeal agar, salts-starch agar, and glycerol-asparagine agar.

D-Glucose, L-arabinose, iso-inositol, and D-fructose are utilized for growth. No growth or only trace of growth on D-xylose, rhamnose, and raffinose. Utilization of sucrose and D-mannitol is doubtful.

Type strain shows the highest sequence similarity to: *S. hiroshimensis*, AB184789, 99.5%; *S. ardus*, AB184864, 99.4%; *S. cinnamoneus*, AB184850, 99.3%; *S. albireticuli*, AB184881, 99.3%; *S. eurocidicus*, AY999790, 99.2%; *S. caeruleus*, EF178675, 99.1%; *S. pseudoechinosporeus*, AB184100, 99.1%; *S. abikoensis*, AB184537, 99%.

Source: not known.

DNA G+C content (mol%): not known.

Type strain: AS 4.1647, ATCC 19731, CBS 470.68, BCRC 13387, CECT 3257, DSM 40029, NBRC 12747, JCM 4184, JCM 4565, NCIMB 9800, NRRL B-5480, NRRL-ISP 5029, RIA 1012.

Sequence accession no. (16S rRNA gene): AY999802.

Further comments: in violation of Rule 33c of the *Bacteriological Code* (1990 Revision), in Validation List no. 38, *Streptomyces blastmyceticus* is proposed as a *nomen revictum* (basonym: "*Streptomyces blastmyceticus*" Watanabe et al. 1957).

74. **Streptomyces bluensis** Mason, Dietz and Hanka 1963b, 608[AL]

blu.en'sis. N.L. masc. adj. *bluensis* (from French adj. *bleu*) belonging to blue, referring to the blue color of the aerial mycelium.

Spore chains in Section *Spirales.* Spirals may be poorly developed on some media and open loops, imperfect spirals, or flexuous spore chains may be common. Observers do not agree on the best medium for spiral formation. Some spirals are reported by different observers on yeast-malt agar, oatmeal agar, salts-starch agar, and glycerol-asparagine agar. Spore chains are short to long, usually with more than 10 spores per chain. Spore surface is spiny.

Color of colony: aerial mass color in the Blue or White color series on yeast-malt agar; White color series on oatmeal agar, salts-starch agar, and glycerol-asparagine agar. Nearest matching color tab in the Blue color series is 19dc, pale blue. Reverse side of colony with no distinctive pigments (light grayish yellow to pale yellow) on yeast-malt agar, oatmeal agar, salts-starch agar, and glycerol-asparagine agar.

Color in medium: melanoid pigments are not formed in peptone-yeast-iron agar, tyrosine agar, or tryptone-yeast broth. No pigment is found in the medium in yeast-malt agar, oatmeal agar, salts-starch agar, or glycerol-asparagine agar.

D-Glucose, L-arabinose, D-xylose, iso-inositol, D-mannitol, D-fructose, rhamnose, sucrose, and raffinose are all utilized for growth.

Type strain shows no sequence similarity over 99%.

Source: not known.

DNA G+C content (mol%): not known.

Type strain: AS 4.1463, ATCC 27420, CBS 239.69, CBS 761.72, DSM 40564, NBRC 13460, JCM 4729, LMG 5969, NCIMB 9754, NRRL 2876, NRRL-ISP 5564, RIA 1421.

Sequence accession no. (16S rRNA gene): X79324.

75. **Streptomyces bobili** (Waksman and Curtis 1916) Waksman and Henrici *in* Breed, Murray and Hitchens 1948, 937[AL] ["*Actinomyces bobili*" Waksman and Curtis 1916, 121; "*Streptomyces bobiliae*" (*sic*) Waksman and Henrici *in* Breed, Murray and Hitchens 1948, 937]

bo.bi'li. N.L. gen. n. *bobili* named for Bobili, the nickname of an individual.

Spore chains in Section *Spirales.* Sporulating aerial mycelium is not produced at 27 or 37°C on yeast-malt agar, oatmeal agar, salts-starch agar, or glycerol-asparagine agar, or on supplementary media (Czapek's agar and potato-glucose agar) used by one observer. Two observers found some spots of sporulating aerial mycelium on salts-starch agar. One collaborator (J.B. Routien) found spiral spore chains with 10 to 30 or more spores per chain on Pridham and Gottlieb's basal salts medium for carbon utilization enriched with xylose, raffinose, or glucose. Incubation was at 37°C. Spirals were most abundant when xylose was carbon source. The first description by Waksman (1916, op. cit) notes the absence of true aerial mycelium or spores on Czapek's medium. The description published in 1919 (Waksman, (1919), op. cit) notes spiral formation on scant white aerial mycelium on glycerin-synthetic agar and traces of aerial mycelium in spots only on starch-agar plates and potato plugs. Aerial mycelium was not produced on the seven other solid media used by Waksman. The present culture seems to conform well to the early descriptions. Spore surface is smooth.

Color of colony: aerial mass color cannot be observed on yeast-malt agar, oatmeal agar, salts-starch agar, or glycerol-asparagine agar. Sparse aerial mycelium, when produced, is white. On the reverse side of the colony, substrate mycelium color may be grayish yellow, or may be modified with red to reddish gray or pink, reddish brown, or reddish orange. The substrate mycelium color is dependent on pH, changing from yellowish pink to violet pink with addition of 0.05 M NaOH and from yellowish pink to yellow orange with 0.05 M HCl.

Color in medium: melanoid pigments are not produced in peptone-yeast-iron agar, tyrosine agar, or tryptone-yeast broth. No pigments are found in medium in yeast-malt agar, oatmeal agar, salts-starch agar, or glycerol-asparagine agar.

D-Glucose, L-arabinose, sucrose, D-xylose, iso-inositol, D-fructose, rhamnose, and raffinose are utilized for growth. No growth or only trace of growth on D-mannitol.

Type strain shows the highest sequence similarity to: *S. galilaeus*, AB045878, 100%; *S. phaeoluteigriseus*, AJ391815, 99.5%; *S. resistomycificus*, AB184166, 99.5%; *S. aureocirculatus*, AB184260, 99.4%; *S. novaecaesareae*, AB184357, 99.3%; *S. chartreusis*, AB184839, 99.3%; *S. pseudovenezuelae*, AB184233, 99.2%; *S. prunicolor*, DQ026659, 99%.

Source: not known.

DNA G+C content (mol%): not known.

Type strain: AS 4.1624, ATCC 23889, ATCC 3310, CBS 419.34, CBS 675.68, BCRC 13671, DSM 40056, HAMBI 1059, NBRC 13199, NBRC 16166, IMET 41372, JCM 4012, JCM 4627, NRRL B-1338, NRRL B-2097, NRRL-ISP 5056, RIA 1116, VKM Ac-1756.

Sequence accession no. (16S rRNA gene): AB249925.

76. **Streptomyces bottropensis** Waksman 1961 182[AL]

bot.trop.en'sis. N.L. masc. adj. *bottropensis* of or pertaining to the German town Bottrop.

Spore chains in Section *Spirales*. Hooks, primitive spirals, and terminal spirals of only one or two turns are common together with open spirals of several turns. Mature spore chains generally have 10–50 spores per chain. This morphology is seen on yeast-malt agar, oatmeal agar, salts-starch agar, and glycerol-asparagine agar, but spirals are best developed on salts-starch agar. Spore surface is smooth.

Color of colony: aerial mass color in the Gray color series on yeast-malt agar, oatmeal agar, salts-starch agar, and glycerol-asparagine agar. Reverse side of colony is yellow to strong brown, reddish brown, or sometimes very dark brown.

Color in medium: melanoid pigments are formed in peptone-yeast-iron agar, tyrosine agar, and tryptone-yeast broth. Brown pigment is found in the medium in yeast-malt agar and glycerol-asparagine agar; this pigment is not pH-sensitive.

D-Glucose, L-arabinose, D-xylose, iso-inositol, D-mannitol, D-fructose, rhamnose, sucrose, and raffinose are utilized for growth, but growth on raffinose is less abundant than on other carbon sources.

Type strain shows no sequence similarity over 99%.

Source: not known.

DNA G+C content (mol%): not known.

Type strain: AS 4.1669, ATCC 25435, CBS 163.64, CBS 667.69, BCRC 12063, CIP 105278, DSM 40262, NBRC 13023, JCM 4459, NRRL-ISP 5262, RIA 1215, VKM Ac-1755.

Sequence accession no. (16S rRNA gene): AB026217.

77. **Streptomyces brasiliensis** (Falcão de Morais, Chaves Batista and Massa et al. 1966) Goodfellow, Williams and Alderson 1986a, 573[VP] (Effective publication: Goodfellow, Williams and Alderson 1986c, 53.) (*Elytrosporangium brasiliense* Falcão de Morais, Chaves Batista and Massa 1966, 170)

bra.si.li.en'sis. N.L. masc. adj. *brasiliensis* of or pertaining to Brazil.

Forms extensively branched substrate and aerial mycelium. The latter bears long spiral spore chains, the former bears occasional short chains of spores (0.8–1.0 µm diameter). The spore surface is smooth. The aerial spore mass is gray to greenish gray; the reverse color is colorless to light cream. Does not form melanin pigments. Esculin, casein, guanine, starch, testosterone, and tyrosine are degraded, but allantoin, arbutin, chitin, elastin, hypoxanthine, lecithin, pectin, urea, and xanthine are not. Nitrate is reduced to nitrite and hydrogen sulfide is produced. Adonitol, L-arabinose, cellobiose, D-fructose, D-galactose, D-glucose, *myo*-inositol, inulin, D-lactose, mannitol, D-mannose, melezitose, melibiose, raffinose, L-rhamnose, salicin, sucrose, trehalose, and D-xylose are used as sole carbon sources, but xylitol is not. Grows on L-arginine, L-cysteine, L-histidine, L-methionine, L-phenylalanine, L-serine, L-threonine, and L-valine, but not on DL-amino-n-butyric acid, L-hydroxyproline, or potassium nitrate, as sole nitrogen sources. Growth occurs at 25–37°C, but not at 10 or 45°C. Tolerant to phenol (0.1%, w/v), sodium azide (0.02%, w/v), and sodium chloride (4%, w/v). Sensitive to rifampin and 7% (w/v) sodium chloride. Does not show antimicrobial activity against *Aspergillus niger* LIV 131, *Bacillus subtilis* NCIB 3610, *Candida albicans* CBS 562, *Escherichia coli* NCIB 9132, *Micrococcus luteus* NCIB 196, *Pseudomonas fluorescens* NCIB 9046^T, *Saccharomyces cerevisiae* CBS 1171^T, or *Streptomyces murinus* ISP 5091. The peptidoglycan contains LL-A_2pm as major diamino acid and is of the A3γ type (Stackebrandt et al., 1981). Contains octahydrogenated menaquinones with nine isoprene units as the predominant isoprenolog.

Type strain shows the highest sequence similarity to: *S. africanus*, AY208912, 99.2%; *S. afghaniensis*, AJ399483, 99.1%.

Source: isolated from soil at Alianca, north of Pernambuco, Brazil.

DNA G+C content (mol%): not known.

Type strain: ATCC 23727, CBS 520.68, DSM 43159, IFM 1210, NBRC 12596, KCC A-0086, KCTC 9071, JCM 3086, NBRC 101283, RIA 911, VKM Ac-1310, VKM Ac-656.

Sequence accession no. (16S rRNA gene): AB249981.

78. **Streptomyces bungoensis** Eguchi, Takada, Nakamura, Tanaka, Makino and Oshima 1993, 797[VP]

bun.go.en'sis. N.L. masc. adj. *bungoensis* of or belonging to Bungo, a region of Japan.

Mature spore chains are spiral, with 20 or more spores per chain. Spores are ellipsoidal, 0.5–0.7 µm in diameter. Spore surface is spiny. Mycelia do not fragment into coccoid or bacilliary structures. Aerial mass on glucose-asparagine agar, glycerol-asparagine agar, inorganic salts-starch agar, and oatmeal agar is grayish brown or gray. Reverse sides of colonies are brownish yellow on glucose-asparagine agar and glycerol-asparagine agar, gray on inorganic salts-starch agar and oatmeal agar, yellow on tyrosine agar, and reddish brown on yeast extract-malt extract agar. The melanoid pigment produced is pale brown on tyrosine agar and brownish black on peptone-yeast extract-iron agar and in tryptone-yeast extract broth. A reddish brown soluble pigment is produced on nutrient agar and a brownish black pigment is produced on yeast extract-malt extract agar. No soluble pigment is produced on sucrose-nitrate agar, glucose-asparagine agar, glycerol-asparagine agar, organic salts-starch agar, oatmeal agar, or Czapek's solution agar. Starch is not hydrolyzed, gelatin is liquefied, and milk is peptonized, but not coagulated. D-Glucose, D-fructose, D-galactose, D-xylose, L-arabinose, and D-mannitol are utilized for growth, but rhamnose, iso-inositol, and salicin are not. Little, if any, growth is observed with raffinose and scant growth is observed with sucrose. The cell-wall composition is chemotype I containing LL-A_2pm and glycine, but no galactose, arabinose, or *meso*-A_2pm.

Type strain shows the highest sequence similarity to: *S. longwoodensis*, AB184580, 99.7%; *S. galbus*, X79852, 99.7%; *S. capoamus*, AB045877, 99.5%; *S. corchorusii*, AB184267, 99.4%; *S. olivaceoviridis*, AB184288, 99.3%; *S. canarius*, AB184396, 99.2%.

Source: not known.

DNA G+C content (mol%): 70.3.

Type strain: AS 4.1653, DSM 41781, FERM 8432, NBRC 15711, JCM 9925, MS16-10G.

Sequence accession no. (16S rRNA gene): AB184696.

79a. **Streptomyces cacaoi subsp. cacaoi** (Waksman *in* Bunting 1932) Waksman and Henrici *in* Breed, Murray and Hitchens 1948, 951[AL] emend. Lanoot, Vancanneyt, Cleenwerck, Wang, Li, Liu and Swings 2002, 828 ("*Actinomyces cacaoi*" Waksman *in* Bunting 1932, 516)

ca.ca'o.i. Mexican Spanish *cacao* the cacao; N.L. gen. n. *cacaoi* of cacao.

Spore chains in Section *Spirales* and poorly developed on all ISP media. Spirals, when formed, are open. Incomplete spirals, loops, hooks, or even flexuous chains may be common when spore chains are short. These are not typical of the large hooks and loops of *Retinaculum-Apertum* cultures. Mature spore chains may contain 10–50 spores per chain on yeast-malt agar, salts-starch agar, and glycerol-asparagine agar, but long chains are not common. Oatmeal agar is not suitable for observation of morphology or aerial mass color. Spore surface is smooth.

Color of colony: aerial mass color in the White color series on yeast-malt agar, salts-starch agar, and glycerol-asparagine agar. Reverse side of colony with no distinctive pigments on yeast-malt agar, oatmeal agar, salts-starch agar, and glycerol-asparagine agar; substrate pigment is not a pH indicator.

Color in medium: melanoid pigments are not formed on peptone-yeast-iron agar, tyrosine agar, or tryptone-yeast extract broth. No pigment is found in medium in yeast-malt agar, oatmeal agar, salts-starch agar, or glycerol-asparagine agar. D-Glucose, L-arabinose, D-xylose, D-mannitol, and D-fructose are utilized for growth. No growth or only trace of growth on iso-inositol and rhamnose. Utilization of sucrose and raffinose is doubtful.

Type strain shows no sequence similarity over 99%.

Source: not known.

DNA G+C content (mol%): 73.0.

Type strain: AS 4.1466, ATCC 19732, ATCC 3082, CBS 471.68, BCRC 12103, DSM 40057, NBRC 12748, IMET 40260, IMRU 3082, JCM 4352, KCTC 9758, LMG 19320, NCIMB 9626, NRRL B-1220, NRRL B-2686, NRRL-ISP 5057, RIA 1013, UNIQEM 123, VKM Ac-733.

Sequence accession no. (16S rRNA gene): AB184115.

Further comments: Streptomyces cacaoi subsp. *cacaoi* (Waksman 1932) Waksman and Henrici 1948 emend. Lanoot et al. 2002 is an earlier heterotypic synonym of *Streptomyces aminophilus* Foster 1961.

79b. **Streptomyces cacaoi subsp. asoensis** Isono, Nagatsu, Kawashima and Suzuki 1965, 853[AL]

aso.en'sis. N.L. masc. adj. *asoensis* probably of or pertaining to Aso, a geographical area in Japan.

Produces the polyoxin complex of selectively specific anti-fungal antibiotics comprised of polyoxins A to L. Polyoxin C apparently exhibits no biological activity.

Type strain shows the highest sequence similarity to: *S. humidus*, DQ442508, 99.5%; *S. rishiriensis*, EF178691, 99.1%; *S. flavovariabilis*, EF178691, 99.1%.

Source: not known.

DNA G+C content (mol%): not known.

Type strain: AS 4.1602, ATCC 19093, CBS 378.69, DSM 41440, NBRC 13813, JCM 4185, KCTC 9700, NCIMB 12769, NRRL B-16592.

Sequence accession no. (16S rRNA gene): DQ026644.

80. **Streptomyces caelestis** DeBoer, Dietz, Wilkins, Lewis and Savage 1955a, 831[AL]

ca.e.les'tis. L. masc. adj. *caelestis* of the sky, heavenly (referring to the blue color of the aerial mycelium and spores).

Spore chains in Section *Spirales* or *Retinaculiaperti*. Spirals are usually poorly developed and show only a few turns; loops and hooks are of small diameter and therefore are not representative of typical *Retinaculum-Apertum* cultures. Short spore chains; usually only 3–10 spores per chain on yeast-malt agar, oatmeal agar, and salts-starch agar. Spore surface is smooth.

Color of colony: aerial mass color in the Blue color series on yeast-malt agar, oatmeal agar, salts-starch agar, and glycerol-asparagine agar. Reverse side of colony with no distinctive pigments on yeast-malt agar, oatmeal agar, salts-starch agar, and glycerol-asparagine agar; substrate pigment is not a pH indicator.

Color in medium: melanoid pigments are formed in peptone-yeast-iron agar and tyrosine agar. Pigments other than melanoids are not formed in yeast-malt agar, oatmeal agar, salts-starch agar, or glycerol-asparagine agar. D-Glucose, L-arabinose, sucrose, D-xylose, iso-inositol, D-fructose, rhamnose, and raffinose are utilized for growth. No growth or only trace of growth on D-mannitol. Type strain shows the highest sequence similarity to: *S. azureus*, EF178674, 99.5%; *S. gancidicus*, AB184660, 99.2%; *S. levis*, AB184670, 99.1%; *S. cellulosae*, DQ442495, 99.1%; *S. capillispiralis*, AB184577, 99.1%; *S. pseudogriseolus*, DQ442541, 99.1%.

Source: not known.

DNA G+C content (mol%): not known.

Type strain: AS 4.1688, ATCC 14924, ATCC 15084, ATCC 19733, CBS 472.68, CBS 967.70, BCRC 13685, DSM 40084, NBRC 12749, IMET 43502, SP 5084, JCM 4218, JCM 4566, LMG 5970, NCIMB 9751, NRRL 2418, RIA 1014, VKM Ac-1822.

Sequence accession no. (16S rRNA gene): X80824.

81. **Streptomyces caeruleus** (Baldacci 1944) Pridham, Hesseltine and Benedict 1958, 60[AL] emend. Lanoot, Vancanneyt, Cleenwerck, Wang, Li, Liu and Swings 2002, 828 ("*Actinomyces caeruleus*" Baldacci 1944, 180)

ca.e.ru'le.us. L. masc. adj. *caeruleus* dark blue, azure.

Spore chains in Section *Rectiflexibiles*. Two observers report poor growth or no sporulating aerial mycelium on yeast-malt agar and oatmeal agar. Mature spore chains with more than 10 spores per chain are usually produced on salts-starch agar, glycerol-asparagine agar, and Hickey and

Tresner medium. Typical spores are not observed; atypical spores are smooth. One observer reports fragmentation of substrate mycelium on Hickey and Tresner agar in 14 d.

Color of colony: aerial mass color in the Gray color series on salts-starch agar, glycerol-asparagine agar, and Hickey and Tresner agar. One observer also records gray aerial mycelium on yeast-malt agar and oatmeal agar, but sporulating growth on these media may be inadequate for color determination. Reverse side of colony with a distinct blackish blue to very dark grayish purple pigment on salts-starch agar, glycerol-asparagine agar, and Hickey and Tresner agar. A dark blue violet to blackish pigment may also be seen in yeast-malt agar or glycerol-asparagine agar when adequate growth occurs. Reverse mycelium pigment is a pH indicator, changing from dark violet or blue to deep green with addition of 0.05 M NaOH.

Color in medium: melanoid pigments probably are not formed in peptone-yeast-iron agar, tyrosine agar, or tryptone-yeast broth; observation may be obscured by dark violet or blue pigments. Brown violet, blue, or blue black pigment is found in the medium in yeast-malt agar, oatmeal agar, salts-starch agar, glycerol-asparagine agar, and Hickey and Tresner agar. This pigment may not be pH-sensitive when tested with 0.05 M NaOH or HCl, or may be somewhat sensitive to 0.05 M NaOH, changing from violet to very deep green.

Growth is generally very poor or absent on Pridham and Gottlieb carbon-utilization medium containing D-glucose or other carbon sources. Carbon utilization cannot be determined accurately on this medium. One observer notes that traces of growth on sucrose and rhamnose are probably equal to growth on D-glucose, but this observation is of doubtful significance.

Type strain shows the highest sequence similarity to: *S. hiroshimensis*, AB184789, 99.7%; *S. abikoensis*, AB184537, 99.5%; *S. lilacinus*, AB184819, 99.3%; *S. cinnamoneus*, AB184850, 99.3%; *S. pseudoechinosporeus*, AB184100, 99.2%; *S. sapporonensis*, AB184508, 99.1%; *S. aureoversilis*, AB184855, 99.1%; *S. blastmyceticus*, AY999802, 99.1%; *S. ardus*, AB184864, 99%.

Source: not known.

DNA G+C content (mol%): not known.

Type strain: ATCC 27421, CBS 645.72, BCRC 13659, DSM 40103, NBRC 13344, IMET 40622, IPV 930, ISP 5103, JCM 4014, JCM 4730, LMG 19399, NRRL B-1623, NRRL B-2194, NRRL-ISP 5103, RIA 1305, RIA 755, VKM Ac-1918.

Sequence accession no. (16S rRNA gene): EF178675.

Further comments: Streptomyces caeruleus (Baldacci 1944) Pridham et al. 1958 emend. Lanoot et al. 2002 is an earlier heterotypic synonym of *Streptomyces niveus* Smith et al. 1956 and an earlier heterotypic synonym of *Streptomyces spheroides* Wallick et al. 1956.

82. **Streptomyces californicus** (Waksman and Curtis 1916) Waksman and Henrici *in* Breed, Murray and Hitchens 1948, 936[AL] ("*Actinomyces californicus*" Waksman and Curtis 1916, 122)

ca.li.for'ni.cus. N.L. masc. adj. *californicus* of or belonging to California, the source of the soil from which the organism was isolated.

Spore chains in Section *Rectiflexibiles*. Mature spore chains generally have 10 to 50 or more spores per chain. This morphology is seen on yeast-malt agar, oatmeal agar, salts-starch agar, and glycerol-asparagine agar. Spore surface is smooth.

Color of colony: aerial mass color in the Red color series on yeast-malt agar, oatmeal agar, salts-starch agar, and glycerol-asparagine agar. One observer recorded both Gray series and Violet series for aerial color on oatmeal agar, salts-starch agar, and glycerol-asparagine agar. This difference may be due to pH. Reverse side of colony is modified by Violet or Red on yeast-malt agar, oatmeal agar, salts-starch agar, and glycerol-asparagine agar; substrate pigment is a pH indicator and is changed from blue-violet to red by addition of 0.05 M HCl, or from red to blue-violet by addition of NaOH.

Color in medium: melanoid pigments not formed in peptone-yeast-iron agar and tyrosine agar; pigment is found in medium in oatmeal agar and tyrosine-asparagine agar. It is pH-sensitive when tested with 0.05 M NaOH or HCl, showing same change as noted for reverse color.

D-Glucose, D-xylose, D-mannitol, and D-fructose are utilized for growth. No growth or only traces of growth on L-arabinose, sucrose, iso-inositol, rhamnose, and raffinose.

Type strain shows the highest sequence similarity to: *S. floridae*, AB184656, 100%; *S. griseinus*, AB184205, 99.9%; *S. badius*, AY999783, 99.9%; *S. pluricolorescens*, DQ442540, 99.9%; *S. sindenensis*, AB184759, 99.9%; *S. mediolani*, AB184674, 99.9%; *S. rubiginosohelvolus*, AB184240, 99.9%; *S. fulvorobeus*, AB184711, 99.8%; *S. alboviridis*, AB184256, 99.8%; *S. praecox*, AB184293, 99.8%; *S. flavofuscus*, AB249935, 99.8%; *S. griseoplanus*, AY999894, 99.8%; *S. lipmanii*, AB184148, 99.8%; *S. albovinaceus*, AB249958, 99.8%; *S. fimicarius*, AY999784, 99.8%; *S. globisporus* subsp. *globisporus*, EF178686, 99.8%; *S. anulatus*, DQ026637, 99.8%; *S. microflavus*, DQ445795, 99.8%; *S. cinereorectus*, AB184646, 99.7%; *S. baarnensis*, EF178688, 99.7%; *S. parvus*, DQ442537, 99.7%; *S. luridiscabiei*, AF361784, 99.7%; *S. acrimycini*, AY999889, 99.7%; *S. cyaneofuscatus*, AB184860, 99.7%; *S. lavendulae* subsp. *lavendulae*, AB184080, 99.7%; *S. cavourensis* subsp. *washingtonensis*, DQ026671, 99.7%; *S. argenteolus*, AB045872, 99.6%; *S. finlayi*, AY999788, 99.6%; *S. bacillaris*, AB184439, 99.6%; *S. halstedii*, EF178695, 99.5%; *S. albolongus*, AB184425, 99.5%; *S. griseobrunneus*, AB249912, 99.5%; *S. celluloflavus*, AB184476, 99.5%; *S. pulveraceus*, AB184806, 99.5%; *S. flavogriseus*, AJ494864, 99.5%; *S. flavovirens*, DQ026635, 99.5%; *S. griseolus*, AB184768, 99.5%; *S. griseus* subsp. *griseus*, AY207604, 99.5%; *S. cavourensis* subsp. *cavourensis*, DQ445791, 99.4%; *S. nitrosporeus*, EF178680, 99.4%; *S. clavifer*, DQ026670, 99.4%; *S. sanglieri*, AB249945, 99.3%; *S. atroolivaceus*, AJ781320, 99.3%; *S. candidus*, DQ026663, 99.3%; *S. olivoviridis*, AB184227, 99.3%; *S. mutomycini*, AB249951, 99.2%; *S. gelaticus*, DQ026636, 99.2%; *S. yanii*, AB006159, 99.2%; *S. atratus*, DQ026638, 99.2%; *S. spiroverticillatus*, AB184814, 99.1%; *S. cremeus*, AB184124, 99%.

Source: isolated from soil from California.

DNA G+C content (mol%): not known.

Type strain: ATCC 19734, ATCC 3312, CBS 125.20, CBS 354.53, CBS 473.68, BCRC 13688, DSM 40058, HUT 6049, IFM 1070, NBRC 12750, NBRC 3386, IMET 40261, JCM 4015, JCM 4567, LMG 19309, NCAIM B.01475, NRRL B-1221, NRRL B-2098, NRRL-ISP 5058, RIA 1015, VKM Ac-575.

Sequence accession no. (16S rRNA gene): AB184755.

83. **Streptomyces calvus** Backus, Tresner and Campbell 1957, 533[AL]

cal'vus. L. masc. adj. *calvus* bald, referring to the sparse formation of aerial mycelium of the organism.

Spore chains in Section *Spirales* on salts-starch agar. Spore chains are short and poorly developed or absent on yeast-malt agar, oatmeal agar, and glycerol-asparagine agar. The original description by Backus, Tresner and Campbell also indicates that aerial mycelium is poorly developed on most media with best sporulation and spiral development on starch containing inorganic salts media. Spore surface is spiny to hairy.

Color of colony: aerial mycelium is inadequate for color determination on most media. When mature sporulating aerial mycelium is formed on yeast-malt agar or salts-starch agar, it is in the Gray color series. Reverse side of colony with no distinctive pigments; colorless or grayish yellowish brown on yeast-malt agar, oatmeal agar, salts-starch agar, and glycerol-asparagine agar.

Color in medium: melanoid pigments are not found in peptone-yeast-iron agar, tyrosine agar, or tryptone-yeast broth. No pigment is found in medium in yeast-malt agar, oatmeal agar, salts-starch agar, or glycerol-asparagine agar.

D-Glucose, L-arabinose, sucrose, D-xylose, iso-inositol, D-mannitol, D-fructose, rhamnose, and raffinose are utilized for growth.

Type strain shows the highest sequence similarity to: *S. asterosporus*, AB184706, 100%; *S. virens*, DQ442554, 99.9%; *S. aureorectus*, AB184710, 99.9%; *S. djakartensis*, AB184657, 99.2%; *S. minutiscleroticus*, EF178696, 99.1%; *S. tuirus*, AB184690, 99.1%; *S. ghanaensis*, AY999851, 99.1%; *S. geysiriensis*, DQ442501, 99.1%; *S. anandii*, AB184402, 99.1%; *S. plicatus*, AB184291, 99%; *S. pilosus*, AB184161, 99%; *S. vinaceusdrappus*, AY999929, 99%; *S. mutabilis*, EF178679, 99%; *S. graminearus* AJ781333, 99%; *S. griseomycini*, AB184137, 99%; *S. griseostramineus*, AB184140, 99%; *S. flavoviridis*, AB184842, 99%; *S. rochei*, AB184237, 99%.

Source: not known.

DNA G+C content (mol%): not known.

Type strain: AS 4.1691, ATCC 13382, ATCC 23890, CBS 350.62, CBS 676.68, BCRC 11859, CECT 3271, DSM 40010, IFM 1093, NBRC 13200, JCM 4326, JCM 4628, NCIMB 12240, NRRL B-2399, NRRL-ISP 5010, RIA 1103, VKM Ac-1185.

Sequence accession no. (16S rRNA gene): AB184329.

84. **Streptomyces canarius** Vavra and Dietz 1965, 76[AL]

ca.na'ri.us. L. masc. adj. *canarius* of or belonging to the Canary Islands; intended to refer to production of a bright canary yellow pigment on a variety of media.

Spore chains in Section *Spirales*; straight to flexuous chains are also seen. Mature spore chains are moderately long with more than 10 spores per chain. This morphology is seen on yeast-malt agar, oatmeal agar, salts-starch agar, and glycerol-asparagine agar. Spore surface is smooth.

Color of colony: aerial mass color in the White to Gray color series (b, white, to 2ge, light olive brown) on yeast-malt agar, oatmeal agar, salts-starch agar, and glycerol-asparagine agar when adequate sporulating aerial mycelium is produced. Aerial mycelium may be inadequate for color determination on some of these media. Reverse side of colony is distinctive greenish yellow to moderate yellow, orange-yellow or strong yellowish brown on yeast-malt agar, oatmeal agar, salts-starch agar, and glycerol-asparagine agar. Reverse mycelium pigment is not a pH indicator.

Color in medium: melanoid pigments are not formed in peptone-yeast-iron agar, tyrosine agar, or tryptone-yeast broth. Yellow to yellow-green pigment may be found in the medium in oatmeal agar and glycerol-asparagine agar. This pigment is not pH-sensitive.

D-Glucose, L-arabinose, D-xylose, iso-inositol, D-mannitol, D-fructose, rhamnose, sucrose, and raffinose are utilized for growth.

Type strain shows the highest sequence similarity to: *S. olivaceoviridis*, AB184288, 100%; *S. corchorusii*, AB184267, 99.8%; *S. capoamus*, AB045877, 99.5%; *S. longwoodensis*, AB184580, 99.3%; *S. bungoensis*, AB184696, 99.2%; *S. galbus*, X79852, 99.1%.

Source: not known.

DNA G+C content (mol%): not known.

Type strain: AS 4.1581, ATCC 27423, CBS 732.72, BCRC 11621, DSM 40528, HAMBI 1014, NBRC 13431, IMET 43539, JCM 4549, JCM 4733, NCIMB 9468, NRRL 2976, NRRL-ISP 5528, RIA 1392.

Sequence accession no. (16S rRNA gene): AB184396.

85. **Streptomyces candidus** (*ex* Krasil'nikov 1941) Sveshnikova 1986, 574[VP] (Effective publication: Sveshnikova *in* Gause, Preobrazhenskaya, Sveshnikova, Terekhova and Maximova 1983.) ("*Actinomyces candidus*" Krasil'nikov 1941)

can'di.dus. L. masc. adj. *candidus* shining white, clear, bright.

Spore chains in Section *Rectiflexibiles*. Mature spore chains are generally long and flexuous, often with more than 50 spores per chain. This morphology is seen on salts-starch agar and glycerol-asparagine agar. Sporulation may be poor on yeast-malt agar and oatmeal agar. Spore surface is smooth.

Color of colony: aerial mass color in the White color series on yeast-malt agar, oatmeal agar, salts-starch agar, and glycerol-asparagine agar, although aerial mycelium may be poorly developed on yeast-malt agar and oatmeal agar. Reverse side of colony with no distinctive pigments (colorless or very pale grayed yellow) on yeast-malt agar, oatmeal agar, salts-starch agar, or glycerol-asparagine agar.

Color in medium: melanoid pigments are not formed in peptone-yeast-iron agar, tyrosine agar, or tryptone-yeast broth. Pigments are not formed in medium in yeast-malt agar, oatmeal agar, salts-starch agar, or glycerol-asparagine agar.

D-Glucose, L-arabinose, D-xylose, D-mannitol, and rhamnose are utilized for growth. No growth or only traces of growth on sucrose, iso-inositol, and raffinose.

Type strain shows the highest sequence similarity to: *S. albolongus*, AB184425, 99.4%; *S. celluloflavus*, AB184476, 99.4%; *S. griseobrunneus*, AB249912, 99.4%; *S. cavourensis* subsp. *cavourensis*, DQ445791, 99.3%; *S. finlayi*, AY999788, 99.3%; *S. spiroverticillatus*, AB184814, 99.3%; *S. floridae*, AB184656, 99.3%; *S. californicus*, AB184755, 99.3%; *S. badius*, AY999783, 99.2%; *S. cremeus*, AB184124, 99.2%;

S. sindenensis, AB184759, 99.2%; *S. pluricolorescens*, DQ442540, 99.2%; *S. mediolani*, AB184674, 99.2%; *S. griseinus*, AB184205, 99.2%; *S. rubiginosohelvolus*, AB184240, 99.2%; *S. clavifer*, DQ026670, 99.2%; *S. anulatus*, DQ026637, 99.1%; *S. parvus*, DQ442537, 99.1%; *S. fimicarius*, AY999784, 99.1%; *S. lipmanii*, AB184148, 99.1%; *S. fulvorobeus*, AB184711, 99.1%; *S. atroolivaceus*, AJ781320, 99.1%; *S. olivoviridis*, AB184227, 99.1%; *S. globisporus* subsp. *globisporus*, EF178686, 99.1%; *S. albovinaceus*, AB249958, 99.1%; *S. alboviridis*, AB184256, 99.1%; *S. praecox*, AB184293, 99.1%; *S. griseoplanus*, AY999894, 99.1%; *S. microflavus*, DQ445795, 99.1%; *S. flavofuscus*, AB249935, 99.1%; *S. cavourensis* subsp. *washingtonensis*, DQ026671, 99.1%; *S. pulveraceus*, AB184806, 99%; *S. cyaneofuscatus*, AB184860, 99%; *S. luridiscabiei*, AF361784, 99%; *S. acrimycini*, AY999889, 99%; *S. lavendulae* subsp. *lavendulae*, AB184080, 99%; *S. cinereorectus*, AB184646, 99%; *S. baarnensis*, EF178688, 99%; *S. mutomycini*, AB249951, 99%.

Source: not known.

DNA G+C content (mol%): not known.

Type strain: AS 4.1664, ATCC 19735, ATCC 19891, ATCC 23891, CBS 677.68, BCRC 13760, DSM 40141, ICMP 12538, NBRC 12846, INA 5855/54, JCM 4629, KCTC 9020, NCIMB 12827, NRRL-ISP 5141, RIA 1131, VKM Ac-1091.

Sequence accession no. (16S rRNA gene): DQ026663.

86. **Streptomyces canescens** Waksman *in* Breed, Murray and Smith 1957, 768[AL] ["*Streptomyces canescus*" (*sic*) Hickey, Corum, Hidy, Cohen, Nager and Kropp 1952, 473]

ca.nes'cens. L. part. adj. *canescens* growing white, whiten.

Spore chains in Section *Rectiflexibiles*. Mature spore chains generally have 10–50 spores per chain. This morphology is observed on yeast-malt agar, oatmeal agar, salts-starch agar, and glycerol-asparagine agar. Spore surface is smooth.

Color of colony: aerial mass color in the Yellow color series on yeast-malt agar, oatmeal agar, salts-starch agar, and glycerol-asparagine agar. Reverse side of colony with no distinctive pigments on yeast-malt agar, oatmeal agar, or glycerol-asparagine agar; grayed greenish yellow on salts-starch agar; substrate is not a pH indicator.

Color in medium: melanoid pigments are not formed in peptone-yeast-iron agar or tyrosine agar. No pigment is found in medium in yeast-malt agar, oatmeal agar, salts-starch agar, and glycerol-asparagine agar.

D-Glucose, D-xylose, and D-mannitol are utilized for growth. No growth or only traces of growth on sucrose, iso-inositol, rhamnose and raffinose. Variable reports on growth with D-fructose and L-arabinose.

Type strain shows the highest sequence similarity to: *S. limosus*, AB184147, 100%; *S. felleus*, AB184129, 100%; *S. daghestanicus*, DQ442497, 100%; *S. albidoflavus*, AB184255, 100%; *S. hydrogenans*, AB184868, 100%; *S. odorifer*, Z76682, 100%; *S. violascens*, AY999737, 100%; *S. griseus* subsp. *solvifaciens*, AB249915, 99.9%; *S. champavatii*, DQ026642, 99.9%; *S. sampsonii*, D63871, 99.7%; *S. koyangensis*, AY079156, 99.6%.

Source: not known.

DNA G+C content (mol%): not known.

Type strain: AS 4.1681, ATCC 19736, CBS 474.68, BCRC 12206, DSM 40001, NBRC 12751, IMET 43077, JCM 4196, JCM 4568, NRRL 2419, NRRL-ISP 5001, RIA 1016, UNIQEM 124, VKM Ac-732.

Sequence accession no. (16S rRNA gene): AB184117.

87. **Streptomyces cangkringensis** Sembiring, Ward and Goodfellow 2001, 1619[VP] (Effective publication: Sembiring, Ward and Goodfellow 2000, 365.)

cang.krin.gen'sis. N.L. masc. adj. *cangkringensis* of or pertaining to Cangkringan, a place in Java, Indonesia.

Spore chains are *Spirales*; spore surface is rugose. On oatmeal agar, the spore mass is gray, the substrate mycelium is grayish-yellow, and the diffusible pigment is yellow. Melanin pigments are not produced. The organism degrades pectin but not adenine and grows at 45°C.

Type strain shows the highest sequence similarity to: *S. rhizosphaericus*, AB249941, 99.8%; *S. asiaticus*, AB249947, 99.8%; *S. indonesiensis*, DQ334783, 99.7%; *S. griseinger*, AJ391818, 99.5%. Type strain shows DNA–DNA similarity to: *S. albiflaviniger* NRRL B-1356[T], 98.4%; *S. geldanamycinus* NRRL 3602[T], 98.9%; *S. griseiniger* NRRL B1865[T], 99.3%; *S. rhizosphaericus* DSM 41760[T], 99.0%; *S. asiaticus* DSM 41761[T], 99.2%; *S. indonesiensis* DSM 41759[T], 98.4%; *S. javensis* DSM 41764[T], 98.4%; *S. yogyakartensis* DSM 41766[T], 98.2%.

Source: isolated from non-rhizosphere soil adjacent to a stand of the tropical legume *Paraserianthes falcataria*.

DNA G+C content (mol%): not known.

Type strain: D13P3, DSM 41769, JCM 11444, NBRC 100775, NCIMB 13684.

Sequence accession no. (16S rRNA gene): AJ391831.

88. **Streptomyces caniferus** (*ex* Krasil'nikov 1970b) Preobrazhenskaya 1986, 574[VP] (Effective publication: Preobrazhenskaya *in* Gause, Preobrazhenskaya, Sveshnikova, Terekhova and Maximova 1983.) ("*Actinomyces caniferus*" Krasil'nikov 1970b)

ca.ni'fe.rus. L. adj. *canus* white; L. adj. *ferus* (from L. v. *fero* to bear) bearing; N.L. masc. adj. *caniferus* bearing (producing) white.

Spore chains are spiral (*Spirales*); spores are smooth. On mineral agar 1 and oatmeal agar: aerial mycelium is gray; substrate mycelium is yellow, dark yellow; no diffusible pigment. On starch-ammonia agar: poor growth; aerial mycelium is gray; substrate mycelium is colorless to yellowish beige; no diffusible pigment. On glycerol-nitrate agar and glycerol-asparagine agar: aerial mycelium is white, whitish gray to gray; substrate mycelium is yellow, dark-yellow; diffusible pigment is yellow, but sometimes absent. On organic agar 2: aerial mycelium is white, later gray; substrate mycelium is colorless, yellow, or dark yellow; no diffusible pigment. Melanoid pigments are not formed. Grows on starch, inositol, mannitol, rhamnose, maltose, sucrose, and glucose; no growth on fructose, arabinose, galactose, xylose, lactose, or raffinose.

Type strain shows the highest sequence similarity to: *S. libani* subsp. *rufus*, AJ781351, 100%; *S. hygroscopius* subsp. *glebosus*, AB184479, 100%; *S. platensis*, AB045882, 99.8%; *S. libani* subsp. *libani*, AB184414, 99.6%; *S. hygroscopius* subsp. *decoyicus*, AY999883, 99.6%; *S. tubercidicus*, AJ621612, 99.5%; *S. nigrescens*, DQ442530, 99.5%; *S. catenulae*, AJ621613, 99.4%; *S. misakiensis*, AB217605, 99.4%; *S. ramulosus*, DQ026662, 99.3%; *S. sioyaensis*, DQ026654, 99.2%; *S. monomycini*, DQ445790, 99%.

Source: not known.

DNA G+C content (mol%): not known.

Type strain: VKM Ac-68, INMI 377, JCM 6914, AS 4.1588, ATCC 43699, DSM 41453, NBRC 15389, NRRL B-16358.

Sequence accession no. (16S rRNA gene): AB184640.

89. **Streptomyces canus** Heinemann, Kaplan, Muir and Hooper 1953, 1239[AL]

ca'nus. L. masc. adj. *canus* white, hoary.

Spore chains in Section *Spirales*, but some spore chains representative of Section *Rectiflexibiles* and *Retinaculiaperti* are also found. Mature spore chains are long, generally 10 to 50 or more spores per chain. This morphology is seen on yeast-malt agar, oatmeal agar, salts-starch agar, and glycerol-asparagine agar. Spore surface is spiny.

Color of colony: aerial mass color in the Gray color series on yeast-malt agar, oatmeal agar, salts-starch agar, and glycerol-asparagine agar. Reverse side of colony with no distinctive pigments on yeast-malt agar, oatmeal agar, salts-starch agar, and glycerol-asparagine agar, except for dark brown on glycerol-asparagine agar in 14–21 d. Substrate is not a pH indicator.

Color in medium: true melanoid pigments are probably not formed in peptone-yeast-iron agar and tyrosine agar. Reports differ on production of melanoid pigments. Each of two observers reported dark pigment, but on different test media; a third collaborator saw no melanoid pigment in peptone-yeast-iron agar or tyrosine agar, but dark brown pigment in the substrate mycelium on glycerol-asparagine agar. An additional test showed no characteristic melanoid pigment. Pigments other than dark pigments noted above are not formed on yeast-malt agar, oatmeal agar, or salts-starch agar.

D-Glucose, L-arabinose, sucrose, D-xylose, iso-inositol, D-mannitol, D-fructose, and rhamnose are utilized for growth. Variable reports on growth with raffinose.

Type strain shows the highest sequence similarity to: *S. ciscaucasicus*, AY508512, 99.8%; *S. pseudovenezuelae*, AB184233, 99.4%; *S. resistomycificus*, AB184166, 99.3%; *S. novaecaesareae*, AB184357, 99%.

Source: not known.

DNA G+C content (mol%): not known.

Type strain: AS 4.1468, ATCC 12237, ATCC 19737, CBS 475.68, BCRC 13652, DSM 40017, IFM 1092, NBRC 12752, JCM 4212, JCM 4569, LMG 19329, NCIMB 9627, NRRL B-1989, NRRL B-3980, NRRL-ISP 5017, RIA 1017, UNIQEM 125, VKM Ac-1011.

Sequence accession no. (16S rRNA gene): AY999775.

90. **Streptomyces capillispiralis** Mertz and Higgens 1982, 123[VP]

ca.pil.li.spi.ra'lis. L. n. *capillus* hair; N.L. adj. *spiralis* (from L. n. *spira* a coil) spiral or spiraled; N.L. masc. adj. *capillispiralis* hairy spiraled.

Spore chains in Section *Spirales*. This morphology is readily observed on all media which support formation of aerial mycelia. Oatmeal agar (ISP 3), ISP 7, tomato paste oatmeal agar, and tap-water agar provide excellent observation of spiral morphology. The spirals are simple, open, loose coils of two to three turns. The sporophores bear chains of 10–50 spores. Spore shape is oval to globose. The spore size ranges from 0.96 to 1.19 μm by 0.54 to 0.71 μm. The mean size is 1.05 by 0.62 μm. Spore surface ornamentation is hairy. Coremia are observed on yeast-malt extract agar (ISP 2). Aerial mycelium is in the Gray color series of Tresner and Backus; light brownish-gray is the predominant shade. Produces a non-fragmenting substrate or primary mycelium which varies from yellow-brown to brownish-black, depending on the medium. A brown, water-soluble pigment is occasionally produced. Does not produce melanoid pigments.

Positive for catalase, phosphatase, and urease; decomposes casein, hypoxanthine, tyrosine, and xanthine. Liquefaction of gelatin, hydrolysis of starch, reduction of nitrate, peptonization of milk, resistance to lysozyme, and decomposition of esculin are negative. Acetate, D-arabinose, melibiose, raffinose, salicin, and sucrose are not utilized. Growth occurs at 10–45°C, with optimum growth at 30°C. Tolerates a pH range of 6–9.5, levels of NaCl up to 6%, and sucrose concentrations up to 35%. Cell analysis demonstrated the presence of LL-A_2pm; no *meso* isomer was detected. Sugar determinations indicated that only glucose and ribose were present. This information indicates a type I cell wall and a type C sugar pattern.

Type strain shows the highest sequence similarity to: *S. pseudogriseolus*, DQ442541, 100%; *S. gancidicus*, AB184660, 100%; *S. cellulosae*, DQ442495, 99.8%; *S. azureus*, EF178674, 99.5%; *S. levis*, AB184670, 99.5%; *S. lusitanus*, AB184424, 99.3%; *S. rubiginosus*, AY999810, 99.3%; *S. carpinensis*, AB184574, 99.3%; *S. matensis*, AB184221, 99.2%; *S. tuirus*, AB184690, 99.2%; *S. albaduncus*, AY999757, 99.1%; *S. caelestis*, X80824, 99.1%; *S. paradoxus*, AB184628, 99.1%; *S. djakartensis*, AB184657, 99%; *S. griseoloalbus*, AB184275, 99%; *S. afghaniensis*, AJ399483, 99%; *S. geysiriensis*, DQ442501, 99%; *S. africanus*, AY208912, 99%; *S. minutiscleroticus*, EF178696, 99%; *S. rochei*, AB184237, 99%; *S. vinaceusdrappus*, AY999929, 99%; *S. mutabilis*, EF178679, 99%; *S. ghanaensis*, AY999851, 99%; *S. plicatus*, AB184291, 99%.

Source: not known.

DNA G+C content (mol%): not known.

Type strain: A49492, DSM 41695, NBRC 14222, JCM 5075, KCTC 1719, NCIMB 12832, NRRL 12279.

Sequence accession no. (16S rRNA gene): AB184577.

91. **Streptomyces capoamus** Gonçalves de Lima, Albert and Gonçalves de Lima 1964, 317[AL]

ca.po.a'mus. Nheêngatû Amazonian dialect *capoama* island; N.L. n. *capoamus* island, referring to Ascension Island, the source of the soil from which the organism was isolated.

Spore chains in Section *Retinaculiaperti*. Terminal spirals, loops, and hooks may be found on moderately long chains of 10–50 spores. Straight to flexuous chains are also common. This morphology is seen on yeast-malt agar, oatmeal agar, and salts-starch agar, but not on glycerol-asparagine agar. Spore surface is smooth.

Color of colony: aerial mass color in the Gray or Red color series on yeast-malt agar and salts-starch agar; Red color series on oatmeal agar. Aerial mycelium is poorly developed or absent on glycerol-asparagine agar. Nearest matching color tabs from the Gray color series are e, medium gray, and 5fe, light grayish reddish brown. Nearest matching tabs from the Red color series are 6ec, grayish yellowish pink, and 7ca, light yellowish pink. Reverse side of colony is red; light reddish brown to grayish yellowish pink on yeast-malt agar, oatmeal agar and salts-starch agar. Reverse mycelium pigment is a pH indicator, changing from red or reddish brown to blue or violet with the addition of 0.05 M NaOH.

Color in medium: melanoid pigments are formed in peptone-yeast-iron agar, tyrosine agar, and tryptone-yeast broth, but the reaction may be weak in tyrosine agar. Yellow, red, or orange may or may not be found in the medium in yeast-malt agar, oatmeal agar, salts-starch agar, or glycerol-asparagine agar. One of three observers reports a pH-sensitive pigment showing essentially the same reactions noted for the reverse mycelium pigments in yeast-malt agar, oatmeal agar, salts-starch agar, and glycerol-asparagine agar.

D-Glucose, L-arabinose, D-xylose, D-mannitol, and D-fructose are utilized for growth. Utilization of raffinose is doubtful. No growth or only traces of growth with iso-inositol, rhamnose, and sucrose.

Type strain shows the highest sequence similarity to: *S. longwoodensis*, AB184580, 99.6%; *S. canarius*, AB184396, 99.5%; *S. corchorusii*, AB184267, 99.5%; *S. olivaceoviridis*, AB184288, 99.5%; *S. galbus*, X79852, 99.5%; *S. bungoensis*, AB184696, 99.5%; *S. regensis*, DQ026649, 99.1%; *S. curacoi*, EF626595, 99%.

Source: isolated from soil from Ascension Island.

DNA G+C content (mol%): not known.

Type strain: AS 4.1696, ATCC 19006, CBS 712.72, BCRC 11860, DSM 40494, NBRC 13411, JCM 4253, JCM 4734, NRRL B-3632, NRRL-ISP 5494, RIA 1372.

Sequence accession no. (16S rRNA gene): AB045877.

92. **Streptomyces carpaticus** Maximova and Terekova 1986, 574[VP] (Effective publication: Maximova and Terekova *in* Gause, Preobrazhenskaya, Sveshnikova, Terekhova and Maximova 1983.)

car.pa'ti.cus. N.L. masc. adj. *carpaticus* of or pertaining to the Carpathians.

Spore chains are spiral (*Spirales*); spores are smooth. On mineral agar 1, glycerol-asparagine agar, oatmeal agar, and glycerol-nitrate agar: aerial mycelium is grayish gray-brown; substrate mycelium and diffusible pigment are dark brown, nearly black with reddish or olive shadow. On starch-ammonia agar: aerial mycelium is light gray; substrate mycelium is colorless; no diffusible pigment. Sometimes substrate mycelium and diffusible pigment are light brown. On organic agar 2 and agar 79: aerial mycelium is grayish gray-brown; substrate mycelium and diffusible pigment are greenish gray-brown, reddish gray-brown; melanoid pigments are not formed.

Rhamnose is utilized for growth; weak growth is observed with sucrose, fructose, glucose, xylose, mannitol, inositol, raffinose, and arabinose. Antibiotic is not produced.

Type strain shows no sequence similarity over 99%.

Source: not known.

DNA G+C content (mol%): not known.

Type strain: AS 4.1621, ATCC 43678, DSM 41468, NBRC 15390, INA 8851, JCM 6915, NRRL B-16359, VKM Ac-1211.

Sequence accession no. (16S rRNA gene): DQ442494.

93. **Streptomyces carpinensis** (Falcão de Morais, Oliveira Da Silva and Machado 1971) Goodfellow, Williams and Alderson 1986a, 574[VP] (Effective publication: Goodfellow, Williams and Alderson 1986c, 53.) (*Elytrosporangium carpinense* Falcão de Morais, Oliveira Da Silva and Machado 1971, 205)

car.pin.en'sis. Gr. n. *karpos* fruit, seed with seed-vessel; N.L. masc. adj. *carpinensis* pertaining to spores within a sporangium.

Forms extensively branched substrate and aerial mycelium. The latter bears long spiral chains and the former bears single spores or short chain of up to seven spores of various sizes (1.0–3.0 µm diameter). The spore surface is smooth. Aerial spore mass is dark gray; the reverse color is black. Does not form melanin pigments. Adenine, esculin, casein, guanine, hypoxanthine, starch, testosterone, and tyrosine are degraded, but allantoin, arbutin, chitin, elastin, lecithin, pectin, urea, and xanthine are not. Hydrogen sulfide is produced, but nitrate is not reduced. L-Arabinose, cellobiose, D-fructose, D-galactose, D-glucose, mannitol, D-mannose, melezitose, melibiose, raffinose, L-rhamnose, salicin, trehalose, and xylitol are used as sole carbon sources, but adonitol, *myo*-inositol, inulin, D-lactose, sucrose, and D-xylose are not. Grows on DL-amino-n-butyric acid, L-arginine, L-cysteine, L-histidine, L-phenylalanine, potassium nitrate, L-serine, L-threonine, and L-valine, but not on L-hydroxyproline or L-methionine, as sole nitrogen source. Growth occurs at 10, 37 and 45°C, but not at 4°C. Tolerant to phenol (0.1%, w/v), sodium azide (0.02%, w/v), and sodium chloride (7%, w/v). Resistant to rifampin but sensitive to sodium chloride at 10% (w/v). Does not show antimicrobial activity against *Aspergillus niger* LIV 131, *Bacillus subtilis* NCIB 3610, *Candida albicans* CBS 562, *Escherichia coli* NCIB 9132, *Micrococcus luteus* NCIB 196, *Saccharomyces cerevisiae* CBS 1171[T], or *Streptomyces murinus* ISP 5091. Contains major amounts of hexa- and octa-hydrogenated menaquinones with nine isoprene units.

Type strain shows the highest sequence similarity to: *S. levis*, AB184670, 99.5%; *S. cellulosae*, DQ442495, 99.4%; *S. pseudogriseolus*, DQ442541, 99.4%; *S. gancidicus*, AB184660, 99.4%; *S. capillispiralis*, AB184577, 99.3%; *S. azureus*, EF178674, 99.1%; *S. spinoverrucosus*, AB184578, 99%; *S. djakartensis*, AB184657, 99%.

Source: isolated from soil in Pernambuco, Brazil.

DNA G+C content (mol%): not known.

Type strain: 70-6-2 (Lab. Microbiol. Inst. Biociencias Univer. Federal Pernambuco, Recife, Brazil), ATCC 27116, DSM 43835, NBRC 14214, IMET 43558, JCM 3301, KCC A-0301, KCTC 9128, NRRL B 16921, RIA 982, VKM Ac-1300, VKM Ac-657.

Sequence accession no. (16S rRNA gene): AB184574.

94. **Streptomyces catenulae** Davisson and Finlay *in* Waksman 1961, 190[AL]

ca.te.nu'la.e. L. dim. n. *catenula* a small chain; L. gen. dim. n. *catenulae* of a small chain.

Spore chains in Section *Rectiflexibiles*. Very short spore chains of 3–10 spores occur in dense clusters so that morphology is difficult to determine. Short chains often form hooks or incomplete spirals. Tight spirals may occur in the dense clusters, but most chains are too short to form true spirals. Short chains are also not typical of *Retinaculiaperti* cultures. This morphology is seen on yeast-malt agar, oatmeal agar, salts-starch agar, and glycerol-asparagine agar. It is representative of morphology described by Davisson and Finlay (op. cit). Spore surface is smooth.

Color of colony: aerial mass color in the Green color series on yeast-malt agar, oatmeal agar, and glycerol-asparagine agar; Gray or Green color series on salts-starch agar. The color on all media is medium gray to olive gray or grayish olive (see tabs 2ih or 2ge in Gray color series and 1½ge in Green color series). Reverse side of colony is colorless or grayish yellow to light grayish olive or olive brown on yeast-malt agar, oatmeal agar, salts-starch agar, and glycerol-asparagine agar. Substrate mycelium pigment is not pH-sensitive.

Color in medium: melanoid pigments are not found in peptone-yeast-iron agar, tyrosine agar, or tryptone-yeast broth; no pigment or only trace of yellow pigment in medium in yeast-malt agar, oatmeal agar, salts-starch agar, and glycerol-asparagine agar.

D-Glucose, D-mannitol, D-fructose, and raffinose are utilized for growth. No growth or only traces of growth on L-arabinose, sucrose, D-xylose, iso-inositol, and rhamnose.

Type strain shows the highest sequence similarity to: *S. misakiensis*, AB217605, 100%; *S. libani* subsp. *libani*, AB184414, 99.6%; *S. nigrescens*, DQ442530, 99.5%; *S. tubercidicus*, AJ621612, 99.5%; *S. caniferus*, AB184640, 99.4%; *S. libani* subsp. *rufus*, AJ781351, 99.3%; *S. hygroscopius* subsp. *glebosus*, AB184479, 99.3%; *S. platensis*, AB045882, 99.2%; *S. sioyaensis*, DQ026654, 99.2%; *S. ramulosus*, DQ026662, 99.1%.

Source: not known.

DNA G+C content (mol%): not known.

Type strain: AS 4.1701, ATCC 12476, ATCC 23893, CBS 679.68, BCRC 12092, DSM 40258, HAMBI 986, NBRC 12848, IMET 42944, JCM 4353, KCTC 9223, NRRL B-2342, NRRL-ISP 5258, RIA 1183, VKM Ac-758.

Sequence accession no. (16S rRNA gene): AJ621613.

95. **Streptomyces caviscabies** Goyer, Faucher and Beaulieu 1996, 638[VP]

ca.vi.sca'bies. L. adj. *cavus* hollow, excavated; L. n. *scabies* scab, mange; N.L. n. *caviscabies* excavated scab, referring to the ability of the organism to cause deep-pitted scab of potatoes.

The cylindrical, smooth spores are 0.87–1.08 × 0.5–0.63 μm. Strains are characterized by a gold mycelium on yeast-malt extract medium and a white mass of spores borne in flexuous chains. Does not produce melanin. All strains grow on proline or methionine as sole nitrogen source. Utilizes raffinose as a sole carbon source. All strains grow in the presence of 4% NaCl, 0.1% phenol, 10 IU/ml penicillin, 10 μg/ml thallous acetate, and 20 μg/ml streptomycin sulfate. Growth is inhibited at pH 4.5. The cell walls contain LL-A_2pm.

For sequence similarity, see type strain of *Streptomyces griseus.*

Source: not known.

DNA G+C content (mol%): 71.0.

Type strain: EF-87, ATCC 51928, CIP 104962, DSM 41811.

Sequence accession no. (16S rRNA gene): AF112160.

Further comments: according to Liu et al. (2005b), *Streptomyces caviscabies* Goyer et al. 1996 is a later heterotypic synonym of *Streptomyces griseus* (Krainsky 1914) Waksman and Henrici 1948 emend. Liu et al. 2005b.

According to Guo et al. (2008), *Streptomyces caviscabies* Goyer et al. (1996) is not a later heterotypic synonym of *Streptomyces griseus* (Krainsky 1914) Waksman and Henrici 1948.

96a. **Streptomyces cavourensis subsp. cavourensis** Skarbek and Brady 1978, 52[AL]

ca.vour.en'sis. N.L. masc. adj. *cavourensis* of or pertaining to Cavour, named for Conti de Cavour (1810–1861), an Italian statesman and hero.

Spore chains in Section *Rectiflexibiles.* Mature spore chains are generally flexuous and very long, with more than 50 spores per chain. This morphology is seen on yeast-malt agar, oatmeal agar, salts-starch agar, and glycerol-asparagine agar. Spore surface is smooth, sometimes with minor surface irregularities.

Color of colony: aerial mass color in the Yellow color series (2db or 2ba, pale yellow) on yeast-malt agar, oatmeal agar, and salts-starch agar; Yellow or White color series on glycerol-asparagine agar. Reverse side of colony with no distinctive pigments (moderate to strong brown on yeast-malt agar; yellow, orange yellow, or olive brown on oatmeal agar, salts-starch agar, and glycerol-asparagine agar).

Color in medium: melanoid pigments are formed in peptone-yeast-iron agar and tryptone-yeast broth, but not in tyrosine agar. No pigment (other than brown) is found in the medium in yeast-malt agar, oatmeal agar, salts-starch agar, or glycerol-asparagine agar.

D-Glucose, L-arabinose, D-xylose, D-mannitol, and D-fructose are utilized for growth. No growth or only traces of growth with iso-inositol, rhamnose, sucrose, and raffinose. Type strain shows the highest sequence similarity to: *S. celluloflavus*, AB184476, 100%; *S. albolongus*, AB184425, 100%; *S. griseobrunneus*, AB249912, 99.9%; *S. bacillaris*, AB184439, 99.4%; *S. californicus*, AB184755, 99.4%; *S. fulvorobeus*, AB184711, 99.3%; *S. floridae*, AB184656, 99.3%; *S. griseinus*, AB184205, 99.3%; *S. parvus*, DQ442537, 99.3%; *S. pluricolorescens*, DQ442540, 99.3%; *S. candidus*, DQ026663, 99.3%; *S. spiroverticillatus*, AB184814, 99.3%; *S. globisporus* subsp. *globisporus*, EF178686, 99.3%; *S. sindenensis*, AB184759, 99.3%; *S. badius*, AY999783, 99.3%; *S. praecox*, AB184293, 99.2%; *S. albovinaceus*, AB249958, 99.2%; *S. microflavus*, DQ445795, 99.2%; *S. fimicarius*, AY999784, 99.2%; *S. flavofuscus*, AB249935, 99.2%; *S. alboviridis*, AB184256, 99.2%; *S. rubiginosohelvolus*, AB184240, 99.2%; *S. lipmanii*, AB184148, 99.2%; *S. mediolani*, AB184674, 99.2%; *S. cremeus*, AB184124, 99.2%; *S. nitrosporeus*, EF178680, 99.1%; *S. anulatus*, DQ026637, 99.1%; *S. acrimycini*, AY999889, 99.1%; *S. griseoplanus*, AY999894, 99.1%; *S. baarnensis*, EF178688, 99.1%; *S. lavendulae* subsp. *lavendulae*, AB184080, 99.1%; *S. cavourensis* subsp. *washingtonensis*, DQ026671, 99.1%; *S. argenteolus*, AB045872, 99%; *S. flavovirens*, DQ026635, 99%; *S. cinereorectus*, AB184646, 99%; *S. cyaneofuscatus*, AB184860, 99%; *S. finlayi*, AY999788, 99%; *S. luridiscabiei*, AF361784, 99%.

Source: not known.

DNA G+C content (mol%): not known.

Type strain: AS 4.1692, ATCC 14889, ATCC 25438, CBS 669.69, DSM 40300, NBRC 13026, JCM 4249, JCM 4298, JCM 4555, NCIMB 8918, NRRL 2740, NRRL-ISP 5300, RIA 1218, VKM Ac-731.

Sequence accession no. (16S rRNA gene): DQ445791.

96b. **Streptomyces cavourensis subsp. washingtonensis** Skarbek and Brady 1978, 52[AL]

wash.ing.ton.en'sis. N.L. masc. adj. *washingtonensis* of or pertaining to Washington.

Strain exhibits straight to flexuous chains of spores on all standard media and thus belongs to the section *Rectiflexibiles.* The spore type is smooth. Malate is moderately solubilized by all strains tested, except one, which gave a weak response. No distinctive pigments other than yellow-brown are found in the reverse mycelium (i.e. reverse side of colony) of cultures. Diffusible, soluble pigments other than brown or black are found only with one strain grown on the standard media. This yellow, soluble pigment is not a pH indicator. All cultures, except that of strain Illinois 205-2M, are found to produce melanoid pigments on at least three of the four media which were utilized. Strain Illinois 205-2M failed to produce melanoid pigments on any of the melanin formation media employed.

Tolerance to heavy metals is as follows. Strain AUW-83 differs from all the other strains in that it produces a soluble, dark brown pigment, modified by gray-green, after 10 d of incubation on the basal agar supplemented with zinc; little soluble pigment, except for a slight yellow-brown exhibited by strain 689, was observed with the remaining strains on the zinc agar.

Type strain shows the highest sequence similarity to: *S. cyaneofuscatus*, AB184860, 100%; *S. lavendulae* subsp. *lavendulae*, AB184080, 100%; *S. alboviridis*, AB184256, 100%; *S. fulvorobeus*, AB184711, 100%; *S. praecox*, AB184293, 100%; *S. anulatus*, DQ026637, 100%; *S. flavofuscus*, AB249935, 100%; *S. cinereorectus*, AB184646, 100%; *S. griseoplanus*, AY999894, 100%; *S. lipmanii*, AB184148, 100%; *S. fimicarius*, AY999784, 100%; *S. microflavus*, DQ445795, 99.9%; *S. sindenensis*, AB184759, 99.9%; *S. rubiginosohelvolus*, AB184240, 99.9%; *S. mediolani*, AB184674, 99.9%; *S. griseinus*, AB184205, 99.9%; *S. acrimycini*, AY999889, 99.9%; *S. pluricolorescens*, DQ442540, 99.9%; *S. argenteolus*, AB045872, 99.9%; *S. badius*, AY999783, 99.9%; *S. floridae*, AB184656, 99.8%; *S. albovinaceus*, AB249958, 99.8%; *S. globisporus* subsp. *globisporus*, EF178686, 99.8%; *S. griseus* subsp. *griseus*, AY207604, 99.8%; *S. griseolus*, AB184768, 99.8%; *S. luridiscabiei*, AF361784, 99.8%; *S. baarnensis*, EF178688, 99.7%; *S. flavovirens*, DQ026635, 99.7%; *S. californicus*, AB184755, 99.7%; *S. parvus*, DQ442537, 99.7%; *S. halstedii*, EF178695, 99.7%; *S. flavogriseus*, AJ494864, 99.6%; *S. olivoviridis*, AB184227, 99.5%; *S. bacillaris*, AB184439, 99.5%; *S. finlayi*, AY999788, 99.5%; *S. pulveraceus*, AB184806, 99.5%; *S. atroolivaceus*, AJ781320, 99.5%; *S. clavifer*, DQ026670, 99.4%; *S. yanii*, AB006159, 99.4%; *S. nitrosporeus*, EF178680, 99.4%; *S. albolongus*, AB184425, 99.3%; *S. sanglieri*, AB249945, 99.3%; *S. mutomycini*, AB249951, 99.2%; *S. atratus*, DQ026638, 99.2%; *S. celluloflavus*, AB184476, 99.2%; *S. gelaticus*, DQ026636, 99.2%; *S. griseobrunneus*, AB249912, 99.2%; *S. spiroverticillatus*, AB184814, 99.1%; *S. cavourensis* subsp. *cavourensis*, DQ445791, 99.1%; *S. cremeus*, AB184124, 99.1%; *S. candidus*, DQ026663, 99.1%.

Source: not known.

DNA G+C content (mol%): not known.

Type strain: AS 4.1635, ATCC 27732, DSM 41423, NBRC 15391, JCM 4967, NRRL B-8030.

Sequence accession no. (16S rRNA gene): DQ026671.

97. **Streptomyces cellostaticus** Hamada 1958, 178[AL]

Etymology of specific epithet is unknown.

Spore chains in Section *Spirales.* Mature spore chains generally have 10–50 spores per chain; longer chains are often observed. This morphology is seen on yeast-malt agar, oatmeal agar, salts-starch agar, and glycerol-asparagine agar. Spore surface is spiny.

Color of colony: aerial mass color in the Red (or Gray) color series on yeast-malt agar, oatmeal agar, salts-starch agar, and glycerol-asparagine agar. Characteristic color is usually between 3ge (light grayish yellowish brown) and 5ge or 5dc (grayish yellowish pink) color tabs of Tresner–Backus color wheels. Reverse side of colony with no distinctive pigment (characteristic grayed yellow or light yellow brown) on yeast-malt agar, oatmeal agar, salts-starch agar, or glycerol-asparagine agar.

Color in medium: melanoid pigments are formed in peptone-yeast-iron agar, tyrosine agar, and tryptone-yeast broth. Pigments other than melanoids or faint traces of yellow pigment not formed in yeast-malt agar, oatmeal agar, salts-starch agar, or glycerol-asparagine agar.

D-Glucose, L-arabinose, sucrose, D-xylose, iso-inositol, D-mannitol, D-fructose, rhamnose, and raffinose are utilized for growth.

Type strain shows the highest sequence similarity to: *S. griseochromogenes*, AB184387, 99.7%; *S. yokosukanensis*, DQ026652, 99.5%; *S. achromogenes* subsp. *achromogenes*, AB184109, 99%; *S. griseoruber*, AB184209, 99%; *S. olivaceoviridis*, AB184288, 99%; *S. corchorusii*, AB184267, 99%.

Source: not known.

DNA G+C content (mol%): not known.

Type strain: AS 4.1637, ATCC 23894, CBS 680.68, DSM 40189, NBRC 12849, IMET 41374, JCM 4183, JCM 4631, NCIMB 9830, NRRL-ISP 5189, RIA 1143, VKM Ac-1222.

Sequence accession no. (16S rRNA gene): AB184192.

98. **Streptomyces celluloflavus** Nishimura, Kimura and Kuroya 1953, 64[AL]

cel.lu.lo.fla'vus. N.L. n. *cellulosum* cellulose; L. adj. *flavus* yellow; N.L. masc. adj. *celluloflavus* cellulose, yellow (intended to refer to the yellow streptomycete that attacks cellulose).

Sparse formation of aerial mycelium. Poor growth on Czapek's solution agar. Produces aureothricin; inhibited by streptomycin.

Type strain shows the highest sequence similarity to: *S. cavourensis* subsp. *cavourensis*, DQ445791, 100%; *S. albolongus*, AB184425, 100%; *S. griseobrunneus*, AB249912, 100%; *S. californicus*, AB184755, 99.5%; *S. bacillaris*, AB184439, 99.4%; *S. fulvorobeus*, AB184711, 99.4%; *S. floridae*,

AB184656, 99.4%; *S. griseinus*, AB184205, 99.4%; *S. pluricolorescens*, DQ442540, 99.4%; *S. candidus*, DQ026663, 99.4%; *S. spiroverticillatus*, AB184814, 99.4%; *S. globisporus* subsp. *globisporus*, EF178686, 99.4%; *S. sindenensis*, AB184759, 99.4%; *S. badius*, AY999783, 99.4%; *S. rubiginosohelvolus*, AB184240, 99.4%; *S. mediolani*, AB184674, 99.4%; *S. praecox*, AB184293, 99.3%; *S. albovinaceus*, AB249958, 99.3%; *S. microflavus*, DQ445795, 99.3%; *S. fimicarius*, AY999784, 99.3%; *S. flavofuscus*, AB249935, 99.5%; *S. alboviridis*, AB184256, 99.3%; *S. lipmanii*, AB184148, 99.3%; *S. cremeus*, AB184124, 99.3%; *S. griseoplanus*, AY999894, 99.3%; *S. anulatus*, DQ026637, 99.3%; *S. parvus*, DQ442537, 99.2%; *S. luridiscabiei*, AF361784, 99.2%; *S. acrimycini*, AY999889, 99.2%; *S. baarnensis*, EF178688, 99.2%; *S. cyaneofuscatus*, AB184860, 99.2%; *S. cavourensis* subsp. *washingtonensis*, DQ026671, 99.2%; *S. cinereorectus*, AB184646, 99.2%; *S. flavovirens*, DQ026635, 99.2%; *S. lavendulae* subsp. *lavendulae*, AB184080, 99.2%; *S. nitrosporeus*, EF178680, 99.1%; *S. argenteolus*, AB045872, 99.1%; *S. finlayi*, AY999788, 99.1%; *S. flavogriseus*, AJ494864, 99.1%; *S. griseolus*, AB184768, 99%; *S. pulveraceus*, AB184806, 99%; *S. clavifer*, DQ026670, 99%; *S. griseus* subsp. *griseus*, AY207604, 99%; *S. halstedii*, EF178695, 99%; *S. cinnamonensis*, AB184707, 99%.

Source: not known.

DNA G+C content (mol%): not known.

Type strain: AS 4.1659, ATCC 29806, CECT 3242, DSM 40839, NBRC 13780, JCM 4126, KCTC 9702, NRRL B-2493.

Sequence accession no. (16S rRNA gene): AB184476.

99. **Streptomyces cellulolyticus** Li 1997, 444[VP]

cel.lu.lo.ly'ti.cus. N.L. n. *cellulosum* cellulose; N.L. masc. adj. *lyticus* (from Gr. masc. adj. *lutikos*) able to dissolve; N.L. masc. adj. *cellulolyticus* decomposing cellulose.

Spore chains are *Rectiflexibiles*, with 20 or more spores per chain. Spores are oval and are 2.1–2.3 × 2.5–2.7 µm. Spore surface is warty. Mycelia do not fragment into coccoid or bacillary structures. The branching substrate mycelium is: yellow on yeast extract-malt extract agar, inorganic salts-starch agar, glucose-asparagine agar, and nutrient agar; brown on glycerol-asparagine agar, tyrosine agar, and Czapek's solution agar; and colorless on oatmeal agar. Aerial spore mass is white to pink. Soluble pigments, including melanin, are not produced. Cellulose is decomposed. D-Glucose, D-fructose, L-arabinose, sucrose, D-xylose, raffinose, and iso-inositol are utilized for growth, but rhamnose is not utilized. L-Asparagine, L-cysteine, and L-threonine can be used as nitrogen sources. Positive for catalase activity and production of H_2S and negative for oxidase activity and indole production. Nitrate is reduced. Starch, casein, and esculin are hydrolyzed. Gelatin is not liquefied. Good growth occurs at pH 7.2 and the optimum temperature is 30°C. Growth occurs in the presence of up to 10% NaCl. Isolate LX^T is susceptible to penicillin G, but not to dimethylchlortetracycline, vancomycin, kanamycin, rifampin, or aminobenzylpenicillin. The cell-wall chemotype is chemotype I and the cell wall contains LL-A_2pm and glycine; no characteristic sugars are detected as whole-cell sugars.

Source: not known.

DNA G+C content (mol%): not known.

Type strain: LX, AS 4.1332.

Sequence accession no. (16S rRNA gene): no sequence available.

100. **Streptomyces cellulosae** (Krainsky 1914) Waksman and Henrici *in* Breed, Murray and Hitchens 1948, 938[AL] ["*Actinomyces cellulosae*" (*sic*) Krainsky 1914, 683]

cel.lu.lo'sa.e. N.L. n. *cellulosum* cellulose; N.L. gen. n. *cellulosae* (*sic*) of cellulose (probably intended to mean the species that degrades cellulose).

Spore chains are typically flexuous. Excellent growth on Czapek's solution agar; exhibits slight anti-bacterial activity.

Type strain shows the highest sequence similarity to: *S. pseudogriseolus*, DQ442541, 99.9%; *S. gancidicus*, AB184660, 99.9%; *S. capillispiralis*, AB184577, 99.8%; *S. carpinensis*, AB184574, 99.4%; *S. azureus*, EF178674, 99.4%; *S. levis*, AB184670, 99.3%; *S. rubiginosus*, AY999810, 99.2%; *S. lusitanus*, AB184424, 99.2%; *S. lavendulocolor*, DQ442516, 99.1%; *S. thermocarboxydus*, U94490, 99.1%; *S. caelestis*, X80824, 99.1%; *S. matensis*, AB184221, 99.1%; *S. djakartensis*, AB184657, 99%; *S. africanus*, AY208912, 99%; *S. viridiviolaceus*, AY999854, 99%; *S. afghaniensis*, AJ399483, 99%; *S. griseoloalbus*, AB184275, 99%; *S. tuirus*, AB184690, 99%; *S. albaduncus*, AY999757, 99%.

Source: not known.

DNA G+C content (mol%): not known.

Type strain: AS 4.1411, ATCC 25439, CBS 122.18, CBS 670.69, BCRC 12087, DSM 40362, NBRC 13027, JCM 4462, KCTC 9703, LMG 19315, NRRL B-2889, NRRL-ISP 5362, RIA 1219, VKM Ac-829.

Sequence accession no. (16S rRNA gene): DQ442495.

101. **Streptomyces champavatii** Uma and Narasimha Rao 1959, 133[AL]

cham.pa.va'ti.i. N.L. gen. n. *champavatii* of Champavathi, named after the Champavathi River in Andhra Pradesh, India.

Forms green vegetative mycelium and diffusible pigment on some media. Poor growth on Czapek's solution agar; produces champamycins A and B (heptaenic antifungal antibiotics) and champavatin, a non-polyenic antifungal antibiotic; produces vitamin B_{12}.

Type strain shows the highest sequence similarity to: *S. limosus*, AB184147, 100%; *S. felleus*, AB184129, 100%; *S. daghestanicus*, DQ442497, 100%; *S. albidoflavus*, AB184255, 100%; *S. odorifer*, Z76682, 100%; *S. violascens*, AY999737, 100%; *S. hydrogenans*, AB184868, 100%; *S. griseus* subsp. *solvifaciens*, AB249915, 99.9%; *S. canescens*, AB184117, 99.9%; *S. koyangensis*, AY079156, 99.5%; *S. sampsonii*, D63871, 99.3%.

Source: not known.

DNA G+C content (mol%): not known.

Type strain: AS 4.1615, BCRC 12231, DSM 40841, NBRC 15392, JCM 5066, NCIMB 12859, NRRL B-5682.

Sequence accession no. (16S rRNA gene): DQ026642.

102. **Streptomyces chartreusis** Leach, Calhoun, Johnson, Teeters and Jackson 1953, 4011[AL] [*"Actinomyces chartreusis"* Preobrazhenskaya 1966, 852]

char.treu'sis. N.L. n. *chartreusum* (from French n. *chartreuse*), a Carthusian monastery famed for a sweet yellow liqueur, hence the color "chartreuse"; N.L. masc. adj. *chartreusis* yellow, referring to color of the diffusible pigment formed by the organism.

Spore chains in Section *Spirales*. Clusters of spiral spore chains sometimes resemble whorls of verticils. Pseudoverticils are not uniformly distributed and do not arise from an acial hypha characteristic of true verticillate forms. Mature spore chains generally have 10–50 spores per chain. This morphology is seen on yeast-malt agar, oatmeal agar, salts-starch agar, and glycerol-asparagine agar. Spore surface is spiny.

Color of colony: aerial mass color in the Blue color series on yeast-malt agar, oatmeal agar, salts-starch agar, and glycerol-asparagine agar (Gray series also reported on oatmeal agar). Reverse side of colony with no distinctive pigments on yeast-malt agar, oatmeal agar, salts-starch agar, and glycerol-asparagine agar; substrate is not a pH indicator.

Color in medium: melanoid pigments formed in peptone-yeast-iron agar and tryptone-yeast broth, but not in tyrosine agar. Pigments other than melanoids are not formed in yeast-malt agar, oatmeal agar, salts-starch agar, and glycerol-asparagine agar.

D-Glucose, L-arabinose, sucrose, D-xylose, iso-inositol, D-mannitol, D-fructose, rhamnose, and raffinose are utilized for growth.

Type strain shows the highest sequence similarity to: *S. resistomycificus*, AB184166, 99.5%; *S. galilaeus*, AB045878, 99.3%; *S. bobili*, AB249925, 99.3%; *S. novaecaesareae*, AB184357, 99.1%; *S. phaeoluteigriseus*, AJ391815, 99.1%; *S. pseudovenezuelae*, AB184233, 99.1%; *S. prunicolor*, DQ026659, 99%; *S. aureocirculatus*, AB184260, 99%.

Source: not known.

DNA G+C content (mol%): not known.

Type strain: AS 4.1639, ATCC 14922, ATCC 19738, CBS 476.68, BCRC 13673, CCT 5005, DSM 40085, NBRC 12753, JCM 4570, KCTC 9704, NRRL 2287, NRRL-ISP 5085, RIA 1018, UNIQEM 126, VKM Ac-1721.

Sequence accession no. (16S rRNA gene): AB184839.

103. **Streptomyces chattanoogensis** Burns and Holtman 1959, 398[AL]

chat.ta.no.o.gen'sis. N.L. masc. adj. *chattanoogensis* of or belonging to Chattanooga, Tennessee, the source of the soil from which the organism was isolated.

Spore chains Section *Spirales*, but spore chains representative of Section *Rectiflexibiles* and Section *Retinaculiaperti* are also found. Mature spore chains generally have 10–50 spores per chain on suitable media. This morphology is seen on yeast-malt agar, oatmeal agar, salts-starch agar, and glycerol-asparagine agar. Spore surface is spiny.

Color of colony: aerial mass color in the Yellow or White color series on yeast-malt agar, oatmeal agar, salts-starch agar, and glycerol-asparagine agar. Yellow, when present, is a very pale yellow. Reverse side of colony with no distinctive pigments on yeast-malt agar, oatmeal agar, salts-starch agar, and glycerol-asparagine agar; substrate is not a pH indicator.

Color in medium: melanoid pigments are not formed in peptone-yeast-iron agar and tyrosine agar. Yellow pigment is found in medium in yeast-malt agar, oatmeal agar, salts-starch agar, and glycerol-asparagine agar; this pigment is not pH-sensitive when tested with 0.05 M NaOH or HCl.

D-Glucose, iso-inositol, D-mannitol, D-fructose, raffinose, and sucrose are utilized for growth. No growth or only traces of growth on D-xylose, rhamnose, and L-arabinose.

Type strain shows the highest sequence similarity to: *S. lydicus*, Y15507, 99.8%; *S. sioyaensis*, DQ026654, 99.3%; *S. rimosus* subsp. *paromomycinus*, AJ621610, 99.2%; *S. chrestomyceticus*, AJ621609, 99.2%; *S. tubercidicus*, AJ621612, 99%; *S. libani* subsp. *libani*, AB184414, 99%; *S. nigrescens*, DQ442530, 99%.

Source: isolated from soil from Chattanooga, Tennessee, USA.

DNA G+C content (mol%): not known.

Type strain: AS 4.1415, ATCC 13358, ATCC 19739, CBS 477.68, BCRC 13655, CECT 3321, DSM 40002, NBRC 12754, JCM 4299, JCM 4571, KCTC 1087, LMG 19339, NCIMB 9809, NRRL B-2255, NRRL-ISP 5002, RIA 1019, VKM Ac-1775.

Sequence accession no. (16S rRNA gene): AJ621611.

104. **Streptomyces cheonanensis** Kim, Lee and Hwang 2006, 474[AL]

che.on.an.en'sis. N.L. masc. adj. *cheonanensis* of or pertaining to Cheonan, Republic of Korea, the geographical origin of the type strain.

Forms extensively branched aerial and substrate hyphae. Short or long, straight to flexuous chains of smooth-surfaced spores are evident on the aerial hyphae. Aerial mycelium is gray to white and the substrate mycelium appears light yellow when grown on ISP 4 agar. Aerial and substrate mycelia grow abundantly on both ISP 3 agar and Bennett's agar. Soluble pigments are generated on ISP3, ISP 3, ISP 5 (glycerol-asparagine agar), and ISP 7. The cell wall contains LL-A_2pm. Predominant cellular fatty acids are 14-methylpentadecanoic acid ($C_{16:0}$ iso; 47.82%), hexadecanoic acid ($C_{16:0}$; 14.44%), and *cis*-9-hexadecenoic acid ($C_{16:1}$ *cis*9; 10.24%). Optimum growth occurs at 29°C. Grows well in yeast extract-malt extract broth adjusted to pH 6.5–8.0. Tolerates NaCl concentrations up to 7%. Capable of utilizing several carbon sources, including adonitol, arabinose, dextran, fructose, *myo*-inositol, mannitol, melezitose, melibiose, raffinose, L-rhamnose, sucrose, xylitol, and xylose. Can also use several nitrogen sources: DL-α-amino-n-butyric acid, L-cysteine, L-histidine, L-hydroxyproline, L-phenylalanine, and L-valine. Resistant to penicillin G, but sensitive to neomycin, oleandomycin, and rifampin. Secretes compounds the inhibit mycelial growth of plant-pathogenic fungi including *Alternaria mali*, *Collectotrichum orbiculare*, *Magnaporthe grisea*, *Fusarium oxysporum* f. sp. *lycopersici*, and *Rhizotonia solani* and the oomycete *Phytophthora capsici*.

Type strain shows no sequence similarity over 99%. Type strain shows DNA–DNA similarity to: *S. thermolineatus* DSM 41451^T, 21.5%; "*S. cattleya*" JCM 4925, 39.8%; *S. macrosporus* DSM 41449^T, 19.8%; *S. acidiscabies* ATCC 49003^T, 60.6%.

Source: not known.

DNA G+C content (mol%): 75.5.

Type strain: VC-A46, KCCM 42119, NBRC 100940.

Sequence accession no. (16S rRNA gene): AY822606.

105. **Streptomyces chibaensis** Suzuki, Nakamura, Okama and Tomiyama 1958, 81^{AL}

chi.ba.en'sis. N.L. masc. adj. *chibaensis* of or belonging to Chiba City, Japan, the source of the soil from which the organism was isolated.

Spore chains in Section *Spirales* or *Rectiflexibiles.* Spirals, when formed, are open and poorly developed. Some spore chains are straight and many are flexuous or curved to form hooks or partial spirals. Most spore chains are too short (3 to 10 or 20 spores per chain) for this culture to be placed in Section *Retinaculiaperti.* This morphology is seen on yeast-malt agar, oatmeal agar, salts-starch agar, and glycerol-asparagine agar. Spore surface is smooth.

Color of colony: mature aerial mass color is usually in the Gray color series on yeast-malt agar, oatmeal agar, and salts-starch agar; Gray or Yellow color series on glycerol-asparagine agar. Color tabs selected from the Gray color series were yellowish gray or grayish yellowish brown. Reverse side of colony with no distinctive pigments (grayed greenish yellow, yellow, or rarely orange-yellow) on yeast-malt agar, oatmeal agar, salts-starch agar, and glycerol-asparagine agar; substrate mycelium pigment is not a pH indicator.

Color in medium: melanoid pigments are not found in peptone-yeast-iron agar, tyrosine agar, or tryptone-yeast broth. Yellow pigment may or may not be found in the medium in yeast-malt agar, oatmeal agar, and glycerol-asparagine agar; this pigment is not pH-sensitive.

D-Glucose, L-arabinose, sucrose, D-xylose, iso-inositol, D-mannitol, D-fructose, rhamnose, and raffinose are all utilized for growth.

For sequence similarity, see type strain of *Streptomyces corchorusii.*

Source: not known.

DNA G+C content (mol%): not known.

Type strain: AS 4.1654, ATCC 23895, CBS 681.68, DSM 40220, IFM 1085, NBRC 12850, JCM 4017, JCM 4632, KCTC 9786, LMG 20456, NRRL B-2904, NRRL-ISP 5220, RIA 1164, VKM Ac-1893.

Sequence accession no. (16S rRNA gene): AB184193.

Further comments: according to Lanoot et al. (2005b), *Streptomyces chibaensis* Suzuki et al. 1958 is a later heterotypic synonym of *Streptomyces corchorusii* Ahmad and Bhuiyan 1958 emend. Lanoot et al. 2005b.

According to Rule 24b of the *Bacteriological Code* (1990 Revision), if two names compete for priority and if both names are listed on the Approved Lists of Bacterial Names, the priority shall be determined by the date of the effective publication of the name before 1 January 1980. *Streptomyces chibaensis* and *Streptomyces corchorusii* are cited in the Approved Lists. The dates of effective publications are 1958. Consequently, to determine priority it is necessary to know the month (and perhaps the day) of the effective publications. *Streptomyces chibaensis* was effectively published in the *Journal of Antibiotics (Tokyo) Series A*, 1958, vol. 11, pp. 81–83, and *Streptomyces corchorusii* in the *Pakistan Journal of Biological and Agricultural Sciences*, 1958, vol. 1, pp. 137–143. It is certainly not easy to know the exact dates of publication of these articles. However, according to Rule 42 of the *Bacteriological Code* (1990 Revision), if the epithets are of the same date, the author who first unites the taxa has the right to choose one of them, and his choice must be followed. Lanoot et al. chose the epithet *corchorusii.*

106. **Streptomyces chrestomyceticus** Canevazzi and Scotti 1959, 248^{AL}

chres.to.my.ce'ti.cus. N.L. n. *chrestomycinum* chrestomycin, name of an antibiotic; L. masc. suff. *-icus* suffix used with the sense of belonging to; N.L. masc. adj. *chrestomyceticus* belonging to chrestomycin.

Spore chains in Section *Spirales.* Spirals are often irregular and may become entangled. Mature spore chains generally are long with 10 to 50 or often more than 50 spores per chain. This morphology is seen on yeast-malt agar, oatmeal agar, salts-starch agar, and glycerol-asparagine agar. Spore surface is smooth. Special morphological characteristics are moist droplets (hygroscopic droplets are sometimes found on oatmeal agar; one observer describes these as similar to sporangia of *Actinosporangium*). Coremia may also be formed on salts-starch agar or glycerol-asparagine agar.

Color of colony: aerial mass color in the Yellow or White color series on yeast-malt agar and glycerol-asparagine agar; White color series on salts-starch agar; Green or White color series on oatmeal agar. Nearest matching color tabs in the Yellow color series are ½ec to 1dc, pale yellow green; 2ba, pale yellow; and 2fb, light yellow. Nearest matching color tab in the Green color series is 1½ge, light grayish olive. Reverse side of colony with no distinctive pigments (colorless to pale yellow or light yellowish brown) on yeast-malt agar, oatmeal agar, salts-starch agar, and glycerol-asparagine agar.

Color in medium: melanoid pigments are not formed in peptone-yeast-iron agar, tyrosine agar, tryptone-yeast broth, or Gause's medium no. 2. No pigment is found in medium in yeast-malt agar, oatmeal agar, salts-starch agar, or glycerol-asparagine agar.

D-Glucose, L-arabinose, D-mannitol, and D-fructose are utilized for growth. Utilization of sucrose, D-xylose, iso-inositol, and raffinose is doubtful. No growth or only traces of growth with rhamnose.

Type strain shows the highest sequence similarity to: *S. rimosus* subsp. *paromomycinus*, AJ621610, 100%; *S. albofaciens*, AB045880, 99.7%; *S. lydicus*, Y15507, 99.3%; *S. erumpens*, AJ621603, 99.3%; *S. chattanoogensis*, AJ621611, 99.2%; *S. sclerotialus*, AJ621608, 99.1%; *S. sioyaensis*, DQ026654, 99%.

Source: not known.

DNA G+C content (mol%): not known.

Type strain: AS 4.1657, ATCC 14947, CBS 745.72, BCRC 12173, DSM 40545, NBRC 13444, JCM 4735, NCAIM B.01478, NCIMB 8995, NRRL B-3293, NRRL B-3310, NRRL B-3672, NRRL-ISP 5545, RIA 1405.

Sequence accession no. (16S rRNA gene): AJ621609.

Further comments: the ISP description of this strain differs from the original description with respect to spiral spore chains versus very short *Rectiflexibiles* chains, Yellow or Green aerial mycelium versus White, and in growth on L-arabinose.

107. **Streptomyces chromofuscus** (Preobrazhenskaya, Blinov and Ryabova *in* Gauze, Preobrazhenskaya, Kudrina, Blinov, Ryabova and Sveshnikova 1957) Pridham, Hesseltine and Benedict 1958, 68[AL] ("*Actinomyces chromofuscus*" Preobrazhenskaya, Blinov and Ryabova *in* Gauze, Preobrazhenskaya, Kudrina, Blinov, Ryabova and Sveshnikova 1957, 176)

chro.mo.fus'cus. Gr. n. *chroma* color; L. adj. *fuscus* dark, tawny; N.L. masc. adj. *chromofuscus* dark or tawny colored.

Spore chains in Section *Spirales*. Mature spore chains are moderately long with 10–50, or often more than 50, spores per chain. This morphology is seen on yeast-malt agar, oatmeal agar, salts-starch agar, and glycerol-asparagine agar. Spore surface is spiny.

Color of colony: aerial mass color in the Gray color series on yeast-malt agar, oatmeal agar, salts-starch agar, and glycerol-asparagine agar. Reverse side of colony with no distinctive pigment (grayed yellow to olive brown) on yeast-malt agar, oatmeal agar, salts-starch agar, and glycerol-asparagine agar.

Color in medium: melanoid pigments are formed in peptone-yeast-iron agar and tryptone-yeast broth, but not in tyrosine agar. Pigments other than melanoids or faint traces of yellow are not found in the medium in yeast-malt agar, oatmeal agar, salts-starch agar, or glycerol-asparagine agar.

D-Glucose, L-arabinose, D-xylose, iso-inositol, D-mannitol, D-fructose, and rhamnose are utilized for growth. Reports vary on utilization of raffinose and iso-inositol. No growth or only traces of growth on sucrose.

Type strain shows the highest sequence similarity to: *S. cinereospinus*, AB184648, 99.1%; *S. coeruleofuscus*, DQ026668, 99.1%.

Source: not known.

DNA G+C content (mol%): not known.

Type strain: AS 4.1451, ATCC 23896, CBS 682.68, DSM 40273, NBRC 12851, INA 13638/58, JCM 4354, LMG 19317, NRRL B-12175, NRRL-ISP 5273, RIA 1191, VKM Ac-974.

Sequence accession no. (16S rRNA gene): AB184194.

108. **Streptomyces chryseus** (Krasil'nikov, Korenyako and Nikitina *in* Krasil'nikov 1965) Pridham 1970, 10[AL] ("*Actinomyces chryseus*" Krasil'nikov, Korenyako and Nikitina *in* Krasil'nikov 1965, 224)

chry'se.us. N.L. masc. adj. *chryseus* (from Gr. masc. adj. *khruseos*) golden.

Spore chains in Section *Spirales* on oatmeal agar or salts-starch agar, but flexuous spore chains suggesting *Rectiflexibiles* morphology may be common on these media and are the predominant morphology on glycerol-asparagine agar. Sporulating aerial mycelium is poorly developed or absent on yeast-malt agar. Mature spore chains are moderately long with 10 to 50 or more spores per chain. Spore surface is smooth.

Color of colony: aerial mass color in the Red color series (5ca, light yellowish pink, or 3ca, pale orange-yellow) on oatmeal agar, salts-starch agar, and glycerol-asparagine agar in 14–21 d. White aerial mycelium may also be seen on these media. Reverse side of colony is pale yellow to light yellowish brown or orange-yellow on yeast-malt agar, oatmeal agar, salts-starch agar, and glycerol-asparagine agar.

Color in medium: melanoid pigments are not formed in peptone-yeast-iron agar, tyrosine agar, and tryptone-yeast broth. Yellow pigment may be found in the medium in yeast-malt agar, oatmeal agar, salts-starch agar, and glycerol-asparagine agar. This pigment is not pH-sensitive when tested with 0.05 M NaOH or HCl.

D-Glucose, L-arabinose, and D-fructose are utilized for growth. Reports vary on utilization of D-xylose and iso-inositol. No growth or only traces of growth with D-mannitol, rhamnose, sucrose, and raffinose.

Type strain shows the highest sequence similarity to: *S. flavidovirens*, AB184270, 100%; *S. helvaticus*, AB184367, 100%; *S. albidochromogenes*, AB249953, 99.9%; *S. enissocaesilis*, AB249930, 99.1%.

Source: not known.

DNA G+C content (mol%): not known.

Type strain: AS 4.1694, ATCC 19829, CBS 678.72, DSM 40420, NBRC 13377, JCM 4737, NCIMB 10041, NRRL B-12347, NRRL-ISP 5420, RIA 1338, VKM Ac-200.

Sequence accession no. (16S rRNA gene): AY999787.

109a. **Streptomyces chrysomallus subsp. chrysomallus** Lindenbein 1952, 369[AL]

chry.so'mal.lus. N.L. masc. adj. *chrysomallus* (from Gr. masc. adj. *khrusomallos*) with golden wool.

Spore chains in Section *Rectiflexibiles*. Mature spore chains generally have 10–50 spores per chain. Typical morphology on yeast-malt agar, oatmeal agar, salts-starch agar, and glycerol-asparagine agar. Spore surface is smooth (some surface irregularities are present, but are less distinct than on characteristic warty spores).

Color of colony: aerial mass color in the Yellow color series on yeast-malt agar, oatmeal agar, salts-starch agar, and glycerol-asparagine agar. Reverse side of colony with no distinctive pigments (yellow to grayed-yellow) on yeast-malt agar, oatmeal agar, salts-starch agar, and glycerol-asparagine agar; substrate pigment is not a pH indicator.

Color in medium: melanoid pigments are not formed in peptone-yeast-iron agar or tyrosine agar. Yellow pigment found in medium in yeast-malt agar and oatmeal agar; traces of yellow pigment may be formed in salts-starch agar and glycerol-asparagine agar. This pigment is not a pH indicator.

D-Glucose, L-arabinose, D-xylose, D-mannitol, D-fructose, and rhamnose are utilized for growth. No growth or only trace of growth on sucrose, iso-inositol, and raffinose.

For sequence similarity, see type strain of *Streptomyces anulatus*.

Source: not known.

DNA G+C content (mol%): not known.

Type strain: AS 4.1676, ATCC 11523, BCRC 11511, DSM 40128, NBRC 15393, IMET 41360, JCM 4296, LMG 20459, NRRL 2250, NRRL 2280, UNIQEM 127.

Sequence accession no. (16S rRNA gene): AB184644.

Further comments: Lanoot et al. (2005b) are of the opinion that *Streptomyces chrysomallus* subsp. *chrysomallus* Lindenbein 1952 is a later heterotypic synonym of *Streptomyces anulatus* (Beijerinck 1912) Waksman 1953. The type of the subspecies *Streptomyces chrysomallus* subsp. *chrysomallus* is automatically the type of *Streptomyces chrysomallus* Lindenbein 1952^{AL}. Consequently, if an author agrees with Lanoot et al., *Streptomyces chrysomallus* must be considered as a later heterotypic synonym of *Streptomyces anulatus*. In expressing that opinion, Lanoot et al. (2005b) have placed the type of *Streptomyces chrysomallus* in a different species. In this case, Rule 37a of the *Bacteriological Code* (1990 Revision) applies and the authors should have dealt with the nomenclature and taxonomic position of *Streptomyces chrysomallus* subsp. *fumigatus* Frommer 1959. Authors who follow the proposal to treat the types of *Streptomyces anulatus* and *Streptomyces chrysomallus* (including the subspecies *Streptomyces chrysomallus* subsp. *chrysomallus*) as synonyms are not at liberty to use the name *Streptomyces chrysomallus* subsp. *fumigatus* and must make a taxonomic proposal for placing this subspecies in another species or subspecies.

109b. **Streptomyces chrysomallus subsp. fumigatus** Frommer 1959, 202^{AL}

fu.mi.ga'tus. L. masc. part. adj. *fumigatus* smoked.

Produces the actinomycin C complex; inhibited by streptomycin; poor growth on Czapek's solution agar.

Type strain shows the highest sequence similarity to: *S. purpureus*, AJ781324, 99.9%; *S. herbaricolor*, AB184801, 99.5%; *S. indigoferus*, AB184214, 99.5%; *S. aburaviensis*, AY999779, 99.5%; *S. xanthocidicus*, AY999858, 99.4%; *S. psammoticus*, AY999862, 99.3%; *S. avellaneus*, AB184413, 99.2%. Type strain shows the highest sequence similarity to following *Kitasatospora*: *Kitasatospora kifunensis*, AB022874, 99.2%.

Source: not known.

DNA G+C content (mol%): not known.

Type strain: AS 4.1589, DSM 41424, NBRC 15394, JCM 3371, KCTC 9705, NRRL B-2289.

Sequence accession no. (16S rRNA gene): AB184645.

Further comments: Lanoot et al. (2005b) are of the opinion that *Streptomyces chrysomallus* subsp. *chrysomallus* Lindenbein 1952 is a later heterotypic synonym of *Streptomyces anulatus* (Beijerinck 1912) Waksman 1953. The type of the subspecies *Streptomyces chrysomallus* subsp. *chrysomallus* is automatically the type of *Streptomyces chrysomallus* Lindenbein 1952^{AL}. So, in expressing that opinion Lanoot et al. have placed the type of *Streptomyces chrysomallus* in a different species. In this case, Rule 37a of the *Bacteriological Code* (1990 Revision) applies and the authors should have dealt with the nomenclature and taxonomic position of *Streptomyces chrysomallus* subsp. *fumigatus* Frommer 1959. Authors who follow the proposal to treat the types of *Streptomyces anulatus* and *Streptomyces chrysomallus* (including the subspecies *Streptomyces chrysomallus* subsp. *chrysomallus*) as synonyms are not at liberty to use the name *Streptomyces chrysomallus* subsp. *fumigatus* and must make a taxonomic proposal for placing this subspecies in another species or subspecies.

110. **Streptomyces cinereorectus** Terekhova and Preobrazhenskaya 1986, 574^{VP} (Effective publication: Terekhova and Preobrazhenskaya *in* Gause, Preobrazhenskaya, Sveshnikova, Terekhova and Maximova 1983.) emend. Lanoot, Vancanneyt, Dawyndt, Cnockaert, Zhang, Huang, Liu and Swings 2005a, 8 (Effective publication: Lanoot, Vancanneyt, Dawyndt, Cnockaert, Zhang, Huang, Liu and Swings 2004, 88.)

ci.ne.re.o.rec'tus. L. adj. *cinereus* similar to ashes, ash-colored; L. adj. *rectus* straight; N.L. masc. adj. *cinereorectus* ash-colored, straight.

Spore chains are straight, short, up to 10 spores per chain (*Rectiflexibiles*); spores are smooth. On mineral agar 1, oatmeal agar, and starch-ammonia agar: moderate or poor growth; aerial mycelium is poor, light gray; substrate mycelium colorless; no diffusible pigment. Glycerol-nitrate agar: abundant growth; aerial mycelium is gray; substrate mycelium grayish yellowish; no diffusible pigment. On glycerol-asparagine agar: moderate growth; aerial mycelium is gray; substrate mycelium colorless; no diffusible pigment. On organic agar 2: no aerial mycelium; substrate mycelium is colorless to yellowish; no diffusible pigment. Melanoid pigments are not formed. Good growth on glucose; moderate growth on fructose and mannitol; poor growth on raffinose; no growth on rhamnose or xylose. Forms antibiotic penicillin N. Contains LL-A_2pm and no diagnostic sugars in whole-cell hydrolysates.

Type strain shows the highest sequence similarity to: *S. argenteolus*, AB045872, 100%; *S. cavourensis* subsp. *washingtonensis*, DQ026671, 100%; *S. lavendulae* subsp. *lavendulae*, AB184080, 100%; *S. cyaneofuscatus*, AB184860, 99.9%; *S. fulvorobeus*, AB184711, 99.9%; *S. lipmanii*, AB184148, 99.9%; *S. griseoplanus*, AY999894, 99.9%; *S. alboviridis*, AB184256, 99.9%; *S. flavofuscus*, AB249935, 99.9%; *S. fimicarius*, AY999784, 99.9%; *S. praecox*, AB184293, 99.9%; *S. griseolus*, AB184768, 99.9%; *S. anulatus*, DQ026637, 99.9%; *S. microflavus*, DQ445795, 99.9%; *S. halstedii*, EF178695, 99.8%; *S. pluricolorescens*, DQ442540, 99.8%; *S. sindenensis*, AB184759, 99.8%; *S. griseinus*, AB184205, 99.8%; *S. badius*, AY999783, 99.8%; *S. mediolani*, AB184674, 99.8%; *S. acrimycini*, AY999889, 99.8%; *S. rubiginosohelvolus*, AB184240, 99.8%; *S. albovinaceus*, AB249958, 99.7%; *S. flavogriseus*, AJ494864, 99.7%; *S. parvus*, DQ442537, 99.7%; *S. griseus* subsp. *griseus*, AY207604, 99.7%; *S. floridae*, AB184656, 99.7%; *S. baarnensis*, EF178688, 99.7%; *S. luridiscabiei*, AF361784, 99.7%; *S. flavovirens*, DQ026635, 99.7%; *S. californicus*, AB184755, 99.7%; *S. globisporus* subsp. *globisporus*, EF178686, 99.7%; *S. yanii*, AB006159, 99.5%; *S. pulveraceus*, AB184806, 99.5%; *S. nitrosporeus*, EF178680, 99.5%; *S. finlayi*, AY999788, 99.4%; *S. olivoviridis*, AB184227, 99.4%; *S. atroolivaceus*, AJ781320, 99.4%; *S. bacillaris*, AB184439, 99.4%; *S. sanglieri*, AB249945, 99.3%; *S. clavifer*, DQ026670, 99.3%; *S. albolongus*, AB184425, 99.2%; *S. gelaticus*, DQ026636, 99.2%; *S. celluloflavus*, AB184476, 99.2%; *S. mutomycini*,

AB249951, 99.2%; *S. griseobrunneus*, AB249912, 99.2%; *S. atratus*, DQ026638, 99.2%; *S. cremeus*, AB184124, 99%; *S. candidus*, DQ026663, 99%; *S. cavourensis* subsp. *cavourensis*, DQ445791, 99%; *S. mauvecolor*, AB184532, 99%; *S. spiroverticillatus*, AB184814, 99%.

Source: not known.

DNA G+C content (mol%): not known.

Type strain: AS 4.1622, ATCC 43679, DSM 41469, NBRC 15395, INA 5202, JCM 6916.

Sequence accession no. (16S rRNA gene): AB184646.

Further comments: according to Lanoot et al. (2004), *Streptomyces cinereorectus* Terekhova and Preobrazhenskaya 1986 is an earlier heterotypic synonym of *Streptomyces cochleatus* Nakagaito et al. 1993b.

111a. **Streptomyces cinereoruber subsp. cinereoruber** Corbaz, Ettlinger, Keller-Schierlein and Zähner 1957b, 331[AL]

ci.ne.re.o.ru'ber. L. adj. *cinereus* similar to ashes, ash-colored; L. adj. *ruber* red; N.L. masc. adj. *cinereoruber* ashy red.

Spore chains in Section *Rectiflexibiles*. Mature spore chains are long, often more than 50 spores per chain. This morphology is found on yeast-malt agar, oatmeal agar, salts-starch agar, and glycerol-asparagine agar. Spore surface is smooth.

Color of colony: aerial mass color in the Gray color series on oatmeal agar; Red series or Gray series on yeast-malt agar, salts-starch agar, and glycerol-asparagine agar. Nearly all color tabs selected by collaborators are near-gray containing some pink or red. Reverse side of colony is the characteristic grayed yellow modified by red on yeast-malt agar or red to violet on oatmeal agar, salts-starch agar, and glycerol-asparagine agar. A change from red to blue reverse color by addition of 0.05 M NaOH is reported by one observer only.

Color in medium: melanoid pigments are formed in peptone-yeast-iron agar and tryptone-yeast broth. Pigments other than melanoids are probably not formed in yeast-malt agar, oatmeal agar, salts-starch agar, and glycerol-asparagine agar.

D-Glucose, L-arabinose, and D-xylose are utilized for growth. No growth or only traces of growth on sucrose, iso-inositol, D-mannitol, rhamnose, and raffinose. Growth on D-fructose is doubtful.

Type strain shows the highest sequence similarity to: *S. violaceorectus*, AB184314, 99.7%; *S. showdoensis*, AB184389, 99.5%; *S. viridobrunneus*, AJ781372, 99.4%; *S. bikiniensis*, X79851, 99.3%; *S. tanashiensis*, AJ781362, 99.1%; *S. nashvillensis*, AB184286, 99.1%; *S. vietnamensis*, DQ311081, 99%; *S. litmocidini*, AB184149, 99%.

Source: not known.

DNA G+C content (mol%): not known.

Type strain: AS 4.1698, ATCC 19740, CBS 479.68, BCRC 11816, DSM 40012, NBRC 12756, JCM 4205, JCM 4572, KCTC 9706, NCIMB 9797, NRRL 2589, NRRL-ISP 5012, RIA 1021, RIA 535, UNIQEM 116, VKM Ac-1860.

Sequence accession no. (16S rRNA gene): AB184121.

111b. **Streptomyces cinereoruber subsp. fructofermentans** Corbaz, Ettlinger, Keller-Schierlein and Zähner 1957b, 331[AL]

fruc.to.fer.men'tans. L. n. *fructus* fruit; L. part. adj. *fermentans* fermenting; N.L. part. adj. *fructofermentans* fruit fermenting (but pertaining to ability of the organism to utilize L-rhamnose, D-fructose, and D-sorbitol).

Moderate growth on Czapek's solution agar; vegetative growth and diffusible pigment in tints and shades of red on some media. Produces cinerobin A and cinerobin B; inhibited by streptomycin.

Type strain shows no sequence similarity over 99%.

Source: not known.

DNA G+C content (mol%): not known.

Type strain: AS 4.1593, DSM 40692, NBRC 15396, JCM 4956, KCTC 9707, NRRL 2588.

Sequence accession no. (16S rRNA gene): AY999758.

112. **Streptomyces cinereospinus** Terekhova, Preobrazhenskaya and Gause 1986, 574[VP] (Effective publication: Terekhova, Preobrazhenskaya and Gause *in* Gause, Preobrazhenskaya, Sveshnikova, Terekhova and Maximova 1983.)

ci.ne.re.o.spi'nus. L. adj. *cinereus* similar to ashes, ash-colored; L. adj. *spineus* spiny; N.L. masc. adj. *cinereospinus (sic)* ash-colored, spiny.

Spore chains are spirals (*Spirales*); spores are spiny, spines are short. On mineral agar 1: aerial mycelium is gray or greenish gray; substrate mycelium is colorless or, after several days, light yellow; no diffusible pigment. On glycerol-nitrate agar: no aerial mycelium; substrate mycelium and diffusible pigment are blue-green to dark gray-green. On oatmeal agar: aerial mycelium is greenish gray; substrate mycelium is colorless or, after several days, spotted green; no diffusible pigment. On starch-ammonia agar: aerial mycelium is gray, greenish gray; substrate mycelium is colorless; no diffusible pigment. On glycerol-asparagine agar: no aerial mycelium; substrate mycelium and diffusible pigment are colorless or light pink to blue-green. On glucose-asparagine agar: no aerial mycelium, substrate mycelium is colorless; diffusible pigment is light pink. On organic agar 2: no aerial mycelium; substrate mycelium colorless; no diffusible pigment. Melanoid pigments are not formed. Moderate growth on glycerin and glucose; poor growth on mannitol, rhamnose, and fructose; no growth on raffinose, inositol, arabinose, xylose, or sugar. Forms antibiotic 1719 from group azotomycin.

Type strain shows the highest sequence similarity to: *S. coeruleofuscus*, DQ026668, 99.2%; *S. chromofuscus*, AB184194, 99.1%.

Source: not known.

DNA G+C content (mol%): not known.

Type strain: AS 4.163, ATCC 43680, DSM 41470, NBRC 15397, INA 1719, JCM 6917, VKM Ac-1215.

Sequence accession no. (16S rRNA gene): AB184648.

113. **Streptomyces cinereus** (Cross, Lechevalier and Lechevalier 1963) Goodfellow, Williams and Alderson 1986a, 574[VP] (Effective publication: Goodfellow, Williams and Alderson 1986a, 53.) (*Microellobosporia cinerea* Cross, Lechevalier and Lechevalier 1963, 428)

ci.ne're.us. L. masc. adj. *cinereus* ash-colored.

Short straight spore chains (2–5); spore surface is smooth; spores are round to oval and borne in both the substrate and aerial mycelium. Spores sizes vary from 1.5 to 3.5 µm (mean diameter 2.5 µm). Extensively branched substrate and aerial mycelium. The aerial spore mass is white; reverse color is red-orange and the pigment is sensitive to pH. Does not form melanin pigments. Adenine, esculin, allantoin, arbutin, casein, hypoxanthine, starch, testosterone, tyrosine, and urea are degraded, but chitin, elastin, guanine, lecithin, pectin, xanthine, and xylan are not. Hydrogen sulfide is produced but nitrate is not reduced. L-Arabinose, cellobiose, D-fructose, D-galactose, D-glucose, *myo*-inositol, inulin, D-lactose, mannitol, D-mannose, melibiose, melezitose, raffinose, L-rhamnose, salicin, sucrose, trehalose, and D-xylose are used as sole carbon sources, but adonitol and xylitol are not. Grows on L-arginine, L-histidine, potassium nitrate, L-threonine, and L-valine, but not on DL-α-amino-n-butyric acid, L-cysteine, L-hydroxyproline, L-methionine, and L-phenylalanine, as sole nitrogen source. Growth occurs at 10–37°C but not at 4 or 45°C. Tolerant to phenol (0.1%, w/v) and sodium chloride (7%, w/v) but not to sodium azide (0.01%, w/v). Resistant to rifampin but sensitive to sodium chloride (10%, w/v). Antimicrobial activity is shown against *Bacillus subtilis* NCIB 3610, but not against *Aspergillus niger* LIV 131, *Candida albicans* CBS 562, *Escherichia coli* NCIB 9132, *Micrococcus luteus* NCIB 196, *Pseudomonas fluorescens* NCIB 9046^T, *Saccharomyces cerevisiae* CBS 1171^T, or *Streptomyces murinus* ISP 5091. The peptidoglycan contains LL-A_2pm as the major diamino acid and is of the A3γ type (Stackebrandt et al., 1981). Contains major amounts of hexa- and octahydrogenated menaquinones with nine isoprene units (Collins et al., 1984).

Type strain shows no sequence similarity over 99%.

Source: not known.

DNA G+C content (mol%): 67.6.

Type strain: AS 4.1672, ATCC 15840, CBS 356.67, BCRC 11616, DSM 43033, IFM 1137, IFM 1237, NBRC 12247, IMET 43557, JCM 3040, KCC A-0040, KCTC 9066, NCIMB 9586, NRRL B-2909, VKM Ac-812.

Sequence accession no. (16S rRNA gene): AB184072.

114. **Streptomyces cinerochromogenes** Miyairi, Tajashima, Shimizu and Sakai 1966, 58^{AL}

ci.ne.ro.chro.mo'ge.nes. L. adj. *cinereus* similar to ashes, ash-colored; Gr. n. *chroma* color; N.L. suff. *-genes* (from Gr. v. *gennaô* to produce) producing; N.L. part. adj. *cinerochromogenes* producing ashy color.

Produces the anti-bacterial antibiotics cineromycin A and cineromycin B. Type strain shows no sequence similarity over 99%.

Source: not known.

DNA G+C content (mol%): not known.

Type strain: Fuji 50, AS 4.162, ATCC 33339, DSM 41651, NBRC 13822, JCM 3385, NRRL B-16928.

Sequence accession no. (16S rRNA gene): AB184507.

115. **Streptomyces cinnabarinus** (Ryabova and Preobrazhenskaya *in* Gauze, Preobrazhenskaya, Kudrina, Blinov, Ryabova and Sveshnikova 1957) Pridham, Hesseltine and Benedict 1958, 62^{AL} ("*Actinomyces cinnabarinus*" Ryabova and Preobrazhenskaya *in* Gauze, Preobrazhenskaya, Kudrina, Blinov, Ryabova and Sveshnikova 1957, 196)

cin.na.ba'ri.nus. N.L. masc. adj. *cinnabarinus* of cinnabar, referring to the vermilion color of vegetative mycelium and diffusible pigment.

Spore chains in Section *Rectiflexibiles* but a very small proportion of strongly flexuous spore chains may suggest *Retinaculiaperti* or *Spiral* morphology on yeast-malt agar or oatmeal agar. Mature spore chains are generally long, often with more than 50 spores per chain. This morphology is seen on yeast-malt agar, oatmeal agar, and salts-starch agar. Sporulating aerial mycelium is poorly developed or absent on glycerol-asparagine agar and yeast-malt agar. Spore surface is smooth; spores of phalangeal type are common.

Color of colony: aerial mass color in the Red color series (3ca, pale orange yellow) on oatmeal agar and salts-starch agar and also on yeast-malt agar when adequate sporulation occurs on this medium. White aerial mycelium may also be seen on these media. Mature sporulating mycelium is inadequate for aerial mass color determination on glycerol-asparagine agar. Reverse side of colony is pale yellow or grayish on yeast-malt agar; yellow is modified by red to yellowish pink or reddish gray on oatmeal agar and to reddish orange, grayish red or reddish brown on salts-starch agar and glycerol-asparagine agar. Substrate pigment is not a pH indicator or is changed only slightly by addition of 0.05 M NaOH or HCl.

Color in medium: melanoid pigments are formed in peptone-yeast-iron agar and tryptone-yeast broth. Red or lavender pigment is found in the medium in oatmeal agar and salts-starch agar. One observer only found this pigment to be pH-sensitive when tested with 0.05 M NaOH and recorded a change from light violet to yellow-colorless. Two observers recorded no change.

D-Glucose, L-arabinose, D-xylose, iso-inositol, D-mannitol, D-fructose, rhamnose, sucrose, and raffinose are utilized for growth.

Type strain shows the highest sequence similarity to: *Streptomyces griseoruber*, AB184209, 99%.

Source: not known.

DNA G+C content (mol%): not known.

Type strain: AS 4.1590, ATCC 23617, ATCC 25440, CBS 671.69, DSM 40467, NBRC 13028, INA 1242, JCM 4463, NRRL B-12382, NRRL-ISP 5467, PCM 2311, RIA 1220, VKM Ac-1904.

Sequence accession no. (16S rRNA gene): AB184266.

116. **Streptomyces cinnamonensis** Okami *in* Maeda, Okami, Kosaka, Taya and Umezawa 1952, 572^{AL}

cin.na.mo.nen'sis. L. n. *cinnamum* cinnamon; N.L. masc. adj. *cinnamonensis* belonging to cinnamon, referring to the color of the aerial mycelium.

Spore chains of typical *Retinaculum-Apertum* type; spores phalangiform. NaCl tolerance >4%, but <7%. Produces actihiazic acid, a biotin antagonist and anti-mycobacterial antibiotic; exhibits anti-fungal activity.

Type strain shows the highest sequence similarity to: *S. pseudoechinosporeus*, AB184100, 99.9%; *S. hiroshimensis*, AB184789, 99.7%; *S. blastmyceticus*, AY999802, 99.3%; *S. caeruleus*, EF178675, 99.3%; *S. aureoversilis*, AB184855, 99.2%; *S. werraensis*, DQ442558, 99.1%; *S. lilacinus*, AB184819, 99.1%; *S. abikoensis*, AB184537, 99.1%; *S. ardus*,

AB184864, 99%; *S. biverticillatus*, AJ781381, 99%.

Source: not known.

DNA G+C content (mol%): not known.

Type strain: 154-T4, NIHJ 35, AS 4.1619, ATCC 12308, CBS 411.63, CECT 3198, DSM 40803, HUT 6050, NBRC 15873, JCM 4019, KCTC 9708, NCIMB 12604, NRRL B-1588, VKM Ac-1912.

Sequence accession no. (16S rRNA gene): AB184707.

117. **Streptomyces cinnamoneus** (Benedict, Dvonch, Shotwell, Pridham and Lindenfelser 1952) Witt and Stackebrandt 1991, 456[VP] (Effective publication: Witt and Stackebrandt 1990, 370.) ["*Streptomyces cinnamoneus*" Benedict, Dvonch, Shotwell, Pridham and Lindenfelser 1952, 591; "*Streptomyces cinnamomeus* forma *cinnamomeus*" (*sic*) Pridham, Shotwell, Stodola, Lindenfelser, Benedict and Jackson 1956, 576; "*Streptoverticillium cinnamomeus* forma *cinnamomeus*" (*sic*) Baldacci 1958, 25; "*Verticillomyces cinnamomeus* forma *cinnamomeus*" (*sic*) Shinobu 1965, 104; *Streptoverticillium cinnamoneum* Baldacci, Farina and Locci 1966, 158]

cin.na.mo′ne.us. L. n. *cinnamum* cinnamon; N.L. adj. *cinnamoneus* cinnamon-colored (after the color of the aerial mycelium).

Spore chains in *Umballate Monoverticillate* (= *Streptomyces* section *Verticillati*, biverticillate). Mature spore chains generally have 3–10, often more than 10, spores per chain. This morphology is seen on yeast-malt agar, oatmeal agar, salts-starch agar, and glycerol-asparagine agar. Spore surface is smooth.

Color of colony: aerial mass color in the Red color series (grayish yellowish pink) on yeast-malt agar, oatmeal agar, salts-starch agar, and glycerol-asparagine agar. Reverse side of colony with no distinctive pigment (grayish yellow to yellow-brown or brown) on yeast-malt agar, oatmeal agar, salts-starch agar, and glycerol-asparagine agar.

Color in medium: melanoid pigments are not formed in peptone-yeast-iron agar, tyrosine agar, or tryptone-yeast broth. No pigment is found in the medium in yeast-malt agar, oatmeal agar, salts-starch agar, or glycerol-asparagine agar. Vegetative growth is generally poor on Pridham and Gottlieb's carbon utilization medium plus D-glucose. Reports form collaborators vary from no growth to slight growth with other carbon sources, but good growth is not observed on any of the following: L-arabinose, sucrose, D-xylose, iso-inositol, D-mannitol, D-fructose, rhamnose, or raffinose. Poor growth on all carbon sources including D-glucose may indicate a requirement for a growth factor not present in the basal medium.

Type strain shows the highest sequence similarity to: *S. cinnamoneus*, X53171, 99.8%; *S. cinnamoneus*, X53165, 99.6%; *S. olivoreticuli*, X53166, 99.6%; *S. cinnamoneus* subsp. *albosporus* DSM 40897[T], 99.6%; *S. cinnamoneus* subsp. *lanosus* DSM 40898[T], 99.6%; *S. cinnamoneus* subsp. *sparsus* DSM 40899[T], 99.6%; *S. roseoverticillatus* DSM 40845[T], 99.6%; *S. olivoreticuli* subsp. *cellulophilus*, X53166, 99.5%; *S. lavendofoliae* DSM 40217[T], 99.5%; *S. parvisporogenes*, DSM 40473[T], 99.5%; *S. baldacii*, X53164, 99.4%; *S. albireticuli*, DSM 40051[T], 99.4%; *S. alboverticillatus*, DSM 41678[T], 99.4%; *S. hiroshimensis*, DSM 40037[T], 99.4%; *S. lilacinus*, DSM 40254[T], 99.4%; *S. blastmyceticus*, DSM 40029[T], 99.3%; *S. gobitricini*, DSM 41701[T], 99.3%; *S. lavendulocolor*, DSM 40216[T], 99.3%; *S. septatus*, DSM 40577[T], 99.3%; *S. eurocidicus*, DSM 40604[T], 99.2%; *S. kashmirensis*, DSM 40336[T], 99.2%; *S. luridus*, DSM 40081[T], 99.2%; *S. kasugaensis*, AB024441, 99.1%; *S. kasugaensis*, AB024442, 99.1%; *Streptomyces* sp., AF012739, 99.1%; *S. salmonis*, X53169, 99.1%; *S. baldacii*, X53164, 99.1%; *S. roseoverticillatus*, X53164, 99.1%; *S. griseoruber*, DSM 40181[T], 99.1%; *S. luteoverticillatus*, DSM 40038[T], 99.1%; *S. celluloflavus*, DSM 40839[T], 99.1%; *S. ehimensis*, DSM 40253[T], 99.1%; *S. mauvecolor*, DSM 41702[T], 99.1%; *S. melanogenes*, DSM 40192[T], 99.1%; *S. michiganensis*, DSM 40015[T], 99.1%; *S. noboritoensis*, DSM 40223[T], 99.1%; *S. roseofulvus*, DSM 40172[T], 99.1%; *S. roseolus*, DSM 40174[T], 99.1%; *S. sapporonensis*, DSM 41675[T], 99.1%; *S. viridoflavum*, DSM 40237[T], 99.1%; *S. xanthochromogenes*, DSM 40111[T], 99.1%; *S. alanosinicus*, DSM 40606[T], 99%; *S. ardus*, DSM 40527[T], 99%; *S. aureoversile*, DSM 40387[T], 99%; *S. filamentosus*, DSM 40022[T], 99%; *S. griseoverticillatus*, DSM 40507[T], 99%; *S. hachijoensis*, DSM 40114[T], 99%; *S. netropsis*, DSM 40259[T], 99%; *S. roseosporus*, DSM 40122[T], 99%; *S. thioluteus*, DSM 40027[T], 99%; *S. varsoviensis*, DSM 40346[T], 99%.

Source: not known.

DNA G+C content (mol%): not known.

Type strain: ATCC 11874, AS 4.1084, AS 4.1706, ATCC 23897, CBS 293.64, CBS 683.68, BCRC 12169, CCUG 11122, CECT 3258, DSM 40005, HAMBI 1067, NBRC 12852, IMET 41381, JCM 4152, JCM 4633, LMG 5971, NCIMB 8851, NRRL B-1285, NRRL-ISP 5005, RIA 1102, RIA 360, VKM Ac-876.

Sequence accession no. (16S rRNA gene): AB184850.

Further comments: in violation of Rule 33c of the *Bacteriological Code* (1990 Revision), on Validation List no. 38, *Streptomyces cinnamoneus* is proposed as a *nomen revictum* (basonym: "*Streptomyces cinnamoneus*" Benedict et al. 1952).

Witt and Stackebrandt proposed to transfer *Streptoverticillium cinnamoneum* (Benedict et al. 1952) Baldacci et al. 1966 to the genus *Streptomyces* as *Streptomyces cinnamoneus* (Benedict et al. 1952) Witt and Stackebrandt 1991. However, Validation List no. 38 does not include formal propositions about *Streptoverticillium cinnamoneum* subsp. *albosporum* Thirumalachar 1968, *Streptoverticillium cinnamoneum* subsp. *cinnamoneum* (Benedict et al. 1952) Baldacci et al. 1966, *Streptoverticillium cinnamoneum* subsp. *lanosum* Thirumalachar 1968 and *Streptoverticillium cinnamoneum* subsp. *sparsum* Thirumalachar 1968.

According to Hatano et al. (2003), *Streptomyces cinnamoneus* (Benedict et al. 1952) Witt and Stackebrandt 1991 is an earlier heterotypic synonym of *Streptomyces griseoverticillatus* (Shinobu and Shimada 1962) Witt and Stackebrandt 1991, of *Streptomyces hachijoensis* (Hosoya et al. 1952) Witt and Stackebrandt 1991, and of *Streptomyces sapporonensis* (Locci and Schofield 1989) Witt and Stackebrandt 1991.

118. **Streptomyces cirratus** Koshiyama, Okanishi, Ohmori, Miyake, Tsukiura, Matsuzaki and Kawaguchi 1963, 65[AL]

cir.ra′tus. L. masc. adj. *cirratus* curled, having ringlets.

Spore chains in Section *Retinaculiaperti*. Spore chains often with terminal loops, primitive spirals, or sometimes well-defined coils of several turns. Mature spore chains are generally long, often with more than 50 spores per chain. This morphology is seen on yeast-malt agar, oatmeal agar, salts-starch agar, and glycerol-asparagine agar. Spore surface is smooth. Special morphological characteristics: knots and moist, nest-like tangles may be seen in long aerial hyphae or moist droplets may form around terminal coils.

Color of colony: aerial mass color usually in the Red color series (5dc or 5cb, grayish yellowish pink, or 5ge, light grayish reddish brown) on yeast-malt agar, oatmeal agar, salts-starch agar, and glycerol-asparagine agar; aerial mycelium sometimes is in the Gray color series (5fe, light grayish reddish brown) on these media. Reverse side of colony with no distinctive pigments (grayish yellow to light yellowish brown on yeast-malt agar; nearly colorless, pale grayish yellow, or light olive brown on oatmeal agar, salts-starch agar, and glycerol-asparagine agar).

Color in medium: melanoid pigments are usually formed in peptone-yeast-iron agar and tryptone-yeast broth, but are formed weakly or not at all in tyrosine agar. Yellow pigment is usually found in the medium in yeast-malt agar, oatmeal agar, salts-starch agar, and glycerol-asparagine agar. This pigment is not pH-sensitive when tested with 0.05 M NaOH or HCl.

D-Glucose, L-arabinose, D-xylose, and D-fructose are utilized for growth. No growth or only traces of growth with iso-inositol, D-mannitol, rhamnose, sucrose, or raffinose.

Type strain shows the highest sequence similarity to: *S. vinaceus*, AB184394, 100%; *S. spororaveus*, AJ781370, 99.9%; *S. nojiriensis*, AJ781355, 99.9%; *S. xanthophaeus*, DQ442560, 99.8%; *S. sporoverrucosus*, DQ442544, 99.7%; *S. goshikiensis*, EF178693, 99.7%; *S. cinnamonensis*, AB184707, 99.6%; *S. avidinii*, AB184395, 99.6%; *S. colombiensis*, DQ026646, 99.6%; *S. subrutilus*, X80825, 99.5%; *S. virginiae*, D85119, 99.4%.

Source: not known.

DNA G+C content (mol%): not known.

Type strain: AS 4.1679, ATCC 14699, CBS 699.72, DSM 40479, NBRC 13398, JCM 4738, KCTC 9709, NRRL B-3250, NRRL-ISP 5479, RIA 1359, VKM Ac-620.

Sequence accession no. (16S rRNA gene): AY999794.

119. **Streptomyces ciscaucasicus** Sveshnikova 1986, 574[VP] (Effective publication: Sveshnikova *in* Gause, Preobrazhenskaya, Sveshnikova, Terekhova and Maximova 1983.)

cis.cau.ca.si'cus. N.L. masc. adj. *ciscaucasicus* of or pertaining to Ciscaucasus.

Spore chains are spiral (*Spirales*); spores are spiny, spines are medium sized. On mineral agar 1, glycerol-nitrate agar, and oatmeal agar: aerial mycelium is light gray, gray; substrate mycelium is yellowish red to red; no diffusible pigment. On starch-ammonia agar: aerial mycelium is gray, sometimes with brown shadow; substrate mycelium is red with yellowish or brownish shadow; no diffusible pigment. On glycerol-asparagine agar: aerial mycelium is light gray to gray; substrate mycelium is brownish red, raspberry red; no diffusible pigment. On organic agar 2: aerial mycelium is white to light gray; substrate mycelium is yellowish, reddish, yellowish reddish, light gray-brownish red; no diffusible pigment. Melanoid pigments are not formed. Good growth on glucose, arabinose, sucrose, xylose, fructose, rhamnose, raffinose, and mannitol. Produces antibiotic pigment of the prodigiosin group. Pigment of substrate mycelium can act as an indicator: yellow under alkaline reaction; dark pink to pinkish red under acidic reaction.

Type strain shows the highest sequence similarity to: *S. canus*, AY999775, 99.8%; *S. pseudovenezuelae*, AB184233, 99.2%; *S. resistomycificus*, AB184166, 99.1%.

Source: not known.

DNA G+C content (mol%): not known.

Type strain: AS 4.1603, ATCC 23626, ATCC 23918, CBS 839.68, DSM 40275, NBRC 12872, IMET 42945, INA 2022/55, JCM 4384, NRRL B-16362, NRRL-ISP 5275, RIA 1193, VKM Ac-1184, VKM Ac-998.

Sequence accession no. (16S rRNA gene): AY508512.

120. **Streptomyces citreofluorescens** (Korenyako, Krasil'nikov, Nikitina and Sokolova 1960) Pridham 1970, 10[AL] ("*Actinomyces citreofluorescens*" Korenyako, Krasil'nikov, Nikitina and Sokolova *in* Rautenshtein 1960, 156)

cit.re.o.flu.o.res'cens. L. n. *citrus* the citrus, the citrontree; N.L. v. *fluoresco* fluoresce; N.L. part. adj. *citreofluorescens* with a yellow fluorescence.

Spore chains in Section *Rectiflexibiles*. Mature spore chains are long, often with more than 50 spores per chain. This morphology is seen on yeast-malt agar, oatmeal agar, salts-starch agar, and glycerol-asparagine agar. Spore surface is smooth.

Color of colony: aerial mass color in the Yellow color series on yeast-malt agar, oatmeal agar, salts-starch agar, and glycerol-asparagine agar. Reverse side of colony with no distinctive pigments (light yellow or grayish yellow to light olive brown) on yeast-malt agar, oatmeal agar, salts-starch agar, and glycerol-asparagine agar.

Color in medium: melanoid pigments are not produced in peptone-yeast-iron agar, tyrosine agar, and tryptone-yeast broth. Yellow pigment is found in medium in glycerol-asparagine agar and yellow or green pigment is found in medium in yeast-malt agar, oatmeal agar, and salts-starch agar. This pigment is not a pH indicator.

D-Glucose, L-arabinose, D-xylose, D-mannitol, D-fructose, and rhamnose are utilized for growth. No growth or only traces of growth on sucrose, iso-inositol, or raffinose.

For sequence similarity, see type strain of *Streptomyces anulatus*.

Source: not known.

DNA G+C content (mol%): not known.

Type strain: AS 4.1652, ATCC 15858, ATCC 23898, CBS 684.68, BCRC 11820, DSM 40265, NBRC 12853, INMI 2292, JCM 4356, KCTC 9710, LMG 20475, NCIMB 9806, NRRL B-3362, NRRL-ISP 5265, RIA 1187, RIA 648, VKM Ac-96.

Sequence accession no. (16S rRNA gene): AB184195.

Further comments: according to Lanoot et al. (2005b), *Streptomyces citreofluorescens* (Korenyako et al. 1960) Pridham 1970 is a later heterotypic synonym of *Streptomyces anulatus* (Beijerinck 1912) Waksman 1953 emend. Lanoot et al. 2005b.

121. **Streptomyces clavifer** (Millard and Burr 1926) Waksman *in* Waksman and Lechevalier 1953, 103^{AL} ("*Actinomyces clavifer*" Millard and Burr 1926, 630)

cla'vi.fer. L. n. *clava* club; L. suff. *-fer* carrying, bearing; N.L. masc. adj. *calvifer* club-bearing.

Poor to fair growth on Czapek's solution agar; NaCl tolerance >10%, but <13%.

Type strain shows the highest sequence similarity to: *S. olivoviridis*, AB184227, 100%; *S. atroolivaceus*, AJ781320, 100%; *S. mutomycini*, AB249951, 99.8%; *S. finlayi*, AY999788, 99.8%; *S. fimicarius*, AY999784, 99.5%; *S. praecox*, AB184293, 99.5%; *S. flavofuscus*, AB249935, 99.5%; *S. anulatus*, DQ026637, 99.5%; *S. badius*, AY999783, 99.4%; *S. griseoplanus*, AY999894, 99.4%; *S. cavourensis* subsp. *washingtonensis*, DQ026671, 99.4%; *S. acrimycini*, AY999889, 99.4%; *S. lavendulae* subsp. *lavendulae*, AB184080, 99.4%; *S. sindenensis*, AB184759, 99.4%; *S. pluricolorescens*, DQ442540, 99.4%; *S. rubiginosohelvolus*, AB184240, 99.4%; *S. griseinus*, AB184205, 99.4%; *S. californicus*, AB184755, 99.4%; *S. mediolani*, AB184674, 99.4%; *S. globisporus* subsp. *globisporus*, EF178686, 99.3%; *S. fulvorobeus*, AB184711, 99.3%; *S. albovinaceus*, AB249958, 99.3%; *S. argenteolus*, AB045872, 99.3%; *S. parvus*, DQ442537, 99.3%; *S. floridae*, AB184656, 99.3%; *S. lipmanii*, AB184148, 99.3%; *S. alboviridis*, AB184256, 99.3%; *S. cinereorectus*, AB184646, 99.3%; *S. baarnensis*, EF178688, 99.3%; *S. cyaneofuscatus*, AB184860, 99.3%; *S. microflavus*, DQ445795, 99.3%; *S. luridiscabiei*, AF361784, 99.2%; *S. candidus*, DQ026663, 99.2%; *S. flavovirens*, DQ026635, 99.2%; *S. griseolus*, AB184768, 99.2%; *S. griseus* subsp. *griseus*, AY207604, 99.2%; *S. flavogriseus*, AJ494864, 99.1%; *S. bacillaris*, AB184439, 99.1%; *S. halstedii*, EF178695, 99.1%; *S. albolongus*, AB184425, 99%; *S. griseobrunneus*, AB249912, 99%; *S. pulveraceus*, AB184806, 99%; *S. celluloflavus*, AB184476, 99%.

Source: not known.

DNA G+C content (mol%): not known.

Type strain: AS 4.1604, CBS 101.27, DSM 40843, NBRC 15398, JCM 5059, NRRL B-2557.

Sequence accession no. (16S rRNA gene): DQ026670.

122. **Streptomyces clavuligerus** Higgens and Kastner 1971, 330^{AL}

cla.vu.li.ge'rus. L. fem. n. *clavula* little club; N.L. suff. *-gerus* bearing; N.L. masc. adj. *clavuligerus* bearing little clubs.

Produces aerial mycelium which is composed of a network of sympodially branched, aerial hyphae that eventually segment into spores. Short, clavate, side branches are formed that usually produce from one to four spores each. Spore chain morphology is classified in the *Rectus-flexibilis* section. Spores are oblong to short-cylindrical averaging 0.64 by 1.53 μm in size, with smooth spore surfaces. Neither fragmentation of hyphae nor formation of spores occurs in the substrate mycelium.

Aerial mycelium is dark grayish green on media with abundant sporulation. The color ranges from white to gray, to light grayish white on other media. Substrate mycelia vary from pale yellow to grayish yellow. No soluble pigment is produced on any of the 11 media used. The culture is assigned to the Gray (GY) and Green (GN) series of Tresner and Backus. The Maerz and Paul (1950) color block most similar to the spore color en masse is 21-B1, and the light grayish olive color of the ISCC-NBS designation method corresponds to this color block. The culture grows over the pH range 5.0–8.5. Growth does not occur at pH 4.0 or 9.0. Sporulation occurs from pH 5.0 to 6.5 and is most abundant at pH 6.0. Whole-cell hydrolysates contain LL-A_2pm, glycine, aspartic acid, alanine, glutamic acid and leucine as major constituents.

Type strain shows no sequence similarity over 99%.

Source: soil.

DNA G+C content (mol%): not known.

Type strain: AS 4.1611, ATCC 27064, CBS 226.75, BCRC 11518, CECT 3125, DSM 40751, DSM 738, NBRC 13307, IMET 43657, JCM 4710, KCTC 9095, NCIMB 12785, NRRL 3585, VKM Ac-602.

Sequence accession no. (16S rRNA gene): AY 999718.

123. **Streptomyces coelescens** (Krasil'nikov, Sorokina, Alferova and Bezzubenkova *in* Krasil'nikov 1965) Pridham 1970, 21^{AL} ("*Actinomyces coelescens*" Krasil'nikov, Sorokina, Alferova and Bezzubenkova *in* Krasil'nikov 1965, 110)

co.e.les'cens. N.L. adj. *coelescens* slightly blue.

Spore chains in Section *Spirales*; spirals are best developed on salts-starch agar or oatmeal agar. Flexuous chains and imperfect spirals suggesting *Retinaculiaperti* morphology are also common on yeast-malt agar, oatmeal agar, salts-starch agar, and glycerol-asparagine agar. Mature spore chains are moderately long with 10–50 spores per chain. Spore surface is smooth.

Color of colony: aerial mass color in the Gray color series (3fe, light brownish gray; 2dc, yellowish gray or 3ge light grayish yellowish brown) on yeast-malt agar, oatmeal agar, salts-starch agar, and glycerol-asparagine agar. Reverse side of colony is dark grayish purple or blackish purple on salts-starch agar and glycerol-asparagine agar; grayish brown to dark brown or reddish black on yeast-malt agar and oatmeal agar. Reverse mycelium pigment is a pH indicator, changing from reddish violet to bluish violet or blue with the addition of 0.05 M NaOH and from bluish violet to reddish violet, brown, or red with the addition of 0.05 M HCl.

Color in medium: melanoid pigments are not formed in peptone-yeast-iron agar, tyrosine agar, or tryptone-yeast broth. Red, violet, or blue pigment, depending on pH, may be found in the medium in yeast-malt agar, oatmeal agar, salts-starch agar, or glycerol-asparagine agar, but pigment is not always found in these media. This pigment is pH-sensitive, showing essentially the same changes noted for the reverse mycelium pigment.

D-Glucose, L-arabinose, D-xylose, iso-inositol, D-mannitol, D-fructose, and rhamnose are utilized for growth. Utilization of sucrose or raffinose is doubtful.

Type strain shows the highest sequence similarity to: *S. humiferus*, AF503491, 100%; *S. violaceolatus*, AF503497, 100%; *S. violaceoruber*, AF503492, 100%; *S. tricolor*, AB184687, 99.9%; *S. anthocyanicus*, AB184631, 99.9%; *S. rubrogriseus*, AB184681, 99.7%; *S. tendae*, D63873, 99.6%; *S. lienomycini*, AJ781353, 99.6%; *S. violaceorubidus*, AJ781374, 99.4%; *S. coelicoflavus*, AB184650, 99.3%; *S. ambofaciens*, M27245, 99%; *S. pactum*, AB184398, 99%.

Source: not known.

DNA G+C content (mol%): not known.

Type strain: AS 4.1594, ATCC 19830, CBS 679.72, DSM 40421, ICSSB 1021, NBRC 13378, INMI 20-41, JCM 4739, NCIMB 10042, NRRL B-12348, NRRL-ISP 5421, RIA 1339, VKM Ac-98.

Sequence accession no. (16S rRNA gene): AF503496.

124. **Streptomyces coelicoflavus** (*ex* Ryabova and Preobrazhenskaya) Terekhova 1986, 574[VP] (Effective publication: Terekhova *in* Gause, Preobrazhenskaya, Sveshnikova, Terekhova and Maximova 1983.) ["*Actinomyces coelicoflavus*" (Ryabova and Preobrazhenskaya) Krasil'nikov 1970b]

co.e.li.co.fla'vus. L. n. *caelum* the sky, heaven; L. adj. *flavus* yellow; N.L. masc. adj. *coelicoflavus* (*sic*) azure, yellow.

Spore chains are spiral (*Spirales*); spores are smooth. On mineral agar 1: aerial mycelium is light gray to gray; substrate mycelium is yellowish to dark blue, dark grayish blue; no diffusible pigment. On starch-ammonia agar: aerial mycelium is absent; substrate mycelium is gray-brownish-blue; no diffusible pigment. On glycerol-nitrate agar: aerial mycelium is light gray to light gray-brownish gray; substrate mycelium is pink to red, later blue, dark blue; no diffusible pigment. On glycerol-asparagine agar: aerial mycelium is light beige; substrate mycelium is yellow, gray; no diffusible pigment. On oatmeal agar: aerial mycelium is light gray, gray; substrate mycelium is colorless to light yellow; no diffusible pigment. On organic agar 2: aerial mycelium is white to light gray; substrate mycelium is first pink to red, later yellow-gray-brown; diffusible pigment is gray-brown or gray-brownish yellow. Melanoid pigments are not formed. Good growth on glucose, sucrose, fructose, rhamnose, arabinose, xylose, mannitol, and inositol.

Type strain shows the highest sequence similarity to: *S. anthocyanicus*, AB184631, 99.5%; *S. tricolor*, AB184687, 99.4%; *S. fragilis*, AY999917, 99.4%; *S. humiferus*, AF503491, 99.3%; *S. violaceolatus* AF503497, 99.3%; *S. violaceoruber*, AF503492, 99.3%; *S. coelescens*, AF503496, 99.3%; *S. rubrogriseus*, AB184681, 99.3%; *S. lienomycini*, AJ781353, 99.2%; *S. diastaticus*, subsp. *ardesiacus*, DQ026631, 99.2%; *S. flaveolus*, AB184764, 99.1%; *S. violaceorubidus*, AJ781374, 99%.

Source: not known.

DNA G+C content (mol%): not known.

Type strain: AS 4.1596, DSM 41471, NBRC 15399, INA 9630, JCM 6918, NRRL B-16363, VKM Ac-1221.

Sequence accession no. (16S rRNA gene): AB184650.

125. **Streptomyces coelicolor** (Müller 1908) Waksman and Henrici *in* Breed, Murray and Hitchens 1948, 935[AL] ("*Streptothrix coelicolor*" Müller 1908, 197; "*Cladothrix coelicolor*" Macé 1913, 758; "*Nocardia coelicolor*" Chalmers and Christopherson 1916, 271; "*Actinomyces coelicolor*" Lieske 1921, 28; "*Corynebacterium coelicolor*" Müller 1950, 274)

co.e.li.co'lor. L. n. *caelum* heaven, sky (blue); L. n. *color* color; N.L. n. *coelicolor* (*sic*) sky (blue) color.

Spore chains in Section *Rectiflexibliles.* Mature spore chains usually have 10–50, or often more than 50, spores per chain. This morphology is seen on yeast-malt agar, oatmeal agar, salts-starch agar, and glycerol-asparagine agar. Spore surface is smooth. Special morphological characteristics: one observer reports substrate conidia on oatmeal agar, Emerson's potato glucose agar, and potato carrot agar. Sclerotia formation on yeast-malt agar and oatmeal agar was reported by one observer.

Color in colony: aerial mass color in the Yellow color series on yeast-malt agar, oatmeal agar, salts-starch agar, and glycerol-asparagine agar. Reverse side of colony is light olive brown to strong brown or very dark brown on yeast-malt agar, salts-starch agar, and glycerol-asparagine agar; brown substrate mycelium pigment on oatmeal agar is pH-sensitive changing from brown to green with addition of 0.05 M NaOH. Potato plug is pigmented dark blue in 4–7 d.

Color in medium: melanoid pigments are not formed in peptone-yeast-iron agar, tyrosine agar, or tryptone-yeast broth. Reddish yellowish brown or greenish pigments are found in the medium in yeast-malt agar, oatmeal agar, salts-starch agar, and glycerol-asparagine agar; yellowish brown pigment is changed to green with addition of 0.05 M NaOH. Liquid surrounding dark blue potato plug is pigmented greenish brown in 4–7 d.

D-Glucose, L-arabinose, D-xylose, D-mannitol, and D-fructose are utilized for growth. No growth or only traces of growth on sucrose, iso-inositol, rhamnose, and raffinose.

Type strain shows the highest sequence similarity to: *S. aurantiogriseus*, AY999793, 99.9%; *S. griseoviridis*, AY999807, 99.3%.

Source: not known.

DNA G+C content (mol%): not known.

Type strain: AS 4.1658, ATCC 23899, CBS 210.27, CBS 795.68, BCRC 12067, CCUG 11110, DSM 40233, NBRC 12854, JCM 4357, NCIMB 9798, NRRL B-2812, NRRL-ISP 5233, PCM 2324, RIA 1173, VKM Ac-738.

Sequence accession no. (16S rRNA gene): DQ442496.

126. **Streptomyces coeruleoflavus** Preobrazhenskaya and Maximova 1986, 574[VP] (Effective publication: Preobrazhenskaya and Maximova *in* Gause, Preobrazhenskaya, Sveshnikova, Terekhova and Maximova 1983.)

co.e.ru.le.o.fla'vus. L. adj. *caeruleus* dark blue, azure; L. adj. *flavus* yellow; N.L. masc. adj. *coeruleoflavus* (*sic*) dark blue, yellow.

Spore chains are spiral (*Spirales*); spores are spiny; spines are medium-sized. On mineral agar 1: aerial mycelium is smooth, color is gray to light blue; substrate mycelium is yellow, dark orange, or brown-yellow; no diffusible pigment. On glycerol-nitrate agar: aerial mycelium is light blue; substrate mycelium and diffusible pigment are yellow-brown to brown. On starch-ammonia agar: aerial mycelium is light blue; substrate mycelium is brown-yellow to brown; diffusible pigment is yellow-gray. On glycerol-asparagine agar: aerial mycelium and diffusible pigment are not found; substrate mycelium is colorless. On oatmeal agar: aerial mycelium is light blue; substrate mycelium and diffusible pigment are dark brown. Melanoid pigments are not formed. Good growth on rhamnose, glucose, mannitol, sucrose, arabinose, xylose, fructose, inositol, and raffinose. Antibiotic: actinomycin D.

Source: not known.

DNA G+C content (mol%): not known.

Type strain: INA 2206.

Sequence accession no. (16S rRNA gene): no sequence available.

127. **Streptomyces coeruleofuscus** (Preobrazhenskaya *in* Gauze, Preobrazhenskaya, Kudrina, Blinov, Ryabova and Sveshnikova 1957) Pridham, Hesseltine and Benedict 1958, 67[AL] ("*Actinomyces coeruleofuscus*" Preobrazhenskaya *in* Gauze, Preobrazhenskaya, Kudrina, Blinov, Ryabova and Sveshnikova 1957, 128)

co.e.ru.le.o.fus'cus. L. adj. *caeruleus* dark blue, azure; L. adj. *fuscus* dark, tawny; N.L. masc. adj. *coeruleofuscus* (*sic*) dark blue, tawny (referring to the bluish aerial mycelium and brownish vegetative mycelium of the organism).

Spore chains in Section *Spirales*. Mature spore chains are long, often more than 50 spores per chain. This morphology is found on yeast-malt agar, oatmeal agar, salts-starch agar, and glycerol-asparagine agar. Spore surface is spiny.

Color of colony: aerial mass color in the Blue color series on yeast-malt agar, oatmeal agar, salts-starch agar, and glycerol-asparagine agar. Reverse side of colony is variable in color; it is yellow-brown modified by red on yeast-malt agar and characteristic grayed yellow or grayed greenish yellow (center of growth) modified by green, red, and brown (borders) on oatmeal agar, salts-starch agar, and glycerol-asparagine agar. Substrate is not a pH indicator; however, one observer reports that on glycerol-asparagine agar, the reverse color is changed from olive green to bluish green by addition of 0.05 M NaOH, and from olive green to very pale red with HCl.

Color in medium: melanoid pigments are formed in peptone-yeast-iron agar, tryptone-yeast broth, and some other media. Pigments other than melanoids are not formed in yeast-malt agar, oatmeal agar, salts-starch agar, or glycerol-asparagine agar.

D-Glucose, L-arabinose, sucrose, D-xylose, iso-inositol, D-mannitol, D-fructose, rhamnose, and raffinose are utilized for growth.

Type strain shows the highest sequence similarity to: *S. cinereospinus*, AB184648, 99.2%; *S chromofuscus*, AB184194, 99.1%.

Source: not known.

DNA G+C content (mol%): not known.

Type strain: AS 4.1667, ATCC 19741, ATCC 23618, CBS 480.68, BCRC 12186, DSM 40144, NBRC 12757, IMET 43574, INA 2922/57, JCM 4358, NRRL B-5417, NRRL-ISP 5144, RIA 1022, UNIQEM 128, VKM Ac-619.

Sequence accession no. (16S rRNA gene): DQ026668.

128. **Streptomyces coeruleoprunus** Preobrazhenskaya 1986, 574[VP] (Effective publication: Preobrazhenskaya *in* Gause, Preobrazhenskaya, Sveshnikova, Terekhova and Maximova 1983.)

co.e.ru.le.o.pru'nus. L. adj. *caeruleus* dark blue, azure; L. n. *prunum* plum; N.L. masc. adj. *coeruleoprunus* (*sic*) dark blue plum.

Spore chains are straight (*Rectiflexibiles*); spores are smooth. On mineral agar 1: aerial mycelium is smooth, color is gray to light blue; substrate mycelium is brown-violet to blue-black or brown-yellow; diffusible pigment is brown-pink to brown. On oatmeal agar: aerial mycelium is light blue; substrate mycelium and diffusible pigment are pink to gray-violet. On starch-ammonia agar: aerial mycelium is light blue; substrate mycelium is colorless to brown; no diffusible pigment. On glycerol-asparagine agar: aerial mycelium color is light to pale blue; substrate mycelium is colorless; no diffusible pigment. On organic agar 2: white to light blue aerial mycelium; substrate mycelium and diffusible pigment are dark gray-brown or brown. Melanoid pigments are not formed. Good growth on glucose, sucrose, rhamnose, mannitol, arabinose, xylose, fructose, inositol, and raffinose. Antibiotic: neomycin.

Type strain shows the highest sequence similarity to: *S. fradiae*, DQ026630, 99.2%; *S. somaliensis*, AJ007403, 99%.

Source: not known.

DNA G+C content (mol%): not known.

Type strain: AS 4.1648, ATCC 43681, DSM 41472, NBRC 15400, INA 1655, JCM 6919, NRRL B-16364, VKM Ac-1208.

Sequence accession no. (16S rRNA gene): AB184651.

129. **Streptomyces coeruleorubidus** (Preobrazhenskaya *in* Gauze, Preobrazhenskaya, Kudrina, Blinov, Ryabova and Sveshnikova 1957) Pridham, Hesseltine and Benedict 1958, 67[AL] ("*Actinomyces coeruleorubidus*" Preobrazhenskaya *in* Gauze, Preobrazhenskaya, Kudrina, Blinov, Ryabova and Sveshnikova 1957, 125)

co.e.ru.le.o.ru'bi.dus. L. adj. *caeruleus* dark blue, azure; L. adj. *rubidus* dark red; N.L. masc. adj. *coeruleorubidus* (*sic*) dark blue, dark red (referring to the bluish aerial mycelium and red vegetative mycelium of the organism on chemically defined media).

Spore chains in Section *Spirales*. Salts-starch agar shows best spiral morphology; incomplete spirals and hooks may also be common, especially on yeast-malt agar, oatmeal agar, and glycerol-asparagine agar. Mature spore chains generally have 10–50 spores per chain. Spore surface is spiny.

Color of colony: aerial mass color is in the Blue color series on oatmeal agar and salts-starch agar; Blue or White color series on yeast-malt agar and glycerol asparagine agar. Reverse side of colony is grayish yellow to yellowish brown on yeast-malt agar; grayed yellow to yellowish green on oatmeal agar and salts-starch agar; both grayed yellowish green and grayed red or orange on glycerol-asparagine agar. These substrate mycelium pigments are not pH indicators when tested with 0.05 M NaOH or HCl.

Color in medium: melanoid pigments are formed in peptone-yeast-iron agar, tyrosine agar, and tryptone-yeast broth. Red pigment may be found in medium in yeast-malt agar, salts-starch agar, and glycerol-asparagine agar after 14 d; it is not pH-sensitive.

D-Glucose, L-arabinose, sucrose, D-xylose, iso-inositol, D-mannitol, D-fructose, rhamnose, and raffinose are utilized for growth.

Type strain shows the highest sequence similarity to: *S. coerulescens*, AY999720, 99.6%; *S. bellus*, AB184849, 99.6%; *S. albogriseolus*, AJ494865, 99.4%; *S. viridodiastaticus*, AY999852, 99.4%; *S. lomondensis*, AB184673, 99.3%; *S. iakyrus*, AB184877, 99.3%; *S. lusitanus*, AB184424, 99.2%;

S. longispororuber, AB184440, 99.2%; *S. olivaceus*, AB184743, 99.2%; *S. purpurascens*, AJ399486, 99.2%; *S. parvulus*, AB184326, 99.2%; *S. pactum*, AB184398, 99.2%; *S. indiaensis*, AB184553, 99%; *S. roseoviolaceus*, AJ399484, 99%; *S. janthinus*, AB184851, 99%; *S. violaceus*, AB184315, 99%; *S. spinoverrucosus*, AB184578, 99%; *S. hawaiiensis*, AB184143, 99%; *S. albosporeus* subsp. *albosporeus*, AJ781327, 99%; *S. atrovirens*, DQ026672, 99%.

Source: not known.

DNA G+C content (mol%): not known.

Type strain: AS 4.1678, ATCC 13740, ATCC 23900, CBS 796.68, BCRC 11463, DSM 40145, NBRC 12855, IMET 42060, INA 12531/54, JCM 4359, KCTC 1922, NCIMB 9620, NRRL B-2569, NRRL-ISP 5145, RIA 1132, VKM Ac-576.

Sequence accession no. (16S rRNA gene): AY999719.

130. **Streptomyces coerulescens** (Preobrazhenskaya *in* Gauze, Preobrazhenskaya, Kudrina, Blinov, Ryabova and Sveshnikova 1957) Pridham, Hesseltine and Benedict 1958, 67[AL] ("*Actinomyces coerulescens*" Preobrazhenskaya *in* Gauze, Preobrazhenskaya, Kudrina, Blinov, Ryabova and Sveshnikova 1957, 120)

co.e.ru.les′cens. L. adj. *caerulus* dark blue, azure; N.L. part. adj. *coerulescens* (*sic*) becoming blue, slightly blue (referring to the color of the aerial mycelium on a chemically defined medium).

Spore chains in Section *Spirales*. Mature spore chains generally have 10–50 spores per chain. Typical morphology is observed on yeast-malt agar, oatmeal agar, salts-starch agar, and glycerol-asparagine agar. Spore surface is spiny.

Color of colony: aerial mass color in the Blue color series on yeast-malt agar, oatmeal agar, salts-starch agar, and glycerol-asparagine agar. Reverse side of colony with no distinctive pigments on yeast-malt agar, oatmeal agar, salts-starch agar, and glycerol-asparagine agar; substrate is not a pH indicator.

Color in medium: melanoid pigments are formed in peptone-yeast-iron agar, and tryptone-yeast broth. Pigments other than melanoids are not formed in yeast-malt agar, oatmeal agar, salts-starch agar, and glycerol-asparagine agar.

D-Glucose, L-arabinose, sucrose, D-xylose, iso-inositol, D-mannitol, D-fructose, rhamnose, and raffinose are utilized for growth.

Type strain shows the highest sequence similarity to: *S. bellus*, AB184849, 100%; *S. coeruleorubidus*, AY999719, 99.6%; *S. lomondensis*, AB184673, 99.4%; *S. purpurascens*, AJ399486, 99.4%; *S. lusitanus*, AB184424, 99.4%; *S. iakyrus*, AB184877, 99.3%; *S. paradoxus*, AB184628, 99.2%; *S. viridodiastaticus*, AY999852, 99.2%; *S. parvulus*, AB184326, 99.2%; *S. matensis*, AB184221, 99.2%; *S. azureus*, EF178674, 99.2%; *S. indiaensis*, AB184553, 99.2%; *S. albogriseolus*, AJ494865, 99.2%; *S. longispororuber*, AB184440, 99.2%; *S. hawaiiensis*, AB184143, 99.2%; *S. thermocarboxydus*, U94490, 99.1%; *S. spinoverrucosus*, AB184578, 99.1%; *S. griseorubens*, AB184139, 99.1%; *S. erythrogriseus*, AJ781328, 99%; *S. griseoincarnatus*, AJ781328, 99%; *S. labedae*, AB184704, 99%; *S. variabilis*, DQ442551, 99%; *S. massasporeus*, AB184152, 99%; *S. roseoviolaceus*, AJ399484, 99%; *S. violaceus*, AB184315, 99%; *S. albosporeus* subsp. *albosporeus*, AJ781327, 99%; *S. janthinus*, AB184851, 99%.

Source: not known.

DNA G+C content (mol%): not known.

Type strain: AS 4.1597, ATCC 19742, ATCC 19896, CBS 481.68, BCRC 11464, DSM 40146, NBRC 12758, IMET 43578, INA 4562, JCM 4360, NCIMB 9615, NRRL B-2701, NRRL-ISP 5146, PCM 2312, RIA 1023, UNIQEM 129, VKM Ac-1843.

Sequence accession no. (16S rRNA gene): AY999720.

131. **Streptomyces collinus** Lindenbein 1952, 380[AL]

col.li′nus. L. masc. adj. *collinus* hilly, mounded.

Spore chains in Section *Spirales* or *Retinaculiaperti*. Spore chains are short resulting in poorly developed spirals of only a few turns; loops and hooks are of small diameter and therefore are not representative of true *Retinaculiaperti* cultures. Distinct spore chains are relatively short, often with only 3–10 spores per chain; longer chains occur, but often coalesce as masses of spores. This morphology is seen on yeast-malt agar, oatmeal agar, salts-starch agar, and glycerol-asparagine agar. Spore surface is smooth. After 14 d, subglobose bodies, composed primarily of masses of spores, can be seen. They appear to originate at the hooked or spiral ends of *Retinaculiaperti* spore chains and to be held together by fluid.

Color of colony: aerial mass color in the Gray color series on yeast-malt agar, oatmeal agar, salts-starch agar, and glycerol-asparagine agar; or (by one observer) in the Red series on salts-starch agar. Reverse side of colony with no distinctive pigments on yeast-malt agar, oatmeal agar, salts-starch agar, and glycerol-asparagine agar.

Color in medium: melanoid pigments formed in peptone-yeast-iron agar and sometimes in tryptone-yeast broth, but not on tyrosine agar. Pigments other than melanoids are not formed in yeast-malt agar, oatmeal agar, salts-starch agar, or glycerol-asparagine agar.

D-Glucose, L-arabinose, sucrose, iso-inositol, D-mannitol, D-fructose, rhamnose, and raffinose are utilized for growth. Two of three observers found good growth with D-xylose.

Type strain shows the highest sequence similarity to: *S. violaceochromogenes*, AY999867, 99.6%; *S. paradoxus*, AB184628, 99.5%; *S. griseoflavus*, AJ781322, 99.5%; *S. viridochromogenes*, DQ442555, 99.3%; *S. iakyrus*, AB184877, 99.3%; *S. matensis*, AB184221, 99.2%; *S. ambofaciens*, M27245, 99.2%; *S. flaveolus*, AB184764, 99.2%; *S. griseorubens*, AB184139, 99.2%; *S. erythrogriseus*, AJ781328, 99.1%; *S. griseoincarnatus*, AJ781328, 99.1%; *S. variabilis*, DQ442551, 99.1%; *S. labedae*, AB184704, 99.1%; *S. violaceorubidus*, AJ781374, 99%.

Source: not known.

DNA G+C content (mol%): not known.

Type strain: AS 4.1623, ATCC 19743, CBS 482.68, BCRC 11465, DSM 2012, DSM 40129, ICMP 12539, NBRC 12759, JCM 4361, KCTC 9713, NRRL B-5412, NRRL-ISP 5129, RIA 1024, UNIQEM 130, VKM Ac-710.

Sequence accession no. (16S rRNA gene): AB184123.

132. **Streptomyces colombiensis** Pridham, Hesseltine and Benedict 1958, 76[AL] emend. Lanoot, Vancanneyt, Dawyndt, Cnockaert, Zhang, Huang, Liu and Swings 2005a, 8 (Effective publication: Lanoot, Vancanneyt, Dawyndt, Cnockaert, Zhang, Huang, Liu and Swings 2004, 88.)

co.lom.bi.en′sis. N.L. masc. adj. *colombiensis* of or belonging to Columbia.

Spore chains in Section *Spirales*. Flexuous spore chains and open loops resembling *Retinaculiaperti* morphology are also present. Mature spore chains are generally long with more than 50 spores per chain. This morphology is seen on yeast-malt agar, oatmeal agar, salts-starch agar, and glycerol-asparagine agar. Spore surface is smooth.

Color of colony: aerial mass color in the Red color series (5dc, 5ec, 5cb, 6ec, grayish yellowish pink; or 5ge, light grayish reddish brown) on yeast-malt agar, oatmeal agar, salts-starch agar, and glycerol-asparagine agar. Reverse side of colony is moderate reddish orange to strong brown on yeast-malt agar; grayish yellow, orange yellow, or yellowish brown on oatmeal agar, salts-starch agar, and glycerol-asparagine agar. Reverse mycelium pigment is not a pH indicator.

Color in medium: melanoid pigments are not formed in tyrosine agar; reports vary on production of melanoid pigments in peptone-yeast-iron agar and tryptone-yeast broth. No pigment is found in medium in yeast-malt agar, oatmeal agar, salts-starch agar, or glycerol-asparagine agar.

D-Glucose and D-fructose are utilized for growth. No growth or only traces of growth with L-arabinose, D-xylose, iso-inositol, D-mannitol, rhamnose, sucrose, and raffinose.

Type strain shows the highest sequence similarity to: *S. goshikiensis*, EF178693, 100%; *S. sporoverrucosus*, DQ442544, 99.9%; *S. nojiriensis*, AJ781355, 99.8%; *S. spororaveus*, AJ781370, 99.8%; *S. xanthophaeus*, DQ442560, 99.7%; *S. vinaceus*, AB184394, 99.7%; *S. cirratus*, AY999794, 99.6%; *S. cinnamonensis*, AB184707, 99.5%; *S. avidinii*, AB184395, 99.5%; *S. subrutilus*, X80825, 99.3%; *S. virginiae*, D85119, 99.1%.

Source: not known.

DNA G+C content (mol%): not known.

Type strain: ATCC 27425, CBS 755.72, DSM 40558, NBRC 13454, JCM 4675, JCM 4740, NRRL B-1990, NRRL-ISP-5558, RIA 739, RIA 1415.

Sequence accession no. (16S rRNA gene): DQ026646.

Further comments: according to Lanoot et al. (2004), *Streptomyces colombiensis* Pridham et al. 1958 is an earlier heterotypic synonym of *Streptomyces distallicus* (Locci et al. 1969) Witt and Stackebrandt 1991.

133. **Streptomyces corchorusii** Ahmad and Bhuiyan 1958, 143[AL] emend. Lanoot, Vancanneyt, Van Shoor, Liu and Swings 2005b, 731

cor.cho.ru′si.i. L. n. *Corchorus* name of a plant and a scientific generic name; L. gen. n. *corchorusii* (*sic*) of Corchorus, because the organism was isolated from soil of a field of jute, *Corchorus capsulatus*.

Spore chains in Section *Spirales* to *Rectiflexibiles*. Incomplete spirals (hooks) and flexuous or straight spore chains are common; spirals usually have only 1–3 turns. Hooks and primitive spirals are on relatively short chains and are therefore not representative of typical *Retinaculiaperti* or *Rectiflexibiles* morphology. Mature spore chains generally have 10–20 spores per chain. This morphology is seen on yeast-malt agar, oatmeal agar, salts-starch agar, and glycerol-asparagine agar. Spore surface is smooth. Electron micrographs show spores of irregular size and shape; some surface irregularities may be present.

Color of colony: aerial mass color in the Gray color series (color tabs 3ge, light grayish yellowish brown and 3fe or 4li, brownish gray) on yeast-malt agar, oatmeal agar, salts-starch agar, and glycerol-asparagine agar. Reverse side of colony with no distinctive pigments (yellow to yellowish brown or olive brown) on yeast-malt agar, oatmeal agar, salts-starch agar, and glycerol-asparagine agar.

Color in medium: melanoid pigments are not formed in peptone-yeast-iron agar, tyrosine agar, or tryptone-yeast broth. No pigment (or only a trace of yellow) is found in the medium in yeast-malt agar, oatmeal agar, salts-starch agar, and glycerol-asparagine agar.

D-Glucose, L-arabinose, D-xylose, iso-inositol, D-mannitol, D-fructose, rhamnose, sucrose, and raffinose are all utilized for growth.

Type strain shows the highest sequence similarity to: *S. olivaceoviridis*, AB184288, 99.9%; *S. canarius*, AB184396, 99.8%; *S. capoamus*, AB045877, 99.5%; *S. longwoodensis*, AB184580, 99.4%; *S. bungoensis*, AB184696, 99.4%; *S. galbus*, X79852, 99.3%; *S. griseochromogenes*, AB184387, 99%; *S. cellostaticus*, AB184192, 99%.

Source: isolated from soil from a field of jute.

DNA G+C content (mol%): not known.

Type strain: AS 4.1592, ATCC 25444, CBS 677.69, BCRC 11821, DSM 40340, NBRC 13032, JCM 4286, JCM 4467, KCTC 9715, LMG 20488, NCIMB 9476, NCIMB 9979, NRRL B-12289, NRRL-ISP 5340, RIA 1224, VKM Ac-1906.

Sequence accession no. (16S rRNA gene): AB184267.

Further comments: according to Lanoot et al. (2005b), *Streptomyces corchorusii* Ahmad and Bhuiyan 1958 emend. Lanoot et al. 2005b is an earlier heterotypic synonym of *Streptomyces chibaensis* Suzuki et al. 1958.

According to Rule 24b of the *Bacteriological Code* (1990 Revision), if two names compete for priority and if both names are listed on the Approved Lists of Bacterial Names, the priority shall be determined by the date of the effective publication of the name before 1 January 1980. *Streptomyces chibaensis* and *Streptomyces corchorusii* are cited in the Approved Lists. The dates of effective publications are 1958. Consequently, to determine priority it is necessary to know the month (and perhaps the day) of the effective publications. *Streptomyces chibaensis* was effectively published in the *Journal of Antibiotics* (*Tokyo*) *Series A*, 1958, vol. 11, pp. 81–83, and *Streptomyces corchorusii* in the *Pakistan Journal of Biological and Agricultural Sciences*, 1958, vol. 1, pp. 137–143. It is certainly not easy to know the exact dates of publication of these articles. However, according to Rule 42 of the *Bacteriological Code* (1990 Revision), if the epithets are of the same date, the author who first unites the taxa has the right to choose one of them, and his choice must be followed. Lanoot et al. chose the epithet *corchorusii*.

134. **Streptomyces costaricanus** Esnard, Potter and Zuckerman 1995, 778[VP]

cos.ta.ri.can'us. N.L. adj. masc. *costaricanus* of or belonging to Costa Rica, the geographic origin of the organism.

Mature spore chains are tightly coiled spirals with 10–50 spores per chain. Spores are smooth and gray-brown in mature colonies. Aerial mycelial mass is grayish brown on yeast extract-malt extract agar, oatmeal agar, inorganic salts-starch agar, and glycerol-asparagine agar and yellow on NZamine medium containing soluble starch and glucose (ATCC medium 172). The substrate mycelium is light yellow on yeast extract-malt extract agar and glycerol-asparagine agar, golden on ATCC medium 172, brown on oatmeal agar, and yellow on inorganic salts-starch agar. A yellow pH-insensitive diffusible pigment is produced on yeast extract-malt extract agar and glycerol-asparagine agar. The pigment color is yellow-orange on ATCC medium 172. No melanoid pigment is produced on peptone-yeast extract-iron agar or tyrosine agar. Color of the reverse side of the colonies is also not sensitive to pH.

D-Fructose, D-glucose, D-mannitol, D-xylose, salicin, and galactose are utilized for growth, but L-arabinose, raffinose, rhamnose, and sucrose are not utilized. Acid is produced from cellobiose, D-glucose, glycerol, maltose, galactose, D-mannitol, and D-xylose but not from L-arabinose, D-fructose, lactose, or sucrose. No growth occurs in the presence of ribitol, galactitol, erythritol, or iso-inositol; 7% NaCl is inhibitory. Cell walls contain LL-A_2pm. The most abundant hydrolyzable fatty acids are $C_{15:0}$ anteiso, $C_{16:0}$, $C_{17:0}$ anteiso, $C_{15:0}$ iso, $C_{16:0}$ iso, and $C_{17:0}$ iso in cells grown on ISP medium 2 agar. The concentration of octadecanoic acid is ninefold higher in ISP medium 2 broth.

Exhibits anti-nematodal activity against *Caenorhabditis elegans* and antibiotic activity against *Rhizoctonia solani* and *Phytium aphanidermatum*.

Type strain shows the highest sequence similarity to: *S. griseofuscus*, AB184206, 100%; *S. murinus*, AB184155, 100%; *S. phaeogriseichromatogenes*, AJ391813, 99.6%.

Source: isolated from a tropical soil.

DNA G+C content (mol%): not known.

Type strain: CR-43, ATCC 55274, DSM 41827, NRRL B-16897, JCM 11306, NBRC 100773, NCIMB 13455.

Sequence accession no. (16S rRNA gene): AB249939.

135. **Streptomyces cremeus** (Kudrina *in* Gauze, Preobrazhenskaya, Kudrina, Blinov, Ryabova and Sveshnikova 1957) Pridham, Hesseltine and Benedict 1958, 66[AL] ("*Actinomyces cremeus*" Kudrina *in* Gauze, Preobrazhenskaya, Kudrina, Blinov, Ryabova and Sveshnikova 1957, 93)

cre'me.us. N.L. masc. adj. *cremeus* cream-white.

Spore chains in Section *Retinaculiaperti*, but with many straight to flexuous spore chains. Open coils and primitive spirals are common. In the original description (Kudrina, ibid), this is characterized as a spiral culture. Mature spore chains are long, often with more than 50 spores per chain. This morphology is observed on yeast-malt agar, oatmeal agar, salts-starch agar, and glycerol-asparagine agar. Spore surface is smooth.

Color of colony: aerial mass color in the Red color series on yeast-malt and salts-starch agar; Yellow series on glycerol-asparagine agar; Yellow or Red series on oatmeal agar. Reverse side of colony with no distinctive pigments on yeast-malt agar, oatmeal agar, salts-starch agar, and glycerol-asparagine agar; substrate is not a pH indicator (one observer reports slight change from pale yellow to pale pink with 0.05 M NaOH).

Color in medium: melanoid pigments are not formed in peptone-yeast-iron agar and tyrosine agar. No pigments or only traces of yellow pigments in yeast-malt agar, oatmeal agar, salts-starch agar, and glycerol-asparagine agar.

D-Glucose, L-arabinose, D-xylose, and D-fructose are utilized for growth. No growth or only traces of growth on sucrose, iso-inositol, D-mannitol, rhamnose, and raffinose.

Type strain shows the highest sequence similarity to: *S. spiroverticillatus*, AB184814, 99.7%; *S. celluloflavus*, AB184476, 99.3%; *S. griseobrunneus*, AB249912, 99.3%; *S. albolongus*, AB184425, 99.3%; *S. cavourensis* subsp. *cavourensis*, DQ445791, 99.2%; *S. candidus*, DQ026663, 99.2%; *S. sindenensis*, AB184759, 99.1%; *S. pluricolorescens*, DQ442540, 99.1%; *S. lavendulae* subsp. *lavendulae*, AB184080, 99.1%; *S. floridae*, AB184656, 99.1%; *S. cavourensis* subsp. *washingtonensis*, DQ026671, 99.1%; *S. mediolani*, AB184674, 99.1%; *S. rubiginosohelvolus*, AB184240, 99.1%; *S. argenteolus*, AB045872, 99.1%; *S. badius*, AY999783, 99.1%; *S. griseinus*, AB184205, 99.1%; *S. anulatus*, DQ026637, 99%; *S. flavogriseus*, AJ494864, 99%; *S. halstedii*, EF178695, 99%; *S. cyaneofuscatus*, AB184860, 99%; *S. parvus*, DQ442537, 99%; *S. albovinaceus*, AB249958, 99%; *S. californicus*, AB184755, 99%; *S. graminofaciens*, AJ781329, 99%; *S. alboviridis*, AB184256, 99%; *S. globisporus* subsp. *globisporus*, EF178686, 99%; *S. lipmanii*, AB184148, 99%; *S. microflavus*, DQ445795, 99%; *S. fulvorobeus*, AB184711, 99%; *S. griseoplanus*, AY999894, 99%; *S. flavovirens*, DQ026635, 99%; *S. praecox*, AB184293, 99%; *S. flavofuscus*, AB249935, 99%; *S. pulveraceus*, AB184806, 99%; *S. griseolus*, AB184768, 99%; *S. fimicarius*, AY999784, 99%; *S. cinereorectus*, AB184646, 99%.

Source: not known.

DNA G+C content (mol%): not known.

Type strain: AS 4.1625, ATCC 19744, ATCC 19897, CBS 483.68, BCRC 11466, DSM 40147, NBRC 12760, IMET 43743, INA 815/54, JCM 4362, NCIMB 10030, NCIMB 9596, NRRL 3241, NRRL B-2583, NRRL-ISP 5147, RIA 1025, UNIQEM 131, VKM Ac-1844.

Sequence accession no. (16S rRNA gene): AB184124.

136. **Streptomyces crystallinus** Tresner, Davies and Backus 1961, 74[AL]

crys.tal.lin'us. L. masc. adj. *crystallinus* made of crystal, crystalline, referring to crystals formed by the organism in some media.

Spore chains are straight to long and flexuous; forms light to dark brown vegetative mycelium and diffusible pigment on many media. Poor growth on Czapek's solution agar; produces hygromycin A and other antibiotics; exhibits anti-fungal activity; inhibited by streptomycin.

Type strain shows the highest sequence similarity to: *S. melanogenes*, AB184222, 99.2%; *S. noboritoensis*, AB184287, 99.2%.

Source: not known.

DNA G+C content (mol%): not known.

Type strain: AS 4.16, DSM 40945, NBRC 15401, JCM 5067, KCTC 9717, NCIMB 12860, NRRL B-3629.

Sequence accession no. (16S rRNA gene): AB184652.

137. **Streptomyces curacoi** Cataldi *in* Trejo and Bennett 1963, 683[AL]

cu.ra'co.i. N.L. gen. n. *curacoi* of Cura-Co in the province of La Pampa, Argentina.

Spore chains in Section *Spirales*. Mature spore chains generally have 10–50 spores per chain; long chains are not common. This morphology is seen on yeast-malt agar, oatmeal agar, salts-starch agar, and glycerol-asparagine agar, but formation of sporulating aerial mycelium is not uniformly good in different laboratories with this strain. Spore surface is spiny.

Color of colony: aerial mass color in the Blue color series on salts-starch agar and on other media (yeast-malt agar, oatmeal agar, glycerol-asparagine agar) when mature spores occur; immature spore chains or non-sporulation aerial mycelium appears white. Reverse side of colony with no distinctive pigments on yeast-malt agar, oatmeal agar, salts-starch agar, and glycerol-asparagine agar; substrate is not a pH indicator.

Color in medium: melanoid pigments are formed in peptone-yeast-iron agar, tyrosine agar, and tryptone-yeast broth. Pigments other than melanoids are not formed in yeast-malt agar, oatmeal agar, salts-starch agar, or glycerol-asparagine agar.

D-Glucose, L-arabinose, sucrose, D-xylose, iso-inositol, D-mannitol, and D-fructose are utilized for growth. Trace of growth is seen on rhamnose or raffinose.

Type strain shows the highest sequence similarity to: *S. longwoodensis*, AB184580, 99%; *S. capoamus*, AB045877, 99%.

Source: not known.

DNA G+C content (mol%): not known.

Type strain: 5828, ATCC 13385, ATCC 19745, CBS 484.68, NBRC 12761, JCM 4219, JCM 4573, NRRL B-2901, NRRL-ISP 5107, RIA 1026, SC 3604, UNIQEM 132.

Sequence accession no. (16S rRNA gene): EF626595.

138. **Streptomyces cuspidosporus** Higashide, Hasegawa, Shibata, Mizumo and Akaike 1966, 2[AL]

cu.spi.do.spo'rus. L. n. *cuspis -idis* point; N.L. n. *spora* a spore; N.L. masc. adj. *cuspidosporus* spore with points or spines.

Forms green to blue to yellowish green diffusible pigment on some media. Excellent growth on Czapek's solution agar; produces sparsomycin, tubercidin, and several other antibiotics.

Type strain shows the highest sequence similarity to: *S. sparsogenes*, AB184301, 99.9%.

Source: not known.

DNA G+C content (mol%): not known.

Type strain: AS 4.1886, ATCC 33340, CBS 192.78, DSM 41425, DSM 41653, NBRC 12378, JCM 4316, KCTC 9718, NRRL B-5620, VKM Ac-599.

Sequence accession no. (16S rRNA gene): AB184090.

139. **Streptomyces cyaneofuscatus** (Kudrina *in* Gauze, Preobrazhenskaya, Kudrina, Blinov, Ryabova and Sveshnikova 1957) Pridham, Hesseltine and Benedict 1958, 58[AL] ("*Actinomyces cyaneofuscatus*" Kudrina *in* Gauze, Preobrazhenskaya, Kudrina, Blinov, Ryabova and Sveshnikova 1957, 85)

cy.a.ne.o.fus.ca'tus. L. adj. *cyaneus* dark blue; L. adj. *fuscus* dark, tawny; L. masc. suff. *-atus* suffix used with the sense of provided with; N.L. masc. adj. *cyaneofuscatus* provided with dark blue, tawny, referring to different pigments formed.

Spore chains are in Section *Rectiflexibiles*. Mature spore chains generally have 10–50 spores per chain. Typical morphology is observed on yeast-malt agar, oatmeal agar, salts-starch agar, and glycerol-asparagine agar. Spore surface is smooth. Coremia may be formed on oatmeal agar, salts-starch agar, and tyrosine agar.

Color of colony: aerial mass color in the Yellow color series on yeast-malt agar, oatmeal agar, salts-starch agar, and glycerol-asparagine agar. Reverse side of colony with no distinctive pigments on yeast-malt agar, oatmeal agar, salts-starch agar, and glycerol-asparagine agar. Reverse color is changed from pale yellowish green to pink by addition of 0.05 M HCl or from pale yellow to yellowish green with NaOH.

Color in medium: melanoid pigments are formed in peptone-yeast-iron agar. Pigments other than a weak yellow pigment are not formed in yeast-malt agar, oatmeal agar, salts-starch agar, or glycerol-asparagine agar. On glucose-mineral salts medium (CPI Kuznetsov: glucose, 2%; KNO_3, 0.1%; $MgSO_4$, 0.05%; NaCl, 0.05%; K_2HPO_4, 0.05%; $CaCO_3$, 0.3%; agar, 2.0%), blue pigment is formed in the medium. Blue pigment is pH-sensitive, changing from blue to pink with 0.05 M HCl.

D-Glucose, D-xylose, D-mannitol, D-fructose, and rhamnose are utilized for growth. No growth or only traces of growth on L-arabinose, sucrose, iso-inositol, and raffinose.

Type strain shows the highest sequence similarity to: *S. cavourensis* subsp. *washingtonensis*, DQ026671, 100%; *S. lavendulae* subsp. *lavendulae*, AB184080, 100%; *S. alboviridis*, AB184256, 99.9%; *S. fulvorobeus*, AB184711, 99.9%; *S. praecox*, AB184293, 99.9%; *S. anulatus*, DQ026637, 99.9%; *S. flavofuscus*, AB249935, 99.9%; *S. cinereorectus*, AB184646, 99.9%; *S. griseoplanus*, AY999894, 99.9%; *S. lipmanii*, AB184148, 99.9%; *S. fimicarius*, AY999784, 99.9%; *S. microflavus*, DQ445795, 99.9%; *S. sindenensis*, AB184759, 99.8%; *S. rubiginosohelvolus*, AB184240, 99.8%; *S. mediolani*, AB184674, 99.8%; *S. griseinus*, AB184205, 99.8%; *S. acrimycini*, AY999889, 99.8%; *S. pluricolorescens*, DQ442540, 99.8%; *S. argenteolus*, AB045872, 99.8%; *S. badius*, AY999783, 99.8%; *S. floridae*, AB184656, 99.7%; *S. albovinaceus*, AB249958, 99.7%; *S. globisporus* subsp. *globisporus*, EF178686, 99.7%; *S. griseus* subsp. *griseus*, AY207604, 99.7%; *S. griseolus*, AB184768, 99.7%; *S. luridiscabiei*, AF361784, 99.7%; *S. baarnensis*, EF178688, 99.7%; *S. californicus*, AB184755, 99.7%; *S. parvus*, DQ442537, 99.7%; *S. halstedii*, EF178695, 99.7%; *S. flavovirens*, DQ026635, 99.6%; *S. flavogriseus*, AJ494864, 99.5%; *S. olivoviridis*, AB184227, 99.4%; *S. bacillaris*, AB184439, 99.4%; *S. finlayi*, AY999788, 99.4%; *S. pulveraceus*, AB184806, 99.4%; *S. atroolivaceus*, AJ781320, 99.4%; *S. yanii*, AB006159, 99.4%; *S. clavifer*,

DQ026670, 99.3%; *S. nitrosporeus*, EF178680, 99.3%; *S. albolongus*, AB184425, 99.2%; *S. sanglieri*, AB249945, 99.2%; *S. mutomycini*, AB249951, 99.2%; *S. gelaticus*, DQ026636, 99.2%; *S. griseobrunneus*, AB249912, 99.2%; *S. celluloflavus*, AB184476, 99.2%; *S. atratus*, DQ026638, 99.1%; *S. spiroverticillatus*, AB184814, 99%; *S. cavourensis* subsp. *cavourensis*, DQ445791, 99%; *S. cremeus*, AB184124, 99%; *S. mauvecolor*, AB184532, 99%; *S. candidus*, DQ026663, 99%.

Source: not known.

DNA G+C content (mol%): not known.

Type strain: AS 4.1612, ATCC 19746, ATCC 23619, CBS 485.68, BCRC 11467, DSM 40148, NBRC 13190, IMET 41583, INA 99/54, JCM 4364, NCIMB 13021, NRRL B-2570, NRRL-ISP 5148, RIA 1027, UNIQEM 133, VKM Ac-752.

Sequence accession no. (16S rRNA gene): AB184860.

140. **Streptomyces cyaneus** (Krasil'nikov 1941) Waksman *in* Waksman and Lechevalier 1953, 42[AL] ("*Actinomyces cyaneus*" Krasil'nikov 1941, 14)

cy'a.ne.us. L. masc. adj. *cyaneus* dark blue.

Spore chains in Section *Spirales.* Compact spirals of 2–6 turns are common, but open spirals, imperfect spirals or flexuous spore chains may also be found. Mature spore chains are moderately long with 10–50 spores per chain. This morphology is seen on yeast-malt agar, oatmeal agar, salts-starch agar, and glycerol-asparagine agar. Spore surface is spiny.

Color of colony: aerial mass color in the Blue or Gray color series on yeast-malt agar, salts-starch agar, and glycerol-asparagine agar; and Gray color series on oatmeal agar. Aerial mycelium may be poorly developed and atypical on glycerol-asparagine agar. The most representative color tabs from the Blue color series are 19fe or dc, pale blue, and the most representative color from the Gray color series is tab e, medium gray. Reverse side of colony is dark grayish blue to dark grayish purple on yeast-malt agar, oatmeal agar, and salts-starch agar; this pigmentation sometimes fails to develop on glycerol-asparagine agar. Reverse mycelium pigment is a pH indicator, changing from violet or purple to blue with the addition of 0.05 M NaOH and from blue to violet or purple with the addition of 0.05 M HCl.

Color in medium: melanoid pigments are formed in peptone-yeast-iron agar, tyrosine agar, and tryptone-yeast broth. Blue or violet pigment is found in the medium in yeast-malt agar, oatmeal agar, and salts-starch agar; it is often not formed in glycerol-asparagine agar. This pigment is pH-sensitive, showing the same changes noted for the reverse mycelium pigment.

D-Glucose, L-arabinose, D-xylose, iso-inositol, D-mannitol, D-fructose, rhamnose, sucrose, and raffinose are all utilized for growth.

Type strain shows the highest sequence similarity to: *S. pseudovenezuelae*, AB184233, 99.1%.

Source: not known.

DNA G+C content (mol%): not known.

Type strain: AS 4.1671, ATCC 14923, CBS 647.72, BCRC 13767, DSM 40108, NBRC 13346, JCM 4220, JCM 4743, KCTC 9719, NRRL B-16305, NRRL B-2296, NRRL-ISP 5108, PCM 2297, RIA 1307, VKM Ac-1712.

Sequence accession no. (16S rRNA gene): AF346475.

141. **Streptomyces cyanoalbus** (Krasil'nikov and Agre in Rautenshtein 1960) Pridham 1970, 13[AL] ("*Actinomyces cyanoalbus*" Krasil'nikov and Agre *in* Rautenshtein 1960, 273)

cy.a.no.al'bus. L. adj. *cyaneus* dark blue; L. adj. *albus* white; N.L. masc. adj. *cyanoalbus* blue, white, referring to formation of colorless (white) or blue vegetative mycelium.

Spore chains in Section *Spirales* or *Rectiflexibiles.* Distinct spirals are rare. Moderately short chains, often with only 10–20 spores, may form strongly wavy chains to open spirals of 1–4 turns. This morphology is seen on yeast-malt agar, oatmeal agar, salts-starch agar, and glycerol-asparagine agar. Spore surface is spiny.

Color of colony: mature aerial mass color (14–21 d) is in the Green color series on yeast-malt agar and salts-starch agar; Green, Gray, or White color series on oatmeal agar and glycerol-asparagine agar. Aerial mycelium is usually poorly developed on oatmeal agar. Reverse side of colony is grayed yellow to grayed yellow-green on yeast-malt and oatmeal agar; grayed yellow-green to blue on salts-starch agar and glycerol-asparagine agar. Substrate mycelium pigment is not a pH indicator.

Color in medium: melanoid pigments are not formed in peptone-yeast-iron agar, tyrosine agar, or tryptone-yeast broth. No pigment is found in medium in yeast-malt agar, oatmeal agar, salts-starch agar, or glycerol-asparagine agar.

D-Glucose, L-arabinose, sucrose, D-xylose, iso-inositol, D-mannitol, D-fructose, and rhamnose are utilized for growth. No growth or only traces of growth on raffinose.

Type strain shows the highest sequence similarity to: *S. hirsutus*, AB184844, 100%; *S. bambergiensis*, AB184869, 99%; *S. prasinus*, DQ026658, 99%.

Source: not known.

DNA G+C content (mol%): not known.

Type strain: AS 4.1426, ATCC 15859, ATCC 23902, CBS 798.68, DSM 40198, HAMBI 1045, NBRC 12857, INMI 414, JCM 4363, LMG 19343, NCIMB 9831, NRRL B-3040, NRRL-ISP 5198, RIA 1150, RIA 662, VKM Ac-585.

Sequence accession no. (16S rRNA gene): AB184882.

142. **Streptomyces daghestanicus** (Sveshnikova *in* Gauze, Preobrazhenskaya, Kudrina, Blinov, Ryabova and Sveshnikova 1957) Pridham, Hesseltine and Benedict 1958, 67[AL] ("*Actinomyces daghestanicus*" Sveshnikova *in* Gauze, Preobrazhenskaya, Kudrina, Blinov, Ryabova and Sveshnikova 1957, 59)

da.ghes.ta'ni.cus. N.L. masc. adj. *daghestanicus* of or belonging to Daghestan, A.S.S.R., the source of the soil from which the organism was isolated.

Spore chains in Section *Retinaculiaperti* but with some well defined spiral spore chains. Mature chains generally have 10–50 spores per chain. Typical morphology is seen on yeast-malt agar, oatmeal agar, salts-starch agar, and glycerol-asparagine agar. Spore surface is smooth with some minor irregularities suggesting wartiness.

Color of colony: aerial mass color in the Red color series on yeast-malt agar, oatmeal agar, salts-starch agar, and glycerol-asparagine agar. Reverse side of colony is grayed yellow to yellow-brown on yeast-malt agar, oatmeal agar, salts-starch agar, and glycerol-asparagine agar; the color is changed only slightly or not at all by addition of 0.05 M NaOH or HCl.

Color in medium: melanoid pigments not formed in peptone-yeast-iron agar or tyrosine agar. No pigment found in medium in yeast-malt agar, oatmeal agar, salts-starch agar, or glycerol-asparagine agar.

D-Glucose, L-arabinose, D-xylose, D-mannitol, D-fructose, and rhamnose are utilized for growth. No growth or only traces of growth on sucrose, iso-inositol, and raffinose.

Type strain shows the highest sequence similarity to: *S. canescens*, AB184117, 100%; *S. felleus*, AB184129, 100%; *S. limosus*, AB184147, 100%; *S. albidoflavus*, AB184255, 100%; *S. hydrogenans*, AB184868, 100%; *S. odorifer*, Z76682, 100%; *S. violascens*, AY999737, 100%; *S. griseus* subsp. *solvifaciens*, AB249915, 100%; *S. champavatii*, DQ026642, 100%; *S. sampsonii*, D63871, 99.8%; *S. koyangensis*, AY079156, 99.7%.

Source: isolated from soil from Daghestan, A.S.S.R.

DNA G+C content (mol%): not known.

Type strain: AS 4.169, ATCC 19747, ATCC 23620, CBS 486.68, BCRC 11468, DSM 40149, NBRC 12762, INA 2656/55, JCM 4365, NRRL B-5418, NRRL-ISP 5149, RIA 1028, UNIQEM 134, VKM Ac-1722, VKM Ac-1862.

Sequence accession no. (16S rRNA gene): DQ442497.

143. **Streptomyces demainii** Goodfellow, Kumar, Labeda and Sembiring 2008, 1[VP] (Effective publication: Goodfellow, Kumar, Labeda and Sembiring 2007, 198.)

de.main'i.i. N.L. gen. masc. n. *demainii* of Demain, named in honor of Arnold Demain, a celebrated actinomycete biologist.

Spore chains in *Spirales*; spore surface is rugose. On oatmeal agar, the aerial spore mass color is gray, becoming black and moist when mature; the reverse side of colony growth is grayish-yellow. Melanin pigments are not formed.

Type strain shows the highest sequence similarity to: *S. sporocinereus*, AB249933, 99.9%; *S. hygroscopius* subsp. *hygroscopius*, AB184428, 99.9%; *S. endus*, AY999911, 99.9%; *S. violaceusniger*, AJ391823, 99.5%; *S. yogyakartensis*, AB249942, 99.5%; *S. albiflaviniger*, AJ391812, 99.3%.

Source: not known.

DNA G+C content (mol%): 71.2.

Type strain: DSM 41600, NRRL B-1478.

Sequence accession no. (16S rRNA gene): DQ334782.

144a. **Streptomyces diastaticus subsp. diastaticus** (Krainsky 1914) Waksman and Henrici *in* Breed, Murray and Hitchens 1948, 939[AL] ("*Actinomyces diastaticus*" Krainsky 1914, 687)

di.a.sta'ti.cus. N.L. masc. adj. *diastaticus* diastatic.

Spore chains in Section *Rectiflexibiles* to *Spirales*. Good spirals are not formed on yeast-malt agar, oatmeal agar, salts-starch agar, or glycerol-asparagine agar, but highly flexuous or crooked spore chains, some of which suggest irregular spirals, are common on these media. Hooks, loops, or irregular spirals are of small diameter and are therefore not representative of true *Retinaculiaperti* morphology. Spore chains are often short, with 3–10 spores per chain, but longer chains are found. Spore surface is smooth.

Color of colony: sporulating aerial mycelium is often poorly developed on yeast-malt agar, oatmeal agar, salts-starch agar, and glycerol-asparagine agar so that aerial mass color is difficult to determine. When sporulating aerial mycelium is formed on these media, it may be in the Gray color series (2dc, yellowish gray) or the aerial color may appear to be in the Yellow color series (2ba, pale yellow; 1cb, pale yellowish green). Reverse side of colony with no distinctive pigments (grayish yellow or pale yellowish green) on yeast-malt agar, oatmeal agar, salts-starch agar, and glycerol-asparagine agar.

Color in medium: melanoid pigments are not formed in peptone-yeast-iron agar, tyrosine agar, or tryptone-yeast broth. No pigment is found in medium in yeast-malt agar, oatmeal agar, salts-starch agar, or glycerol-asparagine agar.

D-Glucose, L-arabinose, D-xylose, D-fructose, sucrose, and D-mannitol are utilized for growth. No growth or only traces of growth with iso-inositol, rhamnose, and raffinose.

Type strain shows the highest sequence similarity to: *S. gougerotii*, AB184742, 100%; *S. rutgersensis* subsp. *rutgersensis*, AB184170, 100%; *S. intermedius*, AB184277, 99.8%; *S. misionensis*, EF178678, 99.2%; *S. phaeoluteichromatogenes*, AJ391814, 99.1%; *S. matensis*, AB184221, 99%; *S. aureoverticillatus*, AY999774, 99%.

Source: not known.

DNA G+C content (mol%): not known.

Type strain: AS 4.1420, ATCC 3315, CBS 126.20, CBS 713.72, CCUG 11116, DSM 40496, NBRC 13412, NBRC 3714, IMET 40274, JCM 4128, JCM 4745, LMG 19322, NRRL B-1241, NRRL B-1270, NRRL-ISP 5496, RIA 104, RIA 1373, VKM Ac-723.

Sequence accession no. (16S rRNA gene): AB184785.

144b. **Streptomyces diastaticus subsp. ardesiacus** (Baldacci, Grein and Spalla 1955) Pridham, Hesseltine and Benedict 1958, 78[AL] ("*Actinomyces diastaticus* subsp. *ardesiacus*" Baldacci, Grein and Spalla 1955, 136)

ar.de.si'a.cus. N.L. masc. adj. *ardesiacus* (from Italian n. *ardesia* slate) intended to mean slate colored.

Excellent growth on Czapek's solution agar. Inhibited by streptomycin. Type strain shows the highest sequence similarity to: *S. coelicoflavus*, AB184650, 99.2%.

Source: not known.

DNA G+C content (mol%): not known.

Type strain: AS 4.1682, CBS 100.56, DSM 40934, NBRC 15402, JCM 5815, NRRL B-1773.

Sequence accession no. (16S rRNA gene): DQ026631.

145. **Streptomyces diastatochromogenes** (Krainsky 1914) Waksman and Henrici *in* Breed, Murray and Hitchens 1948, 941[AL] ("*Actinomyces diastatochromogenes*" Krainsky 1914, 687)

di.a.sta.to.chro.mo'gen.es. Gr. adj. *diastatus* split, divided; Gr. n. *chroma* color; N.L. suff. *-genes* (from Gr. v. *gennaô* to produce) producing; N.L. part. adj. *diastatochromogenes* producing diastatic color, presumably intended to mean producing diastase and color.

Spore chains in Section *Spirales* (or *Retinaculiaperti*). Spirals are open with 4–6 turns, and strongly flexuous to straight spore chains are also common. Mature spore chains are moderately long with 10–50 spores per chain. This morphology is seen on yeast-malt agar, oatmeal agar, and salts-starch agar, but aerial mycelium may be poorly developed or absent on glycerol-asparagine agar. Spore surface is smooth.

Color of colony: aerial mass color in the Gray color series on yeast-malt agar, oatmeal agar, and salts-starch agar; aerial mycelium may be absent on glycerol-asparagine agar, or white aerial mycelium sometimes may be found on this medium, oatmeal agar, and salts-starch agar. Reverse side of colony with no distinctive pigments (light brown or olive brown on yeast-malt agar; pale yellow or grayish yellow on oatmeal agar, salts-starch agar, and glycerol-asparagine agar).

Color in medium: melanoid pigments are formed in peptone-yeast-iron agar, tyrosine agar, tryptone-yeast broth, and Gause's medium no. 2. No pigment is found in medium in yeast-malt agar, oatmeal agar, salts-starch agar, or glycerol-asparagine agar.

D-Glucose, L-arabinose, D-xylose, iso-inositol, D-mannitol, D-fructose, rhamnose, sucrose, and raffinose are all utilized for growth.

Type strain shows no sequence similarity over 99%.

Source: not known.

DNA G+C content (mol%): not known.

Type strain: AS 4.1606, ATCC 12309, CBS 370.58, CBS 690.72, BCRC 13668, CFBP 4540, CIP 105123, DSM 40449, NBRC 13389, NBRC 3337, JCM 4119, JCM 4746, NRRL B-1698, NRRL-ISP 5449, RIA 1350, VKM Ac-1760.

Sequence accession no. (16S rRNA gene): D63867.

146. **Streptomyces distallicus** (Locci, Baldacci and Petrolini Baldan 1969) Witt and Stackebrandt 1991, 456[VP] (Effective publication: Witt and Stackebrandt 1990, 370.) (*Streptoverticillium distallicum* Locci, Baldacci and Petrolini Baldan 1969, 42).

Etymology of specific epithet is unknown.

Spore chains are straight to flexuous and ending in hooks. Reverse colors are darker, after 15 d, and then tend to equal those of *S. kentuckensis*. No great differences in aerial mycelium colors. Aerial mycelium is poorer on Bacto Emerson agar and absent on liquid media. Growth at 37°C is inferior to that at 27°C. Type strain produces distamycins A, B and C, and mycolutein.

For sequence similarity, see type strain of *Streptomyces netropsis*.

Source: not known.

DNA G+C content (mol%): not known.

Type strain: DSM 40846, IMI 72676, JCM 4544, NBRC 15815, NCIB (now NCIMB) 8936, NRRL 2886.

Sequence accession no. (16S rRNA gene): AB184703.

Further comments: in violation of Rule 33c of the *Bacteriological Code* (1990 Revision), on Validation List no. 38, *Streptomyces distallicus* is proposed as a *nomen revictum* (basonym: "*Streptomyces distallicus*" Arcamone et al. 1959.

According to Labeda (1996) and Hatano et al. (2003), *Streptomyces distallicus* (Locci et al. 1969) Witt and Stackebrandt 1991 is a later heterotypic synonym of *Streptomyces netropsis* (Finlay et al. 1951) Witt and Stackebrandt 1991.

According to Lanoot et al. (2004), *Streptomyces distallicus* (Locci et al. 1969) Witt and Stackebrandt 1991 is a later heterotypic synonym of *Streptomyces colombiensis* Pridham et al. (1958).

147. **Streptomyces djakartensis** Huber, Wallhäusser, Fries, Steigler and Weidenmüller 1962, 1191[AL]

dja.kart.en'sis. N.L. masc. adj. *djakartensis* of or belonging to Djakarta, Indonesia, the source of the soil from which the organism was isolated.

Good growth on Czapek's solution agar. Produces niddamycin (3-desacetyl carbomycin B), an anti-bacterial macrolide.

Type strain shows the highest sequence similarity to: *S. ghanaensis*, AY999851, 99.5%; *S. geysiriensis*, DQ442501, 99.5%; *S. minutiscleroticus*, EF178696, 99.5%; *S. mutabilis*, EF178679, 99.4%; *S. rochei*, AB184237, 99.4%; *S. vinaceusdrappus*, AY999929, 99.4%; *S. tuirus*, AB184690, 99.4%; *S. plicatus*, AB184291, 99.4%; *S. calvus*, AB184329, 99.2%; *S. levis*, AB184670, 99.2%; *S. azureus*, EF178674, 99.2%; *S. anandii*, AB184402, 99.1%; *S. asterosporus*, AB184706, 99.1%; *S. flavoviridis*, AB184842, 99.1%; *S. pseudogriseolus*, DQ442541, 99.1%; *S. gancidicus*, AB184660, 99.1%; *S. pilosus*, AB184161, 99.1%; *S. malachitofuscus*, AB184282, 99%; *S. cellulosae*, DQ442495, 99%; *S. aureorectus*, AB184710, 99%; *S. virens*, DQ442554, 99%; *S. carpinensis*, AB184574, 99%; *S. capillispiralis*, AB184577, 99%; *S. brasiliensis*, AB249981, 99%.

Source: isolated from soil from Djakarta, Indonesia.

DNA G+C content (mol%): not known.

Type strain: AS 4.1674, ATCC 13441, DSM 40743, NBRC 15409, JCM 4957, KCTC 9722, NRRL B-12103.

Sequence accession no. (16S rRNA gene): AB184657.

148. **Streptomyces drozdowiczii** Semêdo, Gomes, Linhares, Duarte, Nascimento, Rosado, Margis-Pinheiro, Margis, Silva, Alviano, Manfio, Soares, Linhares and Coelho 2004, 1327[VP]

droz.do.wic'zi.i. N.L. gen. masc. n. *drozdowiczii* of Drozdowicz, named after Adam Drozdowicz, a soil microbiologist who worked in Brazil.

Spore chains in Section *Rectiflexibiles.* Oval spores, borne in chains on the tip of the aerial mycelium. Spore surface is smooth. Branched, gray colored aerial mycelium is produced on inorganic salts-starch agar. The substrate mycelium has no distinctive pigment, but a diffusible yellow-brown pigment is produced. Melanin is produced on tyrosine agar and peptone-yeast extract-iron agar. Cell-wall hydrolysates contain LL-A_2pm. The predominant amino acids in the cell-wall hydrolysate are alanine (major), glycine, glutamic acid, and leucine. The predominant fatty acids found in whole cell methanolysates are $C_{16:0}$ iso (22%), $C_{15:0}$ iso (19%), $C_{15:0}$ anteiso (18%), $C_{17:0}$ iso (10%), $C_{17:0}$ anteiso (7%), $C_{16:0}$ (6%), and $C_{14:0}$ iso (6%). The species description is based on a single strain and hence serves as the type strain description.

Type strain shows no sequence similarity over 99%.

Source: isolated from the Brazilian Atlantic forest soil

(Mendanha Forest. Rio de Janeiro, RJ, Brazil).

DNA G+C content (mol%): not known.

Type strain: M7a, CBMAI 0498, CIP 107837, JCM 13580, NRRL B-24297.

Sequence accession no. (16S rRNA gene): EF654097.

149. **Streptomyces durhamensis** Gordon and Lapa 1966, 754[AL]

dur.ham.en'sis. N.L. masc. adj. *durhamensis* of or belonging to Durham, named after Durham, North Carolina, USA.

Spore chains in Section *Spirales*. Mature spore chains generally contain 10–50 spores per chain. This morphology is seen on yeast-malt agar, oatmeal agar, salts-starch agar, and glycerol-asparagine agar. Spore surface is spiny, but smooth spores may also be seen.

Color of colony: aerial mass color in the Gray or Red color series on yeast-malt agar, oatmeal agar, salts-starch agar, and glycerol-asparagine agar. Nearest matching color tabs in the Gray color series are 3ge, light grayish yellowish brown; 3fe, light brownish gray; and 2dc, yellowish gray. Nearest tab in the Red color series is 5dc, grayish yellowish pink. Reverse side of colony with no distinctive pigments (pale or grayish yellow to yellowish brown) on yeast-malt agar, oatmeal agar, salts-starch agar, and glycerol-asparagine agar.

Color in medium: melanoid pigments are formed in peptone-yeast-iron agar, tyrosine agar, and tryptone-yeast broth. No pigment is found in medium in yeast-malt agar, oatmeal agar, salts-starch agar, or glycerol-asparagine agar.

D-Glucose, L-arabinose, D-xylose, iso-inositol, D-mannitol, D-fructose, sucrose, and raffinose are all utilized for growth. No growth or only traces of growth with rhamnose.

Type strain shows the highest sequence similarity to: *S. filipinensis*, AB184198, 99.8%; *S. puniciscabiei*, AF361785, 99.1%.

Source: not known.

DNA G+C content (mol%): not known.

Type strain: AS 4.1699, ATCC 23194, CBS 742.72, DSM 40539, HAMBI 1064, NBRC 13441, IMET 43359, JCM 4291, JCM 4747, KCTC 9723, NRRL B-3309, NRRL-ISP 5539, RIA 1402, VKM Ac-763.

Sequence accession no. (16S rRNA gene): AY999785.

150. **Streptomyces durmitorensis** Savic, Bratic and Vasiljevic 2007, 2123[VP]

dur.mi.tor.en'sis. N.L. masc. adj. *durmitorensis* of or pertaining to Durmitor, Serbia and Montenegro, where the type strain was isolated.

Spore chains are *Rectiflexibiles*, with 10 or more rod-shaped smooth-surfaced spores (0.5–0.9 × 1–1.5 µm) per chain. Produces a yellowish gray and greenish gray substrate mycelium and a greenish yellow aerial spore-mass on yeast extract-malt extract and glycerol-asparagine agars. Soluble pigments are not formed on oatmeal, yeast extract-malt extract, or glycerol-asparagine agars, while dark gray pigment is formed on inorganic salts-starch agar. Melanoid pigments are not formed on peptone-yeast extract-iron or tyrosine agars. Nitrate is reduced to nitrite, gelatin is liquefied, starch is not hydrolyzed, esculin, and DNA are not degraded. Cellobiose, D(−)-fructose, D(+)-galactose, D(+)-glucose, D(+)-mannose, α-melibiose, D(+)-raffinose, L(+)-rhamnose, D(−)-sucrose, D(+)-xylose, α-trehalose, glycerol, and D(+)-mannitol are utilized for growth, but α-lactose, β-lactose, D(−)-maltose, L(+)-arabinose, D-sorbitol, and *myo*-inositol are not utilized. Acid is produced from glucose. L-Alanine, L-arginine, L-cysteine, L-glycine, L-histidine, L-methionine, L-proline, and L-valine are utilized as sole nitrogen sources, but L-asparagine, L-lysine, L-phenylalanine, ornithine hydrochloride, and thiamine are not utilized. Positive for nitrate reduction, catalase, extracellular protease, haemolysin (β-hemolysis), and urease; lecithinase-negative. H_2S and indole are not produced. Temperature range for growth is 10–37°C, with an optimum between 28 and 32°C. The cell wall contains LL-A_2pm. Grows in the presence of NaCl (9%, w/v) and thallous acetate (0.001%, w/v), but not in sodium-azide (0.01%, w/v), phenol (0.1%, w/v), or potassium tellurite (0.001%, w/v). Susceptible to: apramycin (10 µg/ml), kanamycin (5 µg/ml), gentamicin (5 µg/ml), tetracycline (10 µg/ml), thiostrepton (10 µg/ml), chloramphenicol (35 µg/ml), and spectinomycin(90 µg/ml). Resistant to: ampicillin (100 µg/ml), erythromycin (100 µg/ml), and FK506 (100 µg/ml). Shows antimicrobial activity against *Micrococcus luteus* NCIMB 196 and *Saccharomyces cerevisiae* FAV20, but not against *Bacillus subtilis* NCIMB 3610, *Candida albicans* CBS 562, *Escherichia coli* ATCC 25922, *Pseudomonas aeruginosa* ATCC 27853, *Saccharomyces cerevisiae* FAS20, or *Staphylococcus aureus* ATCC 25923.

Type strain shows the highest sequence similarity to: *S. aureus*, AB249976, 99.6%; *S. kanamyceticus*, DQ442511, 99.2%. Type strain shows DNA–DNA similarity to: *S. aureus* DSM 41785[T], 15.5%; *S. kanamyceticus* DSM 40500[T], 13.3%.

Source: not known.

DNA G+C content (mol%): 72.0.

Type strain: MS405, CIP 108995, DSM 41863.

Sequence accession no. (16S rRNA gene): DQ067287.

151. **Streptomyces echinatus** Corbaz, Ettlinger, Gäumann, Keller-Schierlein, Kradolfer, Neipp, Prelog, Reusser and Zähner 1957a, 203[AL]

e.chi.na'tus. L. masc. adj. *echinatus* set with prickles, prickly.

Spore chain morphology in Section *Retinaculiaperti* or *Spirales*. Tight spirals sometimes occur at tips of sporophores. Mature spore chains generally have 10–50 spores per chain. This morphology is observed on yeast-malt agar, oatmeal agar, salts-starch agar, and glycerol-asparagine agar. Spore surface is spiny.

Color of colony: aerial mass color in the Gray color series on yeast-malt agar, oatmeal agar, salts-starch agar, and glycerol-asparagine agar. Reverse side of colony is grayed yellow modified by green on oatmeal agar, salts-starch agar, and glycerol-asparagine agar. Substrate pigment is not a pH indicator.

Color in medium: melanoid pigments are formed in peptone-yeast-iron agar, tyrosine agar, and tryptone-yeast broth. Pigments other than melanoids are not formed in yeast-malt agar, oatmeal agar, salts-starch agar, or glycerol-asparagine agar.

D-Glucose, L-arabinose, D-xylose, iso-inositol, D-mannitol, D-fructose, rhamnose, and raffinose are utilized for growth (growth on L-arabinose and D-xylose is less than on other sugars). No growth or only traces of growth on sucrose.

Type strain shows no sequence similarity over 99%.

Source: not known.

DNA G+C content (mol%): not known.

Type strain: AS 4.1642, ATCC 19748, ATCC 21133, CBS 409.59, CBS 487.68, BCRC 13656, CECT 3313, DSM 40013, HUT 6090, IFM 1076, NBRC 12763, IMET 40461, JCM 4144, JCM 4574, KCTC 9724, LMG 5972, NCIMB 9598, NCIMB 9799, NRRL 2587, NRRL-ISP 5013, RIA 1029, UNIQEM 135, VKM Ac-762.

Sequence accession no. (16S rRNA gene): AJ399465.

152. **Streptomyces echinoruber** Palleroni, Reichelt, Müller, Epps, Tabenkin, Bull, Schüep and Berger 1981, 382[VP] (Effective publication: Palleroni, Reichelt, Müller, Epps, Tabenkin, Bull, Schüep and Berger 1978, 1224.)

e.chi.no.ru'ber. L. n. *echinus* a hedgehog, urchin; L. adj. *ruber* red; N.L. masc. adj. *echinoruber* spiny red.

Strain shows gray spore mass color, spiral spore chains, and spiny spore surface. Colonies are raised and coarse, with well-defined edges. The most striking character of strain X-14077[T] is the production of the deep cherry-red pigment which freely diffuses into the medium. This pigment is produced in most media tested and it becomes evident after 1 or 2 d of incubation. For a more detailed description of morphological properties, see Palleroni et al. (1978).

Source: not known.

DNA G+C content (mol%): not known.

Type strain: X-14077, AC 4.1707, DSM 41696, NBRC 14238, JCM 5016, KCTC 9725, NCIMB 12831, NRRL 8144.

Sequence accession no. (16S rRNA gene): no sequence available.

153. **Streptomyces ederensis** Wallhäusser, Nesemann, Präve and Steigler 1966, 734[AL]

e.der.en'sis. N.L. masc. adj. *ederensis* of or pertaining to Eder, named for the Eder valley in Germany.

Produces the moenomycin complex of anti-bacterial antibiotics comprised of moenomycins A, B_1, B_2, and C.

Type strain shows the highest sequence similarity to: *S. umbrinus*, AB184305, 100%; *S. phaeochromogenes*, AB184738, 99.6%.

Source: not known.

DNA G+C content (mol%): not known.

Type strain: AS 4.1665, ATCC 15304, CBS 545.70, BCRC 11896, CECT 3212, DSM 40741, NBRC 15410, JCM 4958, KCTC 9726, NRRL B-8146, VKM Ac-845.

Sequence accession no. (16S rRNA gene): AB184658.

154. **Streptomyces ehimensis** corrig. (Shibata, Honso, Tokui and Nakazawa 1954) Witt and Stackebrandt 1991, 456[VP] (Effective publication: Witt and Stackebrandt 1990, 370.) ("*Streptomyces ehimensis*" Shibata, Honso, Tokui and Nakazawa 1954, 168; "*Verticillomyces ehimensis*" Shinobu 1965, 109; *Streptoverticillium ehimense* Locci, Baldacci and Petrolini Baldan 1969, 40).

Etymology of specific epithet is unknown.

Spore chain morphology in Section *Verticillati*. Mature aerial mycelium is usually umbellate-monoverticillate (biverticillate), but sporulating aerial mycelium is very sparse or absent on yeast-malt agar, oatmeal agar, salts-starch agar, and glycerol-asparagine agar. Biverticillate spore chains can be seen on Pridham and Gottlieb carbon utilization medium plus D-xylose or L-arabinose and sometimes can be found on oatmeal agar after 21 d. Spore chains are short, usually with 3–10 spores per chain. Spore surface is smooth.

Color of colony: aerial mass color for mature sporulating aerial mycelium cannot be determined accurately because of poor sporulation on yeast-malt agar, salts-starch agar, and glycerol-asparagine agar. When sporulating aerial mycelium occurs on oatmeal agar, it is in the Gray color series. The spore surface of poorly sporulating growth is generally in the Yellow or Red color series on yeast-malt agar, salts-starch agar, and glycerol-asparagine agar or sparse aerial mycelium may be white. Reverse side of colony is brown to dark brown on yeast-malt agar, but grayish yellow, orange yellow, or yellowish brown on oatmeal agar, salts-starch agar, and glycerol-asparagine agar. Substrate mycelium pigment is not a pH indicator.

Color in medium: melanoid pigments are produced in peptone-yeast-iron agar and tryptone-yeast broth, but only weakly or not at all on tyrosine agar. Yellow or grayish yellowish brown pigment may be found in the medium in yeast-malt agar, oatmeal agar, salts-starch agar, and glycerol-asparagine agar. This pigment is not pH-sensitive.

D-Glucose, iso-inositol, D-mannitol, and D-fructose are utilized for growth. Traces of growth occur on the carbon-free control and on other ISP carbon sources.

Type strain shows the highest sequence similarity to: *S. hygroscopius* subsp. *angustmyceticus*, DQ442509, 99.8%; *S. abikoensis*, AB184537, 99.6%; *S. sapporonensis*, AB184508, 99.4%; *S. lilacinus*, AB184819, 99.3%; *S. luteireticuli*, AB249969, 99.2%; *S. varsoviensis*, DQ026653, 99%.

Source: not known.

DNA G+C content (mol%): not known.

Type strain: AS 4.1668, ATCC 23903, CBS 799.68, BCRC 13319, DSM 40253, HAMBI 1042, NBRC 12858, NBRC 3398, JCM 4162, JCM 4635, KCTC 9727, NRRL B-1967, NRRL-ISP 5253, RIA 1179, VKM Ac-945.

Sequence accession no. (16S rRNA gene): AY999834.

Further comments: the original spelling, *Streptomyces ehimensis* (*sic*), has been corrected in accordance with Rule 61 of the *Bacteriological Code* (1990 Revision).

In violation of Rule 33c of the *Bacteriological Code* (1990 Revision), in Validation List no. 38, *Streptomyces ehimensis* is proposed as a *nomen revictum* (basonym: "*Streptomyces ehimensis*" Shibata et al. 1954.

According to Hatano et al. (2003), *Streptomyces ehimensis* corrig. (Shibata et al. 1954) Witt and Stackebrandt 1991 is a later heterotypic synonym of *Streptomyces abikoensis* (Umezawa et al. 1951) Witt and Stackebrandt 1991.

155. **Streptomyces emeiensis** Sun, Huang, Zhang and Liu 2007, 1638[VP]

e.me.i.en'sis. N.L. masc. adj. *emeiensis* of or pertaining to Emei, a famous mountain in Sichuan Province, southern China, where the sample yielding the type strain was collected.

Aerobic, mesophilic, Gram-stain-positive actinomycete that develops well-branched substrate and aerial mycelium. *Rectiflexibiles* and hooked spore chains of elliptical, spiny-surfaced spores are frequently arranged in a verticillate structure. Diffusible pigments are not formed, nor are melanin pigments produced on peptone-yeast extract-iron or tyrosine agars. Nitrate is reduced. Amylase and gelatinase are not produced. Growth occurs between 15 and 40°C and at pH values from 5.5 to 9.5, but not at pH 4.5 or 10.5. Growth occurs in the presence of 0.1% (w/v) phenol, 5% (w/v) NaCl, and 0.01 (w/v) NaN_3, but not in the presence of 7% (w/v) NaCl. Shows weak antimicrobial activity against strains of *Bacillus subtilis* and *Mycobacterium smegmatis*, but not against strains of *Staphylococcus epidermis*, *Escherichia coli*, *Klebsiella pneumoniae*, *Pseudomonas aeruginosa*, or *Candida albicans*. Sensitive to filter-paper discs soaked in the following (μg/ml): novobiocin (5), streptomycin (10), oxacillin (1), chloramphenicol (30), ciprofloxacin (5), and erythromycin (15). The cell-wall is of type I. Type II phospholipids and menaquinones MK-9(H_6, H_8, H_4) are detected. The fatty acid profile is composed of $C_{15:0}$ anteiso (14.6%), $C_{16:0}$ iso (13.3%), $C_{17:0}$ anteiso (12.6%), $C_{16:0}$ (9.4%), $C_{16:1}\omega 7c$ (8.2%), $C_{17:1}$ anteiso$\omega 9c$ (7.0%), $C_{18:0}$ (3.7%), $C_{17:1}\omega 8c$ (3.6%), iso H-$C_{16:1}$ (3.6%), $C_{15:0}$ (3.1%), $C_{15:0}$ iso (3.0%), $C_{17:1}$ iso$\omega 9c$ (3.0%), $C_{15:0}$ iso (3.0%), $C_{14:0}$ (2.8%), $C_{18:1}\omega 9c$ (2.4%), $C_{18:1}\omega 7c$ (1.6%), $C_{17:0}$ iso (1.3%), $C_{17:0}$ (1.1%), $C_{14:0}$ iso (1.1%), and iso I-$C_{15:1}$ (1.1%).

Type strain shows the highest sequence similarity to: *S. prasinopilosus*, EF626597, 99.4%. Type strain shows DNA–DNA similarity to: *S. prasinopilosus* DSM 40098^T, 62.7%; *S. prasinus* JCM 4603^T, 55.5%; *S. hirsutus* DSM 40095^T, 46.4%; *S. bambergiensis* DSM 40590^T, 31.7%; *S. cyanoalbus* DSM 40198^T, 26.1%.

Source: the type strain was isolated from a soil sample collected from Emei Mountain, Sichuan Province, China.

DNA G+C content (mol%): 70.8.

Type strain: 4776, CGMCC 4.3504, DSM 41884.

Sequence accession no. (16S rRNA gene): DQ462649.

156. **Streptomyces endus** Anderson and Gottlieb 1952, 302[AL]

en'dus. L. praep. *endo* in; N.L. masc. adj. *endus* referring to the site (inside the hyphae) of formation of the antibiotic endomycin.

Spore chains in Section *Spirales* with tightly closed spirals. Individual spores are not easily detached from the spiral chain and compact coils may become hydroscopic (coalesce in black moist droplets). Spore surface is warty.

Color of colony: aerial mass color in the Gray color series after 21 d on yeast-malt agar, oatmeal agar, and salts-starch agar. Aerial mycelium on glycerol-asparagine agar and young aerial mycelium on other media may be white. Older colonies sometimes become moist and black. Reverse side of colony with no distinctive pigments (grayed yellow to yellowish brown or gray) on yeast-malt agar, oatmeal agar, salts-starch agar, and glycerol-asparagine agar. In older cultures, the substrate mycelium sometimes becomes moist and black. Substrate color is not affected by pH change.

Color in medium: melanoid pigments are not formed in peptone-yeast-iron agar, tyrosine agar, or tryptone-yeast broth. No pigments are found in the medium in yeast-malt agar, oatmeal agar, salts-starch agar, or glycerol-asparagine agar.

D-Glucose, L-arabinose, D-xylose, D-mannitol, D-fructose, and rhamnose are utilized for growth. No growth or only traces of growth on sucrose, iso-inositol, and raffinose.

Source: not known.

DNA G+C content (mol%): not known.

Type strain: 9-20, ATCC 23904, CBS 800.68, DSM 40187, NBRC 12859, JCM 4213, JCM 4636, NRRL 2339, NRRL-ISP 5187, RIA 1141.

Sequence accession no. (16S rRNA gene): AY999911.

Type strain shows the highest sequence similarity to: *S. sporocinereus*, AB249933, 100%; *S. hygroscopius* subsp. *hygroscopius*, AB184428, 100%; *S. demainii*, DQ334782, 99.9%; *S. yogyakartensis*, AB249942, 99.5%; *S. violaceusniger*, AJ391823, 99.5%; *S. albiflaviniger*, AJ391812, 99.3%.

157. **Streptomyces enissocaesilis** (*ex* Krasil'nikov 1970b) Sveshnikova 1986, 574[VP] (Effective publication: Sveshnikova *in* Gause, Preobrazhenskaya, Sveshnikova, Terekhova and Maximova 1983.) ("*Actinomyces enissocaesilis*" Krasil'nikov 1970b).

Etymology of specific epithet is unknown.

Spore chains are spiral (*Spirales*); spores are smooth. On mineral agar 1: aerial mycelium is light pink-yellowish-grayish; substrate mycelium is brown; diffusible pigment is brown with grayish violet shadows. On starch-ammonia agar: aerial mycelium is absent or poorly developed, whitish; substrate mycelium is colorless to pale yellowish, growth is poor; no diffusible pigment. On glycerol-nitrate agar: aerial mycelium is white to brownish pink; substrate mycelium and diffusible pigment are brown to dark brown. On glycerol-asparagine agar: aerial mycelium is absent or poorly developed, whitish; substrate mycelium is black-grayish brown; no diffusible pigment. On oatmeal agar: aerial mycelium is poorly developed; substrate mycelium is brown; diffusible pigment is brownish, weak. On organic agar 2: aerial mycelium is light pink-yellowish-grayish; substrate mycelium and diffusible pigment are brown. Melanoid pigment: not extant or poorly developed. Grows on glucose, arabinose, xylose, and mannitol; no growth seen on sucrose, rhamnose, or raffinose.

Type strain shows the highest sequence similarity to: *S. flavidovirens*, AB184270, 99.2%; *S. chryseus*, AY999787, 99.1%; *S. helvaticus*, AB184367, 99.1%; *S. albidochromogenes*, AB249953, 99.1%.

Source: not known.

DNA G+C content (mol%): not known.

Type strain: AS 4.1586, ATCC 43682, DSM 41454, INMI 40-31, JCM 9088, NBRC 100763, NRRL B-16365, VKM Ac-130.

Sequence accession no. (16S rRNA gene): AB249930.

158. **Streptomyces erumpens** Calot and Cercós 1963, 159[AL]

e.rum'pens. L. masc. part. adj. *erumpens* bursting forth.

Poor growth on Czapek's solution agar. Exhibits antibacterial activity; inhibited by streptomycin; produces the tetraenic anti-fungal antibiotic 17732 (tetrins A and B and another polyene).

Type strain shows the highest sequence similarity to: *S. rimosus* subsp. *rimosus*, AB045883, 99.7%; *S. monomycini*, DQ445790, 99.4%; *S. chrestomyceticus*, AJ621609, 99.3%; *S. rimosus* subsp. *paromomycinus*, AJ621610, 99.3%; *S. ochraceiscleroticus*, AB184094, 99.3%; *S. sioyaensis*, DQ026654, 99.2%; *S. hygroscopius* subsp. *decoyicus*, AY999883, 99.2%; *S. purpurogeneiscleroticus*, AJ621604, 99.1%; *S. albofaciens*, AB045880, 99.1%; *S. violens*, AJ621605, 99.1%; *S. sclerotialus*, AJ621608, 99.1%.

Source: not known.

DNA G+C content (mol%): not known.

Type strain: AS 4.1626, ATCC 23266, CBS 252.65, DSM 40941, NBRC 15403, JCM 5060, KCTC 9729, NRRL B-3163.

Sequence accession no. (16S rRNA gene): AJ621603.

Further comments: according to Guo et al. (2008), *Streptomyces erumpens* Calot and Cercós 1963 is a later heterotypic synonym of *Streptomyces griseus* (Krainsky 1914) Waksman and Henrici 1948.

159. **Streptomyces erythraeus** (Waksman *in* Bergey, Harrison, Breed, Hammer and Huntoon 1923) Waksman and Henrici *in* Breed, Murray and Hitchens 1948, 938[AL] ["*Actinomyces erythreus*" (*sic*) Waksman *in* Bergey, Harrison, Breed, Hammer and Huntoon 1923, 370; "*Streptomyces erythreus*" (*sic*) Waksman and Henrici *in* Breed, Murray and Hitchens 1948, 938]

e.ry.thra'e.us. L. masc. adj. *erythraeus* reddish, referring to colony color.

Spore chains in Section *Spirales* or *Retinaculiaperti*. Spore chains are usually short so that imperfect spirals, hooks, or loops of only one turn, and short, straight to flexuous chains are common. Hooks and loops are usually of small diameter and are not representative of true *Retinaculiaperti* morphology. This morphology is seen on yeast-malt agar, oatmeal agar, salts-starch agar, and glycerol-asparagine agar, but aerial mycelium may be poorly developed on glycerol-asparagine agar. Spore surface is spiny.

Color of colony: aerial mass color in the Red color series (4ec, 5cb, 5db, grayish yellowish pink; or 3ca, pale orange yellow) on yeast-malt agar, oatmeal agar, salts-starch agar, and glycerol-asparagine agar. White aerial mycelium may also be seen. Reverse side of colony is brown to grayish reddish brown on yeast-malt agar; orange-yellow to light yellowish brown or grayish yellow on oatmeal agar and salts-starch agar; yellowish brown to strong brown on glycerol-asparagine agar. Reverse mycelium pigment is usually not a pH indicator, but, when pink pigment is present, it may change to pale yellow with the addition of 0.05 M HCl.

Color in medium: melanoid pigments are not formed in peptone-yeast-iron agar, tyrosine agar, or tryptone-yeast broth. Pink or yellow pigment may be found in the medium in yeast-malt agar, oatmeal agar, salts-starch agar, and glycerol-asparagine agar, but pigment is not always present on these media. Pink pigment, when present, becomes colorless with the addition of either 0.05 M HCl or NaOH.

D-Glucose, D-fructose, sucrose, raffinose, and D-mannitol are utilized for growth. Reports vary on utilization of L-arabinose, D-xylose, iso-inositol, and rhamnose. Significant growth is often observed on control plates of carbon-utilization medium without added carbon source.

Source: not known.

DNA G+C content (mol%): not known.

Type strain: ATCC 11635, CBS 727.72, DSM 40517, HUT 6087, IAM 0045, NBRC 13426, JCM 4748, IMRU 3737, NCIB (now NCIMB) 8594, NRRL 2338, NRRL-ISP 5517, RIA 1387, VKM Ac-1189.

Sequence accession no. (16S rRNA gene): no sequence available.

Further comments: the type strain of *Streptomyces erythraeus* (Waksman 1923) Waksman and Henrici 1948 is not a representative of the genus *Streptomyces* Waksman and Henrici 1943 and a new species, *Saccharopolyspora erythraea* Labeda 1987, is created for this strain and other strains having type IV cell walls. Labeda (1987) also proposed an emendation of *Streptomyces erythraeus* [*Streptomyces erythraeus* (Waksman 1923) Waksman and Henrici 1948 emend. Labeda 1987] with a neotype strain NRRL B-5616[T] (NRRL-ISP 5059[T], Sanchez-Marroquin A-24[T]). According to Rules 18c and 18g of the *Bacteriological Code* (1990 Revision), only the Judicial Commission may decide to take action leading to replacement of the type strain and, according to Rule 37a(2) of the *Bacteriological Code* (1990 Revision), retention of a name in a sense which excludes the type can only be effected by conservation and only by the Judicial Commission. Therefore, *Streptomyces erythraeus* (Waksman 1923) Waksman and Henrici 1948 emend. Labeda 1987 is illegitimate and may not be used [Rule 51a of the *Bacteriological Code* (1990 Revision)] and a new species *Streptomyces labedae* Lacey 1987 is provided for this taxon (Lacey, 1987).

Rule 17 of the *Bacteriological Code* (1990 Revision) states that the type determines the application of a taxon if the taxon is subsequently divided or united with another taxon, and Rule 40b of the *Bacteriological Code* (1990 Revision) that the specific epithet must be retained for the species which includes the type strain. Since *Saccharopolyspora erythraea* (Waksman 1923) Labeda 1987 contains the type strain of *Streptomyces erythraeus* (Waksman 1923) Waksman and Henrici 1948, the new taxon must be regarded as a new combination.

160. **Streptomyces erythrogriseus** Falcão de Morais and Dália Maia 1959, 64[AL]

e.ry.thro.gri'se.us. Gr. adj. *euruthros* red; N.L. adj. *griseus* gray; N.L. masc. adj. *erythrogriseus* red, gray (referring to change in color of aerial mycelium from gray to red).

Spore chains in Section *Spirales*, but short spore chains may form imperfect or incomplete spirals or crooked hooks or loops. Spirals are most numerous on salts-starch agar after 14 d and are rare on yeast-malt agar and glycerol-asparagine agar. Mature spore chains are generally short, with 3 to 10 or more spores per chain, but long chains with more than 50 spores are also reported. Spore surface is spiny with short spines. Smooth spores are also found.

Color of colony: aerial mass color usually in the Gray color series on yeast-malt agar and salts-starch agar; Gray, Red, or White color series on glycerol-asparagine agar. Aerial mycelium suitable for color determination is usually not formed on oatmeal agar, but when present it is in the Gray color series. Nearest matching color tabs in the Gray color series are d, light gray, and 5fe, light grayish reddish brown. Nearest matching tabs in the Red color series are 5ge, light grayish reddish brown and 5dc, light yellowish pink. Reverse side of colony is strong brown on yeast-malt agar; orange yellow to yellowish brown on oatmeal agar and salts-starch agar; reddish orange to reddish brown on glycerol-asparagine agar. Reverse mycelium pigment is a pH indicator, changing from yellow to orange with the addition of 0.05 M NaOH and from yellow brown to yellow with the addition of 0.05 M HCl.

Color in medium: melanoid pigments are not formed in peptone-yeast-iron agar, tyrosine agar, or tryptone-yeast broth. Yellow to orange pigment is sometimes found in the medium in yeast-malt agar, oatmeal agar, salts-starch agar, and glycerol-asparagine agar. When this pigment is present, it is pH-sensitive, showing the same changes observed in the reverse mycelium pigment.

D-Glucose, L-arabinose, D-xylose, iso-inositol, D-mannitol, D-fructose, and rhamnose are utilized for growth. No growth or only traces of growth with sucrose and raffinose.

Type strain shows the highest sequence similarity to: *S. griseoincarnatus*, AJ781328, 100%; *S. labedae*, AB184704, 100%; *S. variabilis*, DQ442551, 100%; *S. griseorubens*, AB184139, 99.9%; *S. griseoflavus*, AJ781322, 99.6%; *S. matensis*, AB184221, 99.6%; *S. althioticus*, AY999808, 99.2%; *S. paradoxus*, AB184628, 99.2%; *S. heliomycini*, AB184712, 99.1%; *S. flaveolus*, AB184764, 99.1%; *S. collinus*, AB184123, 99.1%; *S. viridochromogenes*, DQ442555, 99.1%; *S. bellus*, AB184849, 99%; *S. albogriseolus*, AJ494865, 99%; *S. viridodiastaticus*, AY999852, 99%; *S. violaceochromogenes*, AY999867, 99%; *S. malachitofuscus*, AB184282, 99%; *S. violaceorubidus*, AJ781374, 99%; *S. coerulescens*, AY999720, 99%.

Source: not known.

DNA G+C content (mol%): not known.

Type strain: ATCC 27427, CBS 485.74, BCRC 13770, DSM 40116, NBRC 14601, JCM 9650, NRRL B-3808, NRRL-ISP 5116.

Sequence accession no. (16S rRNA gene): AJ781328.

161. **Streptomyces eurocidicus** (Okami, Utahara, Nakamura and Umezawa 1954) Witt and Stackebrandt 1991, 456[VP] (Effective publication: Witt and Stackebrandt 1990, 370.) ("*Streptomyces eurocidicus*" Okami, Utahara, Nakamura and Umezawa 1954, 102; "*Verticillomyces eurocidicus*" Shinobu 1965, 111; *Streptoverticillium eurocidicum* Locci, Baldacci and Petrolini Baldan 1969, 36)

eu.ro.ci'di.cus. N.L. n. *eurocidinum* eurodicin; L. masc. suff. *-icus* suffix used with the sense of belonging to; N.L. masc. adj. *eurocidicus* belonging to eurodicin.

Spore chains in Section Umbellate Monoverticillate (=*Streptomyces* Section Verticillati biverticillate). Mature spore chains contain 3 to 10 or more spores per chain. This morphology is seen on yeast-malt agar, oatmeal agar, salts-starch agar, and glycerol-asparagine agar. Spore surface is smooth.

Color of colony: aerial mass color in the Yellow color series (1ba or 2ba, pale yellow; 1½db, pale greenish yellow; 1db, pale yellow green; 1½fb, light yellow) on yeast-malt agar, oatmeal agar, salts-starch agar, and glycerol-asparagine agar. Reverse side of colony with no distinctive pigments (olive brown to reddish brown on yeast-malt agar; pale grayish yellow to yellowish brown or olive brown on oatmeal agar, salts-starch agar, and glycerol-asparagine agar).

Color in medium: melanoid pigments are formed in peptone-yeast-iron agar and tryptone-yeast broth, but not in tyrosine agar. No pigment is found in the medium in yeast-malt agar, oatmeal agar, salts-starch agar, or glycerol-asparagine agar.

D-Glucose, iso-inositol, and D-fructose are utilized for growth. No growth or only traces of growth with L-arabinose, D-xylose, D-mannitol, rhamnose, sucrose, and raffinose.

Type strain shows the highest sequence similarity to: *S. albireticuli*, AB184881, 99.8%; *S. werraensis*, DQ442558, 99.3%; *S. biverticillatus*, AJ781381, 99.3%; *S. blastmyceticus*, AY999802, 99.2%; *S. stramineus*, AB184720, 99.2%; *S. netropsis*, AB184848, 99.1%.

Source: not known.

DNA G+C content (mol%): not known.

Type strain: AS 4.1086, ATCC 27428, CBS 792.72, BCRC 12424, CECT 3259, DSM 40604, NBRC 13491, IMET 43412, JCM 4029, JCM 4749, NRRL B-1676, NRRL-ISP 5604, RIA 1452, RIA 733, VKM Ac-903.

Sequence accession no. (16S rRNA gene): AY999790.

Further comments: in violation of Rule 33c of the *Bacteriological Code* (1990 Revision), in Validation List no. 38, *Streptomyces eurocidicus* is proposed as a *nomen revictum* (basonym: "*Streptomyces eurocidicus*" Okami et al. 1954.

According to Hatano et al. (2003), *Streptomyces eurocidicus* (Okami et al. 1954) Witt and Stackebrandt 1991 is an earlier heterotypic synonym of *Streptomyces albireticuli* (Nakazawa 1955) Witt and Stackebrandt 1991.

162. **Streptomyces europaeiscabiei** Bouchek-Mechiche, Gardan, Normand and Jouan 2000, 98[VP]

eu.ro.pa.ei.sca'bi.ei. L. adj. *europaeus* european; L. n. *scabies* scab, mange; N.L. gen. n. *europaeiscabiei* of European scab, referring to the European origin of the strains.

Spores are gray and are borne in mature spiral chains. Melanin is produced on tyrosine agar. L-Arabinose, D-fructose, D-glucose, D-mannitol, inositol, raffinose, rhamnose, sucrose, and D-xylose are utilized for growth. Degradation of xanthine differs between the strains studied. All strains are susceptible to 20 µg/ml streptomycin and 0.5 µg/ml crystal violet. They are not susceptible to 25 µg/ml oleandomycin or 10 IU/ml penicillin G. They utilize *trans*-aconitate, D(+)trehalose, ONPG, melibiose, and 5-keto-D-gluconate; most (about 78%) of the strains assimilate gentisate. They do not use betaine, mucate, D-saccharate, DL-lactate, or turanose.

Type strain shows the highest sequence similarity to: *S. scabies*, D63862, 99.8%. Type strain shows DNA–DNA similarity to: *S. stelliscabiei* DSM 41803^T, 42%; *S. reticuliscabiei* DSM 41804^T, 20%.

Source: these strains were isolated from common scab lesions, mostly on potato, but also on carrot and beet, and have been confirmed to be pathogenic on potato cvs Bintje and Urgenta, on carrot cv. Premia and on radish cv. Polka.

DNA G+C content (mol%): 71.3.

Type strain: CFBP 4497, CIP 107062, DSM 41802, ICMP 13714, KACC 20186, NCPPB 4039.

Sequence accession no. (16S rRNA gene): AY207598.

163. **Streptomyces eurythermus** Corbaz, Ettlinger, Gäumann, Keller-Schierlein, Kradolfer, Kyburz, Neipp, Prelog, Reusser and Zähner 1955, 1202^{AL}

eu.ry.ther′mus. Gr. adj. *eurus* wide; Gr. adj. *thermos* hot; N.L. masc. adj. *eurythermus* wide, hot.

Spore chains in Section *Retinaculiaperti* (including many straight and flexuous spore chains and occasional spirals). Mature spore chains generally have 10–50 spores per chain. This morphology is observed on yeast-malt agar, oatmeal agar, salts-starch agar, and glycerol-asparagine agar. Spore surface is smooth.

Color of colony: aerial mass color in the Gray color series on yeast-malt agar, oatmeal agar, salts-starch agar, and glycerol-asparagine agar. Reverse side of colony with no distinctive pigments on yeast-malt agar, oatmeal agar, salts-starch agar, and glycerol-asparagine agar; substrate pigment is not a pH indicator. One observer reports slight change from yellowish brown to reddish brown with addition of 0.05 M NaOH.

Color in medium: melanoid pigments are formed in peptone-yeast-iron agar, tyrosine agar, and tryptone-yeast broth. Pigments other than melanoids are not formed in yeast-malt agar, oatmeal agar, salts-starch agar, or glycerol-asparagine agar.

D-Glucose, L-arabinose, sucrose, D-xylose, D-mannitol, and D-fructose are utilized for growth. Variable reports on growth with iso-inositol, rhamnose, and raffinose.

Type strain shows the highest sequence similarity to: *S. nogalater*, AB045886, 99.2%; *S. fragilis*, AY999917, 99%; *S. tendae*, D63873, 99%.

Source: not known.

DNA G+C content (mol%): not known.

Type strain: AS 4.1697, ATCC 14975, ATCC 19749, CBS 488.68, BCRC 13650, DSM 40014, NBRC 12764, IMET 43078, JCM 4206, JCM 4575, KCTC 9731, NRRL 2539, NRRL-ISP 5014, RIA 1030, VKM Ac-1729.

Sequence accession no. (16S rRNA gene): D63870.

164. **Streptomyces exfoliatus** (Waksman and Curtis 1916) Waksman and Henrici *in* Breed, Murray and Hitchens 1948, 951^{AL} ("*Actinomyces exfoliatus*" Waksman and Curtis 1916, 116)

ex.fo.li.a′tus. L. masc. part. adj. *exfoliatus* stripped of leaves.

Spore chains in Section *Rectiflexibiles*. Mature spore chains generally have 10–50 spores per chain. Typical morphology on oatmeal agar. Some sporulation aerial mycelium with typical morphology on yeast-malt agar, oatmeal agar, salts-starch agar, and glycerol-asparagine agar. Spore surface is smooth. One observer reported fragmentation and spore formation in substrate-mycelium on yeast-malt agar, oatmeal agar, salts-starch agar, and glycerol-asparagine agar in 14 d; another observer reported coremia formation on glycerol-asparagine agar in 14 d.

Color of colony: aerial mass color in the Red color series on oatmeal agar; sporulation aerial mycelium is also in the Red series on yeast-malt agar and salts-starch agar when formed on these media. Reverse side of colony with no distinctive pigments on yeast-malt agar, oatmeal agar, salts-starch agar, and glycerol-asparagine agar; substrate pigment is not a pH indicator.

Color in medium: melanoid pigments are not formed in peptone-yeast-iron agar and tyrosine agar. No pigment found in medium in yeast-malt agar, oatmeal agar, salts-starch agar, and glycerol-asparagine agar.

D-Glucose, L-arabinose, sucrose, D-xylose, D-fructose, rhamnose, and raffinose are utilized for growth. No growth on iso-inositol or D-mannitol.

Type strain shows the highest sequence similarity to: *S. zaomyceticus*, EF178685, 99.9%; *S. venezuelae*, AB045890, 99.9%; *S. lateritius*, AJ781326, 99.7%; *S. wedmorensis*, DQ442557, 99.7%; *S. litmocidini*, AB184149, 99.6%; *S. omiyaensis*, EF178697, 99.6%; *S. yereyanensis*, EF178684, 99.4%; *S. narbonensis*, DQ445794, 99.3%.

Source: not known.

DNA G+C content (mol%): not known.

Type strain: AS 4.1407, ATCC 12627, ATCC 19750, CBS 489.68, BCRC 11469, CCUG 11113, DSM 40060, NBRC 13191, JCM 4366, LMG 19307, NCIMB 12599, NRRL B-1237, NRRL B-2924, NRRL-ISP 5060, PCM 2367, RIA 1031, UNIQEM 137, VKM Ac-767.

Sequence accession no. (16S rRNA gene): AB184324.

165. **Streptomyces felleus** Lindenbein 1952, 374^{AL}

fel′le.us. L. masc. adj. *felleus* of gall, like gall (pertaining to the bitter taste of proactinomycin A).

Spore chains in Section *Rectiflexibiles*. Mature spore chains generally have 10–50 spores per chain. This morphology is observed on yeast-malt agar, oatmeal agar, salts-starch agar, and glycerol-asparagine agar. Spore surface is smooth.

Color of colony: aerial mass color in the Yellow color series on yeast-malt agar, oatmeal agar, salts-starch agar, and glycerol-asparagine agar. Reverse side of colony with no distinctive pigments on yeast-malt agar, oatmeal agar, salts-starch agar, or glycerol-asparagine agar; substrate pigment is not a pH indicator.

Color in medium: melanoid pigments not formed in peptone-yeast-iron agar and tyrosine agar. No pigment in medium or faint yellow to yellowish green color in yeast-malt agar, oatmeal agar, salts-starch agar, and glycerol-asparagine agar.

D-Glucose, L-arabinose, D-xylose, D-mannitol, and D-fructose are utilized for growth. No growth or only traces of growth on sucrose, iso-inositol, rhamnose, and raffinose.

Type strain shows the highest sequence similarity to: *S. limosus*, AB184147, 100%; *S. canescens*, AB184117, 100%; *S. daghestanicus*, DQ442497, 100%; *S. albidoflavus*, AB184255, 100%; *S. hydrogenans*, AB184868, 100%; *S. odorifer*, Z76682, 100%; *S. violascens*, AY999737, 100%; *S. griseus* subsp. *solvifaciens*, AB249915, 100%; *S. champavatii*, DQ026642, 100%; *S. sampsonii*, D63871, 99.8%; *S. koyangensis*, AY079156, 99.7%.

Source: not known.

DNA G+C content (mol%): not known.

Type strain: AS 4.1677, ATCC 19752, CBS 491.68, BCRC 11471, DSM 40130, NBRC 12766, JCM 4368, NCIMB 12974, NRRL-ISP 5130, RIA 1033, UNIQEM 139, VKM Ac-722.

Sequence accession no. (16S rRNA gene): AB184129.

166. **Streptomyces ferralitis** Saintpierre-Bonaccio, Amir, Pineau, Lemriss and Goodfellow 2004, 2064[VP]

fer.ra'li.tis. N.L. gen. n. *ferralitis* of ferralite, denoting the type of soil from which the type strain was isolated.

Forms an extensively branched substrate mycelium and aerial hyphae that differentiate into looped or spiral chains of spores. The spore chains consist of up to 15 barrel-shaped spores with smooth surface. A brown substrate mycelium and a white aerial spore mass are formed on modified Bennett's agar (Jones, 1949). Melanin pigments are not produced on peptone-yeast extract-iron agar. The culture grows well at 20 and 45°C, but does not grow at 10°C. Metabolizes casein, hypoxanthine, L-tyrosine, urea, and xanthine, but not adenine, elastin, gelatin, guanine, starch, Tween 80, or xanthine. D(+)-Galactose, D(+)-glucose, D(+)-mannitol, D(+)-mannose, and D(+)-trehalose are used as sole carbon sources for energy and growth, but adonitol, D-arabinose, D(+)-cellobiose, D(+)-melibiose, and sodium citrate are not. Resistant to penicillin (25 µg/ml), but does not grow in the presence of erythromycin (4 µg/ml), gentamicin sulfate (10 µg/ml), rifampin (6 µg/ml), streptomycin sulfate (5 µg/ml), tetracycline hydrochloride (30 µg/ml), vancomycin hydrochloride (10 µg/ml), crystal violet (0.0002%, w/v), phenol (0.01%, w/v), potassium tellurite (0.005%, w/v), or sodium chloride (5%, w/v). It shows activity against clinical isolates of *Candida albicans*, *Staphylococcus aureus*, *Staphylococcus epidermidis* and a *Corynebacterium* strain, but not against *Fusarium oxysporum*, *Bacillus*, *Erwinia*, *Escherichia coli*, *Klebsiella pneumoniae*, or *Pseudomonas aeruginosa* strains. The species description is based upon a single strain and hence serves as the description of the type strain.

Type strain shows no sequence similarity over 99%.

Source: the type strain was isolated from a ferralitic, oxidic ultramafic soil collected at the southern end of the main island of New Caledonia.

DNA G+C content (mol%): not known.

Type strain: SFOp68, DSM 41836, NCIMB 13954.

Sequence accession no. (16S rRNA gene): AY262826.

167. **Streptomyces fervens** (DeBoer et al. 1959–1960) Witt and Stackebrandt 1991, 456[VP] (Effective publication: Witt and Stackebrandt 1990, 370.) ("*Streptomyces fervens*" DeBoer Dietz, Evans and Michaels 1959–1960), 220; (*Streptoverticillium fervens* Locci, Baldacci and Petrolini Baldan 1969, 23)

fer'vens. L. part. adj. *fervens* boiling hot, referring to its high growth temperature.

Spore chain morphology: Section not determined. Sporulating aerial mycelium is not produced on yeast-malt agar, oatmeal agar, salts-starch agar, or glycerol-asparagine agar. Note that the original descriptions of DeBoer et al. and Locci et al. (op. cit) call attention to the limited sporulation; they report verticillate morphology. Spore surface is smooth.

Color of colony: the thin aerial mycelium, which generally lacks sporophores and spore chains, is in the Red color series on yeast-malt agar, oatmeal agar, salts-starch agar, and glycerol-asparagine agar. Two observers report that an appropriate color tab cannot be found in the red color wheel of Tresner and Backus; one observer identifies nonsporulating aerial mycelium color as 7 ca, light yellowish pink. Reverse side of colony with distinctive red color reported as dark red or eosin red on yeast-malt agar, oatmeal agar, salts-starch agar, and glycerol-asparagine agar. Reverse mycelium pigment is pH-sensitive, changing from deep red to reddish brown with the addition of 0.05 M NaOH and from red to violet red with the addition of 0.05 M HCl.

Color in medium: melanoid pigments are formed in peptone-yeast-iron agar and tryptone-yeast broth, but not in tyrosine agar. No pigment other than melanoids is found in the medium in yeast-malt agar, oatmeal agar, salts-starch agar, or glycerol-asparagine agar.

D-Glucose and iso-inositol are utilized for growth. A trace of growth is seen on L-arabinose, D-xylose, D-mannitol, D-fructose, rhamnose, sucrose and raffinose; utilization of these carbon sources is doubtful.

For sequence similarity, see type strain of *Streptomyces hiroshimensis*.

Source: not known.

DNA G+C content (mol%): not known.

Type strain: ATCC 27429, JCM 4310, JCM 4750, NBRC 13343, NRRL 2755.

Sequence accession no. (16S rRNA gene): AB184871.

Further comments: in violation of Rule 33c of the *Bacteriological Code* (1990 Revision), in Validation List no. 38, *Streptomyces fervens* is proposed as a *nomen revictum* (basonym: "*Streptomyces fervens*" DeBoer et al. 1959–1960).

Witt and Stackebrandt proposed to transfer *Streptoverticillium fervens* (DeBoer et al. 1959–1960) Locci et al. 1969 to the genus. However, Validation List no. 38 does not include formal propositions about *Streptoverticillium fervens* subsp. *fervens* Baldacci and Locci 1974 and *Streptoverticillium fervens* subsp. *melrosporus* Mason et al. 1965.

Labeda (1996) proposes *Streptomyces biverticillatus* (Preobrazhenskaya and Ryabova 1957) Witt and Stackebrandt 1991, *Streptomyces fervens* (DeBoer et al. 1959–1960) Witt and Stackebrandt 1991, and *Streptomyces roseoverticillatus* (Shinobu 1956) Witt and Stackebrandt 1991 as later synonyms of *Streptomyces baldaccii* corrig. (Farina and Locci 1966) Witt and Stackebrandt 1991. This proposal is in violation of Rule 24b(1) of the *Bacteriological Code* (1990 Revision) because the senior epithet is *roseoverticillatus*. *Streptomyces baldaccii*, *Streptomyces biverticillatus*, and *Streptomyces fervens* are therefore to be regarded as later synonyms of *Streptomyces roseoverticillatus*.

According to Hatano et al. (2003), *Streptomyces fervens* (DeBoer et al. 1959–1960) Witt and Stackebrandt 1991 is a later heterotypic synonym of *Streptomyces hiroshimensis* (Shinobu 1955) Witt and Stackebrandt 1991.

168. **Streptomyces filamentosus** Okami and Umezawa *in* Okami, Okuda, Takeuchi, Nitta and Umezawa 1953, 153[AL] emend. Lanoot, Vancanneyt, Dawyndt, Cnockaert, Zhang, Huang, Liu and Swings 2005a, 8 (Effective publication: Lanoot, Vancanneyt, Dawyndt, Cnockaert, Zhang, Huang, Liu and Swings 2004, 88.)

fi.la.men.to'sus. L. n. *filamentum* assembly of threads; L. masc. suff. *-osus* suffix used with the sense of full of, prone to; N.L. masc. adj. *filamentosus* full of threads or filaments.

Spore chains in Section *Rectiflexibiles.* Mature spore chains generally have 10–50 spores per chain. This morphology is found on yeast-malt agar, oatmeal agar, salts-starch agar, and glycerol-asparagine agar. Spore surface is smooth.

Color of colony: aerial mass color in the Red color series on yeast-malt agar, oatmeal agar, salts-starch agar, and glycerol-asparagine agar. Reverse side of colony with no distinctive pigments on yeast-malt agar, oatmeal agar, salts-starch agar, and glycerol-asparagine agar; substrate pigment is not a pH indicator.

Color in medium: melanoid pigments not formed in peptone-yeast-iron agar and tyrosine agar. No pigment found in medium in yeast-malt agar, oatmeal agar, salts-starch agar, or glycerol-asparagine agar.

D-Glucose, L-arabinose, sucrose, D-xylose, and rhamnose are utilized for growth. No growth or only trace of growth on iso-inositol, D-mannitol, and raffinose. Variable reports on utilization of D-fructose.

Type strain shows the highest sequence similarity to: *S. roseolus*, AB184168, 99.4%; *S. roseoviridis*, AB184239, 99%.

Source: not known.

DNA G+C content (mol%): not known.

Type strain: AS 4.1656, ATCC 19753, CBS 492.68, BCRC 13644, DSM 40022, HAMBI 1010, IFM 1180, NBRC 12767, IMET 43562, JCM 4122, JCM 4576, NCIMB 13018, NRRL B-2114, NRRL-ISP 5022, RIA 1034, UNIQEM 140, VKM Ac-1266.

Sequence accession no. (16S rRNA gene): AB184130.

Further comments: according to Lanoot et al. (2004), *Streptomyces filamentosus* Okami and Umezawa 1953 is an earlier heterotypic synonym of *Streptomyces roseosporus* Falcão de Morais and Dália Maia 1961.

169. **Streptomyces filipinensis** Ammann, Gottlieb, Brock, Carter and Whitfield 1955, 559[AL]

fi.li.pi.nen'sis. N.L. masc. adj. *filipinensis* of or belonging to the Philippines, the source of the soil from which the organism was isolated.

Spore chains in Section *Spirales.* Tight spirals or open coils, usually of several turns, occur at the end of moderately long spore chains or 10 to 50 or more spores per chain on yeast-malt agar, oatmeal agar, and salts-starch agar. Spirals may be replaced by *Retinaculiaperti* morphology on glycerol-asparagine agar. Spore surface is spiny.

Color of colony: aerial mass color in the Gray color series on yeast-malt agar, oatmeal agar, salts-starch agar, and glycerol-asparagine agar. Reverse side of colony with no distinctive pigments (colorless to grayish yellow) on yeast-malt agar, oatmeal agar, salts-starch agar, and glycerol-asparagine agar.

Color in medium: melanoid pigments are produced in peptone-yeast-iron agar, tyrosine agar, and peptone-yeast broth. Pigments other than melanoids are not found in the medium in yeast-malt agar, oatmeal agar, salts-starch agar, or glycerol-asparagine agar.

Carbon utilization: D-glucose, L-arabinose, sucrose, D-xylose, iso-inositol, D-mannitol, D-fructose and raffinose are utilized for growth. No growth or only trace of growth on rhamnose.

Type strain shows the highest sequence similarity to: *S. durhamensis*, AY999785, 99.8%.

Source: not known.

DNA G+C content (mol%): not known.

Type strain: AS 4.1452, ATCC 23905, CBS 309.56, CBS 801.68, BCRC 11472, DSM 40112, NBRC 12860, JCM 4369, LMG 19333, NRRL 2437, NRRL-ISP 5112, RIA 1124, VKM Ac-966.

Sequence accession no. (16S rRNA gene): AB184198.

170. **Streptomyces fimbriatus** (Millard and Burr 1926) Waksman *in* Waksman and Lechevalier 1953, 104[AL] ("*Actinomyces fimbriatus*" Millard and Burr 1926, 639)

fim.bri.a'tus. L. masc. adj. *fimbriatus* fibrous, fringed.

Excellent growth on Czapek's solution agar. Produces septacidin, an anti-tumor and anti-fungal purine antibiotic; not inhibited by streptomycin.

Type strain shows no sequence similarity over 99%.

Source: Millard and Burr's original single isolate (no longer extant) was obtained from a case of common potato scab.

DNA G+C content (mol%): not known.

Type strain: AS 4.1598, ATCC 15051, CBS 453.65, DSM 40942, NBRC 15411, JCM 5080, NCIMB 13039, NRRL B-3175, VKM Ac-761.

Sequence accession no. (16S rRNA gene): AY999844.

171. **Streptomyces fimicarius** (Duché 1934) Waksman and Henrici *in* Breed, Murray and Hitchens 1948, 940[AL] ("*Actinomyces fimicarius*" Duché 1934, 346)

fi.mi.ca'ri.us. L. n. *fimus* dung, manure; L. adj. *carus* dear, loving; N.L. masc. adj. *fimicarius* dung-loving.

Spore chains Section *Rectiflexibiles.* Mature spore chains are generally long with 10 to 50 or more spores per chain. This morphology is seen on yeast-malt agar, oatmeal agar, salts-starch agar, and glycerol-asparagine agar. Spore surface is smooth. Fragmentation of the substrate mycelium may be seen on glycerol-asparagine agar in 16 d.

Color of colony: aerial mass color in the Yellow or White color series on yeast-malt agar, oatmeal agar, salts-starch agar, and glycerol-asparagine agar. Reverse side of colony with no distinctive pigments (pale or grayish yellow to yellowish brown) on oatmeal agar, salts-starch agar, and glycerol-asparagine agar; yellow is modified by red (to orange or reddish brown) on yeast-malt agar. This pigment changes from reddish brown to pale brown with addition of 0.05 M HCl.

Color in medium: melanoid pigments are not formed in peptone-yeast-iron agar, tyrosine agar, or tryptone-yeast broth (but according to one observer, some brown pigment is formed in the medium in Gause's medium no. 2). Red (pink to light reddish brown) pigment is found in the medium in yeast-malt agar and oatmeal agar; it may or may not be seen in salts-starch agar and glycerol-asparagine agar. This pigment is pH-sensitive, changing from pink or reddish brown to yellow when tested with 0.05 M HCl.

D-Glucose, L-arabinose, D-xylose, D-mannitol, D-fructose, and rhamnose are utilized for growth. Only traces of growth is seen with iso-inositol, sucrose, and raffinose.

Type strain shows the highest sequence similarity to: *S. cavourensis* subsp. *washingtonensis*, DQ026671, 100%; *S. anulatus*, DQ026637, 100%; *S. badius*, AY999783, 100%; *S. sindenensis*, AB184759, 100%; *S. rubiginosohelvolus*, AB184240, 100%; *S. pluricolorescens*, DQ442540, 100%; *S. griseinus*, AB184205, 100%; *S. acrimycini*, AY999889, 100%; *S. flavofuscus*, AB249935, 100%; *S. mediolani*, AB184674, 100%; *S. praecox*, AB184293, 100%; *S. lavendulae* subsp. *lavendulae*, AB184080, 100%; *S. griseoplanus*, AY999894, 100%; *S. microflavus*, DQ445795, 99.9%; *S. cinereorectus*, AB184646, 99.9%; *S. fulvorobeus*, AB184711, 99.9%; *S. lipmanii*, AB184148, 99.9%; *S. cyaneofuscatus*, AB184860, 99.9%; *S. baarnensis*, EF178688, 99.9%; *S. globisporus* subsp. *globisporus*, EF178686, 99.9%; *S. albovinaceus*, AB249958, 99.9%; *S. alboviridis*, AB184256, 99.9%; *S. californicus*, AB184755, 99.8%; *S. parvus*, DQ442537, 99.8%; *S. argenteolus*, AB045872, 99.8%; *S. halstedii*, EF178695, 99.7%; *S. luridiscabiei*, AF361784, 99.7%; *S. flavogriseus*, AJ494864, 99.7%; *S. griseus* subsp. *griseus*, AY207604, 99.7%; *S. griseolus*, AB184768, 99.7%; *S. floridae*, AB184656, 99.7%; *S. flavovirens*, DQ026635, 99.7%;%; *S. nitrosporeus*, EF178680, 99.5%; *S. pulveraceus*, AB184806, 99.5%; *S. olivoviridis*, AB184227, 99.5%; *S. bacillaris*, AB184439, 99.5%; *S. atroolivaceus*, AJ781320, 99.5%; *S. finlayi*, AY999788, 99.5%; *S. clavifer*, DQ026670, 99.5%; *S. yanii*, AB006159, 99.4%; *S. gelaticus*, DQ026636, 99.3%; *S. atratus*, DQ026638, 99.3%; *S. albolongus*, AB184425, 99.3%; *S. mutomycini*, AB249951, 99.3%; *S. celluloflavus*, AB184476, 99.3%; *S. griseobrunneus*, AB249912, 99.3%; *S. sanglieri*, AB249945, 99.3%; *S. spiroverticillatus*, AB184814, 99.2%; *S. cavourensis* subsp. *cavourensis*, DQ445791, 99.2%; *S. candidus*, DQ026663, 99.1%; *S. mauvecolor*, AB184532, 99%; *S. cremeus*, AB184124, 99%.

Source: not known.

DNA G+C content (mol%): not known.

Type strain: AS 4.1629, ATCC 25449, CBS 420.34, CBS 682.69, BCRC 12245, DSM 40322, NBRC 13037, JCM 4472, NRRL-ISP 5322, RIA 1229, VKM Ac-1724.

Sequence accession no. (16S rRNA gene): AY999784.

172. **Streptomyces finlayi** (Szabó, Marton, Buti and Pártai 1963) Pridham 1970, 35[AL] ("*Actinomyces finlayi*" Szabó, Marton, Buti and Pártai 1963, 209)

fin.lay'i. N.L. gen. masc. n. *finlayi* of Finlay, named for Alexander C. Finlay, discoverer of oxytetracycline.

Spore chains in Section *Rectiflexibiles* to *Spirales*. Short spore chains of 10 or more spores per chain are generally flexuous, crooked, hooked, or in imperfect spirals of only one or two turns; longer chains may form open spirals of three or four turns. Hooks and loops on short spore chains are not characteristic of *Retinaculiaperti* morphology. This morphology is seen on yeast-malt agar, oatmeal agar, salts-starch agar, and glycerol-asparagine agar; spirals are best developed on yeast-malt agar and oatmeal agar. Spore surface is hairy.

Color of colony: aerial mass color in the Gray color series on yeast-malt agar, oatmeal agar, salts-starch agar, and glycerol-asparagine agar. Reverse side of colony is grayed yellow to yellow-brown modified by green (grayish yellow to olive brown on yeast-malt agar and oatmeal agar; pale greenish yellow or yellow-green to moderate or dark olive green on salts-starch agar or glycerol-asparagine agar). Substrate mycelium pigment is not a pH indicator.

Color in medium: melanoid pigments are not produced in peptone-yeast-iron agar, tyrosine agar, or tryptone-yeast broth. No pigments are found in the medium in yeast-malt agar, oatmeal agar, salts-starch agar, or glycerol-asparagine agar.

D-Glucose is utilized for growth. Significant growth also occurs on L-arabinose, D-xylose, and rhamnose, although growth on these carbon sources is less than on D-glucose. Utilization of sucrose is doubtful. No growth or only trace of growth on iso-inositol, D-mannitol, D-fructose, and raffinose.

Type strain shows the highest sequence similarity to: *S. clavifer*, DQ026670, 99.8%; *S. olivoviridis*, AB184227, 99.7%; *S. atroolivaceus*, AJ781320, 99.7%; *S. californicus*, AB184755, 99.6%; *S. mutomycini*, AB249951, 99.6%; *S. pluricolorescens*, DQ442540, 99.5%; *S. rubiginosohelvolus*, AB184240, 99.5%; *S. fulvorobeus*, AB184711, 99.5%; *S. lipmanii*, AB184148, 99.5%; *S. fimicarius*, AY999784, 99.5%; *S. sindenensis*, AB184759, 99.5%; *S. praecox*, AB184293, 99.5%; *S. flavofuscus*, AB249935, 99.5%; *S. badius*, AY999783, 99.5%; *S. mediolani*, AB184674, 99.5%; *S. cavourensis* subsp. *washingtonensis*, DQ026671, 99.5%; *S. anulatus*, DQ026637, 99.5%; *S. griseoplanus*, AY999894, 99.5%; *S. floridae*, AB184656, 99.5%; *S. lavendulae* subsp. *lavendulae*, AB184080, 99.5%; *S. alboviridis*, AB184256, 99.5%; *S. microflavus*, DQ445795, 99.5%; *S. griseinus*, AB184205, 99.5%; *S. cyaneofuscatus*, AB184860, 99.4%; *S. albovinaceus*, AB249958, 99.4%; *S. cinereorectus*, AB184646, 99.4%; *S. globisporus* subsp. *globisporus*, EF178686, 99.4%; *S. baarnensis*, EF178688, 99.4%; *S. acrimycini*, AY999889, 99.4%; *S. luridiscabiei*, AF361784, 99.4%; *S. griseolus*, AB184768, 99.3%; *S. candidus*, DQ026663, 99.3%; *S. griseus* subsp. *griseus*, AY207604, 99.3%; *S. argenteolus*, AB045872, 99.3%; *S. flavovirens*, DQ026635, 99.3%; *S. parvus*, DQ442537, 99.3%; *S. bacillaris*, AB184439, 99.2%; *S. flavogriseus*, AJ494864, 99.2%; *S. halstedii*, EF178695, 99.2%; *S. griseobrunneus*, AB249912, 99.1%; *S. albolongus*, AB184425, 99.1%; *S. celluloflavus*, AB184476, 99.1%; *S. pulveraceus*, AB184806, 99%; *S. nitrosporeus*, EF178680, 99%; *S. cavourensis* subsp. *cavourensis*, DQ445791, 99%.

Source: not known.

DNA G+C content (mol%): not known.

Type strain: AS 4.1436, ATCC 23340, ATCC 23906, CBS 802.68, DSM 40218, HAMBI 1071, NBRC 13201, JCM 4216, JCM 4637, NCIMB 9834, NRRL B-12114, NRRL-ISP 5218, RIA 1162, VKM Ac-967.

Sequence accession no. (16S rRNA gene): AY999788.

173. **Streptomyces flaveolus** (Waksman *in* Bergey, Harrison, Breed, Hammer and Huntoon 1923) Waksman and Henrici *in* Breed, Murray and Hitchens 1948, 936[AL] ("*Actinomyces flaveolus*" Waksman *in* Bergey, Harrison, Breed, Hammer and Huntoon 1923, 368; "*Streptomyces flaveolus* subsp. *flaveolus*" Waksman *in* Pridham, Lyons and Seckinger 1965, 220)

fla.ve'o.lus. L. adj. *flavus* yellow; N.L. dim. masc. adj. *flaveolus* somewhat yellow.

Spore chains in Section *Spirales*, but short spore chains representative of Section *Retinaculiaperti* are also common. Mature spore chains generally have 10–50 spores per chain; shorter chains (3–10 spores) may be found. This morphology is seen on yeast-malt agar, oatmeal agar, salts-starch agar, and glycerol-asparagine agar (although sporulating aerial mycelium may be poorly developed on oatmeal agar or salts-starch agar). Spore surface is hairy with some tendency toward spines; carbon replica method suggests spiny spore surface. Coremia may form on salts-starch agar.

Color of colony: aerial mass color in the Gray color series on yeast-malt agar, oatmeal agar, salts-starch agar, and glycerol-asparagine agar (White series also reported by one observer on salts-starch agar and glycerol-asparagine agar). Reverse side of colony with no distinctive pigments on yeast-malt agar, oatmeal agar, salts-starch agar, and glycerol-asparagine agar; substrate pigment is not a pH indicator.

Color in medium: melanoid pigments not formed in peptone-yeast-iron agar and tyrosine agar. Yellow pigment found in medium in yeast-malt agar, glycerol-asparagine agar, and sometimes in other media. This pigment is not pH-sensitive when tested with 0.05 M NaOH or HCl.

D-Glucose, L-arabinose, sucrose, D-xylose, iso-inositol, D-mannitol, D-fructose, and rhamnose are utilized for growth. No growth or only trace of growth on raffinose.

Type strain shows the highest sequence similarity to: *S. griseoflavus*, AJ781322, 99.5%; *S. fragilis*, AY999917, 99.3%; *S. collinus*, AB184123, 99.2%; *S. viridochromogenes*, DQ442555, 99.2%; *S. griseorubens*, AB184139, 99.2%; *S. ambofaciens*, M27245, 99.1%; *S. glaucescens*, AB184843, 99.1%; *S. paradoxus*, AB184628, 99.1%; *S. matensis*, AB184221, 99.1%; *S. variabilis*, DQ442551, 99.1%; *S. coelicoflavus*, AB184650, 99.1%; *S. griseoincarnatus*, AJ781328, 99.1%; *S. erythrogriseus*, AJ781328, 99.1%; *S. malachitofuscus*, AB184282, 99%; *S. labedae*, AB184704, 99%.

Source: not known.

DNA G+C content (mol%): not known.

Type strain: AS 4.1432, ATCC 19754, ATCC 3319, CBS 128.20, CBS 493.68, CCM 3171, BCRC 12489, CECT 3181, DSM 40061, HAMBI 893, NBRC 12768, NBRC 3408, NBRC 3715, IMET 40233, JCM 4032, JCM 4577, KCTC 9022, LMG 19328, NRRL B-1334, NRRL B-2688, NRRL-ISP 5061, RIA 1035, RIA 485, UNIQEM 141, VKM Ac-965.

Sequence accession no. (16S rRNA gene): AB184764.

174. **Streptomyces flaveus** (Cross, Lechevalier and Lechevalier 1963) Goodfellow, Williams and Alderson 1986a, 574[VP] (Effective publication: Goodfellow, Williams and Alderson 1986a, 53.) (*Microellobosporia flavea* Cross, Lechevalier and Lechevalier 1963, 428)

fla.ve'us. N.L. masc. adj. *flaveus* presumably from *flavus* yellow.

Short straight spore chains (2–5). Spore surface is smooth. Spores borne on both the substrate and aerial mycelium. Forms extensively branched substrate and aerial mycelium. The aerial spore mass is gray; the reverse color is yellow brown. Does not form melanin pigments. Adenine, esculin, casein, hypoxanthine, starch, testosterone, tyrosine, and urea are degraded, but allantoin, arbutin, chitin, elastin, guanine, lecithin, pectin, xanthine, and xylan are not. Hydrogen sulfide is produced but nitrate is not reduced. L-Arabinose, cellobiose, D-fructose, D-galactose, D-glucose, *myo*-inositol, inulin, D-lactose, mannitol, D-mannose, melibiose, raffinose, L-rhamnose, salicin, sucrose, trehalose, and D-xylose are used as sole carbon sources, but adonitol, D-melezitose, and xylitol are not. Grows on DL-α-aminobutyric acid, L-arginine, L-cysteine, L-histidine, L-phenylalanine, potassium nitrate, L-serine, L-threonine, and L-valine, but not on L-hydroxyproline or L-methionine, as sole nitrogen source. Growth occurs at 10 and 37°C, but not at 4 or 45°C. Tolerant to phenol (0.1%, w/v) and sodium chloride (7%, w/v), but not to sodium azide (0.01%, w/v). Sensitive to rifampin and sodium chloride (10%, w/v). Does not show antimicrobial activity against *Aspergillus niger* LIV 131, *Bacillus subtilis* NCIB 3610, *Candida albicans* CBS 562, *Escherichia coli* NCIB 9132, *Micrococcus luteus* NCIB 196, *Pseudomonas fluorescens* NCIB 9046[T], *Saccharomyces cerevisiae* CBS 1171[T], or *Streptomyces murinus* ISP 5091. Wall peptidoglycan contains LL-A_2pm as the major diamino acid. The organism has a type II phospholipid pattern (*sensu* Lechevalier et al., 1977) and contains major amounts of hexa- and octahydrogenated menaquinones with nine isoprene units (Collins et al., 1984).

Type strain shows no sequence similarity over 99%.

Source: not known.

DNA G+C content (mol%): 68–71.

Type strain: ATCC 15332, CBS 355.67, DSM 43153, IFM 1234, NBRC 12190, IMET 43554, LMG 19323, NCIMB 9587, NRRL B-16074, RIA 896, VKM Ac-1295, VKM Ac-633, KCC A-0035, JCM 3035.

Sequence accession no. (16S rRNA gene): DQ026643.

175. **Streptomyces flavidofuscus** Preobrazhenskaya 1986, 574[VP] (Effective publication: Preobrazhenskaya *in* Gause, Preobrazhenskaya, Sveshnikova, Terekhova and Maximova 1983.)

fla.vi.do.fus'cus. L. adj. *flavidus* yellowish; L. adj. *fuscus* dark, tawny; N.L. masc. adj. *flavidofuscus* yellowish, tawny.

Spore chains are spiral (*Spirales*); spores are smooth. On mineral agar 1: starch-ammonia agar: aerial mycelium is velvety, creamy, yellow, sometimes poor; substrate mycelium is colorless; no diffusible pigment. On glycerol-nitrate agar, glycerol-asparagine agar: aerial mycelium is yellow; substrate mycelium is colorless or yellow; no diffusible pigment. On oatmeal agar: aerial mycelium is yellow; substrate mycelium is colorless or pink; no diffusible pigment. On organic agar 2: aerial mycelium is creamy, yellow; substrate mycelium and diffusible pigment are gray-brown. Melanoid pigments are found. Grows on fructose, sucrose, rhamnose, glucose, arabinose, raffinose, mannitol, and inositol. Strains of this species produce the antibiotic echinomycin.

Type strain shows no sequence similarity over 99%.

Source: not known.

DNA G+C content (mol%): 68–71.

Type strain: AS 4.1617, ATCC 43683, DSM 41473, NBRC 15404, INA 15719, JCM 6920, NRRL B-16366, VKM Ac-1209.

Sequence accession no. (16S rRNA gene): AY999914.

Further comments: according to Guo et al. (2008), *Streptomyces flavidofuscus* Preobrazhenskaya 1986 should be transferred to the genus *Nocardiopsis*. However, no formal proposition is made in the paper by Guo et al. (2008).

176. **Streptomyces flavidovirens** (Kudrina *in* Gauze, Preobrazhenskaya, Kudrina, Blinov, Ryabova and Sveshnikova 1957) Pridham, Hesseltine and Benedict 1958, 66[AL] ("*Actinomyces flavidovirens*" Kudrina *in* Gauze, Preobrazhenskaya, Kudrina, Blinov, Ryabova and Sveshnikova 1957, 90)

fla.vi.do.vi'rens. L. adj. *flavidus* yellowish; L. part. adj. *virens* being green; N.L. part. adj. *flavidovirens* being yellowish green.

Spore chains in Section *Retinaculiaperti* or *Rectiflexibiles*. Preponderance of very flexuous spore chains, some of which appear as imperfect or open spirals together with some straight or slightly flexuous chains makes this strain difficult to categorize in respect to spore chain morphology. The illustration accompanying the original description would place this species in Section *Spirales*, but regular spirals were not found by ISP observers. Mature spore chains are moderately long with 10 to 50 or more spores per chain. This morphology is found on yeast-malt agar, oatmeal agar, salts-starch agar, and glycerol-asparagine agar. Spore surface is smooth.

Color of colony: aerial mass color in the Yellow or White color series on yeast-malt agar, oatmeal agar, salts-starch agar, and glycerol-asparagine agar. The nearest matching color tab in the Yellow series is 2ba, pale yellow. Reverse side of colony with no distinctive pigments (colorless to pale yellow or light grayish yellow) on yeast-malt agar, oatmeal agar, salts-starch agar, and glycerol-asparagine agar.

Color in medium: melanoid pigments are not formed in peptone-yeast-iron agar, tyrosine agar, or tryptone-yeast broth. Traces of yellow pigment may or may not be found in the medium in yeast-malt agar, oatmeal agar, salts-starch agar, and glycerol-asparagine agar. This pigment, when present, is not pH-sensitive when tested with 0.05 M NaOH or HCl.

D-Glucose, L-arabinose, and D-xylose are utilized for growth. Reports vary on utilization of iso-inositol, D-fructose, rhamnose, and sucrose. No growth or only trace of growth with D-mannitol and raffinose.

Type strain shows the highest sequence similarity to: *S. helvaticus*, AB184367, 100%; *S. albidochromogenes*, AB249953, 100%; *S. chryseus*, AY999787, 100%; *S. enissocaesilis*, AB249930, 99.2%.

Source: not known.

DNA G+C content (mol%): not known.

Type strain: ATCC 19900, ATCC 25451, CBS 684.69, BCRC 13761, DSM 40150, NBRC 13039, IMET 43744, INA 12287, JCM 4474, NRRL B-2708, NRRL-ISP 5150, RIA 1231, VKM Ac-1771.

Sequence accession no. (16S rRNA gene): AB184270.

177. **Streptomyces flavisclerotícus** (*ex* Pridham 1970) Goodfellow, Williams and Alderson 1986a, 574[VP] (Effective publication: Goodfellow, Williams and Alderson 1986d, 59.)

fla.vi.scle.ro'ti.cus. L. adj. *flavus* yellow; N.L. neut. n. *sclerotium* sclerotium; N.L. masc. adj. *flavisclerotícus* yellow sclerotium, referring to yellow and ability to form sclerotia.

Spore chain morphology: Section not determined. Aerial mycelium is poorly developed or absent on yeast-malt agar, oatmeal agar, salts-starch agar, and glycerol-asparagine agar. Rare spore chains on yeast-malt agar or on glycerol-asparagine agar are reported to be straight or flexuous. The original description by Thirumalachar reports irregular spirals. Spore surface is not determined.

One of three observers reports "globula sporangia" on yeast-malt agar, oatmeal agar, and salts-starch agar, and sclerotia on yeast-malt agar. Fragmentation of substrate mycelium into spore-like bodies is also recorded by this observer.

Color of colony is not determined. Aerial mycelium is not produced on yeast-malt agar, oatmeal agar, salts-starch agar, or glycerol-asparagine agar. Reverse side of colony is olive brown to strong brown on yeast-malt agar; greenish yellow on oatmeal agar; grayish yellow, orange yellow, or moderate yellow on salts-starch agar and glycerol-asparagine agar.

Color in medium: melanoid pigments are not formed in peptone-yeast-iron agar, tyrosine agar, or tryptone-yeast broth. No pigment or only a trace of yellow is found in the medium in yeast-malt agar, oatmeal agar, salts-starch agar, and glycerol-asparagine agar.

D-Glucose, L-arabinose, D-xylose, D-mannitol, D-fructose, and rhamnose are utilized for growth. Utilization of sucrose is doubtful; no growth or only traces of growth with iso-inositol and raffinose.

For sequence similarity, see type strain of *Streptomyces minutisclerotícus*.

Source: not known.

DNA G+C content (mol%): not known.

Type strain: ATCC 19347, AS 4.1071, CBS 658.72, BCRC 12108, CMI 117723, DSM 40270, DSM 43152, NBRC 12998, NBRC 13357, NBRC 14019, NBRC 15148, IMET 43617, JCM 3100, JCM 4751, KCC A-0100, LMG 19886, NCIB (now NCIMB) 11008, NRRL B-12173, NRRL-ISP 5270, PCM 2303, RIA 1318, RIA 883.

Sequence accession no. (16S rRNA gene): AB184634.

Further comments: according to Goodfellow et al. (1986a), the species *Streptomyces flavisclerotícus* (*ex* Pridham 1970) Goodfellow et al. 1986a is a synonym of the species *Chainia flava* Thirumalachar and Sukapure 1964. However, according to Rule 51b (2) of the *Bacteriological Code* (1990 Revision), the transfer of *Chainia flava* Thirumalachar and Sukapure 1964 in the genus *Streptomyces* Waksman and Henrici 1943 as *Streptomyces flavisclerotícus* (*ex* Pridham 1970) Goodfellow et al. 1986a sp. nov., nom. rev. is illegitimate.

According to Lanoot et al. (2005b), *Streptomyces flavisclerotícus* (*ex* Pridham 1970) Goodfellow et al. 1986a is a later heterotypic synonym of *Streptomyces minutisclerotícus* (Thirumalachar 1965) Pridham 1970 emend. Lanoot et al. 2005b.

178. **Streptomyces flavofungini** (*ex* Uri and Békési 1958) Szabó and Preobrazhenskaya 1986, 574[VP] (Effective publication: Szabó and Preobrazhenskaya *in* Gause, Preobrazhenskaya, Sveshnikova, Terekhova and Maximova 1983.) ["*Actinomyces flavofungini*" (Uri and Békési 1958) Szabó and Preobrazhenskaya 1962]

fla.vo.fun.gi'ni. L. adj. *flavus* golden yellow, reddish yellow; L. adj. *funginus* of a mushroom, pertaining to a mushroom; N.L. adj. *flavofungini* yellow and from a mushroom.

Spore chains in Section *Rectiflexibiles*, but sporulation is poor or absent on most media. Absence of sporulation is also mentioned in the original description for this strain. ISP observers record short to long spore chains on oatmeal agar and salts-starch agar but not on yeast-malt agar or glycerol-asparagine agar. Spore chains may also be seen on carbon-utilization media. Spore surface is smooth.

Color of colony: aerial mass color in the White or Yellow color series on salts-starch agar. Nearest matching color tab in the Yellow color series is 1cb, pale yellow green. Reverse side of colony is yellow to greenish yellow or light olive or light olive brown on yeast-malt agar, oatmeal agar, salts-starch agar, and glycerol-asparagine agar.

Color in medium: two of three observers report dark pigments in peptone-yeast-iron agar. No melanoid pigments are seen in tyrosine agar or tryptone-yeast broth. Yellow or pale greenish yellow pigment is sometimes found in the medium in oatmeal agar, salts-starch agar, or glycerol-asparagine agar, or pigment may be absent in these media.

D-Glucose, L-arabinose, D-xylose, iso-inositol, D-mannitol, D-fructose, and raffinose are utilized for growth. Utilization of rhamnose is doubtful. No growth or only traces of growth with sucrose.

Type strain shows the highest sequence similarity to: *S. fulvissimus*, AB184434, 99.7%; *S. alboflavus*, EF178699, 99%.

Source: not known.

DNA G+C content (mol%): not known.

Type strain: SA-IX-3, ATCC 27430, CBS 411.59, CBS 672.72, DSM 40366, NBRC 13371, JCM 4753, NRRL B-12307, NRRL-ISP 5366, RIA 1332, VKM Ac-1179.

Sequence accession no. (16S rRNA gene): AB184359.

179. **Streptomyces flavofuscus** (Kudrina *in* Gauze, Preobrazhenskaya, Kudrina, Blinov, Ryabova and Sveshnikova 1957) Preobrazhenskaya 1986, 574[VP] (Effective publication: Preobrazhenskaya *in* Gause, Preobrazhenskaya, Sveshnikova, Terekhova and Maximova 1983.) ("*Actinomyces globisporus* var. *flavofuscus*" Kudrina *in* Gauze, Preobrazhenskaya, Kudrina, Blinov, Ryabova and Sveshnikova 1957, 81; *Streptomyces globisporus* subsp. *flavofuscus* Pridham, Hesseltine and Benedict 1958, 59)

fla.vo.fus'cus. L. adj. *flavus* yellow; L. adj. *fuscus* dark, tawny; N.L. masc. adj. *flavofuscus* yellow, tawny.

Spore chains are straight (*Rectiflexibiles*); spores are smooth. On mineral agar 1, glycerol-nitrate agar: aerial mycelium is mealy, yellow, green-yellow; substrate mycelium and diffusible pigment are gray-brownish, gray-brown. On oatmeal agar: aerial mycelium is yellow; substrate mycelium is colorless; no diffusible pigment. On synthetic agar Korenjako (Kuchaeva et al., 1960): aerial mycelium is yellow; substrate mycelium is dark green; no diffusible pigment. On organic agar 2: aerial mycelium is yellow; substrate mycelium and diffusible pigment are yellow-gray-brown, gray-brown. Melanoid pigments are not formed. Antibiotic: not produced.

Type strain shows the highest sequence similarity to: *S. rubiginosohelvolus*, AB184240, 100%; *S. praecox*, AB184293, 100%; *S. griseinus*, AB184205, 100%; *S. cavourensis* subsp. *washingtonensis*, DQ026671, 100%; *S. mediolani*, AB184674, 100%; *S. badius*, AY999783, 100%; *S. acrimycini*, AY999889, 100%; *S. griseoplanus*, AY999894, 100%; *S. anulatus*, DQ026637, 100%; *S. flavofuscus*, AB249935, 100%; *S. sindenensis*, AB184759, 100%; *S. fimicarius*, AY999784, 100%; *S. pluricolorescens*, DQ442540, 100%; *S. lavendulae* subsp. *lavendulae*, AB184080, 100%; *S. microflavus*, DQ445795, 99.9%; *S. alboviridis*, AB184256, 99.9%; *S. cyaneofuscatus*, AB184860, 99.9%; *S. globisporus* subsp. *globisporus*, EF178686, 99.9%; *S. baarnensis*, EF178688, 99.9%; *S. lipmanii*, AB184148, 99.9%; *S. fulvorobeus*, AB184711, 99.9%; *S. cinereorectus*, AB184646, 99.9%; *S. albovinaceus*, AB249958, 99.9%; *S. parvus*, DQ442537, 99.8%; *S. californicus*, AB184755, 99.8%; *S. argenteolus*, AB045872, 99.8%; *S. luridiscabiei*, AF361784, 99.7%; *S. halstedii*, EF178695, 99.7%; *S. flavogriseus*, AJ494864, 99.7%; *S. griseus* subsp. *griseus*, AY207604, 99.7%; *S. floridae*, AB184656, 99.7%; *S. flavovirens*, DQ026635, 99.7%; *S. griseolus*, AB184768, 99.7%; *S. bacillaris*, AB184439, 99.5%; *S. olivoviridis*, AB184227, 99.5%; *S. atroolivaceus*, AJ781320, 99.5%; *S. clavifer*, DQ026670, 99.5%; *S. finlayi*, AY999788, 99.5%; *S. pulveraceus*, AB184806, 99.5%; *S. nitrosporeus*, EF178680, 99.5%; *S. yanii*, AB006159, 99.4%; *S. atratus*, DQ026638, 99.3%; *S. gelaticus*, DQ026636, 99.3%; *S. griseobrunneus*, AB249912, 99.3%; *S. celluloflavus*, AB184476, 99.3%; *S. albolongus*, AB184425, 99.3%; *S. mutomycini*, AB249951, 99.3%; *S. sanglieri*, AB249945, 99.3%; *S. cavourensis* subsp. *cavourensis*, DQ445791, 99.2%; *S. spiroverticillatus*, AB184814, 99.2%; *S. candidus*, DQ026663, 99.1%; *S. mauvecolor*, AB184532, 99%; *S. cremeus*, AB184124, 99%.

Source: not known.

DNA G+C content (mol%): not known.

Type strain: ATCC 19908, CBS 121.60, DSM 41426, INA 1565/53, JCM 9766, KCTC 9737, NBRC 100768, NRRL B-2594, RIA 310, VKM Ac-1841.

Sequence accession no. (16S rRNA gene): AB249935.

180. **Streptomyces flavogriseus** (Duché 1934) Waksman *in* Waksman and Lechevalier 1953, 55[AL] ("*Actinomyces flavogriseus*" Duché 1934, 341)

fla.vo.gri'se.us. L. adj. *flavus* yellow; N.L. adj. *griseus* gray; N.L. masc. adj. *flavogriseus* yellowish gray.

Spore chains in Section *Rectiflexibiles*. Mature spore chains have 3–10, or often more than 10 spores per chain; long chains are not common. This morphology is seen on yeast-malt agar, oatmeal agar, salts-starch agar, and glycerol-asparagine agar. Spore surface is smooth.

Color of colony: aerial mass color in the Gray color series on yeast-malt agar, oatmeal agar, salts-starch agar, and glycerol-asparagine agar. Reverse side of colony is strong yellow or orange-yellow on yeast-malt agar, grayish yellow to olive brown on oatmeal agar, greenish yellow on salts-starch agar; reverse mycelium pigment is not a pH indicator.

Color in medium: melanoid pigments are not formed in peptone-yeast-iron agar, tyrosine agar, or tryptone-yeast broth. No pigment, or only a trace of yellow, is found in the medium in yeast-malt agar, oatmeal agar, salts-starch agar, and glycerol-asparagine agar.

D-Glucose, L-arabinose, D-xylose, D-mannitol, D-fructose, and rhamnose are utilized for growth. No growth or only traces of growth with iso-inositol, sucrose, and raffinose.

Type strain shows the highest sequence similarity to: *S. flavovirens*, DQ026635, 100%; *S. nitrosporeus*, EF178680, 99.7%; *S. fimicarius*, AY999784, 99.7%; *S. praecox*, AB184293, 99.7%; *S. cinereorectus*, AB184646, 99.7%; *S. griseolus*, AB184768, 99.7%; *S. flavofuscus*, AB249935, 99.7%; *S. anulatus*, DQ026637, 99.7%; *S. halstedii*, EF178695, 99.6%; *S. pluricolorescens*, DQ442540, 99.6%; *S. sindenensis*, AB184759, 99.6%; *S. mediolani*, AB184674, 99.7%; *S. lavendulae* subsp. *lavendulae*, AB184080, 99.6%; *S. albovinaceus*, AB249958, 99.6%; *S. rubiginosohelvolus*, AB184240, 99.6%; *S. cavourensis* subsp. *washingtonensis*, DQ026671, 99.6%; *S. griseinus*, AB184205, 99.6%; *S. badius*, AY999783, 99.6%; *S. argenteolus*, AB045872, 99.6%; *S. acrimycini*, AY999889, 99.6%; *S. griseoplanus*, AY999894, 99.6%; *S. luridiscabiei*, AF361784, 99.5%; *S. microflavus*, DQ445795, 99.5%; *S. cyaneofuscatus*, AB184860, 99.5%; *S. parvus*, DQ442537, 99.5%; *S. californicus*, AB184755, 99.5%; *S. lipmanii*, AB184148, 99.5%; *S. fulvorobeus*, AB184711, 99.5%; *S. alboviridis*, AB184256, 99.5%; *S. floridae*, AB184656, 99.4%; *S. globisporus* subsp. *globisporus*, EF178686, 99.4%; *S. griseus* subsp. *griseus*, AY207604, 99.4%; *S. baarnensis*, EF178688, 99.4%; *S. pulveraceus*, AB184806, 99.3%; *S. bacillaris*, AB184439, 99.3%; *S. olivoviridis*, AB184227, 99.2%; *S. atroolivaceus*, AJ781320, 99.2%; *S. finlayi*, AY999788, 99.2%; *S. yanii*, AB006159, 99.2%; *S. celluloflavus*, AB184476, 99.1%; *S. griseobrunneus*, AB249912, 99.1%; *S. gelaticus*, DQ026636, 99.1%; *S. spiroverticillatus*, AB184814, 99.1%; *S. sanglieri*, AB249945, 99.1%; *S. clavifer*, DQ026670, 99.1%; *S. albolongus*, AB184425, 99.1%; *S. atratus*, DQ026638, 99.1%; *S. cremeus*, AB184124, 99%.

Source: not known.

DNA G+C content (mol%): not known.

Type strain: ATCC 25452, CBS 101.34, CBS 685.69, BCRC 13440, CECT 3327, DSM 40323, NBRC 13040, IMET 43576, JCM 4475, KCTC 9778, NRRL B-1671, NRRL-ISP 5323, RIA 1232, VKM Ac-1325.

Sequence accession no. (16S rRNA gene): AJ494864.

181. **Streptomyces flavopersicus** (Oliver, Goldstein, Bower, Holper and Otto 1961) Witt and Stackebrandt 1991, 456[VP] (Effective publication: Witt and Stackebrandt 1990, 370.) ("*Streptomyces flavopersicus*" Oliver, Goldstein, Bower, Holper and Otto 1961, 495; *Streptoverticillium flavopersicum* Locci, Baldacci and Petrolini Baldan 1969, 41)

fla.vo.per′si.cus. L. adj. *flavus* yellow; L. n. *persicus* peach; N.L. masc. adj. *flavopersicus* of yellow peach.

Spore chains in Section *Verticillati*. Umbellate monoverticillate (biverticillate); individual spore chains may form hooks, loops, or primitive spirals. Mature spore chains generally have 10–50 spores per chain; but shorter chains of 3–10 spores are also found. This morphology is seen on yeast-malt agar, oatmeal agar, and salts-starch agar. Spore surface is smooth.

Color of colony: aerial mass color in the Red color series on yeast-malt agar, oatmeal agar, and salts-starch agar. Reverse side of colony with no distinctive pigments on yeast-malt agar, oatmeal agar, salts-starch agar, or glycerol-asparagine agar; substrate pigment is not a pH indicator.

Color in medium: melanoid pigments formed weakly after prolonged cultivation in peptone-yeast-iron agar. No pigment found in medium, or very faint color in yeast-malt agar, oatmeal agar, salts-starch agar, and glycerol-asparagine agar.

D-Glucose and iso-inositol are utilized for growth. No growth or only trace of growth on L-arabinose, sucrose, D-xylose, D-mannitol, rhamnose, and raffinose. Utilization of D-fructose is doubtful.

For sequence similarity, see type strain of *Streptomyces netropsis*.

Source: not known.

DNA G+C content (mol%): not known.

Type strain: ATCC 19756, JCM 4307, JCM 4370, NBRC 12769, NRRL 2820, UNIQEM 142.

Sequence accession no. (16S rRNA gene): AB249911.

Further comments: in violation of Rule 33c of the *Bacteriological Code* (1990 Revision), in Validation List no. 38, *Streptomyces flavopersicus* is proposed as a *nomen revictum* (basonym: "*Streptomyces flavopersicus*" Oliver et al. 1961).

According to Labeda (1996) and Hatano et al. (2003), *Streptomyces flavopersicus* (Oliver et al. 1961) Witt and Stackebrandt 1991 is a later heterotypic synonym of *Streptomyces netropsis* (Finlay et al. 1951) Witt and Stackebrandt 1991.

182. **Streptomyces flavotricini** (Preobrazhenskaya and Sveshnikova *in* Gauze, Preobrazhenskaya, Kudrina, Blinov, Ryabova and Sveshnikova 1957) Pridham, Hesseltine and Benedict 1958, 60[AL] ("*Actinomyces flavotricini*" Preobrazhenskaya and Sveshnikova *in* Gauze, Preobrazhenskaya, Kudrina, Blinov, Ryabova and Sveshnikova 1957, 49)

fla.vo.tri.ci′ni. L. adj. *flavus* yellow; Gr. n. *thrix* the hair; N.L. *flavotricini* (*sic*) of yellow hair, probably referring to yellow diffusible pigment and formation of a streptothricin-like antibiotic.

Spore chains in Section *Rectiflexibiles*. Mature spore chains are generally long, often with more than 50 spores per chain. This morphology is seen on yeast-malt agar, oatmeal agar, and salts-starch agar, but not on glycerol-asparagine agar. Spore surface is smooth.

Color of colony: aerial mass color in the Red color series on yeast-malt agar, oatmeal agar, and salts-starch agar. Reverse side of colony with no distinctive pigments on yeast-malt agar, oatmeal agar, salts-starch agar, and glycerol-asparagine agar; substrate pigment is not a pH indicator.

Color in medium: melanoid pigments formed in peptone-yeast-iron agar and tryptone-yeast broth. Pigments other than melanoids not formed in yeast-malt agar, oatmeal agar, salts-starch agar, or glycerol-asparagine agar.

D-Glucose is utilized for growth; doubtful utilization of D-fructose. No growth or only trace of growth on

L-arabinose, sucrose, D-xylose, iso-inositol, D-mannitol, rhamnose, and raffinose.

Type strain shows the highest sequence similarity to: *S. polychromogenes*, AB184292, 99.7%; *S. racemochromogenes*, DQ026656, 99.7%; *S. globosus*, AJ781330, 99.4%; *S. toxytricini*, DQ442548, 99.3%; *S. katrae*, AB184409, 99.2%; *S. cinnamonensis*, AB184707, 99.2%; *S. virginiae*, D85119, 99%; *S. tanashiensis*, AJ781362, 99%.

Source: not known.

DNA G+C content (mol%): not known.

Type strain: ATCC 19757, ATCC 23621, CBS 495.68, BCRC 13762, DSM 40152, NBRC 12770, IMET 42057, INA 11669/58, JCM 4371, NRRL B-5419, NRRL-ISP 5152, RIA 1037, UNIQEM 143, VKM Ac-1277.

Sequence accession no. (16S rRNA gene): AB184132.

183. **Streptomyces flavovariabilis** (*ex* Korenyako and Nikitina *in* Krasil'nikov 1965) Sveshnikova 1986, 574[VP] (Effective publication: Sveshnikova *in* Gause, Preobrazhenskaya, Sveshnikova, Terekhova and Maximova 1983.) ("*Actinomyces flavovariabilis*" Korenyako and Nikitina *in* Krasil'nikov 1965; "*Streptomyces flavovariabilis*" Pridham 1970)

fla.vo.va.ri.a'bi.lis L. adj. *flavus* yellow; L. adj. *variabilis* variable; N.L. masc. adj. *flavovariabilis* yellow, variable.

Spore chains are spiral (*Spirales*); spore surface is spiny or smooth. On mineral agar 1, oatmeal agar: aerial mycelium is whitish, creamy to pale pink; substrate mycelium is grayish green or yellowish green; diffusible pigment weak or not extant. On starch-ammonia agar: aerial mycelium is poor, whitish; substrate mycelium is pale olive; no diffusible pigment. On glycerol-nitrate agar: aerial mycelium is whitish to creamy or grayish; substrate mycelium is greenish gray; diffusible pigment not extant or weak, greenish-grayish. On glycerol-asparagine agar: aerial mycelium is poor, white; substrate mycelium is colorless to pale olive; no diffusible pigment. On organic agar 2: aerial mycelium is pale to creamy; substrate mycelium is dark olive or brown; diffusible pigment is brown. Melanoid pigment is formed.

Good growth on glucose, sucrose, arabinose, rhamnose, xylose, fructose, raffinose, and mannitol. Color of substrate mycelium changes to red under acidic conditions. Blue pigment is very unstable and degenerated quickly, forming dirty yellow, dirty green and other similar pigments which were also found on mycelia of *Streptomyces iakyrus* (Blinov et al., 1975; Machenko et al., 1970). Exhibits anti-bacterial and anti-tumor activity; grows on Czapek's solution agar.

Type strain shows the highest sequence similarity to: *S. pseudovenezuelae*, AB184233, 99.2%; *S. novaecaesareae*, AB184357, 99.1%; *S. cacaoi* subsp. *asoensis*, DQ026644, 99.1%.

Source: not known.

DNA G+C content (mol%): not known.

Type strain: ATCC 43684, DSM 41479, INMI 702, JCM 9089, NBRC 100764, NRRL B-16367, VKM Ac-141.

Sequence accession no. (16S rRNA gene): EF178691.

184. **Streptomyces flavovirens** (Waksman *in* Bergey, Harrison, Breed, Hammer and Huntoon 1923) Waksman and Henrici *in* Breed, Murray and Hitchens 1948, 940[AL] ("*Actinomyces flavovirens*" Waksman *in* Bergey, Harrison, Breed, Hammer and Huntoon 1923, 352)

fla.vo.vi'rens. L. adj. *flavus* golden yellow, reddish yellow; L. v. *vireo* to be green; N.L. part adj. *flavovirens* yellow and becoming green.

Spore chains in Section *Rectiflexibiles*. Mature spore chains are generally long, often with more than 50 spores per chain. This morphology is seen on yeast-malt agar, oatmeal agar, salts-starch agar, and glycerol-asparagine agar. Spore surface is smooth.

Color of colony: aerial mass color in the Gray color series on yeast-malt agar, oatmeal agar, and salts-starch agar. Reverse side of colony is colorless or characteristic grayed yellow modified by green on yeast-malt agar, oatmeal agar, salts-starch agar, and glycerol-asparagine agar; substrate pigment is not a pH indicator.

Color in medium: melanoid pigments not formed in peptone-yeast-iron agar or tyrosine agar. Pigments other than melanoids not formed, or some yellow pigment may be produced on oatmeal agar and glycerol-asparagine agar. This pigment is not pH-sensitive when tested with 0.05 M NaOH or HCl.

D-Glucose, L-arabinose, D-xylose, D-mannitol, and rhamnose are utilized for growth. No growth on sucrose, iso-inositol, or raffinose; trace of growth on D-fructose.

Type strain shows the highest sequence similarity to: *S. flavogriseus*, AJ494864, 100%; *S. sindenensis*, AB184759, 99.7%; *S. pluricolorescens*, DQ442540, 99.7%; *S. fimicarius*, AY999784, 99.7%; *S. cinereorectus*, AB184646, 99.7%; *S. griseinus*, AB184205, 99.7%; *S. badius*, AY999783, 99.7%; *S. halstedii*, EF178695, 99.7%; *S. griseolus*, AB184768, 99.7%; *S. argenteolus*, AB045872, 99.7%; *S. praecox*, AB184293, 99.7%; *S. flavofuscus*, AB249935, 99.7%; *S. nitrosporeus*, EF178680, 99.7%; *S. mediolani*, AB184674, 99.7%; *S. cavourensis* subsp. *washingtonensis*, DQ026671, 99.7%; *S. griseoplanus*, AY999894, 99.7%; *S. rubiginosohelvolus*, AB184240, 99.7%; *S. lavendulae* subsp. *lavendulae*, AB184080, 99.7%; *S. anulatus*, DQ026637, 99.7%; *S. acrimycini*, AY999889, 99.6%; *S. globisporus* subsp. *globisporus*, EF178686, 99.6%; *S. albovinaceus*, AB249958, 99.6%; *S. cyaneofuscatus*, AB184860, 99.6%; *S. fulvorobeus*, AB184711, 99.6%; *S. fulvorobeus*, AB184711, 99.6%; *S. alboviridis*, AB184256, 99.9%; *S. lipmanii*, AB184148, 99.6%; *S. luridiscabiei*, AF361784, 99.5%; *S. baarnensis*, EF178688, 99.5%; *S. griseus* subsp. *griseus*, AY207604, 99.5%; *S. microflavus*, DQ445795, 99.5%; *S. floridae*, AB184656, 99.5%; *S. californicus*, AB184755, 99.5%; *S. parvus*, DQ442537, 99.5%; *S. pulveraceus*, AB184806, 99.4%; *S. bacillaris*, AB184439, 99.4%; *S. yanii*, AB006159, 99.3%; *S. olivoviridis*, AB184227, 99.3%; *S. atroolivaceus*, AJ781320, 99.3%; *S. finlayi*, AY999788, 99.3%; *S. sanglieri*, AB249945, 99.2%; *S. albolongus*, AB184425, 99.2%; *S. spiroverticillatus*, AB184814, 99.2%; *S. griseobrunneus*, AB249912, 99.2%; *S. clavifer*, DQ026670, 99.2%; *S. celluloflavus*, AB184476, 99.2%; *S. atratus*, DQ026638, 99.1%; *S. gelaticus*, DQ026636, 99.1%; *S. cavourensis* subsp. *cavourensis*, DQ445791, 99%; *S. mutomycini*, AB249951, 99%; *S. cremeus*, AB184124, 99%.

Source: not known.

DNA G+C content (mol%): not known.

Type strain: AS 4.575, ATCC 19758, ATCC 3320, CBS 129.20, CBS 279.30, CBS 496.68, CCM 3243, BCRC 13689, DSM 40062, HAMBI 1007, HUT 6019, HUT 6053, NBRC 12771, NBRC 3412, NBRC 3716, IMET 40280, JCM 4035, JCM 4578, LMG 20516, NRRL B-1329, NRRL B-2685, NRRL-ISP 5062, RIA 1038, RIA 635, UNIQEM 144, VKM Ac-1723.

Sequence accession no. (16S rRNA gene): DQ026635.

Further comments: according to Lanoot et al. (2005b), *Streptomyces flavovirens* (Waksman 1923) Waksman and Henrici 1948 emend. Lanoot et al. 2005b is an earlier heterotypic synonym of *Streptomyces nigrifaciens* Waksman (1961).

185. **Streptomyces flavoviridis** (*ex* Preobrazhenskaya *in* Gauze, Preobrazhenskaya, Kudrina, Blinov, Ryabova and Sveshnikova 1957) Preobrazhenskaya 1986, 574[VP] (Effective publication: Preobrazhenskaya *in* Gause, Preobrazhenskaya, Sveshnikova, Terekhova and Maximova 1983.) ("*Actinomyces flavoviridis*" Preobrazhenskaya *in* Gauze, Preobrazhenskaya, Kudrina, Blinov, Ryabova and Sveshnikova 1957)

fla.vo.vi'ri.dis. L. adj. *flavus* yellow; L. adj. *viridis* green; N.L. masc. adj. *flavoviridis* yellow-green.

Spore chains in Section *Spirales* (young spore chains may resemble *Retinaculiaperti*). Mature spore chains generally have 10–50 spores per chain; longer chains are sometimes observed. This morphology is found on yeast-malt agar, oatmeal agar, salts-starch agar, and glycerol-asparagine agar. Spore surface is hairy.

Color of colony: aerial mass color in the Gray or Green series on yeast-malt agar, oatmeal agar, salts-starch agar, and glycerol-asparagine agar; color tabs selected by observers are all light olive gray or light olive brown. Reverse side of colony with no distinctive pigments (grayed yellow to yellow-brown) on yeast-malt agar, oatmeal agar, salts-starch agar, and glycerol-asparagine agar; substrate pigment is not a pH indicator.

Color in medium: melanoid pigments formed in peptone-yeast-iron agar and tryptone-yeast broth. No other pigment in medium in yeast-malt agar, oatmeal agar, salts-starch agar, or glycerol-asparagine agar.

D-Glucose, L-arabinose, D-xylose, iso-inositol, D-mannitol, D-fructose, and rhamnose are utilized for growth. No growth or only traces of growth on sucrose or raffinose.

Type strain shows the highest sequence similarity to: *S. pilosus*, AB184161, 100%; *S. violaceus*, AB184315, 99.3%; *S. albosporeus* subsp. *albosporeus*, AJ781327, 99.3%; *S. janthinus*, AB184851, 99.3%; *S. lomondensis*, AB184673, 99.2%; *S. geysiriensis*, DQ442501, 99.2%; *S. minutiscleroticus*, EF178696, 99.2%; *S. roseoviolaceus*, AJ399484, 99.2%; *S. ghanaensis*, AY999851, 99.2%; *S. luteogriseus*, AB184379, 99.2%; *S. djakartensis*, AB184657, 99.1%; *S. plicatus*, AB184291, 99.1%; *S. rochei*, AB184237, 99.1%; *S. vinaceusdrappus*, AY999929, 99.1%; *S. mutabilis*, EF178679, 99.1%; *S. hawaiiensis*, AB184143, 99.1%; *S. atrovirens*, DQ026672, 99.1%; *S. arenae*, AB249977, 99%; *S. calvus*, AB184329, 99%; *S. tuirus*, AB184690, 99%.

Source: not known.

DNA G+C content (mol%): not known.

Type strain: ATCC 19759, ATCC 19903, CBS 497.68, BCRC 11474, DSM 40153, NBRC 12772, IMET 42058, INA 2314/53, JCM 4372, NRRL-ISP 5153, RIA 1039, UNIQEM 145, VKM Ac-754.

Sequence accession no. (16S rRNA gene): AB184842.

186. **Streptomyces flocculus** (Duché 1934) Waksman and Henrici *in* Breed, Murray and Hitchens 1948, 955[AL] ("*Actinomyces flocculus*" Duché 1934, 300)

floc'cu.lus. L. n. *floccus* a flock of wool; N.L. dim. masc. adj. *flocculus* like a small flock of wool.

Spore chains in Section *Spirales.* Spiral spore chains are usually formed on oatmeal agar; spirals of two or more turns may be formed or short chains of only 3–10 spores may form loops, partial spirals, or hooks. Sporulating aerial mycelium is usually poorly developed on yeast-malt agar, salts-starch agar, and glycerol-asparagine agar so that spirals and hooks may be sparse or absent. The weak growth of this strain on most media was noted by Duché in his original description. Spore surface is smooth.

Color of colony: aerial mass color in the White or Yellow color series on yeast-malt agar, oatmeal agar, salts-starch agar, and glycerol-asparagine agar. The most representative color tab in the Yellow color series is 2ba (pale yellow). Reverse side of colony with no distinctive pigments (colorless or pale grayish yellow) on yeast-malt agar, oatmeal agar, salts-starch agar, and glycerol-asparagine agar.

Color in medium: melanoid pigments are not formed in peptone-yeast-iron agar, tyrosine agar, or tryptone-yeast broth. No pigment found in medium in yeast-malt agar, oatmeal agar, salts-starch agar, or glycerol-asparagine agar.

D-Glucose, L-arabinose, D-xylose, iso-inositol, D-mannitol, D-fructose, and sucrose are utilized for growth. Utilization of rhamnose and raffinose is doubtful.

Type strain shows the highest sequence similarity to: *S. rangoonensis*, AB184295, 99.9%; *S. gibsonii*, AB184663, 99.9%; *S. almquistii*, AB184258, 99.9%; *S. albus* subsp. *albus*, AJ621602, 99.8%.

Source: not known.

DNA G+C content (mol%): not known.

Type strain: ATCC 25453, CBS 686.69, BCRC 12068, DSM 40327, HUT 6615, NBRC 13041, IMET 43522, JCM 4476, NRRL 2960, NRRL B-2465, NRRL B-2843, NRRL-ISP 5327, RIA 1233.

Sequence accession no. (16S rRNA gene): AB184272.

187. **Streptomyces floridae** Bartz, Ehrlich, Mold, Penner and Smith 1951, 4[AL]

flo.ri'da.e. N.L. gen. n. *floridae* of Florida, the source of the soil from which the organism was isolated.

Spore chains are flexuous. Forms dull violet to red-brown vegetative mycelium and diffusible pigment on some media. Poor growth on Czapek's solution agar; produces the viomycin complex; inhibited by streptomycin.

Type strain shows the highest sequence similarity to: *S. californicus*, AB184755, 100%; *S. lipmanii*, AB184148, 99.9%; *S. microflavus*, DQ445795, 99.9%; *S. fulvorobeus*, AB184711, 99.9%; *S. alboviridis*, AB184256, 99.9%; *S.*

cavourensis subsp. *washingtonensis*, DQ026671, 99.8%; *S. lavendulae* subsp. *lavendulae*, AB184080, 99.8%; *S. griseinus*, AB184205, 99.8%; *S. badius*, AY999783, 99.8%; *S. pluricolorescens*, DQ442540, 99.8%; *S. sindenensis*, AB184759, 99.8%; *S. mediolani*, AB184674, 99.8%; *S. rubiginosohelvolus*, AB184240, 99.8%; *S. praecox*, AB184293, 99.7%; *S. flavofuscus*, AB249935, 99.7%; *S. griseoplanus*, AY999894, 99.7%; *S. albovinaceus*, AB249958, 99.7%; *S. fimicarius*, AY999784, 99.7%; *S. globisporus* subsp. *globisporus*, EF178686, 99.7%; *S. anulatus*, DQ026637, 99.7%; *S. cinereorectus*, AB184646, 99.7%; *S. argenteolus*, AB045872, 99.7%; *S. luridiscabiei*, AF361784, 99.7%; *S. cyaneofuscatus*, AB184860, 99.7%; *S. parvus*, DQ442537, 99.7%; *S. griseolus*, AB184768, 99.6%; *S. griseus* subsp. *griseus*, AY207604, 99.6%; *S. baarnensis*, EF178688, 99.6%; *S. acrimycini*, AY999889, 99.6%; *S. finlayi*, AY999788, 99.5%; *S. bacillaris*, AB184439, 99.5%; *S. halstedii*, EF178695, 99.5%; *S. albolongus*, AB184425, 99.5%; *S. flavovirens*, DQ026635, 99.5%; *S. griseobrunneus*, AB249912, 99.4%; *S. celluloflavus*, AB184476, 99.4%; *S. pulveraceus*, AB184806, 99.4%; *S. flavogriseus*, AJ494864, 99.4%; *S. cavourensis* subsp. *cavourensis*, DQ445791, 99.3%; *S. nitrosporeus*, EF178680, 99.3%; *S. clavifer*, DQ026670, 99.3%; *S. atroolivaceus*, AJ781320, 99.3%; *S. candidus*, DQ026663, 99.3%; *S. olivoviridis*, AB184227, 99.3%; *S. sanglieri*, AB249945, 99.2%; *S. yanii*, AB006159, 99.2%; *S. mutomycini*, AB249951, 99.1%; *S. gelaticus*, DQ026636, 99.1%; *S. atratus*, DQ026638, 99.1%; *S. spiroverticillatus*, AB184814, 99.1%; *S. cremeus*, AB184124, 99.1%.

Source: isolated from soil from Florida.

DNA G+C content (mol%): not known.

Type strain: DSM 40938, NBRC 15405, JCM 5068, NCIMB 12830, NRRL 2423.

Sequence accession no. (16S rRNA gene): AB184656.

188. **Streptomyces fluorescens** (Krasil'nikov 1958) Pridham 1970, 15[AL] ("*Actinomyces fluorescens*" Krasil'nikov 1958, 258)

flu.o.res'cens. N.L. v. *fluoresco* to fluoresce; N.L. part. adj. *fluorescens* fluorescing, referring to the yellow fluorescent pigment produced by the organism.

Spore chains in Section *Rectiflexibiles*. Mature spore chains generally have 10–50 spores per chain. This morphology is seen on yeast-malt agar, oatmeal agar, salts-starch agar, and glycerol-asparagine agar. Spore surface is smooth.

Color of colony: aerial mass color in the Yellow color series on yeast-malt agar, oatmeal agar, and salts-starch agar; White or Yellow color series on glycerol-asparagine agar. Reverse side of colony with no distinctive pigments (colorless to grayish yellow on oatmeal agar, salts-starch agar, and glycerol-asparagine agar; grayed orange-yellow to olive brown on yeast-malt agar). Reverse pigment is not pH-sensitive or is changed only slightly from yellow-brown to yellow-green, with addition of 0.05 M NaOH.

Color in medium: melanoid pigments are not produced in peptone-yeast-iron agar, tyrosine agar, or tryptone-yeast broth. A small amount of yellow pigment may be found in the medium in yeast-malt agar, oatmeal agar, salts-starch agar, and tryptone-yeast broth. This pigment, when present, is only slightly pH-sensitive, changing from yellow-brown to yellow-green with addition of 0.05 NaOH.

D-Glucose, L-arabinose, D-xylose, D-mannitol, D-fructose, and rhamnose are utilized for growth. No growth or only traces of growth on sucrose, iso-inositol, and raffinose. The original description for this species states that rhamnose is not utilized, noting that this is one difference between *A. fluorescens* and other species in the fluorescent group. However, all of the three ISP observers recorded good growth on rhamnose, equivalent to growth on D-glucose.

For sequence similarity, see type strain of *Streptomyces anulatus.*

Source: not known.

DNA G+C content (mol%): not known.

Type strain: ATCC 15860, ATCC 23907, CBS 803.68, BCRC 11475, CECT 3130, DSM 40203, NBRC 12861, INMI 592, JCM 4373, LMG 8579, NCIMB 9851, NRRL B-2873, NRRL-ISP 5203, RIA 1154, RIA 647, VKM Ac-147.

Sequence accession no. (16S rRNA gene): AB184199.

Further comments: according to Lanoot et al. (2005b), *Streptomyces fluorescens* (Krasil'nikov 1958) Pridham 1970 is a later heterotypic synonym of *Streptomyces anulatus* (Beijerinck 1912) Waksman 1953 emend. Lanoot et al. 2005b.

189. **Streptomyces fradiae** (Waksman and Curtis 1916) Waksman and Henrici *in* Breed, Murray and Hitchens 1948, 954[AL] emend. Lanoot, Vancanneyt, Dawyndt, Cnockaert, Zhang, Huang, Liu and Swings 2005a, 8 (Effective publication: Lanoot, Vancanneyt, Dawyndt, Cnockaert, Zhang, Huang, Liu and Swings 2004, 88.) ["*Actinomyces fradii*" (*sic*) Waksman and Curtis 1916, 125; "*Streptomyces fradii*" (*sic*) Waksman and Henrici *in* Breed, Murray and Hitchens 1948, 954; "*Streptomyces fradiae* subsp. *fradiae*" Waksman and Curtis *in* Pridham, Lyons and Seckinger 1965, 222]

fra.di'a.e. N.L. gen. n. *fradiae* of Fradia, a patronymic.

Spore chains in Section *Retinaculiaperti* with characteristic range from straight to spiral spore chains. Straight or flexuous spore chains are most common on yeast-malt agar; *Retinaculiaperti* morphology, including open spirals, best developed on oatmeal agar and salts-starch agar. Mature spore chains generally have 10–50 spores per chain; longer chains are sometimes observed. This morphology is seen on yeast-malt agar, oatmeal agar, salts-starch agar, and glycerol-asparagine agar. Spore surface is smooth.

Color of colony: aerial mass color in the Red color series on yeast-malt agar, oatmeal agar, salts-starch agar, and glycerol-asparagine agar. Reverse side of colony with no distinctive pigments on yeast-malt agar, oatmeal agar, salts-starch agar, and glycerol-asparagine agar; substrate pigment is not a pH indicator.

Color in medium: melanoid pigments not formed in peptone-yeast-iron agar and tyrosine agar. No pigment found in medium in yeast-malt agar, oatmeal agar, salts-starch agar, or glycerol-asparagine agar.

D-Glucose, L-arabinose, and D-fructose are utilized for growth. No growth or only traces of growth on sucrose, iso-inositol, D-mannitol, rhamnose, and raffinose. Variable reports on growth with D-xylose.

Type strain shows the highest sequence similarity to: *S. coeruleoprunus*, AB184651, 99.2%.

Source: not known.

DNA G+C content (mol%): not known.

Type strain: ATCC 10745, ATCC 19760, CBS 498.68, CCM 3174, BCRC 12196, CECT 3197, DSM 40063, HAMBI 965, HUT 6095, IFM 1030, NBRC 12773, NBRC 3718, IMET 42051, IMI 061202, JCM 4133, JCM 4579, KCTC 9760, NCIMB 11005, NCIMB 8233, NRRL B-1195, NRRL-ISP 5063, PCM 2330, RIA 1040, RIA 97, UNIQEM 146, VKM Ac-150, VKM Ac-151, VKM Ac-152, VKM Ac-764.

Sequence accession no. (16S rRNA gene): DQ026630.

Further comments: according to Lanoot et al. (2004), *Streptomyces fradiae* (Waksman and Curtis 1916) Waksman and Henrici 1948 is an earlier heterotypic synonym of *Streptomyces roseoflavus* Arai 1951.

190. **Streptomyces fragilis** Anderson, Ehrlich, Sun and Burkholder 1956, 105[AL]

fra'gi.lis. L. masc. adj. *fragilis* fragile.

Spore chains of atypical *Retinaculiaperti* type. Poor growth on Czapek's solution agar; produces *O*-diazoacetyl-L-serine (azaserine), an anti-bacterial, anti-fungal, anti-protozoal, and anti-tumor antibiotic; inhibited by streptomycin; NaCl tolerance: >4%, but <7%.

Type strain shows the highest sequence similarity to: *S. coelicoflavus*, AB184650, 99.4%; *S. flaveolus*, AB184764, 99.3%; *S. matensis*, AB184221, 99%; *S. eurythermus*, D63870, 99%.

Source: not known.

DNA G+C content (mol%): not known.

Type strain: ATCC 23908, CBS 804.68, BCRC 13654, DSM 40044, HAMBI 1083, HAMBI 1090, NBRC 12862, IMET 43575, JCM 4187, JCM 4638, NCIMB 9795, NRRL 2424, NRRL-ISP 5044, RIA 1111, VKM Ac-1773.

Sequence accession no. (16S rRNA gene): AY999917.

191. **Streptomyces fulvissimus** (Jensen 1930) Waksman and Henrici *in* Breed, Murray and Hitchens 1948, 946[AL] ("*Actinomyces fulvissimus*" Jensen 1930, 66)

ful.vis'si.mus. L. sup. masc. adj. *fulvissimus* very yellow.

Spore chains in Section *Rectiflexibiles.* Mature spore chains generally contain 10 to 50 or more spores per chain. This morphology is seen on yeast-malt agar, salts-starch agar, and glycerol-asparagine agar, but sporulating aerial mycelium may be poorly developed or absent on oatmeal agar. Spore surface is smooth.

Color of colony: aerial mass color in the Red color series on yeast-malt agar, oatmeal agar, salts-starch agar, and glycerol-asparagine agar; white aerial mycelium may also be present. Nearest matching color tabs in the Red color series are 3ea, light orange yellow; 5ca and 7ca, light yellowish pink; and 4ea, moderate yellowish pink. Reverse side of colony with distinctive reddish orange pigments on yeast-malt agar, oatmeal agar, salts-starch agar, and glycerol-asparagine agar. Reverse mycelium pigment is not a pH indicator.

Color in medium: melanoid pigments are formed in peptone-yeast-iron agar or tryptone-yeast broth, but not in tyrosine agar. No pigment is found in the medium in yeast-malt agar, oatmeal agar, salts-starch agar, or glycerol-asparagine agar.

Carbon utilization: D-glucose, L-arabinose, iso-inositol, D-mannitol, D-fructose, and raffinose are utilized for growth. Reports vary on the utilization of D-xylose, and the utilization of sucrose and rhamnose is doubtful.

Type strain shows the highest sequence similarity to: *S. flavofungini*, AB184359, 99.7%; *S. alboflavus*, EF178699, 99.3%; *S. variegatus*, AJ781371, 99.1%.

Source: not known.

DNA G+C content (mol%): not known.

Type strain: ATCC 27431, CBS 783.72, BCRC 12172, CCUG 11114, CIP 105783, DSM 40593, NBRC 13482, JCM 4129, JCM 4754, KCTC 9779, LMG 19310, NCIMB 10505, NCIMB 9609, NRRL B-1453, NRRL-ISP 5593, RIA 1443, VKM Ac-994.

Sequence accession no. (16S rRNA gene): AB184434.

192. **Streptomyces fulvorobeus** Vinogradova and Preobrazhenskaya 1986, 574[VP] (Effective publication: Vinogradova and Preobrazhenskaya *in* Gause, Preobrazhenskaya, Sveshnikova, Terekhova and Maximova 1983.)

ful.vo.ro'be.us. L. adj. *fulvus* reddish yellow; L. adj. *robeus* reddish brown; N.L. masc. adj. *fulvorobeus* reddish yellow, reddish brown.

Spore chains are spiral (*Spirales*); spores are smooth. On mineral agar 1, glycerol-nitrate agar: aerial mycelium is creamy, yellow, grayish yellow; substrate mycelium is gray-brownish yellow, yellow-gray-brown. On starch ammonia agar, glycerol-asparagine agar, oatmeal agar: aerial mycelium is white-yellow, yellow; substrate mycelium is colorless; no diffusible pigment. On organic agar 2: aerial mycelium is whitish or absent; substrate mycelium is colorless; no diffusible pigment. Melanoid pigments are not formed. Grows on glucose, maltose, sucrose, and arabinose; no growth on mannitol, xylose, inositol, rhamnose, or galactose.

Type strain shows the highest sequence similarity to: *S. cavourensis* subsp. *washingtonensis*, DQ026671, 100%; *S. microflavus*, DQ445795, 100%; *S. lipmanii*, AB184148, 100%; *S. lavendulae* subsp. *lavendulae*, AB184080, 100%; *S. alboviridis*, AB184256, 100%; *S. luridiscabiei*, AF361784, 99.9%; *S. flavofuscus*, AB249935, 99.9%; *S. cinereorectus*, AB184646, 99.9%; *S. fimicarius*, AY999784, 99.9%; *S. griseoplanus*, AY999894, 99.9%; *S. cyaneofuscatus*, AB184860, 99.9%; *S. floridae*, AB184656, 99.9%; *S. praecox*, AB184293, 99.9%; *S. anulatus*, DQ026637, 99.9%; *S. pluricolorescens*, DQ442540, 99.8%; *S. acrimycini*, AY999889, 99.8%; *S. argenteolus*, AB045872, 99.8%; *S. griseinus*, AB184205, 99.8%; *S. badius*, AY999783, 99.8%; *S. rubiginosohelvolus*, AB184240, 99.8%; *S. sindenensis*, AB184759, 99.8%; *S. mediolani*, AB184674, 99.8%; *S. californicus*, AB184755, 99.8%; *S. globisporus* subsp. *globisporus*, EF178686, 99.8%; *S. parvus*, DQ442537, 99.7%; *S. halstedii*, EF178695, 99.7%; *S. baarnensis*, EF178688, 99.7%; *S. griseolus*, AB184768, 99.7%; *S. griseus* subsp. *griseus*, AY207604, 99.7%; *S. albovinaceus*, AB249958, 99.7%; *S. flavovirens*, DQ026635, 99.6%; *S. finlayi*, AY999788, 99.5%; *S. flavogriseus*, AJ494864, 99.5%; *S. bacillaris*, AB184439, 99.5%; *S. yanii*, AB006159, 99.4%; *S. atroolivaceus*, AJ781320, 99.4%; *S. olivoviridis*, AB184227, 99.4%; *S. pulveraceus*, AB184806, 99.4%; *S. celluloflavus*, AB184476, 99.4%; *S. griseobrunneus*, AB249912, 99.4%; *S. albolongus*, AB184425, 99.4%; *S. clavifer*, DQ026670, 99.3%; *S. nitrosporeus*, EF178680, 99.3%; *S. cavourensis* subsp. *cavourensis*, DQ445791, 99.3%; *S. spiroverticillatus*, AB184814, 99.2%; *S. sanglieri*, AB249945, 99.2%; *S. atratus*, DQ026638, 99.1%; *S. mutomycini*, AB249951, 99.1%; *S. candidus*, DQ026663, 99.1%; *S. gelaticus*, DQ026636, 99.1%; *S. cremeus*, AB184124, 99%.

Source: not known.

DNA G+C content (mol%): not known.

Type strain: DSM 41455, NBRC 15897, INMI 34-280, JCM 9090, VKM Ac-158.

Sequence accession no. (16S rRNA gene): AB184711.

193. **Streptomyces fumanus** (Sveshnikova 1957) *in* Gauze, Preobrazhenskaya, Kudrina, Blinov, Ryabova and Sveshnikova 1957 Pridham, Hesseltine and Benedict 1958, 67[AL] ["*Actinomyces fumanus*" Sveshnikova *in* Gauze, Preobrazhenskaya, Kudrina, Blinov, Ryabova and Sveshnikova 1957, 61; "*Actinomyces fumarius*" (*sic*) Danga and Gottlieb 1959, 43]

fu.ma'nus. N.L. masc. adj. *fumanus* smoky, probably referring to the color of the vegetative mycelium of the organism.

Spore chains in Section *Spirales*. Mature spore chains are moderately long with 10 to 50 or more spores per chain. This morphology is found on yeast-malt agar, oatmeal agar, salts-starch agar, and glycerol-asparagine agar. Spore surface is smooth.

Color of colony: aerial mass color in the Red color series on yeast-malt agar and glycerol-asparagine agar; Red or sometimes Yellow color series on oatmeal agar and salts-starch agar. Reverse side of colony is light grayish yellow may or may not change to dark brown on salts-starch agar and glycerol-asparagine agar or to orange-yellow or strong brown on yeast-malt agar and oatmeal agar.

Color in medium: melanoid pigments are not formed in peptone-yeast-iron agar, tyrosine agar, or tryptone-yeast broth. No pigment (or only a trace of yellow or greenish yellow) is found in the medium in yeast-malt agar, oatmeal agar, salts-starch agar, or glycerol-asparagine agar.

D-Glucose, L-arabinose, D-xylose, D-mannitol, D-fructose, rhamnose, and raffinose are utilized for growth. No growth or only traces of growth on iso-inositol and sucrose.

Type strain shows no sequence similarity over 99%.

Source: not known.

DNA G+C content (mol%): not known.

Type strain: ATCC 19904, ATCC 25454, CBS 687.69, BCRC 12058, DSM 40154, NBRC 13042, INA 10256/54, JCM 4477, NRRL B-3898, NRRL B-5420, NRRL-ISP 5154, RIA 1234, VKM Ac-1845.

Sequence accession no. (16S rRNA gene): AB184273.

194. **Streptomyces fumigatiscleroticus** (*ex* Pridham 1970) Goodfellow, Williams and Alderson 1986a, 59[VP] (Effective publication: Goodfellow, Williams and Alderson 1986d, 59.)

fu.mi.ga.ti.scle.ro'ti.cus. L. part. adj. *fumigatus* smoked; N.L. neut. n. *sclerotium* sclerotium; N.L. masc. adj. *fumigatiscleroticus* smoked sclerotium, referring to smoke color and ability to form sclerotia.

Produces spiral spore chains and shows brownish-black crusty growth; extensively branched substrate and aerial mycelium. Produces sclerotia. Melanin pigments are not formed. Degrades gelatin, starch, urea, and xanthine. Nitrate is reduced but hydrogen sulfide is not produced. L-Arabinose, glucose, lactose, maltose, D-mannitol, mannose, sorbitol, sucrose, and D-xylose are used as sole carbon sources but dulcitol, galactose, glycerol, inulin, *myo*-inositol, L-rhamnose, salicin, and sucrose are not. Acid is produced from D-lactose, D-mannitol, melibiose, methyl α-D-glucoside, raffinose, L-rhamnose, D-sorbitol, and D-xylose but not from adonitol, dulcitol, *meso*-erythritol, *myo*-inositol, or sucrose. Grows at 42°C but not at 10°C. Shows antimicrobial activity against *Bacillus subtilis*, *Escherichia coli*, and *Staphylococcus aureus* but not against *Candida albicans*, *Cryptococcus neoformans*, or *Saccharomyces cerevisiae*.

Type strain shows the highest sequence similarity to: *S. poonensis*, DQ445792, 99.3%; *S spiralis*, EF178683, 99%.

Source: not known.

DNA G+C content (mol%): not known.

Type strain: ATCC 19345, CBS 639.66, BCRC 12344, CMI 117720, DSM 43154, NBRC 12999, KCC A-0101, JCM 3101, NCIMB 11004, NRRL B-3856, RIA 884.

Sequence accession no. (16S rRNA gene): DQ442499.

Further comments: according to Goodfellow et al. (1986d), the species *Streptomyces fumigatiscleroticus* (*ex* Pridham 1970) Goodfellow et al. 1986a is a synonym of the species *Chainia fumigata* Thirumalachar et al. 1966. However, according to Rule 51b(2) of the *Bacteriological Code* (1990 Revision) the transfer of *Chainia fumigata* Thirumalachar et al. 1966 in the genus *Streptomyces* Waksman and Henrici 1943 as *Streptomyces fumigatiscleroticus* (*ex* Pridham 1970) Goodfellow et al. 1986a sp. nov., nom. rev. is illegitimate.

195. **Streptomyces galbus** Frommer 1959, 195[AL]

gal'bus. L. masc. adj. *galbus* greenish yellow.

Spore chains in Section *Spirales* or *Retinaculiaperti*. Spirals may be open coils of several turns, tight spirals at the ends of flexuous spore chains or imperfect spirals and loops at the ends of spore chains suggesting *Retinaculiaperti* morphology. Spore chains are moderately long with 10–50, or often more than 50, spores per chain. Spore surface is smooth to warty. Surface irregularities suggesting very small warts are characteristic.

Color of colony: aerial mass color in the Gray color series on yeast-malt agar and oatmeal agar and in the Yellow color series or Gray color series on salts-starch agar and glycerol-asparagine agar. Gray color is represented by tabs 2fe (medium gray) or 3fe (light brownish gray) on all media; and 1ba or 2ba (pale yellow) or 1½db (pale grayish yellow) on salts-starch agar and glycerol-asparagine agar. Reverse side of colony with no distinctive pigments (yellow to yellow-brown or olive brown) on yeast-malt agar, oatmeal agar, salts-starch agar, and glycerol-asparagine agar.

Color in medium: melanoid pigments are produced in peptone-yeast-iron agar and tryptone-yeast broth, but not in tyrosine agar. Yellow soluble pigment is found in the medium in yeast-malt agar, oatmeal agar, salts-starch agar, and glycerol-asparagine agar. This pigment is not a pH indicator.

D-Glucose, L-arabinose, D-xylose, iso-inositol, D-mannitol, and D-fructose are utilized for growth. Only traces of growth comparable to growth on carbon-free control is found on sucrose, rhamnose, and raffinose.

Type strain shows the highest sequence similarity to: *S. longwoodensis*, AB184580, 99.9%; *S. bungoensis*, AB184696, 99.7%; S. *capoamus*, AB045877, 99.5%; *S. corchorusii*, AB184267, 99.3%; *S. olivaceoviridis*, AB184288, 99.2%; *S. canarius*, AB184396, 99.1%.

Source: not known.

DNA G+C content (mol%): not known.

Type strain: ATCC 23910, CBS 831.68, BCRC 12166, DSM 40089, NBRC 12864, IMET 42937, JCM 4222, JCM 4639, NCIMB 13005, NRRL B-2283, NRRL-ISP 5089, RIA 1121, VKM Ac-165.

Sequence accession no. (16S rRNA gene): X79852.

196. **Streptomyces galilaeus** Ettlinger, Corbaz and Hütter 1958a, 356[AL]

ga.li.la'e.us. L. masc. adj. *galilaeus* of or belonging to Galilee, a Province in Palestine, apparently the source of the soil (Newi Yusha, Israel) from which the organism was isolated.

Spore chains in Section *Spirales.* Open spirals with moderately long chains of 10 to 50 or more spores are found on oatmeal agar, salts-starch agar, and glycerol-asparagine agar; shorter flexuous spore chains are usually found on yeast-malt agar. Spore surface is smooth.

Color of colony: aerial mass color in the Gray color series on yeast-malt agar, oatmeal agar, salts-starch agar, and glycerol-asparagine agar. Nearest matching color tabs for 14- to 21-d-old cultures are 7ih, grayish pink, and 7fe, pale purple. Observers do not agree on the nearest matching color tab from the Gray color series for the aerial mass on salts-starch agar or glycerol-asparagine agar. Reverse side of colony with no distinctive pigments (pale or grayish yellow to yellowish brown or olive brown) on yeast-malt agar, oatmeal agar, salts-starch agar, and glycerol-asparagine agar. Two of three observers note that the yellowish color of the reverse side of the mycelium is changed to yellowish pink or faint pink by the addition of 0.05 M NaOH.

Color in medium: melanoid pigments are formed in peptone-yeast-iron agar, tyrosine agar, and tryptone-yeast broth. No pigment or only a trace of yellow is found in the medium in yeast-malt agar, oatmeal agar, salts-starch agar, and glycerol-asparagine agar. When a faint yellow pigment is found, it shows the same color changes noted for the reverse mycelium pigment when 0.05 M NaOH is added.

D-Glucose, L-arabinose, D-xylose, iso-inositol, D-fructose, rhamnose, sucrose, and raffinose are utilized for growth. No growth or only traces of growth with D-mannitol.

Type strain shows the highest sequence similarity to: *S. bobili*, AB249925, 100%; *S. aureocirculatus*, AB184260, 99.5%; *S. phaeoluteigriseus*, AJ391815, 99.5%; *S. resistomycificus*, AB184166, 99.5%; *S. novaecaesareae*, AB184357, 99.4%; *S. chartreusis*, AB184839, 99.3%; *S. pseudovenezuelae*, AB184233, 99.3%; *S. prunicolor*, DQ026659, 99%.

Source: soil from Palestine.

DNA G+C content (mol%): not known.

Type strain: ATCC 14969, CBS 701.72, BCRC 11828, CCT 4839, DSM 40481, NBRC 13400, JCM 4231, JCM 4757, KCTC 1921, NRRL 2722, NRRL-ISP 5481, RIA 1361, VKM Ac-729.

Sequence accession no. (16S rRNA gene): AB045878.

197. **Streptomyces gancidicus** Suzuki 1957, 538[AL]

gan.ci'di.cus. N.L. n. *gancidinum* gancidin name of an antibiotic; L. masc. suff. *-icus* suffix used with the sense of belonging to; N.L. masc. adj. *gancidicus* belonging to gancidin.

Moderate growth on Czapek's solution agar; produces the gancidin complex (components A and W) effective against Gram-stain-positive bacteria and tumors; inhibited by streptomycin.

Type strain shows the highest sequence similarity to: *S. pseudogriseolus*, DQ442541, 100%; *S. capillispiralis*, AB184577, 100%; *S. cellulosae*, DQ442495, 99.9%; *S. azureus*, EF178674, 99.5%; *S. levis*, AB184670, 99.4%; *S. lusitanus*, AB184424, 99.4%; *S. rubiginosus*, AY999810, 99.4%; *S. carpinensis*, AB184574, 99.4%; *S. albaduncus*, AY999757, 99.2%; *S. matensis*, AB184221, 99.2%; *S. griseoloalbus*, AB184275, 99.2%; *S. caelestis*, X80824, 99.2%; *S. djakartensis*, AB184657, 99.1%; *S. tuirus*, AB184690, 99.1%; *S. paradoxus*, AB184628, 99.1%; *S. afghaniensis*, AJ399483, 99%; *S. geysiriensis*, DQ442501, 99%; *S. viridiviolaceus*, AY999854, 99%; *S. africanus*, AY208912, 99%; *S. lavendulocolor*, DQ442516, 99%; *S. minutiscleroticus*, EF178696, 99%; *S. spinoverrucosus*, AB184578, 99%; *S. thermocarboxydus*, U94490, 99%; *S. malachitofuscus*, AB184282, 99%.

Source: not known.

DNA G+C content (mol%): not known.

Type strain: DSM 40935, IFM 1024, NBRC 15412, JCM 4171, NCIMB 12858, NRRL B-1872.

Sequence accession no. (16S rRNA gene): AB184660.

198. **Streptomyces gardneri** (Waksman *in* Waksman, Horning, Welsch and Woodruff 1942) Waksman 1961, 215[AL] ("*Proactinomyces gardneri*" Waksman *in* Waksman, Horning, Welsch and Woodruff 1942, 289; "*Nocardia gardneri*" Waksman and Henrici *in* Breed, Murray and Hitchens 1948, 914)

gard'ne.ri. N.L. gen. masc. n. *gardneri* of Gardner, named for Professor A.D. Gardner, one of the two who first isolated the organism.

Spore chains in Section *Rectiflexibiles.* Spore chains are moderately long, usually 10–50, or often more than 50, spores per chain. This morphology is seen on yeast-malt agar, oatmeal agar, salts-starch agar, and glycerol-asparagine agar, although aerial mycelium may be poorly developed on oatmeal agar and glycerol-asparagine agar. Spore surface is smooth.

Color of colony: aerial mass color in the Gray color series on salts-starch agar in 14–21 d. Mature aerial mycelium may be inadequate for determination of aerial mass color on yeast-malt agar, oatmeal agar, and glycerol-asparagine agar. One observer obtained good sporulating aerial growth in the Gray color series on potato agar (potato, 200 g; agar, 10 g; tap water, 1 l; pH 6.9–7.1). Reverse side of colony with no distinctive pigment (colorless to grayish yellow or grayed yellow-brown) on yeast-malt agar, oatmeal agar, salts-starch agar, glycerol-asparagine agar, and potato agar.

Color in medium: melanoid pigments are produced in peptone-yeast-iron agar and tryptone-yeast broth, but not in tyrosine agar (or only in trace amounts) in 2–4 d. No pigments are found in the medium in yeast-malt agar, oatmeal agar, salts-starch agar, or glycerol-asparagine agar.

D-Glucose, L-arabinose, sucrose, D-xylose, D-fructose, rhamnose, and raffinose are utilized for growth. No growth or only traces of growth on iso-inositol and D-mannitol.

Type strain shows no sequence similarity over 99%.

Source: not known.

DNA G+C content (mol%): not known.

Type strain: ATCC 23911, ATCC 9604, CBS 832.68, BCRC 12346, BCRC 13687, BCRC 13731, DSM 40064, NBRC 12865, NBRC 13974, NBRC 3385, IMET 7182, JCM 3004, JCM 4375, NCTC 6531, NRRL B-5615, NRRL-ISP 5064, RIA 1117, VKM Ac-1829.

Sequence accession no. (16S rRNA gene): AB184757.

199. **Streptomyces gelaticus** (Waksman *in* Bergey, Harrison, Breed, Hammer and Huntoon 1923) Waksman and Henrici *in* Breed, Murray and Hitchens 1948, 979[AL] ("*Actinomyces gelaticus*" Waksman *in* Bergey, Harrison, Breed, Hammer and Huntoon 1923, 356)

ge.la'ti.cus. L. part adj. *gelatus* congealed, jellied; N.L. masc. adj. *gelaticus* intended to mean resembling hardened gelatin.

Spore chains in Section *Rectiflexibiles.* Aerial mycelium is not produced on most media. Absence of aerial mycelium on most media was also noted in the original description (Waksman and Curtis, 1916) which, however, records open spirals on the scant growth on synthetic agar. ISP observers found wavy spore chains, but no spirals in the thin aerial mycelium on yeast-malt agar and glycerol-asparagine agar. The absence of spirals has also been noted by Anderson et al. (1956), by Waksman in a later description (Waksman, 1961), and by Hütter (1967a). Spore surface is smooth.

Color of colony: aerial mycelium is poorly developed on most media. When an adequate spore mass is produced it is in the Gray color series. This was observed on yeast-malt agar and glycerol-asparagine agar. Additional media used unsuccessfully by collaborators in an effort to get good sporulating growth including Bennett's agar, Anderson's agar, blood agar, Lemko agar, and milk agar. Reverse side of colony with no distinctive pigments (colorless to grayish yellow) on yeast-malt agar, oatmeal agar, salts-starch agar, and glycerol-asparagine agar.

Color in medium: melanoid pigments are not produced in peptone-yeast-iron agar, tyrosine agar, or tryptone-yeast broth. No pigment is found in medium in yeast-malt agar, oatmeal agar, salts-starch agar, or glycerol-asparagine agar.

D-Glucose, sucrose, D-xylose, and rhamnose are utilized for growth. No growth or only traces of growth on L-arabinose, iso-inositol, D-mannitol, and raffinose.

Type strain shows the highest sequence similarity to: *S. pulveraceus*, AB184806, 99.7%; *S. atratus*, DQ026638, 99.6%; *S. sanglieri*, AB249945, 99.5%; *S. yanii*, AB006159, 99.4%; *S. globisporus* subsp. *globisporus*, EF178686, 99.3%; *S. sindenensis*, AB184759, 99.3%; *S. pluricolorescens*, DQ442540, 99.3%; *S. albovinaceus*, AB249958, 99.3%; *S. flavofuscus*, AB249935, 99.3%; *S. griseinus*, AB184205, 99.3%; *S. fimicarius*, AY999784, 99.3%; *S. anulatus*, DQ026637, 99.3%; *S. mediolani*, AB184674, 99.3%; *S. praecox*, AB184293, 99.3%; *S. badius*, AY999783, 99.3%; *S. rubiginosohelvolus*, AB184240, 99.3%; *S. lavendulae* subsp. *lavendulae*, AB184080, 99.2%; *S. californicus*, AB184755, 99.2%; *S. parvus*, DQ442537, 99.2%; *S. cavourensis* subsp. *washingtonensis*, DQ026671, 99.2%; *S. argenteolus*, AB045872, 99.2%; *S. cinereorectus*, AB184646, 99.2%; *S. cyaneofuscatus*, AB184860, 99.2%; *S. griseoplanus*, AY999894, 99.2%; *S. fulvorobeus*, AB184711, 99.1%; *S. acrimycini*, AY999889, 99.1%; *S. microflavus*, DQ445795, 99.1%; *S. griseolus*, AB184768, 99.1%; *S. flavovirens*, DQ026635, 99.1%; *S. halstedii*, EF178695, 99.1%; *S. lipmanii*, AB184148, 99.1%; *S. alboviridis*, AB184256, 99.1%; *S. baarnensis*, EF178688, 99.1%; *S. floridae*, AB184656, 99.1%; *S. griseus* subsp. *griseus*, AY207604, 99.1%; *S. flavogriseus*, AJ494864, 99.1%; *S. luridiscabiei*, AF361784, 99%; *S. nitrosporeus*, EF178680, 99%.

Source: not known.

DNA G+C content (mol%): not known.

Type strain: ATCC 23912, ATCC 3323, CBS 131.20, CBS 369.39, CBS 833.68, BCRC 11477, DSM 40065, NBRC 12866, IMET 40285, JCM 4376, NCIMB 9848, NRRL B-1252, NRRL B-2928, NRRL-ISP 5065, RIA 1118, RIA 89, VKM Ac-1704.

Sequence accession no. (16S rRNA gene): DQ026636.

200. **Streptomyces geldanamycininus** Goodfellow, Kumar, Labeda and Sembiring 2008, 1[VP] (Effective publication: Goodfellow, Kumar, Labeda and Sembiring 2007, 198.)

gel.da.na.my.ci'ni.nus. N.L. neut. n. *geldanamycinum* geldanamycin; L. suff. *-inus* adjectival suffix used with the sense of belonging to or related to; N.L. masc. adj. *geldanamycininus* related to geldanamycin, producing the antibiotic geldanamycin.

Spore chains are *Spirales*; the spore surface is rugose. On oatmeal agar, the aerial spore mass color is grayish-brown and the substrate mycelium grayish-yellow. Melanin pigments are not produced.

Type strain shows the highest sequence similarity to: *S. antimycoticus*, AB184185, 99.8%; *S. sporoclivatus*, AB249934, 99.8%; *S. rutgersensis* subsp. *castelarensis*, AY508511, 99.6%; *S. asiaticus*, AB249947, 99%; *S. melanosporofaciens*, AJ271887, 99%; *S. rhizosphaericus*, AB249941, 99%.

Source: not known.

DNA G+C content (mol%): 70.2.

Type strain: DSM 41894, NRRL B-3602.

Sequence accession no. (16S rRNA gene): DQ334781.

201. **Strept.omyces geysiriensis** Wallhäusser, Nesemann, Präve and Steigler 1966, 734[AL]

gey.si.ri.en'sis. N.L. masc. adj. *geysiriensis* (from Icel. n. *geysir* a geyser) of or belonging to a geyser, referring to the source of the organism (an Iceland geyser).

Moderate growth on Czapek's solution (synthetic) agar; produces the moenomycin complex of anti-bacterial antibiotics (moenomycins A, B_1, B_2, and C).

Type strain shows the highest sequence similarity to: *S. vinaceusdrappus*, AY999929, 100%; *S. ghanaensis*, AY999851, 100%; *S. plicatus*, AB184291, 100%; *S. minutiscleroticus*, EF178696, 100%; *S. rochei*, AB184237, 100%; *S. mutabilis*, EF178679, 99.8%; *S. tuirus*, AB184690, 99.5%; *S. djakartensis*, AB184657, 99.5%; *S. anandii*, AB184402, 99.3%; *S. viola-*

ceorubidus, AJ781374, 99.3%; *S. pilosus*, AB184161, 99.2%; *S. flavoviridis*, AB184842, 99.2%; *S. tendae*, D63873, 99.1%; *S. calvus*, AB184329, 99.1%; *S. azureus*, EF178674, 99.1%; *S. asterosporus*, AB184706, 99.1%; *S. levis*, AB184670, 99%; *S. luteogriseus*, AB184379, 99%; *S. capillispiralis*, AB184577, 99%; *S. pseudogriseolus*, DQ442541, 99%; *S. aureorectus*, AB184710, 99%; *S. virens*, DQ442554, 99%; *S. gancidicus*, AB184660, 99%; *S. naganishii*, DQ442529, 99%.

Source: not known.

DNA G+C content (mol%): not known.

Type strain: ATCC 15303, CBS 546.70, CECT 3209, DSM 40742, NBRC 15413, JCM 4962, NRRL B-12102, VKM Ac-844.

Sequence accession no. (16S rRNA gene): DQ442501.

202. **Streptomyces ghanaensis** Wallhäusser, Nesemann, Präve and Steigler 1966, 734[AL]

gha.na.en'sis. N.L. masc. adj. *ghanaensis* of or belonging to Ghana, the source of the soil from which the organism was isolated.

Moderate growth on synthetic agar; produces the moenomycin complex of anti-bacterial antibiotics (components A, B, B_1, and C).

Type strain shows the highest sequence similarity to: *S. vinaceusdrappus*, AY999929, 100%; *S. geysiriensis*, DQ442501, 100%; *S. plicatus*, AB184291, 100%; *S. minutiscleroticus*, EF178696, 100%; *S. rochei*, AB184237, 100%; *S. mutabilis*, EF178679, 99.8%; *S. tuirus*, AB184690, 99.5%; *S. djakartensis*, AB184657, 99.5%; *S. pilosus*, AB184161, 99.2%; *S. violaceorubidus*, AJ781374, 99.2%; *S. flavoviridis*, AB184842, 99.2%; *S. anandii*, AB184402, 99.2%; *S. calvus*, AB184329, 99.1%; *S. levis*, AB184670, 99%; *S. asterosporus*, AB184706, 99%; *S. tendae*, D63873, 99%; *S. luteogriseus*, AB184379, 99%; *S. capillispiralis*, AB184577, 99%; *S. azureus*, EF178674, 99%.

Source: isolated from soil from Ghana.

DNA G+C content (mol%): not known.

Type strain: ATCC 14672, CBS 544.70, CECT 3210, DSM 40746, NBRC 15414, JCM 4963, KCTC 9882, NRRL B-12104.

Sequence accession no. (16S rRNA gene): AY999851.

203. **Streptomyces gibsonii** (Erikson 1935) Waksman and Henrici *in* Breed, Murray and Hitchens 1948, 963[AL] ["*Actinomyces gibsoni*" (*sic*) Dodge 1935, 722; "*Actinomyces gibsonii*" (*sic*) Erikson 1935, 36; "*Nocardia gibsonii*" Waksman *in* Waksman and Lechevalier 1953, 155]

gib.so'ni.i. N.L. gen. masc. n. *gibsonii* of Gibson, named for A.G. Gibson who first isolated the organism.

Poor growth on Czapek's solution agar; exhibits slight anti-bacterial activity.

Type strain shows the highest sequence similarity to: *S. almquistii*, AB184258, 100%; *S. rangoonensis*, AB184295, 100%; *S. albus* subsp. *albus*, AJ621602, 100%; *S. flocculus*, AB184272, 99.9%.

Source: isolated from a monkey injected with strain NCTC 450 subsequently named *Actinomyces upcottii* Erikson 1935, 36. Originally obtained from the spleen on a case of acholuric jaundice by Dr A.G. Gibson in 1920.

DNA G+C content (mol%): not known.

Type strain: ATCC 6852, CBS 118.60, CBS 119.60, DSM 43284, HUT 6617, NBRC 15415, IMET 7023, JCM 5061, NCTC 4575, NRRL B-1335.

Sequence accession no. (16S rRNA gene): AB184663.

204. **Streptomyces glaucescens** (Preobrazhenskaya *in* Gauze, Preobrazhenskaya, Kudrina, Blinov, Ryabova and Sveshnikova 1957) Pridham, Hesseltine and Benedict 1958, 67[AL] ("*Actinomyces glaucescesens*" Preobrazhenskaya *in* Gauze, Preobrazhenskaya, Kudrina, Blinov, Ryabova and Sveshnikova 1957, 122)

glau.ces'cens. L. adj. *glaucus* bluish gray; N.L. part. adj. *glaucescens* becoming slightly bluish gray, referring to the bluish green color of the aerial mycelium on a chemically defined medium.

Spore chains in Section *Spirales*. Mature spore chains are short with 3–10 spores per chain. Typical morphology on yeast-malt agar, oatmeal agar, and salts-starch agar, but not typical on glycerol-asparagine agar. Spore surface is hairy. Hairs are coarse, showing some tendency towards spines.

Color of colony: aerial mass color in the Blue or Green color series on yeast-malt agar, oatmeal agar, and salts-starch agar. Color tabs selected by observers fall in both series, but all tabs selected are grayish green. Reverse side of colony is grayed yellow modified by red on yeast-malt agar and glycerol-asparagine agar; modified by green on oatmeal agar and salts-starch agar. Substrate pigment is not a pH indicator.

Color in medium: melanoid pigments formed in peptone-yeast-iron agar, tyrosine agar, and tryptone-yeast broth. Pigments other than melanoids are not formed in yeast-malt agar, oatmeal agar, salts-starch agar, or glycerol-asparagine agar.

D-Glucose, L-arabinose, D-xylose, iso-inositol, D-mannitol, D-fructose, and rhamnose are utilized for growth. No growth or only traces of growth on sucrose and raffinose.

Type strain shows the highest sequence similarity to: *S. pharetrae*, AY699792, 99.3%; *S. matensis*, AB184221, 99.2%; *S. paradoxus*, AB184628, 99.2%; *S. flaveolus*, AB184764, 99.1%; *S. misionensis*, EF178678, 99%; *S. viridochromogenes*, DQ442555, 99%; *S. malachitofuscus*, AB184282, 99%.

Source: not known.

DNA G+C content (mol%): not known.

Type strain: AS 4.1408, ATCC 19761, ATCC 23622, CBS 499.68, BCRC 11478, CECT 3133, DSM 40155, NBRC 12774, IMET 43584, INA 8731, JCM 4377, LMG 19330, NCIMB 9619, NCIMB 9844, NRRL B-2706, NRRL-ISP 5155, RIA 1041, UNIQEM 147, VKM Ac-617.

Sequence accession no. (16S rRNA gene): AB184843.

205. **Streptomyces glauciniger** Huang, Li, Wang, Lanoot, Vancanneyt, Rodriguez, Liu, Swings and Goodfellow 2004b, 2088[VP]

glau.ci.ni'ger. L. adj. *glaucus* greenish gray; L. adj. *niger* black; N.L. masc. adj. *glauciniger* greenish black, referring to the color of colony reverse on modified Bennett's agar.

Forms an extensively branched substrate mycelium and aerial hyphae that differentiate into long spiral spore chains with 15–20 cylindrical spores per chain. Spore surface is smooth. Aerial spore mass on oatmeal agar is grayish brown. Soluble pigments are not produced, nor are melanin pigments formed on peptone-yeast extract-iron or tyrosine agars. The organism degrades adenine, casein, hypoxanthine, starch, and xanthine, but not cellulose or elastin. Nitrate is reduced. Gelatin is not liquefied. It uses dextrin, D-galactose, D-glucose (all at 1%, w/v), and sodium acetate, and sodium citrate (both at 0.1%, w/v), but not maltose (1%, w/v), as sole carbon sources for energy and growth. Growth occurs at 10–35°C and pH 5.0–10.0, but not at 40°C or at pH 4.0 or 11.0. Growth also occurs in the presence of phenol (0.1%, w/v) but not in the presence of NaCl (5%, w/v), novobiocin (5 µg/ml), or streptomycin (10 µg/ml). It shows antimicrobial activity against strains of *Bacillus subtilis* and *Candida albicans*, but not against strains of *Aspergillus niger*, *Escherichia coli*, *Klebsiella pneumoniae*, *Pseudomonas aeruginosa*, or *Staphylococcus aureus*. Cell wall is of type I, phospholipid type II and menaquinone MK-9(H_6, H_8). The fatty acid profile is composed of $C_{16:0}$ iso (25.4%), $C_{17:0}$ anteiso (17.4%), $C_{15:0}$ anteiso (16.7%), $C_{15:0}$ iso (9.5%), $C_{16:0}$ (7.7%), $C_{17:0}$ iso (3.8%), $C_{14:0}$ iso (3.5%), $C_{17:1}$ iso ω9*c* (3.3%), $C_{17:1}$ anteiso ω9*c* (3.0%), $C_{16:1}$ iso (2.9%), $C_{15:0}$ (2.8%), and $C_{17:0}$cyclo (2.7%).

Type strain shows no sequence similarity over 99%.

Source: the type strain was isolated from soil in a willow wood collected in Nanning City, Guangxi Province, China.

DNA G+C content (mol%): 67.0.

Type strain: FXJ14, AS 4.1858, JCM 12278, LMG 22082, NBRC 100913.

Sequence accession no. (16S rRNA gene): AB249964.

206. **Streptomyces glaucosporus** (*ex* Krasil'nikov, Agre, Dorokhova and Sokolov 1968) Agre 1986, 574[VP] (Effective publication: Agre *in* Gause, Preobrazhenskaya, Sveshnikova, Terekhova and Maximova 1983.) ("*Actinomyces glaucosporus*" Krasil'nikov, Agre, Dorokhova and Sokolov 1968)

glau.co.spo'rus. L. adj. *glaucus* bluish gray; N.L. n. *spora* a spore; N.L. masc. adj. *glaucosporus* bluish gray spored.

Spore chains are spiral (*Spirales*); spores are warty. On mineral agar 1, oatmeal agar: aerial mycelium poorly or well developed, green-gray; substrate mycelium is dark yellow; no diffusible pigment. On starch-ammonia agar: aerial mycelium is green-gray; substrate mycelium is yellowish; no diffusible pigment. On glycerol-nitrate agar: aerial mycelium is poorly developed, green-gray; substrate mycelium is colorless; no diffusible pigment. On glycerol-asparagine agar: aerial mycelium poorly developed, gray; substrate mycelium is colorless; no diffusible pigment. On organic agar 2: aerial mycelium is green-gray; substrate mycelium and diffusible pigment are dark yellow, brown-yellow. Melanoid pigments are not formed. Good utilization of glucose, maltose, and xylose; weak utilization of sucrose, rhamnose, and inositol.

Type strain shows no sequence similarity over 99%.

Source: not known.

DNA G+C content (mol%): not known.

Type strain: ATCC 25183, DSM 41689, NBRC 15416, INMI 2979, INA G-72, JCM 6921, VKM Ac-1763.

Sequence accession no. (16S rRNA gene): AB184664.

207. **Streptomyces glaucus** (*ex* Lehmann and Schütze 1912) Agre and Preobrazhenskaya 1986, 574[VP] (Effective publication: Agre and Preobrazhenskaya *in* Gause, Preobrazhenskaya, Sveshnikova, Terekhova and Maximova 1983.) ("*Actinomyces glaucus*" Lehmann and Schütze 1912; "*Streptomyces glaucus*" Waksman 1953)

glau'cus. L. masc. adj. *glaucus* bluish gray.

Spore chains are spiral (*Spirales*); spore surface is hairy and rough. On mineral agar 1 and oatmeal agar: aerial mycelium mealy, greenish-bluish; substrate mycelium and diffusible pigment are brown. On glycerol-nitrate agar: aerial mycelium poorly developed, white; substrate mycelium dark brown; diffusible pigment brown. On starch-ammonia agar: aerial mycelium is bluish-greenish; substrate mycelium colorless; diffusible pigment brown. On glycerol-asparagine agar: aerial mycelium is bluish; substrate mycelium is colorless to brownish; diffusible pigment is brownish or absent. On organic agar 2: aerial mycelium poorly developed, white; substrate mycelium dark grayish brown. Melanoid pigments are not formed. Good growth on starch, glucose, lactose, xylose, maltose, rhamnose, glycerol, and mannitol; moderate growth on arabinose and laevulose; poor growth on galactose and inositol; no growth on dulcitol, sorbitol, or sucrose. Cultures belonging to this species are produce antibiotic RA 166 (PA 166-russ).

Type strain shows the highest sequence similarity to: *S. griseomycini*, AB184137, 99.1%; *S. graminearus* AJ781333, 99.1%; *S. griseostramineus*, AB184140, 99%.

Source: not known.

DNA G+C content (mol%): not known.

Type strain: ATCC 43685, DSM 41456, NBRC 15417, INMI 2965, INA G-86, JCM 6922, NRRL B-16368, VKM Ac-803.

Sequence accession no. (16S rRNA gene): AB184665.

208a. **Streptomyces globisporus subsp. globisporus** (Krasil'nikov 1941) Waksman *in* Waksman and Lechevalier 1953, 39[AL] ("*Actinomyces globisporus*" Krasil'nikov 1941, 48)

glo.bi.spo'rus. L. n. *globus* a round body; N.L. n. *spora* a spore; N.L. masc. adj. *globisporus* round spored (as determined by light microscopy).

Spore chains in Section *Rectiflexibiles*. Mature spore chains are predominantly flexuous and moderately long with 10–50, or often more than 50, spores per chain. This morphology is seen on yeast-malt agar, oatmeal agar, salts-starch agar, and glycerol-asparagine agar. Spore surface is smooth.

Color of colony: aerial mass color in the Yellow color series on yeast-malt agar, oatmeal agar, salts-starch agar, and glycerol-asparagine agar. Reverse side of colony with no distinctive pigments (pale yellow to light olive brown) on yeast-malt agar, oatmeal agar, salts-starch agar, and glycerol-asparagine agar.

Color in medium: melanoid pigments are not produced in peptone-yeast-iron agar, tyrosine agar, or tryptone-yeast broth. Yellow to greenish yellow pigment may be found in medium in yeast-malt agar, oatmeal agar, salts-starch agar, and glycerol-asparagine agar. This pigment is not sensitive when tested with 0.05 M HCl or NaOH.

D-Glucose, L-arabinose, D-xylose, D-mannitol, D-fructose, and rhamnose are utilized for growth. No growth or only traces of growth on sucrose, iso-inositol, and raffinose.

Type strain shows the highest sequence similarity to: *S. sindenensis*, AB184759, 100%; *S. pluricolorescens*, DQ442540, 100%; *S. mediolani*, AB184674, 100%; *S. badius*, AY999783, 100%; *S. griseinus*, AB184205, 100%; *S. rubiginosohelvolus*, AB184240, 100%; *S. praecox*, AB184293, 99.9%; *S. flavofuscus*, AB249935, 99.9%; *S. fimicarius*, AY999784, 99.9%; *S. albovinaceus*, AB249958, 99.9%; *S. anulatus*, DQ026637, 99.9%; *S. griseoplanus*, AY999894, 99.9%; *S. californicus*, AB184755, 99.8%; *S. fulvorobeus*, AB184711, 99.8%; *S. parvus*, DQ442537, 99.8%; *S. cavourensis* subsp. *washingtonensis*, DQ026671, 99.8%; *S. acrimycini*, AY999889, 99.8%; *S. lavendulae* subsp. *lavendulae*, AB184080, 99.8%; *S. argenteolus*, AB045872, 99.7%; *S. baarnensis*, EF178688, 99.7%; *S. lipmanii*, AB184148, 99.7%; *S. microflavus*, DQ445795, 99.7%; *S. cyaneofuscatus*, AB184860, 99.7%; *S. floridae*, AB184656, 99.7%; *S. bacillaris*, AB184439, 99.7%; *S. cinereorectus*, AB184646, 99.7%; *S. alboviridis*, AB184256, 99.7%; *S. griseolus*, AB184768, 99.6%; *S. flavovirens*, DQ026635, 99.6%; *S. albolongus*, AB184425, 99.5%; *S. pulveraceus*, AB184806, 99.5%; *S. halstedii*, EF178695, 99.5%; *S. nitrosporeus*, EF178680, 99.5%; *S. atroolivaceus*, AJ781320, 99.4%; *S. olivoviridis*, AB184227, 99.4%; *S. celluloflavus*, AB184476, 99.4%; *S. yanii*, AB006159, 99.4%; *S. finlayi*, AY999788, 99.4%; *S. griseobrunneus*, AB249912, 99.4%; *S. flavogriseus*, AJ494864, 99.4%; *S. cavourensis* subsp. *cavourensis*, DQ445791, 99.3%; *S. atratus*, DQ026638, 99.3%; *S. griseus* subsp. *griseus*, AY207604, 99.3%; *S. luridiscabiei*, AF361784, 99.3%; *S. clavifer*, DQ026670, 99.3%; *S. sanglieri*, AB249945, 99.3%; *S. gelaticus*, DQ026636, 99.3%; *S. mutomycini*, AB249951, 99.1%; *S. candidus*, DQ026663, 99.1%; *S. spiroverticillatus*, AB184814, 99.1%; *S. cremeus*, AB184124, 99%.

Source: not known.

DNA G+C content (mol%): not known.

Type strain: ATCC 15864, ATCC 23913, CBS 834.68, BCRC 11479, CCUG 11107, DSM 40199, NBRC 12867, INMI 2302, JCM 4378, KCTC 9026, NCAIM B.01476, NCIMB 9796, NRRL B-2872, NRRL-ISP 5199, RIA 1151, RIA 335, VKM Ac-179.

Sequence accession no. (16S rRNA gene): EF178686.

208b. **Streptomyces globisporus subsp. caucasicus** (Kudrina *in* Gauze, Preobrazhenskaya, Kudrina, Blinov, Ryabova and Sveshnikova 1957) Pridham, Hesseltine and Benedict 1958, 59[AL] ("*Actinomyces globisporus* var. *caucasicus*" Kudrina *in* Gauze, Preobrazhenskaya, Kudrina, Blinov, Ryabova and Sveshnikova 1957, 79)

cau.ca'si.cus. L. n. *Caucasius* region of the Caucasus; N.L. masc. adj. *caucasicus* belonging to the Caucasus.

Spore chains are typically flexuous; poor growth on Czapek's solution agar. Exhibits anti-bacterial and anti-fungal activity; inhibited by streptomycin; NaCl tolerance >10%, but <13%.

Type strain shows no sequence similarity over 99%.

Source: not known.

DNA G+C content (mol%): not known.

Type strain: ATCC 19907, CBS 120.60, DSM 40814, INA 13195/54, JCM 9867, NBRC 100770, NRRL B-2593, NRRL-ISP 5157, RIA 319, VKM Ac-1846.

Sequence accession no. (16S rRNA gene): EF178676.

208c. **Streptomyces globisporus subsp. flavofuscus** (Kudrina *in* Gauze, Preobrazhenskaya, Kudrina, Blinov, Ryabova and Sveshnikova 1957) Pridham, Hesseltine and Benedict 1958, 59[AL] ("*Actinomyces globisporus* var. *flavofuscus*" Kudrina *in* Gauze, Preobrazhenskaya, Kudrina, Blinov, Ryabova and Sveshnikova 1957, 81)

fla.vo.fus'cus. L. adj. *flavus* yellow; L. adj. *fuscus* dark, tawny; N.L. masc. adj. *flavofuscus* dark yellow.

Spore chains are typically flexuous; poor growth observed on Czapek's solution agar. Exhibits anti-bacterial and anti-fungal activity; inhibited by streptomycin; NaCl tolerance >10%, but <13%.

For sequence similarity, see type strain of *Streptomyces flavofuscus.*

Source: not known.

DNA G+C content (mol%): not known.

Type strain: ATCC 19908, INA 1565/53, JCM 9766.

Sequence accession no. (16S rRNA gene): DQ026648.

Further comments: according to Preobrazhenskaya (1986), *Streptomyces globisporus* subsp. *flavofuscus* (Kudrina 1957) Pridham, Hesseltine and Benedict 1958, is a later heterotypic synonym of *Streptomyces flavofuscus* (Kudrina 1957) Preobrazhenskaya 1986.

209. **Streptomyces globosus** (Krasil'nikov 1941) Waksman *in* Waksman and Lechevalier 1953, 68[AL] ("*Actinomyces globosus*" Krasil'nikov 1941, 58)

glo.bo'sus. L. masc. adj. *globosus* spherical (referring to the shape of the spores when examined with the light microscope).

Forms straight and short chains of spores; excellent growth on Czapek's solution agar; melanin-like chromogenicity is not expressed to great degree with strain IMRU 3763. Exhibits anti-bacterial and anti-fungal activity; inhibited by streptomycin.

Type strain shows the highest sequence similarity to: *S. toxytricini*, DQ442548, 100%; *S. flavotricini*, AB184132, 99.4%; *S. racemochromogenes*, DQ026656, 99.1%; *S. polychromogenes*, AB184292, 99.1%.

Source: not known.

DNA G+C content (mol%): not known.

Type strain: DI-15, ATCC 14979, DSM 40815, IMRU 3736, JCM 13859, NRRL B-2292.

Sequence accession no. (16S rRNA gene): AJ781330.

210. **Streptomyces glomeratus** (*ex* Gause and Sveshnikova 1978) Gause and Preobrazhenskaya 1986a, 574[VP] (Effective publication: Gause and Preobrazhenskaya *in* Gause, Preobrazhenskaya, Sveshnikova, Terekhova and Maximova 1983.) ("*Streptomyces glomeratus*" Gause and Sveshnikova *in* Gause et al. 1978)

glo.me.ra'tus. L. masc. part. adj. *glomeratus*, formed into a ball, conglobated, glomerated.

Spore chains are hooks, loops, or spiral with 1 or 2 turns; spores look like quasi-sporangia. Spores are smooth. On mineral agar 1: aerial mycelium is whitish gray, light gray to greenish gray; substrate mycelium is grayish blue to grayish brown-blue; diffusible pigment is absent or sometimes pale reddish violet colored. On starch-ammonia agar: aerial mycelium is white to gray brownish gray, greenish gray; substrate mycelium is colorless to gray brown and grayish brown-blue; no diffusible pigment. On glycerol-nitrate agar: aerial mycelium poorly developed, whitish, sometimes greenish gray; substrate mycelium is gray brownish; no diffusible pigment. On glycerol-asparagine agar: aerial mycelium is white to gray brownish gray, sometimes with greenish shadow; substrate mycelium is grayish gray-brown, violet gray-brown; no diffusible pigment. On glucose-asparagine agar: aerial mycelium is grayish white, light greenish gray; substrate mycelium is colorless to brown and grayish brown-blue; no diffusible pigment. Oatmeal agar: aerial mycelium is gray-green; substrate mycelium is gray-brown, gray- brown-blue, gray- blue, sometimes with olive shadow; no diffusible pigment. On organic agar: aerial mycelium is grayish green, sometimes absent; substrate mycelium is grayish gray-brown; diffusible pigment is gray brown. Melanoid pigment is formed. Antibiotic: beromycin, nogalamycin (indicator pigment changing color from yellow in acidic medium to red-violet in alkaline medium). Moderate growth on xylose, rhamnose, fructose, and mannitol; poor growth on arabinose, raffinose, and sucrose; no growth on cellulose.

Type strain shows the highest sequence similarity to: *S. bangladeshensis*, AY750056, 99.2%; *S. rameus*, AY999821, 99.1%; *S. achromogenes* subsp. *rubradiris*, AY999846, 99%.

Source: not known.

DNA G+C content (mol%): not known.

Type strain: DSM 41457, NBRC 15898, INA 3980, JCM 9091, VKM Ac-834.

Sequence accession no. (16S rRNA gene): AB249917.

211. **Streptomyces glomeroaurantiacus** (Krasil'nikov and Yuan *in* Krasil'nikov 1965) Pridham 1970, 17[AL] ("*Actinomyces glomeroaurantiacus*" Krasil'nikov and Yuan *in* Krasil'nikov 1965, 50)

glo.me.ro.au.ran.ti'a.cus. L. v. *glomero* form into a ball; N.L. adj. *aurantiacus* orange colored; N.L. masc. adj. *glomeroaurantiacus* (*sic*) orange-colored ball.

Spore chains in Section *Spirales*, but aerial mycelium is poorly developed and typical spore chains are difficult to find on yeast-malt agar, oatmeal agar, salts-starch agar, and glycerol-asparagine agar. Mature spore chains on these media are generally very short or absent. Spore surface: Smooth. Conidia-like spores or fragmentation of the substrate mycelium may be seen on yeast-malt agar, oatmeal agar, salts-starch agar, or glycerol-asparagine agar.

Color of colony: growth of aerial mycelium is not adequate for determination of aerial mass color on yeast-malt agar, oatmeal agar, salts-starch agar, or glycerol-asparagine agar. Reverse side of colony is yellowish pink, orange or reddish orange on yeast-malt agar and glycerol-asparagine agar; no distinctive pigments (grayish yellow, orange yellow or yellowish brown) on oatmeal agar or salts-starch agar. Reverse mycelium pigment is not a pH indicator.

Color in medium: melanoid pigments are not formed in peptone-yeast-iron agar, tyrosine agar, or tryptone-yeast broth. One of three observers records orange to pink or red pigment in the medium in yeast-malt agar, oatmeal agar, and glycerol-asparagine agar. This pigment is not pH-sensitive when tested with 0.05 M NaOH or HCl.

D-Glucose, L-arabinose, D-mannitol, D-fructose, and rhamnose are utilized for growth. Reports vary from doubtful growth to positive growth with D-xylose, iso-inositol, sucrose, and raffinose.

Type strain shows the highest sequence similarity to: *S. aurantiacus*, AJ781383, 100%; *S. tauricus*, AB045879, 99.1%.

Source: not known.

DNA G+C content (mol%): not known.

Type strain: ATCC 15866, DSM 41782, JCM 4677, NBRC 15418, INMI 1464, NRRL B-3375, RIA 683.

Sequence accession no. (16S rRNA gene): AB249983.

212. **Streptomyces gobitricini** (Preobrazhenskaya and Sveshnikova *in* Gauze, Preobrazhenskaya, Kudrina, Blinov, Ryabova and Sveshnikova 1957) Pridham, Hesseltine and Benedict 1958, 67[AL] ("*Actinomyces gobitricini*" Preobrazhenskaya and Sveshnikova *in* Gauze, Preobrazhenskaya, Kudrina, Blinov, Ryabova and Sveshnikova 1957, 34)

go.bi.tri.ci'ni. English n. *Gobi* the Gobi Desert in Mongolia; Gr. n. *thrix* the hair; N.L. *gobitricini* (*sic*) of gobi hair, referring to the Gobi desert, the first source of the soil from which the organism was isolated, and probably to formation of a streptothricin-like antibiotic.

Some spore chains are of atypical *Retinaculum-Apertum* type; moderate growth on Czapek's solution agar. Exhibits anti-bacterial and anti-fungal activity; inhibited by streptomycin; NaCl tolerance >7%, but <10%.

Type strain shows the highest sequence similarity to: *S. lavendofoliae*, AJ781336, 99.8%; *S. luridus*, DQ442523, 99.5%.

Source: isolated from soil from the Gobi desert.

DNA G+C content (mol%): not known.

Type strain: CBS 123.60, DSM 41701, NBRC 15419, JCM 5062, NRRL B-2596.

Sequence accession no. (16S rRNA gene): AB184666.

213. **Streptomyces goshikiensis** Niida *in* Shirling and Gottlieb 1966, 324[AL]

gos.hi.ki.en'sis. etymology unknown.

Spore chains in Section *Spirales*, with open spirals of 4–10 convolutions at the end of long spore chains. Mature spore chains have 10–50 spores per chain; longer chains are often observed. This morphology is seen on yeast-malt agar, oatmeal agar, salts-starch agar, and glycerol-asparagine agar. Spore surface is smooth with minor surface irregularities, but no true spines.

Color of colony: aerial mass color in the Red (or Gray) color series on yeast-malt agar, oatmeal agar, and salts-starch agar. Characteristic color on these media in between 3ge (light grayish yellowish brown) and 5dc or 6ec (grayish yellowish pink) color tabs of Tresner–Backus

Color Wheels. Aerial mycelium may be poorly developed on glycerol-asparagine agar. Reverse side of colony with no distinctive pigments (characteristic grayed yellow on oatmeal agar, salts-starch agar, and glycerol-asparagine agar to light orange yellow or light brown on yeast-malt agar).

Color in medium: melanoid pigments are formed in peptone-yeast-iron agar, tyrosine agar, and tryptone-yeast broth. Pigments other than melanoids or faint traces of yellow are not produced in yeast-malt agar, oatmeal agar, salts-starch agar, and glycerol-asparagine agar.

D-Glucose and possibly D-fructose are utilized for growth. Only traces of growth comparable to growth on carbon-free basal medium seen on L-arabinose, sucrose, D-xylose, iso-inositol, D-mannitol, rhamnose, and raffinose.

Type strain shows the highest sequence similarity to: *S. colombiensis*, DQ026646, 100%; *S. sporoverrucosus*, DQ442544, 100%; *S. nojiriensis*, AJ781355, 99.9%; *S. spororaveus*, AJ781370, 99.9%; *S. xanthophaeus*, DQ442560, 99.8%; *S. vinaceus*, AB184394, 99.7%; *S. cirratus*, AY999794, 99.7%; *S. cinnamonensis*, AB184707, 99.6%; *S. avidinii*, AB184395, 99.6%; *S. subrutilus*, X80825, 99.5%; *S. virginiae*, D85119, 99.4%.

Source: not known.

DNA G+C content (mol%): not known.

Type strain: ATCC 23914, CBS 835.68, BCRC 12330, CCUG 11121, DSM 40190, NBRC 12868, IMET 42067, JCM 4294, JCM 4640, NCIMB 9828, NRRL B-5428, NRRL-ISP 5190, RIA 1144, VKM Ac-1212.

Sequence accession no. (16S rRNA gene): EF178693.

214. **Streptomyces gougerotii** (Duché 1934) Waksman and Henrici *in* Breed, Murray and Hitchens 1948, 947[AL] ["*Actinomyces gougeroti*" (*sic*) Duché 1934, 272; "*Streptomyces gougerotii*" Waksman and Henrici *in* Hütter 1967b, 75]

gou.ge.ro'ti.i. N.L. gen. masc. n. *gougerotii* of Gougerot; named for Professor Gougerot from whom the original culture was obtained.

Spore chains in Section *Rectiflexibiles* on yeast-malt agar; aerial mycelium is very thin on this medium and is usually not produced on oatmeal agar, salts-starch agar, or glycerol-asparagine agar. Spore chains, when produced, are moderately short with 3 to 10 or more spores per chain. Two observers recorded fragmentation of the substrate mycelium and this characteristic is also mentioned in Duché's original description (op. cit). Spore surface is smooth.

Color of colony: aerial mass color in the White or Yellow (2ba, pale yellow) color series on yeast-malt agar. Aerial mycelium is usually thin on yeast-malt agar and is often inadequate for color observation on all other ISP media. Reverse side of colony with no distinctive pigments (colorless to grayish yellow) on yeast-malt agar, oatmeal agar, salts-starch agar, and glycerol-asparagine agar.

Color in medium: melanoid pigments are not formed in tyrosine agar or tryptone-yeast broth; reports vary on production of dark pigment in peptone-yeast-iron agar. No pigment found in medium in yeast-malt agar, oatmeal agar, salts-starch agar, or glycerol-asparagine agar.

D-Glucose, D-xylose, D-mannitol, and D-fructose are utilized for growth. Reports vary on utilization of L-arabinose, iso-inositol, and rhamnose. No growth or only traces of growth on sucrose and raffinose.

Type strain shows the highest sequence similarity to: *S. diastaticus* subsp. *diastaticus*, AB184785, 100%; *S. rutgersensis* subsp. *rutgersensis*, AB184170, 100%; *S. intermedius*, AB184277, 99.8%; *S. misionensis*, EF178678, 99.2%; *S. phaeoluteichromatogenes*, AJ391814, 99.1%%; *S. matensis*, AB184221, 99%%; *S. aureoverticillatus*, AY999774, 99%.

Source: not known.

DNA G+C content (mol%): not known.

Type strain: ATCC 10975, ATCC 25455, CBS 422.34, CBS 688.69, BCRC 12105, DSM 40324, NBRC 13043, NBRC 3198, IMET 40289, JCM 4136, JCM 4478, NRRL B-1344, NRRL B-1903, NRRL-ISP 5324, RIA 1235, VKM Ac-713.

Sequence accession no. (16S rRNA gene): AB184742.

215. **Streptomyces graminearus** Preobrazhenskaya 1986, 574[VP] (Effective publication: Preobrazhenskaya *in* Gause, Preobrazhenskaya, Sveshnikova, Terekhova and Maximova 1983.)

gra.mi.ne.a'rus. N.L. masc. adj. *graminearus* related to grain, isolated from grain.

Spore chains are spiral (*Spirales*); spores are smooth. On mineral agar 1: aerial mycelium is gray or green-gray; substrate mycelium is light yellow; no diffusible pigment. On glycerol-nitrate agar: aerial mycelium is poor, white; substrate mycelium is yellow, citreous; no diffusible pigment. On oatmeal agar: aerial mycelium is gray; substrate mycelium is colorless or light citreous; no diffusible pigment. On starch-ammonia agar: aerial mycelium is grayish gray; substrate mycelium is yellow; no diffusible pigment. On glycerol-asparagine agar: aerial mycelium is poor, white; substrate mycelium is yellow; no diffusible pigment. On organic agar 2: aerial mycelium is white; substrate mycelium is colorless or yellow; no diffusible pigment. Melanoid pigments are not formed. Grows on glucose, fructose, starch, mannitol, sucrose, and arabinose; weak growth on raffinose, rhamnose, and inositol.

Type strain shows the highest sequence similarity to: *S. griseostramineus*, AB184140, 100%; *S. griseomycini*, AB184137, 100%; *S. glaucus*, AB184665, 99.1%; *S. calvus*, AB184329, 99%.

Source: not known.

DNA G+C content (mol%): not known.

Type strain: DSM 41474, NBRC 15420, INA 13982, JCM 6923, NRRL B-16369, VKM Ac-1847.

Sequence accession no. (16S rRNA gene): AJ781333.

216. **Streptomyces graminofaciens** Charney, Fisher, Curran, Machlowitz and Tytell 1953, 1283[AL]

gra.mi.no.fa'ci.ens. L. n. *gramen* grass; L. part. adj. *faciens* producing; N.L. part. adj. *graminofaciens* grass producing, probably refers to production of (strepto)gramin.

Spore chains in Section *Spirales*. Open spirals of four or more turns, irregular spirals and hooks, or tight terminal spirals may be found on various media. Mature spore chains are generally long, sometimes with more than 50 spores per chain. This morphology is seen on yeast-malt agar, oatmeal agar, salts-starch agar, and glycerol-asparagine agar. Spore surface is warty.

Color of colony: aerial mass color in the Gray or White color series on yeast-malt agar, oatmeal agar, salts-starch agar, and glycerol-asparagine agar. Only one of three observers records aerial mycelium in the Red color series (5dc, grayish yellowish pink) on yeast-malt agar in 21 d. Nearest matching color tabs in the Gray color series are d, light gray; 2dc, yellow gray; and 5fe, light grayish reddish brown. Reverse side of colony with no distinctive pigments (grayish yellow to brown on yeast-malt agar; pale grayish or greenish yellow to light yellowish brown on oatmeal agar, salts-starch agar, and glycerol-asparagine agar).

Color in medium: melanoid pigments are not formed in peptone-yeast-iron agar, tyrosine agar, or tryptone-yeast broth. No pigment is found in the medium in yeast-malt agar, oatmeal agar, salts-starch agar, or glycerol-asparagine agar.

D-Glucose, L-arabinose, D-xylose, iso-inositol, D-mannitol, D-fructose, rhamnose, sucrose, and raffinose are all utilized for growth.

Type strain shows the highest sequence similarity to: *S. kurssanovii*, AB184325, 100%; *S. xantholiticus*, AB184349, 99.8%; *S. peucetius*, AB045887, 99.7%; *S. cremeus*, AB184124, 99%.

Source: not known.

DNA G+C content (mol%): not known.

Type strain: ATCC 12705, CBS 756.72, BCRC 12352, CECT 3217, DSM 40559, HAMBI 982, NBRC 13455, IMET 43540, JCM 4157, JCM 4762, NRRL B-2609, NRRL-ISP 5559, RIA 1416, VKM Ac-973.

Sequence accession no. (16S rRNA gene): AJ781329.

217. **Streptomyces griseiniger** Goodfellow, Kumar, Labeda and Sembiring 2008, 1[VP] (Effective publication: Goodfellow, Kumar, Labeda and Sembiring 2007, 198.)

gri.se.i.ni'ger. N.L. adj. *griseus* gray; L. adj. *niger* black; N.L. masc. adj. *griseiniger* gray-black.

Spore chains are *Spriales*; spore surface is rugose. On oatmeal agar, the aerial spore mass color is gray, becoming black and moist when mature; the reverse side of colonial growth is grayish-yellow. Melanin pigments are not formed.

Type strain shows the highest sequence similarity to: *S. rhizosphaericus*, AB249941, 99.8%; *S. asiaticus*, AB249947, 99.8%; *S. cangkringensis*, AJ391831, 99.5%; *S. indonesiensis*, DQ334783, 99.4%. Type strain shows DNA–DNA similarity to: *S. cangkringensis* DSM 41769^T, 59%; *S. hygroscopicus* subsp. *geldanus* NRRL 3602^T, 56%; *S. indonesiensis* DSM 41759^T, 62%; *S. melanosporofaciens* NRRL B-12234^T, 57%.

Source: not known.

DNA G+C content (mol%): 70.2.

Type strain: DSM 41895, NRRL B-1865.

Sequence accession no. (16S rRNA gene): AJ391818.

218. **Streptomyces griseinus** Waksman 1959, 1045[AL]

gri.se.i'nus.N.L. n. *griseinum* grisein, name of an antibiotic; L. masc. suff. *-inus* suffix used with the sense of belonging to; N.L. masc. adj. *griseinus* belonging to grisein, the antibiotic produced by the organism.

Spore chains in Section *Rectiflexibiles*. Mature spore chains are predominantly flexuous and moderately long with 10–50, or often more than 50, spores per chain on yeast-malt agar, oatmeal agar, salts-starch agar, and glycerol-asparagine agar. Spore surface is smooth.

Color of colony: aerial mass color in the Yellow color series on salts-starch agar and glycerol-asparagine agar and in the Yellow or Gray color series on oatmeal agar and yeast-malt agar. Nearest matching color tab from the Gray color series is 2dc, yellowish gray. Reverse side of colony: no distinctive pigments (grayish yellow to yellowish brown or light olive brown) on yeast-malt agar, oatmeal agar, salts-starch agar, or glycerol-asparagine agar. The substrate mycelium pigment may be changed slightly from pale yellow to pale yellow-brown or pale violet with addition of 0.05 M NaOH, or from yellowish brown to pale yellow with 0.05 M HCl.

Color in medium: melanoid pigments are not produced in peptone-yeast-iron agar, tyrosine agar, or tryptone-yeast broth. A trace of yellow to reddish brown pigment may be found in the medium in oatmeal agar. This pigment, when present, is slightly pH-sensitive changing from pale yellow to pale violet with addition of 0.05 M NaOH.

D-Glucose, L-arabinose, D-xylose, D-mannitol, D-fructose, and rhamnose are utilized for growth. No growth or only traces of growth on sucrose, iso-inositol, and raffinose.

Type strain shows the highest sequence similarity to: *S. sindenensis*, AB184759, 100%; *S. albovinaceus*, AB249958, 100%; *S. griseoplanus*, AY999894, 100%; *S. anulatus*, DQ026637, 100%; *S. globisporus* subsp. *globisporus*, EF178686, 100%; *S. pluricolorescens*, DQ442540, 100%; *S. praecox*, AB184293, 100%; *S. flavofuscus*, AB249935, 100%; *S. fimicarius*, AY999784, 100%; *S. mediolani*, AB184674, 100%; *S. badius*, AY999783, 100%; *S. rubiginosohelvolus*, AB184240, 100%; *S. californicus*, AB184755, 99.9%; *S. lavendulae* subsp. *lavendulae*, AB184080, 99.9%; *S. cavourensis* subsp. *washingtonensis*, DQ026671, 99.9%; *S. acrimycini*, AY999889, 99.9%; *S. parvus*, DQ442537, 99.9%; *S. fulvorobeus*, AB184711, 99.8%; *S. lipmanii*, AB184148, 99.8%; *S. microflavus*, DQ445795, 99.8%; *S. cyaneofuscatus*, AB184860, 99.8%; *S. floridae*, AB184656, 99.8%; *S. alboviridis*, AB184256, 99.8%; *S. baarnensis*, EF178688, 99.8%; *S. cinereorectus*, AB184646, 99.8%; *S. griseus* subsp. *griseus*, AY207604, 99.7%; *S. griseolus*, AB184768, 99.7%; *S. luridiscabiei*, AF361784, 99.7%; *S. flavovirens*, DQ026635, 99.7%; *S. argenteolus*, AB045872, 99.7%; *S. bacillaris*, AB184439, 99.6%; *S. pulveraceus*, AB184806, 99.6%; *S. halstedii*, EF178695, 99.6%; *S. flavogriseus*, AJ494864, 99.6%; *S. atroolivaceus*, AJ781320, 99.5%; *S. olivoviridis*, AB184227, 99.5%; *S. finlayi*, AY999788, 99.5%; *S. nitrosporeus*, EF178680, 99.5%; *S. albolongus*, AB184425, 99.4%; *S. clavifer*, DQ026670, 99.4%; *S. griseobrunneus*, AB249912, 99.4%; *S. sanglieri*, AB249945, 99.4%; *S. celluloflavus*, AB184476, 99.4%; *S. yanii*, AB006159, 99.4%; *S. gelaticus*, DQ026636, 99.3%; *S. atratus*, DQ026638, 99.3%; *S. cavourensis* subsp. *cavourensis*, DQ445791, 99.3%; *S. mutomycini*, AB249951, 99.2%; *S. candidus*, DQ026663, 99.2%; *S. spiroverticillatus*, AB184814, 99.1%; *S. cremeus*, AB184124, 99.1%.

Source: not known.

DNA G+C content (mol%): not known.

Type strain: ATCC 23915, CBS 836.68, BCRC 11480, CCUG 11106, DSM 40047, NBRC 12869, JCM 4379, NRRL B-1076, NRRL-ISP 5047, RIA 1113.

Sequence accession no. (16S rRNA gene): AB184205.

219. **Streptomyces griseoaurantiacus** (Krasil'nikov and Yuan *in* Krasil'nikov 1965) Pridham 1970, 17[AL] ("*Actinomyces griseoaurantiacus*" Krasil'nikov and Yuan *in* Krasil'nikov 1965, 52)

gri.se.o.au.ran.ti.a'cus N.L. adj. *griseus* gray; N.L. adj. *aurantiacus* orange colored; N.L. masc. adj. *griseoaurantiacus* orange colored with gray.

Spore chains in Section *Spirales*. Hooks, loops, and primitive spirals suggesting *Retinaculiaperti* morphology are very common. Mature spore chains have 10 to 50 or more spores per chain. This morphology is seen on yeast-malt agar, oatmeal agar, salts-starch agar, and glycerol-asparagine agar. Spore surface is smooth.

Color of colony: aerial mass color in the Gray color series (2dc, yellowish gray; 2fe, medium gray; 3fe, light brownish gray; or 5fe, light grayish reddish brown) on yeast-malt agar, oatmeal agar and salts-starch agar. Gray or Red color series (5dc, grayish yellowish pink) on glycerol-asparagine agar. Reverse side of colony is grayish yellow to orange yellow, yellowish pink, or dark reddish orange on yeast-malt agar; yellowish pink or reddish gray to reddish orange on oatmeal agar, salts-starch agar, and glycerol-asparagine agar. Reverse mycelium pigment on yeast-malt agar, salts-starch agar, and glycerol-asparagine agar is a pH indicator, changing from orange to red with the addition of 0.05 M HCl and from orange to brown with the addition of 0.05 M NaOH.

Color in medium: melanoid pigments are not formed in peptone-yeast-iron agar, tyrosine agar, or tryptone-yeast broth. Yellow orange or pinkish red may be found in the medium in yeast-malt agar, oatmeal agar, salts-starch agar, and glycerol-asparagine agar in 21 d. This pigment, when formed in sufficient quantity, is pH-sensitive, showing the same changes observed for reverse mycelium pigment.

D-Glucose, L-arabinose, D-xylose, iso-inositol, D-mannitol, D-fructose, and rhamnose are utilized for growth. Utilization of raffinose is doubtful. No growth or only traces of growth with sucrose.

Type strain shows the highest sequence similarity to: *S. jietaisiensis*, AY314783, 99.7%.

Source: not known.

DNA G+C content (mol%): not known.

Type strain: ATCC 19840, CBS 682.72, DSM 40430, NBRC 13381, NBRC 15440, JCM 4763, NRRL-ISP 5430, RIA 1342, VKM Ac-1728.

Sequence accession no. (16S rRNA gene): AB184676.

220. **Streptomyces griseobrunneus** Waksman 1961, 220[AL]

gri.se.o.brun'ne.us. N.L. adj. *griseus* gray; N.L. adj. *brunneus* dark brown; N.L. masc. adj. *griseobrunneus* grayish dark-brown colored.

Spore chains in Section *Rectiflexibiles*. Mature spore chains are generally long, often with more than 50 spores per chain. This morphology is seen on yeast-malt agar, oatmeal agar, salts-starch agar, and glycerol-asparagine agar. Spore surface is smooth. Special morphological characteristics: substrate conidia noted by two observers.

Color of colony: aerial mass color in the Yellow color series on yeast-malt agar, oatmeal agar, salts-starch agar, and glycerol-asparagine agar. Reverse side of colony with no distinctive pigments on yeast-malt agar, oatmeal agar, salts-starch agar, or glycerol-asparagine agar; substrate pigment is not a pH indicator.

Color in medium: melanoid pigments formed in peptone-yeast-iron agar. Pigments other than melanoid not formed in yeast-malt agar, oatmeal agar, salts-starch agar, or glycerol-asparagine agar.

D-Glucose, D-xylose, D-mannitol, and D-fructose are utilized for growth. No growth or only trace of growth on L-arabinose, sucrose, iso-inositol, rhamnose, and raffinose.

Type strain shows the highest sequence similarity to: *S. celluloflavus*, AB184476, 100%; *S. albolongus*, AB184425, 100%; *S. cavourensis* subsp. *cavourensis*, DQ445791, 99.9%; *S. californicus*, AB184755, 99.5%; *S. bacillaris*, AB184439, 99.4%; *S. floridae*, AB184656, 99.4%; *S. globisporus* subsp. *globisporus*, EF178686, 99.4%; *S. griseinus*, AB184205, 99.4%; *S. fulvorobeus*, AB184711, 99.4%; *S. pluricolorescens*, DQ442540, 99.4%; *S. candidus*, DQ026663, 99.4%; *S. spiroverticillatus*, AB184814, 99.4%; *S. sindenensis*, AB184759, 99.4%; *S. badius*, AY999783, 99.4%; *S. rubiginosohelvolus*, AB184240, 99.4%; *S. mediolani*, AB184674, 99.4%; *S. praecox*, AB184293, 99.3%; *S. albovinaceus*, AB249958, 99.3%; *S. microflavus*, DQ445795, 99.3%; *S. fimicarius*, AY999784, 99.3%; *S. flavofuscus*, AB249935, 99.5%; *S. alboviridis*, AB184256, 99.3%; *S. lipmanii*, AB184148, 99.3%; *S. cremeus*, AB184124, 99.3%; *S. griseoplanus*, AY999894, 99.3%; *S. anulatus*, DQ026637, 99.3%; *S. cavourensis* subsp. *washingtonensis*, DQ026671, 99.2%; *S. lavendulae* subsp. *lavendulae*, AB184080, 99.2%; *S. parvus*, DQ442537, 99.2%; *S. luridiscabiei*, AF361784, 99.2%; *S. acrimycini*, AY999889, 99.2%; *S. baarnensis*, EF178688, 99.2%; *S. cyaneofuscatus*, AB184860, 99.2%; *S. cinereorectus*, AB184646, 99.2%; *S. flavovirens*, DQ026635, 99.2%; *S. argenteolus*, AB045872, 99.1%; *S. finlayi*, AY999788, 99.1%; *S. flavogriseus*, AJ494864, 99.1%; *S. nitrosporeus*, EF178680, 99%; *S. griseolus*, AB184768, 99%; *S. pulveraceus*, AB184806, 99%; *S. clavifer*, DQ026670, 99%; *S. griseus* subsp. *griseus*, AY207604, 99%; *S. halstedii*, EF178695, 99%; *S. cinnamonensis*, AB184707, 99%.

Source: not known.

DNA G+C content (mol%): not known.

Type strain: ATCC 19762, CBS 500.68, BCRC 13674, CCUG 11105, DSM 40066, HAMBI 1015, NBRC 12775, IMET 42052, JCM 4380, NCIMB 12975, NRRL B-2095, NRRL-ISP 5066, RIA 1042, UNIQEM 148, VKM Ac-753.

Sequence accession no. (16S rRNA gene): AB249912.

221. **Streptomyces griseocarneus** (Benedict, Lindenfelser, Stodola and Traufler 1950) Witt and Stackebrandt 1991, 456[VP] (Effective publication: Witt and Stackebrandt 1990, 370.) ["*Streptomyces griseocarneus*" Benedict, Lindenfelser, Stodola and Traufler 1950; "*Streptoverticillium griseocarneus*" (*sic*) Baldacci 1958; "*Verticillomyces griseocarneus*" Shinobu 1965; *Streptoverticillium griseocarneum* Baldacci, Farina and Locci 1966, 170]

gri.se.o.car'ne.us. L. adj. *griseus* gray; L. adj. *carneus* pertaining to flesh; N.L. masc. adj. *griseocarneus* gray flesh-colored.

Spore chains in Umbellate monoverticillate (=*Streptomyces* Section *Verticillati*, biverticillate). Mature spore chains are short, generally 3–10 spores per chain. This morphology is seen on yeast-malt agar, salts-starch agar, and glycerol-asparagine agar. Mature spores often not produced until 14–21 d. Spore surface is smooth.

Color of colony: aerial mass color in the Red color series on yeast-malt agar, salts-starch agar, and glycerol-asparagine agar; good spore mass not produced on oatmeal agar. Reverse side of colony with no distinctive pigments (grayed yellow to yellow-brown) on yeast-malt agar, salts-starch agar, and glycerol-asparagine agar; substrate pigment is not a pH indicator.

Color in medium: melanoid pigments formed in peptone-yeast-iron agar. Pigments other than melanoids not formed in yeast-malt agar, oatmeal agar, salts-starch agar, or glycerol-asparagine agar.

D-Glucose is utilized for growth. No growth or only traces of growth on sucrose, rhamnose, and raffinose. Variable reports on growth with L-arabinose, D-xylose, iso-inositol, D-mannitol, and D-fructose (but if growth occurs with these carbon sources it is less than with D-glucose).

Type strain shows no sequence similarity over 99%.

Source: not known.

DNA G+C content (mol%): not known.

Type strain: AS 4.1088, ATCC 12628, ATCC 19763, CBS 501.68, CCM 3228, BCRC 13304, CCUG 11123, CECT 3250, DSM 40004, NBRC 12776, NBRC 3387, JCM 4095, JCM 4580, LMG 5973, NCIMB 9623, NRRL B-1068, NRRL B-1350, NRRL-ISP 5004, PCM 2326, PCM 2345, RIA 1043, RIA 132, UNIQEM 149, VKM Ac-881.

Sequence accession no. (16S rRNA gene): X99943.

Further comments: in violation of Rule 33c of the *Bacteriological Code* (1990 Revision), in Validation List no. 38, *Streptomyces griseocarneus* is proposed as a *nomen revictum* (basonym: "*Streptomyces griseocarneus*" Benedict et al. (1950).

According to Hatano et al. (2003), *Streptomyces griseocarneus* (Benedict et al. 1950) Witt and Stackebrandt 1991 is an earlier heterotypic synonym of *Streptomyces alboverticillatus* (Locci and Schofield 1989) Witt and Stackebrandt 1991 and of *Streptomyces septatus* (Locci et al. 1969) Witt and Stackebrandt 1991. Hatano et al. (2003) also propose that *Streptomyces griseocarneus* (Benedict et al. 1950) Witt and Stackebrandt 1991 be a heterotypic synonym of "*Streptomyces tropicalensis*" Gupta 1965b.

222. **Streptomyces griseochromogenes** Fukunaga *in* Fukunaga, Misato, Ishii and Asakawa 1955, 181[AL]

gri.se.o.chro.mo'ge.nes. N.L. adj. *griseus* gray; Gr. n. *chroma* color; N.L. suff. *-genes* (from Gr. v. *gennaô* to produce) producing; N.L. part. adj. *griseochromogenes* producing gray color.

Spore chains in Section *Spirales* to *Retinaculiaperti*. Open loops and hooks as well as well-developed spirals are common. Mature spore chains generally contain 10–50 spores per chain. This morphology is seen on yeast-malt agar, oatmeal agar, salts-starch agar, and glycerol-asparagine agar. Spore surface is spiny.

Color of colony: aerial mass color in the Gray color series (2fe, medium gray, or 3ge, light grayish yellowish brown) on yeast-malt agar, oatmeal agar, salts-starch agar, and glycerol-asparagine agar; white aerial mycelium may also be seen on young cultures or on 21-d-old cultures on glycerol-asparagine agar. Reverse side of colony with no distinctive pigments (grayish yellow to orange yellow on yeast-malt agar; grayish yellow to light olive brown or grayish greenish yellow on oatmeal agar, salts-starch agar, or glycerol-asparagine agar).

Color in medium: melanoid pigments are formed in peptone-yeast-iron agar, tyrosine agar, and tryptone-yeast broth. No pigment is found in the medium in yeast-malt agar, oatmeal agar, salts-starch agar, or glycerol-asparagine agar.

D-Glucose, L-arabinose, D-xylose, iso-inositol, D-mannitol, D-fructose, sucrose, and raffinose are utilized for growth. No growth or only traces of growth with rhamnose.

Type strain shows the highest sequence similarity to: *S. cellostaticus*, AB184192, 99.7%; *S. yokosukanensis*, DQ026652, 99.3%; *S. corchorusii*, AB184267, 99%; *S. olivaceoviridis*, AB184288, 99%.

Source: not known.

DNA G+C content (mol%): not known.

Type strain: ATCC 14511, CBS 714.72, BCRC 11818, DSM 40499, IFM 1229, NBRC 13413, JCM 4039, JCM 4764, KCTC 9027, NRRL B-12423, NRRL-ISP 5499, RIA 1374.

Sequence accession no. (16S rRNA gene): AB184387.

223. **Streptomyces griseoflavus** (Krainsky 1914) Waksman and Henrici *in* Breed, Murray and Hitchens 1948, 948[AL] ("*Actinomyces griseoflavus*" Krainsky 1914, 694)

gri.se.o.fla'vus. N.L. adj. *griseus* gray; L. adj. *flavus* yellow; N.L. masc. adj. *griseoflavus* grayish yellow.

Spore chains in Section *Spirales*. Mature spore chains generally have 10–50 spores per chain, although shorter chains may also be common. This morphology is seen on yeast-malt agar, oatmeal agar, salts-starch agar, and glycerol-asparagine agar. Spore surface is spiny.

Color of colony: aerial mass color in the Gray color series on yeast-malt agar, oatmeal agar, salts-starch agar, and glycerol-asparagine agar (immature cultures may appear to be in the Yellow color series on glycerol-asparagine agar). Reverse side of colony is yellow to orange-yellow on yeast-malt agar, oatmeal agar, salts-starch agar, and glycerol-asparagine agar. Reverse mycelium pigment is not a pH indicator.

Color in medium: melanoid pigments are not formed in peptone-yeast-iron agar, tyrosine agar, or tryptone-yeast broth. No pigment found in medium in yeast-malt agar, oatmeal agar, salts-starch agar, or glycerol-asparagine agar.

D-Glucose, L-arabinose, D-xylose, iso-inositol, D-mannitol, D-fructose, and rhamnose are utilized for growth. No growth or only traces of growth with sucrose and raffinose.

Type strain shows the highest sequence similarity to: *S. griseorubens*, AB184139, 99.7%; *S. griseoincarnatus*, AJ781328, 99.6%; *S. erythrogriseus*, AJ781328, 99.6%; *S. labedae*, AB184704, 99.6%; *S. variabilis*, DQ442551, 99.6%; *S. flaveolus*, AB184764, 99.5%; *S. collinus*, AB184123, 99.5%;

S. matensis, AB184221, 99.5%; *S. heliomycini*, AB184712, 99.3%; *S. paradoxus*, AB184628, 99.3%; *S. malachitofuscus*, AB184282, 99.3%; *S. althioticus*, AY999808, 99.2%; *S. ambofaciens*, M27245, 99.2%; *S. viridochromogenes*, DQ442555, 99.2%; *S. violaceochromogenes*, AY999867, 99.1%; *S. albaduncus*, AY999757, 99%; *S. speibonae*, AF452714, 99%; *S. violaceorubidus*, AJ781374, 99%; *S. longispororuber*, AB184440, 99%; *S. iakyrus*, AB184877, 99%.

Source: not known.

DNA G+C content (mol%): not known.

Type strain: AS 4.1454, ATCC 25456, CBS 409.52, CBS 689.69, BCRC 12232, DSM 40456, NBRC 13044, IMET 43530, JCM 4479, LMG 19344, NRRL B-5312, NRRL-ISP 5456, RIA 1236, VKM Ac-993.

Sequence accession no. (16S rRNA gene): AJ781322.

224. **Streptomyces griseofuscus** Sakamoto, Kondo, Yumoto and Arishima 1962, 98[AL]

gri.se.o.fus′cus. N.L. adj. *griseus* gray; L. adj. *fuscus* dark, tawny; N.L. masc. adj. *griseofuscus* gray, tawny.

Spore chains in Section *Spirales*. Mature spore chains generally have 10–50 spores per chain. This morphology is observed on yeast-malt agar, oatmeal agar, salts-starch agar, and glycerol-asparagine agar. Spore surface is smooth.

Color of colony: aerial mass color in the Red or Gray color series on yeast-malt agar, oatmeal agar, salts-starch agar, and glycerol-asparagine agar. Characteristic color is between 4ge or 5ge (light grayish reddish brown) and 4ig (light grayish brown) color tabs of Tresner–Backus color wheels. Reverse side of colony with no distinctive pigments on yeast-malt agar, oatmeal agar, salts-starch agar, and glycerol-asparagine agar.

Color in medium: melanoid pigments not formed in peptone-yeast-iron agar, tyrosine agar, or tryptone-yeast broth. No pigment found in medium in yeast-malt agar, oatmeal agar, salts-starch agar, or glycerol-asparagine agar.

D-Glucose, L-arabinose, D-xylose, iso-inositol, D-mannitol, and D-fructose are utilized for growth. No growth or only traces of growth on sucrose, iso-inositol, rhamnose, and raffinose.

Type strain shows the highest sequence similarity to: *S. murinus*, AB184155, 100%; *S. costaricanus*, AB249939, 100%; *S. phaeogriseichromatogenes*, AJ391813, 99.6%.

Source: not known.

DNA G+C content (mol%): not known.

Type strain: ATCC 23916, CBS 837.68, BCRC 10483, CECT 3307, DSM 40191, NBRC 12870, IMET 42068, JCM 4276, JCM 4641, NCIMB 9821, NRRL B-5429, NRRL-ISP 5191, RIA 1145, VKM Ac-1707.

Sequence accession no. (16S rRNA gene): AB184206.

225. **Streptomyces griseoincarnatus** (Preobrazhenskaya, Ryabova and Blinov *in* Gauze, Preobrazhenskaya, Kudrina, Blinov, Ryabova and Sveshnikova 1957) Pridham, Hesseltine and Benedict 1958, 69[AL] ("*Actinomyces griseoincarnatus*" Preobrazhenskaya, Ryabova and Blinov *in* Gauze, Preobrazhenskaya, Kudrina, Blinov, Ryabova and Sveshnikova 1957, 169)

gri.se.o.in.car.na′tus. N.L. adj. *griseus* gray; L. part. adj. *incarnatus* flesh-colored; N.L. masc. part. adj. *griseoincarnatus* grayish flesh-colored, referring to changes in color of the aerial mycelium.

Spore chains in Section *Spirales*. Spirals are generally open with 3 to 7 or more turns; flexuous spore chains and imperfect spirals suggesting *Retinaculiaperti* morphology are also common. Spirals are best developed on oatmeal agar and salts-starch agar. Spore chains are moderately long, especially on oatmeal agar and salts-starch agar with 10–50, or often more than 50, spores per chain. Spore surface is spiny.

Color of colony: aerial mass color in the Gray color series on salts-starch agar ater 14–21 d; younger aerial mycelium may be in the Red color series. Aerial mass color in the Gray or Red color series yeast-malt agar, oatmeal agar, and salts-starch agar. Observers did not agree on the selection of individual color tabs representing the aerial color, but all chose tabs ranging form grayish yellowish pink or grayish reddish brown to medium gray. Reverse side of colony is yellow to yellow brown plus red (grayish yellow on salts-starch agar; orange-yellow to reddish brown on yeast-malt agar, oatmeal agar). Substrate mycelium pigment is not a pH indicator.

Color in medium: melanoid pigments are not produced in peptone-yeast-iron agar, tyrosine agar, or tryptone-yeast broth. Red or orange pigment is found in the medium in yeast-malt agar in 7 d and in oatmeal agar in 14–21 d. Red or orange pigment may or may not be found in glycerol-asparagine agar. This pigment is not a pH indicator.

D-Glucose, L-arabinose, sucrose, D-xylose, D-mannitol, D-fructose, and rhamnose are utilized for growth. Reports vary on utilization of iso-inositol and only traces of growth are found on raffinose.

Type strain shows the highest sequence similarity to: *S. erythrogriseus*, AJ781328, 100%; *S. labedae*, AB184704, 100%; *S. variabilis*, DQ442551, 100%; *S. griseorubens*, AB184139, 99.9%; *S. griseoflavus*, AJ781322, 99.6%; *S. matensis*, AB184221, 99.6%; *S. althioticus*, AY999808, 99.2%; *S. paradoxus*, AB184628, 99.2%; *S. heliomycini*, AB184712, 99.1%; *S. collinus*, AB184123, 99.1%; *S. viridochromogenes*, DQ442555, 99.1%; *S. flaveolus*, AB184764, 99.1%; *S. bellus*, AB184849, 99%; *S. albogriseolus*, AJ494865, 99%; *S. viridodiastaticus*, AY999852, 99%; *S. violaceochromogenes*, AY999867, 99%; *S. malachitofuscus*, AB184282, 99%; *S. violaceorubidus*, AJ781374, 99%; *S. coerulescens*, AY999720, 99%.

Source: not known.

DNA G+C content (mol%): not known.

Type strain: AS 4.1409, ATCC 23623, ATCC 23917, CBS 838.68, BCRC 11481, DSM 40274, NBRC 12871, INA 9673/55, JCM 4381, LMG 19316, NCIMB 9825, NRRL B-5313, NRRL-ISP 5274, RIA 1192.

Sequence accession no. (16S rRNA gene): AJ781321.

226. **Streptomyces griseoloalbus** (Kudrina *in* Gauze, Preobrazhenskaya, Kudrina, Blinov, Ryabova and Sveshnikova 1957) Pridham, Hesseltine and Benedict 1958, 58[AL] ("*Actinomyces griseoloalbus*" Kudrina *in* Gauze, Preobrazhenskaya, Kudrina, Blinov, Ryabova and Sveshnikova 1957, 112)

gri.se.o.lo.al′bus. N.L. dim. adj. *griseolus* somewhat gray; L. adj. *albus* white; N.L. masc. adj. *griseoloalbus* somewhat grayish white.

Spore chain in Section *Rectiflexibiles*. Aerial hyphae may be sterile or spores may be poorly defined. Spore surface is smooth. Terminal swellings of two types are sometimes seen on aerial hyphae. Aerial mycelium is best developed on yeast-malt agar and salts-starch agar, but good sporulating aerial mycelium is not found on any of the ISP media. In addition to the terminal swellings noted above, one observer records fragmentation of the substrate mycelium on yeast-malt agar and salts-starch agar.

Color of colony: aerial mass color in the White or Yellow (2ba, pale yellow) color series on yeast-malt agar and salts-starch agar; a white aerial mycelium may or may not be formed on oatmeal agar and glycerol-asparagine agar, but this mycelium is usually sterile. Reverse side of colony is light yellow or grayish yellow to orange-yellow on yeast-malt agar, oatmeal agar, salts-starch agar, and glycerol-asparagine agar.

Color in medium: melanoid pigments are not formed in peptone-yeast-iron agar, tyrosine agar, or tryptone-yeast broth. A trace of yellow to orange-yellow pigment may be found in the medium in yeast-malt agar, oatmeal agar, salts-starch agar or glycerol-asparagine agar. This pigment is not pH-sensitive when tested with 0.05 M NaOH or HCl.

D-Glucose, L-arabinose, D-xylose, iso-inositol, D-mannitol, D-fructose, rhamnose, and sucrose are utilized for growth. Reports vary on the utilization of raffinose.

Type strain shows the highest sequence similarity to: *S. albuduncus*, AY999757, 100%; *S. matensis*, AB184221, 99.2%; *S. gancidicus*, AB184660, 99.2%; *S. paradoxus*, AB184628, 99.%; *S. pseudogriseolus*, DQ442541, 99.1%; *S. capillispiralis*, AB184577, 99%; *S. heliomycini*, AB184712, 99%; *S. malachitofuscus*, AB184282, 99%; *S. cellulosae*, DQ442495, 99%; *S. lusitanus*, AB184424, 99%.

Source: not known.

DNA G+C content (mol%): not known.

Type strain: ATCC 23624, ATCC 25458, CBS 691.69, DSM 40468, NBRC 13046, INA 1875/54, JCM 4480, NRRL B-12383, NRRL-ISP 5468, RIA 1238, VKM Ac-1739.

Sequence accession no. (16S rRNA gene): AB184275.

227. **Streptomyces griseolus** (Waksman 1923) Waksman and Henrici *in* Breed, Murray and Hitchens 1948, 938[AL] ("*Actinomyces griseolus*" Waksman *in* Bergey, Harrison, Breed, Hammer and Huntoon 1923, 369)

gri.se'o.lus. N.L. adj. *griseus* gray; N.L. dim. masc. adj. *griseolus* somewhat gray.

Spore chain in Section *Rectiflexibiles*. Mature spore chains are moderately short with 3–10, or sometimes more than 10, spores per chain (one observer reports 10–50 spores per chain). This morphology is seen on yeast-malt agar; sporulation on oatmeal agar, salts-starch agar, or glycerol-asparagine agar may be poor. Spore surface is smooth.

Color of colony: aerial mass color in the Gray color series on yeast-malt agar; sporulation usually inadequate for color determination on oatmeal agar, salts-starch agar, and glycerol-asparagine agar. Reverse side of colony is yellow-brown to brown on yeast-malt agar and colorless or characteristic grayed yellow to grayed yellow-brown on oatmeal agar, salts-starch agar, and glycerol-asparagine agar. Substrate pigment is not a pH indicator.

Color in medium: melanoid pigments are not formed in peptone-yeast-iron agar and tyrosine agar. No pigment is found in medium in yeast-malt agar, oatmeal agar, salts-starch agar, or glycerol-asparagine agar.

D-Glucose, L-arabinose, D-xylose, and D-fructose are utilized for growth. No growth or only traces of growth with iso-inositol and raffinose. Variable reports on growth with sucrose, D-mannitol, and rhamnose; interpretation is difficult because of significant growth on the carbon-free basal medium.

Type strain shows the highest sequence similarity to: *S. halstedii*, EF178695, 100%; *S. cinereorectus*, AB184646, 99.9%; *S. cavourensis* subsp. *washingtonensis*, DQ026671, 99.8%; *S. lavendulae* subsp. *lavendulae*, AB184080, 99.8%; *S. argenteolus*, AB045872, 99.8%; *S. mediolani*, AB184674, 99.7%; *S. pluricolorescens*, DQ442540, 99.7%; *S. alboviridis*, AB184256, 99.7%; *S. flavofuscus*, AB249935, 99.7%; *S. sindenensis*, AB184759, 99.7%; *S. lipmanii*, AB184148, 99.7%; *S. praecox*, AB184293, 99.7%; *S. microflavus*, DQ445795, 99.7%; *S. griseinus*, AB184205, 99.7%; *S. rubiginosohelvolus*, AB184240, 99.7%; *S. anulatus*, DQ026637, 99.7%; *S. fimicarius*, AY999784, 99.7%; *S. flavovirens*, DQ026635, 99.7%; *S. badius*, AY999783, 99.7%; *S. fulvorobeus*, AB184711, 99.7%; *S. flavogriseus*, AJ494864, 99.7%; *S. griseoplanus*, AY999894, 99.7%; *S. cyaneofuscatus*, AB184860, 99.7%; *S. baarnensis*, EF178688, 99.6%; *S. acrimycini*, AY999889, 99.6%; *S. globisporus* subsp. *globisporus*, EF178686, 99.6%; *S. floridae*, AB184656, 99.6%; *S. griseus* subsp. *griseus*, AY207604, 99.6%; *S. albovinaceus*, AB249958, 99.6%; *S. luridiscabiei*, AF361784, 99.6%; *S. parvus*, DQ442537, 99.5%; *S. californicus*, AB184755, 99.5%; *S. nitrosporeus*, EF178680, 99.5%; *S. pulveraceus*, AB184806, 99.4%; *S. yanii*, AB006159, 99.4%; *S. olivoviridis*, AB184227, 99.3%; *S. finlayi*, AY999788, 99.3%; *S. atroolivaceus*, AJ781320, 99.3%; *S. bacillaris*, AB184439, 99.2%; *S. sanglieri*, AB249945, 99.2%; *S. clavifer*, DQ026670, 99.2%; *S. spiroverticillatus*, AB184814, 99.1%; *S. albolongus*, AB184425, 99.1%; *S. gelaticus*, DQ026636, 99.1%; *S. atratus*, DQ026638, 99.1%; *S. celluloflavus*, AB184476, 99%; *S. cremeus*, AB184124, 99%; *S. mutomycini*, AB249951, 99%; *S. griseobrunneus*, AB249912, 99%.

Source: not known.

DNA G+C content (mol%): not known.

Type strain: ATCC 19764, ATCC 3325, CBS 502.68, BCRC 13677, DSM 40067, HAMBI 1000, HUT 6056, HUT 6099, NBRC 12777, NBRC 3415, NBRC 3719, IMET 42053, JCM 4042, JCM 4043, JCM 4581, KCTC 9028, NCIMB 9606, NRRL B-1062, NRRL B-2925, NRRL-ISP 5067, RIA 1044, RIA 88, UNIQEM 150, VKM Ac-1726.

Sequence accession no. (16S rRNA gene): AB184768.

228. **Streptomyces griseoluteus** Umezawa, Hayano, Maeda, Ogata and Okami 1950, 112[AL]

gri.se.o.lu'te.us. N.L. adj. *griseus* gray; L. adj. *luteus* yellow; N.L. masc. adj. *griseoluteus* grayish yellow.

Spore chains in Section *Rectiflexibiles*. Crooked or hooked spore chains may also be seen. Mature spore chains are generally short, sometimes with less than 10 spores per chain. This morphology is seen on yeast-malt agar, oatmeal agar, salts-starch agar, and glycerol-asparagine agar. Spore surface is smooth. One observer reports sclerotia-like bodies on the substrate mycelium of salts-starch agar and unusual morphology in the sporulating aerial mycelium. Another observer reports short chains of conidia-like spores on the substrate hyphae on yeast-malt agar and salts-starch agar.

Color of colony: aerial mass color in the Gray color series on yeast-malt agar, oatmeal agar, and salts-starch agar. Aerial mycelium may be poorly developed or absent on glycerol-asparagine agar. Reverse side of colony with no distinctive pigments (pale or grayish yellow to yellowish brown or olive brown) on yeast-malt agar, oatmeal agar, and glycerol-asparagine agar. Yellow to yellow brown may or may not be modified by red (orange) on salts-starch agar. Reverse mycelium pigment on salts-starch agar is a pH indicator, changing from red or reddish brown to gray with the addition of 0.05 M NaOH.

Color in medium: melanoid pigments are formed in peptone-yeast-iron agar, tyrosine agar, and tryptone-yeast broth. Red to bluish gray pigment is found in the medium in salts-starch agar. This pigment may show the same change observed for reverse mycelium pigment with the addition of 0.05 M NaOH.

D-Glucose, L-arabinose, D-xylose, D-mannitol, and D-fructose are utilized for growth. No growth or only traces of growth with iso-inositol, rhamnose, sucrose, and raffinose.

Type strain shows the highest sequence similarity to: *S. recifensis*, AB184165, 99.9%; *S. seoulensis*, AB249970, 99.7%.

Source: not known.

DNA G+C content (mol%): not known.

Type strain: AS 4.1440, ATCC 12768, CBS 676.72, DSM 40392, HUT 6058, IFM 1055, NBRC 13375, JCM 4041, JCM 4765, LMG 19356, NRRL B-1315, NRRL-ISP 5392, RIA 1336, VKM Ac-976.

Sequence accession no. (16S rRNA gene): AY999751.

229. **Streptomyces griseomycini** (Preobrazhenskaya, Blinov and Ryabova *in* Gauze, Preobrazhenskaya, Kudrina, Blinov, Ryabova and Sveshnikova 1957) Pridham, Hesseltine and Benedict 1958, 69[AL] ("*Actinomyces griseomycini*" Preobrazhenskaya, Blinov and Ryabova *in* Gauze, Preobrazhenskaya, Kudrina, Blinov, Ryabova and Sveshnikova 1957, 136)

gri.se.o.my.ci'ni. N.L. adj. *griseus* gray; N.L. suff. *-mycinum*, -mycin (antibiotics produced by *Streptomyces* strains); N.L. gen. adj. *griseomycini* of gray, antibiotic (referring to gray aerial mycelium and antibiotic activity).

Spore chains in Section *Spirales* or *Retinaculiaperti*. Sporophores are short and poorly developed but are frequently coiled at the tips; true spirals are rare. Hooks and loops are of small diameter and therefore are not representative of typical *Retinaculiaperti* cultures. Mature spore chains are short, generally 3–10 spores per chain. This morphology is found on yeast-malt agar, oatmeal agar, salts-starch agar, and glycerol-asparagine agar, but sporulation may be poor on oatmeal agar and glycerol-asparagine agar. Spore surface is hairy. Some smooth spores may be observed.

Color of colony: aerial mass color in the Green color series on yeast-malt and salts-starch agar (one observer placed the culture in the Gray series on these two media). Reverse side of colony is grayed yellow modified by green on yeast-malt agar and salts-starch agar; it is colorless, grayed yellow, or grayed greenish yellow on oatmeal agar and glycerol-asparagine agar. Substrate pigment is not a pH indicator.

Color in medium: melanoid pigments formed in peptone-yeast-iron agar. Pigments other than melanoids not formed in yeast-malt agar, oatmeal agar, salts-starch agar, or glycerol-asparagine agar.

D-Glucose, L-arabinose, D-xylose, iso-inositol, D-mannitol, D-fructose, and rhamnose are utilized for growth. No growth or only traces of growth on sucrose and raffinose.

Type strain shows the highest sequence similarity to: *S. graminearus* AJ781333, 100%; *S. griseostramineus*, AB184140, 100%; *S. glaucus*, AB184665, 99.1%; *S. calvus*, AB184329, 99%.

Source: not known.

DNA G+C content (mol%): not known.

Type strain: ATCC 19765, ATCC 23625, CBS 503.68, BCRC 13763, DSM 40159, NBRC 12778, INA 13984, JCM 4382, NCIMB 9845, NRRL B-5421, NRRL-ISP 5159, RIA 1045, UNIQEM 151.

Sequence accession no. (16S rRNA gene): AB184137.

230. **Streptomyces griseoplanus** Backus, Tresner and Campbell 1957, 536[AL]

gri.se.o.pla'nus. N.L. adj. *griseus* gray; L. adj. *planus* flat, level; N.L. masc. adj. *griseoplanus* flat, gray (referring to the restricted, flat, plane growth and grayish spore color *en masse* of the organism).

Spore chains in Section *Spirales*. Spore chains range from flexuous to spiral. Spirals are generally open: Mature spore chains generally have 10–50 spores per chain. This morphology is seen on yeast-malt agar and oatmeal agar; poor growth on salts-starch agar and glycerol-asparagine agar. Spore surface is warty.

Color of colony: aerial mass color in the Gray color series on yeast-malt agar and oatmeal agar. Reverse side of colony with no distinctive pigments (grayed yellow to yellow-brown on yeast-malt agar and colorless or grayed yellow on oatmeal agar, salts-starch agar, and glycerol-asparagine agar). Substrate pigment is not a pH indicator.

Color in medium: melanoid pigments not formed in peptone-yeast-iron agar, tyrosine agar, or tryptone-yeast broth. No pigment found in medium in yeast-malt agar, oatmeal agar, salts-starch agar, or glycerol-asparagine agar.

D-Glucose, L-arabinose, D-xylose, raffinose, and D-fructose are utilized for growth. No growth or only traces of growth on sucrose, iso-inositol, D-mannitol, and rhamnose.

Type strain shows the highest sequence similarity to: *S. fimicarius*, AY999784, 100%; *S. sindenensis*, AB184759, 100%; *S. praecox*, AB184293, 100%; *S. cavourensis* subsp. *washingtonensis*, DQ026671, 100%; *S. rubiginosohelvolus*, AB184240, 100%; *S. lavendulae* subsp. *lavendulae*, AB184080, 100%; *S. anulatus*, DQ026637, 100%; *S. griseinus*, AB184205, 100%; *S. acrimycini*, AY999889, 100%; *S. badius*, AY999783, 100%; *S. pluricolorescens*, DQ442540, 100%; *S. mediolani*, AB184674, 100%; *S. flavofuscus*, AB249935, 100%; *S. albo-*

viridis, AB184256, 99.9%; *S. lipmanii*, AB184148, 99.9%; *S. cinereorectus*, AB184646, 99.9%; *S. baarnensis*, EF178688, 99.9%; *S. albovinaceus*, AB249958, 99.9%; *S. cyaneofuscatus*, AB184860, 99.9%; *S. microflavus*, DQ445795, 99.9%; *S. fulvorobeus*, AB184711, 99.9%; *S. globisporus* subsp. *globisporus*, EF178686, 99.9%; *S. argenteolus*, AB045872, 99.8%; *S. californicus*, AB184755, 99.8%; *S. parvus*, DQ442537, 99.8%; *S. flavovirens*, DQ026635, 99.7%; *S. griseolus*, AB184768, 99.7%; *S. floridae*, AB184656, 99.7%; *S. griseus* subsp. *griseus*, AY207604, 99.7%; *S. luridiscabiei*, AF361784, 99.7%; *S. flavogriseus*, AJ494864, 99.6%; *S. halstedii*, EF178695, 99.6%; *S. bacillaris*, AB184439, 99.5%; *S. pulveraceus*, AB184806, 99.5%; *S. olivoviridis*, AB184227, 99.5%; *S. atroolivaceus*, AJ781320, 99.5%; *S. finlayi*, AY999788, 99.5%; *S. nitrosporeus*, EF178680, 99.4%; *S. clavifer*, DQ026670, 99.4%; *S. yanii*, AB006159, 99.4%; *S. mutomycini*, AB249951, 99.4%; *S. celluloflavus*, AB184476, 99.3%; *S. griseobrunneus*, AB249912, 99.3%; *S. sanglieri*, AB249945, 99.3%; *S. albolongus*, AB184425, 99.3%; *S. gelaticus*, DQ026636, 99.2%; *S. atratus*, DQ026638, 99.2%; *S. spiroverticillatus*, AB184814, 99.1%; *S. candidus*, DQ026663, 99.1%; *S. cavourensis* subsp. *cavourensis*, DQ445791, 99.1%; *S. cremeus*, AB184124, 99%.

Source: not known.

DNA G+C content (mol%): not known.

Type strain: AS 4.1868, ATCC 19766, CBS 504.68, BCRC 13649, DSM 40009, NBRC 12779, JCM 4300, JCM 4582, NCIMB 9811, NRRL B-3064, NRRL-ISP 5009, RIA 1046, UNIQEM 152, VKM Ac-1727.

Sequence accession no. (16S rRNA gene): AY999894.

231. **Streptomyces griseorubens** (Preobrazhenskaya, Blinov and Ryabova *in* Gauze, Preobrazhenskaya, Kudrina, Blinov, Ryabova and Sveshnikova 1957) Pridham, Hesseltine and Benedict 1958, 65[AL] ("*Actinomyces griseorubens*" Preobrazhenskaya, Blinov and Ryabova *in* Gauze, Preobrazhenskaya, Kudrina, Blinov, Ryabova and Sveshnikova 1957, 144)

gri.se.o.ru'bens. N.L. adj. *griseus* gray; L. part. adj. *rubens* blushing, reddening; N.L. part. adj. *griseorubens* gray-reddening, referring to color of the aerial mycelium, vegetative mycelium and diffusible pigment.

Spore chains in Section *Spirales* or *Rectiflexibiles*. Flexuous sporophores are the dominant forms; hooks, loops, and some open spirals are present. On ISP media, this culture does not produce the long spore chains with wide diameter hooks and loops characteristic of typical *Retinaculiaperti* cultures. Mature spore chains are moderately short with 3–10, or sometimes more than 10, spores per chain. This morphology is observed on yeast-malt agar. Sporulation is poor on oatmeal agar, salts-starch agar, and glycerol-asparagine agar. Spore surface is spiny with very short spines or sometimes with smooth spores.

Color of colony: aerial mass color in the Gray color series on yeast-malt and salts-starch agar. Reverse side of colony with no distinctive pigments on yeast-malt agar, oatmeal agar, salts-starch agar, and glycerol-asparagine agar; substrate pigment is not a pH indicator.

Color in medium: melanoid pigments not formed in peptone-yeast-iron agar, tyrosine agar, or tryptone-yeast broth. No pigment found in medium in yeast-malt agar, oatmeal agar, salts-starch agar, or glycerol-asparagine agar.

D-Glucose, D-xylose, D-mannitol, D-fructose, and rhamnose are utilized for growth. No growth or only traces of growth on sucrose and raffinose. Utilization of L-arabinose and iso-inositol is doubtful.

Type strain shows the highest sequence similarity to: *S. labedae*, AB184704, 99.9%; *S. variabilis*, DQ442551, 99.9%; *S. erythrogriseus*, AJ781328, 99.9%; *S. griseoincarnatus*, AJ781328, 99.9%; *S. griseoflavus*, AJ781322, 99.7%; *S. matensis*, AB184221, 99.7%; *S. althioticus*, AY999808, 99.4%; *S. heliomycini*, AB184712, 99.3%; *S. paradoxus*, AB184628, 99.3%; *S. collinus*, AB184123, 99.2%; *S. viridodiastaticus*, AY999852, 99.2%; *S. flaveolus*, AB184764, 99.2%; *S. viridochromogenes*, DQ442555, 99.2%; *S. bellus*, AB184849, 99.1%; *S. violaceochromogenes*, AY999867, 99.1%; *S. violaceorubidus*, AJ781374, 99.1%; *S. malachitofuscus*, AB184282, 99.1%; *S. tendae*, D63873, 99.1%; *S. albogriseolus*, AJ494865, 99.1%; *S. coerulescens*, AY999720, 99.1%; *S. ambofaciens*, M27245, 99%; *S. speibonae*, AF452714, 99%; *S. longispororuber*, AB184440, 99%; *S. iakyrus*, AB184877, 99%.

Source: not known.

DNA G+C content (mol%): not known.

Type strain: ATCC 19767, ATCC 19909, CBS 505.68, BCRC 12104, DSM 40160, NBRC 12780, INA 6124/54, JCM 4383, NCIMB 9846, NRRL B-3982, NRRL-ISP 5160, RIA 1047, UNIQEM 153, VKM Ac-1894.

Sequence accession no. (16S rRNA gene): AB184139.

232. **Streptomyces griseoruber** Yamaguchi and Saburi 1955, 220[AL]

gri.se.o.ru'ber. N.L. adj. *griseus* gray; L. adj. *ruber* red; N.L. masc. adj. *griseoruber* grayish red.

Spore chains in Section *Spirales* or *Retinaculiaperti*. Open or tight spirals are produced at the ends of moderately long spore chains of 10 to 50 or more spores per chain. Spirals are often incomplete, forming only 1 or 2 turns or a hook at the end of the spore chain. Straight to flexuous chains without hooks or spirals are also present. This morphology is usually best on glycerol-asparagine agar, but may also be observed on yeast-malt agar, oatmeal agar, and salts-starch agar. Spore surface is smooth.

Color of colony: aerial mass color in the Red or Gray color series on yeast-malt agar, oatmeal agar, salts-starch agar, and glycerol-asparagine agar. The color is represented by tab 5fe (light grayish reddish brown) from the Gray color series and tab 5ge (also light grayish reddish brown) from the Red color series. Reverse side of colony is yellow to yellow-brown plus red (reddish orange or reddish brown on yeast-malt agar, oatmeal agar, and salts-starch agar; light yellowish pink on glycerol-asparagine agar). Substrate mycelium pigment is a pH indicator changing from reddish orange to purple with addition of 0.05 M NaOH and from reddish orange to yellowish orange with addition of 0.05 M HCl.

Color in medium: melanoid pigments are produced in peptone-yeast-iron agar, tyrosine agar, and tryptone-yeast broth. Yellow to red pigment (depending upon pH) may be found in the medium in yeast-malt agar, oatmeal agar, salts-starch agar, and glycerol-asparagine agar. This pigment is pH-sensitive, changing from orange to purple with addition of 0.05 M NaOH and from orange to yellow with addition of 0.05 M HCl.

D-Glucose, L-arabinose, D-xylose, iso-inositol, D-fructose, and rhamnose are utilized for growth. No growth or only traces of growth on sucrose, D-mannitol, and raffinose.

Type strain shows the highest sequence similarity to: *S. antibioticus*, AY999776, 99.3%; *S. cinnabarinus*, AB184266, 99%; *S. cellostaticus*, AB184192, 99%.

Source: not known.

DNA G+C content (mol%): not known.

Type strain: AS 4.1417, ATCC 23919, CBS 903.68, BCRC 11826, CCUG 11117, DSM 40281, HAMBI 1051, NBRC 12873, JCM 4200, JCM 4642, LMG 19325, NRRL B-1818, NRRL-ISP 5281, RIA 1195, VKM Ac-1900.

Sequence accession no. (16S rRNA gene): AB184209.

233. **Streptomyces griseorubiginosus** (Ryabova and Preobrazhenskaya *in* Gauze, Preobrazhenskaya, Kudrina, Blinov, Ryabova and Sveshnikova 1957) Pridham, Hesseltine and Benedict 1958, 62[AL] ("*Actinomyces griseorubiginosus*" Ryabova and Preobrazhenskaya *in* Gauze, Preobrazhenskaya, Kudrina, Blinov, Ryabova and Sveshnikova 1957, 193)

gri.se.o.ru.bi.gin.o′sus. N.L. adj. *griseus* gray; N.L. adj. *robiginosus* (*sic*) rusty; N.L. masc. adj. *griseorubiginosus* gray, rusty (referring to the gray aerial mycelium and rosy reddish vegetative mycelium and diffusible pigment on a chemically defined medium).

Spore chains in Section *Rectiflexibiles.* The very long spore chains with more than 50 spores per chain may also form a few open loops and terminal hooks suggestive of *Retinaculiaperti* morphology. This morphology is seen on yeast-malt agar, oatmeal agar, salts-starch agar, and glycerol-asparagine agar. Spore surface is smooth; two observers found only smooth spores, a third observer found both smooth and spiny spores. Long aerial hyphae may be entangled, forming knots and sclerotia-like bodies on yeast-malt agar, oatmeal agar, glycerol-asparagine agar, or Czapek's sucrose agar. One observer records *in situ* germination of spores in 7 d on Czapek's sucrose agar.

Color of colony: aerial mass color in the Gray color series (2dc, yellowish gray to 3fe, light brownish gray or 5fe, light grayish reddish brown) on yeast-malt agar, oatmeal agar, salts-starch agar, and glycerol-asparagine agar. Reverse side of colony is brown, grayish reddish brown or dark brown on yeast-malt agar, oatmeal agar, salts-starch agar, and glycerol-asparagine agar. Reverse mycelium pigment is a pH indicator changing from yellowish brown to gray with addition of 0.05 M HCl.

Color in medium: melanoid pigments are formed in peptone-yeast-iron agar, tyrosine agar, or tryptone-yeast broth. Red or reddish brown pigment is found in the medium in yeast-malt agar, oatmeal agar, and glycerol-asparagine agar and yellow or yellowish brown pigment is found in the medium in salts-starch agar. This pigment is pH-sensitive, showing the same changes noted for reverse mycelium pigment.

D-Glucose, L-arabinose, D-xylose, iso-inositol, D-mannitol, D-fructose, rhamnose, sucrose, and raffinose are all utilized for growth.

Type strain shows the highest sequence similarity to: *S. phaeopurpureus*, DQ026666, 100%.

Source: not known.

DNA G+C content (mol%): not known.

Type strain: ATCC 23627, ATCC 25459, CBS 692.69, BCRC 12124, DSM 40469, NBRC 13047, INA 7712, JCM 4481, NRRL B-12384, NRRL-ISP 5469, RIA 1239, VKM Ac-1203.

Sequence accession no. (16S rRNA gene): AJ781339.

234. **Streptomyces griseosporeus** Niida and Ogasawara 1960, 23[AL]

gri.se.o.spo′re.us. N.L. adj. *griseus* gray; N.L. n. *spora* a spore; N.L. masc. adj. *griseosporeus* gray spored.

Spore chains in Section *Retinaculiaperti.* Spore chains may be flexuous or in imperfect spirals, hooks and loops. Flexuous chains are common. Mature spore chains are moderately long with 10–50 or more spores per chain. This morphology is seen on yeast-malt agar, oatmeal agar, salts-starch agar, and glycerol-asparagine agar. Spore surface is smooth.

Color of colony: aerial mass color in the Gray color series (2fe or e, moderate gray; 3fe, light brownish gray; or 5fe, light grayish reddish brown) on yeast-malt agar, oatmeal agar, salts-starch agar, and glycerol-asparagine agar. Reverse side of colony with no distinctive pigments (pale or grayish yellow or sometimes light yellowish brown on yeast-malt agar, oatmeal agar, salts-starch agar, and glycerol-asparagine agar).

Color in medium: melanoid pigments are formed in peptone-yeast-iron agar, tyrosine agar, and tryptone-yeast broth. No pigment is found in the medium in yeast-malt agar, oatmeal agar, salts-starch agar, or glycerol-asparagine agar.

D-Glucose, L-arabinose, D-xylose, iso-inositol, D-mannitol, D-fructose, rhamnose, sucrose, and raffinose are all utilized for growth.

Type strain shows no sequence similarity over 99%.

Source: not known.

DNA G+C content (mol%): not known.

Type strain: ATCC 27435, CBS 137.72, CBS 759.72, DSM 40562, HAMBI 1009, NBRC 13458, IMET 43543, JCM 4766, NRRL B-12498, NRRL-ISP 5562, RIA 1419, VKM Ac-1731.

Sequence accession no. (16S rRNA gene): AB184419.

235. **Streptomyces griseostramineus** (Preobrazhenskaya, Kudrina, Blinov and Ryabova *in* Gauze, Preobrazhenskaya, Kudrina, Blinov, Ryabova and Sveshnikova 1957) Pridham, Hesseltine and Benedict 1958, 65[AL] ("*Actinomyces griseostramineus*" Preobrazhenskaya, Kudrina, Blinov and Ryabova *in* Gauze, Preobrazhenskaya, Kudrina, Blinov, Ryabova and Sveshnikova 1957, 155)

gri.se.o.stra.mi′ne.us. N.L. adj. *griseus* gray; L. adj. *stramineus* of straw, here straw-colored; N.L. masc. adj. *griseostramineus* gray, straw-colored (referring to the gray aerial mycelium and straw-yellow vegetative mycelium on a chemically defined medium).

Spore chains in Section *Spirales.* Short spore chains may form tight spirals of only a few turns on yeast-malt agar and salts-starch agar. Hooks and loops suggestive of Section *Retinaculiaperti* may also be found on these media and on oatmeal agar. Mature spore chains are short, generally 3–10 spores per chain. This morphology is seen on yeast-malt agar and salts-starch agar; sporulation may be poor on oatmeal agar and glycerol-asparagine agar. Spore surface is hairy to spiny: appendages are shorter than characteristic hairs, but longer and more flexuous than typical spines.

Color of colony: aerial mass color in the Green color series on yeast-malt agar and salts-starch agar. Reverse side of colony with no distinctive pigment (grayed yellow or grayed yellow modified by green) on yeast-malt agar, oatmeal agar, salts-starch agar, and glycerol-asparagine agar; substrate pigment is not a pH indicator.

Color in medium: melanoid pigments formed in peptone-yeast-iron agar. Pigments other than melanoids not formed in yeast-malt agar, oatmeal agar, salts-starch agar, or glycerol-asparagine agar.

D-Glucose, L-arabinose, D-xylose, iso-inositol, D-mannitol, D-fructose, and rhamnose are utilized for growth. No growth or only trace of growth on sucrose. Variable reports on growth with raffinose.

Type strain shows the highest sequence similarity to: *S. griseomycini*, AB184137, 100%; *S. graminearus* AJ781333, 100%; *S. calvus*, AB184329, 99%; *S. glaucus*, AB184665, 99%.

Source: not known.

DNA G+C content (mol%): not known.

Type strain: ATCC 19768, ATCC 23628, CBS 506.68, BCRC 12075, CECT 3273, DSM 40161, NBRC 12781, INA 10381, JCM 4385, NRRL B-5422, NRRL-ISP 5161, RIA 1048, UNIQEM 154, VKM Ac-968.

Sequence accession no. (16S rRNA gene): AB184140.

236. **Streptomyces griseoverticillatus** (Shinobu and Shimada 1962) Witt and Stackebrandt 1991, 456[VP] (Effective publication: Witt and Stackebrandt 1990, 370.) ("*Streptomyces griseoverticillatus*" Shinobu and Shimada 1962, 174; "*Verticillomyces griseoverticillatus*" Shinobu 1965; *Streptoverticillium griseoverticillatum* Locci, Baldacci and Petrolini Baldan 1969, 59)

gri.se.o.ver.ti.cil.la'tus. N.L. adj. *griseus* gray; L. masc. n. *verticillus* whorl; N.L. adj. *verticillatus* whorled; N.L. masc. adj. *griseoverticillatus* gray and whorled.

Spore chains in Section *Verticillati*, umbellate monoverticillate (biverticillate). Mature spore chains are short with 3 to 10 or more spores per chain. This morphology is seen on yeast-malt agar, oatmeal agar, salts-starch agar, and glycerol-asparagine agar. Spore surface is smooth.

Color of colony: aerial mass color in the Red color series (usually 4ec, grayish yellowish pink; sometimes 4ge, light grayish reddish brown; 4ie, light brown; or 3ca, pale orange-yellow) on yeast-malt agar, oatmeal agar, salts-starch agar, and glycerol-asparagine agar. Reverse side of colony is yellowish brown or orange-yellow on yeast-malt agar; grayish yellowish pink to light olive brown on oatmeal agar; yellowish brown to strong brown on salts-starch agar; pale or grayish yellow on glycerol-asparagine agar. Reverse mycelium pigment is not a pH indicator.

Color in medium: melanoid pigments are not formed in peptone-yeast-iron agar, tyrosine agar, or tryptone-yeast broth. No pigment is found in the medium in yeast-malt agar, oatmeal agar, salts-starch agar, or glycerol-asparagine agar.

D-Glucose and iso-inositol are utilized for growth. Utilization of D-fructose is doubtful. No growth or only traces of growth with L-arabinose, D-xylose, rhamnose, sucrose, raffinose, and D-mannitol.

For sequence similarity, see type strain of *Streptomyces cinnamoneus.*

Source: not known.

DNA G+C content (mol%): not known.

Type strain: ATCC 27436, CBS 721.72, BCRC 12430, DSM 40507, NBRC 13420, JCM 4202, JCM 4767, NRRL B-12432, NRRL-ISP 5507, PCM 2351, RIA 1381, VKM Ac-883.

Sequence accession no. (16S rRNA gene): AB184862.

Further comments: In violation of Rule 33c of the *Bacteriological Code* (1990 Revision), in Validation List no. 38, *Streptomyces griseoverticillatus* is proposed as a *nomen revictum* (basonym: "*Streptomyces griseoverticillatus*" Shinobu and Shimada (1962).

According to Hatano et al. (2003), *Streptomyces griseoverticillatus* (Shinobu and Shimada 1962) Witt and Stackebrandt 1991 is a later heterotypic synonym of *Streptomyces cinnamoneus* (Benedict et al. 1952) Witt and Stackebrandt 1991.

237. **Streptomyces griseoviridis** Anderson, Ehrlich, Sun and Burkholder 1956, 114[AL]

gri.se.o.vi'ri.dis. N.L. adj. *griseus* gray; L. adj. *viridis* green; N.L. masc. adj. *griseoviridis* gray-green.

Spore chains in Section *Spirales.* Some open spirals have only one or two turns or are poorly developed suggesting *Retinaculiaperti* morphology. Spore chains are moderately long with 10–50 spores per chain. Spore surface is smooth.

Color of colony: aerial mass color in the Red color series (grayish yellowish pink to pale orange yellow) on yeast-malt agar, oatmeal agar, salts-starch agar, and glycerol-asparagine agar. One observer placed this culture in the Yellow color series. The original description (Anderson et al., 1956) included the following statement: "Aerial mycelium is white to pink-tan, ping-gray, or brown or occasionally greenish". The greenish color was not observed on ISP media. Reverse side of colony is grayed yellow to olive brown or dark brown on yeast-malt agar, oatmeal agar, salts-starch agar, and glycerol-asparagine agar. Substrate mycelium pigment is not a pH indicator or is changed only slightly by 0.05 M NaOH or HCl.

Color in medium: melanoid pigments are not found in peptone-yeast-iron agar, tyrosine agar, or tryptone-yeast broth. No pigment in medium, or only trace of yellow, in yeast-malt agar, oatmeal agar, salts-starch agar, or glycerol-asparagine agar.

D-Glucose, L-arabinose, D-xylose, D-mannitol, and rhamnose are utilized for growth. Utilization of D-fructose is doubtful. No growth or only traces of growth on sucrose, iso-inositol, and raffinose.

Type strain shows the highest sequence similarity to: *S. aurantiogriseus*, AY999793, 99.4%; *S. coelicolor*, DQ442496, 99.3%.

Source: not known.

DNA G+C content (mol%): not known.

Type strain: AS 4.1418, ATCC 23920, CBS 904.68, DSM 40229, HAMBI 1086, NBRC 12874, JCM 4250, JCM 4643, KCTC 9780, LMG 19321, NCIMB 9853, NRRL 2427, NRRL-ISP 5229, RIA 1170, VKM Ac-622.

Sequence accession no. (16S rRNA gene): AY999807.

238a. **Streptomyces griseus subsp. griseus** (Krainsky 1914) Waksman and Henrici *in* Breed, Murray and Hitchens 1948, 948[AL] ("*Actinomyces griseus*" Krainsky 1914, 662)

gri'se.us. N.L. masc. adj. *griseus* gray.

Spore chains in Section *Rectiflexibiles.* Mature spore chains are moderately long with 10–50 spores per chain. This morphology is seen on yeast-malt agar, oatmeal agar, salts-starch agar, and glycerol-asparagine agar. Spore surface is smooth.

Color of colony: aerial mass color in the Yellow color series (2db, pale yellow, or 1½db–1½ec, pale greenish yellow) on yeast-malt agar, oatmeal agar, salts-starch agar, and glycerol-asparagine agar. Reverse side of colony with no distinctive pigments (grayed yellow to olive brown or light yellowish brown) on yeast-malt agar, oatmeal agar, salts-starch agar, and glycerol-asparagine agar.

Color in medium: melanoid pigments are formed in tyrosine agar, but not in peptone-yeast-iron agar or tryptone-yeast broth. No pigments other than traces of yellow are found in the medium in yeast-malt agar, oatmeal agar, salts-starch agar, or glycerol-asparagine agar.

D-Glucose, D-xylose, D-mannitol, and D-fructose are utilized for growth. No growth or only traces of growth on L-arabinose, sucrose, iso-inositol, rhamnose, or raffinose.

Type strain shows the highest sequence similarity to: *S. cavourensis* subsp. *washingtonensis*, DQ026671, 99.8%; *S. lavendulae* subsp. *lavendulae*, AB184080, 99.8%; *S. cinereorectus*, AB184646, 99.7%; *S. alboviridis*, AB184256, 99.7%; *S. sindenensis*, AB184759, 99.7%; *S. mediolani*, AB184674, 99.7%; *S. fimicarius*, AY999784, 99.7%; *S. lipmanii*, AB184148, 99.7%; *S. fulvorobeus*, AB184711, 99.7%; *S. anulatus*, DQ026637, 99.7%; *S. praecox*, AB184293, 99.7%; *S. flavofuscus*, AB249935, 99.7%; *S. cyaneofuscatus*, AB184860, 99.7%; *S. pluricolorescens*, DQ442540, 99.7%; *S. griseinus*, AB184205, 99.7%; *S. microflavus*, DQ445795, 99.7%; *S. griseoplanus*, AY999894, 99.7%; *S. rubiginosohelvolus*, AB184240, 99.7%; *S. argenteolus*, AB045872, 99.7%; *S. badius*, AY999783, 99.7%; *S. griseolus*, AB184768, 99.6%; *S. luridiscabiei*, AF361784, 99.6%; *S. floridae*, AB184656, 99.6%; *S. albovinaceus*, AB249958, 99.6%; *S. acrimycini*, AY999889, 99.6%; *S. baarnensis*, EF178688, 99.6%; *S. flavovirens*, DQ026635, 99.5%; *S. halstedii*, EF178695, 99.5%; *S. californicus*, AB184755, 99.5%; *S. parvus*, DQ442537, 99.5%; *S. flavogriseus*, AJ494864, 99.4%; *S. pulveraceus*, AB184806, 99.4%; *S. yanii*, AB006159, 99.4%; *S. finlayi*, AY999788, 99.3%; *S. globisporus* subsp. *globisporus*, EF178686, 99.3%; *S. bacillaris*, AB184439, 99.2%; *S. nitrosporeus*, EF178680, 99.2%; *S. olivoviridis*, AB184227, 99.2%; *S. atroolivaceus*, AJ781320, 99.2%; *S. clavifer*, DQ026670, 99.2%; *S. gelaticus*, DQ026636, 99.1%; *S. atratus*, DQ026638, 99.1%; *S. mutomycini*, AB249951, 99%; *S. sanglieri*, AB249945, 99%; *S. albolongus*, AB184425, 99%; *S. celluloflavus*, AB184476, 99%; *S. griseobrunneus*, AB249912, 99%.

Source: not known.

DNA G+C content (mol%): not known.

Type strain: AS 4.1419, ATCC 23345, ATCC 23921, CBS 905.68, BCRC 13478, CCT 4836, CCUG 11104, CECT 3330, CFBP 4546, CIP 105124, DSM 40236, HAMBI 2315, NBRC 12875, NBRC 15744, JCM 4047, JCM 4644, KACC 20084, KCTC 9080, KCTC 9135, LMG 19302, NCIMB 13023, NCTC 13033, NRRL B-2682, NRRL-ISP 5236, PCM 2331, RIA 1176, VKM Ac-800.

Sequence accession no. (16S rRNA gene): AY207604.

Further comments: according to Liu et al. (2005b), *Streptomyces griseus* (Krainsky 1914) Waksman and Henrici 1948 emend. Liu et al. 2005b is an earlier heterotypic synonym of *Streptomyces argenteolus* Tresner et al. 1961, an earlier heterotypic synonym of *Streptomyces caviscabies* Goyer et al. 1996, and an earlier heterotypic synonym of *Streptomyces setonii* (Millard and Burr 1926) Waksman 1953.

According to Guo et al. (2008), *Streptomyces griseus* (Krainsky 1914) Waksman and Henrici 1948 is an earlier heterotypic synonym of *Streptomyces erumpens* Calot and Cercós 1963.

According to Guo et al. (2008), *Streptomyces argenteolus* Tresner et al. 1961 is not a later heterotypic synonym of *Streptomyces griseus* (Krainsky 1914) Waksman and Henrici 1948.

According to Guo et al. (2008), *Streptomyces caviscabies* Goyer et al. 1996 is not a later heterotypic synonym of *Streptomyces griseus* (Krainsky 1914) Waksman and Henrici 1948.

238b. **Streptomyces griseus subsp. alpha** (Ciferri 1927) Pridham 1970, 37[AL] ["*Actinomyces albus* subsp. alpha" (*sic*) Ciferri 1927, 83]

al'pha. L. n. *alpha*, alpha, first letter of the Greek alphabet.

Spore chains are typically flexuous; grows poorly on Czapek's solution agar. Exhibits slight anti-bacterial activity; inhibited by streptomycin. Reported as a cause of musty odor in cacao beans (*Theobroma cacao* L.).

For sequence similarity, see type strain of *Streptomyces microflavus.*

Source: not known.

DNA G+C content (mol%): not known.

Type strain: CBS 219.25, DSM 40937, NBRC 15421, JCM 5078, LMG 19953, NRRL B-2249.

Sequence accession no. (16S rRNA gene): AB184668.

Further comments: according to Lanoot et al. (2005b), *Streptomyces griseus* subsp. *alpha* (Ciferri 1927) Pridham 1970 is a later heterotypic synonym of *Streptomyces microflavus* (Krainsky 1914) Waksman and Henrici 1948 emend. Lanoot et al. 2005b.

238c. **Streptomyces griseus subsp. cretosus** Pridham 1970, 37[AL] ("*Oospora cretacea*" Krüger 1905, 286; "*Actinomyces cretaceus*" Krasil'nikov 1941, 34; "*Streptomyces cretaceus*" Waksman 1950, 143)

cre.tos'us. L. masc. adj. *cretosus* chalky.

Exhibits slight anti-microbial activity; inhibited by streptomycin; moderate growth observed on Czapek's solution agar.

For sequence similarity, see type strain of *Streptomyces microflavus.*

Source: not known.

DNA G+C content (mol%): not known.

Type strain: ATCC 27903, CBS 137.21, CBS 758.72, DSM 40561, NBRC 13457, JCM 4742, KCTC 9079, LMG 19946, NRRL B-2252, NRRL-ISP 5561, RIA 1418, VKM Ac-712.

Sequence accession no. (16S rRNA gene): AB184418.

Further comments: according to Lanoot et al. (2005b), *Streptomyces griseus* subsp. *cretosus* Pridham 1970 is a later heterotypic synonym of *Streptomyces microflavus* (Krainsky 1914) Waksman and Henrici 1948 emend. Lanoot et al. 2005b.

238d. **Streptomyces griseus subsp. solvifaciens** Pridham 1970, 38[AL]

sol.vi.fa'ci.ens. L. v. *solvo* to loosen; L. v. *facio* to make; N.L. part. adj. *solvifaciens* making loose, dissolving, referring to the lytic activity of actinomycetin.

Spore chains are typically flexuous; poor growth is seen on Czapek's solution agar. Exhibits anti-bacterial and anti-fungal activity; produces actinomycetin, now considered a general term for a number of different lytic enzymic anti-bacterial and anti-viral factors; inhibited by streptomycin.

Type strain shows the highest sequence similarity to: *S. limosus*, AB184147, 100%; *S. felleus*, AB184129, 100%; *S. daghestanicus*, DQ442497, 100%; *S. albidoflavus*, AB184255, 100%; *S. hydrogenans*, AB184868, 100%; *S. odorifer*, Z76682, 100%; *S. violascens*, AY999737, 100%; *S. champavatii*, DQ026642, 99.9%; *S. canescens*, AB184117, 99.9%; *S. sampsonii*, D63871, 99.7%; *S. koyangensis*, AY079156, 99.6%.

Source: not known.

DNA G+C content (mol%): not known.

Type strain: JCM 5079, DSM 40933, NBRC 13689, NRRL B-1561.

Sequence accession no. (16S rRNA gene): AB249915.

Further comments: according to Guo et al. (2008), *Streptomyces griseus* subsp. *solvifaciens* Pridham 1970 should be removed from *Streptomyces griseus.* However, no formal proposition is made in the paper by Guo et al. (2008).

239. **Streptomyces guanduensis** Xu, Wang, Cui, Huang, Liu, Zheng and Goodfellow 2006, 1114[VP]

gu.an.du.en'sis. N.L. masc. adj. *guanduensis* of or belonging to Guandu, the source of the soil from which the type strain was isolated.

Acidophilic streptomycete; forms branched substrate and aerial hyphae. Smooth-surfaced spores are borne in flexuous spore chains. Deep-brown colonies that carry a white to gray aerial spore mass are formed on oatmeal agar, ISP medium 9 supplemented with glucose (1%, w/v), and on yeast extract-malt extract agar. Diffusible pigments are not formed and melanin pigments are not produced on peptone-yeast extract-iron agar or tyrosine agar. Degrades Tween 80, but not adenine, guanine, starch, or xanthine. Cellobiose, D-galactose, D-glucose, D-inulin, D-lactose, D-mannitol, and D-salicin (each at 1%, w/v), and adipic acid and L-phenylalanine (each at 0.1%, w/v) are used as sole carbon sources for energy and growth, but adonitol and D-sorbitol (each at 1%, w/v), and L-alanine, DL-aminobutyric acid, L-arginine, α-L-aspartic acid, L-cysteine, L-valine, sodium acetate, sodium citrate, and sodium oxalate (each at 0.1%, w/v) are not. Growth occurs at temperatures between 20 and 37°C, but not at 15°C, and at pH values from 4.5–7.0, but not at pH 3.5. Does not grow in the presence of 5% (w/v) NaCl. The organism is sensitive to filter-paper discs soaked in the following (μg/ml unless indicated): acetylspiramycin (15), carbenicillin (10), cephalothin (30), ciprofloxacin (5), doxycycline hydrochloride (30), erythromycin (15), josamycin (15), kanamycin sulfate (30), minocyclin hydrochloride (30), and tobramycin sulfate (10), but not to amoxycillin (10), ampicillin (10), azitrhomycin (30), aztreonam (30), penicillin G (10 IU/ml), or sulfamethoxazole (25). Additional properties are shown in Table 264. Chemotaxonomic properties are typical of members of the genus *Streptomyces.*

Type strain shows no sequence similarity over 99%. Type strain shows DNA–DNA similarity to: *S. yeochonensis* NRRL B-24245[T], 18.7%; *S. paucisporeus* JCM 13276[T], 14.9%; *S. rubidus* JCM 13277[T], 21.4%; *S. yanglinensis* JCM 13275[T], 23.1%.

Source: the type strain was isolated from soil from Guandu.

DNA G+C content (mol%): 72.7.

Type strain: 701, CGMCC 4.2022, JCM 13274.

Sequence accession no. (16S rRNA gene): AY876942.

240. **Streptomyces hachijoensis** (Hosoya, Komatsu, Soeda and Sonoda 1952) Witt and Stackebrandt 1991, 456[VP] (Effective publication: Witt and Stackebrandt 1990, 370.) ("*Streptomyces hachijoensis*" Hosoya, Komatsu, Soeda and Sonoda 1952, 508; *Streptoverticillium hachijoense* Locci, Baldacci and Petrolini Baldan 1969, 59)

ha.chi.jo.en'sis. N.L. masc. adj. *hachijoensis* or or belonging to Hachijo (named for the place of origin, Hachijo Jima, a small island in the Pacific Ocean).

Spore chains in Section *Verticillati.* Both monoverticillate and umbellate monoverticillate (biverticillate) spore chains are found. Mature spore chains are short, generally 3–10 spores per chain. This morphology is seen on yeast-malt agar, oatmeal agar, salts-starch agar, and glycerol-asparagine agar. Spore surface is smooth.

Color of colony: aerial mass color in the Red color series on yeast-malt agar, oatmeal agar, salts-starch agar, and glycerol-asparagine agar. Reverse side of colony with no distinctive pigment (grayed yellow to yellow-brown) on yeast-malt agar, oatmeal agar, salts-starch agar, and glycerol-asparagine agar; substrate pigment is not a pH indicator.

Color in medium: melanoid pigments not formed in peptone-yeast-iron agar and tyrosine agar. No pigment in medium, or only trace of yellow, in yeast-malt agar, oatmeal agar, salts-starch agar, and glycerol-asparagine agar.

D-Glucose and iso-inositol are utilized for growth. No growth or only traces of growth on L-arabinose, sucrose, D-xylose, D-mannitol, D-fructose, rhamnose, and raffinose.

For sequence similarity, see type strain of *Streptomyces cinnamoneus*.

Source: isolated from soil from Hachijo Jima, a small island in the Pacific Ocean.

DNA G+C content (mol%): not known.

Type strain: ATCC 19769, CBS 507.68, BCRC 12419, CECT 3260, DSM 2011, DSM 40114, NBRC 12782, JCM 4331, JCM 4583, NRRL B-3106, NRRL-ISP 5114, RIA 1049, UNIQEM 155, VKM Ac-191.

Sequence accession no. (16S rRNA gene): AB184141.

Further comments: in violation of Rule 33c of the *Bacteriological Code* (1990 Revision), in Validation List no. 38, *Streptomyces hachijoensis* is proposed as a *nomen revictum* (basonym: "*Streptomyces hachijoensis*" Hosoya et al. (1952).

According to Hatano et al. (2003), *Streptomyces hachijoensis* (Hosoya et al. 1952) Witt and Stackebrandt 1991 is a later heterotypic synonym of *Streptomyces cinnamoneus* (Benedict et al. 1952) Witt and Stackebrandt 1991.

241. **Streptomyces hainanensis** Jiang, Tang, Wiese, Xu, Imhoff and Jiang 2007, 2697[VP]

hai.nan.en'sis. N.L. masc. adj. *hainanensis* of or pertaining to Hainan, a province of south China, from where the type strain was isolated.

Spore chains are spiral or looped. Spores are elliptical or short rod-shaped. Spore surface is smooth. Aerial mycelia are white and pink white to pink gray. Vegetative mycelia are pale to deep orange yellow. Produces light brown to orange yellowish soluble pigments. Gelatin liquefaction, milk coagulation and peptonization, arginase, phenylalanine deaminase, DNase, and melanin production are negative. Starch hydrolysis, arginine decarboxylase, nitrate reduction, gas production from nitrate, growth on cellulose, and H_2S production are positive. Glucose, cellobiose, starch, esculin, galactoside, and urea are utilized; acids are not produced from the five carbon sources. Galactose, mannose, fructose, arabinose, xylose, ribose, rhamnose, sucrose, lactose, maltose, melibiose, raffinose, turanose, melezitose, sorbin, dextrin, mycose, salicin, adonitol, inositol, mannitol, sorbitol, xylitol, galactitol, erythritol, amygdaloside, sodium citrate, sodium acetate, gluconate, malonate, tartrate, lysine, ornithine, and acetamide are not utilized. Optimal growth occurs at pH 7.0 (range pH 6.0–9.0) and without NaCl (range 0–10% NaCl). Resistant to penicillin G (10 IU), amoxycillin/clavulanic acid (20/10 µg), novobiocin (30 µg), rifampin (5 µg), and ampicillin (10 µg), and sensitive to erythromycin (15 µg IU), gentamicin (10 µg), kanamycin (30 µg), tetracycline (30 µg), vancomycin (30 µg), midecamycin (15 µg), clindamycin (2 µg), sulfamethoxazole/trimethoprim (23.75/1.25 µg), chloramphenicol (30 µg), polymyxin B (300 IU), and norfloxacin (10 µg). Cell wall contains LL-A_2pm, trace *meso*-A_2pm, and glycine. Whole-cell hydrolysates contain galactose and xylose. Main phospholipids are phosphatidylethanolamine and diphosphatidylglycerol (phospholipid type II). The predominant menaquinones are MK-9(H_4) (45.4%), MK-9(H_6) (14.0%), MK-9(H_8) (13.6%), and MK-10(H_0) (27.0%). Fatty acid composition comprises $C_{15:0}$ iso (1.1%), $C_{16:0}$ iso (30.9%), $C_{16:1}\omega7c$/$C_{16:1}\omega6c$ (2.1%), $C_{16:0}$ (13.7%), $C_{17:1}$ iso$\omega9c$ (1.4%), $C_{17:0}$ iso (2.8%), $C_{17:0}$ anteiso (10.8%), $C_{17:1}\omega8c$ (7.1%), cyclo-$C_{17:0}$ (1.4%), $C_{17:0}$ (5.6%), $C_{18:0}$ anteiso/$C_{18:2}\omega6,9c$ (9.7%), $C_{18:1}\omega9c$ (4.9%), $C_{18:0}$ (1.2%), and $C_{17:1}$ iso$\omega9c$/$C_{16:0}$ 10-methyl (1.4%).

Type strain shows no sequence similarity over 99%.

Source: the type strain was isolated from a soil sample collected from evergreen broadleaf forest in Wuzhi Mountain, Hainan Province, China.

DNA G+C content (mol%): 73.4.

Type strain: YIM 47672, CCTCC AA 205017, DSM 41900.

Sequence accession no. (16S rRNA gene): AM398645.

242. **Streptomyces halstedii** (Waksman and Curtis 1916) Waksman and Henrici *in* Breed, Murray and Hitchens 1948, 953[AL] ("*Actinomyces halstedii*" Waksman and Curtis 1916, 124)

hal.ste'di.i. N.L. gen. masc. n. *halstedii* of Halsted, named for Byron David Halsted (1852–1918) of Rutgers University.

Spore chains Section *Rectiflexibiles*. Spore chains are predominantly flexuous, but many hooks and some irregular coils similar to *Rectiflexibiles* morphology are found on yeast-malt agar and glycerol-asparagine agar. Spore chains are short, 3–10 spores per chain, on yeast-malt agar, salts-starch agar, and glycerol-asparagine agar. Sporulation may be poor, especially on oatmeal agar. Since the original characterization by Waksman (1916) describes closed spirals 7–10 µm in diameter, the short flexuous or hooked chains may be atypical. Spore surface is smooth. Fragmentation of substrate mycelium is noted by one observer.

Color of colony: aerial mass color in the Gray color series on yeast-malt agar, salts-starch agar, and glycerol-asparagine agar. Reverse side of colony has no distinctive pigment (grayed yellow to yellow-brown) on yeast-malt agar, oatmeal agar, salts-starch agar, and glycerol-asparagine agar; substrate pigment is not a pH indicator.

Color in medium: melanoid pigments not formed in peptone-yeast-iron agar, tyrosine agar, or tryptone-yeast broth. No pigment found in medium in yeast-malt agar, oatmeal agar, salts-starch agar, or glycerol-asparagine agar.

D-Glucose, L-arabinose, D-xylose, and D-fructose are utilized for growth. No growth or only traces of growth on sucrose, iso-inositol, D-mannitol, rhamnose, and raffinose.

Type strain shows the highest sequence similarity to: *S. griseolus*, AB184768, 100%; *S. cinereorectus*, AB184646, 99.8%; *S. lavendulae* subsp. *lavendulae*, AB184080, 99.7%; *S. argenteolus*, AB045872, 99.7%; *S. cyaneofuscatus*, AB184860, 99.7%; *S. cavourensis* subsp. *washingtonensis*, DQ026671, 99.7%; *S. praecox*, AB184293, 99.7%; *S. anulatus*, DQ026637, 99.7%; *S. flavovirens*, DQ026635, 99.7%; *S. fimicarius*, AY999784, 99.7%; *S. lipmanii*, AB184148, 99.7%; *S. flavofuscus*, AB249935, 99.7%; *S. microflavus*, DQ445795, 99.7%; *S. alboviridis*, AB184256, 99.7%; *S. fulvorobeus*, AB184711, 99.7%; *S. mediolani*, AB184674, 99.6%; *S. badius*, AY999783, 99.6%; *S. acrimycini*, AY999889, 99.6%;

S. griseoplanus, AY999894, 99.6%; *S. sindenensis*, AB184759, 99.6%; *S. pluricolorescens*, DQ442540, 99.6%; *S. rubiginosohelvolus*, AB184240, 99.6%; *S. griseinus*, AB184205, 99.6%; *S. flavogriseus*, AJ494864, 99.6%; *S. baarnensis*, EF178688, 99.5%; *S. luridiscabiei*, AF361784, 99.5%; *S. albovinaceus*, AB249958, 99.5%; *S. parvus*, DQ442537, 99.5%; *S. floridae*, AB184656, 99.5%; *S. californicus*, AB184755, 99.5%; *S. griseus* subsp. *griseus*, AY207604, 99.5%; *S. globisporus* subsp. *globisporus*, EF178686, 99.5%; *S. nitrosporeus*, EF178680, 99.4%; *S. yanii*, AB006159, 99.3%; *S. pulveraceus*, AB184806, 99.3%; *S. finlayi*, AY999788, 99.2%; *S. atroolivaceus*, AJ781320, 99.2%; *S. bacillaris*, AB184439, 99.2%; *S. olivoviridis*, AB184227, 99.2%; *S. gelaticus*, DQ026636, 99.1%; *S. sanglieri*, AB249945, 99.1%; *S. clavifer*, DQ026670, 99.1%; *S. atratus*, DQ026638, 99.1%; *S. albolongus*, AB184425, 99%; *S. spiroverticillatus*, AB184814, 99%; *S. cremeus*, AB184124, 99%; *S. celluloflavus*, AB184476, 99%; *S. griseobrunneus*, AB249912, 99%.

Source: not known.

DNA G+C content (mol%): not known.

Type strain: ATCC 10897, ATCC 19770, CBS 508.68, BCRC 13680, CECT 3328, DSM 40068, HAMBI 993, NBRC 12783, IMET 40322, JCM 4584, NCIMB 9839, NRRL B-1238, NRRL-ISP 5068, RIA 1050, UNIQEM 156, VKM Ac-1768.

Sequence accession no. (16S rRNA gene): EF178695.

243. **Streptomyces hawaiiensis** Cron, Whitehead, Hooper, Heinemann and Lein 1956, 63[AL]

ha.wai.i.en′sis. N.L. masc. adj. *hawaiiensis* of or belonging to Hawaii, the source of the soil from which the organism was isolated.

Spore chains in Section *Spirales*, but on some media short or poorly developed spore chains are flexuous or have hooks, loops or open spirals resembling *Retinaculiaperti* morphology. Mature spore chains are moderately short with 3–10, or sometimes more than 10 spores per chain. This morphology is seen on yeast-malt agar, oatmeal agar, salts-starch agar, and glycerol-asparagine agar. Spore surface is spiny.

Color of colony: aerial mass color in the White color series on salts-starch agar and glycerol-asparagine agar; Yellow or White series on yeast-malt agar and oatmeal agar. Reverse side of colony with no distinctive pigment (grayed yellow to yellow-brown) on yeast-malt agar, oatmeal agar, salts-starch agar, and glycerol-asparagine agar; substrate pigment is not a pH indicator.

Color in medium: melanoid pigments formed in peptone-yeast-iron agar and tryptone-yeast broth; pigments other than melanoids not formed in yeast-malt agar, oatmeal agar, salts-starch agar, or glycerol-asparagine agar.

D-Glucose, L-arabinose, sucrose, iso-inositol, D-mannitol, D-fructose, rhamnose, and raffinose are utilized for growth; utilization of D-xylose is doubtful.

Type strain shows the highest sequence similarity to: *S. massasporeus*, AB184152, 99.7%; *S. arenae*, AB249977, 99.7%; *S. purpurascens*, AJ399486, 99.5%; *S. luteogriseus*, AB184379, 99.4%; *S. indiaensis*, AB184553, 99.3%; *S. janthinus*, AB184851, 99.3%; *S. violaceus*, AB184315, 99.3%; *S. albosporeus* subsp. *albosporeus*, AJ781327, 99.3%; *S. coerulescens*, AY999720, 99.2%; *S. bellus*, AB184849, 99.2%; *S. roseoviolaceus*, AJ399484, 99.2%; *S. flavoviridis*, AB184842, 99.1%; *S. pilosus*, AB184161, 99%; *S. afghaniensis*, AJ399483, 99%; *S. africanus*, AY208912, 99%; *S. lomondensis*, AB184673, 99%; *S. levis*, AB184670, 99%; *S. coeruleorubidus*, AY999719, 99%.

Source: isolated from soil from Hawaii.

DNA G+C content (mol%): not known.

Type strain: ATCC 12236, ATCC 19771, CBS 509.68, BCRC 13653, DSM 40042, IFM 1071, NBRC 12784, IMET 43082, JCM 4172, JCM 4585, LMG 5975, NCIMB 9410, NRRL B-1988, NRRL-ISP 5042, PCM 2315, RIA 1051, UNIQEM 157, VKM Ac-1761.

Sequence accession no. (16S rRNA gene): AB184143.

244. **Streptomyces hebeiensis** Xu, Li, Wu, Wang, Xu and Jiang 2004a, 730[VP]

he.bei.en′sis. N.L. masc. adj. *hebeiensis* of or pertaining to Hebei, a province in northern China where the sample yielding the type strain was collected.

Aerial mycelium and substrate mycelium are well developed. Aerial mycelium at maturity forms long, straight to *Rectiflexibiles* spore chains composed of nonmotile and coccoid spores with a warty surface. Diffusible pigments are produced on several media. The pigment is not a pH indicator or is changed only slightly with addition of 0.05 M HCl in ISP 5 and ISP 6. Colony color is medium-dependent. Casein and xanthine can be metabolized, but adenine and pectin cannot. Tests for gelatin, nitrate reduction, and melanin production are positive and tests for H_2S production and peptonization of milk are negative. Galactose, lactose, mannose, maltose, xylose, sorbitol, sodium citrate, sodium acetate, oxalate, starch, and glycerol are utilized as sole carbon and energy sources, but cellulose and xylan are not. Acid is formed from mannose and starch, but not from arabinose, fructose, galactose, glucose, inositol, lactose, mannitol, maltose, rhamnose, raffinose, sucrose, sorbitol, xylose, sodium citrate, sodium acetate, oxalate, or glycerol. L-Histidine and L-hydroxyproline can be used as sole carbon and nitrogen sources. Grows well at 27, 30, and 37°C but does not grow at 45 or 10°C. Grows in the presence of 4 or 7% NaCl and 0.1% phenol. Diagnostic amino acid of peptidoglycan is LL-A_2pm with trace amounts of *meso*-A_2pm. Whole-cell hydrolysates contain glucose and small quantities of xylose, galactose, and arabinose. The menaquinones are MK9(H_4) (4.6%), MK-9(H_6) (60%), MK-9(H_8) (30.7%), and MK-9(H_{10}) (4.7%) and phosphatidylethanolamine is the diagnostic phospholipid.

Type strain shows no sequence similarity over 99%.

Source: not known.

DNA G+C content (mol%): 71.4.

Type strain: YIM 001, CCTCC AA 203005, CIP 107974, DSM 41837, JCM 12696. *Sequence accession no. (16S rRNA gene):* AY277529.

245. **Streptomyces heliomycini** (*ex* Braznikova, Uspenskaya, Sokolova, Preobrazhenskaya, Gause, Ukholina, Shorin, Rossolimo and Vertogradova 1958) Preobrazhenskaya 1986, 574[VP] (Effective publication: Preobrazhenskaya *in* Gause, Preobrazhenskaya, Sveshnikova, Terekhova and

Maximova 1983.) ("*Actinomyces flavochromogenes* subsp. *heliomycini*" Braznikova, Uspenskaya, Sokolova, Preobrazhenskaya, Gause, Ukholina, Shorin, Rossolimo and Vertogradova 1958)

he.li.o.my.ci'ni. N.L. n. *heliomycinum* heliomycin; N.L. gen. n. *heliomycini* of heliomycin, intended to mean heliomycin producing.

Spore chains are spiral (*Spirales*); spore surface is warty, hairy, and spiny. On mineral agar 1, starch-ammonia agar: aerial mycelium is green-gray; substrate mycelium is yellow, brown-orange; no diffusible pigment. On glycerol-nitrate agar, glycerol-asparagine agar: aerial mycelium is gray; substrate mycelium is dark yellow, brown yellowish, dark gray; no diffusible pigment. On oatmeal agar: aerial mycelium is gray; substrate mycelium is dark yellow, yellow brown; no diffusible pigment. On organic agar 2: aerial mycelium is gray, dark gray, green gray; substrate mycelium is brownish yellow, yellow brown; sometimes no diffusible pigment. Melanoid pigments are not formed. Utilization of glucose, sucrose, xylose, inositol, mannitol, fructose, and rhamnose. Antibiotic: geliomycin.

Type strain shows the highest sequence similarity to: *S. griseorubens*, AB184139, 99.3%; *S. griseoflavus*, AJ781322, 99.3%; *S. atrovirens*, DQ026672, 99.2%; *S. matensis*, AB184221, 99.2%; *S. griseoincarnatus*, AJ781328, 99.1%; *S. erythrogriseus*, AJ781328, 99.1%; *S. labedae*, AB184704, 99.1%; *S. variabilis*, DQ442551, 99.1%; *S. ambofaciens*, M27245, 99.1%; *S. albaduncus*, AY999757, 99%; *S. griseoloalbus*, AB184275, 99%; *S. lienomycini*, AJ781353, 99%.

Source: not known.

DNA G+C content (mol%): not known.

Type strain: DSM 41690, NBRC 15899, INA 2915, JCM 9767, VKM Ac-1778.

Sequence accession no. (16S rRNA gene): AB184712.

246. **Streptomyces helvaticus** (Krasil'nikov, Korenyako and Nikitina *in* Krasil'nikov 1965) Pridham 1970, 18[AL] ("*Actinomyces helvaticus*" Krasil'nikov, Korenyako and Nikitina *in* Krasil'nikov 1965, 224)

helv.va'ti.cus. N.L. n. *Helvetia* Switzerland; N.L. masc. adj. orth. var. *helvaticus* of or belonging to Switzerland.

Spore chains in Section *Retinaculiaperti* to *Spirales*. Straight to flexuous spore chains may also be common, especially when short spore chains are produced. This morphology is seen on yeast-malt agar, oatmeal agar, salts-starch agar, and glycerol-asparagine agar. Sporulating aerial mycelium may be poorly developed or absent on yeast-malt agar. Spore surface is smooth.

Color of colony: aerial mass color probably in the Red color series (3ca, pale orange-yellow to 4ca, light yellowish pink) on oatmeal agar and salts-starch agar when optimum sporulation occurs. Aerial mass color may also appear to be white or yellow (2ba or 2db, pale yellow) on oatmeal agar and salts-starch agar. Aerial mycelium is poorly developed on yeast-malt agar and glycerol-asparagine agar. Reverse side of colony with no distinctive pigment (light yellow or pale yellow) on yeast-malt agar, oatmeal agar, salts-starch agar, and glycerol-asparagine agar.

Color in medium: melanoid pigments are not formed in peptone-yeast-iron agar, tyrosine agar, or tryptone-yeast broth. No pigment or only a trace of yellow is found in the medium in yeast-malt agar, oatmeal agar, salts-starch agar, and glycerol-asparagine agar.

D-Glucose, L-arabinose, D-xylose and iso-inositol are utilized for growth. Utilization of D-fructose is doubtful. No growth or only traces of growth with rhamnose, sucrose, raffinose, and D-mannitol.

Type strain shows the highest sequence similarity to: *S. flavidovirens*, AB184270, 100%; *S. chryseus*, AY999787, 100%; *S. albidochromogenes*, AB249953, 99.9%; *S. enissocaesilis*, AB249930, 99.1%.

Source: not known.

DNA G+C content (mol%): not known.

Type strain: ATCC 19841, CBS 683.72, DSM 40431, NBRC 13382, INMI 1013-B, JCM 4768, NRRL B-12365, NRRL-ISP 5431, RIA 1343, VKM Ac-192.

Sequence accession no. (16S rRNA gene): AB184367.

247. **Streptomyces herbaricolor** Kawato and Shinobu 1959, 114[AL]

her.ba.ri.co'lor. L. n. *herbarius* one skilled in plants, a botanist; L. n. *color* color; N.L. adj. *herbaricolor* grass colored, green (referring to the grass green diffusible pigment produced by the organism on chemically defined media).

Spore chains in Section *Rectiflexibiles*. Mature spore chains generally have 10–50 spores per chain on yeast-malt agar and glycerol-asparagine agar. Sporulation aerial mycelium is poorly developed or absent on oatmeal agar and salts-starch agar. Spore surface is smooth.

Color of colony: aerial mass color in the Gray color series on glycerol-asparagine agar and yeast-malt agar. Aerial mycelium is poorly developed on all ISP media; mass color cannot be determined on oatmeal agar or salts-starch agar. Reverse side of colony: substrate mycelium may be grayish yellow to olive brown or brown or it may contain an additional pH-sensitive pigment with a color range from pale purple or pale purplish pink to green or greenish blue. When this pigment is present, addition of 0.05 M NaOH changes color from violet-red to green or blue and this change is reversed by addition of 0.05 M HCl. The pH-sensitive mycelial pigment was found by one observer in yeast-malt agar, oatmeal agar, salts-starch agar, and glycerol-asparagine agar.

Color in medium: melanoid pigments are not formed or occur only in trace amounts in peptone-yeast-iron agar and tyrosine agar. Pigment may or may not be formed in the medium in yeast-malt agar, oatmeal agar, salts-starch agar, or glycerol-asparagine agar. When formed, this pigment shows the same pH range found in the transient substrate pigment.

D-Glucose, L-arabinose, sucrose, D-xylose, D-fructose, and raffinose are utilized for growth. No growth or only trace of growth on iso-inositol, D-mannitol, and rhamnose.

Type strain shows the highest sequence similarity to: *S. indigoferus*, AB184214, 100%; *S. aburaviensis*, AY999779, 99.6,%; *S. purpureus*, AJ781324, 99.6%; *S. chrysomallus* subsp. *fumigatus*, AB184645, 99.5%; *S. purpeofuscus*, AJ781364, 99.2%; *S. xanthocidicus*, AY999858, 99.2%. Type strain shows the highest sequence similarity to following

Kitasatospora species: *Kitasatospora kifunensis*, AB022874, 99.3%; *Kitasatospora nipponensis*, AY442263, 99%; *Kitasatospora gansuensis*, AY442265, 99%.

Source: not known.

DNA G+C content (mol%): not known.

Type strain: ATCC 23922, CBS 424.61, CBS 906.68, BCRC 13772, DSM 40123, NBRC 12876, NBRC 3932, JCM 4138, JCM 4645, NBRC 3838, NCIMB 9837, NRRL B-3299, NRRL-ISP 5123, RIA 1126, RIA 654, VKM Ac-793.

Sequence accession no. (16S rRNA gene): AB184801.

248. **Streptomyces hiroshimensis** (Shinobu 1955) Witt and Stackebrandt 1991, 456[VP] (Effective publication: Witt and Stackebrandt 1990, 370.) ("*Streptomyces hiroshimensis*" Shinobu 1955; "*Verticillomyces hiroshimensis*" Shinobu 1965; *Streptoverticillium hiroshimense* Farina and Locci 1966, 51)

hi.ro.shim.en′sis. N.L. masc. adj. *hiroshimensis* of or pertaining to Hiroshima.

Spore chains in Umbellate Monoverticillate (=*Streptomyces* Section *Verticillati*, biverticillate). Mature spore chains are short, generally 3–10 spores per chain. This morphology is seen on yeast-malt agar, oatmeal agar, salts-starch agar, and glycerol-asparagine agar in 14–21 d. Spore surface is smooth.

Color of colony: aerial mass color in the Red color series on yeast-malt agar, oatmeal agar, and salts-starch agar. Reverse side of colony is grayed yellow, modified by red on oatmeal agar, salts-starch agar, and glycerol-asparagine agar; or yellow-brown is modified by red on yeast-malt agar. Substrate pigment is not a pH indicator (one observer reports slight change from pink to brown with NaOH on oatmeal agar).

Color in medium: melanoid pigments formed in peptone-yeast-iron agar and tryptone-yeast broth; pigments other than melanoids not formed in yeast-malt agar, oatmeal agar, salts-starch agar, or glycerol-asparagine agar.

D-Glucose, iso-inositol, and D-fructose are utilized for growth. No growth or only traces of growth on L-arabinose, sucrose, D-xylose, D-mannitol, rhamnose, and raffinose.

Type strain shows the highest sequence similarity to: *S. caeruleus*, EF178675, 99.7%; *S. cinnamoneus*, AB184850, 99.7%; *S. abikoensis*, AB184537, 99.5%; *S. blastmyceticus*, AY999802, 99.5%; *S. aureoversilis*, AB184855, 99.5%; *S. pseudoechinosporeus*, AB184100, 99.5%; *S. lilacinus*, AB184819, 99.4%; *S. ardus*, AB184864, 99.2%; *S. biverticillatus*, AJ781381, 99%; *S. albireticuli*, AB184881, 99%; *S. sapporonensis*, AB184508, 99%; *S. werraensis*, DQ442558, 99%.

Source: not known.

DNA G+C content (mol%): not known.

Type strain: ATCC 19772, CBS 510.68, BCRC 13375, CECT 3261, DSM 40037, HUT 6033, NBRC 12785, NBRC 3839, IMET 43546, JCM 4098, JCM 4586, KCTC 9781, NBRC 3720, NCIMB 9838, NRRL B-1823, NRRL B-5484, NRRL-ISP 5037, RIA 1052, RIA 592, UNIQEM 158, VKM Ac-902.

Sequence accession no. (16S rRNA gene): AB184789.

Further comments: according to Hatano et al. (2003), *Streptomyces hiroshimensis* (Shinobu 1955) Witt and Stackebrandt 1991 is an earlier heterotypic synonym of *Streptomyces aureoversilis* corrig. (Locci et al. 1969) Witt and Stackebrandt 1991, of *Streptomyces baldaccii* corrig. (Farina and Locci 1966) Witt and Stackebrandt 1991, of *Streptomyces biverticillatus* (Preobrazhenskaya and Ryabova 1957) Witt and Stackebrandt 1991, of *Streptomyces fervens* (DeBoer et al. 1959–1960) Witt and Stackebrandt 1991, of *Streptomyces rectiverticillatus* (Krasil'nikov and Yuan 1965) Witt and Stackebrandt 1991, of *Streptomyces roseoverticillatus* (Shinobu 1956) Witt and Stackebrandt 1991, of *Streptomyces salmonis* (Baldacci et al. 1966) Witt and Stackebrandt 1991, and of *Streptomyces spitsbergensis* Wieczorek et al. 1993. Hatano et al. (2003) also propose that *Streptomyces hiroshimensis* (Shinobu 1955) Witt and Stackebrandt 1991 be a heterotypic synonym of "*Streptomyces fervens* subsp. *melrosporus*" (NBRC 15920), and of "*Streptoverticillium rubrochlorinum*" (NBRC 14694). In the paper by Hatano et al. (2003), "*Streptomyces fervens* subsp. *melrosporus*" is not in quotes. However, this name has no standing in bacterial nomenclature.

249. **Streptomyces hirsutus** Ettlinger, Corbaz and Hütter 1958a, 344[AL]

hir.su′tus. L. masc. adj. *hirsutus* shaggy, bristly, with stiff hairs.

Spore chains in Section *Retinaculiaperti* or *Spirales*. Flexuous spore chains and chains terminating in hooks and loops are common; well developed spirals of more than 1 or 2 turns are rare. Mature spore chains generally have 10–30 spores per chain. This morphology is seen on yeast-malt agar, oatmeal agar, salts-starch agar, and glycerol-asparagine agar. Spore surface is spiny.

Color of colony: aerial mass color in the Green color series on yeast-malt agar, oatmeal agar, salts-starch agar, and glycerol-asparagine agar. Reverse side of colony with no distinctive pigment (grayed yellow to grayed greenish yellow) on yeast-malt agar, oatmeal agar, salts-starch agar, and glycerol-asparagine agar; substrate pigment is not a pH indicator.

Color in medium: melanoid pigments not formed in peptone-yeast-iron agar or tyrosine agar. No pigment found in medium in yeast-malt agar, oatmeal agar, salts-starch agar, or glycerol-asparagine agar.

D-Glucose, L-arabinose, sucrose, D-xylose, iso-inositol, D-mannitol, D-fructose, rhamnose, and raffinose are utilized for growth.

Type strain shows the highest sequence similarity to: *S. cyanoalbus*, AB184882, 100%; S. *prasinus*, DQ026658, 99.1%; *S. bambergiensis*, AB184869, 99.1%.

Source: not known.

DNA G+C content (mol%): not known.

Type strain: ATCC 19773, CBS 511.68, BCRC 13676, DSM 40095, HAMBI 1003, NBRC 12786, IMET 42054, JCM 4191, JCM 4587, NRRL B-2713, NRRL-ISP 5095, RIA 1053, UNIQEM 159, VKM Ac-623.

Sequence accession no. (16S rRNA gene): AB184844.

250. **Streptomyces humidus** Nakazawa and Shibata *in* Imamura, Hori, Nakazawa, Shibata, Tatsuoka and Miyake 1956, 648[AL]

hu′mi.dus. L. masc. adj. *humidus* wet, damp, moist.

Spore chains in Section *Spirales* with many incomplete spirals resembling *Retinaculiaperti* morphology. Spore chains short to moderately long with 10–50 spores per chain. This morphology is seen on yeast-malt agar, oatmeal agar, salts-starch agar, and glycerol-asparagine agar.

Spore surface is smooth. Although the original description reports production of black, moist (hygroscopic) areas on mature aerial mycelium, this was not recorded on ISP reports.

Color of colony: aerial mass color in the Gray or Red color series (3fe or 3li, brownish gray from the Gray color series to 5dc or 5ch, grayish yellowish pink, in the Red color series). Reverse side of colony with no distinctive pigments (colorless, pale or grayish yellow or light olive gray) on yeast-malt agar, oatmeal agar, salts-starch agar, and glycerol-asparagine agar.

Color in medium: melanoid pigments not produced in peptone-yeast-iron agar, tyrosine agar, or tryptone-yeast broth. No pigment, or only trace of yellow pigment, is found in the medium in yeast-malt agar, oatmeal agar, salts-starch agar, and glycerol-asparagine agar.

D-Glucose, L-arabinose, D-xylose, iso-inositol, D-mannitol, D-fructose, and rhamnose are utilized for growth. No growth or only traces of growth on sucrose and raffinose.

Type strain shows the highest sequence similarity to: *S. cacaoi* subsp. *asoensis*, DQ026644, 99.5%; *S. rishiriensis*, EF178691, 99.3%; *S. novaecaesareae*, AB184357, 99.1%; *S. pseudovenezuelae*, AB184233, 99%.

Source: not known.

DNA G+C content (mol%): not known.

Type strain: ATCC 12760, ATCC 23923, CBS 907.68, BCRC 13707, DSM 40263, NBRC 12877, JCM 4386, NRRL B-3172, NRRL-ISP 5263, RIA 1186, VKM Ac-1703.

Sequence accession no. (16S rRNA gene): DQ442508.

251. **Streptomyces humiferus** Goodfellow, Williams and Alderson 1986a, 574[VP] (Effective publication: Goodfellow, Williams and Alderson 1986e, 63.) (*Actinopycnidium caeruleum* Krasil'nikov 1962, 250)

hu.mi'fer.us. L. n. *humus* ground; L. v. *fero* to bear; N.L. masc. adj. *humiferus* borne of the ground, i.e. soil-borne.

Spore chains are *Spirales*; the spore surface is smooth. Forms extensively branched substrate and aerial mycelium. The aerial spore mass is gray; the reverse is red-orange and the pigment is pH-sensitive; red-orange diffusible pigments which are also pH-sensitive are produced. Does not form melanin pigments. Allantoin, adenine, esculin, arbutin, elastin, gelatin, guanine, hypoxanthine, pectin, starch, testosterone, tyrosine, urea, and xanthine are degraded but chitin and lecithin are not. Hydrogen sulfide is produced and nitrate is reduced. L-Arabinose, cellobiose, D-fructose, D-galactose, D-glucose, *myo*-inositol, D-lactose, mannitol, D-mannose, melibiose, raffinose, L-rhamnose, trehalose, and D-xylose are used as sole carbon sources but adonitol, inulin, melezitose, sucrose, and xylitol are not. Grows on L-arginine, L-cysteine, L-histidine, L-phenylalanine, potassium nitrate, L-threonine, and L-valine but not on DL-amino-n-butyric acid, L-hydroxyproline, L-methionine, or L-serine as sole nitrogen sources. Grows at 10–37°C, but not at 4 or 45°C. Tolerant to phenol (0.01%, w/v) and sodium chloride (10%, w/v) but not to sodium azide (0.01%, w/v). Resistant to rifampin. Antimicrobial activity shown against *Streptomyces murinus* ISP 5091 but not towards *Aspergillus niger* LIV 131, *Bacillus subtilis* NCIB 3610, *Candida albicans* CBS 562, *Escherichia coli* NCIB 9132, *Micrococcus luteus* NCIB 196, *Pseudomonas fluorescens* NCIB 9046[T], and *Saccharomyces cerevisiae* CBS 1171[T]. The peptidoglycan contains LL-A_2pm as the major diamino acid.

Type strain shows the highest sequence similarity to: *S. coelescens*, AF503496, 100%; *S. violaceolatus* AF503497, 100%; *S. violaceoruber*, AF503492, 100%; *S. tricolor*, AB184687, 99.8%; *S. anthocyanicus*, AB184631, 99.8%; *S. rubrogriseus*, AB184681, 99.7%; *S. tendae*, D63873, 99.5%; *S. lienomycini*, AJ781353, 99.5%; *S. violaceorubidus*, AJ781374, 99.3%; *S. coelicoflavus*, AB184650, 99.3%.

Source: not known.

DNA G+C content (mol%): 70.4.

Type strain: AS 4.1070, ATCC 15719, ATCC 15812, DSM 43030, IFM 1139, NBRC 12244, IMET 43409, JCM 3037, KCC A-0037, KCTC 9116, NCIMB 10164, RIA 729, VKM Ac-644.

Sequence accession no. (16S rRNA gene): AF503491.

Further comments: the genus *Actinopycnidium* Krasil'nikov 1962 is reduced to synonymy with *Streptomyces* Waksman and Henrici 1943. For *Actinopycnidium caeruleum* Krasil'nikov 1962, it is necessary to substitute a new specific epithet, because there is a senior homonym, *Streptomyces caeruleus* (Baldacci 1944) Pridham et al. 1958, cited on the Approved Lists of Bacterial Names (Rules 34a and 41a of the *Bacteriological Code* (1990 Revision).

252. **Streptomyces hydrogenans** Lindner, Junk, Nesemann and Schmidt-Thomé 1958, 117[AL]

hy.dro'gen.ans. Gr. n. *hudôr* water; Gr. v. *gennaô* to produce; N.L. part. adj. *hydrogenans* water producing.

Spore chains in Section *Rectiflexibiles*. Mature spore chains are generally long with more than 50 spores per chain. This morphology is seen on yeast-malt agar, oatmeal agar, salts-starch agar, and glycerol-asparagine agar. Spore surface is smooth.

Color of colony: aerial mass color in the Gray color series on yeast-malt agar; White or Yellow color series on oatmeal agar, salts-starch agar, and glycerol-asparagine agar. Nearest matching color tabs in the Gray color series are d, light gray, and 5fe, light grayish reddish brown. Nearest matching tab in the Yellow color series is 2ba, pale yellow. Reverse side of colony with no distinctive pigments (pale or grayish yellow) on yeast-malt agar, oatmeal agar, salts-starch agar, and glycerol-asparagine agar.

Color in medium: melanoid pigments are not formed in peptone-yeast-iron agar, tyrosine agar, or tryptone-yeast broth. No pigment is found in the medium in yeast-malt agar, oatmeal agar, salts-starch agar, or glycerol-asparagine agar.

D-Glucose, L-arabinose, D-xylose, and rhamnose are utilized for growth. No growth or only traces of growth with iso-inositol, D-mannitol, D-fructose, sucrose, and raffinose.

Type strain shows the highest sequence similarity to: *S. limosus*, AB184147, 100%; *S. felleus*, AB184129, 100%; *S. daghestanicus*, DQ442497, 100%; *S. albidoflavus*, AB184255, 100%; *S. violascens*, AY999737, 100%; *S. odorifer*, Z76682, 100%; *S. griseus* subsp. *solvifaciens*, AB249915, 100%; *S. champavatii*, DQ026642, 100%; *S. canescens*, AB184117, 100%; *S. sampsonii*, D63871, 99.8%; *S. koyangensis*, AY079156, 99.7%.

Source: not known.

DNA G+C content (mol%): not known.

Type strain: ATCC 19631, CBS 776.72, BCRC 11855, DSM 40586, HAMBI 405, NBRC 13475, JCM 4771, NRRL B-12091, NRRL-ISP 5586, RIA 1436, VKM Ac-1919.

Sequence accession no. (16S rRNA gene): AB184868.

253a. **Streptomyces hygroscopicus subsp. hygroscopius** (Jensen 1931) Waksman and Henrici *in* Breed, Murray and Hitchens 1948, 953[AL] ("*Actinomyces hygroscopicus*" Jensen 1931, 357)

hy.gro.sco'pi.cus. Gr. adj. *hugros* moist; Gr. n. *skopos* one that watches, watcher; N.L. masc. adj. *hygroscopicus* detecting moisture, covered with moisture, hygroscopic.

Spore chains in Section *Spirales*; tight spirals in dense clusters. Spirals may coalesce as dark, moist masses of spores. Mature spore chains generally contain 10–50 spores per chain. This morphology is seen on yeast-malt agar, oatmeal agar, salts-starch agar, and glycerol-asparagine agar. Spore surface is warty; individual spores are poorly delineated. Moist, black, liquefied (hygroscopic) areas are found in the aerial mycelium in 14–21 d. These are especially common on oatmeal agar and salts-starch agar.

Color of colony: aerial mass color in the Gray color series (3fe, 3li or 5ih, brownish gray; or 5fe, light grayish reddish brown) on yeast-malt agar, oatmeal agar, salts-starch agar, and glycerol-asparagine agar. Dark or medium gray areas are also reported on oatmeal agar and salts-starch agar, and moist black (hygroscopic) areas may be seen on older cultures. Reverse side of colony with no distinctive pigments (colorless to grayish yellow, pale yellow or light olive brown or gray) on yeast-malt agar, oatmeal agar, salts-starch agar, and glycerol-asparagine agar, except that dark discoloration may be seen beneath hygroscopic areas.

Color in medium: melanoid pigments are not formed in peptone-yeast-iron agar, tyrosine agar, or tryptone-yeast broth. No pigment or only a trace of yellow is found in the medium in yeast-malt agar, oatmeal agar, salts-starch agar, and glycerol-asparagine agar.

D-Glucose, D-mannitol, D-fructose, and rhamnose are utilized for growth. Good growth is reported on carbon-free control medium as well as on L-arabinose, D-xylose, and iso-inositol, so that utilization of these carbon sources is doubtful. Sucrose and raffinose are probably not utilized.

Type strain shows the highest sequence similarity to: *S. endus*, AY999911, 100%; *S. sporocinereus*, AB249933, 100%; *S. demainii*, DQ334782, 99.9%; *S. yogyakartensis*, AB249942, 99.5%; *S. violaceusniger*, AJ391823, 99.5%; *S. albiflaviniger*, AJ391812, 99.3%.

Source: not known.

DNA G+C content (mol%): not known.

Type strain: ATCC 27438, CBS 773.72, BCRC 11611, CIP 106840, DSM 40578, NBRC 13472, JCM 4772, LMG 19335, NRRL 2387, NRRL-ISP 5578, RIA 1433, VKM Ac-831.

Sequence accession no. (16S rRNA gene): AB184428.

253b. **Streptomyces hygroscopicus subsp. angustmyceticus** Yüntsen, Ohkuma, Ishii and Yonehara 1956, 200[AL]

an.gust.my.ce'ti.cus. L. adj. *angustus* narrow; Gr. n. *mukês* fungus; N.L. adj. *myceticus* fungus-like; N.L. masc. adj. *angustmyceticus* like a narrow fungus, but referring to the narrow spectrum of anti-bacterial activity of the organism, hence angustmycin.

Excellent growth on Czapek's solution agar; hygroscopic; NaCl tolerance >10%, but <13%. Produces angustmycin A, angustmycin B (adenine) and angustmycin C-anti-mycobacterial antibiotics; exhibits anti-fungal activity; inhibited by streptomycin.

Type strain shows the highest sequence similarity to: *S. ehimensis*, AY999834, 99.8%; *S. abikoensis*, AB184537, 99.7%; *S. sapporonensis*, AB184508, 99.5%; *S. lilacinus*, AB184819, 99.4%; *S. luteireticuli*, AB249969, 99.1%; *S. varsoviensis*, DQ026653, 99.1%; *S. thioluteus*, AB184753, 99%; *S. morookaense*, AJ781349, 99%; *S. mobaraensis*, DQ442528, 99%.

Source: not known.

DNA G+C content (mol%): not known.

Type strain: ATCC 15484, DSM 41683, NBRC 3934, JCM 4053, KCTC 1089, NRRL B-2347, NRRL B-3306.

Sequence accession no. (16S rRNA gene): DQ442509.

253c. **Streptomyces hygroscopicus subsp. decoyicus** Vavra, Dietz, Churchill, Siminoff and Koepsell 1959, 427[AL]

de.co'yi.cus. N.L. n. *decoyininum* decoyinine, name of an antibiotic; L. masc. suff. *-icus* suffix used with the sense of belonging to; N.L. masc. adj. *decoyicus* intended to mean possessing decoyinine.

Excellent growth on Czapek's solution agar; hygroscopic; NaCl tolerance >10%, but <13%. Produces angustmycins A and C; exhibits anti-fungal activity; inhibited by streptomycin.

Type strain shows the highest sequence similarity to: *S. caniferus*, AB184640, 99.6%; *S. libani* subsp. *rufus*, AJ781351, 99.6%; *S. sioyaensis*, DQ026654, 99.6%; *S. hygroscopius* subsp. *glebosus*, AB184479, 99.6%; *S. platensis*, AB045882, 99.4%; *S. monomycini*, DQ445790, 99.4%; *S. erumpens*, AJ621603, 99.2%; *S. libani* subsp. *libani*, AB184414, 99.2%; *S. tubercidicus*, AJ621612, 99.2%; *S. nigrescens*, DQ442530, 99.2%; *S. lydicus*, Y15507, 99%.

Source: not known.

DNA G+C content (mol%): not known.

Type strain: AS 4.1861, CIP 106836, DSM 41427, NBRC 13977, JCM 4550, NCIMB 10502, NCIMB 9752, NRRL 2666.

Sequence accession no. (16S rRNA gene): AY999883.

253d. **Streptomyces hygroscopicus subsp. glebosus** Ohmori, Okanishi and Kawaguchi 1962, 26[AL]

gle.bo'sus. L. masc. adj. *glebosus* full of clods, cloddy.

Ridged spores; excellent growth on Czapek's solution agar; hygroscopic; NaCl tolerance >10%, but <13%. Produces glebomycin; inhibited by streptomycin.

Type strain shows the highest sequence similarity to: *S. caniferus*, AB184640, 100%; *S. libani* subsp. *rufus*, AJ781351, 100%; *S. platensis*, AB045882, 99.9%; *S. libani* subsp. *libani*, AB184414, 99.7%; *S. tubercidicus*, AJ621612, 99.6%; *S. nigrescens*, DQ442530, 99.6%; *S. hygroscopius* subsp. *decoyicus*, AY999883, 99.6%; *S. catenulae*, AJ621613, 99.3%; *S. ramulosus*, DQ026662, 99.3%; *S. misakiensis*, AB217605, 99.3%; *S. sioyaensis*, DQ026654, 99.2%; *S. monomycini*, DQ445790, 99%.

Source: not known.

DNA G+C content (mol%): not known.

Type strain: ATCC 14607, CIP 106832, DSM 40823, NBRC 13786, NBRC 13982, JCM 4954, KCTC 9782, NRRL B-3248.

Sequence accession no. (16S rRNA gene): AB184479.

253e. **Streptomyces hygroscopicus subsp. ossamyceticus** Schmitz, Jubinski, Hooper, Crook, Price and Lein 1965, 87[AL]

os.sa.my.ce'ti.cus. N.L. n. *ossamycinum* ossamycin, name of an antibiotic named for Mount Ossa of Greek mythology; L. masc. suff. *-icus* suffix used with the sense of belonging to; N.L. masc. adj. *ossamyceticus* belonging to ossamycin.

Petri dish cultures grown at 28°C on inorganic salts-starch agar show following morphology. Vegetative mycelium: branched, approx. 0.75–1 µm in diameter, no evidence of fragmentation. Aerial mycelium: branched, approx. 0.75–1 µm in diameter. Sporophore morphology: short side branches located along the main axial hyphae terminate in tight spiral spore chains of two to many turns; sporophores arranged singly, in pairs, or in clusters along the axial hyphae; no evidence of whorl formation. Conidia: catenulate, subglobose to elongated ovoid, most conidia ovoid measuring approx. 0.75 × 1–1.5 µm, smooth walls. For detailed cultural characteristics, see Schmitz et al. (1965). Produces ossamycin.

Type strain shows the highest sequence similarity to: *S. torulosus*, AJ781367, 100%; *S. neyagawaensis*, D63869, 99.4%; *S. ipomoeae*, AB184857, 99.3%.

Source: isolated from a soil sample collected in South America.

DNA G+C content (mol%): not known.

Type strain: ATCC 15420, CIP 106834, DSM 40824, NBRC 13983, JCM 4965, NRRL B-3822.

Sequence accession no. (16S rRNA gene): AB184560.

254. **Streptomyces iakyrus** de Querioz and Albert 1962, 33[AL]

i.a.ky'rus. N.L. masc. adj. *iakryus* [from Amazonian oral aboriginal language (Nheêngatû) adj. *iakryus*] green, referring to the color of the vegetative mycelium and diffusible pigment on some media.

Spore chains in Section *Spirales*. Mature spore chains are generally long with 10 to 50 or more spores per chain. This morphology is seen on yeast-malt agar, oatmeal agar, salts-starch agar, and glycerol-asparagine agar. Spore surface is spiny.

Color of colony: aerial mass color in the Gray or Red color series on oatmeal agar and salts-starch agar; nearest matching color tab in the Gray color series is 5fe, light grayish reddish brown, and in the Red color series 5dc, grayish yellowish pink. These colors as well as white aerial mycelium are reported for yeast-malt agar and glycerol-asparagine agar (a variety of colors reported on yeast-malt agar also includes tabs from the Yellow color series and the Blue color series). Reverse side of colony is greenish yellow, olive gray, olive, or olive brown on yeast-malt agar, oatmeal agar, salts-starch agar, and glycerol-asparagine agar.

Color in medium: melanoid pigments are formed in peptone-yeast-iron agar and tryptone-yeast broth, but only slightly or not at all on tyrosine agar. Yellow to yellow-green pigment is found in the medium in yeast-malt agar, oatmeal agar, salts-starch agar, and glycerol-asparagine agar. This pigment is not pH-sensitive when tested with 0.05 M NaOH or HCl.

D-Glucose, L-arabinose, D-xylose, iso-inositol, D-mannitol, D-fructose, rhamnose, sucrose, and raffinose are all utilized for growth.

Type strain shows the highest sequence similarity to: *S. violaceochromogenes*, AY999867, 99.5%; *S. longispororuber*, AB184440, 99.4%; *S. coerulescens*, AY999720, 99.3%; *S. collinus*, AB184123, 99.3%; *S. coeruleorubidus*, AY999719, 99.3%; *S. bellus*, AB184849, 99.3%; *S. speibonae*, AF452714, 99.3%; *S. lomondensis*, AB184673, 99.1%; *S. purpurascens*, AJ399486, 99.1%; *S. viridodiastaticus*, AY999852, 99.1%; *S. griseorubens*, AB184139, 99%; *S. albogriseolus*, AJ494865, 99%; *S. lusitanus*, AB184424, 99%; *S. griseoflavus*, AJ781322, 99%.

Source: not known.

DNA G+C content (mol%): not known.

Type strain: ATCC 15375, CBS 702.72, BCRC 11930, DSM 40482, NBRC 13401, INMI 15375, JCM 4254, JCM 4773, NRRL B-3317, NRRL B-3634, NRRL-ISP 5482, RIA 1362, VKM Ac-201.

Sequence accession no. (16S rRNA gene): AB184877.

255. **Streptomyces indiaensis** (Gupta 1965a) Kudo and Seino 1987, 243[VP] ("*Streptosporangium indianense*" Gupta 1965a)

in.di.a.en'sis. L. n. *india* India; L. suff. *ensis* indicating origin; N.L. masc. adj. *indiaensis* of or belonging to India.

Spore chains are *Spirales*; spore surface is smooth. Branching substrate and aerial mycelia are formed. Pseudosporangia on aerial mycelia are 2–10 µm in diameter and globular to oval or irregular. Aerial mass color is gray to grayish-violet; the reverse side of the colony is red to violet. Reverse color is changed from violet to red by addition of 0.05 M NaOH. Soluble pigment is red or violet. Melanoid pigment is formed in peptone-yeast extract iron agar, but not in tyrosine agar or tryptone-yeast extract agar. L-Arabinose, D-fructose, D-glucose, iso-inositol, lactose, D-mannitol, sucrose, and D-xylose are utilized for growth, but raffinose, L-rhamnose and salicin are not.

Type strain shows the highest sequence similarity to: *S. purpurascens*, AJ399486, 99.6%; *S. massasporeus*, AB184152, 99.6%; *S. thermocarboxydus*, U94490, 99.3%; *S. hawaiiensis*, AB184143, 99.3%; *S. coerulescens*, AY999720, 99.2%; *S. bellus*, AB184849, 99.2%; *S. lomondensis*, AB184673, 99.1%; *S. coeruleorubidus*, AY999719, 99%; *S. lusitanus*, AB184424, 99%; *S. arenae*, AB249977, 99%.

Source: not known.

DNA G+C content (mol%): 73.7.

Type strain: ATCC 33330, CBS 560.75, DSM 43803, NBRC 13964, JCM 3053, KCC A-0053, KCTC 9489, NCIB (now NCIMB) 9794.

Sequence accession no. (16S rRNA gene): AB184553.

256. **Streptomyces indigoferus** Shinobu and Kawato 1960, 49[AL]

in.di.go.fer'us. N.L. n. *indigo* (from Fr. n. *indigo*, derived from L. n. *indicum*, indigo) the dye indigo; L. suff. *-fer -fera -ferum* (from L. v. *fero*, to bear), bearing; N.L. masc. adj. indigoferus (*sic*) bearing (producing) indigo and referring to production of blue to green diffusible pigments on chemically defined media.

Spore chains in Section *Rectiflexibiles*. Mature spore chains are moderately long with 10–50, or sometimes more than 50, spores per chain. Aerial mycelium is poorly developed on all ISP media and is obscured by extensive coremia formation on yeast-malt agar, salts-starch agar, and glycerol-asparagine agar. Spore surface is smooth.

Color of colony: aerial mass color often cannot be determined because of inadequate aerial growth on oatmeal agar or because of extensive coremia formation on yeast-malt agar, salts-starch agar, and glycerol-asparagine agar. When adequate aerial mycelium is formed, it is in the Gray color series on yeast-malt agar and oatmeal agar. Aerial mass color in the Gray color series was also found on glycerol-asparagine agar and soil extract agar by one observer. Reverse side of colony: substrate mycelium is usually nearly colorless. grayish yellow or dark brown on yeast-malt agar, oatmeal agar, salts-starch agar, and glycerol-asparagine agar. A transient blue to red pigment (depending on pH) may also be found in the mycelium on these media. This pigment changes from blue or colorless to red with addition of 0.05 M HCl.

Color in medium: melanoid pigments are formed in peptone-yeast-iron agar and tryptone-yeast broth, but are not formed in tyrosine agar. Pigments other than melanoids are not found in the medium in yeast-malt agar, oatmeal agar, salts-starch agar, or glycerol-asparagine agar.

D-Glucose and L-arabinose are utilized for growth. Utilization of xylose is doubtful. No growth or only traces of growth on sucrose, iso-inositol, D-mannitol, D-fructose, rhamnose, and raffinose.

Type strain shows the highest sequence similarity to: *S. herbaricolor*, AB184801, 100%; *S. aburaviensis*, AY999779, 99.6%; *S. purpureus*, AJ781324, 99.6%; *S. chrysomallus* subsp. *fumigatus*, AB184645, 99.5%; *S. purpeofuscus*, AJ781364, 99.2%; *S. xanthocidicus*, AY999858, 99.2%. Type strain shows the highest sequence similarity to following *Kitasatospora* species: *Kitasatospora kifunensis*, AB022874, 99.3%; *Kitasatospora nipponensis*, AY442263, 99%; *Kitasatospora gansuensis*, AY442265, 99%.

Source: not known.

DNA G+C content (mol%): not known.

Type strain: ATCC 23924, CBS 908.68, BCRC 13773, DSM 40124, NBRC 12878, NBRC 3868, IMET 42938, JCM 4646, NCIMB 9718, NRRL B-3301, NRRL-ISP 5124, RIA 1127.

Sequence accession no. (16S rRNA gene): AB184214.

257. **Streptomyces indonesiensis** Sembiring, Ward and Goodfellow 2001, 1619[VP] (Effective publication: Sembiring, Ward and Goodfellow2000, 365.)

in.do.ne.si.en'sis. N.L. masc. adj. *indonesiensis* of or pertaining to Indonesia.

Spore chains are *Spirales*; spore surface is rugose. On oatmeal agar, the spore mass is gray, the substrate mycelium grayish-yellow and the diffusible pigment yellow. Melanin pigments are not produced. The strain degrades pectin but not adenine.

Type strain shows the highest sequence similarity to: *S. asiaticus*, AB249947, 99.1%; *S. cangkringensis*, AJ391831, 99.7%; *S. rhizosphaericus*, AB249941, 99.7%; *S. griseinger*, AJ391818, 99.4%. Type strain shows DNA–DNA similarity to: *S. albiflaviniger* NRRL B-1356[T] 98.3%; *S. geldanamycinus* NRRL 3602[T], 98.2%; *S. griseiniger* NRRL B1865[T], 98.7%; *S. rhizosphaericus* DSM 41760[T], 98.3%; *S. asiaticus* DSM 41761[T], 99.1%; *S. javensis* DSM 41764[T], 98.2%; *S. yogyakartensis* DSM 41766[T], 98.1%; *S. cangkringensis* DSM 41769[T], 99.2%.

Source: isolated from the rhizosphere of the tropical legume *Paraserianthes falcataria*.

DNA G+C content (mol%): not known.

Type strain: A4R2, DSM 41759, JCM 11445, NBRC 100776, NCIMB 13673.

Sequence accession no. (16S rRNA gene): DQ334783.

258. **Streptomyces intermedius** (Krüger 1904) Waksman *in* Waksman and Lechevalier 1953, 116[AL] ("*Oospora intermedia*" Krüger 1904, 289; "*Actinomyces intermedius*" Wollenweber 1920, 13)

in.ter.me'di.us. L. masc. adj. *intermedius* intermediate.

Spore chains in Section *Rectiflexibiles*. Flexuous chains of 10–50 spores occur as tufts or clusters on yeast-malt agar, oatmeal agar, salts-starch agar, and glycerol-asparagine agar, although production of sporulating aerial mycelium is not uniformly good in different laboratories with this strain. Spore surface is smooth.

Color of colony: sporulating aerial mycelium is usually inadequate for determination of aerial mass color. When aerial mycelium is formed, it is in the Yellow color series (1bc, pale yellow-green; 2ba, pale yellow) on yeast-malt agar, oatmeal agar, salts-starch agar, and glycerol-asparagine agar (aerial mycelium is described as light green or light gray in the descriptions of Krasil'nikov, 1941, and Waksman, 1953). Reverse side of colony with no distinctive pigments (colorless to pale grayish yellow or light yellowish brown) on yeast-malt agar, oatmeal agar, salts-starch agar, and glycerol-asparagine agar.

Color in medium: melanoid pigments are not formed in peptone-yeast-iron agar, tyrosine agar, or tryptone-yeast broth. No pigment or only a trace of yellow is found in the medium in yeast-malt agar, oatmeal agar, salts-starch agar, and glycerol-asparagine agar.

D-Glucose, L-arabinose, D-xylose, D-mannitol, and D-fructose are utilized for growth. No growth or only traces of growth with iso-inositol, rhamnose, sucrose, and raffinose.

Type strain shows the highest sequence similarity to: *S. gougerotii*, AB184742, 99.8%; *S. diastaticus* subsp. *diastaticus*, AB184785, 99.8%; *S. rutgersensis* subsp. *rutgersensis*, AB184170, 99.8%; *S. koyangensis*, AY079156, 99.1%; *S. misionensis*, EF178678, 99%; *S. phaeoluteichromatogenes*, AJ391814, 99%; *S. aureoverticillatus*, AY999774, 99%; *S. matensis*, AB184221, 99%.

Source: not known.

DNA G+C content (mol%): not known.

Type strain: ATCC 3329, ICMP 12540, AS 4.1467, ATCC 25461, CBS 101.21, CBS 694.69, BCRC 13706, DSM 40372,

ICMP 12540, NBRC 13049, IMET 41384, JCM 4197, JCM 4483, LMG 19304, NRRL B-1327, NRRL B-2670, NRRL-ISP 5372, RIA 1241.

Sequence accession no. (16S rRNA gene): AB184277.

259. **Streptomyces inusitatus** Hasegawa, Yamano and Yoneda 1978, 409[AL]

i.nu.si.ta′tus. L. masc. adj. *inusitatus* unusual.

Spore-bearing hyphae are simply branched and possess open or closed coils of spores. Spores (0.5–0.7 × 0.9–1.4 μm) are ellipsoidal to cylindrical, have smooth spore surfaces, and are arranged in coiled chains. The color of the aerial mycelium is blue-gray to gray-blue on Czapek agar, glucose-asparagine agar, yeast extract-malt extract agar, inorganic salts-starch agar, and tyrosine agar. Reverse side of colony with no distinctive pigments on either chemically defined or organic media. Color in medium: no pigment or only a trace of ivory color is formed on oatmeal agar. Starch is hydrolyzed, gelatin is weakly liquefied, and milk is peptonized. Carbon sources such as maltose, lactose, glycerol, starch, and sodium citrate are utilized for growth. Poor growth is obtained with D-galactose and D-glucose. L-A_2pm acid, glycine, aspartic acid, alanine, leucine, and glutamic acid, but not arabinose and galactose, are contained in cell-wall preparations. Produces oxamicetin.

Type strain shows the highest sequence similarity to: *S. amakusaensis*, AY999781, 99.8%.

Source: not known.

DNA G+C content (mol%): not known.

Type strain: ATCC 33341, CBS 196.78, DSM 41441, NBRC 13601, JCM 4988, NRRL B-16929.

Sequence accession no. (16S rRNA gene): AB184445.

260. **Streptomyces ipomoeae** (Person and Martin 1940) Waksman and Henrici *in* Breed, Murray and Hitchens 1948, 958[AL] [“*Actinomyces ipomoea*” (*sic*) Person and Martin 1940, 923; “*Streptomyces ipomoea*” (*sic*) Waksman and Henrici *in* Breed, Murray and Hitchens 1948, 958]

i.po.mo.e′a.e. N.L. n. *Ipomoea* generic name of the sweet potato; N.L. gen. n. *ipomoeae* of *Ipomoea*, referring to the source of the organism.

Spore chains in Section *Spirales*. Spore chains are usually short with 3–10 spores per chain and may give rise to hooks, loops, incomplete spirals, or spirals of only 1 or 2 turns. This morphology is seen on yeast-malt agar, oatmeal agar, and salts-starch agar, but typical aerial mycelium is usually not found on glycerol-asparagine agar. Spore surface is smooth; wrinkles or folds may be present.

Color of colony: aerial mass color in the Blue color series on yeast-malt agar, oatmeal agar, salts-starch agar, and glycerol-asparagine agar. Reverse side of colony is pale yellow or grayish yellow on yeast-malt agar and glycerol-asparagine agar; yellow may be modified by blue (very pale blue or pale green) on reverse side of mature growth on oatmeal agar and salts-starch agar. This pigment is not a pH indicator.

Color in medium: melanoid pigments are not formed in peptone-yeast-iron agar, tyrosine agar, or tryptone-yeast broth. Traces of yellow or green pigment may or may not be found in the medium in yeast-malt agar, oatmeal agar, salts-starch agar, and glycerol-asparagine agar.

D-Glucose, L-arabinose, D-xylose, iso-inositol, D-mannitol, D-fructose, rhamnose, sucrose, and raffinose are all utilized for growth.

Type strain shows the highest sequence similarity to: *S. hygroscopicus* subsp. *ossamyceticus*, AB184560, 99.3%; *S. torulosus*, AJ781367, 99.2%.

Source: not known.

DNA G+C content (mol%): not known.

Type strain: 9820, ATCC 25462, CBS 695.69, DSM 40383, ICMP 12541, NBRC 13050, KCC S-0484, NRRL-ISP 5383, RIA 1242.

Sequence accession no. (16S rRNA gene): AB184857.

261. **Streptomyces janthinus** (Artamonova and Krasil'nikov 1960) Pridham 1970, 19[AL] (“*Actinomyces janthinus*” Artamonova and Krasil'nikov *in* Rautenshtein 1960, 334)

jan′thi.nus. L. masc. adj. *janthinus* violet-colored, referring to color of vegetative mycelium and diffusible pigment.

Spore chains in Section *Spirales*. Spore chains are moderately long with 10–50, or sometimes more than 50, spores per chain. This morphology is seen on yeast-malt agar, oatmeal agar, salts-starch agar, and glycerol-asparagine agar. Spore surface is spiny.

Color of colony: aerial mass color in the Red or White color series on yeast-malt agar and salts-starch agar; White color series on oatmeal agar, and White or Gray color series on glycerol-asparagine agar. The nearest color tabs in the Red color series are 5dc, 5ca, 5cb, and 6ec (all grayish yellowish pink). Reverse side of colony with no distinctive pigments (colorless to grayed yellow or pale yellow) on yeast-malt agar, oatmeal agar, salts-starch agar, and glycerol-asparagine agar.

Color in medium: melanoid pigments are formed in peptone-yeast-iron agar and tryptone-yeast broth, but not in tyrosine agar. Yellow pigment is found in the medium in salts-starch agar. This pigment is not pH-sensitive when tested with 0.05 M NaOH or HCl.

D-Glucose, L-arabinose, sucrose, D-xylose, iso-inositol, D-mannitol, D-fructose, rhamnose, and raffinose are utilized for growth.

Type strain shows the highest sequence similarity to: *S. roseoviolaceus*, AJ399484, 100%; *S. violaceus*, AB184315, 100%; *S. albosporeus* subsp. *albosporeus*, AJ781327, 100%; *S. luteogriseus*, AB184379, 99.5%; *S. lomondensis*, AB184673, 99.3%; *S. hawaiiensis*, AB184143, 99.3%; *S. flavoviridis*, AB184842, 99.3%; *S. pilosus*, AB184161, 99.2%; *S. arenae*, AB249977, 99.2%; *S. africanus*, AY208912, 99.2%; *S. tuirus*, AB184690, 99.2%; *S. bellus*, AB184849, 99.1%; *S. mutabilis*, EF178679, 99.1%; *S. massasporeus*, AB184152, 99%; *S. afghaniensis*, AJ399483, 99%; *S. levis*, AB184670, 99%; *S. coerulescens*, AY999720, 99%; *S. parvulus*, AB184326, 99%; *S. coeruleorubidus*, AY999719, 99%.

Source: not known.

DNA G+C content (mol%): not known.

Type strain: ATCC 15870, ATCC 23925, CBS 909.68, DSM 40206, NBRC 12879, JCM 4387, INMI 117, KCC S-0387, NRRL B-3365, NRRL-ISP 5206, RIA 659.

Sequence accession no. (16S rRNA gene): AB184851.

262. **Streptomyces javensis** Sembiring, Ward and Goodfellow 2001, 1619[VP] (Sembiring, Ward and Goodfellow 2000, 365)

ja.ven'sis. N.L. masc. adj. *javensis* of or pertaining to Java, Indonesia.

Spore chains are *Spirales*; spore surface is rugose. On oatmeal agar, the spore mass is gray, the substrate mycelium grayish-yellow and the diffusible pigment yellow. Melanin pigments are not produced. The organism degrades xylan and does not grow at 45°C.

Type strain shows the highest sequence similarity to: *S. yogyakartensis*, AB249942, 99.2%; *S. violaceusniger*, AJ391823, 99.1%. Type strain shows DNA–DNA similarity to: *S. albiflaviniger* NRRL B-1356^{T}, 98.9%; *S. geldanamycinus* NRRL 3602^{T}, 98.9%; *S. griseiniger* NRRL B-1865^{T}, 98.6%; *S. rhizosphaericus* DSM 41760^{T}, 98.3%; *S. asiaticus* DSM 41761^{T}, 98.2%; *S. indonesiensis* DSM 41759^{T}, 98.6%; *S. yogyakartensis* DSM 41766^{T}, 98.6%; *S. cangkringensis* DSM 41769^{T}, 98.4%.

Source: isolated from non-rhizosphere soil adjacent to a stand of the tropical legume *Paraserianthes falcataria*.

DNA G+C content (mol%): not known.

Type strain: B22P3, DSM 41764, JCM 11446, NBRC 100777, NCIMB 13679.

Sequence accession no. (16S rRNA gene): AJ391833.

263. **Streptomyces jietaisiensis** He, Li, Huang, Wang, Liu, Lanoot, Vancanneyt and Swings 2005, 1943[VP]

jie.tai.si.en'sis. N.L. masc. adj. *jietaisiensis* of or pertaining to Jietaisi, a place in a suburb of Beijing, where the type strain was isolated.

Spore chains with 10–20 cylindrical spores are *Rectiflexibiles*. The spore surface is smooth. Diffusible pigments are not produced, nor are melanin pigments formed on peptone-yeast extract-iron agar, or tyrosine agars. Adonitol, cellobiose, dextrin, D-galactose, D-glucose, inulin, glycogen, maltose, D-mannitol, D-mannose, melezitose, salicin, trehalose, and D-xylose (all at 1%, w/v), and L-alanine, L-arginine, L-aspartic acid, L-cysteine, L-glutamic acid, L-histidine, L-isoleucine, L-phenylalanine, sodium oxalate, sodium pyruvate, L-threonine, and L-valine (all at 0.1%, w/v) are used as sole carbon sources for energy and growth, but not glycerol, glycine, and xylitol (all at 1%, w/v) or DL-aminobutyric acid (at 0.1%, w/v). L-Alanine, L-arginine, L-aspartic acid, L-glutamic acid, and L-phenylalanine (all at 0.1%, w/v) are metabolized as sole carbon and nitrogen sources, but not L-isoleucine (at 0.1%, w/v). Growth occurs between 10 and 40°C, and between pH 5.0 and 10.0, but not at pH 4.0 or 11.0 or in the presence of streptomycin (10 µg/ml) or novobiocin (5 µg/ml). Cell-wall type I, phospholipid type II, and menaquinone MK-9(H_6,H_8,H_4). The fatty acid profile is composed of $C_{15:0}$ anteiso (35.7%), $C_{17:0}$ anteiso (18.9%), $C_{16:0}$ iso (14.8%), $C_{17:1}$ anteiso ω9*c* (8.1%), $C_{16:0}$ (6.2%), $C_{16:1}$ iso (4.5%), $C_{15:0}$ iso (4.3%), $C_{16:1}$ ω7*c* (2.33%), $C_{17:1}$ iso ω9*c* (1.8%), $C_{17:0}$ iso (1.7%), and $C_{14:0}$ iso (1.7%).

Type strain shows the highest sequence similarity to: *S. griseoaurantiacus*, AB184676, 99.7%. Type strain shows DNA–DNA similarity to: *S. griseoaurantiacus* DSM 40430^{T}, 48.8%.

Source: not known.

DNA G+C content (mol%): 72.3.

Type strain: FXJ46, AS 4.1859, JCM 12279.

Sequence accession no. (16S rRNA gene): AY314783.

264. **Streptomyces kanamyceticus** Okami and Umezawa *in* Umezawa, Ueda, Maeda, Yagishita, Kondō, Okami, Utahara, Ōsato, Nitta and Takeuchi 1957, 183[AL]

ka.na.my.ce'ti.cus. N.L. n. *kanamycinum* kanamycin, name of an antibiotic; L. masc. suff. *-icus* suffix used with the sense of belonging to; N.L. masc. adj. *kanamyceticus* belonging to kanamycin.

Spore chains in Section *Rectiflexibiles*. Sporulating aerial mycelium is usually not abundant or may be absent on yeast-malt agar, oatmeal agar, salts-starch agar, and glycerol-asparagine agar, although one observer reports long spore chains with 50 or more spores per chain on all of these media. Spore surface is smooth. Special morphological characteristics: terminal droplets showing wide variation in size are often found on aerial hyphae. These may contain coalesced masses of spores but they do not appear to be sporangia.

Color of colony: aerial mass color in the Yellow or White color series when sporulating aerial mycelium is present on yeast-malt agar, oatmeal agar, salts-starch agar, and glycerol-asparagine agar. Nearest matching color tabs from the Yellow color series are 1½fb, light yellow; 2db, pale yellow; and 1db, pale yellow-green. Reverse side of colony with no distinctive pigments (colorless to pale yellow or grayish yellow) on yeast-malt agar, oatmeal agar, salts-starch agar, and glycerol-asparagine agar.

Color in medium: melanoid pigments are not formed in peptone-yeast-iron agar, tyrosine agar, or tryptone-yeast broth. No pigment is found in the medium in yeast-malt agar, oatmeal agar, salts-starch agar, or glycerol-asparagine agar.

D-Glucose, L-arabinose, D-xylose, D-fructose, raffinose, and D-mannitol are utilized for growth. No growth or only traces of growth with iso-inositol, rhamnose, or sucrose.

Type strain shows the highest sequence similarity to: *S. aureus*, AB249976, 99.5%; *S. durmitorensis*, DQ067287, 99.2%.

Source: not known.

DNA G+C content (mol%): not known.

Type strain: AS 4.1441, ATCC 12853, CBS 715.72, BCRC 11515, DSM 40500, NBRC 13414, JCM 4433, JCM 4775, KCTC 9225, LMG 5976, LMG 19351, NCIMB 9343, NRRL B-2535, NRRL-ISP 5500, RIA 1375, RIA 690, VKM Ac-837.

Sequence accession no. (16S rRNA gene): DQ442511.

265. **Streptomyces kashmirensis** (*sic*) (Gupta and Chopra 1963b) Witt and Stackebrandt 1991, 456[VP] (Effective publication: Witt and Stackebrandt 1990, 370.) ("*Streptomyces kashmirensis*" Gupta and Chopra 1963a, 112; *Streptoverticillium kashmirense* Locci, Baldacci and Petrolini Baldan 1969, 59)

kash.mir.en'sis. N.L. masc. adj. *kashimirensis* (*sic*) of or pertaining to Kashmir.

Spore chains in Section *Verticillati*, umbellate monoverticillate (biverticillate). Mature spore chains are generally short with 3 to 10 or more spores per chain. This morphology is usually seen on yeast-malt agar, oatmeal agar, salts-starch agar, and glycerol-asparagine agar, although sporulation is not always good on these media. Spore surface is smooth.

Color of colony: aerial mass color in the Red color series (5cb or 5dc, grayish yellowish pink) on yeast-malt agar, oatmeal agar, salts-starch agar, and glycerol-asparagine agar. When aerial mycelium is poorly developed, the surface color may appear to be in the Gray (d, light gray, or 2dc, yellowish gray) or White color series. The original description of Gupta and Chopra (1963b) reports blue aerial mycelium on Emmerson's medium. Reverse side of colony with distinctive red pigments (grayish red or moderate reddish brown on yeast-malt agar and oatmeal agar; light brown to dark brown on salts-starch agar and glycerol-asparagine agar). The reverse mycelium pigment is not a pH indicator.

Color in medium: melanoid pigments are formed in peptone-yeast-iron agar, tyrosine agar, and tryptone-yeast broth in 4 d; reaction may be weak on tyrosine agar. Yellow to orange or reddish pigment is found in the medium in yeast-malt agar, oatmeal agar, salts-starch agar, and glycerol-asparagine agar. This pigment is not pH-sensitive when tested with 0.05 M NaOH or HCl.

D-Glucose and iso-inositol are utilized for growth. Only traces of growth are seen with L-arabinose, D-xylose, D-mannitol, D-fructose, rhamnose, sucrose, and raffinose.

For sequence similarity, see type strain of *Streptomyces lilacinus.*

Source: not known.

DNA G+C content (mol%): not known.

Type strain: ATCC 27439, CBS 665.72, DSM 40336, NBRC 13906, JCM 4776, NRRL B-3103, NRRL-ISP 5336, RIA 1325, VKM Ac-885.

Sequence accession no. (16S rRNA gene): AB184546.

Further comments: in violation of Rule 33c of the *Bacteriological Code* (1990 Revision), in Validation List no. 38, *Streptomyces kashmirensis* is proposed as a *nomen revictum* (basonym: "*Streptomyces kashmirensis*" Gupta and Chopra 1963b).

According to Hatano et al. (2003), *Streptomyces kashmirensis* (sic) (Gupta and Chopra 1963b) Witt and Stackebrandt 1991 is a later heterotypic synonym of *Streptomyces lilacinus* (Nakazawa et al. 1956) Witt and Stackebrandt 1991.

266. **Streptomyces kasugaensis** Hamada, Kinoshita, Hattori, Yoshida, Okami, Higashide, Sakata and Hori 1995b, 879[VP] (Effective publication: Hamada, Kinoshita, Hattori, Yoshida, Okami, Higashide, Sakata and Hori 1995a, 35.)

ka.su.ga.en'sis. N.L. masc. adj. *kasugaensis* of or pertaining to Kasuga, Japan.

Substrate mycelia are well-branched. Aerial mycelia form complete spiral chains of spores. Mature spore chains consist of 10 or more spores. The spores are oval (0.6–0.7 × 0.6–0.8 μm) and the spore surface is smooth. The spores are not motile. No synnemata, sclerotia, or sporangia are observed. Aerial mycelia are light olive gray to light brownish gray, and vegetative mycelia are colorless to pale yellowish brown on various media. Yellow to brownish soluble pigments are produced on various media. Melanoid pigments are not produced on peptone-yeast extract iron agar (ISP 6), tyrosine agar (ISP 7), or in tryptone-yeast extract broth (ISP 1). Starch is not hydrolyzed, gelatin is liquefied, milk is peptonized and coagulated, and nitrate is reduced. D-Glucose, D-fructose, raffinose, and *myo*-inositol are utilized for growth, but L-arabinose, D-xylose, rhamnose, sucrose, and D-mannitol are not utilized. Permissive temperature range for growth is 20–37°C. Whole-cell hydrolysates contain LL-A_2pm, but no detectable arabinose, galactose, or *meso*-A_2pm. Phospholipid pattern type is PII. Mycolic acids are absent. Major menaquinones are MK-9(H_6), MK-9(H_8), and MK-9(H_4). Major components of cellular fatty acids are $C_{15:0}$ anteiso, $C_{16:0}$, $C_{17:0}$ anteiso, and $C_{16:0}$ iso.

Type strain shows no sequence similarity over 99%.

Source: not known.

DNA G+C content (mol%): 70.4–70.9.

Type strain: M338-M1, ATCC 15714, BCRC 12349, DSM 40819, NBRC 13851, JCM 4208, KCTC 1078, KCTC 2113, NCIMB 12239, NCIMB 12718.

Sequence accession no. (16S rRNA gene): AB024441.

267. **Streptomyces katrae** Gupta and Chopra 1963b, 1[AL]

kat'ra.e. N.L. gen. n. *katrae* of Katra, named for Katra, Jammu Province, India, the source of the soil from which the organism was isolated.

Spore chains in Section *Retinaculiaperti.* Long spore chains with terminal primitive spirals, hooks, or loops of wide diameter. Some well-defined spirals are also seen. Mature spore chains are long, often with more than 50 spores per chain. This morphology is seen on yeast-malt agar, oatmeal agar, salts-starch agar, and glycerol-asparagine agar. Spore surface is smooth.

Color of colony: aerial mass color in the Red color series (5dc, 5cb or 4ca, grayish yellowish pink; 5ca, light yellowish pink; or 3ca, pale orange-yellow) on yeast-malt agar, oatmeal agar, salts-starch agar, and glycerol-asparagine agar. Reverse side of colony with no distinctive pigments (yellowish brown to olive brown on yeast-malt agar; grayish yellow to yellowish brown on salts-starch agar, oatmeal agar, and glycerol-asparagine agar).

Color in medium: melanoid pigments are formed in peptone-yeast-iron agar and tryptone-yeast broth but not in tyrosine agar. No pigment is found in the medium in yeast-malt agar, oatmeal agar, salts-starch agar, or glycerol-asparagine agar.

D-Glucose, L-arabinose, D-xylose, D-fructose, and sucrose are utilized for growth. No growth or only traces of growth with iso-inositol, D-mannitol, rhamnose, or raffinose.

Type strain shows the highest sequence similarity to: *S. polychromogenes*, AB184292, 99.5%; *S. racemochromogenes*, DQ026656, 99.5%; *S. flavotricini*, AB184132, 99.2%.

Source: isolated from soil from Katra, Jammu Province, India.

DNA G+C content (mol%): not known.

Type strain: ATCC 27440, CBS 748.72, DSM 40550, NBRC 13447, IMET 43361, JCM 4777, NRRL B-3093, NRRL-ISP 5550, RIA 1408, RIA 794, VKM Ac-1220.

Sequence accession no. (16S rRNA gene): AB184409.

268. **Streptomyces kentuckensis** (Barr and Carman 1956) Witt and Stackebrandt 1991, 456[VP] (Effective publication: Witt and Stackebrandt 1990, 370.) ("*Streptomyces kentuckensis*"

Barr and Carman 1956; "*Verticillomyces kentuckensis*" Shinobu 1965; *Streptoverticillium kentuckense* Baldacci, Farina and Locci 1966, 170)

ken.tuck.en'sis. N.L. masc. adj. *kentuckensis* of or pertaining to Kentucky.

Spore chains in Umbellate Monoverticillate (=*Streptomyces* Section *Verticillati*, biverticillate). Mature spore chains moderately long with 10–50 spores per chain. This morphology is seen on yeast-malt agar, oatmeal agar, salts-starch agar, and glycerol-asparagine agar. Spore surface is smooth.

Color of colony: aerial mass color in the Red color series on yeast-malt agar, oatmeal agar, and salts-starch agar; Red or White color series on glycerol-asparagine agar. Reverse side of colony with no distinct pigments (grayed yellow on oatmeal agar, salts-starch agar, and glycerol-asparagine agar to brown on yeast-malt agar).

Color in medium: melanoid pigments are produced in peptone-yeast-iron agar and tryptone-yeast broth, but not in tyrosine agar. No pigment is found in the medium in yeast-malt agar, oatmeal agar, salts-starch agar, or glycerol-asparagine agar.

D-Glucose, iso-inositol, and fructose are utilized for growth; utilization of D-xylose is doubtful. No growth or only trace of growth on L-arabinose, sucrose, D-mannitol, rhamnose, and raffinose.

For sequence similarity, see type strain of *Streptomyces netropsis*.

Source: not known.

DNA G+C content (mol%): not known.

Type strain: ATCC 12691, ATCC 23926, CBS 910.68, DSM 40052, HAMBI 52, NBRC 12880, IPV 940, IPV 1780, IPV1958, JCM 4153, JCM 4647, KCC S-153, LMG 5977, NRRL B-1831, NRRL-ISP 5052, RIA 1114.

Sequence accession no. (16S rRNA gene): AB184215.

Further comments: in violation of Rule 33c of the *Bacteriological Code* (1990 Revision), in Validation List no. 38, *Streptomyces kentuckensis* is proposed as a *nomen revictum* (basonym: "*Streptomyces kentuckensis*" Barr and Carman 1956).

According to Labeda (1996) and to Hatano et al. (2003), *Streptomyces kentuckensis* (Barr and Carman 1956) Witt and Stackebrandt 1991 is a later heterotypic synonym of *Streptomyces netropsis* (Finlay et al. 1951) Witt and Stackebrandt 1991.

269. **Streptomyces kishiwadensis** (Shinobu and Kayamura 1964) Witt and Stackebrandt 1991, 456[VP] (Effective publication: Witt and Stackebrandt 1990, 370.) ("*Streptomyces kishiwadensis*" Shinobu and Kayamura 1964; "*Verticillomyces kishiwadenis*" Shinobu 1965; *Streptoverticillium kishiwadense* Locci, Baldacci and Petrolini Baldan 1969, 59)

ki.shi.wad.en'sis. N.L. masc. adj. *kishiwadensis* of or belonging to Kishiwada (named after the place of origin, Kishiwada City, Japan).

Spore chains in Section *Verticillati.* Both monoverticillate and umbellate monoverticillate (biverticillate) sporophores are found. Mature sporophores are usually umbellate monoverticillate. Spore chains are short with 3–10 spores per chain. This morphology is seen on yeast-malt agar, oatmeal agar, salts-starch agar, and glycerol-asparagine agar. Spore surface is smooth.

Color of colony: aerial mass color in the Red color series (5ca, light yellowish pink to 5cb, grayish yellowish pink) on yeast-malt agar, oatmeal agar, salts-starch agar, and glycerol-asparagine agar; white to pale yellow or pale orange-yellow (3ca) aerial mass color may also be seen on these media. Reverse side of colony with no distinctive pigments (pale grayish yellow to orange-yellow or yellowish brown) on yeast-malt agar, oatmeal agar, salts-starch agar, and glycerol-asparagine agar.

Color in medium: melanoid pigments are formed in peptone-yeast-iron agar and tryptone-yeast broth, but only weakly or not at all in tyrosine agar. No pigment found in medium with yeast-malt agar, oatmeal agar, salts-starch agar, or glycerol-asparagine agar.

D-Glucose, iso-inositol, D-fructose, and sucrose are utilized for growth. Utilization of D-xylose is doubtful and there is no growth or only traces of growth with L-arabinose, D-mannitol, rhamnose, and raffinose.

For sequence similarity, see type strain of *Streptomyces mashuensis.*

Source: not known.

DNA G+C content (mol%): not known.

Type strain: ATCC 25464, CBS 697.69, DSM 40397, NBRC 13052, JCM 4486, NRRL B-12326, NRRL-ISP 5397, RIA 1244, VKM Ac-931.

Sequence accession no. (16S rRNA gene): AB184858.

Further comments: in violation of Rule 33c of the *Bacteriological Code* (1990 Revision), in Validation List no. 38, *Streptomyces kishiwadensis* is proposed as a *nomen revictum* (basonym: "*Streptomyces kishiwadensis*" Shinobu and Kayamura 1964).

According to Hatano et al. (2003), *Streptomyces kishiwadensis* (Shinobu and Kayamura 1964) Witt and Stackebrandt 1991 is a later heterotypic synonym of *Streptomyces mashuensis* (Sawazaki et al. 1955) Witt and Stackebrandt 1991.

270. **Streptomyces koyangensis** Lee, Lee, Jung and Hwang 2005, 261[VP]

ko.yang.en'sis. N.L. masc. adj. *koyangensis* of or pertaining to Koyang, Republic of Korea, the geographical origin of the type strain.

Spore chains containing 10 or more spores per chain are *Rectiflexibiles.* Spores are spherical (1.2 m in diameter) with a smooth surface. The spore mass is white to gray and the reverse sides of colonies are brown on most agar media. Grows well on yeast extract-malt extract (ISP 2), oatmeal agar (ISP 3), inorganic salts-starch agar (ISP 4), peptone-yeast extract-iron agar (ISP 6), and tyrosine agar (ISP 7). Does not grow well on ISP 5 medium. Aerial mycelia are abundant on most of these media. The color of substrate mycelium is pale brown to dark brown. Production of spores on ISP 4 is prolific. Melanin pigments are produced on ISP 6 and ISP 7. Degrades casein, elastin, esculin, gelatin, starch, tyrosine, and xanthine, but not cellulose. As sole carbon sources, utilizes L-arabinose, D-fructose, mannitol, and xylose for growth, but not adonitol, dextran, *myo*-inositol, melezitose, melibiose, raffinose, L-rhamnose, sucrose, or xylitol. As nitrogen sources, utilizes L-cysteine, L-histidine, L-phenylalanine, and L-valine. It cannot utilize DL-α-amino-n-butyric acid or L-hydroxyproline. Pectin hydrolysis, nitrate reduction, and H_2S production are positive, whereas lecithinase, lipolysis, and hippurate hydrolysis are negative. Grows in the presence of 4, 7, and 10% sodium chloride, but not in 13%. It grows in 0.02% NaN_3 and 0.001% thallous acetate, but not in 0.1% phenol or 0.001% potassium tellurite. Whole-cell hydrolysates contain LL-A_2pm. The predominant cellular fatty acids are $C_{15:0}$ anteiso (16.54%), $C_{16:0}$ iso (28.77%), and $C_{16:0}$ (11.60%). In addition, $C_{17:0}$ anteiso (9.01%), $C_{14:0}$ iso (8.84%), $C_{15:0}$ iso (7.02%), $C_{17:0}$ cyclo (4.54%), $C_{17:1}$ anteiso (3.23%), $C_{17:0}$ iso (1.94%), $C_{14:0}$ (1.33%), $C_{16:1}$ iso (1.86%), and $C_{16:1}$ *cis*9 (2.57) are detected. Resistant to penicillin G, but sensitive to neomycin, rifampin, and oleandomycin. Produces 4-phenyl-3-buteonic acid, which inhibits the mycelial growth of several plant-pathogenic fungi, such as *Alternaria mali, Cladosporium cucumerinum, Colletotrichum gloeosporiuides, Colletotrichum orbiculare, Magnaporthe grisea,* and *Fusarium oxysporum* f. sp. *cucumerinum.*

Type strain shows the highest sequence similarity to: *S. odorifer,* Z76682, 99.7%; *S. hydrogenans,* AB184868, 99.7%; *S. daghestanicus,* DQ442497, 99.7%; *S. limosus,* AB184147, 99.7%; *S. albidoflavus,* AB184255, 99.7%; *S. felleus,* AB184129, 99.7%; *S. violascens,* AY999737, 99.7%; *S. griseus* subsp. *solvifaciens,* AB249915, 99.6%; *S. canescens,* AB184117, 99.6%; *S. champavatii,* DQ026642, 99.5%; *S. sampsonii,* D63871, 99.5%; *S. intermedius,* AB184277, 99.1%. Type strain shows DNA–DNA similarity to: *S. griseus* IFO 12875^T, 68.5%; *S. canescens* DSM 40001^T, 55.2%; *S. coelicolor* DSM 40233^T, 20.8%; *S. sampsonii* ATCC 25495^T, 64.0%; *S. odorifer* DSM 40347^T 34.4%; *S. limosus* DSM 40131^T, 66.8%; *S. felleus* DSM 40130^T, 25.6%; *S. somaliensis* DSM 40267, 57.5%.

Source: not known.

DNA G+C content (mol%): 67.8.

Type strain: VK-A60, KCCM 10555, NBRC 100598.

Sequence accession no. (16S rRNA gene): AY079156.

271. **Streptomyces kunmingensis** (Ruan, Lechevalier, Jiang and Lechevalier 1985) Goodfellow, Williams and Alderson 1986a, 574^{VP} (Effective publication: Goodfellow, Williams and Alderson 1986d, 59.) (*Chainia kunmingensis* Ruan, Lechevalier, Jiang and Lechevalier 1985, 167)

kun.ming.en′sis. N.L. masc. adj. *kunmingensis* of or pertaining to Kunming, a province of South China.

Forms extensively branched substrate and aerial mycelium. Chains of conidia in loose spirals are occasionally borne on the aerial hyphae. Colonies are usually yellowish tan to dark orange yellow, depending on the medium, with or without yellowish white aerial mycelium. Produces sclerotia (5–25 μm) in and on the surface of agar media. Melanin pigments are not formed. Adenine, esculin, casein, hypoxanthine, starch, tyrosine, and xanthine are degraded, but gelatin and urea are not. Phosphatase and nitrate reductase are produced. L-Arabinose, D-mannitol, raffinose, L-rhamnose, and D-xylose are used as sole carbon sources, but *myo*-inositol and sucrose are not.

Acid is formed from L-arabinose, cellobiose, fructose, galactose, glucose, glycerol, *myo*-inositol, D-lactose, maltose, D-mannitol, mannose, melibiose, methyl α-D-glucoside, methyl α-D-xyloside, raffinose, L-rhamnose, salicin, sucrose, trehalose, and D-xylose but not from adonitol, dulcitol, *meso*-erythritol, or D-sorbitol. Grows at 10 and 37°C, but not at 42°C. The wall peptidoglycan contains LL-A_2pm as the major diamino acid. The organism has a type II phospholipid pattern (*sensu* Lechevalier et al., 1977).

Type strain shows no sequence similarity over 99%.

Source: isolated from soil at Dagyanlow's Lake in Kunming, Peoples' Republic of China.

DNA G+C content (mol%): 71.3.

Type strain: 80-3024, ATCC 35682, DSM 41681, NBRC 14463, JCM 7473, NRRL B-16240, VKM Ac-895.

Sequence accession no. (16S rRNA gene): DQ442513.

272. **Streptomyces kurssanovii** (Preobrazhenskaya, Kudrina, Ryabova and Blinov *in* Gauze, Preobrazhenskaya, Kudrina, Blinov, Ryabova and Sveshnikova 1957) Pridham, Hesseltine and Benedict 1958, 69^{AL} ("*Actinomyces kurssanovii*" Preobrazhenskaya, Kudrina, Ryabova and Blinov *in* Gauze, Preobrazhenskaya, Kudrina, Blinov, Ryabova and Sveshnikova 1957, 156)

kurs.sa.nov′i.i. N.L. gen. masc. n. *kurssanovii* of Kursanov, possibly named after L.I. Kursanov, a Russian microbiologist.

Spore chains in Section *Retinaculiaperti.* Spirals are not observed on ISP media. Hooks and loops in the short spore chains are of small diameter and therefore are not truly representative of *Retinaculiaperti* morphology. The original description (Preobrazhenskaya and Ryabova, 1957) characterizes this culture as spiral. Mature spore chains are moderately short with 3–10, or sometimes more than 10, spores per chain. This morphology is seen on yeast-malt agar, oatmeal agar, and salts-starch agar; sporulation may be unsatisfactory on glycerol-asparagine agar. Spore surface is smooth.

Color of colony: aerial mass color in the Gray color series on yeast-malt agar and salts-starch agar. Reverse side of colony with no distinctive pigment (grayed yellow to yellow-brown) on oatmeal agar, salts-starch agar, and glycerol-asparagine agar; yellow-brown is modified by red on yeast-malt agar. Substrate pigment is not a pH indicator.

Color in medium: melanoid pigments formed in peptone-yeast-iron agar; pigments other than melanoids not formed in yeast-malt agar, oatmeal agar, salts-starch agar, or glycerol-asparagine agar.

D-Glucose, L-arabinose, sucrose, D-xylose, D-fructose, and raffinose are utilized for growth. No growth or only traces of growth on iso-inositol, D-mannitol, and rhamnose.

Type strain shows the highest sequence similarity to: *S. graminofaciens,* AJ781329, 100%; *S. xantholiticus,* AB184349, 99.9%; *S. peucetius,* AB045887, 99.8%.

Source: not known.

DNA G+C content (mol%): not known.

Type strain: ATCC 15824, ATCC 19774, ATCC 23629, CBS 512.68, BCRC 12133, CECT 3274, DSM 40162, NBRC

13192, INA 10294, JCM 4388, NCIMB 12788, NRRL B-3366, NRRL-ISP 5162, RIA 1054, UNIQEM 160.

Sequence accession no. (16S rRNA gene): AB184325.

273. **Streptomyces labedae** Lacey 1987, 458[VP]

la.be'da.e. N.L. gen. masc. n. *labedae* of Labeda, named after David P. Labeda, who first recorded the difference between the type strain and second reference strain of *Saccharopolyspora erythraeus* designated by Waksman.

Spore chains are in long, open spirals (*Spirales*). The spore surface is spiny. Spores are moderate gray on most media and reddish gray on glycerol-asparagine or Czapek's agars. Substrate mycelium olivc gray to dark reddish brown. Aerial mycelium white to pinkish, especially on glycerol-asparagine or Czapek's agars. Melanin pigments are not produced on Czapek's agar. Glucose, xylose, rhamnose, fructose, galactose, mannitol, and inositol are assimilated as sole carbon sources; arabinose, salicin, and sucrose are assimilated moderately well; raffinose is poorly assimilated. Wall chemotype I, with LL-A_2pm and glycine.

Type strain shows the highest sequence similarity to: *S. griseoincarnatus*, AJ781328, 100%; *S. erythrogriseus*, AJ781328, 100%; *S. variabilis*, DQ442551, 100%; *S. griseorubens*, AB184139, 99.9%; *S. griseoflavus*, AJ781322, 99.6%; *S. matensis*, AB184221, 99.6%; *S. althioticus*, AY999808, 99.3%; *S. paradoxus*, AB184628, 99.2%; *S. heliomycini*, AB184712, 99.1%; *S. collinus*, AB184123, 99.1%; *S. viridochromogenes*, DQ442555, 99.1%; *S. flaveolus*, AB184764, 99%; *S. bellus*, AB184849, 99%; *S. albogriseolus*, AJ494865, 99%; *S. viridodiastaticus*, AY999852, 99%; *S. violaceochromogenes*, AY999867, 99%; *S. malachitofuscus*, AB184282, 99%; *S. violaceorubidus*, AJ781374, 99%; *S. coerulescens*, AY999720, 99%.

Source: isolated from soil.

DNA G+C content (mol%): not known.

Type strain: Sanchez-Marroquin A-24, DSM 41446, NBRC 15864, IMRU 3737, JCM 9381, NRRL B-5616, NRRL-ISP 5059.

Sequence accession no. (16S rRNA gene): AB184704.

Further comments: Saccharopolyspora erythraeus (Waksman 1923) Waksman and Henrici 1948 emend. Labeda 1987 is illegitimate [Rules 37a, 51a and 53 of the *Bacteriological Code* (1990 Revision)] and the name *Streptomyces labedae* Lacey 1987 is provided for this taxon.

274. **Streptomyces laceyi** Manfio, Atalan, Zakrzewska-Czerwinska, Mordarski, Rodriguez, Collins and Goodfellow 2003b, 1219[VP] (Effective publication: Manfio, Atalan, Zakrzewska-Czerwinska, Mordarski, Rodriguez, Collins and Goodfellow 2003a, 254.)

la'cey.i. N.L. gen. masc. n. *laceyi* of Lacey, named in honor of John Lacey, an English microbiologist.

Forms extensively branched substrate and aerial hyphae. Spiral chains of smooth-surfaced spores are evident on aerial hyphae. Aerial spore mass varies from gray to yellow pink depending on the medium; similarly, the substrate mycelium color varies from yellow to yellow pink. Diffusible pigments are not produced. Degrades xanthine, but shows no activity against hippurate or nitrate. Cleaves methoxysuccinyl-L-alanine-L-lysine-7AMC, L-phenylalanine-7AMC, *N*-acetyl-L-phenylalanine-L-arginine-7AMC (endopeptidase substrates), benzylocarbonyl-L-arginine-7AMC, L-proline-7AMC, L-pyroglutamate-7AMC (exopeptidase substrates), 4MU-β-D-galactopyranoside, 4MU-*N*-acetyl-β-D-glucosaminide, 4MU-α-D-mannopyranoside (glycosides), and 4MU-α-D-pyrophosphate (inorganic ester) but not L-citrulline-7AMC, glutaryl-glycine-L-phenylalanine-7AMC, glutaryl-L-phenylananine-7AMC (endopeptidases), L-glutamine-7AMC, L-iso-leucine-7AMC, L-ornithine-7AMC (exopeptidases), 4MU-α-D-glucoside, 2-deoxy-β-D-glucopyranoside (glucosides), 4MU-elaidate, or 4MU-oleate (organic esters). Dextran, inulin, D(+)mannitol, citrate, and pyruvate are used as sole sources of carbon for energy and growth, but not *myo*-inositol. L-Histidine, potassium nitrate, and L-valine are used as sole carbon and nitrogen sources, but not L-hydroxyproline. Grows from 10–30°C and in the presence of sodium azide (0.01%, w/v), thallous acetate (0.001%, w/v), and neomycin sulfate (50 μg/ml), but shows no activity against *Bacillus subtilis* NCIMB 3610, *Candida albicans* CBS 562, *Micrococcus luteus* NCIMB 196, or *Streptomyces murinus* ISP 5091.

Type strain shows no sequence similarity over 99%.

Source: isolated from soil taken from meadow hay plots.

DNA G+C content (mol%): 69–74.

Type strain: C7654, DSM 41788, JCM 12606, NBRC 100783, NCIMB 13928.

Sequence accession no. (16S rRNA gene): AB249944.

275. **Streptomyces ladakanum** (Hanka, Evans, Mason and Dietz 1966) Witt and Stackebrandt 1991, 456[VP] (Effective publication: Witt and Stackebrandt 1990, 370.) (*Streptoverticillium ladakanus* Hanka, Evans, Mason and Dietz 1966, 620)

la.da.ka'num. N.L. adj. *ladakanum* (*sic*) from Ladislav J. Hanka, who isolated the strain.

Spore chains in Section *Verticillati*, umbellate monoverticillate (biverticillate). Mature spore chains are generally short with 3 to 10 or more spores per chain. This morphology can be seen on yeast-malt agar, oatmeal agar, salts-starch agar, or glycerol-asparagine agar, but sporulation may be poor on these media. Spore surface is smooth.

Color of colony: aerial mass color in the White or Yellow color scrics on yeast-malt agar, oatmeal agar, salts-starch agar, and glycerol-asparagine agar. Abundant aerial mycelium is usually in the White color series; shorter or poorly developed aerial mycelium may be 2ba or 2db, pale yellow. Reverse side of colony is moderate orange to strong brown on yeast-malt agar; light or pale yellow on oatmeal agar; yellowish pink to grayish reddish orange on salts-starch agar, and grayish yellow to reddish brown or strong brown on glycerol-asparagine agar. Reverse mycelium pigment is not a pH indicator.

Color in medium: melanoid pigments are not formed in peptone-yeast-iron agar, tyrosine agar, or tryptone-yeast broth. No pigment or only a trace of yellow is found in the medium in yeast-malt agar, oatmeal agar, salts-starch agar, and glycerol-asparagine agar.

D-Glucose and D-fructose are utilized for growth. No growth or only traces of growth with L-arabinose, D-xylose, iso-inositol, D-mannitol, rhamnose, sucrose, and raffinose.

For sequence similarity, see type strain of *Streptomyces mobaraensis*.

Source: not known.

DNA G+C content (mol%): not known.

Type strain: ATCC 27441, JCM 4778, NBRC 13476, NRRL 3191.

Sequence accession no. (16S rRNA gene): AB184430.

Further comments: according to Labeda (1996), *Streptomyces ladakanum* (Hanka et al. 1966) Witt and Stackebrandt 1991 is a later heterotypic synonym of *Streptomyces mobaraensis* (Nagatsu and Suzuki 1963) Witt and Stackebrandt 1991.

According to Hatano et al. (2003), *Streptomyces ladakanum* (Hanka et al. 1966) Witt and Stackebrandt 1991 is a later heterotypic synonym of *Streptomyces mobaraensis* (Nagatsu and Suzuki 1963) Witt and Stackebrandt 1991.

276. **Streptomyces lanatus** Frommer 1959, 204[AL]

la.na'tus. L. masc. adj. *lanatus* woolly, referring to the nature of the aerial mycelium of the organism.

Spore chains in Section *Spirales*. Mature spore chains generally have 10–50 spores per chain. This morphology is found on yeast-malt agar, oatmeal agar, salts-starch agar, and glycerol-asparagine agar. Spore surface is spiny.

Color of colony: aerial mass color in the Blue color series on yeast-malt agar, oatmeal agar, and salts-starch agar. Reverse side of colony with no distinctive pigment on yeast-malt agar, oatmeal agar, salts-starch agar, and glycerol-asparagine agar; substrate pigment is not a pH indicator.

Color in medium: melanoid pigments formed in peptone-yeast-iron agar, tyrosine agar, and tryptone-yeast broth. Pigments other than yellow or brown not formed in yeast-malt agar, oatmeal agar, salts-starch agar, or glycerol-asparagine agar.

D-Glucose, L-arabinose, sucrose, D-xylose, iso-inositol, D-mannitol, D-fructose, rhamnose, and raffinose are utilized for growth.

Type strain shows no sequence similarity over 99%.

Source: not known.

DNA G+C content (mol%): not known.

Type strain: ATCC 19775, CBS 513.68, BCRC 12060, DSM 40090, NBRC 12787, JCM 4332, JCM 4588, NRRL B-2291, NRRL-ISP 5090, RIA 1055, UNIQEM 161.

Sequence accession no. (16S rRNA gene): AB184845.

277. **Streptomyces lateritius** (Sveshnikova *in* Gauze, Preobrazhenskaya, Kudrina, Blinov, Ryabova and Sveshnikova 1957) Pridham, Hesseltine and Benedict 1958, 67[AL] ("*Actinomyces lateritius*" Sveshnikova *in* Gauze, Preobrazhenskaya, Kudrina, Blinov, Ryabova and Sveshnikova 1957, 70)

la.te.ri'ti.us. L. masc. adj. *lateritius* made or consisting of bricks, intended to mean brick red (referring to the color of the aerial mycelium).

Spore chains in Section *Retinaculiaperti*, but spore chains representative of Section *Rectiflexibiles* are also common. Mature spore chains generally have 10–50 spores per chain; longer chains are sometimes observed. This morphology is seen on salts-starch agar. Aerial mycelium is not produced on glycerol-asparagine agar and may be poorly developed on yeast-malt agar and oatmeal agar. Spore surface is warty.

Color of colony: aerial mass color in the Red color series on salts-starch agar. When sporulating aerial mycelium is formed on yeast-malt agar or oatmeal agar, it is in the Red color series. Reverse side of colony is grayed yellow is modified by violet or blue (see soluble pigment) on oatmeal agar, salts-starch agar, and glycerol-asparagine agar. Reverse color is changed from violet to blue by addition of 0.05 M NaOH, and from violet or blue to red with 0.05 M HCl.

Color in medium: melanoid pigments formed in peptone-yeast-iron agar. Blue or violet pigments found in medium in oatmeal agar, salts-starch agar, and glycerol-asparagine agar. This pigment is pH-sensitive when tested with 0.05 M NaOH or HCl. Violet pigment is changed to blue by 0.05 M NaOH and to red by 0.05 M HCl.

D-Glucose, L-arabinose, D-xylose, D-fructose, and rhamnose are utilized for growth. No growth or only traces of growth on sucrose, D-mannitol, and raffinose. Utilization of iso-inositol is doubtful.

Type strain shows the highest sequence similarity to: *S. venezuelae*, AB045890, 99.9%; *S. zaomyceticus*, EF178685, 99.7%; *S. exfoliatus*, AB184324, 99.7%; *S. wedmorensis*, DQ442557, 99.6%; *S. omiyaensis*, EF178697, 99.5%; *S. litmocidini*, AB184149, 99.5%; *S. yereyanensis*, EF178684, 99.3%; *S. narbonensis*, DQ445794, 99.2%.

Source: not known.

DNA G+C content (mol%): not known.

Type strain: AS 4.1427, ATCC 19776, ATCC 19913, CBS 514.68, BCRC 13774, DSM 40163, NBRC 12788, INA 6993, JCM 4389, NRRL B-5349, NRRL B-5423, NRRL-ISP 5163, RIA 1056, UNIQEM 162, VKM Ac-1849, VKM Ac-577.

Sequence accession no. (16S rRNA gene): AJ781326.

278. **Streptomyces laurentii** Trejo, Dean, Pluscec, Meyers and Brown 1977, 642[AL]

lau.ren'ti.i. N.L. gen. masc. n. *laurentii* derived from latinization of Lawrence, after Lawrence Township, New Jersey, the origin of the soil isolate.

On oatmeal agar, the aerial mycelium is predominantly straight (Rectus) with rare primitive spirals of a single turn; however, on starch-casein agar, hooks and primitive spirals predominate. The spores are smooth as seen by electron microscopy. In shaken culture [18 h at 25°C in tryptone-yeast extract broth (ISP medium 1)], a dusty pink soluble pigment is produced and the whole mycelium fragments into arthrospores and rods of varying length.

On yeast extract malt agar (ISP medium 2): sporulation is scant as a faint pinkish blush on white aerial mycelium; reverse: burnt orange; soluble pigment: light rose. On oatmeal agar (ISP medium 3): sporulation is good and occurs rapidly within 4 d; aerial mycelium is grayish yellow pink; CHM no. 5ec; dusty pink; there is no distinctive reverse color or soluble pigment. On inorganic salts starch agar (ISP medium 4): sporulation is good; CHM no. 4ec; light rose beige; reverse is reddish-orange; no soluble pigment. No melanin is produced on sodium caseinate-tyrosine agar (25 g sodium caseinate, 10 g $NaNO_3$, 1 g L-tyrosine, 1 liter tap water). Carbohydrate utilization was determined on the basal medium of Pridham and Gottlieb (1948) supplemented with individual carbon sources at 1% (w/v). The following carbohydrates, as the sole carbon source, support growth: glucose, xylose, galactose, melibiose, sucrose, and lactose. No growth on mannitol, inositol, sorbitol, arabinose, rhamnose, fructose, raffinose, and trehalose. Cell-wall hydrolysates contain LL-A_2pm.

Type strain shows the highest sequence similarity to: *S. roseofulvus*, AB184327, 99.2%; *S. bikiniensis*, X79851, 99%.

Source: isolated from soil from Lawrence Township, New Jersey, USA.

DNA G+C content (mol%): not known.

Type strain: ATCC 31255, DSM 41684, NBRC 15422, JCM 5063, PCM 2368.

Sequence accession no. (16S rRNA gene): AJ781342.

279. **Streptomyces lavendofoliae** (Kuchaeva, Krasil'nikov, Taptykova and Gesheva 1961) Pridham 1970, 19[AL] ("*Actinomyces lavendofoliae*" Kuchaeva, Krasil'nikov, Taptykova and Gesheva 1961, 120)

la.ven.do.fo.li'a.e. N.L. n. *lavendula* lavender; L. n. *folium* a leaf; N.L. gen. n. *lavendofoliae* (*sic*) of lavender leaf, referring to the color of the aerial mycelium.

Spore chains in Section *Retinaculiaperti*. Long, flexuous spore chains often terminate in spirals or hooks and loops of wide diameter. Mature spore chains have 10–50, or often more than 50, spores per chain. This morphology is seen on yeast-malt agar, oatmeal agar, salts-starch agar, and glycerol-asparagine agar. Spore surface is smooth.

Color of colony: aerial mass color in the Red color series (grayish yellowish pink or yellowish pink) on yeast-malt agar, oatmeal agar, and salts-starch agar. Reverse side of colony with no distinctive pigment (pale yellow to light yellowish brown) on yeast-malt agar, oatmeal agar, salts-starch agar, and glycerol-asparagine agar.

Color in medium: melanoid pigments are formed in peptone-yeast-iron agar and usually in tyrosine agar and tryptone-yeast broth. Pigments other than melanoids are not formed in the medium in yeast-malt agar, oatmeal agar, salts-starch agar, or glycerol-asparagine agar.

D-Glucose, L-arabinose, D-xylose, and iso-inositol are utilized for growth. Utilization of fructose is doubtful. No growth or only traces of growth on sucrose, D-mannitol, rhamnose, and raffinose.

Type strain shows the highest sequence similarity to: *S. gobitricini*, AB184666, 99.8%; *S. luridus*, DQ442523, 99.4%.

Source: not known.

DNA G+C content (mol%): not known.

Type strain: ATCC 15872, ATCC 23928, CBS 912.68, DSM 40217, NBRC 12882, INA 3613, JCM 4391, NCIMB 9823, NRRL B-3371, NRRL-ISP 5217, RIA 1161, RIA 750, VKM Ac-272.

Sequence accession no. (16S rRNA gene): AJ781336.

280a. **Streptomyces lavendulae subsp. lavendulae** (Waksman and Curtis 1916) Waksman and Henrici *in* Breed, Murray and Hitchens 1948, 944[AL] ("*Actinomyces lavendulae*" Waksman and Curtis 1916, 126)

la.ven.du'la.e. N.L. n. *lavendula* lavender; N.L. gen. n. *lavendulae* of lavender color.

Spore chains in Section *Retinaculiaperti* with many *Rectiflexibiles* spore chains as well as coils of wide diameter and some spirals. Mature spore chains generally have 10–50 spores per chain; longer chains are often observed. This morphology is seen on yeast-malt agar, oatmeal agar, salts-starch agar, and glycerol-asparagine agar. Spore surface is smooth.

Color of colony: aerial mass color in the Red (or Gray) color series on yeast-malt agar, oatmeal agar, salts-starch agar, and glycerol-asparagine agar. Characteristic color is between 3ge (light grayish yellowish brown) and 4ec or 5ec (grayish yellowish pink) color tabs of Tresner–Backus Color Wheels. Reverse side of colony with no distinctive pigment on yeast-malt agar, oatmeal agar, salts-starch agar, and glycerol-asparagine agar; substrate pigment is not a pH indicator (one observer reports slight change from yellowish brown to greenish brown by addition of 0.05 M NaOH).

Color in medium: melanoid pigments formed in peptone-yeast-iron agar and usually in tyrosine agar and tryptone-yeast broth. Pigments other than melanoids not formed in yeast-malt agar, oatmeal agar, salts-starch agar, or glycerol-asparagine agar.

D-Glucose is utilized for growth. No growth or only traces of growth on L-arabinose, sucrose, D-xylose, iso-inositol, D-mannitol, rhamnose, and raffinose. Limited growth with D-fructose.

Type strain shows the highest sequence similarity to: *S. cyaneofuscatus*, AB184860, 100%; *S. cavourensis* subsp. *washingtonensis*, DQ026671, 100%; *S. alboviridis*, AB184256, 100%; *S. fulvorobeus*, AB184711, 100%; *S. microflavus*, DQ445795, 100%; *S. praecox*, AB184293, 100%; *S. anulatus*, DQ026637, 100%; *S. flavofuscus*, AB249935, 100%; *S. cinereorectus*, AB184646, 100%; *S. griseoplanus*, AY999894, 100%; *S. lipmanii*, AB184148, 100%; *S. fimicarius*, AY999784, 100%; *S. sindenensis*, AB184759, 99.9%; *S. rubiginosohelvolus*, AB184240, 99.9%; *S. mediolani*, AB184674, 99.9%; *S. griseinus*, AB184205, 99.9%; *S. acrimycini*, AY999889, 99.9%; *S. pluricolorescens*, DQ442540, 99.9%; *S. argenteolus*, AB045872, 99.9%; *S. badius*, AY999783, 99.9%; *S. floridae*, AB184656, 99.8%; *S. albovinaceus*, AB249958, 99.8%; *S. globisporus* subsp. *globisporus*, EF178686, 99.8%; *S. griseus* subsp. *griseus*, AY207604, 99.8%; *S. griseolus*, AB184768, 99.8%; *S. luridiscabiei*, AF361784, 99.8%; *S. baarnensis*, EF178688, 99.8%; *S. flavovirens*, DQ026635, 99.7%; *S. californicus*, AB184755, 99.7%; *S. parvus*, DQ442537, 99.7%; *S. halstedii*, EF178695, 99.7%; *S. flavogriseus*, AJ494864, 99.6%; *S. olivoviridis*, AB184227, 99.5%; *S. bacillaris*, AB184439, 99.5%; *S. finlayi*, AY999788, 99.5%; *S. pulveraceus*, AB184806, 99.5%; *S. atroolivaceus*, AJ781320, 99.5%; *S. clavifer*, DQ026670, 99.4%; *S. yanii*, AB006159, 99.4%; *S. nitrosporeus*, EF178680, 99.4%; *S. albolongus*, AB184425,

99.3%; *S. sanglieri*, AB249945, 99.3%; *S. mutomycini*, AB249951, 99.2%; *S. atratus*, DQ026638, 99.2%; *S. celluloflavus*, AB184476, 99.2%; *S. gelaticus*, DQ026636, 99.2%; *S. griseobrunneus*, AB249912, 99.2%; *S. spiroverticillatus*, AB184814, 99.1%; *S. cavourensis* subsp. *cavourensis*, DQ445791, 99.1%; *S. cremeus*, AB184124, 99.1%; *S. candidus*, DQ026663, 99%.

Source: not known.

DNA G+C content (mol%): not known.

Type strain: ATCC 19777, ATCC 23950, ATCC 8664, CBS 515.68, CCM 3010, CCUG 11120, DSM 2014, DSM 40069, DSM 40213, HUT 6006, IFM 1031, NBRC 12789, NBRC 12903, NBRC 3177, NRRL-ISP 5213, NRRL-ISP 5069, JCM 4055, JCM 4589, JCM 4664, KCTC 1398, NBRC 12343, NCIMB 9840, NRRL B-1230, NRRL B-5617, RIA 1057, RIA 1159, RIA 144, RIA 531, RIA 744, UNIQEM 163, VKM Ac-1278, VKM Ac-624.

Sequence accession no. (16S rRNA gene): D85116.

280b. **Streptomyces lavendulae subsp. grasserius** (Kuchaeva, Krasil'nikov, Taptykova and Gesheva 1961) Pridham 1970, 54[AL] ("*Actinomyces lavendulae grasserius*" Kuchaeva, Krasil'nikov, Taptykova and Gesheva 1961, 119)

gras.se'ri.us. Fr. n. *grasserié* a disease of silkworms; N.L. adj. *grasserius* grasserial (referring to the activity of the organism against silkworm jaundice virus, grasserié).

Spore chains of typical *Retinaculum-Apertum* type; poor growth on Czapek's solution agar. Produces the grasseriomycin complex (a streptothricin complex); inhibited by streptomycin.

Type strain shows no sequence similarity over 99%.

Source: not known.

DNA G+C content (mol%): not known.

Type strain: ATCC 15875, ATCC 25457, CBS 690.69, DSM 40385, NBRC 13045, JCM 4056, JCM 4556, NRRL B-3072, NRRL B-3372, NRRL-ISP 5385, RIA 1237, RIA 746, VKM Ac-1178.

Sequence accession no. (16S rRNA gene): AY999841.

281. **Streptomyces lavenduligriseus** (Locci, Baldacci and Petrolini Baldan 1969) Witt and Stackebrandt 1991, 456[VP] (Effective publication: Witt and Stackebrandt 1990, 370.) (*Streptoverticillium lavenduligriseum* Locci, Baldacci and Petrolini Baldan 1969, 59)

la.ven.du.li.gri'se.us. N.L. n. *lavendula* lavender; N.L. adj. *griseus* gray; N.L. masc. adj. *lavenduligriseus* gray and lavender colored.

Spore chains in Section *Rectiflexibiles* to *Retinaculiaperti* (one observer records monoverticillate sporophores). Clusters of short, flexuous spore chains include hooks, loops, and primitive spirals. Straight spore chains are also seen, and spore chains or clusters of sporophores sometimes seem to be opposite or in whorls. These do not appear to have the uniformity or even spacing characteristic of verticillate cultures. This morphology is seen on yeast-malt agar, oatmeal agar, and salts-starch agar; sporulation is poor or absent on glycerol-asparagine agar. Spore surface is smooth.

Color of colony: aerial mass color in the Gray color series (3fe, light brownish gray; 3ge, light grayish yellowish brown; 3ig, grayish yellowish brown; or 2fe, medium gray) on yeast-malt agar, oatmeal agar, and salts-starch agar. Aerial mycelium is poorly developed or absent on glycerol-asparagine agar. Reverse side of colony with no distinctive pigments (brown to grayish yellow on yeast-malt agar and oatmeal agar; grayish yellow or pale yellow on salts-starch agar and glycerol-asparagine agar).

Color in medium: melanoid pigments are formed in peptone-yeast-iron agar, tyrosine agar, and tryptone-yeast broth. One of three observers reports reddish to orange pigment in the medium in yeast-malt agar and oatmeal agar, and yellow pigment in glycerol-asparagine agar. Two observers report no pigment in ISP media. When soluble pigment is produced, it is reported to be pH-sensitive, changing from red, orange, or yellow to violet with the addition of 0.05 M NaOH.

D-Glucose, L-arabinose, D-xylose, D-fructose, and D-mannitol are utilized for growth. Reports vary on utilization of iso-inositol. Utilization of rhamnose or sucrose is doubtful. No growth or only traces of growth with raffinose.

Type strain shows the highest sequence similarity to: *S. morookaense* AJ781349, 99.6%; *S. thioluteus*, AB184753, 99.2%; *S. abikoensis*, AB184537, 99.1%; *S. luteireticuli* AB249969, 99%; *S. sapporonensis*, AB184508, 99%.

Source: not known.

DNA G+C content (mol%): not known.

Type strain: ATCC 13306, ATCC 29661, CBS 706.72, DSM 40487, NBRC 13405, JCM 4545, JCM 4779, NRRL B-3173, NRRL-ISP 5487, RIA 1366, VKM Ac-1159.

Sequence accession no. (16S rRNA gene): DQ442515.

282. **Streptomyces lavendulocolor** (Kuchaeva, Krasil'nikov, Taptykova and Gesheva 1961) Pridham 1970, 20[AL] ("*Actinomyces lavendulocolor*" Kuchaeva, Krasil'nikov, Taptykova and Gesheva 1961, 120)

la.ven.du.lo.co'lor. N.L. n. *lavendula* the lavender; L. n. *color* color; N.L. *lavendulocolor* intended to mean lavender colored.

Spore chains of typical *Retinaculum-Apertum* type; poor growth on Czapek's solution agar. Said to produce the streptothricin complex; inhibited by streptomycin; said to inhibit some other streptothricin complex-producing strains.

Type strain shows the highest sequence similarity to: *S. cellulosae*, DQ442495, 99.1; *S. gancidicus*, AB184660, 99; *S. pseudogriseolus*, DQ442541, 99%.

Source: not known.

DNA G+C content (mol%): not known.

Type strain: ATCC 15871, ATCC 23927, CBS 911.68, BCRC 12057, DSM 40216, NBRC 12881, INA 4518, JCM 4390, NCIMB 9829, NRRL B-3367, NRRL-ISP 5216, RIA 1160, RIA 749, VKM Ac-215.

Sequence accession no. (16S rRNA gene): DQ442516.

283. **Streptomyces levis** Sveshnikova 1986, 574[VP] (Effective publication: Sveshnikova *in* Gause, Preobrazhenskaya, Sveshnikova, Terekhova and Maximova 1983.)

le'vis. L. masc. adj. *levis* smooth.

Spore chains are spiral (*Spirales*); spores are smooth. On mineral agar 1, oatmeal agar: aerial mycelium is pale pink to pale violet; substrate mycelium is brown; diffusible pigment is brown with grayish violet shadow. On starch-ammonia agar: aerial mycelium is not extant or poor, whitish; substrate mycelium is violet, gray-brown violet or plum black; diffusible pigment is pale violet, weak, or not extant. On glycerol-nitrate agar: aerial mycelium is white, poor; substrate mycelium is colorless to violet; no diffusible pigment. On glycerol-asparagine agar: aerial mycelium is light violet to light pink-violet; substrate mycelium is grayish-brownish yellow later grayish-brownish violet, plum black; no diffusible pigment. On organic agar 2: aerial mycelium is white to pale pink; substrate mycelium is brown to yellow brown; diffusible pigment is brown. Melanoid pigments are not formed. Grows on glucose, arabinose, xylose, raffinose, fructose, and sucrose; moderate growth on mannitol. Antibiotic pigment of group ribromycin-griseorodin having indicator character: alkaline reaction, violet; acidic reaction, red.

Type strain shows the highest sequence similarity to: *S. pseudogriseolus*, DQ442541, 99.5%; *S. capillispiralis*, AB184577, 99.5%; *S. carpinensis*, AB184574, 99.5%; *S. tuirus*, AB184690, 99.4%; *S. gancidicus*, AB184660, 99.4%; *S. azureus*, EF178674, 99.4%; *S. cellulosae*, DQ442495, 99.3%; *S. djakartensis*, AB184657, 99.2%; *S. afghaniensis*, AJ399483, 99.2%; *S. viridiviolaceus*, AY999854, 99.2%; *S. africanus*, AY208912, 99.2%; *S. caelestis*, X80824, 99.1%; *S. spinoverrucosus*, AB184578, 99.1%; *S. mutabilis*, EF178679, 99.1%; *S. misionensis*, EF178678, 99.1%; *S. paradoxus*, AB184628, 99%; *S. ghanaensis*, AY999851, 99%; *S. violaceus*, AB184315, 99%; *S. hawaiiensis*, AB184143, 99%; *S. vinaceusdrappus*, AY999929, 99%; *S. luteogriseus*, AB184379, 99%; *S. albosporeus* subsp. *albosporeus*, AJ781327, 99%; *S. geysiriensis*, DQ442501, 99%; *S. plicatus*, AB184291, 99%; *S. janthinus*, AB184851, 99%; *S. rochei*, AB184237, 99%; *S. minutiscleroticus*, EF178696, 99%; *S. roseoviolaceus*, AJ399484, 99%.

Source: not known.

DNA G+C content (mol%): not known.

Type strain: ATCC 43686, DSM 41458, NBRC 15423, INA 9020, JCM 6924, NRRL B-16370, VKM Ac-835.

Sequence accession no. (16S rRNA gene): AB184670.

284a. **Streptomyces libani subsp. libani** Baldacci and Grein 1966, 196[AL]

li'ba.ni. L. gen. n. *libani* of Lebanon, the source of the soil from which the organism was isolated.

Spore chains in Section *Spirales*. Tight terminal spirals are usually seen on moderately short chains of 10 or more spores. This morphology is found on yeast-malt agar, oatmeal agar, salts-starch agar, and glycerol-asparagine agar. Spore surface is smooth.

Color of colony: aerial mass color in the Gray color series on yeast-malt agar, oatmeal agar, salts-starch agar, and glycerol-asparagine agar. Nearest matching color tabs reported for aerial mass color on yeast-malt agar, oatmeal agar, and glycerol-asparagine agar are d, light gray; e, medium gray; and 2dc and 2bc, yellowish gray. Nearest matching tabs reported for salts-starch agar are 3fe, light brownish gray; 4li, brownish gray; and 3ge, light yellowish brown. Reverse side of colony is gray or pale yellow to orange yellow on yeast-malt agar, oatmeal agar, salts-starch agar, and glycerol-asparagine agar.

Color in medium: melanoid pigments are not formed in peptone-yeast-iron agar, tyrosine agar, or tryptone-yeast broth. No pigment or only a trace of yellow is found in the medium in yeast-malt agar, oatmeal agar, salts-starch agar, and glycerol-asparagine agar.

D-Glucose, D-xylose, iso-inositol, D-mannitol, D-fructose, rhamnose, sucrose, and raffinose are utilized for growth. Utilization of L-arabinose is doubtful.

Type strain shows the highest sequence similarity to: *S. nigrescens*, DQ442530, 100%; *S. tubercidicus*, AJ621612, 100%; *S. hygroscopius* subsp. *glebosus*, AB184479, 99.7%; *S. libani* subsp. *rufus*, AJ781351, 99.7%; *S. caniferus*, AB184640, 99.6%; *S. catenulae*, AJ621613, 99.6%; *S. misakiensis*, AB217605, 99.6%; *S. sioyaensis*, DQ026654, 99.5%; *S. platensis*, AB045882, 99.5%; *S. hygroscopius* subsp. *decoyicus*, AY999883, 99.2%; *S. lydicus*, Y15507, 99.1%; *S. chattanoogensis*, AJ621611, 99%.

Source: isolated from soil from Lebanon.

DNA G+C content (mol%): not known.

Type strain: ATCC 23732, CBS 753.72, DSM 40555, NBRC 13452, JCM 4322, JCM 4781, KCTC 9113, NCAIM B.01474, NCIMB 11012, NRRL B-3446, NRRL-ISP 5555, RIA 1413, VKM Ac-1905.

Sequence accession no. (16S rRNA gene): AB184414.

284b. **Streptomyces libani subsp. rufus** Baldacci and Grein 1966, 197[AL]

ru'fus. L. masc. adj. *rufus* red, reddish.

Moderate growth on Czapek's solution agar; aerial mycelium typically colored in shades of hazel-nut brown (*avellaneus*); hygroscopic; forms reddish violet diffusible pigment with some media. Produces libanomycin; exhibits anti-fungal activity; inhibited by streptomycin.

Type strain shows the highest sequence similarity to: *S. caniferus*, AB184640, 100%; *S. hygroscopius* subsp. *glebosus*, AB184479, 100%; *S. platensis*, AB045882, 99.9%; *S. libani* subsp. *libani*, AB184414, 99.7%; *S. tubercidicus*, AJ621612, 99.6%; *S. nigrescens*, DQ442530, 99.6%; *S. hygroscopius* subsp. *decoyicus*, AY999883, 99.6%; *S. catenulae*, AJ621613, 99.3%; *S. ramulosus*, DQ026662, 99.3%; *S. misakiensis*, AB217605, 99.3%; *S. sioyaensis*, DQ026654, 99.2%; *S. monomycini*, DQ445790, 99%.

Source: not known.

DNA G+C content (mol%): not known.

Type strain: ATCC 23731, CMI 130779, DSM 41230, NBRC 15424, JCM 4325, NCIMB 10976, NRRL B-3445, VKM Ac-600.

Sequence accession no. (16S rRNA gene): AJ781351.

285. **Streptomyces lienomycini** Gause and Maximova 1986a, 574[VP] (Effective publication: Gause and Maximova *in* Gause, Preobrazhenskaya, Sveshnikova, Terekhova and Maximova 1983.)

li.e.no.my.ci'ni. N.L. n. *lienomycinum* lienomycin; N.L. gen. n. *lienomycini* of lienomycin, intended to mean lienomycin producing.

Spore chains are spiral (*Spirales*); spores are smooth. On mineral agar 1: aerial mycelium is white, later gray; substrate mycelium is yellowish or gray brownish yellow; no diffusible pigment. On glycerol-nitrate agar: aerial mycelium is absent; substrate mycelium colorless; no diffusible pigment. On oatmeal agar: aerial mycelium is light gray; substrate mycelium colorless; no diffusible pigment. On starch-ammonia agar: aerial mycelium is grayish white; substrate mycelium yellowish; no diffusible pigment. On glycerol-asparagine agar: aerial mycelium is absent; substrate mycelium yellowish; no diffusible pigment. On organic agar 2: aerial mycelium is absent or whitish; substrate mycelium and diffusible pigment are gray brown. Melanoid pigments are formed. Good growth on xylose, fructose, inositol, sucrose, arabinose, mannitol, rhamnose, and raffinose. Antibiotic: lienomycin.

Type strain shows the highest sequence similarity to: *S. rubrogriseus*, AB184681, 99.9%; *S. violaceorubidus*, AJ781374, 99.7%; *S. anthocyanicus*, AB184631, 99.7%; *S. tricolor*, AB184687, 99.7%; *S. violaceoruber*, AF503492, 99.6%; *S. coelescens*, AF503496, 99.6%; *S. violaceolatus* AF503497, 99.5%; *S. humiferus*, AF503491, 99.5%; *S. tendae*, D63873, 99.5%; *S. coelicoflavus*, AB184650, 99.2%; *S. olivaceus*, AB184743, 99.2%; *S. pactum*, AB184398, 99.2%; *S. ambofaciens*, M27245, 99.1%; *S. matensis*, AB184221, 99.1%; *S. heliomycini*, AB184712, 99%.

Source: not known.

DNA G+C content (mol%): not known.

Type strain: ATCC 43687, DSM 41475, NBRC 15425, INA 478, JCM 6925, VKM Ac-1767.

Sequence accession no. (16S rRNA gene): AJ781353.

286. **Streptomyces lilacinus** (Nakazawa, Tanabe, Shibata, Miyake and Takewaka 1956) Witt and Stackebrandt 1991, 456[VP] (Effective publication: Witt and Stackebrandt 1990, 370.) ("*Streptomyces lilacinus*" Nakazawa, Tanabe, Shibata, Miyake and Takewaka 1956, 81; "*Verticillomyces lilacinus*" Shinobu 1965; *Streptoverticillium lilacinum* Locci, Baldacci and Petrolini Baldan 1969, 59)

li.la'ci.nus. N.L. masc. adj. *lilacinus* lilac colored.

Spore chains in Section *Verticillati*, umbellate monoverticillate (biverticillate). Spore chains are short, usually with 3–10 spores per chain. This morphology is seen on yeast-malt agar, oatmeal agar, salts-starch agar, and glycerol-asparagine agar. Spore surface is smooth.

Color of colony: aerial mass color in the Red color series (grayish yellowish pink) on yeast-malt agar, oatmeal agar, salts-starch agar, and glycerol-asparagine agar. Reverse side of colony is yellow-brown plus red (grayish reddish brown to strong brown) on yeast-malt agar, oatmeal agar, salts-starch agar, and glycerol-asparagine agar; this pigment is somewhat pH-sensitive changing from brown to reddish brown with addition of 0.05 M NaOH and from brown to yellowish brown with addition of 0.05 M HCl.

Color in medium: melanoid pigments are formed in peptone-yeast-iron agar, tyrosine agar, and tryptone-yeast broth. Red pigment is found in the medium in yeast-malt agar, oatmeal agar, salts-starch agar, and glycerol-asparagine agar. This pigment is pH-sensitive when tested with 0.05 M NaOH or HCl, showing the same change noted for substrate color.

D-Glucose is utilized for growth; reports vary on positive utilization of iso-inositol. No growth or only trace of growth on L-arabinose, sucrose, D-xylose, D-mannitol, D-fructose, rhamnose, and raffinose.

Type strain shows the highest sequence similarity to: *S. abikoensis*, AB184537, 99.7%; *S. hygroscopius* subsp. *angustmyceticus*, DQ442509, 99.4%; *S. hiroshimensis*, AB184789, 99.4%; *S. caeruleus*, EF178675, 99.3%; *S. sapporonensis*, AB184508, 99.3%; *S. ehimensis*, AY999834, 99.3%; *S. cinnamoneus*, AB184850, 99.1%; *S. pseudoechinosporeus*, AB184100, 99%.

Source: not known.

DNA G+C content (mol%): not known.

Type strain: ATCC 23930, CBS 914.68, BCRC 12421, CECT 3264, DSM 40254, NBRC 12884, NBRC 3944, JCM 4188, JCM 4648, NRRL B-1968, NRRL-ISP 5254, RIA 1180.

Sequence accession no. (16S rRNA gene): AB184819.

Further comments: in violation of Rule 33c of the *Bacteriological Code* (1990 Revision), in Validation List no. 38, *Streptomyces lilacinus* is proposed as a *nomen revictum* (basonym: "*Streptomyces lilacinus*" Nakazawa et al. 1956).

According to Hatano et al. (2003), *Streptomyces lilacinus* (Nakazawa et al. 1956) Witt and Stackebrandt 1991 is an earlier heterotypic synonym of *Streptomyces kashmirensis* (*sic*) (Gupta and Chopra 1963b) Witt and Stackebrandt 1991.

287. **Streptomyces limosus** Lindenbein 1952, 379[AL]

li.mo'sus. L. masc. adj. *limosus* slimy, referring to the river bank slime from which the organism was isolated.

Spore chains in Section *Rectiflexibiles*. Mature spore chains on suitable media generally 10–50 spores per chain; longer chains are sometimes observed. This morphology is seen on yeast-malt agar, salts-starch agar, and glycerol-asparagine agar. Spore surface is smooth.

Color of colony: aerial mass color in the Yellow color series on yeast-malt agar, oatmeal agar, salts-starch agar, and glycerol-asparagine agar. Reverse side of colony with no distinctive pigment on yeast-malt agar, oatmeal agar, salts-starch agar, or glycerol-asparagine agar; substrate pigment is not a pH indicator.

Color in medium: melanoid pigments not formed in peptone-yeast-iron agar or tyrosine agar; no pigment found in medium in yeast-malt agar, oatmeal agar, salts-starch agar, or glycerol-asparagine agar.

D-Glucose, L-arabinose, D-xylose, D-mannitol, and D-fructose are utilized for growth. No growth or only trace of growth on sucrose, iso-inositol, rhamnose, and raffinose.

Type strain shows the highest sequence similarity to: *S. canescens*, AB184117, 100%; *S. felleus*, AB184129, 100%; *S. daghestanicus*, DQ442497, 100%; *S. albidoflavus*, AB184255, 100%; *S. hydrogenans*, AB184868, 100%; *S. odorifer*, Z76682, 100%; *S. violascens*, AY999737, 100%; *S. griseus* subsp. *solvifaciens*, AB249915, 100%; *S. champavatii*, DQ026642, 100%; *S. sampsonii*, D63871, 99.8%; *S. koyangensis*, AY079156, 99.7%.

Source: isolated from river bank slime.

DNA G+C content (mol%): not known.

Type strain: ATCC 19778, CBS 531.68, BCRC 13700, DSM 40131, NBRC 12790, JCM 4393, KCTC 9033, NCIMB 12976, NRRL B-5413, NRRL-ISP 5131, RIA 1058, UNIQEM 165, VKM Ac-850.

Sequence accession no. (16S rRNA gene): AB184147.

288. **Streptomyces lincolnensis** Mason, Dietz and DeBoer 1963a, 555[AL]

lin.coln.en'sis. N.L. masc. adj. *lincolnensis* of or belonging to Lincoln, referring to the source of the soil from which the organism was isolated, *viz.* Gehring, near Lincoln, Nebraska, USA.

Spore chains in Section *Rectiflexibiles.* Mature spore chains are long and flexuous, often with more than 50 spores per chain. This morphology is seen on yeast-malt agar, oatmeal agar, salts-starch agar, and glycerol-asparagine agar. Spore surface is smooth.

Color of colony: aerial mass color in the Yellow or Red color series on yeast-malt agar, oatmeal agar, salts-starch agar, and glycerol-asparagine agar. Representative color tabs are 2ba (pale yellow) from the Yellow color series; 2ca (pale yellow) and 3ca (pale orange yellow) from the Red color series. Reverse side of colony with no distinctive pigments (pale grayish yellow to orange yellow) on oatmeal agar and salts-starch agar; yellowish brown to brown on yeast-malt agar and olive brown to dark grayish brown on glycerol-asparagine agar.

Color in medium: melanoid pigments are formed in peptone-yeast-iron agar, tyrosine agar, and tryptone-yeast broth. A trace of yellow to greenish yellow pigment is found in the medium in oatmeal agar and glycerol-asparagine agar. This pigment is not pH-sensitive when tested with 0.05 M NaOH or HCl.

D-Glucose, L-arabinose, D-xylose, iso-inositol, D-mannitol, D-fructose, rhamnose, sucrose, and raffinose are all utilized for growth.

Type strain shows no sequence similarity over 99%.

Source: isolated from soil in Lincoln, Nebraska.

DNA G+C content (mol%): not known.

Type strain: ATCC 25466, CBS 630.70, CBS 699.69, BCRC 11173, DSM 2013, DSM 40355, NBRC 13054, JCM 4287, JCM 4488, KCTC 1868, KCTC 9088, KCTC 9089, NCIMB 9413, NRRL 2936, NRRL-ISP 5355, RIA 1246, VKM Ac-727.

Sequence accession no. (16S rRNA gene): AB184279.

289. **Streptomyces lipmanii** (Waksman and Curtis 1916) Waksman and Henrici *in* Breed, Murray and Hitchens 1948, 952[AL] ("*Actinomyces lipmanii*" Waksman and Curtis 1916, 123)

lip.man'i.i. N.L. gen. masc. n. *lipmanii* of Lipman, named for Jacob Goodale Lipman (1874–1939) of the New Jersey Agricultural Experiment Station.

Spore chains in Section *Rectiflexibiles.* Mature spore chains generally have 10–50 spores per chain; longer chains are sometimes observed on suitable media. This morphology is seen on yeast-malt agar and glycerol-asparagine agar. Sporulation may be poor on oatmeal agar and salts-starch agar. Spore surface is smooth.

Color of colony: aerial mass color in the Yellow color series on yeast-malt agar, salts-starch agar and glycerol-asparagine agar. Reverse side of colony with no distinctive pigment on oatmeal agar, salts-starch agar and glycerol-asparagine agar; yellow to yellow-brown on yeast-malt agar. Substrate pigment is not a pH indicator.

Color in medium: melanoid pigments not formed in peptone-yeast-iron agar or tyrosine agar. No pigment found in medium in yeast-malt agar, oatmeal agar, salts-starch agar, or glycerol-asparagine agar.

D-Glucose, D-xylose, D-mannitol, D-fructose, and rhamnose are utilized for growth. No growth or only traces of growth on L-arabinose, iso-inositol, and raffinose. Utilization of sucrose is doubtful.

Type strain shows the highest sequence similarity to: *S. cavourensis* subsp. *washingtonensis*, DQ026671, 100%; *S. fulvorobeus*, AB184711, 100%; *S. microflavus*, DQ445795, 100%; *S. lavendulae* subsp. *lavendulae*, AB184080, 100%; *S. alboviridis*, AB184256, 100%; *S. luridiscabiei*, AF361784, 99.9%; *S. flavofuscus*, AB249935, 99.9%; *S. cinereorectus*, AB184646, 99.9%; *S. fimicarius*, AY999784, 99.9%; *S. griseoplanus*, AY999894, 99.9%; *S. cyaneofuscatus*, AB184860, 99.9%; *S. floridae*, AB184656, 99.9%; *S. praecox*, AB184293, 99.9%; *S. anulatus*, DQ026637, 99.9%; *S. pluricolorescens*, DQ442540, 99.8%; *S. acrimycini*, AY999889, 99.8%; *S. argenteolus*, AB045872, 99.8%; *S. griseinus*, AB184205, 99.8%; *S. badius*, AY999783, 99.8%; *S. rubiginosohelvolus*, AB184240, 99.8%; *S. sindenensis*, AB184759, 99.8%; *S. mediolani*, AB184674, 99.8%; *S. californicus*, AB184755, 99.8%; *S. parvus*, DQ442537, 99.7%; *S. halstedii*, EF178695, 99.7%; *S. baarnensis*, EF178688, 99.7%; *S. griseolus*, AB184768, 99.7%; *S. griseus* subsp. *griseus*, AY207604, 99.7%; *S. globisporus* subsp. *globisporus*, EF178686, 99.7%; *S. albovinaceus*, AB249958, 99.7%; *S. flavovirens*, DQ026635, 99.6%; *S. finlayi*, AY999788, 99.5%; *S. flavogriseus*, AJ494864, 99.5%; *S. yanii*, AB006159, 99.4%; *S. atroolivaceus*, AJ781320, 99.4%; *S. olivoviridis*, AB184227, 99.4%; *S. bacillaris*, AB184439, 99.4%; *S. pulveraceus*, AB184806, 99.4%; *S. celluloflavus*, AB184476, 99.3%; *S. griseobrunneus*, AB249912, 99.3%; *S. albolongus*, AB184425, 99.3%; *S. clavifer*, DQ026670, 99.3%; *S. nitrosporeus*, EF178680, 99.3%; *S. spiroverticillatus*, AB184814, 99.2%; *S. sanglieri*, AB249945, 99.2%; *S. cavourensis* subsp. *cavourensis*, DQ445791, 99.2%; *S. mutomycini*, AB249951, 99.2%; *S. atratus*, DQ026638, 99.1%; *S. candidus*, DQ026663, 99.1%; *S. gelaticus*, DQ026636, 99.1%; *S. cremeus*, AB184124, 99%.

Source: not known.

DNA G+C content (mol%): not known.

Type strain: ATCC 19779, ATCC 3331, CBS 532.68, DSM 40070, HAMBI 1075, HUT 6059, NBRC 12791, NBRC 3410, NBRC 3721, IMET 40336, JCM 4058, JCM 4590, LMG 20047, NCIMB 9841, NRRL B-1229, NRRL B-2100, NRRL-ISP 5070, RIA 1059, RIA 85, UNIQEM 166.

Sequence accession no. (16S rRNA gene): AB184148.

Further comments: according to Lanoot et al. (2005b), *Streptomyces lipmanii* (Waksman and Curtis 1916) Waksman and Henrici 1948 is a later heterotypic synonym of *Streptomyces microflavus* (Krainsky 1914) Waksman and Henrici 1948 emend. Lanoot et al. 2005b.

290. **Streptomyces litmocidini** (Ryabova and Preobrazhenskaya *in* Gauze, Preobrazhenskaya, Kudrina, Blinov, Ryabova and Sveshnikova 1957) Pridham, Hesseltine and Benedict 1958, 65[AL] ("*Actinomyces litmocidini*" Ryabova and Preobrazhenskaya *in* Gauze, Preobrazhenskaya, Kudrina, Blinov, Ryabova and Sveshnikova 1957, 187)

lit.mo.ci.di'ni. N.L. gen. n. *litmocidini* of litmocidin.

Spore chains in Section *Rectiflexibiles* with some straight spore chains and occasional hooks or primitive spirals suggestive of Section *Retinaculiaperti.* Mature spore chains generally have 10–50 spores per chain; longer chains are often observed. This morphology is seen on yeast-malt agar, oatmeal agar, salts-starch agar, and glycerol-asparagine agar. Spore surface is smooth.

Color of colony: aerial mass color in the Gray color series on yeast-malt agar, oatmeal agar, salts-starch agar, and glycerol-asparagine agar. Reverse side of colony is colorless or characteristic grayed yellow on oatmeal agar, but the grayed yellow is modified by violet or red (depending on pH) on salts-starch agar, glycerol-asparagine agar, and sometimes on yeast-malt agar. Reverse color is changed from violet to blue by addition of 0.05 M NaOH, and from violet to red with 0.05 M HCl.

Color in medium: melanoid pigments formed in peptone-yeast-iron agar. Violet, blue, or red pigments found in medium in yeast-malt agar, salts-starch agar, and glycerol-asparagine agar; violet pigment is changed to blue by 0.05 M NaOH and to red by addition of 0.05 M HCl.

D-Glucose and L-arabinose are utilized for growth. No growth or only trace of growth on sucrose, iso-inositol, D-mannitol, rhamnose, and raffinose. Variable reports on growth with D-xylose and D-fructose.

Type strain shows the highest sequence similarity to: *S. yereyanensis* EF178684, 99.8%; *S. exfoliatus*, AB184324, 99.6%; *S. venezuelae*, AB045890, 99.6%; *S. zaomyceticus*, EF178685, 99.6%; *S. omiyaensis*, EF178697, 99.5%; *S. wedmorensis*, DQ442557, 99.5%; *S. lateritius*, AB184324, 99.5%; *S. narbonensis*, DQ445794, 99.1%; *S. cinereoruber* subsp. *cinereoruber*, AB184121, 99%.

Source: not known.

DNA G+C content (mol%): not known.

Type strain: ATCC 19780, ATCC 19914, CBS 533.68, BCRC 11866, DSM 40164, NBRC 12792, INA 1823/55, JCM 4394, NRRL B-3635, NRRL-ISP 5164, RIA 1060, UNIQEM 167, VKM Ac-1887.

Sequence accession no. (16S rRNA gene): AB184149.

291. **Streptomyces lomondensis** Johnson and Dietz 1969, 758[AL]

lo.mond.en'sis. N.L. masc. adj. *lomondensis* possibly pertaining to Loch Lomond, Scotland.

Warty to spiny spores borne on straight to open spiral to spiral sporophores (*Rectiflexibiles, Retinaculiaperti, Spirales*). Spores are poorly differentiated by carbon repligraphy. Blue aerial mycelium. Melanin-positive. The culture may be placed in the Red (R) and Blue (B) color series of Tresner and Backus (1963). Growth of the culture on carbon compounds in a synthetic medium was determined according to the procedure of Pridham and Gottlieb (1948). Growth is good on D-xylose, L-arabinose, rhamnose, D-fructose, D-galactose, D-glucose, D-mannose, maltose, sucrose, lactose, cellobiose, raffinose, dextrin, inulin, soluble starch, glycerol, D-mannitol, inositol, sodium acetate, and sodium succinate; slight on dulcitol and D-sorbitol. No growth on the control, salicin, phenol, cresol, sodium formate, sodium oxalate, sodium tartrate, or sodium salicylate. Grows at temperatures of 18–37°C. It does not grow at 55°C. The optimal temperature is 37°C. Produces the antibiotic lomofungin (lomondomycin, U-24792).

Type strain shows the highest sequence similarity to: *S. lusitanus*, AB184424, 99.4%; *S. bellus*, AB184849, 99.4%; *S. coerulescens*, AY999720, 99.4%; *S. purpurascens*, AJ399486, 99.4%; *S. janthinus*, AB184851, 99.3%; *S. coeruleorubidus*, AY999719, 99.3%; *S. parvulus*, AB184326, 99.3%; *S. roseoviolaceus*, AJ399484, 99.3%; *S. albosporeus* subsp. *albosporeus*, AJ781327, 99.3%; *S. violaceus*, AB184315, 99.3%; *S. luteogriseus*, AB184379, 99.3%; *S. flavoviridis*, AB184842, 99.2%; *S. pilosus*, AB184161, 99.2%; *S. malachitospinus*, AB249954, 99.2%; *S. iakyrus*, AB184877, 99.1%; *S. indiaensis*, AB184553, 99.1%; *S. violaceorubidus*, AJ781374, 99.1%; *S. matensis*, AB184221, 99%; *S. spinoverrucosus*, AB184578, 99%; *S. hawaiiensis*, AB184143, 99%; *S. thermocarboxydus*, U94490, 99%; *S. tuirus*, AB184690, 99%.

Source: soil.

DNA G+C content (mol%): not known.

Type strain: ATCC 25299, BCRC 12208, DSM 41428, NBRC 15426, JCM 4866, NCIMB 10094, NRRL 3252.

Sequence accession no. (16S rRNA gene): AB184673.

292. **Streptomyces longisporoflavus** Waksman *in* Waksman and Lechevalier 1953, 94[AL]

lon.gi.spo.ro.fla'vus. L. adj. *longus* long; N.L. n. *spora* a spore; L. adj. *flavus* yellow; N.L. masc. adj. *longisporoflavus* long-spored, yellow.

Spore chains in Section *Spirales.* Open spirals; spore chains resembling *Retinaculiaperti* morphology are also common. Mature spore chains are long with 10–50 spores per chain; longer chains are often observed. This morphology may be observed on yeast-malt agar, oatmeal agar, salts-starch agar, and glycerol-asparagine agar. Spore surface is smooth.

Color of colony: aerial mass color in the Yellow color series on salts-starch agar. Aerial mycelium is poorly developed or absent on other ISP media. Reverse side of colony with no distinct pigments (colorless, grayish yellow, or light grayish yellowish brown) on yeast-malt agar, oatmeal agar, salts-starch agar, and glycerol-asparagine agar.

Color in medium: melanoid pigments are not formed in peptone-yeast-iron agar, tyrosine agar, or tryptone-yeast broth. No pigment found in medium in yeast-malt agar, oatmeal agar, salts-starch agar, or glycerol-asparagine agar.

D-Glucose, L-arabinose, D-xylose, iso-inositol, D-fructose, and rhamnose are utilized for growth. No growth or only trace of growth on sucrose and raffinose. Utilization of iso-inositol and D-mannitol is doubtful.

No detailed sequence information available.

Source: not known.

DNA G+C content (mol%): not known.

Type strain: AS 4.1453, ATCC 19781, ATCC 19915, ATCC 23932, CBS 915.68, BCRC 13775, DSM 40165, NBRC 12886, IMET 43506, INA 81/53, JCM 4396, LMG 19347, NRRL-ISP 5165, RIA 1133, RIA 312, VKM Ac-1003.

Sequence accession no. (16S rRNA gene): AB184220.

293. **Streptomyces longispororuber** Waksman *in* Waksman and Lechevalier 1953, 99[AL]

lon.gi.spo.ro.ru'ber L. adj. *longus* long; Gr. n. *spora* a seed; N.L. n. *spora* a spore; L. adj. *ruber* red; N.L. adj. *longispororuber* long-spored, red.

Spore chains in Section *Spirales*. Spirals may be found on oatmeal agar or salts-starch agar but are rarely found on yeast-malt agar or glycerol-asparagine agar. Sporulation is generally poor on yeast-malt agar, oatmeal agar, salts-starch agar, and glycerol-asparagine agar. One observer reports only sterile aerial mycelium on these media. Spore surface was not determined. Spores are poorly defined in electron micrographs submitted by ISP observers.

Color of colony: aerial mass color in the Red color series (5ca or 7ca, light yellowish pink; 4ea, moderate yellowish pink; or 5cb or 6ec, grayish yellowish pink) on yeast-malt agar, oatmeal agar, salts-starch agar, and glycerol-asparagine agar. One observer also reports 10ec, grayish purplish pink, in the Violet color series on salts-starch agar. Reverse side of colony with distinctive red pigments (strong or deep reddish orange or purplish pink). Substrate pigment is a pH indicator, changing from yellow orange or orange to pink with the addition of 0.05 M HCl and orange or pink to yellow orange or yellow with the addition of 0.05 M NaOH.

Color in medium: melanoid pigments are formed in peptone-yeast-iron agar and tryptone-yeast broth, but not in tyrosine agar. No pigment is found in the medium in yeast-malt agar, oatmeal agar, salts-starch agar, or glycerol-asparagine agar.

D-Glucose, L-arabinose, iso-inositol, D-mannitol, and D-fructose are utilized for growth. Reports vary on the utilization of D-xylose. No growth or only traces of growth with sucrose, rhamnose, and raffinose.

Type strain shows the highest sequence similarity to: *S. iakyrus*, AB184877, 99.4%; *S. coeruleorubidus*, AY999719, 99.2%; *S. lusitanus*, AB184424, 99.2%; *S. bellus*, AB184849, 99.2%; *S. coerulescens*, AY999720, 99.2%; *S. thermocarboxydus*, U94490, 99.1%; *S. viridodiastaticus*, AY999852, 99.1%; *S. albogriseolus*, AJ494865, 99.1%; *S. speibonae*, AF452714, 99.1%; *S. matensis*, AB184221, 99%; *S. griseorubens*, AB184139, 99%; *S. griseoflavus*, AJ781322, 99%.

Source: not known.

DNA G+C content (mol%): not known.

Type strain: ATCC 27443, CBS 789.72, DSM 40599, NBRC 13488, INA 11668/54, JCM 4784, NCIMB 9629, NRRL B-3736, NRRL B-5761, NRRL-ISP 5599, PCM 2396, RIA 1449, VKM Ac-1735.

Sequence accession no. (16S rRNA gene): AB184440.

294. **Streptomyces longisporus** (Krasil'nikov 1941) Waksman *in* Waksman and Lechevalier 1953, 39[AL] ("*Actinomyces longisporus*" Krasil'nikov 1941, 47)

lon.gi'spo.rus. L. adj. *longus* long; Gr. n. *spora* a seed; N.L. n. *spora* a spore; N.L. adj. *longisporus* long-spored.

Spore chains in Section *Spirales*. Mature spore chains are moderately short, usually 10–50 spores per chain. Spore surface is spiny (one observer found smooth spores).

Color of colony: aerial mass color in the White color series on yeast-malt agar, oatmeal agar, and glycerol-asparagine agar; White or Red (5cb grayish yellowish pink) color series on salts-starch agar. Reverse side of colony with no distinctive pigments (light or grayish yellow on oatmeal agar, salts-starch agar, and glycerol-asparagine agar to olive brown or brown on yeast-malt agar).

Color in medium: melanoid pigments are produced in peptone-yeast-iron agar, tyrosine agar, and tryptone-yeast broth. Pigments other than melanoid are not found in the medium in yeast-malt agar, oatmeal agar, salts-starch agar, or glycerol-asparagine agar.

D-Glucose, L-arabinose, sucrose, D-xylose, iso-inositol, D-mannitol, D-fructose, rhamnose, and raffinose are utilized for growth.

Type strain shows no sequence similarity over 99%.

Source: not known.

DNA G+C content (mol%): not known.

Type strain: ATCC 23931, CBS 916.68, BCRC 13776, CCT 5006, DSM 40166, NBRC 12885, IMET 43090, INA 4417/56, JCM 4395, NRRL B-5336, NRRL-ISP 5166, PCM 2394, RIA 1134, VKM Ac-1896.

Sequence accession no. (16S rRNA gene): AJ399475.

295. **Streptomyces longwoodensis** Prosser and Palleroni 1981, 382[VP] (Effective publication: Prosser and Palleroni 1976, 321.)

long.wood.en'sis. N.L. masc. adj. *longwoodensis* of or belonging to Longwood, named after Longwood Gardens in Kennett Square, Pennsylvania, USA.

Strain produces a substrate mycelium, which does not fragment into spores, and an aerial mycelium, which later forms spore chains. After 14 d of incubation at 28°C, the spore chains appear spiral in form with greater than 10 spores per chain (range: 5 to >50). Spores are smooth and range in size from 1.2 × 0.32 μm to 1.8 × 0.74 μm. Does not hydrolyze hippurate; decomposes adenine, hypoxanthine, and tyrosine but not xanthine; and it slowly peptonizes skim milk. The cell wall of this organism contains LL-A_2pm.

Type strain shows the highest sequence similarity to: *S. galbus*, X79852, 99.9%; *S. bungoensis*, AB184696, 99.7%; *S. capoamus*, AB045877, 99.6%; *S. corchorusii*, AB184267, 99.4%; *S. canarius*, AB184396, 99.3%; *S. olivaceoviridis*, AB184288, 99.2%; *S. curacoi*, EF626595, 99%.

Source: not known.

DNA G+C content (mol%): not known.

Type strain: Roche X-14537, ATCC 29251, BCRC 12034, DSM 41677, NBRC 14251, JCM 4976, KCTC 9783, NRRL B-16923.

Sequence accession no. (16S rRNA gene): AB184580.

296. **Streptomyces lucensis** Arcamone, Bertazzoli, Canevazzi, DiMarco, Ghione and Grein 1957, 119[AL]

lu.cen'sis. L. masc. adj. *lucensis* of or belonging to Lucca, a city in Italy, the source of the soil from which the organism was isolated.

Spore chains in Section *Spirales*. Flexuous spore chains and imperfect spirals are common together with some regular open spirals of 4 or more turns; mature spore chains have 10–50 spores per chain. This morphology is seen on yeast-malt agar, oatmeal agar, salts-starch agar, and glycerol-asparagine agar. Spore surface is hairy to spiny.

Color of colony: aerial mass color in the in the Gray color series (brownish gray) on yeast-malt agar, oatmeal agar, salts-starch agar, and glycerol-asparagine agar. Reverse side of colony with no distinctive pigments (colorless to pale grayish yellow or light yellowish brown) on yeast-malt agar, oatmeal agar, salts-starch agar, and glycerol-asparagine agar.

Color in medium: melanoid pigments are formed in peptone-yeast-iron agar and tryptone-yeast broth, but not in tyrosine agar. No pigment is found in medium in yeast-malt agar, oatmeal agar, salts-starch agar, or glycerol-asparagine agar.

D-Glucose, L-arabinose, D-xylose, D-mannitol, D-fructose, and sucrose are utilized for growth. No growth or only trace of growth with iso-inositol, rhamnose, and raffinose.

Type strain shows the highest sequence similarity to: *S. niveoruber*, DQ445796, 99.5%.

Source: not known.

DNA G+C content (mol%): not known.

Type strain: ATCC 17804, ATCC 25468, CBS 701.69, DSM 40317, NBRC 13056, JCM 4490, NCIMB 12679, NRRL B-16066, NRRL B-5626, NRRL-ISP 5317, RIA 1248, VKM Ac-1737.

Sequence accession no. (16S rRNA gene): DQ442522.

297. **Streptomyces luridiscabiei** Park, Kim, Kwon, Wilson, Yu, Hur and Lim 2003, 2053[VP]

lu.ri.di.sca'bi.ei. L. adj. *luridus* pale yellow; L. n. *scabies* roughness, scabbiness; N.L. gen. n. *luridiscabiei* pale yellow, scab-causing bacteria.

Spores are yellow-white, smooth, and borne in mon-overticillus flexuous spore chains. Melanin is produced on tyrosine and peptone agars. L-Arabinose, D-fructose, D-glucose, D-mannitol, raffinose, rhamnose, sucrose, D-xylose, and iso-inositol are utilized for growth. Minimum pH for growth is 4.5. Sensitive to 5, 6, and 7% (w/v) NaCl, 100 µg/ml thallium acetate, 0.1% (w/v) phenol, 25 and 100 µg/ml oleandomycin, and 20 µg/ml streptomycin, but not to 10 µg/ml thallium acetate, 0.5 µg/ml crystal violet, or 10 IU/ml penicillin.

Type strain shows the highest sequence similarity to: *S. fulvorobeus*, AB184711, 99.9%; *S. lipmanii*, AB184148, 99.9%; *S. alboviridis*, AB184256, 99.9%; *S. lavendulae* subsp. *lavendulae*, AB184080, 99.8%; *S. cavourensis* subsp. *washingtonensis*, DQ026671, 99.8%; *S. microflavus*, DQ445795, 99.8%; *S. cinereorectus*, AB184646, 99.7%; *S. fimicarius*, AY999784, 99.7%; *S. mediolani*, AB184674, 99.7%; *S. rubiginosohelvolus*, AB184240, 99.7%; *S. griseinus*, AB184205, 99.7%; *S. badius*, AY999783, 99.7%; *S. anulatus*, DQ026637, 99.7%; *S. praecox*, AB184293, 99.7%; *S. flavofuscus*, AB249935, 99.7%; *S. argenteolus*, AB045872, 99.7%; *S. californicus*, AB184755, 99.7%; *S. pluricolorescens*, DQ442540, 99.7%; *S. albovinaceus*, AB249958, 99.7%; *S. cyaneofuscatus*, AB184860, 99.7%; *S. griseoplanus*, AY999894, 99.7%; *S. sindenensis*, AB184759, 99.7%; *S. floridae*, AB184656, 99.7%; *S. acrimycini*, AY999889, 99.6%; *S. griseus* subsp. *griseus*, AY207604, 99.6%; *S. griseolus*, AB184768, 99.6%; *S. flavogriseus*, AJ494864, 99.5%; *S. parvus*, DQ442537, 99.5%; *S. halstedii*, EF178695, 99.5%; *S. flavovirens*, DQ026635, 99.5%; *S. finlayi*, AY999788, 99.4%; *S. pulveraceus*, AB184806, 99.3%; *S. atroolivaceus*, AJ781320, 99.3%; *S. globisporus* subsp. *globisporus*, EF178686, 99.3%; *S. olivoviridis*, AB184227, 99.3%; *S. griseobrunneus*, AB249912, 99.2%; *S. nitrosporeus*, EF178680, 99.2%; *S. yanii*, AB006159, 99.2%; *S. cellulofiavus*, AB184476, 99.2%; *S. bacillaris*, AB184439, 99.2%; *S. clavifer*, DQ026670, 99.2%; *S. albolongus*, AB184425, 99.2%; *S. spiroverticillatus*, AB184814, 99.1%; *S. sanglieri*, AB249945, 99.1%; *S. baarnensis*, EF178688, 99.1%; *S. gelaticus*, DQ026636, 99%; *S. mutomycini*, AB249951, 99%; *S. atratus*, DQ026638, 99%; *S. cavourensis* subsp. *cavourensis*, DQ445791, 99%; *S. candidus*, DQ026663, 99%. Type strain shows DNA–DNA similarity to: *S. scabies* ATCC 49173^T, 16%; *S. turgidiscabies* ATCC 700248^T, 15%; *S. acidiscabies* ATCC 49003^T, 10%; *S. bottropensis* DSM 40262^T, 15%; *S. neyagawaensis* DSM 40588^T, 7%; *S. diastatochromogenes* DSM 40449^T, 13%; *S. setonii* DSM 40395^T, 10%; *S. griseus* subsp. *griseus* DSM 40236^T, 7%; *S. sampsonii* DSM 40394^T, 18%; *S. eurythermus* DSM 40014^T, 13%; *S. tendae* DSM 40101^T, 12%; *S. coelicolor* DSM 40233^T 9%; "*S. lividans*" DSM 40434, 10%; *S. ambofaciens* DSM 40053^T, 13%; *S. puniciscabiei* LMG 21391^T, 11%; *S. niveiscabiei* LMG 21392^T, 9%.

Source: isolated from raised corky lesions on potato cv. Atlantic and pathogenic on potato cv. Daeji-ma.

DNA G+C content (mol%): 70.3.

Type strain: S63, KACC 20252, LMG 21390.

Sequence accession no. (16S rRNA gene): AF361784.

298. **Streptomyces luridus** (Krasil'nikov, Korenyako, Meksina, Valedinskaya and Veselov 1957) Waksman 1961, 237[AL] ("*Actinomyces luridus*" Krasil'nikov, Korenyako, Meksina, Valedinskaya and Veselov 1957, 563)

lu'ri.dus. L. adj. *luridus* pale yellow, ghostly pallid (probably referring to the color of the vegetative mycelium).

Spore chains in Section *Retinaculiaperti* with characteristic range of flexuous spore chains, hooks, loops, and occasional spirals (spirals rare). Mature spore chains generally have 10–50 spores per chain; longer chains are often observed. This morphology is seen on yeast-malt agar, oatmeal agar, salts-starch agar, and glycerol-asparagine agar. Spore surface is smooth.

Color of colony: aerial mass color in the Red color series on yeast-malt agar in 14 d. Sporulation aerial mycelium in the Red color series usually develops on oatmeal agar, salts-starch agar, and glycerol-asparagine agar in 21 d. Reverse side of colony with no distinctive pigment on yeast-malt agar, oatmeal agar, salts-starch agar, and glycerol-asparagine agar; substrate pigment is not a pH indicator.

Color in medium: melanoid pigments formed in peptone-yeast-iron agar, tyrosine agar, and tryptone yeast broth. Pigments other than melanoids are not formed in yeast-malt agar, oatmeal agar, salts-starch agar, or glycerol-asparagine agar.

D-Glucose, L-arabinose, D-xylose, and iso-inositol are utilized for growth. No growth or only traces of growth on sucrose, D-mannitol, rhamnose, and raffinose. Utilization of D-fructose is doubtful.

Type strain shows the highest sequence similarity to: *S. gobitricini*, AB184666, 99.5%; *S. lavendofoliae*, AJ781336, 99.4%.

Source: not known.

DNA G+C content (mol%): not known.

Type strain: AS 4.1458, ATCC 19782, CBS 534.68, BCRC 13684, DSM 40081, NBRC 12793, INMI 111, JCM 4591, LMG 19365, NRRL B-5409, NRRL-ISP 5081, RIA 1061, UNIQEM 168, VKM Ac-245.

Sequence accession no. (16S rRNA gene): DQ442523.

299. **Streptomyces lusitanus** Villax 1963, 661[AL]

lu.si.ta'nus. L. masc. adj. *lusitanus* of or belonging to *Lusitania*, the Roman name of Portugal.

Spore chains in Section *Spirales*. Spirals are best developed on salts-starch agar and glycerol-asparagine agar; flexuous chains as well as open spirals and imperfect spirals are common on yeast-malt agar and oatmeal agar. Mature spore chains generally are long with 10 to 50 or more spores per chain. Spore surface is spiny with short spines. Some spores may appear smooth.

Color of colony: aerial mass color in the Gray or White color series on yeast-malt agar, oatmeal agar, salts-starch agar and glycerol-asparagine agar. Nearest matching color tabs in the Gray color series are 3fe, light brownish gray, and 5fe, light grayish reddish brown. Reverse side of colony with no distinctive pigments (pale or grayish yellow on oatmeal agar and glycerol-asparagine agar; olive brown or yellowish brown on yeast-malt agar and salts-starch agar).

Color in medium: melanoid pigments are not formed in peptone-yeast-iron agar, tyrosine agar, or tryptone-yeast broth. No pigment (or only a trace of yellow or pale olive) is found in the medium in yeast-malt agar, oatmeal agar, salts-starch agar, and glycerol-asparagine agar.

D-Glucose, L-arabinose, D-xylose, iso-inositol, D-mannitol, D-fructose, rhamnose, and sucrose are utilized for growth. No growth or only traces of growth with raffinose.

Type strain shows the highest sequence similarity to: *S. thermocarboxydus*, U94490, 99.5%; *S. coerulescens*, AY999720, 99.4%; *S. bellus*, AB184849, 99.4%; *S. lomondensis*, AB184673, 99.4%; *S. gancidicus*, AB184660, 99.4%; *S. pseudogriseolus*, DQ442541, 99.4%; *S. capillispiralis*, AB184577, 99.3%; *S. matensis*, AB184221, 99.3%; *S. longispororuber*, AB184440, 99.2%; *S. coeruleorubidus*, AY999719, 99.2%; *S. purpurascens*, AJ399486, 99.2%; *S. spinoverrucosus*, AB184578, 99.2%; *S. cellulosae*, DQ442495, 99.2%; *S. griseoloalbus*, AB184275, 99.1%; *S. parvulus*, AB184326, 99%; *S. speibonae*, AF452714, 99%; *S. iakyrus*, AB184877, 99%; *S. indiaensis*, AB184553, 99%; *S. viridodiastaticus*, AY999852, 99%; *S. albaduncus*, AY999757, 99%; *S. albogriseolus*, AJ494865, 99%.

Source: not known.

DNA G+C content (mol%): not known.

Type strain: ATCC 15842, ATCC 27444, CBS 765.72, BCRC 11552, DSM 40568, NBRC 13464, JCM 4785, NCIMB 9585, NRRL B-12501, NRRL B-5637, NRRL-ISP 5568, RIA 1425, VKM Ac-1194.

Sequence accession no. (16S rRNA gene): AB184424.

300. **Streptomyces luteireticuli** (*ex* Katoh and Arai 1957) Hatano, Nishii and Kasai 2003, 1528[VP] ("*Streptoverticillium luteoreticuli*" Katoh and Arai 1957)

lu.te.i.re.ti'cu.li. L. adj. *luteus* yellow; L. n. *reticulum* net; N.L. gen. n. *luteireticuli* of a yellow net.

Spore chains in Section Umbellate Monoverticillate (=*Streptomyces Verticillati*, biverticillate). Mature spore chains are generally short with 3 to 10 or more spores per chain. This morphology is seen on oatmeal agar and sometimes on salts-starch agar. Sporulating aerial mycelium is usually thin or absent on yeast-malt agar or glycerol-asparagine agar. Spore surface is smooth.

Color of colony: aerial mass color in the Yellow or Gray color series on oatmeal agar. Nearest matching color tabs in the Yellow color series are 1dc and 1½ec, pale yellow green, and 2db, pale yellow. Nearest matching tabs in the Gray color series are 2dc, yellowish gray, and 2ge, light olive brown. Reverse side of colony with no distinctive pigments (olive brown to dark brown on yeast-malt agar, grayish yellow or yellowish brown to olive brown on oatmeal agar, salts-starch agar, and glycerol-asparagine agar).

Color in medium: melanoid pigments are formed in peptone-yeast-iron agar, tyrosine agar, and tryptone-yeast broth. Yellowish to greenish yellow pigment is found in the medium in yeast-malt agar, oatmeal agar, salts-starch agar, and glycerol-asparagine agar. This pigment is not pH-sensitive when tested with 0.05 M NaOH or HCl.

D-Glucose and iso-inositol are utilized for growth. Utilization of L-arabinose, D-xylose, D-mannitol, D-fructose, rhamnose, sucrose, and raffinose is doubtful since only a small amount of growth is found on these carbon sources.

Type strain shows the highest sequence similarity to: *S. abikoensis*, AB184537, 99.3%; *S. sapporonensis*, AB184508, 99.2%; *S. ehimensis*, AY999834, 99.2%; *S. hygroscopius* subsp. *angustmyceticus*, DQ442509, 99.1%; *S. varsoviensis*, DQ026653, 99.1%; *S. lavenduligriseus*, DQ442515, 99%; *S. thioluteus*, AB184753, 99%.

Source: not known.

DNA G+C content (mol%): not known.

Type strain: ATCC 27446, CBS 723.72, DSM 40509, ISP 5509, JCM 4788, NBRC 13422, RIA 1383.

Sequence accession no. (16S rRNA gene): AB249969.

301. **Streptomyces luteogriseus** Schmitz, Deak, Crook and Hooper 1964, 89[AL]

lu.te.o.gri'se.us. L. adj. *luteus* golden yellow; N.L. adj. *griseus* gray; N.L. masc. adj. *luteogriseus* grayish, golden-yellow, referring to the yellowish gray color of sporulating aerial mycelium on certain media.

Spore chains in Section *Spirales*. Mature spore chains generally contain 10 to 50 or more spores per chain. This morphology is seen on yeast-malt agar, oatmeal agar, salts-starch agar, and glycerol-asparagine agar. Spore surface is smooth.

Color of colony: aerial mass color in the Gray color series on yeast-malt agar, oatmeal agar, salts-starch agar, and glycerol-asparagine agar. Nearest matching color tabs are 5fe, light grayish reddish brown, and 2dc, yellowish gray. One observer also reports aerial mycelium in the Red color series (5dc, grayish yellowish pink) on oatmeal agar and salts-starch agar. Reverse side of colony is strong yellowish brown to dark brown on yeast-malt agar; grayish yellow to olive brown on oatmeal agar, salts-starch agar, and glycerol-asparagine agar.

Color in medium: melanoid pigments are formed in peptone-yeast-iron agar, tyrosine agar, and tryptone-yeast broth, but the reaction may be weak in tyrosine agar. Yellow to brown pigment is found in the medium in yeast-malt agar, oatmeal agar, salts-starch agar, and glycerol-asparagine agar. This pigment is not pH-sensitive when tested with 0.05 M NaOH or HCl.

D-Glucose, L-arabinose, D-xylose, iso-inositol, D-mannitol, D-fructose, rhamnose, sucrose, and raffinose are all utilized for growth.

Type strain shows the highest sequence similarity to: *S. janthinus*, AB184851, 99.5%; *S. albosporeus* subsp. *albosporeus*, AJ781327, 99.5%; *S. roseoviolaceus*, AJ399484, 99.5%; *S. violaceus*, AB184315, 99.5%; *S. hawaiiensis*, AB184143, 99.4%; *S. tuirus*, AB184690, 99.4%; *S. arenae*, AB249977, 99.4%; *S. lomondensis*, AB184673, 99.3%; *S. flavoviridis*, AB184842, 99.2%; *S. parvulus*, AB184326, 99.2%; *S. pilosus*, AB184161, 99.2%; *S. massasporeus*, AB184152, 99.2%; *S. purpurascens*, AJ399486, 99.1%; *S. mutabilis*, EF178679, 99.1%; *S. malachitospinus*, AB249954, 99.1%; *S. rochei*, AB184237, 99%; *S. minutiscleroticus*, EF178696, 99%; *S. geysiriensis*, DQ442501, 99%; *S. bellus*, AB184849, 99%; *S. plicatus*, AB184291, 99%; *S. ghanaensis*, AY999851, 99%; *S. levis*, AB184670, 99%; *S. vinaceusdrappus*, AY999929, 99%.

Source: not known.

DNA G+C content (mol%): not known.

Type strain: ATCC 15072, CBS 703.72, DSM 40483, IFM 1203, NBRC 13402, JCM 4786, NRRL B-12422, NRRL-ISP 5483, RIA 1363, VKM Ac-1913.

Sequence accession no. (16S rRNA gene): AB184379.

302. **Streptomyces luteosporeus** Witt and Stackebrandt 1991, 456[VP] (Effective publication: Witt and Stackebrandt 1990, 370.) (*Streptoverticillium album* Locci, Baldacci and Petrolini Baldan 1969, 59)

lu.te.o.spo're.us. L. adj. *luteus* yellow; Gr. n. *spora* a seed and in biology a spore; N.L. masc. adj. *luteosporeus* yellow spored.

Spore chains are usually shorter. Reverse colors varying from yellow to hazel-nut yellow. Aerial mycelium colors are lighter, tending toward pinkish beige on glucose-asparagine agar, glycerol-asparagine agar, inorganic salts-starch agar, and yeast extract-malt extract agar. Melanin pigment is not produced. L-Methionine, L-proline, and shikimic acid are utilized, but not coumarin, mannitol, melibiose, raffinose, sorbitol, or DL-α-aminobutyric acid. Esculin and L-tyrosine are degraded, but not citrate. H_2S is not produced. Acid is produced from D-ribose and trehalose, but not from D-galactose, D-fructose, or *myo*-inositol. *Aspergillus niger* is not inhibited. Growth is as good at 37°C as at 27°C. Aerial mycelium is better at 37°C. The type strain produces acetopyrrothine (thiolutin).

Type strain shows no sequence similarity over 99%.

Source: not known.

DNA G+C content (mol%): not known.

Type strain: ATCC 33049, JCM 4542, DSM 40833, NBRC 14657, NRRL 2401, VKM Ac-927.

Sequence accession no. (16S rRNA gene): DQ442525.

Further comments: for the transfer of *Streptoverticillium album* Locci et al. 1969 to the genus *Streptomyces* Waksman and Henrici 1943, it is necessary to substitute a new specific epithet to produce *Streptomyces luteosporeus* because there is a senior homonym, *Streptomyces albus* (Rossi Doria 1891) Waksman and Henrici 1943 included on the Approved Lists of Bacterial Names [Rules 34a and 41a of the *Bacteriological Code* (1990 Revision)].

303. **Streptomyces luteoverticillatus** (Shinobu 1956) Witt and Stackebrandt 1991, 456[VP] (Effective publication: Witt and Stackebrandt 1990, 370.) ("*Streptomyces luteoverticillatus*" Shinobu 1956; "*Verticillomyces luteoverticillatus*" Shinobu 1965; *Streptoverticillium luteoverticillatum* Locci, Baldacci and Petrolini Baldan 1969, 59)

lu.te.o.ver.ti.cil.la'tus. L. adj. *luteus* yellow; L. n. *verticillus* a whorl; N.L. adj. *verticillatus* whorled; N.L. masc. adj. *luteoverticillatus* yellow, whorled.

Spore chains in Section *Verticillati*, umbellate monoverticillate (biverticillate). Mature spore chains generally 10 to 20 or more spores per chain after 21 d on salts-starch agar or yeast-malt agar. Aerial mycelium is poorly developed or absent on yeast-malt agar, oatmeal agar, and glycerol-asparagine agar. Absence of good aerial mycelium on various media was also noted by Shinobu in his original description (op. cit). Spore surface is smooth.

Color of colony: mature aerial mycelium in the Red or Yellow color series on salts-starch agar; it is absent or poorly developed (white or yellow) on other ISP media. Reverse side of colony with no distinctive pigments (light brown to grayed yellow) on yeast-malt agar, oatmeal agar, salts-starch agar, and glycerol-asparagine agar.

Color in medium: melanoid pigments are formed in peptone-yeast-iron agar, tyrosine agar, and tryptone-yeast broth. Yellow pigment is found in medium in yeast-malt agar, oatmeal agar, and salts-starch agar after 14–21 d; it is not a pH indicator.

D-Glucose, iso-inositol, and D-fructose are utilized for growth. No growth or only traces of growth on L-arabinose, sucrose, D-xylose, D-mannitol, rhamnose, and raffinose.

For sequence similarity, see type strain of *Streptomyces abikoensis*.

Source: not known.

DNA G+C content (mol%): not known.

Type strain: ATCC 23933, CBS 917.68, BCRC 13323, DSM 40038, NBRC 12887, NBRC 3840, JCM 4099, JCM 4649, NCIMB 9720, NRRL B-1995, NRRL-ISP 5038, RIA 1109, VKM Ac-889.

Sequence accession no. (16S rRNA gene): AB184803.

Further comments: in violation of Rule 33c of the *Bacteriological Code* (1990 Revision), in Validation List no. 38, *Streptomyces luteoverticillatus* is proposed as a *nomen revictum* (basonym: "*Streptomyces luteoverticillatus*" Shinobu 1956).

According to Hatano et al. (2003), *Streptomyces luteoverticillatus* (Shinobu 1956) Witt and Stackebrandt 1991 is a later heterotypic synonym of *Streptomyces abikoensis* (Umezawa et al. 1951) Witt and Stackebrandt 1991.

304. **Streptomyces lydicus** DeBoer, Dietz, Silver and Savage 1955b, 886[AL]

ly'di.cus. L. n. *Lydia* an ancient state in Asia Minor; N.L. masc. adj. *lydicus* of or belonging to Lydia.

Spore chains in Section *Spirales*. Spirals are sometimes at the tips of long spore chains suggesting *Retinaculiaperti* morphology. Mature spore chains generally have 10–50 or sometimes more than 50 spores per chain. This morphology is seen on yeast-malt agar, oatmeal agar, salts-starch agar, and glycerol-asparagine agar. Spore surface is smooth.

Color of colony: aerial mass color in the Gray color series (3fe, light brownish gray, to 3ih, dark gray) on yeast-malt agar and oatmeal agar; Gray or Yellow (2ba, pale yellow) or White color series on salts-starch agar and glycerol-asparagine agar. Reverse side of colony with no distinctive pigments (colorless, pale grayish-yellow or orange-yellow) on yeast No distinctive pigments (colorless, pale grayish-yellow or orange-yellow) on yeast-malt agar, oatmeal agar, salts-starch agar, and glycerol-asparagine agar.

Color in medium: melanoid pigments are not formed in peptone-yeast-iron agar, tyrosine agar, or tryptone-yeast broth, according to two of three observers. The third observer records some darkening of peptone-yeast-iron agar and a trace of dark pigment on tyrosine agar, but not darkening of tryptone-yeast broth. No pigment (or only a trace of yellow) is found in the medium in yeast-malt agar, oatmeal agar, salts-starch agar, and glycerol-asparagine agar.

D-Glucose, L-arabinose, D-xylose, iso-inositol, D-mannitol, D-fructose, sucrose, and raffinose are utilized for growth. Utilization of rhamnose is doubtful.

Type strain shows the highest sequence similarity to: *S. chattanoogensis*, AJ621611, 99.8%; *S. sioyaensis*, DQ026654, 99.3%; *S. rimosus* subsp. *paromomycinus*, AJ621610, 99.3%; *S. chrestomyceticus*, AJ621609, 99.3%; *S. tubercidicus*, AJ621612, 99.1%; *S. libani* subsp. *libani*, AB184414, 99.1%; *S. nigrescens*, DQ442530, 99%; *S. hygroscopius* subsp. *decoyicus*, AY999883, 99%.

Source: not known.

DNA G+C content (mol%): not known.

Type strain: AS 4.1412, ATCC 25470, CBS 703.69, BCRC 11919, CECT 3163, DSM 40461, HAMBI 1063, NBRC 13058, IMET 43531, JCM 4492, LMG 19331, NCIMB 12977, NRRL 2433, NRRL-ISP 5461, RIA 1250, VKM Ac-1869.

Sequence accession no. (16S rRNA gene): Y15507.

305. **Streptomyces macrosporus** (*ex* Krasil'nikov, Agre, Dorokhova and Sokolov 1968) Goodfellow, Lacey and Todd 1988, 328[VP] (Effective publication: Goodfellow, Lacey and Todd 1987a, 3148.) ("*Streptomyces macrosporus*" Krasil'nikov, Agre, Dorokhova and Sokolov 1968)

ma.cro.spo'rus. Gr. adj. *makros* long, large, Gr. n. *spora* seed; N.L. masc. adj. *macrosporus* large spored.

Spores mostly in tight spirals of up to six turns and 50 spores but sometimes in short straight chains of only five spores. Spores appear warty in transmission electron micrographs but are characteristically wrinkled in scanning electron micrographs, 0.7–1.8 × 0.7–1.6 µm, sometimes broader than long, mean 1.12 × 0.96 µm. Growth at 40°C on half-strength nutrient and V-8 juice agars is good; aerial mycelium is in the Green color series near 24ih (Tresner and Backus, 1963), with white flecks, although may be thin and white on half-strength nutrient agar. At 25°C, small colonies only are formed, sometimes with sparse white aerial mycelium. Substrate mycelium is colorless to dark brown with no distinctive pigments but often crystalline deposits are found in the agar. Melanoid pigments are not produced on peptone iron agar. No soluble pigments. Degrades esculin, arbutin, casein, DNA, gelatin, RNA, starch, and L-tyrosine; utilizes D-fructose, *myo*-inositol, D-mannose, L-rhamnose, trehalose, and D-xylose as sole carbon sources, and L-arginine, L-phenylalanine, potassium nitrate, and L-threonine as sole nitrogen sources. Proteolysis and lipolysis evident on egg yolk agar; pectin hydrolyzed; hydrogen sulfide produced but nitrate not reduced.

Type strain shows no sequence similarity over 99%.

Source: isolated from sewage compost and soil.

DNA G+C content (mol%): not known.

Type strain: K44, A1201, ATCC 51533, DSM 41449, HUT 6608, NBRC 14748, INMI 2892, JCM 6305, NCIMB 12473, VKM Ac-777.

Sequence accession no. (16S rRNA gene): Z68099.

306. **Streptomyces malachitofuscus** (*ex* Preobrazhenskaya et al. 1964) Preobrazhenskaya and Terekhova 1986, 574[VP] (Effective publication: Preobrazhenskaya and Terekhova *in* Gause, Preobrazhenskaya, Sveshnikova, Terekhova and Maximova 1983.) ("*Actinomyces malachitofuscus*" Preobrazhenskaya et al. 1964; "*Streptomyces malachitofuscus*" Pridham 1970)

ma.la.chi.to.fus'cus. L. n. *malache* the mallow; L. adj. *fuscus* dark or tawny; N.L. masc. adj. *malachitofuscus* mallow, dark (dark green).

Spore chains in Section *Spirales*. Mature spore chains generally have 10–50 spores per chain on yeast-malt agar, oatmeal agar, and salts-starch agar. Shorter spore chains, especially on glycerol-asparagine agar, may form irregular or incomplete spirals, loops, and hooks. Spore surface is spiny to hairy.

Color of colony: aerial mass color in the Gray color series (3fe, light brownish gray to 5fe, light reddish grayish brown) on yeast-malt agar, oatmeal agar, salts-starch agar, and glycerol-asparagine agar. Reverse side of colony is nearly colorless to yellow or yellowish gray on oatmeal agar and salts-starch agar; grayish olive to olive brown on glycerol-asparagine agar.

Color in medium: melanoid pigments are formed in peptone-yeast-iron agar, tyrosine agar, and tryptone-yeast broth. No pigment is found in medium in salts-starch agar and glycerol-asparagine agar; a trace of yellow or yellow-brown pigment may or may not be found in yeast-malt agar and oatmeal agar.

D-Glucose, L-arabinose, D-xylose, iso-inositol, D-mannitol, D-fructose, rhamnose, and sucrose are utilized for growth. No growth or only traces of growth with raffinose.

Type strain shows the highest sequence similarity to: *S. griseoflavus*, AJ781322, 99.3%; *S. matensis*, AB184221, 99.3%; *S. griseorubens*, AB184139, 99.1%; *S. labedae*, AB184704, 99%; *S. griseoincarnatus*, AJ781328, 99%; *S. erythrogriseus*, AJ781328, 99%; *S. variabilis*, DQ442551, 99%; *S. flaveolus*, AB184764, 99%; *S. violaceorubidus*, AJ781374, 99%; *S. paradoxus*, AB184628, 99%; *S. glaucescens*, AB184843, 99%; *S. albaduncus*, AY999757, 99%; *S. gancidicus*, AB184660, 99%; *S. griseoloalbus*, AB184275, 99%; *S. djakartensis*, AB184657, 99%.

Source: not known.

DNA G+C content (mol%): not known.

Type strain: ATCC 25471, CBS 881.69, DSM 40332, NBRC 13059, INA 739, JCM 4493, KCC S-0493, NRRL B-12273, NRRL-ISP 5332, RIA 1251, VKM Ac-1850.

Sequence accession no. (16S rRNA gene): AB184282.

307. **Streptomyces malachitospinus** (*ex* Preobrazhenskaya et al. 1957) Preobrazhenskaya and Terekhova 1986, 574[VP] (Effective publication: Preobrazhenskaya and Terekhova *in* Gause, Preobrazhenskaya, Sveshnikova, Terekhova and Maximova 1983.) ("*Actinomyces malachitospinus*" Preobrazhenskaya et al. 1957)

ma.la.chi.to.spi'nus. L. n. *malache* the mallow the mallow; L. adj. *spineus* spiny; N.L. masc. adj. *malachitospinus* spiny and mallow green.

Spore chains are spiral (*Spirales*); spores are spiny, spines are long. On mineral agar 1, glycerol-nitrate agar, oatmeal agar: aerial mycelium is light gray; substrate mycelium is green; no diffusible pigment. On starch-ammonia agar: aerial mycelium is absent or light gray; substrate mycelium is green; no diffusible pigment. On glycerol-asparagine agar: aerial mycelium is white to light gray; substrate mycelium is yellowish green; no diffusible pigment. On organic agar 2: aerial mycelium is white or light gray; substrate mycelium is yellowish to greenish; no diffusible pigment. Color of substrate mycelium is a result of green pigment ferroverdin. Melanoid pigments are not formed. Glucose, arabinose, galactose, glycerol, and starch are utilized for growth; no growth with sucrose. Antibiotic: not isolated. Strain INMI 217^T builds a physiologically active substance which stimulates the building of zygotes in Mucorales.

Type strain shows the highest sequence similarity to: *S. parvulus*, AB184326, 99.3%; *S. violaceorubidus*, AJ781374, 99.2%; *S. lomondensis*, AB184673, 99.2%; *S. pactum*, AB184398, 99.2%; *S. luteogriseus*, AB184379, 99.1%; *S. rubrogriseus*, AB184681, 99%; *S. olivaceus*, AB184743, 99%; *S. tendae*, D63873, 99%.

Source: not known.

DNA G+C content (mol%): not known.

Type strain: INMI 217, INA 316, NBRC 101004.

Sequence accession no. (16S rRNA gene): AB249954.

308. **Streptomyces malaysiensis** Al-Tai, Kim, Kim, Manfio and Goodfellow 1999, 1401[VP]

mal.ay.si.en'sis. N.L. masc. adj. *malaysiensis* of or belonging/pertaining to Malaysia, the source of the soil from which the organism was isolated.

Forms extensively branched substrate hyphae (0.3–0.5 μm in diameter) and aerial hyphae which differentiate into tight, spiral spore chains. The appearance of the spores ranges from cylindrical to barrel-shaped (1.3–1.0 × 1.5 μm) and the spore surface is rugose. On inorganic salts-starch agar, the aerial spore-mass color is gray and a yellow diffusible pigment is formed. Xanthine is not degraded. Growth does not occur at 10 or 45°C.

Type strain shows no sequence similarity over 99%.

Source: isolated from a soil sample collected at Tasek Bera, Malaysia's first Ramsar Site.

DNA G+C content (mol%): 72.2.

Type strain: ATB-11, ATCC BAA-13, DSM 41697, NBRC 16446, JCM 10672.

Sequence accession no. (16S rRNA gene): AB249918.

309. **Streptomyces mashuensis** (Sawazaki, Suzuki, Nakamura, Kawasaki, Yamashita, Isono, Anzai, Serizawa and Sekiyama 1955) Witt and Stackebrandt 1991, 456[VP] (Effective publication: Witt and Stackebrandt 1990, 370.) ("*Streptomyces mashuensis*" Sawazaki, Suzuki, Nakamura, Kawasaki, Yamashita, Isono, Anzai, Serizawa and Sekiyama 1955, 46; "*Verticillomyces mashuensis*" Shinobu 1965; *Streptoverticillium mashuense* Locci, Baldacci and Petrolini Baldan 1969, 59)

ma.shu.en'sis. N.L. masc. adj. *mashuensis* of or pertaining to Lake Mashu, Japan.

Spore chains in Section *Verticillati*, umbellate monoverticillate (biverticillate). Spore chains are short with 3–10, or sometimes more than 10, spores per chain. Spore surface is smooth.

Color of colony: aerial mass color in the Red color series on yeast-malt agar, oatmeal agar, salts-starch agar, and glycerol-asparagine agar. Mature aerial mycelium may be poorly developed on glycerol-asparagine agar. Reverse side of colony with no distinctive pigments (grayish yellow to orange yellow on oatmeal agar, salts-starch agar, and glycerol-asparagine agar; olive brown to brown on yeast-malt agar).

Color in medium: melanoid pigments are produced in peptone-yeast-iron agar and tryptone-yeast broth, but not in tyrosine agar. No pigment found in medium in yeast-malt agar, oatmeal agar, salts-starch agar, or glycerol-asparagine agar.

D-Glucose, sucrose, and D-fructose are utilized for growth. Utilization of iso-inositol is doubtful. No growth or only traces of growth on L-arabinose, D-xylose, D-mannitol, rhamnose, and raffinose.

Type strain shows no sequence similarity over 99%.

Source: not known.

DNA G+C content (mol%): not known.

Type strain: ATCC 23934, CBS 279.65, CBS 918.68, BCRC 12420, DSM 40221, IFM 1082, NBRC 12888, IMET 42941, JCM 4059, JCM 4650, NRRL B-3352, NRRL B-8164, NRRL-ISP 5221, RIA 1165, VKM Ac-949.

Sequence accession no. (16S rRNA gene): X79323.

Further comments: in violation of Rule 33c of the *Bacteriological Code* (1990 Revision), in Validation List no. 38, *Streptomyces mashuensis* is proposed as a *nomen revictum* (basonym: "*Streptomyces mashuensis*" Sawazaki et al. 1955).

According to Hatano et al. (2003), *Streptomyces mashuensis* (Sawazaki et al. 1955) Witt and Stackebrandt 1991 is an earlier heterotypic synonym of *Streptomyces kishiwadensis* (Shinobu and Kayamura 1964) Witt and Stackebrandt 1991.

310. **Streptomyces massasporeus** Shinobu and Kawato 1959, 283[AL]

mas.sa.spo're.us. L. n. *massa* mass, lump; N.L. n. *spora* a spore; N.L. masc. adj. *massasporeus* mass, spore, referring to the coalescence of spores into moist masses.

Spore chains in Section *Spirales*, but spore chains representative of Section *Retinaculiaperti* are also present. Mature spore chains generally have 10–50 spores per chain. This morphology is seen on yeast-malt agar, oatmeal agar, salts-starch agar, and glycerol-asparagine agar. Spore surface is smooth. Special morphological characteristics: spores frequently coalesce to form moist masses of spores. Small masses may form from single spiral spore chains of large masses may form by the coalescence of many spirals.

Color of colony: aerial mass color in the Red color series on yeast-malt agar, salts-starch agar, and glycerol-asparagine agar. Red or White series on oatmeal agar. Reverse side of colony is grayed yellow modified by red or violet (see pH) on yeast-malt agar, oatmeal agar, salts-starch agar, and glycerol-asparagine agar. Reverse color is changed from violet to red by addition of 0.05 M HCl and from red to reddish violet to blue violet by addition of 0.05 M NaOH.

Color in medium: melanoid pigments formed in peptone-yeast-iron agar. Red pigments found in medium in oatmeal agar and glycerol-asparagine agar. One observer detected red pigment in yeast-malt agar, oatmeal agar, salts-starch agar, and glycerol-asparagine agar in 14 d and violet pigment in these media at 21 d; changes in color from violet to blue by addition of 0.05 M NaOH and from violet to red by addition of 0.05 M HCl were also noted.

D-Glucose, L-arabinose, sucrose, D-xylose, iso-inositol, D-mannitol, D-fructose, rhamnose, and raffinose are utilized for growth.

Type strain shows the highest sequence similarity to: *S. hawaiiensis*, AB184143, 99.7%; *S. indiaensis*, AB184553, 99.6%; *S. arenae*, AB249977, 99.4%; *S. purpurascens*, AJ399486, 99.3%; *S. luteogriseus*, AB184379, 99.2%; *S. coerulescens*, AY999720, 99%; *S. albosporeus* subsp. *albosporeus*, AJ781327, 99%; *S. janthinus*, AB184851, 99%; *S. roseoviolaceus*, AJ399484, 99%; *S. violaceus*, AB184315, 99%; *S. bellus*, AB184849, 99%.

Source: not known.

DNA G+C content (mol%): not known.

Type strain: AS 4.1433, ATCC 19785, CBS 537.68, BCRC 13647, DSM 40035, NBRC 12796, NBRC 3841, JCM 4139, JCM 4593, LMG 19362, NRRL B-3300, NRRL-ISP 5035, RIA 1064, RIA 652, UNIQEM 171, VKM Ac-578.

Sequence accession no. (16S rRNA gene): AB184152.

311. **Streptomyces matensis** Margalith, Beretta and Timbal 1959, 71[AL]

mat.en'sis. N.L. masc. adj. *matensis* of or belonging to mat (of uncertain derivation).

Spore chains in Section *Spirales*. Open spirals are the dominant forms; spore chains suggesting primitive spirals of *Retinaculiaperti* cultures or flexuous chains of *Rectiflexibiles* are also present. Mature spore chains are moderately long with 10–50 spores per chain. This morphology is seen on yeast-malt agar, oatmeal agar, salts-starch agar, and glycerol-asparagine agar. Spore surface is spiny. Spines are short and inconspicuous; some smooth spores may be seen.

Color of colony: mature aerial mass color in the Gray color series (color tab 3ge, light grayish yellowish brown; to 5fe light grayish reddish brown) on salts-starch agar; Gray or White color series on yeast-malt agar and glycerol-asparagine agar. One observer reports tab 5dc (grayish yellowish pink) from the Red color series for growth on yeast-malt agar, oatmeal agar, and salts-starch agar. This trace of red (or violet) pigment in aerial color is also noted in the original description (Margalith et al., 1959). Reverse side of colony with red pigments in mycelium which modify characteristic substrate colors to dark reddish brown on yeast-malt agar and salts-starch agar; light grayish brown or yellowish pink on oatmeal agar and glycerol-asparagine agar. The substrate mycelium pigment is not a strong pH indicator (one observer reports change from violet to pink with addition of 0.05 M HCl).

Color in medium: melanoid pigments are not found in peptone-yeast-iron agar, tyrosine agar, or tryptone-yeast broth. Red or violet pigments are found in yeast-malt agar, oatmeal agar, and glycerol-asparagine agar. One observer reports a change from violet to red by addition of 0.05 M HCl.

D-Glucose, L-arabinose, D-xylose, iso-inositol, D-mannitol, D-fructose, and rhamnose are utilized for growth. No growth or only traces of growth on raffinose. Utilization of sucrose is doubtful.

Type strain shows the highest sequence similarity to: *S. althioticus*, AY999808, 99.8%; *S. griseorubens*, AB184139, 99.7%; *S. labedae*, AB184704, 99.6%; *S. variabilis*, DQ442551, 99.6%; *S. griseoincarnatus*, AJ781328, 99.6%; *S. paradoxus*, AB184628, 99.6%; *S. erythrogriseus*, AJ781328, 99.6%; *S. griseoflavus*, AJ781322, 99.5%; *S. violaceorubidus*, AJ781374, 99.4%; *S. viridochromogenes*, DQ442555, 99.3%; *S. albaduncus*, AY999757, 99.3%; *S. lusitanus*, AB184424, 99.3%; *S. malachitofuscus*, AB184282, 99.3%; *S. coerulescens*, AY999720, 99.2%; *S. heliomycini*, AB184712, 99.2%; *S. glaucescens*, AB184843, 99.2%; *S. tendae*, D63873, 99.2%; *S. capillispiralis*, AB184577, 99.2%; *S. gancidicus*, AB184660, 99.2%; *S. pseudogriseolus*, DQ442541, 99.2%; *S. griseoloalbus*, AB184275, 99.2%; *S. bellus*, AB184849, 99.2%; *S. collinus*, AB184123, 99.2%; *S. lienomycini*, AJ781353, 99.1%; *S. violaceochromogenes*, AY999867, 99.1%; *S. flaveolus*, AB184764, 99.1%; *S. cellulosae*, DQ442495, 99.1%; *S. longispororuber*, AB184440, 99%; *S. rubrogriseus*, AB184681, 99%; *S. ambofaciens*, M27245, 99%; *S. diastaticus* subsp. *diastaticus*, AB184785, 99%; *S. gougerotii*, AB184742, 99%; *S. rutgersensis* subsp. *rutgersensis*, AB184170, 99%; *S. azureus*, EF178674, 99%; *S. lomondensis*, AB184673, 99%; *S. fragilis*, AY999917, 99%; *S. intermedius*, AB184277, 99%; *S. misionensis*, EF178678, 99%.

Source: not known.

DNA G+C content (mol%): not known.

Type strain: ATCC 23935, CBS 919.68, DSM 40188, HAMBI 1048, NBRC 12889, IMET 42065, JCM 4277, JCM 4268, JCM 4651, NCIMB 9826, NRRL B-2576, NRRL-ISP 5188, RIA 1142, RIA 570.

Sequence accession no. (16S rRNA gene): AB184221.

312. **Streptomyces mauvecolor** Okami and Umezawa *in* Murase, Hikiji, Nitta, Okami, Takeuchi and Umezawa 1961, 117[AL]

mau.ve.co'lor. L. n. *malva* mallow, a plant with violet-colored petals, hence French *mauve*; L. n. *color* color; N.L. adj. *mauvecolor* mauve colored.

Probably grows poorly on Czapek's solution agar. Produces peptimycin, a peptidic anti-tumor antibiotic.

Type strain shows the highest sequence similarity to: *S. xanthochromogenes*, DQ442559, 99.4%; *S. michiganensis*, AB184153, 99.4%; *S. cyaneofuscatus*, AB184860, 99%; *S. flavofuscus*, AB249935, 99%; *S. cinereorectus*, AB184646, 99%; *S. sannanensis*, AB184579, 99%; *S. fimicarius*, AY999784, 99%; *S. anulatus*, DQ026637, 99%; *S. praecox*, AB184293, 99%.

Source: not known.

DNA G+C content (mol%): not known.

Type strain: ATCC 29835, DSM 41702, NBRC 13854, JCM 5002.

Sequence accession no. (16S rRNA gene): AB184532.

313. **Streptomyces mediolani** Bianchi, Grein, Julita, Marnati and Spalla 1970, 243[AL]

me.di.o.la'ni. L. n. *Mediolanum* Milan; L. gen. adj. *mediolani* of Milan.

Sporophores are long and straight; from these, through monopodial branching, the spore bearing hyphae originate as long, straight to slightly flexuous filaments. Spores are cylindrical in shape, and devoid of any ornamentation. Growth is always very good on most common agar media; the substrate mycelium grows as a compact, smooth patina, showing a yolk-yellow to orange-yellow color. The aerial mycelium is very abundant showing a velvety to dusty aspect; its color varies from yellow-vanilla to yellow-rose or yellow-beige, according to the different substrates on which it is grown.

The spore suspension prepared from a culture of *S. mediolani* grown on a solid medium containing yeast extract, glucose, and inorganic salts, was used to inoculate 300 ml Erlenmeyer flasks each containing 60 ml liquid medium having dextrin, casein, corn-steep liquor, calcium carbonate and other inorganic salts as ingredients. After 27 h of incubation at 28°C on a rotary shaker, aliquots of the culture were transferred to 300 ml Erlenmeyer flasks each containing 30 ml of the following production medium: soluble starch, 37.5 g; morsuit (a maltose-rich syrup manufactured by F.R.A.G.D., Italy), 33 g; soybean meal 40 g; NaCl, 1.25 g; KH_2PO_2, 3 g; $MgSO_4 \cdot 7H_2O$, 0.5 g; $CaCO_3$, 3.5 g; $(NH_4)_2SO_4$, 1.5 g; $ZnSO_4 \cdot 7H_2O$, 0.005 g; $FeSO_4 \cdot 7H_2O$, 0.005 g; $CuSO_4 \cdot 5H_2O$, 0.001 g; tap water up to 1000 ml. The flasks were incubated for 8 d on a rotary shaker at 29°C.

The extracts prepared from the mycelium of *S. mediolani* with the aid of different organic solvents contained three main carotenoids, designated A, B, and C, together with other minor pigments. Compounds A and B were epiphasic, while pigment C precipitated at the interfaces in the biphasic system petroleum ether (b.p. 40–60°C)/90% aqueous methanol. Separation of the epiphasic pigments by chromatography on aluminium oxide, using petroleum ether containing increasing amounts of acetone as developing agent, allowed isolation of pigment A [$C_{40}H_{48}$, red crystals m.p. 200–201°C, λ_{max} (petroleum ether) 278, 428 (inflexion), 448, 478 nm] and pigment B [amorphous red power, m.p. 180–185°C, λ_{max} (petroleum ether) 278, 340 (*cis*-peak), 425 (inflexion), 446, 472 nm]. Pigment C was purified by chromatography on silicic acid using 10% acetone in benzene as developing solvent and isolated as a dark-red amorphous solid [m.p. 200°C (dec), λ_{max} (petroleum ether) 278, 340 (*cis*-peak), 425 (inflexion), 445, 475 nm].

Pigment A proved to be identical to isorenieratene (i) by direct comparison with an authentic sample. Pigments C and B were soluble in ethanolic potassium hydroxide and showed a hydroxyl band in the IR at 3590 cm^{-1} (methylene chloride solution). On treatment with acetic anhydride and pyridine, pigments B and C were converted to a monoacetate ($C_{42}H_{50}O_2$, m.p. 156–158°C) and a diacetate ($C_{44}H_{52}O_4$, m.p. 182–184°C), respectively, both derivatives showing UV and visible spectra identical to those displayed by I (no *cis*-peak), and a phenolic acetate band at 1758 cm^{-1} in the IR (KBr pellets). Treatment of the acetyl derivative of pigment B with chromic anhydride in benzene-acetic acid gave crocetindial, 2,3,6-trimethylbenzaldehyde, and 2,3,6-trimethyl-4-acetoxybenzaldehyde, as main breakdown products. All compounds were identified by direct comparison (UV, visible, and IR-spectra, mixed m.p., chromatographic behavior) with authentic samples prepared by synthesis. Similarly, oxidation of the diacetyl derivative of pigment C gave crocetindial and 2,3,6-trimethyl-4-acetoxybenzaldehyde as major reaction products. Structures II and III could therefore be written for pigments B and C.

The new carotenoids B (3-hydroxyisorenieratene) and C (3,3′-dihydroxyisorenieratene) have been subsequently prepared by chemical synthesis. The Wittig reaction of crocetindial with triphenyl-2,3,6-trimethylbenzylphosphonium bromide and triphenyl-[2,3,6-trimethyl-4-(2′-tetrahydropyranyloxy)]-benzylphosphonium bromide as reagents was used. For the preparation of the latter reagent, 2,3,6-trimethyl-4-hydroxybenzaldehyde was converted to the 4-(2′-tetrahydopyranyloxy) derivative, then reduced with lithium aluminium hydride to 2,3,6-trimethyl-4-(2′-tetrahydropyranyloxy)-benzylalcohol (m.p. 85°C), which was in turn converted to the phosphonium bromide by the usual procedure. When crocetindial was allowed to react with triphenyl-2,3,6-trimethyl-4-(2′-tetrahydropyranyloxy)-benzylphosphonium bromide in the presence of *n*-butyl lithium, symmetric 3,3′-ditetrahydropyranyloxyisorenieratene, $C_{50}H_{64}O_4$ (m.p. 185–187°C), was obtained. Removal of the protecting group by acid treatment afforded III (overall yield 60%). When the Wittig reaction was repeated starting with a mixture of both phosphonium bromides, a product was obtained from which, after removal of the tetrahydropyranyl group, 3-hydoxy-isorenieratene (II) was isolated in 30% yield, together with lower amounts of I and III. The synthetic compounds II and II displayed the same visible and IR-spectra, and identical chromatographic behavior, also after acetylation, as natural pigments B and C.

The peptide antibiotic histidomycin was isolated from the filtered broth by absorption-elution on ion exchange resins followed by molecular sieves fractionation. The identification of the antibiotic was performed on the basis of its chemical and biological properties, and of direct comparison with an authentic sample. The substance was responsible for the anti-bacterial activity displayed by culture liquids of *S. mediolani*.

Type strain shows the highest sequence similarity to: *S. sindenensis*, AB184759, 100%; *S. rubiginosohelvolus*, AB184240, 100%; *S. albovinaceus*, AB249958, 100%; *S. griseoplanus*, AY999894, 100%; *S. anulatus*, DQ026637, 100%; *S. globisporus* subsp. *globisporus*, EF178686, 100%; *S. pluricolorescens*, DQ442540, 100%; *S. praecox*, AB184293, 100%; *S. flavofuscus*, AB249935, 100%; *S. fimicarius*, AY999784, 100%; *S. badius*, AY999783, 100%; *S. griseinus*,

AB184205, 100%; *S. californicus*, AB184755, 99.9%; *S. lavendulae* subsp. *lavendulae*, AB184080, 99.9%; *S. cavourensis* subsp. *washingtonensis*, DQ026671, 99.9%; *S. acrimycini*, AY999889, 99.9%; *S. parvus*, DQ442537, 99.9%; *S. fulvorobeus*, AB184711, 99.8%; *S. lipmanii*, AB184148, 99.8%; *S. microflavus*, DQ445795, 99.8%; *S. cyaneofuscatus*, AB184860, 99.8%; *S. floridae*, AB184656, 99.8%; *S. alboviridis*, AB184256, 99.8%; *S. baarnensis*, EF178688, 99.8%; *S. cinereorectus*, AB184646, 99.8%; *S. griseus* subsp. *griseus*, AY207604, 99.7%; *S. griseolus*, AB184768, 99.7%; *S. luridiscabiei*, AF361784, 99.7%; *S. flavovirens*, DQ026635, 99.7%; *S. argenteolus*, AB045872, 99.7%; *S. bacillaris*, AB184439, 99.6%; *S. pulveraceus*, AB184806, 99.6%; *S. halstedii*, EF178695, 99.6%; *S. flavogriseus*, AJ494864, 99.6%; *S. atroolivaceus*, AJ781320, 99.5%; *S. olivoviridis*, AB184227, 99.5%; *S. finlayi*, AY999788, 99.5%; *S. nitrosporeus*, EF178680, 99.5%; *S. albolongus*, AB184425, 99.4%; *S. clavifer*, DQ026670, 99.4%; *S. griseobrunneus*, AB249912, 99.4%; *S. sanglieri*, AB249945, 99.4%; *S. celluloflavus*, AB184476, 99.4%; *S. yanii*, AB006159, 99.4%; *S. gelaticus*, DQ026636, 99.3%; *S. atratus*, DQ026638, 99.3%; *S. mutomycini*, AB249951, 99.3%; *S. candidus*, DQ026663, 99.2%; *S. cavourensis* subsp. *cavourensis*, DQ445791, 99.2%; *S. spiroverticillatus*, AB184814, 99.1%; *S. cremeus*, AB184124, 99.1%.

Source: not known.

DNA G+C content (mol%): not known.

Type strain: ATCC 33021, BCRC 12035, CMI 134886, DSM 41058, DSM 41647, NBRC 15427, JCM 5076, NCIMB 10969, VKM Ac-1917.

Sequence accession no. (16S rRNA gene): AB184674.

314. **Streptomyces megasporus** (*ex* Krasil'nikov, Agre, Dorokhova and Sokolov 1968) Agre 1986, 575[VP] (Effective publication: Agre *in* Gause, Preobrazhenskaya, Sveshnikova, Terekhova and Maximova 1983.) ("*Actinomyces megasporus*" Krasil'nikov, Agre, Dorokhova and Sokolov 1968)

me.ga.spo'rus. Gr. adj. *megas* big; Gr. n. *spora* seed; N.L. masc. adj. *megasporus* large spored.

Spore chains are spirals (*Spirales*); spores have spines and warts. On mineral agar 1: aerial mycelium is pale green-gray; substrate mycelium is colorless; no diffusible pigment. On glycerol-nitrate agar and oatmeal agar: aerial mycelium is white or absent; substrate mycelium is colorless; no diffusible pigment. On starch-ammonia agar: poor growth; aerial mycelium is poor, grayish; substrate mycelium is colorless; no diffusible pigment. On glycerol-asparagine agar: poor growth; aerial mycelium is white; substrate mycelium is colorless; no diffusible pigment. On organic agar 2: aerial mycelium absent; substrate mycelium colorless or yellow; no diffusible pigment. Melanoid pigments are not formed. Growth on glucose, maltose, xylose, rhamnose, and arabinose.

Type strain shows no sequence similarity over 99%.

Source: not known.

DNA G+C content (mol%): not known.

Type strain: ATCC 43688, DSM 41476, HUT 6610, NBRC 14749, INA M-22, INMI 2869, JCM 6926, NRRL B-16372, VKM Ac-1776.

Sequence accession no. (16S rRNA gene): AB184617.

315. **Streptomyces melanogenes** Sugawara and Onuma 1957, 141[AL]

me.la.no'ge.nes. Gr. adj. *melas* black; N.L. suff. *-genes* (from Gr. v. *gennaô* to produce) producing; N.L. part. adj. *melanogenes* producing black.

Spore chains in Section *Rectiflexibiles*. Mature spore chains are long with 10–50, or often more than 50, spores per chain. This morphology is seen on yeast-malt agar, oatmeal agar, salts-starch agar, and glycerol-asparagine agar. Spore surface is smooth.

Color of colony: aerial mass color in the Red color series (5cb, grayish yellowish pink) on oatmeal agar and salts-starch agar; White or Gray color series on yeast-malt agar and salts-starch agar. Reverse side of colony is grayed yellow to yellow-brown is usually modified by red (becoming dark brown to reddish brown) on yeast-malt agar, oatmeal agar, salts-starch agar, and glycerol-asparagine agar; this pigment is not a pH indicator.

Color in medium: melanoid pigments are formed in peptone-yeast-iron agar, tyrosine agar, and tryptone-yeast broth. This pigment is not pH-sensitive when tested with 0.05 M NaOH or HCl.

D-Glucose, L-arabinose, D-xylose, iso-inositol, D-mannitol, D-fructose, and raffinose are utilized for growth. Reports vary on utilization of sucrose. No growth or only traces of growth rhamnose.

Type strain shows the highest sequence similarity to: *S. noboritoensis*, AB184287, 100%; *S. crystallinus*, AB184652, 99.2%.

Source: not known.

DNA G+C content (mol%): not known.

Type strain: ATCC 23937, CBS 921.68, BCRC 12053, DSM 40192, NBRC 12890, JCM 4398, NCIMB 9835, NRRL B-2072, NRRL-ISP 5192, RIA 1146.

Sequence accession no. (16S rRNA gene): AB184222.

316. **Streptomyces melanosporofaciens** Arcamone, Bertazzoli, Ghione and Scotti 1959, 215[AL]

me.la.no.spo.ro.fa'ci.ens. Gr. adj. *melas* black; N.L. n. *spora* a spore; L. part. adj. *faciens* producing; N.L. part. adj. *melanosporofaciens* black spore producing.

Spore chains in Section *Spirales*, but spore chain morphology is sometimes difficult to observe because of accumulations of moist exudate ("hygroscopic" masses) on the sporulating surfaces. Identifiable spore chains are often short with only 3–10 spores per chain and may form loops or incomplete spirals as well as true spirals. This morphology is seen on yeast-malt agar, oatmeal agar, salts-starch agar, and glycerol-asparagine agar. Spore surface is smooth, or with unevenly ridged surface suggesting warty spores on some preparations. In 7–21 d, sporulating surfaces form globose accumulations of liquid exudate. These coalesce and may eventually give rise to a moist black surface containing loose spores.

Color of colony: aerial mass color in the Gray color series on yeast-malt agar and oatmeal agar; poorly sporulating cultures on salts-starch agar and glycerol-asparagine agar may appear to be in the Yellow color series. Areas of sporulation are usually gray at first, but may become moist-black as spore masses coalesce in a liquid exudate on the aerial mycelium. Reverse side of colony is dark grayish yellow to orange-yellow on yeast-malt agar; light yellow to light grayish olive on oatmeal agar; nearly colorless to strong yellow on salts-starch agar and glycerol-asparagine agar. Substrate pigment is not a pH indicator.

Color in medium: melanoid pigments are not formed in peptone-yeast-iron agar, tyrosine agar, or tryptone-yeast broth, but a reddish brown, rose, or yellow non-melanoid pigment may be found in tyrosine agar. No pigment is found in medium in yeast-malt agar, oatmeal agar, salts-starch agar, or glycerol-asparagine agar.

D-Glucose, L-arabinose, D-xylose, iso-inositol, D-mannitol, D-fructose, rhamnose, and raffinose are utilized for growth. Only traces of growth are found with sucrose.

Type strain shows the highest sequence similarity to: *S. sporoclivatus*, AB249934, 99.6%; *S. antimycoticus*, AB184185, 99.6%; *S. rutgersensis* subsp. *castelarensis*, AY508511, 99.4%; *S. geldanamycininus*, DQ334781, 99%.

Source: not known.

DNA G+C content (mol%): not known.

Type strain: ATCC 25473, CBS 883.69, BCRC 12064, DSM 40318, NBRC 13061, JCM 4495, NCIMB 12978, NRRL B-12234, NRRL-ISP 5318, RIA 1253, VKM Ac-1864.

Sequence accession no. (16S rRNA gene): AJ271887.

317. **Streptomyces mexicanus** Petrosyan, García-Varela, Luz-Madrigal, Huitrón and Flores 2003, 272[VP]

mex.i.ca'nus. N.L. masc. adj. *mexicanus* of or belonging to Mexico, the source of the soil from which the organism was isolated.

Forms highly branched substrate mycelium and aerial hyphae which differentiate into long *Rectiflexibiles* chains of 10 or more, green, smooth spores. Spores are spherical, about 0.88 µm in diameter, or slightly oval. Aerial spore mass color is gray with a slight green tint. Substrate mycelium is beige on standard media. Yellowish diffusible pigments are formed on yeast-extract, malt-extract, and oatmeal agar. Melanin pigments are not produced on peptone-iron or tyrosine agars. Positive for H_2S production. Degrades arbutin, starch, xylan, adenine, casein, hypoxanthine, and L-tyrosine, but not guanine or testosterone. Utilizes L-arabinose, fructose, glucose, raffinose, mannitol, mannose, xylose, galactose, maltose, glycerol, lactose, cellobiose, trehalose, and sodium acetate as sole carbon sources, but not *myo*-inositol, melezitose, L-rhamnose, sorbitol, or sucrose. Growth occurs between 20 and 55°C, from pH 4.3–8.0, and in the presence of 6% (w/v) NaCl and ampicillin (100 µg/ml). Growth is inhibited in the presence of chloramphenicol, erythromycin, gentamicin, nalidixic acid sodium salt, kanamycin sulfate, rifampin, tetracycline hydrochloride, and thiostrepton, and in the presence of 7% (w/v) NaCl. No antimicrobial activity is shown against *Escherichia coli* JM 109 or against representative strains of *Aspergillus flavipes, Aureobasidium* sp., *Bacillus subtilis, Pichia pastoris,* "*Sarcina lutea*", or *Trichoderma viridae.*

Type strain shows the highest sequence similarity to: *S. thermoviolaceus* subsp. *apingens*, Z68095, 99%.

Source: isolated from a soil sample obtained at a sugarcane field in the State of Morelos, Mexico.

DNA G+C content (mol%): not known.

Type strain: CH-M-1035, BM-B-384, DSM 41796, JCM 12681, NBRC 100915, NRRL B-24196.

Sequence accession no. (16S rRNA gene): AB249966.

318. **Streptomyces michiganensis** Corbaz, Ettlinger, Keller-Schierlein and Zähner 1957b, 205[AL]

mi.chi.gan.en'sis. N.L. masc. adj. *michiganensis* of or belonging to Michigan, the source of the soil from which the organism was isolated.

Spore chain morphology in Section *Rectiflexibiles*. Mature spore chains are generally long, often with more than 50 spores per chain. This morphology is seen on yeast-malt agar, oatmeal agar, salts-starch agar, and glycerol-asparagine agar. Spore surface is smooth.

Color of colony: aerial mass color in the Yellow color series on yeast-malt agar, oatmeal agar, salts-starch agar, and glycerol-asparagine agar. Reverse side of colony with no distinctive pigment (colorless or characteristic grayed yellow on oatmeal agar, salts-starch agar, and glycerol-asparagine agar).

Color in medium: melanoid pigments formed in peptone-yeast-iron agar and tryptone-yeast broth. Yellow pigment found in medium in yeast-malt agar and oatmeal agar. This pigment is not pH-sensitive when tested with 0.05 M NaOH of HCl.

D-Glucose, D-xylose, iso-inositol, D-mannitol, and D-fructose are utilized for growth. No growth or only trace of growth on L-arabinose, sucrose, and rhamnose. Variable reports on growth with raffinose.

Type strain shows the highest sequence similarity to: *S. xanthochromogenes*, DQ442559, 100%; *S. mauvecolor*, AB184532, 99.4%.

Source: isolated from soil from Michigan.

DNA G+C content (mol%): not known.

Type strain: ATCC 14970, ATCC 19786, CBS 538.68, BCRC 11613, DSM 40015, NBRC 12797, JCM 4594, NRRL B-1940, NRRL-ISP 5015, RIA 1065, UNIQEM 172, VKM Ac-862.

Sequence accession no. (16S rRNA gene): AB184153.

319. **Streptomyces microflavus** (Krainsky 1914) Waksman and Henrici *in* Breed, Murray and Hitchens 1948, 950[AL] ("*Actinomyces microflavus*" Krainsky 1914, 686; "*Micromonospora microflava*" Duché 1934, 29)

mic.ro.fla'vus. Gr. adj. *mikros* small; L. adj. *flavus* yellow; N.L. masc. adj. *microflavus* small, yellow.

Spore chains in Section *Rectiflexibiles.* Mature spore chains generally have 10–50 spores per chain. This morphology is seen on yeast-malt agar, oatmeal agar, salts-starch agar, and glycerol-asparagine agar. Spore surface is smooth.

Color of colony: aerial mass color in the Gray or Yellow color series (2dc, yellowish gray; 2db, pale yellow) on yeast-malt agar and oatmeal agar; Yellow color series (2ba or 2db, pale yellow) on salts-starch agar and glycerol-asparagine agar. Reverse side of colony with no distinctive pigments (grayish yellow to yellowish brown, olive brown or strong brown) on yeast-malt agar, oatmeal agar, salts-starch agar, and glycerol-asparagine agar.

Color in medium: melanoid pigments are not formed in peptone-yeast-iron agar, tyrosine agar, or tryptone-yeast broth. No pigment found in medium in yeast-malt agar, oatmeal agar, salts-starch agar, or glycerol-asparagine agar.

D-Glucose, D-xylose, D-mannitol, D-fructose, and rhamnose are utilized for growth. Reports vary on the utilization of sucrose. Utilization of L-arabinose is doubtful. No growth or only traces of growth with iso-inositol and raffinose.

Type strain shows the highest sequence similarity to: *S. fulvorobeus*, AB184711, 100%; *S. lipmanii*, AB184148, 100%; *S. lavendulae* subsp. *lavendulae*, AB184080, 100%; *S. alboviridis*, AB184256, 100%; *S. cavourensis* subsp. *washingtonensis*, DQ026671, 99.9%; *S. flavofuscus*, AB249935, 99.9%; *S. cinereorectus*, AB184646, 99.9%; *S. fimicarius*, AY999784, 99.9%; *S. griseoplanus*, AY999894, 99.9%; *S. cyaneofuscatus*, AB184860, 99.9%; *S. floridae*, AB184656, 99.9%; *S. praecox*, AB184293, 99.9%; *S. anulatus*, DQ026637, 99.8%; *S. luridiscabiei*, AF361784, 99.8%; *S. pluricolorescens*, DQ442540, 99.8%; *S. acrimycini*, AY999889, 99.8%; *S. argenteolus*, AB045872, 99.8%; *S. griseinus*, AB184205, 99.8%; *S. badius*, AY999783, 99.8%; *S. rubiginosohelvolus*, AB184240, 99.8%; *S. sindenensis*, AB184759, 99.8%; *S. mediolani*, AB184674, 99.8%; *S. californicus*, AB184755, 99.8%; *S. parvus*, DQ442537, 99.7%; *S. halstedii*, EF178695, 99.7%; *S. baarnensis*, EF178688, 99.7%; *S. griseolus*, AB184768, 99.7%; *S. griseus* subsp. *griseus*, AY207604, 99.7%; *S. globisporus* subsp. *globisporus*, EF178686, 99.7%; *S. albovinaceus*, AB249958, 99.7%; *S. flavovirens*, DQ026635, 99.5%; *S. finlayi*, AY999788, 99.5%; *S. flavogriseus*, AJ494864, 99.5%; *S. yanii*, AB006159, 99.4%; *S. atroolivaceus*, AJ781320, 99.4%; *S. olivoviridis*, AB184227, 99.4%; *S. bacillaris*, AB184439, 99.4%; *S. pulveraceus*, AB184806, 99.4%; *S. celluloflavus*, AB184476, 99.3%; *S. griseobrunneus*, AB249912, 99.3%; *S. albolongus*, AB184425, 99.3%; *S. clavifer*, DQ026670, 99.3%; *S. nitrosporeus*, EF178680, 99.3%; *S. spiroverticillatus*, AB184814, 99.2%; *S. sanglieri*, AB249945, 99.2%; *S. cavourensis* subsp. *cavourensis*, DQ445791, 99.2%; *S. atratus*, DQ026638, 99.1%; *S. mutomycini*, AB249951, 99.1%; *S. candidus*, DQ026663, 99.1%; *S. gelaticus*, DQ026636, 99.1%; *S. cremeus*, AB184124, 99%.

Source: not known.

DNA G+C content (mol%): not known.

Type strain: AS 4.1428, ATCC 13231, ATCC 25474, CBS 124.18, CBS 884.69, BCRC 12084, DSM 40331, HAMBI 1019, NBRC 13062, JCM 4496, LMG 19327, NRRL B-2156, NRRL B-2888, NRRL-ISP 5331, RIA 1254, VKM Ac-971.

Sequence accession no. (16S rRNA gene): DQ445795.

Further comments: according to Lanoot et al. (2005b), *Streptomyces microflavus* (Krainsky 1914) Waksman and Henrici 1948 emend. Lanoot et al. 2005b is an earlier heterotypic synonym of *Streptomyces griseus* subsp. *alpha* (Ciferri 1927) Pridham 1970, an earlier heterotypic synonym of *Streptomyces griseus* subsp. *cretosus* Pridham 1970^{AL}, an earlier heterotypic synonym of *Streptomyces lipmanii* (Waksman and Curtis 1916) Waksman and Henrici 1948, and an earlier heterotypic synonym of *Streptomyces willmorei* (Erikson 1935) Waksman and Henrici 1948.

320. **Streptomyces minutiscleroticus** (Thirumalachar *in* Thirumalachar, Rahlkar, Desmukh and Sukapure 1965) Pridham 1970, 41^{AL} emend. Lanoot, Vancanneyt, van Schoor, Liu and Swings 2005b, 731 (*Chainia minutisclerotica* Thirumalachar *in* Thirumalachar, Rahlkar, Desmukh and Sukapure 1965, 7)

mi.nu.ti.scle.ro'ti.cus. L. masc. adj. *minutus* little, small, minute; N.L. n. *sclerotium* sclerotium; L. masc. suff. *-icus* suffix used in adjectives with the sense of belonging to; N.L. masc. adj. *minutiscleroticus* belonging to a small sclerotium.

Spore chains in Section *Spirales*. Mature spore chains generally contain 10 to 50 or more spores per chain. This morphology is seen on yeast-malt agar, oatmeal agar, salts-starch agar, and glycerol-asparagine agar. Spore surface is smooth. Special morphological characteristics: sclerotic granules are produced. Minute sclerotic granules are produced on broad hyphae, often on the agar surface or vegetative substrate. One observer compares these small sclerotia to actinosporangium-like formations, on oatmeal agar and salts-starch agar. Sclerotic granules are reported on yeast-malt agar, salts-starch agar, and glycerol-asparagine agar in 7–21 d.

Color of colony: aerial mass color in the Gray or Yellow color series (2dc, yellowish gray; 24½dc, pale yellow green; or 2ba, pale yellow) on yeast-malt agar and salts-starch agar; Yellow color series (2ba, pale yellow) on oatmeal agar and glycerol-asparagine agar. Reverse side of colony is strong brown on yeast-malt agar; grayish yellow to orange yellow or brown on oatmeal agar and salts-starch agar; yellowish brown, orange yellow or reddish brown on glycerol-asparagine agar. Reverse mycelium pigment is not a pH indicator.

Color in medium: melanoid pigments are not formed in peptone-yeast-iron agar, tyrosine agar, or tryptone-yeast broth. Some yellow pigment may or may not be found in the medium in yeast-malt agar, oatmeal agar, salts-starch agar, and glycerol-asparagine agar. When present, this pigment is not pH-sensitive.

D-Glucose, L-arabinose, D-xylose, iso-inositol, D-mannitol, D-fructose, and rhamnose are utilized for growth. No growth or only traces of growth with sucrose or raffinose.

Type strain shows the highest sequence similarity to: *S. vinaceusdrappus*, AY999929, 100%; *S. geysiriensis*, DQ442501, 100%; *S. plicatus*, AB184291, 100%; *S. ghanaensis*, AY999851, 100%; *S. rochei*, AB184237, 100%; *S. mutabilis*, EF178679, 99.8%; *S. tuirus*, AB184690, 99.5%; *S. djakartensis*, AB184657, 99.5%; *S. anandii*, AB184402, 99.3%; *S. violaceorubidus*, AJ781374, 99.3%; *S. pilosus*, AB184161, 99.2%; *S. flavoviridis*, AB184842, 99.2%; *S. calvus*, AB184329, 99.1%; *S. tendae*, D63873, 99.1%; *S. azureus*, EF178674, 99.1%; *S. asterosporus*, AB184706, 99.1%; *S. levis*, AB184670, 99%; *S. luteogriseus*, AB184379, 99%; *S. aureorectus*, AB184710, 99%; *S. capillispiralis*, AB184577, 99%; *S. pseudogriseolus*, DQ442541, 99%; *S. naganishii*, DQ442529, 99%; *S. virens*, DQ442554, 99%; *S. gancidicus*, AB184660, 99%.

Source: not known.

DNA G+C content (mol%): not known.

Type strain: ATCC 17757, ATCC 19346, CBS 231.65, CBS 662.72, BCRC 12544, CMI 112786, DSM 40301, NBRC

13000, NBRC 13361, JCM 3102, JCM 4790, KCTC 9123, LMG 20062, NCIMB 10996, NRRL B-12202, NRRL-ISP 5301, PCM 2304, RIA 1322, RIA 885.

Sequence accession no. (16S rRNA gene): EF178696.

Further comments: Streptomyces minutiscleroticus (Thirumalachar 1965) Pridham 1970 and *Chainia minutisclerotica* Thirumalachar et al. (1965) have the same type strain and therefore are homotypic synonyms [Rules 24a and 24b (1) of the Bacteriological Code].

According to Lanoot et al. (2005b), *Streptomyces minutiscleroticus* (Thirumalachar 1965) Pridham 1970 emend. Lanoot et al. 2005b is an earlier heterotypic synonym of *Streptomyces flaviscleroticus* (*ex* Pridham 1970) Goodfellow et al. 1986a.

321. **Streptomyces mirabilis** (Kudrina *in* Gauze, Preobrazhenskaya, Kudrina, Blinov, Ryabova and Sveshnikova 1957) Ruschmann 1952, 543[AL] ("*Actinomyces mirabilis*" Kudrina *in* Gauze, Preobrazhenskaya, Kudrina, Blinov, Ryabova and Sveshnikova 1957, 107)

mi.ra'bi.lis. L. masc. adj. *mirabilis* marvellous.

Spore chains in Section *Spirales*. Sterile hyphae may be abundant; these are often flexuous or hooked. Mature spore chains are moderately long, sometimes with more than 10 spores per chain. This morphology is seen on yeast-malt agar, oatmeal agar, salts-starch agar, and glycerol-asparagine agar, but sporulation may be poor, especially on yeast-malt agar and glycerol-asparagine agar. Spore surface is smooth.

Color of colony: aerial mass color in the Gray color series (2fe, medium gray; 3fe, light brownish gray; also 3ig, 4ig, 3li, grayish brown and brownish gray) on yeast-malt agar, oatmeal agar, salts-starch agar, and glycerol-asparagine agar. Reverse side of colony is grayish yellow becoming moderate olive brown to dark olive or dark brown on oatmeal agar, salts-starch agar, and glycerol-asparagine agar; grayish blue or grayish green and moderate yellowish brown on yeast-malt agar. Substrate pigment is not a pH indicator or is modified only slightly from olive brown to dark brown by 0.05 M NaOH on oatmeal agar and glycerol-asparagine agar.

Color in medium: melanoid pigments are formed in peptone-yeast-iron agar and in trace amounts in tryptone-yeast broth but not in tyrosine agar. No pigment is found in the medium in yeast-malt agar, oatmeal agar, salts-starch agar, or glycerol-asparagine agar.

D-Glucose, L-arabinose, iso-inositol, D-mannitol, D-fructose, and rhamnose are utilized for growth. Reports vary on utilization of D-xylose. No growth or only traces of growth with sucrose or raffinose.

Type strain shows the highest sequence similarity to: *S. olivochromogenes*, AY094370, 99.4%.

Source: not known.

DNA G+C content (mol%): not known.

Type strain: ATCC 27447, CBS 751.72, DSM 40553, NBRC 13450, JCM 4551, JCM 4791, NRRL B-2400, NRRL-ISP 5553, RIA 1411.

Sequence accession no. (16S rRNA gene): AB184412.

322. **Streptomyces misakiensis** Nakamura 1961, 86[AL]

mi.sa.ki.en'sis. N.L. masc. adj. *misakiensis* of or belonging to Misaki, referring to Misakicho, Kanagawa Prefecture, Japan, the source of the soil from which the organism was isolated.

Spore chains in Section *Rectiflexibiles*. Mature spore chains are generally long with 10–50, or sometimes more than 50, spores per chain. This morphology is seen on yeast-malt agar, oatmeal agar, salts-starch agar, and glycerol-asparagine agar. Spore surface is smooth.

Color of colony: aerial mass color in the Gray color series on salts-starch agar and glycerol-asparagine agar; Gray or Red color series on yeast-malt agar and oatmeal agar. The most representative color tab from the Gray color-wheel is 2dc (yellowish gray) for all ISP media. One observer selected tab 5dc (grayish yellowish pink) from the Red color-wheel as the most representative color on yeast-malt agar and oatmeal agar. Reverse side of colony is yellow to yellow-brown is modified by red, becoming brown to strong brown on yeast-malt agar, salts-starch agar and glycerol-asparagine agar and orange to light brown on oatmeal agar; this pigment is not a pH indicator.

Color in medium: melanoid pigments are not formed in peptone-yeast-iron agar, tyrosine agar, or tryptone-yeast broth. Red pigment is found in the medium in yeast-malt agar, oatmeal agar, and salts-starch agar. This pigment is not pH-sensitive when tested with 0.05 NaOH or HCl.

D-Glucose, sucrose, D-fructose, and raffinose are utilized for growth. No growth or only trace of growth with L-arabinose, D-xylose, iso-inositol, D-mannitol and rhamnose. In the original description (Nakamura, 1961), L-arabinose and D-mannitol were also reported as supporting growth with Czapek's agar as basal medium.

Type strain shows the highest sequence similarity to: *S. catenulae*, AJ621613, 100%; *S. libani* subsp. *libani*, AB184414, 99.6%; *S. nigrescens*, DQ442530, 99.5%; *S. tubercidicus*, AJ621612, 99.5%; *S. caniferus*, AB184640, 99.4%; *S. libani* subsp. *rufus*, AJ781351, 99.3%; *S. hygroscopius* subsp. *glebosus*, AB184479, 99.3%; *S. platensis*, AB045882, 99.2%; *S. sioyaensis*, DQ026654, 99.2%; *S. ramulosus*, DQ026662, 99.1%.

Source: isolated from soil from Misakicho, Kanagawa Prefecture, Japan.

DNA G+C content (mol%): not known.

Type strain: AS 4.1437, ATCC 23938, CBS 278.65, CBS 922.68, DSM 40222, IFM 1195, NBRC 12891, JCM 4062, JCM 4653, NCIMB 9852, NRRL B-2923, NRRL-ISP 5222, RIA 1166, VKM Ac-625.

Sequence accession no. (16S rRNA gene): AB217605.

323. **Streptomyces misionensis** Cercós, Eilberg, Goyena, Souto, Vautier and Widuczynski 1962, 22[AL]

mi.si.on.en'sis. N.L. masc. adj. *misionensis* of or belonging to Misiones, a province in Argentina, South America, the source of the soil from which the organism was isolated.

Spore chains in Section *Spirales*. Mature spore chains are moderately long, often with more than 50 spores per chain. This morphology is seen on yeast-malt agar, oatmeal agar, salts-starch agar, and glycerol-asparagine agar. Spore surface is smooth or slightly warty.

Color of colony: aerial mass color in the Gray or Red color series (4ig, light grayish brown in Gray series or 4ge, light grayish reddish brown in the Red series) on yeast-malt agar, oatmeal agar, salts-starch agar, and glycerol-asparagine agar. Reverse side of colony is olive brown to moderate orange on yeast-malt agar; grayish yellow, yellowish brown, or olive brown on oatmeal agar, salts-starch agar, and glycerol-asparagine agar.

Color in medium: melanoid pigments not formed in peptone-yeast-iron agar and tyrosine agar. No pigment is found in medium in yeast-malt agar, oatmeal agar, salts-starch agar, or glycerol-asparagine agar.

D-Glucose, L-arabinose, D-xylose, D-mannitol, and D-fructose are utilized for growth. Utilization of iso-inositol, rhamnose, sucrose and raffinose is doubtful.

Type strain shows the highest sequence similarity to: *S. phaeoluteichromatogenes*, AJ391814, 99.8%; *S. rutgersensis* subsp. *rutgersensis*, AB184170, 99.2%; *S. gougerotii*, AB184742, 99.2%; *S. diastaticus* subsp. *diastaticus*, AB184785, 99.2%; *S. levis*, AB184670, 99.1%; *S. paradoxus*, AB184628, 99.1%; *S. glaucescens*, AB184843, 99%; *S. matensis*, AB184221, 99%; *S. intermedius*, AB184277, 99%.

Source: isolated from soil from Misiones, a province in Argentina, South America.

DNA G+C content (mol%): not known.

Type strain: ATCC 14991, ATCC 25475, CBS 885.69, BCRC 12094, DSM 40306, NBRC 13063, JCM 4497, NRRL B-3230, NRRL-ISP 5306, RIA 1255, VKM Ac-626.

Sequence accession no. (16S rRNA gene): EF178678.

324. **Streptomyces mobaraensis** (Nagatsu and Suzuki 1963) Witt and Stackebrandt 1991, 456[VP] (Effective publication: Witt and Stackebrandt 1990, 370.) (*Streptomyces mobaraensis* Nagatsu and Suzuki 1963, 47; *Streptoverticillium mobaraense* Locci, Baldacci and Petrolini Baldan 1969, 42)

mo.ba.ra.en'sis. N.L. masc. adj. *mobaraensis* of or belonging to Mobara (named after the place of origin, Mobara City, Chiba Prefecture, Japan).

Good growth on potato-glucose agar (Baldacci et al., 1954). Color: reverse light beige to greenish yellow; aerial mycelium dirty white with greenish shades to greenish gray. On Bacto Czapek agar and Casamino acids-Czapek agar (1 g/l Difco vitamin-free Casamino acids, replacing sodium nitrate): very poor growth; colorless; traces only of off-white aerial mycelium. On glucose-asparagine agar (ISP medium 5 with 1% glucose replacing glycerol): good growth; color, reverse yellowish; aerial mycelium whitish. On glycerol-asparagine agar: good growth; color, reverse beige to light brown to yellowish brown; aerial mycelium white to off-white beige. On inorganic salts-starch agar: good growth; color, reverse brown-yellowish to dirty greenish yellow; aerial mycelium beige to dirty greenish beige. On yeast extract-malt extract agar: good growth; color: reverse brown-yellow; aerial mycelium white to dirty beige. On Bacto Emerson agar: good growth; color: reverse yellowish; aerial mycelium off-white. On Bennett agar (1% glucose, 0.1% Bacto beef agar, 0.1% yeast extract, 0.2% peptone, 1.5% agar): good growth; color, reverse yellow; aerial mycelium white with pale pink shades in patches. On Oxoid nutrient agar: good growth; color, reverse yellowish; poor, white aerial mycelium.

Grows at 27°C as well as at 37°C. There are also no differences in the amount of aerial mycelium, which is greener in color at 37°C. Greenish shades of the reverse are also more accentuated. No growth at 45°C. The type strain produces piericidin A, piericidin B, and detoxin and exhibits anti-bacterial activity.

Type strain shows the highest sequence similarity to: *S. sapporonensis*, AB184508, 99.1%; *S. hygroscopius* subsp. *angustmyceticus*, DQ442509, 99%.

Source: not known.

DNA G+C content (mol%): not known.

Type strain: ATCC 29032, CBS 199.75, BCRC 12165, DSM 40847, NBRC 13819, JCM 4168, NCIMB 11159, NRRL B-3729, RIA 1627, VKM Ac-928.

Sequence accession no. (16S rRNA gene): DQ442528.

Further comments: in violation of Rule 33c of the *Bacteriological Code* (1990 Revision), in Validation List no. 38, *Streptomyces mobaraensis* is proposed as a *nomen revictum* (basonym: "*Streptomyces mobaraensis*" Kubo et al. (1964).

According to Labeda (1996), *Streptomyces mobaraensis* (Nagatsu and Suzuki 1963) Witt and Stackebrandt 1991 is an earlier synonym of *Streptomyces ladakanum* (Hanka et al. 1966) Witt and Stackebrandt 1991.

According to Hatano et al. (2003), *Streptomyces mobaraensis* (Nagatsu and Suzuki 1963) Witt and Stackebrandt 1991 is an earlier heterotypic synonym of *Streptomyces ladakanum* (Hanka et al. 1966) Witt and Stackebrandt 1991.

325. **Streptomyces monomycini** Gause and Terekhova 1986a, 575[VP] (Effective publication: Gause and Terekhova *in* Gause, Preobrazhenskaya, Sveshnikova, Terekhova and Maximova 1983.)

mo.no.my.ci'ni. N.L. n. *monomycinum* monomycin; N.L. gen. n. *monomycini* of monomycin, intended to mean monomycin producing.

Spore chains are spiral (*Spirales*); spores are smooth. On mineral agar 1: aerial mycelium is white, poor; substrate mycelium is colorless to greenish-gray-brownish; no diffusible pigment. On starch-ammonia agar: no aerial mycelium; colorless substrate mycelium; no diffusible pigment. On glycerol-nitrate agar: aerial mycelium is white; substrate mycelium is gray-brownish-yellow to green-gray-brownish; no diffusible pigment. On glycerol-asparagine agar: aerial mycelium is light gray; substrate mycelium is colorless; no diffusible pigment. On oatmeal agar: aerial mycelium is poorly developed, white; substrate mycelium is greenish yellow; no diffusible pigment. On organic agar 2: aerial mycelium is poorly developed, white; substrate mycelium is gray-brownish yellow to green-gray-brownish; diffusible pigment is absent or weak, colors are similar to substrate mycelium. Melanoid pigments are not formed. Growth on fructose, glucose, xylose, and mannitol, no growth on sucrose, rhamnose, inositol, raffinose, and arabinose. Antibiotic: monomycin.

Type strain shows the highest sequence similarity to: *S. ochraceiscleroticus*, AB184094, 99.5%; *S. hygroscopius* subsp. *decoyicus*, AY999883, 99.4%; *S. niger*, AJ621607,

99.4%; *S. erumpens*, AJ621603, 99.4%; *S. olivaceiscleroticus*, AJ621606, 99.4%; S. *violens*, AJ621605, 99.3%; *S. purpurogeneiscleroticus*, AJ621604, 99.3%; *S. rimosus* subsp. *rimosus*, AB045883, 99.1%; *S. platensis*, AB045882, 99%; *S. sioyaensis*, DQ026654, 99%; *S. libani* subsp. *rufus*, AJ781351, 99%; *S. caniferus*, AB184640, 99%; *S. hygroscopius* subsp. *glebosus*, AB184479, 99%.

Source: not known.

DNA G+C content (mol%): not known.

Type strain: DSM 41801, INA 1465, JCM 9768, NBRC 100769, NRRL B-24309.

Sequence accession no. (16S rRNA gene): DQ445790.

326. **Streptomyces morookaense** (Locci 1985) Witt and Stackebrandt 1991, 456[VP] (Effective publication: Witt and Stackebrandt 1990, 370.) ("*Streptomyces moro-okaensis*" Niida, Hamamoto, Tsuruoka and Hara 1963; "*Streptoverticillium moro-okaense*" Arai 1976; "*Streptoverticillium morookaense*" Locci 1985, 232)

mo.ro.ok.a.en'se. N.L. adj. *morookaense* (*sic*) of or belonging to Moro-oka (possibly the isolation place).

Spore chains in Section *Verticillati*; mature spore chains are predominantly umbellate monoverticillate (biverticillate). Mature spore chains are moderately long, often with more than 10 spores per chain. This morphology is seen on yeast-malt agar, oatmeal agar, salts-starch agar, and glycerol-asparagine agar. Spore surface is smooth.

Color of colony: aerial mass color in the Yellow color series (1dc, 1cb, 1½ec, pale yellow green) on yeast-malt agar, oatmeal agar, salts-starch agar, and glycerol-asparagine agar. Reverse side of colony with no distinctive pigments (pale yellow to pale greenish yellow) on yeast-malt agar, oatmeal agar, salts-starch agar, and glycerol-asparagine agar.

Color in medium: melanoid pigments are not formed in peptone-yeast-iron agar, tyrosine agar, or tryptone-yeast broth. No pigment is found in the medium in yeast-malt agar, oatmeal agar, salts-starch agar, or glycerol-asparagine agar.

D-Glucose, iso-inositol, D-mannitol, D-fructose, and raffinose are utilized for growth. Utilization of rhamnose is doubtful. No growth or only traces of growth with L-arabinose, sucrose, or D-xylose.

Type strain shows the highest sequence similarity to: *S. lavenduligriseus*, DQ442515, 99.6%; *S. thioluteus*, AB184753, 99.5%; *S. abikoensis*, AB184537, 99.1%; *S. hygroscopius* subsp. *angustmyceticus*, DQ442509, 99%.

Source: not known.

DNA G+C content (mol%): not known.

Type strain: ATCC 19166, CBS 717.72, DSM 40503, NBRC 13416, JCM 4673, JCM 4793, NRRL B-12429, NRRL-ISP 55036, RIA 1377, VKM Ac-1916.

Sequence accession no. (16S rRNA gene): AJ781349.

Further comments: in violation of Rule 33c of the *Bacteriological Code* (1990 Revision), in Validation List no. 38, *Streptomyces morookaense* is proposed as a *nomen revictum* (basonym: "*Streptomyces morookaense*" Niida et al. 1963).

According to Hatano et al. (2003), *Streptomyces morookaense* (Locci and Schofield 1989) Witt and Stackebrandt 1991 is a heterotypic synonym of "*Streptomyces aspergilloides*" (NBRC 13461).

327. **Streptomyces murinus** Frommer 1959, 198[AL]

mu.ri'nus. L. masc. adj. *murinus* of mice, mouse-gray; referred to as reddish-gray in original description.

Spore chains in Section *Spirales*. Mature spore chains generally 10–50 spores per chain. This morphology is seen on yeast-malt agar, oatmeal agar, salts-starch agar, and glycerol-asparagine agar. Spore surface is smooth. Air-dried preparation for electron microscopy may show internal contractions that suggest a warty appearance.

Color of colony: aerial mass color in the Red color series on glycerol-asparagine agar; Gray series on salts-starch agar; Gray or Red color series on oatmeal agar and glycerol-asparagine agar. Reverse side of colony is grayed yellow on oatmeal agar and glycerol-asparagine agar; grayed yellow or yellow modified by red on yeast-malt and salts-starch agar. Substrate pigment is not a pH indicator.

D-Glucose, D-xylose, D-mannitol, and D-fructose are utilized for growth. No growth or only trace of growth on L-arabinose, sucrose, rhamnose, and raffinose. Variable reports on growth with iso-inositol.

Type strain shows the highest sequence similarity to: *S. griseofuscus*, AB184206, 100%; *S. costaricanus*, AB249939, 100%; *S. phaeogriseichromatogenes*, AJ391813, 99.6%.

Source: not known.

DNA G+C content (mol%): not known.

Type strain: ATCC 19788, CBS 540.68, BCRC 12061, CECT 3309, DSM 40091, NBRC 12799, NBRC 14802, JCM 4333, JCM 4595, KCTC 9492, LMG 10475, NCIMB 12701, NRRL B-2286, NRRL-ISP 5091, PCM 2369, RIA 1067, UNIQEM 174, VKM Ac-1190.

Sequence accession no. (16S rRNA gene): AB184155.

328. **Streptomyces mutabilis** (Preobrazhenskaya and Ryabova *in* Gauze, Preobrazhenskaya, Kudrina, Blinov, Ryabova and Sveshnikova 1957) Pridham, Hesseltine and Benedict 1958, 69[AL] ("*Actinomyces mutabilis*" Preobrazhenskaya and Ryabova *in* Gauze, Preobrazhenskaya, Kudrina, Blinov, Ryabova and Sveshnikova 1957, 166)

mu.ta'bi.lis. L. masc. adj. *mutabilis* changeable, so named because the organism could not be assigned to any of the species known in the literature.

Spore chains in Section *Spirales* or *Retinaculiaperti*. Spore chains are poorly developed so that hooks, loops, or partial spirals are shorter and of smaller diameter than on typical *Retinaculiaperti* cultures. Spirals, when found, are aberrant. Mature spore chains generally contain only 3–10 spores per chain on yeast-malt agar, oatmeal agar, salts-starch agar, and glycerol-asparagine agar. Spore surface is smooth.

Color of colony: aerial mass color in the Gray or White color series on yeast-malt agar and oatmeal agar; White series on salts-starch agar and glycerol-asparagine agar. Reverse side of colony with no distinctive pigment on yeast-malt agar, oatmeal agar, salts-starch agar, or glycerol-asparagine agar; substrate pigment is not a pH indicator.

Color in medium: melanoid pigments not formed in peptone-yeast-iron agar, tyrosine agar, or tryptone-yeast broth. No pigments found in medium, except for traces of yellow or greenish yellow pigments in yeast-malt agar.

This pigment is not pH-sensitive when tested with 0.05 M NaOH or HCl.

D-Glucose, L-arabinose, D-xylose, iso-inositol, D-mannitol, D-fructose, and rhamnose are utilized for growth. Some growth occurs with sucrose as the carbon source; raffinose is not utilized.

Type strain shows the highest sequence similarity to: *S. rochei*, AB184237, 99.9%; *S. vinaceusdrappus*, AY999929, 99.9%; *S. plicatus*, AB184291, 99.9%; *S. geysiriensis*, DQ442501, 99.8%; *S. ghanaensis*, AY999851, 99.8%; *S. minutiscleroticus*, EF178696, 99.8%; *S. tuirus*, AB184690, 99.6%; *S. djakartensis*, AB184657, 99.4%; *S. violaceorubidus*, AJ781374, 99.2%; *S. flavoviridis*, AB184842, 99.1%; *S. anandii*, AB184402, 99.1%; *S. levis*, AB184670, 99.1%; *S. tendae*, D63873, 99.1%; *S. violaceus*, AB184315, 99.1%; *S. pilosus*, AB184161, 99.1%; *S. luteogriseus*, AB184379, 99.1%; *S. azureus*, EF178674, 99.1%; *S. janthinus*, AB184851, 99.1%; *S. albosporeus* subsp. *albosporeus*, AJ781327, 99.1%; *S. capillispiralis*, AB184577, 99%; *S. roseoviolaceus*, AJ399484, 99%; *S. rubrogriseus*, AB184681, 99%; *S. asterosporus*, AB184706, 99%; *S. africanus*, AY208912, 99%; *S. calvus*, AB184329, 99%.

Source: not known.

DNA G+C content (mol%): not known.

Type strain: ATCC 19789, ATCC 19919, CBS 541.68, DSM 40169, HAMBI 1069, NBRC 12800, IMET 43509, JCM 4400, NRRL-ISP 5169, RIA 1068, UNIQEM 175, VKM Ac-1851.

Sequence accession no. (16S rRNA gene): EF178679.

329. **Streptomyces mutomycini** Gause and Maximova 1986a, 575[VP] (Effective publication: Gause and Maximova *in* Gause, Preobrazhenskaya, Sveshnikova, Terekhova and Maximova 1983.)

mu.to.my.ci'ni. N.L. n. *mutomycinum* mutomycin; N.L. gen. n. *mutomycini* of mutomycin, intended to mean mutomycin producing.

Spore chains are spiral (*Spirales*); spores are spiny, spines are short. On mineral agar 1 and oatmeal agar: aerial mycelium is gray to dark gray; substrate mycelium is colorless, sometimes with gray spots; no diffusible pigment. On glycerol-nitrate agar: aerial mycelium is poor, gray; substrate mycelium and diffusible pigment are yellow-brown. On starch-ammonia agar: poor, gray aerial mycelium; colorless substrate mycelium; no diffusible pigment. On glycerol-asparagine agar: aerial mycelium and diffusible pigment are gray, poor; substrate mycelium is colorless or pinkish gray to yellow-brown. On organic agar 2: no aerial mycelium: substrate mycelium and diffusible pigment are dark yellow or olive-gray. On organic agar 79: no aerial mycelium; substrate mycelium and diffusible pigment are brown. Melanoid pigments are not formed. Moderate growth on sugar and sucrose, fructose, glucose, and mannitol; poor growth on arabinose and xylose. Antibiotic: mutomycin.

Type strain shows the highest sequence similarity to: *S. clavifer*, DQ026670, 99.8%; *S. olivoviridis*, AB184227, 99.7%; *S. atroolivaceus*, AJ781320, 99.7%; *S. finlayi*, AY999788, 99.6%; *S. acrimycini*, AY999889, 99.4%; *S. griseoplanus*, AY999894, 99.4%; *S. fimicarius*, AY999784, 99.3%; *S. flavofuscus*, AB249935, 99.3%; *S. praecox*, AB184293, 99.3%; *S. mediolani*, AB184674, 99.3%; *S. anulatus*, DQ026637, 99.3%; *S. sindenensis*, AB184759, 99.2%; *S. griseinus*, AB184205, 99.2%; *S. rubiginosohelvolus*, AB184240, 99.2%; *S. badius*, AY999783, 99.2%; *S. cavourensis* subsp. *washingtonensis*, DQ026671, 99.2%; *S. cyaneofuscatus*, AB184860, 99.2%; *S. pluricolorescens*, DQ442540, 99.2%; *S. cinereorectus*, AB184646, 99.2%; *S. californicus*, AB184755, 99.2%; *S. lipmanii*, AB184148, 99.2%; *S. lavendulae* subsp. *lavendulae*, AB184080, 99.2%; *S. baarnensis*, EF178688, 99.1%; *S. floridae*, AB184656, 99.1%; *S. albovinaceus*, AB249958, 99.1%; *S. fulvorobeus*, AB184711, 99.1%; *S. microflavus*, DQ445795, 99.1%; *S. alboviridis*, AB184256, 99.1%; *S. globisporus* subsp. *globisporus*, EF178686, 99.1%; *S. griseus* subsp. *griseus*, AY207604, 99%; *S. argenteolus*, AB045872, 99%; *S. griseolus*, AB184768, 99%; *S. candidus*, DQ026663, 99%; *S. bacillaris*, AB184439, 99%; *S. luridiscabiei*, AF361784, 99%; *S. flavovirens*, DQ026635, 99%; *S. parvus*, DQ442537, 99%.

Source: not known.

DNA G+C content (mol%): not known.

Type strain: AS 4.1747, ATCC 43689, DSM 41691, INA 4305, NBRC 100999.

Sequence accession no. (16S rRNA gene): AB249951.

330. **Streptomyces naganishii** Yamaguchi and Saburi 1955, 219[AL]

na.ga.ni'shi.i. N.L. gen. masc. n. *naganishii* of Naganishi, named for Professor H. Naganishi of the University of Hiroshima, Japan.

Spore chains in Section *Spirales*. Mature spore chains are short to moderately long, usually with more than 10 spores per chain. This morphology is best developed on yeast-malt agar. Short spore chains on salts-starch agar and glycerol-asparagine agar may form incomplete spirals or hooks. Sporulation on oatmeal agar is especially poor. Aerial hyphae may terminate in atypical club-like swellings instead of spores. Spore surface is smooth.

Color of colony: aerial mass color in the Gray or Red color series on yeast-malt agar and on carbon utilization medium plus D-glucose. The sporulated aerial growth on oatmeal agar, salts-starch agar, and glycerol-asparagine agar is usually white. Reverse side of colony is pale or grayish yellow to yellow-brown modified in spots or at margins by red (dark pink, purplish pink, or grayish reddish brown). Substrate pigment is not a pH indicator.

Color in medium: melanoid pigments are formed in peptone-yeast-iron agar and tryptone-yeast broth in 2 d, but are produced slowly or not at all in tyrosine agar. A transient red pigment is found in the medium in oatmeal agar and salts-starch agar. This pigment is not pH-sensitive when tested with 0.05 M NaOH or HCl.

D-Glucose, L-arabinose, D-xylose, iso-inositol, D-mannitol, D-fructose, and rhamnose are utilized for growth. Utilization of raffinose is doubtful. No growth or only traces of growth with sucrose.

Type strain shows the highest sequence similarity to: *Streptomyces geysiriensis*, DQ442501, 99; *Streptomyces minutiscleroticus*, EF178696, 99%.

Source: not known.

DNA G+C content (mol%): not known.

Type strain: ATCC 23939, CBS 923.68, DSM 40282, NBRC 12892, JCM 4190, JCM 4654, NRRL B-1816, NRRL-ISP 5282, RIA 1196.

Sequence accession no. (16S rRNA gene): DQ442529.

331. **Streptomyces narbonensis** Corbaz, Ettlinger, Gäumann, Keller-Schierlein, Kradolfer, Kyburz, Neipp, Prelog, Reusser and Zähner 1955, 935[AL] ["*Streptomyces narboensis*" (*sic*) Corbaz, Ettlinger, Gäumann, Keller-Schierlein, Kradolfer, Kyburz, Neipp, Prelog, Reusser and Zähner 1955, 941]

nar.bo.nen'sis. L. masc. adj. *narbonensis* of or belonging to Narbonne, France, the source of the soil from which the organism was isolated.

Spore chains in Section *Rectiflexibiles.* Mature spore chains generally have 10–50 spores per chain; longer chains are sometimes observed. This morphology is seen on yeast-malt agar, oatmeal agar, salts-starch agar, and glycerol-asparagine agar. Spore surface is smooth. Fragmentation of the substrate mycelium was noted by one observer only; unusual fragmentation of the aerial mycelium was recorded by another observer.

Color of colony: aerial mass color in the Gray color series on yeast-malt agar, oatmeal agar, salts-starch agar, and glycerol-asparagine agar. Reverse side of colony with no distinctive pigment (grayed yellow to grayed greenish yellow) on yeast-malt agar, oatmeal agar, salts-starch agar, and glycerol-asparagine agar; substrate pigment is not a pH indicator.

Color in medium: melanoid pigments formed in peptone-yeast-iron agar. Pigments other than melanoids not formed in yeast-malt agar, oatmeal agar, salts-starch agar, or glycerol-asparagine agar.

D-Glucose, L-arabinose, sucrose, D-xylose, D-fructose, rhamnose, and raffinose are utilized for growth. No growth or only traces of growth on iso-inositol and D-mannitol.

Type strain shows the highest sequence similarity to: *S. zaomyceticus*, EF178685, 99.3%; *S. exfoliatus*, AB184324, 99.3%; *S. venezuelae*, AB045890, 99.3%; *S. lateritius*, AB184324, 99.2%; *S. wedmorensis*, DQ442557, 99.2%; *S. litmocidini*, AB184149, 99.1%; *S. omiyaensis*, EF178697, 99.1%.

Source: isolated from soil from Narbonne, a small community near Cannes on the Côte d'Azur, France.

DNA G+C content (mol%): not known.

Type strain: ATCC 19790, CBS 310.55, CBS 542.68, BCRC 13651, DSM 40016, NBRC 12801, JCM 4147, JCM 4596, NRRL B-1680, NRRL-ISP 5016, RIA 1069, RIA 529, UNIQEM 176, VKM Ac-1891.

Sequence accession no. (16S rRNA gene): DQ445794.

332. **Streptomyces nashvillensis** McVeigh and Reyes 1961, 312[AL]

nash.vil.len'sis. N.L. masc. adj. *nashvillensis* of or belonging to Nashville, a city in Tennessee, the source of the soil from which the organism was isolated.

Spore chains in Section *Rectiflexibiles.* Long chains, often with more than 50 spores, are predominantly straight but a very small number of these chains may show *Retinaculiaperti* morphology including some spirals. This morphology is seen on yeast-malt agar, oatmeal agar, salts-starch agar, and glycerol-asparagine agar. Spore surface is smooth.

Color of colony: aerial mass color in the Gray color series (3fe, light brownish gray to 5fe, light grayish reddish brown or 5ih, brownish gray) on yeast-malt agar, oatmeal agar, salts-starch agar, and glycerol-asparagine agar. Reverse side of colony is strong brown to light grayish brown on yeast-malt agar; grayish yellow, light olive gray, or yellowish brown on oatmeal agar, salts-starch agar, and glycerol-asparagine agar. Reverse mycelium pigment is somewhat pH-sensitive changing from yellowish brown to reddish brown with addition of 0.05 M NaOH or from yellowish brown to yellow with addition of 0.05 M HCl.

Color in medium: melanoid pigments are formed in peptone-yeast-iron agar, tyrosine agar, or tryptone-yeast broth, but melanin reaction may be delayed or weak in tyrosine agar. Yellow (to brown) pigment is found in the medium in yeast-malt agar and oatmeal agar and usually salts-starch agar and glycerol-asparagine agar. This pigment is somewhat pH-sensitive when tested with 0.05 M NaOH or HCl, showing the same color change noted for the reverse mycelium pigment.

D-Glucose, L-arabinose, D-xylose, and raffinose are utilized for growth. Utilization of D-fructose is doubtful and there is no growth or only traces of growth with iso-inositol, D-mannitol, rhamnose, or sucrose.

Type strain shows the highest sequence similarity to: *S. tanashiensis*, AJ781362, 100%; *S. violaceorectus*, AB184314, 99.2%; *S. showdoensis*, AB184389, 99.1%; *S. cinereoruber* subsp. *cinereoruber*, AB184121, 99.1%; *S. roseoviridis*, AB184239, 99.1%; *S. racemochromogenes*, DQ026656, 99.1%; *S. viridobrunneus*, AJ781372, 99.1%; *S. roseolus*, AB184168, 99%; *S. polychromogenes*, AB184292, 99%; *S. bikiniensis*, X79851, 99%.

Source: isolated from soil from Nashville, a city in Tennessee.

DNA G+C content (mol%): not known.

Type strain: ATCC 25476, CBS 886.69, BCRC 13625, DSM 40314, NBRC 13064, JCM 4498, NRRL B-2606, NRRL-ISP 5314, RIA 1256, VKM Ac-1766.

Sequence accession no. (16S rRNA gene): AB184286.

333. **Streptomyces netropsis** (Finlay et al. 1951) Witt and Stackebrandt 1991, 456[VP] (Effective publication: Witt and Stackebrandt 1990, 370.) [*Streptoverticillium netropsis* (Finlay et al. 1951) Baldacci, Farina and Locci 1966, 161]

ne.trop'sis. Gr. n. *netron* spindle; Gr. fem. n. *opsis* aspect, appearance; N.L. masc. adj. *netropsis* spindle-like.

Spore chains in Umbellate monoverticillate (=*Streptomyces* Section Verticillati, biverticillate). Whorls may be composed of straight (BIV), flexuous, or spiral (BIV-S) spore chains. Mature spore chains generally have 10–20 spores per chain. This morphology is seen on yeast-malt agar, oatmeal agar, salts-starch agar, and glycerol-asparagine agar. Spore surface is smooth.

Color of colony: aerial mass color in the Red color series on yeast-malt agar, oatmeal agar, salts-starch agar, and glycerol-asparagine agar. Reverse side of colony with no distinctive pigments (grayish yellow to olive brown or brown) on yeast-malt agar, oatmeal agar, salts-starch agar, and glycerol-asparagine agar.

Color in medium: melanoid pigments are formed in

peptone-yeast-iron agar, but not in tyrosine agar. No pigment or only a trace of yellow or brown pigment is found in the medium in yeast-malt agar, oatmeal agar, and salts-starch agar. Pigment, if present, may be slightly pH-sensitive changing from pale yellow to pale pink when tested with 0.05 M NaOH.

D-Glucose, iso-inositol, and D-fructose are utilized for growth. No growth or only trace of growth with L-arabinose, sucrose, D-xylose, D-mannitol, rhamnose, and raffinose.

Type strain shows the highest sequence similarity to: *S. werraensis*, DQ442558, 99.3%; *S. biverticillatus*, AJ781381, 99.3%; *S. stramineus*, AB184720, 99.2%; *S. eurocidicus*, AY999790, 99.1%; *S. albireticuli*, AB184881, 99%.

Source: not known.

DNA G+C content (mol%): not known.

Type strain: ATCC 23010, CBS 921.68, BCRC 13371, CECT 3265, DSM 40259, HUT 6086, IFM 1035, NBRC 12893, NBRC 3723, JCM 4063, JCM 4655, LMG 5979, NCIMB 9592, NRRL 2268, NRRL-ISP 5259, PCM 2251, RIA 1184, RIA 605, VKM Ac-820.

Sequence accession no. (16S rRNA gene): AB184848.

Further comments: in violation of Rule 33c of the *Bacteriological Code* (1990 Revision), in Validation List no. 38, *Streptomyces netropsis* is proposed as a *nomen revictum* (basonym: "*Streptomyces netropsis*" Finlay et al. (1951).

According to Labeda (1996), *Streptomyces netropsis* (Finlay et al. 1951) Witt and Stackebrandt 1991 is an earlier heterotypic synonym of *Streptomyces distallicus* (Locci et al. 1969) Witt and Stackebrandt 1991, an earlier synonym of *Streptomyces flavopersicus* (Oliver et al. 1961) Witt and Stackebrandt 1991, and an earlier synonym of *Streptomyces kentuckensis* (Barr and Carman 1956) Witt and Stackebrandt 1991.

According to Hatano et al. (2003), *Streptomyces netropsis* (Finlay et al. 1951) Witt and Stackebrandt 1991 is an earlier heterotypic synonym of *Streptomyces distallicus* (Locci et al. 1969) Witt and Stackebrandt 1991, of *Streptomyces flavopersicus* (Oliver et al. 1961) Witt and Stackebrandt 1991, of *Streptomyces kentuckensis* (Barr and Carman 1956) Witt and Stackebrandt 1991, and of *Streptomyces syringium* (Konev 1986) Witt and Stackebrandt 1991.

334. **Streptomyces neyagawaensis** Yamamoto, Nakazawa, Horii and Miyake 1960, 286[AL]

ne.ya.ga.wa.en'sis. N.L. masc. adj. *neyagawaensis* of or belonging to Neyagawa City, Japan, near which the soil was obtained from which the organism was isolated.

Spore chains in Section *Spirales* on oatmeal agar and salts-starch agar. Short spore chains may form incomplete spirals (hooks) or flexuous to straight chains on yeast-malt agar and glycerol-asparagine agar. Spore chains are short on some media but long chains of 10–50 or more spores per chain may also be found. Spore chain morphology is often difficult to observe when spore chains and aerial hyphae coalesce in black moist (hygroscopic) masses. Spore surface is smooth. Surface of growth may become moist and black (hygroscopic) on older cultures.

Color of colony: aerial mass color in the Gray color series (3fe, light brownish gray and 5fe, light grayish reddish brown on yeast-malt agar, oatmeal agar, and salts-starch agar; 2dc, yellowish gray or 5ih, brownish gray on glycerol-asparagine agar). Reverse side of colony is dark olive on yeast-malt agar; yellowish gray to light olive gray or brown on oatmeal agar, salts-starch agar, and glycerol-asparagine agar.

Color in medium: melanoid pigments are formed in peptone-yeast-iron agar, tyrosine agar, and tryptone-yeast broth. No pigment, or a trace of gray or olive brown pigment, is found in the medium in yeast-malt agar, oatmeal agar, salts-starch agar, and glycerol-asparagine agar.

D-Glucose, L-arabinose, D-xylose, iso-inositol, D-mannitol, D-fructose, rhamnose, sucrose, and raffinose are all utilized for growth.

Type strain shows the highest sequence similarity to: *S. hygroscopicus* subsp. *ossamyceticus*, AB184560, 99.4%; *S. torulosus*, AJ781367, 99.3%.

Source: isolated from soil near Neyagawa City, Japan.

DNA G+C content (mol%): not known.

Type strain: ATCC 27449, CBS 778.72, DSM 40588, NBRC 13477, NBRC 3784, NRRL B-3092, NRRL-ISP 5588, RIA 1438, VKM Ac-1915, JCM 4796.

Sequence accession no. (16S rRNA gene): D63869.

335. **Streptomyces niger** (Thirumalachar 1955) Goodfellow, Williams and Alderson 1986a, 575[VP] (Effective publication: Goodfellow, Williams and Alderson 1986d, 59.) (*Chainia nigra* Thirumalachar 1955, 935)

ni'ger. L. masc. adj. *niger* black.

Forms extensively branched substrate and aerial mycelium. Spiral spore chains of smooth-surfaced spores are borne on the aerial mycelium. Aerial spore mass is white and the substrate mycelium is green. A green soluble pigment is produced. Sclerotia are formed in 7–14 d on agar media. Melanin pigments are not produced. Adenine, esculin, allantoin, arbutin, casein, elastin, guanine, hypoxanthine, pectin, starch, testosterone, L-tyrosine, xanthine, and xylan are degraded but chitin, lecithin, and urea are not. Nitrate is reduced but hydrogen sulfide is not produced. Adonitol, L-arabinose, cellobiose, D-fructose, D-galactose, D-glucose, *myo*-inositol, inulin, D-lactose, D-mannitol, D-mannose, melibiose, melezitose, raffinose, L-rhamnose, salicin, sucrose, trehalose, xylitol, and D-xylose are all used as sole carbon sources. Acid is formed from adonitol, *meso*-erythritol, *myo*-inositol, D-lactose, D-mannitol, melibiose, methyl α-D-glucoside, raffinose, L-rhamnose, sucrose, and D-xylose but not from dulcitol or D-sorbitol. Grows on DL-α-amino-n-butyric acid, L-arginine, L-histidine, L-phenylalanine, potassium nitrate, L-serine, L-threonine, and L-valine, but not on L-cysteine or L-methionine as sole nitrogen source. Grows at 10, 37, and 45°C but not at 4°C. Tolerant to phenol (0.1%, w/v) and sodium azide (0.02%, w/v) and sodium chloride (10%, w/v). Resistant to rifampin but not to sodium chloride (13%, w/v). Antibiotic activity shown against *Bacillus subtilis* NCIB 3610 and *Saccharomyces cerevisiae* CBS 1171[T], but not against *Aspergillus niger* LIV 131, *Escherichia coli* NCIB 9132, *Micrococcus luteus* NCIB 196, *Pseudomonas fluorescens* NCIB 9046[T], or *Streptomyces murinus* ISP 5091. The organism has a type II phospholipid pattern (*sensu* Lechevalier et al., 1977) and contains major amounts of hexahydrogenated menaquinones with nine isoprene units (Collins et al., 1984).

Type strain shows the highest sequence similarity to: *S. olivaceiscleroticus*, AJ621606, 100%; *S. monomycini*, DQ445790, 99.4%; *S. rimosus* subsp. *rimosus*, AB045883, 99.3%; *S. sclerotialus*, AJ621608, 99.2%; *S. purpurogeneiscleroticus*, AJ621604, 99.1%; *S. ochraceiscleroticus*, AB184094, 99.1%.

Source: not known.

DNA G+C content (mol%): not known.

Type strain: ATCC 17756, CBS 230.65, CBS 663.72, BCRC 11877, CMI 112787, DSM 43049, KCC A-0158, NBRC 13362, NBRC 13902, JCM 3158, NCIMB 10992, NRRL B-3857, NRRL-ISP 5302, PCM 2305, RIA 1323, VKM Ac-1736.

Sequence accession no. (16S rRNA gene): AJ621607.

336. **Streptomyces nigrescens** (Sveshnikova *in* Gauze, Preobrazhenskaya, Kudrina, Blinov, Ryabova and Sveshnikova 1957) Pridham, Hesseltine and Benedict 1958, 70[AL] ("*Actinomyces nigrescens*" Sveshnikova *in* Gauze, Preobrazhenskaya, Kudrina, Blinov, Ryabova and Sveshnikova 1957, 146)

ni.gres′cens. L. part. adj. *nigrescens* becoming black.

Spore chains in Section *Spirales*. Mature spore chains are moderately long with 10–50, or sometimes more than 50, spores per chain. This morphology is seen on yeast-malt agar, oatmeal agar, salts-starch agar, and glycerol-asparagine agar. Spore surface is smooth to warty.

Color of colony: aerial mass color in the Gray color series on yeast-malt agar, oatmeal agar, salts-starch agar, and glycerol-asparagine agar. In the original description (Gauze et al., 1957), emphasis is placed on an autolytic change of gray aerial mycelium to a black shiny mass in ageing cultures. This phenomenon was not observed by ISP cooperators. Reverse side of colony with no distinctive pigments (colorless to grayish yellow) on yeast-malt agar, oatmeal agar, salts-starch agar, and glycerol-asparagine agar.

Color in medium: melanoid pigments are not formed in peptone-yeast-iron agar, tyrosine agar, or tryptone-yeast broth. No pigment is found in the medium in yeast-malt agar, oatmeal agar, salts-starch agar, or glycerol-asparagine agar.

D-Glucose, sucrose, D-xylose, iso-inositol, D-mannitol, D-fructose, and raffinose are utilized for growth. No growth or only trace of growth with L-arabinose and rhamnose.

Type strain shows the highest sequence similarity to: *S. libani* subsp. *libani*, AB184414, 100%; *S. tubercidicus*, AJ621612, 99.9%; *S. hygroscopius* subsp. *glebosus*, AB184479, 99.6%; *S. libani* subsp. *rufus*, AJ781351, 99.6%; *S. caniferus*, AB184640, 99.5%; *S. catenulae*, AJ621613, 99.5%; *S. misakiensis*, AB217605, 99.5%; *S. sioyaensis*, DQ026654, 99.5%; *S. platensis*, AB045882, 99.5%; *S. hygroscopius* subsp. *decoyicus*, AY999883, 99.2%; *S. lydicus*, Y15507, 99%; *S. chattanoogensis*, AJ621611, 99%.

Source: not known.

DNA G+C content (mol%): not known.

Type strain: AS 4.1410, ATCC 23941, CBS 925.68, DSM 40276, NBRC 12894, INA 1800/54, JCM 4401, LMG 19332, NCIMB 9856, NRRL B-12176, NRRL-ISP 5276, RIA 1194, VKM Ac-1705.

Sequence accession no. (16S rRNA gene): DQ442530.

337. **Streptomyces nigrifaciens** Waksman 1961, 247[AL]

ni.gri.fa′ci.ens. L. adj. *niger -gra -grum* black; L. part adj. *faciens* producing; N.L. part adj. *nigrifaciens* producing black pigment.

Spore chains in Section *Rectiflexibiles*. Mature spore chains generally have 10–50 spores per chain; longer chains are sometimes observed. This morphology is seen on yeast-malt agar, oatmeal agar, and salts-starch agar. Spore surface is smooth.

Color of colony: aerial mass color in the Gray color series on yeast-malt agar, oatmeal agar, and salts-starch agar. Reverse side of colony with no distinctive pigment (grayed yellow to yellow-brown on yeast-malt agar, salts-starch agar, glycerol-asparagine agar, and grayed greenish yellow on oatmeal agar). Substrate pigment is not a pH indicator.

Color in medium: melanoid pigments not formed in peptone-yeast-iron agar and tyrosine agar. Some orange to yellow pigment found in medium in yeast-malt agar. In oatmeal agar, salts-starch agar, and glycerol-asparagine agar, yellow pigment is very faint or absent; it is not pH-sensitive.

D-Glucose, L-arabinose, D-xylose, D-mannitol, and rhamnose are utilized for growth. No growth or only trace of growth on sucrose, iso-inositol, and raffinose. Variable reports on growth with D-fructose.

For sequence similarity, see type strain of *Streptomyces flavovirens*.

Source: not known.

DNA G+C content (mol%): not known.

Type strain: ATCC 19791, CBS 543.68, BCRC 13672, DSM 40071, NBRC 12802, JCM 4223, JCM 4597, LMG 20048, NCIMB 13019, NRRL B-2094, NRRL-ISP 5071, RIA 1070, UNIQEM 177, VKM Ac-1888.

Sequence accession no. (16S rRNA gene): AB184158.

Further comments: according to Lanoot et al. (2005b), *Streptomyces nigrifaciens* Waksman 1961 is a later heterotypic synonym of *Streptomyces flavovirens* (Waksman 1923) Waksman and Henrici 1948 emend. Lanoot et al. 2005b.

338. **Streptomyces nitrosporeus** Okami 1952, 477[AL]

ni.tro.spo′re.us. N.L. n. *nitras* nitrate; Gr. n. *spora* a seed; N.L. n. *spora* a spore; N.L. masc. adj. *nitrosporeus* nitrate spored, name is based on rapid spore formation accompanied by vigorous reduction of nitrate.

Spore chains in Section *Rectiflexibiles*. Mature spore chains generally have 10–50 spores per chain. This morphology is seen on yeast-malt agar, oatmeal agar, and salts-starch agar. Spore surface is smooth.

Color of colony: aerial mass color in the Gray color series on yeast-malt agar, oatmeal agar, and salts-starch agar. Reverse side of colony with no distinctive pigment (grayed yellow to yellow-brown) on yeast-malt agar, oatmeal agar, salts-starch agar, and glycerol-asparagine agar; substrate pigment is not a pH indicator.

Color in medium: melanoid pigments not formed in peptone-yeast-iron agar and tyrosine agar. Indistinct yellow, gray, or brownish gray pigment found in medium in yeast-malt agar, oatmeal agar, salts-starch agar, and glycerol-asparagine agar; it is not pH-sensitive.

D-Glucose, L-arabinose, D-xylose, and rhamnose utilized for growth. No growth or only trace of growth on sucrose, iso-inositol, D-mannitol, and raffinose. Variable reports with D-fructose.

Type strain shows the highest sequence similarity to: *S. flavogriseus*, AJ494864, 99.7%; *S. flavovirens*, DQ026635, 99.7%; *S. mediolani*, AB184674, 99.5%; *S. rubiginosohelvolus*, AB184240, 99.5%; *S. badius*, AY999783, 99.5%; *S. cinereorectus*, AB184646, 99.5%; *S. albovinaceus*, AB249958, 99.5%; *S. praecox*, AB184293, 99.5%; *S. flavofuscus*, AB249935, 99.5%; *S. globisporus* subsp. *globisporus*, EF178686, 99.5%; *S. pluricolorescens*, DQ442540, 99.5%; *S. sindenensis*, AB184759, 99.5%; *S. griseinus*, AB184205, 99.5%; *S. fimicarius*, AY999784, 99.5%; *S. anulatus*, DQ026637, 99.5%; *S. griseolus*, AB184768, 99.5%; *S. californicus*, AB184755, 99.4%; *S. parvus*, DQ442537, 99.4%; *S. bacillaris*, AB184439, 99.4%; *S. cavourensis* subsp. *washingtonensis*, DQ026671, 99.4%; *S. argenteolus*, AB045872, 99.4%; *S. griseoplanus*, AY999894, 99.4%; *S. lavendulae* subsp. *lavendulae*, AB184080, 99.4%; *S. acrimycini*, AY999889, 99.4%; *S. halstedii*, EF178695, 99.4%; *S. alboviridis*, AB184256, 99.3%; *S. baarnensis*, EF178688, 99.3%; *S. pulveraceus*, AB184806, 99.3%; *S. fulvorobeus*, AB184711, 99.3%; *S. lipmanii*, AB184148, 99.3%; *S. microflavus*, DQ445795, 99.3%; *S. cyaneofuscatus*, AB184860, 99.3%; *S. floridae*, AB184656, 99.3%; *S. griseus* subsp. *griseus*, AY207604, 99.2%; *S. luridiscabiei*, AF361784, 99.2%; *S. albolongus*, AB184425, 99.1%; *S. celluloflavus*, AB184476, 99.1%; *S. sanglieri*, AB249945, 99.1%; *S. cavourensis* subsp. *cavourensis*, DQ445791, 99.1%; *S. olivoviridis*, AB184227, 99%; *S. griseobrunneus*, AB249912, 99%; *S. atratus*, DQ026638, 99%; *S. finlayi*, AY999788, 99%; *S. atroolivaceus*, AJ781320, 99%; *S. yanii*, AB006159, 99%; *S. gelaticus*, DQ026636, 99%.

Source: not known.

DNA G+C content (mol%): not known.

Type strain: ATCC 12769, ATCC 19792, CBS 544.68, BCRC 13645, DSM 40023, HUT 6032, NBRC 12803, NBRC 3362, JCM 4064, JCM 4598, KCTC 9761, NCIMB 9717, NRRL B-1316, NRRL-ISP 5023, RIA 1071, RIA 503, UNIQEM 178, VKM Ac-1191, VKM Ac-1202.

Sequence accession no. (16S rRNA gene): EF178680.

339. **Streptomyces niveiscabiei** Park, Kim, Kwon, Wilson, Yu, Hur and Lim 2003, 2053[VP]

ni.ve.i.sca'bi.ei. L. adj. *niveus* snow-white; L. n. *scabies* roughness, scabbiness; N.L. gen. n. *niveiscabiei* of white scab, intended to mean white and scab-causing bacteria.

Spores are white, smooth, and borne in simple rectus flexuous spore chains. Melanin is not produced on tyrosine or peptone agars. L-Arabinose, D-fructose, D-glucose, D-mannitol, raffinose, rhamnose, sucrose, D-xylose, and iso-inositol are utilized for growth. Minimum pH for growth is 3.5. Sensitive to 5, 6, and 7% (w/v) NaCl, 20 and 100 μg/ml thallium acetate, 0.5 μg/ml crystal violet, 0.1% (w/v) phenol, 25 and 100 μg/ml oleandomycin, and 20 μg/ml streptomycin, but not to 10 IU/ml penicillin.

Type strain shows no sequence similarity over 99%. Type strain shows DNA–DNA similarity to: *S. scabies* ATCC 49173[T], 16%; *S. turgidiscabies* ATCC 700248[T], 15%; *S. acidiscabies* ATCC 49003[T], 17%; *S. bottropensis* DSM 40262[T], 15%; *S. neyagawaensis* DSM 40588[T], 18%; *S. diastatochromogenes* DSM 40449[T], 19%; *S. setonii* DSM 40395[T], 15%; *S. griseus* subsp. *griseus* DSM 40236[T], 22%; *S. sampsonii* DSM 40394[T], 19%; *S. eurythermus* DSM 40014[T], 17%; *S. tendae* DSM 40101[T], 16%; *S. coelicolor* DSM 40233[T], 12%; "*S. lividans*" DSM 40434, 19%; *S. ambofaciens* DSM 40053[T], 15%; *S. luridiscabiei* LMG 21390[T], 17%; *S. puniciscabiei* LMG 21391[T], 13%.

Source: isolated from raised corky lesions on potato cv. Daeji-ma and pathogenic on potato cv. Daeji-ma.

DNA G+C content (mol%): 70.1.

Type strain: S78, KACC 20254, LMG 21392.

Sequence accession no. (16S rRNA gene): AF361786.

340. **Streptomyces niveoruber** Ettlinger, Corbaz and Hütter 1958a, 350[AL]

ni.ve.o.ru'ber. L. adj. *niveus* snow-white; L. adj. *ruber* red; N.L. masc. adj. *niveoruber* snow-white-red (referring to the white color of the aerial mycelium and red color of the vegetative mycelium).

Excellent growth on Czapek's solution agar; forms red vegetative mycelium on some media. This taxon might also be placed in the White color series. Produces cinerubins A and B.

Type strain shows the highest sequence similarity to: *S. lucensis*, DQ442522, 99.5%; *S. achromogenes* subsp. *achromogenes*, AB184109, 99%.

Source: not known.

DNA G+C content (mol%): not known.

Type strain: ATCC 14971, DSM 40638, NBRC 15428, IMET 43354, JCM 4234, NRRL B-2724.

Sequence accession no. (16S rRNA gene): DQ445796.

341. **Streptomyces niveus** Smith, Dietz, Sokolski and Savage 1956, 135[AL]

ni've.us. L. masc. adj. *niveus* snow-white, referring to the color of the aerial mycelium of the organism.

Spore chains in Section *Spirales*, with as many as 10 or more well-formed volutions of narrow diameter at the ends of long sporophores. Mature spore chains generally have 10–50 spores per chain; longer chains are sometimes observed. This morphology is seen on yeast-malt agar, oatmeal agar, salts-starch agar, and glycerol-asparagine agar. Spore surface is smooth.

Color of colony: aerial mass color in the Yellow color series on yeast-malt agar, oatmeal agar, salts-starch agar, and glycerol-asparagine agar. Reverse side of colony with no distinctive pigment on yeast-malt agar, oatmeal agar, salts-starch agar, and glycerol-asparagine agar; substrate pigment is not a pH indicator.

Color in medium: melanoid pigments are not formed in peptone-yeast-iron agar and tyrosine agar. Pigment, other than traces of yellow, are not formed in yeast-malt agar, oatmeal agar, salts-starch agar, or glycerol-asparagine agar.

D-Glucose, D-xylose, and D-fructose are utilized for growth. No growth or only trace of growth on sucrose, iso-inositol, and raffinose. Variable reports on growth with L-arabinose, D-mannitol, and rhamnose. However, reports from four cooperating laboratories indicate unusual disagreement regarding carbon utilization by this strain. The original description includes sucrose, rhamnose, and raffinose among carbon sources utilized for growth.

For sequence similarity, see type strain of *Streptomyces caeruleus*.

Source: not known.

DNA G+C content (mol%): not known.

Type strain: ATCC 19793, CBS 545.68, BCRC 11514, CCUG 11108, DSM 40088, IFM 1181, NBRC 12804, IMET 43503, JCM 4251, JCM 4599, LMG 5980, LMG 19395, NCIMB 9219, NRRL 2466, NRRL-ISP 5088, RIA 1072, UNIQEM 179.

Sequence accession no. (16S rRNA gene): AB184160.

Further comments: according to Lanoot et al. (2002), *Streptomyces niveus* (Smith et al. 1956) is a later heterotypic synonym of *Streptomyces caeruleus* (Baldacci 1944) Pridham et al. 1958 emend. Lanoot et al. 2002.

342. **Streptomyces noboritoensis** Isono, Yamashita, Tomiyama, Suzuki and Sakai 1957, 21[AL]

no.bo.ri.to.en′sis. N.L. masc. adj. *noboritoensis* of or belonging to noborito, referring to Inada-noborito, Kawasaki City, Kanagawa Prefecture, Japan, the source of the soil from which the organism was isolated.

Spore chains in Section *Rectiflexibiles.* Mature spore chains are moderately long, often with more than 50 spores per chain. This morphology is seen on yeast-malt agar, oatmeal agar, salts-starch agar, and glycerol-asparagine agar, although the aerial mycelium may be thin or poorly developed on some of these media. Spore surface is smooth.

Color of colony: aerial mass color in the Gray color series on yeast-malt agar, oatmeal agar, salts-starch agar, and glycerol-asparagine agar when mature sporulating mycelium is produced in adequate amount. Development of aerial mycelium may not be adequate for spore mass color determination on some or these media. Reverse side of colony is yellow to yellow brown modified to dark reddish brown or very dark brown on yeast-malt agar, oatmeal agar, salts-starch agar, and glycerol-asparagine agar; substrate pigment is not a pH indicator.

Color in medium: melanoid pigments are formed in peptone-yeast-iron agar and tryptone-yeast broth, but may be absent or present only in trace amount in tyrosine agar. Brown pigment is sometimes found in the medium in yeast-malt agar, oatmeal agar, and salts-starch agar or pigment may be absent from these media. The brown pigment is not pH-sensitive when tested with 0.05 M NaOH or HCl.

D-Glucose, L-arabinose, D-xylose, D-mannitol, D-fructose, and raffinose are utilized for growth. Reports vary on the utilization of iso-inositol and utilization of sucrose is doubtful. No growth or only trace of growth with rhamnose.

Type strain shows the highest sequence similarity to: *S. melanogenes*, AB184222, 100%; *S. crystallinus*, AB184652, 99.2%.

Source: isolated from soil from Inada-noborito, Kawasaki City, Kanagawa Prefecture, Japan.

DNA G+C content (mol%): not known.

Type strain: AS 4.1457, ATCC 25477, CBS 887.69, BCRC 11553, DSM 40223, NBRC 13065, JCM 4065, JCM 4557, KCTC 9060, LMG 19337, NRRL B-12152, NRRL-ISP 5223, RIA 1257, VKM Ac-1012.

Sequence accession no. (16S rRNA gene): AB184287.

343. **Streptomyces nodosus** Trejo *in* Waksman 1961, 250[AL]

no.do′sus. L. masc. adj. *nodosus* knotty.

Spore chains in Section *Spirales.* Spirals on short spore chains may be poorly developed showing only a few turns or hooks of small diameter; or tightly knotted spirals may be formed. Spore chains are short, often with only 3–10 spores per chain. This morphology is seen on yeast-malt agar, salts-starch agar, and glycerol-asparagine agar. Spore surface is smooth, but some wrinkles or folds are found on the spore surface.

Color of colony: aerial mass color in the Gray color series on yeast-malt agar, salts-starch agar, and glycerol-asparagine agar. Reverse side of colony is grayed yellow or grayed greenish yellow to dark olive brown or near black on yeast-malt agar, oatmeal agar, salts-starch agar, and glycerol-asparagine agar; substrate mycelium pigment is not a pH indicator.

Color in medium: melanoid pigments not formed in peptone-yeast-iron agar, tyrosine agar, and tryptone-yeast extract broth. Traces of red or yellow orange pigment may be found in the medium in yeast-malt agar, oatmeal agar, and glycerol-asparagine agar. One observer reports this pigment is pH-sensitive changing from red to green by addition of 0.05 M NaOH and from green to blue-red by 0.05 M HCl.

D-Glucose, D-xylose, iso-inositol, D-mannitol, D-fructose, and rhamnose are utilized for growth. No growth or only trace of growth on sucrose and raffinose. Utilization of L-arabinose is doubtful.

Type strain shows no sequence similarity over 99%.

Source: not known.

DNA G+C content (mol%): not known.

Type strain: AS 4.1459, ATCC 14899, ATCC 23942, CBS 926.68, BCRC 13768, DSM 40109, NBRC 12895, JCM 4297, JCM 4656, KCTC 9035, LMG 19340, NCIMB 12816, NRRL B-2371, NRRL-ISP 5109, RIA 1123, RIA 831, VKM Ac-1224.

Sequence accession no. (16S rRNA gene): AF114033.

344. **Streptomyces nogalater** Bhuyan and Dietz 1966, 838[AL]

no.gal.at′er. Spanish n. *nogal* walnut; L. adj. *ater* black; N.L. masc. adj. *nogalater* black walnut, referring to the production (by the organism) on most media of an odor like that of black walnuts.

Spore chains in Section *Rectiflexibiles* to *Retinaculiaperti* or possibly *Spirales.* Spore chains may be straight, strongly flexuous, or in irregular spirals. Straight to flexuous chains are usually seen on yeast-malt agar and glycerol-asparagine agar; strongly flexuous, *Retinaculiaperti* or *Spirales* spore chains may be found on oatmeal agar and salts-starch agar. Spore surface is smooth.

Color of colony: aerial mass color in the Gray color series (3fe or 4ig, light brownish gray; 4li, brownish gray; 3ig, grayish yellowish brown; or 2ih, light olive gray) on yeast-malt agar, oatmeal agar, salts-starch agar, and glycerol-asparagine agar. Reverse side of colony is orange-yellow to light yellowish pink on yeast-malt agar; light yellow to orange yellow on oatmeal agar; pale or grayish yellow on salts-starch agar and glycerol-asparagine agar. One observer reports that the reverse mycelium pigment is a pH indicator changing from orange to dark red or dark violet with the addition of 0.05 M NaOH and from orange to yellow orange with the addition of 0.05 M HCl.

Color in medium: melanoid pigments are not formed in peptone-yeast-iron agar, tyrosine agar, or tryptone-yeast broth. Red or yellow pigment may be found in the medium in yeast-malt agar, oatmeal agar, and glycerol-asparagine agar. One observer reports that this pigment is pH-sensitive, showing the same changes observed with the reverse mycelium pigments when 0.05 M NaOH or HCl is added.

D-Glucose, L-arabinose, D-xylose, D-mannitol, D-fructose, rhamnose, and raffinose are probably utilized for growth, but reports vary for L-arabinose, D-fructose, and rhamnose. No growth or only traces of growth is seen with sucrose or iso-inositol.

Type strain shows the highest sequence similarity to: *Streptomyces eurythermus*, D63870, 99.2%.

Source: not known.

DNA G+C content (mol%): not known.

Type strain: AS 4.1442, ATCC 27451, CBS 238.69, CBS 746.72, BCRC 12316, DSM 40546, HAMBI 951, NBRC 13445, IMET 43360, JCM 4553, JCM 4799, LMG 5981, LMG 19338, NCIMB 9489, NRRL 3035, NRRL-ISP 5546, RIA 1406, VKM Ac-1290.

Sequence accession no. (16S rRNA gene): AB045886.

345. **Streptomyces nojiriensis** Ishida, Kumagai, Niida, Hamamoto and Shomura 1967, 64[AL]

no.ji.ri.en'sis. N.L. masc. adj. *nojiriensis* of or belonging to Nojiri, named for Lake Nojiri at Nagano, Japan, the source of the soil from which the organism was isolated.

Poor growth on Czapek's solution agar. Produces nojirimycin, an anti-bacterial antibiotic.

Type strain shows the highest sequence similarity to: *S. xanthophaeus*, DQ442560, 100%; *S. spororaveus*, AJ781370, 100%; *S. goshikiensis*, EF178693, 99.9%; *S. cirratus*, AY999794, 99.9%; *S. vinaceus*, AB184394, 99.9%; *S. colombiensis*, DQ026646, 99.8%; *S. sporoverrucosus*, DQ442544, 99.8%; *S. cinnamonensis*, AB184707, 99.7%; *S. avidinii*, AB184395, 99.7%; *S. subrutilus*, X80825, 99.7%; *S. virginiae*, D85119, 99.5%; *S. racemochromogenes*, DQ026656, 99%; *S. polychromogenes*, AB184292, 99%.

Source: isolated from soil from Lake Nojiri at Nagano, Japan.

DNA G+C content (mol%): not known.

Type strain: ATCC 29781, DSM 41655, NBRC 13794, JCM 3382, KCTC 9784, NRRL B-16930.

Sequence accession no. (16S rRNA gene): AJ781355.

346. **Streptomyces noursei** Brown, Hazen and Mason 1953, 609[AL]

nour'sei. N.L. gen. masc. n. *noursei* of Nourse, referring to the owner of the farm where soil was obtained from which the organism was isolated.

Poor growth on Czapek's solution agar; NaCl tolerance >7%, but <10%. Exhibits anti-bacterial activity; produces nystatin and cycloheximide.

Type strain shows the highest sequence similarity to: *S. albulus*, AB024440, 100%; *S. yunnanensis*, AF346818, 99.7%.

Source: isolated from soil.

DNA G+C content (mol%): not known.

Type strain: ATCC 11455, CBS 240.57, BCRC 12044, CECT 3240, DSM 40635, NBRC 15452, JCM 4922, KCTC 1083, LMG 5982, NCIMB 8593, NRRL B-1714.

Sequence accession no. (16S rRNA gene): AB184678.

347. **Streptomyces novaecaesareae** Waksman and Henrici *in* Breed, Murray and Hitchens 1948, 951[AL]

no.va.e.ca.e.sa.re'a.e. N.L. gen. n. *novaecaesareae* intended to mean of New Jersey, USA.

Spore chain section undetermined. Aerial mycelium is not found on any of the ISP media. The original description for *Actinomyces violaceus-caeseri* Waksman and Curtis (1916) states that on Czapek agar: "Aerial mycelium produced very late; ... Conidia could not be demonstrated", and little or no aerial mycelium was reported for other media included in the original description. Spore surface undetermined. Color of colony: not determined. Reverse side of colony with distinctive grayish blue or blue green to grayish violet or red violet pigment is produced on oatmeal agar, salts-starch agar, and glycerol-asparagine agar, but the pigment is not always present on these media. Reverse mycelium is usually grayish yellow to grayish olive on yeast-malt agar. Reverse mycelium pigment is a pH indicator changing from pale violet to pale grayish blue with the addition of 0.05 M NaOH and from pale violet to reddish violet or pale grayish red with the addition of 0.05 M HCl.

Color in medium: melanoid pigments are not formed in peptone-yeast-iron agar, tyrosine agar, or tryptone-yeast broth. Blue or violet pigment is sometimes found in the medium in yeast-malt agar, oatmeal agar, salts-starch agar, and glycerol-asparagine agar; it is most consistently present in oatmeal agar. This pigment is pH-sensitive showing the same changes reported for the reverse mycelium pigment.

D-Glucose, L-arabinose, D-xylose, iso-inositol, D-mannitol, D-fructose, rhamnose, sucrose, and raffinose are all utilized for growth.

Type strain shows the highest sequence similarity to: *S. aureocirculatus*, AB184260, 99.5%; *S. pseudovenezuelae*, AB184233, 99.5%; *S. phaeoluteigriseus*, AJ391815, 99.5%; *S. galilaeus*, AB045878, 99.4%; *S. resistomycificus*, AB184166, 99.3%; *S. bobili*, AB249925, 99.3%; *S. humidus*, DQ442508, 99.1%; *S. chartreusis*, AB184839, 99.1%; *S. flavovariabilis*, EF178691, 99.1%; *S. canus*, AY999775, 99%; *S. prunicolor*, DQ026659, 99%.

Source: not known.

DNA G+C content (mol%): not known.

Type strain: AS 4.1471, ATCC 27452, CBS 134.20, CBS 669.72, DSM 40358, NBRC 13368, JCM 4800, NRRL B-1267, NRRL B-3011, NRRL-ISP 5358, RIA 1329, VKM Ac-963.

Sequence accession no. (16S rRNA gene): AB184357.

348. **Streptomyces ochraceiscleroticus** Pridham 1970, 22[AL]

o.chra.ce.i.scle.ro'ti.cus. L. n. *ochra* ochre; N.L. adj. *ochraceus* like-ochre, rust colored; N.L. n. *sclerotium* sclerotium; N.L. masc. adj. *ochraceiscleroticus* sclerotium with rust color.

Spore chains in Section *Spirales*. Mature spore chains are moderately long with 10 to 50 or more spores per chain.

This morphology is seen on yeast-malt agar, oatmeal agar, salts-starch agar, and glycerol-asparagine agar. Spore surface is smooth. Sclerotia-like granules are formed on glycerol-asparagine agar and other media.

Color of colony: aerial mass color in the Yellow or Red color series on yeast-malt agar, oatmeal agar, salts-starch agar, and glycerol-asparagine agar. White or gray aerial mycelium may also be seen on these media. Nearest matching color tab recorded for the Yellow color series is 2ba, pale yellow; nearest tabs in the Red color series are 4ie, light brown, and 3ea, light orange-yellow. Reverse side of colony is strong brown, moderate orange, or orange-yellow on yeast-malt agar and glycerol-asparagine agar; light yellow, yellow or light yellowish brown on oatmeal agar and salts-starch agar. Reverse mycelium pigment is not a pH indicator.

Color in medium: melanoid pigments are not formed in peptone-yeast-iron agar, tyrosine agar, or tryptone-yeast broth. No pigment is found in the medium in yeast-malt agar, oatmeal agar, salts-starch agar, or glycerol-asparagine agar.

D-Glucose, L-arabinose, D-xylose, iso-inositol, D-mannitol, D-fructose, rhamnose, sucrose, and raffinose are all utilized for growth.

Type strain shows the highest sequence similarity to: *S. purpurogeneiscleroticus*, AJ621604, 99.9%; S. *violens*, AJ621605, 99.8%; *S. monomycini*, DQ445790, 99.5%; *S. erumpens*, AJ621603, 99.3%; *S. sioyaensis*, DQ026654, 99.1%; *S. olivaceiscleroticus*, AJ621606, 99.1%; *S. niger*, AJ621607, 99.1%.

Source: not known.

DNA G+C content (mol%): not known.

Type strain: AS 4.1096, ATCC 15814, CBS 168.62, CBS 784.72, BCRC 13310, DSM 40594, DSM 43155, NBRC 12394, NBRC 13483, IMET 43492, JCM 3048, JCM 4801, LMG 19349, NRRL B-3041, NRRL-ISP 5594, PCM 2307, RIA 1444, RIA 710, VKM Ac-651.

Sequence accession no. (16S rRNA gene): AB184094.

Further comments: Streptomyces ochraceiscleroticus Pridham 1970 and *Chainia ochracea* Kuznetsov 1962 have the same type strain and therefore are homotypic synonyms [Rules 24a and 24b (1) of the *Bacteriological Code* (1990 Revision)].

349. **Streptomyces odorifer** (Rullman 1895) Waksman *in* Waksman and Lechevalier 1953, 79[AL] ["*Cladothrix odorifera*" Rullmann 1895, 44; "*Oospora odorifera*" Lehmann and Neumann 1896, 392; "*Actinomyces odoriferus rullmanni I*" (*sic*) Berestnev 1897, 167; "*Cladothrix odoriferus rullmanni I*" (*sic*) Berestnev 1897, 167; "*Actinomyces odorifer*" Lachner-Sandoval 1898, 65; "*Streptothrix odorifera*" Foulerton and Price-Jones 1902, 112; "*Nocardia odorifera*" Castellani and Chalmers 1913, 818]

o.do'ri.fer. L. masc. adj. *odorifer* fragrant.

Spore chains in Section *Rectiflexibiles*. Flexuous chains are common. Mature spore chains generally contain 10 to 50 or more spores per chain. This morphology is seen on yeast-malt agar, oatmeal agar, salts-starch agar, and glycerol-asparagine agar, but aerial mycelium may be poorly developed on salts-starch agar. Spore surface is smooth.

Color of colony: aerial mass color in the Yellow color series (1ba, 2ba, or 2db, pale yellow, or 1dc, pale yellow green) on yeast-malt agar, oatmeal agar, salts-starch agar, and glycerol-asparagine agar. Reverse side of colony with no distinctive pigments (strong brown to olive brown on yeast-malt agar; light yellow to greenish yellow, grayish yellow, or light brown on oatmeal agar, salts-starch agar, and glycerol-asparagine agar).

Color in medium: melanoid pigments are not formed in peptone-yeast-iron agar, tyrosine agar, or tryptone-yeast broth. Yellow pigment is found in the medium in yeast-malt agar and salts-starch agar. One observer also reports yellow pigment on oatmeal agar and glycerol-asparagine agar. This pigment is not pH-sensitive when tested with 0.05 M NaOH or HCl.

D-Glucose, L-arabinose, D-xylose, D-mannitol, D-fructose, and sucrose are utilized for growth. No growth or only traces of growth with iso-inositol, rhamnose, and raffinose.

Type strain shows the highest sequence similarity to: *S. limosus*, AB184147, 100%; *S. felleus*, AB184129, 100%; *S. daghestanicus*, DQ442497, 100%; *S. albidoflavus*, AB184255, 100%; *S. violascens*, AY999737, 100%; *S. hydrogenans*, AB184868, 100%; *S. griseus* subsp. *solvifaciens*, AB249915, 100%; *S. champavatii*, DQ026642, 100%; *S. canescens*, AB184117, 100%; *S. sampsonii*, D63871, 99.9%; *S. koyangensis*, AY079156, 99.7%.

Source: not known.

DNA G+C content (mol%): not known.

Type strain: ATCC 6246, CBS 666.72, BCRC 13704, CECT 3178, DSM 40347, NBRC 13365, IMET 41377, JCM 4198, JCM 4803, NRRL B-1328, NRRL-ISP 5347, RIA 1326, VKM Ac-748.

Sequence accession no. (16S rRNA gene): Z76682.

350. **Streptomyces olivaceiscleroticus** (Thirumalachar and Sukapure 1964) Pridham 1970, 41[AL] (*Chainia olivacea* Thirumalachar and Sukapure 1964, 160)

o.li.va.ce.i.scle.ro'ti.cus. N.L. adj. *olivaceus* olive colored; N.L. n. *sclerotium* sclerotium; N.L. masc. adj. *olivaceiscleroticus* sclerotium with olive color.

Spore chains in Section *Spirales*. Mature spore chains generally contain 10–50 spores per chain. This morphology may be seen on yeast-malt agar, oatmeal agar, salts-starch agar, and glycerol-asparagine agar, or sporulating aerial mycelium may be poorly developed on various media. Spore surface is smooth. Tight spirals may coalesce. One of three observers reports sclerotia formation.

Color of colony: aerial mass color in the Gray color series (2dc, yellowish gray; 3ge light grayish yellowish brown; 3–4li, brownish gray; or 2ge, light olive brown) on yeast-malt agar, oatmeal agar, salts-starch agar, and glycerol-asparagine agar. White or yellow (2ba, pale yellow) aerial mycelium is also reported. Reverse side of colony with no distinctive pigments (grayish yellow to olive brown) on yeast-malt agar, oatmeal agar, salts-starch agar, and glycerol-asparagine agar.

Color in medium: melanoid pigments are not formed in peptone-yeast-iron agar, tyrosine agar, or tryptone-yeast broth. Reddish pigment is sometimes found in the medium in yeast-malt agar, oatmeal agar, and salts-starch agar. Yellowish green pigment may be present in yeast-malt agar and glycerol-asparagine agar, or pigment may be absent in these media; pigments are not pH-sensitive when tested with 0.05 M NaOH or HCl.

D-Glucose, L-arabinose, D-xylose, iso-inositol, D-mannitol, D-fructose, rhamnose, sucrose, and raffinose are all utilized for growth.

Type strain shows the highest sequence similarity to: *S. niger*, AJ621607, 100%; *S. monomycini*, DQ445790, 99.4%; *S. rimosus* subsp. *rimosus*, AB045883, 99.3%; *S. sclerotialus*, AJ621608, 99.2%; *S. purpurogeneiscleroticus*, AJ621604, 99.1%; *S. ochraceiscleroticus*, AB184094, 99.1%.

Source: not known.

DNA G+C content (mol%): not known.

Type strain: ATCC 15722, CBS 296.66, CBS 785.72, BCRC 11608, DSM 40595, NBRC 13484, JCM 3045, JCM 4805, NRRL B-2318, NRRL-ISP 5595, RIA 1445.

Sequence accession no. (16S rRNA gene): AJ621606.

Further comments: Streptomyces olivaceiscleroticus Pridham 1970 and *Chainia olivacea* Thirumalachar and Sukapure 1964 have the same type strain and therefore are homotypic synonyms [Rules 24a and 24b (1) of the *Bacteriological Code* (1990 Revision)].

351. **Streptomyces olivaceoviridis** (Preobrazhenskaya and Ryabova *in* Gauze, Preobrazhenskaya, Kudrina, Blinov, Ryabova and Sveshnikova 1957) Pridham, Hesseltine and Benedict 1958, 65[AL] ("*Actinomyces olivaceoviridis*" Preobrazhenskaya and Ryabova *in* Gauze, Preobrazhenskaya, Kudrina, Blinov, Ryabova and Sveshnikova 1957, 163)

o.li.va.ce.o.vi′ri.dis. N.L. adj. *olivaceus* olive colored; L. adj. *viridis* green; N.L. masc. adj. *olivaceoviridis* olive-green colored, referring to the gray-olive colored aerial mycelium and greenish gray-brown vegetative mycelium on a chemically defined medium.

Spore chains in Section *Retinaculiaperti* or *Spirales*. Hooks, loops, and terminal spirals of only one or two turns are common; flexuous chains and primitive spirals may also be found. Mature spore chains generally contain 10 to 50 or more spores per chain. This morphology is seen on oatmeal agar, salts-starch agar, and glycerol-asparagine agar; spore chains may be short or poorly developed on yeast-malt agar. Spore surface is smooth.

Color of colony: aerial mass color usually in the Gray or Yellow color series on yeast-malt agar, oatmeal agar, salts-starch agar, and glycerol-asparagine agar. The aerial mass color is usually pale yellowish green to grayish yellow or even light grayish olive (tabs 1cb or 1½ec from the Yellow color series; 2dc or 2ge from the Gray color series or 3ec from the Red color series; 1½ge from the Green color series). Immature aerial mycelium may also appear to be in the White color series. Reverse side of colony with no distinctive pigments (pale yellow to orange-yellow on yeast-malt agar; pale yellow to grayish yellow on oatmeal agar, salts-starch agar, and glycerol-asparagine agar).

Color in medium: melanoid pigments are not formed in peptone-yeast-iron agar, tyrosine agar, or tryptone-yeast broth. No pigment, or only a trace of yellow, is found in the medium in yeast-malt agar, oatmeal agar, salts-starch agar, and glycerol-asparagine agar.

D-Glucose, L-arabinose, D-xylose, D-mannitol, D-fructose, rhamnose, sucrose, and raffinose are utilized for growth. Some growth is also reported on iso-inositol, but less than with the other carbon sources noted above.

Type strain shows the highest sequence similarity to: *S. canarius*, AB184396, 100%; *S. corchorusii*, AB184267, 99.9%; *S. capoamus*, AB045877, 99.5%; *S. bungoensis*, AB184696, 99.3%; *S. galbus*, X79852, 99.2%; *S. longwoodensis*, AB184580, 99.2%; *S. cellostaticus*, AB184192, 99%; *S. griseochromogenes*, AB184387, 99%.

Source: not known.

DNA G+C content (mol%): not known.

Type strain: AS 4.1430, ATCC 23630, ATCC 25478, CBS 888.69, DSM 40334, NBRC 13066, IMET 43128, INA 11584, JCM 4499, KCTC 9132, LMG 19324, NCIMB 9982, NRRL B-12280, NRRL-ISP 5334, RIA 1258, VKM Ac-1852.

Sequence accession no. (16S rRNA gene): AB184288.

352. **Streptomyces olivaceus** (Waksman 1923) Waksman and Henrici *in* Breed, Murray and Hitchens 1948, 950[AL] ("*Actinomyces olivaceus*" Waksman 1923, 354)

o.li.va′ce.us. N.L. masc. adj. *olivaceus* olive colored, apparently referring to the color of vegetative mycelium.

Spore chains in Section *Spirales*, with open spirals intergrading through flexuous spore chains suggestive of Section *Rectiflexibiles*. Mature spore chains are generally long, often with more than 50 spores per chain. This morphology is seen on yeast-malt agar, oatmeal agar, salts-starch agar, and glycerol-asparagine agar. Spore surface is smooth.

Color of colony: aerial mass color in the Gray color series on yeast-malt agar, oatmeal agar, salts-starch agar, and glycerol-asparagine agar. Reverse side of colony is usually grayed yellow on yeast-malt agar, oatmeal agar, salts-starch agar, and glycerol-asparagine agar. Production of dark reverse color is variable on these media; substrate pigment is not a pH indicator.

Color in medium: melanoid pigments not formed in peptone-yeast-iron agar and tyrosine agar. No pigment found in medium in yeast-malt agar, oatmeal agar, salts-starch agar, or glycerol-asparagine agar.

D-Glucose, L-arabinose, D-xylose, iso-inositol, D-mannitol, D-fructose, and rhamnose are utilized for growth. No growth or only trace of growth on sucrose and raffinose.

Type strain shows the highest sequence similarity to: *S. pactum*, AB184398, 100%; *S. parvulus*, AB184326, 99.3%; *S. coeruleorubidus*, AY999719, 99.2%; *S. rubrogriseus*, AB184681, 99.2%; *S. lienomycini*, AJ781353, 99.2%; *S. anthocyanicus*, AB184631, 99.1%; *S. tricolor*, AB184687, 99%; *S. malachitospinus*, AB249954, 99%.

Source: not known.

DNA G+C content (mol%): not known.

Type strain: ATCC 19794, ATCC 3335, CBS 546.68, CCM 3188, BCRC 11485, CCUG 11111, DSM 40072, NBRC 12805, INA 3200, JCM 4402, NRRL B-1224, NRRL B-3009, NRRL-ISP 5072, RIA 1073, RIA 481, UNIQEM 180, VKM Ac-254.

Sequence accession no. (16S rRNA gene): AB184743.

353. **Streptomyces olivochromogenes** (Waksman 1923) Waksman and Henrici *in* Breed, Murray and Hitchens 1948, 941[AL] ["*Actinomyces olivochromogenus*" (*sic*) Waksman 1923; "*Streptomyces olivochromogenus*" (*sic*) Waksman and Henrici 1948]

o.li.vo.chro.mo′ge.nes. L. n. *oliva* olive; Gr. n. *chroma* color; N.L. suff. *-genes* (from Gr. v. *gennaô* to produce) producing; N.L. adj. *olivochromogenes* producing an olive color.

Spore chains in Section *Spirales*, but sporulation aerial mycelium is poorly developed and spore chains are generally short (3–10 spores per chain) so that flexuous chains, hooks, loops, and imperfect or irregular spirals are common. This morphology is seen on yeast-malt agar, oatmeal agar, salts-starch agar, glycerol-asparagine agar, Czapek's sucrose agar, and soil agar. The abundant cottony aerial mycelium mentioned in the original description (op. cit) is not seen on ISP media. Spore surface is smooth.

Color of colony: sporulation on yeast-malt agar, oatmeal agar, salts-starch agar and glycerol-asparagine agar is inadequate for accurate spore mass color determination. Reverse side of colony with no distinctive pigments (nearly colorless or grayish yellow to grayish greenish yellow or moderate yellow) on yeast-malt agar, oatmeal agar, salts-starch agar, and glycerol-asparagine agar.

Color in medium: melanoid pigments are formed in peptone-yeast-iron agar, and tryptone-yeast broth; a faint brown pigment may be also produced in tyrosine agar in 4 d. A trace of yellow or greenish pigment may or may not be found in the medium in yeast-malt agar, oatmeal agar, salts-starch agar, and glycerol-asparagine agar.

D-Glucose, L-arabinose, D-xylose, iso-inositol, D-mannitol, D-fructose, rhamnose, and sucrose are utilized for growth. No growth or only trace of growth with raffinose.

Type strain shows the highest sequence similarity to: *S. mirabilis*, AB184412, 99.4%.

Source: not known.

DNA G+C content (mol%): not known.

Type strain: ATCC 25479, ATCC 3336, CBS 889.69, DSM 40451, NBRC 13067, NBRC 3178, IMET 40352, JCM 4163, JCM 4500, KCTC 9064, NRRL B-1341, NRRL-ISP 5451, RIA 1259.

Sequence accession no. (16S rRNA gene): AY094370.

354. **Streptomyces olivomycini** (Gause and Sveshnikova 1986b) Witt and Stackebrandt 1991, 456[VP] (Effective publication: Witt and Stackebrandt 1990, 370.) (*Streptoverticillium olivomycini* Gause and Sveshnikova (Gause and Sveshnikova 1986b)

o.li.vo.my.ci′ni. N.L. n. *olivomycinum* olivomycin; N.L. gen. n. *olivomycini* of olivomycin, intended to mean olivomycin producing.

Spore chains are straight, spores are smooth. On mineral agar 1: aerial mycelium is creamy, pale lilac with a brownish note; substrate mycelium is brownish-greenish, brownish-olive, grayish-greenish; no diffusible pigment. On starch-ammonia agar (ISP 4): aerial mycelium is creamy, poor; substrate mycelium is brownish-gray, weak growth; no diffusible pigment. On glycerol-nitrate agar: aerial mycelium is creamy; substrate mycelium is brownish-greenish, olive green; diffusible pigment is greenish, weak. On glycerol-asparagine agar (ISP 5): aerial mycelium is pale lilac; substrate mycelium is brown-gray greenish; no diffusible pigment. On oatmeal agar: aerial mycelium is creamy, pale lilac; substrate mycelium is olive to brownish; diffusible pigment is light greenish. On organic agar 2: aerial mycelium yellowish to light pink; substrate mycelium is yellowish-brownish to greenish; no diffusible pigment.

Melanoid pigments are not formed. Carbon utilization: weak digestion of sucrose, arabinose, rhamnose, xylose, fructose, raffinose, and mannitol. Antibiotic: olivomycin.

Source: not known.

DNA G+C content (mol%): not known.

Type strain: INA 16749.

Sequence accession no. (16S rRNA gene): no sequence available.

355. **Streptomyces olivoreticuli** (Arai, Nakada and Suzuki 1957) Witt and Stackebrandt 1996, 456[VP] (Effective publication: Witt and Stackebrandt 1990, 370.) ("*Streptomyces olivoreticuli*" Arai, Nakada and Suzuki 1957, 441; *Streptoverticillium olivoreticuli* Baldacci, Farina and Locci 1966, 162)

o.li.vo.re.ti′cu.li. L. n. *oliva* olive; L. n. *reticulum* small net; N.L. gen. n. *olivoreticuli* of olive-colored net, itended to mean having the appearance of olive-colored nets.

Spore chains in Umbellate monoverticillate (=*Streptomyces* Section Verticillati, biverticillate). Mature spore chains are generally short with 3–10 spores per chain. This morphology may be observed on Pridham and Gottlieb carbon utilization medium with glucose or inositol, but good sporulation is usually not seen on yeast-malt agar, oatmeal agar, salts-starch agar, or glycerol-asparagine agar. Spore surface is smooth.

Color of colony: aerial mass color in the White color series on media supporting sporulating aerial growth. In the absence of sporulation, the surface is pale yellow on yeast-malt agar, oatmeal agar, and glycerol-asparagine agar and white on salts-starch agar. Reverse side of colony with no distinctive pigments (colorless to grayish yellowish brown) on yeast-malt agar, oatmeal agar, salts-starch agar, and glycerol-asparagine agar.

Color in medium: melanoid pigments are formed in peptone-yeast-iron agar and tryptone-yeast broth, but not in tyrosine agar. No pigment found in medium in yeast-malt agar, oatmeal agar, salts-starch agar, or glycerol-asparagine agar. Carbon utilization appears to be unusually variable. Reports from three observers differ on utilization of each carbon source tested and results vary on duplicate dishes in the same laboratory. The original description (op. cit.) directs attention to fluctuation of carbon utilization by ultra-violet irridiated strains. This fluctuation appears also to be a characteristic of the type strain.

For sequence similarity, see type strain of *Streptomyces abikoensis.*

Source: not known.

DNA G+C content (mol%): not known.

Type strain: ATCC 23943, CBS 927.68, BCRC 13765, DSM 40105, IFM 1018, NBRC 12896, IMET 43690, JCM 4176, JCM 4657, NRRL B-2091, NRRL-ISP 5105, RIA 1122, VKM Ac-839.

Sequence accession no. (16S rRNA gene): AB184853.

Further comments: in violation of Rule 33c of the *Bacteriological Code* (1990 Revision), in Validation List no. 38, *Streptomyces olivoreticuli* is proposed as a *nomen revictum* (basonym: "*Streptomyces olivoreticuli*" Arai et al. (1957).

Witt and Stackebrandt proposed to transfer *Streptoverticillium olivoreticuli* corrig. (Arai et al. 1957) Baldacci et al. (1966) to the genus *Streptomyces* as *Streptomyces olivoreticuli* (Arai et al. 1957) Witt and Stackebrandt 1991. However, Validation List no. 38 does not include formal propositions about *Streptoverticillium olivoreticuli* subsp. *cellulophilum* corrig. Locci and Schofield 1989 and *Streptoverticillium olivoreticuli* subsp. *olivoreticuli* corrig. (Arai et al. 1957) Baldacci et al. 1966.

According to Hatano et al. (2003), *Streptomyces olivoreticuli* (Arai et al. 1957) Witt and Stackebrandt 1991 is a later heterotypic synonym of *Streptomyces abikoensis* (Umezawa et al. 1951) Witt and Stackebrandt 1991.

356. **Streptomyces olivoverticillatus** (Shinobu 1956) Witt and Stackebrandt 1991, 456[VP] (Effective publication: Witt and Stackebrandt 1990, 370.) ("*Streptomyces olivoverticillatus*" Shinobu 1956, 92; "*Verticillomyces olivoverticillatus*" Shinobu 1965; *Streptoverticillium olivoverticillatum* Baldacci, Farina and Locci 1966, 163)

o.li.vo.ver.ti.cil.la'tus. L. n. *oliva* olive; L. n. *verticillus* a whorl; N.L. adj. *verticillatus* whorled; N.L. masc. adj. *olivoverticillatus* olive, whorled.

Spore chain Section not determined. The original description of Shinobu ((1956), op. cit) states: "The formation of aerial mycelium of this strain was generally not good, even on a suitable media for growth". ISP observers found some aerial mycelium with straight, flexuous, and abortive spiral spore chains on oatmeal agar. Electron microscope preparations from yeast-malt agar, oatmeal agar, and salts-starch agar contained some flexuous to spiral chains of 13 to 20 or more spores. "Whirls" as described by Shinobu ((1956), op. cit) were not found by ISP observers. Spore surface is spiny.

Color of colony: aerial mass color probably in the Yellow or Gray color series (1db, pale yellow green; 3fe light brownish gray) when mature spores occur on yeast-malt agar, oatmeal agar, and salts-starch agar. Thin or immature aerial mycelium is white. Reverse side of colony with no distinctive pigments (nearly colorless, yellow, greenish yellow or olive brown) on yeast-malt agar, oatmeal agar, salts-starch agar, and glycerol-asparagine agar.

Color in medium: melanoid pigments are not formed in peptone-yeast-iron agar, tyrosine agar, or tryptone-yeast broth. Brown or yellow-brown pigment may be found in the medium in yeast-malt agar, oatmeal agar, and salts-starch agar or the medium may remain unpigmented.

D-Glucose and iso-inositol are utilized for growth. Utilization of L-arabinose, D-xylose, D-mannitol, D-fructose, rhamnose, sucrose, and raffinose is doubtful.

Type strain shows the highest sequence similarity to: *S. thioluteus*, AB184753, 99.1%; *S. abikoensis*, AB184537, 99%.

Source: not known.

DNA G+C content (mol%): not known.

Type strain: ATCC 25480, CBS 890.69, BCRC 13610, CECT 3266, DSM 40250, NBRC 13068, NBRC 15273, NBRC 3842, NBRC 3929, JCM 4100, JCM 4501, NCIMB 97148, NRRL B-1994, NRRL-ISP 5250, RIA 1260, RIA 551, VKM Ac-890.

Sequence accession no. (16S rRNA gene): AB184636.

Further comments: in violation of Rule 33c of the *Bacteriological Code* (1990 Revision), in Validation List no. 38, *Streptomyces olivoverticillatus* is proposed as a *nomen revictum* (basonym: "*Streptomyces olivoverticillatus*" Shinobu 1956).

According to Hatano et al. (2003), *Streptomyces olivoverticillatus* (Shinobu 1956) Witt and Stackebrandt 1991 is an earlier heterotypic synonym of *Streptomyces viridiflavus* corrig. (Locci and Schofield 1989) Witt and Stackebrandt 1991.

357. **Streptomyces olivoviridis** (Kuchaeva, Krasil'nikov, Skryabin and Taptykova *in* Rautenshtein 1960) Pridham 1970, 23[AL] ["*Actinomyces olivoviridis*" Kuchaeva, Krasil'nikov, Skryabin and Taptykova *in* Rautenshtein 1960; "*Actinomyces olivovirilis*" (*sic*) Kuchaeva, Krasil'nikov, Skryabin and Taptykova *in* Rautenshtein 1960]

o.li.vo.vi'ri.dis. L. n. *oliva* olive; L. adj. *viridis* green; N.L. masc. adj. *olivoviridis* olive-green.

Spore chains in Section *Rectiflexibiles*. The predominantly flexuous or wavy spore chains may also form occasional hooks and open spirals (rare) resembling *Retinaculiaperti* or Spiral morphology. Mature spore chains are generally long with 10–50 spores per chain. This morphology is seen on yeast-malt agar, oatmeal agar, salts-starch agar, and glycerol-asparagine agar. Spore surface is spiny.

Color of colony: aerial mass color in the Gray color series on yeast-malt agar, oatmeal agar, and salts-starch agar; aerial mycelium is poorly developed or white on glycerol-asparagine agar. Reverse side of colony is yellow to yellow brown is usually modified by green (olive brown to yellow green or grayish green) on yeast-malt agar, oatmeal agar, and salts-starch agar; no distinctive substrate mycelium pigment on glycerol-asparagine agar. Substrate mycelium pigment is not a pH indicator.

Color in medium: melanoid pigments are not formed in peptone-yeast-iron agar or tyrosine agar. Some coloration may appear in medium in tryptone-yeast broth. Yellow or green pigment is found in the medium in yeast-malt agar, oatmeal agar, and salts-starch agar. This pigment is not pH-sensitive when tested with 0.05 M NaOH or HCl.

D-Glucose, L-arabinose, D-xylose, D-mannitol, D-fructose, and rhamnose are utilized for growth. No growth or only trace of growth with sucrose, iso-inositol, and raffinose.

Type strain shows the highest sequence similarity to: *S. clavifer*, DQ026670, 100%; *S. atroolivaceus*, AJ781320, 100%; *S. mutomycini*, AB249951, 99.7%; *S. finlayi*, AY999788, 99.7%; *S. sindenensis*, AB184759, 99.5%; *S. fimicarius*, AY999784, 99.5%; *S. pluricolorescens*, DQ442540, 99.5%; *S. griseoplanus*, AY999894, 99.5%; *S. griseinus*, AB184205, 99.5%; *S. lavendulae* subsp. *lavendulae*, AB184080, 99.5%; *S. mediolani*, AB184674, 99.5%; *S. cavourensis* subsp. *washingtonensis*, DQ026671, 99.5%; *S. flavofuscus*, AB249935, 99.5%; *S. badius*, AY999783, 99.5%; *S. anulatus*, DQ026637, 99.5%; *S. rubiginosohelvolus*, AB184240, 99.5%; *S. praecox*, AB184293, 99.5%; *S. albovinaceus*, AB249958, 99.4%; *S. cinereorectus*, AB184646, 99.4%; *S. lipmanii*, AB184148, 99.4%; *S. baarnensis*, EF178688, 99.4%; *S. fulvorobeus*,

AB184711, 99.4%; *S. acrimycini*, AY999889, 99.4%; *S. microflavus*, DQ445795, 99.4%; *S. alboviridis*, AB184256, 99.4%; *S. cyaneofuscatus*, AB184860, 99.4%; *S. globisporus* subsp. *globisporus*, EF178686, 99.4%; *S. parvus*, DQ442537, 99.3%; *S. griseolus*, AB184768, 99.3%; *S. argenteolus*, AB045872, 99.3%; *S. luridiscabiei*, AF361784, 99.3%; *S. floridae*, AB184656, 99.3%; *S. californicus*, AB184755, 99.3%; *S. flavovirens*, DQ026635, 99.3%; *S. flavogriseus*, AJ494864, 99.2%; *S. halstedii*, EF178695, 99.2%; *S. griseus* subsp. *griseus*, AY207604, 99.2%; *S. candidus*, DQ026663, 99.1%; *S. pulveraceus*, AB184806, 99%; *S. nitrosporeus*, EF178680, 99%; *S. bacillaris*, AB184439, 99%.

Source: not known.

DNA G+C content (mol%): not known.

Type strain: ATCC 15882, ATCC 23944, CBS 928.68, DSM 40211, NBRC 12897, INMI 1475, JCM 4432, JCM 4658, NRRL B-3374, NRRL-ISP 5211, RIA 1157, RIA 661, VKM Ac-259.

Sequence accession no. (16S rRNA gene): AB184227.

358. **Streptomyces omiyaensis** Umezawa and Okami *in* Umezawa, Tazaki, Okami and Fukuyama 1950, 293^{AL}

o.mi.ya.en'sis. N.L. masc. adj. *omiyaensis* of or belonging to Omiya City near Tokyo, Japan, the source of the soil from which the organism was isolated.

Spore chains in Section *Rectiflexibiles.* Mature spore chains are generally long with 10 to 50 or more spores per chain. This morphology is seen on yeast-malt agar, oatmeal agar, salts-starch agar, and glycerol-asparagine agar. Spore surface is smooth.

Color of colony: aerial mass color in the Gray color series (d. light gray, or 2dc, yellowish gray) on yeast-malt agar, oatmeal agar, salts-starch agar, and glycerol-asparagine agar. Reverse side of colony with no distinctive pigments (light yellow, pale yellow, or grayish yellow) on yeast-malt agar, oatmeal agar, salts-starch agar, and glycerol-asparagine agar.

Color in medium: melanoid pigments are not formed in peptone-yeast-iron agar, tyrosine agar or tryptone-yeast broth. No pigment is found in the medium in yeast-malt agar, oatmeal agar, salts-starch agar or glycerol-asparagine agar.

D-Glucose, D-xylose, and rhamnose are utilized for growth. Only traces of growth are seen on L-arabinose or D-fructose; iso-inositol, D-mannitol, raffinose, and sucrose are not utilized.

Type strain shows the highest sequence similarity to: *S. exfoliatus*, AB184324, 99.6%; *S. wedmorensis*, DQ442557, 99.6%; *S. venezuelae*, AB045890, 99.6%; *S. zaomyceticus*, EF178685, 99.6%; *S. lateritius*, AB184324, 99.5%; *S. litmocidini*, AB184149, 99.5%; *S. yereyanensis* EF178684, 99.3%; *S. narbonensis*, DQ445794, 99.1%.

Source: isolated from soil from Omiya City near Tokyo, Japan.

DNA G+C content (mol%): not known.

Type strain: ATCC 27454, CBS 750.72, BCRC 11897, DSM 40552, NBRC 13449, IMET 43362, JCM 4806, NRRL B-1587, NRRL-ISP 5552, RIA 1410, VKM Ac-1903.

Sequence accession no. (16S rRNA gene): EF178697.

359. **Streptomyces orinoci** (Cassinelli, Grein, Orezzi, Pennella and Sanfilippo 1967) Witt and Stackebrandt 1991, 456^{VP} (Effective publication: Witt and Stackebrandt 1990, 370.) (*Streptoverticillium orinoci* Cassinelli, Grein, Orezzi, Pennella and Sanfilippo 1967, 367)

o.ri.no'ci. N.L. gen. n. *orinoci* of Orinoco, named after the Orinoco River, South America, from whose banks it was isolated.

Spore chains in Section Umbellate Monoverticillate (=*Streptomyces* Section Verticillati, biverticillate). Mature spore chains generally contain more than 10 spores per chain. This morphology is usually seen on ISP yeast-malt agar, oatmeal agar, salts-starch agar, and glycerol-asparagine agar, but sporulating aerial mycelium may be observed in the scant growth on carbon-utilization medium enriched with raffinose, D-fructose, rhamnose, or iso-inositol or on half-strength Emerson potato-glucose agar. Spore surface is smooth.

Color of colony: sporulating aerial mycelium is not produced on yeast-malt agar, oatmeal agar, salts-starch agar, or glycerol-asparagine agar. Aerial mass color on carbon-utilization medium plus fructose is in the Gray color series. Reverse side of colony is strong yellow or deep yellow on yeast-malt agar, glycerol-asparagine agar, and half-strength Emerson potato-glucose agar. Colorless to pale or grayish yellow on salts-starch agar and glycerol-asparagine agar.

Color in medium: melanoid pigments are not formed in peptone-yeast-iron agar, tyrosine agar, or tryptone-yeast broth. No pigment is found in the medium in yeast-malt agar, oatmeal agar, salts-starch agar, or glycerol-asparagine agar.

D-Glucose is utilized for growth. No growth or only traces of growth with L-arabinose, D-xylose, iso-inositol, D-mannitol, D-fructose, rhamnose, sucrose, and raffinose.

Type strain shows no sequence similarity over 99%.

Source: isolated from the banks of the Orinoco River, South America.

DNA G+C content (mol%): not known.

Type strain: ATCC 23202, CBS 767.72, CECT 3267, DSM 40571, IFM 1226, NBRC 13466, JCM 4546, JCM 4807, NRRL B-3379, NRRL-ISP 5571, RIA 1427, VKM Ac-929.

Sequence accession no. (16S rRNA gene): AB184866.

360. **Streptomyces pactum** (*sic*) Bhuyan, Dietz and Smith 1962, 185^{AL}

pac'tum. L. adj. *pactum* settled (referring to the compactness of the coiled chains of spores).

Spore chains in Section *Spirales.* Mature spore chains contain 10–50 spores per chain. This morphology is seen on yeast-malt agar, oatmeal agar, salts-starch agar, and glycerol-asparagine agar, but sporulation is usually poor on glycerol-asparagine agar. Spore surface is hairy.

Color of colony: aerial mass color in the Gray color series (d, light gray, or e, medium gray) on oatmeal agar and salts-starch agar; Gray or Blue color series on oatmeal agar. Nearest matching color tab in the Blue color series is 19fe, pale blue. On glycerol-asparagine agar, aerial mycelium is in the White color series. Reverse side of colony with no distinctive pigments (light yellowish brown, grayish yellow or grayish greenish yellow) on yeast-malt agar, oatmeal agar, salts-starch agar, and glycerol-asparagine agar.

Color in medium: melanoid pigments are not formed in peptone-yeast-iron agar, tyrosine agar, or tryptone-yeast broth. Yellow pigment may be found in the medium in yeast-malt agar, oatmeal agar, salts-starch agar, and glycerol-asparagine agar. This pigment is not pH-sensitive when tested with 0.05 M NaOH or HCl.

D-Glucose is utilized for growth and only traces of growth are reported with D-fructose. No growth with L-arabinose, D-xylose, iso-inositol, D-mannitol, rhamnose, sucrose, or raffinose.

Type strain shows the highest sequence similarity to: *S. olivaceus*, AB184743, 100%; *S. parvulus*, AB184326, 99.4%; *S. coeruleorubidus*, AY999719, 99.2%; *S. rubrogriseus*, AB184681, 99.2%; *S. lienomycini*, AJ781353, 99.2%; *S. malachitospinus*, AB249954, 99.2%; *S. anthocyanicus*, AB184631, 99.1%; *S. tricolor*, AB184687, 99.1%; *S. coelescens*, AF503496, 99%; *S. violaceoruber*, AF503492, 99%; *S. violaceorubidus*, AJ781374, 99%.

Source: not known.

DNA G+C content (mol%): not known.

Type strain: AS 4.1443, ATCC 27456, CBS 461.69, CBS 734.72, BCRC 12076, DSM 40530, NBRC 13433, IMET 43357, JCM 4288, JCM 4809, KCTC 9165, LMG 19357, NCIMB 9445, NRRL 2939, NRRL-ISP 5530, RIA 1394, VKM Ac-1911.

Sequence accession no. (16S rRNA gene): AB184398.

361. **Streptomyces paradoxus** (Krasil'nikov and Yuan 1961) Goodfellow, Williams and Alderson 1986a, 575[VP] (Effective publication: Goodfellow, Williams and Alderson 1986e, 62.) (*Actinosporangium violaceum* Krasil'nikov and Yuan 1961, 115)

pa.ra'do.xus. L. masc. adj. *paradoxus* contrary to all expectation.

Spore chains in Section *Retinaculiaperti*. Spore surface is smooth. Forms extensively branched substrate and aerial mycelium. The aerial spore mass is gray; sclerotia are formed. Produces melanin pigments. Allantoin, adenine, esculin, arbutin, gelatin, guanine, hypoxanthine, testosterone, tyrosine, urea, and xanthine are degraded, but chitin, elastin, lecithin, pectin, and starch are not. Hydrogen sulfide is produced but nitrate is not reduced. L-Arabinose, cellobiose, D-fructose, D-galactose, D-glucose, *myo*-inositol, inulin, D-lactose, mannitol, D-mannose, melezitose, melibiose, raffinose, L-rhamnose, sucrose, trehalose, and D-xylose are used as sole carbon sources but adonitol and xylitol are not. Grows on L-arginine, L-cysteine, L-histidine, L-hydroxyproline, L-methionine, potassium nitrate, L-serine, L-threonine, and L-valine, but not on DL-amino-n-butyric acid or L-phenylalanine, as sole nitrogen sources. Grows at 10 and 37°C, but not at 4 or 45°C.Tolerant to phenol (0.1%, w/v) and sodium chloride (4%, w/v), but not sodium azide (0.01%, w/v). Sensitive to rifampin and inhibited by sodium chloride at 7% (w/v). Antimicrobial activity not shown against *Aspergillus niger* LIV 131, *Bacillus subtilis* NCIB 3610, *Candida albicans* CBS 562, *Escherichia coli* NCIB 9132, *Micrococcus luteus* NCIB 196, *Pseudomonas fluorescens* NCIB 9046[T], *Saccharomyces cerevisiae* CBS 171, or *Streptomyces murinus* ISP 5091. The peptidoglycan contains LL-A_2pm as the major diamino acid and is of the A3γ type (Stackebrandt et al., 1981).

Type strain shows the highest sequence similarity to: *S. matensis*, AB184221, 99.6%; *S. collinus*, AB184123, 99.5%; *S. viridochromogenes*, DQ442555, 99.4%; *S. azureus*, EF178674, 99.3%; *S. griseoflavus*, AJ781322, 99.3%; *S. griseorubens*, AB184139, 99.3%; *S. bellus*, AB184849, 99.2%; *S. ambofaciens*, M27245, 99.2%; *S. glaucescens*, AB184843, 99.2%; *S. albaduncus*, AY999757, 99.2%; *S. erythrogriseus*, AJ781328, 99.2%; *S. labedae*, AB184704, 99.2%; *S. griseoincarnatus*, AJ781328, 99.2%; *S. variabilis*, DQ442551, 99.2%; *S. coerulescens*, AY999720, 99.2%; *S. violaceorubidus*, AJ781374, 99.1%; *S. flaveolus*, AB184764, 99.1%; *S. griseoloalbus*, AB184275, 99.1%; *S. capillispiralis*, AB184577, 99.1%; *S. althioticus*, AY999808, 99.1%; *S. pseudogriseolus*, DQ442541, 99%; *S. malachitofuscus*, AB184282, 99%; *S. levis*, AB184670, 99%.

Source: isolated from soil.

DNA G+C content (mol%): 70.4–71.6.

Type strain: ATCC 15813, BCRC 12521, DSM 43350, IFM 1160, NBRC 14887, IMET 43491, INMI 3180, JCM 3052, KCC A-0052, KCTC 9118, NRRL B-3457, NRRL B-3483, PCM 2310, RIA 655, VKM Ac-645.

Sequence accession no. (16S rRNA gene): AB184628.

362. **Streptomyces parvisporogenes** (Locci, Baldacci and Petrolini Baldan 1969) Witt and Stackebrandt 1991, 456[VP] (Effective publication: Witt and Stackebrandt 1990, 370.) (*Streptoverticillium parvisporogenes* Locci, Baldacci and Petrolini Baldan 1969, 37)

par.vi.spo.ro'ge.nes. L. adj. *parvus* little; Gr. n. *spora* a seed and in biology a spore; N.L. suff. *-genes* (from Gr. v. *gennaô* to produce) producing; N.L. part. adj. *parvisporogenes* producing a little number of spores.

Spore chains in Section Verticillate, umbellate monoverticillate (biverticillate). This morphology is seen on yeast-malt agar and oatmeal agar; a sporulating aerial mycelium is usually not produced on salts-starch agar and may also be absent on glycerol-asparagine agar. Spore surface is smooth or with some surface irregularities.

Color of colony: aerial mass color in the Red, White or Yellow color series on yeast-malt agar, oatmeal agar, salts-starch agar, and glycerol-asparagine agar. The yellow color (1ba, 2ba, or 2db, pale yellow) is representative of aerial color in the absence of sporulation. Aerial mass color should be determined on a sporulating surface. Mature (21 d) sporulating aerial mycelium on oatmeal agar and on carbon-utilization medium plus fructose is in the Red color series (3ca, pale orange yellow). Reverse side of colony with no distinctive pigments (pale or grayish yellow on oatmeal agar, salts-starch agar, and glycerol-asparagine agar; yellowish brown or olive brown on yeast-malt agar).

Color in medium: melanoid pigments are formed in peptone-yeast-iron agar and tryptone-yeast broth; weakly or not at all in tyrosine agar. No pigment, or only a trace of yellow, is found in the medium in yeast-malt agar, oatmeal agar, salts-starch agar, and glycerol-asparagine agar.

D-Glucose and iso-inositol are utilized for growth. Reports vary on utilization of D-mannitol, and utilization of D-fructose is doubtful. Only traces of growth are seen on L-arabinose, D-xylose, rhamnose, sucrose, and raffinose.

For sequence similarity, see type strain of *Streptomyces abikoensis*.

Source: not known.

DNA G+C content (mol%): not known.

Type strain: ATCC 12568, CBS 695.72, DSM 40473, NBRC 13907, JCM 4694, JCM 4812, NRRL B-12386, NRRL B-5464, NRRL-ISP 5473, RIA 1355, VKM Ac-878.

Sequence accession no. (16S rRNA gene): AB249913.

Further comments: in violation of Rule 33c of the *Bacteriological Code* (1990 Revision), in Validation List no. 38, *Streptomyces parvisporogenes* is proposed as a *nomen revictum* (basonym: "*Streptomyces parvisporogenes*" *ignotus* 1960).

According to Hatano et al. (2003), *Streptomyces parvisporogenes* (Locci et al. 1969) Witt and Stackebrandt 1991 is a later heterotypic synonym of *Streptomyces abikoensis* (Umezawa et al. 1951) Witt and Stackebrandt 1991.

363. **Streptomyces parvulus** corrig. Waksman and Gregory 1954, 1055[AL]

par'vu.lus. L. masc. adj. *parvulus* very small.

Spore chains in Section *Spirales*. Mature spore chains are generally long, often with more than 50 spores per chain. This morphology is seen on yeast-malt agar, oatmeal agar, salts-starch agar, and glycerol-asparagine agar. Spore surface is smooth.

Color of colony: aerial mass color in the Gray color series on yeast-malt agar, oatmeal agar, salts-starch agar, and glycerol-asparagine agar. Reverse side of colony with no distinctive pigment (grayed yellow to grayed greenish yellow) on yeast-malt agar, oatmeal agar, salts-starch agar, and or glycerol-asparagine agar; substrate pigment is not a pH indicator.

Color in medium: melanoid pigments not formed in peptone-yeast-iron agar and tyrosine agar. No pigment found in medium in yeast-malt agar, oatmeal agar, salts-starch agar, or glycerol-asparagine agar.

D-Glucose, L-arabinose, sucrose, D-xylose, iso-inositol, D-mannitol, D-fructose, and rhamnose are utilized for growth. Variable reports on growth with raffinose.

Type strain shows the highest sequence similarity to: *S. pactum*, AB184398, 99.4%; *S. lomondensis*, AB184673, 99.3%; *S. olivaceus*, AB184743, 99.3%; *S. malachitospinus*, AB249954, 99.3%; *S. coeruleorubidus*, AY999719, 99.2%; *S. bellus*, AB184849, 99.2%; *S. coerulescens*, AY999720, 99.2%; *S. luteogriseus*, AB184379, 99.2%; *S. violaceorubidus*, AJ781374, 99.1%; *S. roseoviolaceus*, AJ399484, 99%; *S. lusitanus*, AB184424, 99%; *S. albosporeus* subsp. *albosporeus*, AJ781327, 99%; *S. spinoverrucosus*, AB184578, 99%; *S. janthinus*, AB184851, 99%; *S. violaceus*, AB184315, 99%; *S. tendae*, D63873, 99%; *S. purpurascens*, AJ399486, 99%.

Source: not known.

DNA G+C content (mol%): not known.

Type strain: ATCC 12434, ATCC 19796, CBS 418.59, CBS 548.68, BCRC 12046, DSM 40048, HUT 6081, ICMP 156, NBRC 13193, IMET 41380, JCM 4068, JCM 4601, LMG 19312, NCIMB 11240, NRRL B-1628, NRRL-ISP 5048, RIA 1075, RIA 507, UNIQEM 182, VKM Ac-1063.

Sequence accession no. (16S rRNA gene): AB184326.

Further comments: the original spelling, *Streptomyces parvullus* (*sic*), was corrected by Hill et al. (1984).

364. **Streptomyces parvus** (Krainsky 1914) Waksman and Henrici *in* Breed, Murray and Hitchens 1948, 939[AL] ("*Actinomyces parvus*" Krainsky 1914, 685; "*Nocardia parva*" Chalmers and Christopherson 1916, 268)

par'vus. L. masc. adj. *parvus* small.

Spore chains in Section *Rectiflexibiles*. Mature spore chains generally contain 10–50 or sometimes more than 50 spores per chain. This morphology is seen on yeast-malt agar, oatmeal agar, salts-starch agar, and glycerol-asparagine agar. Spore surface is smooth.

Color of colony: aerial mass color in the Yellow color series (1ba, 2ba, pale yellow) on yeast-malt agar, oatmeal agar, salts-starch agar, and glycerol-asparagine agar. Reverse side of colony with no distinctive pigments (light yellow to light greenish yellow) on yeast-malt agar, oatmeal agar, salts-starch agar, and glycerol-asparagine agar.

Color in medium: melanoid pigments are not formed in peptone-yeast-iron agar, tyrosine agar, or tryptone-yeast broth. Yellow pigment is found in the medium in yeast-malt agar, oatmeal agar, salts-starch agar, and glycerol-asparagine agar. This pigment is not pH-sensitive when tested with 0.05 M NaOH or HCl.

D-Glucose, L-arabinose, D-xylose, D-mannitol, D-fructose, and rhamnose are utilized for growth. No growth or only traces of growth on sucrose, iso-inositol, and raffinose.

Type strain shows the highest sequence similarity to: *S. rubiginosohelvolus*, AB184240, 99.9%; *S. mediolani*, AB184674, 99.9%; *S. badius*, AY999783, 99.9%; *S. pluricolorescens*, DQ442540, 99.9%; *S. griseinus*, AB184205, 99.9%; *S. sindenensis*, AB184759, 99.9%; *S. flavofuscus*, AB249935, 99.8%; *S. globisporus* subsp. *globisporus*, EF178686, 99.8%; *S. griseoplanus*, AY999894, 99.8%; *S. praecox*, AB184293, 99.8%; *S. fimicarius*, AY999784, 99.8%; *S. albovinaceus*, AB249958, 99.8%; *S. anulatus*, DQ026637, 99.8%; *S. microflavus*, DQ445795, 99.7%; *S. cinereorectus*, AB184646, 99.7%; *S. lipmanii*, AB184148, 99.7%; *S. alboviridis*, AB184256, 99.7%; *S. floridae*, AB184656, 99.7%; *S. lavendulae* subsp. *lavendulae*, AB184080, 99.7%; *S. cyaneofuscatus*, AB184860, 99.7%; *S. cavourensis* subsp. *washingtonensis*, DQ026671, 99.7%; *S. fulvorobeus*, AB184711, 99.7%; *S. acrimycini*, AY999889, 99.7%; *S. californicus*, AB184755, 99.7%; *S. baarnensis*, EF178688, 99.7%; *S. argenteolus*, AB045872, 99.6%; *S. luridiscabiei*, AF361784, 99.5%; *S. flavogriseus*, AJ494864, 99.5%; *S. griseus* subsp. *griseus*, AY207604, 99.5%; *S. halstedii*, EF178695, 99.5%; *S. bacillaris*, AB184439, 99.5%; *S. griseolus*, AB184768, 99.5%; *S. pulveraceus*, AB184806, 99.5%; *S. flavovirens*, DQ026635, 99.5%; *S. nitrosporeus*, EF178680, 99.4%; *S. cavourensis* subsp. *cavourensis*, DQ445791, 99.3%; *S. finlayi*, AY999788, 99.3%; *S. yanii*, AB006159, 99.3%; *S. atroolivaceus*, AJ781320, 99.3%; *S. sanglieri*, AB249945, 99.3%; *S. olivoviridis*, AB184227, 99.3%; *S. clavifer*, DQ026670, 99.3%; *S. albolongus*, AB184425, 99.3%; *S. atratus*, DQ026638, 99.2%; *S. griseobrunneus*, AB249912, 99.2%; *S. gelaticus*, DQ026636, 99.2%; *S. celluloflavus*, AB184476, 99.2%; *S. candidus*, DQ026663, 99.1%; *S. cremeus*, AB184124, 99%; *S. mutomycini*, AB249951, 99%; *S. spiroverticillatus*, AB184814, 99%.

Source: not known.

DNA G+C content (mol%): not known.

Type strain: ATCC 12433, CBS 427.61, DSM 40348, HAMBI 1018, HUT 6062, NBRC 14599, JCM 4069, NCIMB 9608, NRRL B-1255, NRRL B-1455, NRRL-ISP 5348, RIA 610, VKM Ac-725.

Sequence accession no. (16S rRNA gene): DQ442537.

365. **Streptomyces paucisporeus** Xu, Wang, Cui, Huang, Liu, Zheng and Goodfellow 2006, 1113[VP]

pau.ci.spo're.us. L. adj. *paucus* few; N.L. adj. *sporeus* spored; N.L. masc. adj. *paucisporeus* few spored, forming few spores.

Few spores with smooth surfaces are borne in short flexuous spore chains. Forms branched substrate and aerial hyphae. Khaki-colored colonies that carry a white to gray aerial spore mass are formed on acidified oatmeal agar, yeast extract-starch agar, and yeast extract-malt extract agar. Diffusible pigments are not formed, though melanin pigments are produced on tyrosine agar but not on peptone-yeast extract-iron agar. Degrades Tween 80, but not adenine, guanine, starch, or xanthine. Cellobiose, D-galactose, D-glucose, D-mannitol, D-sorbitol (each at 1%, w/v), L-alanine, L-arginine, α-L-aspartic acid, L-cysteine, L-glutamic acid, L-phenylalanine, and L-valine (each at 0.1%, w/v) are used as sole carbon sources for energy and growth, but adonitol, inulin, D-lactose (each at 1%, w/v), adipic acid, DL-aminobutyric acid, sodium acetate, sodium citrate, and sodium oxalate (each at 0.1%, w/v) are not. Does not use L-alanine, L-arginine, L-aspartic acid, L-glutamic acid, L-isoleucine, or L-phenylalanine (each at 0.1%, w/v) as sole carbon and nitrogen sources. Growth occurs at temperatures between 20 and 37°C, but not at 15°C, and at pH values from 4.5–7.5, but not at pH 3.5. Does not grow in the presence of 5% (w/v) NaCl. Sensitive to filter-paper discs soaked in the following (µg/ml unless indicated): azithromycin (30), doxycycline hydrochloride (30), erythromycin (15), josamycin (15), kanamycin sulfate (30), minocycline hydrochloride (30), neomycin sulfate (30), streptomycin sulfate (10), tetracycline hydrochloride (30), and tobramycin sulfate (10), but not to filter-paper discs soaked in acetylspiramycin (15), amoxycillin (10), ampicillin (10), azetreonam (30), carbenicillin (10), penicillin G (10 IU/ml), rifampin (5), or sulfamethoxazole (25). Chemotaxonomic properties are typical of the genus *Streptomyces*.

Type strain shows no sequence similarity over 99%. Type strain shows DNA–DNA similarity to: *S. yeochonensis* NRRL B-24245^{T}, 35.0%; *S. guanduensis* JCM 13274^{T}, 14.9%; *S. rubidus* JCM 13277^{T}, 19.9%; *S. yanglinensis* JCM 13275^{T}, 25.9%.

Source: the type strain was isolated from Dahao pine-forest in Guandu, Yunnan Province, People's Republic of China.

DNA G+C content (mol%): 74.8.

Type strain: 1413, CGMCC 4.2025, JCM 13276.

Sequence accession no. (16S rRNA gene): AY876943.

366. **Streptomyces peucetius** Grein, Spalla, DiMarco and Canevazzi 1963, 109[AL]

peu.ce'ti.us. L. masc. adj. *peucetius* of or pertaining to Peucetia, an ancient name for Central Puglia in Italy, the source of the soil from which the organism was isolated.

Spore chains of atypical *Retinaculiaperti* type. Excellent growth on Czapek's solution agar; reported to form coremia; forms pinkish vegetative mycelium on some media. Produces daunomycin, an anti-bacterial and anti-tumor antibiotic; produces polyenic anti-fungal activity.

Type strain shows the highest sequence similarity to: *S. xantholiticus*, AB184349, 100%; *S. kurssanovii*, AB184325, 99.8%; *S. graminofaciens*, AJ781329, 99.7%.

Source: isolated from soil from Central Puglia, Italy.

DNA G+C content (mol%): not known.

Type strain: ATCC 29050, CCT 4840, CMI 101335, DSM 40754, JCM 9920, KCTC 9199, NBRC 100596, NCIMB 10972.

Sequence accession no. (16S rRNA gene): AB045887.

367. **Streptomyces phaeochromogenes** (Conn 1917) Waksman *in* Breed, Murray and Smith 1957, 778[AL] ["*Actinomyces pheochromogenus*" (*sic*) Conn 1917, 16; "*Streptomyces phaeochromogenus*" (*sic*) Waksman and Henrici 1948, 943; "*Streptomyces pheochromogenus*" Pridham, Hesseltine and Benedict 1958, 74]

pha.e.o.chro.mo'ge.nes. Gr. adj. *phaios* brown; Gr. n. *chroma* color; N.L. suff. *-genes* (from Gr. v. *gennaô* to produce) producing; N.L. part. adj. *phaeochromogenes* producing brown color.

Spore chains in Section *Rectiflexibiles*. Mature spore chains are generally straight chains of 10–50, or sometimes more than 50, spores per chain. This morphology is seen on yeast-malt agar, oatmeal agar, salts-starch agar, and glycerol-asparagine agar. Spirals were reported in Waksman's (1919) description, but not in Conn's original description (Conn, 1917; op. cit). Spore surface is smooth.

Color of colony: aerial mass color usually in the Red color series on yeast-malt agar, oatmeal agar, salts-starch agar, and glycerol-asparagine agar. Most representative color-tabs from the Red color series were 3ca (pale orange yellow) and 5ca or 5cb (light yellowish pink to grayish yellowish pink). Aerial mass color may also be white or pale yellow on oatmeal agar and pale yellow on salts-starch agar and glycerol-asparagine agar. Reverse side of colony with no distinctive pigments (pale yellow or grayish yellow on oatmeal agar and salts-starch agar; light brown to strong brown on glycerol-asparagine agar and yeast-malt agar).

Color in medium: melanoid pigments are formed in peptone-yeast-iron agar, tyrosine agar, and tryptone-yeast broth. No pigment other than trace of yellow is found in the medium in yeast-malt agar, oatmeal agar, salts-starch agar, and glycerol-asparagine agar.

D-Glucose, L-arabinose, sucrose, D-xylose, iso-inositol, D-mannitol, D-fructose, and rhamnose are utilized for growth. Reports vary on utilization of raffinose.

Type strain shows the highest sequence similarity to: *S. ederensis*, AB184658, 99.6%; *S. umbrinus*, AB184305, 99.4%.

Source: not known.

DNA G+C content (mol%): not known.

Type strain: ATCC 23945, ATCC 3338, CBS 282.30, CBS 288.60, CBS 929.68, BCRC 12484, CECT 3070, DSM 40073, IFM 1051, NBRC 12898, NBRC 3180, IMET 40355, JCM 4070, JCM 4659, KCTC 9763, LMG 19348, NCIMB

8505, NRRL B-1248, NRRL B-3010, NRRL-ISP 5073, RIA 1119, RIA 61, VKM Ac-1002.

Sequence accession no. (16S rRNA gene): AB184738.

368. **Streptomyces phaeofaciens** Maeda, Okami, Kosaka, Taya and Umezawa 1952, 327[AL]

pha.e.o.fa'ci.ens. Gr. adj. *phaios* brown; L. part adj. *faciens* producing; N.L. part adj. *phaeofaciens* producing brown, referring to production of brown diffusible pigment.

Spore chains in Section *Spirales*. Mature spore chains are long with 10 to 50 or more spores per chain. This morphology is seen on yeast-malt agar, oatmeal agar, salts-starch agar, and glycerol-asparagine agar. Spore surface is smooth.

Color of colony: aerial mass color usually in the Gray color series (3ih, dark gray; 3ig, grayish yellowish brown; 3ge, light grayish yellowish brown; or 5fe, light grayish reddish brown) on yeast-malt agar, oatmeal agar, salts-starch agar, and glycerol-asparagine agar; the yellow to reddish brown aerial mycelium may also fall in the Red color series as represented by color tab 4ge, light grayish reddish brown. Reverse side of colony with no distinctive pigments (greenish yellow to light olive brown on yeast-malt agar; pale yellow to light olive gray on oatmeal agar, salts-starch agar, and glycerol-asparagine agar).

Color in medium: melanoid pigments are formed in peptone-yeast-iron agar and tryptone-yeast broth, but are formed only weakly or as trace amounts in tyrosine agar. No pigment is found in the medium in yeast-malt agar, oatmeal agar, salts-starch agar, or glycerol-asparagine agar.

D-Glucose, L-arabinose, D-xylose, iso-inositol, D-mannitol, D-fructose, and rhamnose are utilized for growth. No growth or only traces of growth with sucrose or raffinose.

Type strain shows the highest sequence similarity to: *S. phaeofaciens*, AB184360, 100%; *S. rishiriensis*, EF178691, 99.2%.

Source: not known.

DNA G+C content (mol%): not known.

Type strain: ATCC 15034, CBS 426.64, CBS 673.72, DSM 40367, IFM 1177, NBRC 13372, JCM 4125, JCM 4814, NRRL B-1516, NRRL-ISP 5367, RIA 1333, VKM Ac-1865.

Sequence accession no. (16S rRNA gene): AB184360.

369. **Streptomyces phaeogriseichromatogenes** Goodfellow, Kumar, Labeda and Sembiring 2008, 1[VP] (Effective publication: Goodfellow, Kumar, Labeda and Sembiring 2007, 194.)

pha.e.o.gri.se.i.chro.ma.to'ge.nes. Gr. adj. *phaios* brown; N.L. adj. *griseus* gray; Gr. n. *chroma -atos* color; N.L. suff. *-genes* (from Gr. v. *gennaô* to produce) producing; N.L. part. adj. *phaeogriseichromatogenes* producing brown and gray colors.

Spore chains are *Rectiflexibiles*; spore surface is ridged. Aerial spore mass colors range from grayish-white to grayish-brown and substrate mycelium is yellow to yellowish-brown, a yellow diffusible pigment is formed on yeast extract-malt extract agar, but melanin pigments are not produced. Hydrogen sulfide is produced, esculin is hydrolyzed, but not urea. Nitrate is not reduced. Casein, chitin, hypoxanthine, tyrosine, Tween 60, uric acid, and xanthine are degraded, but not adenine, arbutin, guanine, or xylan. L-Arabinose, D-arabitol, butan-1,4-diol, cellobiose, citric acid, dextrin, D-fructose, D-galactose, D-glucose, α-lactose, maltose, D-mannitol, D-mannose, D-ribose, and D-sucrose are used as sole carbon sources, but not L-arabitol, L-fucose, *myo*-inositol, melezitose, methanol, propanol, pyruvic acid, raffinose, L-rhamnose, or salicin. L-Alanine, L-arginine, L-glutamic acid, L-histidine, L-leucine, DL-methionine, DL-norleucine, L-proline, and L-valine are used as sole carbon and nitrogen sources, but not α-alanine, L-aminobutyric acid, L-glycine, L-phenylalanine, L-threonine, or L-tryptophan. Grows from pH 5 to pH 10, at 25 and 37°C, but not at 10 or 40°C. Does not grow in the presence of 13% (w/v) NaCl. Resistant (μg/ml) to cephalosporin (32), cefoxitin (32), cephaloridine (32, 64), doxycycline hydrochloride (4), and rifampin (64), but not to cefoxitin (64), carbenicillin (32, 64), erythromycin (32, 64), oleandomycin phosphate (32, 64), chlortetracycline hydrochloride (4, 8), doxycycline hydrochloride (8), rifampin (32, 64), fusidic acid (8, 16), lincomycin hydrochloride (32, 64), or novobiocin (4, 8).

Type strain shows the highest sequence similarity to: *S. costaricanus*, AB249939, 99.6%; *S. griseofuscus*, AB184206, 99.6%; *S. murinus*, AB184155, 99.6%.

Source: not known.

DNA G+C content (mol%): 71.2.

Type strain: DSM 40710, NRRL 2834.

Sequence accession no. (16S rRNA gene): AJ391813.

370. **Streptomyces phaeoluteichromatogenes** Goodfellow, Kumar, Labeda and Sembiring 2008, 1[VP] (Effective publication: Goodfellow, Kumar, Labeda and Sembiring 2007, 195.)

pha.e.o.lu.te.i.chro.ma.to'ge.nes. Gr. adj. *phaios* brown, dark; L. adj. *luteus* yellow; Gr. n. *chroma -atos* color; N.L. suff. *-genes* (from Gr. v. *gennaô* to produce) producing; N.L. part. adj. *phaeoluteichromatogenes* producing brown and yellow colors.

Spore chains are *Rectiflexibiles*; spore surface is smooth. On oatmeal agar the aerial spore mass color is brown and the substrate mycelium is yellow; a yellow diffusible pigment is produced. Melanin pigments are not formed. Esculin and urea are hydrolyzed. Does not produce hydrogen sulfide or reduce nitrate. Adenine, casein, chitin, hypoxanthine, tyrosine, and Tween 80 are degraded, but not casein, guanine, uric acid, xanthine, or xylan. Uses L-arabinose, D- and L-arabitol. butan-1,4-diol, cellobiose, citric acid, D-fructose, D-galactose, D-glucose, *myo*-inositol, D-mannitol, D-mannose, melezitose, methanol, propanol, and pyruvic acid as sole carbon sources, but not dextrin, L-fucose, α-lactose, maltose, raffinose, L-rhamnose, D-ribose, D-salicin, or sucrose. α-Alanine, L-glutamic acid, L-glycine, L-leucine, L-proline, L-serine, and L-threonine are used as sole carbon and nitrogen sources, but not L-alanine, L-aminobutyric acid, L-histidine, DL-methionine, DL-norleucine, L-phenylalanine, L-tryptophan, or L-valine. Grows from pH 4 to pH 9 and at 25°C and 40°C, but not at 10°C. Growth occurs in the presence of 13% (w/v) NaCl. Resistant (μg/ml) to carbenicillin (32, 64), cephalosporin (32), cefoxitin (32, 64), cephaloridine (32, 64), chlortetracycline hydrochloride (4, 8), doxycycline hydrochloride (4, 8), tetracycline hydrochloride (64), rifampin (32, 64), and novobiocin (4), but sensitive to erythromycin (32, 64), oleandomycin phosphate (32, 64), fusidic acid (8, 16), and lincomycin hydrochloride (32, 64).

Type strain shows the highest sequence similarity to: *S. misionensis*, EF178678, 99.8%; *S. diastaticus* subsp. *diastaticus*, AB184785, 99.1%; *S. rutgersensis* subsp. *rutgersensis*, AB184170, 99.1%; *S. gougerotii*, AB184742, 99.1%; *S. intermedius*, AB184277, 99%.

Source: not known.

DNA G+C content (mol%): 69.8.

Type strain: DSM 41898, NRRL B-5799.

Sequence accession no. (16S rRNA gene): AJ391814.

371. **Streptomyces phaeoluteigriseus** Goodfellow, Kumar, Labeda and Sembiring 2008, 1[VP] (Effective publication: Goodfellow, Kumar, Labeda and Sembiring 2007, 196.)

pha.e.o.lu.te.o.gri'se.us. Gr. adj. *phaios* brown; L. adj. *luteus* yellow; N.L. adj. *griseus* gray; N.L. masc. adj. *phaeoluteigriseus* brown, yellow, and gray-colored.

Spore chains are *Rectiflexibiles*; spore surface is smooth. On oatmeal agar the aerial spore mass color is gray and the substrate mycelium is yellowish-brown; a yellow diffusible pigment is produced, as are melanin pigments. Esculin is hydrolyzed but not urea. Hydrogen sulfide is produced, but nitrate is not reduced. Adenine, arbutin, chitin, hypoxanthine, Tween 80, and xanthine are degraded, but not arbutin, guanine, tyrosine, uric acid, or xylan. Adonitol, L-arabitol, citric acid, and dextrin are used as sole carbon sources, but not L-arabitol, butan-1,4-diol, cellobiose, D-fructose, L-fucose, D-galactose, D-glucose, *myo*-inositol, α-lactose, D-mannitol, D-mannose, melezitose, methanol, propanol, pyruvic acid, raffinose, L-rhamnose, D-ribose, salicin, or sucrose. α-Alanine, L-histidine, and L-threonine are used as sole carbon and nitrogen sources, but not L-alanine, L-aminobutyric acid, L-arginine, L-glutamic acid, L-leucine, DL-methionine, DL-norleucine, L-phenylalanine, L-serine, L-threonine, L-tryptophan, or L-valine. Grows from pH 4 to pH 10 and at 25 and 37°C, but not at 10 or 40°C. Growth occurs in the presence of 13% (w/v) NaCl. Resistant (μg/ml) to carbenicillin (32, 64), cefocitin (32, 64), cephaloridine (32), oleandomycin phosphate (32), chlortetracycline hydrochloride (4), tetracycline hydrochloride (64), fusidic acid (8, 16), and novobiocin (4), but sensitive to cephalosporin (32), cephaloridine (64), erythromycin (32, 64), oleandomycin phosphate (64), chlortetracycline hydrochloride (8), doxycycline hydrochloride (4, 8), rifampin (32, 64), lincomycin hydrochloride (32, 64), and novobiocin (8).

Type strain shows the highest sequence similarity to: *S. bobili*, AB249925, 99.5%; *S. novaecaesareae*, AB184357, 99.5%; *S. galilaeus*, AB045878, 99.5%; *S. pseudovenezuelae*, AB184233, 99.4%; *S. resistomycificus*, AB184166, 99.4%; *S. chartreusis*, AB184839, 99.1%; *S. prunicolor*, DQ026659, 99%.

Source: not known.

DNA G+C content (mol%): 72.2.

Type strain: DSM 41896, ISP 5182, NRRL ISP-5182.

Sequence accession no. (16S rRNA gene): AJ391815.

372. **Streptomyces phaeopurpureus** Shinobu 1957, 63[AL] emend. Lanoot, Vancanneyt, Dawyndt, Cnockaert, Zhang, Huang, Liu and Swings 2005a, 8 (Effective publication: Lanoot, Vancanneyt, Dawyndt, Cnockaert, Zhang, Huang, Liu and Swings 2004, 88.)

pha.e.o.pur.pu're.us. Gr. adj. *phaios* brown; L. adj. *purpureus* purple colored; N.L. masc. adj. *phaeopurpureus* brown, purple colored.

Spore chains in Section *Rectiflexibiles*. Aerial mycelium is usually poorly developed or absent on yeast-malt agar, oatmeal agar, salts-starch agar, and glycerol-asparagine agar. Mature spore chains may be observed on other media including Gause's medium no. 1 where chains of 10 to 50 or more spores may be formed. Spore surface is smooth.

Color of colony: aerial mass color probably in the Gray or Red color series, but aerial sporulation on yeast-malt agar, oatmeal agar, salts-starch agar, and glycerol-asparagine agar is usually inadequate for accurate spore mass color determination. Reverse side of colony is yellow to yellow brown modified by red (light brown, strong brown, or grayish reddish brown on oatmeal agar, salts-starch agar, and glycerol-asparagine agar; moderate to strong reddish brown on yeast-malt agar). Substrate mycelium pigment is not a pH indicator (or is changed only slightly from brown to yellow brown by 0.05 M HCl).

Color in medium: melanoid pigments are formed in peptone-yeast-iron agar, tyrosine agar, and tryptone-yeast broth. Yellow, orange, or red pigment is found in the medium in yeast-malt agar, oatmeal agar, salts-starch agar, and glycerol-asparagine agar. This pigment is slightly pH-sensitive changing from reddish orange or brown to yellow brown with 0.05 M HCl.

D-Glucose, L-arabinose, sucrose, D-xylose, iso-inositol, D-mannitol, D-fructose, rhamnose, and raffinose are utilized for growth.

Type strain shows the highest sequence similarity to: *S. griseorubiginosus*, AJ781339, 100%.

Source: not known.

DNA G+C content (mol%): not known.

Type strain: ATCC 23946, CBS 930.68, BCRC 13754, DSM 40125, HAMBI 950, NBRC 12899, NBRC 3930, JCM 4101, JCM 4660, KCTC 9764, NRRL B-2260, NRRL-ISP 5125, RIA 1128.

Sequence accession no. (16S rRNA gene): DQ026666.

Further comments: according to Lanoot et al. (2004), *Streptomyces phaeopurpureus* Shinobu 1957 is an earlier heterotypic synonym of *Streptomyces phaeoviridis* Shinobu 1957.

373. **Streptomyces phaeoviridis** Shinobu 1957, 63[AL]

pha.e.o.vi'ri.dis. Gr. adj. *phaios* brown; L. adj. *viridis* green; N.L. masc. adj. *phaeoviridis* brown-green.

Spore chains in Section *Rectiflexibiles*. Mature spore chains generally 10–50 spores per chain. This morphology is seen on yeast-malt agar, oatmeal agar and salts-starch agar, but aerial mycelium with mature spore chains is not formed on glycerol-asparagine agar. Spore surface is smooth.

Color of colony: aerial mass color in the Gray color series on yeast-malt agar, oatmeal agar and salts-starch agar. (White also may be seen on these media and on the poorly sporulated surface of glycerol-asparagine agar) Reverse side of colony is yellow to yellow brown is modified slightly by red or orange (light yellowish brown to strong brown, yellowish pink, or grayish reddish orange). Substrate mycelium pigment is not a pH indicator.

Color in medium: melanoid pigments are formed in peptone-yeast-iron agar, tyrosine agar and tryptone-yeast broth. Orange or reddish pigment is found in the medium in yeast-malt agar, oatmeal agar and glycerol-asparagine agar. This pigment is not pH-sensitive when tested with 0.05 M NaOH or HCl. D-Glucose, L-arabinose, sucrose, D-xylose, iso-inositol, D-mannitol, D-fructose, rhamnose and raffinose are utilized for growth.

For sequence similarity, see type strain of *Streptomyces phaeopurpureus.*

Source: not known.

DNA G+C content (mol%): not known.

Type strain: ATCC 23947, CBS 931.68, DSM 40285, NBRC 12900, JCM 4102, JCM 4661, NCIMB 9832, NRRL B-2258, NRRL-ISP 5285, RIA 1199.

Sequence accession no. (16S rRNA gene): AB184230.

Further comments: according to Lanoot et al. (2004), *Streptomyces phaeoviridis* Shinobu 1957 is a later heterotypic synonym of *Streptomyces phaeopurpureus* Shinobu 1957.

374. **Streptomyces pharetrae** Le Roes and Meyers 2005a, 2236^{VP} (Effective publication: Le Roes and Meyers 2005b, 489.)

pha.re'tra.e. L. gen. n. *pharetrae* of a quiver for holding arrows, isolated from soil taken from the base of a Giant quiver tree (*Aloe pillansii*).

Aerobic, Gram-stain-positive actinomycete which forms green aerial mycelium and yellow-blue substrate mycelium on ISP 4 (inorganic salts-starch agar). The color of the substrate mycelium is not pH-sensitive. No fragmentation occurs and no verticils are present. Mature spore chains are of the *Spirales*-type with hairy spore ornamentation. Spores are nonmotile. Good growth is observed on Czapek's solution agar, yeast extract-malt extract agar (ISP 2), and inorganic salts-starch agar (ISP 4). Substrate mycelium color is medium-dependent, but aerial mycelium color is constant. No diffusible pigment is produced on glycerol-asparagine agar (ISP 5). Melanin is produced on peptone-yeast extract-iron agar (ISP 6), but not on tyrosine agar (ISP 7). H_2S production occurs and nitrate is reduced (weak reaction after 14 d). Lecithinase, protease, and lipase activities are observed on egg-yolk agar. Pectin is hydrolyzed but not hippurate. Degrades adenine, esculin, arbutin, casein, cellulose, gelatin, hypoxanthine, L-tyrosine, Tween 80, and urea. Starch hydrolysis is negative at 7 d, but weakly positive when tested at 14 and 21 d. Allantoin, guanine, xanthine, and xylan are not degraded. Uses β-alanine, DL-alanine, ammonium phosphate, L-arginine, L-asparagine, L-cysteine, L-glutamic acid, L-glutamine, L-histidine, L-hydroxyproline, L-isoleucine, L-leucine, L-methionine, L-phenylalanine, potassium nitrate, L-proline, L-serine, urea, and L-valine, but not DL-α-amino-*n*-butyric acid as sole nitrogen sources. Uses dextran, D(−) fructose, D(+)glucose, methyl α-D-glucoside, glycerol, glycogen, D(−)ribose, sodium butyrate, and sodium sorbate as sole carbon sources, but not adonitol, *meso*-erythritol, *myo*-inositol, inulin, maltose, D(+)melibiose, D(+)melezitose, raffinose, salicin, sodium benzoate, sodium formate, sodium lactate, sodium maleate, sodium oxalate, sodium salicylate, sodium L(+)tartrate, trehalose, or xylitol. Weak growth is observed after 21 d on L(+)arabinose, D(−)lactose, D(−)mannitol, D(+)mannose, L(+)rhamnose, sodium acetate, sodium citrate, sodium gluconate, sodium DL-malate, sodium succinate, sucrose, and D(+)xylose. Grows in the presence of 0.3% 2-phenylethanol, 0.0001% crystal violet, 7% NaCl (but not 10%), and 0.1% phenol, but not in the presence of sodium azide (0.01%). Growth is also observed in the presence of cephaloridine (100 μg/ml), lincomycin (100 μg/ml), penicillin G (10 IU/ml), and rifampin (50 μg/ml), but not in the presence of neomycin (50 μg/ml), and streptomycin (100 μg/ml). Growth is observed at 30, 37, and 45°C, but not at 4°C or at pH 4.3. Weak antibiosis is exhibited against *Bacillus coagulans* ATCC 7050^T, *Mycobacterium aurum* A+, and *Acinetobacter calcoaceticus* C91, but there is no antibiosis against *Escherichia coli* ATCC 25922, *Pseudomonas aeruginosa* ATCC 27853, and *Staphylococcus aureus* ATCC 25923.

Type strain shows the highest sequence similarity to: *S. glaucescens*, AB184843, 99.3%.

Source: isolated from soil taken from the base of a Giant quiver tree (*Aloe pillansii*).

DNA G+C content (mol%): 76.0 (±1.4).

Type strain: CZA14, DSM 41856, JCM 13860, NRRL B-24333.

Sequence accession no. (16S rRNA gene): AY699792.

375. **Streptomyces pilosus** Ettlinger, Corbaz and Hütter 1958a, 347^{AL}

pi.lo'sus. L. masc. adj. *pilosus* hairy, shaggy.

Spore chains in Section *Rectiflexibiles* to *Spirales.* Short spore chains include straight or flexuous chains, hooks, loops, and poorly developed spirals. Hooks and loops of small diameter on short sporophores are not typical of the broad hooks and loops of *Retinaculiaperti* cultures. In the original description (Ettlinger et al., 1958a), spore chains are described as open, regular spirals. Mature spore chains are moderately short with 3–10 or sometimes more than 10 spores per chain. This morphology is seen on yeast-malt agar, oatmeal agar, and salts-starch agar; sporulation may be poor on glycerol-asparagine agar. Spore surface is hairy, with a tendency toward spines on some spores.

Color of colony: aerial mass color in the Gray color series on yeast-malt agar, oatmeal agar, and salts-starch agar. Reverse side of colony is grayed yellow to yellow-brown on glycerol-asparagine agar and dark grayed yellow-brown to near black on yeast-malt agar, oatmeal agar, and salts-starch agar; substrate pigment is not a pH indicator.

Color in medium: melanoid pigments formed in peptone-yeast-iron agar and tryptone-yeast extract broth. Pigments other than melanoids are not formed in yeast-malt agar, oatmeal agar, salts-starch agar, or glycerol-asparagine agar.

D-Glucose, L-arabinose, D-xylose, iso-inositol, D-mannitol, D-fructose, and rhamnose are utilized for growth. No growth or only trace of growth on sucrose and raffinose.

Type strain shows the highest sequence similarity to: *S. flavoviridis*, AB184842, 100%; *S. violaceus*, AB184315, 99.2%; *S. albosporeus* subsp. *albosporeus*, AJ781327, 99.2%; *S. janthinus*, AB184851, 99.2%; *S. lomondensis*, AB184673, 99.2%; *S. geysiriensis*, DQ442501, 99.2%; *S. minutiscleroticus*, EF178696, 99.2%; *S. roseoviolaceus*, AJ399484, 99.2%; *S. ghanaensis*, AY999851, 99.2%; *S. luteogriseus*, AB184379, 99.2%; *S. djakartensis*, AB184657, 99.1%; *S. plicatus*, AB184291, 99.1%; *S. rochei*, AB184237, 99.1%; *S. vinaceusdrappus*, AY999929, 99.1%; *S. mutabilis*, EF178679, 99.1%; *S. hawaiiensis*, AB184143, 99%; *S. atrovirens*, DQ026672, 99%; *S. arenae*, AB249977, 99%; *S. calvus*, AB184329, 99%; *S. tuirus*, AB184690, 99%.

Source: not known.

DNA G+C content (mol%): not known.

Type strain: ATCC 19797, CBS 549.68, BCRC 12091, DSM 40097, NBRC 12807, JCM 4403, NRRL 2721, NRRL-ISP 5097, RIA 1076, UNIQEM 183, VKM Ac-1765.

Sequence accession no. (16S rRNA gene): AB184161.

376. **Streptomyces platensis** Tresner and Backus 1956, 244[AL]

plat.en'sis. Gr. adj. *platus* flat, broad, wide; N.L. masc. adj. *platensis* belonging to flat.

Spore chains in Section *Spirales*. Mature spore chains generally have 10–50, or sometimes more than 50, spores per chain. This morphology is seen on yeast-malt agar, oatmeal agar, salts-starch agar, and glycerol-asparagine agar. Spore surface is smooth.

Color of colony: aerial mass color in the Gray color series on yeast-malt agar, oatmeal agar, salts-starch agar, and glycerol-asparagine agar. Spores frequently coalesce to form black, moist (hygroscopic) masses of spores. Sometimes much of the aerial growth will be converted to a black, moist surface. Reverse side of colony is grayish yellow on oatmeal agar and salts-starch agar; sometimes modified by red to grayish yellowish pink or light reddish brown on yeast-malt agar and glycerol-asparagine agar. Reddish substrate mycelium pigment may be changed to colorless or pale yellow by 0.05 M HCl.

Color in medium: melanoid pigments are not formed in peptone-yeast-iron agar, tyrosine agar, or tryptone-yeast broth. Traces of red or yellow pigments are found in the medium in yeast-malt agar, oatmeal agar, salts-starch agar, and glycerol-asparagine agar. The reddish pigment is somewhat pH-sensitive, changing from grayed red to nearly colorless with 0.05 M HCl.

D-Glucose, sucrose, iso-inositol, D-mannitol, D-fructose, and raffinose are utilized for growth. Utilization of L-arabinose and D-xylose is doubtful. No growth or only trace of growth with rhamnose.

Type strain shows the highest sequence similarity to: *S. libani* subsp. *rufus*, AJ781351, 99.9%; *S. hygroscopius* subsp. *glebosus*, AB184479, 99.9%; *S. caniferus*, AB184640, 99.8%; *S. libani* subsp. *libani*, AB184414, 99.5%; *S. nigrescens*, DQ442530, 99.5%; *S. tubercidicus*, AJ621612, 99.5%; *S. hygroscopius* subsp. *decoyicus*, AY999883, 99.4%; *S. misakiensis*, AB217605, 99.2%; *S. catenulae*, AJ621613, 99.2%; *S. ramulosus*, DQ026662, 99.1%; *S. sioyaensis*, DQ026654, 99.1%; *S. monomycini*, DQ445790, 99%.

Source: not known.

DNA G+C content (mol%): not known.

Type strain: ATCC 13865, ATCC 23948, CBS 310.56, CBS 932.68, BCRC 11898, CCUG 11118, DSM 40041, NBRC 12901, JCM 4189, JCM 4662, KCTC 1088, NCAIM B.01481, NCIMB 9607, NRRL 2364, NRRL B-5486, NRRL-ISP 5041, RIA 1110, VKM Ac-1288.

Sequence accession no. (16S rRNA gene): AB045882.

377. **Streptomyces plicatus** Pridham, Hesseltine and Benedict 1958, 65[AL]

pli.ca'tus. L. masc. part. adj. *plicatus* folded, coiled.

Spore chains in Section *Spirales*. Spirals are characteristically open. Wavy spore chains or long flexuous chains terminating in hooks, loops, or incomplete spirals may also suggest *Retinaculiaperti* morphology. Mature spore chains are generally long, often with more than 50 spores per chain. This morphology is seen on yeast-malt agar, oatmeal agar, salts-starch agar, and glycerol-asparagine agar. Spore surface is smooth.

Color of colony: aerial mass color usually in the Gray color series (3fe, light brownish gray or 5fe, light grayish reddish brown) on yeast-malt agar, oatmeal agar, salts-starch agar, and glycerol-asparagine agar, but color similar to 5dc (grayish yellowish pink) from the Red color series may also be seen on yeast-malt agar and salts-starch agar. Reverse side of colony with no distinctive pigments (pale grayish yellow, yellowish brown or occasionally dark grayish brown) on yeast-malt agar, oatmeal agar, salts-starch agar, and glycerol-asparagine agar.

Color in medium: melanoid pigments are not formed in peptone-yeast-iron agar, tyrosine agar, or tryptone-yeast broth. No pigment is found in medium in yeast-malt agar, oatmeal agar, salts-starch agar, or glycerol-asparagine agar.

D-Glucose, L-arabinose, D-xylose, iso-inositol, D-mannitol, D-fructose, and rhamnose are utilized for growth. No growth or only trace of growth with sucrose and raffinose.

Type strain shows the highest sequence similarity to: *S. vinaceusdrappus*, AY999929, 100%; *S. ghanaensis*, AY999851, 100%; *S. geysiriensis*, DQ442501, 100%; *S. minutiscleroticus*, EF178696, 100%; *S. rochei*, AB184237, 100%; *S. mutabilis*, EF178679, 99.9%; *S. tuirus*, AB184690, 99.5%; *S. djakartensis*, AB184657, 99.4%; *S. anandii*, AB184402, 99.2%; *S. violaceorubidus*, AJ781374, 99.2%; *S. pilosus*, AB184161, 99.1%; *S. flavoviridis*, AB184842, 99.1%; *S. tendae*, D63873, 99%; *S. calvus*, AB184329, 99%; *S. azureus*, EF178674, 99%; *S. asterosporus*, AB184706, 99%; *S. levis*, AB184670, 99%; *S. luteogriseus*, AB184379, 99%; *S. capillispiralis*, AB184577, 99%.

Source: not known.

DNA G+C content (mol%): not known.

Type strain: ATCC 25483, CBS 911.69, BCRC 12279, DSM 40319, NBRC 13071, JCM 4504, NCAIM B.01841, NRRL 2428, NRRL-ISP 5319, RIA 1263, VKM Ac-627.

Sequence accession no. (16S rRNA gene): AB184291.

378. **Streptomyces pluricolorescens** Okami and Umezawa *in* Waksman 1961, 259[AL]

plu.ri.co.lor.es'cens. L. comp. adj. *plus*, *pluris* more, many; L. n. *color* color; N.L. part. adj. *pluricolorescens* becoming many-colored or variegated.

Spore chains in Section *Rectiflexibiles.* Mature spore chains are moderately short with 3–10, or sometimes more than 10 spores per chain. This morphology is observed on yeast-malt agar, oatmeal agar, salts-starch agar, and glycerol-asparagine agar. Spore surface is smooth.

Color of colony: aerial mass color in the Yellow or Red color series on yeast-malt agar, oatmeal agar, salts-starch agar, and glycerol-asparagine agar (usually pale yellow in 7 d becoming pale orange-yellow or grayish yellowish pink in 14–21 d). Reverse side of colony is grayed yellow to yellow-brown (modified slightly by red in 14–21 d) on yeast-malt agar, oatmeal agar, salts-starch agar, and glycerol-asparagine agar; substrate pigment is not a pH indicator.

Color in medium: melanoid pigments not formed in peptone-yeast-iron agar and tyrosine agar. No pigment found in medium in yeast-malt agar, oatmeal agar, salts-starch agar, or glycerol-asparagine agar.

D-Glucose, D-xylose, D-mannitol, D-fructose, and rhamnose are utilized for growth. No growth or only trace of growth on sucrose, iso-inositol, and raffinose. Utilization of L-arabinose is doubtful.

Type strain shows the highest sequence similarity to: *S. sindenensis*, AB184759, 100%; *S. albovinaceus*, AB249958, 100%; *S. griseoplanus*, AY999894, 100%; *S. anulatus*, DQ026637, 100%; *S. globisporus* subsp. *globisporus*, EF178686, 100%; *S. praecox*, AB184293, 100%; *S. flavofuscus*, AB249935, 100%; *S. fimicarius*, AY999784, 100%; *S. mediolani*, AB184674, 100%; *S. badius*, AY999783, 100%; *S. griseinus*, AB184205, 100%; *S. rubiginosohelvolus*, AB184240, 100%; *S. californicus*, AB184755, 99.9%; *S. lavendulae* subsp. *lavendulae*, AB184080, 99.9%; *S. cavourensis* subsp. *washingtonensis*, DQ026671, 99.9%; *S. acrimycini*, AY999889, 99.9%; *S. parvus*, DQ442537, 99.9%; *S. fulvorobeus*, AB184711, 99.8%; *S. lipmanii*, AB184148, 99.8%; *S. microflavus*, DQ445795, 99.8%; *S. cyaneofuscatus*, AB184860, 99.8%; *S. floridae*, AB184656, 99.8%; *S. alboviridis*, AB184256, 99.8%; *S. baarnensis*, EF178688, 99.8%; *S. cinereorectus*, AB184646, 99.8%; *S. griseus* subsp. *griseus*, AY207604, 99.7%; *S. griseolus*, AB184768, 99.7%; *S. luridiscabiei*, AF361784, 99.7%; *S. flavovirens*, DQ026635, 99.7%; *S. argenteolus*, AB045872, 99.7%; *S. bacillaris*, AB184439, 99.6%; *S. pulveraceus*, AB184806, 99.6%; *S. halstedii*, EF178695, 99.6%; *S. flavogriseus*, AJ494864, 99.6%; *S. atroolivaceus*, AJ781320, 99.5%; *S. olivoviridis*, AB184227, 99.5%; *S. finlayi*, AY999788, 99.5%; *S. nitrosporeus*, EF178680, 99.5%; *S. albolongus*, AB184425, 99.4%; *S. clavifer*, DQ026670, 99.4%; *S. griseobrunneus*, AB249912, 99.4%; *S. sanglieri*, AB249945, 99.4%; *S. celluloflavus*, AB184476, 99.4%; *S. yanii*, AB006159, 99.4%; *S. gelaticus*, DQ026636, 99.3%; *S. atratus*, DQ026638, 99.3%; *S. cavourensis* subsp. *cavourensis*, DQ445791, 99.3%; *S. mutomycini*, AB249951, 99.2%; *S. candidus*, DQ026663, 99.2%; *S. spiroverticillatus*, AB184814, 99.1%; *S. cremeus*, AB184124, 99.1%.

Source: not known.

DNA G+C content (mol%): not known.

Type strain: ATCC 19798, CBS 550.68, BCRC 13657, DSM 40019, IFM 1101, NBRC 12808, JCM 4302, JCM 4602, NCIMB 9813, NRRL B-2121, NRRL-ISP 5019, RIA 1077, UNIQEM 184, VKM Ac-765.

Sequence accession no. (16S rRNA gene): DQ442540.

379. **Streptomyces polychromogenes** Hagemann, Pénasse and Teillon *in* Hütter 1964, 615[AL]

po.ly.chro.mo'ge.nes. Gr. adj. *polu* many; Gr. n. *chroma* color; N.L. suff. *-genes* (from Gr. v. *gennaô* to produce) producing; N.L. part. adj. *polychromogenes* producing many colors, referring to the characteristic variation of pigmentation.

Spore chains characteristically in Section *Rectiflexibiles* with long, straight spore chains of more than 50 spores per chain on yeast-malt agar, oatmeal agar, salts-starch agar, and glycerol-asparagine agar. A few spore chains may terminate in spirals. Spore surface is smooth. Knots and nest-like tangles may be seen in the aerial mycelium. Some of these tangles fragment into spore-like bodies and one observer reports that spores may also be seen on the substrate mycelium.

Color of colony: aerial mass color in the Red color series (grayish yellowish pink) on yeast-malt agar, oatmeal agar, salts-starch agar, and glycerol-asparagine agar. Reverse side of colony with no distinctive pigments (nearly colorless to grayish yellow or light yellow on oatmeal agar, salts-starch agar, and glycerol-asparagine agar; light grayish reddish brown to strong brown on yeast-malt agar).

Color in medium: melanoid pigments are formed in peptone-yeast-iron agar and tryptone-yeast broth, but only weakly or not at all in tyrosine agar. No pigment, or only a trace of yellow, is found in the medium in yeast-malt agar, oatmeal agar, salts-starch agar, and glycerol-asparagine agar.

D-Glucose, L-arabinose, D-xylose, D-fructose, and sucrose are utilized for growth. Only traces of growth is seen with iso-inositol, D-mannitol, rhamnose, and raffinose.

Type strain shows the highest sequence similarity to: *S. racemochromogenes*, DQ026656, 100. *S. flavotricini*, AB184132, 99.7. *S. katrae*, AB184409, 99.5. *S. cinnamonensis*, AB184707, 99.1. *S. tanashiensis*, AJ781362, 99.1. *S. globosus*, AJ781330, 99.1. *S. nojiriensis*, AJ781355, 99. *S. spororaveus*, AJ781370, 99. *S. toxytricini*, DQ442548, 99. *S. nashvillensis*, AB184286, 99%.

Source: not known.

DNA G+C content (mol%): not known.

Type strain: ATCC 12595, ATCC 25484, CBS 311.56, CBS 912.69, BCRC 11899, DSM 40316, NBRC 13072, JCM 4236, JCM 4505, KCTC 9765, NCIMB 8791, NRRL B-12233, NRRL B-2656, NRRL B-3032, NRRL B-5697, NRRL-ISP 5316, RIA 1264, RIA 362, VKM Ac-1207.

Sequence accession no. (16S rRNA gene): AB184292.

380. **Streptomyces poonensis** (Thirumalachar *in* Kalakoutskii and Krasil'nikov *in* Rautenshtein 1960) Pridham 1970, 42[AL] ("*Chainia poonensis*" Thirumalachar *in* Kalakoutskii and Krasil'nikov *in* Rautenshtein 1960, 45)

po.on.en'sis. N.L. masc. adj. *poonensis* of or belonging to Poona, India, the source of the soil from which the organism was isolated.

Spore chains in Section *Spirales.* Mature spore chains generally contain 3 to 10 or more spores per chain. Aerial mycelium may be absent or spiral spore chains may be seen on yeast-malt agar and salts-starch agar. Straight spore chains are reported on oatmeal agar. Spore surface is smooth. Sclerotia are abundant on yeast-malt agar, oatmeal agar, salts-starch agar, and glycerol-asparagine agar in 14 d. One observer reports LL-A_2pm and no arabinose in the cell wall, placing this strain in cell-wall type I.

Color of colony: aerial mass color in the White color series on yeast-malt agar, oatmeal agar, and salts-starch agar when aerial mycelium is produced. Aerial mycelium is usually not seen on glycerol-asparagine agar and may be absent on oatmeal agar or other ISP media. Reverse side of colony is colorless on glycerol-asparagine agar; yellow to yellow brown is modified by red becoming reddish brown, brownish pink, or purplish pink on yeast-malt agar, oatmeal agar, and salts-starch agar within 21 d. Only one of three observers reports reverse mycelium pigment is pH-sensitive, changing from pink to violet with the addition of 0.05 M NaOH.

Color in medium: melanoid pigments are not formed in peptone-yeast-iron agar, tyrosine agar, or tryptone-yeast broth. Two observers report no pigment in the medium in yeast-malt agar, oatmeal agar, salts-starch agar, and glycerol-asparagine agar. One observer reports pinkish pigment in yeast-malt agar, oatmeal agar, and salts-starch agar; it is reported that this pigment is pH-sensitive, changing from pink to violet with the addition of 0.05 M NaOH.

D-Glucose, L-arabinose, D-xylose, iso-inositol, D-mannitol, and D-fructose are utilized for growth. Reports vary on utilization of rhamnose; no growth or only traces of growth with sucrose or raffinose.

Type strain shows the highest sequence similarity to: *S. fumigatiscleroticus*, DQ442499, 99.3%; *S. ruber*, AB184604, 99.1%.

Source: isolated from soil from Poona, India.

DNA G+C content (mol%): not known.

Type strain: AS 4.1097, ATCC 15723, CBS 295.66, CBS 786.72, BCRC 13311, DSM 40596, HAMBI 987, NBRC 12556, NBRC 13485, IMET 43406, JCM 3071, JCM 3079, JCM 4815, LMG 19326, NRRL B-2319, NRRL B-2951, NRRL-ISP 5596, PCM 2246, RIA 1446, RIA 569, VKM Ac-1715.

Sequence accession no. (16S rRNA gene): DQ445792.

Further comments: Streptomyces poonensis (Thirumalachar 1960) Pridham 1970 and *Chainia poonensis* Thirumalachar 1960 have the same type strain and therefore are homotypic synonyms [Rules 24a and 24b (1) of the *Bacteriological Code* (1990 Revision)].

381. **Streptomyces praecox** (Millard and Burr 1926) Waksman *in* Waksman and Lechevalier 1953, 107[AL] ("*Actinomyces praecox*" Millard and Burr 1926, 633)

pra.e'cox. L. masc. adj. *praecox* premature, precocious.

Spore chains predominantly in Section *Rectiflexibiles* on yeast-malt agar, oatmeal agar, salts-starch agar, and glycerol-asparagine agar. One of three ISP observers records strongly flexuous chains on yeast-malt agar to open spirals on oatmeal agar in 14 d. The original work of Millard and Burr (op. cit) describes "widely open spirals". Spore chains are moderately long with 10–50 or sometimes more than 50 spores per chain. Spore surface is smooth.

Color of colony: mature aerial mass color in the Yellow color series (pale yellow or pale yellow-green) on yeast-malt agar, oatmeal agar, and salts-starch agar; White color series on glycerol-asparagine agar. Reverse side of colony with no distinctive pigments (pale or grayish yellow to light olive-brown) on yeast-malt agar, oatmeal agar, salts-starch agar, and glycerol-asparagine agar.

Color in medium: melanoid pigments are not formed in peptone-yeast-iron agar, tyrosine agar, or tryptone-yeast broth. No pigment, or only a trace of yellow, is found in the medium in yeast-malt agar, oatmeal agar, salts-starch agar, and glycerol-asparagine agar.

D-Glucose, L-arabinose, D-xylose, D-mannitol, D-fructose, and rhamnose are utilized for growth. Utilization of iso-inositol, sucrose, and raffinose is doubtful.

Type strain shows the highest sequence similarity to: *S. flavofuscus*, AB249935, 100%; *S. lavendulae* subsp. *lavendulae*, AB184080, 100%; *S. cavourensis* subsp. *washingtonensis*, DQ026671, 100%; *S. anulatus*, DQ026637, 100%; *S. acrimycini*, AY999889, 100%; *S. griseinus*, AB184205, 100%; *S. fimicarius*, AY999784, 100%; *S. badius*, AY999783, 100%; *S. griseoplanus*, AY999894, 100%; *S. mediolani*, AB184674, 100%; *S. sindenensis*, AB184759, 100%; *S. rubiginosohelvolus*, AB184240, 100%; *S. pluricolorescens*, DQ442540, 100%; *S. globisporus* subsp. *globisporus*, EF178686, 99.9%; *S. alboviridis*, AB184256, 99.9%; *S. lipmanii*, AB184148, 99.9%; *S. fulvorobeus*, AB184711, 99.9%; *S. cyaneofuscatus*, AB184860, 99.9%; *S. albovinaceus*, AB249958, 99.9%; *S. cinereorectus*, AB184646, 99.9%; *S. microflavus*, DQ445795, 99.9%; *S. baarnensis*, EF178688, 99.9%; *S. californicus*, AB184755, 99.8%; *S. parvus*, DQ442537, 99.8%; *S. argenteolus*, AB045872, 99.8%; *S. flavovirens*, DQ026635, 99.7%; *S. griseolus*, AB184768, 99.7%; *S. luridiscabiei*, AF361784, 99.7%; *S. griseus* subsp. *griseus*, AY207604, 99.7%; *S. flavogriseus*, AJ494864, 99.7%; *S. halstedii*, EF178695, 99.7%; *S. floridae*, AB184656, 99.7%; *S. pulveraceus*, AB184806, 99.5%; *S. clavifer*, DQ026670, 99.5%; *S. atroolivaceus*, AJ781320, 99.5%; *S. olivoviridis*, AB184227, 99.5%; *S. nitrosporeus*, EF178680, 99.5%; *S. finlayi*, AY999788, 99.5%; *S. bacillaris*, AB184439, 99.5%; *S. yanii*, AB006159, 99.4%; *S. celluloflavus*, AB184476, 99.3%; *S. griseobrunneus*, AB249912, 99.3%; *S. sanglieri*, AB249945, 99.3%; *S. albolongus*, AB184425, 99.3%; *S. mutomycini*, AB249951, 99.3%; *S. atratus*, DQ026638, 99.3%; *S. gelaticus*, DQ026636, 99.3%; *S. spiroverticillatus*, AB184814, 99.2%; *S. cavourensis* subsp. *cavourensis*, DQ445791, 99.2%; *S. candidus*, DQ026663, 99.1%; *S. mauvecolor*, AB184532, 99%; *S. cremeus*, AB184124, 99%.

Source: not known.

DNA G+C content (mol%): not known.

Type strain: ATCC 25485, ATCC 3374, CBS 104.27, CBS 913.69, DSM 40393, NBRC 13073, IMET 40356, JCM 4506, NRRL B-1586, NRRL B-2551, NRRL-ISP 5393, RIA 1265, RIA 66, VKM Ac-1873.

Sequence accession no. (16S rRNA gene): AB184293.

382. **Streptomyces prasinopilosus** Ettlinger, Corbaz and Hütter 1958a, 345[AL]

pra.si.no.pi.lo'sus. L. adj. *prasinus* green; L. adj. *pilosus* hairy; N.L. masc. adj. *prasinopilosus* green-hairy.

Spore chains in Section *Retinaculiaperti*. Mature spore chains generally have 10–50 spores per chain. This morphology is seen on yeast-malt agar, oatmeal agar, salts-starch agar, and glycerol-asparagine agar. Spore surface is hairy.

Color of colony: aerial mass color in the Green color series on yeast-malt agar, oatmeal agar, salts-starch agar, and glycerol-asparagine agar. Reverse side of colony is grayed yellow or grayed greenish yellow on oatmeal agar, salts-starch agar, and glycerol-asparagine agar, and grayed yellow modified by orange on yeast-malt agar. Substrate pigment is not a pH indicator (one observer reports slight change form orange to brown with NaOH).

Color in medium: melanoid pigments not formed in peptone-yeast-iron agar or tyrosine agar. No pigment found in medium in yeast-malt agar, oatmeal agar, salts-starch agar, or glycerol-asparagine agar.

D-Glucose, L-arabinose, iso-inositol, D-mannitol, D-fructose, and rhamnose are utilized for growth. No growth or only trace of growth on sucrose and raffinose. Variable reports on growth with D-xylose and D-mannitol.

Type strain shows the highest sequence similarity to: *S. emeiensis*, DQ462649, 99.4%.

Source: not known.

DNA G+C content (mol%): not known.

Type strain: ATCC 19799, CBS 551.68, BCRC 13678, DSM 40098, NBRC 12809, JCM 4207, JCM 4404, LMG 19345, NCIMB 9842, NRRL B-2711, NRRL-ISP 5098, RIA 1078, UNIQEM 185, VKM Ac-1740.

Sequence accession no. (16S rRNA gene): EF626597.

383. **Streptomyces prasinosporus** Tresner, Hayes and Backus 1966, 162[AL]

pra.si.no'spo.rus. L. adj. *prasinus* green; N.L. n. *spora* a spore; N.L. masc. adj. *prasinosporus* green-spored.

Spore chains in Section *Spirales* with irregular spirals of a few turns and loops and hooks of small diameter. Spirals are often clumped or tangled. Mature spore chains are generally short with 3 to 10 or more spores per chain. This morphology is seen on yeast-malt agar, oatmeal agar, salts-starch agar, and glycerol-asparagine agar. Spore surface is hairy, with short hairs.

Color of colony: aerial mass color usually in the Green color series when good sporulation occurs on yeast-malt agar, oatmeal agar, salts-starch agar, and glycerol-asparagine agar (1li, olive gray; 1ig, light grayish olive, or 24½ih, dark greenish gray). Aerial mass color may also be in the Gray color series (2ih, light olive gray or, if sporulating aerial mycelium is less developed, g, medium gray or 2dc, yellowish gray). Reverse side of colony with no distinctive pigments (pale yellow to light olive gray on oatmeal agar, salts-starch agar, and glycerol-asparagine agar; grayish greenish yellow on yeast-malt agar).

Color in medium: melanoid pigments are formed in peptone-yeast-iron agar and tryptone-yeast broth, but not in tyrosine agar. No pigment is found in the medium in yeast-malt agar, oatmeal agar, salts-starch agar, or glycerol-asparagine agar.

D-Glucose, L-arabinose, D-xylose, iso-inositol, D-mannitol, D-fructose, and rhamnose are utilized for growth. Utilization of sucrose and raffinose is doubtful.

Type strain shows no sequence similarity over 99%.

Source: not known.

DNA G+C content (mol%): not known.

Type strain: AS 4.1422, ATCC 17918, CBS 720.72, DSM 40506, NBRC 13419, JCM 4816, LMG 19346, NRRL B-12431, NRRL-ISP 5506, RIA 1380, VKM Ac-979.

Sequence accession no. (16S rRNA gene): AB184390.

384. **Streptomyces prasinus** Ettlinger, Corbaz and Hütter 1958a, 343[AL]

pra'si.nus. L. masc. adj. *prasinus* green.

Spore chains in Section *Retinaculiaperti* or *Spirales*. Spirals are poorly developed and show only one to three turns; loops and hooks are of small diameter on short spore chains and therefore are not typical of well developed *Retinaculiaperti* spore chains. Mature spore chains are moderately short with 3–10, or sometimes more than 10 spores per chain. This morphology is seen on yeast-malt agar, oatmeal agar, salts-starch agar, and glycerol-asparagine agar. Spore surface is spiny.

Color of colony: aerial mass color in the Green color series on yeast-malt agar, oatmeal agar, salts-starch agar, and glycerol-asparagine agar. Reverse side of colony with no distinctive pigment (grayed yellow to yellow-brown on yeast-malt agar and oatmeal agar, and grayed yellow-brown modified slightly by green on salts-starch agar and glycerol-asparagine agar). Substrate pigment is not a pH indicator.

Color in medium: melanoid pigments not formed in peptone-yeast-iron agar and tyrosine agar. No pigment found in medium in yeast-malt agar, oatmeal agar, salts-starch agar, or glycerol-asparagine agar.

D-Glucose, sucrose, D-xylose, iso-inositol, D-mannitol, D-fructose, and rhamnose are utilized for growth. Variable reports on growth with L-arabinose and raffinose.

Type strain shows the highest sequence similarity to: *S. bambergiensis*, AB184869, 99.9%; *S. hirsutus*, AB184844, 99.1%; *S. cyanoalbus*, AB184882, 99%.

Source: not known.

DNA G+C content (mol%): not known.

Type strain: ATCC 19800, CBS 552.68, BCRC 13681, DSM 40099, NBRC 12810, JCM 4192, JCM 4603, NRRL B-2712, NRRL-ISP 5099, RIA 1079, UNIQEM 186, VKM Ac-1725.

Sequence accession no. (16S rRNA gene): DQ026658.

385. **Streptomyces prunicolor** (Ryabova and Preobrazhenskaya *in* Gauze, Preobrazhenskaya, Kudrina, Blinov, Ryabova and Sveshnikova 1957) Pridham, Hesseltine and Benedict 1958, 63[AL] ("*Actinomyces prunicolor*" Ryabova and Preobrazhenskaya *in* Gauze, Preobrazhenskaya, Kudrina, Blinov, Ryabova and Sveshnikova 1957, 184)

pru.ni.co'lor. L. n. *prunum* plum; L. n. *color* color; N.L. *prunicolor* plum colored, referring to the color of the vegetative mycelium of the organism.

Spore chains in Section *Rectiflexibiles*. Straight to flexuous spore chains are generally long with 10–50, or sometimes more than 50 spores per chain. This morphology is seen on yeast-malt agar, oatmeal agar, salts-starch agar, and glycerol-asparagine agar. Spore surface is smooth. Long aerial hyphae may become entangled into knots or nest-like bodies.

Color of colony: mature aerial mass color in the Red color series (5cb, grayish yellowish pink) on yeast-malt agar, salts-starch agar, and glycerol-asparagine agar; immature aerial mycelium may be white to yellow on these media and sporulation on oatmeal agar is usually inadequate for accurate spore mass color determination. Reverse side of colony is light brown to reddish brown on yeast-malt agar; grayish yellow, reddish gray, reddish purple, or dark reddish purple on oatmeal agar, salts-starch agar, and glycerol-asparagine agar. Reports vary on the production of the reddish purple reverse pigment. Reverse mycelium pigment is not a pH indicator, or is changed only slightly with addition of 0.05 M HCl.

Color in medium: melanoid pigments are not formed in peptone-yeast-iron agar, tyrosine agar, or tryptone-yeast broth. No pigment, or only a trace of pale brown, is found in the medium in yeast-malt agar, oatmeal agar, salts-starch agar or glycerol-asparagine agar.

D-Glucose, L-arabinose, D-xylose, iso-inositol, D-mannitol, D-fructose, rhamnose, and raffinose are utilized for growth. Reports vary on utilization of sucrose.

Type strain shows the highest sequence similarity to: *S. resistomycificus*, AB184166, 99.1%; *S. galilaeus*, AB045878, 99%; *S. phaeoluteigriseus*, AJ391815, 99%; *S. novaecaesareae*, AB184357, 99%; *S. chartreusis*, AB184839, 99%; *S. bobili*, AB249925, 99%.

Source: not known.

DNA G+C content (mol%): not known.

Type strain: AS 4.1413, ATCC 25487, CBS 915.69, DSM 40335, NBRC 13075, IMET 43129, INA 8805/64, JCM 4508, LMG 19311, NCIMB 9978, NRRL B-12281, NRRL-ISP 5335, PCM 2370, RIA 1267, VKM Ac-992.

Sequence accession no. (16S rRNA gene): DQ026659.

386. **Streptomyces psammoticus** Virgilio and Hengeller 1960, 167[AL]

psam.mo'ti.cus. Gr. n. *psammos* sand; N.L. masc. adj. *psammoticus* sandy.

Spore chains in Section *Rectiflexibiles* to *Retinaculiaperti* with 10 to 50 or more spores per chain on oatmeal agar and glycerol-asparagine agar. Spore chains may be flexuous or irregularly turned in various forms and coremia or sclerotia-like bodies may be formed. Yeast-malt agar and salts-starch agar are not suitable media for observation of spore chain morphology. Spore surface is smooth; spores are irregular in size and shape.

Color of colony: aerial mass color in the Green color series (1½ge, light grayish olive) on oatmeal agar and salts-starch agar, and sometimes in marginal areas on yeast-malt agar or glycerol-asparagine agar. According to one observer the aerial mass color is in the Yellow color series (color tab 1dc, pale yellow-green). Reverse side of colony with no distinctive pigments (nearly colorless to pale yellow or grayish yellow on oatmeal agar, salts-starch agar, and glycerol-asparagine agar; orange-yellow to yellow-brown on yeast-malt agar).

Color in medium: melanoid pigments are not formed in peptone-yeast-iron agar, tyrosine agar, or tryptone-yeast broth. No pigment or a trace of yellow pigment is found in the medium in yeast-malt agar, oatmeal agar, salts-starch agar, and glycerol-asparagine agar.

D-Glucose, sucrose, and D-fructose are utilized for growth. Utilization of L-arabinose is doubtful. No growth or only trace of growth with D-xylose, iso-inositol, D-mannitol, rhamnose, and raffinose.

Type strain shows the highest sequence similarity to: *S. avellaneus*, AB184413, 100%; *S. aureofaciens*, AY207608, 99.6%; *S. chrysomallus* subsp. *fumigatus*, AB184645, 99.3%; *S. purpureus*, AJ781324, 99.3%; *S. aburaviensis*, AY999779, 99.1%; *S. xanthocidicus*, AY999858, 99%.

Source: not known.

DNA G+C content (mol%): not known.

Type strain: AS 4.1465, ATCC 14125, ATCC 25488, CBS 175.61, CBS 299.65, CBS 916.69, BCRC 12241, DSM 40341, NBRC 13877, NBRC 13971, JCM 4434, NRRL B-3291, NRRL B-5753, NRRL-ISP 5341, PCM 2371, RIA 1268, RIA 832, VKM Ac-996.

Sequence accession no. (16S rRNA gene): AY999862.

387. **Streptomyces pseudoechinosporeus** (Konev, Tsyganov, Minbaev and Morogov 1967) Goodfellow, Williams and Alderson 1986a, 575[VP] (Effective publication: Goodfellow, Williams and Alderson 1986a, 52.) ("*Microechinospora grisea*" Konev, Tsyganov, Minbaev and Morogov 1967, 309; *Microellobosporia grisea* Pridham 1970, 17)

pseu.do.e.chi.no.spo're.us. Gr. adj. *pseudes* false; Gr. n. *ekhinos* hedgehog, sea-urchin; Gr. n. *spora* seed; N.L. masc. adj. *pseudoechinosporeus* false spiny spored.

Forms extensively branched substrate and aerial mycelium. Single or chains of up to three spores are formed on both the substrate and aerial mycelium. Spores have smooth surfaces and are unequal in size (1.8–3.5 µm diameter). They were originally thought to be spiny but are now known to be heavily encrusted in needle-like crystals. The aerial spore mass is white to glaucous gray; the reverse color is pink to light violet pink. Does not form melanin pigments. Gelatin is degraded but cellulose is not. Does not reduce nitrate or produce hydrogen sulfide. L-Arabinose, D-fructose, D-galactose, D-glucose, inulin, D-lactose, maltose, mannitol, raffinose, L-rhamnose, starch, sucrose, and D-xylose are used as sole carbon sources but cellulose and sorbitol are not. Grow at 20, 26, and 37°C, but not at 50°C. The wall peptidoglycan contains LL-A_2pm as the major diamino acid.

Type strain shows the highest sequence similarity to: *S. cinnamoneus*, AB184850, 99.9%; *S. hiroshimensis*, AB184789, 99.5%; *S. caeruleus*, EF178675, 99.2%; *S. aureoversilis*, AB184855, 99.1%; *S. blastmyceticus*, AY999802, 99.1%; *S. lilacinus*, AB184819, 99%; *S. abikoensis*, AB184537, 99%.

Source: isolated from sand collected from the south-west part of the Kyzyl-Kum desert, USSR.

DNA G+C content (mol%): 69.0.

Type strain: ATCC 19618, DSM 43035, IFM 1243, NBRC 12518, IMET 43494, KCC A-0066, KCTC 9178, JCM 3066, NCIMB 9918, RIA 554, RIA 897, VKM Ac-1226.

Sequence accession no. (16S rRNA gene): AB184100.

Further comments: for the transfer of *Microellobosporia grisea* (Konev et al. 1967) Pridham 1970 in the genus *Streptomyces* Waksman and Henrici 1943 it is necessary to substitute a new specific epithet to produce *Streptomyces pseudoechinosporeus* because there is a senior homonym, *Streptomyces griseus* (Krainsky 1914) Waksman and Henrici 1948, included on the Approved Lists of Bacterial Names [Rules 34a and 41a of the *Bacteriological Code* (1990 Revision)].

388. **Streptomyces pseudogriseolus** Okami and Umezawa *in* Okami, Utahara, Ōyagi, Nakamura, Umezawa, Yanagisawa and Tunematsu 1955, 128[AL]

pseu.do.gri.se.o'lus. Gr. adj. *pseudes* false; N.L. dim. adj. *griseolus* specific epithet; N.L. dim. adj. *pseudogriseolus* the false *griseolus* referring to resemblance to *Streptomyces griseolus*.

Spore chains in Section *Spirales*. Spore chains suggesting *Retinaculiaperti* morphology may also be common. Mature spore chains are moderately short, often with 3–10 spores per chain, but longer chains are formed on suitable media. This morphology is seen on yeast-malt agar and salts-starch agar, but aerial mycelium is poorly developed on oatmeal agar and glycerol-asparagine agar. Spore surface is smooth.

Color of colony: aerial mass color in the White color series on yeast-malt agar and salts-starch agar; poor sporulation on oatmeal agar and glycerol-asparagine agar. The aerial mycelium is described as grayish buff in the original description by Okami and Umezawa (1955). Reverse side of colony with no distinctive pigments (colorless to grayish yellow or yellowish brown) on yeast-malt agar, oatmeal agar, salts-starch agar, and glycerol-asparagine agar.

Color in medium: melanoid pigments are not formed in peptone-yeast-iron agar, tyrosine agar, or tryptone-yeast broth. No pigment, or only trace of yellow, is found in the medium in yeast-malt agar, oatmeal agar, salts-starch agar, and glycerol-asparagine agar.

D-Glucose, L-arabinose, D-xylose, iso-inositol, D-mannitol, D-fructose, and rhamnose are utilized for growth. No growth or only trace of growth with sucrose and raffinose.

Type strain shows the highest sequence similarity to: *S. gancidicus*, AB184660, 100%; *S. capillispiralis*, AB184577, 100%; *S. cellulosae*, DQ442495, 99.9%; *S. levis*, AB184670, 99.5%; *S. azureus*, EF178674, 99.5%; *S. lusitanus*, AB184424, 99.4%; *S. rubiginosus*, AY999810, 99.4%; *S. carpinensis*, AB184574, 99.4%; *S. albaduncus*, AY999757, 99.2%; *S. matensis*, AB184221, 99.2%; *S. griseoloalbus*, AB184275, 99.1%; *S. djakartensis*, AB184657, 99.1%; *S. tuirus*, AB184690, 99.1%; *S. caelestis*, X80824, 99.1%; *S. paradoxus*, AB184628, 99%; *S. afghaniensis*, AJ399483, 99%; *S. geysiriensis*, DQ442501, 99%; *S. viridiviolaceus*, AY999854, 99%; *S. africanus*, AY208912, 99%; *S. lavendulocolor*, DQ442516, 99%; *S. minutiscleroticus*, EF178696, 99%.

Source: not known.

DNA G+C content (mol%): not known.

Type strain: ATCC 12770, ATCC 23949, CBS 933.68, BCRC 12132, DSM 40026, NBRC 12902, JCM 4071, JCM 4663, NCIMB 9411, NCIMB 9814, NRRL B-3288, NRRL-ISP 5026, RIA 1106, VKM Ac-1859.

Sequence accession no. (16S rRNA gene): DQ442541.

389. **Streptomyces pseudovenezuelae** (Kuchaeva, Krasil'nikov, Taptykova and Gesheva 1961) Pridham 1970, 24[AL] ("*Actinomyces pseudovenezuelae*" Kuchaeva, Krasil'nikov, Taptykova and Gesheva 1961, 114)

pseu.do.ve.ne.zu.e'la.e. Gr. adj. *pseudes* false; N.L. gen. n. *venezuelae* a specific epithet; N.L. gen. n. *pseudovenezuelae* the false venezuelae, referring to resemblance to *Streptomyces venezuelae*.

Spore chains in Section *Rectiflexibiles*. Mature spore chains are generally long, often with more than 50 spores per chain. This morphology is seen on yeast-malt agar, oatmeal agar, salts-starch agar, and glycerol-asparagine agar. Spore surface is smooth.

Color of colony: aerial mass color in the Red color series (5dc, grayish yellowish pink) on yeast-malt agar, oatmeal agar, and salts-starch agar; Gray or White color series on glycerol-asparagine agar (Gray or White is also sometimes reported for yeast-malt agar, oatmeal agar, and salts-starch agar). Reverse side of colony with no distinctive pigments (light grayish yellow to orange yellow or light yellowish brown) on yeast-malt agar, oatmeal agar, salts-starch agar, and glycerol-asparagine agar.

Color in medium: melanoid pigments are formed in peptone-yeast-iron agar, tyrosine agar, and tryptone-yeast broth. Trace of grayish yellow pigment may be found in the medium in yeast-malt agar, oatmeal agar, salts-starch agar, and glycerol-asparagine agar. This pigment is not pH-sensitive when tested with 0.05 M NaOH or HCl.

D-Glucose, L-arabinose, sucrose, D-xylose, iso-inositol, D-mannitol, D-fructose, rhamnose, and raffinose are utilized for growth.

Type strain shows the highest sequence similarity to: *S. resistomycificus*, AB184166, 99.5%; *S. novaecaesareae*, AB184357, 99.5%; *S. phaeoluteigriseus*, AJ391815, 99.4%; *S. canus*, AY999775, 99.4%; *S. galilaeus*, AB045878, 99.3%; *S. bobili*, AB249925, 99.2%; *S. ciscaucasicus*, AY508512, 99.2%; *S. flavovariabilis*, EF178691, 99.2%; *S. chartreusis*, AB184839, 99.1%; *S. cyaneus*, AF346475, 99.1%; *S. aureocirculatus*, AB184260, 99.1%; *S. humidus*, DQ442508, 99%.

Source: not known.

DNA G+C content (mol%): not known.

Type strain: ATCC 23951, CBS 934.68, BCRC 11487, DSM 40212, NBRC 12904, IMET 43512, JCM 4405, JCM 11516, NRRL B-3623, NRRL-ISP 5212, RIA 1158, RIA 742, VKM Ac-1199.

Sequence accession no. (16S rRNA gene): AB184233.

390. **Streptomyces pulveraceus** Shibata, Higashide, Kanzaki, Yamamoto and Nakazawa 1961, 172[AL]

pul.ve.ra'ce.us. L. n. *pulvis, pulveris* powder; N.L. masc. adj. *pulveraceus* powdery.

Probably grows poorly on Czapek's solution agar. Produces neomycins E and F (paromomycin and paromomycin II), zygomycin B, cycloheximide, and naramycin B.

Type strain shows the highest sequence similarity to: *S. atratus*, DQ026638, 99.7%; *S. gelaticus*, DQ026636, 99.7%; *S. sanglieri*, AB249945, 99.7%; *S. griseinus*, AB184205, 99.6%; *S. rubiginosohelvolus*, AB184240, 99.6%; *S. pluricolorescens*, DQ442540, 99.6%; *S. sindenensis*, AB184759, 99.6%; *S. mediolani*, AB184674, 99.6%; *S. badius*, AY999783, 99.6%; *S. albovinaceus*, AB249958, 99.5%; *S. lavendulae* subsp. *lavendulae*, AB184080, 99.5%; *S. californicus*, AB184755, 99.5%; *S. yanii*, AB006159, 99.5%; *S. praecox*, AB184293, 99.5%; *S. griseoplanus*, AY999894, 99.5%; *S. parvus*,

DQ442537, 99.5%; *S. argenteolus*, AB045872, 99.5%; *S. flavofuscus*, AB249935, 99.5%; *S. globisporus* subsp. *globisporus*, EF178686, 99.5%; *S. fimicarius*, AY999784, 99.5%; *S. cavourensis* subsp. *washingtonensis*, DQ026671, 99.5%; *S. anulatus*, DQ026637, 99.5%; *S. cinereorectus*, AB184646, 99.5%; *S. cyaneofuscatus*, AB184860, 99.4%; *S. griseus* subsp. *griseus*, AY207604, 99.4%; *S. baarnensis*, EF178688, 99.4%; *S. griseolus*, AB184768, 99.4%; *S. acrimycini*, AY999889, 99.4%; *S. flavovirens*, DQ026635, 99.4%; *S. alboviridis*, AB184256, 99.4%; *S. microflavus*, DQ445795, 99.4%; *S. lipmanii*, AB184148, 99.4%; *S. floridae*, AB184656, 99.4%; *S. fulvorobeus*, AB184711, 99.4%; *S. flavogriseus*, AJ494864, 99.3%; *S. luridiscabiei*, AF361784, 99.3%; *S. nitrosporeus*, EF178680, 99.3%; *S. halstedii*, EF178695, 99.3%; *S. bacillaris*, AB184439, 99.2%; *S. finlayi*, AY999788, 99%; *S. celluloflavus*, AB184476, 99%; *S. griseobrunneus*, AB249912, 99%; *S. albolongus*, AB184425, 99.%; *S. spiroverticillatus*, AB184814, 99%; *S. olivoviridis*, AB184227, 99%; *S. candidus*, DQ026663, 99%; *S. clavifer*, DQ026670, 99%; *S. atroolivaceus*, AJ781320, 99%; *S. cremeus*, AB184124, 99%.

Source: not known.

DNA G+C content (mol%): not known.

Type strain: ATCC 13875, DSM 41657, NBRC 3855, KCTC 9766, JCM 7545.

Sequence accession no. (16S rRNA gene): AB184806.

391. **Streptomyces puniceus** Patelski *in* Routien and Hofmann 1951, 387[AL]

pu.ni'ce.us. L. adj. *puniceus* reddish, purple.

Spore chains in Section *Rectiflexibiles.* Mature spore chains are generally long, often with more than 50 spores per chain. This morphology is seen on yeast-malt agar, oatmeal agar, salts-starch agar, and glycerol-asparagine agar. Spore surface is smooth.

Color of colony: aerial mass color in the Yellow color series on yeast-malt agar, oatmeal agar, salts-starch agar and glycerol-asparagine agar. One observer places this culture in the Gray series (tabs 2dc, yellowish gray and 3fe, light grayish brown) on these media. Reverse side of colony is violet on yeast-malt agar, oatmeal agar, salts-starch agar, and glycerol-asparagine agar. Reverse color is changed from violet to blue-violet or blue by addition of 0.05 M NaOH and from violet to pink or red with 0.05 M HCl.

Color in medium: melanoid pigments not formed in peptone-yeast-iron agar or tyrosine agar. Traces of blue or violet pigment found in medium in yeast-malt agar and glycerol-asparagine agar after 14–21 d. Pigment, when present, is changed from blue or violet to red-violet by 0.05 M HCl.

D-Glucose, D-xylose, D-mannitol, and D-fructose are utilized for growth. No growth or only trace of growth on L-arabinose, iso-inositol, and rhamnose. Utilization of sucrose is doubtful and reports are variable for growth with raffinose.

Type strain shows the highest sequence similarity to: *S. phaeofaciens*, AB184360, 100%; *S. rishiriensis*, EF178691, 99.1%.

Source: not known.

DNA G+C content (mol%): not known.

Type strain: ATCC 19801, CBS 308.55, CBS 553.68, BCRC 12097, DSM 40083, NBRC 12811, JCM 4406, NRRL B-2895, NRRL-ISP 5083, RIA 1080, UNIQEM 187, VKM Ac-579.

Sequence accession no. (16S rRNA gene): DQ442542.

392. **Streptomyces puniciscabiei** Park, Kim, Kwon, Wilson, Yu, Hur and Lim 2003, 2053[VP]

pu.ni.ci.sca'bi.ei. L. adj. *puniceus* purple or red; L. n. *scabies* roughness, scabbiness; N.L. gen. n. *puniciscabiei* of purple roughness, intended to mean purple or red, scab-causing bacteria.

Spores are pale orange, spiny, and borne in simple rectus flexuous spore chains. Melanin is produced on tyrosine agar, but not on peptone agar. L-Arabinose, D-fructose, D-glucose, D-mannitol, raffinose, rhamnose, sucrose, D-xylose, and iso-inositol are utilized for growth. Minimum pH for growth is 3.5. Sensitive to 10 and 100 µg/ml thallium acetate and 20 µg/ml streptomycin, but not to 5, 6, and 7% (w/v) NaCl, 0.5 µg/ml crystal violet, 0.1% (w/v) phenol, 10 IU/ml penicillin, or 25 or 100 µg/ml oleandomycin.

Type strain shows the highest sequence similarity to: *S. durhamensis*, AY999785, 99.1%. Type strain shows DNA–DNA similarity to: *S. scabies* ATCC 49173^T, 13%; *S. turgidiscabies* ATCC 700248^T, 13%; *S. acidiscabies* ATCC 49003^T, 11%; *S. bottropensis* DSM 40262^T, 12%; *S. neyagawaensis* DSM 40588^T, 6%; *S. diastatochromogenes* DSM 40449^T, 17%; *S. setonii* DSM 40395^T, 13%; *S. griseus* subsp. *griseus* DSM 40236^T, 16%; *S. sampsonii* DSM 40394^T, 42%; *S. eurythermus* DSM 40014^T, 14%; *S. tendae* DSM 40101^T, 15%; *S. coelicolor* DSM 40233^T, 8%; " *S. lividans*" DSM 40434, 29%; *S. ambofaciens* DSM 40053^T, 12%; *S. luridiscabiei* LMG 21390^T, 14%; *S. niveiscabiei* LMG 21392^T, 13%.

Source: isolated from raised corky lesions on potato cv. Daeji-ma and pathogenic on potato cv. Daeji-ma.

DNA G+C content (mol%): 68.3.

Type strain: S77, KACC 20253, LMG 21391.

Sequence accession no. (16S rRNA gene): AF361785.

393. **Streptomyces purpeofuscus** Yamaguchi and Saburi 1955, 207[AL]

pur.pe.o.fus'cus. L. adj. *purpureus* purple; L. adj. *fuscus* dark, tawny; N.L. masc. adj. *purpeofuscus* dark purple, referring to color of vegetative mycelium.

Spore chains in Section *Rectiflexibiles.* Mature spore chains are generally long, often with more than 50 spores per chain. This morphology is seen on yeast-malt agar, oatmeal agar, salts-starch agar, and glycerol-asparagine agar. Spore surface is smooth.

Color of colony: aerial mass color in the Gray color series on yeast-malt agar, oatmeal agar, salts-starch agar, and glycerol-asparagine agar. Reverse side of colony with no distinctive pigments (grayish yellow to olive brown) on yeast-malt agar, oatmeal agar, salts-starch agar, and glycerol-asparagine agar, but dark brown to reddish brown substrate mycelium pigment may be produced. The original description (Yamaguchi and Saburi, 1955) describes purple to brown reverse mycelium color but notes that this coloration may be lost on repeated transfer.

Color in medium: melanoid pigments are formed in peptone-yeast-iron agar, but only traces of brownish black color in tyrosine agar or tryptone-yeast broth. Traces of red, violet, or yellow pigment may be found in the medium in yeast-malt agar, oatmeal agar, and salts-starch agar. This pigment is not pH-sensitive when tested with 0.05 M NaOH or HCl.

D-Glucose, L-arabinose, and D-xylose are utilized for growth. Utilization of fructose is doubtful. No growth or only trace of growth with sucrose, iso-inositol, D-mannitol, rhamnose, and raffinose.

Type strain shows the highest sequence similarity to: *S. indigoferus*, AB184214, 99.2%; *S. herbaricolor*, AB184801, 99.2%; *S. aburaviensis*, AY999779, 99.1%. Type strain shows the highest sequence similarity to following *Kitasatospora* species: *Kitasatospora azatica*, U93312, 99.1%.

Source: not known.

DNA G+C content (mol%): not known.

Type strain: ATCC 23952, CBS 935.68, BCRC 12093, DSM 40283, NBRC 12905, JCM 4156, JCM 4665, NCIMB 9822, NRRL B-1817, NRRL-ISP 5283, RIA 1197.

Sequence accession no. (16S rRNA gene): AJ781364.

394. **Streptomyces purpurascens** Lindenbein 1952, 371^{AL}

pur.pur.as'cens. L. part. adj. *purpurascens* making purple.

Spore chains in Section *Spirales*. Mature spore chains are generally long with 10–50 or often more than 50 spores per chain. This morphology is seen on oatmeal agar, salts-starch agar, and glycerol-asparagine agar, but spiral spore chains may be absent on yeast-malt agar. Spore surface is spiny.

Color of colony: aerial mass color in the Red color series (grayish yellowish pink) or White color series on yeast-malt agar, oatmeal agar, salts-starch agar, and glycerol-asparagine agar (aerial mycelium may be thin and white or absent on yeast-malt agar or on glycerol-asparagine agar). Reverse side of colony is grayish yellowish pink to reddish brown on yeast-malt agar, oatmeal agar, salts-starch agar, and glycerol-asparagine agar. Reverse mycelium pigment is a pH indicator changing from red or purple to blue with addition of 0.05 M NaOH and to red or orange with addition of 0.05 M HCl.

Color in medium: melanoid pigments are not formed in peptone-yeast-iron agar and tryptone-yeast broth and may or may not develop after 4 d in tyrosine agar. Red to violet pigment is sometimes found in the medium in yeast-malt agar, oatmeal agar, salts-starch agar, and glycerol-asparagine agar. When present, this pigment is pH-sensitive showing the same color changes recorded for reverse mycelium pigments.

D-Glucose, L-arabinose, D-xylose, iso-inositol, D-mannitol, D-fructose, rhamnose, sucrose, and raffinose are all utilized for growth.

Type strain shows the highest sequence similarity to: *S. indiaensis*, AB184553, 99.6%; *S. hawaiiensis*, AB184143, 99.5%; *S. bellus*, AB184849, 99.4%; *S. coerulescens*, AY999720, 99.4%; *S. arenae*, AB249977, 99.3%; *S. lomondensis*, AB184673, 99.3%; *S. massasporeus*, AB184152, 99.3%; *S. coeruleorubidus*, AY999719, 99.2%; *S. lusitanus*, AB184424, 99.2%; *S. luteogriseus*, AB184379, 99.1%; *S. iakyrus*, AB184877, 99.1%; *S. parvulus*, AB184326, 99%.

Source: not known.

DNA G+C content (mol%): not known.

Type strain: ATCC 25489, CBS 917.69, BCRC 11872, DSM 40310, NBRC 13077, JCM 4509, NRRL B-12230, NRRL-ISP 5310, PCM 2299, RIA 1269, VKM Ac-755.

Sequence accession no. (16S rRNA gene): AB045888.

395. **Streptomyces purpureus** (Matsumae and Hata 1968) Goodfellow, Williams and Alderson 1986a, 575^{VP} (Effective publication: Goodfellow, Williams and Alderson 1986b, 65.) (*Kitasatoa purpurea* Matsumae and Hata *in* Matsumae, Ohtani, Takeshima and Hata 1968, 617.)

pur.pur'e.us. L. masc. adj. *purpureus* purple colored.

Forms extensively branched substrate and aerial mycelium. Spore chains in Section *Rectiflexibiles*. Spore surface is smooth. Hyphae have been reported to carry club-shaped sporangia enclosing chains of zoospores with a single polar flagellum. The aerial spore mass is either gray or red and a yellow-brown diffusible pigment is produced. Melanin pigments are formed. Adenine, esculin, allantoin, arbutin, casein, elastin, guanine, hypoxanthine, starch, testosterone, tyrosine, urea, and xanthine are degraded but pectin and xylan are not. Nitrate is reduced to nitrite and hydrogen sulfide is produced. L-Arabinose, cellobiose, D-fructose, D-glucose, glycerol, *myo*-inositol, maltose, D-mannose, melezitose, salicin, and trehalose are used as sole carbon sources but adonitol, dulcitol, inulin, D-lactose, mannitol, melibiose, raffinose, L-rhamnose, sucrose, xylitol, and D-xylose are not. Grows on L-arginine and potassium nitrate but not on DL-amino-n-butyric acid, L-cysteine, L-histidine, L-hydroxyproline, L-methionine, L-phenylalanine, L-serine, L-threonine, or L-valine as sole nitrogen source. Growth occurs at 37°C but not at 10 or 45°C. Tolerant to sodium chloride (4%, w/v) but not to phenol (0.1%, w/v) or sodium azide (0.01%, w/v). Resistant to rifampin but sensitive to sodium chloride at 7% (w/v). Some strains show antimicrobial activity against *Aspergillus niger* LIV 131, *Bacillus subtilis* NCIB 3610, *Micrococcus luteus* NCIB 196, and *Streptomyces murinus* ISP 5091, but such activity is not shown against *Candida albicans* CBS 562, *Escherichia coli* NCIB 9132, *Pseudomonas fluorescens* NCIB 9046^T, or *Saccharomyces cerevisiae* CBS 1171^T. The peptidoglycan contains LL-A_2pm as the major diamino acid and is of the A3γ type (Stackebrandt et al., 1981). Have a type II phospholipid pattern (*sensu* Lechevalier et al., 1977) and contain octahydrogenated menaquinones with nine isoprene units as the major isoprenolog. Produces the antibiotic chloramphenicol.

Type strain shows the highest sequence similarity to: *S. chrysomallus* subsp. *fumigatus*, AB184645, 99.9%; *S. herbaricolor*, AB184801, 99.6%; *S. indigoferus*, AB184214, 99.6%; *S. aburaviensis*, AY999779, 99.5,%; *S. xanthocidicus*, AY999858, 99.5%; *S. psammoticus*, AY999862, 99.3%; *S. avellaneus*, AB184413, 99.2%. Type strain shows the highest sequence similarity to following *Kitasatospora* species: *Kitasatospora kifunensis*, AB022874, 99.2%.

Source: isolated from soil in Hawaii and Japan.

DNA G+C content (mol%): not known.

Type strain: ATCC 27787, BCRC 12101, DSM 43362, NBRC 13927, IMET 9041, JCM 3172, KCC A-0172, KCTC

9187, LMG 19368, NCIMB 11311, NRRL B-5403, VKM Ac-1298.

Sequence accession no. (16S rRNA gene): AJ781324.

Further comments: Goodfellow et al. (1986a) proposed that *Kitasatoa diplospora* Matsumae et al. 1968, *Kitasatoa kauaiensis* Matsumae et al. 1968, and *Kitasatoa nagasakiensis* Matsumae and Hata 1968 become later synonyms of *Kitasatoa purpurea* Matsumae and Hata 1968, and that the latter be transferred to the genus *Streptomyces* as *Streptomyces purpureus* (Matsumae and Hata 1968) Goodfellow et al. 1986b, comb. nov.

396. **Streptomyces purpurogeneiscleroticus** (Thirumalachar *in* Thirumalachar and Sukapure 1964) Pridham 1970, 43[AL] ("*Chainia purpurogena*" Thirumalachar *in* Thirumalachar and Sukapure 1964, 160)

pur.pur.o.ge.ni.scle.ro'ti.cus. L. adj. *purpureus* purple colored; N.L. suff. *-genes* (from Gr. v. *gennaô* to produce) producing; N.L. n. *sclerotium* sclerotium; N.L. part. adj. *purpurogeneiscleroticus* sclerotium along with producing purple color.

Spore chains in Section *Spirales* on yeast-malt agar and salts-starch agar. Sporulation may be thin on these media and is generally absent on oatmeal agar and glycerol-asparagine agar. Mature spore chains generally contain 10 or more spores per chain. Spore surface is smooth. Sclerotia are produced on yeast-malt agar, oatmeal agar, salts-starch agar, and glycerol-asparagine agar in 14 d.

Color of colony: aerial mass color in the White color series on yeast-malt agar and salts-starch agar; aerial mycelium is not produced on oatmeal agar or glycerol-asparagine agar. Reverse side of colony is grayish yellow on salts-starch agar; brown or reddish brown on yeast-malt agar, oatmeal agar, and glycerol-asparagine agar. One of three observers reports the reddish brown reverse mycelium pigment is changed from dark brown to light brown with the addition of 0.05 M HCl. The other observers report no change.

Color in medium: melanoid pigments are not formed in peptone-yeast-iron agar, tyrosine agar, or tryptone-yeast broth. Reddish or yellowish pigment is sometimes found in the medium in yeast-malt agar, oatmeal agar, salts-starch agar, or glycerol-asparagine agar. This pigment is pH-sensitive, changing from brown to reddish brown or red with the addition of 0.05 M NaOH and from brown to yellowish brown or yellow with the addition of 0.05 M HCl.

D-Glucose, L-arabinose, D-xylose, iso-inositol, D-mannitol, D-fructose, rhamnose, sucrose, and raffinose are all utilized for growth.

Type strain shows the highest sequence similarity to: *S. ochraceiscleroticus*, AB184094, 99.9%; S. *violens*, AJ621605, 99.7%; *S. monomycini*, DQ445790, 99.3%; *S. olivaceiscleroticus*, AJ621606, 99.1%; *S. sioyaensis*, DQ026654, 99.1%; *S. erumpens*, AJ621603, 99.1%; *S. niger*, AJ621607, 99.1%.

Source: not known.

DNA G+C content (mol%): not known.

Type strain: ATCC 19348, CBS 409.66, CBS 659.72, BCRC 13317, CM 3103, CMI 112722, DSM 43156, HAMBI 1061, NBRC 13001, NBRC 13903, JCM 3080, JCM 4818, NCIMB 10981, NRRL B-2952, NRRL-ISP 5271, PCM 2306, RIA 1319, RIA 886.

Sequence accession no. (16S rRNA gene): AJ621604.

Further comments: Streptomyces purpurogeneiscleroticus Pridham 1970 and *Chainia purpurogena* Thirumalachar and Sukapure 1964 have the same type strain and therefore are homotypic synonyms [Rules 24a and 24b (1) of the *Bacteriological Code* (1990 Revision)].

397. **Streptomyces racemochromogenes** Sugai 1956, 171[AL]

ra.ce.mo.chro.mo'ge.nes. L. n. *racemus* a raceme or cluster of berries; N.L. part. adj. *chromogenes* producing color; N.L. part. adj. *racemochromogenes* raceme, producing color (probably referring to morphology of spore chains and to chromogenicity).

Spore chains in Section *Retinaculiaperti* including many *Rectiflexibiles* spore chains, and some long chains with terminal hooks or primitive spirals of wide diameter. Mature spore chains are generally long with 10–50, or often more than 50, spores per chain. This morphology is seen on yeast-malt agar, oatmeal agar, salts-starch agar, and glycerol-asparagine agar. Spore surface is smooth.

Color of colony: aerial mass color in the Red color series (5ec and 5cb, grayish yellowish pink) on yeast-malt agar, oatmeal agar, salts-starch agar, and glycerol-asparagine agar. Areas of gray or white sporulation may also be seen. Reverse side of colony with no distinctive pigments on oatmeal agar, but yellow to yellow brown substrate mycelium color may be modified in some areas by blue pigment (dark olive green, grayish green or grayish blue) on yeast-malt agar, salts-starch agar, and glycerol-asparagine agar. The blue pigment is not always present. When present, it can be intensified from grayish green to reddish gray by 0.05 M HCl.

Color in medium: melanoid pigments are produced in peptone-yeast-iron agar, tyrosine agar, and tryptone-yeast broth. No pigment other than a trace of yellow is found in the medium in yeast-malt agar, oatmeal agar, salts-starch agar, and glycerol-asparagine agar.

D-Glucose, L-arabinose, and sucrose are utilized for growth. Only traces of growth are found on D-xylose, iso-inositol, D-mannitol, D-fructose, rhamnose, and raffinose.

Type strain shows the highest sequence similarity to: *S. polychromogenes*, AB184292, 100%; *S. flavotricini*, AB184132, 99.7%; *S. katrae*, AB184409, 99.5%; *S. cinnamonensis*, AB184707, 99.1%; *S. tanashiensis*, AJ781362, 99.1%; *S. globosus*, AJ781330, 99.1%; *S. toxytricini*, DQ442548, 99.1%; *S. nashvillensis*, AB184286, 99.1%; *S. nojiriensis*, AJ781355, 99%; *S. spororaveus*, AJ781370, 99%.

Source: not known.

DNA G+C content (mol%): not known.

Type strain: ATCC 23954, CBS 937.68, BCRC 12318, DSM 40194, NBRC 12906, JCM 4407, NRRL B-5430, NRRL-ISP 5194, RIA 1147, VKM Ac-1206.

Sequence accession no. (16S rRNA gene): DQ026656.

398. **Streptomyces radiopugnans** Mao, Tang, Zhang, Wang, Wei, Huang, Liu, Shi and Goodfellow 2007, 2581[VP]

ra.di.o.pug'nans. L. n. *radius* a beam or ray; N.L. pref. *radio-* pertaining to radiation; L. part. adj. *pugnans* fighting or resisting; N.L. part. adj. *radiopugnans* radiation-resisting.

Radiation-resistant actinomycete that forms an extensively branched substrate mycelium which carries aerial hyphae that differentiate into spiral chains of spores with rough to warty surfaces. Moderate to abundant, white to pale gray aerial spore mass is formed on modified Bennett's, Gause's synthetic medium no. 1, inorganic salts-starch, yeast extract-malt extract, and yeast-starch agars. Substrate mycelium is yellowish brown on modified Bennett's, Gause's synthetic medium no. 1, and yeast-starch agars, and light pinkish yellow on inorganic salts-starch and yeast extract-malt extract agars. Diffusible pigments are not formed on any of the media nor are melanin pigments formed on peptone-yeast extract iron or tyrosine agars. L-Arabinose, melezitose, and ribose are used as sole carbon sources for energy and growth, but not cellobiose, lactose, lactulose, melibiose, raffinose, or trehalose (all at 1%, w/v). Similarly, L-cysteine, L-glycine, D-glutamate, sodium azelate, sodium isobutyrate, and sodium malonate are used as sole carbon sources for energy and growth, but not D-glutamic acid, L-hydroxyproline, DL-isoleucine, L-leucine, methyl-D-glucopyranoside, methyl α-D-mannopyranoside, L-phenylalanine, sodium propionate, sodium pyruvate, sodium suberate, or spermidine (all at 0.1%, w/v).

The fatty acid profile consists of $C_{16:0}$ iso (34.5%), $C_{15:0}$ anteiso (15.4%), $C_{16:1}$ iso-H (14.2%), $C_{17:0}$ anteiso (9.1%), $C_{14:0}$ iso (5.6%), and $C_{17:1}$ anteiso ω9*c* (5.2%).

Type strain shows no sequence similarity over 99%.

Source: strain R97^T was isolated from a radiation-contaminated soil sample collected from Yinjiang Province, Northwestern China.

DNA G+C content (mol%): 72.7.

Type strain: R97, CGMCC 4.3519, DSM 41901.

Sequence accession no. (16S rRNA gene): DQ912930.

399. **Streptomyces rameus** Shibata 1959, 398^{AL}

ra'me.us. L. masc. adj. *rameus* pertaining to branches.

Produces streptomycin.

Type strain shows the highest sequence similarity to: *S. bangladeshensis*, AY750056, 99.8%; *S. glomeratus*, AB249917, 99.1%.

Source: not known.

DNA G+C content (mol%): not known.

Type strain: ATCC 21273, DSM 41685, NBRC 16196, JCM 5064, KCTC 9767.

Sequence accession no. (16S rRNA gene): AY999821.

Further comments: in a Request for an Opinion, Hatano (1999) proposed that the type strain of *Streptomyces rameus* Shibata 1959 is strain 43797, NBRC 3782, which was the originally designated type strain. The Judicial Commission supported this request and decided that strain 43797 [NBRC (previously IFO) 3782, ATCC 700861, JCM 11574] has to replace ATCC 21273 (DSM 41685, NBRC 16196, JCM 5064) as given in the Approved Lists.

400. **Streptomyces ramulosus** Ettlinger, Gäumann, Hütter, Keller-Schierlein, Kradolfer, Neipp, Prelog and Zähner 1958b, 217^{AL}

ra.mu.lo'sus. L. masc. adj. *ramulosus* much branched.

Spore chains in Section *Retiflexibiles.* Short dichotomous spore chains often occur in dense clumps of clusters. Short spore chains of 3–10 spores per chain. This morphology is seen on yeast-malt agar, salts-starch agar, and glycerol-asparagine agar. Spore surface is smooth.

Color of colony: aerial mass color in the Gray color series on yeast-malt agar, salts-starch agar, and glycerol-asparagine agar; poor sporulation on oatmeal agar. Reverse side of colony: substrate is colorless on oatmeal agar and grayed reddish brown to dark brown on yeast-malt agar, salts-starch agar, and glycerol-asparagine agar (one observer, only, reports reverse color as changed from grayed red to grayed yellow by 0.05 M NaOH and from grayed red to grayed purplish color by 0.05 M HCl).

Color in medium: melanoid pigments probably not formed in peptone-yeast-iron agar and tyrosine agar (a light orange pigment is formed in peptone-yeast-iron agar). Red pigment found in medium in yeast-malt agar and glycerol-asparagine agar; red to violet pigment found in oatmeal agar and salts-starch agar. One observer, only, reports pigment changed to yellow by 0.05 M NaOH and to violet by 0.05 M HCl (two observers reported no change).

D-Glucose, D-mannitol, and raffinose are utilized for growth. Doubtful growth on D-fructose. No growth or only trace of growth on L-arabinose, sucrose, D-xylose, iso-inositol, and rhamnose.

Type strain shows the highest sequence similarity to: *S. hygroscopius* subsp. *glebosus*, AB184479, 99.3%; *S. libani* subsp. *rufus*, AJ781351, 99.3%; *S. caniferus*, AB184640, 99.3%; *S. catenulae*, AJ621613, 99.1%; *S. misakiensis*, AB217605, 99.1%; *S. platensis*, AB045882, 99.1%; *S. rimosus* subsp. *rimosus*, AB045883, 99%.

Source: not known.

DNA G+C content (mol%): not known.

Type strain: AS 4.1434, ATCC 19802, CBS 554.68, BCRC 12343, DSM 40100, HAMBI 981, NBRC 12812, NBRC 15798, JCM 4193, JCM 4604, KCTC 9768, LMG 19354, NRRL B-2714, NRRL-ISP 5100, RIA 1081, UNIQEM 188, VKM Ac-1001.

Sequence accession no. (16S rRNA gene): DQ026662.

401. **Streptomyces rangoonensis** corrig. (Erikson 1935) Pridham, Hesseltine and Benedict 1958, 61^{AL} ["*Actinomyces rangoon*" (*sic*) Erikson 1935, 37; "*Nocardia rangoonensis*" (*sic*) Waksman and Henrici 1948, 911]

ran.goon.en'sis. N.L. masc. adj. *rangoonensis* of or belonging to Rangoon, Burma.

Spore chains in Section *Spirales.* Flexuous chains and irregular spirals are also common. Mature spore chains generally have 3 to 10 or more spores per chain. This morphology is seen on yeast-malt agar, oatmeal agar, salts-starch agar, and glycerol-asparagine agar. Spore surface is smooth.

Color of colony: aerial mass color in the White color series on yeast-malt agar, oatmeal agar, salts-starch agar, and glycerol-asparagine agar. Reverse side of colony with no distinctive pigments (colorless to pale yellow on oatmeal agar, salts-starch agar, and glycerol-asparagine agar; yellowish brown to light olive brown on yeast-malt agar).

Color in medium: melanoid pigments are not formed in peptone-yeast-iron agar, tyrosine agar, or tryptone-yeast broth. No pigment is found in medium in yeast-malt agar, oatmeal agar, salts-starch agar, or glycerol-asparagine agar.

D-Glucose, D-xylose, D-mannitol, and D-fructose are utilized for growth. Utilization of L-arabinose is doubtful. Only traces of growth or no growth is found on iso-inositol, rhamnose, sucrose, and raffinose.

Type strain shows the highest sequence similarity to: *S. almquistii*, AB184258, 100%; *S. gibsonii*, AB184663, 100%; *S. albus* subsp. *albus*, AJ621602, 100%; *S. flocculus*, AB184272, 99.9%.

Source: not known.

DNA G+C content (mol%): not known.

Type strain: ATCC 25490, ATCC 6860, CBS 918.69, DSM 40452, HUT 6616, NBRC 13078, IMET 41357, JCM 4510, NRRL B-12378, NRRL B-16595, NRRL-ISP 5452, RIA 1270, VKM Ac-1899.

Sequence accession no. (16S rRNA gene): AB184295.

Further comments: the original spelling of the specific epithet, *rangoon* (*sic*), has been corrected by Trüper and De'Clari (1997).

In the opinion of Kilian (1998), the change of the well-established name *Streptomyces rangoon* may be a source of confusion. So, with reference to the first Principle of the *Bacteriological Code* (1990 Revision), Kilian (2001) requested that the original name *Streptomyces rangoon* be conserved. The Judicial Commission 2005 denied this request and no opinion will be issued upon this request.

402. **Streptomyces recifensis** (Gonçalves de Lima, Machado, Araujo, Falcão de Morais and Biermann 1955) Falcão de Morais, Gonçalves de Lima and Maia 1957, 249[AL] ("*Nocardia recifei*" Gonçalves de Lima, Machado, Araujo, Falcão de Morais and Biermann 1955, 26)

re.cif.en'sis. N.L. masc. adj. *recifensis* of or belonging to Recife, Brazil, the source of the soil from which the organism was isolated.

Spore chains in Section *Retinaculiaperti*; spore chains representative of Section *Rectiflexibiles* are also present. Mature spore chains generally have 10–50 spores per chain. This morphology is found on yeast-malt agar, oatmeal agar, salts-starch agar, and glycerol-asparagine agar. Spore surface is smooth.

Color of colony: aerial mass color in the Gray color series on yeast-malt agar, oatmeal agar, salts-starch agar, and glycerol-asparagine agar. Reverse side of colony with no distinctive pigment on oatmeal agar, salts-starch agar, and glycerol-asparagine agar, grayed yellow modified by dark grayish green on yeast-malt agar. Substrate pigment is not a pH indicator.

Color in medium: Melanoid pigments are not formed in peptone-yeast-iron agar and tyrosine agar. No pigment found in medium in yeast-malt agar, oatmeal agar, salts-starch agar, or glycerol-asparagine agar.

D-Glucose, L-arabinose, sucrose, D-xylose, D-mannitol, D-fructose, and raffinose are utilized for growth. No growth or only traces of growth on iso-inositol and rhamnose.

Type strain shows the highest sequence similarity to: *S. griseoluteus*, AY999751, 99.9%; *S. seoulensis*, AB249970, 99.8%.

Source: isolated from soil from Recife, Brazil.

DNA G+C content (mol%): not known.

Type strain: ATCC 19803, CBS 555.68, BCRC 12086, DSM 40115, NBRC 12813, JCM 4408, NRRL B-3811, NRRL-ISP 5115, RIA 1082, UNIQEM 189, VKM Ac-1890.

Sequence accession no. (16S rRNA gene): AB184165.

403. **Streptomyces rectiverticillatus** (Krasil'nikov and Yuan 1965) Witt and Stackebrandt 1991, 456[VP] (Effective publication: Witt and Stackebrandt 1990, 370.) ("*Actinomyces rectiverticillatus*" Krasil'nikov and Yuan 1965, 49; *Streptoverticillium rectiverticillatum* Locci, Baldacci and Petrolini Baldan 1969, 41)

rec.ti.ver.ti.cil.la'tus. L. adj. *rectus* straight; L. masc. n. *verticillus* whorl; N.L. adj. *verticillatus* whorled; N.L. masc. adj. *rectiverticillatus* straight and whorled.

Spore chains in Section *Verticillati*, umbellate monoverticillate. Monoverticillate morphology is also common. Mature spore chains are generally short with 3–10 or sometimes more than 10 spores per chain. This morphology is seen on yeast-malt agar, oatmeal agar, salts-starch agar, and glycerol-asparagine agar. Spore surface is smooth.

Color of colony: aerial mass color in the Red or White color series on yeast-malt agar, oatmeal agar, salts-starch agar, and glycerol-asparagine agar. Reverse side of colony grayish yellowish pink to strong brown or reddish orange on yeast-malt agar, oatmeal agar, salts-starch agar, and glycerol-asparagine agar. Reverse mycelium pigment is not a pH indicator.

Color in medium: melanoid pigments are formed in peptone-yeast-iron agar, tyrosine agar, or tryptone-yeast broth, but may be weak in tyrosine agar. Yellow or red pigment is found in the medium in yeast-malt agar and oatmeal agar. Traces of yellow may also be present in salts-starch agar and glycerol-asparagine agar. This pigment is not pH-sensitive when tested with 0.05 M NaOH or HCl.

D-Glucose, iso-inositol, and D-fructose are utilized for growth. Utilization of sucrose and raffinose is doubtful. No growth or only trace of growth with L-arabinose, D-xylose, D-mannitol, or rhamnose.

For sequence similarity, see type strain of *Streptomyces hiroshimensis*.

Source: not known.

DNA G+C content (mol%): not known.

Type strain: ATCC 19845, ATCC 25491, CBS 951.69, BCRC 13306, CECT 3268, DSM 40436, NBRC 13079, INMI 380, JCM 4511, NRRL B-12369, NRRL-ISP 5436, RIA 1271, VKM Ac-1503.

Sequence accession no. (16S rRNA gene): AB184296.

Further comments: in violation of Rule 33c of the *Bacteriological Code* (1990 Revision), in Validation List no. 38, *Streptomyces rectiverticillatus* is proposed as a *nomen revictum* (basonym: "*Streptomyces rectiverticillatus*" Krasil'nikov and Yuan (1965).

According to Hatano et al. (2003), *Streptomyces rectiverticillatus* (Krasil'nikov and Yuan 1965) Witt and Stackebrandt 1991 is a later heterotypic synonym of *Streptomyces hiroshimensis* (Shinobu 1955) Witt and Stackebrandt 1991.

404. **Streptomyces rectiviolaceus** (*ex* Artamonova *in* Krasil'nikov 1965) Sveshnikova 1986, 575[VP] (Effective

publication: Sveshnikova *in* Gause, Preobrazhenskaya, Sveshnikova, Terekhova and Maximova 1983.) ("*Actinomyces rectiviolaceus*" Artamonova *in* Krasil'nikov 1965, 234; *Streptomyces rectivioleceus* Pridham 1970, 25)

rec.ti.vio.la'ce.us. L. adj. *rectus* straight; L. adj. *violaceus* violet colored; N.L. masc. adj. *rectiviolaceus* straight, violet colored.

Spore chains are straight or flexuous (*Rectiflexibiles*). Spore surface is smooth. Forms violet or red-colored vegetative mycelium on some media. On mineral agar 1: aerial mycelium is white to light pink, poor; substrate mycelium is violet; diffusible pigment is poor, light violet or not extant. Pigment is able to act as an indicator: alkaline reaction, blue; acidic reaction, red. On glycerol-nitrate agar: aerial mycelium is white, good developed; substrate mycelium is reddish violet; no diffusible pigment. On starch-ammonia agar: aerial mycelium is white, poor; substrate mycelium is colorless to light violet; no diffusible pigment. On glycerol-asparagine agar: aerial mycelium is white, good developed; substrate mycelium is pale violet; no diffusible pigment. On glucose-nitrate agar SP-1: aerial mycelium is whitish, poor; substrate mycelium violet; no diffusible pigment. On oatmeal agar: aerial mycelium is white, poor; substrate mycelium is red-violet; no diffusible pigment. On organic agar 2: no aerial mycelium; substrate mycelium is pale violet; no diffusible pigment. Melanoid pigments are not formed. Grows on glucose, arabinose, xylose, rhamnose, sucrose, raffinose, mannitol and inositol. Exhibits anti-bacterial and anti-fungal activity.

Type strain shows no sequence similarity over 99%.

Source: not known.

DNA G+C content (mol%): not known.

Type strain: ATCC 43690, DSM 41459, INMI 563, JCM 9092, NBRC 100765, NRRL B-16374, VKM Ac-282.

Sequence accession no. (16S rRNA gene): DQ026660.

405. **Streptomyces regensis** Gupta, Sobti and Chopra 1963, 15[AL]

reg.en'sis. N.L. masc. adj. *regensis* of or belonging to reg (unknown derivation, possibly referring to a place in India).

Spore chains in Section *Spirales* or *Rectiflexibiles*. Spirals are poorly developed, usually with only one or two turns or with hooks and loops of small diameter on short lateral spore chains. Flexuous or straight chains are also common. Aerial hyphae may be long, but spore chains are often moderately short with 10 or more spores per chain. This morphology is seen on yeast-malt agar, oatmeal agar, salts-starch agar, and glycerol-asparagine agar. Spore surface is smooth or irregularly folded.

Color of colony: aerial mass color in the Gray or Yellow color series on yeast-malt agar, oatmeal agar, salts-starch agar, and glycerol-asparagine agar. Representative color tabs from the Gray color series are 2dc, yellowish green; 2ge, light olive brown; and 3fe, light brownish gray. Representative color chips from the Yellow color series are 1½ec – 1cb, pale yellow green, and 1ba – 2ba, pale yellow. Reverse side of colony is moderate to strong or dark yellow, greenish yellow, or olive on oatmeal agar, salts-starch agar, and glycerol-asparagine agar. Strong yellow or strong brown on yeast-malt agar.

Color in medium: melanoid pigments are formed in peptone-yeast-iron agar, tyrosine agar, and tryptone-yeast broth. Yellow pigment is found in the medium in yeast-malt agar, oatmeal agar, salts-starch agar, and glycerol-asparagine agar. This pigment is not pH-sensitive when tested with 0.05 M NaOH or HCl.

D-Glucose, L-arabinose, D-xylose, iso-inositol, D-mannitol, D-fructose, rhamnose, sucrose, and raffinose are all utilized for growth.

Type strain shows the highest sequence similarity to: *S. capoamus*, AB045877, 99.1%.

Source: not known.

DNA G+C content (mol%): not known.

Type strain: ATCC 27461, CBS 749.72, BCRC 11890, DSM 40551, NBRC 13448, JCM 4820, NRRL B-11479, NRRL-ISP 5551, RIA 1409, VKM Ac-1289.

Sequence accession no. (16S rRNA gene): DQ026649.

406. **Streptomyces resistomycificus** Lindenbein 1952, 376[AL]

re.sis.to.my.ci'fi.cus. L. v. *restisto* to resist; Gr. n. *mukês* fungus; L. masc. suff. *-ficus* (from L. v. *facio* to make) making, producing; N.L. masc. adj. *resistomycificus* making resistant to a fungus; producing resistomycin.

Spore chains in Section *Spirales*, but with many spore chains representative of Section *Retinaculiaperti* (especially when sporophores are immature). Mature spore chains generally have 10–50 spores per chain. This morphology is seen on yeast-malt agar, oatmeal agar, salts-starch agar, and glycerol-asparagine agar. Spore surface is smooth.

Color of colony: aerial mass color in the Gray color series on yeast-malt agar, oatmeal agar, salts-starch agar, and glycerol-asparagine agar. Reverse side of colony dark reddish brown on yeast-malt agar; grayed brown to reddish brown on oatmeal agar and glycerol-asparagine agar; and grayed yellow to faint pinkish yellow on salts-starch agar. Substrate pigment is not a pH indicator.

Color in medium: melanoid pigments formed in peptone-yeast-iron agar, tyrosine agar, and tryptone-yeast extract broth. Pigments other than melanoids are not formed in yeast-malt agar, oatmeal agar, salts-starch agar, or glycerol-asparagine agar.

D-Glucose, L-arabinose, sucrose, D-xylose, iso-inositol, D-mannitol, D-fructose, rhamnose, and raffinose are utilized for growth.

Type strain shows the highest sequence similarity to: *S. pseudovenezuelae*, AB184233, 99.5%; *S. galilaeus*, AB045878, 99.5%; *S. chartreusis*, AB184839, 99.5%; *S. bobili*, AB249925, 99.5%; *S. phaeoluteigriseus*, AJ391815, 99.4%; *S. novaecaesareae*, AB184357, 99.3%; *S. canus*, AY999775, 99.3%; *S. aureocirculatus*, AB184260, 99.1%; *S. prunicolor*, DQ026659, 99.1%; *S. ciscaucasicus*, AY508512, 99.1%; *S. alboniger*, AY845349, 99%.

Source: not known.

DNA G+C content (mol%): not known.

Type strain: ATCC 19804, CBS 556.68, BCRC 13755, DSM 40133, NBRC 12814, JCM 4409, NCIMB 9843, NRRL

2290, NRRL-ISP 5133, PCM 2296, RIA 1083, UNIQEM 190, VKM Ac-1895.

Sequence accession no. (16S rRNA gene): AB184166.

407. **Streptomyces reticuliscabiei** Bouchek-Mechiche, Gardan, Normand and Jouan 2000, 98[VP]

re.ti.cu.li.sca'bi.ei. L. n. *reticulum* reticulum; L. n. *scabies* mange; N.L. gen. n. *reticuliscabiei* referring to the reticulum aspect of the symptoms of the disease.

Spores are light gray and borne in mature flexuous chains. Melanin is not produced on tyrosine agar. Fructose, D-glucose, D-mannitol, inositol, raffinose, rhamnose, sucrose, and D-xylose are used as carbon sources. Utilizes α-D(+)melibiose, mucate, D-saccharate, D(+)trehalose, and 5-keto-D-gluconate. Does not grow in the presence of 0.5 μg/ml crystal violet, 20 μg/ml streptomycin, 100 μg/ml oleandomycin, or 5% (w/v) NaCl. Strains are not susceptible to 10 IU/ml penicillin G and some (about 60%) of the strains are not susceptible to 25 μg/ml oleandomycin. Does not assimilate *trans*-aconitate or ONPG. Most (about 80%) of the strains utilize DL-lactate and turanose. Some strains use betaine.

Type strain shows the highest sequence similarity to: *S. turgidiscabies*, AB026221, 99.1%. Type strain shows DNA–DNA similarity to: *S. europaeiscabiei* DSM 41802^T, 47%; *S. stelliscabiei* DSM 41803^T, 24%.

Source: isolated from netted scab lesions on potato cv. Bintje and have been confirmed to be pathogenic only on potato cv. Bintje.

DNA G+C content (mol%): 69.8.

Type strain: CFBP 4531, CIP 107061, DSM 41804, ICMP 13716, NCPPB 4041.

Sequence accession no. (16S rRNA gene): AJ007428.

Further comments: a numerical analysis of phenotypic characteristics showed that *Streptomyces reticuliscabiei* Bouchek-Mechiche et al. 2000 and *Streptomyces turgidiscabies* Miyajima et al. 1998 belong to the same cluster and share almost all morphological and biochemical traits that are important in the identification of *Streptomyces* species. DNA–DNA hybridization and phylogenetic comparisons of 16S rRNA gene sequences confirm that the two species are genomically closely related. In contrast, pathological data showed that *Streptomyces turgidiscabies* and *Streptomyces reticuliscabiei* cause two distinct diseases. For the pathologist, the fusion of *Streptomyces reticuliscabiei* and *Streptomyces turgidiscabies* under a single species denomination would cause confusion of separate diseases and create a discrepancy between taxonomists and pathologists. Therefore, Bouchek-Mechiche et al. think that the two groups should continue to carry their current denominations, i.e. *Streptomyces reticuliscabiei* Bouchek-Mechiche et al. 2000 for the strains inducing netted scab and *Streptomyces turgidiscabies* Miyajima et al. 1998 for those causing common scab.

408. **Streptomyces rhizosphaericus** corrig. Sembiring, Ward and Goodfellow 2001, 1619[VP] (Effective publication: Sembiring, Ward and Goodfellow 2000, 362.)

rhi.zo.spha.e'ri.cus. Gr. n. *rhiza* root; L. adj. *sphaericus* of or belonging to a ball, spherical; N.L. masc. adj. *rhizosphaericus* belonging to the sphere of the root.

Spore chains are *Spirales*; spore surface is rugose. On oatmeal agar, the spore mass is gray, the substrate mycelium is grayish-yellow, and the diffusible pigment is yellow. Melanin pigments are not produced. Degrades adenine and pectin and grows at 10°C.

Type strain shows the highest sequence similarity to: *S. asiaticus*, AB249947, 100%; *S. cangkringensis*, AJ391831, 99.8%; *S. griseinger*, AJ391818, 99.8%; *S. indonesiensis*, DQ334783, 99.7%; *S. antimycoticus*, AB184185, 99.1%; *S. sporoclivatus*, AB249934, 99.1%; *S. rutgersensis* subsp. *castelarensis*, AY508511, 99.1%; *S. geldanamycininus*, DQ334781, 99%. Type strain shows DNA–DNA similarity to: *S. albiflaviniger* NRRL B-1356^T, 98.1%; *S. geldanamycinus* NRRL 3602^T, 98.7%; *S. griseiniger* NRRL B1865^T, 99.2%; *S. asiaticus* DSM 41761^T, 98.6%; *S. indonesiensis* DSM 41759^T, 98.3%; *S. javensis* DSM 41764^T, 98.3%; *S. yogyakartensis* DSM 41766^T, 97.9%; *S. cangkringensis* DSM 41769^T, 99.0%.

Source: isolated from the rhizosphere of the tropical legume, *Paraserianthes falcataria*.

DNA G+C content (mol%): not known.

Type strain: A10P1, DSM 41760, JCM 11447, NBRC 100778, NCIMB 13674.

Sequence accession no. (16S rRNA gene): AB249941.

Further comments: the original spelling of the specific epithet, *rhizosphaerius* (*sic*), has been corrected on validation according to Rule 61 of the *Bacteriological Code* (1990 Revision).

409a. **Streptomyces rimosus subsp. rimosus** Sobin, Finlay and Kane *in* Waksman and Lechevalier 1953, 47[AL]

ri.mo'sus. L. masc. adj. *rimosus* full of fissures.

Spore chains in Section *Spirales*, but many imperfect spirals, hooks, and loops suggesting *Retinaculiaperti* morphology are also present. Mature spore chains are short to moderately long with 3–10, or often more than 10, spores per chain. This morphology is seen on yeast-malt agar, oatmeal agar, salts-starch agar, glycerol-asparagine agar, and carbon utilization medium plus D-glucose. Spore surface is smooth.

Color of colony: aerial mass color in the Red or White color series on yeast-malt agar and in the Yellow or White color series on oatmeal agar, salts-starch agar, and glycerol-asparagine agar. Representative color tabs are 3ca (pale orange yellow), 5cb (grayish yellowish pink), and 2ca (pale yellow) from the Red color-wheel; 2db and 2ba (pale yellow) from the Yellow color-wheel; and a (white) from the White color-wheel. Reverse side of colony is grayish yellow on oatmeal agar and salts-starch agar; grayish yellow modified slightly by red (moderate to dark orange yellow) on yeast-malt agar and glycerol-asparagine agar. Substrate mycelium pigment is not a pH indicator.

Color in medium: melanoid pigments are not formed in peptone-yeast-iron agar, tyrosine agar, or tryptone-yeast broth. A trace of yellow pigment is found in the medium in yeast-malt agar, oatmeal agar, salts-starch agar, and glycerol-asparagine agar. This pigment is not pH-sensitive when tested with 0.05 M NaOH or HCl.

D-Glucose, L-arabinose, iso-inositol, D-mannitol, D-fructose, and raffinose are utilized for growth. Utilization of

D-xylose is doubtful. No growth or only trace of growth with sucrose and raffinose.

Type strain shows the highest sequence similarity to: *S. erumpens*, AJ621603, 99.7%; *S. sclerotialus*, AJ621608, 99.4%; *S. niger*, AJ621607, 99.3%; *S. olivaceiscleroticus*, AJ621606, 99.3%; *S. monomycini*, DQ445790, 99.1%; *S. ramulosus*, DQ026662, 99%.

Source: not known.

DNA G+C content (mol%): not known.

Type strain: AS 4.1438, ATCC 10970, ATCC 23955, CBS 437.51, CBS 938.68, BCRC 11612, CECT 3144, DSM 40260, HAMBI 1066, HUT 6064, HUT 6100, ICMP 919, IFM 1065, NBRC 12907, IMET 40364, JCM 4073, JCM 4667, KCTC 1077, LMG 5984, LMG 19352, NCIMB 8229, NRRL 2234, NRRL B-2659, NRRL-ISP 5260, RIA 1185, RIA 606, VKM Ac-849.

Sequence accession no. (16S rRNA gene): AB045883.

409b. **Streptomyces rimosus subsp. paromomycinus** Coffey, Anderson, Fisher, Galbraith, Hillegas, Kohberger, Thompson, Weston and Ehrlich 1959, 730[AL] (*Streptomyces rimosus* forma *paromomycinus* Coffey, Anderson, Fisher, Galbraith, Hillegas, Kohberger, Thompson, Weston and Ehrlich 1959, 737)

pa.r.o.mo.my.ci'nus. N.L. n. *paromomycinum* paromomycin; N.L. masc. adj. *paromomycinus* intented to mean paromomycin-producing.

Fair growth on Czapek's solution agar. Produces the aminocyclitol anti-bacterial antibiotics neomycins E and F (paromycins I and II); produces the glutarimide anti-fungal antibiotic (streptimidone); inhibited by streptomycin.

Type strain shows the highest sequence similarity to: *S. chrestomyceticus*, AJ621609, 100%; *S. albofaciens*, AB045880, 99.7%; *S. lydicus*, Y15507, 99.3%; *S. erumpens*, AJ621603, 99.3%; *S. chattanoogensis*, AJ621611, 99.2%; *S. sclerotialus*, AJ621608, 99.1%; *S. sioyaensis*, DQ026654, 99%.

Source: not known.

DNA G+C content (mol%): not known.

Type strain: ATCC 14827, DSM 41429, NBRC 15454, JCM 4541, JCM 4871, NRRL 2455, VKM Ac-605.

Sequence accession no. (16S rRNA gene): AJ621610.

410. **Streptomyces rishiriensis** Kawaguchi, Tsukiura, Okanishi, Miyaki, Ohmori, Fujisawa and Koshiyama 1965, 3[AL]

ri.shi.ri.en'sis. N.L. masc. adj. *rishiriensis* of or belonging to Rishiri, named for Rishiri Island, Hokkaido, Japan, the source of the soil from which the organism was isolated.

Spore chains may be in Section *Retinaculiaperti*. Moderate growth on Czapek's solution agar. Produces the coumermycin complex (coumermycins A_1, A_2, B, C, and D); inhibited by streptomycin.

Type strain shows the highest sequence similarity to: *S. humidus*, DQ442508, 99.3%; *S. phaeofaciens*, AB184360, 99.2%; *S. puniceus*, DQ442542, 99.1%; *S. cacaoi* subsp. *asoensis*, DQ026644, 99.1%.

Source: isolated from soil from Rishiri Island, Hokkaido, Japan.

DNA G+C content (mol%): not known.

Type strain: ATCC 14812, CBS 708.72, BCRC 12333, DSM 40489, NBRC 13407, JCM 4821, NCIMB 11890, NRRL B-3239, NRRL-ISP 5489, RIA 1368, VKM Ac-1188.

Sequence accession no. (16S rRNA gene): EF178682.

411. **Streptomyces rochei** Berger, Jampolsky and Goldberg *in* Waksman and Lechevalier 1953, 40[AL] ("*Actinomyces rochei*" Preobrazhenskaya, Blinov and Ryabova *in* Gauze, Preobrazhenskaya, Kudrina, Blinov, Ryabova and Sveshnikova 1957, 133)

ro'che.i. N.L. gen. masc. n. *rochei* of Roche.

Spore chains in Section *Spirales*. Spirals are usually open, sometimes almost flexuous. Mature spore chains are moderately long with 10–50, or often more than 50, spores per chain. This morphology is seen on yeast-malt agar, oatmeal agar, salts-starch agar, and glycerol-asparagine agar. Spore surface is smooth to warty.

Color of colony: aerial mass color in the Gray color series on yeast-malt agar, oatmeal agar, salts-starch agar, and glycerol-asparagine agar. Reverse side of colony with no distinctive pigments (grayish yellow to yellowish brown) on yeast-malt agar, oatmeal agar, salts-starch agar, and glycerol-asparagine agar.

Color in medium: melanoid pigments are not formed in peptone-yeast-iron agar, tyrosine agar, and tryptone-yeast broth. No pigment is found in the medium in yeast-malt agar, oatmeal agar, salts-starch agar, or glycerol-asparagine agar.

D-Glucose, L-arabinose, D-xylose, iso-inositol, D-mannitol, D-fructose, and rhamnose are utilized for growth. No growth or only trace of growth on sucrose and raffinose.

Type strain shows the highest sequence similarity to: *S. vinaceusdrappus*, AY999929, 100%; *S. ghanaensis*, AY999851, 100%; *S. geysiriensis*, DQ442501, 100%; *S. minutiscleroticus*, EF178696, 100%; *S. plicatus*, AB184291, 100%; *S. mutabilis*, EF178679, 99.9%; *S. tuirus*, AB184690, 99.5%; *S. djakartensis*, AB184657, 99.4%; *S. anandii*, AB184402, 99.2%; *S. violaceorubidus*, AJ781374, 99.2%; *S. pilosus*, AB184161, 99.1%; *S. flavoviridis*, AB184842, 99.1%; *S. tendae*, D63873, 99%; *S. calvus*, AB184329, 99%; *S. azureus*, EF178674, 99%; *S. asterosporus*, AB184706, 99%; *S. levis*, AB184670, 99%; *S. luteogriseus*, AB184379, 99%; *S. capillispiralis*, AB184577, 99%.

Source: not known.

DNA G+C content (mol%): not known.

Type strain: AS 4.1425, ATCC 10739, ATCC 19245, ATCC 23956, CBS 224.46, CBS 939.68, CCUG 11115, CECT 3329, DSM 40231, HAMBI 2114, IFM 1188, NBRC 12908, IMET 41386, JCM 4074, JCM 4668, LMG 19313, NRRL 3533, NRRL B-1559, NRRL B-2410, NRRL-ISP 5231, RIA 1171, VKM Ac-997.

Sequence accession no. (16S rRNA gene): AB184237.

412. **Streptomyces roseiscleroticus** (Thirumalachar *in* Rautenshtein 1960) Pridham 1970, 42[AL] (*Chainia poonensis* Thirumalachar *in* Rautenshtein 1960, 45)

ro.se.i.scle'ro.ti.cus. L. adj. *roseus* rosy; N.L. n. *sclerotium* a sclerotium; N.L. masc. adj. *roseiscleroticus* rosy, belonging to sclerotium.

Spore chains in Section *Spirales*. Mature spore chains are short to long, sometimes with more then 50 spores per chain. This morphology is sometimes seen on yeast-malt agar, oatmeal agar, salts-starch agar, glycerol-asparagine agar, and or Gause's medium no. 1. Sporulating aerial mycelium is often absent on various media. Spore surface is smooth. Special morphological characteristics: sclerotia are produced on yeast-malt agar, oatmeal agar, salts-starch agar, glycerol-asparagine agar, and Gause's medium no. 1 in 7–14 d.

Color of colony: aerial mass color in the White color series when adequate aerial mycelium is produced on yeast-malt agar, oatmeal agar, salts-starch agar, glycerol-asparagine agar, or Gause's medium no. 1. Reports vary on the media supporting good growth of aerial mycelium. Reverse side of colony is yellow to yellow brown is modified by red to strong (reddish) brown on yeast-malt agar and to reddish orange or deep orange on oatmeal agar. Reverse mycelium on salts-starch agar and glycerol-asparagine agar may be colorless, light yellow, orange, or strong brown. No change is reported in reverse mycelium color after the addition of 0.05 M NaOH or HCl.

Color in medium: melanoid pigments are not formed in peptone-yeast-iron agar, tyrosine agar, or tryptone-yeast broth. No pigment or only a trace of brown is found in the medium in yeast-malt agar, oatmeal agar, salts-starch agar, and glycerol-asparagine agar.

D-Glucose, L-arabinose, D-xylose, D-mannitol, D-fructose, and rhamnose are utilized for growth. Utilization of sucrose, iso-inositol, and raffinose is doubtful.

Type strain shows the highest sequence similarity to: *S. ruber*, AB184604, 100%.

Source: not known.

DNA G+C content (mol%): not known.

Type strain: ATCC 17755, CBS 226.65, CBS 664.72, BCRC 12541, CMI 112788, DSM 40303, NBRC 13002, NBRC 13363, IMET 43586, JCM 3104, JCM 4823, NCIMB 11013, NRRL B-3348, NRRL-ISP 5303, RIA 1324, RIA 887, VKM Ac-1718.

Sequence accession no. (16S rRNA gene): AB184251.

Further comments: Streptomyces roseiscleroticus Pridham 1970 and *Chainia rosea* Thirumalachar et al. 1966 have the same type strain and therefore are homotypic synonyms [Rules 24a and 24b (1) of the *Bacteriological Code* (1990 Revision)].

413. **Streptomyces roseodiastaticus** (Duché 1934) Waksman *in* Waksman and Lechevalier 1953, 27[AL] ["*Actinomyces roseodiastaticus*" Duché 1934, 329; *Streptomyces roseodiastaticus* (Wollenweber 1920) Waksman 1961]

ro.se.o.di.a.sta'ti.cus. L. adj. *roseus* rosy; N.L. adj. *diastaticus* diastatic, starch, digesting; N.L. masc. adj. *roseodiastaticus* rosy, diastatic.

Excellent growth on Czapek's solution agar. Inhibited by streptomycin.

For sequence similarity, see type strain of *Streptomyces tricolor*.

Source: not known.

DNA G+C content (mol%): not known.

Type strain: CBS 102.34, DSM 41703, NBRC 15457, JCM 4295, JCM 13861, NRRL B-1906.

Sequence accession no. (16S rRNA gene): AB184683.

Further comments: according to Lanoot et al. (2004), *Streptomyces roseodiastaticus* (Duché 1934) Waksman 1953 is a later heterotypic synonym of *Streptomyces tricolor* (Wollenweber 1920) Waksman 1961.

414. **Streptomyces roseoflavus** Arai 1951, 218[AL] [*Streptomyces fradiae* (Waksman and Curtis 1916, 125) Waksman and Henrici 1948, 954]

ro.se.o.fla'vus. L. adj. *roseus* rosy; L. adj. *flavus* yellow; N.L. masc. adj. *roseoflavus* rose-yellow.

Spore chains in Section *Retinaculiaperti* or *Spirales*. Flexuous spore chains are common; *Retinaculiaperti* and *Spirales* spore chains may not be seen on some media. Mature spore chains of more than 50 spores per chain may be seen if good sporulation occurs, but sporulation is often poor on yeast-malt agar, oatmeal agar, and salts-starch agar, and especially on glycerol-asparagine agar. Spore surface is smooth.

Color of colony: aerial mass color in the Red or Yellow color series on yeast-malt agar, salts-starch agar, and glycerol-asparagine agar. Yellow color series on oatmeal agar. Representative color tabs from the Red color series are 3ca, pale orange yellow; 4ec, grayish yellowish pink; and 5ca, light yellowish pink. Representative chips from the Yellow color series are 1 or 2ba, pale yellow; and 1½ec, pale yellowish green (yellow colors may represent poor sporulation). Reverse side of colony is moderate to deep orange on yeast-malt agar; grayish yellow to orange yellow or yellowish brown on oatmeal agar, salts-starch agar, and glycerol-asparagine agar.

Color in medium: melanoid pigments are not formed in peptone-yeast-iron agar, tyrosine agar, or tryptone-yeast broth. No pigment is found in the medium in yeast-malt agar, oatmeal agar, or glycerol-asparagine agar; a trace of yellow to orange pigment may or may not be present in salts-starch agar.

D-Glucose, L-arabinose, and D-xylose are utilized for growth. Utilization of fructose is doubtful. No growth or only traces of growth with iso-inositol, D-mannitol, rhamnose, sucrose, or raffinose.

For sequence similarity, see type strain of *Streptomyces fradiae*.

Source: not known.

DNA G+C content (mol%): not known.

Type strain: ATCC 13167, CBS 740.72, DSM 40536, NBRC 13439, JCM 4824, NRRL B-1563, NRRL B-2789, NRRL-ISP 5536, RIA 1400, VKM Ac-1907.

Sequence accession no. (16S rRNA gene): no sequence available.

Further comments: according to Lanoot et al. (2004), *Streptomyces roseoflavus* Arai 1951 is a later heterotypic synonym of *Streptomyces fradiae* (Waksman and Curtis 1916) Waksman and Henrici 1948.

415. **Streptomyces roseofulvus** (Preobrazhenskaya *in* Gauze, Preobrazhenskaya, Kudrina, Blinov, Ryabova and Sveshnikova 1957) Pridham, Hesseltine and Benedict 1958, 61[AL] ("*Actinomyces roseofulvus*" Preobrazhenskaya *in* Gauze, Preobrazhenskaya, Kudrina, Blinov, Ryabova and Sveshnikova 1957, 55)

ro.se.o.ful′vus. L. adj. *roseus* rosy: L. adj. *fulvus* deep yellow; N.L. masc. adj. *roseofulvus* rosy, deep yellow, referring to color of aerial mycelium and vegetative mycelium, respectively.

Spore chains in Section *Rectiflexibiles.* Mature spore chains are generally long and straight, often with more than 50 spores per chain. This morphology is seen on yeast-malt agar, oatmeal agar, and salts-starch agar; poor sporulation on glycerol-asparagine agar. Spore surface is smooth.

Color of colony: aerial mass color in the Red color series on yeast-malt agar, oatmeal agar, and salts-starch agar in 21 d. Younger cultures or cultures on media giving poor sporulation may show white aerial mycelium. Reverse side of colony is colorless or pale grayed yellow to pale pinkish yellow on yeast-malt agar, oatmeal agar, salts-starch agar, and glycerol-asparagine agar. Reverse color is changed from colorless or pale yellow to pale pink by addition of 0.05 M NaOH, or from pinkish yellow to pale yellow with 0.05 M HCl.

Color in medium: melanoid pigments not formed in peptone-yeast-iron agar, tyrosine agar, or tryptone-yeast extract broth. Traces of pale grayish yellow or pale brownish gray pigment may be found in yeast-malt agar, oatmeal agar, and salts-starch agar. This faint pigment may be pH-sensitive (one observer only); the change is from pale yellow to pale pink by 0.05 M NaOH.

D-Glucose, L-arabinose, sucrose, D-xylose, D-fructose, rhamnose, and raffinose are utilized for growth. No growth or only trace of growth on iso-inositol and D-mannitol.

Type strain shows the highest sequence similarity to: *S. laurentii*, AJ781342, 99.2%; *S. roseolus*, AB184168, 99.2%.

Source: not known.

DNA G+C content (mol%): not known.

Type strain: ATCC 19805, ATCC 19921, CBS 557.68, BCRC 12051, DSM 40172, NBRC 13194, NBRC 15816, NBRC 15817, INA 14535, JCM 4334, JCM 4605, NRRL B-2729, NRRL-ISP 5172, RIA 1084, UNIQEM 191, VKM Ac-1080.

Sequence accession no. (16S rRNA gene): AB184327.

416. **Streptomyces roseolilacinus** (Preobrazhenskaya and Sveshnikova *in* Gauze, Preobrazhenskaya, Kudrina, Blinov, Ryabova and Sveshnikova 1957) Pridham, Hesseltine and Benedict 1958, 68[AL] ("*Actinomyces roseolilacinus*" Preobrazhenskaya and Sveshnikova *in* Gauze, Preobrazhenskaya, Kudrina, Blinov, Ryabova and Sveshnikova 1957, 35)

ro.se.o.li.la.ci′nus. L. adj. *roseus* rosy; N.L. adj. *lilacinus* lilac colored; N.L. masc. adj. *roseolilacinus* rose, lilac colored, referring to color of the aerial mycelium of the organism.

Spore chains in Section *Retinaculiaperti* to *Spirales.* Spirals are prominent on salts-starch agar, but spore chains representative of Section *Retinaculiaperti* are more common on yeast-malt agar and oatmeal agar; poor sporulation on glycerol-asparagine agar. Mature spore chains generally have 10–50 spores per chain. Spore surface is smooth.

Color of colony: aerial mass color in the Red color series on yeast-malt agar, oatmeal agar, and salts-starch agar. Reverse side of colony with no distinctive pigment on yeast-malt agar, oatmeal agar, salts-starch agar, and glycerol-asparagine agar; substrate pigment is not a pH indicator (or shows only slight change with 0.05 M HCl or NaOH on salts-starch agar and glycerol-asparagine agar).

Color in medium: melanoid pigments not formed in peptone-yeast-iron agar and tyrosine agar. Pigments not formed in medium or found only as traces of pale yellowish or brownish gray, in yeast-malt agar, oatmeal agar, salts-starch agar, and glycerol-asparagine agar.

D-Glucose and L-arabinose are utilized for growth. No growth or only trace of growth on sucrose, D-xylose, iso-inositol, D-mannitol, rhamnose, and raffinose. Doubtful growth with D-fructose.

Type strain shows no sequence similarity over 99%.

Source: not known.

DNA G+C content (mol%): not known.

Type strain: ATCC 19806, ATCC 19922, CBS 558.68, BCRC 12329, DSM 40173, NBRC 12815, INA 14250, JCM 4335, JCM 4606, NRRL B-2699, NRRL-ISP 5173, RIA 1085, UNIQEM 192, VKM Ac-1276.

Sequence accession no. (16S rRNA gene): AB184167.

417. **Streptomyces roseolus** (Preobrazhenskaya and Sveshnikova *in* Gauze, Preobrazhenskaya, Kudrina, Blinov, Ryabova and Sveshnikova 1957) Pridham, Hesseltine and Benedict 1958, 61[AL] ("*Actinomyces roseolus*" Preobrazhenskaya and Sveshnikova *in* Gauze, Preobrazhenskaya, Kudrina, Blinov, Ryabova and Sveshnikova 1957, 37)

ro.se.o′lus. N.L. dim. masc. adj. *roseolus* somewhat rosy, referring to the color of the aerial mycelium.

Spore chains in Section *Rectiflexibiles.* Mature spore chains are generally long with 10–50 spores per chain on yeast-malt agar, oatmeal agar, salts-starch agar, and glycerol-asparagine agar. Spore surface is smooth.

Color of colony: aerial mass color in the Red color series on yeast-malt agar and salts-starch agar, and in the Red or Yellow series on oatmeal agar. Sporulation may be poor on glycerol-asparagine agar. Reverse side of colony with no distinctive pigment on yeast-malt agar, oatmeal agar, salts-starch agar, and glycerol-asparagine agar; substrate pigment is not a pH indicator.

Color in medium: melanoid pigments not formed in peptone-yeast-iron agar and tyrosine agar. No pigment found in medium in yeast-malt agar, oatmeal agar, salts-starch agar, or glycerol-asparagine agar.

D-Glucose, L-arabinose, D-xylose, and rhamnose are utilized for growth. No growth or only trace of growth on sucrose, iso-inositol, D-mannitol, and raffinose. Variable reports of slight growth with D-fructose.

Type strain shows the highest sequence similarity to: *S. filamentosus*, AB184130, 99.4%; *S. roseofulvus*, AB184327, 99.2%; *S. tanashiensis*, AJ781362, 99.1%; *S. nashvillensis*, AB184286, 99%.

Source: not known.

DNA G+C content (mol%): not known.

Type strain: ATCC 23210, CBS 559.68, BCRC 13778, DSM 40174, NBRC 12816, INA 5449/54, JCM 4411, NCIMB 13022, NRRL B-5424, NRRL-ISP 5174, RIA 1086, UNIQEM 193, VKM Ac-848.

Sequence accession no. (16S rRNA gene): AB184168.

418. **Streptomyces roseosporus** Falcão de Morais and Dália Maia 1961, 41[AL] [*Streptomyces venezuelae* subsp. *roseospori* (*sic*) Falcão de Morais, Dália Maia and Souto Maior Genn 1958, 102; *Streptomyces filamentosus* Okami and Umezawa *in* Okami, Okuda, Takeuchi, Nitta and Umezawa 1953, 153]

ro.se.o.spo′rus. L. adj. *roseus* rosy; N.L. n. *spora* a spore; N.L. masc. adj. *roseosporus* rosy-spored.

Spore chains in Section *Rectiflexibiles.* Mature spore chains are generally long, often with more than 50 spores per chain. This morphology is seen on yeast-malt agar, oatmeal agar, salts-starch agar, and glycerol-asparagine agar. Spore surface is smooth.

Color of colony: aerial mass color in the Red color series on yeast-malt agar, oatmeal agar, and salts-starch agar (White, Yellow, or Red color series on glycerol-asparagine agar). Reverse side of colony with no distinctive pigments (grayish yellow to yellowish brown) on yeast-malt agar, oatmeal agar, salts-starch agar, and glycerol-asparagine agar.

Color in medium: melanoid pigments are not formed in peptone-yeast-iron agar, tyrosine agar, or tryptone-yeast broth. No pigment is found in the medium in yeast-malt agar, oatmeal agar, salts-starch agar, or glycerol-asparagine agar.

D-Glucose, L-arabinose, D-xylose, and rhamnose are utilized for growth. Utilization of D-fructose is doubtful. No growth or only trace of growth with sucrose, iso-inositol, D-mannitol, and raffinose.

For sequence similarity, see type strain of *Streptomyces filamentosus.*

Source: not known.

DNA G+C content (mol%): not known.

Type strain: ATCC 23958, CBS 941.68, BCRC 13771, DSM 40122, NBRC 12910, JCM 4412, KCTC 9568, NCIMB 13008, NRRL B-5411, NRRL-ISP 5122, RIA 1125.

Sequence accession no. (16S rRNA gene): AB184238.

Further comments: according to Lanoot et al. (2004), *Streptomyces roseosporus* Falcão de Morais and Dália Maia 1961 is a later heterotypic synonym of *Streptomyces filamentosus* Okami and Umezawa 1953.

419. **Streptomyces roseoverticillatus** (Shinobu 1956) Witt and Stackebrandt 1991, 456[VP] (Effective publication: Witt and Stackebrandt 1990, 370.) ("*Streptomyces roseoverticillatus*" Shinobu 1956; "*Verticillomyces roseoverticillatus*" Shinobu 1965; *Streptoverticillium roseoverticillatum* Farina and Locci 1966, 49)

ro.se.o.vert.i.cil.la′tus. L. adj. *roseus* rosy; L. masc. n. *verticillus* whorl; N.L. adj. *verticillatus* whorled; N.L. masc. adj. *roseoverticillatus* rosy whorled.

Spore chains in Umbellate Monoverticillate (=*Streptomyces* Section Verticillati, biverticillate). Mature spore chains are short, generally 3–10 spores per chain. This morphology is seen on yeast-malt agar, oatmeal agar, salts-starch agar, and glycerol-asparagine agar. Spore surface is smooth.

Color of colony: aerial mass color in the Red color series on yeast-malt agar, oatmeal agar, salts-starch agar, and glycerol-asparagine agar. Reverse side of colony is grayed yellow is modified by red on yeast-malt agar, oatmeal agar, salts-starch agar, and glycerol-asparagine agar. Only slight changes, if any, occur when pH is changed by addition of 0.05 NaOH or HCl.

Color in medium: melanoid pigments formed on peptone-yeast-iron agar and tryptone-yeast extract broth. Traces of yellow, orange, or red pigment may be found in yeast-malt agar and oatmeal agar. This pigment, when present, is not affected by pH or is changed slightly from orange in the presence of 0.05 M NaOH to pink in presence of 0.05 M HCl.

D-Glucose is utilized for growth. No growth or only trace of growth on L-arabinose, D-xylose, D-mannitol, rhamnose, and raffinose. Variable reports of doubtful growth to no growth with sucrose, iso-inositol, and D-fructose.

For sequence similarity, see type strain of *Streptomyces hiroshimensis.*

Source: not known.

DNA G+C content (mol%): not known.

Type strain: ATCC 19807, CBS 560.68, CECT 3269, DSM 40039, NBRC 12817, NBRC 3844, JCM 4103, JCM 4607, NRRL B-1993, NRRL-ISP 5039, PCM 2248, RIA 1087, RIA 552, UNIQEM 194, VKM Ac-880.

Sequence accession no. (16S rRNA gene): AB184169.

Further comments: Streptomyces roseoverticillatus (Shinobu 1956) Witt and Stackebrandt 1991 is an earlier synonym of *Streptomyces baldaccii* corrig. (Farina and Locci 1966) Witt and Stackebrandt 1991, an earlier synonym of *Streptomyces biverticillatus* (Preobrazhenskaya and Ryabova 1957) Witt and Stackebrandt 1991, and an earlier synonym of *Streptomyces fervens* (DeBoer et al. 1959–1960) Witt and Stackebrandt 1991.

Labeda (1996) proposes *Streptomyces biverticillatus* (Preobrazhenskaya and Ryabova 1957) Witt and Stackebrandt 1991, *Streptomyces fervens* (DeBoer et al. 1959–1960) Witt and Stackebrandt 1991, and *Streptomyces roseoverticillatus* (Shinobu 1956) Witt and Stackebrandt 1991 as later synonyms of *Streptomyces baldaccii* corrig. (Farina and Locci 1966) Witt and Stackebrandt 1991. This proposal is in violation of Rule 24b(1) of the *Bacteriological Code* (1990 Revision) because the senior epithet is *roseoverticillatus. Streptomyces baldaccii, Streptomyces biverticillatus,* and *Streptomyces fervens* are therefore to be regarded as later synonyms of *Streptomyces roseoverticillatus.*

According to Hatano et al. (2003), *Streptomyces roseoverticillatus* (Shinobu 1956) Witt and Stackebrandt 1991 is a later heterotypic synonym of *Streptomyces hiroshimensis* (Shinobu 1955) Witt and Stackebrandt 1991.

420. **Streptomyces roseoviolaceus** (Sveshnikova *in* Gauze, Preobrazhenskaya, Kudrina, Blinov, Ryabova and Sveshnikova 1957) Pridham, Hesseltine and Benedict 1958, 68[AL] ("*Actinomyces roseoviolaceus*" Sveshnikova *in* Gauze, Preobrazhenskaya, Kudrina, Blinov, Ryabova and Sveshnikova 1957, 67)

ro.se.o.vi.o.la′ce.us. L. adj. *roseus* rosy; L. adj. *violaceus* violet colored; N.L. adj. *roseoviolaceus* rosy, violet colored, referring to color of aerial mycelium, vegetative mycelium, and diffusible pigment.

Spore chains in Section *Spirales.* Mature spore chains are moderately long with 10 to 50 or more spores per chain.

This morphology is seen on yeast-malt agar, oatmeal agar, salts-starch agar, and glycerol-asparagine agar. Spore surface is spiny.

Color of colony: aerial mass color in the Red color series on oatmeal agar and salts-starch agar; Red or Violet color series on yeast-malt agar and glycerol-asparagine agar. The most representative color-tabs in the Red color series are 5cb or 5ec, grayish yellowish pink and the most representative tab in the Violet color series is 11ca, very pale purple. Reverse side of colony is purplish pink or purplish red on yeast-malt agar, oatmeal agar, salts-starch agar, and glycerol-asparagine agar. Reverse mycelium pigment is a pH indicator changing from red or pink to violet or blue-violet with addition of 0.05 M NaOH and violet to red or pinkish orange with addition of 0.05 M HCl.

Color in medium: melanoid pigments are formed in peptone-yeast-iron agar, tyrosine agar, and tryptone-yeast broth. Red to violet pigment is found in the medium in yeast-malt agar, oatmeal agar, salts-starch agar, and glycerol-asparagine agar. This pigment is pH-sensitive, showing the same changes noted for the reverse mycelium pigment.

D-Glucose, L-arabinose, D-xylose, iso-inositol, D-mannitol, D-fructose, rhamnose, sucrose, and raffinose are all utilized for growth.

Type strain shows the highest sequence similarity to: *S. janthinus*, AB184851, 100%; *S. violaceus*, AB184315, 100%; *S. albosporeus* subsp. *albosporeus*, AJ781327, 100%; *S. luteogriseus*, AB184379, 99.5%; *S. lomondensis*, AB184673, 99.3%; *S. hawaiiensis*, AB184143, 99.2%; *S. flavoviridis*, AB184842, 99.2%; *S. pilosus*, AB184161, 99.2%; *S. arenae*, AB249977, 99.2%; *S. africanus*, AY208912, 99.2%; *S. tuirus*, AB184690, 99.2%; *S. bellus*, AB184849, 99%; *S. mutabilis*, EF178679, 99%; *S. massasporeus*, AB184152, 99%; *S. afghaniensis*, AJ399483, 99%; *S. levis*, AB184670, 99%; *S. coerulescens*, AY999720, 99%; *S. parvulus*, AB184326, 99%; *S. coeruleorubidus*, AY999719, 99%.

Source: not known.

DNA G+C content (mol%): not known.

Type strain: 1020/54, ATCC 25493, CBS 953.69, NBRC 13081, JCM 4513, KCC S-0513, NRRL B-12177, NRRL-ISP 5277, RIA 1273.

Sequence accession no. (16S rRNA gene): AJ399484.

421. **Streptomyces roseoviridis** (Preobrazhenskaya *in* Gauze, Preobrazhenskaya, Kudrina, Blinov, Ryabova and Sveshnikova 1957) Pridham, Hesseltine and Benedict 1958, 61^{AL} ("*Actinomyces roseoviridis*" Preobrazhenskaya *in* Gauze, Preobrazhenskaya, Kudrina, Blinov, Ryabova and Sveshnikova 1957, 57)

ro.se.o.vi'ri.dis. L. adj. *roseus* rosy; L. adj. *viridis* green; N.L. masc. adj. *roseoviridis* rosy green, referring to the rosy aerial mycelium and green vegetative mycelium and diffusible pigment.

Spore chains in Section *Rectiflexibiles*. Mature spore chains generally 10–50 spores per chain. This morphology is seen on yeast-malt agar, oatmeal agar, salts-starch agar, and glycerol-asparagine agar. Spore surface is smooth.

Color of colony: aerial mass color in the Red or Yellow color series on yeast-malt agar, oatmeal agar, salts-starch agar, and glycerol-asparagine agar. Reverse side of colony with no distinctive pigment on yeast-malt agar, oatmeal agar, salts-starch agar, and glycerol-asparagine agar.

Color in medium: melanoid pigments are formed in peptone-yeast-iron agar, tyrosine agar, and tryptone-yeast broth. Pigments other than melanoids are not formed in yeast-malt agar, oatmeal agar, salts-starch agar, or glycerol-asparagine agar.

D-Glucose, L-arabinose, and D-xylose are utilized for growth. No growth or only trace of growth on sucrose, iso-inositol, D-mannitol, rhamnose, and raffinose. Utilization of D-fructose is doubtful.

Type strain shows the highest sequence similarity to: *S. viridobrunneus*, AJ781372, 99.4%; *S. showdoensis*, AB184389, 99.4%; *S. nashvillensis*, AB184286, 99.1%; *S. tanashiensis*, AJ781362, 99.1%; *S. filamentosus*, AB184130, 99%.

Source: not known.

DNA G+C content (mol%): not known.

Type strain: ATCC 23959, CBS 942.68, BCRC 13779, DSM 40175, NBRC 12911, INA 3617, JCM 4414, NCIMB 13012, NRRL B-2730, NRRL-ISP 5175, RIA 1135, VKM Ac-943.

Sequence accession no. (16S rRNA gene): AB184239.

422. **Streptomyces ruber** (Shirling and Gottlieb 1972) Goodfellow, Williams and Alderson 1986a, 575^{AL} (Effective publication: Goodfellow, Willams and Alderson 1986d, 58.) (*Chainia rubra* Shirling and Gottlieb 1972, 347.)

ru'ber. L. masc. adj. *ruber* red.

Forms extensively branched substrate and aerial mycelium. Spiral spore chains containing 10 to over 50 spores are borne on the aerial mycelium. Spore surface is smooth or bears very short spines. Aerial spore mass is white or red and the substrate mycelium red. Sclerotia are formed in 14–21 d on agar medium. Melanin pigments are not produced. Xanthine is degraded but urea is not. Nitrate reductase is not formed. L-Arabinose, D-fructose, D-glucose, D-mannitol, rhamnose, and D-xylose are used as sole carbon sources, but raffinose is not utilized. Acid is formed from D-lactose, D-mannitol, melibiose, methyl α-D-glucoside, L-rhamnose, sucrose, and D-xylose but not from adonitol, dulcitol, *meso*-erythritol, *myo*-inositol, raffinose, or D-sorbitol. Grows at 42°C but not at 10°C. The organism contains major amounts of hexa- and octa-hydrogenated menaquinones with nine isoprene units (Collins et al., 1984).

Type strain shows the highest sequence similarity to: *S. roseiscleroticus*, AB184251, 100%; *S. poonensis*, DQ445792, 99.1%; *S. spiralis*, EF178683, 99%.

Source: not known.

DNA G+C content (mol%): not known.

Type strain: ATCC 17754, CBS 228.65, BCRC 12358, CMI 112789, DSM 40304, NBRC 14600, JCM 3131, KCC A-0131, NCIB (now NCIMB) 10983, NRRL B-5315, NRRL-ISP 5304.

Sequence accession no. (16S rRNA gene): AB184604.

423. **Streptomyces rubidus** Xu, Wang, Cui, Huang, Liu, Zheng and Goodfellow 2006, 1113^{VP}

ru'bi.dus. L. masc. adj. *rubidus* dark red.

Neutrotolerant, acidophilic streptomycete that forms branched substrate and aerial hyphae. Smooth-surfaced spores are borne in flexuous spore chains. Mahogany-colored substrate mycelium, sparse aerial hyphae and diffusible pigments are formed on oatmeal agar and yeast extract-malt extract agar. Melanin pigments are not produced on tyrosine agar or peptone-yeast extract-iron agar. Starch and Tween 80 are degraded, but adenine, guanine, and xanthine are not. Cellobiose, D-galactose, D-glucose, D-lactose, D-mannitol, D-salicin (each at 1%, w/v), L-arginine, and sodium citrate (0.1%, w/v) are used as sole carbon sources for energy and growth, but adonitol, inulin, D-sorbitol (each at 1%, w/v), adipic acid, L-alanine, DL-aminobutyric acid, α-L-aspartic acid, L-cysteine, L-phenylalanine, sodium acetate, sodium oxalate, and L-valine (each at 0.1%, w/v) are not. L-Alanine is used as a sole carbon and nitrogen source, but L-arginine, α-L-aspartic acid, L-glutamic acid, L-isoleucine, and L-phenylalanine (each at 0.1%, w/v) are not. Growth occurs at temperatures between 20 and 37°C, but not at 15°C, and at pH values from 4.5 to7.0, but not at pH 3.5. Does not grow in the presence of 5% (w/v) NaCl. Sensitive to filter-paper discs soaked in the following (μg/ml unless indicated): azithromycin (30), azetreonam (30), carbenicillin (10), cephalothin (30), ciprofloxacin (5), doxycycline hydrochloride (30), erythromycin (15), josamycin (15), kanamycin sulfate (30), minocycline hydrochloride (30), neomycin sulfate (30), streptomycin sulfate (10), tetracycline hydrochloride (30), and tobramycin sulfate (10), but resistant to filter-paper discs soaked in acetylspiramycin (15), amoxycillin (10), ampicillin (10),ofloxacin (5), penicillin G (10 IU/ml), rifampin (5), and sulfamethoxazole (25). Chemotaxonomic properties are typical of the genus *Streptomyces.*

Type strain shows no sequence similarity over 99%. Type strain shows DNA–DNA similarity to: *S. yeochonensis* NRRL B-24245^T, 43.0%; *S. guanduensis* JCM 13274^T, 21.4%; *S. paucisporeus* JCM 13276^T, 19.9%; *S. yanglinensis* JCM 13275^T, 23.8%.

Source: the type strain was isolated from a pine-forest Yanglin, Yunnan Province, People's Republic of China.

DNA G+C content (mol%): 70.6.

Type strain: 13c15, CGMCC 4.2026, JCM 13277.

Sequence accession no. (16S rRNA gene): AY876941.

424. **Streptomyces rubiginosohelvolus** (Kudrina *in* Gauze, Preobrazhenskaya, Kudrina, Blinov, Ryabova and Sveshnikova 1957) Pridham, Hesseltine and Benedict 1958, 59^{AL} ("*Actinomyces rubiginosohelvolus*" Kudrina *in* Gauze, Preobrazhenskaya, Kudrina, Blinov, Ryabova and Sveshnikova 1957, 89)

ru.bi.gi.no.so.hel'vo.lus. L. masc. adj. *rubiginosus* rusty; L. adj. *helvolus* pale yellow, yellowish; N.L. masc. adj. *rubiginosohelvolus* rusty, yellowish.

Spore chains in Section *Rectiflexibiles.* Aerial mycelium is sometimes poorly developed on ISP media. Mature spore chains, when formed, contain 10 to 50 or more spores per chain on yeast-malt agar, oatmeal agar, salts-starch agar, and glycerol-asparagine agar. Spore surface is smooth.

Color of colony: aerial mass color in the Yellow color series on yeast-malt agar, salts-starch agar, and glycerol-asparagine agar. Aerial mycelium is poorly developed on oatmeal agar. One observer also recorded tabs 2ca (pale yellow) and 3ca (pale orange yellow) from the Red color series as a second color on yeast-malt agar and glycerol-asparagine agar. Reverse side of colony is colorless to pale yellow or grayed yellow on yeast-malt agar, oatmeal agar, salts-starch agar, and glycerol-asparagine agar. However, yellowish reverse pigment is a pH indicator changing from yellow to pink with addition 0.05 M NaOH.

Color in medium: melanoid pigments not formed in peptone-yeast-iron agar, tyrosine agar, or tryptone-yeast broth. No pigments or only trace of yellow pigment found in yeast-malt agar, oatmeal agar, salts-starch agar, or glycerol-asparagine agar.

D-Glucose, L-arabinose, D-xylose, D-mannitol, D-fructose, and rhamnose are utilized for growth. No growth or only trace of growth on sucrose, iso-inositol, and raffinose.

Type strain shows the highest sequence similarity to: *S. sindenensis,* AB184759, 100%; *S. albovinaceus,* AB249958, 100%; *S. griseoplanus,* AY999894, 100%; *S. anulatus,* DQ026637, 100%; *S. globisporus* subsp. *globisporus,* EF178686, 100%; *S. pluricolorescens,* DQ442540, 100%; *S. praecox,* AB184293, 100%; *S. flavofuscus,* AB249935, 100%; *S. fimicarius,* AY999784, 100%; *S. mediolani,* AB184674, 100%; *S. badius,* AY999783, 100%; *S. griseinus,* AB184205, 100%; *S. californicus,* AB184755, 99.9%; *S. lavendulae* subsp. *lavendulae,* AB184080, 99.9%; *S. cavourensis* subsp. *washingtonensis,* DQ026671, 99.9%; *S. acrimycini,* AY999889, 99.9%; *S. parvus,* DQ442537, 99.9%; *S. fulvorobeus,* AB184711, 99.8%; *S. lipmanii,* AB184148, 99.8%; *S. microflavus,* DQ445795, 99.8%; *S. cyaneofuscatus,* AB184860, 99.8%; *S. floridae,* AB184656, 99.8%; *S. alboviridis,* AB184256, 99.8%; *S. baarnensis,* EF178688, 99.8%; *S. cinereorectus,* AB184646, 99.8%; *S. griseus* subsp. *griseus,* AY207604, 99.7%; *S. griseolus,* AB184768, 99.7%; *S. luridiscabiei,* AF361784, 99.7%; *S. flavovirens,* DQ026635, 99.7%; *S. argenteolus,* AB045872, 99.7%; *S. bacillaris,* AB184439, 99.6%; *S. pulveraceus,* AB184806, 99.6%; *S. halstedii,* EF178695, 99.6%; *S. flavogriseus,* AJ494864, 99.6%; *S. atroolivaceus,* AJ781320, 99.5%; *S. olivoviridis,* AB184227, 99.5%; *S. finlayi,* AY999788, 99.5%; *S. nitrosporeus,* EF178680, 99.5%; *S. albolongus,* AB184425, 99.4%; *S. clavifer,* DQ026670, 99.4%; *S. griseobrunneus,* AB249912, 99.4%; *S. sanglieri,* AB249945, 99.4%; *S. cellulofiavus,* AB184476, 99.4%; *S. yanii,* AB006159, 99.4%; *S. gelaticus,* DQ026636, 99.3%; *S. atratus,* DQ026638, 99.3%; *S. mutomycini,* AB249951, 99.2%; *S. candidus,* DQ026663, 99.2%; *S. cavourensis* subsp. *cavourensis,* DQ445791, 99.2%; *S. spiroverticillatus,* AB184814, 99.1%; *S. cremeus,* AB184124, 99.1%.

Source: not known.

DNA G+C content (mol%): not known.

Type strain: ATCC 19926, ATCC 23960, CBS 943.68, BCRC 13780, DSM 40176, NBRC 12912, INA 10/53, JCM 4415, NRRL B-5425, NRRL-ISP 5176, RIA 1136, VKM Ac-1072.

Sequence accession no. (16S rRNA gene): AB184240.

425. **Streptomyces rubiginosus** (Preobrazhenskaya, Blinov and Ryabova *in* Gauze, Preobrazhenskaya, Kudrina, Blinov, Ryabova and Sveshnikova 1957) Pridham, Hesseltine and Benedict 1958, 70^{AL} ("*Actinomyces rubiginosus*"

Preobrazhenskaya, Blinov and Ryabova *in* Gauze, Preobrazhenskaya, Kudrina, Blinov, Ryabova and Sveshnikova 1957, 134)

ru.bi.gi.no′sus. L. masc. adj. *rubiginosus* rusty, referring to the red-gray-brown color of vegetative mycelium.

Spore chains in Section *Spirales*. Spirals are best developed on salts-starch agar. Aerial mycelium and spore chains may be poorly developed and atypical on yeast-malt agar, oatmeal agar, and glycerol-asparagine agar. Mature spore chains generally have 10–50 spores per chain. Spore surface is spiny.

Color of colony: aerial mass color in the Gray color series on yeast-malt agar, oatmeal agar, salts-starch agar, and glycerol-asparagine agar. Reverse side of colony with no distinctive pigment (yellowish brown, grayed orange brown or yellowish gray) on yeast-malt agar, oatmeal agar, salts-starch agar, and glycerol-asparagine agar.

Color in medium: melanoid pigments not formed in peptone-yeast-iron agar, tyrosine agar, or tryptone-yeast broth. No pigment found in medium in yeast-malt agar, salts-starch agar, or glycerol-asparagine agar. There may be a trace of red soluble pigment in oatmeal agar; this pigment is not pH-sensitive.

D-Glucose, sucrose, D-xylose, iso-inositol, D-mannitol, D-fructose, and rhamnose are utilized for growth. Growth on sucrose is less abundant than on other carbon sources. No growth or only trace of growth on raffinose.

Type strain shows the highest sequence similarity to: *S. pseudogriseolus*, DQ442541, 99.4%; *S. gancidicus*, AB184660, 99.4%; *S. capillispiralis*, AB184577, 99.3%; *S. cellulosae*, DQ442495, 99.2%.

Source: not known.

DNA G+C content (mol%): not known.

Type strain: ATCC 19927, ATCC 23961, CBS 944.68, DSM 40177, NBRC 12913, INA 11852, JCM 4416, KCTC 9042, NRRL B-3983, NRRL-ISP 5177, RIA 1137, VKM Ac-1089.

Sequence accession no. (16S rRNA gene): AY999810.

426. **Streptomyces rubrogriseus** (*ex* Kurylowicz et al. *in* Prauser 1970) Terekhova 1986, 575[VP] (Effective publication: Terekhova *in* Gause, Preobrazhenskaya, Sveshnikova, Terekhova and Maximova 1983.) ("*Actinomyces rubrogriseus*" Kurylowicz et al. *in* Prauser 1970)

Synonym: "*Actinomyces rubrogriseus*" Kurylowicz *et al. in* Prauser (1970).

ru.bro.gri′se.us. L. adj. *ruber -bra -brum* red; N.L. adj. *griseus* gray; N.L. masc. adj. *rubrogriseus* red, gray.

Spore chains are spiral (*Spirales*); spores are smooth. On mineral agar 1, oatmeal agar: aerial mycelium is light gray, sometimes pink; substrate mycelium is red; no diffusible pigment. On glycerol-nitrate agar, starch ammonia agar, glycerol-asparagine agar: aerial mycelium is white to light gray; substrate mycelium is yellowish to red, raspberry red; no diffusible pigment. On organic agar 2: aerial mycelium is white to light gray; substrate mycelium is yellowish red to raspberry red; no diffusible pigment. Melanoid pigments are not formed. Antibiotic: prodigiosin-pigment; streptothricin.

Type strain shows the highest sequence similarity to: *S. anthocyanicus*, AB184631, 99.9%; *S. lienomycini*, AJ781353, 99.9%; *S. tricolor*, AB184687, 99.9%; *S. humiferus*, AF503491, 99.7%; *S. coelescens*, AF503496, 99.7%; *S. violaceorubidus*, AJ781374, 99.7%; *S. violaceolatus* AF503497, 99.7%; *S. violaceoruber*, AF503492, 99.7%; *S. tendae*, D63873, 99.6%; *S. coelicoflavus*, AB184650, 99.3%; *S. olivaceus*, AB184743, 99.2%; *S. pactum*, AB184398, 99.2%; *S. ambofaciens*, M27245, 99%; *S. mutabilis*, EF178679, 99%; *S. matensis*, AB184221, 99%; *S. malachitospinus*, AB249954, 99%.

Source: not known.

DNA G+C content (mol%): not known.

Type strain: ATCC 43691, DSM 41477, NBRC 15455, INA 2626, JCM 6927, VKM Ac-1216.

Sequence accession no. (16S rRNA gene): AB184681.

427a. **Streptomyces rutgersensis subsp. rutgersensis** (Waksman and Curtis 1916) Waksman and Henrici *in* Breed, Murray and Hitchens 1948, 952[AL] ("*Actinomyces rutgersensis*" Waksman and Curtis 1916, 123)

rut.gers.en′sis. N.L. masc. adj. *rutgersensis* of or belonging to Rutgers; named for Rutgers University, New Brunswick, New Jersey, USA.

Spore chains in Section *Rectiflexibiles* to *Retinaculiaperti*. Short spore chains on this strain are irregularly bent or turned and are not typical of either *Rectiflexibiles* or *Retinaculiaperti* cultures [the characterization of Waksman and Curtis (1916) describes spiral spore chains; atypical spirals are now very rare and reported by only one observer]. Usually only 3–10 spores per chain on yeast-malt agar, oatmeal agar, salts-starch agar, and glycerol-asparagine agar. Spore surface is smooth.

Color of colony: aerial mass color in the Yellow color series on yeast-malt agar, oatmeal agar, salts-starch agar, and glycerol-asparagine agar. Reverse side of colony is grayed yellow modified by green on yeast-malt agar, oatmeal agar, salts-starch agar, and glycerol-asparagine agar; substrate color is not a pH indicator.

Color in medium: melanoid pigments not formed in peptone-yeast-iron agar, tyrosine agar, and tryptone-yeast extract broth. Pigments not formed in significant amounts in yeast-malt agar, oatmeal agar, salts-starch agar, or glycerol-asparagine agar.

D-Glucose, L-arabinose, D-xylose, D-mannitol, and D-fructose are utilized for growth. No growth or only trace of growth on sucrose, iso-inositol, rhamnose, and raffinose.

Type strain shows the highest sequence similarity to: *S. diastaticus* subsp. *diastaticus*, AB184785, 100%; *S. gougerotii*, AB184742, 100%; *S. intermedius*, AB184277, 99.8%; *S. misionensis*, EF178678, 99.2%; *S. phaeoluteichromatogenes*, AJ391814, 99.1%; *S. matensis*, AB184221, 99%; *S. aureoverticillatus*, AY999774, 99%.

Source: not known.

DNA G+C content (mol%): not known.

Type strain: ATCC 19809, ATCC 3350, CBS 562.68, BCRC 13670, BCRC 13701, DSM 40077, HAMBI 1038, HUT 6069, IFM 1033, NBRC 12819, NBRC 3419, NBRC 3727, IMET 43501, JCM 4082, JCM 4608, NRRL B-1256, NRRL B-2102, NRRL-ISP 5077, RIA 1089, RIA 410, UNIQEM 196, VKM Ac-1877.

Sequence accession no. (16S rRNA gene): AB184170.

427b. **Streptomyces rutgersensis subsp. castelarensis** corrig. Cercós 1954, 263[AL]

cas.te.lar.en'sis. N.L. masc. adj. *castelarensis* of or belonging to Castelar, Argentina, South America, the source of the organism (from dust).

Spores are short, cylindrical, phalangiform, and may appear roughened. Probably grows well on Czapek's solution agar. Produces camphomycin (two components); NaCl tolerance >4%, but <7%.

Type strain shows the highest sequence similarity to: *Streptomyces sporoclivatus*, AB249934, 99.8; *Streptomyces antimycoticus*, AB184185, 99.8; *Streptomyces geldanamycininus*, DQ334781, 99.6; *Streptomyces melanosporofaciens*, AJ271887, 99.4; *Streptomyces asiaticus*, AB249947, 99.1; *Streptomyces rhizosphaericus*, AB249941, 99.1%.

Source: isolated from dust from Castelar, Argentina, South America.

DNA G+C content (mol%): not known.

Type strain: ATCC 15191, CBS 309.55, BCRC 11879, DSM 40830, NBRC 15875, INA R-43, JCM 4978, RIA 851, VKM Ac-832.

Sequence accession no. (16S rRNA gene): AY508511.

Further comments: the original spelling, *Streptomyces rutgersensis* subsp. *castelarense* (*sic*), has been corrected by Hill et al. (1984).

428. **Streptomyces salmonis** (Baldacci, Farina and Locci 1966) Witt and Stackebrandt 1996, 836[VP] (Effective publication: Witt and Stackebrandt 1990, 370.) ["*Streptoverticillium salmonicida*" (Rucker 1949) Baldacci, Farina and Locci 1966, 164; *Streptoverticillium salmonis* Locci, Baldacci and Petrolini Baldan 1969, 27]

sal'mo.nis. L. gen. n. *salmonis* of a salmon.

Good growth on potato-glucose agar (Baldacci et al., 1954). Color: reverse, brick red; aerial mycelium, yellowish white; traces of soluble pigment. On Bacto Czapek agar: traces of growth only. On Casamino acids-Czapek agar (1 g/l Difco vitamin-free Casamino acids, replacing sodium nitrate): very poor, colorless growth. On glucose-asparagine agar (ISP medium 5 with 1% glucose replacing glycerol): good growth. Color: reverse, orange-yellow; aerial mycelium, white with traces of pink and yellow. On glucose-asparagine agar: good growth. Color: reverse orange-yellow; aerial mycelium, whitish with shades of yellow. On yeast extract-malt extract agar: good growth. Color: reverse, brick red; aerial mycelium, yellowish white. On Bacto Emerson agar: good growth. Color: reverse, brick red; aerial mycelium, dirty white; brown soluble pigment. On Bennett agar (1% glucose, 0.1% Bacto beef agar, 0.1% yeast extract, 0.2% peptone, 1.5% agar): good growth. Color: reverse, dirty red; aerial mycelium, dirty white. On Oxoid nutrient agar: medium growth. Color: reverse, brown; aerial mycelium, poor pink; brown soluble pigment. The strain grows also at 37°C, very poor or no aerial mycelium being formed. No growth at 45°C. The type strain exhibits anti-bacterial and anti-fungal activity.

For sequence similarity, see type strain of *Streptomyces hiroshimensis*.

Source: not known.

DNA G+C content (mol%): not known.

Type strain: BCRC 12456, DSM 40895, HUT 6085, NBRC 15865, JCM 4083, NRRL B-1472, VKM Ac-944.

Sequence accession no. (16S rRNA gene): X53169.

Further comments: according to Hatano et al. (2003), *Streptomyces salmonis* (Baldacci et al. 1966) Witt and Stackebrandt 1991 is a later heterotypic synonym of *Streptomyces hiroshimensis* (Shinobu 1955) Witt and Stackebrandt 1991.

429. **Streptomyces sampsonii** (Millard and Burr 1926) Waksman *in* Waksman and Lechevalier 1953, 155[AL] ("*Actinomyces sampsonii*" Millard and Burr 1926, 614)

samp.so'ni.i. N.L. gen. masc. n. *sampsonii* of Sampson, a patronymic.

Spore chains in Section *Rectiflexibiles*. Mature spore chains are generally long and flexuous with more than 50 spores per chain. This morphology is seen on yeast-malt agar, oatmeal agar, salts-starch agar, and glycerol-asparagine agar. Spore surface is smooth.

Color of colony: aerial mass color in the Yellow color series (1ba-2ba, pale yellow, or 1db, pale yellowish green) on yeast-malt agar, oatmeal agar, salts-starch agar, and glycerol-asparagine agar; or sometimes in the Gray color series (2dc, yellowish gray) on yeast-malt agar and salts-starch agar. Reverse side of colony with no distinctive pigments (pale yellow to light olive brown on salts-starch agar and glycerol-asparagine agar; strong yellow to orange-yellow or yellowish brown on yeast-malt agar and oatmeal agar).

Color in medium: melanoid pigments are not formed in peptone-yeast-iron agar, tyrosine agar, or tryptone-yeast broth. No pigment is found in medium in yeast-malt agar, oatmeal agar, salts-starch agar, or glycerol-asparagine agar.

D-Glucose, L-arabinose, D-xylose, D-mannitol, and D-fructose are utilized for growth. No growth or only trace of growth with iso-inositol, rhamnose, sucrose, and raffinose.

Type strain shows the highest sequence similarity to: *S. albidoflavus*, AB184255, 99.9%; *S. odorifer*, Z76682, 99.9%; *S. daghestanicus*, DQ442497, 99.8%; *S. felleus*, AB184129, 99.8%; *S. hydrogenans*, AB184868, 99.8%; *S. violascens*, AY999737, 99.8%; *S. limosus*, AB184147, 99.8%; *S. griseus* subsp. *solvifaciens*, AB249915, 99.7%; *S. canescens*, AB184117, 99.7%; *S. koyangensis*, AY079156, 99.5%; *S. champavatii*, DQ026642, 99.3%.

Source: not known.

DNA G+C content (mol%): not known.

Type strain: ATCC 25495, CBS 955.69, BCRC 13705, DSM 40394, NBRC 13083, JCM 4515, NRRL B-12325, NRRL-ISP 5394, RIA 1275, VKM Ac-851.

Sequence accession no. (16S rRNA gene): D63871.

430. **Streptomyces sanglieri** Manfio, Atalan, Zakrzewska-Czerwinska, Mordarski, Rodríguez, Collins and Goodfellow 2003b, 1219[VP] (Effective publication: Manfio, Atalan, Zakrzewska-Czerwinska, Mordarski, Rodríguez, Collins and Goodfellow 2003a, 252.)

san.gli.e'ri. N.L. gen. masc. n. *sanglieri* of Sanglier, named in honor of Jean-Jacques Sanglier, a Belgian biotechnologist.

Aerobe, Gram-stain-positive, mesophilic actinomycetes which form extensively branched substrate and aerial hyphae. Spiral and open loops of smooth-surfaced spores are evident on aerial hyphae. A gray aerial spore mass and a reddish-orange substrate mycelium are formed on modified Bennett's, glycerol-asparagine, oatmeal, and yeast extract-malt extract agars; a reddish-orange pigment is produced on oatmeal agar. Hippurate is hydrolyzed and nitrate is reduced. Cleaves methoxysuccinyl-L-alanine-L-lysine-7AMC, L-citrulline-7AMC, *N*-acetyl-L-phenylalanine-L-arginine-7AMC (endopeptidase substrates), L-cysteine-7AMC (exopeptidase substrates), 4MU-*N*-acetyl-β-D-glucosaminde, 4MU-2-deoxy-β-D-glucopyranoside and 4MU-β-D-xyloside (glycosides), and 4MU-oleate (organic ester), but not L-phenylalanine-7AMC (endopeptidase substrate), L-glutamine-7AMC (exopeptidase substrates), 4MU-α-D-mannopyranoside (glycoside), or 4MU-elaidate (organic ester). L-Phenylalanine is used as sole carbon and nitrogen source but not L-histidine, L-hydroxyproline, or potassium nitrate. Does not use dextran, *myo*-inositol, inulin, mannitol, or pyruvate as sole sources of carbon for energy and growth. Grows between 10 and 35°C, but not at 45°C, or in the presence of sodium azide (0.01%, w/v), sodium chloride (7.0%, w/v), or neomycin sulfate (50 μg/ml) but is sensitive to oleandomycin phosphate (100 μg/ml).

Type strain shows the highest sequence similarity to: *S. atratus*, DQ026638, 100%; *S. yanii*, AB006159, 99.7%; *S. pulveraceus*, AB184806, 99.7%; *S. gelaticus*, DQ026636, 99.5%; *S. badius*, AY999783, 99.4%; *S. pluricolorescens*, DQ442540, 99.4%; *S. sindenensis*, AB184759, 99.4%; *S. mediolani*, AB184674, 99.4%; *S. griseinus*, AB184205, 99.4%; *S. rubiginosohelvolus*, AB184240, 99.4%; *S. albovinaceus*, AB249958, 99.3%; *S. globisporus* subsp. *globisporus*, EF178686, 99.3%; *S. flavofuscus*, AB249935, 99.3%; *S. fimicarius*, AY999784, 99.3%; *S. anulatus*, DQ026637, 99.3%; *S. praecox*, AB184293, 99.3%; *S. lavendulae* subsp. *lavendulae*, AB184080, 99.3%; *S. argenteolus*, AB045872, 99.3%; *S. parvus*, DQ442537, 99.3%; *S. cinereorectus*, AB184646, 99.3%; *S. griseoplanus*, AY999894, 99.3%; *S. californicus*, AB184755, 99.3%; *S. cavourensis* subsp. *washingtonensis*, DQ026671, 99.3%; *S. alboviridis*, AB184256, 99.2%; *S. griseolus*, AB184768, 99.2%; *S. baarnensis*, EF178688, 99.2%; *S. lipmanii*, AB184148, 99.2%; *S. cyaneofuscatus*, AB184860, 99.2%; *S. fulvorobeus*, AB184711, 99.2%; *S. acrimycini*, AY999889, 99.2%; *S. flavovirens*, DQ026635, 99.2%; *S. microflavus*, DQ445795, 99.2%; *S. floridae*, AB184656, 99.2%; *S. halstedii*, EF178695, 99.1%; *S. flavogriseus*, AJ494864, 99.1%; *S. luridiscabiei*, AF361784, 99.1%; *S. nitrosporeus*, EF178680, 99.1%; *S. griseus* subsp. *griseus*, AY207604, 99%; *S. bacillaris*, AB184439, 99%.

Source: isolated from soil taken from meadow hay plots.

DNA G+C content (mol%): 70–76.

Type strain: A46R51, DSM 41791, JCM 12607, NBRC 100784, NCIMB 13929.

Sequence accession no. (16S rRNA gene): AB249945.

431. **Streptomyces sannanensis** Iwasaki, Itoh and Mori 1981, 283^VP^

san.nan.en'sis. N.L. masc. adj. *sannanensis* of or pertaining to Sannan, a town in Japan.

Spore chains in Section *Spirales*. Mature spore chains are generally moderately short, often with 20 or more spores per chain. Spore surface is smooth.

Color of colony: aerial mass color in the gray color series on sucrose-nitrate agar, glucose-asparagine agar, inorganic salts-starch agar, and oatmeal agar. Reverse side of colony with no distinctive pigments (colorless to buff in color) on all media. Diffusible pigment: melanoid pigments are not formed in peptone-yeast extract-iron agar, tyrosine agar, or tryptone-yeast extract broth. No pigment other than a trace of yellow is produced in the media used. Starch is hydrolyzed and gelatin is liquefied. Does not grow in skim milk medium. Melanoid pigment is not produced. No growth or only traces of growth with L-arabinose, D-xylose, D-glucose, D-fructose, sucrose, iso-inositol, L-rhamnose, raffinose, or D-mannitol with Pridham and Gottlieb's basal mineral salts agar. On $CuSO_4$-free Pridham and Gottlieb's basal mineral salts agar and Czapek's solution agar, D-xylose and D-glucose are utilized for growth; D-glucose gives only faint growth. Mesophilic (growth range is from 17–37°C; temperature for optimum growth is between 27 and 34°C). Cell-wall composition: LL-A_2pm occurs in whole-cell hydrolysates. The cell-wall composition is believed to be of Type I. Produces sannamycins.

Type strain shows the highest sequence similarity to: *S. mauvecolor*, AB184532, 99%.

Source: not known.

DNA G+C content (mol%): not known.

Type strain: KC-7038, ATCC 31530, DSM 41705, NBRC 14239, JCM 9651, KCTC 9770.

Sequence accession no. (16S rRNA gene): AB184579.

432. **Streptomyces sapporonensis** (Locci 1985) Witt and Stackebrandt 1996, 836^VP^ (Effective publication: Witt and Stackebrandt 1990, 370.) ["*Streptomyces sapporonensis*" Sakai and Miyoshi 1972; "*Streptoverticillium sapporonense*" (*ex* Arai 1976) Locci 1985, 232]

sap.po.ro.nen'sis. N.L. masc. adj. *sapporoensis* of or pertaining to Sapporo, a town in Japan.

Reddish spore mass, aerial mycelium is cottony. Melanin pigment is not produced. Acid is produced from D-galactose and *myo*-inositol, but not from D-fructose, trehalose, or D-ribose. L-Methionine, shikimic acid, and DL-α-aminobutyric acid are utilized, but not coumarin, mannitol, melibiose, proline, raffinose, or sorbitol. Degrades citrate, but not esculin or L-tyrosine. H_2S production. No NO_3^- reduction. Resistant to carbenicillin (100 μg/ml), cephaloridine (30 μg/ml), cephalotin (30 μg/ml), and colistin (30 μg/ml). *Aspergillus niger* and *Bacillus subtilis* are inhibited, *Candida albicans* is not.

Type strain shows the highest sequence similarity to: *S. hygroscopius* subsp. *angustmycaticus*, DQ442509, 99.5%; *S. varsoviensis*, DQ026653, 99.5%; *S. abikoensis*, AB184537, 99.5%; *S. ehimensis*, AY999834, 99.4%; *S. lilacinus*, AB184819, 99.3%; *S. luteireticuli* AB249969, 99.2%; *S. mobaraensis*, DQ442528, 99.1%; *S. caeruleus*, EF178675, 99.1%; *S. lavenduligriseus*, DQ442515, 99%; *S. hiroshimensis*, AB184789, 99%; *S. thioluteus* AB184753, 99%.

Source: not known.

DNA G+C content (mol%): not known.

Type strain: ATCC 21532, DSM 41675, NBRC 13823, JCM 4934.

Sequence accession no. (16S rRNA gene): AB184508.

Further comments: according to Hatano et al. (2003), *Streptomyces sapporonensis* (Locci and Schofield 1989) Witt and Stackebrandt 1991 is a later heterotypic synonym of *Streptomyces cinnamoneus* (Benedict et al. 1952) Witt and Stackebrandt 1991.

433. **Streptomyces scabiei** corrig. (*ex* Thaxter 1891) Lambert and Loria 1989b, 391[VP] ("*Oospora scabies*" Thaxter 1891; "*Actinomyces scabies*" Güssow 1914; "*Streptomyces scabies*" Waksman and Henrici 1948)

sca'bi.ei. L. gen. n. *scabiei* of mange, in reference to the ability to cause potato scab.

Spore chains in Section *Rectiflexibiles.* Mature spore chains are moderately short with 3–10, or sometimes more than 10, spores per chain. This morphology is seen on yeast-malt agar, oatmeal agar, salts-starch agar, and glycerol-asparagine agar. Spore surface is smooth.

Color of colony: aerial mass color in the Gray color series on yeast-malt agar, oatmeal agar, salts-starch agar, and glycerol-asparagine agar. Reverse side of colony is yellow to yellow brown is usually modified by green to light grayish olive, olive, or olive brown on yeast-malt agar, oatmeal agar, salts-starch agar, and glycerol-asparagine agar. Substrate mycelium pigment is not a pH indicator.

Color in medium: melanoid pigments are not formed in peptone-yeast-iron agar, tyrosine agar, or tryptone-yeast broth. Yellow or greenish yellow pigment may be found in the medium in yeast-malt agar and oatmeal agar, but not in salts-starch agar or glycerol-asparagine agar; this pigment is not pH-sensitive when tested with 0.05 M NaOH or HCl.

D-Glucose, L-arabinose, D-xylose, D-mannitol, D-fructose, and rhamnose are utilized for growth. No growth or only trace of growth with sucrose, iso-inositol, and raffinose.

Type strain shows the highest sequence similarity to: *S. europaeiscabiei*, AY207598, 99.8%.

Source: not known.

DNA G+C content (mol%): not known.

Type strain: RL-34, ATCC 49173, CIP 105438, DSM 41658, ICMP 12542, JCM 7914.

Sequence accession no. (16S rRNA gene): D63862.

Further comments: the original spelling of the specific epithet, *scabies* (*sic*), has been corrected by Trüper and De'Clari (1997).

434. **Streptomyces scabrisporus** Xu, Takahashi, Seino, Iwai and Ōmura 2004b, 580[VP]

sca.bri.spo'rus. L. adj. *scaber -bra -brum* scabby, rough; N.L. n. *spora* spore; N.L. masc. adj. *scabrisporus* referring to the rugose surface of the spores.

Mature spore chains are spiral, with more than 20 spores per chain. Spores are cylindrical in shape, 0.6–0.8 × 1.2–1.6 µm in diameter; the spore surface is rugose. Substrate and aerial mycelia are produced. The reverse sides of colonies are colorless to light ivory. The aerial mass on some agar media, such as water agar, oatmeal agar, and 1/10 V8 juice agar, is gray. Mycelia do not fragment into coccoid or bacillary elements. Melanoid or soluble pigments are not produced on any medium tested. Starch is weakly hydrolyzed, gelatin is not liquefied and milk is weakly coagulated. D-Glucose, D-fructose, D-xylose, *myo*-inositol, and rhamnose are utilized for growth, but D-mannitol, raffinose, and melibiose are not; little if any growth is observed with sucrose, L-arabinose, or salicin. Menaquinone composition is MK-9(H_4), MK-9(H_2), and MK-9(H_6) in the ratio 12:4:3.

Type strain shows no sequence similarity over 99%. Type strain shows DNA–DNA similarity to: *S. chrestomyceticus* ATCC 14947^T, 54%.

Source: the type strain was isolated from a soil sample collected from Ushiku-cho, Ibaraki Prefecture, Japan.

DNA G+C content (mol%): 70.6.

Type strain: KM-4927, JCM 11712, NBRC 100760, NRRL B-24202.

Sequence accession no. (16S rRNA gene): AB030585.

435. **Streptomyces sclerotialus** (Thirumalachar 1955) Pridham 1970, 44[AL] (*Chainia antibiotica* Thirumalachar 1955, 935)

scle.rot.i'a.lus. N.L. neut. n. *sclerotium* sclerotium; N.L. masc. adj. *sclerotialus* pertaining to sclerotia.

Spore chains in Section *Spirales.* Spirals are best developed on salts-starch agar. Aerial mycelium may be poorly developed on yeast-malt agar, oatmeal agar, salts-starch agar, and glycerol-asparagine agar, but some spirals or chains can usually be found on these media. Spore surface is smooth. One of three observers reports sclerotia and fragmentation of substrate mycelium on glycerol-asparagine agar in 21 d. A second observer reports "globular sporangia" on yeast-malt agar, oatmeal agar, and salts-starch agar and "coremia" on yeast-malt agar in 7–14 d. Scanning electron microscopy shows coalesced masses of spores.

Color of colony: aerial mycelium is generally poorly developed on yeast-malt agar, oatmeal agar, salts-starch agar, and glycerol-asparagine agar. Sporulating aerial mycelium adequate for color determination may be formed on salts-starch agar; it is in the White color series. One observer reports aerial mycelium in the Yellow color series (2fb, light yellowish red; 1db, pale yellow green) on yeast-malt agar, oatmeal agar, salts-starch agar, and glycerol-asparagine agar; this may be the color of the vegetative (substrate mycelium) growth. Reverse side of colony with no distinctive pigments (orange yellow to yellowish brown on yeast-malt agar, light grayish yellow or light yellowish brown on oatmeal agar, salts-starch agar, and glycerol-asparagine agar).

Color in medium: melanoid pigments are not formed in peptone-yeast-iron agar, tyrosine agar, or tryptone-yeast broth. No pigment or only a trace of yellow is found in the medium in yeast-malt agar, oatmeal agar, salts-starch agar, and glycerol-asparagine agar.

D-Glucose, L-arabinose, D-xylose, iso-inositol, D-mannitol, D-fructose, rhamnose, sucrose, and raffinose are all utilized for growth.

Type strain shows the highest sequence similarity to: *S. rimosus* subsp. *rimosus*, AB045883, 99.4%; *S. olivaceiscleroticus*, AJ621606, 99.2%; *S. niger*, AJ621607, 99.2%; *S. chresto-*

myceticus, AJ621609, 99.1%; *S. rimosus* subsp. *paromomycinus*, AJ621610, 99.1%; *S. erumpens*, AJ621603, 99.1%.

Source: not known.

DNA G+C content (mol%): not known.

Type strain: ATCC 15721, CBS 167.62, BCRC 13307, DSM 43032, IFM 1141, NBRC 12246, NBRC 13356, NBRC 13904, JCM 3039, JCM 4828, KCTC 9065, NRRL B-2317, NRRL-ISP 5269, RIA 1317, VKM Ac-1909.

Sequence accession no. (16S rRNA gene): AJ621608.

Further comments: Streptomyces sclerotialus Pridham 1970 and *Chainia antibiotica* Thirumalachar 1955 have the same type strain and therefore are homotypic synonyms [Rules 24a and 24b(1) of the *Bacteriological Code* (1990 Revision)].

436. **Streptomyces scopiformis** Li, Lanoot, Zhang, Vancanneyt, Swings and Liu 2002a, 1632[VP]

sco.pi.for'mis. L. fem. n. *scopa* a broom; L. adj. suffix *-formis -is -e* (from L. n. *forma* figure, shape, appearance) -like, in the shape of; N.L. masc. adj. *scopiformis* in the form of a broom, referring to the structure of the spore chains.

Rectiflexibiles chains of roundish, spiny-surfaced spores (0.7–0.8 μm) are arranged in fastigiate form. Spore mass is gray or blue-gray, the reverse is blue to gray-blue. Substrate hyphae are extensively branching, septate, and swollen. Diffusible pigment is not formed. Tests for esculin, starch, dextrin, elastin, nitrate reduction, and gelatin are positive and tests for hippurate, cellulose, and lipolysis are negative. Utilizes L-arabinose, D-fructose, D-galactose, D-glucose, D-maltose, L-rhamnose, D-xylose, D-sucrose, dulcitol, *myo*-inositol, melibiose, trehalose, sodium acetate, and sodium citrate as sole carbon sources but not D-mannitol, D-raffinose, adonitol, methyl α-glucoside, iso-erythritol, or inulin. Grows in presence of penicillin G, biomycin, phenol, and ethanol, at 20–45°C and at pH 5–10 but not in the presence of bacteracin, lysozyme, sodium azide, or methyl violet, or at 10 or 50°C or at pH 4.0 or 11.0. Cell-wall type I, phospholipid type PII and menaquinone MK-9($H_{4,6}$) are detected. The fatty acids are type II. Mycolic acids are not present.

Type strain shows no sequence similarity over 99%. Type strain shows DNA–DNA similarity to: *S. ambofaciens* AS 41528[T], 13±1%; *S. coerulescens* JCM 4358[T], 34±3%; *S. caelestis* JCM 218[T], 23±4%; *S. nogalater* JCM 4799[T], 31±4%; *S. intermedius* JCM 4483[T], 26±3%; *S. albidoflavus* JCM 4446[T], 19±6%.

Source: not known.

DNA G+C content (mol%): 71.0.

Type strain: A25, AS 4.1331, DSM 41825, JCM 12114, LMG 20251, NBRC 100244.

Sequence accession no. (16S rRNA gene): AB249927.

437. **Streptomyces seoulensis** Chun, Youn, Yim, Lee, Kim, Hah and Kang 1997, 495[VP]

se.oul.en'sis. N.L. masc. adj. *seoulensis* of or belonging to Seoul, Republic of Korea, the geographical origin of the species.

Long-chain *Rectiflexibiles* spores with smooth surfaces are produced. Forms a yellow substrate mycelium on glycerol-asparagine agar and a gray aerial mycelium and spores on inorganic salts-starch agar. Verticils are not present. Diffusible pigments are not produced on ISP 5 medium. Melanin is not produced on peptone-yeast extract-iron agar and tyrosine agar. It utilizes L-histidine as sole nitrogen source but not DL-α-amino-*n*-butyric acid, L-cysteine, L-valine, L-phenylalanine, L-hydroxyproline, or potassium nitrate. Uses D-fructose, lactose, mannitol, D-melibiose, raffinose, L-rhamnose, salicin, sodium pyruvate, sucrose, and D-glucose as sole sources of carbon, but not adonitol, L-arabinose, dextran, *myo*-inositol, melezitose, sodium acetate, sodium citrate, sodium propionate, or xylitol. Tests for lecithinase, lipolysis, pectin hydrolysis, and H_2S production are positive but not those for hippurate hydrolysis or nitrate reduction. Degrades allantoin, arbutin, and elastin but not guanine, xanthine, or xylan. Grows in the presence of potassium tellurite, rifampin, and penicillin G but not at 45°C or in the presence of sodium chloride, sodium azide, thallous acetate, neomycin, or oleandomycin. The cell wall contains LL-A_2pm and no diagnostic sugars are present in the cell-wall fraction (chemotype I). Phosphatidylethanolamine, phosphatidylinositol, diphosphatidylglycerol, and phosphatidylinositol mannosides are present in the polar lipid fraction (phospholipid type II). The fatty acids are mainly saturated straight-chain as well as iso- and anteiso-branched fatty acids (fatty acid type 2c). Activity is not exhibited against *Aspergillus niger, Bacillus subtilis, Candida albicans, Micrococcus luteus, Saccharomyces cerevisiae,* or *Streptomyces murinus.*

Type strain shows the highest sequence similarity to: *S. recifensis*, AB184165, 99.8%; *S. griseoluteus*, AY999751, 99.7%.

Source: isolated from Seoul, Korea.

DNA G+C content (mol%): 68.0.

Type strain: IMSNU-1 (deposited in the Institute of Microbiology, Seoul National University, under accession number IMSNU 21266), CIP 105312, NBRC 16255, NBRC 16668, JCM 10116.

Sequence accession no. (16S rRNA gene): AB249970.

438. **Streptomyces septatus** (Locci, Baldacci and Petrolini Baldan 1969) Witt and Stackebrandt 1991, 456[VP] (Effective publication: Witt and Stackebrandt 1990, 370.) (*Streptoverticillium septatum* Locci, Baldacci and Petrolini Baldan 1969, 59)

sep.ta'tus. L. masc. adj. *septatus* surrounded by a bulwark, fenced.

Spore chains in Section Umbellate Monoverticillate (=*Streptomyces Verticillati*, biverticillate). Mature spore chains are generally short, with 3 to 10 or more spores per chain. This morphology is seen on yeast-malt agar, oatmeal agar, salts-starch agar, and glycerol-asparagine agar. Spore surface is smooth.

Color of colony: aerial mass color in the Yellow or Red color series on yeast-malt agar, oatmeal agar, salts-starch agar, and glycerol-asparagine agar (white is also present). Nearest matching color tabs in the Yellow color series are 2db and 2ba, pale yellow, and 1cb, pale yellow green. Nearest matching tabs in the Red color series are 4ec and 5cb, grayish yellowish pink, and 3ca, pale orange yellow. Red aerial mycelium may sometimes occur as scattered tufts of longer aerial hyphae in a yellow or white background.

Reverse side of colony with no distinctive pigments (olive brown to strong brown on yeast-malt agar; pale grayish yellow on oatmeal agar and salts-starch agar; grayish yellow to pale yellowish green on glycerol-asparagine agar).

Color in medium: melanoid pigments are formed in peptone-yeast-iron agar and tryptone-yeast broth, but not in tyrosine agar. No pigment is found in the medium in yeast-malt agar, oatmeal agar, salts-starch agar, and glycerol-asparagine agar in 14 d or a trace of yellow may sometimes be seen in yeast-malt agar, oatmeal agar, and salts-starch agar in 21 d.

D-Glucose and iso-inositol are utilized for growth. Reports vary on growth on D-fructose, but utilization of this sugar is doubtful. No growth or only traces of growth with L-arabinose, D-xylose, D-mannitol, rhamnose, sucrose, and raffinose.

For sequence similarity, see type strain of *Streptomyces griseocarneus.*

Source: not known.

DNA G+C content (mol%): not known.

Type strain: ATCC 27464, CBS 772.72, BCRC 11895, CECT 3251, DSM 40577, NBRC 13471, JCM 4547, JCM 4829, NCIMB 12982, NRRL 2974, NRRL-ISP 5577, RIA 1432, VKM Ac-888.

Sequence accession no. (16S rRNA gene): AB184883.

Further comments: according to Hatano et al. (2003), *Streptomyces septatus* (Locci et al. 1969) Witt and Stackebrandt 1991 is a later heterotypic synonym of *Streptomyces griseocarneus* (Benedict et al. 1950) Witt and Stackebrandt 1991.

439. **Streptomyces setonii** (Millard and Burr 1926) Waksman *in* Waksman and Lechevalier 1953, 107[AL] ("*Actinomyces setonii*" Millard and Burr 1926, 604)

se.to'ni.i. N.L. gen. masc. n. *setonii* of Seton, named for a person, Seton (Probably R.S. Seton of the Harris Insitute, Preston, England).

Spore chains in Section *Rectiflexibiles.* Mature spore chains are generally long, often with more than 50 spores per chain. This morphology is seen on yeast-malt agar, oatmeal agar, salts-starch agar, and glycerol-asparagine agar. Spore surface is smooth.

Color of colony: aerial mass color in the Yellow (2ba, pale yellow) or Gray (2dc, yellowish gray) color series on yeast-malt agar and glycerol-asparagine agar; White color series on oatmeal agar and salts-starch agar. Reverse side of colony with no distinctive pigments (pale or grayish yellow to light yellowish brown) on yeast-malt agar, oatmeal agar, salts-starch agar, and glycerol-asparagine agar.

Color in medium: melanoid pigments are not formed in peptone-yeast-iron agar, tyrosine agar, or tryptone-yeast broth. No pigment is found in medium in yeast-malt agar, oatmeal agar, salts-starch agar, or glycerol-asparagine agar.

D-Glucose, L-arabinose, D-xylose, D-mannitol, D-fructose, and rhamnose are utilized for growth. No growth or only trace of growth with iso-inositol, sucrose, and raffinose.

For sequence similarity, see type strain of *Streptomyces griseus.*

Source: not known.

DNA G+C content (mol%): not known.

Type strain: ATCC 25497, CBS 105.27, CBS 957.69, CFBP 4549, CIP 105279, DSM 40395, ICMP 12543, NBRC 13085, JCM 4226, JCM 4516, KCTC 9144, NRRL B-2555, NRRL-ISP 5395, RIA 1277.

Sequence accession no. (16S rRNA gene): D63872.

Further comments: according to Liu et al. (2005b), *Streptomyces setonii* (Millard and Burr 1926) Waksman 1953 is a later heterotypic synonym of *Streptomyces griseus* (Krainsky 1914) Waksman and Henrici 1948 emend. Liu et al. 2005b.

440. **Streptomyces showdoensis** Nishimura, Mayama, Komatsu, Kato, Shimaoka and Tanaka 1964, 150[AL]

show.do.en'sis. N.L. masc. adj. *showdoensis* of or belonging to Shodo, an island in Kagawa Prefecture, Japan, the source of the soil from which the organism was isolated.

Spore chains in Section *Rectiflexibiles.* Mature spore chains are generally long, with 10 to 50 or more spores per chain. This morphology is seen on yeast-malt agar, oatmeal agar, salts-starch agar, and glycerol-asparagine agar. Spore surface is smooth.

Color of colony: aerial mass color in the Gray or Yellow color series on yeast-malt agar, oatmeal agar, salts-starch agar, and glycerol-asparagine agar. White aerial mycelium may also be seen on these media. Most representative color tabs from the Gray color series are 2dc, yellowish gray, and 3ge, light grayish yellowish brown, and from the Yellow color series, 2db and 2ba, pale yellow. Reverse side of colony with no distinctive pigments (orange yellow or yellowish brown on yeast-malt agar; pale or grayish yellow on oatmeal agar, salts-starch agar, and glycerol-asparagine agar).

Color in medium: melanoid pigments are formed in peptone-yeast-iron agar and tryptone-yeast broth, and weakly in tyrosine agar. No pigment is found in the medium in yeast-malt agar, oatmeal agar, salts-starch agar, or glycerol-asparagine agar.

D-Glucose, D-xylose, and D-fructose are utilized for growth. Utilization of L-arabinose and sucrose is doubtful. No growth or only traces of growth with iso-inositol, D-mannitol, rhamnose, or raffinose.

Type strain shows the highest sequence similarity to: *S. viridobrunneus*, AJ781372, 100%; *S. cinereoruber* subsp. *cinereoruber*, AB184121, 99.5%; *S. roseoviridis*, AB184239, 99.4%; *S. tanashiensis*, AJ781362, 99.1%; *S. nashvillensis*, AB184286, 99.1%; *S. violaceorectus*, AB184314, 99.1%; *S. vietnamensis*, DQ311081, 99%; *S. bikiniensis*, X79851, 99%.

Source: isolated from soil from Shodo, an island in Kagawa Prefecture, Japan.

DNA G+C content (mol%): not known.

Type strain: ATCC 15105, CBS 718.72, BCRC 11868, DSM 40504, NBRC 13417, JCM 4830, NRRL B-12430, NRRL-ISP 5504, RIA 1378, VKM Ac-1219.

Sequence accession no. (16S rRNA gene): AB184389.

441. **Streptomyces sindenensis** Nakazawa and Fujii 1957, 109[AL]

sin.den.en'sis. N.L. masc. adj. *sindenensis* of or belonging to Sinda Village, Osaka Prefecture, Japan, the source of the soil from which the organism was isolated.

Spore chains in Section *Rectiflexibiles*. Mature spore chains are short, often with only 3–10 spores per chain (longer chains may be found on impression mounts or electron micrographs) This morphology is seen on yeast-malt agar, oatmeal agar, salts-starch agar, and glycerol-asparagine agar. Spore surface is smooth.

Color of colony: aerial mass color is difficult to determine because sporulating aerial mycelium is poorly developed on ISP media. When sporulation occurs, it is usually in the White or Red (5cb, grayish yellowish pink) color series on yeast-malt agar and in the White or Yellow color series on oatmeal agar. When present, sporulation aerial mycelium is in the White color series on salts-starch agar and glycerol-asparagine agar. Reverse side of colony with no distinctive pigments (colorless or pale grayish yellow) on yeast-malt agar, oatmeal agar, salts-starch agar, and glycerol-asparagine agar.

D-Glucose, L-arabinose, D-xylose, D-mannitol, and D-fructose are utilized for growth. No growth or only trace of growth with sucrose, iso-inositol, rhamnose, and raffinose.

Type strain shows the highest sequence similarity to: *S. albovinaceus*, AB249958, 100%; *S. griseoplanus*, AY999894, 100%; *S. anulatus*, DQ026637, 100%; *S. globisporus* subsp. *globisporus*, EF178686, 100%; *S. pluricolorescens*, DQ442540, 100%; *S. praecox*, AB184293, 100%; *S. flavofuscus*, AB249935, 100%; *S. fimicarius*, AY999784, 100%; *S. mediolani*, AB184674, 100%; *S. badius*, AY999783, 100%; *S. griseinus*, AB184205, 100%; *S. rubiginosohelvolus*, AB184240, 100%; *S. californicus*, AB184755, 99.9%; *S. lavendulae* subsp. *lavendulae*, AB184080, 99.9%; *S. cavourensis* subsp. *washingtonensis*, DQ026671, 99.9%; *S. acrimycini*, AY999889, 99.9%; *S. parvus*, DQ442537, 99.9%; *S. fulvorobeus*, AB184711, 99.8%; *S. lipmanii*, AB184148, 99.8%; *S. microflavus*, DQ445795, 99.8%; *S. cyaneofuscatus*, AB184860, 99.8%; *S. floridae*, AB184656, 99.8%; *S. alboviridis*, AB184256, 99.8%; *S. baarnensis*, EF178688, 99.8%; *S. cinereorectus*, AB184646, 99.8%; *S. griseus* subsp. *griseus*, AY207604, 99.7%; *S. griseolus*, AB184768, 99.7%; *S. luridiscabiei*, AF361784, 99.7%; *S. flavovirens*, DQ026635, 99.7%; *S. argenteolus*, AB045872, 99.7%; *S. bacillaris*, AB184439, 99.6%; *S. pulveraceus*, AB184806, 99.6%; *S. halstedii*, EF178695, 99.6%; *S. flavogriseus*, AJ494864, 99.6%; *S. atroolivaceus*, AJ781320, 99.5%; *S. olivoviridis*, AB184227, 99.5%; *S. finlayi*, AY999788, 99.5%; *S. nitrosporeus*, EF178680, 99.5%; *S. albolongus*, AB184425, 99.4%; *S. clavifer*, DQ026670, 99.4%; *S. griseobrunneus*, AB249912, 99.4%; *S. sanglieri*, AB249945, 99.4%; *S. celluloflavus*, AB184476, 99.4%; *S. yanii*, AB006159, 99.4%; *S. gelaticus*, DQ026636, 99.3%; *S. atratus*, DQ026638, 99.3%; *S. cavourensis* subsp. *cavourensis*, DQ445791, 99.3%; *S. mutomycini*, AB249951, 99.2%; *S. candidus*, DQ026663, 99.2%; *S. spiroverticillatus*, AB184814, 99.1%; *S. cremeus*, AB184124, 99.1%.

Source: isolated from soil from Sinda Village, Osaka Prefecture, Japan.

DNA G+C content (mol%): not known.

Type strain: ATCC 23963, CBS 946.68, BCRC 11887, DSM 40255, NBRC 12915, NBRC 3399, JCM 4164, JCM 4669, NRRL B-1866, NRRL-ISP 5255, RIA 1181.

Sequence accession no. (16S rRNA gene): AB184759.

442. **Streptomyces sioyaensis** Nishimura, Okamoto, Mayama, Ohtsuka, Nakajima, Tawara, Shimohira and Shimaoka 1961, 257[AL]

si.o.ya.en'sis. N.L. masc. adj. *sioyaensis* of or belonging to Sioya, Kobe, Japan, the source of the soil from which the organism was isolated.

Spore chains in Section *Spirales*. Mature spore chains generally have 10–50 spores per chain; longer chains are sometimes observed. Morphology can be observed on oatmeal agar and glycerol-asparagine agar in 14–21 d. Typical aerial mycelium may develop slowly or fail to develop on yeast-malt agar and salts-starch agar. Spore surface is smooth. One observer reports disintegration of spiral spore chains in dark liquid droplets.

Color of colony: aerial mass color in the Gray color series on yeast-malt agar, oatmeal agar, salts-starch agar, and glycerol-asparagine agar. Reverse side of colony with no distinctive pigments (grayed yellow or grayed yellow modified by green) on yeast-malt agar, oatmeal agar, salts-starch agar, and glycerol-asparagine agar; substrate pigment is not a pH indicator.

Color in medium: melanoid pigments not formed in peptone-yeast-iron agar and tyrosine agar. Trace of yellow pigment found in medium in yeast-malt agar, oatmeal agar, salts-starch agar, and glycerol-asparagine agar; this pigment is not pH-sensitive.

D-Glucose, sucrose, D-xylose, iso-inositol, D-mannitol, D-fructose, and raffinose are utilized for growth. No growth or only trace of growth on L-arabinose and rhamnose.

Type strain shows the highest sequence similarity to: *S. hygroscopius* subsp. *decoyicus*, AY999883, 99.6%; *S. nigrescens*, DQ442530, 99.5%; *S. libani* subsp. *libani*, AB184414, 99.5%; *S. tubercidicus*, AJ621612, 99.5%; *S. lydicus*, Y15507, 99.3%; *S. chattanoogensis*, AJ621611, 99.3%; *S. libani* subsp. *rufus*, AJ781351, 99.2%; *S. misakiensis*, AB217605, 99.2%; *S. caniferus*, AB184640, 99.2%; *S. catenulae*, AJ621613, 99.2%; *S. erumpens*, AJ621603, 99.2%; *S. hygroscopius* subsp. *glebosus*, AB184479, 99.2%; *S. ochraceiscleroticus*, AB184094, 99.1%; *S. purpurogeneiscleroticus*, AJ621604, 99.1%; *S. platensis*, AB045882, 99.1%; *S. chrestomyceticus*, AJ621609, 99%; *S. albospinus*, AB184527, 99%; *S. monomycini*, DQ445790, 99%; *S. auratus*, AJ391816, 99%; *S. rimosus* subsp. *paromomycinus*, AJ621610, 99%.

Source: isolated from soil from Sioya, Kobe, Japan.

DNA G+C content (mol%): not known.

Type strain: ATCC 13989, ATCC 19810, CBS 563.68, BCRC 11878, DSM 40032, NBRC 12820, JCM 4418, KCTC 9043, NRRL B-5408, NRRL-ISP 5032, RIA 1090, UNIQEM 197, VKM Ac-1260.

Sequence accession no. (16S rRNA gene): DQ026654.

443. **Streptomyces sodiiphilus** Li, Zhang, Zhang, Tang, Yu, Xu and Jiang 2005, 1332[AL]

so.di.i'phi.lus. N.L. n. *sodium -i* sodium; N.L. adj. *philus -a -um* (from Gr. adj. *philos -ê -on*) friend, loving; N.L. masc. adj. *sodiiphilus* sodium ion-loving, referring to the characteristic of Na^+-dependent growth.

Long or short chains of spores are straight to flexuous and spores are nonmotile. Both vegetative and aerial hyphae are well-developed and not fragmented. No diffusible pigments are produced except on nutrient agar medium (pale orange-yellow). Positive for gelatin liquefaction and nitrate reduction, but negative for urease, melanin production, starch hydrolysis, H_2S production, milk coagulation, and milk peptonization. Sodium acetate and rhamnose can be used as sole carbon sources for growth, but not most other carbon sources, such as lactose, maltose, fructose, xylose, ribose, arabinose, sucrose, glucose, galactose, sodium citrate, cellobiose, raffinose, mannitol, sorbitol, glycerol, and starch. Grows optimally at 28°C and in ISP medium 2 with 3% NaCl and pH 9.0–10.0. Cell wall contains LL-A_2pm and glycine. Whole-cell hydrolysates mainly contain galactose and glucose and no diagnostic sugars. Predominant menaquinones are MK-9(H_4) (13%), MK-9(H_6) (68%), and MK-9(H_8) (19%), and the diagnostic phospholipid is phosphatidylethanolamine. Major fatty acid components are $C_{15:0}$ anteiso (16.46%), $C_{17:0}$ anteiso (13.30%), and $C_{16:0}$ iso (31.32%).

Type strain shows no sequence similarity over 99%.

Source: the type strain was isolated from soil sample collected from Chaka salt lake, Qinghai Province, China.

DNA G+C content (mol%): 70.5.

Type strain: YIM 80305, CCTCC AA 203015, CIP 107975, JCM 13581.

Sequence accession no. (16S rRNA gene): AY236339.

444. **Streptomyces somaliensis** (Brumpt 1906) Waksman and Henrici *in* Breed, Murray and Hitchens 1948, 965[AL] ("*Indiella somaliensis*" Brumpt 1906, 555)

so.ma.li.en'sis. N.L. masc. adj. *somaliensis* of or pertaining to Somalia.

Spore chains in Section *Rectiflexibiles*; flexuous chains are most common. Mature spore chains generally have 10–50, or sometimes more than 50, spores per chain. This morphology is seen on yeast-malt agar, oatmeal agar, salts-starch agar, and glycerol-asparagine agar. Spore surface is smooth.

Color of colony: aerial mass color in the Yellow color series (2db or 2ba, pale yellow) on yeast-malt agar, oatmeal agar, salts-starch agar, and glycerol-asparagine agar (one observer selected color tab 3ca, pale orange yellow, from the Red color-wheel as the most representative color). Aerial mycelium is often poorly developed on much of the culture surface on ISP media. Good sporulation in the Yellow color series was observed on Czapek's agar. Reverse side of colony with no distinctive pigments (grayish yellow to grayish greenish yellow and light olive brown) on yeast-malt agar, oatmeal agar, salts-starch agar, and glycerol-asparagine agar.

Color in medium: melanoid pigments are not formed in peptone-yeast-iron agar, tyrosine agar, or tryptone-yeast broth. Yellow pigment is found in the medium in yeast-malt agar, oatmeal agar, salts-starch agar, and glycerol-asparagine agar; this pigment is not pH-sensitive when tested with 0.05 M NaOH or HCl.

D-Glucose, L-arabinose, sucrose, D-xylose, D-mannitol, and D-fructose are utilized for growth. No growth or only trace of growth iso-inositol, rhamnose, and raffinose.

Type strain shows the highest sequence similarity to: *S. coeruleoprunus*, AB184651, 99%.

Source: not known.

DNA G+C content (mol%): not known.

Type strain: ATCC 33201, DSM 40738, JCM 12659, IMRU 1274, NCTC 11332.

Sequence accession no. (16S rRNA gene): AJ007403.

445. **Streptomyces sparsogenes** Owen, Dietz and Camiener 1963, 772[AL] emend. Goodfellow, Kumar, Labeda and Sembiring 2008, 5 (Effective publication: Goodfellow, Kumar, Labeda and Sembiring 2007, 192.) ("*Streptomyces sparsogenes* var. *sparsogenes*" Owen, Dietz and Camiener 1963, 772)

spar.so'ge.nes. L. part. adj. *sparsus* scattered; N.L. suff. *-genes* (from Gr. v. *gennaô* to produce) producing; N.L. part. adj. *sparsogenes* scattered producing, probably referring to the sparse formation of aerial mycelium.

Spore chains in Section *Spirales*. Aerial hyphae and spirals may emerge from coremia-like structures. Mature spore chains are moderately long with 10 to 50 or more spores per chain. This morphology may be seen on yeast-malt agar, oatmeal agar, salts-starch agar, and glycerol-asparagine agar, although sporulation is not always uniformly good on these media. One observer notes autolysis or lysis on media in 14 d. Spore surface is spiny; individual spores are indistinct.

Color of colony: aerial mass color in the Gray color series (2dc, yellowish gray or 3fe, light brownish gray to 5fe, light grayish reddish brown) on yeast-malt agar, oatmeal agar, salts-starch agar, and glycerol-asparagine agar; aerial mycelium in the Yellow color series (2ba, pale yellow) may also be seen on yeast-malt agar and glycerol-asparagine agar. Reverse side of colony with no distinctive pigments (light or pale yellow to moderate yellow or orange-yellow) on yeast-malt agar, oatmeal agar, salts-starch agar, and glycerol-asparagine agar.

Color in medium: melanoid pigments are not formed in peptone-yeast-iron agar, tyrosine agar, or tryptone-yeast broth. No pigment is found in medium in yeast-malt agar, oatmeal agar, salts-starch agar, or glycerol-asparagine agar.

D-Glucose, L-arabinose, D-xylose, D-mannitol, D-fructose, rhamnose, sucrose, and raffinose are utilized for growth. Utilization of iso-inositol is doubtful.

Type strain shows the highest sequence similarity to: *S. cuspidosporus*, AB184090, 99.9%. *Source:* not known.

DNA G+C content (mol%): not known.

Type strain: ATCC 25498, CBS 672.69, CBS 958.69, BCRC 12085, DSM 40356, NBRC 13086, JCM 4517, LMG 5985, NCIMB 9449, NRRL 2940, NRRL-ISP 5356, RIA 1278, VKM Ac-1744.

Sequence accession no. (16S rRNA gene): AB184301.

446. **Streptomyces specialis** Kämpfer, Huber, Buczolits, Thummes, Grün-Wollny and Busse 2008, 2605[VP]

spe.ci.a'lis. L. masc. adj. *specialis* particular, special, because of the unusual quinone type and the separate phylogenetic position.

Forms extensively branched substrate mycelium and aerial hyphae that differentiate into spiral chains of spores. Spore chains consist of up to 15 spores. Spores are elliptical or short rods. Good growth on all ISP media (ISP 1 through 5, according to Shirling and Gottlieb,

1966) after 2 weeks incubation at 28°C. On all media, a pinkish white aerial mycelium and a pale yellow substrate mycelium is produced. The isolate produces a black pigment on all tested ISP media. *N*-Acetyl-D-glucosamine, D-glucose, D-gluconate, D-ribose, D-sucrose, adonitol, D-arabitol, D-sorbitol, iso-inositol, propionate, glutarate, butyrate, isobutyrate, iso-valeric acid, L-arginine, and L-asparagine are utilized for growth. The following carbon sources are not utilized: L-arabinose, cellobiose, D-fructose, D-galactose, maltose, D-mannose, melibiose, L-rhamnose, salicin, and D-xylose. Good growth occurs on nutrient agar and medium 65 at 25–30°C; Major fatty acids are $C_{16:0}$ iso, $C_{16:0}$, $C_{17:1}$ anteiso ω9*c*, $C_{17:0}$ anteiso, and $C_{16:1}$ iso H. The quinone system is composed of the major compound MK-10(H_4) and moderate to minor amounts of MK-10(H_6), MK-9(H_4), and MK-9(H_6). In the polar lipid profile, diphosphatidylglycerol is predominant. Phosphatidylethanolamine, phosphatidylglycerol, and phosphatidylinositol are present in moderate amounts and phosphatidylinositol mannosides, an unknown aminolipid, an unknown phospholipid, and three unknown polar lipids are present in minor to trace amounts.

Type strain shows no sequence similarity over 99%.

Source: isolated from soil.

DNA G+C content (mol%): not known.

Type strain: GW41-1564, DSM 41924, CCM 7499.

Sequence accession no. (16S rRNA gene): AM934703.

447. **Streptomyces spectabilis** Mason, Dietz and Smith 1961, 118[AL]

spec.ta'bi.lis. L. masc. adj. *spectabilis* visible, notable, remarkable.

Spore chains in Section *Rectiflexibiles* with straight to flexuous spore chains. Distribution of sporophores sometimes suggests verticillate morphology, but true whorls or verticils are not produced. Mature spore chains are long with 10 to 50 or more spores per chain. This morphology is seen on yeast-malt agar, oatmeal agar, salts-starch agar, and glycerol-asparagine agar. Spore surface is smooth.

Color of colony: aerial mass color in the Red color series (7ca or 5ca, light yellowish pink; 5gc, light reddish brown) on yeast-malt agar, oatmeal agar, salts-starch agar, and glycerol-asparagine agar. White aerial mycelium is sometimes produced. Reverse side of colony with distinctive reddish orange pigments on yeast-malt agar, oatmeal agar, salts-starch agar, and glycerol-asparagine agar. Reverse mycelium pigment is changed only slightly by the addition of 0.05 M NaOH or HCl.

Color in medium: melanoid pigments are formed in peptone-yeast-iron agar and sometimes in tryptone-yeast broth, but not in tyrosine agar. No pigment is found in the medium in yeast-malt agar, oatmeal agar, salts-starch agar, or glycerol-asparagine agar.

D-Glucose, L-arabinose, D-xylose, iso-inositol, D-mannitol, D-fructose, and raffinose are utilized for growth. Utilization of rhamnose is doubtful. No growth or only traces of growth with sucrose.

Type strain shows no sequence similarity over 99%.

Source: not known.

DNA G+C content (mol%): not known.

Type strain: ATCC 27465, CBS 725.72, BCRC 12648, CECT 3146, DSM 40512, NBRC 13424, NBRC 15441, JCM 4308, JCM 4832, KCTC 9218, LMG 5986, NCIMB 9733, NRRL 2494, NRRL-ISP 5512, RIA 1385.

Sequence accession no. (16S rRNA gene): AB184393.

448. **Streptomyces speibonae** Meyers, Porter, Omorogie, Pule and Kwetane 2003, 804[VP]

spei.bo'na.e. L. n. *spes -ei* hope; L. adj. *bonus* good; N.L. gen. n. *speibonae* of good hope, to indicate Cape Town, the Cape of Good Hope, South Africa, the geographical location from which the type strain was isolated.

Spirales-type spore chains with hairy spore sheaths are produced. Forms gray aerial mycelium. The color of the substrate mycelium is not pH-sensitive. Verticils are not present. The mycelium does not fragment. No diffusible pigments are produced on glycerol-asparagine agar or on any other medium. Melanin pigment is produced on both peptone-yeast extract-iron agar and tyrosine agar. Although growth on inorganic salts-starch agar is initially slow, very good growth with profuse sporulation is observed on this medium after 14 d. Very good growth occurs on yeast extract-malt extract agar. Good growth is observed on oatmeal agar and moderate growth on Czapek's solution agar (Atlas, 1993). Growth on glycerol-asparagine agar is poor. The substrate mycelium is blue on yeast extract-malt extract agar and oatmeal agar, but light gray on Czapek and glycerol-asparagine media. Degrades casein, DNA, gelatin, guanine, hypoxanthine, starch (weakly), Tween 80, L-tyrosine, xanthine, and xylan but not adenine, allantoin, or urea. Uses L(+)-arabinose, D(+)-cellobiose, D(−)-fructose, D(+)-galactose, glycerol, *myo*-inositol, lactose, maltose, D-mannitol, D(+)-mannose, L(+)-rhamnose, D(−)-ribose, sodium acetate, sodium butyrate, sodium DL-malate, sodium malonate, sodium propionate, sodium pyruvate, sodium succinate, sucrose (weak growth), trehalose, and D(+)-xylose as sole carbon sources but not adonitol, *meso*-erythritol, inulin, D(+)-melezitose, D(+)-melibiose, methyl α-D-glucoside, raffinose, salicin, sodium benzoate, sodium citrate, sodium formate, sodium maleate, sodium oxalate, sodium salicylate, sodium L(+)-tartrate, L(−)-sorbose, or xylitol. Tests for nitrate reductase and the production of H_2S are positive, but pectin is not hydrolyzed. Lipase and lecithinase are produced on egg-yolk agar, but protease activity is not seen on this medium after the recommended 2 d of incubation (there is weak activity after 6 d). The cell wall contains LL-A_2pm (cell-wall type I). The whole-cell sugar pattern contains no diagnostic sugars. No antibiosis is exhibited against *Escherichia coli* ATCC 25922 or *Pseudomonas aeruginosa* ATCC 27853. Grows in the presence of (μg/ml unless stated otherwise): cefotaxime (100), cephaloridine (100), D-cycloserine (50), lincomycin (100), oleandomycin (100), penicillin G (10 IU/ml), phenol (0.1%), 2-phenylethanol (0.1%), sodium chloride (10%), and vancomycin (50) and at 45°C but not at 4°C, pH 4.3, or in the presence of sodium azide (0.01%), capreomycin (20), gentamicin (100), kanamycin (10), neomycin (50), rifampin (50), streptomycin (100), tobramycin (50), or viomycin (8). Uses DL-α-amino-*n*-butyric acid, 4-amino-*n*-butyric acid, L-arginine, DL-citrulline, L-cysteine,

L-histidine, L-methionine, DL-ornithine, potassium nitrate, L-serine, L-threonine, and L-valine as sole nitrogen sources, but not L-hydoxyproline or L-phenylalanine.

Type strain shows the highest sequence similarity to: *S. iakyrus*, AB184877, 99.3%; *S. longispororuber*, AB184440, 99.1%; *S. griseoflavus*, AJ781322, 99%; *S. viridodiastaticus*, AY999852, 99%; *S. lusitanus*, AB184424, 99%; *S. griseorubens*, AB184139, 99%.

Source: type strain was isolated from Cape Town, the Cape of Good Hope, South Africa.

DNA G+C content (mol%): 73.4.

Type strain: PK-Blue, ATCC BAA-411, CIP 108060, DSM 41797, JCM 12682, KCTC 9973, NRRL B-24240.

Sequence accession no. (16S rRNA gene): AF452714.

449. **Streptomyces speleomycini** Preobrazhenskaya and Szabó 1986, 575[VP] (Effective publication: Preobrazhenskaya and Szabó *in* Gause, Preobrazhenskaya, Sveshnikova, Terekhova and Maximova 1983.)

spe.le.o.my.ci'ni. N.L. n. *speleomycinum* speleomycin; N.L. gen. n. *speleomycini* of speleomycin, intended to mean spelomycin producing.

Spore chains are straight (*Rectiflexibiles*); spores are smooth. On mineral agar 1: aerial mycelium is light yellowish creamy, yellow; substrate mycelium is reddish gray brownish; no diffusible pigment. On glycerol-nitrate agar, oatmeal agar, glucose-nitrate agar, glycerol-asparagine agar: aerial mycelium is mealy, whitish yellow, light yellow, yellow; substrate mycelium is colorless; no diffusible pigment. On starch-ammonia agar, starch agar (Waksman, 1961): aerial mycelium is whitish, light yellow; substrate mycelium is colorless to gray, olive gray; no diffusible pigment. On organic agar 2: aerial mycelium is whitish; substrate mycelium is gray brownish yellow; no diffusible pigment. Melanoid pigments are not formed. Growth on ribose, fructose, xylose, and galactose; no growth on rhamnose, arabinose, or inositol. Antibiotic: speleomycin.

Source: not known.

DNA G+C content (mol%): not known.

Type strain: B-23.

Sequence accession no. (16S rRNA gene): no sequence available.

Further comments: strain B-23 was first described as *Actinomyces erythreus* subsp. *speleomycini* Szabó and Preobrazhenskaya 1962.

450. **Streptomyces spheroides** (Baldacci 1944) Wallick, Harris, Reagan, Ruger and Woodruff 1956, 911[AL] (*Actinomyces caeruleus* Baldacci 1944, 180; *Streptomyces caeruleus* Pridham, Hesseltine and Benedict 1958, 60 emend. Lanoot, Vancanneyt, Cleenwerck,Wang, Li, Liu and Swings 2002, 828)

sphe.ro.i'des. Gr. n. *sphaira* a ball, a sphere; Gr. adj. suff. *eides* similar; N.L. masc. adj. *spheroides* similar to a ball referring to the characteristic compact coils of spores, resembling spheres.

Spore chains in Section *Spirales*. Mature spore chains are generally long, often with more than 50 spores per chain. This morphology is seen on yeast-malt agar, oatmeal agar, salts-starch agar, and glycerol-asparagine agar. Spore surface is smooth.

Color of colony: aerial mass color in the Yellow color series on yeast-malt agar, oatmeal agar, salts-starch agar, and glycerol-asparagine agar. Reverse side of colony with no distinctive pigments (colorless to pale yellow or light yellowish brown) on yeast-malt agar, oatmeal agar, salts-starch agar, and glycerol-asparagine agar.

Color in medium: melanoid pigments are not formed in peptone-yeast-iron agar, tyrosine agar, or tryptone-yeast broth. No pigment found in medium in yeast-malt agar, oatmeal agar, salts-starch agar, or glycerol-asparagine agar.

D-Glucose, L-arabinose, sucrose, D-xylose, D-mannitol, D-fructose, and rhamnose are utilized for growth. Only traces of growth are found on iso-inositol and raffinose. A trace of growth also occurs on the carbon-free basal medium.

For sequence similarity, see type strain of *Streptomyces caeruleus.*

Source: not known.

DNA G+C content (mol%): not known.

Type strain: ATCC 23965, CBS 491.62, CBS 948.68, BCRC 11559, DSM 40292, NBRC 12917, JCM 4252, JCM 4670, LMG 19392, NCIMB 11891, NRRL 2449, NRRL-ISP 5292, RIA 1200, RIA 700.

Sequence accession no. (16S rRNA gene): EF178698.

Further comments: Streptomyces spheroides Wallick et al. 1956 is a later heterotypic synonym of *Streptomyces caeruleus* (Baldacci 1944) Pridham et al. 1958 emend. Lanoot et al. 2002.

451. **Streptomyces spinoverrucosus** Diab and Al-Gounaim 1982, 331[VP]

spi.no.ver.ru.co'sus. L. adj. *spinosus* thorny; L. adj. *verrucosus* warty; N.L. masc. adj. *spinoverrucosus* spiny and warty, referring to the spiny and warty spore surface.

Spore chains are in the form of terminal, closed, or compact spirals with two or more turns. There are 10 to 20 or more spores per chain. Spore surfaces are spiny; some spores have spiny and warty surfaces. Mature aerial mass colors in the green color series on salts starch agar, yeast malt agar, oatmeal agar, and glycerol-asparagine agar (Tresner–Backus color series tabs) were matched to the red, yellow, green, and gray color wheels. T.G. Pridham (who confirmed the color series) stated that strain 163MA represents a good example of the problem involved in use of color as a major criterion in characterization and classification of streptomycetes and streptoverticillia. The reverse sides of the colonies are colorless or faint yellow on salts starch agar and brownish red on yeast malt agar, oatmeal agar, and glycerol asparagine agar. The red color is pH-sensitive.

Color in medium: melanoid pigments are formed in peptone-yeast-iron agar, tyrosine agar, and tryptone yeast broth. The water-soluble pigments produced are brown on yeast malt agar and glycerol-asparagine agar and red or reddish brown on oatmeal agar. An orange water-soluble pigment may be formed in salts-starch agar.

D-Glucose, D-xylose, L-arabinose, iso-inositol, D-mannitol, fructose, rhamnose, sucrose, and raffinose were utilized for growth. Growth on Czapek's solution was good.

A whole-cell hydrolysate analysis revealed the presence of LL-A_2pm.

Type strain shows the highest sequence similarity to: *S. lusitanus*, AB184424, 99.2%; *S. levis*, AB184670, 99.1%; *S. bellus*, AB184849, 99.1%; *S. coerulescens*, AY999720, 99.1%; *S. coeruleorubidus*, AY999719, 99%; *S. carpinensis*, AB184574, 99%; *S. parvulus*, AB184326, 99%; *S. gancidicus*, AB184660, 99%; *S. lomondensis*, AB184673, 99%.

Source: not known.

DNA G+C content (mol%): not known.

Type strain: Diab 163MA, NCIB (now NCIMB) 11666, ATCC 33692, DSM 41648, NBRC 14228, NBRC 14250, JCM 5077, NRRL B-16932.

Sequence accession no. (16S rRNA gene): AB184578.

452. **Streptomyces spiralis** (Falcão de Morais 1970) Goodfellow, Williams and Alderson 1986a, 575[VP] (Effective publication: Goodfellow, Williams and Alderson 1986a, 50.) (*Elytrosporangium spirale* Falcão de Morais 1970, 79)

spi.ra'lis. L. n. *spira* a spiral; N.L. masc. adj. *spiralis* spiraled, referring to spiral spore chains.

Forms extensively branched substrate and aerial mycelium. The latter bears long spirals spore chains and the former occasional chains of spores. The spore surface is smooth. The aerial spore mass is yellowish gray; the reverse color is cream to yellow. Does not form melanin pigments. Adenine, esculin, casein, guanine, hypoxanthine, pectin, starch, testosterone, and tyrosine are degraded but allantoin, arbutin, chitin, elastin, lecithin, urea, and xanthine are not. Hydrogen sulfide is produced but nitrate is not reduced. L-Arabinose, cellobiose, D-fructose, D-galactose, D-glucose, *myo*-inositol, inulin, D-lactose, mannitol, D-mannose, melibiose, raffinose, L-rhamnose, salicin, sucrose, trehalose, and xylitol are used as sole carbon sources but adonitol, melezitose, and D-xylose are not. Grows on L-arginine, L-cysteine, L-histidine, potassium nitrate, L-serine, L-threonine, and L-valine, but not on DL-amino-n-butyric acid, L-hydroxyproline, L-methionine, or L-phenylalanine, as sole nitrogen source. Growth occurs at 37 and 45°C but not at 10°C. Tolerant to phenol (0.1%, w/v), sodium azide (0.02%, w/v), and sodium chloride (7%, w/v). Resistant to rifampin but sensitive to sodium chloride at 10% (w/v). Does not show antimicrobial activity against *Aspergillus niger* LIV 131, *Bacillus subtilis* NCIB 3610, *Candida albicans* CBS 562, *Escherichia coli* NCIB 9132, *Micrococcus luteus* NCIB 196, *Pseudomonas fluorescens* NCIB 9046^T, *Saccharomyces cerevisiae* CBS 1171^T, or *Streptomyces murinus* ISP 5091. Contains octahydrogenated menaquinones with nine isoprene units as the predominant isoprenolog.

Type strain shows the highest sequence similarity to: *S. ruber*, AB184604, 99%; *S. fumigatiscleroticus*, DQ442499, 99%.

Source: isolated from soil, Brazil.

DNA G+C content (mol%): not known.

Type strain: ATCC 25664, DSM 43836, NBRC 14215, KCC A-0302, JCM 3302, NRRL B-16922, VKM Ac-1311.

Sequence accession no. (16S rRNA gene): EF178683.

453. **Streptomyces spiroverticillatus** Shinobu 1958, 93[AL]

spi.ro.ver.ti.cil.la'tus. L. n. *spira* a coil, spiral; L. n. *verticillus* a whorl; N.L. adj. *verticillatus* whorled; N.L. masc. adj. *spiroverticillatus* coiled and whorled.

Spore chains in Section *Retinaculiaperti*. Shinobu's original description makes note of verticils or whorls "near the base of the aerial mycelium". Mature spore chains are generally long, often with more than 50 spores per chain. This morphology is seen on yeast-malt agar, oatmeal agar, salts-starch agar, and glycerol-asparagine agar. Spore surface is smooth. Conidia-like fragments in substrate mycelium (one observer); sclerotia (two observers).

Color of colony: aerial mass color in the White color series on yeast-malt agar and glycerol-asparagine agar; White or Red series on oatmeal agar and salts-starch agar. Reverse side of colony is grayed yellow to yellow-brown is modified by red or orange on yeast-malt agar, oatmeal agar, salts-starch agar, and glycerol-asparagine agar. Reverse color is changed slightly from yellow-brown to reddish brown with the addition of 0.05 M NaOH and from reddish-brown to yellow-brown with 0.05 M HCl.

Color in medium: melanoid pigments are not formed in peptone-yeast-iron agar and tyrosine agar. Pigments other than traces of yellow are not formed in yeast-malt agar, oatmeal agar, salts-starch agar, and glycerol-asparagine agar. One observer reports a trace of yellow pigment that may change to pink with addition of 0.05 M NaOH.

D-Glucose, L-arabinose. D-xylose, and D-fructose are utilized for growth. No growth or only trace of growth on iso-inositol, D-mannitol, rhamnose, and raffinose. Variable reports on growth with sucrose.

Type strain shows the highest sequence similarity to: *S. cremeus*, AB184124, 99.7%; *S. albolongus*, AB184425, 99.4%; *S. celluloflavus*, AB184476, 99.4%; *S. griseobrunneus*, AB249912, 99.4%; *S. candidus*, DQ026663, 99.3%; *S. cavourensis* subsp. *cavourensis*, DQ445791, 99.3%; *S. praecox*, AB184293, 99.2%; *S. fulvorobeus*, AB184711, 99.2%; *S. lipmanii*, AB184148, 99.2%; *S. fimicarius*, AY999784, 99.2%; *S. flavovirens*, DQ026635, 99.2%; *S. anulatus*, DQ026637, 99.2%; *S. flavofuscus*, AB249935, 99.2%; *S. microflavus*, DQ445795, 99.2%; *S. alboviridis*, AB184256, 99.2%; *S. pluricolorescens*, DQ442540, 99.1%; *S. albovinaceus*, AB249958, 99.1%; *S. lavendulae* subsp. *lavendulae*, AB184080, 99.1%; *S. argenteolus*, AB045872, 99.1%; *S. griseolus*, AB184768, 99.1%; *S. mediolani*, AB184674, 99.1%; *S. luridiscabiei*, AF361784, 99.1%; *S. acrimycini*, AY999889, 99.1%; *S. californicus*, AB184755, 99.1%; *S. flavogriseus*, AJ494864, 99.1%; *S. baarnensis*, EF178688, 99.1%; *S. badius*, AY999783, 99.1%; *S. rubiginosohelvolus*, AB184240, 99.1%; *S. floridae*, AB184656, 99.1%; *S. globisporus* subsp. *globisporus*, EF178686, 99.1%; *S. griseinus*, AB184205, 99.1%; *S. sindenensis*, AB184759, 99.1%; *S. cavourensis* subsp. *washingtonensis*, DQ026671, 99.1%; *S. griseoplanus*, AY999894, 99.1%; *S. pulveraceus*, AB184806, 99%; *S. parvus*, DQ442537, 99%; *S. cyaneofuscatus*, AB184860, 99%; *S. cinereorectus*, AB184646, 99%; *S. halstedii*, EF178695, 99%.

Source: not known.

DNA G+C content (mol%): not known.

Type strain: ATCC 19811, CBS 564.68, BCRC 13648, DSM 40036, NBRC 12821, IMET 42050, JCM 4104, JCM 4609, NBRC 3931, NRRL B-2259, NRRL B-5483, NRRL-ISP 5036, RIA 1091, RIA 549, UNIQEM 198, VKM Ac-751.
Sequence accession no. (16S rRNA gene): AB184814.

454. **Streptomyces spitsbergensis** Wieczorek, Mordarska, Zakrzewska-Czerwinska, Gamian and Mordarski 1993, 86[VP]

spits.ber.gen′sis. N.L. masc. adj. *spitsbergensis* of or belonging to Spitsbergen, the source of the soil sample from which the type strain was isolated.

Spore chains are *Rectiflexibiles*. Spore surface is smooth. Color of branching substrate mycelium ranges from red to reddish brown, depending on the growth medium. Aerial mycelium is pink to violet. Soluble pigments (including melanin) are not produced. Hydrolyzes gelatin and starch. Nitrate is not reduced. Grows well at 25–37°C; no growth occurs at 45°C. Type I cell wall and type PII phospholipid pattern (saturated iso and anteiso fatty acids, lack of mycolic acids and major glycolipids). The species produces some prodiginine-like metabolites with immunosuppressive activity.
For sequence similarity, see type strain of *Streptomyces hiroshimensis.*
Source: the type strain was isolated from soil from Spitsbergen.
DNA G+C content (mol%): 71.0.
Type strain: S-2, ATCC 51269, NBRC 15745, JCM 8881, PCM 2404.
Sequence accession no. (16S rRNA gene): AB184700.
Further comments: Hatano et al. (1997) propose *Streptomyces spitsbergensis* Wieczorek et al. 1993 as a later heterotypic synonym of *Streptomyces baldaccii* corrig. (Farina and Locci 1966) Witt and Stackebrandt 1991.
According to Hatano et al. (2003), *Streptomyces spitsbergensis* Wieczorek et al. 1993 is a later heterotypic synonym of *Streptomyces hiroshimensis* (Shinobu 1955) Witt and Stackebrandt 1991.

455. **Streptomyces sporocinereus** (*ex* Krasil'nikov 1970b) Preobrazhenskaya 1986, 575[VP] (Effective publication: Preobrazhenskaya *in* Gause, Preobrazhenskaya, Sveshnikova, Terekhova and Maximova 1983.) ("*Actinomyces sporocinereus*" Krasil'nikov 1970b)

spo.ro.ci.ne′re.us. Gr. n. *spora* seed; L. adj. *cinereus* ash-colored; N.L. masc. adj. *sporocinereus* ash-colored spores.

Spore chains are spiral (*Spirales*); spores are warty. On mineral agar 1 and oatmeal agar: aerial mycelium is gray, dark gray, black; substrate mycelium is colorless; no diffusible pigment. On glycerol-nitrate agar: aerial mycelium is white or beige; substrate mycelium colorless; no diffusible pigment. On starch-ammonia agar: aerial mycelium is gray; substrate mycelium is colorless to yellowish beige; no diffusible pigment. On glycerol-asparagine agar: aerial mycelium is gray; substrate mycelium colorless; no diffusible pigment. On organic agar 2: aerial mycelium is gray; substrate mycelium and diffusible pigment are yellow-brown, brown. Melanoid pigments are not formed. Utilization of glucose, galactose, and starch; no digestion of fructose.

Type strain shows the highest sequence similarity to: *S. endus*, AY999911, 100%; *S. hygroscopius* subsp. *hygroscopius*, AB184428, 100%; *S. demainii*, DQ334782, 99.9%; *S. yogyakartensis*, AB249942, 99.5%; *S. violaceusniger*, AJ391823, 99.5%; *S. albiflaviniger*, AJ391812, 99.3%.
Source: not known.
DNA G+C content (mol%): not known.
Type strain: ATCC 43692, DSM 41460, INMI 32, JCM 9093, NBRC 100766, NRRL B-16376, VKM Ac-312.
Sequence accession no. (16S rRNA gene): AB249933.

456. **Streptomyces sporoclivatus** (*ex* Krasil'nikov 1970b) Preobrazhenskaya 1986, 575[VP] (Effective publication: Preobrazhenskaya *in* Gause, Preobrazhenskaya, Sveshnikova, Terekhova and Maximova 1983.) ("*Actinomyces sporoclivatus*" Krasil'nikov 1970b).

Etymology is not provided.

Spore chains are spiral (*Spirales*); spores are warty. On mineral agar 1, oatmeal agar, starch-ammonia agar: aerial mycelium is whitish gray or gray; substrate mycelium is colorless; no diffusible pigment. On glycerol-nitrate agar: aerial mycelium is white or light yellow; substrate mycelium is dark yellow; no diffusible pigment. On glycerol-asparagine agar: aerial mycelium is white, light gray; substrate mycelium is colorless to yellow; no diffusible pigment. On organic agar 2: aerial mycelium is white; substrate mycelium is yellow; no diffusible pigment. Melanoid pigments are not formed. Grows on glucose, fructose, xylose, galactose, raffinose, and mannitol; no growth on rhamnose or arabinose.
Type strain shows the highest sequence similarity to: *S. antimycoticus*, AB184185, 100%; *S. geldanamycininus*, DQ334781, 99.8%; *S. rutgersensis* subsp. *castelarensis*, AY508511, 99.8%; *S. melanosporofaciens*, AJ271887, 99.6%; *S. rhizosphaericus*, AB249941, 99.1%; *S. asiaticus*, AB249947, 99.1%.
Source: not known.
DNA G+C content (mol%): not known.
Type strain: ATCC 43693, DSM 41461, INMI 97, JCM 9094, NBRC 100767, VKM Ac-315.
Sequence accession no. (16S rRNA gene): AB249934.

457. **Streptomyces spororaveus** (*ex* Krasil'nikov 1970b) Preobrazhenskaya 1986, 575[VP] (Effective publication: Preobrazhenskaya *in* Gause, Preobrazhenskaya, Sveshnikova, Terekhova and Maximova 1983.) ("*Actinomyces spororaveus*" Krasil'nikov 1970b)

spo.ra.ra.ve′us. Gr. n. *spora* seed; N.L. adj. *raveus* gray-yellow, gray; N.L. masc. adj. *spororaveus* gray-yellow, gray spores.

Spore chains are spiral (*Retinaculiaperti* – spore chains are hooks, loops, or spirals with one or two turns); spores are warty and smooth. On mineral agar 1, starch-ammonia agar: aerial mycelium is gray; substrate mycelium is colorless or sometimes light yellow; no diffusible pigment. On glycerol-nitrate agar: aerial mycelium is gray; substrate mycelium is brown to black; diffusible pigment is brown. On oatmeal agar: aerial mycelium is brownish gray; substrate mycelium and diffusible pigment are gray-brown. On glycerol-asparagine agar: aerial mycelium is gray;

substrate mycelium gray-brown; gray brownish or no diffusible pigment. On organic agar 2: aerial mycelium is brown; substrate mycelium and diffusible pigment are black. Melanoid pigments are formed. Grows on glucose, maltose, and starch; no growth on xylose, arabinose, galactose, fructose, raffinose, or rhamnose.

Type strain shows the highest sequence similarity to: *S. nojiriensis*, AJ781355, 100%; *S. xanthophaeus*, DQ442560, 100%; *S. goshikiensis*, EF178693, 99.9%; *S. vinaceus*, AB184394, 99.9%; *S. cirratus*, AY999794, 99.9%; *S. colombiensis*, DQ026646, 99.8%; *S. sporoverrucosus*, DQ442544, 99.8%; *S. cinnamonensis*, AB184707, 99.7%; *S. avidinii*, AB184395, 99.7%; *S. subrutilus*, X80825, 99.7%; *S. virginiae*, D85119, 99.5%; *S. racemochromogenes*, DQ026656, 99%; *S. polychromogenes*, AB184292, 99%.

Source: not known.

DNA G+C content (mol%): not known.

Type strain: ATCC 43694, DSM 41462, NBRC 15456, INMI 101, JCM 6928, NRRL B-16378, VKM Ac-318.

Sequence accession no. (16S rRNA gene): AJ781370.

458. **Streptomyces sporoverrucosus** (*ex* Krasil'nikov 1970b) Preobrazhenskaya 1986, 575[VP] (Effective publication: Preobrazhenskaya *in* Gause, Preobrazhenskaya, Sveshnikova, Terekhova and Maximova 1983.) ("*Actinomyces sporoverrucosus*" Krasil'nikov 1970b)

spo.ro.ver.ru.co'sus. Gr. n. *spora* seed; L. adj. *verrucosus* rough, rugged; N.L. masc. adj. *sporoverrucosus* having rough spores.

Spore chains are spiral; spores are warty. On mineral agar 1, oatmeal agar: aerial mycelium is gray, sometimes grayish beige; substrate mycelium is brownish yellow, yellow-brown; no diffusible pigment. On starch-ammonia agar: aerial mycelium is gray; substrate mycelium colorless; no diffusible pigment. On glycerol-nitrate agar: aerial mycelium is yellow; substrate mycelium is dark yellow; no diffusible pigment. On glycerol-asparagine agar: aerial mycelium is white-gray; substrate mycelium is yellow; no diffusible pigment. On organic agar 2: aerial mycelium is absent or light gray; substrate mycelium and diffusible pigment are brown. Melanoid pigments are formed.

Type strain shows the highest sequence similarity to: *S. goshikiensis*, EF178693, 100%; *S. colombiensis*, DQ026646, 99.9%; *S. nojiriensis*, AJ781355, 99.8%; *S. sporo
raveus*, AJ781370, 99.8%; *S. xanthophaeus*, DQ442560, 99.7%; *S. vinaceus*, AB184394, 99.7%; *S. cirratus*, AY999794, 99.7%; *S. cinnamonensis*, AB184707, 99.5%; *S. avidinii*, AB184395, 99.5%; *S. subrutilus*, X80825, 99.3%; *S. virginiae*, D85119, 99.2%.

Source: not known.

DNA G+C content (mol%): not known.

Type strain: ATCC 43695, DSM 41463, NBRC 15458, INMI 15, JCM 6929, NRRL B-16379, VKM Ac-321.

Sequence accession no. (16S rRNA gene): DQ442544.

459. **Streptomyces stelliscabiei** Bouchek-Mechiche, Gardan, Normand and Jouan 2000, 98[VP]

stel.li.sca'bi.ei. L. n. *stella -ae* a star; L. n. *scabies -ei* mange; N.L. gen. n. *stelliscabei* referring to lesions from which these strains were isolated, which looks like stars.

Spores are gray and are borne in mature spiral chains. Melanin is produced on tyrosine agar. L-Arabinose, D-fructose, D-glucose, D-mannitol, inositol, raffinose, rhamnose, sucrose, and D-xylose are utilized for growth. Most strains studied degrade xanthine. All strains are susceptible to 20 µg/ml streptomycin, 0.5 µg/ml crystal violet, 100 µg/ml oleandomycin, and 5% (w/v) NaCl, but not to 25 µg/ml oleandomycin or 10 IU/ml penicillin G. Utilizes *trans*-aconitate, D(+)trehalose, and α-D(+)melibiose but do not assimilate 5-keto-D-gluconate, ONPG, betaine, mucate, D-saccharate, DL-lactate, gentisate, or turanose.

Type strain shows no sequence similarity over 99%. Type strain shows DNA–DNA similarity to: *S. europaeiscabiei* DSM 41802[T], 47%; *S. reticuliscabiei* DSM 41804[T], 20%.

Source: the type strain was isolated from star-like common scab lesions on potato tubers cv. Belle de Fontenay in France.

DNA G+C content (mol%): 71.7.

Type strain: CFBP 4521, CIP 107060, CIP 107126, DSM 41803, ICMP 13715, NCPPB 4040.

Sequence accession no. (16S rRNA gene): AJ007429.

460. **Streptomyces stramineus** Labeda, Lechevalier and Testa 1997, 752[VP]

stra.mi'ne.us. L. masc. adj. *stramineus* made of straw, of straw, intended to mean straw colored, referring to the color of the aerial mycelium and spore mass.

Aerial mycelium forms verticils on which umbels consisting of straight chains of ovoid, smooth surface, straw yellow spores (0.5 × 1.0 µm) are produced. The substrate mycelium is yellow to yellowish brown. Soluble pigments that range from yellowish to shades of brown are produced on several media, such as yeast extract-malt extract. Casein, esculin, gelatin, hypoxanthine, starch, tyrosine, and xanthine are hydrolyzed; adenine and urea are not hydrolyzed. Phosphatase is produced; nitrate is not reduced. Melanin pigments are not produced from tyrosine. Growth occurs in the presence of glycerol, 5% NaCl, and lysozyme; no growth occurs in the presence of salicylate. Acetate, citrate, malate, propionate, pyruvate, and succinate are assimilated; benzoate, lactate, mucate, oxalate, and tartrate are not assimilated. Acid is produced from adonitol, dextrin, fructose, glucose, glycerol, inositol, maltose, mannitol, mannose, sorbitol, and trehalose; no acid is produced from arabinose, cellobiose, dulcitol, erythritol, galactose, melibiose, methyl α-D-glucoside, raffinose, rhamnose, salicin, sucrose, xylose, or methyl β-D-xyloside. Growth occurs on fructose, glucose, inositol, and mannitol as the sole carbon source but not on galactose, raffinose, rhamnose, salicin, or xylose. Growth occurs at 10 and 42°C, but not 52°C. Produces the antibiotics LL-BO1208α and LL-BO1208β.

Type strain shows the highest sequence similarity to: *S. eurocidicus*, AY999790, 99.2%; *S. netropsis*, AB184848, 99.2%; *S. albireticuli*, AB184881, 99.1%.

Source: isolated from a grassland soil sample from South Dakota.

DNA G+C content (mol%): 75.0.

Type strain: CIP 105398, DSM 41783, NBRC 16131, JCM 10649, NRRL 12292.

Sequence accession no. (16S rRNA gene): AB184720.

461. **Streptomyces subrutilus** Arai, Kuroda, Yamagishi and Katoh 1964, 25[AL]

sub.ru'ti.lus. L. masc. adj. *subrutilus* reddish.

Spore chains in Section *Rectiflexibiles*. Long, straight, or slightly flexuous spore chains often contain more than 50 spores per chain. This morphology is seen on yeast-malt agar, oatmeal agar, salts-starch agar, and glycerol-asparagine agar. Spore surface is smooth.

Color of colony: aerial mass color in the Red color series (4ec, 5ec, 6ec, 5cb, grayish yellowish pink) on yeast-malt agar, oatmeal agar, and salts-starch agar; Red or White color series on glycerol-asparagine agar. Reverse side of colony with no distinctive pigments (moderate orange yellow or moderate orange brown on yeast-malt agar; pale yellow or grayish yellow on oatmeal agar, salts-starch agar, and glycerol-asparagine agar).

Color in medium: melanoid pigments are formed in peptone-yeast-iron agar and tryptone-yeast broth but are formed only weakly or not at all in tyrosine agar. No pigment or only a trace of yellow is found in the medium in yeast-malt agar, oatmeal agar, salts-starch agar, and glycerol-asparagine agar.

D-Glucose and D-fructose are utilized for growth. Reports vary on utilization of sucrose and D-mannitol, but growth is not abundant on these media. No growth or only traces of growth with L-arabinose, D-xylose, iso-inositol, rhamnose, and raffinose.

Type strain shows the highest sequence similarity to: *S. nojiriensis*, AJ781355, 99.7%; *S. spororaveus*, AJ781370, 99.7%; *S. avidinii*, AB184395, 99.7%; *S. xanthophaeus*, DQ442560, 99.6%; *S. vinaceus*, AB184394, 99.5%; *S. goshikiensis*, EF178693, 99.5%; *S. cirratus*, AY999794, 99.5%; *S. cinnamonensis*, AB184707, 99.4%; *S. sporoverrucosus*, DQ442544, 99.3%; *S. virginiae*, D85119, 99.3%; *S. colombiensis*, DQ026646, 99.3%.

Source: not known.

DNA G+C content (mol%): not known.

Type strain: ATCC 27467, CBS 689.72, BCRC 11921, DSM 40445, IFM 1222, NBRC 13388, JCM 4695, JCM 4834, KCTC 9045, NRRL B-12377, NRRL-ISP 5445, RIA 1349, VKM Ac-1210.

Sequence accession no. (16S rRNA gene): X80825.

462. **Streptomyces sulfonofaciens** Miyadoh, Shomura, Ito and Niida 1983, 323[VP]

sul.fon.o.fa'ci.ens N.L. n. *acidum sulfonicum* sulfonic acid; L. part. adj. *faciens* producing; N.L. part. adj. *sulfonofaciens* producing sulfonic acid.

Spores are 0.7–0.8 × 1.0–1.5 μm and oval or cylindrical; Spore surface is smooth. Spore chains are nearly straight and moderately short, bearing 10–30 spores. Fragmentation of mycelium is not observed on agar or in liquid cultures. Neither sclerotia nor sporangia are formed. Colors of the mature aerial masses produced on inorganic salts-starch agar, oatmeal agar, and yeast extract-malt extract agar belong to the Red color series (5cb, grayish yellowish pink). Undersides of colonies are colorless to pale yellowish brown. No diffusible pigments are formed, except for a light rose beige pigment formed on inorganic salts-starch agar. Frequent subculturing of strain SF-2130[T] onto fresh agar media results in the loss of the ability to produce aerial mycelium, and small aerial mycelium tends to be formed and then lysed. Positive physiological properties include gelatin liquefaction, starch hydrolysis, and milk peptonization. Negative properties include production of melanoid pigment on peptone-yeast extract-iron agar and tryptone-yeast extract broth, milk coagulation, and nitrate reduction. On ISP 9 medium, good or moderate growth is obtained with D-glucose, L-arabinose, D-xylose, L-rhamnose, and D-fructose, and poor growth is obtained with D-mannitol and sucrose. No growth is observed with iso-inositol or raffinose. The temperature range for growth is 15–42°C. Optimum growth occurs between 27 and 33°C. Cell Analysis of the cell-wall hydrolysates by paper chromatography demonstrates the presence of LL-A_2pm in addition to glycine. Thus, strain SF-2130[T] can be considered to have cell-wall type I. Produces a new β-lactam antibiotic.

Type strain shows no sequence similarity over 99%.

Source: soil.

DNA G+C content (mol%): not known.

Type strain: SF-2103, ATCC 31892, DSM 41679, NBRC 14260, JCM 5069, NRRL B-16438.

Sequence accession no. (16S rRNA gene): AB249974.

463. **Streptomyces sulphureus** (Gasperini 1894) Waksman *in* Waksman and Lechevalier 1961, 278[AL] ("*Actinomyces sulphureus*" Gasperini 1894, 78; "*Streptothrix sulphurea*" Caminiti 1907, 197; "*Nocardia sulfurea*" Vuillemin 1931, 129)

sul.phu're.us. L. masc. adj. *sulphureus* of sulfur, sulphurous, referring to the bright sulfur-yellow color of the aerial mycelium.

Spore chains in Section *Spirales* or *Retinaculiaperti*. Short, crooked chains form irregular and imperfect spirals, hooks, or loops. These are not representative of the large hooks, loops, and primitive spirals seen in typical *Retinaculiaperti* cultures. This morphology is seen on yeast-malt agar and oatmeal agar and sometimes on glycerol-asparagine agar. Morphology may also be observed on carbon-utilization medium enriched with D-glucose. Spore surface is smooth.

Color of colony: aerial mass color in the Yellow color series (1½fb, light yellow, or 1½db, pale greenish yellow) on yeast-malt agar and oatmeal agar and sometimes on salts-starch agar or glycerol-asparagine agar if adequate aerial mycelium is developed. Reverse side of colony with no distinctive pigments (yellowish brown to olive brown on yeast-malt agar; grayish yellow, grayish greenish yellow, or olive on oatmeal agar, salts-starch agar, and glycerol-asparagine agar).

Color in medium: melanoid pigments are not formed in peptone-yeast-iron agar, tyrosine agar, or tryptone-yeast broth or are formed very weakly in peptone-yeast-iron agar and tryptone-yeast broth. No pigment is usually found in the medium in yeast-malt agar, oatmeal agar, salts-starch agar, or glycerol-asparagine agar, but one of three observers reports a pale violet pigment in oatmeal agar in 14 d. This pigment becomes pale yellowish brown with the addition of 0.05 M HCl.

Growth is poor on carbon-utilization medium enriched with D-glucose. Comparison with growth on D-glucose suggests equal or better utilization of L-arabinose, D-xylose, D-mannitol, D-fructose, and rhamnose, but utilization of all carbon sources should be verified on a basal medium supporting better growth. Sucrose apparently is not utilized, and reports vary on the utilization of raffinose and iso-inositol.

Type strain shows no sequence similarity over 99%.

Source: not known.

DNA G+C content (mol%): not known.

Type strain: ATCC 27468, CBS 646.72, BCRC 13764, DSM 40104, HUT 6080, NBRC 13345, IMET 40623, JCM 4085, JCM 4835, LMG 19355, NRRL B-1627, NRRL B-2195, NRRL-ISP 5104, RIA 1306, VKM Ac-1820.

Sequence accession no. (16S rRNA gene): DQ442546.

464. **Streptomyces synnematoformans** Hozzein and Goodfellow 2007, 2012[VP]

syn.ne.ma.to.for'mans. Gr. prep. *syn* in company with, together with; Gr. n. *nema* thread; N. Gr. n. *synnema* threads wrapping together; L. part. adj. *formans* forming; N.L. part. adj. *synnematoformans* synnemata-forming, referring to the ability of the organism to form synnemata.

Non-acid–alcohol-fast actinomycete; forms extensively branched substrate mycelium that carries aerial hyphae which differentiate into short straight to flexuous chains of smooth-surfaced spores. Forms synnemata-like structures but not sclerotia. Grayish to blackish red aerial spore mass and a dark red to reddish black substrate mycelium are formed on synthetic agar media. Deep grayish red diffusible pigment is produced on inorganic salts-starch agar. Does not hydrolyze arbutin or degrade adenine or uric acid. Cellobiose, D-fructose, D-galactose, D-glucose, maltose, D-mannose, and trehalose are used as sole carbon sources for energy and growth, but not L-arabitol, *meso*-erythritol, melezitose, D-salicin, D-sorbitol, or D-xylitol (all at 1%, w/v). Similarly, pyruvate is used as a sole carbon source, but not acetate, malonate, or succinate (all at 0.1%, w/v). Grows from 10°C to 37°C and from pH 6.5 to 9.5. Grows in the presence of 7% (w/v) sodium chloride.

Type strain shows no sequence similarity over 99%.

Source: the type strain was isolated from a sand dune soil collected at Borg El-Arab in Egypt.

DNA G+C content (mol%): not known.

Type strain: S155, CGMCC 4.2055, DSM 41902.

Sequence accession no. (16S rRNA gene): EF121313.

465. **Streptomyces syringium** (Konev 1986) Witt and Stackebrandt 1996, 836[VP] (Effective publication: Witt and Stackebrandt 1990, 370.) [*Streptoverticillium syringium* (*ex* Konev, Barashkova and Shenin 1974) Konev 1986].

Etymology is not known.

Spore chains in Section *Rectiflexibiles.* Spores are longish with a smooth surface, 0.5–0.6 × 0.8–1.4 μm in size. Aerial mycelium is well developed, white to light pink and violet color. Substrate mycelium is of a brown color. Soluble pigment is brown-greyish. No tyrosine or H_2S production. Utilizes D-glucose, rhamnose, D-fructose, and iso-inositol, but not lactate, sucrose, or raffinose. Weak growth on L-arabinose, D-xylose, and D-mannitol. Antagonistic against Gram-positive bacteria and fungi. Produces polyene.

For sequence similarity, see type strain of *Streptomyces netropsis.*

Source: not known.

DNA G+C content (mol%): not known.

Type strain: DSM 41480, DSM 41502, NBRC 15900, JCM 9948, LIA 0725, VKM Ac-1230.

Sequence accession no. (16S rRNA gene): AJ781375.

Further comments: this species was inadvertently omitted from Validation List no. 38 (Int. J. Syst. Bacteriol. 1991, 41, 456–457); the effective date of validation is that of list no. 38 (July 1991).

According to Hatano et al. (2003), *Streptomyces syringium* (Konev 1986) Witt and Stackebrandt 1991 is a later heterotypic synonym of *Streptomyces netropsis* (Finlay et al. 1951) Witt and Stackebrandt 1991.

466. **Streptomyces tanashiensis** Hata, Ohki and Higuchi 1952, 529[AL]

ta.na.shi.en'sis. N.L. masc. adj. *tanashiensis* of or belonging to Tanashi-machi, a town near Tokyo, Japan, the source of the soil from which the organism was isolated.

Spore chains in Section *Rectiflexibiles.* Mature spore chains generally have 10–50, or sometimes more than 50, spores per chain. This morphology is seen on yeast-malt agar, oatmeal agar, salts-starch agar and glycerol-asparagine agar. Spore surface is smooth.

Color of colony: aerial mass color in the Gray color series on yeast-malt agar, oatmeal agar, salts-starch agar, and glycerol-asparagine agar. Reverse side of colony with no distinctive pigments (pale yellow to light olive brown) on yeast-malt agar, oatmeal agar, salts-starch agar, and glycerol-asparagine agar.

Color in medium: melanoid pigments are formed in peptone-yeast-iron agar and tryptone-yeast broth, but not on tyrosine agar. No pigment is found in medium in yeast-malt agar, oatmeal agar, salts-starch agar, or glycerol-asparagine agar.

D-Glucose, L-arabinose, and D-xylose are utilized for growth. Reports vary on utilization of D-fructose. Only traces of growth are found with sucrose, iso-inositol, D-mannitol, rhamnose, and raffinose.

Type strain shows the highest sequence similarity to: *S. nashvillensis,* AB184286, 100%; *S. violaceorectus,* AB184314, 99.2%; *S. showdoensis,* AB184389, 99.1%; *S. cinereoruber* subsp. *cinereoruber,* AB184121, 99.1%; *S. roseoviridis,* AB184239, 99.1%; *S. racemochromogenes,* DQ026656, 99.1%; *S. viridobrunneus,* AJ781372, 99.1%; *S. roseolus,* AB184168, 99.1%; *S. polychromogenes,* AB184292, 99.1%; *S. bikiniensis,* X79851, 99.1%; *S. vietnamensis,* DQ311081, 99%; *S. flavotricini,* AB184132, 99%.

Source: isolated from soil from Tanashi-machi, a town near Tokyo, Japan.

DNA G+C content (mol%): not known.

Type strain: ATCC 23967, CBS 165.64, CBS 950.68, BCRC 12641, DSM 40195, HUT 6070, NBRC 12919, IMET 42939, JCM 4086, JCM 4671, NRRL B-1692, NRRL-ISP 5195, RIA 1148, VKM Ac-1892.

Sequence accession no. (16S rRNA gene): AJ781362.

467. **Streptomyces tauricus** (*ex* Ivanitskaya, Upiter, Sveshnikova and Gause 1966) Sveshnikova 1986, 575[VP] (Effective publication: Sveshnikova *in* Gause, Preobrazhenskaya, Sveshnikova, Terekhova and Maximova 1983.) ("*Actinomyces tauricus*" Ivanitskaya, Upiter, Sveshnikova and Gause 1966; "*Streptomyces tauricus*" Pridham 1970)

tau'ri.cus. L. masc. adj. *tauricus* of or belonging to the Taurians (a Thracian people, living in what is now Crimea), Albanian.

Spore chains in Section *Spirales* or *Retinaculiaperti*. Terminal spirals of several turns and some hooks or loops are found on long spore chains of 10–50 or more spores per chain. This morphology is seen on yeast-malt agar, oatmeal agar, salts-starch agar, and glycerol-asparagine agar. Spore surface is smooth.

Color of colony: aerial mass color in the Red color series (5cb or 6ec, grayish yellowish pink; 5ca, light yellowish pink; 3ca, pale orange yellow; or 7ca, light yellowish pink) on yeast-malt agar, oatmeal agar, salts-starch agar, and glycerol-asparagine agar. Reverse side of colony with distinctive red pigments. Grayish red, reddish brown, or yellowish pink on yeast-malt agar, oatmeal agar, salts-starch agar, and glycerol-asparagine agar. Reverse mycelium pigment is a pH indicator, changing from orange or red to blue or violet with the addition of 0.05 M NaOH and from brownish red to yellowish red with the addition of 0.05 M HCl.

Color in medium: melanoid pigments are formed in peptone-yeast-iron agar and tryptone-yeast broth but weakly or not at all in tyrosine agar. Red (pink) pigment is sometimes found in the medium in yeast-malt agar, oatmeal agar, and glycerol-asparagine agar. When pigment is present in the medium it is pH-sensitive, showing the same changes observed with the reverse mycelium pigment.

D-Glucose, L-arabinose, D-xylose, iso-inositol, D-mannitol, D-fructose, rhamnose, sucrose, and raffinose are all utilized for growth.

Type strain shows the highest sequence similarity to: *S. glomeroaurantiacus*, AB249983, 99.1%; *S. aurantiacus*, AJ781383, 99%.

Source: not known.

DNA G+C content (mol%): not known.

Type strain: ATCC 27470, CBS 757.72, BCRC 12822, DSM 40560, NBRC 13456, IMET 43541, INA 8173, JCM 4837, NRRL B-12497, NRRL-ISP 5560, RIA 1417, VKM Ac-1853.

Sequence accession no. (16S rRNA gene): AB045879.

468. **Streptomyces tendae** Ettlinger, Corbaz and Hütter 1958a, 351[AL]

ten'da.e. N.L. gen. n. *tendae* of Tende, Germany, the source of the soil from which the organism was isolated.

Spore chains in Section *Retinaculiaperti* or *Spirales*. Straight to flexuous spore chains are most common on yeast-malt agar and spirals are best developed on salts-starch agar. Straight, flexuous, *Retinaculiaperti*, and spiral spore chains are recorded for oatmeal agar and glycerol-asparagine agar. Mature spore chains generally have 10–50 spores per chain, longer chains are sometimes observed. Spore surface is smooth.

Color of colony: aerial mass color in the Gray color series on yeast-malt agar, oatmeal agar, and salts-starch agar. Reverse side of colony is yellow or greenish yellow on yeast-malt agar, oatmeal agar, salts-starch agar, and glycerol-asparagine agar. Reverse color is changed from yellow to orange by addition of 0.05 M HCl.

Color in medium: melanoid pigments not formed in peptone-yeast-iron agar, tyrosine agar, and tryptone-yeast extract broth. Yellow pigment found in medium in glycerol-asparagine agar; traces of yellow pigment may also diffuse into yeast-malt agar, oatmeal agar, and salts-starch agar. The yellow pigment is changed to yellowish orange or orange by 0.05 M HCl.

D-Glucose, L-arabinose, sucrose, D-xylose, iso-inositol, D-mannitol, D-fructose, and rhamnose are utilized for growth. No growth or only trace of growth on raffinose.

Type strain shows the highest sequence similarity to: *S. violaceorubidus*, AJ781374, 99.8%; *S. violaceoruber*, AF503492, 99.6%; *S. coelescens*, AF503496, 99.6%; *S. rubrogriseus*, AB184681, 99.6%; *S. anthocyanicus*, AB184631, 99.5%; *S. humiferus*, AF503491, 99.5%; *S. violaceolatus* AF503497, 99.5%; *S. lienomycini*, AJ781353, 99.5%; *S. tricolor*, AB184687, 99.4%; *S. matensis*, AB184221, 99.2%; *S. griseorubens*, AB184139, 99.1%; *S. mutabilis*, EF178679, 99.1%; *S. geysiriensis*, DQ442501, 99.1%; *S. minutiscleroticus*, EF178696, 99.1%; *S. ghanaensis*, AY999851, 99%; *S. rochei*, AB184237, 99%; *S. eurythermus*, D63870, 99%; *S. parvulus*, AB184326, 99%; *S. malachitospinus*, AB249954, 99%; *S. plicatus*, AB184291, 99%; *S. vinaceusdrappus*, AY999929, 99%.

Source: isolated from soil from Tende, Germany.

DNA G+C content (mol%): not known.

Type strain: AS 4.1460, ATCC 19812, CBS 565.68, BCRC 12167, DSM 40101, IFM 1176, NBRC 12822, IMET 40459, JCM 4149, JCM 4610, LMG 5987, LMG 19314, NCIMB 9614, NRRL B-2313, NRRL-ISP 5101, RIA 1092, RIA 534, UNIQEM 199, VKM Ac-1889.

Sequence accession no. (16S rRNA gene): D63873.

469. **Streptomyces termitum** Duché, Heim and Laboureur *in* Heim 1951, 359[AL]

ter.mi'tum. L. n. *termes -itis* woodworm, termite; L. gen. pl. n. *termitum* of termites, referring to the source of the organism.

Spore chains in Section *Rectiflexibiles*. Mature spore chains generally long and flexuous with 10–50 or often more than 50 spores per chain. This morphology is seen on yeast-malt agar, oatmeal agar, salts-starch agar, and glycerol-asparagine agar. Spore surface is smooth.

Color of colony: aerial mass color in the Red color series (3ca, pale orange-yellow to 5cb, grayish yellowish pink) on yeast-malt agar, oatmeal agar, salts-starch agar, and glycerol-asparagine agar. Reverse side of colony with no distinctive pigments (colorless to pale or grayish yellow) on yeast-malt agar, oatmeal agar, salts-starch agar, and glycerol-asparagine agar.

Color in medium: melanoid pigments are not formed in peptone-yeast-iron agar, tyrosine agar, or tryptone-yeast broth. No pigment is found in medium in yeast-malt agar, oatmeal agar, salts-starch agar, or glycerol-asparagine agar.

D-Glucose and D-xylose are utilized for growth. Utilization of D-fructose is doubtful. Reports vary on utilization of sucrose and rhamnose (two of three observers find no growth on these two carbon sources). No growth or only trace of growth with L-arabinose, iso-inositol, D-mannitol, and raffinose.

Type strain shows no sequence similarity over 99%.

Source: not known.

DNA G+C content (mol%): not known.

Type strain: ATCC 25499, CBS 959.69, BCRC 12592, DSM 40329, NBRC 13087, IMET 43127, NCIMB 9980, NRRL B-3804, NRRL-ISP 5329, RIA 1279, JCM 4518.

Sequence accession no. (16S rRNA gene): AB184302.

470. **Streptomyces thermoalcalitolerans** Kim, Sahin, Minnikin, Zakrzewska-Czerwinska, Mordarski and Goodfellow 1999, 15[VP]

ther.mo.al.ca.li.to'le.rans. Gr. n. *therme* heat; N.L. n. *alcali* (from Arabic article *al* the; Arabic n. *qaliy* ashes of saltwort) alkali; L. part. adj. *tolerans* tolerating, enduring; N.L. part. adj. *thermoalcalitolerans* thermophilic alkali-tolerating.

Spiral chains of warty surfaced spores are borne on aerial hyphae. Forms extensively branched substrate and aerial hyphae. The aerial spore mass is gray; neither distinctive substrate mycelium colors nor diffusible pigments are formed. Melanin pigments are not produced on peptone-iron agar. Casein, DNA, gelatin, starch, testosterone, L-tyrosine, and xylan are degraded, but not adenine, arbutin, elastin, guanine, hypoxanthine, or xanthine. Adonitol, L-arabinose, arabitol, cellobiose, D-fructose, D-galactose, D-glucose, *myo*-inositol, α-lactose, D-mannitol, D-mannose, melezitose hydrate, melibiose, α-L-rhamnose, D-ribose, D-sorbitol, sucrose, trehalose, D-turanose, xylitol, and D-xylose are used as sole carbon sources for energy and growth, but D-raffinose is not. Growth occurs between 25 and 55°C, from pH 6.0 to 11.5 and in the presence of ampicillin (8 μg/ml), bacitracin (16 μg/ml), oleandomycin phosphate (16 μg/ml), penicillin G (15 international units), rifampin (16 μg/ml), streptomycin sulfate (4 μg/ml), tetracycline hydrochloride (16 μg/ml), and tunicamycin (10 μg/ml). In contrast, growth is inhibited in the presence of gentamicin sulfate (8 μg/ml), lincomycin hydrochloride (32 μg/ml), neomycin sulfate (8 μg/ml), novobiocin (4 μg/ml), oleandomycin phosphate (32 μg/ml), polymyxin B phosphate (32 μg/ml), rifampin (32 μg/ml), streptomycin sulfate (16 μg/ml), tetracycline hydrochloride (32 μg/ml), tobramycin sulfate (32 μg/ml), and vancomycin hydrochloride (16 μg/ml).

Type strain shows no sequence similarity over 99%.

Source: isolated from tropical garden soil collected by M. Goodfellow in 1991 from Yogyakarta, Central Java, Indonesia.

DNA G+C content (mol%): 73.0.

Type strain: TA56, DSM 41741, NBRC 16322, JCM 10673.

Sequence accession no. (16S rRNA gene): AB249909.

471. **Streptomyces thermoautotrophicus** Gadkari, Schricker, Acker, Kroppenstedt and Meyer 1991, 456[VP] (Effective publication: Gadkari, Schricker, Acker, Kroppenstedt and Meyer 1990, 3733.)

ther.mo.au.to.tro'phi.cus. Gr. adj. *thermos* hot; Gr. pref. *autos* self; Gr. adj. *throphikos* nursing, tending or feeding; N.L. masc. adj. *thermoautothrophicus* heat-loving self-nourishing, referring to the ability to grow at high temperature at the expense of CO or H_2 plus CO_2.

Forms chains of two to eight oval spores produced in a sheath residing on the substrate and scanty whitish aerial mycelium. No endospores, synnemata, sporangia, or sclerotia are found. Forms relatively stable branching vegetative hyphae with a diameter of 0.2–0.5 μm. Possesses LL-A_2pm and ribose (cell-wall type I). Mycolic acids are absent. Predominant menaquinone is MK-9(H_4); small amounts of MK-9(H_6) also occur. Phospholipid pattern is composed of phosphatidylethanolamine, diphosphatidylglycerol, phosphatidylglycerol, and phosphatidylinositol mannosides (phospholipid type 2). Iso- and anteiso-branched fatty acid patterns are found, with C_{17}, C_{17}, and C_{16} being the predominant fatty acids. Small amounts of cyclopropane fatty acids and traces of 2-hydroxy fatty acids are also present.

Source: not known.

DNA G+C content (mol%): 70.6.

Type strain: UBT1, DSM 41605.

Sequence accession no. (16S rRNA gene): no sequence available.

472. **Streptomyces thermocarboxydovorans** Kim, Falconer, Williams and Goodfellow 1998, 65[VP]

ther.mo.car.bo.xy.do.vo'rans. Gr. n. *therme* heat; N.L. n. *carboxydum* carbon monoxide; L. v. *voro* to eat greedily, devour; N.L. part. adj. *thermocarboxydovorans* thermophilic, carbon monoxide consuming.

Moderately thermophilic, facultatively chemolithotrophic actinomycete with extensively branched substrate and aerial mycelia. *Rectiflexibiles* chains of smooth-surfaced spores are borne on aerial hyphae. The aerial spore mass is gray; distinctive substrate mycelium colors are not produced. Diffusible pigments are not formed. Melanin pigments are not produced on peptone-iron agar. The organism oxidizes hydrogen and can use carbon monoxide as sole source of carbon for energy and growth. Arbutin, DNA, elastin, guanine, hypoxanthine, RNA, starch, L-tyrosine, and xanthine are degraded. Nitrate is reduced and hydrogen sulfide is produced. Neither allantoin nor urea are hydrolyzed. D-Fructose, D-lactose, sodium fumarate, and sodium pyruvate, but not adonitol or xylitol are used as sole carbon sources. L-Arginine, D-aminobutyric acid, L-cysteine, L-histidine, L-methionine, potassium nitrate, L-serine, L-threonine, and L-valine, but not L-hydroxyproline, are used as sole nitrogen sources. Positive activity is seen for chymotrypsin, cystine arylamidase, α-galactosidase and lipase but not for α-glucosidase (API ZYM). Similarly, *N*-acetyl-β-D-glucosamidase is produced but not β-D-cellobiosidase or β-D-xylosidase. Growth occurs between 20°C and 55°C and in the presence of adenine (0.5%, w/v), crystal violet (0.001%, w/v), phenyl ethanol (0.3%, w/v), potassium tellurite (0.001%, w/v), sodium azide (0.01%, w/v), chloramphenicol (2 μg/ml), chlortetracycline (4 μg/ml), gentamicin sulfate (1 μg/ml), novobiocin (0.5 μg/ml), oleandomycin phosphate (1 μg/ml), penicillin G (16 μg/ml), polymyxin B (32 μg/ml), and tobramycin sulfate (1 µg/ml). In contrast, growth was inhibited in the presence of phenol (1%, w/v), sodium azide (0.02%, w/v), rifampin (0.5 µg/ml), and streptomycin sulfate (2 µg/ml).

Type strain shows the highest sequence similarity to: *S. thermospinosisporus*, AF333113, 99.1%.

Source: isolated from soil.

DNA G+C content (mol%): 74.7 ± 0.05.

Type strain: AT52, CIP 105544, DSM 44296, NBRC 16324, JCM 10367.

Sequence accession no. (16S rRNA gene): U94489.

473. **Streptomyces thermocarboxydus** Kim, Falconer, Williams and Goodfellow 1998, 66[VP]

ther.mo.car.bo'xy.dus. Gr. n. *therme* heat; N.L. n. *carboxydum* carbon monoxide; N.L. masc. adj. *carboxydus* pertaining to carbon monoxide.

Moderately thermophilic, facultatively chemolithotrophic actinomycete with extensively branched substrate and aerial mycelia. *Retinaculiaperti* chains of warty-surfaced spores are borne on aerial hyphae. Aerial spore mass is gray; distinctive substrate mycelium colors are not produced. Diffusible pigments are not formed. Melanin pigments are not produced on peptone-iron agar. Organism oxidizes hydrogen and can use carbon monoxide as a sole source of carbon for energy and growth. Esculin, arbutin, DNA, elastin, guanine, hypoxanthine, starch, testosterone, and xanthine are degraded. Nitrate is reduced and hydrogen sulfide is produced. Neither allantoin nor urea are hydrolyzed. D-Fructose, D-glucose, *myo*-inositol, D-lactose, mannitol, trehalose, and sodium propionate, but not adonitol, raffinose, sucrose, xylose, sodium acetate, sodium fumarate, or sodium pyruvate, are used as sole carbon sources. L-Arginine, D-aminobutyric acid, L-cysteine, L-histidine, potassium nitrate, L-serine, L-threonine, and L-valine, but not L-hydroxyproline or L-methionine, are used as sole nitrogen sources. Positive activity is seen for chymotrypsin, cystine arylamidase, α-glucosidase, lipase, and trypsin but not for α-galactosidase (API ZYM). *N*-Acetyl-β-D-galactosamidase, *N*-acetyl-β-D-glucosamidase, and α-L-arabinosidase are produced but not β-D-cellobiosidase, β-D-fucopyranosidase, α-D-mannosidase, or β-D-xylosidase. Growth occurs between 20 and 55°C and in the presence of adenine (0.5%, w/v), crystal violet (0.001%, w/v), phenyl ethanol (0.3%, v/v), potassium tellurite (0.01%, w/v), sodium azide (0.01%, w/v), chloramphenicol (32 µg/ml), chlortetracycline (4 µg/ml), gentamicin sulfate (8 µg/ml), neomycin sulfate (2 µg/ml), novobiocin (2 µg/ml), oleandomycin phosphate (32 µg/ml), penicillin G (16 µg/ml), polymyxin B (32 µg/ml), rifampin (4 µg/ml), streptomycin sulfate (2 µg/ml), tobramycin sulfate (8 µg/ml), and vancomycin (4 µg/ml).

Type strain shows the highest sequence similarity to: *S. lusitanus*, AB184424, 99.5%; *S. indiaensis*, AB184553, 99.3%; *S. longispororuber*, AB184440, 99.1%; *S. bellus*, AB184849, 99.1%; *S. cellulosae*, DQ442495, 99.1%; *S. coerulescens*, AY999720, 99.1%; *S. gancidicus*, AB184660, 99%; *S. lomondensis*, AB184673, 99%.

Source: not known.

DNA G+C content (mol%): 70.9 ± 0.4.

Type strain: AT37, CIP 105545, DSM 44293, NBRC 16323, JCM 10368.

Sequence accession no. (16S rRNA gene): U94490.

474. **Streptomyces thermocoprophilus** Kim, Al-Tai, Kim, Somasundaram and Goodfellow 2000, 506[VP]

ther.mo.co.pro.phi'lus. Gr. n. *therme* heat; Gr. n. *kopros* dung; Gr. adj. *philos* loving; N.L. masc. adj. *thermocoprophilus* dung-loving thermophile.

Moderately thermophilic actinomycete which forms a highly branched substrate mycelium and aerial hyphae which differentiate into long chains of straight spores which are cylindrical and have smooth surfaces (1.1–1.7 × 0.5 µm). Aerial spore mass color is gray. Diffusible pigments are formed on some standard media such as inorganic salts starch agar. Melanin pigments are produced on peptone-iron and tyrosine agars. Casein, starch, xanthine, and xylan are degraded but not adenine. L-Arabinose, D-fructose, D-galactose, D-glucose, *myo*-inositol, maltose, D-mannitol, D-mannose, and D-xylose are used as sole carbon sources for energy and growth but not carboxymethylcellulose, D-raffinose, starch, or sucrose. Growth occurs between 20 and 50°C and in the presence of ampicillin (10 µg/ml), erythromycin (15 µg/ml) and sodium chloride (7%, w/v) but not at 10 or 55°C. Growth is inhibited by chloramphenicol (30 µg/ml), gentamicin sulfate (15 µg/ml), kanamycin sulfate (30 µg/ml), neomycin sulfate (30 µg/ml), streptomycin sulfate (10 µg/ml), and tetracycline hydrochloride (30 µg/ml). Antimicrobial activity is shown against *Bacillus subtilis* NCIB 3610 but not towards *Escherichia coli* NCIB 9132 or *Staphylococcus aureus* ATCC 12600, or against representative strains of *Candida albicans*, *Curvularia lunata*, *Pestalotiopsis gnepini*, *Pyricularia oryzae*, and *Trichoderma viride*.

Type strain shows no sequence similarity over 99%.

Source: isolated from a sample of poultry feces collected from the poultry farm at the University of Malaya.

DNA G+C content (mol%): 68.6.

Type strain: B19, DSM 41700, JCM 10918, NBRC 100771.

Sequence accession no. (16S rRNA gene): AJ007402.

475. **Streptomyces thermodiastaticus** (Bergey, Harrison, Breed, Hammer and Huntoon 1923) Waksman 1953, 102[AL] (*Actinomyces thermodiastaticus* Bergey, Harrison, Breed, Hammer and Huntoon 1923)

ther.mo.di.a.sta'ti.cus. Gr. adj. *thermos* hot; N.L. adj. *diastaticus* starch-hydrolyzing; N.L. masc. adj. *thermodiastaticus* starch-hydrolyzing, heat-loving.

Spore chains in Section *Spirales*. Spirals are often irregular or imperfect; hooks or incomplete spirals and crooked spore chains are common. Mature spore chains are moderately short with 3 to 10 or more spores per chain. This morphology is seen on yeast-malt agar, oatmeal agar, salts-starch agar, and glycerol-asparagine agar. Special incubation temperature: incubation temperatures used by ISP collaborators included 37, 45, 48, and 51°C; typical morphology, color reactions, and physiology are seen within this range. Spore surface is warty or spiny (very small "warts" or very blunt spines).

Color of colony: aerial mass color in the Gray color series on yeast-malt agar, oatmeal agar, salts-starch agar and glycerol-asparagine agar. Nearest matching color tabs in the Gray color series include 3fe, light brownish gray; 3ig, grayish yellowish brown; 3ge, light grayish yellowish brown; and 2ge, light olive brown. Colors are essentially the same at incubation temperatures from 37–51°C. Reverse side of colony is dark brown on yeast-malt agar; grayish yellow, yellowish brown, or olive brown on oatmeal agar, salts-starch agar, and glycerol-asparagine agar. When reddish (brown) pigment is present in the substrate mycelium, it is somewhat pH-sensitive, changing from reddish brown to olive brown or olive with the addition of 0.05 M HCl.

Color in medium: melanoid pigments are not formed in peptone-yeast-iron agar, tyrosine agar, and tryptone-yeast broth except that a grayish yellow or olive yellow pigment is found in tyrosine agar in 2–4 d. Faint reddish (brown) or yellow pigment may be found in the medium in yeast-malt agar, oatmeal agar, salts-starch agar, and glycerol-asparagine agar. When present, this pigment is pH-sensitive, changing from brownish to red with the addition of 0.05 M NaOH and from brownish to olive green with the addition of 0.05 M HCl.

D-Glucose, L-arabinose, D-xylose, iso-inositol, D-mannitol, D-fructose, rhamnose, and raffinose are utilized for growth. No growth or only traces of growth with sucrose. No difference in carbon utilization is seen on cultures incubated at 37 and 51°C.

Type strain shows the highest sequence similarity to: *S. thermospinosisporus*, AF333113, 99%.

Source: not known.

DNA G+C content (mol%): not known.

Type strain: ATCC 27472, CBS 769.72, BCRC 12492, BCRC 12636, DSM 40573, HUT 6606, JCM 4840, NBRC 100020, NRRL B-5316, NRRL-ISP 5573, RIA 1429.

Sequence accession no. (16S rRNA gene): AB018095.

476. **Streptomyces thermogriseus** Xu, Tiang, Zhang, Zhao and Jiang 1998, 1093[VP]

ther.mo.gri′se.us. Gr. adj. *thermos* hot; N.L. adj. *griseus* gray; N.L. masc. adj. *thermogriseus* hot, gray.

Spores are spherical, subspherical, or short rods. Spore surface is smooth. Brown or yellow vegetative hyphae are produced, becoming black after autolysis. Neither diffusible pigment nor melanin is produced. Milk is coagulated and peptonized. Hydrolyzes starch. Reduces nitrate. No H_2S production. No growth on cellulose. No degradation of urea. Utilizes rhamnose and mannitol. May or may not utilize glucose, fructose, xylose, raffinose, ribitol, and cellobiose. Does not utilize inositol. Lecithinase may or may not be produced. No resistance to neomycin or rifampin. No antimicrobial activities are detected. Grows at 65–68°C and in 0.1% phenol. Growth or no growth in 7% NaCl. The cell wall contains LL-A_2pm and glycine. The whole-cell hydrolysate contains galactose.

Type strain shows the highest sequence similarity to: *S. thermovulgaris*, Z68094, 100%.

Source: Strain Y-4027 (=CCTCC AA 97012) was isolated from soil samples of Kunming, Y-5114 (=CCTCC AA97013) was from Chenghai Lake and Y-14046[T] (=CCTCC AA 97014[T]) and Y-14082 (=CCTCC AA97015) were from a hot-spring in Eryuan, Yunnan, China.

DNA G+C content (mol%): not known.

Type strain: Y-14046, CCTCC AA 97014, CIP 105834, DSM 41756, JCM 11269, NBRC 100772.

Sequence accession no. (16S rRNA gene): AB249980.

477. **Streptomyces thermolineatus** Goodfellow, Lacey and Todd 1988, 329[VP] (Effective publication: Goodfellow, Lacey and Todd 1987a, 3147.)

ther.mo.lin.e.a′tus. Gr. fem. n. *therme* heat, L. part. adj. *lineatus* reduced to a straight line, to made straight; N.L. masc. adj. *thermolineatus* heat-loving linear (referring to the spore chains).

Spores in straight or flexuous chains less than 30 spores long. Spores are smooth but ends often prolonged, projecting from the oval spore body and retaining their shape under vacuum when the rest of the spore collapses, to give a phalangiform appearance (Tresner et al., 1966). Spores measure 1.0–2.1 × 0.9–1.3 µm (mean 1.43 × 1.06 µm). Good growth at 40°C on V-8 juice agar, producing abundant aerial mycelium in the green color series, 24ih to 24½ih. Reverse of colony is yellow-brown with no distinctive pigments. No melanoid pigment produced on peptone-iron agar. Degrades casein, gelatin, and starch. Uses trehalose and sometimes cellobiose and mannitol as sole carbon sources, and L-arginine, L-histidine, L-hydroxyproline, L-methionine, L-phenylalanine, potassium nitrate, L-serine, and L-valine as sole nitrogen sources. Reduces nitrate to nitrite and sometimes shows lipolytic and proteolytic activity on egg yolk.

Type strain shows no sequence similarity over 99%.

Source: isolated from sewage compost.

DNA G+C content (mol%): not known.

Type strain: K47, A1484, ATCC 51534, DSM 41451, HUT 6609, NBRC 14750, JCM 6307, NCIMB 12471.

Sequence accession no. (16S rRNA gene): Z68097.

478. **Streptomyces thermonitrificans** Desai and Dhala 1967, 137[AL]

ther.mo.ni.tri′fi.cans. Gr. n. *therme* heat; N.L. part. adj. *nitrificans* nitrifying; N.L. part. adj. *thermonitrificans* heat, nitrifying, referring to thermophily and vigorous nitrate reduction of the organism.

Spore chains in Section *Rectiflexibiles* or *Spirales*. Short spore chains often have only 3–10 spores per chain. On yeast-malt agar, salts-starch agar, and glycerol-asparagine agar, the chains are usually crooked or twisted. Longer chains may form irregular and imperfect spirals, especially on oatmeal agar and occasionally on salts-starch agar. Spore surface is smooth.

Color of colony: aerial mass color in the Gray color series when adequate sporulation occurs on yeast-malt agar, oatmeal agar, and salts-starch agar. Immature aerial mycelium is white; sporulating aerial mycelium is not found on glycerol-asparagine agar. Nearest matching color tabs are 3li and 4li, brownish gray; 3ge, light grayish yellowish brown; and 2ge, light olive brown. Reverse side of colony with no distinctive pigments (grayish olive to dark grayish brown on yeast-malt agar; grayish yellow to grayish yellowish green or grayish olive on oatmeal agar, salts-starch agar, and glycerol-asparagine agar).

Color in medium: melanoid pigments are not formed in peptone-yeast-iron agar, tyrosine agar, or tryptone-yeast broth according to ISP observers [the original description by Desai and Dhala (1967) states that this strain is H_2S- and melanin-positive]. No pigment or only a trace of yellow is found in the medium in yeast-malt agar, oatmeal agar, salts-starch agar, and glycerol-asparagine agar.

D-Glucose, D-xylose, iso-inositol, D-mannitol, D-fructose, and sucrose are utilized for growth. No growth or only traces of growth with L-arabinose, rhamnose, and raffinose.

For sequence similarity, see type strain of *Streptomyces thermovulgaris.*

Source: not known.

DNA G+C content (mol%): not known.

Type strain: ATCC 23385, CBS 774.72, DSM 40579, NBRC 13473, NBRC 16616, IMET 43405, JCM 4841, LMG 19341, NCIM 2007, NRRL-ISP 5579, RIA 1434.

Sequence accession no. (16S rRNA gene): Z68098.

Further comments: according to Kim et al. (1999), *Streptomyces thermonitrificans* Desai and Dhala 1967 is a later heterotypic synonym of *Streptomyces thermovulgaris* Henssen 1957.

479. **Streptomyces thermospinosisporus** corrig. Kim and Goodfellow 2002, 1227[VP]

ther.mo.spi.ni.spo'rus. Gr. adj. *thermos* hot; L. adj. *spinosus* spiny; N.L. n *spora* a spore; N.L. masc. adj. *thermospinisporus* heat-loving, spiny spores.

Moderately thermophilic, facultatively carboxydotrophic actinomycete which forms extensively branched aerial and substrate hyphae. Aerial hyphae differentiate into flexuous chains of spores that show spiny ornamentation. Aerial spore mass is gray, but the substrate mycelium has no distinctive pigments. Soluble pigments are not produced, nor are melanin pigments formed on peptone-yeast extract-iron agar. Utilizes carbon monoxide and carbon dioxide as sole sources of carbon for energy and growth. Nitrate is reduced, and elastin, starch, and L-tyrosine are degraded. Growth is inhibited in the presence of phenol (0.1%, w/v), phenyl ethanol (0.3%, w/v), and sodium azide (0.02%, w/v). *myo*-Inositol, mannitol, raffinose, sucrose and sodium pyruvate are used as sole carbon sources when the organism is grown heterotrophically. Contains major amounts of LL-A_2pm, octahydrogenated menaquinones with nine isoprene units as the predominant isoprenolog, and major amounts of diphosphatidylglycerol and phosphatidylethanolamine.

Type strain shows the highest sequence similarity to: *S. thermocarboxydovorans*, U94489, 99.1%; *S. thermodiastaticus*, AB018095, 99%. Type strain shows DNA–DNA similarity to: *S. thermocarboxydovorans* DSM 44296^T, 72%; *S. thermodiastaticus* DSM 40573^T, 72%.

Source: not known.

DNA G+C content (mol%): 73.6.

Type strain: AT10, DSM 41779, JCM 11756, KCTC 9909, NBRC 100043.

Sequence accession no. (16S rRNA gene): AF333113.

Further comments: the original spelling *Streptomyces thermospinisporus* (*sic*) has been corrected by the List Editor, IJSEM.

480a. **Streptomyces thermoviolaceus subsp. thermoviolaceus** Henssen 1957, 388[AL]

ther.mo.vi.o.la'ce.us. Gr. n. *therme* heat; L. adj. *violaceus* violet colored; N.L. masc. adj. *thermoviolaceus* heat, violet colored (probably referring to the thermophilic nature of the species and the violet color of the aerial mycelium).

Spore chains in Section *Spirales.* Spirals are best developed on carbon-utilization medium plus raffinose. Imperfect spirals or loops may be common on yeast-malt agar, oatmeal agar, salts-starch agar, and glycerol-asparagine agar. Spore surface is smooth.

Color of colony: aerial mass color in the Red color series (5dc, grayish yellowish pink) when adequate sporulating aerial mycelium is present, but sporulation is generally poor on yeast-malt agar, oatmeal agar, salts-starch agar, and glycerol-asparagine agar at 28°C. ISP observers did not incubate at the higher temperatures (40–50°C) recommended by Henssen. Reverse side of colony with no distinctive pigments (olive brown on yeast-malt agar and oatmeal agar; grayish yellow to pale greenish yellow on salts-starch agar and glycerol-asparagine agar). The reverse mycelium pigment is pH-sensitive, changing from yellow or brown to gray or bluish gray with addition of 0.05 M NaOH and becomes reddish with the addition of 0.05 M HCl.

Color in medium: melanoid pigments are not formed in peptone-yeast-iron agar, tyrosine agar, or tryptone-yeast broth. Red or violet pigment is found in the medium in yeast-malt agar, oatmeal agar, and glycerol-asparagine agar. This pigment is pH-sensitive, changing from violet to blue with the addition of 0.05 M NaOH and from violet to red with the addition of 0.05 M HCl.

D-Glucose, D-xylose, D-mannitol, and D-fructose are utilized for growth. Utilization of L-arabinose is doubtful. No growth or only traces of growth with iso-inositol, rhamnose, sucrose, and raffinose.

Type strain shows the highest sequence similarity to: *S. thermoviolaceus* subsp. *apingens*, Z68095, 99.4%.

Source: not known.

DNA G+C content (mol%): not known.

Type strain: AS 4.1471, ATCC 19283, CBS 278.66, CBS 688.72, BCRC 12493, BCRC 12639, DSM 40443, HAMBI 1006, HUT 6604, NBRC 13905, IMET 43353, JCM 4337, JCM 4843, LMG 14943, LMG 19359, NCIMB 10076, NRRL B-12374, NRRL-ISP 5443, RIA 1348, VKM Ac-1857.

Sequence accession no. (16S rRNA gene): Z68096.

480b. **Streptomyces thermoviolaceus subsp. apingens** Henssen 1957, 390[AL]

a.pin'gens. Gr. pref. *a*-denoting negation, not; L. part. adj. *pingens* coloring; N.L. part. adj. *apingens* not coloring.

Aerial hyphae straight or branched, spore chains coiled, 20–40 μm long, spores 10–20, oval, 1.0–1.2 × 1.2–1.6 μm. Substrate mycelium is hardly branched. Colonies are yellowish to ochre brown, aerial mycelium white to violet gray. Physiological characteristics under aerobic and anaerobic conditions at 50°C are as follows. On cellulose-dextrin agar: weak growth, aerial mycelium formed, colonies light brownish. On meat-extract agar: weak growth, aerial mycelium hardly formed, colonies yellow. On casein-glucose agar: weak growth, gray aerial mycelium. On asparagine-glycerol agar: good growth, gray aerial mycelium, colonies yellow. On Czapek's agar: weak growth, gray aerial mycelium. On yeast agar: weak growth, aerial mycelium white to gray. On yeast-glucose agar: good growth, thick, violet-gray aerial mycelium, colonies ochre brown. On potato agar I: good growth, thick, gray aerial mycelium, colonies ochre brown. On potato agar II: good growth, no aerial mycelium, colonies colorless to yellowish. On potato wedge: good growth, aerial mycelium at most marginal, colonies yellowish. Starch is quickly hydrolyzed. Grows on nitrate but does not form nitrite from nitrate. Milk is quickly coagulated and peptonized. Gelatin and nutrient-gelatin are mostly liquefied, colonies are orange yellow. This subspecies grows well at 50°C under exclusion of oxygen. Growth is weak at 28 and 60°C, aerial mycelium is not found; colonies are mostly yellow or brownish.

Type strain shows the highest sequence similarity to: *S. thermoviolaceus* subsp. *thermoviolaceus*, Z68096, 99.4%; *S. mexicanus*, AB249966, 99%.

Source: not known.

DNA G+C content (mol%): not known.

Type strain: ATCC 19994, CBS 140.67, DSM 41392, NBRC 15459, JCM 4312, NCIMB 10077.

Sequence accession no. (16S rRNA gene): Z68095.

481. **Streptomyces thermovulgaris** Henssen 1957, 391[AL]

ther.mo.vul.ga'ris. Gr. n. *therme* heat; L. adj. *vulgaris* common; N.L. masc. adj. *thermovulgaris* heat, common.

Spore chains in Section *Spirales* on yeast-malt agar, oatmeal agar, salts-starch agar, and glycerol-asparagine agar when incubated at 40–50°C for 14 d. *Retinaculiaperti* or flexuous spore chains may also be found. At 28°C, some flexuous spore chains (only) may be found on these media, but aerial mycelium is usually absent or poorly developed at 28°C. Spore surface is smooth. Thermophilic with optimum growth at 40–50°C. Sclerotia-like masses of spore chains are common.

Color of colony: aerial mass color in the Gray color series (usually 3ig or 4ig, light grayish brown; 3fe, light brownish gray or 2dc, yellowish gray) on yeast-malt agar, oatmeal agar, salts-starch agar, and glycerol-asparagine agar. At suboptimal temperatures, the aerial mycelium is white. Reverse side of colony with no distinctive pigments (light yellow or pale grayish yellow to olive-brown) on yeast-malt agar, oatmeal agar, salts-starch agar, and glycerol-asparagine agar.

Color in medium: melanoid pigments are not formed in peptone-yeast-iron agar, tyrosine agar, or tryptone-yeast broth. No pigment is found in medium in yeast-malt agar, oatmeal agar, salts-starch agar, or glycerol-asparagine agar.

D-Glucose, L-arabinose, D-xylose, iso-inositol, D-mannitol, D-fructose, rhamnose, sucrose, and raffinose are all utilized for growth, but growth on rhamnose or raffinose may be less than on the other carbon sources tested.

Type strain shows the highest sequence similarity to: *S. thermogriseus*, AB249980, 100%.

Source: not known.

DNA G+C content (mol%): not known.

Type strain: AS 4.1455, ATCC 19284, ATCC 25501, CBS 276.66, CBS 643.69, BCRC 12488, BCRC 12638, DSM 40444, HUT 6605, NBRC 13089, NBRC 16615, JCM 4338, JCM 4520, NCIMB 10078, NRRL B-12375, NRRL-ISP 5444, RIA 1281, VKM Ac-1745.

Sequence accession no. (16S rRNA gene): Z68094.

Further comments: according to Kim et al. (1999), *Streptomyces thermovulgaris* Henssen 1957 is an earlier heterotypic synonym of *Streptomyces thermonitrificans* Desai and Dhala 1967.

482. **Streptomyces thioluteus** (Okami 1952) Witt and Stackebrandt 1991, 456[VP] (Effective publication: Witt and Stackebrandt 1990, 370.) (*Streptomyces thioluteus* Okami 1952, 30; *Verticillomyces thioluteus* Shinobu 1965, 161; *Streptoverticillium thioluteum* Baldacci, Farina and Locci 1966, 165)

thi.o.lu'te.us. Gr. n. *theion* (L. translit. *thium*) sulfur; L. adj. *luteus* yellow; N.L. masc. adj. *thioluteus* sulfur-yellow in color.

Good growth on potato-glucose agar (Baldacci et al., 1954). Color: reverse, orangish yellow; aerial mycelium, yellowish. On Bacto Czapek agar: very poor growth. Color: colorless; whitish tufts of aerial mycelium. On Casamino acids Czapek agar (1 g/l Difco vitamin-free Casamino acids, replacing sodium nitrate): limited growth. Color: reverse, colorless to yellowish; aerial mycelium, whitish yellow. On glucose asparagine agar (ISP medium 5 with 1% glucose replacing glycerol): good growth. Color: reverse, yellow to brown yellow; poor aerial mycelium, yellowish. On glucose-asparagine agar: good growth. Color: reverse, yellowish brown; aerial mycelium, yellowish. On inorganic salts-starch agar: good growth. Color: reverse, yellowish; aerial mycelium, yellowish. On yeast extract-malt extract agar: good growth. Color: reverse, brown; aerial mycelium, yellowish; soluble pigments present. On Bacto Emerson agar: good growth. Color: reverse, brown; aerial mycelium, greenish yellow; traces of soluble pigment. On Bennett agar (1% glucose, 0.1% Bacto beef agar, 0.1% yeast extract, 0.2% peptone, 1.5% agar): good growth. Color: reverse, yellow to brown-yellow; aerial mycelium, yellowish. On Oxoid nutrient agar: medium growth. Color: aerial mycelium, greenish yellow; reverse, brownish yellow; traces of aerial mycelium and pigment. Grows at 37°C; however, no aerial mycelium is produced in 15 d. No growth at 45°C. Some cultures form traces of H_2S and utilize rhamnose and casein. The type strain produces propiopyrrothine (aureothricin) and aureothin.

Type strain shows the highest sequence similarity to: *S. morookaense* AJ781349, 99.5%; *S. abikoensis*, AB184537, 99.3%; *S. lavenduligriseus*, DQ442515, 99.2%; *S. olivoverticillatus*, AB184636, 99.1%; *S. sapporonensis*, AB184508, 99%; *S. luteireticuli* AB249969, 99%; *S. hygroscopius* subsp. *angustmyceticus*, DQ442509, 99%.

Source: not known.

DNA G+C content (mol%): not known.

Type strain: ATCC 12310, CBS 642.72, BCRC 12428, DSM 40027, DSM 41486, HUT 6071, NBRC 13341, NBRC 3364, JCM 4087, JCM 4844, NRRL B-1667, NRRL-ISP 5027, RIA 1302, VKM Ac-1914.

Sequence accession no. (16S rRNA gene): AB184753.

Further comments: in violation of Rule 33c of the *Bacteriological Code* (1990 Revision), in Validation List no. 38, *Streptomyces thioluteus* is proposed as a *nomen revictum* (basonym: "*Streptomyces thioluteus*" Okami 1952).

483. **Streptomyces torulosus** Lyons and Pridham 1971, 192[AL]

to.ru.lo'sus. L. dim. n. *torulus* a small protuberance; L. masc. suff. *-osus* suffix used with the sense of full of, prone to; N.L. masc. adj. *torulosus* having small protuberances.

Spores are arranged in both dextrorse and sinistrorse coiled chains, with three to five volutions per coil. The color of the aerial mycelium observed with inorganic salts-starch agar after 14 d at 28°C was placed in the Gray series. A brown to black diffusible pigment is formed in peptone-iron agar. Melanin-like chromogenicity: Brown to black diffusible pigments are formed in tryptone-glucose-liver extract-yeast extract (Lyons and Pridham, 1965) broth. D-Glucose, D-xylose, L-arabinose, L-rhamnose, D-fructose, D-galactose, raffinose, D-mannitol, and iso-inositol are utilized; salicin is not. Growth is excellent on Czapek's solution-sucrose agar. The L-form of A_2pm is found in whole-cell hydrolysates. Faint spots for mannose and arabinose also can be detected. Strains are sensitive to streptomycin. The type strain is able to inhibit *Bacillus subtilis, Sarcina lutea, Escherichia coli,* and *Mucor ramannianus. Saccharomyces pastorianus* was very slightly inhibited, and *Candida albicans* was not inhibited.

Type strain shows the highest sequence similarity to: *S. hygroscopicus* subsp. *ossamyceticus*, AB184560, 100%; *S. neyagawaensis*, D63869, 99.3%; *S. ipomoeae*, AB184857, 99.2%.

Source: not known.

DNA G+C content (mol%): not known.

Type strain: ATCC 29340, CBS 801.71, DSM 40894, IFM 1283, NBRC 15460, JCM 4872, NRRL B-3889.

Sequence accession no. (16S rRNA gene): AJ781367.

484. **Streptomyces toxytricini** (Preobrazhenskaya and Sveshnikova *in* Gauze, Preobrazhenskaya, Kudrina, Blinov, Ryabova and Sveshnikova 1957) Pridham, Hesseltine and Benedict 1958, 68[AL] ("*Actinomyces toxytricini*" Preobrazhenskaya and Sveshnikova *in* Gauze, Preobrazhenskaya, Kudrina, Blinov, Ryabova and Sveshnikova 1957, 47)

to.xy.tri.ci'ni. N.L. n. *toxythricinum* toxythricin a suggested, but never used, antibiotic name; N.L. gen. n. *toxytricini* of toxithricin.

Spore chains in Section *Retinaculiaperti* or *Spirales.* Long spore chains of the *Retinaculiaperti* type as well as straight to flexuous sporophores are common on salts-starch agar, glycerol-asparagine agar, yeast-malt agar, and oatmeal agar. Spirals are most numerous as long chains of more than 50 spores on yeast-malt agar and oatmeal agar. Chains of 10 to 50 or more spores per chains are also found on salts-starch agar and glycerol-asparagine agar. Spore surface is smooth.

Color of colony: aerial mass color in the Red color series on yeast-malt agar, oatmeal agar, and salts-starch agar. Reverse side of colony with no distinctive pigment (grayed yellow to yellow-brown) on yeast-malt agar, oatmeal agar, salts-starch agar or glycerol-asparagine agar. Substrate pigment is not a pH indicator.

Color in medium: melanoid pigments formed in peptone-yeast-iron agar and tryptone-yeast extract broth. Pigments other than melanoids not formed in yeast-malt agar, oatmeal agar, salts-starch agar, or glycerol-asparagine agar.

D-Glucose is utilized for growth. Doubtful traces of growth or no growth on L-arabinose, sucrose, D-xylose, iso-inositol, rhamnose, and raffinose. Utilization of D-mannitol and D-fructose is also doubtful.

Type strain shows the highest sequence similarity to: *S. globosus*, AJ781330, 100%; *S. flavotricini*, AB184132, 99.3%; *S. racemochromogenes*, DQ026656, 99.1%; *S. polychromogenes*, AB184292, 99%.

Source: not known.

DNA G+C content (mol%): not known.

Type strain: ATCC 19813, CBS 566.68, BCRC 13472, DSM 40178, NBRC 12823, JCM 4421, NCIMB 9847, NRRL B-5426, NRRL-ISP 5178, RIA 1093, UNIQEM 200, VKM Ac-1279.

Sequence accession no. (16S rRNA gene): DQ442548.

485. **Streptomyces tricolor** (Wollenweber 1920) Waksman 1961, 158[AL] emend. Lanoot, Vancanneyt, Dawyndt Cnockaert, Zhang, Huang, Liu and Swings 2005a, 8 (Effective publication: Lanoot, Vancanneyt, Dawyndt Cnockaert, Zhang, Huang, Liu and Swings 2004, 90.) ("*Actinomyces tricolor*" Wollenweber 1920, 13)

tri'co.lor. L. masc. adj. *tricolor* of three colors.

Forms yellow-, red-, or blue-colored vegetative mycelium on some media; forms blue diffusible pigment on some media.

Type strain shows the highest sequence similarity to: *S. anthocyanicus*, AB184631, 100%; *S. rubrogriseus*, AB184681, 99.9%; *S. violaceoruber*, AF503492, 99.9%; *S. coelescens*, AF503496, 99.9%; *S. violaceolatus* AF503497, 99.8%; *S. humiferus*, AF503491, 99.8%; *S. lienomycini*, AJ781353, 99.7%; *S. violaceorubidus*, AJ781374, 99.5%; *S. tendae*, D63873, 99.4%; *S. coelicoflavus*, AB184650, 99.4%; *S. pactum*, AB184398, 99.1%; *S. olivaceus*, AB184743, 99%.

Source: isolated from flat scab of potato.

DNA G+C content (mol%): not known.

Type strain: CBS 103.21, DSM 41704, NBRC 15461, JCM 5065, NRRL B-16925.

Sequence accession no. (16S rRNA gene): AB184687.

Further comments: according to Lanoot et al. (2004), *Streptomyces tricolor* (Wollenweber 1920) Waksman 1961[AL] is an earlier heterotypic synonym of *Streptomyces roseodiastaticus* (Duché 1934) Waksman 1953.

486. **Streptomyces tubercidicus** Nakamura 1961, 90[AL]

tu.ber.ci'di.cus. L. n. *tuber* nodule; L. v. *caedo* to kill; N.L. masc. adj. *tubercidicus* nodule destroying, referring to antitumor activity of the antibiotic.

Spore chains in Section *Spirales*. Mature spore chains generally have 3 to 10 or more spores per chain. This morphology is seen on yeast-malt agar, oatmeal agar, salts-starch agar, and glycerol-asparagine agar. Spore surface is smooth.

Color of colony: aerial mass color in the Gray color series (d, light gray to 3fe, light brownish gray) on yeast-malt agar, oatmeal agar, and salts-starch agar in 14 d. One observer places 21 d culture in the Red color series (5dc, grayish yellowish pink) on yeast-malt agar, oatmeal agar, and glycerol-asparagine agar. The original description notes a tendency for the aerial mycelium to become moist and black; this hygroscopic characteristic is not recorded by ISP observers. Reverse side of colony is grayish yellow to yellowish brown or olive-brown on oatmeal agar, salts-starch agar, and glycerol-asparagine agar; moderate brown to dark brown on yeast-malt agar. Reverse mycelium pigment is a pH indicator, changing from yellow or yellow-brown to pale pink with addition of 0.05 M NaOH and from pink or grayish brown to yellow with addition of 0.05 M HCl.

Color in medium: melanoid pigments are not formed in peptone-yeast-iron agar, tyrosine agar, or tryptone-yeast broth. Yellow to pink or violet-pink (depending upon pH) pigment is found in the medium in yeast-malt agar, oatmeal agar, salts-starch agar, and glycerol-asparagine agar. This pigment is pH-sensitive showing the same changes noted for the reverse mycelium pigment.

D-Glucose, iso-inositol, D-mannitol, D-fructose, sucrose, and raffinose are utilized for growth. No growth or only trace of growth with L-arabinose, D-xylose, and rhamnose.

Type strain shows the highest sequence similarity to: *S. libani* subsp. *libani*, AB184414, 100%; *S. nigrescens*, DQ442530, 99.9%; *S. hygroscopius* subsp. *glebosus*, AB184479, 99.6%; *S. libani* subsp. *rufus*, AJ781351, 99.6%; *S. caniferus*, AB184640, 99.5%; *S. catenulae*, AJ621613, 99.5%; *S. misakiensis*, AB217605, 99.5%; *S. sioyaensis*, DQ026654, 99.5%; *S. platensis*, AB045882, 99.5%; *S. hygroscopius* subsp. *decoyicus*, AY999883, 99.2%; *S. lydicus*, Y15507, 99.1%; *S. chattanoogensis*, AJ621611, 99%.

Source: not known.

DNA G+C content (mol%): not known.

Type strain: AS 4.1414, ATCC 25502, CBS 644.69, BCRC 11886, CECT 3272, DSM 40261, IFM 1064, NBRC 13090, IMET 43517, JCM 4054, JCM 4558, KCTC 9109, LMG 19361, NRRL B-5440, NRRL-ISP 5261, RIA 1282, VKM Ac-1073.

Sequence accession no. (16S rRNA gene): AJ621612.

487. **Streptomyces tuirus** Albert and Malaquias de Queiroz 1963, 43[AL]

tu'ir.us. Nheêngatû Amazonian dialect *tuira* violet, violet-blue; N.L. masc. adj. *tuirus* violet, violet-blue (referring to the color of vegetative mycelium and diffusible pigment).

Spore chains in Section *Spirales*. Spiral spore chains often arise in verticils from long axial hyphae, suggesting monoverticillate morphology. Spirals are usually open and wavy; hooked spore chains are also present. This morphology is seen on yeast-malt agar, oatmeal agar, salts-starch agar, and glycerol-asparagine agar. Spore surface is smooth.

Color of colony: aerial mass color in the Gray color series (5fe, light grayish reddish brown; 2dc, yellowish gray; d-e, light to medium gray) on yeast-malt agar, oatmeal agar, salts-starch agar, and glycerol-asparagine agar. Reverse side of colony with distinctive red pigments on yeast-malt agar, salts-starch agar, and glycerol-asparagine agar. Reverse color is reddish black on yeast-malt agar and grayish reddish brown, dark pale red, or dark grayish pink on salts-starch agar and glycerol-asparagine agar; grayish brown on oatmeal agar. One of three observers reports that the reverse mycelium pigment is a pH indicator, changing from red to violet with the addition of 0.05 M NaOH and to a more intensive red with the addition of 0.05 M HCl.

Color in medium: melanoid pigments are formed in peptone-yeast-iron agar and tryptone-yeast broth. Melanin reaction may be weak or absent in tyrosine agar. Red, reddish brown, or violet pigment is found in the medium in yeast-malt agar, oatmeal agar, salts-starch agar, and glycerol-asparagine agar. According to one observer, only, this pigment is pH-sensitive, showing the same changes observed in the reverse mycelium pigment.

D-Glucose, L-arabinose, D-xylose, iso-inositol, D-mannitol, D-fructose, rhamnose, sucrose, and raffinose are all utilized for growth.

Type strain shows the highest sequence similarity to: *S. mutabilis*, EF178679, 99.6%; *S. rochei*, AB184237, 99.5%; *S. vinaceusdrappus*, AY999929, 99.5%; *S. plicatus*, AB184291, 99.5%; *S. minutiscleroticus*, EF178696, 99.5%; *S. ghanaensis*, AY999851, 99.5%; *S. geysiriensis*, DQ442501, 99.5%; *S. levis*, AB184670, 99.4%; *S. djakartensis*, AB184657, 99.4%; *S. luteogriseus*, AB184379, 99.4%; *S. capillispiralis*, AB184577, 99.2%; *S. janthinus*, AB184851, 99.2%; *S. albosporeus* subsp. *albosporeus*, AJ781327, 99.2%; *S. roseoviolaceus*, AJ399484, 99.2%; *S. azureus*, EF178674, 99.2%; *S. violaceus*, AB184315, 99.2%; *S. africanus*, AY208912, 99.1%; *S. gancidicus*, AB184660, 99.1%; *S. pseudogriseolus*, DQ442541, 99.1%; *S. afghaniensis*, AJ399483, 99.1%; *S. calvus*, AB184329, 99.1%; *S. virens*, DQ442554, 99%; *S. cellulosae*, DQ442495, 99%; *S. violaceorubidus*, AJ781374, 99%; *S. anandii*, AB184402, 99%; *S. flavoviridis*, AB184842, 99%; *S. lomondensis*, AB184673, 99%; *S. pilosus*, AB184161, 99%; *S. aureorectus*, AB184710, 99%; *S. asterosporus*, AB184706, 99%.

Source: not known.

DNA G+C content (mol%): not known.

Type strain: ATCC 19007, CBS 719.72, BCRC 12217, DSM 40505, NBRC 15617, JCM 4255, JCM 4846, NRRL B-3631, NRRL-ISP 5505, RIA 1379.

Sequence accession no. (16S rRNA gene): AB184690.

488. **Streptomyces turgidiscabies** Miyajima, Tanaka, Takeuchi and Kuninaga 1998, 500[AL]

tur.gi.di.sca'bi.es. L. adj. *turgidus* swollen; L. fem. n. *scabies* scabbiness; N.L. fem. n. *turgidiscabies* a swollen scabbiness.

Spores are 0.5–0.6 × 1.0–1.2 µm, smooth, gray, and are borne in mature flexuous chains containing 8 or more spores. Diffusible pigments and melanin are not produced. Positive for peptonization but not for coagulation of milk, is not susceptible to 25 µg/ml oleandomycin or 10 IU/ml penicillin G and is not positive for utilization of sodium acetate or sodium propionate as sole carbon sources. L-Arabinose, D-fructose, D-glucose, D-mannitol, iso-inositol, rhamnose, sucrose, D-xylose, and raffinose are used as carbon sources. None of the strains studied degrade xanthine or grow at 37°C and all are susceptible to 20 μg/ml streptomycin, 0.5 μg/ml crystal violet, 100 μg/ml potassium tellurite, and 0.1% phenol. Cell wall contains the LL-A_2pm isomer. *S. turgidiscabies* differs from other potato scab pathogens in having gray, smooth, flexuous spore chains.

Type strain shows the highest sequence similarity to: *S. reticuliscabiei*, AJ007428, 99.1%.

Source: the type strain was isolated from raised scab lesions on potato tubers in eastern Hokkaido, Japan, in 1991, and has been confirmed to be pathogenic.

DNA G+C content (mol%): 70.9–72.5.

Type strain: SY9113, ATCC 700248, CIP 105577, NBRC 16080, JCM 10429, NRRL B-24078.

Sequence accession no. (16S rRNA gene): AB026221.

Further comments: a numerical analysis of phenotypic characteristics showed that *Streptomyces reticuliscabiei* Bouchek-Mechiche et al. 2000 and *Streptomyces turgidiscabies* Miyajima et al. 1998 belong to the same cluster and share almost all morphological and biochemical traits that are important in the identification of *Streptomyces* species. DNA–DNA hybridization and phylogenetic comparisons of 16S rRNA gene sequences confirm that the two species are genomically closely related. In contrast, pathological data showed that *Streptomyces turgidiscabies* and *Streptomyces reticuliscabiei* cause two distinct diseases. For the pathologist, the fusion of *Streptomyces reticuliscabiei* and *Streptomyces turgidiscabies* under a single species denomination would cause confusion of separate diseases and create a discrepancy between taxonomists and pathologists. Therefore, Bouchek-Mechiche et al. think that the two groups should continue to carry their current denominations, i.e. *Streptomyces reticuliscabiei* Bouchek-Mechiche et al. 2000 for the strains inducing netted scab and *Streptomyces turgidiscabies* Miyajima et al. 1998 for those causing common scab.

489. **Streptomyces umbrinus** (Sveshnikova *in* Gauze, Preobrazhenskaya, Kudrina, Blinov, Ryabova and Sveshnikova 1957) Pridham, Hesseltine and Benedict 1958, 61[AL] ("*Actinomyces umbrinus*" Sveshnikova *in* Gauze, Preobrazhenskaya, Kudrina, Blinov, Ryabova and Sveshnikova 1957, 62)

um.bri'nus. N.L. masc. adj. *umbrinus* wood brown, the color of the aerial mycelium.

Spore chains in Section *Rectiflexibiles.* Mature spore chains are moderately long with 10 to 50 or more spores per chain. This morphology is seen on yeast-malt agar, oatmeal agar, salts-starch agar, and glycerol-asparagine agar. Spore surface is smooth.

Color of colony: aerial mass color in the Red color series (3ca, pale orange-yellow to 5cb, grayish yellowish pink) on yeast-malt agar, oatmeal agar, salts-starch agar, and glycerol-asparagine agar. Reverse side of colony is reddish brown on yeast-malt agar, light brown to strong brown on oatmeal agar and salts-starch agar, gray to reddish black on glycerol-asparagine agar. Reverse mycelium pigment is not a pH indicator.

Color in medium: melanoid pigments are formed in peptone-yeast-iron agar, tyrosine agar, and tryptone-yeast broth. Yellow pigment may be found in the medium in yeast-malt agar, oatmeal agar, salts-starch agar, and glycerol-asparagine agar. This pigment is not pH-sensitive when tested with 0.05 M NaOH or HCl.

D-Glucose, L-arabinose, D-xylose, iso-inositol, D-mannitol, D-fructose, rhamnose, sucrose, and raffinose are all utilized for growth.

Type strain shows the highest sequence similarity to: *S. ederensis*, AB184658, 100%; *S. phaeochromogenes*, AB184738, 99.4%.

Source: not known.

DNA G+C content (mol%): not known.

Type strain: ATCC 19929, ATCC 25503, CBS 645.69, DSM 40278, NBRC 13091, INA 1703/53, JCM 4521, NRRL B-2572, NRRL-ISP 5278, RIA 1283, VKM Ac-1747.

Sequence accession no. (16S rRNA gene): AB184305.

490. **Streptomyces variabilis** (Preobrazhenskaya, Ryabova and Blinov *in* Gauze, Preobrazhenskaya, Kudrina, Blinov, Ryabova and Sveshnikova 1957) Pridham, Hesseltine and Benedict. 1958, 70[AL] ("*Actinomyces variabilis*" Preobrazhenskaya, Ryabova and Blinov *in* Gauze, Preobrazhenskaya, Kudrina, Blinov, Ryabova and Sveshnikova 1957)

va.ri.a'bi.lis. L. masc. adj. *variabilis* variable.

Spore chains in Section *Spirales.* Open spirals are conspicuous on salts-starch agar, rare on yeast-malt agar. Flexuous spore chains, loops, hooks, and spirals suggestive of Section *Retinaculiaperti* are observed on oatmeal agar, salts-starch agar, and glycerol-asparagine agar. Spore surface is spiny. Spines are short; one observer reports warty to smooth spores.

Color of colony: aerial mass color in the Gray color series on oatmeal agar and salts-starch agar; Gray or Red series on yeast-malt agar. Aerial spore mass is not well-developed on glycerol-asparagine agar. Reverse side of colony with no distinctive pigment on oatmeal agar, salts-starch agar, and glycerol-asparagine agar; reddish brown on yeast-malt agar.

Color in medium: melanoid pigments not formed in peptone-yeast-iron agar or tyrosine agar. No pigment found in medium in yeast-malt agar, oatmeal agar, salts-starch agar, or glycerol-asparagine agar.

D-Glucose, L-arabinose, D-xylose, iso-inositol, D-mannitol, D-fructose, and rhamnose are utilized for growth. No growth or only trace of growth on sucrose and raffinose.

Type strain shows the highest sequence similarity to: *S. erythrogriseus*, AJ781328, 100%; *S. labedae*, AB184704, 100%; *S. griseoincarnatus*, AJ781328, 100%; *S. griseorubens*, AB184139, 99.9%; *S. griseoflavus*, AJ781322, 99.6%; *S. matensis*, AB184221, 99.6%; *S. althioticus*, AY999808, 99.2%; *S. paradoxus*, AB184628, 99.2%; *S. heliomycini*, AB184712, 99.1%; *S. collinus*, AB184123, 99.1%; *S. flaveolus*, AB184764, 99.1%; *S. viridochromogenes*, DQ442555, 99%; *S. bellus*, AB184849, 99%; *S. viridodiastaticus*, AY999852, 99%; *S. violaceochromogenes*, AY999867, 99%; *S. malachitofuscus*, AB184282, 99%; *S. violaceorubidus*, AJ781374, 99%; *S. coerulescens*, AY999720, 99%.

Source: not known.

DNA G+C content (mol%): not known.

Type strain: ATCC 19815, ATCC 19930, CBS 568.68, BCRC 11488, DSM 40179, NBRC 12825, IMET 42059, JCM 4422, NRRL B-3984, NRRL-ISP 5179, RIA 1095, UNIQEM 202, VKM Ac-1854.

Sequence accession no. (16S rRNA gene): DQ442551.

491. **Streptomyces variegatus** Sveshnikova and Timuk 1986, 575[VP] (Effective publication: Sveshnikova and Timuk *in* Gause, Preobrazhenskaya, Sveshnikova, Terekhova and Maximova 1983.)

va.ri.e.ga′tus. L. masc. part. adj. *variegatus* variegated, made of various colors.

Spore chains are spiral (*Spirales*); spores are smooth. On mineral agar 1: aerial mycelium is pink to loud pink; substrate mycelium is raspberry red and green, grayish brownish green. Intensity of building pigments is different, normally first red colored, later covered by green color; no diffusible pigment. Curious color of substrate mycelium is due to pigment types prodigiosin and ferroverdin. On starch-ammonia agar: aerial mycelium grows late, light pink, poor; substrate mycelium is light pink with weak violet shadow; no diffusible pigment. On glycerol-nitrate agar: aerial mycelium is white, poorly developed, sometimes absent; substrate mycelium is green with raspberry colored segments, sometimes brindled green, greenish grayish brown, light yellow, and grayish brownish pink at the same time; no diffusible pigment. On glycerol-asparagine agar: aerial mycelium is whitish to light pink and pink; substrate mycelium is brindled light yellow, orange, raspberry colored, grayish brownish red, and dark green; no diffusible pigment. On glucose-asparagine agar: aerial mycelium whitish to light pink and pink; substrate mycelium is brindled yellow, green, and light raspberry colored; no diffusible pigment. On agar Sp I Krasil'nikov: weak growth; no aerial mycelium; colorless substrate mycelium; no diffusible pigment. On oatmeal agar: aerial mycelium is whitish to light pink; substrate mycelium is yellow, pale orange, pink to raspberry colored; green pigment is weak; no diffusible pigment. On organic agar 2: aerial mycelium is whitish to pink; substrate mycelium is brindled, pale red, pale greenish; no diffusible pigment. Melanoid pigments are not formed. Grows on fructose, xylose, and arabinose; no growth on sucrose, rhamnose, or raffinose. Antibiotic: antibiotic of group α-oxi-keto-Pentaen; pigment with antibiotic character of the prodigiosin group.

Type strain shows the highest sequence similarity to: *S. fulvissimus*, AB184434, 99.1%; *S. alboflavus*, EF178699, 99%.

Source: not known.

DNA G+C content (mol%): not known.

Type strain: ATCC 43696, DSM 41464, NBRC 15462, INA T-511, JCM 6930, VKM Ac-846.

Sequence accession no. (16S rRNA gene): AJ781371.

492. **Streptomyces varsoviensis** Kurylowicz and Woźnicka 1967, 1[AL]

var.so.vi.en′sis. N.L. masc. adj. *varsoviensis* of or pertaining to Varsovia; named for Warsaw, Poland.

Spore chains in Section *Spirales*, *Retinaculiaperti*, or *Rectiflexibiles.* Spiral spore chains are abundant on yeast-malt agar; straight to flexuous chains are most common on glycerol-asparagine agar. Flexuous, spiral, or intermediate (*Retinaculiaperti*) forms may be found on oatmeal agar and salts-starch agar but sporulating aerial mycelium is usually poorly developed on these media. Spore surface is smooth.

Color of colony: aerial mass color in the White or Yellow (2ba, pale yellow) color series on oatmeal agar and salts-starch agar. Sporulation on oatmeal agar and salts-starch agar is usually inadequate for spore mass color determination. Reverse side of colony is yellow to yellow-brown and modified by red (to yellowish pink, orange, grayish reddish orange, or strong brown) on yeast-malt agar, salts-starch agar, and glycerol-asparagine agar. Substrate pigment is not a pH indicator or is modified only slightly by addition of 0.05 M NaOH or HCl.

Color in medium: reports vary on production of melanoid pigments. Some darkening of peptone-yeast-iron agar, tyrosine agar, and tryptone-yeast broth may be seen in 4 d, but usually not in 2 d. Gause's organic medium no. 2 is not darkened in 2–4 d. No pigment is found in medium in yeast-malt agar, oatmeal agar, salts-starch agar, and glycerol-asparagine agar.

D-Glucose and D-mannitol are utilized for growth. Utilization of fructose is doubtful. No growth or only trace of growth with L-arabinose, D-xylose, iso-inositol, rhamnose, sucrose, and raffinose.

Type strain shows the highest sequence similarity to: *S. sapporonensis*, AB184508, 99.5%; *S. abikoensis*, AB184537, 99.2%; *S. hygroscopius* subsp. *angustmyceticus*, DQ442509, 99.1%; *S. luteireticuli* AB249969, 99.1%; *S. ehimensis*, AY999834, 99%.

Source: not known.

DNA G+C content (mol%): not known.

Type strain: AS 4.1431, ATCC 14631c, ATCC 25505, CBS 357.64, CBS 647.69, BCRC 12647, DSM 40346, HAMBI 1046, NBRC 13093, IMET 43351, JCM 4303, JCM 4523, NCIMB 9522, NRRL B-3589, NRRL-ISP 5346, RIA 1285, VKM Ac-1000.

Sequence accession no. (16S rRNA gene): DQ026653.

493. **Streptomyces vastus** Szabó and Marton 1958, 245[AL]

vas′tus. L. masc. adj. *vastus* empty, unoccupied, waste, referring to the occurrence of the organism in the Hortobagy Puszta (eastern Hungary).

Spore chains in Section *Spirales.* Open, terminal spirals of only one to three turns may be seen on yeast-malt agar, oatmeal agar, and salts-starch agar, but sporulating aerial mycelium is usually poorly developed on these media. Sporulation may be absent on glycerol-asparagine agar and Gause's medium no. 1. Poor sporulation of this strain on various media is noted in Szabó and Marton's original description (op. cit.). Spore chains are often short with 3 to 10 or more spores per chain. Spore surface is smooth.

Color of colony: aerial mass color in the White or Gray color series on yeast-malt agar and oatmeal agar when mature sporulating aerial mycelium is formed. Thin, white aerial mycelium may also be seen on salts-starch agar and glycerol-asparagine agar but is inadequate for accurate color determination. Reverse side of colony is grayish green on yeast-malt agar; colorless to pale blue or grayish blue on oatmeal agar and salts-starch agar; colorless to pale yellow on glycerol-asparagine agar. Reverse mycelium pigment is not a pH indicator.

Color in medium: melanoid pigments are not formed in peptone-yeast-iron agar, tyrosine agar, or tryptone-yeast broth. No pigment is found in medium in oatmeal agar or glycerol-asparagine agar. Traces of yellow or green pigment may or may not be found in yeast-malt agar and a trace of blue pigment may be found in salts-starch agar. This pigment, if present, is not a pH indicator when tested with 0.05 M NaOH or HCl.

D-Glucose, L-arabinose, iso-inositol, D-mannitol, D-fructose, rhamnose, sucrose, and raffinose are utilized for growth. Reports vary on utilization of D-xylose.

Type strain shows no sequence similarity over 99%.

Source: not known.

DNA G+C content (mol%): not known.

Type strain: ATCC 25506, CBS 290.60, CBS 648.69, DSM 40309, NBRC 13094, JCM 4524, NRRL B-12232, NRRL-ISP 5309, RIA 1286, VKM Ac-1871.

Sequence accession no. (16S rRNA gene): DQ442552.

494. **Streptomyces venezuelae** Ehrlich, Gottlieb, Burkholder, Anderson and Pridham 1948, 467[AL]

ve.ne.zu.e'la.e. N.L. gen. n. *venezuelae* of Venezuela.

Spore chains in Section *Rectiflexibiles.* Straight spore chains are generally long, often with more than 50 spores per chain. This morphology is seen on oatmeal agar, salts-starch agar, and glycerol-asparagine agar; sporulation may be poor on yeast-malt agar. One observer, only, records fragmentation of substrate mycelium on glycerol-asparagine agar. Spore surface is smooth.

Color of colony: aerial mass color in the Gray color series (2dc, yellowish gray to 5fe, light grayish reddish brown) on oatmeal agar, salts-starch agar, and glycerol-asparagine agar; a good spore is usually not produced on yeast-malt agar. Reverse side of colony with no distinct pigments (grayish yellow) on yeast-malt agar, oatmeal agar, salts-starch agar, and glycerol-asparagine agar.

Color in medium: melanoid pigments are formed in peptone-yeast-iron agar and tryptone-yeast broth, but only weakly or not at all in tyrosine agar. No pigment is found in medium in yeast-malt agar, oatmeal agar, salts-starch agar, or glycerol-asparagine agar.

D-Glucose, L-arabinose, D-xylose, D-fructose, and rhamnose are utilized for growth. A trace of growth is usually found in iso-inositol, D-mannitol, sucrose, and raffinose.

Type strain shows the highest sequence similarity to: *S. zaomyceticus*, EF178685, 99.9%; *S. exfoliatus*, AB184324, 99.9%; *S. lateritius*, AB184324, 99.9%; *S. wedmorensis*, DQ442557, 99.7%; *S. litmocidini*, AB184149, 99.6%; *S. omiyaensis*, EF178697, 99.6%; *S. yereyanensis* EF178684, 99.4%; *S. narbonensis*, DQ445794, 99.3%.

Source: not known.

DNA G+C content (mol%): not known.

Type strain: ATCC 10712, ATCC 25508, CBS 650.69, BCRC 11512, DSM 40230, NBRC 12595, NBRC 13096, IMET 41356, JCM 4526, LMG 19308, NRRL 2277, NRRL-ISP 5230, RIA 1288, RIA 70, VKM Ac-589.

Sequence accession no. (16S rRNA gene): AB045890.

495. **Streptomyces vietnamensis** Zhu, Guo, Yao, Yang, Deng, Phuong, Hanh and Ryan 2007, 1773[VP]

vi.et.nam.en'sis. N.L. masc. adj. *vietnamensis* of or pertaining to Vietnam, the geographical location from where the type strain was isolated.

Straight to flexuous chains of cylindrical spores are produced. Forms a white aerial mycelium and a reddish-brown substrate mycelium. Verticils are not present. The mycelium does not fragment. Diffusible pigments are produced on ISP 2, ISP 3, ISP 4, and ISP 5 media and on Gause's synthetic agar, but not on Czapek's solution agar. Melanin is produced on ISP 7. Although growth on ISP 4 is initially slow, very good growth with profuse sporulation is observed on this medium after 14 d. Very good growth occurs on ISP 2, Gause's synthetic agar, and ISP 3. Moderate growth is observed on ISP 5 but only poor growth on Czapek agar. The substrate mycelium is reddish brown on ISP 2, ISP 5, Gause's synthetic agar, ISP 4, and ISP 3, but grayish orange on Czapek medium. Utilizes melibiose, glucose, sorbinose, sucrose, D-fructose, xylose, D-galactose, rhamnose, and arabinose. Positive for production of H_2S, but pectin is not hydrolyzed. Cell wall contains LL-A_2pm (cell-wall type I). Whole-cell sugar pattern contains diagnostic sugars: mannose, small quantities of ribose and galactose. No antibiosis is exhibited against *Escherichia coli* ATCC 25922, *Pseudomonas aeruginosa* ATCC 6538, *Bacillus subtilis* ATCC 6633, *Candida albicans* ATCC 10231, or *Penicillium citrinum* AS 3.2788.

Type strain shows the highest sequence similarity to: *S. bikiniensis*, X79851, 99.5%; *S. violaceorectus*, AB184314, 99.4%; *S. viridobrunneus*, AJ781372, 99.1%; *S. tanashiensis*, AJ781362, 99%; *S. showdoensis*, AB184389, 99%; *S. cinereoruber* subsp. *cinereoruber*, AB184121, 99%. Type strain shows DNA–DNA similarity to: *S. bikiniensis* ATCC 11062^T, 50.3%.

Source: the type strain was isolated from a forest soil sample in Vietnam.

DNA G+C content (mol%): 73.9.

Type strain: GIMV4.0001, CCTCC M 205143, IAM 15340, JCM 21785.

Sequence accession no. (16S rRNA gene): DQ311081.

496. **Streptomyces vinaceus** Jones 1952, 47[AL] emend. Lanoot, Vancanneyt, Dawyndt, Cnockaert, Zhang, Huang, Liu and Swings 2005a, 8 (Effective publication: Lanoot, Vancanneyt, Dawyndt Cnockaert, Zhang, Huang, Liu and Swings 2004, 90.)

vi.na'ce.us. L. masc. adj. *vinaceus* of or belonging to wine or the grape, referring to the color of the aerial mycelium.

Spore chains in Section *Retinaculiaperti* or *Spirales.* Primitive spirals or open imperfect spirals as well as hooks and loops of wide diameter are common. Mature spore chains are generally long, often with more than 50 spores per chain. This morphology is seen on yeast-malt agar, oatmeal agar, salts-starch agar, and glycerol-asparagine agar. Spore surface is smooth.

Color of colony: aerial mass color in the Red color series (4gc or 4ec, grayish yellowish pink; 5gc, light reddish brown; 5dc, grayish yellowish pink; 5ca, light yellowish pink) on yeast-malt agar, oatmeal agar, salts-starch agar, and glycerol-asparagine agar; white aerial mycelium may be also present on glycerol-asparagine agar. Reverse side of colony is strong brown or orange brown on yeast-malt agar; orange yellow, yellowish brown, or grayish yellowish pink on oatmeal agar and salts-starch agar; grayish yellow to moderate yellowish pink on glycerol-asparagine agar. Reverse mycelium pigment is not a pH indicator.

Color in medium: melanoid pigments are formed in peptone-yeast-iron agar, tryptone-yeast broth, and Gause's medium no. 2, but not in tyrosine agar. No pigment (or only a trace of yellow or pink) is found in the medium in yeast-malt agar, oatmeal agar, salts-starch agar, or glycerol-asparagine agar.

D-Glucose, D-mannitol, and D-fructose are utilized for growth. Utilization of L-arabinose is doubtful or weak. No growth or only traces of growth with D-xylose, iso-inositol, rhamnose, sucrose, and raffinose.

Type strain shows the highest sequence similarity to: *S. cirratus*, AY999794, 100%; *S. spororaveus*, AJ781370, 99.9%; *S. nojiriensis*, AJ781355, 99.9%; *S. xanthophaeus*, DQ442560, 99.8%; *S. sporoverrucosus*, DQ442544, 99.7%; *S. colombiensis*, DQ026646, 99.7%; *S. goshikiensis*, EF178693, 99.7%; *S. cinnamonensis*, AB184707, 99.6%; *S. avidinii*, AB184395, 99.6%; *S. subrutilus*, X80825, 99.5%; *S. virginiae*, D85119, 99.4%.

Source: not known.

DNA G+C content (mol%): not known.

Type strain: ATCC 27476, CBS 726.72, BCRC 11865, DSM 40515, HUT 6082, NBRC 13425, JCM 4090, JCM 4849, KCTC 9771, NRRL 2382, NRRL-ISP 5515, PCM 2366, RIA 1386, RIA 805.

Sequence accession no. (16S rRNA gene): AB184394.

Further comments: according to Lanoot et al. (2004), *Streptomyces vinaceus* Jones 1952 is an earlier heterotypic synonym of *Streptomyces arabicus* Shibata et al. 1957.

497. **Streptomyces vinaceusdrappus** Pridham, Hesseltine and Benedict 1958, 68[AL]

vi.na.ce.us.drap'pus. L. adj. *vinaceus* of or belonging to wine or the grape; L. n. *drappus* a sheet, here referring to the color "drab"; N.L. masc. adj. *vinaceusdrappus* of wine-drab, referring to the drab wine color of the aerial mycelium and spores of the organism.

Spore chains in Section *Spirales*. Mature spore chains are generally long, often with more than 50 spores per chain. This morphology is seen on yeast-malt agar, oatmeal agar, salts-starch agar, and glycerol-asparagine agar. Spore surface is smooth.

Color of colony: aerial mass color in the Red color series (5dc, grayish yellowish pink to 4ge, light grayish reddish brown) on yeast-malt agar, oatmeal agar, salts-starch agar, and glycerol-asparagine agar. One observer places this strain in the Gray color series (5fe, light grayish reddish brown or 3fe, light brownish gray). Reverse side of colony with no distinctive pigments (grayish yellow to yellowish or olive-brown) on yeast-malt agar, oatmeal agar, salts-starch agar, and glycerol-asparagine agar.

Color in medium: melanoid pigments are not formed in peptone-yeast-iron agar, tyrosine agar, or tryptone-yeast broth. No pigment is found in medium in yeast-malt agar, oatmeal agar, salts-starch agar, or glycerol-asparagine agar.

D-Glucose, L-arabinose, D-xylose, iso-inositol, D-mannitol, D-fructose, rhamnose, sucrose, and raffinose are all utilized for growth.

Type strain shows the highest sequence similarity to: *S. rochei*, AB184237, 100%; *S. ghanaensis*, AY999851, 100%; *S. geysiriensis*, DQ442501, 100%; *S. minutiscleroticus*, EF178696, 100%; *S. plicatus*, AB184291, 100%; *S. mutabilis*, EF178679, 99.9%; *S. tuirus*, AB184690, 99.5%; *S. djakartensis*, AB184657, 99.4%; *S. anandii*, AB184402, 99.2%; *S. violaceorubidus*, AJ781374, 99.2%; *S. pilosus*, AB184161, 99.1%; *S. flavoviridis*, AB184842, 99.1%; *S. tendae*, D63873, 99%; *S. calvus*, AB184329, 99%; *S. azureus*, EF178674, 99%; *S. asterosporus*, AB184706, 99%; *S. levis*, AB184670, 99%; *S. luteogriseus*, AB184379, 99%; *S. capillispiralis*, AB184577, 99%.

Source: not known.

DNA G+C content (mol%): not known.

Type strain: ATCC 25511, CBS 653.69, BCRC 12170, DSM 40470, NBRC 13099, JCM 4529, NCIMB 12980, NRRL 2363, NRRL-ISP 5470, RIA 1291, VKM Ac-1902.

Sequence accession no. (16S rRNA gene): AY999929.

498. **Streptomyces violaceochromogenes** (Ryabova and Preobrazhenskaya *in* Gauze, Preobrazhenskaya, Kudrina, Blinov, Ryabova and Sveshnikova 1957) Pridham 1970, 28[AL] ["*Actinomyces violaceus chromogenes*" Krasil'nikov 1949, 55; "*Actinomyces violaceochromogenes*" Ryabova and Preobrazhenskaya *in* Gauze, Preobrazhenskaya, Kudrina, Blinov, Ryabova and Sveshnikova 1957, 183; "*Actinomyces violochromogenes*" (*sic*) Artamonova and Krasil'nikov *in* Rautenshtein 1960, 334]

vi.o.la.ce.o.chro.mo'ge.nes. L. adj. *violaceus* violet; Gr. n. *chroma* color; N.L. suff. *-genes* (from Gr. v. *gennaô* to produce) producing; N.L. part. adj. *violaceochromogenes* producing violet color.

Spore chains in Section *Retinaculiaperti* or *Spirales*. Short chains form incomplete or imperfect spirals, hooks, flexuous chains. These are neither typically spiral nor representative of the long chains with open loops and spirals on true *Retinaculiaperti* cultures. Some longer chains bear terminal spirals suggesting *Retinaculiaperti* morphology. Mature spore chains are generally short with 3–10, or sometimes more than 10, spores per chain. This morphology is seen on yeast-malt agar, oatmeal agar, salts-starch agar, and glycerol-asparagine agar. Spore surface is smooth.

Color of colony: aerial mass color in the Red (5cb, grayish yellowish pink) or Gray (5fe, light grayish reddish brown) color series on yeast-malt agar, oatmeal agar, salts-starch agar, and glycerol-asparagine agar. Reverse side of colony with no distinctive pigments (grayish yellow to strong brown) on yeast-malt agar and salts-starch agar; but yellow to yellow-brown may or may not be modified by red on oatmeal agar and glycerol-asparagine agar. If red reverse, mycelium pigment is present, it is pH-sensitive changing from red to violet or purple with addition of 0.05 M NaOH or from violet to red with addition of 0.05 M HCl.

Color in medium: melanoid pigments are formed in peptone-yeast-iron agar and tryptone-yeast broth, but only weakly or not at all in tyrosine agar. Yellow or yellow-brown pigment is found in the medium in yeast-malt agar and salts-starch agar. This pigment is not pH-sensitive when tested with 0.05 M NaOH or HCl.

D-Glucose, L-arabinose, D-xylose, iso-inositol, D-mannitol, D-fructose, rhamnose, sucrose, and raffinose are all utilized for growth.

Type strain shows the highest sequence similarity to: *S. collinus*, AB184123, 99.6%; *S. iakyrus*, AB184877, 99.5%; *S. griseorubens*, AB184139, 99.1%; *S. paradoxus*, AB184628, 99.1%; *S. griseoflavus*, AJ781322, 99.1%; *S. matensis*, AB184221, 99.1%; *S. variabilis*, DQ442551, 99%; *S. griseoincarnatus*, AJ781328, 99%; *S. labedae*, AB184704, 99%; *S. erythrogriseus*, AJ781328, 99%.

Source: not known.

DNA G+C content (mol%): not known.

Type strain: ATCC 19932, ATCC 25512, CBS 654.69, DSM 40181, NBRC 13100, INA 425, JCM 4530, NRRL B-5427, NRRL-ISP 5181, RIA 1292, VKM Ac-581.

Sequence accession no. (16S rRNA gene): AY999867.

499. **Streptomyces violaceolatus** (Krasil'nikov, Sorokina, Alferova and Bezzubenkova *in* Krasil'nikov 1965) Pridham 1970, 28[AL] ("*Actinomyces violaceolatus*" Krasil'nikov, Sorokina, Alferova and Bezzubenkova *in* Krasil'nikov 1965, 113)

vi.o.la.ce.o.la'tus. L. adj. *violaceus* violet-colored; L. adj. *latus* broad; N.L. masc. adj. *violaceolatus* violet, broad.

Forms blue-, purple-, or red-colored vegetative mycelium and diffusible pigment depending upon pH.

Spore chains in Section *Spirales*. Mature spore chains are long, usually with more than 50 spores per chain. This morphology is seen on yeast-malt agar, oatmeal agar, salts-starch agar, and glycerol-asparagine agar. Spore surface is smooth.

Color of colony: aerial mass color in the Gray color series (3fe, light brownish gray; 5fe, light grayish reddish brown; or 3ge, light grayish yellowish brown) on yeast-malt agar, oatmeal agar, salts-starch agar, and glycerol-asparagine agar. Color may sometimes approach tab 5dc, grayish yellowish pink in the Red color series on glycerol-asparagine agar. Reverse side of colony is blue, purple, or red, depending upon pH and ranging from grayed colors to almost black, depending upon intensity of pigment. Reverse mycelium pigment is not a pH indicator, changing from red or violet to blue with addition of 0.05 M NaOH and from blue or violet to red or pink with addition of 0.05 M HCl.

Color in medium: melanoid pigments are not formed in peptone-yeast-iron agar, tyrosine agar, or tryptone-yeast broth. Blue, violet, or red pigment, depending upon pH, is found in the medium in yeast-malt agar, oatmeal agar, salts-starch agar, and glycerol-asparagine agar. This pigment is pH-sensitive, showing the same changes noted for the reverse mycelium pigment.

D-Glucose, L-arabinose, D-xylose, iso-inositol, D-mannitol, D-fructose, rhamnose, sucrose, and raffinose are all utilized for growth, but growth on sucrose may be somewhat less than on any other carbon sources.

Type strain shows the highest sequence similarity to: *S. coelescens*, AF503496, 100%; *S. humiferus*, AF503491, 100%; *S. violaceoruber*, AF503492, 100%; *S. tricolor*, AB184687, 99.8%; *S. anthocyanicus*, AB184631, 99.8%; *S. rubrogriseus*, AB184681, 99.7%; *S. tendae*, D63873, 99.5%; *S. lienomycini*, AJ781353, 99.5%; *S. violaceorubidus*, AJ781374, 99.3%; *S. coelicoflavus*, AB184650, 99.3%.

Source: not known.

DNA G+C content (mol%): not known.

Type strain: ATCC 19847, ATCC 25513, CBS 655.69, DSM 40438, ICSSB 1022, NBRC 13101, JCM 4531, KCTC 9772, NRRL B-12371, NRRL-ISP 5438, RIA 1293, VKM Ac-582.

Sequence accession no. (16S rRNA gene): AF503497.

500. **Streptomyces violaceorectus** (Ryabova and Preobrazhenskaya *in* Gauze, Preobrazhenskaya, Kudrina, Blinov, Ryabova and Sveshnikova 1957) Pridham, Hesseltine and Benedict 1958, 63[AL] ("*Actinomyces violaceorectus*" Ryabova and Preobrazhenskaya *in* Gauze, Preobrazhenskaya, Kudrina, Blinov, Ryabova and Sveshnikova 1957, 182)

vi.o.la.ce.o.rec'tus. L. adj. *violaceus* violet colored; L. adj. *rectus* straight; N.L. adj. *violaceorectus* violet colored, straight (referring to the color of the vegetative mycelium and diffusible pigment on some media and to the structure of sporophores).

Spore chains in Section *Rectiflexibiles*. Mature spore chains are generally long and straight, often with more than spores per chain. This morphology is seen on yeast-malt agar, oatmeal agar, salts-starch agar, and glycerol-asparagine agar. Spore surface is smooth.

Color of colony: aerial mass color in the Gray or Red color series on yeast-malt agar, oatmeal agar, salts-starch agar, and glycerol-asparagine agar (representative color tab from the Gray color series is 5fe, light grayish reddish brown; representative tabs from the Red color series are 5dc, 5ec, or 6ec, grayish yellowish pink). Reverse side of colony is grayish yellow to brown or grayish reddish brown on yeast-malt agar; yellowish pink or reddish brown to dark purplish pink or dark purplish red on oatmeal agar, salts-starch agar, and glycerol-asparagine agar. Reverse mycelium pigment is a pH indicator, changing from orange or pink to violet or purple with addition of 0.05 M NaOH and from violet to pink or orange with addition of 0.05 M HCl.

Color in medium: melanoid pigments are formed in peptone-yeast-iron agar and tryptone-yeast broth, but not in tyrosine agar. Traces of yellow to pinkish brown pigment may be found in the medium in yeast-malt agar, oatmeal agar, salts-starch agar, and glycerol-asparagine agar. Pink or pinkish brown pigment, when present, is pH-sensitive, changing from pink or brown to pale violet or gray with addition of 0.05 M NaOH.

D-Glucose, L-arabinose, and D-xylose are utilized for growth. Limited growth also may occur on D-fructose and sucrose. No growth or only trace of growth with iso-inositol, D-mannitol, rhamnose, and raffinose.

Type strain shows the highest sequence similarity to: *S. bikiniensis*, X79851, 99.7%; *S. cinereoruber* subsp. *cinereoruber*, AB184121, 99.7%; *S. vietnamensis*, DQ311081, 99.4%; *S. tanashiensis*, AJ781362, 99.2%; *S. nashvillensis*, AB184286, 99.2%; *S. showdoensis*, AB184389, 99.1%; *S. viridobrunneus*, AJ781372, 99%.

Source: not known.

DNA G+C content (mol%): not known.

Type strain: ATCC 25514, CBS 656.69, BCRC 13626, DSM 40279, NBRC 13102, IMET 43520, INA 506, JCM 4532, NRRL B-12181, NRRL-ISP 5279, RIA 1294, VKM Ac-584.

Sequence accession no. (16S rRNA gene): AB184314.

501. **Streptomyces violaceoruber** (Waksman and Curtis 1916) Pridham 1970, 44[AL] ("*Actinomyces violaceus-ruber*" Waksman and Curtis 1916, 127; "*Streptomyces violaceoruber*" Waksman *in* Kutzner and Waksman 1959, 535)

vi.o.la.ce.o.ru'ber. L. adj. *violaceus* violet; L. adj. *ruber* red; N.L. masc. adj. *violaceoruber* violet-red.

Spore chains in Section *Spirales*, but spore chains representative of Sections *Rectiflexibiles* and *Retinaculiaperti* are also reported. Mature spore chains generally have 10–50 spores per chain. Typical morphology on yeast-malt agar, oatmeal agar, salts-starch agar, and glycerol-asparagine agar. Spore surface is smooth.

Color of colony: aerial mass color in the Gray color series on yeast-malt agar, oatmeal agar, salts-starch agar, and glycerol-asparagine agar. Reverse side of colony is blue or violet, depending on pH, on yeast-malt agar, oatmeal agar, salts-starch agar, and glycerol-asparagine agar. Reverse pigment is a pH indicator changing from violet or blue-violet to blue by addition of 0.05 M NaOH and from violet to red-violet or red with 0.05 M HCl.

Color in medium: melanoid pigments not formed in peptone-yeast-iron agar, tyrosine agar, or tryptone-yeast extract broth. Blue or violet pigment found in medium in yeast-malt agar, oatmeal agar, salts-starch agar, and glycerol-asparagine agar. This pigment is pH-sensitive; color changes are identical to changes noted for reverse color.

D-Glucose, L-arabinose, D-xylose, iso-inositol, D-mannitol, D-fructose, and rhamnose are utilized for growth. No growth or only trace of growth on sucrose. Variable reports on growth with raffinose.

Type strain shows the highest sequence similarity to: *S. violaceolatus* AF503497, 100%; *S. coelescens*, AF503496, 100%; *S. humiferus*, AF503491, 100%; *S. tricolor*, AB184687, 99.9%; *S. anthocyanicus*, AB184631, 99.9%; *S. rubrogriseus*, AB184681, 99.7%; *S. tendae*, D63873, 99.6%; *S. lienomycini*, AJ781353, 99.6%; *S. violaceorubidus*, AJ781374, 99.4%; *S. coelicoflavus*, AB184650, 99.3%; *S. ambofaciens*, M27245, 99%; *S. pactum*, AB184398, 99%.

Source: not known.

DNA G+C content (mol%): not known.

Type strain: ATCC 14980, ATCC 19816, ATCC 3355, CBS 569.68, BCRC 11489, DSM 40049, ICSSB 1016, NBRC 12826, JCM 4423, KCTC 9787, NRRL B-12594, NRRL B-2935, NRRL B-3025, NRRL B-3319, NRRL-ISP 5049, RIA 1096, UNIQEM 203, VKM Ac-726.

Sequence accession no. (16S rRNA gene): AF503492.

502. **Streptomyces violaceorubidus** Terekhova 1986, 575[VP] (Effective publication: Terekhova *in* Gause, Preobrazhenskaya, Sveshnikova, Terekhova and Maximova 1983.)

vi.o.la.ce.o.ru'bi.dus. L. adj. *violaceus* violet; L. adj. *rubidus* dark-red; N.L. masc. adj. *violaceorubidus* violet, dark-red.

Spore chains are spiral (*Spirales*); spores are smooth. On mineral agar 1: aerial mycelium is light gray; substrate mycelium and diffusible pigment are light pink, reddish violet; diffusible pigment is weak. On starch-ammonia agar: no aerial mycelium; substrate mycelium is yellow; no diffusible pigment. On glycerol-nitrate agar: aerial mycelium is creamy, light pinkish gray; substrate mycelium and diffusible pigment are yellowish reddish to violet red. On glycerol-asparagine agar: aerial mycelium is white, light gray; substrate mycelium is dark yellow; no diffusible pigment. On oatmeal agar: aerial mycelium is light pinkish gray; substrate mycelium and diffusible pigment are yellowish red to reddish violet; diffusible pigment is weak. On organic agar 2: aerial mycelium not extant or light gray, poor; substrate mycelium is yellowish; diffusible pigment is yellowish or not extant. On organic agar 79: aerial mycelium is light gray; substrate mycelium is yellowish; diffusible pigment is yellowish or not extant. Melanoid pigments are not formed. Good growth on rhamnose, fructose, glucose, xylose, mannitol, inositol, raffinose, and arabinose; sucrose is not utilized. Antibiotic: zinerubin.

Type strain shows the highest sequence similarity to: *S. tendae*, D63873, 99.8%; *S. lienomycini*, AJ781353, 99.7%; *S. rubrogriseus*, AB184681, 99.7%; *S. tricolor*, AB184687, 99.5%; *S. anthocyanicus*, AB184631, 99.5%; *S. matensis*, AB184221, 99.4%; *S. violaceoruber*, AF503492, 99.4%; *S. coelescens*, AF503496, 99.4%; *S. geysiriensis*, DQ442501, 99.3%; *S. minutiscleroticus*, EF178696, 99.3%; *S. humiferus*, AF503491, 99.3%; *S. violaceolatus* AF503497, 99.3%; *S. rochei*, AB184237, 99.2%; *S. ghanaensis*, AY999851, 99.2%; *S. malachitospinus*, AB249954, 99.2%; *S. vinaceusdrappus*, AY999929, 99.2%; *S. mutabilis*, EF178679, 99.2%; *S. plicatus*, AB184291, 99.2%; *S. lomondensis*, AB184673, 99.1%; *S. parvulus*, AB184326, 99.1%; *S. paradoxus*, AB184628, 99.1%; *S. griseorubens*, AB184139, 99.1%; *S. pactum*, AB184398, 99%; *S. malachitofuscus*, AB184282, 99%; *S. griseoflavus*, AJ781322, 99%; *S. tuirus*, AB184690, 99%; *S. variabilis*, DQ442551, 99%; *S. labedae*, AB184704, 99%; *S. griseoincarnatus*, AJ781328, 99%; *S. erythrogriseus*, AJ781328, 99%; *S. coelicoflavus*, AB184650, 99%; *S. viridochromogenes*, DQ442555, 99%; *S. collinus*, AB184123, 99%.

Source: not known.

DNA G+C content (mol%): not known.

Type strain: ATCC 43697, DSM 41478, NBRC 15463, INA 770, JCM 6931, NRRL B-16381, VKM Ac-1292.

Sequence accession no. (16S rRNA gene): AJ781374.

503. **Streptomyces violaceus** (Rossi Doria 1891) Waksman *in* Waksman and Lechevalier 1953, 43[AL] emend. Lanoot, Vancanneyt, Cleenwerk, Wang, Li, Liu and Swings 2002, 828 ["*Streptotrix* (*sic*) *violacea*" Rossi Doria 1891, 411; "*Oospora violacea*" Sauvageau and Radais 1892, 252; "*Actinomyces violaceus*" Gasperini 1894, 84; "*Cladothrix violacea*" Macé 1897, 1032; "*Nocardia violacea*" Chalmers and Christopherson 1916, 270].

vi.o.la'ce.us. L. masc. adj. *violaceus* violet colored, referring to the color of the vegetative mycelium and diffusible pigment of the organism.

Spore chains in Section *Spirales*. Mature spore chains are moderately long with 10 to 50 or more spores per chain. This morphology is seen on yeast-malt agar, oatmeal agar, salts-starch agar, and glycerol-asparagine agar. Spore surface is spiny.

Color of colony: aerial mass color in the Red color series (5cb, grayish yellowish pink or 7ca, light yellowish pink) on salts-starch agar and glycerol-asparagine agar; White or Red on yeast-malt agar; White color series on oatmeal agar. When aerial mycelium is thin, the red to violet color of the substrate mycelium may be evident. Reverse side of colony is reddish orange to purplish pink or pale purple, depending on pH. Reverse mycelium pigment is a pH indicator, changing from red to violet (purple) with addition of 0.05 M NaOH or from violet to red (pink) with addition of 0.05 M HCl.

Color in medium: melanoid pigments are formed in peptone-yeast-iron agar and tryptone-yeast broth, but not in tyrosine agar. Red or violet pigment, depending on pH, is found in the medium in yeast-malt agar, oatmeal agar, salts-starch agar, and glycerol-asparagine agar. This pigment is pH-sensitive showing the same changes noted for the reverse mycelium pigments.

D-Glucose, L-arabinose, D-xylose, iso-inositol, D-mannitol, D-fructose, rhamnose, sucrose, and raffinose are all utilized for growth.

Type strain shows the highest sequence similarity to: *S. roseoviolaceus*, AJ399484, 100%; *S. janthinus*, AB184851, 100%; *S. albosporeus* subsp. *albosporeus*, AJ781327, 100%; *S. luteogriseus*, AB184379, 99.5%; *S. lomondensis*, AB184673, 99.3%; *S. hawaiiensis*, AB184143, 99.3%; *S. flavoviridis*, AB184842, 99.3%; *S. pilosus*, AB184161, 99.2%; *S. arenae*, AB249977, 99.2%; *S. africanus*, AY208912, 99.2%; *S. tuirus*, AB184690, 99.2%; *S. bellus*, AB184849, 99.1%; *S. mutabilis*, EF178679, 99.1%; *S. massasporeus*, AB184152, 99%; *S. afghaniensis*, AJ399483, 99%; *S. levis*, AB184670, 99%; *S. coerulescens*, AY999720, 99%; *S. parvulus*, AB184326, 99%; *S. coeruleorubidus*, AY999719, 99%.

Source: not known.

DNA G+C content (mol%): not known.

Type strain: AS 4.1456, ATCC 15888, ATCC 25515, CBS 657.69, BCRC 11880, CCT 4833, CECT 3235, DSM 40082, NBRC 13103, IMET 43085, INMI 1, JCM 4533, LMG 20257, NRRL B-2869, NRRL-ISP 5082, RIA 1295, RIA 656, VKM Ac-510, VKM Ac-977.

Sequence accession no. (16S rRNA gene): AB184315.

Further comments: Streptomyces violaceus (Rossi Doria 1891) Waksman 1953 emend. Lanoot et al. 2002 is an earlier heterotypic synonym of *Streptomyces violatus* (Artamonova and Krasil'nikov 1960) Pridham 1970.

504. **Streptomyces violaceusniger** corrig. (Waksman and Curtis 1916) Pridham, Hesseltine and Benedict 1958, 63^{AL} emend. Labeda and Lyons 1991b, 400 ["*Actinomyces violaceus-niger*" Waksman and Curtis 1916, 111; "*Streptomyces violaco-niger*" (*sic*) Waksman and Henrici *in* Breed, Murray and Hitchens 1948, 947]

vi.o.la.ce.us.ni'ger. L. adj. *violaceus* violet; L. adj. *niger* black; N.L. masc. adj. *violaceusniger* violet-black.

Spore chains are *Spirales*; the spore surface is smooth to rough. The spore mass is gray, becoming black and moist when it is mature. Reverse side of colonial growth is grayish yellow or light olive green to dark olive to dark grayish green. Melanoid pigments are not formed; soluble pigments are not produced. D-Glucose, L-arabinose, D-xylose, iso-inositol, D-mannitol, D-fructose, rhamnose, sucrose, and raffinose are utilized for growth. Exhibits anti-bacterial and anti-fungal activity; excellent growth on Czapek's solution agar; gray to black vegetative mycelium on some media; hygroscopic; NaCl tolerance >4%, but <7%.

Type strain shows the highest sequence similarity to: *S. yogyakartensis*, AB249942, 100%; *S. albiflaviniger*, AJ391812, 99.7%; *S. endus*, AY999911, 99.5%; *S. sporocinereus*, AB249933, 99.5%; *S. demainii*, DQ334782, 99.5%; *S. hygroscopius* subsp. *hygroscopius*, AB184428, 99.5%; *S. javensis*, AJ391833, 99.1%.

Source: not known.

DNA G+C content (mol%): 71.2.

Type strain: AS 4.1423, ATCC 27477, CBS 760.72, DSM 40563, NBRC 13459, JCM 4850, LMG 19336, NRRL B-1476, NRRL-ISP 5563, RIA 1420, VKM Ac-583.

Sequence accession no. (16S rRNA gene): AJ391823.

Further comments: the original spelling, *Streptomyces violaceoniger* (*sic*), has been corrected by Hill et al. (1984).

505. **Streptomyces violarus** (Artamonova and Krasil'nikov 1960) Pridham 1970, 30^{AL} ("*Actinomyces violarus*" Artamonova and Krasil'nikov *in* Rautenshtein 1960, 334)

vi.o.la'rus. N.L. masc. adj. *violarus* (from L. adj. *violaris*) of or belonging to violets, violet, referring to the color of the vegetative mycelium.

Spore chains in Section *Spirales*. Spirals of three to four turns may be formed or poorly developed irregular spirals, hooks or loops may become entangled. Mature spore chains are moderately long with 10–50 spores per chain. This morphology is seen on yeast-malt agar, oatmeal agar, salts-starch agar, and glycerol-asparagine agar. Spore surface is spiny.

Color of colony: aerial mass color in the Red or Violet color series on yeast-malt agar, oatmeal agar, salts-starch agar, and glycerol-asparagine agar. Nearest matching color tabs are 5cb, 5ec or 6ec, grayish yellowish pink from the Red color series or 11ca, very pale purple, from the Violet color series. Reverse side of colony is reddish orange or reddish brown to purplish pink or purple depending upon pH. Reverse mycelium pigment is a pH indicator, changing from pink or violet to blue-violet or blue with addition of 0.05 M NaOH and from violet to pink or red with addition of 0.05 M HCl.

Color in medium: melanoid pigments are formed in peptone-yeast-iron agar and tryptone-yeast broth, but only weakly or not at all in tyrosine agar. Red or violet pigment, depending upon pH, is found in the medium in yeast-malt agar, oatmeal agar, salts-starch agar, and glycerol-asparagine agar. This pigment is pH-sensitive, showing the same changed noted for reverse mycelium pigment.

D-Glucose, L-arabinose, D-xylose, iso-inositol, D-mannitol, D-fructose, rhamnose, sucrose, and raffinose are all utilized for growth.

Type strain shows no sequence similarity over 99%.

Source: not known.

DNA G+C content (mol%): not known.

Type strain: ATCC 15891, ATCC 25516, CBS 658.69, CCT 5007, CECT 3237, DSM 40205, NBRC 13104, INMI 1212, JCM 4534, KCTC 9788, NRRL B-5432, NRRL-ISP 5205, RIA 1296, RIA 157, UNIQEM 204, VKM Ac-528.

Sequence accession no. (16S rRNA gene): AB184316.

506. **Streptomyces violascens** (Preobrazhenskaya and Sveshnikova *in* Gauze, Preobrazhenskaya, Kudrina, Blinov, Ryabova and Sveshnikova 1957) Pridham, Hesseltine and Benedict 1958, 68[AL] ("*Actinomyces violascens*" Preobrazhenskaya and Sveshnikova *in* Gauze, Preobrazhenskaya, Kudrina, Blinov, Ryabova and Sveshnikova 1957, 41)

vi.o.la'scens. N.L. part. adj. *violascens* becoming violet.

Spore chains in Section *Spirales*. Mature spore chains have 10 to 50 or more spores per chain. This morphology is seen on yeast-malt agar, oatmeal agar, salts-starch agar, and glycerol-asparagine agar. Spore surface is spiny.

Color of colony: aerial mass color in the Violet color series on salts-starch agar and glycerol-asparagine agar; Violet or White color series on oatmeal agar; Violet, White, or Red color series on yeast-malt agar. Reverse side of colony with no distinctive pigment (grayed yellow to yellow-brown) on yeast-malt agar, oatmeal agar, salts-starch agar, and glycerol-asparagine agar.

Color in medium: melanoid pigments are formed in peptone-yeast-iron agar, tyrosine agar, and tryptone-yeast broth. Pigments other than melanoids not formed (or only traces of yellow) in yeast-malt agar, oatmeal agar, salts-starch agar, and glycerol-asparagine agar.

D-Glucose, L-arabinose, D-xylose, D-fructose, and raffinose are utilized for growth. No growth or only trace of growth on D-mannitol and rhamnose. Utilization of sucrose and iso-inositol is doubtful.

Type strain shows the highest sequence similarity to: *S. limosus*, AB184147, 100%; *S. felleus*, AB184129, 100%; *S. daghestanicus*, DQ442497, 100%; *S. albidoflavus*, AB184255, 100%; *S. hydrogenans*, AB184868, 100%; *S. odorifer*, Z76682, 100%; *S. griseus* subsp. *solvifaciens*, AB249915, 100%; *S. champavatii*, DQ026642, 100%; *S. canescens*, AB184117, 100%; *S. sampsonii*, D63871, 99.8%; *S. koyangensis*, AY079156, 99.7%.

Source: not known.

DNA G+C content (mol%): not known.

Type strain: ATCC 23968, CBS 266.66, CBS 951.68, BCRC 12240, CECT 3215, DSM 40183, NBRC 12920, IMET 42061, INA 3959/54, JCM 4424, KCTC 9785, NCIMB 9820, NRRL B-2700, NRRL-ISP 5183, RIA 1138, VKM Ac-1275.

Sequence accession no. (16S rRNA gene): AY999737.

507. **Streptomyces violatus** (Artamonova and Krasil'nikov 1960) Pridham 1970, 30[AL] ("*Actinomyces violatus*" Artamonova and Krasil'nikov *in* Rautenshtein 1960, 334)

vi.o.la'tus. L. masc. adj. *violatus* flavored with violet, referring to the color of the vegetative mycelium and diffusible pigment of the organism.

Spore chains in Section *Spirales*. Mature spore chains are moderately long, with 10 to 50 or more spores per chain. This morphology is seen on yeast-malt agar, oatmeal agar, salts-starch agar, and glycerol-asparagine agar. Spore surface is spiny.

Color of colony: aerial mass color in the White or Red color series on yeast-malt agar, oatmeal agar, salts-starch agar, and glycerol-asparagine agar. Nearest matching color tab in the Red color series is 5cb, grayish yellowish pink. Reverse side of colony is light grayish reddish brown on yeast-malt agar; light reddish purple to dark purplish pink, or yellowish pink on oatmeal agar, salts-starch agar, and glycerol-asparagine agar. Reverse mycelium pigment is a pH indicator, changing from red or pink to violet or blue with the addition of 0.05 M NaOH and from violet to yellowish pink or orange with the addition of 0.05 M HCl.

Color in medium: melanoid pigments are formed in peptone-yeast-iron agar and tryptone-yeast broth, but appear slowly or not at all in tyrosine agar. Red to violet pigment is found in the medium in yeast-malt agar, oatmeal agar, salts-starch agar, and glycerol-asparagine agar. This pigment is pH-sensitive, showing the same changes recorded for the reverse mycelium pigment.

D-Glucose, L-arabinose, D-xylose, D-fructose, rhamnose, sucrose, raffinose, iso-inositol, and D-mannitol are all utilized for growth.

For sequence similarity, see type strain of *Streptomyces violaceus*.

Source: not known.

DNA G+C content (mol%): not known.

Type strain: ATCC 15892, CBS 650.72, DSM 40209, NBRC 13349, INMI 1205, JCM 4237, JCM 4851, NRRL B-2867, NRRL-ISP 5209, RIA 708.

Sequence accession no. (16S rRNA gene): AJ399480.

Further comments: Streptomyces violatus (Artamonova and Krasil'nikov 1960) Pridham 1970 is a later heterotypic synonym of *Streptomyces violaceus* (Rossi Doria 1891) Waksman 1953 emend. Lanoot et al. 2002.

508. **Streptomyces violens** (Kalakoutskii and Krasil'nikov 1960) Goodfellow, Williams and Alderson 1987c, 179[VP] (Effective publication: Goodfellow, Williams and Alderson 1986d, 59.) (*Chainia violens* Kalakoutskii and Krasil'nikov *in* Rautenshtein 1960, 55)

vi.o.lens. L. adj. *violens* raging but probably from L. fem. n. *viola* violet, referring to pink to violet pigment produced by the organism on some media.

Spore chain Section not determined. Aerial mycelium is not produced on yeast-malt agar, oatmeal agar, salts-starch agar, or glycerol-asparagine agar. Spore surface not determined. Sclerotia are produced on yeast-malt agar, oatmeal agar, salts-starch agar, and glycerol-asparagine agar, and the substrate (vegetative) mycelium fragments into rod-shaped or coccoid elements on these media. One observer reports L-A_2pm and the absence of arabinose in the cell wall (cell-wall type I).

Color of colony: aerial mass color not determined; aerial mycelium absent on yeast-malt agar, oatmeal agar, salts-starch agar, and glycerol-asparagine agar. Reverse side of colony with no distinctive pigments (colorless on glycerol-asparagine agar; pale or grayish yellow to light olive on yeast-malt agar, oatmeal agar, and salts-starch agar).

Color in medium: melanoid pigments are not formed in peptone-yeast-iron agar, tyrosine agar, or tryptone-yeast broth. One observer reports that melanin is produced on a peptone-yeast-iron agar of different composition. A pale yellow pigment in oatmeal agar is reported by one observer; and another reports orange or pinkish pigment in glycerol-asparagine agar. The latter is pH-sensitive, changing from orange or rose to violet or light purple with the addition of 0.05 M NaOH.

D-Glucose, L-arabinose, D-xylose, iso-inositol, D-mannitol, D-fructose, rhamnose, sucrose, and raffinose are all utilized for growth.

Type strain shows the highest sequence similarity to: *S. ochraceiscleroticus*, AB184094, 99.8%; *S. purpurogeneiscleroticus*, AJ621604, 99.7%; *S. monomycini*, DQ445790, 99.3%; *S. erumpens*, AJ621603, 99.1%.

Source: not known.

DNA G+C content (mol%): not known.

Type strain: ATCC 15898, CBS 451.65, CBS 787.72, BCRC 12540, DSM 40597, HAMBI 1073, NBRC 12557, NBRC 13486, INMI 1212, JCM 3072, JCM 4852, NCAIM B.01477, NRRL B-3484, NRRL-ISP 5597, PCM 2247, RIA 1447, RIA 565, VKM Ac-586, VKM Ac-653.

Sequence accession no. (16S rRNA gene): AJ621605.

509. **Streptomyces virens** Gause and Sveshnikova 1986a, 575[VP] (Effective publication: Gause and Sveshnikova *in* Gause, Preobrazhenskaya, Sveshnikova, Terekhova and Maximova 1983.)

vi'rens. L. part. adj. *virens* being green.

Spore chains are spiral (*Spirales*); spore surface has spines and warts, warts are situated between hairs and spines. On mineral agar 1: aerial mycelium is light gray to greenish, poorly developed; substrate mycelium is olive, gray brownish olive; diffusible pigment is orange red, moderate. On glycerol-nitrate agar: aerial mycelium is weak developed with white grayish appearance; substrate mycelium is olive, gray-brown olive; diffusible pigment is light brown. On starch-ammonia agar: aerial mycelium is poorly developed, whitish or grayish; substrate mycelium is colorless to light olive; no diffusible pigment. On glycerol-asparagine agar: aerial mycelium is absent or poorly developed with white appearance; substrate mycelium and diffusible pigment are olive. On oatmeal agar: aerial mycelium is absent or whitish; substrate mycelium is olive; no diffusible pigment. On organic agar 2: aerial mycelium is absent or whitish; substrate mycelium is gray brownish; diffusible pigment is reddish. Melanoid pigments are not formed. Moderate growth on arabinose, fructose, and mannitol; no growth on xylose, raffinose, rhamnose, sucrose, or cellulose. Antibiotic: virenomycin.

Type strain shows the highest sequence similarity to: *S. calvus*, AB184329, 99.9%; *S. aureorectus*, AB184710, 99.9%; *S. asterosporus*, AB184706, 99.8%; *S. djakartensis*, AB184657, 99%; *S. geysiriensis*, DQ442501, 99%; *S. tuirus*, AB184690, 99%; *S. anandii*, AB184402, 99%; *S. minutiscleroticus*, EF178696, 99%.

Source: not known.

DNA G+C content (mol%): not known.

Type strain: DSM 41465, NBRC 15901, INA 3831, JCM 9095, NRRL B-24331, VKM Ac-833.

Sequence accession no. (16S rRNA gene): DQ442554.

510. **Streptomyces virginiae** Grundy, Whitman, Rdzok, Rdzok, Hanes and Sylvester 1952, 399[AL]

vir.gi.ni'a.e. N.L. gen. n. *virginiae* of Virginia, referring to the source of the soil (near Roanoke, Virginia) from which the organism was isolated.

Spore chains in Section *Retinaculiaperti*, including flexuous, looped, and spiral forms characteristic of this section (terminal spirals on long spore chains may be common in some areas on mass cultures). Mature spore chains are moderately long with 10–50, or sometimes more than 50, spores per chain. This morphology is seen on yeast-malt agar, oatmeal agar, salts-starch agar, and glycerol-asparagine agar. Spore surface is smooth.

Color of colony: aerial mass color in the Red or Gray color series on yeast-malt agar, oatmeal agar, salts-starch agar, and glycerol-asparagine agar. Color tabs selected by observers: Grayish yellowish pink (5dc, 6ec, Red series), pale purple (7fe, Gray series). Reverse side of colony with no distinctive pigment (grayed yellow to yellow-brown) on yeast-malt agar, oatmeal agar, and glycerol-asparagine agar; dark grayed yellow may be modified by traces of red or violet on salts-starch agar. Substrate pigment is not a pH indicator.

Color in medium: melanoid pigments are not formed in peptone-yeast-iron agar or tryptone-yeast extract broth. Pigments other than melanoids are not formed in yeast-malt agar, oatmeal agar, salts-starch agar, or glycerol-asparagine agar.

D-Glucose and D-fructose are utilized for growth. Trace of growth on L-arabinose, sucrose, D-xylose, iso-inositol, D-mannitol, rhamnose, and raffinose is similar to, or only slight better than growth on basal medium without carbon.

Type strain shows the highest sequence similarity to: *S. cinnamonensis*, AB184707, 99.8%; *S. sporoраveus*, AJ781370, 99.5%; *S. nojiriensis*, AJ781355, 99.5%; *S. xanthophaeus*, DQ442560, 99.5%; *S. vinaceus*, AB184394, 99.4%; *S. goshikiensis*, EF178693, 99.4%; *S. cirratus*, AY999794, 99.4%; *S. avidinii*, AB184395, 99.3%; *S. subrutilus*, X80825, 99.3%; *S. sporoverrucosus*, DQ442544, 99.2%; *S. colombiensis*, DQ026646, 99.1%; *S. flavotricini*, AB184132, 99%.

Source: isolated from soil near Roanoke, Virginia, USA.

DNA G+C content (mol%): not known.

Type strain: ATCC 19817, CBS 291.60, CBS 570.68, BCRC 12069, DSM 40094, NBRC 12827, NBRC 3729, JCM 4425, KCTC 1747, NRRL B-1446, NRRL-ISP 5094, RIA 1097, UNIQEM 205, VKM Ac-1218.

Sequence accession no. (16S rRNA gene): D85119.

511. **Streptomyces viridiflavus** corrig. (Locci and Schofield 1989) Witt and Stackebrandt 1991, 457[VP] (Effective publication: Witt and Stackebrandt 1990, 370.) ["*Streptoverticillium viridoflavum*" (*ex* Waksman and Taber 1953) Locci 1985, 232; *Streptoverticillium viridoflavum* Locci and Schofield *in* Williams, Sharpe and Holt 1989, 2503]

vir.i.di.flav'us. L. adj. *viridis* green; L. adj. *flavus* yellow; N.L. masc. adj. *viridiflavus* green-yellow.

Poor off-white aerial vegetation. White spore mass. Melanin pigment is not produced. Coumarin, L-methionine, L-proline, and shikimic acid are utilized, but not mannitol, melibiose, raffinose, sorbitol, or DL-α-aminobutyric acid. Produces acid from *myo*-inositol, D-ribose, and trehalose, but not from D-galactose or D-fructose. Citrate and hypoxanthine are degraded, but not esculin or L-tyrosine. H_2S is produced. No reduction of NO_3^-. Resistance shown to cephalotin (30 µg/ml), colistin (30 µg/ml), and cephamandole (30 µg/ml). *Aspergillus niger* and *Candida albicans* are inhibited but not *Bacillus subtilis.* The strain shows antifungal activity.

For sequence similarity, see type strain of *Streptomyces olivoverticillatus.*

Source: not known.

DNA G+C content (mol%): not known.

Type strain: Y. E. Konev, ATCC 12631, CBS 652.72, DSM 40237, NBRC 13351, NBRC 15799, JCM 4221, JCM 4857, NRRL B-1548, NRRL-ISP 5237, RIA 1312.

Sequence accession no. (16S rRNA gene): AB184702.

Further comments: in violation of Rule 33c of the *Bacteriological Code* (1990 Revision), in Validation List no. 38, *Streptomyces viridiflavus* corrig. is proposed as a *nomen revictum* (basonym: "*Streptomyces viridoflavum*" Waksman and Taber 1953).

The original spelling of the specific epithet, *viridoflavum* (*sic*), has been corrected by Euzéby (1998).

According to Hatano et al. (2003), *Streptomyces viridiflavus* corrig. (Locci and Schofield 1989) Witt and Stackebrandt 1991 is a later heterotypic synonym of *Streptomyces olivoverticillatus* (Shinobu 1956) Witt and Stackebrandt 1991.

512. **Streptomyces viridiviolaceus** (Ryabova and Preobrazhenskaya *in* Gauze, Preobrazhenskaya, Kudrina, Blinov, Ryabova and Sveshnikova 1957) Pridham, Hesseltine and Benedict 1958, 70^{AL} ("*Actinomyces viridiviolaceus*" Ryabova and Preobrazhenskaya *in* Gauze, Preobrazhenskaya, Kudrina, Blinov, Ryabova and Sveshnikova 1957, 188)

vi.ri.di.vi.o.la'ce.us. L. adj. *viridis* green; L. adj. *violaceus* violet-colored; N.L. masc. adj. *viridiviolaceus* green-violet, referring to the greenish color of the aerial mycelium and the violet color of diffusible pigment.

Spore chains in Section *Spirales.* Mature spore chains are moderately long, with 10–50 spores per chain. This morphology is seen on yeast-malt agar, oatmeal agar, salts-starch agar, and glycerol-asparagine agar. Spore surface is spiny to hairy.

Color of colony: aerial mass color in the Gray color series (3fe, light brownish gray; 5fe or 5ge, light grayish reddish brown; 3ig, grayish yellowish brown) on yeast-malt agar, oatmeal agar, salts-starch agar, and glycerol-asparagine agar. Reverse side of colony is strong brown or reddish brown on yeast-malt agar and glycerol-asparagine agar; yellowish brown or olive brown on oatmeal agar and salts-starch agar. Reverse mycelium pigment is a pH indicator, changing from yellow or orange to violet with the addition of 0.05 M NaOH and from violet or red to orange with the addition of 0.05 M HCl.

Color in medium: melanoid pigments are not formed in peptone-yeast-iron agar, tyrosine agar, or tryptone-yeast broth. Orange or red pigment is found in the medium in yeast-malt agar, salts-starch agar, and glycerol-asparagine agar but not in oatmeal agar. This pigment is pH-sensitive, showing the same changes observed with the reverse mycelium pigment.

D-Glucose, L-arabinose, D-mannitol, and D-fructose are utilized for growth. Utilization of D-xylose, iso-inositol, and sucrose is doubtful. No growth or only traces of growth with rhamnose and raffinose.

Type strain shows the highest sequence similarity to: *S. levis,* AB184670, 99.2%; *S. cellulosae,* DQ442495, 99%; *S. gancidicus,* AB184660, 99%; *S. pseudogriseolus,* DQ442541, 99%.

Source: not known.

DNA G+C content (mol%): not known.

Type strain: ATCC 27478, CBS 660.72, BCRC 12457, DSM 40280, NBRC 13359, INA 5276/56, JCM 4855, NRRL B-12182, NRRL-ISP 5280, RIA 1320, VKM Ac-587.

Sequence accession no. (16S rRNA gene): AY999854.

513. **Streptomyces viridobrunneus** (*ex* Krasil'nikov 1970b) Terekhova 1986, 575^{VP} (Effective publication: Terekhova *in* Gause, Preobrazhenskaya, Sveshnikova, Terekhova and Maximova 1983.) ("*Actinomyces viridobrunneus*" Krasil'nikov 1970b)

vi.ri.do.brun'ne.us. L. adj. *viridis* green; N.L. adj. *brunneus* dark brown; N.L. masc. adj. *viridobrunneus* green dark-brown colored.

Spore chains are straight to flexuous (*Rectiflexibiles*); spores are smooth. On mineral agar 1: aerial mycelium is gray; substrate mycelium is brownish greenish olive gray; diffusible pigment is green to olive. On glycerol-nitrate agar: aerial mycelium is absent or white, poorly developed; substrate mycelium and diffusible pigment are gray brownish olive to olive gray brown. On starch-ammonia agar, glycerol-asparagine agar: aerial mycelium is absent or gray, gray brownish gray; substrate mycelium is gray brownish green; diffusible pigment is green, gray brownish green. On oatmeal agar: aerial mycelium is gray; substrate mycelium is yellowish to olive; no diffusible pigment. On organic agar 2: aerial mycelium is gray; substrate mycelium and diffusible pigment are dark gray brown. Melanoid pigments are formed. Glucose, mannose, sucrose, and raffinose are utilized for growth; no growth on mannitol, rhamnose, inositol, or arabinose. Antibiotic: pigment antibiotic viridomycin. The type strain is characterized by gray aerial mycelium, straight spore chains, building melanoid pigment and having green to olive gray brownish color of substrate mycelium and diffusible pigment under the pigment viridomycin.

Type strain shows the highest sequence similarity to: *S. showdoensis,* AB184389, 100%; *S. cinereoruber* subsp. *cinereoruber,* AB184121, 99.4%; *S. roseoviridis,* AB184239, 99.4%; *S. nashvillensis,* AB184286, 99.1%; *S. tanashiensis,* AJ781362, 99.1%; *S. vietnamensis,* DQ311081, 99.1%; *S. violaceorectus,* AB184314, 99%; *S. bikiniensis,* X79851, 99%.

Source: not known.

DNA G+C content (mol%): not known.

Type strain: ATCC 43698, DSM 41466, NBRC 15902, INMI 300, JCM 9096, VKM Ac-559.

Sequence accession no. (16S rRNA gene): AJ781372.

514. **Streptomyces viridochromogenes** (Krainsky 1914) Waksman and Henrici 1948, 942[AL] ("*Actinomyces viridochromogenes*" Krainsky 1914, 684)

vi.ri.do.chro.mo'ge.nes. L. adj. *viridis* green; Gr. n. *chroma* color; N.L. suff. *-genes* (from Gr. v. *gennaô* to produce) producing; N.L. part. adj. *viridochromogenes* producing green color.

Spore chains in Section *Spirales.* Mature spore chains are generally long, often with more than 50 spores per chain. This morphology is seen on yeast-malt agar, oatmeal agar, salts-starch agar, and glycerol-asparagine agar. Spore surface is spiny.

Color of colony: aerial mass color in the Green color series (24ih, grayish green; 24½h, dark grayish green), Yellow color series (24½dc, pale yellow green) or Gray color series (2dc, yellowish gray) on yeast-malt agar. It is in the Yellow (1dc, pale yellow green), Green (24½li, dark greenish gray), or Blue (22fe, pale green) color series on oatmeal agar. It is in the Blue (19dc, pale blue; 22fe, pale green) color series on salts-starch agar. It is in the Blue (19dc, pale blue; 22fe, pale green) or Yellow (1dc, pale yellow green) color series on glycerol-asparagine agar. Reverse side of colony is olive brown to dark olive on yeast-malt agar; grayish green to grayish olive on oatmeal agar; grayish yellow, olive brown, or greenish yellow on salts-starch agar and glycerol-asparagine agar. Reverse mycelium pigment is a pH indicator, changing from bluish or greenish brown to reddish or purplish brown with the addition of 0.05 M HCl.

Color in medium: melanoid pigments are not formed in peptone-yeast-iron agar or tryptone-yeast broth; tyrosine agar may or may not be darkened in 2–4 d. Greenish pigment is found in the medium in oatmeal agar. This pigment is pH-sensitive, changing from blue green or greenish to pink with the addition of 0.05 M HCl.

D-Glucose, L-arabinose, D-xylose, iso-inositol, D-mannitol, D-fructose, rhamnose, and raffinose are utilized for growth. Utilization of sucrose is doubtful.

Type strain shows the highest sequence similarity to: *S. paradoxus*, AB184628, 99.4%; *S. matensis*, AB184221, 99.3%; *S. collinus*, AB184123, 99.3%; *S. griseoflavus*, AJ781322, 99.2%; *S. griseorubens*, AB184139, 99.2%; *S. flaveolus*, AB184764, 99.2%; *S. griseoincarnatus*, AJ781328, 99.1%; *S. labedae*, AB184704, 99.1%; *S. erythrogriseus*, AJ781328, 99.1%; *S. violaceorubidus*, AJ781374, 99%; *S. glaucescens*, AB184843, 99%; *S. variabilis*, DQ442551, 99%.

Source: not known.

DNA G+C content (mol%): not known.

Type strain: ATCC 14920, CBS 140.20, CBS 648.72, BCRC 13769, CECT 3216, DSM 40110, HAMBI 1023, HUT 6030, NBRC 13347, NBRC 3113, JCM 4265, JCM 4856, NCIMB 9597, NRRL B-1511, NRRL-ISP 5110, RIA 1308, VKM Ac-629.

Sequence accession no. (16S rRNA gene): DQ442555.

515. **Streptomyces viridodiastaticus** (Baldacci, Grein and Spalla 1955) Pridham, Hesseltine and Benedict 1958, 67[AL] ["*Actinomyces virido-diastaticus*" (*sic*) Baldacci, Grein and Spalla 1955, 133]

vi.ri.do.di.a.sta'ti.cus. L. adj. *viridis* green; N.L. adj. *diastaticus*, diastatic, starch, digesting; N.L. masc. adj. *diastaticus* green-diastatic.

Spore chains in Section *Spirales* but abundant open spirals and flexuous chains also suggest *Retinaculiaperti* morphology. Spirals are most abundant on salts-starch agar. Mature spore chains are moderately long with 10–50 spores per chain. This morphology is seen on yeast-malt agar, oatmeal agar, salts-starch agar, and glycerol-asparagine agar. Spore surface is spiny, but spines may not be apparent on some spores.

Color of colony: aerial mass color in the Gray color series (2fe, medium gray to 2ih, light olive-gray or 3ig, grayish yellowish brown) on yeast-malt agar, oatmeal agar, salts-starch agar, and glycerol-asparagine agar. Reverse side of colony with no distinctive pigments (yellowish brown to olive-brown on yeast-malt agar; pale yellow-green or pale yellow and grayish yellow on oatmeal agar, salts-starch agar, and glycerol-asparagine agar).

Color in medium: melanoid pigments are not formed in peptone-yeast-iron agar, tyrosine agar, or tryptone-yeast broth. No pigment is found in medium in yeast-malt agar, oatmeal agar, salts-starch agar, or glycerol-asparagine agar.

D-Glucose, L-arabinose, D-xylose, iso-inositol, D-mannitol, D-fructose, and rhamnose are utilized for growth. Reports vary on utilization of sucrose and raffinose.

Type strain shows the highest sequence similarity to: *S. albogriseolus*, AJ494865, 99.8%; *S. coeruleorubidus*, AY999719, 99.4%; *S. bellus*, AB184849, 99.2%; *S. griseorubens*, AB184139, 99.2%; *S. coerulescens*, AY999720, 99.2%; *S. atrovirens*, DQ026672, 99.2%; *S. iakyrus*, AB184877, 99.1%; *S. longispororuber*, AB184440, 99.1%; *S. variabilis*, DQ442551, 99%; *S. erythrogriseus*, AJ781328, 99%; *S. lusitanus*, AB184424, 99%; *S. labedae*, AB184704, 99%; *S. speibonae*, AF452714, 99%; *S. griseoincarnatus*, AJ781328, 99%.

Source: not known.

DNA G+C content (mol%): not known.

Type strain: ATCC 25518, CBS 660.69, BCRC 12458, DSM 40249, NBRC 13106, JCM 4536, NRRL B-5622, NRRL-ISP 5249, RIA 1298, VKM Ac-1749.

Sequence accession no. (16S rRNA gene): AY999852.

516. **Streptomyces viridosporus** Pridham, Hesseltine and Benedict 1958, 67[AL]

vi.ri.do.spo'rus. L. adj. *viridis* green; N.L. n. *spora* a spore; N.L. masc. adj. *viridosporus* green-spored.

Spore chains in Section *Spirales.* Spirals are sometimes irregular or imperfect when spore chains are short. Mature spore chains are generally short but contain more than 10 spores per chain. This morphology is seen on yeast-malt agar, oatmeal agar, salts-starch agar, and glycerol-asparagine agar. Spore surface is spiny to hairy.

Color of colony: aerial mass color in the Green color series on yeast-malt agar, oatmeal agar, salts-starch agar, and glycerol-asparagine agar. Three observers selected a variety of color tabs as the nearest matching color including: 2ih, light olive; 1½li, olive gray; 1½ge, 1½ig, 1ig, light grayish olive; and 24½ih, dark greenish gray. Reverse side of colony with no distinctive pigments (yellowish brown to olive brown on yeast-malt agar; pale grayish yellow to light yellowish brown or grayish olive on oatmeal agar, salts-starch agar, and glycerol-asparagine agar).

Color in medium: melanoid pigments are not formed in peptone-yeast-iron agar, tyrosine agar, or tryptone-yeast broth in 2–4 d. No pigment is found in the medium in yeast-malt agar, oatmeal agar, salts-starch agar, or glycerol-asparagine agar.

D-Glucose, L-arabinose, D-xylose, iso-inositol, D-mannitol, D-fructose, and rhamnose are utilized for growth. No growth or only traces of growth with sucrose or raffinose.

Type strain shows no sequence similarity over 99%.

Source: not known.

DNA G+C content (mol%): not known.

Type strain: ATCC 27479, CBS 654.72, BCRC 11870, CCUG 37512, DSM 40243, NBRC 13353, IMET 43514, JCM 4859, KCTC 9145, NCIMB 9824, NRRL 2414, NRRL-ISP 5243, RIA 1314, VKM Ac-1769, VKM Ac-618.

Sequence accession no. (16S rRNA gene): DQ442556.

517. **Streptomyces vitaminophilus** corrig. (Shomura, Amano, Yoshida, Ezaki, Ito and Niida 1983) Goodfellow, Williams and Alderson 1986a, 575^{VP} (Effective publication: Goodfellow, Williams and Alderson 1986e, 63.) (*Actinosporangium vitaminophilum* Shomura, Amano, Yoshida, Ezaki, Ito and Niida 1983, 563)

vi.ta.mi.no'phi.lus. N.L. n. *vitaminum* vitamin; N.L. masc. adj. *philus* (from Gr. masc. adj. *philos*) friend, loving; N.L. masc. adj. *vitaminophilus* vitamin-loving.

Spores contained in "pseudosporangia"; the spore surface is smooth. Forms extensively branched substrate mycelium but scant aerial mycelium. The reverse side of colonies is colorless to pale tan or pale grayish yellow. Growth is often enhanced by vitamin B_{12}. Does not form melanin pigments. Gelatin and starch degraded; nitrate is reduced. D-Glucose, glycerol, L-rhamnose, and D-xylose are used as sole carbon sources but L-arabinose, D-fructose, *myo*-inositol, mannitol, raffinose, and sucrose are not. Grows at 15 and 45°C; optimal growth between 25 and 34°C. Wall peptidoglycan contains LL-A_2pm as the major diamino acid; the predominant isoprenologs are hexa- and octa-hydrogenated menaquinones with nine isoprene units. Produces antibiotics of the pyrrolomycin complex.

Type strain shows no sequence similarity over 99%.

Source: isolated from soil.

DNA G+C content (mol%): not known.

Type strain: SF 2080, ATCC 31673, DSM 41686, NBRC 14294, JCM 6054, NRRL B-16933.

Sequence accession no. (16S rRNA gene): AB184589.

Further comments: the original spelling, *Streptomyces vitaminophileus* (*sic*), has been corrected by the Associate Editor, IJSB 1986.

518. **Streptomyces wedmorensis** (*ex* Millard and Burr 1926) Preobrazhenskaya 1986, 575^{VP} (Effective publication: Preobrazhenskaya *in* Gause, Preobrazhenskaya, Sveshnikova, Terekhova and Maximova 1983.) ("*Actinomyces wedmorensis*" Millard and Burr 1926)

wed.mor.en'sis. N.L. masc. adj. *wedmorensis* of or pertaining to Wedmore, a city in England.

Spore chains are straight (*Rectiflexibiles*); spores are smooth. On mineral agar 1, oatmeal agar, and starch-ammonia agar: aerial mycelium is whitish gray to gray; substrate mycelium is colorless; no diffusible pigment. On glycerol-nitrate agar: aerial mycelium is poor, white; substrate mycelium yellow; no diffusible pigment. On organic agar 2: aerial mycelium is poor, white; substrate mycelium colorless; no diffusible pigment. On glycerol-asparagine agar: aerial mycelium is poor, white; substrate mycelium light gray-brown; no diffusible pigment. Melanoid pigments are not formed. Good growth on glucose, xylose, mannitol, and fructose; poor growth on arabinose and rhamnose. Antibiotic: phosphonomycin, antibiotic 280.

Type strain shows the highest sequence similarity to: *S. exfoliatus*, AB184324, 99.7%; *S. zaomyceticus*, EF178685, 99.7%; *S. venezuelae*, AB045890, 99.7%; *S. omiyaensis*, EF178697, 99.6%; *S. lateritius*, AB184324, 99.6%; *S. litmocidini*, AB184149, 99.5%; *S. yereyanensis* EF178684, 99.3%; *S. narbonensis*, DQ445794, 99.2.

Source: not known.

DNA G+C content (mol%): not known.

Type strain: ATCC 21239, BCRC 12667, CECT 3245, DSM 41676, ICMP 12544, NBRC 14062, JCM 4937, NRRL 3426, VKM Ac-1861.

Sequence accession no. (16S rRNA gene): DQ442557.

519. **Streptomyces werraensis** Wallhäusser, Huber, Nesemann, Präve and Zepf 1964, 357^{AL}

wer.ra.en'sis. N.L. masc. adj. *werraensis* of or belonging to River Werra, Germany (referring to "werramycin," the name originally assigned to the antibiotics produced).

Spore chains in Section *Spirales* or *Retinaculiaperti*. Imperfect spirals, flexuous chains, hooks, and loops are common; well-developed spirals are rare. Mature spore chains generally contain 10 to 50 or more spores per chain. This morphology is seen on yeast-malt agar, oatmeal agar, salts-starch agar, and glycerol-asparagine agar. Spore surface is spiny.

Color of colony: aerial mass color in the Gray color series (2dc, yellowish gray; 3fe, light brownish gray; g or e, medium gray) on yeast-malt agar, oatmeal agar, salts-starch agar, and glycerol-asparagine agar. Aerial mycelium is sometimes absent or poorly developed on oatmeal agar, salts-starch agar, and glycerol-asparagine agar. Reverse side of colony with no distinctive pigments (nearly colorless or pale yellowish gray on yeast-malt agar, oatmeal agar, salts-starch agar, and glycerol-asparagine agar).

Color in medium: melanoid pigments are not formed in peptone-yeast-iron agar, tyrosine agar, or tryptone-yeast broth. No pigment is found in the medium in yeast-malt agar, oatmeal agar, salts-starch agar, or glycerol-asparagine agar.

D-Glucose, L-arabinose, D-xylose, iso-inositol, D-mannitol, D-fructose, and rhamnose are utilized for growth. Reports vary on amount of growth on sucrose, but utilization is doubtful. No growth or only traces of growth with raffinose.

Type strain shows the highest sequence similarity to: *S. biverticillatus*, AJ781381, 100%; *S. albireticuli*, AB184881, 99.5%; *S. netropsis*, AB184848, 99.3%; *S. eurocidicus*, AY999790, 99.3%; *S. cinnamoneus*, AB184850, 99.1%; *S. hiroshimensis*, AB184789, 99%.

Source: not known.

DNA G+C content (mol%): not known.

Type strain: ATCC 14424, CBS 437.67, CBS 705.72, BCRC 12038, DSM 40486, NBRC 13404, JCM 4860, NRRL B-5317, NRRL-ISP 5486, RIA 1365.

Sequence accession no. (16S rRNA gene): DQ442558.

520. **Streptomyces willmorei** (Erikson 1935) Waksman and Henrici *in* Breed, Murray and Hitchens 1948, 966[AL] ["*Actinomyces willmorei*" Erikson 1935, 36; *Streptomyces microflavus* (Krainsky 1914) Waksman and Henrici 1948, 950 emend. Lanoot, Vancanneyt, Van Shoor, Liu and Swings 2005b, 731]

will.mo′re.i. N.L. gen. masc. n. *willmorei* of Willmore, named for J.G. Willmore, the surgeon who first isolated the organism.

Spore chains in Section *Rectiflexibiles.* Mature spore chains are short to moderately long, usually with more than 10 spores per chain. This morphology is seen on yeast-malt agar, oatmeal agar, salts-starch agar, and glycerol-asparagine agar. Spore surface is smooth.

Color of colony: aerial mass color in the Gray, Yellow, or White color series on yeast-malt agar, oatmeal agar, and salts-starch agar; White or Yellow color series on glycerol-asparagine agar. Nearest matching color tabs in the Gray color series are d, medium gray, and 2dc, yellowish gray. Nearest matching color tabs in the Yellow color series are 2ba and 2db, pale yellow. Reverse side of colony with no distinctive pigments (light olive brown or yellowish brown on yeast-malt agar; colorless to pale grayish yellow or light grayish olive on oatmeal agar, salts-starch agar, and glycerol-asparagine agar).

Color in medium: melanoid pigments are not formed (or only a very weak discoloration occurs) in peptone-yeast-iron agar, tyrosine agar, or tryptone-yeast broth. No pigment or only a trace of yellow is found in the medium in yeast-malt agar, oatmeal agar, salts-starch agar, and glycerol-asparagine agar.

D-Glucose, D-xylose, D-mannitol, D-fructose, and rhamnose are utilized for growth. Utilization of L-arabinose is doubtful. No growth or only traces of growth with iso-inositol, sucrose, and raffinose.

For sequence similarity, see type strain of *Streptomyces microflavus.*

Source: not known.

DNA G+C content (mol%): not known.

Type strain: ATCC 6867, CBS 372.64, CBS 692.72, BCRC 12640, DSM 40459, NBRC 13391, IMET 41387, JCM 4861, 21046, NCIMB 12984, NCTC 1856, NRRL B-1332, NRRL-ISP 5459, RIA 1352, VKM Ac-1867.

Sequence accession no. (16S rRNA gene): AB184374.

Further comments: according to Lanoot et al. (2005b), *Streptomyces willmorei* (Erikson 1935) Waksman and Henrici 1948 is a later heterotypic synonym of *Streptomyces microflavus* (Krainsky 1914) Waksman and Henrici 1948 emend. Lanoot et al. 2005b.

521. **Streptomyces xanthochromogenes** Arishima, Sakamoto and Sato 1956, 469[AL]

xan.tho.chro.mo′ge.nes. Gr. adj. *xanthos* yellow; Gr. n. *chroma* color; N.L. suff. *-genes* (from Gr. v. *gennaô* to produce) producing; N.L. part. adj. *xanthochromogenes* producing yellow color.

Spore chains in Section *Rectiflexibiles.* Mature spore chains generally have 10–50 spores per chain. This morphology is seen on yeast-malt agar, oatmeal agar, salts-starch agar, and glycerol-asparagine agar. Spore surface is smooth.

Color of colony: aerial mass color in the Yellow color series on yeast-malt agar, oatmeal agar, salts-starch agar, and glycerol-asparagine agar. Reverse side of colony with no distinct pigment (yellow, grayed yellow, or yellow-brown) on yeast-malt agar, oatmeal agar, salts-starch agar, or glycerol-asparagine agar; substrate pigment is not a pH indicator.

Color in medium: melanoid pigments are formed in peptone-yeast-iron agar, tyrosine agar, and tryptone-yeast extract broth. Yellow pigment may be found in oatmeal agar and salts-starch agar or may be absent. Yellow pigment is not pH-sensitive.

D-Glucose, D-xylose, D-mannitol, and D-fructose are utilized for growth. Only trace of growth indicating doubtful utilization of iso-inositol, rhamnose, and raffinose. Variable reports on growth with L-arabinose and sucrose.

Type strain shows the highest sequence similarity to: *S. michiganensis*, AB184153, 100%; *S. mauvecolor*, AB184532, 99.4%.

Source: not known.

DNA G+C content (mol%): not known.

Type strain: AS 4.1435, ATCC 19818, CBS 571.68, BCRC 11876, DSM 2015, DSM 40111, NBRC 12828, JCM 4215, JCM 4612, NRRL B-5410, NRRL-ISP 5111, RIA 1098, UNIQEM 207, VKM Ac-1071.

Sequence accession no. (16S rRNA gene): DQ442559.

522. **Streptomyces xanthocidicus** Nagatsu, Asahi and Suzuki in Asahi, Nagatsu and Suzuki 1966, 196[AL]

xan.tho.ci′di.cus. Gr. adj. *xanthos* yellow; L. v. *caedo* to kill; N.L. masc. adj. *xanthocidicus* pertaining to yellow and to cut, probably referring to the name given the antibiotic produced which, in turn, was probably derived from its activity against *Xanthomonas oryzae.*

Spore chains in Section *Rectiflexibiles.* Mature spore chains generally are long and straight with more than 50 spores per chain. This morphology is seen on yeast-malt agar, oatmeal agar, salts-starch agar, and glycerol-asparagine agar. Spore surface is smooth.

Color of colony: aerial mass color in the Gray color series (5fe, light grayish reddish brown; 3fe, light brownish gray; d-2fe, light to moderate gray) on yeast-malt agar, oatmeal agar, salts-starch agar, glycerol-asparagine agar, and Gause's medium no. 1. Reverse side of colony with no distinctive pigments (colorless to pale grayish yellow on glycerol-asparagine agar; grayish yellow on oatmeal agar and salts-starch agar; light yellowish brown to brown on yeast-malt agar and Gause's medium no. 1).

Color in medium: melanin reaction variable in peptone-yeast-iron agar, tyrosine agar, and tryptone-yeast broth; both weakly positive and negative reactions are recorded by different observers in 2–4 d. No pigment is found in the medium in yeast-malt agar, oatmeal agar, salts-starch agar, or glycerol-asparagine agar.

D-Glucose, L-arabinose, D-xylose, D-fructose, sucrose, and raffinose are utilized for growth. No growth or only traces of growth with iso-inositol, D-mannitol, and rhamnose.

Type strain shows the highest sequence similarity to: *S. purpureus*, AJ781324, 99.5%; *S. aburaviensis*, AY999779, 99.4,%; *S. chrysomallus* subsp. *fumigatus*, AB184645, 99.4%; *S. herbaricolor*, AB184801, 99.2%; *S. indigoferus*, AB184214, 99.2%; *S. aureofaciens*, AY207608, 99.1%; *S. psammoticus*, AY999862, 99%. Type strain shows the highest sequence similarity to following *Kitasatospora* species: *Kitasatospora kifunensis*, AB022874, 99.1%.

Source: not known.

DNA G+C content (mol%): not known.

Type strain: AS 4.1424, ATCC 27480, CBS 770.72, BCRC 11874, DSM 40575, NBRC 13469, JCM 4243, JCM 4862, NRRL B-12504, NRRL-ISP 5575, RIA 1430, VKM Ac-872.

Sequence accession no. (16S rRNA gene): AY999858.

523. **Streptomyces xantholiticus** (Konev and Tsyganov 1962) Pridham 1970, 31[AL] ["*Actinomyces xantholiticus*" (*sic*) Konev and Tsyganov 1962, 1026]

xan.tho.li'ti.cus. Gr. adj. *xanthos* yellow; N.L. masc. adj. *lyticus* (from Gr. masc. adj. *lutikos*), able to loosen, able to dissolve; N.L. masc. adj. *xantholiticus* (*sic*) yellow and soluble (referring to the yellow color of the vegetative mycelium and the tendency of the organism to lyse when maintained on some solid media).

Spore chains in Section *Spirales.* Sporulating aerial mycelium is poorly developed on yeast-malt agar, oatmeal agar, salts-starch agar, and glycerol-asparagine agar. Two observers report compact spirals on salts-starch agar; one observer reports spirals on yeast-malt agar and glycerol-asparagine agar. Spore surface is smooth.

Color of colony: aerial mass color in the Gray color series when adequate sporulating aerial mycelium is formed on salts-starch agar or glycerol-asparagine agar. Color of aerial mycelium cannot be determined on yeast-malt agar or oatmeal agar. Reverse side of colony is pale yellow green to greenish yellow on oatmeal agar, salts-starch agar, and glycerol-asparagine agar; moderate yellow to light olive brown on yeast-malt agar.

Color in medium: melanoid pigments are not formed in peptone-yeast-iron agar, tyrosine agar, or tryptone-yeast broth. Yellow pigment is found in the medium in glycerol-asparagine agar and sometimes in oatmeal agar or salts-starch agar. This pigment is not pH-sensitive when tested with 0.05 M NaOH or HCl.

D-Glucose, D-xylose, and D-fructose are utilized for growth. Utilization of iso-inositol is doubtful; no growth or only traces of growth with L-arabinose, D-mannitol, rhamnose, sucrose, and raffinose.

Type strain shows the highest sequence similarity to: *S. peucetius*, AB045887, 100%; *S. kurssanovii*, AB184325, 99.9%; *S. graminofaciens*, AJ781329, 99.8%.

Source: not known.

DNA G+C content (mol%): not known.

Type strain: ATCC 27481, CBS 655.72, BCRC 12646, DSM 40244, NBRC 13354, JCM 4282, JCM 4863, NCIMB 9857, NRRL B-12153, NRRL-ISP 5244, RIA 1315, VKM Ac-1872.

Sequence accession no. (16S rRNA gene): AB184349.

524. **Streptomyces xanthophaeus** Lindenbein 1952, 378[AL]

xan.tho.pha'e.us. N.L. masc. adj. *xanthophaeus* (from Gr. adj. *xanthopos* golden-looking) shining like gold.

Spore chains in Section *Rectiflexibiles.* Mature spore chains are very long, usually more than 50 spores per chain. This morphology is found on oatmeal agar, salts-starch agar, and glycerol-asparagine agar; poor sporulation on yeast-malt agar. Spore surface is smooth.

Color of colony: aerial mass color in the Red or Gray color series on oatmeal agar, salts-starch agar, and glycerol-asparagine agar. Nearest matching color tabs: grayish yellowish pink (5dc, Red series) to light grayish reddish brown (5fe to 4li, Gray color series). Reverse side of colony is colorless or characteristic grayed yellow to yellow-brown on yeast-malt agar, oatmeal agar, and glycerol-asparagine agar, but very dark reddish gray to near black on salts-starch agar. Reverse color is not a pH indicator.

Color in medium: melanoid pigments are formed in peptone-yeast-iron agar and tryptone-yeast extract broth. Pigments other than traces of yellow are not formed in yeast-malt agar, oatmeal agar, salts-starch agar, and glycerol-asparagine agar. D-Glucose is utilized for growth. No growth or only trace of growth on L-arabinose, sucrose, D-xylose, iso-inositol, D-mannitol, D-fructose, rhamnose, and raffinose.

Type strain shows the highest sequence similarity to: *S. nojiriensis*, AJ781355, 100%; *S. sporoaveus*, AJ781370, 100%; *S. vinaceus*, AB184394, 99.8%; *S. cirratus*, AY999794, 99.8%; *S. goshikiensis*, EF178693, 99.8%; *S. colombiensis*, DQ026646, 99.7%; *S. cinnamonensis*, AB184707, 99.7%; *S. avidinii*, AB184395, 99.7%; *S. sporoverrucosus*, DQ442544, 99.7%; *S. subrutilus*, X80825, 99.6%; *S. virginiae*, D85119, 99.5%.

Source: not known.

DNA G+C content (mol%): not known.

Type strain: ATCC 19819, CBS 572.68, BCRC 13756, DSM 40134, NBRC 12829, JCM 4426, KCTC 9144, KCTC 9220, NRRL B-5414, NRRL-ISP 5134, RIA 1099, UNIQEM 208, VKM Ac-1205, VKM Ac-1823.

Sequence accession no. (16S rRNA gene): DQ442560.

525. **Streptomyces yanglinensis** Xu, Wang, Cui, Huang, Liu, Zheng and Goodfellow 2006, 1114[VP]

yang.lin.en'sis. N.L. masc. adj. *yanglinensis* of or belonging to Yanglin, the source of the soil from which the type strain was isolated.

Neutrotolerant, acidophilic streptomycete that forms branched substrate and aerial hyphae. Smooth-surfaced spores are borne on flexuous spore chains. Gray-colored colonies that carry an abundant white to gray aerial spore mass are formed on acidified oatmeal agar; blackish colonies bearing an abundant white to gray aerial spore mass are formed on ISP 9 agar supplemented with fructose, mannose, or sucrose (each at 1%, w/v). Diffusible pigments are not produced, and melanin pigments are not formed on peptone-yeast extract-iron agar or tyrosine agar. Degrades starch and Tween 80, but not adenine, guanine, or xanthine. Adonitol, cellobiose, D-galactose, D-glucose, D-inulin, D-lactose, D-mannitol, D-salicin, and D-sorbitol (each at 1%, w/v) are used as sole carbon sources for energy and growth, but melezitose (at 1%, w/v), adipic acid, L-alanine, DL-aminobutyric acid, L-arginine, α-L-aspartic acid, L-cysteine, L-phenylalanine, L-valine, sodium acetate, sodium citrate, and sodium oxalate (each at 0.1%, w/v) are not. L-Glutamic acid is used as a sole carbon and nitrogen source, but L-alanine, L-arginine, L-aspartic acid, L-isoleucine, and L-phenylalanine (each at 0.1%, w/v) are not. Growth occurs at temperatures between 20 and 37°C, but not at 15°C, and at pH values between 4.5 and 7.0, but not at pH 3.5. Does not grow in the presence of 5% (w/v) NaCl. Sensitive to filter-paper discs soaked in the following (μg/ml unless indicated): cephalothin (30), doxycycline hydrochloride (30), erythromycin (15), josamycin (15), kanamycin sulfate (30), minocycline hydrochloride (30), neomycin sulfate (30), sulfamethoxazole (25), and tobramycin sulfate (10), but not to filter-paper discs soaked in acetylspiramycin (15), amoxycillin (10), ampicillin (10), azithromycin (30); azetreonam (30), carbenicillin (10), ciprofloxacin (5), ofloxacin (5), penicillin G (10 IU/ml), rifampin, streptomycin sulfate (10), or tetracycline hydrochloride (30).

Type strain shows no sequence similarity over 99%. Type strain shows DNA–DNA similarity to: *S. yeochonensis* NRRL B-24245^T, 21.0%; *S. guanduensis* JCM 13274^T, 23.1%; *S. paucisporeus* JCM 13276^T, 25.9%; *S. rubidus* JCM 13277^T, 23.8%.

Source: the type strain was isolated from a pine-forest soil in Yanglin, Yunnan Province, People's Republic of China.

DNA G+C content (mol%): 74.8.

Type strain: 1307, CGMCC 4.2023, JCM 13275.

Sequence accession no. (16S rRNA gene): AY876940.

526. **Streptomyces yanii** (Yan, Jiang and Zhang 1987) Liu, Shi, Zhang, Zhou, Lu, Li, Huang, Rodríguez and Goodfellow 2005b, 1609^{AL} ("*Microstreptospora cinerea*" Yan, Jiang and Zhang 1987)

ya'ni.i. N.L. gen. masc. n. *yanii* of Yan, named in honor of Yunchu Yan, a Chinese microbiologist.

Forms an extensively branched substrate mycelium and aerial hyphae. The substrate mycelium carries abundant spherical, smooth-surfaced spores (about 2.5 μm in diameter) singly or in short chains of 2–4 spores on short sporophores. Short chains of ellipsoidal smooth-surfaced spores (0.8 × 2.0 μm in diameter) are formed on aerial hyphae. The latter are usually formed only on inorganic media; an abundant gray aerial spore mass is apparent on oatmeal and Krass's No. 1 agars. Forms an abundant gray to black substrate mycelium on Czapek's, Krass's, and oatmeal agars. Gelatin and starch are degraded, but not cellulose or tyrosine. Nitrate is reduced. H_2S is not formed. Milk is coagulated but not peptonized. Good growth occurs at 18–35°C and pH 6–9. Whole-organism hydrolysates contain major proportions of LL-A_2pm, glycine, galactose, and mannose, and a trace of ribose. The muramic acid of the peptidoglycan is *N*-acetylated. The predominant isoprenologs are tetra- and hexa-hydrogenated menaquinones with nine isoprene units and the major phospholipids are diphosphatidylglycerol and phosphatidylinositol mannosides.

Type strain shows the highest sequence similarity to: *S. atratus*, DQ026638, 99.8%; *S. sanglieri*, AB249945, 99.7%; *S. cinereorectus*, AB184646, 99.5%; *S. pulveraceus*, AB184806, 99.5%; *S. griseinus*, AB184205, 99.4%; *S. pluricolorescens*, DQ442540, 99.4%; *S. alboviridis*, AB184256, 99.4%; *S. mediolani*, AB184674, 99.4%; *S. griseoplanus*, AY999894, 99.4%; *S. sindenensis*, AB184759, 99.4%; *S. rubiginosohelvolus*, AB184240, 99.4%; *S. gelaticus*, DQ026636, 99.4%; *S. fimicarius*, AY999784, 99.4%; *S. globisporus* subsp. *globisporus*, EF178686, 99.4%; *S. fulvorobeus*, AB184711, 99.4%; *S. lipmanii*, AB184148, 99.4%; *S. badius*, AY999783, 99.4%; *S. microflavus*, DQ445795, 99.4%; *S. argenteolus*, AB045872, 99.4%; *S. griseolus*, AB184768, 99.4%; *S. acrimycini*, AY999889, 99.4%; *S. lavendulae* subsp. *lavendulae*, AB184080, 99.4%; *S. flavofuscus*, AB249935, 99.4%; *S. anulatus*, DQ026637, 99.4%; *S. griseus* subsp. *griseus*, AY207604, 99.4%; *S. cavourensis* subsp. *washingtonensis*, DQ026671, 99.4%; *S. praecox*, AB184293, 99.4%; *S. cyaneofuscatus*, AB184860, 99.4%; *S. albovinaceus*, AB249958, 99.3%; *S. flavovirens*, DQ026635, 99.3%; *S. parvus*, DQ442537, 99.3%; *S. halstedii*, EF178695, 99.3%; *S. baarnensis*, EF178688, 99.3%; *S. luridiscabiei*, AF361784, 99.2%; *S. californicus*, AB184755, 99.2%; *S. floridae*, AB184656, 99.2%; *S. flavogriseus*, AJ494864, 99.2%; *S. nitrosporeus*, EF178680, 99%; *S. bacillaris*, AB184439, 99%. Type strain shows DNA–DNA similarity to: *S. argenteolus* AS 4.1693^T, 13%; *S. caviscabies* DSM 41811^T, 30%; *S. griseus* AS 4.1419^T, 31%; *S. laceyi* AS 4.1832^T, 14%; *S. sanglieri* AS 4.1831^T, 33%; *S. setonii* AS 4.1774^T, 25%; *S. peucetius* AS 4.1799^T, 14%; *S. purpureus* AS 4.1225^T, 11%; *S. venezuelae* AS 4.1307^T, 16%.

Source: the sole strain was isolated from a mud sample collected from a sewage ditch in Zhanjiang City, Guangdong Province, China.

DNA G+C content (mol%): 69.4.

Type strain: 80-133, AS 4.1146, NBRC 14669, JCM 3331.

Sequence accession no. (16S rRNA gene): AB006159.

Further comments: Streptomyces yanii Liu et al. 2005b was previously known as "*Microstreptospora cinerea*" Yan et al. 1987.

527. **Streptomyces yatensis** Saintpierre, Amir, Pineau, Sembiring and Goodfellow 2003b, 1219^{VP} (Effective publication: Saintpierre, Amir, Pineau, Sembiring and Goodfellow 2003a, 25.)

ya.ten'sis. N.L. masc. adj. *yatensis* pertaining to the Yaté region of New Caledonia, the source of the isolate.

Forms an extensively branched substrate mycelium which carries aerial hyphae that differentiate into spiral spore chains with three or four turns; there are 15–20 cylindrical to barrel-shaped spores per chain. Spore surface is rugose. On oatmeal agar, the spore mass is initially gray, then black; the substrate mycelium is grayish-yellow. Melanin pigments are not produced. Growth occurs at 20, 30, and 37°C, but not at 10°C or 45°C. It also occurs in the presence of ampicillin (10 μg/ml), lysozyme, penicillin (10), and sodium chloride (3%, w/v) but not in the presence of carbenicillin (100), chloramphenicol (30), gentamicin sulfate (10), neomycin sulfate (30), streptomycin sulfate (30), tetracycline hydrochloride (30), or sodium chloride (5% w/v). The organism shows activity against some of the microbial strains isolated from clinical material (the *Bacillus*, *Corynebacterium*, *Candida albicans*, *Klebsiella pneumoniae*, and *Staphylococcus aureus* strains) and against the phytopathogen (*Fusarium oxysporum*) but not against strains of *Erwinia*, *Escherichia coli*, *Pseudomonas aeruginosa*, or *Staphylococcus epidermidis*. 70% toxicity is shown against Kb cells using a concentration of 10 μg/ml of a crude extract.

Type strain shows no sequence similarity over 99%.

Source: the organism was isolated from a ferraltic soil collected at the southern end of New Caledonia.

DNA G+C content (mol%): not known.

Type strain: SFOCin 76, DSM 41771, JCM 13244, NBRC 101000, NRRL B-24116.

Sequence accession no. (16S rRNA gene): AB249962.

528. **Streptomyces yeochonensis** Kim, Seon, Jeon, Bae and Goodfellow 2004, 213[VP]

ye.o.chon.en'sis. N.L. masc. adj. *yeochonensis* of or belonging to Yeochon, a province in Korea, referring to the place where the organism was first isolated.

Aerobic, Gram-stain-positive, nonmotile, neutrotolerant acidophilic streptomycete that forms extensively branched substrate and aerial mycelia. Smooth-surfaced spores are borne in flexuous spore-chains. Aerial spore mass color is gray. Substrate mycelia have no distinctive color; diffusible pigments are not produced. pH range for growth is 4.3–7.3. Casein, gelatin, guanine, starch, and Tween 80 are degraded, but elastin, hypoxanthine, testosterone, Tween 20, tyrosine, and xanthine are not. Good growth occurs between 25 and 37°C, but not at 12 or 45°C. The sugars erythritol, inulin, melezitose, salicin, ribitol, and sorbitol (all at 1%, w/v) are used as sole carbon sources, as are β-hydroxybutyric acid, D-gluconic acid, hippuric acid, α-ketoglutaric acid, 2-keto-D-glucuronic acid, lactic acid, malic acid, malonic acid, oxalic acid, pyruvic acid, and succinic acid (as sodium salts) (all at 0.1%, w/v).

Type strain shows no sequence similarity over 99%.

Source: the type strain was isolated from acidic soil collected in the Yeochon area of the Republic of Korea.

DNA G+C content (mol%): not known.

Type strain: CN732, JCM 12366, KCTC 9926, IMSNU 50114, NBRC 100782, NRRL B-24245.

Sequence accession no. (16S rRNA gene): AF101415.

529. **Streptomyces yerevanensis** (Tsyganov, Zhukova and Timofeeva 1964) Goodfellow, Williams and Alderson 1986a, 575[VP] (Effective publication: Goodfellow, Williams and Alderson 1986a, 52.) ["*Macrospora violaceus*" (*sic*) Tsyganov, Zhukova and Timofeeva 1964, 868; *Microellobosporia violacea* Pridham *in* Buchanan and Gibbons 1974, 844]

ye.re.van.en'sis. N.L. masc. adj. *yerevanensis* of or pertaining to Yerevan, Armenia.

Forms extensively branched substrate and aerial mycelium. Short chains (2–8) of smooth-surfaced spores borne on both the substrate and aerial mycelium. Aerial spore mass is light mouse gray; reverse color is violet and the pigment is pH-sensitive. Violet diffusible pigment, which is also pH-sensitive, is produced. Does not form melanin pigments. Adenine, esculin, arbutin, casein, hypoxanthine, starch, testosterone, and tyrosine are degraded but allantoin, chitin, guanine, lecithin, pectin, urea, xanthine, and xylan are not. Adonitol, L-arabinose, cellobiose, D-fructose, D-glucose, *myo*-inositol, inulin, D-lactose, mannitol, D-mannose, melibiose, melezitose, raffinose, L-rhamnose, salicin, sucrose, and D-xylose are used as sole carbon sources but D-galactose, maltose, trehalose, and xylitol are not. Grows on L-arginine, L-histidine, potassium nitrate, L-serine, L-threonine, and L-valine, but not on DL-α-amino-n-butyric acid, L-cysteine, L-hydroxyproline, L-methionine, or L-phenylalanine, as sole nitrogen source. Growth occurs at 10 and 37°C, but not at 4 or 45°C. Tolerant to phenol (0.1%, w/v) and sodium chloride (4%, w/v), but not to sodium azide (0.01%, w/v). Resistant to rifampin but sensitive to sodium chloride (7%, w/v). Does not show antimicrobial activity against *Aspergillus niger* LIV 131, *Bacillus subtilis* NCIB 3610, *Candida albicans* CBS 562, *Escherichia coli* NCIB 9132, *Micrococcus luteus* NCIB 196, *Pseudomonas fluorescens* NCIB 9046[T], *Saccharomyces cerevisiae* CBS 1171[T], or *Streptomyces murinus* ISP 5091. The peptidoglycan contains LL-A_2pm as the major diamino acid. The organism has a type II phospholipid pattern (Lechevalier et al., 1977) and contains major amounts of tetra-, hexa-, and octahydrogenated menaquinones with nine isoprene units (Alderson et al., 1985; Collins et al., 1984). Produces violacin, a pigmented antibiotic of the rhodomycin-cinerubin-mycetin type.

Type strain shows the highest sequence similarity to: *S. litmocidini*, AB184149, 99.8%; *S. exfoliatus*, AB184324, 99.4%; *S. venezuelae*, AB045890, 99.4%; *S. zaomyceticus*, EF178685, 99.4%; *S. lateritius*, AB184324, 99.3%; *S. wedmorensis*, DQ442557, 99.3%; *S. omiyaensis*, EF178697, 99.3%.

Source: isolated from soil, Armenia, USSR.

DNA G+C content (mol%): 68.0–70.0.

Type strain: AS 4.1464, ATCC 43727, BCRC 11564, DSM 43167, IFM 1151, IFM 1242, NBRC 12517, IMET 43616, JCM 3047, JCM 3065, KCC A-0047, LMG 19363, NCIB (now NCIMB) 9589, NRRL B-16943, RIA 795, VKM Ac-1234.

Sequence accession no. (16S rRNA gene): EF178684.

Further comments: for the transfer of *Microellobosporia violacea* (Tsyganov et al. 1964) Pridham 1974 to the genus *Streptomyces* Waksman and Henrici 1943 it is necessary to substitute a new specific epithet to produce *Streptomyces yerevanensis* because there is a senior homonym, *Streptomyces violaceus* (Rossi Doria 1891) Waksman 1953, included on the Approved Lists of Bacterial Names [Rules 34a and 41a of the *Bacteriological Code* (1990 Revision)].

530. **Streptomyces yogyakartensis** Sembiring, Ward and Goodfellow 2001, 1619[VP] (Effective publication: Sembiring, Ward and Goodfellow 2000, 363.)

yog.ya.kar.ten'sis. N.L. masc. adj. *yogyakartensis* of or pertaining to Yogyakarta, Indonesia.

Spore chains are *Spirales*; spore surface is rugose. On oatmeal agar the spore mass is gray, the substrate mycelium is grayish-yellow, and the diffusible pigment is yellow. Melanin pigments are not produced. The organism grows at 45°C, degrades adenine but does not reduce nitrate.

Type strain shows the highest sequence similarity to: *S. violaceusniger*, AJ391823, 100%; *S. albiflaviniger*, AJ391812, 99.7%; *S. endus*, AY999911, 99.5%; *S. sporocinereus*, AB249933, 99.5%; *S. demainii*, DQ334782, 99.5%; *S. hygroscopius* subsp. *hygroscopius*, AB184428, 99.5%; *S. javensis*, AJ391833, 99.2%. Type strain shows DNA–DNA similarity to: *S. albiflaviniger* NRRL B-1356^T, 98.9%; *S. geldanamycinus* NRRL 3602^T, 98.7%; *S. griseiniger* NRRL B1865^T, 98.3%; *S. rhizosphaericus* DSM 41760^T, 97.9%; *S. asiaticus* DSM 41761^T, 98.1%; *S. indonesiensis* DSM 41759^T, 98.3%; *S. javensis* DSM 41764^T, 98.6%; *S. cangkringensis* DSM 41769^T, 98.2%.

Source: isolated from non-rhizosphere soil adjacent to a stand of the tropical legume, *Paraserianthes falcataria*.

DNA G+C content (mol%): not known.

Type strain: C4R3, DSM 41766, JCM 11448, NBRC 100779, NCIMB 13681.

Sequence accession no. (16S rRNA gene): AB249942.

531. **Streptomyces yokosukanensis** Nakamura 1961, 94^{AL}

yo.ko.su'ka.nen'sis. N.L. masc. adj. *yokosukanensis* (*sic*) of or belonging to Yokosuka City, Kanagawa Prefecture, Japan, the source of the soil from which the organism was isolated.

Spore chains in Section *Spirales* or *Retinaculiaperti*. Tight spirals are common at the ends of long spore chains. Open spirals, hooks, and flexuous chains are also found. Spore surface is spiny according to one observer; another observer finds only smooth spores.

Color of colony: aerial mass color in the Red color series (grayish yellowish pink) on yeast-malt agar, oatmeal agar, salts-starch agar, and glycerol-asparagine agar. Reverse side of colony with no distinctive pigments (grayish yellow, light yellowish brown, olive-brown, or strong brown) on yeast-malt agar, oatmeal agar, salts-starch agar, and glycerol-asparagine agar.

Color in medium: melanoid pigments are formed in peptone-yeast-iron agar, tyrosine agar, and tryptone-yeast broth. No pigment or only a trace of yellow is found in medium in yeast-malt agar, oatmeal agar, salts-starch agar, and glycerol-asparagine agar.

D-Glucose, L-arabinose, D-xylose, iso-inositol, D-mannitol, D-fructose, rhamnose, and raffinose are utilized for growth. No growth or only trace of growth with sucrose.

Type strain shows the highest sequence similarity to: *S. cellostaticus*, AB184192, 99.5%; *S griseochromogenes*, AB184387, 99.3%.

Source: isolated from soil from Yokosuka City, Kanagawa Prefecture, Japan.

DNA G+C content (mol%): not known.

Type strain: ATCC 25520, CBS 662.69, BCRC 11875, DSM 40224, NBRC 13108, JCM 4137, JCM 4559, NRRL B-3353, NRRL-ISP 5224, RIA 1300, VKM Ac-1713.

Sequence accession no. (16S rRNA gene): DQ026652.

532. **Streptomyces yunnanensis** Zhang, Li, Cui, Li, Xu and Jiang 2003, 220^{VP}

yun.nan.en'sis. N.L. masc. adj. *yunnanensis* of or pertaining to Yunnan, a province of south-west China.

Spore chains with many spores are spiral. The spores are rugose with short spines and are short pillar-shaped (0.5–1.0 µm in diameter) and nonmotile. Vegetative and aerial hyphae are abundant and well-developed. Diffusible pigments are not produced and melanin is not produced. Milk is not coagulated but peptonized, starch is hydrolyzed, and H_2S is not produced. Nitrate is not reduced and gelatin is liquefied. Does not hydrolyze cellulose. Utilizes glucose, fructose, rhamnose, inositol, mannitol, arabinose, and raffinose for growth; does not utilize sucrose or xylose. It has antimicrobial activity against *Aspergillus niger* but not against *Bacillus subtilis*. Optimum growth is at 28°C. The cell wall contains LL-A_2pm and glycine (cell-wall chemotype I). Whole-cell hydrolysates contain galactose.

Type strain shows the highest sequence similarity to: *S. albulus*, AB024440, 99.7%; *S. noursei*, AB184678, 99.7%.

Source: the type strain was isolated from red soil of suburb of Kunming in Yunnan, China.

DNA G+C content (mol%): not known.

Type strain: YIM 41004, CGMCC 4.1004, DSM 41793, JCM 12115, NBRC 100781.

Sequence accession no. (16S rRNA gene): AF346818.

533. **Streptomyces zaomyceticus** Hinuma 1954, 134^{AL}

za.o.my.ce'ti.cus. English n. N.L. n. *zaomycinum* zaomycin, an antibiotic named after the Mount Zao, Japan, the source of the soil from which the organism was isolated; N.L. masc. suff. *-icus* suffix used with the sense of belonging to; N.L. masc. adj. *zaomyceticus* belonging to zaomycin.

Spore chains in Section *Rectiflexibiles*. Mature spore chains generally are long, often with more than 50 spores per chain. Long chains are flexuous, but true spirals are not formed. This morphology is seen on yeast-malt agar, oatmeal agar, salts-starch agar, and glycerol-asparagine agar but not on salts-starch agar. Spore surface is smooth.

Color of colony: aerial mass color in the Gray or White color series on yeast-malt agar, oatmeal agar, and glycerol-asparagine agar. A good aerial mycelium is not produced on salts-starch agar. Nearest matching color tabs in the Gray color series are 2dc, yellowish gray, and d, light gray. Reverse side of colony with no distinctive pigments (olive brown to yellowish brown on yeast-malt agar; light grayish yellow to olive brown on oatmeal agar and salts-starch agar; grayish yellow on glycerol-asparagine agar).

Color in medium: melanoid pigments are formed in peptone-yeast-iron agar, tyrosine agar, and tryptone-yeast broth, but the reaction is weak in tyrosine agar. No pigment other than a melanoid pigment is found in the medium in yeast-malt agar, oatmeal agar, salts-starch agar, and glycerol-asparagine agar.

D-Glucose, L-arabinose, and D-xylose are utilized for growth. No growth or only traces of growth with iso-inositol, D-mannitol, D-fructose, rhamnose, sucrose, and raffinose.

Type strain shows the highest sequence similarity to: *S. exfoliatus*, AB184324, 99.9%; *S. venezuelae*, AB045890, 99.9%; *S. wedmorensis*, DQ442557, 99.7%; *S. lateritius*, AB184324, 99.7%; *S. omiyaensis*, EF178697, 99.6%; *S. litmocidini*, AB184149, 99.6%; *S. yereyanensis* EF178684, 99.4%; *S. narbonensis*, DQ445794, 99.3%.

Source: not known.

DNA G+C content (mol%): not known.

Type strain: ATCC 27482, CBS 649.72, BCRC 12317, DSM 40196, NBRC 13348, JCM 4179, JCM 4864, NCIMB 9850, NRRL B-2038, NRRL-ISP 5196, RIA 1309, VKM Ac-1192.

Sequence accession no. (16S rRNA gene): EF178685.

Genus *incertae sedis* I. **Kitasatospora** corrig. Ōmura, Takahashi, Iwai and Tanaka 1983, 672[VP] (Effective publication: Ōmura Takahashi, Iwai and Tanaka 1982, 1014.) emend. Zhang, Wang and Ruan 1997, 1053

Peter Kämpfer

Ki.ta.sa.to.spo′ra. N.L. fem. n. *Kitasatoa* named for Shibasaburo Kitasato (1852–1931), a Japanese bacteriologist; N.L. fem. n. *spora* a seed and, in biology, a spore; N.L. fem. n. *Kitasatospora* Kitasato spore.

Aerobic. Gram-stain-positive, non-acid–alcohol-fast actinomycetes which form an intensively branched, non-fragmenting mycelium. A stable substrate mycelium is as well-developed as that of streptomycetes, and the aerial mycelium bears long spore chains of more than 20 spores. Does not form sporangia. The major constituents of the cell wall are glycine, galactose, and *meso*-A_2pm or LL-A_2pm, depending on the type of cells analyzed. When cells are grown on agar media, aerial spores contain LL-A_2pm, whereas the substrate mycelium contains *meso*-A_2pm. When cells are grown in liquid media, submerged spores which contain LL-A_2pm are formed while the filamentous mycelia contain *meso*-A_2pm. Whole-cell hydrolysates contain galactose, but lack arabinose, madurose, and xylose. Hexa- and octa-hydrogenated menaquinones with nine isoprene units are the predominant isoprenologs, cells are rich in saturated, iso- and anteiso-fatty acids, and have complex polar lipid patterns which contain diphosphatidylglycerol, phosphatidylethanolamine, phosphatidylinositol and phosphatidylinositol mannosides. The glycolate test is negative. The organism is chemo-organotrophic, and grows from 15 to 42°C, and within the pH range 5.5 to 9.0.

Source: not known.

DNA G+C content (mol%): 66–77.

Type species: **Kitasatospora setae** corrig. Ōmura, Takahashi, Iwai and Tanaka 1983, 672 (Effective publication: Ōmura, Takahashi, Iwai and Tanaka 1982, 1014).

Further descriptive information

The genus *Kitasatosporia* (*sic*) was proposed by Ōmura et al. (1982) for actinomycetes that were phenotypically very similar to members of the genus *Streptomyces*, but which contained major amounts of the *meso*-isomer of A_2pm and galactose in their cell walls. Zhang et al. (1997), who revived the genus, corrected the name to *Kitasatospora* albeit without a formal proposal. Some confusion has arisen as to whether the genus name *Kitasatospora* is illegitimate, because no formal proposal was made for the transfer of *Streptomyces setae* to *Kitasatospora setae*. This also has the effect of automatically changing the spelling of all species names validly published under *Kitasatosporia* (*sic*). The problem was handled by the Judicial Commission of the ICSP (De Vos et al., 2005), with the result that no further action is required by the Judicial Commission as all names listed under the genus *Kitasatospora* were validly published in accordance with the Rules of the *Bacteriological Code* (1990 Revision).

The distribution of the two isomers of A_2pm in cells of *Kitasatospora* grown at different stages of differentiation was studied by Takahashi et al. (1983), who found that *Kitasatospora* species contained LL-A_2pm in the aerial mycelium and *meso*-A_2pm in the vegetative mycelium, while *Streptomyces* species contained only LL-A_2pm in both aerial and vegetative mycelia. The relative amounts of the two isomers varied in different experiments, and some *Streptomyces* species were also found to contain various amounts of *meso*-A_2pm, though *Kitasatospora* species consistently had a much higher ratio of *meso*-A_2pm to LL-A_2pm than *Streptomyces* species (Wellington et al., 1992). Strains assigned to the genus *Kitasatospora* were, unlike most *Streptomyces* species, resistant to polyvalent *Streptomyces* phages (Wellington et al., 1992) and formed submerged spores in liquid culture (Ōmura et al., 1982; Takahashi et al., 1983). Wellington et al. (1992) also found that the 16S rRNA gene sequence of *Kitasatospora setae* showed 91.6% similarity to the 16S rRNA gene sequence of *Streptomyces baldaccii*, and that a *Streptomyces*-specific oligonucleotide probe could recognize members of all four valid *Kitasatospora* species. These observations, and the fact that many phenotypic properties are shared by *Kitasatospora* and *Streptomyces* species, led Wellington and her colleagues to propose that all *Kitasatospora* species should be reclassified as *Streptomyces* species. These findings were supported by Ochi and Hiranuma (1994) on the basis of results from an analysis of the N-terminal sequences of ribosomal protein AT-L30.

In contrast, Nakagaito et al. (1992a) had assigned *Kitasatospora* and the *Streptomyces* species to distinct clusters based on DNA–DNA reassociation and phenetic studies. In addition, Kim et al. (1996) conducted a 16S rRNA gene sequence-based analysis of members of many *Streptomyces* species and found that the three *Kitasatospora* species formed a distant lineage in the "*Streptomyces*" tree. Zhang et al. (1997) determined the nucleotide sequences of 16S rRNA genes and the 16S–23S rRNA gene spacers of 12 actinomycete strains which were either previously classified as *Kitasatospora* strains or defined as *Streptomyces* strains, but had been found to contain major amounts of *meso*-A_2pm in their whole-cell hydrolysates. On the basis of their results, Zhang and his colleagues revived the genus *Kitasatospora* Ōmura et al. (1982).

The genus *Kitasatospora* can be distinguished from the genus *Streptomyces* by the ratio of *meso*-A_2pm to LL-A_2pm in whole-cell hydrolysates. The *meso*-A_2pm content is 49–89% in *Kitasatospora* strains and 1–16% in *Streptomyces* strains. Galactose is present in whole-cell hydrolysates of *Kitasatospora* strains, but not in those of *Streptomyces* strains. At the time of writing, the genus encompassed twelve species with validly published names. Representatives of each of these species with validly published names form a distinct line in the 16S rRNA *Streptomycetaceae* gene tree (Figure 339). On the basis of 16S rRNA gene sequence data, all the species are grouped together, showing high similarities. They are very similar to members of the genera *Streptomyces* and *Streptacidiphilus*, i.e. they produce aerial mycelia with long chains of spores, contain hexa- and octa-hydrogenated menaquinones with nine isoprene units as predominant isoprenologs, are rich in saturated, iso- and anteiso-fatty acids, and have complex polar lipid patterns which contain major amounts of diphosphatidylglycerol, phosphatidylethanolamine, phosphatidylinositol, and phosphatidylinositol mannosides. However, *Kitasatospora* and *Streptomyces* are clearly sister taxa as they share so many phenotypic properties.

Enrichment and isolation procedures

Enrichment and isolation procedures are as described for *Streptomyces* species.

Maintenance procedures

Maintenance procedures are as for *Streptomyces* species.

Differentiation of the genus *Kitasatospora* from other genera

Despite some differences, members of the family *Streptomycetaceae* are morphologically and chemically homogeneous (Table 263). Identification of most species is difficult, because in many instances only one (the type) or a few strains have been examined. *Kitasatospora* species may be distinguished from other genera in the family by slight differences in cell-wall sugars and the presence of *meso*-A_2pm in the peptidoglycan (Table 263). Table 273 lists some characters which can be used to distinguish between *Kitasatospora* species.

TABLE 273. Phenotypic properties which separate *Kitasatospora* species (modified from Groth et al., 2004)[a,b]

Test	*K. arboriphila* HKI 0189^{T}	*K. azatica* DSM 41650^{T}	*K. cystarginea* DSM 41680^{T}	*K. gansuensis* sp. nov. HKI 0314^{T}	*K. kifunensis* DSM 41654^{T}	*K. kifunensis* HKI 0316	*K. mediocidica* DSM 43929^{T}	*K. nipponensis* sp. nov. HKI 0315^{T}	*K. paranensis* sp. nov. HKI 0190^{T}	*K. phosalacinea* DSM 43860^{T}	*K. putterlickiae* DSM 44665^{T}	*K. terrestris* sp. nov. HKI 0186^{T}
Spore chain morphology[c]	RF, RA, S	RF	S[d]	RF	RF, S	RF, S	RF, RA	RF, RA, S	RF	RF	RF	RF, RA, S
Formation of melanoid pigment	+	–	–[d]	+	+	+	+	–	+	–	+	+
Degradation of casein	+	+	nd	+	+	–	–	+	+	+	+	+
Liquefaction of gelatin	+	+	–[d]	+	–	–	–	+	+	+	+	+
Hydrolysis of potato starch	+	+	+[d]	+	+	+	+	+	+	+	–	+
Peptonization of milk	+	+	+[d]	+	–	–	–	+	+	+	+	+
Nitrate reduction	+	+	–[d]	+	–	–	–	–	+	+	+	+
Growth on sole carbon sources:												
(+)-L-Arabinose	+	+	±[d]	+	+	–	+	–	+	+[e]	–	+
(+)-D-Fructose	+	+	±[d]	+	–	+	+	(+)	–	+[e]	(+)	+
iso-Inositol	–	–	+[d]	–	–	–	–	–	–	–[e]	–	–
(+)-D-Mannitol	–	–	–[d]	–	+	+	–	–	–	–[e]	–	–
(+)-D-Raffinose	+	–	–[d]	–	–	–	–	–	–	+[e]	–	+
(+)-L-Rhamnose	–	–	–[d]	–	–	–	–	–	+	+[e]	–	(+)
(–)-D-Sucrose	–	+	–[d]	+	+	+	+	(+)	–	+[e]	(+)	(+)
(+)-D-xylose	+	+	–[d]	+	+	–	+	–	+	+[e]	–	+
Enzyme assay (API ZYM):												
N-Acetyl-β-glucosamidase	–	–	–	(+)	–	–	–	+	–	–	–	–
β-Galactosidase	(+)	+	+	+	+	+	–	+	+	+	+	+
α-Glucosidase	+	(+)	+	+	+	(+)	–	–	+	+	–	(+)
β-Glucosidase	–	–	–	+	–	–	–	–	–	–	+	–
α-Mannosidase	–	–	–	–	(+)	+	–	–	–	–	–	–
Naphthol-AS-BI-phosphohydrolase	+	+	+	+	+	+	+	+	+	+	–	+
Growth at:												
6°C	–	–	–[d]	(+)	–	–	(+)	–	–	–	–	–
10°C	–	(+)	–[d]	+	(+)	+	+	+	(+)	+	(+)	–
35°C	+	(+)	+[d]	–	–	+	–	(+)	+	+	+	+
37°C	+	–	+[d]	–	–	+	–	–	+	+	+	+
40°C	(+)	–	+[d]	–	–	–	–	–	–	+	–	+
42°C	–	–	–[d]	–	–	–	–	–	–	+[e]	–	–
Growth in the presence of NaCl (%):												
2.0	+	–	+	+	+	+	+	+	+	+	+	+
2.5	–	–	+	+	+	+	+	(+)	+	+	+	+
3.0	–	–	–	–	–	+	–	–	+	–	+	(+)
3.5	–	–	–	–	–	+	–	–	+	–	+	–

(continued)

TABLE 273. (continued)

Test	*K. arboriphila* HKI 0189[T]	*K. azatica* DSM 41650[T]	*K. cystarginea* DSM 41680[T]	*K. gansuensis* sp. nov. HKI 0314[T]	*K. kifunensis* DSM 41654[T]	*K. kifunensis* HKI 0316	*K. mediocidica* DSM 43929[T]	*K. nipponensis* sp. nov. HKI 0315[T]	*K. paranensis* sp. nov. HKI 0190[T]	*K. phosalacinea* DSM 43860[T]	*K. putterlickiae* DSM 44665[T]	*K. terrestris* sp. nov. HKI 0186[T]
Growth at pH:												
8.0	+	+	+	+	+	+	−	+	+	+	+	+
9.0	−	−	+	+	+	−	−	−	+	+	+	+
9.5	−	−	−	+	−	−	−	−	−	+	+	−
Antibiotic susceptibility:												
Ampicillin (10 μg)	−	+	nd	(+)	(+)	−	−	−	(+)	+	+	−
Lincomycin hydrochloride (2 μg)	−	−	nd	−	−	−	(+)	−	(+)	−	−	−
Methicillin (5 μg)	−	+	nd	−	(+)	−	−	(+)	(+)	−	(+)	−
Norfloxacin (10 μg)	−	(+)	nd	−	−	+	(+)	−	−	−	−	(+)
Novobiocin (5 μg/ml)	+	+	nd	+	+	+	+	+	+	−	+	+
Penicillin G (10 IU)	−	+	nd	−	−	−	−	−	−	+	+	−
Polymyxin B (300 IU)	+	+	nd	−	−	−	(+)	(+)	+	(+)	(+)	+
Sulfonamide (200 μg)	−	+	nd	+	(+)	−	−	(+)	(+)	−	−	−

[a]+, Positive; −, negative; (+), weakly positive; ±, doubtful; nd, not determined.

[b]For the following properties, tests in which strain DSM 41680[T] was not included are indicated by an asterisk. Spores of all of the tested strains are cylindrical with smooth surface. All strains were positive for the production of H_2S*, growth on (+)-D-glucose, and produced acid phosphatase, alkaline phosphatase, esterase (C4), esterase lipase (C8), and leucine arylamidase (API ZYM tests). Good growth occurred at temperatures of 15–32°C and pH 5.0–7.0. All strains were sensitive to chloramphenicol (30 μg)*, ciprofloxacin (5 μg)*, imipenem (10 μg)*, kanamycin sulfate (30 μg)*, nalidixic acid (50 μg/ml agar)*, oxytetracycline (30 μg)*, rifampin (30 μg)*, streptomycin sulfate (10 μg)*, and vancomycin (30 μg)*. They did not use cellulose* as a sole carbon source; did not produce α-chymotrypsin, cystine arylamidase, α-galactosidase, β-glucuronidase, α-fucosidase, lipase (C14), trypsin, or valine arylamidase (API ZYM tests); and did not grow in the presence of NaCl (4%, w/v), at 42°C, or at pH 4.0 or pH 10.

[c]RF, *Rectiflexibiles*; RA, *Retinaculiaperti*; S, *Spirales*.

[d]Data from Kusakabe and Isono (1988).

[e]Data from Takahashi et al. (1984).

List of species of the genus *Kitasatospora*

1. **Kitasatospora setae** corrig. Ōmura, Takahashi, Iwai and Tanaka 1983, 673[VP] (Effective publication: Ōmura, Takahashi, Iwai and Tanaka 1982, 1014.)

se′ta.e. N.L. gen. n. *setae* of Seta, Japan, where the bacterium was isolated.

Spore chain morphology *Rectiflexibiles* type with a smooth spore surface. Color of vegetative mycelia is pale yellow to light ivory on yeast extract-malt extract and glycerol-asparagine agars (Shirling and Gottlieb, 1966). Aerial mass color is white or light gray on yeast extract-malt extract and inorganic salts-starch agars. Forms yellow diffusible pigment on inorganic salts-starch agar. Nitrate is not reduced, milk is coagulated. Raffinose, D-fructose, L-rhamnose, sucrose, melibiose, D-mannitol, iso-inositol, and cellulose are not utilized; D-glucose, L-arabinose, and D-xylose are used as sole carbon sources. Can grow in NaCl at a concentration up to 1.5%. The temperature range for growth is 15–37°C. Cell wall contains both *meso*- and LL-A_2pm. Whole-cell hydrolysates contain galactose, but lack arabinose, madurose, and xylose. Produces setamycin, an antibiotic active against trichomonads and bacteria.

The type strain shows its highest sequence similarities to the following *Kitasatospora* species: *Kitasatospora niigatensis*, AB022876, 99.1%; *Kitasatospora cineracea*, AB022875, 99.1%. It does not show any sequence similarities over 99% to any *Streptacidiphilus* or *Streptomyces* species.

Source: soil.

DNA G+C content (mol%): 73.1.

Type strain: KM-6054, ATCC 33774, DSM 43861, NBRC 14216, JCM 3304, NRRL B-16185, VKM Ac-900.

Sequence accession no. (16S rRNA gene): M55220.

2. **Kitasatospora arboriphila** Groth, Rodríguez, Schüetze, Schmitz, Leistner and Goodfellow 2004, 2125[VP]

ar.bo.ri.phi′la. L. n. *arbor* a tree; N.L. adj. *philus -a -um* (from Gr. adj. *philos -ê -on*) friend, loving; N.L. fem. adj. *arboriphila* tree loving.

Spore chains are long, straight to spiral with hooks and loops with 20 or more cylindrical, smooth-surfaced spores (1.1–1.7 × 1.3–2.4 μm) per chain. Submerged spores are formed sparsely in liquid culture. Produces a yellowish-brown to dark brown or olive substrate mycelium and a gray to dark gray aerial spore mass on glycerol-asparagine, inorganic salts-starch, oatmeal, and yeast extract-malt extract agars. Soluble pigments are not formed, but melanoid pigments are produced on peptone-yeast extract-iron and tyrosine agars. Temperature range for growth is 15–40°C (optimum 28–32°C); growth does not occur at 10°C or above 40°C. pH range for good growth is pH 5.0–8.0; growth does not occur at either pH 4.5 or pH 9.0. The cell wall contains *meso-* and LL-A_2pm; the muramic acid moiety is *N*-acetylated. Whole-organism hydrolysates contain galactose, mannose, glucose, and ribose. The major polar lipids are phosphatidylethanolamine, diphosphatidylglycerol, phosphatidylglycerol, phosphatidylinositol, and phosphatidylinositol mannosides. The predominant fatty acids are $C_{16:0}$ iso (25%), $C_{15:0}$ iso (17%), and $C_{15:0}$ anteiso (12%); mycolic acids are absent. The major menaquinones are MK-9(H_6) (49%) and MK-9(H_8) (28%).

The type strain does not show any sequence similarities over 99% to other *Kitasatospora* species, or to any *Streptacidiphilus* or *Streptomyces* species.

Source: the type strain was isolated from a soil sample collected from the roots of the tree *Maytenus aquifolia* in Ribeirao Preto, Brazil.

DNA G+C content (mol%): not known.

Type strain: 2291-120, HKI 0189, DSM 44785, JCM 13002, NBRC 101834, NCIMB 13973.

Sequence accession no. (16S rRNA gene): AY442267.

3. **Kitasatospora azatica** corrig. (Nakagaito, Yokota and Hasegawa 1992b) Zhang, Wang and Ruan 1997, 1053^VP (*Streptomyces azaticus* Nakagaito, Yokota and Hasegawa 1992b, 118)

a.za'ti.ca. N.L. fem. adj. *azatica* referring to the product azaamino acid antitumor agent.

Mature spore chains on aerial mycelia are long and straight to wavy. Spores are cylindrical with a smooth surface. Melanoid pigment is not produced. Yellow soluble pigment is produced on inorganic salts-starch agar. The color of the vegetative mycelia is pale yellow. The color of the aerial mycelia is white to grayish white. Growth is good, but aerial formation is only slight on yeast extract-malt extract agar, oatmeal agar, glycerol-asparagine agar, and Bennett's agar. The temperature range for growth is 11–34°C. The concentration of NaCl at which growth occurs is less than 3%. Nitrate is reduced, starch is hydrolyzed, gelatin is not liquefied, and milk is peptonized but not coagulated.

D-Glucose, D-fructose, D-xylose, and L-arabinose are utilized, but D-mannitol, rhamnose, raffinose, iso-inositol, and sucrose are not or poorly utilized. The cell wall contains both LL- and *meso*-A_2pm and a trace amount of glycine. Galactose is detected as whole-cell sugar. The phospholipid pattern is type II. MK-9(H_6) and MK-9(H_8) are detected.

Type strain shows the highest sequence similarity to following *Kitasatospora* species: *Kitasatospora nipponensis*, AY442263, 99.1%. The type strain shows its highest sequence similarity to *Streptomyces purpeofuscus*, AJ781364, 99.2%, but does not show any sequence similarities over 99% to any *Streptacidiphilus* species.

Source: not known.

DNA G+C content (mol%): 70.5.

Type strain: ATCC 29755, DSM 41650, NBRC 13803, JCM 8798.

Sequence accession no. (16S rRNA gene): U93312.

4. **Kitasatospora cheerisanensis** Chung, Sung, Mo, Son, Nam, Chun and Bae 1999, 757^VP

che.e.ri.san.en'sis. N.L. fem. adj. *cheerisanensis* of or pertaining to Cheerisan, the name of a mountain in Korea where the species was originated.

Spore chains are *Rectiflexibiles*, with 20 or more rod-shaped smooth-surfaced spores (0.75–0.90 × 1.2–1.5 µm) per chain. Submerged spores are formed in liquid culture. Produces a greenish-yellow substrate mycelium and a whitish-gray aerial mycelium on yeast extract-malt extract agar, inorganic salts-starch agar, tyrosine agar, Bennett's agar, and starch agar. Vegetative mycelia fragment into bacillary structures. Soluble pigments, including melanin, are not produced. Colonies lacking aerial mycelium during early growth stage were pasty, circular, convex, and dull milky. The strain is positive for catalase activity and reduction of nitrate. Acids are produced from arabinose, cellobiose, galactose, glucose, maltose, trehalose, and xylose. Casein, chitin, chitosan, starch, esculin, gelatin, hippurate, and Tweens 40, 60 and 80 are decomposed, but cellulose, hypoxanthine, and tyrosine are not. Growth occurs in the presence of 1% NaCl, but not in 3% NaCl. Good growth occurs at pH 7–8 and the optimum temperature range is 25–30°C. Arabinose, azelate, cellobiose, cronate, fumarate, D-glucose, galactose, iso-butyrate, malonate, maltose, rhamnose, trehalose, and xylose are utilized for growth, but not adonitol, fructose, or glycerol. Mannitol, *myo*-inositol, melezitose, raffinose, sorbitol, suberate, succinate, and sucrose are not utilized. Whole-cell hydrolysates contain glucose and mannose, but lack galactose. Phosphatidylethanolamine, phosphatidylinositol, and diphosphatidylglycerol are contained in the polar lipid fraction. The predominant fatty acids are saturated iso- and anteiso-branched as well as straight-chain fatty acids. The major menaquinone type is MK-9(H_8). This strain is susceptible to ampicillin, chloramphenicol, kanamycin, nalidixic acid, oxytetracycline, rifampin, and streptomycin. The type strain, YC75^T, produces bafilomycin-C1-like anti-fungal compounds.

The type strain does not show any sequence similarities over 99% to other *Kitasatospora* species, or to any *Streptacidiphilus* or *Streptomyces* species.

Source: isolated from Cheerisan, Korea.

DNA G+C content (mol%): 75.8.

Type strain: YC75, KCTC 2395.

Sequence accession no. (16S rRNA gene): AF050493.

5. **Kitasatospora cineracea** Tajima, Takahashi, Seino, Iwai and Ōmura 2001, 1770^VP

ci.ne.ra'ce.a. L. fem. adj. *cineracea* similar to ash, ash-gray, referring to the color of the aerial mycelium.

The spore chains are straight and flexuous, with 20 or more rod-shaped, smooth-surfaced spores (0.9–1.0 × 0.6 µm) per chain. Submerged spores are formed in liquid culture. Produces yellowish brown vegetative mycelium and oyster-white to silver-gray aerial mycelium on yeast extract-malt extract agar, inorganic salts-starch agar, glycerol-asparagine agar, and sucrose-nitrate agar. Soluble pigments, including melanin, are not produced. The strain is positive for peptonization of milk and hydrolysis of starch.

D-Glucose, L-arabinose, D-xylose, raffinose, melibiose, and D-rhamnose are utilized for growth, but D-mannitol, D-fructose, inositol, sucrose, and cellulose are not utilized. The cell wall contains both *meso*- and LL-A_2pm. Whole-cell hydrolysates contain galactose, mannose, and ribose but lack arabinose and xylose. Phosphatidylethanolamine and phosphatidylinositol are contained in the polar lipid fraction. MK-9(H_6) and MK-9(H_8) are predominant menaquinones. The *N*-acyl type is the acetyl type. The temperature range for growth is 15–37°C. The strain is resistant to novobiocin at 100 µg/ml.

The type strain shows its highest sequence similarities to the following *Kitasatospora* species: *Kitasatospora niigatensis*, AB022876, 100%; *Kitasatospora phosalacinea*, AB022869, 99.2%; *Kitasatospora setae*, AB022868, 99.1%; *Kitasatospora griseola*, AB022870, 99.1%. The type strain does not show any sequence similarities over 99% to any *Streptacidiphilus* or *Streptomyces* species.

Source: not known.

DNA G+C content (mol%): 73.7.

Type strain: SK-3255, NBRC 16452, JCM 10915, NRRL B-24134.

Sequence accession no. (16S rRNA gene): AB022875.

6. **Kitasatospora cochleata** corrig. (Nakagaito, Yokota and Hasegawa 1992b) Zhang, Wang and Ruan 1997, 1053^VP^ (*Streptomyces cochleatus* Nakagaito, Yokota and Hasegawa 1992b, 116)

co.chle.a'ta. L. fem. adj. *cochleata* spiral, referring to the formation of spiral aerial mycelia.

Mature spore chains are long, forming hooks and spirals on glycerol-asparagine agar, tyrosine agar, oatmeal agar, glucose-asparagine agar, and water agar. Spores are cylindrical with a smooth surface. The color of vegetative mycelia is grayish brown. The color of aerial mycelia is gray. Melanoid pigments are produced in tyrosine agar. Brown soluble pigments are produced on yeast extract-malt extract agar. Nitrate is not reduced, starch is hydrolyzed, gelatin is not liquefied, and milk is peptonized but not coagulated. D-Glucose and L-arabinose are utilized, but D-fructose, D-mannitol, D-xylose, rhamnose, iso-inositol, and sucrose are poorly utilized or not utilized. The temperature range for growth is 13–38°C. The concentration of NaCl at which growth occurs is less than 3%. The cell wall contains both LL- and *meso*-A_2pm and a small amount of glycine. Galactose and a trace of madurose are detected as whole-cell sugars. The phospholipid pattern is type II. MK-9(H_6) and MK-9(H_8) are detected.

The type strain shows its highest sequence similarity to *Kitasatospora paracochleata*, U93328, 99.5%. The type strain does not show any sequence similarities over 99% to any *Streptacidiphilus* or *Streptomyces* species.

Source: not known.

DNA G+C content (mol%): 72.4.

Type strain: M-5, ATCC 51235, DSM 41652, NBRC 14768, JCM 8799.

Sequence accession no. (16S rRNA gene): U93316.

7. **Kitasatospora cystarginea** corrig. Kusakabe and Isono 1992, 327^VP^ (Effective publication: Kusakabe and Isono 1988, 1758.)

cys.tar.gi'ne.a. N.L. n. *cystarginum* antibiotic cystargin; N.L. fem. adj. *cystarginea* referring to anti-fungal antibiotic cystargin that the organism produces.

Mature spore chains on aerial mycelium are long and spiral. The spores are cylindrical with a smooth surface. The color of vegetative mycelia is light brown. The color of aerial mycelia is gray. Neither melanoid pigment nor soluble pigment is produced. Nitrate is not reduced, starch is hydrolyzed and gelatin is not liquefied. Milk is peptonized but not coagulated. D-Glucose is utilized but D-fructose, D-xylose, L-arabinose, rhamnose, raffinose, sucrose, D-mannitol, and iso-inositol are not utilized or are poorly utilized. The temperature range for growth is 17–40°C. The cell wall contains LL- and *meso*-A_2pm and glycine. Galactose is detected as whole-cell sugar. The phospholipid pattern is type II. MK-9(H_6) and MK-9(H_8) are detected.

The type strain shows its highest sequence similarity to *Kitasatospora paracochleata*, U93328, 99%. The type strain does not show any sequence similarities over 99% to any *Streptacidiphilus* or *Streptomyces* species.

Source: not known.

DNA G+C content (mol%): 70.6.

Type strain: RK-419, ATCC 49931, DSM 41680, FERM P-8006, NBRC 14836, JCM 7356, VKM Ac-2004.

Sequence accession no. (16S rRNA gene): U93318.

8. **Kitasatospora gansuensis** Groth, Rodríguez, Schüetze, Schmitz, Leistner and Goodfellow 2004, 2127^VP^

gan.su.en'sis. N.L. fem. adj. *gansuensis* of or pertaining to Gansu, a province in China, the origin of the soil from which the type strain was isolated.

Spore chains are long, straight to flexuous, with 20 or more cylindrical, smooth-surfaced spores (0.8–1.3 × 1.6–3.0 µm) per chain. Submerged spores and irregular fragments are formed in liquid culture. Produces a yellowish- or orange-brown to dark brown substrate mycelium and a white to beige aerial spore mass on glycerol-asparagine, inorganic salts-starch, oatmeal, and yeast extract-malt extract agars; soluble pigments are not formed on these media but melanoid pigments are produced on peptone-yeast extract-iron and tyrosine agars. Temperature range for growth is 6–32°C (optimum 25–28°C); growth does not occur below 6°C or at 35°C. pH range for good growth is pH 5.0–9.5; growth does not occur at pH 4.5 or above pH 9.5. The cell wall contains *meso*- and LL-A_2pm; the muramic acid moiety is *N*-acetylated and whole organism hydrolysates contain galactose, ribose, mannose, and rhamnose. The polar lipids are phosphatidylethanolamine, diphosphatidylglycerol, phosphatidylglycerol, phosphatidylinositol, phosphatidylinositol mannosides, and an unknown glycolipid. The predominant fatty acids are $C_{15:0}$ anteiso (21%), $C_{16:0}$ (19%), $C_{15:0}$ iso (10%), and $C_{16:0}$ iso (10%); mycolic acids are absent. The major menaquinone is MK-9(H_6) (75%).

The type strain does not show any sequence similarities over 99% to other *Kitasatospora* species, or to any *Streptacidiphilus* species, but shows its highest sequence similarities to the following *Streptomyces* species: *Streptomyces atroaurantiacus*, DQ026645, 99.4%; *Streptomyces indigoferus*, AB184214, 99%; *Streptomyces aburaviensis*, AY999779, 99%; *Streptomyces herbaricolor*, AB184801, 99%.

Source: the type strain was isolated from a sample of forest soil collected in the Lianhua Shan Reservation, Gansu Province, China.

DNA G+C content (mol%): not known.

Type strain: 2050-015, HKI 0314, DSM 44786, JCM 13003, NBRC 101835, NCIMB 13974.

Sequence accession no. (16S rRNA gene): AY442265.

9. **Kitasatospora griseola** corrig. Takahashi, Iwai and Ōmura 1985, 535[VP] (Effective publication: Takahashi, Iwai and Ōmura 1984, 384.)

gri.se.o'la. N.L. dim. fem. adj. *griseola* somewhat gray.

Spore chain morphology *Rectiflexibiles* type. The aerial spores are poorly septated and the spore surface is somewhat wrinkled. Color of vegetative mycelia is golden-olive to parchment on yeast extract-malt extract and glycerol-asparagine agars. Aerial mass color is gray or silvery-gray on most media. Forms a pinkish diffusible pigment on oatmeal (Shirling and Gottlieb, 1966) and glucose-asparagine agars. Can grow in NaCl at a concentration up to 2.0%. The temperature range for growth is 15–37°C; pH range for growth is pH 5.5–9.0. Cell wall contains both *meso-* and LL-A_2pm. Whole-cell hydrolysates contain galactose, but lack arabinose, madurose, and xylose. Produces setamycin, an antibiotic active against trichomonads and bacteria.

The type strain shows its highest sequence similarity to the following *Kitasatospora* species: *Kitasatospora paracochleata*, U93328, 99.2%; *Kitasatospora cineracea*, AB022875, 99.1%; *Kitasatospora niigatensis*, AB022876, 99%. It does not show any sequence similarities over 99% to any *Streptacidiphilus* or *Streptomyces* species.

Source: not known.

DNA G+C content (mol%): 66.0.

Type strain: AM-9660, DSM 43859, NBRC 14371, JCM 3339, NRRL B-16229, VKM Ac-2002.

Sequence accession no. (16S rRNA gene): U93320.

10. **Kitasatospora kifunensis** (Nakagaito, Shimazu, Yokota and Hasegawa 1992a) Groth, Schütze, Boettcher, Pullen, Rodriguez, Leistner and Goodfellow 2003, 2037[VP] (*Streptomyces kifunensis* Nakagaito, Shimazu, Yokota and Hasegawa 1992a, 630)

ki.fu.nen'sis. N.L. fem. adj. *kifunensis* of or belonging to Mount Kifune, Kyoto Prefecture, Japan, the source of the soil from which the organism was isolated.

Spore-chains are straight, hooked to spiral (*Rectiflexibiles* to *Spirales*). Spores are short, cylindrical and smooth-surfaced (1.2–1.3 × 0.6–0.7 μm). Submerged spores are formed sparsely in liquid culture. The color of vegetative mycelia is yellowish brown. The color of aerial mycelia is gray. Produces melanoid pigments on tyrosine agar, but not on peptone-yeast extract-iron agar or in tryptone-yeast extract broth. Nitrate is not reduced to nitrite. Gelatin is not liquefied and milk is not peptonized. Casein is degraded. H_2S is produced. Starch is hydrolyzed. Alkaline phosphatase, esterase (C4), esterase lipase (C8), β-galactosidase, α-glucosidase, leucine arylamidase, naphthol-AS-BI-phosphohydrolase, α-mannosidase (weak), and acid phosphatase are produced, but *N*-acetyl-β-glucosamidase, α-chymotrypsin, cystine arylamidase, α-fucosidase, α-galactosidase, β-glucosidase, β-glucuronidase, lipase (C14), trypsin, and valine arylamidase are not. L-Arabinose, D-galactose, D-glucose, glycerol, maltose, D-mannitol, D-mannose, sucrose, trehalose, and D-xylose are used as sole sources of carbon for energy and growth, but cellulose, chitin, D-fructose, iso-inositol, inulin, D-lactose, raffinose, L-rhamnose, and salicin are not. Temperature range for growth is 10–32°C; the organism does not grow below 10°C or above 32°C. pH range for growth is 5–9; growth does not occur at pH 4.5 or 9.5. Growth is inhibited by ampicillin (10 μg/ml, weak), chloramphenicol (30 μg/ml), ciprofloxacin (5 μg/ml), imipenem (10 μg/ml), kanamycin sulfate (30 μg/ml), methicillin (5 μg/ml), nalidixic acid (50 μg/ml), novobiocin (2.5 μg/ml), oxytetracycline hydrochloride (30 μg/ml), rifampin (30 μg/ml), streptomycin sulfate (10 μg/ml), sulfonamide (200 μg/ml, weak), and vancomycin hydrochloride (30 μg/ml), but not by lincomycin hydrochloride (2 μg/ml), norfloxacin (10 μg/ml), penicillin G (10 IU), or polymyxin B (3400 IU). The strain tolerates 2.5% but not 3.5% (w/v) NaCl. Resistance is shown to the polyvalent *Streptomyces* phage S7. Whole-cell chemistry reveals the presence of both *meso-* and LL-A_2pm; the muramic acid moiety is *N*-acetylated. Whole-organism hydrolysates contain galactose, mannose, glucose, and ribose and the major polar lipids are phosphatidylethanolamine, diphosphatidylglycerol, phosphatidylglycerol, phosphatidylinositol, and phosphatidylinositol mannosides. Predominant fatty acids are $C_{16:0}$, $C_{15:0}$ anteiso, and $C_{15:0}$ iso; mycolic acids are absent. Major menaquinone is hexahydrogenated with nine isoprene units.

The type strain shows its highest sequence similarity to *Kitasatospora nipponensis*, AY442263, 99.1%. It does not show any sequence similarities over 99% to any *Streptacidiphilus* species but shares its highest sequence similarities to the following *Streptomyces* species: *S. herbaricolor*, AB184801, 99.3%; *S. indigoferus*, AB184214, 99.3%; *S. aburaviensis*, AY999779, 99.3%; *S. chrysomallus* subsp. *fumigatus*, AB184645, 99.2%; *S. purpureus*, AJ781324, 99.2%; *S. xanthocidicus*, AY999858, 99.1%.

Source: the type strain was isolated from a soil sample obtained at Mount Kifune, Kyoto Prefecture, Japan.

DNA G+C content (mol%): 71.3.

Type strain: ATCC 51379, DSM 41654, NBRC 15206, JCM 9081.

Sequence accession no. (16S rRNA gene): AY442264.

11. **Kitasatospora mediocidica** corrig. Labeda 1988, 289[VP]

med.i.o.cid'i.ca. N.L. n. *mediocidinum* mediocidin, antbiotic produced by the type strain; L. fem. suff. *-ica* suffix used with the sense of pertaining to; N.L. fem. adj. *mediocidica* posessing mediocidin.

Sporulating aerial mycelium is not produced on any ISP media or on 15 additional media tested by cooperators. Single conidia were observed on the substrate mycelium on glucose-asparagine agar and Waksman's starch agar B. Masses of crystals are formed on the substrate growth on yeast-malt agar. The original description indicates that aerial mycelium was only poorly developed or absent on most media. Color of colony: no sporulation aerial mycelium was found on any of the media tried by the cooperators. Reverse side of colony is yellow to greenish yellow on yeast-malt agar, oatmeal agar, salts-starch agar, and glycerol-asparagine agar. This pigment is not pH-sensitive.

Color in medium: melanoid pigments are formed in peptone-yeast-iron agar and tryptone-yeast broth, but may not be formed in tyrosine agar. Trace of yellow pigment may be found in the medium in yeast-malt agar, oatmeal agar, salts-starch agar, and glycerol-asparagine agar; this pigment is not pH-sensitive.

D-Glucose, iso-inositol, and D-fructose are utilized for growth. No growth or only trace of growth on L-arabinose, sucrose, D-xylose, D-mannitol, rhamnose, and raffinose.

The type strain does not show any sequence similarities over 99% to other *Kitasatospora* species, or to any *Streptacidiphilus* or *Streptomyces* species.

Source: not known.

DNA G+C content (mol%): not known.

Type strain: ATCC 49055, DSM 43929, NBRC 14789, JCM 9868, NRRL B-16109.

Sequence accession no. (16S rRNA gene): U93324.

12. **Kitasatospora niigatensis** Tajima, Takahashi, Seino, Iwai and Ōmura 2001, 1770[VP]

ni.i.gat.en'sis. N.L. fem. adj. *niigatensis* of or belonging to Niigata, the city in Japan where the species originated.

The spore chains are straight and flexuous, with 20 or more rod-shaped, smooth-surfaced spores (1.0–1.1 × 0.5 µm) per chain. Submerged spores are formed in liquid culture. Produces ivory to brownish-gray vegetative mycelium and brownish white to purplish gray aerial mycelium on yeast extract-malt extract agar, oatmeal agar, glycerol-asparagine agar, and sucrose-nitrate agar. Soluble pigments, including melanin, are not produced. The strain is positive for reduction of nitrate, peptonization of milk, and hydrolysis of starch. D-Glucose, L-arabinose, and D-xylose are utilized for growth, but raffinose, melibiose, D-mannitol, D-fructose, D-rhamnose, inositol, sucrose, and cellulose are not utilized. The cell wall contains both *meso-* and LL-A_2pm. Whole-cell hydrolysates contain galactose, mannose, and ribose, but lack arabinose and xylose. Phosphatidylethanolamine and phosphatidylinositol are contained in the polar lipid fraction. MK-9(H_6) and MK-9(H_8) are predominant menaquinones. The *N*-acyl type is the acetyl type. The temperature range for growth is 15–41°C. The strain is resistant to novobiocin at 100 µg/ml.

The type strain shows its highest sequence similarities to the following *Kitasatospora* species: *Kitasatospora cineracea*, AB022875, 100%; *Kitasatospora phosalacinea*, AB022869, 99.1%; *Kitasatospora setae*, AB022868, 99.1%; *Kitasatospora griseola*, AB022870, 99%. It does not show any sequence similarities over 99% to any *Streptacidiphilus* or *Streptomyces* species.

Source: isolated from soil from Niigata, Japan.

DNA G+C content (mol%): 73.5.

Type strain: SK-3406, NBRC 16453, JCM 10916, NRRL B-24135.

Sequence accession no. (16S rRNA gene): AB249960.

13. **Kitasatospora nipponensis** Groth, Rodríguez, Schüetze, Schmitz, Leistner and Goodfellow 2004, 2127[VP]

nip.pon.en'sis. N.L. fem. adj. *nipponensis* of or pertaining to Nippon, the native name for Japan, the origin of the soil from which the type strain was isolated.

Spore chains are open spirals, long straight loops and hooks with 20 or more cylindrical, smooth-surfaced spores (1.1–1.6 × 1.2–2.3 µm) per chain. Submerged spores are formed in liquid culture. Produces a yellowish- or reddish-brown substrate mycelium and a gray aerial spore mass on glycerol-asparagine, inorganic salts-starch, oatmeal, and yeast extract-malt extract agars. A purple soluble pigment is formed on oatmeal agar, but melanoid pigments are not produced on peptone-yeast extract-iron and tyrosine agars. Temperature range for growth is 10–32°C (optimum 25–28°C); growth does not occur at 6 or 37°C. Good growth is observed at pH 5.0–8.0; growth does not occur at pH 4.5 or above pH 8.5. The cell wall contains *meso-* and LL-A_2pm; the muramic acid moiety is *N*-acetylated and whole-organism hydrolysates contain galactose, mannose, ribose, and glucose. The major polar lipids are phosphatidylethanolamine, diphosphatidylglycerol, phosphatidylglycerol, phosphatidylinositol, phosphatidylinositol mannosides, and an unknown phospholipid. The predominant fatty acids are $C_{16:0}$ iso (38%) and $C_{15:0}$ iso (10%); mycolic acids are absent. The major menaquinone is MK-9(H_6) (74%).

The type strain shows its highest sequence similarity to following *Kitasatospora* species: *Kitasatospora kifunensis*, AB022874, 99.1%; *Kitasatospora azatica*, U93312, 99.1%. It does not show any sequence similarities over 99% to any *Streptacidiphilus* species, but shows its highest sequence similarities to the following *Streptomyces* species: *S. herbaricolor*, AB184801, 99%; *S. indigoferus*, AB184214, 99%.

Source: the type strain was isolated from a soil sample collected at Kumagura, Japan.

DNA G+C content (mol%): not known.

Type strain: 2148-013, HKI 0315, DSM 44787, JCM 13004, NBRC 101836, NCIMB 13975.

Sequence accession no. (16S rRNA gene): AY442263.

14. **Kitasatospora paracochleata** corrig. (Nakagaito, Yokota and Hasegawa 1992b) Zhang, Wang and Ruan 1997, 1053[VP] (*Streptomyces paracochleatus* Nakagaito, Yokota and Hasegawa 1992b, 118)

pa.ra.co.chle.a'ta. Gr. prep. *para* beside, alongside of, resembling; L. fem. adj. *cochleata* a specific epithet; N.L. fem. adj. *paracochleata* similar to *Kitasatospora cochleata*.

Mature spore chains on aerial mycelia are long, forming spirals on glycerol-asparagine agar, tyrosine agar, oatmeal agar, glucose-asparagine agar, and water agar. Spores are cylindrical with a smooth surface. The color of vegetative mycelia is grayish brown. The color of aerial mycelia is gray.

Melanoid pigments are produced on tyrosine agar. Brown soluble pigments are produced on yeast extract-malt extract agar. Nitrate is reduced, starch is hydrolyzed, and gelatin is not liquefied. Milk is peptonized but not coagulated. D-Glucose and D-fructose are utilized, but L-arabinose, D-mannitol, D-xylose, rhamnose, raffinose, iso-inositol, and sucrose are not or poorly utilized. The temperature range for growth is 11–39°C. The concentration of NaCl at which growth occurs is less than 3%. The cell wall contains both LL- and *meso*-A_2pm and glycine. Galactose, rhamnose, and a trace of madurose are detected as whole-cell sugars. The phospholipid pattern is type II. MK-9(H_6) and MK-9(H_8) are detected.

The type strain shows its highest sequence similarities to the following *Kitasatospora* species: *Kitasatospora cochleata*, U93316, 99.5%; *Kitasatospora griseola*, AB022870, 99.2%; *Kitasatospora cystarginea*, U93318, 99%. It does not show any sequence similarities over 99% to any *Streptacidiphilus* or *Streptomyces* species.

Source: not known.

DNA G+C content (mol%): 73.1.

Type strain: M-13, ATCC 51236, DSM 41656, NBRC 14769, JCM 8800.

Sequence accession no. (16S rRNA gene): U93328.

15. **Kitasatospora paranensis** Groth, Rodríguez, Schüetze, Schmitz, Leistner and Goodfellow 2004, 2128[VP]

pa.ra.nen'sis. N.L. fem. adj. *paranensis* of or pertaining to Parana, a state of Brazil, the origin of the soil from which the type strain was isolated.

Spore chains are long, straight to flexuous with 20 or more cylindrical, smooth-surfaced spores (1.1–1.4 × 1.2–2.1 μm) per chain. Submerged spores are rarely formed in liquid culture. Produces a yellowish-brown to dark brown substrate mycelium and a gray aerial spore mass on glycerol-asparagine, inorganic salts-starch, oatmeal, and yeast extract-malt extract agars. Soluble pigments are not formed, but melanoid pigments are produced on peptone-yeast extract-iron and tyrosine agars. Temperature range for growth is 10–37°C (optimum 25–28°C); growth does not occur at 6 or 40°C. Good growth is observed at pH 5.0–9.0; growth does not occur at either pH 4.0 or pH 9.5. The cell wall contains both *meso*- and LL-A_2pm; the muramic acid moiety is *N*-acetylated and whole-organism hydrolysates contain galactose, mannose, and glucose. The polar lipids are phosphatidylethanolamine, diphosphatidylglycerol, phosphatidylglycerol, phosphatidylinositol mannosides, phosphatidylserine, and an unknown phospholipid. The predominant fatty acids are $C_{16:0}$ iso (19%), $C_{15:0}$ anteiso (16%), and $C_{16:0}$ (14%); mycolic acids are absent. The major menaquinone is MK-9(H_6) (53%), with minor components MK-9(H_4) (22%) and MK-9(H_2) (14%).

The type strain shows its highest sequence similarities to the following *Kitasatospora* species: *Kitasatospora terrestris*, AY442266, 99.2%. It does not show any sequence similarities over 99% to any *Streptacidiphilus* or *Streptomyces* species.

Source: the type strain was isolated from rhizosphere soil of *Maytenus ilicifolia*, Contenda, Parana State, Brazil.

DNA G+C content (mol%): not known.

Type strain: 2292-041, HKI 0190, DSM 44788, JCM 13005, NBRC 101837, NCIMB 13976.

Sequence accession no. (16S rRNA gene): AY442268.

16. **Kitasatospora phosalacinea** corrig. Takahashi, Iwai and Ōmura 1985, 535[VP] (Effective publication: Takahashi, Iwai and Ōmura 1984, 384.)

pho.sa.la.ci'ne.a. N.L. fem. adj. *phosalacinea* pertaining to phosalacine (an antibiotic produced by the organism).

Spore chain morphology *Rectiflexibiles* type. Color of vegetative mycelia is pale yellowish-brown on most media. Aerial mass color is white or light gray on oatmeal and inorganic salts-starch agars. Forms yellowish-brown diffusible pigment on some media. Can grow in NaCl at a concentration up to 2.0%. The temperature range for growth is 15–42°C. Cell wall contains both *meso*- and LL-A_2pm. Whole-cell hydrolysates contain galactose, but lack arabinose, madurose, and xylose. Produces phosalacine, a herbicidal antibiotic; activity *in vitro* against *Bacillus subtilis* on a synthetic medium, which is reversed by glutamine.

The type strain shows its highest sequence similarities to the following *Kitasatospora* species: *Kitasatospora cineracea*, AB022875, 99.2%; *Kitasatospora niigatensis*, AB022876, 99.1%. It does not show any sequence similarities over 99% to any *Streptacidiphilus* or *Streptomyces* species.

Source: soil.

DNA G+C content (mol%): 66.6.

Type strain: KA-338, DSM 43860, NBRC 14372, JCM 3340, NRRL B-16230, VKM Ac-2006.

Sequence accession no. (16S rRNA gene): M55223.

17. **Kitasatospora putterlickiae** Groth, Schütze, Boettcher, Pullen, Rodriguez, Leistner and Goodfellow 2003, 2037[VP]

put.ter.lic'ki.ae. N.L. gen. n. *putterlickae* of the plant genus *Putterlickia*.

Spore chains are straight to flexuous (*Rectiflexibiles*), with 20 or more cylindrical, smooth-surfaced spores (1.6–2.5 × 1.0–1.5 μm) per chain. Submerged spores are formed in liquid culture. Produces a dark-brown substrate mycelium and a dark-gray aerial spore-mass on glycerol-asparagine and yeast extract-malt extract agars. Brown soluble pigments are formed on oatmeal and yeast extract-malt extract agars and melanoid pigments are formed on peptone-yeast extract-iron and tyrosine agars. Nitrate is reduced to nitrite, gelatin is liquefied, milk is peptonized, casein is degraded, and H_2S is weakly produced. Starch is not hydrolyzed. Alkaline phosphatase, esterase (C4), esterase lipase (C8), leucine arylamidase, acid phosphatase, β-galactosidase and β-glucosidase are produced, but *N*-acetyl-β-glucosamidase, α-chymotrypsin, cystine arylamidase, α-fucosidase, α-galactosidase, α-glucosidase, β-glucuronidase, lipase (C14), α-mannosidase, naphthol-AS-BI-phosphohydrolase, trypsin and valine arylamidase are not. D(+)-fructose (weak), D(+)-glucose, and D(−)-sucrose (weak) are used as sole sources of carbon for energy and growth, but L(+)-arabinose, cellulose, iso-inositol, D(−)-mannitol, D(+)-raffinose, L(+)-rhamnose, and D(+)-xylose are not. Temperature range for growth is 10–37°C (optimum is between 28 and 32°C); growth does not occur at either 6 or 40°C. Good growth occurs at pH 5–9; growth does not occur at pH 4.5 or above 9.5. Growth is inhibited by ampicillin (10 μg/ml), chloramphenicol (30 μg/ml), ciprofloxacin (5 μg/ml), imipenem (10 μg/ml), kanamycin sulfate (30 μg/ml), methicillin (5 μg/ml, weak), novobiocin (2.5 μg/ml), oxytetracycline hydrochloride (30 μg/ml), penicillin G (10 IU), polymyxin B (300 IU, weak), rifampin (30 μg/ml), streptomycin sulfate (10 μg/ml), vancomycin hydrochloride (30 μg/ml), and nalidixic acid (50 μg/ml), but not by lincomycin hydrochloride (2 μg/ml), norfloxacin (10 μg/ml), or sulfonamide (200 μg/ml). NaCl is tolerated up to a concentration of 3.5% (w/v). Resistance is shown to polyvalent *Streptomyces* phage S7. Whole cell chemistry reveals the presence of both *meso*- and LL-A_2pm; the muramic acid moiety is *N*-acetylated. Whole-organism hydrolysates contain galactose, madurose, mannose, and rhamnose, and the major polar lipids are diphosphatidylglycerol, phosphatidylethanolamine, phosphatidylglycerol, phosphatidylinositol, and phosphatidylinositol mannosides. Predominant fatty acids are $C_{16:0}$, $C_{16:0}$, and $C_{15:0}$ iso; mycolic acids are absent. Major menaquinones are tetra- and hexahydrogenated with nine isoprene units.

The type strain does not show any sequence similarities over 99% to other *Kitasatospora* species, or to any *Streptacidiphilus* or *Streptomyces* species.

Source: isolated from the rhizosphere of *Putterlickia verrucosa.*

DNA G+C content (mol%): 66.6.

Type strain: F18-98, DSM 44665, JCM 12393, NBRC 100917, NCIMB 13932.

Sequence accession no. (16S rRNA gene): AY189976.

18. **Kitasatospora sampliensis** Mayilraj, Krishnamurthi, Saha and Saini 2006, 521[VP]

sam.pli.en'sis. N.L. fem. adj. *sampliensis* pertaining to Sampli village, Punjab State, India, where the type strain was isolated.

Spore chain morphology is *Rectiflexibiles* with 10 or more smooth-surfaced spores per chain. Produces pale-gray, dark-gray, or dark-gray-brown substrate mycelium and a pale-gray or dark-gray aerial mycelium on almost all ISP media. Dark-brown or dark-gray soluble pigments are formed on ISP 2, ISP 3, ISP 6, ISP 7, *Streptomyces* agar, actinomycetes isolation agar, and Sabouraud glucose agar. Melanin is produced on peptone-yeast extract-iron and tyrosine agar. Nitrate is reduced to nitrite, casein is not degraded, starch is not hydrolyzed, and gelatin is not liquefied. Positive for utilization of L-arabinose, L-rhamnose, D-fructose, D-mannitol, raffinose, D-sucrose, and D-xylose as sole carbon and energy sources. Can grow in NaCl at a concentration up to 2.5%, but not at 3.0% or above. Optimum growth is observed at pH 7.0–8.0 and at 30°C. Cell wall contains both *meso-* and LL-A_2pm. The *N*-acyl type of muramic acid of the peptidoglycan is acetyl. Major fatty acids are $C_{16:0}$ (16.48%), $C_{15:0}$ iso (20.07%), $C_{16.0}$ (10.94%), $C_{17:0}$ iso (12.54%), $C_{15:0}$ anteiso (9.55%), and $C_{17:0}$ anteiso (9.36%). Whole-cell sugars are galactose, glucose, mannose, and ribose. The polar lipids are phosphatidylinositol, phosphatidylethanolamine, diphosphatidylglycerol, and phosphatidylinositol mannosides. The major menaquinone is MK-9(H6).

The type strain does not show any sequence similarities over 99% to other *Kitasatospora* species, or to any *Streptacidiphilus* or *Streptomyces* species.

Source: the type strain was isolated from a sugar-cane field soil sample collected from Sampli village, Punjab state, India.

DNA G+C content (mol%): 76.5.

Type strain: VT-36, DSM 44898, JCM 13010, NBRC 102069, MTCC 6546.

Sequence accession no. (16S rRNA gene): AY260167.

19. **Kitasatospora terrestris** Groth, Rodriguez, Rodríguez, Schüetze, Leistner and Goodfellow 2004, 2128[VP]

ter.res'tris. L. fem. adj. *terrestris* of the earth, terrestrial.

Spore chains are straight, hooked and spiral with 20 or more cylindrical, smooth-surfaced spores (1.1–1.5 × 1.3–2.8 µm) per chain. Submerged spores are formed in liquid culture. Produces a yellowish-brown to dark brown substrate mycelium and a gray aerial spore mass on glycerol-asparagine, inorganic salts-starch, oatmeal, and yeast extract-malt extract agars. Soluble pigments are not formed. The formation of melanoid pigments is weak on peptone-yeast extract-iron, and tyrosine agars. Temperature range for growth is 15–40°C (optimum 28–32°C); growth does not occur at 10°C or 42°C. pH range for good growth is pH 5.0–9.0; growth does not occur at pH 4.5 or at pH 9.5. The cell wall contains *meso-* and LL-A_2pm; the muramic acid moiety is *N*-acetylated and whole-organism hydrolysates contain galactose, mannose, and glucose. The polar lipids are phosphatidylethanolamine, diphosphatidylglycerol, phosphatidylglycerol (traces), phosphatidylinositol, phosphatidylinositol mannosides, phosphatidylserine, and an unknown glycolipid. The predominant fatty acids are $C_{16:0}$ iso (21%), $C_{15:0}$ iso (16%), and $C_{15:0}$ anteiso (10%); mycolic acids are absent. The major menaquinone is MK-9(H_6) (76%).

The type strain shows its highest sequence similarity to *Kitasatospora paranensis*, AY442268, 99.2%. It does not show any sequence similarities over 99% to any *Streptacidiphilus* or *Streptomyces* species.

Source: the type strain was isolated from a soil sample of the roots of *Maytenus aquifolia,* Ribeirao Preto, Brazil.

DNA G+C content (mol%): not known.

Type strain: 2293-012, HKI 0186, DSM 44789, JCM 13006, NBRC 101838, NCIMB 13977.

Sequence accession no. (16S rRNA gene): AY442266.

20. **Kitasatospora viridis** Liu, Rodríguez, Wang, Cui, Huang, Quintana and Goodfellow 2005a, 709[VP]

vi'ri.dis. L. fem. adj. *viridis* green, referring to the production of a green aerial spore mass.

Non-acid–alcohol-fast, nonmotile actinomycete that forms an extensively branched, light-yellow substrate mycelium and a greenish aerial spore mass on acidified oatmeal agar. Aerial hyphae differentiate into long, spiral chains of smooth-surfaced, cylindrical spores (1.0–1.2 × 0.7–0.8 µm). Starch is degraded, but not adenine, guanine, hypoxanthine, xanthine, or xylan. Adonitol, cellobiose, dextran, D-galactose, D-gluconic acid, D-glucose, inulin, D-lactose, maltose, D-mannose, melezitose, melibiose, D-salicin, D-sorbitol, trehalose, and xylitol are used as sole carbon sources for energy and growth, but not glycerol, *myo*-inositol, or xylan (all at 1%, w/v). Similarly, 2-aminoethanol, α-DL-aminobutyric acid, L-alanine, L-arginine, L-cysteine, L-glutamic acid, L-histidine, L-isoleucine, L-phenylalanine, L-threonine, L-valine, sodium oxalate, and sodium pyruvate are used as sole carbon sources, but not adipic acid or L-aspartic acid (all at 0.1%, w/v). 2-Aminoethanol, L-alanine, L-arginine, L-isoleucine, and L-phenylalanine are used as sole sources of carbon and nitrogen for energy and growth. Growth occurs at 10–37°C, but not at 4 or 45°C. The pH range for growth is pH 4–7.0. Growth occurs in the presence (µg/ml) of amikacin (32), amoxycillin (32), ampicillin (32), cefalexin (32), cephaloridine (64), clindamycin (8), doxycycline hydrochloride (32), fusidic acid (16), gentamicin sulfate (16), kanamycin sulfate (16), lincomycin hydrochloride (16), midecamycin (4), neomycin sulfate

2. **Streptacidiphilus anmyonensis** Cho, Han, Ko and Kim 2008, 1568[VP]

an.myon.en'sis. N.L. masc. adj. *anmyonensis* of or belonging to Anmyon, where the organism was isolated.

Aerial hyphae differentiate into long flexuous chains of spores (0.6 × 0.9 μm); the spore surface is smooth. Forms cream-colored colonies that carry moderate to abundant, white to grayish-white aerial hyphae on acidified oatmeal, inorganic salt-starch, yeast extract-malt extract and modified Bennett's agar plates. The substrate mycelium is brownish gray to brown on acidified oatmeal and yeast extract-malt extract agars, but cream-colored on the other two agars. Soluble pigments are not produced on any of the above-mentioned media. Starch and Tween 80 are degraded, but xanthine, and Tweens 20 and 40 are not. Glycerol, D-gluconic acid, D(+)-glucosamine hydrochloride, *myo*-inositol, melibiose, D-sorbitol, sucrose, D(+)-xylose (all at 1%, w/v), and L-arginine (at 0.1%, w/v) are used as sole carbon sources for energy and growth, but L-aspartic acid and sodium oxalate (all at 0.1%, w/v) are not. L-Isoleucine is used as sole nitrogen source. Growth occurs at pH 3.0–8.0 and between 28 and 35°C. Chemotaxonomic properties are typical of the genus *Streptacidiphilus*. The major fatty acids are $C_{15:0}$ iso (18.4% of the total fatty acid composition), $C_{15:0}$ anteiso (11.4%), $C_{16:0}$ iso (19.1%), *n*-$C_{16:0}$ (14.7%), $C_{17:0}$ iso (7.4%), and $C_{17:0}$ anteiso (7.5%). Contains hexa- and octahydrogenated menaquinones with nine isoprene units [MK-9(H_6) and MK-9(H_8), 24.1% and 55.8% of the total composition, respectively] as predominant isoprenologs. The diamino acid of the peptidoglycan is LL-A_2pm (89% of the total A_2pm composition), although minor amounts of the *meso*-isomer (11%) are also detected.

The type strain shows its highest sequence similarities to the following: *Streptacidiphilus* species: *Streptacidiphilus jiangxiensis*, AB249948, 99.6; *Streptacidiphilus melanogenes*, DQ994689, 99.3%. It does not show any sequence similarities over 99% to any *Kitasatospora* or *Streptomyces* species.

Source: isolated from *Pinus* soils Anmyeon, near coastal areas in Tae-An, Chungnam, Republic of Korea.

DNA G+C content (mol%): not known.

Type strain: AM-11, NBRC 103185, KCTC 19278.

Sequence accession no. (16S rRNA gene): DQ904546.

3. **Streptacidiphilus carbonis** Kim, Lonsdale, Seong and Goodfellow 2003b, 1219[VP] (Effective publication: Kim, Lonsdale, Seong and Goodfellow 2003b, 114.)

car'bo.nis. L. n. *carbo -onis* coal; L. gen. n. *carbonis* of/from coal.

Aerial hyphae differentiate into long flexuous chains of spores (0.6 × 1.0 μm); the spore surface is smooth. Cream-colored colonies form sparse to abundant white aerial hyphae on acidified modified Bennett's agar. The underside of colonies is cream-colored. Elastin is not degraded. L-Arabinose, D-fructose, D-gluconic acid, glycerol, glycogen, *myo*-inositol, inulin, D-raffinose, L-rhamnose (all at 1%, w/v), sodium pyruvate, and sodium succinate (all at 0.1%, w/v) are used as sole carbon sources for energy and growth, but not D-amygdalin, melezitose, methyl α-D-glucoside, D-ribose, D-sorbitol (all at 1%, w/v), sodium adipate, L-isoleucine, or sodium oxalate (all at 0.1%, w/v). L-Alanine, L-arginine, L-aspartic acid, L-glutamic acid, and L-phenylalanine are used as sole carbon and nitrogen sources, but not 2-aminoethanol or L-isoleucine. Grows between 15 and 30°C, at pH 3.5–6.0, and in the presence (all at μg/ml) of gentamicin sulfate (4), lincomycin hydrochloride (16), neomycin sulfate (32), oleandomycin phosphate (4), tobramycin sulfate (4), bismuth citrate (1), cadmium acetate (10), cobalt chloride (10), copper sulfate (10), crystal violet (1), ferrous sulfate (10), lead acetate (50), manganese sulfate (50), thallous acetate (1), and zinc sulfate (100), but not with cephaloridine (2), demeclocycline hydrochloride (2), penicillin G (16), streptomycin sulfate (16), vancomycin hydrochloride (2), cadmium acetate (50), cobalt chloride (50), crystal violet (10), ferrous sulfate (100), lead acetate (100), potassium tellurite (50), or thallous acetate (100).

The type strain does not show any sequence similarities over 99% to other *Streptacidiphilus* species, or to any *Kitasatospora* or *Streptomyces* species.

Source: the type strain was isolated from reclaimed acid coal mine waste.

DNA G+C content (mol%): not known.

Type strain: JL415, DSM 41754, JCM 12364, KCTC 9912, NBRC 100919.

Sequence accession no. (16S rRNA gene): AF074412.

4. **Streptacidiphilus jiangxiensis** Huang, Cui, Wang, Rodriguez, Quintana, Goodfellow and Liu 2005, 1743[VP] (Effective publication: Huang, Cui, Wang, Rodriguez, Quintana, Goodfellow and Liu 2004a, 162.)

ji.ang.xi.en'sis. N.L. masc. adj. *jiangxiensis* of or pertaining to Jiangxi Province, South China, the source of the isolates.

Aerial hyphae differentiate into long *Rectiflexibiles* chains of spores (0.6 × 0.9–1.0 μm) which have smooth surfaces. The organism forms lichenoid colonies that carry moderate to abundant, white to grayish-white aerial hyphae on acidified oatmeal, inorganic salts-starch, modified Bennett's, and yeast extract-malt extract agars. The substrate mycelium is cream-colored on acidified oatmeal agar, stray yellow or Dresden brown on the other three agars. Soluble pigments are not produced on the media cited above, nor are melanin pigments formed on peptone-yeast extract-iron agar. Guanine, hypoxanthine, Tweens 40, 60, and 80, and xanthine are not degraded. Adonitol, D(+)-cellobiose, dextran, D(−)-fructose, D(+)-mannitol, D(+)-mannose, salicin, D(+)-sucrose, xylan, xylitol, D(+)-xylose (all at 1%, w/v), 2-aminoethanol, L-cystine, L-glutamic-acid, L-isoleucine, sodium oxalate, L-threonine, and L-valine (all at 0.1%, w/v) are used as sole carbon sources for energy and growth, but not adipic acid, aminobutyric acid, L-arginine, L-aspartic acid, or L-histidine (all at 0.1%, w/v). L-Alanine, 2-aminoethanol, L-arginine, L-isoleucine, and L-phenylalanine are used as sole carbon and nitrogen sources, but not aspartic acid. Growth occurs between pH 3.5–6.5 and between 15 and 35°C. Sensitive to ampicillin (10 μg/ml), carbenicillin (100), chloramphenicol (30), clarithromycin (15), gentamicin sulfate (10), kanamycin sulfate (30), midecamycin (15), minocycline hydrochloride (30), novobiocin (5), penicillin G (16), rifampin (5), streptomycin sulfate (16), and tobramycin sulfate (10). Chemotaxonomic properties are typical of the genus *Streptacidiphilus*. The predominant fatty acids are $C_{15:0}$ iso (10.8–13.2%), $C_{15:0}$ anteiso (14.7–17.5%), $C_{16:0}$ iso (18.3–22.2%), and $C_{16:0}$ (25.9–30.4%).

The type strain shows its highest sequence similarities to the following *Streptacidiphilus* species: *Streptacidiphilus anmyonensis*, DQ904546, 99.6%; *Streptacidiphilus melanogenes*, DQ994689, 99.1%. It does not show any sequence similarities over 99% to any *Kitasatospora* or *Streptomyces* species.

Source: isolated from rhizosphere soil of wild tea plants growing on the campus of Jiangxi Agricultural University, Jiangxi Province, China.

DNA G+C content (mol%): 70.8–71.7.

Type strain: 33214, AS 4.1857, JCM 12277, NBRC 100920.

Sequence accession no. (16S rRNA gene): AY314780.

5. **Streptacidiphilus melanogenes** Cho, Han, Ko and Kim 2008, 1569[VP]

me.la.no'ge.nes. Gr. adj. *melas -anos* black; N.L. suff. *-genes* (from Gr. v. *gennaô* to produce) producing; N.L. part. adj. *melanogenes* producing black (pigment).

Aerial hyphae differentiate into long flexuous chains of spores (0.6 × 1.0 µm); the spore surface is smooth. Forms cream-colored colonies that carry moderate to abundant white aerial hyphae on acidified oatmeal, inorganic salts-starch, yeast extract-malt extract, and modified Bennett's agar plates. The substrate mycelium is brownish-gray or brown on acidified inorganic salts-starch and yeast extract-malt extract agars, but cream-colored on the other two agars. Brownish gray diffusible pigments are formed on acidified oatmeal and yeast extract-malt extract agars. Soluble pigments are not produced on the two agars. Starch, Tweens 40 and 80, and xanthine are degraded, but Tween 20 is not. Glycerol, D-gluconic acid, D(+)-glucosamine hydrochloride, *myo*-inositol, melibiose, D-sorbitol, sucrose, D(+)-xylose (all at 1%, w/v), L-arginine and sodium oxalate (all at 0.1%, w/v) are used as sole carbon sources for energy and growth, but L-aspartic acid (at 0.1%, w/v) is not. L-Isoleucine is used as nitrogen source. Growth occurs at pH 3.0–8.0 and between 28 and 35°C. Chemotaxonomic properties are typical of the genus *Streptacidiphilus*. The major fatty acids are $C_{15:0}$ iso (20.4% of the total fatty acid composition), $C_{15:0}$ anteiso (6.7%), $C_{16:0}$ iso (14.7%), *n*-$C_{16:0}$ (15.7%), $C_{17:0}$ iso (7.3%), and $C_{17:0}$ anteiso (3.8%). Contains hexa- and octahydrogenated menaquinones with nine isoprene units [MK-9(H_6) and MK-9(H_8), 18.1% and 68.6% of total composition, respectively] as predominant isoprenologs. The diamino acid of the peptidoglycan is LL-A_2pm (100% of the total A_2pm composition).

The type strain shows its highest sequence similarities to the following *Streptacidiphilus* species: *Streptacidiphilus anmyonensis*, DQ904546, 99.3; *Streptacidiphilus rugosus*, DQ904547, 99.2. It does not show any sequence similarities over 99% to any *Kitasatospora* or *Streptomyces* species.

Source: isolated from *Pinus* soils, Sambong, near coastal areas in Tae-An, Chungnam, Republic of Korea.

DNA G+C content (mol%): not known.

Type strain: SB-B34, NBRC 103184, KCTC 19280.

Sequence accession no. (16S rRNA gene): DQ994690.

6. **Streptacidiphilus neutrinimicus** Kim, Lonsdale, Seong and Goodfellow 2003a, 1219[VP] (Effective publication: Kim, Lonsdale, Seong and Goodfellow 2003b, 113.)

neu.tri.ni'mi.cus. L. adj. *neuter -tra -trum* neither or both (here for neutral pH); L. masc. n. *inimicus* an enemy, foe; N.L. n. *neutrinimicus* (nominative in apposition) enemy of the neuter pH.

Aerial hyphae differentiate into long flexuous chains of spores (0.6 × 1.0 µm); the spore surface is smooth. Cream-colored colonies form sparse to abundant white aerial hyphae on acidified modified Bennett's agar. The underside of colonies is either cream-colored or brown. Neither elastin nor xanthine is degraded. L-Arabinose, D-galactose, D-glucose, D-lactose, maltose, melibiose, glycerol, glycogen, raffinose, and trehalose (all at 1%, w/v), and L-alanine, L-arginine, L-aspartic acid, L-glutamic acid, L-histidine, L-phenylalanine, sodium pyruvate, and sodium succinate (all at 0.1%, w/v) are used as sole carbon sources for energy and growth, but not D-amygdalin, *meso*-erythritol, *myo*-inositol, inulin, melezitose, methyl α-D-glucoside, D-ribose, D-sorbitol (all at 1%, w/v), sodium adipate, *p*-hydroxybenzoic acid, L-isoleucine, sodium acetate, sodium hippurate, sodium oxalate, or sodium propionate (all at 0.1%, w/v). L-Arginine, L-aspartic acid, L-glutamic acid, and L-phenylalanine are used as sole carbon and nitrogen sources but not L-isoleucine or sodium hippurate. Growth occurs between 10 and 25°C, and at pH 3.5–5.5. Growth also occurs in the presence (all at µg) of cephaloridine hydrochloride (2), gentamicin sulfate (32), lincomycin hydrochloride (16), neomycin sulfate (32), oleandomycin phosphate (4), penicillin G (16), streptomycin sulfate (16), tobramycin sulfate (32), bismuth citrate (10), cadmium acetate (50), cobalt chloride (10), copper sulfate (10), crystal violet (1), ferrous sulfate (50), lead acetate (100), manganese sulfate (100), potassium tellurite (10), thallous acetate (10), and zinc sulfate (100), but not with chlortetracycline hydrochloride (2), demeclocycline hydrochloride (2), vancomycin hydrochloride (2), bismuth citrate (100), cadmium acetate (50), cobalt chloride (100), crystal violet (10), ferrous sulfate (100), phenol (100), or potassium tellurite (50).

The type strain does not show any sequence similarities over 99% to other *Streptacidiphilus* species, or to any *Kitasatospora* or *Streptomyces* species.

Source: the type strain was isolated from litter and mineral horizons in a spruce forest soil.

DNA G+C content (mol%): not known.

Type strain: JL206, DSM 41755, JCM 12365, KCTC 9911, NBRC 100921.

Sequence accession no. (16S rRNA gene): AF074410.

7. **Streptacidiphilus oryzae** Wang, Huang, Liu, Goodfellow and Rodríguez 2006, 1260[VP]

o.ry'za.e. L. gen. n. *oryzae* of rice, denoting the isolation of the strains from a rice field.

Aerial hyphae differentiate into long flexuous chains of spores (0.7 × 1.0 µm) with smooth surfaces. Forms brown substrate mycelium and abundant grayish-white aerial hyphae on acidified modified Bennett's, inorganic salts-starch, oatmeal, and yeast extract-malt extract agars. Golden brown diffusible pigments are formed on acidified modified Bennett's, oatmeal, and yeast extract-malt extract agars, but not on inorganic salts-starch agar. Degrades adenine, casein, starch, and uric acid, but not elastin, guanine, hypoxanthine, Tween 80, L-tyrosine, xanthine, or xylan. Nitrate is reduced. Esculin, allantoin, and urea are not hydrolyzed. L-Arabinose, D-arabitol, D-cellobiose, D-fructose, D-galactose, D-glucose, glycogen, D-lactose, D-mannitol, D-mannose, raffinose, L-rhamnose, D-salicin (weak), sucrose (weak), trehalose, and D-xylose (each at 1%, w/v), and L-alanine, α-aminobutyric acid, 2-aminoethanol, L-histidine (weak), L-isoleucine (weak), L-phenylalanine, sodium fumarate (weak), sodium pyruvate (weak), L-threonine, and L-valine (each at 0.1%, w/v) are used as sole carbon sources for energy and growth, but not adonitol, dextran, methyl α-D-glucoside, ethanol, and glycine (each at 1%, w/v), or adipic acid, L-arginine, L-aspartic acid, L-cysteine, potassium nitrate, and sodium oxalate (each at 0.1%, w/v). L-Alanine, 2-aminoethanol, L-aspartic acid, L-isoleucine (weak), and L-phenylalanine (each at 0.1%, w/v) are used as sole carbon and nitrogen sources. Growth occurs at pH 3.0–6.5 and at 28 and 37°C and in the presence of (μg/ml) amoxycillin (16), fusidic acid (16), gentamicin sulfate (16, weak), lincomycin hydrochloride (16), and penicillin G (8, weak), but not in the presence of amikacin (32), amoxycillin (32), ampicillin (16, 32), cephalexin (16, 32), cephaloridine hydrochloride (32, 64), clindamycin hydrochloride (4), doxycycline hydrochloride (16), gentamicin sulfate (16), neomycin sulfate (16, 32), novobiocin (8), penicillin G (16), streptomycin sulfate (8, 16), tetracycline hydrochloride (16, 32), or lead acetate (100). Weak growth occurs in the presence of 5% (w/v) NaCl.

The type strain does not show any sequence similarities over 99% to other *Streptacidiphilus* species, or to any *Kitasatospora* or *Streptomyces* species.

Source: the type strain was isolated from a rice-field soil sample collected in Nontaburi Province, Thailand.

DNA G+C content (mol%): not known.

Type strain: TH49, CGMCC 4.2012, JCM 13271.

Sequence accession no. (16S rRNA gene): DQ208700.

8. **Streptacidiphilus rugosus** Cho, Han, Ko and Kim 2008, 1568[VP]

ru.go'sus. L. masc. adj. *rugosus* wrinkled.

Aerial hyphae differentiate into long flexuous chains of spores (0.5 × 1.2 μm); the spore surface is smooth. Forms green-colored, rugose colonies that carry moderate to abundant white aerial hyphae on acidified oatmeal, inorganic salts-starch, yeast extract-malt extract, and modified Bennett's agar plates. The substrate mycelium is cream on acidified inorganic salts-starch, yeast extract-malt extract, and modified Bennett's agars, but yellowish brown or brown on oatmeal agar. Soluble pigments are not produced on any of the above-mentioned media. Starch and Tween 80 are degraded, but xanthine, and Tweens 20 and 40 are not. Glycerol, D-gluconic acid, *myo*-inositol, D-sorbitol, sucrose (all at 1%, w/v), and L-arginine and sodium oxalate (at 0.1%, w/v) are used as sole carbon sources for energy and growth, but D(+)-glucosamine hydrochloride, melibiose, D(+)-xylose (all at 1%, w/v), and L-aspartic acid (at 0.1%, w/v) are not. L-Isoleucine is used as sole nitrogen source. Growth occurs at pH 3.0–8.0 and also between 28 and 35°C. Chemotaxonomic properties are typical of the genus *Streptacidiphilus*. The major fatty acids are $C_{15:0}$ iso (15% of the total fatty acid composition), $C_{15:0}$ anteiso (14.9%), $C_{16:0}$ iso (25.4%), *n*-$C_{16:0}$ (15.6%), $C_{17:0}$ iso (3.0%) and $C_{17:0}$ anteiso (4.7%). Contains hexa- and octahydrogenated menaquinones with nine isoprene units [MK-9(H_6) and MK-9(H_8), 19.8% and 59.6% of total, respectively] as predominant isoprenologs. The diamino acid of the peptidoglycan is LL-A_2pm (94% of the total A_2pm composition), although minor amounts of the *meso*-isomer (6%) are also detected.

The type strain shows highest sequence similarity to *Streptacidiphilus melanogenes*, DQ994689, 99.2%. It does not show any sequence similarities over 99% to any *Kitasatospora* or *Streptomyces* species.

Source: isolated from *Pinus* soil, Anmyeon, near coastal areas in Tea-An, Chungnam, Republic of Korea.

DNA G+C content (mol%): not known.

Type strain: AM-16, NBRC 103186, KCTC 19279.

Sequence accession no. (16S rRNA gene): DQ904547.

References

Adams, M.J. and D.H. Lapwood. 1978. Studies on the lenticel development, surface microflora and infection by common scab (*Streptomyces scabies*) of potato tubers growing in wet and dry soils. Ann. Appl. Biol. *90*: 335–343.

Agre, N.S. 1986. Footnote *f*. *In* Validation of the publication of new names and new combinations previously effectively published outside the IJSB. List no. 22. Int. J. Syst. Bacteriol. *36*: 573–576.

Aharonowitz, Y., G. Cohen and J.F. Martin. 1992. Penicillin and cephalosporin biosynthetic genes: structure, organization, regulation, and evolution. Annu. Rev. Microbiol. *46*: 461–495.

Ahmad, K. and J.A.M. Bhuiyan. 1958. A new antifungal *Streptomyces* species, *Streptomyces corchorusii*. Pak. J. Biol. Agric. Sci. *1*: 137–143.

Ainsa, J.A., N.J. Ryding, N. Hartley, K.C. Findlay, C.J. Bruton and K.F. Chater. 2000. WhiA, a protein of unknown function conserved among gram-positive bacteria, is essential for sporulation in *Streptomyces coelicolor* A3(2). J. Bacteriol. *182*: 5470–5478.

Al-Bari, M.A., M.S. Bhuiyan, M.E. Flores, P. Petrosyan, M. Garcia-Varela and M.A. Islam. 2005. *Streptomyces bangladeshensis* sp. nov., isolated from soil, which produces bis-(2-ethylhexyl)phthalate. Int. J. Syst. Evol. Microbiol. *55*: 1973–1977.

Al-Diwany, L.J. and T. Cross. 1978. Ecological studies on nocardioforms and other actinomycetes in aquatic habitats. Proceedings of the International Symposium on *Nocardia* and *Streptomyces*, 1976, Warsaw, pp. 153–160.

Al-Tai, A., B. Kim, S.B. Kim, G.P. Manfio and M. Goodfellow. 1999. *Streptomyces malaysiensis* sp. nov., a new streptomycete species with rugose, ornamented spores. Int. J. Syst. Bacteriol. *49*: 1395–1402.

Albert, C.A. and V.M. Malaquias de Querioz. 1963. *Streptomyces tuirus* nov. sp., productor do antibiotico tuoromicina. Revista do Instituto de Antibioticos, Universidade do Recife *5*: 43–51.

Alderson, G., M. Goodfellow and D.E. Minnikin. 1985. Menaquinone composition in the classification of *Streptomyces* and other sporoactinomycetes. J. of Gen. Microbiol. *131*: 1671–1679.

Ammann, A., D. Gottlieb, T.D. Brock, H.E. Carter and G.B. Whitfield. 1955. Filipin, an antibiotic effective against fungi. Phytopathology *45*: 559–563.

Anderson, A.S. and E.M. Wellington. 2001. The taxonomy of *Streptomyces* and related genera. Int. J. Syst. Evol. Microbiol. *51*: 797–814.

Anderson, H.W. and D. Gottlieb. 1952. Plant disease control with antibiotics. Economic Bot. *6*: 294–308.

Anderson, L.E., J. Ehrlich, S.H. Sun and P.R. Burkholder. 1956. Strains of *Streptomyces*, the sources of azaserine, elaiomycin, griseoviridin and viridogrisein. Antibiot. Chemother. *6*: 100–115.

Antai, S.P. and D.L. Crawford. 1981. Degradation of softwood, hardwood, and grass lignocelluloses by two *Streptomyces* strains. Appl. Environ. Microbiol. *42*: 378–380.

Antony-Babu, S. and M. Goodfellow. 2008. Biosystematics of alkaliphilic streptomycetes isolated from seven locations across a beach and dune sand system. Antonie van Leeuwenhoek *94*: 581–591.

Antony-Babu, S., J.E. Stach and M. Goodfellow. 2010. Computer-assisted numerical analysis of colour-group data for dereplication of streptomycetes for bioprospecting and ecological purposes. Antonie van Leeuwenhoek *97*: 231–239.

Anzai, Y., T. Okuda and J. Watanabe. 1994. Application of the random amplified polymorphic DNA using the polymerase chain reaction for efficient elimination of duplicate strains in microbial screening. II. Actinomycetes. J. Antibiot. (Tokyo) *47*: 183–193.

Arai, T. 1951. Studies of flavomycin. Taxonomic investigations on the strain, production of the antibiotic and application of cup method to the assay. J. Antibiot. (Tokyo) Ser. A *4*: 215–221.

Arai, T., T. Nakada and M. Suzuki. 1957. Production of viomycin-like substance by a *Streptomyces*. Antibiot. Chemother. *7*.

Arai, T., S. Kuroda, S. Yamagishi and Y. Katoh. 1964. A New Hydroxystreptomycin Source, *Streptomyces subrutilus*. J. Antibiot. (Tokyo) *17*: 23–28.

Arai, T. 1976. Actinomycetes: The Boundary Microorganisms. Toppan, Tokyo, pp. 1–651.

Arcamone, F., C. Bertazzoli, G. Canevazzi, A. Dimarco, M. Ghigne and A. Grein. 1957. La etruscomicina, nuovo antibiotico antifungino prodotto dallo *Streptomyces lucensis*, n. sp. G. Microbiol. *4*: 119–128.

Arcamone, F., C. Bertazzoli, M. Ghione and T. Scotti. 1959. Melanosporin and elaiophylin, new antibiotics from *Streptomyces melanosporus* (sive *melanosporofaciens*) n. sp. G. Microbiol. *7*: 207–216.

Archuleta, J.G. and G.D. Easton. 1981. The cause of deep-pitted scab of potatoes. Am. Potato J. *58*: 385–392.

Arias, M.E., M. Arenas, J. Rodriguez, J. Soliveri, A.S. Ball and M. Hernandez. 2003. Kraft pulp biobleaching and mediated oxidation of a nonphenolic substrate by laccase from *Streptomyces cyaneus* CECT 3335. Appl. Environ. Microbiol. *69*: 1953–1958.

Arishima, M., J.M. Sakamoto and T. Sato. 1956. Studies on an antibiotic *Streptomyces* No. 689 strain. Part I. Taxonomic studies (in Japanese). J. Agric. Chem. Soc. Jpn *30*: 469–471.

Artamonova, O.I. and N.A. Krasil'nikov. 1960. Biology of special groups of actinomycetes. Producers of antibiotics. *In* Transactions of the Institute of Microbiology, vol. 8 (edited by Rautenshtein). Academy of Sciences USSR, pp. 1–344.

Asahi, K., J. Nagatsu and S. Suzuki. 1966. Xanthocidin, a new antibiotic. J. Antibiot. (Tokyo) *19*: 195–199.

Atalan, E., G.P. Manfio, A.C. Ward, R.M. Kroppenstedt and M. Goodfellow. 2000. Biosystematic studies on novel streptomycetes from soil. Antonie van Leeuwenhoek *77*: 337–353.

Atlas, R.M. 1993. Handbook of Microbiological Media. CRC Press, Boca Raton, FL.

August, P.R., L. Tang, Y.J. Yoon, S. Ning, R. Muller, T.W. Yu, M. Taylor, D. Hoffmann, C.G. Kim, X. Zhang, C.R. Hutchinson and H.G. Floss. 1998. Biosynthesis of the ansamycin antibiotic rifamycin: deductions from the molecular analysis of the *rif* biosynthetic gene cluster of *Amycolatopsis mediterranei* S699. Chem. Biol. *5*: 69–79.

Backus, E.J., H.D. Tresner and T.H. Campbell. 1957. The nucleocidin and alazopeptin producing organisms: two new species of *Streptomyces*. Antibiot. Chemother. *7*: 532–541.

Bailey, C.R., C.J. Bruton, M.J. Butler, K.F. Chater, J.E. Harris and D.A. Hopwood. 1986. Properties of *in vitro* recombinant derivatives of pJV1, a multi-copy plasmid from *Streptomyces phaeochromogenes*. J. Gen. Microbiol. *132*: 2071–2078.

Baldacci, E. 1944. Contributo alla sistematica degli attinomiceti: X–XVI - *Actinomyces madurae*; *Proactinomyces ruber*; *Proactinomyces pseudomadurae*, *Proactinomyces polychromogenus*; *Actinomyces violaceus*; *Actinomyces caeruleus*; con un elenco alfabetico delle specie e delle varieta finora studiate. Atti dell'Istituto Botanico della Universita Laboratorio Crittogamico di Pavia Series 5 *3*: 139–193.

Baldacci, E., C. Spalla and A. Grein. 1954. The classification of the *Actinomyces* species (*Streptomyces*). Arch. Mikrobiol. *20*: 347–357.

Baldacci, E., A. Grein and C. Spalla. 1955. Studio di una "Serie" di specie di attinomiceti: *A. diastaticus*. G. Microbiol. *1*: 127–143.

Baldacci, E. 1958. Development in the classification of actinomycetes. G. Microbiol. *6*: 10–27.

Baldacci, E., G. Farina and R. Locci. 1966. Emendation of genus *Streptoverticillium* Baldacci (1958) and revision of some species. G. Microbiol. *14*: 153–171.

Baldacci, E. and A. Grein. 1966. *Streptomyces avellaneus* and *Streptomyces libani*: two new species characterized by a hazel-nut brown (*Avellaneus*) aerial mycelium. G. Microbiol. *14*: 185–198.

Baldacci, E. and R. Locci. 1974. Genus *Streptoverticillium*. *In* Bergey's Manual of Determinative Bacteriology, 8th edn (edited by Buchanan and Gibbons). Williams & Wilkins, Baltimore, pp. 829–842.

Banchio, C. and H.C. Gramajo. 1997. Medium- and long-chain fatty acid uptake and utilization by *Streptomyces coelicolor* A3(2): first characterization of a gram-positive bacterial system. Microbiology *143*: 2439–2447.

Bapteste, E. and Y. Boucher. 2008. Lateral gene transfer challenges principles of microbial systematics. Trends Microbiol *16*: 200–207.

Bapteste, E., M.A. O'Malley, R.G. Beiko, M. Ereshefsky, J.P. Gogarten, L. Franklin-Hall, F.J. Lapointe, J. Dupre, T. Dagan, Y. Boucher and W. Martin. 2009. Prokaryotic evolution and the tree of life are two different things. Biol. Direct *4*: 34.

Barr, F.S. and P.E. Carman. 1956. *Streptomyces kentuckensis*, a new species, the producer of raisnomycin. Antibiot. Chemother. *6*: 286–289.

Bartz, Q.R., J. Ehrlich, J.D. Mold, M.A. Penner and R.M. Smith. 1951. Viomycin, a new tuberculostatic antibiotic. Am. Rev. Tuberc. *63*: 4–6.

Batra, S.K. and B.S. Bajaj. 1965. *Streptomyces anandii* – a new species of *Streptomyces* isolated from soil. Ind. J. Exp. Biol. *3*: 240–242.

Behal, V. 2000. Bioactive products from *Streptomyces*. Adv. Appl. Microbiol. *47*: 113–156.

Beijerinck, M.W. 1912. Mutation bei Mikroben. Folia Mikrobiologiya (Delft) *1*: 4–100.

Benedict, R.G., L.A. Lidenfelser, F.H. Stodola and D.H. Traufler. 1950. Studies on *Streptomyces griseocarneus* and the production of hydroxystreptomycin. J. Bacteriol. *62*: 487–497.

Benedict, R.G., W. Dvonch, O.L. Shotwell, T.G. Pridham and L.A. Lidenfelser. 1952. Cinnamycin, an antibiotic from *Streptomyces cinnamoneus* sp. nov. Antibiot. Chemother. *2*: 591–594.

Benedict, R.G., O.L. Shotwell, T.G. Pridham, L.A. Lidenfelser and W.C. Haynes. 1954. The production of the neomycin complex by *Streptomyces albogriseolus*, nov. sp. Antibiot. Chemother. *4*: 653–656.

Benedict, R.G., T.G. Pridham, L.A. Lindenfelser, H.H. Hall and R.W. Jackson. 1955. Further studies in the evaluation of carbohydrate utilization tests as aids in the differentiation of species of *Streptomyces*. Appl. Microbiol. *3*: 1–6.

Bentley, S.D., K.F. Chater, A.M. Cerdeno-Tarraga, G.L. Challis, N.R. Thomson, K.D. James, D.E. Harris, M.A. Quail, H. Kieser, D. Harper, A. Bateman, S. Brown, G. Chandra, C.W. Chen, M. Collins, A. Cronin, A. Fraser, A. Goble, J. Hidalgo, T. Hornsby, S. Howarth, C.H. Huang, T. Kieser, L. Larke, L. Murphy, K. Oliver, S. O'Neil, E. Rabbinowitsch, M.A. Rajandream, K. Rutherford, S. Rutter, K. Seeger, D. Saunders, S. Sharp, R. Squares, S. Squares, K. Taylor, T. Warren, A. Wietzorrek, J. Woodward, B.G. Barrell, J. Parkhill and D.A. Hopwood. 2002. Complete genome sequence of the model actinomycete *Streptomyces coelicolor* A3(2). Nature *9*: 141–147.

Bérdy, J. 2005. Bioactive microbial metabolites. J. Antibiot. (Tokyo) *58*: 1–26.

Berestnev, N. 1897. Actinomycosis and its Causes. Moscow University, Moscow.

Berger, D.R. and D.M. Reynolds. 1958. The chitinase system of a strain *Streptomyces griseus*. Biochim. Biophys. Acta *29*: 522–534.

Berger, J., L.M. Jampolsky and M.W. Goldberg. 1953. A Guide to the Classification of the Actinomycetes and their Antibiotics (edited by Waksman and Lechevalier). Williams & Wilkins, Baltimore, pp. 1–246.

Bergey, D.H., F.C. Harrison, R.S. Breed, B.W. Hammer and F.M. Huntoon. 1923. Bergey's Manual of Determinative Bacteriology. Williams & Wilkins, Baltimore.

Beyazova, M. and M.P. Lechevalier. 1993. Taxonomic utility of restriction-endonuclease fingerprinting of large DNA fragments from *Streptomyces* strains. Int. J. Syst. Bacteriol. *43*: 674–682.

Beyer, M. and H. Diekmann. 1985. The chitinase system of *Streptomyces* sp. ATCC 11238 and its significance for fungal cell wall degradation. Appl. Microbiol. Biotechnol. *23*: 140–146.

Bhuyan, B.K., A. Dietz and C.G. Smith. 1962. Pactamycin, a new antitumor antibiotic. I. Discovery and biological properties. Antimicrob. Agents Chemother. *1961*: 184–190.

Bhuyan, B.K., S.P. Owen and A. Dietz. 1965. Rubradirin, a new antibiotic. I. Fermentation and biological properties. Antimicrob. Agents Chemother *1964*: 91–96.

Bhuyan, B.K. and A. Dietz. 1966. Fermentaion, taxonomic and biological studies on nogalamycin. Antimicrob. Agents Chemother *1965*: 836–844.

Bianchi, M.L., A. Grein, P. Julita, M.P. Marnati and C. Spalla. 1970. *Streptomyces mediolani* (Arcamone *et al.*) emend. Bianchi *et al.* and its production of carotenoids. Z. Allg. Mikrobiol. *10*: 237–244.

Bibb, M.J., D.H. Sherman, S. Ōmura and D.A. Hopwood. 1994. Cloning, sequencing and deduced functions of a cluster of *Streptomyces* genes probably encoding biosynthesis of the polyketide antibiotic frenolicin. Gene *142*: 31–39.

Bignell, D.E., H. Oskarsson and J.M. Anderson. 1980. Colonization of the epithelial face of the peritrophic membrane and the ectoperitrophic space by actinomycetes in a soil-feeding termite. J. Invertebr. Pathol. *36*: 426–428.

Bignell, D.E., H. Oskarsson and J.M. Anderson. 1981. Association of actinomycetes with soil-feeding termites: a novel symbiotic relationship?. *In Actinomycetes.* Proceedings of the 4th International Symposium on Actinomycete Biology, Cologne, 1979 (edited by Schaal and Pulverer). Gustav Fischer Verlag, Stuttgart, pp. 201–206.

Bignell, D.E. 1984. The arthropod gut as an environment for microorganisms. In Invertebrate–Microbial Interactions (edited by Anderson, Rayner and Walton). Cambridge University Press, Cambridge, UK, pp. 205–227.

Binnie, C., M. Warren and M.J. Butler. 1989. Cloning and heterologous expression in *Streptomyces lividans* of *Streptomyces rimosus* genes involved in oxytetracycline biosynthesis. J. Bacteriol. *171*: 887–895.

Blaak, H. and H. Schrempf. 1995. Binding and substrate specificities of a *Streptomyces olivaceoviridis* chitinase in comparison with its proteolytically processed form. Eur. J. Biochem. *229*: 132–139.

Blanco, G., M.R. Rodicio, A.M. Puglia, C. Mendez, C.J. Thompson and J.A. Salas. 1994. Synthesis of ribosomal proteins during growth of *Streptomyces coelicolor*. Mol. Microbiol. *12*: 375–385.

Blinov, H.O., C.A. Ezoрова and И.В. Machenko. 1975. В кН.: Биология лучистых грибко М: Наука (in Russian). 152–164.

Bormann, C., V. Mohrle and C. Bruntner. 1996. Cloning and heterologous expression of the entire set of structural genes for nikkomycin synthesis from *Streptomyces tendae* Tu901 in *Streptomyces lividans*. J. Bacteriol. *178*: 1216–1218.

Bouchek-Mechiche, K., L. Gardan, P. Normand and B. Jouan. 2000. DNA relatedness among strains of *Streptomyces* pathogenic to potato in France: description of three new species, *S. europaeiscabiei* sp. nov., and *S. stelliscabiei* sp. nov. associated with common scab, and *S. reticuliscabiei* sp. nov. associated with netted scab. Int. J. Syst. Evol. Microbiol. *50*: 91–99.

Boucher, Y. and E. Bapteste. 2009. Revisiting the concept of lineage in prokaryotes: a phylogenetic perspective. Bioessays *31*: 526–536.

Bowen, T., E. Stackebrandt, M. Dorsch and T.M. Embley. 1989. The phylogeny of *Amycolata autotrophica*, *Kibdelosporangium aridum* and *Saccharothrix australiensis*. J. Gen. Microbiol. *135*: 2529–2536.

Braznikova, M.G., T.A. Uspenskaya, L.B. Sokolova, T.P. Preobrazhenskaya, G.F. Gause, R.S. Ukholina, V.A. Shorin, O.K. Rossolimo and T.P. Vertogradova. 1958. A new antivirus antibiotic – heliomycin. Antibiotiki (in Russian) *3*: 29–34.

Brian, P.W. 1957. The ecological significance of antibiotic production. *In* Microbial Ecology (edited by Williams and Spicer). Cambridge University Press, Cambridge, UK, pp. 168–188.

Brown, R., E.L. Hazen and A. Mason. 1953. Effect of fungicidin (nystatin) in mice injected with lethal mixtures of aureomycin and *Candida albicans*. Science *117*: 609–610.

Brown, R.L. and G.E. Peterson. 1966. Cholesterol oxidation by soil actinomycetes. J. Gen. Microbiol. *45*: 441–450.

Brumpt, E. 1906. Les Mycétomes. Arch. Parasitol. *10*: 489–527.

Brüsewitz, G. 1959. Untersuchungen über den Einfluss des Regenwurms auf Zahl, Art und Leistungen von Mikroorganismen im Boden. Arch. Microbiol. *33*: 52–82.

Burg, R.W., B.M. Miller, E.E. Baker, J. Birnbaum, S.A. Currie, R. Hartman, Y.L. Kong, R.L. Monaghan, G. Olson, I. Putter, J.B. Tunac, H. Wallick, E.O. Stapley, R. Oiwa and S. Ōmura. 1979. Avermectins, new family of potent anthelmintic agents: producing organism and fermentation. Antimicrob. Agents Chemother. *15*: 361–367.

Burkholder, P.R., S.H. Sun, J. Ehrlich and L. Anderson. 1954. Criteria of speciation in the genus *Streptomyces*. Ann. N.Y. Acad. Sci. *60*: 102–123.

Burman, N.P., C.P. Oliver and J.K. Stevens. 1969. Membrane filtration techniques for the isolation from water, of coli-aerogenes, *Escherichia coli*, faecal streptococci, *Clostridium perfringens*, actinomycetes and microfungi. *In* Isolation Methods for Microbiologists (edited by Shapton and Gould). Academic Press, London, pp. 127–134.

Burman, N.P. 1973. The occurrence and significance of actinomycetes in water supply. *In Actinomycetales:* Characteristics and Practical Importance (edited by Sykes and Skinner). Academic Press, London, pp. 219–230.

Burns, J. and D.F. Holtman. 1959. Tennecetin: a new antifungal antibiotic. Antibiot. Chemother. *9*: 398–405.

Butler, M.J., P. Bruheim, S. Jovetic, F. Marinelli, P.W. Postma and M.J. Bibb. 2002. Engineering of primary carbon metabolism for improved antibiotic production in *Streptomyces lividans*. Appl. Environ. Microbiol. *68*: 4731–4739.

Calcutt, M.J. and F.J. Schmidt. 1992. Conserved gene arrangement in the origin region of the *Streptomyces coelicolor* chromosome. J. Bacteriol. *174*: 3220–3226.

Calot, L. and A.P. Cercos. 1963. *Streptomyces ornatus*, nov. sp. et *Streptomyces erumpens*, nov. sp. producteurs d'ornamicine et antibiotique 17732. Annales de l'Institut Pasteur (Paris) *105*: 159–161.

Caminiti, R. 1907. Über eine neue Streptothrix species und die Streptothricheen. Allgemeines Zentralbl. Bakteriol. Parasitenk. Infektionskr. Hyg. Abt. Orig. *44*: 193–208.

Canevazzi, G. and T. Scotti. 1959. Descrizione di uno streptomicete (*Streptomyces chrestomyceticus*) sp. nova, produttore del nuovo antibiotico amminosidina. G. Microbiol. *7*: 242–250.

Carvajal, F. 1947. The Production of Spores in Submerged Cultures by Some *Streptomyces*. Mycologia *39*: 426–440.

Carvajal, F. 1953. Phage problems in the streptomycin fermentation. Mycologia *45*: 209–234.

Cassinelli, G., A. Grein, P. Orezzi, P. Pennella and A. Sanfilippo. 1967. New antibiotics produced by *Streptoverticillium orinoci*, n. sp. Arch. Mikrobiol. *55*: 358–368.

Castellani, A. and A.J. Chalmers. 1913. Manual of Tropical Medicine, 2nd edn. Baillière, Tindall and Cox, London.

Cataldi, M. 1963. In Trejo and Bennett (Editors), *Streptomyces* species comprising the blue-spore series. J. Bacteriol. *85*: 676–690.

Cercos, A.P., B.L. Eilberg, J.G. Goyena, J. Souto, E.E. Vautier and I. Widuczynski. 1962. Misionina: antibiotico polienico producido por *Streptomyces misionensis* n. sp. Revista de Investigaciones Agricolas *16*: 5–27.

Cercós, A.P. 1954. *Streptomyces rutgersensis* var. *castelarense* n. var. Nuevas propiedades de la canfomicina. Revista de Investigaciones Agricolas *8*: 263–283.

Challis, G.L. and D.A. Hopwood. 2003. Synergy and contingency as driving forces for the evolution of multiple secondary metabolite production by *Streptomyces* species. Proc. Natl. Acad. Sci. U.S.A. *100*: 14555–14561.

Chalmers, A.J. and J.B. Christopherson. 1916. A Sudanese actinomycosis. Ann. Trop. Med. Parasitol. *10*: 223–282.

Chamberlain, K. and D.L. Crawford. 2000. Thatch biodegradation and antifungal activities of two lignocellulolytic *Streptomyces* strains in laboratory cultures and in golf green turfgrass. Can. J. Microbiol. *46*: 550–558.

Chandramohan, D., S. Ramu and Nataraja.R. 1972. Cellulolytic activity of marine streptomycetes. Curr. Sci. *41*: 245–246.

Charney, J., W.P. Fisher, C. Curran, R.A. Machlowitz and A.A. Tytell. 1953. Streptogramin, a new antibiotic. Antibiot. Chemother. *3*: 1283–1286.

Chater, K.F. 1979. *Streptomycetes. In* Development Biology of Prokaryotes (edited by Parish). Blackwell Scientific Publications, Oxford, pp. 93–114.

Chater, K.F. and C.J. Bruton. 1985. Resistance, regulatory and production genes for the antibiotic methylenomycin are clustered. EMBO J. *4*: 1893–1897.

Chater, K.F. 1986. *Streptomyces* phages and their applications for *Streptomyces* genetics. *In* The Bacteria, Antibiotic-producing *Streptomyces*, vol. 9 (edited by Queener and Day). Academic Press, Orlando, pp. 119–158.

Chater, K.F., N.D. Lomovskaya, T.A. Voeykova, I.A. Sladkova, N.M. Mkrtumian and G.L. Muravnik. 1986. *Streptomyces* ϕC31-like phages: cloning vectors, genome changes and host range. *In* Biological, Biochemical and Biomedical Aspects of Actinomycetes (edited by Szabo, Biro and and Goodfellow). Akademiai Kiado, Budapest, pp. 45–54.

Chater, K.F. and S. Horinouchi. 2003. Signalling early developmental events in two highly diverged *Streptomyces* species. Mol. Microbiol. *48*: 9–15.

Chater, K.F. and G. Chandra. 2006. The evolution of development in *Streptomyces* analysed by genome comparisons. FEMS Microbiol. Rev. *30*: 651–672.

Chen, C.W. 1995. The unstable ends of the *Streptomyces* linear chromosomes: A nuisance without cures? Trends Biotechnol. *13*: 157–160.

Chen, C.W., C.H. Huang, H.H. Lee, H.H. Tsai and R. Kirby. 2002. Once the circle has been broken: dynamics and evolution of *Streptomyces* chromosomes. Trends Genet. *18*: 522–529.

Chesters, C.G.C., A. Apinis and M. Turner. 1956. Studies of the decomposition of seaweeds and seaweed products by microorganisms. Proc. Linn. Soc. Lond. *166*: 87–97.

Cho, S.H., J.H. Han, C.N. Seong and S.B. Kim. 2006. Phylogenetic diversity of acidophilic sporoactinobacteria isolated from various soils. J. Microbiol. *44*: 600–606.

Cho, S.H., J.H. Han, H.Y. Ko and S.B. Kim. 2008. *Streptacidiphilus anmyonensis* sp. nov., *Streptacidiphilus rugosus* sp. nov. and *Streptacidiphilus melanogenes* sp. nov., acidophilic actinobacteria isolated from Pinus soils. Int. J. Syst. Evol. Microbiol. *58*: 1566–1570.

Choulet, F., B. Aigle, A. Gallois, S. Mangenot, C. Gerbaud, C. Truong, F.X. Francou, C. Fourrier, M. Guerineau, B. Decaris, V. Barbe, J.L. Pernodet and P. Leblond. 2006. Evolution of the terminal regions of the *Streptomyces* linear chromosome. Mol. Biol. Evol. *23*: 2361–2369.

Chun, J., H.D. Youn, Y.I. Yim, H. Lee, M.Y. Kim, Y.C. Hah and S.O. Kang. 1997. *Streptomyces seoulensis* sp. nov. Int. J. Syst. Bacteriol. *47*: 492–498.

Chung, Y.R., K.C. Sung, H.K. Mo, D.Y. Son, J.S. Nam, J.S. Chun and K.S. Bae. 1999. *Kitasatospora cheerisanensis* sp. nov., a new species of the genus *Kitasatospora* that produces an antifungal agent. Int. J. Syst. Bacteriol. *49*: 753–758.

Ciferri, R. 1927. Studien über Kakao. Untersuchungen über den muffigen Geruch der Kakaobohnen. Zentralbl. Bakteriol. Parasitenkd. Infektionskr. Hyg. Abt. II *71*: 80–93.

Claessen, D., H.A. Wosten, G. van Keulen, O.G. Faber, A.M. Alves, W.G. Meijer and L. Dijkhuizen. 2002. Two novel homologous proteins of *Streptomyces coelicolor* and *Streptomyces lividans* are involved in the formation of the rodlet layer and mediate attachment to a hydrophobic surface. Mol. Microbiol. *44*: 1483–1492.

Clarke, S.D., D.A. Ritchie and S.T. Williams. 1993. Ribosomal DNA restriction fragment analysis of some closely related *Streptomyces* species. Syst. Appl. Microbiol. *16*: 256–260.

Coffey, G.L., L.E. Anderson, M.W. Fisher, M.M. Galbraith, A.B. Hillegas, D.L. Kohberger, P.E. Thompson and J. Weston Ks Ehrlich. 1959. Biological studies of paromomycin. Antibiot. Chemother. *9*: 730–738.

Collins, M.D. and D. Jones. 1981. Distribution of isoprenoid quinone structural types in bacteria and their taxonomic implication. Microbiol. Rev. *45*: 316–354.

Collins, M.D., M. Faulkner and R.M. Keddie. 1984. Menaquinone composition of some sporeforming actinomycetes. Syst. Appl. Microbiol. *5*: 20–29.

Conn, H.J. 1917. Soil flora studies. V. Actinomycetes in soil. Bull. N.Y. State Agric. Exp. Stat. *60*: 3–25.

Corbaz, R., L. Ettlinger, E. Gäumann, W. Keller, F. Kradolfer, E. Kyburz, L. Neipp, V. Prelog, R. Reusser and H. Zähner. 1955. Stoffwechselprodukte Von Actinomyceten.1. Narbomycin. Helv. Chim. Acta *38*: 935–942.

Corbaz, R., L. Ettlinger, E. Gäumann, W. Kellerschierlein, F. Kradolfer, L. Neipp, V. Prelog, P. Reusser and H. Zähner. 1957a. Stoffwechselprodukte Von Actinomyceten.7. Echinomyccin. Helv. Chim. Acta *40*: 199–204.

Corbaz, R., L. Ettlinger, W. Keller-Schierlein and H. Zähner. 1957b. Zur Systematik der Actinomyceten. 1. Über Streptomyceten mit rhodomycinastigen Pigmenten. Arch. Mikrobiol. *25*: 325–332.

Corke, C.T. and F.E. Chase. 1956. The selective enumeration of Actinomycetes in the presence of large numbers of fungi. Can. J. Microbiol. *2*: 12–16.

Craveri, R. and H. Pagani. 1962. Thermophilic microorganisms among actinomycetes in the soil. Annali di Microbiologia *12*: 115–130.

Crawford, D.L. and E. McCoy. 1972. Cellulases of *Thermomonospora fusca* and *Streptomyces thermodiastaticus*. Appl. Microbiol. *24*: 150–152.

Crawford, D.L. 1978. Lignocellulose decomposition by selected *Streptomyces* strains. Appl. Environ. Microbiol. *35*: 1041–1045.

Crawford, D.L. 1988. Biodegradation of agricultural and urban wastes. *In* Actinomycetes in Biotechnology (edited by Goodfellow, Williams and Mordarski). Academic Press, London, pp. 433–459.

Crawford, D.L., J.D. Doyle, Z. Wang, C.W. Hendricks, S.A. Bentjen, H. Bolton, Jr, J.K. Fredrickson and B.H. Bleakley. 1993. Effects of a lignin peroxidase-expressing recombinant, *Streptomyces lividans* TK23.1, on biogeochemical cycling and the numbers and activities of microorganisms in soil. Appl. Environ. Microbiol. *59*: 508–518.

Crawford, R.L. 1981. Lignin Biodegradation and Transformation. John Wiley & Sons, New York.

Crespi, M., E. Messens, A.B. Caplan, M. van Montagu and J. Desomer. 1992. Fasciation induction by the phytopathogen *Rhodococcus fascians* depends upon a linear plasmid encoding a cytokinin synthase gene. EMBO J. *11*: 795–804.

Cron, M.J., D.F. Whitehead, I.R. Hooper, B. Heinemann and J. Lein. 1956. Bryamycin, a new antibiotic. Antibiot. Chemother. *6*: 63–67.

Crook, P., C.C. Carpenter and P.F. Klens. 1950. The use of sodium propionate in isolating actinomycetes from soils. Science *111*: 656.

Cross, T., M.P. Lechevalier and H. Lechevalier. 1963. A new genus of the *Actinomycetales*: *Microellobosporia* gen. nov. J. Gen. Microbiol. *31*: 421–429.

Cross, T. 1968. Thermophilic actinomycetes. J. Appl. Bacteriol. *31*: 36–53.

Cross, T. 1981a. The monosporic actinomycetes. *In* The Prokaryotes: a Handbook on Habitats, Isolation, and Identification of Bacteria (edited by Starr, Stolp, Trüper, Balows and Schlegel). Springer, New York, pp. 2091–2102.

Cross, T. 1981b. Aquatic actinomycetes: a critical survey of the occurrence, growth and role of actinomycetes in aquatic habitats. J. Appl. Bacteriol. *50*: 397–423.

Cross, T. 1982. Actinomycetes: a continuing source of new metabolites. Dev. Indust. Microbiol. *23*: 1–18.

Cundell, A.M. and A.P. Mulcock. 1975. The biodegradation of vulcanized rubber. Dev. Indust. Microbiol. *16*: 88–96.

Dagan, T., Y. Artzy-Randrup and W. Martin. 2008. Modular networks and cumulative impact of lateral transfer in prokaryote genome evolution. Proc. Natl. Acad. Sci. India *105*: 10039–10044.

Danga, F. and D. Gottlieb. 1959. English translation of G.F. Gauze *et al.*, 1957. Problems in the classification of antagonistic actinomycetes. The American Institute of Biological Sciences, Washington, D.C.

Davisson, J.W. and A.C. Finlay. 1961. The Actinomycetes. Vol. 2. *In* Classification, Identification and Descriptions of Genera and Species (edited by Waksman). Williams & Wilkins, Baltimore, pp. 1–363.

de Queiroz, V.M. and C.A. Albert. 1962. *Streptomyces iakyrus* nov. sp., produtor dos antibióticos Iaquirina I, IIe, III. Revista do Instituto de Antibioticos, Universidade do Recife *4*: 33–46.

De Vos, P., H.G. Trüper and B.J. Tindall. 2005. Judicial Commission of the International Committee on Systematics of Prokaryotes. Xth International (IUMS) Congress of Bacteriology and Applied Microbiology Minutes. Int. J. Syst. Bacteriol. *55*: 525–532.

DeBoer, C., A. Dietz, J.R. Wilkins, C.N. Lewis and G.M. Savage. 1955a. Celesticetin – a new crystalline antibiotic. I. Biologic studies of celesticetin. Antibiot. Ann. *1954*: 831–841.

Deboer, C., A. Dietz, G.M. Savage and W.S. Silver. 1955b. Streptolydigin, a new antimicrobial antibiotic. I. Biologic studies of streptolydigin. Antibiot. Annu. *3*: 886–892.

DeBoer, C., A. Dietz, J.S. Evans and R.M. Michaels. 1959–1960. Fervenulin, a new crystalline antibiotic. I. Discovery and biological activities. Antibiot. Annu. *7*: 220–226.

DeBoer, C., A. Dietz, N.E. Lummis and G.M. Savage. 1961. Porfiromycin, a new antibiotic. I. Discovery and biological activities. Antimicrob. Agents Annu. *1960–1961*: 17–22.

Decker, H. and S. Haag. 1995. Cloning and characterization of a polyketide synthase gene from *Streptomyces fradiae* Tu2717, which carries the genes for biosynthesis of the angucycline antibiotic urdamycin A and a gene probably involved in its oxygenation. J. Bacteriol. *177*: 6126–6136.

Delafield, F.P., M. Doudoroff, N.J. Palleroni, C.J. Lusty and R. Contopoulos. 1965. Decomposition of poly-beta-hydroxybutyrate by pseudomonads. J. Bacteriol. *90*: 1455–1466.

Deobald, L.A. and D.L. Crawford. 1987. Activities of cellulase and other extracellular enzymes during lignin solubilization by *Streptomyces viridosporus*. Appl. Microbiol. Biotechnol. *26*: 158–163.

Desai, A.J. and S.A. Dhala. 1967. *Streptomyces thermonitrificans* sp.n., a thermophilic streptomycete. Antonie van Leeuwenhoek *33*: 137–144.

Develoux, M., M.T. Dieng and B. Ndiaye. 1999. Mycetoma of the neck and scalp in Dakar. J. Mycol. Med. *9*: 179–209.

Dharmatilake, A.J. and K.E. Kendrick. 1994. Expression of the division-controlling gene *ftsZ* during growth and sporulation of the filamentous bacterium *Streptomyces griseus*. Gene *147*: 21–28.

Diab, A. and M.Y. Algounaim. 1982. *Streptomyces spinoverrucosus*, a new species from the air of Kuwait. Int. J. Syst. Bacteriol. *32*: 327–331.

Dietz, A. and J. Mathews. 1971. Classification of *Streptomyces* spore surfaces into five groups. Appl. Microbiol. *21*: 527–533.

Distler, J., A. Ebert, K. Mansouri, K. Pissowotzki, M. Stockmann and W. Piepersberg. 1987. Gene cluster for streptomycin biosynthesis in *Streptomyces griseus:* nucleotide sequence of three genes and analysis of transcriptional activity. Nucleic Acids Res. *15*: 8041–8056.

Dodge, C.W. 1935. Medical mycology. Fungous diseases of men and other mammals. C.V. Mosby Co., St Louis.

Donadio, S., M. Sosio and G. Lancini. 2002. Impact of the first *Streptomyces* genome sequence on the discovery and production of bioactive substances. Appl. Microbiol. Biotechnol. *60*: 377–380.

Doroghazi, J.R. and D.H. Buckley. 2010. Widespread homologous recombination within and between *Streptomyces* species. ISME J. *4*: 1136–1143.

Dosch, D.C., W.R. Strohl and H.G. Floss. 1988. Molecular cloning of the nosiheptide resistance gene from *Streptomyces actuosus* ATCC 25421. Biochem. Biophys. Res. Commun. *156*: 517–523.

Duché, J. 1934. Les *Actinomyces* du groupe albus. Encyclopédie Mycologique *6*: 1–375.

Duché, J., R. Heim and P. Laboureur. 1951. Mémoire sur l'Antennopsis, ectoparasite du termite de Saintonge. IV. *In* Étude du *Streptomyces termitum* n. sp., associé à l'antennopsis, vol. 67 (edited by Heim). Bulletin de la Société Mycologique France, pp. 359–364.

Ducote, M.J., S. Prakash and G.S. Pettis. 2000. Minimal and contributing sequence determinants of the *cis*-acting locus of transfer (*clt*) of streptomycete plasmid pIJ101 occur within an intrinsically curved plasmid region. J. Bacteriol. *182*: 6834–6841.

Duggar, B.M. 1948. Aureomycin; a product of the continuing search for new antibiotics. Ann. N.Y. Acad. Sci. *51*: 177–181.

Dyson, P. 2010. *Streptomyces*: Molecular Biology and Biotechnology. Caister Academic Press.

Eguchi, T., N. Takada, S. Nakamura, T. Tanaka, T. Makino and Y. Oshima. 1993. *Streptomyces bungoensis* sp. nov. Int. J. Syst. Bacteriol. *43*: 794–798.

Ehrlich, J., D. Gottlieb et al. 1948. *Streptomyces venezuelae*, n. sp., the source of chloromycetin. J. Bacteriol. *56*: 467–477.

El-Nakeeb, M.A. and H.A. Lechevalier. 1963. Selective isolation of aerobic actinomycetes. Appl. Microbiol. *11*: 75–77.

Enger, M.D. and B.P. Sleeper. 1965. Multiple cellulase system from *Streptomyces antibioticus*. J. Bacteriol. *89*: 23–27.

Ensign, J.C. 1978. Formation, properties, and germination of actinomycete spores. Annu. Rev. Microbiol. *32*: 185–219.

Epp, J.K., S.G. Burgett and B.E. Schoner. 1987. Cloning and nucleotide sequence of a carbomycin-resistance gene from *Streptomyces thermotolerans*. Gene *53*: 73–83.

Erikson, D. 1935. The pathogenic aerobic organisms of the actinomyces group. Med. Res. Coun. Spec. Rep. Ser. No. *203*: 5–61.

Esnard, J., T.L. Potter and B.M. Zuckerman. 1995. *Streptomyces costaricanus* sp. nov., isolated from nematode-suppressive soil. Int. J. Syst. Bacteriol. *45*: 775–779.

Ettlinger, L., R. Corbaz and R. Hütter. 1958a. Zur Systematik der Actinomyceten. 4. Eine Arteinteilung der Gattung *Streptomyces* Waksman et Henrici. Arch. Mikrobiol. *31*: 326–358.

Ettlinger, L., E. Gäumann, R. Hütter, W. Keller-Schierlein, F. Kradolfer, L. Neipp, V. Prelog and H. Zähner. 1958b. Stoffwechselprodukte von Actimomyceten. 12. Mitteilung über die Isolierung und Characteristierung von Acetomycin. Helv. Chim. Acta *41*: 216–219.

Euzéby, J.P. 1998. Taxonomic note: necessary correction of specific and subspecific epithets according to Rules 12c and 13b of the

International Code of Nomenclature of Bacteria (1990 Revision). Int. J. Syst. Bacteriol. *48*: 1073–1075.

Fahal, A.H. and M.A. Hassan. 1992. Mycetoma. Br. J. Surg. *79*: 1138–1141.

Fahal, A.H. 2004. Mycetoma: a thorn in the flesh. Trans. R. Soc. Trop. Med. Hyg. *98*: 3–11.

Fahal, A.H. 2006. Mycetoma – Clinicopathological Monograph. Khartoum University Press, Khartoum, p. 112.

Fairbairn, D.A., F.G. Priest and J.R. Stark. 1986. Extracellular amylase synthesis by *Streptomyces limosus*. Enzyme Microb. Technol. *8*: 89–92.

Falcão de Morais, J.O., O. Gonçalves de Lima and M.H. Dália Maia. 1957. Novo estudo sôbre *Nocardia recifei* Lima *et al.*, e sua designacao como *Streptomyces recifensis*. Anais da Sociedade de Biologia de Pernambuco *15*: 239–253.

Falcão de Morais, J.O., M.H.D. Maia and M.E.S.M. Genn. 1958. Sôbre uma variedade de *Streptomyces* comum nos solos do Brasil: *Streptomyces venzuelae* var. *roseospori* nov. var. Rev. Inst. Antibiot. Univ. Recife *1*: 99–106.

Falcão de Morais, J.O. and M.H. Dália Maia. 1959. *S. erythrogriseus*: novo *Streptomyces* produtor de antibiotico. Revista do Instituto de Antibioticos, Universidade do Recife *2*: 63–67.

Falcão de Morais, J.O. and M.H. Dália Maia. 1961. Uma contribuição ao estudo toxonômico do gênero *Streptomyces* - Uma tentativa de simplificação. Rev. Inst. Antibiot. Univ. Recife *3*: 33–60.

Falcão de Morais, J.O., A. Chaves Batista and D.M.G. Massa. 1966. *Elytrosporangium*: a new genus of the *Actinomycetales*. Mycopathol. Mycol. Appl. *30*: 161–171.

Falcão de Morais, J.O. 1970. *Elytrosporangium spirale*: nova especie de *Actinoplanaceae* do genero *Elytrosporangium*. Rev. Microbiol. *1*: 79–84.

Falcão de Morais, J.O., J. Da Silva and C. Machado. 1971. Uma Terceira especie de *Actinomycetales* do genero *Elytrosporangium*, *E. carpinense* sp. nov., isolada de solo em Pernambuco. Rev. Microbiol. Brazil *2*: 203–206.

Farina, G. and R. Locci. 1966. Contribution to the study of *Streptoverticillium*: description of a new species (*Streptoverticillium baldacci* sp. nov.) and examination of previously illustrated species (In Italian). G. Microbiol. *14*: 33–52.

Feitelson, J.S. and D.A. Hopwood. 1983. Cloning of a *Streptomyces* gene for an *O*-methyltransferase involved in antibiotic biosynthesis. Mol. Gen. Genet. *190*: 394–398.

Fergus, C.L. 1964. Thermophilic and thermotolerant molds and actinomycetes of mushroom compost during peak heating. Mycologia *56*: 267–284.

Ferguson, E.V., A.C. Ward, J.J. Sanglier and M. Goodfellow. 1997. Evaluation of *Streptomyces* species-groups by pyrolysis mass spectrometry. Zentralbl. Bakteriol. *285*: 169–181.

Festenstein, G.N., J. Lacey, F.A. Skinner, P.A. Jenkins and J. Pepys. 1965. Selfheating of hay and grain in Dewar flasks and the development of farmer's lung antigens. J. Gen. Microbiol. *41*: 389–407.

Finlay, A.C., F.A. Hochstein, B.A. Sobin and F.X. Murphy. 1951. Netropsin, a new antibiotic produced by a *Streptomyces*. J. Am. Chem. Soc. *73*: 341–343.

Fishman, S.E., K. Cox, J.L. Larson, P.A. Reynolds, E.T. Seno, W.K. Yeh, R. Van Frank and C.L. Hershberger. 1987. Cloning genes for the biosynthesis of a macrolide antibiotic. Proc. Natl. Acad. Sci. U.S.A. *84*: 8248–8252.

Flaig, W. and H.J. Kutzner. 1954. Zur Systematik Der Gattung *Streptomyces*. Naturwissenschaften *41*: 287–287.

Flaig, W. and H.J. Kutzner. 1960a. Beitrag Zur Ökologie Der Gattung *Streptomyces* Waksman *et* Henrici. Arch. Mikrobiol. *35*: 207–228.

Flaig, W. and H.J. Kutzner. 1960b. Beitrag Zur Systematik Der Gattung *Streptomyces* Waksman *et* Henrici. Arch. Mikrobiol. *35*: 105–138.

Flowers, T.H. and S.T. Williams. 1977a. Measurement of growth rates of streptomycetes: comparison of turbidimetric and gravimetric techniques. J. Gen. Microbiol. *98*: 285–289.

Flowers, T.H. and S.T. Williams. 1977b. The influence of pH on the growth rate and viabiltiy of neutrophilic and acidophilic streptomycetes. Microbios *18*: 223–228.

Foster, W.J. 1961. Zur Systematik der Actinomyceten. 5. Die Art *Streptomyces albus* (Rossi-Doria emend. Krainsky) Waksman *et* Henrici 1943. Arch. Mikrobiol. *38*: 367–383.

Foulerton, A.G.R. and C. Price-Jones. 1902. On the general characteristics and pathogenic action of the genus *Streptothrix*. Trans. Pathol. Soc. Lond. *53*: 56–127.

Fournier, G.P., J. Huang and J.P. Gogarten. 2009. Horizontal gene transfer from extinct and extant lineages: biological innovation and the coral of life. Philos. Trans. R. Soc. Lond. B. Biol. Sci. *364*: 2229–2239.

Fournier, G.P. and J.P. Gogarten. 2010. Rooting the ribosomal tree of life. Mol. Biol. Evol. *27*: 1792–1801.

Frommer, W. 1959. Zur Systematik Der Actinomycin Bildenden Streptomyceten. Arch. Mikrobiol. *32*: 187–206.

Fukunaga, K. 1955. Blasticidin, a new antiphytopathogenic fungal substance. Part I. Bull. Agric. Chem. Soc. Jpn. *19*: 181–188.

Fulton, T.R., M.C. Losada, E.M. Fluder and G.T. Chou. 1995. rRNA operon restriction derived taxa for *Streptomyces* (RiDiTS). FEMS Microbiol. Lett. *125*: 149–158.

Gadkari, D., K. Schricker, G. Acker, R.M. Kroppenstedt and O. Meyer. 1990. *Streptomyces thermoautotrophicus* sp. nov., a thermophilic CO-oxidizing and H_2-oxidizing obligate chemolithoautotroph. Appl. Environ. Microbiol. *56*: 3727–3734.

Gadkari, D., K. Schricker, G. Acker, R.M. Kroppenstedt and O. Meyer. 1991. *In* Validation of the publication of new names and new combinations previously effectively published outside the IJSB. List no. 38. Int. J. Syst. Bacteriol. *41*: 456–457.

Gasperini, G. 1892. Ricerche morfologiche e biologiche sul genere *Actinomyces* Harz come contributo allo studio delle relative micosi Ann. Ist. d'Igiene, Universitá Roma *2*: 167–231.

Gasperini, G. 1894. Ulteriori ricerche sul senere *Actinomyces*. P.V. Soc. Tosc. Sci. Nat. (Pisa) *9*: 64–89.

Gause, G.F., T.P. Preobrazhenskaya, M.A. Sveshnikova, L.P. Terekhova and T.S. Maximova. 1983. A guide for the determination of actinomycetes. Genera *Streptomyces*, *Streptoverticillium*, and *Chainia*. Nauka, Moscow, URSS.

Gause, G.F. and M.A. Sveshnikova. 1986a. Footnote *f*. *In* Validation of the publication of new names and new combinations previously effectively published outside the IJSB. List no. 22. Int. J. Syst. Bacteriol. *36*: 573–576.

Gause, G.F. and R.S. Sveshnikova. 1986b. Unification of the genera *Streptoverticillium* and *Streptomyces*, and amendation of *Streptomyces* Waksman and Henrici 1943, 339[AL]. Syst. Appl. Microbiol. *13*: 361–371.

Gauze, G.F., T.P. Preobrazhenskaya, E.S. Kudrina, N.O. Blinov, I.D. Ryabova and M.A. Sveshnikova. 1957. Problems in the classification of antagonistic actinomycetes. State Publishing House for Medical Literature (in Russian). Medzig, Moscow.

Genner, C. and E.C. Hill. 1981. Fuels and oils. *In* Microbial Biodeterioration, Econ. Microbiol. 6 (edited by Rose). Academic Press, London, pp. 259–306.

Gerber, N.N. 1979a. Volatile substances from actinomycetes: their role in the odor pollution of water. Crit. Rev. Microbiol. *9*: 191–214.

Gerber, N.N. 1979b. Odorouc substances from actinomycetes. Dev. Indust. Microbiol. *20*: 225–238.

Gladek, A., M. Mordarski, M. Goodfellow and S.T. Williams. 1985. Ribosomal ribonucleic-acid similarities in the classification of *Streptomyces*. FEMS Microbiol. Lett. *26*: 175–180.

Glauert, A.M. and D.A. Hopwood. 1960. The fine structure of *Streptomyces coelicolor* I. The cytoplasmic membrane system. J. Biophys. Biochem. Cytol. *7*: 479–488.

Godden, B., T. Legon, P. Helvenstein and M. Penninckx. 1989. Regulation of the production of hemicellulolytic and cellulolytic enzymes by a *Streptomyces* sp. growing on lignocellulose. J. Gen. Microbiol. *135*: 285–292.

Gonçalves de Lima, V.O., M.P. Machado, L.A. de Araújo, J.O. Falcão de Morais and H. Biermann. 1955. Novo espécie do gênero *Nocardia*: *N. recefei* sua ativadade antagonista. Antibiótico produzido. Anais da Sociedade de Biologia de Pernambuco *13*: 21–36.

Gonçalves de Lima, V.O., C.A. Albert and O. Gonçalves de Lima. 1964. *Streptomyces capoamus* nov. sp., produtor da ciclamicina e das ciclacidinas A e B. Anais da Academia Brazileira de Ciencias *36*: 317–322.

Goodfellow, M. and D. Dawson. 1978. Qualitative and quantitative studies of bacteria colonizing *Picea sitchensis* litter. Soil Biol. Biochem. *10*: 303–307.

Goodfellow, M. and S.T. Williams. 1983. Ecology of actinomycetes. Annu. Rev. Microbiol. *37*: 189–216.

Goodfellow, M. and J.A. Haynes. 1984. Actinomycetes in marine sediments. *In* Biological, Biochemical and Biomedical Aspects of Actinomycetes (edited by Ortiz-Ortiz, Bojalil and Yakoleff). Academic Press, Orlando, pp. 453–472.

Goodfellow, M. and C.H. Dickenson. 1985. Delineation and description of microbial populations using numerical methods. *In* Computer-assisted Bacterial Systematics (edited by Jones and Priest). Academic Press, London, pp. 165–226.

Goodfellow, M. and S.T. Williams. 1986. New strategies for the selective isolation of industrially important bacteria. Biotechnol. Genet. Eng. Rev. *4*: 213–262.

Goodfellow, M., S.T. Williams and G. Alderson. 1986a. *In* Validation of the publication of new names and new combinations previously effectively published outside the IJSB. List no. 22. Int. J. Syst. Bacteriol. *36*: 573–576.

Goodfellow, M., S.T. Williams and G. Alderson. 1986b. Transfer of *Kitasatoa purpurea* Matsumae and Hata to the genus *Streptomyces* as *Streptomyces purpureus* comb. nov. Syst. Appl. Microbiol. *8*: 65–66.

Goodfellow, M., S.T. Williams and G. Alderson. 1986c. Transfer of *Elytrosporangium brasiliense* Falcão de Morais *et al.*, *Elytrosporangium carpinense* Falcão de Morais *et al.*, *Elytrosporangium spirale* Falcão de Morais *et al.*, *Microellobospora cinerea* Cross *et al.*, *Microellobosporia flavea* Cross *et al.*, *Microellobosporia grisea* (Konev *et al.*) Pridham and *Microellobosporia violacea* (Tsyganov *et al.*) Pridham to the genus *Streptomyces* with emended descripton of the species. Syst. Appl. Microbiol. *8*: 48–54.

Goodfellow, M., S.T. Williams and G. Alderson. 1986d. Transfer of *Chainia* species to the genus *Streptomyces* with emended description of species. Syst. Appl. Microbiol. *8*: 55–60.

Goodfellow, M., S.T. Williams and G. Alderson. 1986e. Transfer of *Actinosporangium violaceum* Krasilnikov and Yuan, *Actinosporangium vitaminophilum* Shomura *et al.* and *Actinopycnidium caeruleum* Krasilnikov to the genus *Streptomyces*, with amended descriptions of the species. Syst. Appl. Microbiol. *8*: 61–64.

Goodfellow, M., J. Lacey and C. Todd. 1987a. Numerical classification of thermophilic streptomycetes. J. Gen. Microbiol. *133*: 3135–3149.

Goodfellow, M., C. Lonsdale, A.L. James and O.C. Macnamara. 1987b. Rapid biochemical tests for the characterization of streptomycetes. FEMS Microbiol. Lett. *43*: 39–44.

Goodfellow, M. and K.E. Simpson. 1987. Ecology of streptomycetes. Front. Appl. Microbiol. *2*: 97–125.

Goodfellow, M., S.T. Williams and G. Alderson. 1987c. *In* Validation of the publication of new names and new combinations previously effectively published outside the IJSB. List no. 23. Int. J. Syst. Bacteriol. *36*: 179–180.

Goodfellow, M., J. Lacey and C. Todd. 1988. *In* Validation of the publication of new names and new combinations previously effectively published outside the IJSB. List no. 26. Int. J. Syst. Bacteriol. *38*: 328–329.

Goodfellow, M. 1989. The *Actinomycetes* I. Suprageneric classification of actinomycetes. *In* Bergey's Manual of Systematic Bacteriology, vol. 4 (edited by Williams, Sharpe and Holt). Williams & Wilkins, Baltimore, pp. 2333–2339.

Goodfellow, M. and A.G. O'Donnell. 1993. Roots of bacterial systematics *In* Handbook of New Bacterial Systematics (edited by Goodfellow and O'Donnell). Academic Press, London, pp. 3–56.

Goodfellow, M., Y. Kumar, D.P. Labeda and L. Sembiring. 2007. The *Streptomyces violaceusniger* clade: a home for streptomycetes with rugose ornamented spores. Antonie van Leeuwenhoek *92*: 173–199.

Goodfellow, M., Y. Kumar, D.P. Labeda and L. Sembiring. 2008. *Streptomyces albiflaviniger* sp. nov. *In* List of new names and new combinations previously effectively, but not validly, published. Validation List no. 119. Int. J. Syst. Evol. Microbiol. *58*: 1–2.

Goodfellow, M. and H.P. Fiedler. 2010. A guide to successful bioprospecting: informed by actinobacterial systematics. Antonie van Leeuwenhoek *98*: 119–142.

Gordon, M.A. and E.W. Lapa. 1966. Durhamycin, a pentaene antifungal antibiotic from *Streptomyces durhamensis* sp. n. Appl. Microbiol. *14*: 754–760.

Gottschalk, L.M., R. Nobrega and E.P. Bon. 2003. Effect of aeration on lignin peroxidase production by *Streptomyces viridosporus* T7A. Appl. Biochem. Biotechnol. *105*: 799–807.

Goyer, C., E. Faucher and C. Beaulieu. 1996. *Streptomyces caviscabies* sp. nov., from deep-pitted lesions in potatoes in Quebec, Canada. Int. J. Syst. Bacteriol. *46*: 635–639.

Grantcharova, N., W. Ubhayasekera, S.L. Mowbray, J.R. McCormick and K. Flardh. 2003. A missense mutation in *ftsZ* differentially affects vegetative and developmentally controlled cell division in *Streptomyces coelicolor* A3(2). Mol. Microbiol. *47*: 645–656.

Gravius, B., D. Glocker, J. Pigac, K. Pandza, D. Hranueli and J. Cullum. 1994. The 387 kb linear plasmid pPZG101 of *Streptomyces rimosus* and its interactions with the chromosome. Microbiology *140*: 2271–2277.

Gregory, P.H. and M.E. Lacey. 1963. Mycological examination of dust from mouldy hay associated with farmer's lung disease. J. Gen. Microbiol. *30*: 75–88.

Grein, A., C. Spalla, G. Canevazz and A. Dimarco. 1963. Descrizione e Classificazione di un Attinomicete (*Streptomyces peucetius* sp. nova) Produttore di una Sostanza ad Attivita Antitumorale: la Daunomicina. G. Microbiol. *11*: 109–118.

Greiner-Mai, E., R.M. Kroppenstedt, F. Korn-Wendisch and H.J. Kutzner. 1987. Morphological and biochemical characterization and emended descriptions of thermophilic actinomycetes species. Syst. Appl. Microbiol. *9*: 97–109.

Groth, I., B. Schütze, T. Boettcher, C.B. Pullen, C. Rodriguez, E. Leistner and M. Goodfellow. 2003. *Kitasatospora putterlickiae* sp. nov., isolated from rhizosphere soil, transfer of *Streptomyces kifunensis* to the genus *Kitasatospora* as *Kitasatospora kifunensis* comb. nov., and emended description of *Streptomyces aureofaciens* Duggar 1948. Int. J. Syst. Evol. Microbiol. *53*: 2033–2040.

Groth, I., C. Rodriguez, B. Schütze, P. Schmitz, E. Leistner and M. Goodfellow. 2004. Five novel *Kitasatospora* species from soil: *Kitasatospora arboriphila* sp. nov., *K. gansuensis* sp. nov., *K. nipponensis* sp. nov., *K. paranensis* sp. nov. and *K. terrestris* sp. nov. Int. J. Syst. Evol. Microbiol. *54*: 2121–2129.

Grundy, W.E., A.L. Whitman, E.G. Rdzok, E.J. Rdzok, M.E. Hanes and J.C. Sylvester. 1952. Actithiazic acid. I. Microbiological studies. Antibiot. Chemother. *2*: 399–408.

Gumaa, S.A. and E.S. Mahgoub. 1975. Counterimmunoelectrophoresis in the diagnosis of mycetoma and its sensitivity as compared to immunodiffusion. Sabouraudia *13*: 309–315.

Gumaa, S.A. 1994. The aetiology and epidemiology of mycetoma. Sud. Med. J. *32*: 14–22.

Guo, Y., W. Zheng, X. Rong and Y. Huang. 2008. A multilocus phylogeny of the *Streptomyces griseus* 16S rRNA gene clade: use of multilocus sequence analysis for streptomycete systematics. Int. J. Syst. Evol. Microbiol. *58*: 149–159.

Gupta, K.C. and I.C. Chopra. 1963a. A new whorl-forming species of *Streptomyces*. Hindustan Antibiotics Bulletin *5*: 110–112.

Gupta, K.C. and I.C. Chopra. 1963b. *Streptomyces katrae* - a new species of *Streptomyces* isolated from soil. Ind. J. Microbiol. *3*: 1–4.

Gupta, K.C., R.R. Sobti and I.C. Chopra. 1963. Actinomycin produced by a new species of *Streptomyces*. Hindustan Antibiotics Bulletin *6*: 12–16.

Gupta, K.C. 1965a. A new species of the genus *Streptosporangium* isolated from Indian soil. J. Antibiot. *18*: 125–127.

Gupta, K.C. 1965b. *Streptomyces tropicalensis*, a new whorl-forming species of *Streptomyces*. J. Antibiot. (Tokyo) *18*: 53–55.

Güssow, H.T. 1914. The systematic position of the common potato scab. Science (Washington) *39*: 431–433.

Gyllenberg, H.G. 1976. Application of automation to the identification of streptomycetes. *In* Actinomycetes – The Boundary Microorganisms (edited by Arai). Toppan Co. Ltd, Tokyo, pp. 299–321.

Hagedorn, C. 1976. Influences of soil acidity on *Streptomyces* populations inhabiting forest soils. Appl. Environ. Microbiol. *32*: 368–375.

Hagege, J., J.L. Pernodet, A. Friedmann and M. Guérineau. 1993. Mode and origin of replication of pSAM2, a conjugative integrating element of *Streptomyces ambofaciens*. Mol. Microbiol. *10*: 799–812.

Hagemann, G., L. Pénasse and J. Teillon. 1964. Zur Systematik der Actinomyceten. 9. Streptomyceten mit *cinnamoneus* Luftmycel. *In* Zentralbl. Bakteriol. Parasitenkd. Infektionskr. Hyg., vol. 117 (edited by Hütter). Abteilung II, pp. 603–661.

Hain, T., N. Ward-Rainey, R.M. Kroppenstedt, E. Stackebrandt and F.A. Rainey. 1997. Discrimination of *Streptomyces albidoflavus* strains based on the size and number of 16S-23S ribosomal DNA intergenic spacers. Int. J. Syst. Bacteriol. *47*: 202–206.

Hamada, M., N. Kinoshita, S. Hattori, A. Yoshida, Y. Okami, K. Higashide, N. Sakata and M. Hori. 1995a. *Streptomyces kasugaensis* sp. nov.: a new species of the genus *Streptomyces*. Actinomycetologica *9*: 27–36.

Hamada, M., N. Kinoshita, S. Hattori, A. Yoshida, Y. Okami, K. Higashide, N. Sakata and M. Hori. 1995b. *In* Validation of the publication of new names and new combinations previously effectively published outside the IJSB. List no. 55. Int. J. Syst. Bacteriol. *45*: 879–880.

Hamada, S. 1958. A study of a new antitumor substance, cellostatin. I. On the isolation and some properties of cellostatin. Tohoku J. Exp. Med. *67*: 173–179.

Han, L., K. Yang, E. Ramalingam, R.H. Mosher and L.C. Vining. 1994. Cloning and characterization of polyketide synthase genes for jadomycin B biosynthesis in *Streptomyces venezuelae* ISP5230. Microbiology *140*: 3379–3389.

Hanka, L.J., J.S. Evans, D.J. Mason and A. Dietz. 1966. Microbiological production of 5-azacytidine. I. Production and biological activity. Antimicrob. Agents Chemother. *1966*: 619–624.

Hanka, L.J., P.W. Rueckert and T. Cross. 1985. A method for isolating strains of the genus *Streptoverticillium* from soil. FEMS Microbiol. Lett. *30*: 365–368.

Hanka, L.J. and R.D. Schaadt. 1988. Methods for isolation of streptoverticillia from soils. J. Antibiot. *41*: 576–578.

Harchand, R.K. and S. Singh. 1997. Extracellular cellulase system of a thermotolerant streptomycete: *Streptomyces albaduncus*. Acta Microbiol. Immunol. Hung. *44*: 229–239.

Hasegawa, T., T. Yamano and M. Yoneda. 1978. *Streptomyces inusitatus* sp. nov. Int. J. Syst. Bacteriol. *28*: 407–410.

Hata, T., N. Ohki and T. Higuchi. 1952. Studies on the antibiotic substance "luteomycin". On the strains and the cultural conditions. J. Antibiot. (Tokyo) Ser. *A 5*: 529–534.

Hatano, K., T. Nishii and H. Mordarska. 1997. *Streptomyces spitsbergensis* Wieczorek et al 1993 is a later subjective synonym of *Streptomyces baldaccii* (Farina and Locci 1966) Witt and Stackebrandt 1991. Int. J. Syst. Bacteriol. *47*: 573–574.

Hatano, K. 1999. Replacement of ATCC 21273, the current type strain of *Streptomyces rameus* Shibata 1959, with IFO 3782: Request for an Opinion. Int. J. Syst. Bacteriol. *49*: 931–932.

Hatano, K., T. Nishii and H. Kasai. 2003. Taxonomic re-evaluation of whorl-forming *Streptomyces* (formerly *Streptoverticillium*) species by using phenotypes, DNA–DNA hybridization and sequences of *gyrB*, and proposal of *Streptomyces luteireticuli* (*ex* Katoh and Aral 1957) corrig., sp. nov., nom. rev. Int. J. Syst. Evol. Microbiol. *53*: 1519–1529.

Hayakawa, M. and H. Nonomura. 1987a. Humic acid-vitamine agar, a new medium for the selective isolation of soil actinomycetes. J. Ferment. Technol. *65*: 501–509.

Hayakawa, M. and H. Nonomura. 1987b. Efficacy of artificial humic acid as a selective nutrient in HV agar used for the isolation of soil actinomycetes. J. Ferment. Technol. *65*: 609–616.

He, L., W. Li, Y. Huang, L. Wang, Z. Liu, B. Lanoot, M. Vancanneyt and J. Swings. 2005. *Streptomyces jietaisiensis* sp. nov., isolated from soil in northern China. Int. J. Syst. Evol. Microbiol. *55*: 1939–1944.

Heinemann, B., M.A. Kaplan, R.D. Muir and I.R. Hooper. 1953. Amphomycin, a new antibiotic. Antibiot. Chemother. *3*: 1239–1242.

Helmke, E. 1981. Growth of actinomycetes from marine and terrestrial origin under increased hydrostatic pressure. Proceedings of the 4th International Symposium on Actinomycete Biology, 1979, Cologne, pp. 321–327.

Henssen, A. 1957. Beiträge zur Morphologie und Systematik der thermophilen Actinomyceten. Arch. Mikrobiol. *26*: 373–414.

Herron, P.R. and E.M. Wellington. 1990. New method for extraction of streptomycete spores from soil and application to the study of lysogeny in sterile amended and nonsterile soil. Appl. Environ. Microbiol. *56*: 1406–1412.

Hesketh, A., D. Fink, B. Gust, H.U. Rexer, B. Scheel, K. Chater, W. Wohlleben and A. Engels. 2002a. The GlnD and GlnK homologues of *Streptomyces coelicolor* A3(2) are functionally dissimilar to their nitrogen regulatory system counterparts from enteric bacteria. Mol. Microbiol. *46*: 319–330.

Hesketh, A.R., G. Chandra, A.D. Shaw, J.J. Rowland, D.B. Kell, M.J. Bibb and K.F. Chater. 2002b. Primary and secondary metabolism, and post-translational protein modifications, as portrayed by proteomic analysis of *Streptomyces coelicolor*. Mol. Microbiol. *46*: 917–932.

Hesseltine, C.W., J.N. Porter, N. Deduck, M. Hauck, B. M. and J.H. Williams. 1954. A new species of *Streptomyces*. Mycologia *46*: 16–22.

Heuer, H., M. Krsek, P. Baker, K. Smalla and E.M. Wellington. 1997. Analysis of actinomycete communities by specific amplification of genes encoding 16S rRNA and gel-electrophoretic separation in denaturing gradients. Appl. Environ. Microbiol. *63*: 3233–3241.

Hickey, R.J., C.J. Corum, P.H. Hidy, I.R. Cohen, N. U.F.B. and E. Kropp. 1952. Ascosin, a antifungal antibiotic produced by a streptomycete. Antibiot. Chemother. *2*: 472–483.

Higashide, E., M. Hasegawa, K. Shibata, K. Mizumo and H. Akaike. 1966. Studies on the *Streptomycetes*. *Streptomyces cuspidosporus* nov. sp. and the antibiotics sparsomycin and tubercidin produced thereby. Annual Report of the Takeda Research Laboratory *25*: 1–14.

Higgens, C.E. and R.E. Kastner. 1971. *Streptomyces clavuligerus* sp. nov., a β-lactam antibiotic producer. Int. J. Syst. Bacteriol. *21*: 326–331.

Hill, L.R., V.B.D. Skerman and P.H.A. Sneath. 1984. Corrigenda to the Approved Lists of Bacterial Names. Int. J. Syst. Bacteriol. *34*: 508–511.

Hinuma, Y. 1954. Zaomycin, a new antibiotic from a *Streptomyces* sp. Studies on the antibiotic substances from *Actinomyces*. III. J. Antibiot. (Tokyo) Ser. A *7*: 134–136.

Hiraga, K., T. Suzuki and K. Oda. 2000. A novel double-headed proteinaceous inhibitor for metalloproteinase and serine proteinase. J. Biol. Chem *275*: 25173–25179.

Hirsch, C.F. and D.L. Christensen. 1983. Novel method for selective isolation of actinomycetes. Appl. Environ. Microbiol. *46*: 925–929.

Hofheinz, W. and H. Grisebach. 1965. Die Fettsäuren von *Streptomyces erythreus* und *Streptomyces halstedii*. Z. Naturforsch. *20B*: 43.

Hong, S.T., J.R. Carney and S.J. Gould. 1997. Cloning and heterologous expression of the entire gene clusters for PD 116740 from *Streptomyces* strain WP 4669 and tetrangulol and tetrangomycin from *Streptomyces rimosus* NRRL 3016. J. Bacteriol. *179*: 470–476.

Hopkins, D.W., S.J. Macnaughton and A.G. O'Donnell. 1991. A dispersion and differential centrifugation technique for representatively sampling microorganisms from soil. Soil Biol. Biochem. *23*: 217–225.

Hopwood, D.A. and H.M. Ferguson. 1969. A rapid method for lyophilizing *Streptomyces* cultures. J. Appl. Bacteriol. *32*: 434–436.

Hopwood, D.A., M.J. Bibb, K.F. Chater, T. Kieser, C.J. Bruton, H.M. Kieser, D.J. Lydiate, C.P. Smith, J.M. Ward and S. H. 1985. Genetic Manipulation of *Streptomyces*. A Laboratory Manual. John Innes Foundation, Norwich, UK.

Hopwood, D.A. 2003. *Streptomyces* genes: from Waksman to Sanger. J. Ind. Microbiol. Biotechnol. *30*: 468–471.

Hopwood, D.A. 2007. *Streptomyces* in nature and medcine. *In* The Antibiotic Makers. Oxford University Press.

Hori, H. and S. Osawa. 1987. The rates of evolution in some ribosomal components. J. Mol. Evol. *9*: 191–201.

Horinouchi, S. 2002. Amicrobial hormone, A-factor, as a master switch for morphological differentiation and secondary metabolism in *Streptomyces griseus*. Front. Biosci. *7*: 2045–2057.

Hosoya, S., S. Komatsu, M. Soeda and Y. Sonoda. 1952. Trichomycin, a new antibiotic produced by *Streptomyces hachijoensis* with trichomonadicidal and antifungal activity. Jap. J. Exp. Med. *22*: 505–509.

Hotta, K., N. Saito and Y. Okami. 1980. Studies on new aminoglycoside antibiotics, istamycins, from an actinomycete isolated from a marine environment. I. The use of plasmid profiles in screening antibiotic-producing streptomycetes. J. Antibiot. (Tokyo) *33*: 1502–1509.

Howarth, O.W., E. Grund, R.M. Kroppenstedt and M.D. Collins. 1986. Structural determination of a new naturally occurring cyclic vitamin K. Biochem. Biophys. Res. Commun *140*: 916–923.

Hozzein, W.N. and M. Goodfellow. 2007. *Streptomyces synnematoformans* sp. nov., a novel actinomycete isolated from a sand dune soil in Egypt. Int. J. Syst. Evol. Microbiol. *57*: 2009–2013.

Hsiao, N.H. and R. Kirby. 2008. Comparative genomics of *Streptomyces avermitilis*, *Streptomyces cattleya*, *Streptomyces maritimus* and *Kitasatospora aureofaciens* using a *Streptomyces coelicolor* microarray system. Antonie van Leeuwenhoek *93*: 1–25.

Hsieh, C.J. and G.H. Jones. 1995. Nucleotide sequence, transcriptional analysis, and glucose regulation of the phenoxazinone synthase gene (*phsA*) from *Streptomyces antibioticus*. J. Bacteriol. *177*: 5740–5747.

Hsu, S.C. and J.L. Lockwood. 1975. Powdered chitin agar as a selective medium for enumeration of actinomycetes in water and soil. Appl. Microbiol. *29*: 422–426.

Huang, J., C.J. Lih, K.H. Pan and S.N. Cohen. 2001. Global analysis of growth phase responsive gene expression and regulation of antibiotic biosynthetic pathways in *Streptomyces coelicolor* using DNA microarrays. Genes Dev. *15*: 3183–3192.

Huang, Y., Q. Cui, L. Wang, C. Rodriguez, E. Quintana, M. Goodfellow and Z. Liu. 2004a. *Streptacidiphilus jiangxiensis* sp. nov., a novel actinomycete isolated from acidic rhizosphere soil in China. Antonie van Leeuwenhoek *86*: 159–165.

Huang, Y., W. Li, L. Wang, B. Lanoot, M. Vancanneyt, C. Rodriguez, Z. Liu, J. Swings and M. Goodfellow. 2004b. *Streptomyces glauciniger* sp. nov., a novel mesophilic streptomycete isolated from soil in south China. Int. J. Syst. Evol. Microbiol. *54*: 2085–2089.

Huang, Y., N. Zhao, L. He, L. Wang, Z. Liu, M. You and F. Guan. 2005. *In* Validation of publication of new names and new combinations previously effectively published outside the IJSEM. List no. 105. Int. J. Syst. Evol. Microbiol. *55*: 1743–1745.

Huber, G., K.H. Wallhäeusser, L. Fries, A. Steigler and H.L. Weidenmueller. 1962. Niddamycin ein neues Makrolid-Antibiotikum. Arzneim. Forsch. *12*: 1191–1195.

Huddleston, A.S., J.L. Hinks, M. Beyazova, A. Horan, D.I. Thomas, S. Baumberg and E.M.H. Wellington. 1995. Studies on the diversity of streptomycin-producing streptomycetes. Biotekhnologia *7+8*: 242–253.

Huddleston, A.S., N. Cresswell, M.C. Neves, J.E. Beringer, S. Baumberg, D.I. Thomas and E.M. Wellington. 1997. Molecular detection of streptomycin-producing streptomycetes in Brazilian soils. Appl. Environ. Microbiol. *63*: 1288–1297.

Humm, J.H. and K.S. Shepard. 1946. Three new agar-digesting actinomycetes. Duke University Marine Station Bull. *3*: 76–80.

Hutchinson, M., J.W. Ridgway and T. Cross. 1975. Biodeterioration of rubber in contact with water, sewage and soil. *In* Microbial Aspects of the Deterioration of Materials (edited by Lovelock and Gilbert). Academic Press, London, pp. 187–202.

Hütter, R. 1962. Zur Systematik Der Actinomyceten. 8. Quirlbildende Streptomyceten. Arch. Mikrobiol. *43*: 365–391.

Hütter, R. 1967a. Systematik der Streptomyceten unter besonderer Berücksichtigung der von Ihnen begildeten Antibiotika. Bibl. Microbio. Fasc., vol. 6. Karger AG, Basel, Switzerland, p. 5.

Hütter, R. 1967b. Systematik der Streptomyceten unter besonderer Berücksichtigung der von Ihnen begildeten Antibiotika. *In* Bibl. Microbio. Fasc., vol. 6. Karger, Basel, pp. 1–382.

Hütter, R. and T. Eckhardt. 1988. Genetic Manipulation. *In* Actinomycetes in Biotechnology (edited by Goodfellow, Williams and Mordarski). Academic Press, London, pp. 89–184.

Ikeda, H., J. Ishikawa, A. Hanamoto, M. Shinose, H. Kikuchi, T. Shiba, Y. Sakaki, M. Hattori and S. Ōmura. 2003. Complete genome sequence and comparative analysis of the industrial microorganism *Streptomyces avermitilis*. Nat. Biotechnol. *21*: 526–531.

Ishida, N., K. Kumagai, T. Niida, K. Hamamoto and T. Shomura. 1967. Nojirimycin, a new antibiotic. I. Taxonomy and fermentation. J. Antibiot. (Tokyo) *20*: 62–65.

Isono, K., S. Yamashita, Y. Tomiyama, S. Suzuki and H. Sakai. 1957. Studies on homomycin. II. J. Antibiot. (Tokyo) *10*: 21–30.

Isono, K., J. Nagatsu, Kawashim.Y and S. Suzuki. 1965. Studies on polyoxins, antifungal antibiotics. Part I. Isolation and characterization of polyoxins A and B. Agric. Biol. Chem. *29*: 848–854.

Itoh, T., T. Kudo, F. Parenti and A. Seino. 1989. Amended description of the genus *Kineosporia*, based on chemotaxonomic and morphological studies. Int. J. Syst. Bacteriol. *39*: 168–173.

Ivanitskaya, L.P., G.D. Upiter, M.A. Sveshnikova and G.F. Gauze. 1966. Systematic position, variation and antibiotic properties of the producer of the antitumor antibiotic Tavromycetin (in Russian). Antibiotiki *11*: 973–976.

Iwasaki, A., H. Itoh and T. Mori. 1981. *Streptomyces sannanensis* sp. nov. Int. J. Syst. Bacteriol. *31*: 280–284.

Jagnow, G. 1957. Beiträge zur Ökologie Der *Streptomyceten*. Arch. Mikrobiol. *26*: 175–191.

Jakimowicz, D., J. Majka, W. Messer, C. Speck, M. Fernandez, M.C. Martin, J. Sanchez, F. Schauwecker, U. Keller, H. Schrempf and J. Zakrzewska-Czerwinska. 1998. Structural elements of the *Streptomyces oriC* region and their interactions with the DnaA protein. Microbiology *144*: 1281–1290.

Janshekar, H. and A. Fiechter. 1983. Lignin: biosynthesis, application, and biodegradation. Adv. Biochem. Eng. Biotechnol. *27*: 120–178.

Jendrossek, D., G. Tomasi and R.M. Kroppenstedt. 1997. Bacterial degradation of natural rubber: a privilege of actinomycetes? FEMS Microbiol. Lett. *150*: 179–188.

Jensen, H.L. 1930. Actinomycetes in Danish soils. Soil Sci. *30*.

Jensen, H.L. 1931. Contributions to our knowledge of the *Actinomycetales*. II. The definition and subdivision of the genus *Actinomyces*, with a preliminary account of Australian soil Actinomycetes. Proc. Linn. Soc. N.S.W. *56*: 345–370.

Jeuniaux, C. 1966. Chitinases. Methods Enzymol. *8*: 644–650.

Jiang, Y., S.K. Tang, J. Wiese, L.H. Xu, J.F. Imhoff and C.L. Jiang. 2007. *Streptomyces hainanensis* sp. nov., a novel member of the genus *Streptomyces*. Int. J. Syst. Evol. Microbiol. *57*: 2694–2698.

Johnson, L.E. and A. Dietz. 1969. Lomofungin, a new antibiotic produced by *Streptomyces lomondensis* sp. n. Appl. Microbiol. *17*: 755–759.

Johnstone, D.B. and S.A. Waksman. 1947. Streptomycin II, an antibiotic substance produced by a new species of *Streptomyces*. Proc. Soc. Exp. Biol. Med. *65*: 294–295.

Jones, K.L. 1949. Fresh isolates of actinomycetes in which the presence of sporogenous aerial mycelia is a fluctuating characteristic. J. Bacteriol. *57*: 141–145.

Jones, K.L. 1952. A new *Streptomyces* that produces vitamin B12 actively. Pap. Mich. Acad. Sci. Arts Lett. *37*: 47–48.

Jonsbu, E., B. Christensen and J. Nielsen. 2001. Changes of in vivo fluxes through central metabolic pathways during the production of nystatin by *Streptomyces noursei* in batch culture. Appl. Microbiol. Biotechnol. *56*: 93–100.

Kalakoutskii, L.V. and N.A. Krasil'nikov. 1960. Formation of sclerotia in actinomycetes and the systematic position of genus *Chania*. (En. transl. 1966). *In* Biology of anitbiotic-producing actinomycetes, vol. 8 (edited by Rautenshtein). Akad. Nauk. SSSR Inst. Mikrobiol., pp. 41–51.

Kalakoutskii, L.V. and V.D. Kusnetsov. 1964. A new species of the *Actinoplanes – A. armeniacus* and some peculiarities of its mode of spore formation. Mikrobiologiya *33*: 613.

Kalkus, J., C. Dorrie, D. Fischer, M. Reh and H.G. Schlegel. 1993. The giant linear plasmid pHG207 from *Rhodococcus* sp. encoding hydrogen autotrophy: characterization of the plasmid and its termini. J. Gen. Microbiol. *139*: 2055–2065.

Kämpfer, P. and R.M. Kroppenstedt. 1991. Probabilistic identification of *Streptomycetes* using miniaturized physiological tests. J. Gen. Microbiol. *137*: 1893–1902.

Kämpfer, P., R.M. Kroppenstedt and W. Dott. 1991. A numerical classification of the genera *Streptomyces* and *Streptoverticillium* using miniaturized physiological tests. J. Gen. Microbiol. *137*: 1831–1891.

Kämpfer, P. and D.P. Labeda. 2003. International Committee on Systematics of Prokaryotes. Subcommittee on the taxonomy of *Streptomycetaceae*. Minutes of the meeting, 30 July 2002, Paris, France. Int. J. Syst. Evol. Microbiol. *53*: 925.

Kämpfer, P. 2006. The family *Streptomycetaceae*, Part I: Taxonomy. *In* The Prokaryotes: a Handbook on the Biology of Bacteria, 3rd edn, vol. 3, *Archaea, Bacteria, Firmicutes*, Actinomycetes (edited by Dworkin, Falkow, Rosenberg, Schleifer and Stackebrandt). Springer, New York, pp. 538–604.

Kämpfer, P., B. Huber, S. Buczolits, K. Thummes, I. Grun-Wollny and H.-J. Busse. 2008. *Streptomyces specialis* sp. nov. Int. J. Syst. Evol. Microbiol. *58*: 2602–2606.

Kaneko, M., Y. Ohnishi and S. Horinouchi. 2003. Cinnamate:coenzyme A ligase from the filamentous bacterium *Streptomyces coelicolor* A3(2). J. Bacteriol. *185*: 20–27.

Kataoka, M., T. Seki and T. Yoshida. 1991. Five genes involved in self-transmission of pSN22, a *Streptomyces* plasmid. J. Bacteriol. *173*: 4220–4228.

Kataoka, M., K. Ueda, T. Kudo, T. Seki and T. Yoshida. 1997. Application of the variable region in 16S rDNA to create an index for rapid species identification in the genus *Streptomyces*. FEMS Microbiol. Lett. *151*: 249–255.

Katoh, H. and T. Arai. 1957. On the production of antibiotic substances from the *Streptomyces luteoreticuli*. Annu. Rep. Inst. Food Microbiol. Chiba Univ. *10*: 52–57.

Kawaguchi, H., H. Tsukiura, M. Okanishi, T. Miyaki, T. Ohmori, K. Fujisawa and H. Koshiyama. 1965. Studies on coumermycin, a new antibiotic. I. Production, isolation and characterization of coumermycin A2. J. Antibiot. (Tokyo) *18*: 1–10.

Kawato, M. and R. Shinobu. 1959. On *Streptomyces herbaricolor* sp. nov., supplement: a simple technique for microscopical observation. Osaka Unit. Lib. Arts Educ. B Nat. Sci. *8*: 114–119.

Kebeler, M., E.R. Dabbs, B. Averhoff and G. Gottschalk. 1996. Studies on the isopropylbenzene 2,3-dioxygenase and the 3′-isopropylcatechol 2,3-dioxygenase genes encoded by the linear plasmid of *Rhodococcus erythropolis* BD2. Microbiology *142*: 3241–3251.

Kelly, J., A.H. Kutscher and I.F. Tuot. 1959. Thiostrepton, a new antibiotic: tube dilution sensitivity studies. Oral Surg. Oral Med. Oral Pathol. *12*: 1334–1339.

KenKnight, G. and J.H. Munzie. 1939. Isolation of phytopathogenic actinomycetes. Phytopathology *29*: 1000–1001.

Khan, M.R. and S.T. Williams. 1975. Studies on the ecology of actinomycetes in soil. VIII. Distribution and characteristics of acidophilic actinomycetes. Soil Biol. Biochem. *7*: 345–348.

Khan, M.R., S.T. Williams and M.L. Saha. 1978. Studies on the microbial degradation of jute. Bangladesh J. Jute Fibre Res. *3*: 45–52.

Kieser, T., M.J. Bibb, M.J. Buttner, K.F. Chater and D.A. Hopwood. 2000. Practical *Streptomyces* Genetics. The John Innes Foundation, Norwich, UK.

Kilian, M. 1998. Necessary changes of bacterial names? ASM News *64*: 670.

Kilian, M. 2001. Recommended conservation of the names *Streptococcus sanguis, Streptococcus rattus, Streptococcus cricetus*, and seven other names included in the Approved Lists of Bacterial Names. Request for an Opinion. Int. J. Syst. Evol. Microbiol. *51*: 723–724.

Kim, B., N. Sahin, D.E. Minnikin, J. Zakrzewska-Czerwinska, M. Mordarski and M. Goodfellow. 1999. Classification of thermophilic streptomycetes, including the description of *Streptomyces thermoalcalitolerans* sp. nov. Int. J. Syst. Bacteriol. *49*: 7–17.

Kim, B., A.M. Al-Tai, S.B. Kim, P. Somasundaram and M. Goodfellow. 2000. *Streptomyces thermocoprophilus* sp. nov., a cellulase-free endoxylanase-producing streptomycete. Int. J. Syst. Evol. Microbiol. *50*: 505–509.

Kim, D., J. Chun, N. Sahin, Y.-C. Hah and M. Goodfellow. 1996. Analysis of the thermophilic clades within the genus *Streptomyces* by 16S ribosomal DNA sequence comparisons. Int. J. Syst. Bacteriol. *46*: 581–587.

Kim, H.J., S.C. Lee and B.K. Hwang. 2006. *Streptomyces cheonanensis* sp. nov., a novel streptomycete with antifungal activity. Int. J. Syst. Evol. Microbiol. *56*: 471–475.

Kim, I.S. and K.J. Lee. 1995. Physiological roles of leupeptin and extracellular proteases in mycelium development of *Streptomyces exfoliatus* SMF13. Microbiology *141*: 1017–1025.

Kim, S.B., C. Falconer, E. Williams and M. Goodfellow. 1998. *Streptomyces thermocarboxydovorans* sp. nov. and *Streptomyces thermocarboxydus* sp. nov., two moderately thermophilic carboxydotrophic species from soil. Int. J. Syst. Bacteriol. *48*: 59–68.

Kim, S.B. and M. Goodfellow. 2002. *Streptomyces avermitilis* sp. nov., nom. rev., a taxonomic home for the avermectin-producing streptomycetes. Int. J. Syst. Evol. Microbiol. *52*: 2011–2014.

Kim, S.B., J. Lonsdale, C.N. Seong and M. Goodfellow. 2003a. *In* Validation of publication of new names and new combinations previously effectively published outside the IJSEM. List no. 93. Int. J. Syst. Evol. Microbiol. *53*: 1219–1220.

Kim, S.B., J. Lonsdale, C.N. Seong and M. Goodfellow. 2003b. *Streptacidiphilus* gen. nov., acidophilic actinomycetes with wall chemotype I and emendation of the family *Streptomycetaceae* (Waksman and Henrici 1943^{AL}) emend. Rainey *et al.* 1997. Antonie van Leeuwenhoek *83*: 107–116.

Kim, S.B., C.N. Seong, S.J. Jeon, K.S. Bae and M. Goodfellow. 2004. Taxonomic study of neutrotolerant acidophilic actinomycetes isolated from soil and description of *Streptomyces yeochonensis* sp. nov. Int. J. Syst. Evol. Microbiol. *54*: 211–214.

Kinashi, H., M. Shimaji-Murayama and T. Hanafusa. 1991. Nucleotide sequence analysis of the unusually long terminal inverted repeats of a giant linear plasmid, SCP1. Plasmid *26*: 123–130.

Kirby, R. and E.P. Rybicki. 1986. Enzyme-linked immunosorbent assay (ELISA) as a means of taxonomic analysis of *Streptomyces* and related organisms. J. Gen. Microbiol. *132*: 1891–1894.

Kirk, T.K. and R.L. Farrell. 1987. Enzymatic combustion: the microbial degradation of lignin. Annu. Rev. Microbiol. *41*: 465–505.

Kirsop, B.E. and J.J.S. Snell. 1984. Maintenance of Microorganisms: a Manual of Laboratory Methods. Academic Press, London.

Kluepfel, D. and M. Ishaque. 1982. Xylan-induced cellulolytic enzymes in *Streptomyces flavogriseus*. Dev. Indust. Microbiol. *23*: 389–395.

Kluepfel, D., F. Shareck, F. Mondou and R. Morosoli. 1986. Characterization of cellulase and xylanase activities of *Streptomyces lividans*. Appl. Microbiol. Biotechnol. *24*: 230–234.

Konev, I.E. and V.A. Tsyganov. 1962. A new species in the group of yellow actinomycetes. Mikrobiologiya *31*: 1023–1028.

Konev, I.E., V.A. Tsyganov, R. Minbayev and V.M. Morozov. 1967. New genus of *Actinomycetales*, *Echinospora* gen. nov. Mikrobiologiya *36*: 308–317.

Konev, M. 1986. Footnote *f*. *In* Validation of the publication of new names and new combinations previously effectively published outside the IJSB. List no. 22. Int. J. Syst. Bacteriol. *36*: 573–576.

Konev, Y.E., N.P. Barashkova and Y.D. Shenin. 1974. New verticillate actinomycetes forming hexaene antibiotics. Microbiology (En. transl. from Mikrobiologiya) *43*: 560–565.

Korenyako, A.I., N.A. Krasil'nikov, N.L. Nikitina and A.I. Sokolova. 1960. Biology of antibiotic-producing actinomycetes *In* Transactions of the Institute of Microbiology, vol. 8 (edited by Rautenshtein). Academy of Science, USSR, pp. 1–344.

Kormanec, J., R. Novakova, D. Homerova and B. Rezuchova. 2001. *Streptomyces aureofaciens* sporulation-specific sigma factor sigma(*rpoZ*) directs expression of a gene encoding protein similar to hydrolases involved in degradation of the lignin-related biphenyl compounds. Res. Microbiol. *152*: 883–888.

Korn-Wendisch, F. 1982. Phagentypisierung und Lysogenie bei Actinomyceten. PhD thesis.

Korn-Wendisch, F. and H.J. Kutzner. 1992. The family *Streptomycetaceae*. *In* The Prokaryotes: a Handbook on the Biology of Bacteria: Ecophysiology, Isolation, Identification, Applications, 2nd edn (edited by Balows, Trüper, Dworkin, Harder and Schleifer). Springer, New York, pp. 921–995.

Korn-Wendisch, F. and J. Schneider. 1992. Phage typing: a useful tool in actinomycete systematics. Gene *115*: 243–247.

Korn, F., B. Weingärtner and H.J. Kutzner. 1978. A study of twenty actinophages: Morphology, serological relationships and host range. *In* Genetics of the *Actinomycetales* (edited by Freeksen, Tarnok and Thumin). Gustav Fischer Verlag, Stuttgart, pp. 251–270.

Koshiyama, H., M. Okanishi, T. Ohmori, T. Miyaki, H. Tsukiura, M. Matsuzaki and H. Kawaguchi. 1963. Cirramycin, a new antibiotic. J. Antibiot. (Tokyo) *16*: 59–66.

Kosono, S., M. Maeda, F. Fuji, H. Arai and T. Kudo. 1997. Three of the seven *bphC* genes of *Rhodococcus erythropolis* TA421, isolated from a termite ecosystem, are located on an indigenous plasmid associated with biphenyl degradation. Appl. Environ. Microbiol. *63*: 3282–3285.

Krainsky, A. 1914. Die Aktinomyceten und ihre Bedeutung in der Natur. Zentralbl. Bakteriol. Parasitenkd. Infektionskr. Hyg. Abt. II *41*: 649–688.

Krasil'Nikov, N.A. and C.S. Yuan. 1960. [A new species in the group of *Actinomyces aurantiacus*]. Mikrobiologiia *29*: 482–489.

Krasil'Nikov, N.A. 1962. [A new *Actinomyces* genus *Actinopycnidium* n. gen. from the family *Actinomycetaceae*]. Mikrobiologiia *31*: 250–253.

Krasil'nikov, N.A., N.S. Agre, L.A. Dorokhova and A.A. Sokolov. 1968. [A study of three new species of thermophilic actinomycetes]. Mikrobiologiia *37*: 75–83.

Krasil'nikov, N.A. 1941. Keys to *Actinomycetales* (in Russian). Izvest. Akad. Nauk. SSSR, Moscow.

Krasil'nikov, N.A. 1949. Guide to the bacteria and actinomycetes. Akad. Nauk. SSSR, Moscow.

Krasil'nikov, N.A., A.I. Korenyako, M.M. Meksina, L.K. Valedinskaya and N.M. Veselov. 1957. On the culture of actinomycete no. 111 *Actinomyces luridus* nov. sp. producing an antiviral antibiotic "luridin". Mikrobiologiya *26*: 558–564.

Krasil'nikov, N.A. 1958. The significance of antibiotics as specific characteristics of actinomycetes, and their determination by the method of experimental transformation. Folia Biologica (Praha) *4*: 257–265.

Krasil'nikov, N.A. 1960. Taxonomic principles in the actinomycetes. J. Bacteriol. *79*: 65–71.

Krasil'nikov, N.A., N.L. Nikitina and A.I. Korenyako. 1961. On external features in the taxonomy of actinomycetes. Int. Bull. Bacteriol. Nomencl. Taxon. *11*: 133–159.

Krasil'nikov, N.A. and C.S. Yuan. 1961. *Actinosporangium*, a new genus of the family *Actinoplanaceae*. Izv. Akad. Nauk. SSSR Ser. Biol. *8*: 113–116.

Krasil'nikov, N.A. 1965. Biology of selected groups of actinomycetes (in Russian). Institute of Microbiology, Academy of Science, Publishing Firm "Nauka", Moscow, pp. 1–372.

Krasil'nikov, N.A., E.J. Sorokina, V.A. Alferova and A.P. Bezzubenkova. 1965. Classification of blue actinomycetes. *In* Biology of Selected Groups of Actinomycetes (edited by Krasil'nikov). Institute of Microbiology, Academy of Science, Publishing Firm Nauka, Moscow, pp. 74–123.

Krasil'nikov, N.A. and T. Yuan. 1965. The species composition of orange-colored actinomycetes. *In* Biology of Individual Groups of Actinomycetes (edited by Krasil'nikov), pp. 28–57.

Krasil'nikov, N.A. 1970a. Pigmentation of actinomycetes and its significance in taxonomy. *In* The *Actinomycetales* (edited by Prauser). Gustav Fischer Verlag, Jena, pp. 123–131.

Krasil'nikov, N.A. 1970b. Ray Fungi. Higher Forms. Nauka, Moscow.

Kreimer, A., E. Borenstein, U. Gophna and E. Ruppin. 2008. The evolution of modularity in bacterial metabolic networks. Proc. Natl. Acad. Sci. U.S.A. *105*: 6976–6981.

Kroppenstedt, R. 1977. Untersuchungen zur Chemotaxonomie der Ordnung *Actinomycetales* Buchanan 1917. PhD thesis, University Darmstadt, Germany

Kroppenstedt, R.M., F. Kornwendisch, V.J. Fowler and E. Stackebrandt. 1981. Biochemical and molecular genetic-evidence for a transfer of *Actinoplanes armeniacus* into the family *Streptomycetaceae*. Zentralbl. Bakteriol. Mikrobiol. Hyg. I Abt. Orig. *2*: 254–262.

Kroppenstedt, R.M. 1985. Fatty acid and menaquinone analysis of actinomycetes and related organisms. *In* Chemical Methods in Bacterial Systematics (edited by Goodfellow and Minnikin). Academic Press, London, pp. 173–199.

Kroppenstedt, R.M. 1987. Chemische Untersuchungen an *Actinomycetales* und verwandte Taxa, Korrelation von Chemosystematik und Phylogenie. Habilitationsschrift.

Kroppenstedt, R.M. 1992. The genus *Nocardiopsis*. *In* The Prokaryotes: a Handbook on the Biology of Bacteria: Ecophysiology, Isolation, Identification, Applications, 2nd edn (edited by Balows, Trüper, Dworkin, Harder and Schleifer). Springer, New York, pp. 1139–1156.

Krüger, F. 1904. Untersuchungen über den Gürtelschorf der Zuckerrüben. Arbeiten aus der biologischen Abteilung für Land- und Forstwirtschaft am Kaiserlichen Gesundheitsamt Band IV, Heft 3. Verlagsbuchhandlung Paul Parey, Verlagsbuchhandlung Julius Springer, Berlin, pp. 275–318.

Krüger, F. 1905. Untersuchungen über den Gürtelschorf der Zuckerrüben. Arbeiten aus der Biologischen Abteilung für Land- und Forstwirtschaft am Kaiserlichen Gesundheitsamte Band IV, Heft 3, Verlagsbuchhandlung Paul Parey, Verlagsbuchhandlung Julius Springer, Berlin, pp. 254–318.

Kubo, H., S. Suzuki and S. Tamura. 1964. Process for obtaining a new antibiotic piericidin. Japanese Patent 9443. Japan.

Kuchaeva, A.G., N.A. Krasil'nikov, G.K. Skryabin and S.D. Taptykova. 1960. Biology of antibiotic-producing actinomycetes. *In* Transactions of the Institute of Microbiology, vol. 8 (edited by Rautenshtein). Academy of Science, USSR, pp. 1–344.

Kuchaeva, A.G., N.A. Krasil'nikov, S.D. Taptykova and R.L. Gesheva. 1961. On the systematics of the *Actinomyces* of the *lavendulae* group Izvestiya Na Mikrobiologicheskiya Institut, Bulgarian Academy of Sciences, Class of Biological Sciences, Sofia *13*: 103–124.

Kudo, T. and A. Seino. 1987. Transfer of *Streptosporangium indianense* Gupta 1965 to the genus *Streptomyces* as *Streptomyces indiaensis* (Gupta 1965) comb. nov. Int. J. Syst. Bacteriol. *37*: 241–244.

Kudrina, E.S. 1957. Problems of classification of actinomycetes-antagonists (edited by Gauze, Preobrazhenskaya, Kudrina, Blinov, Ryabova and Sveshnikova). Government Publishing House of Medical Literature, Moscow, pp. 1–398.

Kumar, Y. and M. Goodfellow. 2008. Five new members of the *Streptomyces violaceusniger* 16S rRNA gene clade: *Streptomyces castelarensis* sp. nov., comb. nov., *Streptomyces himastatinicus* sp. nov., *Streptomyces mordarskii* sp. nov., *Streptomyces rapamycinicus* sp. nov. and *Streptomyces ruanii* sp. nov. Int. J. Syst. Evol. Microbiol. *58*: 1369–1378.

Kurylowicz, W. and W. Wóznicka. 1967. *Actinomyces* (*Streptomyces*) *varsoviensis*. I. Taxonomic studies. Medycyna Doswiadczalna i Mikrobiologia *19*: 1–9.

Kurylowicz, W., A. Paszkiewicz, W. Wóznicka, W. Kurtatkowski and T. Szulga. 1975. Numerical taxonomy of streptomycetes. Classification of streptomycetes by different numerical methods. Postepy Hig. Med. Dõsw. *29*: 281–355.

Kusakabe, H. and K. Isono. 1988. Taxonomic studies on *Kitasatosporia cystarginea* sp. nov., which produces a new antifungal antibiotic cystargin. J. Antibiot. (Tokyo) *41*: 1758–1762.

Kusakabe, H. and K. Isono. 1992. *In* Validation of the publication of new names and new combinations previously effectively published outside the IJSB. List no. 41. Int. J. Syst. Bacteriol. *42*: 327–328.

Küster, E. and S.T. Williams. 1964a. Selection of media for isolation of *Streptomycetes*. Nature *202*: 928–929.

Küster, E. and S.T. Williams. 1964b. Production of hydrogen sulfide by *Streptomycetes* and methods for its detection. J. Appl. Microbiol. *12*: 46–52.

Küster, E. 1976. Ecology and predominance of soil streptomycetes. *In* Actinomycetes–the Boundary Microorganisms (edited by Arai). Toppan, Tokyo, pp. 109–121.

Kutzner, H.J. and S.A. Waksman. 1959. *Streptomyces coelicolor* Mueller and *Streptomyces violaceoruber* Waksman and Curtis, two distinctly different organisms. J. Bacteriol. *78*: 528–538.

Kutzner, H.J. 1961a. Effect of various factors on the efficiency of plating and plaque morphology of some *Streptomyces* phages. Pathol. Microbiol. *24*: 30–51.

Kutzner, H.J. 1961b. Specificity of actinophages within a selected group of *Streptomyces*. Pathol. Microbiol. *24*: 170–191.

Kutzner, H.J. 1972. Storage of *Streptomyces* in soft agar and by other methods. Experientia *28*: 1395–1396.

Kutzner, H.J. 1981. The family *Streptomycetaceae*. *In* The Prokaryotes: a Handbook on Habitats, Isolation, and Identification of Bacteria, vol. 2 (edited by Starr, Stolp, Trüper, Balows and Schlegel). Springer, New York, pp. 2028–2090.

Kuznetsov, V.D. 1962. A new species of genus *Chainia*. Mikrobiologiya *31*: 534–539.

Kuznetsov, V.D., T.A. Zajtseva, L.V. Vakulenko and S.N. Flippova. 1992. *Streptomyces albiaxialis* sp. nov.-a new oil hydrocarbon degrading species of thermo- and halotolerant streptomyces. Mikrobiologiya *61*: 84–91.

Kuznetsov, V.D., T.A. Zajtseva, L.V. Vakulenko and S.N. Flippova. 1993. *In* Validation of the publication of new names and new combinations previously effectively published outside the IJSB. List no. 45. Int. J. Syst. Bacteriol. *45*: 398–399.

Labeda, D.P. 1987. Transfer of the type strain of *Streptomyces erythraeus* (Waksman 1923) Waksman and Henrici 1948 to the genus *Saccharopolyspora Lacey* and Goodfellow 1975 as *Saccharopolyspora erythraea* sp. nov., and designation of a neotype strain for *Streptomyces erythraeus*. Int. J. Syst. Bacteriol. *37*: 19–22.

Labeda, D.P. 1988. *Kitasatosporia mediocidica* sp. nov. Int. J. Syst. Bacteriol. *38*: 287–290.

Labeda, D.P. and A.J. Lyons. 1991a. Deoxyribonucleic-acid relatedness among species of the *Streptomyces cyaneus* cluster. Syst. Appl. Microbiol. *14*: 158–164.

Labeda, D.P. and A.J. Lyons. 1991b. The *Streptomyces violaceusniger* cluster is heterogeneous in DNA relatedness among strains: emendation of the descriptions of *S. violaceusniger* and *Streptomyces hygroscopicus*. Int. J. Syst. Bacteriol. *41*: 398–401.

Labeda, D.P. 1993. DNA relatedness among strains of the *Streptomyces lavendulae* phenotypic cluster group. Int. J. Syst. Bacteriol. *43*: 822–825.

Labeda, D.P. 1996. DNA relatedness among verticil-forming *Streptomyces* species (formerly *Streptoverticillium* species). Int. J. Syst. Bacteriol. *46*: 699–703.

Labeda, D.P., M.P. Lechevalier and R.T. Testa. 1997. *Streptomyces stramineus* sp. nov., a new species of the verticillate streptomycetes. Int. J. Syst. Bacteriol. *47*: 747–753.

Labeda, D.P. 1998. DNA relatedness among the *Streptomyces fulvissimus* and *Streptomyces griseoviridis* phenotypic cluster groups. Int. J. Syst. Bacteriol. *48*: 829–832.

Lacalle, R.A., J.A. Tercero and A. Jimenez. 1992. Cloning of the complete biosynthetic gene cluster for an aminonucleoside antibiotic, puromycin, and its regulated expression in heterologous hosts. EMBO J. *11*: 785–792.

Lacey, J. and J. Dutkiewicz. 1976a. Methods for examining the microflora of mouldy hay. J. Appl. Bacteriol. *41*: 13–27.

Lacey, J. and J. Dutkiewicz. 1976b. Isolation of actinomycetes and fungi from mouldy hay using a sedimentation chamber. J. Appl. Bacteriol. *41*: 315–319.

Lacey, J. 1987. Nomenclature of *Saccharopolyspora erythraea* Labeda 1987 and *Streptomyces erythraeus* (Aksman 1923) Waksman and Henrici 1948, and proposals for the alternative epithet *Streptomyces labedae* sp. nov. Int. J. Syst. Bacteriol. *37*: 458–458.

Lacey, J. and M.E. Lacey. 1987. Micro-organisms in the air of cotton mills. Ann. Occup. Hyg. *31*: 1–19.

Lacey, J. 1988. Actinomycetes as biodeteriogens and pollutants of the environment. *In* Actinomycetes in Biotechnology (edited by Goodfellow, Williams and Mordarski). Academic Press, London, pp. 359–432.

Lachner-Sandoval, V. 1898. über Strahlenpilze. Inaugural Dissertation thesis, Universitätsbuchdruckerei Carl Georgi, Bonn, Strassburg.

Lambert, D.H. and R. Loria. 1989a. *Streptomyces acidiscabies* sp. nov. Int. J. Syst. Bacteriol. *39*: 393–396.

Lambert, D.H. and R. Loria. 1989b. *Streptomyces scabies* sp. nov., nom. rev. Int. J. Syst. Bacteriol. *39*: 387–392.

Lange, B.J. and W.J.R. Boyd. 1968. Preservation of fungal spores by drying on porcelain bead. Phytopathology *58*: 1711–1712.

Langham, C.D., S.T. Williams, P.H. Sneath and A.M. Mortimer. 1989. New probability matrices for identification of *Streptomyces*. J. Gen. Microbiol. *135*: 121–133.

Lanoot, B., M. Vancanneyt, I. Cleenwerck, L. Wang, W. Li, Z. Liu and J. Swings. 2002. The search for synonyms among streptomycetes by using SDS-PAGE of whole-cell proteins. Emendation of the species *Streptomyces aurantiacus*, *Streptomyces cacaoi* subsp. *cacaoi*, *Streptomyces caeruleus* and *Streptomyces violaceus*. Int. J. Syst. Evol. Microbiol. *52*: 823–829.

Lanoot, B., M. Vancanneyt, P. Dawyndt, M. Cnockaert, J.L. Zhang, Y. Huang, Z.H. Liu and J. Swings. 2004. BOX-PCR fingerprinting as a powerful tool to reveal synonymous names in the genus *Streptomyces*. Emended descriptions are proposed for the species *Streptomyces cinereorectus*, *S. fradiae*, *S. tricolor*, *S. colombiensis*, *S. filamentosus*, *S. vinaceus* and *S. phaeopurpureus*. Syst. Appl. Microbiol. *27*: 84–92.

Lanoot, B., M. Vancanneyt, P. Dawyndt, M. Crockaert, J. Zhang, Y. Huang, Z. Liu and J. Swings. 2005a. Notification of changes in taxonomic opinion previously published outside the IJSEM. List no. 1. Int. J. Syst. Evol. Microbiol. *55*: 7–8.

Lanoot, B., M. Vancanneyt, A. Van Schoor, Z. Liu and J. Swings. 2005b. Reclassification of *Streptomyces nigrifaciens* as a later synonym of *Streptomyces flavovirens; Streptomyces citreofluorescens, Streptomyces chrysomallus* subsp. *chrysomallus and Streptomyces fluorescens* as later synonyms of *Streptomyces anulatus; Streptomyces chibaensis* as a later synonym of *Streptomyces corchorusii; Streptomyces flaviscleroticus* as a later synonym of *Streptomyces minutiscleroticus;* and *Streptomyces lipmanii, Streptomyces griseus* subsp. *alpha, Streptomyces griseus* subsp. *cretosus* and *Streptomyces willmorei* as later synonyms of *Streptomyces microflavus.* Int. J. Syst. Evol. Microbiol. *55*: 729–731.

Le Roes, M. and P.R. Meyers. 2005a. *Streptomyces pharetrae* sp. nov., isolated from soil from the semi-arid Karoo region. Syst. Appl. Microbiol. *28*: 488–493.

Le Roes, M. and P.R. Meyers. 2005b. *In* Validation of publication of new names and new combinations previously effectively published outside the IJSEM. List no. 106. Int. J. Syst. Bacteriol. *55*: 2235–2238.

Leach, B.E., K.M. Calhoun, L.E. Johnson, C.M. Teeters and W.G. Jackson. 1953. Chartreusin, a new antibiotic produced by *Streptomyces chartreusis* a new species. J. Am. Chem. Soc. *75*: 4011–4012.

Leblond, P. and B. Decaris. 1994. New insights into the genetic instability of *Streptomyces.* FEMS Microbiol. Lett. *123*: 225–232.

Leblond, P., G. Fischer, F.X. Francou, F. Berger, M. Guérineau and B. Decaris. 1996. The unstable region of *Streptomyces ambofaciens* includes 210 kb terminal inverted repeats flanking the extremities of the linear chromosomal DNA. Mol. Microbiol. *19*: 261–271.

Lechevalier, H.A. and C.T. Corke. 1953. The replica plate method for screening antibiotic-producing organisms. Appl. Microbiol. *1*: 110–112.

Lechevalier, H.A. and M.P. Lechevalier. 1970a. A critical evaluation of the genera of aerobic actinomycetes. *In* The *Actinomycetales* (edited by Prauser). Gustav Fischer Verlag, Jena, pp. 395–405.

Lechevalier, H.A., M.P. Lechevalier and N.N. Gerber. 1971. Chemical composition as a criterion in the classification of actinomycetes. Adv. Appl. Microbiol. *14*: 47–72.

Lechevalier, M.P. and H.A. Lechevalier. 1970b. Chemical composition as a criterion in the classification of aerobic actinomycetes. Int. J. Syst. Bacteriol. *20*: 435–443.

Lechevalier, M.P. and H.A. Lechevalier. 1970c. Composition of whole-cell hydrolysates as a criterion in the classification of aerobic actinomycetes. *In* The *Actinomycetales* (edited by Prauser). Gustav Fischer Verlag, Jena, pp. 311–316.

Lechevalier, M.P. 1977. Lipids in bacterial taxonomy - a taxonomist's view. CRC Crit. Rev. Microbiol. *5*: 109–210.

Lechevalier, M.P., C. De Bièvre and H. Lechevalier. 1977. Chemotaxonomy of aerobic actinomycetes: phospholipid composition. Biochem. Syst. Ecol. *5*: 249–260.

Lechevalier, M.P., R.J. Seidler and T.M. Evans. 1980. Enumeration and characterization of standard plate count bacteria in chlorinated and raw water supplies. Appl. Environ. Microbiol. *40*: 922–930.

Lechevalier, M.P., A.E. Stern and H.A. Lechevalier. 1981. Phospholipids in the taxonomy of actinomycetes. Zentralbl. Bakteriol. Parasitenkd. Infektionskr. Hyg. I Abt. Orig. Suppl. *11*: 111–116.

Lechevalier, M.P. 1988. Actinomycetes in agriculture and forestry. *In* Actionmycetes in Biotechnology (edited by Goodfellow, Williams and Mordarski). Academic Press, San Diego, pp. 327–358.

Lee, J.Y., J.Y. Lee, H.W. Jung and B.K. Hwang. 2005. *Streptomyces koyangensis* sp. nov., a novel actinomycete that produces 4-phenyl-3-butenoic acid. Int. J. Syst. Evol. Microbiol. *55*: 257–262.

Lehmann, K.B. and R. Neumann. 1896. Atlas und Grundriss der Bakteriologie und Lehrbuch der speciellen bacteriologischen Diagnostik. Lehmann, München.

Lezhava, A., T. Mizukami, T. Kajitani, D. Kameoka, M. Redenbach, H. Shinkawa, O. Nimi and H. Kinashi. 1995. Physical map of the linear chromosome of *Streptomyces griseus.* J. Bacteriol. *177*: 6492–6498.

Li, W., B. Lanoot, Y. Zhang, M. Vancanneyt, J. Swings and Z. Liu. 2002a. *Streptomyces scopiformis* sp. nov., a novel streptomycete with fastigiate spore chains. Int. J. Syst. Evol. Microbiol. *52*: 1629–1633.

Li, W.J., L.P. Zhang, P. Xu, X.L. Cui, Z.T. Lu, L.H. Xu and C.L. Jiang. 2002b. *Streptomyces beijiangensis* sp. nov., a psychrotolerant actinomycete isolated from soil in China. Int. J. Syst. Evol. Microbiol. *52*: 1695–1699.

Li, W.J., Y.G. Zhang, Y.Q. Zhang, S.K. Tang, P. Xu, L.H. Xu and C.L. Jiang. 2005. *Streptomyces sodiiphilus* sp. nov., a novel alkaliphilic actinomycete. Int. J. Syst. Evol. Microbiol. *55*: 1329–1333.

Li, X.Z. 1997. *Streptomyces cellulolyticus* sp. nov., a new cellulolytic member of the genus *Streptomyces.* Int. J. Syst. Bacteriol. *47*: 443–445.

Liao, D. and P.P. Dennis. 1994. Molecular phylogenies based on ribosomal protein L11, L1, L10, and L12 sequences. J. Mol. Evol. *38*: 405–419.

Lieske, R. 1921. Morphologie und Biologie der Strahlenpilze (Actinomyceten). Borntraeger Bros., Leipzig.

Lin, Y.S., H.M. Kieser, D.A. Hopwood and C.W. Chen. 1993. The chromosomal DNA of *Streptomyces lividans* 66 is linear. Mol. Microbiol. *10*: 923–933.

Lindenbein, W. 1952. über einige chemisch interessante Actinomyceten – stämme und ihre Klassifizierung. Arch. Mikrobiol. *17*: 361–383.

Lindner, F. and K.H. Wallhausser. 1955. Die Arbeitsmethoden Der Forschung Zur Auffindung Neuer Antibiotica. Arch. Mikrobiol. *22*: 219–234.

Lindner, F., R. Junk, G. Nesemann and J. Schmidt-Thomé. 1958. Gewinnung Von 20 β-Hydroxysteroiden Aus 17 α-21-Dihydroxy-20-Ketosteroiden durch Mikrobiologische Hydrierung Mit *Streptomyces hydrogenans.* Hoppe-Seylers Zeitschr. Physiol. Chem. *313*: 117–123.

Lingappa, Y. and J.L. Lockwood. 1962. Chitin media for selective isolation and culture of actinomycetes. Phytopathology *52*: 317–323.

Liu, Z., C. Rodriguez, L. Wang, Q. Cui, Y. Huang, E.T. Quintana and M. Goodfellow. 2005a. *Kitasatospora viridis* sp. nov., a novel actinomycete from soil. Int. J. Syst. Evol. Microbiol. *55*: 707–711.

Liu, Z., Y. Shi, Y. Zhang, Z. Zhou, Z. Lu, W. Li, Y. Huang, C. Rodriguez and M. Goodfellow. 2005b. Classification of *Streptomyces griseus* (Krainsky 1914) Waksman and Henrici 1948 and related species and the transfer of '*Microstreptospora cinerea*' to the genus *Streptomyces* as *Streptomyces yanii* sp. nov. Int. J. Syst. Evol. Microbiol. *55*: 1605–1610.

Lloyd, A.B. 1969. Dispersal of *Streptomycetes* in air. J. Gen. Microbiol. *57*: 35–40.

Locci, R., E. Baldacci and B. Petrolini Baldan. 1969. The genus *Streptoverticillium.* A taxonomic study. G. Microbiol. *17*: 1–60.

Locci, R. and B. Petrolini Baldan. 1971. On the spore formation process in Actinomycetes I. Morphology and development of *Streptoverticillium* species as examined by scanning electron microscopy. Riv. Patol. Veg. Serie IV 7 *(Supplement)*: 3–19.

Locci, R., J. Rogers, P. Sardi and G.M. Schofield. 1981. A preliminary numerical study of named species of the genus *Streptoverticillium.* Ann. Microbiol. Enzimol. *31*: 115–121.

Locci, R. and G.P. Sharples. 1984. Morphology. *In* The Biology of Actinomycetes (edited by Goodfellow, Williams and Mordarski). Academic Press, London, pp. 165–199.

Locci, R. 1985. New combinations and validation of some taxa of the genus *Streptoverticillium.* Ann. Microbiol. Enzimol. *35*: 231–234.

Locci, R. and G.M. Schofield. 1989. Genus *Streptoverticillium. In* Bergey's Manual of Systematic Bacteriology, vol. 4 (edited by Williams, Sharpe and Holt). 2492–2504, Baltimore.

Lombo, F., G. Blanco, E. Fernandez, C. Mendez and J.A. Salas. 1996. Characterization of *Streptomyces argillaceus* genes encoding a polyketide synthase involved in the biosynthesis of the antitumor mithramycin. Gene *172*: 87–91.

Lomovskaya, N.D., K.F. Chater and N.M. Mkrtumian. 1980. Genetics and molecular biology of *Streptomyces* bacteriophages. Microbiol. Rev. *44*: 206–229.

Lonsdale, J.T. 1985. Aspects of the biology of acidophilic actinomycetes. PhD thesis, University of Newcastle, Newcastle upon Tyne.

Loria, R., J. Kers and M. Joshi. 2006. Evolution of plant pathogenicity in *Streptomyces*. Annu. Rev. Phytopathol. *44*: 469–487.

Loria, R., D.R.D. Bignell, S. Moll, J.C. Huguet-Tapia, M.V. Joshi, E.G. Johnson, R.F. Seipke and D.M. Gibson. 2008. Thaxtomin biosynthesis: the path to plant pathogenicity in the genus *Streptomyces*. Antonie van Leeuwenhoek *94*: 3–10.

Ludwig, W. and K.H. Schleifer. 1994. Bacterial phylogeny based on 16S and 23S rRNA sequence analysis. FEMS Microbiol. Rev. *15*: 155–173.

Ludwig, W., O. Strunk, R. Westram, L. Richter, H. Meier, Yadhukumar, A. Buchner, T. Lai, S. Steppi, G. Jobb, W. Forster, I. Brettske, S. Gerber, A.W. Ginhart, O. Gross, S. Grumann, S. Hermann, R. Jost, A. Konig, T. Liss, R. Lussmann, M. May, B. Nonhoff, B. Reichel, R. Strehlow, A. Stamatakis, N. Stuckmann, A. Vilbig, M. Lenke, T. Ludwig, A. Bode and K.H. Schleifer. 2004. ARB: a software environment for sequence data. Nucleic Acids Res. *32*: 1363–1371.

Lutkenhaus, J. 1997. Bacterial cytokinesis: let the light shine in. Curr. Biol. *7*: 573–575.

Lyons, A.J. and T.G. Pridham. 1971. *Streptomyces torulosus* sp. n., an unusual knobby-spored taxon. Appl. Microbiol. *22*: 190–193.

Lyons, A.J., Jr. and T.G. Pridham. 1965. Colorimetric determination of color of aerial mycelium of *Streptomycetes*. J. Bacteriol. *89*: 159–169.

Macé, E. 1897. Traité Pratique de Bactériologie, 4th edn. Baillière, Paris, pp. 1–1144.

Macé, E. 1901. Traité Pratique de Bactériologie, 5th edn. Baillière, Paris, pp. 1–1196.

Macé, E. 1913. Traité Pratique de Bactériologie, 6th edn. Baillière, Paris, pp. 1–918.

Machenko, И.В., С.А. Ezopoвa, H.O. Blinov and H.A. КрасильНиков. 1970. Микробиология 39: 1010 (in Russian).

MacKenzie, C.R., D. Bilous and K.G. Johnson. 1984. Purification and characterization of an exoglucanase from *Streptomyces flavogriseus*. Can. J. Microbiol. *30*: 1171–1178.

Maeda, K., Y. Okami, O. Taya and H. Umezawa. 1952. On new antif ungal substances, moldin and phaeofacin, produced by *Streptomyces* sp. Jpn J. Med. Sci. Biol. *5*: 327–339.

Maerz, A. and M.R. Paul. 1950. A Dictionary of Color, 2nd edn. McGraw Hill, New York.

Mahgoub, E.S. 1985. Mycetoma. Int. J. Dermatol. *24*: 230–239.

Malpartida, F. and D.A. Hopwood. 1984. Molecular cloning of the whole biosynthetic pathway of a *Streptomyces* antibiotic and its expression in a heterologous host. Nature *309*: 462–464.

Malpartida, F., J. Niemi, R. Navarrete and D.A. Hopwood. 1990. Cloning and expression in a heterologous host of the complete set of genes for biosynthesis of the *Streptomyces coelicolor* antibiotic undecylprodigiosin. Gene *93*: 91–99.

Manchester, L., B. Pot, K. Kersters and M. Goodfellow. 1990. Classification of *Streptomyces* and *Streptoverticillium* species by numerical-analysis of electrophoretic protein-patterns. Syst. Appl. Microbiol. *13*: 333–337.

Manfio, G.P., E. Atalan, J. Zakrzewska-Czerwinska, M. Mordarski, C. Rodriguez, M.D. Collins and M. Goodfellow. 2003a. Classification of novel soil streptomycetes as *Streptomyces aureus* sp. nov., *Streptomyces laceyi* sp. nov. and *Streptomyces sanglieri* sp. nov. Antonie van Leeuwenhoek *83*: 245–255.

Manfio, G.P., E. Atalan, J. Zakrzewska-Czerwinska, M. Mordarski, C. Rodriguez, M.D. Collins and M. Goodfellow. 2003b. *In* Validation of publication of new names and new combinations previously effectively published outside the IJSEM. List no. 93. Int. J. Syst. Evol. Microbiol. *53*: 1219–1220.

Mao, J., Q. Tang, Z. Zhang, W. Wang, D. Wei, Y. Huang, Z. Liu, Y. Shi and M. Goodfellow. 2007. *Streptomyces radiopugnans* sp. nov., a radiation-resistant actinomycete isolated from radiation-polluted soil in China. Int. J. Syst. Evol. Microbiol. *57*: 2578–2582.

Margalith, P., G. Beretta and M.T. Timbal. 1959. Matamycin, a new antibiotic. I. Biological studies. Antibiot. Chemother. *9*: 71–75.

Margalith, P. and G. Beretta. 1960. A new antibiotic producing *Streptomyces*: *S. bellus* nov. sp. Mycopathol. Mycol. Appl. *12*: 189–195.

Margolin, W. 2003. Bacterial division: the fellowship of the ring. Curr. Biol. *13*: 16–18.

Márialigeti, K., K. Jager, I.M. Szabo, M. Pobozsny and A. Dzingov. 1984. The faecal actinomycete flora of *Protracheoniscus amoenus* (woodlice; Isopoda). Acta Microbiol. Hung. *31*: 339–344.

Marri, L., E. Barboni, T. Irdani, B. Perito and G. Mastromei. 1997. Restriction enzyme and DNA hybridization analysis of cellulolytic *Streptomyces* isolates of different origin. Can. J. Microbiol. *43*: 395–399.

Martin, J.P., Z. Filip and K. Haider. 1976. Effect of montmorillonite and humate on growth and metabolic activity of some actinomycetes. Soil Biol. Biochem. *8*: 409–413.

Mason, D.J., A. Dietz and R.M. Smith. 1961. Actinospectacin, a new antibiotic. I. Discovery and biological properties. Antibiot. Chemother. *11*: 118–122.

Mason, D.J., A. Dietz and C. De Boer. 1963a. Lincomycin, a new antibiotic. I. Discovery and biological properties. Antimicrob. Agents Chemother. *1962*: 554–559.

Mason, D.J., A. Dietz and L.J. Hanka. 1963b. U-12898, a new antibiotic. I. Discovery, biological properties and assay. Antimicrob. Agents Chemother. *1962*: 607–613.

Mason, D.J., W.L. Lummis and A. Dietz. 1965. U-22956, a new antibiotic. I. Discovery and biological activity. Antimicrob. Agents Chemother. *1964*: *110–113*.

Matsumae, A., M. Ohtani, H. Takeshima and T. Hata. 1968. A new genus of the *Actinomycetales*: *Kitasatoa* gen. nov. J. Antibiot. (Tokyo) *21*: 616–625.

Maximova, T.S. and L.P. Terekhova. 1986. Footnote *f*. *In* Validation of the publication of new names and new combinations previously effectively published outside the IJSB. List no. 22. Int. J. Syst. Bacteriol. *36*: 573–576.

Mayfield, C.I., S.T. Williams, S.M. Ruddick and H.L. Hatfield. 1972. Studies of the ecology of actinomycetes in soil IV. Observation in the form and growth of *Streptomycetes* in soil. Soil. Biol. Biochem. *4*: 79–91.

Mayilraj, S., S. Krishnamurthi, P. Saha and H.S. Saini. 2006. *Kitasatospora sampliensis* sp. nov., a novel actinobacterium isolated from soil of a sugar-cane field in India. Int. J. Syst. Evol. Microbiol. *56*: 519–522.

McCarthy, A.J., M.J. MacDonald, A. Paterson and P. Broda. 1984. Lignocellulose degradation by actinomycetes. J. Gen. Microbiol. *130*: 1023–1030.

McCarthy, A.J. and P. Broda. 1984. Screening for lignin-degrading actinomycetes and characterisation of their activity against ^{14}C-lignin-labelled wheat lignocellulose. J. Gen. Microbiol. *130*: 2905–2913.

McCarthy, A.J., E. Peace and P. Broda. 1985. Studies on the extracellular xylanase activity of some thermophilic actinomycetes. Appl. Microbiol. Biotechnol. *21*: 238–244.

McCarthy, A.J., A. Paterson and P. Broda. 1986. Lignin solubilisation by *Thermomonospora mesophila*. Appl. Microbiol. Biotechnol. *24*: 347–352.

McCue, L.A., J. Kwak, J. Wang and K.E. Kendrick. 1996. Analysis of a gene that suppresses the morphological defect of bald mutants of *Streptomyces griseus*. J. Bacteriol. *178*: 2867–2875.

McKillop, C., P. Elvin and J. Kenten. 1986. Cloning and expression of an extracellular alpha-amylase gene from *Streptomyces hygroscopicus* in *Streptomyces lividans* 66. FEMS Microbiol. Lett. *36*: 3–7.

McNeil, M.M. and J.M. Brown. 1994. The medical important aerobic actinomycetes: epidemiology and microbiology. Clin. Microbiol. Rev. *7*: 357–417.

McVeigh, I. and C.R. Reyes. 1961. A new species of *Streptomyces* and its antibiotic activity. Antibiot. Chemother. *11*: 312–319.

Mehling, A., U.F. Wehmeier and W. Piepersberg. 1995. Application of random amplified polymorphic DNA (RAPD) assays in identifying conserved regions of actinomycete genomes. FEMS Microbiol. Lett. *128*: 119–125.

Menzies, J.D. and C.E. Dade. 1959. A selective indicator medium for isolating *Streptomyces scabies* from potato tubers or soil. Phytopathology *49*: 457–458.

Mertz, F.P. and C.E. Higgens. 1982. *Streptomyces capillispiralis* sp. nov. Int. J. Syst. Bacteriol. *32*: 116–124.

Meyers, P.R., D.S. Porter, C. Omorogie, J.M. Pule and T. Kwetane. 2003. *Streptomyces speibonae* sp. nov., a novel streptomycete with blue substrate mycelium isolated from South African soil. Int. J. Syst. Evol. Microbiol. *53*: 801–805.

Meyers, P.R., C.M. Goodwin, J.A. Bennett, B.L. Aken, C.E. Price and J.M. van Rooyen. 2004. *Streptomyces africanus* sp. nov., a novel streptomycete with blue aerial mycelium. Int. J. Syst. Evol. Microbiol. *54*: 1531–1535.

Mikami, Y., K. Miyashita and T. Arai. 1982. Diaminopimelic acid profiles of alkalophilic and alkaline-resistant strains of *Actinomycetes*. J. Gen. Microbiol. *128*: 1709–1712.

Mikami, Y., K. Miyashita and T. Arai. 1985. Alkalophilic actinomycetes. *In* The *Actinomycetes*, vol. 19, no. 3 (edited by Lechevalier). Rutgers University Publications Department, New Jersey, pp. 176–191.

Mikulik, K., I. Janda, J. Weiser and A. Jiranova. 1982. Ribosomal proteins of *Streptomyces aureofaciens* producing tetracycline. Biochim. Biophys. Acta *699*: 203–210.

Millard, W.A. and S. Burr. 1926. A study of twenty-four strains of *Actinomyces* and their relation to types of common scab of potato. Ann. Appl. Biol. *13*: 580–644.

Millner, P.D. 1982. Thermophilic and thermotolerant actinomycetes in sewage-sludge compost. Dev. Indust. Microbiol. *23*: 61–78.

Minambres, B., E.R. Olivera, R.A. Jensen and J.M. Luengo. 2000. A new class of glutamate dehydrogenases (GDH). Biochemical and genetic characterization of the first member, the AMP-requiring NAD-specific GDH of *Streptomyces clavuligerus*. J. Biol. Chem. *275*: 39529–39542.

Misiek, M. 1955. Comparative studies of *Streptomyces* populations in soils. PhD thesis, Syracuse.

Miyadoh, S., T. Shomura, T. Ito and T. Niida. 1983. *Streptomyces sulfonofaciens* sp. nov. Int. J. Syst. Bacteriol. *33*: 321–324.

Miyairi, N., M. Takashima, K. Shimizu and H. Sakai. 1966. Studies on new antibiotics, cineromycins A and B. J. Antibiot. (Tokyo) Ser. A *19*: 56–62.

Miyajima, K., F. Tanaka, T. Takeuchi and S. Kuninaga. 1998. *Streptomyces turgidiscabies* sp. nov. Int. J. Syst. Bacteriol. *48*: 495–502.

Miyashita, K., T. Fujii and Y. Sawada. 1991. Molecular cloning and characterization of chitinase genes from *Streptomyces lividans* 66. J. Gen. Microbiol. *137*: 2065–2072.

Mordarski, M., J. Wieczorek and B. Jaworska. 1970. On the conditions of amylase production by actinomycetes. Archiwum Immunol. Ther. Experimentalis *18*: 375–381.

Mordarski, M., M. Goodfellow, S.T. Williams and P.H. Sneath. 1986. Evaluation of species groups in the genus *Streptomyces*. *In* Biological, Biochemical and Biomedical Aspects of Actinomycetes (edited by Szabó, Biró and Goodfellow). Akadémiai Kaidó, Budapest, pp. 517–525.

Morita, R.Y. 1985. Starvation and miniaturisation of heterotrophs, with special emphasis on maintenance of the starved viable state. *In* Bacteria in their Natural Environments (edited by Fletcher and Floodgate). Academic Press, London, pp. 111–130.

Morosoli, R., S. Ostiguy and C. Dupont. 1999. Effect of carbon source, growth and temperature on the expression of the *sec* genes of *Streptomyces lividans* 1326. Can. J. Microbiol. *45*: 1043–1049.

Motamedi, H. and C.R. Hutchinson. 1987. Cloning and heterologous expression of a gene cluster for the biosynthesis of tetracenomycin C, the anthracycline antitumor antibiotic of *Streptomyces glaucescens*. Proc. Natl. Acad. Sci. U.S.A. *84*: 4445–4449.

Müller, E. 1950. Medizinische Mikrobiologie - Parasiten, Bakterien, Immunität, 4th edn. Urban and Schwartzenberg, Munich and Berlin.

Müller, R. 1908. Eine Diphtheridee und eine *Streptothrix* mit gleichen blauen Farbstoff sowie Untersuchungen über Streptothrixarten im Allgemeinen. Zentralbl. Bakteriol. Parasitenkd. Infektionskr. Hyg. Abt. II *46*: 195–212.

Murakami, T., S. Anzai, S. Imai, A. Satoh, K. Nagaoka and C.J. Thompson. 1986. The bialalphos biosynthetic genes of *Streptomyces hygroscopicus*: molecular cloning and characterization of the gene cluster. Mol. Gen. Genet. *205*: 42–50.

Muyzer, G., E.C. de Waal and A.G. Uitterlinden. 1993. Profiling of complex microbial populations by denaturing gradient gel electrophoresis analysis of polymerase chain reaction-amplified genes coding for 16S rRNA. Appl. Environ. Microbiol. *59*: 695–700.

Nagatsu, J., S. Suzuki and K. Anzai. 1962. Pathocidin, a new antifungal antibiotic; II. Taxonomic studies on pathocidin-producing organism *Streptomyces albus* var. *pathocidicus*. J. Antibiot. (Tokyo) Ser. A *15*: 103–106.

Nagatsu, J., K. Ansai, S. Suzuki and K. Ohkuma. 1963. Studies on a new antibiotic, tuberin. IV. Taxonomic studies on tuberin producing organism, *Streptomyces amakusaensis*. J. Antibiot. (Tokyo) Ser. A *16*: 207–210.

Nagatsu, J. and S. Suzuki. 1963. Studies on an antitumor antibiotic, cervicarcin. III. Taxonomic studies on cervicarcin producing organism, *Streptomyces ogaensis* nov. sp. J. Antibiot. (Tokyo) Ser. A *16*: 203–206.

Nakagaito, Y., A. Shimazu, A. Yokota and T. Hasegawa. 1992a. Proposal of *Streptomyces atroaurantiacus* sp. nov. and *Streptomyces kifunensis* sp. nov. and transferring *Kitasatosporia cystarginea* Kusakabe and Isono to the genus *Streptomyces* as *Streptomyces cystargineus* comb. nov. J. Gen. Appl. Microbiol. *38*: 627–633.

Nakagaito, Y., A. Yokota and T. Hasegawa. 1992b. Three new species of the genus *Streptomyces*: *Streptomyces cochleatus* sp. nov., *Streptomyces paracochleatus* sp. nov., and *Streptomyces azaticus* sp. nov. J. Gen. Appl. Microbiol. *38*: 105–120.

Nakagaito, Y., A. Shimazu, A. Yokota and T. Hasegawa. 1993a. *In* Validation pf the publication of new names and new combinations previously effectively published outside the IJSB. List no. 46. Int. J. Syst. Bacteriol. *43*: 624.

Nakagaito, Y., A. Shimazu, A. Yokota and T. Hasegawa. 1993b. *In* Validation pf the publication of new names and new combinations previously effectively published outside the IJSB. List no. 44. Int. J. Syst. Bacteriol. *43*: 188–189.

Nakamura, G. 1961. Studies on antibiotic actinomycetes. I. On *Streptomyces* producing a new antibiotic tubermycin. J. Antibiot. (Tokyo) Ser. A *14*: 86–89.

Nakata, K., S. Horinouchi and T. Beppu. 1989. Cloning and characterization of the carbapenem biosynthetic genes from *Streptomyces fulvoviridis*. FEMS Microbiol. Lett. *48*: 51–55.

Nakazawa, K. 1955. *Streptomyces albireticuli* nov. sp. J. Agric. Chem. Soc. Jap. *29*: 647–649.

Nakazawa, K. and M. Shibata. 1956. A new species of *Streptomyces* producing dihydrostreptomycin. Proc. Jpn Acad. Sci., pp. 648–653.

Nakazawa, K., K. Tanabe, M. Shibata, A. Miyake and T. Takewaka. 1956. Studies on streptomycetes. Cladomycin, a new antibiobic produced by *Streptomyces lilacinus* nov. sp. J. Antibiot. (Tokyo) *9*: 81.

Nakazawa, K. and S. Fujii. 1957. Studies on streptomycetes. On *Streptomyces sindenensis* nov. sp. Annual Report of the Takeda Research Laboratory *16*: 109–110.

Naumova, I.B., V.D. Kuznetsov, K.S. Kudrina and A.P. Bezzubenkova. 1980. The occurrence of teichoic acids in streptomycetes. Arch. Mikrobiol. *126*: 71–75.

Nette, I.T., N.V. Pomortzeva and E.I. Kozlova. 1959. Destruction of caoutchouc by microorganisms. Mikrobiologiia *28*: 881–886.

Niida, T. and M. Ogasawara. 1960. Taxonomical study on a new *Streptomyces* producing taitomycin. Scientific Reports of Meiji Seika Kaisha *3*: 23–26.

Niida, T., K. Hamamoto, T. Tsuruoka and T. Hara. 1963. Taxonomic studies on a new *Streptomyces* producing both blasticin S and 8-azaguanine. Sci. Rep. Meiji Seika Kaisha *6*: 27–39.

Niida, T. 1966. Methods for characterization of *Streptomyces* species. *In* Int. J. Syst. Bacteriol., vol. 16 (edited by Shirling and Gottlieb), pp. 313–340.

Nishimura, H. and T. Kimura. 1953. On a yellow crystalline antibiotic, identical with aureothricin isolated from a new species of *Streptomyces*, 39a, and its taxonomic study. J. Antibiot. (Tokyo) Series A *6*: 57–65.

Nishimura, H., T. Kimura, K. Tawara, K. Sasaki, K. Nakajima, N. Shimaoka, S. Okamoto, M. Shimohira and J. Isono. 1957. Aburamycin, a new antibiotic. J. Antibiot. (Tokyo) Series A *10*: 205–212.

Nishimura, H., S. Okamoto, M. Mayama, H. Ohtsuka, K. Nakajima, K. Tawara, M. Shimohira and N. Shimaoka. 1961. Siomycin, a new thiostrepton-like antibiotic. J. Antibiot. (Tokyo) Ser. A *14*: 255–263.

Nishimura, H., M. Mayama, Y. Komatsu, H. Kato, N. Shimaoka and Y. Tanaka. 1964. Showdomycin, a new antibiotic from a *Streptomyces* sp. J. Antibiot. (Tokyo) Ser. A *17*: 148–155.

Nissen, T.V. 1963. Distribution of antibiotic-producing actinomycetes in Danish soils. Experientia *19*: 470–471.

Nolan, R.D. and T. Cross. 1988. Isolation and screening of actinomycetes. *In* Actinomycetes in biotechnology (edited by Goodfellow, Williams and Mordarski). Academic Press, San Diego, pp. 1–32.

Noval, J.J. and W.J. Nickerson. 1959. Decomposition of native keratin by *Streptomyces fradiae*. J. Bacteriol. *77*: 251–263.

Novotna, J., J. Vohradsky, P. Berndt, H. Gramajo, H. Langen, X.M. Li, W. Minas, L. Orsaria, D. Roeder and C.J. Thompson. 2003. Proteomic studies of diauxic lag in the differentiating prokaryote *Streptomyces coelicolor* reveal a regulatory network of stress-induced proteins and central metabolic enzymes. Mol. Microbiol. *48*: 1289–1303.

Nüesch, J. 1965. Isolierung und Selektionierung von *Actinomyceten*. In Symposium ("Anreicherungskultur und Mutantenauslese") Göttingen, April 1964. Zentralbl. Bakteriol. Parasitenkd. Infektionskr. Hyg. Abt. 1 Suppl. *1*: 234–252.

O'Connor, T.J., P. Kanellis and J.R. Nodwell. 2002. The *ramC* gene is required for morphogenesis in *Streptomyces coelicolor* and expressed in a cell type-specific manner under the direct control of RamR. Mol. Microbiol. *45*: 45–57.

Obanye, A.I.C., G. Hobbs, D.C.J. Gardner and S.G. Oliver. 1996. Correlation between carbon flux through the pentose phosphate pathway and production of the antibiotic methylenomycin in *Streptomyces coelicolor* A3(2). Microbiology *142*: 133–137.

Ochi, K. 1989. Heterogeneity of ribosomal proteins among *Streptomyces* species and its application to identification. J. Gen. Microbiol. *135*: 2635–2642.

Ochi, K. 1992. Polyacrylamide gel electrophoresis analysis of ribosomal protein: a new approach for actinomycete taxonomy. Gene *115*: 261–265.

Ochi, K. and H. Hiranuma. 1994. A taxonomic review of the genera *Kitasatosporia* and *Streptoverticillium* by analysis of ribosomal protein AT-L30. Int. J. Syst. Bacteriol. *44*: 285–292.

Ochi, K. 1995. A taxonomic study of the genus *Streptomyces* by analysis of ribosomal protein AT-L30. Int. J. Syst. Bacteriol. *45*: 507–514.

Ogata, S. 1980. Bacteriophage contamination in industrial processes. Biotechnol. Bioeng. *22*: 177–193.

Ogata, S., H. Suenaga and S. Hayashida. 1985. A temperate phage of *Streptomyces azureus*. Appl. Environ. Microbiol. *49*: 201–204.

Oh, C., M. Ahn and J. Kim. 1996. Use of electrophoretic enzyme patterns for streptomycete systematics. FEMS Microbiol. Lett. *140*: 9–13.

Ohmori, T., M. Okanishi and H. Kawaguchi. 1962. Glebomycin, a new member of streptomycin class.III. Taxonomic studies on Strain no. 12096, producer of glebomycin. J. Antibiot. (Tokyo) Ser. A *15*: 21–27.

Ohnuki, T., T. Imanaka and S. Aiba. 1985. Self-cloning in *Streptomyces griseus* of an *str* gene cluster for streptomycin biosynthesis and streptomycin resistance. J. Bacteriol. *164*: 85–94.

Ohta, Y. and M. Ikeda. 1978. Deodorization of pig feces by actinomycetes. Appl. Environ. Microbiol. *36*: 487–491.

Okafor, N. 1966. The ecology of microorganisms on, and the decomoposition of, insect wings in the soil. Plant Soils *25*: 211–237.

Okami, Y. 1952. On an antitubercular antibiotic produced by *Streptomyces cinnamonensis* n. sp. *In* J. Antibiot. (Tokyo), vol. Ser. A 5 (edited by Maeda, Okami, Kosaka, Taya and Umezawa), pp. 572–573.

Okami, Y., T. Okuda, T. Takeuchi, K. Nitta and H. Umezawa. 1953. Studies on antitumor substances produced by microorganisms. IV. Sarkomycin-producing *Streptomyces* and other two *Streptomyces* producing the anti-tumor substance No. 289 and caryomycin. J. Antibiot. (Tokyo) *6*: 153–157.

Okami, Y. and H. Umezawa. 1953. On screening of antiviral substances produced by *Streptomyces* and on an antiviral substance achromoviromycin. *In* Japanese Journal of Medical Science and Biology, vol. 6 (edited by Umezawa, Takeuchi, Okami and Tazaki), pp. 261–268.

Okami, Y., R. Utahara, S. Nakamura and H. Umezawa. 1954. Studies on antibiotic actinomycetes. IX. On *Streptomyces* producing a new antifungal substance mediocidin and antifungal substances of fungicidin-rimocidin-chromin group, eurocidin group and trichomycin-ascosin-candicidin group. J. Antibiot. (Tokyo) *7*: 98–103.

Okami, Y., R. Utahara, H. Oyagi, S. Nakamura, H. Umezawa, K. Yanagi-Sawa and Y. Tunematsu. 1955. The screening of antitoxoplasmic substance produced by streptomycete and anti-toxoplasmic substance No. 534. J. Antibiot. (Tokyo) *8*: 126–131.

Okami, Y. and H. Umezawa. 1957. Production and isolation of a new antibiotic, kanamycin. *In* J. Antibiot. (Tokyo), vol. Ser. A 10 (edited by Umezawa, Ueda, Maeda, Yagashita, Kondo, Osato, Nitta and Takeuchi), pp. 181–188.

Okami, Y. and H. Umezawa. 1961. Peptimycin, a product of *Streptomyces* exhibiting apparent inhibition against Ehrlich carcinoma. *In* J. Antibiot. (Tokyo) vol. Ser. A 14 (edited by Murase, Hikiji, Nitta, Okami, Takeuchi and Umezawa), pp. 113–118.

Okami, Y. and H. Umezawa. 1961. *In* The Actinomycetes, vol. 2, Classification, Identification and Descriptions of Genera and Species (edited by Waksman). Williams & Wilkins, Baltimore, pp. 259–260.

Okami, Y., M. Suzuki and H. Umezawa. 1963. Taxonomical studies on a *Streptomyces* strain producing labilomycin. J. Antibiot. (Tokyo) Ser. A *16*: 152–154.

Okami, Y. and T. Okazaki. 1972. Studies on marine microorganisms. I. *Actinomycetes* in Sagami Bay and their antibiotic substances. J. Antibiot. *25*: 456–460.

Okami, Y., T. Okazaki, T. Kitahara and H. Umezawa. 1976. Studies on marine microorganisms. V: a new antibiotic, aplasmomycin, produced by a streptomycete isolated from shallow sea mud. J. Antibiot. Ser. A *29*: 1019–1025.

Okami, Y. and T. Okazaki. 1978. Actinomycetes in marine environments. Zentralbl. Bakteriol. Parasitenkd. Infektionskr. Hyg. Abt. 1 Suppl. *6*: 145–152.

Okazaki, T. and Y. Okami. 1976. Studies on actinomycetes isolated from shallow sea and their antibiotic substances. *In* Actinomycetes – the

Boundary Microorganisms (edited by Arai). Toppan Co. Ltd, Tokyo, pp. 123–161.

Oliver, T.G., A. Goldstein, R.R. Bower, J.C. Holper and R.H. Otto. 1961. M-141, a new antibiotic. I. Antimicrobial properties, identity with actinospectacin, and production by *Streptomyces flavopersicus* sp. n. Antimicrob. Agents Chemother. 495–502.

Olsen, G.J., H. Matsuda, R. Hagstrom and R. Overbeek. 1994. fast DNAmL: a tool for construction of phylogenetic trees of DNA sequences using maximum likelihood. Comput. Appl. Biosci. *10*: 41–48.

Ōmura, S., Y. Takahashi, Y. Iwai and H. Tanaka. 1982. *Kitasatosporia*, a new genus of the order *Actinomycetales*. J. Antibiot. (Tokyo) *35*: 1013–1019.

Ōmura, S., H. Tanaka, Y. Tanaka, P. Spiri-Nakagawa, R. Oliva, Y. Takahashi, K. Matsuyama and Y. Iwai. 1983. *In* Validation of the publication of new names and new combinations previously effectively published outside the IJSB. List no. 11. Int. J. Syst. Bacteriol. *33*: 673.

Ōmura, S., Y. Takahashi and Y. Iwai. 1989. Genus *Kitasatosporia*. *In* Bergey's Manual of Systematic Bacteriology, vol. 4 (edited by Williams, Sharpe and Holt). Williams & Wilkins, Baltimore, pp. 2594–2598.

Ōmura, S., H. Ikeda, J. Ishikawa, A. Hanamoto, C. Takahashi, M. Shinose, Y. Takahashi, H. Horikawa, H. Nakazawa, T. Osonoe, H. Kikuchi, T. Shiba, Y. Sakaki and M. Hattori. 2001. Genome sequence of an industrial microorganism *Streptomyces avermitilis*: deducing the ability of producing secondary metabolites. Proc. Natl. Acad. Sci. U.S.A. *21*: 12215–12220.

Ottow, J.C. 1972. Rose bengal as a selective aid in the isolation of fungi and actinomycetes from natural sources. Mycologia *64*: 304–315.

Owen, S.P., A. Dietz and G.W. Camiener. 1963. Sparsomycin, a new antitumor antibiotic. I. Discovery and biological properties. Antimicrob. Agents Chemother *1962*: 772–779.

Pagé, N., D. Kluepfel, F. Shareck and R. Morosoli. 1996. Effect of signal peptide alterations and replacement on export of xylanase A in *Streptomyces lividans*. Appl. Environ. Microbiol. *62*: 109–114.

Pahl, A., A. Gewies and U. Keller. 1997. ScCypB is a novel second cytosolic cyclophilin from *Streptomyces chrysomallus* which is phylogenetically distant from ScCypA. Microbiology *143*: 117–126.

Palleroni, N.J., K.E. Reichelt, D. Müeller, R. Epps, B. Tabenkin, D.N. Bull, W. Schüep and J. Berger. 1978. Production of a novel red pigment, rubrolone, by *Streptomyces echinoruber* sp. nov. I. Taxonomy, fermentation and partial purification. J. Antibiot. *31*: 1218–1225.

Palleroni, N.J., K.E. Reichelt, D. Müller, R. Epps, B. Tabenkin, D.N. Bull, W. Schüep and J. Berger. 1981. *In* Validation of the publication of new names and new combinations previously effectively published outside the IJSB, List no. 7. Int. J. Syst. Bacteriol. *31*: 382–383.

Pang, X., Y. Sun, J. Liu, X. Zhou and Z. Deng. 2002a. A linear plasmid temperature-sensitive for replication in *Streptomyces hygroscopicus* 10–22. FEMS Microbiol. Lett. *208*: 25–28.

Pang, X., X. Zhou, Y. Sun and Z. Deng. 2002b. Physical map of the linear chromosome of *Streptomyces hygroscopicus* 10-22 deduced by analysis of overlapping large chromosomal deletions. J. Bacteriol. *184*: 1958–1965.

Paradis, E., C. Goyer, N.C. Hodge, R. Hogue, R.E. Stall and C. Beaulieu. 1994. Fatty acid and protein profiles of *Streptomyces* scabies strains isolated in eastern Canada. Int. J. Syst. Bacteriol. *44*: 561–564.

Paradkar, A., A. Trefzer, R. Chakraburtty and D. Stassi. 2003. *Streptomyces* genetics: a genomic perspective. Crit. Rev. Biotechnol. *23*: 1–27.

Paradkar, A.S., K.A. Aidoo, A. Wong and S.E. Jensen. 1996. Molecular analysis of a beta-lactam resistance gene encoded within the cephamycin gene cluster of *Streptomyces clavuligerus*. J. Bacteriol. *178*: 6266–6274.

Park, D.H., J.S. Kim, S.W. Kwon, C. Wilson, Y.M. Yu, J.H. Hur and C.K. Lim. 2003. *Streptomyces luridiscabiei* sp. nov., *Streptomyces puniciscabiei* sp. nov. and *Streptomyces niveiscabiei* sp. nov., which cause potato common scab disease in Korea. Int. J. Syst. Evol. Microbiol. *53*: 2049–2054.

Park, Y.H., D.G. Yim, E. Kim, Y.H. Kho, T.I. Mheen, J. Lonsdale and M. Goodfellow. 1991. Classification of acidophilic, neutrotolerant and neutrophilic streptomycetes by nucleotide sequencing of 5S ribosomal RNA. J. Gen. Microbiol. *137*: 2265–2269.

Parle, J.N. 1963a. Microorganisms in the intestines of earthworms. J. Gen. Microbiol. *31*: 1–11.

Parle, J.N. 1963b. A microbiological study of earthworm casts. J. Gen. Microbiol. *31*: 13–22.

Patelski, R.A. 1951. *Streptomyces californicus* productor de viomicina. Antibiot. Chemother. *1*: 387–389.

Pathom-aree, W., J.E.M. Stach, A.C. Ward, K. Horikoshi, A.T. Bull and M. Goodfellow. 2006. Diversity of actinomycetes isolated from Challenger Deep sediment (10,898 m) from the Mariana Trench. Extremophiles *10*: 181–189.

Peczynska-Czoch, W. and M. Mordarski. 1988. Actinomycete enzymes. *In* Actinomycetes in Biotechnology (edited by Goodfellow, Williams and Mordarski). Academic Press, London, pp. 219–283.

Pernodet, J.L., J.M. Simonet and M. Guerineau. 1984. Plasmids in different strains of *Streptomyces ambofaciens*: free and integrated form of plasmid pSAM2. Mol. Gen. Genet. *198*: 35–41.

Person, L.H. and W.J. Martin. 1940. Soil rot of sweet potatoes in Louisiana. Phytopathology *30*: 913–926.

Petrosyan, P., M. Garcia-Varela, A. Luz-Madrigal, C. Huitron and M.E. Flores. 2003. *Streptomyces mexicanus* sp. nov., a xylanolytic micro-organism isolated from soil. Int. J. Syst. Evol. Microbiol. *53*: 269–273.

Phillips, L. 1992. The distribution of phenotypic and genotypic characters within streptomycetes and their relationship to antibiotic production. PhD thesis, University of Warwick.

Picardeau, M. and V. Vincent. 1998. Mycobacterial linear plasmids have an invertron-like structure related to other linear replicons in *Actinomycetes*. Microbiology *144*: 1981–1988.

Pinnert-Sindico, S. 1954. Une nouvelle espèce de *Streptomyces* productrice d'antibiotiques; *Streptomyces ambofaciens* n. sp., caractères culturaux. Ann. Inst. Pasteur (Paris) *87*: 702–707.

Polsinelli, M. and P.G. Mazza. 1984. Use of membrane filters for selective isolation of actinomycetes from soil. FEMS Microbiol. Lett. *22*: 79–83.

Pommer, E.H. and G. Lorenz. 1986. The behaviour of polyester and polyether polyurethanes towards microorganisms. *In* Biodeterioration Society Occasional Publications (edited by Seal). International Biodeterioration and Biodegradation Society, Manchester, UK, pp. 77–86.

Porter, J.N., R.L. Hewitt, C.W. Hesseltine, G. Krupka, J.A. Lowery, W.S. Wallace, N. Bohonos and J.H. Williams. 1952. Achromycin: a new antibiotic having trypanocidal properties. Antibiot. Chemother. *2*: 409–410.

Porter, J.N., J.J. Wilhelm and H.D. Tresner. 1960. Method for the preferential isolation of actinomycetes from soils. Appl. Microbiol. *8*: 174–178.

Porter, J.N. and J.J. Wilhelm. 1961. The effect on *Streptomyces* populations of adding various supplements to soil samples. Dev. Indust. Microbiol. *2*: 253–259.

Prauser, H. 1970. Character and genera arrangements in the *Actinomycetales*. *In* The *Actinomycetales* (edited by Prauser). Gustav Fischer Verlag, Jena, pp. 407–418.

Prauser, H. 1984. Phage host ranges in the classification and identification of Gram-positive branched and related bacteria. *In* Biological, Biochemical, and Biomedical Aspects of Actinomycetes (edited by Ortiz-Ortiz, Bojalil and Yakoleff). Academic Press, Orlando, pp. 617–633.

Preobrazhenskaya, T.P. and I.D. Ryabova. 1957. Problems of classification of actinomycetes-antagonists (edited by Gauze, Preobrazhenskaya, Kudrina, Blinov, Ryabova and Sveshnikova). Government Publishing House of Medical Literature, Medgiz, Moscow, pp. 1–398.

Preobrazhenskaya, T.P., T.S. Maksimova and N.O. Blinov. 1964. A study of green pigments from some actinomycetous species by the method of paper chromatography. Antibiotiki *9*: 963–970.

Preobrazhenskaya, T.P. 1966. Characteristics of actinomycetes-antagonists of azureus section Antibiotiki *11*: 849–861.

Preobrazhenskaya, T.P., M.A. Sveshnikova, L.P. Terekhova and N.T. Choromonova. 1978. Selective isolation of soil actinomycetes. *In* *Nocardia* and *Streptomyces* (edited by Mordarski, Kurylowicz and Jeljaszewicz). Gustav Fischer Verlag Stuttgart, New York, pp. 119–123.

Preobrazhenskaya, T.P. 1986. Footnote *f*. *In* Validation of the publication of new names and new combinations previously effectively published outside the IJSB. List no. 22. Int. J. Syst. Bacteriol. *36*: 573–576.

Preobrazhenskaya, T.P. and T.S. Maximova. 1986. Footnote *f*. *In* Validation of the publication of new names and new combinations previously effectively published outside the IJSB. List no. 22. Int. J. Syst. Bacteriol. *36*: 573–576.

Preobrazhenskaya, T.P. and L.P. Terekhova. 1986. Footnote *f*. *In* Validation of the publication of new names and new combinations previously effectively published outside the IJSB. List no. 22. Int. J. Syst. Bacteriol. *36*: 573–576.

Pridham, T.G. and D. Gottlieb. 1948. The utilization of carbon compounds by some *Actinomycetales* as an aid for species determination. J. Bacteriol. *56*: 107–114.

Pridham, T.G., O.L. Shotwell, F.H. Stodola, L.A. Lindenfelser, R.G. Benedict and R.V. Jackson. 1956. Antibiotics against plant disease. II. Effective agents produced by *Streptomyces cinnamoneus forma azacoluta* f. nov. Phytopathology *46*: 575–581.

Pridham, T.G., C.W. Hesseltine and R.G. Benedict. 1958. A guide for the classification of streptomycetes according to selected groups; placement of strains in morphological sections. Appl. Microbiol. *6*: 52–79.

Pridham, T.G., A.J.J.R. Lyons and H.L. Seckinger. 1965. Comparison of some dried holotype and neotype specimens of streptomycetes with their living counterparts. Int. Bull. Bacteriol. Nomencl. Taxon. *15*: 191–237.

Pridham, T.G. 1970. New names and new combinations in the order *Actinomycetales* Buchanan 1917. U.S. Dept. Agric. Tech. Bull. *1424*: 1–55.

Pridham, T.G., A.J. Lyons and B. Phrompatima. 1973. Viability of *Actinomycetales* stored in soil. Appl. Microbiol. *26*: 441–442.

Pridham, T.G. 1974. Genus *Microellobosporia*. *In* Bergey's Manual of Determinative Bacteriology, 8th edn (edited by Buchanan and Gibbons). Williams & Wilkins, Baltimore, pp. 843–845.

Pridham, T.G. and H.D. Tresner. 1974a. Genus I. *Streptomycetaceae* Waksman and Henrici. *In* Bergey's Manual of Systematic Bacteriology, 8th edn (edited by Buchanan and Gibbons). Williams & Wilkins, Baltimore, pp. 747–748.

Pridham, T.G. and H.D. Tresner. 1974b. Family *Streptomycetaceae*. *In* Bergey's Manual of Determinative Bacteriology, 8th edn (edited by Buchanan and Gibbons). Williams & Wilkins, Baltimore, pp. 747–748.

Prosser, B.L.T. and N.J. Palleroni. 1976. *Streptomyces longwoodensis* sp. nov. Int. J. Syst. Bacteriol. *26*: 319–322.

Prosser, B.L.T. and N.J. Palleroni. 1981. *In* Validation of the publication of new names and new combinations previously effectively published outside the IJSB. List no. 7. Int. J. Syst. Bacteriol. *31*: 382–383.

Pruesse, E., C. Quast, K. Knittel, B. Fuchs, W. Ludwig, J. Peplies and F.O. Glöckner. 2007. SILVA: a comprehensive online resource for quality checked and aligned rRNA sequence data compatible with ARB. Nucleic Acids Res. *35*: 7188–7196.

Quintana, E.T., K. Wierzbicka, P. Mackiewicz, A. Osman, A.H. Fahal, M.E. Hamid, J. Zakrzewska-Czerwinska, L.A. Maldonado and M. Goodfellow. 2008. *Streptomyces sudanensis* sp. nov., a new pathogen isolated from patients with actinomycetoma. Antonie van Leeuwenhoek *93*: 305–313.

Ramachandra, M., D.L. Crawford and G. Hertel. 1988. Characterization of an extracellular lignin peroxidase of the lignocellulolytic actinomycete *Streptomyces viridosporus*. Appl. Environ. Microbiol. *54*: 3057–3063.

Rauland, U., I. Glocker, M. Redenbach and J. Cullum. 1995. DNA amplifications and deletions in *Streptomyces lividans* 66 and the loss of one end of the linear chromosome. Mol. Gen. Genet. *246*: 37–44.

Rautenshtein, Y.I. 1960. Biology of antibiotic-producing actinomycetes (in Russian). *In* Transactions of the Institute of Microbiology, vol. 8 (edited by Rautenshtein). Academy of Science USSR, pp. 1–344.

Ravel, J., H. Schrempf and R.T. Hill. 1998. Mercury resistance is encoded by transferable giant linear plasmids in two chesapeake bay *Streptomyces* strains. Appl. Environ. Microbiol. *64*: 3383–3388.

Redenbach, M., F. Flett, W. Piendl, I. Glocker, U. Rauland, O. Wafzig, R. Kliem, P. Leblond and J. Cullum. 1993. The *Streptomyces lividans* 66 chromosome contains a 1 MB deletogenic region flanked by two amplifiable regions. Mol. Gen. Genet. *241*: 255–262.

Redenbach, M., H.M. Kieser, D. Denapaite, A. Eichner, J. Cullum, H. Kinashi and D.A. Hopwood. 1996. A set of ordered cosmids and a detailed genetic and physical map for the 8 Mb *Streptomyces coelicolor* A3(2) chromosome. Mol. Microbiol. *21*: 77–96.

Redenbach, M., J. Scheel, J. Cullum and S. U. 1998. The chromosome of various *Actinomyces* strains is linear (Abstract). *In* 8th International Symposium on the Genetics of Industrial Microorganisms, June 28–July 2, 1998 (edited by Cohen and Aharonowitz), Jerusalem, pp. 69–70.

Ridell, M., G. Wallerström and S.T. Williams. 1986. Immunodiffusion analyses of phenetically defined strains of *Streptomyces*, *Streptoverticillium* and *Nocardiopsis*. Syst. Appl. Microbiol. *8*: 24–27.

Roach, A.W. and J.K.G. Silvey. 1959. The occurrence of marine actinomycetes in Texas Gulf Coast substrates. Am. Midland Naturalist *62*: 482–499.

Robbins, P.W., C. Albright and B. Benfield. 1988. Cloning and expression of a *Streptomyces plicatus* chitinase (chitinase-63) in *Escherichia coli*. J. Biol. Chem. *263*: 443–447.

Rodríguez-García, A., M. Ludovice, J.F. Martín and P. Liras. 1997. Arginine boxes and the *argR* gene in *Streptomyces clavuligerus*: evidence for a clear regulation of the arginine pathway. Mol. Microbiol. *25*: 219–228.

Rossi Doria, T. 1891. Su di alcune specie di "*Streptothrix*" trovate nell'aria studate in rapporto a quelle giá note a specialmente all' "*Actinomyces*". Annali dell'Istituto d'Igiene Sperimentale, Universita Roma *1*: 399–438.

Rothrock, C.S. and D. Gottlieb. 1981. Importance of antibiotic production in antagonism of selected *Streptomyces* species to two soil-borne plant pathogens. J. Antibiot. *34*: 830–835.

Rothrock, C.S. and D. Gottlieb. 1984. Roles of antibiosis in antagonism of *Streptomyces hygroscopicus* var. *geldanus* to *Rhizoctonia solani* in soil. Can. J. Microbiol. *30*: 1440–1447.

Routien, J.B. 1969. Progress in the clarification of the taxonomic and nomenclatural status of some problem actinomycetes. *In* Developments in Industrial Microbiology, vol. 10 (edited by Pridham and Lyons), pp. 183–221.

Ruan, J.S., M.P. Lechevalier, C.L. Jiang and H.A. Lechevalier. 1985. *Chainia kunmingensis*, a new actinomycete species found in soil. Int. J. Syst. Bacteriol. *35*: 164–168.

Rucker, R.R. 1949. A streptomycete pathogenic to fish. J. Bacteriol. *58*: 659–664.

Ruddick, S.M. and S.T. Williams. 1972. Studies on the ecology of actinomycetes in soil. V. Some factors influencing the dispersal and adsorption of spores. Soil Biol. Biochem. *4*: 93–103.

Ruiz-Arribas, A., G.G. Zhadan, V.P. Kutyshenko, R.I. Santamaría, M. Cortijo, E. Villar, J.M. Fernandez-Abalos, J.J. Calvete and V.L. Shnyrov. 1998. Thermodynamic stability of two variants of xylanase (Xys1) from *Streptomyces halstedii* JM8. Eur. J. Biochem. *253*: 462–468.

Rullman, W. 1895. Chemisch bacteriologische Untersuchungen von Zwischendeckenfüllungen mit besonderer Berücksichtung von *Cladotrix oderifera*. Akademische Buchdruckerei von F. Strauv. pp. 1–47, München.

Ruschmann, G. 1952. *Streptomyces mirabilis* und das Miramycin. Pharmazie *7*: 542–550.

Saddler, G.S., M. Goodfellow, D.E. Minnikin and A.G. O'Donnell. 1986. Influence of the growth cycle on the fatty acid and menaquinone composition of *Streptomyces cyaneus* NCIB 9616. J. Appl. Microbiol. *60*: 51–56.

Saddler, G.S., A.G. O'Donnell, M. Goodfellow and D.E. Minnikin. 1987. SIMCA pattern recognition in the analysis of streptomycete fatty acids. J. Gen. Microbiol. *133*: 1137–1147.

Saintpierre-Bonaccio, D., H. Amir, R. Pineau, S. Lemriss and M. Goodfellow. 2004. *Streptomyces ferralitis* sp. nov., a novel streptomycete isolated from a New-Caledonian ultramafic soil. Int. J. Syst. Evol. Microbiol. *54*: 2061–2065.

Saintpierre, D., H. Amir, R. Pineau, L. Sembiring and M. Goodfellow. 2003a. *Streptomyces yatensis* sp. nov., a novel bioactive streptomycete isolated from a New-Caledonian ultramafic soil. Antonie van Leeuwenhoek *83*: 21–26.

Saintpierre, D., H. Amir, R. Pineau, L. Sembiring and M. Goodfellow. 2003b. *In* Validation of publication of new names and new combinations previously effectively published outside the IJSEM. List no. 93. Int. J. Syst. Evol. Microbiol. *53*: 1219–1220.

Sakamoto, J.M., S.I. Kondo, H. Yumoto and M. Arishima. 1962. Bundlins A and B, two antibiotics produced by *Streptomyces griseofuscus* nov. sp. J. Antibiot. (Tokyo) Ser. A *15*: 98–102.

Sanglier, J.J., D. Whitehead, G.S. Saddler, E.V. Ferguson and M. Goodfellow. 1992. Pyrolysis mass spectrometry as a method for the classification, identification and selection of actinomycetes. Gene *115*: 235–242.

Sato, M. and A. Kaji. 1975. Purification and properties of pectate lyase produced by *Streptomyces fradiae* IFO 3439. Agric. Biol. Chem. *39*: 819–824.

Sato, M. and A. Kaji. 1977. Purification and properties of pectate lyase produced by *Streptomyces nitrosporeus*. Agric. Biol. Chem. *41*: 2193–2197.

Sato, M. and A. Kaji. 1980a. Exopolygalacturonate lyase produced by *Streptomyces massasporeus*. Agric. Biol. Chem. *44*: 717–721.

Sato, M. and A. Kaji. 1980b. Another pectate lyase produced by *Streptomyces nitrosporeus*. Agric. Biol. Chem. *44*: 1345–1349.

Sauvageau, C.F. and M. Radais. 1892. Sur les genres *Cladothrix*, *Streptothrix*, *Actinomyces* et description de deux *Streptothrix* nouveaux (sur le genere *Oospora*). Ann. Inst. Pasteur *6*: 242–273.

Savic, M., I. Bratic and B. Vasiljevic. 2007. *Streptomyces durmitorensis* sp. nov., a producer of an FK506-like immunosuppressant. Int. J. Syst. Evol. Microbiol. *57*: 2119–2124.

Sawazaki, T., S. Suzuki, G. Nakamura, M. Kawasaki, S. Yamashita, K. Isono, K. Anzai, Y. Serizawa and Y. Sekiyama. 1955. Streptomycin production by a new strain *Streptomyces mashuensis*. J. Antibiot. *8*: 44–47.

Schäfer, A., R. Konrad, T. Kuhnigk, P. Kämpfer, H. Hertel and H. König. 1996. Hemicellulose-degrading bacteria and yeasts from the termite gut. J. Appl. Bacteriol. *80*: 471–478.

Schleifer, K.H. and O. Kandler. 1972. Peptidoglycan types of bacterial cell walls and their taxonomic implications. Bacteriol. Rev. *36*: 407–477.

Schleifer, K.H. and E. Stackebrandt. 1983. Molecular systematics of prokaryotes. Annu. Rev. Microbiol. *37*: 143–187.

Schmitz, H., S.B. Deak, K.E. Crook, Jr and I.R. Hooper. 1964. Peliomycin, a new cytotoxic agent. I. Production, isolation and characterization. Antimicrob. Agents Chemother. *1963*: 89–94.

Schmitz, H., S.D. Jubinski, I.R. Hooper, K.E. Crook, Jr, K.E. Price and J. Lein. 1965. Ossamycin, a new cytotoxic agent. J. Antibiot. (Tokyo) Ser. A *18*: 82–88.

Schrempf, H., P. Dyson, W. Dittrich, M. Betzler, C. Habiger, B. Mahro, V. Brönneke, A. Kessler and H. Düvel. 1989. Genetic instability in *Streptomyces*. *In* Biology of Actinomycetes '88 (edited by Okami, Beppu and Ogawara). Scientific Press, Tokyo, pp. 145–150.

Schrempf, H. 2006. The family *Streptomycetaceae* - Part II: molecular biology. *In* The Prokaryotes: a Handbook on the Biology of Bacteria, 3rd edn, vol. 3, *Archaea, Bacteria, Firmicutes*, Actinomycetes (edited by Dworkin, Falkow, Rosenberg, Schleifer and Stackebrandt). Springer, New York, pp. 605–622.

Schwecke, T., J.F. Aparicio, I. Molnar, A. Konig, L.E. Khaw, S.F. Haydock, M. Oliynyk, P. Caffrey, J. Cortes, J.B. Lester and *et al.* 1995. The biosynthetic gene cluster for the polyketide immunosuppressant rapamycin. Proc. Natl. Acad. Sci. U.S.A. *92*: 7839–7843.

Sembiring, L., A.C. Ward and M. Goodfellow. 2000. Selective isolation and characterisation of members of the *Streptomyces violaceusniger* clade associated with the roots of *Paraserianthes falcataria*. Antonie van Leeuwenhoek *78*: 353–366.

Sembiring, L., A.C. Ward and M. Goodfellow. 2001. *In* Validation of publication of new names and new combinations previously effectively published outside the IJSEM. List no. 82. Int. J. Syst. Evol. Microbiol. *51*: 1619–1620.

Semêdo, L.T., R.C. Gomes, A.A. Linhares, G.F. Duarte, R.P. Nascimento, A.S. Rosado, M. Margis-Pinheiro, R. Margis, K.R. Silva, C.S. Alviano, G.P. Manfio, R.M. Soares, L.F. Linhares and R.R. Coelho. 2004. *Streptomyces drozdowiczii* sp. nov., a novel cellulolytic streptomycete from soil in Brazil. Int. J. Syst. Evol. Microbiol. *54*: 1323–1328.

Seong, C.N., M. Goodfellow, A.C. Ward and Y.C. Hah. 1993. Numerical classification of acidiphilic actinomycetes isolated from acid soil in Korea. Kor. J. Microbiol. *31*: 355–363.

Seong, C.N., S.K. Park, M. Goodfellow, S.B. Kim and Y.C. Hah. 1995. Construction of probability identification matrix and selective medium for acidophilic actinomycetes usind numerical classification data. J. Microbiol. *33*: 95–102.

Servín-González, L. 1993. Relationship between the replication functions of *Streptomyces* plasmids pJV1 and pIJ101. Plasmid *30*: 131–140.

Servín-González, L., C. Castro, C. Pérez, M. Rubio and F. Valdez. 1997. *bldA*-dependent expression of the *Streptomyces exfoliatus* M11 lipase gene (*lipA*) is mediated by the product of a contiguous gene, *lipR*, encoding a putative transcriptional activator. J. Bacteriol. *179*: 7816–7826.

Sherman, D.H., F. Malpartida, M.J. Bibb, H.M. Kieser and D.A. Hopwood. 1989. Structure and deduced function of the granaticin-producing polyketide synthase gene cluster of *Streptomyces violaceoruber* Tu22. EMBO J *8*: 2717–2725.

Shibata, M., M. Honjo, Y. Tokui and N. Nakazawa. 1954. On a new antifungal and anto-yeast substance candimycin produced by a streptomyces. J. Antibiot. *7*: 168.

Shibata, M., K. Nakazawa, A. Miyake, M. Inoue and A. Okabori. 1957. Studies on streptomycetes. Croceomycin, a new antituberculous substance. Annual Report of the Takeda Research Laboratory *16*: 32–37.

Shibata, M. 1959. On a new streptomycin-producing species. *Streptomyces rameus*, n. sp. J. Antibiot. (Tokyo) Ser. B *12*: 398–400.

Shibata, M., E. Higashide, T. Kanzaki, H. Yamamoto and K. Nakazawa. 1961. Studies on *Streptomycetes* Part I: *Streptomyces pulveraceus* nov. sp., producing new antibiotics zygomycin A and B. Agric. Biol. Chem. *25*: 171–175.

Shibata, M., H. Yamamoto, E. Higashidani and K. Nakazawa. 1962. Studies on streptomycetes. Part I. *Streptomyces atratus* nov. sp., producing new antituberculous antibiotics rufomycin A and B. Agric. Biol. Chem. *26*: 228–233.

Shimo, M., T. Shiga, T. Tomosugi and I. Kamoi. 1959. Studies on taitomycin, a new antibiotic produced by *Streptomyces*, sp. No. 772 (*S. afghaniensis*). I. Studies on the strain and production of taitomycin. J. Antibiot. (Tokyo) Series A *12*: 1–6.

Shinobu, R. 1955. On *Streptomyces hiroshimensis* nov. sp. Seibutsugakkaishi *6*: 43–46.

Shinobu, R. 1956. Three new species of *Streptomyces* forming whirls. Memoirs of the Osaka University of the Liberal Arts and Education *5B*: 84–93.

Shinobu, R. 1957. Two new species of *Streptomyces*. Memoirs of the Osaka University of the Liberal Arts and Education B Natural Science *6*: 63–73.

Shinobu, R. 1958. On *Streptomyces spiroverticillatus* nov. sp. Bot. Mag. (Tokyo) *71*: 87–93.

Shinobu, R. and M. Kawato. 1959. On *Streptomyces massasporeus* nov. sp. Botanical Magazine (Tokyo) *72*: 283–288.

Shinobu, R. and M. Kawato. 1960. On *Streptomyces indigoferus* nov. sp. producing blue to green soluble pigment on some synthetic media. Memoirs of the Osaka University of the Liberal Arts and Education B. Natural Science *9*: 49–53.

Shinobu, R. and Y. Shimada. 1962. On a new whirl-forming species of *Streptomyces*. Bot. Mag. (Tokyo) *75*: 107–175.

Shinobu, R. and Y. Kayamura. 1964. On a new whorl-forming species of *Streptomyces*. Bot. Mag. (Tokyo) *77*: 176–180.

Shinobu, R. 1965. Taxonomy of the whirl-forming *Streptomycetaceae*. Memoirs of the Osaka University of the Liberal Arts and Education B Natural Science *14*: 72–201.

Shirling, E.B. and D. Gottlieb. 1966. Methods for characterization of *Streptomyces* species. Int. J. Syst. Bacteriol. *16*: 313–340.

Shirling, E.B. and D. Gottlieb. 1968a. Cooperative description of type cultures of *Streptomyces*. II. Species description from the first study. Int. J. Syst. Bacteriol. *18*: 69–189.

Shirling, E.B. and D. Gottlieb. 1968b. Cooperative description of type cultures of *Streptomyces*. III. Additional species description from first and second studies. Int. J. Syst. Bacteriol. *18*: 279–392.

Shirling, E.B. and D. Gottlieb. 1969. Cooperative description of type cultures of *Streptomyces*. IV. Species descriptions from the second, third and fourth studies. Int. J. Syst. Bacteriol. *19*: 391–512.

Shirling, E.B. and D. Gottlieb. 1970. Report of the International *Streptomyces* Project. Five years collaborative research. *In* The *Actinomycetales* (edited by Prauser). Gustav Fischer Verlag, Jena, pp. 79–90.

Shirling, E.B. and D. Gottlieb. 1972. Cooperative description of type strains of *Streptomyces*. V. Additional descriptions. Int. J. Syst. Bacteriol. *22*: 265–394.

Shirling, E.B. and D. Gottlieb. 1977. Retrospective evaluation of International *Streptomyces* Project taxonomic criteria. *In* Actinomycetes: the Boundary Microorganisms (edited by Arai). University Park Press, Baltimore pp. 9–41.

Shomura, T., S. Amano, J. Yoshida, N. Ezaki, T. Ito and T. Niida. 1983. *Actinosporangium vitaminophilum* sp. nov. Int. J. Syst. Bacteriol. *33*: 557–564.

Shomura, T., S. Gomi, M. Ito, J. Yoshida, E. Tanaka, S. Amano, H. Watabe, S. Ohuchi, J. Itoh, M. Sezaki and *et al.* 1987. Studies on new antibiotics SF2415. I. Taxonomy, fermentation, isolation, physico-chemical properties and biological activities. J. Antibiot. *40*: 732–739.

Shomura, T., S. Gomi, M. Ito, J. Yoshida, E. Tanaka, S. Amano, H. Watabe, S. Ohuchi, J. Itoh, M. Sezaki, H. Takebe and K. Uatani. 1988. *In* Validation of the publication of new names and new combinations previously effectively published outside the IJSB. List no. 24. Int. J. Syst. Bacteriol. *38*: 136–137.

Siebert, G. and W. Schwartz. 1956. Untersuchungen über das Vorkommen von Mikroorganismen in entstehenden Sedimenten. Arch. Hydrobiol. *52*: 331–366.

Silvestri, L., M. Turri, L.R. Hill and E. Gilardi. 1962. A quantitative approach to the systematics of *Actinomycetales* based on overall similarity. *In* Microbial Classification, Symp. Soc. Gen. Microbiol. vol. 12 (edited by Ainsworth and Sneath), pp. 333–360.

Silvey, J.K.G. and A.W. Roach. 1975. The taste and odor producing aquatic actinomycetes. Crit. Rev. Environ. Control *5*: 233–273.

Sing, P.J. and R.S. Mehrotra. 1980. Biological control of *Rhizoctonia bataticila* on grain by coating seed with *Bacillus* and *Streptomyces* spp. and their influence on plant growth. Plant Soil *56*: 475–483.

Skarbek, J.D. and L.R. Brady. 1978. *Streptomyces cavourensis* sp. nov. (nom. rev.) and *Streptomyces cavourensis* subsp. *washingtonensis* subsp. nov., a chromomycin-producing subspecies. Int. J. Syst. Bacteriol. *28*: 45–53.

Smith, C.G., A. Dietz, W.T. Sokolski and G.M. Savage. 1956. Streptovinicin, a new antibiotic. I. Discovery and biologic studies. Antibiot. Chemother. *6*: 135–142.

Sobin, B.A., A.C. Finlay and J.H. Kane. 1953. A Guide to the Classification of the Actinomycetes and their Antibiotics (edited by Waksman and Lechevalier). Williams & Wilkins, Baltimore, pp. 1–246.

Soh, B.S., P. Loke and T.S. Sim. 2001. Cloning, heterologous expression and purification of an isocitrate lyase from *Streptomyces clavuligerus* NRRL 3585. Biochim. Biophys. Acta *1522*: 112–117.

Sohng, J.K., T.J. Oh, J.J. Lee and C.G. Kim. 1997. Identification of a gene cluster of biosynthetic genes of rubradirin substructures in *S. achromogenes var. rubradiris* NRRL3061. Mol. Cells *7*: 674–681.

Sommer, P., C. Bormann and F. Götz. 1997. Genetic and biochemical characterization of a new extracellular lipase from *Streptomyces cinnamomeus*. Appl. Environ. Microbiol. *63*: 3553–3560.

Song, J., S.-C. Lee, J.-W. Kang, H.-J. Baek and J.-W. Suh. 2003. Phylogenetic analysis of *Streptomyces* spp. isolated from potato scab lesions in Korea on the basis of 16S rRNA gene and 16S–23S rDNA internally transcribed spacer sequences. Int. J. Syst. Evol. Microbiol. *53*: 203–209.

Spicher, G. 1955. Untersuchungen über die Wirkung von Erdextrakt und Spurenelementen auf das Wachstum verschiedener Streptomyzeten. Zentralbl. Bakteriol. Parasitenkd. Infektionskr. Hyg. Abt. 2 *108*: 577–587.

Stackebrandt, E., B. Wunner-Fussl, V.J. Fowler and K.-H. Schleifer. 1981. Deoxyribonucleic acid homologies and ribosomal ribonucleic acid similarities among sporeforming members of the order *Actinomycetales*. Int. J. Syst. Bacteriol. *31*: 420–431.

Stackebrandt, E., W. Liesack, R. Webb and D. Witt. 1991a. Towards a molecular identification of *Streptomyces* species in pure culture and in environmental samples. Actinomycetologica *5*: 38–44.

Stackebrandt, E., D. Witt, C. Kemmerling, R. Kroppenstedt and W. Liesack. 1991b. Designation of streptomycete 16S and 23S ribosomal-RNA-based target regions for oligonucleotide probes. Appl. Environ. Microbiol. *57*: 1468–1477.

Stackebrandt, E., W. Liesack and D. Witt. 1992. Ribosomal RNA and rDNA sequence analyses. Gene *115*: 255–260.

Stackebrandt, E., F.A. Rainey and N.L. Ward-Rainey. 1997. Proposal for a new hierarchic classification system, *Actinobacteria* classis nov. Int. J. Syst. Bacteriol. *47*: 471–491.

Stackebrandt, E., W. Frederiksen, G.M. Garrity, P.A. Grimont, P. Kampfer, M.C. Maiden, X. Nesme, R. Rosselló-Mora, J. Swings, H.G. Trüper, L. Vauterin, A.C. Ward and W.B. Whitman. 2002. Report of the *ad hoc* committee for the re-evaluation of the species definition in bacteriology. Int. J. Syst. Evol. Microbiol. *52*: 1043–1047.

Stackebrandt, E. and P. Schumann. 2006. Introduction to the taxonomy of actinobacteria. *In* The Prokaryotes: a Handbook on the Biology of Bacteria, 3rd edn, vol. 3, *Archaea*, *Bacteria*, *Firmicutes*, Actinomycetes (edited by Dworkin, Falkow, Rosenberg, Schleifer and Stackebrandt). Springer, New York, pp. 297–321.

Stapley, E.O., J.M. Mata, I.M. Miller, T.C. Demny and H.B. Woodruff. 1964. Antibiotic MSD-235. I. Production by *Streptomyces avidinii* and *Streptomyces lavendulae*. Antimicrob. Agents Chemother *1963*: 20–27.

Stindl, A. and U. Keller. 1994. Epimerization of the D-valine portion in the biosynthesis of actinomycin D. Biochemistry *33*: 9358–9364.

Stolp, H. and M.P. Starr. 1981. Principles of isolation, cultivation, and conservation of bacteria. *In* The Prokaryotes: a Handbook on Habitats, Isolation, and Identification of Bacteria (edited by Starr, Stolp, Trüper, Balows and Schlegel). Springer, New York, pp. 135–175.

Stuttard, C. 1982. Temperate phages of *Streptomyces venezuelae*: lysogeny and host specificity shown by phage-SV1 and phage-SV2. J. Gen. Microbiol. *128*: 115–121.

Stutzman-Engwall, K.J. and C.R. Hutchinson. 1989. Multigene families for anthracycline antibiotic production in *Streptomyces peucetius*. Proc. Natl. Acad. Sci. U.S.A. *86*: 3135–3139.

Sugai, T. 1956. New antibiotics 229 and 229B of colorless, water-soluble and basic nature. J. Antibiot. (Tokyo) Ser. B *9*: 170–179.

Suganuma, T., T. Mizukami, K.I. Moori, M. Ohnishi and K. Hiromi. 1980. Studies of the action pattern of an α-amylase from *Streptomyces praecox* Na-273. Biochemistry *88*: 131–138.

Sugawara, A. and M. Onuma. 1957. Melanomycin, a new antitumour substance from *Streptomyces*. II. Description of the strain. J. Antibiot. (Tokyo) Ser. A *10*: 138–142.

Sun, W., Y. Huang, Y.Q. Zhang and Z.H. Liu. 2007. *Streptomyces emeiensis* sp. nov., a novel streptomycete from soil in China. Int. J. Syst. Evol. Microbiol. *57*: 1635–1639.

Suzuki, M. 1957. Studies on an antitumor substance, gancidin. Mycological study on the strain AAK-84 and production, purification of active fractions. J. Chiba Med. Soc. *33*: 535–542.

Suzuki, S., G. Nakamura, K. Okuma and Y. Tomiyama. 1958. Cellocidin, a new antibiotic. J. Antibiot. (Tokyo) Ser. A *11*: 81–83.

Sveshnikova, M.A. 1957. Problems of classification of actinomycetes-antagonists (edited by Gauze, Preobrazhenskaya, Kudrina, Blinov, Ryabova and Sveshnikova). Government Publishing House of Medical Literature, Medgiz, Moscow, USSR, pp. 1–398.

Sveshnikova, M.A. and O.E. Timuk. 1986. Footnote *f*. *In* Validation of the publication of new names and new combinations previously effectively published outside the IJSB. List no. 22. Int. J. Syst. Bacteriol. *36*: 573–576.

Swan, D.G., A.M. Rodriguez, C. Vilches, C. Mendez and J.A. Salas. 1994. Characterisation of a *Streptomyces antibioticus* gene encoding a type I polyketide synthase which has an unusual coding sequence. Mol. Gen. Genet. *242*: 358–362.

Szabó, I. and M. Marton. 1958. A *Streptomyces vastus* és *Streptomyces viridoniger* új sugárgomba fajokról (Adatok a szikestalajok mikrobiologiájahoz). Agrokemia es Talajtan *7*: 243–262.

Szabó, I., M. Marton, I. Buti and G. Partai. 1963. *Actinomyces Finlayi* n. sp. Acta Microbiol. Acad. Sci. Hung. *10*: 207–214.

Szabó, I. and M. Marton. 1964. Zur Frage der spezifischen Bodenmikrofloren. Ein Versuch zur systematischen Bestimmung der Strahlenpilzflora einer mullartigen (Wald-) Rendzina. Zentralbl. Bakteriol. Parasitenkd. Infektionskr. Hyg. Abt. 2 *118*: 265–306.

Szabó, I., M. Marton, L. Ferenczy and I. Buti. 1967. Intestinal microflora of the larvae of St. Mark's fly. II. Computer analysis of instestinal actinomycetes from the larvae of a bibio population. Acta Microbiol. Acad. Sci. Hung. *14*: 239–249.

Szabó, I. and M. Preobrazhenskaya. 1986. Footnote *f*. *In* Validation of the publication of new names and new combinations previously effectively published outside the IJSB. List no. 22. Int. J. Syst. Bacteriol. *36*: 573–576.

Taber, W.A. 1959. Identification of an alkaline-dependent *Streptomyces* as *Streptomyces caeruleus* Baldacci and characterization of the species under controlled conditions. Can. J. Microbiol. *5*: 335–344.

Taber, W.A. 1960. Evidence for the existance of acid-sensitive actinomycetes in soil. Can. J. Microbiol. *6*: 503–514.

Taguchi, S., S. Kojima, K. Miura and H. Momose. 1996. Taxonomic characterization of closely related *Streptomyces* spp. based on the amino acid sequence analysis of protease inhibitor proteins. FEMS Microbiol. Lett. *135*: 169–173.

Taha, A. 1983. A serological survey of antibodies to *Streptomyces somaliensis* and *Actinomadura madurae* in the Sudan enzyme linked immunosorbent assay (ELISA). Trans. R. Soc. Trop. Med. Hyg. *77*: 49–50.

Taig, M.M., N.K. Solov'eva and P.S. Braginskaia. 1969. [Characteristics of the culture-producer of aurenin]. Antibiotiki *14*: 873–876.

Taig, M.M. and N.K. Solovieva. 1986. Footnote *f*. *In* Validation of the publication of new names and new combinations previously effectively published outside the IJSB. List no. 22. Int. J. Syst. Bacteriol. *36*: 573–576.

Tajima, K., Y. Takahashi, A. Seino, Y. Iwai and S. Ōmura. 2001. Description of two novel species of the genus *Kitasatospora* Ōmura *et al.* 1982, *Kitasatospora cineracea* sp. nov. and *Kitasatospora niigatensis* sp. nov. Int. J. Syst. Evol. Microbiol. *51*: 1765–1771.

Takahashi, Y., Y. Iwai and S. Ōmura. 1983. Relationships between cell morphology and the types of diaminopimelic acid in *Kitasatospora setalba*. J. Gen. Appl. Microbiol. *29*: 459–465.

Takahashi, Y., Y. Iwai and S. Ōmura. 1984. Two new species of the genus *Kitasatosporia*, *Kitasatosporia phosalacinea* sp. nov. and *Kitasatosporia griseola* sp. nov. J. Gen. Appl. Microbiol. *30*: 377–387.

Takahashi, Y., Y. Iwai and S. Ōmura. 1985. *In* Validation of the publication of new names and new combinations previously effectively published outside the IJSB. List no. 19. Int. J. Syst. Bacteriol. *35*: 535.

Takahashi, Y., A. Matsumoto, A. Seino, J. Ueno, Y. Iwai and S. Ōmura. 2002. *Streptomyces avermectinius* sp. nov., an avermectin-producing strain. Int. J. Syst. Evol. Microbiol. *52*: 2163–2168.

Taylor, C.F. 1936. A method for isolation of actinomycetes from scab lesions on potato tubers and beet roots. Phytopathology *26*: 287–288.

Tendler, M.D. and P.R. Burkholder. 1961. Studies on the thermophilic actinomycetes. I. Methods of cultivation. Appl. Microbiol. *9*: 394–399.

Terekhova, L.P. 1986. Footnote *f*. *In* Validation of the publication of new names and new combinations previously effectively published outside the IJSB. List no. 22. Int. J. Syst. Bacteriol. *36*: 573–576.

Terekhova, L.P. and T.P. Preobrazhenskaia. 1986. Footnote *f*. *In* Validation of the publication of new names and new combinations previously effectively published outside the IJSB. List no. 22. Int. J. Syst. Bacteriol. *36*: 573–576.

Terekhova, L.P., T.P. Preobrazhenskaia and G.F. Gause. 1986. Footnote *f*. *In* Validation of the publication of new names and new combinations previously effectively published outside the IJSB. List no. 22. Int. J. Syst. Bacteriol. *36*: 573–576.

Thaxter, R. 1891. The potato scab. Conn. Agric. Exp. Sta. Rep. *1890*: 81–95.

Thiemann, J.E. and G. Beretta. 1966. Alanosine, a new antiviral and antitumor antibiotic from *Streptomyces*. J. Antibiot. (Tokyo) Series A *19*: 155–160.

Thirumalachar, M.J. 1955. *Chainia*, a new genus of the *Actinomycetales*. Nature *176*: 934–935.

Thirumalachar, M.J. 1960. Biology of antibiotic-producing actinomycetes *In* Transactions of the Institute of Microbiology, vol. 8 (edited by Kalakoutskii, Krasil'nikov and Rautenshtein). Academy of Science, USSR, pp. 1–344.

Thirumalachar, M.J. and R.S. Sukapure. 1964. Studies on species of the genus *Chainia* from India. Hindustan Antibiot. Bull. *6*: 157–166.

Thirumalachar, M.J. 1965. Production of aburamycin by *Chainia munutisclerotica*, a new species of actinomycete. Hindustan Antibiot. Bull. *8*: 6–9.

Thirumalachar, M.J., P.W. Rahalkar, P.V. Deshmukh and R.S. Sukapure. 1965. Production of aburamycin by *Chainia munutisclerotica*, a new species of actinomycete. Hindustan Antibiot. Bull. *8*: 6–9.

Thirumalachar, M.J., Sukapure, P.W. Rahalkar and K.S. Gopalkrishnan. 1966. Studies on species of the genus *Chainia* from India. II. Hindustan Antibiot. Bull. *9*: 10–14.

Thirumalachar, M.J. 1968. Cultural characteristics and identity of some *Streptoverticillium* species producing polyene antibiotics. *In* Hindustan Antibiot. Bull., vol. 11 (edited by Rahalkar and Thirumalachar), pp. 90–96.

Tindall, B.J., R. Rossello-Mora, H.-J. Busse, W. Ludwig and P. Kämpfer. 2010. Notes on the characterization of prokaryote strains for taxonomic purposes. Int. J. Syst. Evol. Microbiol. *60*: 249–266.

Trejo, W. 1961. The *Actinomycetes*, vol. 2, Classification, Identification and Descriptions of Genera and Species (edited by Waksman). Williams & Wilkins, Baltimore, pp. 1–363.

Trejo, W.H. 1970. An evaluation of some concepts and criteria used in the speciation of streptomycetes. Trans. N.Y. Acad. Sci. Ser. II. *32*: 989–997.

Trejo, W.H., L.D. Dean, J. Pluscec, E. Meyers and W.E. Brown. 1977. *Streptomyces laurentii*, a new species producing thiostrepton. J. Antibiot. (Tokyo) Ser. A *30*: 639–643.

Tresner, H., J.A. Hayes and E.J. Backus. 1966. *Streptomyces prasinosporus* sp. nov. a new green-spored species. Int. J. Syst. Bacteriol. *16*: 161–169.

Tresner, H.D. and E.J. Backus. 1956. A broadened concept of the characteristics of *Streptomyces hygroscopicus*. Appl. Microbiol. *4*: 243–250.

Tresner, H.D., M.C. Davies and E.J. Backus. 1961. Electron microscopy of *Streptomyces* spore morphology and its role in species differentiation. J. Bacteriol. *81*: 70–80.

Tresner, H.D. and E.J. Backus. 1963. System of color wheels for streptomycete taxonomy. Appl. Microbiol. *11*: 335–338.

Tresner, H.D., J.A. Hayes and E.J. Backus. 1967. Morphology of submerged growth of streptomycetes as a taxonomic aid. 1. Morphological development of *Streptomyces aureofaciens* in agitated liquid media. Appl. Microbiol. *15*: 1185–1191.

Tresner, H.D., J.A. Hayes and E.J. Backus. 1968. Differential tolerance of streptomycetes to sodium chloride as a taxonomic aid. Appl. Microbiol. *16*: 1134–1136.

Trolldenier, G. 1966. über die Eignung Erde enthaltender Nährsubstrate zur Zählung und Isolierung von Bodenmikroorganismen auf Membranfiltern. Zentralbl. Bakteriol. Parasitenkd. Infektionskr. Hyg. Abt. 2 *120*: 496–508.

Trujillo, M.E. and M. Goodfellow. 2003. Numerical phenetic classification of clinically significant aerobic sporoactinomycetes and related organisms. Antonie van Leeuwenhoek *84*: 39–68.

Trüper, H.G. and L. De'Clari. 1997. Taxonomic note: necessary correction of specific epithets formed as substantives (nouns) "in apposition". Int. J. Syst. Bacteriol. *47*: 908–909.

Tsao, P.H., C. Leben and G.W. Keitt. 1960. An enrichment method for isolating actinomycetes that produce diffusible antifungi antibiotics. Phytopathology *50*: 88–89.

Tsujibo, H., T. Ohtsuki, T. Iio, I. Yamazaki, K. Miyamoto, M. Sugiyama and Y. Inamori. 1997. Cloning and sequence analysis of genes encoding xylanases and acetyl xylan esterase from *Streptomyces thermoviolaceus* OPC-520. Appl. Environ. Microbiol. *63*: 661–664.

Tsukiura, H., M. Okanishi, H. Koshiyama, T. Ohmori, T. Miyaki and H. Kawaguchi. 1964a. Proceomycin, a new antibiotic. J. Antibiot. (Tokyo) Series A *17*: 223–229.

Tsukiura, H., M. Okanishi, T. Ohmori, H. Koshiyama, T. Miyaki, H. Kitazima and H. Kawaguchi. 1964b. Danomycin, a new antibiotic. J. Antibiot. (Tokyo) Series A *17*: 39–47.

Tsyganov, V.A., R.A. Zhukova and K.A. Timofeeva. 1964. Morphological and biochemical pecularities of a new series, actinomycetes 2732/3. Mikrobiologiia *33*: 863–869.

Uchida, K. and K. Aida. 1977. Acyl type of bacterial cell wall: its simple identification by a colorimetric method. J. Gen. Microbiol. *23*: 249–260.

Ulrich, A. and S. Wirth. 1999. Phylogenetic diversity and population densities of culturable cellulolytic soil bacteria across an agricultural encatchment. Microb. Ecol. *37*: 238–247.

Uma, B.N. and P.L. Narasimha Rao. 1959. Actinomycetes. I. Distribution of streptomycetes in Indian soils. Formation of antifungal antibiotics by *Streptomyces champavati* n. sp. Indian Institute of Sciences Golden Jubilee Research *1909–1959*: 130–141.

Umezawa, H., S. Hayano, K. Maeda, Y. Ogata and Y. Okami. 1950. On a new antibiotic, griseolutein, produced by *Streptomyces*. Jpn. Med. J. *3*: 111–117.

Umezawa, H. and Y. Okami. 1950. On the new source of chloromycetin, *Streptomyces omiyaensis*. J. Antibiot. (Tokyo) Ser. A *3*: 292–296.

Umezawa, H., T. Tazaki and S. Fukuyama. 1951. An antiviral substance, abikoviromycin, produced by *Streptomyces* species. Jpn. Med. J. *4*: 331–346.

Uri, J. and I. Békési. 1958. Flavofungin, a new crystalline antifungal antibiotic: origin and biological properties. Nature (London) *181*: 908.

Uridil, J.E. and P.A. Tetrault. 1959. Isolation of thermophilic streptomycetes. J. Bacteriol. *78*: 243–246.

Van Keulen, G., H.M. Jonkers, D. Claessen, L. Dijkhuizen and H.A. Wosten. 2003. Differentiation and anaerobiosis in standing liquid cultures of *Streptomyces coelicolor*. J. Bacteriol. *185*: 1455–1458.

van Wezel, G.P., E. Vijgenboom and L. Bosch. 1991. A comparative study of the ribosomal RNA operons of *Streptomyces coelicolor* A3(2) and sequence analysis of *rrnA*. Nucleic Acids Res. *25*: 4399–4403.

van Wezel, G.P. and E. Vijgenboom. 2004. Novel aspects of signaling in *Streptomyces* development. Adv. Appl. Microbiol. *56*: 65–88.

Vandamme, P., B. Pot, M. Gillis, P. de Vos, K. Kersters and J. Swings. 1996. Polyphasic taxonomy, a consensus approach to bacterial systematics. Microbiol. Rev. *60*: 407–438.

Vavra, J. and A. Dietz. 1965. U-13,714, a new antiviral agent. I. Discovery and biological properties. Antimicrob. Agents Chemother. *1964*: 75–79.

Vavra, J.J., A. Dietz, B.W. Churchill, P. Siminoff and H.J. Koepsell. 1959. Psicofuranine. III. Production and biological studies. Antibiot. Chemother. *9*: 427–431.

Veldkamp, J. 1955. A study of the aerobic decomposition of chitin by microorganisms. Medelingen van de Landbouwhogeschool. te Wageningen/Nederland Wageningen: H Veenman & Zonen *55*: 127–174.

Ventura, M., C. Canchaya, A. Tauch, G. Chandra, G.F. Fitzgerald, K.F. Chater and D. van Sinderen. 2007. Genomics of *Actinobacteria*: tracing the evolutionary history of an ancient phylum. Microbiol. Mol. Biol. Rev. *71*: 495–548.

Vickers, J.C., S.T. Williams and G.W. Ross. 1984. A taxonomic approach to selective isolation of streptomycetes from soil. *In* Biological, Biochemical and Biomedical Aspects of Actinomycetes (edited by Ortiz-Ortiz, Bojalil and Yakoleff). Academic Press, Orlando, pp. 553–561.

Vickers, J.C. and S.T. Williams. 1987. An assessment of plate inoculation procedures for the enumeration and isolation of streptomycetes. Microbios Lett. *36*: 113–117.

Villax, I. 1963. *Streptomyces lusitanus* and the problem of classification of the various tetracycline-producing streptomyces. Antimicrob. Agents Chemother. *1962*: 661–668.

Vinogradova, K.A. and T.P. Preobrazhenskaya. 1986. Footnote *f. In* Validation of the publication of new names and new combinations previously effectively published outside the IJSB. List no. 22. Int. J. Syst. Bacteriol. *36*: 573–576.

Virgilio, A. and C. Hengeller. 1960. Produzione di Tetraciclina con *Streptomyces psammoticus*. Farmaco, Edizione Scientifica *15*: 164–174.

Virolle, M.J. and M.J. Bibb. 1988. Cloning, characterization and regulation of an alpha-amylase gene from *Streptomyces limosus*. Mol. Microbiol. *2*: 197–208.

Voelskow, H. 1988/89. Methoden der zielorientierten Stammisolierung. *In* Jahrbuch Biotechnologie, Bd. 2. (edited by Präve, Schlingmann, Crueger, Esser, Thauer and Wagner). Carl Hanser Verlag, München, Germany, pp. 343–361.

Vuillemin, P. 1931. Les champignons parasites et les mycoses de l'homm. *In* Encylopédie Mycologique II. Paul Le Chevalier and Sons, Paris, pp. 1–290.

Vujaklija, D., W. Schroder, M. Abramic, P. Zou, I. Lescic, P. Franke and J. Pigac. 2002. A novel streptomycete lipase: cloning, sequencing and high-level expression of the *Streptomyces rimosus* GDS(L)-lipase gene. Arch. Microbiol. *178*: 124–130.

Waksman, S.A. and R.E. Curtis. 1916. The *Actinomyces* of the soil. Soil Sci. *1*: 99–134.

Waksman, S.A. 1919. Cultural studies of species of *Actinomyces.* Soil. Sci. *8*: 71–215.

Waksman, S.A. 1923. Genus *Actinomyces. In* Bergey's Manual of Determinative Bacteriology, 1st edn (edited by Bergey, Harrison, Breed, Hammer and Huntoon). Williams & Wilkins, Baltimore, pp. 339–371.

Waksman, S.A. 1932. *Actinomyces* in cacao-beans. *In* Annals of Applied Biology, vol. 19 (edited by Bunting), pp. 515–517.

Waksman, S.A. and H.B. Woodruff. 1940. The soil as a source of microorganisms antagonistic to disease-producing bacteria. J. Bacteriol. *40*: 581–600.

Waksman, S.A. and H.B. Woodruff. 1941. *Actinomyces antibioticus*, a new soil organism antagonistic to pathogenic and non-pathogenic bacteria. J. Bacteriol. *42*: 231–249.

Waksman, S.A. 1942. Distribution of antagonistic actinomycetes in nature. *In* Soil Science, vol. 54 (edited by Waksman, Horning, Welsch and Woodruff), pp. 281–296.

Waksman, S.A. and A.T. Henrici. 1943. The nomenclature and classification of the actinomycetes. J. Bacteriol. *46*: 337–341.

Waksman, S.A. and A.T. Henrici. 1948. Family III. *Streptomycetaceae* Waksman and Henrici. *In* Bergey's Manual of Determinative Bacteriology, 6th edn (edited by Breed, Murray and Hitchens). Williams & Wilkins, Baltimore, pp. 929–980.

Waksman, S.A. 1950. The Actinomycetes. Their Nature, Occurrence, Activities, and Importance, vol. 9. Chronica Botanica Company, Waltham, MA, pp. 1–230.

Waksman, S.A. 1953. A Guide to the Classification of the Actinomycetes and their Antibiotics (edited by Waksman and Lechevalier). Williams & Wilkins, Baltimore, pp. 1–246.

Waksman, S.A. and W.A. Taber. 1953. Guide to the Classification and Indentification of Actinomycetes and their Antibiotics (edited by Waksman and Lechevalier). Williams & Wilkins, Baltimore, pp. 1–162.

Waksman, S.A. and F.J. Gregory. 1954. Actinomycin-II. Classification of organisms producing different forms of actinomycin. Antibiot. Chemother. *4*: 1050–1056.

Waksman, S.A. 1957. Family *Actinomycetaceae* and family *Steptomycetaceae. In* Bergey's Manual of Determinative Bacteriology, 7th edn (edited by Breed, Murray and Smith). Williams & Wilkins, Baltimore, pp. 744–825.

Waksman, S.A. 1959. Strain specificity and production of antibiotic substances. X. Characterization and classification of species within the *Streptomyces griseus* Group. Proc. Natl. Acad. Sci. U.S.A *45*: 1043–1047.

Waksman, S.A. 1961. The Actinomycetes, Vol. 2. Classification, identification and descriptions of genera and species. Williams & Wilkins, Baltimore, pp. 1–363.

Wallhäusser, K., G. Nesemann, P. Präve and A. Steigler. 1966. Moenomycin, a new antibiotic. I. Fermentation and isolation. Antimicrob. Agents Chemother. *1965*: 734–736.

Wallhäusser, K.H., G. Huber, G. Nesemann, P. Präve and K. Zepf. 1964. Die Antibiotica FF 3582A und B und ihre Identität mit Nonactin und seinen Honologen. Arzneim. Forsch. *14*: 356–360.

Wallick, H., D.A. Harris, M.A. Reagan, M. Ruger and H.B. Woodruff. 1956. Discovery and antimicrobial properties of cathomycin, a new antibiotic produced by *Streptomyces spheroids*, n. sp. Antibiot. Annu. *1955/56*: 909–917.

Wang, E.L., M. Hamada, Y. Okami and H. Umezawa. 1966. A new antibiotic, spinamycin. J. Antibiot. (Tokyo) Series A *19*: 216–221.

Wang, L., Y. Huang, Q. Cui, Q. Xie, Y. Zhang and Z. Liu. 2003. Isolation of acidiphilic and acidoduric streptomycetes using a dispersion and differential centrifugation approach. Microbiologia *30*: 104–106.

Wang, L., Y. Huang, Z. Liu, M. Goodfellow and C. Rodriguez. 2006. *Streptacidiphilus oryzae* sp. nov., an actinomycete isolated from rice-field soil in Thailand. Int. J. Syst. Evol. Microbiol. *56*: 1257–1261.

Wang, Z.M., B.H. Bleakley, D.L. Crawford, G. Hertel and F. Rafii. 1990. Cloning and expression of a lignin peroxidase gene from *Streptomyces viridosporus* in *Streptomyces lividans.* J. Biotechnol. *13*: 131–144.

Warcup, J.H. 1950. The soil-plate method for isolation of fungi from soil. Nature *166*: 117–118.

Watanabe, K., T. Tanaka, K. Fukuhara, N. Miyairi, H. Yonehara and H. Umezawa. 1957. Blastomycin, a new antibiotic from *Streptomyces* sp. J. Antibiot. (Tokyo) *10*: 39–45.

Watson, E.T. and S.T. Williams. 1974. Studies of the ecology of actinomycetes in soil. VII. Actinomycetes in a coastal sand belt. Soil. Biol. Biochem. *6*: 43–52.

Wayne, L.G., D.J. Brenner, R.R. Colwell, P.A.D. Grimont, O. Kandler, M.I. Krichevsky, L.H. Moore, W.E.C. Moore, R.G.E. Murray, E. Stackebrandt, M.P. Starr and H.G. Trüper. 1987. International Committee on Systematic Bacteriology. Report of the *ad hoc* committee on reconciliation of approaches to bacterial systematics. Int. J. Syst. Bacteriol. *37*: 463–464.

Wellington, E.M.H. and S.T. Williams. 1978. Preservation of actinomycete inoculum in frozen glycerol. Microbiol. Lett. *6*: 151–159.

Wellington, E.M.H. and S.T. Williams. 1981a. Transfer of *Actinoplanes armeniacus* Kalakoutskii and Kusnetsov to *Streptomyces*: *Streptomyces armeniacus* (Kalakoutskii and Kusnetsov) comb. nov. Int. J. Syst. Bacteriol. *31*: 77–81.

Wellington, E.M.H. and S.T. Williams. 1981b. Host ranges of phage isolated to *Streptomyces* and other genera. Zentralbl. Bakteriol. Abt. I. Suppl. *11*: 93–98.

Wellington, E.M.H., M. Al-Jawadi and R. Bandoni. 1987. Selective isolation of *Streptomyces* species-groups from soil. Dev. Indust. Microbiol. *28*: 99–104.

Wellington, E.M.H., E. Stackebrandt, D. Sanders, J. Wolstrup and N.O.G. Jorgensen. 1992. Taxonomic status of *Kitasatosporia*, and proposed unification with *Streptomyces* on the basis of phenotypic and 16S ribosomal RNA analysis and emendation of *Streptomyces* Waksman and Henrici 1943, 339AL. Int. J. Syst. Bacteriol. *42*: 156–160.

Welsch, M., R. Corbaz and L. Ettlinger. 1957. Phage typing of streptomycetes. Schweiz. Z. Allgem. Pathol. Bakteriol. *20*: 454–458.

Wenner, T., V. Roth, B. Decaris and P. Leblond. 2002. Intragenomic and intraspecific polymorphism of the 16S–23S rDNA internally transcribed sequences of *Streptomyces ambofaciens.* Microbiology *148*: 633–642.

Weyland, H. 1981a. Characteristics of actinomycetes isolated from marine sediments. *In* Actinomycetes. Proceedings of the 4th International Symposium on Actinomycete Biology, Cologne, 1979 (edited by Schaal and Pulverer). Gustav Fischer Verlag, Stuttgart, pp. 309–314.

Weyland, H. 1981b. Distribution of actinomycetes on the sea floor. Zentrabl. Bakteriol. Mikrobiol. Hyg. I. Abt. Orig. Suppl. *11*: 185–193.

Weyland, H. and E. Helmke. 1988. Actinomycetes in the marine environment. *In* Biology of Actinomycetes '88. Proceedings of the 7th Internatinoal Symposium on Biology of Actinomycetes (edited by Okami, Beppu and Ogawara). Japan Scientific Societies Press, Tokyo, pp. 294–299.

Wieczorek, J., H. Mordarska, J. Zakrzewska-Czerwinska, A. Gamian and M. Mordarski. 1993. *Streptomyces spitsbergensis* sp. nov. Int. J. Syst. Bacteriol. *43*: 84–87.

Wieringa, K.T. 1955. Der Abbau der Pektine; der erste Angriff der organischen Pflanzensubstanz. Pflanzenernährung *69*: 150–155.

Wieringa, K.T. 1966. Solid media with elemental sulphur for detection of sulphur-oxidizing microbes. Antonie van Leeuwenhoek J. Microbiol. Serol. *32*: 183–186.

Wilde, P. 1964. Gezielte Methoden zur Isolierung antibiotisch wirksamer Boden-Actinomyceten. Z. Pflanzenkr. *71*: 179–182.

Wilkin, G.D. and A. Rhodes. 1955. Observations on the morphology of *Streptomyces griseus* in submerged culture. J. Gen. Microbiol. *12*: 259–264.

Williams, S.T. and F.L. Davies. 1965. Use of antibiotics for selective isolation and enumeration of actinomycetes in soil. J. Gen. Microbiol. *38*: 251–261.

Williams, S.T., F.L. Davies and D.M. Hall. 1969. A practical approach to the taxonomy of *Actinomycetes* isolated from soil. *In* The Soil Ecosystem, vol. 8 (edited by Sheals). The Systematics Association, London, pp. 107–117.

Williams, S.T. and T. Cross. 1971. Isolation, purification, cultivation and preservation of actinomycetes. Methods Microbiol. *4*: 295–334.

Williams, S.T., F.L. Davies, C.I. Mayfield and M.R. Khan. 1971. Studies on the ecology of actinomycetes. II. The pH requirements of streptomycetes from two acid soils. Soil Biol. Biochem. *3*: 187–195.

Williams, S.T. and C.I. Mayfield. 1971. Studies on the ecology of actinomycetes in soil. III. The behaviour of neutrophilic streptomycetes in acid soil. Soil. Biol. Biochem. *3*: 197–208.

Williams, S.T., M. Shameemullah, E.T. Watson and C.I. Mayfield. 1972. Studies on the ecology of actinomycetes in soil. VI. The influence of moisture tension on growth and survival. Soil. Biol. Biochem. *4*: 215–225.

Williams, S.T., G.P. Sharples and R.M. Bradshaw. 1973. The fine structure of the *Actinomycetales*. *In Actinomycetales*: Characteristics and Practical Importance (edited by Sykes and Skinner). Academic Press, London, pp. 113–130.

Williams, S.T. and M.R. Khan. 1974. Antibiotics – a soil microbiologist's viewpoint. Postepy Higieny I Medycyny Doswiadczalnej *28*: 395–408.

Williams, S.T. 1978. *Streptomycetes* in the soil ecosystem. Zentralbl. Bakteriol. Parasitenkd. Infektionskr. Hyg. Abt. 1 Suppl. *6*: 137–144.

Williams, S.T. and T.H. Flowers. 1978. The influence of pH on starch hydrolysis by neutrophilic and acidophilic streptomycetes. Microbios *20*: 99–106.

Williams, S.T. and E.M.H. Wellington. 1980. Micromorphology and fine structure of actinomycetes. *In* Microbiological Classification and Identification (edited by Goodfellow and Board). Academic Press, London, pp. 139–165.

Williams, S.T. and C.S. Robinson. 1981. The role of streptomycetes in decomposition of chitin in acidic soils. J. Gen. Microbiol. *127*: 55–63.

Williams, S.T. 1982. Are antibiotics produced in soil? Pedobiologia *23*: 427–435.

Williams, S.T. and E.M.H. Wellington. 1982a. Principles and problems of selective isolation of microbes. *In* Bioactive Microbial Products: Search and Discovery (edited by Bu'lock, Nisbet and Winstanley). Academic Press, London, pp. 9–26.

Williams, S.T. and E.M.H. Wellington. 1982b. Actinomycetes. *In* Methods of Soil Analysis, Part 2, Chemical and Microbiological Properties (edited by Page, Miller and Keeney). American Society of Agronomy and Soil Sciences, Madison, Wisconsin, pp. 969–987.

Williams, S.T., M. Goodfellow, G. Alderson, E.M.H. Wellington, P.H.A. Sneath and M.J. Sackin. 1983a. Numerical classification of *Streptomyces* and related genera. J. Gen. Microbiol. *129*: 1743–1813.

Williams, S.T., M. Goodfellow, E.M.H. Wellington, J.C. Vickers, G. Alderson, P.H.A. Sneath, M.J. Sackin and A.M. Mortimer. 1983b. A probability matrix for identification of some streptomycetes. J. Gen. Microbiol. *129*: 1815–1830.

Williams, S.T., M. Goodfellow and J.C. Vickers. 1984a. New microbes from old habitats? *In* The Microbe1984, Part 2: Prokaryotes and Eukaryotes. Society for General Microbiology Symposium 36 (edited by Kelley and Karr), Cambridge University Press, Cambridge, pp. 219–256.

Williams, S.T., S. Lanning and E.M.H. Wellington. 1984b. Ecology of actinomycetes. *In* The Biology of Actinomycetes (edited by Goodfellow, Mordarski and Williams). Academic Press, London, pp. 481–528.

Williams, S.T., M. Goodfellow and G. Alderson. 1989. Genus *Streptomyces* Waksman and Henrici. *In* Bergey's Manual of Systematic Bacteriology, vol. 4 (edited by Williams, Sharpe and Holt). Williams & Wilkins, Baltimore, pp. 2452–2492.

Wipat, A., M.H. Wellington and V.A. Saunders. 1994. Monoclonal antibodies for *Streptomyces lividans* and their use for immunomagnetic capture of spores from soil. Microbiology *140*: 2067–2076.

Wirth, S. and A. Ulrich. 2002. Cellulose-degrading potentials and phylogenetic classification of carboxymethyl-cellulose decomposing bacteria isolated from soil. Syst. Appl. Microbiol. *25*: 584–591.

Witt, D. and E. Stackebrandt. 1990. Unification of the genera *Streptoverticillum* and *Streptomyces*, and amendation of *Streptomyces* Waksman and Henrici 1943, 339[AL]. Syst. Appl. Microbiol. *13*: 361–371.

Witt, D. and E. Stackebrandt. 1991. *In* Validation of the publication of new names and new combinations previously effectively published outside the IJSB, List no. 38. Int. J. Syst. Bacteriol. *41*: 456–457.

Witt, D. and E. Stackebrandt. 1996. *In* Validation of the publication of new names and new combinations previously effectively published outside the IJSB. List no. 58. Int. J. Syst. Bacteriol. *46*: 836–837.

Wolf, Y.I., I.B. Rogozin, N.V. Grishin and E.V. Koonin. 2002. Genome trees and the tree of life. Trends Genet. *18*: 472–479.

Wollenweber, H.W. 1920. Der Kartoffelschorf. *In* Arbeiten des Forschunginstitutes für Kartoffelbau, no. 2. Verlagsbuchandlung Paul Parey, Berlin, pp. 1–102.

Wood, S., S.T. Williams and W.R. White. 1983. Microbes as a source of earthy flavours in potable water – a review. Int. Biodeterior. Bull. *19*: 83–97.

Xu, C., L. Wang, Q. Cui, Y. Huang, Z. Liu, G. Zheng and M. Goodfellow. 2006. Neutrotolerant acidophilic *Streptomyces* species isolated from acidic soils in China: *Streptomyces guanduensis* sp. nov., *Streptomyces paucisporeus* sp. nov., *Streptomyces rubidus* sp. nov. and *Streptomyces yanglinensis* sp. nov. Int. J. Syst. Evol. Microbiol. *56*: 1109–1115.

Xu, L.-H., Y.-Q. Tiang, Y.-F. Zhang, L.-X. Zhao and C.-L. Jiang. 1998. *Streptomyces thermogriseus*, a new species of the genus *Streptomyces* from soil, lake and hot-spring. Int. J. Syst. Bacteriol. *48*: 1089–1093.

Xu, P., W.J. Li, W.L. Wu, D. Wang, L.H. Xu and C.L. Jiang. 2004a. *Streptomyces hebeiensis* sp. nov. Int. J. Syst. Evol. Microbiol. *54*: 727–731.

Xu, P., Y. Takahashi, A. Seino, Y. Iwai and S. Ōmura. 2004b. *Streptomyces scabrisporus* sp. nov. Int. J. Syst. Evol. Microbiol. *54*: 577–581.

Yamaguchi, H., Y. Nakayama, K. Takeda, K. Tawara, K. Maeda, T. Takeuchi and H. Umezawa. 1957. A new antibiotic, althiomycin. J. Antibiot. (Tokyo) Ser. A *10*: 195–200.

Yamaguchi, T. and Y. Saburi. 1955. Studies on the anti-trichomonal actinomycetes and their classification. J. Gen. Appl. Microbiol. *1*: 201–235.

Yamamoto, H., K. Nakazawa, S. Horii and A. Miyake. 1960. Studies on agricultural antibiotic folimycin, a new antifungal antibiotic produced by *Streptomyces neyagawaensis* nov. sp. J. Agric. Chem. Soc. Jap. *34*: 268–272.

Yan, X., C. Jiang and Y. Zhang. 1987. *Microstreptospora*, a new genus of the order *Actinomycetales*. The Actinomycetes *20*: 89–92.

Yu, T.W., M.J. Bibb, W.P. Revill and D.A. Hopwood. 1994. Cloning, sequencing, and analysis of the griseusin polyketide synthase gene cluster from *Streptomyces griseus*. J. Bacteriol. *176*: 2627–2634.

Yüntsen, H., K. Ohkuma, Y. Ishii and H. Yonehara. 1956. Studies on angustmycin. III. J. Antibiot. (Tokyo) Series A *9*: 195–201.

Zakrzewska-Czerwinska, J. and H. Schrempf. 1992. Characterization of an autonomously replicating region from the *Streptomyces lividans* chromosome. J. Bacteriol. *174*: 2688–2693.

Zhang, Q., W.J. Li, X.L. Cui, M.G. Li, L.H. Xu and C.L. Jiang. 2003. *Streptomyces yunnanensis* sp. nov., a mesophile from soils in Yunnan, China. Int. J. Syst. Evol. Microbiol. *53*: 217–221.

Zhang, Z.S., Y. Wang and J.S. Ruan. 1997. A proposal to revive the genus *Kitasatospora* (Ōmura, Takahashi, Iwai, and Tanaka 1982). Int. J. Syst. Bacteriol. *47*: 1048–1054.

Zhi, X.-Y., W.-J. Li and E. Stackebrandt. 2009. An update of the structure and 16S rRNA gene sequence-based definition of higher ranks of the class *Actinobacteria*, with the proposal of two new suborders and four new families and emended descriptions of the existing higher taxa. Int. J. Syst. Evol. Microbiol. *59*: 589–608.

Zhu, H.H., J. Guo, Q. Yao, S.Z. Yang, M.R. Deng, T.B. Phuong le, V.T. Hanh and M.J. Ryan. 2007. *Streptomyces vietnamensis* sp. nov., a streptomycete with violet blue diffusible pigment isolated from soil in Vietnam. Int. J. Syst. Evol. Microbiol. *57*: 1770–1774.

Order XV. **Streptosporangiales** ord. nov.

Michael Goodfellow

Strep.to.spo.ran.gi'a.les. N.L. neut. n. *Streptosporangium* type genus of the order; suff. -*ales* ending to denote an order; N.L. fem. pl. n. *Streptosporangiales* the *Streptosporangium* order.

Aerobic, Gram-stain-positive, non-acid-fast, chemoorganotrophic actinomycete which have a wall peptidoglycan containing *meso*-diaminopimelic acid, *N*-acetylated muramic acid, and is of the A1γ type. Fatty acid profiles are complex, and mycolic acids absent. **The pattern 16S rRNA signatures consists of nucleotides at positions 127:234 (A–U), 829:857 (G–C), 830:856 (G–C), 953:1228 (U–A), 950:1231 (U–A), 955:1225 (C–G), 986:1219 (A–U) and 987:1218 (A–U)**. The order contains the families *Nocardiopsaceae, Streptosporangiaceae,* and *Thermomonosporaceae.*

DNA G+C content (mol%): 64–77.

Type genus: **Streptosporangium** Couch 1955, 145[AL] emend. Stackebrandt, Kroppenstedt, Jahnke, Kemmering and Gürtler 1994, 268.

Further descriptive information

The order *Streptosporangiales* was formed by elevation of the suborder *Streptosporangineae* Ward-Rainey et al. 1997 emend. Zhi et al. 2009. The families *Nocardiopsaceae, Thermomonosporaceae,* and *Streptosporangiaceae* form distinct clades in the 16S rRNA actinobacterial gene tree and can also be separated from one another and from other actinobacterial families by the pattern of their 16S rRNA signatures (Zhi et al., 2009). *Streptosporangiales* strains produce whole-organism hydrolysates that are rich in *meso*-diaminopimelic acid and which may or may not contain madurose, i.e. they have either a wall chemotype IIIB or IIIC *sensu* Lechevalier and Lechevalier (1970b). They contain complex mixtures of fatty acids but can be assigned to groups based on qualitative and quantitative differences in fatty acid profiles (Kroppenstedt, 1985). *Thermomonosporaceae* and *Streptosporangiaceae* strains are characterized by fatty acid patterns 3a and 3c, respectively; members of the family *Nocardiopsaceae* show greater diversity as some strains exhibit pattern 3a (*Nocardiopsis* and *Streptomonospora*) and others pattern 3e (*Thermobifida*).

Members of the three families can be distinguished on the basis of differences in menaquinone and polar lipid composition (Goodfellow et al., 1988, 1990; Kudo, 2001; Goodfellow and Quintana, 2006; Kroppenstedt and Evtushenko, 2006; Kroppenstedt and Goodfellow, 2006). *Nocardiopsaceae* strains contain menaquinones with 9, 10, or 11 isoprene units (Li et al., 2003; Zhang et al., 1998), whereas those assigned to the other two families have components with nine isoprene units (Kroppenstedt and Goodfellow, 2006; Stackebrandt et al., 1994). *Streptosporangiaceae* and *Thermomonosporaceae* strains can also be distinguished by their menaquinone profiles, as the former is characterized by the presence of major amounts of MK-9(H_2) and MK-9(H_4), saturated at sites III, and III and IV, respectively, and the latter by predominant amounts of MK-9(H_6) saturated at sites II, III, and VIII. Representatives of these taxa typically display polar lipid patterns 1 or 2 (*Thermomonosporaceae*) and IV (*Streptosporangiaceae*), according to the classification of Lechevalier et al. (1977, 1981). In contrast, *Nocardiopsaceae* strains are markedly heterogeneous as they exhibit polar lipid patterns II (*Thermobifida*), III (*Nocardiopsis*), and complex, unusual profiles (*Streptomonospora*) (Cui et al., 2001; Li et al., 2003).

Detection of chemotaxonomic markers

Members of the genera classified in the order *Streptosporangiales* can be distinguished from one another and from sporoactinomycetes assigned to other families based on the discontinuous distribution of chemotaxonomic markers, notably wall amino acids, cellular fatty acids, menaquinones, muramic acid types, sugars, and polar lipids. Standard chromatographic procedures are available for the detection of wall diamino acids (Hancock, 1994; Hasegawa et al., 1983; Staneck and Roberts, 1974), fatty acids (Kroppenstedt et al., 1990; Suzuki and Komagata, 1983) including mycolic acids (Minnikin et al., 1975), menaquinones (Collins, 1994; Kroppenstedt, 1982, 1985; Minnikin et al., 1984), muramic acid residues (Uchida and Aida, 1977; Uchida et al., 1999), and polar lipids (Minnikin et al., 1984; Suzuki et al., 1993).

Primary chemotaxonomic data can be obtained by examining whole-organism hydrolysates for the presence of the isomers of diaminopimelic acid, major diagnostic sugars, and mycolic acids using appropriate standards. The thin-layer chromatographic procedures described by Staneck and Roberts (1974) provide an easy and reliable way of detecting diagnostic amino acids and sugars. These procedures involve the application of unidimensional thin-layer chromatography to establish whether organisms contain LL-, hydroxy- or *meso*-diaminopimelic acid and major diagnostic sugars, including madurose; this sugar has almost the same R_f value as xylose but can be distinguished from the latter as it gives a yellow brown as opposed to a maroon spot. Madurose and other whole-organism sugars can also be detected by gas chromatography (Saddler et al., 1991) and high pressure liquid chromatography (Nakagaito et al., 1993; Yokota and Hasegawa, 1988). Examination of whole-organism methanolysates for the presence of mycolic acids can be achieved using the thin-layer chromatographic technique introduced by Minnikin et al. (1975).

The presence of *meso*-diaminopimelic acid and the absence of mycolic acids and diagnostic sugars other than madurose distinguishes members of the order *Streptosporangiales* from sporoactinomycetes classified in the orders *Catenulisporales, Corynebacteriales, Micromonosporales, Pseudonocardiales,* and *Streptomycetales.*

References

Collins, M.D. 1994. Isoprenoid quinones. *In* Chemical Methods in Prokaryotic Systematics (edited by Goodfellow and O'Donnell). John Wiley & Sons, New York, pp. 265–309.

Couch, J.N. 1955. A new genus and family of the *Actinomycetales* with a revision of the genus *Actinoplanes*. J. Elisha Mitchell Sci. Soc. *71*: 148–155.

Cui, X.L., P.H. Mao, M. Zeng, W.J. Li, L.P. Zhang, L.H. Xu and C.L. Jiang. 2001. *Streptimonospora salina* gen. nov., sp. nov., a new member of the family *Nocardiopsaceae*. Int. J. Syst. Evol. Microbiol. *51*: 357–363.

Goodfellow, M., E. Stackebrandt and R.M. Kroppenstedt. 1988. Chemotaxonomy and actinomycete systematics. *In* Biology of Actinomycetes '88 (edited by Okami, Beppu and Ogawara). Japan Scientific Societies Press, Tokyo, pp. 233–238.

Goodfellow, M., L.J. Stanton, K.E. Simpson and D.E. Minnikin. 1990. Numerical and chemical classification of *Actinoplanes* and some related actinomycetes. J. Gen. Microbiol. *136*: 19–36.

Goodfellow, M. and E.T. Quintana. 2006. The family *Streptosporangiaceae*. *In* The Prokaryotes: a Handbook on the Biology of Bacteria, 3rd edn (edited by Dworkin, Falkow, Rosenberg, Schleifer and Stackebrandt). Springer, New York, pp. 725–753.

Hancock, I.C. 1994. Analysis of cell wall constituents of Gram-positive bacteria. *In* Chemical Methods in Prokaryotic Systematics (edited by Goodfellow and O'Donnell). John Wiley & Sons, Chichester, pp. 63–84.

Hasegawa, T., M. Takizawa and S. Tanida. 1983. A rapid analysis for chemical grouping of aerobic actinomycetes. J. Gen. Appl. Microbiol. *29*: 319–322.

Kroppenstedt, R.M. 1982. Separation of bacterial menaquinones by HPLC using reverse phase (RP 18) and silver loaded ion exchanger as stationary phases. J. Liquid Chromatogr. *5*: 2359–2367.

Kroppenstedt, R.M. 1985. Fatty acid and menaquinone analysis of actinomycetes and related organisms. *In* Chemical Methods in Bacterial Systematics (edited by Goodfellow and Minnikin). Academic Press, London, pp. 173–199.

Kroppenstedt, R.M., E. Stackebrandt and M. Goodfellow. 1990. Taxonomic revision of the actinomycete genera *Actinomadura* and *Microtetraspora*. Syst. Appl. Microbiol. *13*: 148–160.

Kroppenstedt, R.M. and L.I. Evtushenko. 2006. The family *Nocardiopsaceae*. *In* The Prokaryotes: a Handbook on the Biology of Bacteria, 3rd edn (edited by Dworkin, Falkow, Rosenberg, Schleifer and Stackebrandt). Springer, New York, pp. 745–795.

Kroppenstedt, R.M. and M. Goodfellow. 2006. The family *Thermomonosporaceae*: *Actinocorallia*, *Actinomadura*, *Spirillospora* and *Thermomonospora*. *In* The Prokaryotes: a Handbook on the Biology of Bacteria, 3rd edn, vol. 3, *Archaea*, *Bacteria*, *Firmicutes*, Actinomycetes (edited by Dworkin, Falkow, Rosenberg, Schleifer and Stackebrandt). Springer, New York, pp. 682–724.

Lechevalier, M.P. and H.A. Lechevalier. 1970. Chemical composition as a criterion in the classification of aerobic actinomycetes. Int. J. Syst. Bacteriol. *20*: 435–443.

Lechevalier, M.P., C. de Biévre and H.A. Lechevalier. 1977. Chemotaxonomy of aerobic actinomycetes: phospholipid composition. Biochem. Ecol. Systems *5*: 249–260.

Lechevalier, M.P., A.E. Stern and H.A. Lechevalier. 1981. Phospholipids in the taxonomy of actinomycetes. Zentralbl. Bakteriol. Parasitenkd. Infektionskr. Hyg. I Abt. Orig. Suppl. *11*: 111–116.

Li, W.J., P. Xu, L.P. Zhang, S.K. Tang, X.L. Cui, P.H. Mao, L.H. Xu, P. Schumann, E. Stackebrandt and C.L. Jiang. 2003. *Streptomonospora alba* sp. nov., a novel halophilic actinomycete, and emended description of the genus *Streptomonospora* Cui *et al.* 2001. Int. J. Syst. Evol. Microbiol. *53*: 1421–1425.

Minnikin, D.E., L. Alshamaony and M. Goodfellow. 1975. Differentiation of *Mycobacterium*, *Nocardia*, and related taxa by thin-layer chromatographic analysis of whole-organism methanolysates. J. Gen. Microbiol. *88*: 200–204.

Minnikin, D.E., A.G. O'Donnell, M. Goodfellow, G.A. Alderson, M. Athalye, A. Schaal and J.H. Parlett. 1984. An integrated procedure for the extraction of isoprenoid quinones and polar lipids. J. Microbiol. Methods *2*: 233–241.

Nakagaito, Y., Y. Nishii, A. Yokota and T. Hasegawa. 1993. Distribution of madurose, an actinomycete whole-cell sugar, in the genus *Streptomyces*. IFO Res. Commun. *16*: 102–108.

Saddler, G.S., P. Tavecchia, S. Lociuro, M. Zanol, E. Colombo and E. Selva. 1991. Analysis of madurose and other actinomycete whole cell sugars by gas chromatography. J. Microbiol. Methods *14*: 185–191.

Stackebrandt, E., R.M. Kroppenstedt, K.-D. Jahnke, C. Kemmerling and H. Gürtler. 1994. Transfer of *Streptosporangium viridogriseum* (Okuda et al. 1966), *Streptosporangium viridogriseum* subsp. kofuense (Nonomura and Ohara 1969), and *Streptosporangium albidum* (Furumai et al. 1968) to Kutzneria gen. nov. as *Kutzneria viridogrisea* comb. nov., *Kutzneria kofuensis* comb. nov., and *Kutzneria albida* comb. nov., respectively, and emendation of the genus *Streptosporangium*. Int. J. Syst. Bacteriol. *44*: 265–269.

Staneck, J.L. and G.D. Roberts. 1974. Simplified approach to identification of aerobic actinomycetes by thin-layer chromatography. Appl. Microbiol. *28*: 226–231.

Suzuki, K. and K. Komagata. 1983. Taxonomic significance of cellular fatty acid composition in some coryneform bacteria. Int. J. Syst. Bacteriol. *33*: 188–200.

Suzuki, K., M. Goodfellow and A.G. O'Donnell. 1993. Cell envelopes and classification. *In* Handbook of New Bacterial Systematics (edited by Goodfellow and O'Donnell). Academic Press, London, pp. 195–250.

Uchida, K. and K. Aida. 1977. Acyl type of bacterial cell wall: its simple identification by a colorimetric method. J. Gen. Microbiol. *23*: 249–260.

Uchida, K., T. Kudo, K.I. Suzuki and T. Nakase. 1999. A new rapid method of glycolate test by diethyl ether extraction, which is applicable to a small amount of bacterial cells of less than one milligram. J. Gen. Appl. Microbiol. *45*: 49–56.

Ward-Rainey, N.L., F.A. Rainey and E. Stackebrandt. 1997. *In* Stackebrandt, E., F.A. Rainey and N.L. Ward-Rainey. Proposal for a new hierarchic classification system, *Actinobacteria* classis nov. Int. J. Syst. Evol. Microbiol. *45*: 682–692.

Yokota, A. and T. Hasegawa. 1988. The analysis of madurose, an actinomycete whole-cell sugar, by HPLC after enzymatic treatment. J. Gen. Appl. Microbiol. *34*: 445–449.

Zhang, Z., Y. Wang and J. Ruan. 1998. Reclassification of *Thermomonospora* and *Microtetraspora*. Int. J. Syst. Bacteriol. *48*: 411–422.

Zhi, X.-Y., W.-J. Li and E. Stackebrandt. 2009. An update of the structure and 16S rRNA gene sequence-based definition of higher ranks of the class *Actinobacteria*, with the proposal of two new suborders and four new families and emended descriptions of the existing higher taxa. Int. J. Syst. Evol. Microbiol. *59*: 589–608.

Family I. **Streptosporangiaceae** Goodfellow, Stanton, Simpson and Minnikin 1990a, 321[VP] (Effective publication: Goodfellow, Stanton, Simpson and Minnikin 1990b.) emend. Ward-Rainey, Rainey and Stackebrandt 1997, 486 emend. Zhi, Li and Stackebrandt 2009, 600

Michael Goodfellow and Erika T. Quintana

Strep.to.spo.ran.gi.a.ce'a.e. N.L. neut. n. *Streptosporangium* the type genus of the family, suff. *-aceae* ending to denote a family; N.L. fem. pl. n. *Streptosporangiaceae* the *Streptosporangium* family.

Aerobic, Gram-stain-positive, non-acid–alcohol-fast, chemoorganotrophic actinomycetes which form a branched, stable substrate mycelium. When formed, aerial hyphae differentiate either into chains of two or more arthrospores or into spore vesicles that contain one to many spores which may be motile or nonmotile. Spores are borne on the substrate hyphae when aerial mycelium is absent. Wall peptidoglycan contains *meso*-diaminopimelic acid and *N*-acetylmuramic acid and is of the A1γ type. **Lipid profiles typically contain straight, iso-, anteiso-, and 10-methyl branched fatty acids, major proportions of di- and tetrahydrogenated menaquinones with nine isoprene units, and phosphatidylethanolamine and glucosamine containing polar components. Mycolic acids are absent.** The pattern of 16S rRNA signatures consists of nucleotides at positions 440:497 (C–G), 485 (U), 501:544 (C–G), 502:543 (G–C), 833:853 (U–G), and 1355:1367 (A–U). Widely distributed in soil.

DNA G+C content (mol%): 64–77.

Type genus: **Streptosporangium** Couch 1955a, 145[AL] emend. Stackebrandt, Kroppenstedt, Jahnke, Kemmering and Gürtler 1994, 268.

Further descriptive information

Phylogeny. The 11 genera classified in the family *Streptosporangiaceae* form a distinct phyletic line in the 16S rRNA actinobacterial gene tree (Ara and Kudo, 2007; Goodfellow and Quintana, 2006; Goodfellow et al., 2005b). In addition to *Streptosporangium*, the type genus, the family contains the genera *Acrocarpospora* Tamura et al. 2000, *Herbidospora* Kudo et al. 1993, *Microbispora* Nonomura and Ohara 1957, *Microtetraspora* (Thiemann et al. 1968b) Zhang et al. 1998a, *Nonomuria* Zhang et al. 1998a (corrected to *Nonomuraea* by Chiba et al., 1998a), *Planobispora* Thiemann and Beretta 1968c, *Planomonospora* Thiemann et al. 1967, *Planotetraspora* Runmao et al. 1993 emend. Tamura and Sakane 2004, *Sphaerosporangium* Ara and Kudo 2007 (corrected to *Sphaerisporangium* by List Editor, IJSEM, 2007), and *Thermopolyspora* (*ex* Krasil'nikov and Agre 1964) Goodfellow et al. 2005b. Representatives of these taxa form distinct branches in the 16S rRNA *Streptosporangiaceae* gene tree (Figure 360) and share a pattern of 16S rRNA signatures (Zhi et al., 2009), as shown in the description of the family.

Cell morphology. At present, members of the genera classified in the family *Streptosporangiaceae* are distinguished mainly on the basis of morphological criteria (Table 275). Strains that carry one or more spores in chains (*Herbidospora, Microbispora, Microtetraspora, Nonomuraea,* and *Thermopolyspora*) are closely related to organisms that form spores in vesicles (*Acrocarpospora, Planobispora, Planomonospora, Planotetraspora, Sphaerisporangium,* and *Streptosporangium*). Individual spore vesicles contain a coiled chain of arthrospores formed by septation of an unbranched, spiral hypha within an expanded sporophore sheath (Vobis and Kothe, 1985). Spore formation is exogenous, hence the term "spore vesicle" is to be preferred to the original term "sporangium" (Cross, 1970; Sharples et al., 1974). Spores in spore vesicles and in spore chains are formed essentially in the same way, i.e. by a hypha bound by a sheath; sheaths either expand to form the vesicular envelope or remain around the spore chains (Lechevalier et al., 1966b; Sharples et al., 1974; Vobis and Kothe, 1985).

Chemotaxonomy. *Streptosporangiaceae* strains have many chemical markers in common (Kudo, 2001). They have either a wall chemotype IIIB or IIIC *sensu* Lechevalier and Lechevalier (1970, 1970), i.e. they produce hydrolysates that are rich in *meso*-diaminopimelic acid and may or may not contain madurose. They have *N*-acetylated muramic acid and an A1γ peptidoglycan type, are rich in unsaturated, saturated, iso- and anteiso, and 10-methyl branched fatty acids, lack mycolic acids, and typically have a type IV phospholipid pattern and unsaturated, di- and tetrahydrogenated menaquinones with nine isoprene units as the predominant isoprenologues [MK-9, MK-9(H_2), and MK-9(H_4)]. The tetrahydrogenation occurs at the sites of isoprene units III (the third unit from the 2-methyl-1, 4-naphthoquinone moiety) and VIII [MK-9(III, VIII-H_4)] (Ara and Kudo, 2007; Goodfellow et al., 1988; Goodfellow and Quintana, 2006; Kawamoto et al., 1981; Kroppenstedt, 1985, 1990a; Stackebrandt et al., 1994). This chemotaxonomic profile distinguishes members of the family *Streptosporangiaceae* from all of the other families classified in the order *Actinomycetales* including their nearest neighbors, the families *Nocardiopsaceae* and *Thermomonosporaceae*.

The presence of madurose in whole-organism hydrolysates is usually considered to be associated with actinomycetes that have a wall chemotype III, though this sugar has been detected in *Micromonospora rosaria* and several *Streptomyces* species (Nakagaito et al., 1993; Yokota et al., 1989). There is also an unconfirmed report of the presence of madurose in a wall chemotype I actinomycete with a streptomycete-like morphology (Weyland et al., 1982). In addition, 3-*O*-methylgalactosyl (madurosyl) units have been detected in the structure of teichoic acids of a *Nonomuraea roseoviolaceae* subsp. *carminata* strain (formerly *Actinomadura carminata*; Naumova et al., 1986).

Ecology. Members of the family *Streptosporangiaceae* are widely distributed in soil where they are presumably engaged in the turnover of organic matter. The introduction of innovative selective isolation procedures should provide an insight into the occurrence, distribution, numbers, and activities of members of the constituent genera, as exemplified by the multistage procedures introduced for the selective isolation of *Planobispora* and *Planomonospora* strains (Suzuki et al., 2001a, 2001b).

Enrichment and isolation procedures

Streptosporangiaceae strains are generally isolated by plating out suspensions of dry-heat treated, air-dried soil samples onto one or more selective synthetic media or by baiting environmental

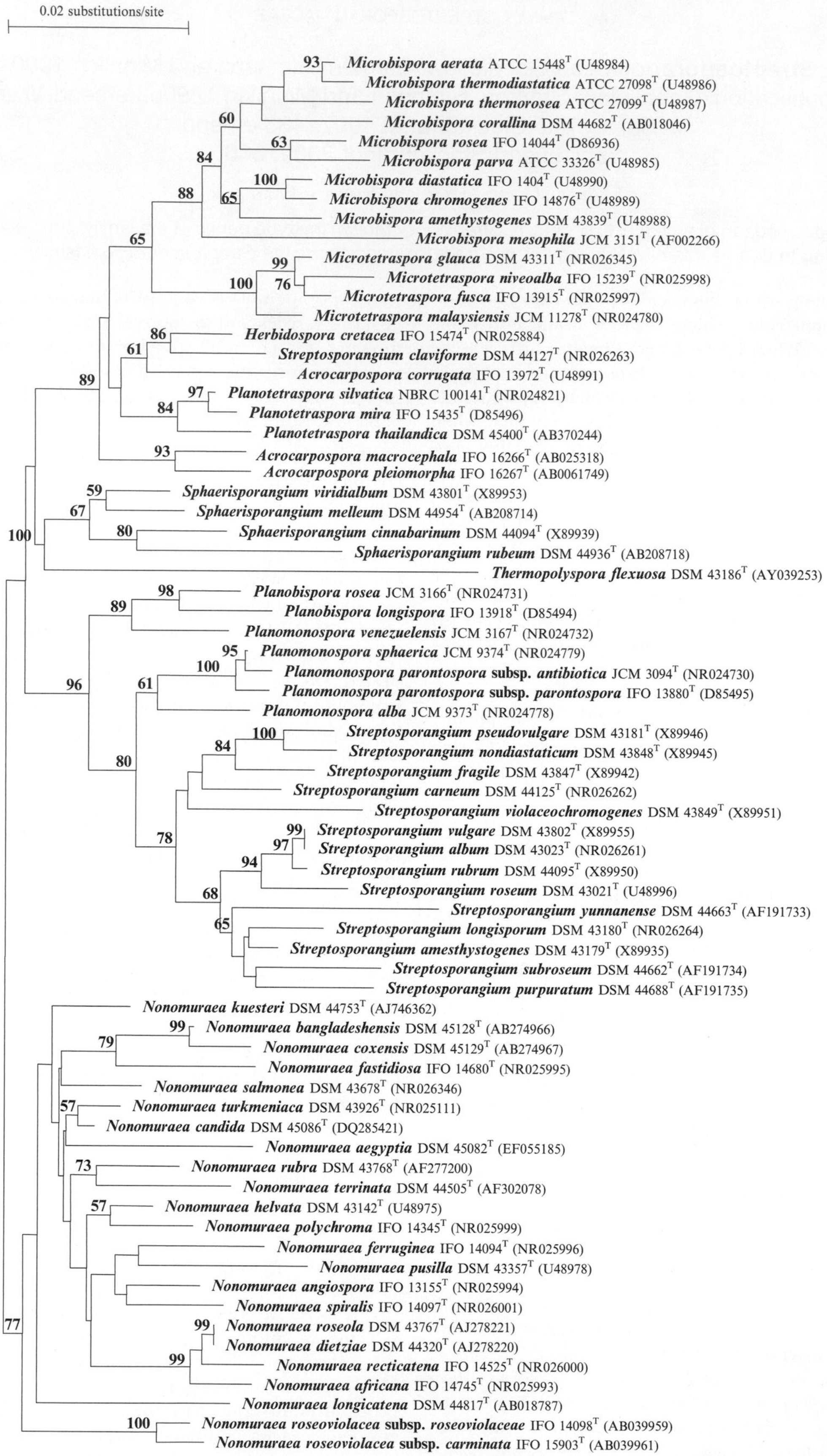

FIGURE 360. Neighbor-joining tree (Saitou and Nei, 1987) based on nearly complete 16S rRNA gene sequences (>1372 nucleotides) showing relationships between representatives of validly named species in the family *Streptosporangiaceae*. The numbers at the nodes indicate the level of bootstrap support (%) based on a neighbor-joining analysis of 1000 re-sampled datasets.^{T}, Type strain.

TABLE 275. Morphological features and chemotaxonomic characteristics of members of the genera classified in the family *Streptosporangiaceae*[a,b]

Characteristic	*Acrocarpospora*	*Herbidospora*	*Microbispora*	*Microtetraspora*	*Nonomuraea*	*Planobispora*	*Planomonospora*	*Planotetraspora*	*Sphaerisporangium*	*Streptosporangium*	*Thermopolyspora*
Vesicle formation	Club or globose spore vesicles on aerial hyphae	Spore chains on aerial hyphae	Spores in characteristic longitudinal pairs on aerial hyphae	Spore chains containing four or more spores on short aerial hyphae	Spore chains or pseudosporangia on aerial hyphae	Cylindrical to clavate spore vesicles containing longitudinal pairs of spores on aerial hyphae	Cylindrical to clavate spore vesicles containing single spores on aerial hyphae	Spore vesicles containing four spores on aerial hyphae	Globose spore vesicles on aerial hyphae	Globose spore vesicles on aerial hyphae	Hooked or irregular spiral chains of 4–10 warty to spiny ornamented spores on aerial hyphae
Aerial mycelium	+	–	+	+	+	+	+	+	+	+	+
Motile spores	–	–	–	–	–	+	+	–		–	–
Cell-wall chemotype[c]	III	III	III	III	III	III	III	III	III	III	III
Peptidoglycan type	A1γ	A1γ	A1γ	A1γ	A1γ	A1γ	A1γ	nd	nd	A1γ	nd
Whole-organism sugar pattern[d]	B,C	B	B,C	B,C	B,C	B	B	B	B	B	C
Fatty-acid type[e]	3c	3c	3c	3c	3c	3c	3c	3d	3c	3c	nr
Menaquinones	MK-9(III, VIII-H_4), MK-9(H_6), MK-9(III-H_2)	MK-10(III, IX-H_4), MK-10(H_6), MK-10	MK-9(III, VIII-H_4), MK-9(III-H_2), MK-9(H_6)	MK-9(III, VIII-H_4), MK-9(III-H_2), MK-9(H_0)	MK-9(III, VIII-H_4), MK-9(III-H_2), MK-9(H_0)	MK-9(III, VIII-H_4), MK-9(III-H_2), MK-9(H_0)	MK-9(III-H_2), MK-9(III VIII-H_4), MK-9(H_0)	MK-9 (III, VIII-H_4), MK-9(III-H_2)	MK-9(III, VIII H_4), MK-9(H_6), MK-9(III), MK-9(H_0), MK-9(H_2)	MK-9(III, VIII-H_4), MK-9(III-H_2), MK-9(H_0)	MK-9(III, VIII-H_4), MK-9(H_0), MK-9(H_2), MK-9 (II, III, VIII-H_6)
Phospholipid type[f]	IV, II	IV	IV	IV	IV	IV	IV	IV	IV	IV	IV
DNA G+C content (mol%)	68–69	69–71	71–73	69–71	64–69	70–71	72	71	70–72	69–71	77
Growth temperature range (°C)	15–30	13–37	25–55	20–45	20–45	28–40	28–37	28–37	20–37	20–50	37–60

[a]Symbols: +, >85% positive; –, 0–15% positive; nr, not reported.

[b]Data from Kroppenstedt et al. (1990); Kudo (2001); Goodfellow et al. (2005b); and Ara and Kudo (2007).

[c]Major constituents: alanine, glutamic acid, glucosamine, and *meso*-A_2pm (Lechevalier and Lechevalier, 1970b).

[d]A, cross-linkage between positions 3 and 4 of adjacent peptide subunits; 1, peptide bridge absent; γ, *meso*-A_2pm at position 3 of the tetrapeptide subunits (Schleifer and Kandler, 1972).

[e]Saturated fatty acids, unsaturated fatty acids, iso-fatty acids (variable), and methyl-branched fatty acids (Kroppenstedt, 1985).

[f]Phospholipid patterns: PII, only phosphatidylethanolamine; PIV, phospholipids containing glucosamine (with phosphatidylmethylethanolamine variable) (Lechevalier et al., 1977, 1981).

samples with natural substrates. Various combinations of pretreatment regimes and selective media have been used to isolate members of the genera *Microbispora* and *Streptosporangium* (Nonomura and Ohara, 1960, 1969a, 1969b), *Microtetraspora* and *Nonomuraea* (Nonomura and Ohara, 1971c, 1971d), and *Nonomuraea* and *Streptosporangium* (Hayakawa and Nonomura, 1987a, 1987b; Nonomura, 1984).

Variations in the approaches mentioned above have involved treating heat-pretreated, air-dried soil suspensions with either benzethonium chloride or chloramine-T prior to plating onto HV agar and examining the incubated plates for the presence of *Streptosporangium* and related taxa (Hayakawa et al., 1991a, 1997). Similarly, *Microbispora* strains can be preferentially isolated by treating suspensions of heat-treated, air-dried soils with phenol, diluting with water, and plating onto HV agar supplemented with nalidixic acid (Nonomura and Hayakawa, 1988). *Microbispora, Planobispora, Planomonospora,* and *Streptosporangium* strains have been isolated by plating centrifuged supernatants of air-dried soils onto HV agar supplemented with streptomycin (Kizuka et al., 1997).

Members of the genera *Microbispora, Microtetraspora, Nonomuraea,* and *Streptosporangium* have been recovered from pretreated soil samples using a range of selective synthetic media supplemented with antibacterial antibiotics including ampicillin, bruneomycin, penicillin, polymixin B, rubromycin, and streptomycin (Hayakawa and Nonomura, 1987a, 1987b; Lavrova et al., 1972; Preobrazhenskaya et al., 1975b; Wang et al., 1999; Whitham et al., 1993). *Microtetraspora* and *Streptosporangium* strains have been isolated following irradiation of soil samples with microwaves, plating out soil suspensions onto Gauze agar 2 (Gauze et al., 1983) either without antibiotics or supplemented with levorin and nalidixic acid to suppress bacterial and fungal growth, respectively, and incubating at 28°C for up to six weeks (Bulina et al., 1997). Baiting environmental samples with pollen or hair has been used to good effect for the isolation of *Planobispora, Planomonospora,* and *Streptosporangium* strains (Couch, 1954, 1955a, 1963). Details of these and other traditional procedures recommended for the selective isolation of *Streptosporangiaceae* strains are given in the chapters dealing with the individual genera.

Taxonomic comments

The family *Streptosporangiaceae* was proposed by Goodfellow et al. (1990b) to provide a home for the genera *Microbispora, Microtetraspora, Planobispora, Planomonospora, Spirillospora,* and *Streptosporangium.* These genera had previously been assigned to higher taxa mainly based on what were seen at the time to be characteristic morphological traits. The family "*Actinosporangiaceae*", which was introduced to accommodate the genus *Streptosporangium* and "sporangiate" actinomycetes belonging to the genus *Actinoplanes* (Couch, 1955a), was subsequently renamed *Actinoplanaceae* (Couch, 1955b). In addition to *Actinoplanes,* the type genus, and *Streptosporangium,* this taxon included other actinomycetes considered to form "sporangia", namely the genera *Amorphosporangium, Ampullariella, Dactylosporangium, Kitasatoa, Pilimelia, Planobispora, Planomonospora,* and *Spirillospora* (Couch and Bland, 1974).

The original members of the family *Actinoplanaceae* were subsequently shown to belong to two distinct DNA–DNA relatedness groups. The genera *Actinoplanes, Ampullariella,* and *Dactylosporangium* were assigned to one group and the genera *Planobispora, Planomonospora,* and *Spirillospora* to the other one (Farina and Bradley, 1970). Organisms in the first group contained *meso*-and/or hydroxydiaminopimelic acid and glycine, i.e. they had a wall chemotype II *sensu* Lechevalier and Lechevalier (1970, 1970), whereas those in the second group contained *meso*-diaminopimelic acid and madurose (wall chemotype IIIB).

The realization that the genera *Actinoplanes* and *Dactylosporangium* had properties in common with the genera *Catellatospora, Couchioplanes, Catenuloplanes, Dactylosporangium, Micromonospora,* and *Pilimelia* led to these taxa being classified in the family *Micromonosporaceae* (Krasil'nikov 1938), a taxon which was emended first by Koch et al. (1996) and then by Stackebrandt et al. (1997). In the meantime, the genera *Amophosporangium* and *Ampullariella* were proposed as subjective synonyms of the genus *Actinoplanes* (Stackebrandt and Kroppenstedt, 1987). Subsequently, the genus *Kitasatoa* became a subjective synonym of the genus *Streptomyces* Goodfellow et al. 1986, and the genus *Spirillospora* became a member of the family *Thermomonosporaceae* (Stackebrandt et al. 1997) emend. Zhang et al. 2001.

Goodfellow and Cross (1984) assigned the oligosporic genera *Actinomadura, Excellospora, Microbispora, Microtetraspora,* and the "sporangiate" genera *Planobispora, Planomonospora, Spirillospora,* and *Streptosporangium* to what they considered to be a somewhat contrived aggregate group, the maduromycetes, a taxon that was recognized in the previous edition of *Bergey's Manual* (Goodfellow, 1989a, 1989b). However, with the exception of the genera *Actinomadura* and *Spirillospora,* these taxa were shown to form a coherent subgeneric group based on 16S rRNA gene cataloging and sequencing (Stackebrandt, 1986). The close relationships shown by the genera in this revised grouping were formalized by the proposal that *Streptosporangium* be recognized as the type genus of the new family *Streptosporangiaceae* Goodfellow et al. 1990b. In addition to the founder members, this family now includes the genera *Acrocarpospora, Herbidospora, Nonomuraea, Planotetraspora, Sphaerisporangium,* and *Thermopolyspora.* Some of these genera, notably *Nonomuraea, Sphaerisporangium,* and *Thermopolyspora* were based on or included taxa that had previously been misclassified with founder members of the family.

The genus *Nonomuraea* was proposed by Zhang et al. (1986a) to accommodate the *Actinomadura pusilla* group (Fischer et al., 1983; Meyer, 1989; Poschner et al., 1985) which had been assigned to the genus *Microtetraspora* (Kroppenstedt et al., 1990a) when it became apparent that the genus *Actinomadura* could be split into two distinct aggregate taxa (see corresponding section on the genus *Actinomadura*). In contrast, *Excellospora rubrobrunea* Agre and Guzeva 1975 was transferred to the revised genus *Actinomadura* as *Actinomadura rubrobrunea* (*ex* Krasil'nikov et al. 1968) Kroppenstedt et al. 1991. The genus *Thermopolyspora* was proposed to provide a taxonomic home for an organism that had initially borne the name "*Thermomonospora flexuosa*" Krasil'nikov and Agre 1964.

The genus *Streptosporangium* was shown to be markedly heterogeneous as it contained several species that had little in common either with one another or with the type species, *Streptosporangium roseum* (Kemmerling et al., 1993; Kudo et al., 1993; Ochi and Miyadoh, 1992). Stackebrandt et al. (1994) classified *Streptosporangium albidum* Furumai et al. 1968, *Streptosporangium viridogriseum* subsp. *kofuense* Okuda et al. 1966, and *Streptosporangium viridogriseum* subsp. *viridogriseum* in a new genus, *Kutzneria*, as *Kutzneria albida*, *Kutzneria kofuensis*, and *Kutzneria viridogrisea*. Similarly, "*Streptosporangium cinnabarinum*" Celmer et al. 1977 and *Streptosporangium viridialbum* Nonomura and Ohara 1960 became initial members of the genus *Sphaerisporangium* as *Sphaerisporangium cinnabarinum* Ara and Kudo 2007 and *Sphaerisporangium viridialbum* (Nonomura and Ohara 1960) Ara and Kudo 2007.

Other changes in the composition of the genus *Streptosporangium* include the transfer of *Streptosporangium corrugatum* Williams and Sharples 1976 to the genus *Acrocarpospora* as *Acrocarpospora corrugata* (Williams and Sharples 1976) Tamura et al. 2000 and *Streptosporangium indianense* Gupta 1965 to the genus *Streptomyces* as *Streptomyces indianensis* (Gupta 1965) Kudo and Seino 1987. The type strains of *Herbidospora cretaceae* Kudo et al. 1993 and *Streptosporangium claviforme* Petrolini et al. 1992 have morphological features in common and belong to the same genomic species; these results led Kudo and his colleagues to propose that the *Streptosporangium claviforme* be seen as a subjective synonym of *Herbidospora cretacea*. However, the two strains form distinct, but related, phyletic lines in the 16S rRNA *Streptosporangiaceae* gene tree (Figure 360).

Differentiation of the genera of the family *Streptosporangiaceae*

The genera classified in the family *Streptosporangiaceae* can be delineated using genotypic and phenotypic markers. They are most readily separated by 16S rRNA gene sequence data including genus specific nucleotide signatures (Ara and Kudo, 2007; Goodfellow and Quintana, 2006), and by morphological criteria (Table 275). The genera *Acrocarpospora*, *Planobispora*, *Planomonospora*, *Planotetraspora*, *Sphaerisporangium*, and *Streptosporangium*, for instance, can be separated on the basis of the shape of spore vesicles, the number and shape of the enclosed spores, and by whether the latter are motile. Similarly, the oligosporic members of the family can be distinguished by the number of spores borne in chains on aerial hyphae and whether spores are formed on substrate hyphae. Microbisporae typically produce chains of two spores, microtetrasporae chains of four spores, whereas nonomuraea form either chains of spores or pseudovesicles. In all cases, the spores are formed on aerial hyphae. *Herbidospora* is unusual as it produces short chains of spores in clusters at the tips of sporophores that arise from substrate hyphae.

It is sometimes necessary to distinguish between genera classified in the family *Streptosporangiaceae* from morphologically similar taxa assigned to other families. Chemotaxonomic markers can be used to distinguish between *Microbispora*, *Microtetraspora* and *Nonomuraea* strains and members of the genus *Actinomadura* which also form chains of spores on aerial hyphae (Kroppenstedt et al., 1990a; Kudo et al., 1993). Actinomadurae typically contain MK-9(H_6), saturated at sites II, III, and VIII as the major isoprenologue, and diphosphatidylglycerol and phosphatidylinositol mannosides but lack nitrogenous phospholipids (phospholipid type I), whereas microbisporae, microtetrasporae, and nonomuraea have major amounts of MK-9(H_4), saturated at points III and VIII, and polar lipid patterns which contain diphosphatidylglycerol, hydroxyethanolamine, uncharacterized glycolipids, and glucosamine-containing phospholipids (phospholipid type IV) (Lechevalier et al., 1977, 1981).

Members of the family *Streptosporangiaceae* which form spore vesicles can be differentiated from morphologically similar strains belonging to the genera *Actinoplanes* and *Dactylosporangium* as the latter have a wall chemotype II and a phospholipid type I or II, whereas the former have a wall chemotype II and a phospholipid pattern IV (Goodfellow and Quintana, 2006; Koch et al., 1996; Vobis, 2006). Care also needs to be taken to distinguish between *Streptosporangium* and *Spirillospora* strains, though only the latter form motile spores (Vobis and Kothe, 1989). The vesicular hyphae within the spore vesicles of spirillosporae are branched, whereas those of streptosporangiae are not. Methods used to detect key chemical markers are cited in the section on the suborder *Streptosporangineae*.

Monciardini et al. (2002) designed a highly specific oligonucleotide primer set for the identification of *Streptosporangiaceae* strains. They also developed selective sets of oligonucleotide primers for PCR-amplification of rDNA for the recognition of members of the families *Micromonosporaceae*, *Streptomycetaceae*, and *Thermomonosporaceae*, and for *Dactylosporangium* strains. The application of these primer sets to environmental samples shows the presence of novel members of these taxa in soil samples.

Genus I. **Streptosporangium** Couch 1955a, 145^{AL} emend. Stackebrandt, Kroppenstedt, Jahnke, Kemmering and Gürtler 1994, 268

Erika T. Quintana and Michael Goodfellow

Strep.to.spo.ran′gi.um. Gr. adj. *streptos* twisted; Gr. n. *spora* a seed; Gr. n. *angium* a vessel; N.L. neut. n. *Streptosporangium* spores coiled within a sporangium.

Aerobic, Gram-stain-positive, non-acid-fast actinomycetes which form a branched, stable, non-fragmenting mycelium bearing aerial hyphae that differentiate into globose or spherical spore vesicles. Oval, spherical, or rod-shaped, nonmotile sporangiospores are formed by septation of a single, coiled, unbranched hypha within the spore vesicle. Grows on a variety of organic and synthetic media. Biochemically versatile and chemically homogeneous. **Whole-organism hydrolysates contain *meso*-diaminopimelic acid and usually madurose. The peptidoglycan is of the A1γ type. Muramic acid moieties are *N*-acetylated. Cells contain major amounts of iso-, anteiso-, saturated, unsaturated, and 10-methyl-branched fatty acids, and MK-9(H_2) and MK-9(III, VIII-H_4) as the predominant isoprenologues. Mycolic acids are absent. Major phospholipids include diphosphatidylglycerol, phosphatidylethanolamine, phosphatidylglycerol, phosphatidylinositol, and glucosamine-containing phospholipids.** The phylogenetic position of the genus *Streptosporangium*, as determined by

16S rRNA gene sequence analysis, is in the family *Streptosporangiaceae*.

DNA G+C content (mol%): 69–71.

Type species: **Streptosporangium roseum** Couch 1955a, 151[AL].

Further descriptive information

Phylogeny. The genus *Streptosporangium* forms a distinct 16S rRNA gene clade within the evolutionary radiation encompassed by genera classified in the family *Streptosporangiaceae* (Ara and Kudo, 2007; Goodfellow et al., 1990b; Kudo, 2001). The genus encompasses 13 validly published species which are most closely related to the genera *Planobispora, Planomonospora,* and *Planotetraspora* (Figure 361). The two most distantly related species, *Streptosporangium violaceochromogenes* and *Streptosporangium yunnanense* share a 16S rRNA gene similarity of 95.2%. In contrast, the type strains of *Streptosporangium album, Streptosporangium roseum,* and *Streptosporangium vulgare* have identical 16S rRNA gene sequences.

Cell morphology. *Streptosporangium* strains characteristically form nonfragmenting, extensively branched substrate hyphae which carry aerial hyphae that differentiate into spore vesicles (Figure 362) which are borne either on short or long sporangiophores (Figure 363, Figure 364, and Figure 365). The spore vesicles may be carried singly or in clusters (Figure 364, Figure 365, and Figure 366). They may be globose or spherical and are usually 4–20 µm in diameter, but may be up to 40 µm. Spore vesicles originate as sac-like structures within which single, unbranched hyphae coil repeatedly and undergo septation giving rise to "sporangiospores" (Locci, 1976; Locci and Petrolini-Baldan, 1971). "Sporangiospores" are formed in the same way as the conidia of streptomycetes and not by cytoplasmic cleavage and nuclear division as in the formation of fungal sporangiospores (Lechevalier et al., 1966b; Sharples et al., 1974; Vobis and Kothe, 1985). Consequently, the use of the term sporangium is inappropriate in this context and should be replaced by the term spore vesicle (Cross, 1970; Sharples et al., 1974). However, unlike streptomycetes, the vesicular sheath does not usually persist around released spores.

As spore vesicles increase in size, the internal arrangement of hyphae and their septation into spores can be seen by the shape of the thin vesicular walls (Figure 364, Figure 365, and Figure 366), though the vesicular membrane of *Streptosporangium fragile* cannot be detected by light microscopy (Shearer et al., 1983). Nonmotile spores are released when spores vesicles are placed in water; the surrounding sheath is ruptured when the intersporal matrices swell and exert pressure on the vesicular membrane. The spores are spherical, oval, or rod shaped and have smooth walls. *Streptosporangium roseum* also forms streptomycete-like chains of arthrospores.

Chemotaxonomy. Streptosporangiae have a wall chemotype 111B (Lechevalier and Lechevalier, 1970a, 1970b), i.e. they contain *meso*-diaminopimelic acid in the wall peptidoglycan and madurose (3-*O*-methyl-D-galactose; Lechevalier and Gerber, 1970) in whole-organism hydrolysates. They have a peptidoglycan of the A1γ type (Schleifer and Kandler, 1972) and *N*-acetylated muramic acid (Kawamoto et al., 1981) and are rich in iso-, anteiso-, saturated, and unsaturated methyl-branched fatty acids (pattern 3c *sensu* Kroppenstedt, 1985; Kudo et al, 1993; Whitham et al., 1993; Stackebrandt et al., 1994; Zhang et al., 2002, 2005), but lack mycolic acids (Goodfellow and Minnikin, 1981; Kudo et al., 1993; Whitham et al., 1993).

Most *Streptosporangium* strains contain di- and tetrahydrogenated menaquinones with nine isoprene units as predominant

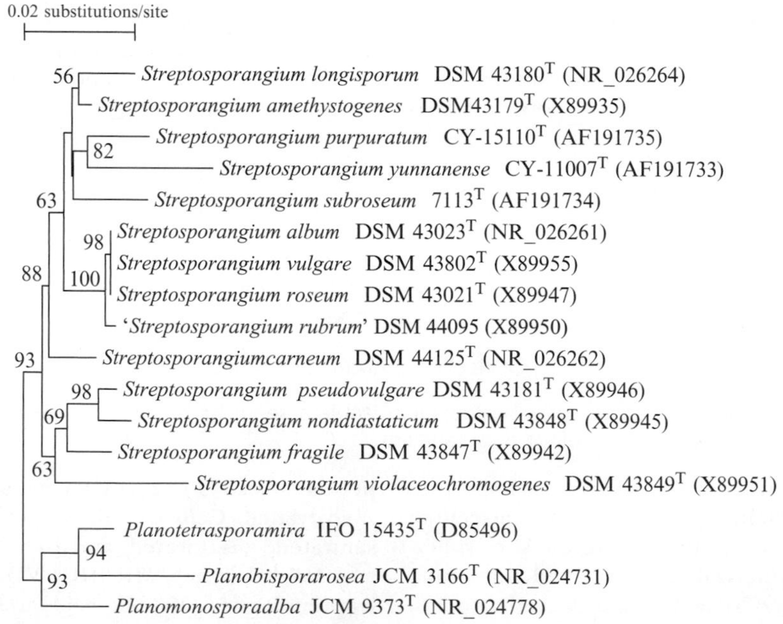

FIGURE 361. Neighbor-joining tree based on almost complete 16S rRNA gene sequences showing relationships of *Streptosporangium* species to one another and to representatives of related genera classified in the family *Streptosporangiaceae*.

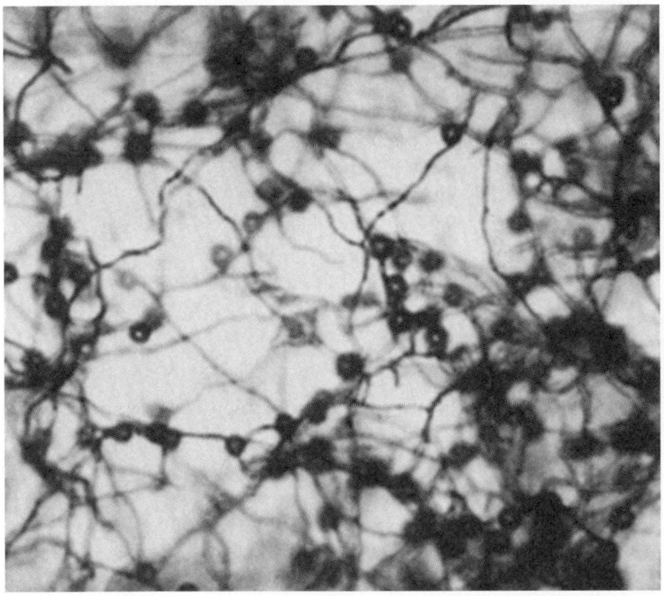

FIGURE 362. Photomicrograph of *Streptosporangium carneum* A84575^T showing profuse spore vesicle formation after growth on inorganic salts-starch agar (ISP medium 4) for 18 d at 30°C. Bright-field illumination. Magnification, 640×. (Reproduced with permission from Mertz and Yao, 1990. Int. J. Syst. Bacteriol. *40*: 247–253.)

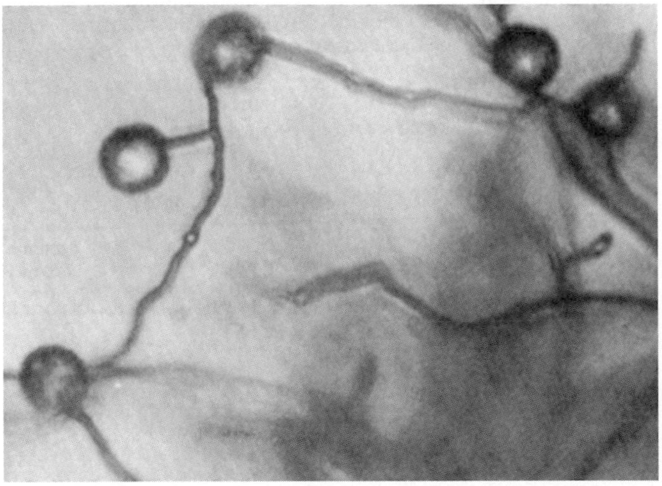

FIGURE 363. Photomicrograph of *Streptosporangium carneum* A84575^T showing sporangiophore formation after growth on inorganic salts-starch agar (ISP medium 4) for 18 d at 30°C. Bright-field illumination, oil immersion. Magnification, 1600×. (Reproduced with permission from Mertz and Yao, 1990. Int. J. Syst. Bacteriol. *40*: 247–253.)

isoprenologues (Kroppenstedt, 1985; Kudo et al., 1993; Stackebrandt et al., 1994; Whitham et al., 1993), though *Streptosporangium purpuratum*, *Streptosporangium subroseum*, and *Streptosporangium yunnanense* also contain substantial proportions of unsaturated menaquinones with nine isoprene units (Zhang et al., 2002, 2005). Streptosporangiae typically have a phospholipid type IV (Lechevalier et al., 1977, 1981); they contain glucosamine-containing lipids with diphosphatidylglycerol, phosphatidylethanolamine, and phosphatidylinositol (Kudo

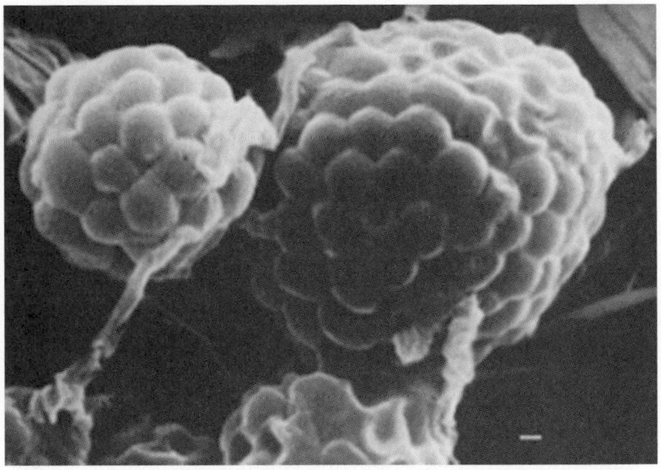

FIGURE 364. Scanning electron micrograph of spore vesicles of *Streptosporangium carneum* A84575^T after 18 d growth at 30°C on inorganic salts-starch agar (ISP medium 4). Bar = 1 μm. (Reproduced with permission from Mertz and Yao, 1990. Int. J. Syst. Bacteriol. *40*: 247–253.)

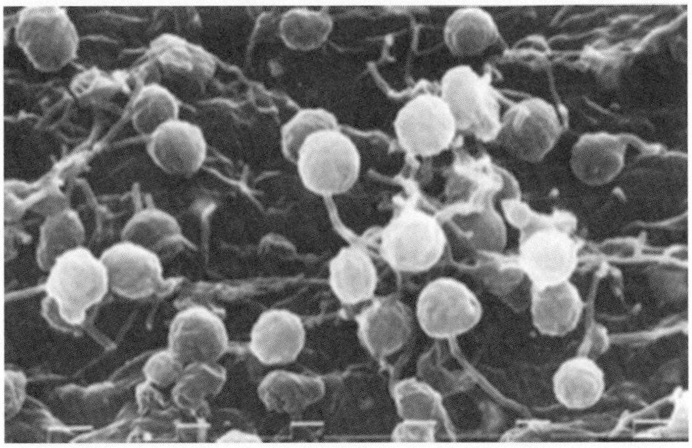

FIGURE 365. Scanning electron micrograph of *Streptosporangium album* CBS 426.61 growing on oatmeal agar. Sporangiophores are short. Bar interval = 10 μm.

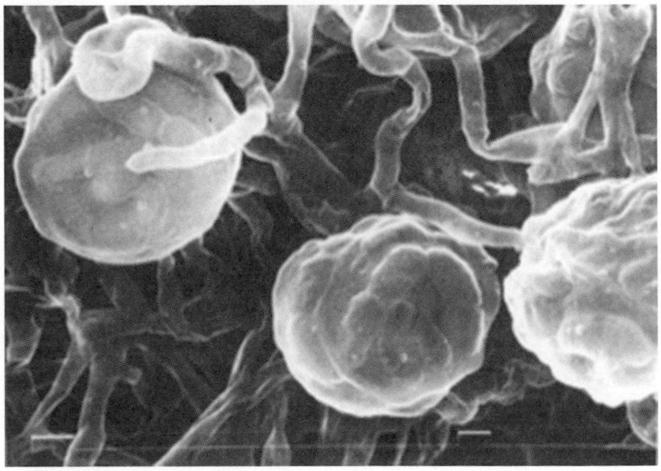

FIGURE 366. Scanning electron micrograph of *Streptosporangium album* CBS 426.61 growing on oatmeal agar. Bar interval = 10 μm. Vesicular walls are thin.

et al., 1993; Stackebrandt et al., 1994; Whitham et al., 1993). The presence of phosphatidylmethylethanolamine is variable (Zhang et al., 2005). *Streptosporangium subroseum* is unusual as it only contains phosphatidylethanolamine (Zhang et al., 2002) and hence has a wall chemotype II *sensu* Lechevalier et al. (1977, 1981). The G+C content of the DNA of streptosporangiae falls within the narrow range of 69.0–71 mol% (Farina and Bradley, 1970; Tsyganov et al., 1966; Yamaguichi, 1967; Zhang et al., 2002, 2005).

Colony morphology. *Streprosporangium* strains form well developed colonies on ISP media that are commonly used to cultivate filamentous actinomycetes (Shirling and Gottlieb, 1966). The substrate mycelium may be pink, red to dark brown, black, or deep purplish red, and the aerial mycelium pinkish-white or yellowish-pink, though the colors can vary between species and are influenced by the cultivation conditions. Some species produce colored diffusible pigments. *Streptosporangium amethystogenes* forms violet crystals of iodinin on oatmeal-YG medium (yeast extract 1 g, glucose 2 g, glycerol 2 g/l).

Nutrition and growth conditions. Streptosporangiae grow well on rich media such as arginine-vitamin agar (Nonomura and Ohara, 1969b, a), glucose-yeast extract agar (Waksman, 1950a), modified Bennett's agar (Jones, 1949), oatmeal agar (Shirling and Gottlieb, 1966), and yeast extract agar (Kudo et al., 1993). They grow particularly well and produce an abundant aerial spore mass on oatmeal-yeast extract (Nonomura and Ohara, 1960; Zhang et al., 2005). Some strains such as *Streptosporangium nondiastaticum* and *Streptosporangium pseudovulgare* require B-vitamins for growth (Zhang et al., 2005). It has also been shown that *Streptosporangium* strains do not have any particular preference for nutrient sources, either for the support of their growth or for their ability to produce anti-*Bacillus subtilis* substances (Platas et al., 1999). Streptosporangiae generally grow well at neutral pH and at 20–30°C; members of some species grow at 42°C (Nonomura, 1989a; Zhang et al., 2002, 2005).

Metabolism and genetics. Little is known about either the metabolism or genetics of *Streptosporangium* strains. In general, streptosporangiae use a broad range of compounds as sole carbon sources (Whitham et al., 1993; Zhang et al., 2002, 2005) and degrade diverse organic substrates (Whitham et al., 1993). *Streptosporangium pseudovulgare*, *Streptosporangium purpuratum*, and *Streptosporangium yunnanenis* degrade cellulose (Zhang et al., 2002, 2005). There is evidence that streptosporangiae can cleave diverse methylumbelliferone conjugated substrates (Whitham et al., 1993). Members of the genus have a chemoorganotrophic metabolism.

Streptosporangiae are a potentially rich source of novel commercially significant products, notably antibiotics (Donadio et al., 2002; Lazzarini et al., 2000), as exemplified by the production of lipopeptides from *Streptosporangium amethystogenes* (Takizawa et al., 1995); glycopeptides from *Streptosporangium carneum* (Michel and Yao, 1991); selenomycin from "*Streptosporangium brasiliense*" (Coronelli and Thiemann, 1969); anthracyclines from *Streptosporangium fragile* (Nash et al., 1981; Shearer et al., 1983); sporamycin from "*Streptosporangium koreanum*" (Celmer et al., 1978); oligopeptides from *Streptosporangium nondiastaticum* (Shokichi et al., 1988); a novel protein (Umezawa and Kamiyama, 1983) and sporamycin from *Streptosporangium pseudovulgare* (Umezawa et al., 1976); an ansa-macrolactam (Hacene et al., 1998), sporangirosomycin (Ghazal and Alb-El-Aziz, 1993), and thiosporamycin (Celmer et al., 1978) from *Streptosporangium roseum*; sibiromycin from "*Streptosporangium sibiricum*" (Brazhnikova et al., 1972; Gauze et al., 1969); platomycins A and B (Takasawa et al., 1975) and victomycin from *Streptosporangium violaceochromogenes* (Kawamoto et al., 1975); and sporacuracins A and B from *Streptosporangium vulgare* (Atsushi et al., 1975).

A new calmodulin antagonist, genistein, has been isolated from *Streptosporangium vulgare* (Goto et al., 1987), a hematopoietic inducer from *Streptosporangium amethystogenes* subsp. *fukuiensis* (Donadio et al., 2002), and endothelin converting enzyme inhibitors from *Streptosporangium roseum* (Tsurumi et al., 1994, 1995). In addition, virginiamycin type peptolides have been isolated from "*Streptosporangium cinnabarinum*" and "*Streptosporangium koreanum*" (Celmer et al., 1977), 1,6- dihydroxy-2-chlorophenazine, a novel phenazine antifungal antibiotic from a *Streptosporangium* strain (Patel et al., 1984), 1-hydroxy-4-methoxy-2-naphthoic acid, a herbicidal compound, from "*Streptosporangium cinnabarinum*" (Pfefferle et al., 1997), a thermostable glucoamylase from a streptosporangial endophyte of maize leaves (Stamford et al., 2002), and pigment-like antibiotics from a *Streptosporangium* strain isolated from an Algerian soil (Boudjella et al., 2007). Other potentially useful substances include inhibitors of lactic acid production by oral flora (Ikeda et al., 1993) and insecticidal compounds (Mishra et al., 1987).

Pfefferle et al. (2000) highlighted the importance of fermentation conditions in the production of secondary metabolites from streptosporangiae. They showed that the optimal conditions for secondary metabolite production were completely different from those of streptomycetes and identified oxygen tension as an important parameter for the optimal production of secondary metabolites from streptosporangiae in submerged culture.

Antibiotic sensitivity. The antibiotic sensitivity profiles of individual species, where known, are given in the species descriptions.

Pathogenicity. There are no proven grounds for believing that streptosporangiae have a role as clinical or veterinary pathogens, though "*Streptosporangium bovinum*" was isolated from infected bovine hooves (Batista et al., 1963).

Ecology. Streptosporangiae have been isolated infrequently from diverse sources, including coastal sediments (Bredholdt et al., 2007), cow dung (Garg et al., 2003), earthworm casts (Mba, 1997), forest soils (Potekhina, 1965; Xu et al. (1996), leaf litter (Van Brummelen and Bent, 1957), plant roots (Coombs and Franco, 2003; de Araujo et al., 2000; Sardi et al., 1992; Solans and Vobis, 2003), and lake sediments (Johnston and Cross, 1976; Lee and Hwang, 2002; Willoughby, 1969a), though organisms labeled *Streptosporangium* type 1 from freshwater streams (Willoughby, 1969b) probably belong to the genus *Actinoplanes*, given their morphological features and ability to form motile spores.

The introduction of an innovative selective isolation procedure showed that streptosporangiae are an integral part of actinomycete communities in soils (Nonomura and Ohara,

1969a) where they presumably have a role in the turnover of organic matter. The number of *Streptosporangium* strains in Japanese soils was estimated at 10^4–10^6 colony forming units per gram dry weight soil (Nonomura, 1984; Nonomura and Ohara, 1969a); lower numbers have been reported from pasture and woodland soils (Whitham et al., 1993). There is evidence that slightly acid, humus rich soils are a good source of streptosporangiae (Hayakawa et al., 1988; Lee and Hwang, 2002; Nonomura and Hayakawa, 1988; Seong et al., 2001).

The application of new selective isolation procedures designed for the recovery of rare actinomycetes have shown that streptosporangiae are present in a range of soil types including chernozem soils (Li et al., 2002), corn and paddy field soils (Nonomura and Hayakawa, 1988) (Masayuki et al., 1991), peaty soils (Zenova et al., 2008), steppe chestnut and desert pale brown soils (Norovsuren et al., 2007; Zhadambaa et al., 2007), and a Saharan soil (Boudjella et al., 2006).

Enrichment and isolation procedures

Isolation and enumeration of streptosporangiae and related actinomycetes, notably microbisporae and microtetrasporae, can be achieved by dry-heat treatment of air-dried soil samples and dilution plate culture with selective synthetic media (Hayakawa and Nonomura, 1984, 1987a, 1987b, 1989; Nonomura and Hayakawa, 1988; Nonomura and Ohara, 1969a, 1969b; Nonomura and Ohara, 1971a, 1971c, 1971d). To this end, soil samples are dried slowly at room temperature for 7 d, passed through a 2 mm sieve, gently ground in a mortar, spread on filter paper, and heated in a hot air oven at 100–120°C for an hour. This procedure greatly reduces the numbers of bacteria and streptomycetes, but leads to an increased isolation frequency of streptosporangiae. There is evidence that pretreatment of soil suspensions with sodium dodecyl sulfate (0.05%, w/v) and yeast extract (6%, w/v) at 40°C for 20 min, followed by dilution in water, activates spores of actinomycetes but kills vegetative cells of other bacteria, thereby leading to increased counts of streptosporangiae and related organisms on isolation plates.

Suspensions of pretreated soil are used to inoculate one or more of the selective synthetic media. Arginine-vitamins (AV) agar*, chitin – V agar†, humic-vitamin (HV)‡, and C-1§ and C-2** have been recommended for the selective isolation of streptosporangiae and related actinomycetes (Hayakawa and Nonomura, 1987a, 1987b; Nonomura and Ohara, 1969a). These media are supplemented with antifungal antibiotics and sometimes with penicillin and polymixin B (Whitham et al., 1993). Inoculated plates are incubated at 30°C for 4–6 weeks and colonies examined for the presence of spore vesicles by using a light microscope fitted with a long working distance objective. The highest counts and cleanest plates are usually obtained with HV-vitamin agar. Growth of streptosporangiae around soil particles can be observed when pretreated soil particles (0.5 g) are sprinkled over selective media that have been incubated for 1 month at 30°C.

Hayakawa et al. (1991a) recommended an improved method for the selective isolation of streptosporangiae from soil. This procedure was based on the ability of streptosporangial spores to withstand dry heat at 120°C for an hour and treatment with benzylthonium chloride (BC), as well as the capacity of streptosporangiae to grow in the presence of leukomycin and nalidixic acid. The dry heat and BC treatments greatly reduce the numbers of bacteria and unwanted actinomycetes, including streptomycetes, from isolation plates. In turn, nalidixic acid in HV agar suppresses the growth of BC-resistant bacteria while leukomycin increases the selectivity of HV agar for streptosporangiae by eliminating unwanted bacteria that remain after the employment of the various pretreatments. In general, streptosporangiae account for about 20% of colonies growing on the resulting isolation plates.

The improved procedure involves several steps. Initially, an air-dried soil sample is ground in a mortar and heated at 120°C for an hour prior to the preparation of a 10^{-1} dilution in water. Half a ml of the 10^{-1} preparation is added to 4.5 ml of sterile 5 mm phosphate buffer (pH 7.0) containing BC at a final concentration of 0.01%, w/v. This preparation is kept at 30°C for 30 min with occasional stirring and a 1 ml fraction diluted in sterile tap water (1:10 or 1:15). Aliquots (0.1 or 0.2 ml) of this preparation are spread over the surfaces of HV agar supplemented with leukomycin in ethanol (1 mg/l) and nalidixic acid (20 mg/l). The inoculated plates are incubated at 30°C for 3–4 weeks then examined as before.

Other procedures based on pretreatment of air-dried soil samples and plating onto selective media have been used to isolate members of genera classified in the family *Streptosporangiaceae*, notably *Herbidospora*, *Microbispora*, *Microtetraspora*, and *Streptosporangium*. Hayakawa et al. (1997) found that spores from representatives of these taxa, unlike those from *Micromonospora*, *Nocardia*, and *Streptomyces* strains, were resistant to toluene-*p*-sulfon sodiumchloroamide trihydrate (chloramine-T). Water suspensions of air-dried soil treated with chloramine-T (1%) and plated onto HV agar supported the growth of the target organisms after incubation at 30°C for 3 weeks. The chlorination treatment specifically inhibits the growth of non-filamentous bacteria and unwanted actinomycetes, thereby facilitating the isolation of the target genera. Isolates presumptively assigned to the genus *Streptosporangium*, on the basis of

*AV agar (Nonomura and Ohara, 1969a): arginine, 0.3 g; glucose, 1 g; glycerol, 1 g, K_2HPO_4, 0.3 g; $MgSO_4 \cdot 7H_2O$, 0.2 g; NaCl, 0.3 g; agar, 15 g; distilled water, 1000 ml; B vitamins (thiamin HCl, riboflavin, niacin, pyridoxine HCl, isositol, calcium pantothenate, *p*-aminobenzoic acid, each at 0.5 mg; biotin, 0.25 mg); trace salts ($CuSO_4 \cdot 5H_2O$, 1 mg; $Fe_2(SO_4)_3$, 10 mg; $MnSO_4 \cdot 7H_2O$, 1 mg; $ZnSO_4 \cdot 7H_2O$, 1 mg); antibiotics (actidione, 50 mg; nystatin, 5 mg; penicillin G, none or 0.8 mg; polymixin B, none or 4 mg); pH 6.4.

†Chitin-V agar (Hayakawa and Nonomura, 1984): colloidal chitin, 2 g (dry w.); K_2HPO_4, 0.15 g; $MgSO_4 \cdot 7H_2O$, 0.2 g; NaCl, 0.3 g; $CaCO_3$, 0.02 g; $FeSO_4 \cdot 7H_2O$, 10 mg; $ZnSO_4 \cdot 7H_2O$, 1 mg; $MnCl_2$, 1 mg; agar, 18 g; distilled water, 1000 ml; B vitamins, as for AV agar; actidione, 50 mg; pH 7.2.

‡HV vitamin agar (Hayakawa and Nonomura, 1984; Nonomura, 1984): humic acid, 1 g (used as an alkaline solution; artificial humic acid prepared from glucose and urea may be used, as may natural humic acids from soil humus, but the pale brown humic acid designated as Rp type gives the best results); Na_2HPO_4, 0.5 g; $MgSO_4 \cdot 7H_2O$, 0.05 mg; KCl, 1.7 g; $FeSO_4 \cdot 7H_2O$, 0.01 g; $CaCO_3$, 0.02 g; agar, 18 g; distilled water, 1000 ml; B vitamins, as for AV agar; actidione, 50 mg; pH 7.2.

§C-1 medium (Nonomura and Ohara, 1969b): casamino acids, 2 g; K_2HPO_4, 0.3 g; $MgSO_4 \cdot 7H_2O$, 0.5 g; NaCl, 0.3 g; agar, 20 g; distilled water, 1000 ml; trace salts ($FeSO_4 \cdot 7H_2O$, 10 mg; $MnSO_4 \cdot 7H_2O$, 1 mg; $CuSO_4 \cdot 5H_2O$, 1 mg); B vitamins (thiamine HCl, riboflavin, niacin, pyridoxine HCl, inositol, calcium pantothenate, *p*-aminobenzoic acid, each 0.5 mg; biotin 0.25 mg); pH 7.2.

**C-2 medium (Nonomura and Ohara, 1969b): casamino acids, 0.5 g; and asparagine, 0.5 g, instead of casamino acids, 2 g, in the composition of C-1 medium. The other components are the same as for C-1 medium.

their ability to form spore vesicles, accounted for 1.8–4.1 × 10^5 c.f.u./g dry weight soil.

Pretreatment of soil suspensions with extreme high frequency (EHF) radiation prior to plating onto Gauze's agar 2 supplemented with nalidixic acid (10 μg/ml) and either levorin (20 μg/ml) or nystatin (50 μg/ml) and incubation at 28°C for 2–6 weeks led to the isolation of increased numbers of rare actinomycetes from soil including streptosporangiae (Bulina et al., 1997; Li et al., 2002). It seems likely that EHF radiation inhibits the growth of unicellular bacteria and breaks the dormancy of actinomycete spores. Streptosporangiae have also been isolated by plating out surface sterilized root material onto selective media (Cao et al., 2005; de Araujo et al., 2000).

An indirect immunomagnetic capture (IMC) method was developed by Mullins et al. (1995) for the selective recovery of *Streptosporangium fragile* from sterile and non-sterile soil. IMC recovery rates for *Streptosporangium fragile* spores from soil were usually in the range of 10–20% of the counts obtained by traditional plating. However, the capture method resulted in cleaner isolation plates and hence could be expected to enhance the successful isolation of the target organism compared with traditional plating methods.

Maintenance procedures

The most convenient method for short-term storage is by serial transfer from agar slants of appropriate media, such as oatmeal-yeast extract agar (Nonomura and Ohara, 1960), every 2 months. The tubes should be tightly closed with cotton plugs dipped in melted paraffin wax and stored at 4°C and room temperature. Longer term preservation of strains can be achieved as frozen stocks in 20% (v/v) aqueous glycerol at −80°C (mechanical freezer) to −172°C (liquid nitrogen vapor) or by using established lyophilization procedures. Strains can also be kept as mycelial suspensions in 20% aqueous glycerol at −20°C to −72°C.

Procedures for testing special characters

Iodinin production is detected after culturing on oatmeal agar supplemented with 0.1% (w/v) yeast extract for 1 month at 23–30°C. Utilization of carbon sources is determined by comparing growth on a given carbon source in basal C-1 or C-2 media with two controls, growth on the basal medium alone, and growth on the basal medium plus 0.5% (w/v) glucose. Most species of *Streptosporangium* do not grow on ISP Carbon Utilization Medium (Shirling and Gottlieb, 1966). Other useful physiological tests are hydrolysis of starch agar supplemented with yeast extract (0.5 g/l), nitrate reduction to nitrite in Bacto-nitrate broth (Difco) supplemented with yeast extract (2 g/l), and gelatin liquefaction (gelatin, 200 g; peptone, 5 g; yeast extract, 2 g; glucose, 2 g; distilled water, 1000 ml).

Differentiation of the genus *Streptosporangium* from other genera

The genus *Streptosporangium* is phylogenetically distinct from all other genera which contain organisms that produce spore vesicles. At present, morphology is the sole criterion used to distinguish between genera classified in the family *Streptosporangiaceae* (see Table 276 in the section on the family). *Streptosporangium* can be readily distinguished from the genera *Herbidospora*, *Microtetraspora*, *Microbispora*, and *Nonomuraea* by its ability to form spore vesicles and from the remaining genera by the shape of spore vesicles, the number of spores they contain, and whether or not the latter are motile. It can be difficult to distinguish between *Streptosporangium fragile*, which has a very thin vesicular membrane, and *Nonomuraea* strains which produce pseudovesicles covered by a slimy material. Similarly, the genus *Spirillospora* may be confused with species of the genus *Streptosporangium* (Vobis and Kothe, 1989). All of the strains belonging to these taxa have multispored, usually spherical, spore vesicles. However, these taxa can be distinguished on the basis of their polar lipid patterns and by the ability of spirillosporae to form motile spores.

Taxonomic comments

Couch (1955a) proposed the genus *Streptosporangium* for sporangiate actinomycetes that formed nonmotile sporangiosphores on an abundant aerial mycelium. Initially, only one species, *Streptosporangium roseum*, was recognized. Additional species were added to the genus primarily based on the capacity of strains to form spore vesicles, as exemplified by proposals for *Streptosporangium album* Nonomura and Ohara 1960, *Streptosporangium amethystogenes* Nonomura and Ohara 1960, *Streptosporangium longisporum* Schäfer 1969, *Streptosporangium nondiastaticum* Nonomura and Ohara 1969b, and *Streptosporangium viridialbum* Nonomura and Ohara 1969b. More exacting studies showed the genus to be heterogeneous, notably on the basis of spore and spore vesicular morphology detected by scanning electron microscopy (Nonomura, 1989a), electrophoretic mobility of ribosomal protein AT-L30 (Ochi and Miyadoh, 1992), partial amino acid sequencing of the N termini of AT-L30 proteins (Ochi et al., 1993), 5S and 16S rRNA gene sequences (Kemmerling et al., 1993; Kudo et al., 1993), and on the discontinuous distribution of chemotaxonomic markers (Stackebrandt et al., 1994).

Stackebrandt and his colleagues assigned members of the genus to two groups based on the discontinuous distribution of chemical markers. Most species, including *Streptosporangium roseum*, the type species, had a phospholipid type IV and predominant menaquinones of the MK-9(H_2) and MK-9(III, VIII-H_4), MK-9 and/or MK-9(H_6) type. The second group, which encompassed *Streptosporangium albidum* and the two subspecies of *Streptosporangium viridugriseum*, had a phospholipid type II and MK-9(II, III, H_4) as the predominant isoprenologues. These results were in excellent agreement with corresponding 16S rRNA gene sequencing data. Consequently, Stackebrandt and his co-workers proposed that *Streptosporangium albidum* Furumai et al. 1968, *Streptosporangium viridogriseum* subsp. *viridogriseum* Nonomura and Ohara 1969b, and *Streptosporangium viridogriseus* subsp. *kofuense* Okuda et al. (1966) be assigned to a new taxon, the genus *Kutzneria*, as *Kutzneria albida* comb. nov., *Kutzneria viridogrisea* comb. nov., and *Kutzneria kofuensis*, respectively, results supported by numerical phenetic data (Whitham et al., 1993).

The position of "*Streptosporangium cinnabarinum*" JCM 3291 and *Streptosporangium viridialbum* DSM 43801^T in the genus

TABLE 276. Characteristics differentiating between *Streptosporangium* species[a,b]

Characteristic	*S. roseum* DSM 43021^T	*S. album* DSM 43023^T	*S. amethystogenes* subsp. *amethystogenes* DSM 43179^T	*S. amethystogenes* subsp. *fukuiense* DSM 44779	*S. carneum* NRRL 18437^T	*S. fragile* DSM 43847^T	*S. longisporum* DSM 43180^T	*S. nondiastaticum* DSM 43848^T	*S. pseudovulgare* DSM 43181^T	*S. purpuratum* DSM 44668^T	*S. subroseum* DSM 44662^T	*S. violaceochromogenes* DSM 43849^T	*S. vulgare* DSM 43802^T	*S. yunnanense* DSM 44663^T
Morphology on oatmeal agar (color of substrate mycelium):	nr	nr	nr	nr	nr	nr	nr	nr	nr	nr	nr	nr	nr	
Brown-black	–	–	–	–	–	+	–	–	–	–	–	–	–	–
Red-orange	+	–	–	–	–	–	+	+	+	+	–	–	+	–
Yellowish brown to brown	+	+	+	+	+	–	–	+	–	–	+	+	+	+
Color of aerial spore mass:	nr	nr	nr	nr	nr	nr	nr	nr	nr	nr	nr	nr	nr	nr
Greenish gray	–	–	–	–	–	–	–	–	–	–	–	–	–	–
Pink	+	–	+	–	+	+	+	+	+	+	+	+	+	+
White	–	+	–	+	–	–	–	–	–	–	–	–	–	–
Soluble pigments:	nr	nr	nr	nr	nr	nr	nr	nr	nr	nr	nr	nr	nr	nr
Brown	–	–	–	–	–	+	–	–	–	–	–	+	–	+
Red-brown to purple brown	+	–	–	–	–	–	–	–	–	–	–	–	–	–
Violet	–	–	–	–	–	–	–	–	–	–	–	+	–	–
Yellow-brown	–	–	–	–	–	–	+	–	–	+	–	–	–	–
Spore vesicle size (μm):	nr	nr	nr	nr	nr	nr	nr	nr	nr	nr	nr	nr	nr	nr
1–5	–	–	–	–	–	–	–	–	–	+	+	–	–	+
6–10	+	+	+	+	+	+	+	–	+	–	–	+	+	–
11–20	(+)	–	–	–	+	+	+	+	–	–	–	–	–	–
21–30	–	–	–	–	(+)	–	–	+	–	–	–	–	–	–
31–50	–	–	–	–	(+)	–	–	–	–	–	–	–	–	–
Sporophore size (μm):	nr	nr	nr	nr	nr	nr	nr	nr	nr	nr	nr	nr	nr	nr
Short (10)	+	+	+	+	–	+	+	+	+	nd	+	+	+	nd
Long (50)	–	–	–	–	+	–	–	–	–	nd	–	–	–	nd
Spore shape:	nr	nr	nr	nr	nr	nr	nr	nr	nr	nr	nr	nr	nr	nr
Spherical to oval	+	+	+	nr	+	+	+	+	+	+	+	+	+	+
Rods	–	–	–	nr	+	–	–	–	–	–	–	–	–	–
Iodinin production	–	–	+	+	–	–	–	–	–	+	–	–	–	–
B vitamins required	+	+	+	+	–	–	–	+	+	–	–	–	+	–
Growth at:	nr	nr	nr	nr	nr	nr	nr	nr	nr	nr	nr	nr	nr	nr
42°C	–	–	–	–	–	+	–	+	+	+	+	–	–	+
50°C	–	–	–	–	–	–	–	–	+	+	–	–	–	–
Biochemical tests:	nr	nr	nr	nr	nr	nr	nr	nr	nr	nr	nr	nr	nr	nr
Nitrate reduction	+	–	+	–	–	+	(+)	+	+	–	+	+	–	+
Degradation tests:	nr	nr	nr	nr	nr	nr	nr	nr	nr	nr	nr	nr	nr	nr
Gelatin hydrolysis	+	+	–	nr	–	–	–	+	+	–	nd	(+)	d	+
Starch hydrolysis	+	–	+	+	–	+	+	–	+	+	–	+	+	+
Sole carbon source utilization (l, w/v):	nr	nr	nr	nr	nr	nr	nr	nr	nr	nr	nr	nr	nr	
Adonitol	+	+	+	–	–	–	+	+	+	nd	nd	nd	+	nd
L(+)-Arabinose	+	+	nd	+	–	+	+	+	+	–	+	–	+	–
D(+)-Galactose	–	+	–	+	+	+	–	+	–	–	+	+	+	–
Glycerol	+	–	nd	–	–	–	–	–	+	nd	nd	+	+	nd
meso-Inositol	–	–	+	–	–	–	–	–	–	–	–	–	+	nd
D(+)-Mannitol	+	+	nd	+	–	+	–	+	+	–	+	–	–	–
L(+)-Rhamnose	–	–	+	+	–	+	–	–	–	–	+	–	+	–
D(+)-Turanose	+	+	nd	nd	nd	+	–	–	+	nd	nd	nd	+	nd

[a]Symbols: +, positive; (+), weak positive; –, negative; d, doubtful; nd, not determined.

[b]Data taken from Nonomura (1989a), Mertz and Yao (1990), Goodfellow (1992), Kudo (2001), Whitham et al. (1993), and Zhang (2002, 2005).

Streptosporangium was questioned by several workers, especially on the basis of 16S rRNA gene sequence data (Stackebrandt et al., 1994; Tamura et al., 2000; Ward-Rainey et al., 1996). These organisms were subsequently found to share genotypic and phenotypic features in common with five soil isolates that were assigned to the new genus *Sphaersporangium* as *Sphaersporangium melleum* and *Sphaersporangium rubeum* (Ara and Kudo, 2007). They also found that "*Streptosporangium cinnabarinum*" and *Streptosporangium viridialbum* could be distinguished from one another and from the other two taxa based on DNA–DNA relatedness and phenotypic data. Consequently, they transferred "*Streptosporangium cinnabarinum*" Celmer et al. 1977 and *Streptosporangium viridialbum* (Nonomura and Ohara 1960) to the genus *Sphaerosporangium* as *Sphaerosporangium cinnabarinum* sp. nov. and *Sphaerosporangium viridialbum* comb. nov., respectively. The genus name *Sphaerosporangium* (*sic*) was corrected on validation [Rule 61 of the *Bacteriological Code* (1990 Revision)] to *Sphaerisporangium* (Lists Editor IJSEM, 2007).

Several other species have been shown to be misplaced in the genus *Streptosporangium*. The type strain of *Streptosporangium claviforme* Petrolini et al. 1992 and *Herbidospora cretacea* Kudo et al. 1993 belong to the same genomic species and form a distinct 16S rRNA gene clade in the *Streptosporangiaceae* gene tree (Tamura et al., 2000), as can be seen in Figure 361 (see section on the family *Streptosporangiaceae*). *Streptosporangium corrugatum* Williams and Sharples 1976 was found to have properties consistent with its transfer to the genus *Acrocarpospora* as *Acrocarpospora corrugata* (Williams and Sharples 1976) Tamura et al. 2000. The proposal to transfer *Streptosporangium indianense* Gupta 1965 to the genus *Streptomyces* as *Streptomyces indianensis* (Gupta 1965) Kudo and Seino 1987 is underpinned by a wealth of phenotypic data, notably by the formation of spiral spore chains and the absence of spore vesicles (Whitham et al., 1993). "*Streptosporangium album* subsp. *thermophilum*" Manachini et al. 1965 is a thermophile, but has been shown to be a member of the genus *Thermoactinomyces* (Goodfellow and Cross, 1984).

The changes outlined above leave the genus *Streptosporangium* as a well-defined taxon that encompasses 13 validly published species. DNA–DNA relatedness studies support the taxonomic integrity of several *Streptosporangium* species (Zhang et al., 2002, 2005) though they do suggest that the two *Streptosporangium amethystogenes* subspecies represent distinct genomic species (Iinuma et al., 1996b), a result underpinned by phenotypic data (Table 276). Further comparative studies are also needed to establish relationships between *Streptosporangium album*, *Streptosporangium roseum*, and *Streptosporangium vulgare* as the type strains of these taxa have identical 16S rRNA gene sequences and share morphological properties. These organisms also have similar polar lipid patterns but can be separated using quantitative fatty acid and menaquinone data (Mertz and Yao, 1990; Stackebrandt et al., 1994). However, there is evidence that the genus is underspeciated (Wang et al., 1999; Whitham et al., 1993). Indeed, Whitham and his colleagues found that isolates from seven environmental samples presumptively assigned to the genus formed 23 centers of variation based on the numerical taxonomic data.

Strains described as "*Streptosporangium brasiliense*" (Coronelli and Thiemann, 1969), "*Streptosporangium koreanum*" (Celmer et al., 1977), and "*Streptosporangium rubrum*" (Potekhina, 1965), share very high 16S rRNA gene similarities with one another and with the type strains of *Streptosporangium album*, *Streptosporangium roseum*, and *Streptosporangium vulgare* (Tamura et al., 2000). Fatty acid date indicate that "*Streptosporangium brasiliense*" and "*Streptosporangium koreanum*" strains are closely related to one another and to the type strain of *Streptosporangium roseum*, whereas a strain labeled "*Streptosporangium sibiricum*" (Potekhina, 1965) formed a distinct single membered group when fatty acid data were expressed as Euclidian distances (Mertz and Yao, 1990). Further comparative studies are needed to determine the relationships between these invalidly named taxa and representatives of validly named *Streptosporangium* species.

Differentiation of species of the genus *Streptosporangium*

Streptosporangium species can be distinguished by using a combination of phenotypic features (Table 276), although a common set of tests has not been used in the description of all species. Most of the type strains can be distinguished on the basis of their 16S rRNA gene sequences, as seen in Figure 361. There is also evidence that these organisms can be differentiated on the basis of their sugar and polar lipid patterns and by qualitative and quantitative differences in fatty acid and menaquinone composition (Kroppenstedt, 1985; Stackebrandt et al., 1994; Whitham et al., 1993; Zhang et al., 2002), though results from individual studies on the same organism do vary, possibly reflecting differences in growth conditions and experimental protocols.

Improved phenotypic tests are needed for the identification of streptosporangiae to the species level. Whitham (1988) generated a probability matrix based on 26 phenotypic properties for the computer-assisted identification of streptosporangiae to established and putatively novel *Streptosporangium* species. Kim (1993), in a continuation of this work, was able to identify known and putatively novel streptosporangiae, notably ones isolated from a range of soil samples.

List of species of the genus *Streptosporangium*

1. **Streptosporangium roseum** Couch 1955a, 151[AL] ("*Angiococcus moliroseus*" Petersen 1959, 169)

ro'se.um. L. neut. adj. *roseum* rose-colored.

Aerial mycelium is pink and the substrate mycelium yellowish brown to orange on oatmeal-yeast extract agar; soluble pigment is reddish to purplish brown on this medium. Spore vesicles are usually 8–10 μm in diameter but larger ones, up to 20 μm, are found in some strains. Spores spherical, short or bent rods, 1.8–2 μm in diameter. Hydrolyzes esculin, but not urea. Nitrate is reduced.

Degrades arbutin, elastin, RNA, and L-tyrosine, but not guanine, hypoxanthine, keratin, testosterone, Tween 20, or xanthine.

Dextrin, inulin, mannose, sucrose, trehalose, and xylose are used as sole carbon sources, but dextran, fructose, lactose, melezitose, melibiose, raffinose, sorbitol, sorbose, stachyose, starch, and xylitol (all at 1%, w/v) are not. Does not use acetate, butyrate, citrate, hippurate, malonate, or pyruvate as sole carbon sources (all at 0.1%, w/v).

Grows in the presence of crystal violet (0.001%, w/v), phenol (0.01%, w/v), phenyl ethanol (0.2%, v/v), potassium tellurite (0.01%, w/v), sodium chloride (2%, w/v), and thallous acetate (0.001%, w/v), but is sensitive to crystal violet (0.01%, w/v), phenol (0.01%, w/v), phenyl ethanol (0.3%, v/v), sodium azide (0.001%, w/v), and sodium chloride (4%, w/v).

Resistant (μg/ml) to amoxycillin (500), azlocillin (500), bactracin (50), cefoxitin (250), cephaloridine (250), cephadrine (500), chloramphenol (25), clavulanic acid (50), flucloxicillin (50), fusidic acid (0.5), keflin (250), kufzol (500), lincomycin (5), mecillinam (500), methicillin (500), oleandomycin phosphate (250), penicillin G (50), pseudomonic acid (500), sulfomethoxazole (25), ticarcillin (50), and tyrothricin (5), but is sensitive to bacitracin (50), clavulanic acid (500), flucloxicillin (250), fusidic acid (0.5), lincomyciin (25), penicillin G (250), rifampin (0.5), tetracycline (5), ticarcillin (500), and vancomycin hydrochloride (5).

Additional phenotypic features are shown in Table 276.

Whole-organism hydrolysates contain galactose and madurose. The fatty profile includes $C_{14:0}$ iso (8.9%), $C_{15:0}$ (2.9%), $C_{16:0}$ iso (43.3%), $C_{16:0}$ (3.7%), $C_{16:0}$ 10-methyl (2.6%), $C_{17:0}$ anteiso (3.1%), $C_{17:0}$ (3.9%), $C_{17:0}$ 10-methyl (18.7%), and $C_{18:0}$ (3.0%). The cellular polar lipid pattern includes diphosphatidylglycerol, hydroxyphosphatidylethanolamine, phosphatidylethanolamine, phosphatidylinositol, and ninhydrin-positive and sugar-positive phospholipids.

Source: a vegetable garden soil.

DNA G+C content (mol%): 70 (HPLC).

Type strain: ATCC 12428, CBS 313.56, DSM 43021, NBRC 3776, JCM 3005, NCIB 10171, NRRL B-2505, VKM Ac-807.

Sequence accession no. (16S rRNA gene): U48996, X70425, X89947.

2. **Streptosporangium album** Nonomura and Ohara 1960, 407[AL]

al'bum. L. neut. adj. *album* white.

Aerial mycelium is white and the substrate mycelium pale yellow on oatmeal-yeast extract agar, but does not form a diffusible pigment on this medium. Melanin pigments are not produced. Spore vesicles are 6–8 μm in diameter and sporangiospores are 1.0–1.3 × 1.5–1.9 μm. Esculin is hydrolyzed, but hippurate and urea are not. Arbutin, L-tyrosine, and Tween 80 are degraded, but guanine, hypoxanthine, keratin, pectin, testosterone, and xanthine are not. Adonitol, dextrin, fructose, inulin, maltose, mannose, raffinose, stachyose, sucrose, turanose, and xylose are used as sole carbon sources, but lactose, melezitose, melibiose, sorbitol, starch, or xylitol (all at 1%, w/v) are not. Acetate and butyrate are used as sole carbon sources, but citrate, hippurate, malonate, propionate, or pyruvate (all at 0.1%, w/v) are not.

Grows in the presence of potassium tellurite (0.005%, w/v), sodium chloride (1%, w/v), and thallous acetate (0.005%, w/v), but is sensitive to crystal violet (0.0005%, w/v), phenol (0.005%, w/v), phenyl ethanol (0.1%, v/v), potassium tellurite (0.01%, w/v), sodium azide (0.005%, v/v), sodium chloride (2%, w/v), and thallous acetate (0.001%, w/v).

Resistant (μg/ml) to amoxycillin (50), cefoxitin (25), cephaloridine (5), chloramphenol (5), clavulanic acid (250), fusidic acid (5), kefzol (50), lincomycin (25), mecillinam (50), neomycin sulfate (0.5), oleandomycin phosphate (5), penicillin G (5), sulfomethoxazole (25), tetracycline (0.5), ticarcillin (50), and tyrothricin (5), but is sensitive to amoxycillin (500), azlocillin (25), bacitracin (25), cefoxitin (50), cephaloridine (25), cephradine (25), chloramphenicol (25), clavulanic acid (500), flucloxicillin (50), fusidic acid (25) gentamicin sulfate (5), keflin (25), kefzol (250), lincomycin (50), mecillinam (250), methicillin (250), neomycin sulfate (5), oleandomycin phosphate (25), penicillin G (25), pseudomonic acid (500), rifampin (0.5), streptomycin sulfate (5), tetracycline (5), and ticarcillin (250).

Additional phenotypic features are shown in Table 276.

Whole-organism hydrolysates contain galactose and a trace of madurose. The fatty profile includes $C_{12:0}$ (8.1%), $C_{14:0}$ iso (4.3%), $C_{15:0}$ (7.3%), $C_{16:0}$ iso (43.3%), $C_{16:0}$ iso (8.0%), $C_{16:1}$ (3.9%), $C_{16:0}$ (6.8%), $C_{16:0}$ 10-methyl (2.6%), $C_{17:0}$ anteiso (21.6%), $C_{17:0}$ (5.3%), $C_{17:0}$ 10-methyl (25.7%), $C_{18:1}$ *cis* (2.8%), and $C_{18:0}$ 10-methyl (3.7%). The cellular polar lipid pattern contains diphosphatidylglycerol, phosphatidylethanolamine, phosphatidylglycerol, phosphatidylinositol mannosides, phosphatidylmethylethanolamine, and several uncharacterized components. The predominant menaquinones are MK-9(H_2) (55.7%) and MK-9(H_4) (42.9%); also contains minor amounts of MK-9(H_0) (1.4%).

Source: a soil sample collected in Japan.

DNA G+C content (mol%): not determined.

Type strain: CBS 429.61, DSM 43023, NRBC 13900, JCM 3025, KCC A-0025, NRRL B-2635, VKM Ac-636.

Sequence accession no. (16S rRNA gene): D85469, X89934.

3. **Streptosporangium amethystogenes** Nonomura and Ohara 1960, 407[AL]

am.e.thys.to'ge.nes. L. adj. *amethystinus* amethyst colored; N.L. suff. *-genes* (from Gr. v. *gennaô* to produce) producing; N.L. part. adj. *amethystogenes* producing violet-colored (crystals).

Aerial mycelium is pink and the substrate mycelium pale brownish gray. Does not produce melanin pigments. Spore vesicles are 6–8 μm in diameter and sporangiospores are 1.0–1.3 × 1.5–1.9 μm.

Esculin is hydrolyzed, but hippurate is not. Degrades arbutin, DNA, elastin, RNA, and L-tyrosine, but not guanine, hypoxanthine, keratin, testosterone, or xanthine. Mannose, trehalose, and xylose are used as sole carbon sources, but dextran, lactose, melezitose, melibiose, raffinose, sorbitol, sorbose, stachyose, starch, and xylitol (all at 1%, w/v) are not. Does not use acetate, butyrate, citrate, hippurate, malonate, propionate, or pyruvate as sole carbon sources (all at 0.1%, w/v).

Grows in the presence of crystal violet (0.0005%, w/v), phenol (0.01%, w/v), phenyl ethanol (0.01%, v/v), potassium tellurite (0.001%, w/v), sodium chloride (1.0%, w/v), and thallous acetate (0.0005%, w/v), but is sensitive to crystal violet (0.01%, w/v), phenol (0.01%, w/v), potassium tellurite (0.01%, v/v), sodium azide (0. 01%, w/v), and sodium chloride (4%, w/v).

Resistant (μg/ml) to gentamicin sulfate (5), sulfomethoxazole (3), tetracycline (0.5), and vancomycin hydrochloride (0.5), but is sensitive to amoxycillin (25), azlocillin (25), bacitracin (25), cefoxitin (25), cephaloridine (50), cephradine (25), chloramphenicol (5), clavulanic acid (25), flucloxicillin (50), fusidic acid (0.5), gentamicin sulfate (25), keflin (25), kefzol (25), lincomycin (25), mecillinam (50), methicillin (250), neomycin sulfate (5), oleandomycin phosphate (25), penicillin G (25), pseudomonic acid (500), rifampin (5), streptomycin sulfate (5), tetracycline (5), ticarcillin (50), tyrothrecin (5), and vancomycin hydrochloride (5).

Additional phenotypic features are shown in Table 276.

Whole-organism hydrolysates contain madurose. The fatty acid profile includes $C_{16:0}$ iso (7.9%), $C_{16:0}$ (13.9%), $C_{17:0}$ anteiso (10.7%), $C_{17:0}$ (4.2%), $C_{18:1}$ *cis* (13.2%), $C_{18:1}$ *trans* (45.1%), and $C_{18:0}$ (3.4%). The cellular polar lipid pattern contains diphosphatidylglycerol, phosphatidylethanolamine, phosphatidylglycerol, phosphatidylinositol, and phosphatidylinositol mannosides. The predominant menaquinones are MK-9(H_2) (49.4%) and MK-9(III, VIII-H_4) (33.5%); also contains MK-9(H_0) (10.1%).

Source: a soil sample collected in Japan.

DNA G+C content (mol%): not determined.

Type strain: ATCC 33327, CBS 430.61, DSM 43179, NRBC 13986, JCM 3026, KCC A-0026, NRRL B-2639, RIA 767, VKM Ac-637.

Sequence accession no. (16S rRNA gene): X89935.

Additional comments: the type strain of *Streptosporangium amethystogenes* KCC A-0026 was incorrectly cited as KCC A-2006 (*sic*) on the Approved Lists of Bacterial Names 1980 (Hill et al., 1984). Subsequently, the species was divided into two subspecies (see below).

3a. **Streptosporangium amethystogenes subsp. amethystogenes** Nonomura and Ohara 1960, 407[AL]

This taxon was automatically created by the valid publication of *Streptosporangium amethystogenes* subspecies *fukuiense* Iinuma et al. (1996b) [Rule 40d of the *Bacteriological Code* (1990 Revision); formerly Rule 46]. The description of *Streptosporangium amethystogenes* subsp. *amethystogenes* is given above.

Source: a soil sample collected in Japan.

DNA G+C content (mol%): not determined.

Type strain: ATCC 33327, CBS 430.61, DSM 43179, NRBC 13986, JCM 3026, KCC A-0026, NRRL B-2639, RIA 767, VKM Ac-637.

Sequence accession no. (16S rRNA gene): X89935.

3b. **Streptosporangium amethystogenes subsp. fukuiense** Iinuma, Yokota and Kanamaru 1996a, 1189[VP] (Effective publication: Iinuma, Yokota and Kanamaru 1996b, 41.)

fu.ku.i.en'se. N.L. neut. adj. *fukuiense* of or belonging to Fukui Prefecture, Japan, source of the soil from which the organism was isolated.

Substrate mycelium is brown to dark brown on oatmeal agar, oatmeal-yeast extract agar, and yeast extract-malt extract agar; sparse white aerial hyphae are formed on oatmeal agar but not on the other two media. Spherical spore vesicles (6–8 μm in diameter) are formed. Sporangiospores are ellipsoidal (0.5–0.8 × 0.8–1.2 μm) and have smooth walls. Temperature growth range is 11–36°C. Iodinin crystals are produced on glycerol-asparagine agar, oatmeal agar, tyrosine agar, and yeast extract-malt extract agar.

Milk is neither coagulated or peptonized. Fructose, glucose, maltose, mannose, trehalose, starch, and xylose are used as sole carbon sources, but cellulose, lactose, raffinose, and sorbitol are not. Does not grow in the presence of lysozyme (0.005%, w/v) or sodium chloride (3%, w/v).

Resistant (μg/ml) to cefoxitin (100), but is sensitive to chloramphenicol (50), gentamicin sulfate (100), kanamycin (100), novobiocin (20), rifampin (50), streptomycin (100), and vancomycin hydrochloride (50).

Produces a hematopoietic cytokine inducer.

Additional phenotypic features are shown in Table 276.

Whole-organism hydrolysates contain arabinose, glucose, madurose, mannose, and xylose. Muramic acid moieties are *N*-acetylated. The fatty profile includes $C_{16:0}$ iso (4.3%), $C_{16:0}$ (18.8%), $C_{17:0}$ (12.3%), $C_{16:1}$ (13.9%), $C_{17:1}$ (11.3%), $C_{18:1}$ (12.9%), $C_{18:0}$ iso (4.3%), $C_{17:0}$ 10-methyl (10.5%), and $C_{18:0}$ 10-methyl (6.3%). The cellular polar lipid pattern contains phosphatidylethanolamine and unknown glucosamine-containing compounds. The predominant menaquinones are MK-9(H_2) (32.5%) and MK-9(H_4) (59.2%); also contains minor amounts of MK-9(H_0) and MK-9(H_2).

Source: a soil sample collected from Fukui Prefecture, Japan in 1990.

DNA G+C content (mol%): not determined.

Type strain: strain AL-23456, CIP 105397, NBRC 15365, DSM 44779, JCM 10083, NBRC 15365.

Sequence accession no. (16S rRNA gene): AB537172.

Additional comments: the 16S rRNA gene sequence of the type strain of *Streptosporangium amethystogenes* subsp. *fukuiense* has still to be determined. The DNA–DNA relatedness values between this strain and the type strains of *Streptosporangium amethystogenes* subsp. *amethystogenes*, *Streptosporangium roseum*, and *Streptosporangium vulgare* are 55, 12, and 23, respectively.

4. **Streptosporangium carneum** Mertz and Yao 1990, 252[VP]

car´ne.um. L. neut. adj. *carneum* of flesh, intended to mean flesh colored, referring to the color of the aerial spore mass.

Produces an orange substrate mycelium and a bright yellow pink aerial mycelium on Emerson agar, glycerol-asparagine agar, inorganic salts-starch agar, tyrosine agar, yeast-glucose agar, and yeast extract-malt extract agar. A reddish-brown diffusible pigment is formed on tyrosine agar and yeast-glucose agar, and a light brown diffusible pigment on Emerson agar. Aerial hyphae bear abundant spore vesicles on chitin agar, Czapek agar, inorganic salts-starch agar, oatmeal agar, tap water agar, and yeast extract-malt extract agar. Spore vesicles range from 3–9 µm in diameter. Sporangiospores are spherical and a mean of 1.3 µm in diameter.

Catalase, hydrogen sulfide, and phosphatase are produced, but does not hydrolyze allantoin, esculin, hippurate, or urea. Nitrate is not reduced. Milk is not peptonized.

Casein, testosterone, and L-tyrosine are degraded, but adenine, calcium malate, elastin, guanine, hypoxanthine, and xanthine are not.

Acid is produced from cellobiose, fructose, galactose, glucose, lactose, mannose, and trehalose, but not from adonitol, D-arabinose, cellulose, dextrin, dulcitol, ethanol, erythritol, glycerol, glycogen, inositol, inulin, maltose, mannitol, melezitose, melibiose, α-methyl-D-glucoside, raffinose, L-rhamnose, ribose, salicin, sorbitol, L-sorbose, sucrose, xylitol, or xylose.

Acetate and pyruvate are used as sole carbon sources, but benzoate, butyrate, citrate, formate, lactate, malate, mucate, oxalate, propionate, succinate, and tartrate are not.

Growth occurs in the presence of 2% (w/v) sodium chloride. Resistant to antibiotic discs containing bacitracin, cephalothin, gentamicin, lincomycin, penicillin, streptomycin, tetracycline, and tobramycin, but is susceptible to neomycin, oleandomycin, rifampin, vancomycin, and lysozyme.

Additional phenotypic features are shown in Table 276.

Whole-organism hydrolysates contain arabinose, madurose, and mannose. The major fatty acids are $C_{14:0}$ (6.9%), $C_{16:0}$ (25.2%), $C_{17:1}$ iso (19.4%), and $C_{18:1}$ iso (6.9%)-like components. The cellular polar lipid pattern contains diphosphatidylglycerol, phosphatidylethanolamine, phosphatidylglycerol, phosphatidylinositol, and unknown glucosamine-containing components. The predominant menaquinone is MK-9(H_4); also contains a minor amount of MK-9(H_2).

Source: soil collected next to the River Tana in Nairobi, Kenya.

DNA G+C content (mol%): not determined.

Type strain: strain A84575, DSM 44125, NBRC 15562, JCM 9926, NRRL 18437, VKM Ac-2007.

Sequence accession no. (16S rRNA gene): X89938.

5. **Streptosporangium fragile** Shearer, Colman and Nash 1983, 364[VP]

fra'gi.le. L. neut. adj. *fragile* fragile, easily broken (referring to vesicular membrane).

Aerial hyphae (0.5–1.0 µm in diameter) bear spore vesicles which are usually 6–12 µm in diameter. Substrate mycelium is light yellowish pink. Aerial mycelium is white to light pink and the substrate mycelium black on oatmeal agar. Soluble light brown pigment is produced. Spore vesicular membrane is so fragile that it is not detected by light microscopy. The temperature range of growth is 15–45°C.

Produces catalase, hydrogen sulfide, reduces nitrate, and peptonizes milk, but does not hydrolyze allantoin, esculin, hippurate, or urea. Arbutin, casein, and L-tyrosine are degraded, but adenine, guanine, hypoxanthine, keratin, pectin, testosterone, and xanthine are not.

Acid is produced from L-arabinose, cellobiose, dextrin, glucose, *meso*-erythritol, fructose, galactose, glycogen, lactose, maltose, mannitol, mannose, α-methyl-D-glucoside, rhamnose, ribose, salicin, starch, trehalose, and xylose, but not from adonitol, dulcitol, inulin, α-methyl-D-mannoside, melibiose, or raffinose. Citrate, lactate, malate, succinate, and pyruvate are utilized, but benzoate and tartrate are not.

Fructose, inulin, maltose, sorbitol, stachyose, starch, sucrose, and xylose are used as sole carbon sources, but dextrin, lactose, melezitose, melibiose, raffinose, and xylitol (all at 1%, w/v) are not. Acetate, butyrate, citrate, hippurate, malonate, and propionate are used as sole carbon sources (all at 0.1%, w/v).

Grows in the presence of crystal violet (0.0005, w/v), phenol (0.01%, w/v), potassium tellurite (0.005%, w/v), and sodium chloride (1%, w/v), but is inhibited by crystal violet (0.001%, w/v), phenol (0.1%, w/v), phenyl ethanol (0.1, v/v), potassium tellurite (0.01%, w/v), sodium azide (0.001%, w/v), sodium chloride (2%, w/v), and thallous acetate (0.0001%, w/v).

Resistant (µg/ml) to chloramphenicol (5), fusidic acid (0.5), gentamicin sulfate (25), lincomycin (5), neomycin sulfate (5), oleandomycin phosphate (5), rifampin (0.5), sulfomethoxazole (25) and tyrothrecin (5), but is sensitive to amoxycillin (25), azlocillin (25), bacitracin (25), cefoxitin (25), cephaloridine (5), cephradine (25), chloramphenicol (25), clavulanic acid (25), flucloxicillin (50), fusidic acid (5), gentamicin sulfate (50), keflin (25), kefzol (25), lincomycin (25), mecillinam (50), methicillin (250), neomycin sulfate (25), oleandomycin phosphate (25), penicillin G (5), pseudomonic acid (500), rifampin (5), streptomycin sulfate (5), ticacillin (50), and vancomycin hydrochloride (0.5).

Additional phenotypic features are shown in Table 276.

Whole-organism hydrolysates contain madurose. The fatty profile includes $C_{16:0}$ (17.3%), $C_{16:0}$ 10-methyl (4.7%), $C_{18:2}$ (26.0%), $C_{18:1}$ *cis* (35.5%), and $C_{18:0}$ (3.5%). The polar lipid pattern contains diphosphatidylglycerol, phosphatidylglycerol, phosphatidylethanolamine, phosphatidylinositol, phosphatidylinositol mannosides, and many uncharacterized components. The predominant menaquinones are MK-9(H_2) (15.0%), MK-9(III, VIII-H_4) (60.0%), and MK-9(H_6) (25.0%).

Source: soil taken from a cultivated field in the Northern Province of Sri Lanka.

DNA G+C content (mol%): not determined.

Type strain: SK&F-BC 2496, ATCC 31519, DSM 43847, NBRC 14311, JCM 6242, NRRL B-16437, VKM Ac-1296.

Sequence accession no. (16S rRNA gene): U48992, X89942.

6. **Streptosporangium longisporum** Schäfer 1969, 368[AL]

lon.gi.spo'rum. L. adj. *longus* long; Gr. n. *spora* a seed; N.L. neut. adj. *longisporum* long spored.

Aerial mycelium is sparse and the substrate mycelium is red on oatmeal agar. Pink aerial mycelium is formed on starch agar. Spherical spore vesicles are usually 3–7 µm in diameter, but may be up to 18 µm in diameter. The vesicular wall is relatively thick. Sporangiospores are rod-shaped (0.6–0.9 × 1.5–3.5 µm).

Esculin is hydrolyzed, but hippurate is not. Degrades arbutin, DNA, elastin, RNA, and L-tyrosine, but not guanine, hypoxanthine, keratin, testosterone, or xanthine. Mannose, trehalose, and xylose are used as sole carbon sources, but dextran, lactose, melezitose, melibiose, raffinose, sorbitol, sorbose, stachyose, starch, and xylitol (all at 1%, w/v) are not. Does not use acetate, butyrate, citrate, hippurate, malonate, propionate, or pyruvate as sole carbon sources (all at 0.1%, w/v).

Grows in the presence of crystal violet (0.0005%, w/v), phenol (0.01%, w/v), phenyl ethanol (0.01%, v/v), potassium tellurite (0.001%, w/v), sodium chloride (1.0%, w/v), and thallous acetate (0.0005%, w/v), but is sensitive to crystal violet (0.01%, w/v), phenol (0.01%, w/v), potassium tellurite (0.01%, w/v), sodium azide (0. 01%, w/v), and sodium chloride (4%, w/v).

Resistant (µg/ml) to gentamicin sulfate (5), sulfomethoxazole (3), tetracycline (0.5) and vancomycin hydrochloride (0.5), but is sensitive to amoxycillin (25), azlocillin (25), bacitracin (25), cefoxillin (25), cephaloridine (50), cephradine (25), chloramphenicol (5), clavulanic acid (25), flucloxicillin (50), fusidic acid (0.5), gentamicin sulfate (25), keflin (25), kefzol (25), lincomycin (25), mecillinam (50), methicillin (250), neomycin sulfate (5), oleandomycin phosphate (25), penicillin G (25), pseudomonic acid (500), rifampin (5), streptomycin sulfate (5), tetracycline (5), ticarcillin (50), tyrothrecin (5), and vancomycin hydrochloride (5).

Additional phenotypic features are shown in Table 276.

Whole-organism hydrolysate contains madurose. The fatty acid profile includes $C_{16:0}$ (11.5%), $C_{18:1}$ *cis* (16.0%), C_{110} *trans* (64.0%), and $C_{18:0}$ (6.0%). The cellular polar lipid pattern contains diphosphatidylglycerol, hydroxyphosphatidylethanolamine, phosphatidylinositol, and ninhydrin-positive and sugar positive phospholipids. The predominant menaquinones are MK-9(H_2) (35.1%) and MK-9(III, VIII-H_4) (45%); also contains MK-9(H_0) (12.7%) and MK-9(H_6) (7.2%).

Source: a soil sample collected in Turkey.

DNA G+C content (mol%): not determined.

Type strain: ATCC 25212, CBS 184.69, DSM 43180, NBRC 13141, JCM 3106, KCC A-0106, NRRL B-16783, VKM Ac-696.

Sequence accession no. (16S rRNA gene): U48993, X89944.

7. **Streptosporangium nondiastaticum** Nonomura and Ohara 1969b, 708[AL]

non.di.as.ta'ti.cum. L. pref. *non-* not; N.L. adj. *diastaticus, -a, -um* diastatic; N.L. neut. adj. *nondiastaticum* not starch digesting.

Aerial mycelium is pink and the substrate mycelium yellowish brown to orange on oatmeal-yeast extract agar; also produces a pale yellow-brown soluble pigment on this medium. Does not produce melanin pigments. Good growth at 42°C. Spore vesicles are 10–15 µm in diameter and sporangiospores 1.3 × 1.5 µm.

Esculin is hydrolyzed, but hippurate and urea are not. Arbutin, Tween 80, and L-tyrosine are degraded, but guanine, hypoxanthine, keratin, pectin, testosterone, and xanthine are not.

Adonitol, dextrin, fructose, inulin, maltose, mannose, surcrose, turanose, and xylose are used as sole carbon sources, but dextran, lactose, melezitose, melibiose, raffinose, sorbitol, stachyose, starch, and xylitol (all at 1%, w/v) are not. Acetate and pyruvate are used as sole carbon sources, but butyrate, citrate, hippurate, malonate, and propionate (all at 0.1%, w/v) are not.

Grows in the presence of crystal violet (0.0005%, w/v), phenol (0.01%, w/v), potassium tellurite (0.01%, w/v), and sodium chloride (2.0%, w/v), but is sensitive to crystal violet (0.001%, w/v), phenol (0. 1%, w/v), phenyl ethanol (0.1%, v/v), sodium azide (0.001%, w/v), sodium chloride (4%, w/v), and thallous acetate (0.0001%, w/v).

Resistant (µg/ml) to amoxycillin (50), azlocillin (50), bactracin (50), cefoxitin (50), cephaloridine (5), clavulanic acid (500), chloramphenicol (5), flucloxicillin (50), fusidic acid (0.5), gentamicin sulfate (5), ketlin (50), kefzol (250), lincomycin (50), mecillinam (500), methicillin (500), neomycin sulfate (5), oleandomycin phosphate (50), penicillin G (50), pseudomonic acid (500), sulfomethoxazole (25), tetracycline (5), ticarcillin (50), and vancomycin hydrochloride (0.5), but is sensitive to amoxycillin (250), azlocillin (250), cefoxitin (250), cephaloridine (25), cephradine (25), chloramphenicol (25), flucloxicillin (250), fusidic acid (5), gentamicin sulfate (25), kefzol (500), oleandomycin phosphate (250), rifampin (0.5), streptomycin sulfate (5), tetracycline (25), ticarcillin (250), and vancomycin hydrochloride (5).

Additional phenotypic features are shown in Table 276.

Whole-organism hydrolysates do not contain madurose. The fatty acid profile includes $C_{14:0}$ iso (2.7%), $C_{15:0}$ iso (1.9%), $C_{15:0}$ (11.2%), $C_{16:0}$ iso (13.2%), $C_{16:1}$ (2.0%), $C_{16:0}$ (5.2%), $C_{16:0}$ 10-methyl (3.6%), $C_{17:0}$ anteiso (19.8%), $C_{17:0}$ (4.4%), $C_{17:0}$ 10-methyl (22.1%), $C_{18:2}$ (5.5%), $C_{18:1}$ *trans* (1.7%), and $C_{18:0}$ 10-methyl (1.5%). The cellular polar lipid pattern contains diphosphatidylglycerol, phosphatidylethanolamine, phosphatidylglycerol, phosphatidylinositol, phosphatidylinositol mannosides, phosphatidylmethylethanolamine, and several uncharacterized components. The predominant menaquinones are MK-9(H_2) (55.0%) and MK-9(III, VIII-H_4) (40%); also contains MK-9(H_0) (2.0%) and MK-9(H_6) (3.0%).

Source: a soil sample collected in Japan.

DNA G+C content (mol%): not determined.

Type strain: ATCC 27101, CBS 800.70, DSM 43848, NBRC 13990, JCM 3114, VKM Ac-1299.

Sequence accession no. (16S rRNA gene): U48994, X70426, X89945.

8. **Streposporangium pseudovulgare** Nonomura and Ohara 1969b, 708[AL]

pseu.do.vul.ga're. Gr. adj. *pseudês* false; L. neut. adj. *vulgare* a specific epithet; N.L. neut. adj. *pseudovulgare* similar in appearance to strain of *Streposporangium vulgare*.

Aerial mycelium is pink and the substrate mycelium yellowish brown to orange on oatmeal-yeast extract agar; a yellowish brown soluble pigment is also produced on this medium. Spore vesicles are 7–10 µm in diameter and sporangiospores 1.2 × 1.5 µm. Good growth at 42°C; also grows at 55°C.

Esculin is hydrolyzed, but hippurate and urea are not. Nitrate is reduced. Degrades arbutin, keratin, pectin, Tween 80, and L-tyrosine, but not guanine, hypoxanthine, testosterone, or xanthine.

Glucose, fructose, inulin, maltose, mannose, turanose, and xylose are used as sole carbon sources, but dextrin, lactose, melezitose, melibiose, raffinose, sorbitol, stachyose, starch, sucrose, and xylitol (all at 1%, w/v) are not. Similarly, pyruvate is metabolized, but acetate, butyrate, citrate, hippurate, malonate, and propionate (all at 0.1%, w/v) are not.

Grows in the presence of crystal violet (0.001, w/v), phenol (0.01%, w/v), phenyl ethanol (0.2%, v/v), potassium tellurite (0.01%, w/v), sodium azide (0.005%, w/v), sodium chloride (2%, w/v), and thallous acetate (0.001%, w/v), but is sensitive to crystal violet (0.01%, w/v), phenol (0.1%, w/v), phenyl ethanol (0.3, v/v), sodium azide (0.02%, w/v), and sodium chloride (4%, w/v).

Resistant (µg/ml) to amoxycillin (250), azlocillin (250), bacitracin (25), cefoxitin (250), cephaloridine (25), cephradine (25), chloramphenicol (5), clavulanic acid (250), flucloxicillin (50), fusidic acid (0.5), gentamicin sulfate (50), keflin (50), kefzol (500), lincomycin (5), mecillinam (500), methicillin (500), neomycin sulfate (5), oleandomycin phosphate (250), penicillin G (25), pseudomonic acid (500), rifampin (25), streptomycin sulfate (5), sulfomethoxazole (25), tetracycline (5), ticarcilliin (500), and vancomycin hydrochloride (0.5), but is sensitive to azlocillin (500), bacitracin (50), cephaloridine (50), cephradine (50), chloramphenicol (25), clavulanic acid (500), flucloxicillin (250), fusidic acid (5), lincomycin (25), penicillin G (50), streptomycin sulfate (25), tetracycline (25), and vancomycin hydrochloride (5).

Additional phenotypic features are shown in Table 276.

Whole-organism hydrolysates contain madurose. The fatty acid profile includes $C_{14:0}$ (1.3%), $C_{15:0}$ (3.5%), $C_{16:0}$ iso (5.6%), $C_{16:1}$ (3.9%), $C_{16:0}$ (25.2%), $C_{17:0}$ anteiso (9.1%), $C_{17:0}$ (4.9%), $C_{17:0}$ 10-methyl (3.2%), $C_{18:1}$ *cis* (11.9%), $C_{18:1}$ *trans* (23.5%), and $C_{18:0}$ (4.7%) The cellular polar lipid pattern contains diphosphatidylglycerol, phosphatidylethanolamine, phosphatidylglycerol, phosphatidylinositol, phosphatidylinositol mannosides, phosphatidylmethylethanolamine, and several uncharacterized components. The predominant menaquinones are MK-9(H_2) (60.4%) and MK-9(III, VIII-H_4) (28.0%); also contains MK-9(H_0) (11.6%).

Source: a soil sample collected in Japan.

DNA G+C content (mol%): not determined.

Type strain: ATCC 27100, CCRC 16308, CCTCC AA 97010, DSM 43181, NBRC 13991, JCM 3115.

Sequence accession no. (16S rRNA gene): U48995, X70428, X89946.

9. **Streptosporangium purpuratum** Zhang, Jiang and Chen 2005, 723[VP]

pur.pu.ra´tum. L. neut. adj. *purpuratum* clad in purple-violet, referring to the colony color.

Substrate mycelium is deep red to deep purplish-red on Bennett's agar, glycerol-asparagine agar, inorganic salts-starch agar, oatmeal agar, oatmeal-yeast extract agar, and yeast extract-malt extract agar; diffusible pigments are not formed on any of these media. Pinkish-white aerial mycelium is formed on most of these media. The exceptions are Bennett's agar, which does not support aerial hyphae, and yeast-extract-malt agar, where the aerial mycelium is sparse. Spherical spore vesicles, usually 2–5 µm in diameter, are formed singly or in clusters on the aerial mycelia on HV and oatmeal agars. Sporangiospores are spherical to oblate. The optimal temperature for growth is 28°C; also grows at 50°C. The optimal pH for growth is 7.2.

Hydrolyzes hippurate. Nitrate is reduced. Cellulose and uric acid are degraded. Cellobiose, fructose, glucose, sucrose, and xylose are used as sole carbon sources, but dextrin, erythrose, lactose, mannose, melibiose, α-methyl-D-glucoside, sorbitol, and sorbose are not.

Additional phenotypic features are shown in Table 276.

Whole-organism hydrolysates contain glucose and madurose. The major fatty acids are $C_{15:0}$ (23.7%), $C_{16:0}$ (42.0%), and $C_{17:0}$ (24.9%). The cellular polar lipid pattern contains diphosphatidylglycerol, phosphatidylethanolamine, phosphatidylinositol, and glucosamic-containing components. The predominant menaquinones are MK-9(H_0), MK-9(H_2), and MK-9(H_4).

Source: a soil sample collected from Yunnan Province, China.

DNA G+C content (mol%): 69.1 (T_m).

Type strain: strain CY-15110, CCRC 16308, CCTCC AA 97010, DSM 44688, JCM 14926.

Sequence accession no. (16S rRNA gene): AF191735.

Additional comments: Streptosporangium purpuratum CY-15110[T] is most closely related to the type strains of *Streptosporangium longisporum* (55.3%), *Streptosporangium pseudovulgare* (45.2%), *Streptosporangium nondiastaticum* (43.6%), and *Streptosporangium roseum* (43.3%) based on DNA–DNA relatedness data (Zhang et al., 2005).

10. **Streptosporangium subroseum** Zhang, Jiang and Chen 2002, 1237[VP]

sub.ro'se.um. N.L. neut. adj. *subroseum* pale rose colored.

Abundant white-pinkish aerial mycelium is formed on Bennett's agar, glycerol-asparagine agar, and oatmeal-yeast extract agar. The substrate mycelium is yellowish-brown. Soluble pigments are not produced. Spherical spore vesicles, commonly 4–10 µm in diameter, are formed on the aerial mycelium on HV agar and oatmeal agar. Sporangiospores are spherical or oval-shaped. The temperature range for growth is 10–42°C; the optimal temperature is 30°C; does not grow at 50°C.

Nitrate is reduced. Degrades cellulose. Cellobiose, fructose, glucose, mannose, raffinose, L-rhamnose, sucrose, and xylose are used as sole carbon sources.

Additional phenotypic features are shown in Table 276.

Whole-organism hydrolysates contain glucose, madurose, ribose, and rhamnose. The major fatty acids are $C_{16:0}$ (5.8%), $C_{17:0}$ (11.4%), $C_{18:0}$ (32.1%), and $C_{19:0}$ (18.6%). The diagnostic polar lipid is phosphatidylethanolamine. The predominant menaquinones are MK-9(H_6), MK-9(H_2), and MK-9(H_4).

Source: a soil sample collected in Yunnan Province, China.

DNA G+C content (mol%): 71.2 (T_m).

Type strain: strain CY-7113, ATCC 21807, BCRC 16302, CCTCC 97008, DSM 44662, JCM 11962, KCC A-0281.

Sequence accession no. (16S rRNA gene): AF191734.

Additional comments: the specific epithet *subroseum* is a N.L. neut. adj. not a L. neut adj. as cited in the paper by Zhang et al. (2002). In the abstract of the paper the type strain BCRC (formerly CCRC) 16302 is erroneously cited as CRC 16302.

11. **Streptosporangium violaceochromogenes** Kawamoto, Takasaka, Okachi, Kohakura, Tkahashi and Nara 1975, 358[AL]

vi.o.la.ce.o.chro.mo'ge.nes. L. adj. *violaceus* violet; Gr. n. *chroma* color; N.L. suff. *-genes* (from Gr. v. *gennaô* to produce) producing; N.L. part. adj. *violaceochromogenes* producing violet color.

Aerial mycelium is pink and the substrate mycelium pale yellow on oatmeal agar. Violet or rose colored pigments are produced on nutrient, Bennett's, and Emerson agars. Spherical spore vesicles (5–9 µm in diameter) with irregular surfaces are borne on long sporangiosphores on the aerial mycelium. Sporangiospores are oval to cylindrical (0.8–0.9 × 1.2–1.6 µm). Does not require biotin or thiamine for growth. Grows from 25–40°C, optimally between 30–37°C, and from pH 6.0–8.5, optimally at pH 7.3.

Reduces nitrate. Milk is coagulated and peptonized. L-Tyrosine is degraded, but cellulose is not.

Fructose, glucose, mannose, starch, sucrose, and xylose are used as sole carbon sources.

Additional phenotypic features are shown in Table 276.

Does not contain madurose in whole-organism hydrolysates. The menaquinones are MK-9(H_0) (19.0%), MK-9(H_2) (50.0%), and MK-9(III, VIII-H_4) (31.0%).

Source: a soil sample collected from a swamp in Yoshioka village, Kitaguninagum, Gunma, Japan.

DNA G+C content (mol%): not determined.

Type strain: ATCC 21807, DSM 43849, NBRC 15560, JCM 3281, KCC A-0281, NRRL B-16784.

Sequence accession no. (16S rRNA gene): U48997, X89951.

12. **Streptosporangium vulgare** Nonomura and Ohara 1960, 407[AL]

vul.ga're. L. neut. adj. *vulgare* common.

Aerial mycelium is pink and the substrate mycelium yellowish-brown to orange on oatmeal-yeast extract agar; a pale yellow to yellow soluble pigment is also formed on this medium. Does not produce melanin pigments. Spore vesicles are 6–8 µm in diameter,- and sporangiospores are 1.0–1.2 × 1.5–1.9.

Hydrolyzes esculin, but not urea. Nitrate is reduced. Degrades arbutin, elastin, RNA, and L-tyrosine, but not guanine, hypoxanthine, keratin, testosterone, Tween 20, or xanthine. Dextrin, inulin, mannose, sucrose, trehalose, and xylose are used as sole carbon sources, but dextran, fructose, lactose, melezitose, melibiose, raffinose, sorbitol, sorbose, stachyose, starch, and xylitol (all at 1%, w/v) are not. Does not use acetate, butyrate, citrate, hippurate, malonate, or pyruvate as sole carbon sources (all at 0.1%, w/v).

Grows in the presence of crystal violet (0.001%, w/v), phenol (0.01%, w/v), phenyl ethanol (0.2%, v/v), potassium tellurite (0.01%, w/v), sodium chloride (2%, w/v), and thallous acetate (0.001%, w/v), but is sensitive to crystal violet (0.01%, w/v), phenol (0. 1%, w/v), phenyl ethanol (0.3% v/v), sodium azide (0.001%, w/v), and sodium chloride (4%, w/v).

Resistant (µg/ml) to amoxycillin (500), azlocillin (500), bacitracin (50), celoxitin (250), cephaloridine (250), cephadrine (500), chloramphenicol (25), clavulanic acid (50), flucloxicillin (50), fusidic acid (0.5), keflin (250), kefzol (500), lincomycin (5), mecillinam (500), methicillin (500), oleandomycin phosphate (250), penicillin G (50), pseudomonic acid (500), sulfomethoxazole (25), ticarcillin (50), and tyrothrecin (5), but is sensitive to bacitracin (50), clavulanic acid (500), flucloxicillin (250), fusidic acid (0.5), lincomycin (25), penicillin G (250), rifampin (0.5), tetracycline (5), ticarcillin (500), and vancomycin hydrochloride (5).

Additional phenotypic features are shown in Table 276.

Whole-organism hydrolysates contain madurose. The fatty acid profile includes $C_{12:0}$ (1.9%), $C_{14:0}$ (6.3%), $C_{15:0}$ (2.1%), $C_{16:0}$ iso (23.6%), $C_{16:0}$ (7.7%), $C_{16:0}$ 10-methyl (2.9%), $C_{17:0}$ (5.4%), $C_{17:0}$ 10-methyl (22.6%), $C_{18:1}$ cis (2.4%), $C_{18:0}$ (2.1%), and $C_{18:0}$ 10-methyl (5.6%). The cellular polar lipid pattern contains diphosphatidylglycerol, hydroxyphosphatidylethanolamine, phosphatidylethanolamine, phosphatidylinositol, and ninhydrin-positive and sugar-positive phospholipids. The predominant menaquinones are MK-9(H_0) (32.2%), MK-9(H_2) (56.7%), and MK-9(III, VIII-H_4) (15.1%).

Source: a soil sample collected from a paddy field in Anjo, Aichi Prefecture, Japan.

DNA G+C content (mol%): not determined.

Type strain: ATCC 33329, CBS 431.61, DSM 43802, NBRC 13985, JCM 3028, KCC A-0028, NRRL B-2633, RIA 765, VKM Ac-641.

Sequence accession no. (16S rRNA gene): U48999, X89955.

13. **Streptosporangium yunnanense** Zhang, Jiang and Chen 2005, 723[VP]

yun.nan.en'se. N.L. neut. adj. *yunnanense* of or pertaining to Yunnan, a province of south-west China.

Aerial mycelium is pale-pink to yellowish-pink and the substrate mycelium is brown to brownish-yellow on Bennett's agar, glycerol-asparagine agar, inorganic salts-starch agar, oatmeal agar, oatmeal-yeast extract agar, and yeast extract-malt extract agar; a pale brown diffusible pigment is formed on all of these media, except Bennett's agar and yeast extract-malt extract agar. Spherical spore vesicles (4–20 μm in diameter) are formed singly or in clusters on HV and oatmeal agars. Sporangiospores are spherical or oblate. The temperature range for growth is 10–42°C, does not grow at 50°C. Optimal pH is 7.2.

Hippurate is hydrolyzed and nitrate reduced. Degrades cellulose and uric acid. Cellobiose, glucose, and sucrose are used as sole carbon sources, but dextrin, erythrose, fructose, inositol, lactose, mannose, mannitol, melibiose, α- methyl-D-glucoside, raffinose, sorbitol, sorbose, and xylose are not.

Additional phenotypic features are shown in Table 276.

Whole-organism hydrolysates contain glucose, madurose, and rhamnose. The major fatty acids are $C_{16:0}$ (14.7%), $C_{17:0}$ (5.0%), $C_{18:0}$ (10.0%), and $C_{19:0}$ (29.6%). The cellular polar lipid pattern contains diphosphatidylglycerol, phosphatidylethanolamine, phosphatidylinositol, phosphatidylmethylethanolamine, and glucosamine-containing components. The predominant menaquinones are MK-9(H_0), MK-9(H_2), and MK-9(H_4).

Source: a soil sample collected from Yunnan Province, China.

DNA G+C content (mol%): 71.1 (T_m).

Type strain: strain CY-11007, CCRC 16307, CCTCC AA 97009, DSM 44663, JCM 14927.

Sequence accession no. (16S rRNA gene): AF191733.

Additional comments: Streptosporangium yunnanense CY-11007[T] is most closely related to the type strains of *Streptosporangium nondiastaticum* (56.8%) and *Streptosporangium pseudovulgare* (53.1%) on the basis of DNA–DNA relatedness data (Zhang et al., 2005).

Species *incertae sedis*

1. **"Streptosporangium rubrum"** Potekhina 1965, 292.

ru'brum. L. neut. adj. *rubrum* red

A pale pink to white substrate mycelium and a white to pale pink aerial mycelium is formed on Czapek agar. Forms spherical spore vesicles (6.0–14 μm in diameter) and branched sporangiophores (0.5–1.2 × 1.6–27 μm). Spores are typically spherical (0.8 × 1.4 μm), but may be ovoid or rod-shaped.

Milk is peptonized and nitrates weakly reduced to nitrites. Cellulose and starch are degraded, but gelatin is not.

Source: a dark gray forest soil.

DNA G+C content (mol%): not determined.

Reference strain: DSM 44095.

Sequence accession no. (16S rRNA gene): X89950.

Genus II. **Acrocarpospora** Tamura, Suzuki and Hatano 2000, 1170[VP]

Tomohiko Tamura

A.cro.car.po.spo'ra. Gr. adj. *akros* uttermost, topmost, highest, at the top, end; Gr. n. *karpos* fruit, harvest; Gr. fem. n. *spora* a seed; N.L. fem. n. *Acrocarpospora* an organism forming spores like fruits on the terminal mycelium

Gram-stain-positive bacterium that is not-acid-fast and forms **branching hyphae**.

Non-fragmentary substrate mycelia are present. Spherical and club shaped structures are borne on the tips of the **aerial mycelium**. These structures contain coiled or straight spore chains. Spores are oval or short rod-like (0.6–0.8 × 0.7–1.0 μm) with a smooth surface and nonmotile. **Strictly aerobic**. Good growth occurs at 20–30°C. Shows good growth on oatmeal agar. In general, the vegetative mycelia are pale yellow and aerial hyphae are white. Cell walls contain glutamate, alanine, and *meso*-diaminopimelate. Wall chemotype is III (Lechevalier and Lechevalier, 1970b), and the peptidoglycan type is presumed to be A1γ (Schleifer and Kandler, 1972). Madurose, glucose, and galactose are detected in whole cell sugars. Major cellular fatty acids are $C_{16:0}$ iso, $C_{17:0}$ 10-methyl, $C_{17:1}$, and $C_{17:0}$. The major menaquinones are MK-9(H_4) and MK-9(H_2), and small amounts of MK-9(H_6) are present. Phosphatidylethanolamine is present as a diagnostic phospholipid. Mycolic acids are absent. The acyl type of cell wall is acetyl. The type species is *Acrocarpospora pleiomorpha.*

DNA G+C content (mol%): 68–69.

Type species: **Acrocarpospora pleiomorpha** Tamura, Suzuki and Hatano 2000, 1170[VP].

Further descriptive information

The sporangium forms two types of structures, club-shaped and various sized spherical bodies. In 7- and 14-d-old cultures, the various sized spherical bodies are commonly found; the club-shaped structures are infrequent.

Although species of *Acrocarpospora* possess similar fatty acids, there are differences in the quantities of some of the most abundant ones (Table 277). For instance, *Acrocarpospora pleiomorpha* contains high levels of $C_{13:0}$, and *Acrocarpospora corrugate* contains low levels of $C_{16:0}$ iso. Similarly, while the physiological

TABLE 277. Cellular fatty acid compositions (%) of the members of the genus *Acrocarpospora*

Fatty acid (%)	*A. pleiomorpha* NBRC 16267^T	*A. corrugata* NBRC 13972^T	*A. macrocephala* NBRC 16266^T
$C_{12:0}$	1.7		
$C_{13:0}$	14.2	0.5	0.9
$C_{14:0}$ iso		4.4	9.8
$C_{14:0}$		0.5	
$C_{15:0}$ iso		3.3	
$C_{15:0}$ anteiso		4.3	
$C_{15:0}$	4.7	4.2	4.1
$C_{16:0}$ iso	43.4	9.2	39.6
$C_{16:1}$ 9*c*		1.2	
$C_{16:0}$	2.5	5.3	2.0
$C_{15:0}$ 2-OH		0.9	
$C_{16:0}$ 10-methyl	1.7	0.8	
$C_{17:0}$ iso		2.8	
$C_{17:0}$ anteiso		5.3	
$C_{17:1}$ 9*c*	4.2	10.6	9.2
$C_{16:0}$ iso 2-OH		1.8	
$C_{17:0}$	6.1	13.2	6.2
$C_{16:0}$ 2-OH		1.0	
$C_{17:0}$ 10-methyl	21.6	19.6	25.0
$C_{18:0}$ iso		0.9	
$C_{18:1}$ 9*c*		2.4	
$C_{17:0}$ iso 2-OH		0.4	
$C_{18:0}$		3.3	
$C_{17:0}$ 2-OH		1.1	
$C_{18:0}$ iso 10-methyl			1.9
TBSA $C_{18:0}$ 10-methyl		3.2	1.4

responses of the species of *Acrocarpospora* are generally similar, some differences are noted (Table 278).

TABLE 278. Physiological characteristics of the members of the genus *Acrocarpospora*[a]

Characteristic	*A. pleiomorpha* NBRC 16267^T	*A. corrugata* NBRC 13972^T	*A. macrocephala* NBRC 16266^T
N-Acetyl-β-glucosaminidase	–	–	–
Acid phosphatase	+	+	+
Akaline phosphatase	+	+	+
Catalase activity	+	+	+
Chymotrypsin	+	–	–
Cystine aminopeptidase	+	w	+
Esculin hydrolysis	+	+	+
Esterase (C4)	+	w	w
Esterase lipase (C8)	+	+	+
α-Fucosidase	–	–	–
α-Galactosidase	–	w	–
β-Galactosidase	–	+	–
β-Glucosidase	w	w	+
Gelatin hydrolysis	+	+	+
Leucine aminopeptidase	+	+	+
Lipase (C14)	w	w	w
α-Mannosidase	–	–	–
Nitrate reduction	–	–	–
Phosphohydrolase	w	+	w
Pyrazinamidase	–	–	–
Pyrrolidonyl arylamidase	–	w	–
Trypsin	+	w	w
Urea hydrolysis	–	+	–
Valine aminopeptidase	+	w	+

[a]Symbols: +, >85% positive; –, 0–15% positive; w, weak reaction.

Enrichment and isolation procedures

Acrocarpospora pleiomorpha strain R-31^T was isolated from soil in Louisiana, USA, and *Acrocarpospora macrocephala* strain R-55^T was isolated from soil in Saitama prefecture, Japan.

The type strain of *Acrocarpospora corrugata* was isolated from beach sand (pH 7.8) at Freshfield, Lancashire, United Kingdom, by using the dilution plate technique on starch-casein medium containing antifungal antibiotics (Williams and Sharples, 1976). In this isolation procedure, most of the actinomycetes appearing on the plates were micromonosporas, and *Streptosporangium corrugatum*-like colonies accounted for less than 1% of the total.

Maintenance procedures

Strains of the genus *Acrocarpospora* are maintained by freezing in water containing 10–30% glycerol at −70°C. Lyophilization of suspensions in 10% skim milk +1% monosodium glutamate and l-drying in 0.01 M potassium phosphate buffer (pH7.0) containing 3% monosodium glutamate are also recommended for long-term preservation.

Differentiation of the genus *Acrocarpospora* from other genera

The genus *Acrocarpospora* forms club-shaped or spherical sporangia at the heads of the aerial mycelia and produces spores by fragmentation of hyphae and the hyphal sheath, which either expands to form the sporangial envelope or remains around the spore chains. Although the morphological characteristics and spore development of isolates are similar in many respects to those of *Streptosporangium* species, there are differences in the irregular shape and variable size of spore vesicles. Under the light microscope, the isolates appear similar to *Nonomuraea pusilla*, which develops pseudosporangia, rather than to the majority of *Streptosporangium* species. The genus *Acrocarpospora* contains MK-9(H_4) as the major menaquinone and is differentiated from the genus *Herbidospora*, which has MK-10(H_4) as a predominant menaquinone.

Taxonomic comments

The phylogenetic tree based on 16S rRNA gene sequence analysis indicates that three species of the genus *Arcocarpospora* form a branch within the family *Streptosporangiaceae* of the order *Streptosporangiales* (Tamura et al., 2000). The genus *Herbidospora* is the most closely related genus to *Acrocarpospora. Acrocarpospora corrugata* was originally proposed as *Streptosporangium corrugatum* by Williams and Sharples (1976) and transferred to the genus *Acrocarpospora* by Tamura et al. (2000).

List of species of the genus *Acrocarpospora*

1. **Acrocarpospora pleiomorpha** Tamura, Suzuki and Hatano 2000, 1170^{VP}

ple.i.o.mor'pha. Gr. adj. *pleios* full; Gr. n. *morphe* form; N.L. fem. adj. *pleiomorpha* pleiomorphic, in various shapes.

Gelatin liquefaction is negative. Hydrolyzes starch. Does not decompose calcium malate. Coagulation of milk is weak. Reduction of nitrate is positive. Glucose, raffinose, rhamnose, and mannitol are utilized, but xylose, arabinose, sucrose, fructose, and inositol are not. As diagnostic phospholipids, phosphatidylethanolamine, ninhydrin positive, and sugar positive phospholipids are present.

Source: soil.

DNA G+C content (mol%): 69 (HPLC).

Type strain: R-31, JCM 10983, NBRC 16267.

Sequence accession no. (16S rRNA gene): AB006174.

2. **Acrocarpospora corrugata** (Williams and Sharples 1976) Tamura, Suzuki and Hatano 2000, 1170^{VP} (*Streptosporangium corrugatum* Williams and Sharples 1976, 45)

cor.ru.ga'ta. L. fem. part. adj. *corrugata* wrinkled.

Aerial hyphae bearing globose vesicles (1.0–5.0 µm diameter) containing coiled chains of about 20 spores and club-shaped structures (3.5–8.0 × 0.75–1.00 µm) containing three to eight spores. Spores from both structures are nonmotile, 0.8–1.0 × 0.6–0.7 µm, with prominent longitudinal ridges and terminal annular ridges. Spore walls become considerably thickened, reaching 150 nm. Aerial mycelium and spore-containing structures are white, the reverse side of the colony is pale buff, and no soluble pigment is produced.

The predominant menaquinones are MK-9(H_2) and MK-9(H_4), and the major cellular fatty acids are $C_{17:1}$ and $C_{17:0}$ (Kudo and Seino, 1987) along with $C_{17:0}$ 10-methyl. The positions of hydrogenation of menaquinone MK-9(H_4) are III and VIII, as in the case of the genus *Streptosporangium* (Collins et al., 1988; Stackebrandt et al., 1994). Galactose and trace amount of madulose are detected in the whole cell sugars (Stackebrandt et al., 1994). In addition to the diagnostic phospholipid, phosphatidylethanolamine, hydroxyphosphatidylethanolamine, diphosphatidylglycerol, and phosphatidylinositol are present (Stackebrandt et al., 1994). Numerical analysis of the phenotypic characteristics were studied by Whitham et al. (1993).

Source: beach sand (pH 7.8).

DNA G+C content (mol%): not available.

Type strain: ATCC 29331, BCRC 12360, DSM 43316, JCM 3181, KCTC 9431, NBRC 13972, NCIMB 11120.

Sequence accession no. (16S rRNA gene): AB188150.

3. **Acrocarpospora macrocephala** Tamura, Suzuki and Hatano 2000, 1170^{VP}

ma.cro.ce'pha.la. Gr. adj. *makrokephalos* long-headed; N.L. fem. n. *macrocephala* large head.

Gelatin liquefaction is negative. Hydrolyzes starch. Does not decompose calcium malate. Coagulates milk. Glucose, raffinose, rhamnose, mannitol, and sucrose are utilized, but xylose, arabinose, fructose, and inositol are not. As diagnostic phospholipids, phosphatidylethanolamine, and ninhydrin positive and sugar positive phospholipids are present.

Source: soil.

DNA G+C content (mol%): 68 (HPLC).

Type strain: R-55, JCM 10982, NBRC 16266.

Sequence accession no. (16S rRNA gene): AB025318.

Genus III. **Herbidospora** Kudo, Itoh, Miyadoh, Shomura and Seino 1993, 327^{VP}

Takuji Kudo

Her.bi.do.spo'ra. L. adj. *herbidus* grassy; Gr. n. *spora* a seed; N.L. fem. n. *Herbidospora* organism forming spores like grass.

Aerobic, stains Gram-positive but non-acid-fast. **Non-fragmented branching vegetative hyphae** are produced, but distinct **aerial hyphae are not formed**. Straight and short chains of nonmotile, smooth-surfaced, oval spores (10–30 spores per chain) are borne on the tips of the sporophores which are derived from the vegetative mycelia in clusters. Cell walls contain *meso*-diaminopimelate and *N*-acetylated muramic acid, but lack a significant amount of glycine (cell-wall type III). Whole-cell hydrolysates contain trace amounts of madurose (whole cell sugar pattern B). Contains phosphatidylethanolamine and its derivatives and glucosamine-containing phospholipids as diagnostic phospholipids (phospholipid type PIV). Cellular fatty acid composition is characterized by $C_{17:0}$ 10-methyl, $C_{17:1}$ ω8c, $C_{16:0}$ iso, and $C_{17:0}$ (fatty acid type 3c). The predominant isoprenoid quinone is a tetrahydrogenated menaquinone with ten isoprene units, and the hydrogenation occurs at units III and IX [MK-10(III,IX-H_4)]. MK-10(H_2), MK-10(H_0), MK-10(H_6), and MK-9(III,VIII-H_4) are also present as minor components. Mycolic acids are absent.

DNA G+C content (mol%): 69–73.

Type species: **Herbidospora cretacea** Kudo, Itoh, Miyadoh, Shomura and Seino 1993, 327^{VP}.

Further descriptive information

The genus *Herbidospora* was first described in 1993 with a single species *Herbidospora cretacea*, and no additional species were proposed for 16 years. From 2009 onward, an additional four species, *Herbidospora osyris* (Li et al., 2009), *Herbidospora yilanensis*, and *Herbidospora daliensis* (Tseng et al., 2010) and *Herbidospora sakaeratensis* (Boondaeng et al., 2011) were described in succession. Based on phylogenetic analysis using 16S rRNA gene

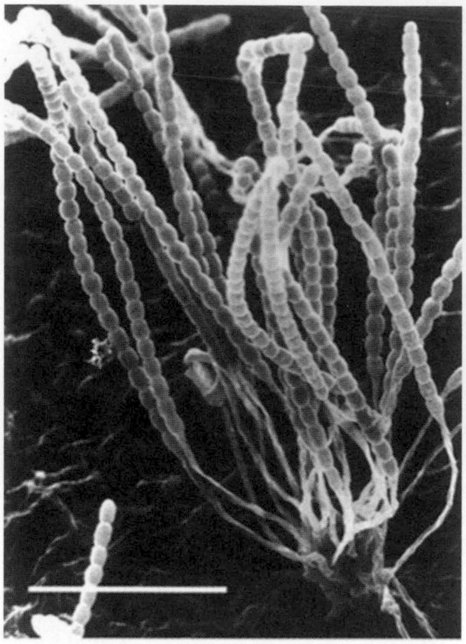

FIGURE 367. Scanning electron micrograph of *Herbidospora cretacea* JCM 8554. Bar = 10 μm.

sequences, the genus *Herbidospora* is monophyletic and within the family *Streptosporangiaceae*. In fact, the morphological and chemotaxonomic characteristics of the members are almost homogeneous and distinct from other genera belonging to the family.

Macroscopically, they produce white or brownish yellow aerial masses on some media, but microscopic observation reveals that relatively long sporophores bearing spore chains develop directly in clusters from the substrate mycelia located at the agar surface (Figure 367). This morphological characteristic is unique among other members of the family *Streptosporangiaceae*.

The chemotaxonomic characteristics of the genus, with the exception of menaquinone composition, are shared by other genera of the family *Streptosporangiaceae*, i.e. cell-wall type III, *N*-acetylated muramic acid in the cell walls, whole cell sugar pattern B, phospholipids type PIV, and cellular fatty acid type 3c. On the other hand, the members of the genus *Herbidospora* contain MK-10(H_4) as a major component, while all of the other genera in the family *Streptosporangiaceae* have menaquinones with nine isoprene units. This difference in the length of the side chain shows the apparent boundary of this genus. The tetrahydrogenation of the side chain of the *Herbidospora* menaquinone occurs at isoprene unit III and IX, while the tetrahydrogenation of those of the other genera belonging to the family *Streptosporangiaceae* occurs at units III and VIII. It is thought that these modes of tetrahydrogenation represent a common mechanism in that the third double bond from the naphthoquinone moiety and the second double bond from the ω-terminus are hydrogenated. Among actinobacteria, this hydrogenation pattern is unique to the family *Streptosporangiaceae*.

Isolation and maintenance procedures

Herbidospora strains have been isolated from soil, the sediment of a river, decayed leaves, the bark of a pine tree, the petals of flowers, and a surface-sterilized tissue of a plant. Humic acid-vitamin agar (Hayakawa and Nonomura, 1987b) was often used as an isolation medium, but there is no information on enrichment. The organisms grow relatively well on general complex agar media used for filamentous actinomycetes such as oatmeal agar (ISP 3), yeast extract-malt extract agar (ISP 2) or yeast-starch agar, but good sporulation occurs on only a few media such as ISP 3, oatmeal-nitrate agar, and ISP 2 agar. As most species of this genus require vitamin B complex, they grow poorly on defined media. They can be stored in 10% glycerol or DMSO solution at −80°C, lyophilized with 10% skim milk, or L-dried with 0.1 M phosphate buffer (pH 7.0) plus 3% monosodium glutamate and 1.5% adonitol for long-term preservation.

Differentiation of the genus *Herbidospora* from other genera

As shown in Figure 368, the genus *Herbidospora* is a member of the family *Streptosporangiaceae*, and the closest relative is the genus *Acrocarpospora*. The genus *Herbidospora* shares chemot-

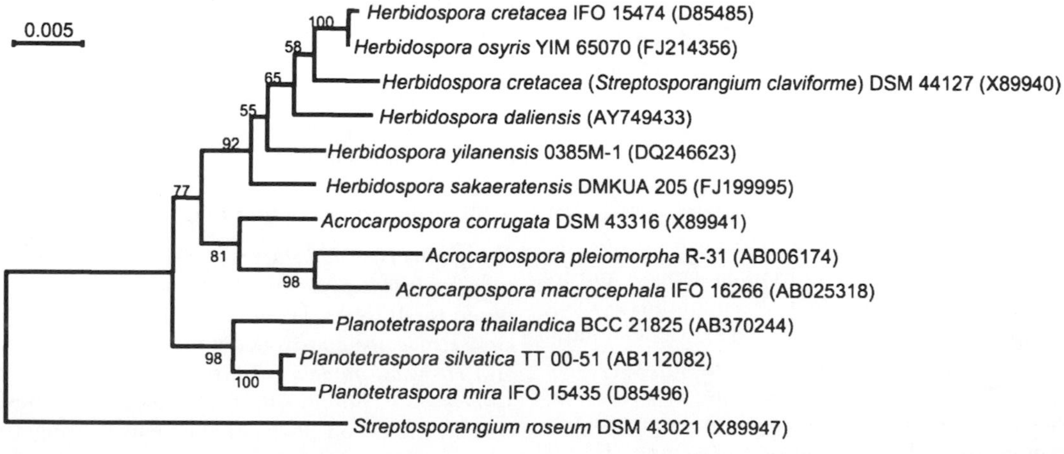

FIGURE 368. Neighbor joining tree based on 16S rRNA gene sequences showing the phylogenetic relationships among *Herbidospora* species and related taxa. Numbers represent bootstrap values for each branch (1000 replicates).

axonomic characteristics with other genera of the family *Streptosporangiaceae*, but it is readily differentiated from them by its distinctive morphology and menaquinone composition. The menaquinones of the genus *Herbidospora* are mainly composed of MK-10(H_4), but other members of the family including the closest genus, *Acrocarpospora*, contain menaquinones with nine isoprene units. Morphologically, the genus *Acrocarpospora* is characterized by formation of spherical and club-shaped sporangium-like structures, but the genus *Herbidospora* does not produce sporangium-like structures. The genus *Actinocorallia* sometimes shows similar morphology to *Herbidospora*, but this genus belongs to the family *Thermomonosporaceae* and has quite different chemotaxonomic characteristics.

Taxonomic comments

Streptosporangium claviforme was described as a sporangium-producing actinomycete (Petrolini et al., 1992), but its morphology is identical with that of the genus *Herbidospora*. In the original description, the authors thought that the chains of spores were enveloped by "sporangial" walls. However, the term "sporangium" has been defined more accurately, and even spores of streptomycetes have sheath-like structures. *Streptosporangium claviforme* also has MK-10(III, IX-H_4) as a major component. Tamura et al. (2000) indicated that the DNA–DNA relatedness between the type strains of *Streptosporangium claviforme* and *Herbidospora cretacea* was 80%, and Boondaeng et al. (2011) reclassified *Streptosporangium claviforme* as a later synonym of *Herbidospora cretacea*.

Differentiation of species of the genus *Herbidospora*

Five species are currently known, and they share morphological and chemotaxonomic characteristics. Thus, differentiation of the species depends on physiological characteristics. However, the accurate boundary of the species cannot be determined physiologically because the descriptions of the species, except for *Herbidospora cretacea*, are based on a single strain. Identification at the species level should be performed by using differential characteristics shown in Table 279 and 16S rRNA gene analysis as well as DNA–DNA hybridization.

TABLE 279. Characteristics differentiating species of the genus *Herbidospora*[a,b]

Characteristic	*H. cretacea*	*H. daliensis*	*H. osyris*	*H. sakaeratensis*	*H. yilanensis*
Decomposition of:					
Esculin	+	–	+	nd	–
Hypoxanthine	+	–	nd	nd	–
Starch	+	+	–	nd	+
Utilization as sole carbon sources:					
L-Arabinose	+	+	–	+	–
Cellobiose	+	–	+	nd	–
Glycerol	–	–	+	nd	+
myo-Inositol	–	–	+	–	+
Maltose	+	+	+	–	+
D-Mannose	+	–	+	+	+
Melezitose	–	+	nd	–	+
Melibiose	w	–	+	nd	+
Raffinose	–	+	+	–	+
L-Rhamnose	v	+	–	–	+
D-Ribose	+	–	+	–	–
Salicin	v	–	+	nd	+
Trehalose	v	–	+	nd	+
D-Xylose	+	–	+	+	–
Utilization of:					
Fumaric acid	+	–	nd	+	+
DL-Lactic acid	+	–	nd	nd	+
L-Malic acid	v	–	+	+	–
Succinic acid	+	–	nd	+	+
Requirement for:					
p-Aminobenzoic acid	–	+	nd	nd	+
myo-Inositol	–	+	nd	nd	+
Riboflavin	–	–	nd	nd	+
Thiamine	+	–	nd	nd	+
NaCl tolerance	<3%	<5%	<3%	<1.5%	<1%

[a]Symbols: +, positive; –, negative; w, weekly positive, v, variable; nd, not determined.

[b]Data from Kudo et al. (1993), Li et al. (2009), Tseng et al. (2010), and Boondaeng et al. (2011).

List of species of the genus *Herbidospora*

1. **Herbidospora cretacea** Kudo, Itoh, Miyadoh, Shomura and Seino 1993, 327[VP]

cre.ta'ce.a. L. fem. adj. *cretacea* chalk-like.

Vegetative mycelia are yellow to brown on most media. When sporulation occurs, the surface of the colony is white or brownish yellow. Good sporulation occurs on oatmeal agar (ISP 3) and oatmeal-nitrate agar. Diffusible pigment and melanoid pigments are not produced. Mesophilic (optimal growth at 28°C) and neutrophilic (optimal growth at pH 7.0–8.0). No growth occurs in the presence of 3% NaCl.

Decomposes casein, deoxyribonucleic acid, esculin, hypoxanthine, and starch, but not adenine, arbutin, guanine, keratin, tyrosine, and xanthine. Utilizes L-arabinose, cellobiose, D-fructose, D-galactose, D-glucose, maltose, D-mannitol, D-mannose, D-ribose, starch, sucrose, and D-xylose as a sole carbon source, but not adonitol, cellulose, dulcitol, *i*-erythritol, glycerol, *myo*-inositol, melezitose, and raffinose. Utilizes fumaric acid, DL-lactic acid, L-malic acid, and succinic acid, but not benzoic acid, citric acid, mucic acid, oxalic acid, propionic acid, and L-tartaric acid. Sensitive to lysozyme, gentamicin, novobiocin, rifampin, streptomycin, and vancomycin. Requires thiamine for growth.

Source: soil, decayed leaves, bark of pine tree, and petal of daisy fleabane; type strain was isolated from soil collected in Saitama, Japan.

DNA G+C content (mol%): 69–70 (HPLC).

Type strain: K-319, ATCC 51904, BCRC 16356, DSM 44071, JCM 8553, KCTC 9335, NBRC 15474, NCIMB 13372, VKM Ac-1997.

Sequence accession no. (16S rRNA gene): D85485.

2. **Herbidospora daliensis** Tseng, Yang and Yuan 2010, 1171[VP]

da.li.en'sis. N.L. fem. adj. *daliensis* of or pertaining to Dali city, Taiwan, where the type strain was isolated.

Vegetative mycelia are yellowish white to yellowish brown on most media. When sporulation occurs, the surface of the colony is white or brownish yellow. Good sporulation occurs on oatmeal agar (ISP 3). Diffusible and melanoid pigments are not produced. Neutrophilic. Growth occurs at 20–40°C. NaCl tolerance is 5%.

Decomposes starch but not adenine, hypoxanthine, tyrosine, and xanthine. L-Arabinose, fructose, galactose, glucose, maltose, mannitol, melezitose, raffinose, rhamnose, and sucrose are utilized. Adonitol, benzoic acid, cellobiose, cellulose, dulcitol, erythritol, fumaric acid, glycerol, *myo*-inositol, DL-lactic acid, L-malic acid, mannose, melibiose, mucic acid, propionic acid, D-ribose, salicin, D-sorbitol, succinic acid, L-tartaric acid, trehalose, and xylose are not utilized. Sensitive to lysozyme. Requires *p*-aminobenzoic acid and *myo*-inositol for growth. Biotin, nicotinic acid, pantothenic acid, pyridoxine, riboflavin, and thiamine are not required for growth.

Morphological and chemical characteristics are given in the description of the genus.

Source: sediment of a river in Dali city, Taiwan.

DNA G+C content (mol%): 70.7 (HPLC).

Type strain: 0385M-1, FIRDI 004, BCRC 16876, LMG 24336, NBRC 106372.

Sequence accession no. (16S rRNA gene): AY749433.

3. **Herbidospora osyris** Li, Zhao, Qin, Zhu, Xu and Li 2009, 3126[VP]

o.sy'ris. L. n. *osyris* a plant (probably the broom-like goosefoot or summer cypress), and also a botanical genus name (*Osirys*); N.L. gen. n. osyris, of *Osyris* the plant genus from which this species was isolated.

Vegetative mycelia are yellowish white to yellowish brown on most media. When sporulation occurs, the surface of the colony is white. Moderate sporulation occurs on yeast-extract-malt extract agar (ISP 2) and YIM 38 agar. A diffusible pigment is not produced. Growth occurs at 10–37°C and pH 6.0–8.0. NaCl tolerance is up to 3%.

Catalase is produced. Negative for the Voges–Proskauer and methyl red tests, for the oxidase reaction, for production of H_2S, for nitrate reduction, and for milk coagulation and peptonization. Tweens 20 and 40 and urea are hydrolyzed, but Tween 80, gelatin, starch, and cellulose are not hydrolyzed. Utilizes amygdalin, arbutin, cellobiose, esculin, D-fructose, D-galactose, glycerol, *myo*-inositol, maltose, D-mannitol, D-mannose, melibiose, raffinose, D-ribose, salicin, sodium DL-malate, D-tagatose, trehalose, turanose, and D-xylose as sole carbon sources. D-Adonitol, D-arabinose, dulcitol, erythritol, D-lactose, L-rhamnose, D-sorbitol, L-sorbose, and xylitol are not utilized. Adenine, L-alanine, L-cysteine, L-cystine, L-histidine, L-lysine, L-phenylalanine, L-proline, L-serine, L-threonine, L-tyrosine, and xanthine can be used as sole nitrogen sources, but L-arginine, L-asparagine, L-glutamic acid, glycine, hypoxanthine, DL-methionine, and L-valine cannot.

Morphological and chemical characteristics are given in the description of the genus.

Source: type strain was isolated from a surface sterilized plant sample, *Osyris wightiana* Wall. *ex* Wight, collected from Yunnan province, south-west China.

DNA G+C content (mol%): 70.4 (HPLC).

Type strain: YIM 65070, CCTCC AA 208019, DSM 45214, JCM 16900, NBRC 106571.

Sequence accession no. (16S rRNA gene): FJ214356.

4. **Herbidospora sakaeratensis** Boondaeng, Suriyachadkun, Ishida, Tamura, Tokuyama and Kitpreechavanich 2011, 779[VP]

sa.ka.e.ra.ten'sis. N.L. fem. adj. *sakaeratensis* of or pertaining to Sakaerat Biosphere Reserve, the source of soil from which the type strain was isolated.

Vegetative mycelia are yellow to yellowish brown. A melanoid pigment is not produced. Growth occurs at 20–40°C and pH 6.0–9.0. NaCl tolerance is up to 1.5%.

Utilizes L-arabinose, D-fructose, D-galactose, D-glucose, D-mannitol, D-mannose, starch, sucrose, and D-xylose as sole carbon sources, but not dulcitol, *myo*-inositol, maltose, melezitose, D-raffinose, L-rhamnose, D-ribose, and D-sorbitol. Utilizes fumaric acid, L-malic acid, and succinic acid, but not benzoic acid and mucic acid. Cellulose degradation, gelatin liquefaction, nitrate reduction, and milk peptonization are negative.

Morphological and chemical characteristics are given in the description of the genus.

Source: a soil sample collected in Sakaerat Biosphere Reserve in Nakhonratchasima, Thailand.

DNA G+C content (mol%): 73 (HPLC).

Type strain: DMKUA 205, BCC 11662, NBRC 102641.

Sequence accession no. (16S rRNA gene): FJ199995.

5. **Herbidospora yilanensis** Tseng, Yang and Yuan 2010, 1170[VP]

yi.lan.en′sis. N.L. fem. adj. *yilanensis* of or pertaining to Yilan county, Taiwan, where the type strain was isolated.

Vegetative mycelia are yellowish white to yellow on most media. When sporulation occurs, the surface of the colony is white or brownish yellow. Moderate sporulation occurs on oatmeal agar (ISP 3), oatmeal-nitrate agar, and glycerol-asparagine agar (ISP 5). Diffusible and melanoid pigments are not produced. Neutrophilic. Growth occurs at 20–40°C. NaCl tolerance is 1%.

Decomposes starch but not adenine, esculin, hypoxanthine, tyrosine, and xanthine. Fructose, fumaric acid, galactose, glucose, glycerol, DL-lactic acid, *myo*-inositol, maltose, mannitol, mannose, melibiose, melezitose, raffinose, rhamnose, salicin, sucrose, succinic acid, and trehalose are utilized. Adonitol, L-arabinose, benzoic acid, cellobiose, cellulose, dulcitol, erythritol, L-malic acid, mucic acid, propionic acid, D-ribose, D-sorbitol, L-tartaric acid, and xylose are not utilized. Sensitive to lysozyme. *p*-Aminobenzoic acid, *myo*-inositol and thiamine are required for growth. Biotin, nicotinic acid, pantothenic acid, and pyridoxine are not required for growth.

Morphological and chemical characteristics are given in the description of the genus.

Source: sediment of a dry lake, Yilan, a county in north Taiwan.

DNA G+C content (mol%): 70.6 (HPLC).

Type strain: 0351M-12, FIRDI 003, BCRC 16875, LMG 24337, NBRC 106371.

Sequence accession no. (16S rRNA gene): DQ246623.

Genus IV. **Microbispora** Nonomura and Ohara 1957, 307[AL] emend. Zhang, Wang and Ruan 1998a, 418

Christopher M. M. Franco

Mi.cro.bi.spo′ra. Gr. adj. *mikros* small; L. adv. num. *bis* twice; Gr. n. *spora* a seed; N.L. fem. n. *Microbispora* the small two-spored (organism).

Aerobic, Gram-stain-positive, non-acid–alcohol-fast, nonmotile, mesophilic, and thermophilic actinobacteria **which form stable, highly branched substrate and aerial mycelia. Spore chains, typically containing two spores, are borne longitudinally on short aerial hyphae. Spores are slightly oval to cylindrical (1.0–1.4 to 1.2–1.7 µm in diameter) and have smooth surfaces. Mesophilic species grow from 20–37°C, with thermophilic species up to 55°C.** Most species require B vitamins, particularly thiamine, for growth. Chemoorganotrophic with an oxidative type of metabolism. **Cell walls contain *N*-acetylated muramic acid and major amounts of *meso*-diaminopimelic acid; peptidoglycan type is A1γ. Whole-cell hydrolysates contain madurose** and complex mixtures of saturated, unsaturated, iso-, anteiso-, and branched chain fatty acids. **Predominant menaquinones have nine isoprene units with hydrogenation at positions III and IV [MK-9(III, IV-H_4), MK-9(H_2), and MK-9(H_0)]. Major phospholipids are phosphatidylcholine and unknown glucosamine-containing compounds, but no phosphatidylglycerol.** Natural habitat is soil, plant litter, and as epiphytes and endophytes of plants.

DNA G+C content (mol%): 68–73.

Type species: **Microbispora rosea** Nonomura and Ohara 1957, 307[AL].

Further descriptive information

Phylogeny. The genus *Microbispora* contains 13 species including *Microbispora rosea, Microbispora aerata, Microbispora amethystogenes, Microbispora chromogenes, Microbispora corallina, Microbispora diastatica, Microbispora indica, Microbispora karnatakensis, Microbispora mesophila, Microbispora parva, Microbispora siamensis, Microbispora thermodiastatica,* and *Microbispora thermorosea.* The type strains of these species form a distinct phyletic line in the *Streptosporangiaceae* 16S rRNA gene tree with *Microtetraspora, Arcocarpospora, Herbidospora,* and *Nonomuraea* (Figure 369) being the most closely related genera. The 16S rRNA gene sequence similarities of the *Microbispora* type strains fall within the range 95.1–99.2% (Wang et al., 1996b). The corresponding DNA–DNA relatedness values range from 27–80% (Nakajima et al., 1999).

Cell morphology. Microbisporae form characteristic pairs of spores on the aerial hyphae or short sporophores; the latter may be short so that the spores appear sessile. Buds formed on the aerial hyphae or tips of a side branch swell and then separate by a side wall. The sporophore of *Microbispora rosea* partially encloses the basal spore giving the appearance of a ball-and-socket joint (Williams, 1970). Mature spores are easily detached from the sporophore and each other when placed in water. Spores from most species of *Microbispora* have a smooth surface and are found along the entire length of the aerial hyphae (Figures 30.11, 30.12 from *Bergey's Manual,* previous edition, Nonomura, 1989b). Single spores have been reported for *Microbispora mesophila* (Zhang et al., 1998a).

Nutrition and growth conditions. Most microbisporae require B-vitamins for growth on synthetic media (Nonomura and Ohara, 1971d) but are readily cultivated on rich media including glucose-yeast extract (Waksman, 1967), oatmeal supplemented with yeast extract (Shirling and Gottlieb, 1966), and yeast extract-malt extract (Shirling and Gottlieb, 1966) agars. Mesophilic microbisporae grow between 20 and 37°C, but the thermophilic strains grow better at 37–45°C.

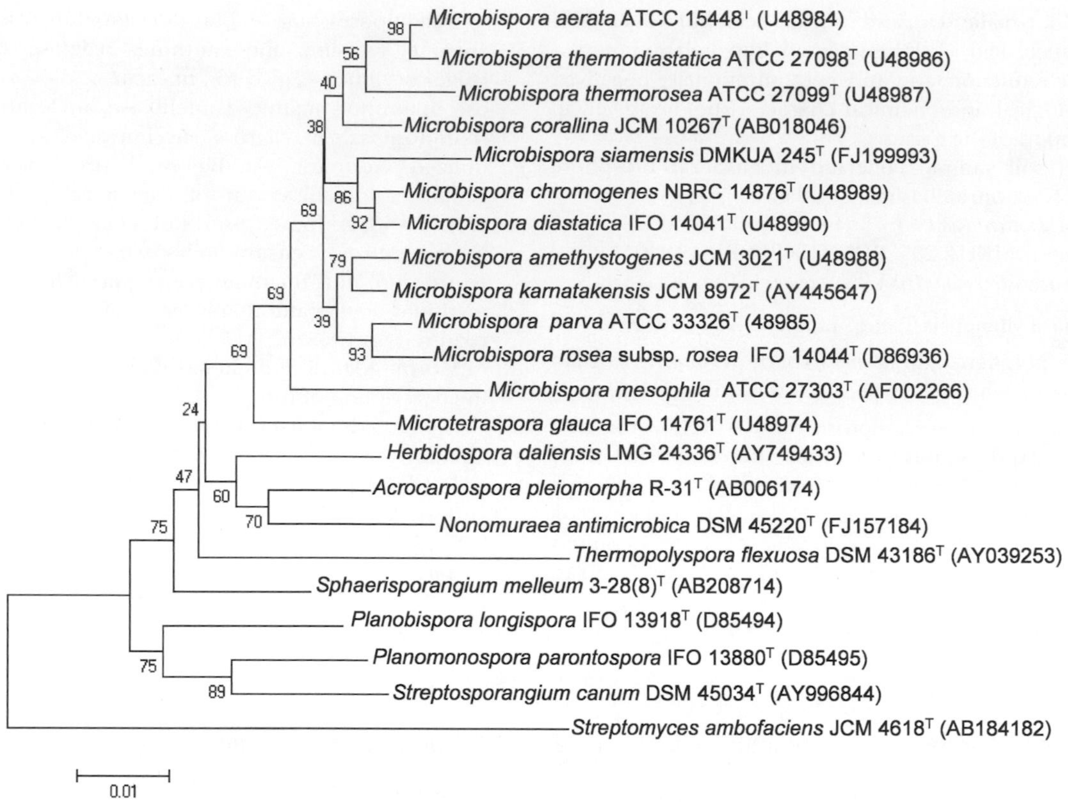

FIGURE 369. 16S rRNA gene-based neighbor-joining tree showing the phylogenetic relationships between *Microbispora* species and between selected microorganisms belonging to the family *Streptosporangiaceae.* The numbers on the branches indicate the percentage bootstrap values of 1000 replicates. *Streptomyces ambofaciens* JCM 9374^T was used as an outgroup.

Cell-wall composition. *Microbispora* strains contain *meso*-diaminopimelic acid as the major wall diamino acid, have a wall chemotype III, and their whole-organism hydrolysates contain madurose (Lechevalier and Lechevalier, 1970b; Nonomura and Ohara, 1971d). The wall peptidoglycan is of the type A1γ and contains *N*-acetylated muramic acid (Nakajima et al., 1999). *Microbispora* contains phosphatidylethanolamine, phosphatidylinositol, diphosphatidylglycerol, phosphatidylinositol mannosides, and ninhydrin and sugar positive phospholipids (phospholipid type IV *sensu* Lechevalier et al., 1977), major proportions of tetrahydrogenated menaquinones with nine isoprene units saturated at sites III and VIII, and substantial proportions of MK-9(H_4, H_2, and H_0) (Kroppenstedt et al., 1990; Nakajima et al., 1999). Members of the genus contain complex mixtures of fatty acids that include major proportions of hexadecanoic acid ($C_{16:0}$), heptadecanoic acid ($C_{17:0}$), 14-methylpentadecanoic ($C_{16:0}$ iso), and 10 methylheptadecanoic ($C_{17:0}$ 10-methyl) acids (Table 281; Kroppenstedt et al., 1990; Nakajima et al., 1999).

Ecology. Microbisporae are common in soils, including soils rich in leaf litter (Nonomura, 1984; Nonomura and Ohara, 1969a), and even desert soils (Takahashi et al., 1996). They have also been isolated in large numbers from surface-sterilized fallen leaves (Matsumoto et al., 1998). Okazaki et al. (1995) collected monocotyledon leaves from 170 plants to obtain 57 *Microbispora* species from the leaf surfaces. The prevalence of strains belonging to *Microbispora* from healthy plant leaves was confirmed by another study by the same group (Kizuka et al., 1998). Endophytic *Microbispora* species were isolated from surface-sterilized wheat roots (Coombs and Franco, 2003) and a range of plants (Taechowisan et al., 2003). Polyethylene-degrading thermophilic strains have been isolated in Taiwan (Hoang et al., 2007; Tseng et al., 2007). Secondary metabolites have been reported from a number of strains; these include a prolyl endopeptidase inhibitor from *Microbispora rosea* and *Microbispora indica* strains (Kimura et al., 1997), tyrosine kinase inhibitors from *Microbispora rosea* strains (Kajiura et al., 1998), antibiotics phenazins and phenoxazinones from *Microbispora aerata* (Gerber and Lechevalier, 1964), glucosylquestiomycin (Igarashi et al., 1998) and bispolides from *Microbispora* species (Okujo et al., 2007), and herbicides from a *Microbispora rosea* strain (Kizuka et al., 1998).

Enrichment and isolation procedures

Microbispora are present in low numbers in soils and constitute a minor component of the actinomycete population (Hayakawa and Nonomura, 1987b). Therefore, selective isolation media such as AV*(Nonomura and Ohara, 1969a), Chitin-V† (Nonomura, 1984), and HV‡ agars (Hayakawa and Nonomura, 1987b) together with pretreatments of samples were developed to yield *Microbispora* species from soils. *Microbispora mesophila* was isolated as "*Thermomonospora mesophila*" using MGA-SE agar§ (Nonomura and Ohara, 1971b). In later improvements to the selective isolation procedures, soils collected from farms in Japan were air-dried, heated to 120°C for 1 h, and treated with 1.5% phenol and 0.03% chlorhexidine gluconate (Hayakawa et al., 1991b) before plating on HV containing 20 mg nalidixic acid. Matsumoto et al. (1998) used glycerol asparagine agar** or water plus 1% proline agar containing the antibiotics nystatin, benomyl, cycloserine, and nalidixic acid to obtain 34 strains of *Microbispora* from a total of 77 actinomycetes obtained from fallen leaves.

Microbispora rosea, *Microbispora amethystogenes*, *Microbispora chromogenes*, *Microbispora diastatica*, and *Microbispora parva* were isolated from soil samples collected in Japan using soil extract agar (Nonomura and Ohara, 1957, 1960). *Microbispora corallina* and *Microbispora siamensis* were isolated from forest soils in Thailand with dilution plating on HV agar (Boondaeng et al., 2009; Nakajima et al., 1999), while *Microbispora indica* and *Microbispora karnatakensis* were isolated from soil samples from India using undisclosed methods (Rao et al., 1987).

Maintenance procedures

Well-sporulated agar slant cultures can be maintained at 4°C for 3–4 months. Recommended media for mesophilic strains are oatmeal agar-YG and glycerol agar (C1 medium†† supplemented with 0.5% (w/v) glycerol) for thermophilic strains. Suspensions of spores and hyphae in glycerol (20%, v/v) can be stored in liquid nitrogen at −80°C. Lyophilization is recommended for long-term preservation by placing the spores and mycelia in skimmed milk.

Differentiation of the genus *Microbispora* from other genera

Microbispora strains can be distinguished from the genera *Microtetraspora* and *Nonomuraea* by their ability to form characteristic longitudinal pairs of spores on short aerial hyphae with smooth spore surfaces, except for *Microbispora mesophila* which is monosporic. *Microtetraspora* species have characteristic chains of four spores on distinct sporophores, whereas *Nonomuraea* species bear chains of spores.

Taxonomic comments

Eleven *Microbispora* species were included in the last edition of *Bergey's Manual of Systematic Bacteriology* (Nonomura, 1989b); these include *Microbispora aerata* Gerber and Lechevalier 1964, *Microbispora amethystogenes* Nonomura and Ohara 1960, *Microbispora bispora* Henssen 1957, *Microbispora chromogenes* Nonomura and Ohara 1960, *Microbispora diastatica* Nonomura and Ohara 1960, *Microbispora echinospora* Nonomura and Ohara 1971b, *Microbispora parva* Nonomura and Ohara 1960, *Microbispora rosea* Nonomura and Ohara 1957, *Microbispora thermodiastatica* Nonomura and Ohara 1969b, *Microbispora thermorosea* Nonomura and Ohara 1969b, and *Microbispora viridis* Miyadoh et al. 1985. Later two others were proposed as valid species, *Microbispora indica* and *Microbispora karnatakensis* (Rao et al., 1987). On the basis of differences in chemotaxonomy and DNA–DNA relatedness, "*Microbispora echinospora*" and "*Microbispora viridis*" were transferred to the genus *Actinomadura* as *Actinomadura echinospora* and *Actinomadura rugatobispora*, respectively (Kroppenstedt et al., 1990; Miyadoh et al., 1990). Miyadoh et al. (1990) further proposed the combination of the 10 extant species, other than *Microbispora bispora*, into two subspecies of *Microbispora rosea* on the basis of DNA–DNA relatedness and chemotaxonomy. "*Microbispora rosea* subsp. *rosea*" contains *Microbispora rosea*, *Microbispora amethystogenes*, *Microbispora chromogenes*, *Microbispora diastatica*, *Microbispora echinospora*, *Microbispora indica*, *Microbispora karnatakensis*, and *Microbispora parva*; whereas "*Microbispora rosea* subsp. *aerata*" contains *Microbispora aerata*, *Microbispora thermodiastatica* and *Microbispora thermorosea*. However, Wang et al. (1996b) called into question the taxonomic status of these combinations, a view supported by Ochi et al. (1991) based on electrophoretic mobility of ribosomal AT-L30 protein. The clear differences observed for *Microbispora bispora* were analyzed by Wang et al. (1996a) who found that, on the basis of 16S rRNA gene sequence dissimilarity, two strains of this species were sufficiently different from the other species of *Microbispora* as well as other members of the family *Streptosporangiaceae* to be transferred to a new genus, *Thermobispora*, as *Thermobispora bispora*. Subsequently, Zhang et al. (1986a) proposed the reclassification of the monosporic *Thermomonospora mesophila* to the genus *Microbispora*, as *Microbispora mesophila*, on the basis of 16S rRNA gene sequence similarity and reported chemotaxonomic properties. Later *Microbispora corallina* was proposed by Nakajima et al. (1999) for two new mesophilic strains isolated from deciduous dipterocarp forest soils in Thailand, and *Microbispora siamensis* was proposed for a thermotolerant strain isolated from a soil sample from the same country (Boondaeng et al., 2009).

Differentiation of species of the genus *Microbispora*

Microbispora species can be distinguished from one another by using a combination of morphological, chemotaxonomic, nutritional, and physiological characteristics (Table 280 and Table 281).

---

*AV agar: arginine, 0.3 g; glucose, 1 g; glycerol, 1 g; K_2HPO_4, 0.3 g; Mg $S0_4 \cdot 7H_2O$, 0.2 g; NaCl, 0.3 g; agar, 15 g; B-vitamins (0.5 mg each of *p*-aminobenzoic acid, calcium pantothenate, inositol, niacin, pyridoxine-HCl, riboflavin, thiamine HCl, and 0.25mg biotin); distilled water 1 liter; antibiotics (Actidione, 50 mg; nystatin, 50 mg; polymixin B, none or 4 mg; penicillin G, none or 0.8 mg); pH 6.4.

† Chitin-V agar: colloidal chitin, 2 g; K_2HPO_4, 0.35 g.; KH_2PO_4, 0.15 g; $MgSO_4 \cdot 7H_2O$, 0.2 g; NaCl, 0.3 g; $CaCO_3$, 0.02 g; Fe $S0_4 \cdot 7H_2O$, 0.01 g;. $ZnS0_4 \cdot 7H_2O$, 1 mg; $MnCl_2$, 1 mg;. agar, 18 g,; β-vitamins, as for AV agar; distilled water 1 liter; Actidione, 50 mg; pH 7.2.

‡ HV agar: humic acid, 1 g; $CaC0_3$, 0.02 g; Fe $S0_4 \cdot 7H_2O$, 0.01 g; KCl, 1.7g; Mg $S0_4 \cdot 7H_2O$, 0.05 g; Na_2HPO_4, 0.5 g.; agar, 15 g; β-vitamins, as for AV agar; distilled water 1 liter; Actidione, 50 mg; pH 7.2.

§ MGA-SE agar:glucose, 2 g; L-asparagine, 1 g; K_2HPO_4, 0.5g; soil extract, 200 ml; agar, 20 g; distilled water 800 ml; antibiotics (Actidione, 50 mg; nystatin, 12.5 mg; polymixin B, 4 mg; penicllin G, 0.8 mg); pH 8.0.

**Glycerol-asparagine agar: L-asparagine, 0.1 g; glycerol, 1 g; K_2HPO_4, 1 mg; $FeSO_4 \cdot 7H_2O$, 0.5 g; $ZnS0_4 \cdot 7H_2O$, 1 mg; $MnCl_2 \cdot 4H_2O$, 1 mg;.agar, 15 g; distilled water 1 liter; antibiotics (Benlate, 20 mg; cycloserine, 50 mg; nalidixic acid, 25 mg; nystatin, 12.5 mg).

†† C1 medium (Nonomura and Ohara, 1969b): casamino acids, 2 g; K_2HPO_4, 0.3 g; $MgSO_4 \cdot 7H_2O$, 0.01 g; NaCl, 0.3 g; Fe $SO_4 \cdot 7H_2O$, 10 mg;. $ZnS0_4 \cdot 7H_2O$, 1 mg; $CuSO_4 \cdot 5H_2O$, 1 mg; B-vitamins (0.5 mg each of *p*-aminobenzoic acid, calcium pantothenate, inositol, niacin, pyridoxine-HCl, riboflavin, thiamine HCl, and 0.25 mg biotin); agar, 20 g; distilled water 1 liter, pH 7.2.

TABLE 280. Characteristics differentiating the type strains of *Microbispora* species[a,b]

Characteristic	*M. rosea*	*M. aerata*	*M. amethystogenes*	*M. chromogenes*	*M. corallina*	*M. diastatica*	*M. indica*	*M. karnatakensis*	*M. mesophila*	*M. parva*	*M. siameensis*	*M. thermodiastatica*	*M. thermorosea*
Aerial mycelium color	Pink	Off-white/ pink	Pale pink	Pink	Pink	Pink	Pinkish white	White	White	White-pale pink	Pale pink	White-pale yellow brown	Pinkish white
Substrate mycelium color	Orange	Yellow-brown	Brown	Orange	Orange-red	Yellow-brown	Violet-orange	Brown-orange	Brown	Yellow-brown	Yellow	Yellow-brown	Yellow-brown
Growth (°C)	17–35	28–55	17–35	17–35	17–35	17–35	28–50	28–50	25–40	17–35	25–50	35–55	35–55
Menaquinone MK-9(H_0) (%)	21	52/40	21	11	+	26	15	15		42	+	8/9	2/5
Menaquinone MK-9(H_2) (%)	49	36/36	54	34	+	51	43	32		45	+	44/33	14/20
Menaquinone MK-9(H_4) (%)	30	9/20	21	50	+	23	40	46		12	+	43/47	57/51
Menaquinone MK-9(H_6) (%)		0/2	2	3			1	5				3/9	22/20
Biochemical tests:													
Biotin requirement	–	+	–	+	–	–	+	+	–	–	–	+	+
Degradation of hypoxanthine	+	+	–	+	–	+	+	+	+	–	nd	+	+
Degradation of testosterone	+	+	–	+	–	+	+	+	nd	+	nd	+	+
Production of iodinin	–	+	+	–	–	–	–	–	–	+	nd	–	–
Reduction of nitrate	++	++	++	++	+	++	++	++	+	–	nd	–	–
Starch hydrolysis	–	+	–	+	+	+	+	–	+	–	nd	+	–
Thiamine requirement	+	+	+	+	+	+	+	+	nd	+	nd	+	+
Growth on sole carbon sources:													
Arabinose	+	–	+	+	nd	+	+	–	–	+	nd	+	+
Glycerol	–	+/–	–	+	++	++	–	–	–	++	–	+/–	+
Lactate	+	+/–	+	+	+	+	+	+	+	+	–	–	–
Malate	+	+	–	–	–	–	+	+	nd	+	nd	–	+
myo-Inositol	–	+/–	–	+	+	+	–	+	–	–	–	–	–
L-Rhamnose							+	+	–		+		
Sorbitol	+/–	+/–	–	+	+	+	+	+		+/–	+	+	+

[a]Symbols: + or ++, positive; –, negative; +/–, variable; nd, not determined.

[b]Data from McCarthy and Cross (1984), Rao et al. (1987), Miyadoh et al. (1990), Kroppenstedt and Goodfellow (1992), Nakajima et al. (1999), Wang et al. (1996a), and Boondaeng et al. (2009).

TABLE 281. Fatty acid profiles of the type strains of *Microbispora* species

Fatty acid	*M. rosea* JCM 3006	*M. aerata*[a] IFO 102581	*M. amethystogenes* JCM 3021	*M. chromogenes* JCM 3022	*M. corallina* JCM 10267	*M. diastatica* JCM 3023	*M. indica* ATCC 35926	*M. karnatakensis* ATCC 35927	*M. mesophila*	*M. parva* JCM 3024	*M. siamensis* BCC 14407	*M. thermodiastatica*[a] IFIO 14046	*M. thermorosea*[a] IFO 14047
Straight-chain fatty acids:													
$C_{15:0}$	2	1	4	3	10	4	3	3	nd	2		7	
$C_{16:0}$	9	7	10	6	8	7	7	8	nd	7		2	7
$C_{17:0}$	10	6	5	7	9	6	4	7	nd	2		2	1
$C_{18:0}$	3	1	3	2	1	2	2	2	nd	2		1	2
Unsaturated fatty acids:													
$C_{16:1}$	3		7	3	2	5	5	3	nd	4			
$C_{18:1}$	3		5	2	1	1	5	3	nd	5			
Branched fatty acids:													
$C_{15:0}$ iso	2		5	4	7	3	2	1	nd	4	+	1	
$C_{16:0}$ iso	43	42	18	35	19	30	32	39	nd	20	+	32	32
$C_{17:0}$ anteiso	9	8	14	6	3	10	11	9	nd	4		21	25
$C_{18:0}$ iso				1	1			4	nd			3	3
10-Methyl fatty acids:													
$C_{17:0}$ iso	2	5	7	9	4	9	8	4	nd	16		5	5
$C_{18:0}$ iso	12	16	15	14	17	17	13	10	nd	15		9	9
$C_{19:0}$ iso	1	5	6	4	2	3	3	3	nd	8		6	6

[a]Cultured at 37°C. Data from Miyadoh et al. (1990), Nakajima et al. (1999), and Boondaeng et al. (2009).

List of species of the genus *Microbispora*

1. **Microbispora rosea** Nonomura and Ohara 1957, 307[AL]

ro′se.a. L. fem. adj. *rosea* rose-colored.

Aerial hyphae white at first, becoming pale pink with the formation of spores on oatmeal yeast extract agar. Spores are oval or round (1.4–1.6 μm) (Figure 370). Good to moderate growth on Emerson's, Bennett's, C.B., nutrient, and oatmeal and soil agars. An orange substrate mycelium is formed on oatmeal agar with 0.1% peptone. B vitamins are required for growth. Grows between 17–35°C, but not at 55°C. Optimal pH range for growth is 6.0–7.0. Degrades casein, hypoxanthine, and testosterone.

Source: soil.

DNA G+C content (mol%): 69.9 (HPLC).

Type strain: ATCC 12950, DSM 43839, JCM 3006, NBRC 14044, NRRL B-2632, VKM Ac-634.

Sequence accession no. (16S rRNA gene): D86936.

2. **Microbispora aerata** (Gerber and Lechevalier 1964) Cross 1974, 859[AL] ("*Waksmania aerata*" Gerber and Lechevalier 1964, 598.)

a.e.ra′ta. L. fem. part. adj. *aerata* covered with bronze.

Poor or moderate growth on oatmeal-yeast extract agar with white to pale pink aerial mycelium, with production of iodinin. Spores are oval or round (1.3–1.5 μm). Grows between 28–55°C. Optimal pH range for growth is 6.0–7.0.

Slight production of melanoid pigments; starch hydrolyzed; nitrites produced from nitrates; slight hydrolysis of gelatin; peptonization of milk. Requires thiamine and low levels of biotin.

DNA G+C content (mol%): 72% (T_m).

Type strain: ATCC 15448, DSM 43176, JCM 3076, NBRC 12581, VKM Ac-1507.

Sequence accession no. (16S rRNA gene): U48984.

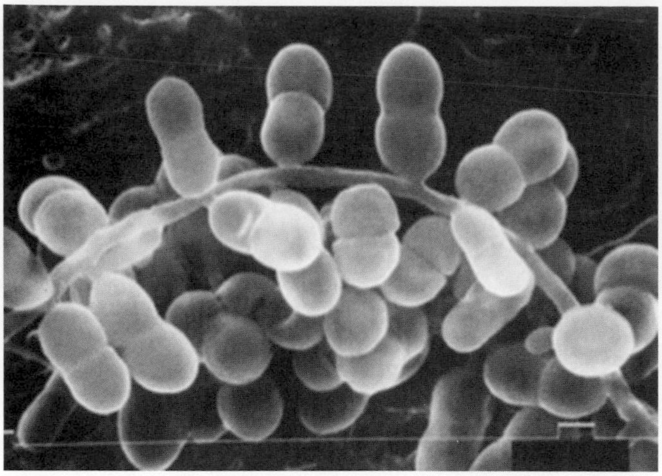

FIGURE 370. Morphology of *Microbispora rosea*: paired spores on hyphae. Scanning electron micrograph. Bar interval = 10 µm.

3. **Microbispora amethystogenes** Nonomura and Ohara 1960, 404[AL]

am.e.thys.to'gen.es. L. adj. *amethystinus* amethyst colored; N.L. suff. *-genes* (from Gr. v. *gennaô* to produce) producing; N.L. part. adj. *amethystogenes* producing violet-colored (crystals).

Aerial mycelium is pink; soluble pigment is pale yellowish brown on oatmeal-yeast extract agar. Abundant violet crystals observed after 30 d. On oatmeal-peptone agar, the substrate mycelium is light brown to brownish gray with scant aerial mycelium. Spores are oval or round (1.3–1.6 µm). Poor growth on glycerine, nutrient, and glycerol-asparagine agar-yeast extract agars with no aerial mycelium. Grows between 17–35°C, but not at 55°C. Optimal pH range for growth is 6.0–7.0.

No melanoid pigments produced; no starch hydrolysis; gelatin liquefied; slight peptonization of milk; nitrites produced from nitrates. Requires thiamine and low levels of biotin.

DNA G+C content (mol%): 70.5 (HPLC).

Type strain: DSM 43164, JCM 3021, NBRC 101907, NRRL B-2637.

Sequence accession no. (16S rRNA gene): U48988.

4. **Microbispora chromogenes** Nonomura and Ohara 1960, 404[AL]

chro.mo'ge.nes. Gr. n. *chroma* color; N.L. suff. *-genes* (from Gr. v. *gennaô* to produce) producing: N.L. part. adj. *chromogenes* producing color.

Aerial mycelium is pink; soluble pigment is dark-purple gray on oatmeal-yeast extract agar. On potato agar, the substrate mycelium is pale orange, aerial mycelium pale pink with pale yellow soluble pigment. A dark green soluble pigment is produced on glycerin agar. Spores are oval or round (1.3–1.6 µm). Good to moderate growth on glycerin, glycerol-asparagine-yeast extract, nutrient, potato, and starch-yeast extract agars. Grows between 17–35°C, but not at 55°C. No melanoid pigments produced; starch hydrolyzed; gelatin liquefied; no peptonization of milk; nitrites produced from nitrates. Requires thiamine and low levels of biotin.

Source: soil.

DNA G+C content (mol%): 70 (HPLC).

Type strain: DSM 43165, JCM 3022, NBRC 14876, NRRL B-2634.

Sequence accession no. (16S rRNA gene): U48989.

5. **Microbispora diastatica** Nonomura and Ohara 1960, 404[AL]

di.a.sta'ti.ca. N.L. fem. adj. *diastatica* starch hydrolyzing.

Aerial mycelium is pink; soluble pigment is pale yellow on oatmeal–yeast extract agar. On oatmeal–peptone agar, the substrate mycelium is pale yellowish brown covered entirely with pale pink aerial mycelium. Spores are oval or round (1.3–1.6 µm). Good growth on glycerine and nutrient agars with no aerial mycelium. Grows between 17–35°C, but not at 55°C. Optimal pH range for growth is 6.0–7.0.

No melanoid pigments produced; starch hydrolyzed; gelatin liquefied; no peptonization of milk; and nitrites not produced from nitrates. Utilizes rhamnose. Requires thiamine and low levels of biotin.

DNA G+C content (mol%): 68.9 (HPLC).

Type strain: JCM 3023, KCC A-0023, NBRC 14041, NBRC 101785, NRRL B-2630.

Sequence accession no. (16S rRNA gene): U48990.

6. **Microbispora mesophila** (Nonomura and Ohara 1971b) Zhang, Wang and Ruan 1998a, 418[VP] (*Thermomonospora mesophila* Nonomura and Ohara 1971b, 899)

me.so.phi'la. Gr. n. *mesos* middle; Gr. fem. adj. *philê* loving; N.L. fem. adj. *mesophila* middle (temperature) loving.

Colonies on agar media are small, entire, raised, with a brown reverse color and abundant white aerial mycelium, with spores either sessile or on short sporophores. Spores with smooth surfaces either oval or round (1.5–2 µm). Good to moderate growth on oatmeal, yeast extract-malt extract, nutrient, starch-yeast extract agars. Very poor growth on glycerol-asparagine agar. Optimum temperature for growth is 35–40°C. Optimum growth at pH 7.5–8.0.

No melanoid pigments produced; starch hydrolyzed; gelatin liquefied weakly; peptonization of milk; nitrites produced from nitrates. Requires thiamine and low levels of biotin.

DNA G+C content (mol%): not determined.

Type strain: ATCC 27303, CIP 105593, DSM 43048, NBRC 14179, JCM 3151, NRRL B-16986, VKM Ac-1953.

Sequence accession no. (16S rRNA gene): AB006170, AF002266.

7. **Microbispora parva** Nonomura and Ohara 1960, 403[AL]

par'va. L. fem. adj. *parva* little, small (growth).

Poor growth on oatmeal agar with a faint yellow soluble pigment; violet crystals observed after 30 d at 30°C. Aerial mycelium is pink; soluble pigment is white to pale pink on oatmeal-yeast extract agar. On oatmeal-peptone agar, the substrate mycelium is pale yellowish brown to light brown with scant aerial mycelium. Spores are oval or round (1.3–1.6 µm). Good to moderate growth on glycerine, nutrient, potato, and glycerol-asparagine agar-yeast extract agars.

Grows between 17–35°C, but not at 55°C. Optimal pH range for growth is 6.0–7.0.

No melanoid pigments produced; no starch hydrolysis; slight hydrolysis of gelatin; slight peptonization of milk. Requires thiamine and low levels of biotin.

DNA G+C content (mol%): 69.8 (HPLC).

Type strain: ATCC 33326, JCM 3024, KCC A-0024, NRRL B-2629.

Sequence accession no. (16S rRNA gene): U48985.

8. **Microbispora siamensis** Boondaeng, Ishida, Tamura, Tokuyama and Kitpreechavanich 2009, 3138[VP]

si.a.men.sis. NL. fem. adj. *siamensis* of or pertaining to Siam, the old name of Thailand, the source of the soil from which the type strain was isolated.

Substrate mycelia are colorless to yellow on most media; yellow and green soluble pigments are produced. Oval spores with smooth surfaces. Good growth on yeast-extract malt extract agar with moderate pink spores.

Temperature growth range is 25–50°C. No growth in the presence of 3% (w/v) NaCl.

Citric acid, L-malic acid, and succinic acid are used, but benzoic acid, fumaric acid, and mucic acid are not. L-Arabinose, D-fructose, D-galactose, D-glucose, D-lactose, D-mannitol, D-mannose, and sucrose are used as sole carbon sources, but dulcitol, inositol, maltose, D-melezitose, D-raffinose, L-rhamnose, and D-sorbitol are not. Cellulose is not degraded; gelatin not liquefied; no nitrites produced from nitrates; no peptonization of milk.

Source: soil.

DNA G+C content (mol%): 68 (HPLC).

Type strain: DMKUA 245, BCC 14407, NBRC 104113.

Sequence accession no. (16S rRNA gene): FJ199993.

9. **Microbispora thermodiastatica** Nonomura and Ohara 1969b, 706[AL]

ther.mo.di.a.sta'ti.ca. Gr. adj. *thermos* hot; N.L. fem. adj. *diastatica* starch-hydrolyzing; N.L. fem. adj. *thermodiastatica* heat-loving, starch-hydrolyzing.

Poor or moderate growth on oatmeal-yeast extract agar with white to pale yellow aerial mycelium at 40°C. Good growth on yeast-starch agar with white to pale yellow brown aerial mycelium. On glucose agar, the aerial mycelium is pale pink with no soluble pigment. Spores are oval or round (1.3–1.5 µm). Grows between 35–55°C. Optimal pH range for growth is 6.0–7.0.

No melanoid pigments produced; starch hydrolyzed; no nitrites produced from nitrates; slight hydrolysis of gelatin; peptonization of milk. Requires thiamine and low levels of biotin.

DNA G+C content (mol%): 70.3 (HPLC).

Type strain: ATCC 27098, JCM 3110.

Sequence accession no. (16S rRNA gene): U48986.

10. **Microbispora thermorosea** Nonomura and Ohara 1969b, 707[AL]

ther.mo.ro'se.a. Gr. adj. *thermos* hot; L. fem. adj. *rosea* rose-colored; N.L. fem. adj. *thermorosea* heat-loving, rose-colored.

Poor or moderate growth on oatmeal agar. Good growth on yeast-starch agar; aerial mycelium is white to pale yellowish brown. Spores are oval or round (1.3–1.6 µm). Grows between 35–55°C. Optimal pH range for growth is 6.0–7.0.

No melanoid pigments produced; no starch hydrolysis; no nitrites produced from nitrates; slight hydrolysis of gelatin; slight peptonization of milk. Requires thiamine and low levels of biotin.

DNA G+C content (mol%): 70.1 (HPLC).

Type strain: ATCC 27099, JCM 3111.

Sequence accession no. (16S rRNA gene): U48987.

Species *incertae sedis*

1. **Microbispora corallina** Nakajima, Kitpreechavanich, Suzuki and Kudo 1999, 1766[VP]

co.ral'li.na. L. fem. adj. *corallina* coral colored.

Substrate mycelia pinkish to brownish red on most media and yellowish soluble pigments produced. Oval spores (0.8~1.2 µm). Good growth on glucose-asparagine agar +1% yeast extract; poor aerial mycelium pink, substrate mycelium russet, yellow diffusible pigment. Good growth on inorganic salts-starch agar +1% yeast extract with cherry pink aerial mycelium, coral substrate mycelium, no diffusible pigment. Good growth on oatmeal agar with scant pink aerial mycelium, yellow diffusible pigment. Good growth on yeast extract-starch agar with poor pink aerial mycelium, dark red substrate mycelium, yellow diffusible pigment. Good growth on yeast-extract-malt extract agar with moderate pink aerial mycelium and deep reddish brown substrate mycelium; no diffusible pigment. Temperature growth range: 17–25°C. No growth in the presence of NaCl 3% (w/v).

Casein and esculin are degraded, but adenine, elastin, guanine, hypoxanthine, keratin, testosterone, tyrosine, xanthine, xylan, and DNA are not. Amylase is produced. Amygdalin, L-arabinose, D-cellobiose, D-fructose, D-galactose, D-glucose, glycerol, *meso*-inositol, D-lactose, maltose, D-mannitol, D-mannose, D-melezitose, D-melibiose, D-ribose, salicin, D-sorbitol, sucrose, D-trehalose, and D- xylose are used as sole carbon sources, but adonitol, D-arabitol, dulcitol, iso-erythritol, inulin, D-raffinose, L-rhamnose, and xylitol are not.

DNA G+C content (mol%): 71.2–71.5 (HPLC).

Type strain: DF-32, = ATCC BAA-20, DSM 44682, NBRC 16416, JCM 10267.

Sequence accession no. (16S rRNA gene): AB018046.

2. **Microbispora indica** Rao, Prabhu, Sridhar, Venkateswarlu and Actor 1987, 184[VP]

in.di.ca. L. fem. adj. *indica* pertaining to India, the source of the soil from which the type strain was isolated.

Substrate mycelia yellow to yellowish brown on most media and orange yellowish soluble pigments produced on oatmeal agar. Oval spores (1.2~1.8 μm). Poor growth on glycerol-asparagine agar with no aerial mycelium, colorless substrate mycelium, no diffusible pigment. Good growth on glycerol-asparagine agar +1% yeast extract with light gray aerial mycelium, yellow brown substrate mycelium, no diffusible pigment. Good growth on oatmeal agar with pink aerial mycelium, violet-orange substrate mycelium, orange-yellow diffusible pigment. Good growth on yeast-extract-malt extract agar with light gray aerial mycelium, yellow brown substrate mycelium, no diffusible pigment. Temperature growth range: 28–50°C. No growth in the presence of NaCl 3% (w/v).

Casein is degraded, but amylase is not produced. L-Arabinose, D-glucose, and L-rhamnose are used as sole carbon sources, but glycerol and *meso*-inositol are not.

DNA G+C content (mol%): not reported

Type strain: SKF-I-101055, ATCC 35926, NBRC 14879, JCM 8971.

Sequence accession no. (16S rRNA gene): not available.

3. **Microbispora karnatakensis** Rao, Prabhu, Sridhar, Venkateswarlu and Actor 1987, 184[VP]

kar.na.ta.ken'sis. N.L. fem. adj. *karnatakensis* of or pertaining to the State of Karnataka, India, the source of the soil from which the type strain was isolated.

Substrate mycelia were orange to yellowish brown on most media, and yellowish orange soluble pigments were produced on most media. Oval spores (1.2–1.8 μm). Good growth on glycerol-asparagine agar, but no aerial mycelium, yellowish brown substrate mycelium, no diffusible pigment. Moderate growth on glycerol-asparagine agar +1% yeast extract with orange substrate mycelium and bright yellow diffusible pigment. Good growth on oatmeal agar with white aerial mycelium, brownish-orange substrate mycelium, deep orange-yellow diffusible pigment. Good growth on yeast-extract-malt extract agar with pale pink aerial mycelium, deep orange substrate mycelium, no diffusible pigment. Temperature growth range is 28–50°C.

Casein is degraded and amylase is produced. D-Glucose, L-rhamnose, and *meso*-inositol are used as sole carbon sources, but L-arabinose and glycerol are not.

DNA G+C content (mol%): not available.

Type strain: SKF-I-58261, ATCC 35927, IMSNU 22065, JCM 8972.

Sequence accession no. (16S rRNA gene): AY445647.

Genus V. **Microtetraspora** Thiemann, Pagani and Beretta 1968b, 296[AL] emend. Zhang, Wang and Ruan 1998a, 420[VP]

Martha E. Trujillo and Michael Goodfellow

Mi.cro.te.tra.spo'ra. Gr. adj. *mikros* small; Gr. adj. *tetra* four; Gr. n. *spora* a seed; N.L. fem. n. *Microtetraspora* the small four-spored (organism).

Aerobic, Gram-stain-positive, non-acid–alcohol-fast, nonmotile, mesophilic actinomycetes **which form stable, highly branched substrate and aerial mycelia. Spore chains, typically contain four spores which are borne on short aerial hyphae. Spores are spherical (1.2–1.5 μm in diameter) or slightly oval to cylindrical (1.0–1.4 to 1.2–1.7 μm in diameter) and have smooth surfaces. Grows from 20–37°C, but not at 40°C**. Some species require B vitamins for growth. Chemoorganotrophic with an oxidative type of metabolism. **Cell walls contain *N*-acetylated muramic acid and major amounts of *meso*-diaminopimelic acid; has a type A1γ peptidoglycan**. Contains complex mixtures of saturated, unsaturated, iso-, anteiso-, and branched chain fatty acids. **Predominant menaquinones have nine isoprene units with hydrogenation at positions III and IV [MK-9(III, IV-H_4)]. Major phospholipids are diphosphatidylglycerol, phosphatidylethanolamine, hydroxyl-phosphatidylethanolamine, and ninhydrin-positive glycophospholipids**. Common in soils.

DNA G+C content (mol%): 69–71.

Type species: **Microtetraspora glauca** Thiemann, Pagani and Beretta 1968b, 296[AL].

Further descriptive information

Phylogeny. The genus *Microtetraspora* contains four species including *Microtetraspora glauca*, *Microtetraspora fusca*, *Microtetraspora malaysiensis*, and *Microtetraspora niveoalba* (Table 282). The type strains of these species form a distinct phyletic line in the *Streptosporangiaceae* 16S rRNA gene tree that is most closely related to the genera *Microbispora* and *Nonomuraea* (Figure 371). The four *Microtetraspora* type strains share 16S rRNA gene similarities within the range 97.9–99.1%. The corresponding DNA–DNA relatedness values range from 21–52% (Nakajima et al., 2003).

Cell morphology. Microtetrasporae form short, sparsely branched aerial hyphae which typically carry chains of four spores (Figure 372, Figure 373, Figure 374, and Figure 375), though chains of two or three spores, and more rarely of five spores, have been reported (Nakajima et al., 2003; Zhang et al., 1998a). Peculiar side branches have been observed on the spore chains of *Microtetraspora niveoalba* (Figure 375). The spores of *Microtetraspora fusca* tend to fuse into a spore mass as the culture ages.

Nutrition and growth conditions. Microtetrasporae are readily cultivated on rich media including modified Bennett's (Jones, 1949), glucose-yeast extract (Gordon and Mihm, 1962), oatmeal (Shirling and Gottlieb, 1966), and yeast extract-malt extract (Shirling and Gottlieb, 1966) agars. *Microtetraspora fusca* and *Microtetraspora glauca* grow well and sporulate on Hickey–Tresner (Hickey and Tresner, 1952) agar. *Microtetraspora malaysiensis* grows well on yeast-malt extract agar (Nakajima et al., 2003). *Microtetraspora niveoalba* requires B vitamins for growth on synthetic media (Nonomura and Ohara, 1971d). Microtetrasporae grow between 20 and 37°C, but not at 40°C.

Cell-wall composition. *Microtetraspora* strains contain *meso*-diaminopimelic acid as the major wall diamino acid (wall

TABLE 282. Characteristics that differentiate the type strains of *Microtetraspora* species[a,b]

Characteristic	*M. glauca*	*M. fusca*	*M. malaysiensis*	*M. niveoalba*
Branched spore chains	–	–	–	+
Aerial spore mass color	Blue-gray/gray	Gray	White-cream	White
Substrate mycelium color	Green-blue	Purplish	Cream-yellow	Pale yellow
Biochemical tests:				
Reduction of nitrate	+	–	nd	+
Urea hydrolysis	–	+	–	+
Degradation of:				
Elastin	–	nd	–	+
Hypoxanthine	+	–	–	+
Starch	+	–	nd	+
Testosterone	+	–	+	+
Tyrosine	–	nd	+	nd
Xanthine	+	–	–	+
Xylan	+	–	–	–
Growth on sole carbon sources:				
L-Arabinose	+	+	–	+
Arbutin	+	+	–	+
D-Fructose	+	–	+	+
D-Galactose	+	–	–	–
Glycerol	+	–	–	+
Inositol	+	–	–	+
D-Mannitol	+	–	+	+
D-Mannose	+	+	–	+
L-Rhamnose	+	–	+	+
Sodium citrate	+	+	+	–
Sodium fumarate	+	–	–	+
Sodium malate	+	–	–	+
Trehalose	+	+	–	+
Xylitol	–	–	–	+
Biotin requirement	+	–	–	–

[a]Symbols:+, positive; –, negative; nd, not determined.

[b]Data from Kroppenstedt et al. (1990a), Najakima et al. (2003), and Nonomura, (1989b).

chemotype III *sensu* Lechevalier and Lechevalier, 1970b). The sugar pattern is variable and may be either B (presence of madurose) or C (no characteristic sugar). The wall peptidoglycan contains *N*-acetylated muramic acid (Kawamoto et al., 1981; Nakajima et al., 2003). Microtetrasporae contain major amounts of diphosphatidylglycerol, phosphatidylethanolamine, hydroxylphosphatidylethanolamine, and ninhydrin and sugar positive phospholipids (phospholipid type IV *sensu* Lechevalier et al., 1977), major proportions of tetrahydrogenated menaquinones with nine isoprene units saturated at sites III and VIII, and substantial proportions of MK-9(H_0) and MK-9(H_2) (Kroppenstedt et al., 1990a; Nakajima et al., 2003). Members of the genus contain complex mixtures of fatty acids that include major proportions of 14-methylpentadecanoic ($C_{16:0}$ iso) and 10-methylheptadecanoic ($C_{17:0}$ 10-methyl) acids (Table 283; Miyadoh et al., 1989; Kroppenstedt et al., 1990a; Nakajima et al., 2003).

Ecology. Microtetrasporae are common in soils including forest soils (Hayakawa et al., 1996, 1988; Lazzarini et al., 2000; Nakajima et al., 2003; Nonomura and Ohara, 1971d; Thiemann et al., 1968b). Using humic acid agar, Nonomura and Hayakawa (1988) recorded mean counts of 3.6×10^4 colony-forming units (c.f.u.) per gram dry weight of soil for soil samples collected from forests in Japan. *Microtetraspora niveoalba* strains were particularly widely distributed, albeit with counts $<10^3$ c.f.u./g dry weight soil (Nonomura and Ohara, 1971d). *Microtetraspora malaysiensis* strains have been isolated from two locations on the Malaysian Peninsula, namely from soil taken from a primary lowland dipterocarp forest at Pasok, Negeri Sembilan and from a dipterocarp forest at the Virgin Jungle Reserve, Gombak, Selangor. *Microtetraspora fusca* and *Microtetraspora glauca* have been isolated from soil samples collected in Brazil, Italy, and Thailand.

Enrichment and isolation procedures

Little is known about the activities of microtetrasporae in natural habitats, though members of the *Microtetraspora glauca* group degrade grass lignocellulose and xylan, but not cellulose (Hayakawa et al., 1996). These workers also showed that some members of the taxon show antimicrobial activity. *Microtetraspora fusca, Microtetraspora glauca,* and *Microtetraspora malaysiensis* were isolated from soil samples using undisclosed methods (Nakajima et al., 2003; Thiemann et al., 1968b). However, a pretreatment of dry heating air-dried soil samples, developed for the isolation of *Microbispora* and *Streptosporangium* (Nonomura and Ohara, 1969a), has proven useful for isolating most *Microtetraspora* species. Spores of microtetrasporae appear to be especially resistant to the use of dry heat at 100–120°C, a practice which significantly reduces associated actinomycetes and bacteria present in soil and allows the slow-growing *Microtetraspora* strains to develop into recognizable colonies on isolation plates. Microtetrasporae have also been isolated by plating out

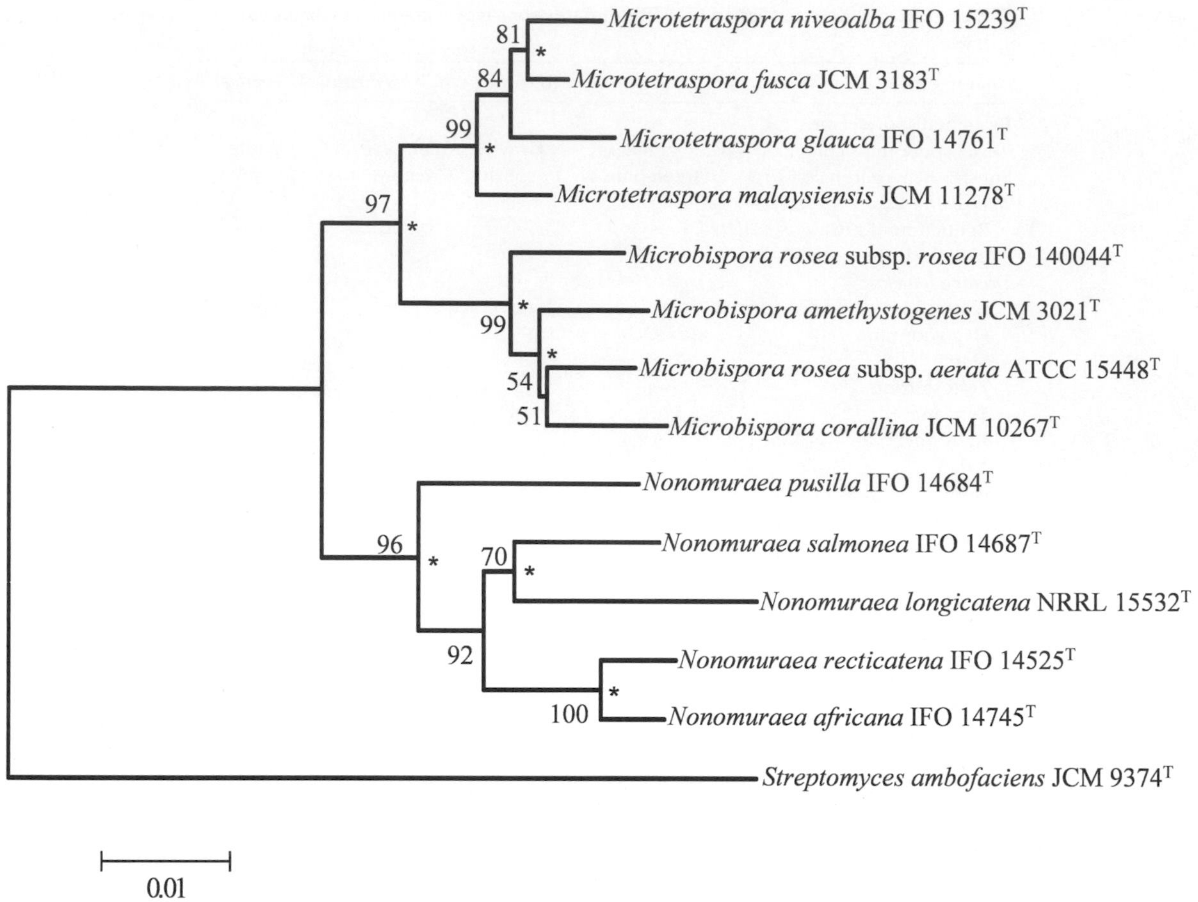

FIGURE 371. Neighbor-joining tree based on 16S rRNA gene sequences showing the phylogenetic relationships between *Microtetraspora* species and selected members of the family *Streptosporangiaceae*. Bootstrap values are shown at branching points. Asterisks indicate that the corresponding nodes were also recovered in maximum parsimony and maximum likelihood trees. *Streptomyces ambofaciens* JCM 9374^T was used as an outgroup. Bar = 1 substitution per 100 nucleotides.

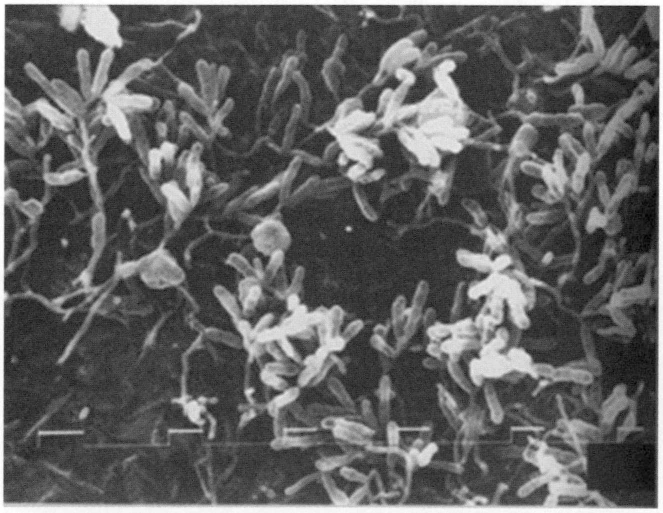

FIGURE 372. Morphology of *Microtetraspora glauca* on oatmeal agar. Scanning electron micrograph. Bar = 10 μm.

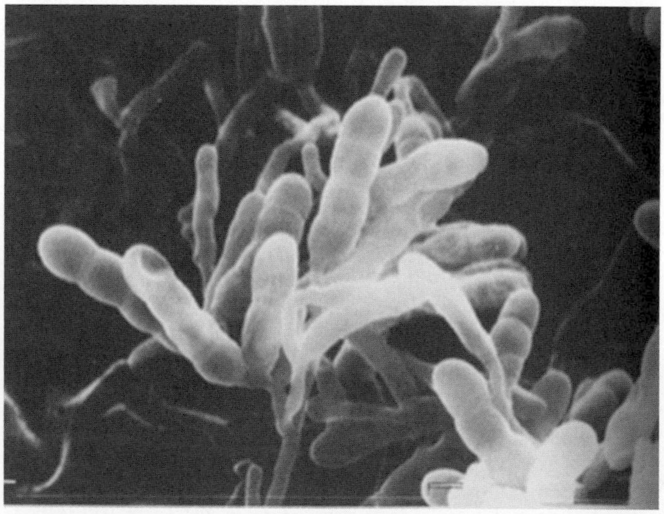

FIGURE 373. Scanning electron micrograph of *Microtetraspora glauca* grown ATCC 27645 on oatmeal agar. Bar = 10 μm.

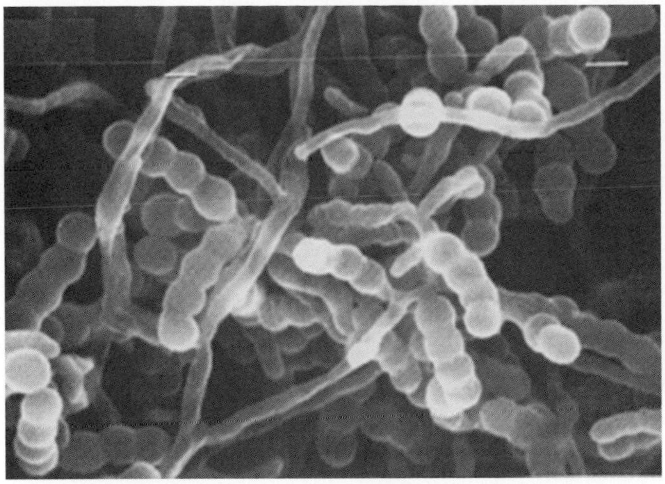

FIGURE 374. Morphology of *Microtetraspora niveoalba* ATCC 27301 on inorganic salts-starch agar. Bar = 10 µm.

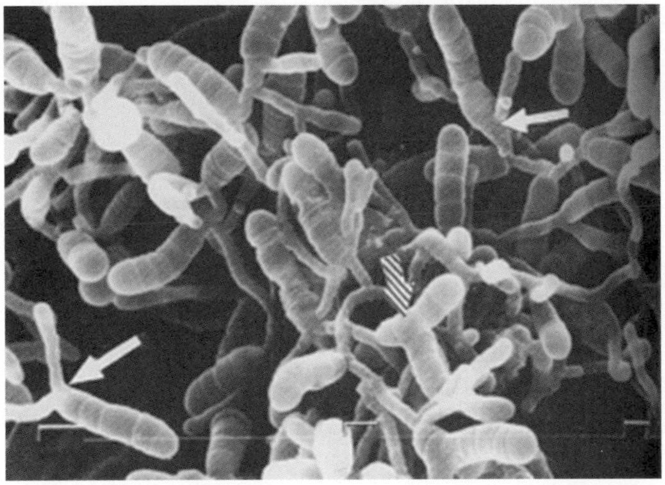

FIGURE 375. Morphology of *Microtetraspora niveoalba*. Striped arrow, branched spore chains; white arrows, branches at broad base of spore chains. Bar = 10 µm.

air-dried and heated soil onto isolation media supplemented with antibiotics such as streptomycin and rifamycin (Li, 1989). Li et al. (2002) selectively isolated actinomycetes, including microtetrasporae, from soil using high frequency radiation.

Microtetraspora niveoalba was isolated from soil which had been dry heated at 120°C for 1 h. Particles of soil were sprinkled sparingly on MGA-SE* agar plates (Nonomura, 1989b; Nonomura and Ohara, 1971d) which were incubated at 38–39°C for 1 month. To obtain pure cultures, fragments of visible colonies that appeared on the primary isolation plates were streaked onto MGA-SE agar. *Microtetraspora glauca* strains have been isolated infrequently on MGA-SE agar plates at 30°C, but *Microtetraspora fusca* was not detected, possibly because of the high pH (8.0) of the medium.

Hayakawa et al. (1996) developed a plate culture procedure for the isolation of members of the *Microtetraspora glauca* group (*Microtetraspora glauca*, *Microtetraspora fusca*, and *Microtetraspora niveoalba*) from soils and sediments. Suspensions of soil in water were treated with benzethonium chloride, plated onto LSV-SE agar† supplemented with kanamycin, norfloxacin, and naladixic acid, and incubated at 32°C for 4 weeks. Members of the *Microtetraspora glauca* group formed large, flat colonies covered with white or whitish gray aerial hyphae which differentiated into linear chains of four spores. Members of the *Microtetraspora glauca* group were isolated from 18 out of 26 environmental samples (16 from field and forest soils, 2 from freshwater sediments) and accounted for 2–27% of the total populations recovered on isolation plates.

Maintenance procedures

Sporulated slant cultures can be maintained at 4°C for 3–4 months; suitable media include glycerol-asparagine, Hickey–Tresner, and oatmeal agars. Suspensions of spores and hyphae in glycerol (20%, v/v) can be stored in liquid nitrogen at −80°C. Lyophilization is recommended for long-term preservation; this involves suspending spores and mycelia in a suitable fluid such as serum plus 7.5% (w/v) glucose or skimmed milk plus 7.5% glucose.

Differentiation of the genus *Microtetraspora* from other genera

Microtetraspora strains can be distinguished from the genera *Actinomadura*, *Microbispora*, and *Nonomuraea* by their ability to form characteristic chains of four spores on distinct sporophores.

Taxonomic comments

Four *Microtetraspora* species were included in the last edition of *Bergey's Manual of Systematic Bacteriology* (Nonomura, 1989b), namely *Microtetraspora glauca* Thiemann et al. 1968b, *Microtetraspora fusca* Thiemann et al. 1968b, *Microtetraspora niveoalba* Nonomura and Ohara 1971d, and "*Microtetraspora viridis*" Nonomura and Ohara 1971d. An additional species, "*Microtetraspora caesia*" Tomita et al. 1980 was cited as a *species incertae sedis*. Miyadoh et al. (1989) transferred "*Microtetraspora viridis*" to the genus *Actinomadura* as *Actinomadura viridis*. The genus *Microtetraspora* provided a temporary home for the *Actinomadura pusilla* group (Fischer et al., 1983; Goodfellow et al., 1988; Kroppenstedt et al., 1990a; Poschner et al., 1985) until it became clear that members of this taxon and those belonging to the *Microtetraspora glauca* group can be separated using DNA–DNA relatedness (Miyadoh et al., 1989), fatty acid (Kroppenstedt et al., 1990a; Miyadoh et al., 1989), numerical taxonomic (Athalye et al., 1985; Goodfellow and Pirouz, 1982), electrophoretic

*MGA-SE agar (Nonomura and Ohara, 1971d): asparagine, 1 g; glucose, 2 g; K_2HPO_4, 0.5 g; soil extract, 200 ml; agar, 20 g; cycloheximide, 50 mg; nystatin, 50 mg; benzyl penicillin, 0.8 mg; polymixin B, 4.0 mg: distilled water, 800 ml; pH 8.0. The soil extract was prepared by autoclaving 1000 g of soil in 1000 ml water for 30 min before decanting and filtering.

†LSV-SE agar: commercial lignin, 1 g, soybean flour, 0.2 g., soil extract, 100 ml, $CaC0_3$, 0.02 g., $FeSO_4 \cdot 7H_2O$, 0.01 g;. KCl, 1.7 g, $MgSO_4 \cdot 7H_2O$, 0.05 g; Na_2 HPO_4, 0.5 g., B-vitamins (0.5 mg each of *p*-aminobenzoic acid, calcium pantothenate, inositol, niacin, pyridoxine-HCl, riboflavin, thiamine HCl, and 0.25 mg biotin), cycloheximide, 50 mg, kanamycin, 20 mg, naladixic acid, 10 mg, norfloxacin, 20 mg, nystatin, 50 mg., distilled water 1 liter.

TABLE 283. Fatty acid profiles (1% of total fatty acids) of the type strains of *Microtetraspora* species[a]

Fatty acid	*M. glauca* JCM 3300	*M. fusca* JCM 3183	*M. malaysiensis* JCM 11278	*M. niveoalba* JCM 3149
Saturated fatty acids:				
$C_{13:0}$	0.2	–	0.1	–
$C_{14:0}$	0.7	0.1	0.5	0.3
$C_{15:0}$	3.8	0.9	2.9	0.5
$C_{15:0}$ 2-OH	0.5	0.3	0.4	–
$C_{16:0}$	5.5	0.3	2.3	1.9
$C_{18:0}$ 2-OH	0.5	0.1	0.2	–
$C_{17:0}$	14.9	0.3	2.8	0.5
$C_{17:0}$ 2-OH	0.3	0.3	–	–
$C_{18:0}$	4.2	–	–	–
$C_{19:0}$	0.2	–	–	–
Unsaturated fatty acids:				
$C_{15:1}$ ($\omega 6c$)	0.1	–	0.3	–
$C_{16:1}$ ($\omega 9c$)	0.3	–	0.5	–
$C_{16:1}$ 2-OH	–	0.2	–	0.2
$C_{17:1}$ $\omega 6c$	3.6	12.4	10.2	9.2
$C_{17:1}$ $\omega 8c$	3.9	0.8	1.7	0.3
$C_{17:1}$ $\omega 9c$	1.0	–	1.4	–
$C_{16:1}$ $\omega 7c$	0.4	–	0.3	–
$C_{18:1}$ $\omega 9c$	1.2	0.1	0.5	–
Branched fatty acids:				
$C_{14:0}$ iso	1.0	0.6	2.9	1.0
$C_{15:0}$ iso	3.3	3.0	4.1	2.7
$C_{15:0}$ anteiso	0.5	0.1	0.7	0.1
$C_{16:1}$ iso 2-OH	0.2	1.0	3.8	0.5
$C_{16:0}$ iso	28.1	56.4	44.0	65.6
$C_{17:0}$ iso	1.8	0.5	0.6	0.3
$C_{17:0}$ anteiso	1.7	0.4	0.9	0.9
$C_{18:1}$ iso 2-OH	0.1	0.3	0.3	–
$C_{18:0}$ iso	4.3	3.6	0.8	3.4
10-Methyl fatty acids:				
$C_{16:0}$ iso	1.5	1.9	1.7	3.3
$C_{17:0}$ iso	13.0	14.6	14.5	6.6
$C_{18:0}$ iso	2.2	1.5	0.8	2.8
$C_{19:0}$ iso	0.1	–	–	–

[a]Data from Kroppenstedt et al. (1990a), Miyadoh et al. (1989), and Nakijama et al. (2003). Quantitative variation between some of the values may be due to differences in cultivation conditions.

mobility of ribosomal AT-L30 protein (Ochi et al., 1991), and 16S rRNA gene sequence data (Wang et al., 1996b). The *Actinomadura pusilla* group was subsequently classified in a new taxon, the genus *Nonomuraea* Zhang et al. 1998a.

"*Microtetraspora tyrrhenii*" was proposed by Tomita et al. (1991) for an organism which formed hooked or spiral spore chains and had other properties consistent with its classification in the *Actinomadura pusilla* group. This organism probably belongs to the genus *Nonomuraea*, but this proposition cannot be tested as the type strain is no longer extant. *Microtetraspora malaysiensis* was proposed for strains isolated from dipterocarp forest soils (Nakajima et al., 2003).

Differentiation of species of the genus *Microtetraspora*

Microtetraspora species can be distinguished from one another by using a combination of morphological, nutritional and physiological characteristics (Table 282).

List of species of the genus *Microtetraspora*

1. **Microtetraspora glauca** Thiemann, Pagani and Beretta 1968b, 296[AL]

glau'ca. L. fem. adj. *glauca* grayish.

Short aerial hyphae typically contain chains of four spores; occasionally chains with two or three spores are formed and very rarely ones with five spores. Spores are smooth and spherical (1.5 µm) to slightly oval (1.4–1.7 µm) (Figure 372). Good growth on modified Bennett's, Hickey–Tresner, oatmeal, and soil agars. A blue-green to yellowish-green substrate mycelium and a blue-gray aerial spore mass are formed on Hickey–Tresner agar. B vitamins are required for growth. Grows between 20 and 37°C, but not at 40°C. Optimal pH growth range 6.0–7.0.

Produces phosphatases. Hydrogen sulfide is produced. Degrades arbutin, casein, chitin, DNA, esculin, gelatin, and Tweens 20 and 40, and xylan, but not adenine, cellulose, guanine, Tweens 60 or 80, or xanthine. Amygdalin, D-arabinose, cellobiose, glucose, glycogen, maltose, α-methyl-D-glucoside, ribose, salicin, starch, xylose, sodium acetate, and sodium butyrate are used as sole carbon sources, but adonitol, dulcitol, *meso*-erythritol, ethanol, inulin, lactose, melezitose, raffinose, sorbitol, and sodium propionate are not.

Tolerant (%, w/v) to brilliant green (0.001), crystal violet (0.0001), sodium chloride (3%), and pyronin (0.1), but sensitive to brilliant green (0.01), crystal violet (0.001), lysozyme (0.005), and sodium chloride (5%). Resistant to (μg/ml) demethylchlortetracycline hydrochloride (500), gentamicin sulfate (50), lincomycin hydrochloride (100), novobiocin (50), rifampin (50), streptomycin sulfate (50), and vancomycin hydrochloride (50), but sensitive to cephaloridine hydrochloride (100), kanamycin sulfate (100), neomycin sulfate (50), tobramycin sulfate (50), and penicillin (10 IU) using freeze-dried filter paper discs soaked in antibiotics at appropriate concentrations.

Additional phenotypic properties are shown in Table 282.

Source: soil.

DNA G+C content (mol%): 69 (HPLC).

Type strain: ATCC 23057, DSM 43311, NBRC 14761, JCM 3300, NRRL B-3735, VKM Ac-663.

Sequence accession no. (16S rRNA gene): D85490, U48974, X97891.

2. **Microtetraspora fusca** Thiemann, Pagani and Beretta 1968b, 296[AL]

fus'ca. L. fem. adj. *fusca* dark, tawny.

Aerial spore mass is gray. Substrate mycelium brown-violet on Hickey–Tresner agar. Good sporulation on glucose-asparagine, Hickey–Tresner, glycerol-asparagine, and soil agars, but does not sporulate on oatmeal agar. Grows between 30–37°C, but not at 40°C. Optimal pH for growth range 6.0–7.0.

Arbutin is used as a sole carbon source.

Additional phenotypic properties are shown in Table 282.

Source: soil.

DNA G+C content (mol%): 70 (HPLC).

Type strain: ATCC 23058, DSM 43841, NBRC 13915, JCM 3183, NRRL B-3628, VKM Ac-662.

Sequence accession no. (16S rRNA gene): U48973.

3. **Microtetraspora malaysiensis** Nakajima, Ho and Kudo 2004, 1[VP] (Effective publication Nakajima, Ho and Kudo 2003, 188.)

mal. ay.si. en'sis. N.L. fem. adj. *malaysiensis* of or pertaining to Malaysia, the source of the soil from which the type strain was isolated.

Straight chains of four spores are formed on sporophores branching from aerial hyphae; occasionally chains with two or three spores are formed and rarely ones with five spores. Spores are oval to cylindrical and have smooth surfaces.

Hickey–Tresner agar: moderate growth; aerial mycelium pearl, substrate mycelium old gold, no diffusible pigment. Inorganic salts-starch agar: poor growth; no aerial mycelium, substrate mycelium yellow, no diffusible pigment. Oatmeal agar: moderate growth; aerial mycelium white, substrate mycelium cream, no diffusible pigment. Oatmeal-nitrate agar: moderate growth; aerial mycelium white, substrate mycelium white, no diffusible pigment. Yeast extract-malt extract agar: good growth; aerial mycelium white, substrate mycelium honey gold, no diffusible pigment. Yeast-extract-starch agar: moderate growth; aerial mycelium oyster white, substrate mycelium cream, no diffusible pigment.

Optimum temperature for growth is 25–30°C; does not grow at 40°C.

Degrades casein, DNA, esculin and testosterone, but not adenine, elastin, guanine, keratin, or xylan. Cellobiose, glucose, lactose, maltose, ribose, salicin, starch, and xylose are used as sole carbon sources, but adonitol, arabitol, arbutin, dulcitol, *meso*-erythritol, *meso*-inositol, inulin, melezitose, melibiose, α-methyl-D-glucoside, raffinose, sorbitol, sorbose, and sucrose are not. Sodium citrate is also used as a sole carbon source, but not sodium benzoate, sodium fumarate, sodium lactate, sodium malate, sodium mucate, sodium oxalate, sodium succinate, or sodium tartrate.

Does not grow in the presence of 3% (w/v) NaCl. Thiamine is needed for growth, but requirement for biotin varies. Whole-organism hydrolysates contain glucose, madurose, mannose, and ribose.

Additional phenotypic properties are shown in Table 282.

Source: dipterocarp forest soils on the Malaysian Peninsula.

DNA G+C content (mol%): 69–70 (HPLC).

Type strain: H47-7, DSM 44579, JCM 11278, NBRC 100735.

Sequence accession no. (16S rRNA gene): AB062383.

4. **Microtetraspora niveoalba** Nonomura and Ohara 1971d, 872[AL]

ni.ve.o.al'ba. L. adj. *niveus* snowy; L. adj. *albus* white; N.L. fem. adj. *niveoalba* snow white.

Mesophilic actinomycete that forms a stable, highly branched substrate mycelium. Substrate mycelium pale yellowish brown on Hickey–Tresner agar. The aerial spore mass color is white. Soluble pigments are not produced.

Good sporulation on oatmeal agar, yeast-starch agar, and on glycerol-asparagine agar supplemented with B vitamins. Branched spore chains are present. Growth occurs between 35–40°C. Optimal pH growth range 7.0–8.0. Whole-organism hydrolysates contain madurose.

Additional phenotypic properties are shown in Table 282.

DNA G+C content (mol%): 71 (HPLC).

Type strain: ATCC 27301, DSM 43174, NBRC 15239, JCM 3149.

Sequence accession no. (16S rRNA gene): U48976.

Species *incertae sedis*

1. **Microtetraspora tyrrhenii** Tomita, Oda, Hoshino, Ohkusa and Chikazawa 1992, 191[AL] (Effective publication: Tomita, Oda, Hoshino, Ohkusa and Chikazawa 1991, 945.)

tyr.re'ni-i. L. gen. n. *tyrrenii* (*sic*), of the Tyrrhenian Sea (part of the Mediterranean Sea) which faces the location from which the soil sample was taken and from which the strain was isolated.

Hooked or spiral spore chains (5–15 spores) formed monopodially on the tips of aerial hyphae. The spores (1.2 × 1.6 ~ 2.5 μm) have vertical folds on the surface.

Colonies on yeast extract-malt extract are convex and crateriform.

Glycerol-asparagine agar: moderate growth; aerial mycelium absent, substrate mycelium dark yellow, no diffusible pigment. Inorganic salts-starch agar: poor growth; scant white aerial mycelium, substrate mycelium light olive brown, no diffusible pigment. Oatmeal agar: moderate growth; no aerial mycelium, substrate mycelium light yellow, no diffusible pigment. Peptone-yeast extract-iron agar: moderate growth; no aerial mycelium, substrate mycelium deep yellowish brown, no diffusible pigment. Tyrosine agar: moderate growth; no aerial mycelium, substrate mycelium dark yellow, no diffusible pigment. Yeast extract-malt extract agar: good growth; aerial mycelium white, substrate mycelium deep yellowish brown, no diffusible pigment.

Temperature growth range: 17–40°C. Grows in the presence of NaCl 3% (w/v).

Adenine, casein, esculin, gelatin, hippuric acid, hypoxanthine, and tyrosine are degraded, but xanthine is not. Amylase is produced. Adonitol, D- and L-arabinose, cellobiose, fructose, galactose, glucose, glycerol, *meso*-inositol, lactose, mannitol, mannose, melibiose, raffinose, rhamnose, ribose, salicin, sucrose, trehalose, and xylose are used as sole carbon sources, but dulcitol, *meso*-erythritol, melezitose, sorbitol, and D-sorbose are not.

DNA G+C content (mol%): not determined.

Type species: Q464-31, ATCC 53931.

2. **"Microtetraspora caesia"** Tomita, Hoshino, Sasahira, Hasegawa and Akiyama 1980

Forms single, pairs or chains of three to eight spores on the aerial mycelium.

Aerial mass color is grayish blue-green. The cell wall contains *meso*-diaminopimelic acid and galactose.

Type strain: E864-61.

Genus VI. **Nonomuraea** corrig. Zhang, Wang and Ruan 1998b, 419[VP]

Peter Kämpfer

No.no.mu.ra'e.a. N.L. fem. n. *Nonomurea* named after Hideo Nonomura, a Japanese taxonomist who has made many contributions to the biology of actinomycetes.

Aerobic, Gram-stain-positive, non-acid-fast bacteria that form **extensively branched substrate and aerial mycelia.** Aerial mycelia bear chains of spores which are hooked, spiral, or straight. The spore surface can be folded, irregular, smooth, or warty. Growth temperature ranges from 20–45°C, in some cases up to 55°C. **Cell walls contain *meso*-diaminopimelic and whole-cell hydrolysates contain madurose as the diagnostic sugar. Contains diphosphatidylglycerol, phosphatidylethanolamine, hydroxylated phosphatidylethanolamine, and ninhydrin and sugar positive phospholipids as predominant phospholipids.. Major menaquinones are MK-9(H_4), MK-9(H_2), and MK-9(H_0).** Major types of fatty acids are C_{17} 10-methyl and C_{16} iso-branched fatty acids.

DNA G+C content (mol%): 64–73.

Type species: **Nonomuraea pusilla** corrig. (Nonomura and Ohara 1971a) Zhang, Wang and Ruan 1998b, 419 (*Actinomadura pusilla* Nonomura and Ohara 1971a, 909).

Further descriptive information

The genus *Nonomuria* (*sic*) was proposed by Zhang et al. (1998b) for species which had been classified in the genus *Actinomadura* (Athalye et al., 1985; Fischer et al., 1983; Poschner et al., 1985) prior to their reclassification to the genus *Microtetraspora* (Kroppenstedt et al., 1990a). Chiba et al. (1999), who pointed out that the name *Nonomuria* (*sic*) was a misnomer according to the *Bacteriological Code* (1990 Revision) (Rule 57a, Appendix 9), introduced the name *Nonomuraea* as the correct spelling of the genus. On the basis of 16S rRNA gene sequence data, this group belongs to the family *Streptosporangiaceae* (Goodfellow et al., 1990b; Goodfellow and Quintana, 2005; Stackebrandt et al., 1997; Zhang et al., 1998b). Interestingly, Goodfellow et al. (1988) were the first to propose the genus *Nonomuria* to accommodate species of the *Actinomadura pusilla* group, but this proposition was not formally published. The taxon encompasses 24 species and 2 subspecies with validly published names. Representatives of each of these validly named species form a distinct line in the 16S rRNA *Streptosporangiaceae* gene tree (Goodfellow and Quintana, 2005; Figure 376). The majority of the species previously assigned to the *Actinomadura pusilla* group, and now part of the genus *Nonomuraea*, originated from soil (Galatenko et al., 1981; Meyer, 1979; Nonomura and Ohara, 1971b).

Only a few reports have been published about the genetics and metabolism of members of the genus despite their potential importance, notably in the discovery of novel bioactive compounds. The gene cluster encoding the biosynthesis of glycopeptide antibiotic A40926 in *Nonomuraea* strain ATCC 39727 has been isolated and characterized by Sosio et al. (2003). This glycopeptide, which is a member of the teichoplanin family of glycopeptides, is the precursor of dalbavancin. Sosio and her colleagues also isolated the novel compound dechloromannosyl-A40926 aglycone following the construction of a *Nonomuraea* mutation by deleting the *dbr* open reading frames 8–10. Subsequently, Alduina et al. (2005) constructed a bacterial artificial chromosomal library of *Nonomuraea* ATCC 39727 using an *Escherichia coli–Streptomyces* artificial chromosome (ESAC) and screened for the presence of dbv genes known to be involved in the biosynthesis of glycopeptide A40926. The heterologous expression of *Nonomuraea* genes in *Streptomyces lividans* was successfully demonstrated by using combined RT-PCR and proteomic approaches. These

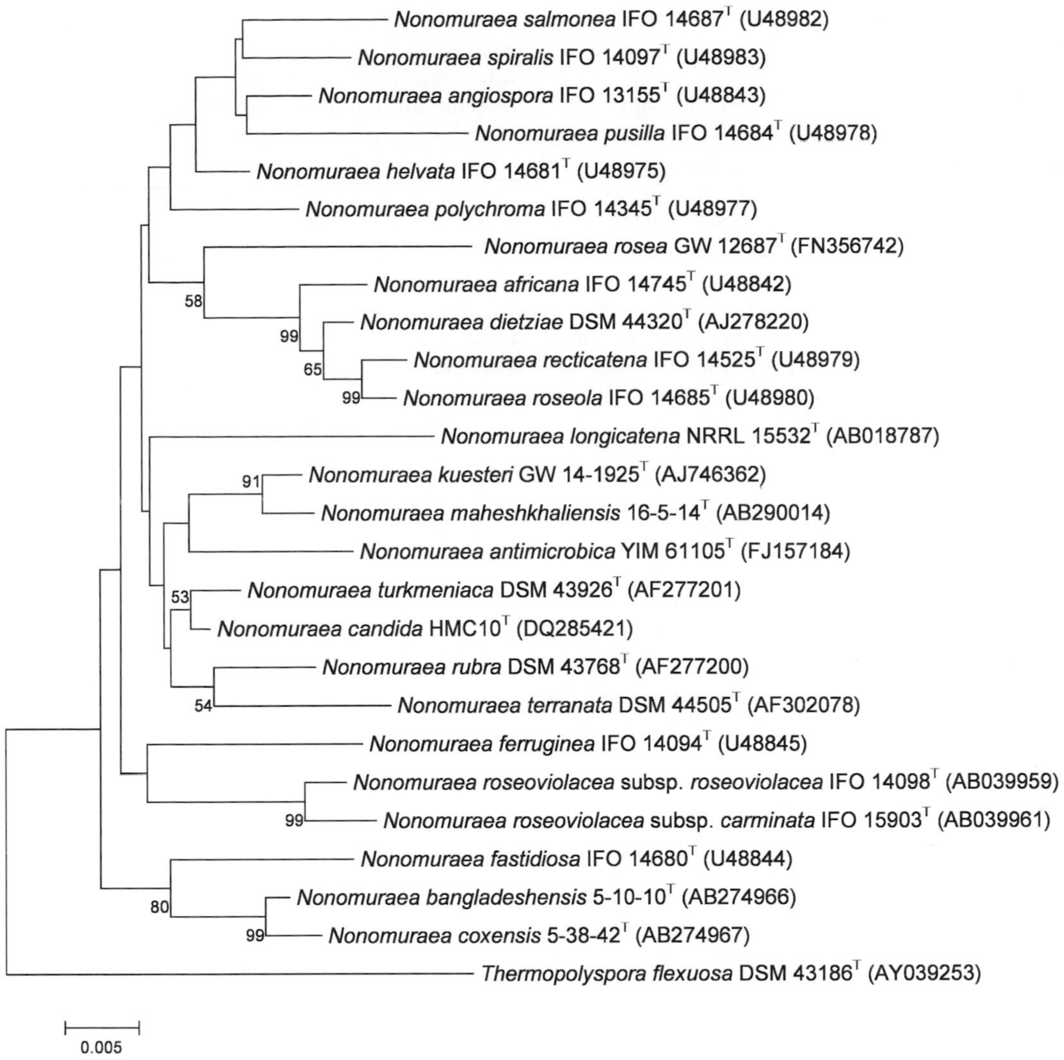

FIGURE 376. Phylogenetic analysis of *Nonomuraea* based on 16S rRNA gene sequences available from the European Molecular Biology Laboratory data library (accession numbers are given in brackets) constructed after multiple alignment of data (distance options according to the Kimura-2 model) and clustering with the neighbor joining method. Bootstrap values based on 1000 replications are listed as percentages at the branching points. Bar = 0.05 nucleotide substitutions per nucleotide position.

results indicate that *Streptomyces lividans* may be a good host for the genetic analysis of *Nonomuraea* strains. It was also demonstrated that *Nonomuraea* ATCC 39727 has a special carbon metabolism. Glucose is primarily metabolized via the Entner–Doudoroff pathway, even though the energetically more favorable Embden–Meyerhof–Parnas pathway is present in this organism (Gunnarsson et al., 2003). In addition, *Nonomuraea* utilizes a PP_i-dependent phosphofructokinase, an enzyme that has been connected with anaerobic metabolism in eukaryotes and higher plants but which has been recognized recently in several actinomycetes. Beltrametti et al. (2004, 2003) showed that the catabolism products of branched chain amino acids provide biosynthetic precursors for the formation of several lipid-containing antibiotics. The effect of valine on the production of glycopeptide antibiotic A40926 was studied in detail. The addition of valine to a minimal medium had a positive effect on the production of A40926; similar results were obtained in a rich production medium.

Members of the family *Streptosporangiaceae*, including *Nonomuraea* species, are an increasingly rich source of commercial products, notably antibiotics and enzymes (Goodfellow and Quintana, 2005). *Nonomuraea rosea* has been shown to produce deoxycephalomycin B (Okazaki and Naito, 1985), *Nonomuraea roseoviolacea* produces carminomicins (Nakagawa et al., 1983, 1989), *Nonomuraea rubra* produces maduromycin (Fleck et al., 1978), *Nonomuraea pusilla* produces actinotiocin (Tamura et al., 1973), and *Nonomuraea spiralis* produces pyralomicin (Naganawa et al., 2002).

Enrichment and isolation procedures

Goodfellow and Quintana (2005) reported that dry heat treatment of air-dried soil samples and dilution plate culture with selective synthetic media is useful for the preferential isolation and enumeration of some members of the family *Streptosporangiaceae*, including members of the genera *Microtetraspora* and *Nonomuraea* (Nonomura and Ohara, 1971b, 1971c, 1971d). In addition, Nonomura and Hayakawa (1988) reported that pretreatment of soil suspensions with yeast extract (6%, w/v) and sodium dodecylsulfate (0.05%, w/v) at 40°C for 20 min, followed by dilution with water, activates actinomycete spores but kills vegetative cells of other soil bacteria in the suspensions, thereby leading to an increase in the counts of actinomycetes on isolation plates.

An improved procedure for the selective isolation of streptosporangiae from soil was reported by Hayakawa et al. (1991a). The method is based on the ability of streptosporangial spores to withstand dry heat and treatment with benzethonium chloride (BC) and the capacity of streptosporangiae to grow in the presence of leukomycin and nalidixic acid. Initially, an air-dried soil sample is ground in a mortar and heated in a hot-air oven for an hour; 0.5 ml of a 10^{-1} dilution in water of the heated sample is transferred to 4.5 ml of sterile 5 mM phosphate buffer (pH 7.0) containing BC at a final concentration of 0.1% (w/v). The resultant preparation is maintained at 30°C for 30 min with occasional stirring, and a portion (1 ml) is then diluted with sterile tap water (1:10 or 1:15). Inocula of 0.1 ml or 0.2 ml of the dilution is then spread over the surface of plates of HV agar supplemented with leukomycin in ethanol (1 mg/l) and nalidixic acid (20 mg/l) and the plates incubated at 30°C for 3–4 weeks. Actinomycetes which appear on the plates are examined by light microscopy (× 600) and assigned to genera on the basis of characteristic morphological properties. This pretreatment procedure has been used to isolate several *Nonomuraea* species including *Nonomuraea helvata*, *Nonomuraea pusilla*, *Nonomuraea roseoviolacea*, and *Nonomuraea spadix* (Nonomura and Ohara, 1971b).

Nonomuraea spores appear to be particularly resistant to dry heat at 100–120°C, thereby allowing the slow-growing nonomuraea to develop into recognizable colonies on dilution plates. Soil dilutions are plated onto various media including AV and MGA-SE agars, and incubated for several weeks at 28–30°C (Nonomura and Ohara, 1971d) (Methods taken from Goodfellow and Quintana, 2005.). Other *Nonomuraea* species such as *Nonomuraea salmonea* and *Nonomuraea roseola* have been isolated from soil on media supplemented with antibiotics. Lavrova et al. (1972) added rubomycin (5, 10, or 20 μg/ml) to medium no. 2 of Gauze et al. (1957); Preobrazhenskaya et al. (1975a) added bruneomycin (0.5, 1, or 2 μg/ml) or streptomycin (0.5, 1, or 2 μg/ml). The use of these antibiotics led to the growth of more *Nonomuraea* colonies on isolation plates while reducing the number of streptomycetes. In contrast, *Nonomuraea ferruginea* and *Nonomuraea spiralis* were isolated by plating soil suspensions onto oatmeal agar or Gauze's no. 1 medium without addition of selective antibiotics (Meyer, 1979).

Nonomuraea species also grow well on rich media including modified Bennett's (Jones, 1949), glucose-yeast extract (Waksman, 1950b), nutrient (BBL), oatmeal (ISP medium 3; Difco 0771), TS (BBL) and yeast extract-malt extract agars [ISP medium 2 (Difco 0770); Shirling and Gottlieb, 1966] as well as on PYES medium (Altenburgera et al., 1996). Most investigators recommend oatmeal-yeast extract agar for the growth of mesophilic strains and glycerol agar for thermophilic strains (Nonomura, 1989c).

Maintenance procedures

Goodfellow and Quintana (2005) reported that the most convenient method for short-term storage is by serial transfer from agar slants of complex media every 2 months (Meyer, 1989); tubes should be tightly closed with cotton plugs dipped in melted paraffin wax. Sporulated spore cultures can be stored at 5°C and at room temperature. Lyophilization, storage in liquid nitrogen, and freezing in glycerol can be used for long-term preservation (Goodfellow and Quintana, 2005; Meyer, 1989). *Nonomuraea* cultures may be lyophilized by procedures commonly used for bacteria. For lyophilization, the spore suspension or vegetative mycelium is suspended in a suitable fluid such as serum plus 7.5% (w/v) glucose or skimmed milk plus 7.5% (w/v) glucose. For storage in liquid nitrogen, the microorganisms are inoculated into small test tubes containing the appropriate medium and incubated until satisfactory growth is visible. Glycerol suspensions are prepared by scraping aerial growth or substrate mycelium or both from heavily inoculated plates and making heavy suspensions in 3 ml of aqueous glycerol in small (e.g. bijoux) bottles, which are stored at −20°C.

Differentiation of the genus *Nonomuraea* from other genera

Despite some differences, members of the family *Streptosporangiaceae* are chemically homogeneous but morphologically diverse (Table 284). *Nonomuraea* strains are distinguished from members of related genera by their ability to form chains of spores or pseudovesicles on aerial hyphae. The species may be distinguished by means of spore chain morphology, spore wall ornamentation, color of mature sporulated aerial mycelium, and substrate mycelium pigmentation. Nevertheless, identification of most species is difficult because, in many instances, only one (the type) or a few strains have been examined. In addition, *Nonomuraea* may be distinguished from other genera in the family *Streptosporangiaceae* by slight differences in cell-wall peptidoglycan, menaquinone type, colony pigmentation, the G+C content of the DNA, and growth at low temperatures (Table 284).

TABLE 284. Chemotaxonomic, morphological, and physiological characteristics of the genera classified in the family *Streptosporangiaceae*[a,b]

Characteristic	*Nonomuraea*	*Acrocarpospora*	*Herbidospora*	*Microbispora*	*Microtetraspora*	*Planobispora*	*Planomonospora*	*Planotetraspora*	*Streptosporangium*	*Thermopolyspora*
Morphology	Spore chains or pseudo-sporangia formed on aerial hyphae	Club or globose spore vesicles on aerial hyphae	Straight chains of smooth-surfaced spores on aerial hyphae	Smooth-surfaced spores in characteristic longitudinal pairs on aerial hyphae	Spore chains typically containing 4 smooth-surfaced spores on aerial hyphae	Cylindrical to clavate spore vesicles containing longitudinal pairs of spores on aerial hyphae	Cylindrical to clavate spore vesicles containing single spores on aerial hyphae	Spore vesicles containing four spores on aerial hyphae	Globose spore vesicles on aerial hyphae	Hooked or irregular spiral chains of 4–10 warty to spiny ornamented spores on aerial hyphae
Motile spores	–	–	–	–	–	+	+	+	–	–
Cell-wall chemotype[c]	III	III	III	III	III	III	III	III	III	III
Major menaquinones[d]	MK-9(H_2, H_4)	MK-9(H_2, H_4, H_6)	MK-10(H_4, H_6, H_8)	MK-9(H_2, H_4)	MK-9(H_2, H_4)	MK-9(H_2, H_4)	MK-9(H_2)	MK-9(H_4)	MK-9(H_2, H_4)	MK-9(H_2, H_4)
Muramic acid type	Acetylated	Acetylated	Acetylated	Acetylated	Acetylated	nd	nd	Acetylated	Acetylated	Acetylated
Phospholipid type	IV	IV, II	IV	IV	IV	IV	IV	nd	IV	IV
Whole-organism sugar pattern[e]	B,C	B,C	B	B,C	B,C	B	B	A,D	B	C
DNA G+C content (mol%)	64–69	68–69	69–71	71–73	69–71	70–71	72	71	69–71	77
Growth temperature range (°C)	20–45	15–30	nd	25–55	20–45	28–40	28–37	28–37	20–50	37–60

[a]Symbols: +, present; –, absent; nd, not determined.

[b]Data were taken from this and previous studies (Goodfellow et al., 1990b; Goodfellow and Quintana, 2005; Greiner-Mai et al., 1987; Kudo et al., 1993; Stackebrandt et al., 2001; Tamura and Sakane, 2004; Tamura et al., 2000).

[c]The major constituents of the cell wall are alanine, glutamic acid, glucosamine, and *meso*-A_2pm (Lechevalier and Lechevalier, 1970a). Diagnostic phospholipid type: II: only phosphatidylethanolamine; IV: glucosamine (with phosphatidylethanolamine and phosphatidylmethylethanolamine variable) (Lechevalier et al., 1977).

[d]Abbreviations exemplified by MK-9(H_4), menaquinone having four of the nine isoprene units hydrogenated.

[e]Whole-organism sugar patterns of actinomycetes containing *meso*-A_2pm: A, arabinose and galactose; B, madurose; C, no diagnostic sugar; D, arabinose and xylose (Lechevalier and Lechevalier, 1970b).

List of species of the genus *Nonomuraea*

1. **Nonomuraea pusilla** corrig. (Nonomura and Ohara 1971a) Zhang, Wang and Ruan 1998b, 419 (*Actinomadura pusilla* Nonomura and Ohara 1971a, 909)

pu.sil'la. L. fem. adj. *pusilla* dwarfish, referring to the aerial mycelium of the organism.

Spore chains in tightly closed spirals forming so-called pseudosporangia, i.e. spores are embedded in a slimy mass (Figure 377). Pseudosporangia are 3–6 μm in diameter. Spore surface smooth.

Glycerol-asparagine agar: good growth, aerial mycelium white, substrate mycelium grayish pink, no diffusible pigment. Inorganic salts-starch agar: scant growth, aerial mycelium microscopically visible, substrate mycelium colorless to brownish, no diffusible pigment. Inorganic salts-starch agar: scant growth, aerial mycelium microscopically visible, substrate mycelium colorless to brownish, no diffusible pigment. Oatmeal agar: abundant growth, surface farinaceous, aerial mycelium white to cream colored, substrate mycelium pale grayish brown, no diffusible pigment. Oatmeal salts-starch agar: good growth, surface granular to farcinoceus, aerial mycelium white, substrate mycelium colorless to white, no diffusible pigment. Yeast extract-malt extract agar: good growth, surface cartilaginous, aerial mycelium only microscopically visible, substrate mycelium dark brown with a red tinge, no diffusible pigment.

Cell hydrolysates contain galactose, madurose, mannose, and ribose. Diphosphatidylglycerol, phosphatidylethanolamine, ninhydrin and sugar positive phospholipids and hydroxylated phosphatidylethanolamine are present. Traces of phosphatidylinositol are found, but phosphatidylglycerol, phosphatidylinostol mannosides, and uncharacterized glycolipids are absent.

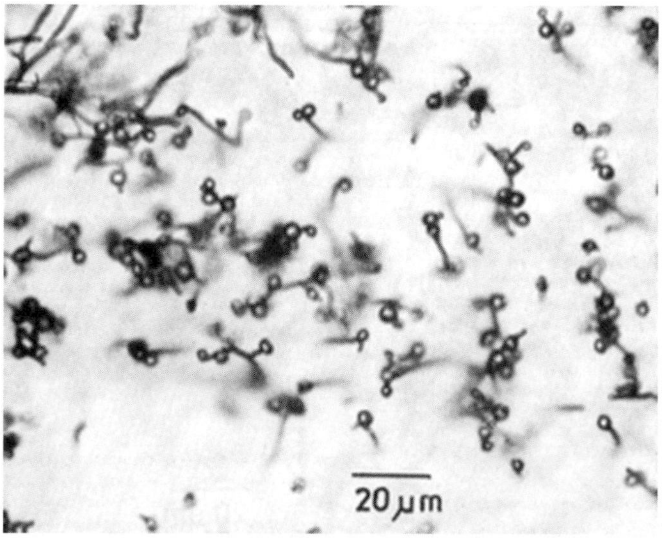

FIGURE 377. *Nonomuraea pusilla*, type strain ATCC 27296. Sporulating aerial mycelium with "pseudosporangia". Grown on yeast extract–malt agar at 28°C for 9 d.

Further chemotaxonomic data are given in Table 284, and the fatty acid pattern of the type strain is shown in Table 285.

Esculin hydrolysis and nitrate reductase tests are positive. Degrades DNA, gelatin, hypoxanthine, and tyrosine. Positive for acetoin, arginine dihydrolase, gelatinase, and urease activities. Further physiological characteristics are given in Table 286. Does not utilize L-arabinose, D-cellobiose, D-fructose, *meso*-inositol, D-mannose, L-rhamnose, or D-xylose as sole carbon sources.

Source: soil.

DNA G+C content (mol%): 68.3 (T_m).

Type strain: ATCC 27296, CBS 262.72, BCRC 11619, CECT 3284, CIP 106954, DSM 43357, NBRC 14684, IMET 9586, JCM 3144, KCTC 9278, NCIMB 11116, NRRL B-16126, VKM Ac-1508.

Sequence accession no. (16S rRNA gene): D85491, U48978.

Additional comments: Nonomuraea pusilla can be easily mistaken for a *Streptosporangium* on superficial microscopic examination; it can only be differentiated by exact proof of a sporangial wall, which indicates a *Streptosporangium* strain. Tamura et al. (1973) isolated a sulfur-containing peptide antibiotic, actinotiocin, from *Nonomuraea pusilla*.

2. **Nonomuraea africana** corrig. (Preobrazhenskaya and Sveshnikova 1974) Zhang, Wang and Ruan 1998b, 419[VP] (*Actinomadura africana* Preobrazhenskaya and Sveshnikova 1974, 865; *Nocardiopsis africana* Preobrazhenskaya and Sveshnikova 1985, 224; *Microtetraspora africana* Kroppenstedt, Stackebrandt and Goodfellow 1990a, 148)

a.fri.ca'na. L. fem. adj. *africana* referring to Africa, the source of the soil sample.

Straight spore chains with 4–10 spores per chain. Spore ornamentation smooth. Growth of aerial mycelium on oatmeal agar is grayish blue, substrate mycelium growth is yellow, and the soluble pigment color yellowish brown.

Esculin is hydrolyzed and nitrate reduced. Casein, DNA, gelatin, hypoxanthine, starch, and tyrosine are degraded. Cell hydrolysates contain madurose. Diphosphatidylglycerol, phosphatidylethanolamine, phosphatidylglycerol, phosphatidylinositol, phosphatidylinositol mannosides, ninhydrin and sugar positive phospholipids, hydroxylated phosphatidylethanolamine, and uncharacterized glycolipids are present.

Further chemotaxonomic data are given in Table 284, and the fatty acid pattern of the type strain is given in Table 285.

L-Arabinose, cellobiose, fructose, mannose, *meso*-inositol, raffinose, rhamnose, sucrose, and xylose are utilized and used as sole carbon sources. Positive for acetoin, arginine dihydrolase, citrate lyase, β-galactosidase, dihydrolase, gelatinase, lysine decarboxylase, ornithine decarboxylase, and urease activities.

Further physiological characteristics are given in Table 286.

DNA G+C content (mol%): 64.0 (T_m).

Type strain: ATCC 35107, DSM 43748, NBRC 14745, INA 1839, JCM 6240, KCTC 9260, NRRL B-16114, RIA 1839, VKM Ac-924.

Sequence accession no. (16S rRNA gene): AJ269555, U48842.

TABLE 285. Cellular fatty acid composition of the type strains of *Nonomuraea* species[a]

Fatty acid	*N. pusilla* DSM 43357	*N. africana* DSM 43748	*N. angiospora* DSM 43173	*N. antimicrobica*[b] DSM 45220	*N. bangladeshensis*[c] JCM 13930	*N. coxensis*[d] JCM 13931	*N. dietziae* DSM 44320	*N fastidiosa* DSM 43674	*N. ferruginea* DSM 43553	*N. flexuosa* DSM 43186	*N. helvata* DSM 43142	*N. kuesteri* DSM 44753	*N. maheshkhaliensis* JCM 13923	*N. longicatena* DSM44817	*N. polychroma* DSM 43925	*N. rosea*[e] DSM 45177	*N. recticatena* DSM 43937	*N. roseola* DSM 43767	*N. roseoviolacea* DSM 43144	*N. rubra* DSM 43768	*N. salmonea* DSM 43678	*N. spiralis* DSM 43555	*N. terrinata* DSM 44505	*N. turkmeniaca* DSM 43926
$C_{14:0}$ iso	1.3	2	0.6	3.9	1.5		3	0.8	0.4	0.8	0.8	0.9	0.4	1.3	0.7	1.3	1.6	1.3		0.6	1.2	3.9	0.5	0.8
$C_{14:0}$	3.6	1.4	1.4	4.2		4.6	1.1	0.7	0.9	0.5	2.3	0.9	3.3	1	0.8	0.5	0.8	2.7	1.5	0.8	0.7	1	1.3	1.6
$C_{15:0}$ iso	10.7	4.1	7.7	6.5	14.7	2.2	3.6	8.1	3	12.5	9.7	7.9	5.5	4.6	7.7	4.3	5.1	2.6	6.3	8.2	7.5	9.4	5.9	9
$C_{15:0}$ anteiso	1	0.4	0.3			2	1.2	1.2		4.3	3.4	0.3	0.5	1.4		0.4	0.7	0.6		0.2	0.6	0.8	0.4	
$C_{15:0}$	7.8	10.7	4.1		1.2	7.5	6.5	3.7	7.7	5.7	8.1	3.8	2.9	7	1.8	3.1	5.4	5.7	3.1	2.3	3.9	10.5	3.6	3.3
$C_{16:1}$ iso	2.3	0.9	4.6	12.7	7.3		4.7	1.7	0.6	0.4	0.5	8.3	3.6	2.2	2.9	6.9	0.9	1.5		4.6	2.9	0.9	2.7	6.9
$C_{16:0}$ iso	19.2	21.1	31.3	32.2	26.4	8.1	13.3	35.8	22.1	13.8	17.6	28.9	13.6	24.8	35.7	43.3	17.5	9.6	22	32.7	40.8	29.9	27.9	26.1
$C_{16:1}$ *cis*	0.7	0.8	1.2			1.3	1.8		1.2	2.8	2	1.8			1.4	0.6		4.1	1.5	2.4	1		1.7	1.4
$C_{15:0}$ iso 2-OH										1.4				1.2		0.4	0.7					2.2		
$C_{16:0}$	5.4	4.7	4.2		1.8	17.4	2.1	2.5	3.5	3.4	12.3	1.8	15.5	1.7	4.4	1.1	3.2	10.7	6.7	5.1	3.9	4.7	4.3	5
$C_{15:0}$ 2-OH	1.1	1.4	0.6				3		1.8		0.5	0.8	0.6	1.7		0.4	2.1	2.1		0.3	0.3	0.8	0.4	
$C_{16:0}$ 10-methyl	5.6	1	3.4		5.8	3.4	1	1.7	1.1	2.4	4.1	2.8	8	4.8	2.3	2.6	0.7	1.5	4.4	4.1	2.3	2.4	2.9	2.5
$C_{17:0}$ iso	2.7	1.1	2.8		10			4.1	1.5	18.7	2.4	1.8	1.8	0.6	1.2	0.5	1.6	0.7	2.9	3.6	2.9	2.2	3.1	0.8
$C_{17:0}$ anteiso	1.2	0.6	2	2.7	1.6	2.4		6.5	0.7	12.4	4.3	1	1	1.5	0.8	0.7	1.2	0.4	0.9	1.4	1.1	0.8	2.8	0.4
$C_{17:1}$ 9*c*	3.4	7	1.5		4.5	2.4	11.3	0.8	9.8	0.5	5.6	4.2		3.1	3.3	2.1	5	13.3	3.2	3.9	2.8	2	5	2.4
$C_{16:0}$ iso 2-OH	4.1	2.8	7.2			1.9	5	7.1	7.4	2.5	4	10.5		10.2	4		6.6	4.2	5.1	7	7.5	5.4	6	4
$C_{17:0}$	5.3	10.3	3.8	2.4		6.7	4.5	2.6	14.3	2	6.4	1.6	1.8	1.8	3.9	1	15.6	10	6	2.6	4.4	7.8	5.7	5.8
$C_{16:0}$ 2-OH	0.5	0.5	0.3			1.9	0.6		0.6		0.4	0.5	3.7		0.3		0.7	4.2	1	0.6	0.4		0.6	
$C_{17:0}$ 10-methyl	20.6	24.1	18.6	17.1	10.5	8.4	33.7	17.6	15.4	6.9	13	15.5	9.6	27.6	18.3	19.8	23.8	13.3	23.3	15.1	13.1	14.7	18	21.2
$C_{17:0}$ iso 2-OH	0.4									3	0.6	0.4					0.5	0.4		0.5			0.4	
$C_{18:0}$ iso	0.5	1	1.3					1.9	2		0.3	0.6	0.2		2.6		1.2		2.2	0.9	1	0.5	1.7	1.8
$C_{18:1}$ 9*c*	0.3	0.9		3	1.2	4.5	0.9	9.5	0.8		0.7	0.6	3.4		1.7		0.4	4	1	0.9	0.4		1.1	0.8
$C_{18:0}$	1.2	1.4	1.9			1.9	0.6	5.4	1.5	2.5	0.7	1.5	1.3		2.6		2.2	2.1	4.1		1.3	0.7	1.9	3.2
$C_{17:0}$ 2-OH	0.3	0.7					0.7		1.6			0.3	0.2				1.8	1.7						
$C_{18:0}$ 10-methyl	0.8	1.3	1.2	5.8	1.6	2.5	1.7	3.2	1		1.1	0.9	4.4	0.6	3.6	0.5	1.2	3.2	5.1	2.7		0.5	2.5	3.7

[a]Also contain traces of $C_{13:0}$ iso, C_{18} iso, and $C_{18:0}$; in DSM 44817 0.9% $C_{18:0}$ iso 10-methyl is found in addition.

[b]Data from Qin et al. (2009).

[c]Data from Ara et al. (2007b).

[d]Data from Ara et al. 2007a).

[e]Data from Kämpfer et al. (2010).

3. **Nonomuraea angiospora** corrig. (Zhukova, Tsyganov and Morozov 1968) Zhang, Wang and Ruan 1998b, 419[VP] (*Micropolyspora angiospora* Zhukova, Tsyganov and Morozov 1968, 728; *Microtetraspora angiospora* Kroppenstedt, Stackebrandt and Goodfellow 1990a, 148)

an.gi.o.spo'ra. Gr. neut. n. *angeion* (L. translit. *angium*) vessel; Gr. n. *spora* a spore; N.L. n. *angiospora* spores enclosed (in capsules).

Spore chains (2–15 spores) are curved or irregularly coiled. Spores spherical to oval (1.5–2.4 × 1.2–2.2 μm); surface spines enveloped in a translucent capsule. White aerial mycelium is formed on soya flour and oatmeal agars; substrate mycelium and colonies white to deep flesh or ochre colored. Grows between 21 and 37°C, and optimally at 27°C.

Cell hydrolysates contain madurose, ribose, and traces of mannose. Diphosphatidylglycerol, phosphatidylglycerol, phosphatidylinositol mannosides, ninhydrin and sugar positive phospholipids, hydroxylated phosphatidylethanolamine, and uncharacterized glycolipids are present. Phosphatidylethanolamine and phosphatidylinositol are absent.

Further chemotaxonomic data are given in Table 284, and the fatty acid pattern of the type strain is shown in Table 285.

Esculin is hydrolyzed; nitrate is not reduced. Positive for acetoin, arginine dihydrolase, gelatinase, and urease activities. Degrades DNA, gelatin, hypoxanthine, and tyrosine. Does not use L-arabinose, cellobiose, D-fructose, *meso*-inositol, D-mannose, D-raffinose, L-rhamnose, sucrose, or D-xylose as sole carbon sources.

Further physiological characteristics are given in Table 286.

DNA G+C content (mol%): not determined.

Type strain: LIA 3479-30, AS 4.1229, BCRC 13334, DSM 43173, NBRC 13155, JCM 3109, KCTC 9191, KCTC 9261, NRRL B-3905.

Sequence accession no. (16S rRNA gene): U48843.

4. **Nonomuraea antimicrobica** Qin, Zhao, Klenk, Li, Zhu, Xu and Li 2009, 2750[VP]

an.ti.mi.cro'bi.ca. Gr. prep. *anti* against; N.L.n. *microbium* microbe; L.adj.suff. *-icus -a -um* suffix used with various meanings; N.L.fem.adj. *antimicrobica* antimicrobial.

Gram-stain-positive, aerobic, non-acid-fast, and non-acid–alcohol-fast actinomycete that forms extensively branched, brown substrate mycelia and white-to-pink aerial mycelia on ISP 2–5 and PDA media. After 14 d of incubation at 28°C, spiral spore chains composed of smooth spores are observed on the aerial mycelium. Temperature range for growth is 15–37°C with optimal growth on 28°C. pH range for growth is 6.0–9.0 with optimal growth at 7.0. No growth is observed with 5% NaCl.

Positive for catalase. Negative for oxidase, milk coagulation, milk peptonization, gelatin liquefaction, cellulose, and starch hydrolysis, H_2S production and nitrate reduction. The diagnostic amino acid of the peptidoglycan is *meso*-diaminopimelic acid. Cell hydrolysates contain ribose, glucose, madurose, and galactose. Polar lipids include diphosphatidylglycerol, phosphatidylglycerol, phosphatidylinositol and phosphatidylinositol mannosides, and the diagnostic phospholipids of phosphatidylethanolamine and an unknown glucosamine-containing phospholipid.

The predominant menaquinone of the type strain is MK-9(H_4); MK-9(H_2), MK-9(H_6), and MK-9 are also present. Major fatty acids (>10%) are $C_{16:0}$ iso, $C_{17:0}$ 10 methyl, $C_{16:1}$ iso G, and $C_{16:0}$.

Source: a surface-sterilized leaf of *Maytenus austroyunnanensis* from the tropical rainforest of Xishuangbanna, Yunnan Province, south-west China.

DNA G+C content (mol%): 69.2 (HPLC).

Type strain: YIM 61105, DSM 45220, CCTCC AA 208016.

Sequence accession no. (16S rRNA gene): FJ157184.

5. **Nonomuraea bangladeshensis** Ara, Kudo, Matsumoto, Takahashi and Ōmura 2007b, 1506[VP]

ban.gla.desh.en'sis. N.L. fem. adj. *bangladeshensis* of or pertaining to Bangladesh, where the strain was isolated.

Forms a well-developed, branched substrate mycelium. Abundant aerial mycelium is present.

Spore chains are spiral with 1 or 2 turns (8–12 spores). Sporangia are not detected. Spores are spherical to cylindrical, and the spore sheath surface is smooth.

Good growth occurs on Bennett's agar, glucose-yeast extract agar, Hickey–Tresner agar, oatmeal agar (ISP-3), oatmeal-nitrate agar, Seino (yeast-starch) agar, tyrosine agar (ISP-7), Waksman agar, and yeast extract-malt extract agar (ISP-2). Moderate growth is found on glycerol-asparagine agar, inorganic salts-starch agar (ISP-4), nutrient agar, and water agar. Poor growth occurs with glucose-asparagine agar and sucrose-nitrate agar. The color of the substrate mycelium is shell to cinnamon brown on various media.

Pale brown aerial mycelium and sporulation occur on ISP-7 agar, ISP-2 agar, ISP-3 agar, oatmeal-nitrate agar, and water agar media after 14 d incubation at 28°C. A soluble pigment is not produced. Aerobic, Gram-stain-positive, and able to grow at pH 5.0–9.0. The temperature range for growth is between 15 and 45°C. No production of melanoid pigments. No growth is observed at 5% NaCl.

The diagnostic diamino acid isomer of the cell-wall peptidoglycan is *meso*-DAP.

Galactose, glucose, madurose, mannose, and ribose are present in whole-cell hydrolysates.

The *N-acyl*-type of the muramic acid is acetyl. Mycolic acids are absent. The polar lipids include PE and OH-PE. Phosphatidylcholine is absent.

The major menaquinone is MK-9(H_4), with minor amounts of MK-9(H_6), MK-9(H_2), and MK-9(H_0). Major fatty acids are $C_{16:0}$ iso (26.4%), $C_{15:0}$ iso (14.7%), and $C_{17:0}$ 10 methyl (10.5%).

Source: sandy soil from Cox's Bazar, Bangladesh.

DNA G+C content (mol%): 72.7 (HPLC).

Type strain: 5-10-10, JCM 13930, MTCC 8089.

Sequence accession no. (16S rRNA gene): AB274966.

6. **Nonomuraea candida** Le Roes and Meyers 2009, 1[VP] (Effective publication: Le Roes and Meyers 2008, 136.)

can´di.da. L. fem. adj. *candida* shining white; pertaining to the light coloring of the substrate and aerial mycelial mass when grown on oatmeal agar (IPS-3).

Gram-stain-positive, non-acid-fast, and non-acid–alcohol-fast actinomycete that forms a leathery, dark brown growth on ISP-2 agar. Tan-colored growth is observed on inorganic salts-starch agar (ISP-4); a yellow diffusible pigment is produced on glycerol-asparagine agar (ISP-5) and beige-colored substrate mycelium and white aerial mycelium is visible. No melanin production is observed on peptone-yeast extract-iron agar (ISP-6) and tyrosine agar (ISP-7). Good growth is observed on ISP-2 agar. Weak antibiosis is exhibited against *Mycobacterium aurum* A+ in agar overlays. Bioautographic analysis of organic solvent extracts of the cell mass shows weak antibiosis against *Mycobacterium aurum* A+, but no activity against *Mycobacterium tuberculosis* H37Rv^T (=ATCC 27294^T).

Growth is observed at 30°C, 37°C, and 45°C, but not at pH 4.3 and 4°C.

The type strain HMC10^T is catalase positive and oxidase negative. It grows in the presence of 0.3% 2-phenylethanol, 3% NaCl, and 0.1% phenol, but not in the presence of 0.01% NaN_3. Growth is observed in the presence of cephaloridine (100 µg/ml) (weak), lincomycin (100 µg/ml), oleandomycin (100 µg/ml), penicillin G (10 i.u./ml), rifampin (50 µg/ml), tobramycin (50 µg/ml) (weak), and vancomycin (50 µg/ml) (weak), but not in the presence of gentamicin (100 µg/ml), kanamycin (100 µg/ml), neomycin (50 µg/ml), and streptomycin (100 µg/ml).

Utilizes DL-α-amino-*n*-butyric acid, L-arginine, L-histidine, L-4-hydroxyproline, L-methionine, L-phenylalanine, potassium nitrate, L-serine, and L-valine as sole nitrogen sources, but not L-casteine. Adonitol, D(–) fructose, D(+) galactose, D(+) glucose, *meso*-inositol, inulin, D(–) lactose, D(–) mannose, D(+) melibiose, raffinose, L(+) rhamnose, D(–) ribose, salicin, sodium acetate (0.1%), sucrose, trehalose, and D(+) xylose are utilized as sole carbon sources. L(+) Arabinose, D(+) cellobiose, D(+) melezitose, and xylitol are weakly utilized, but sodium citrate (0.1%) is not utilized as a sole carbon source. Nitrate is not reduced, but H_2S production occurs. Lecithinase, lipase, and protease activity is observed on egg yolk agar. Hippurate and pectin are hydrolyzed. Degrades adenine, esculin, arbutin, casein, and Tween 80. Gelatin is weakly degraded, while allantoin, guanine, hypoxanthine, starch, L-tyrosine, urea, xanthine, and xylan are not degraded.

Source: South African soil.

DNA G+C content (mol%): 72.7 (HPLC).

Type strain: HMC10, DSM 45086, NRRL B-24552.

Sequence accession no. (16S rRNA gene): DQ285421.

7. **Nonomureae coxensis** Ara, Kudo, Matsumoto, Takahashi and Ōmura 2007b, 1507^{VP}

cox.en'sis. N.L. fem. adj. *coxensis* of or pertaining to Cox's Bazar, Bangladesh, the origin of the soil where the type strain was isolated.

Forms a well-developed, branched substrate mycelium. Abundant aerial mycelium is present.

Spore chains are spiral to hooked with 12–17 spores. Sporangia are not detected. Spores are spherical to oval and the spore surface is smooth.

Good growth on Bennett's agar, glucose-yeast extract agar, Hickey–Tresner agar, oatmeal agar (ISP3), oatmeal-nitrate agar, Seino (yeast-starch) agar, 1/5 yeast-starch agar, Waksman agar, and yeast extract-malt extract agar (ISP2). Moderate growth on inorganic salts-starch agar (ISP4), nutrient agar, and tyrosine agar (ISP7); poor growth on glucose-asparagine agar, glycerol-asparagine agar, and water agar.

Substrate mycelium is light melon yellow to apricot (orange-brown) on various agar media. Pale pink to white aerial mycelium and sporulation occur on ISP-2 agar, ISP-3 agar, Bennett's agar, Hickey–Tresner agar, and 1/5 yeast-starch agar media after 14 d incubation at 28°C. A soluble pigment is not produced. Aerobic; Gram-stain-positive.

Growth occurs at pH 5.0–9.0. The temperature range for growth is between 15 and 45°C.

No production of melanoid pigments. No growth on 4% NaCl. The diagnostic diamino acid isomer of the cell-wall peptidoglycan is *meso*-DAP. Galactose, glucose, madurose, mannose, and ribose are present in whole-cell hydrolysates. The *N*-acyl type of muramic acid is acetyl. Mycolic acids are absent. The polar lipids include PE and OH-PE. Phosphatidylcholine is absent. The major menaquinones are MK-9(H_6) and MK-9(H_4). The fatty acid profile is characterized by the presence of significant amounts of $C_{16:0}$ (17.4%), $C_{17:1}$ ω8c (11.5%), and $C_{17:0}$ 10 methyl (8.4%).

Source: sandy soil from Cox's Bazar, Bangladesh.

DNA G+C content (mol%): 72.3 (HPLC).

Type strain: 5-38-42, JCM 13931, MTCC 8090.

Sequence accession no. (16S rRNA gene): AB274967, AB505224.

8. **Nonomuraea dietziae** Stackebrandt, Wink, Steiner and Kroppenstedt 2001, 1439^{VP}

di.et.zi´a.e. N.L. gen. fem. n. *dietziae* of Dietz, named in honor of the American microbiologist Alma Dietz.

Long spore chains, each containing up to 30 spores, formed on aerial mycelium. Spore chains may be stalked or may form clusters emerging directly from the agar. Sporangia are not detected. Irregular spore chains with spirals of 1–2; sometimes 3–5 turns are seen.

Aerial mycelium develops different colors depending upon the ISP medium used: red-brown (yeast extract malt extract agar, ISP2); white, beige, or pink (oatmeal agar, ISP3); pink (inorganic salts starch agar, ISP4 and tyrosine agar, ISP7); antique rose (glycerol-asparagine agar, ISP5); and white (peptone-yeast extract iron agar, ISP6); after approximately 14 d. Depending upon the ISP medium used, the nonfragmenting substrate mycelium is brown (ISP2), beige (ISP3), pink (ISP3, ISP 7), or rose (ISP5, ISP6) in color. Soluble pigments not produced.

Cell hydrolysates contain galactose, glucose, and ribose, but not arabinose or xylose; madurose occurs in minor amounts. Diphosphatidylglycerol, phosphatidylethanolamine, phosphatidylinositol, ninhydrin and sugar positive phospholipids, hydroxylated phosphatidylethanolamine, and uncharacterized glycolipids are present; Phosphatidylglycerol and phosphatidylinositol mannosides are absent.

Additional chemotaxonomic data are given in Table 284, and the fatty acid pattern of the type strain is shown in Table 285.

TABLE 286. Comparison of the phenotypic properties of *Nonomuraea species*[a,b]

Characteristic	*N. pusilla*	*N. africana*	*N. angiospora*	*N. antimicrobica*	*N. bangladeshensis*	*N. candida*	*N. coxensis*	*N. dietziae*	*N. fastidosa*	*N. ferruginea*	*N. helvata*
Spore chain morphology											
Spore chains[c]	psp	str	sp	Sp	Sp	H	str, h	str, sp	s, sp	h, s	h, psp
Spore ornamentation	Smooth	Smooth	Ridged	Smooth	Smooth	Smooth	Smooth	Cross-ridged, smooth-and-rough	Irregular	Folded	Smooth
Number of spores	>01	4–10	4–15	nd	8–12	nd	12–17	Up to 30	4–10	4–10	4–10
Growth on ISP 3 medium:											
Aerial mycelium	White/ Cream	Grayish blue	White	White/ pink	Pale brown	White	Pink to white	Beige	White/ pink	White/ pink	White
Substrate mycelium	Gray/ brown	Yellow	White/ ochre	Moderate brown	Pale brown	Beige	Orange	Beige	Colorless	Pink	Yellow/ brown
Soluble pigment	None	Yellowish Brown	None	Pink	None	Yellow	None	Yellow	None	None	None
Biochemical tests:											
Esculin hydrolysis	+	+	+	–	nd	+	nd	nd	+	–	+*
Nitrate reductase	+	+	–	–	nd	–		nd	+	+	+
Degradation tests:											
Casein	–	+	+	+	nd	+	nd	nd	+	+	–
DNA	+	+	+	nd	nd	nd	nd	nd	+	+	–
Elastin	–	–	+	nd	nd	nd	nd	nd	+	+	nd
Gelatin	+	+	+	–	nd	+	nd	+	+	–	–
Hypoxanthine	+	+	+	–	nd	–	nd	nd	+	+	–
Starch	–	+	–	–	nd	–	nd	nd	–	+	–
Tyrosine	+	+	+	nd	nd	–	nd	nd	–	+	–
Xanthine	–	–	–	nd	nd	–	nd	nd	–	–	–
Utilization of:											
L-Arabinose	–	+	–	+	w	w	w	w	–	+	–
Cellobiose	–	+	–	+	+	w	+	–	–	–	–
D-Fructose	–	+	–	nd	w	+	w	+	–	+	–
m-Inositol	–	+	+	–	w	+	w	–	+	–	+
D-Mannose	–	+	–	nd	+	+	+	–	–	+	–
D-Raffinose	–	+	+	nd	+	+	w	–	+	–	+
L-Rhamnose	–	+	–	nd	+	+	+	+	–	+	–
Sucrose	–	+	–	–	w	+	+	–	–	+	+
D-Xylose	–	+	+	–	w	+	+	–	+	+	+
Production of:											
Acetoin	+	+	–	nd	nd	nd	nd	+	+	+	+
Arginine dihydrolase	+	+	+	nd	nd	nd	nd	+	+	+	+
Citrate lyase	–	+	–	nd	nd	nd	nd	+	+	+	+
β-Galactosidase	–	+	–	nd	nd	nd	nd	+	+	+	+
Gelatinase	+	+	+	nd	nd	nd	nd	+	+	+	+
Lysine decarboxylase	–	+	+	nd	nd	nd	nd	–	+	+	+
Ornithine decarboxylase	–	+	+	nd	nd	nd	nd	+	+	+	+
Urease	+	+	+	nd	nd	nd	nd	+	+	+	+

[a]Symbols: +, positive; –, negative; w, weak; nd, not determined.

[b]Data were taken from Meyer (1989), Chiba et al. (1999), Stackebrandt et al. (2001), Quintana et al. (2003), Ara et al. (2007a, 2007b), Le Roes and Meyers (2008),

[c]h, hooks, curled; psp, pseudosporingia; s, spirals of 1–2 turns; sp, spirals of 3–5 turns; str, straight.

N. kuesteri	*N. longicatena*	*N. maheshkhaliensis*	*N. polychroma*	*N. recticatena*	*N. rosea*	*N. rosecla*	*N. roseoviolacea*	*N. rubra*	*N. salmonea*	*N. spiralis*	*N. terrnata*	*N. turkmeniaca*
sp	str	sp	nd	str	s	sp, str	psp	h, s, sp	h, s	sp	irregular, psp	sp
nd	Smooth	Folded and rough	nd	Smooth	nd	Folded	Smooth	Smooth	Warty	Folded	Rugose	Smooth
nd	10–30	17–20	nd	4–20	4–10	6–20	4–20	4–20	4–30	4–20	8–15	10–20
Trace	White	White	Trace	White/ cream	White	Pink	Pink/ violet	Trace	Pink	White/ yellow	White	Trace
Yellow	Ochre	Light wheat	Colorless/ brown	Dark yellow/ brown	Pink/ violet	Brown/ red	Violet	Orange/ red	Red	Yellow/ brown	White/ ochre	Violet/ red
None	None	None	None	None	None	None	Violet	Red	None	None	None	Pink/ Violet
+	+	nd	+	+	+	+	+*	–#	+#	+	+	+*
nd	–		–	+	nd	+	+	+	+	+	–	+
–	+	nd	–	–	+	–	–	–	+	–	+	+
nd	–	nd	–	–	nd	–	+	–	+	–	+	–
nd	+	nd	+	+	nd	–	–	+	+	–	–	–
–	–	nd	+	+	+	+	+	+	+	+	–	+
–	+	nd	+	+	+	+	+	+	+	–	+	+
–	+	nd	–	+	+	–	–	+	–	–	+	+
–	–	nd	–	–	+	+	–	+	+	+	–	–
–	–	nd	–	–	+	–	–	–	–	–	–	–
+	+	w	–	+	+	–	+	+	–	–	nd	+
+	–	+	–	–	+	–	–	+	–	–	nd	+
+	nd	+	–	–	+	–	+	+	–	–	nd	+
+	–	w	–	+	+	+	+	+	–	–	nd	+
+	+	+	–	–	+	–	+	+	–	+	nd	+
nd	+	nd	–	–	nd	–	+	+	–	–	nd	+
+	–	+	–	+	+	+	+	+	+	+	nd	+
+	nd	+	–	–	–	+	+	+	+	+	nd	+
+	nd	+	–	–	+	–	–	+	–	+	nd	+
nd	nd	nd	+	+	nd	+	+	+	+	+	nd	+
–	+	nd	+	–	nd	+	+	+	+	–	nd	+
nd	nd	nd	–	–	nd	+	–	+	+	–	nd	+
–	+	nd	–	–	nd	+	–	+	+	+	nd	+
–	–	nd	+	–	nd	+	+	+	+	–	nd	+
–	+	nd	+	–	nd	+	+	+	+	–	nd	+
–	+	nd	+	–	nd	+	+	+	+	–	nd	–
–	+	nd	–	+	nd	+	+	+	+	–	nd	+

Qin et al. (2009), Kämpfer et al. (2010).

Positive for acetoin, arginine dihydrolase, β-galactosidase, citrate lyase, gelatinase, ornithine decarboxylase, and urease activities. D-Fructose and L-rhamnose are utilized as sole carbon sources, but cellobiose, *meso*-inositol, D-mannose, D-rhamnose, sucrose, and D-xylose are not. Utilization of L-arabinose is weak.

Further physiological characteristics are given in Table 286.

Shows no antibacterial effect on *Micrococcus luteus, Streptococcus murinus, Bacillus subtilis, Staphylococcus aureus, Escherichia coli, Pseudomonas aeruginosa, Saccharomyces cerevisiae, Aspergillus niger,* or *Candida albicans.*

Source: soil.

DNA G+C content (mol%): not determined.

Type strain: ATCC 35861, CIP 107127, DSM 44320, NBRC 14309, JCM 3338, NRRL 11111.

Sequence accession no. (16S rRNA gene): AB006156, AJ278220, AJ294350.

9. **Nonomuraea fastidiosa** corrig. (Soina, Sokolov and Agre 1975) Zhang, Wang and Ruan 1998b, 419[VP] (*Actinomadura fastidiosa* Soina, Sokolov and Agre 1975, 883; *Microtetraspora fastidiosa* Kroppenstedt, Stackebrandt and Goodfellow 1990a, 148)

fas.ti.di.o′sa. L. fem. adj. *fastidiosa* fastidious, referring to the difficulties in growing the organism.

Spore chains consist of irregular spirals of 1–2, more rarely 3–4, turns arranged monopodially on long aerial hyphae (Figure 378). Spore surface ornamentation smooth or slightly irregular. Temperature range: 23–55°C, optimum temperature 30–45°C.

Inorganic salts-starch agar (Gauze 1 medium): moderate growth, aerial mycelium sparse and cream colored to pinkish, substrate mycelium colorless, no diffusible pigment. Oatmeal agar: good growth, surface farinaceous, aerial mycelium white to pale pink, substrate mycelium colorless or pale brownish, no diffusible pigment. Yeast extract-malt extract agar: moderate growth, surface granular with spots of whitish aerial mycelium at the edges of colonies, substrate mycelium pale brown, no diffusible pigment.

Cell hydrolysates contain galactose, madurose, mannose, and ribose. Diphosphatidylglycerol, phosphatidylethanolamine, hydroxylated phosphatidylethanolamine, and uncharacterized glycolipids are present; traces of phosphatidylinositol and ninhydrin and sugar positive phospholipids are also found, but phosphatidylglycerol and phosphatidylinositol mannosides are absent.

Further chemotaxonomic data are given in Table 284, and the fatty acid pattern of the type strain is shown in Table 285.

Positive for acetoin, arginine dihydrolase, citrate lyase, β-galactosidase, gelatinase, lysine decarboxylase, ornithine decarboxylase, and urease activities. L-Arabinose, citrate fructose, glucose, and rhamnose are utilized as sole carbon sources, but cellobiose, mannose, *meso*-inositol, rhamnose, sucrose, and xylose are not.

Further physiological characteristics are given in Table 286.

DNA G+C content (mol%): 67 (T_m).

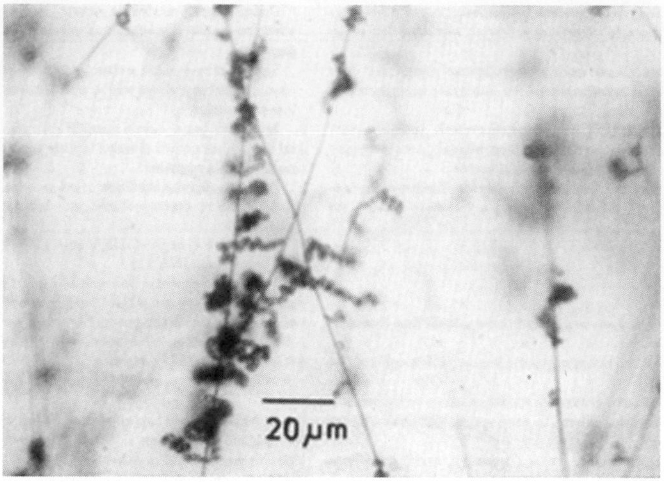

FIGURE 378. *Nonomuraea fastidiosa* strain IMET 9614. Sporulating aerial mycelium. Grown on oatmeal agar at 28°C for 21 d.

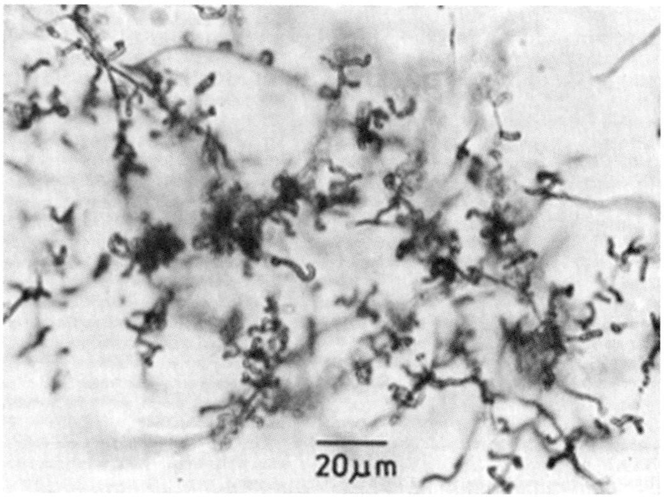

FIGURE 379. *Nonomuraea ferruginea*, type strain IMET 9567. Sporulating aerial mycelium. Grown on oatmeal agar at 28°C for 18 d.

Type strain: ATCC 33516, BCRC 12546, DSM 43674, NBRC 14680, INMI 104, JCM 3321, KCTC 9268, VKM Ac-804.

Sequence accession no. (16S rRNA gene): U48844.

10. **Nonomuraea ferruginea** corrig. (Meyer 1981) Zhang, Wang and Ruan 1998b, 419[VP] (*Actinomadura ferruginea* Meyer 1981, 215; *Microtetraspora ferruginea* Kroppenstedt, Stackebrandt and Goodfellow 1990a, 148)

fer.ru.gi′ne.a. L. fem. adj. *ferruginea* rusty brown, referring to the orange-brown-colored substrate mycelium.

Spore chains short, hooked, or irregular spirals of 1–2 turns, arranged monopodially on long aerial hyphae, 4–9 spores per chain (Figure 379). Spore surface smooth or irregularly folded. Optimum temperature: 28–30°C.

Inorganic salts-starch agar: poor growth, surface leathery, aerial mycelium absent, substrate mycelium colorless to brownish, no diffusible pigment. Oatmeal agar: growth abundant, surface farinaceous, aerial mycelium white to pale pink, substrate mycelium pink, no diffusible pigment. Oatmeal-nitrate agar: growth abundant, surface dusty, aerial mycelium white to pale pink, substrate mycelium orange-pink, no diffusible pigment. Peptone-glucose medium: growth moderate, surface cartilaginous, aerial mycelium absent, substrate mycelium light brown, no diffusible pigment. Yeast extract-malt extract agar: growth abundant, surface farinaceous, aerial mycelium orange-pink, substrate mycelium bright orange-brown, no diffusible pigment.

Cell hydrolysates contain galactose, madurose, and ribose. Diphosphatidylglycerol, phosphatidylinositolmannosides, ninhydrin and sugar positive phospholipids, hydroxylated phosphatidylethanolamine, and uncharacterized glycolipids are present; phosphatidylethanolamine, phosphatidylglycerol, and phosphatidylinositol are absent.

Additional chemotaxonomic data are given in Table 284, and the fatty acid pattern of the type strain is shown in Table 285.

Positive for arginine dihydroloase, acetoin, citrate lyase, β-galactosidase, gelatinase, lysine decarboxylase, ornithine decarboxylase, and urease activities. L-Arabinose, D-fructose, D-mannose, L-rhamnose, sucrose, and D-xylose are utilized as sole carbon sources, but cellobiose, *meso*-inositol, and D-raffinose are not.

DNA G+C content (mol%): 68.1 (T_m).

Type strain: 14094, ATCC 35575, CCM 3424, BCRC 12537, CIP 106925, DSM 43553, NBRC 14094, IMET 9567, JCM 3283, KCTC 9269, NCIMB 11630, NRRL B-16096, VKM Ac-854.

Sequence accession no. (16S rRNA gene): U48845.

11. **Nonomureae helvata** corrig. (Nonomura and Ohara 1971a) Zhang, Wang and Ruan 1998b, 419[VP] (*Actinomadura helvata* Nonomura and Ohara 1971a, 904; *Microtetraspora ferruginea* Kroppenstedt, Stackebrandt and Goodfellow 1990a, 148)

hel.va'ta. N.L. fem. adj. *helvata* honey-yellow, referring to the color of the substrate mycelium.

Spore chains with spirals of 3–5 turns are seen, or hooks and pseudosporangia on long aerial hyphae. About 10 spores per chain. Spore surface smooth.

Inorganic salts-starch agar: moderate growth, surface farinaceous, aerial mycelium white to cream colored, substrate mycelium yellowish brown, no diffusible pigment. Oatmeal agar: moderate growth, surface leathery, aerial mycelium a filmy cover of sterile hyphae, substrate mycelium yellowish brown, no diffusible pigment. Oatmeal-nitrate agar: good growth, surface farinaceous, aerial mycelium white, substrate mycelium cream colored, no diffusible pigment. According to Nonomura and Ohara (1971a), B vitamins are essential for growth. Yeast extract-malt extract agar: good growth, surface farinaceous, aerial mycelium yellowish white, substrate mycelium yellowish brown, no diffusible pigment.

Cell hydrolysates contain galactose, madurose, ribose, and traces of mannose. Diphosphatidylglycerol, phosphatidylglycerol, phosphatidylinositolmannosides, ninhydrin and sugar positive phospholipids, hydroxylated phosphatidylethanolamine, and uncharacterized glycolipids are present; phosphatidylethanolamine and phosphatidylinositol are absent. Esculin is hydrolyzed. Shows arginine dihydrolase, β-galactosidasc, gelatinase, lysine decarboxylase, ornithine decarboxylase, and urease activities, but does degrade casein, gelatin, hypoxynthine, starch, tyrosine, or xanthine. Cellobiose, *meso*-inositol, sucrose, and D-xylose are utilized as sole carbon sources.

Further physiological characteristics are given in Table 286.

Additional chemotaxonomic data are given in Table 284, and the fatty acid pattern of the type strain is shown in Table 285.

DNA G+C content (mol%): not determined.

Type strain: ATCC 27295, CBS 259.72, BCRC 13402, DSM 43142, NBRC 14681, IMET 9584, JCM 3143, KCTC 9274, NCIMB 11115, NRRL B-16123, PCM 2199.

Sequence accession no. (16S rRNA gene): U48975.

12. **Nonomuraea kuesteri** Kämpfer, Kroppenstedt and Grün-Wollny 2005, 848[VP]

ku.es´te.ri. N.L. gen. masc. n. *kuesteri* of Küster, in honor of Eberhard Küster, a German microbiologist, in recognition of his numerous contributions to the taxonomy of actinomycetes.

Spiral spore chains with 3–5 turns. Spore ornamentation not determined. Yellow substrate mycelium, a trace of aerial mycelium, but no soluble pigments formed on oatmeal agar.

Cell hydrolysates contains madurose. Diphosphatidylglycerol, phosphatidylethanolamine, phosphatidylglycerol, phosphatidylinositol, phosphatidylinositolmannosides, ninhydrin-and sugar-positive phospholipids, hydroxylated phosphatidylethanolamine, and uncharacterized glycolipids are present.

Additional chemotaxonomic data are given in Table 284, and the fatty acid pattern of the type strain is shown in Table 285.

Does not produce acid from adonitol, L-arabinose, arabitol, cellobiose, dulcitol, erythritol, glucose, *meso*-inositol, lactose, maltose, D-mannitol, melibiose, methyl D-glucoside, raffinose, L-rhamnose, salicin, sorbitol, sucrose, trehalose, and D-xylose. *N*-Acetyl-D-glucosamine, adonitol, L-aspartate (weak),L-arabinose, D-cellobiose, fumarate (weak), D-fructose, D-galactose, D-glucose, DL-lactate, malate, D-mannitol, D-mannose, D-melibiose, 2-oxoglutarate, L-proline, L-rhamnose, sucrose, D-trehalose, and D-xylose are used as sole carbon sources after 7 d, but acetate, *N*-acetyl-D-galactosamine, *cis*-aconitate, *trans*-aconitate, adipate, 4-aminobutyrate, L-alanine, β-alanine, *p*-arbutin, azelate, citrate, D-gluconate, glutarate, L-histidine, DL-3-hydroxybenzoate, DL-4-hydroxybenzotae, DL-3-hydroxybutyrate, itaconate, *meso*-inositol, L-leucine, maltitol, D-maltose, mesaconate, L-ornithine, L-phenylalanine, L-phenylacetate, putrescine, propionate, pyruvate, D-ribose, salicin, L-serine, sorbitol, suberate, and L-tryptophan are not.

Further physiological characteristics are given in Table 286.

DNA G+C content (mol%): not determined.

Type strain: GW 14-1925, DSM 44753, JCM 13854, NRRL B-24325.

Sequence accession no. (16S rRNA gene): AJ746362.

13. **Nonomuraea longicatena** Chiba, Suzuki and Ando 1999, 1628[VP]

lon.gi.ca.te'na. L. adj. *longus* long; L. n. *catena* chain; N.L. n. *longicatena* long chain, referring to long spore chains.

Straight spore chains with 10–30 spores per chain. Spore ornamentation smooth. No growth at 45°C.

Oatmeal agar: substrate mycelium growth is ochre, white aerial mycelium, no soluble pigment. Yeast extract-malt extract agar: abundant growth, ochre substrate mycelium, white aerial mycelium.

Cell hydrolysates contain madurose. Diphosphatidylglycerol, phosphatidylethanolamine, ninhydrin and sugar positive phospholipids, hydroxylated phosphatidylethanolamine, and uncharacterized glycolipids are present.

Additional chemotaxonomic data are given in Table 284, and the fatty acid pattern of the type strain is shown in Table 285.

Esculin is hydrolyzed; nitrate is not reduced. Degrades casein, elastin, hypoxanthine, and starch.

L-Arabinose, D-mannose, and D-raffinose are utilized as sole carbon sources, but cellobiose, *meso*-inositol, and L-rhamnose are not. Positive for arginine dihydrolase, β-galactosidase, lysine decarboxylase, ornithine decarboxylase, and urease activities. Production of gelatinase is negative; Grows in the presence of up to 3% NaCl.

DNA G+C content (mol%): 71.7 (HPLC).

Type strain: K-252, NBRC 16462, JCM 11136, NRRL 15532.

Sequence accession no. (16S rRNA gene): AB018787.

14. **Nonomuraea maheshkhaliensis** Ara, Kudo, Matsumoto, Takahashi and Ōmura 2007c, 2449[VP] (Effective publication: Ara, Kudo, Matsumoto, Takahashi and Ōmura 2007a, 164.)

ma.hesh.kha.li.en.sis. N.L. fem adj. *maheshkhaliensis* of or pertaining to Maheshkhali, Bangladesh.

Forms a well-developed, branched substrate mycelium. Abundant aerial mycelia are present. Spore chains are spiral with 2–4 turns (17–20 spores). Sporangia are not detected. Spores are spherical to cylindrical and spore surface is folded and rough.

Good growth occurs on Bennett's agar, glucose-yeast extract agar, Hickey–Tresner agar, oatmeal agar (ISP3), oatmeal-nitrate agar, Waksman no. 1 agar, and yeast extract-malt extract agar (ISP2); moderate growth on glycerol-asparagine agar, inorganic salts-starch agar (ISP4), nutrient agar, sucrose-nitrate agar, tyrosine agar (ISP-7) water agar, and yeast extract-malt extract agar; poor growth on glucose-asparagine agar.

Substrate mycelium color is mustard to light wheat color on various agar media.

White aerial mycelia and sporulation occur on ISP3, ISP4 agar, oatmeal-nitrate agar and water agar after 14 d of incubation at 28°C.

A soluble pigment is not produced. Aerobic, Gram-stain-positive, and able to grow at pH 5.0–9.0.

The temperature range for growth is between 15 and 37°C. Production of melanoid pigments is negative. No growth on 4% NaCl.

The isomer of A_2pm was *meso* and galactose, glucose, madurose, mannose, ribose (trace) in the whole-cell hydrolysates. The *N*-acyl type of muramic acid is acetyl. Mycolic acids are absent. The polar lipids include phosphatidylethanolamine and phospholipids of unknown structure containing glucosamine. Phosphatidylcholine is absent.

The menaquinone is MK-9(H_4). Fatty acids with major amounts of $C_{16:0}$ (15.5%), $C_{16:0}$ iso (13.8%), and $C_{17:0}$ 10 methyl (9.6%).

Source: mangrove rhizosphere soil in Maheshkhali, Bangladesh.

DNA G+C content (mol%): 72.0 (HPLC).

Type strain: 16-5-14, JCM 13929, MTCC 8545.

Sequence accession no. (16S rRNA gene): AB290014.

15. **Nonomuraea polychroma** corrig. (Galatenko, Terekhova and Preobrazhenskaya 1981) Zhang, Wang and Ruan 1998b, 419[VP] (*Actinomadura polychroma* Galatenko, Terekhova and Preobrazhenskaya 1981, 803; *Microtetraspora polychroma* Kroppenstedt, Stackebrandt and Goodfellow 1990a, 148)

po.ly.chro'ma. Gr. adj. *polus* many; Gr. n. *chroma* color; N.L. n. *polychroma* intended to mean that the bacterium produces many colors (Note: the epithet *polychroma* is not correct, however, it is not possible to correct it.)

Short spore chains are produced in the form of spirals or pseudosporangia. Spore surface smooth.

Colorless to brown substrate mycelium, traces of white aerial mycelium, but no diffusible pigments formed on oatmeal agar.

The cultures of this species have no antagonistic activity with respect to various test-microbes.

Cell hydrolysates contain madurose and mannose. Diphosphatidylglycerol, phosphatidylethanolamine, ninhydrin and sugar positive phospholipids, hydroxylated phosphatidylethanolamine, and uncharacterized glycolipids are present, as are traces of phosphatidylinositolmannosides.

Additional chemotaxonomic data are given in Table 284, and the fatty acid pattern of the type strain is shown in Table 285.

Does not use L-arabinose, cellobiose, fructose, *meso*-inositol, mannose, raffinose rhamnose, sucrose, or D-xylose as sole carbon sources. Esculin is hydrolyzed, nitrate is not reduced. Elastin, gelatin, and hypoxanthine are degraded. Acetoin, arginine dihydrolase, lysine decarboxylase, and ornithine decarboxylase activities are positive.

DNA G+C content (mol%): not determined.

Type strain: ATCC 49500, DSM 43925, NBRC 14345, IMET 9743, INA 2755, JCM 6834, KCTC 9277, NRRL B-16243, VKM Ac-1084.

Sequence accession no. (16S rRNA gene): U48977.

16. **Nonomuraea recticatena** corrig. (Gauze, Terekhova, Galatenko, Preobrazbenskaya, Borisova and Fedorova 1984) Zhang, Wang and Ruan 1998b (*Actinomadura recticatena* Gauze, Terekhova, Galatenko, Preobrazbenskaya,

Borisova and Fedorova 1984, 3; *Microtetraspora recticatena* Kroppenstedt, Stackebrandt and Goodfellow 1990a, 148)

rec.ti.ca.te'na. L. part. adj. *rectus* straight; L. n. *catena* a chain; N.L. n. *recticatena* (nominative in apposition), a straight chain.

Straight chains of 4–20 spores. Spore ornamentation smooth. The aerial mycelium develops white to cream color on oatmeal agar; substrate mycelium growth is dark yellow to brown; no soluble pigment found.

Cell hydrolysates contain galactose and traces of ribose. Diphosphatidylglycerol, phosphatidylethanolamine, ninhydrin and sugar positive phospholipids, hydroxylated phosphatidylethanolamine, uncharacterized glycolipids, and traces of phosphatidylinositol are present; phosphatidylglycerol and phosphatidylinostol mannosides are absent.

Further chemotaxonomic data are given in Table 284, and the fatty acid pattern of the type strain is shown in Table 285.

Positive for esculin hydrolysis and nitrate reductase. Degrades elastin, gelatin, hypoxanthine, and starch. Positive for acetoin and urease activities. Arginine dihydrolase, citrate lyase, β-galactosidase, lysine decarboxylase, and ornithine decarboxylase are not produced. Further physiological characteristics are given in Table 286. L-Arabinose, *meso*-inositol, and L-rhamnose used as sole carbon sources but cellobiose, D-fructose, D-mannose, D- raffinose, sucrose, and D-xylose are not.

DNA G+C content (mol%): not determined.

Type strain: DSM 43937, NBRC 14525, INA 308, JCM 6835, KCTC 9279, VKM Ac-940.

Sequence accession no. (16S rRNA gene): AJ404230, U48979.

17. **Nonomuraea rosea** Kämpfer, Busse, Tindall, Nimtz and Grün-Wollny 2010, 1123[VP]

ro.se'a. L. fem. adj. *rosea* rose-colored or rosy, referring to the pinkish color of the colonies.

Forms an extensive branched substrate mycelium. A white aerial mycelium is visible on oatmeal agar. Spore chains are spiral; sporangia are not detected. Gram-stain-positive and oxidase positive; shows oxidative metabolism.

Good growth occurs after 3 d incubation on oatmeal agar and nutrient agar (Oxoid) at 25–30°C.

The polar lipids include the major compound diphosphatidylglycerol, moderate amounts of phosphatidylmonomethylethanolamine, phosphatidylethanolamine, hydroxy-phosphatidylethanolamine, hydroxy-phosphatidylmonomethylamine, and an unknown aminophosphoglycolipid. Phosphatidylinositol-mannosides and phosphatidylinositol are present along with small amounts of an unknown phospholipid.

Does not produce acids from the following carbon sources: glucose, lactose, sucrose, D-mannitol, dulcitol, salicin, adonitol, inositol, sorbitol, L-arabinose, raffinose, L-rhamnose, maltose, D-xylose, trehalose, cellobiose, methyl D-glucoside, erythritol, melibiose, and arabitol. The following carbon sources are utilized (after 7 d incubation): *N*-acetyl-D-galactosamine, *N*-acetyl-D-glucosamine, D-cellobiose, D-fructose, D-galactose, D-glucose, D-maltose, D-mannose, L-rhamnose, D-trehalose, D-xylose, adonitol, inositol, 4-aminobutyrate, citrate, fumarate (weak), DL-malate, L-aspartate (weak), and L-proline. The following carbon sources are not utilized: *p*-arbutin, D-gluconate, D-melibiose, D-ribose, sucrose, salicin, maltitol, D-mannitol, sorbitol, putrescine, acetate, propionate, *cis*-aconitate, *trans*-aconitate, adipate, azelate, glutarate, DL-3-hydroxybutyrate, itaconate, mesaconate, pyruvate, DL-lactate, suberate, 2-oxoglutarate, L-alanine, β-alanine, L-histidine, L-leucine, L-ornithine, L-phenylalanine, L-serine, L-tryptophan, DL-3-hydroxybenzoate, DL-4-hydroxybenzoate, and L-phenylacetate.

The fatty acids comprise mainly iso- and 10-heptadecanoic-branched fatty acids. Smaller amounts of unsaturated fatty acids are also detected.

Source: a soil sample.

DNA G+C content(mol%): not determined.

Type strain: GW 12687, CCUG 56107, DSM 45177.

Sequence accession no. (16S rRNA gene): FN356742.

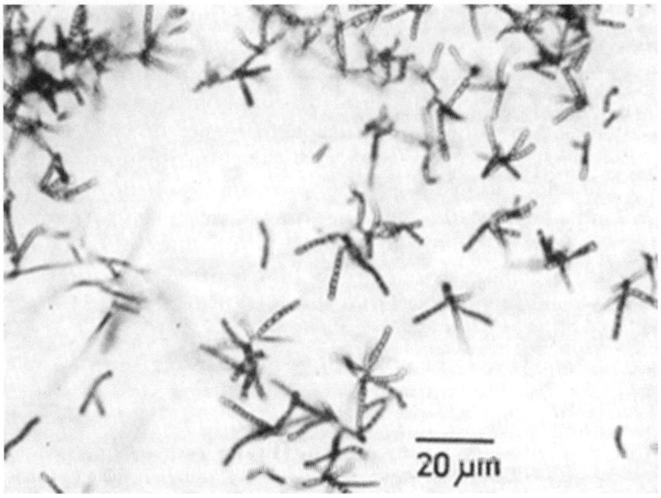

FIGURE 380. *Nonomuraea roseola*, type strain INA 1671. Sporulating aerial mycelium. Grown on oatmeal-nitrate agar at 28°C for 12 d.

18. **Nonomuraea roseola** corrig. (Lavrova and Preobrazhenskaia 1975) Zhang, Wang and Ruan 1998b, 420[VP] (*Actinomadura roseola* Lavrova and Preobrazhenskaya 1975, 483; *Microtetraspora roseola* Kroppenstedt, Stackebrandt and Goodfellow 1990a, 148)

ro.se.o'la. L. adj. *roseus* rose-colored; L. fem. suff. *-ola* diminutive ending; N.L. fem. dim. adj. *roseola* intended to mean with a rosy tinge, referring to the rose-colored aerial mycelium.

Spore chains with spirals of 3–5 turns, or straight chains with 6–30 spores per chain. Spore chains may be stalked or in clusters emerging directly from the agar (Figure 380). Spores elliptical. Spore surface folded.

Czapek sucrose agar: good growth, surface farinaceous, aerial mycelium white, substrate mycelium colorless, no diffusible pigment. Glycerol-asparagine agar: good growth, surface farinaceous, aerial mycelium white to pale pink, substrate mycelium pink to orange colored, no diffusible pigment. Oatmeal agar: abundant growth, surface farinaceous, aerial mycelium pink, substrate mycelium brownish red, no diffusible pigment. Peptone-glucose medium:

good growth, surface leathery, aerial mycelium absent, substrate mycelium yellow to orange colored, no diffusible pigment. Yeast extract-malt extract agar: abundant growth, surface farinaceous, aerial mycelium pink, substrate mycelium rusty brown, no diffusible pigment.

Cell hydrolysates contain galactose, madurose, and ribose; variable for mannose. Diphosphatidylglycerol, phosphatidylethanolamine, ninhydrin and sugar positive phospholipids, hydroxylated phosphatidylethanolamine, and uncharacterized glycolipids are present, as are traces of phosphatidylglycerol; phosphatidylinositol and phosphatidylinositolmannosides are absent.

Additional chemotaxonomic date are given in Table 284, and the fatty acid pattern of the type strain is shown in Table 285.

Esculin is hydrolyzed and nitrate reduced. Degrades gelatin, hypoxanthine, and tyrosine. Positive for arginine dihydrolase, citrate lyase, β-galactosidase, lysine decarboxylase, ornithine decarboxylase, and urease. *meso*-Inositol, L-rhamnose and sucrose are utilized as sole carbon source, but L-arabinose, cellobiose, D-fructose, D-mannose, D-raffinose, and D-xylose are not.

Additional physiological characteristics are given in Table 286.

DNA G+C content (mol%): 66.2 (T_m).

Type strain: ATCC 33579, DSM 43767, NBRC 14685, IMET 9576, INA 1671, JCM 3323, KCTC 9282, VKM Ac-1180.

Sequence accession no. (16S rRNA gene): AJ278221, U48980.

19. **Nonomuraea roseoviolacea** (Nonomura and Ohara 1971a) *Nonomuria roseoviolacea* Zhang, Wang and Ruan 1998b, 420[VP] (*Actinomadura roseoviolacea* Nonomura and Ohara 1971a, 909; *Microtetraspora roseviolacea* Kroppenstedt, Stackebrandt and Goodfellow 1990a, 148)

ro.se.o.vi.o.la'ce.a. L. adj. *roseus* rosy; L. adj. *violaceus* violet colored; N.L. fem. adj. *roseoviolacea* rosy, violet colored, referring to the color of the substrate mycelium (Note: the epithet *roseoviolacea* is not correct, however, it is not possible to correct it.).

Pseudosporangia spore chains with 4–20 spores per chain. Spore ornamentation smooth.

Glucose-asparagine agar: moderate growth, surface slightly farinaceous on the edge of the colonies, substrate mycelium brownish red, pale violet diffusible pigment. Glycerol-asparagine agar: good growth, surface leathery, aerial mycelium absent, substrate mycelium brick red, no diffusible pigment. Inorganic salts-starch agar: good growth, surface farinaceous, aerial mycelium white, substrate mycelium white to pale pink, no diffusible pigment. Oatmeal agar: good growth, surface farinaceous, aerial mycelium pale pink to violet, substrate mycelium pale violet, pale violet diffusible pigment. Yeast extract-malt extract agar: good growth, surface farinaceous to granular, aerial mycelium white to pale pink, substrate mycelium dark purple, no diffusible pigment.

Cell hydrolysates contain madurose, mannose, and ribose. Diphosphatidylglycerol, phosphatidylinositolmannosides, ninhydrin and sugar positive phospholipids, hydroxylated phosphatidylethanolamine, and uncharacterized glycolipids are present; phosphatidylethanolamine, phosphatidylglycerol, and phosphatidylinositol are absent.

Additional chemotaxonomic data are given in Table 284, and the fatty acid pattern of the type strain is shown in Table 285.

Esculin hydrolyzed and nitrate reduced; DNA and hypoxanthine are degraded. L-Arabinose, D-fructose, D-mannose, D-raffinose, L-rhamnose and sucrose, are utilized as sole carbon sources, but cellobiose is not. Positive for acetoin, arginine dihydrolase, lysine decarboxylase, ornithine decarboxylase, and urease activities, but negative for β-galactosidase activity.

DNA G+C content (mol%): 68.5 (T_m).

Type strain: AS 4.1072, ATCC 27297, CBS 260.72, CCM 3491, BCRC 13406, CIP 106924, DSM 43144, NBRC 14098, IMET 9751, JCM 3145, KCTC 9283, NCIMB 11117, NRRL B-16127, VKM Ac-909.

Sequence accession no. (16S rRNA gene): AB039959, AB043101.

Additional comments: in their original description, Nonomura and Ohara (1971a) reported on antibacterial activity of the type strain against *Staphylococcus aureus*. Further investigation of this strain (J. Meyer, unpublished) revealed an active agent similar to daunomycin.

19a. **Nonomuraea roseoviolacea subsp. roseoviolacea** (Nonomura and Ohara 1971a) *Nonomuria roseoviolacea* Zhang, Wang and Ruan 1998b, 420[VP] (*Actinomadura roseoviolacea* Nonomura and Ohara 1971a, 909; *Microtetraspora roseviolacea* Kroppenstedt, Stackebrandt and Goodfellow 1990a, 148)

The original description of *Nonomuraea roseoviolacea* subsp. *roseoviolacea* given by Nonomura and Ohara (1971a) is unchanged. In addition, this subspecies shows reduction of nitrate and liquefies gelatin. Sucrose is not utilized as a sole carbon source. DNA–DNA similarity to *Nonomuraea roseoviolacea* subsp. *carminata* ranges from 49–60%.

DNA G+C content (mol%): 68.5 (T_m).

Type strain: AS 4.1072, ATCC 27297, CBS 260.72, CCM 3491, BCRC 13406, CIP 106924, DSM 43144, NBRC 14098, IMET 9751, JCM 3145, KCTC 9283, NCIMB 11117, NRRL B-16127, VKM Ac-909.

Sequence accession no. (16S rRNA gene): AB039959, AB043101.

19b. **Nonomuraea roseoviolacea subsp. carminata** (Gauze, Sveshnikova, Ukholina, Gavrilina, Filicheva and Gladkikh 1973) Gyobu and Miyadoh 2001, 887[VP] (*Actinomadura carminata* Gauze, Sveshnikova, Ukholina, Gavrilina, Filicheva and Gladkikh 1973, 675)

car.mi.na'ta. N.L. fem. adj. *carminata* (from Fr. *carmin*), carmine.

Substrate hyphae are long, irregularly branched, and do not fragment into short elements. Aerial hyphae branch monopodially and are 0.3–0.4 µm in diameter. Spore chains are formed on the aerial hyphae in tightly closed spirals. Well-developed spore chains are enveloped in a slimy mass, giving the appearance of sporangia (pseudosporangia). Spores are oval to cylindrical in shape and 0.6–1.2 µm in size, with smooth surfaces.

A pink aerial hyphal mass, and a wine-red soluble pigment is formed on yeast-extract–malt extract agar.

Starch is hydrolyzed. Grows at 42°C. Reduction of nitrate, liquefaction of gelatin, peptonization of milk, and production of melanoid pigments are negative. Grows in the presence of 3% NaCl. L-Arabinose, D-fructose, D-galactose D-glucose, L-rhamnose, sucrose, and D-xylose are used as sole carbon sources, but *meso*-inositol, D-mannitol and raffinose are not.

Source: soil.

DNA G+C content (mol%): not determined.

Type strain: DSM 44170, NBRC 15903, INA 4281, JCM 9946, VKM Ac-1780.

Sequence accession no. (16S rRNA gene): AB039961.

20. **Nonomuraea rubra** corrig. (Sveshnikova, Maksimova and Kudrina 1969) Zhang, Wang and Ruan 1998b, 420[VP] (*Micromonospora rubra* Sveshnikova, Maksimova and Kudrina 1969, 883; *Actinomadura rubra* Meyer and Sveshnikova 1974, 167)

ru'bra L. fem. adj. *rubra* red-colored, referring to the color of substrate mycelium.

Spore chains with 4–20 spores per chain form curled hooks, spirals with 1–2 turns or spirals with 3–5 turns. Spore ornamentation smooth. Sporulating aerial mycelium mostly seen only in microscopically visible traces after 4–6 weeks of incubation on oatmeal agar, oatmeal-nitrate agar, and Bennett sucrose agar.

Bennett's sucrose agar: good growth, surface cartilaginous or wrinkled, aerial mycelium only traces and pinkish gray, substrate mycelium red to orange, red diffusible pigment. Oatmeal agar: good growth, surface wrinkled, aerial mycelium very poorly developed and mostly sterile or coremia-like, substrate mycelium bright orange-red, pale red diffusible pigment. Oatmeal-nitrate agar: moderate growth, surface cartilaginous glistening, aerial mycelium only traces and pinkish-gray, substrate mycelium red to orange, no diffusible pigment. Yeast extract-malt extract agar: moderate growth, surface cartilaginous glistening, aerial mycelium only traces of sterile hyphae, substrate mycelium dark red to brown, no diffusible pigment.

Cell hydrolysates contain galactose, madurose, ribose, and traces of mannose. The cell-wall composition of *Nonomuraea rubra* shows a slight difference from type III/B of Lechevalier and Lechevalier (1970b). In addition to *meso*-A$_2$pm, minor amounts of L-A$_2$pm and glycine have been found.

Diphosphatidylglycerol, phosphatidylinositol, phosphatidylinositol mannosides, ninhydrin and sugar positive phospholipids, hydroxylated phosphatidylethanolamine, and uncharacterized glycolipids are present; Phosphatidylethanolamine and phosphatidylglycerol are absent.

Additional chemotaxonomic data are given in Table 284, and the fatty acid pattern of the type strain is shown in Table 285.

Nitrate is reduced but esculin is not hydrolyzed. Degrades elastin, gelatin, hypoxanthine, starch, and tyrosine. Positive for acetoin, arginine dihydrolase, citrate lyase, β-galactosidase, gelatinase, lysine decarboxylase, ornithine decarboxylase, and urease activities. L-Arabinose, cellobiose, D-fructose, *meso*-inositol, D-mannose, D-raffinose, L-rhamnose, sucrose, and xylose are utilized as sole carbon sources.

Produces the antibiotic maduramycin, a red pigment with indicator properties (pH <7.0 yellow, pH >7.0 red) which possesses a strong antimicrobial activity against Gram-stain-positive bacteria (Fleck et al., 1978).

DNA G+C content (mol%): not determined.

Type strain: ATCC 27031, CBS 132.76, BCRC 12591, CIP 107008, DSM 43768, NBRC 14070, NBRC 14686, IMET 8181, INA 325, JCM 3234, JCM 3389, KCTC 9284, NRRL B-16083, VKM Ac-615.

Sequence accession no. (16S rRNA gene): AF277200.

21. **Nonomuraea salmonea** corrig. (Preobrazhenskaya, Lavrova, Ukholina and Nechaeva 1975b) Zhang, Wang and Ruan 1998b, 420[VP] (*Actinomadura salmonea* Preobrazhenskaya, Lavrova, Ukholina and Nechaeva 1975b, 408; *Microtetraspora salmonea* Kroppenstedt, Stackebrandt and Goodfellow 1990a, 148)

sal.mo'ne.a. N.L. fem. adj. *salmonea* from L. n. *salmo* salmon, salmon colored referring to the color of substrate mycelium.

Spore chains with 4–30 spores per chain form curled hooks or spirals with 1–2 turns. Spore ornamentation folded or warty.

Inorganic salts-starch agar: moderate growth, surface wrinkled, aerial mycelium only traces of cream colored mycelium, substrate mycelium pale brown, no diffusible pigment. Oatmeal agar: good growth, surface farinaceous, aerial mycelium pale pink, substrate mycelium dark red, no diffusible pigment. Yeast extract-malt extract agar: moderate growth, surface cartilaginous, aerial mycelium cream colored to pinkish, substrate mycelium light brown, no diffusible pigment.

Cell hydrolysates contain galactose, madurose, mannose, and ribose. Phosphatidylinositol, phosphatidylinositol-mannosides, ninhydrin and sugar positive phospholipids, hydroxylated phosphatidylethanolamine, and uncharacterized glycolipids are present; diphosphatidylglycerol, phosphatidylethanolamine, and phosphatidylglycerol absent.

Additional chemotaxonomic data are given in Table 284, and the fatty acid pattern of the type strain is shown in Table 285.

Esculin is hydrolyzed and nitrate reduced. Degrades casein, DNA, elastin, gelatin, hypoxanthine, and tyrosine. Positive for acetoin, arginine dihydrolase, citrate lyase, β-galactosidase, lysine decarboxylase, ornithine decarboxylase, and urease production activities. L-Rhamnose and sucrose are utilized as sole carbon source, but L-arabinose, cellobiose, *meso*-inositol, D-fructose, D-mannose, D-raffinose, and D-xylose are not.

DNA G+C content (mol%): 66.0 (T_m).

Type strain: ATCC 33580, CIP 107009, DSM 43678, NBRC 14687, IMET 9582, INA 2488, JCM 3324, KCTC 9285, PCM 2201, VKM Ac-913.

Sequence accession no. (16S rRNA gene): U48982, X97892.

22. **Nonomuraea spiralis** corrig. (Meyer 1979) Zhang, Wang and Ruan 1998b, 420[VP] (*Actinomadura spiralis* Meyer 1979, 39; *Microtetraspora spiralis* Kroppenstedt, Stackebrandt and Goodfellow 1990a, 148)

spi.ra'lis. N.L. fem. adj. *spiralis* (from L. n. *spira* coil; and L. suff. *alis -is -e*, suffix denoting pertaining to), coiled, referring to the morphology of the spore chains.

Spore chains in spirals of 2–5 turns, closely packed or more or less loose spirals in pseudoverticillate arrange-

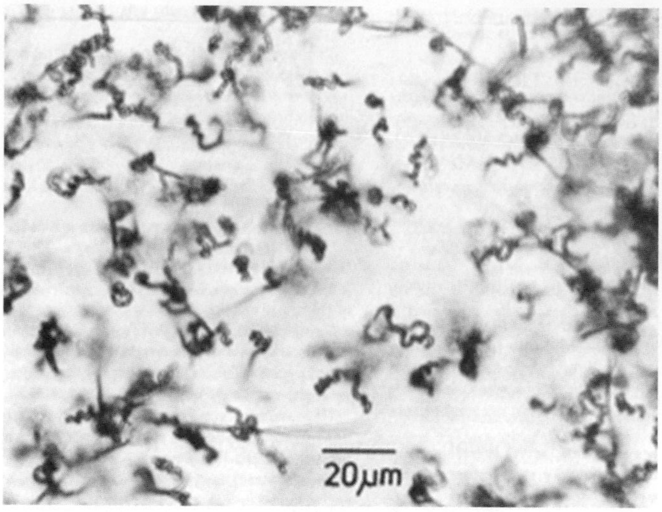

FIGURE 381. *Nonomuraea spiralis*, type strain IMET 9621. Sporulating aerial mycelium. Grown on oatmeal-nitrate agar at 28°C for 18 d.

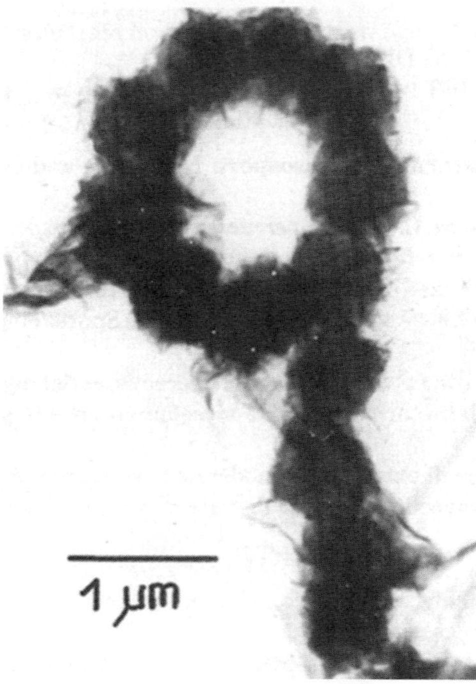

FIGURE 382. *Nonomuraea spiralis*, type strain IMET 9621. Electron micrograph of spore chain.

ment along short straight aerial hyphae (Figure 381 and Figure 382). 10–15 Spores per chain. Spores spherical to subspherical. Spore surface folded.

Inorganic salts-starch agar: poor growth, surface leathery, aerial mycelium absent, substrate mycelium yellow to yellowish brown, no diffusible pigment. Oatmeal agar: good growth, surface woolly to farinaceous, aerial mycelium white to yellowish white, substrate mycelium yellow to yellowish brown, no diffusible pigment. Oatmeal-nitrate agar: good growth, surface farinaceous, aerial mycelium white, substrate mycelium whitish, no diffusible pigment. Peptone-glucose medium: poor growth, surface leathery, aerial mycelium absent, substrate mycelium yellow to yellowish brown, no diffusible pigment. Yeast extract-malt extract agar: moderate growth, surface farinaceous, aerial mycelium white to yellowish, substrate mycelium yellow to yellowish brown; no diffusible pigment.

Cell hydrolysates contain madurose, mannose, and ribose. Diphosphatidylglycerol, phosphatidylinositolmannosides, ninhydrin and sugar positive phospholipids, and hydroxylated phosphatidylethanolamine are present, but phosphatidylethanolamine, phosphatidylglycerol, and phosphatidylinositol are absent.

Additional chemotaxonomic data are given in Table 284, and the fatty acid pattern of the type strain is shown in Table 285.

Esculin is hydrolyzed and nitrate is reduced. Positive for acetoin and β-galactosidase activities, but negative for arginine dihydrolase, citrate lyase, gelatinase, lysine decarboxylase, ornithine decarboxylase, and urease activities. D-Mannose, L-rhamnose, sucrose, and D-xylose are used as sole carbon sources, but L-arabinose, cellobiose, D-fructose, *meso*-inositol, and D-raffinose are not.

DNA G+C content (mol%): 68.1 (T_m).

Type strain: ATCC 35114, CCM 3426, BCRC 13309, CIP 106923, DSM 43555, NBRC 14097, IMET 9621, JCM 3286, KCTC 9286, NCIMB 11633, NRRL B-16098, VKM Ac-853.

Sequence accession no. (16S rRNA gene): U48983.

23. **Nonomuraea turkmeniaca** corrig. (Terekhova, Galatenko and Preobrazhenskaya 1982) Zhang, Wang and Ruan 1998b, 420^{VP} (*Actinomadura turkmeniaca* Terekhova, Galatenko and Preobrazhenskaya 1982, 87; *Microtetraspora turkmeniaca* Kroppenstedt, Stackebrandt and Goodfellow 1990a, 148)

turk.me.ni.a'ca. N.L. fem. adj. *turkmeniaca* pertaining to Turkmen.

Spore chains with spirals of 3–5 turns. Spore ornamentation smooth.

Violet to red substrate mycelium, traces of aerial hyphae, and a pink to violet soluble pigment are formed on oatmeal agar.

Cell hydrolysates contain madurose, mannose, and ribose. Diphosphatidylglycerol, phosphatidylethanolamine, ninhydrin and sugar positive phospholipids, hydroxylated phosphatidylethanolamine, and uncharacterized glycolipids are present, but phosphatidylglycerol, phosphatidylinositol, and phosphatidylinositolmannosides are absent.

Esculin is hydrolyzed and nitrate reduced. Degrades casein, gelatin, hypoxanthine, and starch. Positive for acetoin, arginine dihydrolase, citrate lyase, β-galactosidase, lysine decarboxylase, ornithine decarboxylase, and urease, activities. L-Arabinose, D-fructose, *meso*-inositol, D-mannose, D-raffinose L-rhamnose, sucrose, and D-xylose, are used as sole carbon sources.

DNA G+C content (mol%): not determined.

Type strain: ATCC 49501, CIP 107010, DSM 43926, JCM 6836, NBRC 14348, IMET 9747, INA 3344, KCTC 9287, NRRL B-16246, VKM Ac-852.

Sequence accession no. (16S rRNA gene): AF277201.

Additional organisms

1. **Nonomuraea terrinata** Quintana, Maldonado and Goodfellow 2003, 5[VP]

te.rri.na'ta L. fem. n. *terra* soil; L. part. fem. adj. *nata* born; N.L. fem. part. adj. *terrinata* born from the soil, referring to the source of the isolate.

Spore chains of 8–15 spores, irregular pseudosporangia. Grows well between 20 and 37°C.

A white to ochre substrate mycelium and an abundant aerial mycelium are formed on oatmeal agar, but soluble pigments are not.

Cell hydrolysates contain madurose.

Diphosphatidylglycerol, phosphatidylethanolamine, ninhydrin and sugar positive phospholipids, hydroxylated phosphatidylethanolamine, and uncharacterized glycolipids are present, but phosphatidylglycerol is absent.

Additional chemotaxonomic data are given in Table 284, and the fatty acid pattern of the type strain is shown in Table 285.

Esculin is hydrolyzed, but nitrate is not reduced. Degrades casein, DNA, hypoxanthine, and starch, but not elastin, gelatin, tyrosine, or xanthine.

Source: soil collected in South Korea.

DNA G+C content (mol%): not determined.

Type strain: E626, DSM 44505, NCIMB 13942.

Sequence accession no. (16S rRNA gene): AF302078.

Genus VII. **Planobispora** Thiemann and Beretta 1968, 157[AL]

GERNOT VOBIS, NICOLE LODDERS AND PETER KÄMPFER

Pla.no.bi.spo'ra. Gr. n. *planos* wanderer, tramper; L. adv. num. *bis* twice (double); Gr. fem. n. *spora* a seed, and in biology a spore; N.L. fem. n. *Planobispora* a motile, double-spored organism.

Substrate and aerial mycelia are formed on agar media. Substrate hyphae (0.5–1.0 µm in diameter) are nonfragmenting, branched, and septate. Aerial hyphae (1.0 µm in diameter) are sparsely branched and septate. Gram-stain-positive and non-acid-fast. **Cylindrical to clavate sporangia** (1.0–1.2 µm × 6.0–8.0 µm), each containing a **longitudinal pair of spores**, are developed singly or in bundles on short ramifications of the **aerial hyphae**. The **spores** (zoospores) are oblong and cylindrical with rounded ends, and are **motile** by means of peritrichous flagella. Colonies developed on agar media are flat or occasionally elevated, the **substrate mycelium either without distinctive color or rose colored**. The **aerial mycelium**, which is formed only on certain agar media, is **white or with a light rose tinge**. Strains grow well under aerobic conditions. Chemoorganotrophic and mesophilic, growing well at 28–40°C and pH 6.0–9.0. Cell walls contain *meso*-diaminopimelic acid and whole-cell hydrolysates contain madurose as the diagnostic sugar. Contains diphosphatidylglycerol, phosphatidylethanolamine, hydroxylated phosphatidylethanolamine, and ninhydrin and sugar positive phospholipids as predominant phospholipids. Major menaquinones are MK-9(H_4), MK-9(H_2), and MK-9(H_0). Major types of fatty acids are C_{17} 10-methyl and C_{16} iso-branched fatty acids.

DNA G+C content (mol%): 70–71 (Im).

Type species: **Planobispora longispora** Thiemann and Beretta 1968c, 157[AL].

Further descriptive information

Phylogeny. On the basis of 16S rRNA gene sequence analyses, the genus *Planobispora* is grouped in the family *Streptosporangiaceae.* This is also shown by other 16S rRNA gene sequence studies of this group (Goodfellow et al., 1990b; Goodfellow and Quintana, 2005; Stackebrandt et al., 1997; Zhang et al., 1998b). Within this family, most genera are defined on the basis of distinct chemotaxonomic and morphological properties (Table 275). The genus encompasses two species with validly published names. Representatives of these form a distinct line in the 16S rRNA *Streptosporangiaceae* gene tree (Goodfellow and Quintana, 2005; Figure 383). Their nearest relatives within the family *Streptosporangiaceae* are the genera *Planomonospora, Streptosporangium,* and *Sphaerisporangium.* Similarity values between the genera *Planomonospora* (with the exception of *Planomonospora venezuelensis*) and *Planobispora* range from 97.1–96.2%, between the genera *Planobispora* and *Streptosporangium* from 96.5–93.8%, and between the genera *Planobispora* and *Sphaerisporangium* from 96.2–96.1%. As pointed out in the *Planomonospora* chapter, *Planomonospora venezuelensis* shows a closer relationship to *Planobispora* in the phylogenetic tree than to the genus *Planomonospora.* 16S rRNA gene signature nucleotide positions that can be used to differentiate genera of the family *Streptosporangiaceae* are described by Ara and Kudo (2007). Three of the signature nucleotide positions differ between *Planomonospora* (with the exception of *Planomonospora venezuelensis*) and *Planobispora*: 625 (C and G, respectively), 627 (A and G, respectively), 990:1215 (C–G and C–G/U–G, respectively). For two of the three signature nucleotide positions, *Planomonospora venezuelensis* has the same signature as the genus *Planobispora*: 625 (G) and 627 (G).

Cell morphology. The hyphae of the substrate mycelium are 0.5–1.0 µm in diameter, irregularly branched, and occasionally septate. Fragmentation of hyphae is not apparent either on solid agar media or in liquid cultures (Thiemann and Beretta, 1968). The hyphae of the aerial mycelium are 1.0 µm in diameter, long, slender, and wavy, with few lateral branches, and grow more or less parallel to the surface of the substrate. The aerial mycelium is extremely hydrophobic (Thiemann, 1970).

Cylindrical to clavate sporangia are formed on short side branches of aerial hyphae (Figure 384A). They can be arranged singly or in bundles (Suzuki et al., 2001a; Thiemann and Beretta, 1968; Vobis, 1989). Each sporangium contains a longitudinal pair of spores (Figure 384E, F, G).The mean size of a sporangium is 6.0–8.0 µm × 1.0–1.2 µm (Thiemann, 1974a). The sporangia are connected to the main axis of the sporangiophore by short supporting pedicels (Figure 384E, F), which are only 1.0–3.0 µm

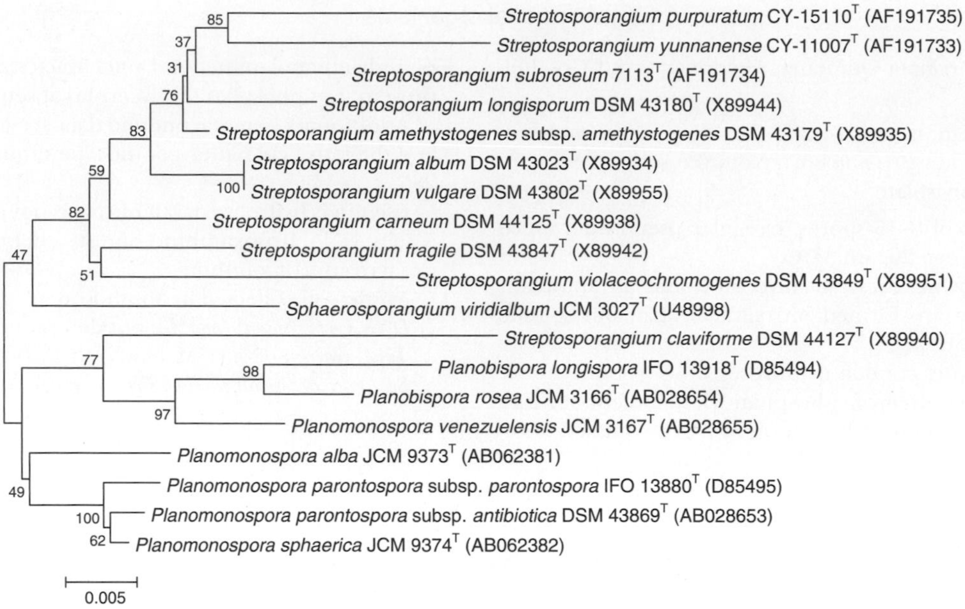

FIGURE 383. Neighbor-joining tree based on 16S rRNA gene sequences of *Planobispora* species and close relatives. The numbers of the nodes indicate the level of bootstrap support (%) based on an analysis of 1000 resampled data sets. The scale bar = 0.5 differences in 16S rRNA gene sequences.

long, very fragile, and collapse easily (Thiemann and Beretta, 1968). As supposed by Bland and Couch (1981), the spores originate from simple transformation of sporogenous hyphae (Figure 384F). The tip of the sporangiophore prolongates until it reaches the length of a ripe sporangium (Figure 384F6). The sporogenous hypha septates from the pedicel and divides into two equal spore-sized segments. The following sporangial primordium continues with the apical outgrowth of the sporangiophore (Figure 384F7). The young spores occupy the sporangial volume (Figure 384G), rounding up later during maturation (Figure 384E, F 1–4). The sporangial envelope is a continuation of the hyphal sheath of the sporangiophore (Figure 384B, C). It is interspersed by longitudinally orientated fibrillar elements, converging like a buttress at the tip of the sporangium (Figure 384B). Between the two spores, a diaphragm is formed transversely, originating from the inner side of the sporangial envelope (Figure 384D, E). In general, the sporangial envelopes are smooth (Thiemann and Beretta, 1968). The studies of Suzuki et al. (2001a) demonstrate various morphological differences in *Planobispora* isolates with clavate to pyriform sporangia, warty sporangial surface, and even three-spored sporangia. The sporangia arranged in bundles form either fan-shaped tufts (Miyadoh et al., 1997), resembling a "half-side" palm leaf pattern like *Planomonospora venezuelensis*, or parallel rows analogous to *Planomonospora parontospora* (Suzuki et al., 2001a).

The spores measure 1.0–1.2 μm × 2.6–4.0 μm, they are oblong with round ends, occasionally slightly curved. The spores (zoospores) are motile by means of peritrichous flagella (Thiemann, 1974a).

The mode of spore liberation was studied by Thiemann (1970). If sporangia of *Planobispora rosea* are dipped into water, the upper spore is released at first through the tip of the sporangium. The lower one remains blocked in the sporangium. Only when the sporangium is broken off at the sporangiophore, the basal spore can be expulsed through the lower end. The central diaphragm (Figure 384D, E), which was called "transverse septum" by Thiemann (1970), occupies an essential function in this process. It separates the two spores by swelling and pushes the spores in opposite directions out of the sporangium. The spores become motile only some time after being dispersed. In spite of good culture conditions, only 2–5% of the spores become motile. The spores germinate with one or two polar germ tubes; occasionally lateral germination is observed (Thiemann and Beretta, 1968).

Cell-wall composition. The peptidoglycan of the cell wall contains *meso*-DAP with madurose (3-*O*-methyl-D-galactose) as the characteristic sugar of whole-cell hydrolysates (Kroppenstedt and Kutzner, 1976, 1978). This chemical cell-wall composition is in accordance with cell-wall type III and sugar pattern B of the classification scheme of Lechevalier and Lechevalier (1970b).

The phospholipids of the cell membranes consist of unknown glucosamine-containing phospholipids, phosphatidylinositol, phosphatidylethanolamine, and diphosphatidylglycerol (Hasegawa et al., 1979), in accordance with phospholipid type IV of Lechevalier et al. (1981).

The fatty acids consist of straight-chain acids, iso- and 10-methyl branched acids, and unsaturated fatty acids. Anteiso-branched fatty acids are not present (Kroppenstedt and Kutzner, 1978), corresponding to fatty acid type 3c of Kroppenstedt (1985).

The two species of *Planobispora* have the menaquinone type 4a of Kroppenstedt (1985) in common, but show variations. *Planobispora rosea* is characterized by di- and tetrahydrogenated menaquinones with nine isoprene units [MK-9(H_2) and MK-9(H_4)] (Whitham et al., 1993), and *Planobispora longispora* by tetra- and dihydrogenated menaquinones with nine units [MK-9(III, VIII-H_4) and MK-9(H_2)], together with MK-9(H_0) (Kudo et al., 1993; Whitham et al., 1993) (Table 287).

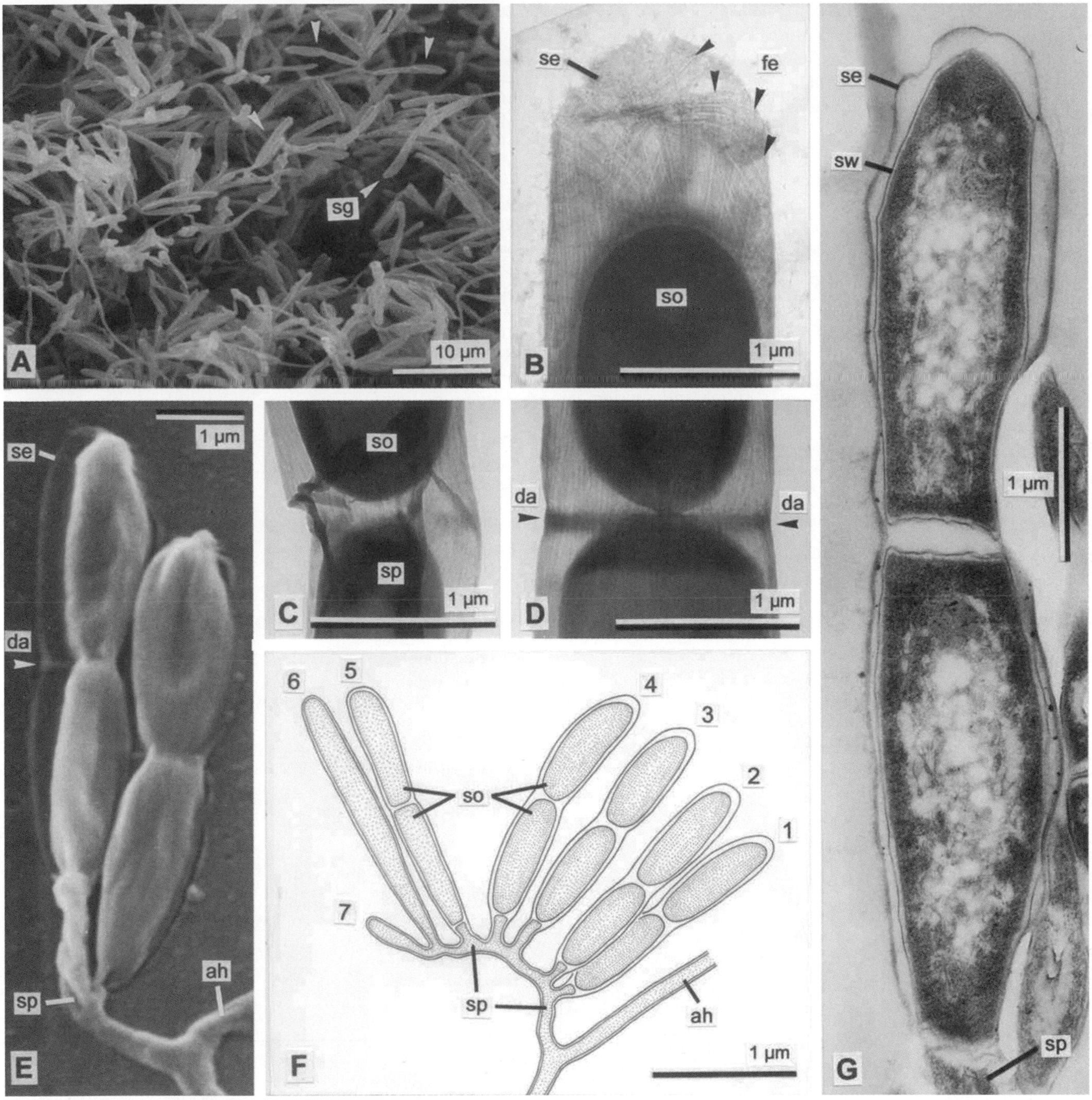

FIGURE 384. Two-spored sporangia of *Planobispora rosea* ATCC 23866 (A, C, D, E, F) and *Planobispora longispora* ATCC 23867 (B, G). A, aerial mycelium with numerous sporangia (SEM); B, tip of the sporangial envelope with fibrous elements (TEM); C, transition from sporangiophore to sporangium (TEM); D, diaphragm in the center (TEM); E, two mature sporangia (SEM); F, sporangiophore with a bundle of 7 sporangia in different stages of development: mature sporangia (1–4), young sporangium (5), immature, still unsegmented sporangium (6) and sporangial primordium (7); G, young sporangium, separated from the sporangiophore and divided in two spore segments (TEM) (see also F 5). Abbreviations: SEM, scanning electron microscope; TEM, transmission electron microscope; ah, aerial hypha; da, diaphragm; fe, fibrous elements; se, sporangial envelope; sg, sporangium; so, spore; sp, sporangiophore; sw, spore wall.

Colony morphology. Substrate and aerial mycelia develop on a variety of agar media. The colonies are mostly flat or occasionally elevated and have a smooth surface. On yeast extract-malt extract agar and on Bennett agar the colonies of *Planobispora longispora* are crusty (Thiemann and Beretta, 1968); *Planobispora rosea* has slightly wrinkled surfaces on Bennett agar and Hickey–Tresner agar (Thiemann, 1970). No distinctive color of the mycelium is formed by *Planobispora longispora* (Thiemann and Beretta, 1968); *Planobispora rosea* produces rose-colored substrate mycelium and white aerial mycelium with rose tinge (Thiemann, 1970). For *Planobispora longispora*, the development of the aerial mycelium and formation of sporangia are promoted by oatmeal, calcium malate, and soil agars (Thiemann and Beretta, 1968). Hickey–Tresner agar and glycerol-asparagine agar have the same effect on *Planobispora rosea* (Thiemann, 1970).

Planobispora rosea strain ATCC 53773 produces the antibiotic GE 2270, a thiazolyl peptide substance inhibiting bacterial protein synthesis (Selva et al., 1995).

Metabolism. Members of *Planobispora* are aerobic, growing well on the various standard culture media recommended by Waksman (1961) and Shirling and Gottlieb (1966). The type strains of *Planobispora longispora* and *Planobispora rosea* can utilize L-arabinose, cellobiose, fructose, glucose, glycogen, inositol, maltose, mannitol, and starch as sole carbon sources. Adonitol, D-arabinose, erythritol, ethanol, glycerol, inulin, lactose, mannose, α-methyl-D-glucoside, raffinose, sorbitol, and sucrose are not utilized (Goodfellow and Pirouz, 1982; Thiemann, 1974a). Acetamide and serine are not utilized as sole carbon and nitrogen sources (Goodfellow and Pirouz, 1982). Further physiological features and degradation abilities are also shown in Table 288.

TABLE 287. Diagnostic characteristics for *Planobispora* species[a]

Characteristic	*P. longispora*	*P. rosea*
Substrate and aerial mycelium rose-colored[b]	−	+
Sensitive to dimethylchlortetracycline[c]	+	−
Sensitive to gentamicin[c]	−	+
Menaquinones:[d,e]		
MK-9(H_0)	+	−
MK-9(H_2)	+	+
MK-9(H_4)	−	+
MK-9(III, VIII-H_4)	+	−

[a]Symbols: +, >85% positive; −, 0–15% positive.

[b]Data from Thiemann (1970).

[c]Data from Goodfellow and Pirouz (1982).

[d]Data from Kudo et al. (1993).

[e]Data from Whitham et al. (1993).

TABLE 288. Characteristics of the *Planobispora* species[a]

Characteristic	*P. longispora*	*P. rosea*
Utilization of sole carbon sources:		
Adonitol	−	−
Amygdalin	+	−
L-Arabinose	+	+
D-Arabinose	−	−
Cellobiose	+	+
Erythritol	−	−
Ethanol	−	−
Fructose	+	+
Galactose	−	+
Glucose	+	+
Glycerol	−	−
Glycogen	+	+
Inositol	+	+
Inulin	−	−
Lactose	−	−
Maltose	+	+
Mannitol	+	−[b],+[c]
Mannose	−	−
Melezitose	−	+
α-Methyl-D-glucoside	−	−
Raffinose	−	−
Rhamnose	+	−
Salicin	−	+
Sorbitol	−	−
Starch	+	+
Sucrose	−	−
Trehalose	−	−
Xylose	+	+
Degradation tests:		
Adenine	−	−
Arbutin	+	+
Calcium malate	−	−
Casein	+	+
Cellulose	−	−
Chitin	−	+
Elastin	+	+
Esculin	−	+
DNA	+	+
Gelatin	+	+
Guanine	−	−
Hippurate	−	−
Hypoxanthine	+	−
Keratin	+	+
RNA	+	+
Starch	+	+
Testosterone	−	−
Tween 20, 40, 60, and 80	+	+
Tyrosine	−[d], +[c]	+
Xanthine	−	−
Xylan	−	−
Additional physiological properties:		
Diffusible pigments produced	−	−
Melanoid pigments	−	nd
H_2S produced	−	nd
Phosphatase produced	+	+
Reduction of nitrate	+	+
Litmus milk:		
Coagulated	+	−
Peptonized	+	−

[a]Symbols: +, >85% positive; −, 0–15% positive; nd, not determined. Data compiled from Thiemann (1970), Goodfellow and Pirouz (1982), and Thiemann and Beretta (1968). Deviations are indicated by superscript letters.

[b]Data from Thiemann (1970).

[c]Data from Goodfellow and Pirouz (1982).

[d]Data from Thiemann and Beretta (1968).

Strains of *Planobispora* are mesophilic. Good growth occurs at 28–40°C, with no growth at 20°C or 45°C (Goodfellow and Pirouz, 1982; Thiemann, 1974a). Cultures grow well at pH 6.0–9.0. As an exception, *Planobispora longispora* grows at pH 5.0 on oatmeal agar (Thiemann and Beretta, 1968). Strains grow in the presence of brilliant green (up to 0.02%, w/v), crystal violet (up to 0.001%, w/v), and pyronine (0.01%, w/v), but they do not tolerate lysozyme (0.005%, w/v) or NaCl (3.0%, w/v) (Goodfellow and Pirouz, 1982). Nitrate is reduced to nitrite and phosphatase is produced. The type strains of *Planobispora longispora* and

Planobispora rosea are able to degrade arbutin, casein, elastin, DNA, gelatin, keratin, RNA, starch, tyrosine, and Tween 20, 40, 60, and 80. The following compounds are not degraded: adenine, cellulose, guanine, hippurate, testosterone, xanthine, and xylan (Goodfellow and Pirouz, 1982) (see Table 288).

Planobispora longispora and *Planobispora rosea* are sensitive to the antibiotics kanamycin, neomycin, novobiocin, tobramycin, and penicillin. They are not sensitive to cephalosporin, lincomycin, rifampin, and streptomycin. *Planobispora longispora* is sensitive to dimethylchlortetracycline, and *Planobispora rosea* is sensitive to gentamicin (Goodfellow and Pirouz, 1982) (Table 287).

Ecology. The genus *Planobispora* was regarded for a long time as a very rare microorganism, known only from a few localities. Several strains, including the type strains, were isolated from two soil samples collected from a river bank in Venezuela. The pH values of the two samples were 5.3 and 7.6 (Thiemann, 1970). Further strains were isolated by D. Schäfer from a soil sample collected in Namibia (Vobis, 1989). Kizuka et al. (1997) recovered *Planobispora* strains from soil originating from arid regions of South Africa. More recently, Suzuki et al. (2001a) tested 1467 soil samples collected from all parts of the world. Only 3.5% of the samples tested contained *Planobispora* strains. The pH value of the samples ranged from 6.2–8.1, nearly 90% in the small range from 7.0–7.9, indicating that the organisms prefer neutral to slightly alkaline environments. The positive samples were collected in Ecuador, Egypt, French Guiana, India, and Madagascar, whereas no isolates were obtained from samples of temperate areas in Europe, North America, and Oceania. The geographical distribution seems to be restricted to tropical and subtropical zones between latitude 35°N and 35°S (Suzuki et al., 2001a).

Enrichment and isolation procedures

Unfortunately, Thiemann and coworkers never published the isolation procedures they used. The traditional baiting technique as described by Bland and Couch (1981) was successfully applied (Vobis, 1989). Kizuka et al. (1997) could isolate *Planobispora* strains by a centrifugation method. Suzuki et al. (2001a) developed an efficient selective isolation method using humic acid–trace salts gellan gum medium (HSG) containing 0.05% nitrohumic acid, 3 mM $CaCl_2$, 5 mM CHES (*N*-cyclohexyl-2-amino-ethanesulfonic acid), and 0.7% gellan gum, with addition of seven antimicrobial agents (μg/ml): trimethoprim (50), nalidixic acid (50), enoxacin (20), sodium ampicillin (2), streptomycin sulfate (1), cycloheximide (50), and nystatin (50). The presence of trace salts ($FeSO_4 \cdot 7H_2O$, $MnCl_2 \cdot 4H_2O$, $ZnSO_4 \cdot 7H_2O$, and $NiSO_4 \cdot 6H_2O$ with 0.0001% concentration of each), the alkaline environment (pH 9.0), and the incubation temperature of 32–37° C stimulate the formation of sporangia. This morphological characteristic is indispensable to recognize *Planobispora* strains under the microscope. Five hundred milligrams of air-dried soil samples are treated in dry heat at 90°C for 60 min. After cooling, each sample is flooded with 2 ml of a solution of 0.1% skim milk (unsterilized), 0.01% Tween 80 in 5 mM CHES (pH 9.0), and incubated with occasional stirring at 35°C for 1 h. After centrifugation (1000 × *g* for 10 min), 800 μl of the supernatant is diluted with sterile saline, and aliquots are spread onto HSG plates. They are incubated for 14–21 d at 32°C. The colonies grown in these enrichment cultures are controlled microscopically. Pure isolates are obtained by streaking onto HSG and tested for zoospore production by flooding with a solution containing 0.1% skim milk in 5 mM CHES (pH 9.0).

Maintenance procedures

For some weeks or even months, cultures can be stored at room temperature in hermetically closed slant culture tubes on agar medium that support good growth of substrate and arial mycelium and production of sporangia. For long-term preservation, strains must be processed as described for other aerobic actinomycetes. Among others, lyophilization is a recommended method.

Differentiation of the genus *Planobispora* from other genera

The genus *Planobispora* may be confused morphologically with members of genera that can produce pairs of spores on aerial and/or substrate mycelia: *Actinobispora*, *Actinomadura*, *Microbispora*, *Microtetraspora*, *Thermobispora* (Miyadoh et al., 1997; Suzuki, 2001; Wang et al., 1996c; Zhang et al., 1998b), and the former genera *Elytrosporangium*, *Kitasatoa*, and *Microellobosporia*, now belonging to the genus *Streptomyces* (Williams et al., 1989). In opposition to *Planobispora*, the spores of those genera are not produced in a sporangium, and their spores are not motile in aquatic habitats. Confusion with strains of the genus *Dactylosporangium*, which produce motile spores within few-spored, cylindrical to clavate sporangia on substrate hyphae, can finally be avoided by determining the cell-wall chemotype. *Planobispora* with cell-wall type III and sugar pattern B distinguishes from *Dactylosporangium* having chemotype II and sugar pattern D.

Taxonomic comments

The majority of the genera classified in the aggregate group "maduromycetes" (Goodfellow, 1989a), including *Planobispora*, were assigned to the family *Streptosporangiaceae* (Goodfellow, 1992; Goodfellow et al., 1990b). Ward-Rainey et al. (1997) emended the description of the family *Streptosporangiaceae* and included the genera *Herbidospora*, *Microbispora*, *Microtetraspora*, *Planobispora*, *Planomonospora*, and *Streptosporangium*. This taxonomical concept was underpinned by Miyadoh et al. (1997). Subsequently, the family *Streptosporangiaceae* was extended by addition of the genera *Acrocarpospora*, *Nonomuraea*, *Planotetraspora*, and *Sphaerisporangium* corrig. (Ara and Kudo, 2007; Miyadoh et al., 2001).

Differentiation of species of the genus *Planobispora*

The two species presently described cannot be clearly differentiated on the basis of morphology. Tests for utilization of 28 carbon sources and degradation of 21 compounds all show only 10% deviation between the two species (Table 288). Besides other distinguishing physiological and cultural characteristics, the colors of substrate and aerial mycelium (Thiemann, 1970) and the menaquinone composition (Kudo et al., 1993; Whitham et al., 1993) are additional features to differentiate the two species (Table 287).

List of species of the genus *Planobispora*

1. **Planobispora longispora** Thiemann and Beretta 1968, 157[AL]

lon.gi.spo'ra. L. adj. *longus* long; Gr. n. *spora* a seed, and in biology a spore; N.L. n. *longispora* (nominative in apposition) the long spore.

Sporangial development is supported by soil, calcium malate, and oatmeal agars. Spores are straight to slightly curved with rounded ends, measuring 1.0–1.2 × 2.6–4.0 μm. They are motile by peritrichous flagella.

No specific color occurs either in the aerial or in the substrate mycelium. The aerial mycelium is white and the substrate mycelium hyaline to creamy colored. No soluble pigments are produced.

Good growth with abundant aerial mycelium occurs on oatmeal agar. On yeast extract-malt extract, Hickey–Tresner, Bennett, and peptone-beef extract agars, colonies also grow well, but aerial mycelium is not developed.

Amygdalin is used for growth; galactose, melezitose, and salicin are not. Hypoxanthine is degraded; esculin and chitin are not degraded. Melanoid pigments are not produced. Litmus milk is coagulated and peptonized.

Contains tetra- and dihydrogenated menaquinones with nine units [MK-9(III,VIII-H_4) and MK-9(H_2)], together with MK-9(H_0)

DNA G+C content (mol%): 71 (T_m).

Type strain: ATCC 23867, DSM 43041, CBS 115.69, JCM 3092, NBRC 13918, VKM Ac-700.

Sequence accession no. (16S rRNA gene): D85494.

2. **Planobispora rosea** Thiemann 1970, 251[AL]

ro'se.a. L. fem. adj. *rosea* rose-colored.

Sporangial development is promoted by all media on which aerial mycelium is formed, e.g. on soil and Hickey–Tresner agars. The spores are elongated and fusiform, with rounded ends, 1.0–1.2 μm × 3.0–3.5 μm. They are motile by peritrichous flagella.

Substrate mycelium in most media is rose colored; if aerial mycelium is developed, it always has a light rose tinge.

On Bennett, peptone-beef extract and potato plug agars the colonies grow well, are slightly wrinkled or flat and rose colored; no aerial mycelium is developed. Good growth occurs also on Hickey–Tresner agar and the colonies are slightly wrinkled and yellow-amber colored with abundant aerial mycelium. Colonies grow well on oatmeal agar; they are smooth and rose colored and aerial mycelium is moderately developed showing a rose tinge. On glycerol-asparagine agar, growth is moderate and the colonies are smooth, flat, and hyaline; abundant aerial mycelium, white with a rose tinge, is formed.

Galactose, melezitose, and salicin are used for growth; amygdalin is not. Esculin and chitin are degraded, hypoxanthine is not hydrolyzed. Litmus milk is neither coagulated nor peptonized.

Contains di- and tetrahydrogenated menaquinones with nine isoprene units [MK-9(H_2) and MK-9(H_4)].

DNA G+C content (mol%): 70 (T_m).

Type strain: ATCC 23866, DSM 43051, JCM 3166, NBRC 15558, NRRL B-8121, VKM Ac-1318.

Sequence accession no. (16S rRNA gene): AB028654.

Genus VIII. **Planomonospora** Thiemann, Pagani and Beretta 1967, 29[AL]

Gernot Vobis, Nicole Lodders and Peter Kämpfer

Pla.no.mo.no.spo'ra. Gr. n. *planos* wanderer, vagabond; Gr. adj. *monos* alone, single; Gr. fem. n. *spora* a seed, and in biology a spore; N.L. fem. n. *Planomonospora* a motile, single spored organism.

Substrate and aerial mycelium developed on various agar media. Substrate hyphae (0.6–1.0 μm in diameter) are nonfragmenting, irregularly branched, and occasionally septated. Aerial hyphae are sparsely branched and rarely septated. Organisms are Gram-stain-positive and non-acid-fast. **Cylindrical to clavate sporangia** (1.0–1.5 μm × 3.5–5.5 μm) **are formed in bundles on the aerial mycelium**, arranged either in narrow parallel rows resembling rows of bananas or in an open palm leaf pattern. **Each sporangium contains a single spore. Spores (zoospores) are oblong to fusiform and motile by means of peritrichous flagella.** Growth under aerobic conditions. Colonies developed on complex agar media are raised or flat with rugose or smooth surfaces. Substrate mycelium may be grayish yellow, light orange, rose, yellowish pink, or violet-brown. Aerial mycelium is white, grayish white, light rose, or pink. Chemoorganotrophic, mesophilic, grows well at 28–37°C and from pH 7.0–8.0.

Cell walls contain *meso*-diaminopimelic acid and whole-cell hydrolysates contain madurose as the diagnostic sugar. Contains diphosphatidylglycerol, phosphatidylethanolamine, hydroxylated phosphatidylethanolamine, and ninhydrin and sugar positive phospholipids as predominant phospholipids. Major menaquinones are MK-9(H_4), MK-9(H_2), and MK-9(H_0). Major types of fatty acids are C_{17} 10 methyl and C_{16} iso branched fatty acids.

DNA G+C content (mol%): 72 (T_m).

Type species: **Planomonospora parontospora** Thiemann, Pagani and Beretta 1967, 29[AL].

Further descriptive information

Phylogeny. 16S rRNA gene sequence analyses place the genus *Planomonospora* in the family *Streptosporangiaceae*. This is confirmed by other 16S rRNA gene sequence studies of this group (Goodfellow et al., 1990b; Goodfellow and Quintana, 2005; Stackebrandt et al., 1997; Zhang et al., 1998b). Within this family, genera are defined on the basis of distinct chemotaxonomic and morphological properties (Table 275). The taxon encompasses four species and two subspecies with validly published names. Representatives of each of these validly described species form a distinct line in the 16S rRNA *Streptosporangiaceae* gene tree (Goodfellow and Quintana, 2005; Figure 383).

TABLE 289. Characteristics of the *Planomonospora* species and subspecies[a,b]

Characteristic	*P. parontospora* subsp. *parontospora*	*P. parontospora* subsp. *antibiotica*	*P. alba*	*P. sphaerica*	*P. venezuelensis*
Utilization of sole carbon sources:					
Adonitol	+	nd	–	–	+
Amygdalin	–	nd	nd	nd	–
D-Arabinose	–	nd	–	–	–
L-Arabinose	+[c], –[d]	+	+	+	+[c], –[d]
Cellobiose	+[c], –[d]	+	+	+	+[c], –[d]
Citrate	+	+	+	+	–
Dextrin	+[e], –[d]	+	+	+	–
Dulcitol	–	nd	–	–	nd
Erythritol	–	nd	–	–	–
Ethanol	–	nd	–	–	–
D-Fructose	+	+	+	+	+[c], d[f], –[d]
D-Galactose	+	+	+	+	+[c], –[d]
D-Glucose	+	+	+	+	+
Glycerol	–	–	+	–	–
Glycogen	+[c], –[d]	+	–	+	+[c], –[d]
Inositol	–	–	–	–	–[c], d[f]
Inulin	+[e], –[c]	nd	–	–	–
Lactose	+	nd	–	–	+
Maltose	–[d,e], +[c]	+	+	+	+[c], –[d]
D-Mannitol	+[c], –[d]	+	+	+	+[c], –[d]
D-Mannose	+[e], –[c,d]	+	+	+	–
Melezitose	+	nd	–	–	–
Melibiose	nd	nd	–	–	nd
α-Methyl-D-glucoside	–	nd	–	–	–
Raffinose	–	–	–	–	–[c], d[g]
L-Rhamnose	+[c,e], –[d]	+[g], –[d]	+	+	+[c,f], –[d]
D-Ribose	–	nd	–	–	nd
Salicin	–	+	–	–	–
D-Sorbitol	–	nd	–	–	–
D-Sorbose	–	nd	–	–	nd
Starch	+[c,e], –[d]	+	+	+	+[c], –[d]
Succinate	–	+	–	+	–
Sucrose	–[d,e], +[c]	+	+	+	+[c], –[d]
Trehalose	+[c], –[d]	+	+	+	+[c], –[d]
D-Xylose	+[c,f], –[d]	+	+	+	+[c,f], –[d]
Sodium salts:					
Acetate	nd	nd	+	+	nd
Benzoate	nd	nd	–	–	nd
Butyrate	nd	nd	+	+	nd
Citrate	nd	nd	–	–	nd
Lactate	nd	nd	+	+	nd
Malate	nd	nd	–	–	nd
Mucate	nd	nd	–	–	nd
Oxalate	nd	nd	–	–	nd
Propionate	nd	nd	–	–	nd
Pyrovate	nd	nd	+	+	nd
Succinate	nd	nd	–	+	nd
Tartrate	nd	nd	–	–	nd
Degradation tests:					
Adenine	–	nd	–	–	–
Arbutin	+	nd	nd	nd	+
Calcium malate	–	–	–	–	–
Casein	+	+	+	+	–[f], +[c]
Cellulose	–	v[g]	–	–	–
Chitin	+	nd	nd	nd	+
DNA	+	nd	+	+	+
Elastin	+	nd	nd	nd	+
Esculin	+	–	–	–	+, –[d]
Gelatin	–[e], +[c]	+	nd	nd	+[c]
Guanine	–	nd	nd	nd	–
Hippurate	–	nd	–	–	–
Hypoxanthine	+, –[f]	–	–	–	+

(continued)

TABLE 289. (continued)

Characteristic	*P. parontospora* subsp. *parontospora*	*P. parontospora* subsp. *antibiotica*	*P. alba*	*P. sphaerica*	*P. venezuelensis*
Keratin	+	nd	nd	nd	+
RNA	+	nd	nd	nd	+
Starch	+	+	nd	nd	+
Testosterone	–	nd	nd	nd	–
Tween (20–80)	+	nd	nd	nd	+
Tyrosine	–[d,e], +[c]	+	+	+	+, –[d]
Xanthine	–	nd	–	–	–
Xylan	–	nd	nd	nd	–
Additional physiological properties:					
H_2S produced	nd	+	nd	nd	+
Gelatinase produced	–	+	+	+	+, –[d]
Litmus milk:					
Coagulated	–	–	nd	nd	–
Peptonized	+	–	nd	nd	–
Melanin produced	–	+	–	–	nd
Nitrate reduced	+[d,e], –[c]	+	–	+	+[d,f], –[c]
Phosphatase produced	+	–	+	+	+
Urease produced	nd	nd	+	+	nd
Growth at:					
15°C	–	+	–	–	–
45°C	–	–	+	–	–
Survives 50°C for 8 h	+	–	+	+	–
Resistance to 5% (w/v) NaCl	–	+	–	–	–

[a]Symbols: +, >85% positive; -, 0–15% positive; nd, not determined; v, strain instability; D, 33–66% of the strains are positive.

[b]Data compiled from [e]Thiemann et al. (1967), [g]Thiemann et al. (1968a), [f]Thiemann, (1970), [c]Goodfellow and Pirouz (1982), and [d]Mertz, (1994). Deviations are indicated by superscript letters.

Planomonospora species and subspecies form a distinct clade in the 16S rRNA gene tree with the exception of *Planomonospora venezuelensis* (Figure 383). Their nearest relatives within the family *Streptosporangiaceae* are the genera *Planobispora*, *Streptosporangium*, and *Sphaerisporangium* corrig. Similarity values between the genera *Planomonospora* (with the exception of *Planomonospora venezuelensis*) and *Planobispora* range from 97.1–96.2%, between the genera *Planomonospora* and *Streptosporangium* from 98.1–95.3%, and between the genera *Planomonospora* and *Sphaerisporangium* from 96.1–95.9%. The 16S rRNA gene sequence of *Planomonospora venezuelensis* differs in 2.2–2.5% from the sequences of the other species of the genus *Planomonospora* (similarity values between the sequences of *Planomonospora venezuelensis* and the other sequences of the genus *Planomonospora* range from 97.5–97.8%). Although similar differences in gene sequences can be found between *Planomonospora venezuelensis* and the genus *Planobispora* (differences of 2.2–2.8%; similarity values range from 97.6–97.8%), *Planomonospora venezuelensis* shows a closer relationship to *Planobispora* in the phylogenetic tree than to the genus *Planomonospora*. 16S rRNA gene signature nucleotide positions that can be used to differentiate genera of the family *Streptosporangiaceae* are described by Ara and Kudo (2007). Three of the signature nucleotide positions differ between *Planomonospora* (with the exception of *Planomonospora venezuelensis*) and *Planobispora*: 625 (C and G, respectively), 627 (A and G, respectively), 990:1215 (C–G and C–G/U–G, respectively). For two of the three signature nucleotide positions, *Planomonospora venezuelensis* has the same signature as the genus *Planobispora*: 625 (G) and 627 (G).

Cell morphology. Strains of *Planomonospora* produce substrate and aerial mycelium on solid media. Substrate hyphae have diameters of 0.6–1.0 µm, and are irregularly branched, septate, and do not fragment on agar media or in liquid-submersed cultures (Mertz, 1994; Thiemann, 1970). Twisting and swelling occurs in *Planomonospora parontospora* (Thiemann et al., 1967, 1968a). Hyphae grow profusely within agar media forming a compact layer on the surface of the substrate mycelium (Thiemann, 1974b). The diameter of aerial hyphae is 0.5–1.0 µm. They are sparsely branched and, in *Planomonospora venezuelensis*, usually long, wavy, and slender (Thiemann, 1970). Hyphae of *Planomonospora sphaerica* develop characteristic sphaerical bodies with diameters up to 5 µm on inorganic salts-starch agar (ISP medium 4) (Mertz, 1994).

The characteristic morphological feature common to all *Planomonospora* species is the monosporous sporangium, which is produced only on the aerial mycelium. Two different morphological arrangements of sporangia can be recognized. In *Planomonospora venezuelensis*, sporangia are bundled like the pinnas of palm leaves ("palm leaf pattern"). The sporangia develop sympodially along the tip of an aerial hypha, which functions as a sporangiophore. The first sporangium originates by differentiation of the final segment of the sporangiophore (Figure 385A) and the primordium of the following sporangium arises as a subterminal protrusion at the base of the last formed sporangium, alternately in each case (Figure 385B). This pattern of development leads to a distichous arrangement of two rows of sporangia on opposite sides (Figure 385C, D). Each sporangium is connected to the common axis by a short stalk. Fibrillar

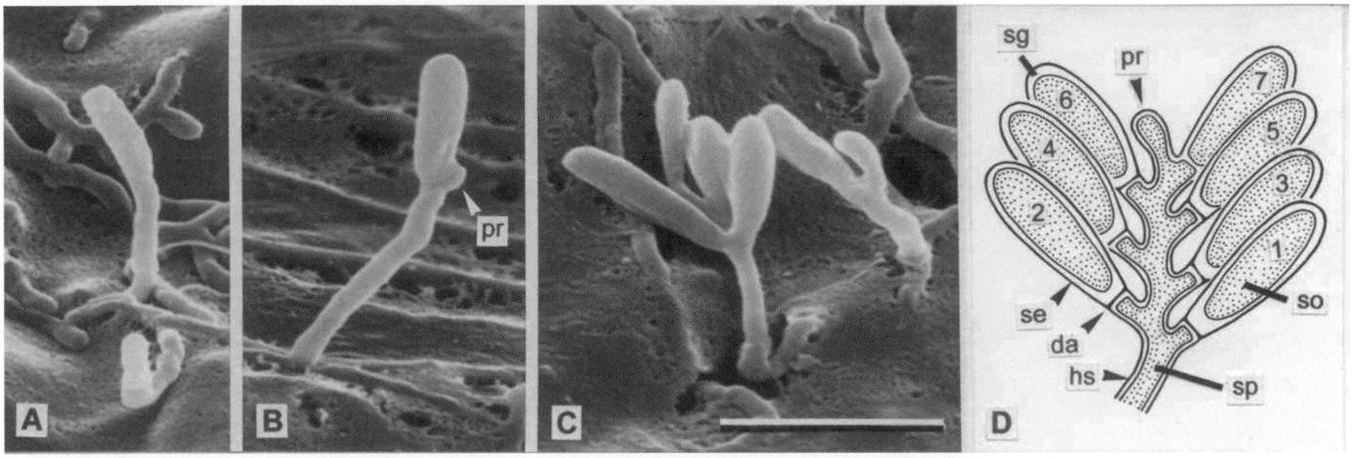

FIGURE 385. Scanning electron micrographs showing ontogenetic development of sporangiophores in *Planomonospora venezuelensis* ATCC 23865 bearing one-spored sporangia in a palm leaf pattern. A, First sporangium produced terminally on an aerial hypha; B, Primordium of the second sporangium protruding at the base of the first sporangium. C, Second sporangium in development (right), sporangiophore with four sporangia (left); D, Sporangiophore with seven sporangia in alternate positions, new sporangial primordium protruding subterminally off the last produced sporangia. Abbreviations: da, diaphragm; hs, hyphal sheath; pr, primordium; se, sporangial envelope; sg, sporangium; so, spore; sp, sporangiophore. Bar = 5 μm.

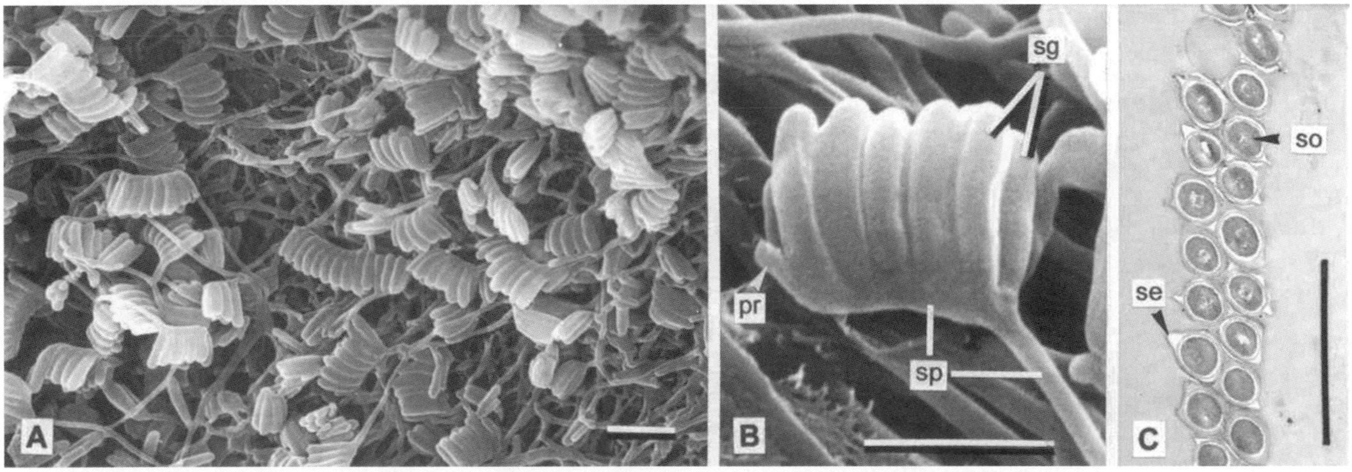

FIGURE 386. Sporangiophores of *Planomonospora parontospora* ATCC 23863. A, Aerial hyphae with numerous sporangial stands; B, sporangiophore bearing a double-row of sessile monosporous sporangia, sporangial primordium protrudes from the sporangiophore apically; C, transverse section of a double row of sporangia, each in alternate position (A, B, scanning electron microscopy; C, transmission electron microscopy). Abbreviations: pr, primordium; se, sporangial envelope; sg, sporangium; sp, sporangiophore. Bars = 5 μm.

elements have been detected on the surface of sporangia (Sharples et al., 1974; Vobis, 1986). The sporangial envelope is a continuation of the hyphal sheath of the sporangiophore (Figure 385D); it retains the hydrophilic character of the aerial hyphae. The morphological appearance of the sporangial strands of *Planomonospora alba*, *Planomonospora parontospora*, and *Planomonospora sphaerica* differ distinctly from those of *Planomonospora venezuelensis*. Sporangia are arranged closely in parallel double rows (Figure 386A). The sporangiophore is slightly bent and can bear up to 60 sporangia, resembling a row of bananas (Mertz, 1994). New sporangia developed from the tip of the sporangiophore (Figure 386B) contribute alternately to each row (Figure 386C). Sporangia are cylindrical with rounded tips, measuring 1.0 μm in diameter and 4.5–5.5 μm in length. Each sporangium is separated from the sporangiophore by a diaphragm (Vobis, 1986). Scanning electron microscope studies show the presence of rostellate structures at the tips of mature and older sporangia, resembling pore-shaped openings or opercula (Locci and Petrolini-Baldan, 1971; Mertz, 1994; Williams, 1970). The fine structure of spore formation was first investigated using transmission electron microscopy by Williams et al. (1973) and Sharples et al. (1974), and reinvestigated by Vobis and Kothe (1985) and Vobis (1986). It can be concluded that sporangial development in *Planomonospora* species follows the same ontogenetical principles as found in other actinomycete genera bearing sporangia on aerial hyphae (Vobis, 1997).

Planomonospora venezuelensis forms fusiform spores, slightly thickened at the terminal end; they are 1.0 μm × 3.0–3.5 μm,

TABLE 290. Diagnostic characteristics for *Planomonospora* species[a]

Characteristic	*P. parontospora*	*P. alba*	*P. sphaerica*	*P. venezuelensis*
Sporangia formed in:				
Parallel rows	+	+	+	–
Palm leaf pattern	–	–	–	+
Production of spherical bodies	–[b]	–	+	–
Color of aerial mycelium:				
White	+[c]	+	–	+
Pink	–	–	+	–
Fatty acids:[d]				
$C_{16:1}$ iso F	–	–	+	–
$C_{16:1}$ iso G	+	+	–	–
$C_{15:0}$ 2-OH	–	+	+	+
$C_{17:1}$	+	+	+	–
$C_{18:2}$ 9c	–	+	+	+

[a]Symbols: +, >85% positive; –, 0–15% positive.

[b]Substrate hyphae with swellings.

[c]With a rose tinge.

[d]Mertz (1994).

filling the sporangium almost completely, and are motile by peritrichous flagella (Thiemann, 1970, 1974b). The spores of *Planomonospora parontospora* are cylindrical to slightly curved or reniform, and are 1.0–1.5 µm × 3.5–4.5 µm. Flagellation is described as lophotrichous (Lechevalier and Lechevalier, 1970a) or peritrichous (Miyadoh et al., 1997; Thiemann, 1974b). Large clavate spores, 10–15 µm long, which move slowly, may be observed (Thiemann et al., 1967). *Planomonospora alba* and *Planomonospora sphaerica* have cylindrical spores, 1.5 × 4.4 µm on mean, which become motile after immersion in water (Mertz, 1994).

The process of spore release begins immediately after sporangia are placed into water. The sporangia change their optical characteristics and become highly opaque (Thiemann, 1970). The spore is pushed upward through the tip, probably due to swelling of material located at the base of the sporangium. The spores become motile by means of peritrichous flagella 30–40 min after they have been expelled (Thiemann et al., 1967). They remain motile for 5–24 h, during which time spore germination may begin (Thiemann, 1970). One or more germ tubes protrude from the spores terminally and subterminally (Miyadoh et al., 1997). In *Planomonospora alba* and *Planomonospora sphaerica*, motility of the sporangiospores has been reported for 30–60 min (Mertz, 1994). The characteristic spherical bodies of *Planomonospora sphaerica* can germinate, producing germ tubes or promycelia (Mertz, 1994); the structures are considered to be atypical-shaped sporangia (Miyadoh et al., 1997).

Cell-wall composition. The peptidoglycan of the cell walls contains *meso*-diaminopimelic acid (Lechevalier and Lechevalier, 1970a; Mertz, 1994) and the characteristic sugar of whole-cell hydrolysates is madurose (3.0-methyl-D-galactose). The sugars galactose and xylose have also been reported (Mertz, 1994) in *Planomonospora alba* and *Planomonospora sphaerica*. *Planomonospora* has a cell-wall chemotype III and a sugar pattern B (Lechevalier and Lechevalier, 1970b).

Planomonospora species show a number of different menaquinone profiles. Di- and tetrahydrogenated menaquinones with nine units [MK-9(H_2) and MK-9(H_4)] are the major components in *Planomonospora parontospora*, whereas tetrahydrogeneted menaquinones with eight isoprene units [MK-8(H_4)] predominate in *Planomonospora venezuelensis* (Collins et al., 1984). In *Planomonospora alba* and *Planomonospora sphaerica*, the major menaquinone detected was MK-9(H_2), though minor amounts of MK-9(H_4) were also present (Mertz, 1994).

The phospholipids of *Planomonospora parontospora* consist of phosphatidylinositol, phosphatidylethanolamine, unknown glucoseamine-containing phospholipids, and diphosphatidylglycerol (DPG) plus lyso-DPG (Hasegawa et al., 1979). The phospholipids present in *Planomonospora alba* and *Planomonospora sphaerica* are diphosphatidylglycerol, phosphatidylglycerol, phosphatidylethanolamine, hydroxyphosphatidylethanolamine, and glucoseamine-containing phospholipids (Mertz, 1994). These polar lipid patterns correspond to phospholipid type IV of Lechevalier et al. (1981).

Planomonospora parontospora and *Planomonospora venezuelensis* have saturated and unsaturated fatty acids; the presence of C_{16} iso and $C_{15/17}$ iso branched fatty acids is variable, but $C_{15/17}$ anteiso is not present (Kroppenstedt, 1985). The presence of 10-methyl branched fatty acids has been demonstrated (Kroppenstedt, 1985; Kroppenstedt and Kutzner, 1978). Differences in fatty acid patterns can be used as diagnostic characteristics for the identification of *Planomonospora* species (Mertz, 1994; Table 290).

Colony morphology. Colonies on agar media are flat or elevated with smooth surfaces, but are occasionally wrinkled or slightly crustose. Abundant aerial mycelium is observed on a few media such as oatmeal agar for *Planomonospora parontospora* (Thiemann et al., 1967) and ATCC medium 172 for *Planomonospora alba* and *Planomonospora sphaerica* (Mertz, 1994). Good or moderately developed aerial mycelium can be observed on Czapek, oatmeal (ISP medium 3), and inorganic salts-starch (ISP medium 4) and potato carrot agar for *Planomonospora alba* and *Planomonospora sphaerica*. In general, *Planomonospora parontospora* and *Planomonospora venezuelensis* develop aerial mycelium moderately or in traces. The color of the aerial mycelium is white with a rose tinge in *Planomonospora parontospora* (Thiemann, 1974b) and white or pink in *Planomonospora sphaerica* (Mertz, 1994). White or grayish white aerial mycelium is a characteristic feature for *Planomonospora alba* (Mertz, 1994) and *Planomonospora venezuelensis* (Thiemann, 1974b).

The substrate mycelium of *Planomonospora parontospora* is light rose to rose color on oatmeal, skim milk, starch, and tyrosine agars. On Bennett's, potato plug, and nutrient agars, the color of the substrate mycelium is yellowish or creamish to light orange; on Hickey–Tresner and Czapek glucose agars it is hyaline (Thiemann et al., 1967, 1968a). The reverse color of colonies of *Planomonospora sphaerica* is pink on ATCC 172 and on ISP media 3 and 4 and on Czapek, potato-carrot, and tap water agars. The colonies of *Planomonospora alba* show grayish or yellow reverse color on these media (Mertz, 1994). The substrate mycelium of *Planomonospora venezuelensis* produces a violet color on oatmeal and skim milk agars and a brown-violet color on Bennett's, glucose-asparagine, Hickey–Tresner, starch, and yeast extract-malt extract agars. On nutrient agar, the color of the substrate mycelium is light brown, on potato plug agar it is gray, and it is hyaline on calcium malate, glycerol-asparagine and tyrosine agars (Thiemann, 1970).

The production of soluble pigments is restricted in *Planomonospora venezuelensis* to traces of a brown-violet pigment on oatmeal agar and an amber to brown pigment on Hickey–Tresner and glucose-asparagine agars (Thiemann, 1970). A very faint yellow pigment is formed by *Planomonospora parontospora* on oatmeal agar (Thiemann et al., 1967). *Planomonospora parontospora* subsp. *antibiotica* produces a light brown diffusible pigment on tyrosine agar (Thiemann et al., 1968a). *Planomonospora parontospora* subsp. *antibiotica* produces an antibacterial agent, sporangiomycin, under submerged conditions (Thiemann et al., 1968a). A protease inhibitor (antipain) has been isolated from a strain of *Planomonospora parontospora* (Wingender et al., 1975). The type strain of *Planomonospora sphaerica* produces a biological substance which has been identified as the antibiotic thiostrepton. The type strain of *Planomonospora alba* produces a biologically active agent against Gram-stain-positive bacteria and algae; it also exhibits antihelminthic activity and shows inhibition to bovine mastitis (Mertz, 1994).

Metabolism. *Planomonospora* species are aerobic, mesophilic to thermotolerant organisms, growing well on various complex and defined media (Mertz, 1994; Thiemann, 1974b). Their type strains can utilize a number of compounds as sole carbon sources, but in a broad distribution pattern (see Table 289). D-Glucose is used as a sole carbon source by all species.

Members of the genus *Planomonospora* grow at 20–40°C, with a temperature optimum at 28–37°C. *Planomonospora alba* can grow at 52°C (Mertz, 1994). Growth of *Planomonospora parontospora* is consistently good at 22–37°C (Goodfellow and Pirouz, 1982; Thiemann, 1974; Thiemann et al., 1967), but growth is not evident at 10°C. pH values from 6.0–9.0 are tolerated by all species, but growth does not occur at pH 5.0. *Planomonospora parontospora* grows sparsely at pH 6.0 and optimally at pH 7.0–8.0 (Goodfellow and Pirouz, 1982; Thiemann, 1974b; Thiemann et al., 1967). *Planomonospora alba* tolerates 2% (w/v) NaCl, *Planomonospora parontospora* subsp. *parontospora* and *Planomonospora venezuelensis* 3% (w/v) NaCl; only *Planomonospora parontospora* subsp. *antibiotica* is resistant to a concentration of 5% (w/v) NaCl (Mertz, 1994). The presence of lysozyme (0.005% w/v) is partially tolerated (Goodfellow and Pirouz, 1982).The type strains of *Planomonospora parontospora* and *Planomonospora venezuelensis* are sensitive to the antibiotics kanamycin, neomycin, and tobramycin, but are resistant to cephaloridine, lincomycin, novobiocin, rifampin, streptomycin, vancomycin, and penicillin. *Planomonospora venezuelensis* is sensitive to demethylchlortetracycline, but *Planomonospora parontospora* is not (Goodfellow and Pirouz, 1982). *Planomonospora alba* and *Planomonospora sphaerica* are resistant to 2.0 µg of kasugamycin per ml (Mertz, 1994). Further physiological features and degradation abilities are shown in Table 289.

Ecology. Members of the genus *Planomonospora* have a worldwide distribution in temperate, arid, and tropical soils. Thiemann (1970, 1974b) isolated 42 strains from 10 out of 454 soil samples (pH values 5.3–7.8) from Argentina, Chile, India, Italy, Peru, and Venezuela. A further 35 strains were isolated by D. Schäfer (personal communication) from soil samples collected in Ceylon, Egypt, France, Greece, Italy, Mexico, Namibia, Turkey, and the United States (Arizona, Florida, and Texas). The type strains of *Planomonospora sphaerica* and *Planomonospora alba* were isolated from soil collected in India and the Sudan, respectively (Mertz, 1994). Suzuki (2001) reported on the distribution of *Planomonospora* strains in 1200 soil samples collected from 28 countries. Strains of the *Planomonospora parontospora* group were present in 10.9% of the samples from 14 countries. Strains of the *Planomonospora venezuelensis* group were less frequently found and were detected only in 13 samples (1.1%). The latter were collected in tropical or temperate areas of Bolivia, Cyprus, Egypt, Greece, India, Japan, New Caledonia, and Turkey. Nearly all strains of *Planomonospora* (94%) were isolated from neutral to slightly alkaline (pH 7.0–9.0) soil samples (Suzuki et al., 2001b).

Enrichment and isolation procedures

Strains of *Planomonospora* can be enriched from soil by using the baiting technique described by Couch (1954) and Bland and Couch (1981). The soil sample (0.5–1.0 g) is placed into a small sterile Petri dish (4 cm diameter) and flooded with sterile distilled water, up to half the level of the dish, so that the material is well covered. After careful stirring, the soil particles are allowed to sediment. The addition of a very small trace of Tween can be helpful. Natural floating baits like pollen grains of *Pinus* (Schäfer, 1973) or grass blades (Mertz, 1994) are exposed on the surface of the water. Enrichment cultures are incubated in darkness at room temperature and examined after 2 and 4 weeks with a dissecting microscope at high magnification (60×). Typical sporangial bundles, i.e. sporangia on sporangiophores in double rows or in palm leaf pattern, can be picked up with a sterile thin pointed needle (Vobis, 1991) and placed onto the surface of an appropriate agar medium such as artificial soil agar (Henssen and Schäfer, 1971). This medium is transparent and promotes the production of aerial mycelium and sporangial development, thereby allowing the detection and examination of new isolates under the microscope (Cross, 1989). Young colonies can be transferred into slant culture tubes containing nutrient rich agar medium.

A new selective isolation technique developed by Suzuki et al. (2001b) employed a humic acid trace salts gellan gum medium (pH 9.0). A combination of the antimicrobial agents ampicillin (2 µg/ml), enoxacin (20 µg/ml), nalidixic acid, and trimethoprim may be added (Suzuki, 2001). The soil samples can be pretreated by dry heat with 100°C for 60 min (Suzuki, 2001). By using a flooding solution containing 0.1% skim milk in 5 mM *N*-cyclohexyl-2-amino-ethanesulfonic acid buffer (pH 9.0), high yields of motile spores may be obtained. The flooded soil samples are incubated at 32°C for 90 min, centrifuged at 100 × g for 10 min, and further incubated at 32°C for 60 min after centrifugation (Suzuki et al., 2001b).

Strains of *Planomonospora* can be detected on isolation plates once micromonosporae and streptomycetes have been removed by using a battery of phages (Kurtböke, 2003).

Maintenance procedures

Cultures on agar slants can be stored at room temperature for several weeks. Long-term preservation can be achieved by using procedures recommended for aerobic actinomycetes.

Differentiation of the genus *Planomonospora* from other genera

Planomonospora is related to other genera with cell-wall chemotype III in the family *Streptosporangiaceae.* With the exception of *Herbidospora* Kudo et al. 1993, all of the strains classified in this family produce spores on aerial mycelia, but the sporulation types are highly variable and can be used to differentiate *Planomonospora* from the other genera. All *Planomonospora* strains form zoospores within sporangia. In contrast, the genera *Herbidospora, Microbispora, Microtetraspora,* and *Nonomuraea* produce nonmotile spores in bisporous, tetrasporous, or oligosporous chains.

The shape of sporangia and the number of spores are of diagnostic value. *Acrosporospora, Sphaerisporangium,* and *Streptosporangium* produce spherical and/or club-shaped sporangia which contain numerous nonmotile spores (see Table 275, family *Streptosporangiaceae*), whereas *Planobispora* and *Planotetraspora* strains form cylindrical sporangia which encase 2 and 4 zoospores, respectively (Runmao et al., 1993; Thiemann, 1970; Thiemann and Beretta, 1968). The sporangia of *Planomonospora* are also cylindrical, but each sporangium contains only a single motile spore. The sporangia of this organism are arranged either in parallel double rows (Figure 386) or in a palm leaf pattern (Figure 385), configurations that are unique among sporangia-forming actinomycetes.

Taxonomic comments

Most of the genera classified in the aggregated group "maduromycetes" (Goodfellow, 1989a), including the genus *Planomonospora,* were assigned to the family *Streptosporangiaceae* (Goodfellow et al., 1990b, 1992). Ward-Rainey et al. (1997) emended the description of the family and included six genera in it. The inclusion of the genera *Herbidospora, Microbispora, Microtetraspora, Planobispora, Planomonospora,* and *Streptosporangium* in the family was underpinned by Miyadoh et al. (1997). Subsequently, the genera *Acrocarpospora, Nonomuraea, Planotetraspora,* and *Sphaerisporangium* corrig. were added to the family (Ara and Kudo, 2007; Miyadoh et al., 2001).

Differentiation of species of the genus *Planomonospora*

Planomonospora species can be distinguished by the morphological arrangement of their sporangia (Figure 385 and Figure 386), by their menaquinone (Collins et al., 1984; Mertz, 1994) and fatty acid composition (Mertz, 1994), and by the characteristic color of the mycelium (Mertz, 1994; Thiemann, 1974b). The diagnostic characteristics of the four species are shown in Table 290. Additional differences are given in Table 289 and in the species descriptions.

List of species of the genus *Planomonospora*

1. **Planomonospora parontospora** Thiemann, Pagani and Beretta 1967, 29[AL]

pa.ron.to.spo'ra. Gr. v. *pareimi* to be by or near one, to be side by side; Gr. n. *spora* a seed, and in biology a spore; N.L. n. *parontospora* (nominative in apposition) spores side by side.

The hyphae of the substrate mycelium (0.6–0.8 µm in diameter) are occasionally branched, septated, and twisted, frequently with swellings. The substrate mycelium is rose to light orange in color. The hyphae of the aerial mycelium (1.0 µm in diameter) are sparsely branched. The aerial mycelium is whitish, always with a light rose tinge, and is abundant on oatmeal, Hickey–Tresner, and soil agars. Growth is very good on Bennett's, Hickey–Tresner, oatmeal, and glucose-asparagine agars.

Monosporous sporangia, developed only on aerial hyphae, are arranged in double parallel rows attached directly to a characteristically bent sporangiophore. Mature sporangia are cylindrical, 1.5 µm × 3.5–4.5 µm. Sporangiospores are motile, fusiform, and slightly curved, measuring 1.0–1.5 × 3.5–4.5 µm. The surfaces of the colonies are smooth, with abundantly to sparsely developed aerial mycelium. Moderate growth occurs on Czapek-glucose, nutrient, and soil agars, but growth is poor on glycerol-asparagine, skim milk, starch, and tyrosine agars. Growth is not apparent on calcium malate, cellulose, or peptone-iron agars.

Melanoid pigments are not produced; litmus milk is peptonized; gelatin is not liquified. Growth occurs at 22–37°C; pH optimum of 7.0–8.0.

DNA G+C content (mol%): 72 (T_m).

Type strain: ATCC 23863, DSM 43177, JCM 3093, NBRC 13880, NRRL B-8120, VKM Ac-664.

Sequence accession no. (16S rRNA gene): D85495.

1a. **Planomonospora parontospora subsp. parontospora** Thiemann, Pagani and Beretta 1967, 29[AL]

The subspecies description is the same as for the species.

DNA G+C content (mol%): 72 (T_m).

Type strain: ATCC 23863, DSM 43177, JCM 3093, NBRC 13880, NRRL B-8120, VKM Ac-664.

Sequence accession no. (16S rRNA gene): D85495.

1b. **Planomonospora parontospora subsp. antibiotica** Thiemann, Coronelli, Pagani, Beretta, Tamoni and Arioli 1968a, 528[AL]

an.ti.bi.o'ti.ca. N.L. fem. adj. *antibioticus* (from Gr. prep. *anti* against; Gr. n. *bios* life; L. suff. *-ticus -a -um* suffix of various meanings, but signifying in general made of or belonging to) related to antibiotic, intended to mean producing antibiotic.

Description as for the species. Good growth occurs on inorganic salts-starch and nutrient agars. A light brown diffusible melanoid pigment is formed on tyrosine agar. Gelatin is liquified, litmus milk is not peptonized, and tyrosine is degraded. H_2S is produced.

Produces the antibiotic sporangiomycin.

DNA G+C content (mol%): not determined.

Type strain: ATCC 23864, DSM 43869, JCM 3094, NBRC 15869.

Sequence accession no. (16S rRNA gene): AB028653.

2. **Planomonospora alba** Mertz 1994, 280^VP^

al'ba. L. fem. adj. *alba* white, referring to the white aerial hyphae.

The hyphae of the substrate mycelium do not fragment and are about 1.0 µm in diameter. Substrate mycelium varies from whitish to yellowish-gray in color. The aerial mycelium is white, and abundant on ATCC medium 172. Aerial hyphae have a diameter of 1.0 µm.

Sporangiophores with long double rows of cylindrical sporangia. Each sporangium contains a single cylindrical to obclavate spore. Spores become motile after 30–60 min immersion in water and are liberated from the sporangial envelope. The mean spore size is 4.4 × 1.5 µm (range 3.5–5.4 × 1.4–1.6 µm).

Abundant growth on ATCC medium 172 and on inorganic salts-starch agar (ISP medium 4); good growth on Aino Henssen agar, Czapek agar, and oatmeal agar (ISP medium 3); and fair growth on potato-carrot agar and tap water agar. Soluble pigments are not produced. Grows at temperatures 20–50°C, tolerates NaCl only at a concentration of 2%, w/v, but is resistant to 2.0 µg of kasugamycin per ml.

Shows biological activity against Gram-stain-positive microorganisms and algae, exhibits antihelminthic activity, and is inhibitory to bovine mastitis.

DNA G+C content (mol%): not determined.

Type strain: A82600, ATCC 51588, DSM 44227, JCM 9373, NRRL 18924.

Sequence accession no. (16S rRNA gene): AB062381.

3. **Planomonospora sphaerica** Mertz 1994, 278^VP^

spha.e'ri.ca. L. fem. adj. *sphaerica* spherical, referring to the presence of spherical bodies.

Hyphae of the substrate mycelium do not fragment and have a diameter of 1.0 µm. The substrate mycelium is yellowish-pink or pink to a distinctive reddish orange color. The aerial mycelium is pink, but is sometimes white, and is abundant on ATCC medium 172.

Aerial hyphae produce sporangiophores bearing double rows of cylindrical sporangia arranged in parallel. Each sporangium contains a single spore which is released after immersed in water for 30–60 min. Spores are cylindrical to subclavate, mean size 4.4 × 1.5 µm (range 3.5–5.4 × 1.4–1.6 µm), and are motile by flagella. Large spherical bodies are formed on inorganic salts-starch agar. These spherical bodies have the capacity to germinate, forming promycelia.

Growth is abundant on ATCC medium 172 and inorganic salts-starch agar (ISP medium 4); good on Aino Henssen agar, Czapek agar, and potato-carrot agar; and fair on tap water agar. Soluble pigments are not produced. Grows at temperatures of 20–42°C; NaCl is tolerated at a concentration of 2%, w/v; is resistant to 2.0 µg of kasugamycin per ml.

Produces the antibiotic thiostrepton.

DNA G+C content (mol%): not determined.

Type strain: A51460, ATCC 51587, DSM 44632, JCM 9374, NRRL 18923.

Sequence accession no. (16S rRNA gene): AB062382.

4. **Planomonospora venezuelensis** Thiemann 1970, 247^AL^

ve.ne.zu.e.len'sis. N.L. fem. adj. *venezuelensis* of or pertaining to Venezuela.

Hyphae of the substrate mycelium are 1.0 µm in diameter and are frequently branched and septate. Substrate mycelium is violet-brown on most agar media. The color of the aerial mycelium is white to grayish white. Aerial hyphae (0.5–0.6 µm in diameter) are sparsely branched, long, wavy, and slender.

Monosporous sporangia are developed at the tips of aerial hyphae, alternately on very short lateral ramifications, arranged in two rows on opposite sides, i.e. in a "palm leaf pattern". Occasionally, sporangia can be produced singly. Sporangia are cylindrical to clavate, 1.0 µm × 4.5–5.5 µm, and are formed abundantly on Hickey–Tresner agar. Spores are motile, fusiform, measuring 1.0 µm in diameter and 3.0–3.5 µm in length.

Good growth on complex media like Bennett's agar, yeast extract-malt extract, nutrient and potato plug agars. Moderate growth on glycerol-asparagine, oatmeal, skim milk, and tyrosine agars. Optimum temperature for growth is 28–37°C. Traces of a brown-violet soluble pigment are produced on oatmeal agar, and an amber to amber-brown pigment on Bennett's and Hickey–Tresner agars. Gelatin stabs are scarcely liquified.

DNA G+C content (mol%): not determined.

Type strain: ATCC 23865, DSM 43178, JCM 3167, NBRC 15590, NRRL B-16603, VKM Ac-699.

Sequence accession no. (16S rRNA gene): AB028655.

Genus IX. **Planotetraspora** Runmao, Guizhen and Junying 1993, 468^VP^ emend. Tamura and Sakane 2004, 2055

Zhi-Heng Liu

Pla.no.te.tra. spo'ra. Gr. n. *planos* wanderer; Gr. adj. *tetra* four; Gr. n. *spora* a seed; N.L. fem. n. *Planotetraspora* four-spored organism.

Aerobic, Gram-stain-positive, non-acid-fast actinomycete which forms a branching, nonfragmenting substrate mycelium. **Long, cylindrical sporangia are formed at the ends of short sporangiophores on aerial hyphae. Each sporangium contains a row of four spores which may be cylindrical, oval, or rod-shaped** (0.4–1.4 × 0.8–1.5 µm), and may be motile. Vegetative hyphae tend to be pale yellow to white. Good growth occurs at 25–30°C. Cell walls contain *meso*-diaminopimelic acid, alanine, and glutamic acid. The peptidoglycan type is A1γ. Muramic acid moieties are *N*-acetylated. Madurose and rhamnose are present in whole-organism hydrolysates. **The major menaquinone is MK-9(H_4), the predominant fatty acid is $C_{18:0}$ 10-methyl, and the diagnostic phospholipid is phosphatidylethanolamine**. Does not contain mycolic acids. Habitat is soil. The phylogenetic position of *Planotetraspora*, as determined by 16S rRNA gene sequence analysis, is in the family *Streptosporangiaceae*.

DNA G+C content (mol%): 71.

Type species: **Planotetraspora mira** Runmao, Guizhen and Junying 1993, 468[VP].

Further descriptive information

The genus *Planotetraspora* contains two species, *Planotetraspora mira* and *Planotetraspora silvatica*, which form a distinct 16S rRNA gene clade within the evolutionary variation encompassed by the family *Streptosporangiaceae*. The closest phylogenetic neighbors of the taxon are the genera *Acrocarpospora* (95.8–97.1%) and *Herbidospora* (95.8–97.1% similarity; Tamura and Sakane, 2004). These workers also found that the type strains of these species share DNA–DNA relatedness levels within the range 38–42%.

Runmao et al. (1993) reported that whole-organism hydrolysates of *Planotetraspora mira* contain arabinose, galactose, mannose, ribose, and xylose, but this was not confirmed in subsequent studies where only madurose and rhamnose were detected as whole-organism sugars (Kudo, 2001; Tamura and Sakane, 2004). In contrast, whole-organism hydrolysates of *Planotetraspora silvatica* NBRC 10041[T] contain galactose, glucose, madurose, 3-*O*-methylmannose, and rhamnose (Tamura and Sakane, 2004).

Planotetraspora mira NBRC 15435[T] grows well and produces aerial mycelia and sporangia on HV agar (Hayakawa and Nonomura, 1989) and on yeast extract-malt extract, oatmeal, inorganic salts-starch, and glycerol-asparagine agars (ISP media 2, 3, 4, and 5, respectively; Shirling and Gottlieb, 1966). The type strain of *Planotetraspora silvatica* also shows good growth on glycerol-asparagine and yeast extract-malt extract agars, and on tyrosine agar (ISP medium 7; Shirling and Gottlieb, 1966).

Enrichment and isolation procedures

The type strain of *Planotetraspora silvatica* was isolated from a sample of forest soil collected from Amami island, Kagashima Prefecture, Japan. The organism was isolated on HV agar (Hayakawa and Nonomura, 1987b) following treatment of the initial dilution with yeast extract and sodium dodecyl sulfate (Hayakawa and Nonomura, 1989). Details were not given on how the *Planotetraspora mira* strains were recovered from a soil sample collected from Wolung village, Sichuan, China (Runmao et al., 1993).

TABLE 291. Phenotypic properties which distinguish between the type strains of *Planotetraspora mira* and *Planotetraspora silvatica*[a,b]

Characteristic	*P. mira* NBRC 15435[T]	*P. silvatica* NBRC 10041[T]
Color of colonies on yeast extract-malt extract agar	White	Yellow
Spores motile	+	–
Acid from:		
Lactose	+	–
Mannitol	+	–
Mannose	+	–
Rhamnose	+	–
Decomposition of:		
Starch	+	+
Xanthine	–	–
Growth on sole carbon sources:		
Glucose	+	–
Mannitol	+	–
Melibiose	–	+
Raffinose	–	+
Xylose	+	–

[a]Symbols: +, >85% positive; –, 0–15% positive.

[b]Data taken from Runmao et al. (1993) and Tamura and Sakane (2004).

Maintenance procedures

Working cultures can be maintained on appropriate standard media such as glucose-asparagine, oatmeal, and yeast extract-malt extract agars. Long-term preservation of strains can be achieved as frozen stocks at −20°C or by using standard lyophilization procedures.

Differentiation of the genus *Planotetraspora* from other genera

Planotetraspora strains can be distinguished from the other genera classified in the family *Streptosporangiaceae* by using a combination of chemotaxonomic and morphological properties (see Table 291 section on the family *Streptosporangiaceae*), notably by their ability to form sporangia which contain a row of four spores. The type strains of the two species form a distinct line of descent within the 16S rRNA *Streptosporangiaceae* gene tree. The signature nucleotides of the 16S rRNA genes of these organisms at positions 502:543 (A–U) and 116:1184 (U–G) are different from those of other members of the family [positions 502:543 (G–C), 1116:1184 (C–G)].

Differentiation of species of the genus *Planotetraspora*

The type strains of *Planotetraspora mira* and *Planotetraspora silvatica* can be distinguished by using a combination of phenotypic properties (Table 291).

List of species of the genus *Planotetraspora*

1. **Planotetraspora mira** Runmao, Guizhen and Jungying 1993, 468[VP]

mi′ra. L. fem. adj. *mira* extraordinary, marvellous.

Forms irregular branching substrate hyphae (0.3–0.6 μm in width) and sparsely branched aerial hyphae (0.2–0.4 μm in width). Cylindrical to clavate sporangia (~2.1–2.7 × 0.6–0.9 μm) are produced on the aerial mycelium either singly or in groups, especially on oatmeal agar; the sporangia are borne on short sporangiophores (0.2–0.8 μm long). Transverse septa divide each sporangium into four spores which at maturity are either cylindrical or spherical (0.8–1.4 × 0.4–0.6 μm). Thirty minutes after being released from the sporangia, spores become motile by means of polar flagella. The sporangiospores are covered by a thin sporangial wall which is readily seen by transmission electron microscopy, but not by light microscopy. Good growth occurs between 28–37°C, and from pH6.0–8.0.

Milk is slowly peptonized and coagulated.

Little or doubtful growth is observed with L-arabitol, L-arabinose, dulcitol, galactose, inulin, maltose, melezitose, rhamnose, D-ribose, salicin, sorbitol, sorbose, sucrose, and trehalose as sole carbon sources for energy and growth.

Additional phenotypic features are shown in Table 291.

Whole-organism hydrolysates contain madurose and rhamnose. Phosphatidylethanolamine is the diagnostic polar lipid. The predominant menaquinone is MK-9(H_4).

Source: a soil sample collected in Wolung village, Suchuan, China.

DNA G+C content (mol%): not determined.

Type strain: NA9211028, SIIA9201, ATCC 51423, DSM 44359, JCM 9131, NBRC 15435, VKM Ac-2000.

Sequence accession no. (16S rRNA gene): D85496.

Additional comment: the sporangia have been considered to have warty surfaces (Runmao et al., 1993).

2. **Planotetraspora silvatica** Tamura and Sakane 2004, 2055[VP]

sil.va'ti.ca. L. fem. adj. *silvatica* of or belonging to a wood.

Forms an extensively branched substrate mycelium and an aerial mycelium. Long cylindrical sporangia are formed at the ends of short sporangiophores borne on aerial hyphae. Each sporangium contains a single row of oval to rod-shaped spores (0.4–1.4 × 0.8–1.5 μm). A brownish soluble pigment is formed on tyrosine agar (ISP medium 7). The optimal temperature for growth is 25–30°C; growth does not occur at 37°C.

Milk is peptonized and coagulated. Gelatin hydrolysis is negative or weakly positive. Does not degrade calcium malate, or grow in the presence of 4% NaCl.

Galactose, lactose, maltose, melibiose, α-methyl-D-glucoside, and rhamnose are used as sole carbon sources for energy and growth, but adonitol, arabinose, dulcitol, and erythritol are not.

Additional phenotypic features are shown in Table 291.

Muramic acid moieties are *N*-acetylated. Whole-organism hydrolysates contain galactose, glucose, madurose, α-O-methylmannose, and rhamnose. The major fatty acids are $C_{18:0}$ 10 methyl and $C_{16:0}$ iso (>14%). Does not contain mycolic acids. Phosphatidylethanolamine and an unidentified phospholipid containing glucosamine are the diagnostic polar lipids. The predominant menaquinone is MK-9(H_4).

Source: a forest soil in Amami Island, Kagashima Prefecture, Japan.

DNA G+C content (mol%): 71 (HPLC).

Type strain: TT 00-51, DSM 44746, JCM 12867, NBRC 100141.

Sequence accession no. (16S rRNA gene): AB112082.

Genus X. **Sphaerisporangium** Ara and Kudo 2007, 2449[VP] (Effective publication: Ara and Kudo 2007c, 18.)

Martha E. Trujillo, Lorena Carro and Michael Goodfellow

Spha.e.ri.spo.ran'gium. L. n. *sphaera* sphere; N.L. neut. n. *sporangium* sporangia; N.L. neut. n. *Sphaerisporangium* an organism with spherical sporangia.

Aerobic, non-acid-fast, Gram-stain-positive actinomycetes which form branched, non-fragmenting substrate and aerial hyphae. **Single or clustered spherical spore vesicles (~1.5–8.0 μm) are produced on aerial hyphae. Vesicles contain coiled chains of non-motile spores which are oval or spherical (0.4–0.9 × 0.6–1.2 μm) with smooth, wrinkled, and prominently ridged surfaces.** Chemo-organotrophic with an oxidative type of metabolism. **The cell wall contains *meso*-diaminopimelic acid. Whole-cell hydrolysates contain galactose, glucose, madurose, mannose, and ribose. The diagnostic phospholipids are phosphatidylethanolamine and ninhydrin-positive phosphoglycolipids. The major fatty acids are $C_{16:0}$ iso and 10-methyl $C_{17:0}$, and the predominant menaquinones, MK-9(H_4) and MK-9(H_6). Phylogenetically related to members of the family *Streptosporangiaceae*.**

DNA G+C content (mol%): 67–72 (HPLC).

Type species: **Sphaerisporangium melleum** corrig. Ara and Kudo 2007c, 2449[VP].

Further descriptive information

Phylogeny. The genus *Sphaerisporangium* forms an independent phyletic branch in the *Streptosporangiaceae* 16S rRNA gene tree (See Figure 360). The species *Thermopolyspora flexuosa* (94.4% sequence similarity) appears to be the closest phylogenetic neighbor forming a deep branch next to *Sphaerisporangium*. The six constituent species form a clade supported by a bootstrap value of 90% (Figure 387). 16S rRNA gene sequence similarity between the six species is 96.3–97.6% and DNA–DNA hybridization values are 32–66% (Cao et al., 2009).

Cell morphology. *Sphaerisporangium* strains form non-fragmenting substrate mycelia and branched aerial hyphae which carry single or clustered spherical spore vesicles (~1.5–8.0 μm) (Figure 388). Spores are nonmotile and are usually formed by the separation of unbranched hyphae within spore vesicles. Spores may be oval, spherical, or pyriform with smooth, wrinkled, and ridged surfaces.

Nutrition and growth conditions. *Sphaerisporangium* strains grow well on standard complex media including Bennett's agar (Jones, 1949), glucose-yeast extract agar (Gordon and Mihm, 1962), oatmeal agar (ISP 3 medium) (Shirling and Gottlieb, 1966), and yeast extract-malt agar (ISP 2 medium). Moderate growth has been observed on glucose-asparagine agar and Hickey–Tresner agar (Hickey and Tresner, 1952), oatmeal-nitrate agar and 1/5 yeast-starch agar. Growth is poor on Czapek's agar (Pridham and Lyons, 1980), glycerol-asparagine agar (ISP 5 medium) (Shirling and Gottlieb, 1966), inorganic salt-starch agar (ISP 4) (Shirling and Gottlieb, 1966), and nutrient agar. All strains are strictly aerobic and chemo-organotrophic with an oxidative metabolism. Most species grow well between 20–37°C and pH 5–9. *Sphaerisporangium cinnabarinum* produces a bright red diffusible pigment.

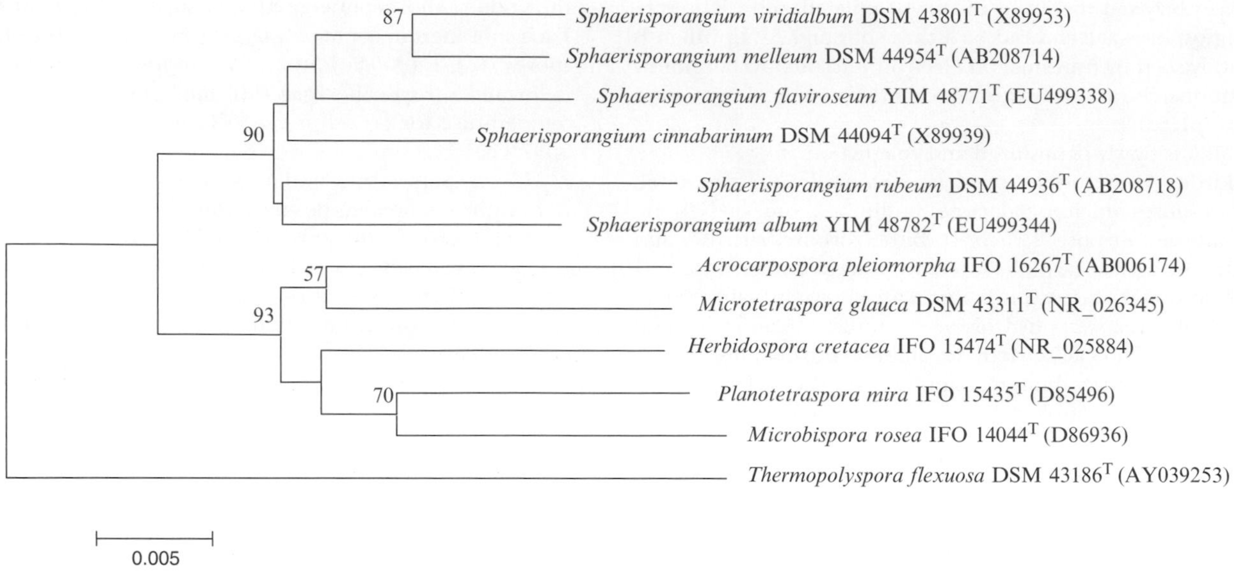

FIGURE 387. Neighbor-joining tree generated from 16S rRNA gene sequences showing relationships between *Sphaerisporangium* species and related taxa. Evolutionary distances were calculated using the Kimura 2-parameter method (Kimura, 1980). Bootstrap values indicated at branching points are expressed as percentages of 1000 replications (only values greater than 50% are shown). Bar = 0.5% sequence divergence.

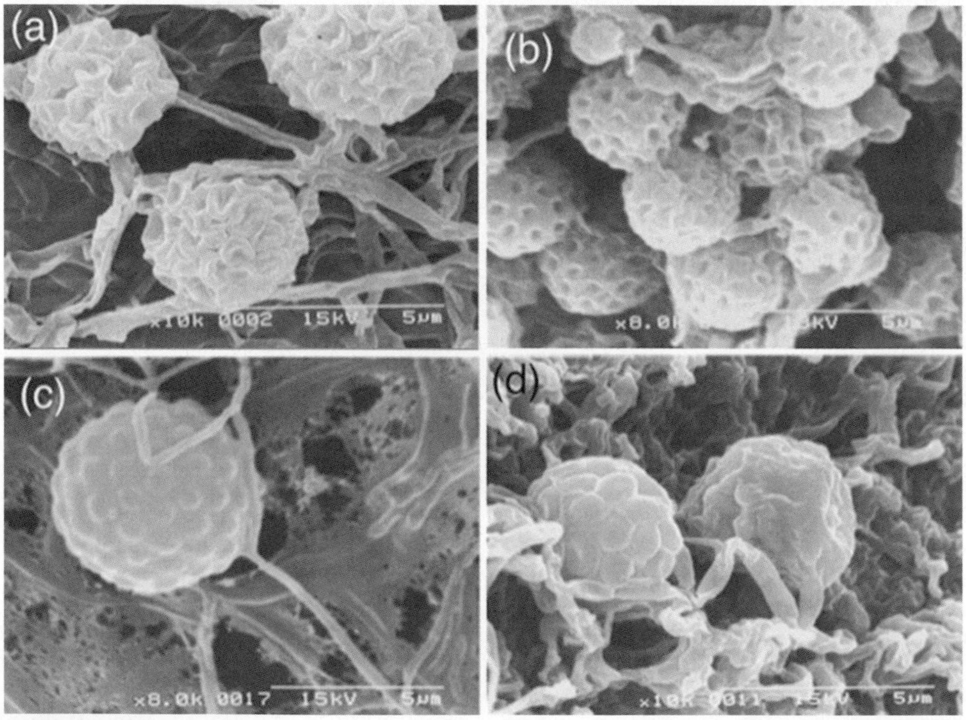

FIGURE 388. Scanning electron micrographs of globose sporangia on aerial mycelia of (a) *Sphaerisporangium melleum* 3-28(8)^{T}; (b) *Sphaerisporangium rubeum* 3D-73(35)^{T}; (c) *Sphaerisporangium cinnabarinum* JCM 3291^{T}, and (d) *Sphaerisporangium viridialbum* JCM 3027^{T} grown on oatmeal-nitrate agar for 21 d at 30°C. (Reproduced with permission from I. Ara and T. Kudo. Society for Actinomycetes, Japan.)

Cell-wall composition. *Sphaerisporangium* species contain *meso*-diaminopimelic acid (*meso*-A_2pm) as the major wall diamino acid and galactose, glucose, madurose, mannose, and ribose which corresponds to a whole-cell sugar pattern B (Lechevalier and Lechevalier, 1970b). The diagnostic sugar madurose is also found in other genera, including *Actinomadura, Dermatophilus, Microbispora, Planobispora, Planomonospora, Spirillospora,* and *Streptosporangium.* The major phospholipids are diphosphatidylglycerol, phosphatidylethanolamine, phosphatidylinositol, phosphatidylinositol mannoside, and ninhydrin-positive phosphoglycolipids mannosides. This pattern corresponds to a phospholipid type IV profile (Lechevalier et al., 1981).

The distribution of menaquinones in the genus *Sphaerisporangium* is rather complex. *Sphaerisporangium cinnabarinum*, *Sphaerisporangium melleum*, *Sphaerisporangium rubeum*, and *Sphaerisporangium viridialbum* contain major amounts of MK-9(H_4) and MK-9(H_6) and small amounts of MK-9(H_2), MK-9(H_0), and MK-9(H_8) (Ara and Kudo, 2007). In the case of *Sphaerisporangium flaviroseum* and *Sphaerisporangium album*, the predominant menaquinones are MK-9(H_4), MK-9(H_2), and MK-9, accounting for ~90% of the total composition (Cao et al., 2009). The MK-9 component in these species is 28.1 and 29.0%, respectively; the remaining *Sphaerisporangium* species contain less than 10% of this molecule. Tetrahydrogenation of the menaquinones occurs at sites of isoprene unit III (the third unit from the 2-methyl-1,4–naphthoquinone moiety) and VIII [MK-9(III, VII-H_4)] (Ara and Kudo, 2007).

The major cellular fatty acids are $C_{16:0}$ iso and $C_{17:0}$ 10-methyl with smaller amounts of the saturated, unsaturated, branched and 10-methyl fatty acids, a profile which corresponds to type 3c after Kroppenstedt (1985). Mycolic acids are absent.

Ecology. The primary habitat of *Sphaerisporangium* strains appears to be soil. *Sphaerisporangium album*, *Sphaerisporangium cinnabarinum*, *Sphaerisporangium flaviroseum*, *Sphaerisporangium melleum*, and *Sphaerisporangium rubeum* have been isolated from this source, and *Sphaerisporangium viridialbum*, from acidic volcanic ash. Janso and Carter (2010) found that 19% of 123 endophytic actinomycetes isolated from tropical plants collected from several locations in Papua New Guinea and Mborokua Island (Solomon Islands), belonged to the *Sphaerisporangium* 16S RNA gene tree suggesting that plants, specifically roots, may be a natural habitat for these micro-organisms.

Isolation procedure. Humic acid-vitamin agar (Hayakawa and Nonomura, 1987b) supplemented with cycloheximide (50 mg/ml), nystatin (50 mg/ml), and nalidixic acid was used for the isolation of several *Sphaerisporangium* species using the dilution plate method (Ara and Kudo, 2007). Isolation plates were incubated under aerobic conditions for 21 d at 30°C. *Sphaerisporangium album* and *Sphaerisporangium flaviroseum* were isolated on glycerol-asparagine agar [per liter: glycerol, 10 g; asparagine, 1 g; $K_2HPO_4 \cdot H_2O$, 1 g; $MgSO_4 \cdot 7H_2O$, 0.5 g; $CaCO_3$, 0.3 g; vitamin mixture powder, 3.7 mg (Hayakawa and Nonomura, 1987b); potassium dichromate, 50 mg; agar, 20 g].

Janso and Carter (2010) isolated endophytic *Sphaerisporangium* strains from stems, roots, and leaves by cutting them into pieces using pruning shears or a scalpel following removal of soil and organic debris by rinsing them thoroughly under running tap water. The tissues were successively cut into 2- by 2-mm pieces with a scalpel, placed inside a tea strainer, and immersed in a series of solutions for surface sterilization. Thin or herbaceous tissues (such as leaves, stems, and petioles), were treated with 70% ethanol for 1 min, 50% Clorox bleach (approx. 3% NaOCl) for 3 min, and 70% ethanol for 0.5 min followed by a rinse in sterile water. The procedure was the same for thicker or woody tissues, such as roots and twigs, except that they were immersed in 50% bleach for 5 min. Tissue samples were aseptically transferred to the surface of arginine-vitamin agar (Nonomura and Ohara, 1969a) supplemented with 3% soil extract (Hayakawa et al., 2000) and cycloheximide (100 µg/ml) and nystatin (50 µg/ml). The soil extract was prepared by mixing 100.0 g organic humus (Jolly Gardener Products Inc., Poland, ME) in 100 ml of tap water; the suspension was autoclaved at 121°C for 30 min, transferred to a 1-liter centrifuge bottle, and centrifuged at 4000 r.p.m. for 20 min. The supernatant was filtered through a 0.2-µm cellulose nitrate filter unit (Nalgene). Inoculated Petri dishes were incubated at room temperature (approx. 23–25°C) for up to 8 weeks.

Maintenance procedures

Sphaerisporangium strains can be maintained on yeast-starch agar [per liter: soluble starch, 15.0 g; yeast extract, 4.0 g; $K_2HPO_4 \cdot 7H_2O$, 0.5 g; agar 15.0 g; (pH 7.2)] or ISP 2 agar (Shirling and Gottlieb, 1966) at 4°C for short-term preservation; mycelial fragments in 20% glycerol (v/v) can be maintained at −25°C. Long-term storage can be achieved by lyophilization.

Differentiation of the genus *Sphaerisporangium* from other genera

16S rRNA gene sequencing and the presence of specific signature nucleotides at position 263 (G), 264 (U), 595 (G), 600:638 (U–G), 602:636 (C–G), 603:635 (C–G), 627 (G), 626 (U), 625 (G), 668:738 (C–G), 669:737 (A–U), 671:735 (G–C), 1012:1017 (G–C), and 1263:1272 (G–U) clearly differentiate *Sphaerisporangium* from other genera classified in the family *Streptosporangiaceae* (Ara and Kudo, 2007). The genera *Herbidospora* and *Sphaerisporangium* can be readily distinguished by their menaquinone profiles.

Taxonomic comments

The genus *Sphaerisporangium* was proposed by Ara and Kudo (2007) to accommodate five strains isolated from soil. They also examined "*Streptosporangium cinnabarinum*" JCM 3291 and *Streptosporangium viridialbum* JCM 3027^T which, from the phylogenetic point of view, were considered to be moderately related to the genus *Streptosporangium* (Kemmerling et al., 1993; Tamura et al., 2000; Ward-Rainey et al., 1996). Ara and Kudo (2007) proposed three new species, *Sphaerosporangium melleum*, the type species, *Sphaerosporangium cinnabarinum*, and *Sphaerosporangium rubeum* and, in addition, transferred *Streptosporangium viridialbum* Nonomura and Ohara 1960, to the genus *Sphaerosporangium* as *Sphaerosporangium viridialbum* Nonomura and Ohara 1960 comb. nov. These taxa were included in Validation List no. 118, under the genus *Sphaerisporangium* corrig. as opposed to the earlier incorrect name *Sphaerosporangium* Ara and Kudo 2007. Two additional species, *Sphaerisporangium album* and *Sphaerisporangium flaviroseum*, were recognized by Cao et al. (2009) who also emended the description of the genus.

Differentiation of species of the genus *Sphaerisporangium*

Sphaerisporangium species can be distinguished from one another by using a combination of morphological, nutritional, and physiological characteristics (Table 292).

TABLE 292. Characteristics that differentiate the type strains of *Sphaerisporangium* species[a,b]

Characteristic	*S. melleum*	*S. album*	*S. cinnabarinum*	*S. flaviroseum*	*S. rubeum*	*S. viridialbum*
Substrate mycelium color (ISP 2 agar)	Honey gold	Pale gray	Bamboo	Deep yellow pink	Coral red	Light tan
Substrate mycelium color (ISP 3 agar)	Mustard gold	Yellow white	Light amber	Soft yellow pink	Light coral red	Bamboo
Reduction of nitrate	−	+	−	−	nd	−
Oxidase activity	−	+	−	−	−	−
Degradation of starch	−	+	−	−	−	−
Growth on sole carbon sources:						
L-Arabinose	−	(+)	+	+	−	+
L-Arginine	−	+	+	(+)	−	+
Cellobiose	+	(+)	+	−	−	+
Dextrin	nd	+	+	+	−	−
D-Fructose	nd	(+)	+	+	+	+
Fucose	+	+	+	+	+	−
D-Galactose	+	+	+	+	−	+
L-Histidine	−	+	−	−	+	+
Inositol	−	+	+	(+)	(+)	(+)
Lactose	nd	+	+	(+)	+	+
L-Lysine	−	+	+	−	−	(+)
Maltose	+	+	+	+	−	+
D-Mannitol	nd	+	−	(+)	+	−
DL-Methionine	(+)	−	(+)	−	−	+
L-Proline	−	+	+	(+)	+	+
Raffinose	nd	+	−	(+)	(+)	−
L-Rhamnose	+	+	−	+	−	+
D-Ribose	+	+	−	+	(+)	+
L- Serine	−	+	+	(+)	+	(+)
Sorbose	(+)	−	+	+	−	−
Sucrose	nd	(+)	(+)	+	+	(+)
L-Tryptophan	−	−	(+)	−	−	(+)
Urea	−	+	−	+	−	−
L-Valine	−	+	+	+	−	+
Xanthine	−	−	+	−	(+)	(+)
D-Xylose	+	(+)	+	(+)	−	+
Major menaquinones	MK-9(H_4), MK-9(H_6)	MK-9(H_4), MK-9(H_2), MK-9	MK-9(H_4), MK-9(H_6)	MK-9(H_4), MK-9(H_2), MK-9	MK-9(H_6), MK-9(H_4)	MK-9(H_4), MK-9(H_2)
Major fatty acids	$C_{16:0}$ iso, $C_{17:0}$ 10-methyl	$C_{16:0}$ iso; $C_{17:0}$ 10-methyl	$C_{16:0}$ iso; $C_{17:0}$ 10-methyl	$C_{16:0}$ iso, $C_{17:0}$ 10-methyl	$C_{16:0}$ iso, $C_{17:0}$ 10-methyl, $C_{15:0}$, $C_{17:0}$	$C_{15:0}$ iso, $C_{17:0}$, $C_{16:0}$ iso, $C_{15:0}$
DNA G+C content (mol%)	71	71	70	67.1	70.4	72

[a]Symbols and abbreviations: +, positive; −, negative; nd, not determined; (+), weak growth; MK-9, MK-9(H_2), MK-9(H_4), and MK-9(H_6) designate menaquinones with nine isoprenoic units in the side chain, of which 0, 1, 2, and 3, respectively, are saturated.

[b]Data from: Ara and Kudo (2007); Cao et al. (2009).

List of species of the genus *Sphaerisporangium*

1. **Sphaerisporangium melleum** corrig. Ara and Kudo 2007c, 2449[VP] (Effective publication: Ara and Kudo 2007, 19.)

mel'le.um. L. neut. adj. *melleum* honey-colored.

Branched, non-fragmenting, substrate and aerial hyphae are formed. The vegetative mycelium is light wheat to honey gold in color and the aerial mycelium is white. Spherical spore vesicles which carry coiled spore chains are formed on the aerial hyphae. Abundant sporulation is obtained on inorganic salts-starch, sucrose-nitrate, tap water, and 1/5 yeast extract-starch agars. Good growth is observed on Bennett's, oatmeal, and yeast extract-starch agars; moderate growth on glucose-asparagine, Hickey–Tresner, and 1/5 yeast extract-starch agars, and poor growth on glycerol-asparagine agar, inorganic salts-starch, nutrient, sucrose-nitrate, tap water and tyrosine agars. Diffusible pigments are not produced on these media. Glycerol, erythritol and *myo*-inositol are weakly utilized as sole carbon sources. Other physiological data are given in Table 292 together with information on menaquinone and fatty acid composition.

Source (type strain): soil collected in Bangladesh.

DNA G+C content (mol%): 71.0 (HPLC).

Type strain: 3-28(8), DSM 44954, JCM 13064.

Sequence accession no. (16S rRNA gene): AB208714.

2. **Sphaerisporangium album** Cao, Jiang, Xu, Jiang 2009, 1682[VP]

al'bum. L. neut. adj. *album* white.

Branched, non-fragmenting, substrate and aerial hyphae are formed. Pale-gray substrate and white aerial mycelia are produced. Spherical and pyriform spore vesicles are borne on aerial mycelia. Good growth is observed on oatmeal and yeast extract-malt extract agars, and moderate growth on Czapek's, glycerol-asparagine and inorganic salt-starch agars. Diffusible pigments are not produced on any of these media. Grows in the presence of 2% (w/v) NaCl. Catalase-positive. H_2S is not produced; milk is not coagulated or peptonized. Does not degrade cellulose. Alkaline phosphatase, esterase (C4), esterase lipase (C8), lipase (C14), leucine arylamidase, valine arylamidase, trypsin, α-chymotrypsin, acid phosphatase, naphthol-AS-BI phosphohydrolase, β-galactosidase, α-glucosidase, β-glucosidase, *N*-acetyl-β-glucosaminidase, and α-mannosidase are produced but not cystine arylamidase, α-fucosidase, α-galactosidase, or β-glucuronidase. L-Arginine, histidine, lysine, ornithine, L-phenylalanine, proline, serine, L-valine, and urea are hydrolyzed, but not glycine, methionine, L-tryptophan, or xanthine. Other physiological data are given in Table 292 together with information on menaquinone and cellular fatty acid composition.

Source (type strain): soil samples collected from virgin forest at Jinbian Rivulet in Hunan Province, China.

DNA G+C content (mol%): 71 (HPLC).

Type strain: CCTCC AA 208026, DSM 45172, YIM 48782.

Sequence accession no. (16S rRNA gene): EU499344.

3. **Sphaerisporangium cinnabarinum** corrig. Ara and Kudo 2007c, 2449[VP] (Effective publication: Ara and Kudo 2007, 19.)

cin.na.ba'ri.num. L. n. *cinnabar -aris*, cinnabar; L. suff. *-inus -a -um*, suffix used with the sense of belonging to; N.L. neut. adj. *cinnabarinum* of cinnabar, referring to the vermilion (bright red) color of the vegetative mycelium and diffusible pigment.

Branched, non-fragmenting, substrate and aerial hyphae are formed. Substrate mycelium is light wheat to melon yellow. White aerial hyphae carry globose spore vesicles containing coiled spore chains. Spores are produced on humic acid-vitamin, oatmeal-nitrate and sucrose-nitrate agars. Good growth is observed on oatmeal and yeast extract-starch; moderate growth on Bennett's, glucose-asparagine, glucose-yeast extract, humic acid-vitamin, inorganic salts-starch, oatmeal-nitrate, and 1/5 yeast extract agars; growth is poor on glycerol-asparagine, Hickey–Tresner, nutrient, tap water, tyrosine, and yeast-extract malt extract agars. Use *myo*-inositol, α-methyl-D-glucoside, and α-D-melibiose as sole carbon sources. Other physiological data are given in Table 292 together with information on menaquinone and cellular fatty acid composition.

Source (type strain): soil collected in the Philippines.

DNA G+C content (mol%): 70 (HPLC).

Type strain: DSM 44094, JCM 3291.

Sequence accession no. (16S rRNA gene): X89939.

4. **Sphaerisporangium flaviroseum** Cao, Jiang, Xu and Jiang 2009, 1682[VP]

fla.vi.ro'se.um. L. adj. *flavus* yellow; L. adj *roseus* rose; N.L. neut. adj. *flaviroseum* yellowish-rose colored.

Branched, non-fragmenting, substrate and aerial hyphae are formed and a yellow-pink substrate and white aerial mycelia produced. Spherical and pyriform spore vesicles are borne on aerial mycelia. Good growth is observed on oatmeal yeast extract-malt extract agars; moderate growth on Czapek's, glycerol-asparagine and inorganic salts-starch agars, ISP 4 agar, and ISP 5 agar. Diffusible pigments are not produced on these media. Grows at pH 8 and in the presence of 1% (w/v) NaCl. Catalase is not produced. Gelatin liquefaction, milk coagulation and peptonization, and production of H_2S are negative. Cellulose is not degraded. Alkaline phosphatase, esterase (C4), esterase lipase (C8), leucine arylmidase, valine arylamidase, trypsin, α-chymotrypsin, acid phosphatase, naphthol-AS-BI-phosphohydrolase, β-galactosidase, α-glucosidase, β-glucosidase, *N*-acetyl-β-glucosaminidase, α-mannosidase, and α-fucosidase are produced, but not cystine arylamidase, α-galactosidase, β-glucuronidase, or lipase (C14) (API ZYM tests). L-Arginine, ornithine, L-phenylalanine, proline, serine, L-valine, and urea are hydrolyzed, but not glycine, histidine, lysine, methionine, L-tryptophan, or xanthine.

Source (type strain): soil collected from Tianzi Mountain in Hunan Province, China.

DNA G+C content (mol%): 67.1 (HPLC).

Type strain: DSM 45170, JCM 16908, KCTC 19393, YIM 48771.

Sequence accession no. (16S rRNA gene): EU499338.

5. **Sphaerisporangium rubeum** corrig. Ara and Kudo 2007c, 2449[VP] (Effective publication: Ara and Kudo 2007, 19.)

ru'be.um. L. neut. adj. *rubeum* red colored.

Branched, non-fragmenting, substrate and aerial hyphae are formed. The vegetative mycelium is light coral pink to coral red and the aerial mycelium is white. Spherical vesicles which carry coiled spore chains are formed on aerial hyphae. Good growth is obtained on yeast extract-starch agar, Bennett's, glucose-yeast extract, Hickey–Tresner, oatmeal and yeast extract-malt extract agars; moderate growth on humic acid-vitamin, inorganic salts-starch, nutrient agar, oatmeal-nitrate, and 1/5 yeast extract-starch agars; growth is poor on glucose-asparagine, glycerol-asparagine, sucrose-nitrate, tap water, and tyrosine agars. Scant sporulation occurs on humic acid-vitamin, oatmeal nitrate, sucrose-nitrate, and water agars.

Adonitol, erythritol, glycerol, and *myo*-inositol are weakly utilized as sole carbon sources. Other physiological data are given in Table 292 together with information on menaquinone and cellular fatty acid composition.

DNA G+C content (mol%): 71 (HPLC).

Type strain: 3D-72(35), DSM 44936, JCM 13067.

Sequence accession no. (16S rRNA gene): AB208718.

6. **Sphaerisporangium viridialbum** (Nonomura and Ohara 1960) corrig. Ara and Kudo 2007c, 2449[VP] (Effective publication: Ara and Kudo 2007, 19.)

vi.ri.di.al'bum. L. adj. *viridis* green; L. neut. adj. *album* white; N.L. neut. adj. *viridialbum* greenish white.

Branched, non-fragmenting, substrate and aerial hyphae are formed. Substrate mycelium is light wheat to bamboo yellow and the aerial mycelium is white. Spherical vesicles which carry coiled spore chains are formed on aerial hyphae. Sporulation occurs on Hickey–Tresner, humic acid-vitamin oatmeal-nitrate, sucrose-nitrate, and 1/5 yeast extract-starch agars. Grows well on Bennett's, glucose-yeast extract, Hickey–Tresner, oatmeal and yeast-extract-starch agars; moderate growth on humic acid-vitamin, nutrient, oatmeal-nitrate, yeast extract-malt extract, and 1/5 yeast extract-starch agars;

growth is poor on glucose-asparagine, glycerol-asparagine, inorganic salts-starch, sucrose-nitrate, tyrosine, and tap water agars. Glycerol and erythritol are weakly utilized as sole carbon sources. Other physiological data are given in Table 292 together with information on menaquinone and cellular fatty acid composition.

Source (type strain): acidic volcanic ash.

DNA G+C content (mol%): 72 (HPLC).

Type strain: ATCC 33328, CBS 432.61, DSM 43801, JCM 3027, KCC A-0027, KCTC 9435, NBRC 13987, NRRL B-2636, VKM Ac-679.

Sequence accession no. (16S rRNA gene): X89953.

Genus XI. **Thermopolyspora** (*ex* Krasil'nikov and Agre 1964) Goodfellow, Maldonado and Quintana 2005, 1980[VP]

Michael Goodfellow

Ther.mo.po.ly.spo'ra. Gr. n. *thermê* heat; Gr. adj. *polu* many; Gr. fem. n. *spora* a seed and, in bacteriology, a spore; N.L. fem. n. *Thermopolyspora* the heat (-loving) many-spored organism.

Aerobic, Gram-stain-positive, non-acid–alcohol-fast, actinomycete which forms non-fragmenting substrate and aerial mycelia. Hooked to irregular spiral chains of 4–10 warty to spiny ornamented spores (1.2–1.5 µm in diameter) are arranged in clusters on long, moderately branched aerial hyphae on potato carrot agar; a light blue aerial mycelium and a brown soluble pigment are also formed on this medium. White to yellowish-white aerial hyphae are borne on a light brown substrate mycelium on oatmeal agar, albeit without the production of diffusible pigments. **Grows from 40–60°C, and optimally between 45 and 55°C. Cell wall contains *meso*-diaminopimelic acid and *N*-acetylated muramic acid. Glucose and ribose are found in whole-organism hydrolysates. The predominant menaquinones are MK-9, MK-9(H_2), and MK-9(H_4), and the major polar lipids are phosphatidylethanolamine, hydroxyphosphatidylethanolamine, phosphatidylinositol mannosides, ninhydrin sugar-positive components, and uncharacterized glycolipids**. The organism is rich in saturated, unsaturated, and branched chain fatty acids, but does not contain mycolic acids. The phylogenetic position of *Thermopolyspora*, as determined by 16S rRNA gene sequence analysis, is in the family *Streptosporangiaceae.*

DNA G+C content (mol%): 77.

Type species: **Thermopolyspora flexuosa** (Meyer 1989) Goodfellow, Maldonado and Quintana 2005, 1982[VP] (*Actinomadura flexuosa* Meyer 1989, 2519).

Further descriptive information

Phylogeny. *Thermopolyspora flexuosa* DSM 43186[T] (formerly *Actinomadura flexuosa*) forms a distinct branch in the *Streptosporangiaceae* 16S rRNA gene tree (see Figure 360 in section on the family *Streptosporangiaceae*). Goodfellow et al. (2005) have shown that this strain produces the PCR product characteristic of members of the family *Streptosporangiaceae* when examined with the family-specific set of oligonucleotide primers developed by Monciardini et al. (2002).

Cell morphology. Strain DSM 43186[T] produces short curved chains of warty to spiny spores on potato carrot agar (Greiner-Mai et al., 1987).

Chemotaxonomy. A comprehensive survey of the chemotaxonomic properties of strain DSM 43186[T] was undertaken by Kroppenstedt et al. (1990a) who found that it produces whole-organism hydrolysates rich in ribose together with a trace of mannose. It also produces complex mixtures of saturated, unsaturated, and branched chain fatty acids and lacks mycolic acids, but contains major amounts of MK-9(H_0), MK-9(H_2), and MK-9(H_4) (points of unsaturation: III and VIII), and minor amounts of MK-9(H_6) (points of unsaturation: II, III and VIII); these components accounted for 41, 31, 25, and 7% of the total menaquinone composition, respectively. The polar lipid pattern was characterized by the presence of major amounts of phosphatidylethanolamine, hydroxylated phosphatidylethanolamine, phosphatidylinositol mannosides, ninhydrin and sugar positive components and uncharacterized glycolipids, and a trace of diphosphatidylglycerol. The diamino acid of the wall peptidoglycan is *meso*-diaminopimelic acid (Becker et al., 1965; Greiner-Mai et al., 1987).

Colony morphology. The organism produces a brown substrate mycelium, a light blue aerial mycelium, and a brown soluble pigment on glucose-yeast extract-malt and potato carrot agars (Greiner-Mai et al., 1987). On oatmeal agar, light brown colonies carry a filmy cover of farinaceous, white to yellowish-white aerial hyphae; sometimes sectors with intensified development of aerial hyphae are seen, but diffusible pigments are not formed (Meyer, 1989). A brown substrate mycelium is formed on peptone-glucose agar, but neither diffusible pigments nor aerial mycelium are produced (Meyer, 1989).

Nutrition and growth conditions. Strain DSM 43186[T] shows good to moderate growth on a range of standard media including glucose-yeast extract-malt extract and oatmeal agars, but grows poorly on Bennett's sucrose and glycerol-asparagine agars (Meyer, 1989). The organism is quite fastidious in its growth temperature and pH requirements; it does not grow below 40°C or above pH 9.0 (McCarthy and Cross, 1984).

Metabolism. Little is known about the metabolic properties of the organism. However, strain DSM 43186[T] was included in a numerical taxonomic study of monosporic actinomycetes and found to be biochemically active with an ability to degrade several complex organic compounds and use a range of sugars as sole carbon sources (McCarthy and Cross, 1984).

Antibiotic sensitivity. The organism is sensitive to a broad range of antibiotics as shown in the species description.

Maintenance procedures

The organism can be maintained on glucose-yeast extract-malt extract agar plates (ISP medium 2; Shirling and Gottlieb, 1966) and as glycerol suspensions of mycelia and spores (20%, v/v) at −20°C. Lyophilization or storage in liquid nitrogen can be used for long-term preservation.

Differentiation of the genus *Thermopolyspora* from other genera

Strain DSM 43186^T can be distinguished from members of other genera classified in the family *Streptosporangiaceae* by using a combination of chemotaxonomic and morphological properties (see Table 275 in section on family *Streptosporangiaceae*), and by 16S rRNA gene sequencing (Goodfellow et al., 2005).

Taxonomic comments

"*Thermopolyspora flexuosa*" was proposed by Krasil'nikov and Agre (1964) for a group of thermophilic actinomycetes which formed short chains of spores on short sporophores. Subsequently, the species had a tortuous taxonomic history as it was included or associated with several taxa, notably the genera *Actinomadura* (Cross and Goodfellow 1973; Lacey et al. 1978), *Micropolyspora* Krasil'nikov et al. 1968, *Nocardia* (Becker et al. 1965; Lechevalier et al. 1966a), and *Nonomuraea* Zhang et al. 1998a. It was classified in the genus *Actinomadura* in the previous edition of *Bergey's Manual of Systematic Bacteriology* (Meyer, 1989). The taxonomic position of the organism was clarified in a polyphasic taxonomic study by Goodfellow et al. (2005) who proposed that the name *Thermopolyspora flexuosa* be revived for the taxon, a proposition advanced by Greiner-Mai et al. (1987) on the basis of chemotaxonomic and morphological data. Goodfellow and his colleagues assigned the genus in the family *Streptosporangiaceae* (Goodfellow et al. 1990b) emend. Ward-Rainey et al. 1997. It is currently classified in the family *Streptosporangiaceae* (Goodfellow et al. 1990b) emend. Zhi et al. 2009.

List of species of the genus *Thermopolyspora*

1. **Thermopolyspora flexuosa** (Meyer 1989) Goodfellow, Maldonado and Quintana 2005, 1982^{VP} (*Actinomadura flexuosa* Meyer 1989, 2519)

fle.xu.o'sa. L. fem. adj. *flexuosa* full of turns or winding, tortuous, flexuous, referring to the morphology of the spore chains.

General chemotaxonomic, colonial, and morphological properties are included in the genus description. Grows well on glucose-yeast extract-malt extract, oatmeal, and peptone-glucose agars. Melanin pigments are formed from tyrosine.

Allantoin and esculin are hydrolyzed, but arbutin and urea are not. Arginine dihydrolase, catalase, citrate lyase, β-galactosidase, β-glucosidase, and oxidase are produced, but acetoin, hydrogen sulfide, lysine decarboxylase, ornithine decarboxylase, and phosphatase are not. Nitrate is reduced.

Agar, casein, DNA, elastin, gelatin, keratin, pectin, starch, testosterone, Tweens 20 and 80, tyrosine, xanthine, and xylan are degraded, but carboxymethylcellulose, chitin, guanine, hypoxanthine, and pectin are not.

L-Arabinose, fructose, galactose, glucose, maltose, mannitol, mannose, sucrose, trehalose, and xylose are used as sole carbon sources for energy and growth, but dulcitol, glycerol, inositol, inulin, lactose, melezitose, melibiose, inositol, raffinose, rhamnose, ribose, sorbitol, and sorbose (all at 1%, w/v) are not. Sodium citrate is used as a sole carbon source, but sodium oxalate (both at 0.1%, w/v) is not.

Alanine, asparagine, arginine, glutamic acid, histidine, tyrosine, and tryptophan are used as sole nitrogen sources, but ammonia salts and nitrates are not.

Grows in the presence of bile salts (0.02%, w/v), crystal violet (0.0,0002%, w/v), potassium tellurite (0.02%, w/v), sodium azide (0.01%, w/v), and thallous acetate (0.001%, w/v), but is sensitive to bile salts (0.5%, w/v), brilliant green (0.0005%, w/v), crystal violet (0.0,0005%, w/v), lysozyme (0.0025%, w/v), sodium azide (0.02%, w/v), sodium chloride (3%, w/v), tetrazolium chloride (0.002%, w/v), and thallous acetate (0.005%, w/v).

Resistant (μg per filter paper disc) to chloramphenicol (25), cephaloridine (5), cephalexin (30), cefoxitin (30), gentamicin sulfate (2), kanamycin sulfate (30), tobramycin sulfate (10), and penicillin (5U), but is sensitive to cephazolin (30), gentamicin sulfate (10), and novobiocin (50 μg/ml).

Major fatty acids (% of total) are $C_{15:0}$ iso (12.5), $C_{15:0}$ (5.7), $C_{16:0}$ iso (13.8), $C_{17:0}$ iso (18.7), $C_{17:0}$ anteiso (12.4), and $C_{17:0}$ 10 methyl-branched (6.9); minor components include $C_{15:0}$ anteiso (4.3), $C_{16:1}$ *cis* (2.8) $C_{15:0}$ iso 2-OH (1.4), $C_{16:0}$ (3.4), $C_{16:0}$ 10 methyl (2.4), $C_{16:0}$ iso 2-OH (2.5), $C_{17:0}$ (2%), $C_{17:0}$ iso 2-OH (3.0), and $C_{18:0}$ (2.5).

Source: soil collected from the Pamir mountains.

DNA G+C content (mol%): 77 (T_m).

Type strain: strain K1132, ATCC 35864, BCRC 12531, CIP 107358, DSM 43186, IMET 9552, JCM 3056, KCC A-0056, KCTC 9270, NBRC 14349, NRRL B-24348.

Sequence accession no. (16S rRNA gene): AY039253.

Additional comments: in the paper by Goodfellow et al. (2005), *Thermopolyspora* is cited as *Thermopolyspora* gen. nov., and *Thermopolyspora flexuosa* as *Thermopolyspora flexuosa* comb. nov., nom. rev. In the abstract of this paper, the strain DSM 43186 is erroneously cited as DSM 41386 [see Erratum. Int. J. Syst. Evol. Microbiol. (2005) *55*:2640]. Basonym: *Actinomadura flexuosa* (*ex* Krasil'nikov and Agre 1964) Meyer 1989, 2519; *Nonomuraea flexuosa* corrig. (Meyer 1989) Zhang et al., 1998a, 420; *Microtetraspora flexuosa* (Meyer 1989) Kroppenstedt et al., 1991, 178; "*Nocardia flexuosa*" (Krasil'nikov and Agre 1964) Becker et al., 1965, 422; "*Thermopolyspora flexuosa*" (Krasil'nikov and Agre 1964) Cross and Goodfellow 1973, 82.

References

Agre, N.S. and L.N. Guzeva. 1975. [New genus of actinomycetes: *Excellospora* gen. nov.]. Mikrobiologiia *44*: 518–523.

Alduina, R., A. Giardina, G. Gallo, G. Renzone, C. Ferraro, A. Contino, A. Scaloni, S. Donadio and A.M. Puglia. 2005. Expression in *Streptomyces lividans* of *Nonomuraea* genes cloned in an artificial chromosome. Appl. Microbiol. Biotechnol. *68*: 656–662.

Altenburger, P., P. Kämpfer, A. Makristathis, W. Lubitz and H.-J. Busse. 1996. Classification of bacteria isolated from a medieval wall painting. J. Biotechnol. *47*: 39–52.

Ara, I. and T. Kudo. 2007. *Sphaerosporangium* gen. nov., a new member of the family *Streptosporangiaceae*, with descriptions of three new species as *Sphaerosporangium melleum* sp. nov., *Sphaerosporangium rubeum* sp. nov. and *Sphaerosporangium cinnabarinum* sp. nov., and transfer of *Streptosporangium viridialbum* Nonomura and Ohara 1960 to *Sphaerosporangium viridialbum* comb. nov. Actinomycetologica *21*: 11–21.

Ara, I., T. Kudo, A. Matsumoto, Y. Takahashi and S. Ōmura. 2007a. *Nonomuraea maheshkhaliensis* sp. nov., a novel actinomycete isolated from mangrove rhizosphere mud. J. Gen. Appl. Microbiol. *53*: 159–166.

Ara, I., T. Kudo, A. Matsumoto, Y. Takahashi and S. Ōmura. 2007b. *Nonomuraea bangladeshensis* sp. nov. and *Nonomuraea coxensis* sp. nov. Int. J. Syst. Evol. Microbiol. *57*: 1504–1509.

Ara, I., T. Kudo, A. Matsumoto, Y. Takahashi and S. Ōmura. 2007c. *In* List of new names and new combinations previously effectively, but not validly, published. Validation List no. 118. Int. J. Syst. Evol. Microbiol. *57*: 2449–2450.

Athalye, M., M. Goodfellow, J. Lacey and R.P. White. 1985. Numerical classification of *Actinomadura* and *Nocardiopsis*. Int. J. Syst. Bacteriol. *35*: 86–98.

Atsushi, T., F. Rizuji and K. Hirotada. 1975. Antibiotic sporocuracin production, Japanese Patent 75,125,094. Japan.

Batista, A.C., S.K. Shome and J.A. De Lima. 1963. *Streptosporangium bovinum* sp. nov. from cattle hoofs. Dermatol. Trop. Ecol. Geogr. *19*: 49–54.

Becker, B., M.P. Lechevalier, R.E. Gordon and H.A. Lechevalier. 1965. Rapid differentiation between *Nocardia* and *Streptomyces* by paper chromatography of whole-cell hydrolysates. Appl. Microbiol. *12*: 421–423.

Beltrametti, F., A. Lazzarini, C. Brunati, E. Selva and F. Marinelli. 2003. Production of demannosyl-A40926 by a *Nonomuraea* sp. ATCC 39727 mutant strain. J. Antibiot. *56*: 310–313.

Beltrametti, F., S. Jovetic, M. Feroggio, L. Gastaldo, E. Selva and F. Marinelli. 2004. Valine influences production and complex composition of glycopeptide antibiotic A40926 in fermentations of *Nonomuraea* sp. ATCC 39727. J. Antibiot. *57*: 37–44.

Bland, C.E. and J.N. Couch. 1981. The family *Actinoplanaceae*. *In* The Prokaryotes: a Handbook on Habitats, Isolation, and Identification of Bacteria (edited by Starr, Stolp, Trüper, Balows and Schlegel). Springer, New York, pp. 2004–2010.

Boondaeng, A., Y. Ishida, T. Tamura, S. Tokuyama and V. Kitpreechavanich. 2009. *Microbispora siamensis* sp. nov., a thermotolerant actinomycete isolated from soil. Int. J. Syst. Evol. Microbiol. *59*: 3136–3139.

Boondaeng, A., C. Suriyachadkun, Y. Ishida, T. Tamura, S. Tokuyama and V. Kitpreechavanich. 2011. *Herbidospora sakaeratensis* sp. nov., isolated from Thailand and reclassification of *Streptosporangium claviforme* as a later synonym of *Herbidospora cretacea*. Int. J. Syst. Evol. Microbiol. *61*: 777–780.

Boudjella, H., K. Bouti, A. Zitouni, F. Mathieu, A. Lebrihi and N. Sabaou. 2006. Taxonomy and chemical characterization of antibiotics of *Streptosporangium* Sg 10 isolated from a Saharan soil. Microbiol. Res. *161*: 288–298.

Boudjella, H., K. Bouti, A. Zitouni, F. Mathieu, A. Lebrihi and N. Sabaou. 2007. Isolation and partial characterization of pigment-like antibiotics produced by a new strain of *Streptosporangium* isolated from an Algerian soil. J. Appl. Microbiol. *103*: 228–236.

Brazhnikova, M.G., N.V. Konstantinova and A.S. Mesentsev. 1972. Sibiromycin: isolation and characterization. J. Antibiot. *25*: 668–673.

Bredholdt, H., O.A. Galatenko, K. Engelhardt, E. Fjaervik, L.P. Terekhova and S.B. Zotchev. 2007. Rare actinomycete bacteria from the shallow water sediments of the Trondheim fjord, Norway: isolation, diversity and biological activity. Environ. Microbiol. *9*: 2756–2764.

Bulina, T.I., I.V. Alferova and L.P. Terekova. 1997. A novel approach to isolation of actinomycetes involving irradiation of soil samples with microwaves. Microbiology *66*: 231–234.

Cao, L., Z. Qiu, J. You, H. Tan and S. Zhou. 2005. Isolation and characterization of endophytic streptomycete antagonists of *Fusarium* wilt pathogen from surface-sterilized banana roots. FEMS Microbiol. Lett. *247*: 147–152.

Cao, Y.R., Y. Jiang, L.H. Xu and C.L. Jiang. 2009. *Sphaerisporangium flaviroseum* sp. nov. and *Sphaerisporangium album* sp. nov., isolated from forest soil in China. Int. J. Syst. Evol. Microbiol. *59*: 1679–1684.

Celmer, W.D., W.P. Cullen, C.E. Moppett, J.B. Routein, M.T. Jefferson, R. Shibakawa and J. Tone. 1977. Mixture of antibiotics produced by new species of *Streptosporangium*. US Patent 4032, 632.

Celmer, W.D., W.P. Cullen, C.E. Moppett, J.B. Routien, P.C. Watts, R. Shibakawa and J. Tone. 1978. Polypeptide antibiotic produced by new subspecies of *Streptosporangium*. US Patent 4,083,963.

Chiba, S., M. Suzuki and K. Ando. 1999. Taxonomic re-evaluation of '*Nocardiopsis*' sp. K-252^T (= NRRL 15532^T): a proposal to transfer this strain to the genus *Nonomuraea* as *Nonomuraea longicatena* sp. nov. Int. J. Syst. Bacteriol. *49*: 1623–1630.

Collins, M.D., M. Faulkner and R.M. Keddie. 1984. Menaquinone composition of some sporeforming actinomycetes. Syst. Appl. Microbiol. *5*: 20–29.

Collins, M.D., R.M. Kroppenstedt, J. Tamaoka, K. Komagata and T. Kinoshita. 1988. Structures of the tetrahydrogenated menaquinones from *Actinomadura angiospora*, *Faenia rectivirgula*, and *Saccharothrix australiensis*. Curr. Microbiol. *17*: 275–279.

Coombs, J.T. and C.M. Franco. 2003. Isolation and identification of actinobacteria from surface-sterilized wheat roots. Appl. Environ. Microbiol. *69*: 5603–5608.

Coronelli, C. and J. Thiemann. 1969. Antibiotic selenomycin from *Streptosporangium brasiliense*. US Patent 2,028,986.

Couch, J.N. 1954. The genus *Actinoplanes* and its relatives. Trans.. N.Y. Acad. Sci. *16*: 315–318.

Couch, J.N. 1955a. A new genus and family of the *Actinomycetales* with a revision of the genus *Actinoplanes*. J. Elisha Mitchell Sci. Soc. *71*: 148–155.

Couch, J.N. 1955b. *Actinosporangiaceae* should be *Actinoplanaceae*. J. Elisha Mitchell Sci. Soc. *71*: 269.

Couch, J.N. 1963. Some new genera and species of the *Actinoplanaceae*. J. Elisha Mitchell Sci. Soc. *79*: 53–70.

Couch, J.N. and C.E. Bland. 1974. The *Actinoplanaceae*. *In* Bergey's Manual of Determinative Bacteriology, 8th edn (edited by Buchanan and Gibbons). Williams & Wilkins, Baltimore, pp. 706–723.

Cross, T. 1970. The diversity of bacterial spores. J. Appl. Bacteriol. *33*: 95–102.

Cross, T. and M. Goodfellow. 1973. Taxonomy and classification of the actinomycetes. *In Actinomycetales:* Characteristics and Practical Importance (edited by Sykes and Skinner). Academic Press, London, pp. 11–112.

Cross, T. 1974. Genus V. *Microbispora*. *In* Bergey's Manual of Determinative Bacteriology, 8th edn (edited by Buchanan and Gibbons). Williams & Wilkins, Baltimore, pp. 856–860.

Cross, T. 1989. Growth and examination of actinomycetes – some guidelines. *In* Bergey's Manual of Systematic Bacteriology, vol. 4 (edited by Williams, Sharpe and Holt). Williams & Wilkins, Baltimore, pp. 2340–2343.

de Araujo, J.M., A.C. Silva and J.L. Azevedo. 2000. Isolation of endophytic actinomycetes from roots and leaves of maize (*Zea mays* L.). Braz. Arch. Biol. Technol. *43*: 447–451.

Donadio, S., P. Monciardini, R. Alduina, P. Mazza, C. Chiocchini, L. Cavaletti, M. Sosio and A.M. Puglia. 2002. Microbial technologies for the discovery of novel bioactive metabolites. J. Biotechnol. *99*: 187–198.

Farina, G. and S.G. Bradley. 1970. Re-association of deoxyribonucleic acids from *Actinoplanes* and other actinomycetes. J. Bacteriol. *102*: 30–35.

Fischer, A., R.M. Kroppenstedt and E. Stackebrandt. 1983. Molecular-genetic and chemotaxonomic studies on *Actinomadura* and *Nocardiopsis*. J. Gen. Microbiol. *129*: 3433–3446.

Fleck, W.F., D.G. Strauss, J. Meyer and G. Porstendorfer. 1978. Fermentation, isolation, and biological activity of maduramycin: a new antibiotic from *Actinomadura rubra*. Z. Allg. Mikrobiol. *18*: 389–398.

Furumai, T., H. Ogawa and T. Okuda. 1968. Taxonomic study on *Streptosporangium albidum* nov. sp. J. Antibiot. (Tokyo) *21*: 179–181.

Galatenko, O.A., L.P. Terekhova and T.P. Preobrazhenskaia. 1981. [New species of *Actinomadura* isolated from soils in Turkmenia and their antagonistic properties]. Antibiotiki *26*: 803–807.

Garg, N., O. Prakash and R.K. Pathak. 2003. Biocontrol of rot of guava by *Streptosporangium pseudovulgare* of cow dung origin. Farm. Sci. J. *12*: 162.

Gauze, G.F., T.P. Preobrazhenskaya, E.S. Kudrina, N.O. Blinov, I.D. Ryabova and M.A. Sveshnikova. 1957. Problems in the classification of antagonistic actinomycetes. State Publishing House for Medical Literature (in Russian). Medzig, Moscow.

Gauze, G.F., T.P. Preobrazhenskaia, L.P. Ivanitskaia and M.A. Sveshnikova. 1969. [Production of the antibiotic sibiromycin by the *Streptosporangium sibiricum* sp. nov.]. Antibiotiki *14*: 963–969.

Gauze, G.F., M.A. Sveshnikova, R.S. Ukholina, D.V. Gavrilina, V.A. Filicheva and K.G. Gladkikh. 1973. Production of antitumor antibiotic carminomycin by *Actinomadura carminata* sp. nov. Antibiotiki *18*: 675–678.

Gauze, G.F., T.P. Preobrazhenskaya, M.A. Sveshnikova, L.P. Terekova and T.S. Maximova. 1983. A Guide to Actinomycetes. Nauka, Moscow.

Gauze, G.F., L.P. Terekhova, O.A. Galatenko, T.P. Preobrazhenskaia and V.N. Borisova. 1984. [New species of *Actinomadura recticatena* sp. nov. and its antibiotic properties]. Antibiotiki *29*: 3–7.

Gerber, N. and M.P. Lechevalier. 1964. Phenazines and Phenoxazinones from *Waksmania aerata* sp. nov. and *Pseudomonas iodina*. Biochemistry *3*: 598–602.

Ghazal, S.A. and Z.K. Alb-El-Aziz. 1993. Sporangiosomycin, a new chromopeptide antibiotic produced by *Streptosporangium roseum* subsp. *antibioticus* subsp. nova. Al-Azhar Bull. Sci. *4*: 265–274.

Goodfellow, M. and D.E. Minnikin. 1981. Introduction to the coryneform bacteria. *In* The Prokaryotes: a Handbook on Habitats, Isolation, and Identification of Bacteria, vol. 2 (edited by Starr, Stolp, Trüper, Balows and Schlegel). Springer, New York, pp. 1811–1826.

Goodfellow, M. and T. Pirouz. 1982. Numerical classification of sporoactinomycetes containing *meso*-diaminopimelic acid in the cell wall. J. Gen. Microbiol. *128*: 503–527.

Goodfellow, M. and T. Cross. 1984. Classification. *In* The Biology of the Actinomycetes (edited by Goodfellow, Mordarski and Williams). Academic Press, London, pp. 7–164.

Goodfellow, M., S.T. Williams and G. Alderson. 1986. Transfer of *Kitasatoa purpurea* Matsumae and Hata to the genus *Streptomyces* as *Streptomyces purpureus* comb. nov. Syst. Appl. Microbiol. *8*: 65–66.

Goodfellow, M., E. Stackebrandt and R.M. Kroppenstedt. 1988. Chemotaxonomy and actinomycete systematics. *In* Biology of Actinomycetes '88 (edited by Okami, Beppu and Ogawara). Japan Scientific Societies Press, Tokyo, pp. 233–238.

Goodfellow, M. 1989a. Maduromycetes. *In* Bergey's Manual of Systematic Bacteriology, vol. 4 (edited by Williams, Sharpe and Holt). Williams & Wilkins, Baltimore, pp. 2509–2510.

Goodfellow, M. 1989b. Suprageneric classification of actinomycetes. *In* Bergey's Manual of Systematic Bacteriology, vol. 4 (edited by Williams, Sharpe and Holt). Williams & Wilkins, Baltimore, pp. 2333–2339.

Goodfellow, M., L. J. Stanton, K.E. Simpson and D.E. Minnikin. 1990a. *In* Validation of the publication of new names and new combinations previously effectively published outside the IJSB. List no. 34. Int. J. Syst. Bacteriol. *40*: 320–321.

Goodfellow, M., L.J. Stanton, K.E. Simpson and D.E. Minnikin. 1990b. Numerical and chemical classification of *Actinoplanes* and some related actinomycetes. J. Gen. Microbiol. *136*: 19–36.

Goodfellow, M. 1992. The family *Streptosporangiaceae*. *In* The Prokaryotes: a Handbook on the Biology of Bacteria: Ecophysiology, Isolation, Identification, Applications, 2nd edn, vol. 2 (edited by Balows, Trüper, Dworkin, Harder and Schleifer). Springer, New York, pp. 1115–1138.

Goodfellow, M., L.A. Maldonado and E.T. Quintana. 2005. Reclassification of *Nonomuraea flexuosa* (Meyer 1989) Zhang 1998 as *Thermopolyspora flexuosa* gen. nov., comb. nov., nom. rev. Int. J. Syst. Evol. Microbiol. *55*: 1979–1983; erratum *55*: 2640.

Goodfellow, M. and E. Quintana. 2005. The family *Streptosporangiaceae*. *In* The Prokaryotes: an Evolving Electronic Resource for the Microbiological Community, 3rd edn (edited by Dworkin, Falkow, Schleifer and Stackebrandt). Springer, New York, pp. 725–753.

Goodfellow, M. and E.T. Quintana. 2006. The family *Streptosporangiaceae*. *In* The Prokaryotes: a Handbook on the Biology of Bacteria, 3rd edn, vol. 3, *Archaea*, *Bacteria*, *Firmicutes*, Actinomycetes (edited by Dworkin, Falkow, Rosenberg, Schleifer and Stackebrandt). Springer, New York, pp. 725–753.

Gordon, R.E. and J.M. Mihm. 1962. Identification of *Nocardia caviae* (Erikson) comb. nov.. Ann. N.Y. Acad. Sci. *98*: 628–636.

Goto, J., K. Matsuda, I. Asano, I. Kawamoto, T. Yasuzawa, K. Shirahata, H. Sano and H. Kase. 1987. K-254-I (Genistein), a new inhibitor of Ca^{2+} and calmodulin-dependent cyclic nucleotide phosphodiesterase from *Streptosporangium vulgare*. Agric. Biol. Chem. *51*: 3003–3009.

Greiner-Mai, E., R.M. Kroppenstedt, F. Kornwendisch and H.J. Kutzner. 1987. Morphological and biochemical characterization and emended descriptions of thermophilic actinomycetes species. Syst. Appl. Microbiol. *9*: 97–109.

Gunnarsson, N., U.H. Mortensen, M. Sosio and J. Nielsen. 2003. Identification of the Entner-Doudoroff pathway in an antibiotic-producing actinomycete species. Mol. Microbiol. *3*: 895–902.

Gupta, K.C. 1965. A new species of the genus *Streptosporangium* isolated from Indian soil. J. Antibiot. *18*: 125–127.

Gyobu, Y. and S. Miyadoh. 2001. Proposal to transfer *Actinomadura carminata* to a new subspecies of the genus *Nonomuraea* as *Nonomuraea roseoviolacea* subsp., carminata comb. nov. Int. J. Syst. Evol. Microbiol. *51*: 881–889.

Hacene, H., F. Boudjellal and G. Lefebvre. 1998. AH7, a non-polyenic antifungal antibiotic produced by a new strain of *Streptosporangium roseum*. Microbios *96*: 103–109.

Hasegawa, T., M.P. Lechevalier and H.A. Lechevalier. 1979. Phospholipid composition of motile actinomycetes. J. Gen. Microbiol. *25*: 209–213.

Hayakawa, M. and H. Nonomura. 1984. HV agar, a new selective medium for isolation of soil actinomycetes. *In* Abstracts of Papers Presented at the Annual Meeting of the Actinomycetologists, Osaka, Japan, p. 6.

Hayakawa, M. and H. Nonomura. 1987a. Efficacy of artificial humic acid as a selective nutrient in HV agar used for the isolation of soil actinomycetes. J. Ferment. Technol. *65*: 609–616.

Hayakawa, M. and H. Nonomura. 1987b. Humic acid-vitamine agar, a new medium for the selective isolation of soil actinomycetes. J. Ferment. Technol. *65*: 501–509.

Hayakawa, M., K. Ishizawa and H. Nonomura. 1988. Distribution of rare actinomycetes in Japanese soils. J. Ferment. Technol. *66*: 367–373.

Hayakawa, M. and H. Nonomura. 1989. A new method for the intensive isolation of actinomycetes from soil. Actinomycetologica *3*: 95–104.

Hayakawa, M., T. Kaihura and H. Nonomura. 1991a. New methods for the highly selective isolation of *Streptosporangium* and *Dactylosporangium* from soil. J. Ferment. Technol. Bioeng. *72*: 327–333.

Hayakawa, M., T. Sadakata, T. Kajiura and H. Nonomura. 1991b. New methods for the highly selective isolation of *Micromonospora* and *Microbispora* from soil. J. Ferment. Bioeng. *72*: 320–326.

Hayakawa, M., Y. Momose, T. Yamazaki and H. Nonomura. 1996. A method for the selective isolation of *Microtetraspora glauca* and related four spore actinomycetes from soil. J. Appl. Bact. *80*: 375–386.

Hayakawa, M., H. Iino, S. Takeuchi and T. Yamazaki. 1997. Application of a method incorporating treatment with chloramine-T for the selective isolation of *Streptosporangiaceae* from soil. J. Ferment. Bioeng. *94*: 599–602.

Hayakawa, M., M. Otoguro, T. Takeuchi, T. Yamazaki and Y. Iimura. 2000. Application of a method incorporating differential centrifugation for selective isolation of motile actinomycetes in soil and plant litter. Antonie van Leeuwenhoek *78*: 171–185.

Henssen, A. 1957. Beiträge zur Morphologie und Systematik der thermophilen Actinomyceten. Arch. Mikrobiol. *26*: 373–414.

Henssen, A. and D. Schäfer. 1971. Emended description of the genus *Pseudonocardia* Henssen and description of the new species *Pseudonocardia spinosa*. Int. J. Syst. Bacteriol. *21*: 29–34.

Hickey, R.J. and H.D. Tresner. 1952. A cobalt-containing medium for sporulation of *Streptomyces* species. J. Bacteriol. *64*: 891–892.

Hill, L.R., V.B.D. Skerman and P.H.A. Sneath. 1984. Corrigenda to the Approved Lists of Bacterial Names. Int. J. Syst. Bacteriol. *34*: 508–511.

Hoang, K.C., M. Tseng and W.J. Shu. 2007. Degradation of polyethylene succinate (PES) by a new thermophilic *Microbispora* strain. Biodegradation *18*: 333–342.

Igarashi, Y., K. Takagi, T. Kajiura and T. Furumai. 1998. Glucosylquestiomycin, a novel antibiotic from *Microbispora* sp. TP-A0184: Fermentation, isolation, structure determination, synthesis and biological activities. J. Antibiot. *51*: 915–920.

Iinuma, S., A. Yokota and T. Kanamura. 1996a. *In* Validation of the publication of new names and new combinations previously effectively published outside the IJSB. List no. 59. Int. J. Syst. Bacteriol. *46*: 1189–1190.

Iinuma, S., A. Yokota and T. Kanamura. 1996b. New subspecies of the genus *Streptosporangium*, *Streptosporangium amethystogenes* subsp. *fukuiense* subsp. nov. Actinomycetologica *10*: 35–42.

Ikeda, T., T. Kurita-Ochial, T. Takizawa and M. Hirasawa. 1993. Isolation and characterisation of the substance isolated from *Streptosporangium* species which inhibits lactic acid production by oral bacteria. Gen. Pharmacol. *24*: 905–910.

Janso, J.E. and G.T. Carter. 2010. Biosynthetic potential of phylogenetically unique endophytic actinomycetes from tropical plants. Appl. Environ. Microbiol. *76*: 4377–4386.

Johnston, D.W. and T. Cross. 1976. The occurrence and distribution of actinomycetes in lakes of the English Lake District. Freshwater Biol. *6*: 457–463.

Jones, K.L. 1949. Fresh isolates of actinomycetes in which the presence of sporogenous aerial mycelia is a fluctuating characteristic. J. Bacteriol. *57*: 141–145.

Kajiura, T., T. Furumai, Y. Igarashi, H. Hori, K. Higashi, K. Ishiyama, M. Uramoto, Y. Uehara and T. Oki. 1998. Signal transduction inhibitors, Hibarimicins A, B, C, D and G produced by *Microbispora*. I. Taxonomy, isolation and physicochemical and biological properties. J. Antibiot. *51*: 394–401.

Kämpfer, P., H.-J. Busse, B.J. Tindall, M. Nimtz and I. Grün-Wollny. 2010. *Nonomuraea rosea* sp. nov. Int. J. Syst. Evol. Microbiol. *60*: 1118–1124.

Kämpfer, P., R.M. Kroppenstedt and I. Grün-Wollny. 2005. *Nonomuraea kuesteri* sp. nov. Int. J. Syst. Evol. Microbiol. *55*: 847–851.

Kawamoto, I., S. Takasawa, R. Okachi, M. Koakura and I. Takahashi. 1975. A new antibiotic victomycin (XK 49-1-B-2). I. Taxonomy and production of the producing organism. J. Antibiot. *28*: 358–365.

Kawamoto, I., T. Oka and T. Nara. 1981. Cell-wall composition of *Micromonospora olivoasterospora*, *Micromonospora sagamiensis*, and related organisms. J. Bacteriol. *146*: 527–534.

Kemmerling, C., H. Gürtler, R.M. Kroppenstedt, R. Toalster and E. Stackebrandt. 1993. Evidence for the phylogenetic heterogeneity of the genus *Streptosporangium*. Syst. Appl. Microbiol. *16*: 369–372.

Kim, H.J. 1993. Selective isolation, characterisation and identification of streptosporangia. PhD thesis, University of Newcastle, Newcastle upon Tyne.

Kimura, K., F. Kanou and M. Yoshihama. 1997. A propyl endopeptidase inhibitor, Propeptin production in the various *Microbispora* sp. Actinomycetologica *11*: 64–68.

Kimura, M. 1980. A simple method for estimating evolutionary rates of base substitutions through comparative studies of nucleotide sequences. J. Mol. Evol. *16*: 111–120.

Kizuka, M., R. Enokita, K. Takahashi and T. Okazaki. 1997. Distribution of actinomycetes in the Republic of South Africa investigated using a newly developed isolation method. Actinomycetologica *11*: 54–58.

Kizuka, M., R. Enokita, K. Takahashi, Y. Okamoto, T. Otsuka, Y. Shigematsu, Y. Inoue and T. Okazaki. 1998. Studies on actinomycetes isolated from plant leaves. Actinomycetologica *12*: 89–91.

Koch, C., R.M. Kroppenstedt, F.A. Rainey and E. Stackebrandt. 1996. 16S ribosomal DNA analysis of the genera *Micromonospora*, *Actinoplanes*, *Catellatospora*, *Catenuloplanes*, *Couchioplanes*, *Dactylosporangium*, and *Pilimelia* and emendation of the family *Micromonosporaceae*. Int. J. Syst. Bacteriol. *46*: 765–768.

Krasil'nikov, N.A., N.S. Agre and G.I. el-Registan. 1968. [New thermophilic species of *Micropolyspora* genus]. Mikrobiologiia *37*: 1065–1072.

Krasil'nikov, N.A. 1938. Ray Fungi and Related Organisms – *Actinomycetales*. Akad. Nauk. S.S.S.R. Moscow.

Krasil'nikov, N.A. and N.S. Agre. 1964. On two new species of *Thermopolyspora*. Hindustan Antibiot. Bull. *6*: 97–107.

Kroppenstedt, R.M. and H.J. Kutzner. 1976. Biochemical markers in the taxonomy of the *Actinomycetales*. Experientia *32*: 318–319.

Kroppenstedt, R.M. and H.J. Kutzner. 1978. Biochemical taxonomy of some problem actinomycetes. Zentralbl. Bakteriol. Parasitenkd. Infektionskr. Hyg. Abt. I Suppl. *6*: 125–133.

Kroppenstedt, R.M. 1985. Fatty acid and menaquinone analysis of actinomycetes and related organisms. *In* Chemical Methods in Bacterial Systematics (edited by Goodfellow and Minnikin). Academic Press, London, pp. 173–199.

Kroppenstedt, R.M., E. Stackebrandt and M. Goodfellow. 1990. Taxonomic revision of the actinomycete genera *Actinomadura* and *Microtetraspora* Syst. Appl. Microbiol. *13*: 148–160.

Kroppenstedt, R.M., E. Stackebrandt and M. Goodfellow. 1991. *In* Validation of the publication of new names and new combinations previously effectively published outside the IJSB. List no. 36. Int. J. Syst. Bacteriol. *41*: 178–179.

Kroppenstedt, R.M. and M. Goodfellow. 1992. The family *Thermomonosporaceae*. *In* The Prokaryotes: a Handbook on the Biology of Bacteria: Ecophysiology, Isolation, Identification, Applications, 2nd edn (edited by Balows, Trüper, Dworkin, Harder and Schleifer). Springer, New York, pp. 1085–1114.

Kudo, H., K.J. Cheng and J.W. Costerton. 1987. Interactions between *Treponema bryantii* and cellulolytic bacteria in the *in vitro* degradation of straw cellulose. Can. J. Microbiol. *33*: 244–248.

Kudo, T. and A. Seino. 1987. Transfer of *Streptosporangium indianense* Gupta 1965 to the genus *Streptomyces* as *Streptomyces indiaensis* (Gupta 1965) comb. nov. Int. J. Syst. Bacteriol. *37*: 241–244.

Kudo, T., T. Itoh, S. Miyadoh, T. Shomura and A. Seino. 1993. *Herbidospora* gen. nov., a new genus of the family *Streptosporangiaceae* Goodfellow et al. 1990. Int. J. Syst. Bacteriol. *43*: 319–328.

Kudo, T. 2001. Family *Streptosporangiaceae*. *In* Identification Manual of Actinomycetes (edited by Miyadoh). Business Center for Academic Societies Japan, Tokyo, pp. 281–291.

Kurtböke, I. 2003. Use of bacteriophages for the selective isolation of rare actinomycetes. *In* Selective Isolation of Rare Actinomycetes (edited by Kurtböke). University of Sunshine Coast, Queensland, pp. 9–54.

Lacey, J., M. Goodfellow and G. Alderson. 1978. The genus *Actinomadura*. *In Nocardia* and *Streptomyces* (edited by Mordarski, Kurylowicz and Jeljaszewicz). Springer, Stuttgart, pp. 107–117.

Lavrova, N.V., T.P. Preobrazhenskaia and M.A. Sveshnikova. 1972. [Isolation of soil actinomycetes on selective media with rubomycin]. Antibiotiki *17*: 965–970.

Lavrova, N.V. and T.P. Preobrazhenskaia. 1975. [Isolation of new species of the genus *Actinomadura* on selective media with rubomycin]. Antibiotiki *20*: 438–438.

Lazzarini, A., L. Cavaletii, G. Toppo and F. Marinelli. 2000. Rare genera of actonomycetes as potential producers of new antibiotics. Antonie van Leeuwenhoek *78*: 399–405.

Le Roes, M. and P.R. Meyers. 2008. *Nonomuraea candida* sp. nov., a new species from South African soil. Antonie van Leeuwenhoek *93*: 133–139.

Le Roes, M. and P.R. Meyers. 2009. *In* List of new names and new combinations previously effectively, but not validly, published. Validation List no. 125. Int. J. Syst. Evol. Microbiol. *59*: 1–2.

Lechevalier, H.A., M.P. Lechevalier and B. Becker. 1966a. Comparison of the chemical composition of cell walls of nocardiae with that of other aerobic actinomycetes. Int. J. Syst. Bacteriol. *16*: 151–160.

Lechevalier, H.A., M.P. Lechevalier and P.E. Holbert. 1966b. Electron microscopic observation of the sporangial structure of strains of *Actinoplanaceae*. J. Bacteriol. *92*: 1228–1235.

Lechevalier, H.A. and M.P. Lechevalier. 1970a. A critical evaluation of the genera of aerobic actinomycetes. *In* The *Actinomycetales* (edited by Prauser). Gustav Fischer Verlag, Jena, pp. 395–405.

Lechevalier, M.P. and H.A. Lechevalier. 1970b. Chemical composition as a criterion in the classification of aerobic actinomycetes. Int. J. Syst. Bacteriol. *20*: 435–443.

Lechevalier, M.P. and N.N. Gerber. 1970. The identity of madurose with 3-O-methyl-D-galactose. Carbohydr. Res. *13*: 451–453.

Lechevalier, M.P., C. de Biévre and H.A. Lechevalier. 1977. Chemotaxonomy of aerobic actinomycetes: Phospholipid composition. Biochemistry and Ecological Systems *5*: 249–260.

Lechevalier, M.P., A.E. Stern and H.A. Lechevalier. 1981. Phospholipids in the taxonomy of actinomycetes. Zentralbl. Bakteriol. Parasitenkd. Infektionskr. Hyg. I Abt. Orig. *Suppl. 11*: 111–116.

Lee, J.Y. and B.K. Hwang. 2002. Diversity of antifungal actinomycetes in various vegetative soils of Korea. Can. J. Microbiol. *48*: 407–417.

Li, G.P. 1989. Isolation of actinomycetes for antibiotic screening. Chin. J. Antibiot. *14*: 452–465.

Li, J., G.-Z. Zhao, S. Qin, W.-Y. Zhu, L.-H. Xu and W.-J. Li. 2009. *Herbidospora osyris* sp. nov., isolated from surface-sterilized tissue of *Osyris wightiana* Wall. *ex* Wight. Int. J. Syst. Evol. Microbiol. *59*: 3123–3127.

Li, Y.V., L.P. Terekhova and M.G. Gapochka. 2002. Isolation of actinomycetes from soil using extremely high frequency radiation. Microbiology *71*: 105–108.

List Editor IJSEM. 2007. *In* List of new names and new combinations previously effectively, but not validly, published. Validation List no. 118. Int. J. Syst. Evol. Microbiol. *57*: 2449–2450.

Locci, R. and B. Petrolini-Baldan. 1971. On the spore formation process in actinomycetes. V. Scanning electron microscopy of some genera of *Actinoplanaceae*. Riv. Pat. Veg. Ser. IV. 81–96.

Locci, R. 1976. Developmental morphology of actinomycetes. *In* Actinomycetes: The Boundary Microorganisms (edited by Arai). Toppan Co. Ltd, Tokyo, pp. 249–297.

Manachini, P.L., A. Ferrari and R. Craveri. 1965. Forme thermofile de *Actinoplanaceae*. Isolamento et caracteristiche di *Streptosporangium album* var. thermophilum.. Annali di Microbiologia et Enzimologia *15*: 129–144.

Masayuki, H., K. Takayuki and H. Nonomura. 1991. New methods for the highly selective isolation of *Streptosporangium* and *Dactylosporangium* from soil. J. Ferm. Bioeng. *72*: 327–333.

Matsumoto, A., Y. Takahashi, M. Mochizuki, A. Seino, Y. Iwai and S. Ōmura. 1998. Characterization of actinomycetes isolated from fallen leaves. Actinomycetologica *12*: 46–48.

Mba, C.C. 1997. Rock phosphate solubilizing *Streptosporangium* isolates from casts of tropical earthworms. Soil Biol. Biochem. *29*: 381–385.

McCarthy, A.J. and T. Cross. 1984. A taxonomic study of *Thermomonospora* and other monosporic actinomycetes. J. Gen. Microbiol. *130*: 5–25.

Mertz, F.P. and R.C. Yao. 1990. *Streptosporangium carneum* sp. nov. isolated from soil. Int. J. Syst. Bacteriol. *40*: 247–253.

Mertz, F.P. 1994. *Planomonospora alba* sp. nov. and *Planomonospora sphaerica* sp. nov., two new species isolated from soil by baiting techniques. Int. J. Syst. Bacteriol. *44*: 274–281.

Meyer, J. and M. Sveshnikova. 1974. [*Micromonospora rubra* Sveshnikova et al. equals *Actinomadura rubra* comb. nov.]. Z. Allg. Mikrobiol. *14*: 167–170.

Meyer, J. 1979. New species of the genus *Actinomadura*. Z. Allg. Mikrobiol. *19*: 37–44.

Meyer, J. 1981. *In* Validation of the publication of new names and new combinations previously effectively published outside the IJSB. List no. 6. Int. J. Syst. Bacteriol. *31*: 215–218.

Meyer, J. 1989. Genus *Actinomadura*. *In* Bergey's Manual of Systematic Bacteriology, 1st edn, vol. 4 (edited by Williams, Sharpe and Holt). Williams & Wilkins, Baltimore, pp. 2511–2526.

Michel, K.H. and R.C. Yao. 1991. New lipoglycopeptide antibiotic A84575 complex-produced by submerged aerobic fermentation of *Streptosporangium carneum* and used to treat and prevent Gram-positive bacterial infections. European Patent 0424051.

Mishra, S.K., J.E. Keller, J.R. Miller, R.M. Helsey, M.G. Nair and A.R. Putnam. 1987. Insecticidal and nematicidal properties of microbial metabolites. J. Ind. Microbiol. *2*: 267–276.

Miyadoh, S., H. Tohyama, S. Amano, T. Shomura and T. Niida. 1985. *Microbispora viridis*, a new species of *Actinomycetales*. Int. J. Syst. Bacteriol. *35*: 281–284.

Miyadoh, S., H. Anzai, S. Amano and T. Shomura. 1989. *Actinomadura malachitica* and *Microtetraspora viridis* are synonyms and should be transferred as *Actinomadura viridis* comb. nov. Int. J. Syst. Bacteriol. *39*: 152–158.

Miyadoh, S., S. Amano, H. Tohyama and T. Shomura. 1990. A taxonomic review of the genus *Microbispora* and a proposal to transfer two species to the genus *Actinomadura* and to combine ten species into *Microbispora rosea*. J. Gen. Microbiol. *136*: 1905–1913.

Miyadoh, S., M. Hamada, K. Hotta, T. Kudo, A. Seino, G. Vobis and A. Yokota. 1997. Atlas of Actinomycetes. Asakura Publishing Co., Ltd., Tokyo.

Miyadoh, S., M. Hamada, K. Hotta, T. Kudo, A. Seino, K. Suzuki and A. Yokota. 2001. Identification Manual of Actinomycetes. Business Center for Academic Societies, Japan.

Monciardini, P., M. Sosio, L. Cavaletti, C. Chiocchini and S. Donadio. 2002. New PCR primers for the selective amplification of 16S rDNA from different groups of actinomycetes. FEMS Microbiol. Ecol. *42*: 419–429.

Mullins, P.H., H. Gürtler and E.M.H. Wellington. 1995. Selective recovery of *Streptosporangium fragile* from soil by indirect immunomagnetic capture. Microbiology *141*: 2149–2156.

Naganawa, H., H. Hashizume, Y. Kubota, R. Sawa, Y. Takahashi, K. Arakawa, S.G. Bowers and T. Mahmud. 2002. Biosynthesis of the aminocyclitol moeity of pyralomicin 1a in *Nonomuraea spiralis* MI178-34F18. J. Antibiot. *55*: 578–584.

Nakagaito, Y., Y. Nishii, A. Yokota and T. Hasegawa. 1993. Distribution of madurose, an actinomycete whole-cell sugar, in the genus *Streptomyces*. IFO Res. Commun. *16*: 102–108.

Nakagawa, M., Y. Hayakawa, H. Kawai, K. Imamura, H. Inoue, A. Shimazu, H. Seto and N. Otake. 1983. A new anthracycline antibiotic *N*-formyl-13-dihydrocarminomycin. J. Antibiot. *36*: 457–458.

Nakagawa, M., Y. Hayakawa, K. Imamura, H. Seto and N. Otake. 1989. Microbial conversion of anthracyclinones to carminomycins by a blocked mutant of *Actinomadura roseoviolacea*. J. Antibiot. *42*: 1698–1703.

Nakajima, Y., V. Kitpreechavanich, K. Suzuki and T. Kudo. 1999. *Microbispora corallina* sp. nov., a new species of the genus *Microbispora* isolated from Thai soil. Int. J. Syst. Bacteriol. *49*: 1761–1767.

Nakajima, Y., C.C. Ho and T. Kudo. 2003. *Microtetraspora malaysiensis* sp. nov., isolated from Malaysian primary dipterocarp forest soil. J. Gen. Appl. Microbiol. *49*: 181–189.

Nakajima, Y., C.C. Ho and T. Kudo. 2004. *In* Validation of publication of new names and new combinations previously effectively published outside the IJSEM. List no. 95. Int. J. Syst. Evol. Microbiol. *54*: 1–2.

Nash, C.H., III, M.C. Shearer, K.M. Snader, J.R. Valenta and D. Cooper. 1981. Anthracycline antibiotics produced by *Streptosporangium fragilis* sp. nov. ATCC 31519. US patent 4,293,546.

Naumova, I.B., N.V. Potekhina, L.P. Terekhova, T.P. Preobrazhenskaya and K. Digimbay. 1986. Wall polyol phosphate polymers of bacteria belonging to the genus *Actinomadura*. *In* Biological, Biochemical and Biomedical Aspects of Actinomycetes (edited by Szabó, Biró and Goodfellow). Akadémai Kiadó, Budapest, pp. 561–566.

Nonomura, H. and Y. Ohara. 1957. Distribution of actinomycetes in the soil. II. *Microbispora*, a new genus of the *Streptomycetaceae*. J. Ferment. Technol. *35*: 307–311.

Nonomura, H. and Y. Ohara. 1960. Distribution of the actinomycetes in soil. IV. The isolation and classification of the genus *Microbispora*. J. Ferment. Technol. *38*: 401–405.

Nonomura, H. and Y. Ohara. 1960. Distribution of actinomycetes in soil. V. The isolation and classification of the genus *Streptosporangium*. J. Ferment. Technol. *38*: 405–409.

Nonomura, H. and Y. Ohara. 1969a. Distribution of actionomycetes in soil. VI. A culture method effective for both preferential isolation and enumeration of *Microbispora* and *Streptosproangium* strains in soil. Part I. J. Ferment. Technol. *47*: 463–469.

Nonomura, H. and Y. Ohara. 1969b. Distribution of actinomycetes in soil. VII. A culture method effective for both preferential isolation and enumeration of *Microbispora* and *Streptosporangium* strains in soil. Part 2. Classification of the isolates. J. Ferm. Technol. *47*: 701–709.

Nonomura, H. and Y. Ohara. 1971a. Distribution of actinomycetes in soil. XI. Some new species of the genus *Actinomadura* Lechevalier et al. J. Ferment. Technol. *49*: 904–912.

Nonomura, H. and Y. Ohara. 1971b. Distribution of actinomycetes in soil. X. New genus and species of monosporic actinomycetes in soil. J. Ferment. Technol. *49*: 895–903.

Nonomura, H. and Y. Ohara. 1971c. Distribution of actinomycetes in soil. VIII. Green-spore group of *Microtetraspora*, its preferential isolation and taxonomic characteristics. J. Ferment. Technol. *49*: 1–7.

Nonomura, H. and Y. Ohara. 1971d. Distribution of actinomycetes in soil. IX. New species of the genus *Microbispora* and *Microtetraspora* and their isolation methods. J. Ferment. Technol. *49*: 887–894.

Nonomura, H. 1984. Design of a new medium for the isolation of soil actinomycetes. The Actinomycetes *18*: 206–209.

Nonomura, H. and M. Hayakawa. 1988. New methods for the selective isolation of soil actinomycetes. *In* Biology of Actinomycetes (edited by Okami, Beppu and Ogawara). Japan Scientific Societies Press, Tokyo, pp. 288–293.

Nonomura, H. 1989a. Genus *Streptosporangium*. *In* Bergey's Manual of Systematic Bacteriology, vol. 4 (edited by Williams, Sharpe and Holt). Williams & Wilkins, Baltimore, pp. 2545–2551.

Nonomura, H. 1989b. Genus *Microbispora*. *In* Bergey's Manual of Systematic Bacteriology, vol. 4 (edited by Williams, Sharpe and Holt). Williams & Wilkins, Baltimore, pp. 2526–2531.

Nonomura, H. 1989c. Genus *Microtetraspora*. *In* Bergey's Manual of Systematic Bacteriology, vol. 4 (edited by Williams, Sharpe and Holt). Williams & Wilkins, Baltimore, pp. 2531–2536.

Norovsuren, Z., R. Baatar and G.M. Zenova. 2007. Actinomycetes of rare genera in plain chestnut soils of Mongolia. Pochvovedenie *3*: 50–52.

Ochi, K., S. Miyadoh and T. Tamura. 1991. Polyacrylamide gel electrophoresis analysis of ribosomal protein AT-L30 as a novel approach to actinomycete taxonomy: application to the genera *Actinomadura* and *Microtetraspora*. Int. J. Syst. Bacteriol. *41*: 234–239.

Ochi, K. and S. Miyadoh. 1992. Polyacrylamide gel electrophoresis analysis of ribosomal protein AT-L30 from an actinomycete genus, *Streptosporangium*. Int. J. Syst. Bacteriol. *42*: 151–155.

Ochi, K., E. Satoh and J. Shima. 1993. Amino acid sequence analysis of ribosomal protein AT-L30 from *Streptosporangium corrugatum* and *Kibdelosporangium aridum*. Syst. Appl. Microbiol. *16*: 13–16.

Okazaki, T. and A. Naito. 1985. Studies in actinomycetes isolated from Australian soils. *In* Biological, Biochemical and Biomedical Aspects of Actinomycetes (edited by Szabó, Biró and Goodfellow). Akadémiai Kiadó, Budapest, pp. 739–741.

Okazaki, T., K. Takahashi, M. Kizuka and R. Enokita. 1995. Studies on actinomycetes isolated from plant leaves. Annual Report of Snakyo Research Laboratories *47*: 97–106.

Okuda, T., Y. Ito, T. Yamaguchi, T. Furumai, M. Suzuki and M. Tsuruoka. 1966. Sporaviridin, a new antibiotic produced by *Streptosporangium viridogriseum* nov. sp. J. Antibiot. *19*: 85–87.

Okujo, N., H. Ilnuma, A. George, K.S. Eim, T.L. Li, N.S. Ting, T.C. Jye, J. Hotta, M. Hatsu, Y. Fukagawa, S. Shibara, K. Numata and S. Kondo. 2007. Bispolides, novel 20-membered ring macrodiolode antibiotics from *Microbispora*. J. Antibiot. *60*: 216–219.

Patel, M., V. Hegde, A.C. Horan, V.P. Gullo, D. Loebenberg, J.A. Marquez, G.H. Miller, M.S. Puar and J.A. Waitz. 1984. A novel phenazine antifungal antibiotic, 1,6-dihydroxy-2-chlorophenazine. Fermentation, isolation, structure and biological properties. J. Antibiot. (Tokyo) *37*: 943–948.

Petersen, J.E. 1959. New species of myxobacter from the bark of living trees. Mycologia *51*: 163–172.

Petrolini, B., S. Quaroni, P. Sardi, M. Saracchi and N. Anterrollo. 1992. A sporangiate actinomycete with unusual morphological features: *Streptosporangium claviforme* sp. nov. Actinomycetes *3*: 45–50.

Pfefferle, C., J. Breinholt, H. Gurtler and H.P. Fiedler. 1997. 1-Hydroxy-4-methoxy-2-naphthoic acid, a herbicidal compound produced by *Streptosporangium cinnabarinum* ATCC 31,213. J. Antibiot. *50*: 1067–1068.

Pfefferle, C., U. Theobald, H. Gurtler and H.P. Fiedler. 2000. Improved secondary metabolite production in the genus *Streptosporangium* by optimization of the fermentation conditions. J. Biotechnol. *80*: 135–142.

Platas, G., F. Pelaez, J. Collado, H. Martinez and M.T. Diez. 1999. Nutritional preferences of a group of *Streptosporangium* soil isolates. J. Biosci. Bioeng. *88*: 269–275.

Poschner, J., R.M. Kroppenstedt, A. Fischer and E. Stackebrandt. 1985. DNA-DNA reassociation and chemotaxonomic studies on *Actinomadura*, *Microbispora*, *Microtetraspora*, *Micropolyspora* and *Nocardiopsis*. Syst. Appl. Microbiol. *6*: 264–270.

Potekhina, L.L. 1965. [*Streptosporangium rubrum* n. sp.–a new species of the genus *Streptosporangium*]. Mikrobiologiia *34*: 245–250.

Preobrazhenskaya, T.P. and M.A. Sveshnikova. 1974. New species of the *Actinomadura* genus. Mikrobiologiya *43*: 864–868.

Preobrazhenskaya, T.P., N.V. Lavrova and N.O. Blinov. 1975a. Taxonomy of *Streptomyces luteofluorescens*. Mikrobiologiya *44*: 524–527.

Preobrazhenskaya, T.P., N.V. Lavrova, R.S. Ukholina and N.P. Nechaeva. 1975b. Isolation of new species of *Actinomadura* on selective media with streptomycin and bruneomycin. Antibiotiki *20*: 404–409.

Preobrazhenskaya, T.P. and M.A. Sveshnikova. 1985. *In* Validation of the publication of new names and new combinations previously effectively published outside the IJSB. List no. 17. Int. J. Syst. Bacteriol. *35*: 223–225.

Pridham, T.G. and A.J. Lyons. 1980. Methodologies for *Actinomycetales* with special reference to streptomycetes and streptoverticillia. *In* Actinomycete Taxonomy, Special Publication no. 6. Society for Industrial Microbiology, Arlington, pp. 153–224.

Qin, S., G.Z. Zhao, H.P. Klenk, J. Li, W.Y. Zhu, L.H. Xu and W.J. Li. 2009. *Nonomuraea antimicrobica* sp. nov., an endophytic actinomycete isolated from a leaf of *Maytenus austroyunnanensis*. Int. J. Syst. Evol. Microbiol. *59*: 2747–2751.

Quintana, E., L. Maldonado and M. Goodfellow. 2003. *Nonomuraea terrinata* sp. nov., a novel soil actinomycete. Antonie van Leeuwenhoek *84*: 1–6.

Rao, V.A., K.K. Prabhu, B.P. Sridhar, A. Venkateswarlu and P. Actor. 1987. Two new species of *Microbispora* from Indian soils - *Microbispora karnatakensis* sp. nov. and *Microbispora indica* sp. nov. Int. J. Syst. Bacteriol. *37*: 181–185.

Runmao, H., W. Guizhen and L. Junying. 1993. A new genus of actinomycetes, *Planotetraspora* gen. nov. Int. J. Syst. Bacteriol. *43*: 468–470.

Saitou, N. and M. Nei. 1987. The neighbor-joining method: a new method for reconstructing phylogenetic trees. Mol. Biol. Evol. *4*: 406–425.

Sardi, P., M. Saracchi, S. Quaroni, B. Petrolini, G.E. Borgonovi and S. Merli. 1992. Isolation of endophytic *Streptomyces* strains from surface-sterilized roots. Appl. Environ. Microbiol. *58*: 2691–2693.

Schäfer, D. 1969. Eine neue *Streptosporangium* Art aus türkischer Steppenerde. Arch. Mikrobiol. *66*: 365–373.

Schäfer, D. 1973. Beiträge zur Klassifizierung und Taxonomie der Actinoplanaceen. Dissertation. Marburg, Germany.

Schleifer, K.H. and O. Kandler. 1972. Peptidoglycan types of bacterial cell walls and their taxonomic implications. Bacteriol. Rev. *36*: 407–477.

Selva, E., P. Ferrari, M. Kurz, P. Tavecchia, L. Colombo, S. Stella, E. Restelli, B.P. Goldstein, F. Ripamonti and M. Denaro. 1995. Components of the GE2270 complex produced by *Planobispora rosea* ATCC 53773. J. Antibiot. *48*: 1039–1042.

Seong, C.N., J.H. Choi and K.S. Baik. 2001. An improved selective isolation of rare actinomycetes from forest soil. J. Microbiol. *39*: 17–23.

Sharples, G.P., S.T. Williams and R.M. Bradshaw. 1974. Spore formation in the *Actinoplanaceae* (*Actinomycetales*). Arch. Microbiol. *101*: 9–20.

Shearer, M.C., P.M. Colman and C.H. Nash. 1983. *Streptosporangium fragile* sp. nov. Int. J. Syst. Bacteriol. *33*: 364–368.

Shirling, E.B. and D. Gottlieb. 1966. Methods for characterization of *Streptomyces* species. Int. J. Syst. Bacteriol. *16*: 313–340.

Shokichi, O., K. Kunihiko, S. Akiko, T. Tamako, V. Junko, A. Shouichi, M. Shinij, M. Yuji, S. Takashi and S. Masaji. 1988. New angiotensis converting enzyme inhibitors SF2513 A, B. and C, produced by *Streptosporangium nondiaslaticum*. Meiji Seika Kenkyu Nenpo *27*: 46–54.

Soina, V.S., A.A. Sokolov and N.S. Agre. 1975. Ultrastructure of mycelium and spores of *Actinomadura fastidiosa* sp. nov. Microbiology (En. transl. from Mikrobiologiya) *44*: 883–887.

Solans, M. and G. Vobis. 2003. Saprophytic actinomycetes associated to the rhizosphere and rhizoplane of *Discaria trinervis*. Ecologia Aust. *13*: 97–107.

Sosio, M., S. Stinchi, F. Beltrametti, A. Lazzarini and S. Donadio. 2003. The gene cluster for the biosynthesis of the glycopeptide antibiotic A40926 by *Nonomuraea* species. Chem. Biol. *10*: 541–549.

Stackebrandt, E. 1986. The significance of "wall types" in phylogenetically based taxonomic studies on actinomycetes. *In* Biological, Biochemical and Biomedical Aspects of Actinomycetes (edited by Szabó, Biró and Goodfellow). Akadémiai Kiadó, Budapest, pp. 497–506.

Stackebrandt, E. and R.M. Kroppenstedt. 1987. Union of the genera *Actinoplanes* Couch, *Ampullariella* Couch, and *Amorphosporangium* Couch in a redefined genus *Actinoplanes*. Syst. Appl. Microbiol. *9*: 110–114.

Stackebrandt, E., R.M. Kroppenstedt, K.D. Jahnke, C. Kemmerling and H. Gürtler. 1994. Transfer of *Streptosporangium viridogriseum* (Okuda et al. 1966), *Streptosporangium viridogriseum* subsp. *kofuense* (Nonomura and Ohara 1969), and *Streptosporangium albidum* (Furumai et al. 1968) to *Kutzneria* gen. nov. as *Kutzneria viridogrisea* comb. nov., *Kutzneria kofuensis* comb. nov., and *Kutzneria albida* comb. nov., respectively, and emendation of the genus *Streptosporangium*. Int. J. Syst. Bacteriol. *44*: 265–269.

Stackebrandt, E., F.A. Rainey and N.L. Ward-Rainey. 1997. Proposal for a new hierarchic classification system, *Actinobacteria* classis nov. Int. J. Syst. Bacteriol. *47*: 479–491.

Stackebrandt, E., J. Wink, U. Steiner and R.M. Kroppenstedt. 2001. *Nonomuraea dietzii* sp. nov. Int. J. Syst. Evol. Microbiol. *51*: 1437–1441.

Stamford, T.L., N.P. Stamford, L.C. Coelho and J.M. Araujo. 2002. Production and characterization of a thermostable glucoamylase from *Streptosporangium* sp. endophyte of maize leaves. Bioresour. Technol. *83*: 105–109.

Suzuki, S. 2001. Establishment and use of gellan gum media for selective isolation and distribution survey of specific rare actinomycetes. Actinomycetologica *15*: 55–60.

Suzuki, S., T. Okuda and S. Komatsubara. 2001a. Selective isolation and study on the global distribution of the genus *Planobispora* in soils. Can. J. Microbiol. *47*: 979–986.

Suzuki, S., T. Okuda and S. Komatsubara. 2001b. Selective isolation and distribution of the genus *Planomonospora* in soils. Can. J. Microbiol. *47*: 253–263.

Sveshnikova, M., T. Maximova and E. Kudrina. 1969. The species belonging to the genus *Micromonospora* Ørskov 1923, and their taxonomy. Mikrobiologiya *38*: 883–893.

Taechowisan, T., J.F. Peberdy and S. Lumyong. 2003. Isolation of endophytic actinomycetes from selected plants and their antifungal activity. World J. Microbiol. Biotechnol. *19*: 381–385.

Takahashi, Y., A. Matsumoto, A. Seino, Y. Iwai and S. Ōmura. 1996. Rare actinomycetes isolated from desert soils. Actinomycetologica *10*: 91–97.

Takasawa, S., I. Kawamoto, I. Takahashi, M. Koakura and R. Okachi. 1975. Platomycins A and B. I. Taxonomy of the producing strain and production, isolation and biological properties of platomycins. J. Antibiot. *28*: 656–661.

Takizawa, M., T. Hida, T. Horiguchi, A. Hiramoto, S. Harada and S. Tanida. 1995. TAN-1511 A, B and C, microbial lipopeptides with G-CSF and GM-CSF inducing activity. J. Antibiot. *48*: 579–588.

Tamura, A., R. Furuta, S. Naruto and H. Ishii. 1973. Actinotiocin, a new sulfur-containing peptide antibiotic from *Actinomadura pusilla*. J. Antibiot. *26*: 343–350.

Tamura, T., S. Suzuki and K. Hatano. 2000. *Acrocarpospora* gen. nov., a new genus of the order *Actinomycetales*. Int. J. Syst. Evol. Microbiol. *50*: 1163–1171.

Tamura, T. and T. Sakane. 2004. *Planotetraspora silvatica* sp. nov. and emended description of the genus *Planotetraspora*. Int. J. Syst. Evol. Microbiol. *54*: 2053–2056.

Terekhova, L.P., O.A. Galatenko and T.P. Preobrazhenskaia. 1982. [New species, *Actinomadura fulvescens* sp. nov. and *Actinomadura turkmeniaca* sp. nov. and their antagonistic properties]. Antibiotiki *27*: 87–92.

Thiemann, J.E., H. Pagani and G. Beretta. 1967. A new genus of the *Actinoplanaceae*: *Planomonospora* gen. nov. G. Microbiol. *15*: 27–38.

Thiemann, J.E., Coronell.C, H. Pagani, G. Beretta, G. Tamoni and V. Arioli. 1968a. Antibiotic production by new form-genera of *Actinomycetales*. I. Sporangiomycin, an antibacterial agent isolated from *Planomonospora parontospora* var. *antibiotica* var. nov. J. Antibiot. *21*: 525–531.

Thiemann, J.E., H. Pagani and G. Beretta. 1968b. A new genus of the *Actinomycetales*: *Microtetraspora* gen. nov. J. Gen. Microbiol. *50*: 295–303.

Thiemann, J.E. and G. Beretta. 1968. A new genus of the *Actinoplanaceae*: *Planobispora* gen. nov. Arch. Mikrobiol. *62*: 157–166.

Thiemann, J.E. 1970. Study of some new genera and species of the *Actinoplanaceae*. *In* The *Actinomycetales* (edited by Prauser). VEB Gustav Fischer Verlag, Jena, pp. 245–257.

Thiemann, J.E. 1974a. Genus *Planobispora* Thiemann and Beretta. *In* Bergey's Manual of Determinative Bacteriology, 8th edn (edited by Buchanan and Gibbons). Williams & Wilkins, Baltimore, pp. 720–721.

Thiemann, J.E. 1974b. Genus *Planomonospora* Thiemann and Beretta. *In* Bergey's Manual of Determinative Bacteriology, 8th edn (edited by Buchanan and Gibbons). Williams & Wilkins, Baltimore, pp. 719–720.

Tomita, K., Y. Hoshino, T. Sasahira, K. Hasegawa, M. Akiyama, H. Tsukiura and H. Kawaguchi. 1980. Taxonomy of the antibiotic Bu-2313-producing organism. *Microtetraspora caesia* sp. nov. J. Antibiot. *33*: 1491–1501.

Tomita, K., N. Oda, Y. Hoshino, N. Ohkusa and H. Chikazawa. 1991. Fluvirucins A1, A2, B1, B2, B3, B4 and B5, new antibiotics active against influenza A virus. IV. Taxonomy on the producing organisms. J. Antibiot. *44*: 940–948.

Tomita, K., N. Oda, Y. Hoshino, N. Ohkusa and H. Chikazawa. 1992. *In* Validation of the publication of new names and new combinations previously effectively published outside the IJSB. List no. 40. Int. J. Syst. Bacteriol. *42*: 191–192.

Tseng, M., K.-C. Hoang, M.-K. Yang, S.-F. Yang and W.S. Chu. 2007. Polyesterdegrading thermophilic actinomycetes isolated from different enironment in Taiwan. Biodegradation *18*: 579–583.

Tseng, M., S.F. Yang and G.F. Yuan. 2010. *Herbidospora yilanensis* sp nov and *Herbidospora daliensis* sp nov., from sediment. Int. J. Syst. Evol. Microbiol. *60*: 1168–1172.

Tsurumi, Y., N. Ohhata, T. Iwamoto, N. Shigematsu, K. Sakamoto, M. Nishikawa, S. Kiyoto and M. Okuhara. 1994. WS79089 A, B and C, new endothelin converting enzyme inhibitors isolated from *Streptosporangium roseum*. No. 79089. Taxonomy, fermentation, isolation, physico-chemical properties and biological activities. J. Antibiot. *47*: 619–630.

Tsurumi, Y., K. Fujie, M. Nishikawa, S. Kiyoto and M. Okuhara. 1995. Biological and pharmacological properties of highly selective new endothelin converting enzyme inhibitor WS 79089B isolated from *Streptosporangium roseum* No. 79089. J. Antibiot. *48*: 169–174.

Tsyganov, V.A., V.P. Namestnikova and N.V. Krasikova. 1966. [DNA composition of various genera of the *Actinomycetales*]. Mikrobiologiia *35*: 92–95.

Umezawa, I., K. Kamiyama, H. Takeshita, J. Awaya and S. Ōmura. 1976. A new antitumour antibiotic, PO-357. J. Antibiot. *29*: 1249–1251.

Umezawa, I. and K. Kamiyama. 1983. Novel protein KUd-PC and preparation thereof. Japanese patent 58,198,422. Japan.

Van Brummelen, J. and J.C. Bent. 1957. *Streptosporangium* isolated from forest litter in the Netherlands. Antonie van Leeuwenhoek *23*: 385–392.

Vobis, G. and H.-W. Kothe. 1985. Sporogenesis in sporangiate actinomycetes. *In* Frontiers in Applied Microbiology, vol. 1 (edited by Mukerji, Pathak and Singh). Print House, Luknow, India, pp. 25–47.

Vobis, G. 1986. Spore development in sporangia-forming actinomycetes. *In* Biological, Biochemical and Biomedical Aspects of Actinomycetes, Part B (edited by Szabó, Biró and Goodfellow). Akadémiai Kiadó, Budapest, pp. 443–452.

Vobis, G. 1989. Genus *Planobispora* Thiemann and Beretta. *In* Bergey's Manual of Systematic Bacteriology, vol. 4 (edited by Williams, Sharpe and Holt). Williams & Wilkins, Baltimore pp. 2536–2539.

Vobis, G. and H.-W. Kothe. 1989. Genus *Spirillospora*. *In* Bergey's Manual of Systematic Bacteriology, vol. 4 (edited by Williams, Sharpe and Holt). Williams & Wilkins, Baltimore, pp. 2543–2545.

Vobis, G. 1991. Morphological appoaches to rapid recognition of sporangiate and non-sporangiate genera. *In* Actinoplanetes and Maduromycetes (Isolation and Characterization), SIM-Workshop on Actinomycetes. University of Wisconsin, Madison, p. 30.

Vobis, G. 1997. Morphology of actinomycetes. *In* Atlas of Actinomycetes (edited by Miyadoh, Hamada, Hotta, Kudo, Seino, Vobis and Yokota). Asakura Publishing, Tokyo pp. 180–191.

Vobis, G. 2006. The genus *Actinoplanes* and related genera. *In* The Prokaryotes: a Handbook on the Biology of Bacteria, 3rd edn (edited by Dworkin, Falkow, Rosenberg, Schleifer and Stackebrandt). Springer, New Yrok, pp. 623–653.

Waksman, S.A. 1950a. The actinomycetes: Their nature, occurrence, activities and importance. Ann. Crypt. Phytopath. *9*: 1–230.

Waksman, S.A. 1950b. The Actinomycetes. Their Nature, Occurrence, Activities, and Importance, vol. 9. Chronica Botanica Company, Waltham, MA, pp. 1–230.

Waksman, S.A. 1961. The Actinomycetes, vol. 2. Classification, Identification and Descriptions of Genera and Species. Williams & Wilkins, Baltimore.

Waksman, S.A. 1967. The Actinomycetes. A Summary of Current Knowledge. Ronald Press, New York.

Wang, Y., Z. Zhang and J. Ruan. 1996a. A proposal to transfer *Microbispora* (Lechevalier 1965) to a new genus, *Thermobispora* gen. nov., as *Thermobispora bispora* comb. nov. Int. J. Syst. Bacteriol. *46*: 933–938.

Wang, Y., Z. Zhang and J. Ruan. 1996b. Phylogenetic analysis reveals new relationships among members of the genera *Microtetraspora* and *Microbispora*. Int. J. Syst. Bacteriol. *46*: 658–663.

Wang, Y., Z.S. Zhang and J.S. Ruan. 1996c. A proposal to transfer *Microbispora bispora* (Lechevalier 1965) to a new genus, *Thermobispora* gen. nov., as *Thermobispora bispora* comb. nov. Int. J. Syst. Bacteriol. *46*: 933–938.

Wang, Y., Z.S. Zhang, J.S. Ruan and S.M. Ali. 1999. Investigation of actinomycete diversity in the tropical rainforests of Singapore. J. Ind. Microbiol. Biotech. *23*: 178–187.

Ward-Rainey, N., F.A. Rainey and E. Stackebrandt. 1996. The phylogenetic structure of the genus *Streptosporangium*. Syst. Appl. Microbiol. *19*: 50–55.

Ward-Rainey, N.L., F.A. Rainey and E. Stackebrandt. 1997. *In* Stackebrandt, E., F.A. Rainey and N.L. Ward-Rainey. Proposal for a new hierarchic classification system, *Actinobacteria* classis nov. Int. J. Syst. Bacteriol. *47*: 479–491.

Weyland, H., E. Helmke, K. Weber and T. Richter. 1982. Madurose in a LL-DAP containing actinomycete. Proceedings of the 5th International Symposium on Actinomycete Biology, Mexico.

Whitham, T.S. 1988. Selective isolation, classification and identification of streptosporangia. PhD thesis, University of Newcastle, Newcastle upon Tyne.

Whitham, T.S., M. Athalye, D.E. Minnikin and M. Goodfellow. 1993. Numerical and chemical classification of *Streptosporangium* and some related actinomycetes. Antonie van Leeuwenhoek *64*: 387–429.

Williams, S.T. 1970. Further investigations of actinomycetes by scanning electron microscopy. J. Gen. Microbiol. *62*: 67–73.

Williams, S.T., G.P. Sharples and R.M. Bradshaw. 1973. The fine structure of the *Actinomycetales*. *In Actinomycetales*: Characteristics and Practical Importance (edited by Sykes and Skinner). Academic Press, London, pp. 113–130.

Williams, S.T. and G.P. Sharples. 1976. *Streptosporangium corrugation* sp. nov., an actinomycete with some unusual morphological features. Int. J. Syst. Bacteriol. *26*: 45–52.

Williams, S.T., M. Goodfellow and G. Alderson. 1989. Genus *Streptomyces* Waksman and Henrici. *In* Bergey's Manual of Systematic Bacteriology, vol. 4 (edited by Williams, Sharpe and Holt). Williams & Wilkins, Baltimore, pp. 2452–2492.

Willoughby, L.G. 1969a. A study of aquatic actinomycetes of Blelham Tarn. Hydrobiologija *34*: 465–483.

Willoughby, L.G. 1969b. A study of aquatic actinomycetes, the allochthonous leaf component. Nova Hedwigia *18*: 45–113.

Wingender, W., H. von Hugo and W. Frommer. 1975. A protease inhibitor isolated from *Planomonospora parontospora*. J. Antibiot. *28*: 611–612.

Xu, L., Q. Li and C. Jiang. 1996. Diversity of soil actinomycetes in Yunnan, China. Appl. Environ. Microbiol. *62*: 244–248.

Yamaguichi, T. 1967. Similarity in DNA of various morphologically distinct actinomycetes. J. Gen. Appl. Microbiol. *13*: 63–71.

Yokota, A., Y. Nakagaito and T. Hasegawa. 1989. *Streptomyces* species with madurose (3-*O*-methyl-D-galactose) as a whole-cell sugar. Arch. Microbiol. *152*: 317–321.

Zenova, G.M., A.A. Gryadunova, A.I. Pozdnyakov and D.G. Zvyagintsev. 2008. Aerobic and microaerophilic actinomycetes of typical agropeat and peat soils. Pochvavedenie *2*: 235–240.

Zhadambaa, N., N.V. Shul'ga-Mikhailova, G.M. Zenova and D.G. Zvyagintsev. 2007. Actinomycetes in soils of Mongolia. Eur. Soil Sci. *35*: 176–182.

Zhang, L.P., C.L. Jiang and W.X. Chen. 2002. *Streptosporangium subroseum* sp. nov., an actinomycete with an unusual phospholipid pattern. Int. J. Syst. Evol. Microbiol. *52*: 1235–1238.

Zhang, L.P., C.L. Jiang and W.X. Chen. 2005. *Streptosporangium yunnanense* sp. nov. and *Streptosporangium purpuratum* sp. nov., from soil in China. Int. J. Syst. Evol. Microbiol. *55*: 719–724.

Zhang, Z., Y. Wang and J. Ruan. 1998a. Reclassification of *Thermomonospora* and *Microtetraspora*. Int. J. Syst. Bacteriol. *48*: 411–422.

Zhang, Z., T. Kudo, Y. Nakajima and Y. Wang. 2001. Clarification of the relationship between the members of the family *Thermomonosporaceae* on the basis of 16S rDNA, 16S–23S rRNA internal transcribed spacer and 23S rDNA sequences and chemotaxonomic analyses. Int. J. Syst. Evol. Microbiol. *51*: 373–383.

Zhang, Z.S., Y. Wang and J.S. Ruan. 1998b. Reclassification of *Thermomonospora* and *Microtetraspora*. Int. J. Syst. Bacteriol. *48*: 411–422.

Zhi, X.-Y., W.-J. Li and E. Stackebrandt. 2009. An update of the structure and 16S rRNA gene sequence-based definition of higher ranks of the class *Actinobacteria*, with the proposal of two new suborders and four new families and emended descriptions of the existing higher taxa. Int. J. Syst. Evol. Microbiol. *59*: 589–608.

Zhukova, R.A., V.A. Syganov and V.M. Morozov. 1968. [A new species of *Micropolyspora* – *Micropolyspora angiospora* sp. nov.]. Mikrobiologiia *37*: 724–728.

Family II. **Nocardiopsaceae** (Rainey, Ward-Rainey, Kroppenstedt and Stackebrandt 1996) emend. Zhi, Li and Stackebrandt 2009, 600^{VP}

Michael Goodfellow and Martha E. Trujillo

No.car.di.op.sa.ce'a.e. N.L. n. *Nocardiopsis* the type genus of the family;suff. *-aceae* ending to denote a family; N.L. fem. pl. n. *Nocardiopsaceae* the *Nocardiopsis* family.

Aerobic, Gram-stain-positive, non-acid-fast actinomycetes which form an extensively branched substrate mycelium which may bear single spores, sometimes in clusters or spore chains which terminate in pseudosporangia. **Aerial hyphae may carry single spores on dichotomously branched sporangia or differentiate into short or long chains of spores.** All spores are nonmotile. **Some strains are moderate or obligate halophiles.** Whole-cell hydrolysates contain *meso*-diaminopimelic acid as the diagnostic diamino acid. **Contains complex mixtures of saturated, unsaturated and branched-chain fatty acids, but lacks mycolic acids. Polar lipid patterns are complex. Menaquinones have nine, ten, and eleven isoprene units with varying degrees of hydrogenation.** The pattern of 16S rRNA signatures consists of nucleotides at positions 440:497 (U–U), 485 (G), 501:544 (G–C), 502:543 (A–U), 833:853 (U–G) and 1355:1367 (G–C).

Widely distributed, but common in soil, especially saline and hypersaline soils.

DNA G+C content (mol%): 64–76.

Type genus: **Nocardiopsis** (Brocq-Rousseu 1904) Meyer 1976, 487^{AL}.

Further descriptive information

Phylogeny. The four genera classified in the family *Nocardiopsaceae* form a distinct phyletic line in the 16S rRNA actinobacterial tree (Cai et al., 2008, 2009; Cui et al., 2001; Tang et al., 2008). In addition to *Nocardiopsis*, the type genus, the family contains the genera *Haloactinospora* Tang et al. (2008), *Streptomonospora* corrig. Cui et al. 2001 and *Thermobifida* (Zhang et al. 1998) Yang et al. 2008b. Representatives of these taxa form distinct branches in the 16S rRNA *Nocardiopsaceae* tree (Figure 389) though the position of *Nocardiopsis arabia* in the *Streptomonospora* clade is more apparent than real. This organism has chemical and morphological properties typical of *Nocardiopsis* strains (Hozzein and Goodfellow, 2008) and its position in the *Nocardiopsaceae* tree needs to be re-examined in light of the acquisition of a full 16S rRNA gene sequence. Members of all four genera share a pattern of 16S rRNA signatures (Zhi et al., 2009), as cited in the family description. The family belongs to the order *Streptosporangiales* which also encompasses the families *Streptosporangiaceae* and *Thermomonosporaceae.*

Morphology. *Nocardiopsaceae* strains typically form an extensively branched substrate mycelium, abundant aerial hyphae and nonmotile spores. However, members of the constituent genera show a range of morphological features of diagnostic value (Table 294). The aerial hyphae of *Nocardiopsis* strains, for instance, are either long and moderately branched, straight and flexuous or irregularly zig-zagged, completely fragmenting into oval to elongated, rod-shaped smooth-surfaced spores (Chen et al., 2008; Hozzein and Goodfellow, 2008; Meyer, 1976, 1989). Short straight to flexuous chains of oval, cylindrical or rod-shaped spores are formed on the aerial mycelium of *Streptomonospora* strains (Cai et al., 2008, 2009; Cui et al., 2001; Li et al., 2003b) and long chains of cylindrical spores with smooth surfaces on the aerial mycelium of *Haloactinospora alba* YIM 90648^T, the sole member of this genus (Tang et al., 2008). In contrast, the aerial mycelium of *Thermobifida* strains carry oval to round, single spores on dichotomously branched sporophores which give the appearance of spore clusters; the latter may also be seen occasionally on the substrate mycelium (Yang et al., 2008b; Zhang et al., 1998). Characteristic chains of wrinkled spores terminating in pseudosporangia are borne on the substrate mycelium of the *Haloactinospora alba* strain (Tang et al., 2008) while single, oval or round spores with wrinkled surfaces are formed on either sporophores or dichotomously branching sporophores on the substrate mycelium of members of the genus *Streptomonospora* (Cai et al., 2008, 2009; Cui et al., 2001; Li et al., 2003b).

Chemotaxonomy. Members of the family contain *meso*-diaminopimelic acid as the wall diamino acid and hence have a wall chemotype III *sensu* Lechevalier and Lechevalier (1970b). *Nocardiopsis* and *Thermobifida* strains lack diagnostic sugars (Kroppenstedt and Evtushenko, 2006) whereas *Streptomonospora* strains contain galactose or arabinose and galactose as a characteristic wall sugar (Cai et al., 2008, 2009; Cui et al., 2001) and *Haloactinospora* galactose and ribose (Tang et al., 2008). Members of the family have complex polar lipid patterns with *Streptomonospora* and *Thermobifida* exhibiting a type II and *Haloactinospora* and *Nocardiopsis* a type III profile after Lechevalier et al. (1977, 1981). All four genera are characterized by complex menaquinone (Table 294) and fatty acid profiles; the latter typically contain large proportions of 14-methylpentadecanoic ($C_{16:0}$ iso) and 14-methylhexadecanoic acid ($C_{17:0}$ anteiso) (Kroppenstedt and Evtushenko, 2006; Tang et al., 2008). The menaquinone composition of some *Streptomonospora* strains is influenced by the growth media (Li et al., 2003b); medium-dependent shifts in menaquinone profiles have been reported for *Micrococcus luteus* (Hiraishi and Komagata, 1989) and age-dependent ones in *Streptomyces cyaneus* (Saddler et al., 1986). Teichoic acids are valuable taxonomic markers for the subgeneric classification of *Nocardiopsis* species (Kroppenstedt and Evtushenko, 2006; Naumova et al., 2001). The cell walls of *Thermobifida fusca* also contain teichoic acids (Potekhina et al., 2003) and a polyglycerophosphate lipoteichoic acid (Rahman et al., 2009).

Differentiation of the genera of the family *Nocardiopsaceae*

The genera classified in the family *Nocardiopsaceae* can be distinguished by using a combination of chemotaxonomic, morphological and physiological criteria (Table 294), by 16S rRNA gene signature nucleotides (Tang et al., 2008) and by comparisons of 16S rRNA gene sequenced data (Figure 389). Standard chemotaxonomic procedures can be used for the detection of diagnostic amino acids and sugars in whole-cell hydrolysates (Hasegawa et al., 1983; Staneck and Roberts, 1974), cellular fatty acids (Kroppenstedt, 1985; Sasser, 1990), and polar lipid patterns (Minnikin et al., 1984). Detailed

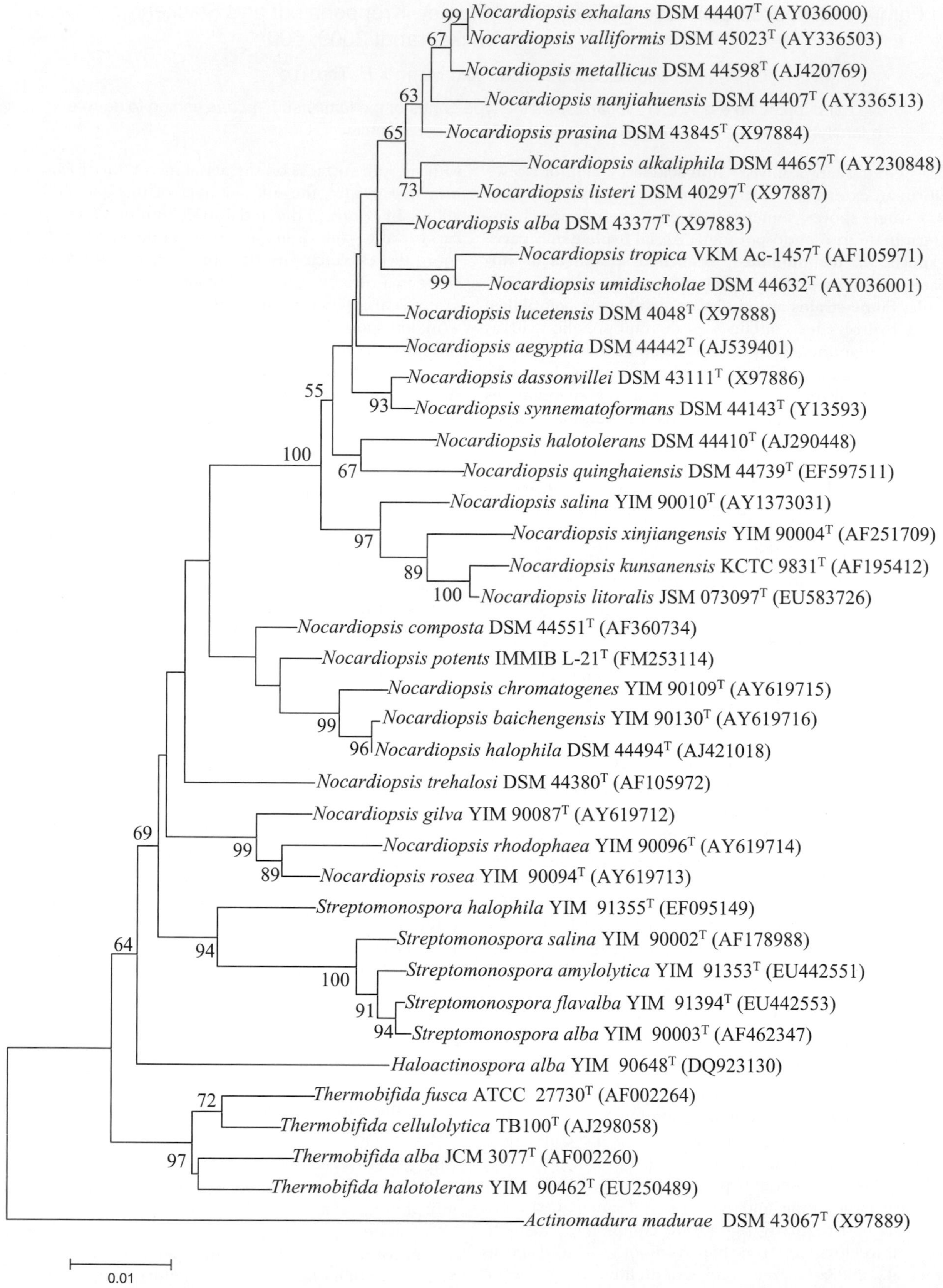

FIGURE 389. Neighbor-joining tree based on 16S rRNA gene sequences showing relationships between taxa classified in the family *Nocardiopsaceae*. Evolutionary distances were calculated using the Kimura two-parameter method of Kimura (1980). Numbers at the nodes are bootstrap values based on 1000 resampled datasets, only values above 50% are given. Bar, 1 nucleotide difference per 100 nucleotides.

TABLE 294. Properties that distinguish between the genera classified in the family *Nocardiopsaceae*[a]

Characteristic	*Nocardiopsis*	*Haloactinospora*	*Streptomonospora*	*Thermobifida*
Spores formed on:				
Aerial mycelium	Long chains	Long chains	Long chains	Single spores on dichotomously branched sporangia
Substrate mycelium	None	Spore chains terminating in pseudosporangia	Single	Occasional single spores borne on dichotomously branched sporangia
Temperature range (°C)	20–45	7–45	20–45	35–60
Growth on NaCl (%)	0–20	9–21	5–25	0–5
Fatty acid type[b]	3d	3d	nd	3e
Phospholipid pattern[c]	DPG, PC, PE, PG, (PI), PME, (PIM)	DPG, PC, PG, PIM	DPG, PC, (PE), PG, PI, (PIMS), (PME), PS)	DPG, PC, (PE) PG, PI, PME
Predominant menaquinones	10(H_2, H_4, H_6), 11(H_2, H_4, H_6, H_8)	11(H_6, H_8)	9(H_2, H_4, H_6, H_8), 10(H_2, H_4, H_6)	10(H_6, H_8)
DNA G+C content (mol%)	64–76	68	69–71	64–69

[a]Data taken from Cui et al. (2001), Cai et al. (2008, 2009), Kroppenstedt and Evtushenko (2006), Li et al. (2003b), Tang et al. (2008) and Zhang et al., (1998).

[b]Fatty acid types according to Kroppenstedt (1985); nd, not determined.

[c]Abbreviations for phospholipids: DPG, diphosphatidylglycerol; PC, phosphatidylcholine; PE, phosphatidylethanolamine; PG, phosphatidylglycerol; PI, phosphatidylinositol; PIM, phosphatidylinositol mannoside(s); PME, phosphatidylmethylmonoethanolamine; PS, phosphatidylserine; (), variable distribution.

analytical procedures are available for establishing menaquinone profiles (Kroppenstedt, 1982, 1985) (Collins, 1994), including the integrated method introduced for the analysis of bacterial isoprenoid quinones and polar lipids (Minnikin et al., 1984). Genus-specific primers are available for the PCR identification of novel *Streptomonospora* strains (Cai et al., 2009; Zhi et al., 2006).

Genus I. **Nocardiopsis** Meyer 1976, 487[AL]

WAEL N. HOZZEIN AND MARTHA E. TRUJILLO

No.car.di.op′sis. N.L. fem. n. *Nocardia* a genus of the order *Actinomycetales*; Gr. fem. n. *opsis* appearance; N.L. fem. n. *Nocardiopsis* that which has the appearance of *Nocardia*.

Gram-stain-positive, aerobic, chemo-organotrophic, **non-acid-fast**, nonmotile filamentous actinomycetes. Substrate mycelium is well developed and hyphae are long and densely branched. **Fragmentation into coccoid and bacillary elements may occur.** Aerial mycelium is well developed and sparse-to-abundant; **aerial hyphae are long**, branched, **straight to flexuous, or irregularly zig-zagged, completely fragmenting into spores of various lengths. Spore surface is smooth. Wall peptidoglycan** contains ***meso*-diaminopimelic acid** and the muramic acid of the peptidoglycan is acetylated. No diagnostic sugars are found in whole-organism hydrolysates. **Mycolic acids are absent.** Major phospholipids are **phosphatidylcholine**, phosphatidylmethylethanolamine, phosphatidylglycerol, and phosphatidylinositol, with small amounts of diphosphatidylglycerol. Menaquinones are predominantly variably hydrogenated with ten isoprene units (**MK-10**), with minor amounts of the **MK-9** and/or **MK-11** series. The main fatty acids are branched and **10-methyl-branched fatty acids, 14-methyl-heptadecanoic acid, and 14-methyl-hexadecanoic acid.** Growth temperature range is 10–45°C. Widely distributed in saline and alkaline soils, and found in compost, vegetable matter, indoor environments, and clinical material of animal and human origin.

DNA G+C content (mol%): 64–69.

Type species: **Nocardiopsis dassonvillei** (Brocq-Rousseu 1904) Meyer 1976, 487[AL].

Further descriptive information

Phylogeny. The genus *Nocardiopsis* is the type genus of the family *Nocardiopsaceae* (Rainey et al., 1996) emend. Zhang et al. (1998), which also includes the genera *Haloactinospora* Tang et al. (2008), *Streptomonospora* (Cui et al., 2001) emend. Li et al. (2003b), and *Thermobifida* (Zhang et al., 1998) emend. Yang et al. (2008b). Phylogenetically, these genera form a distinct and coherent phyletic line in the 16S rRNA gene tree (Cui et al., 2001; Kroppenstedt and Evtushenko, 2006; Tang et al., 2008; Zhang et al., 1998) and can be clearly distinguished from the other two families comprising the suborder *Streptosporangineae* Ward-Rainey et al. 1997, elevated to order *Streptosporangiales* in the present volume, namely the families *Streptosporangiaceae* and *Thermomonosporaceae* (Goodfellow and Quintana, 2006; Gyobu, 2001; Kudo, 2001). The pattern of 16S rRNA signatures consists of nucleotides at positions 440:497 (C–G), 501:544 (C–G), 502:543 (G–C), 831:855 (G–G) 843 (U), 844 (A), and 1355:1367 (A–U) (Zhi et al., 2009).

The genus *Nocardiopsis*, which currently contains 30 species, does not appear to be monophyletic according to 16S rRNA

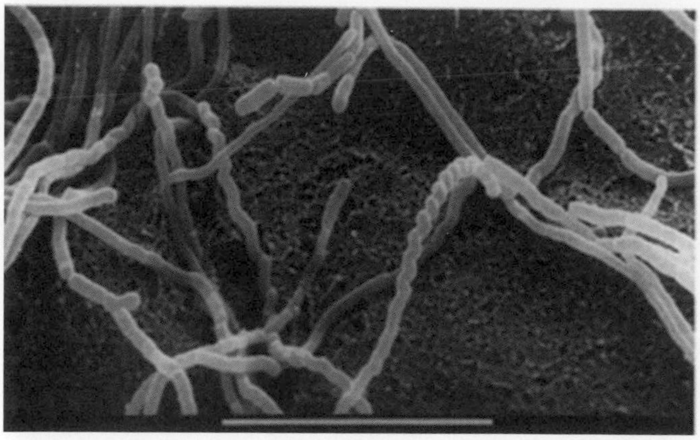

FIGURE 390. Scanning electron micrograph of strain *Nocardiopsis lucentensis* A5-1T^{T}, showing zig-zag hyphae and spore chain with a smooth surface. The culture was grown on ISP medium 4 supplemented with 10% NaCl for 14 d at 35°C. Bar = 1 μm. (Reproduced with permission from Yassin et al., 1993. Int. J. Syst. Bacteriol. *43*: 266–271.)

gene sequence phylogeny (see Figure 389). *Nocardiopsis* species are divided in four phyletic groups, the largest, with 20 species, includes the type species *Nocardiopsis dassonvillei.* On the other hand, a single member cluster is formed by *Nocardiopsis trehalosi.*

Cell morphology. Members of the genus *Nocardiopsis* characteristically form a well-developed and branched substrate mycelium which may fragment into coccoid and bacillary elements. Aerial mycelium production varies from sparse to abundant. The aerial hyphae formed by some strains may not be visible to the unaided eye, requiring careful observation under a microscope. Other cultures are thickly covered with a powdery to velvety aerial mycelium simulating cultures of *Streptomyces griseus* (cf. Gordon and Horan, 1968) because of the yellowish-gray color. Aerial hyphae are long, moderately branched, straight, or flexuous. Initiation of sporulation is often characterized by twisted hyphae, which by examination at higher magnification, reveal a zig-zag arrangement of the developing spores (Figure 390). The elongated spores are smooth and can divide subsequently into smaller spores of irregular size by cross-wall formation. Spores are enclosed within a fibrillar sheath and have thickened polar walls (Williams et al., 1974). *Nocardiopsis synnemataformans* is the only species known to form synnemata from spiral aerial hyphae that wrap together to form long ropes that subsequently fragment into small rod-shaped elements (Figure 391) (Yassin et al., 1997). *Nocardiopsis* strains do not produce sporangia, sclerotia, or motile elements.

The morphology of *Nocardiopsis* resembles that of other arthrospore-forming actinomycetes. However, members of the genus can be presumptively recognized based on various specific macroscopic and microscopic features. Although many *Nocardiopsis* strains resemble some *Streptomyces* strains macroscopically (Gordon and Horan, 1968; Shirling and Gottlieb, 1972), the two genera can be distinguished by direct microscopic comparison of cultures based on their type of life cycle and spore formation. In *Streptomyces*, spores are delimited basipetally almost simultaneously, whereas in *Nocardiopsis*, the cross

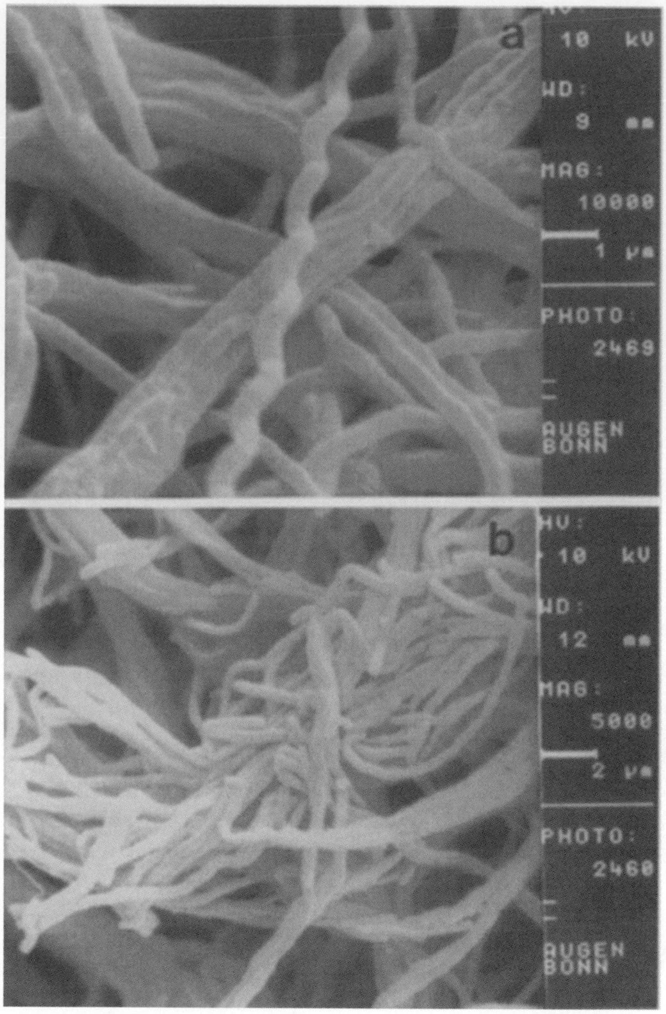

FIGURE 391. Scanning electron micrographs of *Nocardiopsis synnemataformans* strain IMMIB D-1215^{T}, showing spiral hyphae wrapped together to form synnemata (a), which fragment into small rod-shaped elements (b). (Reproduced with permission from Yassin et al., 1997. Int. J. Syst. Bacteriol. *47*: 983–988.)

walls are formed in a relatively uncoordinated manner resulting in spores of various lengths. *Nocardiopsis* species can also differentiated by their nocardioform life cycle.

Cell-wall composition. Members of the genus *Nocardiopsis* have cell-wall chemotype III, which corresponds to peptidoglycan type A1 (Schleifer and Kandler, 1972), i.e. strains contain 2,6-*meso*-diaminopimelic acid (*meso*-DAP), alanine, and glutamic acid in their peptidoglycan. The muramic acid of the peptidoglycan is acetylated (Kroppenstedt, 1987). In whole-cell hydrolyzates, glucose and galactose are detected, but no diagnostic sugars have been found in any of the species analyzed. Cell-wall teichoic acids are present and composed of glycerol and other constituents, like ribitol, glucosamine, galactosamine, succinic, and propionic acids (Naumova et al., 2001). Teichoic acid polymers are considered a valuable chemotaxonomic marker for the intrageneric taxa of *Nocardiopsis* (Kroppenstedt and Evtushenko, 2006; Streshinskaya et al., 1989, 1996, 1998; Tul'skaya et al., 1993, 2000).

Polar lipids. Members of the genus *Nocardiopsis* show a characteristic phospholipid type III profile (Lechevalier et al., 1977) with phosphatidylcholine (PC) as the characteristic polar lipid, plus phosphatidylmethylethanolamine (PME), phosphatidylglycerol (PG), phosphatidylinositol (PI), and small amounts of diphosphatidylglycerol (DPG). Additionally, two spots of glycolipids, identified as monomannosyl diglyceride and monoacetylated glucose, right below the PME spot and two to four unidentified phospholipids above the DPG spot can be detected on two-dimensional thin-layer chromatography plates as additional lipid markers (Grund, 1987; Grund and Kroppenstedt, 1990; Kroppenstedt and Evtushenko, 2006; Minnikin et al., 1977, 1984; Mordarska et al., 1983, 1998).

Menaquinone composition. Members of the genus *Nocardiopsis* show complex menaquinone profiles containing predominant amounts of MK-10(H_0) to MK-10(H_8) and small amounts of the MK-9 and/or MK-11 series (Fischer et al., 1983; Grund and Kroppenstedt, 1990; Minnikin et al., 1978).

Fatty acid profile. *Nocardiopsis* strains are characterized by terminally branched and 10-methyl-branched fatty acids (with chain lengths of 14–18 carbons), but hydroxy fatty acids have never been detected (Grund and Kroppenstedt, 1990; Kroppenstedt and Evtushenko, 2006). Among the terminally branched fatty acids, 14-methyl-heptadecanoic acid ($C_{16:0}$ iso) and 14-methyl-hexadecanoic acid ($C_{17:0}$ anteiso) are the main components. Smaller amounts of the 10-methyl branched tuberculostearic acid, i.e. 10-methyl-octadecanoic acid ($C_{18:0}$10-methyl), and its precursor the unsaturated *cis*9,10 octadecenoic acid ($C_{18:1}$ *cis*) are also found (Grund and Kroppenstedt, 1990). This fatty acid profile belongs to fatty acid type 3d according to Kroppenstedt (1985).

Cultural characteristics. *Nocardiopsis* species tend to grow well on both complex and defined media. Most strains show good to abundant growth on modified Bennett's (Jones, 1949), Czapek–Dox, nutrient (Waksman, 1961b), and glucose-yeast extract-malt extract (Athalye et al., 1981) agars, and on formulations of the International *Streptomyces* Project (ISP media; Shirling and Gottlieb, 1966). Several *Nocardiopsis* species may require NaCl for optimal growth. The halotolerant or moderately halophilic species, *Nocardiopsis halophila, Nocardiopsis kunsanensis, Nocardiopsis salina,* and *Nocardiopsis xinjiangensis* grow optimally on most ISP media in the presence of 10–15% (w/v) NaCl (Al-Tai and Ruan, 1994; Chun et al., 2000; Li et al., 2003a, 2004).

Nutrition and growth conditions. *Nocardiopsis* strains are strictly aerobic and chemo-organotrophic with an oxidative metabolism. Most species grow well at 10–40°C with an optimum growth temperature of 28–30°C; in addition, several *Nocardiopsis* species are considered thermotolerant organisms and may grow at 45–50°C, with an optimal growth temperature of 37°C, namely, *Nocardiopsis baichengensis, Nocardiopsis chromatogenes, Nocardiopsis composta, Nocardiopsis kunsanensis, Nocardiopsis rhodophaea,* and *Nocardiopsis rosea* (Chun et al., 2000; Kämpfer et al., 2002; Li et al., 2006). One thermophilic *Nocardiopsis* isolate, KMD/8, which is able to grow at 65°C with an optimal growth temperature of 50°C, was reported by Kempf (1995).

Nocardiopsis strains can tolerate NaCl concentrations up to 20% (w/v). Halotolerant species include *Nocardiopsis baichengensis, Nocardiopsis chromatogenes, Nocardiopsis composta, Nocardiopsis gilva, Nocardiopsis halotolerans, Nocardiopsis rhodophaea,* and *Nocardiopsis rosea,* which can grow in media supplemented with 15–18% (w/v) NaCl (Al-Zarban et al., 2002; Kämpfer et al., 2002; Li et al., 2006). Only *Nocardiopsis halophila, Nocardiopsis kunsanensis, Nocardiopsis xinjiangensis,* and *Nocardiopsis salina* (Al-Tai and Ruan, 1994; Chun et al., 2000; Li et al., 2003a, 2004) are considered as true halophilic species as NaCl (at least 3%, w/v) is necessary for growth, with an optimal concentration of 10–15% (w/v).

Most *Nocardiopsis* strains are characterized by their alkaliphilic behavior as they prefer mild alkaline conditions and some can even grow at pH 13. Alkaliphilic species include *Nocardiopsis alkaliphila, Nocardiopsis ganjiahuensis, Nocardiopsis kunsanensis, Nocardiopsis litoralis, Nocardiopsis metallicus, Nocardiopsis prasina,* and *Nocardiopsis valliformis* (Chen et al., 2009; Hozzein et al., 2004; Miyashita et al., 1984; Schippers et al., 2002; Yang et al., 2008c; Zhang et al., 2008).

Nocardiopsis strains can utilize a wide range of carbohydrates as carbon sources. They also hydrolyze and degrade diverse compounds. A list of these substrates is given in Table 295.

Pigments. Most *Nocardiopsis* strains do not produce diffusible pigments. However, Gerber (1966) reported that various *Nocardiopsis dassonvillei* strains produced purple colored crystals which were identified as iodinin crystals. Gordon and Horan (1968) also reported the production of yellowish, greenish-yellow, or brown diffusible pigments by strains of the same species. *Nocardiopsis tropica* produces yellow-orange soluble pigments on oatmeal agar or glycerol-yeast extract agar (Evtushenko et al., 2000), whereas *Nocardiopsis chromatogenes* was found to produce a diffusible yellowish-pink pigment on Czapek and ISP 5 agar (Li et al., 2006).

Genetics. Similarity values between the 16S rRNA gene sequences of strains of *Nocardiopsis* species with validly published names range from 93.6 to 99.9%; 41.3% of the similarity values are ≥97%. Although many type strains in the genus share a 16S rRNA gene sequence similarity of 99%, DNA–DNA reassociation values are well below 70%. A value of 55.9% between *Nocardiopsis halophila* and *Nocardiopsis baichengensis* has been reported (Li et al., 2006) and the DNA–DNA values of *Nocardiopsis metallicus* with *Nocardiopsis exhalans* and *Nocardiopsis prasina* are 18.2% and 44.1%, respectively (Schippers et al., 2002).

The *gyrB, sod,* and *rpoB* partial gene sequences of strains of 24 *Nocardiopsis* species with validly published names have been determined; the mean similarities of the *gyrB, sod* and *rpoB* genes of *Nocardiopsis* species were 87.7%, 87.3%, and 94.1% (that of the 16S rRNA gene of *Nocardiopsis* species studied was 96.6%). The topology of the *gyrB* gene tree showed a high consistency with that of the 16S rRNA gene tree suggesting that the *gyrB* gene could be used as a taxonomic phylogenetic marker with a higher degree of discrimination between the species studied (Yang et al., 2007a, 2008a).

Genes encoding industrially important enzymes such as β-1,3-glucanase and chitinases ChiA and ChiB, from two alkaliphic *Nocardiopsis* strains (F96 and *Nocardiopsis prasina* OPC-131, respectively), have been cloned, sequenced, and expressed in *Escherichia coli* (Masuda et al., 2003; Tsujibo et al., 2003). Sequencing of the whole genome of *Nocardiopsis dassonvillei* DSM 43111^T is underway (Project ID: 19709 at DOE Joint Genome Institute).

TABLE 295. Differentiating characters for *Nocardiopsis* species[a]

Character	*N. dassonvillei*	*N. aegyptia*	*N. alba*	*N. alkaliphila*	*N. arabia*	*N. baichengensis*	*N. chromatogenes*	*N. composta*	*N. exhalans*	*N. ganjiahuensis*	*N. gilva*	*N. halophila*	*N. halotolerans*	*N. kunsanensis*	*N. listeri*	*N. litoralis*	*N. lucentensis*	*N. metallicus*	*N. potens*	*N. prasina*	*N. quinghaiensis*	*N. rhodophaea*	*N. rosea*	*N. salina*	*N. synnemataformans*	*N. trehalosi*	*N. tropica*	*N. umidischolae*	*N. valliformis*	*N. xinjiangensis*
Color of:[b]																														
AM	w	w	w	yw	yw	w	w	w	w	gw	py	w	gy	w	–	w	w	w	w	og	w	pi	pi	w	w	w, gy	w	w	yw	w
SM	b	br	py	yb	gy	oy	rb	yb	y	py	b	yr	b	y	b	yw	y	yb	py	lo	py	rb	pi	py	p	lo, yr	oy	y	y	y
Synnemata	–	–	–	–	–	–	–	–	–	–	–	–	–	–	–	–	–	–	–	–	–	–	–	–	+	–	–	–	–	–
Carbon source utilization:																														
L-Arabinose	+	+	–	+	+	+	+	+	+	+	+	+	–	+	+	–	–	nd	+	+	+	+	+	–	–	+	+	+	+	nd
Cellobiose	+	nd	+	+	+	+	+	+	nd	–	+	w	nd	+	+	–	+	+	+	d	+	–	–	–	+	–	+	+	nd	–
D-Galactose	(+)	+	+	w	+	+	+	+	–	nd	+	+	+	v	d	–	–	+	+	–	–	–	–	–	+	+	+	+	nd	–
Inositol	–	+	–	w	nd	–	+	+	–	nd	+	+	nd	v	d	nd	+	nd	+	–	+	+	–	–	–	–	–	–	–	nd
D-Lactose	–	+	–	w	nd	–	+	nd	–	–	+	nd	nd	v	d	–	–	nd	+	–	–	–	+	nd	–	+	–	nd	+	nd
Maltose	(+)	nd	+	+	+	+	+	–	+	nd	–	+	nd	–	–	–	+	+	+	–	–	–	+	–	+	+	+	+	nd	–
Mannitol	+	+	–	–	+	+	+	+	+	nd	+	d	nd	v	d	–	+	+	+	–	+	–	–	–	+	+	+	+	nd	–
Melibiose	–	nd	–	nd	nd	nd	nd	nd	–	nd	nd	nd	+	+	+	–	–	+	nd	–	–	nd	nd	–	nd	+	+	+	nd	nd
L-Rhamnose	+	+	–	w	nd	+	+	–	+	+	–	d	nd	+	+	–	+	+	–	–	–	–	+	–	+	+	+	+	nd	–
Sucrose	(+)	+	–	+	+	+	+	–	+	–	+	+	+	v	d	+	+	+	+	–	+	–	+	+	–	–	+	+	–	+
Trehalose	+	nd	+	nd	+	nd	nd	+	+	nd	nd	+	nd	–	–	–	+	+	+	–	+	nd	nd	–	–	+	+	+	nd	nd
D-Xylose	+	+	–	+	–	–	+	–	+	+	+	+	–	+	+	+	–	+	–	+	nd	–	–	–	+	+	+	+	+	v
Degradation of:																														
Casein	+	nd	+	+	+	nd	nd	nd	nd	nd	nd	+	nd	+	+	–	+	nd	nd	+	+	nd	nd	nd	+	+	–	nd	nd	nd
Tyrosine	(+)	nd	+	+	–	nd	nd	nd	nd	nd	nd	–	nd	+	+	+	+	nd	nd	+	nd	nd	nd	nd	+	–	–	nd	nd	nd
Tween 80	(–)	nd	+	nd	nd	nd	nd	nd	nd	+	nd	+	nd	nd	–	–	–	nd	nd	–	+	nd	nd	nd	nd	+	+	nd	nd	nd
Tween 85	(+)	nd	+	nd	nd	nd	nd	nd	nd	–	nd	nd	nd	nd	–	nd	–	nd	nd	–	nd	nd	nd	nd	nd	–	+	nd	nd	nd
Nitrate reductase	(–)	nd	(–)	nd	nd	–	–	nd	nd	+	+	nd	nd	–	+	–	+	nd	nd	–	–	–	+	+	+	–	–	nd	+	–
Urease	(–)	nd	(+)	nd	nd	–	–	nd	nd	+	–	+	nd	+	–	–	–	nd	nd	–	–	–	–	–	+	+	+	nd	nd	+
Growth at/in:																														
pH (optimal)[c]	8	nd	9	9.5	nd	7.2	7.2	nd	nd	9.0	7.2	nd	nd	9	8	8.5	7.5	8.5	8.5	10	7.0	7.2	7.2	7.2	nd	nd	nd	nd	9.5	7.2
0% NaCl	+	+	+	+	+	+	+	+	+	+	+	d	+	–	+	+	d	+	+	+	+	+	+	–	+	+	+	+	+	nd
10% NaCl	+	–	–	+	+	+	+	+	+	–	+	+	+	+	–	+	–	+	+	–	+	+	+	+	+	–	–	–	–	+
20% NaCl	–	–	–	–	–	–	–	–	–	–	–	+	–	+	–	–	–	–	–	–	–	–	–	+	–	–	–	–	–	nd
10°C	(–)	+	–	+	+	–	–	–	+	+	+	–	+	nd	+	–	+	+	+	–	+	–	–	–	–	–	d	+	–	nd
42°C	(+)	nd	(+)	+	–	+	+	+	–	–	–	–	–	nd	–	–	–	–	–	+	–	+	+	–	–	+	–	–	–	nd
45°C	(–)	–	–	+	–	+	+	+	–	–	–	–	–	nd	–	–	–	–	–	–	–	+	+	–	–	+	–	–	–	nd

[a]Modified from Kroppenstedt and Evtushenko (2006). Data from: Al-Tai and Ruan (1994); Al-Zarban et al. (2002); Chen et al. (2008, 2009); Chun et al. (2000); Evtushenko et al. (2000); Grund and Kroppenstedt (1990); Hozzein et al. (2004); Hozzein and Goodfellow (2008); Kämpfer et al. (2002); Li et al. (2003a, 2006); Peltola et al. (2001); Sabry et al. (2004); Schippers et al. (2002); Yang et al. (2008c); Yassin et al. (1993, 1997, 2009); Zhang et al. (2008). Symbols: +, positive; (+) most strains of the species are positive; –, negative; (–), most strains of the species are negative; nd, not determined; v, variable; d, doubtful result.

[b]AM, aerial mycelium; SM substrate mycelium; –, no aerial mycelium usually formed; b, beige-brown; br, brown; gw, gray-white; gy, gray-yellow; lo, light olive; og, olive-green; oy, orange-yellow; p, pimento; pi, pink; py, pale-yellow; rb, red-brown; w, white; y, yellow; yb, yellow-brown; yr, yellow-red; yw, yellow-white.

Phages. Four *Nocardiopsis* phages were reported by Prauser (1981); however, information about their taxon specificity is still lacking. Clearing effects caused by soil-isolated polyvalent *Streptomyces* phages on *Nocardiopsis dassonvillei* raised speculations about the relationship of these two taxa (Prauser, 1984). However, these two genera can be clearly differentiated and this may only be due to the phage's specificity.

Pathogenicity. The isolation of *Nocardiopsis* strains from human or animal clinical material is well documented (Goodfellow, 1998; Schaal and Beaman, 1984). Like most other actinomycete pathogens, *Nocardiopsis* species are opportunistic rather than invasive pathogens. Liegard and Landrieu (1911) reported the isolation of a strain matching the description of Brocq-Rousseau (1904) from a case of ocular conjunctivitis where they gave it the name *Nocardia dassonvillei*, which later became *Nocardiopsis dassonvillei* (Meyer, 1976). Gordon and Horan (1968) reported that 15 of the 26 *Nocardiopsis* strains in their collection originated from clinical isolates. They speculated that similar cultures from other clinical sources might usually be discarded in laboratories because of their macroscopic resemblance to the genus *Streptomyces*.

The implication of *Nocardiopsis dassonvillei* in skin infections has also been reported (Philip and Roberts, 1984; Singh et al., 1991). The first case of mycetoma caused by *Nocardiopsis dassonvillei* was published by Sindhuphak et al. (1985), who repeatedly isolated *Nocardiopsis dassonvillei* strains from nodules and draining sinuses from the leg of a 39-year-old man. Ajello et al. (1987) confirmed the presence of *Nocardiopsis dassonvillei* in actinomycetoma cases. Furthermore, *Nocardiopsis dassonvillei* is involved in broncho-pulmonary infections (Bernatchez and Lebreux, 1991; Gugnani et al., 1998; Mordarska et al., 1998) and has also been recovered from blood samples (Beau et al., 1999). Although the pathogenicity of *Nocardiopsis synnemataformans* isolated from the sputum of a kidney transplant patient could not be verified, this species is regarded as a potential pathogen.

Antibiotic sensitivity. Very little is known about antibiotic sensitivity of clinical *Nocardiopsis* strains and only a few studies have been carried out. Philip and Roberts (1984) reported that a cutaneous infection caused by *Nocardiopsis dassonvillei* in an elderly man responded to oral treatment with trimethoprim-sulfamethoxazole. Yassin et al. (1997) examined *Nocardiopsis synnemataformans* for its sensitivity by using the agar dilution technique. They reported the following minimum inhibitory concentrations (μg/ml) for: mezlocillin (2.0); amoxycillin plus clavulanic acid (2.0); imipenem (2.0); erythromycin (8.0); clindamycin (>128); tetracycline (≤0.2); vancomycin (≤0.2); gentamicin (≤0.2); tobramycin (≤0.2); amikacin (≤0.2); ciprofloxacin (2.0); and ofloxacin (8.0). In an earlier study, Yassin et al. (1993) also reported that *Nocardiopsis lucentensis* is resistant to lincomycin (128 μg/ml), penicillin G (128 μg/ml), gentamicin (64 μg/ml), streptomycin (64 μg/ml), and neomycin (4 μg/ml), but not to rifampin (128 μg/ml). Similarly, Al-Tai and Ruan (1994) used the paper disk-diffusion method to assess the susceptibility of *Nocardiopsis halophila* and found that it was resistant to amoxycillin, clindamycin, bacitracin, cephalexin, and oxacillin, but susceptible to carbenicillin, chloramphenicol, erythromycin, and novobiocin.

Secondary metabolites. *Nocardiopsis* strains are also known for their potential to produce bioactive metabolites. A disaccharide antibiotic (3-trehalosamine) active against Gram-stain-positive bacteria was reported to be synthesized by *Nocardiopsis trehalosi* (Dolak et al., 1980, 1981). Phenazine antibiotics were obtained from the alkaliphilic *Nocardiopsis* strain OPC-15, which produced different phenazine antibiotics under different culture conditions (Tsujibo et al., 1988). In addition, 1,6-dihydroxyphenazine was obtained from the mycelium after incubation for 6–8 d at 27°C, whereas 1,6-dihydroxyphenazine-5,10-dioxide, known as iodinin, was isolated after incubation for 6 d at 27°C followed by further incubation for 2 d at 4°C. The production of a new indole alkaloid, pendolmycin, by *Nocardiopsis* strain SA 1715, isolated from soil collected in a river near Shanghai, was reported by Yamashita et al. (1988). Members of the genus also produce apoptolidins, new apoptosis inducers in transformed cells (Kim et al., 1997; Wender and Longcore, 2009; Wender et al., 2005).

In the last decade, some bioactive compounds have been discovered from *Nocardiopsis* strains isolated from marine samples. Kahakamides A and B, two new neosidomycin antibiotics (a group of rare indole-*N*-glycosides) were isolated from a *Nocardiopsis dassonvillei* strain (Schumacher et al., 2001). Also, a novel cyclic tetrapeptide was obtained from the culture broth of a marine *Nocardiopsis* isolate (Shin et al., 2003).

A marine-derived actinomycete, *Nocardiopsis* sp. (CMB-M0232), obtained from a sediment sample collected at a depth of 55 m off the coast of Brisbane, Australia, yielded two new FKBP12-binding macrolide polyketides named nocardiopsins A and B (Raju et al., 2010).

As expected, alkalitolerant/alkaliphilic *Nocardiopsis* strains can synthesize many different alkaline enzymes. These compounds have been reported from different *Nocardiopsis* strains (Moreira et al., 2002, 2003). In addition, three chitinases from the alkaliphilic *Nocardiopsis prasina* OPC-131 (Tsujibo et al., 2003), a keratinolytic protease from the alkaliphilic *Nocardiopsis* sp. strain TOA-1 (Mitsuiki et al., 2004), an alkaline protease from *Nocardiopsis* sp. (Monteiro et al., 2005), an endo-1,3-β-glucanase from the alkaliphilic *Nocardiopsis* sp. strain F96 (Fibriansah et al., 2006), and an endo-β-1,4-D-glucanase from the alkalitolerant *Nocardiopsis* sp. SES28 (Walker et al., 2006) have also been isolated. Moreover, a thermostable α-amylase from *Nocardiopsis* sp. was reported by Stamford et al. (2001) and an acid-resistant protease enzyme produced by a strain of *Nocardiopsis alba* was studied by Kelch et al. (2007).

Degradation of organic polymers and transformation of organic compounds have also been reported as applications of some *Nocardiopsis* species. Ghanem et al. (2005) published the first report on the degradation of polyesters, poly(3-hydroxybutyrate) and its copolymers with poly(3-hydroxyvalerate) [poly(3-hydroxybutyrate-co-10–20%-hydroxyvalerate)], by the marine species *Nocardiopsis aegyptia*. Moreover, in a review on transformation of steroids by actinobacteria, Donova (2007) reported that representatives of *Nocardiopsis* are capable of hydroxylating dehydroepiandrosterone.

Ecology. The natural habitat of *Nocardiopsis* strains appears to be the soil where they are frequently isolated together with other actinobacteria (Dolak et al., 1980, 1981; Mikami et al., 1982; Mishra et al., 1987; Wang et al., 1999; Xu et al., 1998). Zitouni et al. (2005) isolated 86 strains from soil samples collected from the Algerian Sahara (desert) and found that 54 isolates belonged to the genus *Nocardiopsis*, whereas Hozzein and colleagues isolated novel *Nocardiopsis* species from desert and sand dune soil samples from Egypt (Hozzein and Goodfellow,

2008; Hozzein et al., 2004). Many *Nocardiopsis* strains have also been found in soils with high salt concentrations (Al-Tai and Ruan, 1994; Al-Zarban et al., 2002; Chen et al., 2008; Chun et al., 2000; Li et al., 2004, 2006; Yassin et al., 1993). Alkaliphilic soils are also a good source for the recovery of *Nocardiopsis* strains. Jiang and Xu (1998) reported that 9 of 49 actinomycetes isolated from alkaline soils were *Nocardiopsis* strains, all grew at pH 12, but six isolates were obligate alkaliphilic strains growing at pH 8–12 but not at pH 7.

In addition, *Nocardiopsis* species have also been reported from other sources including: cotton waste and hay (Lacey, 1977); an Antarctic glacier (Abyzov et al., 1983); deep-sea sediments (Zhang and Zeng, 2008); ovaries of puffer fish (Wu et al., 2005); indoor environments (Peltola et al., 2001); a composting facility (Kämpfer et al., 2002); and a cattle barn (Andersson et al., 1998).

Distribution and abundance of members of the genus *Nocardiopsis* in marine habitats was reported recently (Sabry et al., 2004; Schumacher et al., 2001; Shin et al., 2003). In studies on the diversity of culturable actinobacteria isolated from the marine sponges, *Hymeniacidon perleve*, *Haliclona* sp., and *Iotrochota* sp., *Nocardiopsis* strains were frequently found and it was the second dominant genus after *Streptomyces* in most cases (Jiang et al., 2007, 2008; Zhang et al., 2006).

The presence and isolation of *Nocardiopsis* strains from clinical material including actinomycetomas is well documented (Ajello et al., 1987; Beau et al., 1999; Gugnani et al., 1998; Liegard and Landrieu, 1911; Mordarska et al., 1998; Sindhuphak et al., 1985; Singh et al., 1991; Yassin et al., 1997).

Enrichment and isolation procedures

No specific enrichment or isolation procedures have been recommended for *Nocardiopsis* (Meyer, 1989) and isolation protocols described for streptomycetes are commonly used (Korn-Wendisch and Kutzner, 1992). Most strains grow readily on a variety of media, especially those recommended by the International Streptomycetes Project (ISP); those that have proved to be most effective include yeast extract-malt extract (ISP medium 2), oatmeal (ISP medium 3), glycerol-asparagine (ISP medium 5) (Shirling and Gottlieb, 1966) and glucose-yeast extract-peptone (Naumova et al., 1980) agars and most strains may be recovered from agar plates after incubation at 28–30°C for 7–14 d. Horikoshi (1971) devised a special method for the isolation of alkaliphic *Nocardiopsis* strains: a small amount of soil is suspended in 1 ml sterilized water, 100 µl aliquots of the suspension are spread on dry agar plates, and the plates are incubated for 7–14 d at 27°C. In general, for isolation of alkaliphilic *Nocardiopsis* strains, the pH has to be adjusted to 10 or above with sterile sodium carbonate or sodium hydroxide after autoclaving the isolation medium and the plates should be incubated for 7–14 d at 28°C (Hozzein et al., 2004; Miyashita et al., 1984; Yang et al., 2008c; Zhang et al., 2008).

Various media supplemented with 5–20% (w/v) NaCl have been used to isolate halophilic *Nocardiopsis* strains (Al-Tai and Ruan, 1994; Al-Zarban et al., 2002; Chun et al., 2000; Li et al., 2006; Li et al., 2004; Yassin et al., 1993).

Maintenance procedures

Nocardiopsis strains can be maintained on appropriate media agar slants at 4°C and transferred every 4 months for short-term preservation. To prevent the agar from drying, the tubes should be tightly sealed with silicone stoppers. Medium-term preservation for up to 4 years can be achieved by preparing spore suspensions and/or homogenized mycelia in glycerol (45%, v/v) and keeping them at −25°C (Wellington and Williams, 1978). Storage in liquid nitrogen and lyophilization can be used for long-term storage. For lyophilization, spores and mycelia are suspended in a suitable suspension agent such as skimmed milk (10%, w/v). Storage in liquid nitrogen can be achieved by inoculation of the micro-organisms into small vials containing an appropriate medium which are then incubated until satisfactory growth is visible; the tubes are then tightly sealed and placed in a liquid nitrogen container. Alternatively, a simple, reliable and time-saving method is liquid nitrogen cryopreservation of living cells in small PVC tubes at −196°C (Hoffman, 1989a, 1989b).

Taxonomic comments

The strain that gave its name to the type species of the genus *Nocardiopsis* was originally isolated from mildewed grain and named *Streptothrix dassonvillei* by Brocq-Rosseau (1904), but unfortunately this strain was lost. In 1911, Liegard and Landrieu isolated a micro-organism from a case of ocular conjunctivitis that they found to be similar to the lost strain; however, these authors proposed to include the novel isolate in the genus *Nocardia* as *Nocardia dassonvillei*. Knowledge concerning *Nocardia dassonvillei* was scarce until Gordon and Horan (1968) discovered that the macroscopic appearance and a number of the physiological characters of *Nocardia dassonvillei* were similar to those of *Streptomyces griseus*. Subsequently, Lechevalier and Lechevalier (1970a) transferred *Nocardia dassonvillei* to the genus *Actinomadura* as *Actinomadura dassonvillei*, primarily on the basis of chemical properties.

The genus *Nocardiopsis* was subsequently proposed to accommodate *Actinomadura dassonvillei* as strains of this micro-organism lacked the characteristic whole-organism hydrolysate sugar madurose and formed spores in a distinctive zig-zag formation on aerial hyphae (Meyer, 1976). Additional chemical (Athalye et al., 1984; Collins et al., 1977; Fischer et al., 1983; Lechevalier et al., 1977a; Minnikin et al., 1977; Yamada et al., 1977) and numerical phenetic data (Alderson and Goodfellow, 1979; Athalye et al., 1985; Goodfellow et al., 1979; Goodfellow and Pirouz, 1982) strongly supported the recognition of the genus *Nocardiopsis* with *Nocardiopsis dassonvillei* as the type species.

16S rRNA gene sequence data confirmed the separation between *Nocardiopsis dassonvillei* and *Actinomadura madurae* and showed *Nocardiopsis* to be most closely related to the genera *Microtetraspora* and *Streptomyces* (Goodfellow et al., 1988; Kroppenstedt et al., 1990), whereas the nocardioform soil isolate *Saccharothrix australiensis* (Labeda et al., 1984), which resembles *Nocardiopsis dassonvillei* morphologically and in some chemotaxonomic characters, was found to be closely related to members of the family *Pseudonocardiaceae* (Bowen et al., 1989).

Six of the eight *Nocardiopsis* species with validly published names were subsequently considered to be misclassified (Grund and Kroppenstedt, 1989, 1990). *Nocardiopsis coeruleofusca*, *Nocardiopsis flava*, *Nocardiopsis longispora*, *Nocardiopsis mutabilis*, and *Nocardiopsis syringae* were classified in the genus *Saccharothrix* (Grund and Kroppenstedt, 1989; Labeda and Lechevalier, 1989) as *Saccharothrix coeruleofusca*, *Saccharothrix*

11. **Nocardiopsis gilva** Li, Kroppenstedt, Wang, Tang, Lee, Park, Kim, Xu and Jiang 2006, 1093[VP]

gil′va. L. fem. adj. *gilva* pale yellow.

Aerial mycelium is pale yellow to yellow-white and the substrate mycelium is pale yellow to pale greenish-yellow on media tested. Vegetative hyphae are well developed and fragmented. Spiral spore chains are short and are borne on the aerial hyphae. Spores are smooth-surfaced and nonmotile. No diffusible pigments are produced.

L-Arabinose, cellobiose, D-fructose, D-galactose, D-glucose, glycerol, *myo*-inositol, D-lactose, D-mannitol, raffinose, sodium acetate, sodium citrate, D-sorbitol, starch, sucrose, and D-xylose are used as carbon sources, whereas maltose, D-mannose, L-rhamnose, D-ribose, and D-xylitol are not used. Alanine, arginine, asparagine, glycine, histidine, lysine, proline, and threonine are used as sole nitrogen sources, but adenine, cystine, glutamic acid, hydroxyproline, methionine, phenylalanine, tryptophan, and valine are not used. Milk coagulation, milk peptonization, gelatin liquefaction, starch hydrolysis, H_2S production, urease activity, and melanin production are negative, but nitrate reduction is positive. Optimal growth is at 28–30°C and pH 7.2, with 5–8% (w/v) NaCl. Temperature, pH, and NaCl tolerance ranges are 10–40°C, pH 6–9, and 0–18%, respectively.

The polar lipid pattern contains PME, PC, PI, PG, and DPG. Major menaquinones are MK-11(H_4), MK-11(H_6), and MK-11(H_8).

Source: the type strain was isolated from a saline soil sample in the west of China.

Type strain: CCTCC AA 2040012, DSM 44841, KCTC 19006, YIM 90087.

DNA G+C content (mol%): 68.1 (T_m).

Sequence accession no. (16S rRNA gene): AY619712.

12. **Nocardiopsis halophila** Al-Tai and Ruan 1994, 477[VP]

ha.lo′phi.la. Gr. n. *hals halos* salt; Gr. adj. *philos* loving; N.L. fem. adj. *halophila* loving salt.

Cream yellow to coral red substrate mycelium that rarely fragments. Forms a long and extensively branched substrate mycelium. The aerial mycelium is irregularly branched with very long hyphae in zig-zags which fragment into elongated spores with smooth surfaces. No diffusible pigments are produced.

The following carbon sources are used: arabinose, galactose, glucose, fructose, inositol, maltose, mannitol, mannose, raffinose, rhamnose, ribose, sucrose, and xylose. Acid is produced from arabinose, fructose, inositol, maltose, mannitol, ribose, rhamnose, and xylose. Hydrolyzes urea and Tween 80, but does not degrade carboxymethylcellulose, starch, tyrosine, or xanthine. The optimum growth temperature is 30°C. Tolerates NaCl concentrations up to 20% (w/v).

Resistant to lysozyme, amoxycillin, clindamycin, bacitracin, cephalexin, and oxacillin, but susceptible to carbenicillin, chloramphenicol, erythromycin, and novobiocin.

Predominant phospholipids are PC, lyso-DPG, and DPG. Major menaquinones present are MK-10(H_6) and MK-10(H_8).

Source: the type strain was isolated from a saline soil obtained from Iraq.

Type strain: CCIM A.S.4.1195, DSM 44494, IQ-H3, JCM 9892.

DNA G+C content (mol%): not determined.

Sequence accession no. (16S rRNA gene): AF195411.

13. **Nocardiopsis halotolerans** Al-Zarban, Abbas, Al-Musallam, Steiner, Stackebrandt and Kroppenstedt 2002, 528[VP]

ha.lo.to′le.rans. Gr. n. *hals halos* salt; L. part. adj. *tolerans* tolerating; N.L. part. adj. *halotolerans* salt-tolerating, referring to the ability to tolerate high salt concentrations.

The color of substrate mycelium is beige to brown. Dirty-white to yellow-gray aerial mycelium is formed which shows the typical zig-zag formation prior to sporulation. The long-branched substrate hyphae fragment into nonmotile elements. Diffusible pigments are not produced. Melanin is not observed on either peptone-yeast-iron agar or tyrosine agar (ISP media 6 and 7).

Galactose, glucose, glycerol, D-mannose, melibiose, and sucrose are used as sole carbon sources, but not adonitol, L-arabinose, or D-xylose. Able to degrade feathers. Optimal growth is obtained on starch mineral agar supplemented with 10% (w/v) NaCl at 28°C. Grows at 28–35°C and in NaCl concentrations of 0–15%.

The polar lipid pattern is PC, PI, PG, PME, DPG, and three unknown PLs with high R_f values. The major menaquinones are MK-10, MK-10(H_2), and MK-10(H_4).

Source: the type strain was isolated from salt marsh soil at Al-Khiran, Kuwait.

Type strain: DSM 44410, F100, JCM 11760, NBRC 100347, NRRL B-24124.

DNA G+C content (mol%): 68 (HPLC).

Sequence accession no. (16S rRNA gene): AJ290448.

14. **Nocardiopsis kunsanensis** Chun, Bae, Moon, Jung, Lee and Kim 2000, 1911[VP]

kun.sa.nen′sis. N.L. fem. adj. *kunsanensis* pertaining to Kunsan, a location in Korea where the species was isolated.

Yellow substrate mycelium is formed that bears white aerial hyphae which fragment into elongated nonmotile spores with smooth surfaces. Fragmentation of substrate mycelium has not been observed. Diffusible pigments are not formed.

Utilizes D-glucose and sucrose as sole carbon sources, but not acetate, cellobiose, D-galactose, gluconate, *p*-hydroxybenzoate, *myo*-inositol, maltose, mannitol, paraffin, raffinose, L-rhamnose, or trehalose. L-Alanine is used as sole carbon and nitrogen source, but not gelatin, proline, or serine. Degrades adenine, casein, gelatin, hypoxanthine, starch, tyrosine, urea, and xanthine but not esculin. Nitrate is not reduced to nitrite. H_2S is not produced. Resistant to sodium azide (0.01%), crystal violet (0.0001%), and potassium tellurite (0.0001%); sensitive to phenol (0.1%). Grows optimally at 37°C and pH 9, with 10% (w/v) NaCl.

The polar lipids found are PC, PG, and DPG. Major menaquinone is MK-10(H_8).

Source: the type strain was isolated from a saltern sample collected from Kunsan, Republic of Korea.

Type strain: DSM 44524, HA-9, JCM 10721, KCTC 9831, NBRC 100348.

DNA G+C content (mol%): 71 (T_m).

Sequence accession no. (16S rRNA gene): AF195412.

15. **Nocardiopsis listeri** Grund and Kroppenstedt 1990, 10[VP]

lis'te.ri. N.L. gen. masc. n. *listeri* of Lister, named after Joseph Lister (1827–1912), the father of antiseptic surgery.

Substrate mycelium is colorless and clear in all media tested. Usually no aerial mycelium is formed unless the strain is grown on Hickey–Tresner agar (Hickey and Tresner, 1952), on which a white aerial mycelium is present.

Arabinose, cellobiose, dextrin, fructose, glucose, rhamnose, sucrose, and xylose are used as carbon sources, but not adonitol, *myo*-inositol, lactose, melezitose, melibiose, raffinose, or sorbitol. An alkaline reaction is observed with citrate, lactate, and malate, but not with quinate or malonate. Acid is produced from L-arabinose, galactose, L-rhamnose, and xylose; production from D-lactose and melibiose is variable. Lactate and propionate are decarboxylated, but not oxalate. Adenine, esculin, tyrosine, and xanthine are hydrolyzed, but not calcium oxalate or hypoxanthine. Benzoic acid, salicylic acid, *meta*-hydroxybenzoic acid, and *para*-hydroxybenzoic acid are not degraded.

Growth occurs in the presence of 5% (w/v) NaCl, but not in the presence of lysozyme (50 μg/ml). The optimal growth temperature is 28°C; no growth occurs at 45°C.

Polar lipids present include PC and PME; major menaquinones are MK-10 and MK-10(H_2).

Source: the type strain was isolated from a human clinical specimen.

Type strain: ATCC 27442, CBS 661.72, DSM 40297, IFO 13360, ISP 5297, JCM 4782, KCC S-0782, NCTC 434, RIA 1321, VKM Ac-1881.

DNA G+C content (mol%): not determined.

Sequence accession no. (16S rRNA gene): X97887.

16. **Nocardiopsis litoralis** Chen, Wang, Zhang, Tang, Liu, Xiao, Xu, Cui and Li 2009, 2711[VP]

li.to.ra'lis. L. fem. adj. *litoralis* of or belonging to the seashore.

White aerial mycelium and white to yellow-white substrate mycelium with straight to flexuous hyphae are formed. Substrate hyphae are well developed and fragment with age. Long, straight to flexuous spore chains are borne on aerial hyphae, which fragment into elongated nonmotile spores with smooth surfaces. Good growth occurs on most of the media tested. Diffusible pigments or melanin are not produced.

D-Glucose, sucrose, and xylose are used as carbon sources, but L-arabinose, cellobiose, dextrin, D-fructose, D-galactose, D-lactose, maltose, D-mannose, melezitose, melibiose, raffinose, L-rhamnose, D-ribose, D-salicin, trehalose, adonitol, acetate, citrate, gluconate, L-arginine, L-asparagine, L-glutamic acid, glycine, L-histidine, hydroxy-L-proline, L-isoleucine, L-leucine, L-methionine, L-phenylalanine, L-proline, L-serine, L-threonine, and L-valine are not used as carbon sources. Grows optimally at 25°C and pH 8.5, and in the presence of 5–7% (w/v) total salts. No growth is observed in absence of salts. Positive for catalase, but negative for oxidase. Does not reduce nitrate to nitrite. Hydrogen disulfide is not produced. Adenine, gelatin, hypoxanthine, tyrosine, and xanthine are degraded but not casein, cellulose, chitin, DNA, esculin, starch, Tweens 20, 40, 60, and 80, or urea.

The predominant menaquinones are MK-10(H_4), MK-10(H_6), and MK-10(H_8). Polar lipids comprise DPG, PC, and PG.

Source: the type strain was isolated from homogenates of a sea anemone collected from a tidal flat on Naozhou Island in the South China Sea, near Zhanjiang City, Southern China.

Type strain: DSM 45168, JSM 073097, KCTC 19473.

DNA G+C content (mol%): 70.4 (HPLC).

Sequence accession no. (16S rRNA gene): EU583726.

17. **Nocardiopsis lucentensis** Yassin, Galinski, Wohlfarth, Jahnke, Schaal and Trüper 1993, 268[VP]

lu.cen.ten'sis. N.L. fem. adj. *lucentensis* referring to Lucentum, the ancient Latin name of Alicante, a city in Spain, where the type strain was isolated.

The aerial spore mass color is predominantly white. The reverse side of culture growth is yellow to yellowish brown. An extensive substrate mycelium is produced which fragments. Aerial hyphae are branched, long, and well developed on media supplemented with 5–10% (w/v) NaCl and, at the beginning of sporulation, are more or less zig-zag-shaped. The zig-zag-shaped hyphae subdivide into smaller spores which are elongated and have smooth surfaces (Figure 392). Pigments or melanin are not produced. Cultures grow well on both complex and defined media.

The following carbon sources are used on ISP medium 9 supplemented with 5–10% NaCl: D-fructose, glucose, glycerol, *myo*-inositol, maltose, mannitol, mannose, raffinose, rhamnose, sucrose, and trehalose. On the same medium, the following carbon sources are not used: L-arabinose, D-galactose, lactose, and D-xylose. Acid is produced from glucose, inositol, mannitol, raffinose, and rhamnose. Citrate, malate, succinate, acetate, pyruvate, and propionate are decarboxylated, but lactate and oxalate are not. Adenine, hypoxanthine, tyrosine, and xanthine are degraded; casein, starch, gelatin, and esculin are hydrolyzed, but not allantoin, arbutin, or urea. Produces catalase and phosphatase, but not β-galactosidase or β-glucosidase. Nitrate is reduced to nitrite.

Resistant to (μg/ml) gentamicin (64), lincomycin (128), penicillin G (128), neomycin (4), and streptomycin (64), but not to rifampin (128). The organism is susceptible to lysozyme and tolerates NaCl at levels up to and including 10%.

Phospholipids present are PC, PME, PG, and DPG. Major menaquinones include MK-10(H_8), MK-10(H_6), and MK-10(H_{10}).

Source: the type strain was isolated from a soil sample collected in a salt marsh area near Alicante, Spain.

Type strain: A5-1, ATCC 51300, DSM 44048, IFO 15854, JCM 9420, VKM Ac-1962.

DNA G+C content (mol%): 71 (T_m).

Sequence accession no. (16S rRNA gene): X97888.

18. **Nocardiopsis metallicus** Schippers, Bosecker, Willscher, Spröer, Schumann and Kroppenstedt 2002, 2294[VP]

me.tal'li.cus. L. masc. n. *metallicus* the miner, referring to the ability to mobilize metals from slag.

Yellow-brown substrate mycelium bearing white aerial mycelium with straight spore chains.

Acetate, cellobiose, D-galactose, gelatin, gluconate, D-glucose, maltose, mannitol, proline, L-rhamnose, sucrose, trehalose, and D-xylose are used as carbon sources. The pH range for growth is 7.0–10.5 with optimal growth at pH 8.5. Grows well at 30°C, weakly at 10°C, but not at 40°C. Growth occurs on media supplemented with up to 10% (w/v) NaCl.

The polar lipid profile contains DPG, PG, PE, PME, PI, PIM, PC, and unknown PLs with R_f values above DPG.

Source: the type strain was isolated from an alkaline slag dump associated with metallurgical processing in Germany.

Type strain: DSM 44598, JCM 12409, KBS6, NBRC 101841, NRRL B-24159.

DNA G+C content (mol%): 70.8 (HPLC).

Sequence accession no. (16S rRNA gene): AJ420769.

19. **Nocardiopsis potens** Yassin, Spröer, Hupfer, Siering and Klenk 2009, 2732[VP]

po'tens. L. part. adj. *potens* potent, powerful, pertaining to the metabolic activities of the organism.

Pale yellow to brown-yellow substrate mycelium that carries white aerial hyphae is formed on ISP 2–4 agar. Forms a branched substrate mycelium that fragments into coccoid and rod-shaped elements. Aerial hyphae differentiate into straight to flexuous chains of rod-shaped spores with smooth surfaces. Diffusible pigments or melanoid pigments are not produced.

Hydrolyzes casein, elastin, esculin, guanine, hypoxanthine, testosterone, tyrosine, and urea, but not adenine, gelatin, or xanthine. Assimilates acetate, adonitol, adipate, iso-amyl alcohol, L-arabinose, 2,3-butanediol, cellobiose, citrate, *meso*-erythritol, D-galactose, D-gluconate, D-glucose, *myo*-inositol, L-lactate, lactose, maltose, D-mannitol, melezitose, sucrose, and trehalose as carbon sources, but not *m*-hydroxybenzoate, *p*-hydroxybenzoate, 1,2-propanediol, raffinose, L-rhamnose, D-sorbitol, or D-xylose. Utilizes L-alanine, arginine, gelatin, proline, ornithine, and serine, but not acetamide as simultaneous carbon and nitrogen sources. Grows in the presence of 12% NaCl and at 20–37°C, but not at 10 or 42°C.

Major phospholipids are DPG, PC, PG, PI, PIM, and PE. Major menaquinones are MK-11(H_8), MK-11(H_6), MK-11(H_4), MK-11(H_2), MK-10(H_8), MK-10(H_6), MK-10(H_4), MK-10(H_2), MK-9(H_8), MK-9(H_6), MK-9(H_4), and MK-9(H_2).

Source: the type strain was isolated from household waste.

Type strain: CCUG 56587, DSM 45234, IMMIB L-21.

DNA G+C content (mol%): not determined.

Sequence accession no. (16S rRNA gene): FM253114.

20. **Nocardiopsis prasina** (Miyashita, Mikami and Arai 1984) Yassin, Rainey, Burghardt, Gierth, Ungerechts, Lux, Seifert, Bal and Schaal 1997, 987[VP] (*Nocardiopsis dassonvillei* subsp. *prasina* Miyashita, Mikami and Arai 1984, 408)

pra.si'na. N.L. fem. adj. prasina (from Gr. adj. *prasina*) leek green (referring to the color of the mature aerial mycelium).

The substrate mycelium is colorless and the aerial mycelium is white to pale pink; a greenish shade is occasionally observed. The sporophore morphology is recti-flexibilis and a zig-zag-shaped aerial mycelium is observed at the beginning of sporulation. Excellent growth and abundant aerial mycelium formation appear at pH 9.0 and above.

Arabinose, cellobiose, dextrin, fructose, glucose, and glycerol are used as carbon sources, but not adonitol, *myo*-inositol, lactose, melezitose, melibiose, raffinose, rhamnose, sucrose, sorbitol, or xylose. An alkaline reaction is observed with citrate, lactate, and malate. No alkaline reaction occurs with quinate or malonate as carbon sources. Acid is produced from galactose, mannitol, and sucrose. Lactate, oxalate, and propionate are decarboxylated. Adenine, esculin, hypoxanthine, and xanthine are hydrolyzed, but tyrosine is not. Benzoic acid, salicylic acid, *meta*-hydroxybenzoic acid, and *para*-hydroxybenzoic acid are not degraded. Growth occurs in the presence of 5% (w/v) NaCl, but not in the presence of lysozyme (50 µg/ml). The optimal growth temperature is 28°C; no growth occurs at 45°C.

Diagnostic phospholipid is PC; major menaquinones are MK-10(H_8), MK-10(H_6), and MK-10(H_4).

Source: the type strain was isolated from soil from Japan.

Type strain: ATCC 35940, DSM 43845, IFO 14423, JCM 3336, VKM Ac-1880.

DNA G+C content (mol%): not determined.

Sequence accession no. (16S rRNA gene): X97884.

21. **Nocardiopsis quinghaiensis** Chen, Cui, Kroppenstedt, Stackebrandt, Wen, Xu and Jiang 2008, 703[VP]

quing.hai.en'sis. N.L. fem. adj. *quinghaiensis* pertaining to Qinghai, a province of China in which the sample was collected.

The color of the substrate mycelium is white to pale yellow. Forms white aerial mycelium. Substrate hyphae are well developed and fragment with age. Long spore chains are borne on aerial hyphae. Spores are rod-shaped, smooth, and nonmotile. Diffusible pigments are not produced.

L-Arabinose, cellobiose, citrate, D-fructose, D-glucose, glycerol, *myo*-inositol, D-mannitol, D-mannose, D-ribose, raffinose, sucrose, starch, and trehalose are used as sole carbon and energy sources, whereas acetate, adonitol, D-galactose, D-lactose, maltose, melibiose, D-rhamnose, salicin, D-sorbitol, D-xylitol, and D-xylose cannot be used for growth. L-Alanine, L-asparagine, L-glycine, L-histidine, hydroxy-L-proline, L-methionine, L-proline, L-threonine, and L-tyrosine can be utilized as sole nitrogen sources, whereas adenine, L-arginine, L-cystine, glutamic acid, L-lysine, phenylalanine, L-tryptophan, and L-valine cannot be used. Positive in tests for hydrolysis of casein, chitin, starch, Tween 20, and Tween 80, but negative for hydrolysis of cellulose and gelatin. Nitrate is not reduced to nitrite. H_2S, melanin, and urease are not produced. Optimal growth occurs on marine agar 2216 and Czapek agar with 3% (w/v) NaCl at 28°C and pH 7.0. The temperature, pH, and NaCl concentration ranges for growth are 10–37°C, pH 6.0–8.0, and 0–10%, respectively.

The polar lipid pattern is composed of PC, PG, and DPG. The major menaquinones are MK-10, MK-10(H_2), MK-11(H_2), MK-11, and MK-9(H_4).

Source: the type strain was isolated from a sample of saline soil collected from the Qaidam Basin, Qinghai Province, north-west China.

Type strain: CGMCC 4.3494, DSM 44739, YIM 28A4.

DNA G+C content (mol%): 67.1 (T_m).

Sequence accession no. (16S rRNA gene): EF597511.

22. **Nocardiopsis rhodophaea** Li, Kroppenstedt, Wang, Tang, Lee, Park, Kim, Xu and Jiang 2006, 1094[VP]

rho.do.pha'e.a. Gr. n. *rhodos* the rose; Gr. adj. *phaeos* brown; N.L. fem. adj. *rhodophaea* rose-brown (after the color of the substrate mycelium).

Aerial mycelium is pale pink to light reddish brown and the substrate mycelium is light reddish brown to deep reddish brown on media tested. Vegetative hyphae are well developed and fragmented. Short spore chains are borne on the aerial hyphae. Spores are smooth-surfaced and nonmotile. No diffusible pigments are produced.

L-Arabinose, D-glucose, glycerol, *myo*-inositol, sodium acetate, and D-ribose can be utilized as carbon sources, but not cellobiose, D-fructose, D-galactose, D-lactose, maltose, D-mannose, D-mannitol, raffinose, L-rhamnose, sodium citrate, D-sorbitol, starch, sucrose, D-xylitol, or D-xylose. Alanine, arginine, asparagine, glycine, histidine, proline, and valine are used as sole nitrogen sources, whereas adenine, cystine, glutamic acid, hydroxyproline, lysine, methionine, phenylalanine, threonine, and tryptophan are not utilized. Milk coagulation, gelatin liquefaction, starch hydrolysis, H_2S production, urease activity, nitrate reduction, and melanin production are negative, but milk peptonization is positive. Optimum growth is at 37–40°C and pH 7.2, with 5–8% (w/v) NaCl. Temperature, pH, and NaCl tolerance ranges are 20–60°C, pH 6–9, and 0–18%, respectively.

The polar lipid pattern is composed of PME, PC, PI, PG, DPG, and PIM, together with some unknown PGLs and unknown PLs. Major menaquinones are MK-11(H_6) and MK-11(H_8).

Source: the type strain was isolated from a saline soil sample in the west of China.

Type strain: CCTCC AA 2040014, DSM 44843, KCTC 19049, YIM 90096.

DNA G+C content (mol%): 67.1 (T_m).

Sequence accession no. (16S rRNA gene): AY619714.

23. **Nocardiopsis rosea** Li, Kroppenstedt, Wang, Tang, Lee, Park, Kim, Xu and Jiang 2006, 1094[VP]

ro'se.a. L. fem. adj. *rosea* rose colored.

Aerial mycelium is pink-white to pale pink and the substrate mycelium is pale pink to moderate red on media tested. Vegetative hyphae are well developed and fragmented. Spore chains are borne on the aerial hyphae. Spores are smooth-surfaced and nonmotile. No diffusible pigments are produced.

L-Arabinose, D-fructose, D-glucose, D-lactose, maltose, L-rhamnose, D-ribose, sodium acetate, sucrose, and starch can be utilized as carbon sources, but cellobiose, D-galactose, glycerol, *myo*-inositol, D-mannitol, D-mannose, raffinose, sodium citrate, D-sorbitol, D-xylitol, and D-xylose cannot be utilized. Alanine, arginine, asparagine, glycine, histidine, and proline can be used as sole nitrogen sources, but adenine, cystine, glutamic acid, hydroxyproline, lysine, methionine, phenylalanine, threonine, tryptophan, and valine cannot be utilized. Milk coagulation, milk peptonization, gelatin liquefaction, starch hydrolysis, H_2S production, urease activity, and melanin production are negative; nitrate reduction is positive. Optimum growth is at 37–40°C and pH 7.2, with 5–8% (w/v) NaCl. Temperature, pH, and NaCl tolerance ranges are 20–60°C, pH 6–9, and 0–18%, respectively.

The polar lipid pattern is composed of PME, PC, PI, PG, and DPG, together with some unknown PGLs and unknown PLs. Major menaquinones are MK-11, MK-11(H_2), and MK-11(H_4).

Source: the type strain was isolated from a saline soil sample in the west of China.

Type strain: CCTCC AA 2040013, DSM 44842, KCTC 19007, YIM 90094.

DNA G+C content (mol%): 67.9 (T_m).

Sequence accession no. (16S rRNA gene): AY619713.

24. **Nocardiopsis salina** Li, Park, Tang, Wang, Lee, Xu, Kim and Jiang 2004, 1808[VP] emend. Li, Kroppenstedt, Wang, Tang, Lee, Park, Kim, Xu and Jiang 2006, 1093

sa.li'na..L. fem. adj. *salina* salty, saline.

The color of the aerial mycelium is white on most media tested and the substrate mycelium is pale yellow to light orange-yellow or yellow-white. The vegetative hyphae are long, well developed, and fragmented. Long or short spore chains are borne on the aerial hyphae. Spores (0.4–0.66 × 8–1.2 μm) are rod-shaped, smooth, and nonmotile. No diffusible pigments are produced.

Fructose, raffinose, ribose, sodium acetate, sodium citrate, and sucrose are utilized as carbon sources, whereas arabinose, cellobiose, galactose, glucose, inositol, maltose, mannitol, melibiose, rhamnose, trehalose, xylitol, and xylose and are not used. Adenine, arginine, asparagine, glycine, histidine, hydroxyproline, hypoxanthine, methionine, phenylalanine, proline, serine, threonine, and valine are used as nitrogen sources. Negative in tests for milk coagulation, milk peptonization, starch hydrolysis, H_2S production, urease activity, and melanin production. Doubtful result for gelatin liquefaction; positive for nitrate reduction. Grows optimally at 28°C and pH 7.2, with 10% (w/v) NaCl; the temperature, pH, and NaCl tolerance ranges are 20–40°C, 6–9, and 3–20% (w/v), respectively.

Polar lipids present are PC, PG, PI, DPG, PE, PME, and four small PL spots above DPG. Main menaquinones include MK-9(H_8) and MK-10(H_8).

Source: the type strain was isolated from a saline soil sample in the west of China.

Type strain: CCTCC AA 204009, JCM 13364, KCTC 19003, YIM 90010.

DNA G+C content (mol%): 73.1 (T_m).

Sequence accession no. (16S rRNA gene): AY373031.

25. **Nocardiopsis synnemataformans** Yassin, Rainey, Burghardt, Gierth, Ungerechts, Lux, Seifert, Bal and Schaal 1997, 986[VP]

syn.ne.ma.ta.for'mans. Gr. adv. *syn* together; Gr. n. *nema* thread; N.Gr. n. *synnema* threads wrapping together; L. v.

formare to form; L. pres. part. *formans* forming; N.L. part. adj. *synnemataformans* synnema forming, referring to the ability of the organism to form synnemata.

The substrate mycelium is deep pimento colored, penetrates the agar, and bears aerial mycelia; the aerial mycelium is well developed (0.35–0.45 μm in diameter) with zig-zag or spiral forms and is white with a slight pimento touch, which may be reflected color from the substrate mycelium. Different spirals are wrapped together to form synnemata. The mycelia of a synnema fragment in later stages to form rod-shaped elements. No soluble pigment is produced. Melanoid pigments are not produced on either ISP media 6 or 7.

Cellobiose, citrate, galactose, gluconate, glucose, maltose, mannitol, rhamnose, and xylose are used as carbon sources, but not acetate, adipate, adonitol, arabinose, benzoate, erythritol, *m*-hydroxybenzoate, *p*-hydroxybenzoate, inositol, isoamyl alcohol, lactate, lactose, melezitose, paraffin, raffinose, sorbitol, sucrose, trehalose, 2,3-butanediol, or 1,2-propanediol. Alanine, gelatin, and proline are used as sole carbon and nitrogen sources, but not acetamide or serine. Acid is produced from adonitol, galactose, inositol, D-lactose, mannitol, D-mannose, melibiose, L-rhamnose, and sucrose. Lactate is not decarboxylated. Adenine, casein, elastin, esculin, gelatin, hypoxanthine, tyrosine, urea, and xanthine are hydrolyzed; guanine, keratin, and testosterone are not hydrolyzed. Nitrate reductase, β-glucosidase, β-galactosidase, and phosphatase are produced. Grows in the presence of 10% (w/v) NaCl. Growth occurs at 20, 30, and 37°C but not at 42°C.

Main phospholipids are PC, PE, PG, and DPG. Major menaquinones present are MK-10 and MK-10(H_2).

Source: the type strain was isolated from the sputum of a 35-year-old male who had received a renal transplant.

Type strain: DSM 44143, IMMIB D-1215, JCM 10456, NBRC 102581.

DNA G+C content (mol%): 74.1 (HPLC).

Sequence accession no. (16S rRNA gene): Y13593.

26. **Nocardiopsis trehalosi** nom. rev. (*ex* Dolak, Castle and Laborde 1981) Evtushenko, Taran, Akimov, Kroppenstedt, Tiedje and Stackebrandt 2000, 79[VP] (*Nocardiopsis trehalosei* Dolak, Castle and Laborde 1981)

tre.ha.lo'si. N.L. gen. n. *trehalosi* of trehalose, referring to 3-trehalosamine, an aminoglycoside antibiotic that is produced by the type strain of the species.

The color of the substrate mycelium ranges from pale olive-brownish to pale orange-yellow. The aerial mycelium is white to cream or yellowish gray. Aerial hyphae are zig-zag or twisted-ribbon-like at the beginning of sporulation; spores are irregularly sized (mostly elongated) and have a smooth surface. Long-branched substrate hyphae fragment into nonmotile elements. A light yellow-brownish or light orange-yellow soluble pigment is produced on some media. No melanin is observed on peptone-yeast extract-iron agar (ISP 6).

L-Arabinose, D-fructose, D-galactose, D-glucose, glycerol, lactose, maltose, D-mannitol, D-mannose, L-rhamnose, and D-xylose are used for growth as sole carbon sources, but not adonitol, i-inositol, melibiose, D-sorbitol, or sucrose. Acids are produced from L-arabinose, D-fructose, glycerol, lactose, maltose, D-mannitol, D-mannose, and D-xylose, but not from adonitol, D-galactose, i-inositol, melibiose, L-rhamnose, D-sorbose, or sucrose. Acetate, citrate, formate, fumarate, lactate, malate, malonate, pyruvate, propionate, sebacate, succinate, and tartrate are utilized; no alkaline reaction occurs with aconitate, benzoate, salicylate, or oxalate. Acid is produced from L-arabinose, D-lactose, mannitol, D-mannose, and xylose. Lactate and propionate are decarboxylated, but not oxalate. Casein, calcium oxalate crystals, hypoxanthine, xanthine, Tween 80, and urea are degraded, but not tyrosine, Tween 40, Tween 60, or Tween 85. Esculin hydrolysis is variable. Nitrates are not reduced to nitrites.

Resistant to lysozyme (50 mg/ml). Growth occurs on media supplemented with 5% (w/v) NaCl, 2.5% (v/v) methanol, and 3% (v/v) ethanol, but no growth occurs in 10% (w/v) NaCl or in 0.5% (v/v) butanol. Growth temperature ranges between 18 and 45°C; no growth occurs at 10°C. Optimum growth is at 28–37°C.

Major menaquinones are MK-10(H_4) and MK-10(H_6). Diagnostic phospholipid is PC.

Source: the type strain was isolated from experimental biofilters filled with tree bark compost.

Type strain: CIP 106425, DSM 44380, IFO 14201, JCM 3357, NRRL 12026, VKM Ac-942.

DNA G+C content (mol%): not determined.

Sequence accession no. (16S rRNA gene): AF105972.

27. **Nocardiopsis tropica** Evtushenko, Taran, Akimov, Kroppenstedt, Tiedje and Stackebrandt 2000, 79[VP]

tro'pi.ca. L. fem. adj. *tropica* tropical, of or pertaining to the tropic(s), referring to the tropical region where the type strain was isolated.

Substrate mycelium is colorless to olive-yellow or red-orange on the different media tested. The aerial mycelium is white; however, it has a cream color on Czapek agar. Aerial hyphae are long and zig-zag shaped at the beginning of sporulation, fragmenting into elongated spore-like structures which subsequently subdivide into short rod-shaped spores of irregular size. Spore surface is smooth. Long-branched substrate hyphae fragment into nonmotile elements. Soluble pigments are yellow-orange to yellow on oatmeal agar or glycerol-nitrate agar. Melanin is not produced on peptone-yeast extract-iron agar (ISP 6).

L-Arabinose, D-fructose, D-galactose, D-glucose, glycerol, maltose, mannitol, D-mannose, melibiose, L-rhamnose, sucrose, and D-xylose are used as sole carbon sources for growth; adonitol, dulcitol, i-inositol, lactose, D-ribose, and D-sorbose are not used. Acids are produced from L-arabinose, D-fructose, D-galactose, D-glucose, glycerol, maltose, D-mannitol, melibiose, L-rhamnose, and sucrose, but not from adonitol, dulcitol, i-inositol, lactose, D-ribose, sorbose, or D-xylose. An alkaline reaction is observed with citrate, formate, fumarate, lactate, malate, malonate, oxalate, propionate, pyruvate, sebacate, succinate, and tartrate, but no alkaline reaction occurs with acetate, aconitate, benzoate, or salicylate. Lactate, oxalate, and propionate are decarboxylated. Xanthine, hypoxanthine, calcium oxalate crystals, urea, and Tweens

40, 60, 80, and 85 are degraded; casein and tyrosine are not. Nitrates are not reduced to nitrites. Growth occurs on media supplemented with 10% (w/v) NaCl, 1.5% (v/v) butanol, 2.5% (v/v) methanol, and 3% (v/v) ethanol, but no growth occurs with 12% (w/v) NaCl or lysozyme (50 mg/ml). Growth occurs at 37°C but not at 10°C or 42°C; optimum growth is between 28 and 30°C.

Major menaquinones are MK-10(H_6) and MK-10(H_8). Diagnostic phospholipid is PC.

Source: the type strain was isolated from soil in the rhizosphere of *Casuarina* sp. (Seychelles).

Type strain: CIP 106426, DSM 44381, JCM 10877, VKM Ac-1457.

DNA G+C content (mol%): not determined.

Sequence accession no. (16S rRNA gene): AF105971.

28. **Nocardiopsis umidischolae** Peltola, Andersson, Kämpfer, Auling, Kroppenstedt, Busse, Salkinoja-Salonen and Rainey 2002, 3[VP] (Effective publication: Peltola, Andersson, Kämpfer, Auling, Kroppenstedt, Busse, Salkinoja-Salonen and Rainey 2001, 4303).

u.mi.di.scho'la.e. L. adj. *umidus* moist; L. fem. n. *schola* school; N.L. gen. fem. n. *umidischolae* of a moist school.

The vegetative hyphae are yellowish and penetrate the agar. The aerial hyphae are white, 0.2 µm in diameter, and form thick bundles. Soluble pigments are not produced.

Acetate, *cis*-aconitate, aspartate, β-alanine, L-alanine, arabinose, arbutin, 4-aminobutyrate, 3-hydroxybutyrate cellobiose, citrate, fructose, fumarate, galactose, *N*-acetyl-D-glucosamine, gluconate, glucose, glutarate, histidine, lactate, malate, maltitol, maltose, mannitol, mannose, melibiose, oxoglutarate, phenylacetate, phenylalanine, proline, propionate, pyruvate, rhamnose, ribose, serine, sucrose, trehalose, and xylose are used as carbon sources, but not adipate, adonitol, azelate, inositol, 3-hydroxybenzoate, 4-hydroxybenzoate, itaconate, leucine, mesaconate, ornithine, putrescine, salicin, sorbitol, suberate, or tryptophan. Esculin is not hydrolyzed. Produces α- and β-glucosidases, phosphatase, and peptidases. Grows in the presence of 7.5% (w/v) NaCl. Growth occurs at 10, 28, and 37°C, but not at 50°C.

The polar lipid pattern is composed of PC, PI, PG, PME, and DPG. The menaquinone composition includes MK-10(H_6), MK-10(H_4), MK-10(H_8), and MK-10(H_6).

Source: the type strain was isolated from the indoor dust of a water-damaged school.

Type strain: 66/93, DSM 44362, JCM 11758, NBRC 100349, NRRL B-24122.

DNA G+C content (mol%): not determined.

Sequence accession no. (16S rRNA gene): AY036001.

29. **Nocardiopsis valliformis** Yang, Zhang, Guo, Shi, Lu and Zhang 2008c, 1544[VP]

val.li.for'mis. L. n. *vallum* palisade; L. adj. suffix *-formis -is -e* (from L. n. *forma* figure, shape, appearance) like, in the shape of; N.L. fem. adj. *valliformis* shaped like a palisade, referring to the characteristic mycelium, which is often arranged in a shape like a palisade.

Substrate mycelium is yellow to light-brown and is often arranged in a shape like a fence or palisade. Aerial mycelium is abundant and white to yellowish, and fragments into rod-shaped, smooth-surfaced and nonmotile spores (0.3–0.5 × 1.2–2.5 µm). Melanin is not observed on either tyrosine agar or peptone-yeast-iron agar (ISP media 6 and 7).

L-Arabinose, glycerol, lactose, and D-xylose are used as carbon sources, but not *myo*-inositol or sucrose. Nitrate reduction and gelatin liquefaction are positive; H_2S production is negative. Optimal growth temperature is 28°C; no growth occurs at 10, 42 or 45°C. Optimal pH for growth is pH 9.5–13; has a broad range of growth pH, from pH 8.0 to 14.0. No growth at pH 7.0. Growth occurs in the absence of NaCl and in 1, 3, and 5% (w/v) NaCl; no growth in 10% NaCl.

Major phospholipids include DPG and PC. Major menaquinones are MK-10(H_2), MK-10(H_4), and MK-10(H_6).

Source: the type strain was isolated from a soil sample collected from an alkali lake in Xinjiang, China.

Type strain: CGMCC 4.2135, DSM 45023, HBUM 20028.

DNA G+C content (mol%): 70.6 (HPLC).

Sequence accession no. (16S rRNA gene): AY336503.

30. **Nocardiopsis xinjiangensis** Li, Li, Xu, Cui, Xu and Jiang 2003a, 319[VP]

xin.ji.ang.en'sis. N.L. fem. adj. *xinjiangensis* pertaining to Xinjiang, the province of western China in which the sample was collected.

Fragmented substrate and aerial mycelia are well developed on ISP 4, ISP 5, potato agar, and nutrient agar. Short spore chains are borne on the aerial mycelium; spores are rod-shaped with a smooth surface. Diffusible pigments are not produced.

Alanine, cellobiose, galactose, proline, and serine are used as carbon sources, but not glucose, maltose, mannitol, raffinose, rhamnose, sucrose, or xylose. Gelatin liquefaction, urease activity, and melanin production are positive. Milk coagulation, milk peptonization, starch hydrolysis, nitrate reduction, and H_2S production are negative. Optimum growth temperature is 28°C; optimum salt concentration for growth is 10% (w/v); optimum pH for growth is 7.2.

Predominant menaquinones are MK-10(H_2) and MK-10(H_4). Diagnostic phospholipids are PG and PI.

Source: the type strain was isolated from a saline soil sample in the west of China.

Type strain: CCRC 16285, CCTCC AA 99004, DSM 44589, JCM 12328, YIM 90004.

DNA G+C content (mol%): 74.3 (T_m).

Sequence accession no. (16S rRNA gene): AF251709.

Genus II. **Haloactinospora** Tang, Tian, Zhi, Cai, Wu, Yang, Xu and Li 2008, 2078VP

Martha E. Trujillo and Michael Goodfellow

Ha.lo.ac.ti.no.spo′ra. Gr. n. *hals halos* salt; Gr. n. *aktis -inos* a ray; Gr. n. *spora* a seed, and in biology, a spore; N.L. fem. n. *Haloactinospora* salt-loving and spored ray, referring to a halophilic and spore-forming actinomycete.

Aerobic, Gram-stain-positive, moderately halophilic filamentous actinomycetes which form well-developed, non-fragmented hyphae. **Long chains of cylindrical spores with smooth surfaces are formed on the aerial mycelium. Short spore chains with wrinkled surfaces and terminal pseudosporangia are formed on the substrate mycelium.** Chemo-organotrophic with an oxidative type of metabolism. **Whole-cell hydrolysates contain *meso*-diaminopimelic acid, galactose, and ribose. Cellular fatty acids are rich in branched, straight-chain, and methyl fatty acids. Major polar lipids are diphosphatidylglycerol, phosphatidylglycerol, phosphatidylcholine, and phosphatidylinositol mannoside.** The predominant menaquinones are MK-10(H_8), MK-11(H_6), and MK-11(H_8).

DNA G+C content (mol%): 68 (HPLC).

Type species: **Haloactinospora alba** Tang, Tian, Zhi, Cai, Wu, Yang, Xu and Li 2008, 2078VP.

Further descriptive information

Phylogeny. The genus *Haloactinospora* contains one species, *Haloactinospora alba*. The taxon is classified in the family *Nocardiopsaceae*, which also includes the genera *Nocardiopsis*, *Streptomonospora*, and *Thermobifida*. *Haloactinospora alba* forms an independent branch in the 16S rRNA tree between the genera *Streptomonospora* and *Thermobifida*. 16S rRNA gene sequence similarities between *Haloactinospora alba* and members of the family *Nocardiopsaceae* range from 93.3 to 95%. 16S rRNA gene signature nucleotides are consistent with those described for the family *Nocardiopsaceae* (Zhi et al., 2009) (See Figure 389 in the treatment of the family *Nocardiopsaceae*).

Cell morphology. *Haloactinospora alba* forms a well-developed, non-fragmenting substrate mycelium which carries aerial hyphae. Mature aerial hyphae differentiate into long chains of nonmotile cylindrical spores (0.4–0.6 × 1.0–1.2 µm) with smooth surfaces; spore chains with wrinkled surfaces and terminal pseudosporangia are observed on the substrate mycelium (Figure 393).

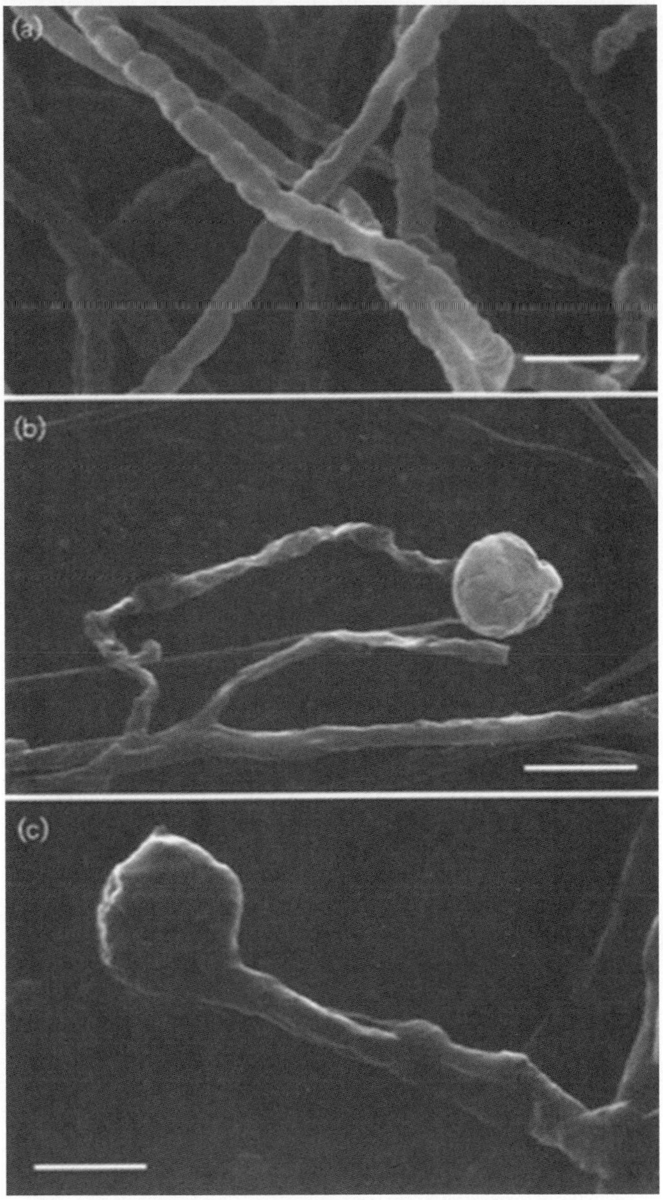

FIGURE 393. Scanning electron micrographs of spore chains and pseudosporangia of *Haloactinospora alba* strain YIM 90648T grown on ISP 2 agar supplemented with 15% (w/v) NaCl for 4 weeks at 37°C. (a) Long spore chains on the aerial mycelium; (b, c) substrate mycelium bearing a short spore chain with a terminal pseudosporangium. Bars, 2 µm (a, b) and 1 µm (c). (Reproduced with permission from Tang et al., 2008. Int. J. Syst. Evol. Microbiol. *58*: 2075–2080.)

Nutrition and growth conditions. *Haloactinospora alba* grows well on Czapek's and yeast extract-malt extract agars (ISP 2 medium; Shirling and Gottlieb, 1966); it shows moderate growth on salts-starch, glycerol-asparagine, and nutrient agars, but grows poorly on oatmeal agar (ISP 3 agar; Shirling and Gottlieb, 1966). The color of the aerial mycelium is white to yellow-white and that of the substrate mycelium is pale yellow to light or deep yellow. Diffusible pigments are not produced on any of these media. Temperature, pH, and NaCl tolerance ranges are 20–45°C (optimum 37°C), pH 6–9 (optimum 7–8), and 9–21% (w/v) NaCl (optimum 15%, w/v), respectively.

Cell-wall composition. *Haloactinospora alba* contains *meso*-diaminopimelic acid as the major wall diamino acid. Whole-cell hydrolyzates also contain galactose, ribose, and minor amounts of glucose. The polar lipid pattern of *Haloactinospora alba* YIM 90648T includes diphosphatidylglycerol, phosphatidylglycerol, phosphatidylcholine, phosphatidylinositol mannoside, and various unidentified lipids (ninhydrin reagent-positive, Dittmer and Lester reagent-negative); hence, it has phospholipid type III *sensu* Lechevalier et al. (1977, 1981). Menaquinone composition is complex and includes a mixture of MK-10(H_8), MK-11(H_4), MK-11(H_6), and MK-11(H_8), which account for about 52% of the total composition. The fatty acid profile contains major amounts of branched fatty acids and minor amounts of straight-chain and methyl fatty acids; the major fatty acids are 14-methyl-heptadecanoic (iso-16:0) and 14-methyl-hexadecanoic (anteiso-17:0) acids. This fatty acid profile places the organism in fatty acid type 3 d according to Kroppenstedt (1985).

Ecology. The only strain currently representing the genus *Haloactinospora* was isolated from a salt lake in China. Like most members of the family *Nocardiopsaceae*, *Haloactinospora alba* is moderately halophilic; it is likely that additional representatives of this taxon are present in saline habitats.

Isolation procedures

Haloactinospora alba YIM 90648^T was isolated from a salt lake sample by inoculating it onto cellulose-casein-multisalts agar (10 g microcrystalline cellulose, 0.3 g casein, 0.2 g KNO_3, 0.5 g K_2HPO_4, 0.02 g $CaCO_3$, 0.01 g $FeSO_4$, 100 g NaCl, 30 g $MgCl_2{\cdot}6H_2O$, 20 g KCl, and 15 g agar, 1000 ml distilled water). The salts solutions (NaCl, KCl, and $MgCl_2$) should be sterilized separately and added to the medium and the pH should be adjusted to 7.5 with 1 M NaOH. Inoculated plates should be incubated for 3 weeks at 37°C.

Maintenance procedures

Strain YIM 90648^T may be maintained on inorganic salts-starch agar slants (ISP 4 agar; Shirling and Gottlieb, 1966) supplemented with 15% (w/v) NaCl at 4°C for short-term preservation; mycelial fragment suspensions in 20% glycerol (v/v) are recommended for freezing at −25°C.

Differentiation of the genus *Haloactinospora* from other genera

A combination of chemotaxonomic, phenotypic, and phylogenetic procedures can be used to differentiate the genus *Haloactinospora* from related genera. The production of pseudosporangia on substrate mycelium separates *Haloactinospora* from the genera *Nocardiopsis*, *Streptomonospora*, and *Thermobifida*. *Haloactinospora* can be distinguished from the genus *Thermobifida* by differences in growth temperature and tolerance to NaCl. Menaquinone and polar lipid patterns are useful markers for distinguishing between *Haloactinospora* and the three other genera classified in the family *Nocardiopsaceae*. 16S rRNA gene sequencing clearly differentiates *Haloactinospora* from other actinobacterial taxa.

Differentiation of the species of the genus *Haloactinospora*

At present, there is only one species in the genus *Haloactinospora*.

List of species of the genus *Haloactinospora*

1. **Haloactinospora alba** Tang, Tian, Zhi, Cai, Wu, Yang, Xu and Li 2008, 2078[VP]

al'ba. L. fem. adj. *alba* white.

Aerial and substrate mycelia are well developed and nonfragmented. The color of the aerial mycelium is white to yellow-white and that of the substrate mycelium is pale yellow to light yellow or even deep yellow. Aerial hyphae differentiate into long chains of cylindrical spores which have smooth surfaces. Chains of wrinkled spores with terminal pseudosporangia are formed on the substrate mycelium. Good growth is observed on media supplemented with 15% (w/v) NaCl, notably on Czapek's, yeast extract-malt extract, and potato agars; moderate growth occurs on inorganic salts-starch, glycerol-asparagine agar, and nutrient agars, but growth is poor on oatmeal agar. Diffusible pigments are not produced on these media.

L-Arabinose, D-arabitol, cellobiose, D-fructose, D-galactose, D-glucose, inulin, raffinose, D-ribose, sucrose, D-tagatose, and D-xylose are used as sole carbon sources, but not D-adonitol, D-arabinose, dulcitol, erythritol, glycogen, inositol, D-lactose, maltose, D-mannitol, D-mannose, melezitose, L-rhamnose, L-sorbose, D-sorbitol, trehalose, xylitol, or L-xylose.

Positive for milk peptonization and coagulation. Tests for gelatin liquefaction, production of H_2S and melanin pigments, and for hydrolysis of cellulose, starch and urea are negative.

Temperature, pH, and NaCl tolerance ranges are 20–45°C, pH 6–9, and 9–21% (w/v) NaCl, respectively. Major fatty acids are $C_{16:0}$ iso (23.8%) and $C_{17:0}$ anteiso (36%).

Source: the type strain was isolated from a salt lake in Xinjiang Province, north-west China.

DNA G+C content (mol%): 68 (HPLC).

Type strain: CCTCC AA 206008, DSM 45015, YIM 90648.

Sequence accession no. (16S rRNA gene): DQ923130.

Genus III. **Streptomonospora** corrig. Cui, Mao, Zeng, Li, Zhang, Xu and Jiang 2001, 362[VP] emend. Li, Xu, Zhang, Tang, Cui, Mao, Xu, Schumann, Stackebrandt and Jiang 2003b, 1424

Xiao-Long Cui

Strep.to.mo.no.spo'ra. Gr. adj. *streptos* pliant, bent; Gr. adj. *monos* single, solitary; Gr. fem. n. *spora* a seed, spore; N.L. fem. n. *Streptomonospora* indicating that this organism forms two types of spores, with wrinkled surfaces, on aerial mycelium and substrate mycelium.

Gram-stain-positive, **aerobic organisms with branching hyphae**, 0.5–0.8 μm in diameter. **Spores are of two types.** The aerial mycelium, at maturity, forms **short chains of nonmotile spores**, which may be oval- to rod-shaped (0.3–1.0 × 0.4–2.0 μm) with wrinkled surfaces. Substrate mycelium is extensively branched with non-fragmenting hyphae. **Single, nonmotile, oval to round spores** (1.4–1.6 μm) are borne on sporophores or dichotomously branched sporophores of substrate hyphae. Optimum growth occurs in media supplemented with **NaCl at a concentration of 10–15% (w/v), at 28–37°C, and at pH 7.0. Peptidoglycan contains *meso*-diaminopimelic acid as diagnostic diamino acid. Cell walls contain galactose or galactose plus arabinose.** The phospholipid pattern is complex, consisting of phosphatidylglycerol, phosphatidylethanolamine, phosphatidylcholine, phosphatidylinositol, and diphosphatidylglycerol; methylphosphatidylethanolamine, phosphatidylinositol mannosides and phosphatidylserine may occur. The menaquinone composition may depend on the growth medium and consists mainly of menaquinones with nine, ten, or eleven isoprenoid chains and various degrees of hydrogenation. The genus *Streptomonospora* belongs to the family *Nocardiopsaceae* (Rainey et al., 1996). **Inhabit saline environments.**

DNA G+C content (mol%): 71.2–74.4 (HPLC).

Type species: **Streptomonospora salina** corrig. Cui, Mao, Zeng, Li, Zhang, Xu and Jiang 2001, 362[VP].

Further descriptive information

Phylogeny. Based on 16S rRNA gene sequence comparisons of members of the family *Nocardiopsaceae*, members of the genus *Streptomonospora* can be distinguished clearly from those of the genera *Haloactinospora*, *Nocardiopsis*, and *Thermobifida*. The five species currently classified in the genus *Streptomonospora* form two 16S rRNA subclusters: one cluster contains the single species *Streptomonospora halophila*, whereas the second cluster includes the species *Streptomonospora alba*, *Streptomonospora amylolytica*, *Streptomonospora flavalba*, and *Streptomonospora salina* (Figure 394). Both clusters are supported by high bootstrap values indicating significant branch stability. The 16S rRNA gene sequence similarities within the type strains of these species range from 96.7 to 99.6%. In addition, DNA–DNA relatedness values between species tested range from 25.1 to 47.3% (Cai et al., 2009). The signature nucleotides of members of the genus *Streptomonospora*, which differentiate them from those of the other genera in the family *Nocardiopsaceae*, are as follows: 81:88 (G–U), 82:87 (G–C), 185:192 (A–G), 196 (C), 197 (A), 207:212 (G–A), 208 (U), 210 (G), 229 (G), 508 (C), 607 (U), 662:743 (C–G), 998:1043 (G–C), 1025:1036 (U–G), 1026:1035 (G–C), 1042 (G).

Cell morphology and cultural characteristics. Species of the genus *Streptomonospora* produce white to pale yellow colonies with well-developed aerial and substrate hyphae (0.5–0.8 µm in diameter) on nutrient agar, potato agar, Czapek's agar, and most ISP media tested (International *Streptomyces* Project medium; Shirling and Gottlieb, 1966). Species do not produce diffusible pigments on any of these media except for *Streptomonospora flavalba* YIM 91394[T], which produces deep yellow diffusible pigments on potato extract agar (pH 7.0) supplemented with 10% (w/v) NaCl. Morphological observation of 7 to 28-d-old cultures grown on glycerol/asparagine agar (ISP medium 5) or inorganic salt/starch agar (ISP medium 4) containing 10–15% (w/v) NaCl revealed that the vegetative hyphae with irregular branches are well developed, but not fragmented (Figure 395). The aerial mycelium, at maturity, forms short chains of spores that are oval- to rod-shaped (0.3–1.0 × 0.4–2.0 µm) with wrinkled surfaces (Figure 395a). Substrate mycelium is extensively branched with non-fragmenting hyphae. Single spores, which are oval to round and 1.4–1.6 µm in diameter, are borne on sporophores of substrate mycelium or dichotomously branched sporophores, and the surfaces of single spores are wrinkled (Figure 395b). Therefore, there are two types of spores, both of which are nonmotile.

Cell-wall composition. All *Streptomonospora* species contain *meso*-diaminopimelic acid (*meso*-DAP) as the diagnostic amino acid; in addition, *Streptomonospora alba* and *Streptomonospora salina* contain alanine, glutamic acid, muramic acid, acetylglucosamine, and di- and tripeptides typical of peptidoglycan type A1γ (Schleifer and Kandler, 1972). In the initial analysis, Cui et al. (2001) reported that the cell wall of *Streptomonospora salina* YIM 90002[T] contained *meso*-DAP, DD-DAP, glycine, and aspartic acid as cell wall amino acids. A second analysis of the cell wall of this micro-organism (Li et al., 2003b) revealed that the cell-wall composition was similar to that of *Streptomonospora alba*, i.e. peptidoglycan of type A1γ. These authors concluded that the additional amino acids reported previously for *Streptomonospora salina* were probably constituents of a protein tightly attached to the peptidoglycan. All *Streptomonospora* species contain galactose as the diagnostic whole-cell sugar. In addition, *Streptomonospora alba* also contains arabinose. *Streptomonospora alba* YIM 90003[T] and *Streptomonospora salina* YIM 90002[T] have been reported to contain glucose as determined by GC-MS analyses (Chen et al., 2000).

Phospholipid composition. The phospholipid composition of *Streptomonospora* strains is quite complex and varies depending on the species. All strains analyzed contain phosphatidylglycerol, phosphatidylcholine, and phosphatidylinositol. *Streptomonospora alba* and *Streptomonospora salina* also contain phosphatidylethanolamine and methylphosphatidylethanolamine (Cai et al., 2008; Cui et al., 2001; Li et al., 2003b). The former also contains phosphatidylserine. An unidentified phospholipid has also been reported for *Streptomonospora alba*, *Streptomonospora halophila*, and *Streptomonospora salina*. In addition, *Streptomonospora amylolytica*, *Streptomonospora flavalba* (Cai et al., 2009), and *Streptomonospora halophila* contain phosphatidylinositol mannosides.

Quinones. The menaquinone composition may depend on the growth medium and consists mainly of menaquinones with nine, ten, or eleven isoprenoid chains and various degrees of hydrogenation: i.e. a combination of one or more representative(s) of the series [MK-9(H_2), (H_4), (H_6), (H_8)], [MK-10(H_2), (H_4), (H_6), (H_8)], and [MK-11(H_4), (H_6), (H_8), (H_{10})]. Following the procedures of Groth et al. (1999), the predominant menaquinones found in *Streptomonospora alba* YIM 90003[T] grown on vitamin-enriched ISP 2 medium were MK-10(H_2), MK-10(H_4), and MK-9(H_8). The composition changed, however, when cells grown on glucose/yeast extract medium were analyzed: the major menaquinone was MK-9(H_4), whereas MK-9, MK-10(H_4), MK-9(H_2), MK-9(H_6) occurred as minor compounds (Li et al., 2003b). Similar results were obtained when *Streptomonospora salina* YIM 90002[T] was analyzed using cultures grown on different media (Li et al., 2003b). The reason for this change has not been investigated, but medium- and age-dependent shifts in menaquinone composition have been reported previously for *Actinobacteria* (Hiraishi and Komagata, 1989; Saddler et al., 1986).

Fatty acid composition. *Streptomonospora* species contain the following major fatty acids: $C_{17:0}$ anteiso, $C_{16:0}$ iso, $C_{15:0}$ iso, $C_{15:0}$ anteiso, $C_{17:0}$ iso, 10-methyl $C_{17:0}$, $C_{16:0}$, $C_{17:0}$, 10-methyl $C_{18:0}$, $C_{18:0}$, $C_{18:0}$ iso, and $C_{14:0}$ iso.

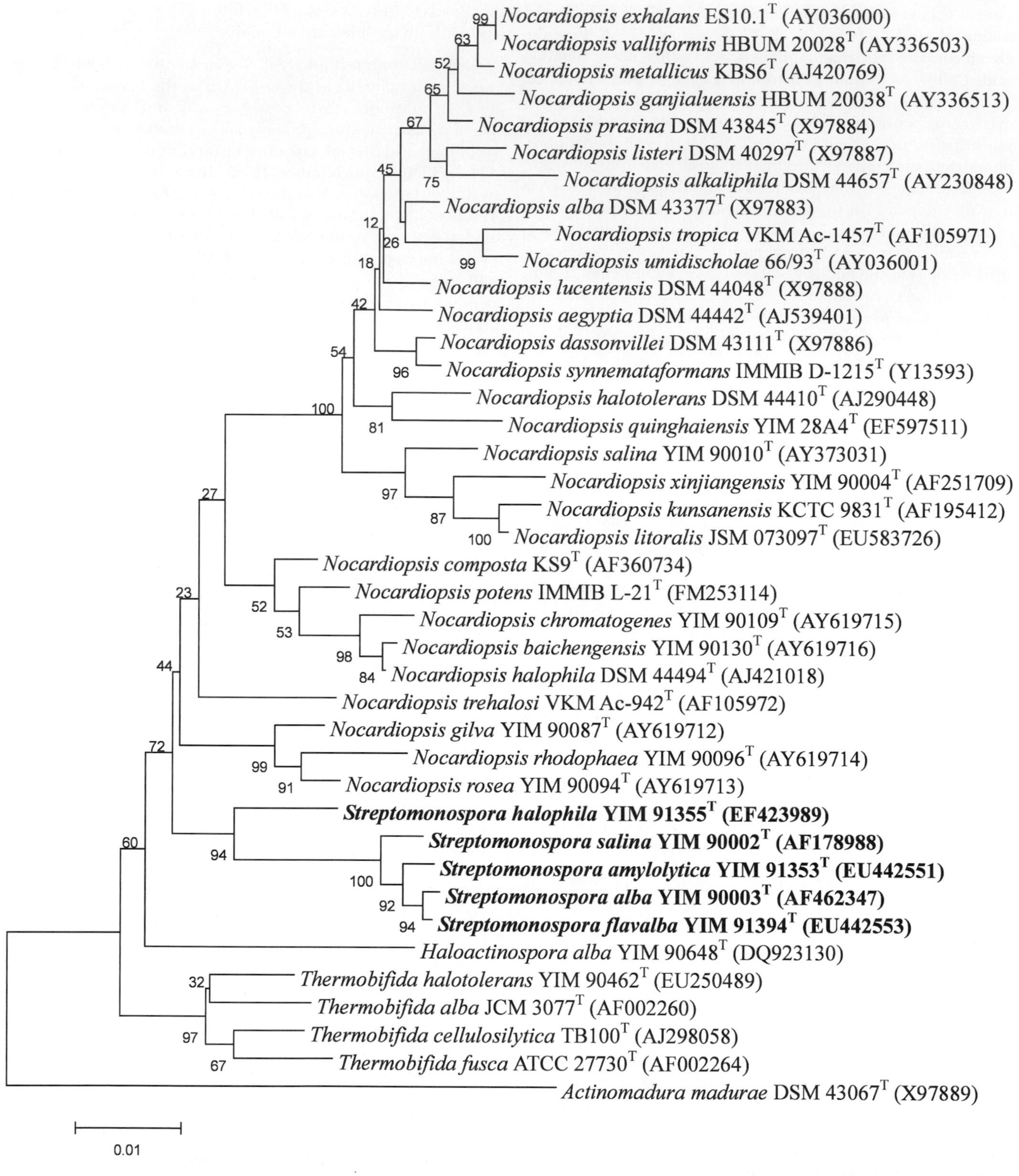

FIGURE 394. Neighbor-joining tree showing the phylogenetic relationships among members of the genera *Streptomonospora*, *Nocardiopsis*, *Haloactinospora*, and *Thermobifida* in the family *Nocardiopsaceae*. *Actinomadura madurae* was used as the outgroup. Bootstrap values from 1000 analyses are shown at the nodes of the tree. Bar = 1 nt substitution per 100 nt of 16S rRNA gene sequence.

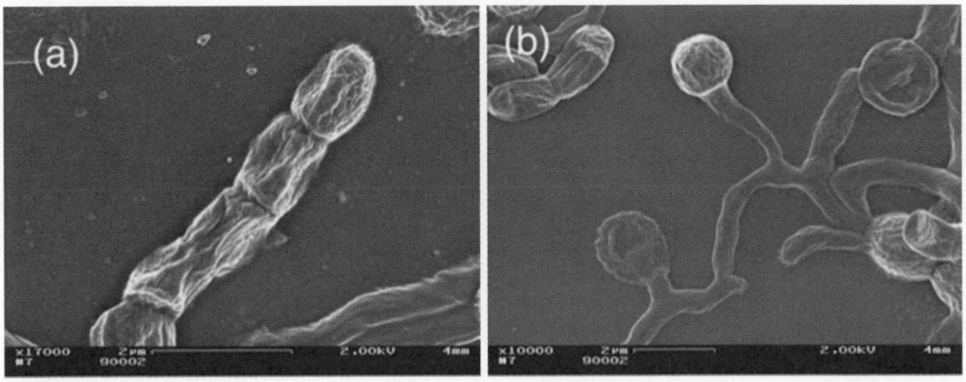

FIGURE 395. Scanning electron micrographs of *Streptomonospora salina* YIM 90002^T grown on ISP medium 5 for 28 d at 28°C, showing a short chain of spores (a) and single spores (b). Bars = 2 μm.

Nutrition and growth conditions. Members of the genus *Streptomonospora* are aerobic, heterotrophic, halophilic actinomycetes that require, for optimal growth, complex media adjusted to a pH of 7.0 and supplemented with 10–15% (w/v) NaCl. The optimal growth temperature for *Streptomonospora alba* and *Streptomonospora salina* is 28°C, whereas *Streptomonospora amylolytica*, *Streptomonospora flavalba*, and *Streptomonospora halophila* show optimal growth at 37°C. Good growth is observed on yeast extract/malt extract agar (ISP 2), inorganic salt/starch agar (ISP 4), and glycerol/asparagine agar (ISP 5), and, in most cases, oatmeal agar (ISP 3). *Streptomonospora alba* YIM 90003^T grows well on ISP 5 medium, nutrient agar, and Czapek's agar. *Streptomonospora amylolytica* YIM 91353^T and *Streptomonospora flavalba* YIM 91394^T show good growth on ISP 3 and 2 media, respectively, while both grow well on nutrient agar. *Streptomonospora halophila* YIM 91355^T grows very well on modified ISP 5 (Cai et al., 2008). *Streptomonospora salina* YIM 90002^T shows good growth on ISP 2 and 5, nutrient agar, Czapek's agar, and potato agar, all supplemented with 15% (w/v) NaCl.

Ecology. All halophilic actinomycetes of the genus *Streptomonospora* have been isolated from hypersaline soil samples from Aiding salt lake in Xinjiang Province in Western China.

Enrichment and isolation procedures

Complex media such as starch/casein agar and modified ISP 5 agar (containing 0.1% K_2HPO_4, 0.1% L-asparagine, 1.0% glycerol, 0.5% yeast extract, 0.5% KNO_3, 10.0% NaCl, 2.0% agar in 1 liter distilled water; pH 7.0) have been used for isolation of *Streptomonospora* strains using dilution plate procedures. Isolation plates are usually incubated at 28 or 37°C for 3 to 4 weeks. Pure cultures may be subsequently obtained on ISP 3 and 4 media, Czapek's agar, potato agar, and nutrient agar, all supplemented with 10–15% (w/v) NaCl.

Maintenance procedures

Streptomonospora cultures may be maintained on ISP 2, ISP 4, and ISP 5 agar slants supplemented with 10–15% (w/v) NaCl at 4°C for short-term preservation. For longer term preservation, all strains can be maintained in 20% (v/v) glycerol suspensions at −80°C. Alternatively, lyophilization is also recommended.

Differentiation of the genus *Streptomonospora* from other genera

Streptomonospora species form two types of spores, which make them easily differentiated from members of the genera *Nocardiopsis*, *Haloactinospora*, and *Thermobifida* of the family *Nocardiopsaceae*. In addition, one set of *Streptomonospora*-specific primers was designed by Zhi et al. (2006, 2007) that can be used to detect/differentiate, by PCR amplification, the large numbers of *Streptomonospora* strains from those of other genera that are obtained from various environments, especially from saline environments.

Taxonomic comments

The genus *Streptomonospora* was first described by Cui et al. (2001) to accommodate a halophilic strain isolated from a salt lake in China; however, the name of the genus was written as "*Streptimonospora*". The original spelling *Streptimonospora* (*sic*) has been corrected by the List Editor, IJSEM (2001), as *Streptomonospora* corrig. Cui et al. 2001, gen. nov.

Differentiation of the species of the genus *Streptomonospora*

Although members of the genus *Streptomonospora* have high 16S rRNA gene sequence similarity values (96.7–99.6%) and share similar morphological properties, low DNA–DNA hybridization values together with differential chemotaxonomic markers, as well as physiological and biochemical characteristics, are useful for differentiating all species (Table 296).

TABLE 296. Phenotypic characteristics that differentiate the type strains of *Streptomonospora* species[a]

Characteristic	*S. salina* YIM 90002[T]	*S. alba* YIM 90003[T]	*S. amylolytica* YIM 91353[T]	*S. flavalba* YIM 91394[T]	*S. halophila* YIM 91355[T]
Media for growth:					
ISP 2	Good	Moderate	Moderate	Good	Good
ISP 3	Poor	Moderate	Good	Moderate	Good
ISP 4	Moderate	Moderate	Moderate	Moderate	Moderate
ISP 5	Good	Good	–	Poor	Good
Czapek's agar	Good	Good	–	Poor	Moderate
Nutrient agar	Good	Good	Good	Good	Poor
Potato agar	Good	Moderate	Moderate	Moderate	Good
Aerial mycelium color	Pale white, white	White	Yellow-white	Yellow-white	Yellow-white
Substrate mycelium color	Mid-yellow	Orange-yellow	Yellow-gray	Pale yellow	Pale yellow
Optimum temperature for growth (°C)	28	28	37	37	37
Diffusible pigment	–	–	–	+ (deep yellow)	–
NaCl concentration for growth (%):					
Range	5–20	5–25	5–20	5–25	5–20
Optimum	15	10–15	10	10	10
Milk peptonization	–	–	+	+	+
Nitrate reduction	–	+	–	–	–
Production of:					
H_2S	–	–	+	–	–
Oxidase	+	–	–	–	–
Melanin	+	–	–	+	–
Amylase	+	–	+	+	–
Cellulase	–	–	–	–	+
Carbon source utilization:					
Arabinose	+	–	+	+	–
Cellobiose	–	–	+	–	–
Glucose	+	+	+	+	–
Inositol	–	–	–	–	+
Maltose	+	–	–	+	+
Rhamnose	–	–	+	+	+
Sorbitol	–	–	+	–	–
Alanine	–	–	–	+	+
Arginine	–	–	+	+	–
Glycine	–	–	–	+	–
Histidine	+	–	–	+	+
Hydroxyproline	–	–	+	–	–
Lysine	–	–	+	–	–
Phenylalanine	–	–	–	+	+
Proline	–	–	+	+	+
Serine	–	–	–	+	–
Threonine	–	–	+	+	+
Tryptophan	–	–	+	+	–
Tyrosine	–	–	–	+	–
Cell-wall sugars	Galactose	Galactose, arabinose	Galactose	Galactose	Galactose
Predominant menaquinones (>10%)	MK-10(H_8) (39.4%), MK-10(H_6) (22.6%), MK-10(H_4) (11.8%)	MK-10(H_8) (50.6%), MK-10(H_6) (15.8%)	MK-10(H_8) (35.8%), MK-10(H_6) (16.4%), MK-9(H_8) (15.3%)	MK-10(H_8) (41.5%), MK-9(H_8) (27.7%), MK-10(H_6) (12.6%)	MK-10(H_8) (43.2%), MK-10(H_6) (17.7%), MK-11(H_8) (10.4%)
Major phospholipids	PG, PI, PC, 2MPE, PL	PG, PE, PI, DPG, MPE, PS, PC, PL	DPG, PC, PG, PI, PIM	DPG, PC, PG, PI, PIM	DPG, PG, PC, PIM, PI, PL
Major fatty acids	$C_{15:0}$ iso (30.1%), $C_{16:0}$ iso (16.5%), 9-methyl $C_{16:0}$ (11.7%), $C_{17:0}$ anteiso (6.7%)	$C_{17:0}$ anteiso (26.0%), $C_{16:0}$ iso (25.1%), $C_{16:0}$ (8.6%), $C_{15:0}$ anteiso (8.0%)	$C_{17:0}$ anteiso (42.4%), $C_{18:0}$ (11.2%)	$C_{17:0}$ anteiso (28.0%), $C_{16:0}$ iso (24.6%)	$C_{16:0}$ iso (39.8%), $C_{17:0}$ anteiso (12.5%), 10-methyl $C_{17:0}$ (10.2%), 10-methyl $C_{18:0}$ (11.1%)
DNA G+C content (mol%)	72.9	74.4	71.2	72.5	72.1

[a]Data are from Cui et al. (2001), Li et al. (2003b) and Cai et al. (2008, 2009). +, Positive; –, negative; DPG, diphosphatidylglycerol; MPE, methylphosphatidylethanolamine; PC, phosphatidylcholine; PE, phosphatidylethanolamine; PG, phosphatidylglycerol; PI, phosphatidylinositol; PIM, PI mannosides; PL, unidentified phospholipid; PS, phosphatidylserine.

List of species of the genus *Streptomonospora*

1. **Streptomonospora salina** corrig. Cui, Mao, Zeng, Li, Zhang, Xu and Jiang 2001, 362[VP] emend. Li, Xu, Zhang, Tang, Cui, Mao, Xu, Schumann, Stackebrandt and Jiang 2003b, 1424

sa.li'na. L. fem. adj. *salina* salted, saline, referring to the saline habitat of the micro-organism.

White aerial mycelium, at maturity, forms short chains of spores that are oval- to rod-shaped (1.5–2 × 1 µm) with wrinkled surfaces. Mid-yellow substrate mycelium is extensively branched with non-fragmenting hyphae. Single spores, which are oval to round (1.4 × 1.6 µm), are borne on sporophores or dichotomously branched sporophores of substrate hyphae; the surfaces of the spores are wrinkled. Both spore types are nonmotile. Grows in 5–20% (w/v) NaCl. Optimum growth occurs in media supplemented with 15% (w/v) NaCl, at 28°C, and at pH 7.0. Diffusible pigments are not produced.

Glucose, sucrose, maltose, arabinose, raffinose, starch, glycerol, mannitol, and histidine are used as sole carbon sources. Positive for catalase, starch hydrolysis, and melanin production, but negative for oxidase, milk coagulation, peptonization, hydrolysis of cellulose, H_2S production, nitrate reduction, and gelatin liquefaction.

The cell wall contains *meso*-DAP, alanine, glutamic acid, muramic acid, acetylglucosamine, and di- and tripeptides typical of peptidoglycan type A1γ; the cell-wall sugar is galactose. The major menaquinones are MK-10(H_8) (39.4%), MK-10(H_6) (22.6%), and MK-10(H_4) (11.8%). The phospholipid pattern consists of phosphatidylglycerol, phosphatidylinositol, phosphatidylcholine, two methylated phosphatidylethanolamines, and an unidentified phospholipid. The major fatty acids are $C_{15:0}$ iso (30.1%), $C_{16:0}$ iso (16.5%), 9-methyl $C_{16:0}$ (11.7%), and $C_{17:0}$ anteiso (6.7%).

Source: the type strain was isolated from a hypersaline environment (a salt lake in China).

DNA G+C content (mol%): 72.9 (HPLC).

Type strain: YIM 90002, CCRC 16284, CCTCC 99003, DSM 44593.

Sequence accession no. (16S rRNA gene): AF178988.

2. **Streptomonospora alba** Li, Xu, Zhang, Tang, Cui, Mao, Xu, Schumann, Stackebrandt and Jiang 2003b, 1424[VP]

al'ba. L. fem. adj. *alba* white, referring to the color of colonies on most media.

Aerial and substrate mycelia are well developed, but not fragmented on most media. White aerial mycelium forms short chains of spores at maturity that are straight to flexuous; spores are oval- to cylindrical-shaped (0.4–0.7 × 0.8–1.6 µm) with wrinkled surfaces and are nonmotile. Single, round to oval spores are borne on substrate mycelium. Color of the substrate mycelium is white (ISP 4, ISP 5, Czapek's agar), gray-white (ISP 3), moderate orange-yellow (ISP 2), deep orange-yellow (potato agar), or brilliant orange-yellow (nutrient agar). Growth occurs in 5–25% (w/v) NaCl (optimum 10–15%), at 28°C, and at pH 7.0. Diffusible pigments are not produced.

Except for glucose, carbon source utilization is not easy to determine because of negative reactions caused by extremely poor growth in basal media. Positive for catalase and nitrate reduction, but negative for oxidase, starch hydrolysis, milk peptonization, H_2S production, hydrolysis of cellulose, and melanin production.

The diagnostic diamino acid of peptidoglycan is *meso*-DAP, with galactose and arabinose as cell-wall sugars. The predominant menaquinones are MK-10(H_8) (50.6%) and MK-10(H_6) (15.8%). Major phospholipids are phosphatidylglycerol, phosphatidylethanolamine, phosphatidylinositol, diphosphatidylglycerol, methylphosphatidylethanolamine, phosphatidylserine, phosphatidylcholine, and an unidentified phospholipid. The major fatty acids are $C_{17:0}$ anteiso (26.0%), $C_{16:0}$ iso (25.1%), $C_{16:0}$ (8.6%), and $C_{15:0}$ anteiso (8.0%).

Source: the type strain was isolated from soil in a hypersaline habitat in Xinjiang Province, western China.

DNA G+C content (mol%): 74.4 (HPLC).

Type strain: YIM 90003, CCTCC AA 001013, DSM 44588, JCM 12680.

Sequence accession no. (16S rRNA gene): AF462347.

3. **Streptomonospora amylolytica** Cai, Tang, Chen, Li, Zhang and Li 2009, 2474[VP]

a.my.lo.ly'ti.ca. Gr. n. *amulon* starch; N.L. adj. *lyticus -a -um* (from Gr. adj. *lutikos -ê -on*) able to loosen, able to dissolve; N.L. fem. adj. *amylolytica* producing lysis of starch.

Aerial and substrate mycelia are not fragmented. White to yellow-white aerial mycelia form spore chains at maturity, with nonmotile, oval to cylindrical-shaped spores (0.3–0.4 × 0.4–0.6 µm). Single, oval spores are borne on sporophores of substrate mycelium. Substrate mycelia are yellow-gray on inorganic salt/starch agar, yeast extract/malt extract agar, and oatmeal agar, pale yellow on potato extract agar, and ivory olive brown on nutrient agar medium. Grows well on oatmeal agar and nutrient agar media supplemented with 10% (w/v) NaCl, but no growth occurs on Czapek's agar or glycerol/asparagine agar supplemented with 10% (w/v) NaCl. Diffusible pigments are not produced on any of the media tested. Growth occurs in 5–20% (w/v) NaCl (optimum 10%), at 20–45°C (optimum 37°C), and at pH 5.0–9.0 (optimum pH 7.0).

Arabinose, cellobiose, glucose, rhamnose, and sorbitol are used as sole carbon and energy sources; arginine, hydroxyproline, lysine, proline, threonine, and tryptophan are used as sole nitrogen, carbon, and energy sources. Positive for catalase, hydrolysis of starch, H_2S production, milk coagulation, and peptonization, but negative for oxidase, gelatin liquefaction, hydrolysis of cellulose and urea, melanin production, and nitrate reduction.

The diagnostic amino acid of the cell-wall peptidoglycan is *meso*-DAP and galactose is the major diagnostic sugar in whole-cell hydrolysates. The predominant menaquinones are MK-10(H_8) (35.8%), MK-10(H_6) (16.4%), and MK-9(H_8) (15.3%). The polar lipid profile consists of diphosphatidylglycerol, phosphatidylcholine, phosphatidylglycerol, phosphatidylinositol, and phosphatidylinositol mannosides. The major fatty acids are $C_{17:0}$ anteiso (42.4%) and $C_{18:0}$ (11.2%).

Source: the type strain was isolated from saline soil collected from a salt lake in Xinjiang Province, north-west China.

DNA G+C content (mol%): 71.2 (HPLC).

Type strain: YIM 91353, CCTCC AA 208048, DSM 45171.

Sequence accession no. (16S rRNA gene): EU442551.

4. **Streptomonospora flavalba** Cai, Tang, Chen, Li, Zhang and Li 2009, 2474[VP]

fla.val'ba. L. adj. *flavus -a -um* yellow; L. adj. *albus -a -um* white; N.L. fem. adj. *flavalba*, yellowish-white, referring to the color of the aerial mycelium.

Aerial and substrate mycelia are not fragmented. Yellow-white aerial mycelia form spore chains at maturity, with nonmotile, oval to cylindrical-shaped spores (0.5–0.6 × 0.7–0.9 μm). Single, oval spores are borne on sporophores of substrate mycelium. Substrate mycelia are: pale-yellow on inorganic salt/starch agar, oatmeal agar, yeast extract/malt extract agar, glycerol/asparagine agar, and nutrient agar; yellow-white on Czapek's agar; and ivory olive brown on potato extract agar. Growth occurs in 5–25% (w/v) NaCl (optimum 10%), at 20–45°C (optimum 37°C), and at pH 5.0–9.0 (optimum pH 7.0). Grows well on yeast extract/malt extract agar and nutrient agar media supplemented with 10% (w/v) NaCl, but poor growth occurs on Czapek's agar and glycerol/asparagine agar supplemented with 10% (w/v) NaCl. A yellow diffusible pigment is produced on potato extract agar (pH 7.0) supplemented with 10% (w/v) NaCl.

Arabinose, glucose, maltose, and rhamnose are used as sole carbon sources; alanine, arginine, glycine, histidine, phenylalanine, proline, serine, threonine, tryptophan, and tyrosine are used as sole nitrogen, carbon, and energy sources. Positive for catalase, hydrolysis of starch, melanin production, milk coagulation, and peptonization; negative for gelatin liquefaction, hydrolysis of cellulose and urea, H_2S production, nitrate reduction, and oxidase.

The diagnostic amino acid of the cell-wall peptidoglycan is *meso*-DAP and galactose is the major diagnostic sugar in whole-cell hydrolysates. The predominant menaquinones are MK-10(H_8) (41.5%), MK-9(H_8) (27.7%), and MK-10(H_6) (12.6%). The polar lipids consist of diphosphatidylglycerol, phosphatidylcholine, phosphatidylglycerol, phosphatidylinositol, and phosphatidylinositol mannosides. The major fatty acids are $C_{17:0}$ anteiso (28.0%) and $C_{16:0}$ iso (24.6%).

Source: the type strain was isolated from saline soil collected from a salt lake in Xinjiang Province, north-west China.

DNA G+C content (mol%): 72.5 (HPLC).

Type strain: YIM 91394, CCTCC AA 208047, DSM 45155.

Sequence accession no. (16S rRNA gene): EU442553.

5. **Streptomonospora halophila** Cai, Zhi, Tang, Zhang, Xu and Li 2008, 1559[VP]

ha.lo'phi.la. Gr. n. *hals halos* salt; N.L. adj. *philus -a -um* (from Gr. adj. *philos -ê -on*) friend, loving; N.L. fem. adj. *halophila* salt-loving, referring to the ability to grow at high NaCl concentrations.

Aerial and substrate mycelia are well developed, but not fragmented. Aerial mycelium forms straight to flexuous spore chains at maturity; spores are oval to cylindrical (0.4–0.5 × 0.8–1.0 μm) and nonmotile. Single, oval to round spores (0.8–1.0 μm diameter) are borne on sporophores of substrate mycelium. Colors of the substrate mycelium are: pale yellow on Czapek's agar, nutrient agar, potato agar, modified ISP 5 medium, and oatmeal agar; pale yellow-white on inorganic salts/starch agar; and yellowish brown on yeast extract/malt extract medium. Diffusible pigments are not produced. Ranges of NaCl concentration (w/v), pH, and temperature for growth are 5–20% (optimum 10%), pH 6.0–9.0 (optimum pH 7.0), and 20–45°C (optimum 37°C), respectively.

Inositol, maltose, and rhamnose are used as sole carbon sources; alanine, histidine, phenylalanine, proline, and threonine are used as sole nitrogen, carbon, and energy sources. Positive for milk peptonization, catalase, and hydrolysis of cellulose, but negative for oxidase, melanin production, amylase, nitrate reduction, and production of H_2S.

The diagnostic amino acid is *meso*-DAP; galactose is the main cell-wall sugar. The predominant menaquinones are MK-10(H_8) (43.2%), MK-10(H_6) (17.7%), and MK-11(H_8) (10.4%). Phospholipids are diphosphatidylglycerol, phosphatidylglycerol, phosphatidylcholine, phosphatidylinositol mannosides, phosphatidylinositol, and an unidentified phospholipid. The major fatty acids are $C_{16:0}$ iso (39.8%), $C_{17:0}$ anteiso (12.5%), 10-methyl $C_{17:0}$ (10.2%), and 10-methyl $C_{18:0}$ (11.1%).

Source: the type strain was isolated from a hypersaline soil in Xinjiang Province, north-west China.

DNA G+C content (mol%): 72.1 (HPLC).

Type strain: YIM 91355, DSM 45075, KCTC 19236.

Sequence accession no. (16S rRNA gene): EF423989.

Genus IV. **Thermobifida** Zhang, Wang and Ruan 1998, 417[VP] emend. Yang, Tang, Zhang, Zhi, Wang, Xu and Li 2008b, 1824

Martha E. Trujillo and Michael Goodfellow

Ther.mo.bi'fi.da. Gr. adj. *thermos* hot; L. adj. *bifidus* cleft; N.L. fem. n. *Thermobifida*, the heat (-loving) cleft (sporophores).

Aerobic, Gram-stain-positive, non-acid-fast, nonmotile, chemoorganotrophic actinomycetes that form an extensively branched, **non-fragmenting substrate mycelium. White aerial hyphae are produced. Single spores, oval to round (0.5–2.0 μm in diameter) are borne on dichotomously branched sporophores, resulting in spore clusters on the aerial mycelium and sometimes on the substrate mycelium. Spores have a smooth or scaly surface and may be heat sensitive. Cultures grow at 35–60°C and at pH 7–9. Many strains degrade cellulose. Cell walls contains *meso*-diaminopimelic acid.** Glucose, galactose, and xylose are found in whole-organism hydrolysates. **The major menaquinones are MK-10(H_6) and MK-10(H_8) and the predominant fatty acids are 14-methylpentadecanoic and 14-methylhexadecanoic acids**, but mycolic acids are not formed. The polar lipid pattern is complex and includes diphosphatidylglycerol, phosphatidylcholine, phosphatidylglycerol, phosphatidylinositol, phosphatidylmethylethanolamine, and an unknown phospholipid; the presence of phosphatidylethanolamine is variable. The phylogenetic position of the genus, as determined by 16S rRNA gene sequence analysis, is in the family *Nocardiopsaceae*.

DNA G+C content (mol%): 66–72 (HPLC, T_m).

Type species: **Thermobifida alba** (Locci, Baldacci and Petrolini 1967) Zhang, Wang and Ruan 1998, 418[VP].

Further descriptive information

Phylogeny. The genus *Thermobifida* belongs to the family *Nocardiopsaceae* together with the genera *Haloactinospora* (Tang et al., 2008), *Nocardiopsis*, and *Streptomonospora* (Cui et al., 2001; Yang et al., 2008b). Based on 16S rRNA gene sequences, the four *Thermobifida* species form a homogeneous cluster supported by a high bootstrap value (see Figure 389 in the treatment of the family *Nocardiopsaceae*). *Thermobifida alba* (previously *Thermomonospora alba*), *Thermobifida cellulosilytica*, *Thermobifida fusca* (previously *Thermomonospora fusca*), and *Thermobifida halotolerans* share 16S rRNA gene sequence similarities between 97.5 and 98.2%. The type strains of the four species show DNA–DNA reassociation values between 33 and 50%.

Cell morphology. *Thermobifida* strains form highly branched, non-fragmenting substrate mycelia and produce white aerial mycelium on nutrient agar. Sporulation occurs by septation of sheathed, branched aerial hyphae and this leads to the formation of single spores (Crawford and Gonda, 1977). The latter, which are oval to round (0.5–2.0 µm in diameter), are borne at the tips of repeatedly branched sporophores, and form spore clusters on the aerial mycelium and sometimes on the substrate mycelium (Figure 396). Spore surfaces are smooth in all species, except for *Thermobifida cellulosilytica* where they have a scale-like appearance resembling pine-cones (Figure 397). Heat-sensitive spores have been reported for *Thermobifida alba*, *Thermobifida cellulosilytica*, and *Thermobifida fusca*.

Nutrition and growth conditions. Members of the genus *Thermobifida* are aerobic and chemo-organotrophic. The four type strains grow well on yeast extract-malt extract agar (ISP medium 2) but poorly on oatmeal, inorganic salts-starch and glycerol-asparagine agars (ISP media 3, 4 and 5; Shirling and Gottlieb, 1966); they also grow well on nutrient and potato agar plates. *Thermobifida alba* and *Thermobifida halotolerans* grow between 20 and 50°C, whereas *Thermobifida cellulosilytica* and *Thermobifida fusca* grow between 28 and 60°C. All of the strains grow between pH 6 and 9. Most thermobifidae grow in the presence of 3% NaCl, but *Thermobifida halotolerans* can grow in media containing 10% (w/v) NaCl (Yang et al., 2008b).

Cell-wall composition. *Thermobifida* strains contain *meso*-diaminopimelic acid (*meso*-A_2pm) as the wall diamino acid, but lack diagnostic sugars (Kroppenstedt and Evtushenko, 2006; Yang et al., 2008b), that is, they have a wall chemotype III *sensu* Lechevalier and Lechevalier (1970b). They contain MK-10(H_6) and MK-10(H_8) as predominant menaquinones and additional components, such as MK-10(H_{10}), MK-11(H_6), and MK-11(H_8), which distinguish between the four species (Kroppenstedt and Goodfellow, 1992; Kukolya et al., 2002; Yang et al., 2008b). Differences are also found between the fatty acid profiles of the four species although, in all cases, the major components are 14-methylpentadecanoic acid ($C_{16:0}$ iso) and 14-methylhexadecanoic acid ($C_{17:0}$ anteiso) (Yang et al., 2008b). The cellular polar lipid patterns include diphosphatidylglycerol, phosphatidylcholine, phosphatidylglycerol,

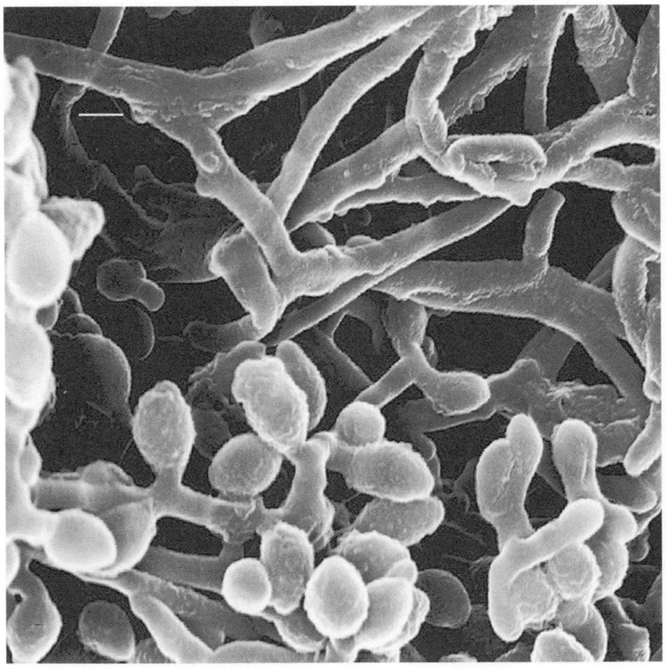

FIGURE 396. Spore formation in *Thermobifida fusca* strain TM51. Bar = 2 µm. (Reproduced with permission from Kukolya et al., 2001. Acta Biol. Hung. *52*: 211–221.)

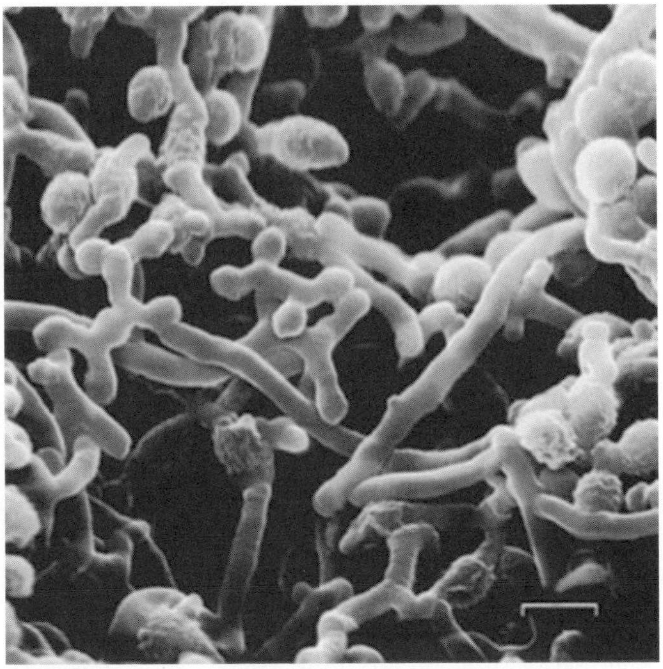

FIGURE 397. Spores of strain *Thermobifida cellulosilytica* TB100[T] formed on aerial hyphae, at the tips of branched sporophores. Bar = 1 µm. (Reproduced with permission from Kukolya et al., 2002. Int. J. Syst. Evol. Microbiol. *52*: 1193–1199.)

phosphatidylinositol, phosphatidylmethylethanolamine, and an unknown phospholipid; *Thermobifida cellulosilytica*, *Thermobifida fusca*, and *Thermobifida halotolerans* also contain phosphatidylethanolamine (Kroppenstedt and Evtushenko, 2006; Yang et al., 2008b). The cell wall of *Thermobifida fusca* contains two teichoic acids, namely, unsubstituted 1,3-poly(glycerol phosphate) and β-glucosylated 1,3-poly(glycerol phosphate) (Potekhina et al., 2003), and a polyglycerophosphate lipoteichoic acid (Rahman et al., 2009).

Metabolism. *Thermobifida* strains are aerobic, catalase-positive, chemo-organotrophic actinomycetes that have an oxidative metabolism. They metabolize a broad range of organic substrates and use diverse sugars as sole carbon sources (Crawford and McCoy, 1972; Kroppenstedt and Evtushenko, 2006; Kukolya et al., 2002; McCarthy and Cross, 1984a; Yang et al., 2008b). Preliminary data suggest that they can cleave a range of 7-amino-4-methylcoumarin and 4-methylumbelliferone-conjugated fluorogenic compounds (Trujillo and Goodfellow, 2003). However, thermobifidae are best known as a source of multiple thermostable enzymes, notably extracellular cellulases (Ball and McCarthy, 1988; Lao et al., 1991; McCarthy, 1987; McGrath and Wilson, 2006; Wilson, 2004) and hemicellulases (McCarthy et al., 1985, 1988).

Thermobifida fusca cellulases have been studied extensively because of their thermostability, broad pH range, and high activity (Beadle et al., 1999; Bellamy, 1973; Irwin et al., 2000; Spiridonov and Wilson, 1998; Wilson, 2004). This organism also has the capacity to degrade plant cell-wall polymers with the exception of lignin. *Thermobifida fusca* YX, a primary degrader of plant cell walls in heated organic matter, grows at 50°C in minimal medium containing cellulose, mannan, starch, or xylan as sole carbon sources and is used as a model system for establishing the complex interactions of glucanases involved in biomass hydrolysis (Warren, 1996).

Extensive studies on *Thermobifida fusca* have led to the identification, purification, and characterization of six extracellular cellulases (Irwin et al., 1993, 1994, 2000, 2003; Wilson et al., 1988; Zhang et al., 1995). In addition, an intracellular β-1,3-glucosidase that degrades cellulose to glucose (Spiridonov and Wilson, 2001), a mannanase (Beki et al., 2003; Hilge et al., 1998, 2001), an extracellular xyloglucanase (Irwin et al., 2003), and the three xylanases (Kim et al., 2004) have been detected, as has the protein CelR, which regulates the induction of cellulases and related enzymes (Wilson, 2004). The structures for three catalytic domains, Cel5Acd, Cel6Acd (Spezio et al., 1993), and Cel9A68 (Sakon et al., 1997), have been elucidated, and the three dimensional structure of *Thermobifida fusca* Cel6A catalytic domain has been determined by X-ray crystallography (Spezio et al., 1993). Components of the *Thermobifida fusca* xylan-degrading system have been purified and characterized (Bachmann and McCarthy, 1989, 1991; Ball and McCarthy, 1989; McCarthy et al., 1985).

The six extracellular enzymes act synergistically to degrade cellulose to cellobiose, traces of cellotriose, and other sugars (Irwin et al., 1993). The released cellobiose is transported into cells where it is converted to glucose by intracellular β-glucosidase (Ball and McCarthy, 1988; Ozaki and Yamada, 1991). Cellulase synthesis in *Thermobifida fusca* is regulated in two ways, induction by cellulose and cellobiose, and carbon source repression (Lin and Wilson, 1987, 1988; Spiridonov and Wilson, 1998). Induction increases cellulase levels about 12-fold and carbon source repression about 9-fold. Other major extracellular proteins induced by cellobiose include a xylanase and two unknown proteins (Chen and Wilson, 2007). *Thermobifida fusca* strains are also very active against arabinoxylan (McCarthy et al., 1985) and can degrade lignocellulosic pulps (Crawford, 1974).

Secreted cellulases are a source of single cell protein (Crawford, 1988; Wood, 1985), while the breakdown of biomass into fermentable sugars has potential for the industrial production of ethanol (Lynd et al., 2002, 2005; Mielenz, 2001; Rubin, 2008). Thermobifidae and other monosporic actinomycetes have been evaluated as a source of single cell protein (Bellamy, 1974, 1977; Crawford, 1988) and ethanol (Ball and McCarthy, 1988; Hägerdahl et al. 1979, 1980; Lee and Humphrey, 1979). Indeed, *Thermobifida fusca* has been used to convert high cellulose, low lignin pulp mill waste into single cell protein (Crawford et al., 1973) that was used as a feed supplement in the diet of chickens (Harkin et al., 1974). The thermostable mannanase produced by this organism has a temperature optimum of 80°C; it hydrolyzes the *O*-glucosidic bonds of mannan and has potential use in pulp and paper production (Hilge et al., 1998; Hilge et al., 2001).

Thermophilic actinomycetes have been used in processes designed to achieve the saccharification of purified cellulose (Bachmann and McCarthy, 1991; Wood, 1985). The cellulases of *Thermobifida fusca* YX, for instance, saccharify finely ground, acid-swollen Avicel, a purified chemically pretreated cellulose (Ferchak, 1980; Su and Paulavicius, 1975). In subsequent studies, Ferchak and Pye (1983) established a process for the saccharification of cellulose using the *Thermobifida fusca* strain and a cellulose/cellobiose mixture; 15–20% glucose syrups were generated from acid-swollen Avicel after 7 d at 50°C. It can be anticipated that the commercial potential of such processes will be improved by the use of cellulase-overproducing *Thermobifida* strains (Meyer and Humphrey, 1982). A composting process for aquatic plants has been developed that involves the use of *Thermobifida fusca* and *Ureibacillus thermosphaericus*, a Gram-stain-negative organism that is unable to ferment sugars (Okuda et al., 2008).

The genome of *Thermobifida fusca* YX has been sequenced (Lykidis et al., 2007), a development that will be of further interest in the metabolism and biotechnological exploitation of this and related organisms. Lykidis and colleagues have shown that the *Thermobifida fusca* genome encodes 45 hydrolytic enzymes, all of the enzymes needed for the glycolytic degradation of monosaccharides and for the biosynthesis of fatty acids, purines and pyrimidines, and enzymes for the *de novo* biosynthesis of all amino acids, apart from arginine. Additionally, glucose utilization is effected through the Entner–Doudoroff pathway. The organism possesses two protein secretion systems, the *sec* general secretion system and the twin-arginine translocation system. Indeed, several of the secreted cellulases have sequence signatures, which indicate that their secretion may be mediated through the twin-arginine translocation system. Extensive transport systems have been detected for import of carbohydrates coupled to transcriptional regulators that control the expression of the transporters and glycosylhydrolases.

Few characterized natural products have been isolated from either archaea or thermophilic bacteria (Donadio et al., 2007).

It is interesting, therefore, that *Thermobifida fusca* YX contains a siderophore biosynthetic gene cluster and has genes that encode a siderophore transport system (Lykidis et al., 2007). This organism also contains a family of structurally related novel nonribosomal peptide siderophores, the fuscachelins, which are products of an orphan gene cluster (Dimise et al., 2008). Fuscachelin A is the first secondary metabolite isolated from *Thermobifida fusca* and the natural product gene cluster is one of only a few detected from a thermophilic micro-organism.

Thermobifida strains produce medically important compounds, as exemplified by the production of topostatin, a novel inhibitor of topoisomerases 1 and 11 (Suzuki et al., 1998a, 1998b). Isoaurastatin, which inhibits topoisomerase 1, has been characterized and shown to be 6,4-dehydroxyisoaurone (Suzuki et al., 2001). A novel non-heme, iron-containing, extracellular peroxidase has been isolated from *Thermobifida fusca* BD25 and its catalytic mechanism has been characterized (Rob et al., 1995, 1996, 1997a, 1997b).

Genetics. In general, genetic studies on *Thermobifida fusca* have been designed to gain an insight into the cellular mechanisms that control the expression and secretion of plant cell wall-degrading enzymes (Wilson, 2004). Genes encoding endoglucanases, exoglucanases, mannosidases, and xylanases have been cloned, sequenced, and expressed in one or more heterologous hosts, namely *Bacillus subtilis* (Ghangas and Wilson, 1987), *Escherichia coli* (Chen and Wilson, 2007; Ghangas et al., 1989; Ghangas and Wilson, 1987, 1988; Irwin et al., 1994, 2000, 2003; Jung et al., 1993; Moser et al., 2008; Spiridonov and Wilson, 2000, 2001), *Streptomyces lividans* (Beki et al., 2003; Ghangas et al., 1989; Ghangas and Wilson, 1988; Irwin et al., 1998, 2000; Zhang et al., 2000; Zhang and Wilson, 1997), and *Pichia pastoris* (Cheng et al., 2005), as a first step towards constructing vehicles for the production of high levels of thermostable enzymes. The xylanase productivity of the *Pichia pastoris* transformant (pPICxXYL) was nearly 70-fold higher than that of the *Thermobifida fusca* strain when cultured in a 5-l fermenter.

Genes encoding two different endo-β-1,4-xylanases have been cloned and expressed in a *Thermobifida alba* strain (Blanco et al., 1997). One of them, *xylA*, was sequenced, subcloned, and overexpressed in *Streptomyces lividans*, and found to encode a protein of 482 amino acids with a deduced molecular mass of 48,456 Da. This protein bound strongly to cellulose and degraded xylans from different origins, producing xylobiose and traces of xylose. There is evidence that chaplin genes are present in *Thermobifida fusca* (Elliot et al., 2003); chaplin amyloids on the surfaces of actinomycetes mediate attachment to hydrophobic surfaces and are responsible for making the surfaces of hyphae and spores hydrophobic (Gebbink et al., 2005).

Thermobifida fusca strains from self-heated organic matter have been found to degrade several diverse biodegradable polyesters, including an aliphatic-aromatic co-polyester (Kleeberg et al., 1998). The gene encoding the *Thermobifida fusca* hydrolase responsible for degrading the co-polyester has been cloned and expressed in *Bacillus megaterium* (Yang et al., 2007b) and *Escherichia coli* (Dresler et al., 2006). In addition, a polyhydroxyalkanoate gene in a *Thermobifida* isolate has been cloned and expressed as a recombinant enzyme (PhaZ-Th) in *Pichia pastoris* (Phithakrotchanakoon et al., 2009). This enzyme is active against *p*-nitrophenyl alkanoates; surface plasmon resonance analysis indicated that PhaZ-Th catalyzed the degradation of poly[(R)-3-hydroxybutyrate] (PHB) films. Surface deterioration of PHB films was observed following exposure to PhaZ-Th by atomic force microscopy.

The genome sequence of *Thermobifida fusca* YX, which consists of a circular chromosome (3.6 Mb), provides the means for elucidating the cellular mechanisms that underpin the metabolism of the organism, including those that control the expression and secretion of cellulases, pectinases, and xylanases (Lykidis et al., 2007). The organism appears to use the TAT secretion system for the export of these enzymes to the extracellular space and produces multiple transcription factors that regulate the expression of glycosylhydrolases and oligo/polysaccharide transport systems. To date, 68% of the 3117 predicted coding sequences have been assigned a function, with 106 genes (3.3%) appearing to be unique to *Thermobifida fusca*. Comparative genome approaches indicate that *Thermobifida fusca* and "*Streptomyces coelicolor*" share a number of developmental genes (Chater and Chandra, 2006; Li et al., 2007; Traag and van Wezel, 2008). *Thermobifida fusca* is one of the few actinobacterial species that does not contain a 5 amino acid insert in glutamyl-tRNA synthase (Gao and Gupta, 2005), an enzyme that plays an essential role in protein synthesis (Woese et al., 2000).

Antibiotic sensitivity. Most *Thermobifida* strains are sensitive to a broad range of antibacterial antibiotics (Kukolya et al., 2002; McCarthy and Cross, 1984a; Trujillo and Goodfellow, 2003) though comparative taxonomic studies on representatives of *Thermobifida* species with validly published names are needed. The antibiotic sensitivity profiles of individual *Thermobifida* species, where known, are provided in the species descriptions.

Pathogenicity. There is no definitive evidence that *Thermobifida* strains, like some other thermophilic actinomycetes, can cause allergic respiratory disorders, though the cause of mushroom workers' disease, a form of allergic alveolitis, is still not settled (Lacey, 1988) It has been claimed that *Thermobifida* aleuriospores may act as allergens (Gusek et al., 1991). In addition, there is evidence that suggests that *Thermobifida alba* and *Thermobifida fusca* may contribute to the occurrence of mushroom workers' lung disease (Van den Bogart et al., 1993).

Ecology. *Thermobifida* strains form an integral part of the authochthonous microflora of overheated plant materials, such as composts, manure heaps, stored hay, and organic household waste (Lacey, 1973; Steger et al., 2007). They are particularly abundant in mushroom composts (Fergus, 1964; Lacey, 1974, 1978; McCarthy and Cross, 1984a, 1984b; McCarthy and Broda, 1984; Song et al., 2001) and in composting anaerobic sludge containing high concentrations of fibers and lipids (Nakasaki et al., 2009). The thermostable, extracellular, cellulolytic (Ball and McCarthy, 1988; McCarthy, 1987), and/or hemicellulolytic (McCarthy et al., 1985, 1988) enzymes produced by *Thermobifida alba* and *Thermobifida fusca* allow them to become dominant in overheated plant materials. *Thermobifida fusca* may be involved in the formation of humic substances in soil (Trigo and Ball, 1994).

Biofilm formation may play a major role in the developmental cycle of *Thermobifida fusca* (Alonso et al., 2008). These workers found that cellulose was specifically colonized by aleuriospores, which germinated, grew, degraded cellulose, and finally differentiated into biofilms encased in a carbohydrate-containing exopolymeric matrix. Cellulose degradation and expression of *celE*, which encodes endoglucanase E5, was similar for *Thermobifida fusca* biofilms and mycelial pellets. *Thermobifida fusca* cells were also found to grow as biofilms attached to both nutritive and non-nutritive surfaces. Alonso and colleagues speculated that biofilm formation might enable *Thermobifida fusca* to adhere to substrates and obtain nutrients, thereby gaining a selective advantage over competitors in composts. They concluded that *Thermobifida fusca* might prove to be an effective model for filamentous bacterial biofilm research.

Enrichment and isolation procedures

Biocomposted paper (Kempf, 1995), mushroom solid waste compost (Resz et al., 1977; Stutzenberger, 1971), mushroom compost (Fergus, 1964), and self-heated grass cuttings (Kutzner, 2000) are rich sources of *Thermobifida* strains. Isolation of strains from such materials can be achieved by suspending 1 g air-dried sample and 10 g sterile plastic beads in 100 ml Ringer's solution followed by shaking at 180 r.p.m. for 30 min to release spores and mycelial fragments from compost particles. Next, 0.1 ml aliquots of a 10-fold dilution series are plated onto modified starch-casein-nitrate agar plates (Kroppenstedt and Evtushenko, 2006); this medium should be predried to reduce swarming of aerobic, endospore-forming bacilli. Sodium propionate (4 g/l) may be added to the medium to reduce the number of unwanted fast-growing bacteria (Lacey, 1974), although this practice may lead to a reduction in the counts of actinomycetes which form aerial mycelia. *Thermobifida* strains can usually be detected after 3–5 d incubation at 50°C.

Thermobifida strains can be isolated using nonselective agar media, but recovery is poor due to the rapid competing growth of thermophilic aerobic endospore-forming bacilli and thermoactinomycetes. The most effective isolation methods are those based on the use of a sedimentation chamber and Andersen air sampler (Lacey and Dutkiewicz, 1976; McCarthy and Broda, 1984). Dried samples are agitated within the sedimentation chamber to create an aerosol of particles which, after 1–2 h of sedimentation, still contains many actinomycete spores, but comparatively few bacteria. *Thermobifida* strains are isolated from this spore suspension using an Andersen sampler loaded with half-strength tryptone soy agar plates supplemented with cycloheximide (50 µg/ml) to prevent growth of fungi. The inoculated plates are incubated at 50°C for 5 d. The isolation of *Thermobifida* strains may be further improved by adjusting the isolation medium to pH 11.0 (Cross, 1981). Cellulolytic isolates can be identified by incorporating cellulose powder or ball-milled straw into the agar (McCarthy and Broda, 1984; Stutzenberger et al., 1970).

Thermobifida halotolerans was isolated from a salt mine sample after incubation for 3 weeks at 45°C on inorganic salts-starch agar (ISP medium 4; Shirling and Gottlieb, 1966) supplemented with 10% NaCl (Yang et al., 2008b). Putatively novel *Thermobifida* strains have been isolated from mushroom compost on R8 agar plates incubated at 50°C for 3–5 d (Song et al., 2001).

Maintenance procedures

Working cultures of *Thermobifida* can be maintained as sporulating cultures at 4°C for up to 4 months on Czapek–Dox-yeast extract-Casamino acids agar at pH 8.0. Drying out of agar can be prevented by tightly sealing tubes with silicone stoppers. Medium-term preservation for up to 4 years can be achieved by preparing spore suspensions or homogenized mycelia in 20% (v/v) glycerol and storing at −20°C (Wellington and Williams, 1978). For long-term preservation, lyophilization of spores and mycelia suspended in 10% skim milk is a convenient method.

Differentiation of the genus *Thermobifida* from other genera

Thermobifida strains can be readily differentiated from members of the three other genera classified in the family *Nocardiopsaceae* by using a combination of chemotaxonomic, growth, morphological, and physiological properties (Table 297, see also

TABLE 297. Differential characteristics of the type strains of members of the genus *Thermobifida*

Character	*T. alba* DSM 43795^T	*T. cellulosilytica* DSM 44535^T	*T. fusca* DSM 43792^T	*T. halotolerans* YIM 90462^T
Aerial mycelium on glucose-yeast extract-malt extract agar	−	+	−	−
Hydrolysis of:				
Arbutin	+	+	+	−
Gelatin	+	+	+	−
Coagulation of milk	+	+	+	−
Peptonization of milk	+	+	+	−
Nitrate reduction	−	−	−	+
Growth on sole carbon sources:				
D-Arabinose	+	−	−	+
D-Fructose	+	−	+	−
Glycerol	+	−	+	+
Lactose	+	−	+	−
Maltose	+	+	+	−
D-Mannose	+	+	+	−
L-Rhamnose	+	+	−	−
D-Ribose	−	+	−	−
L-Sorbose	+	−	−	−
Xylitol	+	−	−	−
Temperature growth range (°C)	20–50	28–55	28–55	20–50
Presence of phosphatidylethanolamine (diagnostic polar lipid)	−	+	+	+
DNA G+C content (mol%)	71.2	70.4	66.2	69.0

treatment of the family *Nocardiopsaceae*) and by 16S rRNA gene sequence studies (see Figure 389 in the treatment of the family *Nocardiopsaceae*). They can be easily separated from their nearest phylogenetic neighbor, the genus *Nocardiopsis*, as members of this taxon form long chains of spores on aerial hyphae and produce a substrate mycelium that tends to fragment. It has also been shown that the genus-specific primers used for PCR identification of *Nocardiopsis* strains do not produce amplification with DNA from *Thermobifida alba* (Salazar et al., 2002).

Taxonomic comments

The taxonomic history of actinomycetes classified in the genus *Thermobifida* as *Thermobifida alba* and *Thermobifida fusca* is complex, confused, and controversial, as is apparent from early reviews on the classification of monosporic actinomycetes (Cross and Lacey, 1970; Cross and Goodfellow, 1973). Indeed, actinomycetes that formed single spores were assigned somewhat unpredictably to a number of taxa, notably to the genera *Actinobifida* Krasil'nikov and Agre 1964, *Micromonospora* Ørskov 1926, *Saccharomonospora* Nonomura and Ohara 1971, *Thermoactinomyces* Tsiklinsky 1899, and *Thermomonospora* Henssen 1957. All of these taxa were essentially morphological in concept. The dangers of over-relying on morphological features was illustrated by Cross and Lacey (1970), who found an almost continuous range of morphological variation between two extremes represented by *Thermomonospora viridis* and *Thermoactinomyces vulgaris*. Between these two extremes, they observed a range of morphological types that overlapped generic boundaries. They also noted that spore arrangement on the aerial mycelium could be influenced by medium composition and incubation temperature.

The rejection of the genus *Actinobifida* as an invalid name (Cross and Goodfellow, 1973; Skerman et al., 1980) and improvements in the classification of the genus *Thermoactinomyces* (Cross and Goodfellow, 1973) helped to define *Thermomonospora* at the genus level. Subsequently, the application of chemotaxonomic and numerical taxonomic methods showed the genus *Thermomonospora* to be a markedly heterogeneous taxon (Kroppenstedt and Evtushenko, 2006; Kroppenstedt and Goodfellow, 1992; McCarthy and Cross, 1984a, 1984b). The detailed taxonomic changes that ensued are presented elsewhere in this volume (see treatment of the family *Thermomonosporaceae*).

Thermomonospora alba DSM 4395^{T}, the first viable member of the genus *Thermomonospora*, was isolated and described by Locci et al. (1967), who classified it as *Actinobifida alba*. The organism was transferred to the genus *Thermomonospora* as it formed heat-sensitive spores on substrate and aerial hyphae (Cross and Goodfellow, 1973). Subsequently, *Thermomonospora mesouviformis* Nonomura and Ohara 1974 was proposed as a synonym of *Thermomonospora alba* McCarthy and Cross 1984a.

Thermomonospora fusca was proposed on the basis of distinctive morphological properties observed in contaminated preparations (Henssen, 1957). The organism was omitted from the Approved Lists of Bacterial Names (Skerman et al., 1980), even though a *Thermomonospora fusca* strain had been isolated and described in detail (Crawford, 1975; Crawford and Gonda, 1977). *Thermomonospora fusca* was later proposed as a validly published name (McCarthy and Cross, 1984a) and added to the Approved Lists of Bacterial Names (Moore et al., 1985).

Kroppenstedt and Goodfellow (1992) assigned *Thermomonospora* strains to three distinct groups, based on the discontinuous distribution of chemical markers, and raised the prospect that *Thermomonospora alba* (including *Thermomonospora mesouviformis*) and *Thermomonospora fusca* merited generic status. Zhang et al. (1998) found that the type strains of *Thermomonospora alba* and *Thermomonospora fusca* were more closely related to *Nocardiopsis* strains than to other *Thermomonospora* species on the basis of 16S rRNA gene sequence data. These results, together with associated chemotaxonomic and phenotypic data, led them to propose the transfer of *Thermomonospora alba* (Locci et al. 1967) Cross and Goodfellow 1973 and *Thermomonospora fusca* (*ex* Henssen 1957) McCarthy and Cross 1984a to the new genus *Thermobifida* as *Thermobifida alba* Locci et al., 1967 comb. nov. and *Thermobifida fusca* McCarthy and Cross 1984a comb. nov. *Thermobifida alba* was designated the type species of the genus as it had been validated before *Thermobifida fusca* (by Henssen, 1957); *Thermobifida cellulosilytica* Kukolya et al. 2002 and *Thermobifida holotolerans* Yang et al. 2008b were subsequently added to the genus.

Differentiation of the species of the genus *Thermobifida*

Thermobifida species can be distinguished by using a combination of chemotaxonomic, morphological, nutritional, and physiological properties (Table 297). The two most studied species, *Thermobifida alba* and *Thermobifida fusca*, have different temperature and pH requirements (Kroppenstedt and Evtushenko, 2006; McCarthy and Cross, 1984a), give different patterns of enzymes in cell-free extracts (Greiner-Mai et al., 1987), and show quantitative differences in fatty acid profiles (Kroppenstedt and Evtushenko, 2006; Yang et al., 2008b). Additionally, unlike *Thermobifida alba*, most *Thermobifida fusca* strains form spores on the substrate mycelium (McCarthy and Cross, 1984a).

List of species of the genus *Thermobifida*

1. **Thermobifida alba** (Locci, Baldacci and Petrolini 1967) Zhang, Wang and Ruan 1998, 418^{VP} [*Actinobifida alba* Locci, Baldacci and Petrolini 1967, 88; heterotypic synonym: *Thermomonospora mesouviformis* Nonomura and Ohara 1974; basonym: *Thermomonospora alba* (Locci, Baldacci and Petrolini 1967) Cross and Goodfellow 1973.]

   al'ba. L. fem. adj. *alba* white.

   Forms an extensively branched substrate mycelium and abundant white aerial hyphae on agar media. Large numbers of single spores are borne on dichotomously branched or unbranched sporophores on aerial hyphae in dense clusters. Does not produce melanin pigments on tyrosine agar. The optimal temperature for growth and sporulation is 40–45°C; growth occurs at 30°C and at pH 10 and 11.

Positive for esculin, catalase, β-galactosidase, β-glucosidase, and phosphatase, but negative for oxidase. Degrades agar, carboxymethylcellulose, casein, DNA, keratin, pectin, RNA, tributyrin, Tween 60, and xylan, but not guanine, hypoxanthine, testosterone, tyrosine, or xanthine. Variable for elastin. Acid is produced from glucose. Glycogen and mannitol are used as sole carbon sources for energy and growth, but not adonitol, L-arabinose, cellobiose, glucosamine, glycerol, *myo*-inositol, mannitol, melezitose, melibiose, raffinose, or salicin (all at 1%, w/v). Similarly, lactic acid is used as a sole carbon source, but not L-arginine, benzamide, L-cystine, ethanolamine, L-glutamine, glycine, L-histidine, L-lysine, L-phenylalanine, sodium hippurate, sodium propionate, sodium pyruvate, or L-valine.

Grows in the presence of [% (w/v) except where indicated]: bile (0.02), phenol (0.01), phenyl ethanol (0.1, v/v), and thallous acetate (0.001), but is inhibited by brilliant green (0.0005), crystal violet (0.00002), phenol (0.1), phenyl ethanol (0.2%, v/v), potassium tellurite (0.0005), sodium azide (0.01), sodium chloride (3), tetrazolium chloride (0.002), and thallous acetate (0.005). Sensitive to (µg per filter paper disc): ampicillin (5), carbenicillin (25), cephaloridine (2), cephalotin (20), cephamandole (20), kanamycin (25), lincomycin (25), neomycin (3), novobiocin (40), oleandomycin (20), rifampin (1), streptomycin (16), tetracycline (20), and tobramycin (5).

Additional phenotypic properties are shown in Table 297. The major fatty acids are $C_{16:0}$ (24.4%) and $C_{17:0}$ anteiso (29.1%).

Source: the type strain was isolated from garden soil.

DNA G+C content (mol%): 71.2 (HPLC).

Type strain: CIP 105591, DSM 43795, IMET 9528, IPV, JCM 3077, KCCA-0077, NBRC 16095.

Sequence accession no. (16S rRNA gene): AF002260.

Additional comments: cleaves a broad range of fluorogenic substrates based on 7-amino-4-methylcoumarin and 4-methylumbelliferone (Trujillo and Goodfellow, 2003). McCarthy (1989) pointed out that the distinction between *Thermobifida alba* and *Thermobifida mesouviformis* had been maintained only because the latter was regarded as a mesophilic species unable to grow at 50–55°C (Cross, 1981; Kurup, 1979; Nonomura and Ohara, 1974). He also noted that the type strain of *Thermobifida mesouviformis* can show poor growth at 50°C; hence, these species should be considered to be synonymous. However, Kroppenstedt and Evtushenko (2006) believe that DNA–DNA hybridization studies are needed to determine the relationship between the two species. *Thermobifida alba* strains do not form a well-delineated species and may be low-temperature variants of *Thermobifida fusca* (see McCarthy and Cross, 1984a; McCarthy, 1989).

2. **Thermobifida cellulosilytica** corrig. Kukolya, Nagy, Láday, Tóth, Oravecz, Márialigeti and Hornok 2002, 1198[VP]

cel.lu.lo.si.ly'ti.ca. N.L. n. *cellulosum* cellulose; N.L. fem. adj. *lytica* (from Gr. fem. adj. *lutikê*) able to loosen, able to dissolve; N.L. fem. adj. *cellulosilytica* cellulose-dissolving.

Forms an extensively branched substrate mycelium and abundant white aerial hyphae on agar media. Sporophores branch repeatedly to form dense spore clusters on the aerial mycelium. Spores have a scale-like surface, are oval (1–1.3 × 0.6 µm) and heat-sensitive (90°C, 10 min). Thermotolerant strains grow between 28 and 45°C and thermophilic ones between 40 and 65°C.

Degrades crystalline cellulose and elastin. Does not produce acid from glucose. Cellobiose, galactose, melezitose, and xylose are used as carbon sources, but not *myo*-inositol, dulcitol, or sorbitol. Cultures grown on lignocellulose substrate produce endoglucanase, cellobiohydrolase, and endoxylanase enzymes and are capable of solubilizing lignocarbohydrates.

Resistant to (µg/ml) erythromycin (20), but sensitive to amoxycillin (10), ampicillin (10), apramycin (2.5), chloramphenicol (10), kanamycin (2.5), novobiocin (10), penicillin (10) streptomycin (2.5), tetracycline (5), and thistrepton (1.25).

Additional phenotypic characteristics are shown in Table 297. The major fatty acids are $C_{16:0}$ (28.7%), $C_{17:0}$ (26.1%), $C_{17:0}$ anteiso (18.3%), and $C_{18:0}$ (12.4%). The cellular polar lipid pattern includes phosphatidylethanolamine, a glycolipid, and unidentified phospholipids. The predominant menaquinones are MK-10(H_6) and MK-10(H_8); the minor components vary, apart from the presence of MK-11(H_6).

Source: isolated from overheated manure compost in Gödöllö, Hungary.

DNA G+C content (mol%): 70–71 (HPLC).

Type strain: DSM 44535, JCM 11767, NCAIM B0, TB100.

Sequence accession no. (16S rRNA gene): AJ298058.

Additional comments: the original spelling *Thermobifida cellulolytica* (*sic*) was corrected to *Thermobifida cellulosilytica* in Validation List no. 37 by the List Editor, IJSEM (2002b).

3. **Thermobifida fusca** (McCarthy and Cross 1984a) Zhang, Wang and Ruan 1998, 418[VP] (McCarthy and Cross 1984a, 22) [Basonym: *Thermomonospora fusca* (*ex* Henssen 1957) McCarthy and Cross 1984a].

fus'ca. L. fem. adj. *fusca* dark, tawny.

Forms an extensively branched substrate mycelium and abundant white aerial hyphae on agar media. Large numbers of single spores in dense clusters are usually produced on the aerial mycelium. The colony reserve pigment on CYC agar is pale yellow. Does not form melanin pigments on tyrosine agar. Grows at pH 10–11 and at 35–53°C, but not at 30°C; some strains grow at 60°C.

Positive for esculin, catalase, β-galactosidase, β-glucosidase, and phosphatase, but negative for oxidase. Degrades agar, carboxymethylcellulose, casein, elastin, DNA, keratin, pectin, RNA, tributyrin, Tween 60, and xylan, but not chitin, guanine, hypoxanthine, testosterone, tyrosine, or xanthine. Produces acid from glucose. Cellobiose, galactose, glycogen, and melezitose are used as carbon sources for energy and growth, but not adonitol, L-arabinose, glucosamine, *myo*-inositol, mannitol, melibiose, raffinose, or xylose (all at 1%, w/v). Similarly, sodium propionate and sodium pyruvate are used as sole carbon sources, but not L-arginine, benzamide, ethanolamine, L-cystine, L-glutamine, glycine, lactic acid, L-histidine, L-lysine, L-phenylalanine, sodium hippurate, or L-valine.

Grows in the presence of (%, w/v) bile (0.05), crystal violet (0.00002), phenol (0.01), phenyl ethanol (0.2%, v/v), potassium tellurite (0.02), sodium azide (0.01), tetrazolium chloride (0.02), and thallous acetate (0.01), but is inhibited by phenol (0.1), phenyl ethanol (0.3%, v/v), sodium azide (0.02), tetrazolium chloride (0.005), and thallous acetate (0.005). Sensitive to (μg per filter paper disc): ampicillin (5), carbenicillin (25), cephaloridine (2), cephalothin (20), cephamandole (20), kanamycin (25), lincomycin (10), neomycin (3), novobiocin (10), oleandomycin (20), rifampin (1), streptomycin (4), tetracycline (20), ticarcillin (70), and tobramycin (0.05).

Additional phenotypic characteristics are shown in Table 297. Major fatty acids are $C_{16:0}$ (38.5%) and $C_{17:0}$ anteiso (15.2%).

Source: isolated from composts and overheated fodders. The type strain was isolated from soil.

DNA G+C content (mol%): 66.2 (HPLC).

Type strain: ATCC 27730, CIP 105594, DSM 43792, IFO (now NBRC) 14071, JCM 3263, NCIB 11185, NRRL B-8184, 190 Th, VKM Ac-.

Sequence accession no. (16S rRNA gene): AF002264.

Additional comments: cleaves a broad range of fluorogenic substrates based on 7-amino-4-methylcoumarin and 4-methylumbelliferone (Trujillo and Goodfellow, 2003). *Thermobifida fusca* was not originally isolated in pure culture (Henssen, 1957) and the unavailability of cultures led to the species being cited as a *nomen dubium* in the 8th edition of the *Manual.* Furthermore, a wholly inaccurate description of *Thermobifida fusca* Henssen, given by Waksman (1961a), led to the use of this name for strains that probably belong to *Thermomonospora chromogena* (Cross and Lacey, 1970; Fergus, 1964; Locci et al., 1967; Nonomura and Ohara, 1969). This conclusion was removed by the subsequent description of what is now the type strain of *Thermobifida fusca* Crawford 1975, although this did not result in validation of the name. Proposals that *Thermobifida fusca* Crawford 1975 be regarded as a synonym of *Thermobifida alba* Cross 1981 or vice versa (Kurup, 1979) are not in agreement with numerical phenetic data (McCarthy and Cross, 1984a), as the respective type strains were recovered in separate clusters.

4. **Thermobifida halotolerans** Yang, Tang, Zhang, Zhi, Wang, Xu and Li 2008b, 1824[VP]

ha.lo.to'le.rans. Gr. n. *hals halos* salt; L. part. adj. *tolerans* tolerating; N.L. part. adj. *halotolerans* referring to the ability to tolerate high salt concentrations.

Forms an extensively branched substrate mycelium and grows well on nutrient, potato, and glucose-yeast extract-malt extract agars, but does not grow on Czapek's agar. Single, smooth-ridged spores are formed at the tips of dichotomously branched sporophores borne on aerial hyphae. Does not form melanin pigments on tyrosine agar. Grows well at 45°C and at pH 7.0 and 8.0. Cellobiose, galactose, glucose, glycerol, raffinose, and xylose are used as sole carbon sources for energy and growth, but not *myo*-inositol or sorbitol.

Additional characteristics are found in Table 297. Diagnostic sugars are glucose, galactose, and xylose. Major fatty acids are $C_{16:0}$ iso (51.1%) and $C_{17:0}$ anteiso (11.7%). The cellular polar lipid pattern contains diphosphatidylglycerol, phosphatidylcholine, phosphatidylethanolamine, phosphatidylglycerol, phosphatidylinositol, phosphatidylmethylethanolamine, and an unknown phospholipid. The predominant menaquinones are MK-9(H_6) and MK-10(H_8).

Source: the type strain was isolated from a saline soil sample collected in Yunnan Province, south-west China.

DNA G+C content (mol%): 69.0 (HPLC).

Type strain: DSM 44931, KCTC 3, YIM 90462.

Sequence accession no. (16S rRNA gene): EU250489.

References

Abyzov, S.S., S.N. Philipova and V.D. Kuznetsov. 1983. *Nocardiopsis antarcticus*, a new species of actinomycetes, isolated from the ice sheet of the central Antarctic glacier. Izv. Akad. Nauk SSSR Ser. Biol. *4*: 559–568.

Abyzov, S.S., S.N. Philipova and V.D. Kuznetsov. 1984. *In* Validation of the publication of new names and new combinations previously effectively published outside the IJSB. List no. 13. Int. J. Syst. Bacteriol. *34*: 91–92.

Ajello, L., J. Brown, E. Macdonald and E. Head. 1987. Actinomycetoma caused by *Nocardiopsis dassonvillei*. Arch. Dermatol. *123*: 426.

Al-Tai, A.M. and J.S. Ruan. 1994. *Nocardiopsis halophila* sp. nov., a new halophilic actinomycete isolated from soil. Int. J. Syst. Bacteriol. *44*: 474–478.

Al-Zarban, S.S., I. Abbas, A.A. Al-Musallam, U. Steiner, E. Stackebrandt and R.M. Kroppenstedt. 2002. *Nocardiopsis halotolerans* sp. nov., isolated from salt marsh soil in Kuwait. Int. J. Syst. Evol. Microbiol. *52*: 525–529.

Alderson, G. and M. Goodfellow. 1979. Classification and identification of *Actinomycetales* causing actinomycosis. Postepy Hig. Med. Dosw. *33*: 109–124.

Alonso, A.N., P.J. Pomposiello and S.B. Leschine. 2008. Biofilm formation in the life cycle of the cellulolytic actinomycete *Thermobifida fusca*. Biofilms *FirstView*: 1–11.

Andersson, M.A., R. Mikkola, R.M. Kroppenstedt, F.A. Rainey, J. Peltola, J. Helin, K. Sivonen and M.S. Salkinoja-Salonen. 1998. The mitochondrial toxin produced by *Streptomyces griseus* strains isolated from an indoor environment is valinomycin. Appl. Environ. Microbiol. *64*: 4767–4773.

Athalye, M., J. Lacey and M. Goodfellow. 1981. Selective isolation and enumeration of actinomycetes using rifampicin. J. Appl. Bacteriol. *51*: 289–297.

Athalye, M., M. Goodfellow and D.E. Minnikin. 1984. Menaquinone composition in the classification of *Actinomadura* and related taxa. J. Gen. Microbiol. *130*: 817–823.

Athalye, M., M. Goodfellow, J. Lacey and R.P. White. 1985. Numerical classification of *Actinomadura* and *Nocardiopsis*. Int. J. Syst. Bacteriol. *35*: 86–98.

Bachmann, S.L. and A.J. McCarthy. 1989. Purification and characterization of a thermostable β-xylosidase from *Thermomonospora fusca*. J. Gen. Microbiol. *135*: 293–299.

Bachmann, S.L. and A.J. McCarthy. 1991. Purification and cooperative activity of enzymes constituting the xylan-degrading system of *Thermomonospora fusca*. Appl. Environ. Microbiol. *57*: 2121–2130.

Ball, A.S. and A.J. McCarthy. 1988. Saccharification of straw by actinomycete enzymes. J. Gen. Microbiol. *134*: 2139–2147.

Ball, A.S. and A.J. McCarthy. 1989. Production and properties of xylanases from actinomycetes. J. Appl. Microbiol. *66*: 439–444.

Beadle, B.M., W.A. Baase, D.B. Wilson, N.R. Gilkes and B.K. Shoichet. 1999. Comparing the thermodynamic stabilities of a related thermophilic and mesophilic enzyme. Biochemistry *38*: 2570–2576.

Beau, F., C. Bollet, T. Coton, E. Garnotel and M. Drancourt. 1999. Molecular identification of a *Nocardiopsis dassonvillei* blood isolate. J. Clin. Microbiol. *37*: 3366–3368.

Beki, E., I. Nagy, J. Vanderleyden, S. Jager, L. Kiss, L. Fulop, L. Hornok and J. Kukolya. 2003. Cloning and heterologous expression of a β-D-mannosidase (EC 3.2.1.25)-encoding gene from *Thermobifida fusca* TM51. Appl. Environ. Microbiol. *69*: 1944–1952.

Bellamy, W.D. 1973. The use of thermophilic microorganisms for the recycling of cellulosic wastes. Am. Inst. Chem. Eng. Symp. *69*: 138–140.

Bellamy, W.D. 1974. Biotechnology report: single cell proteins from cellulosic wastes. Biotechnol. Bioeng *16*: 869–880.

Bellamy, W.D. 1977. Cellulose and lignocellulose digestion by thermophilic actinomycetes for single cell protein production. Dev. Indust. Microbiol. *18*: 249–254.

Bernatchez, H. and E. Lebreux. 1991. *Nocardiopsis dassonvillei* recovered from a lung biopsy and a possible cause of extrinsic alveolitis. Clin. Microbiol. Newsl. *6*: 47–55.

Blanco, J., J.J. Coque, J. Velasco and J.F. Martin. 1997. Cloning, expression in *Streptomyces lividans* and biochemical characterization of a thermostable endo-β-1,4-xylanase of *Thermomonospora alba* ULJB1 with cellulose-binding ability. Appl. Microbiol. Biotechnol. *48*: 208–217.

Bowen, T., E. Stackebrandt, M. Dorsch and T.M. Embley. 1989. The phylogeny of *Amycolata autotrophica*, *Kibdelosporangium aridum* and *Saccharothrix australiensis*. J. Gen. Microbiol. *135*: 2529–2536.

Brocq-Rousseu, D. 1904. Sur un *Streptothrix* cause de l'alteracion des avoines moisies. Res. Bot. *16*: 219–230.

Cai, M., X.Y. Zhi, S.K. Tang, Y.Q. Zhang, L.H. Xu and W.J. Li. 2008. *Streptomonospora halophila* sp. nov., a halophilic actinomycete isolated from a hypersaline soil. Int. J. Syst. Evol. Microbiol. *58*: 1556–1560.

Cai, M., S.K. Tang, Y.G. Chen, Y. Li, Y.Q. Zhang and W.J. Li. 2009. *Streptomonospora amylolytica* sp. nov. and *Streptomonospora flavalba* sp. nov., two novel halophilic actinomycetes isolated from a salt lake. Int. J. Syst. Evol. Microbiol. *59*: 2471–2475.

Chater, K.F. and G. Chandra. 2006. The evolution of development in *Streptomyces* analysed by genome comparisons. FEMS Microbiol. Rev. *30*: 651–672.

Chen, S. and D.B. Wilson. 2007. Proteomic and transcriptomic analysis of extracellular proteins and mRNA levels in *Thermobifida fusca* grown on cellobiose and glucose. J. Bacteriol. *189*: 6260–6265.

Chen, W.-F., L.-Y. Jiang, L.-H. Xu and C.-L. Jiang. 2000. Studies on quantitative of whole-cell sugars in actinomycetes by gas chromatography-mass spectrum. Acta Microbiol. Sin. *27*: 416–420.

Chen, Y.-G., Y.-X. Wang, Y.-Q. Zhang, S.-K. Tang, Z.-X. Liu, H.-D. Xiao, L.-H. Xu, X.-L. Cui and W.-J. Li. 2009. *Nocardiopsis litoralis* sp. nov., a halophilic marine actinomycete isolated from a sea anemone. Int. J. Syst. Evol. Microbiol.: ijs.0.009704–009700.

Chen, Y.G., X.L. Cui, R.M. Kroppenstedt, E. Stackebrandt, M.L. Wen, L.H. Xu and C.L. Jiang. 2008. *Nocardiopsis quinghaiensis* sp. nov., isolated from saline soil in China. Int. J. Syst. Evol. Microbiol. *58*: 699–705.

Cheng, Y.-F., C.-H. Yang and W.-H. Liu. 2005. Cloning and expression of *Thermobifida xylanase* gene in the methylotrophic yeast *Pichia pastoris*. Enzyme and Microbial Technology *37*: 541–546.

Chun, J., K.S. Bae, E.Y. Moon, S.O. Jung, H.K. Lee and S.J. Kim. 2000. *Nocardiopsis kunsanensis* sp. nov., a moderately halophilic actinomycete isolated from a saltern. Int. J. Syst. Evol. Microbiol. *50*: 1909–1913.

Collins, M.D., T. Pirouz, M. Goodfellow and D.E. Minnikin. 1977. Distribution of menaquinones in actinomycetes and corynebacteria. J. Gen. Microbiol. *100*: 221–230.

Collins, M.D. 1994. Isoprenoid quinones. *In* Chemical Methods in Prokaryotic Systematics (edited by Goodfellow and O'Donnell). John Wiley & Sons, New York, pp. 265–309.

Crawford, D.L. and E. McCoy. 1972. Cellulases of *Thermomonospora fusca* and *Streptomyces thermodiastaticus*. Appl. Microbiol. *24*: 150–152.

Crawford, D.L., E. McCoy, J.M. Harkin and P. Jones. 1973. Production of microbial protein from waste cellulose by *Thermomonospora fusca*, a thermophilic actinomycete. Biotechnol. Bioeng. *15*: 833–843.

Crawford, D.L. 1974. Growth of *Thermomonospora fusca* on lignocellulosic pulps of varying lignin content. Can. J. Microbiol. *20*: 1069–1072.

Crawford, D.L. 1975. Cultural, morphological, and physiological characteristics of *Thermomonospora fusca* (strain 190Th). Can. J. Microbiol. *21*: 1842–1848.

Crawford, D.L. and M.A. Gonda. 1977. The sporulation process in *Thermomonospora fusca* as revealed by scanning and transmission electron microscopy. Can. J. Microbiol. *23*: 1088–1095.

Crawford, D.L. 1988. Biodegradation of agricultural and urban wastes. *In* Actinomycetes in Biotechnology (edited by Goodfellow, Williams and Mordarski). Academic Press, London, pp. 433–459.

Cross, T. and J. Lacey. 1970. Studies on the genus *Thermonospora*. *In* The *Actinomycetales* (edited by Prauser). Gustav Fischer Verlag, Jena, pp. 211–219.

Cross, T. and M. Goodfellow. 1973. Taxonomy and classification of the actinomycetes. *In Actinomycetales*: Characteristics and Practical Importance (edited by Sykes and Skinner). Academic Press, London, pp. 11–112.

Cross, T. 1981. The monosporic actinomycetes. *In* The Prokaryotes: a Handbook on Habitats, Isolation, and Identification of Bacteria (edited by Starr, Stolp, Trüper, Balows and Schlegel). Springer, New York, pp. 2091–2102.

Cui, X.L., P.H. Mao, M. Zeng, W.J. Li, L.P. Zhang, L.H. Xu and C.L. Jiang. 2001. *Streptimonospora salina* gen. nov., sp. nov., a new member of the family *Nocardiopsaceae*. Int. J. Syst. Evol. Microbiol. *51*: 357–363.

Dimise, E.J., P.F. Widboom and S.D. Bruner. 2008. Structure elucidation and biosynthesis of fuscachelins, peptide siderophores from the moderate thermophile *Thermobifida fusca*. Proc. Natl. Acad. Sci. U.S.A. *105*: 15311–15316.

Dolak, L.A., T.M. Castle and A.L. Laborde. 1980. 3-Trehalosamine, a new disaccharide antibiotic. J. Antibiot. (Tokyo) *33*: 690–694.

Dolak, L.A., T.M. Castle and L.A. Laborde. 1981. Biologically pure culture of *Nocardiopsis trehalosei* sp. nov. US Patent 4306028 (December 15).

Donadio, S., P. Monciardini and M. Sosio. 2007. Polyketide synthases and nonribosomal peptide synthetases: the emerging view from bacterial genomics. Nat. Prod. Rep. *24*: 1073–1109.

Donova, M. 2007. Transformation of steroids by actinobacteria: a review. Appl. Biochem. Microbiol. *43*: 1–14–14.

Dresler, K., J. van den Heuvel, R.-J. Müller and W.D. Deckwer. 2006. Production of a recombinant polyester-cleaving hydrolase from *Thermobifida fusca* in *Escherichia coli*. Bioprocess Biosyst. Eng. *29*: 169–183.

Elliot, M.A., N. Karoonuthaisiri, J. Huang, M.J. Bibb, S.N. Cohen, C.M. Kao and M.J. Buttner. 2003. The chaplins: a family of hydrophobic cell-surface proteins involved in aerial mycelium formation in *Streptomyces coelicolor*. Genes Dev. *17*: 1727–1740.

Evtushenko, L.I., V.V. Taran, V.N. Akimov, R.M. Kroppenstedt, J.M. Tiedje and E. Stackebrandt. 2000. *Nocardiopsis tropica* sp., nov., *Nocardiopsis trehalosi* sp. nov., nom. rev. and *Nocardiopsis dassonvillei* subsp. *albirubida* subsp. nov., comb. nov. Int. J. Syst. Evol. Microbiol. *50*: 73–81.

Ferchak, J.D. and E.K. Pye. 1983. Effect of cellobiose, glucose, ethanol, and metal ions on the cellulase enzyme complex of *Thermomonospora fusca*. Biotechnol. Bioeng. *25*: 2865–2872.

Ferchak, J.D. and E.K. Pye. 1980. Saccharification of cellulose by the cellulolytic enzyme system of *Thermomonospora* species. I. Stability of cellulolytic activities with respect to time, temperature and pH. Biotechnol. Bioeng. *25*: 2865–2872.

Fergus, C.L. 1964. Thermophilic and thermotolerant molds and actinomycetes of mushroom compost during peak heating. Mycologia *56*: 267–284.

Fibriansah, G., S. Masuda, R. Hirose, K. Hamada, N. Tanaka, S. Nakamura and T. Kumasaka. 2006. Crystallization and preliminary crystallographic analysis of endo-1,3-β-glucanase from alkaliphilic *Nocardiopsis* sp. strain F96. Acta Crystallogr. Sect. F Struct. Biol. Cryst. Commun. *62*: 20–22.

Fischer, A., R.M. Kroppenstedt and E. Stackebrandt. 1983. Molecular-genetic and chemotaxonomic studies on *Actinomadura* and *Nocardiopsis*. J. Gen. Microbiol. *129*: 3433–3446.

Gao, B. and R.S. Gupta. 2005. Conserved indels in protein sequences that are characteristic of the phylum *Actinobacteria*. Int. J. Syst. Evol. Microbiol. *55*: 2401–2412.

Gebbink, M.F.B.G., D. Claessen, B. Bouma, L. Dijkhuizen and H.A.B. Wosten. 2005. Amyloids - a functional coat for microorganisms. Nat. Rev. Microbiol. *3*: 333–341.

Gerber, N.N. 1966. Phenazines and phenoxazines from some novel *Nocardiaceae*. Biochemistry *5*: 4824–3829.

Ghanem, N.B., M.E. Mabrouk, S.A. Sabry and D.E. El-Badan. 2005. Degradation of polyesters by a novel marine *Nocardiopsis aegyptia* sp. nov.: application of Plackett-Burman experimental design for the improvement of PHB depolymerase activity. J. Gen. Appl. Microbiol. *51*: 151–158.

Ghangas, G.S. and D.B. Wilson. 1987. Expression of a *Thermomonospora fusca* cellulase gene in *Streptomyces lividans* and *Bacillus subtilis*. Appl. Environ. Microbiol. *53*: 1470–1475.

Ghangas, G.S. and D.B. Wilson. 1988. Cloning of the *Thermomonospora fusca* endoglucanase E2 gene in *Streptomyces lividans*: affinity purification and functional domains of the cloned gene product. Appl. Environ. Microbiol. *54*: 2521–2526.

Ghangas, G.S., Y.J. Hu and D.B. Wilson. 1989. Cloning of a *Thermomonospora fusca* xylanase gene and its expression in *Escherichia coli* and *Streptomyces lividans*. J. Bacteriol. *171*: 2963–2969.

Goodfellow, M., G. Alderson and J. Lacey. 1979. Numerical taxonomy of *Actinomadura* and related actinomycetes. J. Gen. Microbiol. *112*: 95–111.

Goodfellow, M. and T. Pirouz. 1982. Numerical classification of sporoactinomycetes containing *meso*-diaminopimelic acid in the cell wall. J. Gen. Microbiol. *128*: 503–527.

Goodfellow, M., E. Stackebrandt and R.M. Kroppenstedt. 1988. Chemotaxonomy and actinomycete systematics. *In* Biology of Actinomycetes '88 (edited by Okami, Beppu and Ogawara). Japan Scientific Societies Press, Tokyo, pp. 233–238.

Goodfellow, M. 1998. *Nocardia* and related genera. *In* Topley & Wilson's Microbiology and Microbial Infections, 9th edn, vol. 2 (edited by Balows and Duerden). Arnold, London, pp. 463–489.

Goodfellow, M. and E.T. Quintana. 2006. The family *Streptosporangiaceae*. *In* The Prokaryotes: a Handbook on the Biology of Bacteria, 3rd edn (edited by Dworkin, Falkow, Rosenberg, Schleifer and Stackebrandt). Springer, New York, pp. 725–753.

Gordon, R.E. and A.C. Horan. 1968. *Nocardia dassonvillei*, a macroscopic replica of *Streptomyces griseus*. J. Gen. Microbiol. *50*: 235–240.

Greiner-Mai, E., R.M. Kroppenstedt, F. Korn-Wendisch and H.J. Kutzner. 1987. Morphological and biochemical characterization and emended descriptions of thermophilic actinomycetes species. Syst. Appl. Microbiol. *9*: 97–109.

Groth, I., P. Schumann, B. Schuetze, K. Augsten, I. Kramer and E. Stackebrandt. 1999. *Beutenbergia cavernae* gen. nov., sp. nov., an L-lysine-containing actinomycete isolated from a cave. Int. J. Syst. Bacteriol. *49*: 1733–1740.

Grund, E. 1987. Untersuchungen zur Chernotaxonomie einiger Actinomyceten und coryneformer Bakterien. PhD thesis, Technische Hochschule Darmstadt, Darmstadt, Germany.

Grund, E. and R.M. Kroppenstedt. 1989. Transfer of five *Nocardiopsis* species to the genus *Saccharothrix* Labeda et al. 1984. Syst. Appl. Microbiol. *12*: 267–274.

Grund, E. and R.M. Kroppenstedt. 1990. Chemotaxonomy and numerical taxonomy of the genus *Nocardiopsis*. Int. J. Syst. Bacteriol. *40*: 5–11.

Gugnani, H.C., C. Unaogu, F. Provost and P. Boiron. 1998. Pulmonary infections due to *Nocardiopsis dassonvillei*, *Gordonia sputi*, *Rhodococcus rhodochrous* and *Micromonospora* sp. in Nigeria and literature review. J. Mycol. Med. *8*: 21–25.

Gusek, T.W., R.D. Johnson, M.T. Tyn and J.E. Kinsella. 1991. Effect of agitational shear on growth and protease production by *Thermomonospora fusca*. Biotechnol. Bioeng. *37*: 371–374.

Gyobu, Y. 2001. Family *Nocardiopsaceae*. *In* Identification Manual of Actinomycetes (edited by Miyadoh). Business Center for Academic Societies Japan, Tokyo, pp. 277–280.

Hägerdahl, B.G.R., J.D. Ferchak and E.K. Pye. 1980. Saccharification of cellulose by cellulolytic enzyme system of *Thermomonospora* sp. 1. Stability of cellulolytic activities with respect to time, temperature and pH. Biotechnol. Bioeng *22*: 1515–1528.

Hägerdal, B., H. Harris and E.K. Pye. 1979. Association of β-glucosidase with intact cells of *Thermoactinomyces*. Biotechnol. Bioeng. *21*: 345–355.

Harkin, J.M., D.L. Crawford and E. McCoy. 1974. Bacterial protein from pulp and papermill sludge. TAPPI *57*: 131–134.

Hasegawa, T., M. Takizawa and S. Tanida. 1983. A rapid analysis for chemical grouping of aerobic actinomycetes. J. Gen. Appl. Microbiol. *29*: 319–322.

Henssen, A. 1957. Beiträge zur Morphologie und Systematik der thermophilen Actinomyceten. Arch. Mikrobiol. *26*: 373–414.

Hickey, R.J. and H.D. Tresner. 1952. A cobalt-containing medium for sporulation of *Streptomyces* species. J. Bacteriol. *64*: 891–892.

Hilge, M., S.M. Gloor, W. Rypniewski, O. Sauer, T.D. Heightman, W. Zimmermann, K. Winterhalter and K. Piontek. 1998. High-resolution native and complex structures of thermostable β-mannanase from *Thermomonospora fusca* - substrate specificity in glycosyl hydrolase family 5. Structure *6*: 1433–1444.

Hilge, M., A. Perrakis, J.P. Abrahams, K. Winterhalter, K. Piontek and S.M. Gloor. 2001. Structure elucidation of β-mannanase: from the electron-density map to the DNA sequence. Acta Crystallogr. D Biol. Crystallogr. *57*: 37–43.

Hiraishi, A. and K. Komagata. 1989. Effects of the growth medium composition on the menaquinone homolog formation in *Micrococcus luteus*. J. Gen. Appl. Microbiol. *35*: 311–318.

Hoffman, P. 1989a. Cryopreservation of fungi. World Federation of Culture Collections. Technical Information Sheet No. 5. NESCO/WFCC/Education Committee. Braunschweig, Germany.

Hoffman, P. 1989b. Cryopreservation of basidiomycete cultures: Mushroom Science XII (Part 1). Proceedings of the Proceedings of the Twelth International Congress on the Science and Cultivation of Edible Fungi, 1987, Braunschweig, Germany.

Horikoshi, K. 1971. Production of alkaline enzymes by alkalophilic microorganisms. Part I. Alkaline protease produced by *Bacillus* No. 221. Agric. Biol. Chem. (Tokyo) *35*: 1404–1407.

Hozzein, W.N., W.J. Li, M.I. Ali, O. Hammouda, A.S. Mousa, L.H. Xu and C.L. Jiang. 2004. *Nocardiopsis alkaliphila* sp. nov., a novel alkaliphilic actinomycete isolated from desert soil in Egypt. Int. J. Syst. Evol. Microbiol. *54*: 247–252.

Hozzein, W.N. and M. Goodfellow. 2008. *Nocardiopsis arabia* sp. nov., a halotolerant actinomycete isolated from a sand-dune soil. Int. J. Syst. Evol. Microbiol. *58*: 2520–2524.

Irwin, D., E.D. Jung and D.B. Wilson. 1994. Characterization and sequence of a *Thermomonospora fusca* xylanase. Appl. Environ. Microbiol. *60*: 763–770.

Irwin, D., D.H. Shin, S. Zhang, B.K. Barr, J. Sakon, P.A. Karplus and D.B. Wilson. 1998. Roles of the catalytic domain and two cellulose binding domains of *Thermomonospora fusca* E4 in cellulose hydrolysis. J. Bacteriol. *180*: 1709–1714.

Irwin, D.C., M. Spezio, L.P. Walker and D.B. Wilson. 1993. Activity studies of eight purified cellulases: specificity, synergism, and binding domain effects. Biotechnol. Bioeng. *42*: 1002–1013.

Irwin, D.C., S. Zhang and D.B. Wilson. 2000. Cloning, expression and characterization of a family 48 exocellulase, Cel48A, from *Thermobifida fusca*. Eur. J. Biochem. *267*: 4988–4997.

Irwin, D.C., M. Cheng, B. Xiang, J.K. Rose and D.B. Wilson. 2003. Cloning, expression and characterization of a family-74 xyloglucanase from *Thermobifida fusca*. Eur. J. Biochem. *270*: 3083–3091.

Jiang, C.-L. and L.-H. Xu. 1998. Actinomycete diversity in unusual habitats. *In* Actinomycetes Research (edited by Jiang and Xu). Yunnan University Press, Yunnan, pp. 259–270.

Jiang, S., W. Sun, M. Chen, S. Dai, L. Zhang, Y. Liu, K.J. Lee and X. Li. 2007. Diversity of culturable actinobacteria isolated from marine sponge *Haliclona* sp. Antonie van Leeuwenhoek *92*: 405–416.

Jiang, S., X. Li, L. Zhang, W. Sun, S. Dai, L. Xie, Y. Liu and K. Lee. 2008. Culturable actinobacteria isolated from marine sponge *Iotrochota* sp. Mar. Biol. *153*: 945–952.

Jones, K.L. 1949. Fresh isolates of actinomycetes in which the presence of sporogenous aerial mycelia is a fluctuating characteristic. J. Bacteriol. *57*: 141–145.

Jung, E.D., G. Lao, D. Irwin, B.K. Barr, A. Benjamin and D.B. Wilson. 1993. DNA sequences and expression in *Streptomyces lividans* of an exoglucanase gene and an endoglucanase gene from *Thermomonospora fusca*. Appl. Environ. Microbiol. *59*: 3032–3043.

Kämpfer, P., H.-J. Busse and F.A. Rainey. 2002. *Nocardiopsis compostus* sp. nov., from the atmosphere of a composting facility. Int. J. Syst. Evol. Microbiol. *52*: 621–627.

Kelch, B.A., K.P. Eagen, F.P. Erciyas, E.L. Humphris, A.R. Thomason, S. Mitsuiki and D.A. Agard. 2007. Structural and mechanistic exploration of acid resistance: kinetic stability facilitates evolution of extremophilic behavior. J. Mol. Biol. *368*: 870–883.

Kempf, A. 1995. Untersuchungen über thermophile Actinomyceten: Taxonomie, Ökologie und Abbau von Biopolymeren, PhD thesis. Darmstadt, Germany.

Kim, J.H., D. Irwin and D.B. Wilson. 2004. Purification and characterization of *Thermobifida fusca* xylanase 10B. Can. J. Microbiol. *50*: 835–843.

Kim, J.W., H. Adachi, K. Shin-ya, Y. Hayakawa and H. Seto. 1997. Apoptolidin, a new apoptosis inducer in transformed cells from *Nocardiopsis* sp. J. Antibiot. (Tokyo) *50*: 628–630.

Kimura, M. 1980. A simple method for estimating evolutionary rates of base substitutions through comparative studies of nucleotide sequences. J. Mol. Evol. *16*: 111–120.

Kleeberg, I., C. Hetz, R.M. Kroppenstedt, R.-J. Müller and W.D. Deckwer. 1998. Biodegradation of aliphatic-aromatic copolyesters by *Thermomonospora fusca* and other thermophilic compost isolates. Appl. Environ. Microbiol. *64*: 1731–1735.

Korn-Wendisch, F. and H.J. Kutzner. 1992. The family *Streptomycetaceae*. *In* The Prokaryotes: a Handbook on the Biology of Bacteria: Ecophysiology, Isolation, Identification, Applications, 2nd edn (edited by Balows, Trüper, Dworkin, Harder and Schleifer). Springer, New York, pp. 921–995.

Krasil'nikov, N.A. and N.S. Agre. 1964. A new actinomycete genus – *Actinobifida* n. gen. yellow group – *Actinobifida dichotomica* n. sp. (in Russian). Mikrobiologiya *33*: 935–943.

Kroppenstedt, R.M. 1982. Separation of bacterial menaquinones by HPLC using reverse phase (RP 18) and silver loaded ion exchanger as stationary phases. J. Liquid Chromat. *5*: 2359–2367.

Kroppenstedt, R.M. 1985. Fatty acid and menaquinone analysis of actinomycetes and related organisms. *In* Chemical Methods in Bacterial Systematics (edited by Goodfellow and Minnikin). Academic Press, London, pp. 173–199.

Kroppenstedt, R.M. 1987. Chemische Untersuchungen an *Actinomycetales* und verwandte Taxa, Korrelation von Chemosystematik und Phylogenie. Habilitationsschrift.

Kroppenstedt, R.M., E. Stackebrandt and M. Goodfellow. 1990. Taxonomic revision of the actinomycete genera *Actinomadura* and *Microtetraspora*. Syst. Appl. Microbiol. *13*: 148–160.

Kroppenstedt, R.M. and M. Goodfellow. 1992. The family *Thermomonosporaceae*. *In* The Prokaryotes: a Handbook on the Biology of Bacteria: Ecophysiology, Isolation, Identification, Applications, 2nd edn (edited by Balows, Trüper, Dworkin, Harder and Schleifer). Springer, New York, pp. 1085–1114.

Kroppenstedt, R.M. and L.I. Evtushenko. 2006. The family *Nocardiopsaceae*. *In* The Prokaryotes: a Handbook on the Biology of Bacteria, 3rd edn (edited by Dworkin, Falkow, Rosenberg, Schleifer and Stackebrandt). Springer, New York, pp. 745–795.

Kudo, T. 2001. Family *Streptosporangiaceae*. *In* Identification Manual of Actinomycetes (edited by Miyadoh). Business Center for Academic Societies Japan, Tokyo, pp. 281–291.

Kukolya, J., L. Szabó and L. Hornok. 2001. Surface structures of new and lesser know species of thermobifida as revealed by scanning electron microscopy. Acta Biol. Hung. *52*: 211–221.

Kukolya, J., I. Nagy, M. Laday, E. Toth, O. Oravecz, K. Marialigeti and L. Hornok. 2002. *Thermobifida cellulolytica* sp. nov., a novel lignocellulose-decomposing actinomycete. Int. J. Syst. Evol. Microbiol. *52*: 1193–1199.

Kurup, V. 1979. Characterization of some members of the genus *Thermomonospora* Curr. Microbiol. *2*: 267–272.

Kutzner, H.-J. 2000. Microbiology of composting. *In* Biotechnology, vol. 11c (edited by Kutzner and Reed). Wiley, Weinheim, pp. 35–100.

Labeda, D.P., R.T. Testa, M.P. Lechevalier and H.A. Lechevalier. 1984. *Saccharothrix*: a new genus of the *Actinomycetales* related to *Nocardiopsis*. Int. J. Syst. Bacteriol. *34*: 426–431.

Labeda, D.P. and M.P. Lechevalier. 1989. Amendment of the genus *Saccharothrix* Labeda et al. 1984 and descriptions of *Saccharothrix espanaensis* sp. nov., *Saccharothrix cryophilis* sp. nov., and *Saccharothrix mutabilis* comb. nov. Int. J. Syst. Bacteriol. *39*: 420–423.

Labeda, D.P., K. Hatano, R.M. Kroppenstedt and T. Tamura. 2001. Revival of the genus *Lentzea* and proposal far *Lechevalieria* gen. nov. Int. J. Syst. Evol. Microbiol. *51*: 1045–1050.

Lacey, J. 1973. Actinomycetes in soils, composts and fodders. *In* Actinomycetales: Characteristics and Practical Importance, Society for Applied Bacteriology Symposium Series no. 2 (edited by Sykes and Skinner). Society for Applied Bacteriology, London, pp. 231–251.

Lacey, J. 1974. Allergy in mushroom workers. Lancet Infect. Dis. *1*: 366.

Lacey, J. and J. Dutkiewicz. 1976. Isolation of actinomycetes and fungi from mouldy hay using a sedimentation chamber. J. Appl. Bacteriol. *41*: 315–319.

Lacey, J. 1977. The ecology of actinomycetes in fodders and related substrates. Zentralbl. Bakteriol. Parasitenkd. Infektionskr. Hyg. Abt. I. Orig. *Suppl. 6*: 161–170.

Lacey, J. 1978. Ecology of actinomycetes in fodders and related substances. Zentralbl. Bakteriol. Parasitenkd. Infektionskr. Hyg. Abt. I *Suppl. 6*: 161–170.

Lacey, J. 1988. Actinomycetes as biodeteriogens and pollutants of the environment. *In* Actinomycetes in Biotechnology (edited by Goodfellow, Williams and Mordarski). Academic Press, London, pp. 359–432.

Lao, G., G.S. Ghangas, E.D. Jung and D.B. Wilson. 1991. DNA sequences of three β-1,4-endoglucanase genes from *Thermomonospora fusca*. J. Bacteriol. *173*: 3397–3407.

Lechevalier, H.A. and M.P. Lechevalier. 1970a. A critical evaluation of the genera of aerobic actinomycetes. *In* The *Actinomycetales* (edited by Prauser). Gustav Fischer Verlag, Jena, pp. 395–405.

Lechevalier, M.P. and H.A. Lechevalier. 1970b. Chemical composition as a criterion in the classification of aerobic actinomycetes. Int. J. Syst. Bacteriol. *20*: 435–443.

Lechevalier, M.P., C. de Bièvre and H. Lechevalier. 1977. Chemotaxonomy of aerobic actinomycetes: phospholipid composition. Biochem. Syst. Ecol. *5:* 249–260.

Lechevalier, M.P., A.E. Stern and H.A. Lechevalier. 1981. Phospholipids in the taxonomy of actinomycetes. Zentralbl. Bakteriol. Parasitenkd. Infektionskr. Hyg. I Abt. Orig. Suppl. *11*: 111–116.

Lee, S.E. and A.E. Humphrey. 1979. Use of continuous-culture techniques for determining the growth kinetics of a celluloytic *Thermoactinomyces* sp. Biotechnol. Bioeng. *21*: 1277–1288.

Li, M.G., W.J. Li, P. Xu, X.L. Cui, L.H. Xu and C.L. Jiang. 2003a. *Nocardiopsis xinjiangensis* sp. nov., a halophilic actinomycete isolated from a saline soil sample in China. Int. J. Syst. Evol. Microbiol. *53*: 317–321.

Li, W., J. Wu, W. Tao, C. Zhao, Y. Wang, X. He, G. Chandra, X. Zhou, Z. Deng, K.F. Chater and M. Tao. 2007. A genetic and bioinformatic analysis of *Streptomyces coelicolor* genes containing TTA codons, possible targets for regulation by a developmentally significant tRNA. FEMS Microbiol. Lett. *266*: 20–28.

Li, W.J., P. Xu, L.P. Zhang, S.K. Tang, X.L. Cui, P.H. Mao, L.H. Xu, P. Schumann, E. Stackebrandt and C.L. Jiang. 2003b. *Streptomonospora alba* sp. nov., a novel halophilic actinomycete, and emended description of the genus *Streptomonospora* Cui *et al.* 2001. Int. J. Syst. Evol. Microbiol. *53*: 1421–1425.

Li, W.J., D.J. Park, S.K. Tang, D. Wang, J.C. Lee, L.H. Xu, C.J. Kim and C.L. Jiang. 2004. *Nocardiopsis salina* sp. nov., a novel halophilic actinomycete isolated from saline soil in China. Int. J. Syst. Evol. Microbiol. *54*: 1805–1809.

Li, W.J., R.M. Kroppenstedt, D. Wang, S.K. Tang, J.C. Lee, D.J. Park, C.J. Kim, L.H. Xu and C.L. Jiang. 2006. Five novel species of the genus *Nocardiopsis* isolated from hypersaline soils and emended description of *Nocardiopsis salina* Li *et al.* 2004. Int. J. Syst. Evol. Microbiol. *56*: 1089–1096.

Liegard, H. and M. Landrieu. 1911. Un cas de mycose conjunctivale. Ann. Ocul. *146*: 418–426.

Lin, E. and D.B. Wilson. 1987. Regulation of β-1,4-endoglucanase synthesis in *Thermomonospora fusca*. Appl. Environ. Microbiol. *53:* 1352–1357.

Lin, E.S. and D.B. Wilson. 1988. Transcription of the *celE* gene in *Thermomonospora fusca*. J. Bacteriol. *170*: 3838–3842.

List Editor. 2001. Notification that new names and new combinations have appeared in volume 51, part 2, of the IJSEM. Int. J. Syst. Evol. Microbiol. *51*: 795–796.

List Editor. 2002. Notification that new names and new combinations have appeared in volume 52, part 2, of the IJSEM. Int. J. Syst. Evol. Microbiol. *52*: 691–692.

List Editor. 2002b. Notification that new names and new combinations have appeared in volume 52, part 4, of the IJSEM. Int. J. Syst. Evol. Microbiol. *52*: 1439–1440.

Locci, R., E. Baldacci and B. Petrolini. 1967. Contribution to the study of oligosporic actinomycetes. I. Description of new species of *Actinobifida*: *Actinobifida alba* sp. nov. and revision of the genus. G. Microbiol. *15:* 79–91.

Lykidis, A., K. Mavromatis, N. Ivanova, I. Anderson, M. Land, G. DiBartolo, M. Martinez, A. Lapidus, S. Lucas, A. Copeland, P. Richardson, D.B. Wilson and N. Kyrpides. 2007. Genome sequence and analysis of the soil cellulolytic actinomycete *Thermobifida fusca* YX. J. Bacteriol. *189*: 2477–2486.

Lynd, L.R., P.J. Weimer, W.H. van Zyl and I.S. Pretorius. 2002. Microbial cellulose utilization: fundamentals and biotechnology. Microbiol. Mol. Biol. Rev *66*: 506–577.

Lynd, L.R., W.H. van Zyl, J.E. McBride and M. Laser. 2005. Consolidated bioprocessing of cellulosic biomass: an update. Curr. Opin. Biotechnol. *16*: 577–583.

Masuda, S., K. Endo, T. Hayami, T. Fukazawa, R. Yatsunami and S. Nakamura. 2003. Cloning and expression of *bglF* gene from alkaliphilic *Nocardiopsis* sp. strain F96. Nucleic Acids Res. Suppl.: 317–318.

McCarthy, A.J. and T. Cross. 1984a. A taxonomic study of *Thermomonospora* and other monosporic actinomycetes. J. Gen. Microbiol. *130*: 5–25.

McCarthy, A.J. and T. Cross. 1984b. A taxonomic study of *Thermomonospora* and other monosporic actinomycetes. *In* Biological, Biochemical and Biomedical Aspects of Actinomycetes (edited by Ortiz-Ortiz, Bojalil and Yakoleff). Academic Press, San Diego, pp. 521–536.

McCarthy, A.J. and P. Broda. 1984. Screening for lignin-degrading actinomycetes and characterisation of their activity against ^{14}C-lignin-labelled wheat lignocellulose. J. Gen. Microbiol. *130*: 2905–2913.

McCarthy, A.J., E. Peace and P. Broda. 1985. Studies on the extracellular xylanase activity of some thermophilic actinomycetes. Appl. Microbiol. Biotechnol. *21:* 238–244.

McCarthy, A.J. 1987. Lignocellulose-degrading actinomycetes. FEMS Microbiol. Rev. *46:* 145–163.

McCarthy, A.J., A.S. Ball and S.L. Bachmann. 1988. Ecological and biotechnological implications of lignocellulose degradation by actinomycetes. *In* Biology of Actinomycetes '88 (edited by Okami, Beppu and Ogawara). Japan Scientific Societies Press, Tokyo, pp. 283–287.

McCarthy, A.J. 1989. *Thermomonospora* and related genera. *In* Bergey's Manual of Systematic Bacteriology, vol. 4 (edited by Williams, Sharpe and Holt). Williams & Wilkins, Baltimore, pp. 2552–2559.

McGrath, C.E. and D.B. Wilson. 2006. Characterization of a *Thermobifida fusca* β-1,3-glucanase (Lam81A) with a potential role in plant biomass degradation. Biochemistry *45*: 14094–14100.

Meyer, H.P. and A.E. Humphrey. 1982. Cellulase production by wild and a new mutant strain of *Thermomonospora* sp. Biotechnol. Bioeng. *24*: 1901–1904.

Meyer, J. 1976. *Nocardiopsis*, a new genus of order *Actinomycetales*. Int. J. Syst. Bacteriol. *26*: 487–493.

Meyer, J. 1989. Genus *Nocardiopsis*. *In* Bergey's Manual of Systematic Bacteriology, vol. 4 (edited by Williams, Sharpe and Holt). Williams & Wilkins, Baltimore, pp. 2562–2569.

Mielenz, J.R. 2001. Ethanol production from biomass: technology and commercialization status. Curr. Opin. Microbiol. *4*: 324–329.

Mikami, Y., K. Miyashita and T. Arai. 1982. Diaminopimelic acid profiles of alkalophilic and alkaline-resistant strains of actinomycetes. J. Gen. Microbiol. *128*: 1709–1712.

Minnikin, D.E., T. Pirouz and M. Goodfellow. 1977. Polar lipid composition in the classification of some *Actinomadura* species. Int. J. Syst. Bacteriol. *27*: 118–121.

Minnikin, D.E., M.D. Collins and M. Goodfellow. 1978. Menaquinone patterns in the classification of nocardioform and related taxa. Zentralbl. Bakteriol. Parasitenkd. Infektionskr. Hyg. Abt. 1, Orig. Reihe C. *Suppl. 6*: 85–90.

Minnikin, D.E., A.G. O'Donnell, M. Goodfellow, G.A. Alderson, M. Athalye, A. Schaal and J.H. Parlett. 1984. An integrated procedure for the extraction of isoprenoid quinones and polar lipids. J. Microbiol. Methods *2*: 233–241.

Mishra, S.K., J.E. Keller, J.R. Miller, R.M. Helsey, M.G. Nair and A.R. Putnam. 1987. Insecticidal and nematicidal properties of microbial metabolites. J. Ind. Microbiol. *2*: 267–276.

Mitsuiki, S., M. Ichikawa, T. Oka, M. Sakai, Y. Moriyama, Y. Sameshima, M. Goto and K. Furukawa. 2004. Molecular characterization of a keratinolytic enzyme from an alkaliphilic *Nocardiopsis* sp. TOA-1. Enz. Microb. Technol. *34*: 482–489.

Miyashita, K., Y. Mikami and T. Arai. 1984. Alkalophilic actinomycete, *Nocardiopsis dassonvillei* subsp. *prasina* subsp. nov. isolated from soil. Int. J. Syst. Bacteriol. *34*: 405–409.

Monteiro, T.I.R.C., T.S. Porto, A.M.A. Carneiro-Leão, M.P.C. Silva and M.G. Carneiro-da-Cunha. 2005. Reversed micellar extraction of an extracellular protease from *Nocardiopsis* sp. fermentation broth. Biochem. Eng. J. *24*: 87–90.

Moore, W.E.C., E.P. Cato and L.V.H. Moore. 1985. Index of the bacterial and yeast nomenclatural changes published in the *International Journal of Systematic Bacteriology* since the 1980 Approved Lists of Bacterial Names (1 January 1980 to 1 January 1985). Int. J. Syst. Bacteriol. *35*: 382–407.

Mordarska, H., A. Gamian and J. Carrasco. 1983. Sugar-containing lipids in the classification of representative *Actinomadura* and *Nocardiopsis* species. Arch. Immunol. Ther. Exp. (Warsz) *31*: 135–143.

Mordarska, H., J. Zakrzewska-Czerwinska, M. Pasciak, B. Szponar and S. Rowinski. 1998. Rare, suppurative pulmonary infection caused by *Nocardiopsis dassonvillei* recognized by glycolipid markers. FEMS Immunol. Med. Microbiol. *21*: 47–55.

Moreira, K.A., B.F. Albuquerque, M.F.S. Teixeira, A.L.F. Porto and J.L. Lima Filho. 2002. Application of protease from *Nocardiopsis* sp. as a laundry detergent additive. World J. Microbiol. Biotechnol. *18*: 309–315.

Moreira, K.A., T.S. Porto, M.F.S. Teixeira, A.L.F. Porto and J.L. Lima Filho. 2003. New alkaline protease from *Nocardiopsis* sp.: partial purification and characterization. Process Biochem. *39*: 67–72.

Moser, F., D. Irwin, S. Chen and D.B. Wilson. 2008. Regulation and characterization of *Thermobifida fusca* carbohydrate-binding module proteins E7 and E8. Biotechnol. Bioeng. *100*: 1066–1077.

Nakasaki, K., T.H. Tran le, Y. Idemoto, M. Abe and A.P. Rollon. 2009. Comparison of organic matter degradation and microbial community during thermophilic composting of two different types of anaerobic sludge. Bioresour. Technol. *100*: 676–682.

Naumova, I.B., V.D. Kuznetsov, K.S. Kudrina and A.P. Bezzubenkova. 1980. The occurrence of teichoic acids in streptomycetes. Arch. Mikrobiol. *126*: 71–75.

Naumova, I.B., A.S. Shashkov, E.M. Tul'skaya, G.M. Streshinskaya, Y.I. Kozlova, N.V. Potekhina, L.I. Evtushenko and E. Stackebrandt. 2001. Cell wall teichoic acids: structural diversity, species specificity in the genus *Nocardiopsis*, and chemotaxonomic perspective. FEMS Microbiol. Rev. *25*: 269–284.

Nonomura, H. and Y. Ohara. 1969. Distribution of actinomycetes in soil. VII. A culture method effective for both preferential isolation and enumeration of *Microbispora* and *Streptosporangium* strains in soil. Part 2. Classification of the isolates. J. Ferment. Technol. *47*: 701–709.

Nonomura, H. and Y. Ohara. 1971. Distribution of actinomycetes in soil. X. New genus and species of monosporic actinomycetes in soil. J. Ferment. Technol. *49*: 895–903.

Nonomura, H. and Y. Ohara. 1974. A new species of actinomycetes, *Thermomonospora mesouviformis* sp. nov. J. Ferment. Technol. *53*: 10–13.

Okuda, N., M. Soneura, K. Ninomiya, Y. Katakura and S. Shioya. 2008. Biological detoxification of waste house wood hydrolysate using *Ureibacillus thermosphaericus* for bioethanol production. J. Biosci. Bioeng. *106*: 128–133.

Ørskov, J. 1926. Investigations into the Morphology of the Ray Fungi (The State Serum Institute, Copenhagen. pp 171. Copenhagen: Levin and Munksgaard, 1923.). Arch. Intern. Med. *38*: 412.

Ozaki, H. and K. Yamada. 1991. Isolation of *Streptomyces* sp. producing glucose-tolerant β-glucosidase and properties of the enzyme. Agric. Biol. Chem *55*: 979–987.

Peltola, J.S.P., M.A. Andersson, P. Kämpfer, G. Auling, R.M. Kroppenstedt, H.-J. Busse, M.S. Salkinoja-Salonen and F.A. Rainey. 2001. Isolation of toxigenic *Nocardiopsis* strains from indoor environments and description of two new *Nocardiopsis* species, *N. exhalans* sp. nov. and *N. umidischolae* sp. nov. Appl. Environ. Microbiol. *67*: 4293–4304.

Peltola, J.S.P., M.A. Andersson, P. Kämpfer, G. Auling, R.M. Kroppenstedt, H.-J. Busse, M. Salkinoja-Salonen and F.A. Rainey. 2002. *In* Validation of publication of new names and new combinations previously effectively published outside the IJSEM. List no. 84. Int. J. Syst. Evol. Microbiol. *52*: 3–4.

Philip, A. and G.D. Roberts. 1984. *Nocardiopsis dassonvillei* cellulitis of the arm. 6: 14–15.

Phithakrotchanakoon, C., R. Daduang, A. Thamchaipenet, T. Wangkam, T. Srikhirin, L. Eurwilaichitr and V. Champreda. 2009. Heterologous expression of polyhydroxyalkanoate depolymerase from *Thermobifida* sp. in *Pichia pastoris* and catalytic analysis by surface plasmon resonance. Appl. Microbiol. Biotechnol. *82*: 131–140.

Poschner, J., R.M. Kroppenstedt, A. Fischer and E. Stackebrandt. 1985. DNA–DNA reassociation and chemotaxonomic studies on *Actinomadura, Microbispora, Microtetraspora, Micropolyspora* and *Nocardiopsis*. Syst. Appl. Microbiol. *6*: 264–270.

Potekhina, N.V., A.S. Shashkov, L.I. Evtushenko and I.B. Naumova. 2003. [Teichoic acids in the cell walls of *Microbispora mesophila* Ac-1953t and *Thermobifida fusca* Ac-1952t]. Mikrobiologiia *72*: 189–193.

Prauser, H. 1981. Nocardioform organisms: general characterisation and taxonomic relationships. Zentralbl. Bakteriol. Mikrobiol. Hyg. *11*: 17–24.

Prauser, H. 1984. Phage host ranges in the classification and identification of Gram-positive branched and related bacteria. *In* Biological, Biochemical, and Biomedical Aspects of Actinomycetes (edited by Ortiz-Ortiz, Bojalil and Yakoleff). Academic Press, Orlando, pp. 617–633.

Pridham, T.G. and A.J. Lyons, Jr. 1961. *Streptomyces albus* (Rossi-Doria) Waksman et Henrici: taxonomic study of strains labeled *Streptomyces albus*. J. Bacteriol. *81*: 431–441.

Rahman, O., M. Pfitzenmaier, O. Pester, S. Morath, S.P. Cummings, T. Hartung and I.C. Sutcliffe. 2009. Macroamphiphilic components of thermophilic actinomycetes: identification of lipoteichoic acid in *Thermobifida fusca*. J. Bacteriol. *191*: 152–160.

Rainey, F.A., N. Ward-Rainey, R.M. Kroppenstedt and E. Stackebrandt. 1996. The genus *Nocardiopsis* represents a phylogenetically coherent taxon and a distinct actinomycete lineage: proposal of *Nocardiopsaceae* fam. nov. Int. J. Syst. Bacteriol. *46*: 1088–1092.

Raju, R., A.M. Piggott, M. Conte, Z. Tnimov, K. Alexandrov and R.J. Capon. 2010. Nocardiopsins: new FKBP12-binding macrolide polyketides from an Australian marine-derived actinomycete, *Nocardiopsis* sp. Chemistry *16*: 3194–3200.

Resz, A., J. Schwanbeck and J. Knösel. 1977. Thermophile Actinomyceten und Müllkompost. Temperaturansprüche und proteolytische Aktivität. Forum Städte Hyg *28*: 71–73.

Rob, A., A.S. Ball, M. Tuncer and M.T. Wilson. 1995. Isolation and characterisation of a novel non-haem extracellular peroxidase produced by the thermophilic actinomycete *Thermonospora fusca* BD25. Biochem. Soc. Trans. *23*: 507S.

Rob, A., A.S. Ball, M. Tuncer, G.D. Jones, P.D. Taylor and M.T. Wilson. 1996. Redox reaction of the novel non-haem glycosylated peroxidases from thermophilic actinomycete *Thermomonospora fusca* BD25. Biochem. Soc. Trans. *24*: 455S.

Rob, A., A.S. Ball, M. Tuncer and M.T. Wilson. 1997a. Catalytic mechanism of the novel non-haem iron containing peroxidase produced by the thermophilic actinomycete *Thermomonospora fusca* BD25. Biochem. Soc. Trans. *25*: 64S.

Rob, A., A.S. Ball, M. Tuncer and M.T. Wilson. 1997b. The detection and quantification of novel non-haem extracellular glycosylated peroxidases produced by the thermophilic actinomycete *Thermomonospora fusca* BD25 by means of PAGE-zymogram. Biochem. Soc. Trans. *25*: 37S.

Rubin, E.M. 2008. Genomics of cellulosic biofuels. Nat. Rev. Microbiol. *454*: 841–845.

Sabry, S.A., N.B. Ghanem, G.A. Abu-Ella, P. Schumann, E. Stackebrandt and R.M. Kroppenstedt. 2004. *Nocardiopsis aegyptia* sp. nov., isolated from marine sediment. Int. J. Syst. Evol. Microbiol. *54*: 453–456.

Saddler, G.S., M. Goodfellow, D.E. Minnikin and A.G. O'Donnell. 1986. Influence of the growth cycle on the fatty acid and menaquinone composition of *Streptomyces cyaneus* NCIB 9616. J. Appl. Microbiol. *60*: 51–56.

Sakon, J., D. Irwin, D.B. Wilson and P.A. Karplus. 1997. Structure and mechanism of endo/exocellulase E4 from *Thermomonospora fusca*. Nat. Struct. Biol. *4*: 810–818.

Salazar, O., I. Gonzalez and O. Genilloud. 2002. New genus-specific primers for the PCR identification of novel isolates of the genera *Nocardiopsis* and *Saccharothrix*. Int. J. Syst. Evol. Microbiol. *52*: 1411–1421.

Sasser, M. 1990. Identification of bacteria by gas chromatography of cellular fatty acids. MIDI Technical Note 101, Newark, Delaware, MIDI Inc.

Schaal, K.P. and B.L. Beaman. 1984. Clinical significance of actinomycetes. *In* The Biology of the Actinomycetes (edited by Goodfellow, Mordarski and Williams). Academic Press, London, pp. 389–424.

Schippers, A., K. Bosecker, S. Willscher, C. Spröer, P. Schumann and R.M. Kroppenstedt. 2002. *Nocardiopsis metallicus* sp. nov., a metal-leaching actinomycete isolated from an alkaline slag dump. Int. J. Syst. Evol. Microbiol. *52*: 2291–2295.

Schleifer, K.H. and O. Kandler. 1972. Peptidoglycan types of bacterial cell walls and their taxonomic implications. Bacteriol. Rev. *36*: 407–477.

Schumacher, R.W., B.L. Harrigan and B.S. Davidson. 2001. Kahakamides A and B, new neosidomycin metabolites from a marine-derived actinomycete. Tetrahedron Lett. *42:* 5133–5135.

Shin, J., Y. Seo, H.-S. Lee, J.-R. Rho and S.J. Mo. 2003. A new cyclic peptide from a marine-derived bacterium of the genus *Nocardiopsis*. J. Nat. Prod. *66*: 883–884.

Shirling, E.B. and D. Gottlieb. 1966. Methods for characterization of *Streptomyces* species. Int. J. Syst. Bacteriol. *16*: 313–340.

Shirling, E.B. and D. Gottlieb. 1972. Cooperative description of type strains of *Streptomyces*. V. Additional descriptions. Int. J. Syst. Bacteriol. *22*: 265–394.

Sindhuphak, W., E. Macdonald and E. Head. 1985. Actinomycetoma caused by *Nocardiopsis dassonvillei*. Arch. Dermatol. *121*: 1332–1334.

Singh, S.M., J. Naidu, S. Mukerjee and A. Malkani. 1991. Cutaneous infections due to *Nocardiopsis dassonvillei* (Brocq-Rousseau) Meyer 1976, endemic in members of a family up to fifth degree relatives. Abstr. PS1.91, p. 85. Presented at the XI Congress of the International Society for Human and Animal Mycology.

Skerman, V.B.D., V. McGowan and P.H.A. Sneath. 1980. Approved Lists of Bacterial Names. Int. J. Syst. Bacteriol. *30*: 225–420.

Song, J., H.Y. Weon, S.H. Yoon, D.S. Park, S.J. Go and J.W. Suh. 2001. Phylogenetic diversity of thermophilic actinomycetes and *Thermoactinomyces* spp. isolated from mushroom composts in Korea based on 16S rRNA gene sequence analysis. FEMS Microbiol. Lett. *202*: 97–102.

Spezio, M., D.B. Wilson and P.A. Karplus. 1993. Crystal structure of the catalytic domain of a thermophilic endocellulase. Biochemistry *32*: 9906–9916.

Spiridonov, N.A. and D.B. Wilson. 1998. Regulation of biosynthesis of individual cellulases in *Thermomonospora fusca*. J. Bacteriol. *180*: 3529–3532.

Spiridonov, N.A. and D.B. Wilson. 2000. A *celR* mutation affecting transcription of cellulase genes in *Thermobifida fusca*. J. Bacteriol. *182*: 252–255.

Spiridonov, N.A. and D.B. Wilson. 2001. Cloning and biochemical characterization of BglC, a β-glucosidase from the cellulolytic actinomycete *Thermobifida fusca*. Curr. Microbiol. *42*: 295–301.

Stackebrandt, E., F.A. Rainey and N.L. Ward-Rainey. 1997. Proposal for a new hierarchic classification system, *Actinobacteria* classis nov. Int. J. Syst. Bacteriol. *47*: 479–491.

Stamford, T.L.M., N.P. Stamford, L.C.B.B. Coelho and J.M. Araújo. 2001. Production and characterization of a thermostable α-amylase from *Nocardiopsis* sp. endophyte of yam bean. Bioresour. Technol. *76*: 137–141.

Staneck, J.L. and G.D. Roberts. 1974. Simplified approach to identification of aerobic actinomycetes by thin-layer chromatography. Appl. Microbiol. *28*: 226–231.

Steger, K., A.M. Sjogren, A. Jarvis, J.K. Jansson and I. Sundh. 2007. Development of compost maturity and *Actinobacteria* populations during full-scale composting of organic household waste. J. Appl. Microbiol. *103*: 487–498.

Streshinskaya, G.M., E.M. Tul'skaya, L.P. Terekhova, O.A. Galatenko, I.B. Naumova and T.P. Preobrazhenskaya. 1989. Some chemotaxonomic criteria of the genus *Nocardiopsis* (in Russian). Dokl. Akad. Nauk SSSR. *309*: 477–480.

Streshinskaya, G.M., Yu.I. Kozlova, L.I. Evtushenko, V.V. Taran, A.S. Shashkov and I.B. Naumova. 1996. Cell wall teichoic acid of *Nocardiopsis* subsp. VKM Ac-1457. Biochemistry (Moscow) *61*: 285–288.

Streshinskaya, G.M., E.M. Tul'skaya, A.S. Shashkov, L.I. Evtushenko, V.V. Taran and I.B. Naumova. 1998. Teichoic acids of the cell wall of *Nocardiopsis listeri*, *Nocardiopsis lucentensis*, and *Nocardiopsis trehalosei*. Biochemistry (Moscow) *63*: 230–234.

Stutzenberger, F.J., A.J. Kaufman and R.D. Lossin. 1970. Cellulolytic activity in municipal solid waste composting. Can. J. Microbiol. *16*: 553–560.

Stutzenberger, F.J. 1971. Cellulase production by *Thermomonospora curvata* isolated from municipal solid waste compost. Appl. Microbiol. *22*: 147–152.

Su, T.M. and D. Paulavicius. 1975. Enzymatic saccharification of cellulose by thermophilic actinomycetes. Appl. Polymer Symp. *28*: 221–236.

Suzuki, K., K. Nagao, Y. Monnai, A. Yagi and M. Uyeda. 1998a. Topostatin, a novel inhibitor of topoisomerases I and II produced by *Thermomonospora alba* strain No. 1520. I. Taxonomy, fermentation, isolation and biological activities. J. Antibiot. (Tokyo) *51*: 991–998.

Suzuki, K., S. Yahara, Y. Kido, K. Nagao, Y. Hatano and M. Uyeda. 1998b. Topostatin, a novel inhibitor of topoisomerases I and II produced by *Thermomonospora alba* strain No. 1520. II. Physico-chemical properties and structure elucidation. J. Antibiot. (Tokyo) *51*: 999–1003.

Suzuki, K., S. Yahara, K. Maehata and M. Uyeda. 2001. Isoaurostatin, a novel topoisomerase inhibitor produced by *Thermomonospora alba*. J. Nat. Prod. *64*: 204–207.

Tang, S.K., X.P. Tian, X.Y. Zhi, M. Cai, J.Y. Wu, L.L. Yang, L.H. Xu and W.J. Li. 2008. *Haloactinospora alba* gen. nov., sp. nov., a halophilic filamentous actinomycete of the family *Nocardiopsaceae*. Int. J. Syst. Evol. Microbiol. *58*: 2075–2080.

Traag, B.A. and G.P. van Wezel. 2008. The SsgA-like proteins in actinomycetes: small proteins up to a big task. Antonie van Leeuwenhoek *94*: 85–97.

Trigo, C. and A.S. Ball. 1994. Is the solubilized product from the degradation of lignocellulose by actinomycetes a precursor of humic substances? Microbiology *140*: 3145–3152.

Trujillo, M.E. and M. Goodfellow. 2003. Numerical phenetic classification of clinically significant aerobic sporoactinomycetes and related organisms. Antonie van Leeuwenhoek *84*: 39–68.

Tsiklinsky, P. 1899. On the thermophilic moulds (in French). Ann. Inst. Pasteur *13*: 500–505.

Tsujibo, H., T. Sato, M. Inui, H. Yamamoto and Y. Inamori. 1988. Intracellular accumulation of phenazine antibiotics produced by an alkaliphilic actinomycete. I. Taxonomy, isolation and identification of the phenazine antibiotics. Agric. Biol. Chem. *52*: 301–306.

Tsujibo, H., T. Kubota, M. Yamamoto, K. Miyamoto and Y. Inamori. 2003. Characterization of chitinase genes from an alkaliphilic actinomycete, *Nocardiopsis prasina* OPC-131. Appl. Environ. Microbiol. *69*: 894–900.

Tul'skaya, E.M., G.M. Streshinskaya, I.B. Naumova, A.S. Shashkov and L.P. Terekhova. 1993. A new structural type of teichoic acid and some chemotaxonomic criteria of two species *Nocardiopsis dassonvillei* and *Nocardiopsis antarcticus*. Arch. Microbiol. *160*: 299–305.

Tul'skaya, E.M., A. S. Shashkov, L. I. Evtushenko and I.B. Naumova. 2000. Cell wall teichoic acids of *Nocardiopsis prasina* VKM Ac-1880^{T}. Microbiology (Moscow) *69*: 48–50.

Van den Bogart, H.G.G., G. Van den Ende, P.C.C. Van Loon and L.J.L.D. Van Griensven. 1993. Mushroom worker's lung: Serologic reactions to thermophilic actinomycetes present in the air of compost tunnels. Mycopathologia *122*: 21–28.

Waksman, S.A. 1961a. The Actinomycetes, vol. 2. Classification, Identification and Descriptions of Genera and Species. Williams & Wilkins, Baltimore.

Waksman, S.A. 1961b. The Actinomycetes, vol. 2. Classification, identification and descriptions of genera and species. Williams & Wilkins, Baltimore, pp. 1–363.

Walker, D., P. Ledesma, O.D. Delgado and J.D. Breccia. 2006. High endo-β-1,4-D-glucanase activity in a broad pH range from the alkali-tolerant *Nocardiopsis* sp. SES28. World J. Microbiol. Biotechnol. *22*: 761–764.

Wang, Y., Z. Zhang, J.S. Ruan and S. Ali. 1999. Investigations of actinomycete diversity in the tropical rainforests of Singapore. J. Clin. Microbiol. *23*: 178–187.

Ward-Rainey, N.L., F.A. Rainey and E. Stackebrandt. 1997. *In* Stackebrandt, E., F.A. Rainey and N.L. Ward-Rainey. Proposal for a new hierarchic classification system, *Actinobacteria* classis nov. Int. J. Syst. Bacteriol. *47*: 479–491.

Warren, R.A. 1996. Microbial hydrolysis of polysaccharides. Annu. Rev. Microbiol. *50*: 183–212.

Wellington, E.M.H. and S.T. Williams. 1978. Preservation of actinomycete inoculum in frozen glycerol. Microbiol. Lett. *6*: 151–159.

Wender, P.A., M. Sukopp and K. Longcore. 2005. Apoptolidins B and C: isolation, structure determination, and biological activity. Org. Lett. 7: 3025–3028.

Wender, P.A. and K.E. Longcore. 2009. Apoptolidins E and F, new glycosylated macrolactones isolated from *Nocardiopsis* sp. Org. Lett. *11*: 5474–5477.

Williams, S.T., G.P. Sharples and R.M. Bradshaw. 1974. Spore formation in *Actinomadura dassonvillei* (Brocq-Rousseu) Lechevalier and Lechevalier. J. Gen. Microbiol. *84*: 415–419.

Wilson, D.B., W.A. Wood and S.T. Kellogg. 1988. Cellulases of *Thermomonospora fusca*. *In* Methods in Enzymology, vol. 160. Academic Press, pp. 314–323.

Wilson, D.B. 2004. Studies of *Thermobifida fusca* plant cell wall degrading enzymes. Chem. Rec. *4*: 72–82.

Woese, C.R., G.J. Olsen, M. Ibba and D. Soll. 2000. Aminoacyl-tRNA synthetases, the genetic code, and the evolutionary process. Microbiol. Mol. Biol. Rev. *64:* 202–236.

Wood, W.A. 1985. Useful biodegradation of cellulose. Ann. Proc. Phytochem. Soc. Eur. *26*: 295–309.

Wu, Z., L. Xie, G. Xia, J. Zhang, Y. Nie, J. Hu, S. Wang and R. Zhang. 2005. A new tetrodotoxin-producing actinomycete, *Nocardiopsis dassonvillei*, isolated from the ovaries of puffer fish *Fugu rubripes*. Toxicon *45*: 851–859.

Xu, L.-H., Y.-Q. Tiang, Y.-F. Zhang, L.-X. Zhao and C.-L. Jiang. 1998. *Streptomyces thermogriseus*, a new species of the genus *Streptomyces* from soil, lake and hot-spring. Int. J. Syst. Bacteriol. *48*: 1089–1093.

Yamada, Y., M. Yamashita, Y. Tahara and K. Kondo. 1977. The menaquinone system in the classification of the genus *Actinomadura*. J. Gen. Appl. Microbiol. *23*: 207–219.

Yamashita, T., M. Imoto, K. Isshiki, T. Sawa, H. Naganawa, S. Kurasawa, B.-Q. Zhu and K. Umezawa. 1988. Isolation of a new indole alkaloid, pendolmycin, from *Nocardiopsis*. J. Nat. Prod. *51*: 1184–1187.

Yang, L.-L., X.-Y. Zhi, L.-H. Xu and W.-J. Li. 2008a. Phylogenetic relationships of *Nocardiopsis* species based on partial *gyrB* and 16S rRNA gene sequences. Actinomycetologica *22*: 6–11.

Yang, L.L., X.Y. Zhi and W.J. Li. 2007a. [Phylogenetic analysis of *Nocardiopsis* species based on 16S rRNA, *gyrB*, *sod* and *rpoB* gene sequences]. Wei Sheng Wu Xue Bao (Acta Microbiologica Sinica) *47*: 951–955.

Yang, L.L., S.K. Tang, Y.Q. Zhang, X.Y. Zhi, D. Wang, L.H. Xu and W.J. Li. 2008b. *Thermobifida halotolerans* sp. nov., isolated from a salt mine sample, and emended description of the genus *Thermobifida*. Int. J. Syst. Evol. Microbiol. *58*: 1821–1825.

Yang, R., L.P. Zhang, L.G. Guo, N. Shi, Z. Lu and X. Zhang. 2008c. *Nocardiopsis valliformis* sp. nov., an alkaliphilic actinomycete isolated from alkali lake soil in China. Int. J. Syst. Evol. Microbiol. *58*: 1542–1546.

Yang, Y., M. Malten, A. Grote, D. Jahn and W.D. Deckwer. 2007b. Codon optimized *Thermobifida fusca* hydrolase secreted by *Bacillus megaterium*. Biotechnol. Bioeng. *96*: 780–794.

Yassin, A.F., E.A. Galinski, A. Wohlfarth, K.D. Jahnke, K.P. Schaal and H.G. Trüper. 1993. A new actinomycete species, *Nocardiopsis lucentensis* sp. nov. Int. J. Syst. Bacteriol. *43*: 266–271.

Yassin, A.F., F.A. Rainey, J. Burghardt, D. Gierth, J. Ungerechts, I. Lux, P. Seifert, C. Bal and K.P. Schaal. 1997. Description of *Nocardiopsis synnemataformans* sp. nov., elevation of *Nocardiopsis alba* subsp. *prasina* to *Nocardiopsis prasina* comb. nov., and designation of *Nocardiopsis antarctica* and *Nocardiopsis alborubida* as later subjective synonyms of *Nocardiopsis dassonvillei*. Int. J. Syst. Bacteriol. *47*: 983–988.

Yassin, A.F., C. Spröer, H. Hupfer, C. Siering and H.P. Klenk. 2009. *Nocardiopsis potens* sp. nov., isolated from household waste. Int. J. Syst. Evol. Microbiol. *59*: 2729–2733.

Zhang, H., Y.K. Lee, W. Zhang and H.K. Lee. 2006. Culturable actinobacteria from the marine sponge *Hymeniacidon perleve*: isolation and phylogenetic diversity by 16S rRNA gene-RFLP analysis. Antonie van Leeuwenhoek *90*: 159–169.

Zhang, J.W. and R.Y. Zeng. 2008. Purification and characterization of a cold-adapted alpha-amylase produced by *Nocardiopsis* sp. 7326 isolated from Prydz Bay, Antarctic. Mar. Biotechnol. (NY) *10*: 75–82.

Zhang, S., G. Lao and D.B. Wilson. 1995. Characterization of a *Thermomonospora fusca* exocellulase. Biochemistry *34*: 3386–3395.

Zhang, S. and D.B. Wilson. 1997. Surface residue mutations which change the substrate specificity of *Thermomonospora fusca* endoglucanase E2. J. Biotechnol. *57*: 101–113.

Zhang, S., D.C. Irwin and D.B. Wilson. 2000. Site-directed mutation of noncatalytic residues of *Thermobifida fusca* exocellulase Cel6B. Eur J. Biochem *267*: 3101–3115.

Zhang, X., L.-P. Zhang, R. Yang, N. Shi, Z. Lu, W.X. Chen, C.-L. Jiang and L.-H. Xu. 2008. *Nocardiopsis ganjiahuensis* sp. nov., isolated from a soil from Ganjiahu, China. Int. J. Syst. Evol. Microbiol. *58*: 195–199.

Zhang, Z., Y. Wang and J. Ruan. 1998. Reclassification of *Thermomonospora* and *Microtetraspora*. Int. J. Syst. Bacteriol. *48*: 411–422.

Zhi, X.-Y., S.-K. Tang, W.-J. Li, L.-H. Xu and C.-L. Jiang. 2006. New genus-specific primers for the PCR identification of novel isolates of the genus *Streptomonospora*. FEMS Microbiol. Lett. *263:* 48–53.

Zhi, X.-Y., L.-L. Yang, J.-Y. Wu, S.-K. Tang and W.-J. Li. 2007. Multiplex specific PCR for identification of the genera *Actinopolyspora* and *Streptomonospora*, two groups of strictly halophilic filamentous actinomycetes. Extremophiles *11*: 543–548.

Zhi, X.-Y., W.-J. Li and E. Stackebrandt. 2009. An update of the structure and 16S rRNA gene sequence-based definition of higher ranks of the class *Actinobacteria*, with the proposal of two new suborders and four new families and emended descriptions of the existing higher taxa. Int. J. Syst. Evol. Microbiol. *59*: 589–608.

Zitouni, A., H. Boudjella, L. Lamari, B. Badji, F. Mathieu, A. Lebrihi and N. Sabaou. 2005. *Nocardiopsis* and *Saccharothrix* genera in Saharan soils in Algeria: isolation, biological activities and partial characterization of antibiotics. Res. Microbiol. *156*: 984–993.

Family III. **Thermomonosporaceae** Rainey, Ward-Rainey and Stackebrandt 1997, 486^{VP} emend. Zhang, Wang and Ruan 2001, 381 emend. Zhi, Li and Stackebrandt 2009, 600

Michael Goodfellow and Martha E. Trujillo

Ther.mo.mo.no.spo.ra.ce′a.e. N.L. fem. n. *Thermomonospora* type genus of family; L. suff. *-aceae* ending to denote family; N.L. fem. pl. n. *Thermomonosporaceae*, the *Thermomonospora* family.

Aerobic, Gram-stain-positive, non-acid–alcohol-fast, chemo-organotrophic actinomycete. A stable, branched substrate mycelium carries an aerial mycelium. **Aerial hyphae differentiate into single or short chains of arthrospores or into spore vesicles which release motile spores. Wall peptidoglycan contains *meso*-diaminopimelic acid, *N*-acetylmuramic acid and is of the A1γ type. Major phospholipids are phosphatidylglycerol, phosphatidylinositol, and phosphatidylinositol mannosides. Contains hydrogenated menaquinones with nine isoprene units.** Rich in mixtures of straight chain and branched fatty acids. Lacks mycolic acids. **Whole-organism hydrolysates usually contain madurose.** The pattern of 16S rRNA signatures consists of nucleotides at positions 440:497 (C–G), 501:544 (C–G), 502:543 (G–C), 831:855 (G–G), 843 (U), 844 (A), and 1355:1367 (A–U) (Zhi et al., 2009). A few species are pathogenic for animals, including man. Widely distributed in composts, manures, overheated fodder, and soil.

DNA G+C content (mol%): 66–73.

Type genus: **Thermomonospora** Henssen 1957, 398^{AL} emend. Zhang, Wang and Ruan 1998b, 418.

Further descriptive information

The family *Thermomonosporaceae* (Rainey et al. 1997) emend. Zhang et al. 2001 encompasses the genera *Actinocorallia* (Iinuma et al. 1994) emend. Zhang et al. 2001, *Actinomadura* (Lechevalier and Lechevalier 1970a) emend. Kroppenstedt et al. 1990, *Spirillospora* Couch 1963, and *Thermomonospora* (Henssen 1957) emend. Zhang et al. 2001. Members of these taxa form a distinct phyletic line in the 16S rRNA gene tree (Zhang et al., 1998b, 2001) and can thereby be distinguished from the families *Nocardiopsaceae* and *Streptosporangiaceae* (Goodfellow and Quintana, 2006; Goodfellow et al., 1990; Gyobu, 2001; Kudo, 2001), the other two families that comprise the suborder *Streptosporangineae* Rainey et al. 1997. *Actinomadura*, *Spirillospora*, and *Thermomonospora* strains are phylogenetically intermixed as can be seen in the *Thermomonosporaceae* 16S rRNA gene tree (Figure 398).

Differentiation of the genera of the family *Thermomonosporaceae*

Members of the genera *Actinocorallia*, *Actinomadura*, *Spirillospora*, and *Thermomonospora* can be distinguished from one another using a combination of chemotaxonomic and morphological markers (Table 298).

Taxonomic comments

Cross and Goodfellow (1973) proposed the family *Thermomonosporaceae* for a morphologically diverse group of sporoactinomycetes classified in the genera *Actinomadura*, *Microbispora*, *Microtetraspora*, *Saccharomonospora*, and *Thermomonospora*. They recognized that the taxon was an artificial construct, but also noted that the constituent genera had more in common than the assortment of taxa assigned to the family *Nocardiaceae*, a well known dumping ground for aerobic actinomycetes (Lechevalier, 1976). The new family encompassed mesophilic and thermophilic organisms that produced heat sensitive, motile or nonmotile spores borne singly, in pairs, or as short chains on aerial hyphae and substrate mycelia, or formed spore vesicles as in the genus *Spirillospora*. All but one of the genera had a wall chemotype III, that is, they contained *meso*-diaminopimelic acid (*meso*-A_2pm) but no other characteristic amino acids or sugars (Lechevalier and Lechevalier, 1970, 1970c). The exception, the genus *Saccharomonospora* had a wall chemotype IV, which meant that it had *meso*-A_2pm as the diamino acid of the wall peptidoglycan and a polysaccharide fraction rich in arabinose and galactose. Other wall chemotype III sporoactinomycetes subsequently thought to be related to the genus *Thermomonospora* included the genera *Actinobifida*, *Actinosynnema*, *Nocardiopsis*, *Saccharothrix*, and *Streptoalloteichus* (Goodfellow and Cross, 1984; McCarthy, 1989).

The family *Thermomonosporaceae sensu* Cross and Goodfellow 1973 was not included in the Approved Lists of Bacterial Names (Skerman et al., 1980) nor was it recognized in *Bergey's Manual of Systematic Bacteriology*, Volume 4 (Williams, 1989). Indeed, the difficulties of formally recognizing actinomycete taxa above the genus level led Goodfellow and Cross (1984) to propose "aggregate groups", the composition of which was based on the application of modern taxonomic methods, notably on chemotaxonomic and numerical taxonomic studies. One of the nine "aggregate groups", the maduromycetes, contained the genera *Actinomadura*, *Excellospora*, *Microbispora*, *Microtetraspora*, *Planobispora*, *Planomonospora*, *Spirillospora*, and *Streptosporangium*. The maduromycetes were not conceived as a natural group but were designed to provide a temporary home for organisms that had chemotaxonomic and morphological features in common (Goodfellow, 1989). In the 1st edition of *Bergey's Manual of Systematic Bacteriology*, McCarthy (1989) recognized another artificial group, "*Thermomonospora* and Related Taxa", which encompassed the genera *Actinosynnema*, *Nocardiopsis*, *Streptoalloteichus*, and *Thermomonospora*. These organisms all had a wall chemotype II but were markedly morphologically diverse.

The introduction and application of chemotaxonomic and molecular systematic procedures provided reliable data for establishing relationships between actinomycete taxa above the genus level (Embley and Stackebrandt, 1994; Stackebrandt and Woese, 1981; Stackebrandt et al., 1997; Zhang et al., 2001). It was apparent from such studies that morphological features previously weighted for the assignment of actinomycete genera to families had little predictive value (Goodfellow, 1989; Stackebrandt and Schleifer, 1984). These developments underlined the artificial nature of previously delineated suprageneric groups (Goodfellow and Cross, 1984; Williams, 1989) and provided a baseline for the circumscription of actinomycete taxa above the genus level (Stackebrandt et al., 1997). In addition, the genus *Actinomadura* was seen to encompass two groups of organisms that had little in common (Fischer et al., 1983; Fowler et al., 1985; Goodfellow et al., 1988; Poschner et al., 1985). This division was formally recognized by Kroppenstedt et al. (1990) who proposed that the genus *Actinomadura* be retained for *Actinomadura madurae* and related species and that the *Actinoma-*

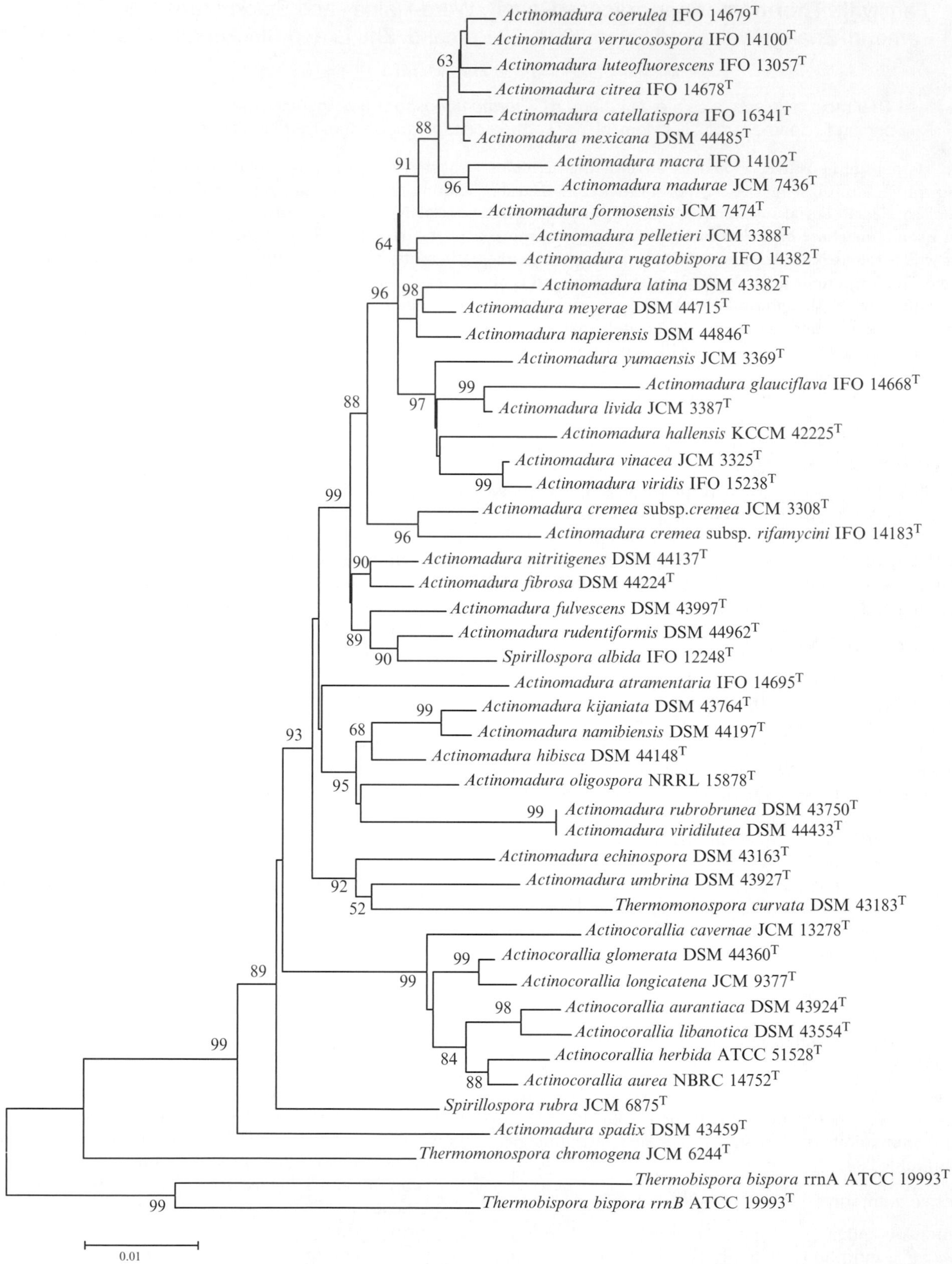

FIGURE 398. Neighbor-joining tree generated from 16S rRNA gene sequences showing relationships between taxa classified in the family *Thermomonosporaceae*. Evolutionary distances were calculated with the Jukes and Cantor method (Jukes and Cantor, 1969). Bootstrap values indicated at branching points are expressed as percentages of 1000 replications (only values greater than 40% are shown). Bar = 1 substitution per 100 nucleotides.

TABLE 298. Phenotypic characteristics separating genera classified in the family *Thermomonosporaceae*

Characteristics	*Thermomonospora*	*Actinocorallia*	*Actinomadura*	*Spirillospora*
Morphology:				
Single spores[a]	+	−	+	−
Chains of spores	+	+	+	−
Spore vesicles	−	−	−	+
Motile spores	−	−	−	+
Chemotaxonomy:				
Phospholipid type[b]	I	II	I(IV)[c]	I/II
Diagnostic sugar	None	Madurose	Madurose	Madurose
Sugar type[d]	C	B	B	B
DNA G+C content (mol%)	nd	66–73	66–73	71–73

[a]Single spores found in *Actinomadura formonsensis* and *Thermomonospora curvata.*

[b]Phospholipid types according to Lechevalier et al. (1977, 1981). Characteristic phospholipids: I, nitrogenous phospholipids absent (phosphatidylglycerol variable); II, only phosphatidylethanolamine; III, phospholipids containing glucosamine (with phosphatidylethanolamine variable) and IV, phospholipids containing glucosamine (with phosphatidylethanolamine variable).

[c]*Actinomadura kijaniata, Actinomadura namibiensis* and *Actinomadura napierensis* have a phospholipid type IV; the phospholipids of *Spirillospora albida* ATCC 14541 were found to be type II (Hasegawa et al., 1979).

[d]Whole-organism sugar patterns of actinomycetes containing *meso*-A_2pm: A, arabinose and galactose; B, madurose (3-*O*-methyl-D-galactose); C, no diagnostic sugars; D, arabinose and xylose (Lechevalier et al., 1970, 1971; Lechevalier and Lechevalier, 1970c).

dura pusilla group be transferred to the genus *Microtetraspora.* The *Actinomadura pusilla* group was subsequently given generic status as *Nonomuraea* Zhang et al. 1998, a proposition raised but never implemented by Goodfellow et al. (1988). Four members of the *Actinomadura madurae* group, *Actinomadura aurantiaca, Actinomadura glomerata, Actinomadura libanotica,* and *Actinomadura longicatena,* were subsequently transferred to the genus *Actinocorallia* by Zhang et al. (2001).

Other genera associated with the genus *Thermomonospora* were assigned to new families. The genera *Microbispora, Microtetraspora, Planobispora,* and *Streptosporangium* were classified in the family *Streptosporangiaceae* Goodfellow et al. 1990, the genera *Actinosynnema* and *Saccharothrix* in the family *Actinosynnemataceae* Labeda and Kroppenstedt 2000, and the genus *Saccharomonospora* in the family *Pseudonocardiaceae* Embley et al. 1988; the genus *Streptoalloteichus* is closely associated with the latter two families (Labeda and Kroppenstedt, 2000). The genus *Nocardiopsis* can be sharply separated from the genus *Thermomonospora* (Fowler et al., 1985; Kroppenstedt et al., 1990) and is now the type genus of the family *Nocardiopsaceae* which also includes the genera *Streptomonospora* Cui et al. 2001 and *Thermobifida* Zhang et al. 1998. The genus *Thermobifida* provided a home for three species previously classified in the genus *Thermomonospora,* namely, *Thermobifida alba, Thermobifida fusca,* and *Thermobifida mesouviformis.* The genus *Excellospora* Agre and Guzeva 1975 was included in the Approved Lists of Bacterial Names with *Excellospora viridilutea* as the type species (Skerman et al., 1980). Members of this taxon have been transferred to the genus *Actinomadura* (Kroppenstedt et al., 1990, 1991; Zhang et al., 2001) thereby underpinning the view that excellosporae should be seen as thermophilic actinomadurae (Meyer, 1989a). These developments leave the genera *Actinocorallia, Actinomadura, Spirillospora,* and *Thermomonospora* as members of the family *Thermomonosporaceae* Rainey et al. 1997 emend. Zhang et al. 2001 emend. Zhi et al. 2009.

Genus I. **Thermomonospora** Henssen 1957, 398[AL] emend. Zhang, Wang and Ruan 1998, 418[VP]

Martha E. Trujillo and Michael Goodfellow

Ther.mo.mo.no.spo′ra. Gr. n. *thermê* heat; Gr. adj. *monos* single, solitary; Gr. fem. n. spora seed; N.L. fem. n. *spora* a spore; N.L. fem. n. *Thermomonospora* the heat (-loving) single-spored (organism).

Aerobic, Gram-stain-positive, non-acid–alcohol-fast, chemo-organotrophic actinomycetes which form branched substrate/aerial mycelia. **Single spores with spiny surfaces are borne at the tips of short sporophores arising from the aerial mycelium**. Substrate mycelium may be yellow-orange to brown. White or light brown aerial mycelium is produced. **Temperature growth range 40–55°C. Cell wall contains *meso*-diaminopimelic acid but no diagnostic sugars**. Predominant menaquinones are MK-9(H_4), MK-9(H_6), and MK-9(H_8). Major phospholipids are diphosphatidylglycerol, phosphatidylglycerol, phosphatidylinositol, and phosphatidylinositol mannoside. **Rich in straight chain, iso- and anteiso- fatty acids, but lack mycolic acids. Isolated from soil, but is common in composts, manures, and overheated fodders**.

DNA G+C content (mol%): not determined for any members of the genus.

Type species: **Thermomonospora curvata** Henssen 1957, 401[VP].

Further descriptive information

Phylogeny. *Thermomonospora* species show a close phylogenetic relationship with members of the genus *Actinomadura.* Comparative 16S and 23S rRNA gene sequence analyses (Zhang et al., 1998, 2001) indicate that *Thermomonospora curvata,* the type species, *Actinomadura echinospora,* and *Actinomadura umbrina* form a moderately stable phyletic line which is separated from other *Actinomadura* species (see Figure 399). The phylogenetic position of *Thermomonospora chromogena* is not clear though it is evident that this organism is not closely related with other taxa in the family *Thermomonosporaceae* (Yap et al., 1999; Zhang et al., 1998). The uncertain phylogenetic relationship between *Thermomonospora chromogena* and related actinomycetes may be a reflection of its two distinct types of rRNA operons (Yap et al., 1999).

Cell morphology. The substrate and aerial mycelia of *Thermomonospora* strains consist of branching nonfragmenting hyphae. Single spores are borne on branched and unbranched sporophores on aerial mycelia. Spore clusters in *Thermomonospora chromogena* are formed by sequential sporulation on incurving hyphae. Spore arrangement on the aerial mycelium can be influenced by medium composition and incubation temperatures (Cross and Lacey, 1970). The spores are heat-sensitive and spore surfaces are spiny.

Nutrition and growth conditions. Members of the genus *Thermomonospora* are strictly aerobic, chemo-organotrophic micro-organisms and are moderately thermophilic as they grow well at 50°C. Some strains grow up to 60°C, but aerial mycelium production and sporulation are often poor at high temperatures. *Thermomonospora chromogena* shows little or no growth below 40°C with optimal growth and sporulation occurring on slightly alkaline nutrient media incubated under aerobic conditions. *Thermomonospora curvata* grows well and sporulates at pH 11.0. Most *Thermomonospora* strains can utilize fructose, glucose, mannose, and trehalose as sole carbon sources, but not dulcitol, inulin, raffinose, sorbitol, or sorbose. Enzymes produced by *Thermomonospora* strains include carboxymethylcellulase, catalase, β-galactosidase, and β-glucosidase; casein, gelatin, and keratin are degraded (McCarthy and Cross, 1984b; Trujillo and Goodfellow, 2003).

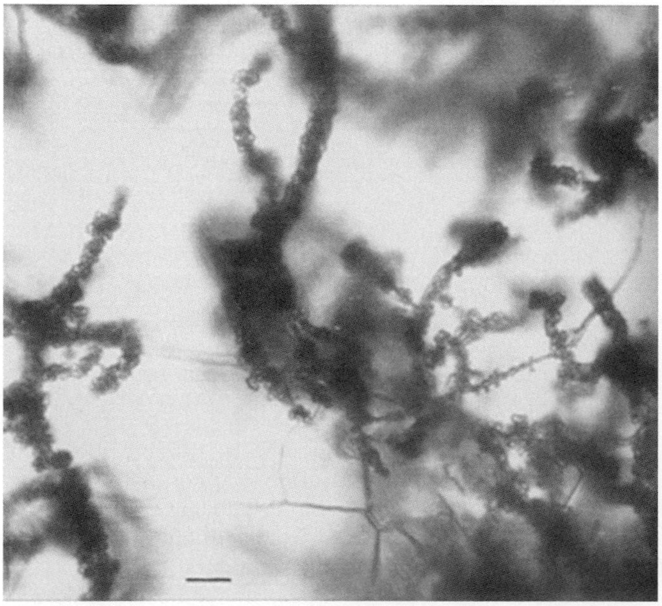

FIGURE 399. *Thermomonospora curvata.* Single spores borne laterally along aerial hyphae on branched and unbranched sporophores. Bar = 10 μm.

Genetics. Genetic studies on *Thermomonospora chromogena* ATCC 43196^T (Yap et al., 1999) show that the genome of this strain contains six rRNA operons (*rrn*), four of which are complete and two incomplete. Comparative sequence analyses demonstrated that five of the operons had almost identical sequences whereas that of the *rrnB* operon displayed a sequence difference of 6–10%. These studies also revealed a close evolutionary relationship between the *rrnB* operon of *Thermomonospora chromogena* and the *rrnA* operon of *Thermobispora bispora.* Based on these data, Yap et al. (1999) raised the possibility that *Thermomonospora chromogena* acquired the *rrnB* operon from either *Thermobispora bispora* or from a related micro-organism via horizontal gene transfer. Plasmids have been isolated from *Thermomonospora chromogena* (McCarthy, 1989) while cellulolytic *Thermomonospora* strains can be infected by various bacteriophages (Lawrence et al., 1986).

Cell-wall composition. The cell-wall peptidoglycan of *Thermomonospora* strains contain *meso*-A_2pm, but no characteristic sugars (wall chemotype III *sensu* Lechevalier and Lechevalier, 1970b). Long-chain fatty-acid composition has been reported to be a mixture of straight-chain and branched-chain *iso-* and *anteiso-* components (Goodfellow and Cross, 1984), that is, a fatty acid pattern type 3a (Kroppenstedt, 1985). The cell wall envelope does not contain mycolic acids. The predominant menaquinones of *Thermomonospora curvata* have side chains with nine isoprene units that are hexa- or octahydrogenated whereas *Thermomonospora chromogena* contains major amounts of MK-9(H_4) (Collins et al., 1982). Thermomonosporae have a phospholipid type I pattern (Lechevalier et al., 1977, 1981), that is, one characterized by the presence of diphosphatidylglycerol, phosphatidylglycerol, phosphatidylinositol, and phosphatidylinositol mannoside.

Ecology. Thermophilic thermomonosporae are found in overheated substrates such as baggasse, composts, fodders, and manures. They are especially abundant in mushroom compost (Fergus, 1964; Lacey, 1974, 1977; McCarthy and Cross, 1981, 1984b). Thermomonosporae secrete thermostable extracellular enzymes which enable them to become established as dominant populations during high temperature composting of plant residues and other wastes (Bernier and Stutzenberger, 1988; Fergus, 1964; George et al., 2001; Stutzenberger, 1971). Furthermore, *Thermomonospora* strains contribute significantly to the degradation of cellulose and lignocellulose residues from agricultural and urban wastes (McCarthy, 1987). *Thermomonospora curvata* is active in the decomposition of municipal waste compost (Stutzenberger and Sterpu, 1971, 1978). The growth of thermophilic actinomycetes in high temperature environments leads to the release of spores than can cause allergic alveolitis. At present, there are no grounds for implicating the spores of *Thermomonospora* species in such respiratory disorders (Lacey, 1988) though the cause of mushroom worker's disease, a form of allergic alveolitis, is still a matter for conjecture.

Isolation procedures

Thermomonospora strains can be isolated from composts and overheated vegetable materials by dilution plating on nonselective media, but recovery is poor owing to the rapid growth of *Bacillus* and *Thermoactinomyces* strains. The most effective isolation methods are those based on the use of a sedimentation chamber and an Andersen air sampler (Lacey and Dutkiewicz, 1976; McCarthy and Broda, 1984; McCarthy and Cross, 1981). Dried environmental samples are shaken within the sedimentation chamber to create an aerosol of propagules which after 1–2 h of sedimentation still contain many actinomycete spores but relatively few bacteria. Actinomycetes are isolated from such spore suspensions using an Andersen sampler loaded with 1/2 strength tryptone soy agar plates supplemented with cycloheximide (50 µg/ml) to prevent the growth of fungi. *Thermomonospora* colonies can usually be recognized after incubation for 3–5 d at 50°C (Kroppenstedt and Goodfellow, 2006; McCarthy, 1989). *Thermomonospora chromogena* is readily isolated on selective media supplemented with either kanamycin (25 µg/ml; McCarthy and Cross, 1981) or rifampin (5 µg/ml; Athalye et al., 1981). Mixtures of organic matter and soil inoculated in partially sealed polyethylene bags yield samples enriched in thermophilic actinomycetes, including thermomonosporae (McCarthy, 1989).

Maintenance procedures

Thermomonospora strains can be maintained as sporulating cultures on Czapek–Dox-yeast extract-Casamino acids agar at pH 8.0, stored at 4°C, and subcultured every 4 weeks. Medium-term preservation for up to four years can be achieved by preparing spore suspensions or homogenized mycelia in 10–20% (v/v) glycerol and storing at –25°C (Zippel and Neigenfind, 1988). Lyophilization of spore and mycelia suspended in 10% skim milk provides a convenient method for long-term storage.

Differentiation of the genus *Thermomonospora* from other genera

The major characteristics that distinguish members of the genus *Thermomonospora* from other genera classified in the family *Thermomonosporaceae* are presented in Table 299, Family *Thermomonosporaceae*. Members of the redefined genus *Thermomonospora* share chemical markers in common with *Actinomadura* strains, i.e. a wall chemotype III, phospholipid type I, menaquinone type 4B2, and fatty acid type 3a (Collins et al., 1982; Kroppenstedt, 1985; Lechevalier et al., 1981, 1977, Lechevalier and H.A. Lechevalier, 1970a). The integrity of the group containing *Thermomonospora chromogena* and *Thermomonospora curvata* is also supported by fatty acid and menaquinone data. Variations in quantitative fatty acid composition and in menaquinone profiles of *Thermomonospora chromogena* and *Thermomonospora curvata* compared with other members of family *Thermomonosporaceae* may be attributed to differences in their temperature requirements for growth (Kroppenstedt et al., 1990). The lack of madurose (sugar type C; Lechevalier and Lechevalier, 1970b) might be an effect of the elevated temperature for growth. As in members of thermophilic *Actinomadura* species (formerly *Excellospora* strains) this sugar is found only in trace amounts.

TABLE 299. Characteristics differentiating between the species of the genus *Thermomonospora*[a, b]

Characteristic	*T. curvata*	*T. chromogena*
Substrate mycelium color	Yellow/orange	Brown
Aerial mycelium color	White	Light brown
Spores on aerial mycelium:		
In clusters	–	+
Single on branched and unbranched sporophores	+	–
Biochemical test:		
Oxidase	–	+
Degradation of:		
Agar	+	–
Cellulose powder (MN300)	+	–
Elastin	–	+
Hypoxanthine	–	+
Starch	+	–
Tyrosine	–	+
Xanthine	–	+
Growth at:		
35°C	+	–
pH 11	+	–
Growth on sole carbon sources (1%, w/v):		
Galactose	–	+
Sucrosc	+	–
Growth in the presence of (µg/ml):		
Kanamycin (25)	–	+

[a]Symbols: +, 90% or more strains positive; –, 10% or less strains positive.

[b]Data from: McCarthy (1989), McCarthy and Cross (1984a, 1984b), and Trujillo and Goodfellow (2003).

Taxonomic comments

The genus *Thermomonospora* was proposed by Henssen (1957) for thermophilic actinomycetes isolated from composted stable manure. Initially the taxon encompassed three species with members that produce single spores on aerial hyphae. All of these organisms formed colorless to pale yellow colonies and a white aerial mycelium, but were distinguished from one another according to aerial mycelium morphology and by the type of branching shown by substrate hyphae. *Thermomonospora curvata*, the only species isolated and maintained in pure culture, subsequently became the type species of the genus (Henssen and Schnepf, 1967). The description of the remaining two species, *Thermomonospora fusca* and *Thermomonospora lineata*, was based on morphological properties in contaminated preparations. Neither of these species were included in the Approved Lists of Bacterial Names (Skerman et al., 1980) despite the fact that *Thermomonospora fusca* had been isolated in pure culture and described in detail (Crawford, 1975; Crawford and Gonda, 1977). *Thermomonospora fusca* was subsequently validated (Moore et al., 1985) but was later transferred to the genus *Thermobifida* based on 16S rRNA gene sequencing data (Zhang et al., 1998).

A mesophilic monosporic actinomycete classified in the genus as *Thermomonospora mesophila* Nonomura and Ohara 1971a was later transferred to the genus *Microbispora* as *Microbispora mesophila* Miyadoh et al. 1990. A second mesophilic species, *Thermomonospora mesouviformis* Nonomura and Ohara 1974, was assigned to the genus *Thermobifida* as a synonym of *Thermobifida alba* Zhang et al. 1998. A third mesophilic species, *Thermomonospora formosensis*, proposed by Hasegawa et al. (1986) was cited as a species *incertae sedis* in *Bergey's Manual of Systematic Bacteriology* (McCarthy, 1989). This species was subsequently transferred to the genus *Actinomadura* as *Actinomadura formosensis* Zhang et al. 1998.

Krasil'nikov and Agre (1964) proposed the genus *Actinobifida* for actinomycetes that formed single spores on dichotomously branched sporophores, but they failed to acknowledge that dichotomous branching had been observed previously in members of the genera *Thermomonospora* (Henssen, 1957) and *Micromonospora* (Jensen, 1930, 1932; Krasil'nikov, 1941). The following year these workers proposed a second species, *Actinobifida chromogena*, thereby implying that all actinomycetes showing dichotomous branching should be transferred to the genus *Actinobifida*. A third species, *Actinobifida alba*, was introduced by Locci et al. (1967). *Actinobifida dichotomica*, the type species of the genus was later transferred to the genus *Thermoactinomyces* as it produced endospores and *Actinobifida alba* to the genus *Thermomonospora* as it formed heat sensitive spores on substrate and aerial hyphae (Cross and Goodfellow, 1973).

A comprehensive numerical taxonomic study of the genus *Thermomonospora* and related organisms (McCarthy and Cross, 1984b, 1984a) confirmed the status of *Thermomonospora curvata* and provided strong evidence for the formal recognition of *Thermomonospora fusca*. In contrast, *Thermomonospora mesouviformis* was considered to be a synonym of *Thermomonospora alba*. These taxa, termed the "white *Thermomonospora* group" because of their white aerial mycelium, were sharply distinguished from organisms which included the type strain of *Actinobifida chromogena* (Krasil'nikov and Agre 1965; "*Thermomonospora falcata*" Henssen 1970, and similar actinomycetes isolated from mushroom compost (McCarthy and Cross, 1981). The "*chromogena*" strains with reddish-brown colonies and a light brown aerial mycelium had been provisionally assigned to the genus *Thermomonospora* Cross 1981 based on wall composition and morphological features. In the meantime, *Thermomonospora viridis* Küster and Locci 1963 had been transferred to the genus *Saccharomonospora* as *Saccharomonospora viridis* Nonomura and Ohara 1971a.

It is clear that the genus *Thermomonospora* has undergone many changes since it was proposed by Henssen (1957). The application of molecular systematic techniques has led to significant improvements in the taxonomy of this taxon, but it seems likely that in future further changes will be necessary. It also seems likely that as more data becomes available, *Thermomonospora chromogena* will be seen to merit generic status.

Differentiation of the species of the genus *Thermomonospora*

Thermomonospora curvata and *Thermomonospora chromogena* can be distinguished using a combination of phenotypic properties as shown in Table 299.

List of species of the genus *Thermomonospora*

1. **Thermomonospora curvata** Henssen 1957, 401[VP]

cur.va'ta. L. v. *curvo* to curve; L. part. fem. adj. *curvata* curved.

Colonies on agar media have a yellow to orange reverse color and bear an abundant white aerial mycelium. Single spores are borne laterally along aerial hyphae on branched and unbranched sporophores (Figure 399). Spores are not formed on the substrate mycelium. Grows in the range 30–53°C, but not at 60°C. Growth observed at pH 11.

Catalase, β-galactosidase and β-glucosidase are produced. Nitrate is reduced. Degrades carboxymethylcellulose, esculin, keratin, Tweens 20 and 80, and xylan, but not chitin, DNA, guanine, or pectin.

Fructose, maltose, mannose, and trehalose are used as sole carbon sources, but not L-arabinose, glycerol, *meso*-inositol, lactose, mannitol, melibiose, or L-rhamnose. Additional phenotypic characteristics are shown in Table 299.

DNA G+C content (mol%): not determined.

Type strain: ATCC 19995, CIP 105592, DSM 43183, HAMBI 1549, IFO (now NBRC) 15933, JCM 3096.

Sequence accession no. (16S rRNA gene): AF002262, D86945.

Sequence accession no. (23S rRNA gene): AF116236.

Sequence accession no. (16S–23S rRNA ITS): AF134112.

2. **Thermomonospora chromogena** (*ex* Krasil'nikov and Agre 1965) McCarthy and Cross, 1984a, 356[VP] (Effective publication: McCarthy and Cross 1984b, 22.)

chro.mo.ge'na. Gr. n. *chroma* color; Gr. v. *gennaio* to produce; N.L. adj. *chromogenes* color producing.

Colonies on agar media are small, entire, raised, and dark reddish brown to light brown. Aerial mycelium white to light brownish, turning blue-gray on prolonged incubation. A dark brown soluble pigment is often produced. Spore clusters on the aerial mycelium are formed by sequential sporulation on incurving hyphae (Figure 400). Spores are not formed on the substrate mycelium. Grows in the range 40–53°C, but not at 60°C. Does not grow at pH 11.

β-Galactosidase and β-glucosidase are produced, but not catalase. Nitrate is reduced. Degrades arbutin, carboxym-

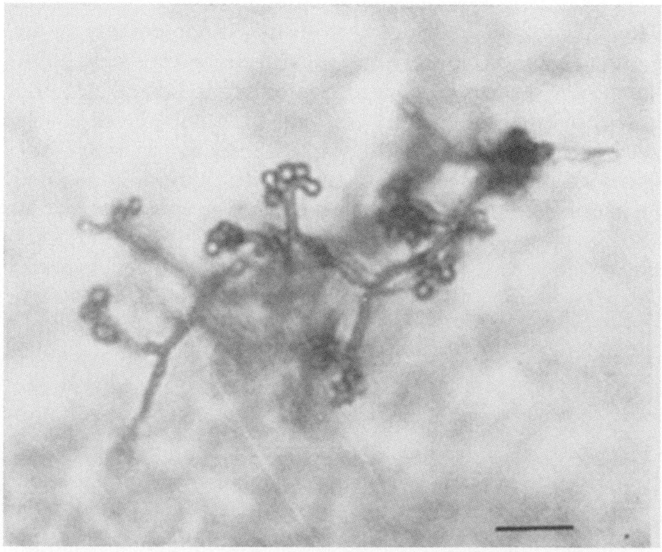

FIGURE 400. *Thermomonospora chromogena*. Single spores produced sequentially on incurving hyphae. Bar = 10 μm.

ethylcellulose, esculin, keratin, pectin, Tweens 20 and 80, tyrosine, and xylan but not chitin or DNA.

Glucose, mannose, and trehalose are used as sole carbon sources, but not L-arabinose, dulcitol, glycerol, inulin, melezitose, melibiose, sorbose, or xylose. Additional phenotypic characteristics are presented in Table 299.

DNA G+C content (mol%): not determined.

Type strain: strain Agre no. 577, ATCC 43196, DSM 43794, NBRC 16096, JCM 6244, NCIB (now NCIMB) 10212, NRRL B-16983.

Sequence accession no. (16S rRNA gene): AF116558.

Genus II. **Actinocorallia** Iinuma, Yokota, Hasegawa and Kanamuru 1994, 233^VP emend. Zhang, Kudo, Nakajima and Wang 2001, 381^VP

Martha E. Trujillo and Michael Goodfellow

Ac.ti.no.co.ral'li.a Gr. n. *actis actinos* a ray; L. n. *corallium* coral; N.L. fem. n. *Actinocorallia* (*sic*) intended to mean a microorganism (actinomycete) that forms sporophores resembling coral.

Aerobic, Gram-stain-positive, nonmotile actinomycete which forms branched non-fragmenting vegetative hyphae. **Spore chains are formed on the aerial mycelium**. Spores are cylindrical or rod-shaped. **Spore surface smooth or warty**. Substrate mycelium pale yellowish to brown. Mature aerial mycelium is pale yellow or pink. Chemo-organotrophic. Temperature growth range is 7–42°C. Melanin pigments are not produced. Does not grow in the presence of 5% (w/v) NaCl. Cell wall contains *meso*-2,6-diaminopimelic acid and *N*-acetylated muramic acid. **Major phospholipids are diphosphatidylglycerol, phosphatidylethanolamine, and phosphatidylinositol**. Predominant menaquinones are MK-9(H_4), MK-9(H_6), and MK-9(H_8). Hexadecanoic, 14-methylpentadecanoic, octadecanoic, and 19-methyloctadecanoic acids are the predominant fatty acids. Mycolic acids are absent.

DNA G+C content (mol%): 66.2–73.0 (T_m, HPLC).

Type species: **Actinocorallia herbida** Iinuma, Yokota, Hasegawa and Kanamuru 1994, 233^VP.

Further descriptive information

Phylogeny. The genus *Actinocorallia* encompasses seven species, *Actinocorallia aurantiaca*, *Actinocorallia aurea*, *Actinocorallia cavernae*, *Actinocorallia glomerata*, *Actinocorallia herbida*, *Actinocorallia libanotica*, and *Actinocorallia longicatena*, which form a distinct phyletic line in the *Thermomonosporaceae* 16S rRNA gene tree (Figure 398). Members of this taxon share 16S rRNA gene similarities within the range 97.3–99.4%. Representatives of the genus contain 16S rRNA signatures characteristic of the family *Thermomonosporaceae* (Stackebrandt et al., 1997). The type strains of *Actinocorallia glomerata* and *Actinocorallia longicatena*, the two most closely related strains in the 16S rRNA phyletic line, share DNA:DNA relatedness values of 36% and 39% in reciprocal crosses (Itoh et al., 1995). The 16S–23S internal transcribed spacer regions of representative strains share high nucleotide sequence similarities and form a well delineated taxon based on partial sequence analyses of 23S rRNA genes (Zhang et al., 2001).

Cell morphology. *Actinocorallia* strains form well developed, branched, nonfragmenting substrate hyphae, features they have in common with members of the genus *Actinomadura*. Aerial hyphae differentiate into long chains of spores after 10–14 d growth on a range of media, including glycerol-asparagine, oatmeal, tyrosine and yeast extract-malt extract agars. *Actinocorallia herbida* produces characteristic coralloid sporophores on glycerol-asparagine, humic acid-vitamins, oatmeal, and yeast extract-malt extract agars. These arise from the substrate mycelium and differentiate into long chains of non-motile spores (more than 30 spores per chain) at the tips; coremia have been observed, albeit rarely, on yeast extract-malt extract agar. The spores of *Actinocorallia aurantiaca* and *Actinocorallia libanotica* are connected by intersporal pads (Figure 401). *Actinocorallia glomerata* forms distinctive globular or amorphous spore masses (pseudosporangia-like structures), 1.5–5.0 μm in diameter, in which hyphae are coiled or intertwined (Figure 402). *Actinocorallia longicatena* is characterized by straight to flexuous chains bearing arthrospores (more than 20 spores) (Figure 403).

Nutrition and growth conditions. *Actinocorallia* strains are aerobic and chemo-organotrophic with an oxidative metabolism. They grow well on oatmeal agar and have no specific requirements for minerals or vitamins. Members of the genus

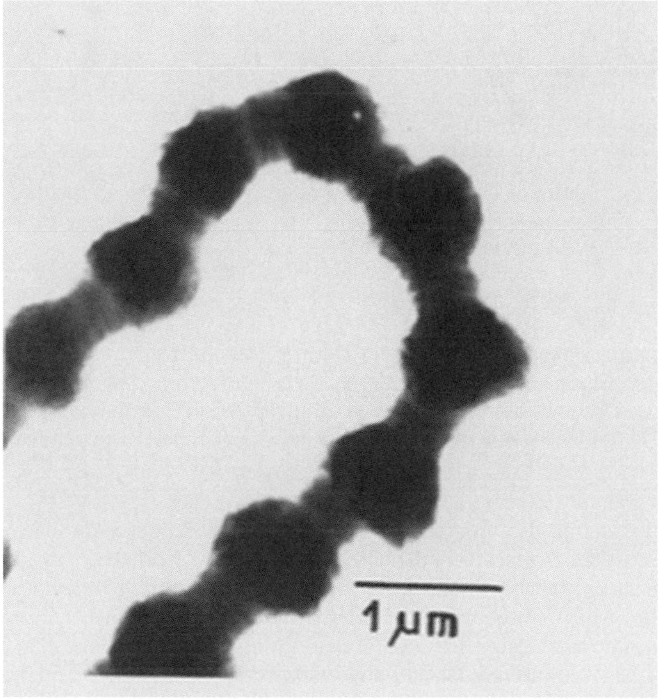

FIGURE 401. *Actinocorallia libanotica* strain IMET 9618. Spore chain with "intersporal pads". Electron micrograph (16,500×).

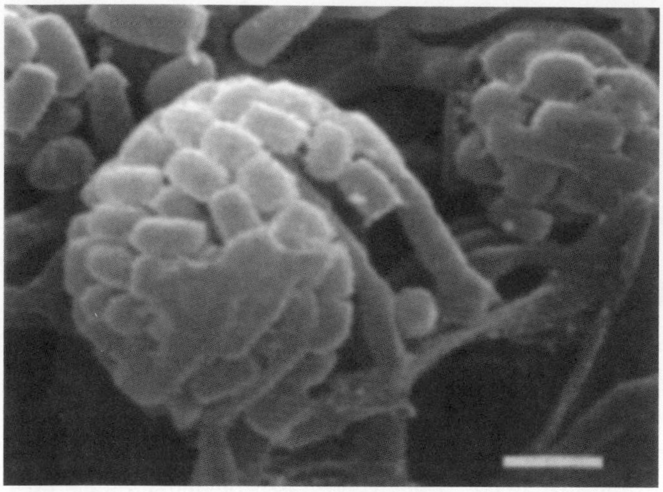

FIGURE 402. *Actinocorallia glomerata* strain I-226. Pseudosporangium-like structures. The micro-organism was incubated on oatmeal agar at 28°C for 2–3 weeks. Electron micrograph (bar = 1 µm). (Reproduced with permission from Itoh et al., 1995. Actinomycetologica *9*: 167.)

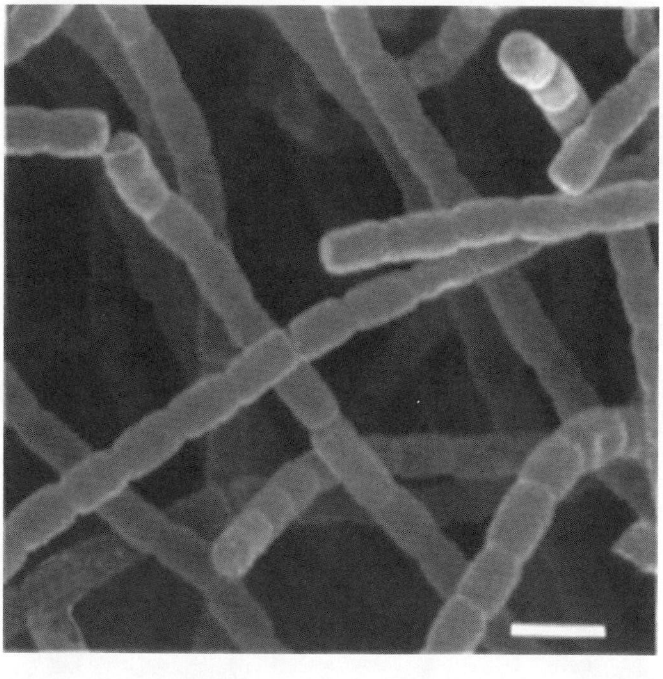

FIGURE 403. *Actinocorallia longicatena* strain I-497. Chains of arthrospores. The micro-organism was incubated on a tenth strength yeast extract-malt extract agar at 28°C for 2–3 weeks. (bar = 1 µm). (Reproduced with permission from Itoh et al., 1995. Actinomycetologica *9*: 167.)

are metabolically active (Table 300), but the descriptions of most species are based on relatively limited studies on single isolates. In contrast, *Actinocorallia aurantiaca* and *Actinocorallia libanotica* strains were included in extensive numerical taxonomic studies on *Actinomadura* and related taxa (Athalye et al., 1985; Trujillo and Goodfellow, 2003).

Cell-wall composition. *Actinocorallia* strains contain *meso*-diaminopimelic acid (*meso*-A_2pm) as the major wall diamino acid and, with the exception of *Actinocorallia cavernae*, contain madurose, that is, they have a wall chemotype III B *sensu* Lechevalier and Lechevalier (1970b). The diagnostic sugar madurose is found in other genera, including *Actinomadura, Dermatophilus, Microbispora, Planobispora, Planomonospora, Spirillospora,* and *Streptosporangium.* Whole-organism hydrolysates of *Actinocorallia glomerata* and *Actinocorallia longicatena* also contain galactose, glucose, mannose, and ribose (Itoh et al., 1995), and those of *Actinocorallia cavernae* contain glucose, mannose, and minor amounts of galactose (Lee, 2006).

The major phospholipids of *Actinocorallia* strains are diphosphatidylglycerol, phosphatidylethanolamine, and phosphatidylinositol, a profile which corresponds to phospholipid pattern type II *sensu* Lechevalier et al. (1981). The predominant menaquinones are MK-9(H_4) and MK-9(H_6) with smaller amounts of MK-9(H_8). *Actinocorallia aurantiaca* also contains a small proportion of MK-9(H_2) and *Actinocorallia libanotica* small amounts of MK-9(H_0) and MK-9(H_2) (Kroppenstedt et al., 1990). *Actinocorallia* strains contain complex mixtures of straight chain saturated, unsaturated, iso-branched, anteiso-branched and 10-methylated branched fatty acids, that is, a type 3a pattern *sensu* Kroppenstedt (1985); the predominant components are hexadecanoic, 14-methylpentadecanoic, octadecenoic, and 14-methyloctadecanoic acids.

Ecology. *Actinocorallia* strains are common in soil (Iinuma et al., 1994; Lavrova and Preobrazhenskaia, 1975; Lee, 2006; Meyer, 1979) where they probably play a role in the turnover of organic matter. They have also been isolated from a decayed leaf of a deciduous tree and from the root of a herbaceous plant (Itoh et al., 1995).

Isolation procedures

Actinocorallia strains can be isolated using methods described for *Actinomadura. Actinocorallia aurantiaca* was isolated from soil by Lavrova and Preobrazhenskaya (1975) using Gauze's medium 2 supplemented with rubromycin (0.5, 1.0, or 2.0 µg/ml), an antibiotic that inhibits the growth of common streptomycetes on isolation plates thereby allowing the development of "rare" actinomycetes. Iinuma et al. (1994) isolated *Actinocorallia herbida* by inoculating serial dilutions of soil onto colloidal chitin-vitamins agar and incubating the inoculated plates at 28°C for up to 3 weeks. *Actinocorallia glomerata* and *Actinocorallia longicatena* were isolated from plant material after washing with sterile water and drying it at 28°C for 7 d. The dried samples were blended in sterile water prior to incorporation into yeast extract agar (0.02%, w/v) supplemented with cycloheximide and nystatin (each at 50 µg/ml) and incubated at 25°C for up to 3 weeks (Itoh et al., 1995).

Maintenance procedures

Short-term storage can be achieved by serial transfer on modified Bennett's or yeast extract-malt extract agars, and long-term storage by lyophilization or storage in liquid nitrogen. For lyophilization, spores and mycelia are suspended in a suitable fluid such as 7.5% (w/v) glucose serum or 10% skimmed milk. Suspensions of spores and mycelium in glycerol (20%, v/v) can be stored in liquid nitrogen at –80°C.

Differentiation of the genus *Actinocorallia* from other genera

Actinocorallia strains can be distinguished from the other three genera classified in the family *Thermomonosporaceae* using a combination of chemotaxonomic, morphological, and

TABLE 300. Characteristics differentiating the species of the genus *Actinocorallia*[a,b]

Characteristic	1. *A. herbida*	2. *A. aurantiaca*	3. *A. aurea*	4. *A. cavernae*	5. *A. glomerata*	6. *A. libanotica*	7. *A. longicatena*
Morphology	Straight coralloid sporophores	Hooks, spirals	Straight to flexuous	Straight or flexuous	Straight, pseudosporangia	Hooks, curled	Straight or flexuous
Spore surface	Smooth	Warty	nd	Smooth	Smooth	Folded, warty	Smooth
Growth on yeast-extract-malt extract agar:							
Degree of growth	Moderate	Poor	Moderate	Good	Good	Abundant	Good
Aerial mycelium	White	White	Green-yellow	Sparse	White	Pale pink	White
Substrate mycelium	Honey gold	Yellowish brown	Light yellow	Olive black	Dull gold	Yellowish-brown	Oak-brown
Esculin hydrolysis	–	–	nd	+	nd	+	nd
Nitrate reductase	–	+	+	–	nd	+	nd
Degradation of:							
Casein	–	–	nd	+	+	–	–
Gelatin	–	+	+	+	–	–	–
Hypoxanthine	nd	–	nd	+	+	–	–
Starch	–	–	+	+	+	+	+
Tyrosine	nd	–	nd	+	–	–	–
Xanthine	nd	–	nd	–	–	–	–
Sole carbon sources:							
Adonitol	nd	+	nd	–	–	+	+
L-Arabinose	–	–	nd	+	+	+	+
Fructose	+	–	nd	+	+	–	+
Glycerol	nd	–	nd	–	+	+	+
Maltose	nd	–	nd	+	+	+	+
Mannose	nd	+	nd	+	+	–	–
Melibiose	nd	–	nd	–	–	–	–
Salicin	nd	–	nd	+	+	+	+
Sucrose	+	–	–	+	+	–	+
Ribose	nd	–	nd	+	–	–	–
Xylose	–	+	nd	+	+	+	+
Tolerance to 1% (w/v) NaCl	–	+	nd	+	+	+	+
Madurose in whole-cell sugars	+	+	+	–	+	+	+

[a]Symbols: +, positive; –, negative; nd, not determined.

[b]Data from: Athalye et al. (1985), Iinuma et al. (1994), Itoh et al. (1995), Lavrova and Preobrazhenskaya (1975), Lee (2006), Meyer (1979), Tamura et al. (2007), Trujillo and Goodfellow (2003) and Zhang et al. (2001).

phylogenetic data (Table 298). They can be separated readily from members of the genus *Actinomadura* as they have a phospholipid type I as opposed to a phospholipid type II pattern (Kroppenstedt and Goodfellow, 2006).

Taxonomic comments

The genus *Actinocorallia* Iinuma et al. 1994 was proposed to accommodate a single species, *Actinocorallia herbida.* The single representative of this taxon produced characteristic coralloid sporophores and was considered to have a wall chemotype III/C, that is, it contained *meso*-A_2pm but no diagnostic sugars. Zhang et al. (1998) found that *Actinocorallia herbida*, *Actinomadura aurantiaca*, and *Actinomadura libanotica* formed a 16S rRNA gene clade that was well separated from one that included the rest of the *Actinomadura* species; these findings raised the question whether the latter two species should be transferred to the genus *Actinocorallia.*

Following an extensive chemotaxonomic and phylogenetic study of members of the family *Thermomonosporaceae*, Zhang et al. (2001) emended the description of the genus *Actinocorallia* in order to accommodate four species previously classified in the genus *Actinomadura.* Thus, *Actinomadura aurantiaca*, *Actinomadura glomerata*, *Actinomadura libanotica*, and *Actinomadura longicatena* were transferred to the genus *Actinocorallia* as *Actinocorallia aurantiaca*, *Actinocorallia glomerata*, *Actinocorallia libanotica*, and *Actinocorallia longicatena*, respectively. Zhang and his colleagues also detected major amounts of diphosphatidylglycerol, phosphatidylethanolamine, and phosphatidylinositol in *Actinocorallia libanotica* and the presence of madurose in *Actinocorallia herbida.* These findings mean that *Actinocorallia libanotica* has a phospholipid type II, not I, as reported by Kroppenstedt et al. (1990) and that *Actinocorallia herbida* has a sugar type B, not C, as reported by Iinuma et al. (Iinuma et al., 1994). It was clear from these findings that six of the seven *Actinocorallia* species contain madurose and have a phospholipid pattern type II *sensu* Lechevalier et al. (1977, 1981). The type strain of *Actinocorallia cavernae* lacks madurose but has a phospholipid pattern type II (Lee, 2006).

Differentiation of the species of the genus *Actinocorallia*

The differential characteristics of the species of *Actinocorallia* are given in Table 300. However, identification of members of the genus is not easy as most of the phenotypic data are based on sole representatives of species and even then there are inconsistencies between data from different studies.

Acknowledgements

The authors are grateful to Dr Olga Genilloud for reading the manuscript and providing helpful comments.

List of species of the genus *Actinocorallia*

1. **Actinocorallia herbida** Iinuma, Yokota, Hasegawa and Kanamuru 1994, 233[VP]

her'bi.da. L. fem. adj. *herbida* grassy, referring to the formation of aerial mycelia like grass.

Spores borne in long straight chains or on coralloid sporophores.

Glycerol-asparagine agar: poor growth, sparse white aerial mycelium, substrate mycelium light ivory, and no diffusible pigment. Inorganic salts-starch agar: poor growth, no aerial mycelium, substrate mycelium light wheat color, and no diffusible pigment. Oatmeal agar: moderate growth, aerial mycelium light yellow, substrate mycelium yellow, and no diffusible pigment. Tyrosine agar: poor growth, no aerial mycelium, substrate mycelium light ivory color, and dusty coral-colored diffusible pigment. Yeast extract-malt extract agar: moderate growth, sparse white aerial mycelium, substrate mycelium honey gold color, and no diffusible pigment. The temperature range for growth is 12–38°C, the optimal range 24–32°C. Glucose is used as a sole carbon source, but not inositol, mannitol, or raffinose. Grows in the presence of 3% but not 5% (w/v) NaCl. Antibiotic resistance (μg/ml) is shown towards cefsulodin (100), but not against gentamicin (100), novobiocin (20), rifampin (50), streptomycin (100), or vancomycin (50). Additional phenotypic properties are shown in Table 300. The predominant fatty acids are $C_{16:0}$ (25.8%), $C_{16:1}$ (15.4%), $C_{17:1}$ (16.0%), and $C_{18:1}$ (26.5%).

Source: soil sample collected in Bangkok, Thailand.

DNA G+C content (mol%): 73.0 (HPLC) (type strain).

Type strain: AL-50780, ATCC 51528, DSM 44254, NBRC 15485, JCM 9647, NCIMB 13337, VKM Ac-1994.

Sequence accession no. (16S rRNA gene): D85473.

Sequence accession no. (23S rRNA gene): AF134086.

Sequence accession no. (16S–23S rRNA ITS): AF134109.

2. **Actinocorallia aurantiaca** (Lavrova and Preobrazhenskaia 1975) Zhang, Kudo, Nakajima and Wang 2001, 381[VP] (*Actinomadura aurantiaca* Lavrova and Preobrazhenskaya 1975, 485[AL])

au.ran.ti'a.ca. N.L. fem. adj. *aurantiaca* orange-colored, referring to the gold-colored substrate mycelium.

Spore chains (of 4–8 spores), hooked or in spirals (one turn), in small clusters emerging directly from the agar. Spore surface warty. Spores connected by intersporal pads. Inorganic salts-starch agar: moderate growth; sparsely scattered dots of sporulating hyphae; aerial mycelium whitish, substrate mycelium cream-colored to orange, and no diffusible pigment. Oatmeal agar: moderate growth; surface farinaceous, aerial mycelium cream-colored to pink; substrate mycelium yellowish white, and no diffusible pigment. Peptone-glucose medium: moderate growth; surface farinaceous, aerial mycelium cream-colored; substrate mycelium yellow to orange, and no diffusible pigment. Yeast extract-malt extract agar: poor growth; sparsely scattered dots of sporulating hyphae; aerial mycelium whitish, substrate mycelium yellow brown, and no diffusible pigment. The temperature range for growth is 10–30°C, the optimal range is 25–28°C. The pH growth range is 6.0–10.0.

Degrades arbutin, DNA, testosterone, and Tweens 20, 40, 60, and 80, but not chitin, elastin, pectin, RNA, or xylan. Galactose, glucose, rhamnose, and trehalose are used as sole carbon sources, but not lactose, mannitol, mannose, melezitose, sorbitol, or xylitol. Antibiotic resistance (µg/ml) is shown towards ampicillin (10), benzylpenicillin (25), carbenicillin (120), cephaloridine (2), cephalothin (20), cephamandole (30), lincomycin (10), oleandomycin (20), and ticarcillin (70), but not against streptomycin (2), tobramycin (0.05), or tetracycline (20). Additional phenotypic properties are shown in Table 300.

Source: soil.

DNA G+C content (mol%): not determined.

Type strain: DSM 43924, IFO (now NRBC) 15554, IMET 9577, INA 1933, JCM 8201, KCTC 9554.

Sequence accession no. (16S rRNA gene): AF134066, AJ293701.

Sequence accession no. (23S rRNA gene): AF134072.

Sequence accession no. (16S–23S rRNA ITS): AF134090.

3. **Actinocorallia aurea** Tamura, Hatano, Suzuki 2007, 2054[VP]

au're.a. L. fem. adj. *aurea* golden.

Straight to flexuous chains of aerial spores are formed on the tip of coralloid sporophores. Spores are short, non-motile rods (0.6–0.9 × 1.0–1.5 mm). Substrate mycelium is yellow on maltose-Bennett's agar, yeast extract-malt-extract agar, inorganic salts-starch agar, glycerol-asparagine agar, and peptone-yeast extract-iron agar, and grayish white on oatmeal and tyrosine agars. The aerial spore mass is yellow on yeast extract-starch agar, maltose-Bennett's agar, and on 1/4 strength oatmeal and yeast extract-malt extract agars. Optimum temperature range for growth is 20–30°C. Grows at 37°C, but not 45°C. Does not grow on 4% (w/v) NaCl. Gelatin is liquefied, starch is hydrolyzed, and nitrate is reduced. Glucose, maltose, rhamnose, and xylose are used as sole carbon sources. Additional phenotypic properties are shown in Table 300. The major fatty acids are $C_{16:0}$ (21%), $C_{18:1}$ (19%), 10-methyl $C_{18:0}$ (10%), and $C_{17:1}$ (13%).

Source: soil.

DNA G+C content (mol%): 71–73 (HPLC).

Type strain: DSM 14752, NBRC 44434.

Sequence accession no. (16S rRNA gene): AB006177.

4. **Actinocorallia cavernae** Lee 2006, 1087[AL]

ca.ver'na.e. L. gen. n. *cavernae* of a cavern, the site from which the type strain was isolated.

Aerial hyphae differentiate into straight to flexuous spores chains. Spore surface smooth. Substrate mycelium is olive black in color on yeast extract-malt extract agar, dark yellow brown on oatmeal agar, and cream on inorganic salts-starch agar. Aerial mycelium is abundant and grayish white on inorganic salts-starch and oatmeal agars, but sparse on yeast extract-malt extract agar. Diffusible pigments are not formed. Grows optimally in the temperature range 25–30°C, but not at 10° or 37°C. Grows in the pH range 5.1–10.1, but not at pH 4.1. Does not degrade DNA or elastin. Urea is hydrolyzed. H_2S is not produced. D-Arabinose, 2,3-butandiol, cellobiose, galactose, glucose, lactose, melezitose, rhamnose, and trehalose are used as sole carbon sources, but not dulcitol, *meso*-erythritol, inositol, inulin, mannitol, methyl-D-glucoside, methyl-D-mannoside, 1,2 propandiol, raffinose, sorbitol, sorbose, or xylitol. Additional phenotypic properties are shown in Table 300. The predominant fatty acids are $C_{16:0}$ (25.9%), $C_{18:1}$ (19.2), $C_{17:0}$ (11.7%), 10-methyl $C_{18:0}$ (tuberculostearic acid, 9.4%), *ai*-$C_{17:0}$ (7.4%), and $C_{18:0}$ (6.7%).

Source: soil sample taken from inside a natural cave on Jeju Island, Republic of Korea.

DNA G+C content (mol%): 70.1 (HPLC).

Type strain: N3-7, DSM 45040, JCM 13278, NRRL B-24429.

Sequence accession no. (16S rRNA gene): AY966427.

5. **Actinocorallia glomerata** (Itoh, Kudo, Oyaizu and Seino 1995) Zhang, Kudo, Nakajima and Wang 2001, 381[VP] (*Actinomadura glomerata* Itoh, Kudo, Oyaizu and Seino 1995, 173)

glo.me.ra'ta. L. fem. part. adj. *glomerata* (from L. v. *glomerare* to form into ball, glomerate) gathered into a mass, referring to the pseudosporangium-like structures.

Rod-shaped spores (0.6–0.8 × 0.4–0.5 µm) are formed in slimy globular, pseudosporangia-like clusters (Figure 402) or occasionally in straight chains. Spore surface smooth. Glycerol-asparagine and tyrosine agars: moderate growth; aerial mycelium white; substrate mycelium pearl-colored, and no diffusible pigment. Inorganic salts-starch agar: moderate growth; no aerial mycelium; substrate mycelium pearl-colored, and no diffusible pigment. Oatmeal agar: good growth; aerial mycelium white; substrate mycelium sand-colored, and no diffusible pigment. Yeast-extract-malt extract agar: good growth; aerial mycelium white; substrate mycelium dull gold color, and no diffusible pigment. Temperature range for growth is 7–41°C, optimal range 28–30°C. Adenine is not degraded. Galactose, glucose, rhamnose, and trehalose are used as sole carbon sources, but not dulcitol, erythritol, inositol, mannitol, raffinose, or sorbitol. Antibiotic resistance (µg/ml) is shown towards novobiocin (20), rifampin (50), streptomycin (50), and vancomycin (50). Additional phenotypic properties are shown in Table 300. Predominant fatty acids are $C_{16:0}$ (28%), $C_{18:1}$ (20%), and 10-methyl $C_{19:0}$ (15%).

Source: a decayed leaf of a deciduous tree collected from a pond in Saitama, Japan.

DNA G+C content (mol%): 71.7 (HPLC) (type strain).

Type strain: DSM 44360, IFO 15960, JCM 9376, NBRC 15960.

Sequence accession no. (16S rRNA gene): AF134068, AJ293704.

Sequence accession no. (23S rRNA gene): AF134077.

Sequence accession no. (16S–23S rRNA ITS): AF134098.

6. **Actinocorallia libanotica** (Meyer 1981) Zhang, Kudo, Nakajima and Wang 2001, 381[VP] (*Actinomadura libanotica* Meyer 1981, 215[VP])

li.ba.no'ti.ca. L. n. *Libanus* Lebanon; L. fem. suff. *-tica*, suffix denoting made of or belonging to; N.L. fem. adj. *libanotica* referring to Lebanon (the country in which the soil sample was taken).

Spores chains in hooks or curled, arranged in clusters on short hyphae on the aerial mycelium, 5–12 spores per chain: spores subspherical, and spore surface folded or warty. Spores are connected by intersporal pads (Figure 401). Inorganic salts-starch agar: moderate growth; only a flimsy cover of aerial mycelium; substrate mycelium grayish, and no diffusible pigment. Oatmeal agar: abundant growth; surface farinaceous; aerial mycelium white to pale pink; substrate mycelium yellowish brown, and no diffusible pigment. Oatmeal-nitrate agar: abundant growth; surface farinaceous; aerial mycelium white to pale pink; substrate mycelium yellowish brown, and no diffusible pigment. Peptone-glucose medium: moderate growth; surface granular; aerial mycelium white; substrate mycelium yellowish brown, and no diffusible pigment. Yeast extract-malt extract agar: abundant growth; surface farinaceous; aerial mycelium pale pink; substrate mycelium yellowish brown, and no diffusible pigment. The temperature range for growth is 10–37°C, and the optimal range is 28–30°C. Does not produce allantoinase or urease. Degrades arbutin, RNA, testosterone, Tweens 20, 40, 60, and 80 and xylan, but not chitin, elastin, guanine, pectin, or xylan. Cellobiose, dextrin, galactose, glycogen, rhamnose, and trehalose are used as sole carbon sources, but not glucosamine, lactose, mannitol, raffinose, sorbitol, or xylitol. Antibiotic resistance (μg/ml) is shown towards ampicillin (10), benzylpenicillin (10), carbenicillin (90), cephaloridine (2), demeclocycline (2), lincomycin (10), neomycin (1), oleandomycin (3), and ticarcillin (30), but not against cephalotin (20), cephamandole (20), gentamicin (4), neomycin (3), rifampin (2), streptomycin (4), tetracycline (20), tobramycin (0.05), or vancomycin (0.25). Additional phenotypic properties are shown in Table 300.

Source: soil.

DNA G+C content (mol%): 66.2 (T_m).

Type strain: ATCC 35576, DSM 43554, IMET 9616, NBRC 14095, JCM 10696, NCIB 11686, NRRL B-16097, VKM Ac-939.

Sequence accession no. (16S rRNA gene): U49007.

Sequence accession no. (23S rRNA gene): AF134078.

Sequence accession no. (16S–23S rRNA ITS): AF134100.

7. **Actinocorallia longicatena** (Itoh, Kudo, Oyaizu and Seino 1995) Zhang, Kudo, Nakajima and Wang 2001, 381[VP] (*Actinomadura longicatena* Itoh, Kudo, Oyaizu and Seino 1995, 175)

lon.gi.ca.te'na. L. adj. *longus* long: L. fem. n. *catena* chain; N.L. fem. n *longicatena* a long chain.

Straight to flexuous chains of 20 or more spores are borne on aerial hyphae. Spores are rod-shaped (0.6–0.9 × 0.4 × 0.5 μm). Spore surface smooth (Figure 403). Glycerol-asparagine agar: growth moderate; sparse white aerial mycelium; substrate mycelium colorless, and no diffusible pigment. Inorganic salts-starch agar: growth poor to moderate; sparse white aerial mycelium; substrate mycelium pearl, and no diffusible pigment. Oatmeal agar: growth good; sparse aerial mycelium; substrate mycelium sandy-colored, and no diffusible pigment. Tyrosine agar: growth moderate; aerial mycelium white; substrate mycelium pearl-colored, and pale brown diffusible pigment. Yeast extract-malt extract agar: good growth; aerial mycelium white; substrate mycelium oak brown, and no diffusible pigment. Temperature range for growth is 7–37°C. Adenine is not degraded. Galactose, glucose, rhamnose, and trehalose are used as sole carbon sources, but not dulcitol, erythritol, inositol, mannitol, raffinose, or sorbitol. Antibiotic resistance (μg/ml) is shown against novobiocin (20), rifampin (50), streptomycin (50), and vancomycin (50). Additional phenotypic properties are shown in Table 300. Predominant fatty acids are $C_{16:0}$ (24%), $C_{18:1}$ (16%), and 10-methyl $C_{19:0}$ (14%).

Source: root of a herbaceous plant (*Bidens* sp.) grown in Saitama, Japan.

DNA G+C content (mol%): 70.9 (HPLC) (type strain).

Type strain: I-497, CIP 105488, DSM 44361, NBRC 15961, JCM 9377.

Sequence accession no. (16S rRNA gene): AF163117, AJ293707.

Sequence accession no. (23S rRNA gene): AF163128.

Sequence accession no. (16S–23S rRNA ITS): AF163139.

Genus III. **Actinomadura** Lechevalier and Lechevalier 1970a, 400[AL] emend. Kroppenstedt, Stackebrandt and Goodfellow 1990, 156

Martha E. Trujillo and Michael Goodfellow

Ac.ti.no.ma.du'ra. Gr. n. *actis actinos* a ray; N.L. n. *Madura* Madura, name of a province in India; N.L. fem. n. *Actinomadura* referring to a micro-organism first described as the causative agent of "Madura foot" disease.

Gram-stain-positive, non-acid–alcohol-fast, nonmotile actinomycetes that form an **extensively branched non-fragmenting, substrate mycelium. Aerial mycelium moderately developed or absent**. When present, aerial hyphae carry up to 50 arthrospores. **Aerial mycelium at maturity forms short or occasionally long chains of arthrospores. Spore chains straight, hooked (open loops), or irregular spirals (1–4 turns). Spore surface folded, irregular, rugose, smooth, spiny, or warty**. Color of mature sporulated aerial mycelium: blue, brown, cream, gray, green, pink, red, white, or yellow. **Colonies have a leathery or cartilaginous appearance** when aerial mycelium is lacking. Aerobic, chemo-organotrophic with an oxidative type metabolism. Temperature growth range 10–60°C. **Cell wall contains *meso*-2,6-diaminopimelic acid as the major diamino acid and *N*-acetylated muramic acid**. Whole-cell hydrolysates contain galactose, glucose, madurose, mannose, and ribose. **The major phospholipids are diphosphatidylglycerol and phosphatidylinositol. Menaquinones are predominantly hexahydrogenated with nine isoprene units saturated at sites II, III, and VIII**. Complex fatty acid profile is rich in branched saturated and unsaturated fatty acids, including tuberculostearic acid. Mycolic acids are absent. Widely distributed in soil. Some strains are pathogenic for animals, including man.

DNA G+C content (mol%): 66–73 (T_m, HPLC).

Type species: **Actinomadura madurae** (Vincent 1894) Lechevalier and Lechevalier 1970a, 400[AL].

Further descriptive information

Phylogeny. The genus *Actinomadura* is classified in the family *Thermomonosporaceae* Rainey et al. 1997 emend. Zhang et al. 2001 emend. Zhi et al. 2009 together with the genera *Actinocorallia* Iinuma et al. 1994 emend. Zhang et al. 2001, *Spirillospora* Couch 1963 and *Thermomonospora* Henssen 1957 emend. Zhang et al. 2001. These taxa form a distinct phyletic line in the 16S rRNA gene tree (Zhang et al., 1998, 2001) and can thereby be distinguished from the other two families that comprise the order *Streptosporangiales*, namely the families *Nocardiopsaceae* and *Streptosporangiaceae* (Goodfellow and Quintana, 2006; Gyobu, 2001; Kudo, 2001). Members of the genera *Actinomadura*, *Spirillospora*, and *Thermomonospora* are phylogenetically intermixed, but can be distinguished using morphological properties (Kroppenstedt and Goodfellow, 2006; Zhang et al. 1998, 2001). The complex topology of the *Thermomonosporaceae* 16S rRNA gene tree is shown in Figure 398.

Cell morphology. *Actinomadura* strains characteristically form nonfragmenting, extensively branched, substrate mycelia and aerial hyphae which carry 1–50 arthrospores. *Actinomadura formosensis* forms single spores on both aerial and substrate mycelia (Hasegawa et al., 1986) whereas *Actinomadura echinospora* and *Actinomadura rugatobispora* usually produce two spores per chain, sometimes three, on aerial hyphae (Miyadoh et al., 1990). Longer spore chains are characteristic of most other *Actinomadura* species though spores have not been detected in *Actinomadura fibrosa* or *Actinomadura latina* strains. Spores are borne in curled, hooked, spiral, or straight chains. Spore sufaces may be folded, irregular, rugose, smooth, shiny, or warty (Table 301).

Some strains, notably those from clinical sources, lack aerial mycelium with colonies exhibiting a cartilaginous or leathery appearance. However, members of most species form a spore-bearing, powdery aerial mycelium on media such as inorganic salts-starch, oatmeal and yeast extract-malt extract agars after cultivation for 10–14 d. At maturity, aerial mycelia may be blue, cream, gray, green, pink, yellow, or white thereby differing little from streptomycetes. Superficial similarity to streptomycetes is also reinforced by the morphology of the sporophores. *Actinomadura spadix* forms pseudosporangia-like structures, i.e. tightly spiraled spore chains embedded in a dry to slimy mass thereby simulating the colonial appearance of *Streptosporangium* strains when examined microscopically at low magnification. Actinomadurae and streptomycetes can be distinguished by direct microscopic comparison of cultures. Most *Actinomadura* strains have spores noticeably larger than the diameter of their hyphae, whereas streptomycetes have spores and hyphae of similar diameter.

Cell-wall composition. The genus *Actinomadura* was defined chemically (Lechevalier and Lechevalier, 1970a) to encompass former *Nocardia* species with cell-wall type III, i.e. with *meso*-A_2pm in the peptidoglycan, but no arabinogalactan polymer, which is in contrast to *Nocardia* species. Members of some species also contain small amounts of the LL-isomer. Whole-organism hydrolysates contain galactose, glucose, mannose, ribose, and madurose (3-*O*-methyl-D-galactose; Lechevalier and Gerber, 1970), though the latter may be found in trace amounts. Madurose also occurs in other wall type III taxa, including *Actinocorallia*, *Dermatophilus*, *Microbispora*, *Nonomuraea*, *Planobispora*, *Planomonospora*, *Spirillospora*, and *Streptosporangium*. Differentiation from these genera is based on molecular and morphological characteristics.

Polar lipid and menaquinone composition. Most *Actinomadura* species, including *Actinomadura madurae*, the type species, have a phospholipid type I pattern (Lechevalier et al., 1977, 1981), which is characterized by the presence of diphosphatidylglycerol, phosphatidylinositol, and phosphatidylinositol mannosides. In contrast, *Actinomadura kijaniata*, *Actinomadura namibiensis*, and *Actinomadura napierensis* have a phospholipid pattern type PIV, which is characterized by the presence of diphosphatidylglycerol, phosphatidylethanolamine, phosphatidylinositol, phosphatidylinositolmannoside, phosphatidylmethylamine, and unknown glucosamine-containing phospholipids (Cook et al., 2005; Wink et al., 2003). *Actinomadura* strains can be differentiated from the genus *Actinocorallia* on the basis of polar lipid profiles. Members of the genus *Actinomadura* contain mainly MK-9(H_6) saturated at sites II, III, and VIII, small amounts of MK-9(H_4) saturated at sites II, III, and VIII, and small amounts of MK-9(H_8) (Kroppenstedt et al., 1990). Differentiation from other wall chemotype III genera, such as *Microbispora* or *Nonomuraea*, is difficult as members of these taxa have similar menaquinone profiles (Athalye et al., 1984; Kroppenstedt et al., 1990; Poschner et al., 1985).

Fatty acid profile. Actinomadurae contain complex mixtures of fatty acids with hexadecanoic (16:0), 14-methylpentadecanoic (iso-16:0), and 10-methyloctadecanoic (tuberculostearic) acids predominating, that is, they have a fatty acid type 3a (Kroppenstedt, 1985). *Actinomadura rubrobrunea*, a thermophilic organism, can be separated from mesophilic actinomadurae by differences in fatty acid composition (Agre and Guzeva, 1975). It contains relatively high proportions of iso-branched fatty acids (high melting point) and low amounts of 10-methyl branched acids (low-melting-point) in line with its thermophilic nature. Hydroxylated fatty acids have been detected in *Actinomadura atramentaria*.

Cultural characteristics. Actinomadurae are slower-growing than streptomycetes and tend to grow well on modified Bennett's (Jones, 1949) and glucose-yeast extract agars (Athalye, 1981), and on formulations that were found to be useful in the International *Streptomyces* Project (ISP; Shirling and Gottlieb, 1966). Most strains show abundant growth on oatmeal agar (ISP 3 medium) at 28–30°C, but *Actinomadura kijaniata* and *Actinomadura macra* grow better on yeast extract- malt extract agar (ISP 2 medium). *Actinomadura vinacea* grows well on glucose-peptone agar but poorly on these media. *Actinomadura spadix* requires vitamin B12 for good growth (Nonomura and Ohara, 1971d). *Actinomadura rubrobrunea* grows well on oatmeal and peptone maize agars at 50°C, and *Actinomadura yumaensis* grows poorly on media containing inorganic nitrogen sources. *Actinomadura latina*, *Actinomadura madurae*, and *Actinomadura pelletieri*, the three recognized pathogenic species, show moderate to good growth on Czapek–Dox Casamino acids (Cross and Attwell, 1974; Trujillo, 1994).

Nutrition and growth conditions. All *Actinomadura* strains are strictly aerobic and chemo-organotrophic with an oxidative metabolism. Most species grow well at 25–40°C, but *Actinomadura formosensis*, *Actinomadura rubrobrunea*, and *Actinomadura viridilutea* are thermophilic, with optimal growth temperatures

TABLE 301. Morphological characters separating *Actinomadura* species[a]

Species	Color on oatmeal agar		Diffusible pigments	Spore chain type	Spore surface type
	Aerial mycelium	Substrate mycelium			
A. madurae	Trace, white specks	Colorless	None	Hooks to spirals	Warty
A. atramentaria	White	Colorless	Inky brown[b]	Straight	Smooth
A. catellatispora	Yellow	Light yellow	None	Straight	Smooth
A. citrea	White to blue	Lemon yellow	Yellow	Hooks or curled	Irregular or warty
A. coerulea	Pink to blue	Colorless to pale pink	None	Hooks to spirals	Warty
A. cremea subsp. *cremea*	White to yellow	Colorless	None	Hooks to spirals	Warty
A. cremea subsp. *rifamicini*	White to yellow	Colorless	None	Hooks to spirals	Warty
A. echinospora	Yellow to pink	Orange	Yellow	Paired spores	Spiny
A. fibrosa	White	Orange	Brown	No spores formed	–
A. formosensis	White	Light apricot	None	Single	Warty
A. fulvescens	White	Red-brown	Red-brown	Spirals	Smooth
A. glauciflava	Bluish green	Yellow to brown	Yellow	Hooks, spirals	Warty
A. hallensis	White	Brown	None	Hooks, spirals	Warty
A. hibisca	Light gray	Red violet	Carmine red	Straight	Smooth
A. kijaniata	Trace, white specks	Pine green	None	Spirals	Smooth
A. latina	Absent	Cream to pink	None	No spores formed	–
A. livida	Trace	Gray to pink	Pale violet	Hooks, spirals	Uneven
A. luteofluorescens	Yellow to blue	Yellow to green	Yellow-green	Hooks, curled	Warty
A. macra	Cream to pink	Cream to pink	None	Hooks, curled	Smooth
A. mexicana	Absent	Cream to yellow	None	Hooks	Warty
A. meyerae	Sparse	Cream to yellow	None	Hooks	Warty
A. namibiensis	None	Salmon pink	None	Spirals	Smooth
A. napierensis	White	Grey	None	Spirals	Smooth
A. nitritigenes	Brown	Colorless to brown	None	Hooks	Smooth
A. oligospora	Trace,white	Gray	None	Straight, hooked	Smooth
A. pelletieri	Absent	Pink to brownish red	None	Hooks to spirals	Warty
A. rubrobrunea	Grayish-blue	Orange	None	Hooks to spirals	Spiny
A. rudentiformis	White	Cream	None	No spores formed	No spores formed
A. rugatobispora	Light to dusty green	Pastel yellow	None	Paired spores	Rugose with vertical ridges
A. spadix	Yellow/brown	Reddish gray/grayish brown	Reddish gray	Pseudosporangia	Smooth
A. umbrina	Scanty, white	Colorless	None	Straight, hooks, spirals	Smooth
A. verrucosospora	Pink to blue	Orange to pink	None	Hooks to spirals	Warty
A. vinacea	Absent	Pink to red	None	Straight	Irregular
A. viridilutea	Blue green	Orange to yellow	None	Hooks to spirals	Spiny
A. viridis	Green	Yellow-brown to green	None	Straight	Smooth
A. yumaensis	Gray to yellow	Gray to yellow	None	Hooks	Smooth

[a]Data from Athalye et al. (1985), Lu et al. (2003), Mertz and Yao (1990), Miyadoh et al. (1985), Meyer (1989a, 1989b), Quintana et al. (2003), Wink et al. (2003), Cook et al. (2005), Lee and Jeong (2006), and Le Roes and Meyers (2007).

[b]Pigment found on tyrosine agar and yeast extract–malt extract agar.

between 45°C and 65°C. *Actinomadura* strains can metabolize a wide range of amino acids and sugars as carbon sources for growth (Table 302); proteolytic activity is shown by the capacity of most strains to hydrolyze arbutin, casein, elastin, gelatin, testosterone, and Tweens 60 and 80 (Kroppenstedt and Goodfellow, 1992; Trujillo and Goodfellow, 2003). On the other hand, xylan is not degraded (Table 303). Pectinase activity has been reported for *Actinomadura mexicana*, *Actinomadura napierensis*, and *Actinomadura verrucosospora* (Cook et al., 2005; Trujillo and Goodfellow, 2003). Pathogenic strains of *Actinomadura madurae* produce collagenolytic enzymes (Rippon, 1968). Most actinomadurae grow in the presence of NaCl 3% (w/v). *Actinomadura atramentaria* tolerates NaCl concentrations of 7% (w/v).

Pigments. *Actinomadura madurae* and *Actinomadura pelletieri* produce prodigiosin-like pigments (Gerber, 1971, 1973; Lechevalier et al., 1971) that are similar to those of *Serratia marcescens*. Members of these species isolated from patients produce prodiginines characterized by a tripyrrole skeleton and identified as cyclononylprodiginine, nonylprodiginine, and undecylprodiginine. *Actinomadura atramentaria* produces an inky brown pigment on several media, especially in submerged cultures (Miyadoh et al., 1987); under alkaline conditions the pigment is dark green-brown and is more soluble in water. In contrast, under acidic conditions, it is dark violet and more soluble in organic solvents.

Genetics. Very little is known about the genetics of *Actinomadura* strains though extensive phylogenetic studies based on the 16S, 23S rRNA, and intergenic spacer regions have been carried out (Wang et al., 1996; Zhang et al., 1998, 2001). 16S rRNA gene sequence similarity of the type strains of all 35 validly named *Actinomadura* species is in the range 93.5–99.6%, suggesting a moderate degree of heterogeneity. Members of the taxon contain the 16S rRNA nucleotide sequence signatures characteristic of members of the family *Thermomonosporaceae* (Stackebrandt et al., 1997).

TABLE 302. Carbon source utilization for *Actinomadura* species[a, b]

Carbon source	*A. madurae*	*A. atramentaria*	*A. catellatispora*	*A. citrea*	*A. coerulea*	*A. cremea*	*A. echinospora*	*A. fibrosa*	*A. formosensis*	*A. fulvescens*	*A. glauciflava*	*A. hallensis*	*A. hibisca*	*A. kijanata*	*A. latina*	*A. livida*	*A. luteofluorescens*	*A. macra*	*A. mexicana*	*A. meyerae*	*A. namibiensis*	*A. napierensis*	*A. nitritigenes*	*A. oligospora*	*A. pelletieri*	*A. rubrobrunea*	*A. rudentiformis*	*A. rugatobispora*	*A. spadix*	*A. umbrina*	*A. verrucosospora*	*A. vinacea*	*A. viridilutea*	*A. viridis*	*A. yumaenesis*
Adonitol	+	−	nd	+	+	−	−	+	nd	+	nd	−	−	+	−	−	+	−	+	+	nd	+	nd	+	−	nd	+	nd	+	nd	+	nd	nd	+	−
Arabinose	d	−	nd	+	−	+	+	+	−	+	nd	+	−	−	+	−	−	−	+	+	+	+	+	+	−	nd	nd	v	+	nd	+	nd	nd	−	−
Cellobiose	+	+	nd	+	+	+	+	+	nd	+	nd	+	+	+	+	+	+	+	+	+	nd	nd	+	+	+	nd	+	nd	+	nd	+	nd	nd	+	+
Fructose	+	−	nd	+	−	+	+	+	−	−	nd	+	nd	+	+	+	+	−	−	+	−	+	+	+	−	nd	+	+	+	nd	+	nd	nd	+	+
Galactose	+	−	nd	−	+	+	−	+	+	+	nd	−	nd	+	+	+	+	−	−	−	nd	+	+	−	−	+	+	nd	+	nd	−	nd	+	+	+
Glucose	+	+	+	+	+	+	+	+	+	+	+	+	+	+	+	+	+	+	+	+	+	nd	nd	+	+	nd	+	+	+	nd	+	nd	nd	+	+
Glycerol	+	+	nd	+	−	+	+	+	−	+	nd	+	−	−	+	−	+	+	+	+	nd	+	+	+	−	+	nd	+	+	nd	+	nd	+	+	+
Glycogen	+	−	nd	+	−	+	+	+	nd	+	nd	nd	nd	+	d	+	+	−	+	+	nd	nd	nd	+	+	nd	nd	nd	+	nd	+	nd	nd	+	+
Lactose	−	+	nd	−	−	−	+	+	−	+	nd	−	−	+	d	−	−	+	−	−	nd	nd	−	+	d	+	+	nd	+	nd	−	nd	+	−	−
Maltose	+	−	nd	+	−	+	+	+	−	−	nd	+	nd	+	+	−	+	−	−	+	nd	+	−	−	d	+	nd	nd	+	nd	+	nd	+	+	+
Mannitol	+	−	nd	+	−	+	+	−	+	−	nd	+	−	−	+	+	+	−	−	+	v	+	nd	−	−	+	+	+	+	nd	+	nd	+	+	−
Mannose	d	+	nd	−	+	+	+	+	−	+	nd	−	nd	+	d	+	−	+	−	−	nd	nd	+	+	−	nd	nd	nd	+	nd	−	nd	nd	−	+
Melezitose	−	−	nd	−	−	−	−	−	nd	−	nd	−	nd	−	+	−	−	−	−	−	nd	+	nd	−	−	nd	+	nd	+	nd	d	nd	nd	−	+
Melibiose	−	+	nd	−	−	−	−	−	nd	−	nd	−	nd	−	−	−	−	−	−	−	nd	−	nd	−	−	nd	+	nd	+	nd	−	nd	nd	−	−
Raffinose	−	−	nd	−	−	−	−	+	−	−	nd	−	nd	−	−	−	−	−	−	−	−	+	nd	−	−	+	+	−	+	nd	−	nd	+	−	−
Rhamnose	d	−	nd	+	−	+	+	+	−	−	nd	−	nd	+	+	−	+	−	+	+	+	+	+	−	d	+	+	+	+	nd	+	nd	+	+	−
Ribose	−	−	nd	+	−	+	−	+	nd	d	nd	nd	nd	+	d	+	−	−	−	+	nd	nd	nd	+	−	nd	+	nd	+	nd	−	nd	nd	+	+
Salicin	−	−	nd	−	−	−	−	−	nd	−	nd	+	nd	+	−	−	−	−	−	−	nd	nd	nd	+	−	nd	+	nd	−	nd	d	nd	nd	−	−
Sorbitol	−	+	nd	−	−	−	−	nd	−	−	nd	−	nd	−	−	−	−	−	−	−	nd	nd	nd	−	−	nd	nd	nd	+	nd	−	nd	nd	−	−
Trehalose	+	+	nd	+	+	+	+	+	−	+	nd	+	nd	+	+	−	+	+	+	+	nd	+	+	+	d	nd	+	nd	+	nd	+	nd	nd	+	+
Xylitol	+	+	nd	−	−	−	−	−	nd	−	nd	−	nd	−	−	−	−	−	−	+	nd	nd	nd	−	−	nd	+	nd	+	nd	−	nd	nd	−	−
Xylose	d	−	nd	+	+	+	+	−	−	+	nd	+	nd	+	+	+	+	−	+	+	v	+	+	+	d	nd	+	−	+	nd	+	nd	nd	+	−
L-Arginine	−	−	nd	+	+	−	−	nd	nd	−	nd	nd	nd	−	d	−	+	−	−	+	nd	nd	nd	−	−	nd	+	nd	−	nd	d	nd	nd	+	+
L-Glutamine	−	−	nd	−	+	−	−	nd	nd	+	nd	nd	nd	+	−	−	+	−	+	−	nd	nd	nd	−	−	nd	nd	nd	−	nd	−	nd	nd	+	+
Glycine	d	−	nd	+	−	−	−	nd	nd	−	nd	nd	nd	+	d	−	−	−	−	−	nd	nd	nd	−	d	nd	nd	nd	−	nd	d	nd	nd	+	−
L-Histidine	d	−	nd	+	+	−	−	nd	nd	+	nd	nd	nd	+	−	+	+	−	−	−	nd	nd	nd	−	d	nd	+	nd	−	nd	d	nd	nd	+	−
L-Leucine	−	+	nd	−	+	−	−	nd	nd	−	nd	nd	nd	+	−	−	+	+	−	−	nd	nd	nd	+	−	nd	nd	nd	−	nd	−	nd	nd	+	−
L-Methionine	−	−	nd	−	−	−	−	nd	nd	−	nd	nd	nd	+	−	−	−	−	−	−	nd	nd	nd	−	−	nd	+	nd	−	nd	−	nd	nd	+	+
L-Phenylalanine	d	−	nd	+	+	−	−	nd	nd	−	nd	nd	nd	+	+	+	+	+	−	−	nd	nd	nd	−	d	nd	+	nd	+	nd	−	nd	nd	−	−
L-Valine	d	+	nd	−	+	−	−	nd	nd	d	nd	nd	nd	+	−	−	−	+	−	−	nd	nd	nd	−	−	nd	+	nd	+	nd	−	nd	nd	+	+

[a]Symbols: +, 90% or more strains positive; −, 10% or less strains positive; d, 11–89% of strains positive; nd, not determined.

[b]Data from Athalye (1981), Athalye et al. (1985), Goodfellow et al. (1979), Horan and Brodsky (1982), Labeda et al. (1985), Lipski and Altendorf (1995), Lu et al. (2003), Mertz and Yao (1986, 1990), Quintana et al. (2003), Trujillo and Goodfellow (1997a, 2003), Kim (1999), Cook et al. (2005), Lee and Jeong (2006), Le Roes and Meyers (2007), and Wink et al. (2003).

TABLE 303. Physiological characteristics differentiating the species of *Actinomadura*[a, b]

Character	*A. madurae*	*A. atramentaria*	*A. catellatispora*	*A. citrea*	*A. coerulea*	*A. cremea*	*A. echinospora*	*A. fibrosa*	*A. formosensis*	*A. fulvescens*	*A. glauciflava*	*A. hallensis*	*A. hibisca*	*A. kijanata*	*A. latina*	*A. livida*	*A. luteofluorescens*	*A. macra*	*A. mexicana*	*A. meyerae*	*A. namibiensis*	*A. napierensis*	*A. nitritigenes*	*A. oligospora*	*A. pelletieri*	*A. rubrobrunea*	*A. rudentiformis*	*A. rugatobispora*	*A. spadix*	*A. umbrina*	*A. verrucosospora*	*A. vinacea*	*A. viridilutea*	*A. viridis*	*A. yumaenesis*
Arbutin hydrolysis	+	+	nd	+	–	+	+	nd	nd	+	nd	nd	nd	+	+	–	+	+	+	+	nd	+	+	+	–	–	+	nd	–	nd	+	nd	nd	+	+
Casein	+	+	–	+	+	+	+	+	nd	+	+	+	+	+	+	+	+	+	+	+	nd	+	+	+	+	nd	+	nd	+	nd	+	+	nd	+	+
Chitin	–	–	nd	–	–	–	–	nd	nd	+	nd	nd	nd	–	–	–	–	–	–	–	nd	nd	nd	–	–	+	nd	nd	–	nd	nd	nd	nd	–	–
DNA	+	+	nd	+	–	–	+	nd	nd	+	nd	+	nd	+	–	+	+	–	nd	nd	nd	nd	nd	+	–	nd	nd	nd	+	nd	+	–	nd	+	+
Esculin hydrolysis	+	+	+	+	+	+	+	+	nd	+	+	+	+	+	+	+	+	+	+	+	nd	+	+	+	–	+	nd	nd	+	nd	+	+	nd	+	+
Elastin	+	+	nd	+	–	+	+	+	+	+	nd	nd	nd	+	d	–	–	+	+	–	nd	nd	+	–	+	nd	nd	nd	+	nd	+	nd	nd	+	+
Gelatin	+	–	+	+	+	+	+	+	+	+	+	+	+	+	+	+	+	+	+	+	–	+	+	+	+	+	+	+	+	nd	+	+	+	+	+
Guanine	+	–	nd	–	+	–	nd	–	nd	nd	nd	nd	nd	–	d	+	+	–	nd	nd	nd	nd	–	–	–	nd	nd	nd	–	nd	+	nd	nd	nd	–
Hypoxanthine	+	+	–	–	–	+	–	+	nd	+	+	–	+	+	–	–	+	+	–	+	nd	+	+	–	+	–	–	nd	–	nd	+	–	nd	–	+
Nitrate reduction	+	+	+	+	+	+	–	+	–	nd	+	–	–	–	nd	+	+	+	nd	nd	nd	nd	+	–	+	+	–	–	+	nd	–	–	+	nd	+
Starch	–	–	–	–	–	+	–	+	–	d	+	+	–	+	+	+	+	–	–	+	nd	+	–	–	–	+	+	+	–	nd	–	–	+	+	+
Testosterone	–	+	nd	+	–	+	–	+	nd	+	nd	nd	nd	+	d	–	+	–	–	nd	nd	nd	+	–	–	nd	nd	nd	–	nd	+	–	nd	–	–
Tributyrin	+	+	nd	+	+	+	–	nd	nd	+	nd	nd	nd	+	+	+	–	–	+	+	nd	nd	nd	+	+	nd	nd	nd	+	nd	+	nd	nd	+	–
Tyrosine	+	nd	–	+	–	+	–	+	nd	–	–	–	+	+	+	–	+	+	–	–	nd	+	d	–	+	–	–	nd	–	nd	+	–	nd	–	+
Urease production	–	–	–	–	–	+	nd	+	nd	nd	nd	+	–	+	nd	–	–	–	–	nd	–	nd	nd	+	–	nd	–	nd	–	nd	–	nd	nd	nd	nd
Xanthine	–	–	–	–	–	–	–	–	nd	–	–	–	+	+	+	–	–	–	–	–	nd	–	–	–	–	–	–	–	+	nd	–	–	nd	+	+

[a]Symbols: +, 90% or more strains positive; –, 10% or less strains positive; d, 11–89% of strains positive; nd, not determined.

[b]Data from Athalye (1981), Athalye et al. (1985), Goodfellow et al. (1979), Horan and Brodsky (1982), Kim (1999), Labeda et al. (1985), Lipski and Altendorf (1995), Lu et al. (2003), Mertz and Yao (1986, 1990), Quintana et al. (2003), Trujillo and Goodfellow (1997a, 2003), Cook et al. (2005), Lee and Jeong (2006), Le Roes and Meyers (2007), and Wink et al. (2003).

Fischer et al. (1983) carried out extensive DNA–DNA hybridization studies on *Actinomadura* species and recorded homology values in the range 18–100%. The highest values were between *Actinomadura madurae* strains (96%) and *Actinomadura pelletieri* strains (85–100%) and between two strains of *Actinomadura verrucosospora* (96%); homology values found between strains of *Actinomadura citrea, Actinomadura madurae, Actinomadura pelletieri*, and *Actinomadura verrucosospora* ranged from 25% to 44%. In a continuation of these studies Poschner et al. (1985) found that homology values between *Actinomadura citrea, Actinomadura coerulea, Actinomadura cremea, Actinomadura madurae*, and *Actinomadura pelletieri* were in the range 27–48%. *Actinomadura glauciflava* shares DNA–DNA relatedness values of 53% with *Actinomadura citrea*, 62% with *Actinomadura luteofluorescens* and *Actinomadura verrucosospora*, and 38% with *Actinomadura coerulea* (Lu et al., 2003).

Recent descriptions of *Actinomadura* species where DNA–DNA hybridization studies were perfomed underline the complex genetic relationships encompassed by members of this taxon. The type strains of *Actinomadura kijaniata* and *Actinomadura namibiensis* share a DNA relatedness value of 72% but can be distinguished using phenotypic data, especially on the basis of diagnostic carbon-utilization and enzyme patterns (Wink et al., 2003). Lee and Jeong (2006) reported that *Actinomadura hallensis* contains an additional extended helix (22 nt in length) between *Escherichia coli* positions 453 and 479 with a unique nucleotide composition suggesting that sequences of this region may prove useful for species identification.

Phages. The occurrence of *Actinomadura* phages has been reported in organic mulches used to support the growth of avocado trees in Australia (Kurtböke et al., 1993).

Pathogenicity. *Actinomadura latina, Actinomadura madurae*, and *Actinomadura pelletieri* cause human actinomycetoma in tropical and subtropical areas, particularly on the African and American continents. The natural habitat of pathogenic actinomadurae is thought to be the environment, namely the surface layers of soil. Actinomadurae from soil invade the human body. The disease, which is often induced by implantation of organisms into tissues by sharp objects, such as thorns or soil contaminated splinters, is characterized by a progressive swelling of the infected area, distortion of the normal anatomy, and multiple draining sinuses and fistulae. Purulent discharge containing the causative agent in the form of granules is characteristic of advanced stages of the disease. Almost 80% of infections enter through the lower extremities of the body. This probably explains the etiology of "Madura foot" described by Gill of Madura in South India in 1842 (Gill, 1842). However, other sites of the human body may be affected such as the back and the head. Non-mycetomic infections caused by *Actinomadura madurae* have been reported (Wust et al., 1990), including one involving an immunocompromised patient (McNeil et al., 1992).

Pathogenic actinomadurae are thought to have a worldwide distribution, but the disease is only a major health problem in tropical and subtropical countries (Fahal, 2006). It is endemic in localities lying between latitudes 15°S and 30°N, an area which includes India, Mali, Mexico, Senegal, Somalia, Sudan, and much of Central and South America (Fahal, 2006; Fahal and Hassan, 1992). There is evidence that the distribution of the causal organisms varies. In India, *Actinomadura madurae* is the predominant causal agent (Klokke et al., 1968; Venugopal and Venugopal, 1991) whereas on the African continent the prevalent organism is *Actinomadura pelletieri* (Develoux et al., 1988; Fahal, 2006; Yu et al., 1993). Infections by *Actinomadura madurae* have been reported from temperate climatic regions, such as Greece, the Netherlands, and the USA (Davis et al., 1999; de Hoog et al., 1993; Ispoglou et al., 2003). Climatic conditions may play a major part in the distribution of the causal agents of actinomycetoma (Develoux et al., 1988; Serrano et al., 1988). The endemic regions have a rainy season of 4–6 months with a relative humidity of 60–80% and temperatures of 30–37°C followed by a dry season of 6–8 months with a relative humidity of 12–30% and daytime temperatures of 45–60°C, which fall to 15–18°C at night (Magaña, 1984).

In general, the disease is four times more frequent in males than females and mainly affects 16–40-year-old adults, but cases involving children and elderly men do occur (El Moghraby, 1971; Lopez, 1993; Serrano et al., 1986). The disease typically affects farmers and herdsmen, though people in occupational activities such as carpentry, housework, and industry can be infected. The lower and upper extremities of the body are the main sites of infection, as the skin in these areas is more likely to be exposed to soil and puncture injuries with contaminated thorns (Goodfellow, 1996; Gumaa et al., 1986; Serrano et al., 1986).

Little information is available on the pathogenicity mechanisms of *Actinomadura*. According to Pulverer and Schaal (1978) it has not been possible to demonstrate the pathogenicity of these actinomadurae for laboratory animals. However, Rippon (1968) reported that virulent strains of *Actinomadura madurae* produce a collagenase that has a significant role in the pathogenicity of the organism.

Antibiotic sensitivity. Cases of actinomycetoma caused by pathogenic actinomadurae have been treated with a range of antibiotics, including amikacin, dapsone, streptomycin, and trimethoprim-sulfamethoxazole (Welsh et al., 1987). Boiron et al. (1992) used the disk diffusion method to assess the susceptibility of *Actinomadura madurae* and *Actinomadura pelletieri* strains against 29 microbial agents and found that they were particularly sensitive to aminocycline, amikacin, imipenem, tobramycin, and vancomycin. Vera-Cabrera et al. (2004) examined 24 *Actinomadura madurae* strains isolated from patients with actinomycetoma for their sensitivity to new quinolones, including garenoxacin, gatifloxacin, moxifloxacin, and two oxazolidinones, linezolid and the compound, DA-7867, and found that all but one of the strains showed high susceptibility to all of these agents; they were most active to DA-7867 showing a minimum inhibitory concentration of 0.06 µg/ml. However, further work needs to be done to determine whether these results can be extrapolated to antibiotic activity in subcutaneous abscesses and granulomas that are formed in cases of mycetoma. Trujillo and Goodfellow (2003) found that most *Actinomadura* strains were resistant to ampicillin and rifampin, and apart from *Actinomadura oligospora* and *Actinomadura yumaensis* were susceptible to tetracycline.

Ecology. Most *Actinomadura* species have been isolated from soil where they probably have a role in organic matter turnover (Galatenko and Preobrazhenskaya, 1981; Horan and Brodsky, 1982; Labeda et al., 1985; Lavrova et al., 1972; Lee and Jeong, 2006; Lu et al., 2003; Meyer, 1979, 1989a, 1989b; Miyadoh et al., 1987, 1989; Nonomura and Ohara, 1971c; Preobrazhenskaya et al., 1975b; Quintana et al., 2003; Wink et al., 2003). Preobrazhenskaya and her colleagues (Chormonova and Preobrazhenskaya, 1981; Galatenko et al., 1987; Galatenko and Preobrazhenskaia, 1981) studied the occurrence and frequency of *Actinomadura* species in different soils. A comparison of chernozem, dark chestnut, and sierozem soils in Kazakhstan and Turkmenistan showed that the number of *Actinomadura* strains was higher in cultivated than in uncultivated soils. The highest numbers of actinomadurae were isolated from chernozem soil (Kazakhstan) though *Actinomadura citrea* was the most prevalent species followed by *Actinomadura cremea* and *Actinomadura verrucosospora.* The dark chestnut soil (cultivated) contained approximately half the number of actinomadurae found in the chernozem soil, but the species diversity of the two soils was similar. The lowest numbers of actinomadurae were isolated from sierozem soil (Turkmenistan). The frequency of actinomadurae appears to depend upon the humus content of the soil. *Actinomadura rubrobrunea* was isolated from Egyptian soils in which maize and rice had been cultivated (Agre and Guzeva, 1975), and *Actinomadura nitritigenes* was isolated from a laboratory scale biofilter that contained tree bark compost as the packing material and which was supplied with ammonia (Lipski and Altendorf, 1995). There is evidence that clinically and ecologically significant actinomadurae are underspeciated (Trujillo and Goodfellow, 2003).

Actinomadura latina, Actinomadura madurae, and *Actinomadura pelletieri* strains have been isolated mainly from clinical material, though there is evidence that members of *Actinomadura madurae* are widespread in organically rich soils. *Actinomadura madurae* strains isolated from environmental samples tend to lack the red endopigment of clinical isolates and sporulate more readily (Gerber, 1971, 1973; Kroppenstedt and Goodfellow, 1992; Lechevalier et al., 1971). *Actinomadura latina* and *Actinomadura pelletieri* have only been found in clinical specimens.

Enrichment and isolation procedures

Many different media have been employed for the isolation of *Actinomadura* strains, especially from soil samples. Those that have proved to be the most effective include egg albumin (Lawson and Davey, 1972), glycerol-asparagine (ISP medium 5; Shirling and Gottlieb, 1966), oatmeal (ISP medium 3; Shirling and Gottlieb, 1966), inorganic salts-starch (ISP medium 4; Shirling and Gottlieb, 1966) and yeast extract-malt extract (ISP medium 2; Shirling and Gottlieb, 1966) agars. Strains may be isolated from agar plates after incubation for up to 6 weeks. Enrichment of actinomadurae from soil can be achieved using pretreatment regimes and selective media. Nonomura and Ohara (1971d) reduced the number of unwanted micro-organisms by air-drying soil, applying dry heat at 100°C for 1 h before plating diluted soil suspensions onto several media, notably arginine-vitamins (AV) and mineral-glucose-asparagine (MGA) agars and incubating for several weeks at 28–30°C. Isolation media should be supplemented with antifungal antibiotics such as cycloheximide to inhibit fungal growth.

Lavrova et al. (1972) and Preobrazhenskaya et al. (1975b) increased the number of actinomadurae isolated from soils by the addition of antibiotics to medium 2 of Gauze et al. (1957); these antibiotics inhibited the growth of bacteria and more frequently occurring streptomycetes thereby providing more favorable conditions for the slow-growing actinomadurae. Bruneomycin (0.5, 1.0, or 2.0 µg/ml), rubomycin (5.0, 10.0, or 20.0 µg/ml), and streptomycin (0.5, 1.0, or 2.0 µg/ml) proved to be the most effective antibiotics. Athalye et al. (1981) combined drying and heat pretreatment regimes with the use of rifampin (5 µg/ml) as the selective agent for the isolation of *Actinomadura* strains from diverse environmental samples. The mean numbers of actinomadurae recovered were higher using this treatment than obtained with untreated samples. Trujillo and Goodfellow (2003) isolated novel *Actinomadura* species from environmental samples collected from diverse geographical locations (Hong Kong, Kenya, Mexico, South Africa, and Venezuela) providing further evidence that the selective isolation procedure recommended by Athalye et al. (1981) is effective.

Pathogen *Actinomadura* species can be isolated from clinical samples, such as pus and biopsy material, using Brain Heart Infusion (Schaal, 1972), Sabouraud glucose (Gordon, 1974) and yeast extract agars (Pridham et al., 1956, 1957). All cultures should be incubated aerobically at 25–27°C and at 36°C for up to 3 weeks and examined both macroscopically and microscopically for growth every 2 d (Schaal, 1984). Actinomadurae can be recognized by their filamentous appearance, leathery colonies, and by the production of red prodiginine pigments. Actinomycetoma granules should be washed in sterile tap water before they are crushed to obtain material for inoculation of culture media.

Maintenance procedures

Heavily sporulating cultures are needed to maintain high viability irrespective of the preservation method. The highest survival rates for nonsporulating cultures are achieved using cells from the exponential growth phase. The culture age of the actinomadurae to be preserved is very important, especially for the preservation of thermophilic strains, as these lyse at the stationary phase of growth. Sporulated actinomadurae can be maintained on oatmeal agar, or other appropriate media, at 4°C and transferred every four months for short-term preservation; tubes should be sealed with silicone stoppers to prevent the agar drying out. Medium-term preservation for up to four years can be achieved by mixing spore suspensions for homogenized mycelia with glycerol (45%, v/v) and storing at –25°C (Zippel and Neigenfind, 1988). Long-term storage can be achieved by lyophilization or storage in liquid nitrogen. For lyophilization, spores and mycelia are suspended in a suitable fluid, such as 7.5% (w/v) glucose serum or 10% skimmed milk. For storage in liquid nitrogen, the micro-organisms are inoculated into small vials containing an appropriate medium, incubated until satisfactory growth is visible when the tubes are tightly sealed and placed in a liquid nitrogen container. An alternative simple and practical method involves nitrogen cryopreservation of material in small polyvinyl chloride tubes ("straws") at –196°C (Hoffmann, 1989a, 1989b).

Differentiation from closely related taxa

Differentiating characteristics of the genus *Actinomadura* and other closely related taxa are presented in Table 298, Family *Thermomonosporaceae.* The genus *Actinomadura* can be readily separated from other members of the family *Thermomonosporaceae* and from other closely related taxa using chemotaxonomic, morphological, and 16S rRNA gene sequence data. The phospholipid pattern is an important criterion for differentiating between members of the genera *Actinomadura* and *Actinocorallia*.

Taxonomic comments

The genus *Actinomadura* was described by Lechevalier and Lechevalier (1970) to accommodate aerobic actinomycetes that had a wall chemotype III and formed a branched substrate mycelium with or without a secondary mycelium that may bear chains of arthrospores. The type species was defined as *Actinomadura madurae* (Vincent 1894) Lechevalier and Lechevalier 1970. *Actinomadura dassonvillei* (Brocq-Rousseu 1904) Lechevalier and Lechevalier 1970, and *Actinomadura pelletieri* (Laveran 1906) Lechevalier and Lechevalier 1970 were also included in the genus.

Actinomadura madurae was first described in 1894 by Vincent as *Streptothrix madurae*, based on isolates from an Algerian case of Madura foot. This combination proved to be illegitimate and Blanchard (1896) transferred the organism to the genus *Nocardia* Trevisan, hence *Nocardia madurae* (Vincent 1894) Blanchard 1896 is the oldest legitimate name of this micro-organism. The organism was subsequently assigned to many genera, including *Streptomyces* (Lacey et al., 1978). The inclusion of *Actinomadura madurae* in either *Nocardia* or *Streptomyces* became untenable when it was shown that whole-organism hydrolysates contained *meso*-A_2pm and a previously unknown sugar, madurose, later identified as 3-*O*-methyl-D-galactose (Becker et al., 1965; Lechevalier, 1968; Lechevalier and Gerber, 1970). In contrast, members of the genus *Nocardia* contained A_2pm, arabinose, and galactose, and *Streptomyces* strains the LL isomer of A_2pm. *Actinomadura dassonvillei* and *Actinomadura pelletieri* have similar taxonomic histories, although *Actinomadura pelletieri* was first placed in the genus *Micrococcus* by Laveran (1906) because the colonies fragmented into cocci suggesting a *Rhodococcus* species. *Actinomadura pelletieri* was subsequently classified in the genus *Nocardia* Pinoy 1912 then in the genus *Streptomyces* Waksman and Henrici 1948. *Actinomadura dassonvillei* was seen to differ from the other two members of the genus *Actinomadura* in its greater vigor and sporulation and lack of madurose. The genus *Nocardiopsis* was subsequently proposed to accommodate *Actinomadura dassonvillei* strains (Meyer, 1976).

Numerical phenetic (Alderson and Goodfellow, 1979; Athalye et al., 1985; Goodfellow, 1971; Goodfellow et al., 1979; Goodfellow and Pirouz, 1982), chemical (Athalye et al., 1984; Collins et al., 1977; Fischer et al., 1983; Lechevalier et al., 1977; Minnikin et al., 1977; Yamada et al., 1977), and 16S rRNA gene sequence data (Goodfellow et al., 1988; Kroppenstedt et al., 1990) confirmed the separation between the genus *Actinomadura* and the genera *Nocardia* and *Nocardiopsis*, and suggested that *Actinomadura madurae* and *Actinomadura pelletieri* were heterogeneous taxa. There was also evidence that the taxon could be separated into two aggregate taxa, the *Actinomadura madurae* and the *Actinomadura pusilla* groups.

Fischer et al. (1983) provided compelling evidence that the genus *Actinomadura* was heterogeneous when they assigned representative strains to two aggregate groups defined using chemotaxonomic and nucleic acid pairing data. rRNA partial oligonucleotide sequencing (Fowler et al., 1985; Goodfellow et al., 1988), menaquinone (Athalye et al., 1984), polar lipid (Agre et al., 1975; Lechevalier et al., 1977), numerical taxonomic (Athalye et al., 1985; Goodfellow and Pirouz, 1982), and DNA homology studies (Poschner et al., 1985) all underlined this division. *Actinomadura madurae* and related species were seen to have a closer affinity to *Thermomonospora curvata* than to *Actinomadura pusilla* and allied taxa that were found to be related to *Streptosporangium roseum* (Fowler et al., 1985).

The division of the genus *Actinomadura* into two aggregate groups was formally recognized by Kroppenstedt et al. (1990) who proposed that the genus *Actinomadura* Lechevalier and Lechevalier 1970 be retained for *Actinomadura madurae* and related species, and that the *Actinomadura pusilla* group be reclassified in the genus *Microtetraspora* Thiemann et al. 1968. This division was supported by a polyacrylamide gel electrophoresis analysis of the ribosomal protein AT-L30 which exhibits electrophoretic mobility that is specific at the genus level (Ochi et al., 1991).

By 1991, the genus *Actinomadura* contained 34 validly described species recognized primarily by chemical and morphological features (Mertz and Yao, 1990; Meyer, 1989a, 1989b; Miyadoh et al., 1987; Terekhova et al., 1987). A phylogenetic analysis based on 16S rRNA gene sequences of *Actinomadura* type strains was carried out by Wang et al. (1996) who supported Kroppenstedt et al. (1990) in classifying members of the *Actinomadura pusilla* group in the genus *Microtetraspora*. These workers also proposed the transfer of *Microbispora echinospora* and *Microbispora viridis* to the genus *Actinomadura* as *Actinomadura echinospora* and *Actinomadura rugatobispora*, respectively (Miyadoh et al., 1990). Subsequently, the genus *Nonomuraea* was proposed for membership in the *Actinomadura pusilla* group following extensive comparative 16S rRNA gene sequencing studies (Zhang et al., 1998), a proposition first raised by Goodfellow et al. (1988).

Zhang and his colleagues also demonstrated that *Thermomonospora formosensis* was related to several *Actinomadura* species on the basis of chemotaxonomic and 16S rRNA gene sequence data (Goodfellow, 1989; Kroppenstedt et al., 1990; Kroppenstedt and Goodfellow, 1992; Kudo, 1997). In a continuation of these studies, Zhang et al. (2001) proposed that *Actinomadura aurantiaca*, *Actinomadura glomerata*, *Actinomadura libanotica*, and *Actinomadura longicatena* be transferred to the new genus *Actinocorallia* as *Actinocorallia aurantiaca*, *Actinocorallia glomerata*, *Actinocorallia libanotica*, and *Actinocorallia longicatena*.

Zhang et al. (2001) considered that *Excellospora viridilutea* should be reclassified in the genus *Actinomadura* as *Actinomadura viridilutea*. They also noted that *Spirillospora albida*, the type species of the genus, was closely related to some *Actinomadura* species on the basis of 16S rRNA gene sequence data. Further studies are needed to unravel relationships between *Actinomadura* and *Spirillospora* species. In contrast, *Actinomadura echinospora* and *Actinomadura umbrina* form a clade with *Thermomonospora curvata* based on 16S and 23S rRNA gene sequence data; further comparative taxonomic investigations are needed to resolve the status of these taxa. Finally, numerical taxonomic (Athalye et al., 1985; Trujillo and Goodfellow, 2003) and 16S and 23S rRNA gene sequence data (Zhang et al., 2001) of *Actinomadura*-related actinomycetes together with phylogenetic data (Zhang et al., 2001) indicate that *Actinomadura spadix* is sharply separated from other *Actinomadura* species, a result that is supported by polyacrylamide gel electrophoresis analysis of the ribosomal protein AT-L30 (Ochi et al., 1991). It seems likely that *Actinomadura spadix* will be found to merit recognition as a new genus.

In summary, the systematics of the genus *Actinomadura* has been markedly improved since the 2nd edition of *Bergey's Manual of Systematic Bacteriology*. The taxon currently encompasses 35 validly published species, though it seems likely that some of the present members will be reclassified in the future.

Differentiation of the species of the genus *Actinomadura*

Published descriptions of *Actinomadura* species are often incomplete since different investigators emphasize some phenotypic features and omit others. This makes identification difficult. Nevertheless, most species can be separated using a combination of morphological and physiological properties (Table 301, Table 302, and Table 303), though in most cases only the type strain has been studied. However, even when several strains have been studied (e.g. *Actinomadura madurae, Actinomadura pelletieri*), the results tend to be variable or inconsistent when those from the literature are compared.

Enzymic substrates based on the fluorophores 4-methylumbelliferone (4-MU) and 7-amino-methylcoumarin (7-AMC) were carried out by Trujillo and Goodfellow (2003) to differentiate *Actinomadura* species. Encouraging results were obtained for differentiating between pathogenic *Actinomadura* species (Table 304) (Goodfellow et al., 1995; Trujillo and Goodfellow, 2003). Enzymic tests were demonstrated by Wink et al. (2003) to differentiate *Actinomadura kijaniata* from *Actinomadura namibiensis*, which share a DNA homology higher than 70%.

TABLE 304. Phenotypic characters separating pathogenic *Actinomadura* species[a,b]

Characteristic	*A. latina*	*A. madurae*	*A. pelletieri*
Degradation tests:			
Arbutin	+	+	–
Esculin	+	+	–
Starch	+	–	+
Cleavage of 7-amino-4-methylcoumarin (7AMC-) substrates:			
Glutaryl-L-phenylalanine-7AMC	–	+	+
L-Pyroglutamide-7AMC	+	+	–
Cleavage of 4-methylumbelliferone (4MU-) substrates:			
4MU-*N*-acetyl-D-galactosoamide	–	d	+
4MU-α-L-arabinopyranoside	+	+	–
4MU-β-D-cellopyranoside	+	+	–
4MU-β-D-galactoside	–	+	–
4MU-β-D-glucoside	+	+	–
4MU-β-D-glucuronide	+	+	–
4MU-β-D-lactoside	–	+	–
4MU-sulfate	–	+	–
4MU-xyloside	+	+	–
Growth on sole carbon sources (1%, w/v):			
Adonitol	–	+	–
Fructose	+	+	–
Galactose	+	+	–
Glycerol	+	+	–
Mannitol	+	+	–
Melezitose	+	–	–

[a]Symbols: +, 90% or more strains positive; –, 90% or more strains negative; d, 11–89% of strains positive.

[b]Data from Trujillo and Goodfellow (1997a, 2003).

Acknowledgements

The authors are particularly indebted to Dr J. Meyer as they drew upon material from the section on *Actinomadura* she wrote for the previous edition of this Manual. They also acknowledge help from Dr L.I. Evtushenko and Dr J. Euzéby for translating some material written in Russian and for revising the etymology of some names, respectively. We also thank Dr P.R. Meyers and Dr E.T. Quintana for reading the manuscript.

List of species of the genus *Actinomadura*

1. **Actinomadura madurae** (Vincent 1894) Kroppenstedt, Goodfellow and Stackebrandt, 1990, 156[AL] (*Streptothrix madurae* Vincent 1894, 132)

ma.du'ra.e. N.L. gen. n. *madurae* of Madura, name of a district in India.

Spore chains short (3–12 spores per chain), hooked or curled, in clusters directly emerging from the agar surface or borne on long aerial hyphae. Spores are elliptical to round. Spore surface warty (Figure 404). Inorganic salts-starch agar*: poor growth; surface granular; aerial mycelium absent; substrate mycelium grayish white, and no diffusible pigment. Oatmeal agar: good growth; surface leathery; aerial mycelium absent or sparse white specks; substrate mycelium colorless center, edge often red, and no diffusible pigment. Peptone-glucose medium: poor growth; surface cartilaginous; aerial mycelium absent; substrate mycelium dark pink to red, and no diffusible pigment. Yeast extract-malt extract agar: moderate growth; surface cartilaginous; aerial mycelium absent; substrate mycelium dark pink to brownish violet, and no diffusible pigment. Growth at 10–45°C, optimally at 28–37°C. Tweens 20, 40, and 60 are degraded. Resistant to ampicillin (5 μg/ml) and carbenicillin (120 μg/ml), but susceptible to lincomycin (10 μg/ml) and tobramycin (1 μg/ml). Additional phenotypic data are shown in Table 302 and Table 303.

Source: clinical specimens (mycetoma) and from soil.

DNA G+C content (mol%): 66.0–68.2 (T_m) (type strain, 68.2).

Type strain: ATCC 19425, CCM 136, CCUG 32944, CECT 3043, CIP 105487, DSM 43067, HAMBI 1926, IAM 14277, IFM 0585, NBRC 14623, JCM 7436, IMET 9585, IMRU 1190, KCTC 9192, NCIMB 13469, NCTC 5654, NRRL B-3843, VKM Ac-809.

*Media used are the following: glycerol-asparagine agar (ISP 5), inorganic salts-starch agar (ISP 4), oatmeal agar (ISP 3), tyrosine agar (ISP 7), and yeast extract-malt extract agar (ISP 2) (Shirling and Gottlieb, (1966)); glucose-asparagine agar (Lindenbein, 1952), inorganic salts-starch agar (Gauze 1) (Gauze et al., 1957); oatmeal-nitrate agar (Prauser and Bergholz, 1974) and peptone-glucose medium (Prauser and Falta, 1968).

FIGURE 404. *Actinomadura madurae*, strain RG 1091. Spore chain, electron micrograph (16,500×).

Sequence accession no. (16S rRNA gene): U58527, X97889.
Sequence accession no. (23S rRNA gene): AF1162290.
Sequence accession no. (16S–23S rRNA ITS): AF134103.

2. **Actinomadura atramentaria** Miyadoh, Amano, Tohyama and Shomura 1987, 345[VP]

a.tra.men.ta'ri.a. N.L. fem. adj. *atramentaria* (from L. n. *atramentum* any black liquid, ink) inky, referring to the inky brown diffusible pigments.

Spores are borne in longitudinal pairs or in straight chains of three or rarely four spores. Chains develop into thick tufts on the aerial hyphae. Straight chains of three to five spores are occasionally observed at the tips of sporulating aerial hyphae. Spores are oval to ellipsoidal (0.6–0.8 × 0.8–1.5 µm). Spore surface smooth (Figure 405). Glycerol-asparagine agar: good growth; moderate white aerial mycelium; substrate mycelium dark brown, and pale brown diffusible pigment. Inorganic salts-starch agar: poor growth; moderate white aerial mycelium; substrate mycelium colorless, and no diffusible pigment. Oatmeal agar: poor growth; thin white aerial mycelium; substrate mycelium colorless, and no diffusible pigment. Tyrosine agar: moderate growth; moderate white aerial mycelium; substrate mycelium dark brown, and inky brown soluble pigment. Yeast extract-malt extract agar: good growth; moderate white aerial mycelium; substrate mycelium dark brown, and inky brown diffusible pigment. Growth at 15–42°C and optimally at 28–37°C. Nitrate is reduced, milk is peptonized. Grows in the presence of NaCl 7%, w/v and at pH 12. Resistant to ampicillin (5 µg/ml), carbenicillin (100 µg/ml) and ticarcillin (80 µg/ml), but susceptible to tetracycline (20 µg/ml). Additional phenotypic data are presented in Table 302 and Table 303.

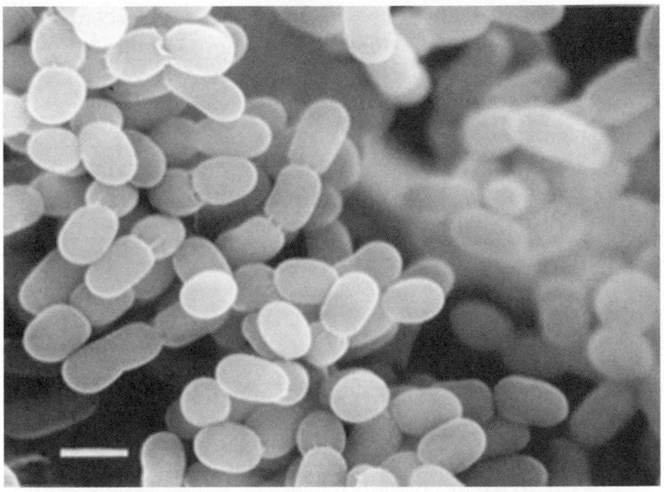

FIGURE 405. Electron micrograph of *Actinomadura atramentaria* strain SF 2197. Bar = 1 µm. (Reproduced with permission from S. Miyadoh, S. Amano, and T. Shomura. Atlas of Actinomycetes, Society for Actinomycetes Japan.)

Source: soil.
DNA G+C content (mol%): 72 (T_m).
Type strain: SF2197, DSM 43919, HUT 6547, NBRC 14695, JCM 6250, KCTC 9620, NCIMB 12618.
Sequence accession no. (16S rRNA gene): U49000.
Sequence accession no. (23S rRNA gene): AF134071.
Sequence accession no. (16S–23S rRNA ITS): AF134089.

3. **Actinomadura catellatispora** Lu, Wang, Zhang, Shi, Liu, Quintana and Goodfellow 2003, 140[VP]

ca.tel.la.ti.spo'ra. L. n. *catella* small chain; Gr. n. *spora* a seed, and in biology a spore; N. L. fem. n. *catellatispora* organism forming small chains of spores.

Short straight chains of spores (0.85 µm diameter) formed on the aerial mycelium. Spore surface smooth. Light-yellow substrate mycelium and a yellow aerial mycelium formed on yeast extract-malt extract agar. Diffusible pigments are not formed. Grows at 18–35°C. Additional phenotypic properties are shown in Table 302 and Table 303.

Source: a mud sample taken from a sewage ditch in Southern China.
DNA G+C content (mol%): 70.8 (T_m).
Type strain: 3.24, AS 4.1522, DSM 44772, NBRC 16341, JCM 10667.
Sequence accession no. (16S rRNA gene): AF154127.

4. **Actinomadura citrea** Lavrova, Preobrazhenskaya and Svesnikova 1972, 967[AL]

ci'tre.a L. fem. adj. *citrea* of or pertaining to the citrus-tree, intended to mean lemon-yellow; referring to the lemon-yellow color of the substrate mycelium.

Spore chains (3–9 spores per chain) short, hooked or curled, in sparse clusters on moderately branched aerial mycelium. Spore surface irregular or warty. Inorganic salts-starch agar: moderate growth; surface leathery; traces of yellow aerial mycelium turning blue with age; substrate mycelium yellowish brown, and diffusible pigment yellow.

Oatmeal agar: abundant growth; surface leathery; traces of yellowish aerial mycelium turning blue with age; substrate mycelium lemon yellow, and diffusible pigment yellow. Peptone-glucose medium: good growth; surface leathery; surface filmy cover of aerial mycelium consisting of sterile hyphae and coremia; substrate mycelium yellow, and no diffusible pigment. Yeast extract-malt extract agar: moderate growth; surface leathery; filmy cover of aerial mycelium consisting of sterile hyphae; substrate mycelium yellowish brown, and no diffusible pigment. Grows at 10–37°C and optimally at 28–30°C. RNA is degraded, but not xylan. Glycogen, maltose, and trehalose are used as carbon sources. Resistant to ampicillin (5 μg/ml), rifampin (10 μg/ml), and ticarcillin (80 μg/ml). Other phenotypic characteristics are shown in Table 302 and Table 303.

Source: soil.

DNA G+C content (mol%): 67.6 (T_m).

Type strain: ATCC 27887, BCRC (formerly CCRC) 13352, DSM 43461, NBRC 14678, IMET 9573, INA 1849, JCM 3295, KCTC 9617, NRRL B-16121, VKM Ac-1119.

Sequence accession no. (16S rRNA gene): U49001, AJ420139.

Sequence accession no. (23S rRNA gene): AF116216.

Sequence accession no. (16S–23S rRNA ITS): AF134091.

5. **Actinomadura coerulea** Preobrazhenskaya, Lavrova, Ukholina and Nechaeva 1975b, 404[AL]

co.e.ru'le.a. L. fem. adj. *coerulea* azure, blue, dark blue, referring to the blue aerial mycelium.

Spore chains curled, hooked or in spirals of one turn, arranged in tufts on long aerial hyphae. Spore surface warty. Inorganic salts-starch agar (Gauze 1): poor growth; surface slightly farinaceous; aerial mycelium pale blue; substrate mycelium colorless, and no diffusible pigment. Oatmeal agar: good growth; surface slightly farinaceous; aerial mycelium pale pink turning blue at maturity; substrate mycelium colorless or pale pink, and no diffusible pigment. Oatmeal-nitrate agar: good growth; surface slightly farinaceous; aerial mycelium white; substrate mycelium white, and no diffusible pigment. Yeast extract-malt extract agar: moderate growth; surface granular with a filmy cover of aerial mycelium consisting of sterile hyphae and coremia; substrate mycelium pale brown, and no diffusible pigment. Grows at 10–37°C and optimally at 28–30°C. RNA is degraded. Arginine and glutamine are used as sole carbon sources. Resistant to ampicillin (10 μg/ml), cephaloridine (10 μg/ml), cephalotin (35 μg/ml), cephamandole (40 μg/ml), and ticarcillin (80 μg/ml), but is susceptible to tetracycline (20 μg/ml). Other phenotypic properties are shown in Table 302 and Table 303.

Source: soil.

DNA G+C content (mol%): 67.0 (T_m).

Type strain: ATCC 33576, DSM 43675, NBRC 14679, IMET 9580, INA 765, JCM 3320, KCTC 9337, VKM Ac-1511.

Sequence accession no. (16S rRNA gene): U49002

Sequence accession no. (23S rRNA gene): AF116217.

Sequence accession no. (16S–23S rRNA ITS): AF134092.

6. **Actinomadura cremea** Preobrazhenskaya, Lavrova, Ukholina and Nechaeva 1975b, 404[AL]

cre'me.a. N.L. fem. adj. *cremea* cream-colored (referring to the color of the aerial mycelium).

Spore chains short, in hooks or spirals of one turn, 3–8 spores per chain arranged in clusters. Spore surface warty. Inorganic salts-starch agar: poor growth; surface granular; aerial mycelium cream-colored to pale pink; substrate mycelium colorless, and no diffusible pigment. Oatmeal agar: moderate growth; surface farinaceous; aerial mycelium white to yellowish white; substrate mycelium colorless, and no diffusible pigment. Peptone-glucose medium: good growth; surface granular; aerial mycelium white to cream-colored; substrate mycelium brown, and no diffusible pigment. Yeast extract-malt extract agar: moderate growth; surface farinaceous; aerial mycelium white; substrate mycelium light brown, and no diffusible pigment. Grows at 20–37°C. Resistant to ampicillin (15 μg/ml), carbenicillin (120 μg/ml), cephaloridine (10 μg/ml), cephalotin (35 μg/ml), cephamandole (40 μg/ml), and ticarcillin (80 μg/ml), but is susceptible to tetracycline (20 μg/ml). Additional phenotypic properties are shown in Table 302 and Table 303.

Source: soil.

DNA G+C content (mol%): 68.0 (T_m).

Type strain: ATCC 33577, BCRC (formerly CCRC) 13394, DSM 43676, NBRC 14182, IMET 9578, INA 292, JCM 3308, NRRL B-16605, VKM Ac-912.

Sequence accession no. (16S rRNA gene): AF134067.

Sequence accession no. (23S rRNA gene): AF134073.

Sequence accession no. (16S–23S rRNA ITS): AF134094.

Additional comments: Gauze et al. (1975) isolated a rifamycin-producing strain (INA 1349) which they described as *Actinomadura cremea* subsp. *rifamycini*. This name was subsequently attributed to Gauze et al. (1987). Euzéby and Kudo (2001) pointed out that this taxon should be attributed to Gauze et al. (1975). However, apart for the production of rifamycin, there are no characteristics that distinguish this strain from *Actinomadura cremea*. Rifamycin is also produced by some *Nocardia* and *Streptomyces* strains.

7. **Actinomadura echinospora** (Nonomura and Ohara 1971c) Miyadoh, Amano, Tohyama and Shomura 1989, 1909[VP] (Basonym: *Microbispora echinospora* Nonomura and Ohara 1971c, 891[AL].)

e.chi.no.spo'ra. L n. *echinus* sea urchin; Gr. n. *spora* a seed, and in biology a spore; N.L. fem. n. *echinospora* spiny spore.

Aerial hyphae differentiate into a non-sporogenous main axis which bears spore-bearing side branches. Spores are formed in clusters. Initially two but sometimes three spores per chain. Spore surface spiny (Figure 406). A yellowish orange to dark yellow substrate mycelium and soluble yellow or yellowish-brown pigments are formed on oatmeal agar. A yellowish pink aerial spore mass and a yellowish brown substrate mycelium are produced on inorganic salts-starch agar-V. The temperature range for optimal growth is 35–40°C; growth is very poor at 25°C. Tween 60 is degraded. Resistant to ampicillin (15 μg/ml), carbenicillin (90 μg/ml), and streptomycin (16 μg/ml), but is susceptible to ticarcillin (70 μg/ml) and tetracycline (20 μg/ml). Requires vitamin B for growth. Additional phenotypic properties are shown in Table 302 and Table 303.

Source: soil.

DNA G+C content (mol%): 74 (T_m).

Type strain: ATCC 27300, BCRC (formerly CCRC) 12547, DSM 43163, HUT 6548, NBRC 14042, JCM 3148, KCTC 9313.

FIGURE 406. Electron micrograph of *Actinomadura echinospora* JCM 3148^T. Aerial hyphae bear lateral branches where spores are produced. Spore surface is spiny. Bar = 1 µm. (Reproduced with permission from S. Amano and S. Miyadoh (A) and M. Hayakawa, H. Iino, and H. Nonomura (B). Atlas of Actinomycetes, Society for Actinomycetes Japan.)

Sequence accession no. (16S rRNA gene): U49004, AJ420135.
Sequence accession no. (23S rRNA gene): AF134075.
Sequence accession no. (16S–23S rRNA ITS): AF134095.

8. **Actinomadura fibrosa** Mertz and Yao 1990, 31^{VP}

fi.bro'sa. L. n. *fibra* a fiber, filament; L. suff. *-osus -a -um* suffix used with the sense of full of, prone to; N.L. fem. adj. *fibrosa* fibrous, referring to the fibrous appearance of the aerial hyphae.

An extensive, reddish brown to reddish orange substrate mycelium is formed. Aerial hyphae are rarely produced except on modified Bennett's, Bennett tomato paste, oatmeal, and yeast extract-malt extract agars. When produced, the aerial mycelium is pink to white. Does not usually form soluble pigments. Grows well on modified Bennett's, glycerol-asparagine, inorganic salts-starch, tyrosine, and yeast extract-malt extract agars. Forms thick fibers. Has not been observed to produce spores. Grows at 20–45°C and optimally at 37°C. Does not degrade guanine. Produces catalase and phosphatase, and liquefies gelatin and skim milk. Acetate, formate, lactate, propionate, pyruvate, and sucrose are used as sole carbon sources. Resistant to cephalotine (30 µg/ml), lincomycin (2 µg/ml), rifampin (5 µg/ml), and lysozyme (50 µg/ml), but is susceptible to bacitracin (10 U), gentamicin (10 µg/ml), and tetracycline (30 µg/ml). Grows in the presence of NaCl (5%, w/v). Produces esterase (C_4), esterase (C_8), leucine arylamidase, naphthol-AS-BI-phosphohydrolase, β-glucosidase, and trypsin, but is negative for cystine arylmidase (API ZYM tests). Additional phenotypic properties are shown in Table 302 and Table 303.

Source: soil collected in West Africa.
DNA G+C content (mol%): not determined.
Type strain: A82810.1, ATCC 49459, DSM 44224, DSM 44224, JCM 9371, NRRL 18348.
Sequence accession no. (16S rRNA gene): AF163114, AJ293702.
Sequence accession no.(23S rRNA gene): AF163136.
Sequence accession no. (16S–23S rRNA ITS): AF163125.

9. **Actinomadura formosensis** (Hasegawa, Tanida and Ono 1986) Zhang, Wang and Ruan 1998, 418^{VP} (*Thermonospora formosensis* Hasegawa, Tanida and Ono 1986, 22^{VP})

for.mo.sen'sis. N.L. fem. adj. *formosensis* of or belonging to Formosa (Republic of China).

Single spores are formed on unbranched sporophores on both aerial and substrate hyphae. The spores (diameter, about 1 µm) are heat sensitive, nonmotile, and globose. Spore surface warty. Glycerol-asparagine agar: moderate growth; aerial mycelium absent; substrate mycelium light apricot, and no diffusible pigment. Inorganic salts-starch agar: poor growth; aerial mycelium absent; substrate mycelium shell-colored, and no diffusible pigment. Oatmeal agar: poor or moderate growth; aerial mycelium poor and white when present; substrate mycelium light apricot, and no diffusible pigment. Peptone-yeast extract-iron agar: moderate growth; aerial mycelium absent; substrate mycelium cinnamon-colored, and no diffusible pigment. Tyrosine agar: moderate growth; aerial mycelium poor and white to flesh pink, and no diffusible pigment. Yeast extract-malt extract agar: aerial mycelium poor and white to flesh pink; substrate mycelium light to dusty orange, and no diffusible pigment. Growth temperature range is 23–41°C. Soluble starch and sucrose are used as sole carbon sources. Growth occurs on media supplemented with up to 2%, w/v NaCl. Susceptible to chloramphenicol, erythromycin, penicillin, tetracycline, and streptomycin. The type strain produces rifamycins O and S and is resistant to rifampin. Additional phenotypic properties are shown in Table 302 and Table 303.

Source: soil.

DNA G+C content (mol%): 72.0 (T_m).

Type strain: C-36820, ATCC 49059, CIP 105595, DSM 43997, NBRC 14204, NBRC 15870, JCM 7474, KCTC 9647, NCIMB 12773, NRRL B-16984, VKM Ac-1954.

Sequence accession no. (16S rRNA gene): AF002263, AJ293703.

Sequence accession no. (23S rRNA gene): AF116218.

Sequence accession no. (16S–23S rRNA ITS): AF134096.

10. **Actinomadura fulvescens** Terekhova, Galatenko and Preobrazhenskaya 1987, 179[VP] (Effective publication: Terekhova, Galatenko and Preobrazhenskaya 1982, 87.)

ful.ves'cens. N.L. part. adj. *fulvescens* becoming reddish-yellow.

Spore chains spiral. Spore surface smooth. Scanty white aerial mycelium, and a colorless or yellowish substrate mycelium is formed on synthetic media and a brownish-yellow substrate mycelium and soluble pigment of the same color on organic media. Grows well on modified Bennett's, Czapek's, oatmeal, and yeast extract-malt extract agars. Grows at 10–45°C. Degrades Tweens 20, 40, 60, and 80. Glutamine and histidine are used as sole carbon sources. Resistant to ampicillin (15 μg/ml), carbenicillin (100 μg/ml), cephalotin (35 μg/ml), and streptomycin (16 μg/ml), but is susceptible to tetracycline (20 μg/ml) and tobramycin (0.5 μg/ml). Additional phenotypic properties are shown in Table 302 and Table 303.

DNA G+C content (mol%): not determined.

Type strain: DSM 43923, NBRC 14347, IMET 9745, INA 3321, JCM 6833, KCTC 9339, NCTC B-16245, NRRL B-16245, VKM Ac-938.

Sequence accession no. (16S rRNA gene): U49005, AJ420137.

Sequence accession no. (23S rRNA gene): AF134076.

Sequence accession no. (16S–23S rRNA ITS): AF134097.

11. **Actinomadura glauciflava** Lu, Wang, Zhang, Shi, Liu, Quintana and Goodfellow 2003, 140[VP]

glau.ci.fla'va. L adj. *glaucus* bluish-green; L. adj. *flavus* yellow; N.L. fem. adj. *glauciflava* bluish-green yellow.

Curled, hooked or spiral spore chains. Spore surface, warty. A yellow-brown, non-fragmenting substrate mycelium is produced. An abundant bluish-green aerial mycelium and a yellow diffusible pigment are formed on yeast extract-malt extract agar. Grows at 18–35°C. Casein, esculin, gelatin, hypoxanthine, and starch are degraded, but not tyrosine or xanthine. Nitrate is reduced.

Source: soil collected in Yunnan Province, China.

DNA G+C content (mol%): 72.0 (T_m).

Type strain: strain 80-60, AS 4.1202, CIP 107902, DSM 44770, DSM 44770, NBRC 14668, JCM 6161, VKM Ac-1273.

Sequence accession no. (16S rRNA gene): AF153881, AB184612.

12. **Actinomadura hallensis** Lee and Jeong 2006, 262[AL]

hal.len'sis. N.L. fem. adj. *hallensis* pertaining to Mt Halla, Jeju Island, Republic of Korea, the origin of the soil sample from which the type strain was isolated.

Short hooked or spiral spore chains. Spore surface warty (Figure 407). Substrate mycelium is grayish reddish purple on inorganic salts-starch agar, and brown on oatmeal agar and blackish purple on yeast-extract-malt extract agar. A moderate amount of white aerial mycelium is formed on inorganic salts-starch and oatmeal agars. Good growth is formed on all of these media. The temperature range for growth is 20–45°C. Produces catalase. H_2S is produced. Does not degrade Tween 80. Sucrose is used as a sole carbon soruce. Other phenotypic properties are shown in Table 302 and Table 303.

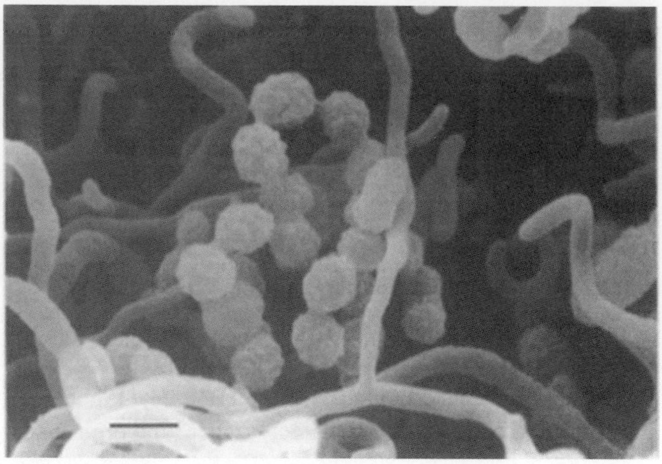

FIGURE 407. Electron micrograph of *Actinomadura hallensis* 647-1[T] grown on ISP 4 medium for 3 weeks. Bar = 0.8 μm. (Reproduced with permission from Lee and Jeong, 2006. Int. J. Syst. Evol. Microbiol. *56*: 259–264.)

Source: soil sample collected from Mt Halla, Jeju Island, Republic of Korea.

DNA G+C content (mol%): 67.5 (HPLC).

Type strain: H647-1, DSM 44987, IMSNU 50760, JCM 13882, KCCM 42245, KCTC 9992, NBRC 102110, NRRL B-24436.

Sequence accession no. (16S rRNA gene): DQ076484.

13. **Actinomadura hibisca** Tomita, Nishio, Saitoh, Yamamoto, Hoshino, Ohkuma, Konishi, Miyaki and Oki 1991, 758[AL] (Effective publication: Tomita, Nishio, Saitoh, Yamamoto, Hoshino, Ohkuma, Konishi, Miyaki and Oki 1990, 758.)

hi'bis.ca. N.L. n. *hibisca* (from Gr. n. *ibiskos* marsh mallow), rose mallow, a plant with reddish flower referring to the production of red diffusible pigments.

Long straight spore-chains (10–50 oblong spores per chain). Spore surface smooth. Coiled spore chains are occasionally formed at the tip of long chains. Pinkish white, deep red, or light brown substrate mycelium is produced on glycerol-asparagine, tyrosine and yeast-extract malt extract agars, respectively. A scant white to light pink aerial mycelium is formed on all of these media. Abundant growth is obtained on glycerol-asparagine agar, and moderate growth on glucose-asparagine and yeast extract-malt extract agars. A reddish pigment due to the production of pradimicins is produced abundantly on peptone-yeast extract-iron and yeast extract-malt extract agars. Produces melanoid pigment on tyrosine agar though growth is poor. Grows at 18–40°C. Adenine is degraded. Acid is produced from cellobiose and glucose. Produces pradimicins A, B, and C.

Enzymic activities (API ZYM system) are positive for cystine arylmidase, esterase (C_4), esterase (C_8), leucine arylamidase, naphthol-AS-BI-phosphohydrolase, β-glucosidase, and trypsin. Additional phenotypic properties are presented in Table 302 and Table 303.

DNA G+C content (mol%): not determined.

Type strain: P157-2, ATCC 53557, DSM 44148, NBRC 15177, JCM 9627, NCIMB 13253.

Sequence accession no. (16S rRNA gene): AF163115, AJ293705.

Sequence accession no. (23S rRNA gene): AF163137.

Sequence accession no. (16S–23S rRNA ITS): AF163126.

14. **Actinomadura kijaniata** Horan and Brodsky 1982, 195[VP]

ki.ja.ni'a.ta. N.L. fem. adj. *kijaniata* derived from "kijani" (the Swahili word for green).

Spore chains in long, open spirals with 10 or more spores per chain. Spores are elliptical (1.0–1.5 µm diameter by 1.5–2.0 µm long). Spore surface smooth. Inorganic salts-starch agar: moderate growth; surface flat and granular; aerial mycelium absent; substrate mycelium center slate-green and periphery light tawny, and faint green diffusible pigment. Oatmeal agar: moderate growth; surface flat to granular; aerial mycelium only white specks; substrate mycelium dark pine-green, and no diffusible pigment. Tyrosine agar: good growth; surface raised and folded; aerial mycelium abundant white; substrate mycelium lead gray, and no diffusible pigment. Yeast extract-malt extract agar: good growth; surface raised and folded; aerial mycelium only white specks; substrate mycelium center dark jade-green and periphery biscuit-colored, and no diffusible pigment. Grows at 10–60°C. RNA and Tweens 20, 40, 60, and 80 are degraded. Glutamine, glycine, histidine, leucine, methionine, phenylalanine, and valine are used as sole carbon sources. Resistant to ampicillin (15 µg/ml), cephaloridine (10 µg/ml), cephamandole (40 µg/ml), neomycin (7 µg/ml), streptomycin (16 µg/ml), rifampin (10 µg/ml), and tobramycin (5 µg/ml), but is susceptible to tetracycline (20 µg/ml). Additional phenotypic data are shown in Table 302 and Table 303.

DNA G+C content (mol%): 69.7 (T_m).

Type strain: 13-363, ATCC 31588, CCRC 13146, DSM 43764, NBRC 14229, IMET 9741, JCM 3306, KCTC 9129, NCIMB 13755, NRRL 12069, NRRL B-16121, SCC 1256, VKM Ac-874.

Sequence accession no. (16S rRNA gene): U49006, X97890.

Sequence accession no. (23S rRNA gene): AF116219.

Sequence accession no. (16S–23S rRNA ITS): AF134099.

Additional comments: Actinomadura kijaniata differs from other *Actinomadura* species as it forms a deep green substrate mycelium and long spore chains on the aerial mycelium. Whole-cell hydrolysates contain *meso*-A_2pm together with a trace of the L-isomer (Horan and Brodsky, 1982). Furthermore, this species produces a complex of acid enol antibiotics, the major component of which was designated kijanimicin. This antibiotic has an unusual *in vitro* spectrum of activity against Gram-stain-positive and anaerobic microorganisms. *In vivo* activity has been shown against *Plasmodium berghei* and *Plasmodium chabaudi* in mice (Waitz et al., 1981).

15. **Actinomadura latina** Trujillo and Goodfellow 1997b, 917[VP] (Effective publication: Trujillo and Goodfellow 1997a, 230.)

la.ti'na. L. fem. adj. *latina* latin; named after *America Latina*, since many clinically significant strains of actinomadurae have been isolated in Latin America.

Moderate growth on Czapek, oatmeal and yeast extract-malt extract agars. Colonies cream to pink, convex, and wrinkled. Aerial mycelium absent or rare. No diffusible pigments. Grows at 10–37°C. Degrades Tweens 20, 40, and 60. Phenylalanine is used as a sole carbon sources. Resistant to rifampin (6 µg/ml), but is susceptible to ticarcillin (70 µg/ml). Additional phenotypic data are shown in Table 302 and Table 303.

Source: a patient with actinomycetoma in the arm.

DNA G+C content (mol%): 67 (T_m).

Type strain: A10, ATCC BAA-277, DSM 43382, IFM 0961, JCM 10674.

Sequence accession no. (16S rRNA gene): AY035998.

16. **Actinomadura livida** Lavrova and Preobrazheskaya 1975, 483[AL]

li'vi.da. L. fem. adj. *livida* livid (referring to the grayish-violet color of the substrate mycelium).

Spore chains in hooks or spirals of one turn. Spore surface irregular. Inorganic salts-starch agar: good growth; surface cartilaginous glistening; aerial mycelium only microscopically visible hyphae; substrate mycelium yellowish brownish, and pale violet diffusible pigment. Oatmeal agar: good growth; surface cartilaginous dull; aerial mycelium microscopically visible hyphae; substrate mycelium pale grayish pink, and pale violet diffusible pigment. Peptone-glucose medium: good growth; surface cartilaginous; aerial mycelium absent; substrate mycelium yellowish brown, and no diffusible pigment. Yeast extract-malt extract agar: moderate growth; surface cartilaginous; aerial mycelium absent; substrate mycelium pale brownish, and no diffusible pigment. Grows at 20–45°C. Histidine and phenylalanine are used as sole carbon sources. Resistant to cephaloridine (10 µg/ml), lincomycin (25 µg/ml), oleandomycin (20 µg/ml), rifampin (10 µg/ml), and ticarcillin (80 µg/ml), but is sensitive to ampicillin (5 µg/ml), benzylpenicillin (25 µg/ml), and tetracycline (20 µg/ml). Additional phenotypic data are shown in Table 302 and Table 303.

Source: soil.

DNA G+C content (mol%): not determined.

Type strain: ATCC 33578, DSM 43677, NBRC 14682, IMET 9575, INA 1678, JCM 3387, VKM Ac-908.

Sequence accession no. (16S rRNA gene): AF163116, AJ293706.

Sequence accession no. (23S rRNA gene): AF163138.

Sequence accession no. (16S–23S rRNA ITS): AF163127.

17. **Actinomadura luteofluorescens** (Shinobu 1962) Preobrazhenskaya and Lavrova in Preobrazhenskaya, Lavrova and Blinov 1975a, 526[AL] (*Streptomyces luteofluorescens* Shinobu 1962, 115)

lu.te.o. flu.o.res'cens. L. adj. *luteus* yellow; N.L. part. adj. *fluorescens* fluorescing; N.L. part. adj. *luteofluorescens* yellow and fluorescing, referring to the greenish tinge of the diffusible pigment produced by the organism.

Short spore chains in hooks or curled, arranged in clusters. Spore surface warty. Inorganic salts-starch agar: poor growth; surface leathery; aerial mycelium microscopically visible hyphae; substrate mycelium orange to brown, and yellow diffusible pigment. Oatmeal agar: good growth; surface farinaceous; aerial mycelium yellowish pink turning blue with age; substrate mycelium pale greenish yellow, and greenish yellow diffusible pigment. Peptone-glucose medium: good growth; surface wrinkled leathery; aerial mycelium absent; substrate mycelium orange to brown, and yellow diffusible pigment. Yeast extract-malt extract agar: moderate growth; surface farinaceous; aerial mycelium yellowish white; substrate mycelium dark yellow, and pale yellow diffusible pigment. Grows at 10–45°C. RNA is degraded. Resistant to ampicillin (15 μg/ml), carbenicillin (120 μg/ml), cephaloridine (10 μg/ml), cephalotin (35 μg/ml), cephamandole (40 μg/ml), rifampin (10 μg/ml), and streptomycin (16 μg/ml), but is susceptible to tetracycline (20 μg/ml). Additional phenotypic data are shown in Table 302 and Table 303.

Source: soil.

DNA G+C content (mol%): not determined.

Type strain: ATCC 25469, CBS 702.69, CGMCC AS 4.1382, CIP 105484, DSM 40398, NBRC 13057, ISP 5398, JCM 4203, JCM 4491, KCC S-0203, KCC S-0491, NRRL B-12327, RIA 1249, VKM Ac-1509.

Sequence accession no. (16S rRNA gene): U49008.

Sequence accession no. (23S rRNA gene): AF134079.

Sequence accession no. (16S–23S rRNA ITS): AF134101.

18. **Actinomadura macra** Huang 1980, 565[VP]

ma'cra. L. fem. adj. *macra* lean (referring to the poor, thin growth of this organism).

Sporulation is extremely rare and delayed. A few short spore chains (flexuous, hooked, or straight; 4–15 spores) have been observed after 5 weeks of incubation on isolation agar (Jensen, 1930) and Czapek sucrose agar. Spores oval to elliptical (0.8–1.0 × 1.2–2.0 μm). Spore surface smooth. Czapek sucrose-nitrate agar: poor growth; surface smooth; aerial mycelium scant and pale cream-colored; substrate mycelium colorless, and no diffusible pigment. Inorganic salts-starch agar: very scanty growth; surface smooth; aerial mycelium absent; substrate mycelium colorless or pale grayish, and no diffusible pigment. Oatmeal agar: moderate growth; surface smooth; aerial mycelium scant and cream to faint pink; substrate mycelium cream to faint pink, and no diffusible pigment. Yeast extract-malt extract agar: good growth; surface raised and wrinkled; aerial mycelium white to grayish; substrate mycelium black, and brown diffusible pigment. Grows at 10–45°C. Tweens 20, 40, 60, and 80 are degraded. Lactose, phenylalanine, and valine are used as sole carbon sources. Resistant to ampicillin (15 μg/ml), cephalotin (30 μg/ml), and carbenicillin (100 μg/ml), but is susceptible to benzylpenicillin (25 μg/ml), ticarcillin (70 μg/ml), rifampin (2 μg/ml), and tetracycline (20 μg/ml).

Source: soil.

DNA G+C content (mol%): not determined.

Type strain: Pfizer FD 25934, ATCC 31286, CIP 105532, BCRC (formerly CCRC) 13378, DSM 43862, NBRC 14102, IMET 9754, JCM 3287, KCC A-0287, KCTC 9342, NRRL B-16124.

Sequence accession no. (16S rRNA gene): U49009.

Sequence accession no. (23S rRNA gene): AF134080.

Sequence accession no. (16S–23S rRNA ITS): AF134102.

19. **Actinomadura mexicana** Quintana, Trujillo and Goodfellow 2004, 307[VP] (Effective publication: Quintana, Trujillo and Goodfellow 2003, 514.)

me.xi.ca'na. N.L. fem. adj. *mexicana*, referring to Mexico, the country of origin of the soil sample from which the organism was isolated.

Aerial hyphae are rare but when present may differentiate into hooked chains of spores (1 μm). Spore surface warty. Extensively branched non-fragmenting, light yellow substrate mycelium and yellow diffusible pigment formed on yeast extract-malt extract agar. Pink to palid red, convex colonies with a wrinkled morphology formed on modified Bennett's agar. Grows at 10–37°C. L-Glutamine is used as sole carbon sources. Resistant to ampicillin (15 μg/ml), benzylpenicillin (25 μg/ml), and lincomycin (25 μg/ml), but is susceptible to neomycin (1 μg/ml), streptomycin (2 μg/ml), cephalotine (20 μg/ml), and tetracycline (20 μg/ml). Additional phenotypic data are shown in Table 302 and Table 303.

Source: soil collected in Mexico.

DNA G+C content (mol%): not determined.

Type strain: A290, DSM 44485, JCM 13236, NRRL B-24203.

Sequence accession no. (16S rRNA gene): AF277195.

20. **Actinomadura meyerae** Quintana, Trujillo and Goodfellow 2004, 307[VP] (Effective publication: Quintana, Trujillo and Goodfellow 2003, 515.)

me.ye'ra.e. N.L. gen. fem. n. *meyerae* of Meyer, named after the German microbiologist Jutta Meyer in recognition of her contributions to actinomycete systematics.

Aerial hyphae are rare but when present may differentiate into hooks and spiral chains of spores. Spore surface warty. Cream to light yellow, convex colonies with a wrinkled morphology formed on modified Bennett's agar, and a light yellow substrate mycelium on yeast extract-malt extract agar. Diffusible pigments not produced. Grows at 25–45°C and pH 7–9. RNA is degraded, but not pectin. Arginine is used as a carbon source. Resistant to ampicillin (15 μg/ml), benzylpenicillin (25 μg/ml), lincomycin (25 μg/ml), and oleandomycin (20 μg/ml), but is susceptible to rifampin (2 μg/ml) and tetracycline (20 μg/ml). Additional phenotypic data are shown in Table 302 and Table 303.

Source: garden soil collected in Mexico.

DNA G+C content (mol%): not determined.

Type strain: A288, DSM 44715, JCM 13237, NRRL B-24247.

Sequence accession no. (16S rRNA gene): AY273787.

21. **Actinomadura namibiensis** Wink, Kroppenstedt, Seibert and Stackebrandt 2003, 723[VP]

na.mi.bi.en'sis. N.L. fem. adj. *namibiensis* of or belonging to Namibia from where the type strain was isolated.

Spiral spore chains of about 20 spores are formed after about 20 d. Spore surface smooth. Substrate mycelium is salmon pink on glycerol-asparagine, inorganic salts-starch, oatmeal, tyrosine, and yeast extract-malt extract agars and colorless on peptone-yeast extract-iron agar. Soluble pigments are not produced on any of these media. White aerial hyphae are formed on glycerol-asparagine inorganic salts-starch and tyrosine agars. Enzymic activities (API ZYM system) are shown for cystine arylmidase, leucine arylamidase, naphthol-AS-BI-phosphohydrolase, β-glucosidase, and trypsin, but not for esterases (C_4) and (C_8). Other phenotypic properties are shown in Table 302 and Table 303.

Source: soil sample collected in Namibia.

DNA G+C content (mol%): not determined.

Type strain: HAG 010767, CIP 108365, DSM 44197, FH-A 1198, JCM 13238, NRRL B-24153.

Sequence accession no. (16S rRNA gene): AJ420134.

22. **Actinomadura napierensis** Cook, le Roes and Meyers 2005, 705[VP]

na.pi.e.ren'sis. N.L. fem. adj. *napierensis* of or belonging to Napier in South Africa, the source of the type strain.

Spiral spore chains are formed. Spore surfaces smooth. Grey aerial and brown-gray substrate mycelia formed on Czapek, Middlebrook 7H9, and yeast extract-malt extract agars. White aerial and substrate mycelia produced on inorganic salts-starch agar. A faint blue diffusible pigment is produced on tyrosine agar. Tween 80 is degraded, but not allantoin, guanine, hippurate, or xylan. Shows a weak H_2S reaction. Dextrin is used as a sole carbon source. Grows in the presence of 2% but not 5% (w/v) NaCl.

Source: soil in Napier, Western State Province, South Africa.

DNA G+C content (mol%): 70 (T_m).

Type strain: B60, DSM 44846, JCM 13850, NRRL B-24319.

Sequence accession no. (16S rRNA gene): AY568292.

23. **Actinomadura nitritigenes** Lipski and Altendorf 1995, 722[VP]

ni.tri.ti'ge.nes. N.L. n. *nitris -itis* nitrite; N.L. suff. *-genes* (from Gr. v. *gennaô* to produce) producing; N.L. adj. *nitritigenes* nitrite producing.

Oval to ellipsoidal spores produced in straight or hook-like chains that are up to seven spores long. Spore surface smooth. Substrate mycelium is colorless to brown on glycerol-asparagine, inorganic salts-starch, oatmeal, peptone-yeast extract-iron, tyrosine, and yeast extract-malt extract agars. Good growth is obtained on all these media except for peptone-yeast extract-iron agar. Diffusible pigments are not produced. Aerial mycelium is either white or brown. Grows at 45°C. Tweens 20 and 80 are degraded, but not adenine, guanine, or xylan. H_2S is not produced. Starch and sucrose are used as sole carbon sources, but not alanine, propionate, pyruvate, or serine. Resistant to neomycin (3 μg/ml), but is susceptible to gentamicin (4 μg/ml) and streptomycin (4 μg/ml). Enzymic activity (API ZYM system) shown for cystine arylmidase, esterase (C_8), β-glucosidase, leucine arylamidase, naphthol-AS-BI-phosphohydrolase, and trypsin, but not for esterase (C_4). Other phenotypic properties are shown in Table 302 and Table 303.

Source: experimental biofilters filled with tree bark compost.

DNA G+C content (mol%): 74 (HPLC).

Type strain: L46, DSM 44137, NBRC 15918, JCM 10104, NCIMB 13456.

Sequence accession no. (16S rRNA gene): AY035999.

24. **Actinomadura oligospora** Mertz and Yao 1986, 180[VP]

o.li.go.spo'ra. Gr. adj. *oligos* few; Gr. n. *spora* seed; N.L. n. *oligospora* organism forming few spores, referring to the relative absence of sporophores in this organism.

Aerial mycelia are absent except for trace amounts produced on inorganic salts-starch and sodium butyrate agars. Substrate mycelia give rise to occasional sparse twisted oyster white aerial hyphae that differentiate into flexuous or hooked chains of 10–50 spores. Spores are oblong (0.5–0.7 × 0.9–1.3 μm). Spore surface smooth. Colonies grow slowly (10–20 d) and are convex with filamentous margins. Growth is best on complex media. Colony surfaces and reverse surfaces are yellowish gray to brown. Soluble pigments are produced on inorganic salts-starch and yeast-glucose agars. Grows at 20–37°C. Does not degrade adenine, calcium malate, guanine, hippurate, or keratin. Leucine is used as a sole carbon source, but not dextran or inulin. Resistant to bacitracin (10 U), cephaloridine (10 μg/ml), cephalotin (35 μg/ml), cephamandole (40 μg/ml), rifampin (10 μg/ml), and tetracycline (20 μg/ml), but is susceptible to gentamicin (10 μg/ml), streptomycin (10 μg/ml), tobramycin (10 μg/ml), and vancomycin (30 μg/ml). Produces a polyether antibiotic. Does not grow at pH 8. Additional phenotypic data are shown in Table 302 and Table 303.

Source: soil collected in Karnataka, India.

DNA G+C content (mol%): not determined.

Type strain: A80190.1, ATCC 43269, JCM 10648, NRRL 15878.

Sequence accession no. (16S rRNA gene): AF163118

Sequence accession no. (23S rRNA gene): AF163140.

Sequence accession no. (16S–23S rRNA ITS): AF163129.

25. **Actinomadura pelletieri** (Laveran 1906) Lechevalier and Lechevalier 1970a, 400[AL] (*Micrococcus pelletieri* Laveran 1906, 341)

pel.le.ti'e.ri. N.L. gen. masc. n. *pelletieri* of Pelletier, named after T. Pelletier, who first isolated this species.

Spore chains short (2–6 spores per chain), hooked or in spirals of two to three turns. Spores subspherical. Spore surface warty. Oatmeal agar: moderate growth; surface cartilaginous; aerial mycelium absent; substrate mycelium pink to brownish red, and no diffusible pigment. Oatmeal-nitrate agar: moderate growth; surface cartilaginous; traces of sporulating hyphae on aerial mycelium; substrate mycelium colorless, and no diffusible pigment. Peptone-glucose medium: moderate growth; surface cartilaginous; aerial mycelium absent; substrate mycelium brownish red, and no diffusible pigment. Yeast extract-malt extract agar: moderate growth; surface cartilaginous; traces of sterile hyphae on aerial mycelium; substrate mycelium pink to brownish red, and no diffusible pigment. Grows at 25–37°C. Tweens 20, 40, and 60 are degraded, but not adenine. Starch hydrolysis is variable. Collagenase is produced and nitrate reduced.

Cellobiose, glucose, and glycogen are used as sole carbon sources, but not adonitol, arabinose, erythritol, inositol, raffinose, or sorbitol. Does not grow in the presence of 4%, w/v, NaCl. Additional phenotypic properties are shown in Table 302 and Table 303.

Source: actinomycetoma of the arm. Most strains classified as *Actinomadura pelletieri* have been isolated from clinical samples.

DNA G+C content (mol%): 65.5–67.3 (T_m).

Type strain: ATCC 33385, CCUG 38891, CIP 105483, DSM 43383, HUT 6549, IAM 12634, IMET 9693, JCM 3388, KCTC 9110, NBRC 103052, NCTC 4162, NRRL B-3997.

Sequence accession no. (16S rRNA gene): AF163119, AJ293710.

Sequence accession no. (23S rRNA gene): AF163141.

Sequence accession no. (16S–23S rRNA ITS): AF163130.

Additional comments: Cummins (1962), who was the first to isolate *meso*-A_2pm from whole-organism hydrolysates of members of this species, also found a small amount of L-isomer and glycine in some strains.

26. **Actinomadura rubrobrunea** (*ex* Krasil'nikov, Agre and El-Registan 1968) Kroppenstedt, Goodfellow and Stackebrandt 1991, 178[VP] (*Excellospora rubrobrunea* Agre and Guzeva 1975, 522)

ru.bro.bru'ne.a. L. adj. *ruber, bra brum* red; N.L. adj. *bruneus a um* brown; N.L. fem. adj. *rubrobrunea* reddish brown.

Coiled to spiral chains of 2–20 spores, some simple. Spore surface spiny. Reddish-brown substrate mycelium. Grows at 37–65°C and optimally at 45–55°C. Does not degrade adenine or cellulose. Additional phenotypic properties are shown in Table 302 and Table 303.

Source: soil.

DNA G+C content (mol%): not determined.

Type strain: ATCC 49883, CIP 105486, DSM 43750, NBRC 15178, IMET 9705, INMI 2991, JCM 7345, KCTC 9493, VKM Ac-1470.

Sequence accession no. (16S rRNA gene): AF134069, EU637008.

Sequence accession no. (23S rRNA gene): AF134081.

Sequence accession no. (16S–23S rRNA ITS): AF134104.

27. **Actinomadura rudentiformis** le Roes and Meyers 2007, 48[AL]

ru. den. ti. for'mis. L. masc. n. *rudens* rope; L. adj. suff. *-formis* like, in the shape of; N.L. fem. adj. *rudentiformis* shaped like a rope.

Spores have not been detected. Cream-colored substrate mycelium with a sparse white aerial mycelium formed on oatmeal agar. Wrinkled, cream-colored substrate mycelium, but no sporulation on yeast extract-malt extract agar. White substrate mycelium but does not sporulate on inorganic salts-starch agar. Pigments are not produced on glycerol-asparagine, peptone-yeast extract-iron, or tyrosine agars. Rope-like or fiber-like growth visible under scanning electron microscopy. Growth occurs at 30, 37, and 45°C, but not at 4°C or pH 4.3. Degrades adenine, Tween 80, and xylan. Guanine and starch are weakly degraded. Hydrolyzes hippurate, but not allantoin, pectin, or urea. Lecithinase, lipase, and proteolytic activities are not observed on egg-yolk agar. Does not produce H_2S or reduce nitrate. Inositol, inulin, melezitose (weak), ribose (weak), salicin (weak), sodium acetate, and sucrose are used as sole carbon sources, but not sodium citrate. DL-α-Amino-*n*-butyric acid, L-arginine, L-cysteine, L-histidine, L-hydroxyproline, L-methionine, L-phenylalanine, potassium nitrate, L-servine, L-threonine, and L-valine are used as sole carbon and nitrogen sources. Resistant to kanamycin (10 μg/ml), neomycin (30 μg/ ml), oleandomycin (100 μg/ml), penicillin G (10 IU), rifampin (50 μg/ml), and tobramycin (50 μg/ml), but is sensitive to cephaloridine (100 μg/ml), gentamicin (100 μg/ml), lincomycin (100 μg/ml), streptomycin (100 μg/ml), and vancomycin (50 μg/ml). Additional phenotypic properties are shown in Table 302 and Table 303.

Source: soil taken from the banks of the Gamka River in the Swartberg Nature Reserve, Western Cape Province, South Africa.

DNA G+C content (mol%): not determined.

Type strain: HMC1, DSM 44962, JCM 14907, NRRL B-24458.

Sequence accession no. (16S rRNA gene): DQ285420.

28. **Actinomadura rugatobispora** (Miyadoh, Tohyama, Amano, Shomura and Niida 1985) Miyadoh, Amano, Tohyama and Shomura 1990, 1909[VP] (*Microbispora viridis* Miyadoh, Tohyama, Amano, Shomura and Niida 1985, 281[VP])

ru.ga.to.bi.spo'ra. L. part. adj. *rugatus* wrinkled; Gr. pref. *bi* twice; Gr. n. *spora* a seed and, in biology, a spore; N.L. n. *rugatobispora* organism forming wrinkled paired-spores, referring to spore morphology.

Aerial hyphae differentiate into a non-sporogenous main axis which bears spore-bearing side branches. Spores are oval (1.9–1.2 × 1.4–2.0 μm). Usually two but sometimes three spores per chain. Spore surface rugose with vertical ridges (Figure 408). Aerial mycelium is grayish green and the substrate mycelium pastel yellow on oatmeal agar. The substrate mycelium is pale yellowish brown on inorganic salts-starch agar. Grows at 25–35°C. Does not require vita-

FIGURE 408. SEM of *Actinomadura rugatobispora* SF 2240[T] (IFO 14382[T]). Usually two but sometimes three spores per chain. Spore surface rugose with vertical ridges. Bar = 1 μm. (Reproduced with permission from S. Miyadoh, S. Amano, and T. Shomura. Atlas of Actinomycetes, Society for Actinomycetes Japan.)

min B for growth. Additional phenotypic properties are shown in Table 302 and Table 303.

Source: soil.

DNA G+C content (mol%): 73 (T_m).

Type strain: SF2240, ATCC 51643, CIP 105482, DSM 44130, NBRC 14382, JCM 3366, NRRL B-16566.

Sequence accession no. (16S rRNA gene): U49010.

Sequence accession no. (23S rRNA gene): AF134082.

Sequence accession no. (16S–23S rRNA ITS): AF134105.

29. **Actinomadura spadix** Nonomura and Ohara 1971d, 911[AL]

spa′dix. L. fem. adj. *spadix* chestnut-colored (referring to the color of the substrate mycelium).

Short spore chains with 5–10 spores in small, round spore masses (pseudosporangia) formed on long aerial hyphae. Sporulation observed on soil extract agar and partly on oatmeal-yeast-glucose agar. Spore surface smooth (Figure 409). Glycerol-asparagine agar: aerial mycelium absent; substrate mycelium grayish brown, and diffusible pigment reddish-gray or grayish brown. Oatmeal agar: aerial mycelium light grayish-yellowish brown; substrate mycelium grayish brown, and diffusible pigment reddish-gray or grayish brown. Yeast extract-malt extact agar: aerial mycelium absent; substrate mycelium dark brown, and diffusible pigment brown. Grows at 30°C, but not at 25 or 37°C. Does not degrade Tweens 20, 40, 60, and 80. Phenylalanine and valine are used as sole carbon sources, but not arginine or glutamine. Resistant to rifampin (10 μg/ml), but is susceptible to ampicillin (5 μg/ml), benzylpenicillin (25 μg/ml), carbenicillin (90 μg/ml), tobramycin (0.05 μg/ml), and tetracycline (20 μg/ml). Additional phenotypic properties are shown in Table 302 and Table 303.

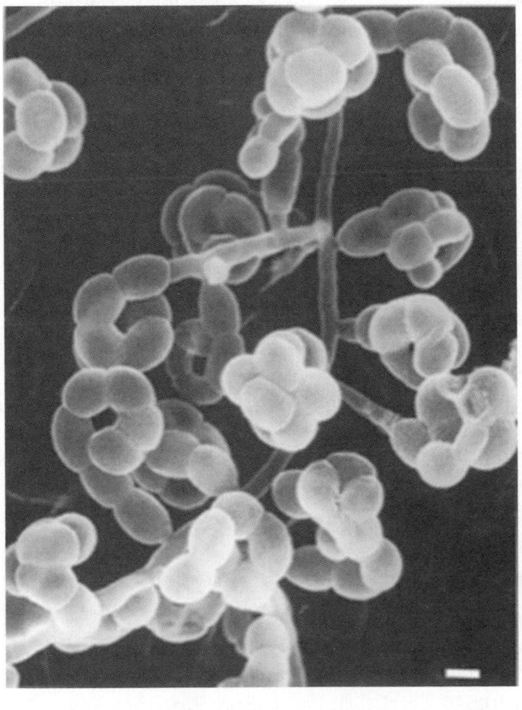

FIGURE 409. SEM of *Actinomadura spadix* MH 193-16 F4. Short spore chains with 5–10 spores in small, round spore masses (pseudosporangia) formed on long aerial hyphae. Spore surface smooth. Bar = 1 μm. (Reproduced with permission from M. Hamada and T. Shomura. Atlas of Actinomycetes, Society for Actinomycetes Japan.)

Source: soil.

DNA G+C content (mol%): 66.4 (T_m).

Type strain: ATCC 27298, CBS 261.72, BCRC (formerly CCRC) 13386, CIP 105479, DSM 43459, NBRC 14099, IMET 9752, JCM 3146, KCC A-0146, KCTC 9252, NCIB 11118, NRRL B-16128.

Sequence accession no. (16S rRNA gene): AF163120, AB364581.

Sequence accession no. (23S rRNA gene): AF163142.

Sequence accession no. (16S–23S rRNA ITS): AF163131.

Additional comments: according to Nonomura and Ohara (1971c), Actinomadura spadix requires B vitamins for growth. Phylogenetic data (16S and 23S rRNA) suggests that this strain is misclassified in the genus *Actinomadura* and may represent a new genus though additional data are needed to clarify its taxonomic status.

30. **Actinomadura umbrina** Galatenko, Terekhova and Preobrazhenskaya 1987, 179[VP] (Effective publication: Galatenko, Terekhova and Preobrazhenskaya 1981, 803.)

um.bri′na. N.L. fem. adj. *umbrina* wood brown.

Short spore chains may be straight, hooked or spiral, often branching. Smooth spore surface. Substrate mycelium is brown or black as are soluble pigments. White scanty aerial mycelium produced. Good growth occurs on mineral (Gauze 1), Czapek's, glycerol-nitrate, glycerol asparagine, and inorganic salts-starch agars. Does not grow on carbon utilization agar supplemented with different sugars, including glucose. Additional phenotypic properties are shown in Table 302 and Table 303.

Source: soil.

DNA G+C content (mol%): not determined.

Type strain: ATCC 49502, CIP 105485, DSM 43927, NBRC 14346, IMET 9746, INA 2309, JCM 6837, NRRL B-16244, VKM Ac-1086.

Sequence accession no. (16S rRNA gene): AF163121, AJ293713.

Sequence accession no. (23S rRNA gene): AF123143.

Sequence accession no. (16S–23S rRNA ITS): AF163132.

31. **Actinomadura verrucosospora** Nonomura and Ohara 1971d, 908[AL]

ver.ru.co.so.spo′ra. L. fem. adj. *verrucosa* warty; Gr. n. *spora* seed; N.L. n. *verrucosospora* organism forming warty spores, referring to the warty surface of the spores.

Spore chains (5–12 spores per chain) hooked or curled or spirals of one turn, arranged in tufts on long aerial hyphae. Spores elliptical. Spore surface warty. Glycerol-asparagine agar: moderate growth; aerial mycelium absent; substrate mycelium bright red, and no diffusible pigment. Inorganic salts-starch agar: poor growth; surface farinaceous to granular; aerial mycelium white; substrate mycelium yellowish white, and no diffusible pigment. Oatmeal agar: good growth; surface farinaceous; aerial mycelium pink turning blue with maturity; substrate mycelium orange to pink, and no diffusible pigment. Oatmeal-nitrate agar: good growth; surface slightly farinaceous; aerial mycelium white to grayish; substrate mycelium white to pink, and no diffusible pigment. Yeast extract-malt extract agar: moderate to good growth; surface farinaceous; aerial mycelium white; substrate mycelium yellowish, and no diffusible pigment. Grows at 10–37°C. Does not degrade RNA or xylan. Resistant to ampicillin (10 μg/ml), benzylpencillin (25 μg/ml), carbenicillin (120 μg/ml), and rifampin (10 μg/ml), but is susceptible to lincomycin (10 μg/ml) and tetracycline (20 μg/ml). Additional phenotypic properties are shown in Table 302 and Table 303.

Source: soil in Japan.

DNA G+C content (mol%): 69.0 (T_m).

Type strain: ATCC 27299, CBS 258.72, CCM 3492, BCRC (formerly CCRC) 13408, CIP 105480, DSM 43358, DSM 43550, IMET 9588, JCM 3147, KCC A-0147, NCIMB 11119, NRRL B-16129, RIA 1503, VKM Ac-668.

Sequence accession no. (16S rRNA gene): U49011.

Sequence accession no. (23S rRNA gene): AF134083.

Sequence accession no. (16S–23S rRNA ITS): AF134106.

32. **Actinomadura vinacea** Lavrova and Preobrazhenskaya 1975, 486[AL]

vi.na'ce.a. L. fem. adj. *vinacea* of or belonging to wine, referring to the brownish-red color of the substrate mycelium.

Short, straight spore chains, arranged in small clusters on moderately branched hyphae of the aerial mycelium. Spore surface irregular. Czapek-sucrose agar: poor growth; aerial mycelium pink or absent; substrate mycelium brownish red, and diffusible pigment pale lilac to red. Inorganic salts-starch agar (Gauze 1): poor growth; surface slightly velvety; aerial mycelium white to gray; substrate mycelium colorless, and diffusible pigment pale pink. Oatmeal agar: poor growth; surface dull leathery; aerial mycelium absent; substrate mycelium pink to red, and no diffusible pigment. Peptone glucose medium: poor to moderate growth; surface cartilaginous; aerial mycelium absent; substrate mycelium brownish red, and diffusible pigment dark red. Yeast extract-malt extract agar: poor growth; surface leathery; aerial mycelium absent; substrate mycelium dark brown-red, and diffusible pigment red. Additional phenotypic properties are shown in Table 302 and Table 303.

Source: soil.

DNA G+C content (mol%): not determined.

Type strain: ATCC 33581, CIP 105481, DSM 43765, NBRC 14688, IMET 9574, IMET 9673, INA 1682, JCM 3325, KCC A-0325, KCTC 9344.

Sequence accession no. (16S rRNA gene): AF134070.

Sequence accession no. (23S rRNA gene): AF134084.

Sequence accession no. (16S–23S rRNA ITS): AF134107.

33. **Actinomadura viridilutea** (Agre and Guzeva 1975) Zhang, Kudo, Nakajima and Wang 2001, 381[VP] (*Excellospora viridilutea* Agre and Guzeva 1975)

vi.ri.di.lu'te.a. L. adj. *viridis* green; L. adj. *luteus* yellow; N.L. fem. adj. *viridilutea* green-yellow.

Spore chains hooked to spiral. Spore surface spiny. Substrate mycelium is yellow; aerial mycelium is bluish green. Grows at 37–65°C and optimally at 45–55°C. Does not degrade cellulose. The following carbon sources were used when assayed in Czapek's agar without sucrose as basal medium: acetate, arabinose (weak), citrate (weak), dulcitol (weak), galactose, glucose, glycerol, inositol, lactose, maltose, mannitol, raffinose, rhamnose, starch, sucrose, and xylose (weak). Asparagine, NH_4, urea, and tyrosine are used as nitrogen sources. Does not show antibacterial or antifungal activities against *Bacillus mycoides, Bacillus subtilis, Escherichia coli, Sarcina lutea, Serratia marcescens, Staphylococcus aureus, Streptomyces globisporus, Streptomyces fluorescens, Streptomyces griseus,* or *Streptomyces levoris* or against *Aspergillus niger, Fusarium oxysporum, Penicillium notatum,* and *Saccharomyces cerevisiae.* Additional phenotypic properties are shown in Table 302 and Table 303.

DNA G+C content (mol%): not determined.

Type strain: ATCC 33925, BCRC (formerly CCRC) 13638, DSM 44433, NBRC 14480, IMET 9742, INMI 187, JCM 3398, JCM 7346.

Sequence accession no. (16S rRNA gene): D86943.

Sequence accession no. (23S rRNA gene): AF134087.

Sequence accession no. (16S–23S rRNA ITS): AF134110.

Additional comments: Actinomadura viridilutea can be differentiated from *Actinomadura rubrobrunea* by the solubility, light absorbance, and pH color change of its pigments, by its antimicrobial response, and the color of its substrate mycelium.

34. **Actinomadura viridis** (Nonomura and Ohara 1971b) Miyadoh, Anzai, Amano and Shomura 1989, 156[VP] (*Microtetraspora viridis* Nonomura and Ohara 1971b, 5[AL])

vi'ri.dis. L. fem. adj. *viridis* green.

Spores in straight chains. Spore surface smooth. Substrate mycelium pale yellowish-brown on Hickey–Tresner agar. Good sporulation on inorganic salts-starch and MGA agars, but none on Czapek oatmeal or soil extract agars. Grows at 20–45°C. Resistant to neomycin (1 μg/ml), streptomycin (4 μg/ml), cephalotin (35 μg/ml), ampicillin (15 μg/ml), and rifampin (10 μg/ml), but is susceptible to benzylpenicillin (25 μg/ml) and tetracycline (20 μg/ml). Additional phenotypic properties are shown in Table 302 and Table 303.

Source: soil.

DNA G+C content (mol%): 72 (T_m).

Type strain: ATCC 27103, CBS 833.70, BCRC (formerly CCRC) 13398, CIP 105478, DSM 43175, NBRC 15238, IMET 9546, JCM 3112, KCC A-0112, KCTC 9290, VKM Ac-1315.

Sequence accession no. (16S rRNA gene): D85467, AJ420141.

Sequence accession no. (23S rRNA gene): AF134085.

Sequence accession no. (16S–23S rRNA ITS): AF134108.

35. **Actinomadura yumaensis** Labeda, Testa, Lechevalier and Lechevalier 1985, 333[AL]

yu.ma.en'sis. N.L. fem. adj. *yumaensis* of or belonging to Yuma County, Arizona, USA, the source of the soil sample from which the type strain was isolated.

Spore chains are short (mean, 20 spores per chain), loosely coiled and usually borne on branched, and almost verticillate aerial sporophores. Spores are ovoid and measure 0.6–0.8 × 1.0–1.4 μm. Spore surface smooth (Figure 410). Bennett-sucrose agar: good growth; aerial mycelium absent; convoluted substrate mycelium yellowish gray to grayish brown; substrate mycelium dark grayish-yellowish brown, and diffusible pigment orange. Inorganic salts-starch agar: poor growth; white aerial mycelium; substrate

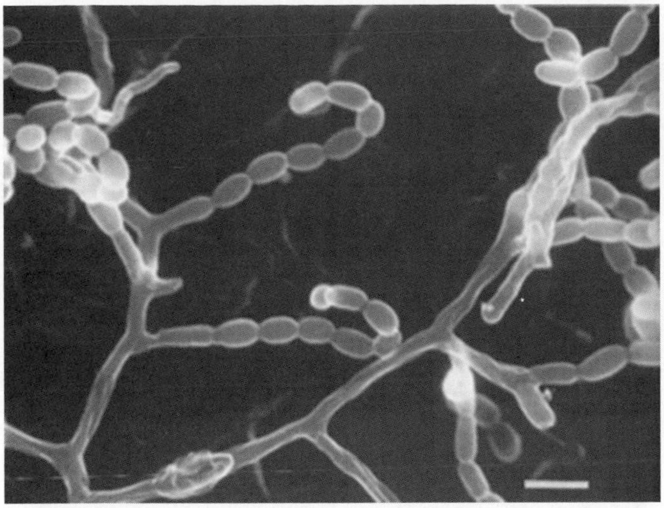

FIGURE 410. Electron micrograph of *Actinomadura yumaensis* IFO 14689^T. Spore chains are coiled and usually borne on branched, almost verticillate aerial sporophores. Bar = 2 μm. (Reproduced with permission from S. Miyadoh and S. Amano. Atlas of Actinomycetes, Society for Actinomycetes Japan.)

mycelium colorless, and no diffusible pigment. Oatmeal agar: moderate growth; surface flat waxy; grayish yellow; moderate white aerial mycelium; substrate mycelium grayish yellow, and no diffusible pigment. Yeast-extract-malt extract agar: good growth; surface raised waxy; convoluted colonies yellowish gray to grayish-yellowish brown; aerial mycelium absent; substrate mycelium dark yellowish brown, and diffusible pigment orange. Grows at 10–45°C. Arginine and sucrose are used as sole carbon sources. Resistant to ampicillin (15 μg/ml), benzylpenicillin (25 μg/ml), cephaloridine (10 μg/ml), cephalotin (35 μg/ml), cephamandole (30 μg/ml), neomycin (7 μg/ml), rifampin (10 μg/ml), streptomycin (16 μg/ml), and tetracycline (40 μg/ml). Additional phenotypic features are shown in Table 302 and Table 303.

Source: soil sample collected in Yuma County, Arizona.

DNA G+C content (mol%): not determined.

Type strain: LL-C23024, ATCC 43060, CIP 105436, NBRC 14689, JCM 3369, KCTC 9495, NRRL 12515.

Sequence accession no. (16S rRNA gene): AF163122.

Sequence accession no. (23S rRNA gene): AF163144.

Sequence accession no. (16S–23S rRNA ITS): AF163133.

Species *incertae sedis*

a. **"Actinomadura azurea"** Nakamura and Isono 1983, 1468

Produces cationomycin.

Type strain: ICM 2033.

b. **"Actinomadura luzonensis"** Tomita, Hoshino, Sasahira and Kawaguchi 1980, 1098

Produces the antitumor antibiotic complex BBM-928.

Type strain: ATCC 31491 (G455–101).

c. **"Actinomadura pulveracea"** Iwami, Kiyoto, Nishikawa, Terano, Kohsaka, Aoki and Imanaka 1985, 835

Produces antitumor antibiotics FR-900405 and FR-900406.

Type strain: No. 6049.

d. **"Actinomadura albolutea"** Tohyama, Miyadoh, Ito, Shomura, Ito and Ishikawa 1984, 1144

The placement of this species in the genus *Actinomadura* is doubtful. Morphologically, Actinomadura albolutea resembles *Nocardiopsis* however, the presence of maduroe in whole-cell hydrolysates, phospholipids of type PIV, and MK-9(H_4) as major menaquinones indicate that this micro-organism is clearly differentiated from *Nocardiopsis*. Produces an indole-*N*-glycoside antibiotic SF-2140.

Type strain: FERM-BP 386 (SF2140).

Genus IV. **Spirillospora** Couch 1963, 61^{AL}

Michael Goodfellow and Martha E. Trujillo

Spi.ril. lo.spo′ra. L. n. *spira* a spiral; N.L. dim. neut. n. *spirillum* a short spiral; Gr. n. *spora* a seed and, in biology, a spore; N.L. fem. n. *Spirillospora* an organism with spores in spirals.

Aerobic, Gram-stain-positive, chemo-organotrophic, mesophilic actinomycetes which **form spherical to vermiform spore vesicles** (5.0–24.0 μm in diameter) **on aerial mycelia. Spore vesicles contain numerous spores arranged in coiled and branched chains. Spores are rod-shaped or curved** (0.5–0.7 × 2.0–6.0 μm) **and are motile by means of one to seven subpolar flagella**. The hyphae of the substrate and aerial mycelia are 0.2–1.0 μm thick, branched, and septate. **The substrate mycelium is white to pale yellow or pale buffy pink to red; the aerial mycelium is usually white. Cell-wall peptidoglycan contains *meso*-diaminopimelic acid; madurose is the characteristic sugar of whole-organism hydrolysates.** Grows at 18–35°C. Predominant menaquinones are MK-9(H_4) and MK-8 (H_4).

DNA G+C content (mol%): 71.0–73.0 (T_m, Bd).

Type species: **Spirillospora albida** Couch 1963, 61^{AL}.

Further descriptive information

Phylogeny. Members of the genus *Spirillospora* were originally classified in the family *Streptosporangiaceae*, but it is evident from chemotaxonomic, nucleic acid reassociation, and comparative 16S rRNA sequencing studies that the genus is a *bona fide* member of the family *Thermomonosporaceae* (Stackebrandt et al., 1997, 1981; Zhang et al., 2001). However, the type strains of the two constituent species, *Spirillospora albida* and *Spirillospora rubra*, are well separated in thc *Thermomonosporaceae* 16S rRNA gene tree

(Figure 398 in *Thermomonosporaceae* chapter). These organisms share a 16S rRNA gene similarity of 96.8%, a value which corresponds to 58 differences out of a total of 1444 nucleotides. *Spirillospora albida* IFO 12248^T and *Spirillospora rubra* JCM 6875^T share the highest 16S rRNA gene similarity with *Actinomadura rudentiformis* DSM 44962^T (97.7% and 97.2%, respectively). The type strains of *Spirillospora albida* and *Spirillospora rubra* have also been shown to be distantly related in an analysis of 23S rRNA gene sequences (Zhang et al., 2001). In addition, the type strain of *Spirillospora albida* shares a 70% 16S–23S rRNA internal transcribed spacer (ITS) sequence identity with several *Actinomadura* species, a value much higher than those (>40%) reported between many *Actinomadura* species (Zhang et al., 2001). In contrast, Zhang and his colleagues found that the type strain of *Actinomadura rubra* exhibits only 30–40% 16S–23S ITS sequence similarity to those of all other species in the family *Thermomonosporaceae*.

Cell morphology. *Spirillospora* strains form substrate and aerial mycelia. The hyphae are branched and septate and are 0.2–1.0 µm in diameter (Couch and Bland, 1974). Spore vesicles (sporangia) produced on aerial mycelia are usually spherical with a diameter of 5.0–24.0 µm; the mean diameter is 10 µm (Figure 411). Subspherical to elongated or club-shaped and vermiform spore vesicles are also formed (Couch, 1963). At the initiation of spore vesicle development, the end of an aerial hypha winds into a coil that is enclosed in a common sheath (Bland and Couch, 1981; Lechevalier et al., 1966; Locci and Petrolini-Baldan, 1971; Vobis, 1985). In some cases it appears that the first coils are temporarily free (Vobis and Kothe, 1985). The coiled sporogenous hyphae are branched and fragment into oblong segments that differentiate into spore-size segments (Lechevalier et al., 1966; Vobis, 1985).

The spores are short to long rods, and are frequently curved (0.5–0.7 × 2–7 µm) (Couch and Bland, 1974). A subpolarly inserted tuft of one to seven flagella gives the spores a slight motility which can become more vigorous in the presence of an energy source (Higgins et al., 1967). In addition to the spores enclosed in spore vesicles, free, exposed spores in regular or irregular coils may be found among the aerial hyphae. When flooded with water, the coils break up into rod-shaped to curved spores that subsequently become motile. Conidia-like structures in moniliform arrangement may be produced by the substrate mycelium (Couch, 1963). The septa involved in spore formation are double-layered [cross wall type 2 of Williams et al. (1973)]. Aerial hyphae are additionally covered with a thick sheath from which the spore vesicular envelope originates (Lechevalier et al., 1966; Vobis and Kothe, 1985). *Spirillospora albida* forms spore vesicles on Czapek, peptone-Czapek, and oatmeal agars and *Spirillospora rubra* on cornmeal-soil agar (Vobis and Kothe, 1989) and artificial soil agar (Henssen and Schäfer, 1971).

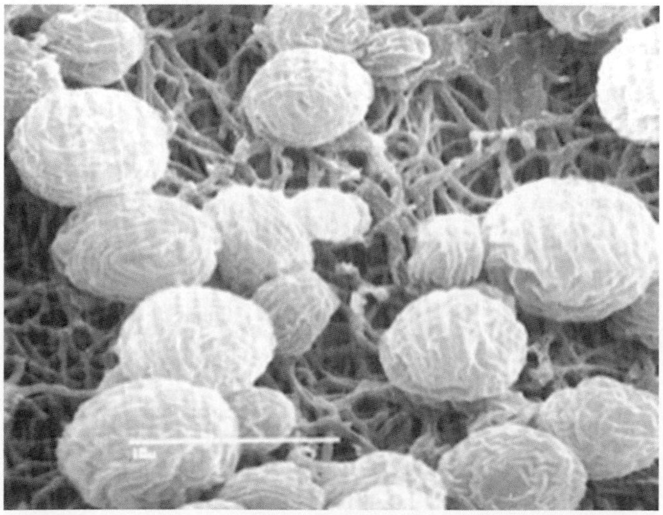

FIGURE 411. *Spirillospora albida* ATCC 15331. Scanning electron micrograph of spherical spore vesicles formed on the aerial mycelium after incubation for 15 d (3200×). Bar = 10 µm.

Nutrition and growth conditions. Spirillosporae can be cultured on various complex media. They grow well on Czapek, peptone-Czapek, and oatmeal agars and moderately on casein and tyrosine agars. Colonies are compact and elevated, sometimes with protruberances on casein, peptone-Czapek, and tyrosine agars. Flattish or confluent colonies are formed on Czapek and oatmeal agars. Some strains produce a blue confluent pigment called spirillomycin, which shows antibiotic activity against some Gram-stain-positive bacteria (Domnas, 1968; McInnis and Domnas, 1970). The pigment producing strains use amylopectin, galactose, glucose, maltose, and sucrose as carbon sources for growth (Domnas, 1970). *Spirillospora* grows well at 18–35°C (Couch and Bland, 1974).

Cell-wall composition. The cell-wall peptidoglycan of *Spirillospora* strains contain *meso*-diaminopimelic acid, and whole-organism hydrolysates contain the diagnostic sugar madurose (Lechevalier and Lechevalier, 1970c; Yamaguichi, 1965); this means that they have a wall chemotype III and a sugar pattern B according to the scheme of Lechevalier and Lechevalier (1970b). The type strain of *Spirillospora albida* contains diphosphatidylglycerol, phosphatidylinositol, and phosphatidylinositol mannosides as major polar lipids (Zhang et al., 2001) and hence has a type 1 phospholipid pattern (Lechevalier et al., 1981) though previous workers reported a type II pattern for this organism (Hasegawa et al., 1978). The predominant menaquinones are MK-9(H_4) and MK-9(H_6) (Collins et al., 1984). Spirillosporae contain major amounts of hexadecanoic ($C_{16:0}$; 20–26% of total fatty acids), octadecenoic ($C_{18:1}$; 21–27%), 14-methylpentadecanoic ($C_{16:0}$ iso; 5–11%), and 10-methyloctadecanoic (tuberculostearic) acids (Zhang et al., 2001), that is, they have a fatty acid type 3a pattern *sensu* Kroppenstedt (1985).

Ecology. *Spirillospora* strains have been isolated from soil, albeit infrequently, by using baits (pollen for the isolation of *Spirillospora albida* and hair for the recovery of *Spirillospora rubra*). However, these organisms accounted for less than 1% of the "sporangiate" actinomycetes isolated from soil using baits (Schäfer, 1973). One part of the life cycle of spirillosporae is adapted to the aquatic milieu. Motile zoospores are able to colonize pollen grains that float on the surface of water and constitute their natural substrate. Mycelia are developed, forming spore vesicles when in contact with air. The spore vesicles are considered to be resistant stages against desiccation. When flooded with water, the zoospores are released from the spore vesicles through either a rupture in the envelope or through a large, irregular pore (Couch, 1963).

Isolation procedures

Spirillosporae have been isolated from soil by baiting with natural substrates (Bland and Couch, 1981; Couch, 1954; Schäfer, 1973). A small amount of soil, approximately one level teaspoonful, is placed in a sterile Petri dish and flooded with sterile water (distilled water or filtered soil or charcoal water extracts may be used). Various types of pollen [from members of the genera *Liquidamber, Pinus and Sparanium*; (Schäfer, 1973)] or other natural substrates such as hair, snakeskin, or boiled grass are added as baits and after incubation for 1–4 weeks the baits are examined with a binocular dissecting microscope (×100). Aerial mycelium and spore vesicles can be recognized at the surface of the water by their glistening appearance. Pollen baits bearing spore vesicles are picked up with a thin needle and transferred to the surface of agar media in small Petri dishes. Individual spore vesicles are separated from the baits and rolled on the agar surface to free them from contaminating bacteria. The cleaned spore vesicles are used as an inoculum for pure cultures. Various complex media support the growth of *Spirillospora* strains. *Spirillospora albida* grows well on Czapek, peptone-Czapek, and oatmeal agars whereas *Spirillospora rubra* prefers half-concentrated corn meal agar (Difco) supplemented with sterilized garden soil (50 g/liter) or half concentrated skim milk agar (Difco; Schäfer, 1973).

Maintenance procedures

Subcultures should be made after a period of 12 weeks. For long-term preservation, the organisms must be lyophilized by using standard procedures employed for aerobic actinomycetes, preferably using well-sporulating cultures. A simple, reliable, and quick method involves nitrogen cryopreservation of living cells in small polyvinyl tubes ("straws") at −196°C (Hoffman, 1989a, 1989b).

Differentiation of the genus *Spirillospora* from other genera

Spirillospora strains can be distinguished from the other three genera classified in the family *Thermomonosporaceae* using a range of chemotaxonomic, morphological and phylogenetic data (Table 298, see chapter on Family *Thermomonosporaceae*). Members of the genus can be easily confused on morphological grounds with those of some species of *Streptosporangium*. All strains belonging to these taxa have multispored, usually spherical, spore vesicles borne on aerial hyphae. However, members of the two genera can be distinguished on the basis of menaquinone and polar lipid composition (Collins et al., 1984; Zhang et al., 2001). In addition, spirollosporae, unlike streptosporangiae, produce zoospores.

Taxonomic comments

Spirillospora albida, the type species of the genus, is closely related to many *Actinomadura* species (Figure 398). The 16S rRNA and 23S rRNA gene sequences of *Spirillospora albida* and some *Actinomadura* strains exhibit much higher similarity than the sequence similarity between most representatives of *Actinomadura* species, as is the case with 16S–23S ITS sequences (Zhang et al., 2001). These data strongly suggest that *Spirillospora albida* has a very close evolutionary relationship with some *Actinomadura* species. In contrast, corresponding data on *Spirillospora rubra* shows that this organism is only distantly related to *Spirillospora albida* and to most other taxa classified in the family *Thermomonosporaceae*, which suggests that this species may merit generic status. However, further comparative taxonomic studies are needed on additional *Spirillospora* strains to clarify their relationships with one another and with members of other genera assigned to the family *Thermomonosporaceae*, notably the genus *Actinomadura*.

Differentiation of species of the genus *Spirillospora*

The two *Spirillospora* species can be distinguished by the characteristic colors of their substrate mycelia. The colonies of *Spirillospora albida* are white to pale yellow whereas those of *Spirillospora rubra* are red to reddish brown.

List of species of the genus *Spirillospora*

1. **Spirillospora albida** Couch 1963, 65[AL]

al'bi. da. L. fem. adj. *albida* whitish.

Spherical spore vesicles (mean, 10 μm in diameter; range, 5.0–24.0 μm) are formed at the tips of aerial hyphae on Czapek, peptone-Czapek, and oatmeal agars. Subspherical to elongated or club-shaped and vermiform spore vesicles are also formed. Spores are rod-shaped, frequently curved (0.5–0.7 × 2.0–6 μm), and weakly motile.

The color of the substrate mycelium is white to pale yellow or buffy pink; the aerial mycelium is white. After 8 weeks, colonies reach diameters up to 13 mm on Czapek, peptone-Czapek, and oatmeal agars, and diameters 5–10 mm on casein and tyrosine agars. A pale yellowish soluble pigment is produced on casein agar and a clay-colored pigment on tyrosine agar. Some strains produce a blue soluble pigment on peptone-Czapek agar. Good growth occurs at 18–35°C and optimum growth at 25°C. Degrades casein and L-tyrosine.

Source: soil.

DNA G+C content (mol%): 70.0–72.0 (T_m) for the type strain and 72.9 (B_d) for strain UNCC 761.

Type strain: ATCC 15331, CBS 291.04, DSM 43034, NBRC 12248, IMET 9031, JCM 3041, NRRL B-3350, VKM Ac-926.

Sequence accession no. (16S rRNA gene): D85498.

Sequence accession no. (23S rRNA gene): AF134088.

Sequence accession no. (16S–23S rRNA ITS): AF134111.

2. **Spirillospora rubra** (*ex* Schäfer 1973) nom. rev. *Spirillospora rubra* Schäfer 1973, 199[AL]

ru'bra. L. fem. adj. *rubra* red.

Hyphae of substrate mycelium are branched (0.4–0.9 μm in diameter), as are aerial hyphae (0.6–1.2 μm in diameter). Spherical spore vesicles, with a diameter of 10.0–25.0 μm, are formed at the tips of aerial hyphae on artificial soil agar and on cornmeal-soil agar.* Rod-shaped spores are frequently slightly curved (0.8 × 1.8–2.8 μm), and weakly motile.

Substrate mycelium is red to reddish brown; the aerial mycelium is white. Good growth occurs on casein and yeast extract-starch agars, and moderate growth on oatmeal-yeast extract, peptone, starch, and tyrosine agars. Grows well at 20–37°C. Degrades casein, but not L-tyrosine. Nitrate is not reduced.

Source: soil.

DNA G+C content (mol%): not determined.

Type strain: CBS 571.75, JCM 6875.

Sequence accession no. (16S rRNA gene): AF163123

Sequence accession no. (23S rRNA gene): AF163145.

Sequence accession no. (16S–23S rRNA ITS): AF163134.

*Difco cornmeal agar: cornmeal, 8.5 g, agar, 7.5 g, sterile garden soil, 50 g, distilled water, 1000 ml.

References

Agre, N.S., T.P. Efimova and L.N. Guzeva. 1975. [Heterogeneity of the genus *Actinomadura* Lechevalier a. Lechevalier]. Mikrobiologiia *44*: 253–257.

Agre, N.S. and L.N. Guzeva. 1975. [New genus of actinomycetes: *Excellospora* gen. nov.]. Mikrobiologiia *44*: 518–523.

Alderson, G. and M. Goodfellow. 1979. Classification and identification of *Actinomycetales* causing actinomycosis. Postepy Hig. Med. Dosw. *33*: 109–124.

Athalye, M. 1981. Classification and isolation of actinomadurae. PhD thesis, University of Newcastle, Newcastle upon Tyne.

Athalye, M., J. Lacey and M. Goodfellow. 1981. Selective isolation and enumeration of actinomycetes using rifampicin. J. Appl. Bacteriol. *51*: 289–297.

Athalye, M., M. Goodfellow and D.E. Minnikin. 1984. Menaquinone composition in the classification of *Actinomadura* and related taxa. J. Gen. Microbiol. *130*: 817–823.

Athalye, M., M. Goodfellow, J. Lacey and R.P. White. 1985. Numerical classification of *Actinomadura* and *Nocardiopsis*. Int. J. Syst. Bacteriol. *35*: 86–98.

Becker, B., M.P. Lechevalier and H.A. Lechevalier. 1965. Chemical composition of cell-wall preparations from strains of various form-genera of aerobic actinomycetes. Appl. Microbiol. *13*: 236–243.

Bernier, R. and F. Stutzenberger. 1988. Extracellular and cell-associated of β-glucosidase in *Thermomonospora curvata*. Lett. Appl. Microbiol. *7*: 103–107.

Blanchard, R. 1896. Parasites végetaux à l'exclusion des bactéries. *In* Traité de Pathologie Générale, vol. II (edited by Bouchard). G. Masson, Paris, pp. 811–932.

Bland, C.E. and J.N. Couch. 1981. The family *Actinoplanaceae*. *In* The Prokaryotes: a Handbook on Habitats, Isolation, and Identification of Bacteria (edited by Starr, Stolp, Trüper, Balows and Schlegel). Springer, New York, pp. 2004–2010.

Boiron, P., I. Medici de Jugo, M. Trujillo, F. Provost and M. Goodfellow. 1992. *In vitro* antibiotic susceptibility testing of agents of actinomycetoma. Med. Microbiol. Lett. *1*: 38–42.

Brocq-Rousseu, D. 1904. Sur un *Streptothrix* cause de l'alteracion des avoines moisies. Res. Bot. *16*: 219–230.

Chormonova, N.T. and T.P. Preobrazhenskaya. 1981. Occurrence of *Actinomadura* in Kazakhstan soils. Antibiotiki *26*: 341–345.

Collins, M.D., T. Pirouz, M. Goodfellow and D.E. Minnikin. 1977. Distribution of menaquinones in actinomycetes and corynebacteria. J. Gen. Microbiol. *100*: 221–230.

Collins, M.D., A.J. McCarthy and T. Cross. 1982. New highly saturated members of the vitamin K2 series from *Thermomonospora*. Zentralbl. Bakteriol. Hyg. I. Abt. Orig. C *3*: 358–363.

Collins, M.D., M. Faulkner and R.M. Keddie. 1984. Menaquinone composition of some sporeforming actinomycetes. Syst. Appl. Microbiol. *5*: 20–29.

Cook, A.E., M. Roes and P.R. Meyers. 2005. *Actinomadura napierensis* sp. nov., isolated from soil in South Africa. Int. J. Syst. Evol. Microbiol. *55*: 703–706.

Couch, J.N. 1954. The genus *Actinoplanes* and its relatives. Trans.. N.Y. Acad. Sci. *16*: 315–318.

Couch, J.N. 1963. Some new genera and species of the *Actinoplanaceae*. J. Elisha Mitchell Sci. Soc. *79*: 53–70.

Couch, J.N. and C.E. Bland. 1974. Genus 1. *Actinoplanes*. *In* Bergey's Manual of Determinative Bacteriology, 8th edn (edited by Buchanan and Gibbons). Williams & Wilkins, Baltimore, pp. 708–710.

Crawford, D.L. 1975. Cultural, morphological, and physiological characteristics of *Thermomonospora fusca* (strain 190Th). Can. J. Microbiol. *21*: 1842–1848.

Crawford, D.L. and M.A. Gonda. 1977. The sporulation process in *Thermomonospora fusca* as revealed by scanning and transmission electron microscopy. Can. J. Microbiol. *23*: 1088–1095.

Cross, T. and J. Lacey. 1970. Studies on the genus *Thermonospora*. *In* The *Actinomycetales* (edited by Prauser). Gustav Fischer Verlag, Jena, pp. 211–219.

Cross, T. and M. Goodfellow. 1973. Taxonomy and classification of the actinomycetes. *In Actinomycetales:* Characteristics and Practical Importance (edited by Sykes and Skinner). Academic Press, London, pp. 11–112.

Cross, T. and R.W. Attwell. 1974. Recovery of viable thermoactinomycete endospores from deep mud cores. *In* Spore Research 1973 (edited by Barker, Gould and Wolf). Academic Press, London, pp. 11–20.

Cross, T. 1981. The monosporic actinomycetes. *In* The Prokaryotes: a Handbook on Habitats, Isolation, and Identification of Bacteria (edited by Starr, Stolp, Trüper, Balows and Schlegel). Springer, New York, pp. 2091–2102.

Cui, X.L., P.H. Mao, M. Zeng, W.J. Li, L.P. Zhang, L.H. Xu and C.L. Jiang. 2001. *Streptimonospora salina* gen. nov., sp. nov., a new member of the family *Nocardiopsaceae*. Int. J. Syst. Evol. Microbiol. *51*: 357–363.

Cummins, C.S. 1962. Chemical composition and antigenic structure of cell walls of *Corynebacterium*, *Mycobacterium*, *Nocardia*, *Actinomyces* and *Arthrobacter*. J. Gen. Microbiol. *28*: 35–50.

Davis, J.D., P.A. Stone and J.J. McGarry. 1999. Recurrent mycetoma of the foot. J. Foot Ankle Surg. *38*: 55–60.

de Hoog, G.S., A. Buiting, C.S. Tan, A.B. Stroebel, C. Ketterings, E.J. de Boer, B. Naafs, R. Brimicombe, M.K. Nohlmans-Paulssen, G.T. Fabius and et al. 1993. Diagnostic problems with imported cases of mycetoma in The Netherlands. Mycoses *36*: 81–87.

Develoux, M., J. Audoin, J. Treguer, J.M. Vetter, A. Warter and A. Cenac. 1988. Mycetoma in the Republic of Niger: clinical features and epidemiology. Am. J. Trop. Med. Hyg. *38*: 386–390.

Domnas, A. 1968. Pigments of the *Actinoplanaceae*. 1. Pigment production by *Spirillospora* 1655. J. Elisha Mitchell Sci. Soc. *84*: 163–123.

Domnas, A. 1970. Pigment production in the *Actinoplanaceae* as affected by cultural conditions. *In* The *Actinomycetales* (edited by Prauser). VEB Gustav Fischer Verlag, Jena, pp. 259–263.

El Moghraby, I.M. 1971. Mycetoma in Gezira. Sudan Med. J. *9*: 77–89.

Embley, M.T., J. Smida and E. Stackebrandt. 1988. The phylogeny of mycolateless wall chemotype-IV actinomycetes and description of *Pseudonocardiaceae* fam. nov. Syst. Appl. Microbiol. *11*: 44–52.

Embley, T.M. and E. Stackebrandt. 1994. The molecular phylogeny and systematics of the actinomycetes. Annu. Rev. Microbiol. *48*: 257–289.

Euzéby, J.P. and T. Kudo. 2001. Corrigenda to the Validation Lists. Int. J. Syst. Evol. Microbiol. *51*: 1933–1938.

Fahal, A.H. and M.A. Hassan. 1992. Mycetoma. Br. J. Surg. *79*: 1138–1141.

Fahal, A.H. 2006. Mycetoma - Clinicopathological Monograph. Khartoum University Press, Khartoum, p. 112.

Fergus, C.L. 1964. Thermophilic and thermotolerant molds and actinomycetes of mushroom compost during peak heating. Mycologia *56*: 267–284.

Fischer, A., R.M. Kroppenstedt and E. Stackebrandt. 1983. Molecular-genetic and chemotaxonomic studies on *Actinomadura* and *Nocardiopsis*. J. Gen. Microbiol. *129*: 3433–3446.

Fowler, V.J., W. Ludwig and E. Stackebrandt. 1985. Ribosomal ribonucleic acid cataloguing in bacterial systematics: the phylogeny of *Actinomadura*. *In* Chemical Methods in Bacterial Systematics (edited by Goodfellow and Minnikin). Academic Press, London, pp. 17–40.

Galatenko, O.A. and T.P. Preobrazhenskaia. 1981. [*Actinomadura* of the sierozem soils of Turkmenia and their antagonistic properties]. Antibiotiki *26*: 723–727.

Galatenko, O.A., L.P. Terekhova and T.P. Preobrazhenskaia. 1981. [New species of *Actinomadura* isolated from soils in Turkmenia and their antagonistic properties]. Antibiotiki *26*: 803–807.

Galatenko, O.A., L.P. Terekova and T.P. Preobrazhenskaya. 1987. *In* Validation of the publication of new names and new combinations previously effectively published outside the IJSB. List no. 23. Int. J. Syst. Bacteriol. *37*: 179–180.

Gauze, G.F., T.P. Preobrazhenskaya, E.S. Kudrina, N.O. Blinov, I.D. Ryabova and M.A. Sveshnikova. 1957. Problems in the classification of antagonistic actinomycetes. State Publishing House for Medical Literature (in Russian). Medzig, Moscow.

Gauze, G.F., T.P. Preobrazhenskaia, N.V. Lavrova, R.S. Ukholina and N.V. Kochetkova. 1975. [*Actinomadura cremea* var. rifamycini, a producer of rifamycin O]. Antibiotiki *20*: 963–966.

Gauze, G.F., L.P. Terekhova, O.A. Galatenko, T.P. Preobrazhenskaia and V.N. Borisova. 1984. [New species of *Actinomadura recticatena* sp. nov. and its antibiotic properties]. Antibiotiki *29*: 3–7.

Gauze, G.F., T.P. Preobrazhenskaia, N.V. Lavrova, R.S. Ukholina and N.V. Kochetkova. 1987. *In* Validation of the publication of new names and new combinations previously effectively published outside the IJSB. List no. 23. Int. J. Syst. Bacteriol. *37*: 179–180.

George, S.P., A. Ahmad and M.B. Rao. 2001. A novel thermostable xylanase from *Thermomonospora* sp.: influence of additives on thermostability. Bioresour. Technol. *78*: 221–224.

Gerber, N.N. 1971. Prodigiosin-like pigments from *Actinomadura* (*Nocardia*) *pelletieri*. J. Antibiot. (Tokyo) *24*: 636–640.

Gerber, N.N. 1973. Minor prodiginine pigments from *Actinomadura madurae* and *Actinomadura pelletieri*. J. Heterocycl. Chem. *10*: 925.

Gill. 1842. Indian Naval Medical Reports - quoted by Ghosh LM et al. Madura foot (mycetoma). Indian Medical Gazette 1950 *85*: 288–291.

Goodfellow, M. 1971. Numerical taxonomy of some nocardioform bacteria. J. Gen. Microbiol. *69*: 33–80.

Goodfellow, M., G. Alderson and J. Lacey. 1979. Numerical taxonomy of *Actinomadura* and related actinomycetes. J. Gen. Microbiol. *112*: 95–111.

Goodfellow, M. and T. Pirouz. 1982. Numerical classification of sporoactinomycetes containing *meso*-diaminopimelic acid in the cell wall. J. Gen. Microbiol. *128*: 503–527.

Goodfellow, M. and T. Cross. 1984. Classification. *In* The Biology of the Actinomycetes (edited by Goodfellow, Mordarski and Williams). Academic Press, London, pp. 7–164.

Goodfellow, M., E. Stackebrandt and R.M. Kroppenstedt. 1988. Chemotaxonomy and actinomycete systematics. *In* Biology of Actinomycetes '88 (edited by Okami, Beppu and Ogawara). Japan Scientific Societies Press, Tokyo, pp. 233–238.

Goodfellow, M. 1989. Maduromycetes. *In* Bergey's Manual of Systematic Bacteriology, vol. 4 (edited by Williams, Sharpe and Holt). Williams & Wilkins, Baltimore, pp. 2509–2510.

Goodfellow, M., L.J. Stanton, K.E. Simpson and D.E. Minnikin. 1990. Numerical and chemical classification of *Actinoplanes* and some related actinomycetes. J. Gen. Microbiol. *136*: 19–36.

Goodfellow, M., M.E. Trujillo and G. Alderson. 1995. Approaches towards the identification of sporoactinomycetes that cause mycetoma. Biotechnologia *7–8*: 271–286.

Goodfellow, M. 1996. *Actinomycetes: Actinomyces, Actinomadura, Nocardia, Streptomyces* and related taxa. *In* Mackie and McCartney Practical Medical Microbiology (edited by Collee, Fraser, Marmion and Simmons). Churchill Livingstone, Edinburgh, pp. 343–359.

Goodfellow, M. and E.T. Quintana. 2006. The family *Streptosporangiaceae*. *In* The Prokaryotes: a Handbook on the Biology of Bacteria, 3rd edn, vol. 3, *Archaea, Bacteria, Firmicutes*, Actinomycetes (edited by Dworkin, Falkow, Rosenberg, Schleifer and Stackebrandt). Springer, New York, pp. 725–753.

Gordon, M.A. 1974. Aerobic pathogenic *Actinomycetaceae*. *In* Manual of Clinical Microbiology, 4th edn (edited by Lennette, Spauling and Truant). American Society for Microbiology, Washington, D.C., pp. 175–188.

Gumaa, S.A., E.S. Mahgoub and M.A. el Sid. 1986. Mycetoma of the head and neck. Am. J. Trop. Med. Hyg. *35*: 594–600.

Gyobu, Y. 2001. Family *Nocardiopsaceae*. *In* Identification Manual of Actinomycetes (edited by Miyadoh). Business Center for Academic Societies Japan, Tokyo, pp. 277–280.

Hasegawa, T., M.P. Lechevalier and H.A. Lechevalier. 1978. New genus of *Actinomycetales*: *Actinosynnema* gen. nov. Int. J. Syst. Bacteriol. *28*: 304–310.

Hasegawa, T., M.P. Lechevalier and H.A. Lechevalier. 1979. Phospholipid composition of motile actinomycetes. J. Gen. Microbiol. *25*: 209–213.

Hasegawa, T., S. Tanida and H. Ono. 1986. *Thermomonospora formosensis* sp. nov. Int. J. Syst. Bacteriol. *36*: 20–23.

Henssen, A. 1957. Beiträge zur Morphologie und Systematik der thermophilen Actinomyceten. Arch. Mikrobiol. *26*: 373–414.

Henssen, A. and E. Schnepf. 1967. [On the knowledge of thermophilic actinomycetes]. Arch. Mikrobiol. *57*: 214–231.

Henssen, A. 1970. Spore formation in thermophilic actinomycetes. *In* The *Actinomycetales* (edited by Prauser). Gustav Fischer Verlag, Jena, pp. 205–210.

Henssen, A. and D. Schäfer. 1971. Emended description of the genus *Pseudonocardia* Henssen and description of the new species *Pseudonocardia spinosa*. Int. J. Syst. Bacteriol. *21*: 29–34.

Higgins, M.L., M.P. Lechevalier and H.A. Lechevalier. 1967. Flagellated actinomycetes. J. Bacteriol. *93*: 1446–1451.

Hoffman, P. 1989a. Cryopreservation of basidiomycete cultures: mushroom (Part 1). Proceedings of the Twelfth International Congress on the Science and Cultivation of Edible Fungi (1987), Braunschweig, Germany.

Hoffman, P. 1989b. Cryopreservation of fungi. World Federation of Culture Collections. Technical Information Sheet No. 5. NESCO/WFCC/Education Committee. Braunschweig, Germany.

Horan, A.C. and B.C. Brodsky. 1982. A novel antibiotic-producing *Actinomadura*, *Actinomadura kijaniata* sp. nov. Int. J. Syst. Bacteriol. *32*: 195–200.

Huang, L.H. 1980. *Actinomadura macra* sp. nov., the producer of antibiotics Cp-47,433 and Cp-47,434. Int. J. Syst. Bacteriol. *30*: 565–568.

Iinuma, A., A. Yokota, T. Hasegawa and T. Kanamura. 1994. *Actinocorallia* gen. nov., a new genus of the order *Actinomycetales*. Int. J. Syst. Bacteriol. *44*: 235–245.

Ispoglou, S.S., A. Zormpala, A. Androulaki and N.V. Sipsas. 2003. Madura foot due to *Actinomadura madurae*: imaging appearance. Clin. Imaging *27*: 233–235.

Itoh, T., T. Kudo, H. Oyaizu and A. Seino. 1995. Two new species in the genus *Actinomadura*: *A. glomerata* sp. nov., and *A. longicatena* sp. nov. Actinomycetologica *9*: 164–177.

Iwami, M., S. Kiyoto, M. Nishikawa, H. Terano, M. Kohsaka, H. Aoki and H. Imanaka. 1985. New antitumor antibiotics, FR-900405 and FR-900406. I. Taxonomy of the producing strain. J. Antibiot. (Tokyo) *38*: 835–839.

Jensen, H.L. 1930. The genus *Micromonospora* Ørskov, a little known group of soil microorganisms. Proc. Linnean Soc. N.S.W. *55*: 231–248.

Jensen, H.L. 1932. Contribution to our knowledge of *Actinomycetales*. III. Further observations on the genus *Micromonospora*. Proc. Linnean Soc. N.S.W. *57*: 173–180.

Jones, K.L. 1949. Fresh isolates of actinomycetes in which the presence of sporogenous aerial mycelia is a fluctuating characteristic. J. Bacteriol. *57*: 141–145.

Jukes, T.H. and C. Cantor. 1969. Evolution of protein molecules. *In* Mammalian Protein Metabolism (edited by Murano). Academic Press, New York pp. 21–132.

Kim, B. 1999. Polyphasic taxonomy of thermophilic actinomycetes. PhD thesis, Newcastle upon Tyne.

Klokke, A.H., G. Swamidasan, R. Anguli and A. Verghese. 1968. The causal agents of mycetoma in South India. Trans. R. Soc. Trop. Med. Hyg. *62*: 509–516.

Krasil'nikov, N.A., N.S. Agre and G.I. el-Registan. 1968. [New thermophilic species of *Micropolyspora* genus]. Mikrobiologiia *37*: 1065–1072.

Krasil'nikov, N.A. 1941. Keys to *Actinomycetales* (In Russian). Izvest. Akad. Nauk SSSR, Moscow.

Krasil'nikov, N.A. and N.S. Agre. 1964. A new actinomycete genus – *Actinobifida* n. gen. yellow group – *Actinobifida dichotomica* n. sp. (in Russian). Mikrobiologiya *33*: 935–943.

Krasil'nikov, N.A. and N.S. Agre. 1965. The brown group of *Actinobifida chromogena* n. sp. (in Russian). Mikrobiologiya *34*: 284–291.

Kroppenstedt, R.M. 1985. Fatty acid and menaquinone analysis of actinomycetes and related organisms. *In* Chemical Methods in Bacterial Systematics (edited by Goodfellow and Minnikin). Academic Press, London, pp. 173–199.

Kroppenstedt, R.M., E. Stackebrandt and M. Goodfellow. 1990. Taxonomic revision of the actinomycete genera *Actinomadura* and *Microtetraspora* Syst. Appl. Microbiol. *13*: 148–160.

Kroppenstedt, R.M., E. Stackebrandt and M. Goodfellow. 1991. *In* Validation of the publication of new names and new combinations previously effectively published outside the IJSB. List no. 36. Int. J. Syst. Bacteriol. *41*: 178–179.

Kroppenstedt, R.M. and M. Goodfellow. 1992. The family *Thermomonosporaceae*. *In* The Prokaryotes: a Handbook on the Biology of Bacteria: Ecophysiology, Isolation, Identification, Applications, 2nd edn (edited by Balows, Trüper, Dworkin, Harder and Schleifer). Springer, New York, pp. 1085–1114.

Kroppenstedt, R.M. and M. Goodfellow. 2006. The family *Thermomonosporaceae*: *Actinocorallia*, *Actinomadura*, *Spirillospora* and *Thermomonospora*. *In* The Prokaryotes: a Handbook on the Biology of Bacteria, 3rd edn, vol. 3, *Archaea*, *Bacteria*, *Firmicutes*, Actinomycetes (edited by Dworkin, Falkow, Rosenberg, Schleifer and Stackebrandt). Springer, New York, pp. 682–724.

Kudo, T. 1997. Family *Thermomonosporaceae*. *In* Atlas of Actinomycetes (edited by Miyadoh). Asakura Publishing, Tokyo, pp. 82–100.

Kudo, T. 2001. Family *Streptosporangiaceae*. *In* Identification Manual of Actinomycetes (edited by Miyadoh). Business Center for Academic Societies Japan, Tokyo, pp. 281–291.

Kurtböke, D.I., C.R. Wilson and K. Sivasithamparam. 1993. Occurrence of *Actinomadura* phage in organic mulches used for avocado plantations in Western Australia. Can. J. Microbiol. *39*: 389–394.

Küster, E. and R. Locci. 1963. Transfer of *Thermoactinomyces viridis* Schuurmans et al., 1956 to the genus *Thermomonospora* as *Thermomonospora viridis* comb. nov.. Int. Bull. Bacteriol. Nomencl. Taxon. *13*: 214–216.

Labeda, D.P., R.T. Testa, M.P. Lechevalier and H.A. Lechevalier. 1985. *Actinomadura yumaensis* sp. nov. Int. J. Syst. Bacteriol. *35*: 333–336.

Labeda, D.P. and R.M. Kroppenstedt. 2000. Phylogenetic analysis of *Saccharothrix* and related taxa: proposal for *Actinosynnemataceae* fam. nov. Int. J. Syst. Evol. Microbiol. *50*: 331–336.

Lacey, J. 1974. Allergy in mushroom workers. Lancet Infect. Dis. *1*: 366.

Lacey, J. and J. Dutkiewicz. 1976. Isolation of Actinomycetes and fungi from mouldy hay using a sedimentation chamber. J. Appl. Bacteriol. *41*: 315–319.

Lacey, J. 1977. The ecology of actinomycetes in fodders and related substrates. Zentralbl. Bakteriol. Parasitenkd. Infektionskr. Hyg. Abt. I. Orig. Suppl. *6*: 161–170.

Lacey, J., M. Goodfellow and G. Alderson. 1978. The genus *Actinomadura*. *In Nocardia* and *Streptomyces* (edited by Mordarski, Kurylowicz and Jeljaszewicz). Springer, Stuttgart, pp. 107–117.

Lacey, J. 1988. Actinomycetes as biodeteriogens and pollutants of the environment. *In* Actinomycetes in Biotechnology (edited by Goodfellow, Williams and Mordarski). Academic Press, London, pp. 359–432.

Laveran, M. 1906. Tumeur provoquée par un microcoque rose en zooglées. C. R. Hebd. Soc. Biol. *2*: 340–341.

Lavrova, N.V., T.P. Preobrazhenskaia and M.A. Sveshnikova. 1972. [Isolation of soil actinomycetes on selective media with rubomycin]. Antibiotiki *17*: 965–970.

Lavrova, N.V. and T.P. Preobrazhenskaia. 1975. [Isolation of new species of the genus *Actinomadura* on selective media with rubomycin]. Antibiotiki *20*: 438–438.

Lawrence, H.M., H. Merivuori, J.A. Sands and K.A. Pidcock. 1986. Preliminary characterization of bacteriophages infecting the thermophilic actinomycete *Thermomonospora*. Appl. Environ. Microbiol. *52*: 631–636.

Lawson, E.N. and L.M. Davey. 1972. A waterborne actinomycete resembling strains causing mycetoma. J. Appl. Bacteriol. *35*: 389–394.

le Roes, M. and P.R. Meyers. 2007. *Actinomadura rudentiformis* sp. nov., isolated from soil. Int. J. Syst. Evol. Microbiol. *57*: 45–50.

Lechevalier, H.A., M.P. Lechevalier and P.E. Holbert. 1966. Electron microscopic observation of the sporangial structure of strains of *Actinoplanaceae*. J. Bacteriol. *92*: 1228–1235.

Lechevalier, H.A. and M.P. Lechevalier. 1970a. A critical evaluation of the genera of aerobic actinomycetes. *In* The *Actinomycetales* (edited by Prauser). Gustav Fischer Verlag, Jena, pp. 395–405.

Lechevalier, H.A., M.P. Lechevalier and N.N. Gerber. 1971. Chemical composition as a criterion in the classification of actinomycetes. Adv. Appl. Microbiol. *14*: 47–72.

Lechevalier, M.P. 1968. Identification of aerobic actinomycetes of clinical importance. J. Lab. Clin. Med. *71*: 934–944.

Lechevalier, M.P. and H.A. Lechevalier. 1970b. Chemical composition as a criterion in the classification of aerobic actinomycetes. Int. J. Syst. Bacteriol. *20*: 435–443.

Lechevalier, M.P. and H.A. Lechevalier. 1970c. Composition of whole-cell hydrolysates as a criterion in the classification of aerobic actinomycetes. *In* The *Actinomycetales* (edited by Prauser). Gustav Fischer Verlag, Jena, pp. 311–316.

Lechevalier, M.P. and N.N. Gerber. 1970. The identity of madurose with 3-*O*-methyl-D-galactose. Carbohydr. Res. *13*: 451–453.

Lechevalier, M.P. 1976. The taxonomy of the genus *Nocardia*: Some light at the end of the tunnel? *In* The Biology of the Nocardiae (edited by Goodfellow, Brownell and Serrano). Academic Press, London, pp. 1–38.

Lechevalier, M.P., C. De Bièvre and H. Lechevalier. 1977. Chemotaxonomy of aerobic actinomycetes: phospholipid composition. Biochem. Syst. Ecol. *5*: 249–260.

Lechevalier, M.P., A.E. Stern and H.A. Lechevalier. 1981. Phospholipids in the taxonomy of actinomycetes. Zentralbl. Bakteriol. Parasitenkd. Infektionskr. Hyg. I Abt. Orig. Suppl. *11*: 111–116.

Lee, S.D. 2006. *Actinocorallia cavernae* sp. nov., isolated from a natural cave in Jeju, Korea. Int. J. Syst. Evol. Microbiol. *56*: 1085–1088.

Lee, S.D. and H.S. Jeong. 2006. *Actinomadura hallensis* sp. nov., a novel actinomycete isolated from Mt. Halla in Korea. Int. J. Syst. Evol. Microbiol. *56*: 259–264.

Lindenbein, W. 1952. Über einige chemisch interessante Actinomyceten – stämme und ihre Klassifizierung. Arch. Mikrobiol. *17*: 361–383.

Lipski, A. and K. Altendorf. 1995. *Actinomadura nitritigenes* sp. nov., isolated from experimental biofilters. Int. J. Syst. Bacteriol. *45*: 717–723.

Locci, R., E. Baldacci and B. Petrolini. 1967. Contribution to the study of oligosporic actinomycetes. I. Description of new species of *Actinobifida*: *Actinobifida alba* sp. nov. and revision of the genus. G. Microbiol. *15*: 79–91.

Locci, R. and B. Petrolini-Baldan. 1971. On the spore formation process in actinomycetes. V. Scanning electron microscopy of some genera of *Actinoplanaceae*. Riv. Pat. Veg. Ser. IV.: 81–96.

Lopez, R. 1993. Aislamiento de *Nocardia brasiliensis* y su patogenicidad. *In* Actinomicetos Universidad Autónoma Metropolitana-Xochimilco (edited by Sandoval-Trujillo), Mexico, pp. 193–200.

Lu, Z., L. Wang, Y. Zhang, Y. Shi, Z. Liu, E.T. Quintana and M. Goodfellow. 2003. *Actinomadura catellatispora* sp. nov. and *Actinomadura glauciflava* sp. nov., from a sewage ditch and soil in southern China. Int. J. Syst. Evol. Microbiol. *53*: 137–142.

Magaña, M. 1984. Mycetoma. Int. J. Dermatol. *23*: 221–236.

McCarthy, A.J. and T. Cross. 1981. A note on a selective isolation medium for the thermophilic actinomycete *Thermomonospora chromogena*. J. Appl. Bacteriol. *51*: 299–302.

McCarthy, A.J. and P. Broda. 1984. Screening for lignin-degrading actinomycetes and characterisation of their activity against ^{14}C-lignin-labelled wheat lignocellulose. J. Gen. Microbiol. *130*: 2905–2913.

McCarthy, A.J. and T. Cross. 1984a. A taxonomic study of *Thermomonospora* and other monosporic actinomycetes. *In* Biological, Biochemical and Biomedical Aspects of Actinomycetes (edited by Ortiz-Ortiz, Bojalil and Yakoleff). Academic Press, San Diego, USA, pp. 521–536.

McCarthy, A.J. and T. Cross. 1984b. A taxonomic study of *Thermomonospora* and other monosporic actinomycetes. J. Gen. Microbiol. *130*: 5–25.

McCarthy, A.J. 1987. Lignocellulose-degrading actinomycetes. FEMS Microbiol. Rev. *46*: 145–163.

McCarthy, A.J. 1989. *Thermomonospora* and related genera. *In* Bergey's Manual of Systematic Bacteriology, vol. 4 (edited by Williams, Sharpe and Holt). Williams & Wilkins, Baltimore, pp. 2552–2559.

McInnis, T.M., Jr and A. Domnas. 1970. Pigments of the *Actinoplanaceae*. 3. A spirillomycin-type pigment from *Spirillospora* 1309-b. Z. Allg. Mikrobiol. *10*: 129–136.

McNeil, M.M., J.M. Brown, G. Scalise and C. Piersimoni. 1992. Nonmycetomic *Actinomadura madurae* infection in a patient with AIDS. J. Clin. Microbiol. *30*: 1008–1010.

Mertz, F.P. and R.C. Yao. 1986. *Actinomadura oligospora* sp. nov., the producer of a new polyether antibiotic. Int. J. Syst. Bacteriol. *36*: 179–182.

Mertz, F.P. and R.C. Yao. 1990. *Actinomadura fibrosa* sp. nov. isolated from soil. Int. J. Syst. Bacteriol. *40*: 28–33.

Meyer, J. 1976. *Nocardiopsis*, a new genus of order *Actinomycetales*. Int. J. Syst. Bacteriol. *26*: 487–493.

Meyer, J. 1979. New species of the genus *Actinomadura*. Z. Allg. Mikrobiol. *19*: 37–44.

Meyer, J. 1981. *In* Validation of the publication of new names and new combinations previously effectively published outside the IJSB. List no. 6. Int. J. Syst. Bacteriol. *31*: 215–218.

Meyer, J. 1989a. Genus *Actinomadura*. *In* Bergey's Manual of Systematic Bacteriology, 1st edn, vol. 4 (edited by Williams, Sharpe and Holt). Williams & Wilkins, Baltimore, pp. 2511–2526.

Meyer, J. 1989b. Genus *Nocardiopsis*. *In* Bergey's Manual of Systematic Bacteriology, vol. 4 (edited by Williams, Sharpe and Holt). Williams & Wilkins, Baltimore, pp. 2562–2569.

Minnikin, D.E., T. Pirouz and M. Goodfellow. 1977. Polar lipid composition in the classification of some *Actinomadura* species. Int. J. Syst. Bacteriol. *27*: 118–121.

Miyadoh, S., H. Tohyama, S. Amano, T. Shomura and T. Niida. 1985. *Microbispora viridis*, a new species of *Actinomycetales*. Int. J. Syst. Bacteriol. *35*: 281–284.

Miyadoh, S., S. Amano, H. Tohyama and T. Shomura. 1987. *Actinomadura atramentaria*, a new species of the *Actinomycetales*. Int. J. Syst. Bacteriol. *37*: 342–346.

Miyadoh, S., H. Anzai, S. Amano and T. Shomura. 1989. *Actinomadura malachitica* and *Microtetraspora viridis* are synonyms and should be transferred as *Actinomadura viridis* comb. nov. Int. J. Syst. Bacteriol. *39*: 152–158.

Miyadoh, S., S. Amano, H. Tohyama and T. Shomura. 1990. A taxonomic review of the genus *Microbispora* and a proposal to transfer two species to the genus *Actinomadura* and to combine ten species into *Microbispora rosea*. J. Gen. Microbiol. *136*: 1905–1913.

Moore, W.E.C., E.P. Cato and L.V.H. Moore. 1985. Index of the bacterial and yeast nomenclatural changes published in the *International Journal of Systematic Bacteriology* since the 1980 Approved Lists of Bacterial Names (1 January 1980 to 1 January 1985). Int. J. Syst. Bacteriol. *35*: 382–407.

Nakamura, G. and K. Isono. 1983. A new species of *Actinomadura* producing a polyether antibiotic, cationomycin. J. Antibiot. (Tokyo) *36*: 1468–1472.

Nonomura, H. and Y. Ohara. 1971a. Distribution of actinomycetes in soil. X. New genus and species of monosporic actinomycetes in soil. J. Ferment. Technol. *49*: 895–903.

Nonomura, H. and Y. Ohara. 1971b. Distribution of actinomycetes in soil. VIII. Green-spore group of *Microtetraspora*, its preferential isolation and taxonomic characteristics. J. Ferment. Technol. *49*: 1–7.

Nonomura, H. and Y. Ohara. 1971c. Distribution of actinomycetes in soil. IX. New species of the genus *Microbispora* and *Microtetraspora* and their isolation methods. J. Ferment. Technol. *49*: 887–894.

Nonomura, H. and Y. Ohara. 1971d. Distribution of actinomycetes in soil. XI. Some new species of the genus *Actinomadura* Lechevalier et al. J. Ferment. Technol. *49*: 904–912.

Nonomura, H. and Y. Ohara. 1974. A new species of actinomycetes, *Thermomonospora mesouviformis* sp. nov. J. Ferment. Technol. *53*: 10–13.

Ochi, K., S. Miyadoh and T. Tamura. 1991. Polyacrylamide gel electrophoresis analysis of ribosomal protein AT-L30 as a novel approach to actinomycete taxonomy: application to the genera *Actinomadura* and *Microtetraspora*. Int. J. Syst. Bacteriol. *41*: 234–239.

Pinoy, E. 1912. Isolement et culture d'une nouvelle oospora pathogene. In Thiroux et Pelletier. Mycetome a grains rouges de la paroi thoracique. Bull. Soc. Path. Exot. *5*: 585–589.

Poschner, J., R.M. Kroppenstedt, A. Fischer and E. Stackebrandt. 1985. DNA-DNA reassociation and chemotaxonomic studies on *Actinomadura, Microbispora, Microtetraspora, Micropolyspora* and *Nocardiopsis*. Syst. Appl. Microbiol. *6*: 264–270.

Prauser, H. and R. Falta. 1968. [Phage sensitivity, cell wall composition and taxonomy of actinomyctes]. Z. Allg. Mikrobiol. *8*: 39–46.

Prauser, H. and M. Bergholz. 1974. Taxonomy of actinomycetes and screening for antibiotic substances. Postepy Hig. Med. Dosw. *28*: 441–457.

Preobrazhenskaya, T.P., N.V. Lavrova and N.O. Blinov. 1975a. Taxonomy of *Streptomyces luteofluorescens*. Mikrobiologiya *44*: 524–527.

Preobrazhenskaya, T.P., N.V. Lavrova, R.S. Ukholina and N.P. Nechaeva. 1975b. Isolation of new species of *Actinomadura* on selective media with streptomycin and bruneomycin. Antibiotiki *20*: 404–409.

Pridham, T.G., O.L. Shotwell, F.H. Stodola, L.A. Lindenfelser, R.G. Benedict and R.V. Jackson. 1956. Antibiotics against plant disease. II. Effective agents produced by *Streptomyces cinnamoneus* forma azacoluta f. nov. Phytopathology *46*: 575–581.

Pridham, T.G., P. Anderson, C. Foley, L.A. Lindenfelser, C.W. Hesseltine and R.G. Benedict. 1957. A selection of media for maintenance and taxonomic study of *Streptomyces*. Antibiot. Annu. 947–953.

Pulverer, G. and K.P. Schaal. 1978. Pathogenicity and medical importance of aerobic and anaerobic actinomycetes. *In Nocardia* and *Streptomyces* (edited by Mordarski, Kurylowicz and Jeljaszewicz). Gustav Fischer Verlag, New York, pp. 417–428.

Quintana, E.T., M.E. Trujillo and M. Goodfellow. 2003. *Actinomadura mexicana* sp. nov. and *Actinomadura meyerii* sp. nov., two novel soil sporoactinomycetes. Syst. Appl. Microbiol. *26*: 511–517.

Quintana, E.T., M.E. Trujillo and M. Goodfellow. 2004. *In* Validation of publication of new names and new combinations previously effectively published outside the IJSEM. List no. 96. Int. J. Syst. Evol. Microbiol. *54*: 307–308.

Rainey, F.A., N.L. Ward-Rainey and E. Stackebrandt. 1997. *In* Stackebrandt, E., F.A. Rainey and N.L. Ward-Rainey. Proposal for a new hierarchic classification system, *Actinobacteria* classis nov. Int. J. Syst. Bacteriol. *47*: 479–491.

Rippon, J.W. 1968. Extracellular collagenase produced by *Streptomyces madurae*. Biochim. Biophys. Acta *159*: 147–152.

Schaal, K.P. 1972. Zur mikrobiologisher Diagnostik der Nocardiose. Zentralbl. Bakteriol. Parasitenkd. Infektionskr. Hyg. I. Abt. Orig. *22*: 242–246.

Schaal, K.P. 1984. Laboratory diagnosis of actinomycete diseases. *In* The Biology of the Actinomycetes (edited by Goodfellow, Mordarski and Williams). Academic Press, London, pp. 425–456.

Schäfer, D. 1973. Beitrage zur Klassifizierung und Taxonomic der *Actinoplanaceae*. PhD dissertation, Marburg, Germany.

Serrano, J.A., B.L. Beaman, T.E. Viloria, M.A. Mejia and R. Zamora. 1986. Histological and ultrastructural studies on human actinomycetomas. *In* Biological, Biochemical and Biomedical Aspects of Actinomycetes (edited by Szabó, Biró and Goodfellow). Adakémiai Kiadó, Budapest, pp. 647–662.

Serrano, J.A., B. Beaman, M.A. Mejia, J.E. Viloria and R. Zamora. 1988. The actinomycetoma in Venezuela: a ten year study (1976–1986). Rev. Inst. Med. Trop. Sao Paulo *30*: 297–304.

Shinobu, R. 1962. A new *Streptomyces* species producing fluorescent-yellow soluble pigment. Mem. Osaka Univ. Liv. Arts Educ. Ser. B Nat. Sci. *11*: 115–122.

Shirling, E.B. and D. Gottlieb. 1966. Methods for characterization of *Streptomyces* species. Int. J. Syst. Bacteriol. *16*: 313–340.

Skerman, V.B.D., V. McGowan and P.H.A. Sneath. 1980. Approved Lists of Bacterial Names. Int. J. Syst. Bacteriol. *30*: 225–420.

Stackebrandt, E., B. Wunner-Fussl, V.J. Fowler and K.-H. Schleifer. 1981. Deoxyribonucleic acid homologies and ribosomal ribonucleic acid similarities among sporeforming members of the order *Actinomycetales*. Int. J. Syst. Bacteriol. *31*: 420–431.

Stackebrandt, E. and C.R. Woese. 1981. The evolution of prokaryotes. *In* Molecular and Cellular Aspects of Microbial Evolution (edited by Carlile, Collins and Moseley). University Press, Cambridge, UK., pp. 1–31.

Stackebrandt, E. and K.H. Schleifer. 1984. Molecular systematics of actinomycetes and related organisms. *In* Biological, Biochemical and Biomedical Aspects of *Actinomycetales* (edited by Ortiz-Ortiz, Bojalil and Yakoleff). Academic Press, Orlando, pp. 485–504.

Stackebrandt, E., F.A. Rainey and N.L. Ward-Rainey. 1997. Proposal for a new hierarchic classification system, *Actinobacteria* classis nov. Int. J. Syst. Bacteriol. *47*: 479–491.

Stutzenberger, F. and I. Sterpu. 1978. Effect of municipal refuse metals on cellulase production by *Thermomonospora curvata*. Appl. Environ. Microbiol. *36*: 201–204.

Stutzenberger, F.J. 1971. Cellulase production by *Thermomonospora curvata* isolated from municipal solid waste compost. Appl. Microbiol. *22*: 147–152.

Tamura, T., K. Hatano and K. Suzuki. 2007. Classification of '*Sarraceniospora aurea*' Furihata et al. 1989 as *Actinocorallia aurea* sp. nov. Int. J. Syst. Evol. Microbiol. *57*: 2052–2055.

Terekhova, L.P., O.A. Galatenko and T.P. Preobrazhenskaia. 1982. [New species, *Actinomadura fulvescens* sp. nov. and *Actinomadura turkmeniaca* sp. nov. and their antagonistic properties]. Antibiotiki *27*: 87–92.

Terekhova, L.P., O.A. Galatenko and T.P. Preobrazhenskaya. 1987. *In* Validation of the publication of new names and new combinations previously effectively published outside the IJSB. List no. 23. Int. J. Syst. Bacteriol. *37*: 179–180.

Thiemann, J.E., H. Pagani and G. Beretta. 1968. A new genus of the *Actinomycetales*: *Microtetraspora* gen. nov. J. Gen. Microbiol. *50*: 295–303.

Tohyama, H., S. Miyadoh, M. Ito, T. Shomura, T. Ito, T. Ishikawa and M. Kojima. 1984. A new indole *N*-glycoside antibiotic SF-2140 from an *Actinomadura*. I. Taxonomy and fermentation of producing microorganism. J. Antibiot. (Tokyo) *37*: 1144–1148.

Tomita, K., Y. Hoshino, T. Sasahira and H. Kawaguchi. 1980. BBM-928, a new antitumor antibiotic complex. II. Taxonomic studies on the producing organism. J. Antibiot. (Tokyo) *33*: 1098–1102.

Tomita, K., M. Nishio, K. Saitoh, H. Yamamoto, Y. Hoshino, H. Ohkuma, M. Konishi, T. Miyaki and T. Oki. 1990. Pradimicins A, B and C: new antifungal antibiotics. I. Taxonomy, production, isolation and physico-chemical properties. J. Antibiot. (Tokyo) *43*: 755–762.

Tomita, K., M. Nishio, K. Saitoh, H. Yamamoto, Y. Hoshino, H. Ohkuma and M. Konishi. 1991. *In* Validation of the publication of new names and new combinations previously effectively published outside the IJSB. List no. 23.. Int. J. Syst. Bacteriol. *41*: 179–180.

Trujillo, M.E. 1994. Taxonomic revision of the genus *Actinomadura* and related taxa using rapid methods. PhD thesis, University of Newcastle, Newcastle upon Tyne.

Trujillo, M.E. and M. Goodfellow. 1997a. Polyphasic taxonomic study of clinically significant actinomadurae including the description of *Actinomadura latina* sp. nov. Zentralbl. Bakteriol. *285*: 212–233.

Trujillo, M.E. and M. Goodfellow. 1997b. *In* Validation of the publication of new names and new combinations previously effectively published outside the IJSB. List no. 62. Int. J. Syst. Bacteriol. *47*: 915–916.

Trujillo, M.E. and M. Goodfellow. 2003. Numerical phenetic classification of clinically significant aerobic sporoactinomycetes and related organisms. Antonie van Leeuwenhoek *84*: 39–68.

Venugopal, P.L. and T.L. Venugopal. 1991. *Actinomadura madurae* causing mycetomas in Madras. Indian J. Pathol. Microbiol. *34*: 119–125.

Vera-Cabrera, L., E.Y. Ochoa-Felix, G. Gonzalez, R. Tijerina, S.H. Choi and O. Welsh. 2004. In vitro activities of new quinolones and oxazolidinones against *Actinomadura madurae*. Antimicrob. Agents Chemother. *48*: 1037–1039.

Vincent, H. 1894. Étude sur le parasite du pied le madura. Ann. Inst. Pasteur *8*: 129–151.

Vobis, G. 1985. Spore development in sporangia-forming actinomycetes. *In* Biological, Biochemical and Biomedical Aspects of Actinomycetes (edited by Szabó, Biró and Goodfellow). Académiai Kiadó, Budapest, pp. 443–452.

Vobis, G. and H.-W. Kothe. 1985. Sporogenesis in sporangiate actinomycetes. *In* Frontiers in Applied Microbiology, vol. 1 (edited by Mukerji, Pathak and Singh). Print House, Luknow, India, pp. 25–47.

Vobis, G. and H.-W. Kothe. 1989. Genus *Spirillospora*. *In* Bergey's Manual of Systematic Bacteriology, vol. 4 (edited by Williams, Sharpe and Holt). Williams & Wilkins, Baltimore, pp. 2536–2539.

Waitz, J.A., A.C. Horan, M. Kalyanpur, B.K. Lee, D. Loebenberg, J.A. Marquez, G. Miller and M.G. Patel. 1981. Kijanimicin (Sch 25663), a novel antibiotic produced by *Actinomadura kijaniata* SCC 1256. Fermentation, isolation, characterization and biological properties. J. Antibiot. (Tokyo) *34*: 1101–1106.

Waksman, S.A. and A.T. Henrici. 1948. Family III. *Streptomycetaceae* Waksman and Henrici. *In* Bergey's Manual of Determinative Bacteriology, 6th edn (edited by Breed, Murray and Hitchens). Williams & Wilkins, Baltimore, pp. 929–980.

Wang, Y., Z.S. Zhang and J.S. Ruan. 1996. Phylogenetic analysis reveals new relationships among members of the genera *Microtetraspora* and *Microbispora*. Int. J. Syst. Bacteriol. *46*: 658–663.

Welsh, O., E. Sauceda, J. Gonzalez and J. Ocampo. 1987. Amikacin alone and in combination with trimethoprim-sulfamethoxazole in the treatment of actinomycotic mycetoma. J. Am. Acad. Dermatol. *17*: 443–448.

Williams, S.T., G.P. Sharples and R.M. Bradshaw. 1973. The fine structure of the *Actinomycetales*. *In Actinomycetales*: Characteristics and Practical Importance (edited by Sykes and Skinner). Academic Press, London, pp. 113–130.

Williams, S.T. 1989. Bergey's Manual of Systematic Bacteriology, vol. 4 (edited by Sharpe and Holt). Williams & Wilkins, Baltimore, pp. 2299–2648.

Wink, J., R.M. Kroppenstedt, G. Seibert and E. Stackebrandt. 2003. *Actinomadura namibiensis* sp. nov. Int. J. Syst. Evol. Microbiol. *53*: 721–724.

Wust, J., H. Lanzendorfer, A. von Graevenitz, H.J. Gloor and B. Schmid. 1990. Peritonitis caused by *Actinomadura madurae* in a patient on CAPD. Eur. J. Clin. Microbiol. Infect. Dis. *9*: 700–701.

Yamada, Y., M. Yamashita, Y. Tahara and K. Kondo. 1977. The menaquinone system in the classification of the genus *Actinomadura*. J. Gen. Appl. Microbiol. *23*: 207–219.

Yamaguichi, T. 1965. Comparison of the cell wall composition of morphologically distinct actinomycetes. J. Bacteriol. *89*: 444–453.

Yap, W.H., Z. Zhang and Y. Wang. 1999. Distinct types of rRNA operons exist in the genome of the actinomycete *Thermomonospora chromogena* and evidence for horizontal transfer of an entire rRNA operon. J. Bacteriol. *181*: 5201–5209.

Yu, A.M., S. Zhao and L.Y. Nie. 1993. Mycetomas in northern Yemen: identification of causative organisms and epidemiologic considerations. Am. J. Trop. Med. Hyg. *48*: 812–817.

Zhang, Z., Y. Wang and J. Ruan. 1998. Reclassification of *Thermomonospora* and *Microtetraspora*. Int. J. Syst. Bacteriol. *48*: 411–422.

Zhang, Z., T. Kudo, Y. Nakajima and Y. Wang. 2001. Clarification of the relationship between the members of the family *Thermomonosporaceae* on the basis of 16S rDNA, 16S-23S rRNA internal transcribed spacer and 23S rDNA sequences and chemotaxonomic analyses. Int. J. Syst. Evol. Microbiol. *51*: 373–383.

Zhang, Z.S., Y. Wang and J.S. Ruan. 1998. Reclassification of *Thermomonospora* and *Microtetraspora*. Int. J. Syst. Bacteriol. *48*: 411–422.

Zippel, M. and M. Neigenfind. 1988. Preservation of streptomycetes. J. Gen. Appl. Microbiol. *34*: 7–14.

Order XVI. **Incertae sedis**

Previously classified within the *Pseudonocardiaceae* (Garrity et al., 2005), this placement of *Thermobispora* is not consistent with chemotaxonomic markers or subsequent rRNA gene analyses. The cell-wall diamino acid is *meso*-diaminopimelic acid; whole-cell hydrolysates contain the sugars madurose and galactose; the major menaquinone is MK-9(H_0); and the phospholipid pattern is type PIV. These properties distinguish it from the *Pseudonocardiaceae*. Moreover, in analyses of the 16S rRNA genes, *Thermobispora* appears as a deep lineage within the *Streptosporangiales* but without a clear association to any of the described families (Ludwig et al., 2012). Given the ambiguities, this genus is reclassified in an order *incertae sedis* in this volume.

References

Garrity, G., J.A. Bell and T. Lilburn. 2005. The revised roadmap to the *Manual*. *In* Bergey's Manual of Systematic Bacteriology, 2nd edn, vol. 2A, The *Proteobacteria*, Introductory Essays. Springer, New York, pp. 159–220.

Ludwig, W., J. Euzéby and W. Whitman. 2012. Taxonomic outline of the phylum *Actinobacteria*. *In* Bergey's Manual of Systematic Bacteriology, 2nd edn, vol. 5, The *Actinobacteria* (edited by Goodfellow, Kämpfer, Busse, Trujillo, Suzuki, Ludwig and Whitman). Springer, New York, pp. 29–31.

Genus I. **Thermobispora** Wang, Zhang and Ruan 1996, 937[VP]

Seung Bum Kim

Ther.mo.bi.spo'ra. Gr. adj. *thermos* hot; Gr. adj. *bis* two; Gr. fem. n. *spora* seed; N.L. fem. n. *Thermobispora* high temperature, two-spored organism.

Gram-stain-positive, aerobic, non-acid-fast, **thermophilic** organisms that produce substrate hyphae (0.5–0.8 µm in diameter). **The aerial mycelia branch monopodially and bear longitudinal pairs of spores**, usually arranged alternately on side branches arising from the main hyphae. **The smooth-surfaced spores are oval to spherical and nonmotile**. Fragmentation of mycelia is not observed on agar or in liquid media. Neither sclerotia, sporangia, nor any other special structures are formed. The cell wall contains a major amount of ***meso*-diaminopimelic acid**, and **the whole-cell hydrolysates contain madurose and galactose**. **MK-9(H_0) is the major menaquinone**. **The phospholipid pattern is type PIV**, including phosphatidylethanolamine and a small amount of glucosamine-containing phospholipids. The major fatty acid is 14-methylpentadecanoic acid ($C_{16:0}$ iso).

DNA G+C content (mol%): 71.

Type species: **Thermobispora bispora** (Henssen 1957) Wang, Zhang and Ruan 1996, 937[VP].

Further descriptive information

Thermobispora is classified with an order *incertae sedis* within this volume. The spores of the type and only species, *Thermobispora bispora*, are usually 1.2–2.0 µm in diameter, but in liquid media spores with a diameter of 3 µm have been observed. The aerial mycelium is white, and the substrate mycelium yellow or yellowish brown on ISP 4 agar medium. Soluble pigments are not produced. The predominant menaquinone of *Thermobispora bispora* is MK-9(H_0), with MK-9(H_2) and MK-9(H_4) present in smaller amounts. The strains of *Thermobispora bispora* can grow at 65°C, but not at 35°C or lower temperatures. The 16S rRNA gene sequences of two *Thermobispora bispora* strains (strains ATCC 19993^T and JCM 3082) were determined by Wang et al. (1996).

Enrichment and isolation procedures

Thermobispora strains have been isolated from (decaying) manure samples. Colonies can be obtained using selective media for actinobacteria and incubation at 50°C or higher temperatures.

Maintenance procedures

Thermobispora bispora can be maintained at temperatures between 50 and 65°C on ISP medium 4, Bennett's agar, oatmeal agar or rolled oats mineral medium (DSM medium no. 84). For long-term preservation, storage of the mycelial or spore suspensions in 20% (v/v) glycerol at −20°C, or lyophilization is recommended.

Differentiation from closely related taxa

The production of aerial spores in longitudinal pairs from monopodially branching aerial mycelia distinguishes *Thermobispora* from the related genera of the family *Pseudonocardiaceae*. The presence of menaquinone MK-9(H_0) in *Thermobispora* is also a useful marker to distinguish the genus from other genera of the family. Unlike most genera of the family *Pseudonocardiaceae* that have type II or III phospholipids, *Thermobispora* has a type IV phospholipid profile. In addition, strains of *Thermobispora* are obligate thermophiles, growing at temperatures between 50 and 65°C, which is comparable to the growth range (20–62.5°C) of another thermophilic genus within the family, *Thermocrispum*. *Thermobispora* can also be differentiated from related genera using the 16S rRNA gene restriction fragment patterns, as suggested by Cook and Meyers (2003).

Taxonomic comments

Thermobispora bispora was originally described as "*Thermopolyspora bispora*" Henssen 1957, and later transferred to *Microbispora* as "*Microbispora bispora*" Lechevalier 1965. However, the strains classified as "*Microbispora bispora*" differed in many ways from the other validly named species of *Microbispora*, in the levels of DNA homology, growth temperature, predominant menaquinones, phospholipid type, major fatty acid profiles and the electrophoretic pattern of ribosomal proteins (Miyadoh et al., 1990; Ochi et al., 1993). Finally, Wang et al. (1996) sequenced the 16S rRNA gene of strains "*Microbispora bispora*" JCM 3082 and "*Microbispora bispora*" ATCC 19993, and combined these results with chemotaxonomic and DNA–DNA hybridization data to classify the two strains in the genus *Thermobispora*. In addition, Wang and colleagues (1996) highlighted the thermophilic lifestyle of *Thermobispora bispora* to support the reclassification of these strains.

Previously, *Thermobispora* was assigned to the family *Pseudonocardiaceae*, with *Pseudonocardia* and *Actinopolyspora* as neighboring genera (Garrity et al., 2005). However, this assignment was not supported by chemotaxonomic markers and analyses of the 16S rRNA genes performed for this volume suggested that *Thermobispora* appeared as a deep lineage within the *Streptosporangiales*, but without a clear association to any of the described families (Ludwig et al., 2012). Because this assignment was not strongly supported by chemotaxonomic markers, the genus is placed in an order *incertae sedis*.

Miscellaneous comments

Thermobispora bispora ATCC 19993^T was reported to have four copies of rRNA operons, containing two distinct types of 16S rRNA genes, both transcriptionally active (Wang et al., 1997). Considerable numbers of nucleotide differences between the two gene sequences exist, but the invariable or rarely invariable regions of both genes remain unchanged.

List of species of the genus *Thermobispora*

1. **Thermobispora bispora** (Henssen 1957) Wang, Zhang and Ruan 1996, 937^{VP} [*Thermopolyspora bispora* Henssen 1957, 395; *Microbispora bispora* (Henssen 1957) Lechevalier 1965, 141]

bi.spo′ra. Gr. adj. *bis* two; Gr. fem. n. *spora* seed; N.L. fem. adj. *bispora* two spores.

Substrate and aerial mycelia are formed, and longitudinal pairs of spores are borne on the branches of the aerial mycelium. The diameter of the spores is usually 1.2–2.0 µm, but occasionally spores with a diameter of 3 µm are observed in liquid culture. The aerial mycelium is white and the substrate mycelium is yellow or yellowish brown on ISP medium 4 and IFO328 medium. Soluble pigments have not been observed. Starch is not hydrolyzed. Negative for iodinin production and nitrate reduction. Inositol and rhamnose are utilized for growth, but not arabinose or glycerol. The type strain grows between 50 and 65°C but not at 35°C.

Habitat: decaying manure.

DNA G+C content (mol%): 71 (T_m).

Type strain: ATCC 19993, CBS 139.67, NBRC 14880.

Sequence accession no. (16S rRNA gene): U58523 (ATCC 19993).

References

Cook, A.E. and P.R. Meyers. 2003. Rapid identification of filamentous actinomycetes to the genus level using genus-specific 16S rRNA gene restriction fragment patterns. Int. J. Syst. Evol. Microbiol. *53*: 1907–1915.

Garrity, G., J.A. Bell and T. Lilburn. 2005. The revised roadmap to the *Manual*. *In* Bergey's Manual of Systematic Bacteriology, 2nd edn, vol. 2A, The *Proteobacteria*, Introductory Essays. Springer, New York, pp. 159–220.

Henssen, A. 1957. Beiträge zur Morphologie und Systematik der thermophilen Actinomyceten. Arch. Mikrobiol. *26*: 373–414.

Lechevalier, H.A. 1965. Priority of the generic name *Microbispora* over *Waksmania* and *Thermopolyspora*. Int. Bull. Bacteriol. Nomencl. Taxon. *15*: 139–142.

Ludwig, W., J. Euzéby and W. Whitman. 2012. Taxonomic outline of the phylum *Actinobacteria*. *In* Bergey's Manual of Systematic Bacteriology, 2nd edn, vol. 5, The *Actinobacteria* (edited by Goodfellow, Kämpfer, Busse, Trujillo, Suzuki, Ludwig and Whitman). Springer, New York, pp. 29–31.

Miyadoh, S., S. Amano, H. Tohyama and T. Shomura. 1990. A taxonomic review of the genus *Microbispora* and a proposal to transfer two species to the genus *Actinomadura* and to combine ten species into *Microbispora rosea*. J. Gen. Microbiol. *136*: 1905–1913.

Ochi, K., K. Haraguchi and S. Miyadoh. 1993. A taxonomic review of the genus *Microbispora* by analysis of ribosomal protein AT-L30. Int. J. Syst. Bacteriol. *43*: 58–62.

Wang, Y., Z.S. Zhang and J.S. Ruan. 1996. A proposal to transfer *Microbispora bispora* (Lechevalier 1965) to a new genus, *Thermobispora* gen. nov., as *Thermobispora bispora* comb. nov. Int. J. Syst. Bacteriol. *46*: 933–938.

Wang, Y., Z. Zhang and N. Ramanan. 1997. The actinomycete Thermobispora bispora contains two distinct types of transcriptionally active 16S rRNA genes. J. Bacteriol. *179*: 3270–3276.

Class II. **Acidimicrobiia** class. nov.

Paul R. Norris

A.ci.di.mi.cro.bi′i.a. N.L. neut. n. *Acidimicrobium* type genus of the type order; suff. *-ia* ending to denote a class; N.L. pl. neut. n. *Acidimicrobiia* the *Acidimicrobiales* class.

The class *Acidimicrobiia* was delineated on the basis of 16S rRNA gene sequence analyses and taxon-specific 16S rRNA signature nucleotides as for the subclass *Acidimicrobidae* Stackebrandt et al. 1997. The pattern of 16S rRNA signature nucleotides was revised as 242:284 (U–G), 291:309 (U–A), 316:337 (C–G), 819 (U), 952:1229 (C–G), and 1115:1185 (U–G) (Zhi et al., 2009). The class contains a single order, the type order *Acidimicrobiales*. The nomenclatural type is the genus *Acidimicrobium* Clark and Norris 1996, 1189^{VP} (Effective publication: Clark and Norris 1996, 790.)

References

Clark, D.A. and P.R. Norris. 1996. *Acidimicrobium ferrooxidans* gen. nov., sp. nov.: mixed culture ferrous iron oxidation with *Sulfobacillus* species. Microbiology *142*: 785–790.

Clark, D.A. and P.R. Norris. 1996. *In* Validation of the publication of new names and new combinations previously effectively published outside the IJSB. List no. 59. Int. J. Syst. Bacteriol. *46*: 1189–1190.

Stackebrandt, E., F.A. Rainey and N.L. Ward-Rainey. 1997. Proposal for a new hierarchic classification system, *Actinobacteria* classis nov. Int. J. Syst. Bacteriol. *47*: 479–491.

Zhi, X.-Y., W.-J. Li and E. Stackebrandt. 2009. An update of the structure and 16S rRNA gene sequence-based definition of higher ranks of the class *Actinobacteria*, with the proposal of two new suborders and four new families and emended descriptions of the existing higher taxa. Int. J. Syst. Evol. Microbiol. *59*: 589–608.

Order I. **Acidimicrobiales** Stackebrandt, Rainey and Ward-Rainey 1997, 483[VP]

Paul R. Norris

A.ci.di.mi.cro.bi'a.les. N.L. neut. n. *Acidimicrobium* type genus of the order; suff. *-ales* ending to denote an order; N.L. fem. pl. n. *Acidimicrobiales* the *Acidimicrobium* order.

In 2009, the emended description of the order *Acidimicrobiales* (Zhi et al., 2009) noted a single family, the *Acidimicrobiaceae*, with the pattern of 16S rRNA signature nucleotides as given for the subclass *Acidimicrobidae*, now the class *Acidimicrobiia*. Also in 2009, a second family of the order was proposed, the *Iamiaceae* (Kurahashi et al., 2009). Nucleotides U–A at positions 952:1229 in the type species of *Iamia*, the single genus in the family *Iamiaceae*, expand the pattern of 16S rRNA signature nucleotides given for the class *Acidimicrobiia* and the order *Acidimicrobiales*, which also includes C–G at positions 952:1229 (Zhi et al., 2009).

A large number of uncultured bacteria of the class *Actinobacteria* have been designated potential members of the *Acidimicrobiia* and *Acidimicrobiales* through analysis of their cloned 16S rRNA genes. Three genera have been described (*Acidimicrobium*, *Ferrimicrobium*, and *Ferrithrix*), which contain species from acidic mine waters or geothermal sites. The rRNA gene sequences of these species are relatively closely related among more divergent sequences found in some of the uncultured *Acidimicrobiia*. Another acidophilic, iron-oxidizing species, related to *Acidimicrobium ferrooxidans* has been isolated and will require placement in a new genus ("*Acidithiomicrobium*" sp.). These genera share their acidophilic phenotype (optimum growth at about pH 2) and a capacity for ferrous iron oxidation. In contrast, the type species of the genus *Iamia* (*Iamia majanohamensis*) is a neutrophile (optimum growth at pH 7; pH range 6–9). It was isolated from the epidermis of sea cucumber and is unlikely to oxidize ferrous iron. The 16S rRNA gene sequence of *Iamia majanohamensis* indicates a phylogenetic placement between the acidophiles of the *Acidimicrobiales* and representatives of the relatively closely related orders of the *Actinobacteria* (sequences from the type species of the type genera of the type families of the *Rubrobacteriales*, *Bifidobacteriales*, and *Coriobacteriales* were used in phylogenetic tree construction; Figure 412). Further consideration of the content, limits, and possible subdivision of the *Acidimicrobiales*, and any concurrence of these aspects with phenotype, will be facilitated by isolation and characterization of some of the large number of uncultured "*Acidimicrobiia*" that are phylogenetically peripherally related to the classified acidophiles of the order.

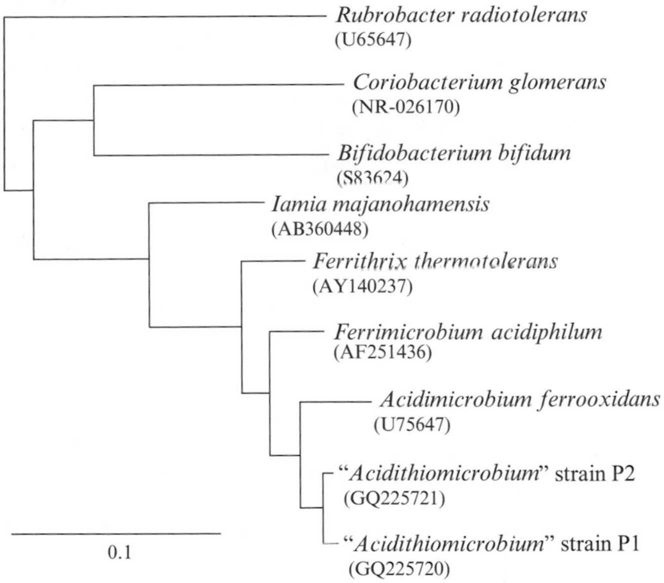

FIGURE 412. Phylogenetic relationships of members of the *Acidimicrobiales* and the most closely related orders of the *Actinobacteria*. Maximum-likelihood evolutionary tree of 16S rRNA gene sequences (GenBank accession nos in parentheses) from type species of the type genera of the orders *Rubrobacteriales* (the outgroup), *Bifidobacteriales* and *Coriobacteriales*, from the type species of the type genus of the family *Iamia*, and from the cultured, ferrous iron-oxidizing actinobacteria of the *Acidimicrobiales*. Bar = 0.1 substitutions per site. Bootstrap values from 100 replicates are shown.

References

Kurahashi, M., Y. Fukunaga, Y. Sakiyama, S. Harayama and A. Yokota. 2009. *Iamia majanohamensis* gen. nov., sp. nov., an actinobacterium isolated from sea cucumber *Holothuria edulis*, and proposal of *Iamiaceae* fam. nov. Int. J. Syst. Evol. Microbiol. *59*: 869–873.

Stackebrandt, E., F.A. Rainey and N.L. Ward-Rainey. 1997. Proposal for a new hierarchic classification system, *Actinobacteria* classis nov. Int. J. Syst. Bacteriol. *47*: 479–491.

Zhi, X.-Y., W.-J. Li and E. Stackebrandt. 2009. An update of the structure and 16S rRNA gene sequence-based definition of higher ranks of the class *Actinobacteria*, with the proposal of two new suborders and four new families and emended descriptions of the existing higher taxa. Int. J. Syst. Evol. Microbiol. *59*: 589–608.

Family I. **Acidimicrobiaceae** Stackebrandt, Rainey and Ward-Rainey 1997, 483[VP]

Paul R. Norris

A.ci.di.mi.cro.bi.a.ce'a.e. N.L. neut. n. *Acidimicrobium* type genus of the family; suff. *-aceae* ending to denote a family; N.L. fem. pl. n. *Acidimicrobiaceae* the *Acidimicrobium* family.

Small rods, sometimes occur as filaments. Acidophilic, mesophilic, or moderately thermophilic, with the capacity for ferrous iron oxidation. All species are capable of heterotrophic growth. Some species are also autotrophic. The genera of this family (*Acidimicrobium, Ferrimicrobium,* and *Ferrithrix*) have single species representatives that, along with the unclassified "*Acidithiomicrobium* sp.", have the pattern of 16S rRNA signatures proposed for the class *Acidimicrobiia.*

Type genus: **Acidimicrobium** Clark and Norris 1996, 1189[VP] (Effective publication: Clark and Norris 1996, 790.).

Genus I. **Acidimicrobium** Clark and Norris 1996, 1189[VP] (Effective publication: Clark and Norris 1996, 790.)

A.ci.di.mi.cro'bi.um. N.L. neut. n. *acidum* (from L. adj. *acidus* sour) an acid; Gr. adj. *mikros* small; Gr. masc. n. *bios* life; N.L. neut. n. *microbium*, a microbe; N.L. neut. n. *Acidimicrobium* referring to a small bacterium from acidic environments.

Thermotolerant or **moderately thermophilic, acidophilic bacterium. Small, rod-shaped cells**, which may be in filaments of variable length. **Optimum growth occurs at 45–50°C** and approximately pH 2. Heterotrophic growth occurs on yeast extract, during which cells are motile. **Autotrophic growth occurs on ferrous iron. Most rapid growth is observed aerobically with ferrous iron and yeast extract.** Found in warm, acidic, mineral sulfide-rich environments.

DNA G+C content (mol%): 67–68 (T_m and HPLC).

Type species: **Acidimicrobium ferrooxidans** Clark and Norris 1996, 1189[VP] (Effective publication: Clark and Norris 1996, 790.).

Further descriptive information

The phylogenetic relationships of *Acidimicrobium* and related species are illustrated in Figure 412. The rod-shaped cells (0.35–0.4 × 1–1.5 µm) are sometimes found in long filaments, depending on the strain and growth conditions (Figure 413). The major fatty acids of total cell hydrolysates are $C_{16:0}$ iso (83%) and $C_{17:0}$ anteiso (8%) and the major menaquinone is MK-9(H_8) (Kurahashi et al., 2009).

Autotrophic growth occurs with ferrous iron as substrate in the presence of low concentrations of reduced sulfur compounds such as tetrathionate, which might provide a source of reduced sulfur for biosynthesis. Heterotrophic growth occurs with yeast extract (0.02%, w/v). Most rapid growth occurs in the presence of ferrous iron and yeast extract. Pyrite is rapidly oxidized during growth in the presence of yeast extract, but growth on elemental sulfur has not been described. Anaerobically, reduction of ferric iron occurs in the presence of organic substrates (yeast extract and glycerol).

Found in warm, acidic, iron-, sulfur- or mineral-sulfide-rich environments, including industrial commercial copper leach dumps, coal mine drainage, and natural geothermal springs.

Enrichment and isolation procedures

The type strain of the type species (*Acidimicrobium ferrooxidans* strain ICP[T]) was isolated from enrichment cultures grown in a mineral salts medium which contained (g/l): $MgSO_4 \cdot 7H_2O$ (0.4), $(NH_4)_2SO_4$ (0.2), K_2HPO_4 (0.1), $FeSO_4 \cdot 7H_2O$ (13.9), and yeast extract (0.2). The phosphate concentration was reduced (K_2HPO_4, 0.01 g/l) in medium solidified with phytagel (0.4%, w/v) for single colony isolation.

Maintenance procedures

Cultures can be maintained for several months at room temperature without shaking in medium (initially pH 2) containing pyrite (50–100 µm diameter particle size) and yeast extract (0.02%, w/v). The type strain can also be maintained vacuum dried and at −70°C in the presence of 7% (v/v) DMSO (Hallberg and Johnson, 2007).

Differentiation of the genus *Acidimicrobium* from other genera

The DNA G+C content and 16S rRNA gene sequence differentiate bacteria of the *Acidimicrobium* and related genera of iron-oxidizing actinobacteria (Table 305, Figure 412), whereas most other features are common to more than one genus. The single species representatives of the genera *Acidimicrobium, Ferrimicrobium,* and *Ferrithrix* (and the unclassified "*Acidithiomicrobium*" species) have the pattern of 16S rRNA signatures proposed for the *Acidimicrobiia, Acidimicrobiales,* and *Acidimicrobiaceae,* with nucleotides at positions 242:284 (U–G), 291:309 (U–A), 316:337 (C–G), 819 (U), 952:1229 (C–G), and 1115:1185 (U–G).

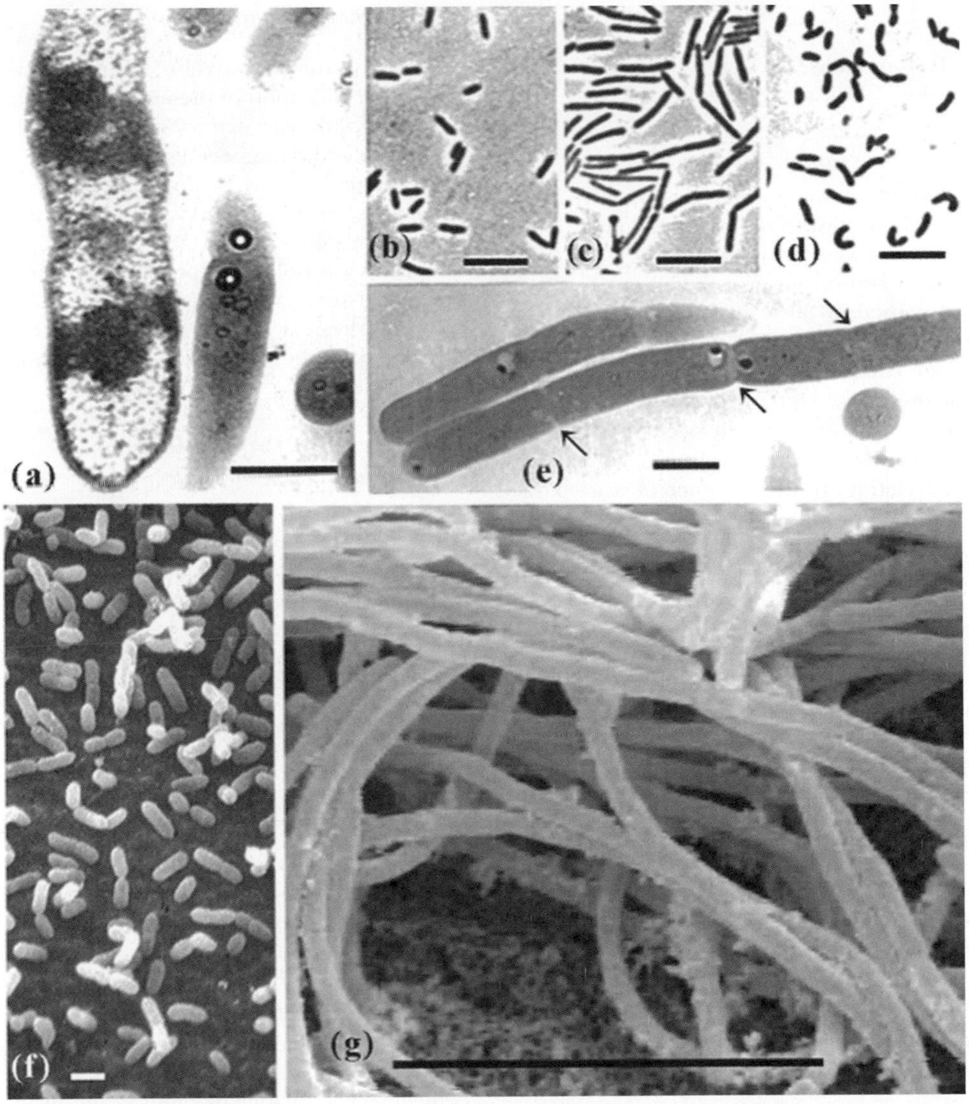

FIGURE 413. Morphology of ferrous-iron-oxidizing actinobacteria. (a) Electron micrograph of thin sections of a mixed culture of iron-oxidizing moderate thermophiles from which the type strain of *Acidimicrobium ferrooxidans* (strain ICP^T) was isolated, showing small *Acidimicrobium*-like bacteria with a larger *Sulfobacillus* sp. (bar = 0.5 µm). Light microscopy of the type strain of *Acidimicrobium ferrooxidans* grown (b) autotrophically on ferrous iron, (c) on ferrous iron and yeast extract, and (d) heterotrophically on yeast extract (bars = 5 µm). (e) Thin sections of *Acidimicrobium ferrooxidans* strain TH3 grown on ferrous iron and yeast extract, with arrows indicating the cross walls in long filaments of cells (bar = 0.5 µm). Scanning electron micrographs of the type strains of (f) *Ferrimicrobium acidiphilum* (bar = 1 µm) and (g) *Ferrithrix thermotolerans* (bar = 5 µm), each grown on ferrous iron and yeast extract. [(a–e) reproduced with permission from Clark and Norris, 1996. Microbiology *142:* 785–790; (f–g) kindly provided by D.B. Johnson, University of Wales, Bangor, UK.]

TABLE 305. Differential characteristics of *Acidimicrobium ferrooxidans* and related, acidophilic, ferrous-iron-oxidizing bacteria[a]

Characteristic	*Acidimicrobium ferrooxidans*	*Ferrimicrobium acidiphilum*	*Ferrithrix thermotolerans*	"*Acidithiomicrobium*" strains P1 and P2
DNA G+C content (mol%)	68	55	50	52–55
16S rRNA identity to *Acidimicrobium ferrooxidans* (%)	100	93	91	95
Motility	+	+	–	+
Growth temperature (°C)[b]	48/57	35/37	43/50	50/59
Sulfur oxidation	–	–	–	+
Autotrophy	+	–	–	+

[a]+, Positive; –, negative.

[b]Optimum growth temperature/maximum growth temperature.

List of species of the genus *Acidimicrobium*

1. **Acidimicrobium ferrooxidans** Clark and Norris 1996, 1189[VP] (Effective publication: Clark and Norris 1996, 790.)

fer.ro.o'xi.dans. L. n. *ferrum* iron; N.L. v. *oxido* to oxidize; N.L. part. adj. *ferrooxidans* iron-oxidizing.

Cells are small and rod-shaped and may be in filaments of variable length. In contrast to the type strain, ICP[T], strain TH3 generally grows in filaments (Figure 413).

The optimum temperature for growth is about 48°C, with good activity maintained at 30°C and the maximum temperature for growth is about 57°C. Autotrophic growth on ferrous iron is weak compared to that of most autotrophic iron-oxidizing acidophiles. Heterotrophic growth occurs with yeast extract. Most rapid growth occurs in the presence of ferrous iron and yeast extract, where iron oxidation rapidly goes to completion. In the presence of yeast extract, pyrite is extensively oxidized during growth. In contrast, only weak acidification of the medium has been observed during growth with yeast extract and sulfur, and a capacity for sulfur oxidation has not been established.

Source: the type strain, ICP[T], was isolated from a pyrite enrichment culture established with a sample from an Icelandic geothermal site. Previously, strain TH3 of the same species was isolated from a copper mine leach dump in the USA (Norris and Barr, 1985). The TH3 designation was used previously for the original isolate of the species which came from the same source, but was not retained in culture (Brierley, 1978).

DNA G+C content (mol%): 67–68 (T_m and HPLC).

Type strain: ICP, DSM 10331.

Sequence accession no. (16S rRNA gene): U75647.

Genus II. **Ferrimicrobium** Johnson, Bacelar-Nicolau, Okibe, Thomas and Hallberg 2009, 1087[VP]

Paul R. Norris

Fer.ri.mi.cro'bi.um. L. n. *ferrum* iron; N.L. neut. n. *microbium* microbe; N.L. neut. n. *Ferrimicrobium* iron microbe, referring to its capacity for ferrous iron oxidation.

Motile rods, Gram-stain-negative. **Acidophilic**, mesophilic, and obligately heterotrophic. Capable of **oxidation of ferrous iron** and pyrite, and reduction of ferric iron. Grows with yeast extract or a limited range of organic compounds.

DNA G+C content (mol%): 55 (T_m).

Type species: **Ferrimicrobium acidiphilum** Johnson, Bacelar-Nicolau, Okibe, Thomas and Hallberg 2009, 1087[VP].

Further descriptive information

All information, including 16S rRNA phylogeny (Figure 412), pertains to the type strain of *Ferrimicrobium acidiphilum*, T23[T] (Johnson et al., 2009).

Colonies of the type species (strain T23[T]) display a "fried egg"-like morphology on solid medium (Johnson et al., 2009). Cells are rod-shaped and motile, with chains of three to five cells common during exponential growth. *meso*-Diaminopimelic acid is present in the peptidoglycan, which is of type A1γ. The major fatty acids are $C_{16:0}$ iso and $C_{14:0}$ iso (acetone extraction of whole-cell lysate). The dominant menaquinone is MK-8(H_{10}).

Optimum growth is observed at 35°C (no growth at 45°C) and pH 2.0 (minimum pH 1.4). Heterotrophic growth occurs on yeast extract and a limited range of defined organic substrates, including glycerol, citric acid, and glutamic acid.

16S rRNA gene sequences almost identical to that of *Ferrimicrobium acidiphilum* have been cloned from DNA extracted from various acidic mine waters or soils worldwide, including from the UK, the USA, Spain, and China.

Enrichment and isolation procedures

Ferrimicrobium acidiphilum T23[T] was isolated from pyrite enrichment cultures inoculated with mine water from the abandoned Cae Coch sulfur mine, North Wales, UK, by plating on a ferrous iron-containing overlay solid medium (Johnson and Hallberg, 2007).

Maintenance procedures

Ferrimicrobium acidiphilum T23[T] is routinely subcultured in liquid media containing 10 mM ferrous sulfate and 0.02% (w/v) yeast extract at pH 2.0 and 30°C. Cultures can be maintained for several months at room temperature without shaking in medium (initially pH 2) containing pyrite (50–100 μm particle size diameter) and yeast extract (0.02%, w/v). The type strain can also be maintained at −70°C in the presence of 7% (v/v) DMSO (Hallberg and Johnson, 2007).

Differentiation of the genus *Ferrimicrobium* from other genera

Among the single species-containing genera of iron-oxidizing actinobacteria, *Ferrimicrobium acidiphilum* is the only mesophile and can also be distinguished by its DNA composition and 16S rRNA gene sequence (Table 305).

List of species of the genus *Ferrimicrobium*

1. **Ferrimicrobium acidiphilum** Johnson, Bacelar-Nicolau, Okibe, Thomas and Hallberg 2009, 1087[VP]

a.ci.di'phi.lum. N.L. neut. n. *acidum* (from L. adj. *acidus* sour) an acid; N.L. adj. *philus -a -um* (from Gr. adj. *philos -ê -on*) friend, loving; N.L. neut. adj. *acidiphilum* acid-loving.

Cells are rod-shaped, 1–3 μm in length and 0.5 μm in diameter, and motile. Gram-stain-negative. Forms gelatinous colonies of approximately 1–3 mm diameter and with ferric iron deposits in the center on ferrous iron overlay plates. Mesophilic, optimum growth is at 35°C (maximum 37°C), and acidophilic, optimum growth is at pH 2 (minimum pH 1.4). Obligately heterotrophic with growth on yeast extract, glycerol, citric acid, and glutamic acid. Capable of oxidation of ferrous iron and pyrite, and, in the presence of glycerol, reduction of ferric iron. *meso*-Diaminopimelic acid is present in the peptidoglycan, which is of type A1γ. The major fatty acids are $C_{16:0}$ iso (64%), $C_{14:0}$ iso (11%), $C_{17:1}$ 6*c* (5.97%), $C_{15:0}$ anteiso (5.14%), $C_{15:0}$ iso (5.03%), $C_{17:0}$ anteiso (3.18%), $C_{16:1}$ iso H (2.49%), and $C_{14:0}$ iso 3-OH (1.20%). The dominant menaquinone is MK-8(H_{10}), with minor amounts of MK-8 and MK-8(H_{10}) containing one or two methyl groups (Johnson et al., 2009).

Source: found in acidic iron-rich environments associated with the oxidation of sulfide minerals, such as acid mine drainage waters.

DNA G+C content (mol%): 55 (T_m).

Type strain: T23, DSM 19497, ATCC BAA-1647.

Sequence accession no. (16S rRNA gene): AF251436.

Genus III. **Ferrithrix** Johnson, Bacelar-Nicolau, Okibe, Thomas and Hallberg 2009, 1088[VP]

Paul R. Norris

Fer.ri.thr'ix. L. n. *ferrum* iron; Gr. fem. n. *thrix* hair, thread; N.L. fem. n. *Ferrithrix* iron thread, referring to filamentous nature and capacity for ferrous iron oxidation.

Grows as filaments (occasionally as single rods). Gram-stain-negative. **Acidophilic, thermotolerant**, or moderately thermophilic, with optimum growth at 43°C. **Obligately heterotrophic**. Grows on yeast extract, glycerol, or ethanol. Capable of **ferrous iron oxidation** and ferric iron reduction. Unknown growth factor(s) from yeast extract is (are) required for growth.

DNA G+C content (mol%): 50 (T_m).

Type species: **Ferrithrix thermotolerans** Johnson, Bacelar-Nicolau, Okibe, Thomas and Hallberg 2009, 1088[VP].

Further descriptive information

All information (see list of species) relates to the type strain (Y005[T]) of the single species (Johnson et al., 2009). Its phylogenetic relationship to other acidophilic, iron-oxidizing actinobacteria is illustrated in Figure 412.

Ferrithrix thermotolerans has not yet been isolated from outside of geothermal regions of Yellowstone National Park, USA, but closely related 16S rRNA gene sequences have been cloned from acid mine drainage in the UK and Spain.

Enrichment and isolation procedures

Ferrithrix thermotolerans was isolated from overlay solid medium containing ferrous iron that was streaked with a sample of a ferrous iron/yeast extract enrichment culture established with an acidic sample from the Beryl Spring/Gibbon River area of Yellowstone National Park (Johnson et al., 2003).

Maintenance procedures

Cultures can be maintained for at least 1 year at room temperature in medium (initially pH 2) containing pyrite (50–100 μm diameter particle size) and yeast extract (0.02%, w/v). The type strain can also be maintained at −70°C in the presence of 7% (v/v) DMSO (Hallberg and Johnson, 2007).

Differentiation of the genus *Ferrithrix* from other genera

Among the single species-containing genera of iron-oxidizing actinobacteria, *Ferrithrix thermotolerans* can be distinguished by its optimum growth temperature, which is between that of the mesophilic *Ferrimicrobium acidiphilum* and the moderately thermophilic *Acidimicrobium ferrooxidans*, and by its DNA composition and 16S rRNA gene sequence (Table 305).

List of species of the genus *Ferrithrix*

1. **Ferrithrix thermotolerans** Johnson, Bacelar-Nicolau, Okibe, Thomas and Hallberg 2009, 1088[VP]

ther.mo.to'le.rans. Gr. n. *thermê* heat; L. part. adj. *tolerans* tolerating; N.L. part. adj. *thermotolerans* heat-tolerating, able to tolerate high temperatures.

Grows as filaments (Figure 413), which form visible flocs of approximately 1–3 mm diameter, and occasionally as nonmotile, rod-shaped cells of 3–4 × 0.5 μm. Gram-stain-negative. Colonies are small (1–2 mm in diameter), rhizoidal, and iron-encrusted on ferrous iron, ferrous iron/tetrathionate solid overlay media. The peptidoglycan type is A1γ. The major fatty acid is $C_{16:0}$ iso, with other fatty acids present in minor amounts. Optimal growth is at 43°C (maximum 50°C). Acidophilic, with optimal growth at pH 1.8 (minimum pH 1.6). Obligately heterotrophic, capable of growth with yeast extract, glycerol, and ethanol. An unknown growth factor(s) supplied by yeast extract is (are) required for growth. Also able to oxidize ferrous iron and reduce ferric iron.

Source: found in geothermal springs.

DNA G+C content (mol%): 50 (T_m).

Type strain: Y005, DSM 19514, ATCC BAA-1645.

Sequence accession no. (16S rRNA gene): AY140237.

Other bacteria

An *Acidimicrobium*-like bacterium, which was found in a pyrite-enrichment mixed culture with *Acidimicrobium ferrooxidans*, was initially referred to as "*Acidimicrobium*" species 2 (Cleaver et al., 2007). This bacterium was subsequently obtained in pure culture, but it has not yet been officially named. A closely related bacterium was isolated from an acidic, geothermal soil in Greece. These moderately thermophilic acidophiles have been described as "*Acidithiomicrobium*" strains P1 and P2, respectively (Norris et al., 2011; 16S rRNA gene sequence accession nos GQ225720 and GQ225721, respectively). They are noted here as close relatives of *Acidimicrobium ferrooxidans* and as species which increase the metabolic diversity of this family of actinobacteria. They grow strongly autotrophically on ferrous iron or pyrite in contrast to *Acidimicrobium ferrooxidans* and grow strongly autotrophically with elemental sulfur as the sole substrate (Davis-Belmar and Norris, 2009). They grow at slightly higher temperatures than *Acidimicrobium ferrooxidans* (Table 305). These features and their phylogenetic relationship to the other iron-oxidizing actinobacteria are noted (Table 305, Figure 412).

References

Brierley, J.A. 1978. Thermophilic iron-oxidizing bacteria found in copper leaching dumps. Appl. Environ. Microbiol. *36*: 523–525.

Clark, D.A. and P.R. Norris. 1996. *Acidimicrobium ferrooxidans* gen. nov., sp. nov.: mixed culture ferrous iron oxidation with *Sulfobacillus* species. Microbiology *142*: 785–790.

Clark, D.A. and P.R. Norris. 1996. *In* Validation of the publication of new names and new combinations previously effectively published outside the IJSB. List no. 59. Int. J. Syst. Bacteriol. *46*: 1189–1190.

Cleaver, A.A., N.P. Burton and P.R. Norris. 2007. A novel *Acidimicrobium* species in continuous culture of moderately thermophilic, mineral sulfide-oxidizing acidophiles. Appl. Environ. Microbiol. *73*: 4294–4299.

Davis-Belmar, C.S. and P.R. Norris. 2009. Ferrous iron and pyrite oxidation by "*Acidithiomicrobium*" species. Adv. Mat. Res. *71–73*: 271–274.

Hallberg, K.B. and D.B. Johnson. 2007. Isolation, enumeration, growth and preservation of acidophilic prokaryotes, pp. 1155–1165. *In* Manual of Environmental Microbiology, 3rd edn (edited by Hurst, Crawford, Garland, Lipson, Mills and Stetzenbach), American Society for Microbiology, Washington, D.C.

Johnson, D.B., N. Okibe and F.F. Roberto. 2003. Novel thermo-acidophilic bacteria isolated from geothermal sites in Yellowstone National Park: physiological and phylogenetic characteristics. Arch. Microbiol. *180*: 60–68.

Johnson, D.B. and K.B. Hallberg. 2007. Techniques for detecting and identifying acidophilic mineral-oxidising microorganisms. *In* Biomining (edited by Rawlings and Johnson). Springer, Heidelberg, pp. 237–261.

Johnson, D.B., P. Bacelar-Nicolau, N. Okibe, A. Thomas and K.B. Hallberg. 2009. *Ferrimicrobium acidiphilum* gen. nov., sp. nov. and *Ferrithrix thermotolerans* gen. nov., sp. nov.: heterotrophic, iron-oxidizing, extremely acidophilic actinobacteria. Int. J. Syst. Evol. Microbiol. *59*: 1082–1089.

Kurahashi, M., Y. Fukunaga, Y. Sakiyama, S. Harayama and A. Yokota. 2009. *Iamia majanohamensis* gen. nov., sp. nov., an actinobacterium isolated from sea cucumber *Holothuria edulis*, and proposal of *Iamiaceae* fam. nov. Int. J. Syst. Evol. Microbiol. *59*: 869–873.

Norris, P.R. and D.W. Barr. 1985. Growth and iron oxidation by acidophilic moderate thermophiles. FEMS Microbiol. Lett. *28*: 221–224.

Norris, P.R., C.S. Davis-Belmar, C.F. Brown and L.A. Calvo-Bado. 2011. Autotrophic, sulfur-oxidizing actinobacteria in acidic environments. Extremophiles *15*: 155–163.

Stackebrandt, E., F.A. Rainey and N.L. Ward-Rainey. 1997. Proposal for a new hierarchic classification system, *Actinobacteria* classis nov. Int. J. Syst. Bacteriol. *47*: 479–491.

Family II. **Iamiaceae** Kurahashi, Fukunaga, Sakiyama, Harayama and Yokota 2009, 872[VP]

Paul R. Norris

I.a.mi.a.ce′a.e. N.L. fem. n. *Iamia* type genus of the family; suff. *-aceae* ending to denote a family; N.L. fem. pl. n. *Iamiaceae* the *Iamia* family.

The family *Iamiaceae* was created for the type species *Iamia majanohamensis* in recognition of the evolutionary distance of its 16S rRNA gene sequence from that of *Acidimicrobium ferrooxidans*, the type species of the type genus of the *Acidimicrobiaceae*.

The pattern of 16S rRNA gene sequence signature nucleotides for the family consists of 408:434 (G–C), 722:733 (G–G), 1118:1155 (U–A), 443:491 (U–A), 1165:1171 (G–C), and 1263:1272 (A–U).

Type genus: **Iamia** Kurahashi, Fukunaga, Sakiyama, Harayama and Yokota 2009, 871[VP].

Genus I. **Iamia** Kurahashi, Fukunaga, Sakiyama, Harayama and Yokota 2009, 871[VP]

Paul R. Norris

I.a'mi.a. N.L. fem. n. *Iamia* arbitrary name formed from the acronym of the Institute of Applied Microbiology at the University of Tokyo, which has made significant contributions to microbiology.

Gram-stain-positive rods. Aerobic. Oxidase- and catalase-positive. Mesophilic. The predominant menaquinone is MK-9(H_6). The major whole-cell fatty acids are $C_{17:0}$, $C_{17:1}$ ω8*c*, $C_{15:0}$, and $C_{16:0}$. Isolated from the epidermis of a sea cucumber and grown on marine agar.

DNA G+C content (mol%): 74.

Type species: **Iamia majanohamensis** Kurahashi, Fukunaga, Sakiyama, Harayama and Yokota 2009, 871[VP].

Further descriptive information

The phylogenetic relationship of the single species of *Iamia* to those of the *Acidimicrobiaceae*, the other family of the *Acidimicrobiales*, is illustrated in Figure 412. Further descriptive information is as for the single species, *Iamia majanohamensis* (Kurahashi et al., 2009).

Enrichment and isolation procedures

Iamia majanohamensis was obtained from the abdominal epidermis of a sea cucumber, *Holothuria edulis*, which was collected off the coast of Aka Island, Okinawa prefecture, Japan (Kurahashi and Yokota, 2004). Cultures were maintained routinely on marine agar at 25°C.

List of species of the genus *Iamia*

1. **Iamia majanohamensis** Kurahashi, Fukunaga, Sakiyama, Harayama and Yokota 2009, 871[VP]

ma.ja.no.ha.men'sis. N.L. fem. adj. *majanohamensis* pertaining to Majanohama, the site on the coast of Aka Island, Japan, where the type strain was isolated.

Gram-stain-positive rods, 1.2–1.7 by 0.3–0.5 µm. Aerobic. Oxidase- and catalase-positive. Colonies on marine agar (which contains peptone and yeast extract) are small, white, circular, convex, smooth, shiny, and 0.2–0.3 mm in diameter after 3 weeks incubation at 30°C. Optimal growth temperature is 28–30°C. No growth occurs below 10°C or above 45°C. Hydrolysis of gelatin is observed. Positive for arginine dihydrolase, lysine decarboxylase, ornithine decarboxylase, production of acetoin and gelatinase, and reduction of nitrate to N_2, but negative for β-galactosidase, urease, tryptophan deaminase, utilization of citrate, production of H_2S and indole, and assimilation of amygdalin, arabinose, D-glucose, inositol, D-mannose, melibiose, rhamnose, sucrose, and sorbitol.

Source: the type strain was isolated from the ventral epidermis of the sea cucumber *Holothuria edulis* at Aka Island, Okinawa, Japan.

DNA G+C content (mol%): 74.4 (HPLC).

Type strain: F12, NBRC 102561, DSM 19957.

Sequence accession no. (16S rRNA gene): AB360448.

References

Kurahashi, M. and A. Yokota. 2004. *Agarivorans albus* gen. nov., sp. nov., a γ-proteobacterium isolated from marine animals. Int. J. Syst. Evol. Microbiol. *54*: 693–697.

Kurahashi, M., Y. Fukunaga, Y. Sakiyama, S. Harayama and A. Yokota. 2009. *Iamia majanohamensis* gen. nov., sp. nov., an actinobacterium isolated from sea cucumber *Holothuria edulis*, and proposal of *Iamiaceae* fam. nov. Int. J. Syst. Evol. Microbiol. *59*: 869–873.

Class III. **Coriobacteriia** class. nov.

Helmut König

Co.ri.o.bac.te.ri'i.a. N.L. neut. n. *Coriobacterium* type genus of the type order; suff. *-ia* ending to denote a class; N.L. pl. neut. n. *Coriobacteriia* the *Coriobacteriales* class.

The class is formed by elevation of the subclass *Coriobacteridae* Stackebrandt, Rainey and Ward-Rainey 1997. The subclass *Coriobacteridae* was circumscribed on the basis of phylogenetic analysis of the 16S rRNA gene sequences (Figure 414). The 16S rDNA/rRNA signature pattern is as that of the family *Coriobacteriaceae*.

Type order: **Coriobacteriales** Stackebrandt, Rainey and Ward-Rainey 1997, 483[VP].

Reference

Stackebrandt, E., F.A. Rainey and N.L. Ward-Rainey. 1997. Proposal for a new hierarchic classification system, *Actinobacteria* classis nov. Int. J. Syst. Bacteriol. *47*: 479–491.

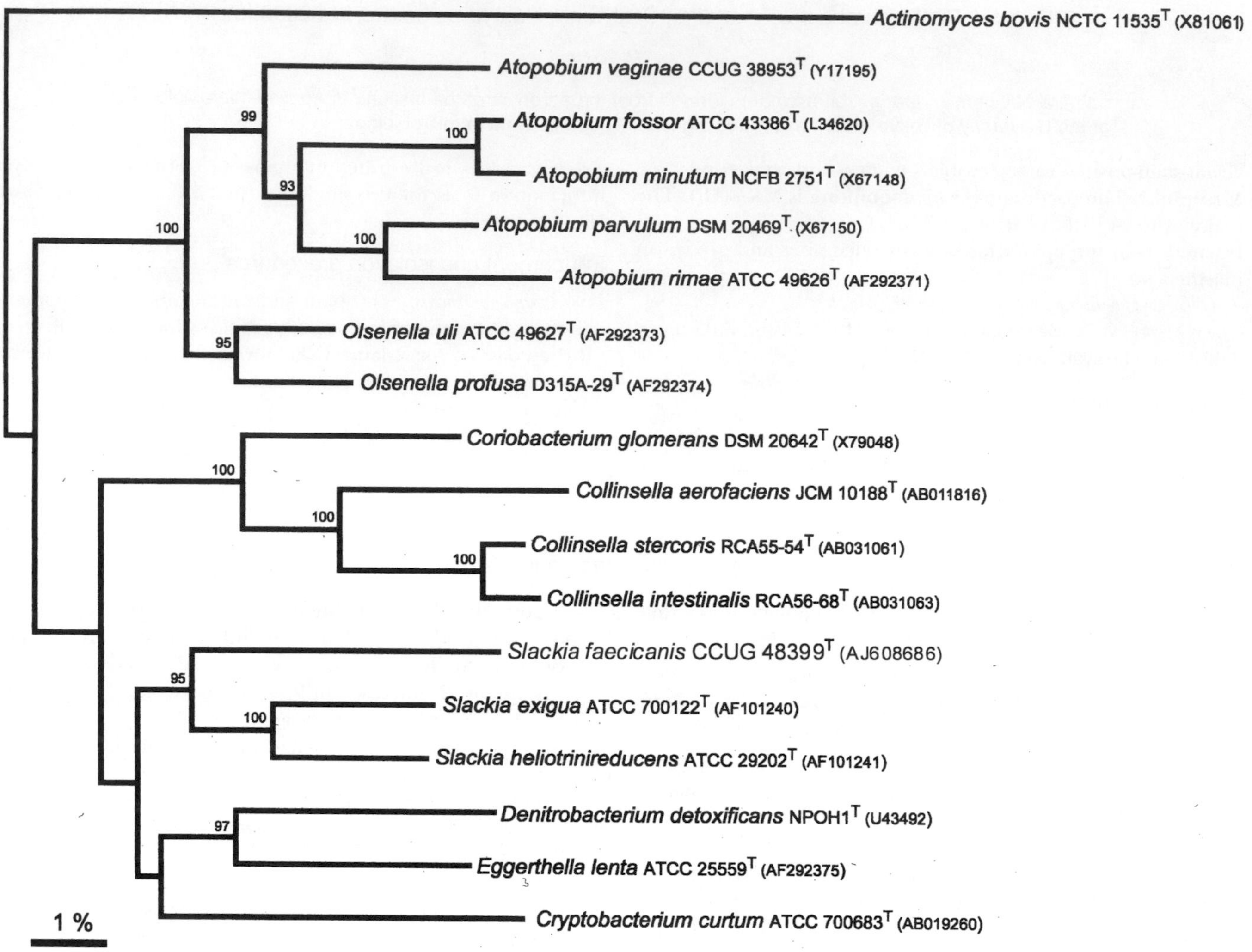

FIGURE 414. Unrooted tree of the species of the family *Coriobacteriaceae*. The tree was constructed using the neighbor-joining method based on a comparison of approximately 1330 nucleotides. Bootstrap values, expressed as a percentage of 1000 replications, are given at branching points. Bar = 1% sequence divergence. (Reprinted with permission from Lawson et al., 2005. Int. J. Syst. Evol. Microbiol. *55*: 1243–1246.)

Order I. **Coriobacteriales** Stackebrandt, Rainey and Ward-Rainey 1997, 483VP

Helmut König

Co.ri.o.bac.te.ri.a′les. N.L. neut. n. *Coriobacterium*, type genus of the order, suff. *-ales* ending to denote an order; N.L. fem. pl. n. *Coriobacteriales* the *Coriobacterium* order.

The order contains the type family *Coriobacteriaceae*. The 16S rDNA/rRNA gene signature pattern is as that of the family *Coriobacteriaceae*. The order *Coriobacteriales* was circumscribed on the basis of phylogenetic analysis of the 16S rRNA gene sequences.

Type genus: **Coriobacterium** Haas and König 1988, 383VP.

References

Haas, F. and H. König. 1988. *Coriobacterium glomerans* gen. nov., sp. nov. from the intestinal tract of the red soldier bug. Int. J. Syst. Bacteriol. *38*: 382–384.

Stackebrandt, E., F.A. Rainey and N.L. Ward-Rainey. 1997. Proposal for a new hierarchic classification system, *Actinobacteria* classis nov. Int. J. Syst. Bacteriol. *47*: 479–491.

Family I. **Coriobacteriaceae** Stackebrandt, Rainey and Ward-Rainey 1997, 483[VP]

Helmut König

Co.ri.o.bac.te.ri.a.ce′a.e. N.L. neut. n. *Coriobacterium* type genus of the family: suff. *-aceae* ending to denote a family; N.L. fem. pl. n. *Coriobacteriaceae* the *Coriobacterium* family.

The pattern of 16S rDNA/rRNA signature nucleotides (base positions relative to *Escherichia coli*) of members of the family consists of 113:314 (C–G), 294:303 (G–C), 295:302 (U–A), 407:435 (C–G), 613:627 (G–C), 670:736 (G–C), 771:808 (U–A), 772:807 (A–U), 823:877 (A–U), 941:1342 (A–U), 950:1231 (U–G), 1120:1153 (G–C), 1148 (C), 1165:1171 (C–G), 1242:1295 (G–C), 1313:1324 (G–C), 1321 (C), 1410:1490 (A–U), 1415:1485 (C–G) and 1416:1484 (C–G). The family contains the genera *Coriobacterium*, *Atopobium*, *Collinsella*, *Cryptobacterium*, *Denitrobacterium*, *Eggerthella*, *Olsenella*, and *Slackia*. The genera were circumscribed on the basis of phylogenetic analysis of the 16S rDNA/rRNA gene sequences.

Rod-shaped or coccoid cells occur singly, in pairs or as rods. Spores and flagella are absent. Gram-stain-positive. Strictly or facultatively anaerobic growth. Sugars may or may not be fermented. The family *Coriobacteriaceae* was circumscribed on the basis of phylogenetic analysis of the 16S rDNA/rRNA gene sequences (Figure 414).

Type genus: **Coriobacterium** Haas and König 1988, 383[VP].

Genus I. **Coriobacterium** Haas and König 1988, 383[VP]

Helmut König

Co.ri.o.bac.ter′i.um. Gr. fem. n. *koris* bug; Gr. neut. n. *bakterion* a small rod; N.L. neut. n. *Coriobacterium*, rodlet associated with bugs.

The cells form chains of **pear-shaped** to irregularly shaped rods. The bacteria are **obligately anaerobic**. The cells stain **Gram-positive**. Spores and **flagella are absent**. Under strictly anaerobic conditions, the **fermentation products** of glucose are **acetic acid**, **L-lactic acid**, and **ethanol**. The **peptidoglycan** belongs to the **Lys–Asp type** (A4α). The bacteria occur in the **intestinal tracts** of insects.

DNA G+C content (mol%): 60–61.

Type species: **Coriobacterium glomerans** Haas and König 1988, 383[VP].

Further descriptive information

The fermentation products from glucose are L-lactic acid, acetic acid, and ethanol under strict anaerobic conditions with N_2/CO_2 as gas phase (using the culture technique of Balch et al. (1979). Under certain conditions in TPY medium (Scardovi, 1981), CO_2 and H_2 are also produced.

Enrichment and isolation procedures

A Gram-stain-positive anaerobic bacterium was regularly found in great numbers in histological sections and anaerobic cultures of the third bulbous mid-gut portion of the red soldier ant *Pyrrhocoris apterus* (L.). It formed long chains of irregular pear-shaped cells with large spherical involutions. All histological sections showed bundles of Gram-stain-positive chain-forming bacteria attached to the epithelium of the intestine at the beginning of the third bulbous portion of the mid-gut. The chains could reach more than 150 μm into the lumen of the gut.

Flattened, dry-looking colonies of chain-forming rod-like bacteria, with filamentous and irregular edges, could be detected when the gut contents were incubated anaerobically on blood-agar or supplemented Schaedler agar plates after 5–10 d at an incubation temperature of 25–37°C. The bacteria were further purified on TPY agar plates. The colonies consisted of long filaments, bent into hairpins, curls, and loops. In fluid TPY medium, bacteria formed a flocculent, woolly sediment with a clear supernatant.

Maintenance procedures

The bacteria can be maintained on TPY agar plates at 37°C. For longer periods, cultures can be stored in liquid nitrogen in the presence of 10% glycerol.

Differentiation of the genus *Coriobacterium* from other genera

Coriobacterium glomarans, the only species of this genus, is distinguished from members of the genus *Lactobacillus* by its higher G+C content (strain PW2[T], 60–61 mol%; *Lactobacillus*, 32–53 mol%) and by the molar ratio of the fermentation products.

Coriobacterium glomarans shares some characteristics with the genus *Bifidobacterium* (e.g. G+C content of the DNA 55–67 mol%) and it possesses the type 6 murein of this genus (Scardovi, 1981). Based on its ability to ferment carbohydrates, it does not correspond to any species in this group. It differs from the genus *Bifidobacterium* by its lack of a positive test for fructose-6-phosphate phosphoketolase and by production of hydrogen. The high amount of ethanol produced cannot be explained by the hexose fermentation pathway in bifidobacteria.

Strain PW2[T] is distinguished from *Eubacterium* by the G+C mol% content of its DNA (*Eubacterium*, 30–55 mol%), usually by its fermentation products, and by its murein composition. Several members of the genus *Eubacterium* produce hydrogen, and acetic and lactic acids as the sole major acid products, but among these species, where the G+C content of the DNA is known, it is about 45 mol%; also, those species of *Eubacterium* that have been investigated possess the quite different type B murein.

The genus *Lachnospira* has been defined as containing curved-to-helical subterminally flagellated rods.

Coriobacterium glomarans is immotile, and flagella have not been observed by electron microscopy. *Lachnospira* but not *Coriobacterium glomarans* forms formic acid from glucose.

Taxonomic comments

Coriobacterium glomerans Haas and König 1988, 383[VP] is the only species so far described in the genus *Coriobacterium*, but other strains/clones have been described (strain CCUG 33917, AJ131149; strain CCUG 33918, AJ131150; strain EKSO3, AJ245921; strain 3WC8.1, AJ586811; strain SG1214, AY669285; clone CC1015, AY669270).

List of species of the genus *Coriobacterium*

1. **Coriobacterium glomerans** Haas and König 1988, 383[VP]

glo'me.rans. L. part. adj. *glomerans* agglomerating.

The cells form flocculent, wooly sediments with a clear supernatant in fluid media. They are regular to pear-shaped, forming cell chains under strictly anaerobic conditions (using the culture technique of Balch et al. (1979). They form irregularly shaped rods in older cultures or after culture in the GasPak Anaerobic System. The bacteria are strictly anaerobic. The cells are Gram-stain-positive (Kopeloff modification of the Gram stain).

The cells have an electron-dense Gram-stain-positive cell wall, 40 nm wide. The peptidoglycan belongs to the Lys–Asp type (A4α). The peptide subunit is composed of Asp, Glu, Gly, Ala, and Lys in a molar ratio of 0.91:1.00:0.45:1.96:1.01.

The length of the cells varies (0.44–1.80 µm). Spherical involution forms with a diameter of up to 198 µm are common. The bacteria form chains more than 150 µm long (Figure 415).

The colonies consist of long filamentous chain-forming bacteria, bent into hairpins, curls, and loops. The colony size is about 0.6 mm. The organisms grow on blood agar (Columbia agar base [BBL]), supplemented Schaedler agar (BBL), and TPY agar at 25°C and 30°C. When grown in TPY medium, the bacteria produce acetic acid, L-lactic acid, and ethanol from glucose in a molar ratio of 1.16:1.00:0.95. CO_2 and H_2 are also produced. D-Lactic acid, formic acid, volatile short-chain alcohols, or other volatile fatty acids are not formed. Under more stringent anaerobic conditions with N_2–CO_2 (80:20) as the gas phase, but lacking H_2, the molar ratio of acetic acid to L-lactic acid to ethanol is 1.20:1.00:0.23. This indicates that in the presence of the higher concentrations of molecular hydrogen (produced by the commonly used GasPack anaerobic system) strain PW2T forms increased amounts of ethanol.

FIGURE 415. Electron micrograph of *Coriobacterium glomerans* strain PW2. Platinum shadowing of a cell chain showing irregular cells.

The cells ferment L-arabinose, cellobiose, glucose, maltose, mannitol, melibiose, raffinose, mannose, ribose, salicin, sucrose, and D-xylose. Inositol, inulin, lactose, melibiose, raffinose, and starch are not fermented.

Source: the third bulbous midgut portion of all stages of the red soldier ant (*Pyrrhocoris apterus*; Heteroptera: Pyrrhocoridae), except the egg stage. The filamentous cell chains are attached to the epithelia of the intestines.

DNA G+C content (mol%): ca. 61 (T_m and nucleotide analysis)

Type strain: PW2, ATCC 49209, DSM 20642, JCM 10262.

DNA G+C content (mol%): X79048 (EMBL).

Genus II. **Atopobium** Collins and Wallbanks 1993, 188[VP] (Effective publication: Wallbanks and Collins 1992, 239.)

Paul A. Lawson

A.to.po'bi.um. Gr. adj. *atopos* having no place, strange; Gr. neut. part. used as a noun *bion* living thing; N.L. neut. n. *Atopobium* strange living thing.

Cells consist of short rods, often with central swellings, or small cocci that may appear to be elliptical. They occur singly, in pairs, and in short chains. Gram-stain-positive, does not form spores, and is nonmotile. **Obligately or facultatively anaerobic and catalase-negative.** Nitrate is not reduced. **The major fermentation products from glucose are lactic acid together with acetic and formic acids; trace amounts of succinic acid may be formed.** Hydrogen is not produced. Growth is stimulated by

TABLE 306. Differential characteristics among members of the genus *Atopobium*[a,b]

Characteristic	*A. minutum*	*A. fossor*	*A. parulum*	*A. rimae*	*A. vaginae*
Acid produced from:					
Cellobiose	–	–	+	+	nd
Glucose	+	+	+	+	–
Lactose	–	–	+	–	–
Maltose	–	–	+	+	–
Mannose	+	+	+	+	–
Ribose	–	–	–	+	–
Sucrose	–	–	+	+	–
Enzyme production:					
Acid phosphatase	–	nd	+	+	+
Alanine arylamidase	–	nd	+	–	–
Arginine dihydrolase	+	nd	–	–	+
Arginine arylamidase	+	nd	+	–	+
Histidine arylamidase	–	nd	–	–	+
β-Galactosidase	–	nd	+	–	–
Leucine arylamidase	–	nd	+	–	+
Proline arylamidase	+	nd	–	–	+
Pyroglutamic acid arylamidase	d	nd	+	+	–
Glycine arylamidase	–	nd	+	–	+
Serine arylamidase	–	nd	–	–	+
Tyrosine arylamidase	–	nd	+	–	–
Fermentation products from glucose	L, A, F, s	L, A	L, A, F, s	L, A, F, s	nd
DNA G+C content (mol%)	44	44	39	45	44
Source	Human blood, abdominal wounds, pelvis abscesses	Horse pharynxes and root abscesses	Human gingival crevices	Human gingival crevices, periodontal pockets	Human vagina

[a]Symbols and abbreviations: –, negative; + positive; d, differences between strains; nd, not determined; a, acetic acid; f, formic acid; l, lactic acid; s, succinic acid; minor products are indicated by lower-case letters.

[b]Data from Rodriguez et al. (1999) and Kageyama et al. (1999b).

Tween 80. Growth may occur in the presence of 6.5% w/v NaCl. Gelatin is not liquefied, meat is not digested and indole is not produced. Isolated from human and animal sources.

DNA G+C content (mol%): 35–46 (T_m).

Type species: **Atopobium minutus** (Hauduroy, Ehringer, Urbain, Guillot and Magrou 1937) Collins and Wallbanks 1993, 188[VP] (Effective publication: Wallbanks and Collins 1992, 239.) (*Lactobacillus minutus* Hauduroy, Ehringer, Urbain, Guillot and Magrou 1937; Moore and Holdeman 1972, 63[AL] emend. Olsen, Johnson, Moore and Moore 1991).

Further descriptive information

In addition to being Gram-stain-positive and non-sporeforming rods to cocci, cells may contain central swellings. A variety of peptidoglycan types are evident, *Atopobium minutus* contains L-Orn–L-Ser–D-Glu (A4β), *Atopobium parvulum* L-Lys–D-Asp (A4α), and *Atopobium fossor* and *Atopobium rimae* L-Ala–D-Glu–LL-Dpm-Gly (A3γ). The cell-wall structure of *Atopobium vaginae* has not been determined. *Atopobium* species possess some unifying traits, but can be distinguished by a number of tests that include carbohydrate fermentation patterns, the presence or absence of metabolic enzymes, and the end products of glucose metabolism (see Table 306).

Atopobium spp. have been isolated from a number of human and animal sources. *Atopobium parulum* and *Atopobium rimae* were originally isolated from the human gingival crevice and *Atopobium vaginae* from the human vagina. Numerous studies employing both culture and culture-independent methods based on the sequencing of 16S rRNA genes have shown that *Atopobium* and other members of the family *Coriobacteriaceae* contribute towards the normal flora of the gastrointestinal tract and are readily recovered from human feces (Harmsen et al., 2000; Holdeman et al., 1976). Harmsen and colleagues developed a specific probe to two clusters that include *Coriobacterium* and *Collinsella* (S-*-Cor-0653-a-A18 (COR653) 5′ CCTCCC(/C) TACCGGACCC) and *Atopobium* S-*-Ato-0291-a-A-17 (ATO291) 5′ GGTCGGTCTCTCAACCC) which also includes the *Coriobacterium* group. These studies suggested that the diversity of *Atopobium* strains present in feces appeared to increase with the age of the individual. Although considered to be part of the normal flora of the GI tract, *Atopobium* may under particular conditions act as an opportunistic pathogen and contribute towards disease processes. *Atopobium rimae* and *Atopobium parvulum*, for example, have been identified as agents of chronic periodontitis (Kumar et al., 2005; Olsen et al., 1991).

Kumar et al. (2005) used a quantitative 16S clonal analysis to investigate biofilms associated with chronic periodontitis and periodontal health. Their data demonstrated that in addition to a number of other genera, *Atopobium* phylotypes were elevated in disease tissue. Furthermore, Angelakis et al. (2009)

reported an unusual case of human bacteremia caused by *Atopobium rimae* that was directly responsible for the septicemia.

Perhaps the most compelling evidence for the association of *Atopobium* with disease comes from investigations involving *Atopobium vaginae* with bacterial vaginosis (BV). Studies employing specific PCR primers towards *Atopobium* suggest that as many as 70% of women with BV harbor *Atopobium vaginae* and that postmenopausal women are particularly susceptible to BV (Burton et al., 2004). Although the exact etiology is still not clear, a possible explanation is that the available fermentable carbohydrate glycogen from vaginal epithelial cells decreases during the onset of menopause (Boskey et al., 1999; Burton et al., 2004; Cauci et al., 2002) promoting the use of other metabolic sources (i.e. proteins) for energy.

Atopobium fossor, the only non-human species of this genus, was originally isolated from horses; it is often recovered as a common component of the bacterial mixed-culture of the pharynx in addition to tooth root abscesses. Moreover, in a recent study of the lower respiratory tract of horses diagnosed with pneumonia, Racklyeft and Love (2000) found that 32% of the animals harbored *Atopobium fossor* using culture-based methods. Although they found a number of obligate anaerobes in co-culture, these authors were unable to suggest an association between any particular species and disease.

Isolation procedures

Complex media recommended for anaerobes by the Anaerobe Laboratory at the Virginia Polytechnic Institute (Holdeman et al., 1976) and the Wadsworth-KTL Anaerobe Laboratory Manual (Jousimies-Somer et al., 2002b) are frequently used as basal media. Supplements that include hemin and vitamin K are routinely added along with Tween 80, blood, or serum to enhance growth. There is no enrichment or selective medium for these organisms. *Atopobium minutum*, *Atopobium parvulum*, *Atopobium rimae*, and *Atopobium vaginae* have been recovered from human sources while *Atopobium fossor* has only been isolated from horses.

Maintenance procedures

For short-term storage, *Atopobium* strains can be maintained in chopped meat medium and DSMZ medium 104 PYG-medium (http://www.dsmz.de). For long-term preservation, strains can be maintained in a medium containing 15–20% glycerol at −70°C or lyophilized.

Taxonomic comments

Based on phylogenetic considerations and associated phenotypic features, the novel genus *Atopobium* was proposed by Collins and Wallbanks (1992) to accommodate misclassified *Lactobacillus* and *Streptococcus* species isolated from either human gingival crevices or from abscesses, blood, and wounds (Olsen et al., 1991). The genus originally contained three species, namely *Atopobium minutum*, *Atopobium parvulum*, and *Atopobium rimae*. Kageyama et al. (1999b) transferred *Eubacterium fossor* to the genus *Atopobium* based on phylogenetic and phenotypic evidence. This organism was recovered from the normal flora of the pharynx and from necrotic diseases of tissue surrounding the teeth of horses.

Some initial confusion existed on the precise phylogenetic position of *Atopobium*; the genus was originally described in a study that included organisms with phenotypic traits of lactic acid producing bacteria and other low G+C Gram-stain-positive organisms. However, Stackebrandt and Ludwig (1994) in a more comprehensive study showed that *Atopobium* formed a phyletic line in the *Actinomycetes* line of descent. In addition, Rainey et al. (1994) demonstrated the genera *Atopobium* and *Coriobacterium* were close phylogenetic neighbors which shared a number of phenotypic characteristics; members of these taxa were Gram-stain-positive, nonmotile, and non-spore-forming and gave end products of metabolism that included acetic and lactic acids. The close relationship between the two genera was underpinned in the hierarchic classification system of Stackebrandt et al. (1997) who classified these taxa in the family *Coriobacteriaceae* as they shared 16S rDNA/rRNA signature nucleotides, namely 113:314 (C–G), 294:303 (G–C), 295:302 (U–A), 670:736 (G–C), 771:808 (U–A), 772:807 (A–U), 823:877 (A–U), 941:1342 (A–U),950:1231 (U–G), 1120:1153 (G–C), 1148 (C), 1165:1171 (C–G), 1242:1295 (G–C), 1313:1324 (G–C), and 1410:1490 (A–U). Furthermore, the 16S rDNA loop 1357–1364 of these genera consists of only eight nucleotides while all other taxa in the *Actinomycetes* line of descent contain an additional nucleotide (Rainey et al., 1994).

Differentiation of the genus *Atopobium* from other genera

In addition to differentiation by 16S rRNA gene sequence analysis (Figure 416), *Atopobium* can be readily distinguished from its closest phylogenetic relatives using a combination of biochemical and chemotaxonomic criteria (Table 307).

List of species of the genus *Atopobium*

1. **Atopobium minutum** (Hauduroy, Ehringer, Urbain, Guillot and Magrou 1937) Collins and Wallbanks 1993, 188[VP] (Effective publication: Wallbanks and Collins, 1992, 239.) (*Lactobacillus minutus* Hauduroy, Ehringer, Urbain, Guillot and Magrou 1937; Moore and Holdeman 1972, 63[AL] emend. Olsen, Johnson, Moore and Moore 1991)

mi.nu'tum. L. neut. adj. *minutum* minute, small.

Strictly anaerobic, small, elliptical, nonmotile, Gram-stain-positive rods (approx. 0.6–1.0 × 0.8–3.0 μm) occur singly, in pairs, and in short chains. Colonies on brain heart blood agar are pinpoint to 1 mm in diameter, raised or low convex, entire, translucent to transparent, and buff, gray, light yellow, or white after 5 d of anaerobic incubation at 37°C. Growth is stimulated by 0.02 (v/v) Tween 80 and may be enhanced when 10% (v/v) rabbit serum is added to culture media. Six out of 10 tested strains grew at 45°C. Four out of 11 strains grew in the presence of 6.5% (w/v) NaCl.

The fermentation end products from glucose are lactic acid (0.2–4.5 meq/100 ml of culture; mean for 21 strains, 2.3 meq/100 ml), acetic acid (0–0.9 meq/100 ml; mean for

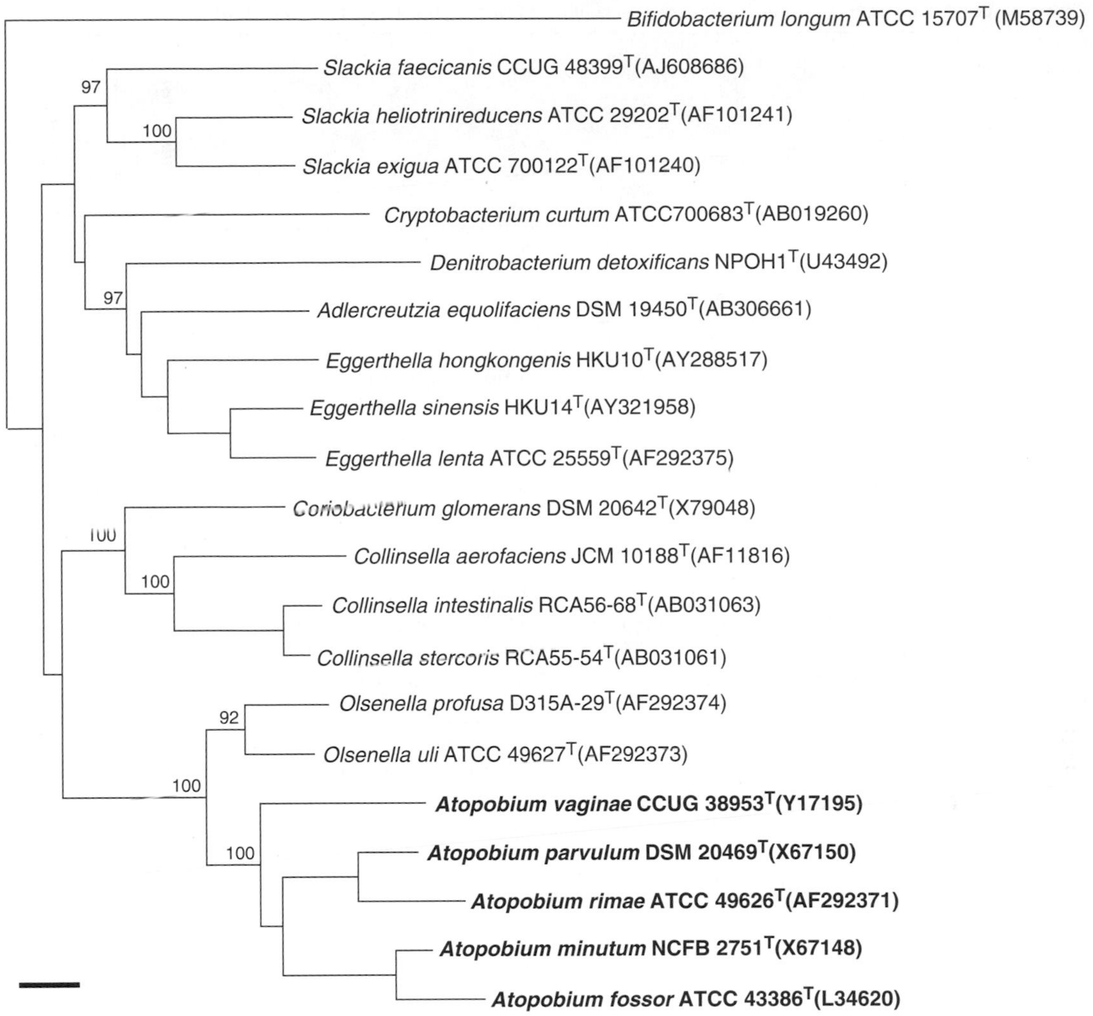

FIGURE 416. Unrooted neighbor-joining tree depicting the estimated phylogenetic relationships of members of the genus *Atopobium* and other members of the family *Coriobacteriaceae. Bifidobacterium longum* was used as the outgroup. The numbers on the branches refer to bootstrap values determined from 1000 replications. Only values about 90% are shown. Bar = 1% sequence divergence.

21 strains, 0.3 meq/100 ml), formic acid (0–0.7 meq/100 ml), and occasionally trace amounts of succinic acid. Little, if any, gas is detected in deep agar cultures. Hydrogen is not produced. The terminal pH range of PYG-Tween 80 broth cultures is 4.5–5.8; the mean terminal pH for 20 tested strains was 5.1.

A few strains (3 out of 27) weakly ferment starch (pH 5.5–5.7), but not amygdalin, erythritol, esculin, glycerol, glycogen, inositol, mannitol, melezitose, melibiose, raffinose, rhamnose, sorbitol, and xylose.

Gelatin is not liquefied, meat is not digested, indole is not produced, and nitrate is not reduced. Reactions are negative for bile-esculin, catalase, DNase, and hippurate hydrolysis.

Susceptibility to antibiotics varies: 1 out of 14 strains were resistant (μg per ml) to chloramphenicol (12), clindamycin (1.6), and erythromycin (3); 2 out of 14 strains were resistant to penicillin G (2U), and 3 out of 14 strains to tetracycline (6).

The major cellular fatty acid derivatives are $C_{18:1}$ *cis*9 fatty acid methyl ester and $C_{18:1}$ dimethylacetyl.

Source: abdominal wounds, blood, pelvic abscesses, and similar human infections.

DNA G+C content (mol%): 44 (T_m).

Type strain: ATCC 33267, CCUG 31167, DSM 20586, JCM 1118, LMG 9439, NCIMB 702751 (formerly NCFB 2751), VPI 9428.

Sequence accession no. (16S rRNA gene): M59059, X67148.

2. **Atopobium fossor** Kageyama, Benno and Nakase 1999d, 1325[VP] (Effective publication: Kageyama, Benno and Nakase 1999b, 394.) (*Eubacterium fossor* Bailey and Love 1986, 386[VP])

fos'sor. L. n. *fossor* a digger, delver.

TABLE 307. Characteristics that are useful in distinguishing the genera of the *Coriobacteriaceae* and *Adlercreutzia*[a,b]

Characteristic	1. *Atopobium*	2. *Adlercreutzia*	3. *Collinsella*	4. *Coriobacterium*	5. *Cryptobacterium*	6. *Eggerthella*	7. *Denitrobacterium*	8. *Olsenella*	9. *Slackia*
End products of metabolism	a, L, (s)	None	a, f, H, l	a, l	None	(a, l, s)[c]	nd	a, L, (s)	(a, iv)
Glucose fermentation	+	–	+	+	–	–	–	+	–
Growth stimulated by:									
Tween 80	+	–	+	nd	–	–	nd	+	–
Arginine	–	+	nd	nd	+	+	nd	–	+
Growth in 20% bile	–	–	nd	nd	–	+[b]	nd	d	–
Hydrogen production	–	–	+	+	nd	–	+	d	nd
Peptidoglycan type	A3γ, A4α, A4β	A1γ	A4	A4α	nd	A4γ	nd	A4β, A4α	nd
Major cellular fatty acid(s)	$C_{18:1}$ *cis9*	$C_{18:1}$ *cis9*	nd	nd	nd	$C_{14:0}$ br, $C_{15:0}$ br[c,d]	$C_{14:0}$ br, $C_{16:0}$ DMA	$C_{18:1}$ *cis9*[e]	$C_{18:0}$ br, $C_{18:1}$
DNA G+C content (mol%)	35–46	64.1–66.5	60–65	60–61	50–51	61.1–64.9	56–60	63–64	60–64
Source	Human oral cavity, blood, abdominal wounds. Horse pharynxes and root abscesses	Human and rat intestine	Human intestine	Red soldier beetle intestine	Human oral cavity	Human intestine and blood culture	Bovine rumen	Human oral cavity	Human oral cavity and dog intestine

[a]Abbreviations: a, acetic acid; f, formic acid; h, hydrogen; iv, isovaleric acid; l, lactic acid; s, succinic acid; minor products are indicated by lower-case letters, products in parentheses may or may not be formed; nd, not determined; br, branched. Symbols: –, negative; + positive; d, differences between strains.

[b]Data from Dewhirst et al. (2001) and Maruo et al. (2008).

[c]Determined for *Eggerthella lenta* but not for *Eggerthella hongkongensis* or *Eggerthella sinensis*.

[d]Maruo et al. (2008) found to be $C_{16:0}$ br.

[e]Determined for *Olsenella uli* but not for *Olsenella profuse*.

Obligately anaerobic, Gram-stain-positive, nonmotile cells (0.5–0.9 × 0.3 µm) that vary from coccobacillary to rod-shaped with parallel sides and rounded ends. Polar pili are present. On SBA, colonies are minute after 24 h, enlarging to 1 mm after 48 h when they appear to be set in small pits (0.3 mm deep and 3 mm wide) in the agar. After 4 d, colonies have umbonate to raised centers (1 mm) with either a roughened depressed region or a series of concentric rings radiating outward 1 to 2 mm from the center of colonies. Hemolysis does not occur. Growth in broth is improved by the addition of Tween 80 (0.75%, v/v) or serum (5%, v/v). Dense growth with uniform turbidity and a fine granular deposit is evident when organisms are grown in either prereduced peptone-yeast extract-glucose broth or cooked meat-glucose broth.

Acetic and lactic acids (lactic acid > acetic acid) are the major fermentation products. Gas is not detected in PYT-glucose deep agar cultures, although ammonia, hydrogen, and sulfide are produced.

Acid is formed from mannose but not from adonitol, amygdalin, arabinose, cellobiose erythritol, esculin, fructose, glycogen, inositol, lactose, maltose, mannitol, melezitose, melibiose, raffinose, rhamnose, ribose, salicin, sorbitol, starch, sucrose, trehalose, or xylose.

Most strains, including the type strain, ferment glucose weakly. Acetylmethylcarbinol, catalase, indole, lecithinase, lipase, and urease are not produced nor are esculin, starch, Tween 20, Tween 40, or Tween 80 hydrolyzed. Nitrate is not reduced. There is no reaction in milk and no digestion of gelatin. Bile neither stimulates nor inhibits growth. Cells do not survive heating for 10 min at 80°C.

The organism is susceptible (µg per ml) to ampicillin (4), carbenicillin (100), doxycycline (6), chloramphenicol (12), erythromycin (3), and to penicillin (2U per ml).

Source: the pharynx of a healthy horse.

DNA G+C content (mol%): 44 (T_m).

Type strain: ATCC 43386, CIP 106638, DSM 15642, JCM 9981, NCTC 11919, VPB 2127.

Sequence accession no. (16S rRNA gene): AB015945, L34620.

3. **Atopobium parvulum** (*ex* Weinberg, Nativelle and Prévot 1937) Collins and Wallbanks 1993, 188[VP] (Effective publication: Collins and Wallbanks 1992, 239.) (*Streptococcus parvulus ex* Weinberg, Nativelle and Prévot 1937, Cato 1983, 83)

par'vu.lus. L. neut. dim. adj. *parvulus* somewhat small.

Obligately anaerobic, nonmotile, nonspore-forming, Gram-stain-positive cocci (approx. 0.3–0.6 µm in diameter) that occur singly, in pairs, in clumps, and in short chains. Occasional elongate cells, some with a central swelling, may occur in stained preparations prepared from growth in media without Tween 80. Colonies on brain heart blood agar are pinpoint to 1 mm in diameter, raised or low convex, circular, entire, translucent to transparent, and colorless to gray after 2–5 d of anaerobic incubation at 37°C. Growth is stimulated (30 out of 30 tested strains) in culture media supplemented with 0.02% (v/v) Tween 80 and with 10% (v/v) rabbit serum. Of 72 strains tested, 58 grew at 45°C, and 6 out of 67 strains grew in the presence of 6.5% (w/v) NaCl.

The fermentation products from glucose are lactic acid (0.7–6.3 meq/100 ml of culture; mean for 40 strains, 4.2 meq/100 ml), acetic acid (0.1–0.6 meq/100 ml; mean for 40 strains, 0.3 meq/100 ml), formic acid (0–0.3 meq/100 ml) and occasionally trace amounts of succinic acid. Little, if any, gas is detected in deep agar cultures. Hydrogen is not produced. The terminal pH range in PYG-Tween 80 broth cultures is 4.0–5.1; the mean terminal pH for 32 strains was 4.4.

A total of 11 out of 82 tested strains fermented esculin; 1 out of 68, glycerol; 1 out of 82, melibiose; and 16 out of 82, xylose. Amygdalin, erythritol, glycogen, inositol, mannitol, melezitose, raffinose, rhamnose, sorbitol, and starch are not fermented.

Starch is not hydrolyzed in the absence of added serum (which contains amylase). Gelatin is not liquefied, meat is not digested, indole is not produced, and nitrate is not reduced. Reactions are usually negative for bile-esculin (none out of 68 tested strains were positive), DNase (1 out of 68 strains positive), and hippurate hydrolysis (none of 68 strains positive). Catalase activity is not detected.

Susceptibility to antibiotics varies; 1 out of 82 tested strains were resistant (μg per ml) to chloramphenicol (12), 10 out of 82 to clindamycin (1.6), 5 out of 82 to erythromycin (3), 6 out of 82 to penicillin G (2U), and 1 out of 82 to tetracycline (6).

The major cellular fatty acid derivatives are $C_{18:1}$ *cis*9 fatty acid methyl ester and $C_{18:1}$ dimethylacetyl.

Source: human gingival crevices and, according to Prévot et al. (1967), from patients with pulmonary gangrene, tonsillitis, sinusitis, otitis, and dental and thoracic abscesses.

DNA G+C content (mol%): 39 (T_m).

Type strain: ATCC 33793, CCUG 32760, CIP 102970, DSM 20469, IPP 1246, JCM 10300, VPI 0546.

Sequence accession no. (16S rRNA gene): S44206, X67150.

4. **Atopobium rimae** (Olsen, Johnson, Moore and Moore 1991) Collins and Wallbanks 1993, 188[VP] (Effective publication: Wallbanks and Collins 1992, 239; *Lactobacillus rimae* Olsen, Johnson, Moore and Moore 1991, 264.)

ri'mae. L. gen. n. *rimae* of a fissure, pertaining to the gingival crevice.

Strictly anaerobic, small, elliptical, nonmotile, Gram-stain-positive rods occur singly, in pairs, and in short chains. Central swellings may occur, particularly in cells grown on solid medium. Colonies on brain heart blood agar are 1–2 mm in diameter, raised or low convex, entire, and translucent to transparent after 5 d of anaerobic incubation at 37°C. Growth is stimulated by 0.02% (v/v) Tween 80 and usually when 10% (v/v) rabbit serum (45 out of 53 strains) is added to culture media. Of 33 tested strains, 21 grew at 45°C. There is no growth in the presence of 6.5% (w/v) NaCl.

The fermentation products from glucose are lactic acid (1–10 meq/100 ml of culture; mean for 33 strains, 4.7 meq/100 ml), acetic acid (0.2–0.6 meq/100 ml; mean for 33 strains, 0.3 meq/100 ml), formic acid (0–0.4 meq/100 ml; mean, 0.05 meq/100 ml), and occasionally trace amounts of succinic acid. Little, if any, gas is detected in deep agar cultures. Hydrogen is not produced. The range of terminal pH of PYG-Tween 80 broth cultures is 3.8–4.9; the mean terminal pH for 30 strains was 4.3.

Glycogen is fermented only when serum (which contains amylase) is added to the medium. In the absence of serum, starch is seldom fermented (1 out of 66 strains positive) or hydrolyzed (11 out of 66 strains positive). Amygdalin, erythritol, and esculin are rarely fermented (1 out of 76 strains positive), glycerol, inositol, mannitol, melezitose, melibiose, raffinose, rhamnose, sorbitol, and xylose are not fermented.

Gelatin is not liquefied, meat is not digested, indole is not produced, and nitrate is not reduced. Reactions are negative for bile-esculin, DNase, and hippurate hydrolysis. Catalase activity is not detected.

Susceptibility to antibiotics: none of the 70 tested strains were resistant (μg per ml) to chloramphenicol (12), clindamycin (1.6), erythromycin (3), or tetracycline (6); 5 out of 70 strains were resistant to penicillin G (2U).

The major cellular fatty acid derivatives are $C_{18:1}$ *cis*9 fatty acid methyl ester and $C_{18:1}$ dimethylacetyl.

Source: human gingival crevices, especially in patients with periodontitis.

DNA G+C content (mol%): 45 (T_m).

Type strain: ATCC 49626, CCUG 31168, DSM 7090, NBRC 15546, NCFB 2896, JCM 10299, LMG 11476, VPI D140H-11A.

Sequence accession no. (16S rRNA gene): S44205, AF292371.

Additional comments: formerly known as *Lactobacillus* DO2.

5. **Atopobium vaginae** Rodriguez-Jovita, Collins, Sjödén and Falsen 1999, 1575[VP]

va.gi'nae. L. n. *vagina* vagina; L. gen. n. *vaginae* of the vagina.

Facultatively anaerobic, small-elongated Gram-stain-positive cocci occur singly, in pairs, or short chains. Tiny pin-head colonies are formed on blood agar at 37°C.

Acid is not produced from mannose or raffinose.

Positive for arginine dihydrolase, arginine arylamidase, acid phosphatase, glycine arylamidase, histidine, arylamidase, leucine arylamidase, proline arylamidase and serine arylamidase, but negative for alanine arylamidase, α-arabinosidase, cystine arylamidase, chymotrypsin, ester lipase C8, esterase C4, α-fucosidase, α- and β-galactosidase, β-galactosidase 6-phosphate, α- and β-glucosidase, glutamic acid decarboxylase, glutamylglutamic acid arylamidase, β-glucuronidase, α-mannosidase, β-*N*-acetylglucosaminidase, lipase C14, pyroglutamic acid arylamidase, tyrosine arylamidase, trypsin, and urease. Weak activity for phenylalanine arylamidase may be detected. Indole-negative. Nitrate is not reduced. Gelatin and esculin are not hydrolyzed.

Source: vaginal flora of a healthy person. Habitat is unknown.

DNA G+C content (mol%): 44 (T_m).

Type strain: ATCC BAA-55, CCUG 38953, CIP 106431, DSM 15829.

Sequence accession no. (16S rRNA gene): Y17195.

Genus III. **Collinsella** Kageyama, Benno and Nakase 1999c, 564[VP]

AKIKO KAGEYAMA AND YOSHIMI BENNO

Col.lin.sel'la. N.L. fem. dim. ending *-ella*; N.L. fem. n. *Collinsella* named to honor Matthew D. Collins, a contemporary English microbiologist, for his outstanding contributions to microbial systematics.

Gram-stain-positive, obligatory anaerobic organism which forms chains of rods (0.3–0.7 × 1.2–4.3 μm). **Does not form spores or flagella**. Colonies are produced in 2 d at 37°C on EG agar in an anaerobic jar containing 100% CO_2. **Fermentation products of glucose are ethanol, formate, hydrogen, and lactate**. Cell wall contains an A4 type peptidoglycan. The 16S rDNA sequence analysis shows that the organism is a member of the family *Coriobacteriaceae*, which also includes the genera *Atopobium*, *Coriobacterium*, *Cryptobacterium*, *Denitrobacterium*, *Eggerthella*, *Olsenella*, and *Slackia*.

DNA G+C content (mol%): 60–65 (HPLC).

Type species: **Collinsella aerofaciens** (Eggerth 1935) Kageyama, Benno and Nakase 1999c, 564[VP].

Further descriptive information

16S rRNA gene sequence data show that the genus *Collinsella* is closely related to the genus *Coriobacterium* of the family *Coriobacteriaceae*, class *Coriobacteriales* Stackebrandt et al. 1997 (Figure 417). Almost complete 16S rRNA gene sequences (positions 28–1491; *Escherichia coli* numbering system) have been determined for *Collinsella* strains. The range of sequence similarity values is 96.6–97.8% for *Collinsella aerofaciens* strains, and 98.2–99.5% for *Collinsella stercoris* strains; *Collinsella intestinalis* strains share a 99.6% sequence similarity.

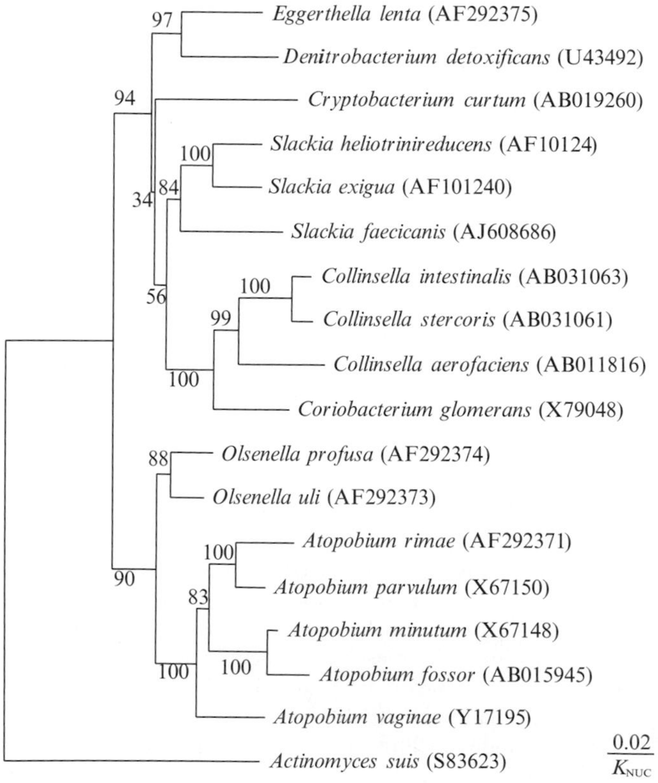

FIGURE 417. Phylogenetic tree derived from 16S rRNA gene sequences of the members of the family *Coriobacteriaceae* and constructed using the neighbor-joining method and K_{nuc} values. Numbers at branching points are bootstrap values (1000 resamplings). The tree was unrooted and *Actinomyces suis* was used as an outgroup.

Species-specific primers have been designed for the three *Collinsella* species. *Collinsella aerofaciens*-specific primers AERO-F (5′-CTTTCAGCAGGGAAGAGTCAA-3′; positions 436–466; *Escherichia coli* numbering system) and AERO-R (5′-AGCCATGCACCACCTGTATGG-3′; positions 1060–1039; *Escherichia coli* numbering system) generate an amplicon of about 590 bp; optimum PCR results are obtained at 63°C after 25 cycles. Similarly, *Collinsella intestinalis*-specific primers INTE-F (5′-CTTACCAGGGCTTGACATGA-3′; positions 980–999; *Escherichia coli* numbering system) and INRE-R (5′-TTAGGCGCCTCCCCCCAGAT-3′; positions 1469–1450; *Escherichia coli* numbering system) yield an amplicon of about 486 bp; optimum PCR results are obtained at 58°C after 25 cycles. Finally, *Collinsella stercoris*-specific primers STER-F (5′-TGCAAGTCGAACGGCACCCG-3′; positions 56–75; *Escherichia coli* numbering system) and STER-R (5′-CCGTCTGGGCTTTGCCGGCC-3′; positions 223–204; *Escherichia coli* numbering system) produce an amplicon of about 158 bp; optimum PCR results are obtained at 58°C and after 25 cycles.

Collinsella strains have an A4-type peptidoglycan (Schleifer and Kandler, 1972). *Collinsella aerofaciens* has an A4β-type peptidoglycan with a (L-Ala)–D-Glu–L-Orn–L-Asp peptide subunit and an interpeptide bridge which only contains L-asparagine moieties; *Collinsella intestinalis* an A4α-type peptidoglycan with a (L-Ala)–D-Glu–L-Lys–D-Glu peptide subunit and interpeptide bridge consisting of D-glutamic acid moieties, and *Collinsella stercoris* an A4β-type peptidoglycan with a (L-Ala)–D-Glu–L-Orn–D-Asp peptide subunit and an interpeptide bridge containing D-asparagine moieties.

The antibiotic sensitivity pattern of *Collinsella* strains, including 32 strains of *Collinsella aerofaciens*, eight strains of *Collinsella stercoris*, and two strains of *Collinsella intestinalis*, has been determined (Table 308). *Collinsella* strains are sensitive to various antibiotics, but not to ceftazidime or cefotetan (MIC: 12.5–25 μg/ml), or to aztreonam or carumonam (MIC: <100 μg/ml).

Collinsella aerofaciens is the most abundant bacterium in the human intestine [about 10^{10} cells (g feces)/min], and is found in more than 90% of human intestines (Benno et al., 1986; Finegold and Sutter, 1978; Moore and Holdeman, 1974). *Collinsella intestinalis* and *Collinsella stercoris* are also present in the human intestine. None of these three species exhibit pathogenicity.

Enrichment and isolation procedures

Strains of *Collinsella* species can be isolated from human feces of patients with colon cancer, Crohn's disease, and ulcerative colitis, and from healthy individuals on EG agar medium* and

*Premixed EG agar (Eiken Chemical Co., Ltd.) supplemented with 5% horse blood contains 3 g of beef extract, 5 g of yeast extract, 10 g of peptone, 1.5 g of glucose, 0.5 g of L-cysteine HCl, 0.2 g of L-cystine, 4 g of Na_2HPO_4, 0.5 g of soluble starch,, 0.5 g of polysorbate 80, 0.5 g of silicone, and 15 g of agar in 1000 ml of distilled water, pH 7.7.

TABLE 308. Susceptibilities of *Collinsella aerofaciens*, *Collinsella intestinalis*, and *Collinsella stercoris* to antimicrobial agents

	MIC (mg/ml)			
Organism and antimicrobial agent	Range	50%	90%	MIC (mg/ml) for type strain
C. aerofaciens (32 strains)				
Ampicillin	0.025–0.78	0.2	0.2	0.2
Erythromycin	0.05–0.39	0.05	0.1	0.1
Tetracycline	0.39–50	0.78	1.56	0.78
Cephalothin	0.39–6.25	0.78	1.56	3.13
Cefuroxime	0.78–3.13	1.56	3.13	3.13
Ceftazidime	3.13–25	12.5	25	25
Cefotetan	6.25–50	12.5	25	50
Aztreonam	12.5–100	50	50	100
Carumonam	50–400	100	200	200
C. stercoris (8 strains)				
Ampicillin	0.05	0.05	0.05	0.05
Erythromycin	0.05	0.05	0.05	0.05
Tetracycline	0.39–0.78	0.39	0.78	0.78
Cephalothin	1.56	1.56	1.56	1.56
Cefuroxime	0.78–3.13	1.56	3.13	1.56
Ceftazidime	12.5	12.5	12.5	12.5
Cefotetan	25	25	25	25
Aztreonam	100–200	200	200	200
Carumonam	100–200	100	200	200
C. intestinalis (2 strains)				
Ampicillin	0.2	0.2	0.2	0.2
Erythromycin	0.78	0.78	0.78	0.78
Tetracycline	0.78	0.78	0.78	0.78
Cephalothin	12.5	12.5	12.5	12.5
Cefuroxime	1.56–3.13	1.56	3.13	1.56
Ceftazidime	12.5	12.5	12.5	12.5
Cefotetan	50	50	50	50
Aztreonam	12.5	12.5	12.5	12.5
Carumonam	100	100	100	100

in EGF broth.* Organisms are cultivated for 2 d at 37°C in an anaerobic chamber using these media.

Maintenance procedures

Collinsella strains can be maintained for several weeks on EG agar plates in an anaerobic jar containing 100% CO_2. Long-term storage can be achieved at −80°C in 10% glycerol or by lyophilization in the presence of stabilizers such as 10% skim milk.

Differentiation of the genus *Collinsella* from other genera

A number of characteristics that can be used to distinguish the genus *Collinsella* from other genera classified in family *Coriobacteriaceae* are shown in Table 307. These properties include fermentation end products. Specific primers can also be used to distinguish *Collinsella* strains from the other genera.

Taxonomic comments

Eubacterium aerofaciens was reclassified as *Collinsella aerofaciens* by Kageyama et al. (1999c). Moore et al. (1971) had reported that these organisms belonged to the genus *Eubacterium* because they were Gram-stain-positive, obligatorily anaerobic, non-spore-forming rods that did not produce propionic acid as a major product (*Propionibacterium*); lactic acid as a sole major acid (*Lactobacillus*); large amounts of succinic (in the presence of CO_2) and lactic acids, sometimes with acetic and formic acids (*Actinomyces*); or acetic and lactic acids (acetic > lactic), with or without formic acid, as sole major acid products (*Bifidobacterium*). The genus *Eubacterium* had a broad definition and acted as a repository for large numbers of phenotypically diverse species (Andreesen, 1992).

*Contains 2.4 g of Lab-Lemco powder (Oxoid), 10 g of proteose peptone No.3 (Difco), 5 g of yeast extract (Difco), 4 g of Na_2HPO_4, 40 ml of sterilized horse blood, 5 g of glucose, 0.5 g of soluble starch (Sigma) and 0.5 g of L-cycteine HCl in 960 ml of distilled water, pH 7.6.

The type species of the genus *Eubacterium* is *Eubacterium limosum*. It was evident from several studies that *Eubacterium barkeri*, *Eubacterium callanderi*, and *Eubacterium limosum* (Mountfort et al., 1988) formed the nucleus of the redefined genus *Eubacterium*. Based on the characteristics of this group, a preliminary working definition of *Eubacterium sensu stricto* was proposed: Gram-stain-positive, rod-shaped nonmotile obligatory anaerobic organisms which may form endospores, are saccharoclastic, produce acetate, butyrate, lactate, and H_2, and formate or CO_2 as products of glucose fermentation. The cell wall has a type B peptidoglycan and the G+C content of the DNA is within the range 45–47 mol% (Willems and Collins, 1996).

Some species classified in the genus *Eubacterium*, but not assigned to *Eubacterium sensu stricto*, have been transferred to established or new genera (Kageyama et al., 1999b, 1999c; Ludwig et al., 1992; Wade et al., 1999; Willems et al., 1997). *Eubacterium aerofaciens* was reclassified to the novel genus *Collinsella* based on phylogenetic characteristics and cell-wall peptidoglycan type (Kageyama et al., 1999c). Subsequently, two new *Collinsella* species were proposed, *Collinsella intestinalis* and *Collinsella stercoris*, and the description of genus *Collinsella* was emended (Kageyama and Benno, 2000).

Differentiation of species of the genus *Collinsella*

Properties which can be used to distinguish between *Collinsella* species are shown in Table 309 and Table 310. Species specific primers also can be used for species differentiation.

TABLE 309. Differences in antibiotic susceptibilities between *Collinsella* species

	MIC_{90} (mg/ml)[a]		
	C. aerofaciens	*C. stercoris*	*C. intestinalis*
Erythromycin	0.1	0.05	0.78
Cephalothin	1.56	1.56	12.5
Cefotetan	25	25	50
Aztreonam	50	200	12.5

[a]MICs at which 90% of strains are inhibited.

TABLE 310. Differential characteristics of *Collinsella* species[a,b]

	C. aerofaciens	*C. stercoris*	*C. intestinalis*
Acid from:			
Lactose	+	+	–
Maltose	+	+	–
Enzyme activity:			
Alkaline phosphatase	–	w	w
Acid phosphatase	w	+	+
Leucine arylamidase	w	+	+
N-Acetyl-β-glucosaminidase	–	+	+

[a]Data based on reactions of six strains of *Collinsella aerofaciens*, two strains of *Collinsella intestinalis*, and eight strains of *Collinsella stercoris*.

[b]Symbols: +, positive; w, weakly positive; –, negative.

List of species of the genus *Collinsella*

1. **Collinsella aerofaciens** (Eggerth 1935) Kageyama, Benno and Nakase 1999c, 564[VP] [*Eubacterium aerofaciens* (Eggerth 1935) Prévot 1938, 295[AL]]

ae.ro.fa′ci.ens. Gr. n. *aer, aeros* air, gas; L. v. *facere* to make, to produce; N.L. part. adj. *aerofaciens* gas-producing.

Gram-stain-positive, obligatory anaerobic organism which occur in chains of 6–120 cells (0.3–0.7×1.2–4.3 μm). Spores and flagella are absent. Colonies are white in the center with clear surroundings and exhibit protuberances on EG agar. Can be cultivated in 2 d at 37°C on EG agar in an anaerobic jar containing 100% CO_2. Acid is produced from fructose, galactose, glucose, lactose, maltose, and mannose, but not from arabinose, erythritol, glycogen, inositol, mannitol, melezitose, soluble starch, sorbitol, or xylose. Has an A4β-type peptidoglycan with a (L-Ala)–D-Glu–L-Orn–L-Asp peptide subunit and interpeptide bridge composed of L-asparagine moieties.

Source: human feces.

DNA G+C content (mol%): 61 (HPLC).

Type strain: JCM 10188, ATCC 25986, CCUG 28087, DSM 3979, NCTC 11838, VPI 1003.

Sequence accession no. (16S rRNA gene): AB011816.

2. **Collinsella intestinalis** Kageyama and Benno 2000, 1773[VP]

in′test.in.alis. n. *intestinum* gut, intestine; L. suff. *-alis* suffix denoting pertaining to; N.L. masc. adj. *intestinalis* pertaining to the intestine.

Gram-stain-positive, obligate anaerobic organism that occur in chains of 2–20 cells (0.3–0.5 × 1.2–2.2 μm). Colonies are white or gray in the center with clear margins on EG agar. Spores and flagella are absent. Grows on EG agar after 2 d at 37°C in an anaerobic jar containing 100% CO_2. Acid is produced from cellobiose, fructose, galactose, glucose, mannose, ribose, and salicin, but not from amygdalin, L-arabinose, erythritol, esculin, glycogen, inositol, lactose, maltose, mannitol, melezitose, raffinose, rhamnose, sorbitol, starch, sucrose, trehalose, or D-xylose. Positive for leucine arylamidase, acid phosphatase, naphthol-AS-BI-phosphohydrolase, and *N*-acetyl-β-glucosaminidase, but negative for chymotrypsin, cystine arylamidase, α-fucosidase, α-galactosidase, β-glucuronidase, lipase (C4), α-mannosidase, trypsin, and valine arylamidase. Has an A4α type peptidoglycan with a (L-Ala)–D-Glu–L-Lys–D-Glu peptide subunit, and interpeptide bridge composed of D-glucose moieties.

Source: human feces.

DNA G+C content (mol%): 64.4 (HPLC).

Type strain: JCM 10643, DSM 13280, RCA56-68, CCUG 45296, CIP 106914.

Sequence accession no. (16S rRNA gene): AB031063.

3. **Collinsella stercoris** Kageyama and Benno 2000, 1773[VP]

ster′co.ris. L. n. *stercus -oris* dung, excrement, feces; L. gen. n. *stercoris* of feces, referring to source of the isolate.

Gram-stain-positive, obligate anaerobic organism which occurs in chains of 2–20 cells (0.3–0.4 × 1.3–2.4 μm). Colonies are white or gray in the center with clear margins on EG agar. Spores and flagella are absent. Grows on EG agar in 2 d at 37°C in an anaerobic jar containing 100% CO_2.

Acid is produced from cellobiose, fructose, galactose, glucose, lactose, maltose, mannose and salicin, but not from amygdalin, L-arabinose, erythritol, esculin, glycogen, inositol, mannitol, melezitose, raffinose, rhamnose, ribose, sorbitol, starch, sucrose, trehalose or D-xylose. Positive for acid phosphatase, β-galactosidase, leucine arylamidase, naphthol-AS-BI-phosphohydrolase, and *N*-acetyl-β-glucosaminidase-positive, but negative for chymotrypsin, cystine arylamidase, α-fucosidase, α-galactosidase, β-glucuronidase, lipase (C4), α-mannosidase, trypsin, or valine arylamidase. Has an A4β type peptidoglycan with a (L-Ala)–D-Glu–L-Orn–D-Asp peptide subunit, and interpeptide bridge consisting only of D-asparagine moieties.

Source: human feces.

DNA G+C content (mol%): 61.2 (HPLC).

Type strain: JCM 10641, DSM 13279, RCA55-54, CCUG 45295, CIP 106913.

Sequence accession no. (16S rRNA gene): AB031061.

Genus IV. **Cryptobacterium** Nakazawa, Poco, Ikeda, Sato, Kalfas, Sundqvist and Hoshino 1999, 1199[VP]

Futoshi Nakazawa

Crypt.o.bac.te′ri.um. Gr. n. *kruptos* hidden; Gr. n. *bakterion* a small rod; N.L. neut. n. *Cryptobacterium* a hidden rod-shaped bacterium.

Cells are very short rods, 0.4 × 0.8–1.0 µm. Gram-stain-positive, occasionally Gram-stain-negative in stationary phase, having a cell wall typical of Gram-stain-positive bacteria (Figure 418). **Nonmotile, non-sporing, and obligately anaerobic. Catalase-negative. Asaccharolytic and no volatile end product in peptone-yeast extract-glucose broth**. Positive for arginine hydrolysis, negative for nitrate reduction and inert in most conventional biochemical tests. Isolated from the periodontal pocket of an adult human with periodontal disease, and from necrotic dental pulp.

DNA G+C content (mol%): 50–51 (T_m).

Type species: **Cryptobacterium curtum** Nakazawa, Poco, Ikeda, Sato, Kalfas, Sundqvist and Hoshino 1999, 1199[VP].

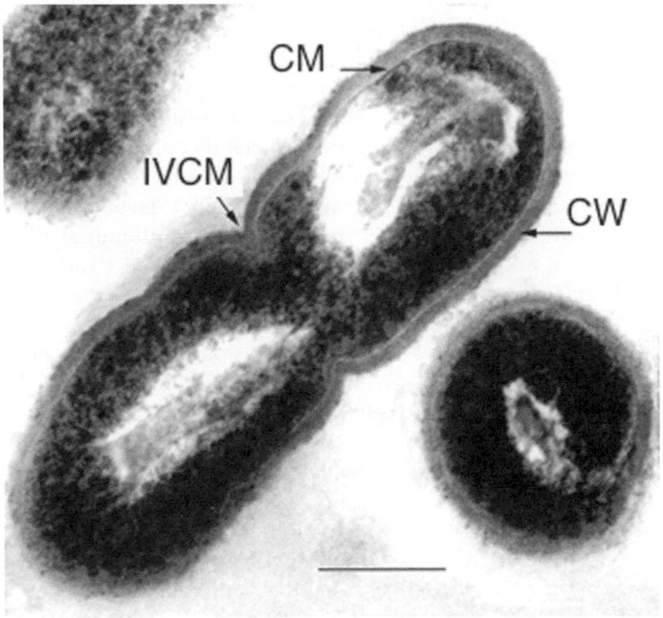

FIGURE 418. Transmission electron microphotograph of *Cryptobacterium curtum*. CM, cytoplasmic membrane; CW, cell wall, IVCM, invagination of cytoplasmic membrane.

Enrichment and isolation procedures

Growth in broth media is very poor with or without carbohydrates. *Cryptobacterium* forms circular, convex, and translucent colonies on BHI-blood agar plates, which are about 0.3–0.5 mm in diameter. Colonies are less than 1 mm in diameter even after prolonged incubation. Hemolysis does not occur around colonies on BHI-blood agar plates.

Differentiation of the genus *Cryptobacterium* from other genera

According to the results of 16S rRNA gene sequence analyses, the genus *Cryptobacterium* belongs to the family *Coriobacteriaceae* which contains anaerobic, asaccharolytic, Gram-stain-positive rods. DNA–DNA hybridization studies indicate that the genus *Cryptobacterium* shows low levels (1–2%) of DNA relatedness against closely related taxa such as the genera *Collinsella*, *Eggerthella* (Nakazawa and Hoshino, 2004), *Mogibacterium* (Nakazawa et al., 2000, 2002), and *Slackia*. The genus *Cryptobacterium* can also be distinguished by restriction fragment length polymorphism (RFLP) analysis of 16S rDNA using the endonucleases *Alu*I and *Smf*. SDS-PAGE analysis of whole-cell proteins and Western blotting reactions with rabbit antisera also allow the genus *Cryptobacterium* to be distinguished from related genera. In addition, the genus can be differentiated from the genera *Collinsella*, *Eggerthella*, and *Slackia* by metabolic end products, arginine hydrolysis, and nitrate reduction.

Taxonomic comments

Initially organisms now classified as *Cryptobacterium* species were isolated from the human oral cavity as anaerobic, asaccharolytic, *Eubacterium*-like isolates. However, these isolates were inert in most conventional biochemical tests and could not be assigned to previously established bacterial species. The genus *Cryptobacterium*, with *Cryptobacterium curtum* as the type strain, was described on the basis of a polyphasic taxonomic study, notably by using 16S rRNA gene sequence and DNA–DNA hybridization data.

List of species of the genus *Cryptobacterium*

1. **Cryptobacterium curtum** Nakazawa, Poco, Ikeda, Sato, Kalfas, Sundqvist and Hoshino 1999, 1199[VP]

cur'tum. L. neut. adj. *curtum* shortened (a shortened cell of this organism).

Cells are very short, Gram-stain-positive, obligately anaerobic, nonmotile, non-sporing rods which occur singly or in masses, and which sometimes stain Gram-negative in older cultures. On BHI-blood agar plates they form minute circular, convex, translucent colonies less than 1 mm in diameter, even after prolonged incubation in an anaerobic glove box. Growth in broth media is poor with or without carbohydrates. Hemolysis does not occur on BHI-blood agar plates. They are inert in most biochemical tests. Neither esculin nor starch are hydrolyzed. Does not reduce nitrate or liquefy gelatin. Negative for catalase, indole, and urease. Ammonia is produced from arginine. Nonfermentative and do not utilize adonitol, amygdalin, arabinose, cellobiose, erythritol, esculin, fructose, galactose, glucose, glycogen, inositol, lactose, maltose, mannitol, mannose, melezitose, melibiose, rhamnose, ribose, salicin, sorbitol, starch, sucrose, trehalose, or xylose. No metabolic end product is detected in peptone-yeast extract medium supplemented with glucose or peptone-yeast extract-glucose broth.

Source: the periodontal pocket of an adult human with periodontal disease, and from necrotic dental pulp.

DNA G+C content (mol%): 50–51 (HPLC).

Type strain: 12-3, ATCC 700863.

Sequence accession no. (16S rRNA gene): AB019260.

Genus V. **Denitrobacterium** Anderson, Rasmussen, Jensen and Allison 2000, 633[VP]

Robin C. Anderson and Thaddeus B. Stanton

De.nit.ro.bac.te'ri.um. L. pref. *de* from; L. n. *nitro* nitrocompound; Gr. neut. dim. n. *bakterion* a small rod; N.L. neut. n. *Denitrobacterium* named for a nitrocompound-reducing rod-shaped bacterium.

Rod-shaped cells (0.5–1.0 × 1.0–1.5 μm). Cells may possess irregular, bulbous ends. **Gram-stain-positive, non-sporulating,** and **nonmotile**. Strictly **anaerobic** chemo-organotroph; growth is supported by suitable electron acceptors and is proportional to amount of acceptor reduced. **Nonfermentative.** Growth occurs in anaerobic media containing clarified ruminal fluid, peptone, and a suitable electron acceptor. **Dimethyl sulfoxide, nitroethane, 3-nitropropan-1-ol, 2-nitropropan-1-ol, 3-nitropropionate, trimethyl amine-oxide and, by some strains, nitrate, may be used as electron acceptors and support good growth. Nitroethanol, 2-nitrobutane, and 1-nitropropane also may be used as electron acceptors, but growth with these compounds is marginal, presumably due to the accumulation of toxic end products. Azide, chlorate, fumarate, nitrite, 2-nitrobenzene, perchlorate, sulfate, and sulfite do not support growth.** Little if any acid is produced during growth in medium with hydrogen or formate as electron donor. **Hydrogen, formate, and (DL)-lactate can be used as electron donors**; acetate is produced when grown with lactate. **Gelatin is not hydrolyzed, indole and H_2S are not produced**. Palmitic acid ($C_{14:0}$) is the predominant cellular fatty acid of cells grown in modified brain heart infusion medium, although an abundance of other straight and branched chain fatty acids, aldehydes, and dimethyl acetals are present. Cells contain a *c*-type cytochrome. Isolated from bovine ruminal contents. A single species of *Denitrobacterium*, *Denitrobacterium detoxificans*, has been characterized. The results of 16S rRNA gene sequence analysis indicate *Denitrobacterium* belongs to the *Actinobacteria* (Figure 419).

DNA G+C content (mol%): 56–61 (T_m).

Type species: **Denitrobacterium detoxificans** Anderson, Rasmussen, Jensen and Allison 2000, 633[VP].

Further descriptive information

Based on 16S rRNA gene sequence analysis, the genus *Denitrobacterium* falls within class *Actinobacteria*, subclass *Coriobacteridae*, order *Coriobacteriales*, suborder *Coriobacterineae*, family *Coriobacteriaceae*. A single species, *Denitrobacterium detoxificans*, has been named. *Denitrobacterium* is most closely related to the genus *Eggerthella* (Figure 419).

Cells of *Denitrobacterium detoxificans* are rod-shaped (0.5–1.0 × 1.0–1.5 μm), stain Gram-positive, and do not react with 3% KOH to yield a slimy interface typical of Gram-stain-negative cell types (Gregersen, 1978). Electron micrographs of thin sections reveal Gram-stain-positive cell-wall structures and the presence of unidentified inclusion bodies (Anderson et al., 1997). Stains for spores or known types of inclusion bodies such as polyphosphate granules, glycogen, starch, or poly β-hydroxybutyrate are all negative. Cells do not survive treatment for 10 min at 80°C (Anderson et al., 1997). Flagella are not observed (Anderson et al., 1997).

Palmitic acid ($C_{14:0}$) is the predominant fatty acid present in cells grown in modified brain heart infusion medium; however, considerable amounts of other straight and branched chain fatty acids (C_{12} to C_{18}), as well as aldehydes and dimethyl acetals, are also present (Anderson et al., 1997, 2000). Cells contain a *c*-type cytochrome (Anderson et al., 1997).

Denitrobacterium detoxificans obtains energy for growth exclusively via anaerobic respiration, coupling the oxidation of hydrogen, formate, or lactate to the reduction of suitable anaerobic electron acceptors such as dimethyl sulfoxide, nitroethane, 3-nitropropan-1-ol, 2-nitropropan-1-ol, 3-nitropropionate, nitroethanol, 1-nitropropane, 2-nitrobutane, and trimethyl amine-oxide. Nitrate is also used as an electron acceptor and is reduced to ammonia by the type strain *Denitrobacterium*

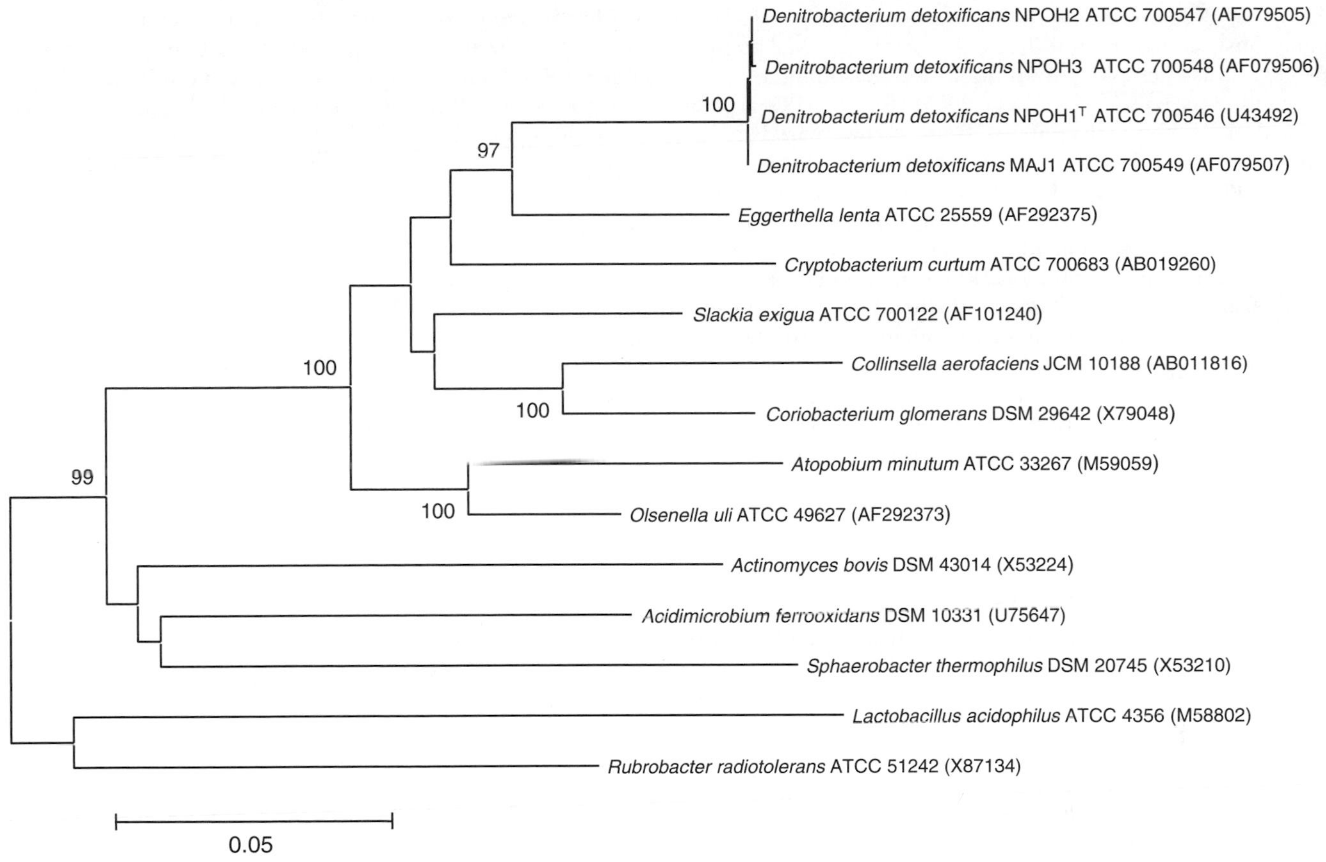

FIGURE 419. Phylogenetic tree based on 16S rRNA gene sequences of *Denitrobacterium detoxificans* strains and type strains of their closest relatives. Phylogenetic and molecular evolutionary analyses were conducted using MEGA version 2.1 (Kumar et al., 2001). Sequence alignments were based on 1592 nucleotide positions and compared by Jukes–Cantor analysis. The tree was constructed using the neighbor-joining method. Scale bar represents a 5% difference in nucleotide sequence. Bootstrap values greater than 95 and based on 1000 replicates are indicated near branch points.

detoxificans NPOH1^T, but not by any of the other three strains tested, NPOH2, NPOH3, and MAJ1 (Anderson et al., 2000).

Denitrobacterium strains are routinely cultured in anaerobic medium B* (Anderson et al., 1996) supplemented with an appropriate source of reductant and electron acceptor. Reductant (provided as 16 mM sodium formate) and acceptor (provided as 3-nitropropionate; generally 5–10 mM) support good growth and are convenient to prepare, filter-sterilize, and to add to previously prepared medium. Dimethyl sulfoxide, nitroethane, 3-nitropropan-1-ol, trimethyl amine-oxide, and, in the case of strain NPOH1^T, nitrate also support good growth, whereas nitroethanol, 2-nitrobutane, and 1-nitropropane do not. Reductant can also be provided as H_2 in the gas phase or as lactate (Anderson et al., 1997). Undefined nutritional factors in phytone are important for maximum growth of all four strains tested.

Cells of *Denitrobacterium detoxificans* NPOH1^T exhibit a positive reaction for arginine aminopeptidase and a weakly positive alkaline phosphatase reaction in the AN-IDENT test system; they exhibit positive reactions for phosphatase acid and naphthol-AS-BI-phosphorohydrolase and a weak positive reaction for lipase-C14 in the API ZYM test system (Anderson et al., 1997). Reactions to all other AN-IDENT and API ZYM tests are negative.

Of the four strains of *Denitrobacterium detoxificans* analyzed, all share >99% 16S rRNA gene sequence similarity.

Denitrobacterium detoxificans strain NPOH1^T exhibits sensitivity to 6 μg/ml cephalothin, 12 μg/ml, chloramphenicol, 1.6 μg/ml clindamycin, and 3 μg/ml erythromycin, and is resistant to 4 μg/ml ampicillin, 100 μg/ml carbenicillin, 2 U/ml penicillin, and 6 μg/ml tetracycline (Anderson et al., 1997).

Denitrobacterium detoxificans strains were isolated from bovine ruminal contents enriched for enhanced rates of 3-nitropropionic acid or 3-nitropropan-1-ol metabolism (Anderson et al.,

*Medium B contains clarified ruminal fluid at 8% (v/v) and (mg/liter): K_2HPO_4, 225.0; KH_2PO_4, 225.0; $(NH_4)_2SO_4$, 450.0; NaCl, 450.0; $MgSO_4 \cdot 7H_2O$, 45.0; $CaCl_2$, 22.5; thiamine, 2.0, pantothenate, 2.0; nicotinamide, 2.0; pyridoxine-HCl, 2.0; riboflavin, 2.0; *p*-aminobenzoic acid, 1.0; biotin, 0.05; folic acid, 0.05; lipoic acid, 0.05; vitamin B-12, 0.02; resazurin, 1.0; cysteine-HCl, 500.0; Na_2CO_3, 4000.0; phytone peptone, 8000.0. Clarified ruminal fluid is prepared by autoclaving freshly collected ruminal fluid that had been strained through cheese cloth or nylon paint strainers and then centrifuging at least twice at 10,000 *g* to remove particulate matter. Cysteine-HCl and Na_2CO_3 are added after the medium containing the other ingredients is adjusted to pH 6.8. The complete medium is boiled and then cooled on ice while being flushed with oxygen-free CO_2 or H_2CO_2 (50:50). When cooled, the medium is anaerobically distributed to culture tubes (Bryant, 1972) which are immediately closed with stoppers and autoclaved.

1996, 1997). 3-Nitropropionic acid or 3-nitropropan-1-ol are toxic constituents of some forages, notably many *Astragalus* species (milkvetches), known to cause poisoning of grazing ruminants (Anderson et al., 2005). Strain NPOH1^T was isolated from ruminal contents from a cow reared at the National Animal Disease Center (NADC) in Ames, Iowa; strains NPOH2 and NPOH3 were isolated likewise from ruminal contents from a different cow reared at the same facility and strain MAJ1 was isolated from ruminal contents from a cow reared in British Columbia. Rates of ruminal nitrotoxin metabolism are markedly increased when cattle are fed sublethal amounts of nitro-containing forage or nitroethane, a nontoxic analog, thus implicating a selection mechanism to enrich for bacteria like *Denitrobacterium detoxificans*, which can confer protection to the host (Anderson, 1998; Majak, 1992b, 1992a; Majak et al., 1986). *Denitrobacterium detoxificans*-like 16S rDNA has been detected in a carious dental sample (Chhour et al., 2005).

Enrichment and isolation procedures

Denitrobacterium can be isolated by enriching anaerobic nitro-respiring bacteria. *Denitrobacterium detoxificans* was isolated from ruminal contents enriched from 3×10^4 cells/ml ruminal fluid to $>10^7$ cells/ml during consecutive (24 h intervals at 39°C) batch culture in a minimal ruminal fluid medium supplemented with gradually increasing concentrations (5–10 mM) of 3-nitropropionic acid or 3-nitropropan-1-ol (Anderson et al., 1996). Concentrations of other bacteria decreased concurrently during the enrichment to approximately 10^7 cells/ml. Anaerobic techniques for both broth and agar (2%) medium (Bryant, 1972) were used during enrichment and colony isolation. Hydrogen (50% in CO_2) in the gas phase served as reductant during the isolations although 16 mM formate, which supports nearly equal amounts of growth (Anderson et al., 1997), may work equally well. Isolates picked at random from roll tubes containing solidified medium B were cultured in nitro-supplemented medium B broth. The isolates that grew and reduced nearly all of the added 3-nitropropionic acid or 3-nitropropan-1-ol, as determined via colorimetric analysis of spent culture fluid (Anderson et al., 1993), were indicative of *Denitrobacterium detoxificans.*

Maintenance procedures

All strains of *Denitrobacterium detoxificans* can be maintained by serial transfer, at 3–4 d intervals, in anaerobically prepared medium B (Anderson et al., 1996) supplemented with 16 mM formate and 9 mM 3-nitropropionic acid. Anaerobic *Denitrobacterium detoxificans* cultures preserved as slants or broths in 20% glycerol can be stored at −80°C for at least 10 years.

Differentiation of the genus *Denitrobacterium* from other genera

Denitrobacterium species can be differentiated from other anaerobic respiring organisms by their obligate respiratory metabolism and their ability to respire on nitroalkane compounds such as 3-nitropropan-1-ol, 2-nitropropan-1-ol, 3-nitropropionic acid, nitroethanol, nitroethane, 1-nitropropane, and 2-nitrobutane but not on other acceptors such as O_2, chlorate, fumarate, perchlorate, sulfate, or sulfite. *Denitrobacterium* species can also be differentiated from other genera by 16S rRNA gene sequence analysis.

Differentiation of species of the genus *Denitrobacterium*

The genus contains a single species, *Denitrobacterium detoxificans.*

List of species of the genus *Denitrobacterium*

1. **Denitrobacterium detoxificans** Anderson, Rasmussen, Jensen and Allison 2000, 633^{VP}

de.tox.if'i.cans. L. pref. *de* from; L. n. *toxicum* poison; L. neut. n. *detoxificans* poison reducer.

The species description is as stated above for the genus. Isolated from ruminal contents enriched for enhanced rates of 3-nitropropan-1-ol metabolism, the toxic aglycone of miserotoxin (3-nitro-1-propyl-β-D-glucopyranoside). All four strains examined share close (>99%) 16S rRNA gene sequence similarity, and have been deposited in the American Type Culture Collection. The type strain differs from the others (strain NPOH2, ATCC 700547; strain NPOH3, ATCC 700548 and MAJ1, and ATCC 700549) in that it has the ability to reduce nitrate.

DNA G+C content (mol%): 56–61 (T_m).

Type strain: NPOH1, ATCC 700546, CCUG 5674.

Sequence accession no. (16S rRNA gene): U43492.

Genus VI. **Eggerthella** Wade, Downes, Dymock, Hiom, Weightman, Dewhirst, Paster, Tzellas and Coleman 1999, 599^{VP}

William G. Wade and Floyd E. Dewhirst

Egg.er.thel'la. N.L. fem. n. named to honor Arnold H. Eggerth (1889–1962), who first described the organism later named *Eubacterium lentum.*

Gram-stain-positive obligately **anaerobic bacilli, which** are **nonmotile** and do not produce endospores. **Growth** is **stimulated by arginine. Acid is not formed** from sugars. The cell wall contains an A4γ-type peptidoglycan with an (L-Ala)–D-Glu–*m*-Dpm–D-Glu peptide subunit with an interpeptide bridge that consists only of D-Glu.

DNA G+C content (mol%): 61–65 (HPLC).

Type species: **Eggerthella lenta** Wade, Downes, Dymock, Hiom, Weightman, Dewhirst, Paster, Tzellas and Coleman 1999, 599^{VP}.

Further descriptive information

Members of this genus and related genera tend to be unreactive in the biochemical tests commonly used for identification and therefore have historically been a difficult group to characterize. The phenotypic description of *Eggerthella lenta*, in particular, has long been known to define a heterogeneous group of organisms isolated from the large intestine of humans and animals. In one study investigating this heterogeneity, 32 strains were divided into 4 groups on the basis of growth stimulation by arginine, production of H_2S, catalase, and bile acid-degrading enzymes (MacDonald et al., 1979). Groups A–C degraded steroids and related molecules, while members of Group D did not display this activity and had a cellular fatty acid composition distinct from the other groups (Verhulst et al., 1987). The type strain was subsequently shown to belong to the Group A–C cluster (Mosca et al., 1998), and the colonies of most strains were shown to fluoresce red under UV light (Mosca et al., 1993). Whether the three groups in this cluster correspond to the three currently validly published species remains to be established. Given the marked phenotypic differences between Group D and Groups A–C, it is likely that Group D strains do not belong to this genus. In addition, the use of a commercially available identification kit has frequently resulted in the mis-identification of oral *Slackia exigua* isolates as *Eggerthella lenta* (Wade et al., 1990). A systematic polyphasic study of a large collection of strains identified as *Eggerthella* by a combination of sequence analysis of 16S rRNA genes and the presence of the characteristic peptidoglycan described above is required. Once this has been done, it will be possible to construct more precise descriptions of the three *Eggerthella* species described to date. It is also likely that additional novel species, and perhaps novel genera, would be detected by such an approach.

Eggerthella species are found in the human large intestine and in clinical specimens from a wide variety of infections including those of the blood and postoperative wounds and various kinds of abscesses found in different parts of the body such as the brain and associated with the rectum, scrotum, and pelvis (Chan and Mercer, 2008; Lattuada et al., 2005; Lau et al., 2004c; Moore et al., 1971; Schwiertz et al., 2000). *Eggerthella lenta* is one of a number of Gram-stain-positive anaerobic bacilli, including *Collinsella aerofaciens*, another member of the *Coriobacteriaceae*, capable of inducing arthritis in a rat model (Severijnen et al., 1990).

Enrichment and isolation procedures

Strains of all species grow readily on horse blood agar forming colonies at least 0.5 mm in diameter after a 48-h anaerobic incubation at 37°C. The inclusion of 0.5% arginine in agar and broth media stimulates the growth of all species.

Maintenance procedures

Strains can be maintained on blood agar incubated anaerobically at 37°C and subcultured weekly. Lyophilization of cultures in the early stationary phase of growth is recommended for long-term storage of most strains. Strains can also be stored at –70°C in Brain Heart Infusion broth supplemented with 10% glycerol.

Procedures for testing special characters

The general methods described for the characterization of anaerobes in the VPI *Anaerobe Laboratory Manual* (Holdeman et al., 1977b) and the *Wadsworth-KTL Anaerobic Bacteriology Manual* (Jousimies-Somer et al., 2002a) are suitable for the study of members of this genus. The RapidID 32A anaerobe identification kit (bioMérieux) includes tests useful for the differentiation of the species of this genus. However, as mentioned above, identification by 16S rRNA gene sequencing is the most reliable method for this group and will also detect as yet un-named taxa.

Differentiation of the genus *Eggerthella* from other genera

Key characteristics of *Eggerthella* and related genera are shown in Table 307.

Taxonomic comments

Eggerthella lenta was originally described as a member of the genus *Eubacterium*, but was shown by 16S rRNA gene sequence analysis to belong to the family *Coriobacteriaceae* within the phylum *Actinobacteria*, while *Eubacterium limosum*, the type species of *Eubacterium*, is in the *Firmicutes* (Kageyama et al., 1999a; Wade et al., 1999). A new genus, *Eggerthella*, was proposed to accommodate this species (Wade et al., 1999) and two additional species were subsequently described (Lau et al., 2004b). The phylogenetic relationship between the genus *Eggerthella* and related genera within the family *Coriobacteriaceae* can be seen in the tree constructed from 16S rRNA gene comparisons in Figure 420.

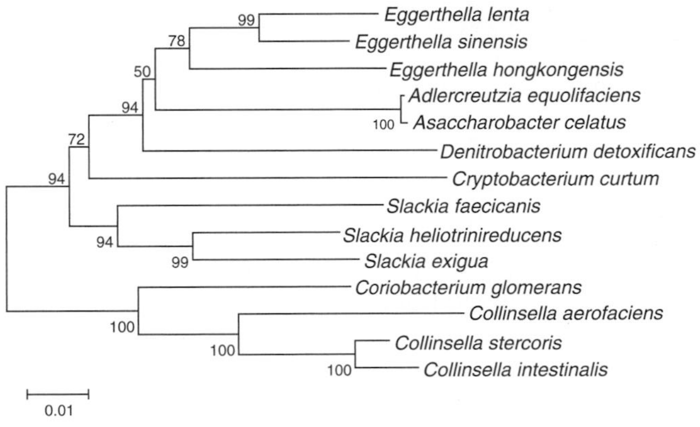

FIGURE 420. Neighbor-joining tree based on 16S rRNA gene sequence comparisons over 1367 aligned bases showing relationships between *Eggerthella* species and between them and related genera. Numbers at the nodes represent bootstrap values for each branch based on data for 100 trees; only values greater than 50% are shown. Bar = 0.01 substitutions per nucleotide position.

Differentiation of species of the genus *Eggerthella*

A scheme for the differentiation of *Eggerthella* species is shown in Table 311.

TABLE 311. Phenotypic characteristics of *Eggerthella lenta*, *Eggerthella hongkonensis*, and *Eggerthella sinensis*[a,b,c]

Chaacteristic	*E. lenta*	*E. hongkongensis*	*E. sinensis*
Arginine arylamidase	–	v	+
β-Glucosidase	–	+	–
Leucine arylamidase	–	v	–
Nitrate reduction	+	–	–

[a]Symbols: +, present; –, absent; v, variable.
[b]Data from Maruo et al. (2008).
[c]Tests performed using the Rapid ID 32A kit (bioMérieux).

Further reading

Moore W.E.C., E.P. Cato and L.V. Holdeman. 1971. *Eubacterium lentum* (Eggerth) Prévot 1938: emendation of description and designation of the neotype strain. Int. J. Syst. Bacteriol. *21*: 299–303.

List of species of the genus *Eggerthella*

1. **Eggerthella lenta** Wade, Downes, Dymock, Hiom, Weightman, Dewhirst, Paster, Tzellas and Coleman 1999, 599[VP]

len'ta. L. fem. adj. *lenta* slow.

Cells from PYG broth cultures are 0.2–0.4 × 0.2–2.0 μm and occur singly and in pairs and short chains. Colonies on horse blood agar are 0.5–2.0 mm, circular, entire to erose, raised to low convex, translucent to semiopaque, dull to shiny, smooth, and sometimes with a mottled appearance. Most strains grow at 30 and 45°C; some grow at 25°C. Arginine enhances growth. Gas is not detected in PYG agar deep cultures. Ammonia is produced from arginine. Acid is not produced from sugars. Hydrogen sulfide is produced in the butt of TSI (triple-sugar iron) slants incubated anaerobically, but not in SIM medium. Catalase is produced by strains grown on agar medium containing 1%, w/v arginine. Hippurate is not hydrolyzed.

The metabolic end products from glucose are acetate with trace amounts of lactate or succinate. H_2 is not produced.

Source: human large intestine, feces, blood and clinical infections, primarily wounds and abscesses.

DNA G+C content (mol%): 61 (HPLC) for the type strain.

Type strain: ATCC 25559, CCUG 17323 A, CIP 106637, DSM 2243, JCM 9979, NCAIM B.01418, NCTC 11813.

Sequence accession no. (16S rRNA gene): AF292375.

Additional comments: arginine is degraded via the arginine dihydrolase pathway (Sperry and Wilkins, 1976a). The type strain contains cytochromes *a*, *b*, and *c* and a carbon monoxide-binding pigment (Sperry and Wilkins, 1976b).

Eggerthella lenta has been shown to transform a wide range of compounds, including steroids and lignans, primarily via dehydrogenases and dehydroxylase. Bile acids such as cholic and chenedeoxycholic acids are degraded by the action of 3α-, 7α-, and 12α- dehydrogenases; the 7α- activity is only demonstrable under anaerobic conditions (Hirano and Masuda, 1981; MacDonald et al., 1977). *Eggerthella lenta* strains also degrade corticosteroids (Bokkenheuser et al., 1977, 1979), linoleic acid, and related unsaturated long-chain fatty acids (Verhulst et al., 1986). The biological activity of the cardiac glycosides digitoxin and digoxigenin are destroyed by *Eggerthella lenta*-mediated reduction of a critical double bond (Dobkin et al., 1982; Robertson et al., 1986). Significant destruction of digoxigenin is thought to occur in 10% of treated patients owing to carriage of high concentrations of *Eggerthella lenta* in their gut microbiota (Chandrasekaran et al., 1987). *Eggerthella lenta* can participate in the metabolism of phyto-estrogens such as secoisolariciresinol diglucoside to produce the estrogen-like compounds enterodiol and enterolactone, which may play a role in the prevention of cancer, atherosclerosis, and diabetes (Clavel et al., 2005, 2006; Jin et al., 2007).

The gene encoding the tetracycline efflux pump, TetK, conferring resistance to this antimicrobial, has been detected in *Eggerthella lenta* (Roberts et al., 1991)

2. **Eggerthella hongkongensis** Lau, Woo, Woo, Fung, Wong, Chan, Tam and Yuen 2006, 2025[VP] (Effective publication: Lau 2004b, 255.)

hong.kong'en'sis. N.L. fem. adj. *hongkongensis* pertaining to Hong Kong, the city where the bacterium was discovered.

Cells are obligately anaerobic, nonsporing, nonmotile Gram-stain-positive coccobacilli which occur in chains. Colonies on blood agar are gray-white and 0.5 mm in diameter after a 48-h anaerobic incubation at 37°C. Catalase, arginine dihydrolase and β-glucosidase are produced, but production of alanine arylamidase, arginine arylamidase, leucine arylamidase, and lysine arylamidase is variable. Alkaline phosphatase, indole, and urease are not produced and nitrate is not reduced. Acid is not produced from glucose or other sugars.

Source: blood.

DNA G+C content (mol%): 61 (HPLC) for the type strain.

Type strain: HKU10, DSM 16106, JCM 14552, LMG 22124.

Sequence accession no. (16S rRNA gene): AY288517.

3. **Eggerthella sinensis** Lau, Woo, Woo, Fung, Wong, Chan, Tam and Yuen 2006, 2025[VP] (Effective publication: Lau 2004b, 255.)

sin'en'sis. N.L. gen. n. *sinae* of China; N.L. fem. adj. *sinensis* pertaining to China, the country where the bacterium was discovered.

Cells are obligately anaerobic, non-sporing, nonmotile, Gram-stain-positive, coccobacilli which occur in chains. Colonies on blood agar are gray-white, and 0.5 mm in diameter after a 48-h anaerobic incubation at 37°C. Catalase, arginine dihydrolase, arginine arylamidase and lysine arylamidase are produced but not alkaline phosphatase, indole, urease, or β-glucosidase. Nitrate is not reduced and acid is not produced from sugars.

Source: blood.

DNA G+C content (mol%): 65 (HPLC) for the type strain.

Type strain: HKU14, DSM 16107, JCM 14551, LMG 22123.

Sequence accession no. (16S rRNA gene): AY321958.

Genus VII. **Olsenella** Dewhirst, Paster, Tzellas, Coleman, Downes, Spratt and Wade 2001, 1802^{VP}

Floyd E. Dewhirst and William G. Wade

Ol.sen.el'la. L. fem. dim. ending *-ella*; N.L. fem n. *Olsenella* of Olsen, named to honor Ingar Olsen, a contemporary Norwegian microbiologist who first described *Lactobacillus uli.*

Cells are small, **elliptical rods** which occur singly, in pairs or short chains. Nonsporeforming. Nonmotile. Gram-stain-positive. **Obligate anaerobes**. Glucose fermented to **acetate and lactate**. Does not produce catalase, urease or indole, or reduce nitrate.

DNA G+C content (mol%): 63–64 (HPLC).

Type species: **Olsenella uli** (Olsen, Johnson, Moore and Moore 1991) Dewhirst, Paster, Tzellas, Coleman, Downes, Spratt and Wade 2001, 1802^{VP} (*Lactobacillus uli* Olsen, Johnson, Moore and Moore 1991, 263.).

Further descriptive information

The major fatty acid of the cell membrane is $C_{18:1}$ (*cis*9). Strains have been isolated from the human oral cavity (Dewhirst et al., 2001; Munson et al., 2004; Olsen et al., 1991; Siqueira et al., 2005), human bacteremia (Lau et al., 2004a), and cow rumen (GenBank AJ251324).

Enrichment and isolation procedures

Strains of *Olsenella* species have been recovered using agar media commonly used for the cultivation of anaerobes, such as Brain Heart Infusion, Peptone Yeast Glucose, and Fastidious Anaerobe Agar, supplemented with 5% horse or sheep blood. Specific enrichment procedures have not been reported for the genus. The growth of *Olsenella uli*, but not that of *Olsenella profusa*, is markedly stimulated by addition of Tween 80 (0.02%, v/v).

Maintenance procedures

Stock cultures of *Olsenella* species can be maintained under anaerobic conditions by weekly transfer on Blood Agar plates. Cultures can be stored by lyophilization or freezing at −80°C in Brain Heart Infusion Broth supplemented with 10% glycerol.

Differentiation of the genus *Olsenella* from other genera

Characters that differentiate *Olsenella* from closely related genera are shown in Table 307. *Olsenella* can be differentiated from *Atopobium*, *Cryptobacterium*, and *Denitrobacterium* by G+C content of the DNA, from *Adlercreutzia*, *Denitrobacterium*, *Eggerthella*, and *Slackia* by sugar fermentation, and from *Collinsella*, *Coriobacterium*, and *Eggerthella* by site of origin. Molecular methods, such as sequencing of the 16S rRNA gene, provide the most definitive identification.

Taxonomic comments

Prior to the use of 16S rRNA-based phylogenetic analysis, strains of species in the genus *Olsenella* were misclassified in the genera *Eubacteria* and *Lactobacillus* of the phylum *Firmicutes.* The genus *Olsenella* is now recognized as one of nine genera classified in the family *Coriobacteriaceae* in the phylum *Actinobacteria* (Dewhirst et al., 2001; Wade et al., 1999) as shown in Figure 421.

Using 16S rRNA gene sequence analysis, examination of oral strains in the Virginia Polytechnic Institute Anaerobe

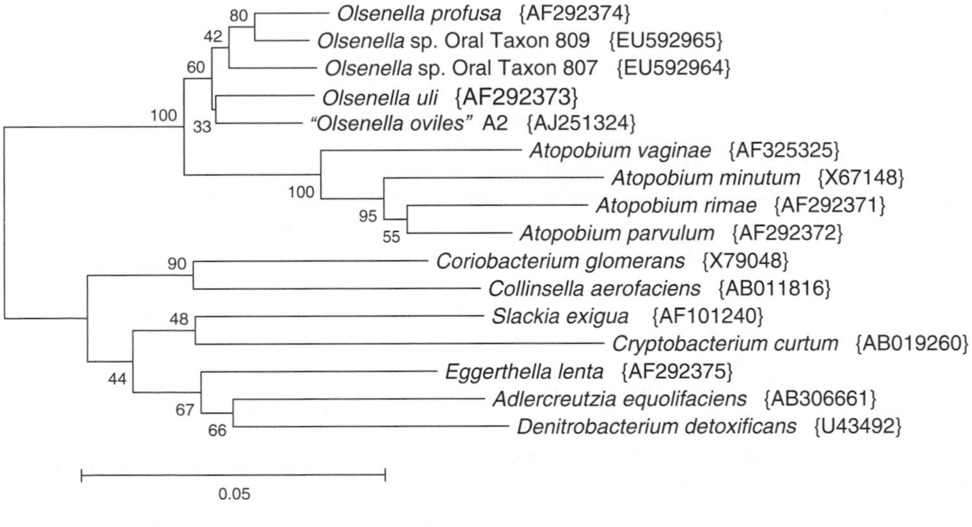

FIGURE 421. Neighbor-joining tree (Saitou and Nei, 1987) showing relationships between *Olsenella* species and representatives of other genera classified in the family *Coriobacteriaceae.* The optimal tree with the sum of branch length = 0.581 is shown. The percentage of replicate trees in which the associated taxa clustered together in the bootstrap test (1000 replicates) is shown next to the branches (Felsenstein, 1985). The tree is drawn to scale, with branch lengths in the same units as those of the evolutionary distances used to infer the phylogenetic tree. The evolutionary distances were computed using the Jukes–Cantor method (Jukes and Cantor, 1969) and are expressed as the number of base substitutions per site. All positions containing alignment gaps and missing data were eliminated only in pairwise sequence comparisons (pairwise deletion option). There were a total of 1491 positions in the final dataset. Phylogenetic analyses were conducted in MEGA4 (Tamura et al., 2007).

TABLE 312. Differential characteristics of *Olsenella uli* and *Olsenella profusa*[a]

Test	*O. uli* (48 strains)	*O. profusa* (2 strains)
Rapid ID 32A profile	2012 0337 05	4516 0537 05
Growth in PYG broth	Poor (1+)	Good (3–4+)
Growth stimulation by Tween 80	Marked	Slight
Arginine hydrolysis	+	–
Fermentation of:		
Arabinose	–	+
Mannitol	–	+
Melibiose	–	+
Raffinose	–	+
Colony morphology[b]	Raised, grey, semi-translucent	Pyramidal, cream, opaque

[a]Data from Olsen et al. (1991) and Dewhirst et al. (2001).

[b]After incubation for 7 d on Fastidious Anaerobe Agar.

Laboratory collection of W.E.C. Moore and L.V. Holderman Moore has identified two additional *Olsenella* species (F.E. Dewhirst, unpublished studies). These strains were designated Coccus D49 and *Lactobacillus* D19 by the Moores, and are now designated *Olsenella* sp. Oral Taxon 807 and *Olsenella* sp. Oral Taxon 809 as shown in Figure 421 and described in the Human Oral Microbiome Database (www.homd.org; Dewhirst et al., 2008). A rumen strain was isolated and informally called "*Atopobium oviles*" in a GenBank submission (AJ251324), but the name has not been formally published. Because the rumen strain has 97% mean similarity to *Olsenella uli* and *Olsenella profusa*, and only 93.5% mean similarity to *Atopobium* species, we would suggest the informal designation should be "*Olsenella oviles*". Several dozen clone sequences in GenBank, from a variety of sources, are greater than 90% similar to species in the genera *Olsenella* and *Atopobium*. Some of these clones come from environmental sources, but may represent bacteria shed from mammalian hosts.

Acknowledgements

Floyd E. Dewhirst was supported by NIH grant DE106937 from the National Institute of Dental and Craniofacial Research.

Differentiation of species of the genus *Olsenella*

Biochemical characteristics useful in distinguishing the species of genus are listed in Table 312. The two species are easily differentiated from one another by using Rapid ID32A anaerobe identification kits (bioMérieux) and on the basis of 16S rRNA gene sequences.

List of species of the genus *Olsenella*

1. **Olsenella uli** (Olsen, Johnson, Moore and Moore 1991) Dewhirst, Paster, Tzellas, Coleman, Downes, Spratt and Wade 2001, 1802^{VP} (*Lactobacillus uli* Olsen, Johnson, Moore and Moore 1991, 263)

u'li. Gr. n. *oulon* the gums; N.L. gen. n. *uli* of the gum.

This description comes primarily from Olsen et al. (1991) with additional information from Dewhirst et al. (2001). Cells are small elliptical rods, which occur singly, in pairs, or short chains and may show central swellings. Nonmotile. Colonies on brain heart blood agar are 1–2 mm in diameter, raised or low convex, and entire after 5 d of anaerobic incubation.

Obligate anaerobe. Grows in Peptone Yeast extract Glucose Hemin vitamin K (PYG-HK) broth (Holdeman et al., 1977a). Growth is stimulated by Tween 80 (0.02%, v/v).

The major fermentation products from glucose are acetate and lactate with occasional minor amounts of succinate. Acid is produced from fructose, glucose, maltose, mannose, salicin, and sucrose, but not from amygdalin, erythritol, esculin, glycerol, inositol, mannitol, melezitose, melibiose, raffinose, rhamnose, sorbitol, or xylose. Indole is not produced and nitrate is not reduced. Negative reaction for bile-esculin, DNase, and hippurate hydrolysis. The Rapid ID32A profile is 2012 0337 05. The major cell wall fatty acid is $C_{18:1}$ (*cis*9).

Source: human gingival crevices of patients with periodontitis (Olsen et al., 1991) and from patients with endodontic infections (Siqueira et al., 2005). Has been recovered from human bacteremia (Lau et al., 2004a).

DNA G+C content (mol%): 64 (HPLC) for type strain.

Type strain: VPI D76D-27C, ATCC 49627, CCUG 31166, DSM 7084, JCM 12494, LMG 11480.

Sequence accession no. (16S rRNA gene): AF292373.

Additional comments: the species was previously designated *Lactobacillus* D01 in the Virginia Polytechnic Institute Anaerobe Laboratory collection of W.E.C. Moore and L.V. Holdeman Moore and initially named *Lactobacillus uli* by Olsen et al. (1991). The phenotypic description of the species needs validation as some strains used in the description of *Lactobacillus uli* have been identified recently as belonging to other *Olsenella* species and to *Atopobium* species by 16S rRNA gene sequence analysis (Dewhirst, unpublished observation).

2. **Olsenella profusa** Dewhirst, Paster, Tzellas, Coleman, Downes, Spratt and Wade 2001, 1802^{VP}

pro.fus'a. L. adj. *profusus* profuse, referring to the good growth of the organism.

The description is based primarily on Dewhirst et al. (2001). Cells are short rods 0.6 × 0.8–2.0 µm, which occur singly or in chains. Nonmotile. After incubation for 7 d on FAA plates, colonies are approximately 1–1.3 mm in diameter, circular, entire, pulvinate to pyramidal, cream colored, and opaque when viewed under a plate microscope. Growth in broth media is good, and addition of Tween 80 does not significantly enhance growth.

Acetate and lactate are major end products of metabolism. Arabinose, cellobiose, fructose, glucose, lactose, maltose, mannitol, mannose, melibiose, raffinose, salicin, sucrose, and trehalose are fermented, but not melezitose, or rhamnose. Esculin is hydrolyzed, but not arginine or starch. Gelatin is not liquified meat is not digested. Does not produce catalase, indole, or H_2S or reduce nitrate.

Source: human subgingival plaque of patients with periodontitis (Dewhirst et al., 2001) and dental caries (Munson et al., 2004). Detected in endodontic infections (Rôças and Siqueira, 2005).

DNA G+C content (mol%): 64 (HPLC) for the type strain.

Type strain: VPI D315A-29, DSM 13989, CCUG 45371, CIP 106885, DSM 13989, JCM 14553.

Sequence accession no. (16S rRNA gene): AF292374.

Additional comments: the species was previously designated *Eubacterium* D52 in the Virginia Polytechnic Institute Anaerobe Laboratory Collection of W.E.C. Moore and L.V. Holderman Moore.

Genus VIII. **Slackia** Wade, Downes, Dymock, Hiom, Weightman, Dewhirst, Paster, Tzellas and Coleman 1999, 598^{VP}

William G. Wade and Floyd E. Dewhirst

Slack'ia. N.L. fem. n. named to honor Geoffrey Slack, distinguished British microbiologist and dental researcher.

Cells are Gram-stain-positive **cocci, coccobacilli, or short bacilli**, which are obligately **anaerobic**, nonmotile, and do not produce endospores. **Arginine is hydrolyzed**; catalase, urease, and indole are not produced and esculin is not hydrolyzed. **Acid is not produced from sugars**. Metabolic end products from glucose are either **acetate or not detectable**.

DNA G+C content (mol%): 60–64 (HPLC).

Type species: **Slackia exigua** Wade, Downes, Dymock, Hiom, Weightman, Dewhirst, Paster, Tzellas and Coleman 1999, 598^{VP}.

Further descriptive information

The predominant long chain cellular fatty acids in *Slackia faecicanis* have been shown to be $C_{16:0}$, $C_{16:1}$ (*cis*9), $C_{18:0}$, $C_{18:1}$ (*cis*9), and $C_{19:1}$ iso (*cis*9) (Lawson et al., 2005) but comparable data are not available for *Slackia exigua* or *Slackia heliotrinireducens*. *Slackia exigua* degrades arginine by the arginine deaminase pathway, as does *Cryptobacterium curtum*, another member of the family *Coriobacteriaceae* (Uematsu et al., 2006).

Strains corresponding to *Slackia exigua* were referred to as *Eubacterium* D6 in the Virginia Polytechnic Institute Anaerobe Laboratory collection of W.E.C. Moore and L.V. Holdeman Moore, "*Eubacterium* Cluster 2" (Wade et al., 1990) and "*Eubacterium* S-group" (Sato et al., 1993) prior to the formal description of the species, initially as "*Eubacterium exiguum*" (Poco et al., 1996). *Slackia exigua* is part of the human oral microbiota and has been associated with periodontitis (Moore et al., 1983), endodontic infections (Hashimura et al., 2001; Sato et al., 1993), and dentoalveolar abscesses (Downes et al., 2001; Wade et al., 1994). In patients with periodontitis, the species is particularly associated with deep pockets suggesting that this is its preferred habitat (Booth et al., 2004). In addition, patients with rapidly progressive periodontitis and refractory adult periodontitis have been shown to have raised antibody titers to *Slackia exigua*, suggesting a role in the pathogenesis of the disease (Smith and Wade, 1999). *Slackia exigua* has also been isolated from a case of bacteremia as part of a polymicrobial infection (Woo et al., 2004).

Ingestion of ragwort by grazing animals can lead to death due to hepatic cirrhosis up to 2 years after ingestion (Muth, 1968). However, sheep are resistant to the toxic substances in ragwort while horses and cattle are susceptible. It is the presence of *Slackia heliotrinireducens* and other as yet uncharacterized bacteria in the sheep rumen that is thought to confer resistance by degrading the pyrrolizidine alkaloids responsible for toxicity (Craig et al., 1992). *Slackia heliotrinireducens* can degrade heliotrine and lasiocarpine but not jacobine or seneciphylline (Hovermale and Craig, 2002).

Enrichment and isolation procedures

Slackia exigua has been successfully grown on Brain Heart Infusion agar and Fastidious Anaerobe Agar, both supplemented with 5% horse or sheep blood, while *Slackia faecicanis* can be cultured on Bacteroides agar (Holdeman et al., 1977a). Strains of all species are slow-growing, but *Slackia exigua* is particularly slow, forming colonies no greater than 1 mm in diameter, even after prolonged incubation.

Enrichment for *Slackia heliotrinireducens* can be achieved by culture in sheep rumen contents supplemented with heliotrine (2 mg/ml) and chloral hydrate (0.2 mg/ml) in an atmosphere of H_2-CO_2 (4:1) for 24–30 h, followed by successive subcultures in a medium containing 30% rumen fluid, 0.2% heliotrine, and salts. Dilutions from the final enrichment can be cultured in roll tubes in medium containing 30% rumen fluid, 0.1% heliotrine, salts, and 1% (w/v) Ionagar (Oxoid no. 2). After incubation for 7–8 d at 38°C, colonies resembling those of *Slackia heliotrinreducens* can be selected and subcultured (Lanigan, 1976). Methods for assay of the alkaloids and their metabolites are given by Lanigan and Smith (1970).

Maintenance procedures

Strains can be maintained on Blood Agar plates, or tryptone-yeast-mineral salts (TYM; Lanigan, 1976) for *Slackia heliotrinireducens*, incubated anaerobically at 37°C, and subcultured weekly. Lyophilization of cultures is recommended for long-term storage, although strains have been successfully stored at −70°C in Brain Heart Infusion Broth + 10% glycerol.

Differentiation from closely related taxa

A simple scheme for the differentiation of the three *Slackia* species, primarily based on a commercial kit for the identification of anaerobic bacteria, is shown in Table 313. Given the slow growth and general unreactivity of members of this group and related *Actinobacteria*, molecular methods such as sequencing of the 16S rRNA gene are recommended for unambiguous identification.

Taxonomic comments

The phylogenetic relationship between the genus *Slackia* and related genera within the family *Coriobacteriaceae* can be seen in the tree constructed from 16S rRNA gene comparisons in Figure 420.

TABLE 313. Differential characteristics of *Slackia* species[a,b]

Characteristic	*S. faecicanis*	*S. exigua*	*S. heliotrinireducens*
Cell shape	Rods	Rods	Cocci
Nitrate reduction	+	−	+
Fermentation products from glucose	nd	nd	Acetate (trace)
API Rapid ID32A system tests:			
Alanine arylamidase	−	+	+
Arginine arylamidase	−	+	−
Glycine arylamidase	−	+	+
Histidine arylamidase	−	+	+
Proline arylamidase	−	+	+
Serine arylamidase	−	+	+
Tyrosine arylamidase	−	+	+

[a]Symbols and abbreviations: +, present; −, absent; nd, none detected.
[b]Data from Lawson et al. (2005).

Acknowledgements

Martin Slayne, Sarah Cheeseman, Sarah Hiom, Dave Dymock, and Julie Downes are thanked for their scientific and technical contributions to the work described in this chapter. Paul Lawson is thanked for discussions clarifying characteristics of *Slackia faecicanis*.

List of species of the genus *Slackia*

1. **Slackia exigua** (*Eubacterium exiguum* Poco, Nakazawa, Ikeda, Sato, Sato and Hoshino 1996) Wade, Downes, Dymock, Hiom, Weightman, Dewhirst, Paster, Tzellas and Coleman 1999, 598[VP]

ex.igu'a. L. fem. adj. *exigua* scanty, small, referring to the scanty or poor growth of the organism.

Cells are short Gram-stain-positive rods that are obligately anaerobic, nonmotile, and nonsporing. Individual cells are 0.5 × 1.0 mm, and the cells occur singly or in clumps. Sometimes cells from older cultures stain Gram-negative. On BHI-blood agar plates, the cells form minute, circular, convex, translucent colonies that are less than 1 mm in diameter even after prolonged incubation in an anaerobic glove box. Growth in broth media is poor with or without carbohydrates but is moderately enhanced in the presence of 5% bovine serum, 0.2% lysine, or 0.2% arginine. Hemolysis does not occur around colonies on BHI blood agar plates. The cells are inert in most biochemical tests. Starch and esculin are not hydrolyzed, and nitrate is not reduced. Does not liquefy gelatin. Catalase-, indole- and urease-negative. Ammonia is produced from arginine. All strains are nonfermentative and do not utilize adonitol, amygdalin, arabinose, cellobiose, erythritol, esculin, fructose, galactose, glucose, glycogen, inositol, lactose, maltose, mannitol, mannose, melezitose, melibiose, rhamnose, ribose, salicin, sorbitol, starch, sucrose, trehalose, or xylose. No detectable metabolic end products are produced in peptone-yeast extract medium supplemented with glucose or peptone-yeast extract-glucose broth.

Source: human necrotic pulp samples, periapical infections, and acute dento-alveolar abscesses.

DNA G+C content (mol%) of the type strain: 60 (HPLC).

Type strain: S-7, ATCC 700122, CIP 105133, JCM 11022.

Sequence accession no. (16S rRNA gene): AF101240.

2. **Slackia heliotrinireducens** Wade, Downes, Dymock, Hiom, Weightman, Dewhirst, Paster, Tzellas and Coleman 1999, 599[VP]

he.li.o.trin.i.re.duc.ens. N.L. n. *heliotrinum* derived from heliotrine, a pyrrolizidine alkaloid; L. part. adj. *reducans* leading back, bringing back and in chemistry converting to a different oxidation state; N.L. part. adj. *heliotrinireducens* referring to the organism's ability to bring about oxidative cleavage of the heliotrine molecule.

Cells are cocci (0.5–0.7 µm in diameter), that occur in pairs, in small clusters and short chains. Subsurface colonies in TYM agar are lenticular with an entire margin and smooth surface. After incubation for 7–10 d, colonies are 0.6–0.8 mm in diameter, grayish-white, and translucent. Surface colonies are 1–2 mm in diameter, effuse, colorless, and transparent with an entire edge. Optimum growth temperature is between 38°C and 42°C with some growth between 30°C and 46°C. Maximal growth occurs between pH 6.5 and 7.0 with some growth between pH 6.2 and 7.2. The presence of 2% NaCl is inhibitory to growth in TYM medium. Growth is enhanced by arginine (10–25 mM), but not by 1% (v/v) serum or 1% (v/v) Tween 80, while glycine (40 mM) is inhibitory. Addition of alanine, glutamate, histidine, ornithine, proline, serine, threonine, and combinations of alanine plus ornithine or alanine plus proline do not affect growth. Ammonia is formed from adenine, arginine, tryptone, yeast extract, and uracil. Nitrates are completely reduced to ammonia if sufficient electron donor (H_2 or formate) is present. Sulfates are not reduced. Creatinine is not hydrolyzed. Does not form gas in TYM-agar medium.

Only a trace of acetate is detected in PY-glucose culture by gas chromatographic analysis. Small amounts of H_2 and CO_2 are produced. Fumarate is reduced to succinate if formate or H_2 is present. Lactate, malate, and pyruvate are not metabolized. Pyrrolizidine alkaloids (europine, heleurine, heliotrine, lasiocarpine, and supinine) are reduced to 1-methylene derivatives in the presence of H_2 or formate (electron donors). Some macrocyclic diesters are metabolized, but more slowly than the monoesters. Anacrotine, cynaustraline, heliotridine, retrorsine, and sarracine are not metabolized. A *c*-type cytochrome is present in ultrasonic extracts of cells. Growth is inhibited in media containing 10 units/ml penicillin.

Source: sheep rumen.

DNA G+C content (mol%) of the type strain: 61 (HPLC).

Type strain: RHS1, ATCC 29202, NCTC 11029, CCUG 47954, JCM 14554, DSM 20476.

Sequence accession no. (16S rRNA gene): AF101241.

3. **Slackia faecicanis** Lawson, Greetham, Gibson, Giffard, Falsen and Collins 2005, 1245[VP]

fae.ci.ca'nis. L. n. *faex, faecis* feces; L. gen. n. *canis* dog; N.L. gen. n. *faecicanis* from dog feces.

Cells are short Gram-stain-positive rods (0.5 × 1–2 μm). Colonies after a 48-h anaerobic incubation at 37°C are 1–2 mm in diameter, translucent to gray, with an uneven surface with irregular edges. Strains are obligate anaerobes. Nitrate is reduced to nitrite. Catalase and indole are not produced. Acid is not produced from glucose, lactose, maltose, mannose, mannitol, melibiose, ribose, starch, or sucrose. Using the API Rapid ID32A system, only arginine dihydrolase is positive. The predominant long-chain cellular fatty acids consist of $C_{16:0}$, $C_{16:1}$ (*cis*9), $C_{18:0}$, $C_{18:1}$ (*cis*9), and $C_{19:1}$ iso (*cis*9).

Source: the primary habitat is unknown but is probably the intestinal tract of the dog.

DNA G+C content (mol%) of the type strain: 61 (HPLC).

Type strain: 5WC12, CCUG 48399, CIP 108281, JCM 14555.

Sequence accession no. (16S rRNA gene): AJ608686.

References

Anderson, R.C., M.A. Rasmussen and M.J. Allison. 1993. Metabolism of the plant toxins nitropropionic acid and nitropropanol by ruminal microorganisms. Appl. Environ. Microbiol. *59*: 3056–3061.

Anderson, R.C., M.A. Rasmussen and M.J. Allison. 1996. Enrichment and isolation of a nitropropanol-metabolizing bacterium from the rumen. Appl. Environ. Microbiol. *62*: 3885–3886.

Anderson, R.C., M.A. Rasmussen, A.A. DiSpirito and M.J. Allison. 1997. Characteristics of a nitropropanol-metabolizing bacterium isolated from the rumen. Can. J. Microbiol. *43*: 617–624.

Anderson, R.C., M.A. Rasmussen, N.S. Jensen and M.J. Allison. 2000. *Denitrobacterium detoxificans* gen. nov., sp. nov., a ruminal bacterium that respires on nitrocompounds. Int. J. Syst. Evol. Microbiol. *50*: 633–638.

Anderson, R.C., W. Majak, M.A. Rassmussen, T.R. Callaway, R.C. Beier, D.J. Nisbet and M.J. Allison. 2005. Toxicity and metabolism of the conjugates of 3-nitropropanol and 3-nitropropionic acid in forages poisonous to livestock. J. Agric. Food Chem. *53*: 2344–2350.

Anderson, R.C., W. Majak, M.A. Rasmussen and M.J. Allison. 1998. Detoxification potential of a new species of ruminal bacteria that metabolize nitrate and naturally occurring nitrotoxins. *In* Toxic Plants and Other Natural Toxicants (edited by Garland and Barr). CAB International, New York, pp. 154–158.

Andreesen, J.R. 1992. The genus *Eubacterium*. *In* The Prokaryotes: a Handbook on the Biology of Bacteria: Ecophysiology, Isolation, Identification, Applications, 2nd edn, vol. 2 (edited by Balows, Trüper, Dworkin, Harder and Schleifer). Springer, New York, pp. 1914–1924.

Angelakis, E., V. Roux, D. Raoult and M. Drancourt. 2009. Human case of *Atopobium rimae* bacteremia. Emerg. Infect. Dis. *15*: 354–355.

Bailey, G.D. and N. Love. 1986. *Eubacterium fossor* sp. nov., an agar-corroding organism from normal pharynx and oral and respiratory tract lesions of horses. Int. J. Syst. Bacteriol. *36*: 383–387.

Balch, W.E., G.E. Fox, L.J. Magrum, C.R. Woese and R.S. Wolfe. 1979. Methanogens: reevaluation of a unique biological group. Microbiol. Rev. *43*: 260–296.

Benno, Y., K. Suzuki, K. Suzuki, K. Narisawa, W.R. Bruce and T. Mitsuoka. 1986. Comparison of the fecal microflora in rural Japanese and urban Canadians. Microbiol. Immunol. *30*: 521–531.

Bokkenheuser, V.D., J. Winter, P. Dehazya and W.G. Kelly. 1977. Isolation and characterization of human fecal bacteria capable of 21-dehydroxylating corticoids. Appl. Environ. Microbiol. *34*: 571–575.

Bokkenheuser, V.D., J. Winter, S.M. Finegold, V.L. Sutter, A.E. Ritchie, W.E.C. Moore and L.V. Holdeman. 1979. New markers for *Eubacterium lentum*. Appl. Environ. Microbiol. *37*: 1001–1006.

Booth, V., J. Downes, J. Van den Berg and W.G. Wade. 2004. Gram-positive anaerobic bacilli in human periodontal disease. J. Periodont. Res. *39*: 213–220.

Boskey, E.R., K.M. Telsch, K.J. Whaley, T.R. Moench and R.A. Cone. 1999. Acid production by vaginal flora *in vitro* is consistent with the rate and extent of vaginal acidification. Infect. Immun. *67*: 5170–5175.

Bryant, M.P. 1972. Commentary on the Hungate technique for culture of anaerobic bacteria. Am. J. Clin. Nutr. *25*: 1324–1328.

Burton, J.P., E. Devillard, P.A. Cadieux, J.-A. Hammond and G. Reid. 2004. Detection of *Atopobium vaginae* in postmenopausal women by cultivation-independent methods warrants further investigation. J. Clin. Microbiol. *42*: 1829–1831.

Cato, E.P. 1983. Transfer of *Peptostreptococcus parvulus* (Weinberg, Nativelle and Prévot 1937) Smith 1957 to the genus *Streptococcus* - *Streptococcus parvulus* (Weinberg, Nativelle and Prévot 1937) comb. nov., nom. rev. emend. Int. J. Syst. Bacteriol. *33*: 82–84.

Cauci, S., S. Driussi, D. De Santo, P. Penacchioni, T. Iannicelli, P. Lanzafame, F. De Seta, F. Quadrifoglio, D. de Aloysio and S. Guaschino. 2002. Prevalence of bacterial vaginosis and vaginal flora changes in peri- and postmenopausal women. J. Clin. Microbiol. *40*: 2147–2152.

Chan, R.C. and J. Mercer. 2008. First Australian description of *Eggerthella lenta* bacteraemia identified by 16S rRNA gene sequencing. Pathology *40*: 409–410.

Chandrasekaran, A., L.W. Robertson and R.H. Reuning. 1987. Reductive inactivation of digitoxin by *Eubacterium lentum* cultures. Appl. Envrion. Microbiol. *53*: 901–904.

Chhour, K.L., M.A. Nadkarni, R. Byun, F.E. Martin, N.A. Jacques and N. Hunter. 2005. Molecular analysis of microbial diversity in advanced caries. J. Clin. Microbiol. *43*: 843–849.

Clavel, T., G. Henderson, C.-A. Alpert, C. Philippe, L. Rigottier-Gois, J. Doré and M. Blaut. 2005. Intestinal bacterial communities that produce active estrogen-like compounds enterodiol and enterolactone in humans. Appl. Environ. Microbiol. *71*: 6077–6085.

Clavel, T., G. Henderson, W. Engst, J. Doré and M. Blaut. 2006. Phylogeny of human intestinal bacteria that activate the dietary lignan secoisolariciresinol diglucoside. FEMS Microbiol. Ecol. *55*: 471–478.

Collins, M.D. and S. Wallbanks. 1992. Comparative sequence analyses of the 16S rRNA genes of *Lactobacillus minutus*, *Lactobacillus rimae* and *Streptococcus parvulus*: proposal for the creation of a new genus *Atopobium*. FEMS Microbiol. Lett. *95*: 235–240.

Collins, M.D. and S. Wallbanks. 1993. *In* Validation of the publication of new names and new combinations previously effectively published outside the IJSB. List no. 44. Int. J. Syst. Bacteriol. *43*: 188–189.

Craig, A.M., C.J. Latham, L.L. Blythe, W.B. Schmotzer and O.A. O'Connor. 1992. Metabolism of toxic pyrrolizidine alkaloids from tansy ragwort (*Senecio jacobaea*) in ovine ruminal fluid under anaerobic conditions. Appl. Envrion. Microbiol. 58: 2730–2736.

Dewhirst, F.E., B.J. Paster, N. Tzellas, B. Coleman, J. Downes, D.A. Spratt and W.G. Wade. 2001. Characterization of novel human oral isolates and cloned 16S rDNA sequences that fall in the family *Coriobacteriaceae*: description of *Olsenella* gen. nov., reclassification of *Lactobacillus uli* as *Olsenella uli* comb. nov. and description of *Olsenella profusa* sp. nov. Int. J. Syst. Evol. Microbiol. *51*: 1797–1804.

Dewhirst, F.E., J. Izard, B.J. Paster, A.C. Tanner, W.G. Wade, W.-H. Yu and T. Chen 2008, posting date. The Human Oral Microbiome Database. http://www.homd.org. [Online.]

Dobkin, J.F., J.R. Saha, V.P. Butler, Jr, H.C. Neu and J. Lindenbaum. 1982. Inactivation of digoxin by *Eubacterium lentum*, an anaerobe of the human gut flora. Trans. Assoc. Am. Physic. *95*: 22–29.

Downes, J., M.A. Munson, D.A. Spratt, E. Kononen, E. Tarkka, H. Jousimies-Somer and W.G. Wade. 2001. Characterisation of *Eubacterium*-like strains isolated from oral infections. J. Med. Microbiol. *50*: 947–951.

Eggerth, A.H. 1935. The gram-positive non-spore-bearing anaerobic bacilli of human feces. J. Bacteriol. *30*: 277–290.

Felsenstein, J. 1985. Confidence limits on phylogenies: an approach using the bootstrap. Evolution *39*: 783–791.

Finegold, S.M. and V.L. Sutter. 1978. Fecal flora in different populations with special reference to diet. Am. J. Clin. Nutr. *27*: 1456–1469.

Gregersen, T. 1978. Rapid method for distinction of gram-negative from gram-positive bacteria. Appl. Microbiol. Biotechnol. *5*: 123–127.

Haas, F. and H. König. 1987. Characterization of an anaerobic symbiont and the associated aerobic bacterial flora of *Pyrrhocoris apterus* (Heteroptera: Pyrrhocoridae). FEMS Microbiol. Ecol. *45*: 99–106.

Haas, F. and H. König. 1988. *Coriobacterium glomerans* gen. nov., sp. nov. from the intestinal tract of the red soldier bug. Int. J. Syst. Bacteriol. *38*: 382–384.

Harmsen, H.J.M., A.C.M. Wildeboer-Veloo, J. Grijpstra, J. Knol, J.E. Degener and G.W. Welling. 2000. Development of 16S rRNA-based probes for the *Coriobacterium* group and the *Atopobium* cluster and their application for enumeration of *Coriobacteriaceae* in human feces from volunteers of different age groups. Appl. Environ. Microbiol. *66*: 4523–4527.

Hashimura, T., M. Sato and E. Hoshino. 2001. Detection of *Slackia exigua*, *Mogibacterium timidum* and *Eubacterium saphenum* from pulpal and periradicular samples using the Polymerase Chain Reaction (PCR) method. Int. Endodont. J. *34*: 463–470.

Hass, H. and H. König. 1988. *Coriobacterium glomerans* gen. nov., sp. nov. from the intestinal tract of the red soldier bug. Int. J. Syst. Bacteriol. *38*:382–384.

Hauduroy, A., G. Ehringer, A. Urbain, G. Guillot and J. Magrou. 1937. Dictionnaire des Bactéries Pathogènes. Masson et Cie, Paris.

Hirano, S. and N. Masuda. 1981. Transformation of bile acids by *Eubacterium lentum*. Appl. Envrion. Microbiol. *42*: 912–915.

Holdeman, L.V., I.J. Good and W.E. Moore. 1976. Human fecal flora: variation in bacterial composition within individuals and a possible effect of emotional stress. Appl. Environ. Microbiol. *31*: 359–375.

Holdeman, L.V., E.P. Cato and W.E.C. Moore (editors). 1977a. Anaerobe Laboratory Manual, 4th edn. Anaerobe Laboratory, Virginia Polytechnic Institute and State University, Blacksburg, VA.

Holdeman, L.V.H., E.P. Cato and W.E.C. Moore. 1977b. Anaerobe Laboratory Manual, 4th edn. Virginia Polytechnic Institute and State University, Blacksburg, VA.

Hovermale, J.T. and A.M. Craig. 2002. Metabolism of pyrrolizidine alkaloids by *Peptostreptococcus heliotrinreducens* and a mixed culture derived from ovine ruminal fluid. Biophys. Chem. *101–102*: 387–399.

Jin, J.S., Y.F. Zhao, N. Nakamura, T. Akao, N. Kakiuchi, B.S. Min and M. Hattori. 2007. Enantioselective dehydroxylation of enterodiol and enterolactone precursors by human intestinal bacteria. Biol. Pharm. Bull. *30*: 2113–2119.

Jousimies-Somer, H., P. Summanen, D.M. Citron, E.J. Baron, H.M. Wexler and S.M. Finegold. 2002a. Wadsworth Anaerobic Bacteriology Manual, 6th edn. Star Publishing, Belmont, CA.

Jousimies-Somer, H.R., P. Summanen, D.M. Citron, E.J. Baron, H.M. Wexler and S.M. Finegold. 2002b. Wadsworth-KTL Anaerobic Bacteriology Manual, 6th edn (edited by Finegold and Jousimies-Somer). Star Publishing Company, Belmont, CA.

Jukes, T.H. and C. Cantor. 1969. Evolution of protein molecules. *In* Mammalian Protein Metabolism (edited by Murano). Academic Press, New York pp. 21–132.

Kageyama, A., Y. Benno and T. Nakase. 1999a. Phylogenetic evidence for the transfer of *Eubacterium lentum* to the genus *Eggerthella* as *Eggerthella lenta* gen. nov., comb. nov. Int. J. Syst. Bacteriol. *49*: 1725–1732.

Kageyama, A., Y. Benno and T. Nakase. 1999b. Phylogenic and phenotypic evidence for the transfer of *Eubacterium fossor* to the genus *Atopobium* as *Atopobium fossor* comb. nov. Microbiol. Immunol. *43*: 389–395.

Kageyama, A., Y. Benno and T. Nakase. 1999c. Phylogenetic and phenotypic evidence for the transfer of *Eubacterium aerofaciens* to the genus *Collinsella* as *Collinsella aerofaciens* gen. nov., comb. nov. Int. J. Syst. Bacteriol. *49*: 557–565.

Kageyama, A., Y. Benno and T. Nakase. 1999d. *In* Validation of publication of new names and new combinations previously effectively published outside the IJSB. List no. 71. Int. J. Syst. Bacteriol. *49*: 1325–1326.

Kageyama, A. and Y. Benno. 2000. Emendation of genus *Collinsella* and proposal of *Collinsella stercoris* sp. nov. and *Collinsella intestinalis* sp. nov. Int. J. Syst. Evol. Microbiol. *50*: 1767–1774.

Kumar, P.S., A.L. Griffen, M.L. Moeschberger and E.J. Leys. 2005. Identification of candidate periodontal pathogens and beneficial species by quantitative 16S clonal analysis. J. Clin. Microbiol. *43*: 3944–3955.

Kumar, S., K. Tamura, I.B. Jakobsen and M. Nei. 2001. MEGA2: molecular evolutionary genetics analysis software. Bioinformatics *17*: 1244–1245.

Lanigan, G.W. 1976. *Peptococcus heliotrinreducans*, sp. nov. cytochrome-producing anaerobe which metabolizes pyrrolizidine alkaloids. J. Gen. Microbiol. *94*: 1–10.

Lanigan, G.W., and L. W. Smith. 1970. Metabolism of pyrrolizidine alkaloids in the ovine rumen. I. Formation of 7-hydroxy-1-methyl-8-pyrrolizidine from heliotrine and lasiocarpine. Aust. J. Agric. Res. *21*: 493–500.

Lattuada, E., A. Zorzi, M. Lanzafame, D. Antolini, R. Fontana, S. Vento and E. Concia. 2005. Cutaneous abscess due to *Eubacterium lentum* in injection drug user: a case report and review of the literature. J. Infect. *51*: E71–72.

Lau, S.K., P.C. Woo, A.M. Fung, K.M. Chan, G.K. Woo and K.Y. Yuen. 2004a. Anaerobic, non-sporulating, Gram-positive bacilli bacteraemia characterized by 16S rRNA gene sequencing. J. Med. Microbiol. *53*: 1247–1253.

Lau, S.K., P.C. Woo, G.K. Woo, A.M. Fung, M.K. Wong, K.M. Chan, D.M. Tam and K.Y. Yuen. 2004b. *Eggerthella hongkongensis* sp. nov. and *Eggerthella sinensis* sp. nov., two novel *Eggerthella* species, account for half of the cases of *Eggerthella* bacteremia. Diagn. Microbiol. Infect. Dis. *49*: 255–263.

Lau, S.K.P., P.C.Y. Woo, A.M.Y. Fung, K.-m. Chan, G.K.S. Woo and K.-y. Yuen. 2004c. Anaerobic, non-sporulating, Gram-positive bacilli bacteraemia characterized by 16S rRNA gene sequencing. J. Med. Microbiol. *53*: 1247–1253.

Lau, S.K.P., P.C.Y. Woo, G.K.S. Wood, A.M.Y. Fung, M.K.M. Wong, K. Chan, D.M.W. Tam and K. Yuen. 2006. *In* List of new names and new combinations previously effectively, but not validly, published. Validation List no. 111. Int. J. Syst. Evol. Microbiol. *56*: 2025–2027.

Lawson, P.A., H.L. Greetham, G.R. Gibson, C. Giffard, E. Falsen and M.D. Collins. 2005. *Slackia faecicanis* sp. nov., isolated from canine faeces. Int. J. Syst. Evol. Microbiol. *55*: 1243–1246.

Ludwig, W., G. Kirchhof, M. Weizenegger and N. Weiss. 1992. Phylogenetic evidence for the transfer of *Eubacterium suis* to the genus *Actinomyces* as *Actinomyces suis* comb. nov. Int. J. Syst. Bacteriol. *42*: 161–165.

MacDonald, I., D. Mahony, J. Jellet and C. Meier. 1977. NAD-dependent 3α- and 12α-hydroxysteroid dehydrogenase activities from *Eubacterium lentum* ATCC no. 25559. Biochim. Biophys. Acta *489*: 466–476.

MacDonald, I.A., J.F. Jellett, D.E. Mahony and L.V. Holdeman. 1979. Bile salt 3α- and 12α-hydroxysteroid dehydrogenases from *Eubacterium lentum* and related organisms. Appl. Environ. Microbiol. *37*: 992–1000.

Majak, W., K.J. Cheng and J.W. Hall. 1986. Enhanced degradation of 3-nitropropanol by ruminal microorganisms. J. Anim. Sci. *62*: 1072–1080.

Majak, W. 1992a. Metabolism and absorption of toxic glycosides by ruminants. J. Range Man. *45*: 67–70.

Majak, W. 1992b. Further enhancement of nitropropanol detoxification by ruminal bacteria in cattle. Can. J. Anim. Sci. *72*: 863–870.

Maruo, T., M. Sakamoto, C. Ito, T. Toda and Y. Benno. 2008. *Adlercreutzia equolifaciens* gen. nov., sp. nov., an equol-producing bacterium isolated from human faeces, and emended description of the genus *Eggerthella*. Int. J. Syst. Evol. Microbiol. *58*: 1221–1227.

Moore, W.E. and L.V. Holdeman. 1974. Human fecal flora: the normal flora of 20 Japanese-Hawaiians. Appl. Microbiol. *27*: 961–979.

Moore, W.E., L.V. Holdeman, E.P. Cato, R.M. Smibert, J.A. Burmeister and R.R. Ranney. 1983. Bacteriology of moderate (chronic) periodontitis in mature adult humans. Infect. Immun. *42*: 510–515.

Moore, W.E.C., E.P. Cato and L.V. Holdeman. 1971. *Eubacterium aerofaciens* (Eggerth) Prévot 1938: Emendation of description and designation of the neotype strain. Int. J. Syst. Bacteriol. *21*: 307–310.

Moore, W.E.C. and L.V. Holdeman. 1972. Anaerobe Laboratory Manual (edited by Moore and Holdeman). Virginia Polytechnic Institute Anaerobe Laboratory, Blacksburg, VA.

Mosca, A., C.A. Strong and S.M. Finegold. 1993. UV red fluorescence of *Eubacterium lentum*. J. Clin. Microbiol. *31*: 1001–1002.

Mosca, A., P. Summanen, S.M. Finegold, G. De Michele and G. Miragliotta. 1998. Cellular fatty acid composition, soluble protein profile, and antimicrobial resistance pattern of *Eubacterium lentum*. J. Clin. Microbiol. *36*: 752–755.

Mountfort, D.O., W.D. Grant, R. Clarke and R.A. Asher. 1988. *Eubacterium callanderi* sp. nov. that demethoxylates *O*-methoxylated aromatic acids to volatile fatty acids. Int. J. Syst. Bacteriol. *38*: 254–258.

Munson, M.A., A. Banerjee, T.F. Watson and W.G. Wade. 2004. Molecular analysis of the microflora associated with dental caries. J. Clin. Microbiol. *42*: 3023–3029.

Muth, O.H. 1968. Tansy ragwort (*Senecio jacobaea*), a potential menace to livestock. J. Am. Vet. Med. Assoc. *153*: 310–312.

Nakazawa, F., S.E. Poco, T. Ikeda, M. Sato, S. Kalfas, G. Sundqvist and E. Hoshino. 1999. *Cryptobacterium curtum* gen. nov., sp. nov., a new genus of Gram-positive anaerobic rod isolated from human oral cavities. Int. J. Syst. Bacteriol. *49*: 1193–1200.

Nakazawa, F., M. Sato, E.S. Poco, T. Hashimura, T. Ikeda, S. Kalfas, G. Sundqvist and E. Hoshino. 2000. Description of *Mogibacterium pumilum* gen. nov., sp. nov. and *Mogibacterium vescum* gen. nov., sp. nov., and reclassification of *Eubacterium timidum* (Holdeman *et al.* 1980) as *Mogibacterium timidum* gen. nov., comb. nov. Int. J. Syst. Evol. Microbiol. *50*: 679–688.

Nakazawa, F., S.E. Poco, M. Sato, T. Ikeda, S. Kalfas, G. Sundqvist and E. Hoshino. 2002. Taxonomic characterization of *Mogibacterium diversum* sp. nov. and *Mogibacterium neglectum* sp. nov., isolated from human oral cavities. Int. J. Syst. Evol. Microbiol. *52*: 115–122.

Nakazawa, F. and E. Hoshino. 2004. DNA-DNA relatedness and phylogenetic positions of *Slackia exigua*, *Slackia heliotrinireducens*, *Eggerthella lenta* and other related bacteria. Oral Microbiol. Immunol. *19*: 343–346.

Olsen, I., J.L. Johnson, L.V.H. Moore and W.E.C. Moore. 1991. *Lactobacillus uli* sp. nov. and *Lactobacillus rimae* sp. nov. from the human gingival crevice and emended descriptions of *Lactobacillus minutus* and *Streptococcus parvulus*. Int. J. Syst. Bacteriol. *41*: 261–266.

Poco, S.E., F. Nakazawa, T. Ikeda, M. Sato, T. Sato and E. Hoshino. 1996. *Eubacterium exiguum* sp. nov., isolated from human oral lesions. Int. J. Syst. Bacteriol. *46*: 1120–1124.

Prévot, A.R. 1938. Études de systématique bactérienne. III. Invalidité du genre *Bacteroides* Castellani et Chalmers démembrement et reclassification. Ann. Inst. Pasteur *20*: 285–307.

Racklyeft, D.J. and D.N. Love. 2000. Bacterial infection of the lower respiratory tract in 34 horses. Aust. Vet. J. *78*: 549–559.

Rainey, F.A., N. Weiss and E. Stackebrandt. 1994. *Coriobacterium* and *Atopobium* are phylogenetic neighbors within the *Actinomycetes* line of descent. Syst. Appl. Microbiol. *17*: 202–205.

Roberts, M.C., B.J. Moncla and S.L. Hillier. 1991. Characterization of unusual tetracycline-resistant gram-positive bacteria. Antimicrob. Agents Chemother. *35*: 2655–2657.

Robertson, L.W., A. Chandrasekaran, R.H. Reuning, J. Hui and B.D. Rawal. 1986. Reduction of digoxin to 20R-dihydrodigoxin by cultures of *Eubacterium lentum*. Appl. Environ. Microbiol. *51*: 1300–1303.

Rôças, I.N. and J.F. Siqueira, Jr. 2005. Species-directed 16S rRNA gene nested PCR detection of *Olsenella* species in association with endodontic diseases. Lett. Appl. Microbiol. *41*: 12–16.

Rodriguez-Jovita, M.R., M.D. Collins, B. Sjödén and E. Falsen. 1999. Characterization of a novel *Atopobium* isolate from the human vagina: description of *Atopobium vaginae* sp. nov. Int. J. Syst. Bacteriol. *49*: 1573–1576.

Saitou, N. and M. Nei. 1987. The neighbor-joining method: a new method for reconstructing phylogenetic trees. Mol. Biol. Evol. *4*: 406–425.

Sato, T., E. Hoshino, H. Uematshu and T. Noda. 1993. Predominant obligate anaerobes in necrotic pulps of human deciduous teeth. Microb. Ecol. Health Dis. *6*: 269–275.

Scardovi, V. 1981. The genus *Bifidobacterium*. *In* The Prokaryotes: a Handbook on Habitats, Isolation, and Identification of Bacteria, vol. 2 (edited by Starr, Stolp, Trüper, Balows and Schlegel). Springer, New York, pp. pp. 1951–1961.

Schleifer, K.H. and O. Kandler. 1972. Peptidoglycan types of bacterial cell walls and their taxonomic implications. Bacteriol. Rev. *36*: 407–477.

Schwiertz, A., G. Le Blay and M. Blaut. 2000. Quantification of different *Eubacterium* spp. in human fecal samples with species-specific 16S rRNA-targeted oligonucleotide probes. Appl. Environ. Microbiol. *66*: 375–382.

Severijnen, A.J., R. van Kleef, M.P. Hazenberg and J.P. van de Merwe. 1990. Chronic arthritis induced in rats by cell wall fragments of *Eubacterium* species from the human intestinal flora. Infect. Immun. *58*: 523–528.

Siqueira, J.F., I.N. Rôças, C.D. Cunha and A.S. Rosado. 2005. Novel bacterial phylotypes in endodontic infections. J. Dent. Res. *84*: 565–569.

Smith, A.J. and W.G. Wade. 1999. Serum antibody response against oral *Eubacterium* species in periodontal disease. J. Periodont. Res. *34*: 175–178.

Sperry, J.F. and T.D. Wilkins. 1976a. Arginine, a growth-limiting factor for *Eubacterium lentum*. J. Bacteriol. *127*: 780–784.

Sperry, J.F. and T.D. Wilkins. 1976b. Cytochrome spectrum of an obligate anaerobe, *Eubacterium lentum*. J. Bacteriol. *125*: 905–909.

Stackebrandt, E. and W. Ludwig. 1994. The importance of using outgroup reference organisms in phylogenetic studies: the *Atopobium* case. Syst. Appl. Microbiol. *17*:39–43.

Stackebrandt, E., F.A. Rainey and N.L. Ward-Rainey. 1997. Proposal for a new hierarchic classification system, *Actinobacteria* classis nov. Int. J. Syst. Bacteriol. *47*: 479–491.

Tamura, K., J. Dudley, M. Nei and S. Kumar. 2007. MEGA4: Molecular Evolutionary Genetics Analysis (MEGA) software version 4.0. Mol. Biol. Evol. *24*: 1596–1599.

Uematsu, H., N. Sato, A. Djais and E. Hoshino. 2006. Degradation of arginine by *Slackia exigua* ATCC 700122 and *Cryptobacterium curtum* ATCC 700683. Oral Microbiol. Immunol. *21*: 381–384.

Verhulst, A., G. Parmentier, G. Janssen, S. Asselberghs and H. Eyssen. 1986. Biotransformation of unsaturated long-chain fatty acids by *Eubacterium lentum*. Appl. Environ. Microbiol. *51*: 532–538.

Verhulst, A., H. Van Hespen, F. Symons and H. Eyssen. 1987. Systematic analysis of the long-chain components of *Eubacterium lentum*. J. Gen. Microbiol. *133*: 275–282.

Wade, W.G., M.A. Slayne and M.J. Aldred. 1990. Comparison of identification methods for oral asaccharolytic *Eubacterium* species. J. Med. Microbiol. *33*: 239–242.

Wade, W.G., M.A. Lewis, S.L. Cheeseman, E.G. Absi and P.A. Bishop. 1994. An unclassified *Eubacterium* taxon in acute dento-alveolar abscess. J. Med. Microbiol. *40*: 115–117.

Wade, W.G., J. Downes, D. Dymock, S.J. Hiom, A.J. Weightman, F.E. Dewhirst, B.J. Paster, N. Tzellas and B. Coleman. 1999. The family *Coriobacteriaceae*: reclassification of *Eubacterium exiguum* (Poco *et al.* 1996) and *Peptostreptococcus heliotrinreducens* (Lanigan 1976) as *Slackia exigua* gen. nov., comb. nov. and *Slackia heliotrinireducens* gen. nov., comb. nov., and *Eubacterium lentum* (Prévot 1938) as *Eggerthella lenta* gen. nov., comb. nov. Int. J. Syst. Bacteriol. *49*: 595–600.

Weinberg, M., R. Nativelle and A.R. Prévot. 1937. Les Microbes Anaérobies. Masson et Cie, Paris.

Willems, A. and M.D. Collins. 1996. Phylogenetic relationships of the genera *Acetobacterium* and *Eubacterium sensu stricto* and reclassification of *Eubacterium alactolyticum* as *Pseudoramibacter alactolyticus* gen. nov., comb. nov. Int. J. Syst. Bacteriol. *46*: 1083–1087.

Willems, A., W.E.C. Moore, N. Weiss and M.D. Collins. 1997. Phenotypic and phylogenetic characterization of some *Eubacterium*-like isolates containing a novel type B wall murein from human feces: description of *Holdemania filiformis* gen. nov., sp. nov. Int. J. Syst. Bacteriol. *47*: 1201–1204.

Woo, P.C., H. Tse, K.M. Chan, S.K. Lau, A.M. Fung, K.T. Yip, D.M. Tam, K.H. Ng, T.L. Que and K.Y. Yuen. 2004. "*Streptococcus milleri*" endocarditis caused by *Streptococcus anginosus*. Diagn. Microbiol. Infect. Dis. *48*: 81–88.

Class IV. **Nitriliruptoria** class. nov.

WOLFGANG LUDWIG, JEAN EUZÉBY AND WILLIAM B. WHITMAN

Ni.tri.li.rup.tor′i.a. N.L. masc. n. *Nitriliruptor* type genus of the type order; suffix *-ia* ending to denote a class; N.L. pl. neut. n. *Nitriliruptoria* the *Nitriliruptorales* class.

This class is formed by elevation of the subclass *Nitriliruptoridae* Kurahashi et al. 2010, which was delineated on the basis of 16S rRNA gene sequence analyses and the distinctive phyletic lineage of the type order.

Type order: **Nitriliruptorales** Sorokin, van Pelt, Tourova and Evtushenko 2009, 252^{VP}.

Reference

Kurahashi, M., Y. Fukunaga, Y. Sakiyama, S. Harayama and A. Yokota. 2010. *Euzebya tangerina* gen. nov., sp. nov., a deeply branching marine actinobacterium isolated from the sea cucumber *Holothuria edulis*, and proposal of *Euzebyaceae* fam. nov., *Euzebyales* ord. nov. and *Nitriliruptoridae* subclassis nov. Int. J. Syst. Evol. Microbiol. *60*: 2314–2319.

Order I. **Nitriliruptorales** Sorokin, van Pelt, Tourova and Evtushenko 2009, 252^VP

THE EDITORIAL BOARD

Ni.tri.li.rup.tor.a′les. N.L. masc. n. *Nitriliruptor* type genus of the order; suffix *-ales* ending to denote an order; N.L. fem. pl. n. *Nitriliruptorales* the order of the genus *Nitriliruptor*.

Description is the same as the family *Nitriliruptoraceae*.

Type genus: **Nitriliruptor** Sorokin, van Pelt, Tourova and Evtushenko 2009, 251^VP.

Reference

Sorokin, D.Y., S. van Pelt, T.P. Tourova and L.I. Evtushenko. 2009. *Nitriliruptor alkaliphilus* gen. nov., sp. nov., a deep-lineage haloalkaliphilic actinobacterium from soda lakes capable of growth on aliphatic nitriles, and proposal of *Nitriliruptoraceae* fam. nov. and *Nitriliruptorales* ord. nov. Int. J. Syst. Evol. Microbiol. *59*: 248–253.

Family I. **Nitriliruptoraceae** Sorokin, van Pelt, Tourova and Evtushenko 2009, 251^VP

THE EDITORIAL BOARD

Ni.tri.li.rup.to.ra.ce′a.e. N.L. masc. n. *Nitriliruptor* type genus of the family; suffix *-aceae*, ending to denote a family; N.L. fem. pl. n. *Nitriliruptoraceae* the family of the genus *Nitriliruptor*.

Rods that stain Gram-positive. **Aerobic heterotrophs with the ability to degrade organic nitriles** as well as organic acids and sugars. **Alkaliniphilic and moderately salt-tolerant**. The description is otherwise the same as the type genus.

Type genus: **Nitriliruptor** Sorokin, van Pelt, Tourova and Evtushenko 2009, 251^VP.

Genus I. **Nitriliruptor** Sorokin, van Pelt, Tourova and Evtushenko 2009, 251^VP

THE EDITORIAL BOARD

Ni.tri.li.rup′tor. N.L. n. *nitrilum* nitrile, nitrile group; L. masc. n. *ruptor* breaker; N.L. masc. n. *nitriliruptor* nitrile-breaker.

Short rods that **stain Gram-positive**. Aerobes that **utilize short-chain organic acids, amides, and aliphatic nitriles** as energy and carbon sources. The cell envelopes contain *meso*-diaminopimelate and saturated C_{14}–C_{16} fatty acids.

DNA G+C content (mol%): 70.8 (T_m).

Type species: **Nitriliruptor alkaliphilus** Sorokin, van Pelt, Tourova and Evtushenko 2009, 251^VP.

Further descriptive information

This description of the genus is based upon a single isolate of a single species and must necessarily be preliminary. Cells are short rods (0.4 × 1.5–3.0 µm) and nonmotile. Colonies develop very slowly over a period of 1 month and are flat, colorless, and spreading. The cells stain Gram-positive and contain alanine, glutamate, and *meso*-diaminopimelate in a 2:1:1 molar ratio. Small amounts of other amino acids as well as the sugars glucose, galactose and glycerol are also present. Fatty acids include $C_{16:0}$ (19%), C_{14} (17%), $C_{16:1}$ ω7 (16%), $C_{14:0}$ (13%), C_{16} (8%), $C_{17:1}$ ω8 (5%), and $C_{18:1}$ ω9 (4%).

Growth is aerobic, and anaerobic growth on sugars with or without nitrate as an electron acceptor is not observed. Isobutyronitrile is completed degraded as a sole carbon and nitrogen source. Other nitriles utilized include propionitrile, butyronitrile, valeronitrile, and capronitrile, but not acetonitrile. Simple monocarboxylic and dicarboxylic organic acids, pyruvate, malate, fumarate, and citrate are also utilized. Sugars and sugar alcohols are utilized, including D-glucose, D-fructose, maltose, D-mannose, melezitose, α,α-trehalose, sucrose, D-arabinose, inositol, cellobiose, and glycerol. Yeast extract and peptone also support slow growth.

Isolated from a soda lake, the pH optimum and range are 9.0–9.4 and 8.4–10.4, respectively. The NaCl optimum and range are 0.2–0.3 M and 0.1–2.0 M, respectively. A mesophile, the temperature optimum is 32°C.

The type species was isolated from an isobutyronitrile-degrading consortium initiated from soda lake sediments of the Kulunda Steppe, Altai, Russia in pH 10 medium containing a total Na concentration of 0.6 M (Sorokin et al., 2007). Environmental clones with ~95 % sequence similarity have been isolated from sediments of other soda lakes and from saline and other soils.

List of species of the genus *Nitriliruptor*

1. **Nitriliruptor alkaliphilus** Sorokin, van Pelt, Tourova and Evtushenko 2009, 251[VP]

al.ka.li'phi.lus. N.L. n. *alkali* soda ash; Gr. adj. *philos* loving; N.L. adj. *alkaliphilus* alkali-loving.

The description is the same as for the genus.

DNA G+C content (mol%): 70.8 (T_m).

Type strain: ANL-iso2, DSM 45188, NCCB 100119, UNIQEM U239.

Sequence accession no. (16S rRNA gene): EF422408.

References

Sorokin, D.Y., S. van Pelt, T.P. Tourova and G. Muyzer. 2007. Microbial isobutyronitrile utilization under haloalkaline conditions. Appl. Environ. Microbiol. *73*: 5574–5579.

Sorokin, D.Y., S. van Pelt, T.P. Tourova and L.I. Evtushenko. 2009. *Nitriliruptor alkaliphilus* gen. nov., sp. nov., a deep-lineage haloalkaliphilic actinobacterium from soda lakes capable of growth on aliphatic nitriles, and proposal of *Nitriliruptoraceae* fam. nov. and *Nitriliruptorales* ord. nov. Int. J. Syst. Evol. Microbiol. *59*: 248–253.

Order II. **Euzebyales** Kurahashi, Fukunaga, Sakiyama, Harayama and Yokota 2010, 2318[VP]

The Editorial Board

Eu.ze.by'a.les. N.L. fem. n. *Euzebya* type genus of the family; *-ales* ending to denote an order; N.L. fem. pl. n. *Euzebyales* the order of the genus *Euzebya*.

The description of the order is the same as for the genus and based upon the distinct 16S rRNA gene phyletic lineage.

Type genus: **Euzebya** Kurahashi, Fukunaga, Sakiyama, Harayama and Yokota 2010, 2318[VP].

Reference

Kurahashi, M., Y. Fukunaga, Y. Sakiyama, S. Harayama and A. Yokota. 2010. *Euzebya tangerina* gen. nov., sp. nov., a deeply branching marine actinobacterium isolated from the sea cucumber *Holothuria edulis*, and proposal of *Euzebyaceae* fam. nov., *Euzebyales* ord. nov. and *Nitriliruptoridae* subclassis nov. Int J Syst Evol Microbiol *60*: 2314–2319.

Family I. **Euzebyaceae** Kurahashi, Fukunaga, Sakiyama, Harayama and Yokota 2010, 2318[VP]

The Editorial Board

Eu.ze.by.a.ce′a.e. N.L. fem. n. *Euzebya* type genus of the family; *-aceae* ending to denote a family; N.L. fem. pl. n. *Euzebyaceae* the family of the genus *Euzebya*.

The description of the family is the same as for the genus and based upon the distinct 16S rRNA gene phyletic lineage.

Type genus: **Euzebya** Kurahashi, Fukunaga, Sakiyama, Harayama and Yokota 2010, 2318[VP].

Genus I. **Euzebya** Kurahashi, Fukunaga, Sakiyama, Harayama and Yokota 2010, 2318[VP]

The Editorial Board

Eu.ze′by.a. N.L. fem. n. *Euzebya* named for Jean Paul Marie Euzéby, a French microbiologist who has contributed significantly to microbial systematics, including the Latinization of microbial names.

Rods that stain Gram-positive and do not form endospores. Aerobic chemo-organotrophs, oxidase- and catalase-positive. **Sodium chloride is required** for growth. **MK-9(H$_4$)** is the predominant menaquinone. The cell envelopes contain ***meso*-diaminopimelate, rhamnose, and galactose.** The major fatty acids are $C_{16:1}$ ω7*c*, $C_{16:0}$, and $C_{17:1}$ ω8*c*. The major polar lipid is phosphatidylglycerol.

DNA G+C content (mol%): 68.3 (HPLC).

Type species: **Euzebya tangerina** Kurahashi, Fukunaga, Sakiyama, Harayama and Yokota 2010, 2318[VP].

Further descriptive information

This description of the genus is based upon a single isolate of a single species and must necessarily be preliminary. Cells are rods (0.6–0.8 × 1.5–6.0 μm) and nonmotile. They stain Gram-positive and do not form endospores. Colonies develop very slowly over a period of 5 weeks and are pulvinate with an entire edge, tangerine in color, nearly opaque, and have a hard texture. The cells wall contains alanine, glutamate, and *meso*-diaminopimelate in a 2:1:1 molar ratio as well as the sugars rhamnose and galactose. Fatty acids include $C_{16:1}$ ω7*c* (35%), $C_{16:0}$ (19%), $C_{17:1}$ ω8*c* (18%), $C_{18:1}$ ω9*c* (8%), $C_{17:0}$ (5%), and 10-methyl $C_{17:0}$ (4%). The predominant menaquinone is MK-9(H$_4$).

Growth is aerobic, and nitrate is not reduced. Grows on complex media with peptone and yeast extract as carbon sources, such as marine agar or broth 2216 and TYSW medium (2 g tryptone and 1 g yeast extract in 1 l artificial seawater). Also assimilates L-arabinose and melibiose, but not amygdalin, D-glucose, inositol, D-mannose, rhamnose, sucrose, and sorbitol. Positive for hydrolysis of gelatin and urea and production of catalase, oxidase and acetoin. Negative for the production of β-galactosidase, arginine dihydrolase, lysine decarboxylase, ornithine decarboxylase, tryptophan deaminase, H_2S, and indole. Unable to utilize citrate.

Isolated from a marine source, the NaCl range for growth is 0.5–12%. The temperature optimum and range are 20–28°C and >10 and <40°C, respectively. The pH range is 7–9. Stock cultures are maintained in marine broth 2216 with 5% DMSO at −80°C.

The type species was isolated from the abdominal epidermis of a sea cucumber, *Holothuria edulis*, which had been collected from the coast of Aka Island, Okinawa prefecture, Japan, at a depth of 6 m (Kurahashi and Yokota, 2004; Valenzuela-Encinas et al., 2009). Environmental clones with greater than ~98% sequence similarity have been isolated from dolomite rock (Horath and Bachofen, 2009), haloalkaline soil (Valenzuela-Encinas et al., 2009), and anthracene-contaminated soils.

List of species of the genus *Euzebya*

1. **Euzebya tangerina** Kurahashi, Fukunaga, Sakiyama, Harayama and Yokota 2010, 2318[VP]

tan.ge.ri′na. N.L. fem. adj. *tangerina* tangerine-colored, referring to the colony color.

The description is the same as for the genus.

DNA G+C content (mol%): 68.3 (HPLC).

Type strain: F10, NBRC 105439, KCTC 19736.

Sequence accession no. (16S rRNA gene): AB478418.

References

Horath, T. and R. Bachofen. 2009. Molecular characterization of an endolithic microbial community in Dolomite rock in the Central Alps (Switzerland). Microb. Ecol. *58*: 290–306.

Kurahashi, M. & A. Yokota. 2004. *Agarivorans albus* gen. nov., sp. nov., a γ-proteobacterium isolated from marine animals. Int. J. Syst. Evol. Micobiol. *54*: 693–697.

Kurahashi, M., Y. Fukunaga, Y. Sakiyama, S. Harayama and A. Yokota. 2010. *Euzebya tangerina* gen. nov., sp. nov., a deeply branching marine actinobacterium isolated from the sea cucumber *Holothuria edulis*, and proposal of *Euzebyaceae* fam. nov., *Euzebyales* ord. nov. and *Nitriliruptoridae* subclassis nov. Int. J. Syst. Evol. Microbiol. *60*: 2314–2319.

Valenzuela-Encinas, C., I. Neria-González, R.J. Alcántara-Hernández, I. Estrada-Alvarado, F.J. Zavala-Díaz de la Serna, L. Dendooven and R. Marsch. 2009. Changes in the bacterial populations of the highly alkaline saline soil of the former lake Texcoco (Mexico) following flooding. Extremophiles *13*: 609–621.

Class V. **Rubrobacteria** class. nov.

Ken-ichiro Suzuki

Ru.bro.bac.te'ri.a. N.L. masc. n. *Rubrobacter* type genus of the type order; suff. *-ia* ending to denote a class; N.L. pl. neut. n. *Rubrobacteria* the *Rubrobacterales* class.

The subclass *Rubrobacteridae* was established by Rainey et al. (1997) for the phylogenetic construction of the hierarchical structure of the class *Actinobacteria*. At the time, only the genus *Rubrobacter* was known, and the subclass contained the single order *Rubrobacterales*, which was represented by the family *Rubrobacteraceae* for the genus *Rubrobacter*. Although the genus *Thermoleophilum* had already been described (Zarilla and Perry, 1984), it was not included in the study. Later, the genus *Thermoleophilum* was reported to be the member of the subclass *Rubrobacteridae* (Yakimov et al., 2003). Almost at the same time, the new genera *Conexibacter* (Monciardini et al. 2003) and *Solirubrobacter* (Singleton et al. 2003) were described. Although their classification at the higher ranks was discussed, a formal recommendation was not made. In 2005, the *Manual* tentatively assigned the genera *Conexibacter*, *Solirubrobacter*, and *Thermoleophilum* to the family *Rubrobacteraceae*, which was the only family validly published at the time (Appendix 2 of Garrity et al., 2005). Stackebrandt (2004, 2005) reanalyzed the relationship of these genera to the subclass *Rubrobacteridae* and proposed the families *Conexibacteraceae*, *Solirubrobacteraceae* and *Thermoleophilaceae* with the emendation of the subclass *Rubrobacteridae* and the family *Rubrobacteraceae*. The results of Stackebrandt (2004, 2005) were considered in the later study for reconstruction of the class *Actinobacteria* based on the 16S rRNA gene sequences (Zhi et al., 2009). Almost at the same time, Reddy and Garcia-Pichel (2009) proposed the orders *Thermoleophilales* and *Solirubrobacterales*, which reflected the low sequence similarity of the 16S rRNA genes between these families and *Rubrobacter* spp.

Lastly, Ludwig et al. (2012) found little phylogenetic or phenotypic evidence for an association between *Rubrobacter* and the orders *Thermoleophilales* and *Solirubrobacterales*. For this reason, they proposed the classification of these groups into the novel classes *Rubrobacteria* and *Thermoleophilia*.

The rRNA gene signatures for the families of the classes *Rubrobacteria* and *Thermoleophilia* are given in Table 314 and differential signatures are given in Table 315.

Type order: **Rubrobacterales** Rainey, Ward-Rainey and Stackebrandt 1997, 483[VP] emend. Zhi, Li and Stackebrandt 2009, 593.

TABLE 314. Signature sequences of the members of the classes *Rubrobacteria* and *Thermoleophilia*[a]

	Rubrobacterales	*Solirubrobacterales*			*Thermoleophilales*
Position(s)	*Rubrobacteraceae*	*Solirubrobacteraceae*	*Conexibacteraceae*	*Patulibacteraceae*	*Thermoleophilus*
52:359	G–C	C–G	U–A	C–G	C–G
63:104	**C–G**	G–C	G–C	G–C	G–C
70:98	**A–U**	G–C	G–C	G–C	G–C
127:234	G–C	G–C	G–C	G–C	G–C
139:224	**U–A**	A–U	G–C	G–C	G–C
144:178	**G–C**	C–G	U–A	C–G	C–G
291:309	U–A	U–A	U–A	U–A	U–A
370:391	C–G	C–G	C–G	C–G	**G–C**
408:434	G–C	G–C	A–U	G–C	G–C
580:776	U–A	U–A	U–A	U–A	**C–G**
590:649	C–G	C–G	U–A	U–A	C–G
600:638	U–G	C–G	U–G	U–G	C–G
657:749	**G–C**	U–A	U–A	U–A	U–A
670:736	A–U	A–U	A–U	A–U	**G–C**
681:709	C–G	**U–A**	**U–A**	**U–A**	C–G
823:877	G–C	G–C	G–C	A–U	G–C
906	A	A	A	A	A
941:1342	A–U	A–U	A–U	A–U	**G–C**
953:1228	**U–A**	G–C	G–C	G–C	G–C
954:1226	**C–G**	G–C	G–C	G–C	G–C
955:1225	U–A	U–A	U–A	U–A	U–A
999:1041	U–A	U–A	G–U	U–A	**A–U**
1051:1207	**C–G**	G–C	G–C	G–C	G–C
1115:1185	C–G	C–G	C–G	C–G	C–G
1118:1155	C–G	**U–A**	**U–A**	**U–A**	C–G
1311:1326	A–U	A–U	A–U	A–U	**G–C**
1410:1490	U–A	U–A	U–A	A–U	U–A

[a]The order-characteristic signatures are indicated in bold letters (Reddy and Garcia-Pichel, 2009).

TABLE 315. Signature sequences that differentiate the families of the classes *Rubrobacteria* and *Thermoleophilia* (modified from Zhi et al., 2009)

	52:359	63:104	144:178	145:177	242:284	291:309	293:304	316:337	370:391	377:386
Rubrobacteraceae		C–G	G–C	G–C	C–G	U–A	G–U	C–G		G–C
Conexibacteraceae			U–A	U–A			G–C			C–G
Patulibacteraceae	C–G		C–G				G–C			C–G
Solirubrobacteraceae	C–G		C–G	C–G			G–C			
Thermoleophilaceae	C–G		C–G	C–G			G–U		G–C	C–G
	408:434	409:433	418:425	502:543	590:649	600:638	657:749	670:736	681:709	722:733
Rubrobacteraceae		C–G					G–C		C–G	G–G
Conexibacteraceae	A–U		U–A		U–A	U–G				
Patulibacteraceae				G–C	U–A	U–G				
Solirubrobacteraceae					C–G	C–G				
Thermoleophilaceae					C–G	C–G		G–C	C–G	
	819	941:1342	952:1229	953:1228	954:1226	1051:1207	1115:1185	1311:1326	1313:1324	1354:1368
Rubrobacteraceae	A		U–A	U–A	C–G	C–G	C G		U–A	
Conexibacteraceae										
Patulibacteraceae										U–G
Solirubrobacteraceae										
Thermoleophilaceae		G–C						G–C		

References

Garrity, G., J.A. Bell and T. Lilburn. 2005. The revised roadmap to the *Manual*. *In* Bergey's Manual of Systematic Bacteriology, 2nd edn, vol. 2A, The *Proteobacteria*, Introductory Essays. Springer, New York, pp. 159–220.

Ludwig, W., J. Euzéby and W. Whitman. 2012. Taxonomic outline of the phylum *Actinobacteria*. *In* Bergey's Manual of Systematic Bacteriology, 2nd edn, vol. 5, The *Actinobacteria* (edited by Goodfellow, Kämpfer, Busse, Trujillo, Suzuki, Ludwig and Whitman). Springer, New York, pp. 29–31.

Monciardini, P., L. Cavaletti, P. Schumann, M. Rohde and S. Donadio. 2003. *Conexibacter woesei* gen. nov., sp. nov., a novel representative of a deep evolutionary line of descent within the class *Actinobacteria*. Int. J. Syst. Evol. Microbiol. *53*: 569–576.

Rainey, F.A., N.L. Ward-Rainey and E. Stackebrandt. 1997. *In* Stackebrandt, E., F.A. Rainey and N.L. Ward-Rainey. Proposal for a new hierarchic classification system, *Actinobacteria* classis nov. Int. J. Syst. Bacteriol. *47*: 479–491.

Reddy, G.S. and F. Garcia-Pichel. 2009. Description of *Patulibacter americanus* sp. nov., isolated from biological soil crusts, emended description of the genus *Patulibacter* Takahashi *et al.* 2006 and proposal of *Solirubrobacterales* ord. nov. and *Thermoleophilales* ord. nov. Int. J. Syst. Evol. Microbiol. *59*: 87–94.

Singleton, D.R., M.A. Furlong, A.D. Peacock, D.C. White, D.C. Coleman and W.B. Whitman. 2003. *Solirubrobacter pauli* gen. nov., sp. nov., a mesophilic bacterium within the *Rubrobacteridae* related to common soil clones. Int. J. Syst. Evol. Microbiol. *53*: 485–490.

Stackebrandt, E., F.A. Rainey and N.L. Ward-Rainey. 1997. Proposal for a new hierarchic classification system, *Actinobacteria* classis nov. Int. J. Syst. Bacteriol. *47*: 479–491.

Stackebrandt, E. 2004. Will we ever understand? The undescribable diversity of the prokaryotes. Acta Microbiol. Immunol. Hung. *51*: 449–462.

Stackebrandt, E. 2005. In Validation of publication of new names and new combinations previously effectively published outside the IJSEM. List no. 102. Int. J. Syst. Evol. Microbiol. *55*: 547–549.

Yakimov, M.M., H. Lunsdorf and P.N. Golyshin. 2003. *Thermoleophilum album* and *Thermoleophilum minutum* are culturable representatives of group 2 of the *Rubrobacteridae* (*Actinobacteria*). Int. J. Syst. Evol. Microbiol. *53*: 377–380.

Zarilla, K.A. and J.J. Perry. 1984. *Thermoleophilum album* gen. nov. and sp. nov., a bacterium obligate for thermophily and normal-alkane substrates. Arch. Microbiol. *137*: 286–290.

Zhi, X.-Y., W.-J. Li and E. Stackebrandt. 2009. An update of the structure and 16S rRNA gene sequence-based definition of higher ranks of the class *Actinobacteria*, with the proposal of two new suborders and four new families and emended descriptions of the existing higher taxa. Int. J. Syst. Evol. Microbiol. *59*: 589–608.

Order I. **Rubrobacterales** Rainey, Ward-Rainey and Stackebrandt 1997, 483^VP emend. Zhi, Li and Stackebrandt 2009, 593

Ken-ichiro Suzuki

Ru.bro.bac.te'ra.les. N.L. masc. n. *Rubrobacter* type genus of the order; suff. *-ales* ending to denote an order; N.L. fem. pl. n. *Rubrobacterales* the *Rubrobacter* order.

The description of the order is the same as the family *Rubrobacteraceae*. It is defined on the basis of the 16S rRNA gene sequences (see Table 314 and Table 315) and supported by similarities in phenotypic properties.

Type genus: **Rubrobacter** Suzuki, Collins, Iijima and Komagata 1989, 44^VP.

References

Rainey, F.A., N.L. Ward-Rainey and E. Stackebrandt. 1997. *In* Stackebrandt, E., F.A. Rainey and N.L. Ward-Rainey. Proposal for a new hierarchic classification system, *Actinobacteria* classis nov. Int. J. Syst. Bacteriol. *47*: 479–491.

Suzuki, K., M.D. Collins, E. Iijima and K. Komagata. 1989. *In* Validation of the publication of new names and new combinations previously effectively published outside the IJSB. List no. 28. Int. J. Syst. Bacteriol. *39*: 93–94.

Zhi, X.-Y., W.-J. Li and E. Stackebrandt. 2009. An update of the structure and 16S rRNA gene sequence-based definition of higher ranks of the class *Actinobacteria*, with the proposal of two new suborders and four new families and emended descriptions of the existing higher taxa. Int. J. Syst. Evol. Microbiol. *59*: 589–608.

Family I. **Rubrobacteraceae** Rainey, Ward-Rainey and Stackebrandt 1997, 483[VP] emend. Zhi, Li and Stackebrandt 2009, 593

Ken-ichiro Suzuki

Ru.bro.bac.te.ra.ce′a.e. N.L. masc. n. *Rubrobacter* type genus of the family; suff. *-aceae* ending to denote a family; N.L. fem. pl. n. *Rubrobacteraceae* the *Rubrobacter* family.

The description of the family is the same as the genus *Rubrobacter*. It is defined on the basis of the 16S rRNA gene sequences (see Table 314 and Table 315) and supported by similarities in phenotypic properties.

Type genus: **Rubrobacter** Suzuki, Collins, Iijima and Komagata 1989, 44[VP].

Genus I. **Rubrobacter** Suzuki, Collins, Iijima and Komagata 1989, 93[VP] (Effective publication: Suzuki, Collins, Iijima and Komagata 1988, 38.)

Ken-ichiro Suzuki

Ru.bro.bac′ter. L. adj. *ruber* red; N.L. masc. n. *bacter* equivalent of the Gr. neut. n. *baktron* a rod; N.L. masc. n. *Rubrobacter* red rod.

Gram-stain-positive irregular rods. Nonmotile and non-sporforming. Cells are generally 0.8–1.0 by 1.0–4.0 µm. Cells in old culture present shorter coccoid form. Aerial mycelia are not formed. **Obligately aerobic**. Catalase and cytochrome oxidase reactions are positive. Nitrite is produced from nitrate.

Where known, the **diamino acid of the cell-wall peptidoglycan is L-lysine** (cell-wall type A3α). However, there is no data for the species *Rubrobacter taiwanensis*. Predominant respiratory quinone is menaquinone with eight isoprene units (**MK-8**). The polar lipids consist of phospholipids including diphosphatidylglycerol, phosphatidylglycerol and one phosphoglycolipid. Some additional phospholipids and glycolipids are also found. Predominant cellular fatty acids are internal branched acids such as **12-methyl hexadecanoic acid and/or 14-methyl octadecanoic acid**. Mycolic acid is not present.

DNA G+C content (mol%): 65–69 (HPLC, T_m).

Type species: **Rubrobacter radiotolerans** (Yoshinaka, Yano and Yamaguchi 1973) Suzuki, Collins, Iijima and Komagata 1989, 93[VP] (Effective publication: Suzuki, Collins, Iijima and Komagata 1988, 38.).

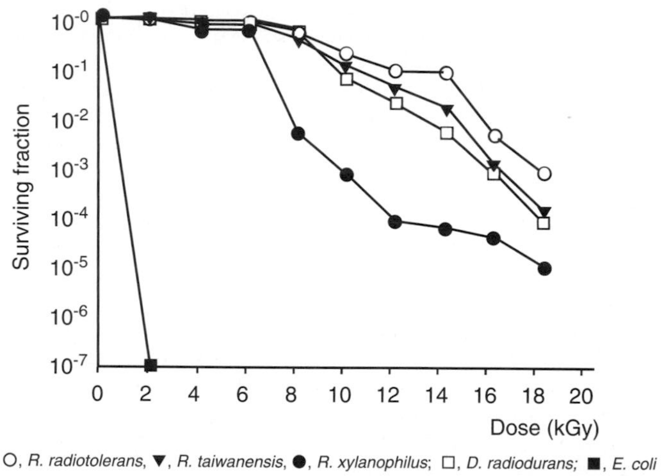

FIGURE 422. Gamma-radiation resistance of species of the genus *Rubrobacter* (from Chen et al., 2004). *Rubrobacter radiotolerans* DSM 5868 (○); *Rubrobacter taiwanensis* BCRC 17173 (▼); *Rubrobacter xylanophilus* DSM 9441 (●): *Deinococcus radiodurans* DSM 20539 (□); *Escherichia coli* K-12 (■).

Further descriptive information

Tolerance to γ-irradiation is the most distinctive feature of the genus *Rubrobacter*, although this characteristic is not specified in the original description of the genus (Yoshinaka et al., 1973). All the described species of the genus *Rubrobacter* show resistance against gamma-radiation (Chen et al., 2004). In addition, the species described so far are moderate thermophiles growing optimally at 46–60°C.

Resistance to γ-irradiation. The type strain of *Rubrobacter radiotolerans*, the type species of the genus, was isolated from a hot spring with high background radiation, and the sample was exposed to γ-radiation prior to isolation. Later, the strain was found to be extremely radiotolerant, comparable to that of species of the genus *Deinococcus* which is also strongly radiatino resistant. For *Rubrobacter radiotolerans*, the shoulder dose, or the dose at which killing is first observed, of exponentially growing cells is 6 kGy, and the D_{10}, or the dose necessary for 10% survival, is 10.0 kGy in phosphate buffer on aerated condition (Ito et al., 1983). The strains of the other species are also extremely radiotolerant, similar to that of *Deinococcus* species whose radiotolerance is well known (Figure 422, from Chen et al., 2004; Ferreira et al., 1999; Saito et al., 1994).

Cellular fatty acids. Strains of the genus *Rubrobacter* contain a large proportion of fatty acids which have methyl branch at a characteristic position in the middle of acyl chain. Such branched fatty acids widely distributed in actinobacteria are 10-methyl acids. 10-Methyl acids are known as the derivatives of corresponding monounsaturated fatty acids such as oleic acid

($C_{18:0}$ ω9). The predominant fatty acids of the genus *Rubrobacter* are 12-methyl hexadecanoic acid (12-Me $C_{16:0}$) and 14-methyl octadecanoic acid (14-Me $C_{18:0}$). Their methyl branch is at a unique position, the 5th carbon from the omega or apolar end of the fatty acid. They are possibly biosynthesized by methylation of corresponding unsaturated acids, namely $C_{16:0}$ ω5 and $C_{18:0}$ ω5, respectively (Suzuki et al., 1993). These unsaturated acids are of the same series of fatty acids with double bonds between the 4th and 5th carbons from the omega end of the acyl chain. Furthermore, 12-Me $C_{16:0}$ is the predominant fatty acid for *Rubrobacter radiotolerans* (Suzuki et al., 1988), and 14-Me $C_{18:0}$ is that for *Rubrobacter taiwanensis* and *Rubrobacter xylanophilus*. The optimal growth temperature of *Rubrobacter radiotolerans* is 45°C, in contrast to 60°C for the latter two species. It is expected that strains growing at higher temperatures have longer cellular fatty acids. Suzuki et al. (1988) reported that $C_{19:0}$ anteiso and $C_{17:0}$ 2-OH were also present in the cells of *Rubrobacter radiotolerans*,. However, these fatty acids were not found by subsequent investigators (Carreto et al., 1996; Chen et al., 2004). It will be interesting if there is found to be some relationship between these distinctive fatty acids and radiation resistance.

Enrichment and isolation procedures

The three species of the genus *Rubrobacter* are characterized by their strong resistance to γ-irradiation. The type strain of *Rubrobacter radiotolerans* was isolated from a sample pre-treated with 1.7 kGy irradiation by Cs^{137} for 5 h (Yoshinaka et al., 1973). Furthermore, the samples for isolation were collected at a radioactive hot spring. Although *Rubrobacter radiotolerans* is a moderate thermophile, it grows slowly. The characteristic pink colonies were picked up after 1-week cultivation at 37°C on an agar plate on which the irradiated sample was spread. Some faster-growing bacteria appeared after only a few days prior to the small pink colonies of the genus *Rubrobacter*.

The strains of the other two species were isolated from samples without radiation pretreatment and background radioactivity. These *Rubrobacter* strains were isolated from samples incubated at 50°C for growth of moderate thermophiles. In general, the isolation procedure involved spreading samples on agar plates and incubating at 50°C for 5–7 d. Pink colonies that appear slowly are picked (Carreto et al., 1996; Chen et al., 2004).

The strains of the genus *Rubrobacter* grow on various media, such as nutrient agar, tryptic soy broth, R medium, and *Thermus* medium. *Thermus* medium gives the best growth. The strains of the genus *Rubrobacter* are enriched by strong γ-irradiation and aerobic cultivation at 45–60°C.

Maintenance procedures

The strains of the genus *Rubrobacter* are preserved by freezing at lower than −70°C suspensions in *Thermus* medium containing 15% glycerol. Lyophilization in 10% skimmed milk containing 1% monosodium glutamate is also useful for long-term preservation and shipment.

Differentiation of the genus *Rubrobacter* from other genera

The genus *Deinococcus*, which also shows very strong radiation resistance, is classified in the phylum *Deinococci* and clearly differentiated by 16S rRNA gene sequences.

Taxonomic comments

Rubrobacter radiotolerans was firstly proposed by Yoshinaka et al. (1973) as a new species of the genus *Arthrobacter* because of the pleomorphic cell morphology, absence of acid production from sugars, and cell-wall peptidoglycan containing L-lysine. Later, Suzuki et al. (1988) proposed a new genus *Rubrobacter* for the species on the basis of the distinctive chemotaxonomic features in addition to the unusual physiological characteristics. The distinctiveness of this organism was further supported by phylogenetic analyses of the 16S rRNA gene sequences of the class *Actinobacteria* by Stackebrandt et al. (1997). The genus *Rubrobacter* was found to form a deep lineage, and it was classified outside of the subclass *Actinobacteridae*, which accommodated the order *Actinomycetales* in the class *Actinobacteria*. Stackebrandt et al. (1997) classified the single-genus taxon in the subclass *Rubrobacteridae*, order *Rubrobacteriales*, and family *Rubrobacteriaceae*. Later, the genera *Thermoleophilum* Yakimov et al. 2003 and *Solirubrobacter* Singleton et al. 2003 were classified in this subclass. Although the genus *Conexibacter* was closely related to *Solirubrobacter*, it was not included (Monciardini et al., 2003). Subsequently, these genera were reclassified into independent families, namely *Conexibacteraceae*, *Solirubrobacteraceae*, and *Thermoleophilaceae* (Stackebrandt 2004; Zhi et al. 2009), and novel orders, *Solirubrobacterales* and *Thermoleophilales*. The order *Solirubrobacterales* contains the families *Conexibacteraceae*, *Patulibacteraceae*, and *Solirubrobacter* (Reddy and Garcia-Pichel 2009). The order *Thermoleophilales* accommodates the family *Thermoleophilaceae*. In the absence of clear evidence of relatedness, these genera were finally reclassified in a novel class, *Thermoleophilia*, as well (Ludwig et al., 2012).

Currently, three species have been described in the genus *Rubrobacter*, and they share common chemotaxonomic, biochemical and physiological features. Because only one or two strains have been described for each species, the variability within species is not known. The species classification has also been confirmed by DNA relatedness studies (Chen et al., 2004).

Differentiation of the species of the genus *Rubrobacter*

Three species share many phenotypic and chemotaxonomic features, such as isoprenoid quinones and cell-wall peptidoglycan type (Table 316). Likewise, the cellular fatty acid profiles are very similar. Features that distinguish the species include the optimum growth temperature and a number of other physiological and biochemical characteristics (Table 316). The DNA relatedness and 16S rRNA gene sequence differences also strongly support the differentiation.

The common biochemical and physiological characteristics for the three species of the genus *Rubrobacter* are as follows. Positive reaction: catalase, cytochrome oxidase, β-glucosidase, β-galactosidase, growth in 5% NaCl and utilization of D-cellobiose, D-raffinose, D-trehalose, D-arabinose, D-fructose, D-mannose, lactose, L-glutamate, and pyruvate. Negative reaction: β-glucosidase, β-galactosidase, hydrolysis of starch, casein, cellulose, and tributyrin, and utilization of D-sorbitol. Characteristics variable among species are shown in Table 316.

TABLE 316. Phenotypic characteristics of *Rubrobacter* species[a,b]

Characteristic	1. *R. radiotolerans*[a]	2. *R. taiwanensis*[b]	3. *R. xylanophilus*[c]
Colony color	Reddish pink	Light pink	Light pink
Cell size (μm)	0.8–1.0 × 1.0–4.0	0.9–1.0 × 1.0–4.0	0.9–1.0 × 1.0–3.0
Optimum growth temperature (°C)	46–48	60	60
Cell-wall peptidoglycan	L-Lys–L-Ala	nd	L-Lys–L-Ala
Menaquinones	MK-8	MK-8	MK-8
Cellular fatty acids (at cultivation temp.)	12Me-$C_{16:0}$ (71%), $C_{19:0}$ anteiso (12%)*, and 2OH-$C_{17:0}$ (17%)* (at 37°C)	14Me-$C_{18:0}$ anteiso (31–33%), 12Me-$C_{17:0}$ (13%) and 12Me-$C_{16:0}$ (12%) (at 60°C)	14Me-$C_{18:0}$ (53%), $C_{18:0}$ (20%),12Me-$C_{16:0}$ (12%), and others (at 60°C)
DNA G+C content (mol%)	67.9 (HPLC)	67.9–68.5 (HPLC)	67.6 (T_m)
Nitrate reduction to nitrite	+	nd	+
Ammonium as a sole N source	+	nd	+
Resistance to γ-irradiation	+++	++	+
Hydrolysis of:			
Gelatin	–	+	+
Xylan	–	–	+
DNA	–	+	–
Esculin	+	–	+
Utilization of:			
D-Glucose	+	d	+
D-Galactose	–	+	+
D-Xylose	–	+	+
D-Melibiose	–	+	+
Rhamnose	+	d	+
Glycerol	+	–	–
Galactitol	–	nd	w
D-Mannitol	+	–	–
Ribitol	+	–	–
myo-Inositol	–	+	+
Acetate	–	nd	w
Malate	+	–	+
Succinate	–	–	+
L-Asparagine	+	d	+
L-Glutamine	–	+	w
L-Serine	–	+	–
Acetamide	–	nd	+

[a]Symbols : +, positive; –, negative; d, different in strains; nd, no data.

[b]The data are those of type strains except *Rubrobacter taiwanensis* (two strains) cited from Yoshinaka et al. (1973), Suzuki et al. (1988), Carreto et al. (1996), and Chen et al. (2004).

List of species of the genus *Rubrobacter*

1. **Rubrobacter radiotolerans** (Yoshinaka, Yano and Yamaguchi 1973, 2273[AL]) Suzuki, Collins, Iijima and Komagata 1989, 93[VP] (Effective publication: Suzuki, Collins, Iijima and Komagata 1988, 38.) (Basonym: *Arthrobacter radiotolerans* Yoshinaka, Yano and Yamaguchi 1973, 2273[AL]).

ra.di.o.to'le.rans. L. n. *radius* a beam or ray; N.L. pref. *radio-* pertaining to radiation; L. pres. part. *tolerans* tolerating; N.L. part. adj. *radiotolerans* (γ-ray) radiation-tolerating.

Gram-stain-positive irregular rod. Cells are 0.8–1.0 × 1.0–4.0 μm in size. Coccoid cells are often found in older cultures. Nonmotile. Colonies on nutrient agar are circular, smooth, entire convex or umbonate, glistening, opaque and reddish-pink in color. Colonies are 0.5–1.0 mm in diameter after 2 weeks, increasing in size to 4–5 mm after 4 weeks.

Xylose, arabinose, glucose, fructose, maltose and glycerol are utilized for growth. Galactose and mannose are weakly utilized. Sucrose, lactose, sorbitol and citrate are not utilized. Acid is not produced from any of carbon sources tested. Gelatin is not liquefied. Neither indole nor acetylmethylcarbinol is produced. Methyl red test negative. Ammonium sulfate is utilized as a sole nitrogen source.

Optimal temperature for growth is 46–48°C. Optimal pH for growth is 7.0–7.4. Growth is observed at 6% NaCl and lower but not at 10% NaCl in the culture medium.

The peptidoglycan structure is L-Lys–L-Ala. The cellular fatty acids are composed of more than 70% of 12-methyl branched octadecanoic acid, 12% of 16-methyl branched octadecanoic acid and 17% of 2-hydroxy heptadecanoic acid. Instead of these two acids, Carreto et al. (1996) report the presence of 14-Me $C_{18:0}$ and some long-chain alcohol.

The type strain was isolated from the radioactive hot spring at Misasa, Tottori, Japan after treatment of samples with γ-irradiation. Highly resistance to γ-irradiation. Shoulder dose of logarithmic growing cells is 6×10^5 rad and D_0 is 1×10^6 rad in phosphate buffer on aerated conditions.

Source of type strain: radioactive hot spring at Misasa, Tottori, Japan

DNA G+C content (mol%): 68 (HPLC).

Type strain: Yoshinaka strain P-1 = ATCC 51242 = CIP 106991 = DSM 5868 = IAM 12072 = IFO (now NBRC) 14777 = JCM 2153 = VKM Ac-1989.

Sequence accession no. (16S rRNA gene): D45058.

2. **Rubrobacter taiwanensis** Chen, Wu, Lin, Lu, Lin, Chang and Tsay 2004, 1853[VP]

tai.wan.en'sis. N.L. masc. adj. *taiwanensis* of Taiwan, where the micro-organism was first isolated.

Cells are irregular short rods, 0.9–1.0 μm wide to 1.0–3.0 μm long. Colonies on *Thermus* medium grown at 60°C for 7 d are 1.6–2.2 mm in diameter, circular, convex, smooth, opaque, and light pink. The temperature range for growth is 30–70°C. Optimal growth is at 60°C. The pH range for growth is 6–11. Optimal growth is at pH 8.0. Growth occurs at NaCl concentrations of 5% and below in *Thermus* medium. Cytochrome oxidase and β-galactosidase are positive. Gelatin and DNA are hydrolyzed. The carbon sources utilized are shown in Table 316. 14-Methyl octadecanoic acid is the predominant cellular fatty acid (31–33%). Other fatty acids present are 12-methyl heptadecanoic acid and 12-methyl hexadecanoic acid. The strains of this species show strong resistance to γ-irradiation.

Source of type strain: thermally heated water/soil/mud samples of Lu-shan hot springs, Nantou, Taiwan.

DNA G+C content (mol%): 67.9–68.5 (HPLC).

Type strain: strain LS-293 = ATCC BAA-406 = BCRC 17173 = JCM 12932.

Sequence accession no. (16S rRNA gene): AF465803 (type strain).

3. **Rubrobacter xylanophilus** Carreto, Moore, Nobre, Wait, Riley, Sharp and Da Costa 1996, 463[VP]

xy.la.no'phi.lus. N.L. n. *xylanum* xylan; Gr. adj. *philos* liking, friendly to; N.L. masc. adj. *xylanophilus* liking xylan.

Forms pleomorphic short rod-shaped cells that are 0.9–1.0 μm wide and 1.0–3.0 μm long; coccoid cells are also present. Colonies on *Thermus* medium incubated at 60°C for 7 d are 1.5–2.0 mm in diameter, circular, convex, smooth, opaque and light pink. Grows optimally at 60°C, and no growth is observed below 40°C and above 70°C. Optimal pH for growth is between 7.5 and 8.0, and no growth occurs below 6.0 and above 10.0. Growth occurs at NaCl concentrations of 6% and below in *Thermus* medium. β-Galactosidase-positive. Xylan is hydrolyzed. Growth occurs on various sugars, organic acids and amino acids (Table 316). The cell-wall peptidoglycan is L-Lys–L-Ala (type A3α). The major fatty acid is 14-methyl octadecanoic acid. Octadecanoic acid and by 12-methyl hexadecanoic acid are also present in cells grown at either 60°C or 45°C.

Source of type strain: thermally polluted run-off (temperature 50°C) from a carpet factory, Wilton, Wiltshire, UK.

DNA G+C content (mol%): 67.6 (T_m) for type strain.

Type strain: strain PRD-1 = CIP 105412 = DSM 9941 = NBRC 16129 = JCM 11954 = NBRC 100952.

Sequence accession no. (16S rRNA gene): CP000386.

References

Carreto, L., E. Moore, M.F. Nobre, R. Wait, P.W. Riley, R.J. Sharp and M.S. da Costa. 1996. *Rubrobacter xylanophilus* sp. nov.: a new thermophilic species isolated from a thermally polluted effluent. Int. J. Syst. Bacteriol. *46*: 460–465.

Chen, M.Y., S.H. Wu, G.H. Lin, C.P. Lu, Y.T. Lin, W.C. Chang and S.S. Tsay. 2004. *Rubrobacter taiwanensis* sp. nov., a novel thermophilic, radiation-resistant species isolated from hot springs. Int. J. Syst. Evol. Microbiol. *54*: 1849–1855.

Ferreira, A.C., M.F. Nobre, E. Moore, F.A. Rainey, J.R. Battista and M.S. da Costa. 1999. Characterization and radiation resistance of new isolates of *Rubrobacter radiotolerans* and *Rubrobacter xylanophilus*. Extremophiles *3*: 235–238.

Ito, H., H. Watanabe, M. Takehisa and H. Iizuka. 1983. Isolation and identification of radiation-resistant cocci belonging to the genus *Deinococcus* from sewage sludges and animal feeds. Agric. Biol. Chem. *47*: 1239–1247.

Ludwig, W., J. Euzéby and W. Whitman. 2012. Taxonomic outline of the phylum *Actinobacteria*. *In* Bergey's Manual of Systematic Bacteriology, 2nd edn, vol. 5, The *Actinobacteria* (edited by Goodfellow, Kämpfer, Busse, Trujillo, Suzuki, Ludwig and Whitman). Springer, New York, pp. 29–31.

Monciardini, P., L. Cavaletti, P. Schumann, M. Rohde and S. Donadio. 2003. *Conexibacter woesei* gen. nov., sp. nov., a novel representative of a deep evolutionary line of descent within the class *Actinobacteria*. Int. J. Syst. Evol. Microbiol. *53*: 569–576.

Rainey, F.A., N.L. Ward-Rainey and E. Stackebrandt. 1997. *In* Stackebrandt, E., F.A. Rainey and N.L. Ward-Rainey. Proposal for a new hierarchic classification system, *Actinobacteria* classis nov. Int. J. Syst. Bacteriol. *47*: 479–491.

Reddy, G.S. and F. Garcia-Pichel. 2009. Description of *Patulibacter americanus* sp. nov., isolated from biological soil crusts, emended description of the genus *Patulibacter* Takahashi *et al.* 2006 and proposal of *Solirubrobacterales* ord. nov. and *Thermoleophilales* ord. nov. Int. J. Syst. Evol. Microbiol. *59*: 87–94.

Saito, T., H. Terato and A. Yamamoto. 1994. Pigments of *Rubrobacter radiotolerans*. Arch. Microbiol. *162*: 414–421.

Singleton, D.R., M.A. Furlong, A.D. Peacock, D.C. White, D.C. Coleman and W.B. Whitman. 2003. *Solirubrobacter pauli* gen. nov., sp. nov., a mesophilic bacterium within the *Rubrobacteridae* related to common soil clones. Int. J. Syst. Evol. Microbiol. *53*: 485–490.

Stackebrandt, E., F.A. Rainey and N.L. Ward-Rainey. 1997. Proposal for a new hierarchic classification system, *Actinobacteria* classis nov. Int. J. Syst. Bacteriol. *47*: 479–491.

Stackebrandt, E. 2004. Will we ever understand? The undescribable diversity of the prokaryotes. Acta Microbiol. Immunol. Hung. *51*: 449–462.

Suzuki, K., M.D. Collins, E. Iijima and K. Komagata. 1988. Chemotaxonomic characterization of a radiotolerant bacterium, *Arthrobacter radiotolerans*: description of *Rubrobacter radiotolerans* gen. nov., comb. nov. FEMS Microbiol. Lett. *52*: 33–39.

Suzuki, K., M.D. Collins, E. Iijima and K. Komagata. 1989. *In* Validation of the publication of new names and new combinations previously effectively published outside the IJSB. List no. 28. Int. J. Syst. Bacteriol. *39*: 93–94.

Suzuki, K., M. Goodfellow and A.G. O'Donnell. 1993. Cell envelopes and classification. *In* Handbook of New Bacterial Systematics (edited by Goodfellow and O'Donnell). Academic Press, London, pp. 195–250.

Yakimov, M.M., H. Lunsdorf and P.N. Golyshin. 2003. *Thermoleophilum album* and *Thermoleophilum minutum* are culturable representatives of group 2 of the *Rubrobacteridae* (*Actinobacteria*). Int. J. Syst. Evol. Microbiol. *53*: 377–380.

Yoshinaka, T., K. Yano and Yamaguch.H. 1973. Isolation of highly radioresistant bacterium, *Arthrobacter radiotolerans* nov. sp. Agric. Biol. Chem. *37*: 2269–2275.

Zhi, X.-Y., W.-J. Li and E. Stackebrandt. 2009. An update of the structure and 16S rRNA gene sequence-based definition of higher ranks of the class *Actinobacteria*, with the proposal of two new suborders and four new families and emended descriptions of the existing higher taxa. Int. J. Syst. Evol. Microbiol. *59*: 589–608.

Class VI. **Thermoleophilia** class. nov.

KEN-ICHIRO SUZUKI AND WILLIAM B. WHITMAN

Ther.mo.le.o.phi'li.a. N.L. neut. n. *Thermoleophilum* type genus of the type order; suff. *-ia* ending to denote a class; N.L. pl. neut. n. *Thermoleophilia* the *Thermoleophilales* class.

The class is defined on the basis of the 16S rRNA gene sequences (Table 314 and Table 315) and is supported by similarities in phenotypic properties. Although initial phylogenetic analyses described an affiliation of the families *Thermoleophilaceae*, *Conexibacteraceae*, *Patulibacteraceae*, and *Solirubrobacteraceae* within the order *Rubrobacterales* (Zhi et al., 2009), more recent analyses do not provide strong support for this association (Ludwig et al., 2012) (Figure 423). For this reason, these families were reclassified in their own class.

Type order: **Thermoleophilales** Reddy and Garcia-Pichel 2009, 91[VP].

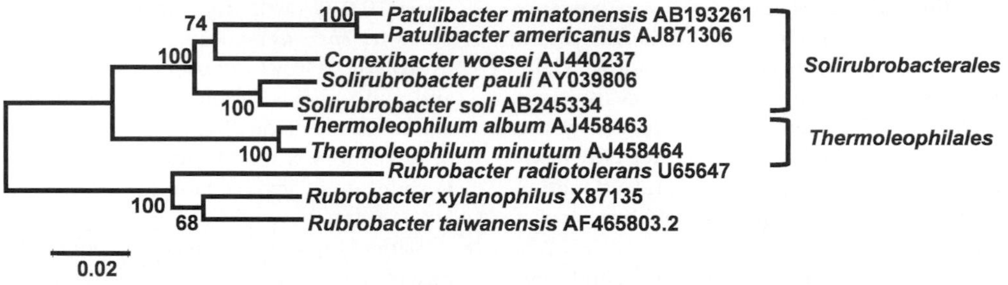

FIGURE 423. Phylogeny of the 16S rRNA genes of the representatives of the class *Thermoleophilia*. Members of the class *Rubrobacteria* serve as an outgroup. The phylogenetic tree was calculated with the Minimum Evolution algorithm of MEGA4. Evolution distance, calculated by the Kimura two-parameter model, is shown on the scale bar. Numbers represent the percentage bootstrap support for the indicated nodes based upon 1000 replicates.

References

Ludwig, W., J. Euzéby and W. Whitman. 2012. Taxonomic outline of the phylum *Actinobacteria*. *In* Bergey's Manual of Systematic Bacteriology, 2nd edn, vol. 5, The *Actinobacteria* (edited by Goodfellow, Kämpfer, Busse, Trujillo, Suzuki, Ludwig and Whitman). Springer, New York, pp. 29–31.

Reddy, G.S. and F. Garcia-Pichel. 2009. Description of *Patulibacter americanus* sp. nov., isolated from biological soil crusts, emended description of the genus *Patulibacter* Takahashi *et al.* 2006 and proposal of *Solirubrobacterales* ord. nov. and *Thermoleophilales* ord. nov. Int. J. Syst. Evol. Microbiol. *59*: 87–94.

Zhi, X.-Y., W.-J. Li and E. Stackebrandt. 2009. An update of the structure and 16S rRNA gene sequence-based definition of higher ranks of the class *Actinobacteria*, with the proposal of two new suborders and four new families and emended descriptions of the existing higher taxa. Int. J. Syst. Evol. Microbiol. *59*: 589–608.

Order I. **Thermoleophilales** Reddy and Garcia-Pichel 2009, 91[VP]

KEN-ICHIRO SUZUKI

Ther.mo.le.o.phi'la.les. N. L. neut. n. *Thermoleophilum* type genus of the order; suff. *-ales* ending to denote an order; N. L. fem. pl. n. *Thermoleophilales* the *Thermoleophilum* order.

Type genus: **Thermoleophilum** Zarilla and Perry 1986, 355[VP] (Effective publication: Zarilla and Perry 1984, 290.).

The order is defined on the basis of 16S rRNA gene sequences (Table 314) and supported by similarities in phenotypic properties. The properties are the same as for the family *Thermoleophilaceae*.

References

Reddy, G.S. and F. Garcia-Pichel. 2009. Description of *Patulibacter americanus* sp. nov., isolated from biological soil crusts, emended description of the genus *Patulibacter* Takahashi *et al.* 2006 and proposal of *Solirubrobacterales* ord. nov. and *Thermoleophilales* ord. nov. Int. J. Syst. Evol. Microbiol. *59*: 87–94.

Zarilla, K.A. and J.J. Perry. 1984. *Thermoleophilum album* gen. nov. and sp. nov., a bacterium obligate for thermophily and normal-alkane substrates. Arch. Microbiol. *137*: 286–290.

Zarilla, K.A. and J.J. Perry. 1986. *In* Validation of the publication of new names and new combinations previously effectively published outside the IJSB. List no. 20. Int. J. Syst. Bacteriol. *36*: 354–356.

Family I. **Thermoleophilaceae** Stackebrandt 2005, 548[VP] (Effective publication: Stackebrandt 2004.) emend. Zhi, Li and Stackebrandt 2009, 594

Ken-ichiro Suzuki

Ther.mo.le.o.phi.la.ce′a.e. N.L. neut. n. *Thermoleophilum* type genus of the family; suff. *-aceae* ending to denote a family; N.L. fem. pl. n. *Thermoleophilaceae* the *Thermoleophilum* family.

Type genus: **Thermoleophilum** Zarilla and Perry 1986a, 355[VP] (Effective publication: Zarilla and Perry 1984, 290).

The family is defined on the basis of the 16S rRNA gene sequences (see Table 314) and supported by similarities in phenotypic properties. Members of the family are small, nonmotile rods which stain Gram-positive. Described species are moderately thermophilic aerobes that utilize *n*-alkanes, such as *n*-heptadecane, and a few other compounds as sole carbon sources.

Genus I. **Thermoleophilum** Zarilla and Perry 1986a, 355[VP] (Effective publication: Zarilla and Perry 1984, 290.)

Jerome J. Perry*

Ther.mo.le.o′phi.lum. Gr. n. *therme* heat; L. n. *oleum* oil; Gr. adj. *philos* loving; N.L. neut. n. *Thermoleophilum* heat- and oil-loving microbe.

Very small rods, 0.3–0.4 µm in diameter and 0.7–1.5 µm in length. Regular in shape. **No resting stage or endospores are observed,** Gram-stain-negative. **The major diamino acid in the cell wall is diaminopimelic acid.** Obligately **aerobic.** Nonmotile. **Optimum temperature for growth is 55–62°C** (minimum 45°C, maximum 70°C). pH optimum is 6–7.5. **Chemo-organotrophic, utilizing *n*-alkanes from 13 to 20 carbons in length; no other growth substrates are utilized.** Oxygen is terminal electron acceptor. Catalase-positive. Unpigmented. Growth is not stimulated by increased oxygen tension (shaking). **Generation time is 6–9 h.** Isolated from both thermal and non-thermal environments.

DNA G+C content (mol%): 68.8–70.4 (T_m).

Type species: **Thermoleophilum album** Zarilla and Perry 1986a, 355 (Effective publication: Zarilla and Perry 1984, 290.).

Further descriptive information

The sequence of bases in the 16S rRNA of *Thermoleophilum album* strain NM indicated that this micro-organism should be placed in a separate "phylum" within the *Eubacteria* (C. Woese, personal communication). Further study (Brown and Haas, 1997) provided evidence that the genus *Thermoleophilum* is related to and should be grouped with the green non-sulfur bacteria. A more recent study analyzing the morphology and genotype of the three strains of *Thermoleophilum album* and *Thermoleophilum minutum* indicated that the genus should be placed in the *Rubrobacter* subdivision of the *Actinobacteria* (Yakimov et al., 2003). These authors suggest that *Thermoleophilum* is a member of the group 2 subclass as the sole known culturable genus. On the basis of 16S rRNA gene sequence analysis, the type strains of *Thermoleophilum album* and *Thermoleophilum minutum* form a distinct lineage within the class *Actinobacteria.* Figure 424 shows a maximum-likelihood tree of selected members of the class *Actinobacteria.* The closest relatives of *Thermoleophilum album* and *Thermoleophilum minutum* are *Conexibacter woesei* (sequence similarities of 90.1% and 90.2%, respectively), *Solirubrobacter soli* (90% sequence similarity), *Solirubrobacter pauli* (sequence similarities of 89.8% and 89.9%, respectively), and *Patulibacter minatonensis* (90% sequence similarity). The genus *Thermoleophilum* is currently the only member of the family *Thermoleophilaceae.*

Six strains of *Thermoleophilum* have been isolated and all are small Gram-stain-negative rods that are regular in shape and occur singly. There is no evidence of flagella or motility (Zarilla and Perry, 1984). The major diamino acid in the cell walls is diaminopimelic acid, with lesser amounts of lysine or ornithine in some strains (Merkel et al., 1978a). Electron micrographs of thin sections reveal internal structures (Kennedy and Finnerty, 1975) that may be involved in alkane uptake and storage (Figure 425). Growth on agar surfaces is inconsistent and single colonies are rare. Growth is generally confluent, whitish, and flat. The optimum growth temperature is 60°C at pH 7.0. All strains are strict aerobes and grow solely on *n*-alkanes from *n*-tridecane to *n*-eicosane; optimal growth is on *n*-heptadecane (Zarilla and Perry, 1984, 1986b). Growth occurs on a mineral salts medium with *n*-alkane added at 0.1% (v/v) with NH_4 as preferred nitrogen source (Perry, 1992). Addition of growth factors (B vitamins, amino acids, etc.) does not increase growth rate or total cell yield. The total cell yield under putatively optimal conditions is 0.3–0.6 g/l. The generation time varies among strains and is 6–9 h. Addition of ^{14}C-labeled acetate to the growth medium (mineral salts + *n*-heptadecane) results in little incorporation of radiolabel into the resultant cell mass (<5% of total carbon). *Thermoleophilum* strains lack 2-oxoglutarate dehydrogenase and apparently do not utilize the tricarboxylic acid cycle in energy generation. They do possess the enzymes of the glyoxylate cycle (Weaver et al., 1987).

The DNA G+C content is 68.8–70.4 mol%. Thermal elution profiles of *Thermoleophilum* duplexes affirm that there are two species within the genus. Reassociation experiments with DNA from *Thermoleophilum* strains confirm that this genus has no relationship to *Thermus* sp., *Thermomicrobium roseum,* or thermophilic hydrocarbon-utilizing bacilli (Zarilla and Perry, 1987). *Thermoleophilum* has a novel respiratory quinone, a tetrahydrogenated menaquinone, 2-methyl-3-VI,VII-tetrahydroheptaprenyl-1,4-naphthoquinone (Collins et al., 1986). Although these organisms are strict aerobes, they do not respond favorably to increased aeration (Allgood and Perry, 1985). Examination of the enzymes involved in defense against toxic products of oxygen reduction, i.e. superoxide dismutase, peroxidase, and catalase, indicate that only catalase responds favorably to increased

*Deceased 17 May 2011.

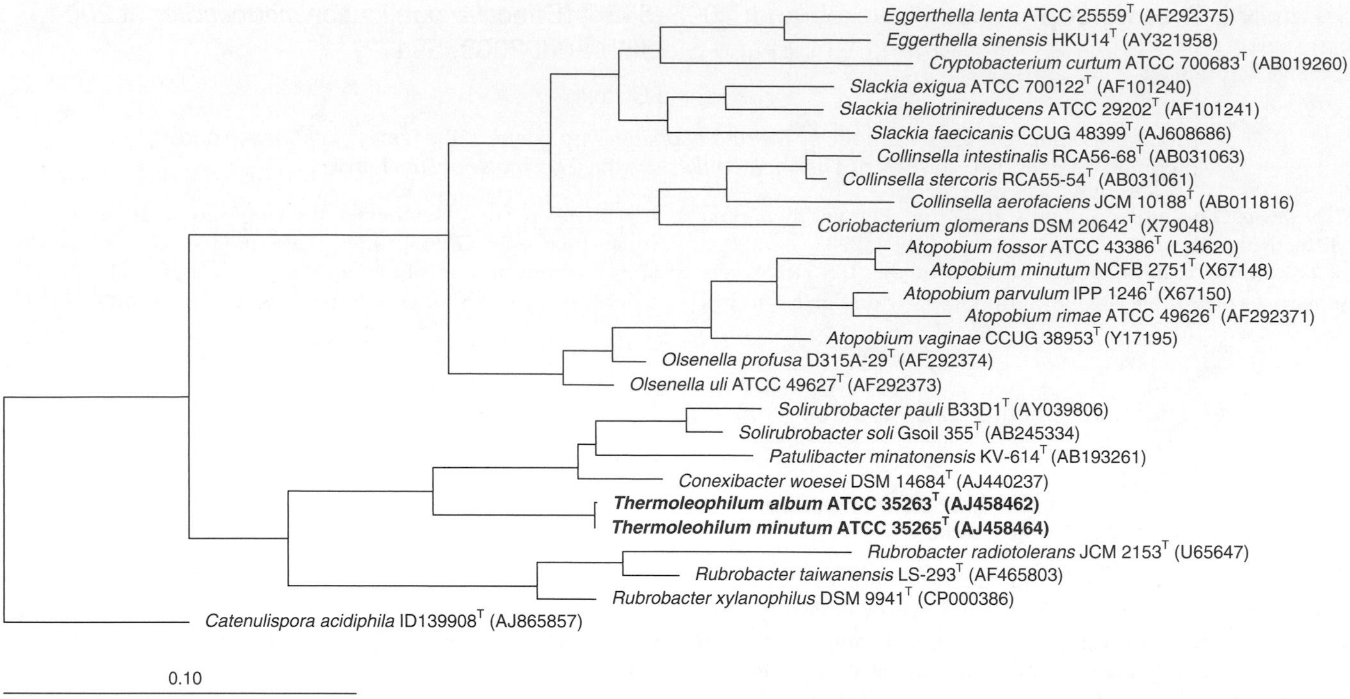

FIGURE 424. Phylogenetic analysis based on 16S rRNA gene sequences available from the EMBL data library (accession numbers are given in parentheses). The phylogenetic tree was constructed using the ARB software package (December 2007 version; Ludwig et al., 2004) and the corresponding SILVA SSURef 95 database (July 2008 version; Pruesse et al., 2007). Tree building was performed using the maximum-likelihood method with fastDNAml (Olsen et al., 1994) and no conservatory filter. Bar = 0.10 nucleotide substitutions per nucleotide position.

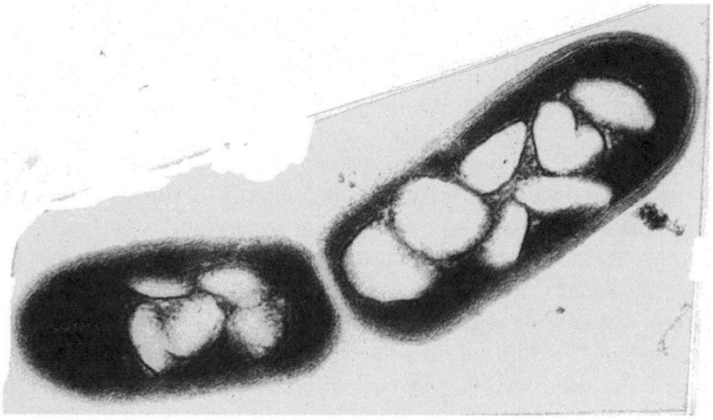

FIGURE 425. Electron micrograph of a thin section of *Thermoleophilum album* with clear areas containing stored *n*-alkane. Magnification = 31,000×.

oxygenation. Catalase was recovered and purified from *Thermoleophilum album* NM and is a heat-stable, manganese-containing enzyme with an M_r of 141,000 (Allgood and Perry, 1986). The manganese catalase gene from *Thermoleophilum album album* NM has been cloned and the nucleotide sequence has been determined (Phucharoen et al., 2001). The gene consists of 88 bp and encodes 294 amino acids with a molecular mass of 32,500 Da.

Electrophoretic mobility of selected enzymes on polyacrylamide gel indicated that enzymes from *Thermoleophilum* strains differ in mobility from equivalent enzymes in other species. An enzyme from each *Thermoleophilum* species was purified and characterized: malate dehydrogenase from *Thermoleophilum album* and isocitrate dehydrogenase from *Thermoleophilum minutum* (Novotny and Perry, 1990, 1991). The results indicate that the difference in mobility of these enzymes was related to some differences in amino acid composition and to the conformation of the protein. P RNA from *Thermoleophilum album* NM has been characterized and the RNase P is a type A ribonuclease (Brown and Haas, 1997).

The minimum antibiotic concentrations (μg/ml) that were inhibitory to all strains are as follows: chlortetracycline (25–50), neomycin (5–10), chloramphenicol (5–10), penicillin (5–25), and novobiocin (5–10).

Enrichment and isolation procedures

Six strains of *Thermoleophilum* have been obtained in axenic culture. These strains can be separated into two species: *Thermoleophilum album* and *Thermoleophilum minutum*. The six strains were isolated by enrichment from soil or mud obtained from across the USA (Table 317). Mud was added to a basal salts medium

TABLE 317. Source of the soil sample from which the obligately thermophilic strains of *Thermoleophilum* were isolated

Species	Source (in the USA)	Strain
T. album	Hot Springs, AR	HS-5^T
	Faywood Hot Springs, NM	NM
	Yellowstone National Park, WY	YS-3
	Roanoke Rapids, NC	RR-D
T. minutum	Yellowstone National Park, WY	YS-4^T
	Beaufort, NC	PTA-1

(Leadbetter and Foster, 1958) with *n*-heptadecane as sole carbon source (Merkel et al., 1978b). The enrichment was incubated at 60°C for 1–2 weeks and a transfer was made from media having visible turbidity to a sterile medium of the same composition. After several transfers in liquid medium, axenic cultures were obtained by streaking on a solid medium (2.0% agar) with *n*-heptadecane (0.2 ml) added in the cover of the inverted plate. *Thermoleophilum*, as the sole culturable genus in the subclass, serves as a prime example of the exacting culture conditions required for the isolation of a fastidious species from nature.

An exceedingly limited substrate range, high growth temperature, and sparse growth on a solid substrate offer ample evidence that culturing many "unculturable" microbes from nature is a difficult task.

Differentiation of the genus *Thermoleophilum* from other genera

The distinct and readily determinable characteristic of members of the genus *Thermoleophilum* is their restriction to growth with *n*-alkanes as substrate. Several characteristics that separate *Thermoleophilum* species from other Gram-stain-negative thermophiles are listed in Table 318. Under phase-contrast microscopy, *Thermomicrobium* cells appear pleomorphic, whereas *Thermus* cells are long, thin, regular rods and the cells of *Thermoleophilum* species are very small and regular in shape. The only other thermophilic hydrocarbon-utilizing aerophiles known are members of the genus *Bacillus* and can be distinguished by Gram-stain and production of endospores.

TABLE 318. Differential characteristics of Gram-stain-negative thermophilic rods

Characteristic	*Thermus* strains	*Thermomicrobium roseum*	*Thermoleophilum* strains
Peptidoglycan major diamino acid	Ornithine	None	Diaminopimelic acid
Cell morphology	Rods, 5–10 µm	Pleomorphic rods, 3–6 µm	Short rods, 0.7–1.5 µm
n-Alkane utilization	–	–	+
Growth on complex media	+	+	–
Growth on glucose	+	–	–
Generation time	20–60 min	5.5 h	6–9 h
Pigment	Cream, yellow, orange, pink	Pink	None
Respiratory quinone	MK-8	MK-8	MK-7(H_4)
Temperature range (°C)	60–70	70–75	55–65

List of species of the genus *Thermoleophilum*

1. **Thermoleophilum album** Zarilla and Perry 1986a, 355[VP] (Effective publication: Zarilla and Perry 1984, 290).

al'bum. L. neut. adj. *album* white.

Rod-shaped cells, 0.9 µm in length, nonmotile, nonsporeforming, Gram-stain-negative, aerobic, catalase-positive, and non-pigmented. Forms very small translucent to white colonies on agar surfaces. Grows solely with *n*-alkanes from 13 to 20 carbons in length at temperatures from 45 to 70°C. Addition of growth factors or increased aeration do not result in an increase in the growth rate or total cell yield.

Source: mud samples taken from both thermal and nonthermal environments.

DNA G+C content (mol%): 70.4 (T_m).

Type strain: HS-5, ATCC 35263.

Sequence accession no. (16S rRNA gene): AJ458462.

Further comments: additional strains of this species are NM (= ATCC 35266), YS-3 (= ATCC 35264), and RR-D (= ATCC 35267).

2. **Thermoleophilum minutum** Zarilla and Perry 1986b, 16[VP]

mi.nu'tum. L. neut. adj. *minutum* small, referring to cell size.

Rod-shaped cells, 1–1.5 µm in length. Nonmotile, nonsporeforming. Gram-stain-negative, aerobic, and catalase-positive. Nonpigmented. Forms small translucent to white colonies on agar surfaces. Grows solely with *n*-alkanes from 13 to 20 carbons in length at temperatures from 45 to 70°C. Addition of growth factors or increased aeration do not result in an increase in the growth rate or total cell yield. This species can be differentiated from *Thermoleophilum album* solely by thermal elution profile of DNA/DNA duplexes (Zarilla and Perry, 1986b).

Source: the type strain was isolated from a hot spring in Yellowstone National Park, USA.

DNA G+C content (mol%): 70.0 (T_m).

Type strain: YS-4, ATCC 35265.

Sequence accession no. (16S rRNA gene): AJ458464.

Further comments: an additional strain of this species is PTA-1 (= ATCC 35268), isolated from a mud sample from Beaufort, NC, USA.

References

Allgood, G.S. and J.J. Perry. 1985. Oxygen defense systems in obligately thermophilic bacteria. Can. J. Microbiol. *31*: 1006–1010.

Allgood, G.S. and J.J. Perry. 1986. Characterization of a manganese-containing catalase from the obligate thermophile *Thermoleophilum album*. J. Bacteriol. *168*: 563–567.

Brown, J.W. and E.S. Haas. 1997. RNase P RNAs of green non-sulfur bacteria. *In* Nucleic Acids Symposium Series no. 36. Oxford University Press, London, pp. 73–75.

Collins, M.D., O.W. Howarth and J.J. Perry. 1986. A new respiratory quinone, 2-methyl-3-VI,VII-tetrahydroheptaprenyl-1,4-naphthoquinone isolated from *Thermoleophilum album*. FEMS Microbiol. Lett. *34*: 167–171.

Kennedy, R.S. and W.R. Finnerty. 1975. Microbial assimilation of hydrocarbons. II. Intracytoplasmic membrane induction in *Acinetobacter* sp. Arch. Microbiol. *102*: 85–90.

Leadbetter, E.R. and J.W. Foster. 1958. Studies on some methane-utilizing bacteria. Arch. Mikrobiol. 30: 91–118.

Ludwig, W., O. Strunk, R. Westram, L. Richter, H. Meier, Yadhukumar, A. Buchner, T. Lai, S. Steppi, G. Jobb, W. Forster, I. Brettske, S. Gerber, A.W. Ginhart, O. Gross, S. Grumann, S. Hermann, R. Jost, A. Konig, T. Liss, R. Lussmann, M. May, B. Nonhoff, B. Reichel, R. Strehlow, A. Stamatakis, N. Stuckmann, A. Vilbig, M. Lenke, T. Ludwig, A. Bode and K.H. Schleifer. 2004. ARB: a software environment for sequence data. Nucleic Acids Res. *32*: 1363–1371.

Merkel, G.J., S.S. Stapleton and J.J. Perry. 1978a. Isolation and peptidoglycan of Gram-negative hydrocarbon-utilizing thermophilic bacteria. J. Gen. Microbiol. *109*: 141–148.

Merkel, G.J., W.H. Underwood and J.J. Perry. 1978b. Isolation of thermophilic bacteria capable of growth solely in long-chain hydrocarbons. FEMS Microbiol. Lett. *3*: 81–83.

Novotny, J.F. and J.J. Perry. 1990. Characterization of the malate dehydrogenase from *Thermoleophilum album* NM. Arch. Microbiol. *154*: 304–307.

Novotny, J.F., Jr and J.J. Perry. 1991. Characterization of a heat-stable NADP-dependent isocitrate dehydrogenase from the obligate thermophile *Thermoleophilum minutum* YS-4. Appl. Microbiol. Biotechnol. *35*: 461–465.

Olsen, G.J., H. Matsuda, R. Hagstrom and R. Overbeek. 1994. fastDNAmL: a tool for construction of phylogenetic trees of DNA sequences using maximum likelihood. Comput. Appl. Biosci. *10*: 41–48.

Perry, J.J. 1992. The genus *Thermoleophilum In* The Prokaryotes: a Handbook on the Biology of Bacteria: Ecophysiology, Isolation, Identification, Applications, 2nd edn, vol. IV (edited by Balows, Trüper, Dworkin, Harder and Schleifer). Springer, New York, pp. 3780–3784.

Phucharoen, K., Y. Takenaka and T. Shinozawa. 2001. Molecular cloning and sequence analysis of the manganese catalase gene from *Thermoleophilum* album NM. DNA Seq. *12*: 413–417.

Pruesse, E., C. Quast, K. Knittel, B. Fuchs, W. Ludwig, J. Peplies and F.O. Glöckner. 2007. SILVA: a comprehensive online resource for quality checked and aligned rRNA sequence data compatible with ARB. Nucleic Acids Res. *35*: 7188–7196.

Stackebrandt, E. 2004. Will we ever understand? The undescribable diversity of the prokaryotes. Acta Microbiol. Immunol. Hung. *51*: 449–462.

Stackebrandt, E. 2005. *In* Validation of publication of new names and new combinations previously effectively published outside the IJSEM. List no. 102. Int. J. Syst. Evol. Microbiol. *55*: 547–549.

Weaver, K.L., K.C. Terlesky and J.J. Perry. 1987. Metabolism in the obligately hydrocarbonoclastic genus *Thermoleophilum*. Proceedings of the Abstracts of the 87th General Meeting of the American Society for Microbiology, Washington, D.C., p. 195.

Yakimov, M.M., H. Lunsdorf and P.N. Golyshin. 2003. *Thermoleophilum album* and *Thermoleophilum minutum* are culturable representatives of group 2 of the *Rubrobacteridae* (*Actinobacteria*). Int. J. Syst. Evol. Microbiol. *53*: 377–380.

Zarilla, K.A. and J.J. Perry. 1984. *Thermoleophilum album* gen. nov. and sp. nov., a bacterium obligate for thermophily and normal-alkane substrates. Arch. Microbiol. *137*: 286–290.

Zarilla, K.A. and J.J. Perry. 1986a. *In* Validation of the publication of new names and new combinations previously effectively published outside the IJSB. List no. 20. Int. J. Syst. Bacteriol. *36*: 354–356.

Zarilla, K.A. and J.J. Perry. 1986b. Deoxyribonucleic acid homology and other comparisons among obligately thermophilic hydrocarbonoclastic bacteria, with a proposal for *Thermoleophilum minutum* sp. nov. Int. J. Syst. Bacteriol. *36*: 13–16.

Zarilla, K.A. and J.J. Perry. 1987. *Bacillus thermoleovorans* sp. nov., a species of obligately thermophilic hydrocarbon utilizing endospore-forming bacteria. Syst. Appl. Microbiol. *9*: 258–264.

Zhi, X.-Y., W.-J. Li and E. Stackebrandt. 2009. An update of the structure and 16S rRNA gene sequence-based definition of higher ranks of the class *Actinobacteria*, with the proposal of two new suborders and four new families and emended descriptions of the existing higher taxa. Int. J. Syst. Evol. Microbiol. *59*: 589–608.

Order II. **Solirubrobacterales** Reddy and Garcia-Pichel 2009, 91[VP]

William B. Whitman and Ken-ichiro Suzuki

So.li.ru.bro.bac'te.ra.les. N.L. masc. n. *Solirubrobacter* type genus of the order; suff. *-ales* ending to denote an order; N.L. fem. pl. n. *Solirubrobacterales* the *Solirubrobacter* order.

The order *Solirubrobacterales* contains three families, *Solirubrobacteraceae*, *Conexibacteraceae*, and *Patulibacteraceae*. The order is defined on the basis of 16S rRNA gene sequence (see Table 314) and phenotypic similarities (Table 319). Although only a few genera have been described, cloning and sequencing of environmental DNA suggest that many more related taxa exist in nature. Currently described species are all Gram-stain positive, mesophilic and sometimes psychrotolerant, and contain *meso*-diaminopimelate as the diamino acid in their peptidoglycan.

Type genus: **Solirubrobacter** Singleton, Furlong, Peacock, White, Coleman and Whitman 2003, 489[VP].

TABLE 319. Characteristics of the genera in the order *Solirubrobacterales*[a]

Properties	*Solirubrobacter*[b]	*Conexibacter*[c]	*Patulibacter*[d]
Morphology (length in μm)	Rod (1.4)	Rod (0.9–1.2)	Rod (1.2–1.5)
Motility	Nonmotile	Motile	Motile
Temperature optimum (°C)	28–30	28–37	24–27
pH range	6–7.7	7–7.5	6–8
Oxidase	Negative	Positive	Negative
Major fatty acid	$C_{16:0}$ iso	$C_{18:1}$ ω9*c*	$C_{18:1}$ ω9*c*
Diamino acid in peptidoglycan	nd	*meso*-DAP	*meso*-DAP
Isoprenoid quinone	MK-7(H_4)	MK-7(H_4)	DMK-7, MK-7(H_4)

[a]Abbreviations: *meso*-DAP, *meso*-diaminopimelate; iso, iso methyl-branched; DMK, demethylmenaquinone; MK, menaquinone; nd, not determined.

[b]Data from Singleton et al. (2003) and Kim et al. (2007).

[c]Data from Monciardini et al. (2003).

[d]Data from Takahashi et al. (2006) and Reddy and Garcia-Pichel (2009).

References

Kim, M.K., J.R. Na, T.H. Lee, W.T. Im, N.K. Soung and D.C. Yang. 2007. *Solirubrobacter soli* sp. nov., isolated from soil of a ginseng field. Int. J. Syst. Evol. Microbiol. *57*: 1453–1455.

Monciardini, P., L. Cavaletti, P. Schumann, M. Rohde and S. Donadio. 2003. *Conexibacter woesei* gen. nov., sp. nov., a novel representative of a deep evolutionary line of descent within the class *Actinobacteria*. Int. J. Syst. Evol. Microbiol. *53*: 569–576.

Reddy, G.S. and F. Garcia-Pichel. 2009. Description of *Patulibacter americanus* sp. nov., isolated from biological soil crusts, emended description of the genus *Patulibacter* Takahashi *et al.* 2006 and proposal of *Solirubrobacterales* ord. nov. and *Thermoleophilales* ord. nov. Int. J. Syst. Evol. Microbiol. *59*: 87–94.

Singleton, D.R., M.A. Furlong, A.D. Peacock, D.C. White, D.C. Coleman and W.B. Whitman. 2003. *Solirubrobacter pauli* gen. nov., sp. nov., a mesophilic bacterium within the *Rubrobacteridae* related to common soil clones. Int. J. Syst. Evol. Microbiol. *53*: 485–490.

Takahashi, Y., A. Matsumoto, K. Morisaki and S. Ōmura. 2006. *Patulibacter minatonensis* gen. nov., sp. nov., a novel actinobacterium isolated using an agar medium supplemented with superoxide dismutase, and proposal of *Patulibacteraceae* fam. nov. Int. J. Syst. Evol. Microbiol. *56*: 401–406.

Family I. **Solirubrobacteraceae** Stackebrandt 2005, 548[VP] emend. Zhi, Li and Stackebrandt 2009, 594

William B. Whitman and Ken-ichiro Suzuki

So.li.ru.bro.bac'te.ra.ce.ae. N. L. masc. n. *Solirubrobacter* type genus of the family; suff. *-aceae* ending to denote a family; N. L. fem. pl. n. *Solirubrobacteraceae* the *Solirubrobacter* family.

Type genus: **Solirubrobacter** Singleton, Furlong, Peacock, White, Coleman and Whitman 2003, 489[VP].

The family is defined on the basis of 16S rRNA gene sequences (see Table 314) and supported by similarities in phenotypic properties. Members of the family are nonmotile rods which stain Gram-positive. Described species are aerobes that utilize many sugars and a few other compounds as sole carbon sources. For a comparison to related families, see the order description.

Genus I. **Solirubrobacter** Singleton, Furlong, Peacock, White, Coleman and Whitman 2003, 489[VP]

William B. Whitman

So.li.ru.bro.bac'ter. L. n. *solum* soil; L. adj. *ruber* red; N.L. n. *bacter* the masculine equivalent of the Gr. neut. n. *bakterion* a rod; N.L. masc. n. *Solirubrobacter* a soil red rod.

Gram-stain positive, **rod** of medium length. **Nonmotile.** Spores are not formed. **Aerobe** and mesophile. Grows well on common sugars, a few amino acids and other compounds as sole carbon source. The **major phospholipid fatty acids (PFLAs) are $C_{16:0}$ iso and $C_{18:1}$ ω9*c*,** and **MK-7(H_4) is the predominant menaquinone** in the only species that has been tested. Some species are sensitive to desiccation.

DNA G+C content (mol%): 71–72 (Lc).

Type species: **Solirubrobacter pauli** Singleton, Furlong, Peacock, White, Coleman and Whitman 2003, 489[VP].

Further descriptive information

This genus is represented by two species, each one represented by a single strain, and the description must necessarily be preliminary. Clones of 16S rRNA genes with high sequence similarity to members of this genus are frequently encountered in clone libraries of agricultural, pasture and forest soils; insect guts; plant-associated bacteria; and aerosols (Brodie et al., 2007; Fall et al., 2007; Gremion et al., 2003; Joseph et al., 2003; Singleton et al., 2003; Wang et al., 2008; Youssef et al., 2009; Yu et al., 2008). Presumably, this genus is representative of a group of abundant but largely uncharacterized microorganisms.

For the type strain of *Solirubrobacter pauli* B33D1[T], cells are short rods, especially in older cultures (Singleton et al., 2003). Much longer cells are occasionally observed in fresh cultures. In liquid culture, cells often grow in long chains that wrap around each other producing large aggregates (Figure 426). Even in rich medium, only a low cellular density (maximum turbidity, near 0.1 at 600 nm) is obtained. Colonies are usually pink in color. Plates incubated at higher temperatures (≥28°C) often produce less pigment initially, although a deep pink color appears over time. The color resembles that of *Rubrobacter* species, and the pigments like those of *Rubrobacter* species are not easily extracted using traditional methods. However, methanol can extract a small amount of pigment, which has absorption maxima at 466, 493, and 526 nm. These maxima are similar to those reported for pigments extracted from *Rubrobacter radiotolerans* (Saito et al., 1994).

The menaquinone composition has only been examined in *Solirubrobacter soli*, where it was found to be MK-7(H_4) (Kim et al., 2007). The major phospholipid fatty acids (PFLAs) are $C_{16:0}$ iso and $C_{18:1}$ ω9*c*. Smaller amounts of $C_{18:3}$ ω6*c*, $C_{17:1}$ ω6*c*, $C_{19:0}$, $C_{18:0}$ iso, and $C_{16:0}$ may also be present. Related organisms *Conexibacter woesei*, *Patulibacter minatonensis*, and *Patulibacter americanus* also

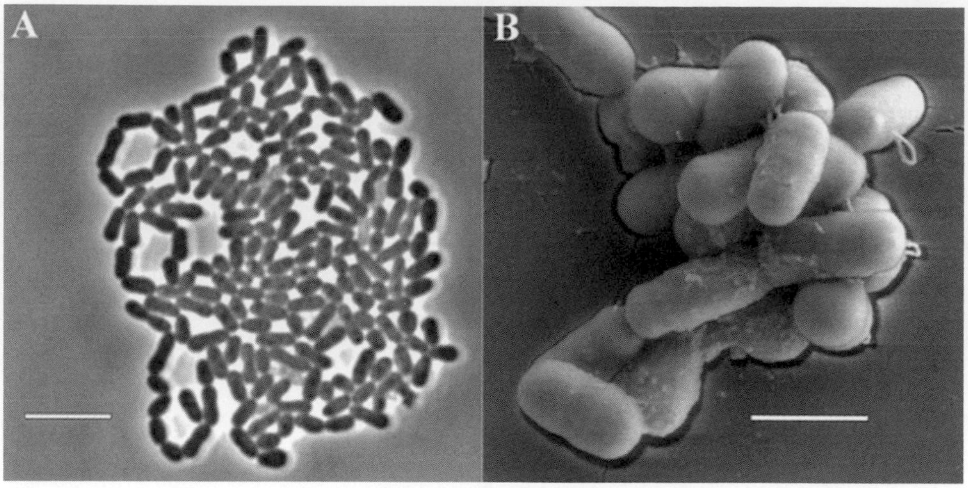

FIGURE 426. Morphology of *Solirubrobacter pauli*. Photomicrographs of strain B33D1 grown in nutrient broth. (A) Phase-contrast image of an aggregate of cells. Scale bar = 5 μm. (B) Scanning electron micrograph of an aggregate. Scale bar = 1 μm. (Reprinted with permission from Singleton et al., 2003. Int. J. Syst. Evol. Microbiol. *53*: 485–490.)

possess high levels of $C_{18:1}$ ω9*c* (Monciardini et al., 2003; Reddy and Garcia-Pichel, 2009; Takahashi et al., 2006).

Enrichment and isolation procedures

The type strain B33D1^{T} of *Solirubrobacter pauli* was isolated from a burrow of the epigeic earthworm *Lumbricus rubellus* in an agricultural soil in Athens, GA, USA (Furlong et al., 2002). Fresh soil was diluted to extinction and plated on medium composed of 50% Nutrient Broth (Difco, pH 7.0, ~23°C). Isolates were allowed to grow for 2 weeks before colonies were picked. A single pink colony was selected and maintained on the same medium. It was the only representative of this taxon of the more than 200 isolates screened from these soils. At present, no evidence suggests that its isolation from an earthworm burrow has significance. Subsequent studies indicated that spread plates greatly underestimate the viability of strain B33D1, and that pour plates containing 1.5% agar enhance plating efficiency.

The type strain Gsoil 355^{T} of *Solirubrobacter soli* was isolated from soil of a ginseng field in Daejon, Korea (Kim et al., 2007). Soil suspensions were serially diluted and plated on R2A medium (Difco). After 5 d of incubation at 30°C, colorless colonies were transferred to fresh medium.

Maintenance procedures

The organism has been stored at −70°C in medium containing 15% glycerol.

Taxonomic comments

Upon its isolation, *Solirubrobacter* was affiliated with the 16S rRNA group 2 of the *Rubrobacteridae* subclass of the *Actinobacteria* phylum, as defined by Holmes et al. (2000). Aside from the thermophilic and radiation-resistant species of *Rubrobacter*, group 2 at that time was based entirely on gene sequences obtained directly from soil DNA. Concurrent with the description of *Solirubrobacter*, another mesophile *Conexibacter woesei* isolated from soil and the moderate thermophiles *Thermoleophilum album* and *Thermoleophilum minutum* were also recognized as members of group 2 (Monciardini et al., 2003; Yakimov et al., 2003). Although originally described in the 1980s, the relatedness of *Thermoleophilum* to the *Actinobacteria* was not known until sequencing of its 16S rRNA genes. Subsequently, additional related genera and species were isolated from soil, including *Patulibacter minatonensis*, *Patulibacter americanus*, and *Solirubrobacter soli* (Kim et al., 2007; Reddy and Garcia-Pichel, 2009; Takahashi et al., 2006).

Based upon the low degree of sequence similarity of their 16S rRNA genes and phenotypic differences, the genera *Thermoleophilum*, *Solirubrobacter*, *Conexibacter*, and *Patulibacter* were classified in separate families (Stackebrandt, 2004; Takahashi et al., 2006; Zhi et al., 2009). In recognition of the differences between *Thermoleophilum* and other members of the group, the families were subsequently classified into two orders, *Thermoleophilales* and *Solirubrobacterales* (Reddy and Garcia-Pichel, 2009). More recently, phylogenetic analyses of the 16S rRNA genes failed to provide strong support for the relatedness of these groups to *Rubrobacter* (Ludwig et al., 2012). For that reason, they are currently classified within the class *Thermoleophilia*. Although represented by only a few cultivated bacteria, the large number of genes cloned from mesophilic soils suggests that these are abundant and diverse groups in nature. However, it is very speculative to infer the properties of the uncultivated members at this time.

Differentiation of species of the genus

The descriptions of *Solirubrobacter pauli* and *Solirubrobacter soli* were based upon descriptions of single strains with growth properties measured under somewhat different growth conditions. Therefore, the reliability of the reported differences in phenotype is not certain. With that caveat, *Solirubrobacter pauli* but not *Solirubrobacter soli* utilizes acetate, D-sorbitol, and L-alanine as sole carbon sources. *Solirubrobacter soli* but not *Solirubrobacter pauli* grows in 1% (w/v) NaCl.

List of species in the genus *Solirubrobacter*

1. **Solirubrobacter pauli** Singleton, Furlong, Peacock, White, Coleman and Whitman 2003, 489[VP]

pau'li. N.L. gen. n. *pauli* of *Paulus*; named for the prominent soil microbiologist Eldor Alvin Paul.

Cells are rods (ca. 1.4 μm × 0.7 μm). Much longer cells are occasionally observed in fresh cultures, and older cultures generally have uniformly shorter rods. In liquid culture, cells often grow in long chains that wrap around each other to produce large aggregates. A capsule is not present. Colonies grown on Difco Nutrient Broth agar plates are round, convex, with entire edges, and usually pink in color. Catalase-positive and oxidase-negative. Casein and Tween 80 are not hydrolyzed. No hemolysis is observed on blood agar. In minimal medium, a variety of sugars serve as sole carbon sources including fructose, galactose, glucose, lactose, mannose, sorbitol, sucrose, and xylose. Cellobiose and mannitol do not support growth. Among the organic acids, pyruvate supports good growth. Acetate supports weak growth. Citrate, malate, and succinate do not support growth. Casamino acids provide good growth. When tested individually, only the amino acids alanine, arginine, and lysine support growth. Glycerol is the only alcohol tested that supports growth. Alcohols that do not support growth are methanol, ethanol, 1-propanol, 2-propanol, butanol, isobutyl alcohol, and iso-amyl alcohol. Of the aromatic compounds tested, only chlorogenic acid supports growth. Anthranilic acid, benzoic acid, catechol, protocatechuic acid, *p*-coumaric acid, gentisic acid, ferulic acid, *p*-hydroxybenzoic acid, syringic acid, and vanillic acid do not produce observable growth after 2 weeks. Ammonium will serve as a sole nitrogen source, and Casamino acids will serve as a sole carbon and nitrogen source. Nitrate and urea are not nitrogen sources. Optimal growth at 28–30°C and pH 6.0–6.5. The temperature and pH ranges for growth are 19–38°C and 6–7.5. It fails to grow anaerobically and requires at least 1% (v/v) air for growth. No growth is observed in the presence of 1% NaCl. Sensitive to polymyxin, ampicillin, tetracycline, and streptomycin.

DNA G+C content (mol%): 71.8 (Lc).

Type strain: B33D1, ATCC BAA-492, DSM 14954, JCM 13025.

Sequence accession no. (16S rRNA gene): AY039806.

2. **Solirubrobacter soli** Kim, Na, Lee, Im, Soung and Yang 2007, 1454[VP]

so'li. L. neut. gen. n. *soli* of the soil, the source of the type strain.

Cells are short rods, 1–3 μm in length. Colonies are not pigmented following growth at 30°C on R2A agar. Weakly positive for oxidase. Gelatin is hydrolyzed. Sugars serve as sole carbon sources, including L-arabinose, L-fucose, D-glucose, D-maltose, D-melibiose, L-rhamnose, D-ribose, sucrose, and glycogen. Other utilized compounds include adipate, gluconate, *myo*-inositol, L-proline, *N*-acetyl-D-glucosamine, and salicin. Compounds that do not support growth as sole carbon source include 2-ketogluconate, 3-hydroxybenzoate, 4-hydroxybenzoate, 3-hydroxybutyrate, 5-ketogluconate, acetate, caprate, citrate, itaconate, lactate, L-malate, malonate, phenylacetate, propionate, suberate, *N*-valerate, D-mannitol, D-sorbitol, L-alanine, and L-histidine. In the API test systems, the following reactions were positive: *N*-acetyl-β-glucosaminidase, acid phosphatase, alkaline phosphatase, cystine arylamidase, esterase (C4), esterase (C8), α-glucosidase, β-glucosidase, β-galactosidase, leucine arylamidase, and valine arylamidase. Nitrate is not reduced to nitrite or N_2. Optimal growth at 30°C, slow growth at 15°C, and no growth at 37°C. Growth is observed in 0–1.5% (w/v) NaCl.

DNA G+C content (mol%): 71.5 (Lc).

Type strain: Gsoil 355, JCM 14923, KCTC 12628, LMG 23485.

Sequence accession no. (16S rRNA gene): AB245334.

References

Brodie, E.L., T.Z. DeSantis, J.P. Parker, I.X. Zubietta, Y.M. Piceno and G.L. Andersen. 2007. Urban aerosols harbor diverse and dynamic bacterial populations. Proc. Natl. Acad. Sci. U.S.A. *104*: 299–304.

Fall, S., J. Hamelin, F. Ndiaye, K. Assigbetse, M. Aragno, J.L. Chotte and A. Brauman. 2007. Differences between bacterial communities in the gut of a soil-feeding termite (*Cubitermes niokoloensis*) and its mounds. Appl. Environ. Microbiol. *73*: 5199–5208.

Furlong, M.A., D.R. Singleton, D.C. Coleman and W.B. Whitman. 2002. Molecular and culture-based analyses of prokaryotic communities from an agricultural soil and the burrows and casts of the earthworm *Lumbricus rubellus*. Appl. Environ. Microbiol. *68*: 1265–1279.

Gremion, F., A. Chatzinotas and H. Harms. 2003. Comparative 16S rDNA and 16S rRNA sequence analysis indicates that *Actinobacteria* might be a dominant part of the metabolically active bacteria in heavy metal-contaminated bulk and rhizosphere soil. Environ. Microbiol. *5*: 896–907.

Holmes, A.J., J. Bowyer, M.P. Holley, M. O'Donoghue, M. Montgomery and M.R. Gillings. 2000. Diverse, yet-to-be-cultured members of the *Rubrobacter* subdivision of the *Actinobacteria* are widespread in Australian arid soils. FEMS Microbiol. Ecol. *33*: 111–120.

Joseph, S., J.P. Hugenholtz, P. Sangwan, C.A. Osborne and P.H. Janssen. 2003. Laboratory cultivation of widespread and previously uncultured soil bacteria. Appl. Environ. Microbiol. *69*: 7210–7215.

Kim, M.K., J.R. Na, T.H. Lee, W.T. Im, N.K. Soung and D.C. Yang. 2007. *Solirubrobacter soli* sp. nov., isolated from soil of a ginseng field. Int. J. Syst. Evol. Microbiol. *57*: 1453–1455.

Ludwig, W., J. Euzéby and W. Whitman. 2012. Taxonomic outline of the phylum *Actinobacteria*. *In* Bergey's Manual of Systematic Bacteriology, 2nd edn, vol. 5, The *Actinobacteria* (edited by Goodfellow, Kämpfer, Busse, Trujillo, Suzuki, Ludwig and Whitman). Springer, New York, pp. 29–31.

Monciardini, P., L. Cavaletti, P. Schumann, M. Rohde and S. Donadio. 2003. *Conexibacter woesei* gen. nov., sp. nov., a novel representative of a deep evolutionary line of descent within the class *Actinobacteria*. Int. J. Syst. Evol. Microbiol. *53*: 569–576.

Reddy, G.S. and F. Garcia-Pichel. 2009. Description of *Patulibacter americanus* sp. nov., isolated from biological soil crusts, emended description of the genus *Patulibacter* Takahashi *et al.* 2006 and proposal of *Solirubrobacterales* ord. nov. and *Thermoleophilales* ord. nov. Int. J. Syst. Evol. Microbiol. *59*: 87–94.

Saito, T., H. Terato and A. Yamamoto. 1994. Pigments of *Rubrobacter radiotolerans*. Arch. Microbiol. *162*: 414–421.

Singleton, D.R., M.A. Furlong, A.D. Peacock, D.C. White, D.C. Coleman and W.B. Whitman. 2003. *Solirubrobacter pauli* gen. nov., sp. nov., a mesophilic bacterium within the *Rubrobacteridae* related to common soil clones. Int. J. Syst. Evol. Microbiol. *53*: 485–490.

Stackebrandt, E. 2004. Will we ever understand? The undescribable diversity of the prokaryotes. Acta Microbiol. Immunol. Hung. *51*: 449–462.

Stackebrandt, E. 2005. *In* Validation of publication of new names and new combinations previously effectively published outside the IJSEM. List no. 102. Int. J. Syst. Evol. Microbiol. *55*: 547–549.

Takahashi, Y., A. Matsumoto, K. Morisaki and S. Ōmura. 2006. *Patulibacter minatonensis* gen. nov., sp. nov., a novel actinobacterium isolated using an agar medium supplemented with superoxide dismutase, and proposal of *Patulibacteraceae* fam. nov. Int. J. Syst. Evol. Microbiol. *56*: 401–406.

Wang, H.X., Z.L. Geng, Y. Zeng and Y.M. Shen. 2008. Enriching plant microbiota for a metagenomic library construction. Environ. Microbiol. *10*: 2684–2691.

Yakimov, M.M., H. Lunsdorf and P.N. Golyshin. 2003. *Thermoleophilum album* and *Thermoleophilum minutum* are culturable representatives of group 2 of the *Rubrobacteridae* (*Actinobacteria*). Int. J. Syst. Evol. Microbiol. *53*: 377–380.

Youssef, N., C.S. Sheik, L.R. Krumholz, F.Z. Najar, B.A. Roe and M.S. Elshahed. 2009. Comparison of species richness estimates obtained using nearly complete fragments and simulated pyrosequencing-generated fragments in 16S rRNA gene-based environmental surveys. Appl. Environ. Microbiol. *75*: 5227–5236.

Yu, H., Z. Wang, L. Liu, Y. Xia, Y. Cao and Y. Yin. 2008. Analysis of the intestinal microflora in *Hepialus gonggaensis* larvae using 16S rRNA sequences. Curr. Microbiol. *56*: 391–396.

Zhi, X.-Y., W.-J. Li and E. Stackebrandt. 2009. An update of the structure and 16S rRNA gene sequence-based definition of higher ranks of the class *Actinobacteria*, with the proposal of two new suborders and four new families and emended descriptions of the existing higher taxa. Int. J. Syst. Evol. Microbiol. *59*: 589–608.

Family II. **Conexibacteraceae** Stackebrandt 2005, 547[VP] (Effective publication: Stackebrandt 2004, 457.) emend. Zhi, Li and Stackebrandt 2009, 594

Peter Schumann

Co.ne.xi.bac.te.ra.ce′a.e. N.L. masc. n. *Conexibacter* type genus of the family; suff. *-aceae* ending to denote a family; N.L. fem. pl. n. *Conexibacteraceae* the *Conexibacter* family.

The pattern of 16S rRNA gene sequence signatures consists of nucleotides at positions: 144:178 (U–A), 145:177 (U–A), 293:304 (G–C), 377:386 (C–G), 408:434 (A–U), 418:425 (U–A), 590:649 (U–A), and 600:638 (U–G) (Zhi et al., 2009). The family *Conexibacteraceae* is a member of the order *Solirubrobacterales* Reddy and Garcia-Pichel (2009).

Type genus: **Conexibacter** Monciardini, Cavaletti, Schumann, Rohde and Donadio 2003, 574[VP].

Taxonomic comments

The family *Conexibacteraceae* Stackebrandt 2005 contains the genus *Conexibacter* Monciardini et al., 2003 and belongs, together with the families *Solirubrobacteraceae* Stackebrandt 2005 emend. Zhi, Li and Stackebrandt 2009 and *Patulibacteraceae* Takahashi, Matsumoto, Morisaki and Ōmura 2006 emend. Zhi, Li and Stackebrandt 2009, to the order *Solirubrobacterales* Reddy and Garcia-Pichel 2009 (Reddy and Garcia-Pichel, 2009; Stackebrandt, 2004; Takahashi et al., 2006; Zhi et al., 2009). The members of the order *Solirubrobacterales* represent a separate cluster in the phylogenetic tree (http://www.arb-silva.de/projects/living-tree/) and are coherent with respect to the set of 16S rRNA gene signature nucleotides, the positive Gram-staining, the occurrence of *meso*-diaminopimelic acid in the peptidoglycan and seven isoprenoid units in the side chain of the respiratory quinones as well as to the psychrotolerant to mesophilic growth (Reddy and Garcia-Pichel, 2009). The affiliation of the family *Conexibacteraceae* to the order *Rubrobacterales* Rainey et al. 1997 emend. Zhi, Li and Stackebrandt 2009 as proposed by Stackebrandt et al. (1997) and Zhi et al. (2009) is neither supported by the topology of the phylogenetic tree (http://www.arb-silva.de/projects/living-tree/) nor by phenotypic characteristics. The families *Conexibacteraceae*, *Rubrobacteraceae*, *Solirubrobacteraceae*, *Thermoleophilaceae*, and *Patulibacteraceae* contain only few culturable strains and the family *Conexibacteraceae* is based on a single strain only. However, many 16S rRNA gene sequences of clones from soils of different regions of the world (e.g. Rheims et al., 1996; Rheims and Stackebrandt, 1999; Yakimov et al., 2003) and of unidentified soil isolates (Joseph et al., 2003; Sangwan et al., 2005) show high similarities to cultured organisms of this group (see Figure 427). It is expected that additional strains will be isolated after elaboration of appropriate cultivation techniques, which will give a better insight into the phenotypic diversity of members of the orders *Rubrobacterales*, *Thermoleophilales*, and *Solirubrobacterales*.

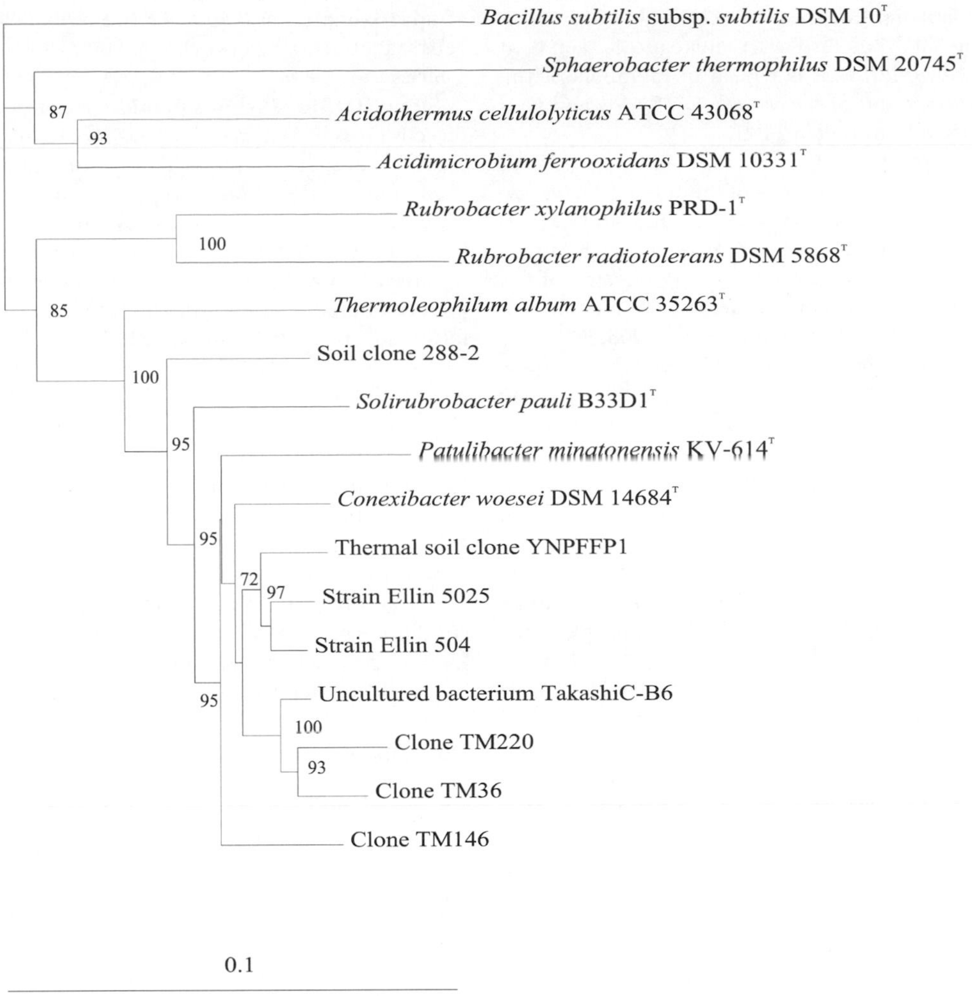

FIGURE 427. 16S rRNA gene sequence neighbor-joining tree (Felsenstein, 1993) of *Conexibacter woesei* DSM 14684^T among representatives of the class *Rubrobacteria*, including species with validly published names, unidentified strains, and clones obtained from environmental DNA. Bootstrap values (>70%) of 1000 resamplings (Felsenstein, 1985) are indicated at nodes. *Bacillus subtilis* subsp. *subtilis* DSM 10^T served as root. Bar = 10 nt substitution per 100 nt.

Genus I. **Conexibacter** Monciardini, Cavaletti, Schumann, Rohde and Donadio 2003, 574^{VP}

Peter Schumann

Co.nex.i.bac′ter. L. part. adj. *conexus* bound, tied; N.L. masc. n. *bacter* rod; N.L. masc. n. *Conexibacter* a rod that is bound.

Cells are small rods (0.6–0.7 × 0.9–1.2 µm), occurring singly or in pairs. **Gram-stain-positive** and **non-sporulating**. **Motile by long, peritrichous flagella.** Aerobic. **Catalase- and oxidase-positive**. The **peptidoglycan** is of the **A1γ type** (based on ***meso*-diaminopimelic acid**, direct cross-linkage). Mycolic acids are absent. The major isoprenoid quinone is menaquinone **MK-7(H$_4$)**. The **polar lipid** pattern consists of **phosphatidylinositol** and an unidentified phospholipid; amino-functional lipids and glycolipids are absent. The **fatty acid profile** is dominated by **oleic, 14-methyl-pentadecanoic, hexadecanoic, and ω6*c*-heptadecenoic acids**. Phylogenetically, the genus is a member of the family *Conexibacteraceae* Stackebrandt 2005 emend. Zhi, Li and Stackebrandt 2009, order *Solirubrobacterales* Reddy and Garcia-Pichel 2009.

DNA G+C content (mol%): 71 (HPLC).

Type species: **Conexibacter woesei** Monciardini, Cavaletti, Schumann, Rohde and Donadio 2003, 574^{VP}.

Further descriptive information

The genus *Conexibacter* was established to accommodate an isolate from forest soil that showed only remote 16S rRNA gene sequence similarity to validly named taxa (Monciardini et al.,

2003). *Conexibacter woesei* shows the 16S rRNA signature nucleotide A at position 906, but displays U instead of A or C at position 955, a substitution that is typical of members of the sublasses *Rubrobacteridae* and *Sphaerobacteridae* (Reddy and Garcia-Pichel, 2009; Stackebrandt et al., 1997).

An insertion of approximately 100 nt within domain III of the 23S rRNA genes found in members of the subclass *Actinobacteridae* but absent in *Atopobium minutum* (Embley and Stackebrandt, 1994) was also not detected in *Conexibacter woesei* strain ID131577^T, suggesting that the deep-branching lineages of the class *Actinobacteria* might differ in this feature from members of the subclass *Actinobacteridae* (Monciardini et al., 2003).

The closest phylogenetic neighbors of *Conexibacter woesei* ID131577^T with validly published names are *Solirubrobacter soli* Gsoil 355^T (Kim et al., 2007), *Patulibacter americanus* (Reddy and Garcia-Pichel, 2009), *Patulibacter minatonensis* JCM 12834^T (Takahashi et al., 2006) and *Solirubrobacter pauli* ATCC BAA-492^T (Singleton et al., 2003) with 94.3, 93.5, 93.5, and 93.3% 16S rRNA gene sequence similarities, respectively (according to a search using the EzTaxon server; Chun et al., 2007). Only moderate 16S rRNA gene sequence similarity values are shown between *Conexibacter woesei* ID131577^T and strains of other members of the subclass *Rubrobacteridae* Rainey et al. 1997 emend. Stackebrandt 2004 (Stackebrandt, 2004; Stackebrandt et al., 1997) (see Figure 427), e.g. 90.2 and 90.1% similarity, respectively, with *Thermoleophilum minutum* and *Thermoleophilum album* (Zarilla and Perry, 1984, 1986), and 84.5% with *Rubrobacter xylanophilus* (Carreto et al., 1996). The 16S rRNA gene sequences of the isolate and its phylogenetic neighbors *Solirubrobacter soli*, *Solirubrobacter pauli*, *Patulibacter americanus*, and *Patulibacter minatonensis* fell into a cluster consisting of clones of hitherto-uncultured bacteria obviously occurring worldwide in different soils (Rheims et al., 1996; Rheims and Stackebrandt, 1999; Yakimov et al., 2003) and strains of the subclass *Rubrobacteridae* isolated from pasture soil (Ellinbank, Victoria, Australia) by cultivation on modified VL55 medium (Sait et al., 2002) and incubation at 25°C for up to 3 months (Joseph et al., 2003; Sangwan et al., 2005). The highest binary similarity value (95.8%) was found between *Conexibacter woesei* ID131577^T and strain Ellin 504 (Sangwan et al., 2005).

The isolation and cultivation of representatives of this phylogenetic cluster offered the opportunity to gain an insight into the phenotypic characteristics of these organisms. Colonies of strain ID131577^T are smooth, mucoid, and of whitish-creamy color on Todd–Hewitt agar. They are sticky and difficult to disintegrate. Cells of *Conexibacter woesei* ID131577^T are short, Gram-stain-positive rods (0.6–0.7 × 0.9–1.2 µm), occurring singly or in pairs. Cells are motile by long peritrichous flagella (Figure 428). After approximately 1 week of cultivation in liquid media, rigid nonmotile spiral bodies are observed by phase-contrast microscopy (shown by Monciardini et al., 2003). Electron microscopic investigation revealed that these bodies were formed by self-aggregation of flagella (Figure 428) in a way that cells may become entangled in flagellar networks (Figure 428). Discarded flagella give rise to the formation of undulating or spiral bodies with a regular periodicity of approximately 2–3 µm (Figure 428). This aggregation of cells via a flagellar network appears to be a unique feature of *Conexibacter woesei*.

Strain ID131577^T shows peptidoglycan of type A1γ based on directly cross-linked *meso*-diaminopimelic acid, whereas mycolic acids are absent. A unique feature of strain ID131577^T is the occurrence of the tetrahydrogenated menaquinone MK-7(H_4), which has not been found as a single respiratory quinone in other bacteria before. Phosphatidylinositol and an additional unknown phospholipid are components of the polar lipid pattern, but not amino-functional lipids or glycolipids. The fatty acid profile of strain ID131577^T is composed of $C_{18:1}$ ω9*c* (41.4%), $C_{16:0}$ iso (16.3%), $C_{17:1}$ ω6*c* (13.9%), $C_{16:0}$ (12.7%), $C_{16:1}$ ω7*c* (1.9%), $C_{18:0}$ (1.6%), $C_{17:1}$ ω8*c* (1.5%), $C_{19:1}$ ω6*c* (1.5%) and $C_{17:0}$ iso (1.2%) (fatty acids representing less than 1% of the total fatty acids are not reported). The DNA G+C content of the type strain of *Conexibacter woesei* is 71 mol%.

Strain ID131577^T grows on brain-heart infusion (BHI; Difco), trypticase soy-yeast extract medium (medium no. 92, www.dsmz.de), and Todd–Hewitt agar at 28–37°C and pH 7–7.5 under aerobic conditions. No growth occurs under anaerobic conditions or at NaCl concentrations higher than 2% (w/v). Physiological features and susceptibility to antibiotics are given in the species description.

Isolation and maintenance procedures

Strain ID131577^T was detected as a contaminant during the isolation of filamentous actinomycetes from a soil sample of a wooded area in Gerenzano, Italy. Soil was plated on half-strength HV medium (Hayakawa and Nonomura, 1987) following dilution in water. Colonies were serially transferred on new half-strength HV medium, ISP3 medium (Shirling and Gottlieb, 1966), and Todd–Hewitt medium (Difco) until pure colonies were obtained.

Serial transfers at 4-week intervals followed by maintenance at 4°C and storage of cells as 20% (w/v) glycerol suspensions at −20°C and at −80°C were suitable provisions for preservation. Long-term preservation methods are freeze-drying in skim milk and maintenance in liquid nitrogen at −196°C.

Differentiation of the genus *Conexibacter* from related genera

Signature nucleotides that differentiate the genus *Conexibacter* from members of the families *Patulibacteraceae* and *Solirubrobacteraceae* are the following: 52:359 (U–A), 144:178 (U–A), 408:434 (A–U) and 999:1041 (G–U) (Reddy and Garcia-Pichel, 2009). The genus *Conexibacter* can be differentiated from other culturable members of the classes *Thermoleophilia* and *Rubrobacteridae* by 16S rRNA gene sequence similarity values of <95%. The phenotypic characteristics of *Conexibacter woesei* ID131577^T and its closest phylogenetic relatives *Solirubrobacter pauli* B33D1^T, *Solirubrobacter soli* Gsoil 355^T, *Patulibacter americanus* CP177-2^T and *Patulibacter minatonensis* KV-614^T are compiled for comparison in Table 320.

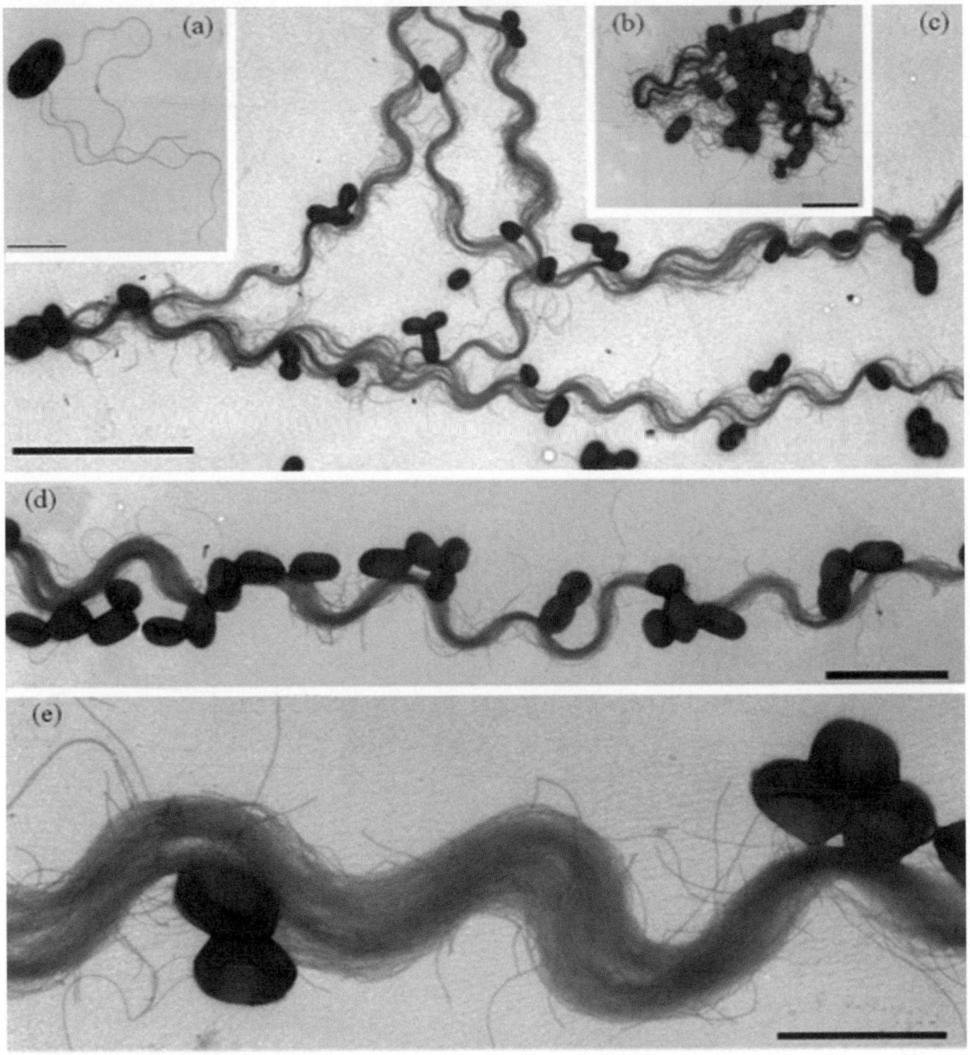

FIGURE 428. Electron micrographs of negatively stained cells of strain ID131577^{T}. Individual bacteria exhibit peritrichous long flagella (a). After a longer period of growth, bacteria form aggregates (b) and, due to self-aggregation of flagella, they exhibit an undulating network-like structure with entangled bacteria (c–e). Bars: (a, e), 1 µm; (b, d), 3 µm; (c) 5 µm. (Reprinted with permission from Monciardini et al., 2003. Int. J. Syst. Evol. Microbiol. *53*: 569–576.)

List of species of the genus *Conexibacter*

1. **Conexibacter woesei** Monciardini, Cavaletti, Schumann, Rohde and Donadio 2003, 574[VP]

woe'se.i. N.L. gen. masc. n. *woesei* of Woese, named to honor Carl R. Woese for his pioneering work on the use of 16S rRNA in phylogenetic analysis.

In addition to the properties described for the genus, colonies on Todd–Hewitt agar are smooth, mucoid to sticky, and of white to cream color. Nitrate is reduced to nitrite. Gelatin and esculin are hydrolyzed. Urea is not decomposed. NaCl is not tolerated at concentrations of 2% (w/v) or above. The type strain is able to utilize the following substrates: glycerol, L-arabinose, D-ribose, D-xylose, acetic acid, α-ketovaleric acid, propionic acid, pyruvic acid (Biolog GP microplate), methylpyruvate, β-hydroxybutyric acid, α-ketoglutaric acid, and α-ketovaleric acid (Biolog GN microplate). The type strain shows the following enzyme activities: esterase for 2-naphthyl caprylate and 2-naphthyl butyrate, leucine arylamidase, acid phosphatase, and naphthol-AS-BI-phosphohydrolase (API ZYM test). The following tests of the Biolog substrate panels, API 20NE gallery, and API ZYM enzyme assay are negative: mannan, α-cyclodextrin, dextrin, glycogen, Tween 40, Tween 80, *N*-acetyl-D-galactosamine, *N*-acetyl-D-glucosamine, adonitol, D-arabitol, cellobiose, iso-erythritol, D-fructose, L-fucose, D-galactose, gentiobiose, α-D-glucose, *myo*-inositol, α-lactose, α-D-lactose, lactulose, maltose, D-mannitol, D-mannose, melibiose, methyl β-D-glucoside, raffinose, L-rhamnose, D-sorbitol, sucrose, trehalose, turanose, xylitol, monomethyl succinate, *cis*-aconitic acid, citric acid, formic acid, D-galactonic acid lactone, D-galacturonic acid, D-gluconic acid, D-glucosaminic acid, D-glucuronic acid, α-hydroxybutyric acid, γ-hydroxybutyric acid, *p*-hydroxyphenylacetic acid, itaconic acid, α-ketobutyric acid, DL-lactic acid, malonic acid, quinic

TABLE 320. Differential phenotypic properties of *Conexibacter woesei* DSM 14684[T], *Patulibacter minatonensis* KV-614[T], *Patulibacter americanus* CP177-2[T], *Solirubrobacter pauli* DSM 14954[T] and *Solirubrobacter soli* Gsoil 355[Ta]

Characteristic	*C. woesei* DSM 14684[T]	*P. minatonensis* KV-614[T]	*P. americanus* CP177-2[T]	*S. pauli* DSM 14954[T]	*S. soli* Gsoil 355[T]
Cell size (μm)	0.6–0.7 × 0.9–1.2	0.6–0.7 × 1.2–1.5	nd	0.7 × 1.4	1.0-3.0 (length)
Cell arrangement	Singly or in pairs	Singly	Singly	Long chains	Singly
Motility	Motile by long peritrichous flagella	Motile by long flagella	Variable (jerking)	None	None
Optimal growth temperature (°C)	28–37	24–27	25	28–30	30
pH range	7.0–7.5	6–8	5–9, optimal 7	6.0–7.7, optimal 6.5	nd
Oxidase	Positive	Negative	Negative	Negative	Weakly positive
Substrates utilized	Glycerol, C_2–C_5 acids, pentoses	Broad range of sugars, alcohols, acids and few amino acids	Few sugars, arginine	Common sugars, few amino acids	Common sugars
Major fatty acids (>10%)	$C_{18:1}$ ω9*c*, $C_{16:0}$ iso, $C_{17:1}$ ω6*c*, $C_{16:0}$	$C_{18:1}$ ω9*c*, $C_{15:0}$ anteiso	$C_{18:1}$ ω9*c*, $C_{15:0}$ anteiso	$C_{16:0}$ iso, $C_{18:1}$ ω9*c*	$C_{16:0}$ iso, $C_{18:1}$ ω9*c*
Diamino acid of the peptidoglycan	*meso*-A_2pm	*meso*-A_2pm	*meso*-A_2pm	nd	nd
Isoprenoid quinone	MK-7(H_4)	DMK-7	MK-7(H_2)	nd	MK-7(H_4)
DNA G+C content (mol%)	71	72	72	71.8	71.5
Isolation site	Temperate forest soil	Soil	Biological soil crusts	Burrow of earthworm	Soil

[a]All strains are Gram-stain-positive, aerobic, catalase-positive and do not form endospores. Data from Monciardini et al. (2003), Takahashi et al. (2006), Reddy and Garcia-Pichel (2009), Singleton et al. (2003) and Kim et al. (2007). A_2pm, 2,6-Diaminopimelic acid; nd, not determined.

acid, D-saccharic acid, sebacic acid, succinic acid, bromosuccinic acid, succinamic acid, glucuronamide, alaninamide, D-alanine, L-alanine, L-alanyl glycine, L-asparagine, L-aspartic acid, L-glutamic acid, glycyl L-aspartic acid, glycyl L-glutamic acid, L-histidine, hydroxy-L-proline, L-leucine, L-ornithine, L-phenylalanine, L-proline, L-pyroglutamic acid, D-serine, L-serine, L-threonine, DL-carnitine, γ-aminobutyric acid, urocanic acid, inosine, uridine, thymidine, phenylethylamine, putrescine, 2-aminoethanol, 2,3-butanediol, DL-α-glycerol phosphate, glucose 1-phosphate, glucose 6-phosphate, β-cyclodextrin, inulin, *N*-acetylmannosamine, amygdalin, arbutin, lactulose, maltotriose, melezitose, methyl α-D-galactoside, 3-methyl glucose, methyl α-D-glucoside, methyl α-D-mannoside, palatinose, salicin, sedoheptulosan, stachyose, D-tagatose, lactamide, D-lactic acid methyl ester, L-lactic acid, D-malic acid, L-malic acid, methyl succinate, *N*-acetyl-L-glutamic acid, adenosine, 2′-deoxyadenosine, AMP, TMP, UMP, fructose 6-phosphate, caprate, adipate, phenylacetate, degradation of tryptophan, fermentation of glucose, arginine dihydrolase, urease, alkaline phosphatase, lipase, valine arylamidase, cystine arylamidase, trypsin, chymotrypsin, α-galactosidase, β-galactosidase, β-glucuronidase, α-glucosidase, β-glucosidase, *N*-acetyl-β-glucosaminidase, α-mannosidase, and α-fucosidase. Good growth occurs at pH 7.0–7.5 and at 28–37°C. The type strain is susceptible to amikacin (30 μg), gentamicin (10 μg), nitrofurantoin (300 μg), novobiocin (30 μg), polymyxin B (300 IU), and teicoplanin (30 μg) and only weakly susceptible to chloramphenicol (30 μg), erythromycin (15 μg), tetracycline (30 μg), and vancomycin (30 μg). Cells are resistant to ampicillin (10 μg), aztreonam (100 μg), ceftazidime (30 μg), ciprofloxacin (5 μg), clindamycin (2 μg), kanamycin (30 μg), methicillin (5 μg), norfloxacin (10 μg), oxacillin (1 μg), rifampin (30 μg), streptomycin (10 μg), trimethoprim (5 μg), and tobramycin (10 μg).

Source: temperate forest soil.

DNA G+C content (mol%): 71 (HPLC).

Type strain: ID131577, DSM 14684, JCM 11494.

Sequence accession no. (16S rRNA gene): AJ440237.

References

Carreto, L., E. Moore, M.F. Nobre, R. Wait, P.W. Riley, R.J. Sharp and M.S. da Costa. 1996. *Rubrobacter xylanophilus* sp. nov: a new thermophilic species isolated from a thermally polluted effluent. Int. J. Syst. Bacteriol. *46*: 460–465.

Chun, J., J.H. Lee, Y. Jung, M. Kim, S. Kim, B.K. Kim and Y.W. Lim. 2007. EzTaxon: a web-based tool for the identification of prokaryotes based on 16S ribosomal RNA gene sequences. Int. J. Syst. Evol. Microbiol. *57*: 2259–2261.

Embley, T.M. and E. Stackebrandt. 1994. The molecular phylogeny and systematics of the actinomycetes. Annu. Rev. Microbiol. *48*: 257–289.

Felsenstein, J. 1985. Confidence limits on phylogenies: an approach using the bootstrap. Evolution *39*: 783–791.

Felsenstein, J. 1993. PHYLIP (Phylogeny Inference Package), 3.5c edn. Department of Genetics, University of Washington, Seattle.

Hayakawa, M. and H. Nonomura. 1987. Humic acid-vitamine agar, a new medium for the selective isolation of soil actinomycetes. J. Ferment. Technol. *65*: 501–509.

Joseph, S.J., P. Hugenholtz, P. Sangwan, C.A. Osborne and P.H. Janssen. 2003. Laboratory cultivation of widespread and previously uncultured soil bacteria. Appl. Environ. Microbiol. *69*: 7210–7215.

Kim, M.K., J.R. Na, T.H. Lee, W.T. Im, N.K. Soung and D.C. Yang. 2007. *Solirubrobacter soli* sp. nov., isolated from soil of a ginseng field. Int. J. Syst. Evol. Microbiol. *57*: 1453–1455.

Monciardini, P., L. Cavaletti, P. Schumann, M. Rohde and S. Donadio. 2003. *Conexibacter woesei* gen. nov., sp. nov., a novel representative of a deep evolutionary line of descent within the class *Actinobacteria*. Int. J. Syst. Evol. Microbiol. *53*: 569–576.

Rainey, F.A., N.L. Ward-Rainey and E. Stackebrandt. 1997. *In* Stackebrandt, E., F.A. Rainey and N.L. Ward-Rainey. Proposal for a new hierarchic classification system, *Actinobacteria* classis nov. Int. J. Syst. Bacteriol. *47*: 479–491.

Reddy, G.S. and F. Garcia-Pichel. 2009. Description of *Patulibacter americanus* sp. nov., isolated from biological soil crusts, emended description of the genus *Patulibacter* Takahashi *et al.* 2006 and proposal of *Solirubrobacterales* ord. nov. and *Thermoleophilales* ord. nov. Int. J. Syst. Evol. Microbiol. *59*: 87–94.

Rheims, H., C. Spröer, F.A. Rainey and E. Stackebrandt. 1996. Molecular biological evidence for the occurrence of uncultured members of the actinomycete line of descent in different environments and geographical locations. Microbiology *142 (Pt 10)*: 2863–2870.

Rheims, H. and E. Stackebrandt. 1999. Application of nested polymerase chain reaction for the detection of as yet uncultured organisms of the class *Actinobacteria* in environmental samples. Environ. Microbiol. *1*: 137–143.

Sait, M., P. Hugenholtz and P.H. Janssen. 2002. Cultivation of globally distributed soil bacteria from phylogenetic lineages previously only detected in cultivation-independent surveys. Environ. Microbiol. *4*: 654–666.

Sangwan, P., S. Kovac, K.E. Davis, M. Sait and P.H. Janssen. 2005. Detection and cultivation of soil *Verrucomicrobia*. Appl. Environ. Microbiol. *71*: 8402–8410.

Shirling, E.B. and D. Gottlieb. 1966. Methods for characterization of *Streptomyces* species. Int. J. Syst. Bacteriol. *16*: 313–340.

Singleton, D.R., M.A. Furlong, A.D. Peacock, D.C. White, D.C. Coleman and W.B. Whitman. 2003. *Solirubrobacter pauli* gen. nov., sp. nov., a mesophilic bacterium within the *Rubrobacteridae* related to common soil clones. Int. J. Syst. Evol. Microbiol. *53*: 485–490.

Stackebrandt, E., F.A. Rainey and N.L. Ward-Rainey. 1997. Proposal for a new hierarchic classification system, *Actinobacteria* classis nov. Int. J. Syst. Bacteriol. *47*: 479–491.

Stackebrandt, E. 2004. Will we ever understand? The undescribable diversity of the prokaryotes. Acta Microbiol. Immunol. Hung. *51*: 449–462.

Stackebrandt, E. 2005. *In* Validation of publication of new names and new combinations previously effectively published outside the IJSEM. List no. 102 Int. J. Syst. Evol. Microbiol. *55*: 547–549.

Takahashi, Y., A. Matsumoto, K. Morisaki and S. Ōmura. 2006. *Patulibacter minatonensis* gen. nov., sp. nov., a novel actinobacterium isolated using an agar medium supplemented with superoxide dismutase, and proposal of *Patulibacteraceae* fam. nov. Int. J. Syst. Evol. Microbiol. *56*: 401–406.

Yakimov, M.M., H. Lunsdorf and P.N. Golyshin. 2003. *Thermoleophilum album* and *Thermoleophilum minutum* are culturable representatives of group 2 of the *Rubrobacteridae* (*Actinobacteria*). Int. J. Syst. Evol. Microbiol. *53*: 377–380.

Zarilla, K.A. and J.J. Perry. 1984. *Thermoleophilum album* gen. nov. and sp. nov., a bacterium obligate for thermophily and normal-alkane substrates. Arch. Microbiol. *137*: 286–290.

Zarilla, K.A. and J.J. Perry. 1986. Deoxyribonucleic acid homology and other comparisons among obligately thermophilic hydrocarbonoclastic bacteria, with a proposal for *Thermoleophilum minutum* sp. nov. Int. J. Syst. Bacteriol. *36*: 13–16.

Zhi, X.-Y., W.-J. Li and E. Stackebrandt. 2009. An update of the structure and 16S rRNA gene sequence-based definition of higher ranks of the class *Actinobacteria*, with the proposal of two new suborders and four new families and emended descriptions of the existing higher taxa. Int. J. Syst. Evol. Microbiol. *59*: 589–608.

Family III. **Patulibacteraceae** Takahashi, Matsumoto, Morisaki and Ōmura 2006, 405^{VP}

Yōko Takahashi

Pa.tu.li.bac.te.ra.ce'a.e. N.L. masc. n. *Patulibacter* type genus of the family; suff. *-aceae* ending to denote a family; N.L. fem. pl. n. *Patulibacteraceae* the *Patulibacter* family.

Cells are rods, Gram-stain-positive, non-endospore-forming. Cell-wall peptidoglycan is based upon *meso*-diaminopimelic acid. The main isoprenoid quinone is demethylmenaquinone. Mycolic acids are absent. The signature nucleotides in the 16S rRNA gene sequence (*Escherichia coli* sequence; Brosius et al., 1978) of members of the family are: 52:359 (C–G), 98 (A), 139 (G), 144:178 (C–G), 370:391 (C–G), 580:761 (U–A), 590:649 (U–A), 600:638 (U–G), 670:736 (A–U), 953:1228 (G–C), 954:1226 (G–C), 999:1041 (U–A), 657:749 (U–A), 681:709 (U–A), 941:1342 (A–U), 1051:1207 (G–C), and 1311:1326 (A–U). Phylogenetically, the family is a member of the phylum *Actinobacteria*, class *Thermoleophilia*, order *Solirubrobacterales*.

DNA G+C content (mol%): 72.0 (HPLC).

Type genus: **Patulibacter** Takahashi, Matsumoto, Morisaki and Ōmura 2006, 405^{VP} emend. Reddy and Garcia-Pichel 2009, 91.

Further descriptive information

In the present volume, the order *Solirubrobacterales* Reddy and Garcia-Pichel 2009 contains three families, *Conexibacteraceae*, *Solirubrobacteraceae*, and *Patulibacteraceae* (Takahashi et al., 2006). The family *Patulibacteraceae* can be distinguished within the order *Solirubrobacterales* on the basis of 16S rRNA signature nucleotides (Stackebrandt et al., 1997; Reddy and Garcia-Pichel, 2009).

The family *Patulibacteraceae* comprises the sole genus *Patulibacter*.

Genus I. **Patulibacter** Takahashi, Matsumoto, Morisaki and Ōmura 2006, 405^{VP} emend. Reddy and Garcia-Pichel 2009, 91

Yōko Takahashi

Pa.tu.li.bac'ter. L. adj. *patulus* spreading; N.L. masc. n. *bacter* from Gr. n. *baktron* a rod; N.L. masc. n. *Patulibacter*, spreading growth-bacterium.

Cells are rods, Gram-stain-positive, and non-endospore-forming. Bacterial respiration is aerobic. Cell-wall peptidoglycan contains ***meso*-diaminopimelic acid (DAP)**, alanine, and glutamic acid. Cell-wall type A1γ (based on *meso*-DAP, direct cross-linkage). The acyl type of the peptidoglycan is acetyl. The fatty acid profile is dominated by $\mathbf{C_{18:1}}$ **ω9*c***, $C_{15:0}$ anteiso, and $C_{17:0}$ anteiso. Mycolic acids are absent. The predominant isoprenoid quinone is **demethylmenaquinone with seven isoprene units (DMK-7).**

DNA G+C content (mol%): 72.0 (HPLC).

Type species: **Patulibacter minatonensis** Takahashi, Matsumoto, Morisaki and Ōmura 2006, 405^{VP}.

Further descriptive information

Patulibacter minatonensis forms flat and nearly transparent colonies with a white color on one-fifth strength nutrient agar (1/5 NA, Difco), ISP 3 medium (Shirling and Gottlieb, 1966), and R2A agar (Difco) consisting of 0.05% yeast extract, 0.05% proteose peptone, 0.05% Casamino acids, 0.05% glucose, 0.05% soluble starch, 0.03% sodium pyruvate, 0.03% dipotassium phosphate, 0.005% magnesium sulfate, and 1.5% agar, and with a pale yellow color on heart infusion agar and Todd–Hewitt agar (Difco) consisting of 0.9% beef heart digest, 1.1% pancreatic digest of casein, 0.3% soy bean peptone, 0.2% glucose, 0.25% sodium carbonate, 0.05% monosodium phosphate, 0.2% sodium chloride, and 1.5% agar. Bacterial cells are rod-shaped (1.2–1.5 × 0.6–0.7 μm), Gram-stain-positive, show negative oxidase reaction and reduce nitrate to nitrite. Cells possess motility due to the presence of long flagella (Figure 429). The strain does not grow on media containing ≥2% (w/v) NaCl. The pH range for growth is 6–8. The temperature range and optimum temperature for growth is 16–28°C and 24–27°C, respectively.

Enrichment and isolation procedures

Patulibacter minatonensis KV-614 was isolated from a soil sample using an agar medium supplemented with superoxide dismutase (SOD) following the method of Takahashi et al. (2003). Glucose-peptone-meat extract agar (GPM agar) medium consisting of 1% D-glucose (Wako Pure Chemical Industry), 0.5% peptone (Kyokuto Seiyaku Co.), 0.5% meat extract (Kyokuto), 0.3% NaCl and 1.2% agar (Wako) was used for isolation. Solid GPM agar medium supplemented with SOD (Sigma) was prepared by spreading SOD (300 U/plate) on the surface. Then 100-μl samples of soil suspensions diluted with sterilized water were spread on dishes of these solid media, which were incubated at 27°C for 14 d. The strain was cultured on 1/5 NA, ISP 3 medium, heart infusion agar, R2A agar, Todd–Hewitt agar (Difco), and yeast extract-glucose agar medium containing 1.0% yeast extract (Difco), 1.0% glucose, and 1.2% agar.

The appropriate cultivation conditions have not been found for many unknown micro-organisms. While searching for factors that promote bacterial colony growth, it was found that using isolation agar medium supplemented with SOD increases the number of colonies from a soil sample and that addition of SOD and catalase increases it still further (Takahashi et al., 2003).

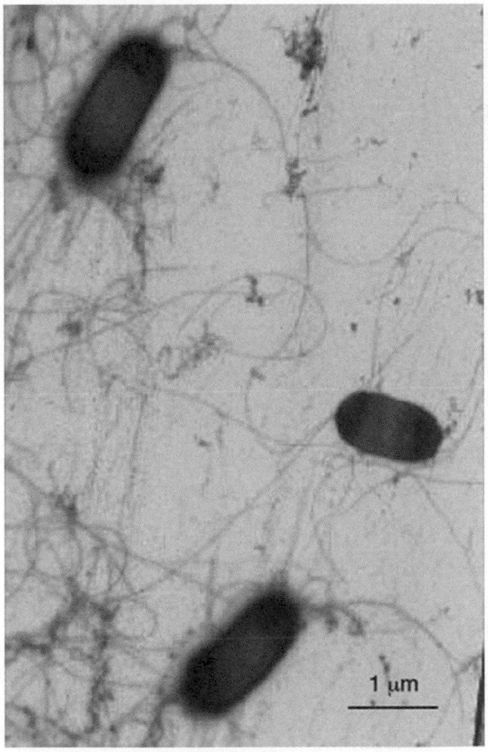

FIGURE 429. Transmission electron micrograph of negatively stained cells with long flagella of *Patulibacter minatonensis* KV-614. (Reprinted with permission from Takahashi et al., 2006. Int. J. Syst. Evol. Microbiol. *56*: 401–406.)

Maintenance procedures

Cultures can be maintained in the laboratory by transfer to the same media used for isolation and stored for a short period. One-fifth strength nutrient agar medium is recommended for maintenance. Recommended conditions for long-term preservation are storage at −80°C with 20% glycerol or lyophilization in the presence of 10% skim milk as stabilizer. The lyophilized ampoules should be stored at 5°C.

Differentiation of the genus *Patulibacter* from other genera

Figure 430 shows a phylogenetic tree constructed by the neighbor-joining method on the basis of 16S rRNA gene sequences of *Patulibacter minatonensis* KV-614 and closely related taxa. *Patulibacter* forms a distinct branch within the order *Rubrobacterales* (Stackebrandt et al., 1997), and is associated with the genera *Conexibacter* (Monciardini et al., 2003) and *Solirubrobacter* (Singleton et al., 2003). The 16S rRNA gene sequence similarity of the type strain KV-614 to *Conexibacter woesei* DSM 14684 is 92.4% and to *Solirubrobacter pauli* B33D1 is 89.7%. The genus *Conexibacter* has one species. The new species of *Solirubrobacter*, *Solirubrobacter soli* Gsoil 355, reported in 2007, also has a low 16S rRNA gene sequence similarity (91.8%) to the type strain (Kim et al., 2007).

Table 320 shows the phenotypic characteristics of *Patulibacter minatonensis* and two related genera, *Conexibacter* and *Solirubrobacter*. *Conexibacter woesei* also contains *meso*-DAP in the cell wall peptidoglycan and produces long flagella, but clearly differs from *Patulibacter minatonensis* in the major menaquinone MK-7(H_4) and predominant fatty acids. *Solirubrobacter pauli* and *Solirubrobacter soli* are also different from *Patulibacter minatonensis* by being non-motile cells and having different fatty acid composition.

Miscellaneous comments

The signature nucleotides in the 16S rRNA gene sequence of members of the subclass *Rubrobacteridae* (Stackebrandt, 2004), which has been elevated to class *Rubrobacteria*, 127:234 (G–C), 291:309 (U–A), 955:1225 (U–A), 1115:1185 (C–G), and 1410:1490 (A–U) (*Escherichia coli* sequence; Brosius et al., 1978) is also present in *Patulibacter*. An A residue at position 906 and a U residue at position 955 are characteristics of members of the class *Rubrobacteridae* (Stackebrandt et al., 1997).

Figure 431 shows a phylogenetic tree constructed with 16S rRNA gene sequences of strain *Patulibacter minatonensis* KV-614 and those of its closest clonal and cultured relatives retrieved from the GenBank database. Clones prefixed "TM" were isolated from a peat bog (Rheims et al., 1999, 1996), and strains prefixed "Ellin" were isolated from Australian soils by culture on simple solid media at 25°C for up to 3 months (Joseph et al., 2003; Sait et al., 2002). The sequences of clones YNPFFP1 and 1959-7 are available from public databases. The closest sequence is that of the bacterium Ellin 5025 (93.2%). This reveals that the strain is closely related to a phylogenetic cluster consisting of uncultured bacteria from terrestrial environments and unidentified bacterial strains.

DMK-7 detected in *Patulibacter minatonensis* KV-614 is a unique isoprenoid quinone and has been known to occur in *Enterococcus faecalis* (Hiraishi, 1988), *Pasteurella*, and *Haemophilus* (Kroppenstedt and Mannheim, 1989). Demethylmenaquinones hitherto have been detected in two genera *Patulibacter* and *Demequina* (Yi et al., 2007) in the class *Actinobacteria*.

Since this chapter was prepared, a second species, *Patulibacter americanus*, was published (Reddy and Garcia-Pichel, 2009); see Table 320 in *Conexibacteraceae*, above, for distinguishing characteristics.

Acknowledgements

The author thanks Atsuko Matsumoto for help with phylogenetic tree construction, and also Ken-ichiro Suzuki for reading the manuscript.

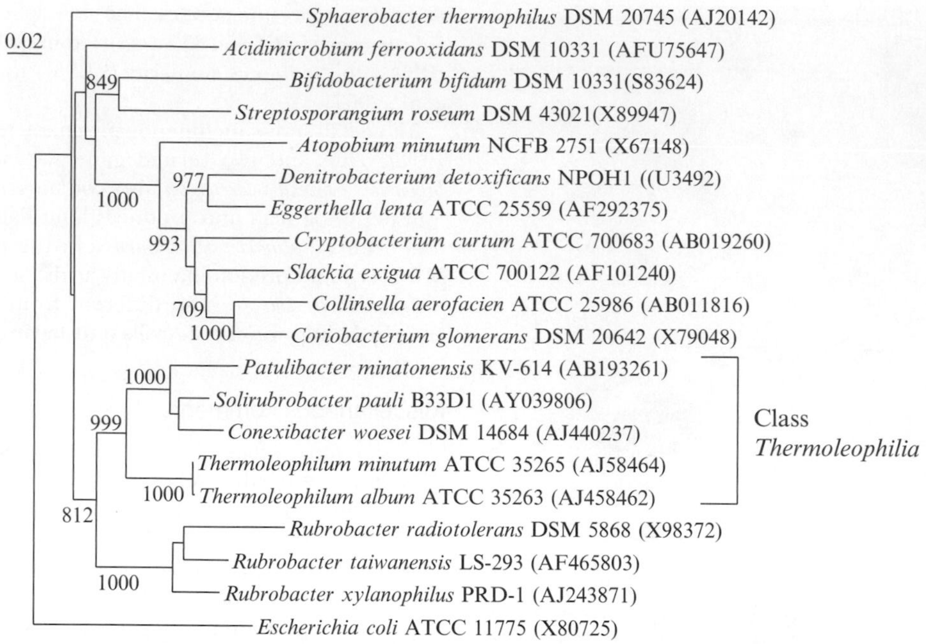

FIGURE 430. Phylogenetic tree showing the position of *Patulibacter* and the other members of the order *Rubrobacterales* based on 16S rRNA gene sequences. Numbers at nodes indicate the level of bootstrap support based on neighbor-joining analysis of 1000 resampled datasets. Only values higher than 700 are shown. Bar = 2 inferred nucleotide substitutions per 100 nt. The tree was rooted with *Escherichia coli.*

List of species of the genus *Patulibacter*

1. **Patulibacter minatonensis** Takahashi, Matsumoto, Morisaki and Ōmura 2006, 405[VP]

mi.na.to.nen'sis. N.L. masc. adj. *minatonensis* pertaining to Minato-ku, the ward of Tokyo, Japan, where the species originated.

Patulibacter minatonensis forms flat and nearly transparent colonies with whitish or pale yellow color. Cells are 1.2–1.5 × 0.6–0.7 µm in size, motile by long flagella, aerobic, Gram positive, and non-endospore-forming, catalase-positive, and oxidase-negative. Cell-wall peptidoglycan contains *meso*-DAP, alanine, and glutamic acid. The acyl type is acetyl. Mycolic acids are absent. The predominant isoprenoid quinone is DMK-7. Cellular fatty acid components are ω9c octadecenoic acid (oleic acid, $C_{18:1}$ ω9*c*, 63%), 12-methyl tetradecanoic acid ($C_{15:0}$ anteiso, 10%), 14-methyl hexadecanoic acid ($C_{17:0}$ anteiso, 8%), hexadecanoic acid ($C_{16:0}$, 5%), and octadecanoic acid ($C_{18:0}$, 4%). The strain does not grow on media containing 2% (w/v) NaCl. The pH range for growth is 6–8. The temperature range for growth is 16–28°C.

The type strain is able to utilize the following substrates: dextrin, inulin, mannan, *N*-acetyl-D-mannosamine, L-arabinose, D-arabitol, cellobiose, D-fructose, L-fucose, D-galactose, D-galacturonic acid, α-D-glucose, *m*-inositol, maltose, maltotriose, D-mannitol, D-mannose, D-melezitose, α-methyl-D-galactoside, β-methyl-D-glucoside, α-methyl-D-glucoside, α-methyl-D-mannoside, D-psicose, L-rhamnose, salicin, sedoheptulosan, turanose, α-hydroxybutyric acid, lactamide, D-lactic acid methyl ester, L-lactic acid, L-serine, 2,3-butanediol, adenosine, inosine, uridine, adenosine-5′-monophosphate, fructose-6-phosphate, glucose-1-phosphate, glucose-6-phosphate, DL-α-glycerol phosphate, D-ribose, D-tagatose, D-xylose, acetic acid, propionic acid, D-alanine, putrescine, maltose, potassium gluconate, adipic acid, and L-malic acid. The type strain shows the following enzyme activities: alkaline phosphatase, elastase (C4), elastase lipase (C8), lipase (C4), leucine arylamidase, acid phosphatase, and naphthol-AS-BI-phosphohydrolase.

In the disc susceptibility test, cells are susceptible to tetracycline (30 µg/disc), polymyxin B (300 IU/disc), amikacin (30 µg/disc), chloramphenicol (30 µg/disc), erythromycin (15 µg/disc), ciprofloxacin (5 µg/disc), tobramycin (10 µg/disc), vancomycin (30 µg/disc), kanamycin (30 µg/disc), gentamicin (10 µg/disc), ampicillin (10 µg/disc), and imipenem (10 µg/disc). Conversely, cells are resistant to aztreonam (30 µg/disc), ceftazidime (30 µg/disc), nalidixic acid (1 µg/disc), norfloxacin (10 µg/disc), and oxacillin (1 µg/disc).

DNA G+C content (mol%): 72.0 (HPLC).

Type strain: KV-614, NRRL B-24346, JCM 12834, NBRC 100761.

Sequence accession no. (16S rRNA gene): AB193261.

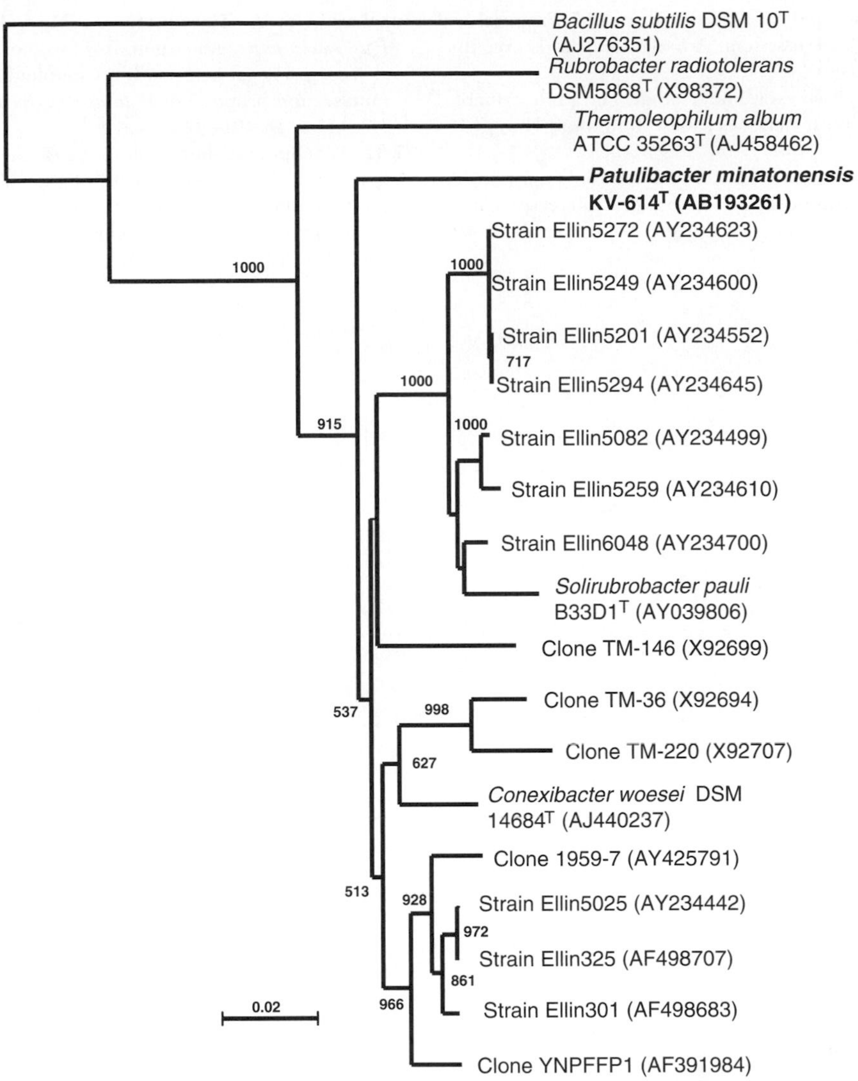

FIGURE 431. Phylogenetic tree constructed based on 16S RNA gene sequences of *Patulibacter minatonensis* KV-614 and its closest clonal and cultured relatives. Numbers at nodes are bootstrap values based on 1000 resamplings; only values higher than 500 are shown. Bar = 2 inferred nucleotide substitutions per 100 nt. The tree was rooted with *Bacillus subtilis*. (Reprinted with permission from Takahashi et al., 2006. Int. J. Syst. Evol. Microbiol. *56*: 401–406.)

References

Brosius, J., M.L. Palmer, P.J. Kennedy and H.F. Noller. 1978. Complete nucleotide sequence of a 16S ribosomal RNA gene from *Escherichia coli*. Proc. Natl. Acad. Sci. U.S.A. *75*: 4801–4805.

Hiraishi, A. 1988. High-performance liquid chromatographic analysis of demethylmenaquinone and menaquinone mixtures from bacteria. J. Appl. Bacteriol. *64*: 103–105.

Joseph, S.J., P. Hugenholtz, P. Sangwan, C.A. Osborne and P.H. Janssen. 2003. Laboratory cultivation of widespread and previously uncultured soil bacteria. Appl. Environ. Microbiol. *69*: 7210–7215.

Kim, M.K., J.R. Na, T.H. Lee, W.T. Im, N.K. Soung and D.C. Yang. 2007. *Solirubrobacter soli* sp. nov., isolated from soil of a ginseng field. Int. J. Syst. Evol. Microbiol. *57*: 1453–1455.

Kroppenstedt, R.M. and W. Mannheim. 1989. Lipoquinones in members of the family *Pasteurellaceae*. Int. J. Syst. Bacteriol. *39*: 304–308.

Monciardini, P., L. Cavaletti, P. Schumann, M. Rohde and S. Donadio. 2003. *Conexibacter woesei* gen. nov., sp. nov., a novel representative of a deep evolutionary line of descent within the class *Actinobacteria*. Int. J. Syst. Evol. Microbiol. *53*: 569–576.

Reddy, G.S. and F. Garcia-Pichel. 2009. Description of *Patulibacter americanus* sp. nov., isolated from biological soil crusts, emended description of the genus *Patulibacter* Takahashi *et al.* 2006 and proposal of *Solirubrobacterales* ord. nov. and *Thermoleophilales* ord. nov. Int. J. Syst. Evol. Microbiol. *59*: 87–94.

Sait, M., P. Hugenholtz and P.H. Janssen. 2002. Cultivation of globally distributed soil bacteria from phylogenetic lineages previously only detected in cultivation-independent surveys. Environ. Microbiol. *4*: 654–666.

Shirling, E.B. and D. Gottlieb. 1966. Methods for characterization of *Streptomyces* species. Int. J. Syst. Bacteriol. *16*: 313–340.

Singleton, D.R., M.A. Furlong, A.D. Peacock, D.C. White, D.C. Coleman and W.B. Whitman. 2003. *Solirubrobacter pauli* gen. nov., sp. nov., a mesophilic bacterium within the *Rubrobacteridae* related to common soil clones. Int. J. Syst. Evol. Microbiol. *53*: 485–490.

Stackebrandt, E., F.A. Rainey and N.L. Ward-Rainey. 1997. Proposal for a new hierarchic classification system, *Actinobacteria* classis nov. Int. J. Syst. Bacteriol. *47*: 479–491.

Stackebrandt, E. 2004. Will we ever understand? The undescribable diversity of the prokaryotes. Acta Microbiol. Immunol. Hung. *51*: 449–462.

Takahashi, Y., S. Katoh, N. Shikura, H. Tomoda and S. Ōmura. 2003. Superoxide dismutase produced by soil bacteria increases bacterial colony growth from soil samples. J. Gen. Appl. Microbiol. *49*: 263–266.

Takahashi, Y., A. Matsumoto, K. Morisaki and S. Ōmura. 2006. *Patulibacter minatonensis* gen. nov., sp. nov., a novel actinobacterium isolated using an agar medium supplemented with superoxide dismutase, and proposal of *Patulibacteraceae* fam. nov. Int. J. Syst. Evol. Microbiol. *56*: 401–406.

Yi, H., P. Schumann and J. Chun. 2007. *Demequina aestuarii* gen. nov., sp. nov., a novel actinomycete of the suborder *Micrococcineae*, and reclassification of *Cellulomonas fermentans* Bagnara *et al.* 1985 as *Actinotalea fermentans* gen. nov., comb. nov. Int. J. Syst. Evol. Microbiol. *57*: 151–156.

Author index

Index of scientific names of *Archaea* and *Bacteria*

Key to the fonts and symbols used in this index:

Nomenclature

Lower case, Roman	Genera, species, and subspecies of bacteria. Every bacterial name mentioned in the text is listed in the index. Specific epithets are listed individually and also under the genus.*
CAPITALS, ROMAN:	Names of taxa higher than genus (tribes, families, orders, classes, phyla, domains).

Pagination

Roman:	Pages on which taxa are mentioned.
Boldface:	Indicates page on which the description of a taxon is given.†

* Infrasubspecific names, such as serovars, biovars, and pathovars, are not listed in the index.

† A description may not necessarily be given in the *Manual* for a taxon that is considered as *incertae sedis* or that is listed in an addendum or note added in proof; however, the page on which the complete citation of such a taxon is given is indicated in boldface type.

Index of scientific names of *Archaea* and *Bacteria*